Geologic Time Scale

ERA	PERIOD		EPOCH	TIME (BEGINNING)
Cenozoic	Quaternary		Holocene	10,000 ybp
			Pleistocene	2.6 mya
	Tertiary	Neogene	Pliocene	5.3 mya
			Miocene	23 mya
		Paleogene	Oligocene	33.9 mya
			Eocene	56 mya
			Paleocene	66 mya
Mesozoic	Cretaceous			145 mya
	Jurassic			201 mya
	Triassic			252 mya
Paleozoic	Permian			299 mya
	Carboniferous		Pennsylvanian	323 mya
			Mississippian	359 mya
	Devonian			419 mya
	Silurian			444 mya
	Ordovician			485 mya
	Cambrian			541 mya
Precambrian	Ediacaran			635 mya
				4.57 bya

ybp = years before present; mya = million years ago; bya = billion years ago

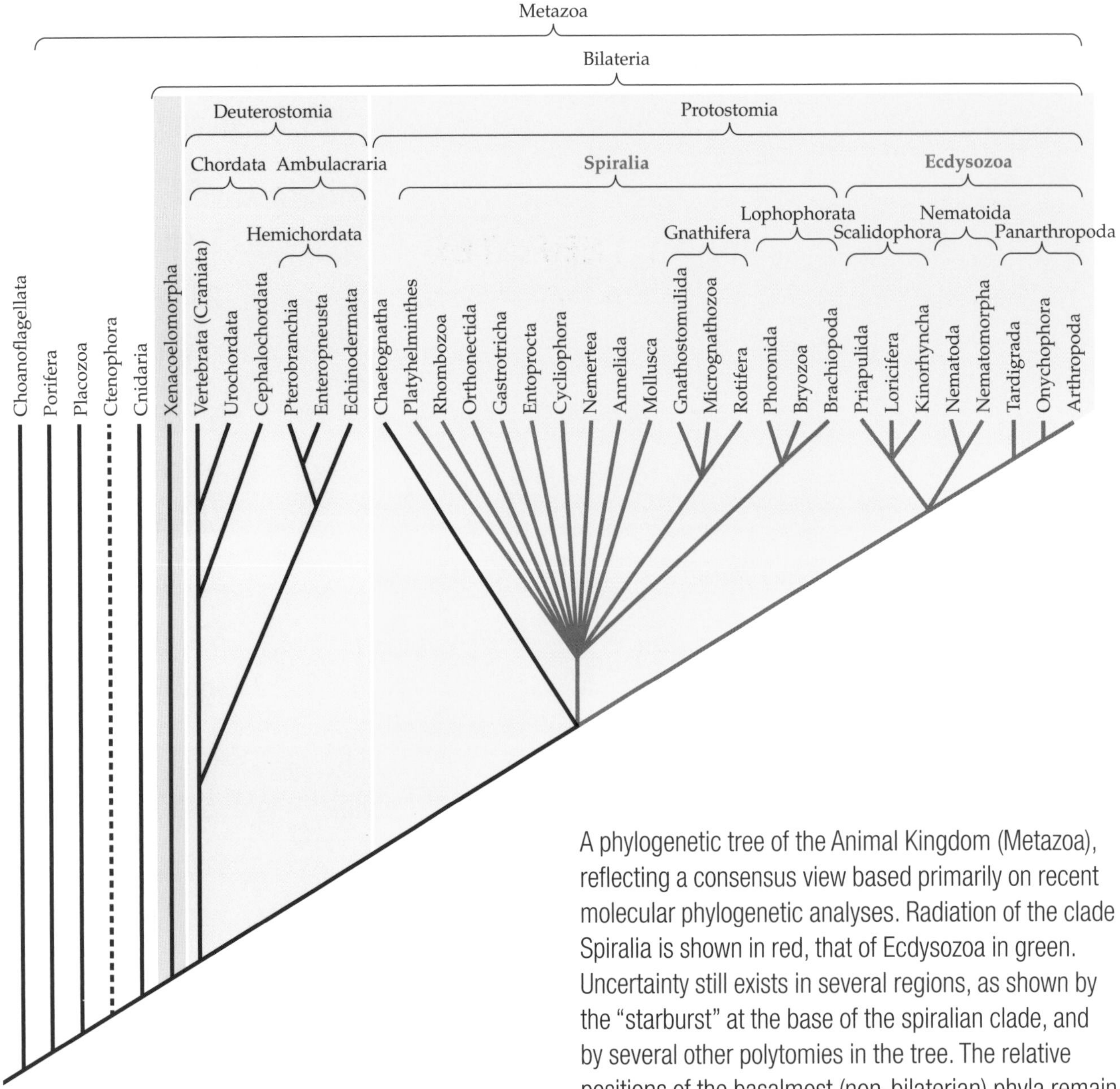

A phylogenetic tree of the Animal Kingdom (Metazoa), reflecting a consensus view based primarily on recent molecular phylogenetic analyses. Radiation of the clade Spiralia is shown in red, that of Ecdysozoa in green. Uncertainty still exists in several regions, as shown by the "starburst" at the base of the spiralian clade, and by several other polytomies in the tree. The relative positions of the basalmost (non-bilaterian) phyla remain uncertain, especially Ctenophora. Further resolution of metazoan phylogeny is enthusiastically anticipated in the coming years. See Chapter 28 for additional details.

INVERTEBRATES

Third Edition

INVERTEBRATES

Third Edition

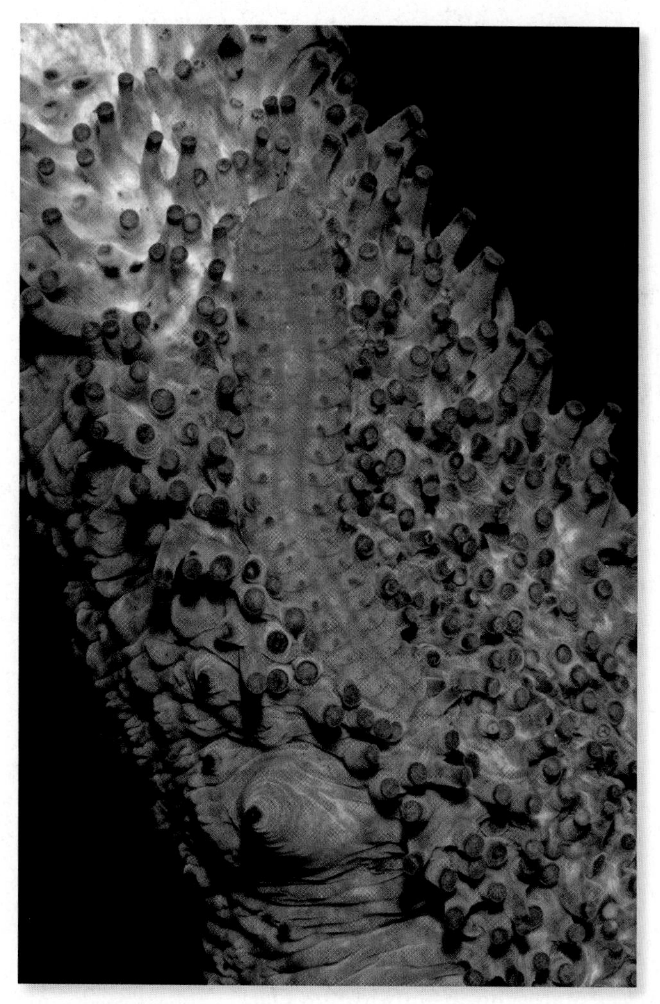

RICHARD C. BRUSCA, PhD
Executive Director Emeritus,
Arizona-Sonora Desert Museum
Research Scientist, Department of Ecology and Evolutionary Biology,
University of Arizona

WENDY MOORE, PhD
Assistant Professor and Curator,
Department of Entomology,
University of Arizona

STEPHEN M. SHUSTER, PhD
Professor of Invertebrate Zoology and Curator,
Department of Biological Sciences,
Northern Arizona University

with illustrations by
NANCY HAVER

Sinauer Associates, Inc., Publishers
Sunderland, Massachusetts USA

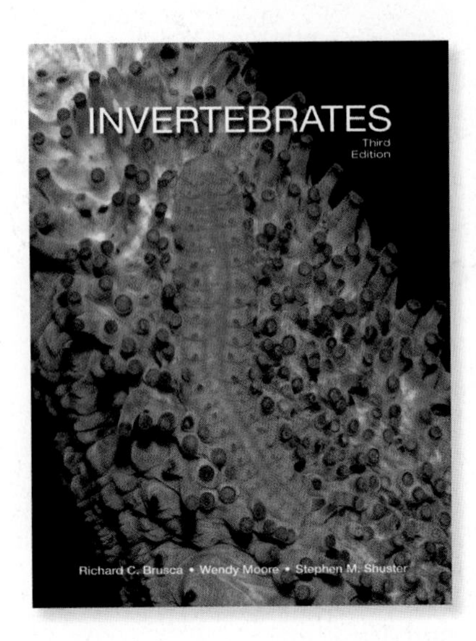

About the cover

The scale worm *Arctonoe pulchra* on the warty sea cucumber (*Apostichopus parvimensis*). Photo by Larry Jon Friesen. Dr. Friesen's spectacular photographs were also part of the Second Edition of *Invertebrates* (including its cover), and over 130 of his photographs grace the pages of this Third Edition. The success of this project has been due, in no small part, to his contributions.

Larry Jon Friesen, PhD, completed his graduate research at the University of California, Santa Barbara, in Animal Communication. Dr. Friesen is a Professor of Biological Sciences at Santa Barbara City College, teaching Cell Biology and Animal Diversity and incorporating his life-long passion for nature photography.

Invertebrates, Third Edition

For information or to order, address:
Sinauer Associates
P.O. Box 407
Sunderland, MA 01375 USA
Fax: 413-549-1118
E-mail: publish@sinauer.com
www.sinauer.com

Library of Congress Cataloging-in-Publication Data

Names: Brusca, Richard C., author. | Moore, Wendy (Entomologist), author. | Shuster, Stephen M., 1954- , author.
Title: Invertebrates / Richard C. Brusca, Wendy Moore, Stephen M. Shuster.
Description: Third edition. | Sunderland, Massachusetts U.S.A. : Sinauer Associates, Inc., Publishers, 2016.
Identifiers: LCCN 2015038708 | ISBN 9781605353753
Subjects: LCSH: Invertebrates.
Classification: LCC QL362 .B924 2016 | DDC 592--dc23
LC record available at http://lccn.loc.gov/2015038708

Printed in the USA
5 4 3 2 1

We dedicate this book to all our fellow teachers
and students of Invertebrate Zoology around the world.

Brief Contents

CHAPTER 1 ■ **Introduction** 1

CHAPTER 2 ■ **Systematics, Phylogeny, and Classification** 35

CHAPTER 3 ■ **The Protists:** Kingdom Protista 55

CHAPTER 4 ■ **Introduction to the Animal Kingdom:** Animal Architecture and Body Plans 135

CHAPTER 5 ■ **Introduction to the Animal Kingdom:** Development, Life Histories, and Origin 183

CHAPTER 6 ■ **Two Basal Metazoan Phyla:** Porifera and Placozoa 213

CHAPTER 7 ■ **Phylum Cnidaria:** Anemones, Corals, Jellyfish, and Their Kin 265

CHAPTER 8 ■ **Phylum Ctenophora:** The Comb Jellies 327

CHAPTER 9 ■ **Introduction to the Bilateria and the Phylum Xenacoelomorpha:** Triploblasty and Bilateral Symmetry Provide New Avenues for Animal Radiation 345

CHAPTER 10 ■ **Phylum Platyhelminthes:** The Flatworms 373

CHAPTER 11 ■ **Four Enigmatic Protostome Phyla:** Rhombozoa, Orthonectida, Chaetognatha, Gastrotricha 413

CHAPTER 12 ■ **Phylum Nemertea:** The Ribbon Worms 435

CHAPTER 13 ■ **Phylum Mollusca** 453

CHAPTER 14 ■ **Phylum Annelida:** The Segmented (and Some Unsegmented) Worms 531

CHAPTER 15 ■ **Two Enigmatic Spiralian Phyla:** Entoprocta and Cycliophora 603

CHAPTER 16 ■ **The Gnathifera:** Phyla Gnathostomulida, Rotifera (including Acanthocephala), and Micrognathozoa 613

CHAPTER 17 ■ **The Lophophorates:** Phyla Phoronida, Bryozoa, and Brachiopoda 635

CHAPTER 18 ■ **The Nematoida:** Phyla Nematoda and Nematomorpha 669

CHAPTER 19 ■ **The Scalidophora:** Phyla Kinorhyncha, Priapula, and Loricifera 693

CHAPTER 20 ■ **The Emergence of the Arthropods:** Tardigrades, Onychophorans, and the Arthropod Body Plan 709

CHAPTER 21 ■ **Phylum Arthropoda:** Crustacea: Crabs, Shrimps, and Their Kin 761

CHAPTER 22 ■ **Phylum Arthropoda:** The Hexapoda: Insects and Their Kin 843

CHAPTER 23 ■ **Phylum Arthropoda:** The Myriapods: Centipedes, Millipedes, and Their Kin 895

CHAPTER 24 ■ **Phylum Arthropoda:** The Chelicerata 911

CHAPTER 25 ■ **Introduction to the Deuterostomes and the Phylum Echinodermata** 967

CHAPTER 26 ■ **Phylum Hemichordata:** Acorn Worms and Pterobranchs 1007

CHAPTER 27 ■ **Phylum Chordata:** Cephalochordata and Urochordata 1021

CHAPTER 28 ■ **Perspectives on Invertebrate Phylogeny** 1047

Contents

CHAPTER 1

Introduction 1

Keeping Track of Life 4
Prokaryotes and Eukaryotes 7
Where Did Invertebrates Come From? 10
- The Dawn of Life 10
- The Ediacaran Period and the Origin of Animals 10
- The Paleozoic Era (541–252 Ma) 12
- The Mesozoic Era (252–66 Ma) 16
- The Cenozoic Era (66 Ma–present) 16

Where Do Invertebrates Live? 17
- Marine Habitats 17
- Estuaries and Coastal Wetlands 20
- Freshwater Habitats 22
- Terrestrial Habitats 22
- A Special Type of Environment: Symbiosis 23

Biodiversity Patterns 24
New Views of Invertebrate Phylogeny 25
Some Comments on Evolution 26
A Final Introductory Message to the Reader 32

CHAPTER 2

Systematics, Phylogeny, and Classification 35

Biological Classification 36
Nomenclature 36
Systematics 39
Monophyly, Paraphyly, and Polyphyly 40
Characters and the Concept of Homology 41
Phylogenetic Trees 43
Pleisiomorphy and Apomorphy 44
Constructing Phylogenies and Classifications 44
Molecular Phylogenetics 51

CHAPTER 3

The Protists: Kingdom Protista 55

Taxonomic History and Classification 59
Overviews of the Major Clades, or Groups 62
The General Protistan Body Plan 64
The Protist Phyla 70
Group 1: Amoebozoa 70
Phylum Amoebozoa: Amebas 70
Group 2: Chromalveolata 76
Phylum Dinoflagellata: Dinoflagellates 76
Phylum Apicomplexa: Gregarines, Coccidians, Haemosporidians, and Their Kin 81
Phylum Ciliata: The Ciliates 87
Phylum Stramenopila: Diatoms, Brown Algae, Golden Algae, Slime Nets, Oomycetes, etc. 97
Phylum Haptophyta: Coccolithophores 101
Phylum Cryptomonada: Cryptomonads 103

Group 3: Rhizaria 104
Phylum Chlorarachniophyta: Chlorarachniophyte Algae 104
Phylum Granuloreticulosa: Foraminiferans and Their Kin 105
Phylum Radiolaria: Radiolarians 108
Phylum Haplosporidia: Haplosporidians 111
Group 4: Excavata 112
Phylum Parabasalida: Trichomonads, Hypermastigotes, and Their Kin 112
Phylum Diplomonadida: Diplomonads 115
Phylum Heterolobosea: Heterolobosids 117
Phylum Euglenida: Euglenids 118
Phylum Kinetoplastida: Trypanosomes, Bodonids, and Their Kin 120
Group 5: Opisthokonta 124
Phylum Choanoflagellata: Choanoflagellates 124

Protist Phylogeny 125
The Origin of the Protista 125
Relationships among the Protists 127

CHAPTER 4

Introduction to the Animal Kingdom: Animal Architecture and Body Plans **135**

Body Symmetry 136
Cellularity, Body Size, Germ Layers, and Body Cavities 138
Locomotion and Support 140
Feeding and Digestion 148
Excretion and Osmoregulation 157
Circulation and Gas Exchange 162
Nervous Systems and Sense Organs 168
Bioluminescence 174
Nervous Systems and Body Plans 174
Hormones and Pheromones 175
Reproduction 176
Asexual Reproduction 176
Sexual Reproduction 177
Parthenogenesis 179

CHAPTER 5

Introduction to the Animal Kingdom: Development, Life Histories, and Origin **183**

Evolutionary Developmental Biology—EvoDevo 184
Developmental Tool Kits 184
The Relationship Between Genotype and Phenotype 185
The Evolution of Novel Gene Function 185
Gene Regulatory Networks 185
Eggs and Embryos 187
Eggs 187
Cleavage 187
Orientation of Cleavage Planes 188
Radial and Spiral Cleavage 188
Cell Fates 191
Blastula Types 192
Gastrulation and Germ Layer Formation 193
Mesoderm and Body Cavities 195
Life Cycles: Sequences and Strategies 197
Classification of Life Cycles 197
Indirect Development 197
Settling and Metamorphosis 199

Direct Development 199
Mixed Development 200
Adaptations to Land and Fresh Water 200
Parasite Life Cycles 200
The Relationships Between Ontogeny and Phylogeny 201
The Origin of the Metazoa 203

CHAPTER 6

Two Basal Metazoan Phyla: Porifera and Placozoa 213

Phylum Placozoa 215
Phylum Porifera: The Sponges 216
Taxonomic History and Classification 220
The Poriferan Body Plan 222
Some Additional Aspects of Sponge Biology 249
Distribution and Ecology 249
Biochemical Agents 250
Growth Rates 250
Symbioses 251
Poriferan Phylogeny 254
The Origin of Sponges 254
Evolution within the Porifera 255

CHAPTER 7

Phylum Cnidaria: Anemones, Corals, Jellyfish, and Their Kin 265

Taxonomic History and Classification 268
The Cnidarian Body Plan 274
Cnidarian Phylogeny 317
Ediacaran Cnidaria? 317
Cnidarian Origins 318
Relationships within Cnidaria 319

CHAPTER 8

Phylum Ctenophora: The Comb Jellies 327

Taxonomic History and Classification 328
The Ctenophoran Body Plan 332
Ctenophoran Phylogeny 341

CHAPTER 9

Introduction to the Bilateria and the Phylum Xenacoelomorpha: Triploblasty and Bilateral Symmetry Provide New Avenues for Animal Radiation 345

The Basal Bilaterian 346
Protostomes and Deuterostomes 347
Phylum Xenacoelomorpha 349
Classification of Phylum Xenacoelomorpha 35
Class Acoela 351
The Acoel Body Plan 353
Class Nemertodermatida 360
The Nemertodermatid Body Plan 362
Subphylum Xenoturbellida 366
The Xenoturbellid Body Plan 367

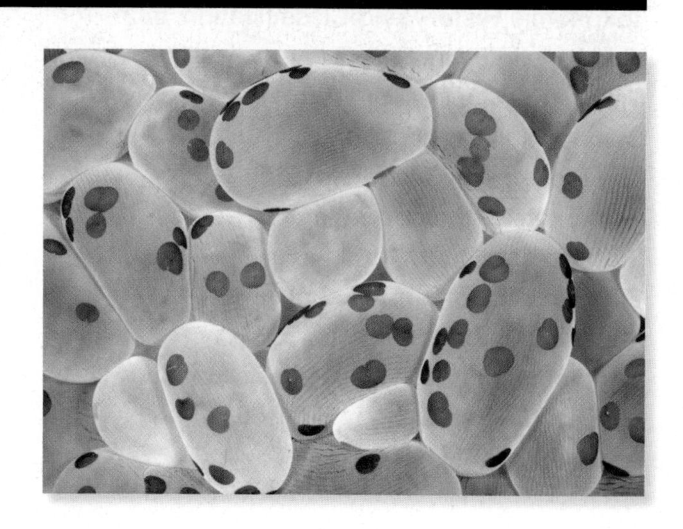

CHAPTER 10

Phylum Platyhelminthes: The Flatworms 373

Taxonomic History and Classification 374
The Platyhelminth Body Plan 380
Platyhelminth Phylogeny 405

CHAPTER 11

Four Enigmatic Protostome Phyla: Rhombozoa, Orthonectida, Chaetognatha, Gastrotricha 413

Phylum Rhombozoa 414
The Dicyemida 414
The Heterocyemida 417
Phylum Orthonectida 418
Phylum Chaetognatha 420
Chaetognath Classification 422
The Chaetognath Body Plan 423
Phylum Gastrotricha: The Gastrotrichs 428
Gastrotrich Classification 429
The Gastrotrich Body Plan 429

CHAPTER 12

Phylum Nemertea: The Ribbon Worms 435

Taxonomic History and Classification 436
The Nemertean Body Plan 438
Nemertean Phylogeny 450

CHAPTER 13

Phylum Mollusca 453

Taxonomic History and Classification 454
The Molluscan Body Plan 472
Molluscan Evolution and Phylogeny 521

CHAPTER 14

Phylum Annelida: The Segmented (and Some Unsegmented) Worms 531

Taxonomic History and Classification 532
The Annelid Body Plan 541
Sipuncula: The Peanut Worms 572
Sipunculan Classification 574
The Sipunculan Body Plan 575
Echiuridae: The Spoon Worms 579
Siboglinidae: Vent Worms and Their Kin 584
Siboglinid Taxonomic History 588
The Siboglinidae Body Plan 588
Hirudinoidea: Leeches and Their Relatives 591
The Hirudinoidean Body Plan 592
Annelid Phylogeny 597

CHAPTER 15

Two Enigmatic Spiralian Phyla: Entoprocta and Cycliophora 603

Phylum Entoprocta: The Entoprocts 603
Entoproct Classification 605
The Entoproct Body Plan 606
Phylum Cycliophora: The Cycliophorans 609

CHAPTER 16

The Gnathifera: Phyla Gnathostomulida, Rotifera (including Acanthocephala), and Micrognathozoa 613

Phylum Gnathostomulida: The Gnathostomulids 615
Gnathostomulid Classification 615
The Gnathostomulid Body Plan 616
Phylum Rotifera: The Free-Living Rotifers 616
Rotifer Classification 618
The Rotifer Body Plan 618
Phylum Rotifera, Subclass Acanthocephala: The Acanthocephalans 624
The Acanthocephalan Body Plan 624
Phylum Micrognathozoa: The Micrognathozoans 626

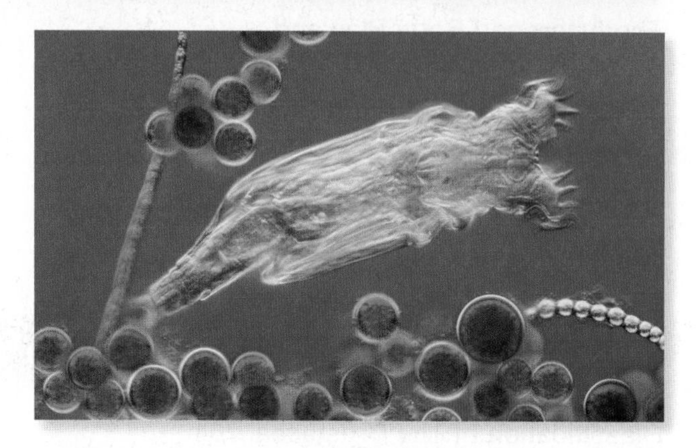

The Micrognathozoan Body Plan 628

CHAPTER 17

The Lophophorates: Phyla Phoronida, Bryozoa, and Brachiopoda 635

Taxonomic History of the Lophophorates 636
The Lophophorate Body Plan 637
Phylum Phoronida: The Phoronids 638
The Phoronid Body Plan 638
Phylum Bryozoa: The Moss Animals 644
The Bryozoan Body Plan 646
Phylum Brachiopoda: The Lamp Shells 657
The Brachiopod Body Plan 660

CHAPTER 18

The Nematoida: Phyla Nematoda and Nematomorpha 669

Phylum Nematoda: Roundworms and Threadworms 671
Nematode Classification 672
The Nematode Body Plan 673
Life Cycles of Some Parasitic Nematodes 682
Phylum Nematomorpha: Horsehair Worms and Their Kin 686
The Nematomorphan Body Plan 687

CHAPTER 19

The Scalidophora: Phyla Kinorhyncha, Priapula, and Loricifera 693

Phylum Kinorhyncha: The Kinorhynchs 695
Kinorhynch Classification 696
The Kinorhynch Body Plan 696
Phylum Priapula: The Priapulans 698
Priapulan Classification 699
The Priapulan Body Plan 699
Phylum Loricifera: The Loriciferans 701

CHAPTER 20

The Emergence of the Arthropods: Tardigrades, Onychophorans, and the Arthropod Body Plan 709

Phylum Tardigrada 711
The Tardigrade Body Plan 715
Phylum Onychophora 718
The Onychophoran Body Plan 722
An Introduction to the Arthropods 728
Taxonomic History and Classification 728
The Arthropod Body Plan and Arthropodization 730
The Evolution of Arthropods 751
The Origin of Arthropods 751
Evolution within the Arthropoda 751

CHAPTER 21

Phylum Arthropoda: Crustacea: Crabs, Shrimps, and Their Kin 761

Classification of The Crustacea 764
Synopses of Crustacean Taxa 767
The Crustacean Body Plan 798
Crustacean Phylogeny 831

CHAPTER 22

Phylum Arthropoda: The Hexapoda: Insects and Their Kin 843

Hexapod Classification 847
The Hexapod Body Plan 859
Hexapod Evolution 887

CHAPTER 23

Phylum Arthropoda: The Myriapods: Centipedes, Millipedes, and Their Kin 895

Myriapod Classification 897
The Myriapod Body Plan 899
Myriapod Phylogeny 908

CHAPTER 24

Phylum Arthropoda: The Chelicerata 911

Chelicerate Classification 915
The Euchelicerate Body Plan 927
The Class Pycnogonida 955
The Pycnogonid Body Plan 958
Chelicerate Phylogeny 961

CHAPTER 25

Introduction to the Deuterostomes and the Phylum Echinodermata 967

Phylum Echinodermata 968
Taxonomic History and Classification 969
The Echinoderm Body Plan 975
Echinoderm Phylogeny 1000
First Echinoderms 1000
Modern Echinoderms 1001

CHAPTER 26

Phylum Hemichordata: Acorn Worms and Pterobranchs 1007

Hemichordate Classification 1008
The Hemichordate Body Plan 1009
Enteropneusta (Acorn Worms) 1009
Pterobranchs 1015
Hemichordate Fossil Record and Phylogeny 1018

CHAPTER 27

Phylum Chordata: Cephalochordata and Urochordata 1021

Chordate Classification 1022
Phylum Chordata, Subphylum Cephalochordata: The Lancelets (Amphioxus) 1023
The Cephalochordate Body Plan 1041
Phylum Chordata, Subphylum Urochordata: The Tunicates 1027
The Tunicate Body Plan 1041
Chordate Phylogeny 1041

CHAPTER 28

Perspectives on Invertebrate Phylogeny 1047

Illustration Credits 1053
Index 1061

Preface to the Third Edition

For this edition of *Invertebrates*, Wendy Moore and Stephen M. Shuster joined as co-authors. In addition, 22 other contributing authors graciously agreed to revise selected chapters or chapter sections. And two dozen reviewers were kind enough to critically read various chapters of the book. There is likely no way *Invertebrates,* Third Edition could have the depth and accuracy it has without the help of these wonderful professionals and specialists, and we are deeply indebted to them.

An information explosion has occurred since the Second Edition of this book, especially in the fields of molecular biology and phylogenetics. Just as the Second Edition of *Invertebrates* was going into production, the beginning framework of a new metazoan phylogeny was starting to appear in the scientific literature, although at that time it was based almost entirely on ribosomal gene trees and considerable disagreement existed. In the intervening decade, this new phylogeny was refined—although many details still remain to be worked out. Most importantly, Protostomia and Deuterostomia have been redefined, and the long-standing Articulata group (based upon a hypothesized sister-group relationship between Annelida and the Panarthropoda) has been disarticulated with annelids now being placed among the Spiralia, and arthropods among the Ecdysozoa. The phyla Echiura and Sipuncula have been subsumed into Annelida. The basal, diploblastic phyla have been shuffled about near the base of the Metazoa tree, and they might continue to shuffle about for a bit longer; as this edition goes to press. We predict that, by the Fourth Edition of *Invertebrates*, the phylogenetic positions of all (or at least most) of the metazoan phyla will be stabilized—a lofty goal long sought by zoologists.

As in the Second Edition, important new terms are printed in boldface when first defined (and these are noted in the Index). Specific gene names, like species names, are italicized (though note that names for classes of genes, e.g., Hox and ParaHox genes, are not italicized). We have again included the protists in the book, because instructors teaching Invertebrate Zoology usually cover the "kingdom Protista" and have asked for it. Our knowledge of protistan biology and phylogeny has expanded so much since the Second Edition that the amount of new information, even briefly presented, is substantial. The ICZN (International Code for Zoological Nomenclature) eschews use of diacritical marks in formal taxonomic names, and we follow that recommendation. However, for other terms we generally retain the original spellings and diacritics. So for example, there are archoöphoran and neoöphoran flatworms (the terms describing modes of egg development), but there are also the (now largely abandoned) taxa Archoophora and Neoophora.

Much of the art for this edition has been updated. However, we continue to include diagrams that will be useful to students in the laboratory, including for animal dissections. We also continue to provide rather detailed classifications and taxonomic synopses within each phylum. We don't expect these to read in the same way as the rest of the chapter, but rather to be used as a reference to look up taxonomic names, understand the traits that distinguish groups, or get an overall sense of the scope of the higher taxa in a phylum.

To say this book is a "labor of love" would be an understatement. Without a deep passion for invertebrates, on the part of all the contributors, it would not have been possible. Hopefully this book elevates in its readers their own passion and enthusiasm for that 96% of the animal kingdom that has so successfully flourished without backbones.

R.C.B.
Tucson, Arizona
December 2015

Preface to the Second Edition

It is not now nor will it ever be given to one man to observe all the things recounted in the following pages.

WALDO L. SCHMITT
Crustaceans, 1965

During the revision of *Invertebrates* my brother Gary passed away. For a while the project was stalled. But, buoyed by the support of family, friends, and colleagues, I eventually returned to the task, which, at times, seemed overwhelming. The field of invertebrate biology is so vast, and cuts across so many disciplinary lines, that even in a book of this size it is necessary to generalize about some topics and to slight others. As university instructors, my brother and I realized early on that the teaching of invertebrate zoology should not be compartmentalized. Thus, in planning this book we were concerned about two potential dangers. First, the text might become an encyclopedic list of "facts" about one group after another, the sort of "flash-card" approach that we wanted to avoid. Second, the book might be a rambling series of stories or vignettes about randomly selected animals (or "model organisms") and their ways of life. The first book would be dull, would encourage rote memorization instead of understanding, and might give the misconception that there is little left to discover. The second book might be full of interesting "gee whiz" stuff but would seem disorganized and without continuity or purpose to the serious student. Either approach could fail to present the most important aspects of invertebrates—their phenomenal diversity, their natural history, and their evolutionary relationships. We also held to the belief that what we *know* about these animals is not as important as how we *think* about them. You should be prepared to assimilate much new material, but you should also be prepared for a great deal of uncertainty and mystery, as much remains to be discovered.

To avoid the pitfalls noted above, and to establish threads of continuity in our discussions about invertebrates, we developed our book around two fundamental themes: **unity** and **diversity**. The first theme we approach by way of functional body architecture, or what we call the bauplan concept. The second theme we approach through the principles of phylogenetic biology. Our hope was that weaving the book tightly to these themes would provide a meaningful flow as readers move from one phylum to the next. The first five chapters provide background for these themes and thus provide an important foundation upon which the rest of the book rests. Please read these chapters carefully and refer back to them throughout your study.

The bulk of this book is devoted to a phylum-by-phylum discussion of invertebrates. Fairly detailed classifications or taxonomic synopses for each phylum are included in separate sections of each chapter to serve as references. A consistent organization is maintained throughout each chapter, although we did yield to the important and sometimes different lessons to be learned by investigating the special attributes of each group of animals. In addition, because of their size and diversity, some taxa receive more attention than others—although this does not mean that such groups are more "important" biologically than smaller or more homogeneous ones. (Five chapters are devoted to the arthropods and their kin.) In certain chapters more than one phylum is covered. In some cases the phyla covered are thought to be closely related to one another; in other cases the phyla merely represent a particular grade of complexity and their inclusion in a single chapter facilitates our comparative approach.

Certain aspects of this book have, of course, been influenced by our own biases; this is especially true of the discussions on phylogeny. We use a combination of phylogenetic trees (cladograms) and narrative discussions to talk about animal evolution. Cladograms are used when appropriate, because they provide the least ambiguous statements that can be made about animal relationships. We always knew that some of you, professors and students both, would disagree with our methods and ideas to various degrees—at least we *hoped* that you would. Never placidly accept what you see in a textbook, or anyplace else for that matter, but try to be critical in your reading.

The book's final chapter is a phylogenetic summary of the animal kingdom. It reinforces the point that much remains to be explored and learned about the evolutionary relationships of invertebrates. Like all scientific knowledge, we are dealing here with provisional, transient "truths" that always remain open to challenge and revision. And, of course, scientists disagree. It is this disagreement and the constant challenging of hypotheses that enliven the field and push the frontiers of knowledge forward.

There are a few other things you should know about this book. A brief historical review of the classification of each major group is provided. We felt this material was not only interesting but also served to imbue students with a sense of the dynamic nature of taxonomy and the development of our understanding of each group. Unless otherwise indicated, the Classification section in each chapter deals only with extant taxa. Descriptions of taxa in these annotated classifications are written in somewhat telegraphic style to save space; we never expected these sections to be "read"—they are for reference. Important new words, when first defined, are set in boldface type. These boldfaced terms are also indicated by boldfaced page references in the index; thus the index can also be used as a glossary. We tried hard to be consistent in our usage of zoological terminology, but the existence of similar terms for entirely different structures in certain groups is notoriously troublesome—these are noted in the text.

For this Second Edition of *Invertebrates*, of course, we have tried to be as current as possible with the research literature, but even as this book goes into production important new publications appear daily. It has been estimated that the volume of scientific information is doubling about every 10 years (or faster). A half-million nonclinical biology papers are published annually. As Professor George Bartholomew noted, "If one equates ignorance with the ratio between what one knows and what is available to be known...each biological investigator becomes more ignorant with every passing day." My goal has been to provide sufficient reference material to lead the interested student quickly into the heart of the relevant literature. Most of the references cited in the text will be found at the end of the corresponding chapter. However, to conserve space and eliminate redundancy, in a number of cases (especially in figure citations) references of a general nature may be listed only once, usually in the introductory chapters. You will also notice citations of a fair number of references that are quite old, some from the nineteenth century. These are included not out of whimsy, but because many of these are benchmark research papers or they stand out as some of the best available descriptions for the subject at hand. (It is surprising how many of the illustrations in modern biology texts can be traced back to origins in nineteenth-century publications.) It is distressing to see how commonplace it has become for researchers to ignore the excellent (and important) work of past decades. For example, many phylogenetic research papers completely ignore 150 years of careful embryological research that was published, largely in the German and American literature, in the nineteenth and twentieth centuries. For some scientists, biological research seems to be little more than "sound bites" from the past decade. Sadly, today, this "sound bite research culture" is often imbued in graduate students—a shocking and dangerous trend that encourages dilettantes. To understand animals requires a thorough understanding of their overall biology, and the dedication of a career, not just dabbling.

Since the first edition of this book, there has been an explosion of research in the field of molecular biology. Much of this has been in molecular phylogenetics, but huge strides are also being made in the area of molecular developmental biology. Papers in these fields now appear at such a pace that it is difficult to write about them in a textbook, for fear the ideas will be obsolete in six months. There have been many new phylogenetic hypotheses proposed on the basis of DNA sequence analyses since the fist edition of this book. Many of the molecular phylogenetic trees that were published before 2000 were quirky and troublesome, due to the simple fact that the field is still new and emerging. Because most of these trees are relatively new and still await rigorous testing with independent data, we do not discuss them all. However, we do discuss molecular-based hypotheses that have a growing body of support or have received widespread attention. But, in general we have taken a conservative approach in this regard—we are only just beginning to discover which genes are appropriate for different levels of phylogenetic analysis, and how best to analyze them.

These things being said, I hope you are now ready to forge ahead in your study of invertebrates. The task may at first seem daunting, and rightly so. I hope that this book will make this seemingly overwhelming task a bit more manageable. If I succeed in enhancing your enjoyment and appreciation of invertebrates, then my efforts will have been worthwhile.

R.C.B.
Tucson, Arizona
December 2002

Acknowledgments

This edition of *Invertebrates* has again benefitted greatly from conscientious reviews provided by many specialists, and we extend to these wonderful professionals our utmost gratitude. A very special note of appreciation goes to Gonzalo Giribet, who not only revised several chapters and sections for this edition, but also graciously took the time to review several other chapters; a more informed invertebrate zoologist does not exist today and we deeply appreciate the help he provided. Rebecca Rundell read the entire text—a daunting undertaking—providing critical scientific and instructor-oriented feedback; we were fortunate that she was willing to undertake this formidable task and we thank her profusely! Other colleagues who went above and beyond in providing assistance include Larry Jon Friesen, Jens Høeg, Reinhardt Kristensen, Brian Leander, Sally Leys, Claus Nielsen, and Martin Sørensen, to whom we owe a great debt of gratitude. Larry Friesen's photographic contributions to this edition of *Invertebrates* tremendously enhanced the book.

Invertebrates is in four languages and enjoys a broad readership, especially in Europe and Latin America. Many students have written over the years expressing their support and encouragement and sending photographs or other material, especially from Mexico and South America, but none has been as creative or inspirational as Lorena Viana and her friends from the University of São Paulo. *Muito obrigado, Lorena.*

Most of the original artwork in this text was done from our own sketches or from other sources by the award-winning scientific illustrator Nancy Haver, supported by our publisher, Sinauer Associates. We have been incredibly fortunate to have Nancy working with us on all three editions of *Invertebrates*. The highly talented production staff at Sinauer has always been a joy to work with, and for this edition we once again had the skills of Janice Holabird, Martha Lorantos, David McIntyre, Marie Scavotto, and Chris Small. We are especially grateful to Production Editor Martha Lorantos, who led the team for this edition, doing so with extraordinary technical skill, patience, and good humor. Martha somehow managed to keep this locomotive on the tracks no matter how many twists, turns, and switches it took. Andy Sinauer has been part of this project since the mid-1980s, and he has been unwavering in his wise council, patience, and great insight. We are deeply indebted to Andy for his consistent support and personal interest in this text throughout its history. As a book publisher (and book lover) Andy's dedication to quality and professionalism is unequaled.

Guest Contributors

Jesús Benito, Hemichordata (with Fernando Pardos), Universidad Complutense, Madrid, Spain

C. Sarah Cohen, Urochordata, California State University at San Francisco, California, USA

Gonzalo Giribet, Onychophora, Nemertea, Chelicerata (with Gustavo Hormiga), Annelida: Sipuncula, Metazoan Phylogeny (with Richard C. Brusca), Museum of Comparative Zoology, Harvard University, Cambridge, Massachusetts, USA

Rick Hochberg, Gastrotricha, University of Massachusetts Lowell, Lowell, Massachusetts, USA

Gustavo Hormiga, Chelicerata (with Gonzalo Giribet), The George Washington University, Washington, DC, USA

Reinhardt Møbjerg Kristensen, Tardigrada (with Richard C. Brusca), Loricifera, Micrognathozoa (with Katrine Worsaae), Natural History Museum of Denmark, University of Copenhagen, Copenhagen, Denmark

David Lindberg, Mollusca (with Winston Ponder and Richard C. Brusca), University of California, Berkeley, California, USA

Carsten Lüter, Brachiopoda, Museum für Naturkunde, Berlin, Germany

Joel W. Martin, Crustacea (with Richard C. Brusca), Natural History Museum of Los Angeles County, Los Angeles, California, USA

Alessandro Minelli, Myriapoda, University of Padova, Padova, Italy

Rich Mooi, Echinodermata, California Academy of Sciences, San Francisco, California, USA

Ricardo Cardoso Neves, Cycliophora, Biozentrum, University of Basel, Basel, Switzerland

Claus Nielsen, Entoprocta, Bryozoa, Natural History Museum of Denmark, University of Copenhagen, Copenhagen, Denmark

Fernando Pardos, Hemichordata (with Jesús Benito), Universidad Complutense, Madrid, Spain

Winston Ponder, Mollusca (with David Lindberg and Richard C. Brusca), Australian Museum, Sydney, Australia

Greg Rouse, Annelida: non-Sipuncula, Scripps Institution of Oceanography, University of California, San Diego, California, USA

Scott Santagata, Phoronida, Long Island University, Greenvale, New York, USA

Andreas Schmidt-Rheasa, Nematomorpha, Zoological Museum, University of Hamburg, Germany

George Shinn, Chaetognatha, Truman State University, Kirksville, Missouri, USA

Martin Vinther Sørensen, Kinorhyncha, Priapula, Gnathostomulida, Rotifera, Natural History Museum of Denmark, University of Copenhagen, Copenhagen, Denmark

S. Patricia Stock, Nematoda, University of Arizona, Tucson, Arizona, USA

Katrine Worsaae, Micrognathozoa (with Reinhardt Møbjerg Kristensen), University of Copenhagen, Denmark

Chapter reviewers for the Third Edition include the following:

Nicole Boury-Esnault, Vlaams Instituut voor de Zee, Oostende, Belgium

Jose Luis Carballo, Universidad Nacional Autónoma de México, Estación Mazatlán, México

Allen Collins, Smithsonian Institution, Washington, D.C.

Alexander V. Ereskovsky, French National Center for Scientific Research, Institut Méditerranéen de Biodiversité et d'Ecologie Marine et Continentale (IMBE), Marseille, France.

Daphne G. Fautin, Professor Emerita, University of Kansas, Lawrence, USA

Gonzalo Giribet, Harvard University, Cambridge, Massachusetts, USA

Gordon Hendler, Natural History Museum of Los Angeles County, Los Angeles, USA

Jens Høeg, University of Copenhagen, Denmark

Matthew Hooge, University of Maine, Orono, USA

Michael N. Horst, Mercer University, Georgia, USA

Michelle Kelly, NIWA (National Institute of Water and Atmospheric Research), New Zealand

Kevin Kocot, University of Alabama, Tuscaloosa, USA

Reinhardt Kristensen, University of Copenhagen, Denmark

Christopher Laumer, Harvard University, Cambridge, Massachusetts, USA

Brian Leander, University of British Columbia, Vancouver, Canada

Sally Leys, University of Alberta, Edmonton, Canada

Renata Manconi, Università degli Studi di Sassari (UNISS), Italy

Mark Q. Martindale, Director, Whitney Laboratory and Seahorse Key Marine Laboratory, and Professor of Biology, University of Florida, Gainesville, USA

Rick McCourt, Academy of Natural Sciences, Philadelphia, USA

Catherine S. McFadden, Harvey Mudd College, Claremont, California, USA

Claus Nielsen, University of Copenhagen, Denmark

David Pawson, National Museum of Natural History, Smithsonian Institution, Washington, DC, USA

Hilke Ruhberg, University of Hamburg, Germany

Rebecca Rundell, State University of New York-ESF, Syracuse, New York, USA

Alastair Simpson, Dalhousie University, Halifax, Nova Scotia, Canada

Christiane Todt, University of Bergen, Norway

Jean Vacelet, Université de la Méditerranée Aix-Marseille, France

R. W. M. van Soest, University of Amsterdam, The Netherlands

Media and Supplements to accompany *Invertebrates*, Third Edition

eBook

Invertebrates, Third Edition is available for purchase as an eBook, in several different formats, including VitalSource, Yuzu, BryteWave, and RedShelf. The eBook can be purchased as either a 180-day rental or a permanent (non-expiring) subscription. All major mobile devices are supported. For details on the eBook platforms offered, please visit www.sinauer.com/ebooks.

For the Instructor

(Available to qualified adopters)

Instructor's Resource Library

The *Invertebrates,* Third Edition Instructor's Resource Library includes an extensive collection of visual resources for use in preparing lectures and other course materials. The IRL includes the following:

Textbook Figures and Tables: All of the textbook's figures and tables are included as both high- and low-resolution JPEGs, for easy use in presentation software, learning management systems, and assessments. New for the Third Edition, this now includes all of the textbook's photographs.

Supplemental Photo Collection: Expanded for the Third Edition, this collection of over 1,000 photographs depicts organisms that span the entire range of phyla covered in the textbook.

PowerPoint Presentations: Two ready-to-use PowerPoint presentations are provided for each chapter of the textbook: one that contains all of the textbook figures and tables, and one that contains all of the relevant photos from the supplemental photo collection.

CHAPTER 1 Introduction

The incredible array of extant (= living) invertebrate species on Earth is the outcome of hundreds of millions of years of evolution. Indirect evidence of the first life on Earth, prokaryotic organisms, has been found in some of the oldest sedimentary rocks on the planet, suggesting that life first appeared in Earth's seas almost as soon as the planet cooled enough for it to exist. The Earth is 4.57 billion years old, and the oldest rocks found so far are about 4.3 billion years old. Although the precise date of the first appearance of life on Earth remains debatable, there are tantalizing 3.8-billion-year-old trace fossils from Australia that resemble prokaryotic cells—although these have been challenged, and opinion is now split on whether they are traces of early bacteria or simply mineral deposits. However, good evidence of prokaryotic life has been found in pillow lava that formed on the seabed 3.5 billion years ago, now exposed in South Africa. And 3.4-billion-year-old fossil cells (probably sulfur bacteria) have been found among cemented sand grains on an ancient beach in Australia.[1]

The next big step in biological evolution came about when prokaryotic cells began taking in guests. Around 2 to 2.5 billion years ago, one of these primitive cells took in a free-living bacterium that established permanent residency, giving rise to the organelles we call mitochondria—and this was the origin of the eukaryotic cell. Mitochondria, you will recall, generate energy for their host cells by oxidizing sugars, and in this case they also equipped early life to survive in Earth's increasing oxygen levels. Evidence suggests that mitochondria evolved just once, from a

[1]There are three popular theories on how life first evolved on Earth. The classic "primeval soup" theory, dating from Stanley Miller's work in the 1950s, proposes that self-replicating organic molecules first appeared in Earth's early atmosphere and were deposited by rainfall into the ocean, where they reacted further to make nucleic acids, proteins, and other molecules of life. More recently, the idea of the first synthesis of biological molecules by chemical and thermal activity at deep-sea hydrothermal vents has been suggested. Hydrothermal vents also spew out compounds that could have been incorporated into the first life forms. The third proposal is that organic molecules, or even prokaryotic life itself, first arrived on Earth from another planet (recently Mars has been at the forefront), or from deep space, on comets or meteorites. Meteorites that fall to Earth contain amino acids and organic carbon molecules such as formaldehyde. Clearly, raw materials were not the issue—the trick was assembling the organic compounds to create a living, reproducing system.

TABLE 1.1 Numbers of Described Living Species in the 32 Animal Phyla (~1,382,402 total; ~1,324,402, or 96% are invertebrates)[a]

Taxon	Number of Described Species	Percent of Total Described Animal Species
Phylum Porifera	9000	0.65
Phylum Placozoa	1 (or 2)	0.0001
Phylum Cnidaria	13,200	0.95
Phylum Ctenophora	100	0.007
Phylum Xenacoelomorpha	400	0.03
Phylum Platyhelminthes	26,500	1.92
Phylum Chaetognatha	130	0.009
Phylum Gastrotricha	800	0.06
Phylum Rhombozoa (= Dicyemida)	70	0.005
Phylum Orthonectida	21	0.002
Phylum Nemertea	1300	0.09
Phylum Mollusca	80,000	5.79
Phylum Annelida	20,000	1.45
Phylum Entoprocta	200	0.014
Phylum Cycliophora	2	0.0001
Phylum Gnathostomulida	100	0.007
Phylum Micrognathozoa	1	0.0001
Phylum Rotifera	2000	0.14
Phylum Phoronida	11	0.0008
Phylum Bryozoa (= Ectoprocta)	6000	0.43
Phylum Brachiopoda	400	0.03
Phylum Nematoda (= Nemata)	25,000	1.81
Phylum Nematomorpha	360	0.03
Phylum Kinorhyncha	200	0.014
Phylum Priapula	20	0.001
Phylum Loricifera	35	0.003
Phylum Tardigrada	1200	0.087
Phylum Onychophora	200	0.014
Phylum Arthropoda		
Subphylum Crustacea	70,000	5.06
Phylum Arthropoda		
Subphylum Hexapoda	926,990	67.06
Phylum Arthropoda		
Subphylum Chelicerata	113,335	8.20
Phylum Arthropoda		
Subphylum Myriapoda	16,360	1.18
Phylum Arthropoda TOTAL	1,126,685	81.48

symbiotic α-proteobacterium, and then subsequently diversified broadly. Modern free-living relatives of this bacterium harbor about 2,000 genes across several million bases, but their mitochondrial descendants have far fewer, sometimes as few as three genes. And human mitochondrial DNA harbors only about 16,000 bases. On the other hand, some plants have greatly expanded their mitochondrial genome, the largest so far discovered in the genus *Silene*, with around 11 million bases. Another prokaryotic intracellular guest, a cyanobacterium, became the ancestor of chloroplasts through the same symbiogenic process; chloroplasts, of course, are the photosynthesizing organelles that made plants and algae possible. In some plant and algae lines, the original chloroplast was lost, and a new one was picked up when a host cell took in an alga and co-opted its chloroplast in another kind of symbiogenic event (see Chapter 3 for a detailed discussion of these symbiogenic events).

Controversial hydrocarbon biomarkers suggest that the first eukaryotic cells might have appeared as early as 2.7 billion years ago (late Archean), although the earliest fossils that have been proposed to be eukaryotes—based on cell surface features and their large size—are about 1.8 billion years old (early Proterozoic). Multicelled algae (protists) date as far back as 1.2 billion years. Eukaryote-like microfossils have been described from 1-billion-year-old freshwater deposits, suggesting that the eukaryotes might have left the sea and invaded the terrestrial realm long ago. Even though the eukaryotic condition appeared early in Earth's history, it took a few hundred million more years for multicellular organisms to first evolve.

Disputed trace fossils have led some to suggest that the earliest

TABLE 1.1 (Continued)

Taxon	Number of Described Species	Percent of Total Described Animal Species
Phylum Echinodermata	7300	0.53
Phylum Hemichordata	135	0.01
Phylum Chordata		
Subphylum Cephalochordata	30	0.002
Phylum Chordata		
Subphylum Urochordata	3000	0.22
Phylum Chordata		
Subphylum Vertebrata	58,000	4.2
Phylum Chordata TOTAL	**61,030**	**4.41**

[a]Estimated numbers of described (living) species of Prokaryota = 10,300; Protista = 200,000; Plantae = 315,000; and Fungi = 100,000.

animals (Metazoa) might have originated a billion years ago, but recent molecular clock estimates put the origin of Metazoa at 875 to 650 million years ago. The oldest generally accepted metazoan fossils are from the Ediacaran period, found in the Fermeuse Formation of Newfoundland (~560 Ma) and the Doushantuo Formation of southern China (600–580 Ma). A 560-million-year-old likely cnidarian (named *Haootia quadriformis*) has been described from Newfoundland, with quadraradial symmetry and clearly preserved bundled muscular fibers. *Haootia* appears to be a nearly 6-cm-long polyp, or perhaps an attached medusa—it resembles modern species of Staurozoa. Cnidarians and other apparent diploblastic animals have been reported from the Doushantuo deposits, although these have been met with skepticism in some quarters. However, in 2015, a seemingly reliable 600-million-year-old fossil sponge (*Eocyathispongia qiania*) was described from the Doushantuo Formation. In 2009, Jun-Yuan Chen and colleagues reported on embryos of reputed bilaterians (triploblasts) in the Doushantuo deposits (dated 600–580 Ma)—32-cell stage embryos with micromeres and macromeres, apparent anterior-posterior and dorsoventral patterning, and ectoderm-like cells around part of their periphery. This finding was challenged, and the fossils were variously declared as prokaryotes or protists by other workers. However, further discoveries of additional embryos seemed to support the view of these being bilaterian embryos and, in some cases, perhaps diapause embryos ("resting eggs") of bilaterians. Good trace fossils (tracks) of a minute wormlike bilaterian animal, possibly with legs, have also been described from 585-million-year-old rocks in Uruguay. These fossil records put the appearance of "higher metazoans" (i.e., bilaterians) millions of years before the beginning of the Cambrian period.

There is no argument that Metazoa are monophyletic (i.e., a clade), and the animal kingdom is defined by numerous synapomorphies, including: gastrulation and embryonic germ layer formation; unique modes of oogenesis and spermatogenesis; a unique sperm structure; mitochondrial gene reduction; epidermal epithelia with septate junctions, tight junctions, or zonula adherens; striate myofibrils; actin-myosin contractile elements; type IV collagen; and the presence of a basal lamina/basement membrane beneath epidermal layers (of course, some of these features have been secondarily lost in some groups). Evidence is strong that Metazoa arose out of the protist group Choanoflagellata, or a common ancestor, and the two comprise sister groups in almost all recent analyses. They are, in turn, part of a larger clade known as Opisthokonta that also includes the fungi and several small protist groups (see Chapter 3).

The three great lineages of life on Earth—Bacteria, Archaea, Eukaryota—are very different from one another. Bacteria and Archaea have their DNA dispersed throughout the cell, whereas in Eukaryota the DNA is enclosed within a membrane-bound nucleus. The cell lineages that gave rise to the Eukaryota are still unknown. The many millennia between the origin of Eukaryota and the explosive radiation that apparently began in the Ediacaran is sometimes called the "boring billion years," but the fossil record is fairly sparse for that time period, so we're not sure how "boring" it actually was. A popular hypothesis suggests that oxygen levels were too low during that time for larger organisms to evolve (see below).

Today, there are an estimated 2,007,702 described and named living species. About 58,000 of these are vertebrates and 1,324,402 are invertebrates (Table 1.1). In addition, about 200,000 protists have been described, 315,000 plants (290,000 seed plants), and 100,000 fungi. And 15,000 to 20,000 new species are described every year. It seems likely that a significant portion of Earth's biodiversity, at the level of both genes and species, resides in the "invisible" prokaryotic world, and we have come to realize how little we know about this hidden world. About 10,300 species of prokaryotes have been described, but there are an estimated 10 million (or perhaps more) undescribed prokaryote species on Earth. An estimated 135,000 more plant species remain to be described. Overall, estimates of undescribed eukaryotes range from lows of 3–8 million to highs of 100 million or more.

Keeping Track of Life

For a gentleman should know something of invertebrate zoology, call it culture or what you will, just as he ought to know something about painting and music and the weeds in his garden.

Martin Wells
Lower Animals, 1968

How can we possibly keep track of all these species names and information about each of them, and how do we organize them in a meaningful way? We do so with classifications. **Classifications** are lists of species, ranked in a subordinated fashion that reflects their evolutionary relationships and phylogenetic history. Classifications summarize the overarching aspects of the tree of life. At the highest level of classification, we can recognize two superkingdoms: Prokaryota (containing the kingdoms Archaea and Bacteria) and Eukaryota (containing the kingdoms Protista, Fungi, Plantae, and Animalia/Metazoa). Because "Protista" is not a monophyletic group, the protists are sometimes broken up into several kingdoms, or other classificatory ranks, but the relationships among the protists are still being debated (see Chapter 3).

One of the earliest and best-known evolutionary trees of life published from a Darwinian (genealogical) perspective was by Ernst Haeckel in 1866 (Figure 1.1). Haeckel coined the term "phylogeny," and his famous trees codified what became a tradition of depicting phylogenetic hypotheses as branching diagrams, a tradition that has persisted since that time. However, a hand-drawn sketch in Charles Darwin's field notebook (1837) clearly depicts his view of South American mammal evolution in a branching tree of extant and fossil species. And in his book, *On the Origin of Species* (1859), Darwin presented an abstract branching diagram of a theoretical tree of species as a way of illustrating his concept of descent with modification. Jean Baptiste Lamarck probably presented the first historical trees of animals in his *Philosophie Zoologique* in 1809, and the French botanist Augustin Augier published a tree showing the relationships among plants in 1801 (perhaps the first tree ever published)—although both Lamarck's and Augier's trees were produced before the modern concept of evolution had been clearly articulated. We discuss various ways in which phylogenetic trees are developed in Chapter 2.

Since Haeckel's day, many names have been coined for the branches that sprout from these trees, and in recent years a glut of new names has been introduced to label various molecularly based clades nested within the tree of life. We will not burden you with all of these names, but a few of them need to be defined here before we launch into our study of the invertebrates. Some of these names refer to groups of organisms that are thought to be natural phylogenetic lineages (i.e., groups that include all the descendants of a stem species, known as **monophyletic** groups, or **clades**). Examples of such natural, or monophyletic groups are the superkingdom Eukaryota, kingdom Metazoa (the animals), and kingdom Plantae (the lower and higher plants).[2] All three of these large groupings are thought to have had a single origin, and they each include all of the species descended from that original ancestor. Monophyletic groups comprise a cluster of terminal branches (with a single origin) embedded within a much larger tree. Some other named groups are natural, having a single evolutionary origin, but the group *does not contain all of the members* of the lineage. Such groups are said to be **paraphyletic**, and they are often the basal or ancestral lineages in a much larger clade. Paraphyletic groups comprise some, but not all descendants of a stem species. The Protista are paraphyletic because the grouping excludes three large multicelled lineages that evolved out of it (e.g., Metazoa, Plantae, Fungi). Another well-known paraphyletic group is Crustacea (which excludes the Hexapoda/Insecta, a clade that evolved out of it). The clade that includes both Crustacea and Hexapoda is known as Pancrustacea. Classifications of life are derived from evolutionary or phylogenetic trees, and thus generally include only monophyletic groups. However, sometimes paraphyletic taxa are also used because, if they are unambiguous, they can be important in facilitating meaningful communication among scientists and between the scientific community and society (e.g., Protista and Crustacea).

Some names refer to unnatural, or composite, groupings of organisms, such as "microbes" (i.e., any organism that is microscopic in size, such as bacteria, archaeans, yeasts, unicellular fungi, and some protists). These unnatural groups are **polyphyletic**. For example, yeasts are unicellular fungi that evolved several times independently from multicellular filamentous ancestors; today they are assigned to one of three higher fungal phyla, so the concept of "yeast" represents a polyphyletic, or unnatural grouping. The name "slugs"

[2]For decades, taxonomists have debated the boundary between protists and Plantae. We accept the view that it should be placed just prior to the evolutionary origin of chloroplasts and that Plantae should comprise all eukaryotes with plastids directly descending from the initially enslaved cyanobacterium, i.e., Rhodophyta (red algae), Glaucophyta (glaucophyte algae), and Viridiplantae ("green plants"), but exclude those like chromists that obtained their chloroplasts from plants secondarily by subsequent eukaryote-to-eukaryote lateral transfers. The structure of plastid genomes and the derived chloroplast protein-import machinery support a single origin of these closely related groups. Thus, Plantae is a monophyletic clade containing two subkingdoms, Biliphyta (phyla Glaucophyta and Rhodophyta) and Viridiplantae (the phyla Chlorophyta, Charophyta, Anthocerotophyta, Bryophyta, Marchantiophyta, and Tracheophyta). In the past, some workers have restricted Plantae to land plants (embryophytes, or higher plants) and included the other Viridiplantae with the Protista in a larger group called Protoctista (which also included the lower fungi).

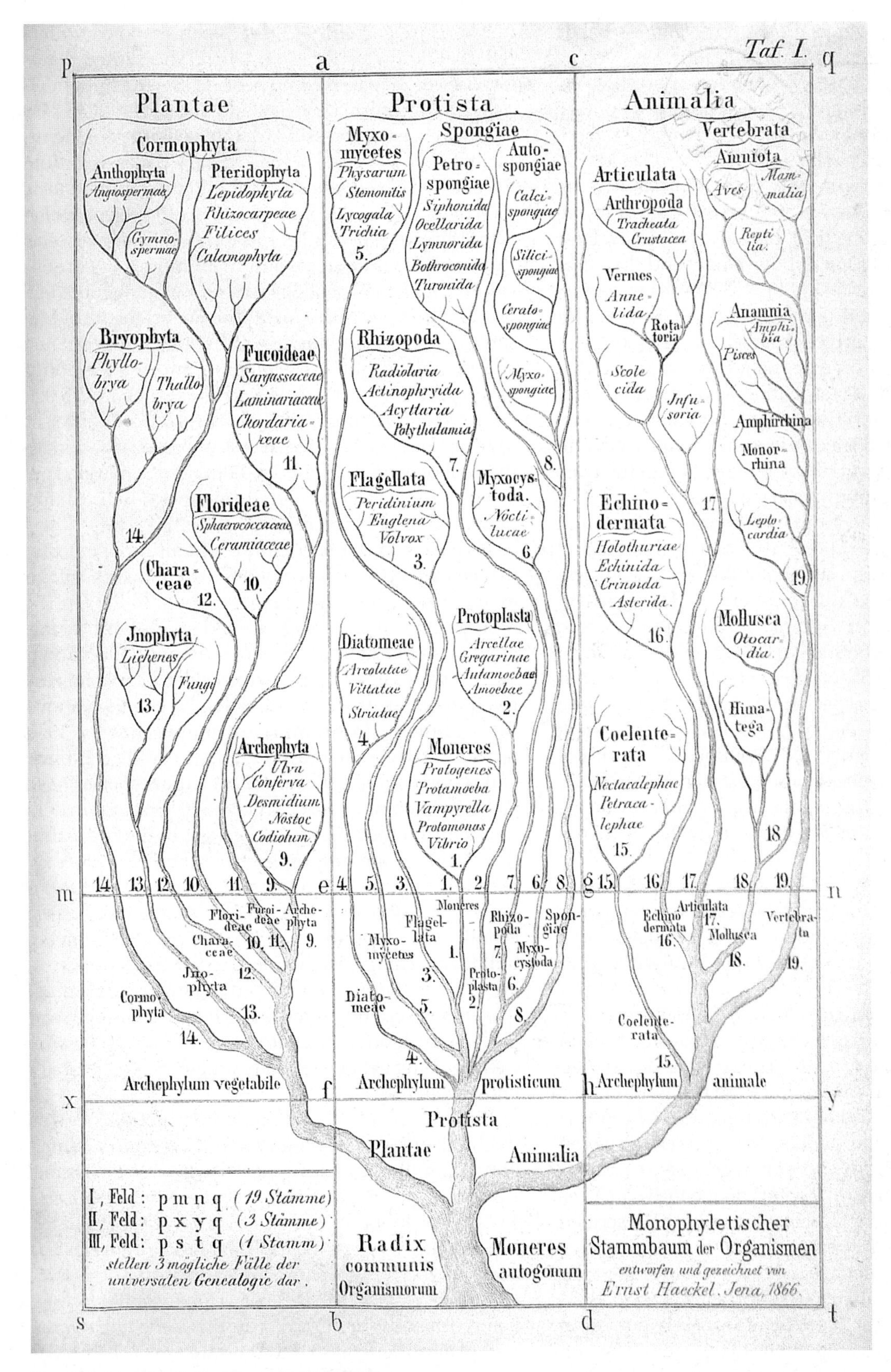

Figure 1.1 Haeckel's Tree of Life (1866).

(or "sea slugs") also refers to a group of animals that do not share a single ancestry (the slug form has evolved many times among gastropod molluscs), so slugs are polyphyletic (see chapter opener photo of a Spanish shawl slug, *Flabellina iodinea*). We explore these concepts more fully in Chapter 2.

We know approximately how many genes are in organisms from yeast (about 6,000 genes) to humans (about 25,000 genes), but we don't know how many living species inhabit our planet and the range in estimates is surprisingly broad. How many undescribed species are lingering out there, waiting for names? However derived, predictions of global species diversity rely on extrapolations from existing real data. Methods of estimation have included rates of past species descriptions, expert opinion, the fraction of undescribed species in samples collected, and ratios between taxa in the taxonomic hierarchy. Each method has its limitations. Two recent estimates of undescribed marine animals (Mora et al. 2011 and Appeltans et al. 2012) reached the conclusion that 91% vs. 33–67% (respectively) of the world's eukaryotic marine fauna is still undescribed. More recently, a large research program sampling the Western Australian upper continental slope for Crustacea and Polychaeta found 95% of the species to be undescribed (with the rate of new species obtained by the sampling program not even leveling off). Given the vast extent of the poorly-sampled world's continental slopes, not to mention the deep sea, rainforests, and other little-sampled habitats, these data suggest that estimates of over 90% undescribed eukaryotes on Earth are not unreasonable.

Our great uncertainty about how many species of living organisms exist on Earth is unsettling and speaks to the issue of priorities and funding in biology. At our current rate of species descriptions, it might take us 10,000 years or more to describe just the rest of Earth's eukaryotic life forms. Not all of the species remaining to be described are invertebrates—between 1990 and 2002 alone, 38 new primate species were discovered and named. And if prokaryotes are thrown into this mix, the numbers become even larger. Recent gene-sequence surveys of the world's oceans (based largely on DNA "barcodes"—16S ribosomal gene sequences for Bacteria and Archaea, 18S ribosomal gene sequences for eukaryotes) have revealed a massive undescribed biota of microbes in the sea. Similar discoveries have been made with genetic searches for soil microbes. For example, there are about 30,000 formally named bacterial varieties that are in pure culture, but estimates of undescribed species range from 10 million to a billion or more! And we now know that thousands of bacterial species inhabit the human body, almost all of which are not yet even named and described. Viruses still lack a universal molecular identifier, and the world scope of viral biodiversity is essentially unknown.

There are currently several attempts to compile a list of all known species on Earth. The United States Geological Service (USGS) hosts ITIS—the Integrated Taxonomic Information System. The goal of ITIS is to create an easily accessible database with reliable information on species names and their classification. Recently, ITIS and several other initiatives turned their data over to the Catalogue of Life (CoL) project, which is building the species list (up to 1.5 million species as this book went to press) and maintaining a "consensus classification" of all life (see www.catalogueof life.org, and Ruggiero et al. 2015). The Encyclopedia of Life (EOL) project is building a website that offers not just species names, but also ecological information about each species; it currently contains more than 175,000 vetted species pages. WoRMS (The World Register of Marine Species), an open-access online database with the goal of listing all described eukaryotic marine species, predicts the completed inventory will catalog 222,000 to 230,000 species.

However, at our current rate of anthropogenically driven extinction a majority of Earth's species will go extinct before they are ever described. In the United States alone, at least 5,000 named species are threatened with extinction, and an estimated 500 known species have already gone extinct since people first arrived in North America. Globally, the United Nations Environment Programme estimates that by 2030 nearly 25% of the world's mammals could go extinct, and recent counts indicate 322 vertebrate species have already become extinct since 1500. Some workers now refer to the time since the start of the Industrial Revolution as the Anthropocene—a period marked by humanity's profound global transformation of the environment. More than half of Earth's terrestrial surface is now plowed, pastured, fertilized, irrigated, drained, bulldozed, compacted, eroded, reconstructed, mined, logged, or otherwise converted to new uses. Human-driven deforestation removes 15 billion trees per year. E. O. Wilson once estimated that about 25,000 species are going extinct annually on Earth (we just don't know what they are!).

Pimm et al. (2014) calculated extinction rates as fractions of species going extinct over time—extinctions per million species-years (E/MSY). For example, 1,230 bird species have been described since 1900, and 13 of these are now extinct. This cohort accumulated 98,334 species-years, meaning an average species has been known for 80 years. The extinction rate is thus $13/98{,}334 \times 10^6 = 132$ E/MSY. They calculated that, before *Homo sapiens* arrived on the scene, the overall background animal extinction rate was 0.1 E/MSY; today it is about 100 E/MSY (a thousand times higher). Proportionally speaking, larger animals (e.g., vertebrates), higher in the food chain and fewer in number, are more likely to go extinct than invertebrates. This means invertebrates (and protists), which already run the world, will play even more important ecosystem roles in the future.

Even though invertebrates make up 96% of the described animal kingdom (Table 1.1), they account for a mere 38% of the 500 or so species now under protection by the U.S. Endangered Species Act. NatureServe has argued that more than 1,800 invertebrate species need protection, while the IUCN Red List of Threatened Species documents the extinction risk of nearly 50,000 species of animals and plants. In 2002, the U.N. Convention on Biological Diversity committed nations to significantly reduce rates of biodiversity loss by 2010, and in 2010 this call was renewed with a set of specific targets for 2020. However, several recent studies have shown that the convention has so far failed and rates of biodiversity loss do not appear to be slowing at all. Further, the Convention has crippled scientific fieldwork around the world by instilling in many nations a fear of outside researchers "stealing their genetic biodiversity."

The single greatest threat to species survival for the past 200 years has been habitat loss. Although we hear mostly about deforestation, 30 to 50% of the Earth's coastal environments have been degraded during past decades, at rates exceeding those of tropical forest loss. Looking forward, the damaging effects of habitat loss will likely be matched (and escalated) by anthropogenically-driven global climate change. The concentration of carbon dioxide (CO_2) in Earth's atmosphere has risen by about 38% since the start of the industrial era as a result of fossil fuel burning and land use change, and a quarter to a third of all the CO_2 emitted through human activities has been absorbed by the ocean, resulting in acidification of surface waters. In fact, the oceans overall are now about 30% more acidic than they were 100 years ago. The drop in ocean pH is creating hardships on animals with calcium carbonate skeletons, and damage has been documented in everything from corals to sea butterflies (pteropod molluscs). In May 2013, the concentration of CO_2 in the atmosphere reached 400 ppm, the highest it has been over the past 2 million years. Never before has such a rapid global-scale increase in CO_2 (and temperature) occurred during the history of human civilization. Rising concentrations of greenhouse gasses in the atmosphere are leading to increasing global temperatures and changes in precipitation regimes, and these changes are impacting the distribution of biota across the planet. Globally, average air temperatures have risen about 0.8°C since 1880, mean land surface temperature has warmed 0.27°C per decade since 1979, and projections from global climate models predict global atmospheric temperatures to increase by about 4°C by the end of this century. Melting polar ice and glaciers, combined with expansion of warming ocean waters, are driving up sea level, which is expected to be as much as four feet higher by the end of this century.

Prokaryotes and Eukaryotes

The discovery that organisms with a cell nucleus constitute a natural (monophyletic) group divided the living world neatly into two categories, the **prokaryotes** (Archaea and Bacteria: those organisms lacking membrane-enclosed organelles and a nucleus, and without linear chromosomes), and the **eukaryotes** (those organisms that do possess membrane-bound organelles, a nucleus, and linear chromosomes). Investigations by Carl Woese and others, beginning in the 1970s, led to the discovery that the prokaryotes themselves comprise two distinct groups, called **Bacteria** (= Eubacteria) and **Archaea** (= Archaebacteria), both quite distinct from eukaryotes (Box 1A). Bacteria correspond more or less to our traditional understanding of bacteria. Archaea strongly resemble Bacteria, but they have genetic and metabolic characteristics that make them quite unique. For example, Archaea differ from both Bacteria and Eukaryota in the composition of their ribosomes, in the construction of their cell walls, and in the kinds of lipids in their cell membranes. Some Bacteria conduct chlorophyll-based photosynthesis, a trait that is never present in Archaea (photosynthesis is the harvesting of light to produce energy/sugars and oxygen). Current thinking favors the view that prokaryotes ruled Earth for about a billion years before the eukaryotic cell appeared.

As the prokaryotes evolved, they adapted to colonize every conceivable environment on Earth. During the early evolution of prokaryotes, Earth's air had almost no oxygen, consisting primarily of CO_2, methane, and nitrogen. The metabolism of the earliest prokaryotes relied on hydrogen, methane, and sulfur, and did not produce oxygen as a byproduct. It was the appearance of the first oceanic photosynthesizing prokaryotes that led to increased atmospheric oxygen concentrations, setting the stage for the evolution of complex multicellular life. And aquatic species were probably able to colonize land only because the oxygen helped create the ozone layer that shields against the sun's ultraviolet radiation. Just *when* oxygen-producing photosynthesis began is still being debated, but when it happened, most of the early chemoautotrophic prokaryotes were likely poisoned by the "new gas" in the environment. A large body of evidence points to a sharp rise in the concentration of atmospheric oxygen between 2.45 and 2.32 billion years ago (this is sometimes called the "great oxidation event"), around the same time the eukaryotic cell first appeared. This evidence includes red beds or layers tinged by oxidized iron (i.e., rust) and oil biomarkers that may be the remains of Cyanobacteria (true Bacteria). However, in western Australia, thick shale deposits that are 3.2 billion years old have bacterial remains that hint at oxygen-producing photosynthesis. These ancient oxygen levels might have reached around 40% of present atmospheric levels. There is

BOX 1A The Six Kingdoms of Life*

PROKARYOTES (Superkingdom Prokaryota)*

Kingdom Bacteria (= Eubacteria)
The "true" bacteria, including Cyanobacteria (blue-green algae), Proteobacteria, Spirochaetae, and numerous other phyla; never with membrane-enclosed organelles or nuclei, or a cytoskeleton; none are methanogens; some use chlorophyll-based photosynthesis; with peptidoglycan in cell wall; with a single known RNA polymerase

Kingdom Archaea (= Archaebacteria)
Anaerobic or aerobic, largely methane-producing microorganisms; never with membrane-enclosed organelles or nuclei, or a cytoskeleton; none use chlorophyll-based photosynthesis; without peptidoglycan in cell wall; with several RNA polymerases.

EUKARYOTES (Superkingdom Eukaryota)

Cells with a variety of membrane-enclosed organelles (e.g., mitochondria, lysosomes, peroxisomes) and with a membrane-enclosed nucleus. Cells gain structural support from an internal network of fibrous proteins called a cytoskeleton.

Kingdom Protista
Largely unicellular eukaryotes that do not undergo tissue formation through the process of embryological layering. A paraphyletic grouping of many phyla, including euglenids, diatoms and some other brown algae, ciliates, dinoflagellates, foraminiferans, amoebae, and others (Chapter 3). Protists descended from bacteria by the acquisition of a nucleus, endomembrane, cytoskeleton, and mitochondria. The 200,000 or so described living species probably represent about 10% of the actual protist diversity on Earth today.

Kingdom Fungi
The fungi. Probably a monophyletic group that includes molds, mushrooms, yeasts, and others. Saprobic, heterotrophic, multicellular organisms. We view the demarcation between Protists and Fungi as lying immediately before the origin of the chitinous wall around vegetative fungal cells and the associated loss of phagotrophy. The earliest fossil records of fungi are from the Middle Ordovician, about 460 million years ago. The 100,000 described species are thought to represent only a small percent of the actual diversity, and estimates of total fungal biodiversity range from 3 to 10 million.

Kingdom Plantae (= Archaeplastida)
Unicellular and multicellular, photosynthetic, chlorophyll-bearing eukaryotes with plastids directly descended from an initially enslaved cyanobacterium. Includes Glaucophyta (glaucophyte algae), Rhodophyta (red algae), and Viridiplantae (green plants). Viridiplantae includes the phyla Chlorophyta (green algae), Charophyta (freshwater green algae), Anthocerotophyta (hornworts and their kin), Marchantiophyta (liverworts), Bryophyta (other non-vascular plants), and Tracheophyta (the vascular plants, about 260,000 of which are flowering plants). Tracheophytes develop through embryonic tissue layering, in a manner analogous to animal embryogeny. The described species of Plantae are thought to represent about half of Earth's actual plant diversity.

Kingdom Animalia (= Metazoa)
The multicellular animals. A monophyletic taxon, containing 32 phyla of ingestive, heterotrophic, multicellular organisms that develop by tissue layering during embryogenesis. About 1,382,402 species of living metazoans have been described; estimates of the number of undescribed animal species range from lows of 3–8 million to highs of over 100 million.

*Portions of the old "Kingdom Monera" are now included in the Bacteria (Eubacteria) and the Archaea. Viruses (about 5,000 described "species") and subviral parasites such as viroids and prions, which are all regarded as laterally transmissible parasitic genetic elements, are not included in this classification. Prions are infectious proteins, devoid of a nucleic acid genome but subject to mutation and thus evolution. Viruses comprise an ancient, polyphyletic group of stripped-down parasitic genetic fragments. Recent work suggests that they might have played important roles in major evolutionary transitions, such as the invention of DNA and DNA replication mechanisms, the formation of the major "superkingdoms" of life, and perhaps even the origin of the eukaryotic nucleus. Viral classification is standardized by the ICTV (International Committee on Taxonomy of Viruses).

also evidence in the geological record that atmospheric oxygen did not steadily increase, but fluctuated wildly, dropping at times to a mere 0.1% of current levels. It may not have been until around 800 million years ago that high oxygen levels stabilized.

Terrestrial photosynthesis has little effect on atmospheric O_2 because it is nearly balanced by the reverse processes of respiration and decay. By contrast, marine photosynthesis is a net source of O_2 because a small fraction (~0.1%) of the organic matter synthesized in the oceans is buried in sediments. It is this small "leak" in the marine organic carbon cycle is responsible for most of our atmospheric O_2. Cyanobacteria are thought to have been largely responsible for the initial rise of atmospheric O_2 on Earth, and even today *Prochlorococcus* can be the numerically dominant phytoplankter in tropical and subtropical oceans, accounting for 20 to 48% of the photosynthetic biomass and production in some regions. Today most marine photosynthesis is performed by Cyanobacteria and single-celled protists, such as diatoms and coccolithophores. Cyanobacteria are nearly unique among the prokaryotes in performing oxygenic photosynthesis, often together with nitrogen fixation, and thus they are major primary producers in both marine and terrestrial ecosystems.

Many Archaea live in extreme environments, and this pattern is often interpreted as a refugial lifestyle—in other words, such creatures tend to live in places where they have been able to survive without confronting dangerous environments or competition from more highly derived life forms. Many of these "extremophiles" are anaerobic chemoautotrophs, and they have been found in a variety of habitats, such as deep-sea hydrothermal vents, benthic marine cold seeps, hot springs, saline lakes, sewage treatment ponds, in subglacial lakes beneath the ice of Antarctica (where the water is liquid due to a combination of geothermal heating and pressure), and in the guts of humans and other animals. One of the most astonishing discoveries of the 1980s was that extremophile Archaea (and some fungi) are widespread in the deep rocks of Earth's crust. Since then, a community of hydrogen-eating Archaea has been found living in a geothermal hot spring in Idaho, 600 feet beneath Earth's surface, relying on neither sunshine nor organic carbon. Archaea are known to grow at temperatures up to 121°C in deep-sea hydrothermal vents, and one species has been shown to fix nitrogen at temperatures up to 92°C. Archaea have been found at depths as great as 2.8 km, living in igneous rocks with temperatures as high as 75°C. Recently, a diverse microbial biota (Bacteria, Archaea, and even Eukaryotes) has been found to live as deep as 2.5 km beneath the seafloor, in ancient buried sediments, and they are remarkably abundant in coal-bed layers. And in the early part of this century, using new high-throughput sequencing technologies, it was discovered that there are massively more species of prokaryotes in the environment than had been thought. Even deep within sediments, at least 800 m below the seafloor, huge populations of prokaryotes have been discovered.

We are now realizing that the 10,000 or so described species of Bacteria and Archaea are barely the tip of the iceberg. Extremophiles include **halophiles** (which grow in the presence of high salt concentrations, in some cases as high as 35% salt), **thermophiles** and **psychrophiles** (which live at very high or very low temperatures), **acidiphiles** and **alkaliphiles** (which are optimally adapted to acidic or alkaline pH environments), and **barophiles** (which grow best under pressure). Molecular phylogenetic studies now suggest that some of these extremophiles, particularly the thermophiles, may be very similar to the "universal ancestor" of all life on Earth.[3]

Courses and texts on invertebrates often include discussions of two eukaryotic kingdoms: the Animalia (= Metazoa) and certain protist phyla. Following this tradition, we treat 32 phyla of Metazoa and 17 phyla of protists in this book. In today's world of diminishing "-ology" courses many students will find this to be their only detailed exposure to this very important group of organisms. The vast majority of kinds (species) of living organisms that have been described are animals. The kingdom Animalia, or **Metazoa**, is usually defined as the ingestive, heterotrophic, sexual, multicellular eukaryotes that undergo embryonic tissue formation.[4] The process of embryonic tissue formation takes place through a major reorganization and differentiation process called **gastrulation**. However, its members possess other unique attributes as well, such as an acetylcholine/cholinesterase-based nervous system, special types of cell–cell junctions, and a unique family of connective tissue proteins called collagens. Metazoans have also been shown to possess a distinctive set of insertions and deletions in genes coding for proteins in the mitochondrial genome. Among the Metazoa are some species that possess a backbone (or vertebral column), but most do not. Those that possess a backbone constitute the subphylum Vertebrata of the phylum Chordata, and account for only about 4% (about 58,000 species) of all described animals. Those that do not possess a backbone (the remainder of the phylum Chordata, plus 31 additional animal phyla) constitute the invertebrates. Thus we can see that the division of animals into invertebrates and vertebrates is based more on tradition and convenience, reflecting a dichotomy of interests among zoologists, than it is on the recognition of natural biological groupings. About 10,000 to 12,000 new species are named and described by biologists each year, most of them invertebrates (mainly insects).

Not only are invertebrates diverse and numerous, they span over 6 orders of magnitude in size. Many species are microscopic (even though they may have thousands of cells in their bodies). Some of the smallest are cycliophorans (350 µm, or .35 mm), loriciferans (as small as 85 µm, or .085 mm), and the coral reef nematode *Greeffiella minutum*, that is only 80 µm (.08 mm) long. However, species with bodies smaller than one millimeter occur in many other phyla. Invertebrates can also be quite large. Some jellyfish reach 25 m including the tentacles, the giant squid (*Architeuthis dux*) reaches 13 m including the tentacles, a sperm whale nematode parasite (*Placentonema gigantisima*) can exceed 8 m, and some earthworms can reach 3 m in length. There is a record of a nemertean reaching 60 m, but this lacks reliable verification. Giant clams (*Tridacna*) can exceed 400

[3]One of the most striking examples of a thermophile is *Pyrolobus fumarii*, a chemolithotrophic archaean that lives in oceanic hydrothermal vents at temperatures of 90°–113°C. (Chemolithotrophs are organisms that use inorganic compounds as energy sources.) On the other hand, *Polaromonas vacuolata* grows optimally at 4°C. *Picrophilus oshimae* is an acidiphile whose growth optimum is pH 0.7 (*P. oshimae* is also a thermophile, preferring temperatures of 60°C). The alkaliphile *Natronobacterium gregoryi* lives in soda lakes where the pH can rise as high as 12. Halophilic microorganisms abound in hypersaline lakes such as the Dead Sea, Great Salt Lake, and solar salt evaporation ponds, and the green alga *Dunaliella salina* lives in the Dead Sea at salinities of 23% salt. Such lakes are often colored red by dense microbial communities (e.g., *Halobacterium*). *Halobacterium salinarum* lives in the saltpans of San Francisco Bay and colors them red. Barophiles have been found living at all depths in the sea, and one unnamed species from the Mariana Trench (the deepest part of the ocean) has been shown to require at least 500 atmospheres of pressure in order to grow.

[4]Heterotrophic organisms are those that derive their nutritional requirements from complex organic substances (i.e., by consuming other organisms or organic materials as food). In contrast, autotrophs are able to form nutritional organic substances from simple inorganic matter such as carbon dioxide (e.g., plants).

kg in weight, and the terrestrial coconut crab (*Birgus latro*) can exceed 4 kg—perhaps the heaviest land invertebrate.

Where Did Invertebrates Come From?

Evolutionary analyses confirm that the ancestors of plants and animals (and fungi) were protists, and the phenomenon of multicellularity arose independently in these three groups, although in different ways. Genetic and developmental data confirm that the basic mechanisms of pattern formation and cell–cell communication during development were independently derived in animals and in plants. In animals, segmental identity is established by the spatially specific transcriptional activation of an overlapping series of master regulatory genes, the homeobox (Hox) genes. The master regulatory genes of plants are not members of the homeobox gene family, but belong to the MADS box family of transcription factor genes. There is no evidence that the animal homeobox and MADS box transcription factor genes are homologous.

Although the fossil record is rich with the history of many early animal lineages, many others have left few or no fossils. Many ancient animals were very small, many were soft-bodied and did not fossilize well, and others lived where conditions were not suitable for the formation of fossils. However, groups such as the echinoderms (sea stars, urchins), molluscs (clams, snails), arthropods (crustaceans, insects), corals, ectoprocts, brachiopods, and vertebrates have left rich fossil records. In fact, for some groups (e.g., echinoderms, brachiopods, ectoprocts, molluscs), the number of extinct species known from fossils exceeds the number of known living forms. Representatives of nearly all of the extant animal phyla were present in the Cambrian period. Life on land, however, did not appear until later, and terrestrial radiations probably began only about 470 million years ago. The following account briefly summarizes the early history of life and the rise of the invertebrates.

The Dawn of Life

It used to be thought that the Proterozoic eon, 2.5 billion years ago to 541 million years ago, was a time of only a few simple kinds of life; hence the name. However, recent discoveries have shown that life on Earth began early and had a very long history throughout the Proterozoic. As noted above, some of the oldest rocks known on Earth already have markings suspected to represent anaerobic sulfate-reducing prokaryotes and perhaps even cyanobacterial stromatolites. Undisputed cyanobacterial fossils 2.9 billion years of age are known. The first actual fossil traces of eukaryotic life (benthic algae) are 1.7 to 2 billion years old, whereas the first certain eukaryotic fossils (phytoplankton) are 1.4 to 1.7 billion years old. Together, these prokaryotes and protists appear to have formed diverse communities in shallow marine habitats throughout the Proterozoic. Living stromatolites (compact layered colonies of Cyanobacteria and mud) are still with us, and can be found in certain high evaporation/high-salinity coastal environments in places as Shark Bay (Western Australia), Scammon's Lagoon (Baja California), salt ponds on islands in the Sea of Cortez, the Persian Gulf, the Paracas coast of Peru, the Bahamas, and Antarctica. Living stromatolites also occur in some isolated inland waters, such as the famous Cuatro Cienegas Oasis of Chihuahua, Mexico.

The end era of the Proterozoic (the Neoproterozoic, 1 billion years ago to 541 Ma) was a different world from today. Atmospheric oxygen levels were much lower. The deep ocean was especially oxygen poor. And three ice-covered Earth events are hypothesized to have occurred during the Neoproterozoic Era, based on the occurrence of multiple glaciations at sea level in low latitudes in the geological record. These "Snowball Earth" events, which might have each lasted up to 10 million years, are thought to have been interrupted by periods of rapid warming and global greenhouse conditions. The causes of these wide climatic swings during the Neoproterozoic are still unclear although there is evidence that the world oceans lacked the $CaCO_3$-based carbon stabilization chemistry we have today. Indeed, major carbon precipitators such as foraminiferans and coccolithophores had probably not yet evolved in the world's oceans. It seems likely that buildup of atmospheric CO_2 generated by massive volcanic eruptions drove the warming events. All these changes were taking place during the breakup of a supercontinent known as Rodinia (about 750 Ma)—the supercontinent that preceded Pangaea. It was against this backdrop that the Metazoa first arose, some time between 650 million years ago and 1 billion years ago (the exact timing is still uncertain). Around 800 million years ago, atmospheric oxygen seems to have stabilized at a relatively high level, setting the stage for the more rapid evolution of eukaryotes and the Bilateria. Oxygen, of course, was critical to the diversification of large and metabolically active, complex life.

The prevailing view is that Metazoa arose in the Neoproterozoic (latest Precambrian) and began to diversify, and then radiated rapidly in the Cambrian—the "Cambrian explosion." However, during the Neoproterozoic these early animal lineages must have survived some extreme climate swings on Earth, including profound changes in ocean and atmospheric chemistry and widespread global glacial events.

The Ediacaran Period and the Origin of Animals

One of the most perplexing unsolved mysteries in biology is the origin and early radiation of Metazoa (the animal kingdom). Fossil (and molecular clock) evidence suggests that by the end-Proterozoic Ediacaran period, a worldwide marine invertebrate fauna was already well established. Although the animals that existed during this time left few records, the fauna of the Ediacaran (635–541 Ma) contains the first evidence of many modern phyla. Living phyla thought to be represented among the Ediacaran fauna include Porifera, Cnidaria, Mollusca, possibly Annelida (including possible echiurids and pogonophorans), and others. Some authors have even suggested the presence of Onychophora, Arthropoda, and Echinodermata, although these are controversial. The Ediacaran *Dickinsonia* has been said to represent a "placozoan-grade animal," *Kimberella* may be a mollusc, *Eoandromeda* a ctenophore, and *Thectardis* a sponge (although all of these classifications are debatable). Over 100 Ediacaran animal genera have been described from around the world. However, many Ediacaran animals cannot be unambiguously assigned to any living phylum, and these animals may represent phyla or other high-level taxa that went extinct at the Proterozoic–Cambrian transition.

Ediacaran fossils were first reported from sites in Newfoundland and Namibia, but the name is derived from the superb assemblages of these fossils discovered at Ediacara in the Flinders Ranges of South Australia. Most of the Ediacaran organisms were preserved as shallow-water impressions on sandstone beds, but some of the 25 or so worldwide sites probably represent deep-water and continental slope communities. The Ediacaran fauna was mostly soft bodied, and there have been no heavily shelled creatures reported from these deposits. Even the alleged molluscs and arthropod-like creatures from this fauna are thought to have had soft (unmineralized or lightly mineralized) skeletons. A few chitinous structures developed during this time, such as the jaws of some annelid-like creatures (and the chitinous sabellid-like tubes of others) and the radulae of early molluscs.[5] Siliceous spicules of hexactinellid sponges have been reported from Australian and Chinese Ediacaran deposits. In 2014, Ediacaran-age metazoan reefs were discovered in Namibia that were built by skeletonized creatures named *Cloudina*. Resembling cnidarians, these are the oldest known structural reef-building organisms although their phylogenetic affinities are uncertain.

Many of these Ediacaran animals appear to have lacked complex internal organ structures. Many were small and possessed radial symmetry. However, at least by the late Ediacaran, large animals with bilateral symmetry seem to have appeared, and some almost certainly had well-developed internal organs (e.g., the segmented, sheetlike *Dickinsonia*, which reached a meter in length; Figure 1.2). Some workers have questioned whether or not Ediacaran life even included any animals, suggesting that it might represent only protists, algae, fungi, lichens, and microbial colonies and mats. However, some Ediacaran fossils seem to be unmistakably Porifera and Cnidaria, and others (e.g., *Kimberella*) show strong similarity to molluscs. *Kimberella* has been associated with scratch marks reminiscent of radula scrapings. Still others are thought to possess bilateral symmetry, an idea once controversial but now gaining traction. Among the bilaterian fossils are *Kimberella*, the annelid- or arthropod-like *Spriggina*, "small shelly fossils" (e.g., the enigmatic *Cloudina*), and bilaterian embryos from the Doushantuo Formation of south China, although these bilaterian records have their detractors. Other evidence of bilaterians in these deposits include what appear to be predatory bore holes in the small shelly fossils, and a variety of horizontal and vertical burrows that were perhaps made by motile animals. An entire group of species called Proarticulata, appear to possess both segmentation and bilateral symmetry (e.g., *Dickinsonia*, *Vendia*, *Onega*, *Praecambridium*).

The beginning of the Ediacaran period coincides with the end of the last "Snowball Earth" event of the Precambrian, when most of Earth's land, and perhaps much of its ocean, was frozen over. And at the end of the Ediacaran (the end of the Precambrian) many living creatures seem to have gone extinct.[6] The Ediacaran period was followed by the Cambrian period (541–485 Ma) and the great "explosion" of skeletonized metazoan life associated with that time (see below). Life quickly got very interesting and very complicated in the Cambrian! Why large, skeletonized animals appeared at that particular time, and in such great profusion, remains a mystery, but the fossil record clearly informs us that by the early Cambrian most of the major animal phyla we recognize today had come into being. Certainly, great changes were occurring at the end of the Precambrian and the start of the early Cambrian—the breakup of the supercontinents, rising sea levels, a possible nutrient crisis, fluctuations in atmospheric composition (including oxygen and CO_2 levels), changes in ocean chemistry—and these likely

[5]Chitin is a cellulose-like family of compounds that is widely distributed in nature, especially in invertebrates, fungi, and many protists, but it is apparently uncommon in deuterostome animals and higher plants, perhaps due to the absence of the chitin synthase enzyme.

[6]The largest mass extinctions recorded in the fossil record occurred at the ends of the Precambrian (the Ediacaran–Cambrian transition), Ordovician, Devonian, and Permian, and in the Early Triassic, Late Triassic, and end-Cretaceous. Most of these extinction events were experienced by both marine and terrestrial organisms. The Triassic events and the end-Devonian event have recently been shown to be less severe extinction events than once thought. And a relatively little-studied mass extinction event in the late Permian, about 260 million years ago, has recently begun to emerge as a much larger extinction event than previously thought, with perhaps 56% of plant species and 58% of marine invertebrate genera disappearing.

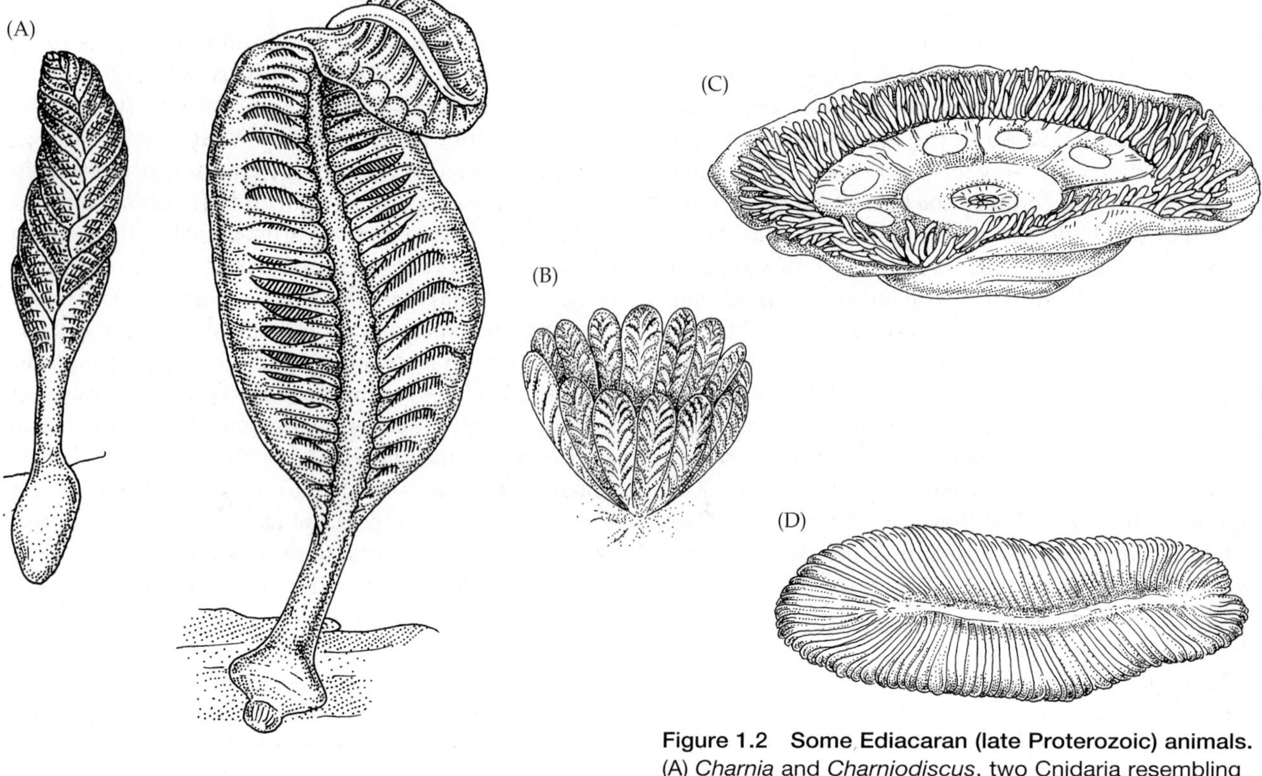

Figure 1.2 Some Ediacaran (late Proterozoic) animals. (A) *Charnia* and *Charniodiscus*, two Cnidaria resembling modern sea pens (Anthozoa, Pennatulacea). (B) A bush-like fossil of uncertain affinity (suggestive of a cnidarian). (C) *Ediacara*, a cnidarian medusa. (D) *Dickinsonia*, possibly an annelid. (E) One of the numerous soft-bodied trilobites known from the Ediacaran period (some of which also occurred in the early Cambrian).

played a role the demise of the Ediacaran fauna and the rise of Cambrian life.

The Paleozoic Era (541–252 Ma)

The Phanerozoic eon (the time of abundant life on Earth that encompasses the Paleozoic, Mesozoic and Cenozoic eras) was ushered in with the almost simultaneous appearance in the lower Cambrian of well-developed, large, calcareous body skeletons in numerous groups, such as archaeocyathans (coral-like organisms that were probably early sponges), molluscs, bryozoans, brachiopods, crustaceans, and trilobites. Shelled animals (molluscs, etc.) first appear among the "small shelly fossils" of the earliest Cambrian, and chaetognaths (represented by protoconodonts) are known from the same period. The appearance of well-mineralized animal skeletons thus defines the beginning of the Cambrian, and it was an event of fundamental importance in the history of life. The newly skeletonized animals radiated quickly and filled a multitude of ecological roles in shallow-water marine environments, and by the early Cambrian most living phyla possessing hard parts had appeared.

Evidence suggests that in the early Cambrian the world ocean increased its calcium content dramatically, which coincided with a proliferation of skeletonized animals. Climates were warm, and shallow epicontinental seas covered broad regions of Paleozoic continents. The explosion of bilaterally symmetrical animals, the Bilateria, also began around the Proterozoic–Cambrian transition. Seed ferns and other gymnosperms also arose during the Paleozoic, perhaps about 400 million years ago, whereas angiosperms may have arisen around 200 million years ago, as suggested by molecular data (although the oldest undisputed angiosperm fossils arose about 130 Ma).

Did all these metazoan lineages arise quickly, during the first 10 million years or so of the Cambrian? Or were many (or most) of them already getting established in the Ediacaran, but we have yet to find evidence of this, perhaps because they lacked hard skeletons or were very small. This latter idea, sometimes called the "phylogenetic fuse" hypothesis, is supported by the discovery of undisputed and modern-appearing crustaceans in the earliest Cambrian, as well as many creatures from the Ediacaran that resemble modern phyla. Given that advanced phyla like arthropods were already well established in the earliest Cambrian, it seems likely that earlier-derived phyla of ecdysozoans must have been alive and well during the Ediacaran. Indeed, apparent phosphatized bilaterian animal embryos from the Ediacaran provide evidence of the cryptic roots of Metazoa and the likelihood that the evolutionary fuse of modern animal phyla was lit well before the Cambrian began. A recent, large phylogenetic analysis by Omar Rota-Stabelli and colleagues used 67 fossil calibration points to date a molecular timetree of ecdysozoan phyla, concluding that the clade originated in the Ediacaran.

Geological evidence tells us that Earth's earliest atmosphere lacked free oxygen, and the radiation of the animal kingdom could not have begun under those conditions. Free oxygen probably accumulated over many millions of years as a by-product of photosynthetic activity in the oceans, particularly by the cyanobacterial (blue-green algae) stromatolites. A longstanding hypothesis suggests that the beginning of an oxygenated atmosphere was around 2.4 billion years ago, and it was then that "modern life" began to evolve (at the beginning of the Proterozoic). However, evidence for free oxygen levels in the Proterozoic is still unclear, and there are some data that suggest an oxygenated atmosphere might have evolved even earlier than that. Proterozoic seas might have been oxic near the surface, but anoxic in deep waters and on the bottom. Some workers suggest that the absence of metazoan life in the early fossil record is due to the simple fact that the first animals were small, lacked skeletons, and did not fossilize well. The discovery of highly diverse communities of metazoan meiofauna in the Proterozoic strata of south China and in deposits from the middle and upper Cambrian (e.g., the Swedish Orsten fauna) lends support to the idea that many of the first animals were microscopic.[7] However, large animals also are not uncommon among the Ediacaran and early Cambrian faunas.

It has also been proposed that the advent of predatory lifestyles, early in the Cambrian, was the key that favored the first appearance of animal skeletons (as defensive structures), leading to the "Cambrian explosion." The rapid appearance and spread of diverse metazoan skeletons in the early Cambrian certainly heralded the beginning of the Phanerozoic (i.e., all time after the Precambrian; the Paleozoic–Mesozoic–Cenozoic eras). The Ediacaran fauna seems to have included primarily passive suspension and detritus feeders; very few of these animals appear to have been active carnivores or herbivores. Early Cambrian animal communities, on the other hand, included most of the trophic roles found in modern marine communities, including giant predatory arthropods.

Much of what we know about earliest Cambrian life comes from the lower Cambrian Chengjiang fossil deposits of the Yunnan Province in southern China, and similarly aged (although less well preserved) deposits spread across China and the Siberian Platform. The Chengjiang deposits are the oldest Cambrian occurrences of well-preserved soft-bodied and hard-bodied animals, and although they are dominated by arthropods they also include a rich assemblage of exquisitely preserved onychophorans, medusae (Cnidaria), and brachiopods, many of which appear closely related to Ediacaran species.

In the middle Cambrian (e.g., the Burgess Shale fauna of western Canada and similar deposits elsewhere; Figures 1.3 and 1.4), annelids and tardigrades made their first positive appearance in the fossil record, and the first complete echinoderm skeletons appear. Twelve million years younger than the Chengjiang fauna, the Burgess Shale material is also dominated by arthropods (and their allies), including the infamous *Opabinia* and *Anomalocaris*. Most of the Burgess Shale fauna can also be assigned to living phyla. There are numerous other (less famous) Cambrian fossil sites, such as the Sirius Passet fauna in Greenland, the Emu Bay Shale in Australia, the Sinsk biota in Russia, and the Kaili and Guanshan biotas in China. In the upper Cambrian (e.g., the Orsten deposits of southern Sweden and similar strata), the first pentastomid Crustacea and the first agnathan fishes make their appearances. By the end of the Cambrian, nearly all of the major, modern animal phyla had appeared. The Burgess Shale-like assemblage in Lower Ordovician rocks of Morocco tells us that many of the Cambrian lineages survived for tens of millions of years (Figure 1.4).

The driving force for this Cambrian explosion has perplexed scientists since Darwin's day. In fact, it has been referred to as "Darwin's dilemma," because he could not reconcile such rapid origination and diversification of major animal groups with his view of gradual evolution driven only by natural selection. Some analyses have suggested that rates of phenotypic and genomic evolution might have been many times faster during the Cambrian than in the rest of the Paleozoic. One popular notion is that atmospheric oxygen reached a critical level around 580 million years ago, in the Ediacaran, that allowed for larger animals and skeletonized animals to begin to evolve. This might have led to the appearance of large carnivores, which rely on oxygen for high-energy pursuits. Once the carnivores

[7]"Orsten"-type preservation entails the phosphatization of cuticles with almost no deformation. It preserves the finest details of organisms, including setae, and such deposits have yielded three-dimensional fossils at the scale of 0.1–2.0 mm. First discovered in Sweden, Orsten-type deposits are now known from several continents, from the early Cambrian (520 Ma) to the middle Cretaceous (100 Ma). Most of the earliest "Orsten fossils" are undisputed arthropods.

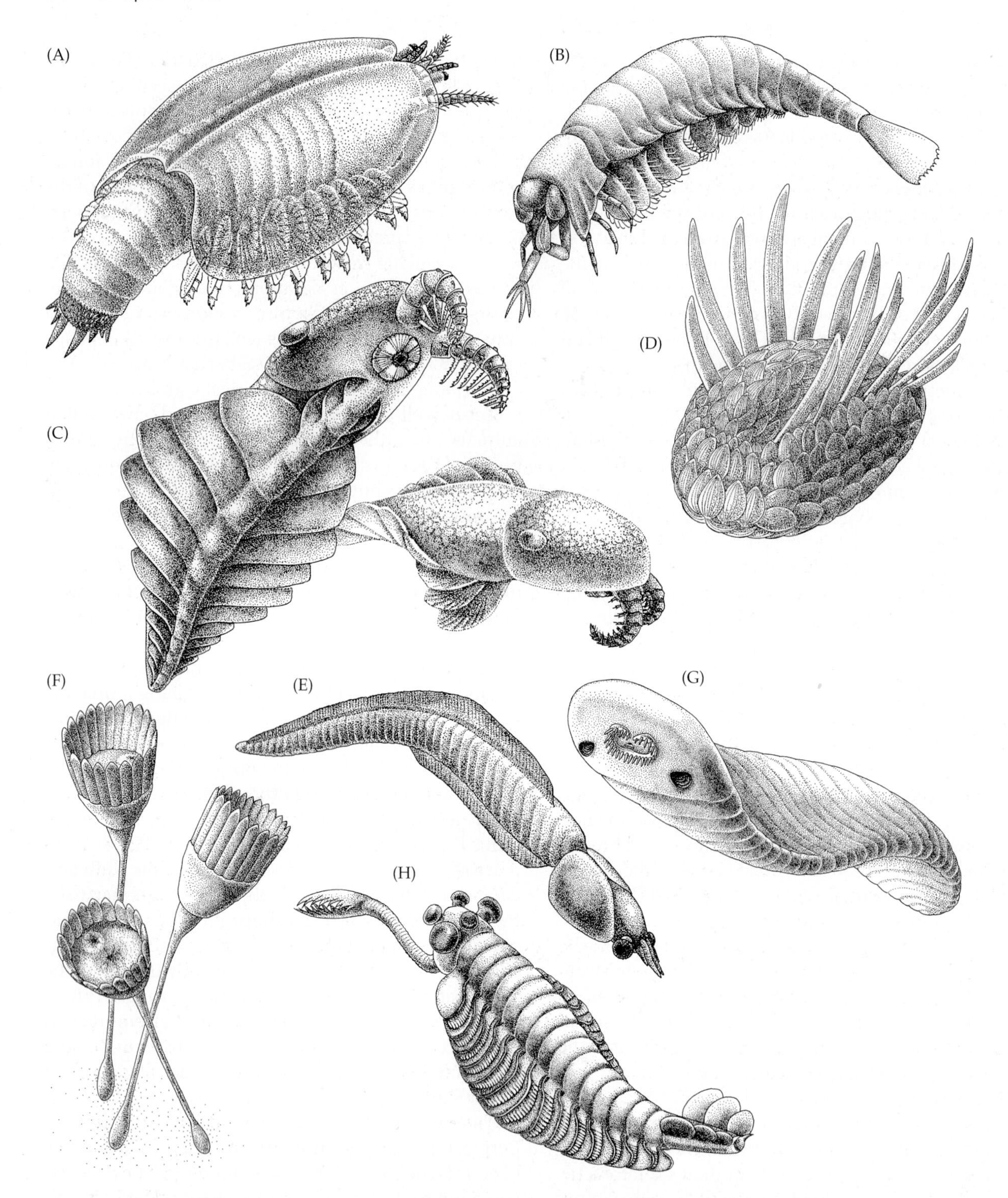

Figure 1.3 Some Cambrian life forms from the Burgess Shale deposits of Canada. (A) *Canadaspis*, an early malacostracan crustacean. (B) *Yohoia*, an arthropod of uncertain classification. (C) Two species of *Anomalocaris*, *A. nathorsti* (above) and *A. canadensis*. Anomalocarids were once thought to represent an extinct phylum of segmented animals, but are now regarded by many workers as primitive arthropods dating back to the Ediacaran. (D) *Wiwaxia*, a Burgess Shale animal with no clear affinity to any known metazoan phylum (although some workers regard it as a free-living annelid). (E) *Nectocaris*, another creature that has yet to be classified into any known phylum (despite its strong chordate-like appearance). (F) *Dinomischus*, a stalked creature with a U-shaped gut and with the mouth and anus both placed on a radially symmetrical calyx. Although superficially resembling several extant phyla, *Dinomischus* is now thought to belong to an unnamed extinct phylum of sessile Cambrian animals. (G) The elusive *Odontogriphus*, an appendageless flattened vermiform creature of unknown affinity. (H) One of the more enigmatic of the Burgess Shale animals, *Opabinia*; this segmented creature was probably an ancestral arthropod. Notice the presence of five eyes, a long prehensile "nozzle," and gills positioned dorsal to lateral flaps.

Figure 1.4 "Fauna of the Burgess Shale," Carel Brest von Kempen. Acrylic on illustration board. Depicted are stromatolites, *Leptomitus, Vauxia, Billingsella, Hallucigenia, Aysheaia, Anomalocaris, Opabinia, Lejopyge, Olenoides, Asaphiscus, Elrathia, Modocia, Naraoia, Habellia, Burgessia, Plenocaris, Sarotrocercus, Odaraia, Pseudoarctolepis, Canadaspis, Marrella, Branchiocaris, Ottoia, Hyolithes, Canadia, Gogia, Pikaia, Wiwaxia, Dinomischus,* and *Amiskwia.*

proliferated, a predator–prey arms race ensued, driving the rapid evolution of never-ending new body forms. However, some other evidence suggests that high oxygen levels existed well before that date.

Another explanation for the explosive radiation of Metazoa in the Cambrian is a rapid increase in nutrient supplies to the world's oceans, especially phosphorus (P) and potassium (K). This is hypothesized to have occurred as the Earth cooled enough for subducting mantle to hydrate with seawater about 600 million years ago, thus moving water from the oceans into the lower mantle and lowering sea level, which in turn exposed great swaths of new landmasses that underwent erosion of P and K, which in turn were transported to the sea as fertilizers. In addition, massive uplift of mountains and plume-driven dome-up of landmasses also might have added new nutrient-erosion sources around the Precambrian–Cambrian transition.

The early Paleozoic also saw the first xiphosurans, eurypterids, trees, and teleost fishes (in the Ordovician). The first land animals (arachnids, centipedes, myriapods) appeared in the Upper Silurian. By the middle Paleozoic (the Devonian period), life on land had begun to proliferate. Forest ecosystems became established and began reducing atmospheric CO_2 levels (eventually terminating an earlier Paleozoic greenhouse environment). The first insects also appeared in the early Paleozoic era (see chapter 22). Insects probably developed flight in the Early Devonian (~406 million years ago), and they began their long history of coevolution with plants shortly thereafter (at least by the mid-Carboniferous, when tree fern galls first appeared in the fossil record).[8] During the Carboniferous period, global climates were generally warm and humid, and extensive coal-producing swamps existed.

The late Paleozoic witnessed the formation of Earth's most recent supercontinent Pangaea, in the Permian period (about 270 Ma). The end of the Paleozoic era/Permian period (252 Ma) was marked by the largest mass extinction known, in which about 90% of Earth's marine species (and 70% of the terrestrial vertebrate genera) were lost over a brief span of a few million years. The Paleozoic reef corals (Rugosa and Tabulata) went extinct, as did the once dominant trilobites, never to be seen again. The driving force of the massive Permian extinction is a hotly debated subject, and hypotheses range from rapid global warming to rapid global cooling! Either might have been driven by a huge asteroid impact, perhaps coupled with

[8]Coevolution is the reciprocal adaptation that occurs over time between closely interacting species. One classic example is plants and their pollinators, which have evolved in lockstep since the late Paleozoic era to form remarkably finely tuned anatomies, physiologies, and behaviors, among willing (or unwilling) partners.

massive Earth volcanism, and perhaps also degassing of methane from stagnant ocean basins. Evidence suggests that toxic waters decimated shallow bottom marine communities at that time. All of these might have created a large and rapid global warming (or cooling) event, depending on how things played out. The trillions of tons of carbon released into the atmosphere and oceans from enormous volcanism likely also led to ocean acidification and a drop in dissolved oxygen in the oceans, further possible causes of oceanic extinction events. This volcanism was likely the one that created the massive flood basalts known as the Siberian Traps in Asia, an event coincident in time that could have led to atmospheric "pollution" in the form of dust and sulfur particles that cooled Earth's surface and/or massive gas emissions that led to a prolonged greenhouse warming and ocean acidification.

The Mesozoic Era (252–66 Ma)

The Mesozoic era is divided into three broad periods: the Triassic, Jurassic, and Cretaceous. The Triassic began with the continents joined together as Pangaea. The land was high, and few shallow seas existed. Global climates were warm, and deserts were extensive. Although the Triassic is perhaps best known for the emergence of the dinosaurs, vertebrate diversity in general exploded during this period, as the first land mammals made their appearance. In Triassic seas, modern-looking scleractinian corals appeared, and the diversity of predatory invertebrates and fishes increased dramatically, although the paleogeological data suggest that deeper marine waters might have been too low in oxygen to harbor much (or any) multicellular life. The end of the Triassic witnessed a global extinction event that resulted in the disappearance of around half of all the living species. This was perhaps driven by the combination of asteroid impact and widespread volcanism that created the Central Atlantic Magmatic Province of northeastern South America 200 million years ago, although there is considerable disagreement about the extent and cause of this extinction event. Recent studies suggest elevated atmospheric CO_2 was the culprit behind the end-Triassic extinction event, possibly caused by high rates of magmatic CO_2 degassing

The Jurassic saw a continuation of warm, stable climates, with little latitudinal or seasonal variation and probably little mixing between shallow and deep oceanic waters. Pangaea split into two large landmasses, a northern Laurasia and southern Gondwana, separated by a circumglobal tropical seaway known as the **Tethys Sea**. Many tropical marine families and genera today are thought to be direct descendants of inhabitants of the pantropical Tethys Sea. On land, modern genera of many gymnosperms and advanced angiosperms appeared, and the first birds begin to evolve (*Archaeopteryx*, ~150 Ma). Leaf-mining insects (lepidopterans) appeared by the late Jurassic (150 Ma), and other leaf-mining orders appeared through the Cretaceous, coincident with the radiation of the vascular plants.

In the Cretaceous, large-scale fragmentation of Gondwana and Laurasia took place, resulting in the formation of the Atlantic and Southern Oceans. During this period, landmasses subsided and sea levels were high; the oceans sent their waters far inland, and great epicontinental seas and coastal swamps developed. As land masses fragmented and new oceans formed, global climates began to cool and oceanic mixing began to move oxygenated waters to greater depths in the sea. The oldest fossil of modern birds dates from the early Cretaceous (*Archaeornithura meemannae*, ~130 Ma).

The end of the Cretaceous was marked by the Cretaceous–Paleocene mass extinction, in which an estimated 50% of Earth's species were lost, including the (nonavian) dinosaurs and all of the sea's rich Mesozoic ammonite diversity. There is strong evidence that this extinction event was driven by a combination of two factors: massive Earth volcanism associated with the great flood basalts of western India known as the Deccan Traps, combined with a major asteroid impact (documented by the Chicxulub Crater in the Yucatán region of modern Mexico). The Deccan Traps flood basalts comprise an almost unimaginable 1.3 million cubic kilometers of erupted lava, 3,000 m thick in some places. The main eruptions initiated about 250,000 years before the Cretaceous-Paleocene boundary, and overall the lava flowed for over 750,000 years. However, an estimated 90% of the eruptive flow took place rapidly, over less than 1 million years, coincident with the Cretaceous–Paleocene boundary. The sulfur dioxide gas injected into the atmosphere from this massive volcanic event would have converted to sulfate aerosols that caused climate cooling, and when these aerosols washed out (as acid rain) they would have acidified the oceans. The scale of biological turnover between the Cretaceous and Paleocene is nearly unprecedented in Earth history. All the nonavian dinosaurs, marine reptiles, ammonites, and rudistid clams went extinct. Planktonic foraminiferans and land plants were devastated. There is recent evidence suggesting that the full strength of the end-Cretaceous mass extinctions might not have kicked in until about 300,000 years after the Chicxulub impact. If comet impacts are beginning to sound like a recurring theme in the history of life on Earth, it's because so many large impact craters have been found (at least 70 that are larger than 6 km in diameter), and many of these coincide with major transitions in the fossil record.

The Cenozoic Era (66 Ma–present)

The Cenozoic era dawned with a continuing worldwide cooling trend. As South America decoupled from Antarctica, the Drake Passage opened to initiate the circum-Antarctic current, which eventually drove the formation of the Antarctic ice cap, which in turn led to our modern cold ocean bottom conditions (in the Miocene). India moved north from Antarctica and collided with southern Asia (in the early Oligocene). Africa collided with western Asia (late Oligocene/early Miocene), separating the Mediterranean Sea from the Indian Ocean and breaking up the circumtropical Tethys Sea. Around 56 million years ago, the Earth warmed rapidly as greenhouse gasses spiked in the atmosphere and global oceans warmed all the way down to the deep sea. The cause of this Paleocene–Eocene Thermal Maximum is uncertain, but massive volcanic eruptions that could have cooked CO_2 out of organic seafloor sediments have been suggested, as have wildfires burning through Paleocene peat deposits, and a comet impact (that could have released large deposits of methane hydrate from the seafloor). Relatively recently (in the Pliocene), the Arctic ice cap formed, and the Panama isthmus rose, separating the Caribbean Sea from the Pacific and breaking up the last remnant of the ancient Tethys Sea about 3 million years ago (although recent work has proposed a much earlier date for the closure of the Panama Seaway, perhaps 15–13 Ma). Modern coral reefs (scleractinian-based reefs) appeared early in the Cenozoic, reestablishing the niche once held by the rugose and tabulate corals of the Paleozoic.

This textbook focuses primarily on invertebrate life at the very end of the Cenozoic, in the Quaternary period (Pleistocene + Holocene). However, evaluation of the present-day success of animal groups also involves consideration of the deeper history of modern lineages, the diversity of life over time (numbers of species and higher taxa), and the abundance of life in various environments. The predominance of certain kinds of invertebrates today is unquestionable. For example, of the 1,382,402 or so described species of animals (1,324,402 of which are invertebrates), 81.5% are arthropods (and 82% of those are insects). It would be hard to argue that insects are not the most successful group of animals on Earth today. And the fossil record tells us that arthropods have always been key players in the biosphere, even before the appearance of the insects. Table 1.1 conveys a general idea of the levels of diversity among the animal phyla today.

Where Do Invertebrates Live?

Marine Habitats

The global ocean is Earth's largest biome. In fact, it can be said that Earth is a marine planet—salt water covers 71% of its surface. The vast three-dimensional world of the seas contains 99% of Earth's inhabited space. Life almost certainly evolved in the sea, and the major events described above leading to the diversification of invertebrates occurred in late Proterozoic and early Cambrian shallow seas. Many aspects of the marine world minimize physical and chemical stresses on organisms. The challenge of evolving gas exchange and osmotic regulatory structures that can function in freshwater and terrestrial environments are formidable, and relatively few lineages were able to do so and escape their marine origins. Thus, is not surprising to find that the marine environment continues to harbor the greatest diversity of higher taxa and major body plans—14 of the 32 living animal phyla are strictly marine (e.g., Placozoa, Ctenophora, Chaetognatha, Rhombozoa, Orthonectida, Cycliophora, Gnathostomulida, Phoronida, Brachiopoda, Kinorhyncha, Priapula, Loricifera, Echinodermata, Hemichordata) and many others have barely penetrated the terrestrial or freshwater realm. Productivity in the world's oceans is very high, and this also probably contributes to the high diversity of animal life in the sea (the total primary productivity of the seas is about 48.7×10^9 metric tons of carbon per year). Perhaps the most significant factor, however, is the special nature of seawater itself.

Water is a very efficient thermal buffer. Because of its high heat capacity, it is slow to heat up or cool down. Large bodies of water, such as oceans, absorb and lose great amounts of heat with little change in actual water temperature. In fact, the oceans store over 90% of the excess heat accumulating in the world's atmospheric climate system today. Thus, oceanic temperatures are very stable in comparison with those of freshwater and terrestrial environments. Short-term temperature extremes occur only in intertidal and estuarine habitats, and invertebrates living in such areas must possess behavioral and physiological adaptations that allow them to survive these temperature changes, which are often combined with aerial exposure during low tide periods.

The saltiness, or salinity, of seawater averages about 3.5% (usually expressed as parts per thousand, 35‰). This property, too, is quite stable, especially in areas away from shore and the influence of freshwater runoff. The salinity of seawater gives it a high density, which enhances buoyancy, thereby minimizing energy expenditures for flotation. Furthermore, the various ions that contribute to the total salinity occur in fairly constant proportions. These qualities result in a total ionic concentration in seawater that is similar to that in the body fluids of most animals, minimizing the problems of osmotic and ionic regulation (see Chapter 4). The pH of seawater is also quite stable throughout most of the ocean. Naturally occurring carbonate compounds participate in a series of chemical reactions that buffer seawater at about pH 7.5–8.5. However, today's anthropogenic-driven increase in atmospheric CO_2 threatens to alter the carbonate buffering capacity of the world's seas. A large fraction (over 25%) of the CO_2 added to the atmosphere by burning of fossil fuels enters the ocean. Seawater reacts with CO_2 to form carbonic acid, decreasing the pH of ocean surface waters.

The lowering of sea surface pH disrupts formation of calcium carbonate skeletons in both animals and protists.

In shallow and nearshore waters, CO_2, various nutrients, and sunlight are generally available in quantities sufficient to allow high levels of photosynthesis, either seasonally or continuously (depending on latitude and other factors). Dissolved oxygen levels rarely drop below those required for normal respiration, except in stagnant waters such as might occur in certain estuarine or ocean basin habitats, or where anthropogenic activities have created eutrophic conditions that can result in oxygen minimum zones (OMZs), hypoxic regions, or even anoxic seafloor conditions. Hypoxia occurs when dissolved oxygen falls below < 2 ml of O_2/liter. The size and number of OMZs, defined by O_2 concentrations below 20 μM and sometimes called "dead zones," in the world oceans have been growing rapidly in recent decades as a result of humans releasing huge quantities of organic waste and fertilizers into the sea. Deep-sea ecosystems contain the largest hypoxic and anoxic regions of the Biosphere. Permanently anoxic conditions in the oceans are present in the subsurface seafloor and, among other areas, in the interior of the Black Sea[9] and in the deep (> 3,000 m) hypersaline anoxic basins of the Mediterranean Sea, where it has long been thought that no metazoans could live. However, recently several species of Loricifera were discovered living in the anoxic Mediterranean basins (perhaps the only animals able to live permanently in anoxic conditions).

Because the marine realm is home to most of the animals discussed in this book, some terms that describe the subdivisions of that environment and the categories of animals that inhabit them will be useful. Figure 1.5 illustrates a generalized cross section through an ocean. The shoreline marks the littoral region, where sea, air, and land meet and interact (Figure 1.6A). Obviously, this region is affected by the rise and fall of the tides, and we can subdivide it into zones or shore elevations relative to the tides. The **supralittoral zone**, or splash zone, is rarely covered by water, even at high tide, but it is subjected to storm surges and spray from waves. The **eulittoral zone**, or true intertidal zone, lies between the levels of the highest and lowest tides. It can be subdivided by its flora and fauna, and by mean monthly hours of aerial exposure, into high, mid-, and low intertidal zones. The **sublittoral zone**, or subtidal zone, is never uncovered, even at very low tides, but it is influenced by tidal action (e.g., by changes in turbulence, turbidity, and light penetration).

Organisms that inhabit the world's littoral regions are subjected to dynamic and often demanding conditions, and yet these areas commonly are home to exceptionally high numbers of species. Most animals and plants are more or less restricted to particular elevations along the shore, a condition resulting in the phenomenon of zonation. Such zones are visible as distinct bands or communities of organisms along the shoreline. The upper elevational limit of an intertidal organism is commonly established by its ability to tolerate conditions of exposure to air (e.g., desiccation, temperature fluctuations), whereas its lower elevational limit is often determined by biological factors (competition with or predation by other species). There are, of course, many exceptions to these generalizations.

Extending seaward from the shoreline is the **continental shelf**, a feature of most large landmasses. The continental shelf may be only a few kilometers wide, or it may extend up to 1,000 km from shore (50 to 100 km is average for most areas). It usually reaches a depth of 150–200 m. These nearshore shelf areas are among the most productive environments of the ocean, being rich in nutrients and shallow enough to permit photosynthesis from surface to seafloor.

The outer limit of the continental shelf—called the **continental edge**—is indicated by a relatively sudden increase in the steepness of the bottom contour. The "steep" parts of the ocean floor, the **continental slopes**, actually have slopes of only 4–6% (although the slope is much steeper around volcanic islands). The continental slope continues from the continental edge to the deep ocean floor, which forms the expansive, relatively flat **abyssal plain**. The abyssal plain is an average of about 4 km below the sea's surface, but it is interrupted by a variety of ridges, seamounts, mountain ranges, trenches, and other formations. The bottoms of some deep-sea trenches exceed 10 km in depth.

Organisms that inhabit the water column are known as **pelagic** organisms, whereas those living on the sea bottom anywhere along the entire contour shown in Figure 1.5 are referred to as **benthic** organisms. Organisms living *near* the seafloor are **demersal**. Both the variety and the abundance of life tend to decrease with increasing depth, from the rich littoral and continental shelf environments to the deep abyssal plain. However, an overgeneralization of this relationship can be misleading. For example, although pelagic biomass declines exponentially with depth, both diversity and biomass increase again near the bottom, in a thick layer of resuspended sediments called the **benthic boundary layer**. Also, shelf and slope habitats in temperate regions are often characterized by low animal density but high species diversity. In many areas, benthic diversity increases abruptly below the continental edge (100–300 m depth), peaks at 1,000 to 2,000 m depth, and then decreases gradually. Species diversity in the benthic abyssal region itself may be surprisingly high. The first impression of early marine scientists—that the deep seabed was an environment able to sustain only a

[9]Despite its large surface area (423,500 km^2), the Black Sea has only a thin surface layer that supports eukaryotic life. The water mass below about 175 m is devoid of dissolved oxygen, making this the largest anoxic body of water in the world.

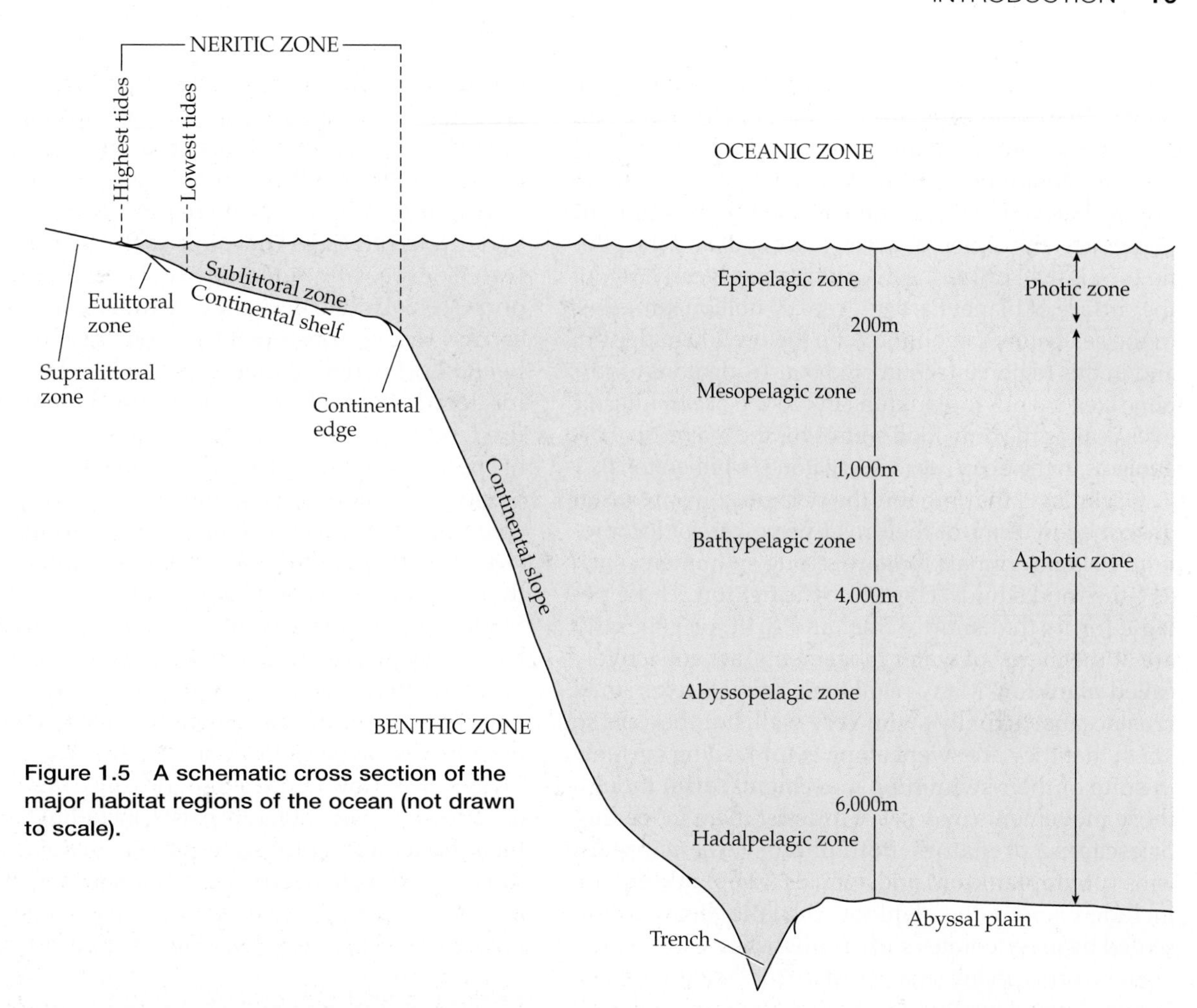

Figure 1.5 A schematic cross section of the major habitat regions of the ocean (not drawn to scale).

few species in impoverished simple communities—was long ago shown to be incorrect, and deep-sea biodiversity is now known to be quite high, in some areas even rivaling tropical rain forests. Additionally, the deep ocean covers more than half the planet!

Benthic animals may live on the surface of the substratum (**epifauna**, or **epibenthic** forms, such as most sea anemones, sponges, many snails, and barnacles) or burrow within soft substrata (**infauna**). Infaunal forms include many relatively large invertebrates, such as clams and various crustaceans and worms, as well as some specialized, very tiny forms that inhabit the spaces between sand grains, termed **interstitial** organisms (the smallest of which are **meiofauna**, usually defined as animals smaller than 0.5 mm). Six phyla of metazoans are exclusively meiobenthic in the sea: Gastrotricha, Gnathostomulida, Kinorhyncha, Loricifera, Micrognathozoa, and Tardigrada. Benthic animals may also be categorized by their locomotor capabilities. Animals that are generally quite motile and active are described as being **errant** (e.g., crabs, many worms), whereas those that are firmly attached to the substratum are **sessile** (e.g., sponges, corals, barnacles). Others are unattached or weakly attached, but generally do not move around much (e.g., crinoids, solitary anemones, most clams); these animals are said to be **sedentary**.

The region of water extending from the surface to near the bottom of the sea is called the **pelagic zone**. The pelagic region over the continental shelf is called the **neritic zone**, and that over the continental slope and beyond is called the **oceanic zone**. The pelagic region can also be subdivided into increments on the basis of water depth (Figure 1.5) or the depth to which light penetrates. The latter factor is, of course, of paramount biological importance. Only within the **photic zone** does enough sunlight penetrate that photosynthesis can occur, and (except in a few special circumstances) all life in the deeper, **aphotic zone** depends ultimately upon organic input from the overlying sunlit layers of the sea. Notable exceptions are the restricted deep-sea hydrothermal vent and benthic cold seep communities in which sulfur-fixing microorganisms serve as the basis of the food chain.[10] The photic zone can be up to 200 m deep in the clear waters of the open ocean, decreasing to about 40 m over continental shelves and to as little

[10]In addition to deep-sea hydrothermal vents, nonphotosynthetic chemoautotroph-based communities have recently been discovered in a cave, Movile Cave in Romania. The base of the food chain in this unique cave ecosystem is autotrophic microorganisms (bacteria and fungi) thriving in thin mats in and near geothermal waters that contain high levels of hydrogen sulfide. This community sustains dozens of microbial and invertebrate species. It is thought that the hydrogen sulfide originates from a deep magmatic source, similar to that seen in deep-sea vents.

as 15 m in some coastal waters. Phytoplankton in the photic zone of the world's oceans account for about half the production of organic matter on Earth. Note that some oceanographers restrict the term aphotic zone to depths below 1,000 m, where absolutely no sunlight penetrates; the region between this depth and the photic zone is then called the **disphotic zone**. Nearly 64% of the surface of planet Earth—over 200 million km^2—lies in the sea below the photic zone (below 200 m depth), and in this region it is estimated that 16 gigatons of carbon fixed by phytoplankton sink to the ocean interior ever year as the only food source for the majority of organisms in the deep sea (one gigaton is a billion tons).

Organisms that inhabit the pelagic zone are often described in terms of their relative powers of locomotion. Pelagic animals that are strong swimmers, such as fishes and squids, constitute the **nekton**. Those pelagic forms that simply float and drift, or generally are at the mercy of water movements, are collectively called **plankton**. Many planktonic animals (e.g., small crustaceans) actually swim very well, but they are so small that they are swept along by prevailing currents in spite of their swimming movements, even though those movements may serve to assist them in feeding or escaping predators. Both photosynthetic organisms (**phytoplankton**) and animals (**zooplankton**) are included among the plankton, the latter being represented by invertebrates such as jellyfishes, comb jellies, arrow worms, many small crustaceans, and the pelagic larvae of many benthic adults. Planktonic animals that spend their entire lives in the pelagic realm are called **holoplanktonic** animals; those whose adult stage is benthic are called **meroplanktonic** animals.

The oceans hold some unique habitats. Perhaps the most well known are coral reefs, which comprise one of the ocean's most diverse ecosystems. In 1997, Marjorie Reaka-Kudla estimated there were about 93,000 described animal species living in the world ocean's coral reefs, but that this was only 10% of the actual diversity (due to all of the undescribed species). Since then, some workers have suggested the number of coral reef animals might be three times, or more, than Reaka-Kudla estimated.

Another well-known ocean ecosystem is the hydrothermal vent communities. These are abundant in the world's oceans, and tend to occur in areas where tectonic plates are moving apart and near plate hotspots (the land equivalents are geysers, fumaroles and hot springs). Oceanic hydrothermal vents have in common an abundance of reduced chemicals, such as sulfides and methane. Chemosynthetic bacteria and archaea form the base of the food chain in these areas. One of the signature animals of hydrothermal vents is the tubeworm annelid *Riftia pachyptila*, which has no mouth, gut or anus and cannot feed by normal means (see Chapter 14). Instead, *Riftia* depends on intracellular chemoautotrophic symbionts—which fill a large internal organ called the trophosome—for nutrition. The symbionts are Proteobacteria, which are functionally analogous to plant chloroplasts in that they generate organic carbon as a food source for their worm hosts.

Another unique marine habitat consists of large vertebrate carcasses that sink to the bottom in deep waters, especially cetaceans (whales, dolphins, and porpoises). Unlike in shallower waters, where a large carcass will be consumed by scavengers over a relatively short period of time, these "whale falls" can last for decades and allow for the establishment of an entire food web, characterized by a specific assemblage of species, especially crustaceans, annelids, and fishes. Some of the more famous members of the whale fall community are species of *Osedax*, the so-called zombie worms, which consume the bones of dead cetaceans in these communities (see Chapter 14).

One of the most unusual marine environments inhabited by invertebrates are the pockets of concentrated brines that are encased in the ice matrix of Earth's polar regions (brines are generally considered to be 5% or more dissolved salts; ocean water is 3.4–3.5% salt). Living at very low temperatures (to –20°C) and light levels, a food web in miniature exists, including photosynthetic bacteria and protists (especially diatoms), heterotrophic protists, flatworms, small crustaceans, etc. These are highly-adapted, normally planktonic species that get trapped, and survive, when winter sea ice forms.

Estuaries and Coastal Wetlands

Estuaries usually occur along low-lying coasts and are created by the interaction of fresh and marine waters, typically where rivers meet the sea. Here one finds an unstable blending of freshwater and saltwater conditions, moving water, tidal influences, and drastic seasonal fluctuations. Estuaries receive high concentrations of nutrients from terrestrial runoff in their freshwater sources and are typically highly productive environments. Temperature and salinity vary greatly with tidal activity and with season. Depending on tides and turbulence, the waters of estuaries may be relatively well mixed and more or less homogeneously brackish, or they may be distinctly stratified, with freshwater floating on the denser salt water below.

The amount of dissolved oxygen in an estuary may also change markedly throughout a 24-hour cycle as a function of temperature and the metabolism of autotrophic organisms. In many cases, hypoxic (very low oxygen) conditions may occur on a daily basis, especially in the early morning hours. Animals inhabiting these areas must be capable of migrating to regions of higher oxygen levels, be able to store oxygen bound to certain body fluid pigments, or be able to switch temporarily to metabolic processes that do not require oxygen-based respiration. Furthermore, vast amounts of silt borne by freshwater runoff are carried into the waters of estuaries; most of this silt settles out and creates

Figure 1.6 A few of Earth's major ecosystems. (A) Exposed rocks and algae in the intertidal zone, northern California. (B) A tidal flat in a salt marsh, New York. (C) A mangrove swamp at low tide, in Mexico. (D) A freshwater stream in a tropical wet forest ("rain forest"), Costa Rica. (E) Flowering trees in a tropical dry forest, Costa Rica. (F) The Sonoran desert in Arizona. Fully a third of the land on our planet is desert or semi-desert, and this ecosystem is predicted to grow with global warming.

extensive tidal flats or deltaic regions (Figure 1.6B). In addition to the natural stresses common to estuarine existence, the inhabitants of estuaries are also subject to stresses resulting from human activity—pollution, thermal additions from power plants, dredging and filling, excessive siltation resulting from coastal and upland deforestation and development, and storm drain discharges are some examples.

Most coastal wetlands and estuaries, such as salt marshes and mangrove swamps, are characterized by stands of halophytes (flowering plants that flourish in saline conditions; Figure 1.6B,C). Salt marshes and mangrove swamps are alternately flooded and uncovered by tidal action within the estuary, and are thus subjected to the fluctuating conditions described above. The dense halophyte stands and the mixing of waters of different salinities create an efficient nutrient trap. Instead of being swept out to sea, most dissolved nutrients entering an estuary (or generated within it) are utilized there, yielding some of the most productive regions in the world. This great productivity does eventually enter the sea in two principal ways: as plant detritus (mainly from halophyte debris), and via the nektonic animals that migrate in and out of the estuary. The contribution of estuaries to general coastal productivity can hardly be exaggerated. The organic matter produced by plants of the Florida Everglades, for example, forms the base of a major detritus food web that culminates in the rich fisheries of Florida Bay. Furthermore, 60–80% of the world's commercial marine fishes rely on estuaries directly, either as homes for migrating adults or as protective nurseries for the young. Estuaries and other coastal wetlands are also of prime importance to both resident and migratory populations of water birds.

A large number of invertebrates have adapted to life in these dynamic environments. In general, animals have but two alternatives when encountering stressful conditions: either they migrate to more favorable environments, or they remain and tolerate (accommodate to) the changing conditions. Many animals migrate into estuaries to spend only a portion of their life cycle, whereas others move in and out on a daily basis with the tides. Other species remain in estuaries throughout their lives, and these species show a remarkable range of physiological adaptations to the environmental conditions with which they must cope (Chapter 4).

Freshwater Habitats

Because bodies of fresh water are so much smaller than the oceans, they are much more readily and drastically influenced by extrinsic environmental factors, and thus are relatively unstable environments (Figure 1.6D). Changes in temperature and other conditions in ponds, streams, and lakes may occur quickly and be of a magnitude never experienced in most marine environments. Seasonal changes are even more extreme, and may include complete freezing during the winter and complete drying in the summer. Ponds that hold water for only a few weeks during and after rainy seasons are called **ephemeral pools** (or vernal pools). They typically contain a unique and highly specialized invertebrate fauna capable of producing resting, or diapause, stages (usually eggs or embryos) that can survive for months or even years without water. As stressful as this sounds, ephemeral pools contain rich communities of plant and animal life, especially endemic species of crustaceans. **Diapause** is a form of dormancy in which invertebrates in any stage of development before the adult, including the egg stage, cease their growth and development. Diapause is genetically determined. Some species are programmed to enter diapause when certain environmental conditions provide the proper cues (often a combination of temperature and length of daylight). Hibernation and aestivation are two other types of dormancy, but they are not genetically programmed and may occur irregularly, or not at all, during any stage of an animal's development. **Hibernation** is a temporary response to cold, while **aestivation** is a temporary response to heat.

The very low salinity of fresh water (rarely more than 1‰) and the lack of constant relative ion concentrations subject freshwater inhabitants to severe ionic and osmotic stresses. These conditions, along with other factors such as reduced buoyancy, less stable pH, and rapid nutrient input and depletion produce environments that support far less biological diversity than the ocean does. Nonetheless, many different invertebrates do live in fresh water and have solved the problems associated with this environment. Special adaptations to life in fresh water are summarized in Chapter 4 and discussed in relation to the groups of invertebrates that have such adaptations in later chapters.

Stygobionts are aquatic obligates in subterranean ground waters, including streams and underground lakes that often connect to the sea. Aquatic creatures that inhabit subterranean alluvial waters and karst ecosystems also fall into this category.

Freshwater habitats are some of the most threatened environments on Earth. Throughout the United States, people destroy 100,000 acres of wetlands annually. Rare aquatic habitats such as ephemeral pools and subterranean rivers are disappearing faster than they can be studied. Underground, or hypogean, habitats are often aquatic, and these habitats are quickly being destroyed by pollution and groundwater overdraft.

Terrestrial Habitats

Life on land is in many ways even more rigorous than life in fresh water. Temperature extremes are usually encountered on a daily basis, water balance is a critical problem, and just physically supporting the body requires major expenditures of energy. Water provides a medium for support, for dispersing gametes, larvae, and adults, and for diluting waste products, and is a source of dissolved materials needed by animals. Animals living in terrestrial environments do not enjoy these benefits of water, and must pay the price.

Relatively few phyla have successfully invaded the terrestrial world. Invertebrate success on land is exemplified by the arthropods, notably the terrestrial isopods, insects, and spiders, mites, scorpions, and other arachnids. These arthropod groups include truly terrestrial species that have invaded even the most arid environments (Figure 1.6F). Except for some snails and nematodes, all other land-dwelling invertebrates, including such familiar animals as earthworms, are largely restricted to relatively moist areas.

Terrestrial environments are commonly described in terms of moisture availability. **Xeric** habitats are dry, **mesic** environments have a moderate amount of moisture, and **hydric** habitats are very wet. Adaptations of terrestrial invertebrates to these various conditions are described in Chapter 4 and, for individual taxa, in subsequent chapters.

A Special Type of Environment: Symbiosis

Many invertebrates live in intimate association with other animals or plants. And, of course all animals share an ancient genetic relationship with prokaryotes. An intimate association between two different species is termed a symbiotic relationship, or **symbiosis**. Symbiosis was first defined in 1879 by German mycologist H. A. DeBary as "unlike organisms living together." In most symbiotic relationships, a larger organism (called the host) provides an environment (its body, burrow, nest, etc.) on or within which a smaller organism (the symbiont) lives. Some symbiotic relationships are rather transient—for example, the relationship between ticks or lice and their vertebrate host—whereas others are more or less permanent. Some symbionts are opportunistic (**facultative**), whereas others cannot survive without their host (**obligatory**). And, importantly, some symbioses evolve such intimacy that, over time, the genes of the smaller symbiont get incorporated into the genome of the "host."

Symbiotic relationships can be subdivided into several categories based on the nature of the interaction between the symbiont and its host (although in many cases the exact nature of the relationship is unknown). Perhaps the most familiar type of symbiotic relationship is **parasitism**, in which the symbiont (a parasite) receives benefits at the host's expense. Parasites may be external (**ectoparasites**), such as lice, ticks, and leeches; or internal (**endoparasites**), such as liver flukes, some roundworms, and tapeworms. Other parasites may be neither strictly internal nor strictly external; rather, they may live in a body cavity or area of the host that communicates with the environment, such as the gill chamber of a fish or the mouth or anus of a host animal (**mesoparasites**). Some parasites live their entire adult lives in association with their hosts and are permanent parasites, whereas temporary, or intermittent parasites, such as bedbugs, only feed on the host and then leave it. There are even nest parasites, such as ant-nest beetles (Carabidae: Paussini) that inhabit the nests of host ants and feed on them, apparently by tricking the ants into thinking they are one of them. In some of these cases, the line between parasitism and predation becomes blurred. Temporary parasites, such as mosquitoes and aegiid isopods, are often referred to as **micropredators**, in recognition of the fact that they usually "prey" on several different host individuals (that happen to be much larger than themselves).

Parasites that parasitize other parasites are **hyperparasitic**. **Parasitoids** are insects, usually flies or wasps, whose immature stages feed on their hosts' bodies, usually other insects, and ultimately kill the host. A **definitive host** is one in which the parasite reaches reproductive maturity. An **intermediate host** is one that is required for parasite's development, but in which the parasite does not reach reproductive maturity. Multiple hosts are more common than not in the world of parasites. Even most human pathogens circulate in animals (or else originated in nonhuman hosts), such as influenza, plague, and trypanosomiasis, which all transmit from animals to humans. Over half of all human pathogens are **zoonotic** (have animals hosts, other than humans, in their life cycle). Every species probably serves as host to several (or many) parasites. It has been suggested that parasitism is the most popular lifestyle on Earth! In fact, it has been estimated that 50 to 70% of the world's species are parasitic, making parasitism the most common way of life. Since insects are the most diverse group of organisms on Earth, and since all insects harbor numerous parasites, it is fair to say that the most common mode of life on Earth is that of an insect parasite. Most parasites have yet to be described, and as species go extinct, so, often, do their parasites (when the passenger pigeon was eliminated in 1914, it took two species of parasitic lice with it).

A few groups of invertebrates are predominantly or exclusively parasitic, and almost all invertebrate phyla have at least some species that have adopted parasitic lifestyles. Many texts and courses on parasitology pay particular attention to the effects of these animals on humans, crops, and livestock. Here we also try to focus on parasitism from "the parasite's point of view," that is, as a particular lifestyle suited to a specific environment, requiring certain adaptations and conferring certain advantages.

Mutualism is another form of symbiosis that is generally defined as an association in which both host and symbiont benefit. Such relationships may be extremely intimate and important for the survival of both parties; for example, the bacteria in our own large intestine are important in the production of certain vitamins and in processing material in our gut. In fact, beneficial associations with specific bacterial symbionts characterize many, if not all, animal species, although most of these relationships have not been well studied. Another example is the relationship between termites and certain protists that inhabit their digestive tracts and are responsible for the breakdown of cellulose into compounds that can be assimilated by the insect hosts. Other mutualistic relationships may be less binding on the organisms involved. Cleaner shrimps, for example, inhabit coral reef environments, where they establish "cleaning stations" that are visited regularly by reef-dwelling fishes that present themselves to the shrimps for the removal of parasites. Obviously, even this rather loose association results in benefits for the shrimps (a meal) as well as for the fishes (removal of parasites). The mutualistic relationships between plants and their pollinators are essential to the survival of most flowering plants and their insect partners (and, in some cases, their bird or nectar-feeding bat partners). Mutualisms are some of the most threatened ecological relationships on earth. For example, as we loose pollinator species the plants they rely on suffer and begin to decline. In some cases, plants rely on single species for pollination. Imagine what will happen to the giant fig trees of the tropics (*Ficus* is the most widespread plant genus in the tropics) as they loose the single parasitic wasp varieties that pollinate each of the 900 or so species. Of immediate concern is Colony Collapse Disorder (CCD), a syndrome that is killing honeybees worldwide. Bees pollinate 71% of the crops that provide 90% of human food; but since 2006, U.S. beekeepers have seen honeybee colony loss rates increase to 35% per year. The causes for CCD have yet to be elucidated, but overuse of pesticides in the environment appears to play a key role.

A third type of symbiosis is called **commensalism**. This category is something of a catchall for associations in which neither significant harm nor mutual benefit is obvious. Commensalism is usually described as an association that is advantageous to one party (the symbiont) but leaves the other (the host) unaffected. For instance, among invertebrates there are numerous examples of one species inhabiting the tube or burrow of another (**inquilism**); the former obtains protection, food, or both with little or no apparent effect on the latter. A special type of commensalism is **phoresis**, wherein the two symbionts "travel together," but there is no physiological or biochemical dependency on the part of either participant. Usually one phoront is smaller than the other and is mechanically carried about by its larger companion.

There is a good deal of overlap among the categories of symbiosis described above, and many animal relationships have elements of two or even of all the categories, depending on life history stage or environmental conditions. Taken in its broad sense, the concept of symbiosis has profound implications for understanding Earth's biodiversity. It has been said that at least half the planet's species are symbionts and that all species have many symbiotic partnerships—concepts suggesting that every individual is an ecosystem.

Biodiversity Patterns

Biodiversity has been defined as the variety of genes, species, higher taxa, and ecosystems that constitute life on Earth. There are all kinds of biodiversity patterns on the Earth. At the species level, perhaps the best-known pattern is the **latitudinal biodiversity gradient** ranging from low at the poles to high at the equator. This pattern is so general across so many taxa (though not all taxa) that it suggests the existence of a general explanation. Many hypotheses have been proposed to explain this pattern, but none have proven flawless. Some hypotheses focus on area relationships—for example, the tropics (roughly between the Tropics of Cancer and Capricorn) comprise the largest and most benign landscape-scale region on the Earth and over time this might result in a great accumulation of species. Other hypotheses focus on biology and ecology. For example, because of their latitude the tropics receive more sunlight and increased solar energy can drive higher productivity, which in turn can presumably support more species. One of the most studied hypotheses to explain biodiversity patterns is the humped-back model (HBM; in reference to a curve often revealed when diversity is plotted against productivity), which suggests that plant diversity peaks in communities at intermediate productivity levels. At low productivity few species can tolerate the environmental stresses, and at high productivity a few highly competitive species dominate. Support for HBM has waxed and waned over the years; currently, it seems rather well supported by the evidence, at least in terrestrial environments. There are also ideas that suggest there may be faster speciation rates in warmer (tropical) climates leading to an accumulation of species. Other hypotheses focus on historical factors, such as climate stability in the tropics over Phanerozoic time. Still others explain the latitudinal biodiversity gradient on the basis of evolutionary ecology, suggesting that species competition, predation, and symbioses are stronger in the tropics and these interactions promote species coexistence and specialization, leading to greater speciation and/or niche specialization in the tropics. Paleontological studies of oceanic reef deposits support the idea that animals living in tropical reefs have high speciation rates (i.e., they are "cradles of evolution") and tend to export species to other, less biodiverse regions.

Within the marine realm today, the tropical West Pacific (especially the Indo-Australian Archipelago) has far and away the greatest biodiversity, especially associated with the coral habitats (both reefs and non-reefs) that dominate shallow seas in that region. This extraordinary diversity has been attributed to three possible mechanisms: (1) the coral habitats have served as refugia, preserving species from extinction during global cooling episodes, such as the 30 or so glacial cycles of the Pleistocene (i.e., these habitats are "species sinks" where species accumulate over time); (2) these habitats have been sources of recruits and recolonization during favorable periods (i.e., they are "species pumps"); and (3) these habitats have exceptionally high speciation rates (i.e., they are "evolutionary cradles"). Of course, any or all of these explanations could be correct. The paleontological record tells us that the West Pacific biodiversity hotspot is much older than the Pleistocene, and also that over Earth's history, marine hotspots have moved about, probably as a result of changes in ocean basin geography.

Interestingly, research shows that there is a correlation between species diversity and the size of the areas being measured. For terrestrial species, some studies show that precipitation is most influential at small spatial scales, but cloud cover and area are more important at larger scales. At the local-to-landscape scale, fire, storm, and hurricane history often correlate to regional diversity. At global scales, processes such as tectonic plate movements and variations in sea level can account for differences in higher-level diversity (e.g., mammal families).

New Views of Invertebrate Phylogeny

Significant changes in our view of animal phylogeny have come about over the past two decades, primarily from the rapidly expanding field of molecular phylogenetics, and also from new paleontological work and new ultrastructural and embryological studies. Broadly speaking, the 32 phyla of Metazoa are now divided into the four basal, or non-bilaterian phyla (Porifera, Cnidaria, Ctenophora, Placozoa) and the Bilateria (the triploblastic phyla). The Bilateria comprise two long-recognized clades, Protostomia and Deuterostomia, but four phyla once regarded as deuterostomes are now recognized as protostomes (Chaetognatha, Phoronida, Bryozoa, Brachiopoda), leaving only three phyla of Deuterostomia—Echinodermata, Hemichordata, Chordata. The newly-created phylum Xenacoelomorpha, housing taxa formerly allied with the Platyhelminthes (Acoela, Nemertodermatida, and *Xenoturbella*), remains enigmatic but most analyses place this group as basal bilaterians.

Among the protostomes, molecular phylogenetic research identifies two major clades, Ecdysozoa and Spiralia. However, five protostome phyla still remain elusive in their affinities: Chaetognatha, Platyhelminthes, Rhombozoa, Orthonectida, and Gastrotricha. And, despite great efforts to date, the specific relationships of the phyla within Ecdysozoa and Spiralia also remain somewhat evasive, although some strongly supported internal clades have been identified. For example, the Ecdysozoa comprises three well-supported clades—Scalidophora (Priapula, Loricifera, Kinorhyncha), Nematoida (Nematoda, Nematomorpha), and Panarthropoda (Onychophora, Tardigrada, Arthropoda). Relationships among the Spiralia have been more difficult to tease apart, although one well-supported internal clade is recognized, the Gnathifera, containing the phyla Gnathostomulida, Micrognathozoa, and Rotifera. Also, recent analyses suggest that the Lophotrochozoa might be a clade within Spiralia, containing (as the name suggests) the lophophorate phyla (Phoronida, Bryozoa/Ectoprocta, Brachiopoda) plus Nemertea, Annelida, and Mollusca (and possibly also Cycliophora, Entoprocta, and Platyhelminthes); however, this needs further testing. Some recent studies have suggested that ctenophores, not sponges, could be basal Metazoa, but this idea might be based on methodological artifacts (Pisani et al 2015).

Some phyla recognized in the last edition of this book are now known to be specialized lineages of other phyla, hence Echiura and Sipuncula are now viewed as specialized clades within the Annelida, and Acanthocephala are now seen as a parasitic lineage of Rotifera. Hexapoda (insects and their kin) now seem certain to have arisen from within the Crustacea, making the latter a paraphyletic group. And the Myxozoa, once thought to be protists, are now known to be highly derived and extraordinarily specialized, parasitic cnidarians.

Among the deuterostomes, Echinodermata and Hemichordata now appear to form a well-supported clade (Ambulacraria), and this is the sister group to Chordata. Within Chordata, the long-standing idea that lancelets (Cephalochordata) are the sister group to the Vertebrata (Craniata) has been overturned, and strong evidence now supports the tunicates (Urochordata) as the sister group to the vertebrates. The many phylogenetic ideas about these phyla and clades are presented in the following chapters. A summary classification is provided on the front inside cover, and a geologic time scale is on the back inside cover.

Not only have the phyla assigned to the old groups Protostomia and Deuterostomia been shuffled, but the meaning of these old names now rings only partly true. New developmental research, especially gene-expression studies, has revealed that many (perhaps most) members of Protostomia do not have "protostomous" development, but instead have idiosyncratic patterns or deuterostomous patterns.

Teasing apart the evolutionary relationships of the metazoan phyla has been challenging because of their ancient roots, and it has required finding informative phylogenetic markers for groups that are hundreds of millions of years old. The main obstacle has been that different regions of the genome have experienced different evolutionary histories, thus producing conflicting phylogenetic signals. Not only has this required using multiple genes, and even genome-level analyses, it has also led to the development of sophisticated algorithms that model the evolution of individual amino acids. Those phyla still stubbornly resisting phylogenetic understanding will probably require new and even more sophisticated analysis protocols. One of the important things that has been revealed over the past couple decades through the work of molecular phylogenetics and developmental biology is that there is far more homoplasy in the animal world than we suspected, even in such complex features as segmentation and nervous systems. For example, one surprising new discovery is that the annelids (now considered spiralians) and the Panarthropoda (Onychophora, Tardigrada, Arthropoda—now placed within Ecdysozoa) are far more distantly related than was previously thought. It might be hard to imagine the highly complex, multilevel process of segmentation shared by annelids and arthropods evolving independently, and it seems more likely that they share some ancient bilaterian genes that predisposed them to a nearly identical embryological segmental patterning process. We still have much to learn about this surprising new twist in animal phylogeny. Even spiral cleavage, once thought to be a largely immutable aspect of animal development (due to "developmental constraints"), has been shown to be flexible—to the point of being greatly altered or possibly even lost in some lineages. Further, we now know that many genes once thought to be specific to particular innovations actually appeared in the tree of life much earlier than did the features themselves; so we can no longer necessarily expect to find novel genes associated with novel morphological or developmental features. For example, cellular adhesion and transcriptional regulation genes essential to animal multicellularity also occur in the protistan ancestral line leading to Metazoa. And, we now know that many genes once thought to be specific to vertebrates can be found as far down the tree of life as Cnidaria (but have been lost in many lineages in-between). Many novel phenotypes have arisen by way of modification of gene function and by interactions between existing genes.

Legacy Names

As you might have guessed by now, some of the names in use today for higher taxa were created before biologists achieved their current understanding of animal phylogeny, and due to more recent reassignments of phyla these names are no longer fully descriptive. As noted above, the two great clades of Bilateria have long been called Protostomia and Deuterostomia. The name Protostomia was created for those animal phyla in which the blastopore gives rise to the mouth during embryogenesis (*proto* = first, *stoma* = mouth). In contrast, the name Deuterostomia was created for those animals that form the mouth not from the blastopore but elsewhere (*deutero* = second); in many of these groups, the the blastopore gives rise to the anus. However, over the past few years, we have discovered that some phyla with deuterostomous development actually belong in the clade called Protostomia (e.g., Chaetognatha, Phoronida, Bryozoa, Brachiopods, Priapula, Arthropoda). Thus both names have been rendered less-than-perfectly descriptive. Similarly, the name Spiralia was created by Waldemar Schleip in 1929, after the stereotypical spiral cleavage seen in most member phyla, but even this name is not fully descriptive because several phyla of Spiralia lack this developmental pattern (e.g., Bryozoa, Brachiopoda, Gastrotricha, Orthonectida). These kinds of names for higher taxa, that are no longer fully descriptive based on their original intent, are **legacy names**. To fully appreciate them, one needs to understand a bit of their history, otherwise they can seem illogical.

Phylogenetics and Classification Schemes

As noted above, the rapidly growing number of molecular phylogenetic studies has changed many of our views on animal relationships and classifications. This work is also generating large and highly detailed phylogenetic trees, with long branching patterns depicting the history of life on Earth. Such detail creates challenges for biologists who like to produce classifications that accurately reflect phylogeny, because the traditional Linnean hierarchical ranks are too few in number to capture the great depth and detail of these trees. Many of the new multigene or genomic trees have scores of branching points, or even hundreds of branches that appear as long comblike topologies. But the standard Linnean ranks number only 30 or so. There are a few solutions to this dilemma, such as using unranked classifications, but none of the solutions is perfect. We briefly discuss these in Chapter 2. In this book, we mostly use ranked classifications that mirror, or at least do not conflict with the phylogeny of the groups in question. In a few instances, we have had to use fully or partly unranked subordinated classifications (e.g., annelids and molluscs). Again, this is all explained in more detail in Chapter 2. But the upshot is, things are changing and the "traditional" classification schemes many students are used to seeing are beginning to look rather different these days.

Some Comments on Evolution

**Fitness By Any Other Name
Would Be As Loose**

A group inept

Might better opt

To be adept

And so adopt

Ways more apt

To wit, adapt.

John Burns
Biograffiti, 1975

This book takes evolution as its central theme. Although evolutionary biology has expanded rapidly over the past few decades, at its core remains the seminal and galvanizing concept of Charles Darwin and Alfred Russell Wallace—that natural selection and descent with modification are the agent and manifestation, respectively, of evolutionary change. Since the "molecular biology revolution" began in the 1980s, the paradigms that have guided evolutionary biology have expanded greatly. Molecular biology has produced dramatic discoveries and will no doubt continue to do so for many decades to come.

There are three fundamental patterns we see when we examine evolutionary history: anagenesis, speciation, and extinction. **Anagenesis** is the process by which a genetic or phenotypic character changes within a species over time, whether the change is random or nonrandom, slow or rapid. Anagenesis seems to be driven by those neo-Darwinian processes often referred to as microevolution—the within-species, generation-by-generation evolution of populations and groups of populations over the "lifetime" of a species. Natural selection and adaptation are powerful driving forces at this level. **Speciation** is the "birth" of a species, and **extinction** is the "death" (termination) of a species. Speciation and extinction engage processes outside the natural selection/adaptation paradigm—processes often referred to as macroevolution. The mechanisms that initiate and sculpt each of these processes differ. Most college courses today focus primarily on microevolution, or anagenesis, and most students reading this book already know a great deal about population genetics and natural selection. Population genetics focuses on vertical transmission of genetic information. However, the view that all of evolution can be understood solely on the basis of microevolutionary phenomena is being reexamined in light of new ideas regarding evolutionary change. Consequently, we would like to introduce readers to some ideas with which they might be less familiar. We will do so by first discussing within-species processes (presented here under the term "microevolution"), and then speciation and extinction (grouped under the heading "macroevolution").

Microevolution

The neo-Darwinian evolutionary model, or so-called "modern synthesis," that resulted from the integration of Mendelian genetics into Darwinian natural selection theory dominated evolutionary biology through the twentieth century. Basically, the neo-Darwinian view holds that all evolutionary changes result from the action of natural selection on variation within populations (see John Burns's poem on page 26). This view has been called the "adaptationist paradigm." The theory focuses on adaptation and deals primarily with genes and changes in allelic frequencies within populations. These genetic variations come about primarily by recombination and mutation, although the random phenomena of genetic drift and founder effect are also part of the neo-Darwinian synthesis.

Evolution by natural selection can be viewed as a deterministic process, even though certain elements of chance are accepted within the theory (e.g., mutation, random mating, the founder effect). The theory of natural selection implies that, given a complete understanding of the environment and genetics, evolutionary outcomes should be largely predictable. The theory of natural selection further implies that virtually all of the characteristics animals possess are products of adaptations leading to increased fitness (ultimately, to increased reproductive success). An adaptationist view might lead one to assume that every aspect of an animal's phenotype is the product of natural selection working to increase the fitness of a species in a particular environment. Microevolution is thus seen as a deterministic, within-species phenomenon that affects population genetics on a generation-to-generation basis to produce changes and patterns in gene frequencies within and among populations.

Macroevolution

Macroevolution is the focus of some of the most interesting debates among biologists today. Macroevolutionary phenomena can include the origin of new species (cladogenesis), "explosive" adaptive radiations that appear to be linked to the opening up of new ecological arenas or niches, transgenic events, major shifts in developmental processes that might result in new body plans, various karyotypic alterations (e.g., polyploidy and polyteny), and mass extinction events (and the subsequent new biotic proliferations). One of the best examples of macroevolution is the origin and rise of birds from their massive-bodied theropod dinosaur ancestors. The lineage leading to modern birds underwent rapid sustained miniaturization, with species evolving at a much faster rate than seen in other theropod lineages. Birds also rapidly developed new ecological and morphological innovations linked to smaller size, including a rate of skeletal adaptations four times faster than other dinosaurs. In part, the rapid radiation of birds was driven by the innovation of flight. Macroevolution asks the question, "what drives evolutionary innovation?"

Mass extinction events in Earth's history have played major roles in reshaping the directions of animal evolution in unpredictable ways. The largest of these extinction events wiped out a majority of life forms on Earth. In the Permian–Triassic extinction event, an estimated 90% of all marine species went extinct (although no phylum is known to have gone extinct since the start of the Cambrian) resulting in a worldwide reorganization of life. Mass extinctions are profound macroevolutionary events that can abruptly (in geological time) terminate millions of species and lineages.

In contrast to microevolution, macroevolution is evolutionary change, often rapid, that produces phylogenetic pattern formation above the species level (e.g., the patterns depicted on the phylogenetic trees in this book). The fossil record suggests that speciation events (one species giving rise to one or more new species) tend to be rapid, or geologically instantaneous. Analysis of the fossil record also shows that the number of species has increased gradually since the end of the Proterozoic, with this diversification periodically interrupted by mass extinctions. And mass extinctions have always been followed by periods of rapid speciation and radiation at higher taxonomic levels (i.e., macroevolution).

Newer views suggest that speciation might not be initiated by natural selection, but rather by processes outside the natural selection paradigm—perhaps most frequently by stochastic processes. Microevolution can be thought of as a within-species process that maintains genomic continuity and continually "fine-tunes" populations and species to their changing environment. A reasonable analogy might be the basic metabolic activities that keep your own body "fine-tuned" to the environment—a background process that is always at work maintaining a level of homeostasis (within your body, or within a species' gene pool). A macroevolutionary event, on the other hand, is typically a process that disrupts that genomic, or reproductive, continuity in a species and may thus initiate speciation events. Following the above analogy, macroevolutionary events disrupt the homeostasis of species' gene pools. Some examples of stochastic events that can lead to macroevolutionary change are described below.

The geneticist Richard Goldschmidt, the paleontologist Otto Schindewolf, and the zoologists René Jeannel and Claude Cuénot all maintained until the 1950s that neither evolution within species nor simple allopatric speciation could fully explain macroevolution. They advanced an idea called saltation theory—the sudden origin of wholly new types of organisms—the "hopeful monsters" of Goldschmidt—in great leaps of change. It has been proposed that one way such rapid changes might occur is through transgenic events, involving the lateral transfer of genetic material from one species to another. Proposed mechanisms of lateral genetic transfer include transposable genetic elements and symbiogenesis. The endosymbiont bacteria that gave rise to mitochondria and chloroplasts have been one major source of bacterial genes in eukaryotic nuclear genomes, and their ancestral lineages are the α-Proteobacteria and Cyanobacteria, respectively. Even in the human genome, it has been shown that dozens of genes have probably been transferred from bacteria to humans (or to one of our vertebrate ancestors) over the course of evolution.

Probably all phyla have some DNA in their genome that was "adopted" from other species in their distant past, especially from Eubacteria and Archaea. Two phyla that have been shown to have notably large percentages of their DNA derived from other life forms via horizontal transfer are rotifers and tardigrades. In the latter case, as much as one-sixth of their DNA might have been acquired by horizontal transfer (Chapter 20). A recent study by Thomas Boothby and others (2015) suggested that tardigrades (and rotifers, and perhaps other invertebrates) might be prone to integrating foreign genes into their genomes because they famously have the ability to survive extreme environmental stress. These researchers speculated that in conditions of extreme stress, such as desiccation, a tardigrade's DNA breaks into tiny pieces and then, when the cells rehydrate, the cell membranes and nucleus become temporarily leaky and DNA molecules can pass through, allowing an opportunity for the "host" to stitch in foreign DNA from the environment.

Transposable elements (TEs) are specialized DNA segments that move (transpose) from one location to another, either within a cell's DNA, between individuals in a species, or even between species. They were discovered in maize (*Zea mays*) by the Nobel laureate Barbara McClintock in the 1950s, but little was known about them until recently. With the growth of molecular genetics, hundreds of TEs now have been identified—over 40 different ones are known from the laboratory fruit fly *Drosophila melanogaster* alone. The mechanisms of TE transfer between organisms are not yet well understood. However, the transfer of genetic elements from one species to another is suspected to be by way of viruses, bacteria, arthropod parasites, or other vectors. There is strong evidence, for example, that parasitic mites have been responsible for the lateral transfer of genetic elements among *Drosophila* species.

The movement of a transposable element within a genome is mediated by a TE-encoded protein called a "transposase," probably interacting in complex ways with certain cellular factors. A transposase recognizes the ends of the TE, breaks the DNA at these ends to release the TE from its original position, and joins the ends to a new target sequence. The transposition of some TEs from bacteria to bacteria, and from bacteria to plant cells, is partially understood, and we know that the introduction of these DNA segments can contribute powerful mutagenic qualities to the new host's genome. Recent work suggests that a great deal of such "gene swapping" took place during the early evolution of the prokaryotes. TEs have been best studied in prokaryotes, but they have been found in most organisms that have been examined, including insects, mammals, flowering plants, sponges, and flatworms. Although we lack specific evidence, there is reason to suspect that TEs could have been responsible for some of the major genetic innovations that have taken place in the history of life.

Aside from transposable elements, the duplication of a chromosomal segment that then becomes separated from the original segment, ending up in a different chromosomal location, is now known to be a fairly common occurrence. A number of human genetic disorders are known to be associated with the increased expression of genes contained within such duplications. Duplicate genes provide a rich new substrate on which evolution can work. One member of the duplicate pair could take on a new function, or two duplicate genes could divide the multiple functions of the ancestral gene between them with natural selection then refining each copy separately to a more restricted set of tasks. Even single genes can duplicate, giving rise to redundancy that can provide the fodder for rapid evolution of new gene functions. For example, in the water flea (Crustacea: Branchiopoda: Diplostraca: *Daphnia*), tandem gene duplication has been extensive and is likely responsible for the extreme phenotypic plasticity of these creatures, as well as their extreme ecophysiological adaptability.

Rapid speciation by way of successful hybridization between two species is another example of macroevolution. Although most common in plants, and probably microbes, hybrid speciation does occur in animals as well, and it has been documented in African cichlids, cyprinid fishes, *Rhagoletis* fruit flies, and *Heliconius* butterflies. The Appalachian tiger swallowtail, *Papilio appalachiensis* evolved from mixing between the eastern tiger swallowtail, *P. glaucus*, and the Canadian tiger swallowtail, *P. canadensis*. The Appalachian species rarely reproduces with its parental species. Studies suggest the Appalachian tiger evolved about 100,000 years ago, and its genome is a mixture of the parental genomes.

Another way in which evolutionary novelties can arise is through symbiosis. The Russian biologist Konstantin Mereschkovsky (1855–1921) developed the "two-plasm" (cell within a cell) theory, claiming that chloroplasts originated from blue-green algae (Cyanobacteria). For this process, he invented the term symbiogenesis (today, this is often called the **endosymbiotic theory**). In Chapter 3, we describe the symbiogenic origin of the eukaryotic cell, which probably arose by way of incorporation of once free-living prokaryotes that came to be what we recognize today as mitochondria, chloroplasts, cilia, flagella, and other organelles. Although symbiogenesis is an old idea, Lynn Margulis and others vigorously championed it in the twentieth century. Symbiogenesis is the permanent merger of two organisms from phylogenetically distant lineages into one radically more complex organism. Once thought to be rare, we now know that this phenomenon is common throughout life. Three examples are exceptionally important with regards to the evolution of the animal kingdom: (1) intracellular enslavement by an early eukaryote of an α-proteobacterium by host protein insertion to make the first mitochondria, (2) later capture of a cyanobacterium by a heterotrophic protist to create the first chloroplast, thereby launching the kingdom Plantae, and (3) secondary enslavement of a red alga to yield more complex membrane topology in the phagophototrophic protist group known as the Chromalveolata. It has been estimated that over 50% of all formally described protists are chromalveolates. In addition to the enslavement of a red alga to give rise to the plastid ancestor of the Chromalveolata event, green algae have also been endosymbiotically captured, at least twice, to begin other new lineages of protists.

Beyond the origin of the eukaryotic cell, symbiogenesis has also been at work in many other systems, but we are just beginning to understand how pervasive this has been among Metazoa. In extremely intimate symbiotic partnerships, the two symbionts can have profound effects on each other's genetic evolution. Such partnerships are major sources of evolutionary innovation and they have driven rapid diversification of organisms, allowed hosts to harness new forms of energy, and resulted in profound modifications of Earth's nutrient cycles and geochemistry. Recent genomic studies have revealed the ubiquity of intimate symbioses. Many invertebrates are involved in such relationships, including the corals and other cnidarians that serve as hosts for symbiotic dinoflagellates (called zooxanthellae) that live within their tissues. Various animals that harbor (and exploit) tetrodotoxin-secreting bacteria (many chaetognaths, the blue-ringed octopus, a sea star, a horseshoe crab, and certain tetraodontid fishes), squids with luminous bacteria, and lichens (an intimate association between fungi and Cyanobacteria or green algae) are other examples. Many insects harbor endosymbionts—bacteria that live within the host's cells. Although separate organisms, they function as a metabolic unit. For example, species of mealy bugs depend on bacteria endosymbionts for nutrient provisioning, and the endosymbiont can in turn harbor its own endosymbiont. Endosymbionts can even speciate within their hosts, as has been shown in cicadas. That symbionts can affect the evolution of their hosts in unexpected ways can also be seen in parasites that enhance their own chances for survival by altering aspects of their host's lives—for example, parasites that increase the likelihood that their intermediate host will fall prey to their definitive host by changing the intermediate host's size, color, biochemistry, or behavior in ways that make it more vulnerable to predation. One common outcome of host-symbiont integration is a reduction in symbiont genome size. A species of leaf hopper, *Macrosteles quadrilineatus*, harbors the endosymbiont *Nasuia deltocephalinicola*, the smallest bacterial genome sequenced to date.

Another revelation in our thinking about macroevolution has come from the discovery of homeobox (Hox) genes. These master regulatory genes encode proteins that regulate the expression of other genes. They modulate other sets of developmental genes and in doing so "select" the developmental pathways that are followed by dividing cells. Hox genes have two functions in the early development of embryos: (1) they encode short regulatory proteins that bind to particular sequences of bases in DNA and either enhance or repress gene expression, and (2) they encode proteins that are expressed in complex patterns that determine the basic geometry of the organism. The term "Hox genes" refers specifically to those genes that are clustered in an array on the chromosome and function primarily in establishing regional or segmental identities. In all animal phyla that have been examined, regional or segmental specialization is controlled by the spatially localized expression of these genes, which play crucial roles in determining body patterns. They underlie such fundamental attributes as anterior–posterior differentiation (in both invertebrates and vertebrates) and the positioning of body wall outgrowths (e.g., limbs). The pivotal role played in evolution by developmental genes and the processes they regulate has led to the emerging field of evolutionary developmental biology, or "EvoDevo" for short. EvoDevo is concerned with the generation of form within an individual and how developmental mechanisms evolve over time—that is, how evolutionary innovation comes about.

Hox genes have been conserved to a remarkable degree throughout the animal kingdom, and they are now known from all animal phyla that have been examined. Hox proteins regulate the genes that control the cellular processes involved in morphogenesis. In doing so, they demarcate relative positions in animals—they do not specify the precise nature of particular structures. For example, in arthropods, Hox genes regulate where body appendages form, and they can either suppress limb development or modify it (in concert with other regulatory genes) to create unique appendage morphologies. Mutations in Hox genes, and other developmental genes, can create gross mutations (homeotic mutations or homeosis).

Only a small fraction of all genes, fewer than 1%, are devoted to the construction and patterning of animal bodies during their development from fertilized egg to adult—the process of a single egg becoming a complex, multibillion- (or multitrillion-) celled organism that looks and functions properly. The rest are involved in the everyday tasks of cells within various organs and tissues. Thus, mutations or other changes in relatively few genes, a "toolkit" common to all animals, can have large embryological outcomes. There is a growing body of evidence suggesting that Hox genes (and other "master control genes") have played major roles in the evolution of new body plans among the Metazoa. The evolutionary potential of Hox genes lies in their hierarchical and combinatorial nature. We now know that a single Hox gene can modulate the expression of dozens of interacting downstream genes, the products of which determine developmental outcomes. Variation in the output of these multigene networks can arise at many levels simply through changes in the relative timing of developmental gene expression (i.e., by heterochrony; see Chapter 5), or through interactions between genes in the regulatory network. To understand the profound potential of Hox genes to drive evolutionary change, consider that within the genome of *Drosophila* 85–170 different genes are regulated by the product of the Hox gene *Ultrabithorax* (*Ubx*) alone. Changes in the *Ubx* protein could potentially alter the regulation of all these genes! In some families of sea spiders (Chelicerata: Pycnogonida), Hox gene mutations appear to have produced spurious segment/leg duplications, creating polymerous lineages.

There are now many examples of the extraordinary potential, and flexibility, of developmental genes. The potential of Hox genes is seen in the abdominal limbs of insects. Abdominal limbs ("prolegs") occur on larvae of various insects in several orders, and they are ubiquitous in the Lepidoptera (e.g., caterpillars). These limbs were probably present in insect ancestors, hence prolegs may have reappeared through the de-repression of an ancestral limb developmental program (i.e., they are a Hox gene mediated atavism). Proleg formation appears to involve a change in the regulation and expression of a single gene (*abd-A*) during embryogenesis. Another fundamental feature of animals is eyes, and these come in an enormous variety of styles, from the sophisticated camera-type eyes of vertebrates and cephalopods, to the simple light-sensitive eyespots of flatworms, to the compound faceted eyes of arthropods. It has been estimated the eyes have been "invented" independently dozens of times in Metazoa. Yet we now know that the gene required for eye formation in fruit flies (the *Pax-6* gene) is the exact counterpart of the gene required for eye formation in humans and squids and many other animals. Major innovations, macroevolution, can come about by "teaching old genes new tricks." And, most recently, it has been discovered that new genes can arise *de novo*, from ancestrally nongenic sequences. Since early in the twenty-first century, *de novo* genes have been identified in *Drosophila*, rodents, rice, yeast and humans.

In summary, the processes of microevolution (e.g., natural selection) act on individuals and populations, maintain genomic continuity, and create anastomosing patterns of relationship over time (Figure 1.7). Macroevolutionary processes (e.g., speciation and extinction), on the other hand, act on species and lineages, disrupt genomic continuity, and create ascending, bifurcating patterns of relationship over time (Figure 1.7). In a cladogram of species, the line segments represent the places where anagenesis (microevolution) is taking place within a given species. The nodes in the cladogram represent macroevolutionary events, speciation, and extinction. Although Darwin titled his book *On the Origin of Species*, he dealt primarily with the maintenance of adaptations. In fact, the nature of

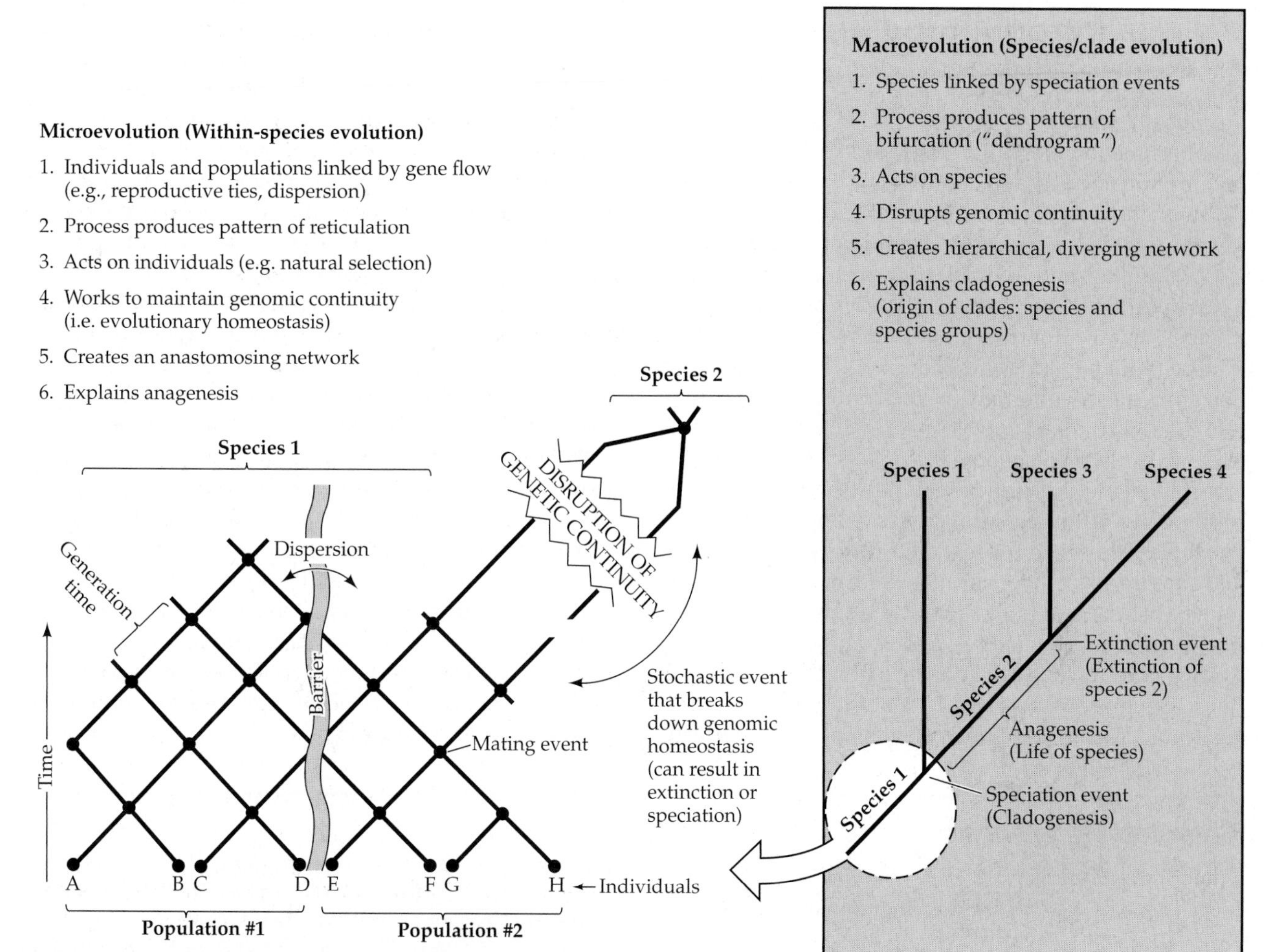

Figure 1.7 Microevolution and macroevolution depicted graphically. The highlighted portion of the cladogram (on the right) is shown in detail in the drawing to the left.

the relationship between anagenesis and cladogenesis is still not well understood. Evidence for the disengagement of natural selection and speciation comes from the fossil record, which suggests that most species do not change significantly throughout their existence; rather, they remain phenotypically stable for millions of years, then undergo a rapid change in which they essentially "replace themselves" with one or more new and different species. These new species, in turn, remain phenotypically static for millions more years. The fossil record suggests that most species of marine invertebrates persist more or less unchanged for 5–10 million years, whereas the time required for significant anatomical change seems to be only a few thousand years or less. The pattern of speciation in rapid bursts, sandwiched between long periods of species stasis, was presented in the famous punctuated equilibrium model of Niles Eldredge and Stephen Jay Gould (1972).

Biologists are still a long way from understanding all the causes and mechanisms of the evolutionary process, although we are developing excellent methods for analyzing the patterns of the history of evolution (e.g., molecular phylogenetics). The current debates concern the process—the nature of the evolutionary mechanisms themselves. It seems probable that different processes, working at different levels, have created the patterns we see in the world today. Despite the many evolutionary questions currently being discussed, and despite whatever evolutionary processes are at work, biologists are able to continue their efforts at reconstructing the evolutionary history of life on Earth, because the processes of evolution result in new organisms that are distinct by virtue of the unique new characters or attributes that they have acquired. Their descendants retain these attributes and in time acquire still others, which are retained by their descendants. In this fashion, the living world provides us with an analyzable hierarchical pattern consisting of nested sets of features recognizable both in fossils and in living organisms. Those features, in turn, are the data (i.e., the "characters") with which we can reconstruct a history of the ascent of life. We will have much more to say regarding this reconstruction process in the following chapter, because understanding what characters are and how they are evaluated is fundamental to comparative biology and to an appreciation of the invertebrate world.

A Final Introductory Message to the Reader

Because of our comparative approach, it is critical that you become familiar with Chapters 1, 2, 4, and 5 before attempting to study and comprehend the sections dealing with individual animal groups. These four chapters are designed to accomplish several goals: (1) to define some basic terminology, (2) to introduce a number of important concepts, and (3) to describe in detail the themes that we use throughout the rest of the book.

The fundamental theme of this book is evolution, and we approach invertebrate evolution primarily through the field of comparative biology. In Chapter 2 we provide an explanation of how biologists derive evolutionary schemes and classifications, how theories about the phylogeny of animal groups grow and change, and how the information presented in this text has been used to construct theories on how life evolved on Earth. In Chapters 4 and 5 we lay out the fundamental anatomical and morphological designs and developmental strategies of metazoans. Like the features of organisms, these designs and strategies are not random, but form patterns. Recognition and analysis of these patterns constitute the basic building blocks of this book. We then proceed in the "animal chapters" to explore the evolution of the invertebrates in light of various combinations of these basic functional body plans and lifestyles. With this background, you should be able to follow the evolutionary changes and branchings among the invertebrate phyla, their body systems, and their various pathways to success on Earth. Note that chapters treating large phyla (e.g., molluscs, annelids, the arthropods) have rather lengthy taxonomic synopses sections. We don't expect students to read every word of these synopses; they are provided more as a source reference for readers to look up taxa within the context of our current state of knowledge for the groups.

Through our approach, we hope to add continuity to the massive subject of invertebrate zoology, which is often covered (in texts and lectures) by a sort of "flash-card" method, in which the primary goal is to have the student memorize animal names and characteristics and keep them properly associated, at least until after the examination. Thus, we urge you to look back frequently at these first few chapters as you read ahead and explore how invertebrates are put together, how they live, and how they evolved.

Selected References

General References

Appeltans, W. and 119 others. 2012. The magnitude of global marine species diversity. Curr. Biol. 22: 1–14.

Boothby, T. C. and 12 others. 2015. Evidence for extensive horizontal gene transfer from the draft genome of a tardigrade. PNAS. doi: 10.1073/pnas.1510461112

Boyce, D. G., M. R. Lewis and B. Worm. 2010. Global phytoplankton decline over the past century. Nature 466: 591–596.

Bray, P. S. and K. B. Anderson. 2009. Identification of Carboniferous (320 million years old) Class Ic amber. Science 326: 132–134.

Buchsbaum, R., M. Buchsbaum, J. Pearse and V. Pearse. 1987. *Animals without Backbones*, 3rd Ed. University of Chicago Press, Chicago. [This classic book provides delightful reading and many excellent photographs.]

Butchart, S. H. M. and 44 others. 2010. Global biodiversity: indicators of recent declines. Science 328: 1164–1168.

Butlin, R. K., J. R. Bridle and D. Schulter (eds.). 2009. *Speciation and Patterns of Diversity*. Cambridge University Press, Cambridge.

Carroll, S. B. 2005. *Endless Forms Most Beautiful. The New Science of Evo Devo*. W. W. Norton & Company, New York.

Carroll, S. B. 2007. *The Making of the Fittest. DNA and the Ultimate Forensic Record of Evolution*. W. W. Norton & Company,, New York.

Combes, C. 2001. *Parasitism: The Ecology and Evolution of Intimate Interactions*. University Chicago Press, Chicago. [Brings a unifying approach to a subject usually presented in a fragmented fashion.]

Curtis, T. P., W. T. Sloan and J. W. Scannell. 2002. Estimating prokaryotic diversity and its limits. Proc. Natl. Acad. Sci. U.S.A. 99(16): 10494–10499. [Also see comment by B. Ward, same issue, 10234–10236.]

Danovaro, R., A. Dell'Anno, A. Pusceddu, C. Gambi, I. Heiner and R. M. Kristensen. 2010. The first Metazoa living in permanently anoxic conditions. BMC Biol. 8: 30–39.

Dirzo, R., H. S. Young, M. Galetti, G. Ceballos, N. J. B. Isaac and B. Collen. 2014. Defaunation in the Anthropocene. Science 345: 401–406

Dykhuizen, D. 2005. Species numbers in bacteria. Proc. Calif. Acad. Sci. 56, Suppl. I(6): 62–71.

Ekman, S. 1953. *Zoogeography of the Sea*. Sedgwick and Jackson, London. [Excellent review of marine invertebrate distributions; dated, but a benchmark work.]

Erwin, T. L. 1991. How many species are there? Revisited. Conserv. Biol. 5: 1–4. [See also F. Ødegaard. 2000. How many species of arthropods? Erwin's estimate revised. Biol. J. Linn. Soc. 71: 583–597.]

Fraser, L. H. and 61 others. 2015. Worldwide evidence of a unimodal relationship between productivity and plant species richness. Science 349(6245): 302–305.

Fredrickson, J. K. and T. C. Onstott. 1996. Microbes deep inside the Earth. Sci. Am. 275: 68–73.

Gilbert, S. F. 2014. *Developmental Biology*, 10th Ed. Sinauer Associates, Sunderland, MA.

Hardy, A. C. 1956. *The Open Sea*. Houghton Mifflin, Boston. [Still the best introduction to the world of plankton.]

Harrison, F. W. (ed.). 1991–1997. *Microscopic Anatomy of Invertebrates*. Wiley-Liss, New York. [A 20-book series providing up-to-date, detailed treatments of anatomy, histology, and ultrastructure.]

Hoppert, M. and F. Mayer. 1999. Prokaryotes. Amer. Sci. 87: 518–525.

Hyman, L. H. 1940–1967. *The Invertebrates. Vols. 1–6*. McGraw-Hill, New York. [This series has ended. Naturally, some of the material in early volumes is out of date, but they still remain among the best comparative anatomy references available.]

Kaestner, A. 1967–1970. *Invertebrate Zoology. Vols. 1–3.* Wiley-Interscience, New York. [Translated from German.]

King, J. L., M. A. Simovich and R. C. Brusca. 1996. Species richness, endemism and ecology of crustacean assemblages in northern California vernal pools. Hydrobiologia 328: 85–116. [A detailed look at a rare and threatened environment.]

Little, C. T. S. 2010. The prolific afterlife of whales. Sci. Am., Feb: 78–84.

Madigan, M. 2000. Extremophilic bacteria and microbial diversity. Ann. Missouri Bot. Garden 87(1): 3–12.

Maldonado, M. 2004. Choanoflagellates, choanocytes, and animal multicellularity. Invert. Biol. 123: 1–22

McClain, C. 2010. An empire lacking food. Am. Sci. 98: 470–477. [An excellent review of deep-sea biodiversity.]

Moore, R. C. (ed.). 1952–present. *Treatise on Invertebrate Paleontology.* Geological Society of America and University of Kansas Press, Lawrence. [Detailed coverage of fossil forms; many volumes still pending.]

Mora, C., D. P. Tittensor, S. Adl, A. G. B. Simpson and B. Worm. 2011. How many species are there on Earth and in the ocean? PLoS Biol. 9(8). doi: 10:1371/pbio.10011127

Pellissier, L. and 9 others. 2014. Quaternary coral reef refugia preserved fish diversity. Science 344: 1016–1019.

Penny, A. M., R. Wood, A. Curtis, F. Bowyer, R. Tostevin and K.-H. Hoffman. 2014. Ediacaran metazoan reefs from the Nama Group, Namibia. Science 344: 1504–1506.

Pimm, S. and 8 others. 2014. The Biodiversity of species and their rates of extinction, distribution, and protection. Science 334(6187): 987.

Piper, R. 2013. *Animal Earth. The Amazing Diversity of Living Creatures.* Thames & Hudson, London.

Poore, G. C. B. and 17 others. 2014. Invertebrate diversity of the unexplored marine western margin of Australia: taxonomy and implications for global biodiversity. Mar. Biodivers. doi: 10.1007/s1252 04-0255-y

Porter, S. M. 2007. Seawater chemistry and early carbonate biomineralization. Science 316: 1302.

Prosser, C. L. (ed.). 1991. *Environmental and Metabolic Animal Physiology.* Wiley-Liss, New York. [One of the best accounts of comparative physiology.]

Putnam, N. H. 18 others. 2007. Sea anemone genome reveals ancestral eumetazoan gene repertoire and genomic organization. Science 317: 86–94.

Reaka-Kudla, M. L. 1997. The global biodiversity of coral reefs: a comparison with rain forests. Pp. 83–108 in M. Reaka-Kudla, D. E. Wilson and E. O. Wilson (eds.), *Biodiversity II: Understanding and Protecting Our Biological Resources.* Washington, DC: Joseph Henry Press.

Roberts, L. S. and J. Janovy, Jr. 1996. *Foundations of Parasitology,* 5th Ed. Wm. C. Brown, Dubuque, IA. [One of the better of a generally disappointing field of textbooks on parasitology.]

Ruggiero, M. A. and 8 others. A higher level classification of all living organisms. PLoS ONE 10(4):e0119248. doi:10.1371/journal.pone/0119248

Rundell, R. J. and B. S. Leander. 2010. Masters of miniaturization: Convergent evolution among interstitial eukaryotes. BioEssays 32: 430–437.

Ryan, J. F. 17 others. 2013. The genome of the ctenophore *Mnemiopsis leidyi* and its implications for cell type evolution. Science 342:1242592.

Sarbu, S. M. and T. C. Kane. 1995. A subterranean chemoautotrophically based ecosystem. NSS Bull. 57: 91–98.

Schoene, B. and 7 others. 2014. U-Pb geochronology of the Deccan Traps and relation to the end-Cretaceous mass extinction. Science 347(6218): 182–184.

Science [Magazine]. 2002. This special issue of Science has numerous excellent articles on environmental microbiology—the world of microbes (pp. 1055–1081).

Shuster, S. M. 2008. Mating systems. Pp. 2266–2273 in S. E. Jorgensen, *Behavioral Ecology.* Elsevier, Oxford.

Shuster, S. M. 2009. Sexual selection and mating systems. PNAS 106, Suppl. 1: 10009–10016.

Small, A. M., W. H. Adey and D. Spoon. 1998. Are current estimates of coral reef biodiversity too low? The view through the window of a microcosm. Atoll Res. Bull. 458: 1–20.

Urban, M. C. 2015. Accelerating extinction risk from climate change. Science 348: 571–573.

Wagner, G. P. 2014. *Homology, Genes and Evolutionary Innovation.* Princeton University Press.

Wilson, E. O. 1992. *The Diversity of Life.* Belknap Press, Harvard University Press, Cambridge, MA. [Outstanding writing by one of the greatest living American naturalists.]

Phylogeny and Paleontology

Antcliffe, J. B. 2013. Questioning the evidence of organic compounds called sponge biomarkers. Palaeontology 56: 917–925.

Bengtson, S. and Y. Zhao. 1997. Fossilized metazoan embryos from the earliest Cambrian. Science 277: 1645–1648.

Bengtson, S. and G. Budd. 2004. Comment on "Small bilaterian fossils from 40 to 55 million years before the Cambrian." Science 306: 1291a.

Bengtson, S., J. A. Cunningham, C. Yin and P. C. J. Donoghue. 2012. A merciful death for the "earliest bilaterian," *Vernanimalcula.* Evol. Dev. 14(5): 421–427.

Borner, J. P. Rehm, R. O. Schill, I. Ebersberger and T. Burmester. 2014. A transcriptome approach to ecdysozoan phylogeny. Mol. Phylogenet. Evol. 80: 79–87.

Briggs, D. E. G., D. E. Erwin and F. J. Collier. 1994. *The Fossils of the Burgess Shale.* Smithsonian Institution Press, Washington, D.C.

Chen, J-Y., P. Oliveri, E. H. Davidson and D. J. Bottjer. 2004. Small bilaterian fossils from 40 to 55 million years before the Cambrian. Science 305: 218–222.

Chen, J-Y., P. Oliveri, E. H. Davidson and D. J. Bottjer. 2004. Response to comment on "Small bilaterian fossils from 40 to 55 million years before the Cambrian." Science 306: 1291b.

Chen, J-Y. and 10 others. 2009. Complex embryos displaying bilaterian characters from Precambrian Doushantuo phosphate deposits, Weng'an, Guizhou, China. PNAS 106(45): 19056–19060.

Chen, L., S. Xiao, K. Pang, C. Zhou and X. Yuan. 2014. Cell differentiation and germ-soma separation in Ediacaran animal embryo-like fossils. Nature. doi: 10.1038/nature13766

Cohen, P. A., A. H. Knoll and R. B. Kodner. 2009. Large spinose microfossils in Ediacaran rocks as resting stages of early animals. Proc. Natl. Acad. Sci. 1006(16): 6519–6524.

Conway Morris, S. 1999. *The Crucible of Creation: The Burgess Shale and the Rise of Animals.* Oxford Univ. Press, Oxford.

Cunningham, J. A. and 7 others. 2014. Distinguishing geology from biology in the Ediacaran Doushantuo biota relaxes constraints on the timing of the origin of bilaterians. Proc. R. Soc. B, doi: 10.1098/rspb.2011.2280

Cunningham, J. A. and 9 others. 2015. Critical appraisal of tubular putative eumetazonas from the Ediacaran Weng'an Doushantuo biota. Proc. R. Soc. B, 282. doi: 10.1098/rspb.2015.1169

Dunn, C. W., G. Giribet, G. D. Edgecombe and A. Hejnol. 2014. Animal phylogeny and its evolutionary implications. Ann. Rev. Ecol. Evol. Syst. 45: 371–395.

Erwin, D. H., and numerous others. 2011. The Cambrian conundrum: early divergence and later ecological success in the early history of animals. Science 334: 1091–1097.

Erwin, D. H., M. Laflamme, S. M. Tweedt, E. A. Sperling, D. Pisani, and K. J. Peterson. 2011. The Cambrian conundrum: early divergence and later ecological success in the early history of animals. Science 334: 1091–1097.

Erwin, D. H. and J. W. Valentine 2013. *The Cambrian Explosion. The Construction of Animal Biodiversity.* Roberts and Co., Greenwood Village, CO.

Gáspar, J., J. Paps and C. Nielsen. 2015. The phylogenetic position of ctenophores and the origin(s) of nervous systems. EvoDevo 6: 1. doi: 10.1186/2041-9139-6-1

Hejnol, A. and J. M. Martín-Durán. 2015. Getting to the bottom of anal evolution. Zool. Anz. doi: 10.1016/j. jcz.2015.02.006

Huldtgren, T.,J. A. Cunningham, C. Yin, M. Stampanoni, F. Marone, P. C. J. Donoghue and S. Bengtson. 2011. Fossilized nuclei and germination structures identify Ediacaran "animal embryos" as encysting protists. Science 334: 1696–1699.

Kiessling, W., C. Simpson and M. Foote. 2010. Reefs as cradles of evolution and sources of biodiversity in the Phanerozoic. Science 327: 196–198.

Kouchinsky, A., S. Bengtson, B. Runnegar, C. Skovsted, M. Steiner and M. Vendrasco. 2012. Chronology of Early Cambrian biomineralization. Geol. Mag. 149: 221–251.

Laumer, C. E. and 10 others. 2015. Spiralian phylogeny informs the evolution of microscopic lineages. Curr. Biol. 25: 1–6.

Lee, M. S. Y., A. Cau, D. Naish and G. J. Dyke. 2014. Sustained miniaturization and anatomical innovation in the dinosaurian ancestors of birds. Science 345: 562–566.

Lee, M. S. Y., J. Soubrier and G. D. Edgecombe. 2013. Rates of phenotypic and genomic evolution during the Cambrian Explosion. Curr. Biol. 23: 1889–1895.

Liu, P., S. Xiao, C. Yin, S. Chen, C. Zhou and M. Li. 2014. Ediacaran acanthomorphic acritarchs and other microfossils from chert nodules of the Upper Doushantuo Formation in the Yangtze Gorges area, South China. J. Paleontol. 88: 1–139.

Love, G. D. and numerous others. 2009. Fossil steroids record the appearance of Demospongiae during the Cryogenian Period. Nature 457: 718–721.

Misof, B. and 100 others. 2014. Phylogenomics resolves the timing and pattern of insect evolution. Science 346: 763–767.

Osigus, H.-J., M. Eitel, M. Bernt, A. Donath and B. Schierwater. 2013. Mitogenomics at the base of Metazoa. Mol. Phylogenet. Evol. 69: 339–351.

Osigus, H.-J., M. Eitel and B. Schierwater. 2013. Chasing the urmetazoon: striking a blow for quality data? Mol. Phylogenet. Evol. 66: 551–557.

Peterson, K. J. and N. J. Butterfield. 2005. Origin of the Eumetazoa: testing ecological predictions of molecular clocks against the Proterozoic fossil record. Prod. Natl. Acad. Sci. 102: 9547–9552.

Pisani, D. and 7 others. 2015. Genomic data do not support comb jellies as the sister group to all other animals. PNAS, doi/10.1073/pnas.1518127112

Planavsky, N. J. and 8 others. 2014. Low Mid-Proterozoic atmospheric oxygen levels and the delayed rise of animals. Science 346: 635–638.

Rota-Stabelli, O., A. C. Daley and D. Pisani. 2013. Molecular timetrees reveal a Cambrian colonization of land and a new scenario for ecdysozoan evolution. Current Biology 23: 392-398.

Sánchez-Villagra, M. 2012. *Embryos in Deep Time. The Rock Record of Biological Development.* University of California Press, Berkeley.

Schiffbauer, J. D., S. Xiao, K. Sen Sharma and G. Wang. 2012. The origin of intracellular structures in Ediacaran metazoan embryos. Geology 40: 223–226.

Schmidt-Rhaesa, A. 2007. *The Evolution of Organ Systems.* Oxford Univ. Press, England.

Schopf, J. W. (ed.) 2002. *Life's Origin. The Beginnings of Biological Evolution.* University of California Press, Berkeley.

Seilacher, A., P. K. Bose and F. Pflüger. 1998. Triploblastic animals more than one billion years ago: Trace fossil evidence from India. Science 282: 80–83.

Shen, X. and 7 others. 2015. Phylomitogenomic analyses strongly support the sister relationship of the Chaetognatha and Protostomia. Zool. Scr. doi: 10.1111/zsc.12140

Suga, H. and 18 others. 2013. The *Capsaspora* genome reveals a complex unicellular prehistory of animals. Nat. Commun. 4: 23–25.

Telford, M. J. and D. T. J. Littlewood (eds.). 2009. *Animal Evolution: Genes, Genomes, Fossils and Trees.* Oxford University Press, Oxford.

Torruella, G. and 6 others. 2011. Phylogenetic relationships within the Opisthokonta based on phylogenomic analyses of conserved single-copy protein domains. Mol. Biol. Evol. 29: 531–544.

Vargas, P. and R. Zardoya (eds.). 2014. *The Tree of Life. Evolution and Classification of Living Organisms.* Sinauer Associates, Sunderalnd, MA.

Vickers-Rich, P. and P. Komarower (eds.). 2007. *The Rise and Fall of the Ediacaran Biota.* The Geological Society, Bath, UK.

Wägele, J. W. and T. Bartolomaeus (eds.). 2014. *Deep Metazoan Phylogeny: The Backbone of the Tree of Life. New Insights from Analyses of Molecules, Morphology, and Theory of Data Analysis.* De Gruyter, Berlin.

Waloszek, D. 2003. The "Orsten"-window—a three-dimensionally preserved Upper Cambrian meiofauna and its contribution to our understanding of the evolution of Arthropoda. Paleontol. Res. 7: 71–88.

Whelan, N. V., K. M. Kocot, L. L. Moroz and K. M. Halanych. 2015. Error, signal, and the placement of Ctenophora sister to all other animals. PNAS 112(18): 5773–5778.

Xiao, S., C. Zhou, P. Liu, D. Wang and X. Yuan. 2014. Phosphatized acanthomorphic acritarchs and related microfossils from the Ediacaran Doushantuo Formation at Weng'an (South China) and their implications for biostratigraphic correlation. J. Palentol. 88: 1–67.

Yin, Z., P. Liu, G. Li, P. Tafforeau and M. Zhu. 2014. Biological and taphonomic implications of Ediacaran fossil embryos undergoing cytokinesis. Gondwana Res. 25: 1019–1026.

Yin, Z., Z. Maoyan, E. H. Davidson, D. J. Bottjer, F. Zhao and P. Tafforeau. 2015. Sponge grade body fossil with cellular resolution dating 60 Myr before the Cambrian. Proc. Natl. Acad. Sci. doi: 10.1073/pnas.1414577112

Yin, Z., M. Zhu, P. Tafforeau, J. Chen, P. Liu and G. Li. 2011. Early embryogenesis of potential bilaterian animals with polar lobe formation from the Edicaran Weng'an Biota, South China. Precambrian Res. 225: 44–57

Zhuravlev, A. and R. Riding (eds.). 2001. *The Ecology of the Cambrian Radiation.* Columbia University Press, New York.

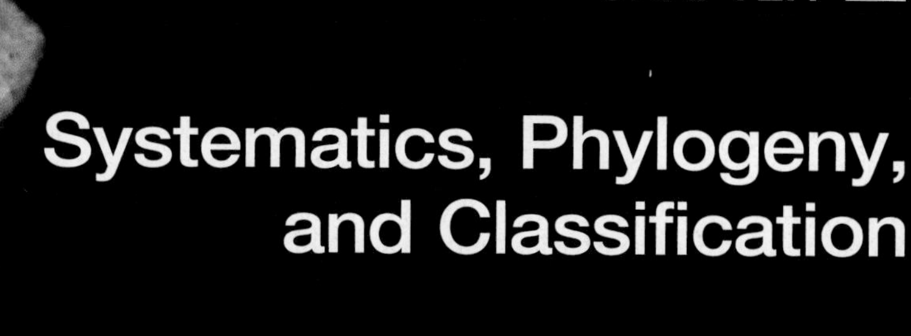

CHAPTER 2

Systematics, Phylogeny, and Classification

This book deals with the field of **comparative biology**, or what may be called the science of the diversity of life. Scientists may use comparative biology for many reasons, but evolutionary biologists use it to study the characteristics of organisms in ways that allow them to estimate the history of life. Biologists have been undertaking comparative studies of anatomy, morphology, embryology, physiology, and behavior for over 150 years. And for the past 20 years or so, comparative molecular phylogenetics and evolutionary-developmental biology ("EvoDevo") have played critically important roles. Because we cannot directly observe the history of life (aside from the paleontological record), we must rely on the strength of the scientific method to reconstruct it, or infer it. This chapter provides an overview of this process. Comparative biology, then, in its attempt to understand diversity in the living world, deals with three distinguishable elements: (1) descriptions of organisms, particularly in terms of similarities and differences in their attributes (including their genetic characteristics); (2) the phylogenetic history of organisms through time; and (3) the distributional history of organisms in space. Many comparative biologists think of themselves as systematists. **Systematics** is the science of documenting Earth's biological diversity, reconstructing the history of that biodiversity, and developing natural classifications that reflect its evolutionary history.

The field of biological systematics has experienced a revolution in its theory and application in the past 40 years, especially with regard to phylogenetic reconstruction. Some philosophical aspects and operating principles of this exciting field are described in this chapter. It is essential that biology students have a basic grasp of how classifications are developed and phylogenetic relationships inferred, and we urge you to reflect carefully on the ideas presented in this introductory chapter.

Today, our classifications of life are built upon careful phylogenetic analyses of morphological, developmental, and genetic characteristics of species that we hypothesize reflect their evolutionary history. More and more,

---

The chapter opener photo shows representatives of two phyla that are only distantly related, deeply in time: sponges (Porifera) and crabs (Arthropoda). Both are ancient phyla that arose in Precambrian seas more than 550 million years ago.

we rely on molecular sequence data from genes as our basis for these analyses. But this process is not always straightforward. Genes (and morphology) can trick us in many ways. Similarities can come about in indirect ways, characters can be lost or transformed in ways that are not always obvious, and many long-held ideas about animal development have recently been altered (or even overturned). In this chapter, we will help you understand the underpinnings of comparative biology as it relates to phylogenetic analysis and the construction of classifications.

Biological Classification

And you see that every time I made a further division, up came more boxes based on these divisions until I had a huge pyramid of boxes. Finally you see that while I was splitting the cycle up into finer and finer pieces, I was also building a structure. This structure of concepts is formally called a hierarchy and since ancient times has been a basic structure for all Western knowledge.

Robert M. Pirsig
Zen and the Art of Motorcycle Maintenance, 1974

The term **biological classification** has two meanings. First, it means the *process* of classifying, which consists of delimiting, ordering, and ranking organisms in groups. Second, it describes the *product* of this process, or the classification scheme itself. The living world has an objective structure that can be empirically documented and described. One goal of biology is to discover and describe this structure, and classifications are one way of doing this. Carrying out the process of biological classification constitutes one of the principal tasks of the systematist.

The construction of a classification may at first appear straightforward; basically, the process consists of analyzing patterns in the distribution of characters among organisms. On the basis of such analyses, specimens are grouped into species (the word "species" is both singular and plural); related species are grouped into genera (singular, genus); related genera are grouped to form families; and so forth. The grouping process creates a system of subordinated, or nested, taxa (singular, taxon) arranged in a hierarchical fashion following basic set theory. If the taxa are properly grouped according to their degree of shared similarity (that is, on the basis of shared derived characteristics), the hierarchy will reflect patterns of evolutionary descent—the "descent with modification" of Darwin and Wallace.

The concept of similarity is fundamental to taxonomy, the classificatory process, and comparative biology as a whole. Similarity, evaluated on the basis of characteristics shared among organisms, is generally accepted by biologists to be a measure of biological (evolutionary) relatedness among individuals and taxa. The concept of relatedness, or genealogical kinship, lies at the core of systematics and evolutionary biology. Patterns of relatedness are usually displayed by biologists in branching diagrams called **trees** (e.g., phylogenetic, genealogical, or evolutionary trees). Once constructed, such trees can then be converted into classification schemes, which are a dynamic way of representing our understanding of the history of life on Earth. Thus, trees and classifications are actually hypotheses of the evolution of life and the natural order it has created.

Classifications are necessary for several reasons, not the least of which is to efficiently catalog the enormous number of species of organisms on Earth. Nearly two million species of prokaryotes and eukaryotes have been named and described (and a great many more remain undescribed). The insects alone comprise nearly a million named species, and over 350,000 of those are beetles! Classifications provide a detailed system for storage and retrieval of these names. Second, and most important to evolutionary biologists, classifications serve a descriptive function. This function is served not only by the descriptions that define each taxon, but also, as noted above, by the detailed hypotheses of evolutionary relationships among the organisms that inhabit Earth. In other words, classifications are (or should be) constructed from evolutionary relationships; that is, from the patterns of ancestry and descent depicted in phylogenetic trees.

If biological classification schemes summarize the hypotheses defined by phylogenetic trees, then classifications, like other hypotheses in science, have a third function—prediction. And like all hypotheses, the more precise and less ambiguous the classification, the greater its predictive value. Predictability is another way of saying testability, and it is testability that places an endeavor in the realm of science rather than in the realm of art, faith, or rhetoric. Like other hypotheses (or theories), classifications are always subject to refutation, refinement, and growth as new data become available. These new data may be in the form of newly discovered species or characteristics of organisms, new tools for the analysis of characters, or new ideas regarding how characteristics are evaluated. Changes in classifications reflect changes in our view and understanding of the natural world.

Nomenclature

The names employed within classifications are governed by rules and recommendations that are analogous to the rules of grammar that govern Western languages. The most fundamental goals of **biological nomenclature** are the creation of classifications in which (1) any single kind of organism has one and only one correct name, and (2) no two kinds of organisms bear the same name. All of the codes of nomenclature for the various groups of life address these two fundamental requirements. Nomenclature is an important tool of biologists that facilitates communication and stability.[1]

Prior to the mid-1700s, animal and plant names consisted of one to several words or often simply a descriptive phrase. In 1735, at the age of just 28, the great Swedish naturalist Carl von Linné (Carolus Linnaeus, in the Latinized form he preferred) established a system of naming organisms now referred to as **binomial nomenclature**. Linnaeus's system required that every organism have a two-part scientific name, that is, be a **binomen**. The two parts of a binomen are the generic, or genus name and the specific name, or specific epithet. For example, the scientific name for one of the common Pacific coast sea stars is *Pisaster giganteus*. These two names together constitute the binomen; *Pisaster* is the animal's generic (genus) name, and *giganteus* is its **specific epithet**. The specific epithet is never used alone, but must be preceded by the generic name, and the animal's "species name" is thus the complete binomen. Use of the first letter of a genus name preceding the specific epithet is also acceptable once the name has appeared spelled out on the page or in a short article (e.g., *P. giganteus*).

The 1758 version of Linnaeus's system was the tenth edition of his famous *Systema Naturae*, in which he listed all animals known to him at that time and included critical guidelines for classifying organisms. Linnaeus distinguished and named over 4,400 species of animals, including *Homo sapiens*. Linnaeus's *Species Plantarum* (in which he named over 8,000 species) did the same for the plants in 1753. Linnaeus was one of the first naturalists to emphasize the use of similarities among species (or among other taxa) in constructing a classification, rather than using differences between them. In doing so, he unknowingly began classifying organisms by virtue of their genetic, and hence evolutionary, relatedness. Linnaeus produced his *Systema Naturae* 100 years prior to the appearance of Darwin and Wallace's theory of evolution by natural selection (1859), and thus his use of similarities in classification foreshadowed the subsequent emphasis by biologists on evolutionary relationships among taxa. Linnaeus was granted nobility in 1761 (and became Carl von Linné); he died in 1778.

[1]We generally avoid using common, or vernacular, names in this book, simply because they are frequently misleading. Most invertebrates have no specific common name, and those that do typically have more than one name. For example, several dozen different species of sea slugs are known as "Spanish dancers." All manner of creatures are called "bugs," most of which are not true bugs (Hemiptera) at all, e.g., "ladybugs," "sowbugs," "potato bugs," and so on. Recently, there has been a movement to codify common names, in an attempt to establish a single preferred vernacular for any given species. This movement is taking place mainly among vertebrate specialists, and there is, as yet, no widely accepted initiative to do this for invertebrates.

Binomens are Latin (or Latinized) because of the custom followed in Europe prior to the eighteenth century of publishing scientific papers in Latin, the universal language of the educated people of the time. For several decades after Linnaeus, names for animals and plants proliferated, and there were often several names for any given species (different names for the same species are called **synonyms**). The name in common use was usually the most descriptive one, or often it was simply the one used by the preeminent authority of the time. In addition, some generic names and specific epithets were composed of more than one word each. This lack of nomenclatural uniformity led, in 1842, to the adoption of a code of rules formulated under the auspices of the British Association for the Advancement of Science, called the Strickland code. In 1901 the newly formed International Commission on Zoological Nomenclature adopted a revised version of the Strickland code, called the **International Code of Zoological Nomenclature (ICZN)**. Botanists had adopted a similar code for plants in 1813, the Théorie Elémentaire de la Botanique, which became in 1930 the International Code of Botanical Nomenclature (there is also a separate, but complementary code for cultivated plants). Since then, it has been revised as the International Code of Nomenclature for Algae, Fungi, and Plants. There is also an International Code of Nomenclature of Bacteria.

The ICZN established January 1, 1758 (the year the tenth edition of Linnaeus's *Systema Naturae* appeared) as the starting date for modern zoological nomenclature. Any names published the same year, or in subsequent years, are regarded as having appeared after the *Systema*. The ICZN also slightly changed the description of Linnaeus's naming system, from binomial nomenclature (names of two parts) to **binominal nomenclature** (names of two names). However, one still sees the former designation in common use. This subtle change implies that the system must be truly binary; that is, both generic and trivial names can be only one word each. Although the system is binary, it also accepts the use of subspecies names, creating a **trinomen** (three names) within which is contained the mandatory binomen. For example, the sea star *Pisaster giganteus* is known to have a distinct form occurring in the southern part of its range, which is designated as a subspecies, *Pisaster giganteus capitatus*.

All codes of biological nomenclature share the following five basic principles:

1. Botanical, bacterial, and zoological codes are independent of each other. It is therefore permissible, although not recommended, for a plant genus and an animal genus to bear the same name (e.g., the name *Cannabis* is used for both a plant genus and a bird genus).

2. A taxon can bear one and only one correct name.
3. No two genera within a given code can bear the same name (i.e., generic names are unique); and no two species within one genus can bear the same name (i.e., binomens are unique).
4. The correct or valid name of a taxon is based on priority of publication (first usage).
5. For the categories of superfamily in animals and order in plants, and for all categories below these, taxon names must be based on type specimens, type species, or type genera.[2]

When strict application of a code results in confusion or ambiguity, problems are referred to the appropriate commission for a "legal" decision. Rulings of the International Commission on Zoological Nomenclature are published regularly in its journal, the Bulletin of Zoological Nomenclature. Note that the international commissions rule only on nomenclature or "legal" matters, not on questions of scientific or biological interpretation; these latter problems are the business of systematists. Obviously, the correct name of a species (and any future changes in that name) has great importance to all fields of biology (e.g., ecology, conservation biology, physiology) because scientists must know the correct names of their study organisms in order to communicate about them.

The hierarchical categories recognized by the ICZN are as follows:

Kingdom
 Phylum
  Class
   Cohort
    Order
     Family
      Tribe
       Genus
        Species
         Subspecies

The above names represent **categories**; the actual animal group that is placed at any particular categorical level forms a **taxon**. Thus, the taxon Echinodermata is placed at the hierarchical level corresponding to the category phylum—Echinodermata is the taxon; phylum is the category. All categories (and taxa) above the species level are referred to as the higher categories (and higher taxa), as distinguished from the species group categories (species and subspecies). Taxonomic ranks may have super-, and sub-, and infracategories as well, e.g., superorder, suborder, infraorder.

The common Pacific sea star *Pisaster giganteus* is classified as follows:

Category	Taxon
Phylum	Echinodermata
Subphylum	Asterozoa
Class	Asteroidea
Order	Forcipulatida
Family	Asteriidae
Genus	*Pisaster*
Species	*Pisaster giganteus* (Stimpson, 1857)

Notice that a person's name follows the species name in this classification. This is the name of the author of that species—the person who first described the species and gave it its name. In this particular case the author's name is in parentheses, which indicates that this species is now placed in a different genus than originally assigned by Professor Stimpson. Authors' names usually follow the first usage of a species name in the primary literature (i.e., articles published in professional scientific journals). In the secondary literature, such as textbooks and popular science magazines, authors' names are rarely used.

The names given to animals and plants are usually descriptive in some way, or perhaps indicative of the geographic area in which the species occurs. Others are named in honor of persons for one reason or another. Occasionally one runs across purely whimsical names, or even names that seem to have been formulated for seemingly diabolical reasons.[3]

The **biological species definition** (or genetical species concept), as codified by Ernst Mayr, defines species as groups of interbreeding (or potentially interbreeding) natural populations that are reproductively isolated from other such groups. Obviously, this definition fails to accommodate nonsexual species. George Gaylord Simpson and Edward O. Wiley developed the **evolutionary species concept**, which states that a species is a single lineage of ancestor–descendant populations that maintains its identity separate from other such lineages and that has its own evolutionary tendencies and historical fate. In reality, of course, biologists rely heavily on morphological aspects of organisms (and increasingly on gene sequence data) as surrogates in gauging these conceptual views of species. That is, we conceive of species as genetic or evolutionary entities, but we recognize them primarily by their phenotypic (or gene sequence) characters. Hence, an understanding of such characters is of great importance; read on.

Higher taxa (categories and taxa above the species level) are natural groups of species (or lineages) chosen by biologists for naming in order to reflect our state of knowledge regarding their evolutionary relationships.

[2]When a biologist first names and describes a new species, he or she takes a "typical" or representative individual, declares it a type specimen, and deposits it in a safe repository such as a large natural history museum. If later workers are ever uncertain about whether they are working with the same species described by the original author, they can compare their material to the type specimen. Although of substantially less value, the designation of a "typical" or type species for a genus, or a type genus for a family, serves a somewhat similar purpose in establishing a "typical" species or genus upon which a genus or family is based.

Higher taxa, if correctly constructed, represent ancestor–descendant lineages (or clades) that, like species, have an origin, a common ancestry and descent, and eventually a death (extinction of the lineage); thus they too are evolutionary units with definable boundaries. There are no rules for how many species should make up a genus—only that it be a natural group. Nor are there rules about how many genera constitute a family, or whether any group of genera should be recognized as a family, or a subfamily, or an order, or any other categorical rank. What matters is simply that the named group (the taxon) be a natural group. Hence, it is incorrect to assume that families of insects are in some way evolutionarily comparable to families of molluscs, or orders of worms comparable to orders of crustaceans. Nor are there any rules about categorical rank and geological or evolutionary age. These aspects of higher taxa are often misunderstood. Interestingly, this being said, family-level taxa often tend to be the most stable taxonomic groupings, usually recognizable even to laypersons—think, for example, of cats (Felidae), dogs (Canidae), abalone (Haliotidae), ladybird beetles (Coccinellidae), mosquitoes (Culicidae), octopuses (Octopodidae), or weevils (Cuculionidae). This stability seems to be an artifact of the history of taxonomy, but it nonetheless makes families convenient higher taxa to study and discuss. However, biologists err when they compare equally ranked higher in ways that presuppose them to be somehow equivalent.

Systematics

The science of systematics is the oldest and most encompassing of all fields of biology. To paraphrase the eminent biologist G. G. Simpson, systematics is the study of the diversity of life on Earth, and of any and all relationships among species. The modern systematist is a natural historian of the first order. His or her training is broad, cutting across the fields of zoology and botany, genetics, paleontology, biogeography, geology, historical biology, ecology, and even ethology, chemistry, philosophy, and cellular and molecular biology. Ernst Mayr said that the field of systematics can be thought of as a continuum, from the routine naming and describing of species (a process known

[3]Among the many clever names given to animals are *Agra vation* (a tropical beetle that was extremely difficult for Dr. Terry Erwin to collect) and *Lightiella serendipida* (a small crustacean; the generic name honors the famous Pacific naturalist S. F. Light, 1886–1947, while the species epithet is taken from "serendipity," a word coined by Walpole in allusion to the tale of "The Three Princes of Serendip," who in their travels were always discovering, by chance or sagacity, things they did not seek—the term is said to aptly describe the circumstances of the initial discovery of this species). There are actually over 500 described species of *Agra* (those carabid beetles known as "elegant canopy beetles"), including *Agra eponine*, named after the street urchin in *Les Miserables* who, in the Broadway version of the story, personified tragic beauty ("such is the state of the tropical forests where these beetles live," according to Dr. Erwin, who also named this species). Another of Erwin's names is *Agra ichabod*, referring to the fact that the holotype is missing its head, the allusion referring to the frightened schoolteacher Ichabod Crane's phantom nemesis, the Headless Horseman, in "The Legend of Sleepy Hollow." The nineteenth-century British naturalist W. E. Leach erected numerous genera of isopod crustaceans whose spellings were anagrams of the name Caroline. Exactly who Caroline was (and the nature of her relationship with Professor Leach) is still being debated, but the prevailing theory implicates Caroline of Brunswick, who was in the public eye at this time in history. It is said that Caroline was badly treated by her husband (the Prince Regent, later George IV), and that she was herself a lady of questionable fidelity. Leach, from Devon, may have taken the side of support for Caroline by honoring her with a long series of generic names, including *Cirolana, Lanocira, Rocinela, Nerocila, Anilocra, Conilera, Olincera*, and others.

A light-hearted attitude toward naming organisms has not always been without Freudian overtones, as there also exist *Thetys vagina* (a large, hollow, tubular pelagic salp), *Succinea vaginacontorta* (a hermaphroditic snail whose vagina twists in corkscrew fashion), *Phallus impudicus* (a slime-covered mushroom), and *Amanita phalloides* and *Amanita vaginata* (two species of highly toxic mushrooms around which numerous aboriginal ceremonies and legends exist). *Humbert humberti* is a wasp named after Vladimir Nabokov's Humbert Humbert, the narrator in the great novel *Lolita* who was obsessed with his 12-year old, soon-to-be stepdaughter. *Crepidula fornicata* is a hermaphroditic slipper shell (gastropod) that forms stacks of alternating male-functioning/female-functioning individuals (males on top turn into females as they grow). Injecting a lyrical dose of sexual innuendo into taxonomy is not new. Linnaeus himself incorporated a few good zingers into his writings and, in fact, drew parallels between plant sexuality and human love. In 1729 he wrote of flower petals, "[These] serve as bridal beds which the Creator has so gloriously arranged, adorned with such noble bed curtains, and perfumed with so many soft scents, that the bridegroom with his bride might there celebrate their nuptials with so much the greater solemnity." Such sexually explicit writing (in the early eighteenth century) did not go uncriticized, and Linnaeus had his detractors. The German botanist Johann Siegesbeck (a Demonstrator at the Botanical Garden at St. Petersburg) called it "loathsome harlotry" and commented, "Who would have thought that bluebells, lilies, and onions could be up to such immorality?" Linnaeus had his revenge, however, when he named a small, ugly, foul-smelling, mud-inhabiting European weed (St. Paul's wort) *Siegesbeckia*.

Other fun names include the hoopoe (a bird), *Upupa epops*, euphoniously named for its call, and the fish *Zappa confluentus*, which was named by a fan of Frank Zappa's. The Grateful Dead have a fly named in their honor (*Dicrotendipes thanatogratus*). And there is the vampire squid *Vampyroteuthis infernalis* (the "vampire squid from hell"), a bivalve named *Abra cadabra*, a blood-sucking spider *Draculoides bramstokeri*, and a wasp *Aha ha*. Even Linnaeus created a curious name for a common ameba, *Chaos chaos*. And, in a stroke of whimsy, the entomologist G. W. Kirkaldy created the bug genera *Polychisme* ("Polly kiss me"), *Peggichisme, Marichisme, Dolychisme*, and *Florichisme*. There are fish genera named *Zeus, Satan, Zen, Batman*, and *Sayonara*. There are insect genera named *Cinderella, Aloha, Oops*, and *Euphoria*. Some other clever binomens include *Leonardo davincii* (a moth), *Phthiria relativitae* (a fly), and *Ba humbugi* (a snail). A few biologists have gone overboard in erecting names for new animals, and many binomens exceed 30 letters in length, including those of the chaetognath *Sagitta pseudoserratadentatoides* (31 letters) and the common North Pacific sea urchin *Strongylocentrotus droebachiensis* (31 letters). *Lagenivaginopseudobenedenia* is a 27-letter genus name for a group of monogenean flukes. Amphipod crustaceans probably win the grand prize in the longest overall name category, with *Siemienkiewicziechinogammarus siemienkiewitschii* (47 letters) and *Cancelloidokytodermogammarus* (*Loveninsuskytodermogammarus*) *loveni* (61 letters, including the subgenus name)—these are cases in which the journal editor simply was not doing his or her job properly!

as **taxonomy**) through the compilation of large faunal compendia and monographs, to more synthetic studies such as the fitting of these species into classifications that depict evolutionary relationships, generating phylogenetic hypotheses based on morphological or molecular genetic data, biogeographic analyses, studies of population biology and genetics, and evolutionary and speciation studies. Mayr designated three stages of study within this continuum, which he called alpha, beta, and gamma, corresponding to the three general levels of complexity he perceived in systematics. When a group of organisms is first discovered or is in a poorly known state, work on that group is necessarily at the alpha level (e.g., the describing of new species, or taxonomy). It is only when most, or at least many, species in a taxon become known that the systematist is able to work at the beta or gamma levels within that group (e.g., to perform evolutionary studies). These stages in systematic research overlap and cycle back on themselves in a highly iterative fashion. In sum, the role of systematics is to document and understand Earth's biological diversity, to reconstruct the history of that biodiversity, and to develop natural (evolutionary) classifications of living organisms.

Systematists use a great variety of tools to study the relationships among taxa. These tools include not only the traditional and highly informative techniques of comparative and functional anatomy, but also the methods of embryology, serology, physiology, immunology, biochemistry, population and molecular genetics, and molecular gene sequencing (which now works at the genomic level). A sound classification lies at the root of any study of evolutionary significance. Without systematics, the science of biology would grind to a halt, or worse yet, would drift off into pockets of isolated reductionist or deterministic schools with no conceptual framework or continuity.

Monophyly, Paraphyly, and Polyphyly

One of the concepts most crucial to our understanding of biological systematics and evolution in general is monophyly. A **monophyletic group** (or monophyletic taxon) is a group of species that includes all of the descendants of a common ancestor—that is, a natural group (Figure 2.1). Species belonging to a monophyletic group are related to one another through a unique history of descent (with modification) from a common ancestor—a single evolutionary lineage. Another name for a monophyletic group is a **clade**.

A group whose member species are all descendants of a common ancestor, but that does *not* contain all the species descended from that ancestor, is called a **paraphyletic group**. Paraphyly implies that for some reason (e.g., lack of knowledge, purposeful manipulation of a classification) one or more members of a natural group have been separated out and placed in a different group. As we will see below, a great many paraphyletic taxa exist within animal classifications today. Some biologists consider paraphyly a subset of monophyly, using the term **holophyly** to denote the strictly monophyletic groups. However, this alternative use of these terms is not common and we do not use it in this book.

A third possible kind of taxon is a **polyphyletic group**—a group comprising species that arose from two or more, different immediate ancestors. Such

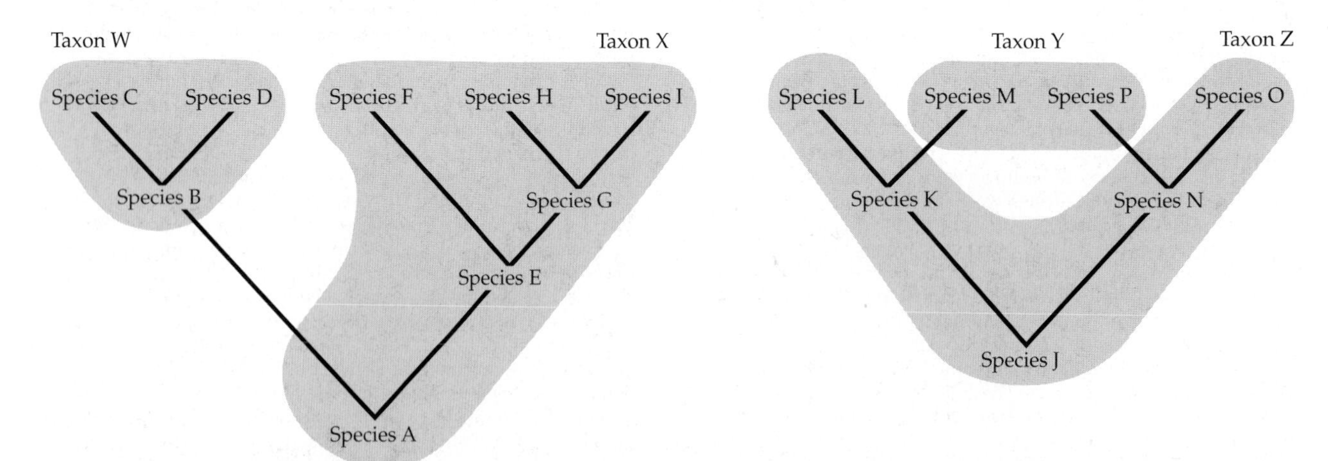

Figure 2.1 Two trees (dendrograms), illustrating three kinds of taxa. Taxon W, comprising three species, is monophyletic because it contains an ancestor (species B) and all its descendants (species C and D). Taxon X is paraphyletic because it includes an ancestor (species A), but only some of its descendants (species E through I, leaving out species B, C, and D). Taxon Y is polyphyletic because it contains taxa that are not derived from an immediate common ancestor; species M and P may look very much alike as a result of evolutionary convergence or parallelism, and therefore may have been mistakenly placed together in a single taxon. Taxon Z is paraphyletic. In this case, further systematic research should eventually reveal the correct relationships among these taxa, resulting in species M being classified with species K and L, and species P with species N and O.

composite taxa would have been established because of insufficient knowledge concerning the species in question. One of the principal goals of systematists is to discover such polyphyletic or "artificial" taxa and, through careful study, reclassify their members into appropriate monophyletic taxa. These three kinds of taxa or species groups are illustrated diagrammatically in Figure 2.1.

There are many examples of known or suspected polyphyletic taxa in the zoological literature. For example, the old phylum Gephyrea contained species that we now classify into three distinct taxa that are only distantly related to one another—Sipuncula, Echiura, and Priapula. Another example is the old group Radiata, which included all animals possessing radial symmetry (e.g., cnidarians, ctenophores, and echinoderms). Most recently, the group called Articulata, containing the annelids and the arthropods (and their kin), has been shown by molecular phylogenetics to be polyphyletic. Polyphyletic taxa usually are established because the features or characters used to recognize and diagnose them are the result of evolutionary convergence in different lineages, as discussed below. Convergent evolution can often be understood only by careful comparative embryological or anatomical studies, sometimes requiring the efforts of several generations of specialists. The complex nature of the probable convergence between annelid and arthropod segmental development, for example, is yet to be fully worked out.

Characters and the Concept of Homology

Characters are the attributes, or features, of organisms or groups of organisms (clades, taxa) that biologists rely on to indicate their relatedness to other similar organisms and to distinguish them from other groups. Characters are the observable features and expressions of the genotype, and they can be anything from the actual amino acid sequences of the genes themselves to the phenotypic expressions of the genotype. A character can be any genetically based trait that taxonomists can examine and measure; it can be a morphological, anatomical, developmental, or molecular feature of an organism, its chromosomal makeup (karyotype) or biochemical "fingerprint," or even a physiological, or ethological (behavioral) attribute. A large number of biochemical and molecular techniques for measuring similarity and inferring relationships among organisms have been developed over the past few decades. Thus, a variety of kinds of data are available that provide systematists with the characters needed to define and compare species and infer phylogenies.

The fundamental basis for comparative biology is the concept of **homology**. Characters that share descent from a common ancestor are called homologues. In other words, **homologues** are characters that are present in two or more taxa, but are traceable phylogenetically and ontogenetically (i.e., they share genetic and developmental ancestry) to the same character in the common ancestor of those taxa. In order to compare characters among different organisms or groups of organisms, it must be established that the characters being compared are homologous. Our ability to recognize anatomical homologues often depends on developmental or embryological evidence and on the relative position of the anatomical structure in adults (Chapter 5).

Homology is an absolute relationship: characters either are or are not homologous. Homology is also completely independent of function. The functions of homologous structures may be similar or different, but this has no bearing on the underlying homology of the structures involved. Genes, like anatomical structures, may be homologous if they are derived from a common ancestral gene either by duplication (which generates **paralogous genes**) or as simple copies passed on via reproduction and descent, including through speciation events (**orthologous genes**). The process of evolutionary descent with modification has produced a hierarchical pattern of homologies that can be traced through lineages of living organisms. It is this pattern that we use to reconstruct the history of life.

Homology is a concept that is applicable to anatomical structures, to genes, and to developmental processes. However, homology at one of these levels does not necessarily indicate homology at another. Biologists should always be clear regarding the level at which they are inferring homology: genes, their expression patterns, their developmental roles, or the structures to which they give rise. Researchers sometimes assume similar patterns of regulatory gene expression are also evidence of homology among structures. This is a mistake because it ignores the evolutionary histories of the genes and of the structures in which they are expressed. The functions of homologous genes (orthologues or paralogues), just like those of homologous structures, can diverge from one another over evolutionary time. Similarly, the functions of nonhomologous genes can converge over time. Therefore, similarity of function is not a valid criterion for the determination of homology of either genes or structures. For example, the phenomenon of gene recruitment (co-option) can lead to situations in which truly orthologous genes are expressed in nonhomologous structures during development. Most regulatory genes play several distinct roles during development, and homologous genes can be independently recruited to superficially similar roles. A classic example is the regulatory gene *Distal-less*, which is expressed in the distal portion of appendages of many animals during their embryogeny (e.g., arthropods, echinoderms, chordates). Although the domains of *Distal-less* gene expression might reflect a homologous role in specifying proximodistal axes of appendages, the appendages themselves are clearly not homologous.

Attempts to relate two taxa by comparing nonhomologous characters will result in errors. For example, the hands of chimpanzees and humans are homologous characters (i.e., homologues) because they have the same evolutionary and developmental origin; the wings of bats and butterflies, although similar in some ways, are not homologous characters because they have completely different origins. In a strict sense, the concept of homology has nothing to do with similarity or degree of resemblance. Some homologous features look very different in different taxa (e.g., the pectoral fins of whales and the arms of humans; the forewings of beetles and of flies). Again, the concept of homology is related to the level of analysis being considered. The wings of bats and birds are homologous as tetrapod forelimbs, but they are not homologous as "wings," because wings evolved independently in these two groups (i.e., the wings of bats and birds do not share a common ancestral wing). Homology is a powerful concept, but it is important to remember that homologies are really hypotheses, open to testing and possible refutation.

Through the phenomenon of **convergent evolution**, similar-appearing (but nonhomologous) structures may evolve in distantly related groups of organisms in quite different ways; that is, they have separate genetic and developmental origins. For example, early biologists were misled by the superficial similarities between the vertebrate eye and the cephalopod eye, the bivalve shells of molluscs and of brachiopods, and the sucking mouthparts of true bugs (Hemiptera) and of mosquitoes (Diptera). Structures such as these, which appear superficially similar but that have arisen independently and have separate genetic and phylogenetic origins, are called convergent characters. There are both ecological and genomic explanations for the evolution of morphological similarity. Through the phenomenon of convergent evolution, similar-appearing structures have arisen independently, with separate genetic and developmental origins, in response to the same ecological factors. Convergent traits in animals and plants have been recognized at nearly all levels of biological organization, ranging from molecules to morphology to behaviors.

One of the most interesting cases of convergent evolution is the recently discovered analogies between voice and vocal learning in some mammals and birds (vocal learning is the ability to imitate sounds). Not only have the vocal areas of certain bird and mammal brains converged in their anatomy, but more than 50 genes have contributed to their convergent specialization—convergent behavior and neural circuits for vocal learning are accompanied by convergent molecular changes of multiple genes in species separated by millions of years from a common ancestor.

Convergence is often confused with **parallelism**. Parallel characters are similar features that have arisen more than once in different species, but that share a common genetic and developmental basis. Parallel evolution is the result of "distant" or underlying homology; for parallel evolution to occur, the genetic potential for certain features must persist within a group, thus allowing the feature to appear and reappear in various related species or groups of species. Parallelism might be thought of as a kind of "evolutionary redundancy."[4] Failure to recognize convergences (and parallelisms) among different groups of organisms has led to the creation of "unnatural," or polyphyletic, taxa in the past. For example, the intracellular parasites known as Myxozoa were long thought to be protists, but have recently been shown to be highly specialized parasitic cnidarians. These and some other groups (e.g., yeasts) mistakenly classified among the Protista made protists a polyphyletic group; removal of those non-protistan taxa has left the Protista a paraphyletic group (a natural group, with a single origin, but from which animals, plants and fungi are excluded). Parallelism is commonly encountered in characters of morphological "reduction," such as reduction in the number of segments, spines, fin rays, and so on in many different kinds of animals.

A third phenomenon in this general category is **evolutionary reversal**, wherein a feature reverts back to a previous, ancestral condition. Together, these three evolutionary processes (convergence, parallelism, reversal) constitute the phenomenon known as **homoplasy**—the recurrence of similarity in evolution (Figure 2.2). As you might guess, for systematists, homoplasy can be both fascinating and frustrating!

When comparing homologues among species, one quickly sees that variation in the expression of a character is the rule, rather than the exception. The various conditions of a homologous character are often referred to as its **character states**. A character may have only two contrasting states, or it may have several different states within a taxon. Polymorphic species are those that show a range of phenotypic or genetic variation as a result of the presence of numerous character states for the features being examined. A simple example is hair color in humans; black, brown, red, and blond are all states of the character "hair color." Not only can characters vary within a species, they also typically have several states among groups of species within higher taxa, such as patterns of body hair among various primates or the spine patterns on the legs of crustaceans.[5]

---

[4]Parallelism in this context is not to be confused with the evolution of species (or characters within species) "in parallel," that is, when two species (or characters) change more or less together over time. Host–parasite coevolution is an example of "evolution in parallel."

[5]In practical usage the terms "character" and "character state" are often used interchangeably. This practice can be a bit confusing. When the term "character" is used in a discussion of two or more homologues, it is typically being used in the same sense as "character state."

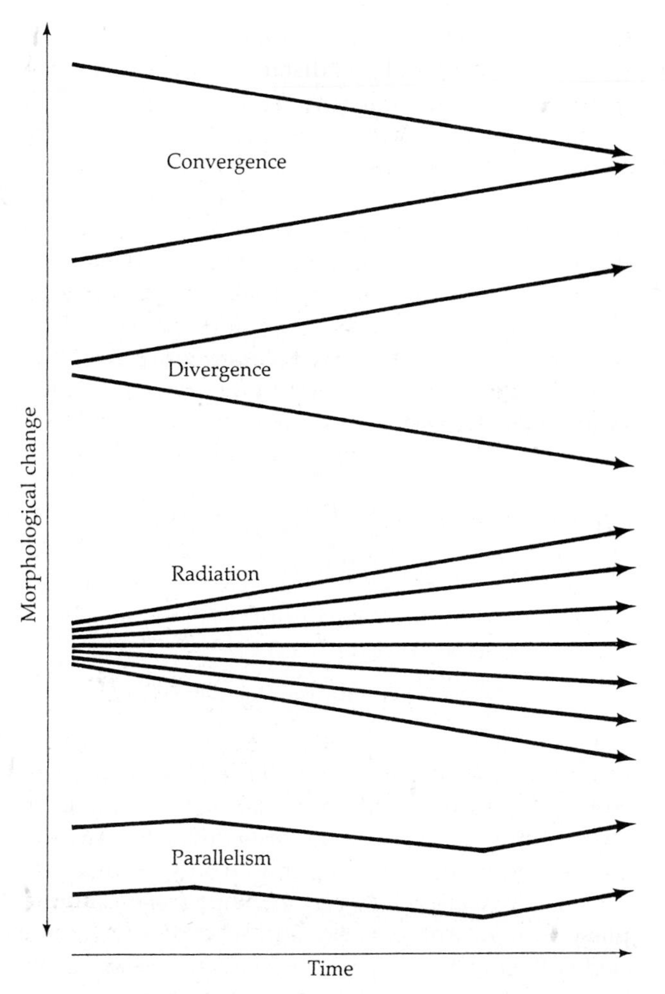

Figure 2.2 **Common patterns of evolution displayed by independent lineages.** Convergence occurs when two or more lineages (or characters) evolve independently toward a similar state. Convergence generally refers to very distantly related taxa and to characters sharing no common genetic (phylogenetic or ontogenetic) basis. Divergence occurs when two or more lineages (or characters) evolve independently to become less similar. Radiations are multiple divergences from a common ancestor that result in more than two descendant lineages. Parallel evolution occurs when two or more species (or lineages) change similarly so that, despite evolutionary activity, they remain similar in some ways. Parallelism generally refers to closely related taxa, usually species, within which the characters or structures in question share a common genetic basis.

It is important to understand that what we designate a "character" is really a hypothesis—that two attributes that appear different in different organisms are simply alternative states of the same feature (i.e., they are homologues). Note that convergences are not homologies, whereas parallelisms and reversals do represent an underlying genetic homology. In other words, some kinds of homoplastic characters are homologues, and others are not. The recognition and selection of proper characters is clearly of primary importance in systematics and phylogenetics, and a great deal has been written on this subject. Systematics is, to a great extent, a search for the homologues that define natural evolutionary lineages. Most recently, great efforts have been made to define homologous gene sequences that can serve as reliable characters to infer relationships among species.

Phylogenetic Trees

Another important concept in systematics and comparative biology is the **phylogenetic tree**. A phylogenetic tree is a branching diagram depicting the relationships among groups of organisms. It is a graphical means of expressing relationships among species or other taxa. Most trees are intended to depict genealogical or evolutionary relationships, with the base representing the oldest (earliest) ancestors and the higher branches indicating successively more recent divisions of evolutionary lineages. Most modern phylogenetic trees, whether based on morphology or gene sequence data, also have an implied (or explicit) time axis, although not all do.

When examining trees and classifications derived from them, it is important to understand the concept of grades and clades. As depicted in Figure 2.3, a **clade** is a monophyletic group or branch of a tree, which may undergo very little or a great deal of diversification. A clade, in other words, is a group of species related by direct descent—a natural group. A **grade**, on the other

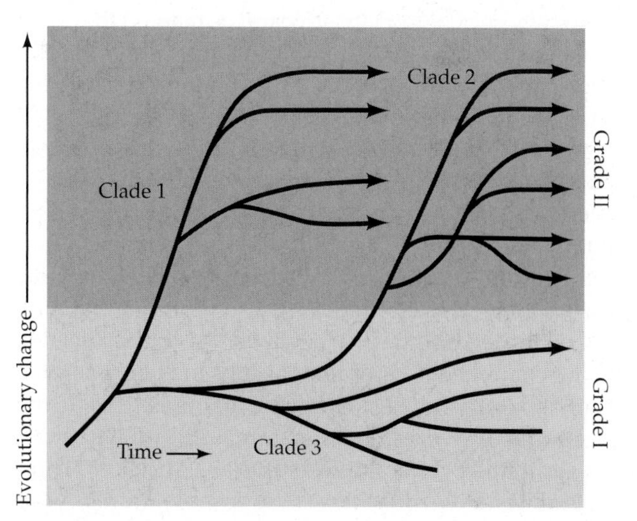

Figure 2.3 **Clades and grades.** Clades are monophyletic branches in a tree. Grades are groups of organisms classified together on the basis of levels of functional or morphological complexity. Grades may be monophyletic, paraphyletic, or polyphyletic. In this figure, grade I is monophyletic, encompassing only a single clade (clade 3); grade II is polyphyletic, because the associated level of complexity has been achieved independently by two separate lineages, clades 1 and 2. An example of a polyphyletic animal clade is "slugs," which comprise a cluster of groups of gastropods that have all lost their shells independently.

hand, is a group of species (or higher taxa) defined by somewhat more abstract measures. In fact, it is a group defined by a particular level of functional or morphological complexity. Thus, a grade can be polyphyletic, paraphyletic, or monophyletic (in the latter case, it is also a clade). A good example of a grade is the large group of gastropod taxa that have achieved shellessness. These "slugs," however, do not constitute a clade, because shell loss has occurred independently in several different lineages of gastropods; thus the grade of shellessness we recognize as "slugs" is a polyphyletic grouping. An example of a monophyletic grade is the subphylum Vertebrata (animals with backbones).

Pleisiomorphy and Apomorphy

One last concept important to our understanding of systematics is that of primitive versus advanced character states. In the most general sense, **primitive character states** are attributes of species that are relatively "old" and have been retained from some remote ancestor; in other words, they have been around for a long time, geologically or genealogically speaking. Character states of this kind are sometimes referred to as **ancestral character states. Advanced character states,** on the other hand, are attributes of species that are of relatively recent origin—often called **derived character states**. Within the phylum Chordata, for example, the possession of hair, milk glands, and three middle ear bones are derived character states whose evolutionary appearance marked the origin of the mammals (thus distinguishing them from all other chordates). Within a subset of the Mammalia, however, such as the primates, these same features represent retained ancestral features, whereas possession of an opposable thumb is a defining, derived trait.

It should be apparent from the preceding paragraph that the designations "primitive" and "advanced" are relative, and that any given character state or attribute can be viewed as either ancestral or derived, depending on the level of the phylogenetic tree or classification being examined. Opposable thumbs may be a derived trait defining primates within the mammal lineage, but it is not a derived character state within the primate line itself (all primates have opposable thumbs). Thus, in the primate genus *Homo,* "opposable thumbs" is a primitive (ancestral) feature, and certain features of the nervous system that distinguish humans from the "lower apes" would be considered derived (such as Broca's center in the human brain). Thus it behooves us to more precisely define the concepts of primitive and advanced. The most unambiguous way to describe and use these important concepts is to define the exact place in the phylogenetic history of a group of organisms at which a character actually undergoes an evolutionary transformation from one state to another. At the specific point on a phylogenetic tree where such a transformation takes place, the new (derived) character state is called an **apomorphy** and the former (ancestral) state a **plesiomorphy**. Thus the appearance of an apomorphy denotes the specific place in the tree of life where a "primitive" or "ancestral" character state changes to become an "advanced" or "derived" character state. Use of these terms thus implies a precise phylogenetic placement of the character in question, and this placement constitutes a testable phylogenetic hypothesis in and of itself. Apomorphies shared by two or more taxa are termed **synapomorphies;** plesiomorphies shared by two or more taxa are called **symplesiomorphies**.

Constructing Phylogenies and Classifications

Our classifications will come to be, as far as they can be so made, genealogies.

Charles Darwin, *The Origin of Species,* 1859

From what you have read so far in this chapter, it should be evident that comparative biologists, particularly systematists, spend a great deal of their time seeking to identify and unambiguously define two natural entities, homologues and monophyletic groups (i.e., clades). In fact, modern phylogenetic trees are composed primarily of these two entities: homologues and clades. Biologists may present their ideas on such matters of relationship in the form of trees, classifications, or narrative discussions (evolutionary scenarios). In all three contexts, these presentations represent sets of evolutionary hypotheses—hypotheses of common ancestry (or ancestor–descendant relationships).

The least ambiguous (most testable) way to present evolutionary hypotheses is in the form of a dendrogram, or branching tree. Although classification schemes are ultimately derived from such dendrograms, they do not always reflect precisely the arrangement of natural groups in the tree. Discrepancies between phylogenetic trees and classifications derived from them most commonly occur when biologists purposely choose to establish or recognize paraphyletic taxa. Thus, to recognize the protists as a distinct taxon (the kingdom Protista) would be to recognize a paraphyletic group (because it excludes three large lineages that descended from it—animals, plants, and fungi). Whereas most systematists advocate that only monophyletic taxa be recognized in a formal classification, many paraphyletic taxa persist in animal classification, for convenience or tradition, or because they are simply not yet known to be paraphyletic (there are probably thousands of taxa for which we do not know whether they are monophyletic or paraphyletic). For example, the long-recognized group Reptilia is paraphyletic because it excludes one of that group's most distinct lineages, the birds. The subphylum Crustacea is paraphyletic because it omits the insects, which evolved out of the crustaceans long ago. And as we will see in Chapter 14, the class Polychaeta is also paraphyletic. And of course, the group "invertebrata" is a paraphyletic group—it is Metazoa excluding the vertebrates. Even the Prokaryota is a paraphyletic group (the Eukaryota evolved out of it). In fact, taxonomic groups at all levels are likely to have been derived from *within* other taxonomic groups, leaving the latter paraphyletic, and we are beginning to discover that paraphyly abounds in the Linnean hierarchy of life that has been built over the past century. The issue of how to deal with such long-standing, well-known paraphyletic taxa in classification schemes is still being debated. One way of doing this might be to indicate their paraphyletic status by a code in the classification scheme (e.g., some type of notation beside the name). This code would inform readers that to view the precise phylogenetic relationships of such taxa, they must look to the phylogenetic tree.

Biologists today use a method known as **phylogenetic systematics**, or cladistics, when construing biological dendrograms. Phylogenetic systematics had its origin in 1950 in a book by the German biologist Willi Hennig; the English translation (with revisions) appeared in 1966. Its popularity has grown steadily since that time. Through the years, cladistics has evolved well beyond the framework Hennig originally proposed. Its detailed methodology has been formalized and expanded and will certainly continue to be elaborated for some time to come. (For good discussions of the philosophical underpinnings of phylogenetic systematics see Eldredge and Cracraft 1980, Nelson and Platnick 1981, and Wiley and Lieberman 2011.) The goal of phylogenetic systematics is to produce explicit and testable hypotheses of genealogical relationships among monophyletic groups of organisms. As a systematic methodology, cladistics is based entirely on recency of common descent (i.e., genealogy). The dendrograms used by phylogenetic systematists are called cladograms, or phylogenetic trees, and they are constructed to depict only genealogy, or ancestor–descendant relationships. The term **cladogenesis** refers to splitting; in the case of biology, this means the splitting of one species (or lineage) into two or more species (or lineages). It is this splitting process that produces genealogical (ancestor–descendant) relationships.

Phylogenetic systematists rely heavily on the concept of ancestral versus derived character states discussed earlier. They identify these homologies in the strict sense, as plesiomorphies and apomorphies. An apomorphy restricted to a single species is referred to as an **autapomorphy**, whereas an apomorphic character state that is shared between two or more species (or other taxa) is called a synapomorphy. Identifying synapomorphies (also known as shared derived characters, or evolutionary novelties) is the systematist's most powerful means of recognizing close evolutionary (genealogical) relationships. Because synapomorphies are shared homologues inherited from an immediate common ancestor, all homologues may be considered synapomorphies at one (but only one) level of phylogenetic relationship, and they therefore constitute symplesiomorphies at all lower levels. As noted earlier, hair, milk glands, and so forth are synapomorphies uniquely defining the appearance of the mammals within the vertebrates, but these are symplesiomorphies *within* the group Mammalia. Jointed legs are a synapomorphy of the Arthropoda, but *within* the arthropods jointed legs are a symplesiomorphy. The keystone of phylogenetic systematics is the recognition that all homologues define monophyletic groups at some level. The challenge is, of course, recognizing the level at which each character state is a unique synapomorphy. Generally speaking, synapomorphies are either structural or genetic features. However, in the broadest sense, and in the context of the biological species definition, reproductive isolation can be thought of as a synapomorphy for any given species. And thus, incomplete reproductive isolation (successful hybridization) could be viewed as a symplesiomorphy shared among the species involved.

Numerous methods and criteria have been used to determine which is the apomorphic and which is the plesiomorphic form of two character states—a process sometimes referred to as character state polarity analysis. No method is foolproof, but some may be better than others under specific circumstances. Only three methods appear to have a strong evolutionary basis and provide a reasonably powerful means for recognizing the relative place of origin of a synapomorphy on a tree: out-group analysis (seeking clues to ancestral character states in groups thought to be earlier derived than the study group), developmental studies (ontogenetic analysis, or seeking clues to ancestral character states in the embryogeny of the study group), and study of the fossil record. Out-group analysis helps identify the states of the characters in question in taxa that are closely related to the study group, but are not part of it. Ontogenetic analysis identifies character changes that occur during the development of a species (see the discussion of ontogeny and phylogeny in Chapter 5). And the use of fossils and associated dating and stratigraphic techniques provides direct historical information (e.g., providing critical calibration points for molecular-based evolutionary studies). These techniques of polarity analysis are not discussed in detail here; we refer those with a serious interest in systematics, evolution, and comparative biology to the readings listed at the end of this chapter.

A cladistic analysis often comprises four steps: (1) identifying homologous characters among the organisms being studied (whether they be based on anatomy, gene sequences, gene expression, etc.), (2) assessing the direction of character change or character evolution (character state polarity analysis), (3) constructing a phylogenetic tree of the taxa possessing the characters analyzed, and (4) testing the tree with new data (new taxa, new characters, new character interpretations, new statistical tests, etc.). Phylogenetic trees depict only one kind of event—the origin or sequence of appearances of unique derived character states (synapomorphies). Hence, phylogenetic trees may be thought of in the most fundamental sense as nested synapomorphy patterns. However, biologists define and categorize taxa by the character states they possess. Thus, in a larger sense, the sequential branching of nested sets of synapomorphies in a cladogram creates a "family tree"—an evolutionary pattern of hypothesized monophyletic lineages.

Phylogenetic systematists generally adhere to the principle of logical parsimony[6] and thus generally prefer the tree containing the smallest number of evolutionary transformations (character state changes). Typically this will also be the tree with the least evolutionary redundancy (= homoplasy). Although parsimony is the only inference method currently used for analyses of non-molecular data, the use of gene sequence data has spawned a new family of model-based methods that incorporate hypotheses of nucleotide evolution. In these methods (i.e., maximum likelihood and distance methods, Bayesian analyses), DNA nucleotide sequences from organisms in the study group are analyzed within a framework of assumptions based on how we believe nucleotides operate and change over time.

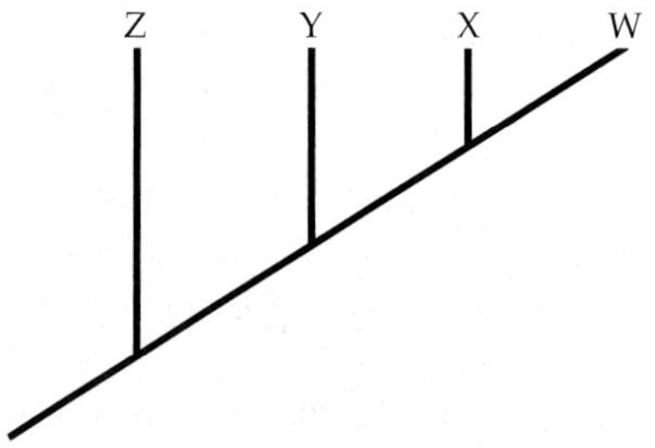

Z Y X W

Figure 2.4 A cladogram of four taxa, illustrating the concept of sister groups. Taxon W is the sister of taxon X; taxon W + X is the sister group of taxon Y; taxon W + X + Y is the sister group of taxon Z.

Construction of a cladogram can be a time-consuming process. The number of mathematically possible trees (of all branching patterns) for more than a few species is enormous—for three taxa there are only four possible trees, but for ten taxa there are about 280 million possible trees, 34 million of which are fully dichotomous. Needless to say, such analyses are not possible without the aid of a computer. Algorithms for computer-assisted tree construction began appearing in the late 1970s, and today the development of such programs comprises an entire field of research. But, all of these programs generate trees by clustering taxa into clades on the basis of nested sets of synapomorphies.

By identifying the precise points at which synapomorphies occur, phylogenetic trees unambiguously define monophyletic lineages. Hence, these trees are called **explicit phylogenetic hypotheses**. Being explicit, they can be tested (and potentially falsified) by anyone. The synapomorphies are markers that identify specific places in the tree where new monophyletic taxa arise. For phylogenetic systematists, a phylogeny consists of a genealogical branching pattern expressed as a phylogenetic tree. Each split or dichotomy within the tree produces a pair of newly derived taxa called **sister taxa**, or sister groups (for example, sister species). Sister groups always share an immediate common ancestor. In Figure 2.4, set W is the sister of set X; set W + X is the sister group of set Y; and set W + X + Y is the sister group of set Z. This nested-set pattern of hierarchical relationships results from the fact that cladogenesis is a historical process. As depicted on a phylogenetic tree, the product of cladogenesis (or the splitting of a taxon) is two (or more) new lineages that constitute sister groups. Another way of stating this is to say that the two (or more) subsets of any set defined by a synapomorphy constitute sister groups.

Like all scientific hypotheses, phylogenetic analyses and their resulting trees are tested by the discovery of new data. As new characters or new species are identified and their character states elucidated, new data matrices are developed, and new analyses are undertaken.

[6]Parsimony is a method of logic in which economy in reasoning is sought. The principle of parsimony, also known as Ockham's razor, has strong support in science. William of Ockham (Occam), the fourteenth-century English philosopher, stated the principle as, "Plurality must not be posited without necessity." Modern renderings would read, "An explanation of the facts should be no more complicated than necessary," or, "Among competing hypotheses, favor the simplest one." Scientists in all disciplines follow this rule daily, and it can be viewed as a consequence of deeper principles that are supported by statistical inferences. Thus, parsimonious solutions or hypotheses are those that explain the data in the simplest way. Evolutionary biologists rely on the principle of logical parsimony for the same reason other scientific disciplines rely on it: doing so presumes the fewest ad hoc assumptions and produces the most testable (i.e., the most easily falsified) hypotheses. If evidential support favored only one hypothesis, we would have little need for parsimony as a method. The reason we must rely on parsimony in science is that there is virtually always more than one hypothesis that can explain our data. Parsimony considerations come into play most strongly when a choice must be made among equally supported hypotheses.

In phylogenetic reconstruction, any given data set can be explained by a great number of possible trees. A three-taxon data set has 3 possible dichotomous (all lines divide into just two branches) trees that explain it. A four-taxon data set has 15 possible dichotomous trees, a five-taxon data set has 105 possible dichotomous trees, and so on. Thus, the evidence alone does not sufficiently narrow the class of admissible hypotheses, and some extra-evidential criterion (parsimony) is required. The virtue of choosing the shortest (i.e., most parsimonious) tree among a universe of possible trees lies in its testability. William of Ockham, by the way, also denied the existence of universals except in the minds of humans and in language. This notion resulted in a charge of heresy from the Church, after which he fled to Rome and, alas, died of the plague.

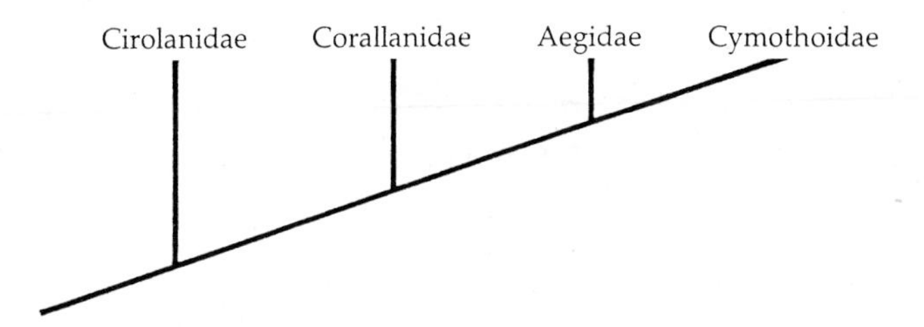

Classification scheme A

Family Aegidae
Family Cirolanidae
Family Corallanidae
Family Cymothoidae

Classification scheme B

Family Cirolanidae
Family Corallanidae
- Subfamily Corallaninae
- Subfamily Aeginae
  - Tribe Aegini
  - Tribe Cymothoini

Classification scheme C

Family Cirolanidae
Family Corallanidae
Family Aegidae
Family Cymothoidae

Figure 2.5 A phylogenetic tree of four closely related families of isopod crustaceans (marine pillbugs; see Chapter 21). In this example, the four families constitute an interesting "evolutionary series," from the free-living carnivorous Cirolanidae through micropredators and temporary fish parasites (Corallanidae and Aegidae) to obligatory fish parasites (Cymothoidae). Classification scheme A simply lists the families in alphabetical order, and there is no way to recover their phylogenetic history from this classification. Scheme B arranges the taxa in a subordinated (hierarchical) classification and precisely reflects the structure of the tree (but requires assigning the taxa to new nomenclatural ranks). Scheme C utilizes the phylogenetic sequencing convention to arrange the taxa in the exact sequential order in which they appear on the tree, but without subordination; it also precisely reflects the structure of the phylogenetic tree.

The first molecular phylogenetic trees were based on a single gene. But as techniques improved, these trees were tested with multiple gene sets, and eventually with entire genomic data sets. Hypotheses (branches of the tree) that consistently resist refutation are said to be highly corroborated. For example, the clade called Arthropoda has been examined in thousands of analyses using a great variety of data, and it has consistently been shown to constitute a monophyletic group (i.e., it is a highly corroborated phylogenetic hypothesis).

The final step in a phylogenetic analysis is often the conversion of the tree into a classification scheme. There is more than one way to make this conversion. Figure 2.5 shows a phylogenetic tree of four families of marine isopods, with three possible ways to classify the taxa. In this tree, these four families show an evolutionary trend from free-living (the Cirolanidae) to parasitic lifestyles (the Cymothoidae). The Cymothoidae (a family of isopods that are obligatory parasites on fishes) is the sister group of Aegidae (a family of temporary fish parasites); together they constitute a sister group of the Corallanidae (micropredators on fishes); and all three constitute a sister group of the Cirolanidae (carnivorous predators and scavengers). Each of these nested sister-group pairs shares one or more unique synapomorphies that defines them. In Figure 2.5, the synapomorphies that define the sister group Cirolanidae + Corallanidae + Aegidae + Cymothoidae become symplesiomorphies higher in the cladogram (i.e., for each of the separate families). Sister groups are monophyletic by definition.

As illustrated in Figure 2.5 (classification scheme B), some phylogenetic systematists early on suggested that every lineage depicted in a tree should be designated by a formal name and categorical rank, and that each member of a sister-group pair must be of the same categorical rank. A moment's thought reveals that giving names to every branching point in a cladogram would result in an impossible proliferation of names and ranks. Other systematists have proposed a method of avoiding such name proliferation, called the phylogenetic sequencing convention. When this convention is used, linear sequences of taxa can all be given equal categorical designations (e.g., they can all be classified as genera, or all as families, and so on), so long as they are listed in the classification scheme in the precise sequence in which the branches appear on the cladogram (classification scheme C). Either method of creating a classification scheme allows one to convert the classification scheme directly back into a cladogram—that is, to visualize the phylogenetic branching pattern it depicts. Classification scheme A does not reflect the phylogeny of these four families at all (it is simply alphabetical), so the only information it provides is that these families are "somehow" related to one another within a monophyletic clade.

The adoption of cladistic methodologies (the empirical construction of phylogenetic trees with precise branching patterns based on synapomorphies) in the 1980s led to an explosion of new phylogenetic work. This resulted in a proliferation of new phylogenetic trees, which in turn led to new classifications. A similar explosion of the field of molecular phylogenetics (using cladistic principals) shortly thereafter gave evolutionary biology another big boost, and has further changed some of our views on animal relationships and classifications. And just since the turn of the century another revolution in phylogenetics has begun as we have learned to sequence not just one gene at a time, but thousands of genes or even entire genomes. This latest development has begun to shine a very bright light on our understanding of animal evolution. It is also generating large and highly detailed phylogenetic trees, with long branching patterns depicting the history of life on earth (Figure 2.6, for example). Such detail creates challenges for biologists who like to produce classifications that accurately reflect phylogeny, because the traditional Linnean hierarchical ranks are too few in number to capture the great depth and detail

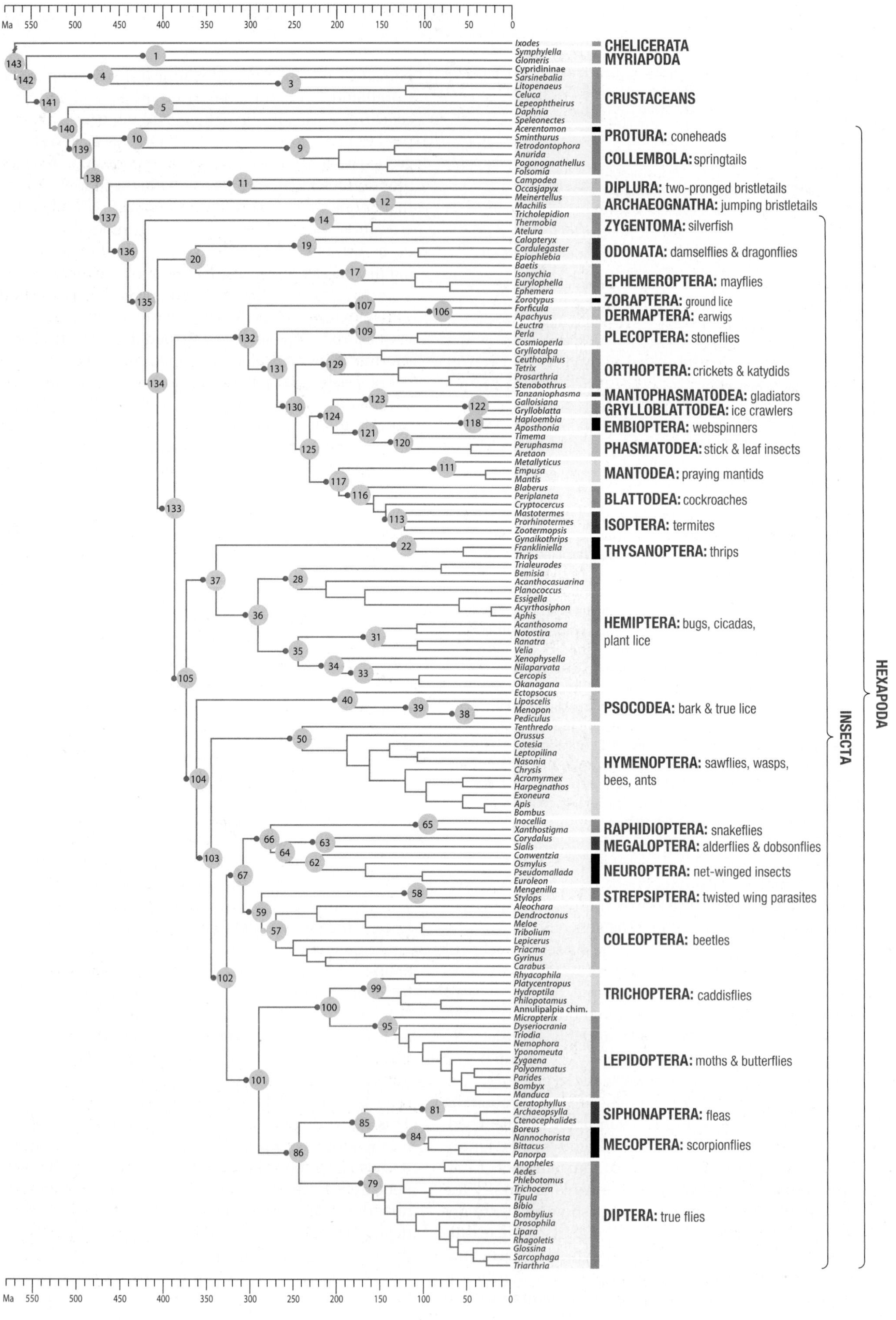

Ma 550 500 450 400 350 300 250 200 150 100 50 0

Ixodes — CHELICERATA
Symphylella, Glomeris — MYRIAPODA
Cypridininae, Sarsinebalia, Litopenaeus, Celuca, Lepeophtheirus, Daphnia, Speleonectes — CRUSTACEANS
Acerentomon — PROTURA: coneheads
Sminthurus, Tetrodontophora, Anurida, Pogonognathellus, Folsomia — COLLEMBOLA: springtails
Campodea, Occasjapyx — DIPLURA: two-pronged bristletails
Meinertellus, Machilis — ARCHAEOGNATHA: jumping bristletails
Tricholepidion, Thermobia, Atelura — ZYGENTOMA: silverfish
Calopteryx, Cordulegaster, Epiophlebia — ODONATA: damselflies & dragonflies
Baetis, Isonychia, Eurylophella, Ephemera — EPHEMEROPTERA: mayflies
Zorotypus — ZORAPTERA: ground lice
Forficula, Apachyus — DERMAPTERA: earwigs
Leuctra, Perla, Cosmioperla — PLECOPTERA: stoneflies
Gryllotalpa, Ceuthophilus, Tetrix, Prosarthria, Stenobothrus — ORTHOPTERA: crickets & katydids
Tanzaniophasma — MANTOPHASMATODEA: gladiators
Galloisiana, Grylloblatta — GRYLLOBLATTODEA: ice crawlers
Haploembia, Aposthonia — EMBIOPTERA: webspinners
Timema, Peruphasma, Aretaon — PHASMATODEA: stick & leaf insects
Metallyticus, Empusa, Mantis — MANTODEA: praying mantids
Blaberus, Periplaneta, Cryptocercus — BLATTODEA: cockroaches
Mastotermes, Prorhinotermes, Zootermopsis — ISOPTERA: termites
Gynaikothrips, Frankliniella, Thrips — THYSANOPTERA: thrips
Trialeurodes, Bemisia, Acanthocasuarina, Planococcus, Essigella, Acyrthosiphon, Aphis, Acanthosoma, Notostira, Ranatra, Velia, Xenophysella, Nilaparvata, Cercopis, Okanagana — HEMIPTERA: bugs, cicadas, plant lice
Ectopsocus, Liposcelis, Menopon, Pediculus — PSOCODEA: bark & true lice
Tenthredo, Orussus, Cotesia, Leptopilina, Nasonia, Chrysis, Acromyrmex, Harpegnathos, Exoneura, Apis, Bombus — HYMENOPTERA: sawflies, wasps, bees, ants
Inocellia, Xanthostigma — RAPHIDIOPTERA: snakeflies
Corydalus, Sialis — MEGALOPTERA: alderflies & dobsonflies
Conwentzia, Osmylus, Pseudomallada, Euroleon — NEUROPTERA: net-winged insects
Mengenilla, Stylops — STREPSIPTERA: twisted wing parasites
Aleochara, Dendroctonus, Meloe, Tribolium, Lepicerus, Priacma, Gyrinus, Carabus — COLEOPTERA: beetles
Rhyacophila, Platycentropus, Hydroptila, Philopotamus, Annulipalpia chim. — TRICHOPTERA: caddisflies
Micropterix, Dyseriocrania, Triodia, Nemophora, Yponomeuta, Zygaena, Polyommatus, Parides, Bombyx, Manduca — LEPIDOPTERA: moths & butterflies
Ceratophyllus, Archaeopsylla, Ctenocephalides — SIPHONAPTERA: fleas
Boreus, Nannochorista, Bittacus, Panorpa — MECOPTERA: scorpionflies
Anopheles, Aedes, Phlebotomus, Trichocera, Tipula, Bibio, Bombylius, Drosophila, Lipara, Rhagoletis, Glossina, Sarcophaga, Triarthria — DIPTERA: true flies

INSECTA
HEXAPODA

Ma 550 500 450 400 350 300 250 200 150 100 50 0

Figure 2.6 This phylogenetic tree of hexapod relationships (Misof et al. 2014) illustrates the challenges of converting large molecular-based phylogenies into classifications. The tree includes over 130 hexapod species and nearly as many branches. In order to erect a classification based on this tree, decisions must be made as to which nodes (clades) to name, whether or not to limit the named clades to the 30 or so traditional Linnean ranks or use unranked names, and whether to make the classification perfectly match the phylogeny. To see one way of dealing with these classification issues, see Chapter 22 (Hexapoda). Note that the seven Crustacea included in the analysis are shown to be a paraphyletic cluster at the base of hexapod clade (i.e., "Crustacea" is a paraphyletic grouping; Chapter 21), and that the sister group of Hexapoda is Remipedia. Note also that this is a dated phylogeny (the time scale, in millions of years, is shown at the top and bottom of the figure).

of these trees. Many of the new multigene or genomic trees have scores of branching points, or even hundreds of branches that appear as long comblike topologies. But the standard Linnean ranks number only nine or ten, or about four times that at best, when the prefixes super-, sub-, and infra- are also used.

There are a few solutions to this dilemma, but none of them is perfect (Figure 2.7). The most straightforward solution is to not use ranks at all in the classification, instead just indenting the subordinated groups in ways that reflect the phylogeny—an **unranked classification**. Another solution is the phylogenetic sequencing convention, described above. A third solution (and probably still the most frequently seen) is to create classifications that *do not* precisely mirror the phylogeny, referring readers to the tree if they want to understand the precise relationships of the taxa in a classification. Unranked classifications are useful when the phylogeny of a group is still quite poorly known (e.g., Chapter 14, Annelida). Fully ranked classifications

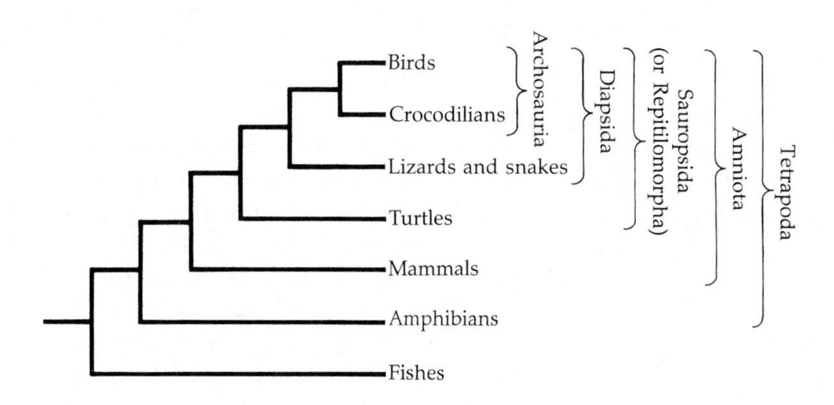

Classification scheme A

Phylum Chordata
  Subphylum Vertebrata
    Class Pisces (fishes)
    Class Amphibia (amphibians)
    Class Reptilia (turtles, crocodilians, snakes, lizards)
    Class Aves (birds)
    Class Mammalia (mammals)

Classification scheme B

Phylum Chordata
  Vertebrata
    Tetrapoda
      Lissamphibia (amphibians)
      Amniota
        Mammalia (mammals)
        Sauropsida (Reptilomorpha)
          Anapsida (turtles)
          Diapsida
            Lepidosaura (snakes, lizards)
            Archosauria
              Crocodilia (crocodilians)
              Aves (birds)

Classification scheme C

Phylum Chordata
  Subphylum Vertebrata
    Class Pisces
    Class Amphibia
    Class Mammalia
    Class Anapsida (turtles)
    Class Lepidosaura (snakes, lizards)
    Class Crocodilia (crocodilians)
    Class Aves (birds)

Figure 2.7 Sometimes genealogy and overall morphological similarity/dissimilarity can lead to conflicting approaches to a classification. This is exemplified by the birds and reptiles. The cladogram in this figure depicts the generally accepted view of the relationships among the major groups of living vertebrates. Classification scheme A depicts a traditional classification of the vertebrates, in which crocodilians are classified with lizards, snakes, and turtles in the taxon Reptilia, while birds are retained as a separate taxon, Aves. In this classification, class Reptilia is paraphyletic (because it excludes the birds). Classification schemes B and C strictly reflect the phylogenetic tree. Scheme B reflects the branching pattern of the cladogram with subordinated taxa (unranked in this case); thus, the reptiles are broken into separate taxa in recognition of their genealogical relationships—the snakes and lizards (Lepidosaura) are classified together as a separate sister group to the birds and crocodilians. Scheme C also strictly mirrors the tree, but uses the phylogenetic sequencing convention rather than subordinated sets of sister-groups. But in both B and C, all taxa are monophyletic. Notice that scheme C, using the sequencing convention, requires four fewer taxonomic names than scheme B.

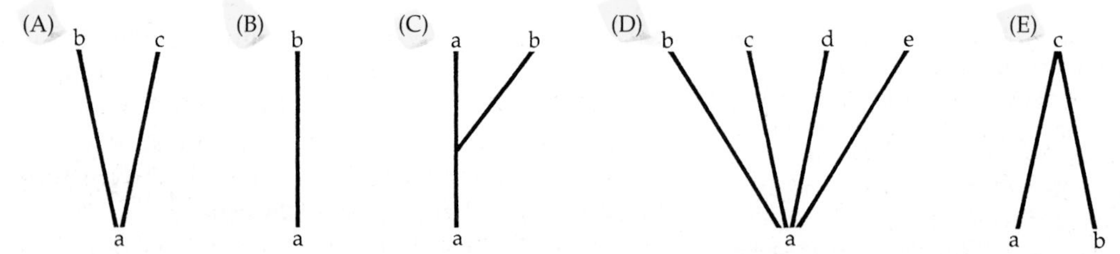

Figure 2.8 Common models of speciation. (A) One species splits into two new species. (B) One species is transformed into another over time. This type of speciation may be viewed as either gradual or rapid. (C) One species remains unchanged, while an isolated peripheral population evolves into a distinct new species. (D) "Explosive adaptive radiation," in which one species suddenly splits into many new species. Speciation events represented by this model are predicted to occur when a species is suddenly confronted with a vast new array of habitats or "unfilled niches" to exploit, resulting in rapid specialization and reproductive isolation as the new niches are filled. Explosive radiation might also occur when the range of a widespread species is fragmented into numerous smaller, isolated populations. Classic examples of explosive radiation include cichlid fishes in Africa's Rift Lakes and ant nest beetles in Madagascar (and elsewhere). (E) A new species is "created" by hybridization of two other species; this type of speciation appears to be rare and may occur primarily in plants and protists, although examples are also known from the Metazoa (e.g., whiptail lizards).

can get cumbersome when "pushed to the limit" of available taxonomic categories (e.g., Chapter 10, Platyhelminthes). The upshot is, things are changing and the "traditional" classification schemes many students are used to seeing are beginning to look rather different these days. In this book, we mostly use ranked classifications that mirror, or at least do not conflict with the phylogeny of the groups in question. In a few instances, we have had to use fully or partly unranked classifications (e.g., annelids and molluscs).

Further confusion is added when names that have been used for many decades are difficult to discard. Some of these **legacy names** have been redefined to represent monophyletic groups, or clades, such as Protostomia and Deuterostomia. Others, however, are now known to represent nonmonophyletic groups, yet the names have not yet disappeared from either textbooks or the scientific literature, for example, the gastropod group "heteropods" (which is a polyphyletic group of pelagic gastropods with laterally-compressed translucent bodies and reduced or lost shells). In this book, if we use a name that is not indicative of a natural (i.e., monophyletic) group, we point this out.

One criticism of phylogenetic systematics occasionally heard is that it always depicts the speciation process as the splitting of an ancestral species into two sister species, despite the probability that numerous other speciation modes exist (Figure 2.8). In a phylogenetic tree, once a new species appears, a "split" must be placed on the tree, and the two branches represent sister groups, whether or not the original species has in fact changed at all. Some biologists have claimed that this practice is misleading. This criticism is unfounded, and it derives from simple lack of understanding. First of all, trees need not always be completely dichotomous; they can (theoretically) have branching points that are trichotomous or even polytomous (Figure 2.8D). Second, a terminal taxon on a cladogram may lack any defining synapomorphies, thus indicating that it is not only the sister group of its adjacent lineage, but also the actual ancestor of that lineage. A phylogenetic tree can express any kind of speciation event; it simply does so in a restricted way—by way of branches depicting a pattern of nested synapomorphies.

The methods of phylogenetic systematics force the systematist to be explicit about groups and characters. The method is also largely independent of the biases of the discipline in which it is applied. In its fundamental principles, it is not restricted to biology, but is applicable to a variety of fields in which the relations that characterize groups are comparable to the homology concept and possess a hierarchical nature. Thus, cladistic analyses have been applied to other historical systems, such as linguistics and textual criticism (in which the "homologues" are shared tongues or texts), and even to the classification of musical instruments. Cladistic analysis has also used in biogeographic analyses, wherein taxa are replaced by their appropriate areas of endemism, and the "homologues" are thus sister groups shared by geographic regions. Although the information stored in a phylogenetic tree is restricted to genealogy, such trees are often used to test other kinds of hypotheses, such as modes of speciation, historical relationships among geographic areas, and coevolution in host–parasite lineages.

As stressed earlier, the concept of similarity plays a central role in phylogenetic systematics. There are really only three kinds of evolutionary similarity expressed among organisms: (1) shared evolutionary novelties inherited from an immediate common ancestor (i.e., apomorphies); (2) similarity inherited from some more remote ancestor (any number of descendant taxa may retain such similarity); and (3) similarity due to evolutionary convergent evolution. Phylogenetic systematists accept only the first kind of similarity (synapomorphies) as valid evidence of close affinity (common ancestry) between two taxa. The second kind of similarity reflects plesiomorphies, attributes that originated before the appearance of the study group and thus, while being informative, do not provide specific evidence relevant to branching points in phylogenetic trees. The third kind of similarity (convergence) holds no value at all in phylogenetic analyses, although it can lead to interesting questions about evolutionary biology.

It is worth noting that the concept of shared derived characters has been around for many decades, and a careful review of the work produced by the most critical systematists through time will reveal that most were striving to delimit monophyletic taxa and construct phylogenetic trees based on nested sets of synapomorphies. However, many existing older classifications are still based in part on symplesiomorphies rather than solely on synapomorphies, and these classifications are destined to be revised as modern studies are accomplished.

Molecular Phylogenetics

On February 12, 1988 (by appropriate coincidence, Charles Darwin's birthday), a paper published by Katherine Field, Rudy Raff, and others presented the first credible molecular analysis of metazoan phylogeny based on sequences from the small subunit ribosomal RNA gene (SSU). This work initiated a paradigm shift in phylogenetic analysis, and today the field of molecular phylogenetics is rooted in the methods pioneered in that important paper. In 1997, Anna Marie Aguinaldo and colleagues also published a revolutionary paper, proposing a radical new view of animal phylogeny—one that hypothesized the Protostomia to comprise two distinct clades, a "molting clade" (called Ecdysozoa) and a nonmolting clade (called Spiralia in this book). With the advent of technology for rapid amplification of DNA fragments and sequencing of the nucleic acids, the field of systematics has undergone dramatic change. Questions of evolutionary relationship, which had traditionally been addressed only by comparative morphology, can now be tested by a new and independent source of data. Since the 1990s, **molecular phylogenetics** has swept the literature with evolutionary trees built by analyses of gene sequences. Molecular matrices are DNA sequence data, constructed of the four nucleotides of the genetic code: adenine (A), thymine (T), cytosine (C), and guanine (G). Molecular trees have corroborated some relationships previously inferred by analyses of morphological traits, but they also provided support for novel relationships never predicted.

The first molecular phylogenies were constructed from analyses of ribosomal genes, which code for RNA that forms the 3D structure of ribosomes (the large subunit 28S, and small subunit 18S).[7] However, not long after the first molecular-based trees were published, it was recognized that analyses of molecular sequence data are prone to predicting erroneous relationships under certain circumstances. Rapidly evolving lineages were inferred to be closely related, regardless of their true evolutionary relationships, due to a phenomenon known as **long branch attraction** (LBA). Since there are only four possible character states in molecular sequence data (the four nucleotides), when DNA substitution rates are high, there is a high probability that two lineages will *independently* evolve the same nucleotide at the same site by chance alone. In these circumstances, phylogenetic algorithms (especially parsimony methods) erroneously interpret these convergences to be signs of shared ancestry (synapomorphies) and therefore misinterpret taxa on long branches to be close relatives.

Using phylogenetic algorithms that incorporate **models of evolution** can minimize the LBA problem. These models include three components: (1) models of DNA substitution, which describe the rates at which one nucleotide replaces another over evolutionary time, (2) the relative nucleotide base frequency in a dataset, and (3) the relative rates at which sites in an alignment evolve in a dataset. Popular evolutionary-model-based algorithms include **maximum likelihood** and **Bayesian methods** of phylogenetic estimation.

The number of genes used to infer phylogenies increased quickly to include both nuclear and mitochondrial protein coding genes. Inferring phylogeny from protein coding genes offers several advantages. Protein sequences from distant relatives are easier to align to one another than ribosomal genes. And since a three-nucleotide codon in a nucleic acid sequence specifies an amino acid according to the genetic code, protein-coding genes can be analyzed at either the nucleotide or the amino acid level.[8] Whereas analyses of single genes were the standard only a few years ago, most molecular phylogenetic analyses today use multiple, preferably unlinked, genes concatenated together in one super matrix of 5 to 10 or more genes. In fact, new and relatively inexpensive DNA sequencing technology, known as **next generation sequencing**, is allowing for molecular phylogenies to be constructed from larger portions of the genome, up to tens of thousands of genes. Specific regions of the genome can be targeted a priori through methods such as anchored hybrid enrichment, or novel genes of phylogenetic significance can be discovered after shotgun sequencing of all DNA molecules (genomes) or RNA molecules

[7]The genetic information in the nucleic acid DNA (deoxyribonucleic acid) is transcribed to the nucleic acid RNA (ribonucleic acid), and this information is then translated from RNA to protein.

[8]The human genome contains about 21,000 genes, and protein-coding regions account for only around 1.5% of the genome. Many other complex species get by with far fewer genes.

(transcriptomes). No mater what type of genetic data is selected or how it is sequenced, these methods result in rich multigene datasets for phylogenetic inference. Resulting trees are therefore inferred from a larger portion of the genome than previous methods that relied on only a handful of genes, whose history may or may not precisely reflect the history of the taxa being analyzed.

Equally exciting are techniques being developed to extract DNA from fossilized bone, tissue, or dung. In 2003, scientists managed to extract DNA from Siberian permafrost sediments and soils of caves in New Zealand. The Siberian sediments yielded the oldest reliable ancient DNA up to that time, from plants as much as 400,000 years old (angiosperms, gymnosperms, and mosses) as well as from numerous animals, including both living and extinct species up to 30,000 years old. In 2008, the entire genome of the extinct woolly mammoth (*Mammuthus primigenius*) from Siberia was sequenced, showing a sequence identity of over 98 percent with modern African elephants (the two diverged from one another about 6 million years ago). Also in 2008, DNA was sequenced from lice (Insecta: Phthiraptera: *Pediculus humanus*) preserved in the scalps of 1000-year old Peruvian mummies. The lice belonged to a subtype of head and body lice found all over the world, thus proving that human lice were in the New World well before Columbus. Since then, scientists have begun sequencing DNA from even older animals—the oldest so far is a 700,000-year-old horse, perhaps pushing the limits of how long DNA can actually survive. By extracting fragments of DNA from ancient soils and sediments (known as "environmental DNA") in North America, scientists have begun to reconstruct Pleistocene and early Holocene ecosystems. The extraction of DNA from 2000- to 4000-year-old cores from the Arctic region has yielded a variety of plants, bison, horses, bears, mammoths, and lemmings. Environmental DNA comes from urine, feces, hair, skin, eggshells, fathers, and even the saliva of animals, as well as from the decaying leaves and fine rootlets of plants. Ancient DNA has led to the discovery of new types of ancient humans and revealed interbreeding between our ancestors and our archaic cousins, which left a genetic legacy that shapes who we are today. Much of the genome of Neanderthal has now been sequenced. As a result, we now know that modern *Homo sapiens* carry a remnant of Neanderthal DNA from interbreeding events that have been postulated to have occurred as humans migrated out of Africa and into Eurasia, at least 80,000 years ago.

Research suggests that in closely-related species (e.g., sister species) that have recently diverged from one another there are key genes that diverge rapidly and produce reproductive incompatibility, thus reducing the likelihood of viable or fertile hybrid offspring. Other key genes that evolve quickly during the origin of a new species appear to be those that confer a special advantage to the new population. For example, the genomes of humans and chimpanzees (*Pan troglodytes*) are about 99 percent identical. But among the genes that comprise that one percent are two small stretches of bases known as the HAR1 (the human accelerated region 1) and the *FOXP2*. Before humans appeared, HAR1 appears to have evolved extremely slowly, but in *Homo sapiens* it underwent abrupt evolution. It turns out that HAR1 is active during embryogenesis in a type of neuron that plays a key role in the pattern and layout of the developing cerebral cortex of the brain. *FOXP2* is involved in speech. Thus, it appears that the rapid burst of substitutions in these genetic regions may have altered our brains significantly from those of chimpanzees. A small amount of rapid gene change can have a major impact in species divergence and evolution. Interestingly, the *FOXP2* sequence of Neanderthals is very similar to that of modern humans, suggesting that they may have had speech similar to ours.

Selected References

Aguinaldo, A. M. A., J. M. Turbeville, L. S. Linford, M. C. Rivera, J. R. Garey, R. A. Raff and J. A. Lake. 1997. Evidence for a clade of nematodes, arthropods, and other moulting animals. Nature 387: 489–494.

Baum, D. A. and S. D. Smith. 2013. *Tree Thinking: An Introduction to Phylogenetic Biology*. Roberts and Company, Greenwood Village, CO.

Bieler, R. and 19 others. 2014. Investigating the bivalve tree of life-an exemplar-based approach combining molecular and novel morphological characters. Invertebr. Syst. 28: 32–115.

Blackwelder, R. E. 1967. *Taxonomy: A Text and Reference Book*. Wiley, NY. [A pleasant little text on practical taxonomy, identification of specimens, curatorial practices, the use of names, the use of taxonomic literature, and publication; a considerable portion of the text is devoted to the intricacies of nomenclature and the rules for name publication, changes, etc.]

Bull, J. J., J. P. Huelsenbeck, C. W. Cunningham, D. L. Swofford and P. J. Waddell. 1993. Partitioning and combining data in phylogenetic analysis. Syst. Biol. 42: 384–397.

Cracraft, J. and N. Eldredge (eds.). 1979. *Phylogenetic Analysis and Paleontology*. Columbia University Press, NY. [Somewhat dated, but still with excellent discussions of phylogenetic reconstruction; an eclectic overview of systematics and evolution; good reading.]

Croizat, L. 1958. *Panbiogeography*. Published by the author. Caracas, Venezuela.

Croizat, L. 1964. *Space, Time, Form: The Biological Synthesis*. Published by the author. Caracas, Venezuela. [Croizat's two books initiated a paradigm shift in the field of biogeography.]

Croizat, L., G. Nelson, and D. E. Rosen. 1974. Centers of origins and related concepts. Syst. Zool. 23: 265–287.

Cummings, M. P., S. P. Otto and J. Wakeley. 1995. Sampling properties of DNA sequence data in phylogenetic analysis. Mol. Biol. Evol. 12(5): 814–822.

Deans, A. R. and 72 others. 2015. Finding our way through phenotypes. PLoS Biol. 13, e1002033.

Donoghue, P. C. and M. J. Benton. 2007. Rocks and clocks: calibrating the tree of life using fossils and molecules. Trends Ecol. Evol. 22: 424–431.

Dopazo, H., J. Santoyo and J. Dopazo. 2004. Phylogenomics and the number of characters required for obtaining an accurate phylogeny of eukaryote model species. Bioinformatics 20 (Suppl. 1): I116–I121.

Eldredge, N. and J. Cracraft. 1980. *Phylogenetic Pattern and the Evolutionary Process: Method and Theory in Comparative Biology.* Columbia University Press, NY. [Still one of the best texts on the theory of cladistic analysis and classification.]

Felsenstein, J. 1983. Parsimony in systematics: Biological and statistical issues. Annu. Rev. Ecol. Syst. 14: 313–333.

Felsenstein, J. 1985. Phylogenies and the comparative method. Am. Nat. 126: 1–25.

Field, K. G. J. and 7 others. 1988. Molecular phylogeny of the Animal Kingdom. Science 239: 748–753.

Frizzell, D. L. 1933. Terminology of types. Am. Midland Nat. 14(6): 637–668. [A listing of every kind of "type" designation Frizzell could locate, the vast majority of which have no particular nomenclatural validity.]

Giribet, G. 2015. Morphology should not be forgotten in an era of genomics—a phylogenetic perspective. Zool. Anz. 256: 96–103.

Giribet, G. 2015 New animal phylogeny: future challenges for animal phylogeny in the age of phylogenomics. Org. Divers. Evol. doi: 10.1007/s13127-015-0236-4.

Gould, S. J. and R. C. Lewontin. 1979. The spandrels of San Marco and the Panglossian paradigm: A critique of the adaptationist programme. Proc. R. Soc. Lond. Ser. B 205: 581–598. [A classic; Gould at his best.]

Hall, B. K. 1994. *Homology: the Hierarchical Basis of Comparative Biology.* Academic Press, SanDiego, CA.

Hall, B. K. 2011. *Phylogenetic Trees Made Easy, Fourth Ed. A How-To Manual.* Sinauer Associates, Sunderland, MA.

Harvey, P. H. and M. D. Pagel. 1991. *The Comparative Method in Evolutionary Biology.* Oxford University Press, NY. [A detailed elucidation of the method.]

Hennig, W. 1979. *Phylogenetic Systematics.* University of Illinois Press, Urbana. [The "third edition" of Hennig's original 1950 text on cladistic classification theory; be aware that the philosophy and methodology of cladistics has changed and grown a great deal since Hennig's original ideas.]

Hillis, D., C. Mortiz and B. Mable. 1996. *Molecular Systematics, 2nd Ed.* Sinauer Associates, Sunderland, MA.

Huelsenbeck, J. P. and J. J. Bull. 1996. A likelihood ratio test to detect conflicting phylogenic signal. Syst. Biol. 45: 92–98.

Huelsenbeck, J. P. and B. Rannala. 1997. Phylogenetic methods come of age: Testing hypotheses in an evolutionary context. Science 276: 227–232.

International Commission on Zoological Nomenclature. 2000. *International Code of Zoological Nomenclature,* 4th Ed. The International Trust for Zoological Nomenclature, London. [The nomenclatural rule book.]

Jablonski, D. and D. Bottjer. 1991. Environmental patterns in the origins of higher taxa: The post-Paleozoic fossil record. Science 252: 1831–1833.

Jefferys, W. H. and J. O. Berger. 1992. Ockham's razor and Bayesian analysis. Am. Sci. 80: 64–72.

Lemey, P., M. Salemi and A.-M. Vandamme (eds.). 2009. *The Phylogenetic Handbook. A Practical Approach to Phylogenetic Analysis and Hypothesis Testing.* Cambridge University Press, NY.

Lemmon A.R., S.A. Emme and E.M. Lemmon. 2012. Anchored hybrid enrichment for massively high-throughput phylogenomics. Syst. Biol. 61(5): 727–744.

Lomolino, M. V., B. R. Riddle, R. J. Whittaker and J. H. Brown. 2010. *Biogeography,* 4th Ed. Sinauer Associates, Sunderland, MA.

Lomolino, M. V., D. F. Sax and J. H. Brown (eds.). 2004. *Foundations of Biogeography. Classic Papers with Commentaries.* University of Chicago Press, Chicago, IL.

Maddison, D. R. 1991. The discovery and importance of multiple islands of most-parsimonious trees. Syst. Zool. 40: 315–328.

Maddison, W. P. 1997. Gene trees in species trees. Syst. Biol. 46: 523–536.

Maddison, W. P., M. J. Donoghue and D. R. Maddison. 1984. Outgroup analysis and parsimony. Syst. Zool. 33(1): 83–103.

Mayr, E. and P. D. Ashlock. 1991. *Principles of Systematic Zoology,* 2nd Ed. McGraw-Hill, NY. [The "bible" of systematics, prior to the establishment of cladistic phylogenetics.]

McCormack, J.E., S. M. Hird, A. J. Zellmer, B. C. Carstens and R. T. Brumfield. 2013. Applications of next-generation sequencing to phylogeography and phylogenetics. Mol. Phylog. Evol. 66: 526–538.

Minelli, A. 2015. Taxonomy faces speciation: the origin of species or the fading out of the species? Biodiversity Journal 6(1): 123–138.

Minelli, A. 2015. Grand challenges in evolutionary developmental biology. Front. Ecol. Evol. doi: 10.3389/fevo.2014.00085

Minelli, A. 2015. EvoDevo and its significance for animal evolution and phylogeny. Pp. 1–23 in A. Wanninger (ed.), *Evolutionary Developmental Biology of Invertebrates 1: Introduction, Non-Bilateria, Acoelomorpha, Xenoturbellida, Chaetognatha.* Springer-Verlag, Vienna.

Misof, B. and 100 others. Phylogenomics resovles the timing and pattern of insect evolution. Science. 346 (6210): 763–767.

Nelson, G. and N. I. Platnick. 1981. Systematics and biogeography: Cladistics and vicariance. Cladistics 5: 167–182.

Nelson, G. and N. I. Platnick. 1981. *Systematics and Biogeography: Cladistics and Vicariance.* Columbia University Press, NY. [Excellent review of the history and development of systematics and biogeography, as well as a thorough, although theoretical, treatment of cladistics and vicariance biogeography.]

Nosenko, T. and numerous others. 2013. Deep metazoan phylogeny: when different genes tell different stories. Mol. Phylog. Evol. 67: 223–233.

Parham, J. F. and 24 others. 2012. Best practices for justifying fossil calibrations. Syst. Biol. 61: 346–359.

Patterson, C. 1982. Morphological characters and homology. Pp. 21–74 in K. Joysey and A. Friday, *Problems of Phylogenetic Reconstruction.* Systematics Association Special Volume, No. 21. Academic Press, NY.

Patterson, C. 1990. Reassessing relationships. Nature 344: 199–200.

Peterson, K. J.,l M. A. McPeek and D. A. D. Evans. 2005. Tempo and mode of early animal evolution: inferences from rocks, Hox, and molecular clocks. Paleobiology 31(2): 36-55.

Philippe, H., A. Chenuil and A. Adoutte. 1994. Can the Cambrian explosion be inferred through molecular phylogeny? Development (Suppl.): 15–25.

Pfenning, A. R. and 25 others. 2014. Convergent transcriptional specializations in the brains of humans and song-learning birds. Science 346: 1333. doi: 10.1126/science.1256846

Pollard, K. S. 2009. What makes us human? Sci. Am. May 2009: 44-49.

Raff, R. A. 1996. *The Shape of Life.* University of Chicago Press, Chicago.

Richter, S. and C. S. Wirkner. 2014. A research program for evolutionary morphology. J. Zool. Syst. Evol. Res. 52: 338–350.

Rose, M. R. and G. V. Lauder. 1996. *Adaptation.* Academic Press, San Diego, CA.

Rosen, D. E. 1985. Geological hierarchies and biogeographic congruence. Ann. Missouri Bot. Garden 72: 636–659.

Sanderson, M. J. 1990. Flexible phylogeny reconstruction: A review of phylogenetic inference packages using parsimony. Syst. Zool. 39: 414–420.

Sanderson, M. J. 1995. Objections to bootstrapping phylogenies: A critique. Syst. Biol. 44: 299–320.

Sanderson, M. J. 2008. Phylogenetic signal in the eukaryotic tree of life. Science 321: 121–123.

Sanderson, M. J. and M. J. Donoghue. 1989. Patterns of variation in levels of homoplasy. Evolution 43: 1781–1795.

Sanderson, M. J. and L. Hufford. 1996. *Homoplasy. The Recurrence of Similarity in Evolution.* Academic Press, San Diego, CA.

Schuh, R. T. 2000. *Biological Systematics: Principles and Applications.* Comstock Publishing Associates, Ithaca, NY.

Simpson, G. G. 1961. *Principles of Animal Taxonomy.* Columbia University Press, NY. [Although this classic text is aging fast, it still provides a sound look at non-cladistic thinking on evolution, the species concept, and traditional classification methods.]

Slater, G. J., L. J. Harmon and M. E. Alfaro. 2012. Integrating fossils with molecular phylogenies improves inference of trait evolution. Evolution 66: 3931–3944.

Stanley, S. M. 1982. Macroevolution and the fossil record. Evolution 36: 460–473.

Stevens, P. F. 1980. Evolutionary polarity of character states. Annu. Rev. Ecol. Syst. 11: 333–358.

Watrous, L. E. and Q. D. Wheeler. 1981. The out-group comparison method of character analysis. Syst. Zool. 30(1): 1–11.

Wiley, E. O. and B. S. Lieberman.2011. *Phylogenetics: Theory and Practice of Phylogenetic Systematics,* 2nd Ed. Wiley-Blackwell, NY.

Wiley, E. O. 1988. Vicariance biogeography. Annu. Rev. Ecol. Syst. 19: 513–542.

Wilkins, A. S. 2002. *The Evolution of Developmental Pathways.* Sinauer Associates, Sunderland, MA.

Willerslev, E. and 9 others. 2003. Diverse plant and animal genetic records from Holocene and Pleistocene sediments. Science 300: 791–795.

Winston, J. E. 1999. *Describing Species: Practical Taxonomic Procedure for Biologists.* Columbia University Press, NY.

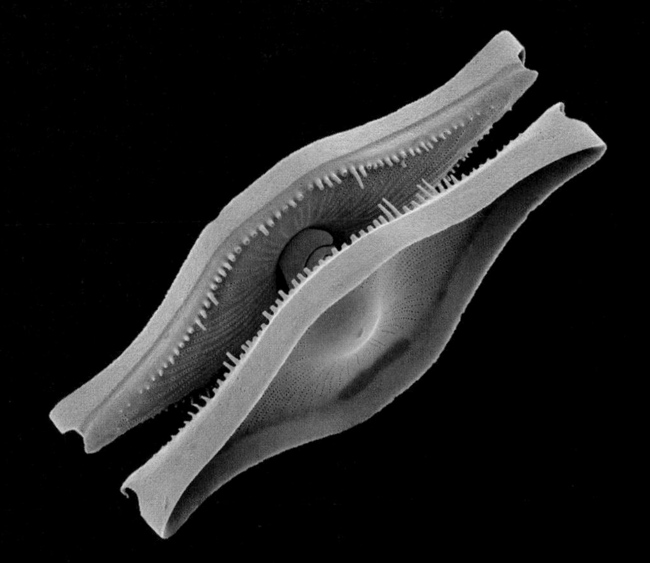

CHAPTER 3

The Protists

Kingdom Protista

It has long been a tradition in invertebrate zoology courses to cover the protists, at least the most common groups, and today well over half of the invertebrate courses taught in North America still do so. And in today's world of diminishing "-ology" courses many students will find this to be their only detailed exposure to this very important assemblage of organisms. For this reason we have chosen to retain the protist chapter in this edition of *Invertebrates*, despite the challenges of presenting this highly diverse collection of phyla in a succinct fashion. The formal taxon "Protozoa" was established by Richard Owen in 1858, the same year Charles Darwin and Alfred Russel Wallace published their theory of natural selection. In its most common use it has always been a paraphyletic taxon, although before the Myxozoa, yeasts, and a few others were removed it was a polyphyletic grouping. However, the name "Protozoa" has been used for a plethora of different groupings of eukaryotes over the past 150 years, and it is probably best laid to rest as a formal taxonomic name today.

Because the protists are a paraphyletic group, they are not easy to define. Like other paraphyletic groups, they are defined primarily by what they lack. One definition is, "eukaryotic, largely unicellular organisms that do not undergo tissue formation through the process of embryological layering." The assemblage we call protists is paraphyletic because it excludes numerous multicellular descendant lineages (Metazoa, Fungi, Plantae), as well as the Glaucophyta, and the red and green algae. Thus protists are the "soup" out of which the three great multicellular kingdoms evolved. Some protists (such as the choanoflagellates) are more closely related to animals than to other protists; and four phyla once regarded as protists are now usually classified in the kingdom Plantae (=Archaeplastida): Glaucophyta (glaucophyte algae), Rhodophyta (red algae), Chlorophyta (green algae), and Charophyta (stoneworts).[1] Recently, some workers have subdivided the protists into

[1]Plants form a monophyletic group, comprising unicellular and multicellular, photosynthetic, chlorophyll-bearing eukaryotes with chloroplasts directly descended from an initially enslaved cyanobacterium (i.e., the "primary plastid"). These chloroplasts are surrounded by two membranes, and they may contain chlorophyll *a* and/or *b*. Plantae contains two subkingdoms: Biliphyta (phyla Glaucophyta and Rhodophyta) and Viridiplantae (the phyla Chlorophyta, Charophyta, Anthocerotophyta, Bryophyta, Marchantiophyta, and Tracheophyta). The two-layered chloroplast membrane, structure of plastid genomes, and derived chloroplast protein-import machinery all suggest a single origin for the Plantae.

two groups—Protozoa and Chromista. This arrangement, championed by Thomas Cavalier-Smith, separates out the Chromista as those **chromophtye algae** (those with chlorophyll *a* and *c*, but not *b*) thought to have evolved by symbiogenetic enslavement of a red alga, as well as all heterotrophic protists descended from them by loss of photosynthesis or entire plastids. In this scheme, the "kingdom Chromista" would include such phyla as Haptophyta, Ciliophora, and Cercozoa; and the "kingdom Protozoa" would contain phyla such as Euglenozoa, Amoebozoa, and Choanozoa (see Ruggiero et al. 2015 for a version of this classification). However, multigene analyses of protists are still ambiguous on the precise nature of this dichotomy and whether Chromista is monophyletic, and we feel the redefined use of the now highly ambiguous term "Protozoa" is confusing. Thus we have not adopted this classification for this edition of *Invertebrates*. In fact, classification of the protists is still in a significant state of flux as new multigene phylogenies continue to discover, sort out, and rearrange the many clades. Because the protists are no more than a paraphyletic assemblage, and because brown algae (Phaeophyta), like red and green algae, arose from a separate protistan ancestor, some workers would exclude the Phaeophyta from the protists.

About 200,000 species of protists have so far been described, but taxonomists have only scratched the surface and it is almost certain that species diversity in this assemblage of phyla exceeds that of Metazoa. For example, recent large-scale molecular screening techniques have revealed an unimagined diversity of minute (picoplankton, 0.2–2 μm) protists in the world's oceans, most of which have not even begun to be classified. Most protists are unicellular, but there are also colonial and multicellular species, although the multicellular forms do not undergo embryonic tissue formation as seen in the plants and animals (thus distinguishing them from the kingdoms Plantae and Metazoa). Some protists are exclusively asexual; others can also reproduce sexually, or at least employ the sexual processes of meiosis and syngamy. Some protists are **photoautotrophs** containing plastids, organelles with light-capturing pigments for photosynthesis. Some are **heterotrophs**, absorbing organic molecules or ingesting larger food particles (phagotrophy). Still others are **mixotrophs**, combining photosynthesis and heterotrophic nutrition. In years past, protists were divided into three large groupings on the basis of these nutritional modes: protozoans (the ingestive heterotrophs, or animal-like protists); algae (the photosynthetic, or plantlike protists); and the funguslike (absorptive) protists. However, we now know that these represented no more than loose ecological groupings.

The prokaryotic and eukaryotic phytoplankton, mainly Cyanobacteria ("blue-green algae") and photosynthetic protists such as the diatoms, dinoflagellates, and coccolithophores, are critically important to oceanic food chains, the global carbon cycle, and marine ecosystem functioning. Even though they account for less than 1% of Earth's standing photosynthetic biomass, they are responsible for about half of Earth's annual production of atmospheric oxygen (the rest coming from terrestrial plants, that, unlike marine primary producers, are much longer-lived and channel more of their energy to standing biomass). And, of course, much of the **petroleum** we consume at an ever-increasing pace came from phytoplankton that settled on the ocean floor and was buried and "cooked" over past millennia. This took place on a warmer Earth that had less dissolved oxygen in the sea, and less decay of organic matter as it sank to the seafloor. However, it has been estimated that only about 2% of all oceanic plankton makes it to the ocean floor in modern seas, the rest being recycled before it reaches the benthic realm. The smallest oceanic protists (below 1 mm body size), such as diatoms, are, like bacteria, completely at the mercy of the currents and tend to disperse far and wide. For this reason, species of small protists tend to disperse everywhere, and the local environment selects where they can survive.

The "big three" eukaryotic primary producers that are conspicuous in the world's oceans today (diatoms, dinoflagellates, coccolithophores) came to prominence during the Mesozoic era, although they first arose much earlier than that. Chemical evidence of lipid biosignatures from dinoflagellates has been found in Neoproterozoic and Paleozoic strata. Evidence of coccolithophores has been found in Late Triassic strata. The siliceous remains of diatoms have left a somewhat substantial fossil record, the earliest clear evidence (fossil frustules) being Jurassic (remnants of their silica shells make up the porous rock called diatomite). However, molecular clock analyses suggest an origin for the diatoms around the Permian-Triassic boundary. Interestingly, all three groups contain plastids derived from an ancestral red alga by secondary (or even higher-level) symbioses, and all three groups began their major radiations subsequent to the End-Permian mass extinction event.[2]

Photosynthetic protists contain a variety of types of chlorophyll and have differently constructed chloroplasts, reflecting their many separate lineages. Like mitochondria, plastids have an intricate internal structure of folded membranes. But the plastid's membranes are not continuous with the inner membrane of the chloroplast envelope. Instead, the internal membranes lie in flattened disclike sacs called **thylakoids**. Each thylakoid consists of an outer thylakoid membrane surrounding an inner space. Thylakoids are piled up like plates. Most protists have thylakoids that form stacks two or

[2]Some dinoflagellates, and other groups, obtained their plastids through tertiary or higher-order symbioses (i.e., acquisition of a eukaryotic algal cell and retention of the plastid as well as other pieces of cell machinery).

three units thick. Green algae and land plants have many thylakoids stacked into piles called **grana** (sing. granum), and a chloroplast may contain many grana. Red algae (Rhodophyta) have unstacked thylakoids.

The body plans of protists demonstrate a remarkable diversity of nonmetazoan form, function, and survival strategies. Most, but not all, are unicellular. In any case, they carry out all of life's functions using only the organelles found in a "typical" eukaryotic cell. Many of the fundamentally unicellular protist phyla also contain species that form colonies. Others are multicellular, though lacking the embryonic tissue-layering processes and differentiated organs and tissues seen in animals, plants, and fungi. Multicellularity in a number of protist groups (e.g., the formation of cellular slime molds in dictyostelids; the formation of multicellular thalli in different groups of brown algae) is but one example of rampant convergent evolution seen among these creatures. Different lineages of free-living protists have, for instance, independently evolved a wide variety of cytoskeletal extensions (e.g., flagella, axopods, haptonemes), feeding apparatuses (e.g., rods and gullets), secreted cell coverings (e.g., scales, loricas, frustules, spines), and intracellular armor (e.g., thecae, pellicles).

Protists include an awesome array of shapes and functional types. Figure 3.1 illustrates some of this variety. Most unicellular protists are microscopic, although a few, such as the foraminiferans, are commonly visible to the naked eye. Many protists are actually larger than the smallest metazoans (e.g., some gastrotrichs, kinorhynchs, nematodes, loriciferans, and others). Protists include marine, freshwater, terrestrial, and symbiotic species, and the last category includes many serious pathogens. Humans are hosts to over 30 species of protistan symbionts, many of which are pathogenic.

Protist diversity, indeed the diversity of the entire superkingdom Eukaryota, had its origin in a remarkable and critically important evolutionary event that occurred 2–2.7 billion years ago, when a free-living proteobacterium took up residence in an anaerobic prokaryote, first becoming an endosymbiont, and then being fully assimilated into the cytoplasm and cellular machinery of its host to become the first mitochondrion. This **primary endosymbiosis** origin of mitochondria is believed to have occurred only once (Figure 3.2), and it represents the origin of the monophyletic lineage called Eukaryota.

Some time later, a cyanobacterium became endosymbiotic in a protist and, as with the proteobacterium, became enslaved and incorporated into its host. As the cyanobacterium became a permanent fixture in this host, the photosynthetic apparatus of its plastid (or chloroplast) continued to function and thus serve its new master. From this singular endosymbiotic event, arose the kingdom Plantae, or Archaeplastida. The plastids in these phyla are called **primary plastids**. The bulk of the genes present in the genome of the cyanobacterial progenitor of plastids were either lost or transferred to the nuclear genome of the eukaryotic host, where they are now expressed and their protein products targeted back to the organelle in which they were originally encoded by a dedicated plastid protein import system.[3]

Subsequent to the ancient primary endosymbiotic event, many other protist clades adopted chloroplasts secondarily by entering into a symbiosis with a green or red alga, and then gradually reducing the symbionts until all that remained was the chloroplast—a process known as **secondary endosymbiosis**. At least three independent secondary endosymbioses between unrelated host cells and red and green algal endosymbionts probably occurred, one involving a red alga and two involving green algae. From the red algal secondary endosymbiotic event arose the cryptomonads, haptophytes, heterokonts, dinoflagellates, and probably apicomplexans (the apicomplexans having lost their chloroplasts as they took up their parasitic lifestyle). From the green algal line, euglenids and chlorarachniophytes each evolved independently. And we now know of tertiary, and perhaps even quaternary endosymbiotic events that transferred photosynthetic capabilities to newly evolving protist clades. Indeed, one of the most stunning achievements of protists has been their ability to engage in endosymbiotic relationships with one another—a trait seen in nearly every major lineage. Some fundamentally heterotrophic groups have become specialists at developing symbioses with a variety of autotrophic protists, such as the forams, which may harbor endosymbiotic diatoms, dinoflagellates, or red or green algae. Diatoms themselves are a highly chimeric group, with about 10% of their nuclear genes being of foreign algal origin (from ancient, secondary endosymbiotic events with red and green algae).

---

[3]Photosynthetic plastids (those with chlorophyll) are called chloroplasts, and they contain their own DNA (cpDNA). Plastids without chlorophyll are usually called leucoplasts (and generally contain lpDNA). Both cpDNA and lpDNA are transcribed and translated just like nuclear DNA, except these organelle genes are effectively haploid. Chloroplasts are the solar-powered, energy-generating, oxygen-producing organelles of plants and algae (photosynthetic protists). All are descendants of once-free-living cyanobacteria. Chlorophyll (a green pigment) is found in most plants, algae, and Cyanobacteria. Chlorophyll absorbs light most strongly in the blue and red wavelengths, and only poorly in the green wavelengths (hence the reflected green color). Chlorophyll molecules are arranged in and around pigment protein complexes called **photosystems**, which are embedded in the thylakoid membranes of chloroplasts. Besides the very common chlorophyll *a* (which may be universal in photosynthetic eukaryotes), there are other pigments, called accessory pigments, which occur in the photosystems. These include other forms of chlorophyll, such as chlorophyll *b* in green algae and higher plants, and chlorophyll *c* or *d* in other algae. There are also nonchlorophyll accessory pigments, such as carotenoids and phycobilins (phycobiliproteins), which also absorb light and transfer that energy to the photosystem's chlorophyll. The different chlorophyll and nonchlorophyll pigments all have different absorptive spectra (e.g., the phycobilins in red algae, and typically in Cyanobacteria, and some cryptomonads, can absorb green light relatively well).

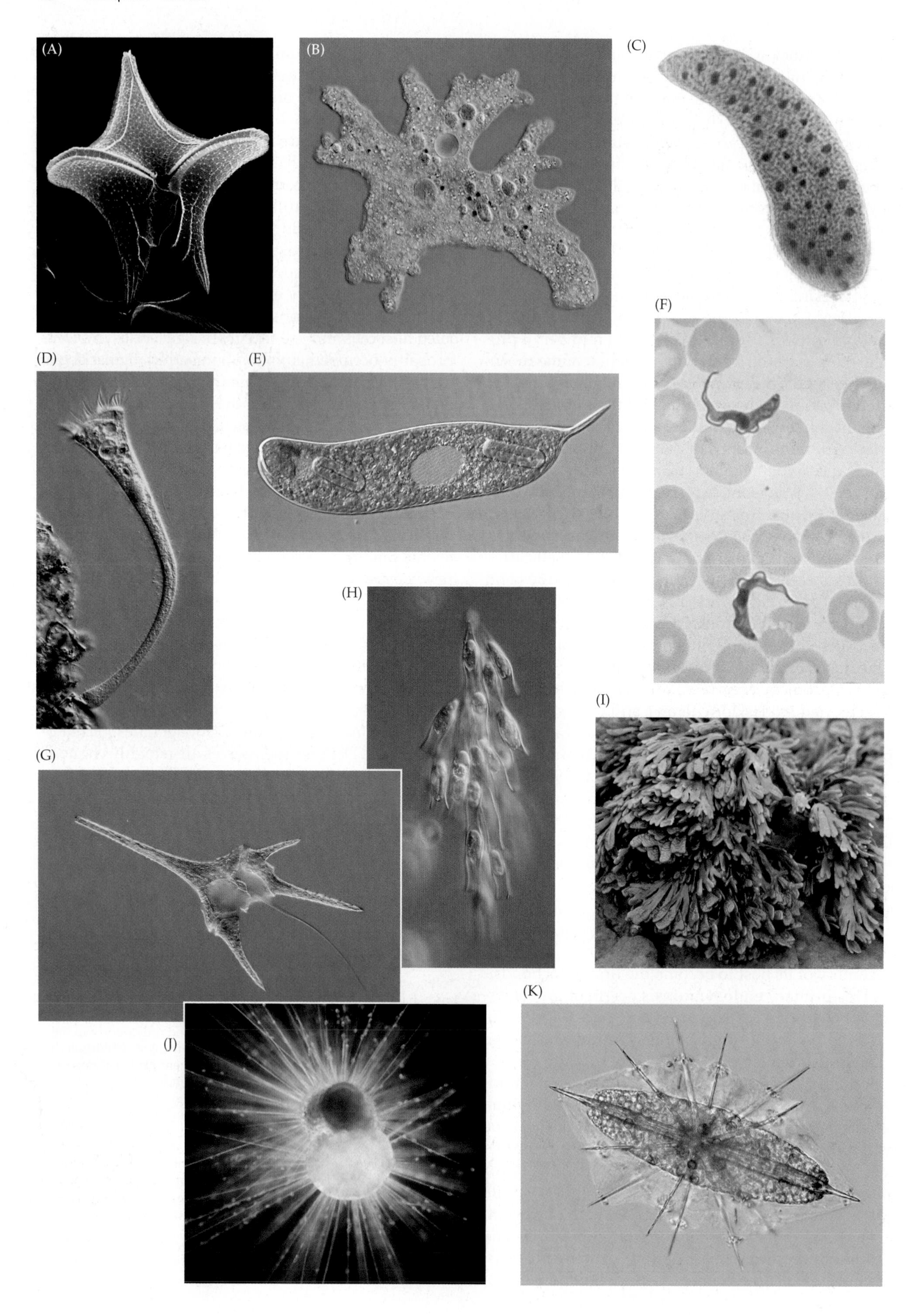
(A)
(B)
(C)
(D)
(E)
(F)
(G)
(H)
(I)
(J)
(K)

◀ **Figure 3.1 Protist diversity.** (A) Phylum Dinoflagellata, *Peridinium*. (B) Phylum Amoebozoa, *Amoeba proteus* (formerly *Chaos diffluens*); a large water expulsion (contractile) vacuole, green food vacuoles, and lobopodia can be seen. (C) Phylum Stramenopila, *Opalina*. (D) Phylum Ciliata, *Stentor*. (E) Phylum Euglenida, *Lepocinclis*. (F) Phylum Kinetoplastida, *Trypanosoma* in blood smear. (G) Phylum Dinoflagellata, *Ceratium hirudinella*, a freshwater dinoflagellate. (H) Phylum Stramenopila, *Dinobryon*, a colonial golden alga. (I) Phylum Stramenopila, *Fucus* (a brown alga). (J) Phylum Granuloreticulosa, *Globigerinella*, a foram (note the calcareous spines that radiate out from the body). (K) Phylum Radiolaria, *Amphilonche heteracantha*.

Cryptomonads and chlorarachniophytes have been the subject of intense research in recent years, because they each possess four genomes—two nuclear genomes, an endosymbiont-derived plastid genome, and a mitochondrial genome derived from the host cell. Like mitochondrial and plastid genomes, the persistent genome of the endosymbiont nucleus, called a **nucleomorph**, has been retained in these two groups. Much of the nucleomorph genome has been lost in cryptomonads and chlorarachniophytes, reduced through the combined effects of gene loss and intracellular gene transfer to the "host" nucleus. In fact, nucleomorph genomes are the smallest genomes known in the Eukaryota. It was an ultrastructural understanding of nucleomorphs, largely led by the work of A. D. Greenwood in the 1970s, that showed them to be degenerate nuclei and provided the "smoking gun" of the secondary endosymbiotic theory. The still unanswered question is, why have the ancestral endosymbiont genomes been retained in cryptomonads and chlorarachniophytes, when they have been lost in all other secondary plastid-containing organisms? One answer is, they are still gradually disappearing as their genes are lost or transferred to the "host" nucleus.

Taxonomic History and Classification

Antony van Leeuwenhoek is generally credited with being the first person to report seeing protists, in about 1675. In fact, Leeuwenhoek was the first to describe a number of microscopic aquatic life forms (e.g., rotifers), referring to them as "animalcules" (little animals). For nearly 200 years, protists were classified along with a great variety of other microscopic life forms under various names (e.g., Infusoria). The name protozoon (Greek, *proto*, "first"; *zoon*, "animal") was coined by Goldfuss in 1818 as a subgrouping of a huge assemblage of animals known at that time as the Zoophyta (protists, sponges, cnidarians, rotifers, and others). Ernst Haeckel's Protista (in 1866) did not include the green algae or ciliates (placed among the animals) but did include the sponges. Following the discovery of cells in 1839, the distinctive nature of protists became apparent. On the basis of this distinction, Karl von Siebold, in 1845, restricted the name protozoa to apply

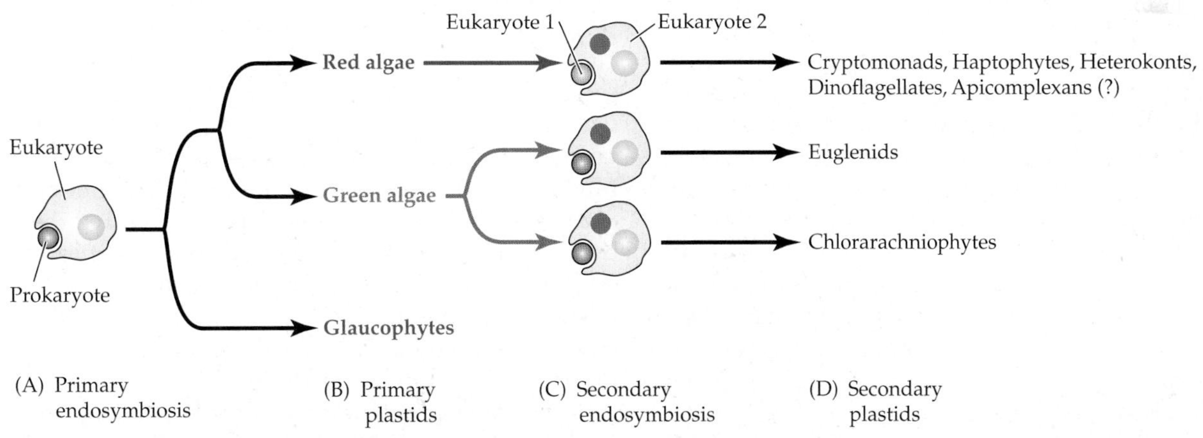

Figure 3.2 The origin and spread of plastids in eukaryotic cells. (A) Primary endosymbiosis between a heterotrophic eukaryote and a photosynthetic prokaryote. (B) Three modern-day eukaryotic lineages (red and green algae, glaucophytes) possess primary plastids whose ancestry traces directly back to the primary endosymbiosis; the green algae spawned multicellular lineages such as land plants. (C) At least three independent secondary (i.e., eukaryote-eukaryote) endosymbiosis events between unrelated host cells and red and green algal endosymbionts have probably occurred. (D) Diversity of eukaryotes harboring secondary plastids; cryptomonads, haptophytes, heterokonts, many dinoflagellates, and (probably) apicomplexans harbor plastids derived from a red algal endosymbiont, while euglenids and chlorarachniopytes acquired their plastids from green algae, probably in separate events. The cryptomonads and chlorarachniopytes are the only known secondary plastid-containing algae to still possess the nucleus and nuclear genome of their algal endosymbionts. Some dinoflagellate lineages have replaced their ancestral red algal secondary plastid with a tertiary plastid through the uptake of cryptomonad, heterokonts, or haptophytes algae. The apicomplexans are an entirely nonphotosynthetic lineage of parasites, some of which have completely lost their plastid.

to all unicellular forms of animal life. It was the great naturalist Ernst Haeckel who united the algae and Protozoa into a single group, the Protista.

Throughout most of the twentieth century, a relatively standard classification scheme was used for the protists, or "Protozoa." This scheme, which has its roots in the work of the great nineteenth century German zoologist Otto Bütschli, was based on the idea that the different groups could be classified primarily by their modes of locomotion and nutrition. Thus the protozoans were divided into the Mastigophora (locomotion with flagella), Ciliata (locomotion with cilia), Sarcodina (locomotion with pseudopodia), and Sporozoa (parasites, with no obvious locomotory structures, but forming resistant spores for transmission between hosts). Flagellated protists were further divided into the zooflagellates (heterotrophs) and phytoflagellates (photosynthetic autotrophs). While these divisions might accurately describe protists' roles in ecosystems, we now know that they do not accurately reflect evolutionary relatedness. Pseudopodia and flagella are present in many different kinds of cells (including plant and animal cells) and their presence does not necessarily indicate unique relatedness. Flagella (and cilia) are clearly shared primitive features, or symplesiomorphies, whereas pseudopodia come in many different forms that represent cases of independent evolution. Modern molecular phylogenetics and comparative ultrastructural studies have completely reorganized the classification of protists, and will continue to make adjustments for some time come.

Although there continues to be much debate over how these enigmatic organisms are related to each other, there are now dozens of well-defined clades of protists, many of which cluster into larger hypothesized clades. However, readers are cautioned that the field of protist systematics is highly dynamic, and major changes continue to appear at a rapid pace. In fact, protistan classification is probably the most unstable of any classification seen in eukaryote systematics today, and nearly every book that treats the subject uses a somewhat different classification scheme. This instability simply reflects the exciting, rapid state of ongoing discovery among the protists. Certainly, by the time this book is printed protistan classification will have already undergone numerous revisions, and it will continue to do so for the foreseeable future.

One of the more surprising recent discoveries is that the former protist phylum Myxozoa comprises a group of highly modified cnidarians, parasitic in certain invertebrates and vertebrates (see Chapter 7). This revelation was made possible through DNA analyses, and by the discovery of certain metazoan and cnidarian features (e.g., collagen, nematocysts) in myxozoans. Also, the Microsporidia, once thought to be basal protists, are now thought to be highly atypical fungi that dramatically reduced the mitochondria during their evolution. Microsporidia are an enigmatic group of parasites, and some workers retain them among the Protista because they are phagotrophic and lack the chitinous wall around vegetative cells that characterize Fungi. And we now know that the amebas do not constitute a monophyletic group, but are spread through many distantly related protist taxa (i.e., pseudopodia are very different in structure and function in different protist groups, and the basic ameba cell form has evolved may times independently). The kingdom Fungi is now known to have been polyphyletic as once defined, and many of its former members are now spread through several protistan phyla, with the remaining "true fungi" sometimes given the name Eumycota (all true fungi, or Eumycota, have chitin in their cell walls, except the Microsporidia). There are many protist groups that remain quite enigmatic, and we still are not certain about their phylogenetic relationships.

Depending on which specialists you follow, protists can be divided into as few as a half-dozen, or as many as 50+ phyla, and phylogenetic research is clustering these into a half-dozen or so larger clades (Box 3A). The deep relationships among protists are now beginning to emerge, and the completion of genome sequencing projects on various species is shedding critical new light on protist relationships. In fact, new data are appearing so rapidly, and so fluid are many of the new hypotheses, that many workers are now opting not to use formal taxonomic names at all, and instead refer to hypothesized "higher clades" simply as "groups" (a practice we follow, in part, in this book). Other workers consider these major clades worthy of "kingdom" or some other status, and still others prefer to simply use vernacular or formal names that have no categorical taxonomic standing at all ("rankless taxa"). The protist phyla we treat in this chapter are the most commonly encountered ones. Although we have organized them by current phylogenetic thinking, this classification scheme will surely be modified as new studies emerge. At the time this book went to press, a conservative view based on molecular phylogenetics (with support from ultrastructural studies) groups the eukaryotes largely into six groups, or putative clades: (1) Amoebozoa, (2) Chromalveolata, (3) Rhizaria, (4) Excavata, (5) Opisthokonta (Choanoflagellata, Metazoa, and Fungi), and (6) Plantae (= Archaeplastida).

CLASSIFICATION OF THE PROTIST PHYLA TREATED IN THIS CHAPTER

GROUP 1 AMOEBOZOA

PHYLUM AMOEBOZOA Examples include: *Acanthamoeba, Amoeba, Arcella, Chaos, Centropyxis, Difflugia, Endolimax, Entamoeba, Euhyperamoeba, Flabellula, Hartmanella,*

BOX 3A Classification of Eukaryota, Including the 17 Protist Phyla Covered in this Book

(nonprotist taxa are shown in blue)

Kingdom Protista*

GROUP 1 AMOEBOZOA

Phylum Amoebozoa Lobate-pseudopod amebas, myxomycetes, dictyostelids, myxogastrids (plasmodial slime molds), and dictyostelids (cellular slime molds or "social amoebae") (200+ described species)

GROUP 2 CHROMALVEOLATA

Phylum Dinoflagellata Dinoflagellates.(2,000 described species)

Phylum Apicomplexa Gregarines, coccidians, haemosporidians, and their kin (5,000+ described species)

Phylum Ciliata (= Ciliophora) Ciliates (10,000–12,000 described species)

Phylum Stramenopila Bacillariophytes (photosynthetic diatoms), Phaeophyta (brown algae), Chrysophytes (golden algae), the funguslike nonphotosynthetic Oomycetes (water molds or Oomycota, downy mildews, etc.), and certain parasitic (opalines and blastocystids) and free-living (some heliozoans and flagellates) groups (9,000 described species)

Phylum Haptophyta (= Prymnesiophyta) Coccolithophores and their relatives

Phylum Cryptomonada Cryptomonads

GROUP 3 RHIZARIA

Phylum Chlorarachniophyta Chlorarachniophytes

Phylum Granuloreticulosa Foraminiferans and their kin (4,000+ described species)

Phylum Radiolaria Radiolarians (2,500 described species)

Phylum Haplosporidia Haplosporidians

*Paraphyletic group

GROUP 4 EXCAVATA

Phylum Parabasalida Trichomonads, hypermastigotes, etc. (300 described species)

Phylum Diplomonadida Diplomonads (100 described species)

Phylum Euglenida Euglenids (1,000 described species)

Phylum Kinetoplastida Trypanosomes, bodonids, and their kin (600 described species)

Phylum Heterolobosea (Naegleria, Stephanopogon, etc.)

GROUP 5 OPISTHOKONTA

Phylum Choanoflagellata Choanoflagellates (150 described species)

Kingdom Metazoa

Kingdom Fungi

Kingdom Plantae (= Archaeplastida)

Subkingdom Biliphyta

Phylum Glaucophyta Glaucophytes

Phylum Rhodophyta Red algae

Subkingdom Viridiplantae

Phylum Chlorophyta Green algae

Phylum Charophyta Freshwater green algae; includes the stoneworts

Subgroup Embryophyta

Phylum Anthocerotophyta Hornworts

Phylum Bryophyta Other nonvascular plants

Phylum Marchantiophyta Liverworts

Phylum Tracheophyta Vascular plants. Includes clubmosses, horsetails, ferns, gymnosperms, and angiosperms (flowering plants)

Iodamoeba, Mayorella, Pamphagus, Pelomyxa, Thecamoeba, Vannella, Dictyostelium (cellular slime mold), *Fuligo* (plasmodial slime mold), *Physarum* (plasmodial slime mold)

GROUP 2 CHROMALVEOLATA

PHYLUM DINOFLAGELLATA (OR DINOZOA) Examples include: *Amphidinium, Ceratium, Haplozoon, Kofoidinium, Gonyaulax, Nematodinium, Nematopsides, Noctiluca, Peridinium, Perkinsus, Pfiesteria, Polykrikos, Protoperidinium, Symbiodinium, Syndinium, Zooxanthella*

PHYLUM APICOMPLEXA Gregarines, coccidians, haemosporidians, and their kin (e.g., *Cryptosporidium, Diaplauxis, Didymophyes, Eimeria, Gregarina, Haemoproteus, Lankesteria, Lecudina, Leucocytozoon, Plasmodium, Pterospora, Selenidium, Strombidium, Stylocephalus, Toxoplasma*)

PHYLUM CILIATA (OR "CILIOPHORA") The ciliates (e.g., *Balantidium, Coleps, Colpidium, Colpoda, Didinium, Euplotes, Halteria, Laboea, Oxytricha, Paramecium, Podophrya, Stentor, Tetrahymena, Tintinnidium, Vorticella*)

PHYLUM STRAMENOPILA Brown algae (= Phaeophyta), Chrysophytes ("golden algae"), the funguslike nonphotosynthetic Oomycetes (water molds or Oomycota, downy mildews), and certain parasitic (opalines and blastocystids) and free-living (some heliozoans and flagellates) groups (e.g., *Actinophrys, Actinosphaerium, Dinobryon, Fucus, Macrocystis, Opalina, Poteriochromas, Protopalina, Saprolegnia, Synura*); and the photosynthetic diatoms [Bacillariophyta] (e.g., *Actinoptychus, Chaetoceros, Coscinodiscus, Didymosphenia, Melosira, Navicula, Nitzschia, Pseudonitzschia, Thalassiosira*)

PHYLUM HAPTOPHYTA (= PRYMNESIOPHYTA) Coccolithophores and their relatives; placement of Haptophyta within the Chromalveolata is only weakly supported (e.g., *Emiliania, Pavlova*)

PHYLUM CRYPTOMONADA Placement of Cryptomonada within the Chromalveolata is only weakly supported (e.g., *Cryptomonas* [= *Chilomonas*], *Goniomonas, Guillardia*)

GROUP 3 RHIZARIA

PHYLUM CHLORARACHNIOPHYTA Examples include: *Bigeloiella, Bigelowiella, Chlorarachnion, Gymnochlora, Lotharella*

PHYLUM GRANULORETICULOSA Foraminiferans and their kin (e.g., *Allogromia, Ammonia, Astrorhiza, Arachnula, Biomyxa, Chitinosiphon, Elphidium, Glabratella, Globigerina, Globigerinella, Gromia, Iridia, Lenticula, Microgromia, Nummulites, Rhizoplasma, Rotaliella, Technitella, Tretomphalus*)

PHYLUM RADIOLARIA Examples include: *Acanthodesmia, Acanthosphaera, Arachnosphaera, Artopilium, Challengeron, Dendrospyris, Heliodiscus, Helotholus, Lamprocyclas, Peridium, Phormospyris, Sphaerostylus*

PHYLUM HAPLOSPORIDIA Examples include: *Bonamia, Haplosporidium, Marticella, Minchinia, Urosporidium*

GROUP 4 EXCAVATA

PHYLUM PARABASALIDA Trichomonads, hypermastigotes (e.g., *Dientamoeba, Histomonas, Monocercomonas, Pentatrichomonas, Trichomonas, Trichonympha, Tritrichomonas*)

PHYLUM DIPLOMONADIDA Examples include: *Enteromonas, Giardia, Hexamita, Octomitis, Spironucleus, Trimitus*

PHYLUM HETEROLOBOSEA Examples include: *Naegleria* and *Stephanopogon*

PHYLUM EUGLENIDA (= EUGLENOZOA) Examples include: *Ascoglena, Calkinsia, Colacium, Entosiphon, Euglena, Lepocinclis, Menoidium, Peranema, Phacus, Rapaza, Strombomonas, Trachelomonas*

PHYLUM KINETOPLASTIDA Trypanosomes, bodonids, and their kin (e.g., *Bodo, Cryptobia, Dimastigella, Leishmania, Leptomonas, Procryptobia, Rhynchomonas, Trypanosoma*)

GROUP 5 OPISTHOKONTA

PHYLUM CHOANOFLAGELLATA Examples include: *Codosiga, Monosiga, Proterospongia*

The Opisthokonta also includes the taxa Nucleariida, Ichthyosporea (= Mesomycetozoea), and the kingdoms Metazoa and Fungi.

Overviews of the Major Clades, or Groups

GROUP 1: AMOEBOZOA

The Amoebozoa include most of the well-known amebas with lobate (rather than threadlike) pseudopodia, as well as the bizarre slime molds. Although lacking clear-cut morphological synapomorphies, most are unicellular and most produce lobate pseudopodia. Amoebozoa is well supported by molecular phylogenetic studies. The group contains only a single nominate phylum (Amoebozoa), and includes such familiar groups as the gymnamoebas, entamoebas, and plasmodial and cellular slime molds. Most amoebozoans are free-living heterotrophs that feed by engulfing other cells with their pseudopodia. Also included are some mitochondria-lacking organisms (pelobionts and entamoebas) and several facultative or obligate parasites. Current evidence suggests both the plasmodial slime molds and cellular slime molds (the social amebas) belong to this clade. Cellular slime molds (e.g., *Dictyostelium*) are amebas that periodically congregate to form an asexual, spore-producing phase called a fruiting body. Plasmodial slime molds (e.g., *Fuligo*) don't have congregating cells, but do have sexual fruiting bodies that produce aggregated cell masses.

GROUP 2: CHROMALVEOLATA

This chapter treats six of the phyla included in this large and diverse assemblage: Dinoflagellata, Apicomplexa, Ciliata, Stramenopila, Haptophyta, and Cryptomonada. Photosynthetic species in these groups have plastids that contain chlorophyll *c*, in addition to chlorophyll *a*. Chromalveolates are predominately unicellular, and they may be photosynthetic or nonphotosynthetic. They are united by the "chromalveolate hypothesis," which states that a single secondary endosymbiosis with a red alga gave rise to a plastid ancestor of all chromalveolates. This plastid was secondarily lost or reduced in some lines, and there was a tertiary reacquisition of a plastid in others. Support for this group is largely based on plastid-related characters, and no single character- or molecular phylogeny has been able to unite all of its hypothesized members.

The phyla Dinoflagellata, Apicomplexa, and Ciliata form a monophyletic clade, the Alveolata, uniquely characterized by a system of flattened, membrane-bounded sacs or cavities, called **alveoli**, that lie just beneath the outer cell membrane. The function of the alveoli is unknown; researchers hypothesize that they may help stabilize the cell surface or regulate the cell's water and ion content. The alveolate clade is also strongly supported by gene sequence phylogenetics.

All three alveolate phyla contain predatory and parasitic species, but only dinoflagellates (and an unusual lineage called *Chromera*) are known to contain fully integrated and photosynthetic plastids.

Closely related to the alveolates is a large and diverse cluster of protists called the Stramenopila, now usually given phylum status. The Stramenopila were first identified through molecular phylogenetic studies and later received confirmation from comparative anatomical work. "Stramenopila" (Latin *stramen*, straw; *pilos*, hair) refers to a flagellum lined with numerous, fine, tubular hairs—a distinctive feature of these protists. In most stramenopiles, this "hairy" flagellum is paired with a smooth (nonhairy) flagellum. In some stramenopile groups, the only flagellated cells are motile reproductive cells. The Stramenopila include: photosynthetic diatoms; brown algae (formerly placed in their own phylum, Phaeophyta); Chrysophytes (the "golden algae"); the funguslike, nonphotosynthetic Oomycetes (water molds or Oomycota, downy mildews, etc.); and certain parasitic (opalines and blastocystids) and free-living (some heliozoans and flagellates) groups. The opalines and diatoms are thought to have secondarily lost the unique hollow hairs (although in centric diatoms the male gametes have hairy flagella). The largest known eukaryotes are stramenopiles—brown algae known as kelp (e.g., *Macrocystis, Nereocystis, Egregia*). Recent work suggests that many stramenopiles also contain nuclear genes from an ancient secondary endosymbiotic event with a green alga, perhaps even predating the red algal gene acquisition that now dominates the photosynthetic machinery in this phylum.

Two other chromalveolate groups, Haptophyta (or Prymnesiophyta; coccolithophores and their relatives) and Cryptomonada (e.g., *Cryptomonas*, = *Chilomonas*) have plastids that contain chlorophyll *a* and *c*, suggesting they also might belong to the alveolate assemblage. However, the precise nature of their affinity with stramenopiles and alveolates has yet to be established. Recent molecular analyses suggest the Stramenopiles, Alveolata, and Rhizaria (the "SAR group") might share a common ancestry to the exclusion of the haptophytes and cryptomonads, which would leave these three taxa with an uncertain classification. There is also reasonable support for Haptophyta being sister to the SAR group.

GROUP 3: RHIZARIA

The clade called Rhizaria contains the mixotrophic, green chloroplast-containing phylum Chlorarachniophyta, the parasitic Haplosporidia, and the phyla Granuloreticulosa and Radiolaria, as well as some fungilike plant parasites such as the plasmodiophorids. Most amebas (including radiolarians) that have threadlike pseudopodia are part of Rhizaria, although these filopodia may vary from simple, to branching or anastomosing. The relationships of the major rhizarian clades remain unresolved, and even the number and placement of the "phyla" is highly unstable. Some recent molecular evidence suggests that Radiolaria and Granuloreticulosa are basal groups, the remaining groups forming the subgroup, or clade, sometimes called Cercozoa (e.g., chlorarachniophytes, plasmodiophorids, haplosporids). There are numerous flagellated cercozoans, many of which use pseudopodia for feeding, if not movement. Although poorly studied, it is known that many small-to-medium sized, heterotrophic cercozoans (e.g., *Cercomonas, Heteromita, Euglypha*), both flagellates and amebas, are important members of benthic and soil microbial communities. Chlorarachniophytes (e.g., *Chlorarachnion, Cryptochlora, Gymnochlora, Lotharella*) are unusual within the cercozoan clade in having chloroplasts. Most studies suggest that Radiolaria is the deepest branch, and that Granuloreticulosa and Haplosporidia branch next to, or even within Cercozoa, but it has also been proposed that Granuloreticulosa and Radiolaria are a clade, or that Granuloreticulosa and Cercozoa are sister groups. The group Rhizaria is united only by molecular phylogenetics; other kinds of synapomorphies are as yet unknown. Recent molecular analyses link the rhizarians closely to the chromalveolates.

GROUP 4: EXCAVATA

The excavates comprise a somewhat loose amalgamation of protists whose relationships to one-another are just beginning to be understood. These are unicellular eukaryotes that share an array of cytoskeletal features as well as a distinctive ventral excavation that functions as a feeding groove (a suspension-feeding cytostome capturing small particles from a feeding current generated by a posteriorly directed flagellum), plus forms that have apparently lost some of these features. Overall, the clade Excavata is only weakly supported by molecular data. Most excavates are heterotrophic flagellates, and many have greatly modified mitochondria.

Currently included within the Excavata are the following protist groups: the phyla Parabasalida (trichomonads and hypermastigotes; e.g., *Dientamoeba, Histomonas, Monocercomonas, Pentatrichomonas, Trichomonas, Trichonympha, Tritrichomonas*), Diplomonada (e.g., *Enteromonas, Giardia, Hexamita, Octomitis, Spironucleus, Trimitus*), and Heterolobosea (e.g., *Naegleria, Stephanopogon*), as well as the groups Jakobida (e.g., *Reclinomonas*), Oxymonada, Retortamonada, Euglenozoa, and a few others. Recent multigene analyses of excavate taxa identify three clades. One clade comprises diplomonads, parabasalids, and the free-living amitochondriate protist *Carpediemonas*. The second clade consists of two other amitochondriate groups, oxymonads and *Trimastix*. The third clade is composed of Euglenozoa, Heterolobosea, and Jakobida. Several of the excavate taxa were long thought to lack mitochondria, but recent evidence suggests these groups have highly reduced or modified mitochondria (e.g., Parabasalida, Diplomonada). The jakobids have the most primitive (bacteria-like) mitochondrial genomes known.

The phylum Heterolobosea is more closely related to euglenids and kinetoplastids than to other amebas. *Naegleria fowleri* (= *N. aerobia*) is the major agent of a disease called primary amebic meningoencephalitis (PAM), or simply "amebic meningitis." PAM is an acute, fulminant, rapidly fatal illness usually affecting young people who have been exposed to water harboring the free-living trophozoites, most commonly in lakes and swimming pools (but this ameba has even been isolated from bottled mineral water in Mexico). It is thought that the amebas are forced into the nasal passages when the victim dives into the water. Once in the nasal passages, they migrate along the olfactory nerves, through the cribiform plate, and into the cranium. Death from brain destruction is rapid. They do not form cysts in the host. *Naegleria* infections are rare, but usually fatal. Only a few hundred cases have been documented since its discovery in Australia in 1960s, including a few dozen in the United States. As environments warm, *Naegleria* is expected to spread to higher latitudes.

The excavate subclade Euglenozoa contains the "plantlike" phylum Euglenida, the phylum Kinetoplastida (trypanosomes, bodonids, and their kin), and a few other odds and ends (e.g., diplonemids). Euglenozoans comprise a diverse group of flagellate predatory heterotrophs, photosynthetic autotrophs, and pathogenic parasites. Two principal anatomical features distinguish euglenozoans: (1) a spiral, or crystalline rod inside each of their two flagella, which insert into an anterior pocket; and (2) disk-shaped mitochondrial cristae. Parasitic and commensal euglenozoans have evolved independently several times among the kinetoplastids. The Euglenozoa clade is also supported by molecular phylogenetic studies. Both the Euglenida and the Kinetoplastida were formerly classified in the old "protozoan" phylum Sarcomastigophora.

The excavates are mostly heterotrophic flagellates, and even within the parasitic groups there are heterotrophic members. Many excavates have highly modified mitochondria not used for oxidative phosphorylation, and these are common in low-oxygen habitats (including animal guts). Broad-pseudopod forming amebas have evolved in one group (Heterolobosea), independently of the Amoebozoa, and even include their own group of slime molds (the acrasids). The euglenids appear to have originated via secondary endosymbiosis between a predatory euglenid and a green alga prey.

GROUP 5: OPISTHOKONTA

The clade known as Opisthokonta includes the kingdoms Metazoa and Fungi (and its likely sister group, Nucleariida), the Choanoflagellata, and a few other small protist groups. Two spore-forming groups once allied with protists in this group are now known to be animals (Myxozoa) and fungi (Microsporidia). In this chapter, we treat the choanoflagellates, which comprise the sister group, and probable direct ancestors, of the Metazoa. Together, Choanoflagellata and Metazoa (along with a few enigmatic unicellular eukaryotes that predated choanoflagellates) comprise a clade called Holozoa. The clades Opisthokonta and Holozoa are both strongly supported by molecular phylogenetics.

The General Protistan Body Plan

While realizing that the protists do not represent a monophyletic group, it is still advantageous to examine them together from the standpoint of the strategies and constraints of a unicellular, or at least a nontissue-level, eukaryotic body plan. Protists represent the "most primitive," or more accurately, the most ancient, living eukaryotes, yet, within the limitations imposed by their unicellularity, these creatures still must accomplish all of the basic life functions common to the Metazoa. Recall that the Eukaryota is distinguished from the other two major clades of life (prokaryotic Bacteria and Archaea) by the structural complexity of the cells—characterized by internal membranes and by having many functions segregated into semi-autonomous regions (organelles)—and by the cytoskeleton. Fundamentally unique to the Eukaryota, and evidence of their singular origin, is the double membrane-bound nucleus with its linear chromosomes (*eukaryote*, "true nuclei").

Body Structure, Excretion, Gas Exchange, and Single-Celledness

Most life processes are dependent upon activities associated with surfaces, notably with cell membranes. Even in the largest multicellular organisms, the regulation of exchanges across cell membranes and the metabolic reactions along the surfaces of various cell organelles are the phenomena on which all life ultimately depends. Consequently, the total area of these important surfaces must be great enough relative to the volume of the organism to provide adequate exchange and reaction sites. Nowhere is the "lesson" of the surface area-to-volume ratio more clearly demonstrated than among the protists, where it reveals the impossibility of massive, 100 kg amebas (1950s horror movies notwithstanding). Lacking both an efficient mechanism for circulation within the body and the presence of membrane partitions (multicellularity) to enhance and regulate exchanges of materials, protists must remain relatively small (with a few notably unique groups, such as the brown algae). The diffusion distances between protists' cell membranes (their "body surface") and the innermost parts of their bodies can never be so great that it prevents adequate movement of materials from one place to another within the cell. Certainly there are structural elements (e.g., microtubules, endoplasmic reticula) and various processes (e.g., protoplasmic streaming, active transport) that supplement passive phenomena. But the fact is, unicellularity mandates that a high surface area-to-volume ratio be maintained by restricting shape and size. This is the principle behind the fact that the largest protists (other than certain colonies, or colony-like species) assume shapes that are elongate, thin, flattened, or hollow—shapes that maintain small diffusion distances.

The formation of membrane-bounded pockets, or vesicles, is common in protists, and these structures help maintain a high surface area for internal reactions and exchanges. The elimination of metabolic wastes and excess water, especially in freshwater forms living in hypotonic environments, is facilitated by water expulsion vesicles (Chapter 4, Figure 4.22). As explained in Chapter 4, these vesicles (frequently called contractile vacuoles) release their contents to the outside in a more or less controlled fashion, often counteracting the normal diffusion gradients between the cell and the environment.

Support and Locomotion

The cell surface is critical not only in providing a means of exchange of materials with the environment but also in providing protection and structural integrity to the cell. The plasma membrane itself serves as a mechanical and chemical boundary to the protist "body," and when present alone (as in the "naked amebas"), it allows great flexibility and plasticity of shape. However, many protists maintain a more or less constant shape (spherical, radial, or even bilaterally symmetrical) by thickening the cell membrane to form a rigid or semirigid **pellicle**, by secreting scales or a shell-like covering called a **test** (usually of cellulose, $CaCO_3$ or SiO_2), by accumulating particles from the environment, or by other skeletal arrangements described below.[4]

The cytoskeleton is a complex array of proteins that provides the structural framework for protist (indeed, for all eukaryote) cells and its components and organelles. Locomotory capabilities are also ultimately provided by interactions between the cell surface and the surrounding medium. Pseudopodia, cilia, and flagella provide the means by which many protists push or pull themselves along.

Pseudopodia come in a variety of forms. **Lobopods** (lobopodia) are broad and blunt-tipped. **Filopods** (filopodia) are thin and tapering; they can be simple, branching or anastomosing. **Axopods** (axopodia) are also thin and tapering, but supported by microtubules. **Reticulopods** are thin and anastomosing and supported by microtubules.

Nutrition

Various types of nutrition occur among protists, but fundamentally they may be either autotrophic or heterotrophic, and many are both. Photosynthetic protists have plastids and are capable of photosynthesis, although not all use the same pigments, and they may differ in plastid structure (Figure 3.3). All heterotrophic protists acquire food through some interaction between the cell surface and the environment. Heterotrophic forms may be saprobic, taking in dissolved organics by diffusion, active transport, or pinocytosis. Or they may be holozoic, taking in solid foods—such as organic detritus or whole prey (e.g., bacteria, smaller protists)—by phagocytosis. Many heterotrophic protists are symbiotic on or within other organisms. Those protists that engage in pinocytosis or phagocytosis rely on the formation of membrane-bounded vesicles called food vacuoles (Figure 3.4). These structures may form at nearly any site on the cell surface, as they do in the amebas, or at particular sites associated with some sort of "cell mouth," or **cytostome**, as they do in most protists with more or less fixed shapes. The cytostome may be associated with further elaborations of the cell surface that form permanent invaginations or feeding structures (discussed in more detail below, under specific taxa).

Once a food vacuole has formed and moved into the cytoplasm, it begins to swell as various enzymes and other chemicals are secreted into it. The vacuole first becomes acidic, and the vacuolar membrane develops numerous inwardly directed microvilli (Figure 3.4). As digestion proceeds, the pH of the vacuolar fluid shifts to become increasingly alkaline. The cytoplasm just inside the vacuolar membrane takes on a distinctive appearance from the products of digestion. Then the vacuolar membrane forms tiny vesicles that pinch off and carry these products into the cytoplasm. Much of this latter activity resembles cell-surface pinocytosis. The result is numerous, tiny, nutrient-carrying vesicles offering a greatly increased surface area for absorption of the digested products into the cell's cytoplasm. During this period of activity, the original vacuole gradually shrinks and undigested materials eventually are expelled from the cell. In some protists (e.g., many amebas), the spent vacuole may discharge anywhere

[4]Cellulose is the most abundant organic polymer on Earth. It is a polysaccharide consisting of a linear chain of several hundred to many thousands of $\beta(1 \rightarrow 4)$ linked D-glucose units.

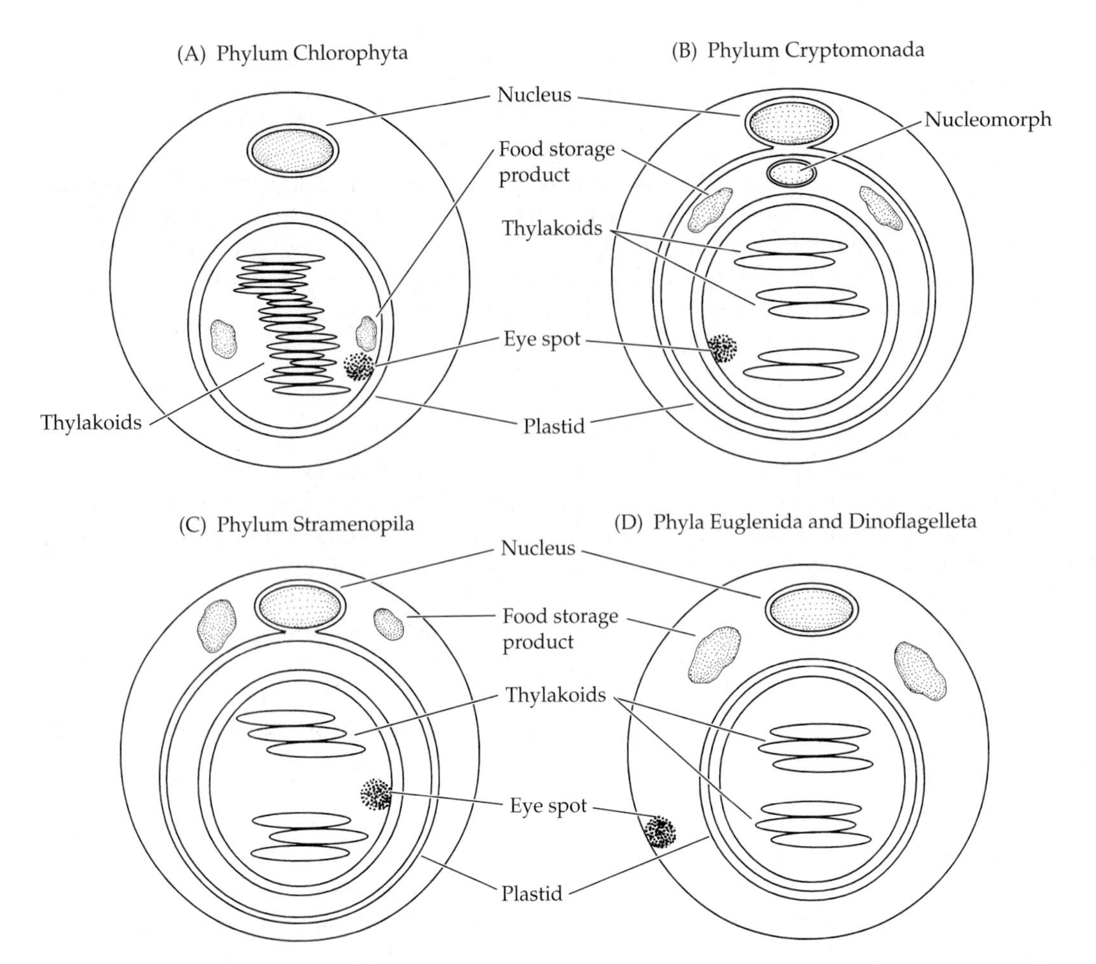

Figure 3.3 Variations in protist (and Chlorophyta) chloroplast anatomy. (A) Phylum Chlorophyta. As in land plants, the chloroplast in Chlorophyta is surrounded by two membranes and the thylakoids are arranged in irregular stacks, or grana. Also as in land plants, the primary photosynthetic pigments in chlorophytes are chlorophylls *a* and *b*, and food reserves are stored as starch inside the chloroplast. (B) Phylum Cryptomonada. In cryptomonads, the chloroplast is surrounded by four membranes and the thylakoids occur in stacks of two. The inner two membranes enclose the thylakoids and eyespot; the outer two membranes also enclose the storage product granules and nucleomorph. The outermost membrane of the four is also continuous with the nuclear envelope. The nucleomorph is thought to be the nucleus of an ancient endosymbiont that eventually became the chloroplast. Food reserves are stored as starch and oils, and the primary photosynthetic pigments are chlorophylls *a* and c_2; accessory pigments include phycobilins and alloxanthin. (C) Phylum Stramenopila. In photosynthetic stramenopiles, the chloroplast is surrounded by four membranes and the thylakoids occur in stacks of three. In many stramenopiles, the outermost membrane is continuous with the inner envelope. Food reserves are stored as liquid polysaccharide (usually laminarin) and oils, which are located in the cytoplasm. The primary photosynthetic pigments are chlorophylls *a*, c_1, and c_2. (D) Phyla Euglenida and Dinoflagellata. In both of these phyla, the chloroplasts are surrounded by three membranes and the thylakoids are arranged in stacks of three. Also in both phyla, the food storage products (starch and oils) and the eyespots are located outside of the chloroplast. The primary photosynthetic pigments in euglenids are chlorophylls *a* and *b*. Food reserves are stored as paramylon. In dinoflagellates, the photosynthetic pigments include chlorophylls *a* and c_2; accessory pigments include the xanthophyll peridinin, which is unique to dinoflagellates. Note that in some dinoflagellates, the eyespot is located inside the chloroplast rather than in the cytoplasm. The food storage products are starch and oils.

on the cell surface. But in ciliates and others in which a relatively impermeable covering exists around the cell, the covering bears a permanent pore (**cytoproct**) through which the vacuole releases material to the outside. Anything left in the food vacuole when it reaches the cytoproct is discharged.

In most protists, as in other eukaryotic organisms, the organelles responsible for most ATP production are the mitochondria. The mitochondria of protists, like all mitochondria, have two membranes, but the inner membranes, or **cristae**, have different forms—tubular, discoidal, and lamellar (Figure 3.5). However, in several groups of protists the mitochondria have been profoundly modified to generate ATP using alternative non-oxygen–dependent pathways, or no longer have a known energy-generating role at all.

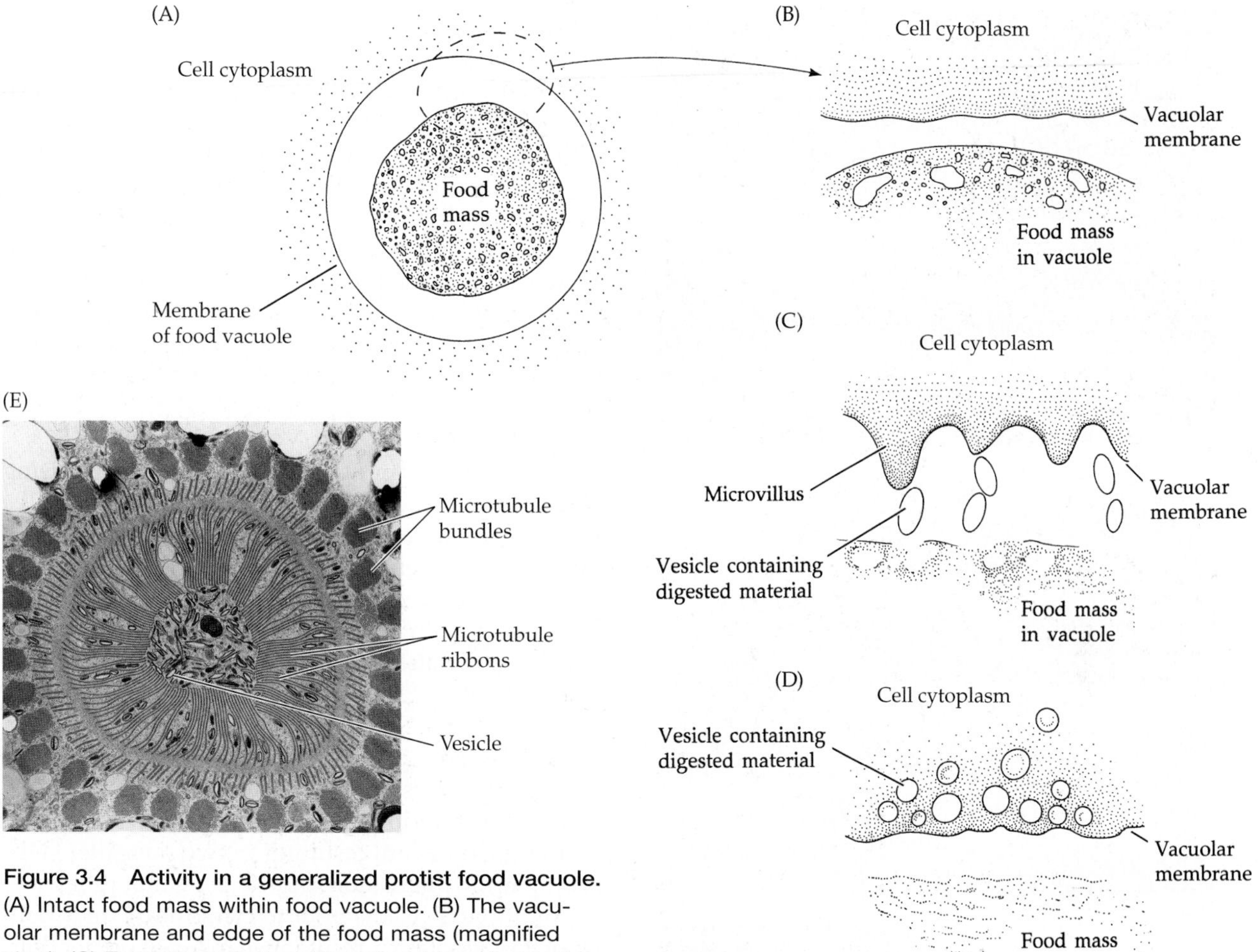

Figure 3.4 Activity in a generalized protist food vacuole. (A) Intact food mass within food vacuole. (B) The vacuolar membrane and edge of the food mass (magnified view). (C) Formation of microvilli and vesicles of vacuolar membrane. (D) Uptake of vesicles containing products of digestion into the cytoplasm. (E) Cross section through the cytostome of the ciliate *Helicoprorodon*, showing the area of food vacuole formation at center. Microtubules provide support to the mouth.

Activity and Sensitivity

Many protists display remarkable degrees of sensitivity to environmental stimuli and are capable of surprisingly complex behaviors. But, unlike that of animals, protists' entire stimulus–response circuit lies within the confines of a single cell. Response behavior may be a function of the general sensitivity and conductivity of protoplasm, or it may involve special organelles. Sensitivity to touch often involves distinctive locomotor reactions in motile protists and avoidance responses in many sessile forms. Cilia and flagella are touch-sensitive; when mechanically stimulated, they typically stop beating or beat in a pattern that moves the organism away from the point of stimulus. These responses are most dramatically expressed by sessile stalked ciliates, which display very rapid reactions when the cilia of the cell body are touched. Contractile elements within the stalk shorten, pulling the animal's body away from the source of the stimulus.

Many protists have **extrusomes**, membrane-bound (exocytotic) organelles containing various chemicals. Extrusomes have a variety of functions (e.g., protection, food capture, secretion), but they have one feature in common: they readily, and sometimes explosively, discharge their contents when subjected to stimuli. The best-known extrusome is the **trichocyst** of ciliates such as *Paramecium*, but about ten different types are known among protists.

Thermoreception is known to occur in many protists but is not well understood. Under experimental conditions, most motile protists will seek optimal temperatures when given a choice of environments. This behavior probably is a function of the general sensitivity of the organism and not of special receptors. Some evidence suggests that thermoreception in protists may be under electrophysiological control. Chemotactic responses are probably similarly induced. Most protists react positively or negatively to various chemicals or concentrations of chemicals. For example, amebas are able to distinguish food from nonfood items and quickly eject the latter from their vacuoles. Many ciliates, especially predators, have specialized patches of sensory cilia that aid in finding prey, and even filter feeders use

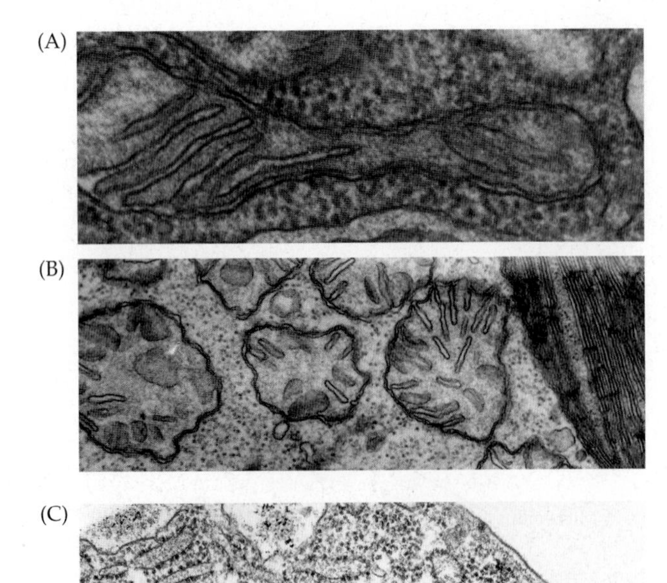

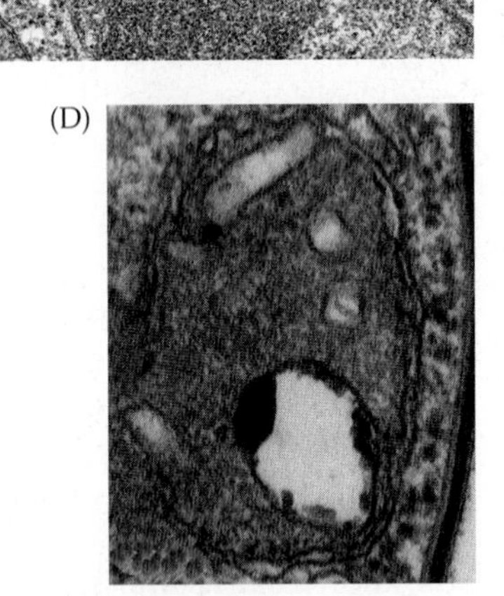

Figure 3.5 Protist mitochondria, showing variation in the inner membrane (i.e., cristae), plus the chlorophyte *Pteromonas lacerata*. (A) Lamellar cristae from the mitochondrion of the choanoflagellate *Stephanocea* (× 80,000). (B) Discoidal cristae from the mitochondrion of the euglenid *Euglena spirogyra* (× 40,000). (C) Dilated tubular cristae from the mitochondrion of *Apusomonas proboscidea*, an enigmatic flagellate of uncertain affinity (× 97,000). (D) Tubular cristae from the mitochondrion of the chlorophyte *Pteromonas lacerata* (× 27,000).

cilia located around the cytostome to "taste" and then accept or reject food items.

Photosynthetic protists typically show a positive taxis to low- or moderate light intensities, an obviously advantageous response for these creatures. They usually become negatively phototactic in very strong light. Specialized light-sensitive organelles are known among many flagellates, especially the photosynthetic ones. These organisms often have pigmented eyespots, or **stigmata** (sing. stigma) that are usually associated with the flagellar apparatus. The stigmata are oriented to shade the light-sensitive region (often part of the flagellum itself) from some directions, and the flagella thus behave differently depending on whether or not the sensitive region is receiving direct light. Some stigmata are actually contained inside a lobe of the chloroplast. Eyespots vary in complexity, ranging from very simple pigment spots to complex, lens-like structures.

Reproduction

A major aspect of protist success is their surprising range of reproductive strategies. Most protists have been able to capitalize on the advantages of both asexual and sexual reproduction, although some apparently reproduce only asexually. Many of the complex life cycles seen in certain protists (especially parasitic forms) involve alternation between sexual and asexual processes, with a series of asexual divisions between brief sexual phases.

Protists undergo a variety of strictly asexual reproductive processes including binary fission, multiple fission, and budding. **Binary fission** involves a single mitotic division, resulting in two daughter cells. During **multiple fission**, the nucleus undergoes several multiple divisions prior to **cytokinesis** (partitioning of the cytoplasm), resulting in many daughter cells. Some protists engage in a process called **plasmotomy**, considered by some to be a form of budding, in which a multinucleate adult simply divides into two multinucleate daughter cells. Still others undergo a type of internal budding called **endopolygeny**, during which daughter cells actually form within the cytoplasm of the mother cell.

The advantage of sexual reproduction has long been viewed to be the generation and maintenance of genetic variation within populations and species. Protists have evolved a variety of methods that achieve this, not all of which result in the immediate production of additional individuals. If we expand our traditional definition of meiosis to include any nuclear process that results in a haploid condition, then meiosis can be considered a general eukaryotic phenomenon. This disclaimer is necessary because protist "meiosis" is more variable than that seen in animals, and it is certainly less well understood. Nonetheless, reduction division does occur, and haploid cells or nuclei of one kind or another are produced, and then fuse to restore the diploid condition. The production and subsequent fusion of gametes in protists is called **syngamy**. However, not all protists are diploid. In the diploid forms, gametes are produced by mitosis, or simply by an existing cell starting to behave as a gamete—when two gametes from haploid individuals fuse to form a

BOX 3B Six Categories of Mitosis in Protists

Open mitosis

1. *Open orthomitosis.* The nuclear envelope breaks down completely; the spindle is symmetrical and bipolar; an equatorial plate occurs.

Semi-open mitosis

2. *Semi-open orthomitosis.* The nuclear envelope persists except for small fenestrae through which the spindle microtubules enter the nucleus; the spindle is symmetrical and bipolar; an equatorial plate occurs.
3. *Semi-open pleuromitosis*. The nuclear envelope persists except for small fenestrae through which the spindle microtubules enter the nucleus; the spindle is asymmetrical; an equatorial plate does not form.

Closed mitosis with an intranuclear spindle

4. *Intranuclear orthomitosis.* Nuclear envelope persists throughout mitosis; spindle is symmetrical, bipolar, and forms inside the nucleus; an equatorial plate usually forms.
5. *Intranuclear pleuromitosis.* Nuclear envelope persists throughout mitosis; spindle is asymmetrical and forms inside the nucleus; an equatorial plate does not form.

Closed mitosis with an extranuclear spindle

6. *Extranuclear pleuromitosis.* Nuclear envelope persists throughout mitosis; spindle is asymmetrical and forms outside of the nucleus; an equatorial plate does not form.

diploid cell, the fusion is followed by meiosis to form the next generation of nonreproductive cells (e.g., in apicomplexans). There are also taxa that will live asexually in either the diploid or haploid state for extended periods (e.g., haptophytes), and in these cases the two forms might look quite different.

Protist cells responsible for the production of gametes are often called **gamonts**. Syngamy may involve gametes that are all similar in size and shape (**isogamy**), or the more familiar condition of gametes of two distinct types (**anisogamy**). Thus, as in the Metazoa, both haploid and diploid phases can be produced in the life histories of sexual protists. As noted above, the meiotic process may immediately precede the formation and union of gametes (prezygotic reduction division), or it may occur immediately after fertilization (postzygotic reduction division), as it also does in haploid protists and in many plants. Other sexual processes that result in genetic mixing by the exchange of nuclear material between mates (**conjugation**) or by the re-formation of a genetically "new" nucleus within a single individual (**autogamy**) are best known among the ciliates and are discussed below for that phylum.

There is also considerable variability in mitosis among protists (Box 3B). Different mitotic patterns are primarily distinguished on the basis of persistence of the nuclear membrane (= envelope), and the location and symmetry of the spindle (Figure 3.6). The terms "open," "semi-open," and "closed" refer to the persistence of the nuclear envelope. If mitosis is open, the nuclear membrane breaks down completely; if it is semi-open, the nuclear envelope remains intact except for small holes

		Pleuromitoses	Orthomitosis
Closed	Extranuclear		—
	Intranuclear		
Semiopen			
Open		—	

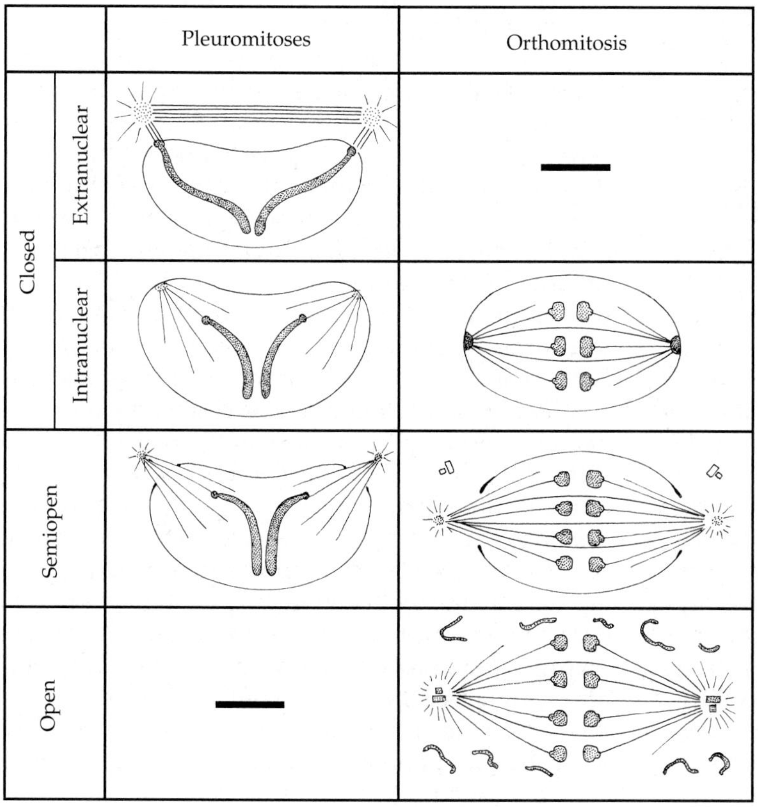

Figure 3.6 **Mitosis in protists.** In open mitosis the nuclear membrane breaks down completely. In semi-open mitosis the nuclear envelope remains intact except for small holes where the spindle microtubules penetrate the nuclear envelope. In closed mitosis the nuclear envelope remains completely intact throughout mitosis, and the spindle forms either inside the nucleus (intranuclear mitosis) or outside the nucleus (extranuclear mitosis). Orthomitosis refers to the spindle being bipolar and symmetrical, usually with the formation of an equatorial plate. Pleuromitosis occurs when the spindle is asymmetrical and an equatorial plate does not form.

(fenestrae) where the spindle microtubules penetrate the nuclear envelope; if it is closed, the nuclear envelope remains completely intact throughout mitosis. The terms "orthomitosis" and "pleuromitosis" refer to the symmetry of the spindle. During **orthomitosis**, the spindle is bipolar and symmetrical, and an equatorial plate usually forms. During **pleuromitosis**, the spindle is asymmetrical, and an equatorial plate does not form. The terms intranuclear and extranuclear refer to the location of the spindle. During **intranuclear mitosis**, the spindle forms inside of the nucleus; during **extranuclear mitosis**, the spindle forms outside of the nucleus.

Protist nuclei also show remarkable diversity. The most common type of nucleus is the **vesicular nucleus**, and is characterized as being between 1–10 μm in diameter, round (usually), with a prominent nucleolus, and uncondensed chromatin. **Ovular nuclei** are characterized as being large (up to 100 μm in diameter), with many peripheral nucleoli, and uncondensed chromatin. Chromosomal nuclei are characterized by the tendency of the chromosomes to remain condensed during interphase, and for there to be one nucleolus that is associated with a chromosome. Ciliates are unique in their possession of two different kinds of nuclei: small **micronuclei** (with no nucleoli and dispersed chromatin), and large **macronuclei** (with many prominent nucleoli and compact chromatin). In summary, protist diversity and success is reflected by the tremendous variation within the unicellular body plan. The following accounts of protist phyla explore this variation in some detail.

BOX 3C Characteristics of the Phylum Amoebozoa

1. Mostly free living, but some are endo- or ectosymbiotic, including some pathogens
2. Cell surrounded by plasma membrane (naked amebas), which may be coated with layer of glycoprotein (the glycocalyx); some also form external test (testate amebas)
3. With temporary extensions of cytoplasm (pseudopodia) for feeding and locomotion; pseudopodia are broad and blunt (lobopods)
4. Some species harbor intracellular symbionts (e.g., algae, bacteria).
5. Mitochondria with tubular cristae
6. Most with single vesicular nucleus
7. Mitotic patterns variable
8. Asexual reproduction by binary fission and multiple fission
9. Sexual reproduction reported for slime molds, otherwise not confirmed
10. Without plastids; strictly heterotrophic
11. Some store glycogen (e.g., *Pelomyxa*), but most do not appear to store carbohydrates

THE PROTIST PHYLA

GROUP 1: AMOEBOZOA

Phylum Amoebozoa: Amebas

The phylum Amoebozoa, although small, comprises a singular clade of protists consisting of just over 200 species. Most are free-living, but some endosymbiotic groups are known, including some pathogenic forms. The most obvious characteristic of amoebozoans is that they form temporary extensions of the cytoplasm, or pseudopodia (described in Chapter 4, used in feeding and locomotion (Figure 3.7; see also Figures 3.1B and 3.10) (Box 3C). Amoebozoans are ubiquitous creatures that can be found in nearly any moist or aquatic habitat: in soil or sand, on aquatic vegetation, on wet rocks, in lakes, streams, glacial meltwater, tidepools, bays, estuaries, on the ocean floor, and afloat in the open ocean. Many are ectocommensals on aquatic organisms and some are parasites of diatoms, fishes, molluscs, arthropods, or mammals. Some amoebozoans harbor intracellular symbionts such as algae, bacteria, and viruses, although the nature of these relationships is not well understood. Amoebozoans have often been used in laboratories as experimental organisms for studies of cell locomotion (*Amoeba proteus*), nonmuscle contractile systems (*Acanthamoeba*), and the effects of removing and transplanting nuclei.

Although most amoebozoans are harmless free-living creatures, some are endosymbiotic, and many of these are considered to be parasites. They occur most commonly in arthropods, annelids, and vertebrates (including humans). Three cosmopolitan species are commensals in the large intestines of humans: *Endolimax nana*, *Entamoeba coli*, and *Iodamoeba buetschlii*. All three species feed on other microorganisms in the gut. *Entamoeba coli* infection levels reach 100% of the population in some areas of the world. *Entamoeba coli* often coexists with the more problematic *E. histolytica*; transmission (via cysts) is by the same method, and the trophozoites of the two species are difficult to differentiate. *Iodamoeba buetschlii* infects humans, other primates, and pigs. *Entamoeba gingivalis* was the first ameba of humans to be described. Like *E. coli*, it is a largely harmless commensal, residing only on the teeth and gums, in gingival pockets near the base of the teeth, and occasionally in the crypts of the tonsils. It also occurs in dogs, cats, and other primates. No cyst is formed and transmission is direct, from one person to another. An estimated 50% of the human population with healthy mouths harbors this ameba.

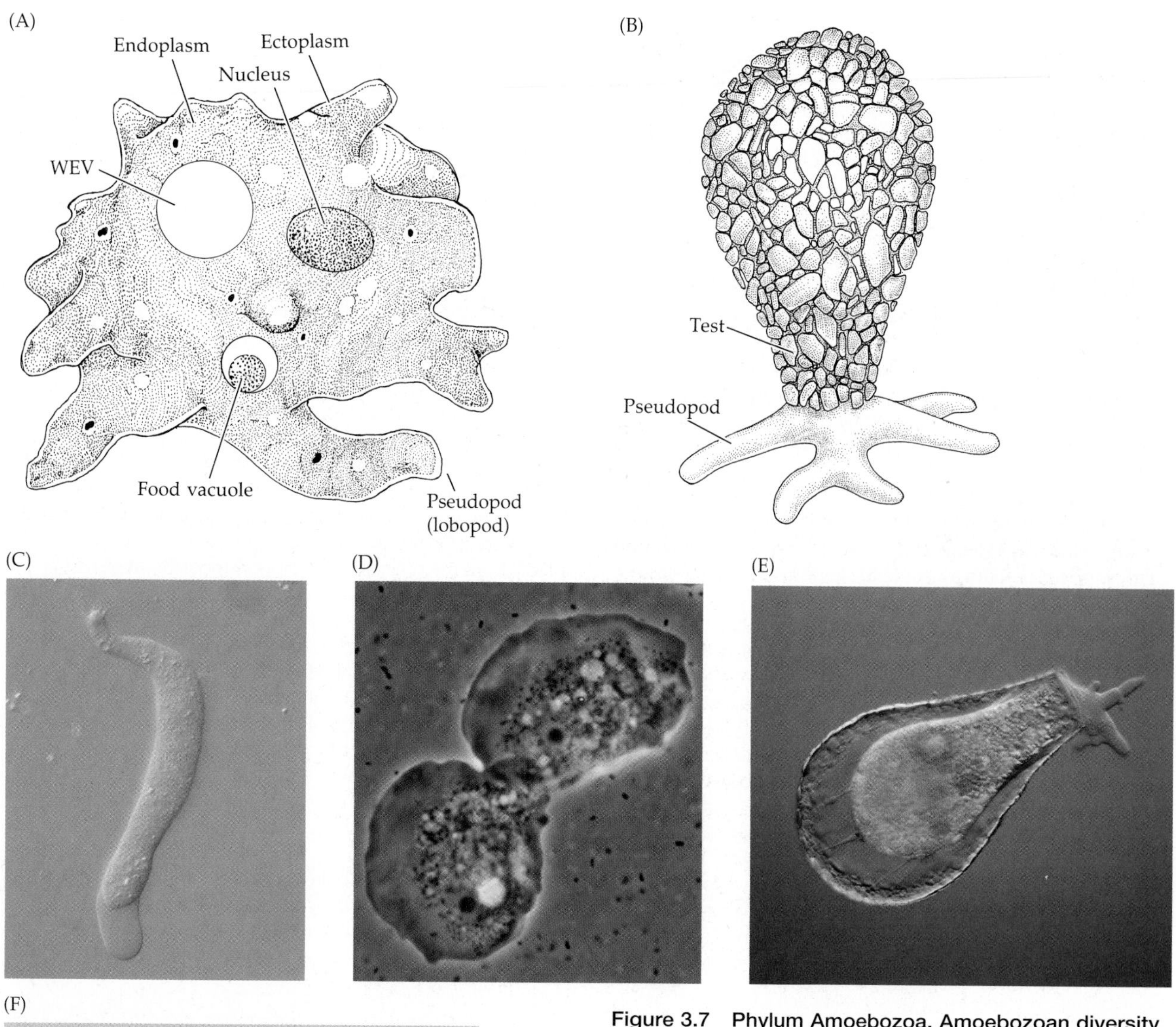

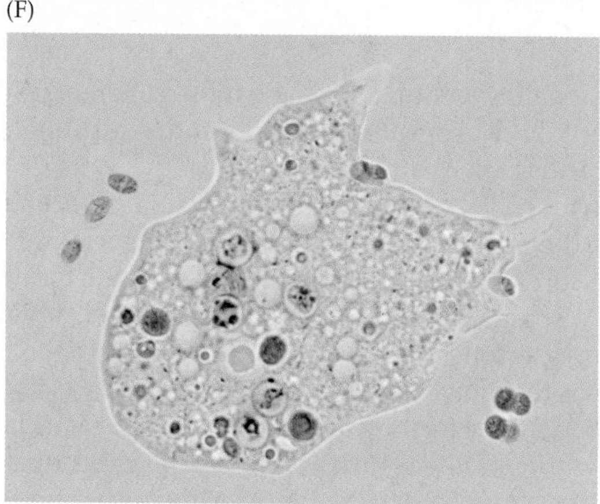

Figure 3.7 Phylum Amoebozoa. Amoebozoan diversity. (A) Anatomy of a "naked" ameba; note the multiple lobopods. (B) A testate ameba, *Difflugia* (test of microscopic mineral grains). (C) *Hartmanella*, with single fingerlike lobopod. (D) *Vannella*, with fan-shaped pseudopod. (E) *Nebela collaris*, a testate ameba. Thick pseudopods (lobopods) extend from the test aperture, and the ameba's body is attached to the interior of the shell by fine pseudopods (filopods). (F) *Mayorella*, a "naked" ameba.

Under severe stress, symbiotic gut amoebozoans (e.g., *Entamoeba coli*) that are normally harmless can increase to abnormally high numbers and cause temporary mild gastrointestinal distress in people. *Entamoeba histolytica*, however, is a serious pathogen in humans (Box 3D). This species causes amebic dysentery, an intestinal disorder resulting in destruction of cells lining the gut. The parasite is usually ingested in its cyst stage, and is acquired by way of fecal contamination. Emergence of individuals in the active (motile) stage (i.e., the trophozoites) takes place quickly once in the host's gut, and it is this stage that releases the histolytic enzymes that break down the epithelium of the large intestine and rectum.

One of the most enigmatic groups of amoebozoans is the social amebas, or cellular slime molds. These odd creatures, and their close relatives the plasmodial (acellular) slime molds (Myxogastrida), were once thought to be allied with the fungi, but are now classified as amoebozoans. So bizarre are the lives of these creatures that they inspired one of the all-time great 1950s science fiction movies, *The Blob* (with Steve McQueen).

BOX 3D Histolytic (Amebic) Dysentery

Entamoeba histolytica is the etiological agent of amebic dysentery, a disease that has plagued humans throughout all of recorded history. *E. histolytica* is the third most common cause of parasitic death in the world. About 500 million people in the world are infected at any one time, with up to 100,000 deaths annually. Interestingly, *E. histolytica* comes in two sizes. The smaller race (trophozoites 12–15 μm in diameter, cysts 5–9 μm wide) is nonpathogenic, and some workers consider it a separate species (*E. hartmanni*). The larger form (trophozoites 20–30 μm in diameter, cysts 10–20 μm wide) is sometimes pathogenic and other times not. Another species of *Entamoeba*, *E. moshkovskii*, is identical in morphology to *E. histolytica*, but is not a symbiont; it lives in sewage and is often mistaken for *E. histolytica*.

When swallowed, the cysts of *E. histolytica* pass through the stomach unharmed. When they reach the alkaline medium of the small intestine, they break open and release trophozoites that are swept to the large intestine. They can survive both anaerobically and in the presence of oxygen. Trophozoites of *E. histolytica* may live and multiply indefinitely within the crypts of the mucosa of the large intestine, apparently feeding on starches and mucous secretions. In order to absorb food in this setting, they may require the presence of certain naturally occurring gut bacteria. However, they can also invade tissues by hydrolyzing mucosal cells of the large intestine, and in this mode they need no help from their bacterial partners to feed. *E. histolytica* produces several hydrolytic enzymes, including phosphatases, glycosidases, proteinases, and an RNAse. They erode ulcers into the intestinal wall, eventually entering the bloodstream to infect other organs such as the liver, lungs, or even the skin. Cysts form only in the large intestine and pass with the host's feces. Cysts can remain viable and infective for many days, or even weeks, but are killed by desiccation and temperatures below 5°C and above 40°C. The cysts are resistant to levels of chlorine normally used for water purification.

Symptoms of amebiasis vary greatly, due to the strain of *E. histolytica* and the host's resistance and physical condition. Commonly, the disease develops slowly, with intermittent diarrhea, cramps, vomiting, and general malaise. Some infections may mimic appendicitis. Broad abdominal pain, fulminating diarrhea, dehydration, and loss of blood are typical of bad cases. Acute infections can result in death from peritonitis, the result of gut perforation, or from cardiac failure and exhaustion. Hepatic amebiasis results when trophozoites enter the mesenteric veins and travel to the liver through the hepato-portal system; they digest their way through the portal capillaries and form abscesses in the liver. Pulmonary amebiasis usually develops when liver abscesses rupture through the diaphragm. Other sites occasionally infected are the brain, skin, and penis (possibly acquired through sexual contact).

Although amebiasis is most common in tropical regions, where up to 40% of the population may be infected, the parasite is firmly established from Alaska to Patagonia. Transmission is via fecal contamination, and the best prevention is a sanitary lifestyle. Filth flies, particularly the common housefly (*Musca domestica*), and cockroaches are important mechanical vectors of cysts, and houseflies' habit of vomiting and defecating while feeding is a key means of transmission. Human carriers (cyst passers) handling food are also major sources of transmission. The use of human feces as fertilizer in Asia, Europe, and South America contributes heavily to transmission in those regions. Although humans are the primary reservoir of *E. histolytica*, dogs, pigs, and monkeys have also been implicated

Among the cellular slime molds, each organism begins its life as a unicellular ameba, primarily feeding on soil bacteria. However, when food is insufficient they aggregate to form a multicellular fruiting body. The fruiting body (or "fructification") takes several forms, depending on the species. This process has been best described for the "model organism" *Dictyostelium discoideum*. In *D. discoideum*, the aggregate of up to 100,000 cells first transforms into a finger-shaped structure, the "slug." The "head" region of the slug senses environmental stimuli such as temperature and light and directs the slug toward the soil's upper surface, where spores will be dispersed. The slug then stands up to form the asexual fruiting body, or **sorocarp** (Figure 3.8). The cells in the head region move into a prefabricated cellulose tube and differentiate into stalk cells that ultimately die. The remaining "body" cells then crawl up the stalk and encapsulate to form spores. Recent work with *Dictyostelium* suggests the spores carry with them soil bacterial "seeds," which are used to inoculate the new ameba's new location. Thus the Dictyostelida display characteristics akin to multicellularity, such as cell-cell signaling, cellular specialization, coherent cell movement, and programmed cell death—although embryonic tissue layering, diagnostic of the Metazoa, obviously does not occur. The dictyostelids' ability to cultivate their own "food crop" of soil bacteria puts these creatures in a unique class of "invertebrate farmers" that also includes fungus-growing ants, termites, and ambrosia beetles.

Whereas the dictyostelids aggregate to from asexual fruiting bodies, the fruiting bodies of the Myxogastrida (e.g., *Physarum, Fuligo*) function in sexual reproduction. Further, myxogastrids don't form fruiting bodies by aggregation. Instead, two haploid cells fuse by syngamy. The resulting diploid ameba then grows without undergoing further cell division to form a multinucleate plasmodium. This super-cell eventually grows fruiting bodies containing uninucleate spores. The process is sexual because there is a meiotic step in the production of the spore cells in the fruiting bodies, so the spores, and the amebas that hatch from them, are haploid. Thus the life cycle of myxogastrids contains critical haploid and diploid components.

Figure 3.8 Phylum Amoebozoa. Cellular slime molds (social amebas) and plasmodial slime molds. (A) Fruiting bodies of the cellular slime mold (Dictyostelida) *Dictyostelium discoideum*. (B) The wood-rot slime mold *Stemonitis*, with its distinctive brown sporangia supported on slender stalks. (C) Large, multinucleate (but acellular) plasmodium of the plasmodial slime mold (Myxogastrida) *Fuligo septica*, the "scrambled-egg slime mold." (D–F) The highly variable slime mold *Physarum polycephalum*.

In North America, the famous scrambled-egg slime mold, or "dog-vomit slime mold" (Myxogastrida: *Fuligo septica*) is often encountered (Figure 3.8C). The gigantic (up to 40 cm across), yellow, cushiony crusts of *F. septica* grow on dead or living trees, and on wood mulches in landscaped areas. The plasmodia slowly creep over the surface, engulfing bacteria, plant and fungal spores, other protists, and organic detritus. When the large, multinucleate (but acellular) plasmodium converts into a spore-bearing structure (an aethalium) it becomes a spongy mass that is the largest spore-producing structure known among the Amebozoa. The spores of *F. septica* are dispersed by beetles of the family Lathridiidae.

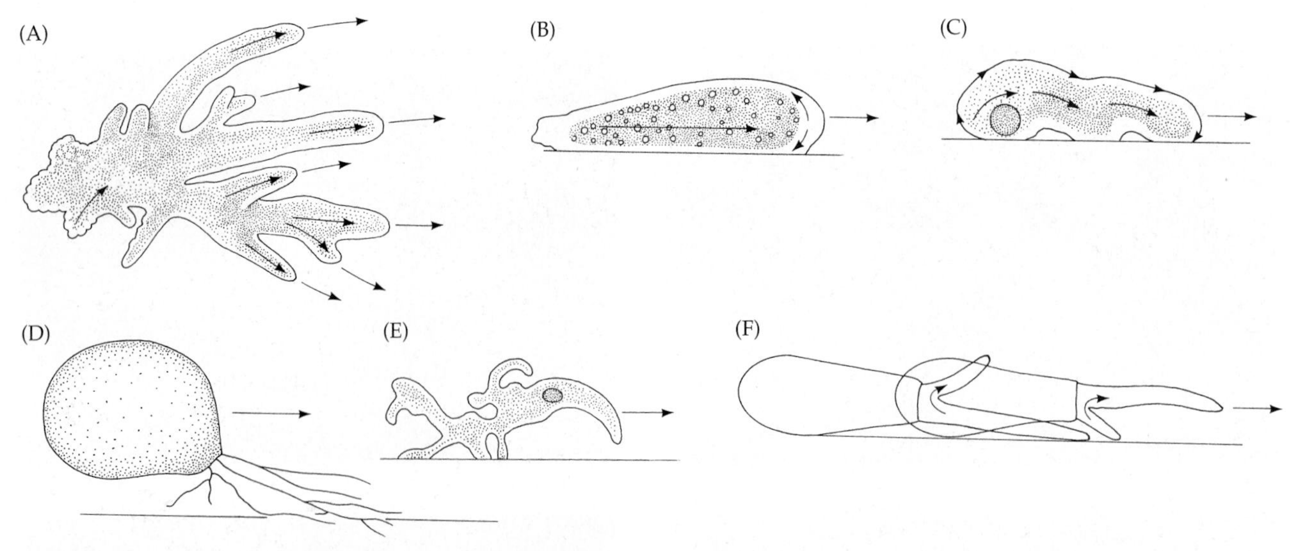

Figure 3.9 Phylum Amoebozoa. Locomotion in amebas. (A) Typical ameboid movement by lobopods in *Amoeba proteus*. (B) Creeping, "limax" form of movement. (C) Rolling, treadlike movement. (D) Filopodial creeping in *Chlamydophorus*. (E) "Walking" locomotion in certain naked amebas. (F) "Bipedal-stepping" in *Difflugia*.

Support and Locomotion

Most amoebozoans are surrounded only by a plasma membrane. These are the so-called **naked amebas** (see Figures 3.7A–C and 3.8). Others, known as **testate amebas**, have the plasma membrane covered by some sort of test (see Figure 3.7D–E). The tests of amoebozoans may be composed primarily of particulate material either gathered from the environment (e.g., *Difflugia*) or secreted by the cell itself (e.g., *Arcella*). A few species produce an external covering of very small scales. Some of the naked amebas (e.g., the genus *Amoeba*) may secrete a mucopolysaccharide layer, called the **glycocalyx**, on the outside of the plasma membrane. Sometimes there may be flexible, sticky structures protruding from the glycocalyx, which are thought to aid in the capture and ingestion of bacteria during feeding.

Amoebozoans use pseudopodia for feeding and locomotion. Although pseudopodia vary in shape among protists, there are two primary types—**lobopods** and **filopods** (sometimes called rhizopods). Amoebozoans have only lobopods, which are blunt and rounded at the tip (filopods are thin and tapering) (Figure 3.9A,E). Some amoebozoans produce several pseudopodia, or lobopods, that extend in different directions at the same time. Probably the most familiar organism that produces multiple lobopods is *Amoeba proteus* (see Figure 3.9A). Similar pseudopodia are formed by some testate amoebozoans such as *Arcella*, *Centropyxis*, and *Difflugia*. Some amoebozoans that produce multiple lobopods also produce subpseudopodia on the surfaces of their lobopods. This condition is found in the genus *Mayorella*, which forms fingerlike subpseudopodia, and *Acanthamoeba*, which forms thin subpseudopodia called **acanthopods**.

Some amoebozoans produce only a single lobopod. One such group is the so-called **limax amoebozoans**, which form a single fingerlike "anterior" lobopod (giving the organism a sluglike, or *Limax*, appearance) (Figure 3.9B). Limicine locomotion is commonly found in amoebozoans that dwell in soil (e.g., *Chaos*, *Euhyperamoeba*, *Hartmanella*, *Pelomyxa*). Other amoebozoans that produce a single lobopod include the genera *Thecamoeba* and *Vannella*. In *Vannella*, the lobopod is shaped such that it gives the body a fanlike appearance (see Figure 3.7C), while in *Thecamoeba*, the lobopod has a somewhat indefinite shape and creates the impression that the cell rolls like the tread of a tractor or tank, the leading surface adhering temporarily to the substratum as the organism progresses (Figure 3.9C).

As described in Chapter 4, the physical processes involved in pseudopodial movement are still not fully understood. It is likely that more than one method of pseudopod formation occurs among the different amoebozoans; certainly the gross mechanics involved in the actual use of pseudopodia vary greatly (see Figure 3.9). Recall the typical differentiation of the cytoplasm into ectoplasm (plasmagel) and endoplasm (plasmasol), the latter being much more fluid than the former. The formation of lobopods results from the streaming of the inner plasmasol into areas where the constraints of the plasmagel have been temporarily relieved.

While many amoebozoans move by "flowing" into their pseudopodia or by "creeping" with numerous filopodia, some engage in more bizarre methods of getting from one place to another. Some hold their bodies off the substratum by extending pseudopodia downward; leading pseudopodia are then produced and extended sequentially, pulling the organism along in a sort of "multi-legged" walking fashion. Some of the shelled amoebozoans (e.g., *Difflugia*) that possess a single pylome extend two pseudopodia through the

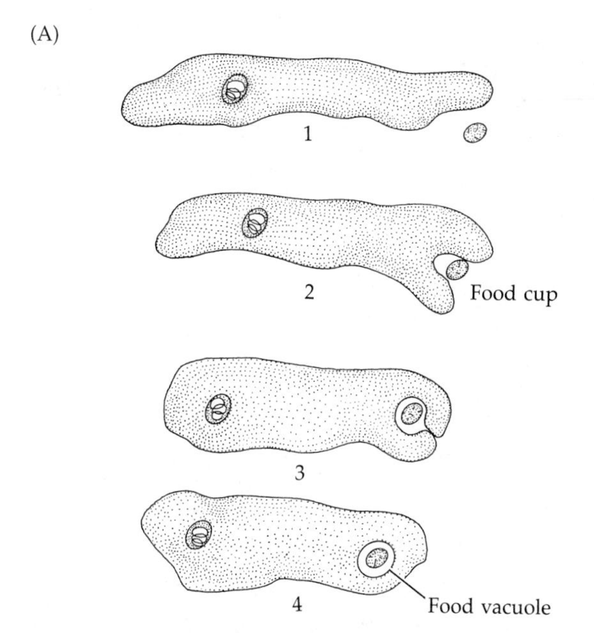

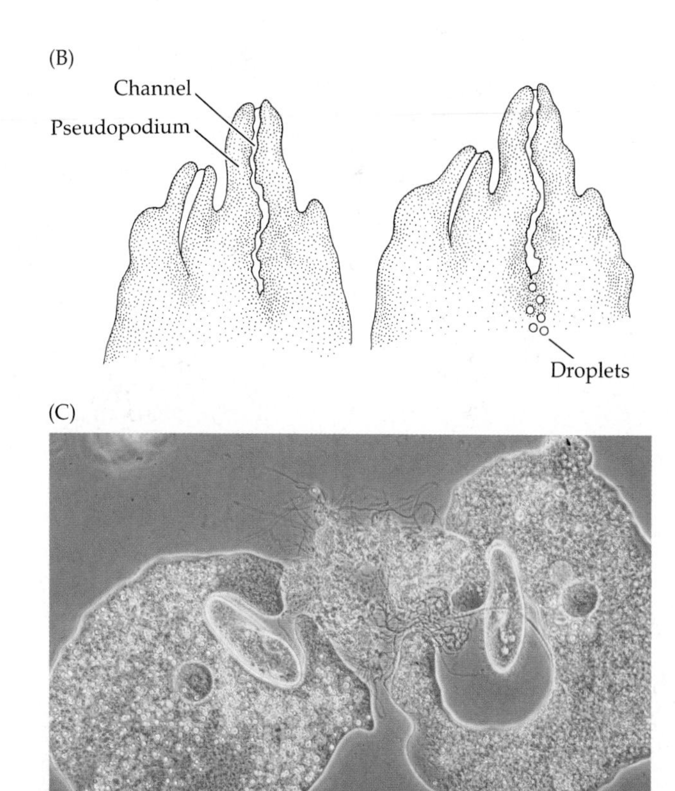

Figure 3.10 Phylum Amoebozoa. Feeding in amebas. (A) Sequence of events during which a lobopodium engulfs a food particle. (B) Uptake of dissolved nutrients through a pinocytotic channel in *Amoeba*. (C) Two soil amebas, *Vahlkampfia*, ingesting ciliates by phagocytosis.

aperture (Figure 3.9F; see also Figure 3.7B). By alternately extending and retracting these pseudopodia, the organism "steps" forward. During locomotion, one pseudopodium is extended and used to "pull" the organism along, trailing the other pseudopodium behind the cell.

Nutrition

While there is little doubt that amoebozoans take up dissolved organics directly across the cell membrane, the most common mechanisms of ingestion are pinocytosis and phagocytosis (Figure 3.10). The size of the food vacuoles varies greatly, depending primarily on the size of the food material ingested. Generally, ingestion can occur anywhere on the surface of the body, there being no distinct cytostome. Most large amoebozoans are carnivores and are frequently predators, whereas the smallest species mostly eat prokaryotes. Some, such as *Pelomyxa*, inhabit soils or muds and are predominantly herbivorous, but they are known to ingest nearly any sort of organic matter in their environment. As explained earlier, a food vacuole forms from an invagination in the cell surface—sometimes called a food cup—that pinches off and drops inward. This process, sometimes called **endocytosis**, occurs in response to some stimulus at the interface between the cell membrane and the environment. Vacuole formation in amoebozoans may be induced by either mechanical or chemical stimuli; even nonfood items may be incorporated into food vacuoles, but they are soon egested.

Not only the size of a food item, but also the amount of water taken in during feeding determine the size of the food vacuole. Frequently, pseudopodia that form a food cup do not actually contact the food item; thus a packet of the environmental medium is taken in with the food. In other cases, the walls forming the vacuole press closely against the food material; thus, little water is included in the vacuole. Food vacuoles move about the cytoplasm and sometimes coalesce. If live prey have been ingested, they generally die within a few minutes from the paralytic and proteolytic enzymes present in the vacuole. Undigested material that remains within the vacuole is eventually expelled from the cell when the vacuole wall reincorporates into the cell membrane. In most amoebozoans this process of cell defecation may occur anywhere on the body, but in some active forms it tends to take place at or near the trailing end of the moving cell.

Feeding in amoebozoans with skeletal elements varies with the form of the test and the type of pseudopodia. Amebas with a relatively large single aperture or opening, such as *Arcella* and *Difflugia*, feed much as described above. By extending lobopods through the aperture, they engulf food in typical vacuoles.

Reproduction

Simple binary fission is the most common form of asexual reproduction, differing only in minor details among the different groups (Figure 3.11). In the naked amoebozoans, nuclear division occurs first and then cytoplasmic division follows. During cytoplasmic division, the

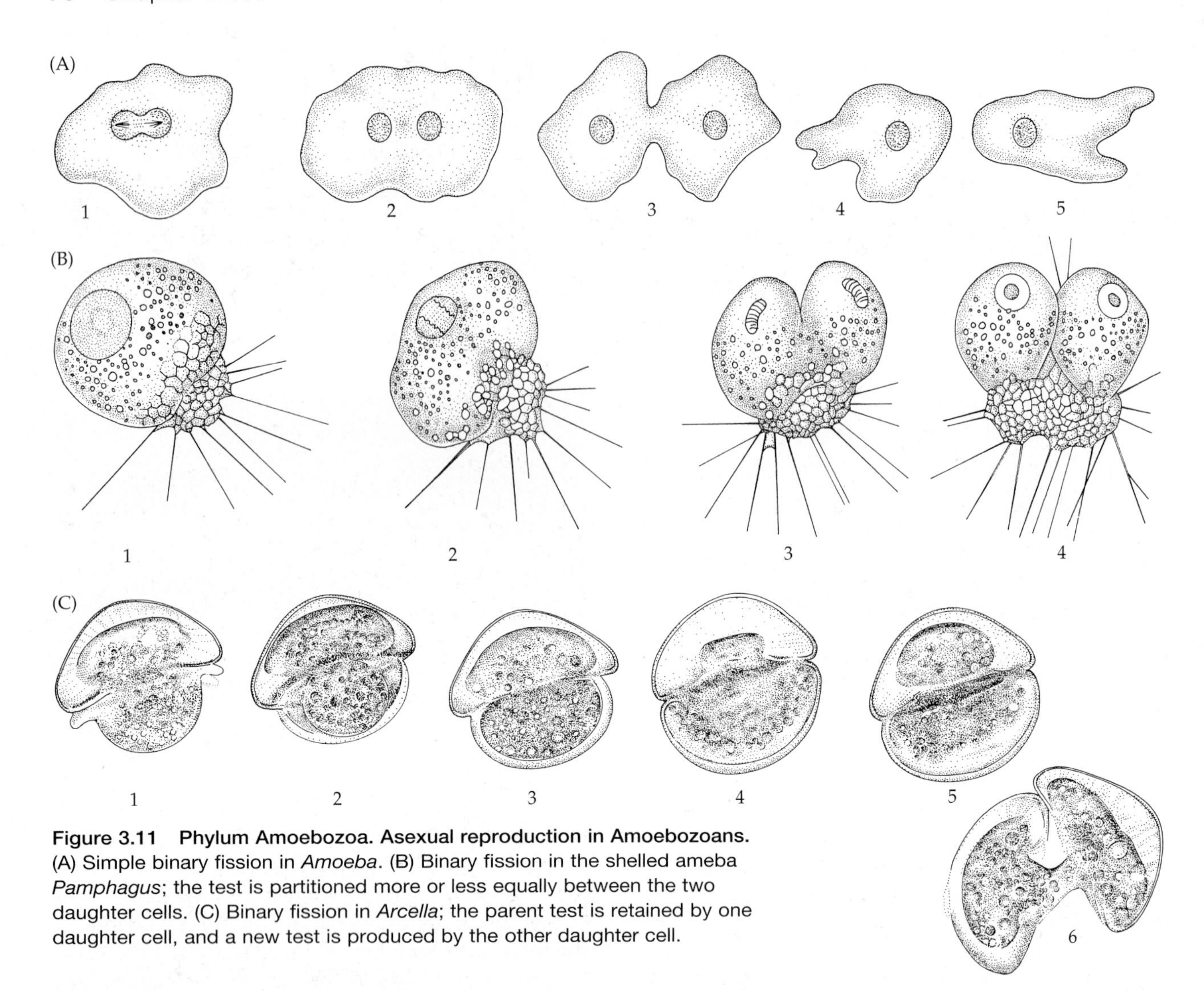

Figure 3.11 Phylum Amoebozoa. Asexual reproduction in Amoebozoans. (A) Simple binary fission in *Amoeba*. (B) Binary fission in the shelled ameba *Pamphagus*; the test is partitioned more or less equally between the two daughter cells. (C) Binary fission in *Arcella*; the parent test is retained by one daughter cell, and a new test is produced by the other daughter cell.

two potential daughter cells form locomotor pseudopodia and pull away from each other. In species with an external test, the shell itself may divide more or less equally in conjunction with the formation of daughter cells (e.g., *Pamphagus*); or, as occurs more frequently, the shell may be retained by one daughter cell, the other producing a new shell (e.g., *Arcella*). Multiple fission is also known among many amoebozoans. Certain endosymbiotic naked species, including *Entamoeba histolytica*, produce cysts in which multiple fission takes place.

Cyst formation during unfavorable environmental conditions is well developed in some amoebozoans, including all testate amebas, most soil amebas, and the parasitic amebas. In the parasitic amebas (e.g., *Entamoeba*), cysts protect the organism as it passes through the digestive tract of the host.

Mitotic patterns in amoebozoans vary and have been used as a criterion for classification within the phylum. In most species, mitosis is characterized as open orthomitosis without centrioles; in some, the breakdown of the nucleus and nucleolus is delayed. Closed intranuclear orthomitosis with a persistent nucleolus occurs in some amoebozoans, such as *Entamoeba*.

Although there have been suggestions that sexual reproduction might occur in amoebozoans, there has been little evidence to support this possibility, other than in the slime molds.

GROUP 2: CHROMALVEOLATA

Phylum Dinoflagellata: Dinoflagellates

The Dinoflagellata comprise about 4,000 described species, including the known extinct forms. Although unquestionable fossil dinoflagellates date back to the Triassic (240 Ma), evidence from organic markers in early Cambrian rocks suggests that they might have been abundant as early as 540 million years ago. The fossil record suggests that dinoflagellate biodiversity was low after the End-Permian mass extinction event, recovered by the early Eocene (~55 Ma), and then subsequently began a long-term decline that continues to the present (also see Chapter 1). The historical patterns of diversity of dinoflagellates and coccolithophores are roughly concordant, but contrast

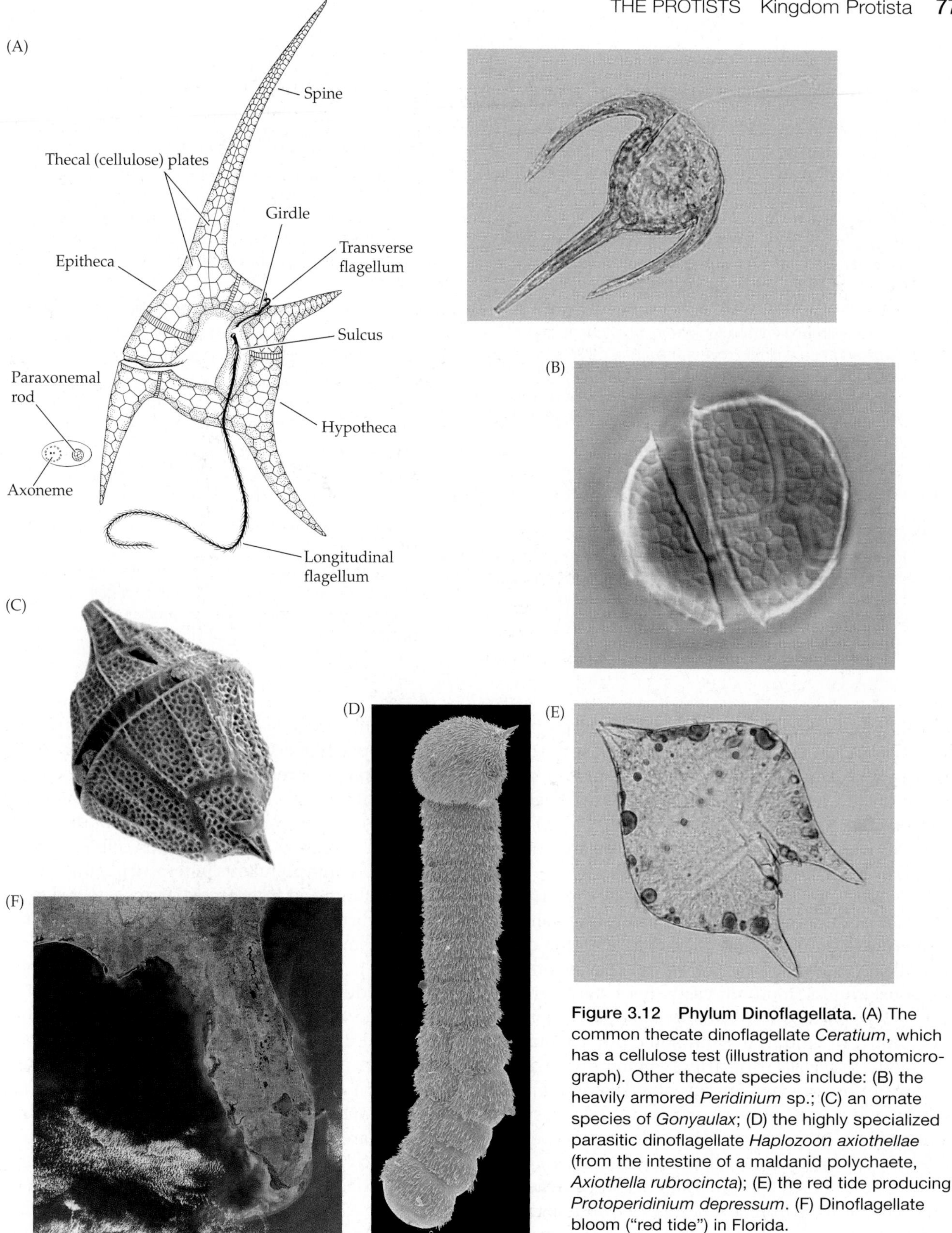

Figure 3.12 Phylum Dinoflagellata. (A) The common thecate dinoflagellate *Ceratium*, which has a cellulose test (illustration and photomicrograph). Other thecate species include: (B) the heavily armored *Peridinium* sp.; (C) an ornate species of *Gonyaulax*; (D) the highly specialized parasitic dinoflagellate *Haplozoon axiothellae* (from the intestine of a maldanid polychaete, *Axiothella rubrocincta*); (E) the red tide producing *Protoperidinium depressum*. (F) Dinoflagellate bloom ("red tide") in Florida.

with that of diatoms—the former peaked in diversity in Mesozoic seas, whereas diatoms began their broad evolutionary radiation much more recently, in the early Cenozoic, corresponding to the onset of major polar ice caps and increased oceanic thermohaline circulation.

Dinoflagellates are common in all aquatic environments, but about 90% of the described species are planktonic in the world's seas (Figure 3.12). Approximately half of the living species of dinoflagellates are photosynthetic, and these are important primary producers in many aquatic environments. They can be quite

BOX 3E Characteristics of the Phylum Dinoflagellata

1. Shape of cell maintained by pellicle consisting of alveolar vesicles beneath the plasma membrane; in some species (thecate species) alveoli may be filled with polysaccharides, usually cellulose; species with "empty" alveoli are said to be naked (athecate)
2. Most species are unicellular; some form filamentous colonies. Endosymbiotic marine species (zooxanthellae) occur as coccoid cells inside their invertebrate or protist hosts, but produce motile cells periodically.
3. Usually with two flagella for locomotion: one is transverse and has a single row of hairs, the other is longitudinal and has two rows of hairs; both flagella are supported by a paraxonemal rod. Flagella usually oriented in longitudinal groove and equatorial groove (a diagnostic feature of this group)
4. Mitochondria with tubular cristae
5. With both autotrophic and heterotrophic species (and many can switch back and forth). Most photosynthetic species (about half of all species in the phylum) have chlorophylls *a* and c_2. Accessory pigments (phycobilins, carotenoids, xanthophylls) often give cells a brownish color. Predatory species often with extrusomes (e.g., trichocysts, mucocysts, nematocysts)
6. Many dinoflagellates are capable of bioluminescence (using a luciferin–luciferase system).
7. Thylakoids occur in stacks of three, with three surrounding membranes; the outer membrane is not continuous with nuclear membrane. Various species show evidence of plastid origin via secondary and even tertiary endosymbiosis with a variety of other protists.
8. Food reserves stored as starch and oils
9. Many with unique system of pusules for osmoregulation, excretion, or buoyancy regulation
10. Nuclei contain permanently condensed chromosomes; few or no histone proteins associated with DNA
11. Nuclear division occurs by closed extranuclear pleuromitosis without centrioles; no obvious organizing center for mitotic spindle
12. Asexual reproduction by binary fission along the longitudinal plane
13. Sexual reproduction occurs in some. Meiosis involves two divisions, one just after nuclei from a pair of gametes fuse, and one after cell undergoes period of dormancy.

beautiful and many are capable of bioluminescence (e.g., *Gonyaulax*) using a luciferin–luciferase system. Although most are unicellular, some form filamentous, multicellular colonies. Dinoflagellates have two flagella, positioned such that they whirl or spin as they swim (see Figures 3.1A,G), the attribute for which they are named (Greek *dinos*, "whirling, turning") (Box 3E).

Endosymbiotic, photosynthetic, marine dinoflagellates that occur as coccoid cells when inside their invertebrate or protist hosts, but produce motile cells periodically, are called zooxanthellae. They belong to the poorly understood but vastly important genera *Zooxanthella* (symbionts of radiolarians), *Symbiodinium* (symbionts of cnidarians and some other metazoans), and *Zoochlorella* (primarily symbionts of various freshwater organisms). Species of *Symbiodinium* are best known as critically important mutualistic symbionts of hermatypic corals. There are several species of *Symbiodinium* in corals. All are photosynthetic and provide nutrients to the corals and help create the internal chemical environment necessary for the coral to secrete its calcium carbonate skeleton. Zooxanthellae also occur in many cnidarians other than scleractinian corals, such as milleporinids, chondrophorans, sea anemones, and various medusae. *Symbiodinium* has even been found as an endosymbiont in certain ciliates (as have green algal symbionts of the genus *Chlorella*).

Some planktonic dinoflagellates occasionally undergo periodic bursts of population growth to produce red tides. Red tides have nothing to do with actual tides, and they are only rarely red. A red tide is simply a streak or patch of ocean water discolored, generally a pinkish orange or reddish-brown, by the presence of trillions of dinoflagellates (and occasionally diatoms or other algae). During a red tide, densities of these protists may be as high as 10 to 100 million cells per liter of seawater. Organic pollutants in terrestrial runoff (e.g., from agriculture and from animal farms) are tied to increased occurrences of red tides worldwide. Many red tide organisms are also bioluminescent, so observers are treated to the spectacular demonstrations of their abundance during both day and night observations!

Many red tide organisms manufacture toxic substances, and some are among the strongest known poisons. One group of toxins produced by dinoflagellate species such as *Gymnodinium catenatum*, *Pyrodinium bahamense*, and *Alexandrium* spp., is called **saxitoxins**. Saxitoxins are tasteless, odorless, and water-soluble, with a toxicity similar to that of the biological-weapon poison ricin. Saxitoxins block the sodium–potassium pump of nerve cells and prevent normal impulse transmission. When suspension feeders such as mussels and clams eat these dinoflagellates, they store the toxins in their bodies. Extremely high concentrations of toxic dinoflagellates will even kill suspension feeders and occasionally also fish caught in the thick of the bloom. The shellfish feeding on the protists become toxic to animals that eat them. In humans, the result is a disease known as paralytic shellfish poisoning (PSP). Extreme cases of PSP result in muscular paralysis and respiratory failure. Over 300 human deaths worldwide have been documented from PSP, and this number is growing as red tides become increasingly frequent around the world (linked to anthropogenic disturbances of coastal environments).[5]

The dinoflagellate *Karenia brevis* (formerly known as *Gymnodinium breve*) releases a family of toxins called **brevetoxins** that result in neurotoxic shellfish poisoning (NSP). Humans consuming animals that have this toxin accumulated in their tissues experience uncomfortable gastrointestinal side effects such as diarrhea, vomiting, and abdominal pain, and also neurological problems, including dizziness and an odd reversal of temperature sensation. Though temporarily incapacitating, no human deaths have been reported from NSP. Ocean spray containing *K. brevis* toxins can blow ashore and cause temporary health problems for seaside residents and visitors (skin, eye, and throat problems). *Karenia brevis* is responsible for producing devastating red tides that have produced massive kills of fish, sea birds, sea turtles, manatees, and dolphins all along the coastline of the Gulf of Mexico. One of the worst blooms of *K. brevis* ever recorded was from January 2005 to January 2006, along the coast of west-central Florida, an event that resulted in widespread hypoxia and the death of tens of thousands of fish, turtles, and marine mammals.

In recent years, two newly discovered species of dinoflagellate called *Pfiesteria piscicida* and *P. shumwayae* have been creating havoc in the coastal areas of the eastern United States. *Pfiesteria piscicida* normally exists in a benign state, and it reportedly can even utilize photosynthesis if it consumes certain other protists that contain chloroplasts it can co-opt. However, with the proper stimulation, which is believed to be high levels of fish oils or fish excrement in the water, *P. piscicida* becomes a voracious predator. It first produces a toxin that causes fish to become lethargic and then releases other toxins that cause open sores to form on the fish's body, exposing the tissues on which it feeds. The toxins of *P. piscicida* have been reported to affect humans but have caused no known deaths. In the Carolinas, it has been claimed that outbreaks of *Pfiesteria* might be linked to large-scale hog farming in North Carolina. The industry dumps hundreds of millions of gallons of untreated hog feces and urine into earthen lagoons along the coast that often leak or collapse. In 1995, 25 million gallons of liquid swine manure (more than twice the size of the Exxon Valdez oil spill) flowed into the New River when a lagoon was breached.

In 2009 red tides caused by the dinoflagellate *Akashiwo sanguinea* in Monterey Bay (California) caused a mass stranding and high mortality of seabirds, not due to toxins but due to their sheer numbers. The birds' plumage became coated in a sticky green froth exuded from the algae that contained surfactant mycosporine-like amino acids, which acted like a detergent to strip the feathers of their natural waterproofing oils.

About half the known dinoflagellates are not photosynthetic (i.e., they are plastid-free), whereas most with plastids have one basic form—a peridinin-containing plastid surrounded by three membranes. These plastids are presumed to have derived from a singular event of secondary endosymbiosis. A minority of plastid-containing dinoflagellates have plastids derived from other sources. One group has a plastid derived from an endosymbiotic haptophyte, thus representing a tertiary symbiosis because the haptophyte plastid itself was the product of a secondary endosymbiosis (*Karenia,* the agent of neurotoxic shellfish poisoning, is one of these). There are also groups with cryptomonad-derived or chlorophyte-derived plastids. It is thought likely that these various atypical plastids might represent plastid replacements in lineages that already had the peridinin-containing plastid, rather than original plastid acquisitions by heterotrophic lineages.

Parasitic dinoflagellates have a broad range of hosts, morphological diversity, and life histories. The genus *Haplozoon* is a small group of intestinal parasites occurring in marine worms with a highly unusual organization of differentiated cells, and a body plan so bizarre that early workers classified some of them as mesozoans and gregarine apicomplexans. Species of *Haplozoon* have long been viewed as "multicellular" or "colonial" in their organization. However, recent studies of species in this genus have revealed a unique cellular organization in which the entire organism is bounded by a single continuous membrane, suggesting that these protists are not "multicelled" but rather syncytial, composed of cell-like compartments separated by sheets of alveoli.

Support and Locomotion

The shape of dinoflagellates appears to be at least partly maintained by a system of flattened sacs, called **alveoli**, beneath the cell's outer (plasma) membrane, plus a layer of supporting microtubules. In some, the alveoli are filled to form plates of polysaccharides, typically cellulose, and these dinoflagellates are said to be **thecate**, or armored (e.g., *Protoperidinium*, *Ceratium*). Dinoflagellates that have empty alveoli are said to be **athecate**, or naked (e.g., *Noctiluca*). The part of the theca above the girdle is called the **epitheca** in armored species and the **epicone** in naked species; the part below the girdle is the **hypotheca** in armored species and the **hypocone** in naked species (see Figure 3.12).

Dinoflagellates possess two flagella that enable their locomotion. A transverse flagellum with a row

[5]Pilots who flew U-2 spy missions over the Soviet Union were reportedly given tiny pellets of saxitoxin extracted from dinoflagellates and instructed to take the suicide capsules if they were shot down.

of slender hairs wraps around the cell in a groove, or **girdle** (see Figure 3.12). When it beats, this flagellum spins the cell around, effectively pushing it through the water like a screw. The transverse flagellum provides much, if not most of the forward propulsion. The second, longitudinal flagellum has two rows of hairs and also lies in a groove on the cell's surface, called the **sulcus**. It extends posteriorly behind the cell and its beat adds to the forward propulsion. Both flagella are supported by a **paraxonemal rod** of uncertain function but similar to that found in the kinetoplastids, euglenids, and others.[6]

Osmoregulation

Most freshwater and some marine dinoflagellates have a unique system of double-membrane–bound tubules called **pusules**, which open to the outside via a canal. The two membranes of the pusules distinguish them from water expulsion vesicles, but apparently these membranes have a similar function—osmoregulation.

Nutrition

Dinoflagellates exhibit wide variation in feeding habits. Approximately half of the living species are photosynthetic, but most of these also are heterotrophic to some extent, and some dinoflagellates with functional chloroplasts can switch entirely to heterotrophy in the absence of sufficient light.

In most photosynthetic species the chloroplasts are surrounded by three membranes, and thylakoids are arranged in stacks of three (see Figure 3.3). Some contain eyespots (stigmata) that can be very simple pigment spots or more complex organelles with lens-like structures that apparently focus the light. Photosynthetic pigments include chlorophylls *a* and c_2, phycobilins, carotenoids (e.g., β-carotene), and also xanthophyll pigments known as **peridinin**, found only in dinoflagellates (neoperidinin, dinoxanthin, neodinoxanthin). These xanthophylls mask the chlorophyll pigments and account for the golden or brown color that is commonly seen in dinoflagellates.

Some dinoflagellates always lack chloroplasts and are obligate heterotrophs. Most of these are free living, but some parasitic species are known. The feeding mechanisms of heterotrophic dinoflagellates are quite diverse. Both free-living and endoparasitic dinoflagellates that live in environments rich in dissolved organic compounds take in dissolved organic nutrients by saprotrophy. Other dinoflagellates ingest food particles by phagocytosis. Many, in fact, are voracious predators that ingest other protists and microinvertebrates or use specialized cellular appendages to pierce prey and suck out their cytoplasmic contents.

Some dinoflagellates (e.g., *Kofoidinium* and *Noctiluca*) have a permanent cell mouth or cytostome supported by sheets of microtubules. The cytostome is often surrounded by extrusomes, of three types—trichocysts, mucocysts, and nematocysts. The most common are **trichocysts** (similar to those found in ciliates), which are fired in defense or to capture and bind prey. Sac-like **mucocysts** secrete sticky mucous material onto the surface of the cell. This may aid in attachment to substrata (e.g., *Amphidinium*) or may help to capture prey (e.g., *Noctiluca*). Other dinoflagellates (e.g., *Nematodinium*, *Nematopsides*, *Polykrikos*) have "**nematocysts**" that resemble, but probably are not homologous to, the stinging organelles of cnidarians of the same name.

Reproduction

The nuclei of dinoflagellates have three unusual features: (1) they contain five to ten times the amount of DNA that is found in most eukaryotic cells; (2) the five histone proteins that are typically associated with the DNA of other eukaryotic cells are absent; and (3) the chromosomes of dinoflagellates remain condensed and the nucleolus remains intact during interphase and mitosis. Most dinoflagellates (but not *Noctiluca*) spend much of their lives as haploid cells, called vegetative cells to distinguish them from haploid gametes.

Nuclear division is by closed extranuclear pleuromitosis. No centrioles are present, and the organizing center for the mitotic spindle is not obvious. Asexual reproduction occurs by oblique, longitudinal fission, beginning at the posterior end of the cell. The thecate forms may divide the thecal plates between the two daughter cells (e.g., *Ceratium*), or they may shed the thecal plates prior to cell division (Figure 3.13A). In the former case, each daughter cell synthesizes the missing plates; in the latter case, each daughter cell synthesizes all of the thecal plates anew.

Sexual reproduction begins when the haploid vegetative cells divide by mitosis to produce two flagellated daughter cells, which act as gametes. When a pair of gametes fuses to form a zygote, a fertilization tube develops beneath the basal bodies of its flagella. The nucleus from each gamete enters the tube where they fuse. The first meiotic division follows shortly after nuclear fusion. Over the next few weeks, the zygote grows in size and then enters a resting stage, or cyst. The cyst develops a resistant outer wall and remains dormant for an indefinite period of time. Eventually the second meiotic division occurs, all but one of the nuclei disintegrate, and a haploid vegetative cell emerges from the cyst (Figure 3.13B).

[6]Paraxonemal rods are internal, solid or hollow, proteinaceous rods that extend nearly the entire length of a flagellum/cilium. They are located between the axoneme and flagellar membrane, and usually are connected to the axoneme and membrane by specific links. Although they are seen in dinoflagellates, euglenids, kinetoplastids, silicoflagellates, and others, they differ in ultrastructure and biochemical composition and their homology seems unlikely. They also are known as "flagellar rods," "paraxial rods," and "paraflagellar rods." Similar structures (also called "paraxonemal rods") have even been described from the flagella of certain vertebrate spermatozoa.

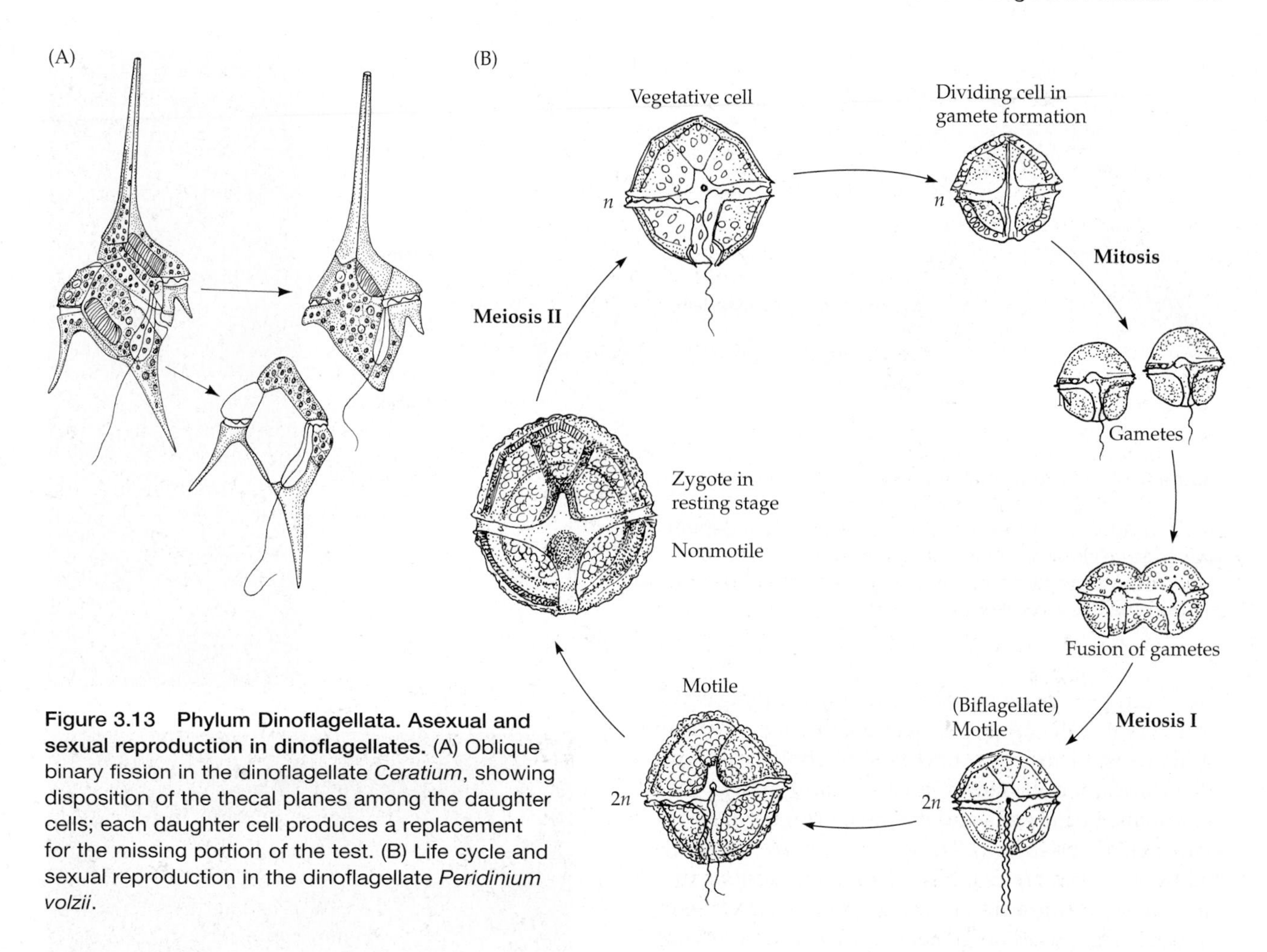

Figure 3.13 Phylum Dinoflagellata. Asexual and sexual reproduction in dinoflagellates. (A) Oblique binary fission in the dinoflagellate *Ceratium*, showing disposition of the thecal planes among the daughter cells; each daughter cell produces a replacement for the missing portion of the test. (B) Life cycle and sexual reproduction in the dinoflagellate *Peridinium volzii*.

Phylum Apicomplexa: Gregarines, Coccidians, Haemosporidians, and Their Kin

The phylum Apicomplexa, with more than 5,000 species, is arguably the most successful group of specialist parasites on Earth. This phylum includes the gregarines, coccidians, haemosporidians, and piroplasms. Among these is the vastly important human pathogen *Plasmodium* (the causative agent of malaria), as well as the human disease causing genera *Toxoplasma* and *Cryptosporidium*. All apicomplexans are parasitic, and the group is characterized by the presence of a unique combination of organelles at the anterior end of the cell called the **apical complex** (Figure 3.14A). The apical complex apparently attaches the parasite to a host cell and releases a substance that causes the host cell membrane to invaginate and draw the parasite into its cytoplasm in a vacuole. Some apicomplexans also possess a bizarre, vestigial plastid, the **apicoplast**. This DNA-containing organelle is the remnant of a secondary endosymbiosis involving a red algal "prey" cell and an apicomplexan. In these species, the apicoplast appears to be necessary for the survival of the "host" cell, and functions in fatty acid, heme, and isoprynoid synthesis (Box 3F).

Gregarines comprise a large group of apicomplexans that occupy the guts, reproductive systems, and other body cavities of numerous phyla of invertebrates, including annelids, molluscs, nemerteans, sipunculans, tunicates, echinoderms, phoronids, hemichordates, appendicularians, and arthropods. The genus *Gregarina* itself has nearly a thousand described species, mostly insect parasites. The trophozoites (motile cells) of gregarines are often very large cells. Unlike most other Apicomplexa, gregarines are largely extracellular, rather than intracellular parasites. Their apical complex is generally specialized for attachment to the host's epithelia, rather than for host cell invasion, and thus is typically equipped with hooks or suckers (Figure 3.14B–D). Although both cellular and molecular data indicate gregarines diverged early within the radiation of Apicomplexa, these are highly-derived/specialized parasites that have many novel ultrastructural and behavioral adaptations, and many of the marine species have become giants among single-celled organisms.

The coccidians are parasites of several groups of animals, mostly vertebrates. They typically reside within the epithelial cells of their host's gut, at least during some stages, and many are pathogenic. Some coccidians pass their entire life cycle within a single host;

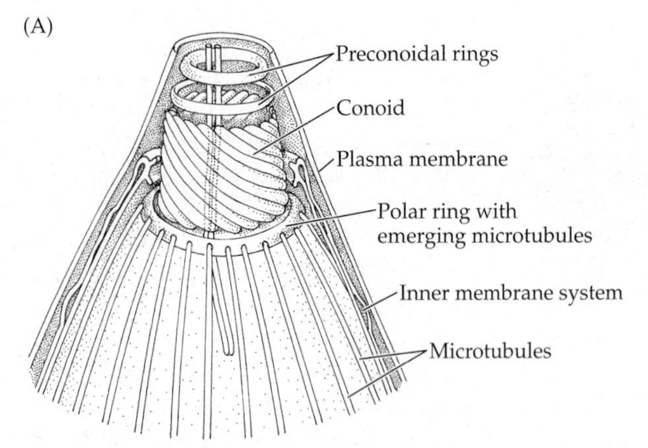

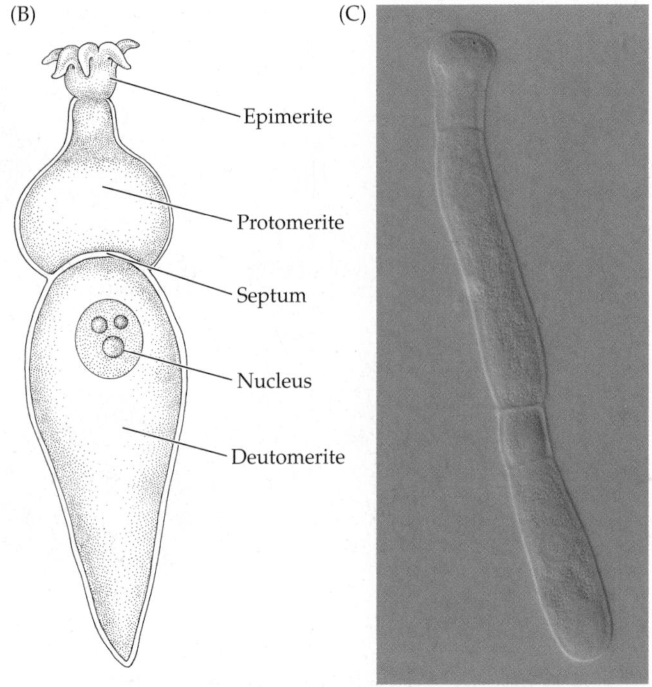

Figure 3.14 Phylum Apicomplexa. (A) Microstructure of the apical complex. (B) The body of a gregarine is commonly divided into three recognizable regions. (C) Septate gregarines (gamonts) undergoing syzygy. (D) *Pterospora floridiensis*, a gregarine that lives in the coelom of maldanid polychaetes (*Axiothella rubrocincta*).

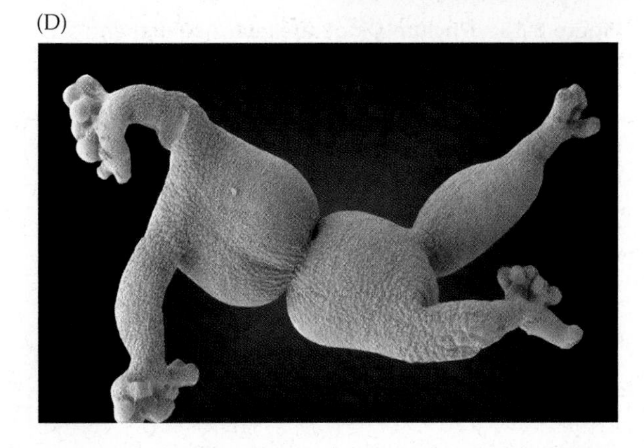

many others require an intermediate host that serves as a vector. Coccidians are responsible for a variety of diseases, including coccidiosis in rabbits, cats, and birds, and toxoplasmosis in people and many other mammals. Piroplasms and haemosporidians are both parasites of vertebrates. The piroplasms are transmitted by ticks and are responsible for some serious diseases of domestic animals, including red-water fever in cattle. The coccidian genus *Eimeria* causes a disease known as cecal coccidiosis in chickens. This disease was estimated by the USDA to cost American poultry farmers more than $600 million dollars annually in lost animals, medication, and additional labor.

Haemosporidians are blood parasites of vertebrates that are transmitted by biting insects and include the organisms that cause malaria and malaria-like ailments in humans, birds, and other vertebrates. Human malaria, which affects people in over 100 countries, is caused by four species of the genus *Plasmodium* (Box 3G). More people have died of malaria than from any other disease in history. Although malaria was greatly reduced worldwide during the 1960s, it is making an alarming comeback, and it is one of the most prevalent and severe health problems in the developing world. Nearly a quarter-million new cases of malaria are contracted each year, and over one-half million deaths occur annually due to the disease, most of them children and 90% of them in Africa where *Anopheles gambiae* mosquitoes are the major vectors of *Plasmodium falciparum*. Although *Anopheles* mosquitoes are the only insects capable of harboring the human malaria parasite, the complex life cycle of this parasite challenges researchers (some *A. gambiae* strains are completely resistant to *Plasmodium* infections). Individuals of the *Plasmodium* parasite differentiate into a series of morphologically distinct forms in both the vertebrate and mosquito hosts. They alternate between morphologically related invasive stages (sporozoites, merozoites, ookinetes) and replicative stages (pre-erythrocytic, erythrocytic-schizont, oocyst) interposed by a single phase of sexual development that mediates transmission from the human host to the anopheline vector. In all, in its two hosts, human *Plasmodium* parasites undergo 10 morphological transitions in five different host tissues. Recent genomic studies have revealed that *Plasmodium* has many genes with no known homologs in other species, indicating its machinery for interacting with the host environment is highly specialized. Rodent malaria parasite species provide model systems that facilitate research not easily accomplished in humans, and three species are in common laboratory use (*Plasmodium berghei*, *P. chabaudi*, *P. yoelii*). Other malaria-like genera (e.g., *Haemoproteus*, *Leucocytozoon*) are parasites of birds and reptiles but have life cycles similar to *Plasmodium* and also utilize mosquitoes as vectors. In the Old World, migrating birds can pick up various species of malaria parasites while overwintering in the tropics. A

BOX 3F Characteristics of the Phylum Apicomplexa

1. Shape of cell maintained by pellicle consisting of alveolar vesicles beneath plasma membrane, and by supporting microtubules
2. Locomotion characterized as gliding, but cilia/flagella absent (although some species produce flagellated or ameboid gametes)
3. With unique system of organelles—the apical complex—in anterior region of cell. These function in host cell attachment and/or host cell invasion.
4. All species are parasitic.
5. Carbohydrates stored as paraglycogen (= amylopectin)
6. Some species with nonphotosynthetic plastid—the apicoplast—which fulfills essential biosynthetic functions
7. Mitochondria with vesicular cristae
8. With single vesicular nucleus
9. Nuclear division occurs by semi-open pleuromitosis in all except gregarines, which show much diversity in their mitoses: mitotic spindle organizers are disks, plugs, or electron-dense crescents on nuclear envelope in the location of fenestrae. In coccidians, centrioles are associated with spindle organizers; in haemosporidians and piroplasms, centrioles absent.
10. Asexual reproduction by binary fission, multiple fission, or endopolygeny
11. Sexual reproduction is gametic; gametes either isogametous or anisogametous; meiosis involves single division after formation of zygote

recent study of migrant great reed warblers, which nest in Sweden and overwinter in Africa, found that birds infected with species of *Plasmodium* and *Haemoproteus* laid fewer eggs in their lifetime, were less successful at rearing offspring, and died at a younger age (probably due to telomere damage).

Toxoplasma gondii is extraordinarily prevalent in the developed world, and it is a source of congenital neurological birth defects in children whose mother happen to be exposed during pregnancy. In adults, *Toxoplasma* usually produces no symptoms, or only mild symptoms, but it and another apicomplexan, *Cryptosporidium*, have been increasingly problematic in AIDS patients and other immunosuppressed people. Transmission of *Toxoplasma* parasites is thought to be via raw or undercooked meat (beef, pork, lamb), through fecal contamination from pet cats, or by way of flies and cockroaches (which can carry the *T. gondii* cysts from a cat's litter box to the table). In 1982 Martina Navratilova lost the U.S. Open Tennis Championship (and a half-million dollars) when she had toxoplasmosis.

Support and Locomotion

The fixed shape of apicomplexans is maintained by a pellicle composed of inner membrane sacs, or alveoli, which lie just beneath the plasma membrane. Microtubules originate at the apical complex and run beneath the alveoli, providing additional support. Apicomplexa do not have cilia, flagella or pseudopodia (except in some microgametes). Nevertheless, they can be observed flexing and gliding along surfaces, and this movement is produced by microtubules and microfilaments underneath the alveoli. The motion involves attachment proteins that are released to the cell membrane from vesicles in the apical complex, and actin-myosin interactions, that is the same fundamental motor system that powers muscle cells in Metazoa.

Nutrition

The alveoli are interrupted at both the anterior and posterior ends, and at tiny invaginations of the cell membrane called **micropores**, which have been implicated in feeding. Nutrient ingestion is thought to occur primarily by pinocytosis at the microspores. In the haemosporidians, ingestion of the host's cytoplasm through the microspores has been observed. Absorption of nutrients has also been reported in some gregarines at the point where the parasite attaches to the host's cell.

Reproduction and Life Cycles

Asexual reproduction in Apicomplexa is by binary fission, multiple fission, or endopolygeny. During multiple fission, the nucleus undergoes several divisions prior to cytokinesis (partitioning of the cytoplasm), resulting in many (e.g., 32) daughter cells. Some engage in a process called **plasmotomy**, considered by some to be a form of budding, in which a multinucleate adult simply divides into two multinucleate daughter cells. Other members of the Apicomplexa undergo a type of internal budding called **endopolygeny**, during which daughter cells actually form within the cytoplasm of the mother cell.

Mitosis is a semi-open pleuromitosis in all apicomplexans except for some gregarines. Gregarines undergo a variety of mitoses, depending on the species. For example, *Diaplauxis hatti* and *Lecudina tuzetae* undergo semi-open orthomitosis, *Monocystis* sp. and *Stylocephalus* sp. undergo open orthomitosis, and *Didymophyes gigantea* undergoes closed intranuclear orthomitosis. Sexual reproduction occurs by a union of haploid gametes, which can be the same size (isogametous) or different sizes (anisogametous), and may be flagellated or form pseudopodia.

The life cycle varies somewhat between the different groups of apicomplexan protists, but can be divided into three general stages: (1) **gamontogony** (the sexual phase), (2) **sporogony** (the spore-forming stage), and (3) **merogony** or **schizogony** (the growth phase).

BOX 3G Malaria

Human malaria has been known since antiquity. Egyptian mummies show signs of malaria, and descriptions of the disease are found in Egyptian papyruses and temple hieroglyphs (outbreaks followed the annual flooding of the Nile). Alexander the Great likely died of it, leading to the unraveling of the Greek Empire. Malaria might have been what stopped the armies of both Attila the Hun and Genghis Khan. Malaria was almost certainly brought to the New World by the conquering Spaniards and their African slaves. At least four popes died of it. George Washington, Abraham Lincoln, and Ulysses S. Grant all suffered from it. During World War II, there were more casualties in the Pacific theater from malaria than from combat. Some scientists believe that one out of every two people who have ever lived have died of malaria. In 1946, the Centers for Disease Control was founded in Atlanta specifically to combat malaria. Other forms of vertebrate malaria are probably hundreds of millions of years old—some evidence suggests dinosaurs might have had malaria. Mice, birds, porcupines, lemurs, monkeys and apes, bats, snakes and flying squirrels all have their own forms of malaria. In 2002, scientists sequenced the genome of *Plasmodium falciparum*, the most deadly species of malaria parasite, and in 2010 they concluded that this parasite was descended from a single lineage (*P. reichenowi*) that jumped from gorillas or chimpanzees a few million years ago. And in 2014, Neafsey et al. published an analysis of the genomes of 16 species of *Anopheles* mosquitoes, providing detailed insight into the evolution of these insects.

The relationship between malaria in humans and the smell of nearby swampland led to the belief that it could be contracted by breathing "bad air" (from the Italian, *mal aria*)—hence the origin of the name.. It was not until 1897 that Britain's Sir Ronald Ross discovered that the malaria parasite is transmitted by the *Anopheles* mosquito. By the time the Panama Canal was built, malaria (and yellow fever) was well established in the New World, and until the twentieth century the disease occurred as far north as the American Midwest. The presence of malaria (and yellow fever) was the principal reason the French failed in their attempt to construct the Panama Canal. William Gorgas, the medical officer in charge during the U.S. phase of the canal's construction, became a hero when his mosquito-control efforts allowed for the canal's successful completion by U.S. engineers. The president made Gorgas Surgeon General, Oxford University gave him an honorary doctorate, and the King of England knighted him.

The first effective remedy for human malaria—the bark of the cinchona tree (a close cousin of coffee)—was discovered in Peru and Ecuador. This remedy came to be distributed worldwide as quinine. The cinchona tree was eventually established on plantations in India and elsewhere. Quinine disrupts the parasite's reproductive process; however it is short-acting, and if taken too frequently can cause serious side effects, including hearing loss. In the 1940s synthetic malaria medicine was developed—initially a drug named chloroquine, which worked as both a prophylactic and a partial remedy. Today, however, most strains of *Plasmodium falciparum* have developed chloroquine resistance, apparently through a simple mutation in a single gene. In the mid-twentieth century DDT (dichlorodiphenyltrichloroethane) was invented by the Swiss chemist Paul Müller, and this potent insecticide quickly became the centerpiece of global mosquito eradication. Today, however, the use of DDT is illegal almost everywhere.. Most recently, the discovery of the potent antimalarial properties of artemisinin has helped helped turn the tide against malaria, especially when used in combination therapies and in association with insecticide-treated bednets. Chinese scientists isolated artemisinin in the 1970s from sweet wormwood (*Artemisia annua*), a plant used to treat fever for centuries. However, it may be only a matter of time until the highly adaptable *P. falciparum* parasite develops a resistance to artemisinin.

Among the many symptoms of malaria are cyclical paroxysms, wherein the patient feels intensely cold as the hypothalamus (the body's thermostat) is activated; body temperature then rises rapidly to 104°–106° F. Nausea and vomiting are usual. Copious perspiration signals the end of the hot stage, and the temperature drops back to normal within 2 or 3 hours; the entire paroxysm is over within 8 to 12 hours. It is thought that these episodes are stimulated by the appearance of the waste products of parasites feeding on erythrocytes, which are released when the blood cells lyse. Secondary symptoms include anemia due to the red blood cell destruction. In extreme cases of falciparum malaria, massive lysis of erythrocytes results in high levels of free hemoglobin and various breakdown products that circulate in the blood and urine, resulting in a darkening of these fluids, hence the condition called blackwater fever.

One of the principal causes of the recent malaria resurgence is the dramatic rise in the numbers of pesticide-resistant strains of *Anopheles* mosquitoes, the insect vector for *Plasmodium*. The development of a malaria vaccine continues to elude researchers; in fact, vaccines do not exist for any human metazoan parasitic disease. By 1968, 38 strains or species of *Anopheles* in India alone had been identified as largely pesticide-resistant; and between 1965 and 1975 the incidence of malaria in Central America tripled. Researchers have recently discovered that *Plasmodium* species, like *Trypanosoma*, have the ability to avoid detection by the human immune system by switching among as many as 150 genes that code for different versions of the protein that coats the surface of their cells (it is this protein coat that the human immune system relies on as a recognition factor). This trick is known as **antigenic variation**. Furthermore, because the parasite sequesters itself inside red blood cells, it is largely protected from most drugs. Recent evidence also suggests that one effect the parasite may have on its mosquito host is inducing it to bite more frequently than uninfected mosquitoes.

At least one of the four malaria-causing *Plasmodium* species occurs on every continent save Antarctica, and the disease threatens nearly half the world's population. The less deadly *Plasmodium vivax* occurs mainly in South America and Asia. Today, sub-Saharan Africa is not only the largest remaining sanctuary of the deadly *P. falciparum*, but also the home of *Anopheles gambiae*, the most aggressive of the more than 60 mosquito species that transmit malaria to people. Falciparum malaria can be devastating, and infected individuals commonly suffer seizures, coma, heart, and lung failure. Those who survive can suffer mental or physical handicaps or chronic debilitation. Most of the genome of *Plasmodium falciparum*, and its host mosquito *Anopheles gambiae*, have now been sequenced, opening doors to possible new control measures for this deadly disease. The battle against malaria got a big boost in 2005 when the Bill and Melinda Gates Foundation announced grants totaling $258 million to support development of new drugs and improved mosquito-control methods.

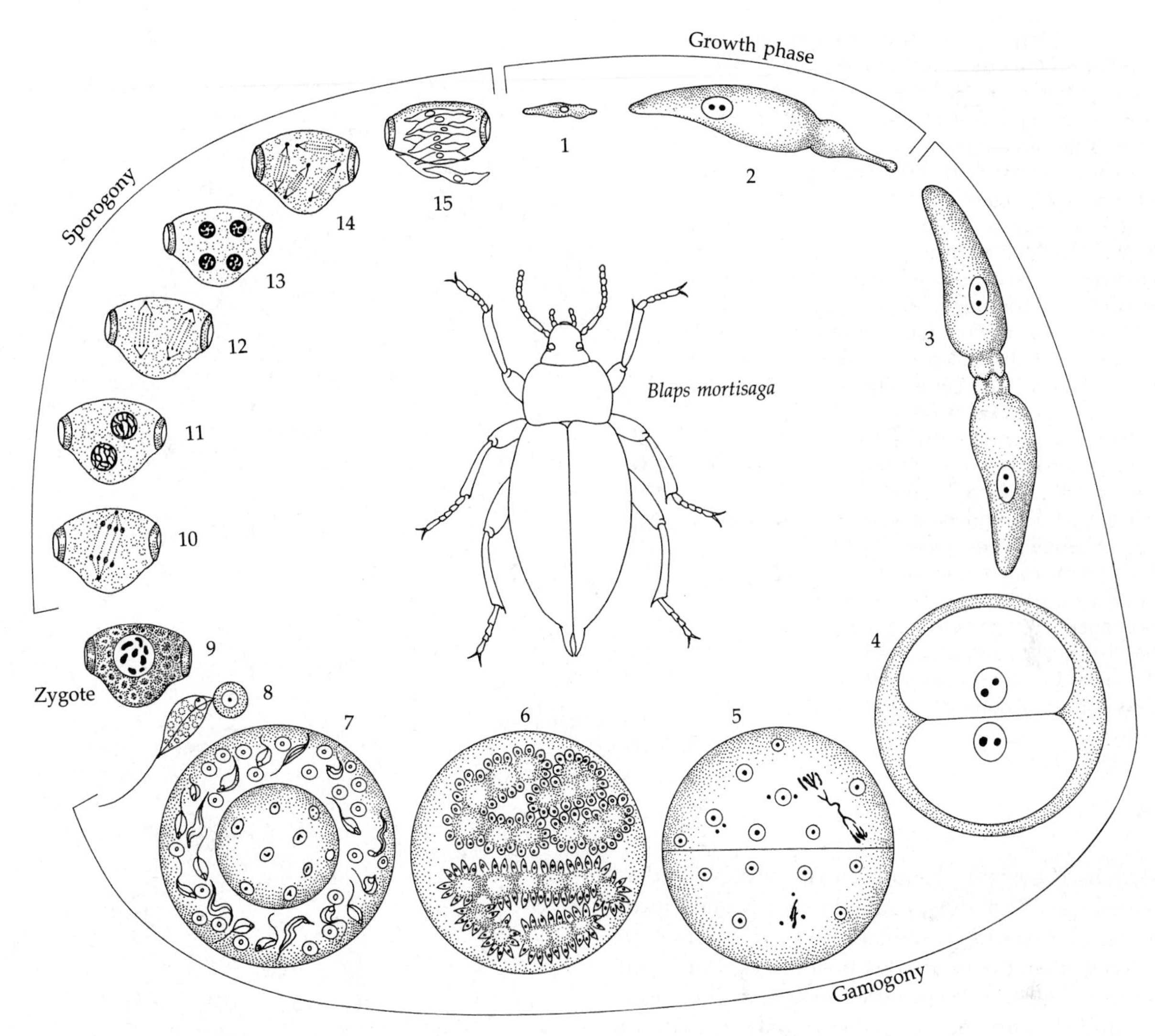

Figure 3.15 Phylum Apicomplexa. Life cycle of the gregarine *Stylocephalus longicollis*, a gut parasite of the Old World "churchyard beetle" (Tenebrionidae: *Blaps mortisaga*). Stages 1–4 take place within the host, 5–15 outside the host. The spores (15) are ingested by the beetle and release sporozoites within the gut lumen. Each sporozoite grows into a gamont (2); the gamonts subsequently mate (3–4), becoming enclosed within a mating cyst, which leaves the host with the feces. Repeated mitotic divisions within the cyst produce anisogametes (5–7); these ultimately fuse (8) to produce a zygote (9), which eventually becomes a spore. The first divisions of the spore cell are meiotic (10), so all subsequent stages leading back to gamete fusion are haploid.

A great deal has been written about the life cycles of these protists and a full description is beyond the scope of this text, however, it is worth noting that typical Apicomplexa are haploid for most of their life cycle. The life cycles of the gregarine *Stylocephalus* and the haemosporidan *Plasmodium* are given as examples to illustrate the basic themes and variations in apicomplexan reproduction.

The life cycle of gregarines is usually **monoxenous**—it involves only one host. Some of the best-studied gregarines are those found in coleopterans (beetles), and the life cycles of these are well understood (the life cycle of *Stylocephalus longicollis* is diagrammed in Figure 3.15). Once two trophozoites join together for sex, a process called **syzygy**, the cells are called **gamonts**. A wall forms around the gamonts forming a **gametocyst**, within which several rounds of mitosis produce hundreds of gametes. Both isogamy and anisogamy are known among different gregarines. Zygotes (the fleeting diploid stage) are formed after one gamete from each gamont fuse together. Each zygote that is formed by the fusion of two gametes becomes a thick-walled **sporocyst**, which undergoes meiosis to produce four or more sporozoites. Sporozoite-filled gametocysts are released into the environment by way of host feces, host gametes, or host disintegration. New hosts inadvertently consume sporocysts in their environment, and these find their way to the appropriate body cavity and penetrate the host cells, completing the cycle.

The life cycle of haemosporidians is **heteroxenous**, which means it involves two hosts, usually a vertebrate

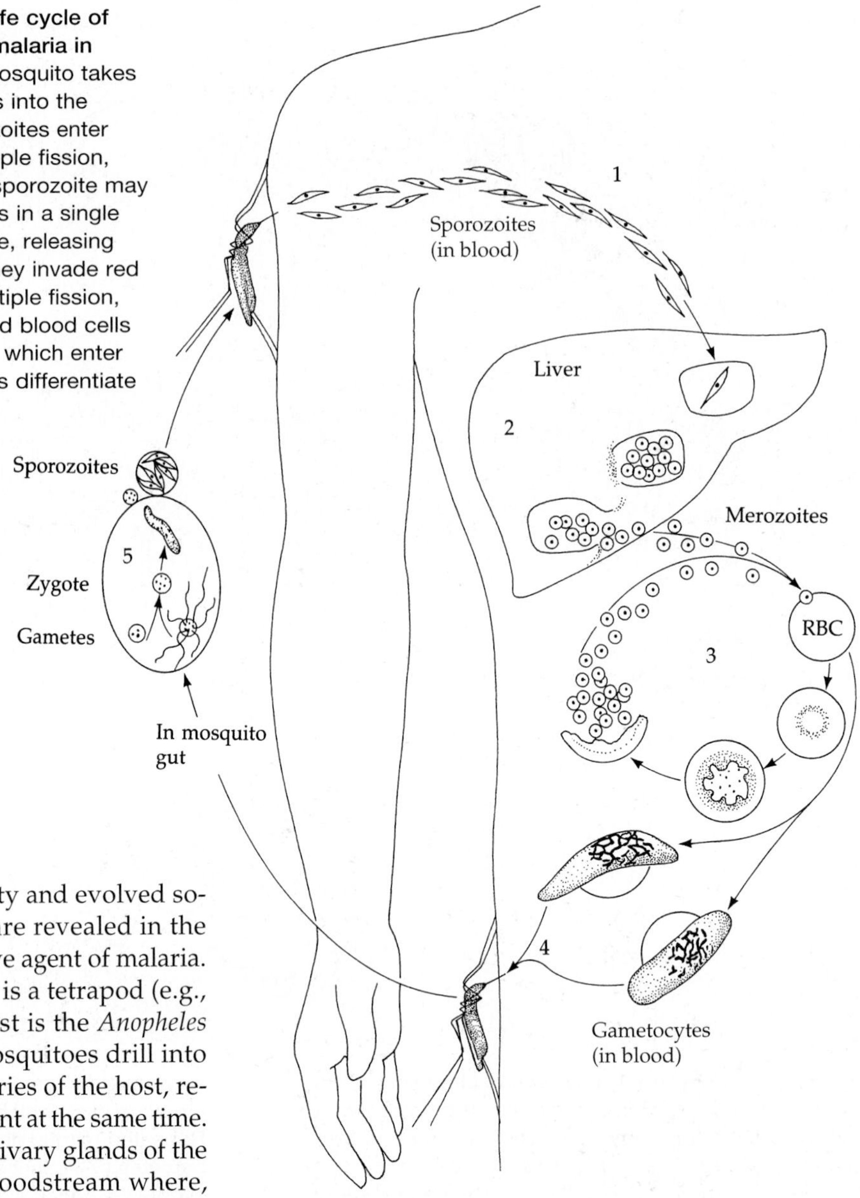

Figure 3.16 Phylum Apicomplexa. Life cycle of *Plasmodium*, the causative agent of malaria in humans. When a female anopheline mosquito takes a blood meal, she releases sporozoites into the victim's bloodstream (1). These sporozoites enter the host's liver cells and undergo multiple fission, producing many merozoites (2); each sporozoite may produce as many as 20,000 merozoites in a single liver cell. The infected liver cells rupture, releasing the merozoites into the blood where they invade red blood cells (3). Through continued multiple fission, more merozoites are produced. The red blood cells eventually burst, releasing merozoites, which enter other red blood cells. Some merozoites differentiate to become gametocytes (4), which are picked up by mosquitoes. The female gametocyte forms a single macrogamete; the male gametocyte typically undergoes multiple fission to produce several motile, flagellated microgametes within the gut of the mosquito (5). After fertilization occurs, the zygote migrates to the mosquito's salivary glands and divides to form numerous sporozoites, thereby completing the life cycle.

and an invertebrate. The complexity and evolved sophistication of haemosporidians are revealed in the lifecycle of *Plasmodium,* the causative agent of malaria. In *Plasmodium,* the vertebrate host is a tetrapod (e.g., humans), and the invertebrate host is the *Anopheles* mosquito (Figure 3.16). Female mosquitoes drill into the network of blood-filled capillaries of the host, releasing an anticoagulant and lubricant at the same time. Sporozoites, which reside in the salivary glands of the mosquito, are released into the bloodstream where, within minutes, they migrate to the liver and enter the hepatocytes, or liver cells. A single successful sporozoite is enough to initiate the disease. Once in the liver cells, the sporozoites undergo multiple fission (schizogony) for a week or so, until the host cell has been entirely digested and is bulging with parasites, upon which the cell ruptures to release the merozoites. The merozoites immediately enter red blood cells (erythrocytes) and transform into trophozoites. In the erythrocytes, the trophozoites undergo schizogony again to form more merozoites (or, occasionally, they transform into gamonts), eventually bursting the erythrocyte. The released merozoites infect other red blood cells and the population of parasites grows rapidly.

The destruction of red blood cells by human *Plasmodium* parasites, and the release of metabolic byproducts, are responsible for the characteristic chills, fever, and anemia that are common symptoms of malaria. So many oxygen-carrying red cells are destroyed that the lungs must fight for air and the heart struggle to pump – and the blood acidifies. In some cases of falciparum malaria, infected red blood cells pass through the capillaries of the brain and the infection turns into cerebral malaria, the most feared manifestation of the disease. Sickle-cell anemia confers partial resistance to *P. falciparum.* If the trophozoite becomes a gamont, it will be morphologically distinguishable as either a macrogamont or a microgamont. The life cycle of gamonts proceeds no further unless they are taken in by a mosquito during its blood meal (and, remember, only female mosquitoes bite). Once in the gut of the mosquito, the macrogamont becomes a spherical macrogamete, while the microgamont undergoes three nuclear divisions and develops eight projections (microgametes), which each receive a nucleus. The microgametes break away and each fertilizes a single macrogamete to form a diploid zygote called an ookinete (this cell is the only diploid stage in the entire life cycle). The ookinete then actively burrows through the mosquito's stomach and secretes a covering around itself on the outside of the stomach, forming an oocyst. Inside the oocyst, the zygote undergoes a meiotic reduction division followed by schizogony to form haploid sporozoites that are released from the oocyst into the gut, to migrate to the salivary gland where they remain until the next time the insect feeds.

The ability of *Plasmodium* to develop within vertebrate red blood cells is a challenge met by very few intracellular parasites. Normally, "diseased" and senescent erythrocytes are cleared by the spleen, but in *Plasmodium*-infected erythrocytes the parasite modifies the host cell by exporting its own proteins into the cytoplasm and plasma membrane of the red cell as an immune evasion mechanism.

Phylum Ciliata: The Ciliates

There are an estimated 10,000 to 12,000 described species of ciliates. They derive their name from the Latin word for "eyelash" ("*cilia*"). Ciliates are very common in benthic and planktonic communities in marine, brackish, and freshwater habitats, as well as in damp soils. A number are also ecto- or endosymbiotic, and some are even obligate or opportunistic parasites. Members of at least one genus, *Collinia*, are marine parasitoids that infect euphausid crustaceans (krill), killing their host in order to achieve transmission and complete their life cycle. Species of *Collina* have been reported causing mass euphausid mortality events in the North Pacific. Most ciliates occur as single cells, and some get quite large (~2 mm long). Branching and linear colonies also occur in several species. The ciliates shown in Figures 3.17 and 3.1D illustrate some of the variety of cell forms within this large and complex group of protists.

Some ciliates are important mutualistic endosymbionts of ruminants such as goats, sheep, and cattle. They are found by the millions in the digestive tracts, feeding on plant matter ingested by the host and converting it into a form that can be metabolized by the ruminant. Some ciliates parasitize fish or invertebrates (e.g., American lobster), and at least one (*Balantidium coli*) is known to be an occasional endoparasite of the human digestive tract. Ciliates such as *Tetrahymena* and *Colpidium* have been used as models in cell biology, genomics, and proteomics (especially studies of cilia). Others are widely used as indicators of water quality, and some have been used to clarify water in sewage treatment plants (Box 3H). Ribozymes (catalytic RNA molecules) and the enzyme telomerase, which is involved in creating the ends of eukaryotic chromosomes, were discovered through studies of model ciliates.

BOX 3H Characteristics of the Phylum Ciliata

1. Common in virtually all aquatic and wet environments, freshwater, saltwater or soil
2. Shape of cell maintained by pellicle consisting of alveolar membrane vesicles and underlying proteinaceous layer (the epiplasm), associated with microtubules, all beneath plasma membrane. Some secrete calcium carbonate plates within their alveoli; some others secrete external skeletons (loricae)
3. With cilia for locomotion. Associated with the basal bodies (kinetosomes) of cilia are two microtubular roots and one fibrous root; these roots plus basal bodies are collectively known as infraciliature. The diagnostic rows of cilia are called kineties.
4. With extrusomes and mitochondria (with tubular cristae)
5. Without plastids
6. Carbohydrates sorted as glycogen
7. Most with buccal cavity and cytostome ("mouth"); often also a cytoproct ("anus")
8. With two distinct types of nuclei—hyperpolyploid macronuclei and a diploid micronuclei
9. Micronucleus divides by closed intranuclear orthomitosis (in most) without centrioles. Electron-dense bodies inside nucleus act as organizing centers for mitotic spindle. Macronucleus divides amitotically by simple constriction.
10. Asexual reproduction by transverse binary fission (homothetogenic)
11. Sexual reproduction by conjugation: pair of ciliates bond together and exchange micronuclei through a cytoplasmic connection at point of joining

Support and Locomotion

The fixed cell shape of ciliates is maintained by the alveolar membrane system and an underlying proteinaceous layer called the **epiplasm**, or cortex (Figure 3.18). A few types (e.g., tintinnids) secrete external skeletons, or loricae, which have been documented in the fossil record as early as the Ordovician (500 Ma). Another common group (*Coleps* and its relatives) has calcium carbonate plates in their alveoli.

The cilia are in rows called **kineties** (sing. kinety), and the different patterns these rows create are used as taxonomic characters for identification and classification. Associated with the basal body (or centriole), which is called a **kinetosome** in ciliates, are three important cytoskeletal structures—two microtubular roots, the postciliary microtubules and the transverse

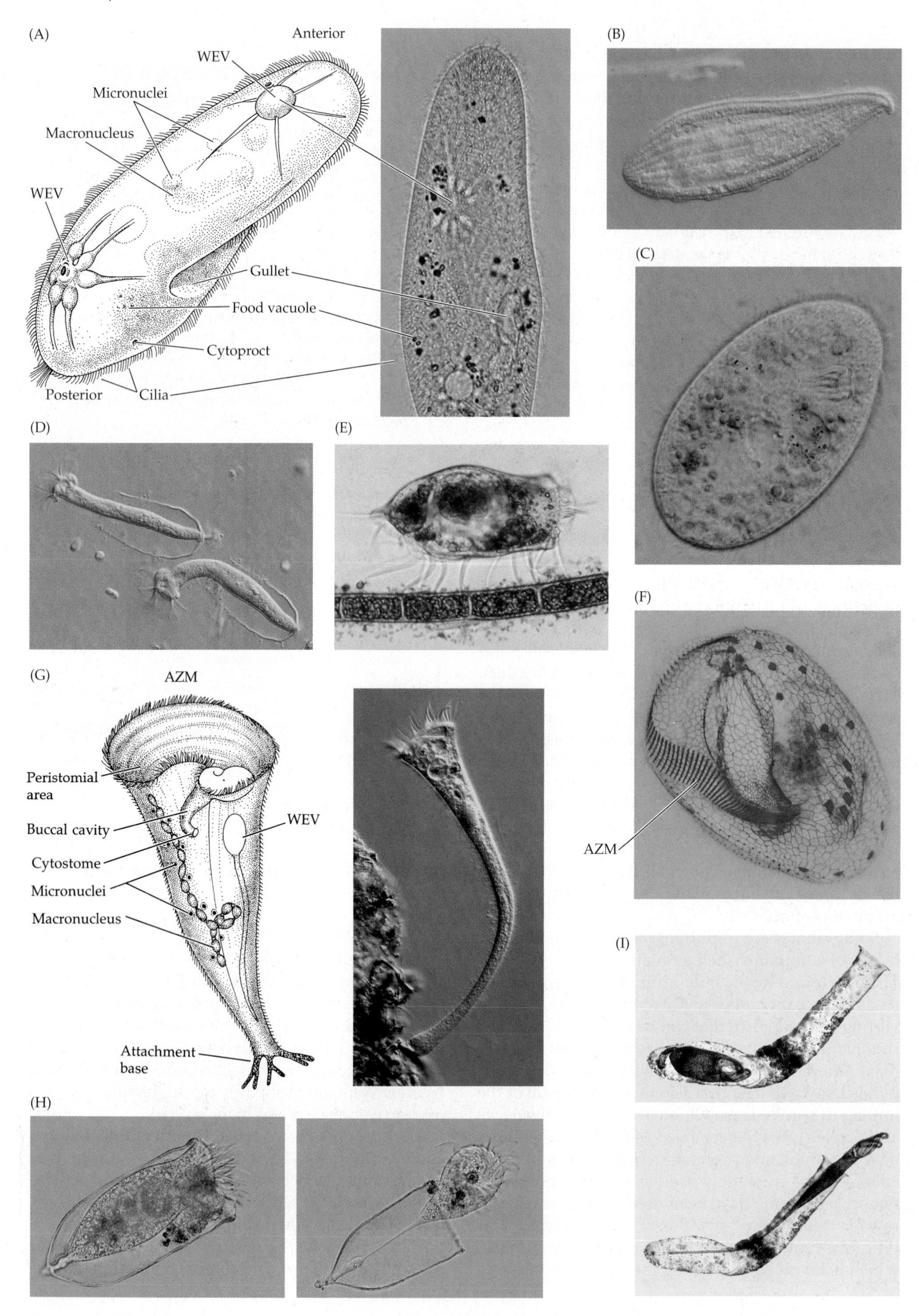
(A)
Anterior
WEV
Micronuclei
Macronucleus
WEV
Gullet
Food vacuole
Cytoproct
Posterior
Cilia
(B)
(C)
(D)
(E)
(F)
AZM
(G)
AZM
Peristomial area
Buccal cavity
WEV
Cytostome
Micronuclei
Macronucleus
Attachment base
(H)
(I)

◀ **Figure 3.17 Phylum Ciliata. Representative ciliates.** (A) *Paramecium* (a peniculine); illustration and photomicrograph. (B) *Loxophyllum* (a haptorid). (C) *Nassula* (a nassulid). (D) *Vaginacola* (a loricated peritrich). (E) *Euplotes* (a hypotrich), walking on an algal filament. (F) *Euplotes*; note the AZM. (G) *Stentor* (a heterotrich), illustration and photomicrograph. (H) *Flavella*, retracted and extended from its lorica. (I) *Folliculina*, retracted and extended from its lorica. Species of Folliculinidae often attach their bottle-shaped lorica to various invertebrates or seaweeds. The lorica can reach several millimeters in length and these ciliates are easily mistaken for small animals.

microtubules, and one fibrous root, the **kinetodesmal fiber**. These roots, together called the **infraciliature**, anchor the cilium to its neighbors and provide additional support for the cell surface (Figure 3.19). The kinetosomes of ciliates are homologous to the centrioles of other eukaryotic cells, being a tubular cylinder of nine triplets of microtubules.

The cilia can be grouped into two functional and two structural categories. Cilia associated with the cytostome and the surrounding feeding area comprise the **oral ciliature**, whereas cilia of the general body surface are the **somatic ciliature**. The cilia in both of these categories may be single (**simple cilia**), or the kinetosomes may be grouped close together to form **compound ciliature** (e.g., cirri, membranelles). Ciliatologists have developed a detailed and complicated terminology for talking about their favorite creatures. This special language reaches almost overwhelming proportions in matters of ciliature, and we present here only a necessary minimum of new words in order to adequately describe these organisms. Beyond this, we offer the reference list at the end of this chapter, especially the extensive and illustrated glossary in J. O. Corliss' *The Ciliated Protozoa* (1979).

The cilia are also, of course, the locomotory organelles of ciliate protists. Their structural similarities and homology to flagella are well known; but ciliates do not move like protists with flagella. The differences are due in large part to the facts that cilia are much more numerous and densely distributed than flagella, and the patterns of ciliation on the body are extremely varied and thus allow a range of diverse locomotor behaviors not possible with just one or a few flagella. Cilia can also play a role in feeding, sensation, and even attachment.

As discussed in Chapter 4, each individual cilium undergoes an effective (power) stroke as it beats. The cilium does not move on a single plane, but describes a distorted cone as it beats (Figure 3.19A,B). The beating of a ciliary field occurs in metachronal waves that pass over the body surface (Figure 3.19C). The coordination of these waves is apparently due largely to hydrodynamic effects generated as each cilium moves. Microdisturbances created in the water by the action of one cilium stimulate movement in the neighboring cilium, and so on over the cell surface. In ciliates, the tip of the cilium usually follows a counterclockwise path during the recovery stroke, resulting in a spiraling motion as the organism swims (Figure 3.19G). Many ciliates (e.g., *Didinium*, *Paramecium*) can vary the direction of ciliary beating and metachronal waves. In such forms, complete reversal of the body's direction of movement is possible by simply reversing the ciliary beat and wave directions.

Perhaps more than any other protist group, the ciliates have been studied for their complex locomotory behavior. *Paramecium*, a popular laboratory protist, has received most of the attention of protozoological behaviorists. When a swimming *Paramecium* encounters a mechanical or chemical environmental stimulus of sufficient intensity, it begins a series of rather intricate response activities. The protist first initiates a reversal of movement, effectively backing away from the source of the stimulus. Then, while the posterior end of the body remains more or less stationary, the anterior end swings around in a circle. This action is appropriately called the cone-swinging phase. The *Paramecium* then proceeds forward again, usually along a new pathway.

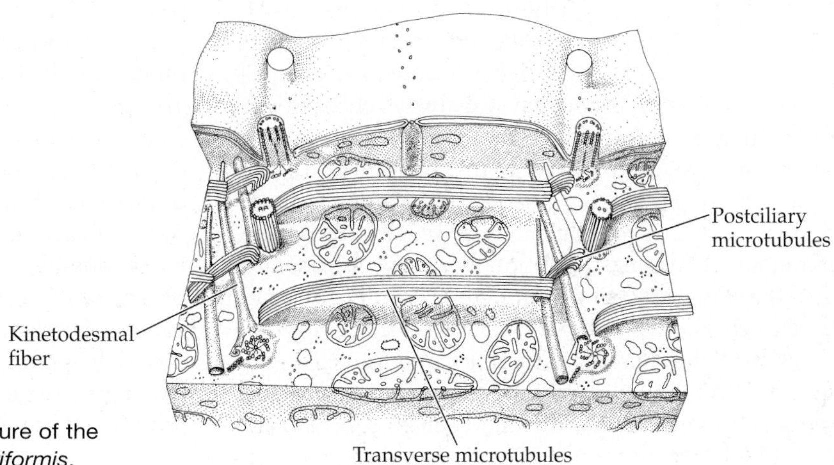

Figure 3.18 Phylum Ciliata. Fine structure of the epiplasma, or cortex of *Tetrahymena pyriformis*.

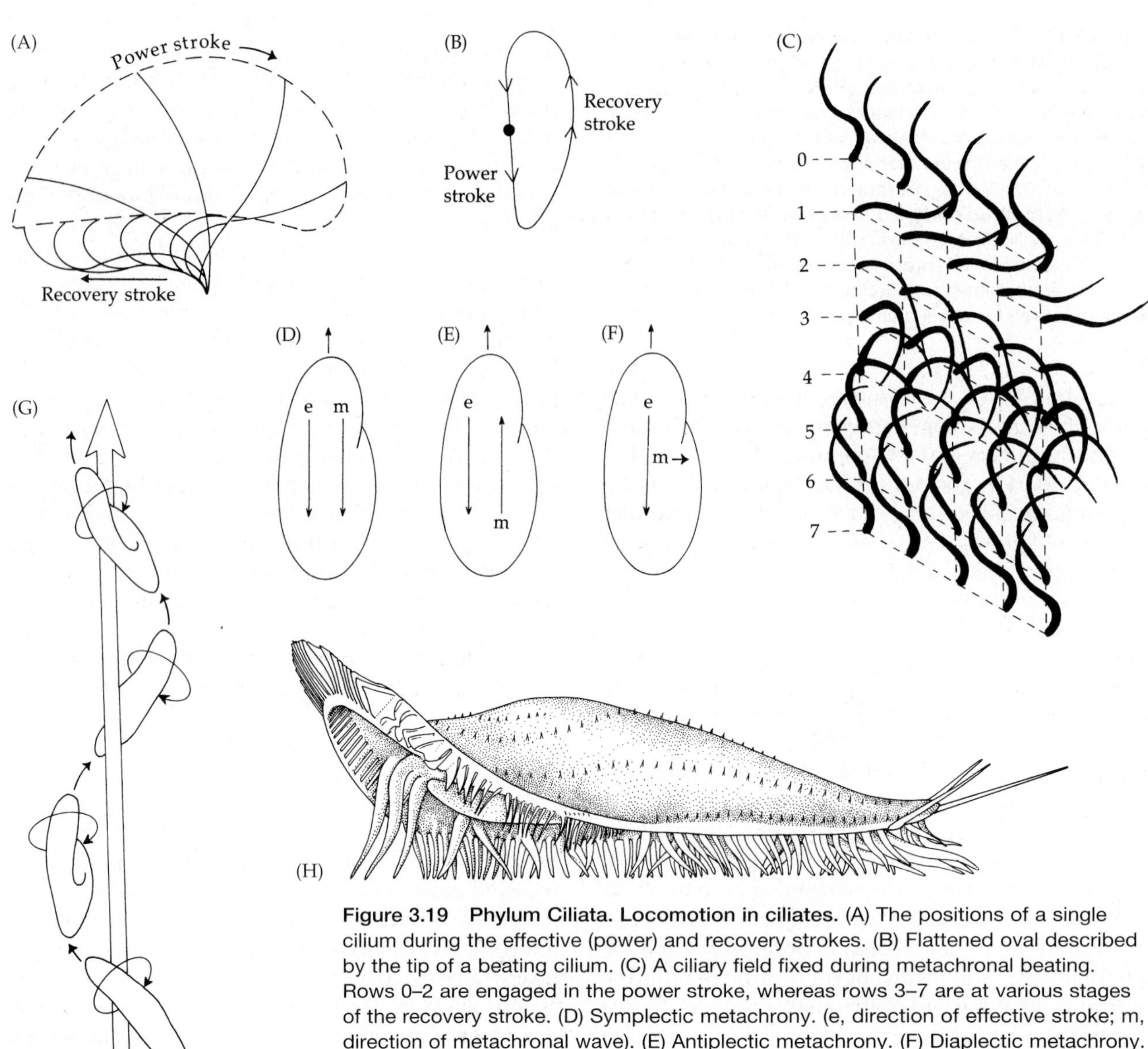

Figure 3.19 Phylum Ciliata. Locomotion in ciliates. (A) The positions of a single cilium during the effective (power) and recovery strokes. (B) Flattened oval described by the tip of a beating cilium. (C) A ciliary field fixed during metachronal beating. Rows 0–2 are engaged in the power stroke, whereas rows 3–7 are at various stages of the recovery stroke. (D) Symplectic metachrony. (e, direction of effective stroke; m, direction of metachronal wave). (E) Antiplectic metachrony. (F) Diaplectic metachrony. (G) Helical pattern of forward movement of *Paramecium*. (H) *Stylonychia* uses cirri for "walking."

Much of the literature refers to this behavior in terms of simple "trial and error," but the situation is not so easily explained. The response pattern is not constant, because the cone-swinging phase may not always occur; sometimes the cell simply changes direction in one movement and swims forward again. Furthermore, the cone-swinging phase occurs even in the absence of recognizable stimuli and thus may be regarded as a phenomenon of "normal" locomotion. Recent studies suggest that *Paramecium* swimming behavior is governed by the cell's membrane potential. When the membrane is "at rest," the cilia beat posteriorly and the cell swims forward. When the membrane becomes depolarized, the cilia beat in a reverse direction (ciliary reversal) and the cell backs up. In *Paramecium* genetic mutations are known to result in abnormal behavior characterized by prolonged periods of continuous ciliary reversal (the so-called "paranoiac *Paramecium*").

An interesting form of locomotion in ciliates is exhibited by the hypotrich ciliates (e.g., *Euplotes*). In this group, the somatic cilia are arranged in bundles called cirri, which they use to "crawl" or "walk" over surfaces (Figure 3.19H and 3.17E). Sessile ciliates are also capable of movement in response to stimuli. The attachment stalk of many peritrichs (e.g., *Vorticella*) contains contractile **myonemes**, fibrillar contractile elements in the cytoplasm, which serve to pull the cell body against the substratum. Similar myonemes are found in the cell walls of other ciliates (e.g., *Stentor*), and they are capable of contracting and extending the entire cell. Other ciliates (e.g., *Lacrymaria*) use sliding microtubules to contract.

Figure 3.20 Phylum Ciliata. Formation of and digestion within a food vacuole in *Paramecium caudatum*. The sequence of digestive events may be followed by staining yeast cells with Congo red dye and allowing the stained cells to be ingested by the protist. The changes in color from red to red-orange to blue-green reflects the change to an acid condition within the food vacuole and thus the initial stage of the digestive process. The change back to red-orange occurs as the vacuole subsequently becomes more alkaline. The pattern of movement of the food vacuole (arrows) is typical of this organism or cell and is often termed cyclosis.

Nutrition

The ciliates include many different feeding types. Some are filter feeders, others capture and ingest other protists or small invertebrates, many eat algal filaments or diatoms, some graze on attached bacteria, and a few are saprophytic parasites. In almost all ciliates, feeding is restricted to a specialized oral area containing the cytostome, or "cell mouth." Food vacuoles are formed at the cytostome and then are circulated through the cytoplasm as digestion occurs (Figure 3.20). Because of the different ciliate feeding types, however, there are a variety of structures associated with, and modifications of, the cytostome.

Holozoic ciliates that ingest relatively large food items usually possess a nonciliated tube, called the cytopharynx, which extends from the cytostome deep into the cytoplasm. The walls of the cytopharynx are often reinforced with stiff rods called **nematodesmata**, composed of microtubules. In a few forms, most notably *Didinium*, the cytopharynx is normally everted to form a projection that sticks to prey and then inverts back into the cell, thus pulling the prey into a food vacuole. In this way, *Didinium* can engulf its relatively gigantic prey, *Paramecium* and other ciliates (Figure 3.21A). Other ciliates, such as the hypostomes, have complex nematodesmal baskets in which microtubules work together to draw filaments of algae into the cytostome, reminiscent of the way a human sucks up a piece of spaghetti (Figure 3.21B). In most of these ciliates, the cilia around the mouth are relatively simple.

(A)

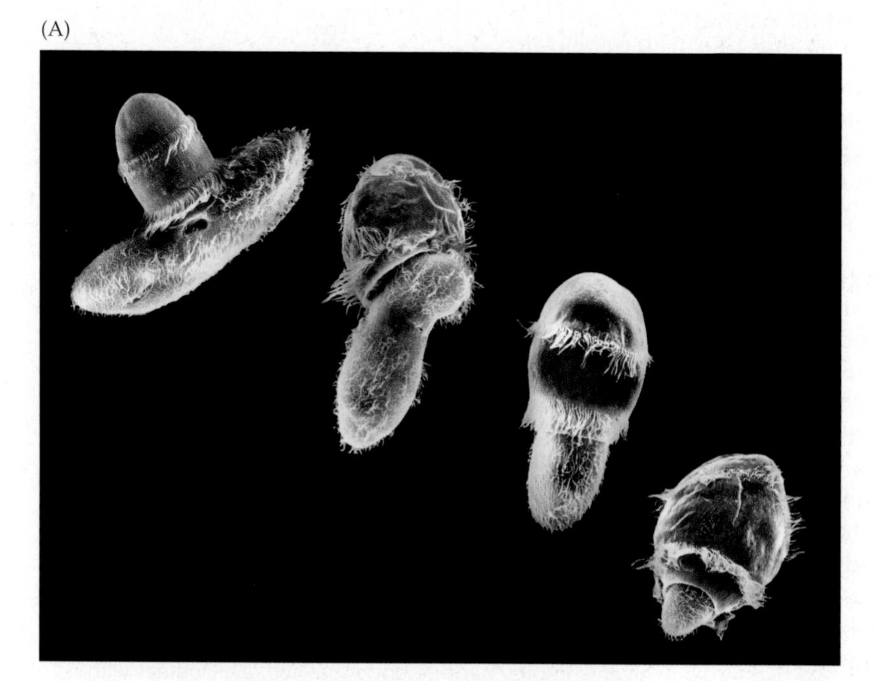

(B)

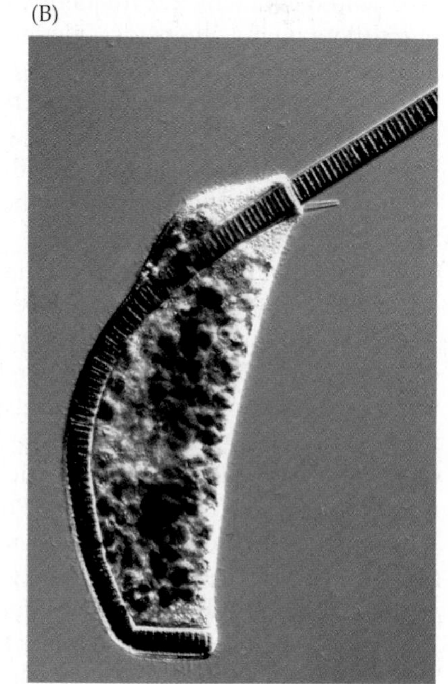

Figure 3.21 Phylum Ciliata. Holozoic feeding in ciliates. (A) The predatory ciliate *Didinium nasutum*, attacking and ingesting a *Paramecium* (composite of SEM photographs). (B) *Nassulopsis* ingesting blue-green algae.

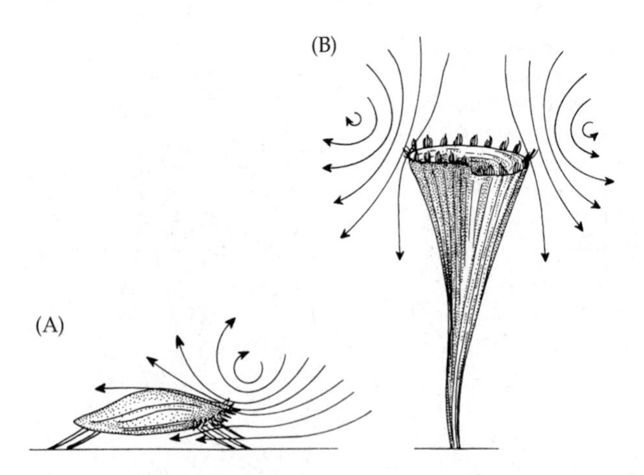

Figure 3.22 Phylum Ciliata. Feeding currents produced by two ciliates. (A) *Euplotes*. (B) *Stentor*. The ciliary currents bring suspended food to the cell, where it can be ingested.

Other ciliates, including many of the more familiar forms (e.g., *Stentor*) are suspension feeders (Figure 3.22). These often lack or have reduced cytopharynxes. Instead, they have elaborate specialized oral cilia for creating water currents, and filtering structures or scraping devices. Their cytostomes often sit in a depression on the cell surface, the **buccal cavity**. The size of the food eaten by such ciliates depends on the nature of the feeding current and, when present, the size of the depression. The oral ciliature often consists of compound ciliary organelles, called the **adoral zone of membranelles**, or simply the **AZM** (see Figure 3.17E,F) on one side of the cytostome, and a row of closely situated paired cilia which is frequently called the **paroral membrane** on the other side. Ciliates that feed like this include such common genera as *Euplotes*, *Stentor*, and *Vorticella*. Many hypotrichs (e.g., *Euplotes*) that move about the substratum with their oral region oriented ventrally use their specialized oral ciliature to swirl settled material into suspension and then into the buccal cavity for ingestion.

Among the most specialized ciliate feeding methods are those used by the suctorians, which lack cilia as adults and instead have knobbed feeding tentacles (Figure 3.23). A few suctorians have two types of tentacles, one form for food capture and another for ingestion. The swellings at the tips of the tentacles contain extrusomes called **haptocysts**, which are discharged upon contact with a potential prey. Portions of the haptocyst penetrate the victim and hold it to the tentacle. Sometimes prey are actually paralyzed after contact with haptocysts, presumably by enzymes released during discharge. Following attachment to the prey, a temporary tube forms within the tentacle, and the contents of the prey are sucked into the tentacle and incorporated into food vacuoles (Figure 3.23B–D).

In addition to haptocysts, several other types of extrusomes are present in ciliates. Some predatory ciliates have tubular extrusomes, called **toxicysts**, in the oral region of the cell (Figure 3.24A). During prey capture, the toxicysts are extruded and release their contents, which apparently include both paralytic and digestive enzymes. Active prey are first immobilized and then partially digested by the discharged chemicals; this partially digested food is later taken into food vacuoles. Some ciliates have extrusomes called **mucocysts** located just beneath the pellicle (Figure 3.24B). Mucocysts discharge mucus onto the surface of the cell as a protective coating; they may also play a role in cyst formation. Others have **trichocysts**, which contain nail-shaped structures that can be discharged through the pellicle. Most specialists suggest that these structures

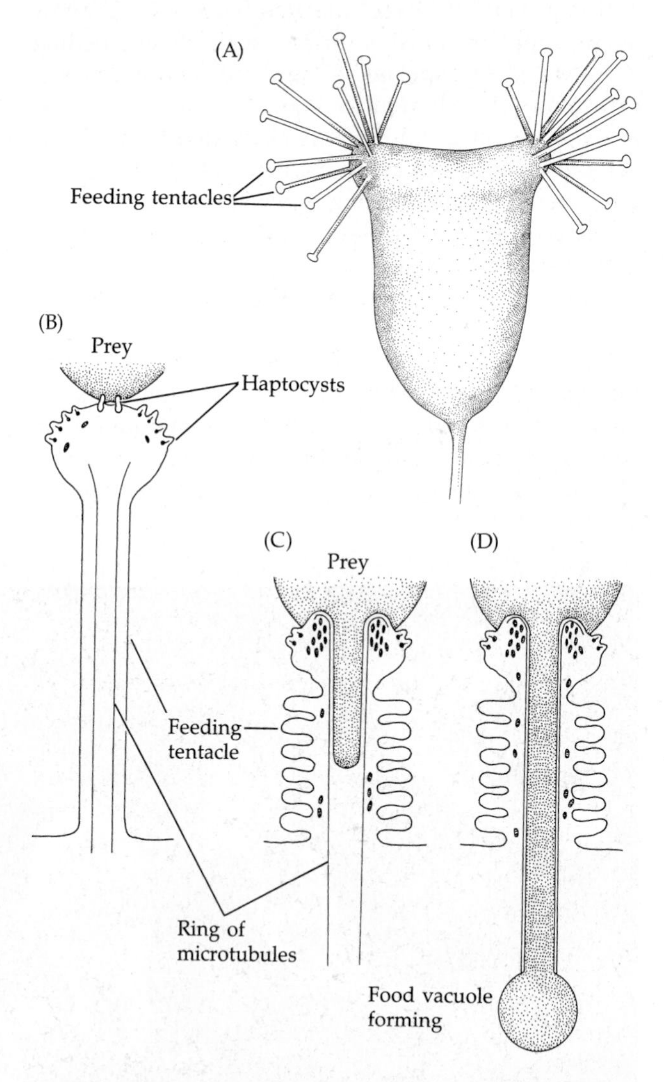

Figure 3.23 Phylum Ciliata. Feeding in the suctorian ciliate Acineta. (A) *Acineta* has capitate feeding tentacles; note the absence of cilia. (B–D) Schematic drawings of enlarged feeding tentacles, showing the sequence of events in prey capture and ingestion. (B) Contact with prey and firing of haptocysts into prey. (C) Shortening of tentacle and formation of a temporary feeding duct within a ring of microtubules. (D) Drawing of contents of prey into duct and formation of food vacuole.

(A)
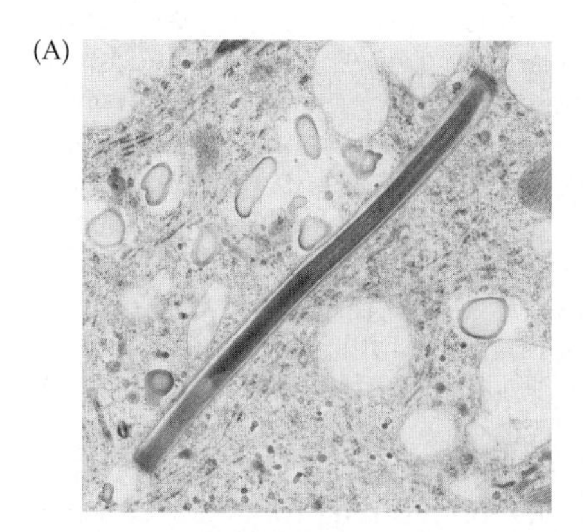
(B)
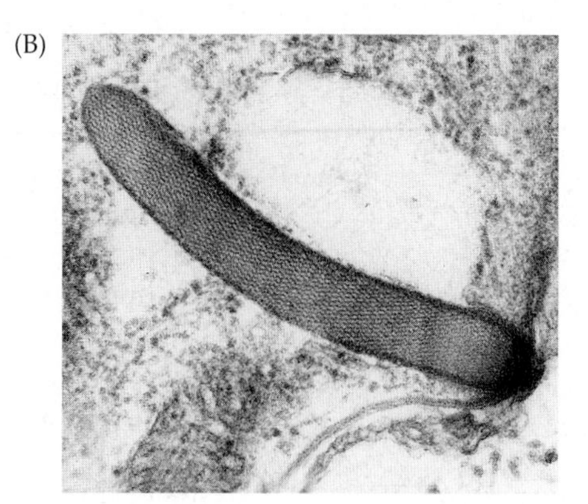
(C)
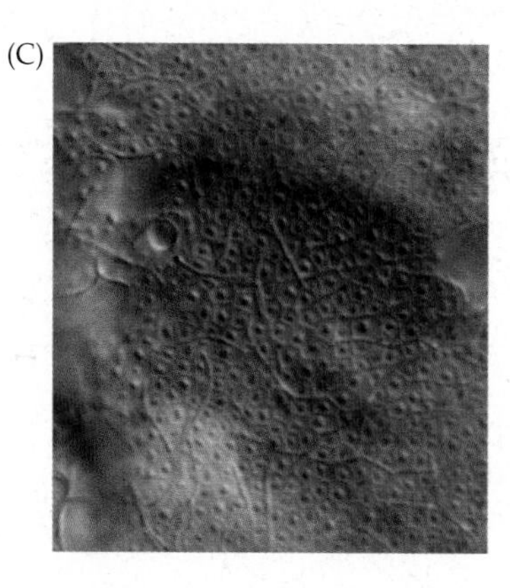

Figure 3.24 Phylum Ciliata. Extrusomes in ciliates. (A) Toxicyst (longitudinal section) from *Helicoprorodon*. (B) Mucocyst (longitudinal section) from *Colpidium*. (C) The pellicle of *Nassulopsis elegans*, showing mucocysts (raised dots) just below the surface.

are not used in prey capture, but serve a defensive function.

A number of ciliates are ecto- or endosymbionts associated with a variety of vertebrate and invertebrate hosts. In some cases these symbionts depend entirely upon their hosts for food. Some suctorians, for example, are true parasites, occasionally even living within the cytoplasm of other ciliates. A number of hypostome ciliates are ectoparasites on freshwater fishes and may cause significant damage to their hosts' gills. *Balantidium coli*, a large vestibuliferan ciliate, is common in pigs and occasionally is acquired by humans, where it can cause intestinal lesions. The rumen of ungulates contains whole communities of ciliates, including species that break down the grasses eaten by the host, bactivorous species, and even predators preying on the other ciliates. Members of the order Chonotrichida are mostly ectosymbiotic on crustaceans (and occasionally on whales). Chonotrichs are sessile, attaching to their hosts by a stalk produced from a special adhesive organelle. Other ciliates are symbiotic on a variety of hosts, including bivalve and cephalopod molluscs, polychaete worms, and perhaps mites.

Although plastids have not been definitely demonstrated in ciliates, several independent lineages in this group harbor photosynthetic symbionts that are intermittently replenished by feeding, especially planktonic forms (e.g., *Laboea*, *Mytridium*, *Strombidium*) that sequester photosynthetically functional plastids from ingested algae. The chloroplasts lie free in the cytoplasm, beneath the pellicle, where they actively contribute to the ciliate's carbon budget. This unusual practice has also been documented in foraminiferans. The cellular mechanisms by which the prey chloroplasts are removed, sequestered, and maintained are not known. Other ciliates maintain entire algal cells as endosymbionts. During the summer, in some areas, "photosynthetic ciliates" of one sort or the other can comprise a majority of the freshwater planktonic ciliate fauna. The common freshwater ciliate *Paramecium bursaria* typically harbors hundreds of symbiotic green algae (*Chlorella* sp.) in its cytoplasm, in a mutualistic relationship. The algal symbionts can be experimentally removed from the host, in which case growth rates decline (in the *Paramecium*). The host can regain its symbionts simply by ingesting them from the environment and incorporating them into its cytoplasm alive and well.

Reproduction

Ciliates are unique among the Eukaryota in possessing two different kinds of nuclei in each cell. The larger type—the **macronucleus**—controls the general operation of the cell. The macronucleus is usually hyperpolyploid (containing many sets of chromosomes) and may be compact, ribbon-like, beaded, or branched. The smaller type—the **micronucleus**—has a reproductive function, synthesizing the DNA associated with reproduction. It is usually diploid. In a process that is utterly unique among eukaryotes, the macronucleus is actually generated from the micronucleus by amplification of the genome (along with some heavy genetic editing). Division of the macronucleus occurs by "amitosis," the segregation of the chromosomes, by a process whose mechanism is not yet well known. This process is not perfect, and after about 200 generations the cell shows signs of aging. Thus periodically the macronuclei must be regenerated and this is accomplished during the process of conjugation (see below).

Asexual reproduction in ciliates is usually by binary fission, although multiple fission and budding are also known (Figure 3.25). Binary fission in ciliates is usually transverse. The micronucleus is the reservoir of genetic material in ciliates. As such, each micronucleus within the cell (even when there are many) forms an internal mitotic spindle during fission, thus distributing daughter micronuclei equally to the progeny of division. Macronuclear division is highly variable, although the nuclear envelope never seems to break down. The large, sometimes multiple, macronuclei usually condense into a single macronucleus which divides by constriction. Some macronuclei have internal

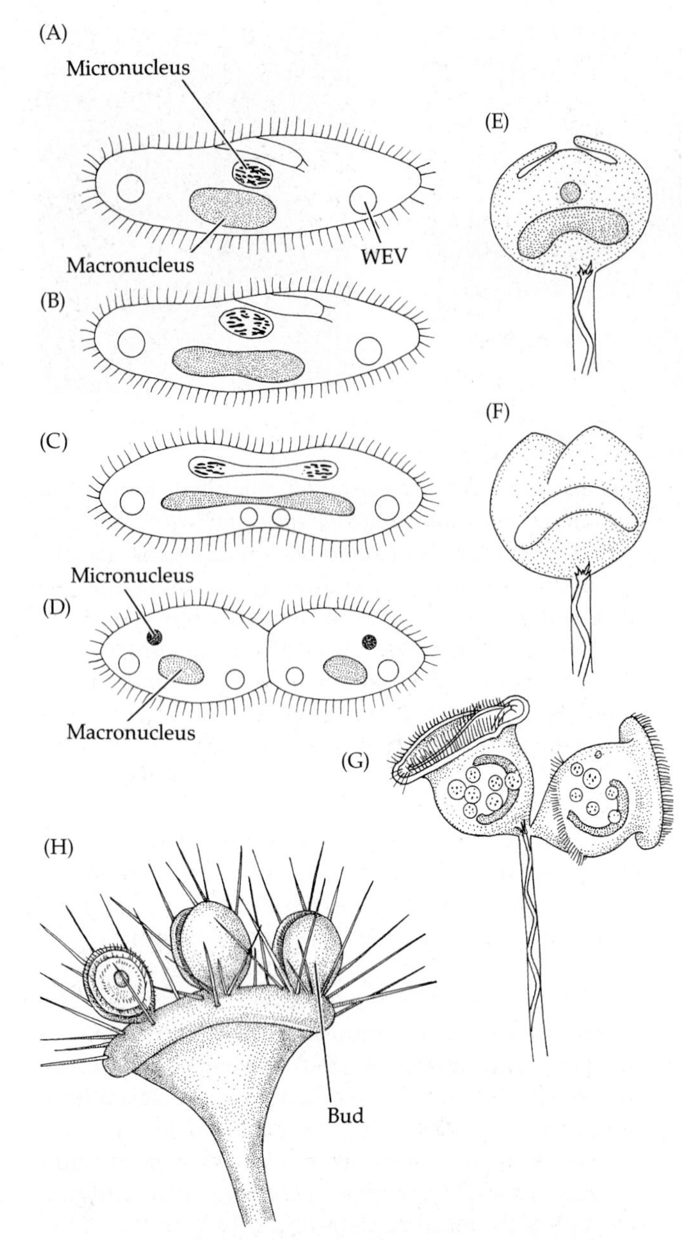

Figure 3.25 Phylum Ciliata. Asexual reproduction in ciliates. (A–D) Transverse binary fission in *Paramecium*; the micronucleus divides mitotically, whereas the macronucleus simply splits. (E–G) Binary fission in *Vorticella*. (H) Budding in the suctorian *Ephelota gigantea*.

microtubules that appear to push daughter nuclei apart, but there is never a clear, well-organized spindle. Since many ciliates are anatomically complex and frequently bear structures that are not centrally or symmetrically placed on the body (especially structures associated with the cytostome), a significant amount of reconstruction must occur following fission. Such reformation of parts or special ciliary fields does not take place haphazardly; it apparently is controlled, at least in part, by the macronucleus.

Binary fission is typical of the colonial and solitary peritrichs. In colonial species, the division is equal, with both daughter cells remaining attached to the growing colony, but in solitary species divisions may be unequal and may involve a swimming phase. A type of unequal division frequently referred to as budding occurs in a variety of sessile ciliates, including chonotrichs and suctorians. In these cases the ciliated bud is released as a so-called swarmer that swims about before adopting the adult morphology and lifestyle. In some cases, several buds are formed and released simultaneously.

True multiple fission is known in a few groups of ciliates and typically follows the production of a cyst by the prospective parent. Repeated divisions within the cyst produce numerous offspring, which are eventually released with breakdown of the cyst coating.

Sexual reproduction (or more precisely, genetic recombination) by ciliates is usually by conjugation, less commonly by autogamy. Conjugation is perhaps most easily understood by first describing it in *Paramecium*. As with any sexual process, the functional advantage of the activity is genetic mixing or recombination, and it is accomplished during conjugation by an exchange of micronuclear material. The following account (Figure 3.26) is of *Paramecium caudatum*—details vary in other species of the genus.

As paramecia move about and encounter one another, they recognize compatible "mates" (i.e., members of another clone). After making contact at their anterior ends, the "mates"—called **conjugants**—orient themselves side by side and attach to each other at their oral areas. In each conjugant, the micronucleus undergoes two divisions that are equivalent to meiosis and reduce the chromosome number to the haploid condition. Three of the daughter micronuclei in each conjugant disintegrate and are incorporated into the cytoplasm; the remaining haploid micronucleus in each cell divides once more by mitosis. The products of this postmeiotic micronuclear division are called **gametic nuclei**. One gametic nucleus in each conjugant remains in its "parent" conjugant while the other is transferred to the other conjugant via a cytoplasmic connection formed at the point of joining. Thus each conjugant sends a haploid micronucleus to the other, thereby accomplishing the exchange of genetic material. Each migratory gametic nucleus then fuses with the stationary micronucleus of the recipient, producing a diploid nucleus, or synkaryon, in each conjugant. This process is analogous to mutual cross-fertilization in metazoan invertebrates (e.g., land slugs).

Following nuclear exchange, the cells separate from each other and are now called **exconjugants**. The process is far from complete, however, for the new genetic combination must be incorporated into the macronucleus if it is to influence the organism's phenotype. This is accomplished as follows. The macronucleus of each exconjugant has disintegrated during the meiotic and exchange processes. The newly formed diploid

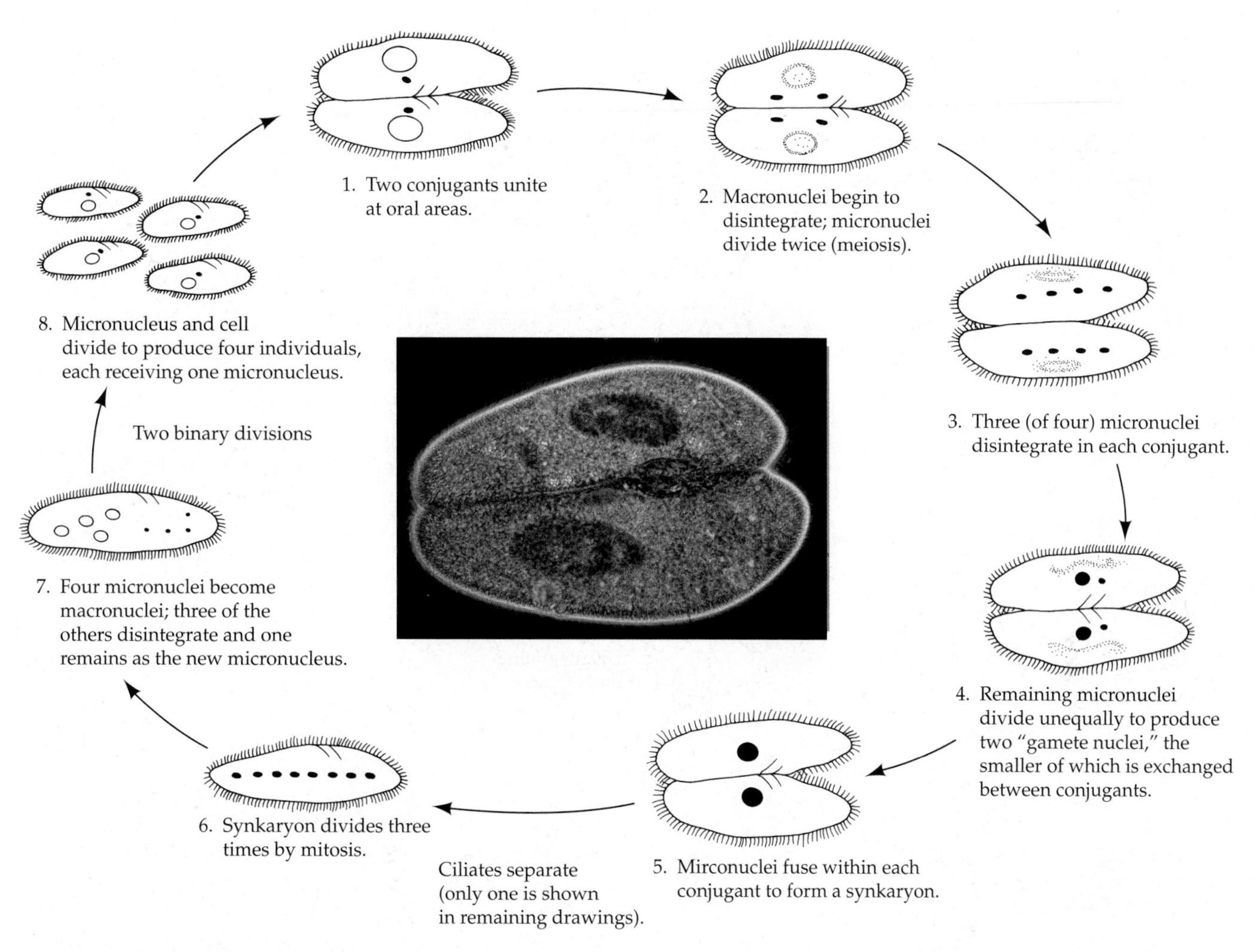

Figure 3.26 Phylum Ciliata. The photograph shows a pair of conjugating *Paramecium caudatum*.

synkaryon divides mitotically three times, producing eight small nuclei (all, remember, containing the combined genetic information from the two original conjugants). Four of the eight nuclei then enlarge to become macronuclei. Three of the remaining four small nuclei break down and are absorbed into the cytoplasm. Then the single remaining micronucleus divides twice mitotically as the entire organism undergoes two binary fissions to produce four daughter cells, each of which receives one of the four macronuclei and one micronucleus. Thus, the ultimate product of conjugation and the subsequent fissions is four new diploid daughter organisms from each original conjugant.

Variations on the sequence of events described above for *Paramecium* include differences in the number of divisions, which seem to be determined in part by the normal number of micronuclei present in the cell. Even when two or more micronuclei are present, they typically all undergo meiotic divisions. All but one disintegrate, however, and the remaining micronucleus divides again to produce the stationary and migratory gametic nuclei.

In most ciliates the members of the conjugating pair are indistinguishable from each other in terms of size, shape, and other morphological details. However, some species, especially in the Peritrichida, display distinct and predictable differences between the two conjugants, particularly in size. In such cases, the members of the mating pair are called the **microconjugant** and the **macroconjugant** (Figure 3.27). The formation of the microconjugant generally involves one or a series of unequal divisions, which may occur in a variety of ways. The critical difference between conjugation of similar mates and that of dissimilar mates is that in the latter case, there is often a one-way transfer of genetic material. The microconjugant alone contributes a haploid micronucleus to the macroconjugant; thus only the larger individual is "fertilized." Following this activity, the entire microconjugant usually is absorbed into the cytoplasm of the macroconjugant (Figure 3.27E). A similar process occurs in most chonotrichs. One conjugant appears to be swallowed into the cytostome of another, after which nuclear fusion and reorganization take place. There are several other modifications on this complex sexual process in ciliates, but all have the same fundamental result of introducing genetic variation into the populations.

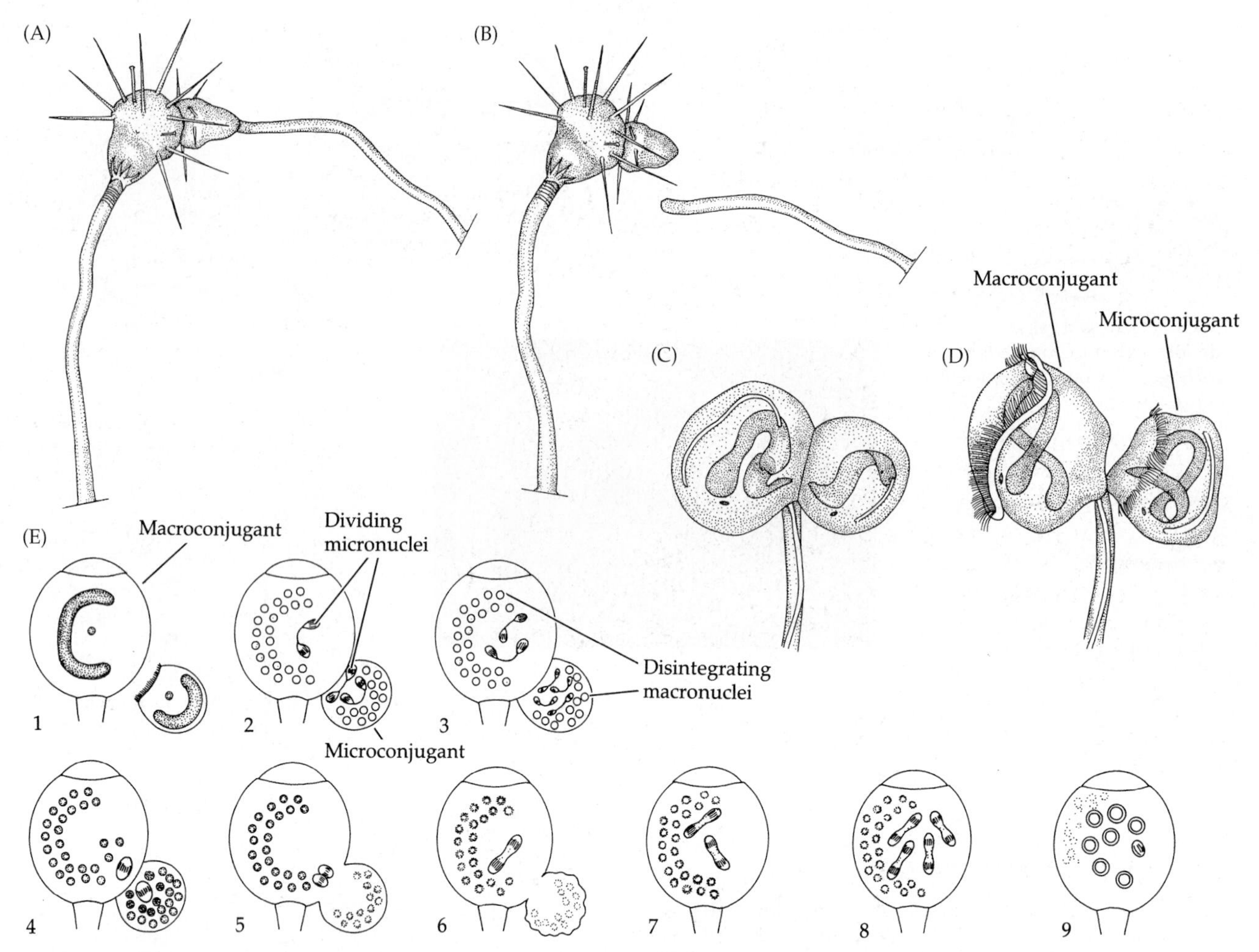

Figure 3.27 Phylum Ciliata. Sexual processes in ciliates. (A,B) *Ephelota gemmipara* (a suctorian); two mating partners of unequal sizes are attached to each other, apparently following chemical recognition. Both have undergone nuclear meiosis. The smaller mate detaches from its stalk and is absorbed by the larger one, then the gametic nuclei fuse. Subsequent nuclear divisions produce the multimicronuclear and macronuclear components of the normal individual. (C,D) Unequal divisions of *Vorticella campanula* result in macro- and microconjugants; conjugation follows. (E) Schematic diagrams of sexual activities in certain peritrichs. Unequal divisions result in macro- and microgamonts; the latter detach from their stalks and become free-swimming organisms; eventually the free-swimming microgamont attaches itself to a sessile macrogamont (1–2). The macronuclei begin to disintegrate (2) and ultimately disappear (9). The micronucleus of the macrogamont divides twice (2–3) and the micronucleus of the microgamont divides three times (2–3). All but one of the micronuclei in each gamont disintegrate, and the remaining micronucleus of the microgamont moves to fuse with the micronucleus of the macrogamont (4–5). As the zygotic nucleus (synkaryon) begins to divide, the microgamont is absorbed into the cytoplasm of the macrogamont. The synkaryon divides three times (6–8); one of the daughter nuclei becomes the micronucleus and the others eventually form the new macronucleus (9). It should be noted that the sequence of nuclear activities and numbers of divisions vary among different peritrichs.

One other aspect of conjugation that deserves mention is that of mating types. Individuals of the same genetic mating type (e.g., members of a clone produced by binary fission) cannot successfully conjugate with one another. In other words, conjugation is not a random event but can occur only between members of different mating types, or clones. This restriction presumably ensures good genetic mixing among individuals.

The second basic sexual process in ciliates is autogamy. Among ciliates in which it occurs (e.g., certain species of *Euplotes* and *Paramecium*), the nuclear phenomena are similar if not identical to those occurring in conjugation. However, only a single individual is involved. When the point is reached at which the cell contains two haploid micronuclei, these two nuclei fuse with one another, rather than one being transferred to a mate. Autogamy is known in relatively few ciliates, although it may actually be much more common than demonstrated thus far. Its significance in terms of genetic variation is not clear.

Phylum Stramenopila: Diatoms, Brown Algae, Golden Algae, Slime Nets, Oomycetes, etc.

The stramenopiles comprise a highly diverse taxon of some 9,000 species including diatoms, brown algae (Phaeophyta), golden algae (Chrysophytes, e.g., *Dinobryon*), some heterotrophic and parasitic flagellates, labyrinthulids (slime nets), oomycetes and hyphochytridiomycetes (the latter two groups formerly classified as fungi) (see Figures 3.28 and 3.1C,H,I) (Box 3I). Each of these groups has been placed in their own phylum in the past, but we herein follow the current convention of classifying them together in the phylum Stramenopila. A few stramenopiles, including brown seaweeds (Phaeophyta) such as the kelps are complex, differentiated organisms, and the Phaeophyta in particular may reach extremely large sizes. Brown algae are always multicellular, marine, photoautotrophs, and are predominantly marine. The largest brown algae, the kelps, may reach a hundred feet in length and resemble true plants. However, they have no vascular tissues (every cell fending for itself), the blades are not homologous to plant leaves, they do not undergo gastrulation nor produce tissues by way of embryological layering. The diversity of form in Stramenopila is staggering, and at first glance it may be difficult to imagine diatoms, kelp, and oomycete "fungi" being closely related. However, their relatedness is hypothesized on the basis that almost all have unique, complex, three-part tubular hairs on the flagella at some stage in the life cycle. The name Stramenopila (Latin *stamen*, "straw"; *pilus*, "hair") refers to the appearance of these hairs (Figure 3.29). Molecular phylogenetics also support this clade. While it might be tempting to think of some species in this clade (e.g., giant kelps) as multicellular in the sense of green plants, one key difference is that higher plants derive their adult tissues through a process of embryological tissue formation, similar though convergent to that seen in the Metazoa. Protists, including kelps, never have an embryogenic tissue-layering phase in their life history.

Stramenopiles are found in a variety of habitats, and most are very small (< 10 microns) biflagellated bacteriovores. Freshwater and marine plankton are rich in diatoms and chrysophytes, and these can also occur in moist soils, sea ice, snow, and glaciers. They have even been found alive in clouds in the atmosphere! Heterotrophic free-living stramenopiles are also found in marine, estuarine, and freshwater habitats. A few are symbiotic on other algae in marine or estuarine environments. Many produce calcite or silicon scales, shells, cysts, or tests, which are preserved in the fossil record. The oldest of these fossils are from the Cambrian/Precambrian boundary, about 550 million years ago.

BOX 3I Characteristics of the Phylum Stramenopila

1. A highly diverse group found in virtually every habitat on Earth.
2. Cell surrounded by plasma membrane, which may be supported by silica (silicon dioxide), calcium carbonate, or proteinaceous shells, scales, or tests
3. Almost all species with unique, complex, three-part tubular hairs on the flagella at some stage in life cycle. Most exhibit heterokont flagellation (i.e., with two flagella, one directed anteriorly, one directed posteriorly). Sometimes only the reproductive cells are flagellated and the trophic cells lack any obvious mode of locomotion. Opalinids are thought to have lost their flagellation and replaced it with rows of cilia (unlike those of the Ciliata).
4. Most are heterotrophic, but others are photosynthetic, saprophytic, or parasitic.
5. Photosynthetic forms have chlorophylls *a*, c_1, and c_2. Thylakoids in stacks of three; four membranes surrounding the chloroplast with the outermost membrane continuing around nucleus. Yellow and brown xanthophylls give them a brownish-green color that has earned them the common name "golden algae."
6. Food reserves stored as liquid polysaccharide (usually laminarin) or oils
7. Mitochondria with short tubular cristae
8. With single vesicular nucleus
9. Nuclear division occurs by open pleuromitosis without centrioles.
10. Asexual reproduction by binary fission
11. Sexual reproduction usually gametic; gametes usually isogametous (not the case for many diatoms and brown algae)

Diatoms (e.g., *Chaetoceros*, *Thalassiosira*) are key components of marine ecosystems and, along with dinoflagellates and coccolithophores, contribute greatly to oceanic primary productivity. In today's ocean, diatoms might be responsible for up to 40% of the net primary production (and upward of 50% of the organic carbon exported to oceanic ecosystems). When conditions are right, oceanic diatoms can multiply at astonishing rates, creating "blooms" that are sometimes toxic. It has been estimated that diatoms produce 19 billion tons of organic carbon annually (giving them a key roll in processing CO_2 into solid matter to mitigate climate warming). Recent estimates suggest diatoms alone account for up to 20% of global carbon fixation. Some species of diatoms can form extensive blooms. Because their shells (**frustules**) are composed of silica, they are also extremely important for the biogeochemical cycling of silicon.

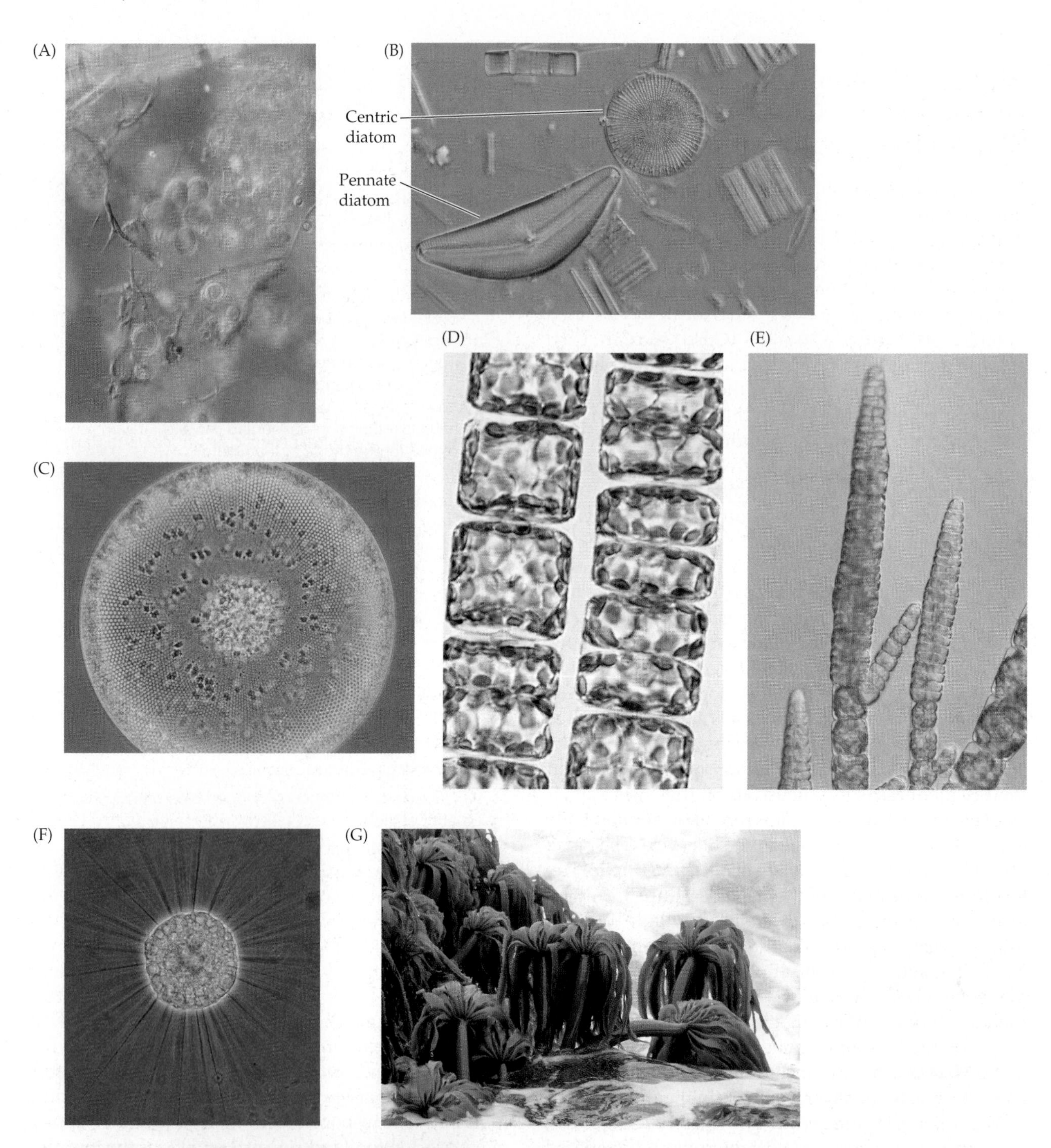

Figure 3.28 Phylum Stramenopila. A diversity of stramenopiles. (A) *Synura*, a colonial golden alga. (B) Centric and pennate diatoms in a plankton sample. (C) The centric diatom *Coscinodiscus* (the yellow patches are chloroplasts). (D) *Melosira,* a chain-forming diatom; the valves of adjacent individuals attached by mucilage pads. (E) *Ectocarpus*, a filamentous brown alga. (F) A heliozoan, *Actinophrys*. (G) *Postelisa*, a brown seaweed. (H) A kelp bed in California waters (*Macrocystis*).

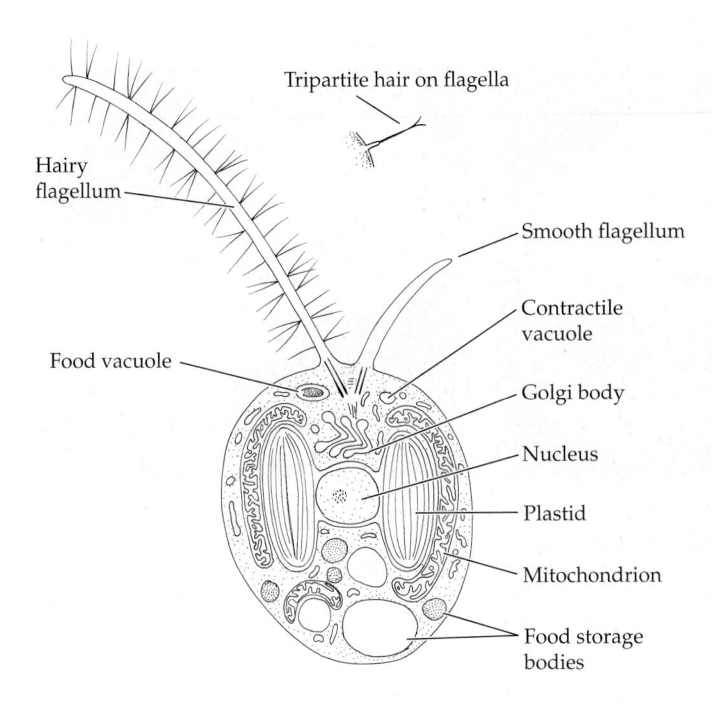

Figure 3.29 Phylum Stramenopila. Generalized anatomy (*Ochromonas*). Note the tripartite hairs on the flagellum.

A few marine diatoms produce toxins similar in potency to those seen in some dinoflagellates. Of particular concern are increasing occurrences of *Pseudonitzschia australis* blooms. In southern California, blooms of this diatom have caused deaths of sea lions, seabirds, fish and shellfish. Diatoms in the genus *Pseudonitzschia* produce **domoic acid** that accumulates in the food chain. In humans, domoic acid poisoning causes severe symptoms, including gastrointestinal disorders and memory loss (amnesic shellfish poisoning, or ASP), and it can be fatal. Cases of ASP have been reported with increasing frequency worldwide since its discovery late in the twentieth century.

The photosynthetic, freshwater diatom *Didymosphenia geminata*, native to cool oligotrophic waters of northern Asia, Europe and North America, has recently become an invasive species, both within its range and elsewhere in the world (e.g., New Zealand, Chile). Individuals attach to the bottom by a stalk, and colonies of the diatom can from thick brown mats in lakes and rivers that smother benthic communities and deplete bottom oxygen as they decompose. The flowing "tails" from these brownish mats look like tissue paper or cotton waving in the water, resulting in the common name of "rock snot." Massive blooms of this diatom have had a major negative impact on freshwater ecosystems since the mid-1980s. It isn't known why populations of this species have become invasive, although reduced stream flow is one important factor (the diatom cannot withstand rapid water flow).

Unlike dinoflagellates and coccolithophores, pelagic (mostly centric) diatoms have a large nutrient storage vacuole that occupies about 40% of the cell volume. This storage organelle allows diatoms to take advantage of short-term pulses of environmental nutrients, storing reserves for rapid reproduction or to survive periods of nutrient depletion. Their nutrient storage capacity also adapts diatoms for fluctuating environments. In addition, the vacuole in pelagic species aids in buoyancy due to its light osmolite content.

The siliceous cell walls (frustules) of diatoms do not preserve well in deep marine sediments because the silica dissolves. Although the earliest certain fossils are Mesozoic, molecular clock estimates suggest that they originated near the Permian-Triassic boundary. The fossil record indicates diatoms experienced two major bursts of radiation during the Cenozoic era, one at the Eocene/Oligocene boundary, and one through the middle and late Miocene. It has been suggested that the rise of diatoms in the Cenozoic stems from their unique, absolute requirement for a large supply of silica in the world's oceans. The theory suggests that the rise of land plants, especially grasses and the grazing ungulates that consume them, led to greatly increased removal of silica from soils and it's eventual transport to the sea, setting the stage for diatoms to radiate and flourish.

Brown seaweeds form an integral base to many coastal food webs, especially on temperate coasts. **Algin**, extracted from certain brown algae (kelp), is used as an emulsifier in everything from paint to baby food to cosmetics. This material has many industrial uses (e.g., in paint as a spreader; as filtration material in food production and water purification). Silicon deposits produced by diatoms and other stramenopiles are used in geology and limnology as markers of different stratigraphic layers of the Earth. Benthic deposits of the siliceous shells of dead marine diatoms can, over geologic time, result in massive uplifted land formations that are mined as diatomaceous earth.

Most Stramenopila are heterotrophic, but a few are serious parasites or agents of disease. Because some secrete fishy-smelling aldehydes, they may become a nuisance when they occur in great quantities, but stramenopiles only rarely cause fish kills or foul drinking water. However, the Oomyceta contain many plant parasites, including many that attack domestic crops. The devastating Irish (European, actually) potato blight disease and resulting famine of the nineteenth century was caused by an oomycete, *Phytophthora infestans*.

Chrysophytes, or "golden algae," are very common in fresh water; more than 1,000 species have been described. Most are photosynthetic, but some species are colorless and strictly heterotrophic. In fact, most of the autotrophic species will become facultatively heterotrophic in the absence of adequate light, or in

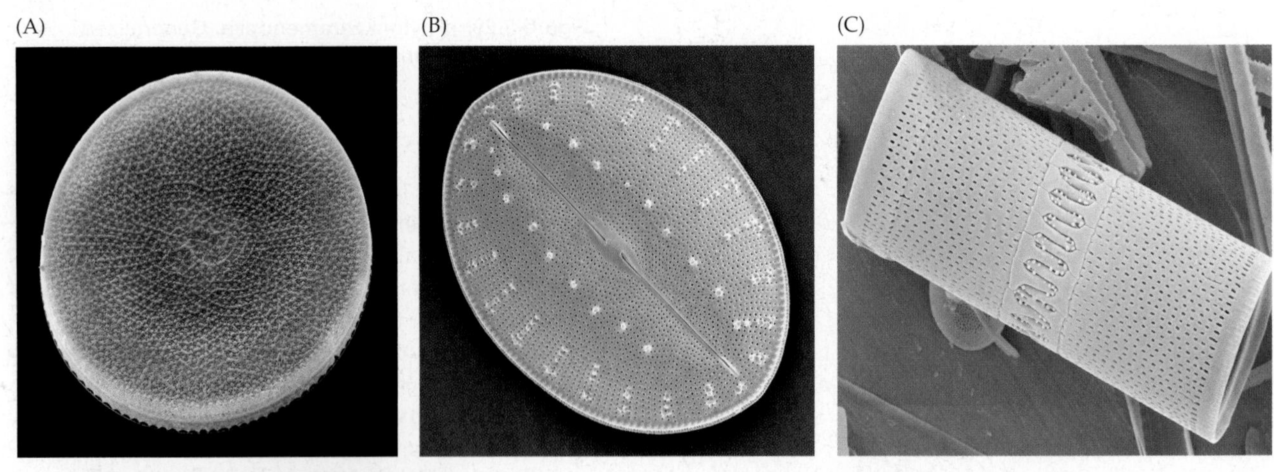

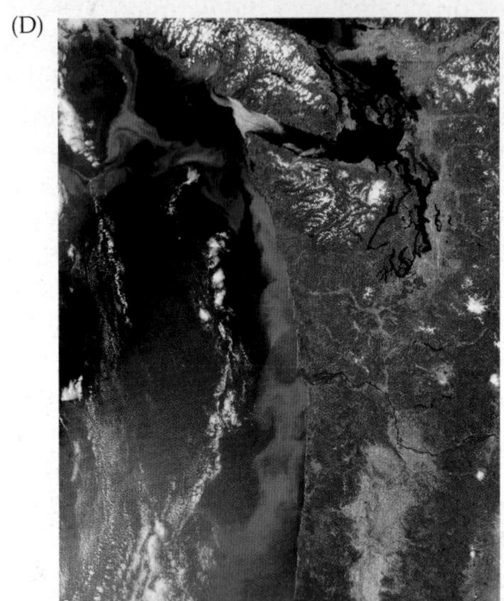

Figure 3.30 Phylum Stramenopila. Stramenopile skeletons. (A) Colorized SEM of the silicious skeleton of the centric diatom *Actinocyclus* sp. (B) Colorized SEM of the pennate diatom *Navicula* sp. (C) Tubular diatom *Aulacoseira italica*. (D) Large diatom bloom in the northeast Pacific (coastal Washington State), July 2014.

the presence of plentiful food. When this occurs, the chloroplasts (called **chrysoplasts** in this group) atrophy and the algae may turn into predators, feeding on bacteria or other protists, including diatoms. The photosynthetic species are important primary producers in lakes. Although most are unicellular, there are also filamentous and colonial forms. In the well-known *Dinobryon* (a freshwater genus), the individual cells are surrounded by vase-shaped loricae, composed of chitin fibrils and other polysaccharides. The colonies grow as branched or unbranched chains. Species of *Synura* form spherical colonies covered by silica scales. The oldest known Chrysophytes are from Cretaceous deposits, and the group has a fairly complete fossil record because of the siliceous resting cysts that freshwater species tend to form.

The enigmatic group Blastocystida includes *Blastocystis*, a highly prevalent, fecal-transmitted parasite that infects the gastrointestinal tract of humans and other animals, including many vertebrates and a few insects (e.g., cockroaches). Infection with *Blastocystis*, blastocystosis, causes abdominal pain, constipation and/or diarrhea. Most of the species found in mammals and birds are also able to cause infections in humans. These protists were first classified as yeasts, later as sporozoans, and then shown by molecular analyses to be stramenopiles.

Support and Locomotion

Stramenopile support structures are highly varied. In addition to their cell membrane, many also possess shells, tests, and other support structures, resulting in many different shapes and appearances. Like most protists, they rarely have chitin in the cell wall (further distinguishing them from true fungi, or Eumycota, which rely heavily on chitin for support). Some chrysophytes produce small discs of calcite, protein, or even silicon in their cells. These are then packaged in endoplasmic reticulum vesicles and secreted onto the surface of the cell to form a layer of distinctive scales—much the same mechanism seen in other scaly protists, including Haptophyta. In some stramenopiles, these can be quite elaborate and beautiful (Figure 3.30). The silicoflagellates have a distinctive internal skeleton of tubular pieces of silica associated with a central nucleus, and a complex lobed body that contains many chloroplasts.

Diatoms also secrete silica in the form of an external test, or frustule, which consists of two parts, called valves. Beneath the test is the cell membrane, enclosing the nucleus, chloroplasts, and the rest of the cytoplasm. There are two different diatom forms: **centric diatoms** have radially symmetrical frustules, and since one valve is slightly larger than the other, they resemble a petri dish (Figures 3.30A and 3.28); **pennate diatoms** are bilaterally symmetrical and often have longitudinal grooves on the valves (Figure 3.30B).

Stramenopiles generally exhibit **heterokont flagellation**. That is, they possess two flagella, one directed anteriorly, the other usually extended posteriorly. The anteriorly directed flagellum has a bilateral array of tripartite, tubular hairs, while the posterior is either smooth or has a row of fine, filamentous hairs (see Figure 3.29). The tripartite, tubular hairs are stiff and reverse the direction of the thrust of the flagellum so that, even though the flagellum is beating in front of the cell, the cell is still drawn forward.

Labyrinthulids are commonly called "slime nets," and because of their unique lifestyle and locomotion, in the past they have been classified as a distinct phylum. However, because they can produce cells with two heterokont flagella they are now classified as Stramenopila. The nonflagellated stage of the labyrinthulid's life cycle forms complex organisms consisting of numerous spindle-shaped bodies, each containing a nucleus, which glide rapidly within a common membrane-bound ectoplasmic network. This network contains a calcium-dependent contractile system of actinlike proteins that is responsible for shuttling the cells through the net. The net spreads out over decaying material, or acts as a pathogen of living plants. They are associated in particular with marine and brackish habitats.

Nutrition

As you have no doubt already guessed, stramenopiles exhibit a wide variety of nutritional habits. Some are photosynthetic, others are ingesting heterotrophs, and still others are saprophytic. Those forms that are photosynthetic have: chlorophylls a, c_1, and c_2; thylakoids in stacks of three; and four membranes surrounding the chloroplast (as in haptophytes and cryptomonads, the outermost membrane is often continuous with the nuclear envelope) (see Figure 3.3). Yellow and brown accessory pigments (primarily xanthophylls such as fucoxanthin, but carotenoids are present as well) give many of them a brownish-green color that has earned them the common name "golden algae." There is usually an eyespot associated with the region of the chloroplast near the basal bodies.

Many of the heterotrophic stramenopiles use the anteriorly directed flagellum with tripartite hairs to capture food particles, which are engulfed by small pseudopodia or a hoop-shaped cytostome near the base of the flagella. Other heterotrophic forms feed saprophytically by excreting enzymes that digest food items outside the cell and then absorbing nutrients through small pores on the cell surface. This mode of nutrition is similar to that of true fungi (Eumycota), and it is the reason that the labyrinthulids, oomycetes ("water molds"), and hyphochytridiomycetes were once classified as fungi. However, the presence of a heterokont-flagella stage in these organisms makes it clear that they are stramenopiles. The oomycetes produce hyphae that absorb nutrients much like fungi, and are mostly coenocytic. Included among the oomycetes are the water molds, white rusts, and downy mildews.

Reproduction

Mitosis in most stramenopiles is characterized as open pleuromitosis. During division, the basal bodies of the two flagella separate and a spindle forms adjacent to them or adjacent to the striated root at the base of each. In those forms with scales, the scaly armor appears to be added to the surface of the daughter cells as division proceeds. In diatoms, each daughter cell gets one of the silica valves and makes a new second valve to complete the frustule.

Sexual reproduction is poorly studied in most forms but appears to almost always occur by the production of haploid gametes, which fuse to form a zygote. In many, the gametes are undifferentiated, but in a few (such as diatoms), one of the gametes is flagellated and motile, and the other is stationary. Many of the brown algae (Phaeophyta) have alternation of generations.

The opalinids, once classified as "protociliates," then as zooflagellates, then as a separate phylum of uncertain relationship, is included here in the Stramenopila based primarily on analyses of DNA sequence data. Their numerous oblique rows of cilia clearly differ from the rows in ciliates in that they lack the kinetidal system. During asexual reproduction, the fission plane parallels the oblique ciliary rows; thus it is longitudinal (as it is in flagellates) rather than transverse (as it is in ciliates). Some opalinids are binucleate, others multinucleate, but all are homokaryotic (i.e., the nuclei are all identical). There are about 150 species of opalinids, in several genera, almost all being endosymbiotic in the hindgut of anurans (frogs and toads) where they ingest dissolved material anywhere on their body surface. Sexual reproduction is by syngamy and asexual reproduction is by binary fission and plasmotomy, the latter involving cytoplasmic divisions that produce multinucleate offspring. *Opalina* and *Protopalina* are two genera commonly encountered (Figure 3.31). Opalinids are often found in routine dissections of frogs in the classroom; their large size and graceful movements through the frog's rectum make them a pleasant discovery for students.

Phylum Haptophyta: Coccolithophores

Haptophyta (also known as Prymnesiophyta) comprise a group of single-celled, phytoplanktonic protists most of which are known as coccolithophores (or coccolithophorids) (Figure 3.32). Haptophytes are distinguished by their possession of two normal flagella plus a third flagellum-like, or peglike structure called a

Figure 3.31 The enigmatic *Opalina*, once classified in its own phylum, now generally regarded as a member of the phylum Stramenopila.

BOX 3J Characteristics of the Phylum Haptophyta

1. A small group of phytoplanktonic protists, almost exclusively marine, most of which are known as coccolithophores
2. Distinguished by two normal flagella, plus a third flagellum-like, or peglike haptonema. The haptonema is usually coiled and differs completely from the true flagella.
3. With two golden-brown chloroplasts that have diadinoxanthin and fucoxanthin accessory pigments
4. Cells usually covered by scales (coccoliths) composed of carbohydrates, calcium carbonate, or silicon
5. Reproduction is poorly understood for this group, but evidence suggests that both asexual and sexual methods occur.

haptonema. The haptonema is generally coiled, and it differs from the flagella proper in its internal structure, basal attachment, and function. During the nonmotile phase of the life cycle, the flagella disappear, but the haptonema often remains. Its primary function is not entirely clear and may vary from species to species. It is used as a food collection device in at least some species, but otherwise it appears to have attachment and/or mechanosensory functions. The cell itself contains a nucleus and two golden-brown chloroplasts that carry the yellow-brown accessory pigments diadinoxanthin and fucoxanthin. The cells are usually covered by scales, either small scales made of carbohydrate, or, in the case of some or all life stages of coccolithophores, small or large scales—**coccoliths**—composed predominantly of calcium carbonate (see below). Some may have siliceous scales. The scales show a broad variety and complexity of shape—they can resemble buckets, donuts, muffins, pentagons, or even be trumpetlike (Box 3J).

Although almost exclusively marine, there are a few reported freshwater and "terrestrial" species. Coccolithophores live in huge numbers throughout the photic zone of the world's oceans, and they are one of the main primary producers in this environment. One of the most abundant species in the world's oceans is *Emiliania huxleyi* (Figure 3.32). The secreted plates, or coccoliths, are held in place on the surface of the cell by an organic coating, the **coccosphere**. The coccoliths are usually composed of calcium carbonate ($CaCO_3$), but unmineralized (carbohydrate compounds) and even siliceous plates are also known from a few species. The function of the coccosphere is presumably either protection or buoyancy. A single coccolithophore will be covered by 10 to over 100 coccoliths, and these vary wildly in shape between species. As might be expected, coccoliths play a central role in haptophyte taxonomy.

Figure 3.32 Phylum Haptophyta. Colorized SEM of the coccolithophore *Emiliania huxleyi*.

It is estimated that coccolithophores produce several million tons of calcite annually. Most of this ends up on the sea floor as calcareous ooze that eventually contributes to the Earth's sedimentary record. In fact, these "calcareous nannofossils" (generally less than 30 μm in diameter) have become the preferred tool for quick accurate stratigraphic age determination in post-Paleozoic calcareous sequences. The earliest uncontested calcareous nannofossils are from the Late Triassic, thought to correspond to the first time open-ocean planktonic organisms utilized calcareous skeletons and exported calcium carbonate to the deep oceans. Fossilizable Haptophytes reached their greatest abundance in the Late Cretaceous, experiencing a mass extinction at the end of the Cretaceous (two-thirds of the ~50 known genera disappeared). Many new groups appear in the Paleocene fossil record. Exposed ancient calcareous coccolithophore sediments are recognized as limestone, or "chalk," mostly Mesozoic in age. The White Cliffs of Dover (England) is one of the more well-known coccolithophore-formed limestone exposures.

The life history of coccolithophores is only partly understood, but the existence of both haploid and diploid phases is inferred by DNA studies, with mitotic reproduction occurring in both stages. In at least some species, such as *Emiliania huxleyi*, the coccosphere-bearing phase is diploid and capable of asexual reproduction, allowing for rapid population growth during periods of optimum conditions when "coccolithophore blooms" may occur. Some of these blooms, especially those of *Emiliania huxleyi*, have been measured at densities of 1.5 million cells per liter! Motile naked haploid gametes may be produced by meiosis, and nonmotile benthic stages also are known to be produced. Syngamy has not been observed but is assumed to occur. Both single and double fission occur, sometimes accompanied by a swarm spore stage. Large blooms can be problematic because of the mucilage surrounding the cells that can clog fish gills or render them permeable to dissolved toxins.

Phylum Cryptomonada: Cryptomonads

Cryptomonads include several genera of photosynthetic, and sometimes heterotrophic flagellates that occur in marine and freshwater habitats. One heterotrophic form, previously called *Chilomonas* but now classified in *Cryptomonas*, is a commonly used research tool in biological laboratories. Cryptomonads are biflagellated cells with a large flagellar pocket, a semirigid cell surface supported by proteinaceous plates (called the **periplast**), and a single large mitochondrion with cristae that appear to be flattened tubes. The plastids of cryptomonads are surrounded by four membranes and contain chlorophylls *a* and *c*. As with chlorarachniophytes, the plastid of cryptomonads includes a small pocket of cytoplasm between the inner and outer pair of membranes, and also contains ribosomes and a **nucleomorph**—the highly reduced nucleus of the eukaryotic, photosynthetic endosymbiont that became the plastid (Figure 3.33). Unlike chlorarachniophytes, however, the cryptomonad plastid is of red algal origin (rather than green algal origin), and starch is synthesized and stored within the plastid's cytoplasmic compartment. Thus cryptomonads and chlorarachniophytes stem from two independent and unrelated endosymbiotic events. The nucleomorph of cryptomonads consists of three small chromosomes that primarily encode only those genes necessary for the maintenance of the nucleomorph itself. This "enslaved" and reduced algal nucleus now encodes only around 20 of the proteins necessary for the maintenance of the plastid, and extensive lateral gene transfer has moved most of the essential algal genes to the host nucleus. The cryptomonad "model organism" *Bigelowiella natans* has a nucleomorph genome of only 373,000 base pairs, one of the smallest eukaryotic genomes known (Box 3K).

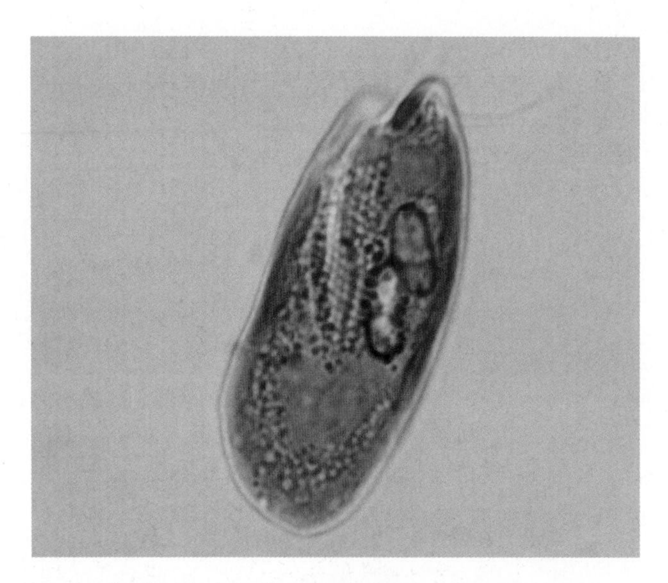

Figure 3.33 Phylum Cryptomonada. *Cryptomonas.*

BOX 3K Characteristics of the Phylum Cryptomonada

1. A small group of primarily photosynthetic flagellates that occurs in marine and freshwater habitats
2. Cells biflagellate; with semirigid cell surface supported by proteinaceous places (the periplast)
3. Single large mitochondrion with cristae that appear as flattened tubes
4. Plastids surrounded by four membranes that contain chlorophylls *a* and *c*. Plastid has small pocket of cytoplasm between inner and outer pair of membranes, and also contains ribosomes and a nucleomorph (the nucleomorph is a highly reduced nucleus of the red algal symbiont that became the plastid).
5. Starch is synthesized and stored within the plastid's cytoplasmic compartment.

(A)

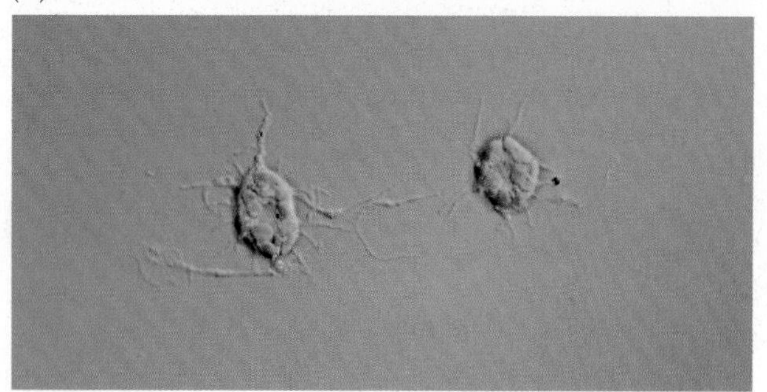

(B)

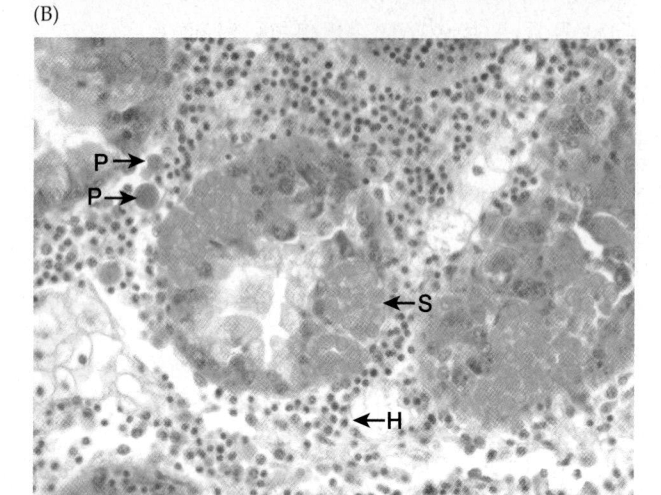

Figure 3.34 Two rare rhizarian phyla, Chlorarachniophyta (A) and Haplosporidia (B). (A) *Chlorarachnion reptans*; the green algal endosymbionts are visible within the cell, each retaining its own nucleus. (B) *Haplosporidium nelsoni*, the causative disease agent in MSX, in the Pacific oyster (*Crassostrea gigas*). The vascular spaces between the digestive gland tubules contain numerous hemocytes, or blood cells, elevated in number as a response to the infection. Note the extensive sporulation within the epithelial cells of the digestive gland tubules, as well as multinucleate plasmodial stages and hemocyte infiltration within the vascular spaces of the digestive gland. Host nuclei are dark purple/blue, cytoplasm and cell membranes are pink. H: hemocytes; S: sporulation/spores; P: plasmodia stages. (5 μm-section stained with haematoxylin and eosin)

BOX 3L Characteristics of the Phylum Chlorarachniophyta

1. A small group of uncommon protists in tropical and subtropical seas
2. Normally small, unicellular, and ameboid with branching cytoplasmic extensions (pseudopodia) that sometimes connect several cells together in netlike fashion
3. Some form uniflagellated zoospores, and coccoid cells (thought to be cysts).
4. Mixotrophic, feeding on bacteria and other protists, or photosynthesizing with green chloroplasts containing chlorophylls *a* and *b* (and a prominent projecting pyrenoid). Chloroplasts with four envelope membranes and a prominent pyrenoid (food storage structure)
5. With nucleomorph (reduced nucleus of the endosymbiotic green algal cell that became the chloroplast) present in space between second and third envelope membranes of each chloroplast
6. Thylakoids often loosely stacked in threes
7. Mitochondrial cristae tubular
8. Asexual reproduction by normal mitotic cell division or zoospore formation. Sexual reproduction apparently rare, but both anisogamy and isogamy have been reported

GROUP 3: RHIZARIA

Phylum Chlorarachniophyta: Chlorarachniophyte Algae

The Chlorarachniophyta (or Chlorarachniophytes) comprise a recently recognized, small, and fascinating group of single-celled algae that have become a major focus of evolutionary biologists (Figure 3.34A). They occur, apparently with rarity, in tropical and subtropical seas. The species are typically mixotrophic, ingesting bacteria and smaller protists as well as conducting photosynthesis. They have green chloroplasts but have a cell structure completely different from that of green algae (phylum Chlorophyta). Normally they have the form of small amebas, with branching cytoplasmic extensions (reticulopods) that capture prey and sometimes connect several cells together in a netlike fashion. They may also form flagellated zoospores, which characteristically have a single subapical flagellum that spirals backwards around the cell body. Some species also form walled coccoid cells (thought to be cysts) (Box 3L).

Chlorarachniophytes are distinguished by the following characteristics: always unicellular, although cells may anastomose their pseudopodia to form "nets"; photosynthetic, with green chloroplasts, with four envelope membranes, and that contain chlorophylls *a* and *b* and a prominent projecting pyrenoid[7]; nucleomorph present in space between the second and third envelope membranes of each chloroplast; thylakoids often loosely stacked in threes; mitochondrial

[7]Pyrenoids are protein-rich structures found inside some types of chloroplasts. They are thought to function in the carbon concentration/carbon fixation portions of photosynthesis, and are often closely associated with accumulations of storage materials, such as starch.

cristae tubular; fundamentally amoeboid cells with long, thin (filose) pseudopodia. Species that have coccoid stages in their life history have been found from coastal areas, while those with flagellated stages (zoospores) tend to occur in oceanic waters (as picoplankton). The patterns of the life cycle vary among species. Coccoid cells tend to be regarded as cysts. Asexual reproduction is carried out by either normal mitotic cell division or zoospore formation. Sexual reproduction has been reported from two species: *Chlorarachnion reptans* and *Cryptochlora perforans*. In the former, two different types of cells, amoeboid and coccoid, fuse to form a zygote (anisogamy), while in *C. perforans* the fusion occurs between two amoeboid cells (isogamy).

The chlorarachniophytes are one of several protist groups that acquired their chloroplasts via secondary endosymbiosis, in which a nonphotosynthetic eukaryote engulfed a eukaryotic alga and co-opted it as a symbiont, which over evolutionary time was reduced to a photosynthetic organelle. In at least two cases of secondary endosymbiosis, Chlorarachniophyta and Cryptomonada, nuclear material (and a small amount of cytoplasm) has been retained from the assimilated photosynthetic endosymbiont (a green and a red alga, respectively), along with the plastid to the present day by the "host" lineage. In both of these phyla the symbiont's nuclei are greatly reduced and termed nucleomorphs. The nucleomorphs, and particularly the genes they contain, have been a key to understanding the phenomenon of secondary endosymbiosis.

The nucleomorph in chlorarachniophytes is the vestigial nucleus of the eukaryotic green algal endosymbiont that gave rise to the chloroplast itself. In the one species studied to date, the nucleomorph genome is just 380 kb—one of the smallest eukaryotic genomes known. The genome consists of three linear chromosomes that encode about 300 genes arranged in a highly compacted manner. Genes encoded by the nucleomorph genome are mostly "housekeeping genes," genes for maintaining its own replication and expression systems. Only a few chloroplast-targeted proteins are encoded by the nucleomorph, suggesting that the majority of nuclear-encoded chloroplast genes have been co-opted by the host's nuclear genome.

Like other DNA-containing organelles, the nucleomorph is semiautonomous. Nucleomorph division takes place just before chloroplast division by the sequential infolding of the inner and outer membranes of the nucleomorph envelope. No spindle formation has been observed during the division. How the chromosomes are correctly segregated into two daughter nucleomorphs is not clear.

Despite the striking similarity in structure and genome organization between the chlorarachniophytes and the cryptomonads, their origins are clearly different. Cryptomonad nucleomorphs originated from a red algal endosymbiont, whereas the chlorarachniophytes nucleomorphs came from a green algal endosymbiont. Nucleomorphs are thus an excellent example of convergent evolution through endosymbiosis (see phylogeny section below). Molecular phylogenetic studies suggest that the "host" clade of Chlorarachniophyta is monophyletic, and that its ancestry lies in a heterotrophic protist that we would place within the clade Cercozoa.

Phylum Granuloreticulosa: Foraminiferans and Their Kin

The phylum Granuloreticulosa contains about 40,000 described living and fossil species (Box 3M). Members of this phylum are nearly ubiquitous in all aquatic habitats from the poles to the equator, and at all depths in the world's oceans. The phylum consists of two major groups: the Athalamida and Foraminifera (including the monothalamids). Athalamids are found in fresh water, soil, and marine environments and are distinguished from the forams in that they lack a test and the pseudopodia can emerge any place on the body.

The Foraminifera (e.g., *Globigerina*), also called Foraminiferida or forams, are the most common and well-known members of the Granuloreticulosa (Figures 3.35 and 3.36; see also Figure 3.1J). They are

BOX 3M Characteristics of the Phylum Granuloreticulosa

1. Common protists, ubiquitous in all aquatic habitats and all depths
2. Cell surrounded by plasma membrane, which may by supported by organic, agglutinated, or calcareous test; test situated outside plasma membrane
3. Locomotion and feeding involves long thin pseudopodia called reticulopodia, which branch and fuse to from a network.
4. Heterotrophic and without plastids. However, many species host symbiotic, photoautotrophic protists within their cell (e.g., diatoms, dinoflagellates, red algae, green algae).
5. Mitochondria with tubular cristae
6. Nuclei either ovular or vesicular; many are multinucleate; some exhibit nuclear dualism
7. Nuclear division by closed intranuclear pleuromitosis
8. Asexual reproduction by budding and/or multiple fission
9. Sexual reproduction known in most. The life cycle is usually complex, involving alternation of an asexual form (agamont) and a sexual form (gamont).

(A)

(B)

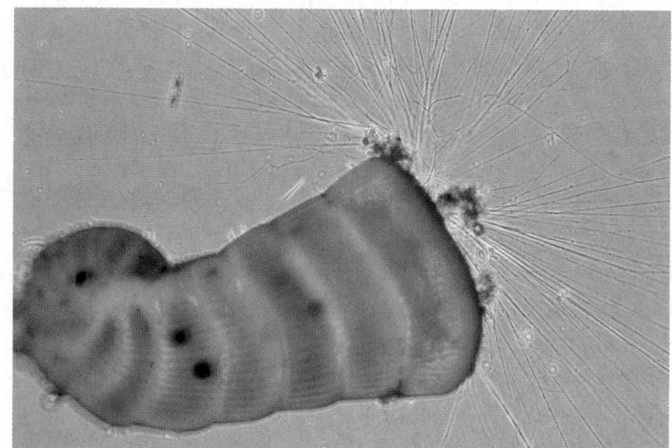

Figure 3.35 Phylum Granuloreticulosa. (A) Foraminiferan shells from the Red Sea. (B) The foraminiferan *Spirolina* sp., with its extended reticulopodia.

most frequently found in marine and brackish water and are characterized by the presence of a test with one to multiple chambers and distinctive reticulopodia—long, thin pseudopodia (supported by microtubules) that branch and anastomose, and house a rapid bidirectional intracellular transport mechanism. Most forams are benthic and have flattened tests. A small number, however, are pelagic and may have calcareous spines that have been implicated in prey capture (e.g., *Globigerinella*; see Figure 3.1J). Foraminiferans (noncalcareous species) have even been found living in the deepest ocean trenches, such as the Challenger Deep (10,896 m). Trench and abyssal forams primarily are simple, single-chambered species belonging to ancient lineages that long ago gave rise to the more complex multichambered groups that dominate shallower seas.

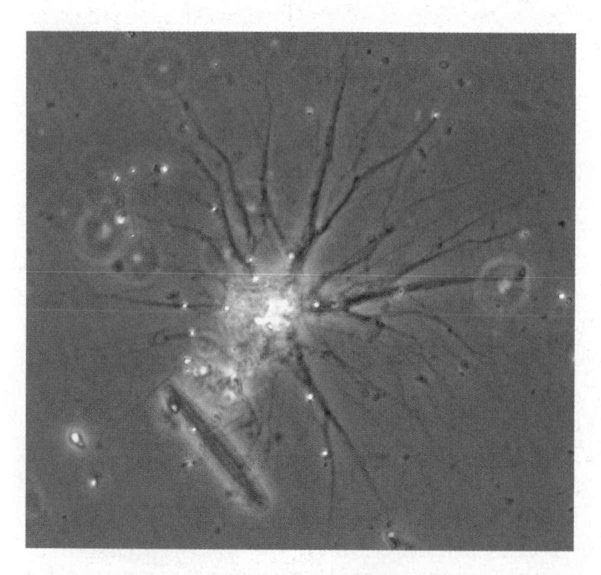

Figure 3.36 Phylum Granuloreticulosa. An unidentified athalamid with reticulopodia.

Members of these single-chambered lineages are also the only forams to have invaded freshwater and terrestrial habitats. Thus the great ocean depths and terrestrial ecosystems serve as refugia for ancient foram lineages.

The tests of forams have left an excellent fossil record dating back to at least the Lower Cambrian, and perhaps the Precambrian. The tests of planktonic forams are used by geologists as paleoecological and biostratigraphic indicators, and deposits of benthic foram tests are often used by petroleum geologists in their search for oil. The tests of some species are surprisingly durable. On the island of Bali, the tests of one species are mined and used as gravel in walks and roads! Much of the world's chalk, limestone, and marble is composed largely of foraminiferan tests or the residual calcareous material derived from the tests. Most of the stones used to build the great pyramids of Egypt are foraminiferan in origin (it's engaging and instructive to think of the great pyramids as being built of protist skeletons). Before they get buried on the sea floor, the tests of foraminiferans function as homes and egg-laying sites for many minute metazoan species, such as small sipunculans, polychaetes, nematodes, copepods, isopods, and others.

Support and Locomotion

Although little is known about locomotion in the Granuloreticulosa, the reticulopodia are involved (see Figure 3.36). Most species have tests covering their plasma (cell) membrane. However, the Althalamida lack a test and instead are covered by a thin, fibrous envelope. The tests of forams are usually constructed as a series of inter-connected chambers of increasing size, with main openings, or apertures, in the largest chamber from which the reticulopodia emerge. The connecting pores between the chambers are called **foramina** (hence the group's name) and they actually constitute the former apertures. There may be one large aperture, or several smaller ones in the largest (youngest) chamber. Cytoplasm emerging from the aperture(s) forms the reticulopodial network and also often forms a layer covering the outside of the test.

There are three types of tests in the Foraminifera: (1) organic, (2) agglutinated, and (3) calcareous. The nature of the test is a taxonomic feature used to classify forams. Organic tests are composed of complexes of proteins and mucopolysaccharides. These tests are flexible and allow the organisms that secrete them (e.g., *Allogramia*) to change shape rapidly.

Agglutinated tests are composed of materials gathered from the environment (e.g., sand grains, sponge spicules, diatoms, etc.) that are embedded in a layer of mucopolysaccharide secreted by the cell. The test may be made rigid by calcareous and iron salts. Some forams that have agglutinated tests are highly selective about the building materials used to build their tests (e.g., *Technitella*), while others are not (e.g., *Astrorhiza*).

Calcareous tests are composed of an organic layer reinforced with calcite ($CaCO_3$) secreted by the forams themselves. The arrangement of calcite crystals gives the tests a characteristic appearance, and three major categories of calcareous tests are recognized: (1) porcelaneous, (2) hyaline, and (3) microgranular. Porcelaneous tests appear shiny and white, and these are probably the most familiar to students (see Figure 3.35). These tests generally lack perforations and the reticulopodia emerge from a single aperture. Hyaline tests have a glasslike appearance in reflected light and often are perforated with tiny holes. Microgranular tests have a sugary (granular) appearance in reflected light. Planktonic foraminiferans can occur in such high numbers that the calcareous tests of dead individuals constitute a major portion of the sediments of ocean basins. In some parts of the world, these sediments—called foraminiferan ooze—are hundreds of meters thick. Such sediments are restricted to depths shallower than 3,000 to 4,000 m, however, because $CaCO_3$ dissolves under high pressure.

Nutrition

All Granuloreticulosa are heterotrophic and feed by phagocytosis. The prey varies depending on the species. Some are herbivores, other carnivores, and still others are omnivores or detritivores. The life cycles of many herbivorous planktonic species are tuned to the bloom of certain algae, such as diatoms or chlorophytes, and they graze heavily on the phytoplankton at those times. All species probably use their reticulopodia to trap food. Vesicles at the tip of the reticulopodia secrete a sticky substance to which prey adhere upon contact. Benthic species trap prey by spreading their reticulopodia out on the lake or ocean bottom. The prey are eventually transported to the cell's plasma membrane where they are engulfed in food vacuoles.

Shallow-water benthic and planktonic forams living near the water's surface often harbor endosymbiotic algae such as diatoms, dinoflagellates, or red or green algae, which can migrate along the reticulopodia to expose themselves to more sunlight. These forams are particularly abundant in warm tropical seas. Studies suggest that nutrient and mineral recycling occurs between the forams and their algal symbionts. Furthermore, it has been shown that the symbionts may enhance the test-building capacities of forams and that their presence often allows their hosts to grow to very large sizes (e.g., the Eocene foram *Nummulites gizehensis* reached 12 cm in diameter), even in nutrient-poor waters. "Giant" forams, such as *Nummulites*, are much more common in fossil deposits than they are today.

Reproduction and Some Life Cycles

The life cycles of granuloreticulosans are frequently complex, and most are incompletely understood. These cycles often involve an alternation of sexual and asexual phases (Figure 3.37). However, some smaller species apparently only reproduce asexually, by budding and/or multiple fission. Nuclear division in both sexually and asexually reproducing species occurs by intranuclear pleuromitosis. In those that reproduce sexually, it is not uncommon to find individuals of the same foraminiferan species differing greatly in size and shape at different phases of the life cycle. The size difference is generally determined by the size of the initial shell chamber (the **proloculum**), produced following a particular life-cycle event. Often the proloculum that is formed following asexual processes is significantly larger than one formed after syngamy. Individuals with large prolocula are called the **macro-** or **megaspheric generation**; individuals with small prolocula are the **microspheric generation**.

During the sexual phase of the life cycle, the haploid individuals (gamonts) undergo repeated divisions to produce and release bi- or triflagellated isogametes, which pair and fuse to form the asexual individuals. Asexual, diploid individuals (called agamonts) undergo meiosis and produce haploid gamonts—the sexual individuals. The means of return to the diploid condition varies. In many foraminiferans (*Elphidium*, *Iridia*, *Tretomphalus*, and others), flagellated gametes are produced and released; fertilization occurs freely in the seawater to produce a young agamont. In others, such as *Glabratella*, two or more gamonts come together and temporarily attach to one another. The gametes, which may be flagellate or amoeboid, fuse within the chambers of the paired tests. The shells eventually separate, releasing the newly formed agamonts. True autogamy occurs in *Rotaliella*: each gamont produces gametes that pair and fuse within a single test, and the zygote is then released as an agamont.

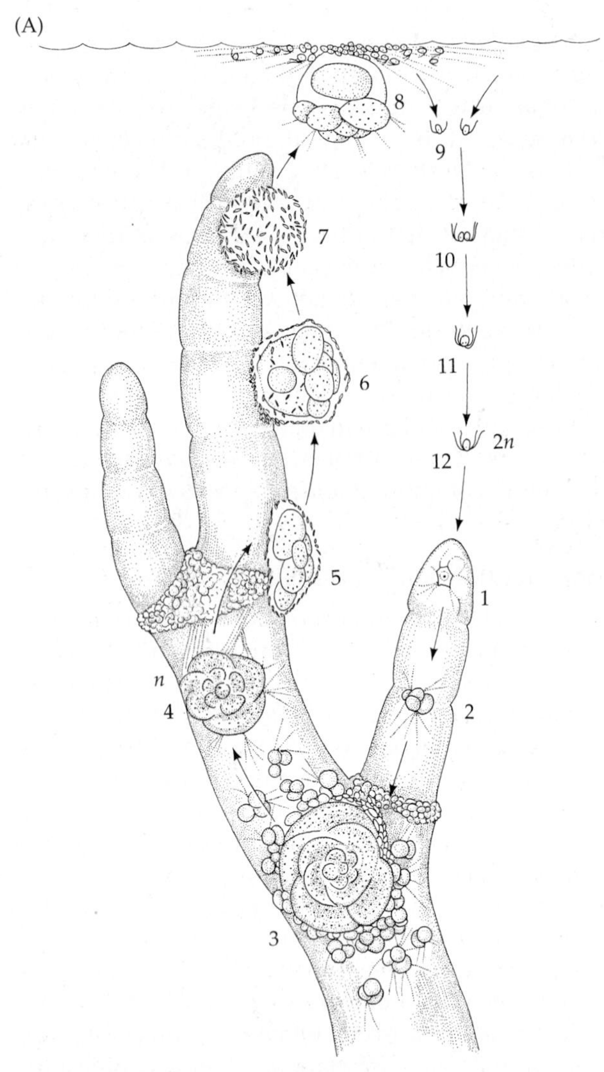

Figure 3.37 Phylum Granuloreticulosa. (A) Life cycle of the foraminiferan *Tretomphalus bulloides*. (1) The shell-less, ameboid zygote settles on the surface of an alga or sea grass (e.g., *Thalassia testudinum*). (2) The cell grows and matures as an agamont, which (3) asexually produces young gamonts. Each mature gamont (4) accumulates particles of detritus (5–6) to produce a flotation chamber (7). (8) The gamont floats to the surface and produces and releases gametes (9), which fuse to produce a swimming zygote (10–12). (B) A common habitat for *Tretomphalus bulloides* in the Caribbean is sea grass beds (*Thalassia testudinum*).

BOX 3N Characteristics of the Phylum Radiolaria

1. Common marine planktonic protists in the world's oceans; most common in the tropics
2. Cytoplasm divided into two regions, endoplasm and ectoplasm (=calymma), separated by a capsular wall composed (usually) of mucoprotein. Cell surrounded by plasma membrane, which may be supported by skeleton secreted by the cell and usually internal; skeletons of variable composition
3. Locomotion mostly passive; some movement can be accomplished with axopods
4. Mitochondria with tubular cristae (in most)
5. Heterotrophic, capturing prey with axopods and extrusomes. Although without plastids, some species harbor symbiotic, photoautotrophic protists (e.g., chlorophytes, dinoflagellates).
6. Most with single vesicular nucleus; some have single ovular nucleus; some have multiple nuclei
7. Nuclear division occurs by closed intranuclear pleuromitosis. Electron-dense plaques act as organizers for mitotic spindle; pair of centrioles are located outside nucleus and situated near plaques. Amorphous structures called polar caps located in cytoplasm act as mitotic spindle organizers.
8. Asexual reproduction by binary fission, multiple fission, or budding
9. Meiosis involves two divisions prior to formation of gametes.

Phylum Radiolaria: Radiolarians

The phylum Radiolaria comprises about 2,500 described species whose threadlike pseudopodia appear to ally them with chlorarachniophytes and many heliozoans—the three forming a clade known as the Cercozoa (Box 3N). Three groups are typically classified in the Radiolaria: Polycystina, Acantharia, and Phaeodaria, although it is likely that none of these represents a monophyletic group (Figure 3.38). Most have internal siliceous (SiO_2) skeletons that preserve well as fossils, although acantharians have a skeleton of strontium sulfate (celestite, $SrSO_4$) that is more water soluble and less frequently found in the fossil record. Radiolarians are all planktonic; they are found exclusively in marine habitats and are most abundant in warm waters (26°–37°C).

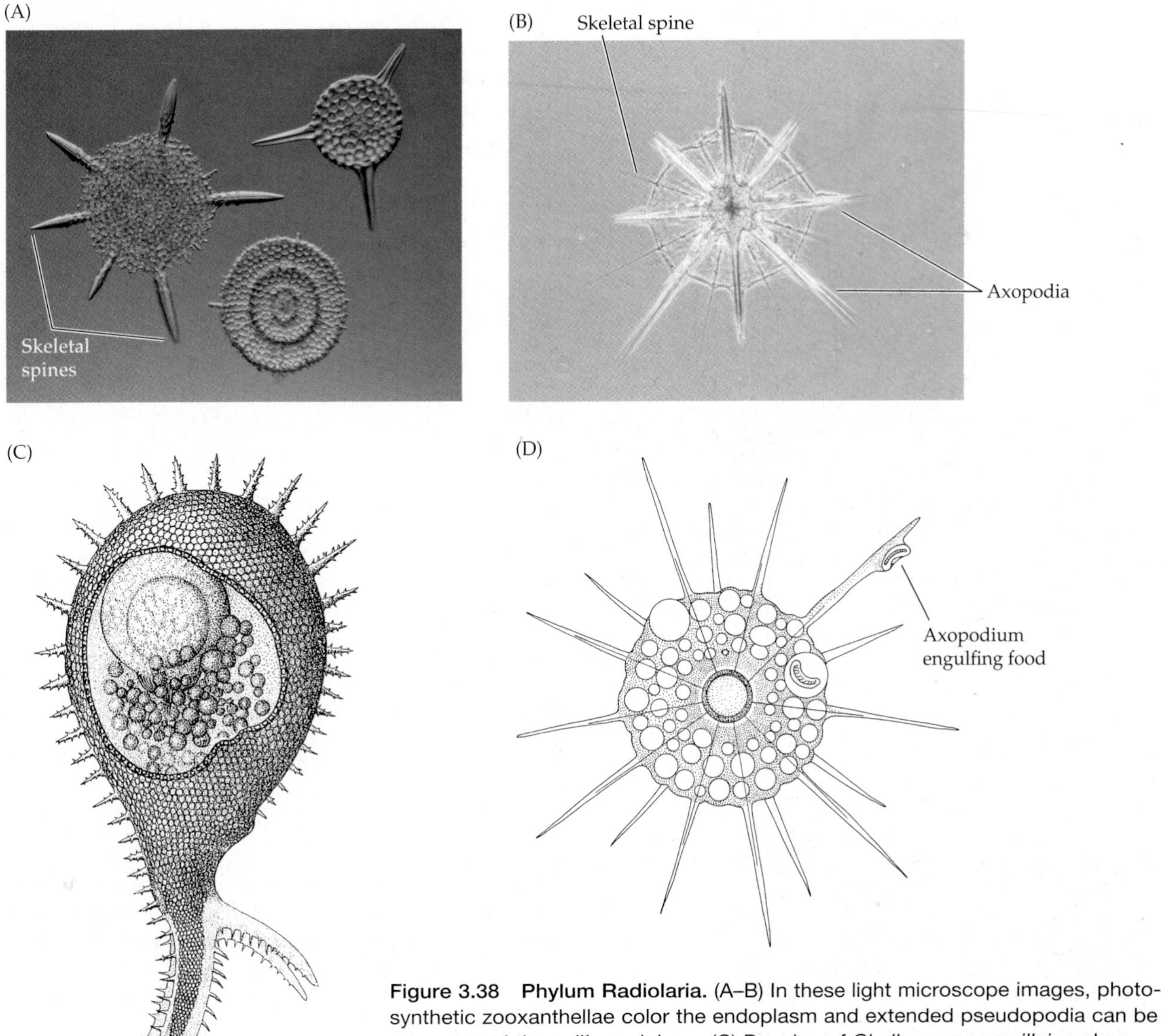

Figure 3.38 Phylum Radiolaria. (A–B) In these light microscope images, photosynthetic zooxanthellae color the endoplasm and extended pseudopodia can be seen around the cell's periphery. (C) Drawing of *Challengeron wyvillei*, a phaeodarian. (D) Food capture by an axopodium.

The axopods, which radiate from the bodies of these beautiful protists, are slender pseudopodia supported by an inner core of microtubules. In many groups, the axopods extend from a central region of the cell called the **axoplast** (Figure 3.39). The pattern of the microtubule arrangement within the axopods varies and is an important taxonomic feature. Axopods function primarily in feeding and, in some cases, locomotion. The cytoplasm exhibits a characteristic bidirectional movement (like the Granuloreticulosa), circulating substances in the cytoplasm between the pseudopodia and the main body of the cell. One of the most useful aspects of radiolarians for humans is related to the nature of their skeletons—the strontium sulfate skeletons of acantharians have been used by scientists to measure amounts of natural or anthropogenic radioactivity in marine environments. The siliceous skeletons of polycystines and phaeodarians do not dissolve under great pressure and therefore accumulate, along with diatom tests, as deposits called siliceous ooze on the floors of deep ocean basins (3,500–10,000 m deep). These skeletons date back to the Cambrian and have been used as paleoenvironmental indicators.

Support and Locomotion

The cytoplasm of radiolarians is divided into two regions, the endoplasm and the ectoplasm, which are separated by a capsular wall that is composed (usually) of mucoprotein. The central endoplasm is granular and dense, and contains most of the organelles: nucleus, mitochondria, Golgi apparatus, pigmented granules, digestive vacuoles, crystals, and the axoplast. Axopods emerge from the axoplast in the endoplasm through pores in the capsule wall. The pore pattern is

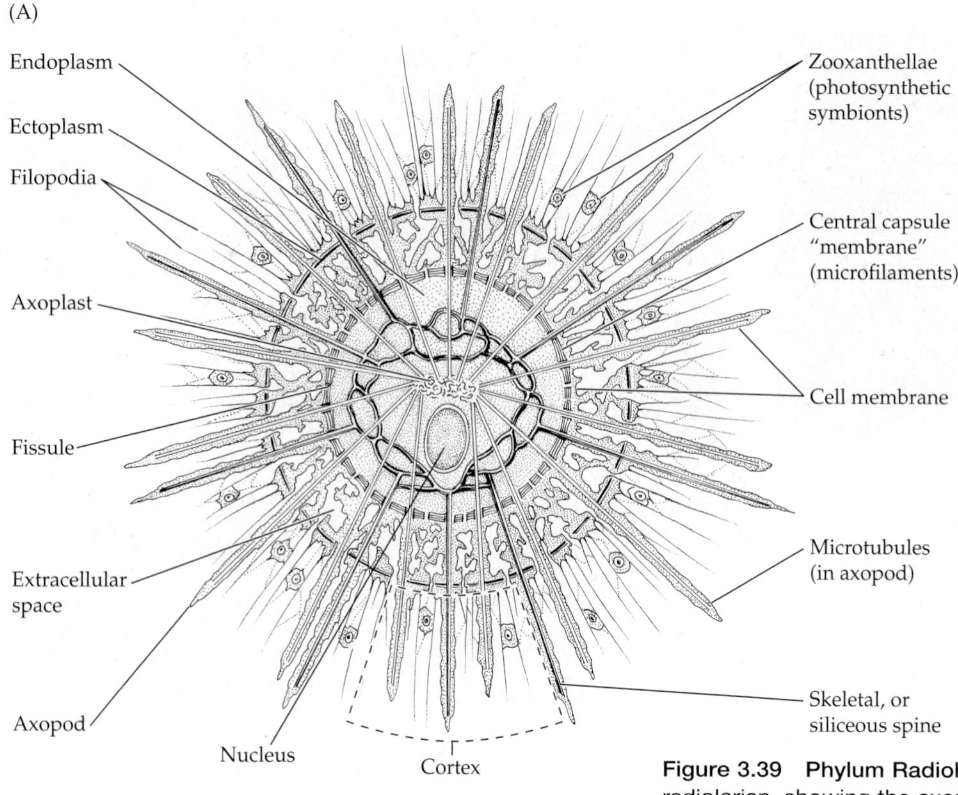

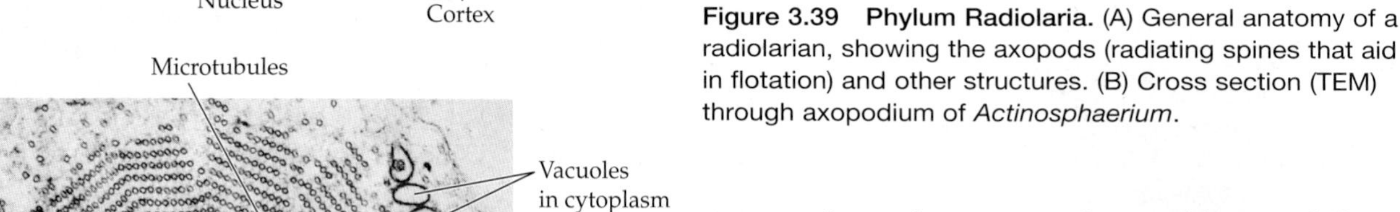

Figure 3.39 Phylum Radiolaria. (A) General anatomy of a radiolarian, showing the axopods (radiating spines that aid in flotation) and other structures. (B) Cross section (TEM) through axopodium of *Actinosphaerium*.

variable. In polycystines, for example, there are many pores in the capsule wall, all of which are associated with collarlike structures called **fusules**. In phaeodarians, there are only three pores in the capsule wall. The largest pore, the **astropyle**, is associated with fusules. Axopods emerge from the two smaller pores, the **parapyles**, which are not associated with fusules.

Most radiolarians have skeletons for support. The skeleton is formed and housed within the endoplasm and is therefore internal. In Polycystina and Phaeodaria, the skeleton is composed primarily of siliceous elements that are solid in polycystines and hollow in phaeodarians. In acantharians, the skeleton is composed of strontium sulfate embedded in a proteinaceous matrix. These skeletons vary greatly in construction and ornamentation, and frequently bear radiating spines that aid in flotation. In the Acantharia, there is a strict arrangement of 20 radial spicules, which is a diagnostic feature of this group.

The ectoplasm, often called the **calymma**, lies outside the capsule wall and contains mitochondria, large digestive vacuoles, extrusomes, and (in some) algal symbionts. The calymma has a rather foamy appearance due to the presence of a large number of vacuoles (see Figures 3.38 and 3.39). The vacuoles, some of which house oil droplets and other low-density fluids, aid in flotation in free-living species. When surface water conditions become rough and potentially dangerous to these delicate protists, the calymma expels some of its contents and the creature sinks to calmer depths. Eventually the cell replaces the oils and other fluids, and the organism rises toward the surface again. Unique to the ectoplasm of phaeodarians are balls of waste products called **phaeodium**, after which this group was named. The ectoplasm of acantharians is covered with a netlike cortex that is anchored to the apex of the spicules by contractile myonemes. Both the cortex and the presence of myonemes in association with the skeletal spicules are distinctive features of the acantharians.

Locomotion in radiolarians is limited. Most radiolarians drift passively in the water column using the axopods, skeletal spines (if present), and ectoplasmic vacuoles as flotation devices. In some cases, however, the axopods and spines play a more active role in locomotion. For example, the axopods may also help these organisms maintain their position in the water column by the expansion and contraction of vacuoles between the axopods. This has been suggested because it has been observed in polycystines that when the ectoplasm and axopods are lost during cell division, the organisms sink. In at least one genus, *Sticholonche*, the axopods appear to be used as tiny oars. In acantharians, it is thought that contraction of the myonemes that are attached to the spicules may somehow regulate buoyancy.

All radiolarians are heterotrophic, obtaining food by phagocytosis, and many are voracious predators. Prey items include bacteria, other protists (e.g., ciliates, diatoms, flagellates), and even small invertebrates (e.g., copepods). Radiolarians use their axopods as traps for prey. The Radiolaria are usually equipped with extrusomes such as mucus-producing mucocysts, and **kinetocysts**, which eject barbed, threadlike structures. Prey items adhere to mucus (discharged by mucocysts) that covers the extended axopods, or they become attached to the axopods by the discharged kinetocysts.

The size and motility of the prey determines the particular feeding mechanism used. Small prey are engulfed in food vacuoles directly, whereas large prey may be partially digested extracellularly by the action of secretory lysosomes in the mucous coat or broken into pieces by the action of large pseudopodia. The extracellular food is drawn toward the cell body by cytoplasmic streaming, eventually enclosed within food vacuoles, and completely digested in the central portion of the cell. In polycystines, it has been observed that when fast-moving or large prey (especially those with skeletons, such as diatoms) contact the axopods, the axopods actually collapse, drawing the prey into the cell body where it is engulfed by thin filopodia and then enclosed in a food vacuole (see Figure 3.38D). The collapse of the axopods is thought to involve microtubule disassembly.

An interesting feeding arrangement is found in the phaeodarians. As mentioned earlier, they have only three openings in the capsule wall: two parapyles and the single astropyle. Prey become trapped on the axopods. Then a large pseudopod, formed from the astropyle, engulfs the prey item into a food vacuole where it is digested in the ectoplasm. Because of this behavior, some workers have referred to the astropyle as a cytostome.

Many polycystines and acantharians live near the water's surface. These protists often have algal symbionts, including chlorophytes and dinoflagellates, which presumably provide them with additional nutrients. Phaeodarians do not have algal symbionts, which is not surprising since they tend to be found in water depths unsuitable for photosynthesis.

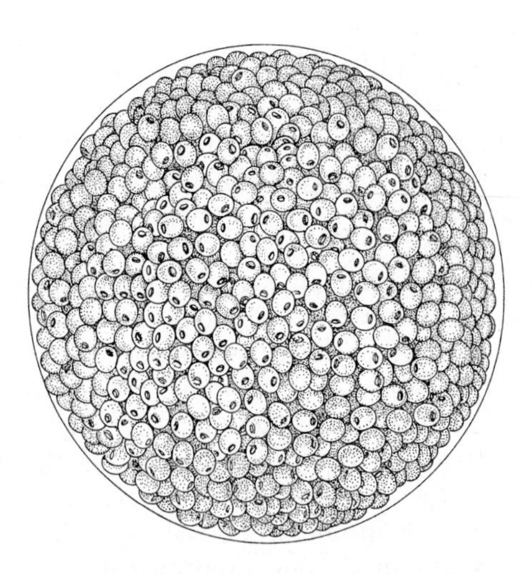

Figure 3.40 Phylum Radiolaria. Mass of swarmers produced by multiple fission within the central capsule of the radiolarian *Thalassophysa*.

Reproduction

Asexual reproduction occurs by binary fission, multiple fission, or budding. In the Polycystina and various shelled forms, however, division occurs along planes predetermined by body symmetry and skeletal arrangement. The same basic mode of multiple fission is seen in all groups. A polyploid nucleus results from numerous mitotic divisions. The nucleus fragments, producing many biflagellate individuals called swarmers, which eventually lose their flagella and develop into adults (Figure 3.40). In polycystines, swarmers have a crystal of strontium sulfate in their cytoplasm. In most polycystine and acantharian species, multiple fission is the only mode of asexual reproduction.

Sexual reproduction is poorly documented in radiolarians. Autogamy is usually triggered by either a lack of food or, conversely, it follows heavy feeding. First the cell encysts and undergoes a mitotic division to produce two gamonts. Each gamont nucleus divides by meiosis without cytokinesis. All but two of the nuclei disintegrate. The two surviving haploid nuclei fuse while still inside the cyst, forming a diploid zygote that later emerges from the cyst when environmental conditions become more favorable.

Nuclear division occurs by closed intranuclear pleuromitosis. Electron-dense plaques, located on the inner surface of the nuclear envelope, act as organizers for the mitotic spindle. A pair of centrioles, located outside of the nucleus is found near the plaques.

Phylum Haplosporidia: Haplosporidians

The haplosporidians comprise a small phylum of parasitic protists that typically (though not always, e.g., *Bonamia*) form uninucleate spores without polar capsules or polar filaments; the spore wall has an orifice at one pole (see Figure 3.34B). In *Urosporidium*, the orifice is "closed" by an internal diaphragm; in the other genera there is an external hinged operculum. Tubular, filamentous, or ribbon-like strands of material have often been observed on the external spore surface, but this ornamentation is structurally diverse and not well understood. There is, in fact, no single ultrastructural feature that characterizes haplosporidians, and they are usually recognized on the basis of a combination of features: a uninucleate stage, usually with a central nucleus, which develops to a diplokaryotic stage. In the later stage, two nuclei lie close and in contact with one another after division, with further nuclear division producing multinucleate plasmodia, from which operculate spores may develop. Haplosporidians are exclusively intracellular parasites of certain invertebrates, including flatworms (Turbellaria), annelids, crustaceans, and especially molluscs. They occur worldwide, primarily in marine environments and to a lesser extent in freshwater habitats. Causative agents of MSX disease, haplosporidians (*Haplosporidium* and *Bonamia*) have been known to decimate coastal oyster populations. We have little knowledge of haplosporidian life histories, and species are characterized on the basis of spore structure and gene sequencing.

The largest genus in the phylum, *Haplosporidium*, contains about two dozen species, most of which are parasites on bivalve molluscs, although some species infest commercially important crabs and shrimps (e.g., *H. littoralis* infects the common European shore crab *Carcinus maenas*).

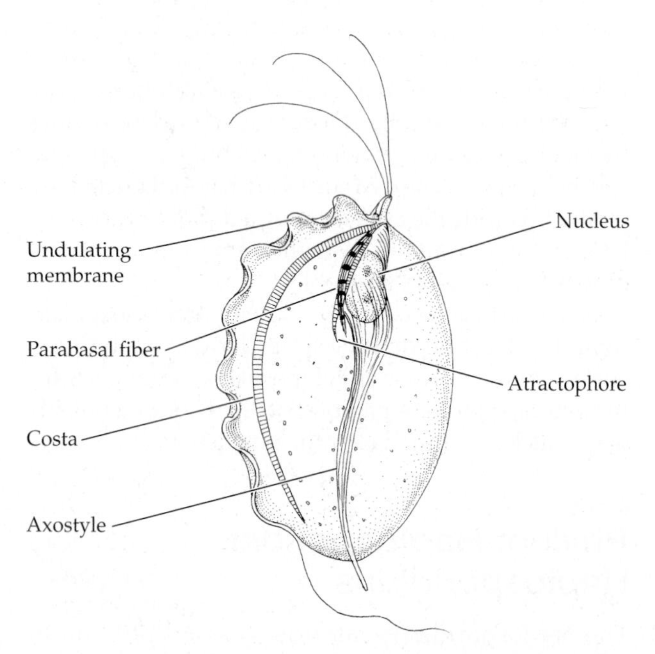

Figure 3.41 Phylum Parabasalida. *Trichomonas murius*, a trichomonad inhabiting the large intestine of mice.

Haplosporidians have been a troublesome group for taxonomists. Historically, the taxon was treated as a catchall for any spore-forming parasites with multinucleated naked cells (plasmodia) in their life cycles that could not be easily classified elsewhere. The Haplosporidia were first erected (in 1899) as an order of the phylum Sporozoa. In 1979 the Haplosporidia (and the Paramyxea) were separated from the "other Sporozoa" and placed in a new phylum called Ascetospora, not all of which have a spore stage in their life history. Recently, the Ascetospora was abandoned and the Haplosporidia and Paramyxea were each elevated to phylum rank. The phylum Paramyxea also contains several important parasites of oysters and other bivalves, including species of *Marteilia* from the Indo-Pacific region. Molecular phylogeneticists have begun trying to unravel the relationships of these groups, but so far with little success. Some protistologists now regard haplosporidians as an order within a recently erected phylum called Retaria, while others place them in a phylum called Cercozoa.

GROUP 4: EXCAVATA

Phylum Parabasalida: Trichomonads, Hypermastigotes, and Their Kin

The phylum Parabasalida contains about 300 species of heterotrophic flagellate protists, all of which are endosymbionts (mostly parasites) of animals (Figure 3.41). There are two major subgroups of parabasalids: the trichomonads and the hypermastigotes (Box 3O). The hypermastigotes (e.g., *Trichonympha*) are obligate mutualists in the digestive tracts of wood-eating insects such as termites and wood roaches. The obligate mutualism between hypermastigotes and termites and wood roaches is well studied. Although these insects eat wood, they lack the enzymes necessary to break it down. The hypermastigotes produce the enzyme cellulase, which breaks the cellulose in wood down to a form that the insect can metabolize.

Trichomonads are symbionts in the digestive, reproductive, and respiratory tracts of vertebrates, including humans. There are four species of trichomonads found in humans, three that are normally harmless commensals (*Dientamoeba fragilis, Pentatrichomonas hominis, Trichomonas tenax*), and one that is an extremely prevalent, sexually transmitted pathogen (*T. vaginalis*). *Pentatrichomonas hominis* is found in the gut of humans, other primates, dogs, and cats. It is generally present in less than 2% of the population, although in many developing countries the prevalence is much higher (e.g., 32% in Mexico). It is directly associated with poor hygiene since the parasite is transmitted by the oral-fecal route via contaminated food, water, and filth insects such as flies and roaches.

BOX 3O Characteristics of the Phylum Parabasalida

1. Small phylum of heterotrophic flagellates, without plastids, all of which are endosymbionts (mostly parasites) of animals. Hypermastigotes are obligate mutualists in the digestive tracts of wood-eating insects. Trichomonads are symbionts in the digestive, reproductive, and respiratory tracts of vertebrates.
2. Body surrounded only by plasma membrane; some rigidity provided by cytoskeleton, associated with the flagella
3. With flagella for locomotion (almost always); number of flagella can vary between one to thousands; fibrous roots (parabasal fiber, atractophore) and microtubular structures (axostyle, pelta) associated with basal bodies of the flagella
4. Highly modified mitochondria present in the form of hydrogenosomes (lacking many normal mitochondrial features, including DNA and oxidative phosphorylation)
5. Hypermastigotes possess either a single chromosomal or vesicular nucleus with a prominent nucleolus. Trichomonads possess a single vesicular nucleus with a minute nucleolus.
6. Nuclear division occurs by closed extranuclear pleuromitosis without centrioles.
7. Asexual cell division by longitudinal binary fission
8. Sexual reproduction occurs in some hypermastigotes, but is unknown in trichomonads. In hypermastigotes, sexual reproduction varies, occurs by gametogamy, gamontogamy, or autogamy.

The parabasalids get their name from a fiber, the **parabasal fiber**, which extends from the basal bodies to the Golgi apparatus. Several other unusual cytoskeletal fibers, basically rodlike or sheetlike bundles of microtubules, are associated with the basal bodies (an **atractophore**, an **axostyle**, and a **pelta**) and their presence, along with the parabasal fiber, is a diagnostic feature of this group. Most species also have an undulating membrane. Parabasalids have highly modified mitochondria called **hydrogenosomes**. These generate energy (ATP) in the absence of oxygen, releasing hydrogen gas as one of the waste products. Similar organelles are found in some other protists that live in oxygen-poor environments (e.g., many ciliates).

Trichomonads have been the focus of much research because there are four species found as symbionts in humans and several species that parasitize domestic stock and fowl. *Trichomonas vaginalis* is a cosmopolitan species found in the vagina and urethra of women, and in the prostate, seminal vesicles, and urethra of men. It is transmitted primarily by sexual intercourse, although it has been found in newborn infants. Its occasional presence in very young children suggests that the infection can also be contracted from shared washcloths, towels, or clothing. Most strains of *T. vaginalis* are of such low pathogenicity that the victim is virtually asymptomatic, although urethritis and prostatitis are not uncommon. However, other strains cause intense inflammation with itching and a copious greenish-white discharge (**leukorrhea**) that is swarming with the parasite. *Trichomonas vaginalis* is basically a highly voracious and indiscriminant predator, feeding on bacteria, vaginal epithelial cells, erythrocytes and leukocytes, and cell exudates. An estimated 170 million new cases of *T. vaginalis* occur annually.

O. F. Müller discovered *Trichomonas tenax* in 1773, when he examined a culture of tartar from his own teeth. *T. tenax* is generally regarded as a harmless commensal of the human mouth, its prevalence ranging from 4% to 53% of the population. However, a rare, but serious disease known as pulmonary trichomoniasis can be contracted by aspirating *T. tenax*. *Dientamoeba fragilis* is a fairly common parasite of human intestinal tracts, where it lives in the large intestine and feeds mainly on debris. Although traditionally considered a harmless commensal, recent studies suggest that infections by this protist routinely result in abdominal stress (e.g., diarrhea, abdominal pain). *Tritrichomonas foetus* is a parasite in cattle and other large mammals and is one of the leading causes of abortion in these animals; it is common in the United States and Europe. *Histomonas meleagridis* is a cosmopolitan parasite of gallinaceous fowl (i.e., domestic poultry and game birds of the order Galliformes). Histomoniasis in chickens and turkeys causes over a million dollars in losses annually.

Trichomonads are members of the parabasalid lineage that lacks "normal" mitochondria and peroxisomes, but contains hydrogenosomes. Analogous organelles have been identified in ciliates and some other eukaryotes. *Trichomonas vaginalis* is known to use carbohydrate as a main energy source via fermentative metabolism under aerobic and anaerobic conditions.

Support and Locomotion

In parabasalids, the cell body is surrounded only by plasma membrane, but some rigidity is provided by a system of supporting fibers and microtubules that are associated with the kinetosomes. There are two striated fibrous roots (a parabasal fiber and an atractophore) and two microtubular roots (an axostyle and a pelta) (Figure 3.41). The number of parabasal fibers is variable. In small trichomonads, there are only a few, whereas in hypermastigotes, such as *Trichonympha*, there can be over a dozen. The atractophore extends toward the nucleus from the basal bodies. The axostyle is a rodlike bundle of microtubules that originates near the basal bodies and curves around the nucleus as it extends to the posterior region of the cell. The pelta is a sheet of microtubules that encloses the flagellar bases. In trichomonads, an additional striated fiber called the costa is present. This fiber originates at the bases of the flagella and extends posteriorly beneath the undulating membrane. These fibers, along with the flagella and the nucleus, comprise the karyomastigont system (similar to those found in diplomonads). Trichomonads are characterized by protoplasmic plasticity, so that various shapes may be assumed. However, they are usually pear-shaped and possess 3 to 5 anterior flagella, with a recurrent anterior flagellum attached to the body as an undulating membrane.

Locomotion is accomplished by the beats of flagella. In *Trichomonas vaginalis*, for example, four free flagella form a tuft in the anterior region of the cell. The fifth flagellum is attached to the cell body at regular attachment sites so that when it beats, the cell membrane in that region of the body is pulled up into a fold, forming an undulating membrane. As in trypanosome kinetoplastids, the flagellum-undulating membrane complex seems to be efficient in moving the organism through viscous media. Hypermastigotes usually have dozens or even hundreds of flagella occurring all over the body. In these protists, the basal bodies of the flagella are arranged in parallel rows and are connected by microfibrils. The beat of the flagella is synchronized (as in ciliates, the synchronization is imposed by hydrodynamic effects), forming metachronal waves.

Some trichomonads (e.g., *Dientamoeba fragilis*, *Histomonas meleagridis*, *Trichomonas vaginalis*) also form pseudopodia. These pseudopodia function primarily in phagocytizing food particles, but they can also aid in locomotion.

Nutrition

All parabasalids are heterotrophic but lack a distinct cytostome (although claims have been made for the presence of a rudimentary, anteriorly positioned cytostome). In some trichomonads (e.g., *Tritrichomonas*), fluid is taken up by pinocytosis in depressions on the cell surface. Trichomonads also form pseudopodia that engulf bacteria, cellular debris, and leukocytes. Most parabasalids, however, take in particulate matter by phagocytosis. In hypermastigotes, pseudopodia formed in a sensitive region at the posterior end of the cell engulf wood particles. The chloroplasts of phototrophic euglenids originated through a secondary endosymbiotic relationship between eukaryovorous euglenids and green algae.

Reproduction

Asexual reproduction is by longitudinal binary fission (Figure 3.42A). Nuclear division occurs by closed extranuclear pleuromitosis with an external spindle. The atractophores are thought to act as microtubular organizing centers.

Sexual reproduction is unknown for the trichomonads but it occurs in some hypermastigotes, where it is well understood. A variety of sexual processes are exhibited by hypermastigotes, including **allogamy**, or cross-fertilization (e.g., union of two single-celled gametes to form a zygote) and **autogamy**, or self-fertilization (the fusion of two nuclei originating from a single cell). Hypermastigotes spend most of their lives as haploids in the digestive tract of wood-eating insects, dividing asexually by mitosis. Sexual reproduction is stimulated when the host insect molts and produces ecdysone, the molting hormone.

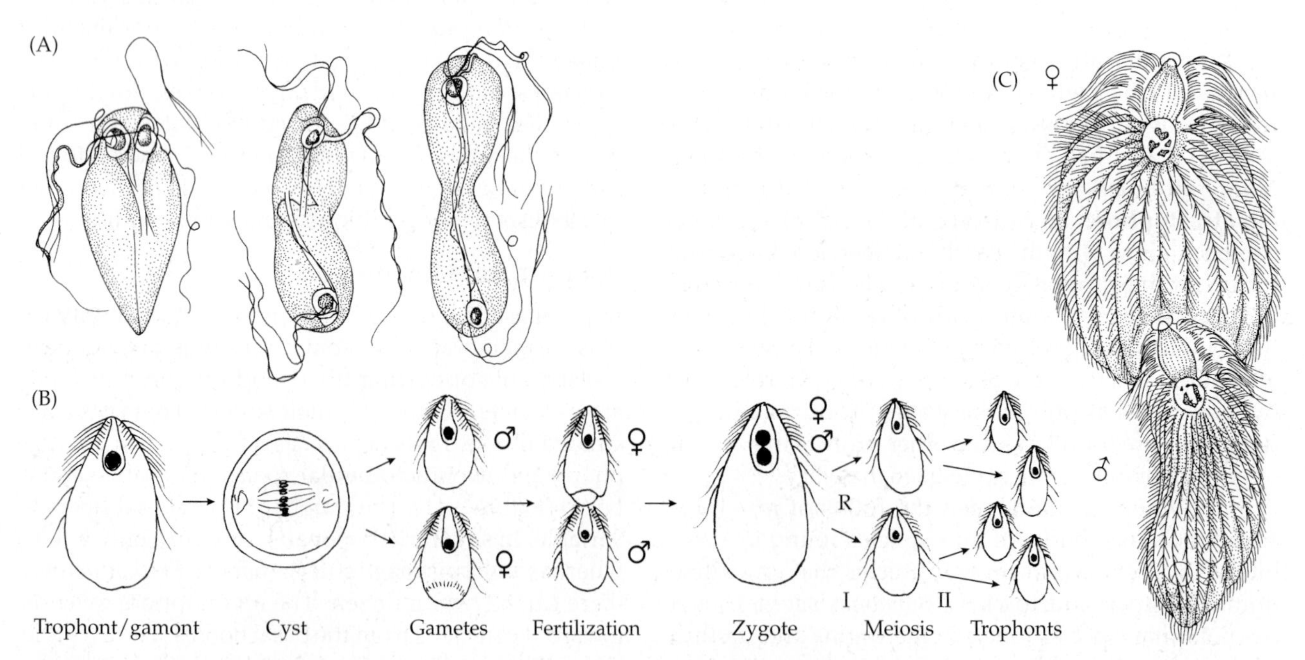

Figure 3.42 Phylum Parabasilida. Reproduction in parabasalids. (A) Longitudinal binary (asexual) fission in the trichomonad *Devescovina*. (B) Sexual reproduction in *Trichonympha* (a hypermastigote). (C) Mating activity (fertilization) in *Eucomonympha* (a hypermastigote), in which individuals act as gametes.

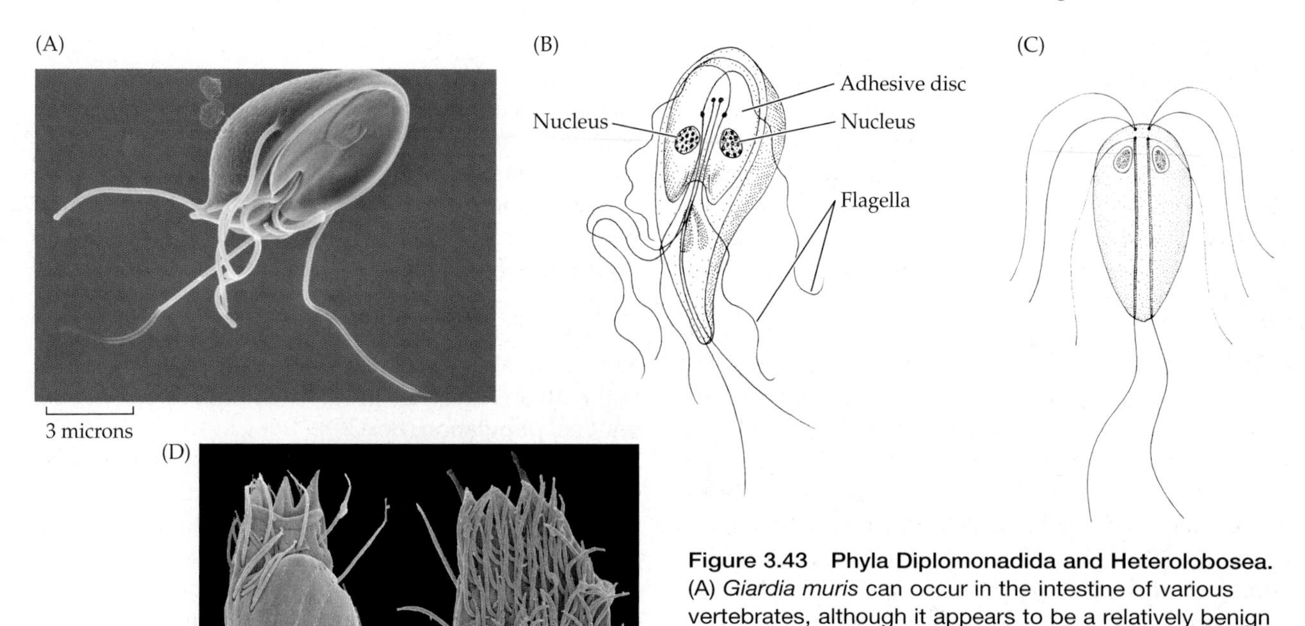

Figure 3.43 Phyla Diplomonadida and Heterolobosea. (A) *Giardia muris* can occur in the intestine of various vertebrates, although it appears to be a relatively benign parasite. The ventral adhesive disk and four pairs of flagella can be seen in this SEM photograph. Schematic drawings of *Giardia* (B) and *Hexamita* (C), illustrating the paired nuclei and numerous flagella. (D) Two views (SEMs) of *Stephanopogon minuta* (Heterolobosea), which has 8 rows of flagella on one side.

An example of a life cycle involving gametogamy is seen in *Trichonympha* (Figure 3.42B). In this group, the gametes are anisogametous, the male gamete being smaller than the female gamete. In some other species that undergo gametogamy, the gametes are isogametous. The haploid individual encysts and transforms into a gamont. While still encysted, the gamont divides by mitosis to produce a pair of flagellated gametes, one male and one female, which escape from the cyst. The posterior end of the female gamete is modified to form a fertilization cone through which the male gamete enters the cell. Once the male gamete enters, it is entirely absorbed by the female gamete, and nuclear fusion produces a diploid zygote. Within a few hours, the zygote undergoes meiosis, resulting in four haploid cells. Because *Trichonympha* are obligate anaerobes in the guts of insects, encysting prior to host molting may allow the insects to maintain their protist symbionts.

Phylum Diplomonadida: Diplomonads

The diplomonads were one of the first protist groups ever to be observed. Antony van Leeuwenhoek described a diplomonad protist, now known as *Giardia intestinalis* (=*G. lamblia*), from his own diarrheic stool as early as 1681 (Figure 3.43A). About 100 species of diplomonads are known today (Box 3P). This is a group of predominantly symbiotic flagellates, but a few free-living genera are known. Free-living species tend to be found in organically rich, oxygen-poor waters or sediments. Most diplomonads live as harmless commensals within the digestive tracts of animals, but a few are serious pathogens.

BOX 3P Characteristics of the Phylum Diplomonadida

1. Predominantly symbiotic flagellates, most of which are harmless commensals in the guts of animals (some are serious pathogens)
2. Heterotrophic; without plastids
3. Body surrounded only by plasma membrane; some rigidity provided by up to three microtubular roots associated with the flagella
4. With flagella for locomotion. Number of flagella varies (typically eight); usually divided into two equal clusters; one flagellum from each cluster usually directed posteriorly
5. Most possess two vesicular nuclei (one associated with each vesicular cluster) with minute nucleoli.
6. Mitochondria represented by highly aberrant organelles called mitosomes (which stem from an ancient endosymbiotic event)
7. Nuclear division occurs by semi-open orthomitosis, synchronous between the two nuclei. Replicated basal bodies (rather than separate centrioles) act as organizing centers for the mitotic spindle.
8. Asexual cell division by longitudinal binary fission
9. Meiosis and sexual reproduction unknown

The diplomonads are so-named because the first species described from this group had a two-fold symmetry defined by a pair of karyomastigont systems (Figure 3.43B). It was later discovered that some genera (the enteromonads) have only one system. Each "karyomastigont" consists of a nucleus and a set of flagella connected to it, plus a system that consists of fibers (mostly microtubular) that originate at the basal bodies of the flagella. The pair of anteriorly located nuclei (one from each karyomastigont), along with their nucleoli, makes the protist seem to have eyes that peer up at the observer (these are the eyes van Leeuwenhoek saw looking at him in 1681). The mitochondria are highly unusual; for a long time they were thought to be completely absent, but in 2003 tiny organelles called **mitosomes** were identified as probable mitochondrial homologues. The mitosomes stem from the endosymbiotic event that gave rise to ancestral mitochondria, but they are now highly reduced and no longer function in cellular energy production. They are, however, known to function in at least one biosynthetic pathway.

As noted above, some diplomonads are pathogenic. *Hexamita salmonis*, a parasite of fish, causes many deaths in salmon and trout hatcheries. *Giardia intestinalis* is a common and nearly ubiquitous intestinal parasite in humans that causes diarrhea, dehydration, and intestinal pain. Although it is not fatal if treated promptly, giardiasis is one of the top ten most common parasite diseases in the world today. There are literally hundreds of millions of *Giardia* infections each year, mostly in the developing world, resulting mainly from sewage contamination of drinking water. In the United States, *G. intestinalis* is the most prevalent parasitic protist, with an incidence possibly as high as 0.7% of the national population (Box 3Q).

Most of the genome of *Giardia intestinalis* was recently sequenced. Unusual features of this enigmatic protist include the presence of two diploid nuclei and the absence of mitochondria. The two nuclei, present in the trophozoite stage, both seem to be fully functional and essentially indistinguishable, containing evidential copies of the genome. The absence of typical mitochondria have led some biologists to speculate that this genus is an ancient clade that evolved prior to the endosymbiotic event that first generated mitochondria in eukaryotes. Analysis of sequence data from the

BOX 3Q *Giardia*

The genus *Giardia* is notable among human parasites in lacking mitochondria, conspicuous Golgi bodies, and lysosomes. For many years, these had been interpreted as primitive traits that placed *Giardia* and other diplomonads near the point of divergence between prokaryotes and eukaryotes (hence the genus has been referred to as a "missing link"). However, recent studies suggest that the absence of mitochondria represents a secondary loss. There are probably five valid species in the intensely studied genus *Giardia*: *G. lamblia* (= *G. intestinalis*, = *G. duodenalis*) and *G. muris* from mammals, *G. ardeae* and *G. psittaci* from birds, and *G. agilis* from amphibians. The closely related genus *Hexamita* has no human parasites, but *H. meleagridis* is a common parasite of the guts of young galliform birds (e.g., turkey, quail, pheasant), and it causes millions of dollars in losses to the U.S. turkey industry annually. The genus *Spironucleus* includes species that cause serious diseases in fishes, including farmed salmon.

Giardia lamblia is a cosmopolitan species that occurs most commonly in developing countries. It is the most common flagellated protist of the human digestive tract—there are hundreds of millions of infections in the developing world each year, thanks to human-to-human transmission and sewage contamination of drinking water. However, it is also present in natural water bodies, inducing in North America, and it can be contracted by drinking nonpurified, or poorly purified water while camping and hiking. Over 30,000 cases of giardiasis are reported annually in the United States, where animal reservoirs of *G. lamblia* include beavers, dogs, cats, and sheep. Treatment with quinacrine or metronidazole ("Flagyl") usually effects complete cure within a few days.

Giardia lamblia is a teardrop-shaped organism, dorsoventrally flattened with the ventral surface bearing a concave bilobed adhesive disc with which the cell adheres to the host tissue. Eight flagella arise from kinetosomes located between the anterior portions of the two nuclei. The flagella facilitate rapid swimming. Members of this genus also possess a unique pair of large, curved, dark-staining median bodies lying posterior to the adhesive discs; their function is unknown. In severe infections, the free surface of nearly every cell in the infected portion of the gut is covered by parasites. A single diarrheic stool can contain up to 14 billion parasites, facilitating the rapid spread of this very common protist. Some infections show no evidence of disease, whereas others cause severe gastritis and associated symptoms, no doubt due to differences in host susceptibility and strains of the parasite. The dense coating of these protists on the intestinal epithelium interferes with absorption of fats and other nutrients. Stools are fatty, but never contain blood. The parasite does not lyse host cells, but appears to feed on mucous secretions. Some protective immunity can apparently be acquired.

Lacking mitochondria, the tricarboxylic acid cycle and cytochrome system are absent in *Giardia*, but the organisms avidly consume oxygen when it is present. Glucose is apparently the primary substratum for respiration, and the parasites store glycogen. However, they also multiply when glucose is absent. Trophozoites divide by binary fission. As with trypanosomes and *Plasmodium*, *G. lamblia* exhibits antigenic variation, with up to 180 different antigens being expressed over 6 to 12 generations.

ribosomal RNA and elongation factor genes tend to place *Giardia* as a basal eukaryote, whereas other genes position it well within the diplomonads as one of many eukaryotic lineages that diverged nearly simultaneously with the opisthokonts and plants. However, genomic analysis has revealed the presence of a mitochondrion-like *cpn60* gene and a mitosome, which implies that the absence of typical (respiring) mitochondria in *Giardia* may reflect adaptation to a microaerophilic lifestyle, rather than divergence before the endosymbiotic event with the mitochondrial ancestor. The genome of *G. intestinalis* is small, compact, and distributed on just five chromosomes. DNA synthesis, transcription, RNA processing, and cell cycle machinery are also highly simplified. In fact, *Giardia* has fewer nucleotide sugar transporters than any other known eukaryotic genome. There are essentially no homologues for Krebs cycle enzymes and, except for scavenging pathways, no evidence of vestigial genes associated with purine and pyrimidine biosynthesis. The genome contains a single actin gene, but does not encode other classical microfilament proteins. It lacks myosins (as does the closely related genus *Trichomonas*), suggesting that either unknown novel proteins or altered cytoskeletal dynamics must be present. When attached to the surface of the intestinal mucosa, *Giardia* trophozoites have ample opportunity to pick up genes from bacteria and to scavenge products of host and bacterial metabolism. Hence, it is not surprising that, like both *Trichomonas* and *Entamoeba*, *Giardia*'s genome contains many genes that appear to have been gained secondarily via lateral gene transfer. In fact, one reason is has been so difficult to resolve *Giardia*'s evolutionary history is probably because so many genes may have been derived from horizontal transfer.

Support and Locomotion

The cell is surrounded by a plasma membrane, but some rigidity is provided by three microtubular roots that are associated with the basal bodies. These roots include a supranuclear fiber that passes over or in front of the nuclei, an infranuclear fiber that extends beneath or behind the nuclei, and a band of microtubules that parallels the posteriorly directed flagellum. Some genera have additional fibrous structures that are associated with the basal bodies. For example, the genus *Giardia* attaches to the host's intestinal epithelium with an adhesive disc that is constructed in part from the microtubular bands of the cytoskeleton. The disc is delimited by a ridge or lateral crest that is composed of actin and is used to "bite" into the host's tissue. The contractile proteins myosin, actin, and tropomyosin have all been reported around the periphery of the disc and may be involved with attaching to the host. The supranuclear fiber in *Giardia* is composed of a single ribbon of microtubules that connects to the plasma membrane of the disc. Each microtubule of the supranuclear fiber is associated with a ribbon of protein that extends into the cytoplasm.

Each karyomastigont system typically has four flagella. The flagella are of at least two kinds, although in *Giardia* each of the four flagella that form each karyomastigont has a distinct appearance. Locomotion is accomplished by the coordinated actions of the eight flagella (although not all play a direct locomotory role in all species). It has been suggested that the flagella may also be involved with creating a suction force beneath the adhesive disc in *Giardia*, enabling it to attach to its host.

Nutrition

Most diplomonads are phagotrophic and feed on bacteria. These forms have a cytostome through which bacteria are engulfed via endocytosis. In some (e.g., *Spironucleus*, *Hexamita*) the two intracellular channels in which the posterior flagella lie function as cytostomes. Other genera such as *Giardia* and *Octomitis* lack cytostomes and are saprozoic, feeding upon mucus secretions of the host's intestinal tissue through simple pinocytosis.

Reproduction

Asexual reproduction is the only reproductive mode known to occur in diplomonads, and division occurs along the longitudinal plane. Nuclear division involves semi-open orthomitosis and is synchronous between the two nuclei (if there are two). Replicated basal bodies act as organizing centers for the mitotic spindle. Most symbiotic diplomonads form cysts at some point during their life cycle, thus alternating between a motile trophozoite form and a dormant encysted form. *Giardia intestinalis*, for example, will form a thick protective covering that resists desiccation as it passes from the host's small intestine into the large intestine, where it is prone to dehydration. Once it leaves the digestive system through the anus, it must be swallowed by another host, where it will travel through the digestive system until it reaches the duodenum of the small intestine and excyst.

Phylum Heterolobosea: Heterolobosids

The Heterolobosea comprise a small and enigmatic protist group, first described in 1985, that seems to be most closely related to the phyla Euglenida and Jakobida. Many heterolobosids can transform between amoeboid, flagellate, or cyst forms, although the amoeboid stage is most frequently seen. Flagellate stages typically have 2 or 4 flagella and possess the typical Excavata feeding groove. The ameboid stage does not form true pseudopods. Instead, they move with "eruptive waves" which bulge from one end of the cell. All have mitochondria. Ameboid stages seem to be mainly

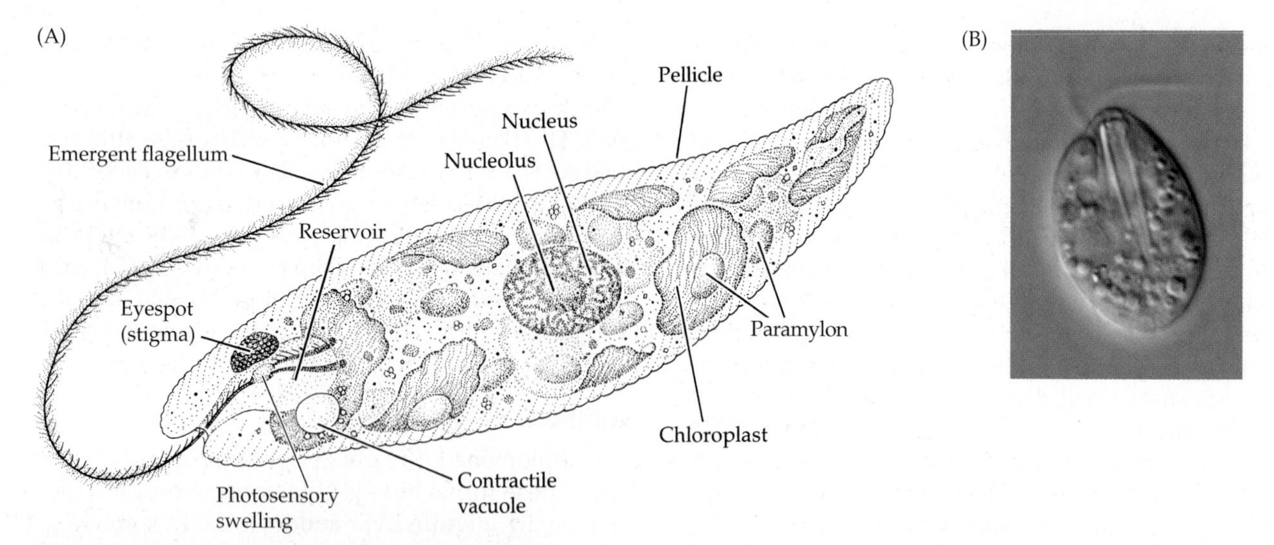

Figure 3.44 Phylum Euglenida. (A) Anatomy of *Euglena*. (B) *Entosiphon*. (C) A variety of *Euglena* species and body shapes.

feeding; flagellate stages, many locomotory. Most heterolobosids live as bacteriovores in soil, but they are also found in freshwater and organic waste (including feces). A few species are marine and some species are parasitic, including *Naegleria fowleri*, which can become pathogenic in humans and is mostly fatal.

Phylum Euglenida: Euglenids

The Euglenida include about 1,000 described species, mostly from fresh water but marine and brackish-water species are also common. Most are noncolonial, but some colonial forms exist (e.g., *Colacium*). Euglenids come in a wide variety of shapes (e.g., elongate, spherical, ovoid, peltate) (Figures 3.44 and 3.1F). The familiar genus *Euglena* has been used extensively in research

BOX 3R Characteristics of the Phylum Euglenida

1. Mostly marine, although the majority of the described species are freshwater
2. Shape of cell maintained by pellicle, formed by interlocking strips of protein beneath the cell membrane that are associated with linked microtubules arranged in regular pattern (also beneath the cell membrane). Some species secrete a mucous lorica.
3. With two flagella of unequal length for locomotion, supported by paraxonemal rods (one of the flagella may be greatly reduced); each flagellum usually with hairs; one or both flagella arise from an anterior pocket, the reservoir
4. Single mitochondrion has discoidal cristae
5. With single chromosomal nucleus
6. Heterotrophic or autotrophic. Photosynthetic forms have chlorophylls *a* and *b*; usually appear grass-green in color. Thylakoid membranes arranged in stacks of three; three membranes surround the chloroplast; outermost membrane not continuous with nuclear membrane
7. Food reserves stored in cytoplasm as the unique starchlike carbohydrate, paramylon
8. Nuclear division occurs by closed intranuclear pleuromitosis without centrioles; organizing center for mitotic spindle is not obvious
9. Asexual reproduction by longitudinal binary fission
10. May be strictly asexual—neither meiosis nor sexual reproduction has been confirmed

laboratories for decades, and it is commonly studied in introductory biology and invertebrate zoology courses (Box 3R).

The phyla Euglenida and Kinetoplastida are closely related, even though some euglenids are photosynthetic and some kinetoplastids (see below) are parasitic heterotrophs. Shared morphological features include: linked microtubules underlying the cell membrane; discoidal mitochondrial cristae; flagella containing a latticelike, spiral, or crystalline supportive rod (the paraxonemal rod); an anterior pocket from which the two flagella arise; and a similar pattern of mitosis. Molecular studies have also corroborated the close relationship of these two groups. Most recently, the odd, low-oxygen, deep-sea genus *Calkinsia* has also been shown to belong to the Euglenozoa clade, sharing the same paraxonemal rods, microtubular root system, and extrusomes seen in Euglenida and Kinetoplastida.

Euglenids are commonly found in bodies of water rich in decaying organic matter. As such, some of them are useful indicator organisms of water quality (e.g., *Leocinclis, Phacus, Trachelomonas*). Some species of *Euglena* have been used in experiments for wastewater treatment and have been reported to extract heavy metals such as magnesium, iron, and zinc from sludge. Other euglenids, however, are environmental pests, and some have been shown to produce toxic substances associated with diseases in trout fry. Still others are responsible for toxic blooms, which have caused destruction of fishes and molluscs in Japan. Many species are phagotrophic, hunting particulate food items such as other small protists and bacteria.

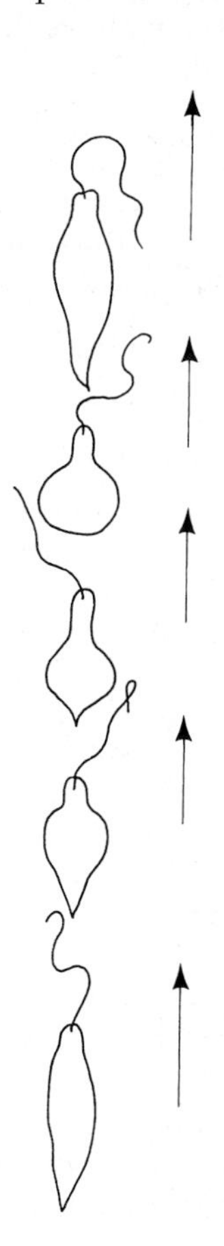

Support and Locomotion

The shape of euglenids is maintained by a pellicle consisting of interlocking longitudinally or helically arranged strips of protein that articulate along their lateral margins. The stripes that can sometimes be seen on a euglenid are the seams between the long protein strips winding around the cell. The pellicle is also supported by regularly arranged microtubules lying just underneath each strip. The rigidity of the pellicle is variable. Some (e.g., *Menodium, Rhobdomonas*) have protein strips that are fused together into a rigid pellicle, while others (e.g., *Euglena*) have protein strips that are articulated to produce a flexible pellicle. Those euglenids with a flexible pellicle undergo euglenoid movement, or **metaboly**, in which the cell undulates as it rapidly extends and contracts (Figures 3.44 and 3.45). Although this type of movement is not fully understood, it is accomplished by the microtubular-driven sliding of adjacent protein strips against one another (Figure 3.46).

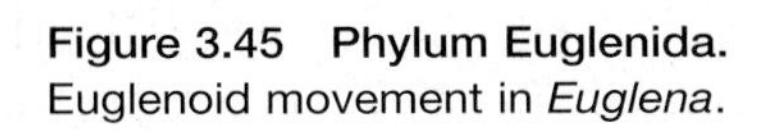

Figure 3.45 Phylum Euglenida. Euglenoid movement in *Euglena*.

A few euglenids (e.g., *Ascoglena, Colacium, Strombomonas, Trachelomonas*) secrete a lorica, or envelope, exterior to the cell membrane. The lorica is formed by the mucus secretions of small vesicles called **mucocysts**, which are located under the cell membrane along the seams between the protein strips of the pellicle. Secretions of mucocysts are also used to form protective coverings when environmental conditions become unfavorable.

Locomotion in euglenids is primarily by flagella. They have two flagella, but one may be very short or represented by just a kinetosome. The flagella originate in an invagination at the anterior end of the cell, called a **reservoir** (= flagellar pocket). In photosynthetic species, the longer, anteriorly directed flagellum (the emergent flagellum) propels the cell through the water or across surfaces. The shorter flagellum either trails behind or does not emerge from the reservoir at all. In many heterotrophs (e.g., *Entosiphon*), the posterior flagellum is actually the longer of the two and is involved in gliding locomotion, while the anterior flagellum is probably primarily for food detection and other tactile functions. Both flagella have a single row of hairs on their surface and a latticelike supporting rod, called the paraxonemal rod, lying adjacent to the microtubules within the shaft. In phagotrophic euglenids, the flagella (with associated paraxonemal rods and hairs) are used to glide along substrates.

Nutrition

Euglenids are quite variable in their nutrition. About one-third of the Euglenids have chloroplasts and are photoautotrophic. These species are positively phototactic and have a swelling near the base of the anterior

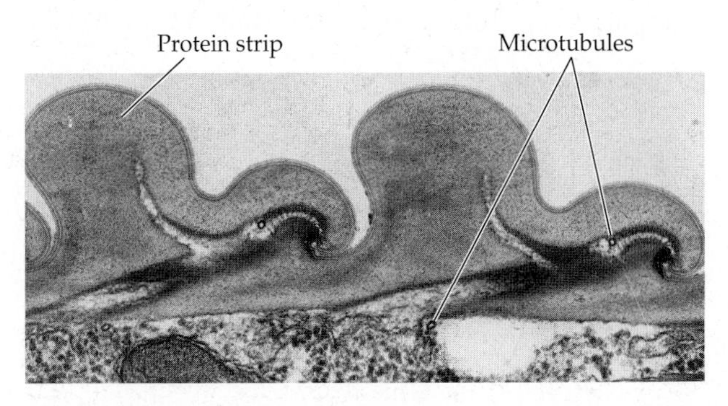

Figure 3.46 Phylum Euglenida. Cross section (TEM) through the pellicle of *Euglena* showing the protein strips and microtubules.

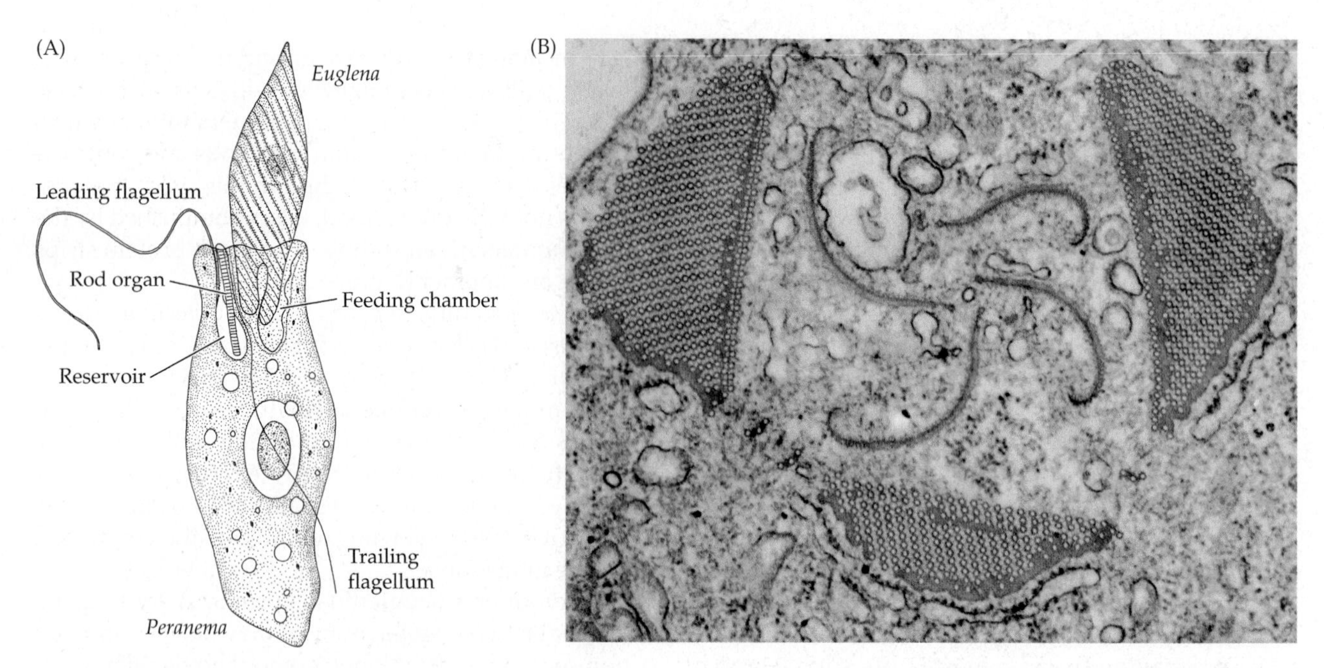

Figure 3.47 Phylum Euglenida. (A) *Paranema* feeding on *Euglena*. *Paranema* possesses an expandable feeding pocket separate from the reservoir in which the flagella arise. The rod organ can be extended to pierce prey and either pull it into the feeding apparatus or hold it while the contents are sucked out. (B) TEM through the cytopharyngeal rods and vanes of the euglenid *Entosiphon*.

flagellum that acts as a photoreceptor. The chloroplast is surrounded by three membranes and has thylakoids that are arranged in stacks of three (see Figure 3.2). The photosynthetic pigments include: chlorophylls *a* and *b*, phycobilins, β-carotene, and the xanthophylls neoxanthin and diadinoanthin.

Approximately two-thirds of the described species of euglenids lack chloroplasts and are thus obligate heterotrophs, and even phototrophic forms can lose their chloroplasts and switch to heterotrophy. A few parasitic species have been reported in invertebrates and frog tadpoles, but these reports remain questionable. Most euglenids also take in dissolved organic nutrients by saprotrophy, and this is generally restricted to parts of the cell not covered by the pellicle (e.g., the reservoir). Some euglenids also ingest particulate food items by phagocytosis of relatively large (sometimes comparatively huge) food materials. These have a cytostome located near the base of the flagella where food vacuoles form (e.g., *Peranema*; Figure 3.47A). The cytostome typically leads to a tube (the "cytopharynx") that extends deep into the cytoplasm. The walls of the cytopharynx are often reinforced by highly organized bundles of microtubules (e.g., *Entosiphon*, *Peranema*) (Figure 3.47B). Extrusomes are often found near the cytostome and presumably aid in prey capture. Euglenids typically store food reserves in the form of starches and a unique molecule called **paramylon**, which is a starchlike carbohydrate.

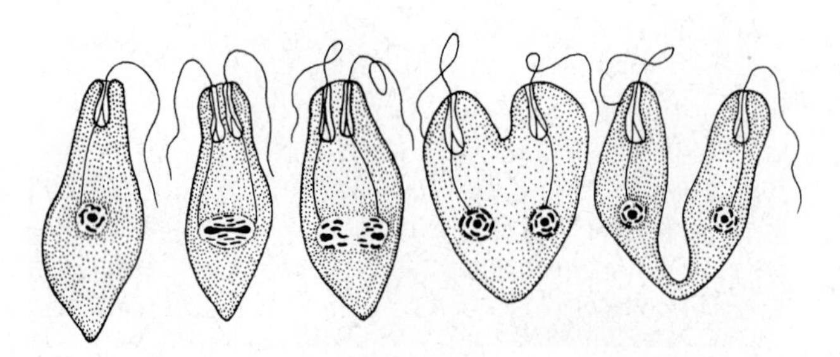

Figure 3.48 Phylum Euglenida. Asexual reproduction. Longitudinal fission in *Euglena*, in which the flagella and reservoir duplicate prior to cell division.

Reproduction

Asexual reproduction in euglenids primarily is by longitudinal cell division (Figure 3.48). Nuclear division occurs by closed intranuclear pleuromitosis. During mitosis, the nucleolus remains distinct and no obvious microtubular organizing center is evident. Sexual reproduction has been reported in one species, but this has not been confirmed.

Phylum Kinetoplastida: Trypanosomes, Bodonids, and Their Kin

There are about 600 described species of kinetoplastids. The phylum includes two major groups: the bodonids and the trypanosomes (*Trypanosoma*, *Leptomonas*, *Leishmania*, etc.) (Figures 3.49 and 3.1F). The bodonids are primarily free living in marine and freshwater en-

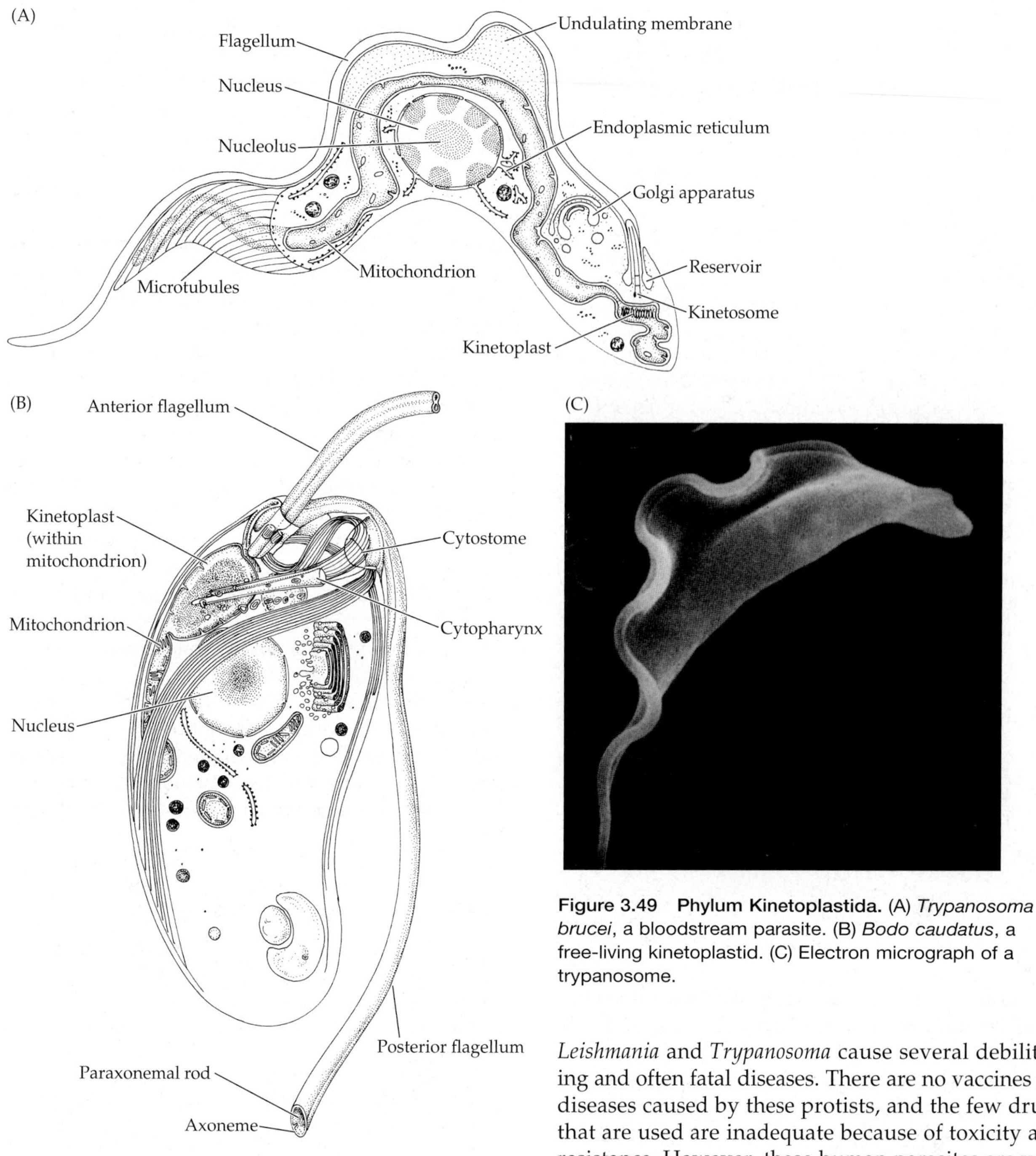

Figure 3.49 Phylum Kinetoplastida. (A) *Trypanosoma brucei*, a bloodstream parasite. (B) *Bodo caudatus*, a free-living kinetoplastid. (C) Electron micrograph of a trypanosome.

vironments, especially those rich in organic material. Like phagotrophic euglenids, they are often associated with surfaces, but generally being smaller they tend to take smaller prey and consequently are important consumers of bacteria in these habitats. Because bodonids have strict oxygen preferences they often aggregate at a particular distance from the water surface. The trypanosomes (or trypanosomatids) are exclusively parasitic, and they occur in the digestive tracts of invertebrates, phloem vessels of certain plant species, and the blood of vertebrates. *Leptomonas* exhibits the simplest life cycle, in which an insect is the sole host and transmission occurs by way of an ingested cyst. In humans, *Leishmania* and *Trypanosoma* cause several debilitating and often fatal diseases. There are no vaccines for diseases caused by these protists, and the few drugs that are used are inadequate because of toxicity and resistance. However, these human parasites are very closely related, sharing a conserved core proteome of about 6,200 genes, suggesting that a genetic-based scientific attack on these protists might be possible as we come to better understand their gene expression patterns (Box 3S).

Kinetoplastids are best known as agents of disease in humans and domestic animals. Species of *Leishmania* cause a variety of ailments collectively called leishmaniasis, and these include kala-azar (a visceral infection that particularly affects the spleen), oriental sore, and "Baghdad boil" (characterized by open, slow-healing, sometimes disfiguring skin boils; parasites that spread to internal organs and can cause death if left untreated), and several other skin and mucous membrane

BOX 3S Characteristics of the Phylum Kinetoplastida

1. Bodonids are primarily free-living heterotrophs in marine and freshwater environments; trypanosomes are strictly parasitic in the digestive tracts of invertebrates.
2. Shape of cell maintained by a pellicle, consisting of the cell membrane and a corset of microtubules beneath it.
3. With one (trypanosomes) or two (bodonids) flagella for locomotion. Flagella contain a paraxonemal rod and arise from an anterior pocket. Flagellum of trypanosomes usually forms an undulating membrane.
4. Single, large, elongate mitochondrion has discoidal cristae and conspicuous discoidal concentration of mDNA (the kinetoplast). Shape of cristae can change as organism progresses through life cycle, but predominantly discoidal.
5. With single vesicular nucleus; prominent nucleolus typically evident.
6. Nuclear division occurs by closed intranuclear pleuromitosis without centrioles. Plaques on inside of nuclear envelope may act as organizing centers for mitotic spindle.
7. Asexual reproduction by longitudinal binary fission.
8. Neither meiosis nor sexual reproduction has been confirmed, although indirect evidence suggests sex might occur in at least some kinetoplastids.
9. Without plastids.
10. Mitochondrial DNA forms aggregates, collectively known as the kinetoplast, readily seen with the light microscope.

infections. Leishmaniasis strikes over one million humans annually but due to effective treatment, it only kills about 1,000 people each year (although many survivors are badly scarred). Leishmaniasis occurs in the tropics and subtropics, and it is transmitted almost exclusively by the bite of sand flies (Diptera: Psychodidae: Phlebotominae). The Neotropical species *Leishmania mexicana* first appeared in the United States (in Texas) in 2007—one of many tropical parasites that are beginning to show up in the United States as climates warm. Both *Trypanosoma* and *Leishmania* were long thought to be asexual and clonal, but in the 1980s solid evidence began to appear that they were also capable of genetic exchange (Box 3T).

More serious diseases are caused by members of the genus *Trypanosoma*, all of which are parasites of vertebrates. *Trypanosoma brucei* is a debilitating parasite that lives in the bloodstream of African hoofed animals, in which it causes a disease called nagana (which kills about three million farm animals annually). It also attacks domestic livestock, including horses, sheep, and cattle; in the latter case it is often fatal, making it impossible to raise livestock on more than 4.5 million square miles of the African continent (an area larger than the United States). Two other African species (sometimes considered subspecies of *T. brucei*) are *T. gambiense* and *T. rhodesiense*, both of which cause sleeping sickness in humans. These parasites are introduced into the blood of humans from the salivary glands of the blood-sucking tsetse fly (*Glossina*). From the blood, trypanosomes can enter the lymphatic system and ultimately the cerebrospinal fluid. It is estimated that 300,000 to 500,000 people annually contract sleeping sickness, and once it enters the brain it is invariably fatal if untreated (it kills about 100,000 people annually). Tsetse flies have been called "vampires of the insect world," because of their voracious appetite for blood. Whereas only female mosquitoes sip blood, among tsetse flies both sexes will drink nearly their own weight in blood with each meal. Infected flies are known to drink more frequently, favoring transmission of the trypanosome parasite to more vertebrate hosts.

Chagas disease (common in Central and South America, and Mexico) is caused by *Trypanosoma cruzi* and is transmitted to humans by cone-nosed hemipteran bugs (also known as assassin or kissing bugs; family Reduviidae, subfamily Triatominae). These bugs feed on blood and often bite sleeping humans. They commonly bite around the mouth, hence the vernacular name. After feeding they leave behind feces that contain the infective stage, which invades through mucous

BOX 3T Leishmaniasis

Species of *Leishmania* are difficult to differentiate morphologically, and their taxonomy is unsettled (although now being unraveled by molecular genetics). Most widespread are *Leishmania infantum* and *L. major*, which occur in Africa and southern Asia and are transmitted by species of the sand fly genus *Phlebotomus* (Psychodidae). These species produce the cutaneous ulcers variously known as oriental sore, cutaneous leishmaniasis, Jericho boil, Aleppo boil, and Delhi boil. *Leishmania donovani* is endemic to southern Asia but also occurs in low levels in Latin America and the Mediterranean region; it is the etiological agent of Dum-Dum fever, or kala-azar. Kala-azar can result in extreme and even grotesque skin deformations. *Leishmania braziliensis* is endemic to Brazil, where it causes espundia, or uta, which often leads to such severe destruction of the skin and associated tissues that complete erosion of the lips and gums ensues. *Leishmania mexicana* occurs in northern Central America, Mexico, Texas, and probably some Caribbean islands, where it mostly affects agricultural or forest laborers. Infections of *L. mexicana* cause a cutaneous disease called chiclero ulcer, because it is so common in "chicleros," men who harvest the gum of chicle trees.

membranes or the wound caused by the insect's bite. Occasionally the bugs bite around the eyes of the sleeping victims, and subsequent rubbing leads to conjunctivitis and swelling of a particular lymph node, a symptom known as Romaña's Sign. The parasites migrate to the bloodstream, where they circulate and invade other tissues. In chronic human infections, *T. cruzi* can cause severe tissue destruction, including the enlargement and thinning of walls of the heart. In Central and South America the incidence of Chagas disease is high, and an estimated 15 to 20 million persons are infected at any given time with an annual death rate recognized at ~21,000 people. A study in Brazil attributed a 30% mortality rate to Chagas disease. In the United States, at least 14 species of mammals may serve as reservoirs for *Trypanosoma cruzi* (including dogs, cats, opossums, armadillos, and wood rats). However, the U.S. strain of *T. cruzi* is less pathogenic than the Mexican and Central and South American strains, and the species of *Triatoma* in the U.S. tend not to defecate when they bite, thus leading to a lower incidence of the disease north of Mexico. In recent decades, *T. cruzi* has also been found to spread by way of blood and organ donations.

The Kinetoplastida have a single, large, elongate mitochondrion with a uniquely conspicuous dark-staining concentration of mitochondrial DNA (mDNA) called the **kinetoplast** (hence the phylum name). Generally the kinetoplast is found in the part of the mitochondrion lying close to the kinetosomes, although there is no known relationship between these two structures. The size, shape, and position of the kinetoplast are important in the taxonomy of trypanosomes and bodonids, and it is used in distinguishing between different stages in the life cycle. Kinetoplast DNA (kDNA) in trypanosomes is organized into a network of linked circles, quite unlike the DNA in mitochondria of other organisms. Kinetoplastids share many ultrastructural and molecular features with the euglenids, their putative sister group (see Euglenida, above).

Support and Locomotion

The shape of the cells is maintained by a pellicle consisting of the cell membrane and a supporting layer of microtubules. In some smaller bodonids, the pellicular microtubules consist of three microtubular bands, whereas in other bodonids, and in trypanosomes the pellicular microtubules are evenly spaced and form a more-or-less complete corset that envelops the entire body. In trypanosomes, a layer of glycoprotein (12 to 15 μm thick) coats the outside of the cell and acts as a protective barrier against the host's immune system. The composition of the glycoprotein coat is changed cyclically; as a result, the trypanosome is able to avoid the host's immune system. This has been well studied in pathogenic trypanosomes such as *Trypanosoma brucei* (the causative agent of African sleeping sickness). When the trypanosome enters the host's body, the immune system recognizes the glycoprotein as foreign (an antigen) and specific antibodies are made against it. Although most of the trypanosome population is destroyed, a few cells are able to evade the immune system by changing their glycoprotein coat so that the new coat is unrecognizable to the host's antibodies. Once a new antibody is produced by the host, another new glycoprotein is produced by the trypanosome, and so on. About 1,000 genes contain information that encodes surface glycoproteins, although it appears that only one of these genes is expressed at a time. The ability of trypanosomes to change their glycoprotein coat makes treatment of trypanosome infections difficult.

Both bodonids and trypanosomes move using flagella, which, like those of euglenids, usually emerge from an inpocketing and contain a paraxonemal rod. Trypanosomes have two kinetosomes, but only one has a flagellum. In many forms this flagellum lies against the side of the cell and its outer membrane is attached to the cell body's membrane. When the flagellum beats, the membrane of the cell is pulled up into a fold and looks like a waving or undulating membrane (see Figure 3.49). This arrangement appears to be relatively efficient in moving the cell through viscous media (e.g., blood). Although trypanosomes can change the direction of flagellar beat in response to chemical or physical stimuli, usually the beat begins at the tip of the flagellum and proceeds toward the kinetosome. This is the reverse of the way the flagella of other eukaryotes beat (from base to tip). Bodonids usually have two flagella: one is extended anteriorly while the other trails behind and may be partially attached to the body in some species (e.g., *Dimastigella, Procryptobia*). Although most bodonids swim, they also employ various mechanisms to move efficiently across surfaces, including a gliding type of locomotion.

Nutrition

All kinetoplastids are heterotrophic. Free-living bodonids capture particulate food, primarily bacteria, with the aid of their anterior flagellum and ingest through a permanent cytostome. Most are raptorial feeders, ingesting prey items (especially attached bacteria) one at a time. The cytostome leads to a cytopharynx, which is supported by microtubules. At the base of the cytopharynx, food is enclosed in food vacuoles by endocytosis.

Little is known about feeding mechanisms in trypanosomes, all of which are parasitic. Some trypanosomes have a cytostome-cytopharyngeal complex through which proteins are ingested. The proteins are taken into food vacuoles by pinocytosis at the base of the cytopharynx. It has also been reported that some trypanosomes can take in proteins by pinocytosis from the membrane lining the flagellar pocket or by some sort of cell membrane-mediated mechanism.

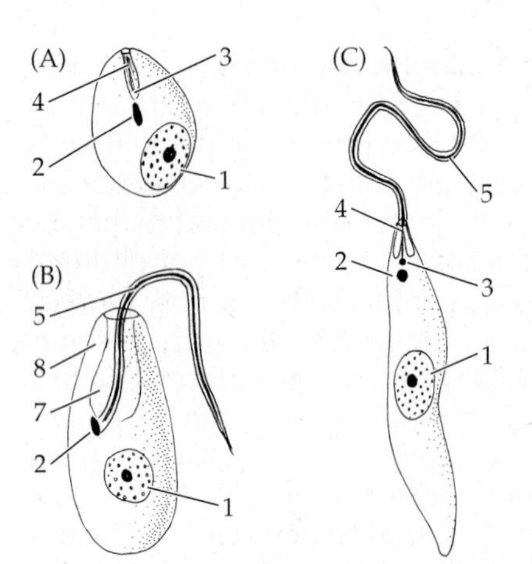

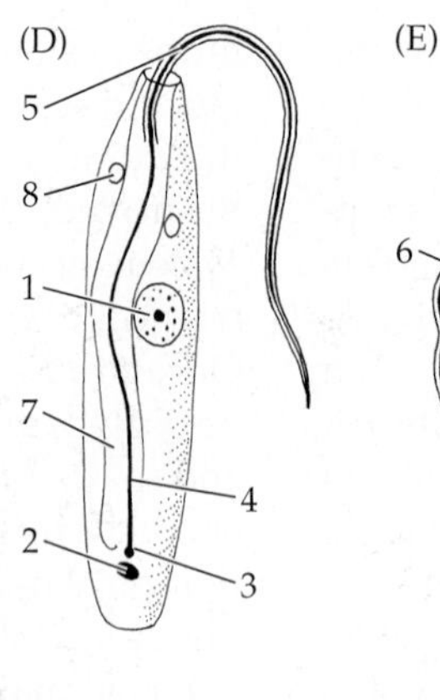

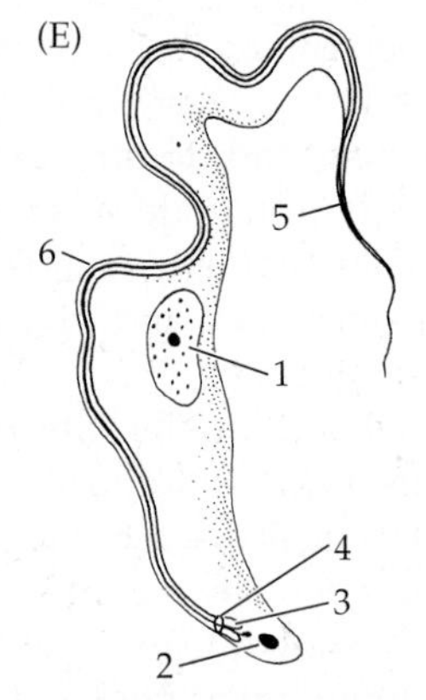

Key
1. Nucleus
2. Kinetoplast
3. Kinetosome
4. Axoneme
5. Flagellum
6. Undulating membrane
7. Flegellar pocket
8. Water expulsion vehicle

Figure 3.50 Phylum Kinetoplastida. Body plans of various trypanosomes. (A) *Leishmania* (amastigote form). (B) *Crithidia* (choanomastigote form). (C) *Leptomonas* (promastigote form). (D) *Herpetomonas* (opisthomastigote form). (E) *Trypanosoma* (trypomastigote form).

Reproduction and Life Cycles

Although sexual reproduction has never been observed in kinetoplastids, there is indirect genetic evidence that it occurs. Asexual reproduction occurs by longitudinal binary fission, as in *Euglena*. Nuclear division is by closed intranuclear pleuromitosis. During pleuromitotic mitosis, the nucleolus remains distinct and plaques on the inside of the nuclear envelope appear to organize the spindle (centrioles are absent). An unusual feature of kinetoplastid mitosis is that condensed chromosomes cannot be identified when the nucleus is dividing, even though they are typically conspicuous during interphase.

The life cycles of trypanosomes are complex and involve at least one host, but usually more. Monoxenous trypanosomes (those with one host) usually are found infecting the digestive tracts of arthropods or annelids. Most heteroxenous forms (those with more than one host) live part of their life cycle in the blood or organs of vertebrates and the remaining part of their life cycle in the digestive tracts of blood-sucking invertebrates, usually insects, sometimes leeches. As a trypanosome progresses through its life cycle, the shape of the cell undergoes different body form changes, depending on the phase of the cycle and the host it is parasitizing. Not all of these forms (Figure 3.50) occur in all genera. Body forms can differ in shape, position of the kinetosome and kinetoplast, and in the development of the flagellum.

GROUP 5: OPISTHOKONTA

Phylum Choanoflagellata (Choanoflagellates)

The choanoflagellates are stalked, sessile cells existing singly or in colonies (Figure 3.51). They are distinctive in that they are seemingly identical to choanocytes, the flagellated feeding cells of sponges. Like choanocytes, they have a single flagellum that is encircled by a basketlike transparent collar of retractile microvilli that contain actin filaments. The microvilli are sometimes called "collar tentacles"—perhaps to differentiate them from microvilli that are not retractile and lack a special cytoskeleton of microtubules connected to the basal body of the cilium. The central flagellum has a pair of basal, lateral vanes, just as in sponge choanocytes. Also, in both choanoflagellates and sponges a true ciliary rootlet is absent.[8] The collar acts as a food-catching net; feeding is accomplished when food particles are swept into the collar by the beating of the flagellum, pressed down against the cell surface, and engulfed by small pseudopodia. The choanoflagellates have long been viewed as a transitional link between flagellated protists and the sponges, or more specifically, as the actual ancestors of the Porifera, and thus the Metazoa. In addition to the structure of the collar itself, some choanoflagellates secrete a skeleton of siliceous pieces

[8]Rootlets do occur in monociliated cells of sponge larvae and in most metazoan monociliated cells.

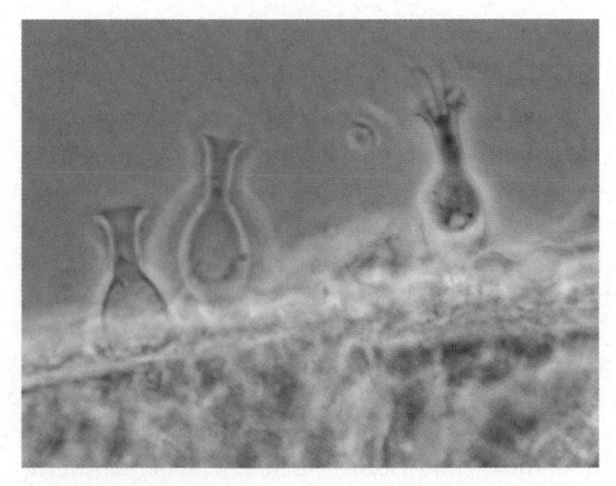

Figure 3.51 Phylum Choanoflagellata. The choanoflagellate *Salpingoeca*.

similar to sponge spicules. DNA data support this hypothesis, although extant choanoflagellates appear to be a monophyletic group and thus sister to Metazoa. On the other hand, some protozoologists in the past have suggested the possibility that, because they are not obviously related to any other protist group, choanoflagellates might actually be highly reduced sponges!

About 150 species of choanoflagellates, in three families, have been described. Most are marine, but a few freshwater species are also known. Codosigidae (=Monosigidae) are naked cells, or cells with thin organic vestments (e.g., *Codonosiga*, *Sphaeroeca*). Salpingoecidae have cellulose theca (e.g., *Salpingoeca*, *Stelexomonas*). Acanthoecidae produce extracellular loricae made of tiny strips of silica, usually forming an open basketlike structure (e.g., *Bicosta*, *Stephanoeca*). The lorica of acanthoecids may be several times larger than the cell. Salpingoecids and acanthoecids are restricted to marine and brackish waters and are mostly planktonic. They are present in huge numbers in the world's oceans, and are thought to be among the most important groups of bacteria-consuming organisms in marine systems, and therefore of major ecological significance.

Protist Phylogeny

The Origin of the Protista

We can do no more than touch upon the myriad questions and interesting ideas concerning the origin and evolution of the protists. Beyond the fast-moving field of protist phylogeny, we are faced here with questions about the very origin of eukaryotic life, as well as the ancestry of the multicellular eukaryotic kingdoms—Plantae, Metazoa, and Fungi. The origin of the Protista (and the Eukaryota) probably took place 2.0 to 2.5 billion years ago. Although there are over 30,000 known fossil species of protists, they are of little use in establishing the origin or subsequent evolution of the various protist lineages. Only those with hard parts have left us much of a fossil record, and only the foraminiferans and radiolarians have substantial records in Precambrian rocks. However, there are some deposits inferred to be testate amoebae in ~750 million-year-old rocks, and there are also some isolated "algal" fossils scattered between 750 to 1,200 million-year-old period (these are probably red algae/Rhodophyta). The origin of the eukaryotic condition was, of course, a pivotal event in the biological history of our planet, for it enabled life to escape from the severe limitations of the prokaryotic body plan by providing the various subcellular units that have formed the basis of specialization among the protists as well as the tissue-forming multicellular kingdoms of life. In addition to the eukaryotic cellular condition, the origin of the protists also ushered in the origin of sex, enabling related cells to fuse and pool their genetic resources in new and creative ways. Prevailing consensus is that the eukaryotic condition evolved only once, making the Eukaryota monophyletic, and protists paraphyletic.

It is now broadly agreed that the origin of modern eukaryotic diversity involved a series of endosymbiotic events, described as the **serial endosymbiotic theory** (or **SET** for short). The SET theory is one of the most fascinating ideas in biology, and a quick historical review is worthwhile. By the late nineteenth century workers had observed that both chloroplasts and mitochondria behave like independent, autonomous organisms that grow in number by division, and that mitochondria had the same staining properties as bacteria. In 1905, the brilliant Russian biologist C. Mereschkowsky hypothesized that this independent behavior of chloroplasts is because they are the evolutionary descendants of endosymbiotic Cyanobacteria-like organisms. And, in 1927, I. Wallin hypothesized that mitochondria also evolved from once free-living bacteria. Thus was born the SET theory. However, the theory remained controversial and sidelined until it was revived by Lynn Margulis in 1970 (it was Margulis who actually first used the phrase "serial endosymbiosis theory," in 1979).

The premise in SET theory is that eukaryotes arose first through an intimate symbiotic relationship between two prokaryotic cells in which one came to live inside the other in a type of permanent endosymbiosis (Figure 3.52). Over time, the symbiont became dependent upon (and integrated with) its host until it eventually became no longer obviously recognizable as a separate organism. It appeared, to all intents and purposes, to be an organelle of the host cell—an organelle we now call the mitochondrion. A second, later event involving a different prokaryotic symbiont (a photosynthetic cyanobacterium) gave rise to the first chloroplast organelle. This "horizontal" mode-of-origin of an organelle is called **symbiogenesis** (as distinct from "autogenesis," or origin by vertical descent with modification).

Unlike prokaryotic cells, all eukaryotic cells contain several kinds of membrane-bounded organelles that harbor distinct genetic systems. The membranes of these organelles are there *because* the organelles originated as membrane-bound, free-living, prokaryotic cells (e.g., the two membranes that surround chloroplasts are derived from the inner and outer membranes of the original gram-negative cyanobacterium). Hence, we are faced with the fascinating proposition that the functions now performed by these various eukaryotic organelles must have evolved long before the eukaryotic cell itself evolved. We are also faced with the reality that evolution has proceeded not only by the splitting and divergence of lineages (as classically depicted on phylogenetic trees), but also by merging of distantly related lineages into evolutionary chimaeric cells.

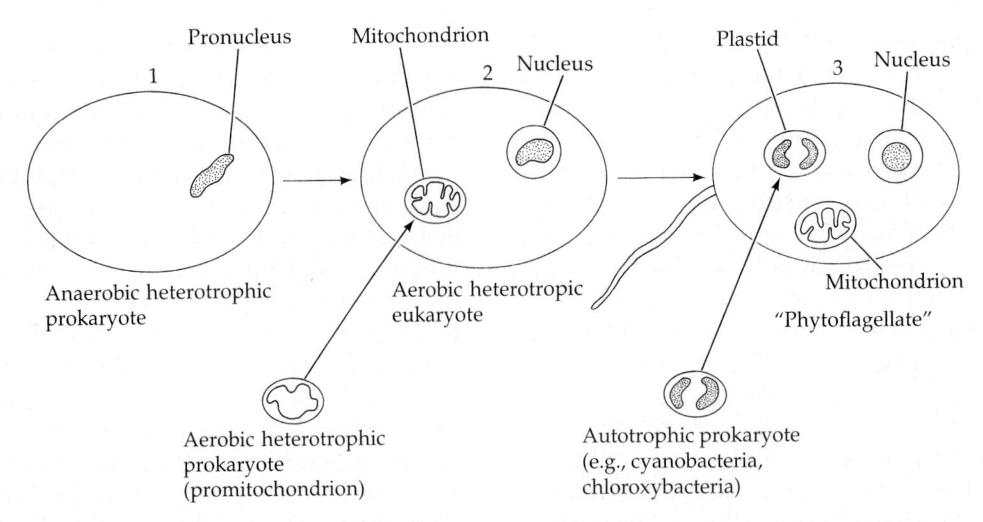

Figure 3.52 A simple model of the origin of eukaryotic cells by symbiosis (the serial endosymbiotic theory). The major events depicted are: acquisition of an aerobic heterotrophic prokaryote (origin of mitochondrion) and acquisition of an autotrophic prokaryote (origin of plastid).

Both ultrastructural and molecular genomic (DNA) evidence suggest that the eukaryotic mitochondrion evolved by way of a symbiotic relationship with a prokaryote that was similar (if not identical) to modern α-proteobacteria. This event is estimated to have transpired between 2 and 2.5 billion years ago. This particular association provided a means for early eukaryotic cells, previously limited to anaerobic metabolism, to carry out aerobic respiration. Once oxygen could be used as a terminal electron acceptor, the energy derived from ingested food increased by a factor of almost 20.

DNA data also indicate that, later in eukaryotic history, one lineage of heterotrophic eukaryotes acquired a photosynthetic cyanobacterium symbiont that then evolved into plastids (e.g., chloroplasts), thus setting the stage for the origin of red and green algae, perhaps between 1.2 and 1.5 billion years ago. Even today, the DNA of plastid genes in red and green algae closely resembles the DNA of Cyanobacteria. Thus the two membranes that surround the plastids in red and green algae correspond to the inner and outer membranes of the original gram-negative Cyanobacteria endosymbionts.

Plastids and mitochondria, which have retained large fractions of their prokaryotic biochemistry, contain only a few remnants of the protein-coding genes that their ancestors possessed. Studies have shown that most of their genes have been lost or have been transferred from the organelles to the nucleus of the eukaryotic "host cell." Today, more than 90% of the proteins required for the operation of any mitochondrion or plastid are encoded by the nuclear genome, rather than the organellar genome.

Support for the SET theory also comes from evidence of eukaryote–eukaryote **secondary endosymbiosis** in numerous photoautotrophic protist clades. It seems that plastids from both red algae and green algae have been secondarily co-opted by various heterotrophic protists to secondarily incorporate photosynthesis as an option in their nutritional arsenal. This process appears to have taken place many times, and most of the important groups of algae in the modern ocean are actually the products of these secondary symbiotic events. For example, diatoms (Stramenopiles), coccolithophores (Haptophytes), peridinin-containing dinoflagellates, and cryptomonads all probably acquired their chloroplasts from red algae, while chlorarachniophytes and euglenids obtained their chloroplasts from two different green algal species. Perhaps the most unusual group to have acquired a plastid from another protist is the parasitic apicomplexans, whose bizarre apicoplast is apparently a vestigial plastid from a photosynthetic ancestor with a red algal, secondary endosymbiont. In most cases little remains of the engulfed alga apart from the plastids themselves, but in two groups, the cryptomonads and chlorarachniophytes, a small remnant nucleus (and a bit of cytoplasm) of the once-engulfed alga remains to this day. These tiny nuclei, called nucleomorphs, are the smallest and most compact eukaryotic genomes known. These kinds of chimeric organisms complicate but make more exciting the study of protists.

Interestingly, endosymbiosis is not restricted to protists, but is also fairly common among animals. The hydrothermal vent clam *Calyptogena magnifica*, for example, harbors a sulfur-oxidizing proteobacterium in specialized cells of its gills. The clam depends on these symbiotic bacteria for its nutrition, and the bacteria are transmitted, like mitochondria, via the eggs of the animal. And, the bacteria have apparently lost their ability to live freely in the marine environment. Many, well-known examples of endosymbiosis are seen among the insects. One of the best examples is the bacterium *Buchnera amphidicola*, a mutualist with aphids. Aphids suck phloem sap that is rich in many nutrients but deficient in amino acids that are provided by *Buchnera*, which are intracellular and restricted to the cytoplasm of one aphid cell type. These endosymbionts are maternally inherited via the aphid ovary. Thus the relationship is mutualistic and obligatory. Molecular phylogenetic studies indicate that the *Buchnera*-aphid relationship is hundreds of millions of years old and is extraordinarily fine-tuned and successful. The intracellular *Buchnera* have a highly reduced genome and resemble endosymbiotic bacteria at the proto-organelle grade of evolution.

Another fascinating invertebrate example of endosymbiosis occurs in certain green sea slugs. These animals feed by evacuating the cellular contents of siphonaceous green algae (e.g., *Vaucheria*) and transferring the metabolically active chloroplasts into their bodies. The chloroplasts are then distributed throughout the slug's body and become lodged only one cell layer beneath the epidermis, where light can reach them. By this means, the animals are capable of photoautotrophic CO_2 fixation. The chloroplasts remain active for a limited amount of time and eventually new ones must be acquired by the sea slug. Analogous processes are common in certain protists (e.g., see the section on ciliates).

In addition to secondary endosymbiosis, many protists maintain close commensalistic relationships with other, photoautotrophic protists. Many species of foraminiferans, for example, harbor symbiotic diatoms, dinoflagellates, or red or green algae. These cases exemplify potential early stages of secondary endosymbiotic evolutionary events.

Relationships among the Protists

We are still a ways from understanding how all protists are interrelated and how they should be classified. However, there are a number of emerging, higher-level protistan clades, identified by combinations of molecular genetics and ultrastructural/biochemical studies. Most of these identified clades aren't given (and might never be given) standard categorical rankings, although some workers refer to them as kingdoms, groups, superphyla and such. Below, we briefly describe the six, major, well-supported clades. Taxonomic summaries and descriptions of these groups were provided earlier in this chapter. One current view of their relatedness is shown in Figure 3.53.

The clade Amoebozoa is well supported by molecular phylogenetic analyses, although there have been no unique ultrastructural synapomorphies identified to distinguish this group. All species (at least primitively) have lobose pseudopodia, although lobopods are not unique to Amoebozoa.

The clade Chromalveolata is one of the most diverse protist assemblages, and it includes dinoflagellates, apicomplexans, ciliates, stramenopiles, and some other small groups—perhaps also Haptophyta and Cryptomonada. With the exception of the last two groups, the clade is fairly strongly supported by molecular data, and in addition, photosynthetic species in these groups share the feature of plastids containing chlorophyll *c*, as well as chlorophyll *a*. Mounting evidence supports the hypothesis that all chromalveolatans are descended from one secondary endosymbiotic event, probably involving a red algal symbiont (the "chromalveolate hypothesis"). Three phyla—Dinoflagellata, Apicomplexa, Ciliata—comprise a well-supported subclade called the Alveolata, and molecular analyses suggest these phyla may be part of a larger monophyletic group that also includes the phylum Stramenopila. The alveolates are uniquely characterized by the presence of alveoli beneath the outer cell membrane. The Stramenopila (treated as a phylum in this text) was first identified through

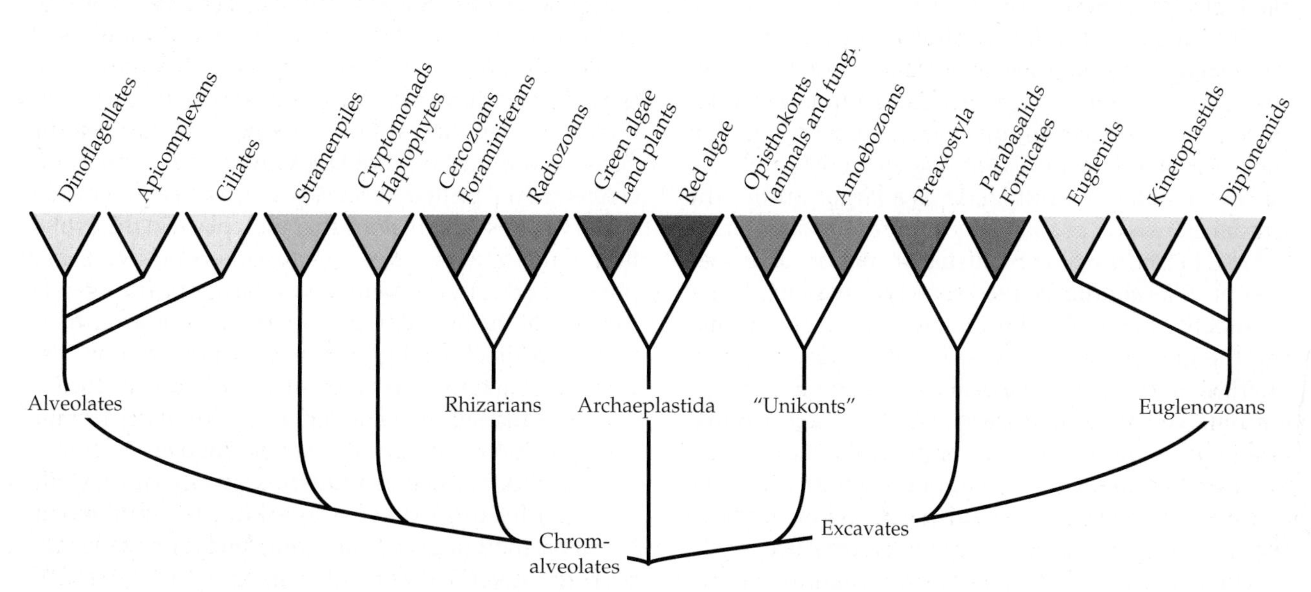

Figure 3.53 A phylogeny of the Eukaryota, showing current understandings of protist relationships.

molecular phylogenetic studies and later received confirmation from comparative anatomical work, notably the shared character of flagella lined with fine, tubular (hollow) hairs. The opalines and diatoms are usually included within the Stramenopila, and they are thought to have secondarily lost the hollow hairs. The Haptophyta (coccolithophores and their relatives) and the Cryptomonada (e.g., *Cryptomonas*) have plastids that contain chlorophyll *a* and *c*, suggesting they also belong to the chromalveolate assemblage. Multigene phylogenetic analyses indicate that these two phyla are closely related to each other, and perhaps these in turn allied with stramenopiles and alveolates.

The clade called Rhizaria contains the mixotrophic, green chloroplast-containing phylum Chlorarachniophyta, and the phyla Granuloreticulosa (forams and their kin), Radiolaria, Haplosporidia, and a few others. Amoeboid species in this clade have filopods or axopods (filopods supported by microtubules), neither of which are unique to this group. Rhizaria was discovered and delineated primarily by analyses of molecular sequence data, but the relationships of the major rhizarian clades remain unresolved. Some recent molecular evidence suggests that Radiolaria may be the basal group, with the remaining groups forming the subgroup called Cercozoa (e.g., forams, chlorarachniophytes, plasmodiophorids, haplosporids, and a few other odd groups), or with those that are sister to the Granuloreticulosa, called cercozoans (i.e., granuloreticulosans and cercozoans form a sister group, that is in turn sister to Radiolaria). However, it has also been proposed that Granuloreticulosa and Radiolaria are a clade. Chlorarachniophytes are unusual among the Rhizaria in their possession of chloroplasts, which they acquired via a secondary endosymbiotic event with a green alga (Chlorophyta) symbiont. Like the Rhizaria itself, the clade Cercozoa has no known unique morphological synapomorphies.

Growing evidence from molecular phylogenetics is suggesting an alliance of Rhizaria with most of the Chromalveolata, as the SAR group (stramenopiles, alveolates, rhizarians). Some workers feel that the molecular data are also strong enough to unite the Chromalveolata and Rhizaria as a larger group, the Chromista, a term proposed by Thomas Cavalier-Smith in 1981. In fact, Cavalier-Smith has championed the idea that protists should be reassigned to two kingdoms: "Protozoa" (a redefined version of this taxon containing the groups Amoebozoa, Excavata and the choanoflagellates) and Chromista (containing the groups Chromalveolata and Rhizaria, and a few other odds-and-ends). However, molecular phylogenetic analyses have yet to provide strong support for the Chromista; thus we have not used it in our classification, nor have we used the proposed SAR clade as a formal taxon.

The clade Excavata is supported mainly by cellular ultrastructural features and only modestly by molecular phylogenetic studies. Excavates typically have a suspension-feeding cytostome of the "excavate" type (i.e., a feeding groove used for capture and ingestion of small particles from a feeding current generated by a posteriorly directed flagellum; margin and floor of groove supported by the microtubular root). The cytostome is presumed to be secondarily lost in many taxa. Currently included within the Excavata are Parabasalida (trichomonads, hypermastigotes, etc.), Diplomonada, Heterolobosea, Jakobida, Oxymonada, Retortamonada, and a few others. Molecular and cell structure evidence also support placing the phyla Euglenida and Kinetoplastida (trypanosomes, bodonids, and their kin) within the Excavata, the two forming a clade known as Euglenozoa which is in turn thought to be closely related to the Heterolobosea and Jakobida. Two principal anatomical features distinguish the euglenozoans: (1) a spiral, or crystalline rod inside each of their two flagella, which inserts into an anterior pocket; and (2) disk-shaped mitochondrial cristae.

The jakobids have the most primitive (bacteria-like) mitochondrial genomes known. The phylum Heterolobosea is more closely related to euglenids and kinetoplastids than to other amebas. Broad-pseudopod-forming amebas have evolved in one group (Heterolobosea), independently of the Amoebozoa (which have similar pseudopods), and heterolobosids even include their own clade of "slime molds" (the acrasids).

Two excavate taxa, Parabasalida and Diplomonada, were long thought to lack mitochondria, but recent evidence suggests these groups simply have highly reduced or modified mitochondria lacking DNA. Parabasalida and Diplomonada also lack plastids, electron transport chains, and the enzymes that are normally needed for the citric acid cycle. Also, most species are found in anaerobic environments. For these reasons, and due to early molecular phylogenetic analyses, it was long thought that these two phyla might represent the oldest (earliest) surviving branches in the eukaryote tree. However, there is now growing evidence that the absence of mitochondrial DNA and electron transport chains are not primary losses, but instead represent secondary losses—reductions that took place in the evolution of these two groups. Most workers now consider parabasalids' and diplomonads' earlier positioning at the base of the protist tree in molecular studies to have been an artifact of long-branch attraction. The two are closely related to each other, and could derive from a common ancestor that had already secondarily lost the classic features of mitochondria noted above.

The clade known as Opisthokonta comprises the protist phylum Choanoflagellata, the kingdoms Metazoa and Fungi (= Eumycota), and a few other obscure groups. The group Opisthokonta is well supported by molecular sequence studies, and its members also share the attributes of flattened mitochondrial cristae and a single posterior cilium/flagellum on the male reproductive cells. As we noted above, the choanoflagellates comprise the sister group called Holozoa, and Choanoflagellata is the probable direct ancestor of Metazoa (although see Chapter 6). Choanoflagellates, most animal sperm, and the zoospores of chytrids (the only fungi with flagella) all swim with their single, unadorned flagellum emerging from their posterior end. Surprisingly, this arrangement is nearly unique, and it appears to have been inherited from the common ancestor of opisthokonts. The true Fungi include the well-known groups Basidiomycota, Ascomycota, Saccharomycetes, Microsporidia and Chytridiomycetes (the chytrids being implicated in the global die-off of amphibians). Fungi are heterotrophic (not phagotrophic), with cell walls (when present) containing β-glucan and usually chitin; plastids and tubular mastigonemes are absent.

We recognize the clade Plantae, or Archaeplastida, to include all organisms that contain chlorophylls *a* and *b*, store their photosynthetic products as starch (inside the double-membrane–bounded chloroplasts in which it is produced), and typically have cell walls made of cellulose. These are the primary plastid-containing lineages (hence the descriptive name, "Archaeplastida"), although some groups have secondarily lost or reduced their plastids. As this book went to press, two major groupings of plants were recognized, the Biliphyta (phyla Glaucophyta and Rhodophyta) and the Viridiplantae (including the phyla Chlorophyta and Charophyta, plus the embryophytes—phyla Anthocerotophyta, Bryophyta, Marchantiophyta and Tracheophyta). Glaucophytes comprise a small group of microscopic freshwater algae. Rhodophytes are the red algae, and chlorophytes are the green algae. Charophytes are an obscure group of freshwater green algae that include the Charales, or stoneworts). Anthocerotophytes are the hornworts, flattened green plants (gametophytes) that produce hornlike sporophyte structures. Bryophytes are mosslike ground-hugging plants. Although the name "bryophyte" was long used to denote any nonvascular plant, today it is restricted to a specific clade of them. Marchantiophytes are the liverworts. Tracheophytes are the vascular, or "higher" plants (including the clubmosses, horsetails, ferns, gymnosperms, and angiosperms). Anthocerotophyta, Bryophyta, Marchantiophyta, and Tracheophyta (the vascular plants) are often classified together under the name Embryophyta, a descriptive term referring to the origination of tissue layers during embryogeny in this group. The plastids (chloroplasts) of Plantae/Archaeplastida appear to be monophyletic—that is, to be descended from a single, original primary endosymbiotic event.

Although the Chlorophyta may rest at the base of the clade leading to the land plants, the precise phylogenetic relationships of Glaucophyta, Rhodophyta, Chlorophyta, Charophyta and the embryophytes have not yet been resolved. The Charophyta, especially the Charales (e.g., *Chara* and Coleochaetales), appear to be the most closely related to the embryophytes (the "land plants"), with which they share several synapomorphies: rose-shaped complexes for cellulose synthesis, peroxisome enzymes (the specialized enzymes of peroxisomes help minimize the loss of organic products as a result of photorespiration), the structure of the flagellated sperm, and formation of a phragmoplast (an alignment of cytoskeletal elements and Golgi-derived vesicles across the midline of dividing cells during cell division). The phylogenetic branching order of the green algal groups that gave rise to the land plants remains uncertain.

Recent biochemical and molecular phylogenetic analyses also place the phylum Rhodophyta (red algae) in the Archaeplastida. The oldest fossil rhodophytan, *Bangiomorpha pubescens*, strongly resembles the modern *Bangia* but occurs in rocks dated at 1.2 billion years.

On the microscopic level, embryophyte cells remain similar to those of green algae, although they lack flagella and centrioles, except in certain gametes. Embryophytes probably evolved out of green algae (Chlorophyta) in the Paleozoic era. The algaelike stoneworts (Charales) are perhaps the best living illustration of that early evolutionary stage. In the first embryophytes, the sporophytes were very small and dependent on the parent for their entire brief life—these are the nonvascular plants. During the Silurian and Devonian, land plants radiated rapidly and during this time the Tracheophyta, or vascular plants emerged. Vascular plants have vascular tissues that transport water throughout the body. Some time in the Devonian (~385 million years ago), desiccation-resistant capsules, called seeds, appeared, a feature distinguishing that group of tracheophytes called the Spermatophyta. Five groups of spermatophytes are recognized: Cycadophyta (cycads), Ginkgophyta (ginkos), Pinophyta (conifers), Gnetophyta (gnetaens), and Magnoliophyta (flowering plants). The first four groups are the gymnosperms. The angiosperms were the last to evolve, probably some time in the Jurassic period, spreading globally and rapidly in the Cretaceous. Categorical names and taxonomic assignments vary considerably among botanists and textbooks.

Thus we see that the origins of the three familiar multicelled kingdoms (Metazoa, Plantae, Fungi) lie in three different protist ancestors and embryogenic tissue-layering processes evolved independently in each line. Other than the embryonic formation of tissues, there are few obvious differences between the ancestral protists and their earliest multicelled descendants. Thus most unicellular green algae strongly resemble primitive green plants (embryophytes), and choanoflagellates are strikingly similar to sponges. In addition to embryonic tissue layering, metazoans (animals) are distinguished by: cells that are typically held together by intercellular junctions; an extracellular matrix (the basement membrane, or basal lamina) with fibrous proteins, typically collagens, between two dissimilar epithelia; sexual reproduction with the generation of eggs cells that are fertilized by smaller, monociliated, sperm cells; phagotrophy; and, lack of a cell wall (typical of plants).

Of course, most unicellular clades are not directly related to any of the three multicellular kingdoms. As the protistan lineages come into focus, it becomes clear that the commonly used six-kingdom classification scheme is inadequate for describing the true nature of diversity among the Eukaryota. New kingdoms, or groups, based on very different criteria than have traditionally been used, are now beginning to be recognized, as noted above.

Selected References

General References

Anderson, D. M., A. W. White and D. G. Baden. 1986. *Toxic Dinoflagellates*. Elsevier Publishing.

Anderson, O. R. 1983. *Radiolaria*. Springer-Verlag, New York.

Armbrust, E. V. and 44 others. 2004. The genome of the diatom *Thalassiosira pseudonana*: ecology, evolution, and metabolism. Science 306: 79–86.

Bates, S. S. and 16 others. 1989. Pennate diatom *Nitzschia pungens* as the primary source of domoic acid, a toxin in shellfish from eastern Prince Edward Island, Canada. Can. J. Fish. Aquat. Sci. 46: 1203–1215.

Be, A. 1982. Biology of planktonic Foraminifera. University of Tennessee Studies in Geology 6: 51–92.

Beutlich, A. and R. Schnetter. 1993. The life cycle of *Cryptochlora perforans* (Chlorarachniophyta). Bot. Acta 106: 441–447.

Bonner, J. Tyler. 2009. *The Social Amoebae: The Biology of Cellular Slime Molds*. Princeton University Press. [Highly recommended; a great natural history read.]

Brock, D. A., T. E. Douglas, D. C. Queller and J. E. Strassman. 2011. Primitive agriculture in a social amoeba. Nature 469: 393–396.

Buetow, D. E. (ed.) 1968, 1982. *The Biology of Euglena*. Vols. 1–2 and 3. Academic Press, New York.

Capriulo, G. M. (ed.). 1990. *Ecology of Marine Protozoa*. Oxford University Press, New York.

Carey, P. 1991. *Marine Interstitial Ciliates: An Illustrated Key*. Chapman & Hall.

Cavalier-Smith, T. 1993. The protozoan phylum Opalozoa. J. Euk. Microbiol. 40: 609–1615.

Cavalier-Smith, T. 1995. Cell cycles, diplokaryosis and the archezoan origin of sex. Arch. Protistenkd. 145: 189–207.

Cavalier-Smith, T. 2003. The excavate protozoan phyla Metamonada Grassé emend. (Anaeromonadea, Parabasalia, *Carpediemonas*, Eopharyngia) and Loukozoa emend. (Jakobea, *Malawimonas*): their evolutionary affinities and new higher taxa. Int. J. Syst. Evol. Micr. 53: 1741–1758.

Cavalier-Smith, T. 2013. Symbiogenesis: mechanisms, evolutionary consequences, and systematic implications. Ann. Rev. Ecol. Evol. Syst. 44: 145–172.

Clark, C. G. and A. J. Roger. 1995. Direct evidence for secondary loss of mitochondria in *Entamoeba histolytica*. Proc. Natl. Acad. Sci. 92: 6518–6521.

Coats, D. W. 1999. Parasitic life styles of marine dinoflagellates. J. Euk. Microbiol. 46: 402–409.

Cole, K. M. and R. G. Sheath (eds.). 1990. *Biology of the Red Algae*. Cambridge Univ. Press, Cambridge.

Curds, C. R. 1992. *Protozoa in the Water Industry*. Cambridge Univ. Press, Cambridge.

Dick, M. W. 2001. *Straminipilous Fungi. Systematics of the Peronosporomycetes, Including Accounts of the Marine Straminipilous Protists, the Plasmodiophorids and Similar Organisms*. Kluwer Academic Publishers, Boston.

Dolan, J. R. 1991. Microphagous ciliates in mesohaline Chesapeake Bay waters: estimates of growth rates and consumption by copepods. Mar. Biol. 111: 303–309.

Dolan, J. R., D. J. S. Montagnes, S. Agatha, D. W. Coats and D. Stoecker. 2012. *The Biology and Ecology of Tintinnid Ciliates. Models for Marine Plankton*. Wiley-Blackwell, London.

Druehl, L. 2000. *Pacific Seaweeds. A Guide to the Common Seaweeds of the West Coast*. Harbour Publishing, Madeira Park, British Columbia.

Falkowski, P. G. and A. H. Knoll (eds.). 2007. *Evolution of Primary Producers in the Sea*. Academic Press (Elsevier), Burlington, Massachusetts. [A contemporary, comprehensive, and authoritative view of the origins of photosynthesis, eukaryotic life, and geobiology of the Earth.]

Fenchel, T. 1980. Suspension feeding in ciliated protozoa: structure and function of feeding organelles. Arch. Protistenkd. 123: 239–260. [A fine review of the subject.]

Fenchel, T. 1987. *Ecology of Protozoa: The Biology of Free-Living Phagotrophic Protists*. Springer-Verlag, New York.

Fenchel, T. and B. J. Finlay. 1995. *Communities and Evolution in Anoxic Worlds*. Oxford Univ. Press, Oxford.

Fensome, R. A., F. J. R. Taylor, G. Norris, W. A. S. Sarjeant, D. I. Wharton and G. L. Williams. 1993. A classification of living and fossil dinoflagellates. Micropaleontology Special Pubs. 7. Sheridan Press, Hanover, PA.

Foissner, W. 1987. Soil protozoa: fundamental problems, ecological significance, adaptations in ciliates and testaceans, bioindicators, and guide to the literature. Prog. Protistol. 2: 69–212.

Gilson, P. R. and G. I. McFadden. 1995. The chlorarachniophyte: a cell with two different nuclei and two different telomeres. Chromosoma 103: 635–641.

Gilson, P. R. and G. I. McFadden. 1997. Good things come in small packages: the tiny genomes of chlorarachniophyte endosymbionts. BioEssays 19: 167–173.

Gilson, P. R., U.-G. Maier and G. I. McFadden. 1997. Size isn't everything: lessons in genetic miniturisation from nucleomorphs. Curr. Opinion Genet. Dev. 7: 800–806.

Gojdics, M. 1953. *The Genus Euglena*. University Wisconsin Press, Madison. [Includes keys, descriptions, and figures of all species known at the time.]

Gómez-Gutiérrez, J., W. T. Peterson, A. de Robertis and R. D. Brodeur. 2003. Mass mortality of krill caused by parasitoid ciliates. Science 301: 339.

Gooday, A. 1984. Records of deep-sea rhizopod tests inhabited by metazoans in the northeast Atlantic. Sarsia 69: 45–53. [Animals that use protist skeletons as homes!]

Grain, J. 1986. The cytoskeleton of protists. Int. Rev. Cytol. 104: 153–249.

Grell, K. B. 1990. Some light microscope observations on *Chlorarachnion reptans* Geitler. Arch. Protistenkd. 138: 271–290.

Gupta, B. K. S. (ed). 2002. *Modern Foraminifera*. Kluwer Academic.

Hallegraeff, G. M. and 8 others. 2010. *Algae of Australia: Phytoplankton of Temperate Coastal Waters(Algae of Australia Series)*. CSIRO Publishing. Canberra, Australia.

Harrison, F. W. and J. O. Corliss. 1991. Protozoa. In F. W. Harrison (ed.), *Microscopic Anatomy of Invertebrates*, Vol. 1. Wiley-Liss, New York.

Hausmann, K. and P. C. Bradbury. 1996. *Ciliates: Cells as Organisms*. Gustav Fischer, Stuttgart.

Hausmann, K. and N. Hülsmann (eds.). 1993. *Progress in Protozoology*. Gustav Fischer Verlag, Stuttgart, Germany.

Hausmann, K. and N. Hülsmann. 1996. *Protozoology*, 2nd Ed. Georg Thieme Medical Publishers, Inc., New York.

Hausmann, K. and R. Peck. 1979. The mode of function of the cytopharyngeal basket of the ciliate *Pseudomicrothorax dubius*. Differentiation 14: 147–158.

Hedley, R. H. and C. G. Adams (eds.). 1974. *Foraminifera*. Academic Press, New York.

Hyman, L. H. 1940. *The Invertebrates. Vol. 1, Protozoa through Ctenophora*. McGraw-Hill, New York. [Obviously dated, but still with a wealth of basic anatomical information.]

Ishida, K., B. R. Green and T. Cavalier-Smith. 1999. Diversification of a chimaeric algal group, the Chlorarachniophytes: phylogeny of nuclear and nucleomorph small-subunit rRNA genes. Mol. Biol. Evol. 16(3): 321–331.

Jeon, K. W. (ed.). 1973. *The Biology of Amoeba*. Academic Press, New York. [21 of the world's ameba specialists contributed to this fine book dealing with all aspects of biology of free-living amebas.]

Jones, A. R. 1974. *The Ciliates*. St. Martin's Press, New York.

Jurand, A. and G. G. Selman. 1969. *The Anatomy of Paramecium aurelia*. Macmillan, London, and St. Martin's Press, New York. [All you ever wanted to know about the anatomy of *Paramecium*; many fine illustrations and micrographs.]

Kappe, S. H. I., A. M. Vaughan, J. A. Boddey and A. F. Cowman. 2010. That was then but this is now: malaria research in the time of an eradication agenda. Science 328: 862–865.

Kemp, P. F., J. J. Cole, B. F. Sherr and E. B. Sherr (eds). 1993. *Handbook of Methods in Aquatic Microbial Ecology*. Lewis Publishers/CRC Press, Boca Raton, FL.

Kreier, J. P. 1991–1994. *Parasitic Protozoa* [eight volumes], 2nd Ed. Academic Press, New York.

Laybourn-Parry, J. 1985. *A Functional Biology of Free-Living Protozoa*. University of California Press, Berkeley.

Leander, B. S. 2007. Marine gregarines: evolutionary prelude to the apicomplexans radiation? Cell. doi: 10.1016/j.pt.2007.11.005: 60–67

Leander, B. S., H. J. Esson and S. A. Breglia. 2007. Macroevolution of complex cytoskeletal systems in euglenids. BioEssays 29(10): 987–1000.

Leander, B. S., J. F. Saldarriaga and P. J. Keeling. 2002. Surface morphology of the marine parasite *Haplozoon axiothellae* Siebert (Dinoflagellata). Eur. J. Protistol. 38: 287–297.

Lee, J. J. and O. R. Anderson. 1991. *Biology of Foraminifera*. Academic Press, London.

Lee, J. J., G. F. Leedale and P. Bradbury (eds.). 2000. *An Illustrated Guide to the Protozoa*. 2nd ed. Society of Protozoologists, Lawrence, Kansas.

Lembi, C. A. and J. Waaland. 1988. *Algae and Human Affairs*. Cambridge Univ. Press, Cambridge.

Levine, N. D. 1988. *The Protozoan Phylum Apicomplexa*. Vols. 1–2. CRC Press, Boca Raton, Florida. [Provides diagnoses of all genera and species.]

Lopez-Garcia, P., F. Rodriguez-Valera, C. Pedros-Alio and D. Moreira. 2001. Unexpected diversity of small eukaryotes in deep-sea Antarctic plankton. Nature 409: 603–607.

Lumsden, W. H. R. and D. A. Evans (eds.). 1976, 1979. *Biology of the Kinetoplastida*. Vols. 1–2. Academic Press, New York.

Lüning, K. 1990. *Seaweeds: Their Environment, Biogeography, and Ecophysiology*. Wiley, New York.

Lynn, D. H. 2010. *The Ciliated Protozoa: Characterization, Classification and Guide to the Literature*, 3rd Ed. Springer Verlag, Berlin.

Mackinnon, M. J. and K. Marsh. 2010. The selection landscape of malaria parasites. Science 328: 866–871.

Margulis, L., J. O. Corliss, M. Melkonian and D. J. Chapman (eds.). 1989. *Handbook of Protoctista*. Jones and Bartlett, Boston.

Melkonian, M., R. A. Anderson and E. Schnepf (eds). 2001. *The Cytoskeleton of Flagellate and Ciliate Protists*. Springer Verlag, Berlin.

Møestrup, Ø. 1982. Flagellar structure in algae: a review with new observations particularly on the Chrysophyceae, Phaeophyceae, Euglenophyceae and Reckertia. Phycologia. 21: 427–528.

Mondragon, J. and J. Mondragon. 2003. *Seaweeds of the Pacific Coast. Common Marine Algae from Alaska to Baja California*. Sea Challengers, Monterey, California.

Muravenko, O. V., I. O. Selyakh, N. V. Kononenko and I. N. Stadnichuk. 2001. Chromosome numbers and nuclear DNA contents in the red microalgae *Cyanidium caldarium* and three *Galdieria* species. Eur. J. Phycol. 36: 227–232.

Murray, J. 2006. *Ecology and Applications of Benthic Foraminifera*. Cambridge University Press, New York.

Neafsey, D. E. and 120 others. 2015. Highly evolvable malaria vectors: the genomes of 16 *Anopheles* mosquitoes. Science. doi: 10.1126/science.1258522

Nigrini, C. and T. C. Moore. 1979. *A Guide to Modern Radiolaria*. Special Publ. No. 16, Cushman Foundation for Foraminiferal Research, Washington, DC.

Nisbet, B. 1983. *Nutrition and Feeding Strategies in Protozoa*. Croom Helm Publishers, London.

Ogden, C. G. and R. H. Hedley. 1980. *An Atlas of Freshwater Testate Amoebae*. Oxford University Press, Oxford. [Magnificent SEM photographs of ameba shells accompanied by descriptions of the species.]

Olive, L. S. 1975. *The Mycetozoans*. Academic Press, New York.

Patterson, D. J. 1996. *Free-Living Freshwater Protozoa. A Colour Guide*. John Wiley & Sons, NY.

Patterson, D. J. 1999. The diversity of eukaryotes. Am. Nat. 154 (suppl.): S96–S124.

Patterson, D. J. and J. Larsen. 1991. *The Biology of Free-Living Heterotrophic Flagellates*. Clarendon Press, Oxford.

Pickett-Heaps, J. D. 1975. *Green Algae: Structure, Reproduction and Evolution in Selected Genera*. Sinauer Assoc., Sunderland, MA.

Pickett-Heaps, J. D. 2004. *Diatoms. Life in Glass Houses*. [DVD] Sinauer Assoc., Sunderland, MA.

Pickett-Heaps, J. D. and J. Pickett-Heaps. 2006. *The Kingdom Protista. The Dazzling World of Living Cells*. [DVD] Sinauer Associates, Sunderland, MA.

Polin, M., I. Tuval, K. Drescher, J. P. Gollub and R. E. Goldstein. 2009. *Chlamydomonas* swims with two "gears" in a eukaryotic version of run-and-tumble locomotion. Science 325: 487–490.

Poxleitner, M. K., M. L. Carpenter, J. J. Mancuso, C.-J. R. Wang, S. C. Dawson and W. Z. Cande. 2008. Evidence for karyogamy and exchange of genetic material in the binucleate intestinal parasite *Giardia intestinalis*. Science 319: 1530–1533.

Ragan, M. A. and R. R. Gutell. 1995. Are red algae plants? Bot. Jour. Linnean Soc. 118: 81–105.

Raikov, I. B. 1994. The diversity of forms of mitosis in protozoa: a comparative review. Eur. J. Protistol. 30: 253–259.

Roberts, L. S., J. Janovy, Jr. and S. Nadler. 2012. *Foundations of Parasitology*, 9th Ed. McGraw-Hill.

Round, F. E., R. M. Crawford and D. G. Mann. 1990. *The Diatoms. Biology & Morphology of the Genera*. Cambridge University Press, New York.

Ruggiero, M. A. and 8 others. 2015. A higher level classification of all living organisms. PLoS ONE. 10(4): e0119248. doi: 10.1371/journal.pone.0119248

Schaap, P. and 11 others. 2006. Molecular phylogeny and evolution of morphology in the social amoebas. Science 314: 661–663.

Seliger, H. H. (ed.). 1979. *Toxic Dinoflagellate Blooms*. Elsevier/ North Holland, NY.

Sleigh, M. A. (ed.). 1973. *Cilia and Flagella*. Academic Press, London.

Sleigh, M. A. 1989. *Protozoa and Other Protists*. 2nd Ed. Edward Arnold, London.

Stentiford, G. D., K. S. Bateman, N. A. Stokes and R. B. Carnegie. 2013. *Haplosporidium littoralis* sp. nov.: a crustacean pathogen within the Haplosporidia (Cercozoa, Ascetospora). Dis. Aquat. Organ. 105: 243–252.

Stoecker, D. K., M. W. Silver, A. E. Michaels and L. H. Davis. 1988. Obligate mixotrophy in *Laboea strobila*, a ciliate which retains chloroplasts. Mar. Biol. 99: 415–423.

Sturm, A. and 9 others. 2006. Manipulation of host hepatocytes by the malaria parasite for delivery into liver sinusoids. Science 313: 1287–1290.

Tarnita, C. E., A. Washburne, R. Martinez-Garcia, A. E. Sgro and S. A. Levin. 2014. Fitness tradeoffs between spores and nonaggregating cells can explain the coexistence of diverse genotypes in cellular slime molds. PNAS 112(9): 2276–2781.

Tartar, V. 1961. *The Biology of Stentor*. Pergamon Press, New York.

Taylor, F. J. R. 1987. *The Biology of Dinoflagellates*. Blackwell Scientific Publications, Oxford.

Thomas, D. 2002. *Seaweeds*. Life Series. Natural History Museum, London.

Todo, Y., H. Kitazato, J. Hashimoto and A. J. Gooday. 2005. Simple foraminifera flourish at the ocean's deepest point. Science 307: 689.

Tomas, Carmelo R. 1997. *Identifying Marine Phytoplankton*. Academic Press, San Diego.

Trench, R. K. 1980. Uptake, retention and function of chloroplasts in animal cells. Pp. 703–730 in W. Schwemmler and H. Schenk (eds.), *Endocytobiology*. Vol. I. Walter de Gruyter, Berlin.

van den Hoek, C., D. G. Mann and H. M. Jahns. 1995. *Algae. An Introduction to Phycology*. Cambridge Univ. Press, Cambridge.

Vroom, P. S. and C. M. Smith. 2001. The challenge of siphonous green algae. Am Sci. 89: 525–531.

Wehr, J. D. and Sheath, R. G. 2003. *Freshwater Algae of North America*. Academic Press, Boston.

Wichterman, R. 1986. *The Biology of Paramecium*, 2nd Ed. Plenum, New York.

Williams, A. G. and G. S. Coleman. 1992. *The Rumen Protozoa*. Springer Verlag, Berlin.

Yubuki, N., V. P. Edgcomb, J. M. Bernhard and B. S. Leander. 2009. Ultrastructure and molecular phylogeny of *Calkinsia aureus*: cellular identity of a novel clade of deep-sea Euglenozoans with epibiotic bacteria. BMC Microbiol. 9(16): 1–22.

Protist Phylogeny

Adams, K. L. and J. D. Palmer. 2003. Evolution of mitochondrial gene content: gene loss and transfer to the nucleus. Mol. Phylogenet. Evol. 29: 380–395.

Adl, S. M. and 27 others. 2005. The new higher level classification of eukaryotes with emphasis on the taxonomy of protists. J. Euk. Microbiol. 53(5): 399–451.

Anderson, J. O., S. W. Sarchfield and A. J. Roger. 2005. Gene transfers from Nanoarchaeota to an ancestor of diplomonads and parabasalids. Mol. Biol. Evol. 22: 85–90.

Angiosperm Phylogeny Group. 2009. An update of the Angiosperm Phylogeny Group classification for the orders and families of flowering plants. Bot. J. Linn. Soc. 161: 105–121.

Archibald, J. M. 2007. Nucleomorph genomes: structure, function, origin and evolution. BioEssays 29: 392–402.

Arisue, N., M. Hasegawa and T. Hashimoto. 2005. Root of the Eukaryota tree as inferred from combined maximum likelihood analyses of multiple molecular sequence data. Mol. Biol. Evol. 22(3): 409–420.

Baldauf, S. L. 2003. The deep roots of eukaryotes. Science 300: 1703–1706.

Baldauf, S. L. and W. F. Doolittle. 1997. Origin and evolution of the slime molds (Mycetozoa). Proc. Natl. Acad. Sci. 94: 12007–12012.

Baldauf, S. L. and J. D. Palmer. 1993. Animals and fungi are each other's closest relatives: Congruent evidence from multiple proteins. Proc. Natl. Acad. Sci. 90: 11558–11562.

Baldauf, S. L., A. J. Roger, I. Wenk-Siefert and W. F. Doolittle. 2000. A kingdom-level phylogeny of eukaryotes based on combined protein data. Science 290: 972–977.

Banks, J. A. and 102 others. 2011. The *Selaginella* genome identifies genetic changes associated with the evolution of vascular plants. Science 332: 960–963.

Bass, D. and 7 others. 2005. Polyubiquitin insertions and the phylogeny of Cercozoa and Rhizaria. Protist 156: 149–161.

Bhattacharya, D. (ed.). 1997. *Origins of Algae and Their Plastids*. Springer-Verlag, New York.

Borchiellini, C., N. Boury-Esnault, J. Vacelet and Y. Le Parco. 1998. Phylogenetic analysis of the Hsp 70 sequences reveals the monophyly of Metazoa and specific phylogenetic relationships between animals and fungi. Mol. Biol. Evol. 15: 647–655.

Burki, F., K. Shalchian-Tabrizi, M. Minge, A. Skjaeveland, S. I. ikolaev, K. S. Jakobsen and J. Pawlowski. 2007. Phylogenomics reshuffles the eukaryotic supergroups. PLoS ONE. (*e790*): 1–6.

Butterfield, N. J. 2000. *Bangiomorpha pubescens* n. gen., n. sp.: implications for the evolution of sex, multicellularity, and the Mesoproterozoic/Neoproterozoic radiation of eukaryotes. Paleobiology 26(3): 386–404.

Cavalier-Smith, T. 1999. Principles of protein and lipid targeting in secondary symbiogenesis: Euglenoid, dinoflagellate, and sporozoan plastid origins of the eukaryote family tree. J. Euk. Microbiol. 46: 347–366.

Cavalier-Smith, T. 2002. The phagotrophic origin of eukaryotes and phylogenetic classification of Protozoa. Int. J. Syst. Evol. Micro. 52: 297–354.

Cavalier-Smith, T. 2004. Only six kingdoms of life. Proc. Roy. Soc. Lond., B 271: 1251–1262.

Cavalier-Smith, T. 2010. Kingdoms Protozoa and Chromista and the eozoan root of the eukaryotic tree. Biol. Lett. 6: 342–345.

Cavalier-Smith, T., M. Allsopp and E. E.-Y. Chao. 1994. Thraustochytrids are chromists, not fungi: 18S rRNA signatures of Heterokonta. Phil. Trans. Royal Soc. London, Ser. B, 346: 387–397.

Cavalier-Smith, T. and E. E.-Y. Chao. 2003. Phylogeny of Choanozoa, Apusozoa and other Protozoa and early eukaryote evolution. J. Mol. Evol. 56: 540–563.

Cavalier-Smith, T. and E. E.-Y. Chao. 2004. Protalveolate phylogeny and systematics and the origins of Sporozoa and dinoflagellates (phylum Myzozoa nom. Nov.). Eur. J. Protistol. 40: 185–212.

Cavalier-Smith, T. and E. E.-Y. Chao. 2006. Phylogeny and megasystematics of phagotrophic heterokonts (Kingdom Chromista). J. Mol. Evol. 62: 388–420.

Cavalier-Smith, T. and S. von der Heyden. 2007. Molecular phylogeny, scale evolution and taxonomy of centrohelid Heliozoa. Mol. Phylogenet. Evol. 44: 1186–1203.

Chase, M. W. and J. L. Reveal. 2009. A phylogenetic classification of the land plants to accompany APG III. Botanical J. Linnean Soc. 161: 122–127.

Clark, C. G. and A. J. Roger. 1995. Direct evidence for secondary loss of mitochondria in *Entamoeba histolytica*. Proc. Natl. Acad. Sci. 92: 6518–6521.

Falkowski, P. G., M. E. Katz, A. H. Knoll, A. Quigg, J. A. Raven, O. Schofield and F. J. R. Taylor. 2004. The evolution of modern eukaryotic phytoplankton. Science 305: 354–360.

Fast, N. M., J. M. Logsdon and W. F. Doolittle. 1999. Phylogenetic analysis of the TATA box binding protein (TBP) gene from *Nosema locustae*: evidence for a microsporidia-fungi relationship and spliceosomal intron loss. Mol. Biol. Evol. 16: 1415–1419.

Freshwater, D. W., S. Fredericq, B. S. Butler, M. H. Hommersand and M. W. Chase 1994. A gene phylogeny of the red algae (Rhodophyta) based on plastid rbcL. Proc. Natl. Acad. Sci. 91: 7281–7285. [Also see Ragan et al. 1994.]

Funes, S., E. Davidson, A. Reyes-Prieto, S. Magallón, P. Herion, M. P. King and D. González-Halphen. 2002. A green algal apicoplast ancestor. Science 298: 2155.

Gajadhar, A. A., W. C. Marquardt, R. Hall, J. Gunderson, E. V. A. Carmona and M. L. Sogin. 1991. Ribosomal RNA sequences of *Sarcocystis muris*, *Theileria annulata*, and *Crypthecodinium cohnii* reveal evolutionary relationships among apicomplexans, dinoflagellates, and ciliates. Mol. Biochem. Parisit. 45: 147–154.

Gray, M. W., B. F. Lang and G. Burger. 2004. Mitochondria of protists. Annu. Rev. Genet. 38: 477–524.

Hampl, V. D. S. Horner, P. Dyal, J. Kulda, J. Flegr, P. Foster and T. M. Embley. 2005. Inference of the phylogenetic position of oxymonads based on nine genes: support for Metamonada and Excavata. Mol. Biol. Evol. 22: 2508–2518.

Hampl, V., L. Hug, J. W. Leigh, J. B. Dacks, B. F. Lang, A. G. B. Simpson and A. J. Roger. 2009. Phylogenomic analyses support the monophyly of Excavata and resolve relationships among eukaryotic "supergroups." Proc. Natl. Acad. Sci. 106: 3859–3864.

Harper, J. T. and P. J. Keeling. 2003. Nucleus-encoded, plastid-targeted glyceraldehyde-3-phosphate dehydrogenase (GAPDH) indicates a single origin for chromalveolate plastids. Mol. Biol. Evol. 20: 1730–1735.

Harper, J. T., E. Waanders and P. J. Keeling. 2005. On the monophyly of chromalveolates using a six-protein phylogeny of eukaryotes. Int. J. Syst. Evol. Micr. 55: 487–496.

Hirt, R. P. and D. Horner (eds.). 2004. *Organelles, Genomes and Eukaryote Evolution.* Taylor & Francis, London.

Hirt, R. P., J. M. Logsdon, Jr., B. Healy, M. W. Dorey, W. F. Doolittle and E. M. Embley. 1999. Microsporidia are related to fungi: evidence from the largest subunit of RNA polymerase II and other proteins. Proc. Natl. Acad. Sci. 96: 580–585.

Ishida, K., B. R. Green and T. Cavalier-Smith. 1999. Diversification of a chimaeric algal group, the Chlorarachniophytes: phylogeny of nuclear and nucleomorph small subunit rRNA genes. Mol. Biol. Evol. 16(3): 321–331.

James, T. Y. and 69 others. 2006. Reconstructing the early evolution of Fungi using a six-gene phylogeny. Nature 443: 818–822.

John, P. and F. W. Whatley. 1975. *Paracoccus dentrificans*: A present-day bacterium resembling the hypothetical free-living ancestor of the mitochondrion. Sym. Soc. Exp. Biol. 29: 39–40.

Katz, L. A. 1999. The tangled web: gene genealogies and the origin of eukaryotes. Am. Nat. 154 (suppl.): S137–S145.

Keeling, P. J. 1998. A kingdom's progress: Archeozoa and the origin of eukaryotes. BioEssays 20: 87–95.

Keeling, P. J. 2009. Chromalveolates and the evolution of plastids by secondary endosymbiosis. J. Euk. Microbiol. 56(1): 1–8.

Keeling, P. J. and 7 others. 2005. The tree of eukaryotes. Trends Ecol. Evol. 20: 670–676.

Keeling, P. J., M. A. Luker and J. D. Palmer. 2000. Evidence from beta-tubulin phylogeny that Microsporidia evolved from within the fungi. Mol. Biol. Evol. 17: 23–31.

Keeling, P. J. and J. D. Palmer. 2000. Phylogeny—Parabasalian flagellates are ancient eukaryotes. Nature 405: 635–637.

Köhler, S. and 7 others. 1997. A plastid of probable green algal origin in apicomplexan parasites. Science 275: 1485–1489.

Kutschera, U. and K .J. Niklas. 2005. Endosymbiosis, cell evolution, and speciation. Theor. Biosci. 124: 1–24.

Lane, C. E. and J. M. Archibald. 2008. The eukaryotic tree of life: endosymbiosis takes its TOL. Trends Ecol. Evol. 23: 268–275.

Lang, B. F., M. W. Gray and G. Burger. 1999. Mitochondrial genome evolution and the origin of eukaryotes. Ann. Rev. Genet. 33: 351–397.

Lang, B .F., C. J. O'Kelly, T. A. Nerad, M. W. Gray and G. Burger. 2002. The closest unicellular relatives of animals. Curr. Biol. 12: 1773–1778.

Leander, B. S. and P. J. Keeling. 2004. Early evolutionary history of dinoflagellates and apicomplexans (Alveolata) as inferred form hsp90 and actin phylogenies. J. Phycol. 40: 341–350.

Leander, B. S., S. A. J. Lloyd, W. Marshall and S. C. Landers. 2006. Phylogeny of marine gregarines (Apicomplexa)—*Pterospora*, *Lithocystis* and *Lankesteria*—and the origin(s) of coelomic parasitism. Protist 157: 45–60.

Leipe, D. D. and 7 others. 1994. The stramenopiles from a molecular perspective: 16S-like rRNA sequenced from *Labyrinthuloides minuta* and *Cafeteria roenbergensis*. Phycol. 33: 369–377.

Lewis, L. A. and R. M. McCourt. 2004. Green algae and the origin of land plants. Am. J. Bot. 91(10): 1535–1556.

Lukes, J., B. S. Leander and P. J. Keeling. 2009. Cascades of convergent evolution: the corresponding evolutionary histories of euglenozoans and dinoflagellates. Proc. Natl. Acad. Sci.106, Suppl. 1: 9963–9970.

Lutzoni, F. and 43 others. 2004. Assembling the fungal tree of life: progress, classification, and evolution of subcellular traits. Am. J. Bot. 91: 1446–1480.

Maldonado, M. 2004. Choanoflagellates, choanocytes, and animal multicellularity. Invertebr. Biol. 123: 1–22.

Margulis, L. 1981. *Symbiosis in Cell Evolution*. W. H. Freeman, San Francisco. [An assessment of the serial endosymbiotic theory and a review of the evolution of life on Earth; dated, but still a benchmark summary.]

McCourt, R. M., C. F. Delwiche and K. G. Karol. 2004. Charophyte algae and land plant origins. Trends Ecol. Evol. 19(12): 661–666.

McFadden, G. I., P. R. Gilson and R. Waller. 1995. Molecular phylogeny of chlorarachniophytes based on plastid rRNA and rbcL sequences. Arch. Protistenkd. 145: 231–239.

McFadden, G. I., P. R. Gilson and C. J. Hofmann. 1997. Division Chlorarachniophyta. Pp. 175–185 in D. Bhattacharya (ed.), *Origins of Algae and their Plastids*. Springer-Verlag, New York.

Melkonian, M. and B. Surek. 1995. Phylogeny of the Chlorophyta: Congruence between ultrastructural and molecular evidence. Bull. Soc. Zool. France 120: 191–208.

Moreira, D., H. Le Guyader and H. Phillippe. 2000. The origin of red algae and the evolution of chloroplasts. Nature 405: 69–72.

Morin, L. 2000. Long-branch attraction effects and the status of "basal eukaryotes": phylogeny and structural analysis of the ribosomal RNA gene cluster of the free-living diplomonad *Trepomonas agilis*. J. Euk. Microbiol. 47: 167–177.

Moustafa, A., B. Beszteri, U. G. Maier, C. Bowler, K. Valentin and D. Bhattacharya. 2009. Genomic footprints of a cryptic plastid endosymbiosis in diatoms. Science 324: 1724–1726.

Nikolaev, S. I. and 8 others. 2004. The twilight of Heliozoa and rise of Rhizaria, an emerging supergroup of amoeboid eukaryotes. Proc. Natl. Acad. Sci. 101: 8066–8071.

Nozaki, H. and 9 others. 2003. The phylogenetic position of red algae revealed by multiple nuclear genes from mitochondria-containing eukaryotes and an alternative hypothesis on the origin of plastids. J. Mol. Evol. 56(4): 485–497.

Nozaki, H., K. Misawa, T. Kajita, M. Kato, S. Nohara and M. Watanabe. 2000. Origin and evolution of the colonial Volvocales (Chlorophyceae) as inferred from multiple chloroplast gene sequences. Mol. Phylogenet. Evol. 17: 256–268.

O'Kelly, C. J. 1993. The jakobid flagellates: structural features of *Jakoba*, *Reclinomonas* and *Histiona* and implications for the early diversification of eukaryotes. J. Euk. Microbiol. 40: 627–636.

Polet, S., C. Berney, J. Fahrni and J. Pawlovski. 2004. Small-subunit ribosomal RNA gene sequences of *Phaeodaria* challenge the monophyly of Haeckel's Radiolaria. Protist 155: 53–63.

Ragan, M. A., C. J. Bird. E. L. Rice, R. R. Gutell, C. A. Murphy and R. K. Singh. 1994. A molecular phylogeny of the marine red algae (Rhodophyta) based on the nuclear small-subunit rRNA gene. Proc. Natl. Acad. Sci. 91: 7276–7280. [Also see Freshwater et al. 1994.]

Round, F. E., R. M. Crawford and D. G. Mann. 1990. *The Diatoms. Biology and Morphology of the Genera*. Cambridge Univ. Press, Cambridge.

Saldarriaga, J. F., F. J. R. Taylor, P. J. Keeling and T. Cavalier-Smith. 2001. Dinoflagellate nuclear SSU rDNA phylogeny suggests multiple plastid losses and replacements. J. Mol. Evol. 53: 204–213.

Saunders, G. W. and M. H. Hommersand. 2004. Assessing red algal supraordinal diversity and taxonomy in the context of contemporary systematic data. Am. J. Bot. 91: 1494–1507.

Schlegel, M. 2003. Phylogeny of eukaryotes recovered with molecular data: highlights and pitfalls. Eur. J. Protistol. 39: 113–122.

Shalchian-Tabrizi, K. and 11 others. 2006. Telonemia, a new protist phylum with affinity to chromist lineages. Proc. Roy. Soc. Lond., B 273: 1833–1842.

Shalchian-Tabrizi, K., H. Kauserud, R. Massana, D. Klaveness and K. S. Jakobsen. 2007. Analysis of environmental 18S ribosomal RNA sequences reveals unknown diversity of the cosmopolitan phylum Telonemia. Protist 158: 173–180.

Sierra, R. and 7 others. 2013. Deep relationships of Rhizaria revealed by phylogenomics: a farewell to Haeckel's Radiolaria. Mol. Phylogenet. Evol. 67: 53–59.

Simpson, A. G. B. 2003. Cytoskeletal organization, phylogenetic affinities and systematics in the contentious taxon Excavata (Eukaryota). Int. J. Syst. Evol. Microbiol. 53: 1759–1777.

Simpson, A. G. B., Y. Inagaki and A. J. Roger. 2006. Comprehensive multigene phylogenies of excavate protists reveal the evolutionary positions of "primitive" eukaryotes. Mol. Biol. Evol. 23(3): 615–625.

Simpson, A. G. B., J. Lukes and A. J. Roger. 2002. The evolutionary history of kinetoplastids and their kinetoplasts. Mol. Biol. Evol. 19: 2071–2083.

Simpson, A. G. B. and A. J. Roger. 2002. Eukaryotic evolution: getting to the root of the problem. Curr. Biol. 12: R691–R693.

Simpson, A. G. B. and A. J. Roger. 2004. The real 'kingdoms' of eukaryotes. Curr. Biol. 14(17): 693–696.

Simpson, A. G. B. and A. J. Roger. 2004. Excavata and the origin of amitochondriate eukaryotes. Pp. 27–54 in R. P. Hirt and D. S. Horner (eds.). *Organelles, Genomes and Eukaryote Phylogeny: An Evolutionary Synthesis in the Age of Genomics*. CRC Press, Boca Raton, FL.

Simpson, A. G. B. and 7 others. 2002. Evolutionary history of 'early diverging' eukaryotes: the excavate taxon *Carpediemonas* is closely related to *Giardia*. Mol. Biol. Evol. 19: 1782–1791.

Soltis, D. E., M. J. Moore, J. G. Burleigh, C. D. Bell and P. S. Soltis. 2010. Assembling the angiosperm tree of life: progress and future prospects. Ann. Mo. Bot. Gard. 97: 514–526.

Spoon, D. M., C. J. Hogan and G. B. Chapman. 1995. Ultrastructure of a primitive, multinucleate, marine cyanobacteriophagous amoeba (*Euhyperamoeba biospherica* n. sp.) and its possible significance in the evolution of the eukaryotes. Invertebr. Biol. 114 (3): 189–201.

Stiller, J. W. and B. D. Hall. 1997. The origin of red algae: implications for plastic evolution. Proc. Natl. Acad. Sci. 94: 4520–4525.

Stiller, J. W., J. Riley and B. D. Hall. 2001. Are red algae plants? A critical evaluation of three key molecular data sets. J. Mol. Evol. 52(6): 527–539.

Van de Peer, Y., A. Ben Ali and A. Meyer. 2000. Microsporidia: accumulating molecular evidence that a group of amitochondriate and suspected primitive eukaryotes are just curious fungi. Gene 246: 1–8.

Von der Heyden, S., E. E. Chao, K. Vickerman and T. Cavalier-Smith. 2004. Ribosomal RNA phylogeny of bodonids and diplonemid flagellates and the evolution of Euglenozoa. J. Euk. Microbiol. 51: 402–416.

Wainright, P. O., G. Hinkle, M. L. Sogin and S. K. Stickel. 1993. Monophyletic origin of the Metazoa: an evolutionary link with fungi. Science 260: 340–342.

Walsh, D. A. and F. W. Doolittle. 2005. The real "domains" of life. Curr. Biol. 15: R237–R240.

Wegener Parfrey, L., E. Barbero, E. Lasser, M. Dunthorn, D. Bhattacharya, D. Petterson and L. A. Katz. 2006. Evaluating support for the current classification of eukaryotic diversity. PLoS Genet. 2 (e220): 2062–2073.

Yoon, H. S., J. D. Hackett, G. Pinto and D. Bhattacharya. 2002. The single ancient origin of chromist plastids. Proc. Natl. Acad. Sci. 99: 15507–15512.

Yoon, H.-S. K. M. Müller, R. G. Sheath, F. D. Ott and D. Bhattacharya. 2006. Defining the major lineages of red algae (Rhodophyta). J. Phycol. 42: 482–492.

Yubuki, N. and B. S. Leander. 2008. Ultrastructure and molecular phylogeny of *Stephanopogon minuta*: An enigmatic microeukaryote from marine interstitial environments. Eur. J. Protistol. 44: 241–253.

Zettler, L. A. A., T. A. Nerad, C. J. O'Kelly and M. L. Sogin. 2000. The nucleariid amoeba: more protists at the animal-fungi boundary. J. Euk. Microbiol. 48: 293–297.

CHAPTER 4

Introduction to the Animal Kingdom

Animal Architecture and Body Plans

Animal bodies are marvelous things—so complex, yet so consistent and true to form within a species. All the parts seem to work together in perfect harmony, like the architecture of a beautiful building, which is how it should be after millions of years of evolutionary fine-tuning. This chapter is about the architecture of animals. The German language includes a wonderful word that expresses the essence of animal architecture—*bauplan* (pl. *baupläne*). The word means, literally, "a structural plan or design," but a direct translation is not entirely adequate. The concept of a *bauplan* captures in a single word the essence of structural range and architectural limits, as well as the functional aspects of a design. The term "body plan" is a close English equivalent. If an organism is to "work," all of its body components must be both structurally and functionally compatible. The entire organism encompasses a definable body plan, and the specific organ systems themselves also encompass structural plans; in both cases the structural and functional components of the particular plan establish both capabilities and limits. Thus, body plans determine the major constraints that operate at both the organismic and the organ system levels.

The diversity of form in the biological world is dazzling, yet there are real limits to what may be successfully molded by evolutionary processes. All animals must accomplish certain basic tasks in order to survive and reproduce. They must acquire, digest, and metabolize food and distribute its usable products throughout their bodies. They must obtain oxygen for cellular respiration, while at the same time ridding themselves of metabolic wastes and undigested materials. The strategies employed by animals to maintain life are extremely varied, but they rest upon relatively few biological, physical, and chemical principles, regardless of what body plan they have. But within the constraints imposed by particular body plans, animals have a limited number of options available to accomplish life's tasks. For this reason, a few recurring fundamental themes are apparent. This chapter is a general review of these themes: the structural/functional aspects of invertebrate body plans and the basic survival strategies employed within each. It is a description of how invertebrates are put together and how they manage to survive and reproduce. Each subject discussed here reflects fundamental principles of animal mechanics, physiology, and adaptation. We realize that much of

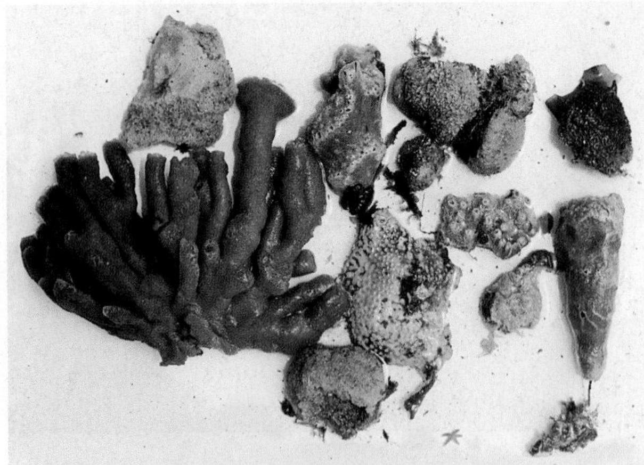

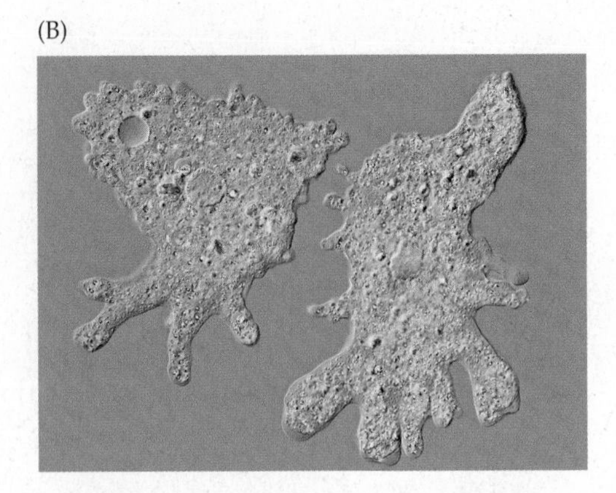

Figure 4.1 Examples of asymmetrical animals and protists. (A) An assortment of sponges. (B) Two amebas.

this chapter presents material that students learn in introductory biology courses, but here we provide specific viewpoints that set the stage for topics covered in more detail in each "animal" chapter of this book.

Keep in mind that even though this chapter is organized on the basis of what might be called the "components" of animal structure, whole animals are integrated functional combinations of these components. Furthermore, there is a strong element of predictability in the concepts discussed in this chapter. For example, given a particular type of symmetry, one can make reasonable guesses about other aspects of an animal's structure that should be compatible with that symmetry—some combinations work, others do not. Herein are explained many of the concepts and terms used throughout this book, and we encourage you to become familiar with this material now as a basis for understanding the remainder of the text.

Body Symmetry

A fundamental aspect of an animal's body plan is its overall shape or geometry. In order to discuss invertebrate architecture and function, we must first acquaint ourselves with a basic aspect of body form: symmetry. Symmetry refers to the regular arrangement of body structures relative to the axis of the body. Animals that can be bisected or split along at least one plane, so that the resulting halves are similar to one another, are said to be symmetrical. For example, a shrimp can be bisected vertically through its midline, head to tail, to produce right and left halves that are mirror images of one another. Some animals have no body axis and no plane of symmetry, and are said to be **asymmetrical**. Many sponges, for example, have an irregular growth form and lack any clear plane of symmetry. Similarly, many protists, particularly the ameboid forms, are asymmetrical (Figure 4.1).

One form of symmetry is **spherical symmetry**. It is seen in creatures whose bodies lack an axis and have the form more or less of a sphere, with the body parts arranged concentrically around, or radiating from, a central point (Figure 4.2). A sphere has an infinite number of planes of symmetry that pass through its center to divide it into like halves. Spherical symmetry is rare in nature; in the strictest sense, it is found only in certain protists. Organisms with spherical symmetry share an important functional attribute with asymmetrical organisms, in that both groups lack polarity. That is, there exists no clear differentiation along an axis. In all other forms of symmetry, some level of polarity has been achieved; and with polarity comes specialization of body regions and structures.

A body displaying **radial symmetry** has the general form of a cylinder, with one main axis around which the various body parts are arranged (Figure 4.3). In a body displaying perfect radial symmetry, the body

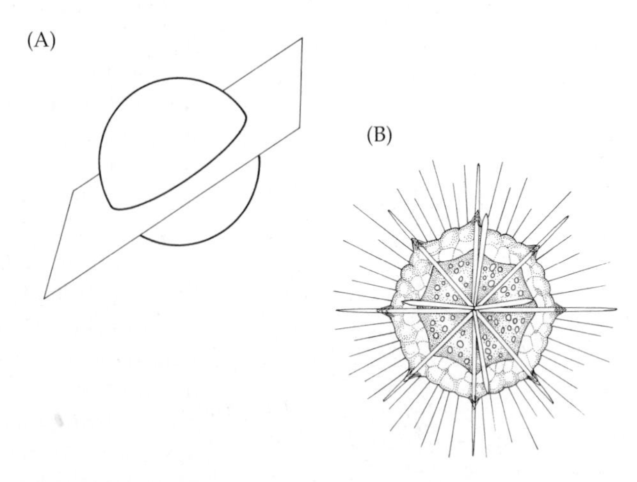

Figure 4.2 Spherical symmetry in animals. (A) An example of spherical symmetry; any plane passing through the center divides the organism into like halves. (B) A radiolarian (protist).

Figure 4.3 Radial symmetry in invertebrates. The body parts are arranged radially around a central oral–aboral axis. (A) Representation of perfect radial symmetry. (B) The barrel-like sponge *Xestospongia*. (C) The sea anemone *Epiactis*, whose mouth alignment and internal organization produce biradial symmetry. (D) The hydromedusa *Scrippsia*, with quadriradial symmetry. (E) The starfish *Patiria*, with pentaradial symmetry. (F) The sea biscuit, *Clypeaster*, with pentaradial symmetry.

parts are arranged equally around the axis, and any plane of sectioning that passes along that axis results in similar halves (rather like a cake being divided and subdivided into equal halves and quarters). Nearly perfect radial symmetry is rare in nature but occurs in some sponges and cnidarian polyps (Figure 4.3A,B). But most radially symmetrical animals have evolved modifications on this theme. **Biradial symmetry**, for example, occurs where portions of the body are specialized and only two planes of sectioning can divide the animal into perfectly similar halves. Common examples of biradial organisms are ctenophores and many sea anemones (Figure 4.3C).

Further specializations of the basic radial body plan can produce nearly any combination of multiradiality. For example, many jellyfishes possess **quadriradial symmetry** (Figure 4.3D). Most echinoderms have **pentaradial symmetry** (Figure 4.3E,F), although many multiarmed starfish are also known. In fact, the presence in starfish of certain organs (e.g., the madreporite) allows for only one plane by which perfectly matching halves exist, and thus starfish actually possess a form of pentaradial bilaterality. But this is splitting hairs. The adaptive significance of body symmetry operates at a much coarser level than organ position, and in this regard most echinoderms, including starfish, are functionally radially symmetrical although their pentaradiality is unique in the animal kingdom and should not be ignored (see Chapter 25).

A radially symmetrical animal has no front or back end; rather it is organized about an axis that passes through the center of its body, like an axle through a wheel. When a gut is present, this axis passes through the mouth-bearing (oral) surface to the opposite (aboral) surface. Radial symmetry is most common in

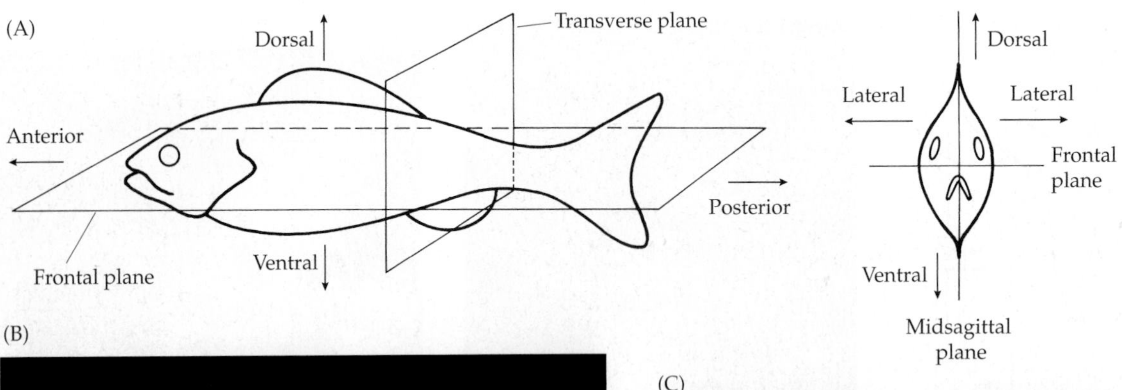

(B)

(C)

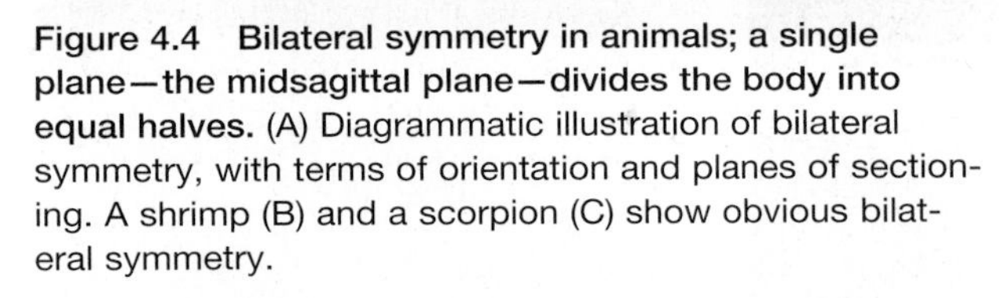

Figure 4.4 Bilateral symmetry in animals; a single plane—the midsagittal plane—divides the body into equal halves. (A) Diagrammatic illustration of bilateral symmetry, with terms of orientation and planes of sectioning. A shrimp (B) and a scorpion (C) show obvious bilateral symmetry.

sessile and sedentary animals (e.g., sponges, starfish, and sea anemones) and drifting pelagic species (e.g., jellyfishes and ctenophores). Given these lifestyles, it is clearly advantageous to be able to confront the environment equally from a variety of directions. In such creatures the feeding structures (tentacles) and sensory receptors are distributed at equal intervals around the periphery of the organisms, so that they contact the environment more or less equally in all directions. Furthermore, many bilaterally symmetrical animals have become functionally radial in certain ways associated with sessile lifestyles. For example, their feeding structures may be in the form of a whorl of radially arranged tentacles, an arrangement allowing more efficient contact with their surroundings.

The body parts of **bilaterally symmetrical** animals are oriented about an axis that passes from the front (anterior) to the rear (posterior) end. A single plane of symmetry—the **midsagittal plane** (or median sagittal plane)—passes along the axis of the body to separate right and left sides. Any longitudinal plane passing perpendicular to the midsagittal plane and separating the upper (dorsal) from the underside (ventral) is called a frontal plane (Figure 4.4). Any plane that cuts across the body perpendicular to the main body axis and the midsagittal plane is called a **transverse plane** (or, simply, a cross section). In bilaterally symmetrical animals the term lateral refers to the sides of the body, or to structures away from (to the right and left of) the midsagittal plane. The term medial refers to the midline of the body, or to structures on, near, or toward the midsagittal plane.

Whereas spherical and radial symmetry are typically associated with sessile or drifting animals, bilaterality is generally found in animals with highly directional mobility. In these animals, the anterior end of the body confronts the environment first. Associated with bilateral symmetry and unidirectional movement is a concentration of feeding and sensory structures at the anterior end of the body. The evolution of a specialized "head," containing those structures and the nervous tissues that innervate them, is called **cephalization** (from the Greek *kephalos*, "head"). Furthermore, the surfaces of the animal differentiate as dorsal and ventral regions, the latter becoming locomotory and the former being specialized for protection. A variety of secondary asymmetrical modifications of bilateral (and radial) symmetry have occurred, for example, the spiral coiling of snails and hermit crabs.

Cellularity, Body Size, Germ Layers, and Body Cavities

One of the main characteristics used to define grades of animal complexity is the presence or absence of true tissues. Tissues are aggregations of morphologically and physiologically similar cells that perform a specific function. As we saw in Chapter 3, protists do not possess tissues, but occur only as single cells or as simple colonies of cells. In a sense, they are all at a unicellular grade of construction. Beyond the protists is the vast array of multicelled animals, the Metazoa. The Metazoa are often divided into two major levels, or grades—the non-bilateria (diploblastic animals) and the bilateria (= triploblastic animals), as described below. These names do not represent formal taxa, but may be used to group the Metazoa by their level of overall structural complexity. The non-bilateria comprise a nonmonophyletic grouping; the bilateria is almost certainly a clade.

Each of these two grades of body complexity is associated with inherent constraints and capabilities, and within each grade there are obvious limits to size. As the British biologist D'Arcy Thompson wrote, "Everything has its proper size...men and trees, birds and fishes, stars and star-systems, have...more or less narrow ranges of absolute magnitudes." As a cell (or an organism) increases in size, its volume increases at a rate faster than the rate of increase of its surface area (surface area increases as the square of linear dimensions; volume increases as the cube of linear dimensions). Because a cell ultimately relies on transport of material across its plasma membrane for survival, this disparity quickly reaches a point at which the cytoplasm can no longer be adequately serviced by simple cellular diffusion. Some unicellular forms develop complexly folded surfaces or are flattened or threadlike in shape. Such creatures can be quite large, but eventually a limit is reached; thus we have no meter-long protists.

To increase in size, ultimately the only way around the **surface-to-volume dilemma** is to increase the number of cells constituting a single organism; hence the Metazoa. But size increase in the Metazoa is also limited. Those Metazoa lacking complex specializations of tissues and organs must rely on diffusion into and out of the body, and this is inadequate to sustain life unless a majority of the body's cells are near or in contact with the external environment. In fact, diffusion is an effective method of oxygenation only when the diffusion path is less than about 1.0 mm. So here, too, there are limits. An animal simply cannot increase indefinitely in volume when most of its cells must lie close to the body surface. Some animals solve this problem to some degree by arranging their cellular material so that diffusion distances from cell to environment are comfortably short. One method of accomplishing this is to pack the internal bulk of the body with nonliving (or largely nonliving) material, such as the jelly-like mesoglea of medusae and ctenophores. Another is to assume a body geometry that maximizes the surface area. Increase in one dimension leads to a vermiform body plan, like that of cestid ctenophores (Ctenophora; Figure 8.1E) or ribbon worms (Nemertea). Increase in two dimensions results in a flat, sheetlike body like that of flatworms (Platyhelminthes). In these cases the diffusion distances are kept short. Sponges effectively increase their surface area by a process of complex branching and folding of the body, both internally and externally. This folding keeps most of the body cells close to the environment.

If these were the only solutions to the surface-to-volume dilemma, the natural world would be filled with tiny, thin, flat animals and convoluted, spongelike creatures. However, many organisms increase in size by one to several orders of magnitude during their ontogeny, and life forms on Earth span about 19 orders of magnitude (in mass). Thus, another solution arose during the course of animal evolution that allowed for increases in body size. This solution was to bring the "environment" functionally closer to each cell in the body by the use of internal transport and exchange systems with large surface areas. A significant three-dimensional increase in body size thus necessitated the development of sophisticated internal transport mechanisms (e.g., circulatory systems) for nutrients, oxygen, waste products, and so on. These evolving transport structures became the organs and organ systems of higher animals. For example, the body volume of humans is so large that we require a highly branched network of gas exchange surfaces (our lungs) to provide an adequate surface area for gas diffusion. This network has about 1,000 square feet of surface—as much area as half a tennis court! The same constraints apply to food absorption surfaces; hence the evolution of very long, highly folded, or branched guts.

The embryonic tissue layers of eumetazoans are called **germ layers** (from the Latin *germen*, "a sprout, bud, or embryonic primordium"), and it is from these germ layers that all adult structures develop. Chapter 5 provides an overview of germ layer formation and other aspects of metazoan developmental patterns. Here we need only point out that the germ layers initially form as outer and inner sheets or coherent masses of embryonic cells, termed **ectoderm** and **endoderm** (or entoderm), respectively. In the embryogeny of the radiate phyla Cnidaria and Ctenophora, only these two germ layers develop (or if a middle layer does develop, it is produced by the ectoderm, is largely noncellular, and is not considered a true germ layer). These animals are regarded as **diploblastic** (Greek *diplo*, "two"; *blast*, "bud" or "sprout"). In the embryogeny of most animals, however, a third cellular germ layer, the **mesoderm**, arises between the ectoderm and the endoderm; these metazoans are said to be **triploblastic**.

The evolution of a mesoderm greatly expanded the evolutionary potential for animal complexity. As we shall see, the triploblastic phyla have achieved many more highly sophisticated body plans than are possible within the confines of a diploblastic body plan. Simply put, a developing triploblastic embryo has more building material to work with than does a diploblastic embryo.

One of the major trends in the evolution of the triploblastic Metazoa has been the development of a fluid-filled cavity between the outer body wall and the digestive tube; that is, between the adult derivatives of the ectoderm and the endoderm. The evolution of this space created a radically new architecture, a tube-within-a-tube design in which the inner tube (the gut and its associated organs) was largely freed from the constraint of being attached to the outer tube (the body wall). The fluid-filled cavity not only served as a mechanical buffer between these two largely independent tubes, but also allowed for the development and expansion of new structures within the body, served as a storage chamber for various body products (e.g., gametes), provided a medium for circulation, and was in itself an incipient hydrostatic skeleton. The nature of this cavity (or the absence of it) is associated with the formation and subsequent development of the mesoderm, as discussed in detail in Chapter 5.

Three major grades of construction are recognizable among the triploblastic Metazoa: **acoelomate**, **blastocoelomate**, and **eucoelomate**. The acoelomate grade (Greek *a*, "without"; *coel*, "hollow, cavity") occurs in numerous triploblastic phyla: Xenacoelomorpha, Platyhelminthes, Entoprocta, Cycliophora, Gnathostomulida, Micrognathozoa, Nematomorpha, and Gastrotricha. In these animals, the mesoderm forms a more or less solid mass of tissue, sometimes with small open spaces (lacunae), between the gut and body wall (Figure 4.5A). In nearly all other triploblastic animals, an actual space develops as a fluid-filled cavity between the body wall and the gut. In many phyla (e.g., annelids and echinoderms), this cavity arises within the mesoderm itself and is completely enclosed within a thin lining called the **peritoneum**, which is derived from the mesoderm. Such a cavity is called a true **coelom** (eucoelom). Notice that the organs of the body are not actually free within the coelomic space itself, but are separated from it by the peritoneum (Figure 4.5C). Peritoneum is usually a squamous epithelial layer, at least that portion of it covering the gut and internal organs.

Several groups of triploblastic Metazoa (e.g., Rotifera, Nematoda, Loricifera, Priapula, Tardigrada, and some Kinorhyncha) possess small or large body cavities that are neither formed from the mesoderm nor fully lined by peritoneum or any other form of mesodermally derived tissue. Such a cavity used to be called a "pseudocoelom" (Greek *pseudo*, "false"; *coel*, "hollow, cavity") (Figure 4.5B). The organs of these animals actually lie free within the body cavity and are bathed directly in its fluid. In most cases the space represents persistent remnants of the embryonic blastocoel, and since there is nothing "false" about it, we use the more descriptive term **blastocoelom** in this text.

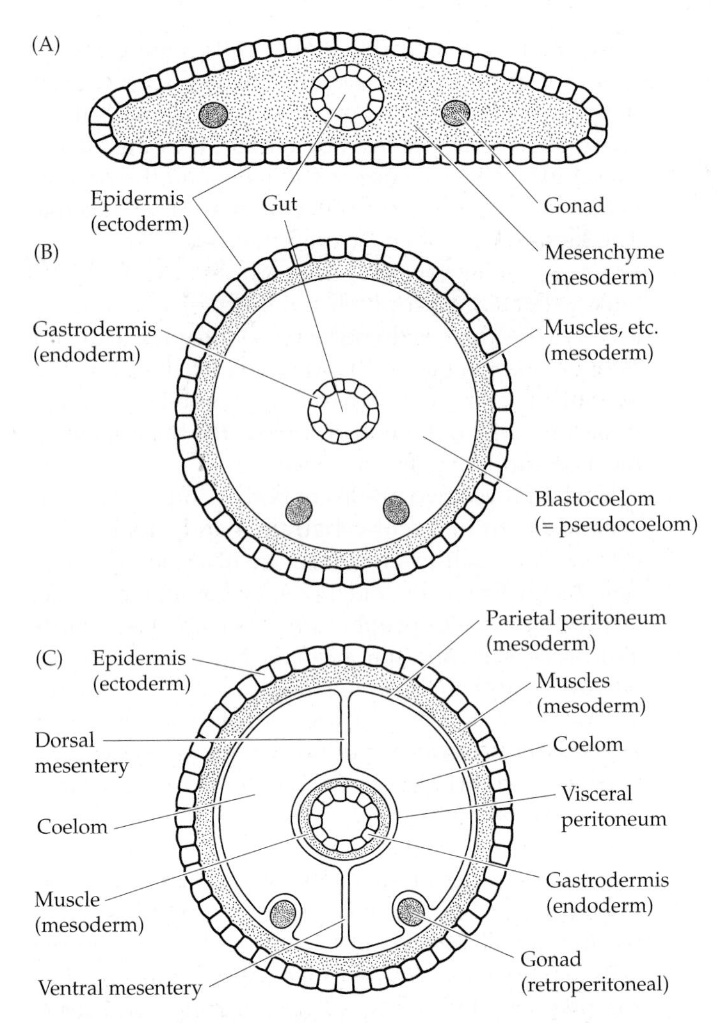

Figure 4.5 Principal body plans of triploblastic Metazoa (diagrammatic cross sections). (A) The acoelomate body plan. (B) The blastocoelomate body plan. (C) The coelomate (= eucoelomate) body plan.

The organization of the body cavity shows only a loose correlation to phylogeny. However, within the constraints inherent in each of the basic body organizations discussed above, animals have evolved a multitude of variations on these themes. Throughout the remainder of this chapter we describe the fundamental organizational plans of major body systems as they have evolved within these basic body plans. In subsequent chapters, we describe how members of the various phyla have modified these basic plans through their own particular evolutionary program or direction.

Locomotion and Support

As eukaryotic life progressed from the single-celled stage to multicellularity, body size increased dramatically. And this increase in body size, coupled with directed movement, was accompanied by the evolution of a variety of support structures and locomotor mechanisms. Because these two body systems evolved mutually and usually work in a complementary fashion, they are conveniently discussed together. And of course, there was a rapid evolutionary boost in body size and support in the Cambrian, approximately 540 to 485 million years ago—an event known as the Cambrian Explosion. The reasons for this rapid radiation of larger animals are still being debated (Chapter 1).

There are four fundamental locomotor patterns in protists and Metazoa: ameboid movement, ciliary and flagellar movement, hydrostatic propulsion, and locomotor limb movement. There are three basic kinds of support systems: structural endoskeletons, structural exoskeletons, and hydrostatic skeletons. In this section we briefly describe the basic architecture and mechanics of the various combinations of these systems.

Reynolds Number

Most invertebrate lineages inhabit water, and aquatic environments present obstacles and advantages to support and locomotion that are quite different from those of terrestrial environments. Just staying in one place in the face of swiftly moving water, without being damaged or dislodged, can require both support and flexibility. Animals moving through water (or moving water over their bodies—the effect is much the same) face problems of fluid dynamics created by the interaction between a solid body and a surrounding liquid. What happens during this interaction is tied to the concept of Reynolds number, a unitless value based on the experiments of Osborne Reynolds (1842–1912). Reynolds number represents a ratio of inertial force to viscous force. At higher Reynolds numbers, inertial force predominates and determines the behavior of water flow around an object. At lower Reynolds numbers, viscous force predominates and determines the behavior of the water flow. The importance of this concept is being increasingly recognized and applied to biological systems. Although there is still a great deal to be done in this area, some interesting generalizations can be made about locomotion of aquatic animals and, as we discuss later, aquatic suspension feeding. Reynolds number is expressed by the following equation:

$$R_e = \frac{plU}{v}$$

where p equals the density of fluid, l is some measurement of the size of the solid body, U equals the relative velocity of the fluid over the body surface, and v is the viscosity of the fluid. The formula was derived by Reynolds to describe the behavior of cylinders in water. Of course, since animals' bodies are not perfect cylinders, the size variable (l) is difficult to standardize. Nonetheless, meaningful relative values can be derived and applied to living creatures in water.

Without belaboring this issue beyond its importance here, it turns out that the problems of a large animal swimming through water are very different from those of a small animal. Large animals such as fishes, whales, or even humans, by virtue of their size or high velocity or both, move in a world of high Reynolds numbers. With increased body size, fluid viscosity becomes less and less significant as far as the animal's energy output during locomotion is concerned. At the same time, however, inertia becomes more and more important. A large animal must expend more energy than a small animal does to put its body in motion. But, by the same token, inertia works in favor of the moving large animal by carrying it forward when the animal stops swimming. When large animals move at high Reynolds numbers, the effect of inertia also imparts motion to the water around the animal's body. Thus, as the Reynolds number increases, a point is reached at which the flow of water changes from laminar to turbulent, decreasing swimming efficiency.

Small organisms generally move in a world of low Reynolds numbers. For example, a larva 1 mm in diameter, moving at a speed of 1 mm/sec, has a Reynolds number of about 1.0. Inertia and turbulence are virtually nonexistent, but viscosity becomes important—increasingly so as body size and velocity decrease (i.e., as the Reynolds number decreases). Small organisms swimming through water have been likened to a human swimming through liquid tar or thick molasses. The effect of this situation is that tiny creatures, such as ciliate and flagellate protists and many small Metazoa, start and stop instantaneously, and the motion of the water set up by their swimming also ceases immediately if the animal stops moving. Thus, small creatures neither pay the price nor reap the benefits of the effects of inertia. The organism only moves forward when it is expending energy to swim; as soon as it stops moving its cilia, or flagella, or appendages, it stops—and so does the fluid surrounding it. Tiny organisms swimming at low Reynolds numbers (i.e., < 1.5) must expend an incredible amount of energy to propel themselves through their "viscous" surroundings.

Ameboid Locomotion

Ameboid movement is used principally by certain protists and by numerous kinds of ameboid cells that occur within the bodies of most Metazoa. Ameboid cells possess a gel-like **ectoplasm**, which surrounds a more fluid **endoplasm** (Figure 4.6). Movement is facilitated by changes in the states of these regions of the cell. At one (or several) points on the cell surface, **pseudopods** (or pseudopodia) develop; and as endoplasm flows into a growing pseudopod (or pseudopodium), the cell creeps in that direction. This seemingly simple process actually involves complex changes in cell fine structure, chemistry, and behavior. The innermost endoplasm moves "forward" while the outermost endoplasm takes on a granular appearance and remains fairly stable. The advancing portion of endoplasm pushes

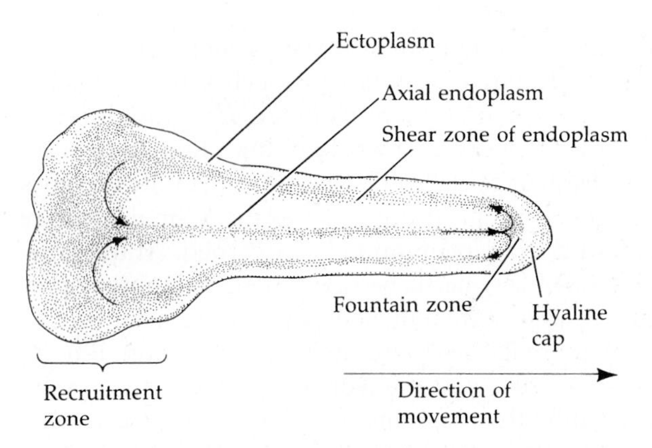

Figure 4.6 Ameboid locomotion by way of pseudopod formation in an ameba.

forward and then becomes semirigid ectoplasm at the tip of the advancing pseudopodium. Concurrently, endoplasm is recruited from the trailing end of the cell, from whence it streams forward to join in the "growing" pseudopod. Many protists, and certain cells of eukaryotes, have wide, blunt-tipped pseudopods called **lobopods** (or lobopodia). In many protist groups long, slender, tapering pseudopods occur—**filopods** (or filopodia). The formation of filopodia appears to involved actin polymerization. **Reticulopods** (or reticulopodia) are filopods that are very long and anastomose in a netlike fashion. Evidence suggests that reticulopods are supported by microtubules. Overall, it appears that various forms of pseudopods have evolved multiple times within the protists.

The molecular basis of ameboid movement may be essentially the same as that of vertebrate muscle contraction, involving actin, myosin, and ATP. Two principal theories exist to explain the process. Perhaps the more popular of the two ideas has the actin molecules floating freely in the endoplasm, polymerizing into their filamentous form at the point of active pseudopodium growth, where they interact with myosin molecules. The resultant contraction literally pulls the streaming endoplasm forward, while at the same time converting it to the ectoplasm that rings the forward-streaming pseudopodium. The second theory suggests that the actin–myosin interaction takes place at the rear of the cell, where it produces a contraction of the ectoplasm. The contraction squeezes the cell like a tube of toothpaste, causing the endoplasm to stream forward and create a pseudopodium directly opposite the point of ectoplasmic contraction. Some modifications of pseudopodial movement were discussed in Chapter 3.

Cilia and Flagella

Cilia or flagella or both occur in virtually every animal phylum (with the qualified exception of the Arthropoda). Structurally, cilia and flagella are nearly identical (and clearly homologous), but the former are shorter and tend to occur in relatively larger numbers (in patches or tracts), whereas the latter are long and generally occur singly or in pairs. During cell development, each new flagellum or cilium arises from an organelle called the **kinetosome** (sometimes called a basal body or a blepharoplast), to which it remains anchored.

The movement of cilia and flagella creates a propulsive force that either moves the organism through a liquid medium or, if the animal (or cell) is anchored, creates a movement of fluid over it. Such action always occurs at very low Reynolds numbers. When the animal is large, the viscosity may be increased by secretion of mucus, which lowers the Reynolds number. The general structure of a flagellum or cilium consists of a long, flexible rod, the outer covering of which is an extension of the plasma membrane of the cell (Figure 4.7A). Inside is a circle of nine paired microtubules (often called doublets) that runs the length of the flagellum or cilium. One microtubule of each doublet bears two rows of projections, the **dynein arms**, directed toward the adjacent doublet. Flagella and cilia move as the microtubules slide up or down against one another, bending the flagellum or cilium in one direction or another. Microtubule sliding is driven by the dynein arms, particularly by protein complexes called **radial spokes** that arise from the arms. The radial spokes attach to each doublet microtubule immediately adjacent to the inner row of dynein arms and project centrally. Down the center of the doublet circle is an additional pair of **microtubules**. This familiar 9+2 pattern is characteristic of nearly all flagella and cilia (Figure 4.7B).[1] Flagellar/ciliary microtubules are modified hollow tubules similar to those present in the matrix of most cells. The principal function of these cellular tubules appears to be support. Just as the ectoplasm helps retain the shape and integrity of a protozoan cell (acting as a type of rudimentary "exoskeleton"), so the cytoplasmic microtubules act as a sort of simple "endoskeleton" that help protists (and other cells) retain their shape. Microtubules are also components of the spindle and help distribute the chromosomes during cell division.

In addition to the locomotor function seen in some protists and small Metazoa, cilia and flagella have an enormous variety of functions in many other animals. For example, they create feeding and gas exchange currents, they line digestive tracts and facilitate food movement, and they propel sex cells and larvae. They also form sensory structures of many kinds. Here we focus on their use as locomotor structures.

Analysis by high-speed photography reveals that the movement of these structures is complex and differs among taxa, and even at different locations on the same organism. Some flagella beat back and forth,

[1]Dyneins are a family of adenosine triphosphatases that cause microtubule sliding in ciliary and flagellar axonemes.

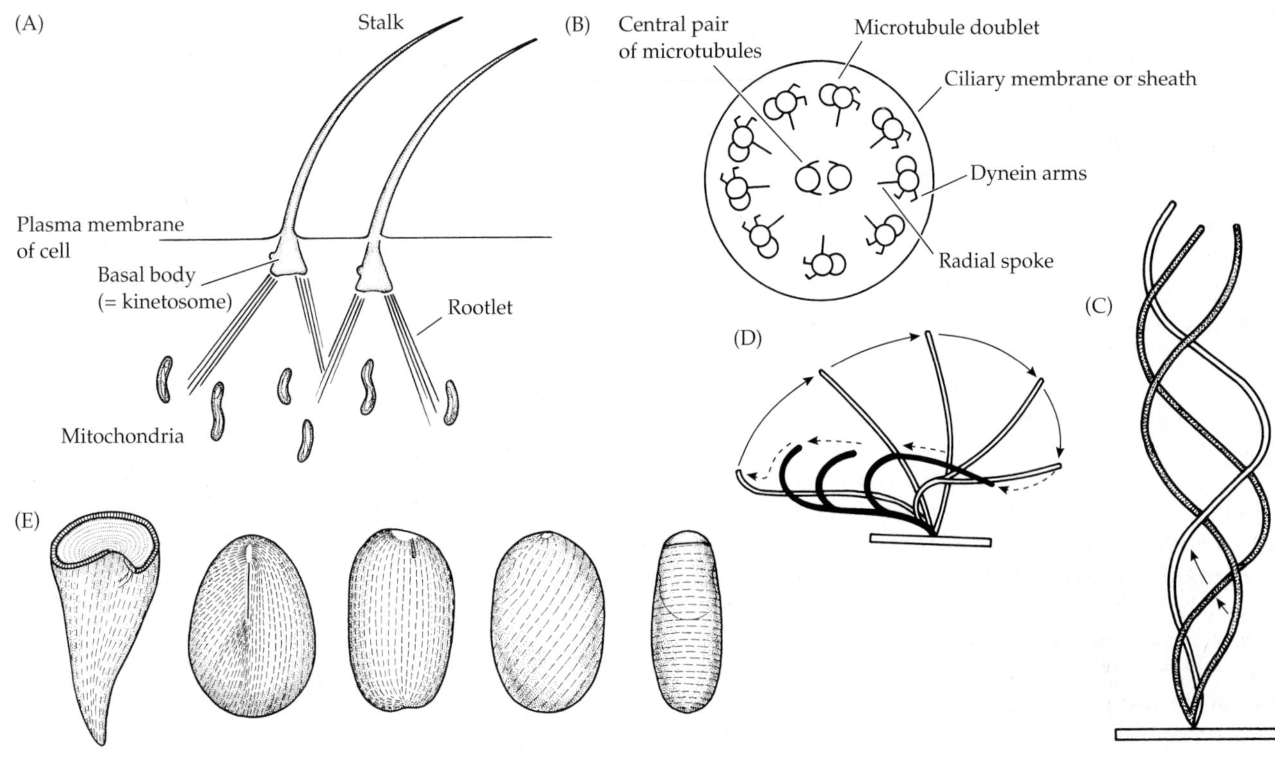

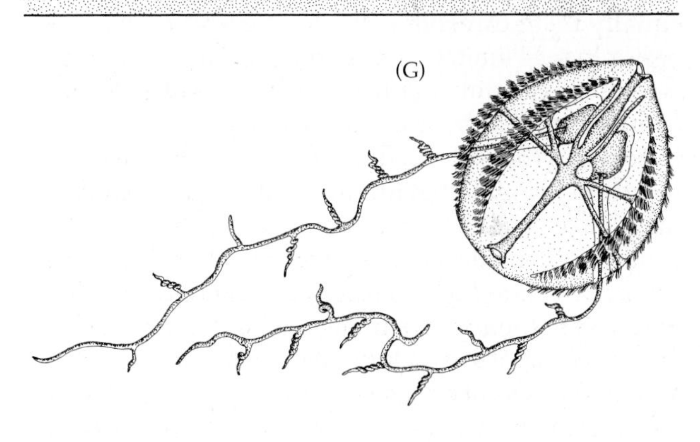

Figure 4.7 Cilia and flagella. (A) Structures of two adjacent cilia. (B) Cross section of a cilium. (C) Three successive stages in the undulatory movement of a flagellum. (D) Successive stages in the oarlike action of a cilium. The power stroke is shown in white, the recovery stroke in black. (E) Examples of ciliary tract patterns in various ciliate protists (tracts indicated by dashed lines). (F) Appearance of metachronal waves of a line of cilia. (G) The comb jellies (ctenophores) are the largest animals known to rely primarily on cilia for locomotion. Shown here is the rather small ctenophore, *Pleurobrachia bachei* (about two cm in diameter).

while others beat in a helical rotary pattern that drives flagellate protist cells something like the propeller of an outboard motor (Figure 4.7C). Depending on whether the undulation moves from base to tip or from tip to base, the effect will be, respectively, to push or pull the cell along.

Some flagella possess tiny, hairlike side branches called **mastigonemes** that increase the effective surface area and thus improve the propulsive capability. The beat of a cilium is generally simpler, consisting of a power stroke and a relaxed recovery stroke (Figure 4.7D). When many cilia are present on a cell, they often occur in distinct tracts, and their action is integrated, with beats usually moving in metachronal waves over the cell surface (Figure 4.7E,F). Since at any one time some cilia are always performing a power stroke, metachronal coordination ensures a uniform and continuous propulsive force.

It was once suggested that ciliary tracts on individual cells were coordinated by a primitive sort of cellular "nervous system," but this hypothesis was never confirmed. Current thinking suggests that the coordinated beating of cilia is probably due to hydrodynamic constraints imposed on them by the interference effects of the surrounding water layers and by the simple mechanical stimulation of moving, adjacent cilia. Nevertheless, some ciliary responses in animals are clearly under neural control, for example, reversal of power stroke direction.

Ciliated protists are the swiftest of the single-celled organisms. Flagellated protists are the next most rapid, and amebas are the slowest. Most amebas move at rates around 5 μm /sec (about 2 cm/hr), or about 100 times slower than most ciliates. Cilia are also used for locomotion by members of several metazoan phyla (including Rhombozoa, Orthonectida, Ctenophora, Platyhelminthes, Rotifera, and some gastropod molluscs), and by the larval stages of a great many taxa.

Muscles and Skeletons

Almost all animals have some sort of a skeleton, the major functions of which are to maintain body shape, provide support, serve as attachments for muscles, transmit the forces of muscle contraction to perform work, and extend relaxed muscles. These functions may be attained either by hard tissues or secreted skeletons, or even by the turgidity of body fluids or tissues under pressure. Muscles, skeletons, and body form are closely integrated, both developmentally and functionally. When rigid skeletal elements are present, they can serve as fixed points for muscle attachment. For example, the rigid and jointed exoskeleton of arthropods allows for a complex system of levers that results in very precise and restricted limb movements. Many invertebrates lack hard skeletons and can change their body shape by alternate contraction and relaxation of various muscle groups attached to tough connective tissues or to the inside of the body wall. These "soft-bodied" invertebrates usually have a hydrostatic skeleton.

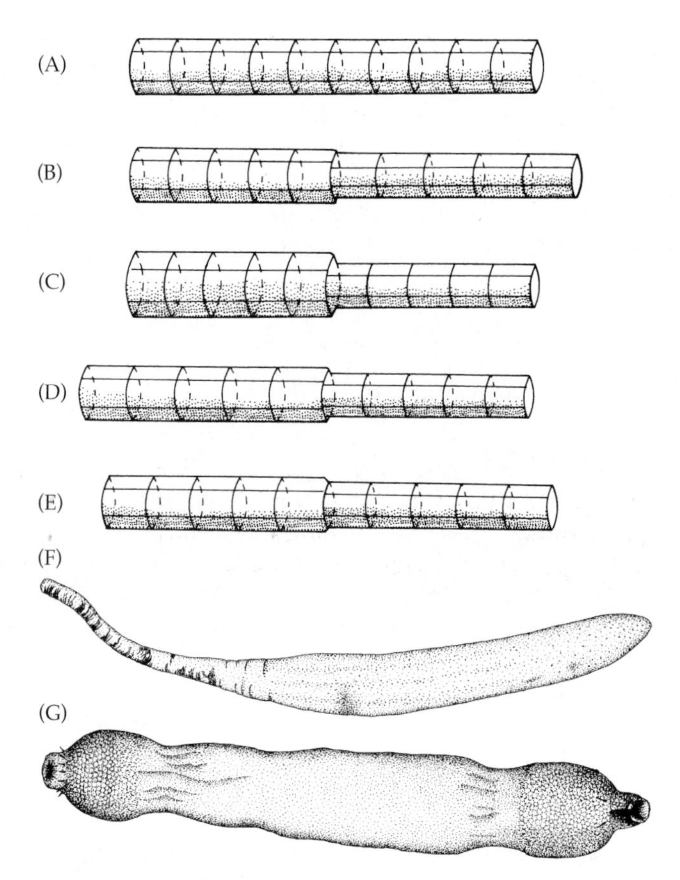

Figure 4.8 The hydrostatic skeleton. (A–E) The initial state and the four possible results of contraction of the circular muscles at one end of a cylindrical animal with a hydrostatic skeleton. (A) The muscles are all relaxed. (B) The circular muscles of the right-hand end have contracted and this end has elongated; the left-hand end has remained unaltered. (C) The length of the right-hand end has remained the same but the diameter of the left-hand end has increased. (D) The length of the right-hand end and the diameter of the left-hand end have remained the same, but the length of the left-hand end has increased. (E) The lengths of both ends have increased, but their respective diameters have remained the same. (F,G) Two animals that rely on hydrostatic skeletons for support and locomotion. (F) The sipunculan worm *Phascolosoma*. (G) The echiurid worm *Urechis caupo*.

The hydrostatic skeleton The performance of a hydrostatic skeleton is based on two fundamental properties of liquids: their incompressibility and their ability to assume any shape. Because of these features, body fluids transmit pressure changes rapidly and equally in all directions. It is important to realize a basic physical limitation concerning the action of muscles—they can only perform work by getting shorter (contracting).[2] In hydrostatic skeletons, a body region is extended (or protruded) by the contractile force of a muscle being imparted to a fluid-filled body compartment, creating a hydrostatic pressure that displaces the wall of the compartment. Such indirect muscle actions can be compared to squeezing a rubber glove filled with water, thereby extending and stiffening the fingers. The enclosure of a fluid-filled chamber (e.g., a coelom) within sets of opposing muscle layers establishes a system in which muscles in one part of the body can contract, forcing body fluids into another region of the body, where the muscles relax; the body is thus extended or otherwise changed in shape (Figure 4.8).

In the most common plan, two muscle layers surround a fluid-filled body cavity, and the fibers of the layers run in different directions (e.g., a circular muscle layer and a longitudinal muscle layer). A soft-bodied invertebrate can move forward by using its hydrostatic skeleton in the following way. The circular muscles at the posterior end of the animal contract; the hydrostatic pressure generated then pushes anteriorly to extend the relaxed longitudinal muscles of the front of the body. Then contraction of the posterior longitudinal muscles pulls the rear end of the body forward. This sequence of muscle contractions results in a directed and controlled movement forward. Such movement requires that the posterior end be anchored when the anterior end is extended, and that the anterior end be anchored when the posterior end is pulled forward. This system is commonly used for locomotion by many worms that generate posterior-to-anterior metachronal waves of muscle action, resulting in what is called peristalsis. A similar hydrostatic system can be used to temporarily or intermittently extend selected parts of the body, such as the feeding proboscis of worms, tube feet of echinoderms, and siphons of clams.

The contraction of the circular muscles at one end of a vermiform animal may actually have four possible effects: the contracting end may elongate, the opposite end may elongate or thicken, or both ends may

[2]Bear in mind that muscles can contract isometrically and do no work.

elongate. The event that transpires depends not on the contraction of the circular muscles of the contracting end, but instead on the state of contraction of the longitudinal and circular muscles in other parts of the body (Figure 4.8). Such combinations of muscle contraction and relaxation create a versatile movement system based on very simple mechanical principles.

Reliance on only circular and longitudinal muscles could result in twists and kinks when a hydrostatic system engages itself against the resistance of the substratum. Hence, most animals that rely on hydrostatic movement also have helically wound, diagonal muscle fibers—a left- and right-handed set, intersecting at an angle between 0 and 180 degrees. The diagonal muscles allow extension and contraction, even at a constant volume, without stretching and while preventing kinking and twisting. A good analogy is the children's toy—the helically woven straw cylinder into which one youngster convinces another to insert his two index fingers; pushing your fingers together increases the diameter of the cylinder (and decreases the length), pulling them apart decreases the diameter (and increases the length). All of this is accomplished without an appreciable stretching or compression of the straw fibers (or the diagonal muscles of an invertebrate). When a cylinder is extended, the fiber angle decreases; when it is compressed, the angle increases.

The volume of the working fluid in a hydrostatic skeleton should remain constant, thus any leakage should not be greater than the rate at which the fluid can be replaced. Body fluids must be retained despite "holes" in the body wall, such as the excretory pores of many coelomate animals or the mouth openings of cnidarians. Such openings are often encircled by sphincter muscles that can close and control the loss of body fluids.

One way in which movement by a hydrostatic skeleton can be made more precise is to divide an animal up into a series of separate compartments. For example, in annelid worms, partitioning of the coelom and body muscles into segments with separate neural control enables body expansions and contractions to be confined to a few segments at a time. By manipulating particular sets of segmental muscles, most annelids not only can move forward and backward but can turn and twist in complex maneuvers.

The rigid skeleton In "hard-bodied" invertebrates, a fixed or rigid skeleton prevents the gross changes in body form seen in soft-bodied invertebrates. This trade-off in flexibility gives hard-bodied animals several advantages: the capacity to grow larger (an advantage that is especially useful in terrestrial habitats, which lack the buoyancy provided by aquatic environments), more precise or controlled body movements, better defense against predators, and often greater speed of movement.

Hard skeletons can be broadly classed as either endoskeletons or exoskeletons. Endoskeletons are generally derived from mesoderm, whereas exoskeletons are derived from ectoderm; both usually have organic and inorganic components. It has been hypothesized that rigid skeletons may have originated by chance, as byproducts of certain metabolic pathways. By sheer accident (preadaptation or exaptation), for example, the accumulation of nitrogenous wastes and their incorporation into complex organic molecules might have resulted in the evolution of the chitinous exoskeleton so common among invertebrates. Similar speculation suggests that a metabolic system that originally functioned to eliminate excess calcium from the body might have produced the first calcareous shell of molluscs. In any event, marine invertebrates are capable of forming, through their various biological activities, a vast array of minerals, some of which cannot be formed inorganically in the biosphere. Indeed, the ever-increasing amounts of these biominerals have radically altered the character of the biosphere since the origin of hard skeletons in the earliest Cambrian. Most common among these biominerals are various carbonates, phosphates, halides, sulfates, and iron oxides.

Invertebrate skeletons may be of the articulating type (e.g., the exoskeletons of arthropods, clams, and brachiopods and the endoskeleton of some echinoderms), or they may be of the nonarticulating type, as seen in the simple one-piece exoskeletons of snails and the rigid endoskeletons composed of interlocking fused plates of sea urchins and sand dollars. Animal endoskeletons may be as simple as the microscopic calcareous or siliceous spicules embedded in the body of a sponge, cnidarian, or sea cucumber, or they may be as complex as the bony skeleton of vertebrates (Figure 4.9). Hard skeletons of calcium carbonate have evolved in many animal (and some algal) phyla, but the use of silica in a mineralized skeleton occurs only in sponges (Porifera). Vertebrate skeletal tissues include a calcium phosphate-collagen matrix. In invertebrates, collagen often forms a substratum upon which calcareous spicules or other skeletal structures form, but with a single exception (certain gorgonians) collagen is never incorporated directly into the calcareous skeletal material.[3]

In the broadest sense, virtually every group of invertebrates has developed an exoskeleton of sorts (Figure 4.10). Even cells of protists possess a semirigid ectoplasm, and some have surrounded themselves with a test comprising bits of sand or other foreign matter glued together. Other protists build a test made from chemicals that they either extract from seawater or produce themselves.

[3]Collagen, of course, also is the primary component of basement membrane that underlies epithelia in all Metazoa, providing structural integrity to tissues.

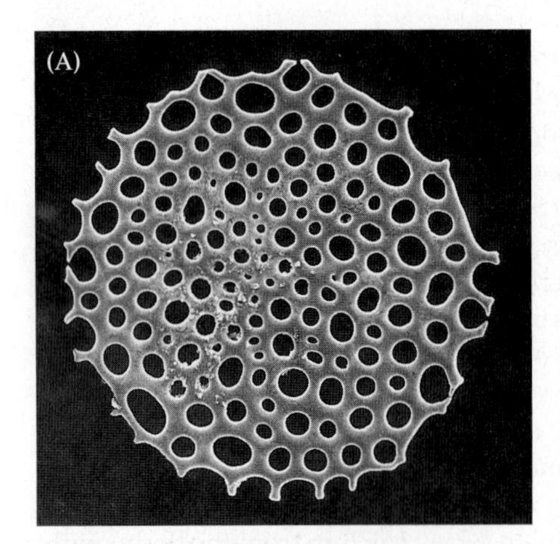

Figure 4.9 Some invertebrate endoskeletons. (A) an ossicle (skeletal element) from a sea cucumber (Echinodermata). (B) Isolated sponge spicules. (C) A deep-water glass sponge (Hexactinellida) from the eastern Pacific; the long siliceous spicules can be seen protruding from the body. (D) The rigid test of a sea urchin, in which the calcareous plates are sutured together by connective tissue and calcite interdigitations

From their epidermis, many Metazoa secrete a nonliving external layer called the cuticle, which serves as an exoskeleton. The cuticle varies in thickness and complexity, but it often has several layers of differing structure and composition. In the arthropods, for example, the cuticle is a complex combination of the polysaccharide chitin and various proteins.[4] This skeleton may be strengthened by the formation of internal cross-linkages (a process called **tanning**) and by the addition of calcium and pigments. The chitinous layer of most invertebrates is the first line of defense against infective microbes and desiccation. In most insects, the outermost layer is impregnated with wax, which decreases its permeability to water. The cuticle is often ornamented with spines, tubercles, scales, or striations; frequently it is divided into rings or segments, a feature lending flexibility to the body. Other examples of exoskeletons are the calcareous shells of many molluscs and the casings of corals (Figure 4.10).

[4]The term chitin refers to a family of closely related chemical compounds, which, in various forms, are produced by and incorporated into the cuticles of many invertebrates. Certain types of chitin are also produced by some fungi and diatoms. Chitins are high-molecular-weight, nitrogenous polysaccharide polymers that are tough yet flexible (Figure 4.10F). In addition to its supportive and protective functions in the formation of exoskeletons, chitin is also a major component of the teeth, jaws, and grasping and grinding structures of a wide variety of invertebrates. That chitin is one of the most abundant macromolecules on earth is evidenced by the estimated 10^{11} tons produced annually in the biosphere—most of it in the ocean.

Most skeletons act as body elements against which muscles operate and by which muscle action is converted to body movement. Because muscles cannot elongate by themselves, they must be stretched by antagonistic forces—usually other muscles, hydrostatic forces, or elastic structures. In animals possessing rigid but articulated skeletons, antagonistic muscles often appear in pairs, for example, flexors and extensors. These muscles extend across a joint and are used to move a limb or other body part (Figure 4.11). Most muscles have a discrete origin, where the muscle is anchored, and an insertion, which is the point of major body or limb movement. A classic vertebrate example of this system is the biceps muscle of the human arm, in which the origin is on the scapula and the insertion is on the radius bone of the forearm; contraction of the biceps causes flexion of the arm by decreasing the angle between the upper arm and the forearm. Movement of a limb toward the body is brought about by **flexor muscles**, of which the biceps is an example (Figure 4.11A,B). The muscle antagonistic to the biceps is the triceps, an **extensor muscle** whose contraction extends the forearm away from the body. Other common sets of antagonistic muscles and actions are **protractors** and **retractors**, which respectively cause anterior and posterior movement of entire limbs at their place of juncture with the body; and **adductors** and **abductors**, which move a body part toward or away from a particular point of reference. Although vertebrates have endoskeletons and arthropods have exoskeletons, most muscles of arthropods are arranged in antagonistic sets

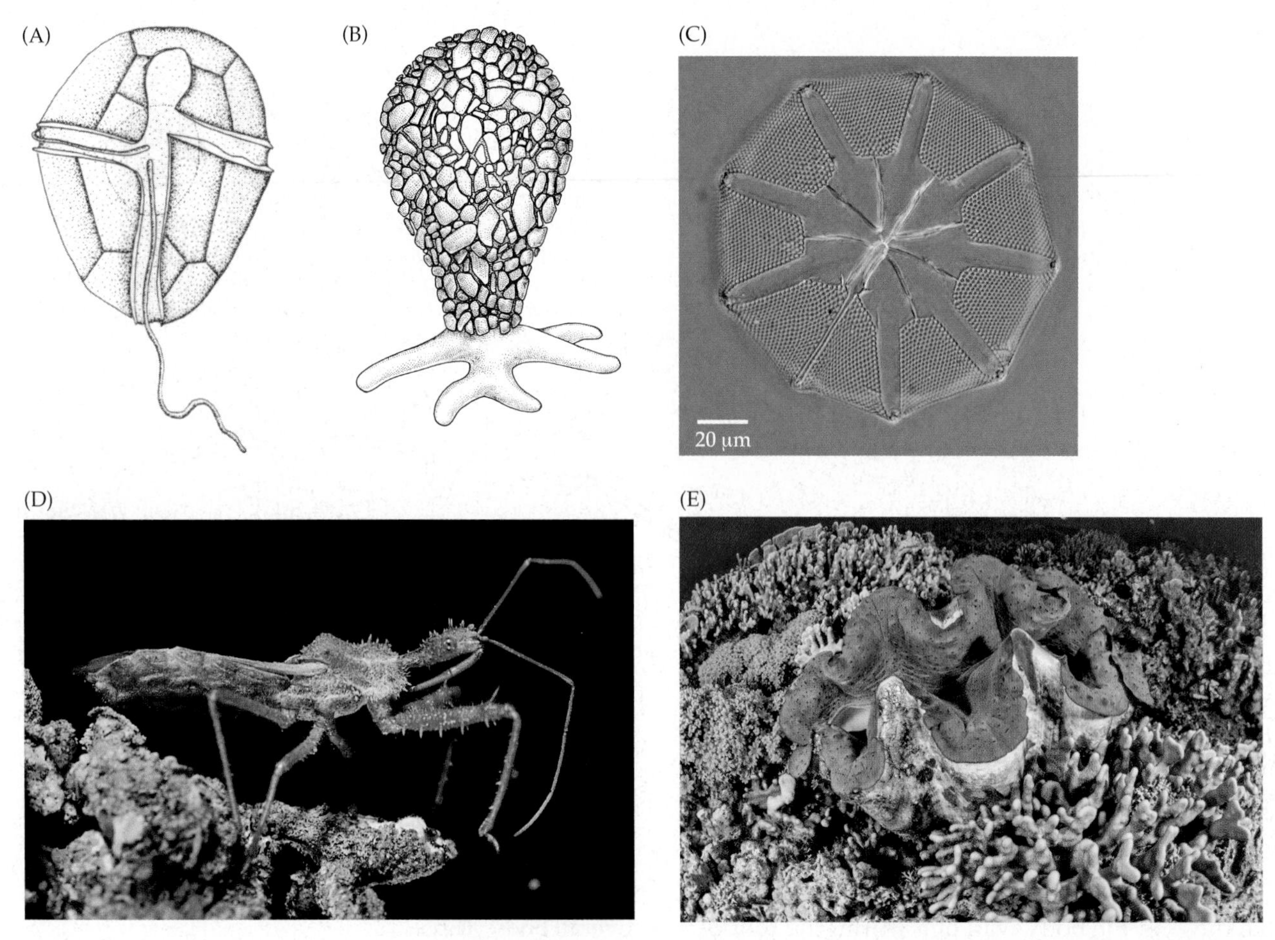

Figure 4.10 Some exoskeletons. (A) The dinoflagellate protist *Gonyaulax*, encased in cellulose plates. (B) The testate ameba *Difflugia*, with a test of minute sand grains. (C) The foraminiferan *Cyclorbiculina* (a granuloreticulosan protist), with a calcareous, multichambered shell. (D) An assassin bug, with a jointed, chitinous exoskeleton. (E) The giant clam, *Tridacna*, among corals. These two very different animals both have calcareous exoskeletons. (F) Chemical structure of the polysaccharide chitin.

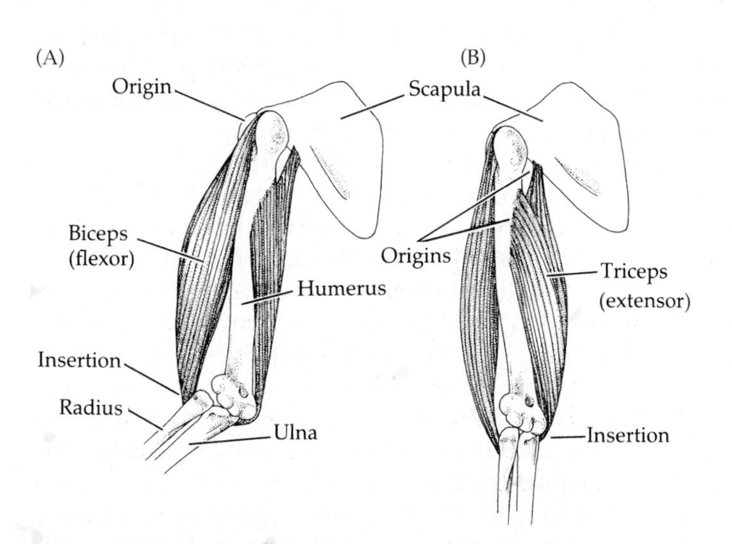

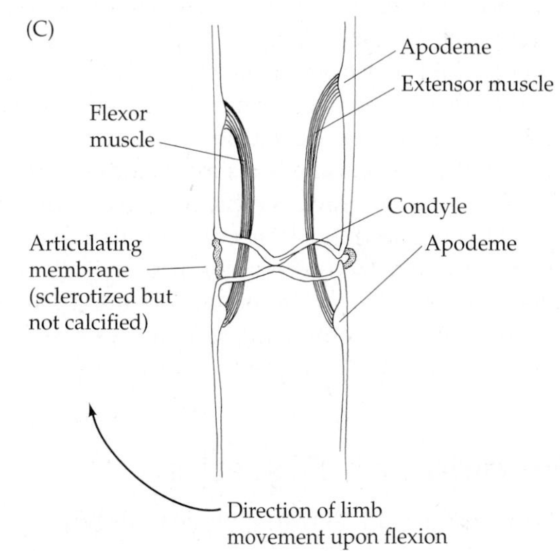

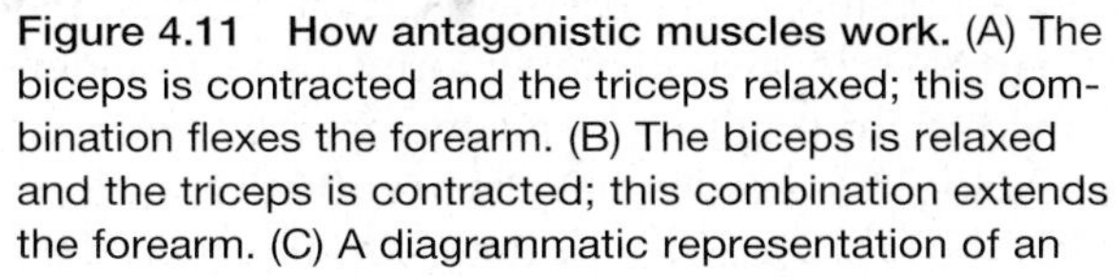

Figure 4.11 How antagonistic muscles work. (A) The biceps is contracted and the triceps relaxed; this combination flexes the forearm. (B) The biceps is relaxed and the triceps is contracted; this combination extends the forearm. (C) A diagrammatic representation of an arthropod joint, illustrating a similar relationship between flexor and extensor. In arthropods, however, the muscles attach to the inside of the skeleton.

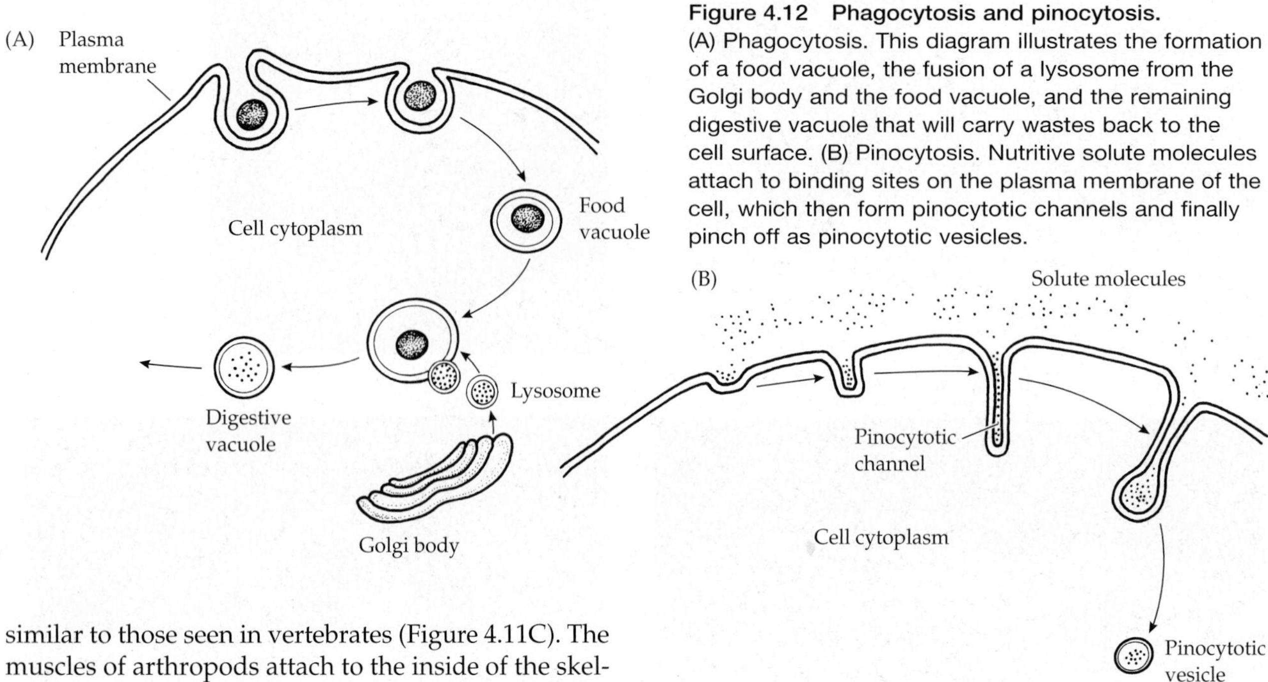

Figure 4.12 Phagocytosis and pinocytosis. (A) Phagocytosis. This diagram illustrates the formation of a food vacuole, the fusion of a lysosome from the Golgi body and the food vacuole, and the remaining digestive vacuole that will carry wastes back to the cell surface. (B) Pinocytosis. Nutritive solute molecules attach to binding sites on the plasma membrane of the cell, which then form pinocytotic channels and finally pinch off as pinocytotic vesicles.

similar to those seen in vertebrates (Figure 4.11C). The muscles of arthropods attach to the inside of the skeletal parts, whereas those of vertebrates attach to the outside, but they both operate systems of levers.

Not all muscles attach to rigid endo- or exoskeletons. Some form masses of interlacing muscle fibers, like those in the body wall of a worm, the foot of a snail, or the muscle layers in the walls of "hollow" organs (like those surrounding a gut tube or uterus). In these cases, the muscles have no definite origin and insertion but act on each other and the surrounding tissues and body fluids to affect changes in the shape of the body or body parts.

The basic physiology and biochemistry of muscle contraction is the same in vertebrates and invertebrates, although a variety of specialized variations on the basic model have evolved. For example, the adductor muscle of a clam (the muscle that holds the shell closed) is divided into two parts. One part is heavily striated and used for rapid shell closure (the phasic, or "quick" muscle); the other is smooth, or tonic, and is used to hold the shell closed for hours or even days at a time (the "catch" muscle). Brachiopods have a similar adductor muscle specialization—a good example of convergent evolution. Other specializations are found in crustacean muscle innervation, which differs from that typically seen in other invertebrates, and in certain insect flight muscles that are capable of contracting at frequencies far higher than can be induced by nerve impulses alone.

Feeding and Digestion

Intracellular and Extracellular Digestion

Virtually all Metazoa and heterotrophic protists must locate, select, capture, ingest, and finally digest and assimilate food. Although the physiology of digestion is similar at the biochemical level, considerable variation exists in the mechanisms of capture and digestion as a result of constraints placed on organisms by their overall body plans.

Digestion is the process of breaking down food by hydrolysis into units suitable to the nutrition of cells. When this breakdown occurs outside the body, it is called extracorporeal digestion; when it occurs in a gut chamber of some sort, it is referred to as extracellular digestion; and when the process occurs within a cell, it is called intracellular digestion. Regardless of the site of digestion, all organisms are ultimately faced with the fundamental challenge of cellular capture of nutritional products (food, digested or not). This cellular challenge is met by the process of **phagocytosis** (literally, "eating by cells") and **pinocytosis** ("drinking by cells"). These processes, collectively called **endocytosis**, are mechanically simple and involve the engulfment of food "particles" at the cell surface.

In 1892 the great comparative anatomist Elie Metchnikoff made a discovery that led to his receiving the Nobel Prize 16 years later. Metchnikoff discovered the process by which certain ameboid cells in the coelomic fluid of starfish engulf and destroy foreign matter such as bacteria. He named this process phagocytosis. In phagocytosis, extensions of a cell's plasma membrane encircle the particle to be captured (whether it be food or a foreign microbe), form an inpocketing on the cell surface, and then pinch off the pocket inside the cell (Figure 4.12A). The resultant intracellular membrane-bounded structure is called a food vacuole. Because the food particle is inside a chamber formed and bounded by a piece of the original plasma membrane of the cell, some biologists consider that it is not actually "inside" the cell, but this point is of little consequence. The plasma membrane surrounding the food vacuole is, of course, no longer part of the cell's outer membrane and in this sense it and whatever is in the vacuole are now "inside" the cell, and the subsequent digestive processes that take place are considered intracellular, not extracellular. However, food inside the food vacuole is not actually incorporated into the cell's cytoplasm until it is digested and the resultant molecules are released.

Protists and sponges rely on phagocytosis as a feeding mechanism, and the digestive cells of metazoan guts take up food particles in the same fashion. Once a cell has phagocytosed a food particle and intracellular digestion has been completed, any remaining waste particles may be carried back to the cell surface by what remains of the old food vacuole, which fuses with the plasma membrane to discharge its wastes in a sort of reverse phagocytosis, called **exocytosis**.

Pinocytosis can be thought of as a highly specialized form of phagocytosis, in which molecule-sized particles are taken up by the cell. Such molecules are always dissolved in some fluid (e.g., a body fluid, or sea water). During pinocytosis, minute invaginations (pinocytotic channels) form on the cell surface, fill with liquid from the surrounding medium (which includes the dissolved nutritional molecules), and then pinch off to enter the cytoplasm as pinocytotic vesicles (Figure 4.12B). Pinocytosis generally occurs in cells lining some body cavity (e.g., the gut) in which considerable extracellular digestion has already taken place and nutritive molecules have been released from the original food source. In some cases, however, nutritional molecules may be taken up directly from seawater, and there is growing evidence that many invertebrates rely substantially on the direct uptake of dissolved organic matter (DOM) from their environment.

Some animals have no true digestive tract (e.g., Porifera, Placozoa, tapeworms, acanthocephalans), but most metazoans possess some sort of dedicated, internal chamber into which food is moved for processing. In some (e.g., Cnidaria, Ctenophora, Xenacoelomorpha, Platyhelminthes) there is only one opening through which food is ingested and undigested materials eliminated. These animals are said to have an **incomplete gut** (or blind gut). Most other Metazoa have both mouth and anus (a **complete gut**, or **through gut**), an arrangement that allows the one-way flow of food and the specialization of different gut regions for functions such as grinding, secretion, storage, digestion, and absorption. As the noted biologist Libbie Hyman so aptly put it, "The advantages of an anus are obvious."

The overall anatomy and physiology of an animal's gut are closely tied to the type and quality of food consumed. In general, the guts of herbivores are long and often have specialized chambers for storage, grinding, and so on because vegetable matter is difficult to digest and requires long residence times in the digestive system. Carnivores tend to have shorter, simpler guts; the animal foods they consume are higher quality and easier to digest.

Feeding Strategies

Just as body architecture influences and limits the digestive modes of invertebrates, it is also intimately associated with the processes of food location, selection, and ingestion. Animals and animal-like protists are generally defined as **heterotrophic** organisms (as opposed to autotrophs and saprophytes); they ingest organic material in the form of other organisms, or parts thereof. However, in several groups of protists (e.g., many euglenoids and chlorophytans), both photosynthesis and heterotrophy can occur as nutritional strategies. In addition, many nonphotosynthetic invertebrate groups have developed intimate symbiotic relationships with single-celled algae, especially with certain species of dinoflagellates. These invertebrates use photosynthetic by-products as an accessory (or occasionally as the primary) food source. Notable in this regard are reef-building corals, giant clams (*Tridacna*), and certain flatworms, sea slugs, hydroids, ascidians, sea anemones, freshwater sponges, and even some species of the protist *Paramecium*. However, the overwhelming majority of invertebrates lead strictly heterotrophic lives.

Biologists classify heterotrophic feeding strategies in a number of ways. For example, organisms can be considered herbivores, carnivores, or omnivores; or they can be classed as grazers, predators, or scavengers. Organisms can also be classified as microphagous or macrophagous by the comparative size of their food or prey, or they can be classified by the environmental source of their food as suspension feeders, deposit feeders, or detritivores. In the remainder of this section we define some important feeding-strategy terms and explain some common themes of feeding.

Few invertebrates are strictly herbivores or carnivores, even though most show a clear preference for either a vegetable or a meat diet. For example, the Atlantic purple sea urchin *Arbacia punctulata* usually feeds on micro- and macroalgae. However, in certain portions of its range, where algae may become seasonally scarce, epifaunal animals constitute the bulk of this urchin's diet. Omnivores, of course, must have the anatomical and physiological capability to capture, handle, and digest both plant and animal material. Among invertebrates, there are two large categories of feeding strategies in which omnivory prevails: suspension feeding and deposit feeding.

Suspension feeding

Suspension feeding is the removal of suspended food particles from the surrounding medium by some sort of capture, trapping, or filtration mechanism. Lest you think suspension feeding is a sideline strategy in the animal kingdom, let us remind you that the largest living animals utilize it—baleen whales and several lineages of sharks and rays. This mode of feeding has three basic steps: transport of water past the feeding structures, removal of particles from the water, and transport of the captured particles to the mouth. It is a major mode of feeding in sponges, ascidians, appendicularians, brachiopods, ectoprocts, entoprocts, phoronids, most bivalves, and many crustaceans, polychaetes, and gastropods. The main food selection criterion is particle size, and the size limits of food are determined by the nature of the particle-capturing device. In some cases potential food particles may also be sorted on the basis of their specific gravity, or even their perceived nutritional quality.

Suspension-feeding invertebrates generally consume bacteria, phytoplankton, zooplankton, and some detritus. All suspension feeders probably have optimal ranges of particle size; but some are capable, experimentally, of preferentially selecting "enriched" artificial food capsules over nonenriched (nonfood) capsules, an observation suggesting that chemosensory selectivity may occur in situ as well. To capture food particles from their environment, suspension feeders either must move part or all of their body through the water, or water must be moved over their feeding structures. As with locomotion in water, the relative motion between a solid and liquid during suspension feeding creates a system that behaves according to the concept of Reynolds numbers. Virtually all suspension-feeding invertebrates capture particles from the water at low Reynolds numbers. The flow rates in such systems are very low and the feeding structures are small (e.g., cilia, flagella, setae).

Recall that at low Reynolds numbers viscous forces dominate, and water flow over small feeding structures is laminar and nonturbulent and ceases instantaneously when energy input stops. Thus, in the absence of inertial influence, suspension feeders that generate their own feeding currents expend a great deal of energy. Some suspension feeders conserve energy by depending to various degrees on prevailing ambient water movements to continually replenish their food supplies (e.g., barnacles on wave-swept shores and mole crabs in the wash zone on sandy beaches). For most organisms, however, the effort expended for feeding is a major part of their energy budget.

Only a relatively few suspension feeders are true filterers. Because of the principles outlined above, it is energetically extremely costly to drive water through a fine-meshed filtering device. For small animals, this is somewhat analogous to moving a fine-mesh filter through thick syrup. Such actual sieving does occur, most notably in many bivalve molluscs, many tunicates, some larger crustaceans, and some worms that produce mucous nets. However, most suspension feeders employ a less expensive method of capturing particles from the water, one that does not involve continuous filtration. Many invertebrates simply expose a sticky surface, such as a coating of mucus, to flowing water. Suspended particles contact and adhere to the surface and then are moved to the mouth by ciliary tracts (as in crinoids), setal brushes (as in certain crustaceans), or by some other means of transport. Other "contact suspension feeders" living in still water may simply expose a sticky surface to the rain of particulate material settling downward from the water above, thus letting gravity do much of the work of food-getting. Some oysters are suspected of this feeding strategy, at least on a part-time basis. Several other contact methods of suspension feeding may occur, but all eliminate the costly activity of actual sieving in the highly viscous world of low Reynolds numbers.

Another nonfiltering suspension feeding method is called "scan-and-trap." The general strategy here is to move water over part or all of the body, detect suspended food particles, isolate the particles in a small parcel of water, and process only that parcel by some method of particle extraction. The animal thus avoids the energetic expense of continuously driving water over the feeding surface at low Reynolds numbers. The precise methods of particle detection, isolation and capture vary among invertebrates that use the scan-and-trap technique, and this basic strategy is probably employed by certain crustaceans (e.g., planktonic copepods), many bryozoans, and a variety of larval forms.

The trick of removing small food particles from the surrounding environment is achieved through four fundamentally different mechanisms. Because there are a limited number of ways in which animals can suspension feed, it is not surprising that a great deal of evolutionary convergence has appeared among their feeding mechanisms.

Among some crustaceans, certain limbs are equipped with rows of feather-like setae adapted for removing particles from the water (Figure 4.13). The size of the particles captured is often directly proportional to the "mesh" size of the interlaced setae on the food-capture structure. In sessile crustaceans such as barnacles, the feeding appendages are swept through the water or held taut against moving water. In either case, sessile animals are dependent upon local currents to continually replenish their food supply. Motile setal-net feeders, like many larger planktonic crustaceans and certain benthic crustaceans (e.g., porcelain crabs), may have modified appendages that generate a current across the feeding appendages that bear the capture setae. Sometimes these same appendages serve simultaneously for locomotion. In cephalocarid and many branchiopod crustaceans, for example, complex coordinated movements of the highly setose thoracic legs propel the animal forward and also produce a constant current of water (Figure 4.13D). These appendages simultaneously capture food particles from the water and collect them in a median ventral food groove at the leg bases, where they are passed forward to the mouth region.

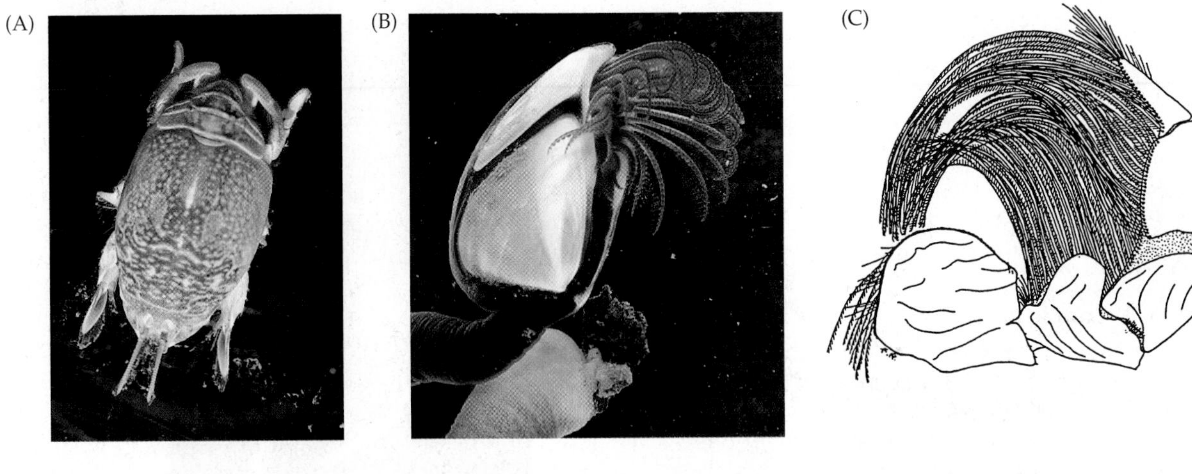

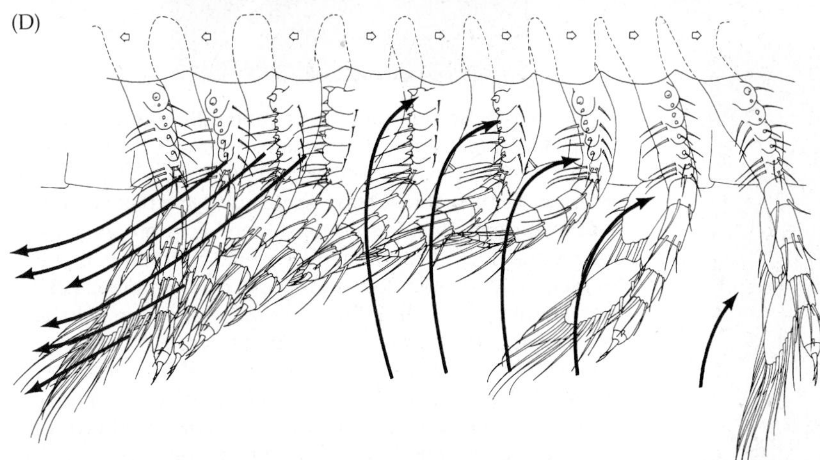

Figure 4.13 Some setal-net suspension-feeding invertebrates. (A) The sand crab *Emerita*. (B) A goose barnacle, *Pollicipes*, with feeding appendages extended. (C) The third maxilliped of the porcelain crab *Petrolisthes elegans*. Note the long, dense setae used in feeding. (D) A portion of the trunk (sagittal view) of a cephalocarid crustacean during the metachronal cycle of the feeding limbs. The arrows indicate the direction of water currents; the arrow above each trunk limb indicates the limb's direction of movement.

A second suspension-feeding device is the mucous net, or mucous trap, wherein patches or a sheet of mucus are used to capture suspended food particles. Most mucous-net feeders consume their net along with the food and recycle the chemicals used to produce it. Again, sessile and sedentary species often rely largely on local currents to keep a fresh supply of food coming their way. Some, however, especially benthic burrowers, actively pump water through their burrow or tube, where it passes across or through the mucous sheet. A classic example of mucous-net feeding is seen in the annelid worm *Chaetopterus* (Figure 4.14A). This animal lives in a U-shaped tube in the sediment and pumps water through the tube and through a mucous net. As the net fills with trapped food particles, it is periodically manipulated and rolled into a ball, which is then passed to the mouth and swallowed. An example of mucous-trap feeding is seen in the tube-building gastropods (family Vermetidae). These wormlike snails construct colonies of meandering calcareous tubes in the intertidal zone. Each animal secretes a mucous trap that is deployed just outside the opening of the tube, until nearly the entire colony surface is covered with mucus. Suspended particulate matter settles and becomes trapped in the mucus. At periodic intervals, each animal withdraws its mucous sheet and swallows it, whereupon a new sheet is immediately constructed.

Another type of suspension feeding is the ciliary-mucous mechanism, in which rows of cilia carry a mucous sheet across some structure while water is passed through or across it. Ascidians (sea squirts; Figure 4.14B) move a more or less continuous mucous sheet across their sievelike pharynx, while at the same time pumping water through it. Fresh mucus is secreted at one side of the pharynx while the food-laden mucus at the other side is moved into the gut for digestion. Several polychaete groups also make use of the ciliary-mucous feeding technique (Figure 4.14C). For example, some species of tube-dwelling fan worms feed with a crown of tentacles that are covered with cilia and mucus and bear ciliated grooves that slowly move captured food particles to the mouth. Many sand dollars capture suspended particles, especially diatoms, on their mucus-covered spines; food and mucus are transported by the tube feet and ciliary currents to food tracts, and then to the mouth.

Still another kind of suspension feeding is tentacle or tube feet suspension feeding. In this strategy, some sort of tentacle-like structure captures larger food particles, with or without the aid of mucus. Food particles captured by this mechanism are generally larger than those captured by setal or mucous traps or sieves. Examples of tentacle or tube feet suspension feeding are most commonly encountered in the echinoderms

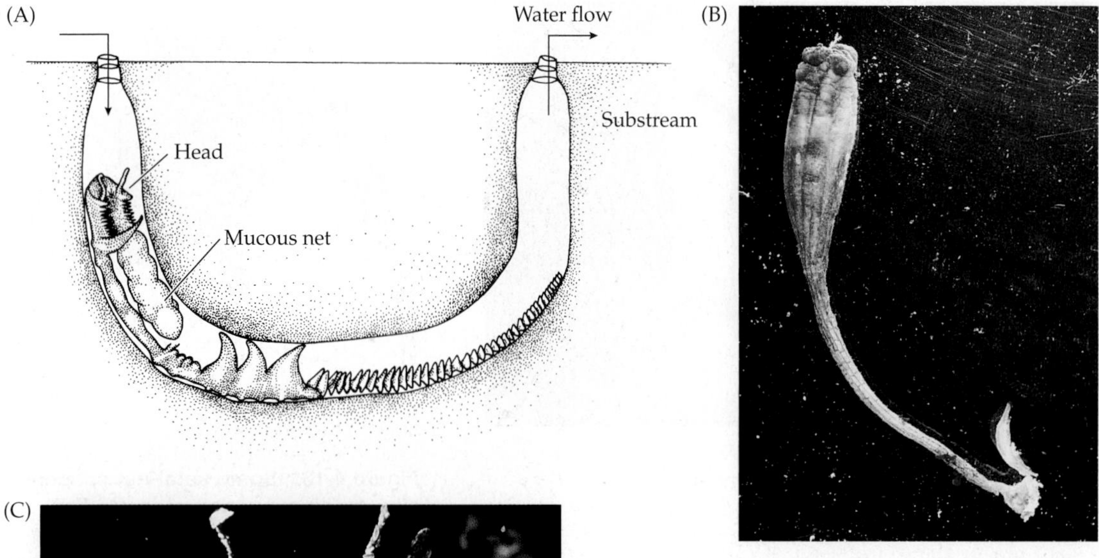

(C)

Figure 4.14 Some mucous-net and ciliary-mucous suspension feeders. (A) The annelid worm *Chaetopterus* in its burrow. Note the direction of water flow through its mucous net. (B) The solitary ascidian *Styela* has incurrent and excurrent siphons through which water enters and leaves the body. Inside, the water passes through a sheet of mucus covering holes in the wall of the pharynx. (C) A maldanid polychaete, *Praxillura maculata*. This animal constructs a membranous tube that bears 6–12 stiff radial spokes. A mucous web hangs from these spokes and passively traps passing food particles. The worm's head is seen sweeping around the radial spokes to retrieve the mucous web and its trapped food particles.

(e.g., many brittle stars and crinoids) and cnidarians (e.g., certain sea anemones and corals; Figure 4.15).

Much research has been done on suspension feeding, and we now know what size range of particles many animals feed on and what kinds of capture rates they have. In general, feeding rates increase with food particle concentration to a plateau, above which the rate levels off. At still higher particle concentrations, entrapment mechanisms may become overtaxed or clogged and feeding is inhibited or simply ceases. In sessile and sedentary suspension feeders, for example, pumping rates decrease quickly as the amount of suspended inorganic sediment (mud, silt, and sand) increases beyond a given concentration. For this reason, the amount of sediment in coastal waters limits the distribution and abundance of certain invertebrates such as clams, corals, sponges, and ascidians. Many tropical coral reefs are dying as a result of increased coastal sediment loads generated by run-off from land areas subjected to deforestation or urban development.

Deposit feeding The **deposit feeders** make up another major group of omnivores. These animals obtain nutrients from the sediments of soft-bottom habitats (muds and sands) or terrestrial soils, but their techniques for feeding are diverse. Direct deposit feeders

Figure 4.15 Tube feet suspension feeding. Food-particle capture in the brittle star *Ophiothrix fragilis*. The photographs show two views of a captured food particle being transported by the arm tentacles to the mouth.

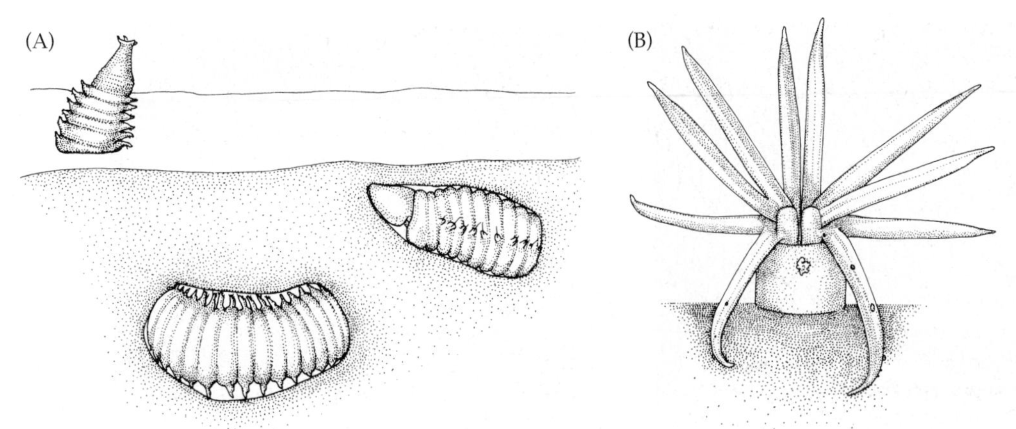

Figure 4.16 Some deposit feeding invertebrates. (A) A lumbrinerid polychaete burrowing in the sediment. This worm is a subsurface deposit feeder. (B) The sabellid polychaete *Manayunkia aestuarina* in its feeding posture. A pair of branchial filaments are being used to feed. The large particle falling in front of the tube has just been expelled from the branchial crown by a rejection current. (C) A surface deposit-feeding holothurian (*Synapta*).

simply swallow large quantities of sediment—mud, sand, soil, organic matter, everything. They may consume up to 500 times their body weight daily. The usable organics are digested and the unusable materials passed out the anus. The resultant fecal material is essentially "cleaned dirt." This kind of deposit feeding is seen in many polychaete annelids (Figure 4.16A), some snails, some sea urchins, and most earthworms. Some deposit feeders utilize tentacle-like structures to consume sediment, such as some sea cucumbers, most sipunculans, certain clams, and several types of polychaetes (Figure 4.16B,C). Tentacle-utilizing deposit feeders preferentially remove only the uppermost deposits from the sediment surface and thus consume a far greater percentage of living (especially bacteria, diatoms, and protists) and detrital organic material that accumulates there than do the burrowing deposit feeders. These animals are generally called **selective deposit feeders**. Aquatic deposit feeders may also rely to a significant extent on fecal material that accumulates on the bottom, and many will actively consume their own fecal pellets (**coprophagy**), which may contain some undigested or incompletely digested organic material as well as microorganisms. Studies have shown that only about half of the bacteria ingested by marine deposit feeders is digested during passage through the gut. In all cases, deposit feeders are microphagous.

The ecological role of deposit feeding in sediment turnover is a critical one. When burrowing deposit feeders are removed from an area, organic debris accumulates, subsurface oxygen is depleted by bacterial decomposition, and anaerobic sulfur bacteria eventually bloom. On land, earthworms and other burrowers are important in maintaining the health of agricultural and garden soils.

Herbivory The following discussion deals with macroherbivory, or the consumption of macroscopic plants. **Herbivory** is common throughout the animal kingdom. It is most dramatically illustrated when certain invertebrate herbivores undergo a temporary population explosion. Famous examples are outbreaks of locust, which can destroy virtually all plant material in their path of movement. In a similar fashion, herbivory by extremely high numbers of the Pacific sea urchin *Strongylocentrotus* results in the wholesale destruction of kelp beds. Unlike suspension- and deposit-feeding herbivory, in which mostly single-celled and microscopic plant matter is consumed, macroherbivory requires the ability to "bite and chew" large pieces of vegetable matter. Although the evolution of biting and chewing mechanisms has taken place within the architectural framework of a number of different invertebrate lineages, it is always characterized by the development of hard (usually calcified or chitinous) "teeth," which are manipulated by powerful muscles. Members of a number of major invertebrate taxa have evolved macroherbivorous lifestyles, including molluscs, polychaetes, arthropods, and sea urchins.

Most molluscs have a unique structure called a **radula**, which is a muscularized, belt-like rasp armed with chitinous teeth. Herbivorous molluscs use the radula to scrape algae off rocks or to tear pieces of algal fronds or the leaves of terrestrial plants. The radula acts like a curved file that is drawn across the feeding surface

(A)

(B)

(C)

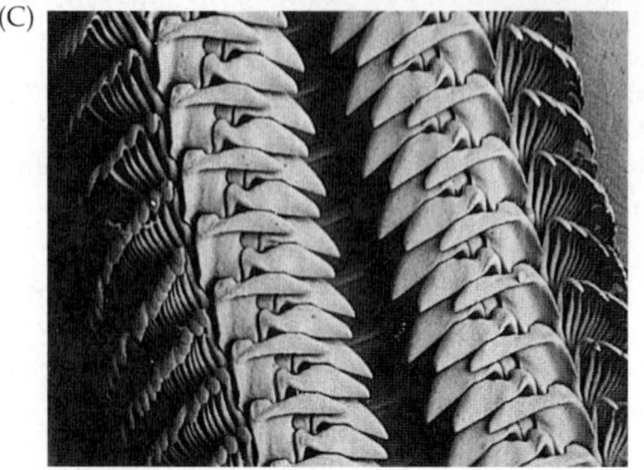

(D)

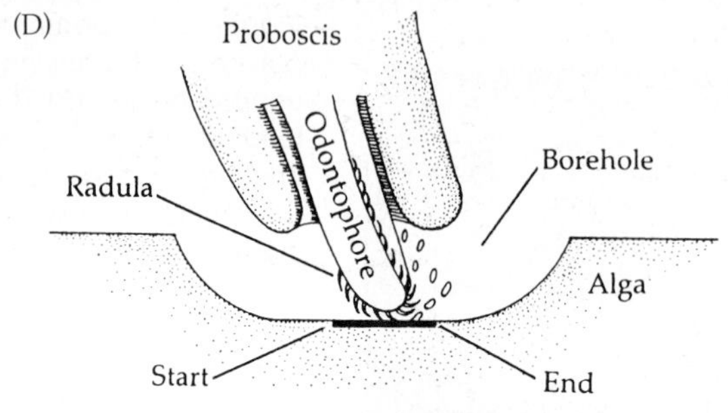

(E)

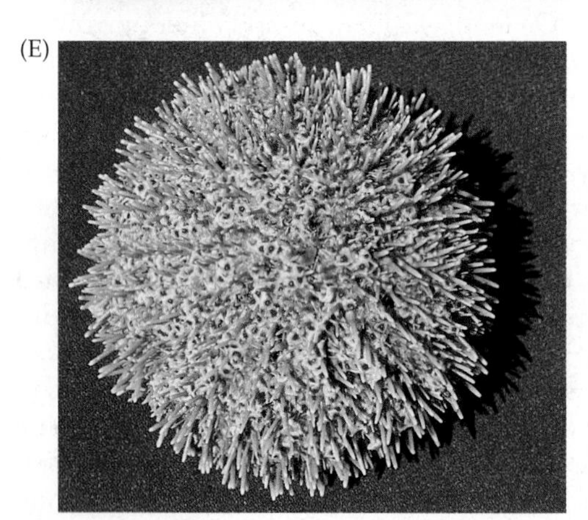

Figure 4.17 Some herbivorous invertebrates. (A) The common land snail *Cornu* (formerly *Helix*), munching on some foliage. (B) The red abalone *Haliotis rufescens*. (C) The radula, or rasping organ, of *H. rufescens*. (D) Diagram of a radula removing food particles from an alga (sagittal section). (E) The tropical Pacific sea urchin *Toxopneustes roseus*.

(Figure 4.17C,D). Some polychaetes such as nereids (family Nereidae) have sets of large chitinous teeth on an eversible pharynx or proboscis. The proboscis is protracted by hydrostatic pressure, exposing the teeth, which by muscular action tear or scrape off pieces of algae that are swallowed when the proboscis is retracted. As might be expected, the toothed pharynx of polychaetes is also suited for carnivory, and many primarily herbivorous polychaetes can switch to meat-eating when algae are scarce.

Macroherbivory in arthropods is best illustrated by certain insects and crustaceans. Both of these large groups have powerful mandibles capable of biting off pieces of plant material and subsequently grinding or chewing them before ingestion. Some macroherbivorous arthropods are able to temporarily switch to carnivory when necessary. This switching is rarely seen in the terrestrial herbivores because it is almost never necessary; terrestrial plant matter can almost always be found. In marine environments, however, algal supplies may at times be very limited. Some herbivorous invertebrates cause serious damage to wooden man-made structures (like homes, pier pilings, and boats) by burrowing through and consuming the wood (Figure 4.18).

Carnivory and scavenging The most sophisticated methods of feeding are those that require the active capture of live animals, or predation.[5] Most carnivorous predators will, however, consume dead or dying animal matter when live food is scarce. Only a few generalizations about the many kinds of predation are presented here; detailed discussions of various taxa are presented in their appropriate chapters.

Active predation often involves five recognizable steps: prey location (predator orientation), pursuit (usually), capture, handling, and finally ingestion. Prey location usually requires a certain level of nervous system sophistication in which specialized sense organs

[5]Although in the broad sense, herbivory is a form of predation (on plants), for clarity of discussion we restrict the use of these terms to vegetable eating and animal eating/carnivory, respectively.

Figure 4.18 Shipworms. Wood from the submerged part of an old dock piling, split open to show the work of the wood-boring bivalve shipworm *Teredo navalis* (Mollusca). The shell valves are so reduced that they can no longer enclose the animal; instead they are used as "auger blades" in boring. The walls of the burrow are lined with a smooth, calcareous, shell-like material.

are present (discussed later in this chapter). Many carnivorous invertebrates rely primarily on chemosensory location of prey, although many also use visual orientation, touch, and vibration detection. Chemoreceptors tend to be equally distributed around the bodies of radially symmetrical carnivores (e.g., jellyfish) but, coincidentally with cephalization, most invertebrates have their gustatory and olfactory receptors ("tasters" and "smellers") concentrated in the head region. A number of insects rely on CO_2 sensing to locate their food source, including fruit flies, mosquitoes, and moths, but the sensing mechanisms are not yet fully characterized.

Predators may be classified by how they capture their prey—as motile stalkers, lurking predators (ambushers), sessile opportunists, or grazers (Figure 4.19). Stalkers actively pursue their prey; they include members of such disparate groups as ciliate protists, polyclads, nemerteans, polychaete worms, gastropods, octopuses and squids, crabs, and starfish. In all these groups, chemosensation is highly important in locating potential prey, although some cephalopods are known to be the most highly visual of all the invertebrate predators.

Lurking predators are those that sit and wait for their prey to come within capture distance, whereupon they quickly seize the victim. Many lurking predators, such as certain species of mantis shrimps (stomatopods), crabs, snapping shrimp (Alpheidae), spiders, and polychaetes, live in burrows or crevices from which they emerge to capture passing prey. There are even ambushing planarian flatworms, which produce mucous patches that form sticky traps for their prey. The cost of building traps is significant. Ant lions, for example, may increase their energy consumption as much as eightfold when building their sand capture pits, and energy lost in mucus secretion by planarians may account for 20% of the worm's energy. Predatory invertebrates, especially lurking predators, are frequently more or less territorial.

Sessile opportunists operate in much the same fashion as lurking predators do, but they lack the mobility of the latter. The same may be said for drifting opportunists, such as jellyfishes. Many sessile predators, such as some protists, barnacles, and cnidarians, are actually suspension feeders with a strong preference for live prey.

Grazing carnivores move about the substratum picking at the epifauna. Grazers may be indiscriminate, consuming whatever happens to be present, or they may be fairly choosy about what they eat. In either case, their diet consists largely of sessile and slow-moving animals, such as sponges, ectoprocts, tunicates, snails, small crustaceans, and worms. Most grazers are omnivorous to some degree, consuming plant material along with their animal prey. Many crabs and shrimps are excellent grazers, continuously moving across the bottom and picking through the epifauna for tasty morsels. Sea spiders (pycnogonids) and some carnivorous sea slugs can also be classed as grazers on hydroids, ectoprocts, sponges, tunicates, and other sessile epifauna. Ovulid snails (family Ovulidae) inhabit, and usually mimic, the gorgonians and corals upon which they slowly crawl about, nipping off polyps as they go.

One special category of carnivory is cannibalism, or intraspecific predation. Gary Polis (1981) examined over 900 published reports describing cannibalism in about 1,300 different species of animals. In general, he found that species of large animals (and also larger individuals in any given species) are the most likely to be cannibals. By far, the majority of the victims are juveniles. However, in a number of invertebrate groups the tables turn and cannibalism occurs when smaller individuals band together to attack and consume a larger individual. Furthermore, females tend generally to be more cannibalistic than males, and males tend to be eaten far more often than females. In many species, filial cannibalism is common, in which a parent eats its dying, deformed, weak, or sick offspring. Polis

Figure 4.19 Some predatory invertebrates. (A) Most octopuses are active hunting predators; this one is a member of the genus *Eledone*. (B) The crown-of-thorns starfish, *Acanthaster*, feeds on corals. (C) The moon snail, *Polinices*, drills holes in the shells of bivalve molluscs to feed on the soft parts. (D) A mantis shrimp (stomatopod); the two drawings (E) depict its raptorial strike to capture a passing fish. (F) The predatory flatworm *Mesostoma* attacking a mosquito larva. (G) A cone snail (*Conus*) eating a fish. (H) *Acanthina*, a predatory gastropod feeding on small barnacles.

concluded that cannibalism is a major factor in the biology of many species and may influence population structure, life history, behavior, and competition for mates and resources. He went so far as to point out that *Homo sapiens* may be "the only species capable of worrying whether its food is intra- or extraspecific."

Dissolved organic matter The total living biomass of the world's oceans is estimated to be about 2×10^9 tons of organic carbon (roughly 500 times the amount of organic carbon in the terrestrial environment). Furthermore, an additional 20×10^9 tons of particulate organic matter is estimated to occur in the seas, and another 200×10^9 tons of organic carbon (C) may occur in the seas as **dissolved organic matter (DOM)**. Thus, at any moment in time, only a small fraction of the organic carbon in the world's seas actually exists in living organisms. Amino acids and carbohydrates may be the most common dissolved organics. Typical oceanic values of DOM range from 0.4 to 1.0 mg C/liter, but may reach 8.0 mg C/liter near shore. Pelagic and benthic algae release copious amounts of DOM into the environment, as do certain invertebrates. Coral mucus, for example, is an important fraction of suspended and dissolved organic material over reefs, and it contains significant amounts of energy-rich and nitrogen-rich compounds, including mono- and polysaccharides and amino acids. Other sources of DOM include decomposing tissue, detritus, fecal material, and metabolic by-products discharged into the environment.

The idea that DOM may contribute significantly to the nutrition of marine invertebrates has been around for over a hundred years. Marine microorganisms are known to use DOM, but the relative role of dissolved organic matter in the nutrition of aquatic Metazoa is uncertain. Available data strongly suggest that members of all marine taxa (except perhaps arthropods and vertebrates) are capable of absorbing DOM to some extent, and in the case of ciliary-mucous suspension feeders, marine larvae, many echinoderms, and mussels, the ability to rapidly take up dissolved free amino acids from a dilute external medium is well established. But because of the complex chemical nature of dissolved organics, and the difficulty of measuring their rates of influx and loss, we still lack strong evidence of the actual use, or relative nutritional importance, of DOM to invertebrates.

Evidence from numerous studies indicates that absorption of DOM occurs directly across the body wall of invertebrates, as well as via the gills. Also, inorganic particles of colloidal dimensions provide a surface on which small organic molecules are concentrated by adsorption, to be captured and utilized by suspension-feeding invertebrates. Interestingly, most freshwater organisms seem incapable of removing small organic molecules from solution at anything like the rates characteristic of marine invertebrates. In fresh water, the uptake of DOM is probably retarded by the processes of osmoregulation. Also, with the exception of the aberrant hagfish, marine vertebrates seem not to utilize DOM to any significant extent.

Chemoautotrophy A special form of autotrophy that occurs in certain bacteria relies not on sunlight and photosynthesis as a source of energy to make organic molecules from inorganic raw materials (**photoautotrophy**), but rather on the oxidization of certain inorganic substances. This is called **chemoautotrophy**. Chemoautotrophs use CO_2 as their carbon source, obtaining energy by oxidizing hydrogen sulfide (H_2S), ammonia (NH_3), methane (CH_4), ferrous ions (Fe^{2+}), or some other chemical, depending on the species. These prokaryotes are not uncommon in aerated soils, and certain species live as symbionts in the tissues of a few marine invertebrates.

Some of the most interesting of these chemoautotrophic organisms derive their energy from the oxidation of hydrogen sulfide released at hot water vents on the deep-sea floor—where, in fact, they are the sole primary producers in the ecosystem. In this environment, chemoautotrophic bacteria inhabit the tissues of certain mussels, clams, and vestimentiferan tube worms, where they produce organic compounds that are utilized by their hosts. Similar invertebrate–bacteria relationships have been discovered in shallow cold-water petroleum and salt (brine) seeps, where the chemoautotrophic microorganisms live off the methane- and hydrogen sulfide-rich waters associated with such sea floor phenomena. In all these cases, the bacteria actually live within the cells of their hosts. In bivalves, the bacteria inhabit the gill cells and extract methane or other chemicals from the water that flows over those structures. In the case of the tube worms, the host must transport the H_2S to their bacterial partners, which live in tissues deep within the animals' body. The worms have a unique type of hemoglobin that transports not only oxygen (for the worm's metabolism) but sulfide as well.

Excretion and Osmoregulation

Excretion is the elimination from the body of metabolic waste products, including carbon dioxide and water (produced primarily by cellular respiration) and excess nitrogen (produced as ammonia from deamination of amino acids). The excretion of respiratory CO_2 is generally accomplished by structures that are separate from those associated with other waste products and is discussed in the section that follows.

The excretion of nitrogenous wastes is usually intimately associated with osmoregulation—the regulation of water and ion balance within the body fluids—so these processes are considered together here.

Excretion, osmoregulation, and ion regulation serve not only to rid the body of potentially toxic wastes, but also to maintain concentrations of the various components of body fluids at levels appropriate for metabolic activities. As we shall see, these processes are structurally and functionally tied to the overall level of body complexity and construction, the nature of other physiological systems, and the environment in which an animal lives. We again emphasize the necessity of looking at whole animals, the integration of all aspects of their biology and ecology, and the possible evolutionary histories that could have produced compatible and successful combinations of functional systems.

Nitrogenous Wastes and Water Conservation

The source of most of the nitrogen in an animal's system is amino acids produced from the digestion of proteins. Once absorbed, these amino acids may be used to build new proteins, or they may be deaminated and the residues used to form other compounds (Figure 4.20). The excess nitrogen released during deamination is typically liberated from the amino acid in the form of ammonia (NH_3), a highly soluble but quite toxic substance that either must be diluted and eliminated quickly or converted to a less toxic form. The excretory products of vertebrates have been studied much more extensively than those of invertebrates, but the available data on the latter allow some generalizations. Typically, one nitrogenous waste form tends to predominate in a given species, and the nature of that chemical is generally related to the availability of environmental water.

The major excretory product in most marine and freshwater invertebrates is ammonia, since their environment provides an abundance of water as a medium for rapid dilution of this toxic substance. Such animals are said to be **ammonotelic**. Being highly soluble, ammonia diffuses easily through fluids and tissues, and much of it is lost straight across the body walls of some ammonotelic animals. Animals that do not possess definite excretory organs (e.g., sponges, cnidarians, and echinoderms) are more or less limited to the production of ammonia and thus are restricted to aquatic habitats.

Terrestrial invertebrates (indeed, all land animals) have water conservation challenges. They simply cannot afford to lose much body water in the process of diluting their wastes. These animals convert their nitrogenous wastes to more complex but far less toxic substances. These compounds are energetically expensive to produce, but they often require relatively little or no dilution by water, and they can be stored within the body prior to excretion.

There are two major metabolic pathways for the detoxification of ammonia: the urea pathway and the uric acid pathway. The products of these pathways, **urea** and **uric acid**, are illustrated in Figure 4.20, along with

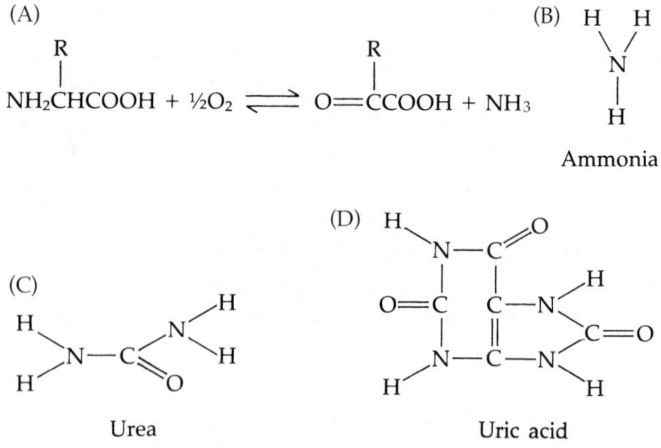

(A)

$$NH_2CHRCOOH + \tfrac{1}{2}O_2 \rightleftharpoons O{=}CRCOOH + NH_3$$

(B) Ammonia

(C) Urea

(D) Uric acid

Figure 4.20 Nitrogenous waste products. (A) The general reaction for deamination of an amino acid producing a keto acid and ammonia. (B–D) The structures of three common excretory compounds. (B) Ammonia. (C) Urea. (D) Uric acid.

ammonia for comparison. **Ureotelic animals** include amphibians, mammals, and cartilaginous fishes (sharks and rays); urea is a relatively rare and insignificant excretory compound among invertebrates. On the other hand, the ability to produce uric acid is critically associated with the success of certain invertebrates on land. **Uricotelic animals** have capitalized on the relative insolubility (and very low toxicity) of uric acid, which is generally precipitated and excreted in a solid or semisolid form with little water loss. Most land-dwelling arthropods and snails have evolved structural and physiological mechanisms for the incorporation of excess nitrogen into molecules of uric acid. We emphasize that various combinations of these and other forms of nitrogen excretion are found in most animals. In some cases, individual animals can actually vary the proportion of these compounds they produce, depending on short-term environmental changes affecting water loss.

Osmoregulation and Habitat

In addition to its relationship to excretion, osmoregulation is directly associated with environmental conditions. As mentioned in Chapter 1, the composition of seawater and that of the body fluids of most invertebrates is very similar, in terms of total concentration *and* the concentrations of many ions. Thus, the body fluids of many marine invertebrates and their habitats are close to being isotonic. We hasten to add, however, that probably no animal has body fluids that are exactly isotonic with sea water, and therefore all are faced with the need for some degree of ionic and osmoregulation. Nonetheless, marine invertebrates certainly do not face the extreme osmoregulatory problems encountered by land and freshwater forms.

As shown in Figure 4.21, the body fluids of freshwater animals are strongly hypertonic with respect to their environment, and thus they face serious problems

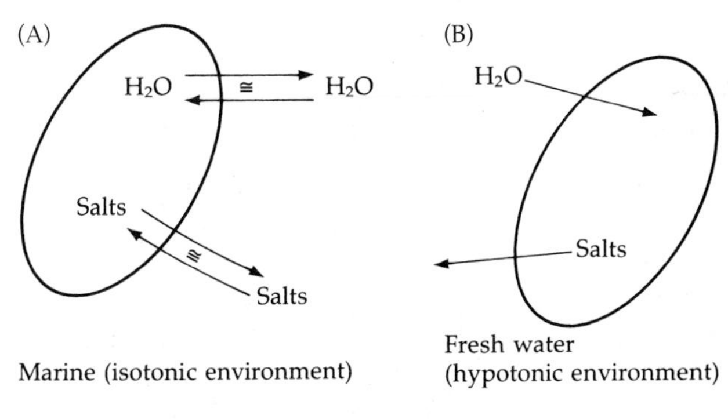

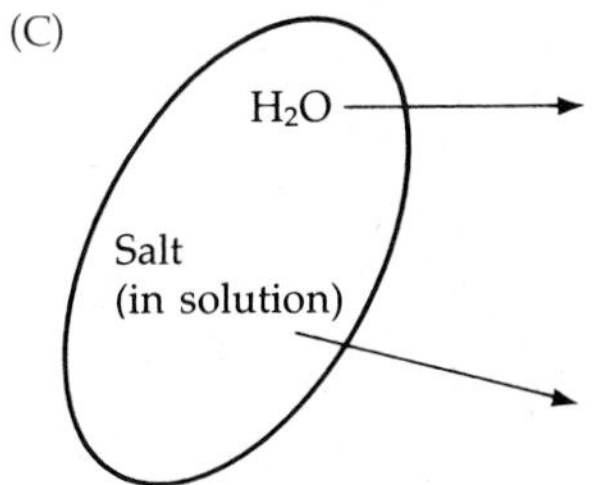

Figure 4.21 Relative osmotic and ionic conditions existing between marine, freshwater, and terrestrial invertebrates and their environments. The arrows indicate the directions in which water and salts move passively in response to concentration gradients. Remember that in each of these cases movement occurs in both directions, but it is the potential net movement along the gradient that is important and against which freshwater and terrestrial animals must constantly battle. For marine invertebrates, the body fluids and the environment are nearly isotonic to one another and there is little net movement in either direction. (A) The organism is isotonic to its environment. (B) The organism is hypertonic. (C) The organism is hypotonic.

of water influx as well as the potential loss of precious body salts. Terrestrial animals are exposed to air and thus to problems of water loss. The evolutionary invasion of land and fresh water was accompanied by the development of mechanisms that solved these problems, and only a relatively small number of invertebrate groups have managed to do this. Animals inhabiting freshwater and terrestrial habitats generally have excretory structures that are responsible for eliminating or retaining water as needed, and they often possess modifications of the body wall to reduce overall permeability. The most successful invertebrate body plans on land, and in some ways of all environments, are those of the arthropods and gastropods. Their effective excretory structures and thickened exoskeletons provide them with physiological osmoregulatory capabilities plus a barrier against desiccation.

Osmoregulatory problems of aquatic animals are, of course, determined by the salinity of the environmental water relative to the body fluids (Figure 4.21). Organisms respond physiologically to changes in environmental salinities in one of two basic ways. Some, such as most freshwater forms (certain crustaceans, protists, and oligochaetes), maintain their internal body fluid concentrations regardless of external conditions and are thus called **osmoregulators**. Others, including a number of intertidal and estuarine forms (mussels and some other bivalves, and a variety of soft-bodied animals), allow their body fluids to vary with changes in environmental salinities; they are appropriately called **osmoconformers**. Again, even the body fluids of marine, so-called osmoconformers are not exactly isotonic with respect to their surroundings; thus, these animals must osmoregulate slightly. Neither of these strategies is without limits, and tolerance to various environmental salinities varies among different species. Those that are restricted to a very narrow range of salinities are said to be **stenohaline**, while those that tolerate relatively extensive variations, such as many estuarine animals, are **euryhaline**.

Although the preceding discussion may seem clear-cut, it is an oversimplification. Experimental data from whole animals tell only part of the story of osmoregulation. When a whole marine animal is placed in a hypotonic medium, it tends to swell (if it is an osmoconformer) or to maintain its normal body volume (if it is an osmoregulator). Even at this gross level, most invertebrates usually show evidence of both conforming and regulating. For instance, an osmoconformer generally swells for a period of time in a lowered salinity environment and then begins to regulate. Its swollen volume will decrease, although probably not to its original size. The same is true of most osmoregulators when faced with a decrease in environmental salinity, but the degree of original swelling is much reduced. In both cases, the swelling of the body is a result of an influx of environmental water into the extracellular body fluids (blood, coelomic fluids, and intercellular fluids). Within limits, this excess water is handled by excretory organs and various surface epithelia of the gut and body wall. However, the second part of the osmoregulatory phenomenon takes place at the cellular level.

As the tonicity (relative concentrations) of the body fluids drops with the entrance of water, the cells in contact with those fluids are placed in conditions of stress—they are now in hypotonic environments. These stressed cells swell to some degree because of the diffusion of water into their cytoplasm, but not to the degree one might expect given the magnitude of the osmotic gradient to which they are subjected. Cellular-level osmoregulation is accomplished by a loss of dissolved materials from the cell into the surrounding intercellular fluids. The solutes released from these cells include both inorganic ions and free amino acids. Thus, osmoconformers are not passive animals that inactively tolerate extremes of salinities. Nor are marine invertebrates free from osmotic problems, even though we read statements that they are "98% water" or other such comments.

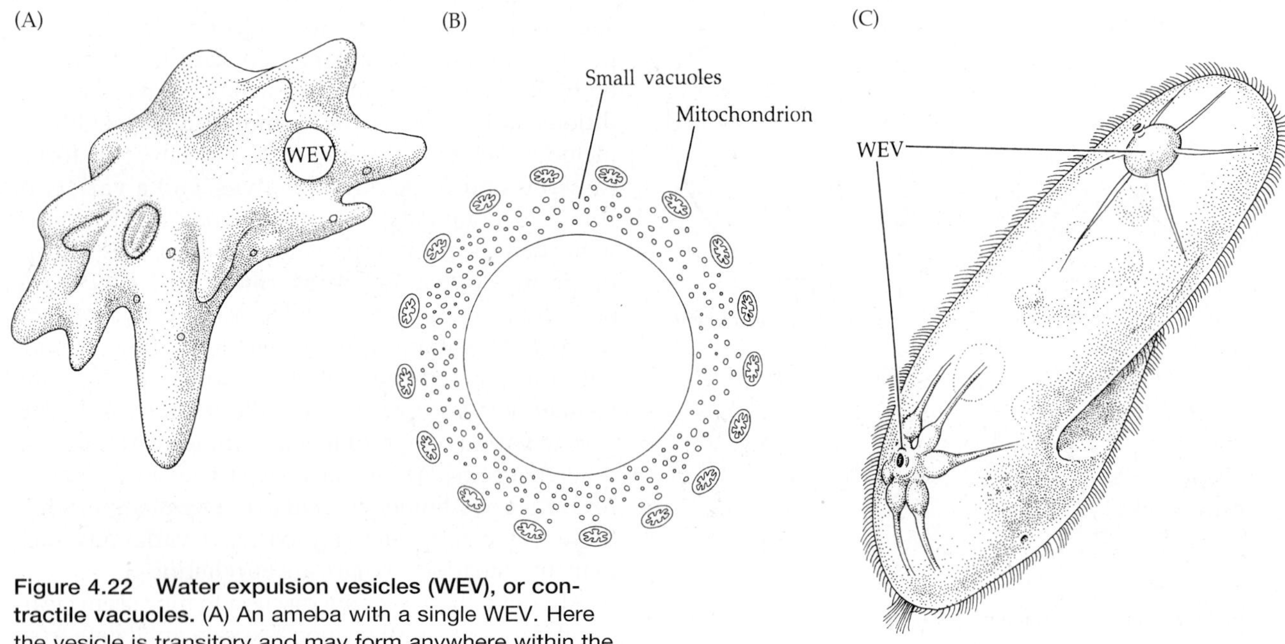

Figure 4.22 Water expulsion vesicles (WEV), or contractile vacuoles. (A) An ameba with a single WEV. Here the vesicle is transitory and may form anywhere within the cell. (B) The WEV of an ameba, and its association with mitochondria. The numerous small vacuoles accumulate water and then contribute their contents to the main WEV. (C) *Paramecium*. Note the positions of two fixed WEV surrounded by arrangements of collecting canals that pass water to the vesicle. (D) The WEV of *Paramecium*, infilled (bottom) and emptied (top) conditions. Enlarged areas show details of a collecting canal surrounded by cytoplasmic tubules that accumulate cell water. The water is passed into the main vesicle, which is collapsed by the action of contractile fibrils, thereby expelling the water through a discharge channel to the outside.

(D)

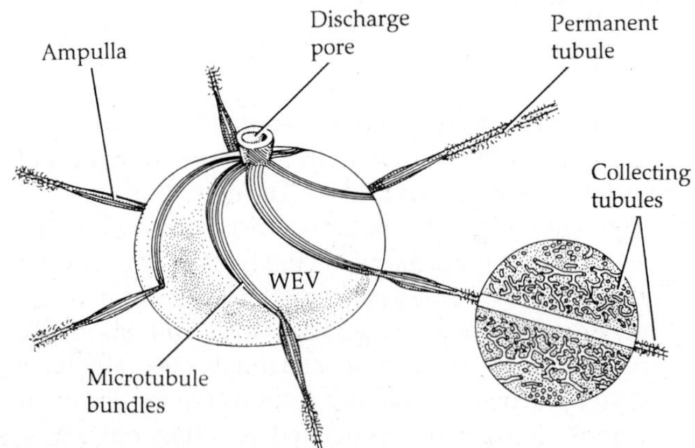

Excretory and Osmoregulatory Structures

Water expulsion vesicles The form and function of organs or systems associated with excretion and osmoregulation are related not only to environmental conditions, but also to body size (especially the surface-to-volume ratio) and other basic features of an organism's body plan. In very small creatures, notably the protists, most metabolic wastes diffuse easily across the body covering because these organisms have sufficient body surface (environmental contact) relative to their volume. However, this high surface area-to-volume ratio presents a distinct osmoregulatory problem, particularly for freshwater forms. Freshwater protists (and even some marine species) typically possess specialized organelles called **contractile vacuoles**, or **water expulsion vesicles** (WEVs), which actively excrete excess water (Figure 4.22). These structures accumulate cytoplasmic water and expel it from the cell. Both of these activities apparently require energy, as suggested in part by the large numbers of mitochondria typically associated with WEVs. The idea that WEVs are primarily osmoregulatory in function is supported by a good deal of evidence. Most convincing is the fact that their rates of filling and emptying change dramatically when the cell is exposed to different salinities. For example, the marine flagellate *Chlamydomonas pulsatilla* lives in supralittoral tidal pools and is exposed to low salinities during rainy periods, at which times it regulates its cell volume and internal osmotic pressure via the action of WEVs (which increase in activity as the salinity of their rock pool drops). Interestingly, WEVs also occur in freshwater sponges, where they probably perform similar osmoregulatory functions.

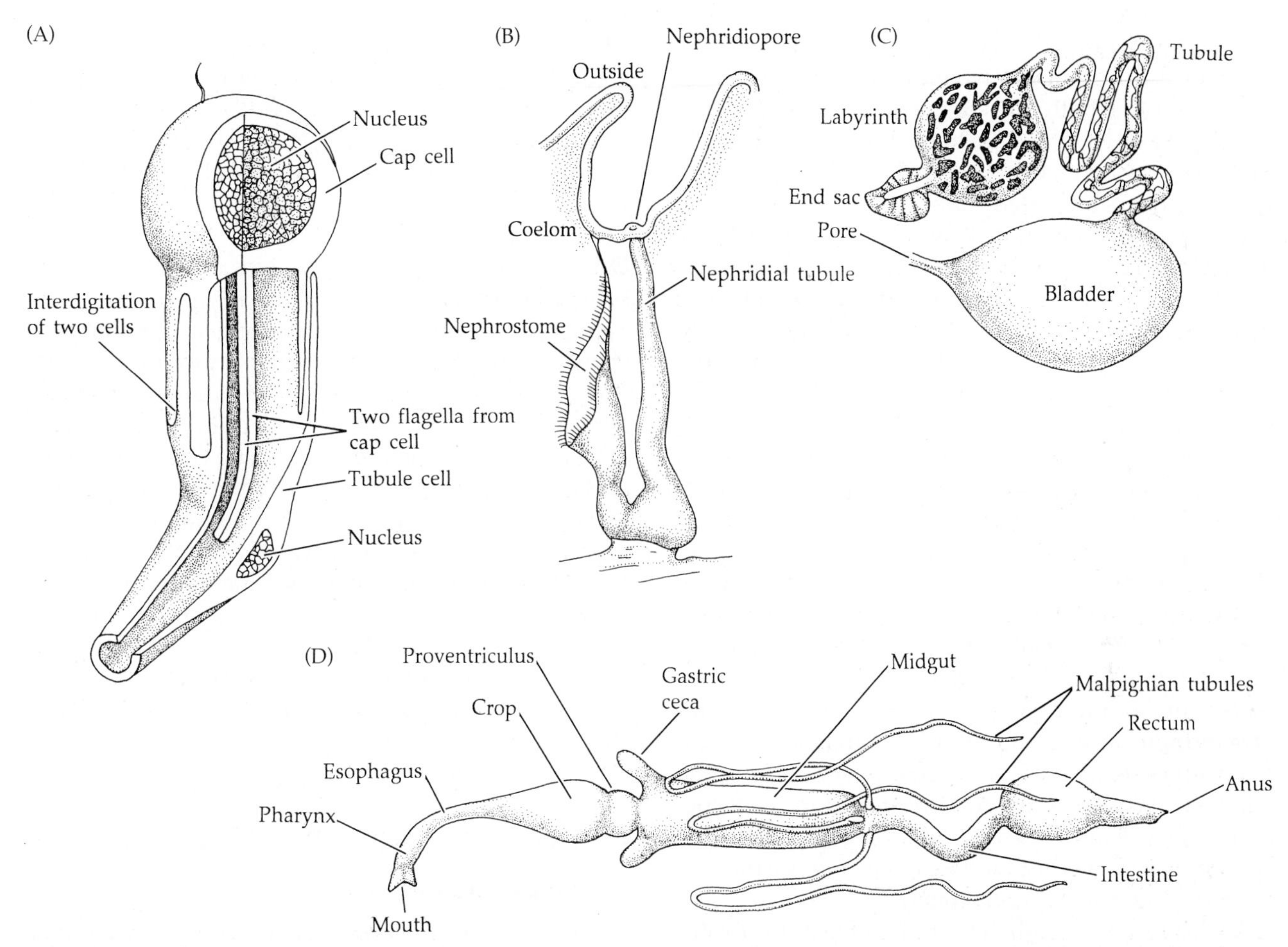

Figure 4.23 Some invertebrate excretory structures. (A) A single protonephridium, with the cap cell and tubule cell (cutaway view). (B) A simple metanephridium from a marine annelid. The nephrostome opens to the coelom, and the nephridiopore opens to the exterior. (C) The internally closed nephridium (antennal gland) of a crustacean. (D) An insect's digestive tract. Excretory Malpighian tubules extract wastes from the hemocoel and empty them into the gut.

Nephridia Although certain metazoan invertebrates possess no known excretory structures, most have some sort of ectodermally derived nephridia that serve for excretion or osmoregulation, or both. The evolution of various types of invertebrate nephridia and their relationships to other structures were discussed by E. S. Goodrich in 1945 in a classic paper, "The Study of Nephridia and Genital Ducts since 1895."

Probably the earliest type of nephridium to appear in the evolution of animals was the **protonephridium** (Figure 4.23A). Protonephridial systems are characterized by a tubular arrangement opening to the outside of the body via one or more nephridiopores and terminating internally in closed unicellular units. These units are the cap cells (or terminal cells) and may occur singly or in clusters. Each cell is folded into a cup shape, creating a concavity leading to an excretory duct (nephridioduct) and eventually to the nephridiopore. Two generally recognized types of protonephridia are **flame bulbs**, bearing a tuft of numerous cilia within the cavity, and **solenocytes**, usually with only one or two flagella. There is some evidence that several different types of flame bulb protonephridia have been independently derived from solenocyte precursors, but the details of nephridial evolution are still controversial.

The cilia or flagella drive fluids down the nephridioduct, thereby creating a lowered pressure within the tubule lumen. This lowered pressure draws body fluids, carrying wastes, across the thin cell membranes and into the duct. Selectivity is based primarily on molecular size. Protonephridia are common in adult acoelomates, many blastocoelomates, and some annelids, but are rare among most adult coelomates (although they occur frequently in various larval types). Protonephridia are probably more important in osmoregulation than in excretion. In most of these animals, nitrogenous wastes are expelled primarily by diffusion across the general body surface.

A second and probably more advanced type of excretory structure among invertebrates is the

metanephridium (Figure 4.23B). There is a critical structural difference between protonephridia and metanephridia: both open to the outside, but metanephridia are open internally to the body fluids as well. Metanephridia are also multicellular. The inner end typically bears a ciliated funnel (nephrostome), and the duct is often elongated and convoluted and may include a bladder-like storage region. Metanephridia function by taking in large amounts of body fluid through the open nephrostome and then selectively absorbing most of the reclaimable components back into the body fluids through the walls of the bladder or the excretory duct.

In very general terms, we can relate the structural and functional differences between proto- and metanephridia to the body plans with which they are commonly associated. Whereas protonephridia can adequately serve animals that have solid bodies (acoelomates), body cavities of small volume (blastocoelomates), or very small bodies (e.g., larvae), metanephridia cannot. Open funnels would be ineffective in acoelomates, and would quickly drain small blastocoelomates of their limited body fluids. Conversely, protonephridia are generally not capable of handling the relatively large body and fluid volumes typical of coelomate invertebrates. Thus, in many large coelomate animals (e.g., annelids, molluscs) one or more pairs of metanephridia are typically found.

We have very broadly interpreted the terms protonephridia and metanephridia in the above discussion, and we use them as explained above throughout this text unless specified otherwise. However, there are more complications than our simple usage suggests. For example, there is a frequent association of nephridia, especially metanephridia, with structures called coelomoducts. **Coelomoducts** are tubular connections arising from the coelomic lining and extending to the outside via special pores in the body wall. Their inner ends are frequently funnel-like and ciliated, resembling the nephrostomes of metanephridia. Coelomoducts may have arisen evolutionarily as a means of allowing the escape of gametes to the outside; they are, in fact, considered homologous to the reproductive ducts of many invertebrates. Primitively, the coelomoducts and nephridia were separate units; however, through evolution they have in many cases fused in various fashions to become what are called nephromixia.

Generally speaking, there are three types of **nephromixia**. When a coelomoduct is joined with a protonephridium and they share a common duct, the structure is called a **protonephromixium**. When a coelomoduct is united with a metanephridium, the result is either a **metanephromixium** or **mixonephridium**, depending on the structural nature of the union. Whereas coelomoducts originate from the coelomic lining, the nephridial components arise from the outer body wall, so nephromixia are a combination of mesodermally and ectodermally derived parts. Obviously there is some confusion at times about which term applies to a particular "nephridial" type if the precise developmental origin is not clear. We do not wish to belabor this point, so we leave it here to be resurrected periodically in later chapters.

Other organs of excretion Not all Metazoa possess excretory organs that are clearly proto- or metanephridia. In some taxa (e.g., sponges, echinoderms, chaetognaths, cnidarians), no definite excretory structures are known. In such cases wastes are eliminated across the surface of the skin or gut lining, perhaps with the aid of ameboid phagocytic cells that collect and transport these products. Other groups possess excretory organs that may represent highly modified nephridia or secondarily derived ("new") structures. For example, the antennal and maxillary glands of crustaceans appear to be derived from metanephridia, whereas the Malpighian tubules of insects and spiders arose independently (Figure 4.23C,D). The details of these structures are discussed in appropriate later chapters.

Circulation and Gas Exchange

Internal Transport

The transport of materials from one place to another within an organism's body depends on the movement and diffusion of substances in body fluids. Nutrients, gases, and metabolic waste products are generally carried in solution or bound to other soluble compounds within the body fluid itself or sometimes in loose cells (such as blood cells) suspended in fluid. Any system of moving fluids that reduces the functional diffusion distance that these products must traverse may be referred to as a circulatory system, regardless of its embryological origin or its ultimate design. The nature of the circulatory system is directly related to the size, complexity, and lifestyle of the organism in question. Usually the circulatory fluid is an internal, extracellular, aqueous medium produced by the animal. There are, however, a few instances in which circulatory functions are accomplished at least partly by other means. For instance, in most protists the protoplasm itself serves as the medium through which materials diffuse to various parts of the cell body, or between the organism and the environment. Sponges and most cnidarians utilize water from the environment as a circulatory fluid, sponges by passing the water through a series of channels in their bodies, and cnidarians by circulating water through the gut (Figure 4.24A,B).

In all Metazoa, the intercellular tissue fluids play a critical role as a transport medium. Even where complicated circulatory plumbing exists, tissue fluids are still necessary to bring dissolved materials in contact with

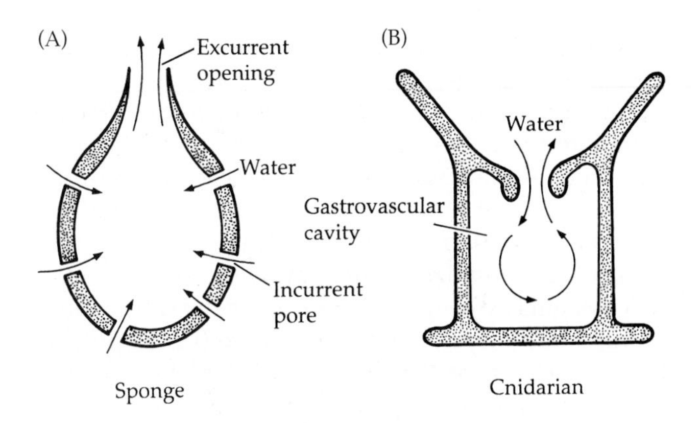

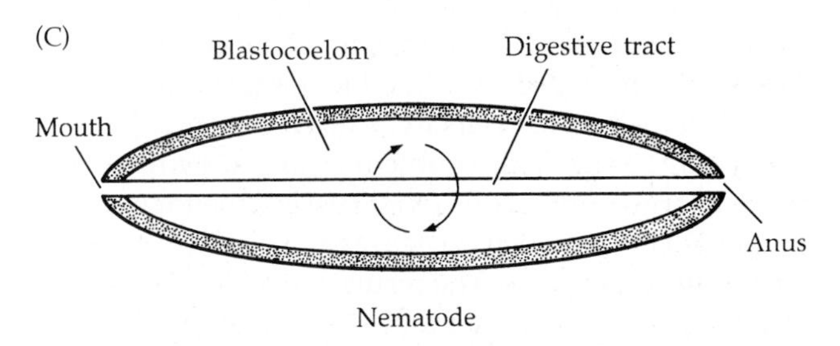

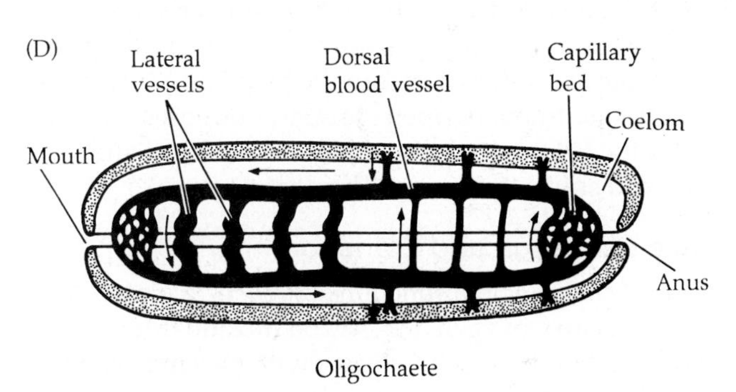

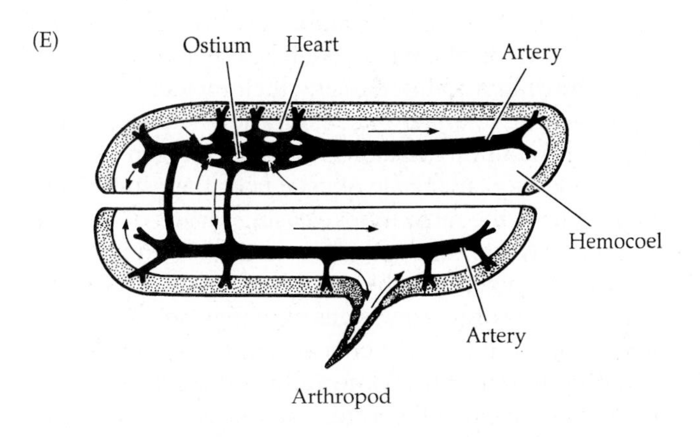

Figure 4.24 Invertebrate circulatory systems. Sponges (A) and cnidarians (B) utilize environmental water as their circulatory fluid. (C) Blastocoelomates (e.g., rotifers and nematodes) use their body cavity fluid for internal transport. (D) The closed circulatory system of an earthworm contains blood that is kept separate from the coelomic fluid. (E) Arthropods are characterized by an open circulatory system, in which the blood and body cavity (hemocoelic) fluid are one and the same.

cells, a vital process for life support. In some animals (e.g., flatworms), there are no special chambers or vessels for body fluids other than the gut and intercellular spaces through which materials diffuse on a cell-to-cell level. This condition limits these animals to relatively small sizes or to shapes that maintain low diffusion distances. Most animals, however, have some specialized structure to facilitate the transport of various body fluids and their contents. This structure may include the body cavities themselves or actual circulatory systems of vessels, chambers, sinuses, and pumping organs. Actually, many animals employ both their body cavity and a circulatory system for internal transport.

Blastocoelomate invertebrates use the fluids of the body cavity for circulation (Figure 4.24C). Most of these animals (e.g., rotifers and roundworms) are quite small, or are long and thin, and adequate circulation is accomplished by the movements of the body against the body fluids, which are in direct contact with internal tissues and organs. Several types of cells are generally present in the body fluids of blastocoelomates. These cells may serve in activities such as transport and waste accumulation, but their functions have not been well studied. A few coelomate invertebrates (e.g., sipunculans and most echinoderms) also depend largely on the body cavity as a circulatory chamber.

Circulatory Systems

Beyond the relatively rudimentary circulatory mechanisms discussed above, there are two principal designs or structural plans for accomplishing internal transport (exceptions and variations are discussed under specific taxa). These two organizational plans are closed and open circulatory systems, both of which contain a circulatory fluid, or blood. In **closed circulatory systems** the blood stays in distinct vessels and perhaps in lined chambers; exchange of circulated material with parts of the body occurs in special areas of the system such as capillary beds (Figure 4.24D). Since the blood itself is physically separated from the intercellular fluids, the exchange sites must offer minimal resistance to diffusion; thus one finds capillaries typically have membranous walls that are only a single cell-layer thick. Closed circulatory systems are common in animals with well developed or spacious coelomic compartments (e.g., annelids, phoronids, vertebrates). Such arrangements facilitate the transport from one body area to another of materials that might otherwise be isolated by the mesenteries or peritoneum of the body cavity. In such situations the blood and coelomic fluid may be quite different from one another, both in composition and in function. For example, blood may transport nutrients and gases, while coelomic fluid may accumulate metabolic wastes for removal by nephridia and also serves as a hydrostatic skeleton.

It takes power to keep a fluid moving through a plumbing system. Many invertebrates with closed systems rely on body movements and the exertion of coelomic pressure on vessels (often containing one-way valves) to move their blood. These activities are frequently supplemented by muscles of the blood vessel walls that contract in peristaltic waves. In addition, there may be special heavily muscled pumping areas along certain vessels. These regions are sometimes referred to as hearts, but most are more appropriately called contractile vessels.

Open circulatory systems are associated with a reduction of the adult coelom, including a secondary loss of most of the peritoneal lining around the organs and inner surface of the body wall. The circulatory system itself usually includes a distinct heart as the primary pumping organ and various vessels, chambers, or ill-defined sinuses (Figure 4.24E). The degree of elaboration of such systems depends primarily on the size, complexity, and to some extent the activity level of the animal. This kind of system, however, is "open" in that the blood, often called the **hemolymph**, empties from vessels into the body cavity and directly bathes the organs. The body cavity is called a **hemocoel**. Open circulatory systems are typical of arthropods and non-cephalopod molluscs, and such animals are sometimes referred to as being hemocoelomate.

Just because the open circulatory system seems a bit sloppy in its organization, it should not be viewed as poorly "designed" or inefficient. In fact, in many groups this type of system has assumed a variety of functions beyond circulation. For example, in bivalves and gastropods, the hemocoel functions as a hydrostatic skeleton for locomotion and certain types of burrowing activities. In aquatic arthropods, it also serves a hydrostatic function when the animal molts and temporarily loses its exoskeletal support. In large terrestrial insects, the transport of respiratory gases has been largely assumed by the tracheal system, and one of the primary responsibilities taken on by the open circulatory system appears to be thermal regulation. In most spiders, the limbs are extended by forcing hemolymph into the appendages.

Hearts and Other Pumping Mechanisms

Circulatory systems, open or closed, generally have structural mechanisms for pumping the blood and maintaining adequate blood pressures. Beyond the influence of general body movements, most of these structures fall into the following categories: contractile vessels (as in annelids); ostiate hearts (as in arthropods); and chambered hearts (as in molluscs and vertebrates). The method of initiating contraction of these different pumps (the pacemaker mechanisms) may be intrinsic (originating within the musculature of the structure itself) or extrinsic (originating from motor nerves arising outside the structure). The first case describes the myogenic hearts of molluscs and vertebrates; the second describes the neurogenic hearts of most arthropods and, at least in part, the contractile vessels of annelids.

Blood pressure and flow velocities are intimately associated not only with the activity of the pumping mechanism but also with vessel diameters. Energetically, it costs a good deal more to maintain flow through a narrow pipe than through a wide pipe. This cost is minimized in animals with closed circulatory systems by keeping the narrow vessels short and using them only at sites of exchange (i.e., capillary beds), and by using the larger vessels for long-distance transport from one exchange site to another. In the human circulatory system, for example, arteries have an average radius of 2.0 mm, veins 2.5 mm, and capillaries 0.006 mm. But reducing the diameter of a single vessel increases flow velocity, which poses problems at an exchange site. This problem is solved by the presence of large numbers of small vessels, the total cross-sectional area of which exceeds that of the larger vessel from which they arise. The result is that blood pressure and total flow velocity actually decrease at capillary exchange sites. A drop in blood pressure and a relative rise in blood osmotic pressure along the capillary bed facilitate exchanges between the blood and surrounding tissue fluids. In open systems, both pressure and velocity drop once the blood leaves the heart and vessels and enters the spacious hemocoel.

Gas Exchange and Transport

One of the principal functions of most circulatory fluids is to carry oxygen and carbon dioxide through the body and exchange these gases with the environment. With few exceptions, oxygen is necessary for cellular respiration. Although a number of invertebrates can survive periods of environmental oxygen depletion—either by dramatically reducing their metabolic rate or by switching to anaerobic respiration—most cannot; they depend upon a relatively constant oxygen supply.

All animals can take in oxygen from their surroundings while at the same time releasing carbon dioxide, a metabolic waste product of respiration. We define the uptake of oxygen and the loss of carbon dioxide at the surface of the organism as gas exchange, reserving the term respiration for the energy-producing metabolic activities within cells. Some authors distinguish these two processes with the terms external respiration and cellular (internal) respiration.

Gas exchange in nearly all animals operates according to certain common principles regardless of any structural modifications that serve to enhance the process under different conditions. The basic strategy is to bring the environmental medium (water or air) close to the appropriate body fluid (blood or body cavity fluid) so that the two are separated only by a wet membrane across which the gases can diffuse. The system must be moist because the gases must be in solution in order to diffuse across the membrane. The diffusion process

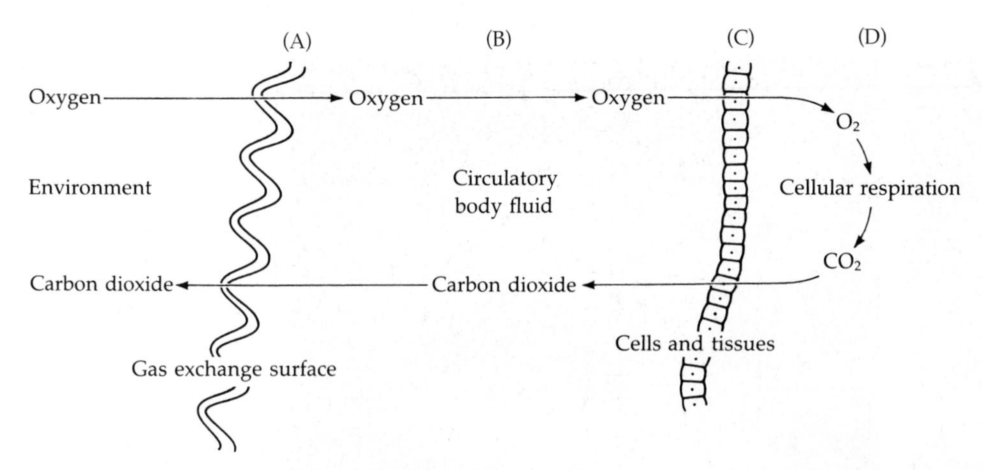

Figure 4.25 Gas exchange in animals. Oxygen is obtained from the environment at a gas exchange surface, such as an epithelial layer (A), and is transported by a circulatory body fluid (B) to the body's cells and tissues (C), where cellular respiration occurs (D). Carbon dioxide follows the reverse path. See text for details.

depends on the concentration gradients of the gases at the exchange site; these gradients are maintained by the circulation of internal fluids to and away from these areas (Figure 4.25).

Gas exchange structures Protists and a number of invertebrates lack special gas exchange structures. In such animals gas exchange is said to be integumentary or cutaneous, and occurs over much of the body surface. Such is the case in many tiny animals with very high surface-to-volume ratios and in some larger soft-bodied forms (e.g., cnidarians and flatworms). Most animals with integumentary gas exchange are restricted to aquatic or damp terrestrial environments where the body surface is kept moist. Integumentary gas exchange also supplements other methods in many animals, even certain vertebrates (e.g., amphibians).

Most marine and many freshwater invertebrates possess gills (Figure 4.26A–C,G), which are external organs or restricted areas of the body surface specialized for gas exchange. Basically, gills are thin-walled processes, well supplied with blood or other body fluids, which promote diffusion between this fluid and the environment. Gills are frequently highly folded or digitate, increasing the diffusive surface area. A great number of nonhomologous structures have evolved as gills in different taxa, and they often serve other functions in addition to gas exchange (e.g., sensory input and feeding). By their very nature, gills are permeable surfaces that must be protected during times of osmotic stress, such as occur in estuaries and intertidal environments. In these instances, the gills may be housed within chambers or be retractable.

A few marine invertebrates employ the lining of the gut as the gas exchange surface. Water is pumped in and out of the hindgut, or a special evagination thereof, in a process called **hindgut irrigation**. Many sea cucumbers and echiurid worms use this method of gas exchange (Figure 4.26F).

As you can imagine, protruding gills would not work on dry land. Here, the gas exchange surfaces must be internalized to keep them moist and protected and to prevent body water loss through the wet surfaces. The lungs of terrestrial vertebrates are the most familiar example of such an arrangement. Among the invertebrates, the arthropods have managed to solve the problems of "air-breathing" in two basic ways. Spiders and their kin possess book lungs, and most insects, centipedes, and millipedes possess tracheae (Figure 4.26D,E). **Book lungs** are blind inpocketings with highly folded inner linings across which gases diffuse between the hemolymph and the air. **Tracheae**, however, are branched, usually anastomosed invaginations of the outer body wall and are open both internally and externally.

The tracheae of most insects allow diffusion of oxygen from air directly to the tissues of the body; the blood plays little or no role in gas transport. Rather, intercellular fluids extend part way into the tracheal tubes as a solvent for gases. Atmospheric pressure tends to prevent these fluids from being drawn too close to the external body surface where evaporation is a potential problem. In addition, the outside openings (**spiracles**) of the tracheae are often equipped with some mechanism of closure. In many insects, especially large ones, special muscles ventilate the tracheae by actively pumping air in and out. Terrestrial isopod crustaceans (e.g., sowbugs and pillbugs) have invaginated gas exchange structures on some of their abdominal appendages. These inpocketings are called **pseudotrachea**, but are probably not homologous to the trachea or the book lungs of insects and spiders.

The only other major group of terrestrial invertebrates whose members have evolved distinct air-breathing structures are the land snails and slugs (Figure 4.26H). The gas exchange structure here is a lung that opens to the outside via a pore called the **pneumostome**. This lung is derived from a feature common to molluscs in general, the mantle

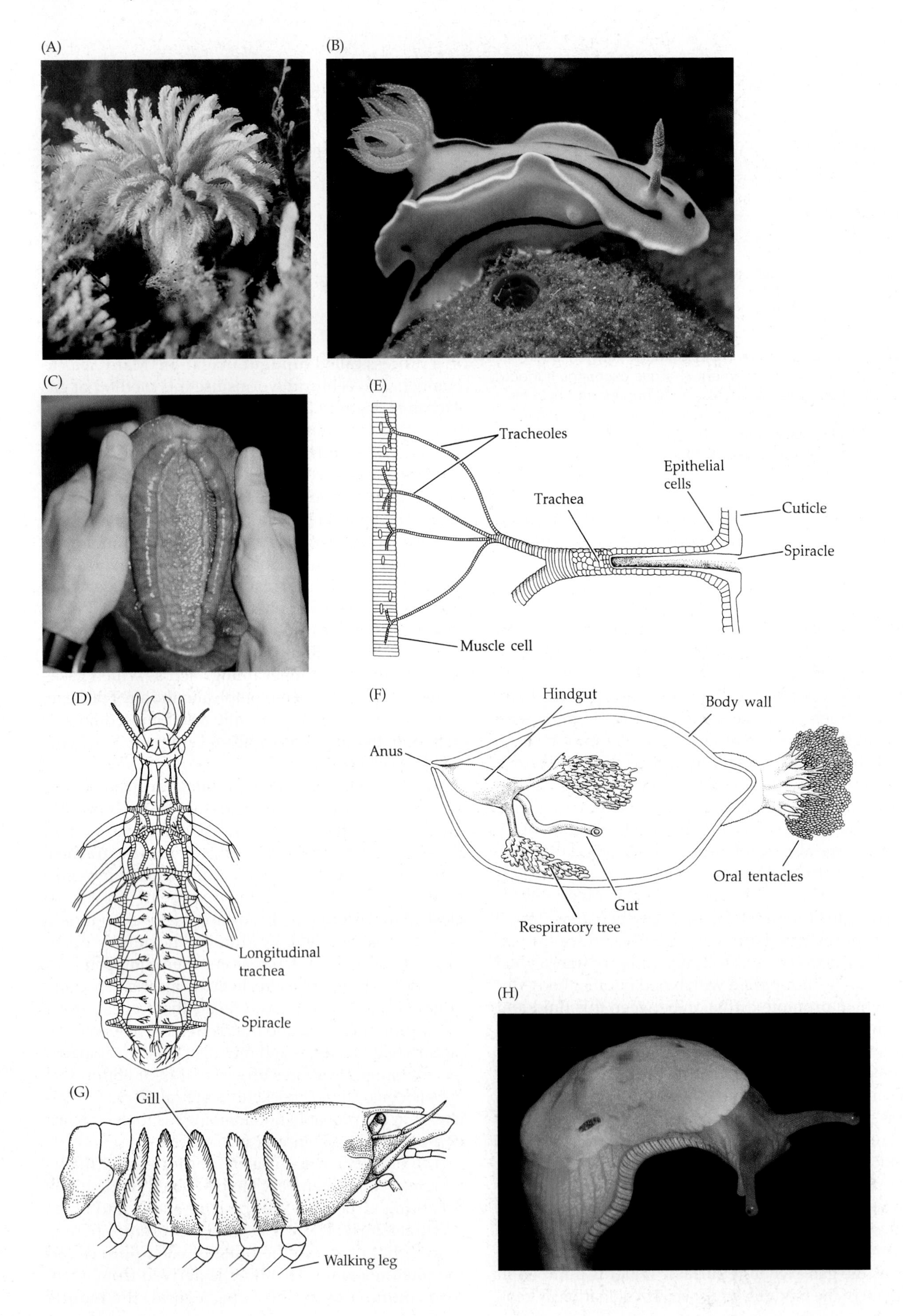
(A)
(B)
(C)
(E)
Tracheoles
Epithelial cells
Trachea
Cuticle
Spiracle
Muscle cell
(D)
Longitudinal trachea
Spiracle
(F)
Hindgut
Body wall
Anus
Oral tentacles
Gut
Respiratory tree
(G)
Gill
Walking leg
(H)

Figure 4.26 Some gas exchange structures in invertebrates. (A) The tube-dwelling polychaete worm *Eudistylia*, with its feeding–gas exchange tentacles extended. (B) A sea slug (nudibranch) displaying its branchial plume. (C) The gills of the giant gumboot chiton (*Cryptochiton stelleri*) are visible along the right side of its foot. (D) A general plan of the tracheal system of an insect. (E) A single insect trachea and its branches (tracheoles), which lead directly to a muscle cell. (F) A sea cucumber dissected to expose the paired respiratory trees, which are flushed with water by hindgut irrigation. (G) The placement of gills beneath the flaps (carapace) of the thorax in a crustacean (lateral view). (H) A terrestrial banana slug has a pneumostome that opens to the air sac, or "lung."

cavity, which in other molluscs houses the gills and other organs.

Gas transport As illustrated in Figure 4.25, oxygen must be transported from the sites of environmental gas exchange to the cells of the body, and carbon dioxide must get from the cells where it is produced to the gas exchange surface for release. Generally, groups displaying marked cephalization circulate freshly oxygenated blood through the "head" region first, and secondarily to the rest of the body.

Invertebrates vary considerably in their oxygen requirements. In general, active animals consume more oxygen than sedentary ones. In slow-moving and sedentary invertebrates, oxygen consumption and utilization are quite low. For example, no more than 20% oxygen withdrawal from the gas exchange water current has ever been demonstrated in sessile sponges, bivalves, or tunicates. The amount of oxygen available to an organism varies greatly in different environments. The concentration of oxygen in dry air at sea level is uniformly about 210 ml/liter, whereas in water it ranges from near zero to about 10 ml/liter. This variation in aquatic environments is due to such factors as depth, surface turbulence, photosynthetic activity, temperature, and salinity (oxygen concentrations drop as temperature and salinity increase). With the exception of certain areas prone to oxygen depletion (e.g., muds rich in organic detritus), most habitats provide adequate sources of oxygen to sustain animal life. Also, the relatively low capacity of body fluids to carry oxygen in solution is greatly increased by binding oxygen with complex organic compounds called respiratory pigments.

Respiratory pigments differ in molecular architecture and in their affinities for oxygen, but all have a metal ion (usually iron, sometimes copper) with which the oxygen combines. In most invertebrates, these pigments occur in solution within the blood or other body fluid, but in some invertebrates (and virtually all vertebrates), they may be in specific blood cells. In general, the pigments respond to high oxygen concentrations by "loading" (combining with oxygen), and to low oxygen concentrations by "unloading" or dissociating from oxygen (releasing oxygen). The loading and unloading qualities are different for various pigments in terms of their relative saturations at different levels of oxygen in their immediate surroundings, and are generally expressed in the form of dissociation curves. Respiratory pigments load at the site of gas exchange, where environmental oxygen levels are high relative to the body fluid, and unload at the cells and tissues, where surrounding oxygen levels are low relative to the body fluid. In addition to simply carrying oxygen from the loading to the unloading sites, some pigments may carry reserves of oxygen that are released only when tissue levels are unusually low. Other factors, such as temperature and carbon dioxide concentration, also influence the oxygen-carrying capacities of respiratory pigments.

Hemoglobin is among the most common respiratory pigments in animals. There are actually a number of different hemoglobins. Some function primarily for transport, whereas others store oxygen and then release it during times of low environmental oxygen availability. Hemoglobins are reddish pigments containing iron as the oxygen-binding metal. They are found in a variety of invertebrates and, with the exception of a few fishes, in all vertebrates. Among the major groups of invertebrates, hemoglobin occurs in many annelids, some crustaceans, some insects, and a few molluscs and echinoderms. Interestingly, hemoglobin is not restricted to the Metazoa; it is also produced by some protists, certain fungi, and in the root nodules of leguminous plants. Among animals, hemoglobin may be carried within red blood cells (erythrocytes), in coelomic cells called hemocytes (in a few echinoderms), or it may simply be dissolved in the blood or coelomic fluid.

Hemocyanins are the most commonly occurring respiratory pigments in molluscs and arthropods, and they occur only in members of these two phyla. Among arthropods, hemocyanin occurs in chelicerates, a few myriapods, and the "higher Crustacea." There is indirect evidence that it also occurred in trilobites. Hemocyanin has been found in most classes of molluscs. Although hemocyanins, like hemoglobins, are proteins, they display significant structural differences, contain copper rather than iron, and tend to have a bluish color when oxygenated. The oxygen-binding site on a hemocyanin molecule is a pair of copper atoms linked to amino acid side chains. Unlike most hemoglobins, hemocyanins tend to release oxygen easily and provide a ready source of oxygen to the tissues as long as there is a relatively high concentration of available environmental oxygen. Hemocyanins are always found in solution, never in cells, a characteristic probably related to the necessity for rapid oxygen unloading. Hemocyanins often give a bluish tint to the hemolymph of arthropods, although the presence of

TABLE 4.1 Properties of oxygen-carrying respiratory pigments

Pigment	Molecular weight	Metal	Ratio of metal to O_2	Metal associate
Hemoglobin	65,000	Fe	1:1	Porphyrin
Hemerythrin	40,000–108,000	Fe	2:1	Protein chains
Hemocyanin	40,000–9,000,000	Cu	2:1	Protein chains
Chlorocruorin	3,000,000	Fe	1:1	Porphyrin

carotenoid pigments (beta-carotene and related molecules) commonly impart a brown or orange coloration.

Two other types of respiratory pigments occur incidentally in certain invertebrates; these are **hemerythrins** and **chlorocruorins**, both of which contain iron. The former is violet to pink when oxygenated; the latter is green in dilute concentrations but red in high concentrations. Chlorocruorins generally function as efficient oxygen carriers when environmental levels are relatively high; hemerythrins function more in oxygen storage. Chlorocruorin is structurally similar to hemoglobin and may have been derived from it. Chlorocruorin occurs in several families of polychaete worms; hemerythrin is known from sipunculans, at least one genus of polychaetes, and some priapulans and brachiopods.

Table 4.1 gives some of the basic properties of oxygen-carrying pigments. There seems to be no obvious phylogenetic rhyme or reason to the occurrence of these pigments among the various taxa. Their sporadic and inconsistent distribution suggests that some of them may have evolved more than once, through parallel or convergent evolution. Respiratory pigments are rare among insects and are known only from the occurrence of hemoglobin in chironomid midges, some notonectids, and certain parasitic flies of the genus *Gastrophilus*. The absence of respiratory pigments among the insects reflects the fact that most of them do not use the blood as a medium for gas transport, but employ extensive tracheal systems to carry gases directly to the tissues. In those insects without well-developed tracheae, oxygen is simply carried in solution in the hemolymph.

Respiratory pigments raise the oxygen-carrying capacity of body fluids far above what would be achieved by transport in simple solution. Similarly, carbon dioxide levels in body fluids (and in sea water) are much higher than would be expected strictly on the basis of its solubility. The enzyme carbonic anhydrase greatly accelerates the reaction between carbon dioxide and water, forming carbonic acid:

$$CO_2 + H_2O \rightleftharpoons H_2CO_3$$

Furthermore, carbonic acid ionizes to hydrogen and bicarbonate ions, so a series of reversible reactions takes place:

$$CO_2 + H_2O \rightleftharpoons H_2CO_3 \rightleftharpoons H^+ + HCO_3^-$$

By "tying up" CO_2 in other forms, the concentration of CO_2 in solution is lowered, thus raising the overall CO_2-carrying capacity of the blood. This set of reactions responds to changes in pH, and in the presence of appropriate cations (e.g., Ca^{2+} and Na^+) it shifts back and forth, serving as a buffering mechanism by regulating hydrogen ion concentration.

Nervous Systems and Sense Organs

All living cells respond to some stimuli and conduct some sort of "information," at least for short distances. Thus, even when no real nervous system is present—the condition found in protists and sponges—coordination and reaction to external stimulation do occur. The regular metachronal beating of cilia in ciliate protists and the responses of certain flagellates to varying light intensities are examples. In addition, most protists are known to respond to gradients of various environmental factors by moving to or away from areas of high concentration. For example, when subjected to conditions of low oxygen concentration (hypoxia), paramecia move to regions of lower water temperature, thus lowering their metabolic rate and presumably their oxygen need. But the integration and coordination of bodily activities in Metazoa are in large part due to the processing of information by a true nervous system. The functional units of nervous systems are **neurons**: cells that are specialized for high-velocity impulse conduction.

The generation of an impulse within a true nervous system usually results from a stimulus imposed on the nervous elements. The source of stimulation may be external or internal. A typical pathway of events occurring in a nervous system is shown in Figure 4.27. A stimulus received by some receptor (e.g., a sense organ) generates an impulse that is conducted along a **sensory nerve** (**afferent nerve**) via a series of adjacent neurons to some coordinating center or region of the system. The information is processed and an appropriate response is "selected." A **motor nerve** (**efferent nerve**) then conducts an impulse from the central processing center to an effector (e.g., a muscle), where the response occurs. Once an impulse is initiated within the system, the mechanism of conduction is essentially

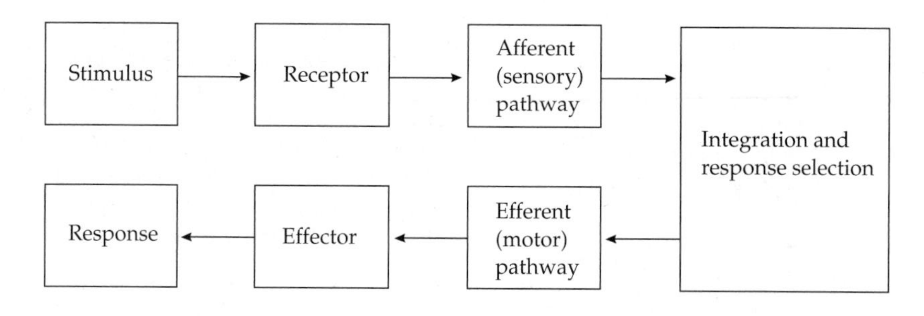

Figure 4.27 A generalized pathway within the nervous system. A stimulus initiates an impulse within some sensory structure (the receptor); the impulse is then transferred to some integrative portion of the nervous system via sensory nerves. Following response selection, an impulse is generated and transferred along motor nerves to an effector (e.g., muscle), where the appropriate response is elicited.

the same in all neurons, regardless of the stimulus. The wave of depolarization along the length of each neuron and the chemical neurotransmitters crossing the synaptic gaps between neurons are common to virtually all nervous conduction. How then is the information interpreted within the system for response selection? The answer to this question involves three basic considerations.

First is the occurrence of a point called a threshold, which corresponds to the minimum intensity of stimulation necessary to generate an impulse. Receptor sites consist of specialized neurons whose thresholds for various kinds of stimuli are drastically different from one another because of structural or physiological qualities. For example, a sense organ whose threshold for light stimulation is very low (compared with other potential stimuli) functions as a light sensor, or photoreceptor. In any such specialized sensory receptor, the condition of differential thresholds essentially screens incoming stimuli so that an impulse normally is generated by only one kind of information (e.g., light, sound, heat, or pressure). Second is the nature of the receptor itself. Receptor units (e.g., sense organs) are generally constructed in ways that permit only certain stimuli to reach the impulse-generating cells. For example, the light-sensitive cells of the human eye are located beneath the eye surface, where stimuli other than light would not normally reach them.

And third, the overall "wiring" or circuitry of the entire nervous system is such that impulses received by the integrative (response-selecting) areas of the system from any particular nerve will be interpreted according to the kind of stimulus for which that sensory pathway is specialized. For example, all impulses coming from a photoreceptor are understood as being light induced. Threshold and circuitry can be demonstrated by introducing false information into the system by stimulating a specialized sense organ in an inappropriate manner: if photoreceptors in the eye are stimulated by electricity or pressure, the nervous system will interpret this input as light. Remember that an impulse can be generated in any receptor by nearly any form of stimulation if the stimulus is intense enough to exceed the relevant threshold. A blow to the eye often results in "seeing stars," or flashes of light, even when the eye is closed. In such a situation, the photoreceptor's threshold to mechanical stimulation has been reached. By the same token, the application of extreme cold to a heat receptor may feel hot.

Nervous systems in general operate on the principles outlined above. However, this description applies largely to nervous systems that have structural centralized regions. Following a discussion below of the basic types of sense organs (receptor units), we discuss centralized and noncentralized nervous systems and their relationships to general body architecture.

Sense Organs

Invertebrates possess an impressive array of receptor structures through which they receive information about their internal and external environments. An animal's behavior is in large part a function of its responses to that information. These responses often take the form of some sort of movement relative to the source of a particular stimulus. A response of this nature is called a **taxis** and may be positive or negative depending on the reaction of the animal to the stimulus. For example, many animals tend to move away from bright light and are thus said to be negatively phototactic.

The activities of receptor units represent the initial step in the usual functioning of the nervous system; they are a critical link between the organism and its surroundings. Consequently, the kinds of sense organs present and their placement on the body are intimately related to the overall complexity, mode of life, and general body plan of any animal. The following general review provides some concepts and terminology that serve as a basis for more detailed coverage in later chapters. The first five categories of sense organs may all be viewed as mechanoreceptors, in that they respond to mechanical stimuli (e.g., touch, vibrations, and pressure). The last three are sensitive to nonmechanical input (e.g., chemicals, light, and temperature). In addition, a few invertebrates have been shown to possess a magnetic compass. For example, during their migrations between North America and central Mexico, monarch butterflies probably navigate using a combination of the sun and the inclination angle component of the Earth's magnetic field to guide their flights, as has been shown for most vertebrate migrators.

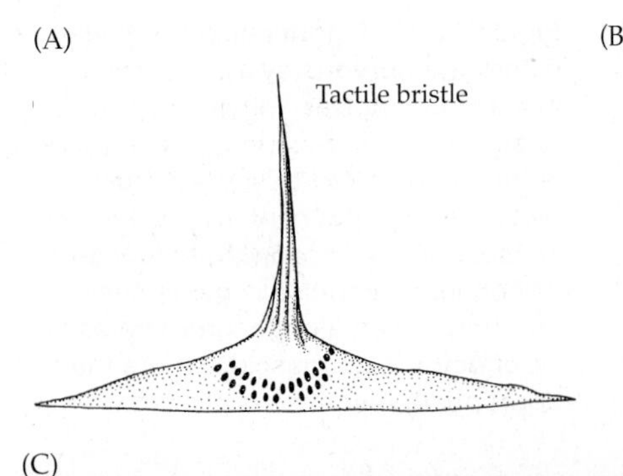

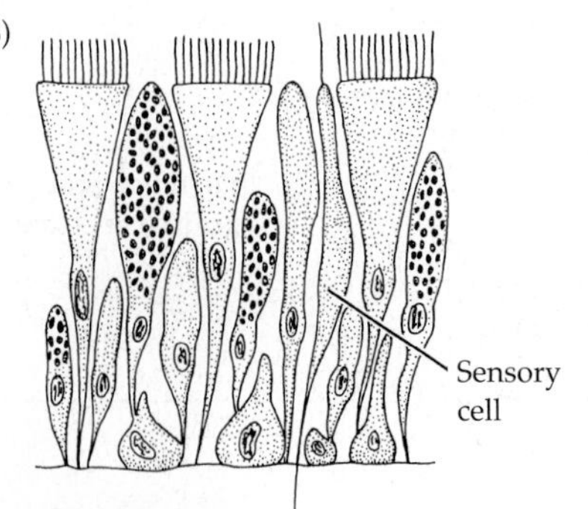

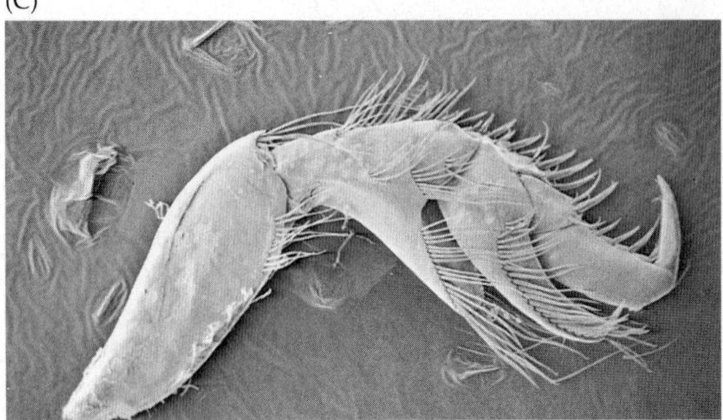

Figure 4.28 Some invertebrate tactile receptors. (A) Tactile organ of *Sagitta bipunctata* (an arrow worm, phylum Chaetognatha). (B) A sensory epithelial cell of a nemertean worm. (C) Long, touch-sensitive setae (and stout grasping setae) on the leg of the isopod, *Politolana* (SEM).

Tactile receptors Touch or tactile receptors are generally derived from modified epithelial cells associated with sensory neurons. The nature of the epithelial modifications depends a great deal on the structure of the body wall. For instance, the form of a touch receptor in an arthropod with a rigid exoskeleton must be different from that in a soft-bodied cnidarian. Most such receptors, however, involve projections from the body surface, such as bristles, spines, setae, tubercles, and assorted bumps and pimples (Figure 4.28). Objects in the environment with which the animal makes contact move these receptors, thereby creating mechanical deformations that are imposed upon the underlying sensory neurons to initiate an impulse.

Virtually all animals are touch-sensitive, but their responses are varied and often integrated with other sorts of sensory input. For example, the gregarious nature of many animals may involve a positive response to touch (positive thigmotaxis) combined with the chemical recognition of members of the same species. Some touch receptors are highly sensitive to mechanically induced vibrations propagated in water, loose sediments, through solid substrata, or other materials. Such vibration sensors are common in certain tube-dwelling polychaetes that retract quickly into their tubes in response to movements in their surroundings. Some crustacean ambush-predators are able to detect the vibrations induced by nearby potential prey animals, and web-building spiders quickly and accurately sense prey in their webs through vibrations of the threads. Some spiders have highly sensitive tactile setae on their appendages, called **trichobothria**, that sense airborne vibrations of prey, such as wing beats and perhaps even some sound frequencies.

Georeceptors Georeceptors respond to the pull of gravity, giving animals information about their orientation relative to "up and down." Most georeceptors are structures called **statocysts** (Figure 4.29). Statocysts usually consist of a fluid-filled chamber containing a solid granule or pellet called a **statolith**. The inner lining of the chamber includes a touch-sensitive epithelium from which project bristles or "hairs" associated with underlying sensory neurons. In aquatic invertebrates, some statocysts are open to the environment and thus are filled with water. In some of these the statolith is a sand grain obtained from the animal's surroundings. Most statoliths, however, are secreted within closed capsules by the organisms themselves.

Because of the resting inertia of the statolith within the fluid, any movement of the animal results in a change in the pattern or intensity of stimulation of the sensory epithelium by the statolith. Additionally, when the animal is stationary, the position of the statolith within the chamber provides information about the organism's orientation to gravity. The fluid within statocysts of at least some invertebrates (especially certain crustaceans) also acts something like the fluid of the semicircular canals in vertebrates. When the animal moves, the fluid tends to remain stationary—the relative "flow" of the fluid over the sensory epithelium provides the animal with information about its linear and rotational acceleration relative to its environment.

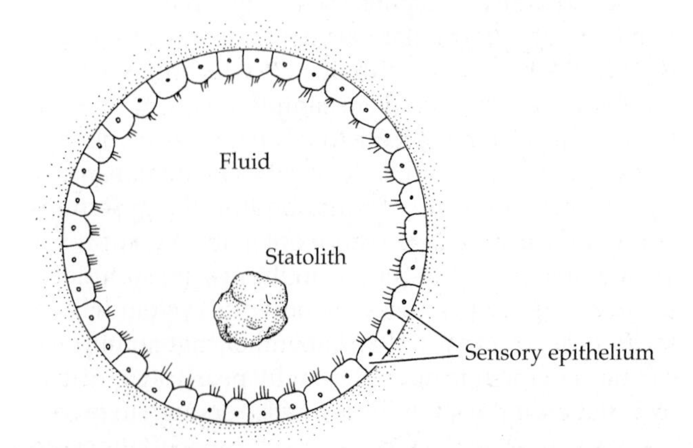

Figure 4.29 A generalized statocyst, or georeceptor (section).

Whether stationary or in motion, animals utilize the input from georeceptors in different ways, depending on their habitat and lifestyle. The information from these statocysts is especially important under conditions where other sensory reception is inadequate. For example, burrowing invertebrates cannot rely on photoreceptors for orientation when moving through the substratum, and some employ statocysts for that purpose. Similarly, planktonic animals face orientation problems in their three-dimensional aqueous environment, especially in deep water and at night; many such creatures possess statocysts.

There are a few exceptions to the standard statocyst arrangements described above. For example, a number of aquatic insects detect gravity by using air bubbles trapped in certain passageways (e.g., tracheal tubes). The bubbles move according to their orientation to the vertical, much like the air bubble in a carpenter's level, and stimulate sensory bristles lining the tube in which they are located.

Proprioceptors Internal sensory organs that respond to mechanically induced changes caused by stretching, compression, bending, and tension are called **proprioceptors**, or simply stretch receptors. These receptors give the animal information about the movement of its body parts and their positions relative to one another. Proprioceptors have been most thoroughly studied in vertebrates and arthropods, where they are associated with appendage joints and certain body extensor muscles. The sensory neurons involved in proprioception are associated with and attached to some part of the body that is stretched or otherwise mechanically affected by movement or muscle tension. These parts may be specialized muscle cells, elastic connective tissue fibers, or membranes that span joints. As these structures are stretched, relaxed, and compressed, the sensory endings of the attached neurons are distorted accordingly and thus stimulated. Some of these receptor arrangements can detect not only changes in position but also in static tension.

Phonoreceptors General sensitivity to sound— phonoreception—has been demonstrated in a number of invertebrates (certain annelid worms and a variety of crustaceans), but true auditory receptors are known only in a few groups of insects and perhaps some arachnids and centipedes. Crickets, grasshoppers, and cicadas possess phonoreceptors called **tympanic organs** (Figure 4.30). A rather tough but flexible tympanum covers an internal air sac that allows the tympanum to vibrate when struck by sound waves. Sensory neurons attached to the tympanum are stimulated directly by the vibrations. Most arachnids possess structures called **slit sense organs**, which, although poorly studied, are suspected to perform auditory functions; at least they appear to be capable of sensing sound-induced vibrations. Certain centipedes bear so-called **organs of Tömösváry**, which some workers believe may be sensitive to sound.

Figure 4.30 An arthropod phonoreceptor, or auditory organ, of the fork-tailed katydid, *Scudderia furcata*. Note the position of the right-side tympanum on the tibia of the first walking leg.

Baroreceptors The sensitivity of invertebrates to pressure changes—baroception—is not well understood, and no structures for this purpose have been positively identified. However, behavioral responses to pressure changes have been demonstrated in several pelagic invertebrates including medusae, ctenophores, cephalopods, and copepod crustaceans, as well as in some planktonic larvae. Aquatic insects also sense changes in pressure, and may use a variety of methods to do so. Some intertidal crustaceans coordinate daily migratory activities with tidal movements, perhaps partly in response to pressure as water depth changes.

Chemoreceptors Many animals have a general chemical sensitivity, which is not a function of any definable sensory structure but is due to the general irritability of protoplasm itself. When they occur in sufficiently high concentrations, noxious or irritating chemicals can induce responses via this general chemical sensitivity. In addition, most animals have specific chemoreceptors.

Chemoreception is a rather direct sense in that the molecules stimulate sensory neurons by contact, usually after diffusing in solution across a thin epithelial covering. The chemoreceptors of many aquatic invertebrates are located in pits or depressions, through which water may be circulated by ciliary action. In arthropods, the chemoreceptors are usually in the form of hollow "hairs" or other projections, within which are chemosensory neurons. While chemosensitivity is a universal phenomenon among invertebrates, a wide range of specificities and capabilities exists.

The types of chemicals to which particular animals respond are closely associated with their lifestyles. Chemoreceptors may be specialized for tasks such as general water analysis, humidity detection, sensitivity to pH, prey tracking, mate location, substratum analysis, and food recognition. Probably all aquatic organisms leak small amounts of amino acids into their environment through the skin and gills as well as in their urine and feces. These released amino acids form an organism's "body odor," which can create a chemical picture of the animal that others detect to identify such characteristics as species, sex, stress level, distance and direction, and perhaps size and individuality. Amino acids are widely distributed in the aquatic environment, where they provide general indicators of biological activity. Many aquatic animals can detect amino acids with much greater sensitivity than our most sophisticated laboratory equipment.

Photoreceptors Nearly all animals are sensitive to light, and most have some kind of identifiable photoreceptors. Although members of only a few of the metazoan phyla appear to have evolved eyes capable of image formation (Cnidaria, Mollusca, Annelida, Arthropoda, and Chordata), virtually all animal photoreceptors share structurally similar light receptor molecules that probably predate the origin of discrete structural eyes. Thus, the structural photoreceptors of animals share the common quality of possessing light-sensitive pigments. These pigment molecules are capable of absorbing light energy in the form of photons, a process necessary for the initiation of any light-induced, or photic, reaction. The energy thus absorbed is ultimately responsible for stimulating the sensory neurons of the photoreceptor unit.

Beyond this basic commonality, however, there is an incredible range of variation in complexity and capability of light-sensitive structures. Arthropods, molluscs, and some polychaete annelids possess eyes with extreme sensitivity, good spatial resolution, and, in some cases, multiple spectral channels. Most classifications of photoreceptors are based upon grades of complexity, and the same categorical term may be applied to a variety of nonhomologous structures, from simple pigment spots (found in protists) to extremely complicated lensed eyes (found in squids and octopuses). Functionally, the capabilities of these receptors range from simply perceiving light intensity and direction to forming images with a high degree of visual discrimination and resolution.

Certain protists, particularly flagellates, possess subcellular organelles called **stigmata**, which are associated with simple spots of light-sensitive pigment (Figure 4.31A). The simplest metazoan photoreceptors are unicellular structures scattered over the epidermis or concentrated in some area of the body. These are usually called **eyespots**. Multicellular photoreceptors may be classified into three general types, with some subdivisions. These types include **ocelli** (sometimes called simple eyes or eyespots), **compound eyes** (found in many arthropods), and **complex eyes** (the "camera" eyes of cephalopod molluscs and vertebrates). In multicellular ocelli, the light-sensitive (retinular) cells may face outward; these ocelli are then said to be direct. Or the light-sensitive cells may be inverted. The inverted type is common among flatworms and nemerteans and is made up of a cup of reflective pigment and retinular cells (Figure 4.31B). The light-sensitive ends of these neurons face into the cup. Light entering the opening of the pigment cup is reflected back onto the retinular cells. Because light can enter only through the cup opening, this sort of ocellus gives the animal a good deal of information about light direction as well as variations in intensity.

Compound eyes are composed of a few to many distinct units called **ommatidia** (Figure 4.31C). Although eyes of multiple units occur in certain annelid worms and some bivalve molluscs, they are best developed and best understood among the arthropods. Each ommatidium is supplied with its own nerve tract leading to a large optic nerve, and apparently each has its own discrete field of vision. The visual fields of neighboring ommatidia overlap to some degree, with the result that a shift in position of an object within the total visual field causes changes in the impulses reaching several ommatidial units; based in part on this phenomenon, compound eyes are especially suitable for detecting movement. Compound eyes are described in more detail in Chapter 20.

The complex eyes of squids and octopuses (Figure 4.31D) are probably the best image-forming eyes among the invertebrates. Cephalopod eyes are frequently compared with those of vertebrates, but they differ in many respects. The eye is covered by a transparent protective cornea. The amount of light that enters the eye is controlled by the iris, which regulates the size of the slitlike pupil. The lens is held by a ring of ciliary muscles and focuses light on the retina, a layer of densely packed photosensitive cells from which the neurons arise. The receptor sites of the retinal layer face in the direction of the light entering the eye. This direct eye arrangement is quite different from the indirect eye condition in vertebrates, where the retinal layer is inverted. Another difference is that in many vertebrates, focusing is accomplished by the action of muscles that change the shape of the lens, whereas in cephalopods it is achieved by moving the lens back and forth with the ciliary muscles and by compressing the eyeball.

A good deal of work suggests that metazoan photoreceptors evolved primarily along two lines. On one hand are photoreceptor units derived from or closely associated with cilia (e.g., in cnidarians, echinoderms, and chordates). These types of eyes are called **ciliary eyes**. On the other hand are photoreceptors derived from microvilli or microtubules and referred to as **rhabdomeric eyes** (e.g., in flatworms, annelids, arthropods, and molluscs). All animal photoreceptors may

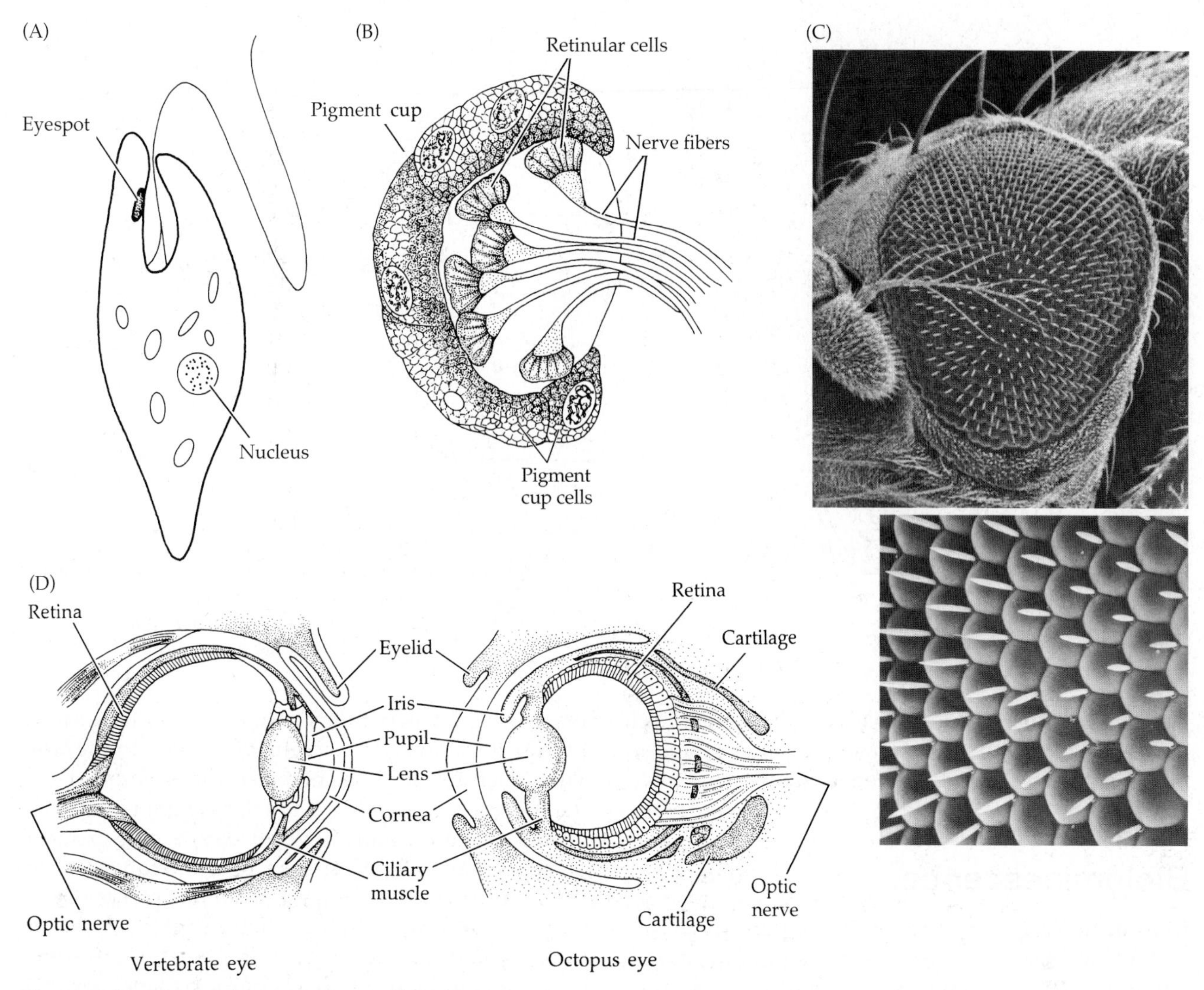

Figure 4.31 Some photoreceptors. (A) A protist, *Euglena*. Note the position of the stigma. (B) An inverted pigment-cup ocellus of a flatworm (section). (C) An insect's compound eye. A single unit is called an ommatidium. (D) A vertebrate eye (left) and a cephalopod eye (right) (vertical sections).

share a deep developmental homology with the *Pax-6* gene, which is known to initiate eye development (even structurally non-homologous eyes) in numerous distantly related phyla of both Protostomia and Deuterostomia.

Thermoreceptors The influence of temperature changes on all levels of biological activity is well documented. Every student of general biology has learned about the basic relationships between temperature and rates of metabolic reactions. Furthermore, even the casual observer has noticed that many organisms' activity levels range from lethargy at low temperatures to hyperactivity at elevated temperatures, and that thermal extremes can result in death. The problem is determining whether the organism is simply responding to the effects of temperature at a general physiological level, or whether discrete thermoreceptor organs are also involved.

There is considerable circumstantial evidence that at least some invertebrates are capable of directly sensing differences in environmental temperatures, but actual receptor units are for the most part unidentified. A number of insects, some crustaceans, and the horseshoe crab (*Limulus*) apparently can sense thermal variation. The only nonarthropod invertebrates that have received much attention in this regard are certain leeches, which apparently are drawn to warm-blooded hosts by some heat-sensing mechanism. Other ectoparasites (e.g., ticks) of warm-blooded vertebrates may also be able to sense the "warmth of a nearby meal," but little work has been done on this subject.

Independent Effectors

Independent effectors are specialized sensory response structures that not only receive information from the environment but also elicit a response to the stimulus directly, without the intervention of the nervous system per se. In this sense, independent effectors are like closed circuits. As discussed in later chapters,

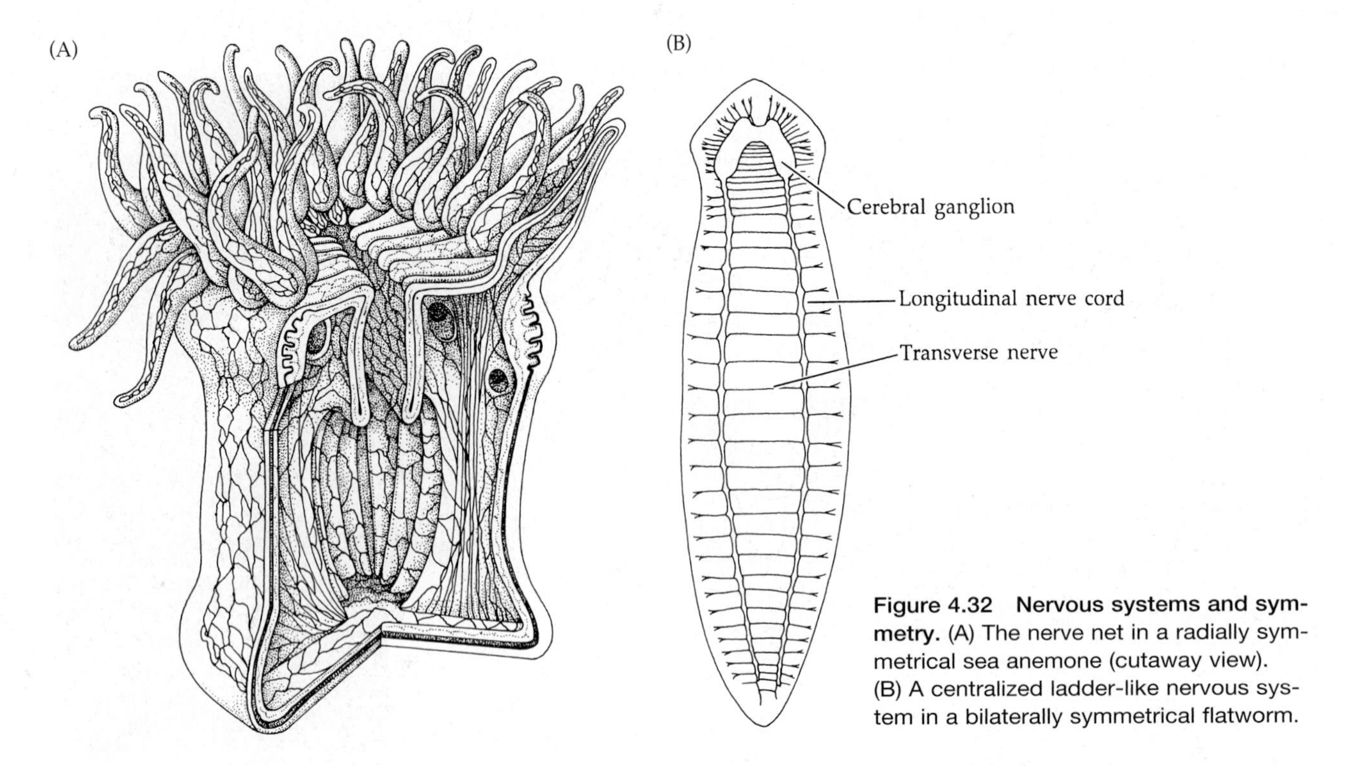

Figure 4.32 Nervous systems and symmetry. (A) The nerve net in a radially symmetrical sea anemone (cutaway view). (B) A centralized ladder-like nervous system in a bilaterally symmetrical flatworm.

the stinging capsules (nematocysts) of cnidarians and the adhesive cells of ctenophores are, at least under most circumstances, independent effectors.

Bioluminescence

Bioluminescence, the production of light by living creatures, occurs in a great variety of organisms in the sea and in some land animals, but it is, curiously, rare in freshwaters. On land, it is seen in a number of beetle groups, as well as some flies and springtails, centipedes and millipedes, a few earthworms, and at least one snail (as well as some fungi). In freshwater it has been reported from only a few insect larvae and a freshwater limpet. In the sea, bioluminescence plays a significant role in animal communication, and it has been recorded from various protists, cnidarians, ctenophores, chaetognaths, annelids, molluscs, numerous arthropods, echinoderms, hemichordates, tunicates, at least one nemertean, and, of course, numerous fishes (as well as bacteria and fungi). Most bioluminescence commonly seen in the sea is produced by dinoflagellates emitting rapid (one-tenth of a second) flashes. But the patient nighttime observer will also discover that flashes of light are produced by some species of medusae, ctenophores, copepods, benthic ostracods, brittle stars, sea pansies and sea pens, chaetopterid and syllid polychaetes, limpets, clams, tunicates, and others.

Luminescence is the emission of light without heat. It involves a special type of chemical reaction in which the energy, instead of being released as heat as occurs in most chemical reactions, is used to excite a product molecule that releases energy as a photon. In all cases, the reaction involves the oxidation of a substrate called **luciferin**, catalyzed by an enzyme called **luciferase**. The structures of these chemicals differ among taxa, but the reaction is similar. The color of light varies from deep blue (shrimp and dinoflagellates) to blue-green or green (certain millipedes, ostracods, and tunicates), to yellow and even red (fireflies). Bioluminescence serves several functions, including offense, defense, prey attraction, and intraspecific communication. In some cases, the luminescent organs of metazoans (particularly fishes) are not intrinsic but are symbiotic colonies of microorganisms.

Nervous Systems and Body Plans

The nervous system is always receiving information via its associated receptors, processing this information, and eliciting appropriate responses. We limit our discussion at this point to those conditions in which distinct systems of identifiable neurons exist, leaving the special situations in protists and sponges for later chapters.

The structure of the nervous system of any animal is related to its body plan and mode of life. Consider first a radially symmetrical animal with limited powers of locomotion, such as a planktonic jellyfish or a sessile sea anemone. In such animals the major receptor organs are more or less regularly (and radially) distributed around the body; the nervous system itself is a noncentralized, diffuse meshwork generally called a **nerve net** (Figure 4.32A). Radially symmetrical animals tend to be able to respond equally well to stimuli coming from any direction—a useful ability for creatures with either sessile or free-floating lifestyles. Interestingly, at least in cnidarians, there are both polarized and non-polarized synapses within the nerve net. Impulses can travel in either direction across the nonpolarized synapses because the neuronal processes on both sides are capable of releasing synaptic transmitter chemicals. This capability, coupled with the gridlike form of the nerve net, enables impulses to travel in all directions from a point of stimulation. From this brief description, it might be assumed that such a simple and "unorganized" nervous system would not provide enough integrated information to allow complex behaviors and coordination. In the absence of a structurally recognizable integrating center, the nerve net does not fit well with our earlier description of the sequence of events from stimulus to response. But many cnidarians are in fact capable of fairly intricate behavior, and the system works, often in ways that are not yet fully understood. In any case, symmetry, sense organ distribution, nervous system organization, and lifestyles are clearly correlated to one another.

The tremendous evolutionary success of bilateral symmetry and unidirectional locomotion must have depended in large part on associated changes in the organization of the nervous system and the distribution of sense organs. The evolutionary trend among animals has been to centralize and concentrate the major coordinating elements of the nervous system. This central nervous system is usually made up of an anteriorly located neuronal mass (ganglion) from which arise one or more longitudinal nerve cords that often bear additional ganglia (Figure 4.32B). The anterior ganglion is referred to by a variety of names. Many authors have abandoned the term brain for such an organ because of the multifaceted implications of that word and adopt the more neutral term **cerebral ganglion** (or cerebral ganglia) for the general case. In many instances, a term of its relative position to some other organ is applied. For example, the cerebral ganglion commonly lies dorsal to the anterior portion of the gut and is thus a supraenteric (or supraesophageal, or suprapharyngeal) ganglion.

In addition to the cerebral ganglion, most bilaterally symmetrical animals have many of the major sense organs placed anteriorly. The concentration of these organs at the front end of an animal is called **cephalization**—the formation of a head region. Even though cephalization may seem an obvious and predictable outcome of bilaterality and mobility, it is nonetheless extremely important. It simply would not do to have information about the environment gathered by the trailing end of a motile animal, lest it enter adverse and potentially dangerous conditions unawares. Hunting, tracking, and other forms of food location are greatly facilitated by having the appropriate receptors placed anteriorly—toward the direction of movement.

Longitudinal nerve cords receive information through peripheral sensory nerves from whatever sense organs are placed along the body, and they carry impulses from the cerebral ganglion to peripheral motor nerves to effector sites. Additionally, nerve cords and peripheral nerves often serve animals in reflex actions and in some highly coordinated activities that do not depend on the cerebral ganglion. The most primitive centralized nervous system may have been similar to that seen today in xenacoelomorph worms and some free-living flatworms, where pairs of longitudinal cords may attached to one another by transverse connectives in a ladder-like fashion (Figure 4.32B). Among those Metazoa that have developed active lifestyles (e.g., errant polychaetes, most arthropods, cephalopod molluscs, vertebrates), the nervous system has become increasingly centralized through a reduction in the number of longitudinal nerve cords. However, a number of invertebrates (e.g., ectoprocts, tunicates, echinoderms) have secondarily taken up sedentary or sessile modes of existence. Within these groups there has been a corresponding decentralization of the nervous system and a general reduction in and dispersal of sense organs.

Hormones and Pheromones

We have stressed the significance of the integrated nature of the parts and processes of living organisms and have discussed the general role of the nervous system in this regard. Organisms also produce and distribute within their bodies a variety of chemicals that regulate and coordinate biological activities. This very broad description of what may be called chemical coordinators obviously includes almost any substance that has some effect on bodily functions. One special category of chemical coordinators is the **hormones**. This term refers to any chemicals that are produced and secreted by some organ or tissue, and are then carried by the blood or other body fluid to exert their influence elsewhere in the body. In vertebrates, we associate this type of phenomenon with the endocrine system, which includes well known glands as production sites. For our purposes we may subdivide hormones into two types. First are **endocrine hormones**, which are produced by more or less isolated glands and released into the circulatory fluid. Second are **neurohormones**, which are produced by special neurons called neurosecretory cells.

Much remains unknown concerning hormones in invertebrates. Most of our information comes from studies on insects and crustaceans, although hormonal activity has been demonstrated in a few other taxa and is suspected in many others. Among the arthropods, hormones are involved in the control of growth, molting, reproduction, eye pigment migration, and probably other phenomena; in at least some other taxa (e.g., annelids), hormones influence growth, regeneration, and sexual maturation.

Hormones do not belong to any particular class of chemical compounds, nor do they all produce the same effects at their sites of action: some are excitatory, some are inhibitory. Because endocrine hormones are carried in the circulatory fluid, they reach all parts of an animal's body. The site of action, or target site, must be able to recognize the appropriate hormone(s) among the myriad other chemicals in its surroundings. This recognition usually involves an interaction between the hormone and the cell surface at the target site. Thus, under normal circumstances, even though a particular hormone is contacting many parts of the body, it will elicit activity only from the appropriate target organ or tissue that recognizes it.

In a general sense, pheromones are substances that act as "interorganismal hormones." These chemicals are produced by organisms and released into the environment, where they have an effect on other organisms. Most pheromone research has been on intraspecific actions, especially in insects, where activities such as mate attraction are frequently related to these airborne chemicals. We may view intraspecific pheromones as coordinating the activities of populations, just as hormones help coordinate the activities of individual organisms. There is also a great deal of evidence for the existence of interspecific pheromones. For example, some predatory species (e.g., some starfish) release chemicals into the water that elicit extraordinary behavioral responses on the part of potential prey species, generally in the form of escape behavior. We discuss examples of various pheromone phenomena for specific animal groups throughout the book.

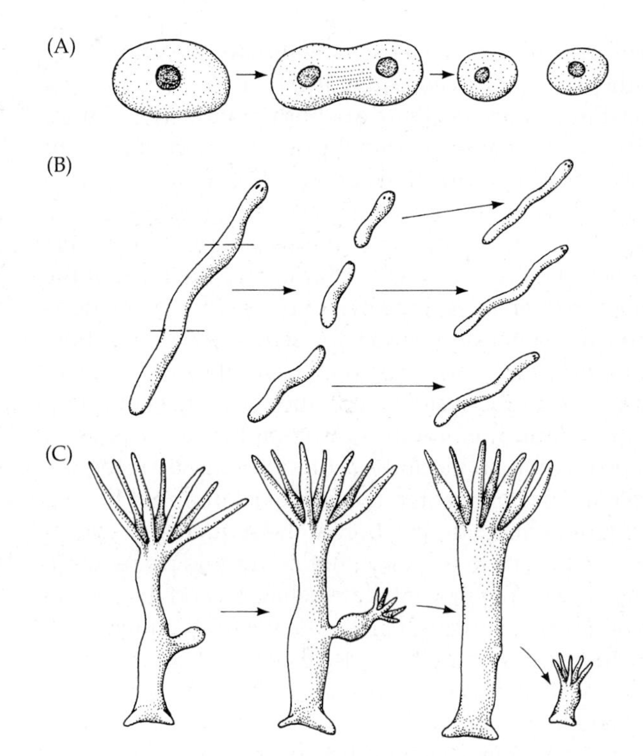

Figure 4.33 Some common asexual reproductive processes. (A) Simple mitotic binary fission; this process occurs in most protists. (B) Fragmentation, followed by regeneration of lost parts. This process occurs in a number of vermiform invertebrates. (C) Budding may produce separate solitary individuals, as it does in *Hydra* (shown here); or it may produce colonies (see Figure 4.34).

Reproduction

The biological success of any species depends upon its members staying alive long enough to reproduce themselves. The following account includes a discussion of the basic methods of reproduction among invertebrates and leads to the account of embryology and developmental strategies provided in Chapter 5.

Asexual Reproduction

Asexual reproductive processes do not involve the production and subsequent fusion of haploid cells, but rely solely on vegetative growth through mitosis. Cell division itself is a common form of asexual reproduction among the protists, and many invertebrates engage in various types of body fission, budding, or fragmentation, followed by growth to new individuals (Figure 4.33). These asexual processes depend largely on the organism's "reproductive exploitation" of its ability to regenerate (regrow lost parts). Even wound healing is a form of regeneration, but many animals have far more dramatic capabilities. The replacement of a lost appendage in familiar animals such as starfish and crabs is a common example of regeneration. However, these regenerative abilities are not "reproduction" because no new individuals result, and their presence does not imply that an animal capable of replacing a lost leg can necessarily reproduce asexually. Examples of organisms that possess regenerative abilities of a magnitude permitting asexual reproduction include protists, sponges, many cnidarians (corals, anemones, and hydroids), many colonial animals, and certain types of worms.

In many cases asexual reproduction is a relatively incidental process and is rather insignificant to a species' overall survival strategy. In others, however, it is an integral and even necessary step in the life cycle. There are important evolutionary and adaptive aspects to asexual reproduction. Organisms capable of rapid asexual reproduction can quickly take advantage of favorable environmental conditions by exploiting temporarily abundant food supplies, newly available living space, or other resources. This competitive edge is frequently evidenced by extremely high numbers of asexually produced individuals in disturbed or unique habitats, or in other unusual conditions. In addition, asexual processes are often employed in the production

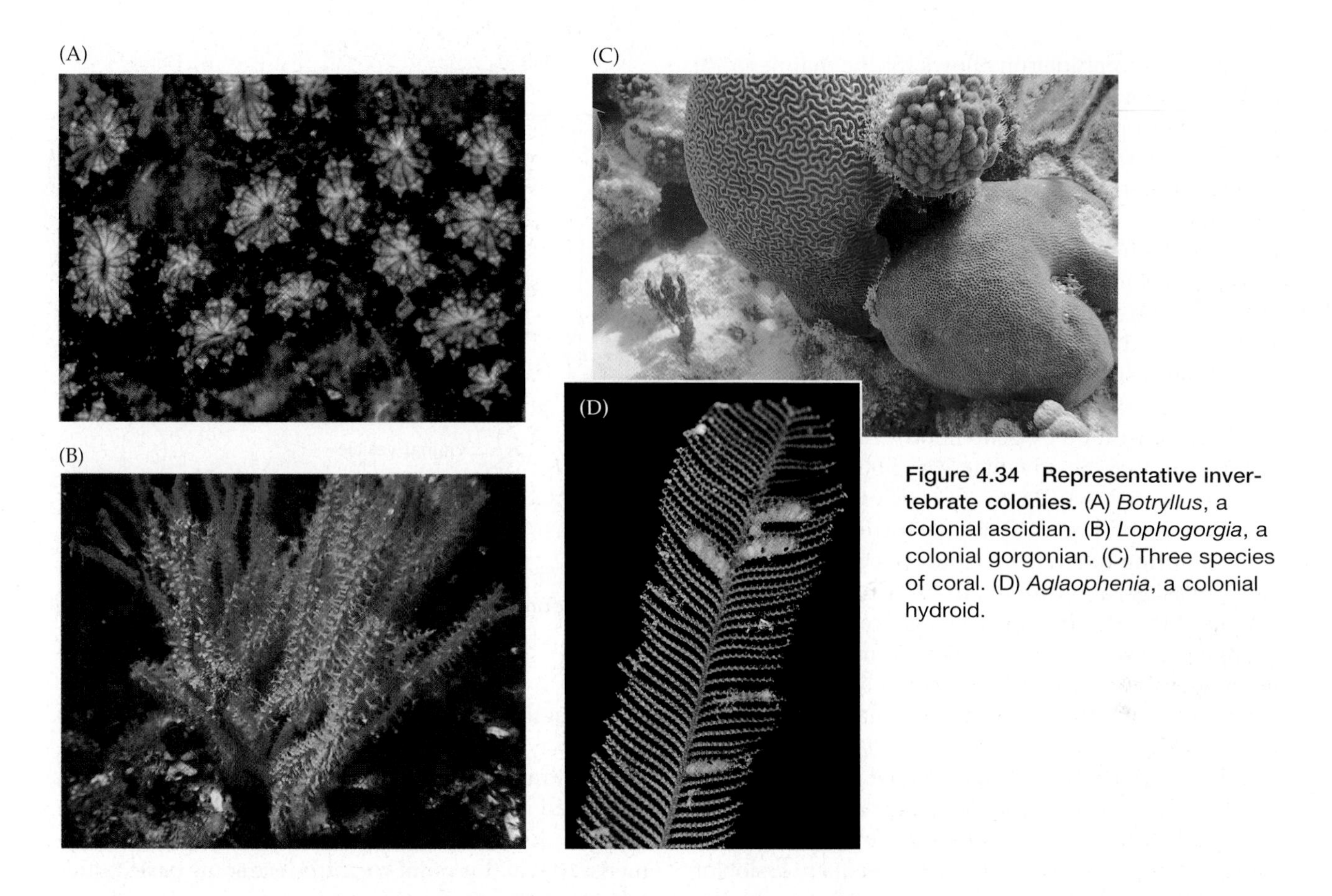

Figure 4.34 Representative invertebrate colonies. (A) *Botryllus*, a colonial ascidian. (B) *Lophogorgia*, a colonial gorgonian. (C) Three species of coral. (D) *Aglaophenia*, a colonial hydroid.

of resistant cysts or overwintering bodies, which are capable of surviving through periods of harsh environmental conditions. When favorable conditions return, these structures grow to new individuals.

A word about colonies A frequent result of asexual reproduction, particularly some forms of budding, is the formation of **colonies**. This phenomenon is especially common in certain taxa (e.g., cnidarians, ascidians, bryozoans) (Figure 4.34). The term colony is not easy to define. It may initially bring to mind ant or bee colonies, or even groups of humans; but these examples are more appropriately viewed as social units rather than as colonies, at least in the context of our discussions. For our purposes, we define colonies as associations in which the constituent individuals are not completely separated from each other, but are organically connected together, either by living extensions of their bodies, or by material that they have secreted. While occasional mixed colonies may exist (usually due to fusion of two separate colonies, as in some tunicates), most colonies are composed of genetically identical individuals. We describe the nature of numerous examples of colonial life in later chapters.

The formation of colonies not only may enhance the benefits of asexual reproduction in general, but also produces overall functional units that are much greater in size than mere individuals; thus this growth habit may be viewed as a partial solution to the surface-to-volume dilemma. Increased functional size through colonialism can result in a number of advantages for animals—it can increase feeding efficiency, facilitate the handling of larger food items, reduce chances of predation, increase the competitive edge for food, space, and other resources, and allow groups of individuals within the colony to specialize for different functions.

Sexual Reproduction

Although reproduction is critical to a species' survival, it is the one major physiological activity that is not essential to an individual organism's survival. In fact, when animals are stressed, reproduction is usually the first activity that ceases. Sexual reproduction is especially energy costly, yet it is the characteristic mode of reproduction among multicellular organisms.[6]

Given the advantages of asexual reproduction, one might wonder why all animals do not employ it and abandon sexual activities entirely. The most frequently given explanation for the popularity of sexual reproduction (aside from anthropomorphic views, of course) focuses on the long-term benefits of genetic

[6]In thinking of animals, we typically view "sex" and "reproduction" as one in the same. However, at the cellular level these two processes are opposites: reproduction is the division of one cell to form two, whereas the sexual process incudes two cells fusing to form one!

variation. Recombination allows for the maintenance of high genetic heterozygosity in individuals and high polymorphism in populations. Through regular meiosis and recombination, a level of genetic variation is maintained generation after generation, within and among populations; thus species are thought to be more "genetically prepared" for environmental changes, including both shifts in the physical environment and the changing milieu of competitors, predators, prey, and parasites.

Although this advantage must surely be real, does it satisfactorily explain the role of sex in short-term selection (i.e., generation by generation)? Presumably even in the short term an advantage lies in the maintenance of genetic variability. That is, genetic variability in both individuals and populations may increase their chances of adapting to environmental fluctuations, predators, parasites, and disease. In 1973, Leigh Van Valen proposed the idea that in order just to "keep up" with changing environments, populations must continually access new and different gene combinations through the process of natural selection—a notion called the "red queen hypothesis" after the Red Queen in Alice in Wonderland, who commanded her courtiers to run continuously just to stay in the same place.

Sexual reproduction involves the formation of haploid cells through meiosis and the subsequent fusion of pairs of those cells to produce a diploid zygote (Figure 4.35). The haploid cells are **gametes**—sperm and eggs—and their fusion is the process of fertilization, or **syngamy**. (Exceptions to these general terms and processes are common among protists as discussed in Chapter 3.) The production of gametes is accomplished by the gonads—**ovaries** in females and **testes** in males—or their functional equivalents. The gonads are frequently associated with reproductive systems that may include various arrangements of ducts and tubes, accessory organs such as yolk glands or shell glands,

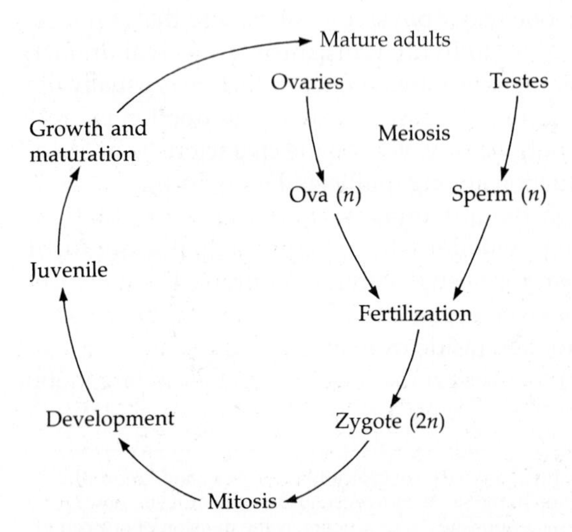

Figure 4.35 A generalized metazoan life cycle.

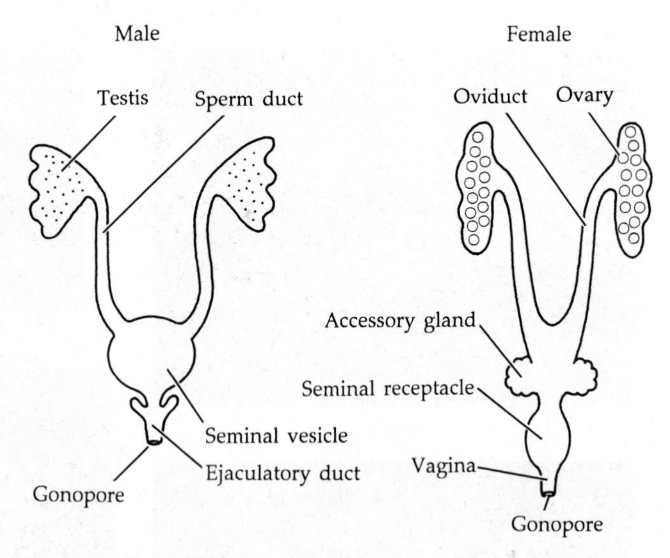

Figure 4.36 Schematic and generalized male and female reproductive systems. See text for explanation.

and structures for copulation. The different levels of complexity of these systems are related to the developmental strategies used by the organisms in question, as discussed in Chapter 5 and described in the coverage of each phylum. The variation in such matters is immense, but at this point we introduce some basic terminology of structure and function.

Many invertebrates simply release their gametes into the water in which they live (**broadcast spawning**), where external fertilization occurs. In such animals the gonads are usually simple, often transiently occurring structures associated with some means of getting the eggs and sperm out of the body. This release is accomplished through a discrete plumbing arrangement (coelomoducts, metanephridia, or gonoducts—sperm ducts and oviducts), or by temporary pores in or rupture of the body wall. In such animals, synchronous spawning is critical, and marine species rely largely on this synchrony and the water currents to achieve fertilization. Water temperature, light, phytoplankton abundance, lunar cycle, and the presence of conspecifics have all been implicated in synchronized spawning events of invertebrates.

On the other hand, invertebrates that pass sperm directly from the male to the female, where fertilization occurs internally, must have structural features to facilitate such activities. Figure 4.36 illustrates stylized male and female reproductive systems. A general scenario leading to internal fertilization in such systems is as follows. Sperm are produced in the testes and transported via the **sperm duct** to a precopulatory storage area called the **seminal vesicle**. Prior to mating, many invertebrates incorporate groups of sperm cells into sperm packets, or **spermatophores**. Spermatophores provide a protective casing for the sperm and facilitate transfer with minimal sperm loss. In addition, many spermatophores are themselves motile, acting as independent sperm carriers. Some sort of male copulatory or intromittent organ (e.g., penis, cirrus, gonopod) is inserted through the female's gonopore and into the vagina. Sperm are passed through the male's ejaculatory duct directly, or by way of a copulatory organ, into the female system, where they are received and often stored by a **seminal receptacle**.

In the female, eggs are produced in the ovaries and transported into the region of the **oviducts**. Sperm eventually travel into the female's reproductive tract, where they encounter the eggs; fertilization often takes place in the oviducts. Among invertebrates, the sperm may move by flagellar or ameboid action, or by locomotor structures on the spermatophore packet; they may be aided by ciliary action of the lining of the female reproductive tract. Various accessory glands may be present both in males (such as those that produce spermatophores or seminal fluids) and in females (such as those that produce yolk, egg capsules, or shells). This simple sequence is typical (although with many elaborations) of most invertebrates that rely on internal fertilization.

Animals in which the sexes are separate, each individual being either male or female, are termed **gonochoristic**, or dioecious. However, many invertebrates are hermaphroditic, or monoecious: each animal contains both ovaries and testes and thus is capable of producing both eggs and sperm (though not necessarily at the same time).[7] Although self-fertilization may seem to be a natural advantage in this condition, such is not the case. In fact, with some exceptions, self-fertilization in hermaphrodites is usually prevented. Fertilizing one's self would be the ultimate form of inbreeding and would presumably result in a dramatic decrease in potential genetic variation and heterozygosity. The rule for many hermaphroditic invertebrates is mutual cross fertilization, wherein two individuals function alternately or simultaneously as males and exchange sperm, and then use the mate's sperm to fertilize their own eggs. The real advantage of hermaphroditism now becomes clear: a single sexual encounter results in the impregnation of two individuals, rather than only one as in the gonochoristic condition.

A common phenomenon among hermaphroditic invertebrates is **protandric hermaphroditism**, or simply **protandry** (Greek *proto*, "first"; *andro*, "male"), where an individual is first a functional male, but later in life changes sex to become a functional female. The less common reverse situation, female first and then male, is called **protogynic hermaphroditism**, or simply **protogyny** (Greek *gynos*, "female"). At least some invertebrates alternate regularly between being functional males and females, as explained by Jerome Tichenor (*Poems in Contempt of Progress*, 1974):

> *Consider the case of the oyster,*
> *Which passes its time in the moisture;*
> *Of sex alternate,*
> *It chases no mate,*
> *But lives in self-contained cloister.*

In addition to the clever oysters immortalized by Professor Tichenor, some other taxa in which the hermaphroditic condition is common include barnacles, arrow worms (Chaetognatha), flatworms, clitellate annelids, tunicates, many advanced gastropods, and cymothoid isopods (Crustacea).

The sexual conditions of colonial animals include myriad variations on the themes described above. Colonies may include only one sex, both sexes, or the individuals may be hermaphroditic.

Parthenogenesis

Parthenogenesis (Greek *partheno*, "virgin"; *genesis*, "birth") is a special reproductive strategy in which unfertilized eggs develop into viable adult individuals. Parthenogenetic species are known in many invertebrate (and vertebrate and plant) groups, including gastrotrichs, rotifers, tardigrades, nematodes, gastropods, certain insects, and various crustaceans. The taxonomic distribution of parthenogenesis is spotty; it is rare to find a whole genus, let alone any higher taxon, that is wholly parthenogenetic. Some higher taxa that are largely parthenogenetic (e.g., aphids, cladocerans) are cyclically parthenogenetic, and they punctuate their life histories with sex.[8] There are also a number of protist higher taxa for which sex has yet to be described. Among the invertebrates, parthenogenesis usually occurs in small-bodied species that are parasites, or are free-living but inhabit extreme or highly variable habitats such as temporary freshwater ponds. There is a general trend for parthenogenesis to become more prevalent as one moves toward higher latitudes or into harsher environments. Overall, it appears that parthenogenetic taxa arise from time to time and succeed in the short run due to certain immediate advantages, but in the long run they might be condemned to extinction through competition with their sexual relatives.

---

[7]Hermaphoditus, the beautiful son of Hermes and Aphrodite, was united with a water nymph at the Carian fountain. Thus his body became both male and female.

[8]A number of fishes and amphibians are parthenogenetic, but none seems to have overcome the need for their egg to be penetrated by a sperm in order to initiate development. The parthenogenetic females usually mate with a male of another species, providing sperm that trigger development (a behavior called **pseudogamy**). A few lizards apparently have no need for a sperm to trigger parthenogenetic development. No parthenogenetic wild birds or mammals have been documented.

In most species that have been studied, parthenogenetic periods alternate with periods of sexual reproduction. In temperate freshwater habitats, parthenogenesis often occurs during summer months, with the population switching to sexual reproduction as winter approaches. In some species, parthenogenesis takes place for many generations, or several years, eventually to be punctuated by a brief period of sexual reproduction. In some rotifers, parthenogenesis predominates until the population attains a certain critical size, at which time males appear and a period of sexual reproduction ensues. Cladocerans switch from parthenogenesis to sexual reproduction under a number of conditions, such as overcrowding, adverse temperature, food scarcity, or even when the nature of the food changes. Many parasitic species alternate between a free-living sexual stage and a parasitic parthenogenetic one; this arrangement is seen in some nematodes, thrips (Thysanoptera), gall wasps, aphids, and certain other hemipterans.

One of the most interesting examples of parthenogenesis occurs in honeybees; in these animals the queen is fertilized by one or more males (drones) at only one period of her lifetime, in her "nuptial flight." The sperm are stored in her seminal receptacles. If sperm are released when the queen lays eggs, fertilization occurs and the eggs develop into females (queens or workers). If the eggs are not fertilized, they develop parthenogenetically into males (drones).

The question of the existence or prevalence of purely parthenogenetic species has been debated for decades. Many species once thought to be entirely parthenogenetic have proved, upon closer inspection, to alternate between parthenogenesis and brief periods of sexual reproduction. In some species purely parthenogenetic populations apparently exist only in some localities. In other species, parthenogenetic lineages have been traced to sexual ancestral populations occupying relictual habitats. Nevertheless, for some parthenogenetic animals, males have yet to be found in any population, and these may indeed be purely clonal species. One cannot help but wonder how long such species can exist in the face of natural selection without the benefits of any genetic exchange. One would predict that, as with any form of asexual reproduction, obligatory parthenogenesis would eventually lead to genetic stagnation and extinction. There may, however, be some as yet unexplained genetic mechanisms to avoid this, because some parthenogenetic animals (e.g., some earthworms, insects, and lizards) are capable of inhabiting a wide range of habitats. Presumably they either have a significant level of genetic adaptability or possess "general purpose genotypes."

Selected References

Adiyodi, K. G. and R. G. Adiyodi (eds.). 1983–2002. *Reproductive Biology of Invertebrates. Vols. 1–11*. Wiley, New York. [Detailed examinations of all aspects of invertebrate reproduction.]

Autrum, H., R. Jung, W. R. Loewenstein, D. M. Mackay and H. L. Teuber (eds.). 1972–1981. *Handbook of Sensory Physiology. Vols. 1–7*. Springer-Verlag, New York. [This multivolume work includes the efforts of over 400 authors.]

Bartolomaeus, T. and P. Ax. 1992. Protonephridia and metanephridia—their relation within the Bilateria. Z. Zool. syst. Evolut.-forsch. 30: 21–45.

Bejan, A. and J. H. Marden. 2006. Constructing animal locomotion from new thermodynamics theory. Am. Sci. 94: 342–349.

Biewener, A. A. 2003. *Animal Locomotion*. Oxford Univ. Press, Oxford.

Cronin, T. W. 1986. Photoreception in marine invertebrates. Am. Zool. 26: 403–415.

Denny, M. 1996. *Air and Water: The Biology and Physics of Life's Media*. Princeton University Press, Ewing, NJ. [A great read.]

DeSimone, D. W. and A. R. Horwitz. 2014. Many modes of motility. Science 345: 1002–1003.

Emlet, R. B. and R. R. Strathmann. 1985. Gravity, drag, and feeding currents of small zooplankton. Science 228: 1016–1017.

Fauchald, K. and P. A. Jumars. 1979. The diet of worms: A study of polychaete feeding guilds. Oceanogr. Mar. Biol. Annu. Rev. 17: 193–284.

Fernald, R. D. 2006. Casting a genetic light on the evolution of eyes. Science 313: 1914–1918.

Fretter, V. and A. Graham. 1976. *A Functional Anatomy of Invertebrates*. Academic Press, New York. [Dated, but still a benchmark reference for the serious student.]

Giese, A. C. (ed.). 1964–1973. *Photophysiology, Vols. 1–8*. Academic Press, New York.

Giese, A. C., J. S. Pearse and V. B. Pearse. 1974–1991. *Reproduction of Marine Invertebrates. Vols. 1-9*. Academic Press, New York, and Boxwood Press, Pacific Grove, CA.

Goodrich, E. S. 1945. The study of nephridia and genital ducts since 1895. Q. J. Micros. Sci. 86: 113–392. [Goodrich's classic paper on the evolution of coelomoducts, gonoducts, and nephridia.]

Gould, S. J. 1977. *Ontogeny and Phylogeny*. The Belknap Press of Harvard University Press, Cambridge, MA. [A classic.]

Gould, S. J. 1980. The evolutionary biology of constraint. Daedalus 109: 39–52.

Gould, S. J. 1992. Constraint and the square snail: Life at the limits of a covariance set. The normal teratology of *Cerion disforme*. Biol. J. Linnean Soc. 47: 4077–437.

Greenspan, R. J. 2007. *An Introduction to Nervous Systems*. Cold Spring Harbor Laboratory Press, Cold Spring Harbor, N.Y.

Guerra, P. A., R. J. Gegear and S. M. Reppert. 2014. A magnetic compass aids monarch butterfly migration. Nat. Comm. 5. doi: 10.1038/ncomms5164

Haddock, S. H. D., M. A. Moline and J. F. Case. 2010. Bioluminescence in the sea. Ann. Rev. Mar. Sci. 2: 443–493.

Hall, B. K. 1996. Baupläne, phylotypic stages, and constraint. Why are there so few types of animals? Pp. 215–261 in M. K. Hecht et al. (eds.), *Evolutionary Biology, Vol. 29*. Plenum, New York.

Harrison, F. W. (ed.). 1991–1997. *Microscopic Anatomy of Invertebrates*. Wiley-Liss, New York.

Hickman, C. S. 1988. Analysis of form and function in fossils. Amer. Zool. 28: 775–793.

Hill, R. W., G. A. Wyse and M. Anderson. 2012. *Animal Physiology*, 3nd Ed. Sinauer Associates, Sunderland.

Hughes, R. N. (ed.). 1993. Diet Selection: An Interdisciplinary Approach to Foraging Behaviour. Blackwell Science, Boston.

Hyman, L. H. 1940, 1951. *The Invertebrates. Vol. 1, Protozoa through Ctenophora; Vol. 2, Platyhelminthes and Rhynchocoela: The Acoelomate Bilateria.* McGraw-Hill, New York. [These two volumes of Hyman's series are especially useful in their discussion of body architecture in "lower Metazoa."]

Jackson, J. B. C., L. W. Buss and R. E. Cook (eds.). 1985. *Population Biology and Evolution of Clonal Organisms.* Yale University Press, New Haven, CT.

Jeffrey, D. J. and J. D. Sherwood. 1980. Streamline patterns and eddies in low Reynolds number flow. J. Fluid Mech. 96: 315–334.

Jorgensen, C. B., T. Kiorboe, J. Mohlenberg and H. U. Riisgard. 1984. Ciliary and mucus-net filter feeding, with special reference to fluid mechanical characteristics. Mar. Ecol. Prog. Ser. 15: 283–292.

Keegan, B. F., P. O. Ceidigh and P. J. S. Boaden (eds.). 1971. *Biology of Benthic Organisms.* Pergamon Press, New York.

Koehl, M. A. R. 1984. How do benthic organisms withstand moving water? Am. Zool. 24: 57–70.

Koehl, M. A. R. and J. R. Strickler. 1981. Copepod feeding currents: Food capture at low Reynolds number. Limnol. Oceanogr. 26: 1062–1093.

LaBarbera, M. 1984. Feeding and particle capture mechanisms in suspension feeding animals. Am. Zool. 24: 71–84.

LaBarbera, M. and S. Vogel. 1982. The design of fluid transport systems in organisms. Am. Sci. 70: 54–60.

Lockwood, G. and B. R. Rosen (eds.). 1979. *Biology and Systematics of Colonial Animals,* Special Vol. No. 11. Academic Press, New York. Published for The Systematics Association.

Lowenstam, H.A. 1981. Minerals formed by organisms. Science 211: 1126–1131.

McMahon, B. R. and L. E. Burnett. 1990. The crustacean open circulatory system. A reexamination. Physiol. Zool. 63: 35–71.

Maynard Smith, J. 1978. *The Evolution of Sex.* Cambridge University Press, Cambridge.

Morin, J. G. 1986. "Firefleas" of the sea: Luminescent signaling in marine ostracode crustaceans. Fla. Entomol. 69: 105–121.

Morse, A. N. C. 1991. How do planktonic larvae know where to settle? Am. Sci. 79: 154–167.

Nelson, R. J. 2011. *An Introduction to Behavioral Endocrinology.* 4th Ed. Sinauer Associates, Sunderland.

Nicol, J. A. C. 1960. *The Biology of Marine Animals.* Putnam, New York. [One of the best summaries of ecological physiology ever written; a shame it was never revised.]

North, G. and R. J. Greenspan (eds.). 2007. *Invertebrate Neurobiology.* Cold Spring Harbor Laboratory Press, Cold Spring Harbor, N.Y.

Polis, G. A. 1981. The evolution and dynamics of intraspecific predation. Ann. Rev. Ecol. Syst. 12: 225–251.

Prosser, C. L. (ed.). 1991. *Environmental and Metabolic Animal Physiology* (the 4th edition of Prosser's classic *Comparative Animal Physiology* text). John Wiley and Sons, New York.

Ratcliffe, N. A. and A. F. Rowley (eds.). 1981. *Invertebrate Blood Cells. Vols. 1–2.* Academic Press, New York. [For a good introduction to the literature on respiratory pigments, see Am. Zool. 1980, 20(1), "Respiratory Pigments."]

Ruppert, E. E. and P. R. Smith. 1988. The fundamental organization of filtration nephridia. Biol. Rev. 6: 231–258.

Rundell, R. J. and B. S. Leander. 2010. Masters of miniaturization: Convergent evolution among interstitial eukaryotes. Bioessays 32: 430-437.

Sandeman, D. C. and H. L. Atwood (eds.). 1982. *The Biology of Crustacea. Vol. 4, Neural Integration and Behavior.* Academic Press, New York.

Schmidt-Nielsen, K. 1984. *Scaling. Why is Animal Size so Important.* Cambridge University Press, New York.

Schmidt-Nielsen, K. 1997. *Animal Physiology. Adaptation and Environment,* 5th Ed. Cambridge University Press, New York.

Schmidt-Nielsen, K., L. Bolis, and S. H. P. Maddrell (eds.). 1978. *Comparative Physiology: Water, Ions and Fluid Mechanics.* Cambridge University Press, New York.

Schmidt-Rhaesa, A. 2007. *The Evolution of Organ Systems.* Oxford University Press, Oxford.

Sellers, J. R. and B. Kachar. 1990. Polarity and velocity of sliding filaments: Control of direction by actin and of speed by myosin. Science 249: 406–408.

Simmons, P. and D. Young. 2010. *Nerve Cells and Animal Behaviour.* Cambridge Univ. Press, Cambridge.

Stossel, T. P. 1990. How cells crawl. Am. Sci. 78: 408–423.

Strathmann, R. R. 1978. The evolution and loss of feeding larval stages of marine invertebrates. Evolution 32: 894–906.

Tennekes, H. 1996. The Simple Science of Flight: From Insects to Jumbo Jets. M.I.T Press, Cambridge, MA.

Thompson, D'Arcy. 1942. *On Growth and Form, Rev. Ed.* Macmillan, New York. [A delightful read.]

Vincent, J. 2012. *Structural Biomaterials,* 3rd Ed. Princeton Univ. Press, Princeton.

Vogel, S. 1981. Life in Moving Fluids: The Physical Biology of Flow, 2nd Ed. Princeton University Press, Princeton.

Vogel, S. 1988. How organisms use flow-induced pressures. Amer. Sci. 76: 28–34.

Vogel, S. 1988. *Life's Devices. The Physical World of Animals and Plants.* Princeton University Press, Princeton.

Vogel, S. 1998. *Cats' Paws and Catapults: Mechanical Worlds of Nature and People.* W. W. Norton & Co., New York.

Vogel, S. 2002. Prime Mover. *A Natural History of Muscle.* W. W. Norton & Co., New York.

Vogel, S. 2003 *Comparative Biomechanics. Life's Physical World.* Princeton Univ. Press, Princeton.

Walsh, P. and P. Wright (eds.). 1995. *Nitrogen Metabolism and Excretion.* CRC Press, Boca Raton, FL.

Watson, S.-A., L. S. Peck, P. A. Tyler, P. C. Southgate, K S. Tan, R. W. Day and S. A. Morley. 2012. Marine invertebrate skeleton size varies with latitude, temperature and carbonate saturation: Implications for global change and ocean acidification. Global Change Biology 18(10): 3026–3038.

Widder, E. A. 2010. Bioluminescence in the ocean: Origins of biological, chemical, and ecological diversity Science 328: 704–708.

Wigglesworth, V. B. 1984. *Insect Physiology,* 7th Ed. Chapman and Hall, London.

Wilbur, K. M. and C. M. Yonge (eds.). 1964, 1967. *Physiology of Mollusca, Vols. 1 and 2.* Academic Press, New York.

Wray, G. A. and R. R. Strathmann. 2002. Stasis, change, and functional constraint in the evolution of animal body plans, whatever they may be. Vie Milieu 52: 189–199.

Yeates, D. K. 1995. Groundplans and exemplars: Paths to the tree of life. Cladistics 11: 343–357.

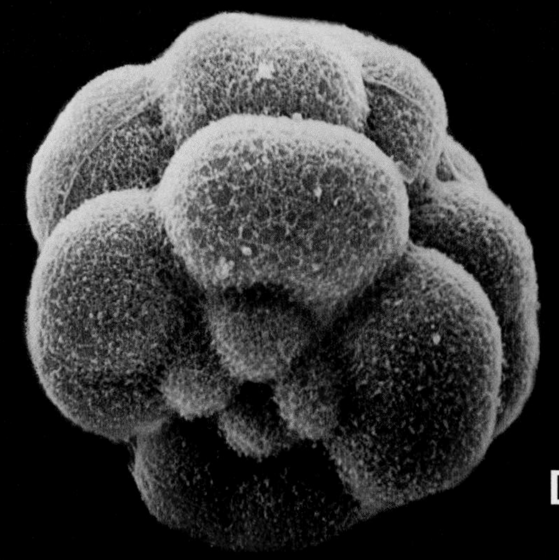

CHAPTER 5

Introduction to the Animal Kingdom

Development, Life Histories, and Origin

The process by which unicellular zygotes transform themselves into multicellular individuals and eventually into reproducing adults, is called **ontogeny**. At the core of metazoan ontogeny is **embryogenesis**—the growth and development of the embryo. The embryo is that stage of an organism between fertilization and birth. It is the principal phase of life history that mediates between genotype and phenotype.

As we saw in Chapter 4, the cells of animals are organized into functional units, generally as tissues and organs, with specific roles that support the whole animal. These different cell types are interdependent and their activities are coordinated in predictable patterns and relationships. The tissues and organs develop through a series of events early in an organism's embryogeny. The embryonic tissues, or **germ layers**, form the framework upon which metazoan body plans are constructed. Thus, the cells of animals (Metazoa) are specialized, interdependent, coordinated in function, and develop through a genetically-orchestrated tissue layering process during embryogeny—a combination of features that is absent among the protists.

The influence of yolk on early zygotic divisions, as well as on the differentiation of embryonic tissues, is one of the most well-established principles of zoology. So it is not surprising that a large body of detailed descriptive research on animal embryos has accumulated, leading to a strong focus on structure and physical processes in the field of developmental biology. Because particular developmental schemes characterize different animal lineages, developmental biology has compiled detailed descriptions of cell divisions, cell fates, embryonic germ layers, and larval or adult structures in the search for shared evolutionary patterns among animal phyla. Among the most detailed studies in biology are the meticulous accounts of early metazoan development by Wilhelm Roux, Hans Driesch, Edmund Beecher Wilson, Hans Spemann, and others. The apparent tendency for progressive stages of embryonic development to reveal the evolutionary history of animals is what inspired Ernst Haeckel, Karl von Baer, Walter Garstang, and others to formulate their versions of the **recapitulation hypothesis**, or "**biogenetic law**" (see the end of this chapter). And, despite some ongoing philosophical debate, from a morphological perspective ontogeny does appear to recapitulate many aspects of animal phylogeny.

However, the recent emergence of *molecular* developmental biology has changed some long-standing embryological principles in subtle and compelling ways. Haeckel's biogenetic law is no longer seen to govern the development of metazoan embryos in the ways embryologists once thought that it did. Taxon-specific patterns, and deviations of those patterns in animal development, have revealed more complicated interactions than could ever have been anticipated from morphological descriptions alone. For these reasons, developmental biology can no longer consist primarily of descriptions of ova, early cell division, the formation of increasingly complex balls of cells and cell layers, although these classic patterns are still a major part of this field. A modern approach to the study of animal development includes molecular as well as morphological studies.

Evolutionary Developmental Biology—EvoDevo

Molecular genetics has had an explosive impact on the study of developmental biology and animal phylogenetics and evolution. "**EvoDevo**," the catchy abbreviation for the field of evolutionary developmental biology, is used for the burgeoning field of comparative developmental genetics—the evolutionary study of the spatial and temporal expression of genes that control body architecture among metazoans. Much of EvoDevo focuses on the roles of **transcription factors** (**TFs**), which are the products of the genes involved in controlling early development. These genes and TFs define the primary embryonic body axes (e.g., the anterior–posterior and dorsal–ventral axes), the directionality of structures developing along these axes, and the appearance of particular structures or organizations such as body cavities, segments, or appendages. EvoDevo investigates the genetic underpinnings of nearly every process described within classical studies of developmental biology. The focus of developmental biology has thus shifted, from the ontogenetic study of animal structural organization, to the comparative ontogenetic study of gene functions and their roles in body organization.

As genomic methods become less expensive and more widely explored, the number of animal species whose full or partial genomic sequences are known is dramatically increasing. Each year, new "model organisms" in the genetic sense are identified, and as the number of species considered within these frameworks grows, opportunities for evolutionary comparison of the different patterns and mechanisms become increasingly possible. The use of developmental genomics has therefore become more than the mere determination of whether ontogeny recapitulates phylogeny, an idea that is now seen to be an oversimplification. Instead, the comparison of developmental similarities and differences, as well as how the differences are played out, has led to a much more subtle and discerning understanding of how animals are organized and how that organization unfolds within individual lifespans. Although the breadth of these changes is beyond the scope of this book, some general patterns are now apparent and we summarize these briefly below.

Developmental Tool Kits

If animals arose from among the Protista, a basic set of genetic factors are likely to have existed in that ancestry that supported multicellularity (which arose several times in the Eukaryota), facilitated its expression, and allowed these traits to be favored by selection. Likely genetic factors may have included a tendency for cells to aggregate rather than disperse after mitosis, the biochemical ability for cells to remain together as a group, and the movement or differentiation of cells bearing one set of specializations in response to proximity with cells bearing another set of specializations.

The possibility that such traits existed implies they were mediated in some way by the transcription products of genes borne by the protistan ancestors of animals. Sets of functional genes that control ontogenetic processes have been called **developmental tool kits**, and with respect to the evolution of multicellularity, they are likely to have been associated with three processes: (1) the adhesion of cells to one another, (2) the transduction of biochemical signals within and between cell types (i.e., **cell signaling pathways**), and (3) differentiation of cells from primordial to specialized states.

The transition to multicellularity from a unicellular state was a fundamental event in metazoan evolution. However, until recently, there were few avenues available for investigating this transition. The possibility that developmental tool kits underlying multicellularity existed at the origin of the Metazoa suggests that recognizable homologous gene sequences in different species, or **orthologs**, and their transcription factors might be found in predictable ways within the ancestors of metazoans, as well as throughout the metazoans themselves.

The structural organization of choanoflagellates (see Chapter 3), as well as their similarity to metazoans such as sponges, have made these creatures compelling candidates as metazoan ancestors, and molecular phylogenetic methods have allowed this hypothesis to be directly tested. Many analyses of nuclear and mitochondrial genes from diverse metazoan lineages strongly support the idea of animals and choanoflagellates having a common ancestor, and as predicted, a majority of choanoflagellate-expressed genes have homologous sequences (orthologs) in animal genotypes. Noteworthy too are observations that choanoflagellates express a surprising diversity of adhesion and cell signaling homologues. Thus, by identifying genes shared among animals and their apparent most recent common ancestor, tests of specific developmental hypotheses regarding genome evolution in early animals have become increasingly possible. However, things aren't always as easy as they seem.

The Relationship Between Genotype and Phenotype

A search for genetically conserved tool kits underlying particular events or morphological landmarks in animal development seems exciting indeed. Unfortunately, evidence now suggests that a direct correspondence between genotype and phenotype in metazoan development is rare—things are usually more complicated. Early studies were lured in this direction by the striking degree to which **Hox clusters** (groups of homeotic genes that control the body plan and limb organization of developing embryos along their anterior–posterior axis) seemed to exhibit organizational and functional consistency throughout the Metazoa. The discovery in the late 1970s that animals as diverse as vertebrates and arthropods had structurally and functionally related Hox genes led to the assumption that all animals would contain similar gene clusters. And some gene sequences, including Hox clusters and other similar genes, confirmed this assumption. The developmental gene *Brachyury* (*bra*) is a good example. The transcription products of *bra* define the midline of bilaterians, and the gene also is expressed in the notochord of species belonging to the phylum Chordata. Its expression pattern confirms the homologies suggested by morphology, in that its expression in the notochord of chordates provides an example of a homologous gene with a homologous function in a homologous morphological character—the notochord (which is a synapomorphy of the phylum).

The use of gene sequences and gene products for phylogenetic analysis has proven to be more complicated than expected. For example, Hox genes that are apparently homologous in their sequences may not be identical or even necessarily similar in their expression in different taxa, even among closely related taxa. Gene sequence homology, it seems, does not guarantee that the specific functions of the genes will be similar. And, Hox genes have been found to be linearly arranged only in certain taxa, despite their linear representation in most review papers and textbooks (Figure 5.1).

Despite their presumed role as an evolutionary "magic bullet," developmental genes now seem every bit as likely as morphological characters are to exhibit homoplasy (e.g., convergent evolution). While developmental genetic programs may at first appear to contain phylogenetic information, observed patterns are likely to represent patterns shaped by the needs of the developmental program in different lineages. Because such difficulties are widespread, using particular genes to establish homology in different metazoan lineages can be challenging.

The Evolution of Novel Gene Function

New functional roles appear to routinely evolve in developmental genetic systems. Developmental genes, despite being under strong stabilizing selection to perform precise functions, can evolve in unexpected ways. This tendency is called **developmental system drift** or **DSD**. The existence of DSD can be useful for tracking the evolution of particular structures (e.g., visual elements) but can cause confusion because DSD itself often leads to morphological divergences.

DSD appears to occur in two ways. First, a gene that performs a function in one taxon may be co-opted for another use in a different taxon. The process often begins with **gene duplication**, when replication errors create multiple copies of functional genes. Redundancy in regulatory function allows continuation of normal metabolism, but also allows duplicated gene products to be used elsewhere. For example, the gene *Pax6* is expressed in a range of both eyed and eyeless bilaterians suggesting that this patterning gene has been coopted for multiple functions in the regulation of eye development.

A second DSD mechanism occurs when differences in gene expression arise due to genetic interactions, or epistasis, occurring within different genetic backgrounds. Identical sequences can produce different phenotypic effects in different species, or even among members of the same species. Although the magnitude of these effects can become increasingly pronounced as species divergence increases, the complexities of epistatic interactions are seldom predictable.

Gene Regulatory Networks

A major contribution of genomic analyses to developmental biology is the conclusion that individual genes do not control cellular or development function. That is, there is no one gene "for" a particular phenotype or structure. Instead, groups of genes, usually called **gene regulatory networks**, are responsible for producing functional traits. Four major groups of these regulatory clusters exist: (1) cell differentiation networks, which appear to allow groups of cells to differentiate in particular ways; (2) subcircuits, developmental gene networks that are repeatedly used in general cell function; (3) switches, regulatory gene networks that turn cell functions off or on to regulate the timing of particular developmental events; and (4) kernels, complex and tightly conserved gene networks designed to specify the fields of cells from which particular body parts will eventually arise.

Because kernels appear to be the type of network most important for organizing lineage-specific

(A) Conventional representation of Hox gene clusters

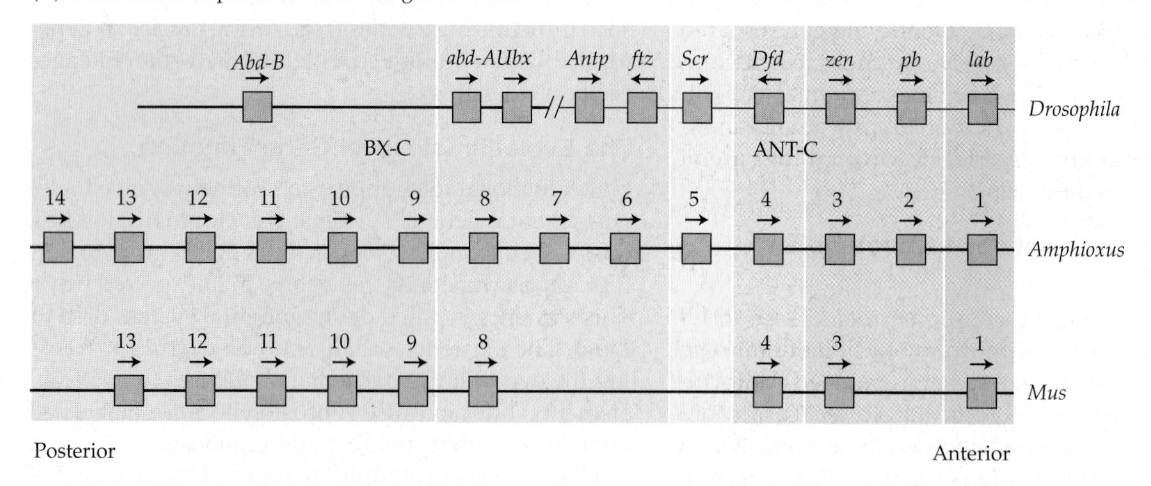

Posterior Anterior

(B) Structural representation of Hox gene clusters

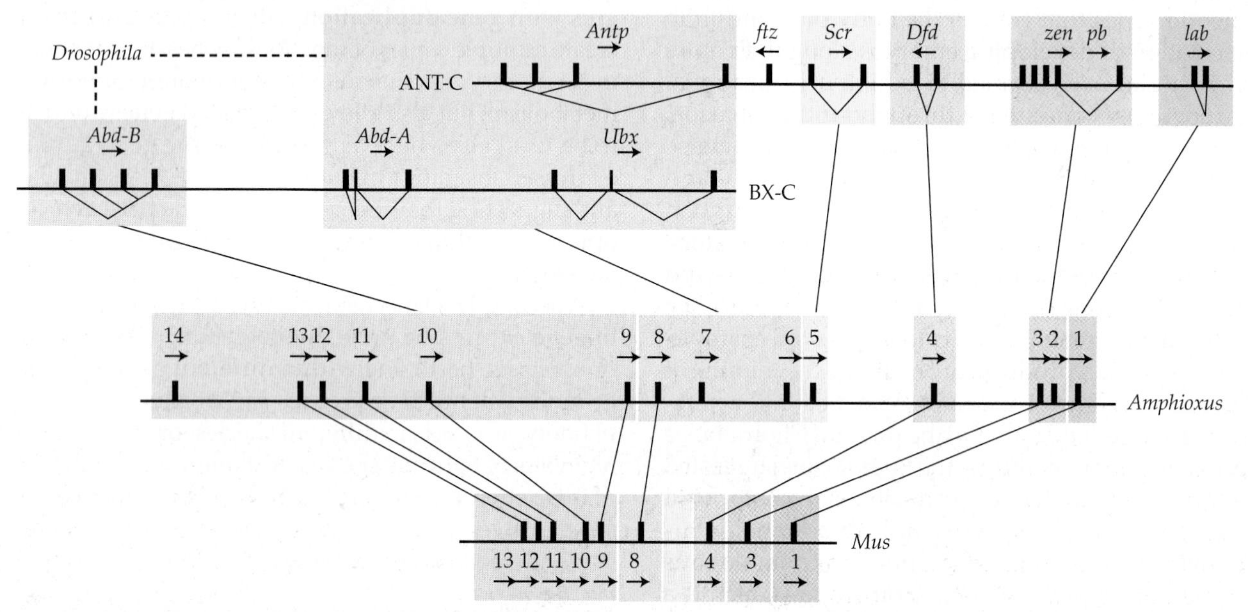

Figure 5.1 Representations of Hox genes in textbooks are often oversimplified and thus can lead to erroneous views of metazoan phylogenetic relationships. In this figure, the horizontal lines represent chromosomes. The colored boxes represent Hox gene "clusters," those genes that are related in their sequences and thus appear to be derived from a common ancestral Hox gene, in an insect (*Drosophila*), a cephalochordate (*Amphioxus*), and a vertebrate (mouse, *Mus*). The arrows represent gene orientation, that is, the direction in which transcription of the DNA coding strand proceeds. (A) Hox "clusters" 1 through 14 as they are often represented in textbooks and some scientific articles. In such oversimplified representations, Hox genes, including BX-C (the bithorax complex of *Drosophila* that controls development of abdominal and posterior thoracic segments) and ANT-C (the antennipedia complex of *Drosophila* that controls the formation of legs) are incorrectly shown as being located in close proximity, with clusters that are similar in nucleotide sequence length, and with gene clusters that are arranged in a linear order on chromosomes. (B) A more accurate representation of Hox clusters as they actually exist within *Drosophila*, *Amphioxus* and *Mus*, shown to scale with regard to the actual size of the genes (the number of nucleotides involved including non-coding sequences), the relative distances between the genes (the degree to which genes form "clusters"), and clearer spatial and gene orientation arrangements of Hox elements with respect to one another on the chromosome. Note that in *Drosophila*, the BX-C and ANT-C complexes are "split" (spatially separated as indicated by the dashed line) on chromosome III, compared with less spatially dispersed arrangements characteristic of chordates (*Amphioxus* and *Mus*). Note also that while *Drosophila* gene orientations are variable (indicated by arrows), they are consistently unidirectional in *Amphioxus* and *Mus*. While Hox clusters in chordates (*Amphioxus* and *Mus*) tend to be more spatially condensed than in other metazoans, they are especially "organized" in this way in vertebrates (e.g., *Mus*). Accurate physical representations of Hox genes are needed to inform the discussion on the structural and functional evolution of Hox clusters.

development (e.g., there are likely to be bilaterian, protostome, spiralian, and ecdysozoan kernels) these are considered most likely useful for phylogenetic analyses of the kind originally conceived of for Hox clusters. Nevertheless, it is still possible for gene regulatory networks to undergo substantial evolution within lineages. Thus, as with all such analyses, care must be used in choosing which networks to include.

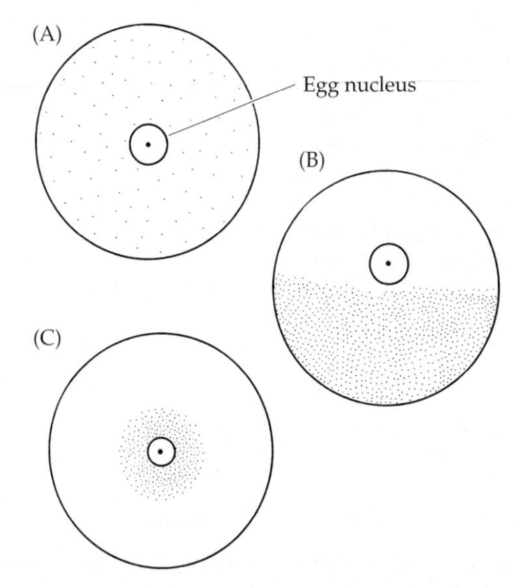

Figure 5.2 Types of ova. Stippling denotes the distribution and relative concentration of yolk within the cytoplasm. (A) An isolecithal ovum has a small amount of yolk distributed evenly. (B) The yolk in a telolecithal ovum is concentrated toward the vegetal pole. The amount of yolk in such eggs varies greatly. (C) A centrolecithal ovum has yolk concentrated at the center of the cell.

Eggs and Embryos

The physical (and physiological) attributes that distinguish the Metazoa are the result of their embryonic development. Stated differently, adult phenotypes result from specific sequences of developmental stages. Therefore, both animal unity and diversity are as evident in patterns of development as they are in the body architecture of adults. The patterns of development discussed below reflect this unity and diversity, and serve as a basis for understanding the sections on embryology in later chapters.

Eggs

Biological processes in general are cyclical. Successive generations illustrate this generality, as the term "life cycle" implies, and where the description of a process begins is only a matter of convenience. Here, we begin with the egg, or **ovum**, a single remarkable cell capable of developing into a new individual. Once fertilized, all of the different cell types of an adult animal are derived during embryogenesis from this single totipotent entity. A fertilized ovum contains not only the genetic information necessary to direct development but also some quantity of nutrient material called **yolk**, which sustains the early stages of life.

Eggs are polarized along an **animal–vegetal axis**. This polarity may be apparent in the egg itself, or it may be recognizable as development proceeds. The **vegetal pole** is associated with the formation of nutritive organs (e.g., the digestive system), whereas the **animal pole** produces other regions of the embryo. Animal ova are categorized primarily by the amount and location of yolk within the cell (Figure 5.2), two factors that greatly influence certain aspects of development. **Isolecithal** eggs contain a relatively small amount of yolk that is more or less evenly distributed throughout the cell. Ova in which the yolk is concentrated at one end (toward the vegetal pole) are termed **telolecithal** eggs; those in which the yolk is concentrated in the center are called **centrolecithal** eggs.

The actual amount of yolk in telolecithal and centrolecithal eggs is highly variable. Yolk production (**vitellogenesis**) is typically the longest phase of egg production, although its duration varies by orders of magnitude among species. Rates of yolk production depend on the specific vitellogenic mechanism used. In general, opportunistic (so-called *r*-selected) species have evolved vitellogenic pathways for the rapid conversion of food into egg production, while more specialist (so-called *K*-selected) species utilize slower pathways.

Cleavage

The penetration of an ovum by a sperm cell and the subsequent fusion of the male and female nuclei initiate the transformation of an ovum into a **zygote** or fertilized egg. The initial cell divisions of a zygote are called **cleavage**, and the resulting cells are called **blastomeres**. Certain aspects of the pattern of early cleavage are determined by the amount and placement of yolk, whereas other features are inherent in the genetic programming of the particular organism. Isolecithal and weakly to moderately telolecithal ova generally undergo **holoblastic** cleavage. That is, the cleavage planes pass completely through the cell, producing blastomeres that are separated from one another by thin cell membranes (Figure 5.3A). Whenever very large amounts of yolk are present (as in strongly telolecithal eggs), the cleavage planes do not pass readily through the dense yolk, so the blastomeres are not fully separated from one another by cell membranes. This pattern of early cell division is called **meroblastic** cleavage (Figure 5.3B). The pattern of cleavage in centrolecithal eggs is dependent on the amount of yolk and varies from holoblastic to various modifications of meroblastic.

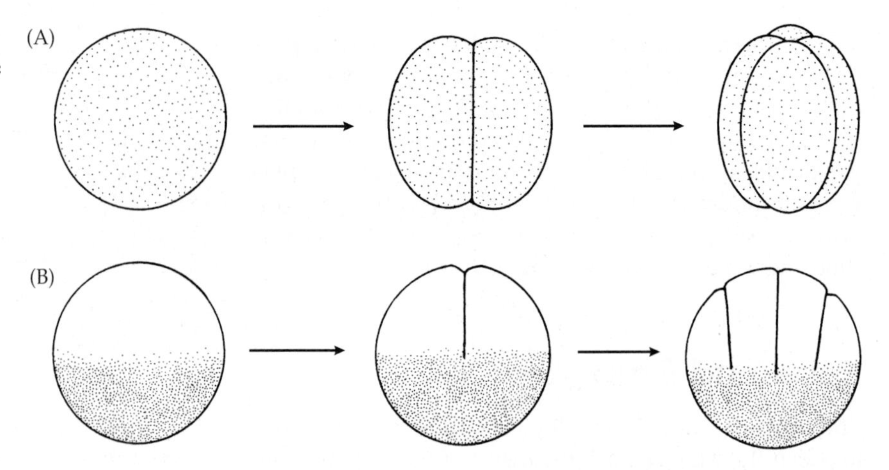

Figure 5.3 Types of early cleavage in developing zygotes. (A) Holoblastic cleavage. The cleavage planes pass completely through the cytoplasm. (B) Meroblastic cleavage. The cleavage planes do not pass completely through the yolky cytoplasm. The stippling represents yolk distribution in the egg and early zygote.

Orientation of Cleavage Planes

A number of terms describe the relationship of cleavage planes to the animal–vegetal axis of the egg and the relationships of the resulting blastomeres, one to another (Figure 5.4). Cell divisions during cleavage may be **equal** or **unequal**, indicating the comparative sizes of groups of blastomeres. The term **subequal** is used when blastomeres are only slightly different in size. When cleavage is distinctly unequal, the larger cells lying at the vegetal pole are called **macromeres.** The smaller cells located at the animal pole are called **micromeres**.

Cleavage planes that pass through or parallel to the animal–vegetal axis produce **longitudinal** (= meridional) divisions; those that pass at right angles to the axis produce **transverse** divisions. Transverse divisions may be equatorial, when the embryo is separated equally into animal and vegetal halves, or simply latitudinal, when the division plane does not pass through the equator of the embryo.

Radial and Spiral Cleavage

Most invertebrates display one of two cleavage patterns defined on the basis of the orientation of the blastomeres about the animal–vegetal axis. These patterns are called **radial** cleavage and **spiral** cleavage and are illustrated in Figure 5.5. Radial cleavage involves strictly longitudinal and transverse divisions. Thus, the blastomeres are arranged in rows either parallel or perpendicular to the animal–vegetal axis. The placement of the blastomeres shows a radially symmetrical pattern in polar view. The chapter opening photo shows the 16-cell stage embryo of a radially-cleaving echinoderm (*Lytechinus pictus*), with four large macromeres behind four smaller micromeres (and eight "mesomeres" behind the macromeres).

Spiral cleavage is quite another matter. Although not inherently complex, it can be difficult to describe. The first two divisions are longitudinal, generally equal or subequal. Subsequent divisions, however, displace the blastomeres laterally so that they lie within the furrows between previously divided cells. This condition is a result of the formation of the mitotic spindles at oblique angles rather than parallel to the axis of the embryo; hence the cleavage planes are neither perfectly longitudinal nor perfectly transverse. The division from four to eight cells involves a displacement of the cells near the animal pole in a clockwise (**dextrotropic**) direction (viewed from the animal pole). The next division, from eight to sixteen cells, occurs with a displacement in a counterclockwise (**levotropic**) direction; the

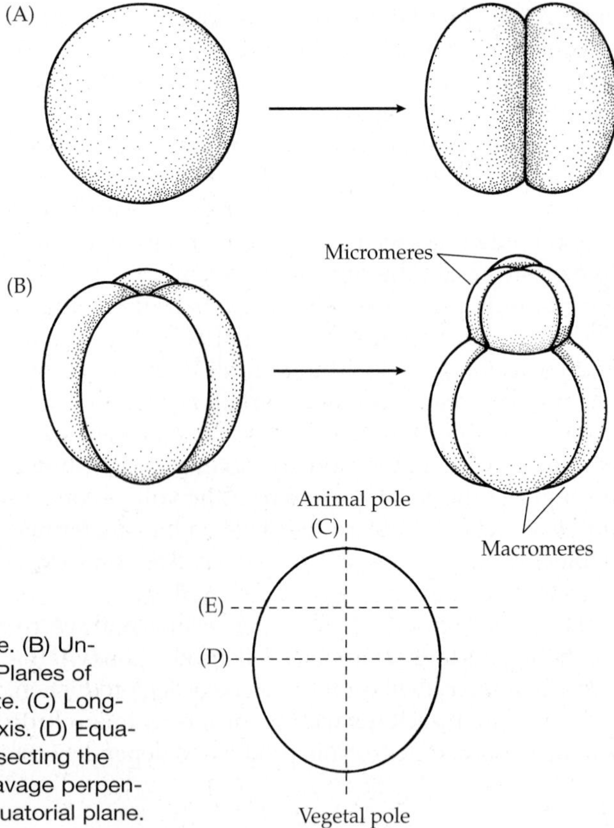

Figure 5.4 Planes of holoblastic cleavage. (A) Equal cleavage. (B) Unequal cleavage produces micromeres and macromeres. (C–E) Planes of cleavage relative to the animal–vegetal axis of the egg or zygote. (C) Longitudinal (= meridional) cleavage parallel to the animal–vegetal axis. (D) Equatorial cleavage perpendicular to the animal–vegetal axis and bisecting the zygote into equal animal and vegetal halves. (E) Latitudinal cleavage perpendicular to the animal–vegetal axis but not passing along the equatorial plane.

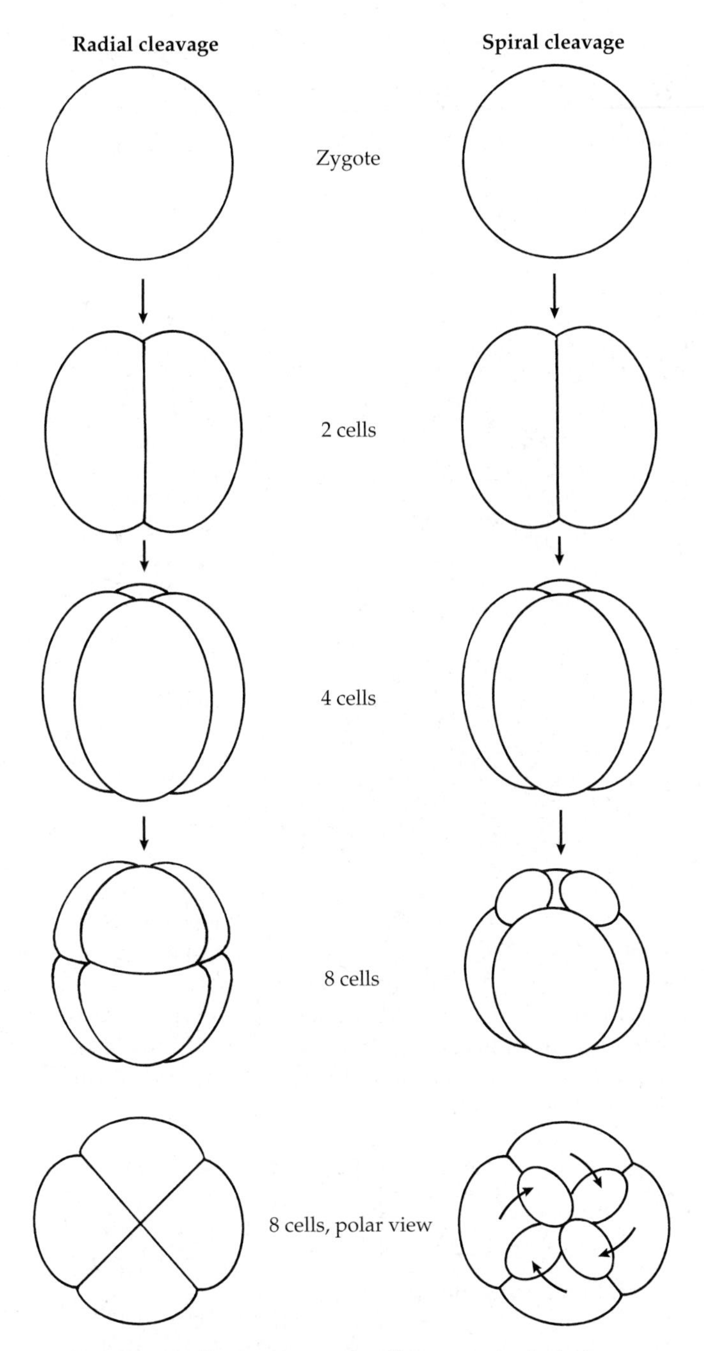

Figure 5.5 Comparison of radial versus spiral cleavage through the 8-cell stage. During radial cleavage, the cleavage planes all pass either perpendicular or parallel to the animal–vegetal axis of the embryo. Spiral cleavage involves a tilting of the mitotic spindles, commencing with the division from 4 to 8 cells. The resulting cleavage planes are neither perpendicular nor parallel to the axis. The polar views of the resulting 8-cell stages illustrate the differences in blastomere orientation.

next is clockwise, and so on—alternating back and forth until approximately the 64-cell stage. We hasten to add that divisions are frequently not synchronous; not all of the cells divide at the same rate. Thus, a particular embryo may not proceed from four cells to eight, to sixteen, and so on, as neatly as in our generalized description.

During his extensive studies on the polychaete worm *Neanthes succinea*, conducted at the Marine Biological Laboratory at Woods Hole, E. B. Wilson (1892) devised an elegant coding system for spiral cleavage, one that allows us to follow the developmental lineage of every embryonic cell. Wilson's system is usually applied to spiral cleavage in order to trace cell fates and compare development among species. Our account of spiral cleavage is a general one, but it provides a point of reference for later consideration of the patterns in different groups of animals.

At the 4-cell stage, following the initial longitudinal divisions, the cells are given the codes of A, B, C, and D, and are labeled clockwise in that order when viewed from the animal pole (Figure 5.6A). These four cells are referred to as a **quartet** of macromeres, and they may be collectively coded as simply Q. The next division is more or less unequal, with the four cells nearest the animal pole being displaced in a dextrotropic fashion, as explained above. These four smaller cells are called the first quartet of micromeres (collectively the 1q cells) and are given the individual codes of 1a, 1b, 1c, and 1d. The numeral "1" indicates that they are members of the first micromere quartet to be produced; the letters correspond to their respective macromere origins. The capital letters designating the macromeres are now preceded with the numeral "1" to indicate that they have divided once and produced a first micromere set (Figure 5.6B). We may view this 8-celled embryo as four pairs of daughter cells that have been produced by the divisions of the four original macromeres as follows:

$$A \begin{matrix} \nearrow 1a \\ \searrow 1A \end{matrix} \quad B \begin{matrix} \nearrow 1b \\ \searrow 1B \end{matrix} \quad C \begin{matrix} \nearrow 1c \\ \searrow 1C \end{matrix} \quad D \begin{matrix} \nearrow 1d \\ \searrow 1D \end{matrix}$$

Note that although the macromeres and micromeres are sometimes similar in size, these terms are nonetheless always used in describing spiral cleavage. Much of the size discrepancy depends upon the amount of yolk present at the vegetal pole in the original egg; this yolk tends to be retained primarily in the larger macromeres.

The division from 8 to 16 cells occurs levotropically and involves cleavage of each macromere and micromere. Notice that the only code numbers that are changed through subsequent divisions are the prefix numbers of the macromeres. These are changed to indicate the number of times these individual macromeres have divided, and to correspond to the number of micromere quartets thus produced. So, at the 8-cell stage, we can designate the existing blastomeres as the 1Q (= 1A, 1B, 1C, 1D) and the 1q (= 1a, 1b, 1c, 1d). The macromeres (1Q) divide to produce a second quartet of micromeres (2q = 2a, 2b, 2c, 2d), and the prefix numeral of the daughter macromeres is changed to "2." The first micromere quartet also divides and now comprises eight cells, each of which is identifiable not only by the letter corresponding to its

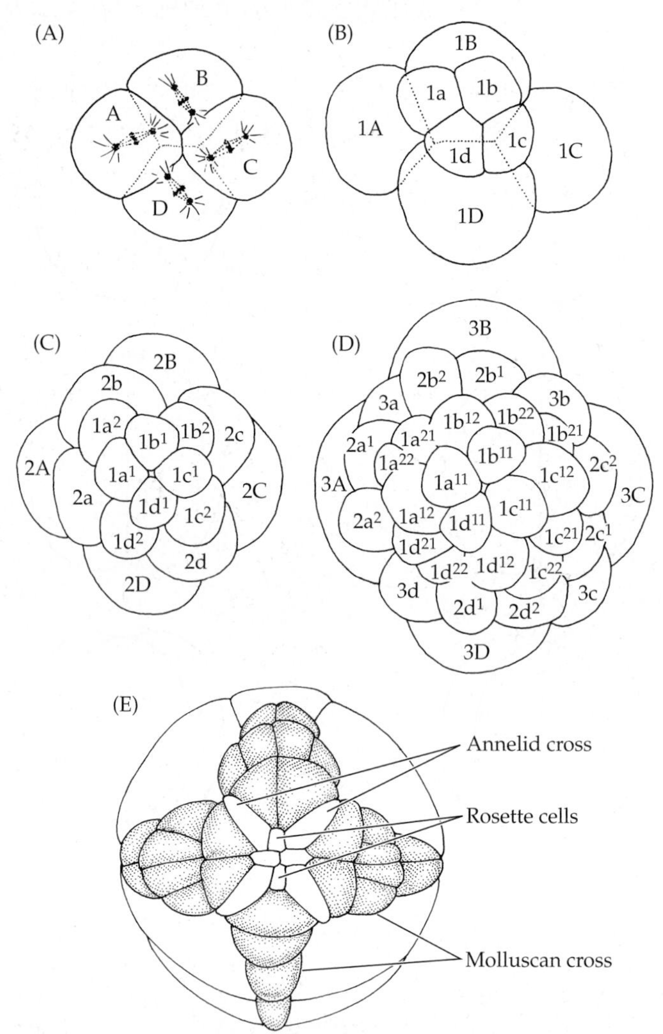

Figure 5.6 Spiral cleavage. (A–D) Spiral cleavage through 32 cells (assumed synchronous) labeled with E. B. Wilson's coding system (all diagrams are surface views from the animal pole). (E) Schematic diagram of a composite embryo at approximately 64 cells showing the positions of the rosette, annelid cross, and molluscan cross.

parent macromere but now by the addition of superscript numerals. For example, the 1a micromere (of the 8-cell embryo) divides to produce two daughter cells coded the 1a1 and the 1a2 cells. The cell that is physically nearer the animal pole of the embryo receives the superscript "1," the other cell the superscript "2." Thus, the 16-cell stage (Figure 5.6C) includes the following cells:

Derivatives of the 1q
- $1a^{1}$ $1b^{1}$ $1c^{1}$ $1d^{1}$
- $1a^{2}$ $1b^{2}$ $1c^{2}$ $1d^{2}$

Derivatives of the 1Q
- 2q = 2a 2b 2c 2d
- 2Q = 2A 2B 2C 2D

The next division (from 16 to 32 cells) again involves dextrotropic displacement. The third micromere quartet (3q) is formed, the daughter macromeres are now given the prefix "3" (3Q), and all of the 12 existing micromeres divide. Superscripts are added to the derivatives of the first and second micromere quartets according to the rule of position as stated above. Thus, the $1b^{1}$ cell divides to yield the $1b^{11}$ and $1b^{12}$ cells; the $1a^{2}$ cell yields the $1a^{21}$ and $1a^{22}$ cells; the 2c yields the $2c^{1}$ and $2c^{2}$, and so on. Do not think of these superscripts as double-digit numbers (i.e., "twenty-one" and "twenty-two"), but rather as two-digit sequences reflecting the precise lineage of each cell ("two-one" and "two-two").

The elegance of Wilson's system is that each code tells the history as well as the position of the cell in the embryo. For instance, the code $1b^{11}$ indicates that the cell is a member (derivative) of the first quartet of micromeres, that its parent macromere is the B cell, that the original 1b micromere has divided twice since its formation, and that this particular cell rests uppermost in the embryo relative to its sister cells. The 32-cell state (Figure 5.6D) is composed of the following:

Derivatives of the 1q
- $1a^{11}$ $1b^{11}$ $1c^{11}$ $1d^{11}$
- $1a^{12}$ $1b^{12}$ $1c^{12}$ $1d^{12}$
- $1a^{21}$ $1b^{21}$ $1c^{21}$ $1d^{21}$
- $1a^{22}$ $1b^{22}$ $1c^{22}$ $1d^{22}$

Derivatives of the 2q
- $2a^{1}$ $2b^{1}$ $2c^{1}$ $2d^{1}$
- $2a^{2}$ $2b^{2}$ $2c^{2}$ $2d^{2}$

Derivatives of the 2Q
- 3q = 3a 3b 3c 3d
- 3Q = 3A 3B 3C 3D

The division to 64 cells follows the same pattern, with appropriate coding changes and additions of superscripts. The displacement is levotropic and results in the following cells:

Derivatives of the 1q
- $1a^{111}$ $1b^{111}$ $1c^{111}$ $1d^{111}$
- $1a^{112}$ $1b^{112}$ $1c^{112}$ $1d^{112}$
- $1a^{121}$ $1b^{121}$ $1c^{121}$ $1d^{121}$
- $1a^{122}$ $1b^{122}$ $1c^{122}$ $1d^{122}$
- $1a^{211}$ $1b^{211}$ $1c^{211}$ $1d^{211}$
- $1a^{212}$ $1b^{212}$ $1c^{212}$ $1d^{212}$
- $1a^{221}$ $1b^{221}$ $1c^{221}$ $1d^{221}$
- $1a^{222}$ $1b^{222}$ $1c^{222}$ $1d^{222}$

Derivatives of the 2q
- $2a^{11}$ $2b^{11}$ $2c^{11}$ $2d^{11}$
- $2a^{12}$ $2b^{12}$ $2c^{12}$ $2d^{12}$
- $2a^{21}$ $2b^{21}$ $2c^{21}$ $2d^{21}$
- $2a^{22}$ $2b^{22}$ $2c^{22}$ $2d^{22}$

Derivatives of the 3q
- $3a^{1}$ $3b^{1}$ $3c^{1}$ $3d^{1}$
- $3a^{2}$ $3b^{2}$ $3c^{2}$ $3d^{2}$

Derivatives of the 3Q
- 4q = 4a 4b 4c 4d
- 4Q = 4A 4B 4C 4D

Notice that no two cells share the same code, so exact identification of individual blastomeres and their lineages is always possible.

Late in the spiral cleavage of certain animals, distinctive cell patterns appear, formed by the orientation of some of the apical first-quartet micromeres (Figure 5.6E). The topmost cells ($1q^{111}$ micromeres) lie at the embryo's apex and form the **rosette**. In some groups (e.g., annelids), other micromeres ($1q^{112}$ micromeres) produce an **annelid cross** roughly at right angles to the rosette cells. In molluscs, the annelid cross may appear (often called peripheral rosette cells in these groups), but an additional **molluscan cross** forms from the $1q^{12}$ cells and their derivatives. The arms of the molluscan cross lie between the cells of the annelid cross (Figure 5.6E), and this configuration is not known to occur in any other metazoan phylum. Some phylogenetic significance has been given to the appearance of these crosses, as we discuss in later chapters.

Cell Fates

Tracing the fates of cells through development has been a popular and productive endeavor of embryologists for over a century. Such studies have played a major role in allowing researchers to describe development as well as establish homologies among the attributes in different animals. The cells of embryos eventually become established as functional parts of tissues or organs, but before they do there is much variation in the timing and degree to which cell fates become firmly fixed. Although under normal conditions their functions are specialized, even in adults, the cells of some animals (e.g., sponges) retain the ability to change their structure and function. Other animal taxa have remarkable power to regenerate lost parts, wherein cells dedifferentiate and then generate new tissues and organs. In still other taxa, cell fates are relatively fixed and cells are able only to produce more of their own kind.

By carefully watching the development of any animal, it becomes clear that certain cells predictably form certain structures. Here too, the emerging field of molecular developmental biology has shown that many molecular components of development are also widely conserved throughout the animal kingdom. For example, some transcription factors and cell signaling systems from widely divergent phyla are clearly homologous and evidently operate in much the same way. On the other hand, these highly conserved molecular components can also be used in diverse ways by embryos. The pattern of orthologous gene expression in early metazoan embryos illustrates both aspects of this relationship. Even such basic developmental features as adult body axis formation and cleavage geometry differ among the metazoan phyla (Figure 5.7). Such fundamental developmental variations appear to have been essential in fabricating the highest levels of animal body plans.

In some cases, cell fates are determined very early during cleavage—as early as the 2- or 4-cell stage. If one experimentally removes a blastomere from the early embryo of such an animal (as Roux did), then that embryo will fail to develop normally; the fates of the cells have already become fixed, and the missing cell cannot be replaced. Animals whose cell fates are established very early are said to have **determinate cleavage**. On the other hand, the blastomeres of some animals can be separated at the 2-cell, 4-cell (as Driesch did), or even later stages (as Spemann did), and each separate cell will develop normally; in these cases the fates of the cells are not fixed until relatively late in development. Such animals are said to have **indeterminate cleavage**. Eggs that undergo determinate cleavage are often called **mosaic ova**, because the fates of regions of undivided cells can be mapped. Eggs that undergo indeterminate cleavage are called **regulative ova**, in that they can "regulate" to accommodate lost blastomeres and thus cannot easily be predictably mapped prior to division.

In any case, formation of the basic body plan is generally determined by the time the embryo comprises about 104 cells (usually after one or two days). By this time, all available embryonic material has been apportioned into specific cell groups, or "founder regions." These regions are relatively few, each forming a territory within which still more intricate developmental patterns unfold. As these zones of undifferentiated tissue are established, the unfolding genetic code drives them to develop into their "preassigned" body tissues, organs, or other structures. Graphic representations of these regions are called **fate maps**, although such devices are rarely used by developmental biologists any longer.

In the past, mosaic eggs and determinate cleavage have been equated with spirally cleaving embryos, and regulative ova and indeterminate cleavage with radially cleaving embryos. However, surprisingly few actual tests for determinacy have been performed, and what evidence is available suggests that there are many exceptions to this generalization. That is, some embryos with spiral cleavage appear indeterminate, and some with radial cleavage appear determinate.

In spite of the variations and exceptions, there is a remarkable underlying consistency in the fates of blastomeres among embryos that develop by typical spiral cleavage. Many examples of these similarities are discussed in later chapters, but we illustrate the point by noting that the germ layers of spirally cleaving embryos tend to arise from the same groups of cells. The first three quartets of micromeres and their derivatives give rise to **ectoderm** (the outer germ layer), the 4a, 4b, 4c, and 4Q cells to **endoderm** (the inner germ layer), and the 4d cell to **mesoderm** (the middle germ layer). Many students of embryology view this uniformity of cell fates as strong evidence that taxa sharing this

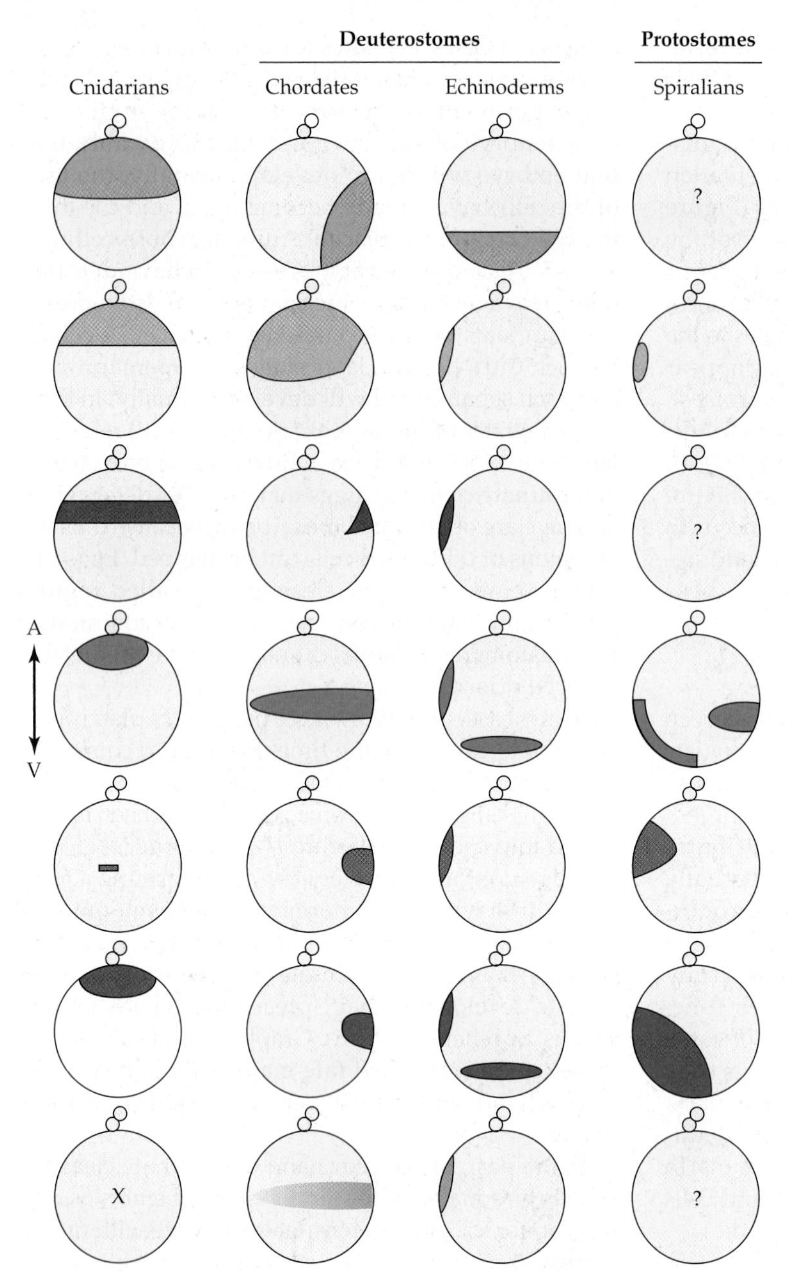

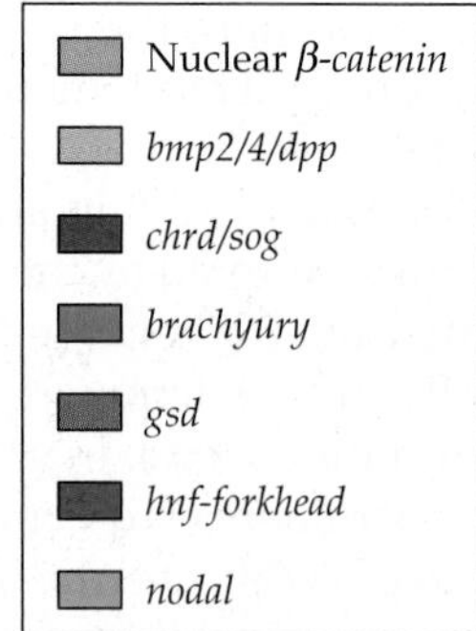

Figure 5.7 Locations and patterns of expression in some genes at the beginning of gastrulation in diverse metazoan embryos (cnidarians, chordates, echinoderms, spiralians). A↔V represents the animal-vegetal axis. All but one of the genes shown (*nodal*) are present in cnidarians as well as in chordates. While some genes (e.g., nuclear β-*catenin*) seem to track the changing location of gastrulation across taxa, other genes (e.g., *chordin* [*chrd*] and *goosecoid* [*gsd*] in echinoderms) are associated with gastrulation in some species and disassociated from such activity in others. The two circles at the animal pole represent polar bodies, X indicates that the gene is not present in the genome of a lineage, and ? indicates that the gene has not been recovered in a lineage. Gene abbreviations: *bmp*, bone morphogenetic protein; *dpp*, decapentaplegic; *sog*, short gastrulation; *hnf-forkhead*, hepatocyte nuclear factor, a forkhead homolog.

pattern are related to one another in some fundamental way and that they share a common evolutionary heritage. We will have much more to say about this idea throughout this book.

EvoDevo analyses, for their part, provide an objective means for assessing germ layer homology by identifying genes that are transcribed at different developmental times, rather than identifying pools of cells by eventual fate (Figure 5.7). Genes that have proven useful for such analyses include: *GATA 4*, *5*, and *6*, genes associated with mucus production in endodermal tissue; *twist*, a gene associated with mesodermal development; *snail*, a repressor of E-cadherin, and thus important for downregulating ectodermal genes within mesoderm and allowing mesenchymal development; and *brachyury*, a gene important in defining the midline of many bilaterians. However this approach is complicated by the fact that even among closely related taxa, similar structures may be derived from different germ layers (e.g., Malpighian tubules are derived from ectoderm in insects but arise from endoderm in chelicerates). As explained above, traits used for evolutionary comparative analyses, even at the molecular level, must be selected with caution.

Blastula Types

The product of early cleavage is called the **blastula**, which may be defined developmentally as the embryonic stage preceding the formation of embryonic germ layers. Several types of blastulae are recognized among invertebrates. Holoblastic cleavage generally results in either a hollow or a solid ball of cells. A **coeloblastula**

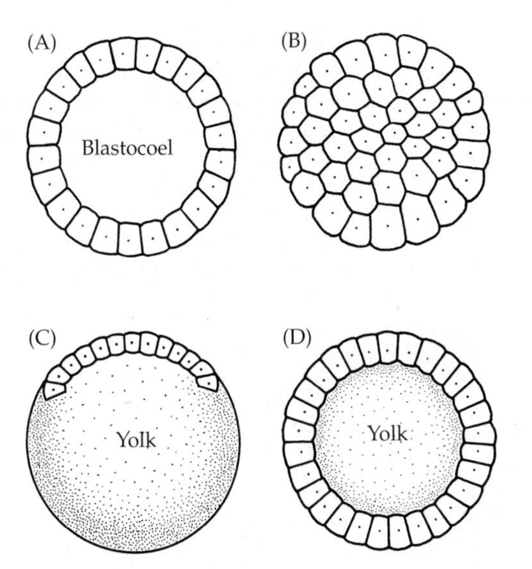

Figure 5.8 Types of blastulae. These diagrams represent sections along the animal–vegetal axis. (A) Coeloblastula. The blastomeres form a hollow sphere with a wall one cell layer thick. (B) Stereoblastula. Cleavage results in a solid ball of blastomeres. (C) Discoblastula. Cleavage has produced a cap of blastomeres that lies at the animal pole, above a solid mass of yolk. (D) Periblastula. Blastomeres form a single cell layer enclosing an inner yolky mass.

(Figure 5.8A) is a hollow ball of cells, the wall of which is usually one cell-layer thick. The space within the sphere of cells is the **blastocoel**, or primary body cavity. A **stereoblastula** (Figure 5.8B) is a solid ball of blastomeres; obviously there is no blastocoel at this stage. Meroblastic cleavage sometimes results in a cap or disc of cells at the animal pole over an uncleaved mass of yolk. This arrangement is appropriately termed a **discoblastula** (Figure 5.8C). Some centrolecithal ova undergo odd cleavage patterns to form a **periblastula**, similar in some respects to a coeloblastula that is centrally filled with noncellular yolk (Figure 5.8D).

Gastrulation and Germ Layer Formation

Through one or more of several methods the blastula develops toward a multilayered form, a process called **gastrulation** (Figure 5.9). The structure of the blastula dictates to some degree the nature of the process and the form of the resulting embryo, the **gastrula**. Gastrulation is the formation of the embryonic germ layers, the tissues on which all subsequent development eventually depends. In fact, we may view gastrulation as the embryonic analogue of the transition from protistan to metazoan grades of complexity. It achieves separation of those cells that must interact directly with the environment (i.e., locomotor, sensory, and protective functions) from those that process materials ingested from the environment (i.e., nutritive functions).

The initial inner and outer sheets of cells are the **endoderm** and **ectoderm**, respectively; in most animals a third germ layer, the **mesoderm**, is produced between the ectoderm and the endoderm. One striking example of the unity among the Metazoa is the consistency of the fates of these germ layers. For example, ectoderm always forms the nervous system, the outer skin and its derivatives; endoderm forms the main portion of the gut and associated structures; mesoderm forms the coelomic lining, the circulatory system, most of the internal support structures, and the musculature. The process of gastrulation, then, is a critical one in establishing the basic materials and their locations for building the whole organism.

Coeloblastulae often gastrulate by **invagination**, a process commonly used to illustrate gastrulation in general zoology classes. The cells in one area of the surface of the blastula (frequently at or near the vegetal pole) pouch inward as a sac within the blastocoel (Figure 5.9A). These invaginated cells are now called the endoderm, and the sac thus formed is the embryonic gut, or **archenteron**; the opening to the outside is the **blastopore**. The outer cells are now called ectoderm, and a double-layered hollow coelogastrula has been formed. The blastopore may become the primordial mouth or anus depending on the animal lineage. If this structure becomes the mouth, the ectoderm lining the interior of the blastopore is considered **stomodeal**. If the blastopore becomes the anus, the ectoderm lining the interior of this structure is considered **proctodeal**. Note that the diagrams in Figure 5.9 represent 3-dimensional embryos. Thus, the **coelogastrula** (Figure 5.9A) actually resembles a balloon with an invisible finger poking into it.

The coeloblastulae of many cnidarians undergo gastrulation processes that result in a solid gastrula (**stereogastrula**). Usually the cells of the blastula divide such that the cleavage planes are perpendicular to the surface of the embryo. Some of the cells detach from the wall and migrate into the blastocoel, eventually filling it with a solid mass of endoderm. This process is called **ingression** (Figure 5.9B) and may occur only at the vegetal pole (**unipolar** ingression) or more or less over the whole blastula (**multipolar** ingression). In a few instances (e.g., certain hydroids), the cells of the blastula divide with cleavage planes that are parallel to the surface, a process called **delamination** (Figure 5.9C). This process produces a layer or a solid mass of endoderm surrounded by a layer of ectoderm.

Stereoblastulae that result from holoblastic cleavage generally undergo gastrulation by **epiboly**. Because there is no blastocoel into which the presumptive endoderm can migrate by any of the above methods, gastrulation of this form involves a rapid growth by a sheet of presumptive ectodermal cells around the presumptive endoderm (Figure 5.9D). Cells of the animal pole proliferate rapidly, growing down and over the vegetal cells to enclose them as endoderm. The blastopore forms where the edges of this ectodermal sheet

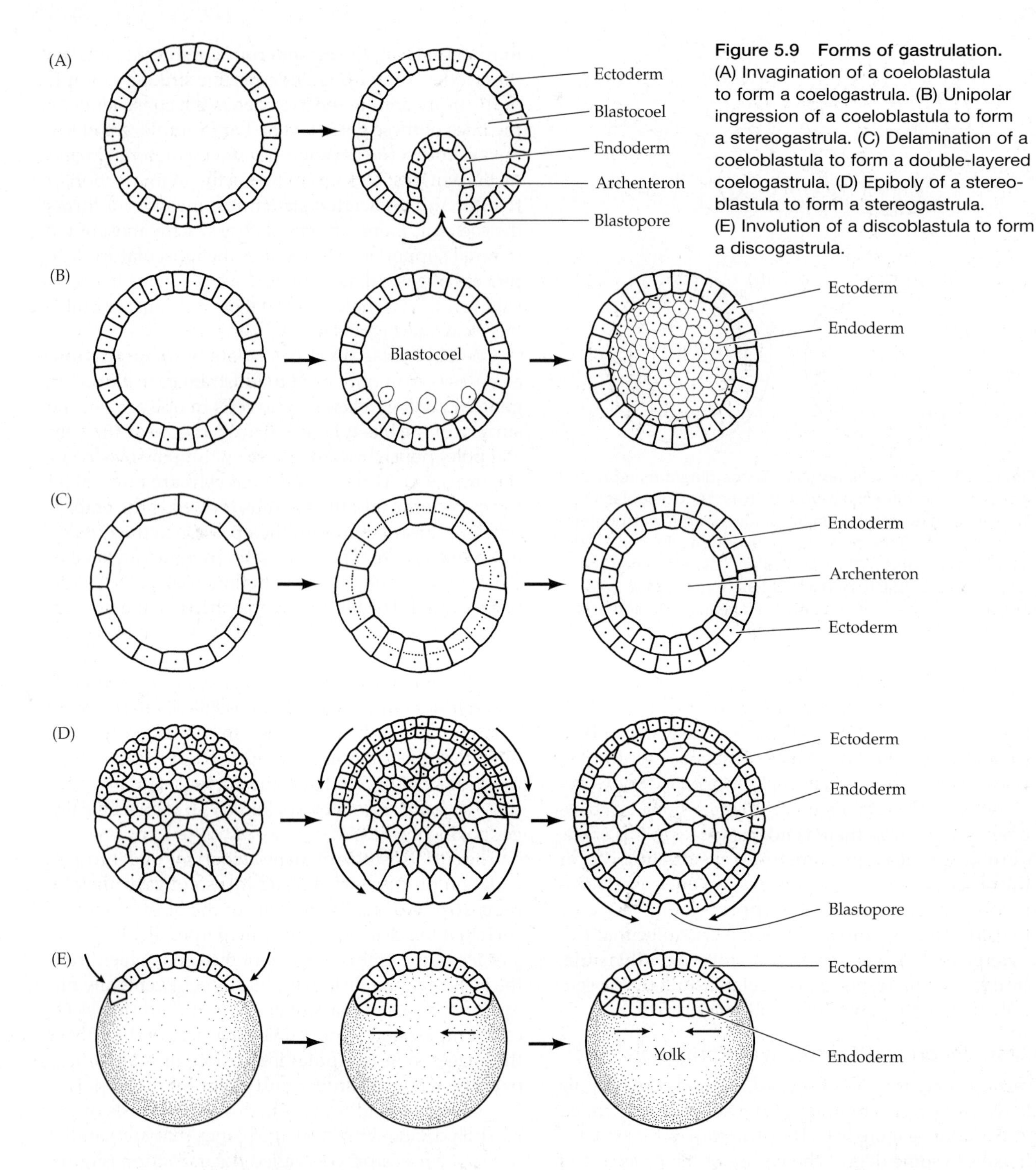

Figure 5.9 Forms of gastrulation. (A) Invagination of a coeloblastula to form a coelogastrula. (B) Unipolar ingression of a coeloblastula to form a stereogastrula. (C) Delamination of a coeloblastula to form a double-layered coelogastrula. (D) Epiboly of a stereoblastula to form a stereogastrula. (E) Involution of a discoblastula to form a discogastrula.

converge from all sides upon a single point at the vegetal pole. The archenteron typically forms secondarily as a space within the developed endoderm.

Figure 5.9E illustrates gastrulation by **involution**, a process that usually follows the formation of a discoblastula. The cells around the edge of the disc divide rapidly and grow beneath the disc, thus forming a double-layered gastrula with ectoderm on the surface and endoderm below. There are several other types of gastrulation, mostly variations or combinations of the above processes. These gastrulation methods are discussed in later chapters.

During gastrulation, subtle shifts in the timing of regulatory gene expression, the timing of cell fate specification, or in the movement of cells relative to one another, can generate distinct developmental pathways. Such developmental divergences may dramatically shift larval or even adult formation within a lineage. For example, sea urchin larvae appear to have switched from **planktotrophy** (feeding larvae) to **lecithotrophy** (nonfeeding larvae) at least 20 times within the history of this echinoderm clade. Among nonfeeding larvae, egg size is often greater, cleavage is significantly altered, and the average larval life span is shorter.

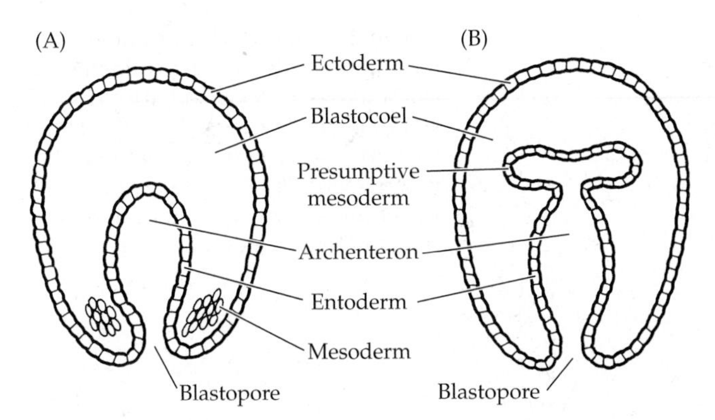

Figure 5.10 Methods of mesoderm formation in late gastrulae (frontal sections). (A) Mesoderm formed from derivatives of a mesentoblast. (B) Mesoderm formed by archenteric pouching.

Mesoderm and Body Cavities

During or soon after gastrulation, a middle layer forms between the ectoderm and the endoderm. This middle layer may be derived from ectoderm, as it is in members of the diploblastic phylum Cnidaria, or from endoderm, as it is in members of the triploblastic phyla. In the first case the middle layer is said to be **ectomesoderm**, and in the latter case **endomesoderm** (or "true mesoderm"). Thus, the triploblastic condition, by definition, includes endomesoderm. In this text, and most others, the term mesoderm in a general sense refers to endomesoderm rather than ectomesoderm. Although endomesoderm is characteristic of triploblastic metazoans, in many lineages some ectomesoderm is also produced.

In diploblastic and certain triploblastic phyla (the acoelomates), the middle layer does not form thin sheets of cells; rather it produces a more-or-less solid but loosely organized mesenchyme consisting of a gel matrix (the **mesoglea**) containing various cellular and fibrous inclusions. In a few cases (e.g., the hydrozoans) a virtually noncellular mesoglea lies between the ectoderm and endoderm (see Chapter 7).

In most animals, the area between the inner and outer body layers includes a fluid-filled space. As discussed in Chapter 4, this space may be either a **blastocoelom**, a cavity not completely lined by mesoderm, or a true **coelom**, a cavity fully enclosed within thin sheets of mesodermally derived tissue. Endomesoderm generally originates in one of two basic ways, as described below (Figure 5.10); modifications of these processes are discussed in later chapters. In most phyla that undergo spiral cleavage (e.g., flatworms, annelids, molluscs), a single micromere—the 4d cell, called the **mesentoblast**—proliferates as mesoderm between the developing archenteron (endoderm) and the body wall (ectoderm) (Figure 5.10A). The other cells of the 4q (the 4a, 4b, and 4c cells) and the 4Q cells generally contribute to endoderm. In some other taxa (e.g., echinoderms and chordates) the mesoderm arises from the wall of the archenteron itself (that is, from preformed endoderm), either as a solid sheet or as pouches (Figure 5.11B).

In addition to giving rise to other structures (such as the muscles of the gut and body wall), in coelomate animals mesoderm is intimately associated with the formation of the body cavity. In those instances where mesoderm is produced as solid masses derived from a mesentoblast, the body cavity arises through a process called **schizocoely**. Normally in such cases, bilaterally paired packets of mesoderm gradually enlarge,

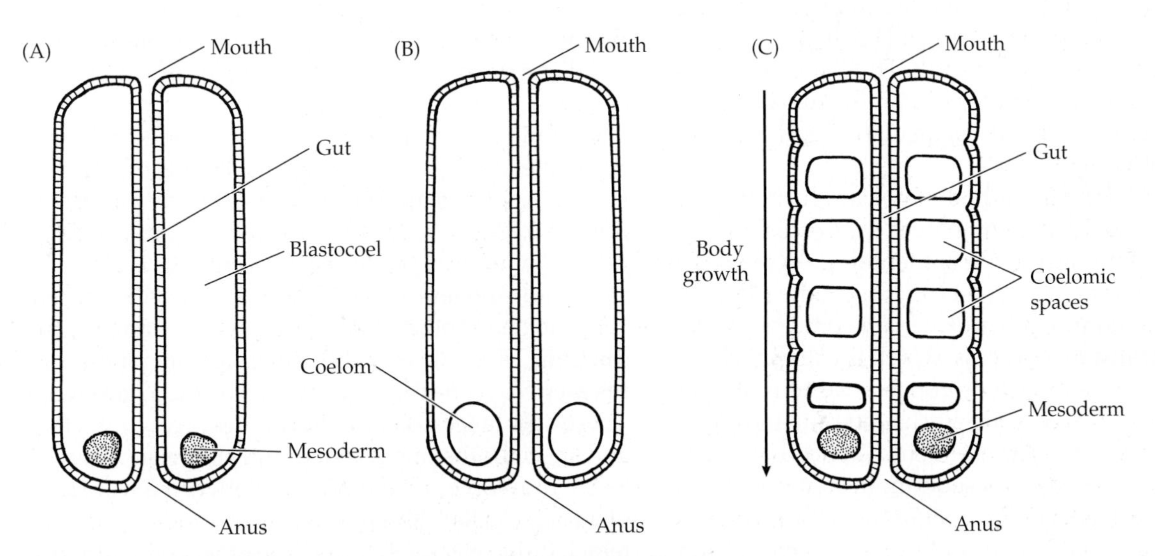

Figure 5.11 Coelom formation by schizocoely (frontal sections). (A) Precoelomic conditions with paired packets of mesoderm. (B) Hollowing of the mesodermal packets to produce a pair of coelomic spaces. (C) Progressive proliferation of serially arranged pairs of coelomic spaces. This process occurs in metameric annelids.

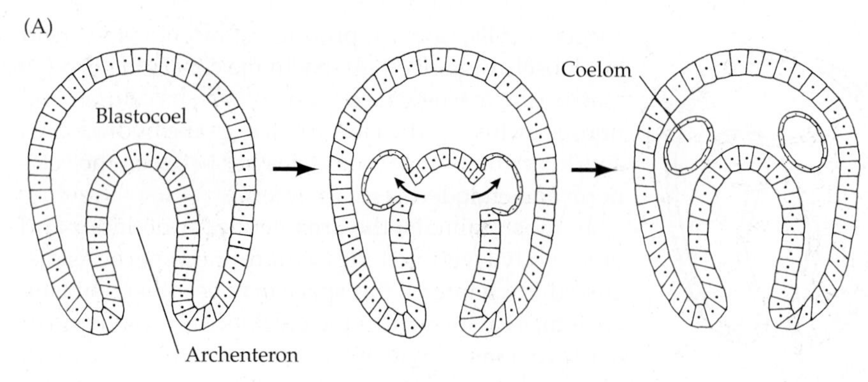

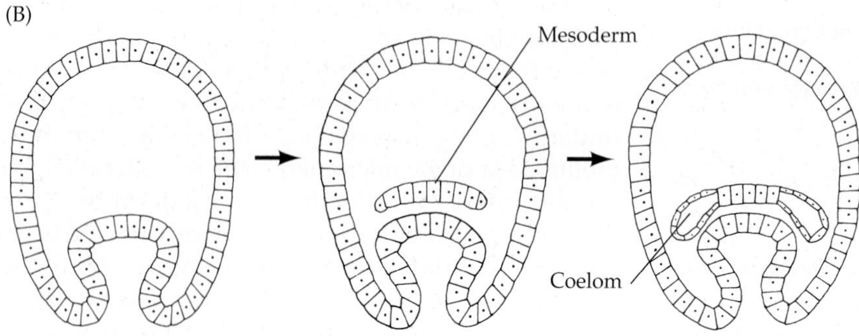

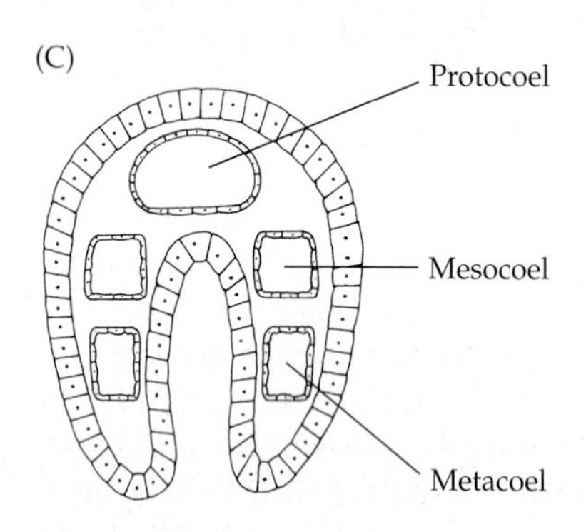

Figure 5.12 Coelom formation by enterocoely (frontal sections). (A) Archenteric pouching. (B) Proliferation and subsequent hollowing of a plate of mesoderm from the archenteron. (C) The typical tripartite arrangement of coeloms in a deuterostome embryo.

split internally, and then expand to simultaneously line the body wall, support viscera and create thin-walled coelomic spaces (Figure 5.11A,B). The number of such paired coeloms varies among different animals and is frequently associated with segmentation, as it is in annelid worms (Figure 5.11C).

The other general method of coelom formation is called **enterocoely**; it accompanies the process of mesoderm formation from the archenteron. In the most direct sort of enterocoely, mesoderm production and coelom formation are one and the same process. Figure 5.12A illustrates this process, which is called **archenteric pouching**. A pouch or pouches form in the gut wall. Each pouch eventually pinches off from the gut as a complete coelomic compartment. The walls of these pouches are defined as mesoderm. In some cases the mesoderm arises from the wall of the archenteron as a solid sheet or plate that later becomes bilayered and hollow (Figure 5.12B). Some authors consider this process to be a form of schizocoely (because of the "splitting" of the mesodermal plate), but it is in fact a modified form of enterocoely. Enterocoely frequently results in a tripartite arrangement of the body cavities, which are designated **protocoel**, **mesocoel**, and **metacoel** (Figure 5.12C).

Following germ tissue establishment, cells begin to specialize and sort themselves out to form the organs and tissues of the body—a poorly understood process known as **morphogenesis**. Cell movements are an essential part of morphogenesis. In addition, in order to sculpt the organs and systems of the body, cells need to "know" when to stop growing and even die. For example, in nematode worms the vas deferens first develops with a closed end; the cell that blocks the end of this tube helps the vas deferens link up to the cloaca. But once the connection has been made, this terminal cell dies and disassociates, creating the opening to the cloaca.

Recent research suggests that the same families of molecules that guide the earliest stages of embryogenesis—setting up the elements of body patterning (Figure 5.7)—also play vital roles during morphogenesis. Communication among adjacent cells is also critical to morphogenesis, and there are three ways cells "talk" to one another during this process, known as **induction**. The first is via diffusible signaling molecules released from one cell and detected by the adjacent cell. These substances include hormones, growth factors, and special substances called **morphogens**. A second method involves actual contact between the surfaces of adjacent cells, allowing cell surface molecules to interact. Cells selectively recognize other cells, adhering to some and migrating over others. A third method involves the movement of substances through gap junctions between cells. Of all the stages of ontogeny, we know least about morphogenesis.

Life Cycles: Sequences and Strategies

The patterns of early development described above are not isolated sequences of events, but are related to the mode of sexual reproduction, the presence or absence of larval stages in the life cycle, and the ecology of the adult. Efforts to classify various invertebrate life cycles and to explain the evolutionary forces that gave rise to them have produced a large number of publications and a great deal of controversy. Most of these studies concern marine invertebrates, on which we center our attention first. We then present some comments on the special adaptations of terrestrial and freshwater forms.

Classification of Life Cycles

Our discussion of life cycles focuses on sexually reproducing animals. Sexual reproduction with some degree of gamete dimorphism is nearly universal among eukaryotes. Male and female gametes may be produced by the same individual (**hermaphroditism**, cosexuality, or in plants, monoecy) or by separate individuals (**gonochory**, or in plants, dioecy). Most terrestrial animals are gonochoristic, but hermaphroditism is widespread among marine invertebrates (as is monoecy among land plants). Mechanisms of sex determination are diverse; in certain arthropods, females may be diploid and males haploid, a system known as **haplodiploidy**. Other forms of sex determination involve structurally distinct sex chromosomes. In **male heterogamety**, males carry X and Y sex chromosomes and females are XX, as in some vertebrates. In **female heterogamety**, females are ZW and males ZZ, as seen in many crustaceans. There is typically little or no recombinational exchange between X and Y chromosomes (or between Z and W) because there is almost no genetic homology between the sex chromosomes. Most of the Y (or W) chromosome is devoid of functional gene loci, other than a few RNA genes and some genes required for male (or female) fertility and sex determination. In any event, it is the fusion of male and female gametes that initiates the process of ontogeny and a new cycle in the life history of an organism.

A number of classification schemes for life cycles have been proposed over the past five decades (see papers by Thorson, Mileikovsky, Chia, Strathmann, Jablonsky, Lutz, and McEdward). We have generalized from the works of various authors and suggest that most animals display some form of one of the three following basic patterns (Figure 5.13).

1. **Indirect development** The life cycle includes free spawning of gametes followed by the development of a free larval stage (usually a swimming form), which is distinctly different from the adult and must undergo a more or less drastic metamorphosis to reach the juvenile or young adult stage. The equivalent in terrestrial invertebrates is seen in insects with holometabolous development. In aquatic groups, two basic larval types can be recognized.
   a. Indirect development with **planktotrophic** larvae. The larva survives primarily by feeding, usually on plankton. (The feeding larvae of some deep-sea species are demersal and feed on detrital matter, never swimming very far off the bottom.)
   b. Indirect development with **lecithotrophic** larvae. The larva survives primarily on yolk supplied to the egg by the mother.
2. **Direct development** The life cycle does not include a free larva. In these cases the embryos are cared for by the parents in one way or another (generally by brooding or encapsulation) until they emerge as juveniles. The equivalent in terrestrial invertebrates is seen in insects with ametabolous or hemimetabolous development.
3. **Mixed development** The life cycle involves brooding or encapsulation of the embryos at early stages of development and subsequent release of free planktotrophic or lecithotrophic larvae. The initial source of nutrition and protection is the adult.

Not every species can be conveniently categorized into just one of the above developmental patterns. For example, some species have free larvae that depend on yolk for a time, but begin to feed once they develop the ability to do so. Some species actually display different developmental strategies under different environmental conditions—convincing evidence that embryogenies, like so many other aspects of a species or population, are subject to selection pressures and can readily evolve.

These life cycle patterns provoke three basic questions. First, how do different developmental sequences relate to other aspects of reproduction such as egg types and mating or spawning activities? Second, how do overall developmental sequences relate to the survival strategies of larvae and adults? Third, what evolutionary mechanisms are responsible for the patterns seen in any given species? Given the large number of interacting factors to be considered, these are complex questions, and our understanding of them is still incomplete. However, by first examining cases of direct and indirect development, we can illustrate some of the principles that underlie their relationships to different ecological situations. Then we will briefly address some ideas about mixed development.

Indirect Development

Consider first a life cycle with planktotrophic larvae (Figure 5.13A). The metabolic expense incurred on the part of the adults involves only the production and

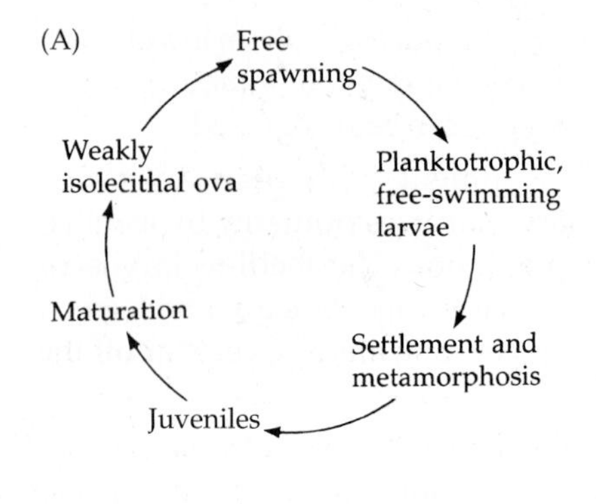

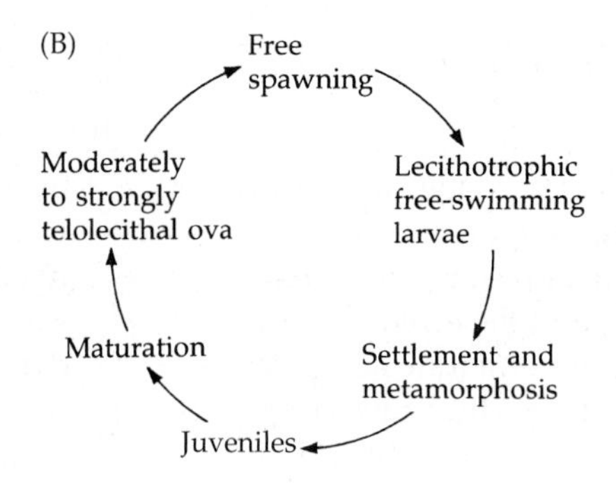

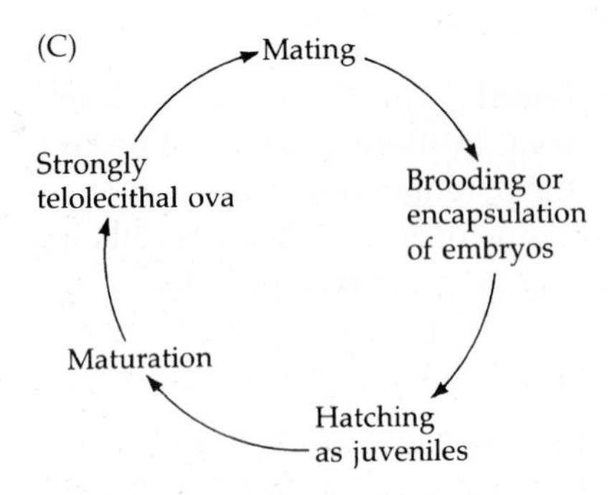

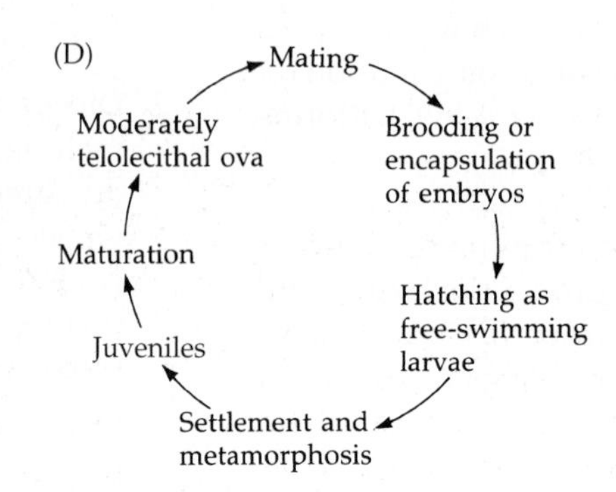

Figure 5.13 Some generalized invertebrate life cycle strategies. (A) Indirect development with planktotrophic larvae. (B) Indirect development with lecithotrophic larvae. (C) Direct development. (D) Mixed life cycle.

release of gametes. Animals with fully indirect development generally do not mate; instead, they shed their eggs and sperm into the water, thus divorcing the adults from any further responsibility of parental care. Such animals typically undergo **synchronous (epidemic) broadcast spawning** of large numbers of gametes, thereby ensuring some level of successful fertilization. This pattern of development is relatively common in opportunistically settling and colonizing (*r*-selected) marine species that make use of tides or ocean currents to disperse their progeny, and are capable of rapid production of high numbers of gametes.

The eggs of animals expressing such traits are usually isolecithal and individually inexpensive to produce. The overall cost—and it is a significant one to each potential parent—is in the production of very large numbers of eggs. Being supplied with little yolk, the embryos must develop quickly into feeding larvae to survive. Mortalities among the embryos and larvae are extremely high and can result from a variety of factors, including lack of food, predation, or adverse environmental conditions. Each successful larva must accumulate enough nutrients from feeding to provide for its immediate survival, as well as for the processes of settling and metamorphosis from larva to juvenile or subadult. That is, they must feed to excess as they prepare for a new lifestyle as a juvenile. Survival rates from zygote to settled juvenile are often less than one percent. Such high mortalities are offset by the initial high production of gametes. But by the same token, high larval mortalities offset the high production of gametes—if all of these zygotes survived, the Earth would quickly be covered by the offspring of animals with indirect development.

What are the advantages and limitations of such a life history, and under what circumstances might it be successful? This sort of planktotrophic development is most common among benthic marine invertebrates in relatively shallow water and the intertidal zones of tropical and warm temperate seas. Here the planktonic food sources are more consistently available (although often in low concentration) than they are in colder or deeper waters, thus reducing the danger of starvation of the larvae. Such **meroplanktonic life cycles** allow animals to take advantage of two distinct resources (plankton in the upper water column as larvae; benthos and bottom plankton as adults). This arrangement reduces or eliminates competition between larvae and adults. Indirect development also provides a mechanism for dispersal, a particularly important benefit to species that are sessile or sedentary as adults. There is good evidence to suggest that animals with free-swimming larvae are likely to recover more quickly from damage to the adult population than those engaging in direct development. A successful set of larvae is a ready-made new population to replace lost adults.

The disadvantages of planktotrophic development result from the unpredictability of larval success. Excessive larval deaths can result in poor recruitment and the possibility of invasion of suitable habitats by competitors. Conversely, unusually high survival rates of larvae can lead to overcrowding and intraspecific competition upon settling.

Animals that produce fully lecithotrophic larvae (Figure 5.13B) must produce yolky and thus more metabolically expensive eggs. This built-in nutrient supply releases the larvae from dependence on environmental food supplies and generally results in reduced mortalities. It is not surprising that these animals produce somewhat fewer ova than those with planktotrophic larvae. The eggs are either spawned directly into the water or are fertilized internally and released as zygotes. Again, the adults' parental responsibility ends with the release of gametes or zygotes into the environment. Although survival rates of lecithotrophic larvae are generally higher than those of planktotrophic types, they are low compared with those of embryos that undergo direct development.

Marine invertebrates that live in relatively deep benthic environments tend to produce lecithotrophic larvae. Here, some of the advantages of indirect development are realized, but larvae do not require environmental food supplies and therefore avoid the intense predation commonly encountered in surface water. The trade-off is clear: In deeper water fewer, more expensive zygotes are produced, but they can survive where more numerous, less expensive planktotrophic larvae cannot.

Settling and Metamorphosis

Of particular importance to the successful completion of animal life cycles with free larval stages are the processes of settlement and metamorphosis. These events are crucial and dangerous times in an animal's life cycle because they often require rapid and dramatic changes in individual habitats and lifestyle. Free-swimming larvae usually metamorphose into benthic juveniles, a process that involves the shedding of larval structures and the rapid growth or mobilization of juvenile ones. Surviving this transformation in form and function and adopting a new mode of life requires adequate stored resources, appropriate responses to internal and external conditions, and considerable luck.

Throughout their free-swimming lives larvae have been "preparing" for these events, until they reach a condition in which they are physiologically capable of metamorphosis. Such larvae are termed **competent**. The duration of the free-swimming period varies greatly among metazoan larvae and depends on factors such as original egg size, yolk content, and the availability of food for planktotrophic forms. Once a larva becomes competent, it generally begins to respond to certain environmental cues that induce settling behavior. Metamorphosis is often preceded by settling, although some species metamorphose prior to settling and still others engage in both processes simultaneously. In any case, larvae typically become negatively phototactic and/or positively geotactic and move toward the bottom to settle. In species that are planktonic both as larvae and as adults (**holoplanktonic species**), the larvae obviously do not settle on the benthos.

Once contact with a substratum is made, a larva tests it, to determine its suitability as a habitat. This act of substratum selection may involve processing physical, chemical, and biological information. A number of studies show that important factors include: substratum texture, composition, and particle size; presence of conspecific adults (or dominant competitors); presence of key chemical cues; presence of appropriate food sources; and the nature of bottom currents or turbulence. Contact with the substrate includes risks. Previously settled planktivores and predators are likely to be common in many potentially suitable habitats. Once again, larval mortalities at this stage are high.

Many invertebrate larvae touch down on the bottom for a few minutes, and then launch themselves back up into the current again and again until a suitable substratum is found. Assuming an appropriate situation is encountered, metamorphosis is induced and proceeds to completion. Interestingly, some feeding larvae are able to postpone metamorphosis and resume planktonic life if they initially encounter an unsuitable substratum. In such cases, however, the larvae become gradually less selective; eventually, metamorphosis ensues regardless of the availability of a proper substratum. The ability to prolong the larval period until conditions are favorable for settlement has obvious survival advantages, and invertebrates differ greatly in this capability. Those that can postpone settlement may do so by several hours, days, or even months (based on laboratory experiments).

Direct Development

Direct development avoids some of the disadvantages but also misses some of the advantages of indirect development. A typical scenario involves the production of relatively few, very yolky eggs, followed by some sort of mating activity and internal fertilization (Figure 5.13C). The embryos receive prolonged parental care, either directly (by brooding in or on the parent's body) or indirectly (by encapsulation in egg cases provided by the parent). Animals that simply deposit their fertilized eggs, either freely or in capsules, are said to be **oviparous**. A great number of invertebrates as well as some vertebrates (amphibians, many fishes, reptiles, and birds) display oviparity. Animals that brood their embryos internally and nourish them directly, such as placental mammals or peracarids crustaceans, are described as **viviparous**. **Ovoviviparous** animals brood their embryos internally but rely on the yolk within the eggs to nourish their developing young. Most internally brooding invertebrates are ovoviviparous.

The large, yolky eggs of most invertebrates with direct development are metabolically expensive to produce. But while only a few eggs are possible, the investment is protected though parental effort and survival rates are relatively high. The dangers of planktonic larval life and metamorphosis are avoided and the embryos eventually hatch as juveniles.

What sorts of environments and lifestyles might result in selection for such a developmental sequence? At the risk of overgeneralizing, we can say that there is a tendency for specialist (e.g., *K*-selected) species to display direct development. Another situation in which direct development occurs is when the adults have no dispersal problems. We find, for example, that holoplanktonic species with pelagic adults (e.g., arrow worms, phylum Chaetognatha; pelagic gastropods) often undergo direct development, either by brooding or by producing floating egg cases. A second situation is one in which critical environmental factors (e.g., food, temperature, water currents) are highly variable. There is a trend among benthic invertebrates to switch from planktotrophic indirect development to direct development at increasingly higher latitudes. The relatively harsh conditions and strongly seasonal occurrence of planktonic food sources in polar and subpolar areas partially explain this tendency.

In addition to avoiding some of the danger of larval life, direct development has another distinct advantage. The juveniles hatch in suitable habitats where the adults brooded them or deposited the eggs in capsules. Thus, there is a reasonable assurance of appropriate food sources and other environmental factors for the young.

Mixed Development

As defined earlier, mixed life histories involve some period of brooding prior to release of a free larval stage. Costly, yolky zygotes are protected for some time and then are released as larvae, exploiting the advantages of dispersal. This developmental pattern is often ignored when classifying life histories, but in fact it is widespread among gastropods, insects, crustaceans, sponges, cnidarians, and a host of other animal groups. Some workers view mixed development as either the "best" or the "worst" of both worlds (i.e., fully indirect or direct). Others suggest that such sequences are evolutionarily unstable, and that local environmental pressures are driving them toward direct or indirect development. There are, however, other possible explanations. It may very well be that under some environmental situations a brooding period followed by a larval phase is adaptive and stable.

Furthermore, at least some species show population variability in the relative lengths of time embryos exist in a brooded versus a free larval phase. If this variability responds to local environmental pressures, then clearly such a species might adapt quickly to changing conditions, or even exploit this ability by extending its geographic range to live under a variety of settings. In this regard, mixed life histories may represent **developmental polymorphisms**, in which the frequency and intensity of particular environmental cues influences the proportion of the population that expresses or does not express a particular larval phenotype. Such **phenotypic plasticity** in life history expression is an area in need of further investigation.

Our short description of life history strategies certainly does not explain all observable patterns in nature. The historical and evolutionary forces acting on invertebrates (and their larvae) are highly complex. For example, larvae are subject to all manner of oceanographic variables (e.g., diffusion, lateral and vertical transport, sea floor topography, storms) as well as their self-directed vertical movements, seasonality, and biotic factors (predators, prey, competition, nutrient availability). Life history predictions based strictly on environmental conditions do not always hold true. Invertebrates living in the deep sea and at the poles do not always brood (as was once thought). We now know that all life history strategies occur in these regions, and many deep-sea and polar species release free-swimming larvae, and even planktotrophic larvae. Even some invertebrates of deep-sea hydrothermal vent communities produce free-swimming larvae. In many cases, this may be due to evolutionary constraints: vent gastropods, for example, belong to lineages that are almost strictly lecithotrophic, regardless of latitude or habitat. Thus, vent gastropods are apparently constrained by their phylogenetic histories. Other vent species that release free larvae, however, are not so constrained: mytilid bivalves, for example, possess a wide range of reproductive modes, and tend to release planktotrophic larvae in deep-sea and vent environments. Furthermore, reproductive cycles in many abyssal invertebrates appear to be seasonal, perhaps cued by annual variations in surface water productivity. There is still much to be learned.

Adaptations to Land and Fresh Water

The foregoing account of life cycle strategies applies largely to marine invertebrates. Many invertebrates, however, have invaded land or fresh water, and their success in these habitats requires not only adaptation of the adults to special problems, but also adaptation of the developmental forms. As discussed in Chapter 1, terrestrial and freshwater environments are more rigorous and unstable than the sea, and they are generally unsuitable for reproductive strategies that involve free spawning of gametes or the production of delicate larval forms. Most groups of terrestrial and freshwater invertebrates have adopted internal fertilization followed by direct development, while their marine counterparts often exhibit external fertilization and produce free-swimming larvae. The insects, flatworms and nematodes are notable exceptions, in which a wide range of mixed development life histories have evolved. In these cases, larvae are highly adapted to their freshwater, terrestrial or parasitic environments with unique traits that are unlikely to have existed in any marine larval ancestor.

Parasite Life Cycles

The evolutionary success of parasites is clear. Every animal species examined for symbionts appears to provide habitat for at least one, and usually many associated species. These symbionts often draw benefits from their host at their host's expense, and thus are parasites. Most parasites have rather complicated life cycles, and specific examples are given in later chapters. For now, we will examine parasitic lifestyles in a general way to understand their central features, and to introduce some basic terminology.

As outlined in Chapter 1, parasites may be classified as **ectoparasites** (living upon the host), **endoparasites** (living internally, within the host), or **mesoparasites** (living in some cavity of the host that opens directly to the outside, such as the oral, nasal, anal, or gill cavities). While associated with a host, a parasite may engage in sexual or asexual reproduction, but the eggs or embryos are usually released to the outside via some avenue from the host's body. The problems at this point are very similar to those encountered during indirect larval development: some mechanism must be provided to ensure adequate survival through the developmental stages, and some sequence of events must bring the parasite back to an appropriate host (the proper "substratum") for maturation and reproduction. As explained earlier, habitat transitions are risky. Thus, many parasites are **parthenogenetic**—a form of reproduction in which the ovum undergoes embryonic development and produces a new individual without fertilization. Parthenogenesis produces offspring that are genetically identical to their parent. Other parasites may be capable of asexual reproduction by way of fission or budding. The production of asexual progeny appears to be one mechanism by which parasites offset the high mortality that attends transitions from one host to the next.

Parasites exploit at least two different habitats in their life cycles. This practice is essential because when hosts die, their parasites usually die with them. Thus, the developmental period from zygote to adult parasite involves either the invasion of another host, or a free-living period between host invasions. When more than one host species is utilized for the completion of the life cycle, the organism harboring the adult parasite is called the **primary** or **definitive** host. Hosts in which developmental or larval forms reside are called **intermediate** hosts. The completion of complex life cycles often requires elaborate methods of transfer from one host to the other, and again, surviving the changes from one habitat to another can be problematic. Losses are routinely high. Thus, we find that many parasites enjoy some of the benefits of indirect development (e.g., dispersal and exploitation of multiple resources) while being subjected to accompanying high mortalities and the dangers of very specialized lifestyles.

We emphasize again that the above discussions of life cycles are generalities to which there are many exceptions. But given these basic patterns, you should recognize and appreciate the adaptive significance of life history patterns of the different invertebrate groups discussed later. You might also be able to predict the sorts of sequences that would be likely to occur under different conditions. For example, given a situation in which a particular species is known to produce very high numbers of free-spawned, isolecithal ova, what might you predict about cleavage pattern, blastula and gastrula type, presence or absence of a larval stage, type of larva, adult lifestyle, and ecological settings in which such a sequence would be advantageous? We hope you will develop the habit of asking these kinds of questions and thinking in this way about all aspects of your study of invertebrates.

The Relationships Between Ontogeny and Phylogeny

Of the many fields of study from which we draw information used in phylogenetic investigations, embryology has been one of the most important. The construction of phylogenies may be accomplished and subsequently tested by several different methods (Chapter 2). But regardless of method, one of the principal problems of phylogeny reconstruction—in fact, central to the process—is separating true homologies from similar character traits that are the result of evolutionary convergence. Even when these problems involve comparative adult morphology, one must often seek answers in studies of the development of the organisms and structures in question. The search is for developmental processes or structures that are homologues and thus demonstrate relationships between ancestors and descendants. Changes that take place in developmental stages are not trivial evolutionary events. It has been effectively argued that developmental phenomena may themselves provide the evolutionary mechanisms by which entire new lineages originated (Chapter 1). As Stephen Jay Gould (1977) has noted,

> *Evolution is strongly constrained by the conservative nature of embryological programs. Nothing in biology is more complex than the production of an adult . . . from a single fertilized ovum. Nothing much can be changed very radically without discombobulating the embryo.*

Indeed, the persistence of distinctive body plans throughout the history of life is testimony to the resistance to change of complex developmental programs. (See Hall 1996 for an excellent analysis of these issues.)

Although few workers would argue against a significant relationship between ontogeny and phylogeny, the exact nature and extent of the relationship have historically been subjects of considerable controversy, a good deal of which continues today. (Gould 1977 presents a fine analysis of these debates.) Central to much of the controversy is the concept of recapitulation.

The Concept of Recapitulation

In 1866 Ernst Haeckel, a physician who found a higher calling in zoology and never practiced medicine, introduced his **law of recapitulation** (or the **biogenetic law**), most commonly stated as "ontogeny recapitulates phylogeny." Haeckel suggested that a species' embryonic development (ontogeny) reflects the adult forms of that species' evolutionary history (phylogeny). According to Haeckel, this was no accident, but a result of a close mechanistic relationship between the two processes: phylogenesis is the actual cause of embryogeny. Restated, animals have an embryogeny *because* of their evolutionary history. Evolutionary change over time has resulted in a continual adding on of morphological stages to the developmental process of organisms. The implications of Haeckel's proposal are immense. Among other things, it means that to trace the phylogeny of an animal, one need only examine its development to find therein a sequential or "chronological" parade of the animal's adult ancestors.

Ideas and disagreement concerning the relationship between ontogeny and phylogeny were by no means new even at Haeckel's time. Over 2,000 years ago Aristotle described a sequence of "souls" or "essences" of increasing quality and complexity through which animals pass in their development. He related these conditions to the adult "souls" of various lower and higher organisms, a notion suggestive of a type of recapitulation.

Descriptive embryology flourished in the nineteenth century, stimulating vigorous controversy regarding the relationship between development and evolution. Many of the leading developmental biologists of the time were in the thick of things, each proposing his own explanation (Meckel 1811; Serres 1824; von Baer 1828; and others). It was Haeckel, however, who really stirred the pot with his discourse on the "law" of recapitulation. He offered a focal point around which biologists argued pro or con for 50 years; sporadic skirmishes still erupt periodically. Walter Garstang critically examined the biogenetic law and gave us a different line of thinking. His ideas, presented in 1922, are reflected in many of his poems (published posthumously in 1951). Garstang made clear what a number of other biologists had suggested: that evolution must be viewed not as a succession of ancestral adult forms, but as a succession of ontogenies. Each animal is a result of its own developmental processes, and any change in an adult must represent a change in its ontogeny. So what we see in the embryogeny of a particular species are not tiny replicas of its adult ancestors, but rather an evolved pattern of development in which clues or traces of ancestral ontogenies, and thus phylogenetic relationships to other organisms, may be found.

Arguments over these matters did not end with Garstang, and they continue today in many quarters. In general, we tend to agree with the approach (if not all of the details) of Gosta Jägersten in *Evolution of the Metazoan Life Cycle* (1972). Recapitulation per se should not categorically be accepted or dismissed as an "always" or "never" phenomenon. The term must be clearly defined in each case investigated, not locked in to Haeckel's original definition and implications. For instance, similar, distinctive, homologous larval types within a group of animals reflect some degree of shared ancestry (e.g., crustacean nauplii or molluscan veligers). And we may speculate on such matters at various taxonomic levels, even when the adults are quite different from one another (e.g., the similar trochophore larvae of polychaetes and molluscs). These phenomena may be viewed as developmental evidence of relatedness through shared ancestry, and thus they are examples of "recapitulation" in a broad sense.

Jägersten's example of vertebrate gill slits is particularly appropriate because, to him, it provides a case in which Haeckel's strict concept of recapitulation is manifest. In writing of this feature Jägersten (1972) stated,

> *The fact remains . . . that a character which once existed in the adults of the ancestors but was lost in the adults of the descendants is retained in an easily recognizable shape in the embryogenesis of the latter. This is my interpretation of recapitulation (the biogenetic 'law').*

Hyman (1940) perhaps put it most reasonably when she wrote,

> *Recapitulation in its narrow Haeckelian sense, as repetition of adult ancestors, is not generally applicable; but ancestral resemblance during ontogeny is a general biological principle. There is no need to quibble over the word recapitulation; either the usage of the word should be altered to include any type of ancestral reminiscence during ontogeny, or some new term should be invented.*

Other authors, however, are not comfortable with such flexibility and have made great efforts to categorize and define the various possible relationships between ontogeny and phylogeny, of which strict recapitulation is considered only one (see especially Chapter 7 of Gould 1977). Although much of this material is beyond the scope of this book, we discuss a few commonly used terms here because they bear on topics in later chapters. We have drawn on a number of sources cited in this chapter to mix freely with our own ideas in explaining these concepts.

Heterochrony and Paedomorphosis

When comparing two ontogenies, one often finds that some features appear earlier or later in one sequence than in the other. Such temporal displacement during development is called **heterochrony**. When comparing suspected ancestral and descendant embryogenies, for example, we may find the very rapid (accelerated) development of a particular feature and thus its relatively early appearance in a descendant species or lineage. Conversely, the development of a trait may be slower (retarded) in a descendant than in an ancestor and thus appear later in the descendant's ontogeny. This retardation may be so pronounced that a structure may never develop to more than a rudiment of its ancestral condition. (For excellent reviews of heterochrony and its impact on phylogeny see Gould 1977, and McKinney and McNamara 1991.)

Particular types of heterochrony result in a condition known as **paedomorphosis**, wherein sexually mature adults possess features characteristically found in early developmental stages of related forms (i.e., juvenile or larval features). Paedomorphosis results when adult reproductive structures develop before completion of the development of all the adult nonreproductive (somatic) structures. Thus, we find a reproductively functional animal retaining what in the ancestor were certain embryonic, larval, or juvenile characteristics. This condition can result from two different heterochronic processes. These are **neoteny**, in which somatic development is retarded, and **progenesis**, in which reproductive development is accelerated. These two terms are frequently used interchangeably because it is not always possible to know which process has given rise to a particular paedomorphic condition. Recognition of paedomorphosis may play a significant role in examining evolutionary hypotheses concerning the origins of certain lineages. For example, the evolution of precocious sexual maturation of a planktonic larval stage (that would "normally" continue developing to a benthic adult) might result in a new diverging lineage in which the descendants pursue a fully pelagic existence. Such a scenario, for example, may have been responsible for the origin of some small planktonic crustaceans. Paedomorphosis has also played major roles in theories regarding the origin of the vertebrates.

Myriad questions about the role of embryogenesis in evolution and the usefulness of embryology in constructing and testing phylogenies persist. As the following accounts show, different authors continue to hold a variety of opinions about these matters.

The Origin of the Metazoa

One theme we develop throughout this book is the evolutionary relationships within and among the invertebrate taxa. Life has probably existed on this planet for nearly 4 billion years; humans have been observing it scientifically for only a few hundred years, and evolutionarily for only about 150 years. Thus, the thread of evolutionary continuity we actually see around us today looks a bit like frazzled ends, representing the many successful animal lineages that survive today, but omitting the legions of extinct species and lineages whose identities could provide a clearer understanding of the history of life on Earth. It is only through conjecture, study, inference, and the testing of hypotheses that we are able to trace phylogenetic strands back in time, joining them at various points to produce hypothetical pathways of evolution. We do not operate blindly in this process, but use rigorous scientific methodology to draw upon information from many disciplines in attempts to make our evolutionary hypotheses meaningful and (we hope) increasingly closer to the truth—to the actual biotic history of Earth (Chapter 2).

In Chapter 1 we briefly reviewed the history of life, in part inferred from the fossil record, and in Chapter 28 we present a phylogenetic tree of the animal kingdom. However, many workers have not been satisfied to develop phylogenetic analyses based solely upon known (extant and extinct) animal groups, but have felt compelled to speculate on hypothetical ancestors that might have occurred along the evolutionary road to modern life. A variety of evolutionary stories have been proposed to describe these sequences of hypothetical metazoan ancestors. We discuss some of these below, and some key works are cited in the references at the end of this chapter and Chapter 28.

Origin of the Metazoan Condition

The origin of the metazoan condition has received attention for more than a century. One of the most spectacular phenomena in the fossil record is the abrupt diversification of nearly all of the metazoan phyla living today in a brief span of 30 million years, at the Precambrian–Cambrian transition (approximately 570–600 million years ago). There is now little doubt that animals—the Metazoa—arose as a monophyletic group from a protist ancestor, 650 million years ago or earlier (Chapter 1). The debates now concern which protist group was ancestral to the first Metazoa, what these first animals were like, what environments they inhabited, and how the changes from unicellularity to multicellularity took place.

Historical Perspectives on Metazoan Origins

What intermediate forms might have linked protists and metazoans? Some authors have chosen to design logical but hypothetical creatures for this purpose. Others rummage among extant types, arguing the advantages of using "real" organisms. Although it is probable that the actual precursor of the Metazoa is long extinct, the existence of modern-day forms that combine protist and metazoan traits keeps this debate alive. These organisms include enigmatic multicellular animals of uncertain position, imagined and real colonial flagellates, and hypothetical multinucleate ciliates. Figure 5.14 illustrates some of these creatures for comparative purposes.

Before molecular tools convincingly linked protists to metazoan ancestry, several theories of metazoan evolution enjoyed support. In 1892, Johannes Frenzel

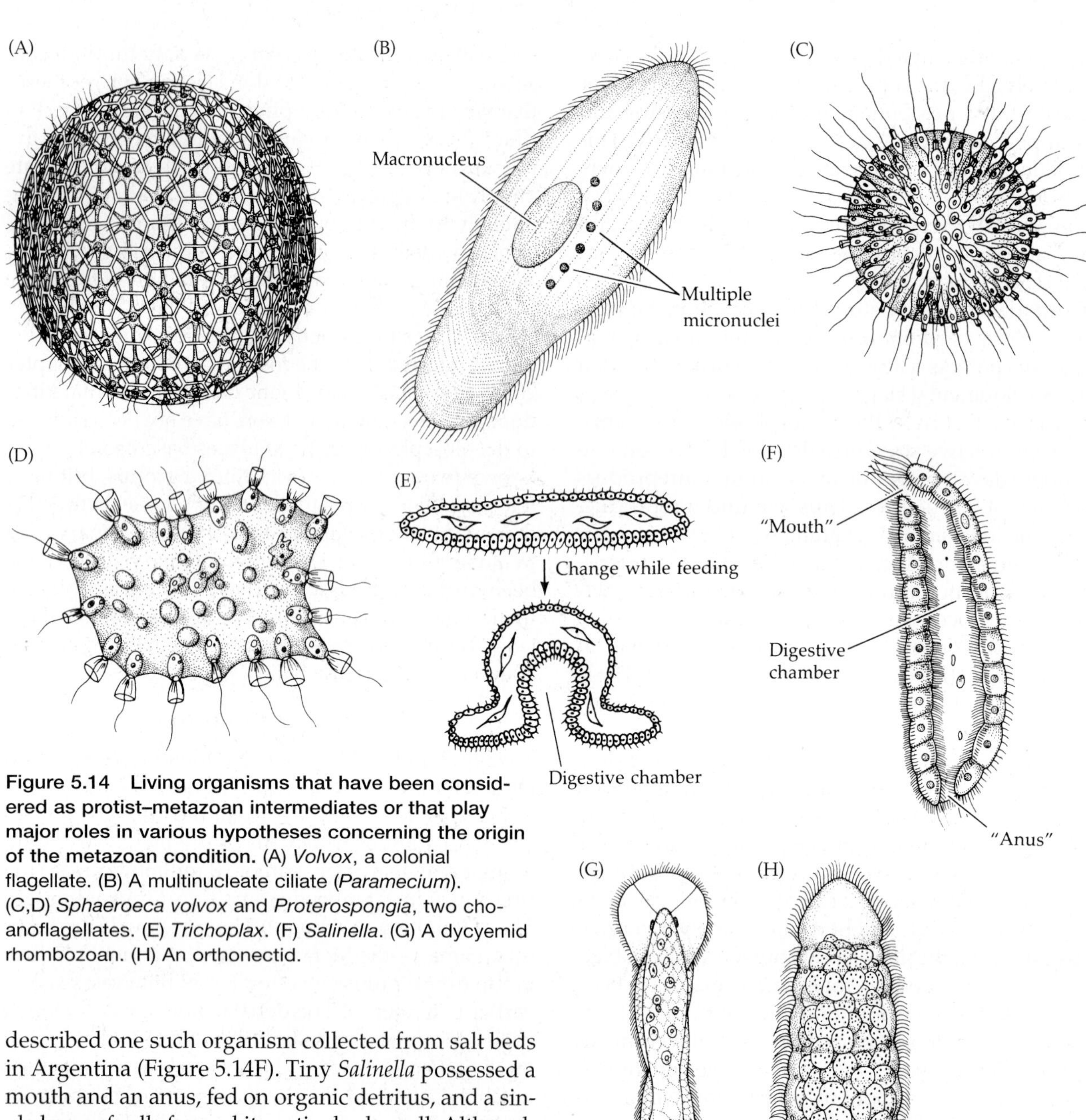

Figure 5.14 Living organisms that have been considered as protist–metazoan intermediates or that play major roles in various hypotheses concerning the origin of the metazoan condition. (A) *Volvox*, a colonial flagellate. (B) A multinucleate ciliate (*Paramecium*). (C,D) *Sphaeroeca volvox* and *Proterospongia*, two choanoflagellates. (E) *Trichoplax*. (F) *Salinella*. (G) A dycyemid rhombozoan. (H) An orthonectid.

described one such organism collected from salt beds in Argentina (Figure 5.14F). Tiny *Salinella* possessed a mouth and an anus, fed on organic detritus, and a single layer of cells formed its entire body wall. Although this creature lacked the layered cellular construction of the Metazoa, it displayed a higher level of organization than colonial protists, and the phylum Monoblastozoa was erected for it. Sadly, *Salinella* has not been seen since the original report, and many zoologists suspect that Frenzel seriously misinterpreted whatever creature he saw. Other so-called "mesozoan" phyla, Rhombozoa and Orthonectida (Figure 5.14G,H), are also structurally simple, but these animals are endoparasites of invertebrates and have complex life cycles. While possibly resembling early metazoans, most workers consider their body organization and life cycles, and their phylogenetic position to be more derived than ancestral.

The **colonial theory** of metazoan evolution was first expressed by Ernst Haeckel (1874), who proposed that a colonial flagellated protist gave rise to a planuloid metazoan ancestor (the planula is the basic larval type of cnidarians; see Chapter 7). The ancestral protist in this theory was a hollow sphere of flagellated cells that developed anterior–posterior locomotor orientation, and specialization of cells into separate somatic and reproductive functions. As we explain in Chapter 3, similar conditions are common in living colonial protists, including freshwater, colonial, photosynthetic flagellates such as *Volvox* (Figure 5.14A). Haeckel called this hypothetical protometazoan ancestor a **blastea** (Figure 5.15A) and supported its validity by noting the widespread occurrence of coeloblastulae among modern animals.

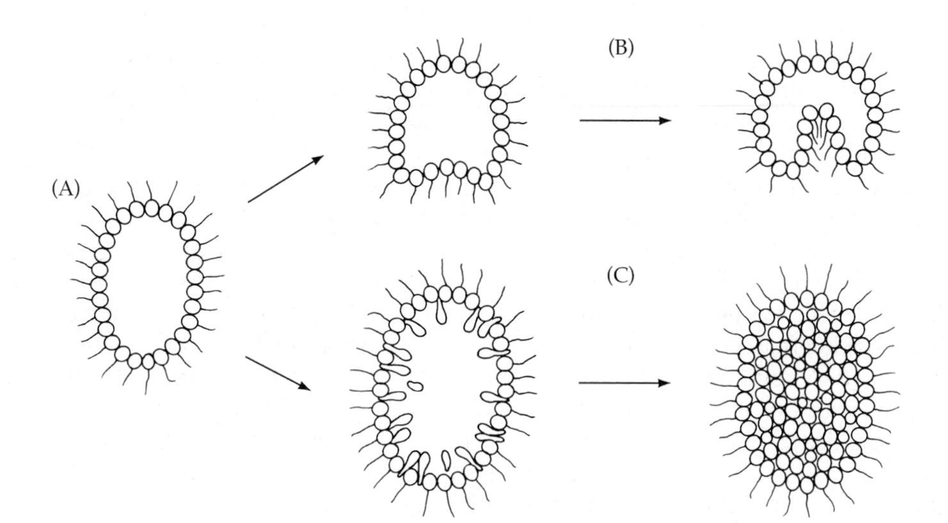

Figure 5.15 Two versions of the colonial theory of the origin of the Metazoa. (A) The hypothetical colonial flagellate ancestor, Haeckel's "blastea" (section). (B) According to Haeckel, the transition to a multicellular condition occurred by invagination, a developmental process that resulted in a hollow "gastrea." (C) According to Metschnikoff, the formation of a solid "gastrea" occurred by ingression.

In Haeckel's scenario, the first Metazoa arose by invagination of the blastea; the resulting animals had a double-layered, gastrula-like body (a **gastrea**) with a blastopore-like opening to the outside (Figure 5.15B) similar to the gastrulae of many modern animals. Haeckel believed that these ancestral creatures (the blastea and gastrea) were recapitulated in the ontogeny of modern animals, and the gastrea was viewed as the metazoan precursor to the cnidarians. It has been said that the monociliated cells of the body wall of Porifera and Cnidaria support this hypothesis. Haeckel's original ideas were somewhat modified over the years by various authors (e.g., Elias Metschnikoff, Libbie Hyman). Some have argued that the transition to a layered construction occurred by ingression rather than by invagination, and that the original Metazoa were solid, not hollow, based in large part on the view that ingression is the primitive form of gastrulation among cnidarians (Figure 4.15C).

In 1883, Otto Bütschli presented another variant of the colonial theory, a bilaterally symmetrical, flattened creature consisting of two cell layers, which fed by crawling over its food, and using its ventral layer as a digestive surface. Bütschli called this creature a **plakula**. In amazing support of the plakula hypothesis, a tiny, flagellated, multicellular creature was discovered in a marine aquarium in the early twentieth century. *Trichoplax adhaerens* was placed in its own phylum, the Placozoa (see Chapter 6), and like Bütschli's plakula has an outer, partly flagellated epithelium surrounding an inner mesenchymal cell mass. Its body margins are irregular, its cells show some specialization for somatic and reproductive function, and when feeding, *Trichoplax* "hunches up" to form a temporary digestive chamber on its underside (Figure 5.14E)—producing a form strikingly similar to Bütschli's hypothetical creature. While this hypothesis is compelling, molecular phylogenetic analyses do not place *Trichoplax* at the base of the metazoan tree.

In the 1950s and 60s J. Hadzi and E. D. Hanson envisioned the metazoan ancestor as a multinucleate, bilaterally symmetrical, benthic ciliate, crawling about with its oral groove directed toward the substratum. This **syncytial theory**, proposed that a cellular epidermis surrounding an inner syncytial mass could form if this creature's surface nuclei partitioned themselves off from one another with cell membranes, producing acoel worm-like creature. Arguments in support of this hypothesis rested upon similarities between modern ciliates and acoels (Chapter 9), including shape, symmetry, mouth location, surface ciliation and size; large ciliates are larger than small acoels. However, objections to this hypothesis were more convincing. Acoels undergo a complex embryonic development; nothing of this sort occurs in ciliates. Acoel guts are cellular, not syncytial. And molecular phylogenetics has shown acoels to be basal bilaterians, not primitive metazoans. Not surprisingly, the syncytial theory enjoys little support today.

The Origin of Multicellularity

Molecular phylogenetic studies have revealed that multicellularity likely evolved in at least a dozen or more eukaryotic clades, and has led to monophyletic lineages of such disparate groups as plants, animals, several different groups of amoebas, and others. Conditions favoring unicellularity persisted for protists over 1.5 billion years by most accounts until two events occurred. First, atmospheric oxygen of sufficient concentration to support multicellular organization became available due to the activities of photosynthetic algae. Second, predation pressure from heterotrophic protists, capable of phagocytizing or otherwise devouring other unicellular individuals, appears to have favored aggregation of cells after mitosis.

Once a tendency to aggregate arose, there appears to have been competition within individuals for certain functions. If, as appears likely, the first multicellular animals were flagellated, these individuals faced a tradeoff between the ability to swim and the ability to engage in mitotic division. The cellular machinery for both functions appear to compete, as is evidenced even today by the fact that animal cells bearing flagella or cilia never replicate until they have retracted and inactivated their flagellar or ciliary apparatus. In xenacoelomorphs, worn out ciliated epidermal cells are simply reabsorbed (Chapter 9). A balance may have arisen between the ability to move and the ability to replicate cells, favoring a tendency toward cellular specialization. If selection favored a shift in the location of non-flagellated cells toward the interior of the individual, with flagellated cells remaining outside, further specialization of internal cells may have become possible, necessitating the evolution of layers of cells with flexible ontogenetic fates, as well as biochemical mechanisms that distinguished or allowed particular cellular interactions.

Most evidence today points to the protist phylum Choanoflagellata as the likely ancestral group from which the Metazoa arose. Choanoflagellates possess collar cells essentially identical to those found in sponges. Choanoflagellate genera such as *Proterospongia*, *Sphaeroeca*, and others are animal-like colonial protists (Figure 5.14C,D) and are commonly cited as typifying a potential metazoan precursor. However, some recent molecular evidence suggests that ctenophores, not sponges, lie at the base of the metazoan tree. Clearly, debate on the emergence of Metazoa from their protist ancestor will continue for some time to come (Chapter 28).

The Origin of the Bilateral Condition and the Coelom

We discussed the functional significance of bilaterality briefly in Chapter 4. The evolution of an anterior–posterior body axis, unidirectional movement, and cephalization almost certainly coevolved to some degree, and probably coincided with the invasion of benthic environments and the development of creeping locomotion. Furthermore, the origin of the triploblastic condition likely took place soon after the appearance of the first bilateral forms. Among modern-day invertebrates, bilaterality and triploblasty generally co-occur.

Various hypotheses concerning the origin of coelom are summarized in R. B. Clark's fine book *Dynamics in Metazoan Evolution* (1964). Clark's personal approach was a functional one that emphasized the adaptive significance of the coelom as the central criterion for evaluating ideas concerning its origin. When early soft-bodied, bilaterally symmetrical animals larger than a few millimeters assumed a benthic, crawling, or burrowing lifestyle, a fluid (hydrostatic) skeleton was essential for certain types of movement. The evolution of a body cavity filled with fluid against which muscles could operate would have offered a tremendous locomotory advantage in addition to providing a circulatory medium and space for organ development. How might such spaces have originated?

Most of the ideas concerning the evolutionary origin of the coelom were developed from the mid-nineteenth to early twentieth century, during the heyday of comparative embryology. Most of these hypotheses shared the premise of monophyly—that the coelomic condition arose only once. The inherent problem with a monophyletic approach is the difficulty of relating existing coelomate animals to a single common coelomate ancestor. Considering the advantages of possessing a coelom, the very different methods of embryonic development (schizocoely and various forms of enterocoely), and the variety of adult coelomic body plans, it may be more biologically reasonable to suggest that the coelomic condition arose twice. There are several current ideas about how this might have happened, and a number of others have mostly been discarded as being incompatible with existing evidence or with our standard definition of the coelom.

The coelom may have originated by the pinching off and isolation of embryonic gut diverticula as occurs in the development of many extant enterocoelous animals (Figure 5.16). This so-called **enterocoel theory** (in several versions) enjoyed relatively strong support by many authors since it was originally proposed by Sir E. Ray Lankester in 1877. An obvious point in favor of this general idea is that enterocoely does occur in many living animals, thus retaining the hypothetical ancestral process. In addition, various authors cite examples

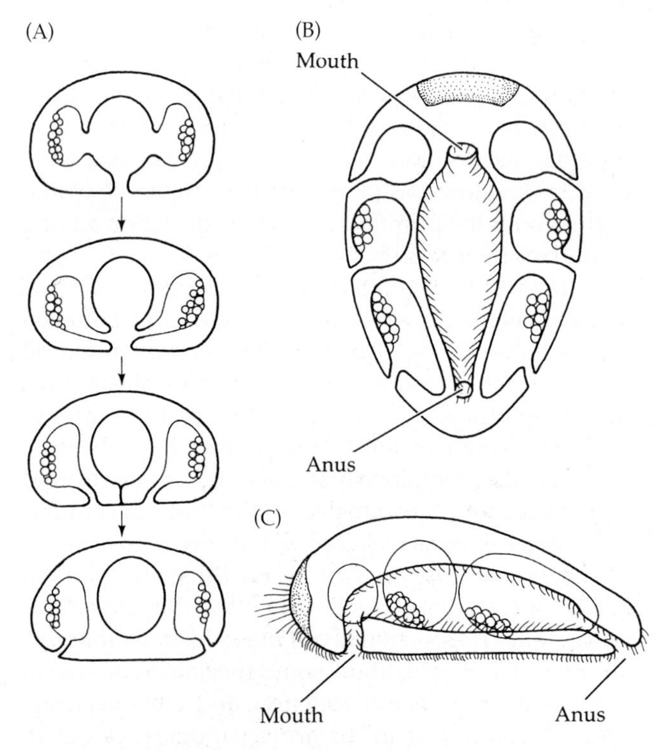

Figure 5.16 Jägersten's bilaterogastrea theory, according to which the coelomic compartments arise by enterocoelic pouching. (A) The formation of paired coeloms from the wall of the archenteron. The slitlike blastopore of the bilaterogastrea closes midventrally, leaving mouth and anus at opposite ends (B). (B,C) The tripartite coelomic condition in Jägersten's hypothetical early coelomate animal (ventral and lateral views).

of noncoelomate animals (anthozoans and flatworms) in which gut diverticula exist in arrangements that resemble possible ancestral patterns.

Another popular idea concerning coelom origin is the **gonocoel theory** (see publications by Bergh, Hatschek, Meyer, Goodrich and others). This hypothesis suggests that the first coelomic spaces arose by way of mesodermally derived gonadal cavities that persisted subsequent to the release of gametes (Figure 5.17). The serial arrangement of gonads, as seen in animals such as flatworms and nemerteans, could have resulted in serially arranged coelomic spaces and linings such as occurs in annelids, where they often still produce and store gametes. A major argument against this hypothesis is that in no modern-day coelomate animals do gonads develop before coelomic spaces. As we have seen, however, heterochrony can account for such turnabouts.

Another idea on coelom origin is called the **nephrocoel theory** (see publications by Lankester, Ziegler, Faussek, Snodgrass, and others). The association between the coelom and excretion has prompted different versions of this hypothesis through about 85 years of moderate support. One idea is that the protonephridia of flatworms expanded to coelomic cavities, arguing that the coelom first arose from ectodermally derived structures. Another view is that coelomic spaces arose as cavities within the mesoderm and served as storage areas for waste products. Certainly the coelomic cavities of many animals are related to excretory functions, but there is no convincing evidence that this relationship was the primary selective force in the origin of the coelomate condition.

Clark (1964) speculated that schizocoely, as we know it today, could have evolved by the formation of spaces within the solid mesoderm of acoelomate animals and then have been retained in response to the positive selection for the resulting hydrostatic skeleton. This is a very straightforward view, in part because, like the enterocoel theory, it accommodates a real developmental process.

As we mentioned earlier, these hypotheses share the fundamental constraint of arguing a monophyletic origin to all coelomate animals. The basic developmental differences between the two great clades of coelomate animals (the Protostomia and the Deuterostomia) suggest that the coelom may have arisen separately in these two lineages. Given the strong similarities between the coelomate Protostomia and acoelomorph worms, it is easy to envision the protostome clade arising from a triploblastic acoelomate ancestor. Hollowing of the mesoderm in such a precursor to produce fluid-filled hydrostatic spaces can be easily explained both developmentally (modern-day schizocoely) and functionally (peristaltic burrowing, increased size, and so on).

To derive the Deuterostomia and Protostomia from an immediate coelomate ancestor creates a complicated scenario. The simplest hypothesis might be to view the deuterostome ancestor as a diploblastic animal, perhaps a planuloid form, in which enterocoely occurred. Deriving the Deuterostomia separately from the evolution of spiral cleavage and the other features of protostomes avoids many of the complications inherent in a monophyletic view of coelom origin. Imagine a hollow, invaginated, gastrula-like metazoan swimming with its blastopore trailing, as do the planula larvae of some cnidarians. Enterocoely may have accompanied a tendency toward benthic life, giving the animal a peristaltic burrowing ability. The archenteron may have then opened anteriorly as a mouth, and the new coelomate creature adopted a deposit-feeding lifestyle. If such a story began at the level of diploblastic Metazoa (e.g., cnidarians), then the radial cleavage seen today in the Bilataria was also present in the ancestor to that group.

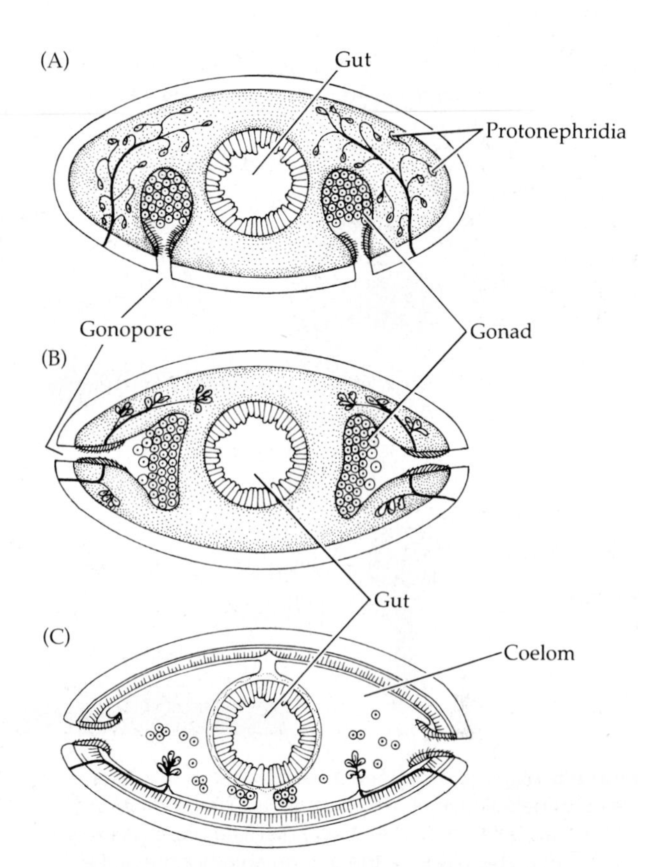

Figure 5.17 A version of the gonocoel theory (schematic cross sections). (A) The condition in flatworms, which have mesodermally derived gonads leading to ventral gonopores. (B) The condition in nemerteans, which have serially arranged gonadal masses leading to laterally placed gonopores. (C) The condition in polychaetes, in which the linings of the gonads have expanded to produce coelomic spaces with coelomoducts to the outside.

The Trochaea Theory

The Danish zoologist Claus Nielsen has envisioned the two major bilaterian clades, Protostomia and Deutero-

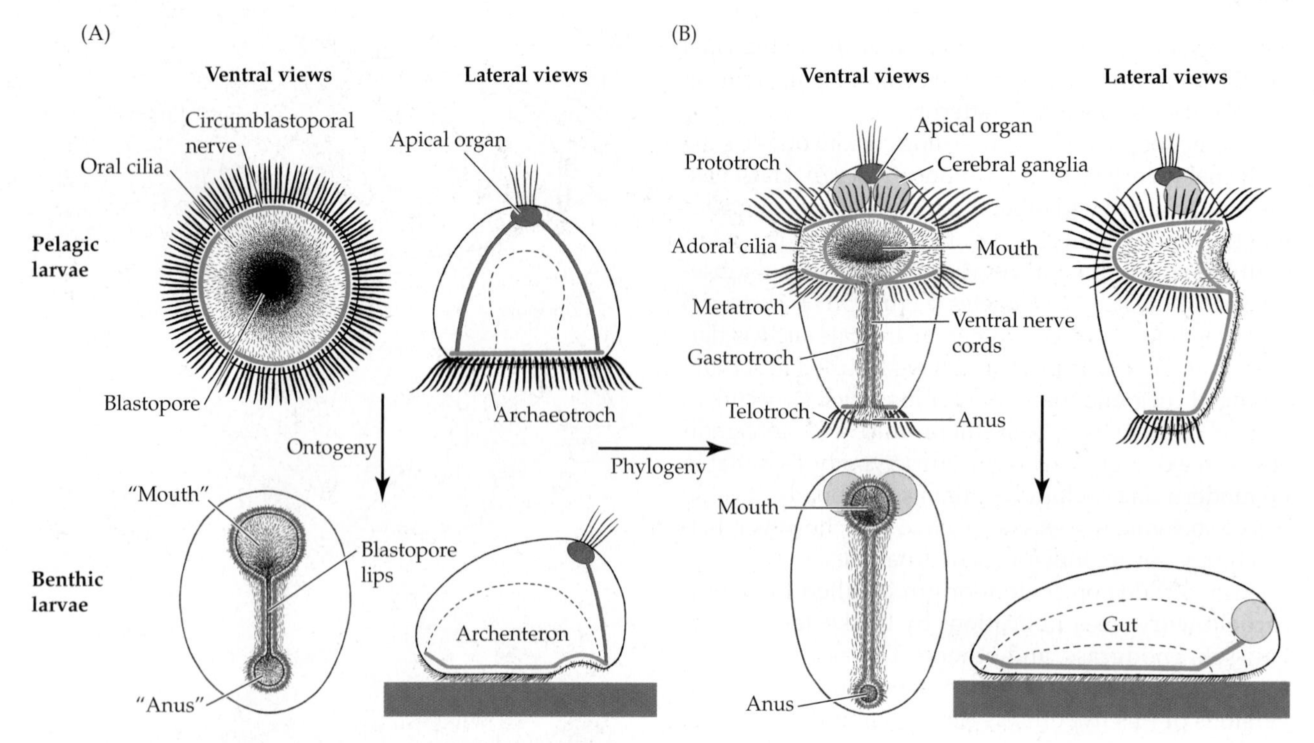

Figure 5.18 The Trochaea Theory. Ventral and lateral views of pelagic larvae and benthic adults predicted by Claus Nielsen's Trochaea Theory. (A) The upper drawings show the morphology of the holopelagic trochaea; the lower drawings illustrate the pelago-benthic life cycle of an early protostomian ancestor. (B) The upper drawings show the pelagic phase of the life cycle of the fully differentiated ancestral protostomian with a trochophore larva, the lower drawings show the benthic form of this animal. Apical organ red; cerebral ganglia yellow; blastoporal nervous system green

stomia, arising from an ancient common ancestor that conforms to Haeckel's radially symmetrical gastrea (see References section). Nielsen's theory proposes that Protostomia arose by way of at least two hypothetical ancestral forms, called the **trochaea** and the **gastroneuron**. The deuterostome line was originally believed to have led to the Deuterostomia by way of a hypothetical **notoneuron** ancestor. (The names gastroneuron and notoneuron referred to the ventral versus dorsal positions of the major nerve cords in most protostomes and deuterostomes, respectively.)

The theory provided a scenario of the evolution of the ancestral Protostomia, which possessed a trochophore larva and a ventral nervous system. The early ancestor in this model was a holopelagic planktotrophic gastraea with a ring of compound cilia (the **archaeotroch**) around the blastopore, which was used in swimming and particle collection by the downstream method (Figure 5.18A). After settling, the adult form of this animal was presumed to creep on the bottom, collecting detritus using monociliated cells around the blastopore. An anterior-posterior axis evolved along with the establishment of a creeping lifestyle. Transport of food particles into and out of the archenteron may have become enhanced by compression of the lateral blastopore lips, which were fused in the adult leaving an anterior mouth and a posterior anus (i.e., a through gut). This fusion of the blastopore lips may soon have become established in the larval stage. The archaeotroch was lost in the creeping adult but retained in the pelagic larva.

The anterior part of the archaeotroch around the mouth could have become laterally extended, with the anterior region becoming the **prototroch** and the posterior region the **metatroch**, bordering a lateral extension of the perioral ciliary area, the **adoral ciliary zone** (Figure 5.18B). Over evolutionary time, this may have created the characteristic trochophore ciliary feeding and swimming structures seen in modern protostomes, wherein the posterior part of the archaeotroch became the **telotroch**. The ciliary bands of the trochophores rely on downstream ciliary feeding, in which the larvae capture food particles from the water on the downstream side of the ciliary feeding bands, and these particles then are transported to the mouth by the adoral ciliary band. The lateral blastopore closure may have resulted in a differentiation of a circumblastoporal ring nerve into an anterior loop around the mouth, the paired (or secondarily fused) ventral nerve cords, and a small loop around the anus (in both the trochophore larvae and the adult). The brain of the trochophore and the adult ancestor consisted of the anteriormost part of the perioral nerve loop and a new paired structure, the large cerebral ganglion developing from the **episphere** of the larva, i.e., from the area in front of the prototroch.

Owing to the realization that Deuterostomia have the neural tube morphologically ventral, and that deuterostomy occurs in several phyla of Protostomia, Nielsen has revised his views on the origin of the former, and in the latest version of his theory the gastroneuron is seen as the latest common ancestor of *all* bilaterians.

As you can see, when one attempts to describe hypothetical ancestors, evolutionary analysis at the level of phyla can be convoluted and problematical. Many different viewpoints of the same phenomena will inevitably arise. We trust, however, that you have gained some insights not only into the particular hypotheses discussed here, but also into evolutionary speculation. A fundamental caveat should be kept in mind: any number of evolutionary pathways can be proposed and made to appear convincing on paper by imagining appropriate hypothetical ancestors or intermediates, but one must always ask whether these marvelous hypothetical creatures would have worked as functional organisms, and whether rigorous phylogenetic analyses support the hypotheses. Clark (1964) spends a good deal of time on this point and emphasizes it in his conclusion with the following passage (p. 258):

> *The most important and least considered of these [principles] is that hypothetical constructs which represent ancestral, generalized forms of modern groups, or stem forms from which several modern phyla diverge, must be possible animals. In other words, they must be conceived as living organisms, obeying the same principles that we have discovered in existing animals.*

In such terms, evolutionary hypotheses can be evaluated. From a phylogenetic point of view, it may be best to avoid initial speculation on what a hypothetical ancestor might have looked like, and instead rely on the analysis of known taxa to establish genealogical relationships or branching patterns. Once a tree has been constructed, the pattern of features associated with the taxa on the tree will themselves predict the nature (character combination) of the ancestor for each branch. This method attempts to avoid the potential problem of circular reasoning, in which a hypothetical ancestor is established first and hence constrains and foretells the nature of the taxa descended from it. In either case, for the hypotheses to be truly scientific, they must be testable with new data gathered outside the framework of that used in their initial formulation.

Selected References

General Invertebrate Embryology

Adiyodi, K. G. and R. G. Adiyodi (eds.). 1983–1998. *Reproductive Biology of Invertebrates. Vols. 1–8*. Wiley, New York.

Conn, D. B. 1991. *Atlas of Invertebrate Reproduction*. Wiley-Liss, New York. [Includes photographs of developmental stages of most major groups.]

Duboule, D. 2007. The rise and fall of Hox gene clusters. Development 124: 2549–2560.

Eckelbarger, K. J. 1994. Diversity of metazoan ovaries and vitellogenic mechanisms: Implications for life history theory. Proc. Biol. Soc. Wash. 107: 193–218.

Giese, A. C. and J. S. Pearse (and V. B. Pearse, Vol. 9) (eds.). 1974–1987. *Reproduction of Marine Invertebrates. Vols. 1–5, 9*. Blackwell Scientific, Palo Alto, CA. *Vol. 6. Echinoderms and Lophophorates*, Boxwood Press, Pacific Grove, CA. [An outstanding series of volumes containing reviews of the invertebrate phyla.]

Gilbert, S. F. 2014. *Developmental Biology*. 10th Ed. Sinauer Associates, Sunderland, MA.

Gilbert, S. F. and A. M. Raunio (eds.). 1997. *Embryology: Constructing the Organism*. Sinauer Associates, Sunderland, MA. [Includes chapters on the development of most major invertebrate phyla.]

Haag, E. S. 2014. The same but different: Worms reveal the pervasiveness of developmental system drift. PLoS Genet. 10(2): e1004150. doi: 10.1371/journal.pgen.1004150

Hall, B. K. 1992. *Evolutionary Developmental Biology*. Chapman & Hall, New York.

Harrison, F. W. and R. R. Cowden (eds.). 1982. *Developmental Biology of Freshwater Invertebrates*. A. R. Liss, New York.

Heffer, A., J. Xiang and L. Pick. 2013. Variation and constraint in *Hox* gene evolution. Proc. Nat. Acad. Sci. 110: 2211–2216.

King, N. 2004. The unicellular ancestry of animal development. Develop. Cell 7: 213–325.

Marthy, H. J. (ed.). 1990. *Experimental Embryology in Aquatic Plants and Animals*. Plenum, New York.

Martindale, M. Q. 2005. The evolution of metazoan axial properties. Nature Rev. Genet. 6: 917–927.

Minelli, A. 2015. EvoDevo and its significance for animal evolution and phylogeny. In A. Wanninger (ed.), *Evolutionary Developmental Biology of Invertebrates 1: Introduction, Non-Bilateria, Acoelomorpha, Xenoturbellida, Chaetognatha*. Springer-Verlag, Vienna.

Raff, R. A. 1996. *The Shape of Life: Genes, Development, and the Evolution of Animal Form*. University of Chicago Press, Chicago.

Richards, G. S. and B. M. Degnan. 2012. The expression of *Delta* ligands in the sponge *Amphimedon queenslandica* suggests an ancient role for *Notch* signaling in metazoan development. EvoDevo 3: 1–15.

Sánchez-Villagra, M. 2012. *Embryos in Deep Time. The Rock Record of Biological Development*. University of California Press, Berkeley.

Sawyer, R. H. and R. M. Showman (eds.). 1985. *The Cellular and Molecular Biology of Invertebrate Development*. University of South Carolina Press, Columbia.

Strathmann, M. F. 1987. *Reproduction and Development of Marine Invertebrates of the Northern Pacific Coast*. University of Washington Press, Seattle.

Technau, U. and C. B. Scholz. 2003. The origins and evolution of endoderm and mesoderm. Int. J. Deve. Biol. 47: 531–539.

Willmore, K. E. 2010. Development influences evolution. American Scientist 98: 220–227.

Wilson, E. B. 1892. The cell lineage of *Nereis*. J. Morphol. 6: 361–480. [Wilson's classic work establishing the coding system for spiral cleavage.]

Wilson, W. H., S. Stricker, and G.L. Shinn (eds.). 1994. *Reproduction and Development of Marine Invertebrates*. Johns Hopkins University Press, Baltimore, MD. [A collection of recent work on a host of developmental topics.]

Life Histories

Ayal, Y. and U. Safriel. 1982. r-Curves and the cost of the planktonic stage. Am. Nat. 119: 391–401.

Cameron, R. A. (ed.). 1986. Proceedings of the Invertebrate Larval Biology Workshop held at Friday Harbor Laboratories, University of Washington, 26–30 Mar. 1985. Bull. Mar. Sci. 39: 145–622. [Thirty-seven papers on larval biology.]

Caswell, H. 1978. Optimal life histories and the age-specific cost of reproduction. Bull. Ecol. Soc. Am. 59: 99. [This paper and the two below include interesting discussions of the adaptive qualities of various life cycle patterns, especially those that do not fit the typical direct or indirect definitions.]

Caswell, H. 1980. On the equivalence of maximizing fitness and maximizing reproductive value. Ecology 61: 19–24.

Caswell, H. 1981. The evolution of "mixed" life histories in marine invertebrates and elsewhere. Am. Nat. 117(4): 529–536.

Charlesworth, B. 1991. The evolution of sex chromosomes. Science 251: 1030–1033.

Charnov, E. L., J. M. Smith, and J. J. Bull. 1976. Why be an hermaphrodite? Nature 263: 125–126.

Chia, F. S. 1974. Classification and adaptive significance of developmental patterns in marine invertebrates. Thalassia Jugosl. 10: 121–130.

Chia, F. S. and M. Rice (eds.). 1978. *Settlement and Metamorphosis of Marine Invertebrate Larvae*. Elsevier/North-Holland, New York.

Christiansen, F. and T. Fenchel. 1979. Evolution of marine invertebrate reproductive patterns. Theor. Pop. Biol. 16: 267–282.

Crisp, D. 1974. Energy relations of marine invertebrate larvae. Thalassia Jugosl. 10: 103–120.

Crisp, D. 1974. Factors influencing the settlement of marine intertidal larvae. Pp. 177–265 in P. T. Grant and A. M. Macie (eds.), *Chemoreception in Marine Organisms*. Academic Press, New York.

Davidson, E. H. and M. S. Levine. 2008. Properties of developmental gene regulatory networks. Proc. Nat. Acad. Sci. 105: 20063–20066.

Dawydoff, C. 1928. Traité d´Embryologie Comparée des Invertébrés. Masson et Cie, Libraires de l´Academie de Médicine, Paris. [In many ways out of date, yet still a useful benchmark work.]

Eckelbarger, K. J. 1994. Diversity of metazoan ovaries and vitellogenic mechanisms: Implications for life history theory. Proc. Biol. Soc. Wash. 107: 193–218.

Eckelbarger, K. J. and L. Watling. 1995. Role of phylogenetic constraints in determining reproductive patterns in deep-sea invertebrates. Invert. Biol. 114(3): 256–269.

Emlet, R. B. and E. E. Ruppert (eds.). 1994. Symposium: Evolutionary morphology of marine invertebrate larvae and juveniles. Am. Zool. 34: 479–585.

Erwin, D. H. 2009. Early origin of the bilaterian developmental toolkit. Phil. Trans. Roy. Soc. B 364: 2253–2261.

Erwin, D. H. 2015. Was the Ediacaran-Cambrian radiation a unique evolutionary event? Paleobiology 41: 1–15.

Gilbert, L. I. and E. Frieden (eds.). 1981. Metamorphosis: A Problem in Developmental Biology, 2nd Ed. Plenum, New York. [A largely biochemical approach.]

Grosberg, R. K. 1981. Competitive ability influences habitat choice in marine invertebrates. Nature 290: 700–702.

Hadfield, M. G. 1978. Metamorphosis in marine molluscan larvae: An analysis of stimulus and response. Pp. 165–175 in F. S. Chia and M. E. Rice (eds.), *Marine Natural Products Chemistry*. Plenum, New York.

Hadfield, M. G. 1984. Settlement requirements of molluscan larvae: New data on chemical and genetic roles. Aquaculture 39: 283–298.

Jablonsky, D. and R. A. Lutz. 1983. Larval ecology of marine benthic invertebrates: Paleobiological implications. Biol. Rev. 58: 21–89. [An excellent review of larval ecology from an evolutionary perspective.]

Jeffrey, W. R. and R. A. Raff (eds.). 1982. *Time, Space, and Pattern in Embryonic Development*. Alan R. Liss, New York.

Kohn, A. J. and F. E. Perron. 1994. *Life History and Biogeography: Patterns in* Conus. Clarenton Press, Oxford.

Lutz, R. A., D. Jablonski and R. D. Turner. 1984. Larval development and dispersal at deep-sea hydrothermal vents. Science 226: 1451–1454.

McEdward, L. R. (ed.). 1995. *Ecology of Marine Invertebrate Larvae*. CRC Press, Boca Raton, FL.

Mileikovsky, S. 1971. Types of larval development in marine bottom invertebrates, their distribution and ecological significance: A re-evaluation. Mar. Biol. 10: 193–213. [An excellent treatment of the subject.]

Perron, F. and R. Carrier. 1981. Egg size distribution among closely related marine invertebrate species: Are they bimodal or unimodal? Am. Nat. 118: 749–755. [Explains, in part, how egg size varies with other aspects of the developmental pattern.]

Rokas, A. 2008. The origins of multicellularity and the early history of the genetic toolkit for animal development. Ann. Rev. Genet. 42: 235–251.

Sammarco, P. W. and M. L. Heron (eds.). 1994. *The Bio-Physics of Marine Larval Dispersal*. Am. Geophysical Union, Wash. DC. [An excellent blending of physical oceanography and biology.]

Starr, M., J. H. Himmelman and J. C. Therriault. 1990. Direct coupling of marine invertebrate spawning with phytoplankton blooms. Science 247: 1071–1074.

Steidinger, K. A. and L. M. Walker (eds.). 1984. *Marine Plankton Life Cycle Strategies*. C.R.C. Press, Boca Raton, FL, pp. 93–120.

Strathmann, R. 1977. Egg size, larval development and juvenile size in benthic marine invertebrates. Am. Nat. 111: 373–376.

Strathmann, R. 1978. The evolution and loss of feeding larval stages of marine invertebrates. Evolution 32(4): 894–906.

Strathmann, R. 1985. Feeding and nonfeeding larval development and life history evolution in marine invertebrates. Ann. Rev. Ecol. Syst. 16: 339–361.

Strathmann, R. and M. Strathmann. 1982. The relationship between adult size and brooding in marine invertebrates. Am. Nat. 119: 91–101.

Thorson, G. 1946. Reproduction and larval development of Danish marine bottom invertebrates with special reference to the planktonic larvae in the South (Oresund). Medd. Danm. Fisk., Havunders., Ser. Plankton: 4.

Thorson, G. 1950. Reproduction and larval ecology of marine bottom invertebrates. Biol. Rev. 25: 1–45. [These two works by G. Thorson laid the foundation for modern studies concerning the classification of invertebrate life cycles and their significance.]

Todd, C. D. and R. W. Doyle. 1981. Reproductive strategies of marine benthic invertebrates: A settlement-timing hypothesis. Mar. Ecol. Prog. Ser. 4: 75–83.

Wray, G. A. and R. A. Raff. 1991. The evolution of developmental strategy in marine invertebrates. Trends Ecol. Evol. 6: 45–50.

Young, C. M. 1990. Larval ecology of marine invertebrates: A sesquicentennial history. Ophelia 32: 1–48.

Young, C. M. and K. J. Eckelbarger (eds.). 1994. *Reproduction, Larval Biology, and Recruitment of the Deep-Sea Benthos*. Columbia University Press, New York.

Phylogeny and the Origins of Major Clades

Alberch, P., S. J. Gould, G. F. Osta, and D. B. Wake. 1979. Size and shape in ontogeny and phylogeny. Paleobiology 5(3): 296–317.

Bergh, R. S. 1885. Die Exkretionsorgane der Würmer. Kosmos, Lwow 17: 97–122.

Bergstrom, J. 1989. The origin of animal phyla and the new phylum Procoelomata. Lethalia 22: 259–269.

Bütschli, O. 1883. Bemerkungen zur Gastrea Theorie. Morph. Jahrb. 9.

Carter, G. S. 1954. On Hadzi's interpretations of animal phylogeny. Syst. Zool. 3: 163–167. [An analysis, sometimes quite pointed, of Hadzi's views.]

Clark, R. B. 1964. *Dynamics in Metazoan Evolution*. Oxford University Press, New York. [A fine functional approach to metazoan evolution, especially concerning the origin of the coelom and metamerism.]

Dougherty, E. C. (ed.). 1963. *The Lower Metazoa: Comparative Biology and Phylogeny*. University of California Press, Berkeley.

Eaton, T. H. 1953. Paedomorphosis: An approach to the chordate– echinoderm problem. Syst. Zool. 2: 1–6.

Faussek, V. 1899. Über die physiologische Bedeutung des Cöloms. Trav. Soc. Nat. St. Petersberg 30: 40–57.

Faussek, V. 1911. Vergleichend–embryologische Studien. (Zur Frage über die Bedeutung der Cölom-hölen). Z. Wiss. Zool. 98: 529–625. [Faussek's works include his views on the nephrocoel theory.]

Frenzel, J. 1892. *Salinella*. Arch. Naturgesch. 58, Pt. 1.

Garstang, W. 1922. The theory of recapitulation. J. Linn. Soc. Lond. Zool. 35: 81–101. [Garstang's revolutionary ideas on Haeckel's recapitulation concept.]

Garstang, W. 1985. *Larval Forms and Other Zoological Verses*. University of Chicago Press, Chicago. [A wonderful collection of prose and poetry by Walter Garstang, published after his death. The biographical sketch by Sir Alister Hardy and the Foreword by Michael LaBarbera chronicle many of Garstang's contributions to our understanding of the relationships between ontogeny and phylogeny and serve as a delightful introduction to the 26 poems in this little volume. This newer edition of the original (1951) version also includes Garstang's famous address on "The Origin and Evolution of Larval Forms."]

Goodrich, E. S. 1946. The study of nephridia and genital ducts since 1895. Q. J. Microsc. Sci. 86: 113–392. [One of the great classics concerning the origin of the coelom and related evolutionary matters.]

Gould, S. J. 1977. *Ontogeny and Phylogeny*. Harvard University Press, Cambridge, MA. [A scholarly coverage of ideas concerning recapitulation and other interactions between development and evolution.]

Grell, K. G. 1971. *Trichoplax adhaerens* F. E. Schulze, und die Entstehung der Metazoen. Naturwiss. Rundsch. 24(4): 160–161.

Grell, K. G. 1971. Embryonalentwicklung bei *Trichoplax adhaerens* F. E. Schulze. Naturwiss. 58: 570.

Grell, K. G. 1972. Formation of eggs and cleavage in *Trichoplax adhaerens*. Z. Morphol. Tiere 73(4): 297–314.

Grell, K. G. 1973. *Trichoplax adhaerens* and the origin of the Metazoa. Actualite's Protozooligiques. IVe. Cong. Int. Protozoologie. Paul Couty, Clermont–Ferrand.

Grell, K. G. and G. Benwitz. 1971. Die Ultrastruktur von *Trichoplax adhaerens* F. E. Schulze. Cytobiologie 4(2): 216–240.

Gutman, W. F. 1981. Relationships between invertebrate phyla based on functional–mechanical analysis of the hydrostatic skeleton. Am. Zool. 21: 63–81.

Hadzi, J. 1963. *The Evolution of the Metazoa*. Macmillan, New York. [Overkill. But then, any book that begins with the sentence, "It was in 1903, 58 years ago, that I, then a young man who had just left the classical grammar school at Zagreb, went to Vienna to study natural sciences and above all my beloved Zoology at Vienna University," can't be all bad!]

Haeckel, E. 1866. *Generelle Morphologie der Organismen: Allgemeine Grundzüuge der organischen Formen-Wissenschaft mechansch begrüundet durch die von Charles Darwin reformierte Descendenz-Theorie. Vols. 1–2*. George Reimer, Berlin.

Haeckel, E. 1874. The gastrea-theory, the phylogenetic classification of the animal kingdom and the homology of the germ-lamellae. Q. J. Microscop. Sci. 14: 142–165; 223–247. [Haeckel's concepts of recapitulation and blastea-gastrea idea of metazoan origin. A translation of the original German paper that introduced the colonial theory of metazoan origin (Jena. Z. Naturwiss. 8: 1–55).]

Hall, B. K. 1996. Baupläne, phylotypic stages, and constraints. Why are there so few types of animals? Pp. 215–261 in, M. K. Hecht, et al. (eds.), *Evolutionary Biology, Vol. 29*. Plenum, New York.

Hanson, E. D. 1958. On the origin of the eumetazoa. Syst. Zool. 7: 16–47. [Support for Hadzi's views.]

Hanson, E. D. 1977. *The Origin and Early Evolution of Animals*. Wesleyan University Press, Middletown, CT.

Hatschek, B. 1877. Embryonalentwicklung und Knospung der Pedicellina echinata. Z. Wiss. Zool. 29: 502–549. [Some early thoughts on the gonocoel theory.]

Hatschek, B. 1878. Studien üuber Entwicklungsgeschichte der Anneliden. Ein Beitrag zur Morphologie der Bilaterien. Arb. Zool. Inst. Wien 1: 277–404.

Hejnol, A. and J. M. Martín-Durán. 2015. Getting to the bottom of anal evolution. Zool. Anz. doi:10.1016/j.jcz.2015.02.006

Hyman, L. H. 1940–1967. The Invertebrates. Vols. 1–6. McGraw-Hill, New York. [All volumes include especially fine discussions on embryology of the included taxa. Volumes 1 and 2 include the author's views on the origin of the Metazoa, bilaterality, and coelom, and other related matters.]

Inglis, W. G. 1985. Evolutionary waves: Patterns in the origins of animal phyla. Aust. J. Zool. 33: 153–178.

Ivanova-Kazas, O. M. 1982. Phylogenetic significance of spiral cleavage. Soviet J. Mar. Biol. 7(5): 275–283.

Jablonski, D. and D. J. Bottjer. 1991. Environmental patterns in the origins of higher taxa: The post-Paleozoic fossil record. Science 252: 1831–1833.

Jägersten, G. 1955. On the early phylogeny of the Metazoa. The bilaterogastrea theory. Zool. Bidr. Uppsala 30: 321–354.

Jägersten, G. 1959. Further remarks on the early phylogeny of the Metazoa. Zool. Bidr, Uppsala 33: 79–108.

Jägersten, G. 1972. *Evolution of the Metazoan Life Cycle*. Academic Press, London. [The phylogeny of the Metazoa according to Jägersten, based in part on his bilaterogastrea hypothesis. Included are some of the author's thoughts on recapitulation.]

Jefferies, R. P. S. 1986. *The Ancestry of the Vertebrates*. British Museum (Natural History), London.

Lang, A. 1881. Der Bau von Gunda segmentata und die Verwandtschaft der Platyhelminthen mit Coelenteraten und Hirundineen. Mitt. Zool. Sta. Neapel. 3: 187–251.

Lang, A. 1903. Beitrüage zu einer Trophocoltheorie. Jena. Z. Naturw. 38: 1–373. [Lang's 1881 paper was in support of the enterocoel theory, suggesting that the coelom arose from pinched-off gut diverticula in flatworms; this opinion was based upon his study of the turbellarian *Gunda* (now *Procerodes*). However, Lang eventually switched his allegiance to the gonocoel theory (1903).]

Lankester, E. R. 1874. Observations on the development of the pond snail (*Lymnaea stagnalis*), and in the early stages of other Mollusca. Q. J. Microsc. Sci. 14: 365–391. [Thoughts on the origin of the coelom.]

Lankester, E. R. 1877. Notes on the embryology and classification of the animal kingdom; comprising a revision of speculations relative to the origin and significance of the germ layers. Q. J. Microsc. Sci. 17: 399–454. [In addition to the ambitious title, this work includes thoughts about the gonocoel theory.]

Margulis, L. 1981. *Symbiosis in Cell Evolution: Life and Its Environment on the Early Earth*. W. H. Freeman, San Francisco.

Marlow, H. and 6 others. 2014. Larval body patterning and apical organs are conserved in animal evolution. BMC Biol. 12: 7.

Martindale, M. Q. and A. Hejnol. 2009. A developmental perspective: Changes in the position of the blastopore during bilaterian evolution. Dev Cell 17: 162–174.

Masterman, A. 1897. On the theory of archimeric segmentation and its bearing upon the phyletic classification of the Coelomata. Proc. R. Soc. Edinburgh 22: 270–310. [Masterman was generally a proponent of the enterocoel theory.]

McKinney, M. L. and K. J. McNamara. 1991. *Heterochrony: The Evolution of Ontogeny*. Plenum Press, NY.

Meckel, J. 1811. Entwurf einer Darstellung der zwischen dem Embryozustande der höheren Tiere und dem Permanenten der niedere stattfindenen Parallele: Beitrüage zur vergleichenden Anatomie, Vol. 2. Carl Heinrich Reclam., Leipzig, pp. 1–60.

Meckel, J. 1811. Über den Charakter der allmüahligen Vervollkommung der Organisation, oder den Unterschied zwischen den höheren und niederen Bildungen: Beytrüage zur vergleichenden Anatomie, Vol. 2. Carl Heinrich Reclam., Leipzig, pp. 61–123. [Works by Meckel contain interesting pre-Haeckelian concepts of relationships between development and evolution as it was understood before Darwin.]

Metschnikoff, E. 1883. Untersuchungen über die intracellulare Verdauung bei wirbellosen Thieren. Arb. Zool. Inst. Wien. 5: 141–168. [Translated into English and published as, "Researches on the intracellular digestion of invertebrates," Q. J. Microsc. Sci. (1884) 24: 89–111. This paper includes some of the studies that led Metschnikoff and eventually others to conclude that ingression was the original form of gastrulation.]

Meyer, E. 1890. Die Abstimmung der Anneliden. Der Ursprung der Metamerie und die Bedeutung des Mesoderms. Biol. Cbl. 10: 296–308. [An English translation appeared in Am. Natur. 24: 1143–1165.]

Meyer, E. 1901. Studien üuber den Körperbau der Anneliden. V. Das Mesoderm der Ringelwüurmer. Mitt. Zool. Sta. Neapel. 14: 247–585. [The two papers by Meyer include coverage of the gonocoel theory.]

Morris, S. C. J. D. George, R. Gibson and H. M. Platt (eds.). 1985. *The Origins and Relationships of Lower Invertebrates*. Clarenton Press, Oxford. Published for the Systematics Association, Special Vol. 28.

Nielsen, C. 1985. Animal phylogeny in light of the trochaea theory. Biol. J. Linn. Soc. London 25: 243–299.

Nielsen, C. 1987. Structure and function of metazoan ciliary bands and their phylogenetic significance. Acta Zool. 68: 205–262.

Nielsen, C. 1994. Larval and adult characters in animal phylogeny. Am. Zool. 34: 492–501.

Nielsen, C. 2012. How to make a protostome. Invertebrate Systematics 26: 25–40

Nielsen, C. 2012. *Animal Evolution: Interrelationships of the Living Phyla*, 3rd Ed. Oxford University Press, Oxford.

Nielsen, C. 2013. Life cycle evolution: was the eumetazoan ancestor a holopelagic planktotrophic gastraea? BMC Evolutionary Biology 13: 171.

Nielsen, C. 2015. Evolution of deuterostomy—and origin of the chordates. Biological Reviews, doi: 10.1111/brv.12229

Nielsen, C. 2015. Larval nervous systems: True larval and precocious adult. J. Exper. Biol. doi: 10.1242/jeb.109603

Nielsen, C. and A. Nørrevang. 1985. The trochea theory: An example of life cycle phylogeny. Pp. 28–41 in Morris, S. C. J. D. George, R. Gibson and H. M. Platt (eds.). 1985. *The Origins and Relationships of Lower Invertebrates*. Clarenton Press, Oxford. Published for the Systematics Association, Special Vol. 28.

Patterson, C. 1990. Reassessing relationships. Nature 344: 199–200.

Popkov. D. V. 1993. Polytrochal hypothesis of origin and evolution of trochophora type larvae. Zool. Zh. 72: 1–17.

Raff, R. A. 2008. Origins of the other metazoan body plans: the evolution of larval forms. Phil. Trans. Royal Soc. B: Biol. Sci. 363: 1473–1479.

Raff, R. A. and T. C. Kaufman. 1983. *Embryos, Genes, and Evolution*. Macmillan, New York.

Rieger, R. M. 1994. The biphasic life cycle—A central theme of metazoan evolution. Am. Zool. 484–491.

Salvini-Plawen, L. 1980. Was ist eine Trochophora? Eine Analyse der Larventypen mariner Protostomier. Zool. Jb., Anat. 103: 389–423.

Salvini-Plawen, L. V. 1982. A paedomorphic origin of the oligomerous animals? Zool. Scr. 11: 77–81.

Sarvaas, A. E. du Marchie. 1933. La theorie du coelome. Thesis, University of Utrecht. [Some ideas on the schizocoel theory that never quite took hold.]

Schleip, W. 1929. Die Determination der Primitiventwicklung. Akad. Verlags, Leipzig. [The origin of the concept of the "Spiralia."]

Sedgwick, A. 1884. On the nature of metameric segmentation and some other morphological questions. Q. J. Microsc. Sci. 24: 43–82. [This work provided the main driving force behind the idea that the coelom arose (via enterocoely) from cnidarian gut pouches rather than by a pinching off of the diverticula in flatworm digestive tracts.]

Serres, E. R. A. 1824. Explication de systéme nerveux des animaux invertébrés. Ann. Sci. Nat. 3: 377–380.

Serres, E. R. A. 1830. Anatomie transcendante—Quatrieme mémoire: Loi de symétrie et de conjugaison du systéme sanguin. Ann. Sci. Nat. 21: 5–49.

Siewing, R. 1980. Das Archichelomatenkonzept. Zool. Jahrb. Abt. Anat. Ontog. Tiere 8, 103: 439–482.

Simonetta, A. M. and S. Conway Morris (eds.). 1989. *The Early Evolution of Metazoa and the Significance of Problematic Taxa*. Cambridge Univ. Press, Cambridge.

Valentine, J., S. M. Awramik, P. S. Signor and P. M. Sadler. 1991. The biological explosion at the Precambrian–Cambrian boundary. Pp. 279–356 in M. K. Hecht, B. Wallace and R. J. Macintyre (eds.), *Evolutionary Biology, Vol. 25*. Plenum, New York.

Vecchia, G. L., R. Valvassori and M. D. C. Carnevali (eds.). 1995. Body cavities: Function and phylogeny. Proceedings of the International Symposium on Body Cavities, Varese. Collana U.Z.I. Selected Symposia and Monographs No. 8. Mucchi Editore, Modena, Italy.

von Baer, K. E. 1828. Entwicklungsgeschichte der Thiere: Beobachtung und Reflexion. Borntrager, Konigsberg.

Wilson, E. B. 1898. Considerations in cell-lineage and ancestral reminiscence. Ann. N. Y. Acad. Sci. 11: 1–27.

Ziegler, H. E. 1898. Über den derzeitigen Stand der Colomfrage. Verh. Dtsch. Zool. Ges. 8: 14–78.

Ziegler, H. E. 1912. Leibeshöhle. Handwörterbuch Naturwiss. 6: 148–165.

CHAPTER 6

Two Basal Metazoan Phyla

Porifera and Placozoa

Chapters 1 through 5 provide a detailed introduction to the Metazoa (or Animalia). Metazoans are a monophyletic clade of eukaryotes—those creatures whose cells contain membrane-enclosed organelles and have a membrane-enclosed nucleus. However, they differ from other eukaryotes (i.e., fungi, plants, and the myriad protist clades) in their combination of multicellularity, heterotrophic and ingestive nutrition, and unique style of tissue formation through embryonic germ layering. **Metazoans are heterotrophic multicellular eukaryotes that undergo embryogenesis by way of tissue layering.** The formation of embryonic germ layers takes place through a process called gastrulation, and even the most primitive metazoans (e.g., sponges) undergo this process—metazoan gastrulation is a hallmark of this kingdom. Gastrulation is a process that achieves separation of those cells that must interact directly with the environment (e.g., locomotor, sensory, protective) from those that process materials obtained from the environment (e.g., nutritive functions).

As we noted in Chapter 1, Metazoa almost certainly constitutes a monophyletic clade, defined by numerous synapomorphies, including: gastrulation; unique modes of oogenesis and spermatogenesis; a unique sperm structure; mitochondrial gene reduction; epidermal epithelia with septate, tight, or zona adherens junctions; striate myofibrils; actin–myosin contractile elements; type IV collagen; and the presence of a basal lamina or basement membrane beneath epidermal layers (of course, some of these features have been secondarily lost in some groups). Evidence is strong that Metazoa are descended from the choanoflagellate protist group, or a common ancestor, and the two clades comprise sister groups in almost all recent phylogenetic analyses.

However, despite these fundamental shared similarities, there are four phyla of Metazoa that are so ancient and possess such a simple body construction that their relationships to other animals have so far eluded our understanding—these are the four non-bilaterian phyla: Porifera, Placozoa, Cnidaria, and Ctenophora. The first two (sponges and placozoans) are covered in this chapter,

Classification of The Animal Kingdom (Metazoa)

Non-Bilateria*
(a.k.a. the diploblasts)
- **PHYLUM PORIFERA**
- **PHYLUM PLACOZOA**
- PHYLUM CNIDARIA
- PHYLUM CTENOPHORA

Bilateria
(a.k.a. the triploblasts)
- PHYLUM XENACOELOMORPHA

Protostomia
- PHYLUM CHAETOGNATHA

SPIRALIA
- PHYLUM PLATYHELMINTHES
- PHYLUM GASTROTRICHA
- PHYLUM RHOMBOZOA
- PHYLUM ORTHONECTIDA
- PHYLUM NEMERTEA
- PHYLUM MOLLUSCA
- PHYLUM ANNELIDA
- PHYLUM ENTOPROCTA
- PHYLUM CYCLIOPHORA

Gnathifera
- PHYLUM GNATHOSTOMULIDA
- PHYLUM MICROGNATHOZOA
- PHYLUM ROTIFERA

Lophophorata
- PHYLUM PHORONIDA
- PHYLUM BRYOZOA
- PHYLUM BRACHIOPODA

ECDYSOZOA

Nematoida
- PHYLUM NEMATODA
- PHYLUM NEMATOMORPHA

Scalidophora
- PHYLUM KINORHYNCHA
- PHYLUM PRIAPULA
- PHYLUM LORICIFERA

Panarthropoda
- PHYLUM TARDIGRADA
- PHYLUM ONYCHOPHORA
- PHYLUM ARTHROPODA
  - SUBPHYLUM CRUSTACEA*
  - SUBPHYLUM HEXAPODA
  - SUBPHYLUM MYRIAPODA
  - SUBPHYLUM CHELICERATA

Deuterostomia
- PHYLUM ECHINODERMATA
- PHYLUM HEMICHORDATA
- PHYLUM CHORDATA

*Paraphyletic group

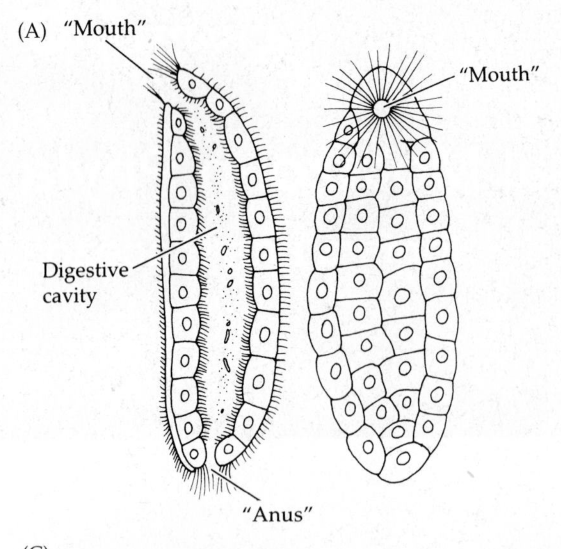

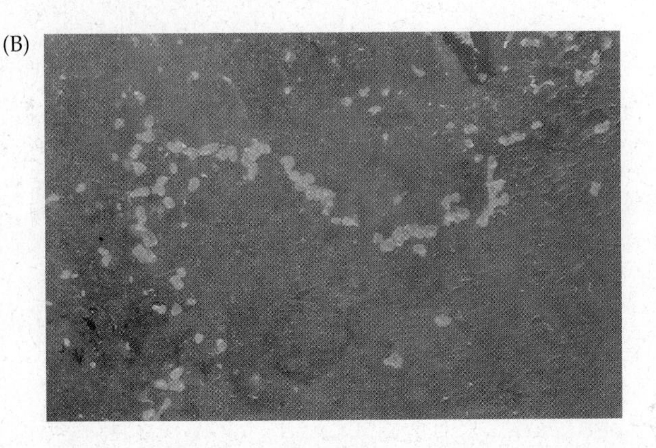

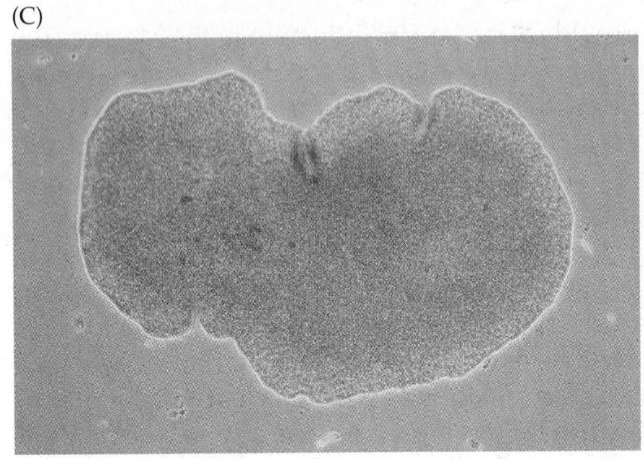

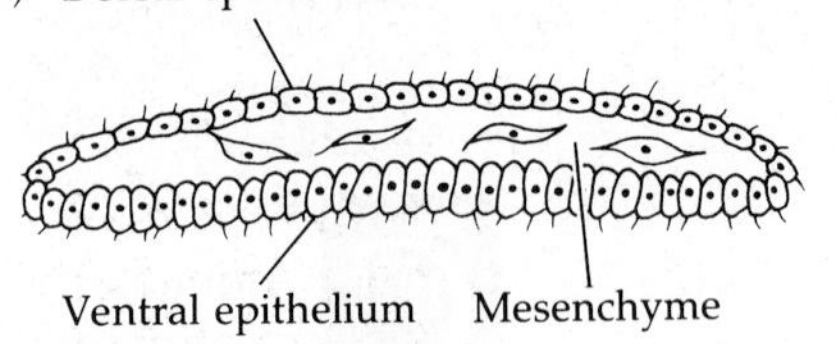

Figure 6.1 (A) The mysterious *Salinella* (Monoblastozoa) (sagittal section and lower view). (B) Several specimens of *Trichoplax adhaerens* (Placozoa) on an algal mat. (C) A single *Trichoplax*. (D) Section through *Trichoplax adhaerens*.

the next two in the following chapters. Phylogenetic analyses inform us that these four phyla are basal to all other Metazoa, and that Porifera is likely the oldest of all living animal phyla. In addition to lacking bilateral symmetry, these four phyla are generally considered to lack true mesoderm development, that is, they are diploblastic (rather than triploblastic) metazoans. Sponges and placozoans appear at the base of the animal tree in most analyses, and many workers consider these two phyla to lack true tissues of any kind, and also to lack a permanent digestive cavity, true nerves, and muscles. The Cnidaria, Ctenophora, and Bilateria are sometimes considered a monophyletic lineage called "Eumetazoa," but this does not seem like a very useful categorization.

One other enigmatic "phylum" was based on a microscopic creature named *Salinella salve,* from salt beds in Argentina. It was described by the German biologist Johannes Frenzel in the nineteenth century. Over the course of time, *Salinella* was treated as a protist and a larval stage of an unknown metazoan, but it was eventually assigned to its own monotypic phylum—Monoblastozoa. The single described species of *Salinella* has not been seen since its reported discovery in 1892, and there is serious question about the accuracy of the original description of this very odd beast. Many attempts have been made to rediscover *Salinella* without success. Yet, Frenzel was a serious scholar and a meticulous artist, leaving us with quite a mystery. One recent, unsuccessful expedition in search of this mystery animal, by another German scholar, Michael Schrödl, discovered that Frenzel had not actually collected the soil samples from which he cultured *Salinella* himself, but that they have been given to him by a geologist friend, adding further to the enigma.

According to Frenzel's description, the body wall of *Salinella* consists of only one or two layers of cells. The inner borders of the cells line a cavity, which is open at both ends (Figure 6.1A). The openings were said to function as an anterior "mouth" and posterior "anus," both of which are ringed by bristles. The rest of the body, inside and out, is densely ciliated. The animal was said to move by ciliary gliding, much like ciliated protists, small flatworms, and xenacoelomorphans (one of which Frenzel might have had under his microscope). *Salinella* was thought to feed by ingesting organic detritus through the "mouth" and digesting it in the internal cavity, undigested material being moved by cilia to the "anus" for expulsion. Asexual reproduction was said to take place by transverse fission of the body, and sexual reproduction was suspected to occur as well. The true nature of this animal, including its very existence, remains elusive. And the combination of characters he presented does not match up very well with anything we can imagine.

Phylum Placozoa

Trichoplax adhaerens was discovered by F. E. Schulze in 1883 in a seawater aquarium at the Graz Zoological Institute in Austria. Specimens have subsequently been found throughout the world's tropical and subtropical seas. A second species, *Treptoplax reptans*, was described in 1896, but has not been seen since. These creatures remained a mystery until the great German protozoologist Karl Grell, Director of the Zoology Institute at Tübingen, began working on them. Molecular genetics of *T. adhaerens* from around the world has recently suggested that this species could be a cryptic amalgamation of many species, yet to be distinguished morphologically. Box 6A lists the major characteristics of Placozoa.

BOX 6A Characteristics of the Phylum Placozoa

1. Minute, flattened metazoans comprised of ciliated upper and lower cell layers (epithelial layers?), with fibrous cells in between; adults asymmetrical. Only four somatic cell types have been identified.
2. Cells with desmosome cell–cell connections
3. With unique shiny spheres, possibly defensive structures, in upper cell layer
4. Without a structural nervous system, muscles, or digestive system

The body of *Trichoplax* is only 1 to 3 mm in diameter, although it consists of a few thousand cells, of only a few types, arranged as a simple double-layered plate (Figure 6.1B–D). It lacks anterior–posterior polarity, symmetry, a mouth or gut, nervous system, muscles, or extra-cellular matrix. However, the cells of the upper (top) and lower (bottom) layers differ in shape, and there is a consistent orientation of the body relative to the substratum. Because the concepts of dorsal and ventral are generally taken to be characteristics of triploblastic animals, we will use the terms upper and lower cells. The upper layer cells are flattened and monociliate, epithelium-like, and they have curious extracellular structures, called **shiny spheres**, unique to placozoans. Long thought to be lipid inclusions, they appear to be easily dislodged and fall out of the cell layer. In 2007, Vicki Buchsbaum Pearse and Oliver Voight reported that gastropods, flatworms, and sabellid polychaetes recoil upon contact with placozoans, suggesting they might possess a chemical deterrent to predators. In 2009, Alexis Jackson and Leo Buss tested this hypothesis, finding that when individual *Trichoplax* were fed to the hydrozoan *Podocoryna carnea*, the polyps became paralyzed, suggesting that the shiny spheres might indeed be anti-predator devices.

Most of the lower layer cells are also monociliate, but they are all more columnar and lack distinct shiny spheres; "gland cells" may also occur on the lower surface. Fiber cells can be seen between the upper and lower cell layers. These internal cells have thin extensions that connect to each other in a network. Cellular material such as microtubules and microfilaments traverse the extensions from fiber cell to fiber cell. It has been suggested that this network plays a role in coordinating movement of the animal. Importantly, desmosomes, cell–cell junctions of extracellular cell-adhesion proteins, have been found in placozoans, a feature not seen in Porifera, but present in all higher Metazoa.

The lower cell layer can be temporarily invaginated, presumably for feeding. This observation supports the notion that there are functional as well as structural differences between the two cell layers. Between these two cell layers (or, epithelial sheets) is a mesenchymal layer of stellate ameboid cells embedded in a supportive gel matrix. Grell (1982) considered *Trichoplax* to be a true diploblastic metazoan and suggested that the upper and lower cell layers were true epithelia and might be homologous to ectoderm and endoderm, respectively. However, a basement membrane has not yet been identified beneath either layer, which suggests that *Trichoplax* may be closer to the Porifera in organization than it is to Cnidaria, Ctenophora, or the triploblastic eumetazoa. No fossil placozoans have been found.

Trichoplax moves by ciliary gliding along a solid surface, aided by irregular, ameba-like shape changes along the body edges. Very small, presumably young individuals can swim, while larger individuals seem to always crawl. *Trichoplax* apparently feeds by phagocytosis of organic detritus using the lower surface, which can contract to form a "feeding chamber." Although there is no evidence of extracellular digestion, it is possible that *Trichoplax* secretes digestive enzymes onto its food within the lower digestive pocket. It is not known what these animals eat in nature, but lab cultures can be sustained on diets of flagellated protists (e.g., *Cryptomonas*, *Chlorella*).

Trichoplax reproduces asexually by fission of the entire body into two new individuals, and also by a budding process that yields numerous multicellular flagellated "swarmers," each of which forms a new individual. It is also capable of regeneration from damage to the body. Sexual reproduction is also known, followed by a developmental period of holoblastic cell division and growth. Eggs have been observed within the mesenchyme, but their origin is unknown.

With only four somatic cell types and a lack of defined symmetry or constant body axis, *Trichoplax* has the simplest body of any known metazoan. Over the years, it has been suggested that *Trichoplax* might be a secondarily reduced cnidarian (or a sister group to the Cnidaria). However, the more highly-derived cnidarians (Scyphozoa, Hydrozoa, Cubozoa) all have a linear mtDNA molecule—a unique synapomorphy of that clade within Cnidaria—in contrast to the circular mtDNA of Anthozoa and all other Metazoa. *Trichoplax* also has circular mtDNA. Further, the secondary structure morphology of the 16S mitochondrial genome of *Trichoplax* differs markedly from that seen in Cnidarians. Most recently, phylogenomic analyses have suggested that Placozoa are not cnidarians, but do lie near the base of the metazoan tree, among the other non-bilaterian phyla. The exact position among these four basal phyla was, at the time this book was being written, still unresolved, with the two strongest arguments favoring it branching off between a basal Porifera and the Cnidaria + Ctenophora, or positioning it as a sister group to the Bilateria. Surprisingly, a 2008 genomic study showed that *Trichoplax* has many of the genes responsible for guiding the development of body shape and organ development in higher metazoans, as has also been shown for many cnidarians. And, curiously, a shift from circular to linear mitochondrial genomes also occurred in a clade of calcareous sponges.

Phylum Porifera: The Sponges

The phylum Porifera (Latin *porus*, "pore"; *ferre*, "to bear") comprises those odd but fascinating animals called sponges. At first glance, sponges may seem difficult to reconcile within the animal kingdom—adults lack a gut, conventional muscles, nerves and conventional neuronal signaling systems, typical metazoan organs, gap junctions between cells, an obvious anterior–posterior polarity (except in larvae), and some of the key metazoan developmental genes. In addition, they have cross-striated ciliary rootlets in larval cells and choanocytes—a feature characteristic of many protists. However, they do possess the metazoan-defining attributes of multicellularity derived by embryonic layering, specialized junctions between cells, actin–myosin contractile elements, and type IV collagen. In addition, recent genomic analyses of *Amphimedon queenslandica* (class Demospongiae) and *Oscarella carmela* (class Homoscleromorpha) reveal the presence of certain key homeobox genes, and representatives of most higher metazoan molecules involved in cell–cell communication, signaling pathways, complex epithelia, and immune recognition. They also undergo typical animal-like sexual reproduction, and the development of embryos through a structured series of cellular divisions (cell cleavages) that result in a spatially organized larva with multiple cell layers and sensory capabilities. Most larvae have an obvious anterior–posterior symmetry, and many adult sponges possess an apical–basal symmetry, or polarity, defined by the presence of a large osculum at one end (although the positions of oscula are often dictated purely by hydrodynamic forces in the environment). Others show such polarity by virtue of their pedunculate or pinnate growth form, often even with stems/stalks and root-like structures. On the other hand, many sponges have no symmetry whatsoever as adults. Molecular genetic analyses indicate Porifera is monophyletic and clearly within Metazoa. In fact, sponge genes have been recently discovered that are implicated in regulating anterior–posterior polarity and specifying particular tissues during the development of other basal metazoans, supporting the contention that sponges undergo true gastrulation during embryogenesis. Figure 6.2 illustrates a variety of sponge body forms and some sponge anatomy. Box 6B lists the major characteristics of sponges.

Sponges are sessile, primarily suspension-feeding, multicellular animals that utilize flagellated cells called

BOX 6B Characteristics of the Phylum Porifera

1. Metazoa partly at the cellular grade of construction, with simple tissues, but with a high degree of cellular pluripotency; adults asymmetrical or with a distinct apical-basal axis (often superficially appearing as radially symmetrical); larvae usually with anterior-posterior symmetry
2. Cells with adhaerens junctions in some species, but without gap junctions
3. With unique flagellated cells—choanocytes—that drive water through canals and chambers constituting the aquiferous system
4. Adults are primarily sessile suspension feeders; larval stages are motile and usually lecithotrophic.
5. Type IV collagen basement membranes occur in most Homoscleromorpha, and also (to a lesser extent) in the other classes.
6. Middle layer—the mesohyl—is variable, but always includes motile cells and usually some skeletal material.
7. Skeletal elements, when present, are composed of calcium carbonate or silicon dioxide (typically in the form of spicules), and/or collagen fibers.
8. Without neurons; only true sense organ is the osculum, which utilizes primary cilia to detect water flow rates
9. Ciliated cells of adult sponges bear only a single cilium (largely lacking the rootlet system seen in higher metazoans); some larvae with cilia that have rootlet systems; some larvae with bi-ciliated cells on surface (postulated by some workers to be products of defective cell division)
10. Porifera is one of the few animal phyla to possess both cilia (e.g., larval epithelial cells, adult oscular epithelium) and flagella (e.g., adult choanocytes) in somatic tissues.

choanocytes to circulate water through a unique system of water canals. Most rely on an internal skeleton of calcium carbonate or silicon dioxide spicules to support their body, which can be quite large. It was long thought that Porifera lacked distinct embryological germ layering that leads to definable tissues, a condition sometimes referred to as "the parazoan grade of body construction." However, we now know that sponges undergo distinct gastrulation events from which the adult tissues derive, and the old concept of "parazoa" is probably best laid to rest. However, some of the adult tissues in sponges are somewhat transmutable and not fixed, due to a degree of cellular pluripotency—most cells are capable of changing form and function, and some are kept in a totipotent state to be recruited "on demand" (although pinacocytes and sclerocytes cannot do so). So, despite the fact that sponges are large-bodied multicellular animals typically supported by an internal skeleton of spicules or stiffened collagen (spongin), in some ways they function like organisms at the unicellular grade of complexity. In fact, as you will discover in this chapter, their nutrition, gas exchange, and response to environmental stimuli are all very protist-like. So, superficially, sponges might be viewed as tight consortiums of semi-autonomous cells, and thus quite simple animals. But, appearances can be deceiving. Read on.

Despite their seeming simplicity, sponges have experimented with various aspects of higher metazoan body organization, and they have developed primitive tissues, a sparse basement membrane, and in some species even predatory behaviors, and other features typical of the higher Metazoa. Some might argue that poriferans are "caught between two worlds"—the world of protists and the world of higher metazoans—while others would argue they are Metazoans in every sense of the word.

One of the most remarkable attributes of sponges is their tendency to maintain symbiotic relationships with a variety of heterotrophic and autotrophic Bacteria, Archaea, and Protista. Some of these intimate relationships have developed to the point that more biomass is actually contributed by the symbiont than by the sponge, and in these species microscopic examination of the sponge reveals mostly cells of microbes! We are just beginning to explore this community hidden within sponges, but already hundreds of symbiotic species, in over a dozen bacterial and archaean phyla (and several protistan groups) have been documented. As the role of microorganisms in sponges begins to be better understood, the emerging evidence suggests strong mutualism in many cases. Sponges of different types, in different ocean basins, seem to host strikingly similar microbial communities, suggesting the symbiotic relationships are very old. And some of these microbes appear to be transported in the eggs, nurse cells, and even sperm of sponges.

Sponges produce the largest and most diverse storehouse of secondary metabolites of any animal phylum—compounds that function to deter predators, prevent fouling of the sponge's surface, screen ultraviolet radiation, and nurture their symbiotic partnerships. Some sponges even "walk" over rocks, using lobelike extensions of the body that grow and elongate and then disappear, sometimes leaving separate living pieces—progeny—in their wake. At least one lineage of sponges, possibly more, has taken a dramatic evolutionary turn and become predatory carnivores; instead of filter feeding, these magnificent creatures are carnivores that capture and engulf small prey that are caught on highly specialized Velcro-like surfaces. Well over 100 species of carnivorous sponges have been described, mainly in the deep-sea family Cladorhizidae and two other small carnivorous families.

As of this writing, nearly 9,000 living species of sponges have been described, all but about 220 (the freshwater species) being restricted to benthic marine environments. Freshwater species occur at all latitudes, from deserts to equatorial rainforests, and from sea level to alpine lakes and even subterranean habitats. About 60 new sponge species are described each year. It has been estimated that less than half of the living species have so far been described. Sponges occur at all depths, but are most abundant in unpolluted littoral and tropical reef habitats, cold temperate continental shelf regions, and Antarctic seas. However, deep-water "sponge grounds" are also important components of deep-sea ecosystems. Most littoral sponges grow as thick or thin layers, or as erect structures, on hard surfaces. Sponges that live on soft substrata typically are upright and tall, or possess funnel-like structures on top of a buried basal body, thus avoiding burial by the shifting sediments of their environment.

Some sponges reach considerable size (up to 2 m in height on Caribbean reefs, and even larger in Antarctica) and may constitute a significant portion of the benthic structure and biomass. In Antarctic seas, sponges can make up almost 75% of the total benthic biomass at a depth of 100–200 m. Areas of the deep Antarctic shelf have been called "sponge kingdoms," and here over 300 species have been recorded with high biomass and density. Subtidal and deeper water species that do not confront strong tidal currents or surge are often large and exhibit a stable, even symmetrical (radial), external form. The deeper water hexactinellid sponges often assume unusual shapes, many being delicate glasslike structures, others round and massive, and still others growing in a ropelike fashion. Siliceous sponge reefs have been documented from several periods in Earth history, and they culminated in the Late Jurassic when they formed a discontinuous deep-water reef belt extending more than 7,000 km. This reef system was the largest biotic structure ever built on Earth (at 2,000 km, the Great Barrier Reef

Figure 6.2 Representative sponges. Class Calcarea: (A) *Leucilla nuttingi*; (B) *Sycon* (= *Scypha*), a syconoid sponge; (C) the unusual *Clathrina clathrus* (Mediterranean Sea), which often grows on sea cave walls and ceilings. Class Demospongiae: (D) *Aplysina archeri* (Caribbean); (E) *Agelas* sp. (Belize); (F) a yellow encrusting *Haliclona* sp. (Gulf of Aden, Djibouti); (G) *Speciospongia confoederata* (close-up of pinacoderm showing dermal pores and oscula); (H) *Tethya aurantia*, close-up showing oscula protected by long spicules. (I,J) Tree sponges! Freshwater sponges from all three New World families occur in the rivers of the Amazon Basin—shown here is *Drulia* (?) that lives 5–10 meters above the dry season low water line (when these photos were taken). (K) The freshwater *Spongilla* (Minnesota, USA). (L) The massive calcareous base of a coralline sponge. (M) A red poecilosclerid sponge growing on the back of a decorator crab makes it nearly invisible (Antipodes Island, New Zealand). Hexactinellida: (N) Three specimens of deep-sea glass sponges (from the eastern Pacific) with silica rope stalks; (O) *Euplectella aspergillum* (Venus's flower basket). (P) Close-up of *Euplectella* skeleton showing arrangement of spicule bundles.

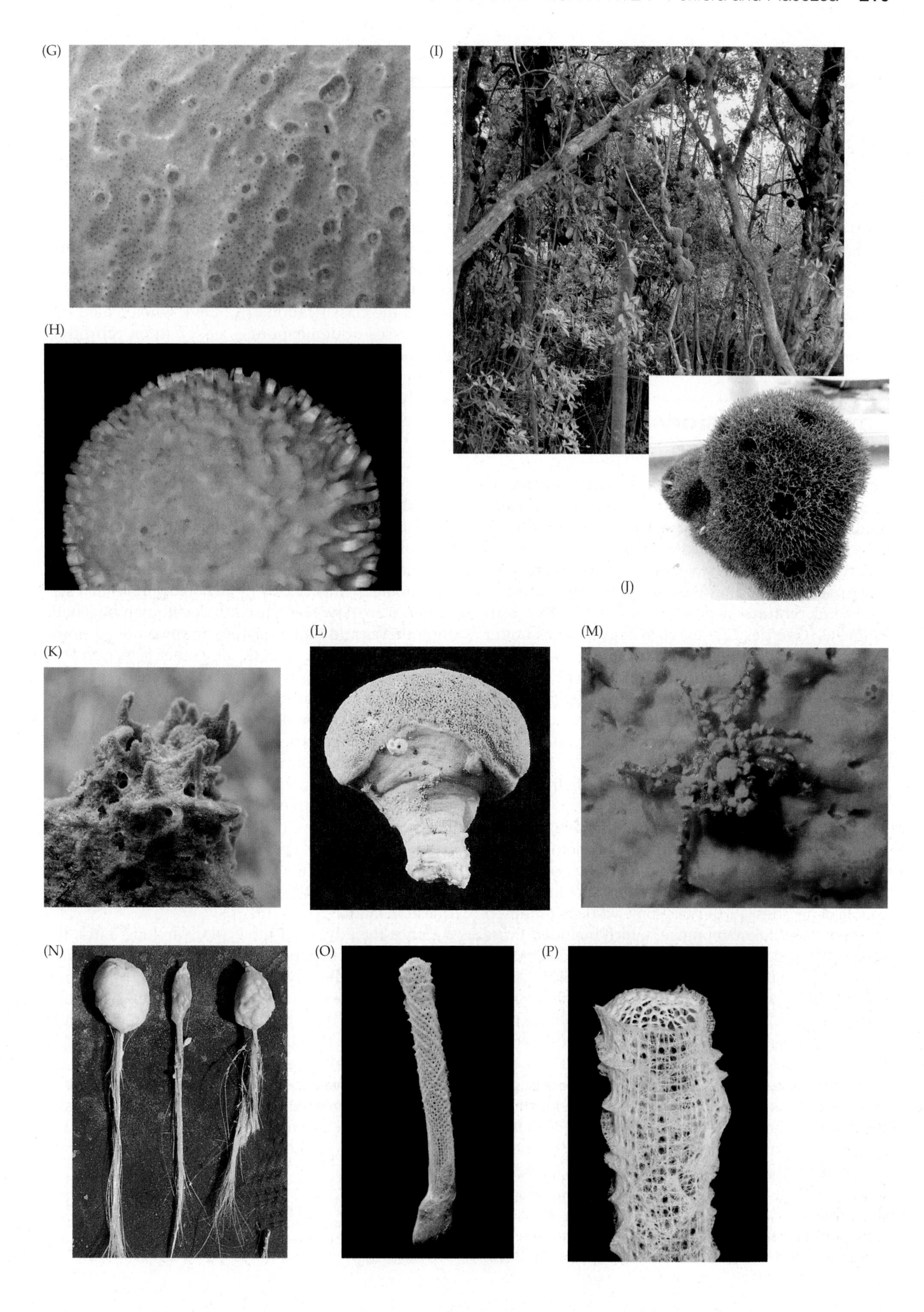
(G)
(I)
(H)
(J)
(K)
(L)
(M)
(N)
(O)
(P)

of Australia is relatively small compared to the Jurassic sponge reef belt).

Sponges display nearly every color imaginable, including bright lavenders, blues, yellows, crimsons, and pure white. In many species it is the symbiotic bacteria or algae that give color to their host's body, especially in the tropics. And, sponges are the only phylum of animals that utilize silica, rather than calcium, in their mineral skeleton (in Demospongiae, Homoscleromorpha, and Hexactinellida). In one of the four classes of sponges, Calcarea, the skeleton is composed not of siliceous spicules but of calcium carbonate spicules (although a few species in other sponge classes are known to secrete a firm calcium carbonate base, upon which the siliceous skeleton rests).

Taxonomic History and Classification

The sessile nature of sponges, and their often amorphous or asymmetrical growth form convinced early naturalists that they were plants. It was not until 1765, when the nature of their internal water currents was described, that sponges were recognized as animals. The great naturalists of the late eighteenth and early nineteenth centuries (e.g., Jean-Baptiste Lamarck, Karl Linnaeus, Georges Cuvier) classified sponges under Zoophytes or Polypes, regarding them as allied to anthozoan cnidarians. Throughout much of the nineteenth century they were placed with cnidarians under the name Coelenterata or Radiata. The morphology and physiology of sponges were first adequately understood by R. E. Grant. Grant created for them the name Porifera, although other names were frequently used (e.g., Spongida, Spongiae, Spongiaria).

Historically, the classes of Porifera have been defined by the nature of their internal skeletons. Until recently, three classes of sponges had long been recognized: Calcarea, Hexactinellida, Demospongiae. For a couple of decades (1970–1990) some workers proposed another class, Sclerospongiae, which included those species that produce a solid, calcareous, rocklike matrix (in addition to the spicular skeleton) on which the living animal grows. Over a dozen species of living sclerosponges, also known as coralline sponges, have been described. Late in the twentieth century, ultrastructural work and DNA analyses showed that the class Sclerospongiae was actually a polyphyletic grouping and it was thus abandoned, its members relegated to the Calcarea and Demospongiae. Today, the "coralline sponges" are known to be the last survivors of otherwise extinct stromatoporids, sphinctozoans, and chaetetids—ancient reef building sponges that were highly diversified in Paleozoic and Mesozoic seas. These ancient coralline sponges were probably among the first metazoans to produce a carbonate skeleton. In 2010, the distinctive nature of the Homoscleromorpha (formerly included within the Demospongiae) was seen to merit elevating this group to class status, thus establishing a fourth class of living Porifera.

Demospongiae is the largest sponge class, comprising 81% of the living species. Because of its size and morphological variability, the class Demospongiae presents the greatest number of problems for taxonomists. The only synapomorphy that distinguishes the class Demospongiae is the presence of a spongin-based skeleton, and yet not all species of demosponges possess this feature. For many years, spongologists followed the classification of C. Lévi, who created two subclasses of demosponges based upon reproductive modes, Tetractinomorpha and Ceractinomorpha. However, by the turn of the century these were widely recognized as polyphyletic. In the early twenty-first century, molecular phylogenetics showed both that the class is monophyletic (exclusive of Homoscleromorpha), and that it can be divided into three or four distinct subclasses as noted below.

Although the mainstay of sponge taxonomy has traditionally been the chemical composition, shape, ornamentation, dimensions, and localization of the spicules, additional kinds of information, including secondary metabolite chemistry, and more notably molecular systematics are now being used to develop phylogenetic hypotheses and higher classifications. Indeed, some sponge species lack spicules altogether (e.g., *Oscarella, Hexadella, Halisarca*), whereas many spicules types appear to be homoplasious among sponges (e.g., asters, acanthostyles, sigmata). Sponge specialists also are now using embryological, biochemical, histological, and cytological methods to diagnose and analyze the Porifera. In the past, the considerable difficulty in precisely setting the limits of some sponge species made poriferan taxonomy challenging. Sponges are famous for their paucity of reliable taxonomic characters, and even the great sponge taxonomist Arthur Dendy was known to frequently end a species diagnosis with a question mark.[1]

Since the advent of molecular phylogenetics, the fascinating world of poriferology has grown more tractable, and modern systematics is beginning to build a robust framework for this phylum. A few molecular phylogenetic studies of sponges suggested they might be paraphyletic, but that work considered a relatively small number of nuclear genes and larger studies have not supported that idea. In addition to resolving long-standing phylogenetic questions, molecular work has let to the discovery that many "cosmopolitan species" are, in fact, clusters of closely-related but distinct species.

[1]The term a "sleeze" of sponges was coined by Ristau (1978) to describe an aggregation of sponges; the usage is comparable to other such collective nouns that define animal groups (e.g., flock, herd, gaggle).

In addition, since the 1970s important bioactive compounds have been discovered in sponges, many having potential pharmacological significance (e.g., antibacterial, antiviral, anti-inflammatory, antitumor, and cytotoxic compounds, as well as channel blockers and antifouling chemicals). The discovery of these natural products in sponges also has led to a renewed interest in the group.

PHYLUM PORIFERA

CLASS CALCAREA Calcareous sponges (Figure 6.2A–C). Spicules of mineral skeleton composed entirely of calcium carbonate laid down as calcite, secreted extracellularly within a collagenous sheath (but with no axial filament); skeletal elements often not differentiated into megascleres and microscleres; spicules usually 1-, 3-, or 4-rayed; body with asconoid, syconoid, or leuconoid construction; many species exhibit superficial radial symmetry around a long axis, but in others there is no visible axial symmetry; early cleavage is total and equal; embryonic cleavage patterns probably fundamentally radial; all studied species are viviparous. All marine; occurring at all latitudes. About 685 described species. Although embryogenesis and larval morphology of the two subclasses differ profoundly, molecular phylogenetics provides evidence that the class and both subclasses are monophyletic.

SUBCLASS CALCINEA Free-living larvae are hollow, flagellated "coeloblastulas" (calciblastula); choanocyte nuclei located basally, and flagellum arises independent of nucleus (a presumptive synapomorphy of this subclass); with free, regular, triradiate spicules. (e.g., *Clathrina*, *Dendya*, *Leucascus*, *Leucetta*, *Soleneiscus*, the coralline genus *Murrayona*)

SUBCLASS CALCARONEA Free-living larvae are unique, partly flagellated amphiblastulas, these typically forming by way of eversion of earlier stomoblastula "prelarvae" (which are held internally)—a presumptive synapomorphy of this subclass; choanocyte nuclei apical, and flagellum arises directly from nucleus; spicules free or fused (e.g., *Amphoriscus*, *Grantia*, *Leucilla*, *Leucosolenia*, *Sycon* [= *Scypha*], the coralline genus *Petrobiona*)

CLASS HEXACTINELLIDA Glass sponges (Figure 6.2N,O,P). Skeleton composed of a large array of siliceous spicules of various shapes and sizes, secreted intracellularly around a square proteinaceous axial filament; spicules with fundamental 3-axon or 6-rayed symmetry (triaxonic); both megascleres and microscleres always present. Entire sponge formed by a single continuous syncytial tissue, the trabecular reticulum, which stretches from the outside or dermal membrane, to the inside or atrial membrane, enclosing the cellular components of the animal. Body wall cavernous, filled primarily with trabecular syncytium, connected via open and plugged cytoplasmic bridges to choanosome, with its flagellated chambers; external pinacoderm absent and replaced by a noncellular dermal membrane; choanosome with anucleate choanocytes embedded in a trabecular syncytium. All studied species are viviparous; some produce a uniquely hexactinellid larva, the trichimella. Long-lived, exclusively marine, usually vase- or tube-shaped (never encrusting), primarily deep-water sponges (maximum diversity is at 300 to 600 m depth); 690 described species. Many species harbor Archaea-dominated microbial communities within their bodies. The hexactinellid body plan is perhaps the most unusual in the entire animal kingdom because nearly all the tissues in adults consist of a giant multinucleated syncytium that forms the inner and outer layers of the sponge, joined by cytoplasmic bridges to limited uninucleate cellular regions. Two subclasses

SUBCLASS AMPHIDISCOPHORA Body anchored in soft sediments by a basal tuft or tufts of spicules; megascleres are discrete spicules, never fused into a rigid network; with birotulate microscleres, never hexasters (e.g., *Hyalonema*, *Monorhaphis*, *Pheronema*)

SUBCLASS HEXASTEROPHORA Usually attached to hard substrata, but sometimes attached to sediments by a basal spicule tuft or mat; microscleres are hexasters; megascleres free, or can be fused into a rigid skeletal framework, in which case sponge may assume large and elaborate morphology (e.g., *Aphrocallistes*, *Caulophacus*, *Euplectella*, *Hexactinella*, *Leptophragmella*, *Lophocalyx*, *Rosella*, *Sympagella*)

CLASS DEMOSPONGIAE Demosponges (Figure 6.2D–M). With siliceous spicules and/or an organic skeleton (or, occasionally, with neither) or in some groups, a solid calcitic skeleton; spicules secreted intra- or extracellularly around a triangular or hexagonal axial filament; spicules never 6-rayed (i.e., not triaxons); organic skeleton a collagenous network ("spongin"); most produce parenchymella larvae (Figure 6.19); viviparous or oviparous (oviparous species occur in the subclasses Myxospongiae, Haplosceromorpha and Heteroscleromorpha); marine, brackish, or freshwater sponges occurring at all depths. Many species exhibit a mesohyl community of Eubacteria, mainly Proteobacteria and gram-positive bacteria, as well as species of Archaea. About 7,400 described species. The two subclasses Tetractinomorpha and Ceractinomorpha, long recognized as polyphyletic, have recently been abandoned, and four new subclasses have been proposed: Keratosa, Myxospongiae, Haploscleromorpha, and Heteroscleromorpha. Molecular analyses to date support the monophyly of these new subclasses, although not all four have been uniquely defined morphologically.

SUBCLASS KERATOSA Skeleton of spongin fibers only, or with a hypercalcified skeleton (*Vaceletia*). Spongin fibers either homogenous or pithed, and strongly laminated with pith grading into bark. Reproduction typically viviparous, larvae are parenchymella. All commercial sponges belong to this subclass (e.g., *Spongia*, *Hippospongia*, *Coscinoderma*, *Rhopaleoides*). Includes the two orders Dendroceratida (Darwinellidae: e.g., *Aplysilla*, *Darwinella*, *Dendrilla*; Dictyodendrillidae: e.g., *Dictyodendrilla*, *Spongionella*) and Dictyoceratida (Spongiidae: e.g., *Spongia*, *Hippospongia*, *Rhopaleoides*; Thorectidae: *Cacospongia*; *Thorecta*, *Phyllospongia*, *Carteriospongia*;

Irciniidae: e.g., *Ircinia*, *Sarcotragus*; Dysideidae: e.g., *Dysidea*, *Pleraplysilla*; Verticillitidae: *Vaceletia*)

SUBCLASS MYXOSPONGIAE (= VERONGIMORPHA) Without a skeleton or with skeleton of spongin fibers only (with a laminated bark and finely fibrillar or granular pith); one genus with skeleton of siliceous asters (*Chondrilla*). All demosponges lacking a skeleton belong to this subclass (e.g., *Chondrosia*, *Halisarca*, *Hexadella*, *Thymosiopsis*). Reproduction oviparous. Two or three orders are now recognized (e.g., *Aplysina, Aplysinella, Chondrosia, Chondrilla, Halisarca, Hexadella, Ianthella, Suberea, Thymosia, Thymosiopsis, Verongula*).

SUBCLASS HAPLOSCLEROMORPHA With an isodictyal anisotropic or isotropic choanosomal skeleton; spicules diactinal megascleres (oxeas or strongyles); microscleres, if present, sigmas and/or toxas, microxeas, or microstrongyles. Reproduction typically viviparous, with the exception of some petrosid genera that are oviparous. Larvae typically parenchymella. Includes Calcifibrospongia, previously assigned to Sclerospongiae. Includes only one order, Haplosclerida. Most of the families and genera are probably polyphyletic, and this subclass itself is sometimes considered part of Heteroscleromorpha. Among the most common genera are: *Haliclona*, *Callyspongia*, *Petrosia*, *Xestospongia*, *Niphates*, *Amphimedon*, *Siphonochalina*, and also *Calcifibrospongia*, *Dendroxea* and *Janulum*.

SUBCLASS HETEROSCLEROMORPHA Skeleton composed of siliceous spicules that can be monaxones and/or tetraxones and, when present, highly diversified microscleres. Reproduction mostly viviparous, but oviparous species occur in some genera (e.g., *Agelas*, *Axinella*, *Raspailia*, *Suberites*). This subclass contains most of the Demospongiae (about 5,000 species), usually organized into eight orders. Most of the sponges once assigned to Sclerospongiae are in this subclass, including stromatoporoids (e.g., *Astrosclera*), tabulates (*Merlia*, *Acanthochaetetes*), and ceratoporellids (e.g., *Ceratoporella*, *Goreauiella*, *Hispidopetra*, and *Stromatospongia*). The order Tetractinellida (*Astrophorina*, *Spirophorina* and most of the lithistids) includes all sponges with tetractine spicules (e.g., *Geodia*, *Penares*, *Stelletta*, *Tetilla*, *Cinachyra*, *Discodermia*). The freshwater sponge order Spongillida is here, which includes 220 species in six families (e.g., *Spongilla*, *Ephydatia*, *Lubomirskia*, *Metania*, *Potamolepis*). Other orders include: Poecilosclerida (e.g., *Clathria*, *Hymedesmia*, *Mycale*, *Myxilla*, *Desmacella*, *Asbestopluma*), Agelasida (e.g., *Agelas*, *Astrosclera*, *Hymerhabdia*, *Acanthostylotella*), Axinellida (e.g., *Axinella*, *Higginsia*, *Stelligera*, *Raspailia*, *Eurypon*, *Myrmekioderma*), Biemnida (e.g., *Biemna*, *Neofibularia*, *Sigmaxinella*), Halichondrida (e.g., *Halichondria*, *Hymeniacidon*), and Hadromerida (e.g., *Cliona*, *Spirastrella*, *Tectitethya*, *Tethya*).

CLASS HOMOSCLEROMORPHA The most recently proposed class of sponges, homoscleromorphs have been shown to be distinct (and monophyletic) on the basis of molecular phylogenetics, anatomy, and embryology (Figure 6.20B). Spongin skeleton always absent; rigid skeleton almost always absent, but when present composed of small (mostly <100 μm) tetraxon (four often unequal rays) siliceous spicules called calthrops,[2] similar but distinct from certain tetraxon spicules seen in demosponges, or peculiar irregular kinked oxeas (2 uneven rays); all spicules are of a similar size, with no differentiation into megascleres and microscleres (hence the class name); with ciliated exopinacocytes and endopinacocytes; a type IV collagen basement membrane underlies both choanoderm and pinacoderm (this type of collagen has also been found in several species of demosponges and calcareans); zonula adherens cell junctions in adult and larval epithelia; spermatozoids with an acrosome; choanocyte chambers oval to spherical, with large choanocytes. To date, septate junctions have not been found in this class. Homoscleromorphs are viviparous (incubate their embryos), form a hollow blastula by multipolar egression, and develop a unique cinctoblastula larva. Although some work had suggested this group might be more closely allied with higher metazoans than with other sponges, recent studies based on complete mitochondrial genome sequences strongly support its inclusion within a monophyletic Porifera, perhaps most closely related to the Calcarea. About 85 known species of exclusively marine sponges, divided into two monophyletic families, Plakinidae (spiculate species) and Oscarellidae (aspiculate species). Most species inhabit hard bottoms of the continental shelf, but some are known from more than 1,000 m depth. The fossil record, poor as it is, dates from the early Carboniferous. About 90 species (e.g., *Corticium, Oscarella, Placinolopha, Plakina, Plakinastrella, Plakortis, Pseudocorticium*).

The Poriferan Body Plan

A mind-boggling diversity exists in the shape, color, and size of sponges. Increases in size and surface area are accomplished by folding of the body wall into a variety of patterns, and also from different growth patterns that appear in response to environmental conditions. This plasticity, plus the fact that sponges maintain certain cells in a pluripotent state, compensate in part for the absence of true organs. Two unique organizational attributes define sponges and have played major roles in poriferan success: first is the water current channel system, or **aquiferous system**, and its unique all-purpose cells (pumping, feeding, gas-exchange, and waste-expulsion) known as **choanocytes**.

[2]A "caltrop" or "calthrop" was an evil-looking medieval landmine, consisting mainly of four short, sharpened spikes radiating from a common center at equal angles. These were scattered in front of defensive positions and covered under a thin layer of dirt, where they tended to inconvenience charging knights. Homoscleromorph tetractine spicules often have the same general shape.

Second is the highly pluripotent nature of sponge cells in general. The aquiferous system brings water through the sponge and close to the cells responsible for food gathering and gas exchange. At the same time, excretory and digestive wastes are expelled by way of the water currents. The volume of water moving through a sponge's aquiferous system is remarkable. A single 1 × 10-cm individual of the complex sponge *Leuconia* pumps about 23 liters of water through its body daily. Researchers have recorded sponge pumping rates that range from 0.002 to 0.84 ml of water per second per cubic centimeter of sponge body. Large sponges filter their own volume of water every 10 to 20 seconds.

The Hexactinellida ("glass sponges") might be the most unusual creatures in the entire animal kingdom. Their triaxonal spicules, with a square proteinaceous internal filament, distinguish them from other siliceous sponges. Their syncytial body, which arises by fusion of early embryonic cells, is unique among Metazoa. Both the larvae and adults have elegant combinations of multinucleate and cellular cytoplasmic regions unlike anything seen in any other animal. The continuity of this tissue allows food to be transported symplastically (analogous to plants), and it also allows electrical signals to travel throughout the sponge body (analogous to the nervous system of higher Metazoa).

Sponges are not colonial animals. Any and all sponge material bounded by a continuous outer covering (the pinacoderm) constitutes a single individual. The growth of each individual is dictated by a combination of genetics (e.g., most larvae have anterior–posterior symmetry, many adults have a radial–axial growth form) and environmental factors (e.g., water flow dynamics, contours of the substratum). For most sponges, changes in body form can arise anywhere in or on the organism in response to external factors.

Body Structure and the Aquiferous System

The outer surface cells of a sponge make up the **pinacoderm** and are called **pinacocytes**, typically unciliated flattened cells that are also known as "pavement cells." The pinacoderm can be considered a true epithelium. Most of the inner surfaces comprise the **choanoderm** and are composed of flagellated cells called **choanocytes**. Both of these layers are a single cell thick. Between these two thin cellular sheets is the **mesohyl**, which may be very thin in some encrusting sponges, or massive and thick in larger species (Figure 6.3). The pinacoderm is perforated by small holes called **dermal pores** or **ostia** (sing. ostium), depending on whether the opening is surrounded by several cells or one cell, respectively (Figure 6.3). Water is pulled through these openings and is driven across the choanoderm by the beating of the choanocyte flagella. The choanocytes pump large volumes of water through the sponge body, thus enabling the aquiferous system. The cost of sponges pushing so much water through the body has been estimated to be about one-third of its overall metabolism.

Sponge choanocytes are thought to be fundamentally identical to cells of choanoflagellate protists (see Chapter 3). They originate during metamorphosis from the ciliated epithelial cells of larvae, and they can also derive from archaeocytes found in gemmules—in both cases, these precursor cells differentiate into choanoblasts that, in turn, grow collars to become choanocytes. Their long flagellum has a basal, bilaterally-arranged pair of wing-shaped projections called **vanes**. The flagellar vanes probably lend support to the flagellum. The flagellum is surrounded by a **collar** of microvilli containing actin filaments, the microvilli being connected to one another via a thin membrane composed of a meshwork of extracellular glycoprotein (the **glycocalyx mesh**) (Figure 6.8 F–H).[3]

A cuticle, or layer of coherent collagen, may cover the pinacoderm in some species. The pinacoderm itself is a simple external sheet (**exopinacoderm**) of cells (**exopinacocytes**), and it also lines some of the internal cavities of the aquiferous system where choanocytes do not occur. Pinacocytes that line internal canals form **endopinacoderm** and are called **endopinacocytes**. The choanoderm can be simple and continuous, or folded and subdivided in various ways. The mesohyl varies in thickness and plays vital roles in digestion, gamete production, secretion of the skeleton, and transport of nutrients and waste products by special ameboid cells. The mesohyl includes a noncellular colloidal mesoglea in which are embedded collagen fibers, spicules, and various cells; as such, it is really a type of mesenchyme. A great number of cell types may be found in the mesohyl. Most of these cells are able to change from one type to another as required, but some differentiate irreversibly, such as those that commit themselves to reproduction or to skeleton formation.

Some sponges actually move from one place to another—ameboid cells along the base of the sponge "crawl," as others carry forward spicules to support the leading edge of the sponge. Studies suggest that some amoebocytes can actually break free from a sponge and move about on their own for a time, eventually returning to the parent sponge body. This "locomotion" in sponges obviously is not sufficient to provide them with a quick escape mechanism from predators; we're talking about millimeters a day here!

As a sponge grows, the pinacoderm and choanoderm remain only one cell thick. But by increasing

[3]Monociliated cells bearing a collar of microvilli surrounding the cilium have been found in almost all metazoan phyla. However, in other metazoans the cilium is usually immobile, lacks vanes, has a striated rootlet, and is frequently involved in sensory processes. It is therefore thought that such cells are not homologues to sponge choanocytes and choanoflagellate protists.

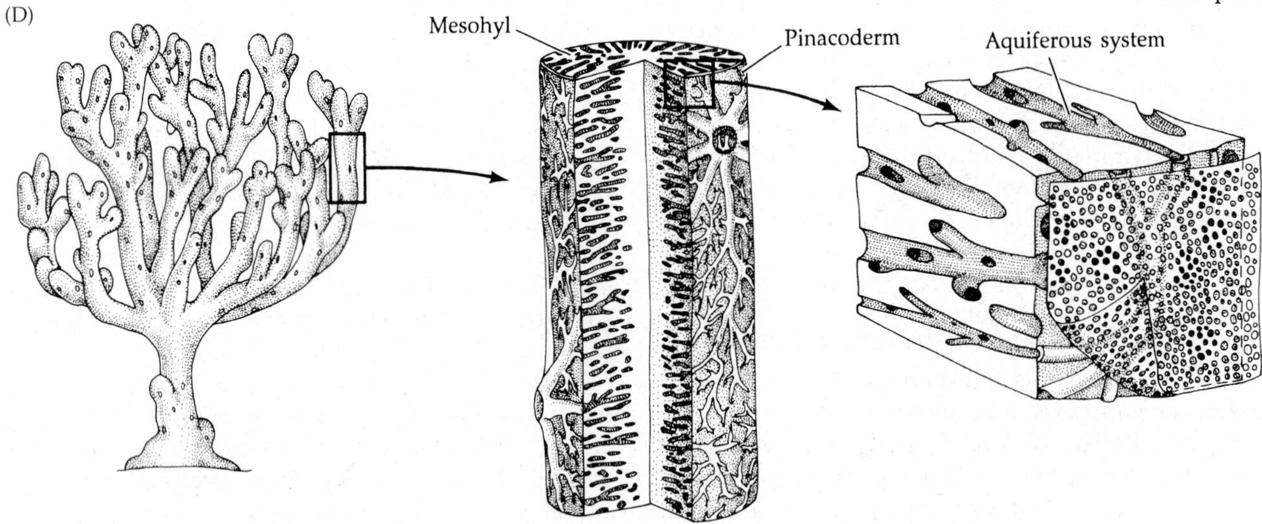

Figure 6.3 Sponge body forms. (A) The unusual demosponge *Coelosphaera hatchi* (height in life 27 mm). (B) The coralline sponge *Merlia normani* (vertical section) has a basal calcareous matrix within which individual compartments are filled by secondary deposition. The superficial soft tissue contains the choanocyte chambers and is supported by tracts of siliceous spicules. (C) The demosponge *Haliclona* sp., a sponge with a tubular type of architecture; three successive levels of magnification are shown, from left to right. (D) *Clathria prolifera*, a demosponge with a more solid type of architecture; three successive levels of magnification are shown, from left to right.

folding as the mesohyl volume increases, these layers maintain a surface area-to-volume ratio sufficient to sustain adequate nutrient and waste exchange throughout the whole individual. Unique to the class Calcarea, some species go through an ontogenetic sequence in which an increasingly complex body architecture develops; adults may express the most complex structure, or they can retain one of the simpler architectures. Thus, some adult Calcarea may retain a largely unfolded, simple, and continuous choanoderm (the **asconoid condition**; Figure 6.4A), or the choanoderm may become folded (the **syconoid condition**; Figure 6.4B,C), or it may become both folded and subdivided into separate flagellated chambers (the **leuconoid condition**; Figure 6.4D). Asconoid and syconoid anatomies occur only in the class Calcarea.

The asconoid condition is found in the early growth stage (olynthus) of newly settled calcareous sponges and in some adult, radially symmetrical calcareous sponges (e.g., *Clathrina*, *Leucosolenia*) (Figure 6.5). Asconoid and syconoid sponges rarely exceed a few cm in height and remain as simple, vase-shaped (apical–basal radial symmetry) tubular units, or meandering networks of these vase-shaped units, or occasionally

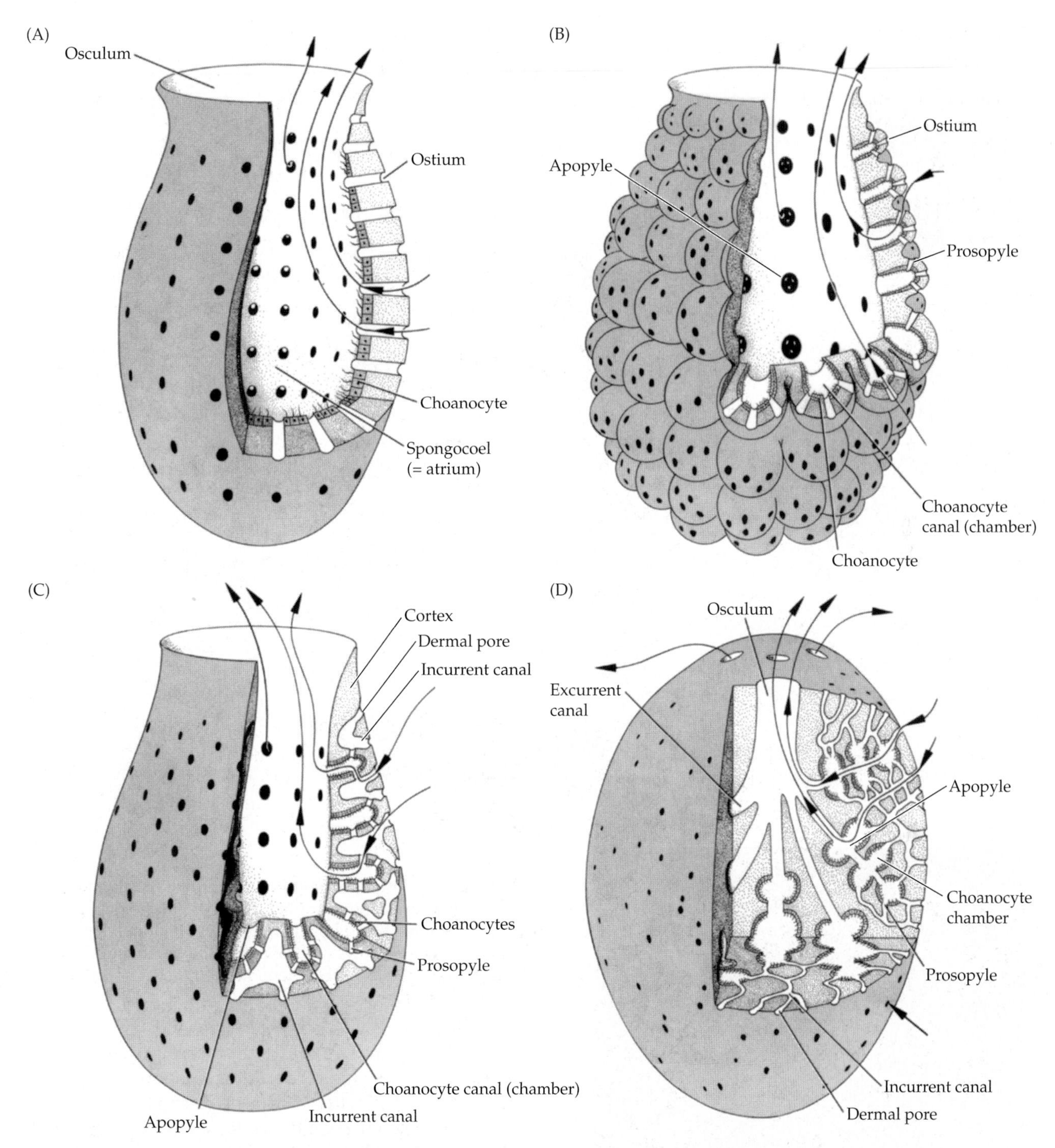

Figure 6.4 Body complexity in sponges (arrows indicate flow of water). (A) The asconoid condition. (B) A simple syconoid condition. (C) A complex syconoid condition with ectosomal growth. (D) A leuconoid condition. Asconoid and syconoid anatomies occur only in the class Calcarea.

even stalked sponges (e.g., some *Clathrina*). The thin walls enclose a central cavity called the **atrium** (= **spongocoel**), which opens to the outside via a single **osculum** (pl. oscula). The pinacoderm of asconoid, and very simple syconoid sponges rely on **ostia** as their incurrent pores. Ostia develop during embryogeny as specialized cells (**porocytes**) that elongate and roll to form cylindrical tubes. Each porocyte extends all the way through the pinacoderm, the thin mesohyl, and the choanoderm, and thus into the atrium where it emerges between adjacent choanocytes like a porthole. (Porocytes also occur in a few demosponge groups, particularly the freshwater species.) The choanoderm is a simple, unfolded layer of choanocytes lining the entire atrium. Water moving through an asconoid sponge thus flows through the following structures: ostium → atrium/spongocoel (over the choanoderm) → osculum.

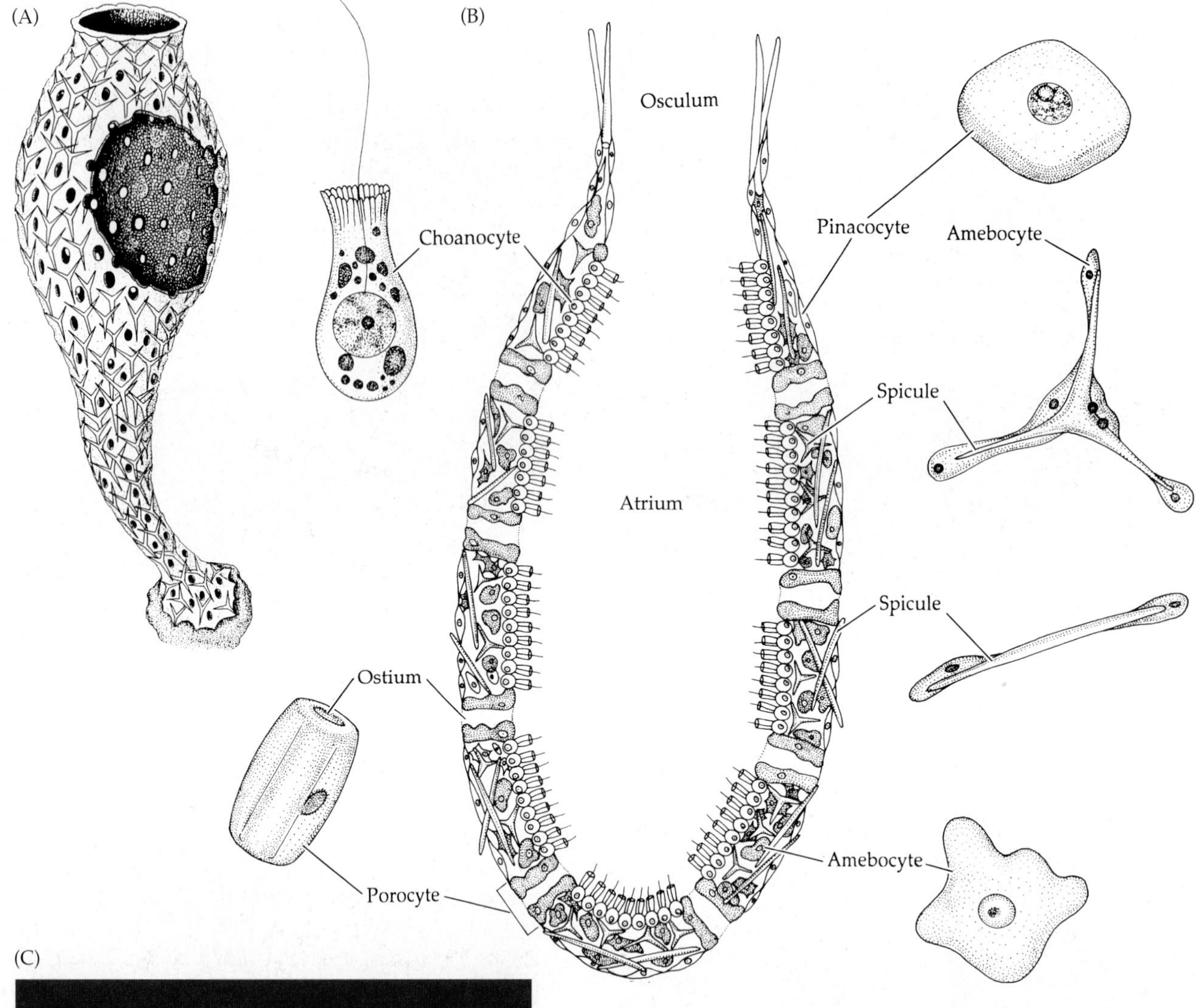

Figure 6.5 The asconoid condition (class Calcarea). (A) An olynthus, the asconoid form that follows larval settlement in calcareous sponges. (B) Major cell types in an asconoid sponge. (C) The simple calcareous sponge *Leucosolenia* shows the asconoid body form and skeleton of $CaCO_3$ spicules.

As stated above, in calcareous sponges, simple folding of the pinacoderm and choanoderm produces the syconoid condition, within which several levels of complexity are possible (Figure 6.4B,C). As complexity increases, the mesohyl may thicken and appear to have two layers. The outer layer (formerly referred to as the cortical region or cortex) is called the **ectosome**. The ectosome contains an accumulation of spicules that are different from those found in the interior portion of the mesohyl. In calcareous sponges with an ectosome, the incurrent openings are lined by several cells (not formed by a single porocyte) and are referred to as **dermal pores**. In the syconoid condition, choanocytes are restricted to specific chambers or diverticula of the atrium called **choanocyte chambers** (also known as flagellated chambers, or radial canals). Each choanocyte chamber opens to the atrium by a wide aperture called an **apopyle**. Syconoid sponges with a thick ectosome possess a system of channels or incurrent canals that lead from the dermal pores through the mesohyl to the choanocyte chambers. The openings from these channels to the choanocyte chambers are called **prosopyles**. In such a complex syconoid sponge, water moving from the surface into the body flows along the following route: dermal (incurrent) pore → incurrent canal → prosopyle → choanocyte chamber → apopyle → atrium → osculum. Syconoid construction is found in many calcareous sponges, including the well-known genera *Grantia* and

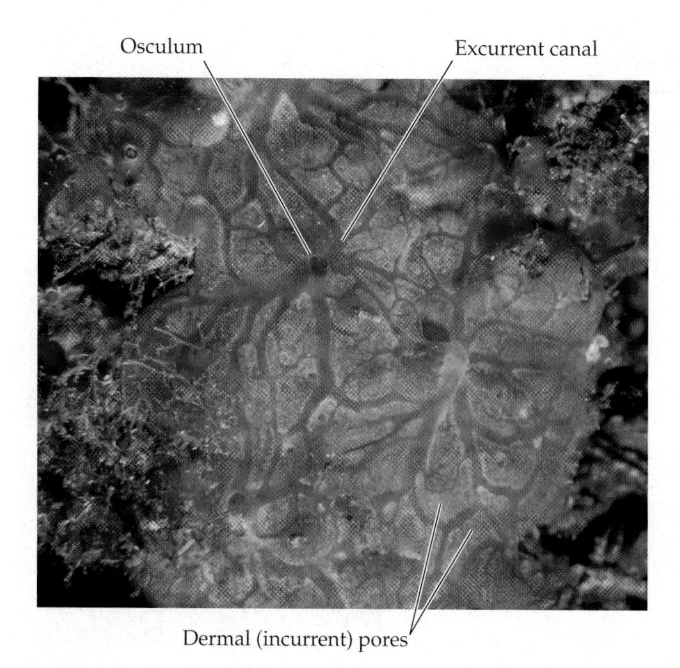

Figure 6.6 The surface of a living demosponge (*Clathria*). The complex system of ostia opens into underlying incurrent canals, and large oscula receive several excurrent canals, all visible through the thin pinacoderm.

Sycon (formerly known as *Scypha*), and such construction often yields vase-shaped adults, as in many asconoid Calcarea. Some syconoid sponges express a radial symmetry externally, but their complex internal organization is largely asymmetrical.

The leuconoid condition is found in all four sponge classes. Here, the choanoderm is much more highly folded and there is a thickening of the mesohyl by growth of the ectosome. Associated with this is a subdivision of the flagellated surfaces into discrete oval choanocyte chambers (Figure 6.4D). In the leuconoid condition, one finds an increase in number and a decrease in size of the choanocyte chambers, which typically cluster in groups in the thickened mesohyl. The atrium is often reduced to a series of **excurrent canals** (or "exhalent" canals) that carry water from the choanocyte chambers to the oscula (Figure 6.6). The flow of water through a leuconoid sponge is: dermal pore → incurrent canal → prosopyle → choanocyte chamber → apopyle → excurrent canal → (atrium) → osculum. Leuconoid organization is typical of all noncalcareous sponges, and many calcareous sponges (e.g., *Leucilla*).

It is important to realize that flow rate is not uniform through the various parts of the aquiferous system. Functionally, it is critical that water be moved very slowly over the choanoderm, allowing time for exchanges of nutrients, gases, and wastes between the water and the choanocytes. The changes in water flow velocity through this plumbing system are a function of the effective accumulated cross-sectional diameters of the channels through which the water moves (see Chapter 4, or your old physics notes). Water flow velocity decreases as the cross-sectional diameter increases—liquids move slower in larger pipes—or when a single pipe divides into numerous smaller pipes. Thus, in a sponge, velocities are lowest over the choanoderm, which has the largest cross-sectional area. However, water leaving the osculum must be carried far enough away to prevent it from fouling or being immediately recycled by the sponge, so smaller accumulated cross-sectional diameters (and greater water velocity) are found in excurrent canal–oscula system.

The recognition of the various levels of organization and complexity among calcareous sponges used to be thought to provide important evolutionary clues. However, there is no evidence that the asconoid plan is necessarily the most phylogenetically primitive, or that all sponge lineages have moved through these three levels of complexity during their evolution. Nor do most sponges pass through these three stages during their development. Calcareous sponges of the leuconoid condition usually do pass through asconoid and syconoid stages as they grow, but it is only in this class that all three organizational body plans clearly occur.

The hexactinellid sponges differ strikingly from other sponge classes (Figure 6.7). The bodies of hexactinellid sponges often display a considerable degree of superficial radial symmetry. There is not a typical pinacoderm in hexactinellids. Instead, a **dermal membrane** is present, but it is extremely thin and is continuous with all the other parts of the sponge; no discrete or continuous cellular structure supports it. Incurrent pores are simple holes in this dermal membrane. The main tissue of the hexactinellid is called the **trabecular syncytium**—it forms a syncytial, trabecular network stretching across interconnecting internal cavities (the subdermal lacunae) near the surface of the sponge and penetrating into and forming the supporting framework for the **choanosome** (Figure 6.7A). The trabecular network resembles a three-dimensional cobweb. The thimble-shaped flagellated chambers of the choanosome are arranged in a single layer and are supported within the trabecular network. Both the trabecular network and the walls of the flagellated chambers are syncytial. In these sponges, the choanoblasts produce new collars but do not produce new nuclei, thus creating a single-nucleate syncytium with many collar bodies. As new collars "bud off," the expanding, single-cell body of the choanoblast cell recesses itself beneath the trabecular syncytium, leaving just the collars exposed to do the work. Water enters incurrent pores in the dermal membrane, passes into the subdermal lacunae, and from there enters the syncytial choanosome chambers via the prosopyles. The soft tissue of the syncytium serves to deliver food and other metabolites to all parts of a hexactinellid sponge by cytoplasmic streaming.

The unique structure of hexactinellids is so striking that some workers have even suggested the

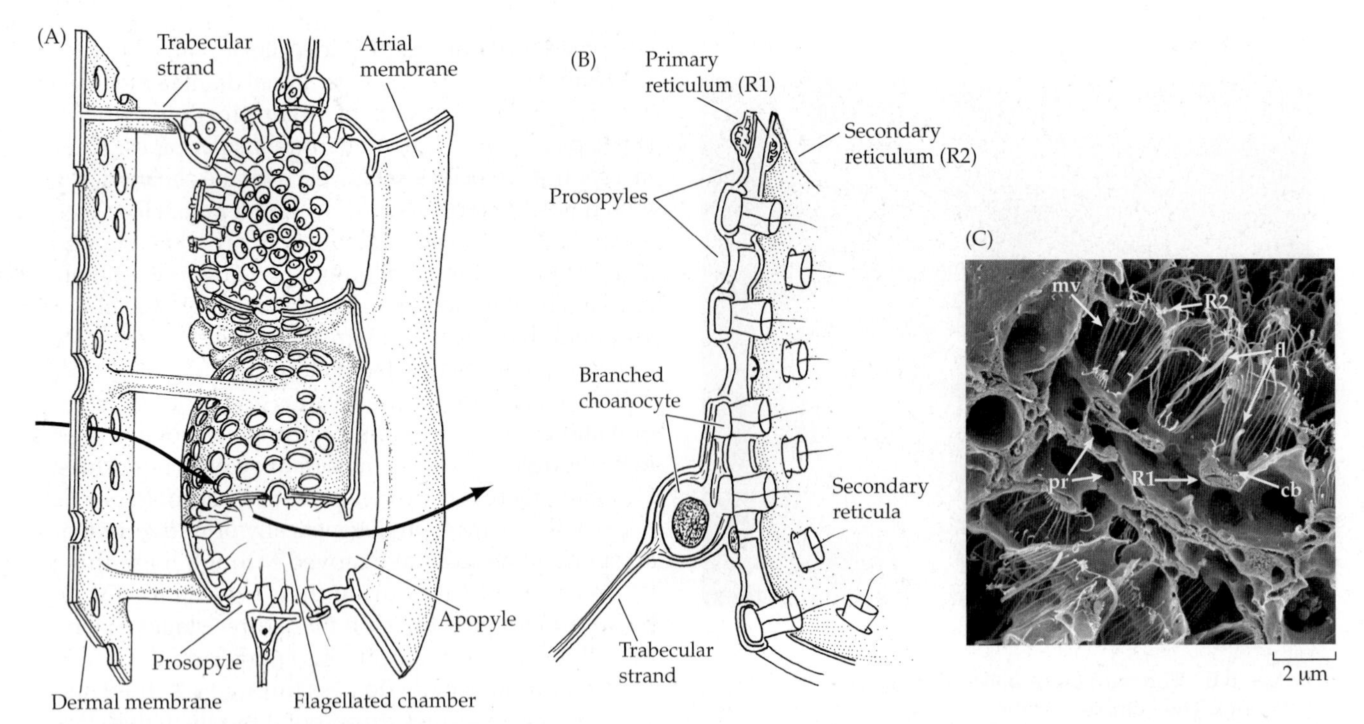

Figure 6.7 Internal anatomy of Hexactinellida showing the trabecular network. (A) Two flagellated chambers connected to the dermal membrane by trabecular strands; arrow shows water flow (*Farrea occa*). (B) Section through wall of a flagellated chamber, based on *Rhabdocalyptus dawsoni*. Water (arrows) is drawn through prosopyles in the primary reticulum and through choanocyte collars by the beating of the flagella. The secondary reticulum surrounds the collars, forcing water through the collar microvilli. (C) Canal system and tissue structure in *Aphrocallistes vastus* (scanning electron micrograph); primary (R1) and secondary (R2) reticula, extensions of the trabecular reticulum, incurrent pores, prosopyles (pr), collar bodies (cb) branching from choanocytes, with collar microvilli (mv) and flagellum (fl).

hexactinellids might be regarded as a separate subphylum or phylum (the "Symplasma"), distinct from the other sponge classes (the "Cellularia"). However, as explained in Chapter 2, phylogenetic relationships are best sought in similarities among groups (i.e., shared derived features), not in differences. By this reasoning, and through molecular phylogenetics, we know without doubt that hexactinellids are poriferans (and probably closely related to demosponges). Nevertheless, hexactinellids are striking and bizarre in that no other metazoan group possesses such extensive syncytial tissues.[4]

More on Sponge Cell Types

Prior to the 1970s, texts generally recognized only a few basic kinds of poriferan cells. However, subsequent detailed histochemical and ultrastructural studies have revealed a broad variety of cell types. These discoveries make succinct "textbook classification" of their cells difficult. We present below a highly abridged version of sponge cell classification.

Cells that line surfaces As noted above, the pinacoderm forms a continuous layer on the external surface of sponges and also lines all incurrent and excurrent canals. The pinacocytes that make up this layer are usually flattened and usually overlapping (Figure 6.8A,B). The external surface, or exopinacoderm, of demosponges has an extracellular matrix (ECM) beneath it, and in the Calcarea a loose collagenous mesohyl underlies the pinacoderm. In homoscleromorph sponges, and some species in other classes, a true basement membrane underlies the pinacoderm. In all cases, this layer of secreted material helps maintain the positional

Figure 6.8 Cells that line sponge surfaces. (A) A pinacocyte from the surface of the demosponge *Halisarca* (drawn from an electron micrograph). The outer surface is covered with a polysaccharide-rich glycocalyx coat. The cell is fusiform and overlaps adjacent pinacocytes. (B) Pinacoderm from a calcareous sponge (section). T-shaped pinacocytes alternate with fusiform pinacocytes. (C,D) A porocyte from the calcareous asconoid sponge *Leucosolenia*. (C) Cross section. (D) Side view. (E) *Myocytes* surrounding a prosopyle. (F) A section of choanoderm, showing three choanocytes; arrows indicate direction of water current. (G) A choanocyte. (H) Ultrastructure of a choanocyte (longitudinal section, drawn from an electron micrograph). (I) A choanocyte chamber opening into an excurrent canal in a demosponge.

[4]Syncytial tissues are seen in some other metazoans. For example, the epithelia of some cnidarians and the skeletogenic tissues of echinoids are syncytial, by way of incomplete cytokinesis. The giant neurons in squid, and vertebrate striated muscle cells (your biceps, for example) are multinucleate by fusion of separate cells.

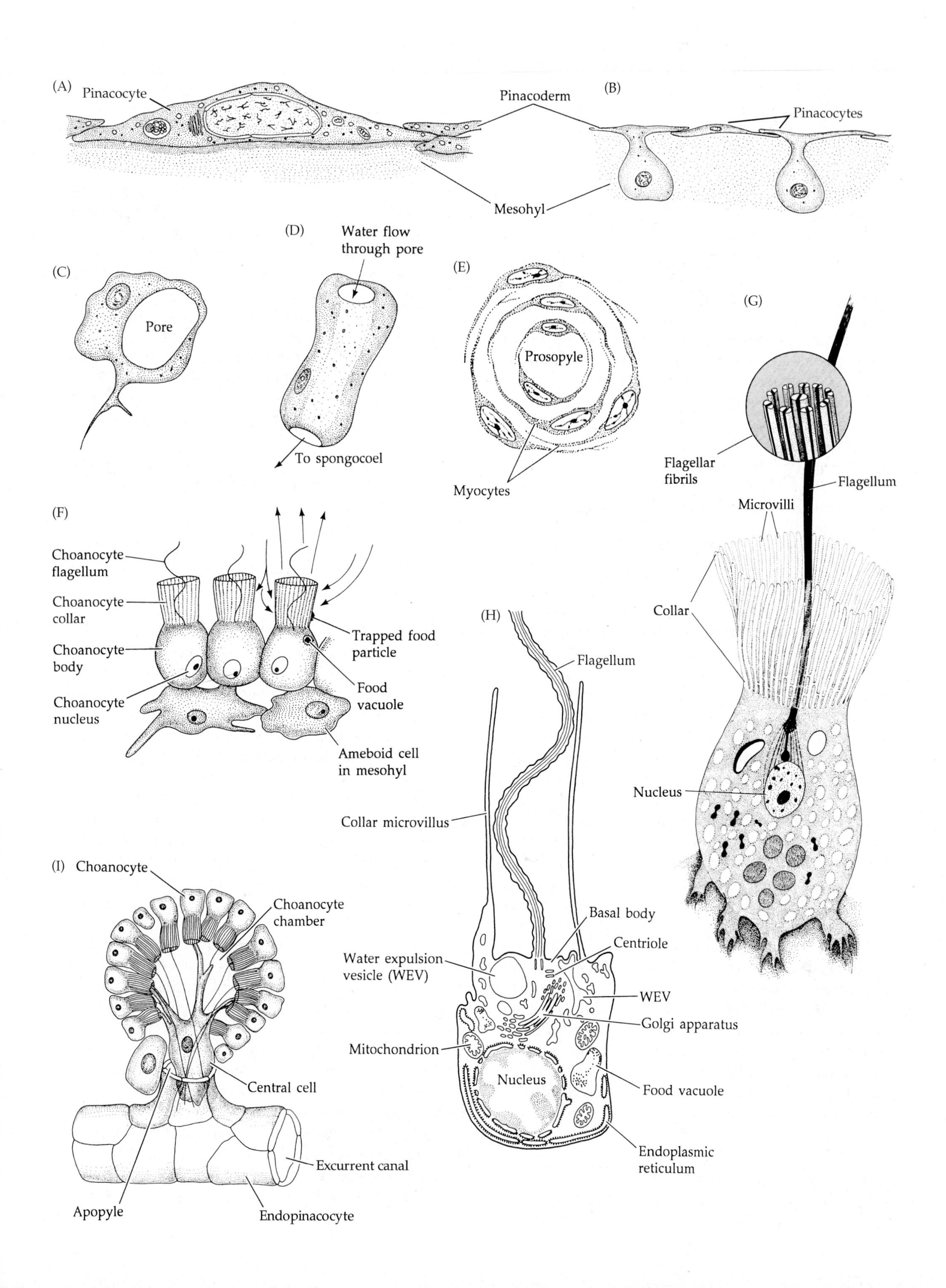
(A)
Pinacocyte
Pinacoderm
(B)
Pinacocytes
Mesohyl
(C)
Pore
(D)
Water flow through pore
To spongocoel
(E)
Prosopyle
Myocytes
(G)
Flagellar fibrils
Flagellum
Microvilli
Collar
Nucleus
(F)
Choanocyte flagellum
Choanocyte collar
Choanocyte body
Choanocyte nucleus
Trapped food particle
Food vacuole
Ameboid cell in mesohyl
(H)
Flagellum
Collar microvillus
Basal body
Centriole
Water expulsion vesicle (WEV)
WEV
Golgi apparatus
Mitochondrion
Nucleus
Food vacuole
Endoplasmic reticulum
(I) Choanocyte
Choanocyte chamber
Central cell
Excurrent canal
Apopyle
Endopinacocyte

integrity of pinacocytes. Basal membranes are sheet-like complexes of extracellular matrix proteins that are secreted by the epithelial layer; they are highly structured and contain type IV collagen. Such membranes typically underlie epithelial (and endothelial) tissues in the Metazoa, but not in fungi or plants. They provide an added mechanical function as supporting structures, and also play a biological role as molecular sieves. The pinacoderm of Homoscleromorpha is especially well organized, forming a well-developed epithelium (with an underlying basal membrane). However, the apparent absence of a basal membrane in most Porifera distinguishes sponge pinacoderm from the well-developed tissue epithelia of the higher Metazoa.[5]

Internal, canal-lining pinacocytes (**endopinacocytes**) are usually more fusiform in shape and have less overlap than outer **exopinacocytes**. Although the endopinacoderm is epithelial in function, it is probably phagocytic as well. External cells of the basal or attaching region of a sponge surface are called **basopinacocytes**. These flattened cells are responsible for secreting a fibrillar collagen–polysaccharide complex by which sponges attach to the substratum. In freshwater sponges, the basopinacocytes are active in feeding and extend ameba-like "filopodia" to engulf bacteria. Freshwater sponge basopinacocytes also play an active role in osmoregulation and contain large numbers of water expulsion vesicles, or contractile vacuoles.

As noted earlier, porocytes are cylindrical or flat, tubelike cells of the pinacoderm that form ostia in some sponge species (Figure 6.8C,D). They are contractile and can open and close the pore and regulate the ostial diameter; however, no microfilaments have been observed in them and their precise method of contraction and expansion is unknown. Some species are capable of producing a diaphragm-like cytoplasmic membrane across the ostial opening that also regulates pore size.

Choanocytes are the flagellated collar cells that make up the choanoderm and create the currents that drive water through the aquiferous system (Figure 6.8F–H). Choanocytes are not coordinated in their beating, not even within a given chamber. However, they are aligned such that the flagella, which beat from base to tip, are directed toward the apopyle. The long flagellum is always surrounded by the choanocyte collar, which is made up of 20 to 55 cytoplasmic microvilli (= villi). The villi have microfilament cores and are connected to one another by anastomosing mucous strands (a mucous reticulum) that form a glycocalyx band around the collar. Choanocytes rest on the mesohyl, held in place by interdigitation of adjacent basal surfaces. In keeping with their central role in phagocytosis and pinocytosis, choanocytes are highly vacuolated.

[5]Type IV collagen, diagnostic of Metazoa, is a long triple helix molecule; it is one of several collagen types found in sponges and other metazoans, and it occurs in all four sponge classes.

Cells that secrete the skeleton There are several types of ameboid cells in the mesohyl that secrete elements of sponge skeletons. In almost all sponges, the entire supportive matrix is built on a framework of fibrillar collagen. The cells that secrete this material are called collencytes, lophocytes, and spongocytes. **Collencytes** are morphologically nearly indistinguishable from pinacocytes, whereas **lophocytes** are large, highly motile cells that can be recognized by a collagen tail they typically trail behind them (Figure 6.9C). The primary function of both cell types is to secrete the dispersed fibrillar collagen found intercellularly in virtually all sponges. **Spongocytes** produce the fibrous supportive collagen referred to as **spongin** (Figure 6.12A). Spongocytes operate in groups and are always found wrapped around a spicule or spongin fiber (Figure 6.9D).

Sclerocytes are responsible for the production of calcareous and siliceous sponge spicules (Figure 6.9A,B). They are active cells that possess abundant mitochondria, cytoplasmic microfilaments, and small vacuoles. Numerous types of sclerocytes have been described; these cells always disintegrate after spicule secretion is complete.

Contractile cells Contractile cells in sponges, called **myocytes**, are found in the mesohyl (Figure 6.8E). They are usually fusiform and grouped concentrically around oscula and major canals. Myocytes are distinguished by the great numbers of microtubules and microfilaments contained in their cytoplasm, and they rely on the classic actin/myosin structure seen in all animals. Because of the nature of their filament arrangement, it has been suggested that myocytes are homologous with the smooth muscle cells of higher invertebrates. However, myocytes are independent effectors with a slow response time, and, unlike neurons and true muscle fibers, they are insensitive to electrical stimuli.

Other cell types **Archaeocytes** are highly versatile, largely totipotent ameboid cells that are capable of quickly differentiating to give rise to virtually any other cell type. These large, highly motile cells also play a major role in digestion and food transport (Figure 6.10). They possess a variety of digestive enzymes (e.g., acid phosphatase, protease, amylase, lipase) and can accept phagocytized material from the choanocytes. They also phagocytize material directly through the pinacoderm of water canals. Archaeocytes carry out much of the digestive, transport, and excretory activities in sponges. And, as cells of maximum totipotency, archaeocytes are essential to the developmental program of sponges

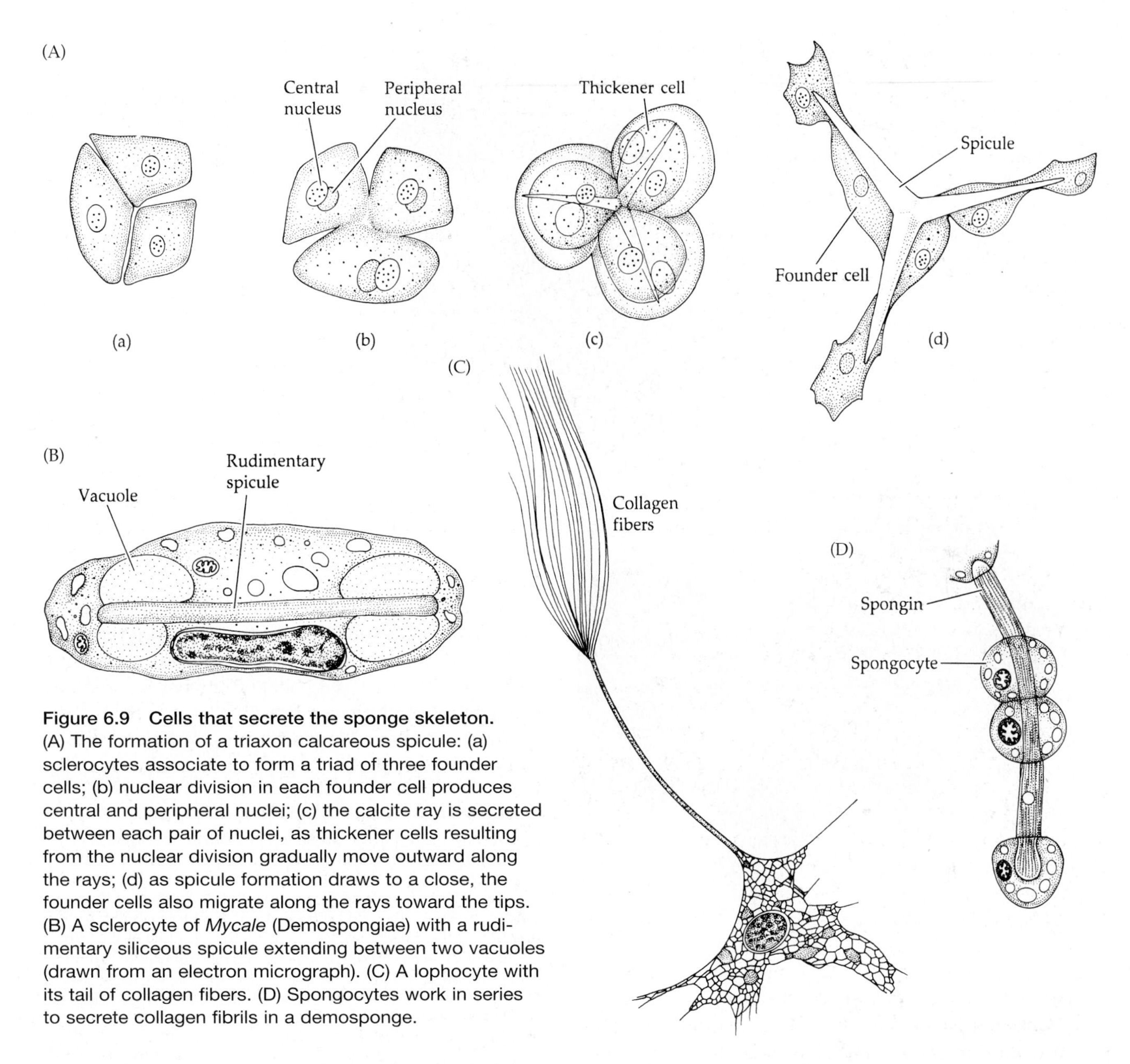

Figure 6.9 Cells that secrete the sponge skeleton. (A) The formation of a triaxon calcareous spicule: (a) sclerocytes associate to form a triad of three founder cells; (b) nuclear division in each founder cell produces central and peripheral nuclei; (c) the calcite ray is secreted between each pair of nuclei, as thickener cells resulting from the nuclear division gradually move outward along the rays; (d) as spicule formation draws to a close, the founder cells also migrate along the rays toward the tips. (B) A sclerocyte of *Mycale* (Demospongiae) with a rudimentary siliceous spicule extending between two vacuoles (drawn from an electron micrograph). (C) A lophocyte with its tail of collagen fibers. (D) Spongocytes work in series to secrete collagen fibrils in a demosponge.

and to various asexual processes (e.g., gemmule formation). **Spherulous cells** are large mesohyl cells containing various chemical inclusions.

Cell reaggregation Around the beginning of the twentieth century, H. V. Wilson first demonstrated the remarkable ability of sponge cells to reaggregate after being mechanically dissociated. Although this discovery was interesting in itself, lending insight into the plasticity of cellular organization in sponges, it also foreshadowed more far-reaching cytological research that has since shed light on basic questions about how cells self-recognize, adhere, segregate, and specialize. Many sponges that are dissociated and maintained under proper conditions will form aggregates, and some will eventually reconstitute their aquiferous system. For example, when pieces of the Atlantic "red beard sponge" (*Clathria prolifera*) are pressed through fine cloth, the separated cells immediately begin to reorganize themselves by active cell migration. Within 2 to 3 weeks, a functional sponge reforms and the original cells return to their respective functions. Furthermore, if cell suspensions of two different sponge species are mixed, the cells sort themselves out and reconstitute individuals of each separate species—a remarkable display of cellular self-recognition. The controversial sponge biologist M. W. de Laubenfels described the situation in 1949 in slightly different terms: "Sponges endure mutilation better than any other known animal." The discovery that dissociated sponge cells will reaggregate to form a functional organism was the basis for the establishment of sponge cell cultures that have been used as a model for the study of fundamental processes in developmental

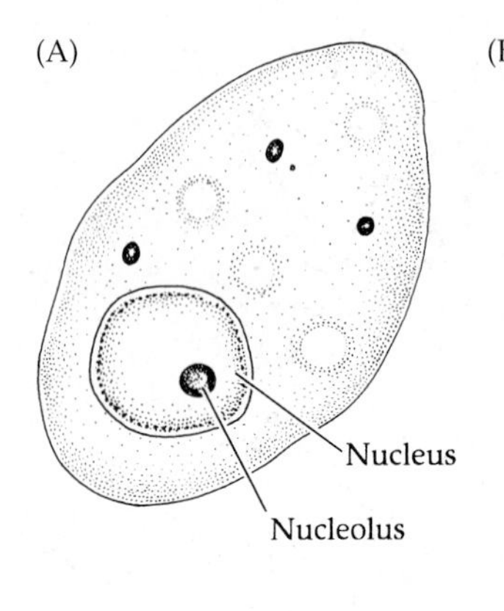

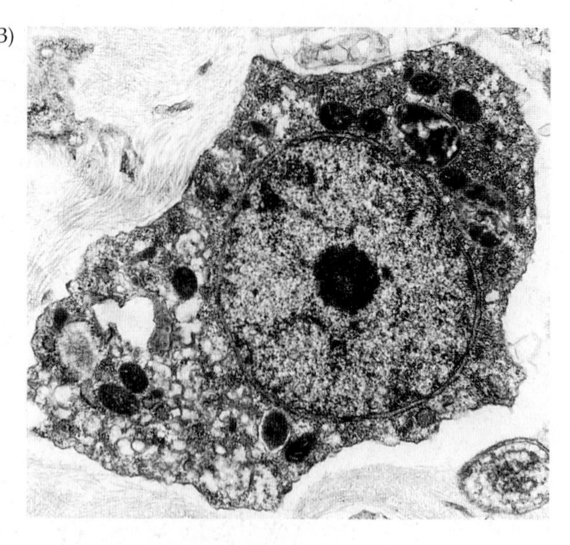

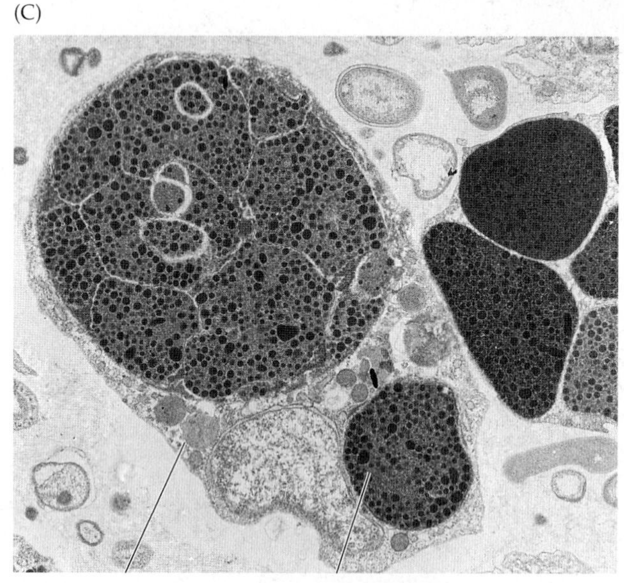

Figure 6.10 Archaeocytes. (A) A typical archaeocyte with a large nucleus and a prominent nucleolus. (B) Photo of a typical archaeocyte. (C) An archaeocyte engages in phagocytosis.

biology and immunology. We now know that sponge cells (like other animal cells) have surface markers that allow for self- versus non-self recognition.

Support

The skeletal elements of sponges are of two types, organic and inorganic. The former is always collagenous and the latter either siliceous (hydrated silicon dioxide) or calcareous—calcium carbonate in the form of calcite or aragonite. Sponges are the only animals that use hydrated silica as a skeletal material. Calcite is a common crystalline form of natural calcium carbonate, $CaCO_3$, that is the basic constituent of many protist and animal skeletons, as well as the fossiliferous sedimentary rocks known as limestone, marble, and chalk.[6]

Collagen is the major structural protein in invertebrates; it is found in virtually all metazoan connective tissues. In higher Metazoa, about 20 different kinds of collagen have been identified, but in Porifera only two are known so far: fibrillar collagen and type IV collagen (the latter, a key component of basal membranes, has to date been found primarily in the Homoscleromorpha, but also in a number of species of demosponges and calcareans).

In sponges, fibrillar collagen is either dispersed as thin fibrils in the intercellular matrix or organized as a fibrous framework called spongin in the mesohyl. Although the dispersed collagen fibers seen in many sponges are sometimes referred to as "spongin," true spongin (fibrillar collagen forming a skeletal framework in the mesohyl) is found only in members of the

[6]Calcite and aragonite are closely related calcium carbonate ($CaCO_3$) minerals representing the two most common, naturally occurring, crystalline forms. Both are formed by biological and physical processes—primarily precipitation in marine and freshwater environments—and both are used by animals in the construction of their skeletons. Calcite is the more stable of the polymorphs of calcium carbonate. Calcite crystals are trigonal-rhombohedral, though actual calcite rhombohedra are rare as natural crystals. In animal skeletons, calcite occurs as lamellar or compact deposits, or occasionally in a fibrous form. Calcite is transparent to opaque. Single calcite crystals display an optical property called birefringence (double refraction) that causes objects viewed through a clear piece of calcite to appear doubled. Although calcite is fairly insoluble in cold water, acidity can cause its dissolution (a big problem for animals and protists living in an acidifying global ocean). Calcite exhibits an unusual characteristic called retrograde (or inverse) solubility in which it becomes less soluble in water as the temperature increases. Aragonite's crystal lattice differs from that of calcite, resulting in a different crystal shape, an orthorhombic system with needlelike crystals. Aragonite may be columnar or fibrous. Aragonite is thermodynamically unstable at standard temperature and pressure, and tends to alter to calcite on scales of 107 to 108 years. Because calcite is more stable than aragonite, and dissolves more slowly than aragonite in water, it is more likely to fossilize. Thus, the fossil record for Paleozoic corals is better than the fossil record for Cenozoic corals, the former being made of calcite, the latter of aragonite.

Calcite skeletons occur in most invertebrates that have hard skeletons, including sponges, brachiopods, echinoderms, most bryozoans, and most bivalves. Coccoliths and planktonioc foraminiferans also have calcitic skeletons, and certain red algae (coralline red algae) also produce it. Because calcite is so stable, larger calcite shells in the fossil record (e.g., bivalves) thus tend to the "real thing," not mineral replacements. Purely aragonite-based shells do exist, such as in the molluscan class Polyplacophora (the chitons), but they are rare.

Figure 6.11 Mediterranean bath sponges (*Spongia officinalis*) for sale in an open-air market in Provence, France.

classes Demospongiae (Figure 6.12A). The amount of this fibrillar collagen varies greatly from species to species—in hexactinellids it is quite sparse, whereas in demosponges it is abundant and may form dense bands in the ectosome. Unlike the ectosome of calcareous sponges, which is fundamentally a layer of concentrated spicules, the ectosome of demosponges is a feature of the outer mesohyl and is basically a well-developed collagen layer lacking choanocyte chambers. The network often contains very thick fibers, and may incorporate siliceous spicules into its structure. Spongin often cements siliceous spicules together at their points of intersection. The encysting coat of the asexual gemmules of freshwater (and some marine) sponges is also composed largely of spongin.

Mineral skeletons of either silica or calcium are found in almost all sponges, except a few species of demosponges and homoscleromorphs. Sponges lacking mineral skeletons possess only fibrous collagen networks, and these are still used as bath sponges despite the prevalence nowadays of synthetic "sponges." Several demosponge and homoscleromorph genera lack both spongin and a spicule skeleton (e.g., *Chondrosia*, *Halisarca*, *Hexadella*, *Oscarella*).

Sponges have been harvested for millennia. Evidence of an active Mediterranean sponge trade is found at least as early as 3,000 BCE (Egyptians), and later in the ancient Phoenician, Greek and Roman civilizations (Figure 6.11). Homer and other ancient Greek writers mention a thriving Mediterranean sponge trade. Prior to the 1950s, active natural sponge fisheries existed in south Florida, the Bahamas, and the Mediterranean. The industry peaked in 1938, when the world's annual sponge catch (including cultivated sponges) exceeded 2.6 million pounds, 700,000 pounds of which came from the United States and the Bahamas. Almost all commercial sponges belong to the genera *Hippospongia* and *Spongia*, but these sponges have now been largely "fished out" in the traditional sponge hunting grounds of the Mediterranean and Florida. In addition, they are prone to current-transmitted epidemic diseases (three such events wiped out bath sponge populations in both the Old World and the New World in 1938, 1947, and in the late 1980s).

Mineralized sponge spicules (Figure 6.12) are produced by special mesohyl cells called sclerocytes, which are capable of accumulating calcium or silicate and depositing it in an organized way. In some cases, one sclerocyte produces one spicule; in others, several sclerocytes work together cooperatively to produce a single spicule, often two cells per spicule ray (Figure 6.9A–D).

The construction of a siliceous spicule begins with the secretion of an organic axial filament within a vacuole in a sclerocyte. As the axial filament elongates at both ends, hydrated silica is secreted into the vacuole and deposited around the filament core. Recent work suggests that, at least in some homoscleromorphan species (e.g., *Corticium candelabrum*) pinacocytes might also be capable of siliceous spicule production. About 92% of all living sponge species are siliceous. One might wonder how a skeleton made of glass (as in the Hexactinellida) could be strong enough to provide support while not being fragile as a windowpane. The trick is that glass sponges have evolved a highly sophisticated, multilayered biomechanical structure to their skeleton. They construct their glass skeleton by first consolidating nanometer-scaled silica spheres, arranged in microscopic concentric rings separated from one another by alternating layers of an organic matrix-glue to form laminated spicules. The spicules are then bundled together by an organic, silica-based cement, resulting in the formation of micron-scale, multilayered "beams." The beams are then assembled into the recognizable square-lattice cagelike structure we see in glass sponges, such as *Euplectella*. The lattice itself is further reinforced by diagonal ridges, giving the whole skeleton remarkable strength plus a resilient flexibility. In species of *Euplectella*, seven hierarchical structural levels have been identified in the skeleton—a textbook example in biomechanical engineering, and perhaps one day an inspiration for glass fiber manufacturers. Even in Demospongiae and Homoscleromorpha, the siliceous spicules exhibit exceptional flexibility and toughness due to their composite, layered construction.

Unlike siliceous spicules, calcium carbonate spicules do not have an organic axial core. Calcareous spicules are produced extracellularly, in intercellular spaces bounded by a number of sclerocytes. Each spicule is essentially a single crystal of calcite or aragonite.

Considerable taxonomic weight has been given to spicule morphology, and an elaborate nomenclature exists to classify these skeletal structures. Spicules are termed either **microscleres** or **megascleres**. The former are small to minute, reinforcing (or packing) spicules; the latter are large structural spicules. The demosponges and hexactinellids have both types; calcareous sponges often have only megascleres (although the spicules can differ considerably in size). Descriptive terms that designate the number of *axes* in a spicule end in the suffix *-axon* (e.g., monaxon, triaxon). Terms that designate the number of *rays* end in the suffixes *-actine* or *-actinal* (e.g., monactinal, hexactinal, tetractinal). In addition, there is a detailed nomenclature specifying shape and ornamentation of various spicules (Figure 6.12).

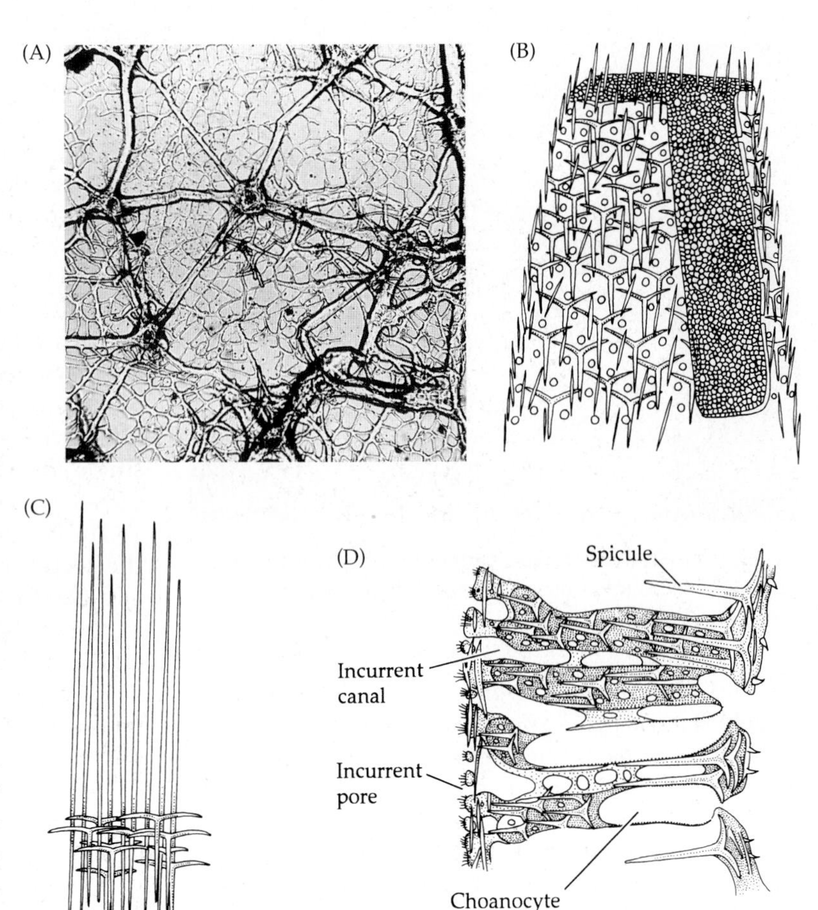

A spicular skeleton may be viewed as a supplemental supporting structure. If the amount of inorganic material is increased in relation to organic material, the sponge becomes increasingly solid until the texture approaches that of a rock, as it may in some members of the demosponge order Tetractinellida and a few others (e.g., the "lithistid" sponges). In contrast to discrete spicules, the massive calcareous skeletons of some species (the coralline, or sclerosponges) have a polycrystalline microstructure; they are composed of needles ("fibers") of either calcite or aragonite embedded in an organic fibrillar matrix. The advantage of incorporating organic matter into the calcareous framework has been compared to lathe-and-plaster, or reinforced concrete. The mix of organic and inorganic materials probably yields fibrous calcites and aragonites that are less prone to fracture while also producing substances that are more easily molded by the organism. In some demosponges (e.g., Lithistida) and many Hexactinellida, the spicules may be linked or fused into such a rigid framework that it is capable of fossilizing.[7]

Nutrition, Excretion, and Gas Exchange

Although sponges lack the complex organs and organ systems seen in the higher Metazoa, they are nevertheless a hugely successful group of animals. As noted above, their success seems partly due to their very "ancientness"—the flexibility inherent in their developmental programming, cellular pluripotency, highly versatile aquiferous system, and the general plasticity of their body form.

Unlike most Metazoa, nearly all sponges rely on intracellular digestion, and thus on phagocytosis and pinocytosis as means of food capture. The aquiferous system has already been described; with it, sponges more or less continuously circulate water through their bodies, bringing with it the microscopic food particles upon which they feed. They are size-selective particle feeders, and the arrangement of the aquiferous system creates a series of "sieves" of decreasing mesh size (e.g., inhalant ostia or dermal pores → canals → prosopyles → choanocyte villi → intertentacular mucous reticulum). The upper limit of the diameter of incurrent openings is usually around 50 μm, so larger particles do not enter the aquiferous system. A few species have larger incurrent pores, reaching diameters of 150 to 175 μm, but in most species the incurrent openings range from 5 to 50 μm in diameter. Internal particle capture in the 2 to 10 μm range (e.g., bacteria, small protists, unicellular algae, organic detritus) is by choanocytes and by phagocytic motile archaeocytes that move to the lining of the incurrent canals. As in the choanoflagellate protists, the propagation of the flagellar wave in sponge choanocytes is thought to draw water in through the microvilli of the collar, which then slows or blocks food particles, allowing them to be phagocytized. The smallest particles, large organic molecules and small bacteria in the 0.1–0.2 μm

[7]The order Lithistida has long been recognized as polyphyletic, but kept mainly as a matter of convenience, essentially because it is used by paleontologists. However, as with the former taxon Sclerospongiae, it is clearly time to abandon the name "Lithistida" and begin reassigning its component species.

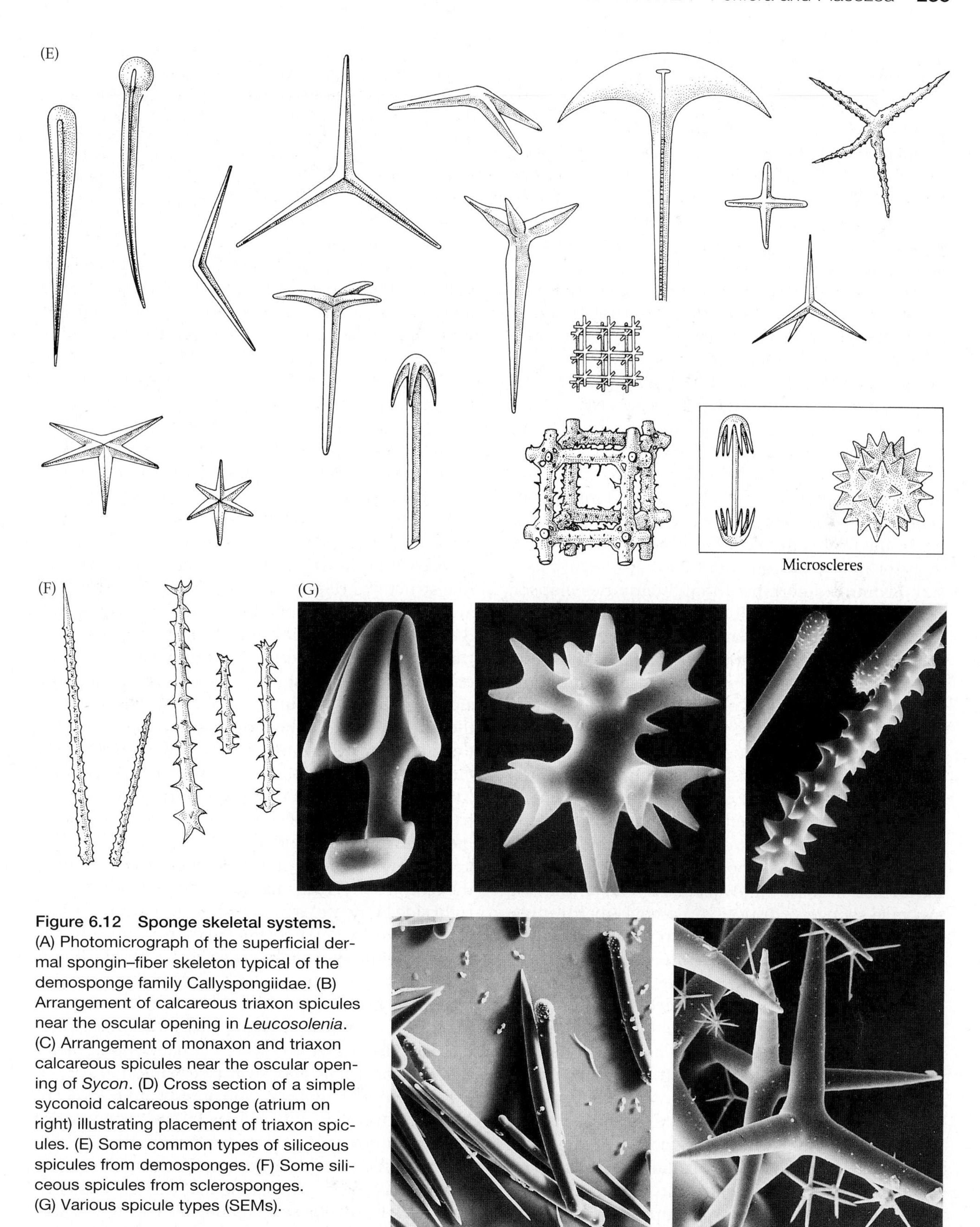

Figure 6.12 Sponge skeletal systems. (A) Photomicrograph of the superficial dermal spongin–fiber skeleton typical of the demosponge family Callyspongiidae. (B) Arrangement of calcareous triaxon spicules near the oscular opening in *Leucosolenia*. (C) Arrangement of monaxon and triaxon calcareous spicules near the oscular opening of *Sycon*. (D) Cross section of a simple syconoid calcareous sponge (atrium on right) illustrating placement of triaxon spicules. (E) Some common types of siliceous spicules from demosponges. (F) Some siliceous spicules from sclerosponges. (G) Various spicule types (SEMs).

range (the distance between adjacent villi is ~1.5 μm or less), may be trapped on the collar itself. Larger particles, up to ~3 μm, adhere to the body of the choanocyte only to be engulfed by "pseudopodial" extensions (phagocytosis) that may be longer than the collar villi themselves. Undulations of the collar might move some trapped food particles down to the choanocyte cell body, or stalled food might be captured by long phagocytizing pseudopods that form from the collar microvilli and migrate from the base of the collar (as in Choanoflagellates) or by pseudopods from the choanocyte cell body itself.

Interestingly, it would seem that food particle capture can take place almost anywhere on the surface of a sponge. Even pinacocytes are known to be capable of phagocytizing particles as large as 6 μm on the surface of a sponge. It has been estimated that the pinacocytes lining canals can take particles as large as 50 μm.

In the case of archaeocyte phagocytosis, digestion takes place in the food vacuole formed at the time of capture. In the case of choanocyte capture, food particles are partly digested in the choanocytes and then quickly passed on to a mesohyl archaeocyte (or other wandering amebocyte) for final digestion. In both cases, the mobility of the mesohyl cells assures transport of nutrients throughout the sponge body.

The efficiency of food capture and digestion was dramatically shown in an elegant study long ago, by Schmidt (1970), using fluorescence-tagged bacteria fed to the freshwater sponge *Ephydatia fluviatilis*. By monitoring the movement of the fluorescent material, Schmidt determined that 30 minutes elapsed from the onset of feeding until the bacteria had been captured by choanocytes and moved to the base of the cells. Transfer of the fluorescent material to the mesohyl commenced 30 minutes later. Twenty-four hours later, fluorescent wastes began to be discharged into the water, and no fluorescent material remained in the sponges after 48 hours. Additional studies on this same species led to an estimate of 7,600 choanocyte chambers per cubic millimeter of sponge body, each chamber pumping a phenomenal 1,200 times its own volume of water daily. More complex leuconoid sponges have as many as 18,000 choanocyte chambers per cubic millimeter. In some thin-walled asconoid and simple syconoid sponges, a distinctive mesohyl is hardly present. In these sponges the choanocytes assume both capture and digestive/assimilative functions. More recent studies have shown that sponges are capable of removing up to 95% of the bacteria and heterotrophic protists from the water they filter.

Many sponges appear to also take up significant amounts of dissolved organic matter (DOM) by pinocytosis from the water within the aquiferous system, and there is some evidence that this is needed by the microbial symbionts within the sponge. Studies have shown that 80% of the organic matter taken in by some shallow-water marine sponges can be of a size below that resolvable by light microscopy. The other 20% comprises primarily bacteria and dinoflagellates. On the other hand, some sponges appear to rely very little on DOM for nutrition (e.g., some hexactinellids that have been studied). Recent studies show that at least some sponges form simple fecal pellets. Experiments on the widespread North Atlantic species *Halichondria panicea* have revealed that undigested material is expelled as discrete pellets coated with a thin layer of mucus. Exactly how these pellets are formed is unclear.

Figure 6.13 These remarkable SEMs and color photographs show predation in the carnivorous sponge *Asbestopluma* (family Cladorhizidae). (A–D) *Asbestopluma* in ambush posture (A), followed by capture of a mysid. (E) Fifteen minutes after capture of a mysid on its tentacle-like feeding filaments. (F–H) The mysid prey has been partly engulfed by the sponge. (I) The prey is entirely engulfed.

Although the phylum Porifera is characterized by filter feeding, members of the deep-sea demosponge family Cladorhizidae display an entirely different and unique mode of feeding. Species in this group have lost most or all of the characteristic choanocyte-lined aquiferous system and instead feed as macrophagous carnivores; in fact, they are passive suspension-feeding predators, spending a minimal amount of energy during long periods between rare feeding opportunities! They feed by trapping small prey on hook-shaped spicules that protrude from the surfaces of tentacle-like structures (Figure 6.13 and 6.14). Trapped prey are gradually enveloped by migrating feeding cells that undertake digestion and absorption. Most cladorhizids live at great ocean depths, and *Asbestopluma occidentalis* is the deepest known sponge. Most cladorhizids probably remain undescribed and, unfortunately, they are common on the tops of seamounts—a habitat threatened by fishing and mineral extraction interests. However, one species of *Asbestopluma* (*A. hypogea*) lives in shallow caves in the Mediterranean, where it has been the subject of considerable study. One of the more bizarre cladorhizids is *Chondrocladia lyra* (the lyre sponge), which has a series of vanes extending from the central body, each vane in turn with a series of upright side branches, creating a lyre- or harp-like structure that works for passive suspension capture of small zooplankters, especially crustaceans. Cladorhizids apparently lack oscula, ostia, and a canal system, and some may even lack choanocytes! Another of the remarkable cladorhizid sponges, *Cladorhiza methanophila*, has been discovered to harbor methanotrophic bacterial symbionts in its cells, such as seen in many animals inhabiting hydrothermal vents and cold seeps. The sponge thus feeds both by predation and by direct consumption of its microbial symbionts. Recently, species in several other deep-sea families have shown evidence of carnivory, suggesting that this habit has perhaps evolved several times among the Demospongiae.

Sponges continuously excrete waste from their body via the osculum. Occasionally, they also undergo a stereotypic series of whole-body contractions, sometimes called a "sneeze." It appears that these contractions are stimulated by a reduction in water flow through the oscula, which would typically be caused by obstruction

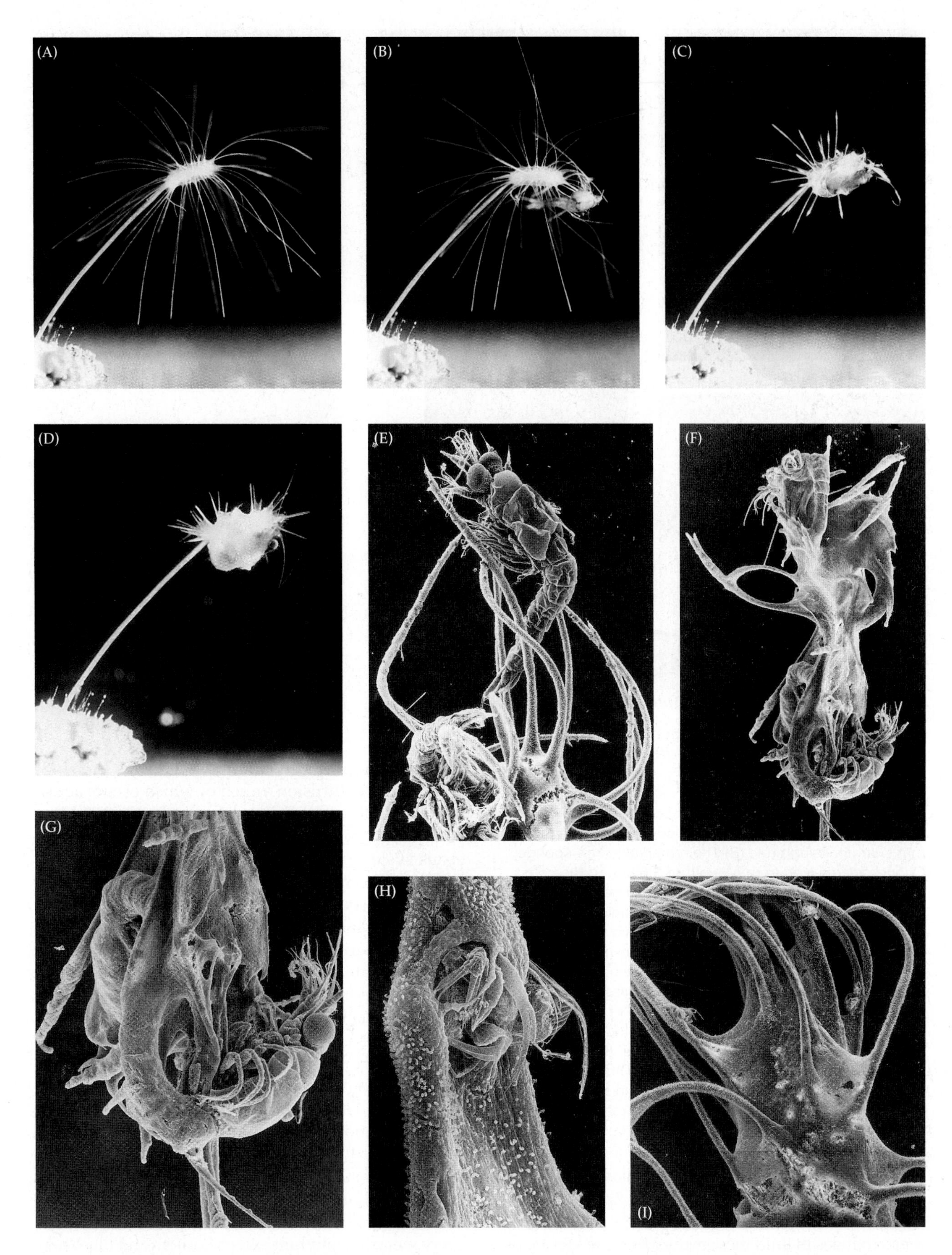

from an excessive accumulation of particulates that can clog the aquiferous system. The decrease in oscular flow is registered by special short cilia that line the inner osculum epithelium. Unlike most cilia, that possess the classic 9+2 anatomy, these oscular cilia lack the central pair of microtubules and are nonmotile. Their

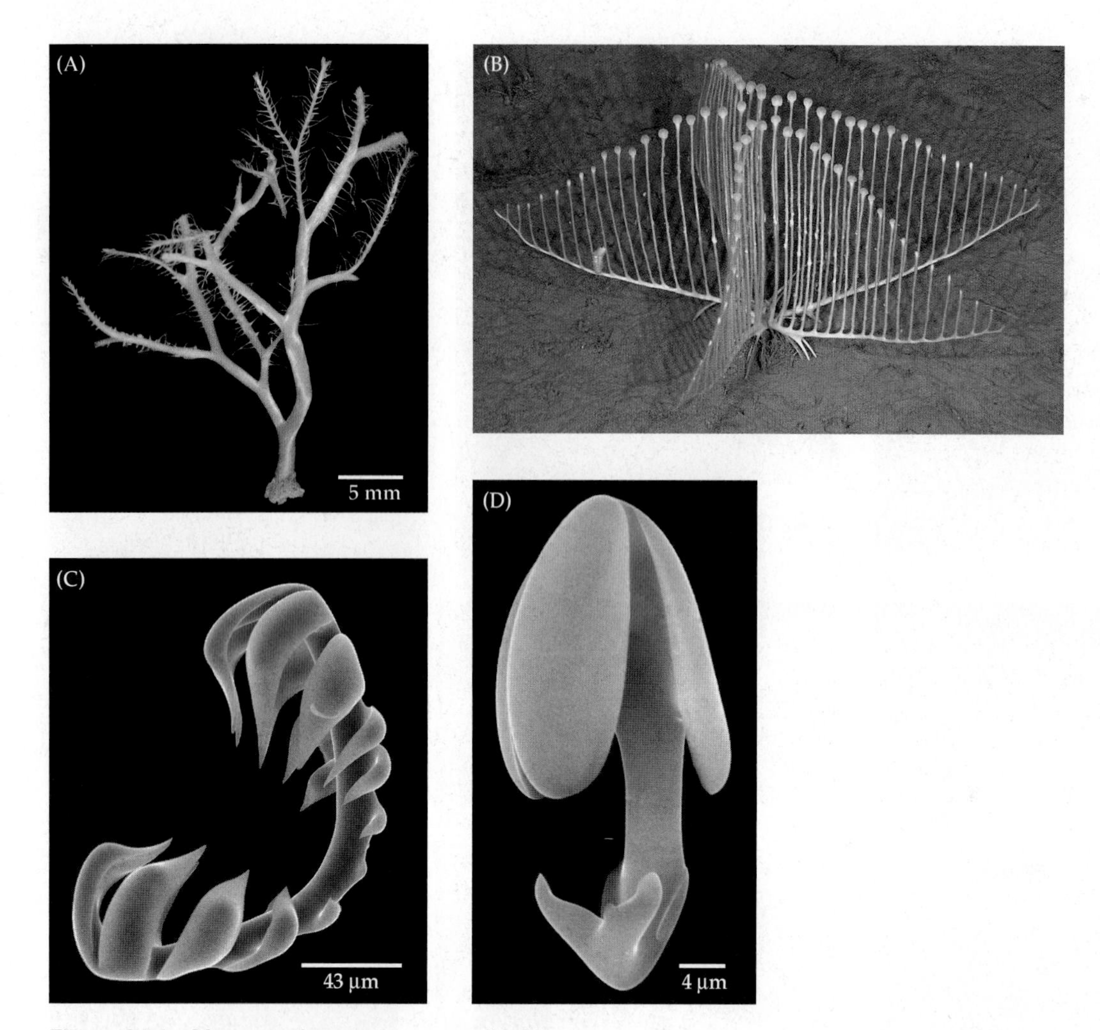

Figure 6.14 More carnivorous sponges (family Cladorhizidae). (A) *Asbestopluma desmophora*, a tree-like carnivorous sponge from New Zealand. Large clear megascleres are aligned vertically along the main "trunk," twisting around it to give a silky sheen to the creature. The small, thin, side branches are covered in microscleres that snare passing prey. (B) The remarkable lyre sponge, *Chondrocladia lyra*, from a depth of 3.5 km off the coast of Monterey, California, reaches 36 cm in height and has up to six long vanes that grow from the main body, each vane with a parallel series of side branches that are used for passive suspension feeding of small invertebrates, particularly crustaceans, using Velcro-like barbed hooks (microscleres). (C) A microsclere from *Abyssocladia carcharias*, named for the remarkable shape of the "abyssochelae" microscleres, which resemble the jaws of the great white shark *Carcharodon carcharias*. (D) Microsclere ("anisochelae") of *Asbestopluma agglutinans*. In carnivorous sponges, microscleres are involved in snaring prey and typically line the sponge's surface.

function seems strictly to be the detection of flow rates through the oscula. Such cilia, called "primary cilia" are known from other metazoans, where they typically sense changes in fluid movement. Thus, the sponge osculum can be viewed as a true sense organ—the only one known in this phylum! Experimentally de-osculated sponges cannot be triggered to sneeze.

Excretion (primarily ammonia) and gas exchange are by simple diffusion, much of which occurs across the choanoderm. We have already seen how folding of the body, combined with the presence of an aquiferous system, overcomes the surface-to-volume dilemma posed by an increase in size. The efficiency of the poriferan body plan is such that diffusion distances never exceed about 1.0 mm, the distance at which gas exchange by diffusion becomes notably inefficient. In addition, water expulsion vesicles (contractile vacuoles) occur in freshwater sponges and presumably aid in osmoregulation.

Activity and Sensitivity

Sponges do not possess neurons, a nervous system, or discrete adult sense organs (except for the oscula, see above), and nothing resembling the synaptic connections (or gap junctions) of higher Metazoa is known in these animals. Neuronal synapses, which provide for chemically-mediated and directional nerve impulse transmission, do not occur in sponges (these occur in every metazoan phylum except Porifera and Placozoa).

Despite the absence of a nervous system, sponges are capable of responding to a variety of environmental stimuli by closure of the ostia or oscula, canal constriction, backflow, body contractions that flush the aquiferous system ("sneezing"), and reorganization of flagellated chambers. Acetylcholinesterase, catecholamines, and serotonin have all been shown to be present in sponges and these probably play roles in coordinating "tissue" contractions, although electrophysiological evidence of a conducting mechanism is still lacking. Sponge larvae show both positive and negative phototaxis, depending on whether they are in their planktonic or settling phase. In the demosponge *Reniera*, larval photoreceptors are thought to be a posterior ring of columnar monociliated epithelial cells that possess cilia and pigment-filled protrusions. In addition, the ability to conduct electrical impulses (when stimulated by an electrical probe) has been shown to exist in some Hexactinellida, where the absence of cellular membranes within the syncytium presumably allows action currents to spread in any direction from sites of depolarization. When the action currents reach choanocytes within the trabecular network, choanocyte activity is suppressed. In all sponges, the "default" behavior seems to be active pumping by the choanocytes to keep the aquiferous system functioning; and the usual effect of environmental stimuli is to reduce or stop the flow of water through the aquiferous system. For example, when suspended particulates become too large or too concentrated, sponges will respond by closing the incurrent openings and immobilizing the choanocyte flagella. Indeed, sponges have long been known to contract their ostia and oscula, and portions of their canal system, although the rates of propagation are very slow. Direct physical stimulation will also elicit this reaction, which is easily observed by simply running one's finger across a sponge surface and observing the dermal pore or oscular contractions with a hand lens or low-power microscope. In addition, the mesohyl of some species has been shown to react to mechanical stimulation by stiffening.

Choanocyte pumping activity also varies with certain endogenous factors. For example, during a major growth phase, such as canal or chamber reorganization, choanocyte-pumping activity typically decreases. Periods of reproductive activity also cause a substantial decrease in water pumping, in part because many choanocytes are expended in the reproduction process (see next section). Even under normal conditions, variations in pumping rates occur. Some sponges cease pumping activity periodically, for a few minutes or for hours at a time; others cease activity for several days at a time—reasons for these changes in activity are not always apparent.

The switch from full pumping activity to complete cessation requires at least several minutes; considering the organism, however, this is a fairly fast response time. The spread of stimulation-and-response in most sponges appears to be by simple mechanical action (stimulation from one cell to the adjacent cells) and perhaps also by diffusion of certain chemical messengers (hormones, or other kinds of signaling molecules) released by cells. The process by which cells communicate with other cells by secreting specific chemicals released into the surrounding extracellular matrix is called a **paracrine signaling**, or a **paracrine system**, and some workers consider sponges to use this kind of cell–cell signaling.

The contractile myocytes of sponges act as independent effectors; they are organized into a network formed by contacts between filopodial extensions of adjacent myocytes and pinacocytes. Response time of myocytes is relatively slow. Latency periods average 0.01 to 0.04 seconds, and conduction velocities are typically less than 0.04 cm/sec (except in the hexactinellids, where velocities of 0.30 cm/sec have been recorded).[8] Conduction is unpolarized and diffuse, and it may rely on both hormones (or other signaling molecules) and direct mechanical action of one cell on the next. Considerable research once focused on the myocytes in attempts to shed light on the possible presence of a sponge nervous system analogous or homologous to that in higher Metazoa. But, in spite of these efforts, there has been no verification of such a system.

Reproduction and Development

All sponges are capable of sexual reproduction, and several types of asexual processes are also common. Many of the details of these processes are unknown, however, largely because sponges lack distinct or localized gonads (gametes and embryos can generally occur throughout the mesohyl). All hexactinellids, homoscleromorphans, and calcareous sponges are probably viviparous, as are most demosponges.

Asexual reproduction Probably all sponges are capable of regenerating viable adults from fragments. Many species "pinch off" buds of various types (Figure 6.15) by a process of cellular reorganization, whereas in others (especially branching species) pieces of the sponge body are simply broken off by storm surges. Either way, the dislodged pieces fall off and regenerate into new individuals. This regenerative ability was once commonly used by commercial sponge farmers, who propagated their sponges by attaching "cuttings" to submerged rocks or cement blocks. Even today, in Pohnpei (Federated States of Micronesia) sponge farmers string cuttings of *Coscinoderma mathewsi* on lines in lagoons, growing them into large bath sponges. Additional asexual processes of poriferans include formation of gemmules and reduction bodies, budding, and possibly formation of asexual larvae.

[8]It has been suggested that, in hexactinellids, the syncytium serves as an electrical conduction pathway, propagating impulses along its membranes that trigger flagellar arrests in the flagellated chambers.

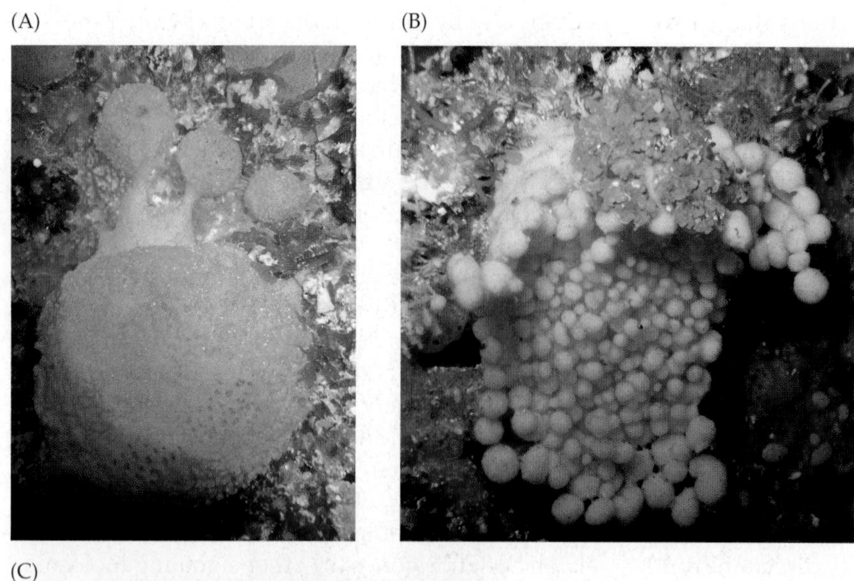

Figure 6.15 Budding in three demosponge species from New Zealand. (A) *Tethya burtoni*. (B) *Stelletta* sp. (C) *Tethya bergquistae*.

In freshwater sponges, small spherical structures called **gemmules** are produced at the onset of winter (Figure 6.16). These dormant, over-wintering bodies are invested with a thick collagenous coating in which supportive siliceous microscleres (**gemmoscleres**) are embedded. Gemmules are highly resistant to both freezing and drying. The gemmules of some species can withstand exposure to –70°C for up to an hour, whereas others experience mortality at –10°C. Experiments suggest that gemmules of at least of some species can also withstand seasonal anoxic conditions. Fossil gemmules have been found as far back as the Cretaceous, and these ancient forms are very similar to modern ones. Gemmule morphology is highly specific and is used diagnostically at the family, genus, and even species levels. Three of the six freshwater sponge families do not produce gemmules: Lubomirskiidae and Metschinkowiidae (both known only from the Caspian Sea and Lake Baikal), and Malawispongiidae (from the ancient lakes of the Great African Rift Valley, through the Middle East, and to Sulawesi Island in the Indian Ocean).

The formation and eventual growth of gemmules are remarkable examples of poriferan cell totipotency. As winter approaches, archaeocytes aggregate in the mesohyl and undergo rapid mitosis. "Nurse cells," called trophocytes, stream to the archaeocyte mass and are engulfed by phagocytosis. The result is a mass of archaeocytes containing food reserves stored in elaborate vitelline platelets. This entire mass eventually becomes surrounded by a three-layered spongin covering. Developing gemmoscleres, **amphidisc spicules** (spicules having a stellate disk at each end) in most species, are transported by their parent cells to the growing gemmule and incorporated into the spongin envelope. The final bit of the gemmule to be enclosed by the spongin case is covered only by a single layer of spongin that is devoid of spicules; this single-layered patch is the micropyle. Thus formed, hibernation of the gemmule commences, while the parent sponge usually dies and disintegrates—such species can be thought of as "annuals."

When environmental conditions are again favorable, the micropyle opens and the first archaeocytes begin to flow out (Figure 6.16C). They immediately flow onto the substratum, whereupon they begin to construct a framework of new pinacoderm and choanoderm. The second wave of archaeocytes to leave the gemmule colonizes this framework. In the course of gemmule "hatching," and in an impressive example of cellular totipotency, archaeocytes quickly give rise to every cell type of the adult sponge. Gemmule dormancy appears to be of two types, a quiescence and a true diapause. Quiescence is imposed by generally

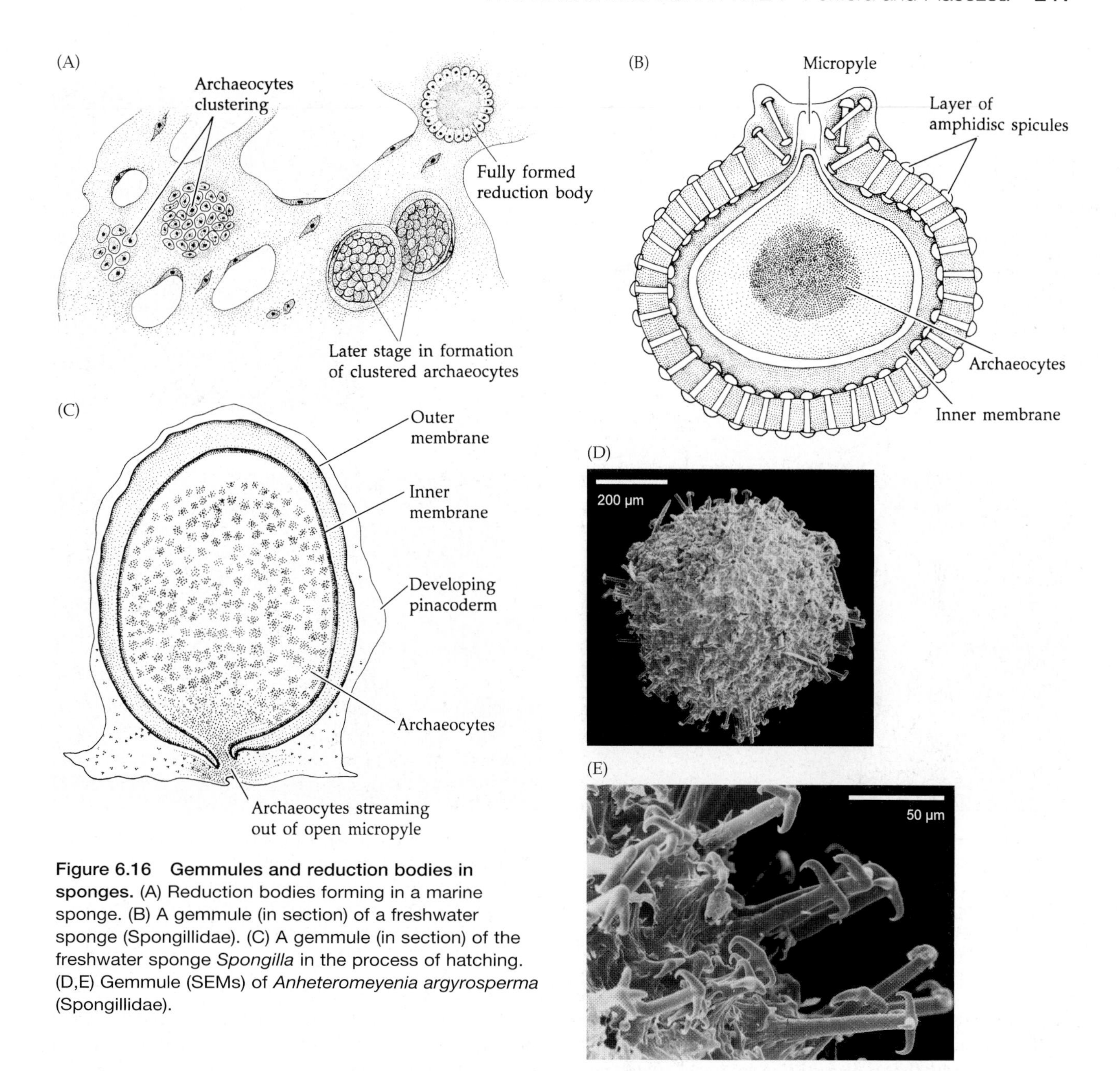

Figure 6.16 Gemmules and reduction bodies in sponges. (A) Reduction bodies forming in a marine sponge. (B) A gemmule (in section) of a freshwater sponge (Spongillidae). (C) A gemmule (in section) of the freshwater sponge *Spongilla* in the process of hatching. (D,E) Gemmule (SEMs) of *Anheteromeyenia argyrosperma* (Spongillidae).

unfavorable conditions, including low temperatures, and it ends when suitable conditions return. Diapause, on the other hand, typical of annual species, is imposed by a combination of endogenous mechanisms and adverse environmental conditions. The breaking of a diapause state typically requires exposure to very low temperatures for a prescribed number of days.

Many marine species produce asexual reproductive bodies (called **reduction bodies**) that are roughly similar to freshwater gemmules, though much simpler in design and cell wall structure, and these incorporate a variety of amebocytes. And, in several genera of marine sponges (e.g., *Haliclona, Chalinula*) asexual reproductive bodies very similar to gemmules are formed, with a spongin coat containing spicules (oxeas).

Sexual processes The ancient origin of sponges is reflected in their extreme variability of reproductive and developmental strategies. Many are hermaphroditic, but they typically produce eggs and sperm at different times. This sequential hermaphroditism may take the form of protogyny or protandry, and the sex change may occur only once, or an individual may repeatedly alternate between male and female. In some species, individuals appear to be permanently male or female. In still other species, some individuals are permanently gonochoristic, whereas other individuals in the same population are hermaphroditic. In all cases, cross-fertilization is probably the rule.

Although sponge embryogenesis shares many cardinal features with higher Metazoa, including the

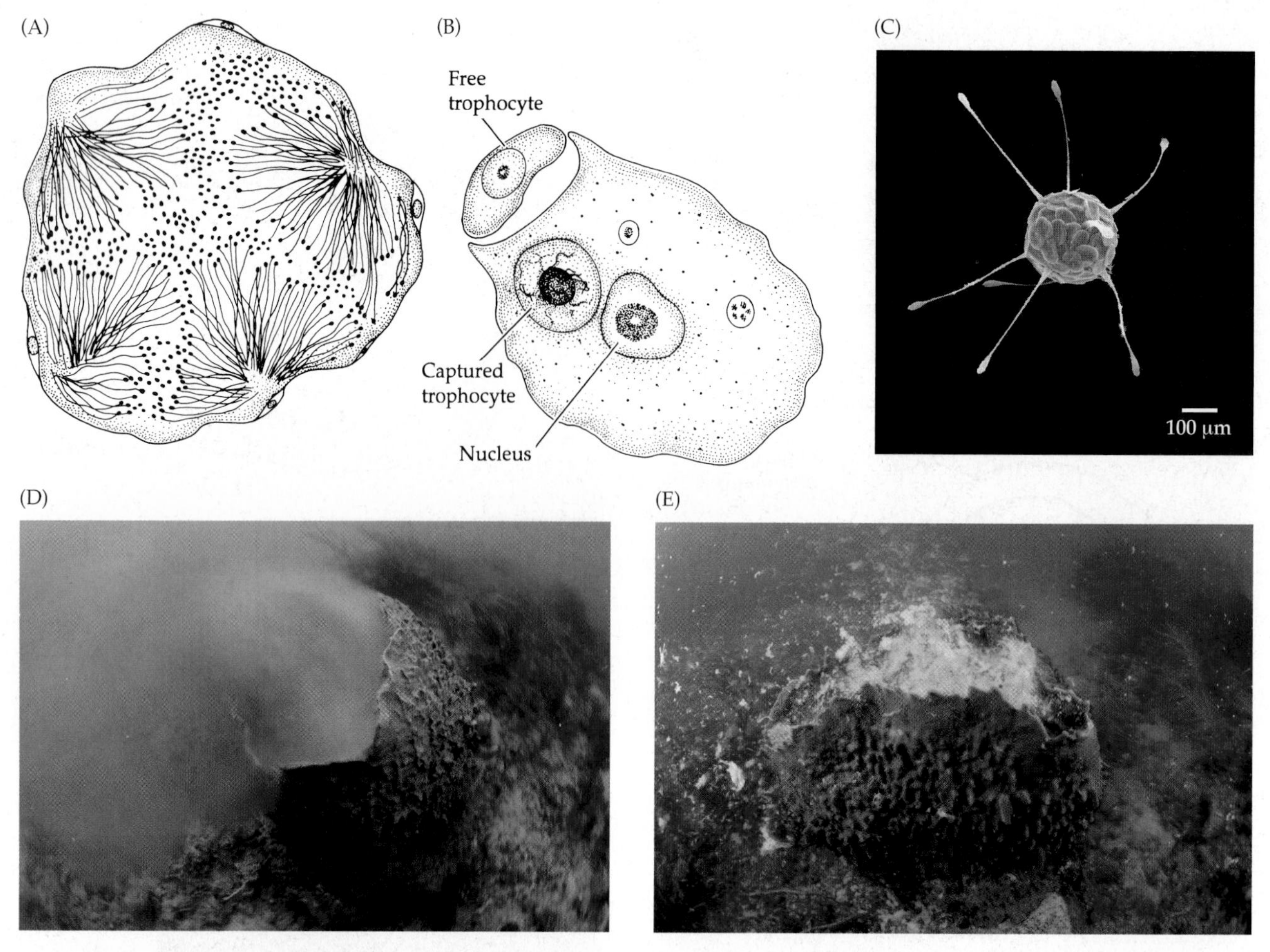

Figure 6.17 Sexual reproduction in Demospongia. (A) Sperm follicle (in section) containing mature spermatozoa. (B) An oocyte (in section) of *Ephydatia fluviatilis* is phagocytizing a trophocyte. Inside the oocyte is a trophocyte that was recently ingested. (C) The unusual armored "hoplitomella larva" typical of boring sponges in the genus *Thoosa*. The armoring consists of flat plates that are actually highly modified spicules (called discotrienaenes). The long fibers emerging from the larva are also modified spicules that presumably aid in flotation. (D) Sperm release from *Xestospongia muta*, Florida. (E) Oocyte release in *Xestospongia muta*, Florida.

uniquely metazoan process of programmed cleavage and tissue layering through the process of embryonic morphogenesis that we regard as true gastrulation, it also expresses great variability. In fact, the range of ontogenetic adaptation among sponges is so broad that it is difficult to generalize about their development. The same cleavage pattern and blastula type may lead to several different larval types, and the same larval types may develop from different cleavage patterns. Overall, sponges appear to have remained at an "experimental grade" of development, in which fixed cleavage patterns and modes of gastrulation are not yet well established, and a variety of developmental processes are exploited among various lineages. This extreme developmental variation is described below.

Sperm appear to arise primarily from choanocytes; eggs from choanocytes or archaeocytes. Spermatogenesis usually occurs in distinct **spermatic cysts** (= **sperm follicles**), which form either when all the cells of a choanocyte chamber are transformed into spermatogonia, or when transformed choanocytes migrate into the mesohyl and aggregate there (Figure 6.17A). During oogenesis, solitary oocytes typically develop within cysts surrounded by a layer of follicle cells and, sometimes, nurse cells (**trophocytes**) (Figure 6.17B).

In Demospongiae and Calcarea, mature sperm and (in the case of demosponge oviparity) oocytes are released into the environment through the aquiferous system. The rapid release of sperm from sponge oscula can be highly visible and quite dramatic, and such individuals are often referred to as "smoking sponges" (Figure 6.17D). Sperm release may be synchronized in a local population or restricted to certain individuals. Fertilization takes place in the water (oviparity) leading to planktonic larvae, or within the female sponge itself (viviparity). In the viviparous coral-boring sponge *Cliona vermifera*, each zygote inside the sponge body becomes encapsulated; these are eventually released into the water in a series of rapid pulsations of the aquiferous system. In viviparous species, sperm

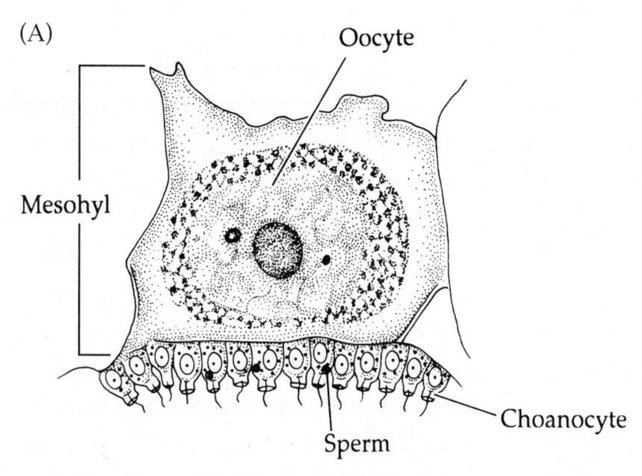

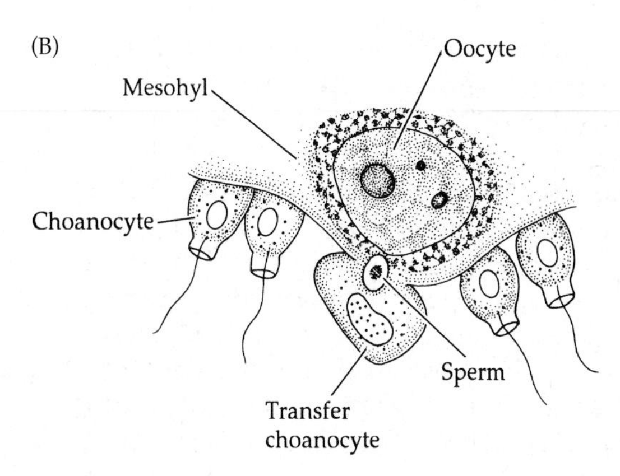

Figure 6.18 Fertilization in the calcareous sponge. (A) Sperm are trapped by choanocytes; an egg is lying in the mesohyl adjacent to the choanoderm. (B) A transfer choanocyte gives up its sperm to the egg; note that the egg lies next to the choanoderm and that the choanocyte has lost its flagellum.

(probably from a different individual) are taken into the aquiferous system and then must then cross the cellular barrier of the choanoderm, enter the mesohyl, locate the oocytes, penetrate the follicular barrier, and finally fertilize the egg. In all studied species of the subclass Calcaronea, this impressive feat involves sperm capture by choanocytes and enclosure in an intracellular vesicle (somewhat like the formation of a "benign food vacuole"). The choanocyte then loses its collar and flagellum and migrates through the mesohyl as an ameboid cell, transporting the sperm to the oocyte (Figure 6.18A). The migratory choanocyte is called a **carrier cell**, or **transfer choanocyte** (Figure 6.18B). Choanocytes no doubt regularly consume and digest the unlucky sperm of different species of sponges and other benthic invertebrates but, by some as yet undescribed recognition mechanism, they respond with a remarkably different behavior to sperm of their own kind. Presumably this process is directed by cell surface proteins/glycoproteins.

In viviparous species, embryos are typically released as mature swimming larvae via the excurrent plumbing of the aquiferous system, or through a rupture in the parent's body wall. Larvae may settle directly, they may swim about for several hours or a few days before settling, or they may simply move about the substratum until ready to attach. In all sponge species that have been studied, the larvae are lecithotrophic. In general, littoral sponges tend to produce planktonic larvae, whereas the larvae of subtidal species tend to settle directly or move about on the ocean floor for a few days before beginning to grow into a new adult individual.

Embryogenesis and larvae Until fairly recently, larval development in sponges was surrounded by a great deal of mystery, partly because it had not been well studied and partly because huge variation exists. The broad array of blastulation/gastrulation processes and larval types seen in sponges is mind boggling, and many different, but distinct larval types have been described. In fact, a number of embryogenic pathways occur in sponges that do not occur in any other Metazoa (e.g., multipolar egression in Homoscleromorpha, polarized delamination in some demosponges). Since the 1990s, significant new work has altered our view of development in this phylum although many questions still remain. The diversity of sponge developmental pathways is remarkable, and in some instances the same cleavage pattern and blastula type may be characteristic of several different larval types. On the other hand, the same larval type may arise from very different cleavage patterns. For example, the parenchymella larva of *Reniera* (Demospongiae: Haplosclerida) develops as a result of chaotic (seemingly unpatterned) cleavage and multipolar delamination, whereas the parenchymella larva of *Halisarca* (Demospongiae: Halisarcida) arises by polyaxial cleavage and multipolar ingression.

Much of the confusion surrounding sponge embryogenesis has centered on the question of when gastrulation occurs. Many older studies had reported that, during larval development, the external ciliated cells lost their cilia and migrated internally, whereupon they differentiated into adult choanocytes. These studies suggested that larvae formed in such a way represented the blastula stage, with later gastrulation coinciding with larval metamorphosis. Such an embryonic processes, producing blastular larvae, would make sponges unique among metazoans. However, recent work suggests that this migration/inversion of cell layers during larval development was probably a misinterpretation in most if not all cases, and in fact sponge larvae probably represent post-gastrulation stage individuals (as with most other metazoan larvae). Cell migrations such as these do occur in some poriferans, but only *after* embryogenesis has formed a fully differentiated bi- or tri-layered larva, and such cell movements are thus not related to gastrulation. It now seems that most, if not all, sponge larvae derive from cellular reorganization that can be equated (though not necessarily directly homologized) with gastrulation processes seen in other metazoans. Thus, gastrulation precedes the larval stage, and probably no sponge larvae represent blastula stages.[9]

Specialists have long wondered what the original, or ancestral cleavage pattern in sponges (and Metazoa) might be—most researchers would probably vote for radial cleavage, and indeed this form of cleavage is seen in many demosponges, although it is not known from any other sponge class. At least three other forms of holoblastic cleavage have been described: chaotic (also known as "anarchic") cleavage, table palintomy, and polyaxial cleavage. **Chaotic cleavage** (seemingly random, or unpatterned) is characteristic of many demosponges. In **table palintomy**, the cleavage furrows pass obliquely relative to the egg's animal–vegetal axis; this form of cleavage occurs in some calcareous sponges (Calcaronea). It is also reminiscent of the green algal genus *Volvox* and several other colonial protists. However, the Calcaronea also undergo an utterly unique process of inversion of the blastula, and this has led some workers to suggest that palintomy is an aberrant cleavage process. **Polyaxial cleavage** is characteristic of the genus *Halisarca* and possibly also the Calcinea (Calcarea). It is characterized by cleavage furrows that form perpendicular to the surface of the embryo during the transition from the 8 to 16-cell stage, with axes of symmetry that radiate at certain angles from the center of the embryo; however, after the 16-cell stage division becomes chaotic.

Although sponge embryogeny is highly variable, all species go through an ordered sequence of blastomere divisions and movements leading to morphogenesis of larval tissue structures (which we consider to be gastrulation) and formation of a larval axis that is often referred to as an anterior–posterior axis. Spongologists have so far recognized at least seven types of sexual development in sponges, recognized by their resultant larval forms, but continued research may reveal even more: (1) trichimella larva (Hexactinellida), (2) calciblastula larva (Calcarea: Calcinea), (3) amphiblastula larva (Calcarea: Calcaronea), (4) cinctoblastula larva (Homoscleromorpha), (5) disphaerula larva (Demospongiae: Halisarcidae), (6) direct development (Demospongiae, Tetractinellida, *Tetilla*), and (7) parenchymella larvae (most Demospongiae). These seven developmental pathways are depicted in Figure 6.19 and described below. The anterior flagellated pole of swimming sponge larvae (e.g., amphiblastula, cinctoblastula, parenchymella; Figure 6.20) putatively corresponds to the animal pole of other metazoan larvae, and it is this pole that gives rise to the internal choanoderm, while ameboid cells at the posterior pole generate the outer layer (pinacoderm) of the adult sponge.

Homologues of metazoan developmental genes are just beginning to be described from Porifera, but many more should be revealed soon with the recent completion of a "complete genome" for the demosponge *Amphimedon queenslandica*. Already, a diversity of metazoan homeobox genes have been identified in sponges, and in *A. queenslandica* a broad range of transcription factor gene classes, many of which appear to be metazoan specific, are expressed during development. However, the possible homologies of adult sponge tissues to metazoan endoderm and ectoderm remain quite uncertain, and for this reason the terms "gastrodermis" and "epidermis" are used for sponges to avoid implications of embryological homology (and the inner mesohyl layer certainly does not appear to

---

[9]The question of whether-or-not, and when, gastrulation occurs in sponges has long been debated. In fact, it depends on how one defines "gastrulation." Some workers define gastrulation simply as the epithelialization of embryos, thus clearly indicating sponges undergo gastrulation. Other workers define it as the embryogenic process through which metazoan ectoderm and endoderm, and a gut are derived; but, of course, sponges have no guts and their inner and outer cell layers appear not to be homologous to the epithelia of higher metazoans, and on this basis one could argue that gastrulation does not occur in Porifera. Spongologists following this line of logic refer to "embryonic morphogenesis" rather than "gastrulation." Still others have defined poriferan gastrulation as the process of metamorphosis from larva to adult, when the choanocyte chambers are formed. In this book, we define gastrulation (in the general sense) as the reorganization of cells of the blastula to form multiple embryonic germ layers—the tissues upon which all subsequent development depends. Gastrulation, in most cases, creates the separation of those cells that must interact directly with the environment, and those that process materials taken in from the environment. Thus, gastrulation is a keystone, defining feature of the Metazoa. By our definition then, gastrulation in sponges takes place during formation of the larva, when the first germ layers are laid down, regardless of the process. It is worth noting, however, that the absence of solid cell–cell adhesion, and the ability of sponge cells to become mobile, might argue against true gastrulation in sponges because there is not necessarily conservation of cells in specific germ layers.

Interestingly, the term "gastrulation" derives from the name given by Ernst Haeckel (in 1872) to a stage in the development of calcareous sponges, his **gastrula** stage, a ciliated egg-shaped larva that settles to the benthos and, according to Haeckel, formed a "mouth and a gut." Haeckel claimed that the gastrula stage could be found in the development of all animals, and thus represents the recapitulation of the ancestral metazoan, a hypothetical ancestor he called the **gastraea**—a diploblastic animal with a ciliated gut. This recapitulation, he argued, was proof that the Metazoa are monophyletic. Until recently, only one other observation of Haeckel's "gastrula" stage in calcareous sponges had been made, by the German philosopher and biologist Ernst Hammer, in 1908. And yet, "gastrulation by invagination" has been widely conceived, for over a hundred years, to be the ancestral mode of germ layer formation in animals. The conundrum, of course, is that we now know sponges don't form a typical metazoan gut, so the transient cavity formed by invagination during settlement cannot represent a future gut. An outstanding sponge biologist by the name of Sally Leys has recently shown that what Haeckel (and later Hammer) witnessed was a very brief transitory stage in which the anterior cells of a settled calcareous sponge larva invaginate into the posterior half of the larva. However, the "invagination hole" closes completely and it is not until some days later that the sponge forms an osculum (not a gut!) at its apical pole. Hence, there is no reason to assume that this transitory "gastrula" stage is a precursor for formation of a gut via invagination in higher animals. Further, it is the earlier formation of the two cellular regions of the larva that comprises the actual gastrulation event (whereas metamorphosis upon settlement involves the reorganization of these already differentiated regions).

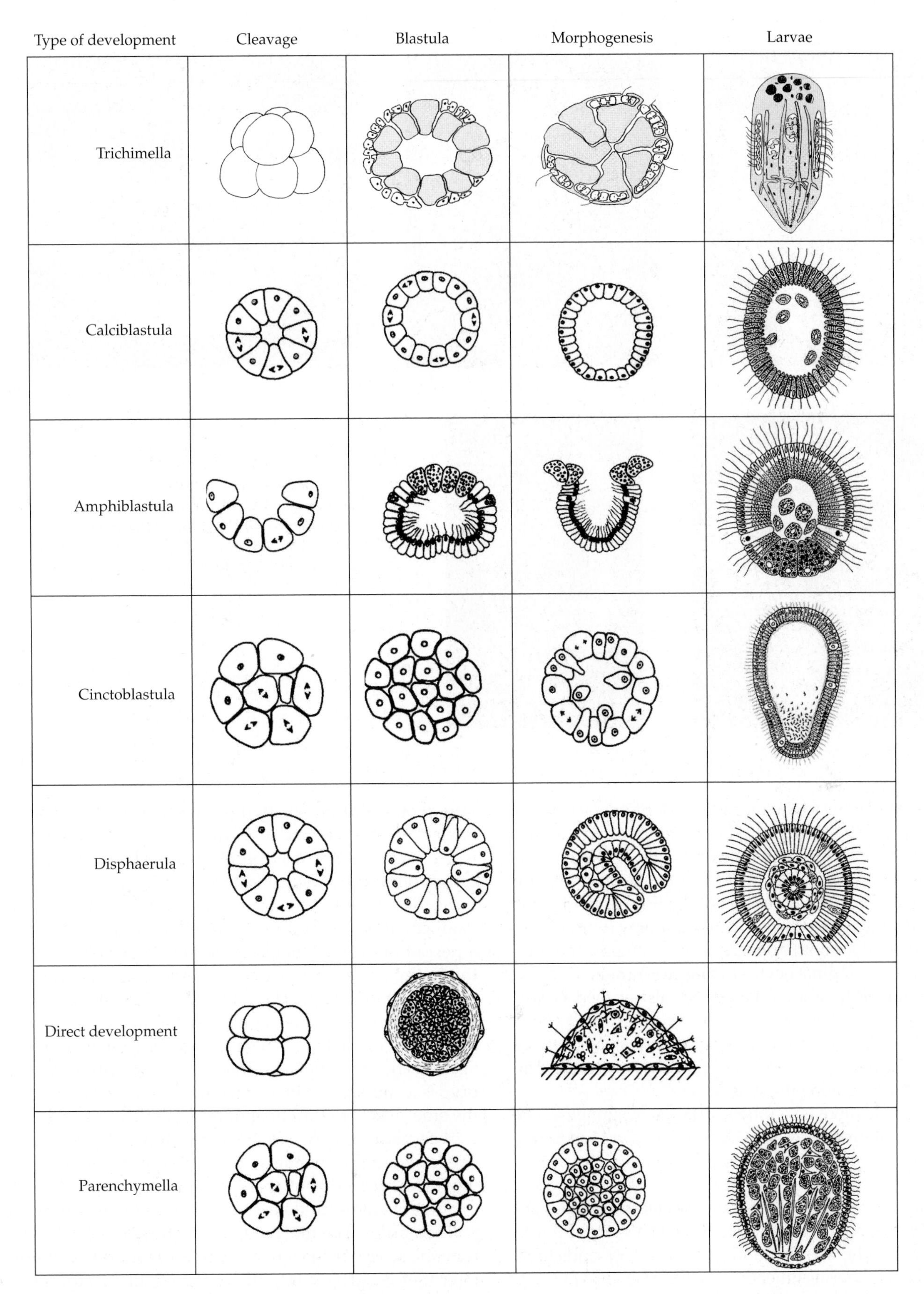

Figure 6.19 Schematic diagram summarizing the various types of development and their resultant larvae in different groups of Porifera; see text for details.

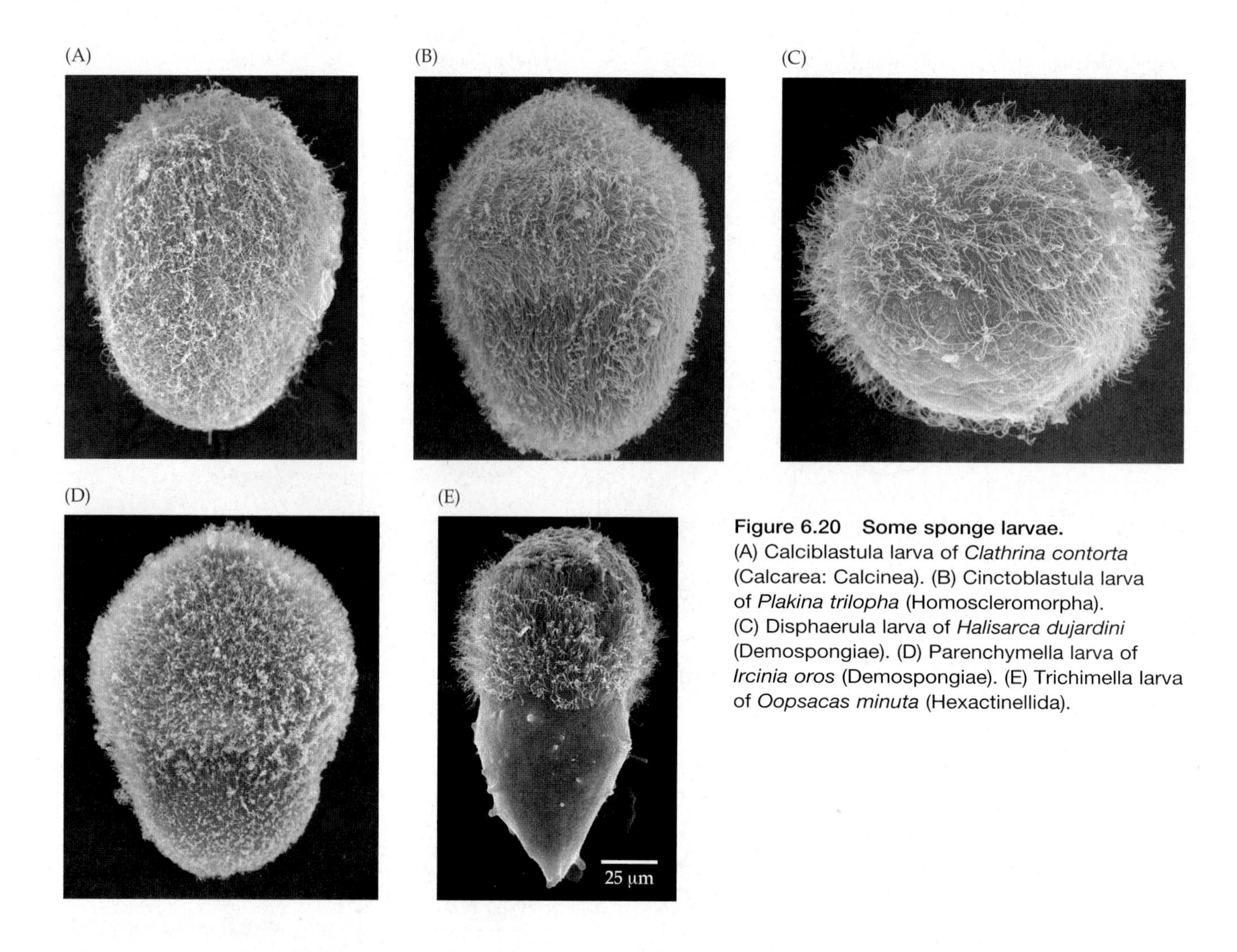

Figure 6.20 Some sponge larvae. (A) Calciblastula larva of *Clathrina contorta* (Calcarea: Calcinea). (B) Cinctoblastula larva of *Plakina trilopha* (Homoscleromorpha). (C) Disphaerula larva of *Halisarca dujardini* (Demospongiae). (D) Parenchymella larva of *Ircinia oros* (Demospongiae). (E) Trichimella larva of *Oopsacas minuta* (Hexactinellida).

be homologous to the mesoderm of higher Metazoa). This inability to directly homologize sponge tissues with those of higher Metazoa is one of the reasons some workers reject using the term "gastrulation" for Porifera. With the exception of the homoscleromorph demosponges, typical epithelial anatomy (with underlying basal lamina) appears to be lacking in sponges. The ability of pinacocytes to migrate into the mesohyl and convert to other cell types has also argued against considering this layer to be true ectodermally-derived epithelium (the motility of pinacocytes has been ascribed to the general absence of specialized desmosomal junctions among these cells).

Because there is so much variation in sponge developmental processes, each class is reviewed separately below.

Calcarea So far as is known, all calcareous sponges are viviparous. In the subclass Calcinea, oocytes develop within the mesohyl and grow in size (by phagocytizing neighboring amebocytes). Embryos also develop in the mesohyl and cleavage is total and equal to produce a hollow blastula larva (coeloblastula). The coeloblastula of some species has two types of cells: ciliated outer cells, and one or two posterior granular cells. After larval release, and just prior to larval attachment to the substrate, in some species, some cells loose their cilia and ingress into the blastocoel via ingression.

In the subclass Calcaronea, oocytes differentiate from choanocytes and move into the mesohyl. After a period of growth they migrate to the periphery of the sponge. Cleavage is total and equal, and the first three are meridional. The fourth division (at least in *Leucosolenia*) is oblique, almost exactly like the cleavage in the colonial green alga *Volvox*—a type of cleavage referred to as table palintomy. The resulting embryo is a cup-shaped blastula, with a small opening at the side closest to the choanoderm. While most of the cells continue to cleave in the same manner, forming a single-layered epithelium, a few that lie directly under the choanoderm do not; these remain much larger than the other cells, with their cytoplasm filled with large yolk granules. The most unusual feature of the calcaronean embryo is that cilia, which differentiate on the micromeres, project into the center of the blastocoel (inverted, compared with those in all other sponge embryos); this stage is called a **stomoblastula** (Figure 6.21B–D). To reach the final orientation, the embryo

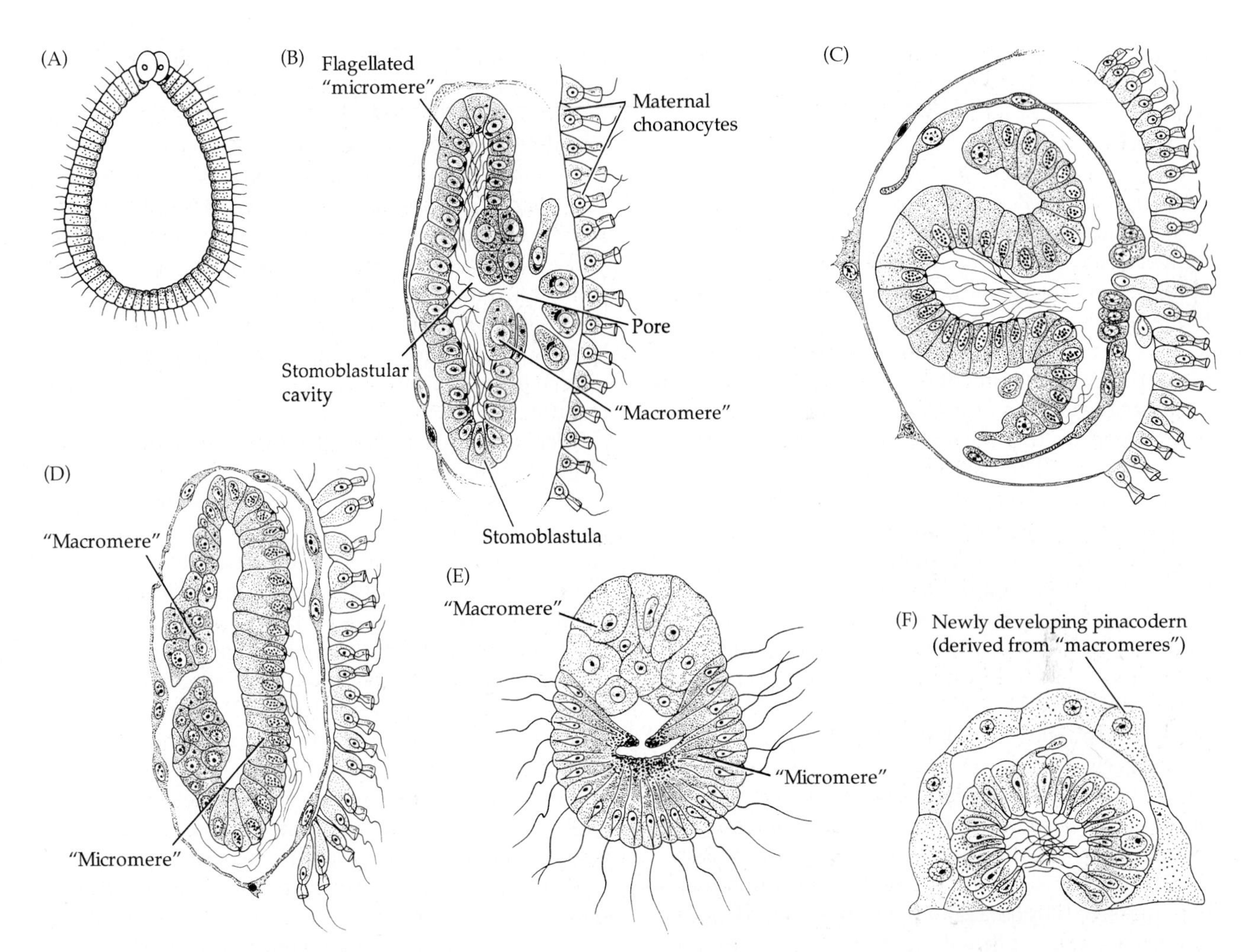

Figure 6.21 "Coeloblastula" and amphiblastula larvae (in section). (A) Typical "coeloblastula" larva with its posterior "macromeres." (B–D) During the remarkable process of inversion in *Sycon* (= *Scypha*), the stomoblastula turns itself inside out to form an amphiblastula larva with externally directed flagella. (E) A typical amphiblastula larva (*Sycon*). (F) Settled young sponge (*Sycon*) after invagination of flagellated cells.

actually turns itself inside out. To do this, the ciliated blastomeres move upwards and through the opening as a "single epithelium," into the choanocyte chamber (much as a bag is turned inside out). The resulting larva is an **amphiblastula** (Figure 6.21E), composed of anterior ciliated cells, posterior granular cells, and inner nutritive amebocytes.

The free-swimming amphiblastula larvae of *Sycon* (class Calcaronea) exit the parent sponge via the osculum and settle within 12 hours, attaching by their cilia, which are thought to be adhesive (probably via glycoproteins on the cell surface). In some cases, larvae settle on their anterior end, on the hole left from the earlier invagination event. The hole is thus unrelated to the future osculum (nor is it a primitive "mouth" of any kind); it is, in fact, engulfed by the newly forming epithelium of the metamorphosing larva. Thus, the transient cavity that is formed by invagination is not the future osculum of the sponge, as Ernst Haeckel long-ago intimated, and this invagination does not represent gastrulation. Once settled, they quickly undergo metamorphosis into a juvenile (sexually immature) sponge.[10] During metamorphosis, the anterior cells loose their cilia and migrate in to form an inner cell mass; some of these cells differentiate into choanocytes, while others remain amoeboid. Metamorphosis of these larvae seems to involve a rapid proliferation of the "macromeres" to form pinacoderm that overgrows the flagellated hemisphere. The flagellated cells pocket inward to form a chamber lined with cells destined to become choanocytes (Figure 6.21F). An osculum breaks through, and the tiny asconoid-like sponge becomes capable of circulating water and feeding. This

[10]This is the stage that Ernst Haeckel focused upon as the root of gastrulation. Haeckel, however, mistook the amphiblastula for a bilayered planula larva and claimed a number of observations that have never been repeated. It is likely that the microscopical tools that Haeckel had were simply insufficient to resolve the rapidly unfolding changes during embryogenesis and metamorphosis in *Sycon*.

initial functional stage is called an **olynthus** (Figure 6.5A). After further growth, it will become an asconoid, syconoid, or leuconoid adult, depending upon the species.

Hexactinellida Because reproductive individuals in this class are rarely encountered, and most species live in deep water, almost all of our knowledge on development comes from a few small cave-dwelling species that are accessible by SCUBA. Although based upon studies of only a few species, a general overview of glass sponge development is as follows. Hexactinellids are probably all viviparous. Gametes arise from groups of archaeocytes that are suspended within the trabecular reticulum. Each archaeocytes cluster is called a **congery**, and all the cells are connected by cytoplasmic bridges (although mature oocytes are independent cells). It is not yet known how sperm find oocytes or how fertilization occurs. Cleavage to the 32-cell stage is total, equal, and asynchronous. The first cleavage is generally meridional, and the second either equatorial or rotational. By the 16- and 32-cell stage, the embryo is a hollow blastula. The following cleavages are unequal, producing smaller micromeres that lie on the outside and larger yolk-rich macromeres on the inside. Even the earliest micromeres are connected to one another by cytoplasmic bridges. The macromeres divide unequally, gradually filling the center of the blastocoel, and eventually they envelop the micromeres with massive filopodia. Uniquely, these cells then fuse to form a single multinucleated giant tissue, the new trabecular syncytium, which completely envelops the micromeres. In this way, the embryo gains its outer epithelium. The formation of micromeres was originally considered to represent gastrulation by cellular delamination. However, the epithelialization of the larva—the enveloping of the micromeres by the incipient trabecular reticulum—is probably better viewed as the true gastrulation event. The fully differentiated hexactinellid larva is a **trichimella** (Figure 6.19). The major tissue of the larva is the syncytial trabecular reticulum, continuous throughout the larva including its surface "epithelium." After swimming for one to several days, the larva settles on its anterior, lipid-filled pole and undergoes metamorphosis. The flagellated cells of the trichimella are multiciliated cells, which is unique among the sponges.

Demospongiae The class Demospongiae is large and diverse—without an easily characterized or uniform developmental pattern. Detailed recent embryological studies have been accomplished for only a dozen or so species. Both viviparity and oviparity occur in this class. Development is highly varied among the orders, families, genera, and in some cases even within single species. Spermatocytes may develop in special cysts within the mesohyl, and in most demosponges, notably in freshwater species, these derive from choanocytes. Oocytes may be isolecithal or telolecithal, and they usually arise from archaeocytes in the mesohyl—although in the family Lubomirskiidae (freshwater sponges in Lake Baikal) they are thought to derive from choanocytes—and they may be accompanied by yolk-rich nurse cells, or follicular cells, that form a thick coating around them, the yolk being incorporated into the cytoplasm of the oocytes. Spawning of eggs and sperm has been linked to lunar cycles in some species, as in many higher invertebrates. Fertilization is usually external.

Cleavage appears to be complete, equal or unequal, often chaotic, and (at least in some species) produce micromeres and macromeres. The former develop a cilium, and eventually both types segregate out until the ciliated micromeres cover the periphery of the embryo (a process termed multipolar delamination). At this point, the embryo may be a solid blastula (stereoblastula), or a coeloblastula (although the central cavity may be filled with bacteriocytes and small granular cells of the follicle layer). Many species appear to produce parenchymella larvae that are polarized and possess a ciliated outer layer. Settlement seems to occur quickly, and in some species the eggs are released in a mucus mass and a planktonic larval stage is absent (larvae develop upon the substratum near the parent sponge).

Among the Demospongiae, the genus *Halisarca* stands out in several ways. It completely lacks a skeleton, and has mixed developmental pathways (that have been studied for over 100 years). Most, or all species are gonochoristic and viviparous. Spermatocytes originate from choanocytes via spermatic cysts. Oocytes also derive from choanocytes, and during vitellogenesis symbiotic bacteria characteristic of the mesohyl of adults are incorporated into the oocyte. Cleavage is total and equal, producing a hollow blastula. In many cases, cells invade the blastocoel by both multipolar and unipolar ingression. Cilia appear on the external cells of the 32- to 64-cell stage blastula, while internal cells differentiate to become archaeocytes. At this point, things get strange, because three very different kinds of larvae can now develop, even within the same individual parent sponge. The first is a coeloblastula, with a single layer of surface cells surrounding a small lumen. The second is a parenchymella that has an outer layer surrounding an inner cell mass of amoeboid cells. The third has two layers of epithelial cells, one external and the other internal, that line a small lumen; this type of larva is called a disphaerula. All three types are polarized by external patterns of ciliation, and all swim by rotating in a left-handed direction.

Homoscleromorpha The eggs of homoscleromorphs are isolecithal, rich in yolk inclusions, and completely surrounded by a follicle formed of parental endopinacocytes. Cleavage is holoblastic and equal. At the third division, cleavage becomes irregular and asynchronous, and proceeds to form a solid morula of undifferentiated yolk-rich blastomeres. At about the 64-cell stage, the outer layer of the morula begins to differentiate. The blastomeres close to the surface divide more actively, while the internal cells migrate to the periphery of the embryo to gradually form a monolayered larva with a central cavity that contains symbiotic bacteria and maternally secreted cells. The resulting embryo is a hollow coeloblastula, and as the outer layer differentiates the cells grow cilia. So far as is known, the centrifugal migration of cells from the center to the periphery of the morula, to form the coeloblastula, is unique not only within Porifera but for all of Metazoa. This unusual process has been termed **multipolar egression**. The ciliated cells of the outermost layer become elongate (columnar) and are closely linked by desmosome-like junctions.

Loose collagen fibrils fill most of the internal cavity as the coeloblastula matures into a larva. However, immediately below the outermost cell layer there develops a tough extracellular matrix of consolidated collagen fibrils, and beneath that there is a second layer of more loosely consolidated collagen fibrils. These two layers of collagen constitute a basement membrane similar to the one found below the pinacoderm and choanoderm of adult homoscleromorphs. (A similar network of extracellular matrix has been described from beneath the larval epithelium of one demosponge larva.) The outermost cell layer of homoscleromorph larvae is a true epithelium, homologous to that seen in adult sponges and in other Metazoa, its cells being columnar and producing a basement membrane/basal lamina.

As development continues, the central cavity is progressively filled by collagen fibrils (along with the symbiotic bacteria) and the larval epithelium becomes regionally differentiated, giving the larva a symmetry that most regard as truly anterior-posterior. The homoscleromorph larva was originally called an "amphiblastula" because it was thought to possess two fairly distinct regions, one ciliated and the other not. However, morphogenesis in sponges of this group is not at all like that of sponges in the Calcaronea (a group for which the amphiblastula larva is characteristic), and the larvae are in fact completely ciliated. Thus it was suggested that the homoscleromorph larva be called a **cinctoblastula**, a term which more accurately describes its distinctive feature of paracrystalline intranuclear inclusions that from a belt around the posterior pole. The free-swimming cinctoblastula larva is ovoid or pear shaped, wider at the anterior pole than the posterior pole, and the entire surface is ciliated. All larvae that have been observed swim with a left-handed rotation.

Some Additional Aspects of Sponge Biology

Some basic sponge ecology has been presented in the previous sections of this chapter. However, because sponges play such important roles in so many marine habitats, and because they are intrinsically interesting in their phylogenetic "primitiveness," we add here some additional aspects of their natural history.

Distribution and Ecology

Certain distributional patterns are evident among the four classes of sponges. Calcareous sponges (and coralline demosponges) are far more abundant in shallow waters (less than 200 m), although they are not uncommon at slope depths, and a few species (particularly *Sycon*; Figure 6.21) have even been reported from depths to 3,800 m. Hexactinellids, which were common in shallow seas of past eras, are now largely restricted to depths below 200 m, except in extremely cold environments (such as western Canada and Alaska, New Zealand, and Antarctica), where some species occur in shallow waters. Demosponges live at all depths, whereas homoscleromorphs occur from the continental shelf to around 1,000 m depth. Calcareous sponges may be restricted largely to shallow waters because they require a firm substratum for attachment. On the other hand, many demosponges and hexactinellids grow on soft sediments, attaching by means of rootlike spicule tufts or mats. The coralline sponges, once a predominant group on shallow tropical reefs, are now largely restricted to shaded crevices and caves, to subreef depths, or to warm-temperate waters where their potential competitors (the hermatypic corals) cannot grow (Figure 6.2L). They are thought to be relicts of major reef-constructing sponge clades of Mesozoic and Paleozoic seas.

Although sponges are very sensitive to suspended sediment in their environment, they seem to be quite resistant to hydrocarbon and heavy metal contamination. Many species can actually accumulate these contaminants without apparent harm, and may be seen, for example, in the ferric granules on the spongin skeleton of fibrous bath sponges from intertidal regions. The capacity of certain species to accumulate metals at far higher levels than that of the ambient environment has been suggested as a possible defense mechanism (anti-predation, antifouling). Detergents also do not appear to affect many sponges, and in fact may even serve as a source of nutrition for these amazingly adaptable animals.

Sponges are the dominant animals in a great many benthic marine habitats. Most rocky littoral regions harbor enormous numbers of sponges, and they often occur in vast numbers (and large sizes) around Antarctica. Although many animals prey on sponges, the amount of serious damage they do is usually slight. Some tropical fishes (e.g., some angelfish) and turtles (e.g., hawksbill turtles) crop certain kinds of sponges, a number of sea stars and shelled gastropods are sponge predators, and smaller predators (mainly opisthobranchs that closely match the prey sponge's color) consume limited amounts of living sponge material in both warm and temperate seas. Overall, however, sponges appear to be very stable and long-lived animals, perhaps in part due to their spicules and toxic or distasteful compounds that discourage would-be predators.

Biochemical Agents

Seashore explorers and SCUBA divers quickly notice that sponges are just about everywhere, except on shallow sandy habitats. Most grow on open rock or occasionally mud surfaces where they are obviously exposed to potential predation. Clearly, some mechanism(s) must be working to prevent these animals from being cropped excessively by predators. The primary defense mechanisms in sponges are mechanical (skeletal structures) and biochemical. Studies have also show that sponges manufacture a surprisingly broad spectrum of biotoxins, some of which are quite potent. A few, such as *Tedania* and *Neofibularia*, can cause painful skin rashes in humans.

Research in sponge biochemistry has also revealed the widespread occurrence of antimicrobial agents in sponges. Sponges appear to use "chemical warfare" not only to reduce predation and prevent infections, but also to compete for space with other invertebrates such as ectoprocts, ascidians, and even other sponges. Different species have evolved chemicals (allelochemicals) that may be species-specific deterrents or actually lethal weapons for use against competing sessile and encrusting organisms. For example, the coral-inhabiting sponge *Siphonodictyon* releases a toxic chemical into the mucus exuded from its oscula, thus preventing potential crowding by maintaining a zone of dead coral polyps around each osculum (Figure 6.22).

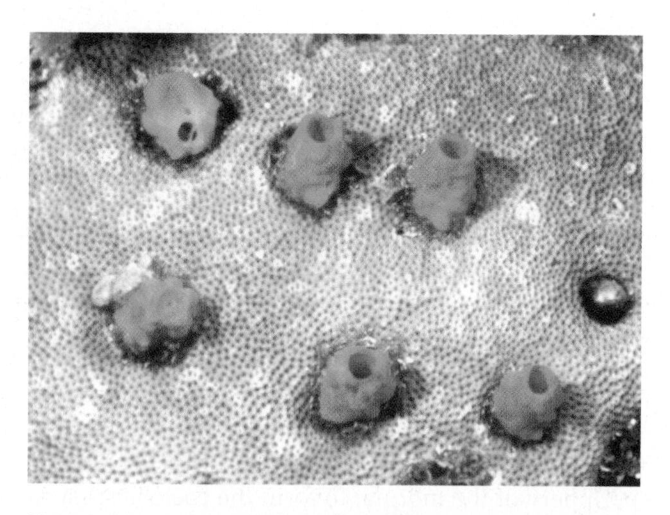

Figure 6.22 *Siphonodictyon coralliphagum* infests the hermatypic coral *Siderastrea siderea* on a Caribbean reef. Note the "dead zone" between the oscular chimneys of the sponge and the coral polyps.

Many of the chemicals produced by sponges and other marine invertebrates are being closely studied by natural products chemists and biologists interested in their potential as pharmaceutical agents. Compounds with respiratory, cardiovascular, gastrointestinal, anti-inflammatory, antitumor, cytotoxic, and antibiotic activities have been identified from many marine sponges. One study found that 87% of the temperate sponges, and 58% of the tropical species examined in New Zealand produced extracts with specific antibacterial activity. One New Zealand sponge (*Halichondria moorei*) has long been used by native Maoris to promote wound healing and was recently discovered to contain remarkably high concentrations (10% of the sponge dry weight) of the potent anti-inflammatory agent potassium fluorosilicate. In the 1950s, compounds belonging to a class of chemicals called arabinosides, that are active against viruses, were found in the tropical sponge *Tectitethya* (= *Tecthya*) *crypta*. This led to the development of a number of important drugs that are in use today, including Ara-C (Citarabina®), sold under the trade name Cytosar-U®, by Pharmacia & Upjohn. Cytosar-U® was approved for treatment of certain leukemias in 1969, making it the first such approved marine-derived drug for use in cancer chemotherapy. Other drugs, also derived from natural products in *T. crypta*, include AZT (Zidovudine), manufactured under the trade name Retrovir® by the drug company GlaxoSmithKline, which was the first drug licensed for the treatment of HIV infection, and Ara-A (sold as Acyclovir/Zovirax®), an antiviral compound commonly prescribed for treating herpes.

Some sponges, including the west Pacific species *Luffariella variabilis*, produce a remarkable terpenoid compound called manoalide that is not only an extremely powerful antibacterial compound but also acts as both an analgesic and anti-inflammatory agent. Sponges of the genera *Halichondria* and *Pandaros* are known to produce potent antitumor compounds belonging to a group of chemicals called halichondrins. For example, eribulin mesylate (sold under the trade name Halaven®), a powerful treatment for metastatic breast cancer, is a synthetic analogue of halichondrin B, a product isolated from the marine sponge *Halichondria okadai*.

Growth Rates

Little is known regarding growth rates in sponges, but rates appear to vary widely among species. Some species are annuals (especially small-bodied calcareous sponges of colder waters); hence, they grow from larvae or reduction bodies to reproductive adulthood in a matter of months. Most, however, appear to be perennials and often grow so slowly that almost no change can be seen from one year to the next; this growth pattern is especially true of tropical and polar demosponges. Until recently, age approximations of perennial species ranged from 20 to 100 years. However, research on the Caribbean giant barrel sponge *Xestospongia muta*, which can exceed two meters in height, suggests that this species might be capable of living more than 2,000 years. Radiocarbon dating of several hexactinellid *Rossella* species from Antarctica recently revealed growth rates of about 3 mm per year, aging the sponges at about 440 years old. Even more impressive, the west Pacific species *Monorhaphis chuni* has giant SiO_2 spicules that can reach lengths of nearly 3 m, representing the largest biogenic silica structures in the animal kingdom; the lifespan of the spicule is estimated to be 11,000 (±3,000) years, making this the longest-lived animal species on Earth!

Some sponges are capable of very rapid growth, and they regularly overgrow neighboring flora and fauna. For example, the tropical encrusting sponge *Terpios* is known to overgrow both living and nonliving substrata. In Guam, this sponge grows at rates averaging 23 mm per month, over almost every live coral species in the area as well as over hydrocorals, mollusc shells, and many algae. Experiments have shown that *Terpios* is toxic to living corals, and presumably to many other animals. However, in most cases of observed sponge overgrowth of living corals, the corals have been stressed (by temperature, sediment, pollution) or damaged and presumably weakened.

Still another physiological trick of some sponges is the ability to rapidly produce copious amounts of mucus when disturbed. On the west coast of North America, the beautiful red-orange *Antho* (= *Plocamia*) *karykina* covers itself with a thick layer of mucus when injured or disturbed. Yet the little red sea slug *Rostanga pulchra* has evolved the ability to live and feed inconspicuously on this and other red-colored sponges, and even lays its camouflaged red egg masses on the sponge's exposed surface without eliciting the mucous reaction.

Symbioses

Symbiosis is common among sponges of all kinds. It would be difficult to find a marine sponge that is not utilized by at least some smaller invertebrates and often by fishes (e.g., gobies and blennies) as refuge. The porous nature of sponges makes them ideally suited for habitation by opportunistic crustaceans, ophiuroids, molluscs (e.g., scallops, mussels), and various worms. A single specimen of *Spheciospongia vesparia* from Florida marine waters was found to have over 16,000 alphaeid shrimps living in it, a study from the Gulf of California found nearly 100 different species of plants and animals on and in a 15 × 15 cm *Geodia mesotriaena*, and an examination of eight sponge species off the coast of South Carolina and Georgia (at depths of 18 to 875 m) recorded 236 symbiotic invertebrate species. The last study concluded that sponge-symbiont assemblages can constitute legitimate ecological communities harboring complete food webs as well as gravid and juvenile individuals. Even freshwater sponges host abundant symbionts, and most freshwater invertebrate phyla have been recorded with sponges (e.g., Hydrozoa, Nematoda, Oligochaeta, Polychaeta, Gastropoda, Bivalvia, Isopoda, Amphipoda, Ostracoda, Copepoda, Hydracarina, Bryozoa).

Most symbionts of sponges use their hosts only for space and protection, but some rely on the sponge's water current for a supply of suspended food particles. A classic example of this phenomenon is the male–female pair of shrimp (*Spongicola*) that inhabit the globally distributed, deep-water hexactinellid sponges known as Venus's flower baskets (*Euplectella*; Figure 6.2O,P). The shrimp enter the sponge when they are young, only to become trapped in their host's glasslike case as they grow too large to escape. Here they spend their lives as "prisoners of love." Appropriately, this sponge (with its guests) is a traditional wedding gift in Japan—a symbol of the lifetime bond between two partners.

Other, far more intimate symbiotic relationships with sponges are also common. Some snails and clams characteristically have specific sponges encrusting their shells, and many species of crabs (hermits and brachyurans) collect certain sponges and cultivate them on their shell or carapace (Figure 6.2M). Demosponges, such as *Suberites*, are commonly involved in these commensalistic relationships. The sponge serves primarily as protective camouflage for its host, and it perhaps benefits by being carried about to new areas. The sponge also no doubt feeds off small bits of organic material dislodged during the feeding activities of its host. Evidence suggests that scallops and oysters with their shells overgrown by sponges (frequently *Halichondria panicea*) experience reduced predation from seastars. Certain dromiid crabs carry sponges on their carapace, probably also as a predator avoidance adaptation, and numerous species of decorator crabs stick pieces of sponge on their carapace as camouflage. A particularly odd symbiosis occurs in the Mediterranean Sea, where most colonies of the bryozoan *Smittina cervicornis* are overgrown by the thinly encrusting sponge *Halisarca harmelini*; somehow, the feeding currents actually appear to be strengthened for both partners by this collaboration. On the other hand, the Indonesian sponge *Mycale vansoesti* seems to rely on a coralline alga (*Amphiroa* sp.) as its skeleton; the alga completely invests the sponge and holds it erect, the sponge itself having almost no spicular skeleton of its own.

Other spectacular examples of poriferan symbioses are sponge–Bacteria/Archaea and sponge–algae associations, some of which appear to be mutualistic whereas others are not. For example, a typical member of the demosponge order Verongida contains a mesohyl bacterial population accounting for some 38% of its body's volume, far exceeding the actual sponge-cell volume of only 21%. Presumably, the sponge matrix provides a rich medium for bacterial growth, and the host benefits by being able to conveniently phagocytize the bacteria for food. Similar relationships are common between poriferans and various Cyanobacteria. The Indian Ocean demosponge *Tethya orphei*, and other haplosclerid sponge species, have been shown to harbor an ectosome permeated by filamentous cyanobacteria (probably *Oscillatoria spongeliae*). Vertical transmission of symbiotic bacteria (from the parent sponge to the next generation) is a specific trait of poriferan development. The bacteria are transmitted to the buds or gemmules when the sponge reproduces asexually. During sexual reproduction the bacteria are transmitted through eggs (in oviparous species) or through larvae (in ovoviviparous forms). It has been suggested that in *Tethya seychellensis* the green alga *Ostreobium* sp. grows exclusively along the siliceous spicule bundles in order to capture sunlight via these natural optical fibers, and other evidence from *T. aurantium* also suggests that silicious spicules act like optical fibers to channel light to photosynthetic microbes living deep inside their bodies. Recent evidence suggests that some products of normal cyanobacterial metabolism (e.g., glycerol and certain organic phosphates) are translocated directly to the sponge for nutrition. In many sponges, both regular bacteria and Cyanobacteria co-occur, the former in deeper cellular regions, the latter closer to the surface where light is available. On the far-offshore regions of the Great Barrier Reef, 80% of the sponge individuals harbor commensal cyanobacteria. In one remarkable study, C. R. Wilkinson (1983) showed that 6 of the 10 most common sponge species on the forereef slope of Davies Reef (Great Barrier Reef) are actually net primary producers, with three times more oxygen produced by photosynthesis (by their symbionts) than consumed by respiration. Such relationships also exist with certain dinoflagellates, filamentous green algae, and red algae.

Recent studies of the Caribbean giant barrel sponge (*Xestospongia muta*) have provided evidence of two very different cyanobacterial symbionts in the *Synechococcus* group, both of which are implicated in "sponge bleaching" events in this species (similar to coral bleaching). There is no evidence that either species of *Synechococcus* maintains a mutualistic relationship with *Xestospongia muta*. One species seems to be a commensal, benefiting from the relationship but doing nothing positive or negative to the sponge. This species dies back on a cyclical basis (as do about 25% of the sponges off southern Florida), probably in association with abnormally warm seawater temperatures, and the cyanobacterial death causes a temporary bleaching (whitening) of the host sponge, but it does not kill it. The other species of *Synechococcus*, which also dies back with warmer than normal waters (and perhaps under other conditions), may be a pathogen and it kills its host sponge as it dies. Similar deaths, which are often associated with a distinctive orange "death band" that moves across the sponge body, have been recorded for several species of *Aplysina* in the Caribbean.

In some areas of the Caribbean and Great Barrier Reef, sponges are second only to corals in overall biomass, and they appear to owe their rapid growth to the presence of large populations of symbiotic Cyanobacteria. Most freshwater spongillids maintain similar relationships with zoochlorellae (symbiotic green algae, or Chlorophyta). These sponges grow larger and more rapidly than specimens of the same species that are kept in dark conditions. Some marine sponges (e.g., the boring sponges *Cliona* and *Spheciospongia*) harbor commensal zooxanthellae similar to those of corals.

Many species of sponges prefer to grow on the roots of mangrove trees in littoral habitats, and experiments have shown that sponge-infested roots elongate much faster than those without sponges. Stable isotope studies have suggested transfer of dissolved inorganic nitrogen from sponge to mangrove, and transfer of carbon from mangrove to sponge! However, this unusual sponge–plant symbiosis is not well understood.

One of the most intriguing examples of poriferan symbiosis is the rare case of intimate association between two different sponge species. In North America, for example *Halichondria poa* is almost always overgrown by *Hymeniacidon sanguinea*, while in Europe *Haliclona cratera* almost always overgrows *Ircinia oros*. And, in the Gulf of California (Mexico) *Haliclona sonorensis* overgrows *Geodia media*. It is not known how the covered species of sponge obtains sufficient water flow to survive, or what the cost/benefit relationships are of these unusual symbioses.

The boring demosponges (e.g., *Cliona, Spheciospongia*) excavate complex galleries in calcareous material such as corals and mollusc shells (Figure 6.23). This phenomenon, known as bioerosion, causes significant damage to commercial oysters as well as to natural coral, bivalve, and gastropod populations. In fact, in the Caribbean the boring sponge *Cliona delitrix* has been shown to undercut entire coral heads by their excavations, resulting in a collapse of the head. The active boring process involves chemical and mechanical removal of fragments or chips of the calcareous coral or shell material by specialized archaeocytes called **etching cells**, which release acid phosphatase. The coarse silt-sized excavated chips are expelled in the excurrent canal system and can actually contribute significantly to local sediments. Many species in the remarkable boring genus *Cliona* (family Clionaidae) have two or three different growth forms (e.g., *C. californiana* of the

(A)

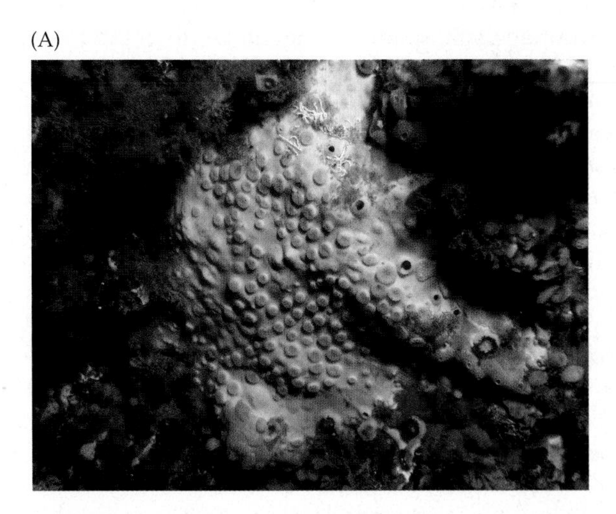

(B)

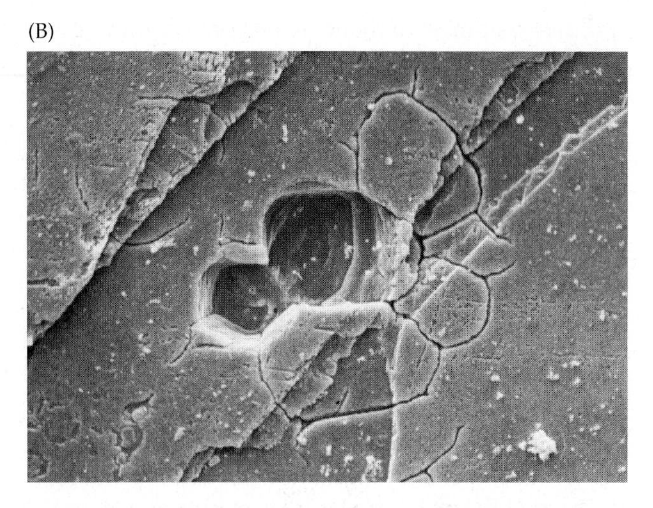

Figure 6.23 Boring sponges. (A) Surface of a coral infected by the yellow boring sponge *Cliona*. (B) A close view (SEM) of the surface of a clam shell, showing six eroded "chips," two of which have been entirely removed and four that are only partly etched by *Cliona*.

eastern Pacific region). The alpha stage (Figure 6.24A) is purely boring (living internally within the "host"), the beta stage (Figure 6.24B) is eruptive (having "escaped" to overgrow the surface of the calcareous substratum), and the gamma stage is "free living" (i.e., has given up the boring stage of its life history) (Figure 6.24C,D).

Sponge bioerosion has a significant impact on coral reefs. Perhaps even more important than actual erosion is the weakening of attachment regions of large corals. This action can result in considerable coral loss during heavy tropical storms, and it is increasing as corals are weakened by bleaching events and ocean acidification resulting from increased atmospheric CO_2. Boring sponges do not appear to gain any direct nourishment from their host coral; rather, they use it as a protective space in which to reside. If you carefully examine the shells of dead bivalves along any shoreline, you will

(A) (B)

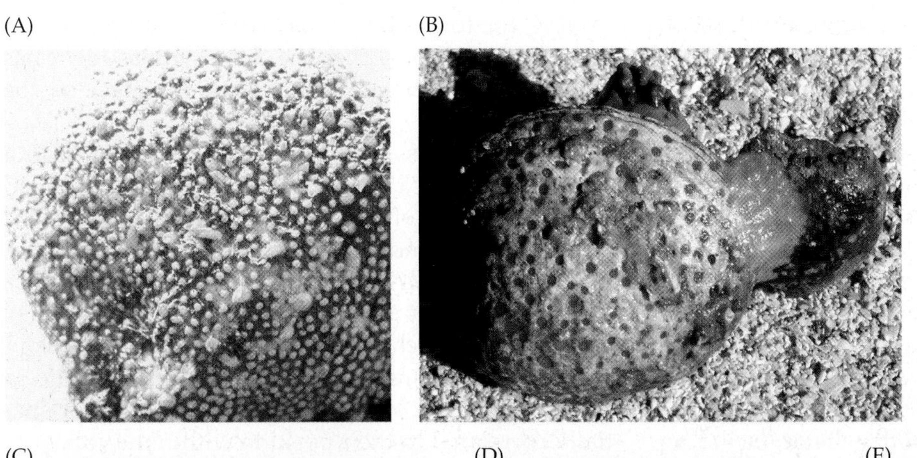

(C)

(D)

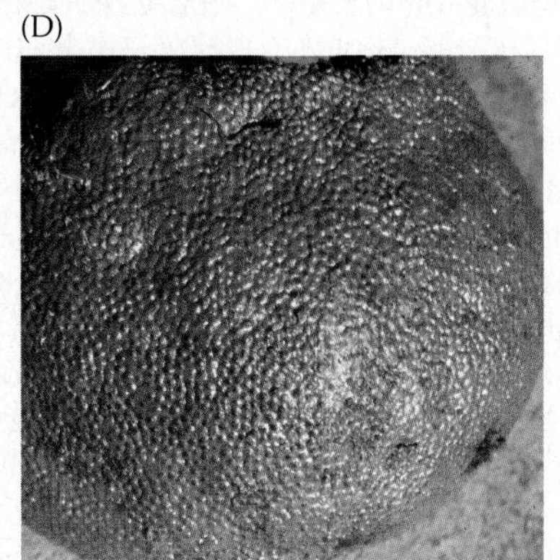

(E)

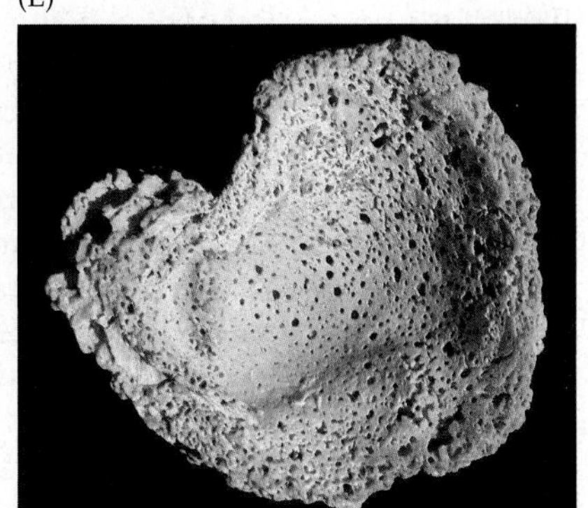

Figure 6.24 The three life history stages of the Pacific boring sponge *Cliona californiana*. (A) In the alpha-stage, the sponge lives primarily within the shell (or other calcareous structure) in which it bores. (B) In the beta-stage, the sponge erupts from its calcareous home and begins to overgrow the substratum. (C,D) In the gamma-stage, the sponge is entirely free living. (E) On most beaches in the world, cast-ashore shells show evidence of boring sponge activity.

discover that most of them are perforated with minute holes and galleries from boring sponges (Figure 6.24E). Bioeroding sponges are responsible for a major portion of the initial breakdown of such calcareous structures, and thus they set the stage for their eventual decomposition and recycling through Earth's biogeochemical cycle.

Poriferan Phylogeny

The Origin of Sponges

Being the living descendants of what might have been the first multicellular animals on earth, sponges are a key group in which to look for innovations that form the basis of the metazoan body plan. But sponges are an ancient group, and the important events in their origin and early evolution lie hidden in the Proterozic eon of the Precambrian. Both the ancient and the unique nature of the poriferan body plan are strikingly revealed by the aquiferous system, cellular pluripotency, and reproductive flexibility; and by the lack of a gut, fixed reproductive organs, nervous system, muscles, cell gap junctions, or strong adult body polarity. These features, along with the choanoflagellate-like choanocytes of sponges, suggest a protistan ancestry, as also shown in many (though not all) molecular phylogenetics. Generally speaking, the poriferans seem to share as many similarities with protists as they do with the higher metazoans. At the same time, sponges clearly stand apart from all other Metazoa in their possession of such features as an aquiferous system and choanocytes with flagella bearing paired vanes (nearly identical to those of the protist phylum Choanoflagellata) (Box 6B).

Understanding sponge evolution is thus closely tied to understanding the transition from unicellular to multicellular life—a phenomenon that we actually know surprisingly little about. Molecular phylogenetic dating estimates indicate that the Porifera likely arose from a common ancestor with the protist phylum Choanoflagellata more than 600 million years ago, perhaps over 700 million years ago.

Although subtle ultrastructural differences exist, the collar cells (choanocytes) of sponges are nearly identical to the collar cells of choanoflagellates, including the unique possession of two distinct winglike projections (the vanes) that arise on either side of the flagellum, and a glycocalyx sheath on the collar that joins the microvilli. In choanoflagellates, the collars are funnel shaped and the sheath forms an uninterrupted fibrous layer on the cell surface. In sponge choanocytes, the collars tend to be tubular and the glycocalyx forms a mesh that both holds the microvilli together and connects adjacent collar microvilli. The vanes, in both groups, are also formed from horizontal fibers of glycocalyx that extend from the flagellum. The vanes in sponge choanocytes appear to be wider than that described from Choanoflagellata, spanning the full width of the collar with the edges of the two wings lying against the inside of the collar (and perhaps attached to it) in sponges.

Somewhat similar collar cells (but without flagellar vanes and generally with a nonmotile cilium) have also been found in some other metazoan phyla (e.g., in certain echinoderm larvae and adults, in some corals, in the trunk epithelium of the enteropneust *Harrimania kupfferi*). The homology of these other metazoan cells to sponge choanocytes is questionable however. In addition, they do little to diminish the force of the argument for a direct relationship between Porifera and Choanoflagellata (an idea that dates back to at least 1866). Molecular data support the monophyly of both the Choanoflagellata and the Porifera.

So similar are choanoflagellates and sponges that periodically through the eighteenth and nineteenth centuries the proposal came up that sponges are no more than highly-organized colonial choanoflagellates. However, the presence of distinctly animal features, including metazoan developmental genes, in the Porifera argues against this idea (e.g., gastrulation during embryogenesis, spermatogenesis, adhaerens junctions in some species, type IV collagen, etc.). And, recent molecular developmental work suggests that certain genes found in sponges and other metazoans probably played critical roles in the transition from unicellularity to multicellularity. Homologues of many typical metazoan genes have now been identified in sponges, such as those encoding proteins involved in immune responses, myosin production, developmental patterning, the formation of the extracellular matrix, and other vital functions. Even the chitin synthase gene has been found in sponges, although it does not seem to be active.

The major class of transcription-factor-encoding genes that regulate animal development is called Antennapedia (ANTP). The ANTP group includes the Hox, ParaHox, and NK genes, all of which are paralogues (i.e., they have arisen in different animals from a shared ancestor as a result of gene-duplication events). These genes, which are involved in multiple developmental processes, are often found in clusters, and in some animals their expression is temporally or spatially correlated to their position within the cluster (a phenomenon known as colinearity). Until recently, no ANTP genes had been found in the phylum Porifera, leading to the suspicion that this gene family did not evolve until the emergence of a line subsequent to sponges—the name "ParaHoxozoa" was given to that lineage of phyla with ANTP class genes. But when the first sponge genome was sequenced (from a species of Demospongiae) a group of gene clusters was found, which are known to be neighbors of Hox and ParaHox genes in higher metazoans. This led to a proposal in 2012 that these developmental genes had been present in the common ancestor of all animals, but had been secondarily lost in sponges—an idea called the "ghost locus" hypothesis. A test of this hypothesis came in 2014, when researchers looked closely at sponge genomes in the class Calcarea, and it was discovered that this sponge class contained both NK and ParaHox genes remarkably similar to those of higher animals. In fact, they found that ParaHox expression in the choanoderm cell layer was very similar to that seen in the endoderm of bilaterians. Although homology in developmental genes does not necessarily equate to homology of organs, the discovery provided support for the ghost locus hypothesis, and it suggests that the first Metazoa probably had a ParaHox gene repertoire. The discovery of these ANTP genes in calcarean sponges adds further data to the recognized, considerable genetic divergence among the sponge classes, which is not too surprising given that Calcarea and Demospongiae are estimated to have diverged from one another at least 600 million years ago.

Recently, genes belonging to the cadherin family have also been identified in sponges. The cadherin family of genes comprise critical mediators of metazoan cell adhesion and signaling and provide the structural basis for vital developmental processes, including tissue morphogenesis and maintenance, cell sorting, and cell polarization. Plants and fungi lack cadherins, which are so far known only from metazoans and choanoflagellates. Cadherin activity has been identified in choanoflagellates in the actin-filled microvilli that comprise the apical collar, at the basal pole of the cell, and in certain cellular bodies of unknown function. The localization of cadherins on the collar suggests that actin filaments and cadherins have been associated since before the origin of the Metazoa. In metazoan epithelial cells, recruitment of β catenin facilitates essential interactions between classical cadherins and the actin cytoskeleton to establish and maintain cell shape and polarity.

This brings us to the interesting proposal that choanoflagellates might be nothing more than "reduced" sponges. This idea derives partly from the inability, so far, to connect the Choanoflagellata phylogenetically with any other protist group. In addition, the mitochondria of choanoflagellates (which are characterized by non-discoidal flattened cristae) and their basal bodies (which bear accessory centrioles) are two other features that are shared with metazoans. On the other hand, comparison of the mitochondrial genomes of protists and metazoans yields at least 13 genes in choanoflagellates that are absent from animal mitochondria, in support of their protist roots. It is often noted that sponges, and all other metazoans, produce true gametes and have embryogenesis—hallmarks of the Metazoa that are lacking in choanoflagellates. However, several choanoflagellates with a sessile trophic stage also produce free-swimming cells, called **zoospores**. They have no collar, and swim vigorously with the flagellum propelling the cell from behind, just as in animal sperm! Choanoflagellates and sponges are clearly closely related and reside near the point of origin of the Metazoa.[11]

The solutions that poriferans evolved to survive environmental challenges created a group of animals unlike any other. Sponges achieved multicellularity and large body size without such typically metazoan traits as fixed organs, nervous and muscular systems, extracellular digestion, or excretory organs. Taken together, these and other poriferan attributes support the notion that sponges are the earliest evolving metazoans for which living representatives exist today. However, despite this evidence several phylogenetic studies have suggested that Ctenophora, not Porifera, could be the basalmost metazoan, although those studies have been seriously questioned.

Evolution within the Porifera

Sponges comprise such an ancient and enigmatic phylum that their phylogeny largely eluded scientists until fairly recently. Although the Porifera have nearly consistently appeared at the base of metazoan molecular and morphological trees, data have been slow to resolve the relationships of sponges to the other basal metazoan phyla (e.g., Cnidaria, Ctenophora, Placozoa), and as this book went to press those relationships were still unsettled, although overall evidence still supported them as basal metazoans. Recent genomic-level studies strongly suggest that Placozoa arose between sponges and all other metazoans, thus clearly separating the Porifera from Cnidaria and Ctenophora. And only since the early twenty-first century have there been robust phylogenetic hypotheses of relationships among the four poriferan classes. Although fossils provide a poor record of noncalcareous sponges, first-occurrence data support a Precambrian origin of the Hexactinellida and Demospongiae, and an early Cambrian origin of the Calcarea (Figure 6.25). Estimated rates of molecular evolution place the origin of the Hexactinellida over 600 million years ago. Even considering some preservation biases in the fossil record, well over 1,000 fossil genera have been described, about 20% of which are still extant.

Sponges first appear in the fossil record in the Precambrian, and the oldest known fossils are from about 600 million years ago, although a well-preserved

[11] Several other enigmatic, unicellular organisms that appear to fall on the line leading to Metazoa are *Capsaspora owczaraki*, *Ministeria vibrans*, and the group Ichthyosporea. *Capsaspora owczarzaki* is a symbiont in the hemolymph of the freshwater snail *Biomphalaria glabrata*. *Ministeria vibrans* is a free-living species that feeds on bacteria. Phylogenetic analyses suggest that these two species might form a sister group to the clade Choanoflagellata + Metazoa. The large group Ichthyosporea, which are symbionts with various animals, might be sister to that group, and this larger clade has been called Holozoa. Holozoa, in turn, is the likely sister group to Fungi, a clade known as Opisthokonta. However, the status of these creatures might well change as more information about them becomes available in the near future.

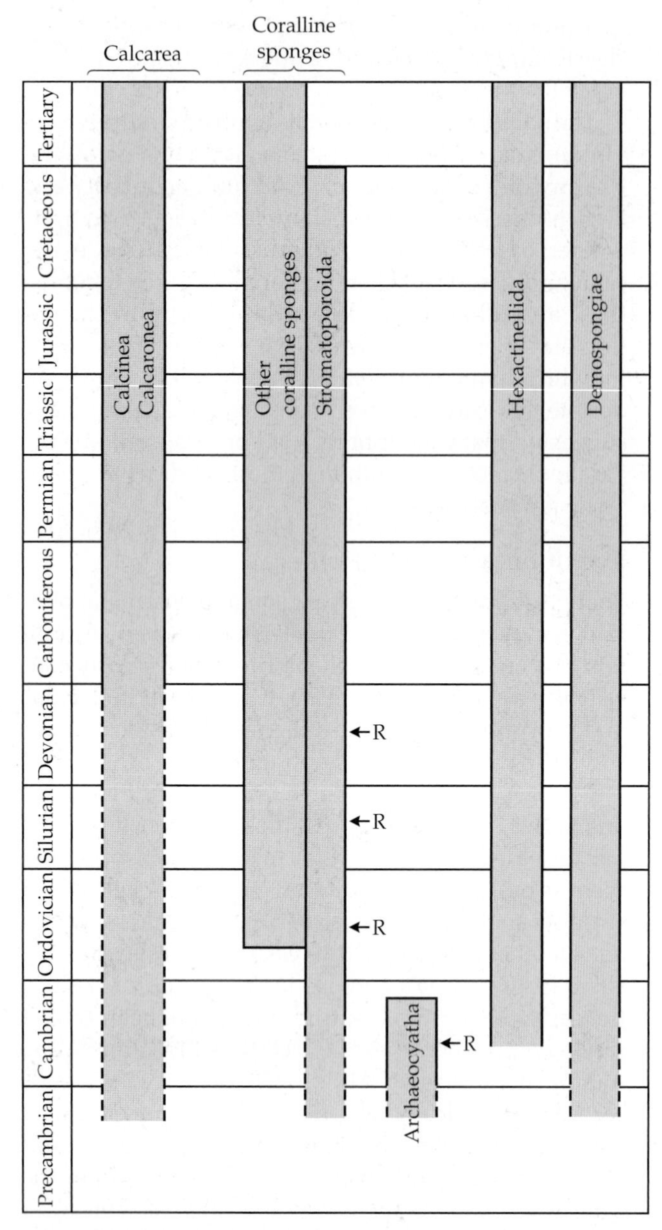

Figure 6.25 Fossil record of three sponge classes, the Archaeocyatha, and the other coralline sponges. Dashed lines indicate suggested occurrence, even though fossils have not yet been found. "R" indicates the times when the group in question is known to have been an important marine reef builder.

600-million-year-old fossil of an Ediacaran sponge was described in 2015 from the Doushantuo Formation of South China. This fossil sponge, *Eocyathispongia qiania*, is only just over one millimeter wide but composed of hundreds of thousands of cells, and it has a structure consisting of several hollow tubes sharing a common base. The external surface is covered with flattened cells resembling pinacocytes, punctuated with small pores.

The fossil record suggests that sponges might have had their ecological heyday in the Paleozoic and Mesozoic when massive tropical reefs developed composed largely of four groups: the archaeocyathans, stromatoporoids, sphinctozoans, and chaetetids. The oldest of these, the coralline Archaeocyatha (Figure 6.26), had a short life within the Cambrian (550–500 Ma). Sphinctozoans and stromatoporids also appeared in the Cambrian (about 540 Ma), while chaetetids first appeared in the Ordovician (about 480 Ma). The affinities of these four coral-like groups have been debated for a hundred years, and various alliances with Cyanobacteria, red algae, ectoprocts, cnidarians, and foraminiferans have been proposed. However, the discovery of living coralline sponges has led most workers to believe that the majority of species in these four groups were primitive, but true, sponges. Research suggests that the calcium-carbonate secreting sponges utilize the same α-carbonic anhydrase genetic pathway as seen in higher animals that biomineralize. And it was exploitation of this gene group that led the explosion of mineralized skeletons among the Metazoa during their rapid Cambrian radiation, made possible by unique genes inherited from their poriferan "roots."

Unlike the coralline sponges, which have decreased in abundance and diversity since the Mesozoic, the Calcarea and Demospongiae appear to have increased in diversity throughout their history. The demosponges were well established by the mid-Cambrian, and there is biochemical evidence that demosponges might have been living as much as 1.8 billion years ago (e.g., sponge-specific biochemicals associated with ancient stromatolites). Several recent molecular analyses,

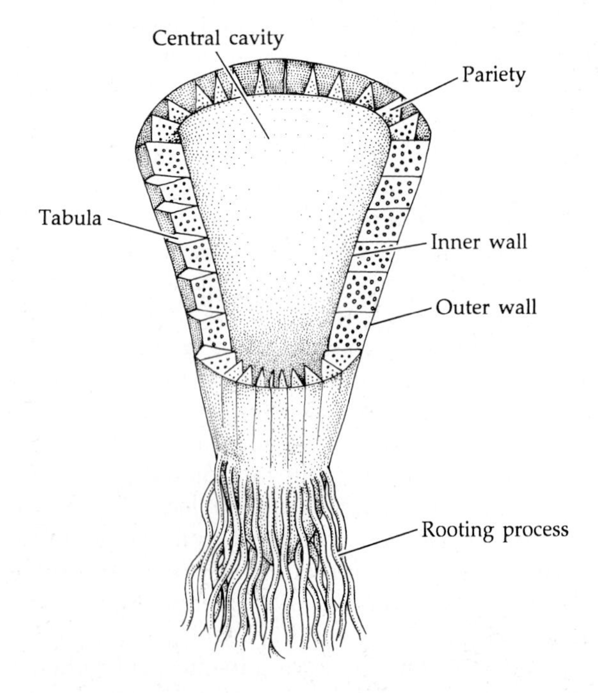

Figure 6.26 A typical archaeocyathan. A vertical section has been partly cut away to show the structure between the inner and outer walls (i.e., vertical parieties and horizontal tabula).

including an analysis of protein kinase C in sponges and some higher metazoans, favor the hypothesis that Hexactinellida were the first of the three sponge classes to appear. The Late Devonian extinction event brought many demosponge clades to an end, and the modern demosponge lineages largely appeared subsequent to that event. All known orders of modern demosponges are found in Cretaceous rocks.

Hexactinellids were most diverse and abundant during the Cretaceous. The oldest hexactinellid fossils, of the early Cambrian, were all thin-walled, saclike sponges with a dispersed surface spicule layer that probably could not support a thick body wall. During the Paleozoic, hexactinellids were common in shallow-water environments. Since then, however, they have become restricted largely to deeper oceanic regions.

Until recently, two fundamentally opposing views of sponge phylogeny existed. One hypothesis suggested that the living sponge classes that produce siliceous spicules comprise a monophyletic lineage (e.g., Demospongiae, Homoscleromorpha, Hexactinellida), with the Calcarea being phylogenetically more distant. The other view proposed that the classes of cellular sponges (Demospongiae, Homoscleromorpha, Calcarea) comprise a monophyletic clade, with the syncytial Hexactinellida standing apart from them.

The first hypothesis was proposed by de Laubenfels in 1955, who reduced the siliceous groups to a single class, Hyalospongiae. Subsequent authors renamed the two groups Silicea (or Silicispongia) and Calcarea (or Calcispongia). An obvious objection to this concept had been the profoundly different anatomies and spicule geometrys of Hexactinellida and Demospongiae/ Homoscleromorpha. However, the same process of secretion of the siliceous spicules within sclerocytes around an axial filament occurs in all three groups, and that itself might be viewed as a synapomorphy of the Silicea. In addition, these three groups lack cross-striated rootlets, which are present in Calcarea and other metazoans (as well as in choanoflagellates and some other protists). Thus, *loss* of this structure may be another synapomorphy of the Silicea.

The second hypothesis of class-level relationships in Porifera suggests that the Hexactinellida stand apart from the other sponges by their unique syncytial body plan. This view recognized the two groups Symplasma (or Nuda) for hexactinellids, and Cellularia (or Gelatinosa) for demosponges and calcareous sponges. One of the arguments against this view has been that hexactinellid sponges begin their lives as cellular embryos and partly-cellular larvae, only later transitioning to the syncytial adult body form (by fusion of individual blastomeres), thus suggesting that Hexactinellida may have arisen from a demosponge-like ancestry.

As it turns out, neither of these views of sponge phylogeny is likely fully correct. DNA analyses since 2009 strongly favor of the first hypothesis, but with the Homoscleromorphans removed from the demosponges (Figure 6.27). It now appears that Demosponges and Hexactinellids comprise a sister group, as possibly do Calcarea and Homoscleromorpha. All four classes appear to be monophyletic groups.

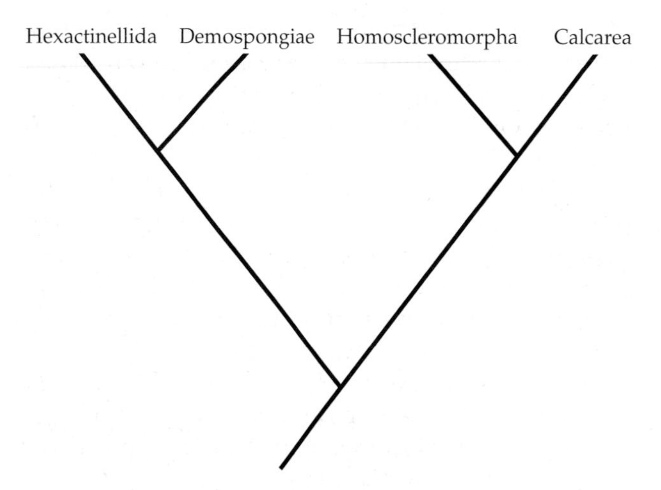

Figure 6.27 A modern view of poriferan phylogeny, based on molecular sequence analyses—18S and 28S rDNA, plus mitochondrial genes. Synapomorphies of the classes discussed in the text include the following: Hexactinellida with unique syncytial body anatomy, triaxonic spicules, and a square axial filament; Homoscleromorpha never with a spongin skeleton; Calcarea with monocrystalline calcareous spicules.

Hexactinellid sponges, with their Precambrian record, might be considered the basal extant sponges. In fact, hexactinellids are the earliest known metazoans that can be related to an extant animal group. Hexactinellids formed enormous reefs in the Jurassic and their fossilized remains crop up all round the world, but they declined in the Cretaceous and have largely disappeared. However, in 1987 the Geological Survey of Canada was engaged in routine surveys of the seafloor off British Columbia. Here, at around 200 m depth in ancient glacially-scoured seafloor troughs, they discovered massive hexactinellid sponge reefs. Since that discovery, teams of scientists have documented the extent and composition of these living sponge reefs, which appear to be about 9,000 years old. The reefs cover roughly 700 km^2, and they are composed of fewer than a dozen hexactinellid species. They are largely identical to the massive hexactinellid-lithistid sponge reefs of Mesozoic times, known to paleontolgists as "reef mounds" or "mud mounds," which culminated in the Late Jurassic when a 7,000 km-long reef existed on the northern shelf of the Tethys Sea—the largest bioconstruction ever built on Earth. Hexactinellid reefs declined after the Cretaceous.

The long fossil record of hexactinellids comprises two principal clades, the Amphidiscophora and the

Hexasterophora, differing in shape and type of their microscleres. Hexactinellid synapomorphies are many, including their unique cellular and syncytial tissue organization, possession of triaxonic spicules, and an axial filament with a square section.

The key synapomorphy of the Calcarea is their monocrystalline calcareous spicules (unique among sponges). DNA analyses also support the monophyly of the two classes of Calcarea—Calcinea and Calcaronea. An important proposed synapomorphy of the Calcinea is the basal position of nuclei in choanocytes, with no relation to the flagella. A proposed synapomorphy of the Calcaronea is the formation of the amphiblastula larva through the original process of eversion of the stomoblastula.

Overall, relationships within the Demospongiae remain uncertain, although there is growing support for the four subclasses used herein (although some researchers would subsume the Haploscleromorpha within the Heteroscleromorpha). For many years, Homoscleromorpha was viewed as the most primitive group of living Porifera (and deeply embedded within the Demospongiae) owing to its seemingly simple anatomic organization. One species, *Oscarella lobularis*, was even suggested as the prototype of all sponges because of its reduced mesohyl (and appearance of having only two layers, choanoderm and pinacoderm) and lack of a mineral skeleton. For a brief period of time, some workers were proposing the Homoscleromorpha be removed from the Porifera altogether, and allied with the Eumetazoa (a grouping dubbed Epitheliozoa). However, just the opposite view now prevails. The recent discovery that Homoscleromorpha are not demosponges, but should stand alone as a monophyletic fourth class of living sponges (probably most closely allied to Calcarea, on the basis of DNA sequence data and the unique cross-striated rootlet in larval ciliated cells in these two groups) was based on anatomical data, as well as 18S and 28S rDNA and complete mitochondrial DNA sequence data. In fact, Homoscleromorphs have a number of distinctly metazoan-like attributes, such as apical cell junctions, flagellated exopinacoderm, a well-developed basement membrane beneath both the pinacoderm and the choanoderm (in larvae and adults), acrosomes in their spermatozoa (also known from a few other sponges), and a basal apparatus of flagellated larval cells with true cross-striated rootlets (as also seen in some Calcarea).

What is one to make of these higher metazoan-like aspects of the Homoscleromorpha, especially given that other features (including their mitochondrial genome) place them squarely within the phylum Porifera? One explanation that has been offered is that these features represent retained ancestral characteristics that have been lost in most other sponges. This would suggest that sponges are actually far more "evolved" than conventional wisdom holds, and that they have undergone a massive simplification since their origin, traces of which can be seen in the Homoscleromorpha. A simpler interpretation might be that the genetic potential for such features as basement membrane, acrosome-bearing spermatozoa, and cross-striated ciliary rootlets has existed since the dawn of the Metazoa, and was simply unevenly expressed in early animal evolution. Certainly other features of sponges, such as the wide variation in embryogenesis and larval forms, support the idea of Porifera being an ancient group that was one of the first to begin experimenting with metazoan body plans.

Selected References

Placozoa

Buchholz, K. and A. Ruthmann. 1995. The mesenchyme-like layer of the fiber cells of *Trichoplax adhaerens* (Placozoa), a syncytium. Zeitschrift fuer Naturforschung, Sec. C Biosciences 50: 282–285.

Da Silva, F. B., V. C. Muschner and S. L. Bonatto. 2007. Phylogenetic position of *Placozoa* based on large subunit (LSU) and small subunit (SSU) rRNA genes. Genet. Mol. Biol. 30: 127–132.

Ender, A. and B. Schierwater. 2003. Placozoa are not derived cnidarians: evidence from molecular morphology. Mol. Biol. Evol. 20(1): 130–134.

Grell, K. 1971a. Embryonalentwicklung bei *Trichoplax adhaerens* F.E. Schulze. Naturwiss. 58: 570.

Grell, K. 1971b. *Trichoplax adhaerens* F.E. Schulze und die Entstehung der Metazoen. Naturwiss. Rundsch. 24: 160–161.

Grell, K. 1972. Eibildung und Furchung von *Trichoplax adhaerens* F.E. Schulze (Placozoa). Z. Morphol. Tiere 73: 297–314.

Grell, K. 1973. *Trichoplax adhaerens* and the origin of the Metazoa. Actualites Protozoologiques. Ive Cong. Int. Protozoologie. Paul Couty, Clermont-Ferrand.

Grell, K. 1982. Placozoa. P. 639 in S. Parker (ed.), *Synopsis and Classification of Living Organisms*. McGraw-Hill, New York.

Grell, K. and G. Benwitz. 1971. Die Ultrastruktur von *Trichoplax adhaerens* F.E. Schulze. Cytobiologie 4: 216–270.

Grell, K. and G. Benwitz. 1974. Elektronenmikroskopische Beobachtungen über das Wachstum der Eizelle und die Bildung der "Befruchtungsmembran" von *Trichoplax adhaerens* F. E. Schulze (Placozoa). Z. Morphol. Tiere 79: 295–310.

Grell, K. and A. Ruthmann. 1991. Placozoa, in F. W. Harrison and W. J. A. (eds.), *Microscopic Anatomy of Invertebrates. Vol. 2. Placozoa, Porifera, Cnidaria and Ctenophora*. Wiley-Liss, New York.

Harrison, F. and J. Wesfall (eds.). 1991. *Microscopic Anatomy of Invertebrates, Vol. 2: Placozoa, Porifera, Cnidaria, and Ctenophora.* Wiley-Liss, New York.

Jackson, A. M. and L. W. Buss. 2009. Shiny spheres of placozoans (*Trichoplax*) function in anti-predator defense. Invert. Biol. 128(3): 205–212.

Pearse, V. B. 1989. Growth and behavior of *Trichoplax adhaerens*: first record of the phylum Placozoa in Hawaii. Pac. Sci. 43: 117–121.

Pearse, V. B. and O. Voight. 2007. Field biology of placozoans (*Trichoplax*): distributoin, diversity, biotic interactions. Integr. Comp. Biol. 47: 677–692

Philippe, H. and 19 others. 2009. Phylogenomics revives traditional views on deep animal relationships. Curr. Biol. 19: 1–7.

Pick, K. S. and 10 others. 2010. Improved phylogenomic taxon sampling noticeably affects nonbilaterian relationships. Mol. Biol. Evol. 27(9): 1983–1987.

Ruthmann, A. 1977. Cell differentiation, DNA content, and chromosomes of *Trichoplax adhaerens* F. E. Schulze. Cytobiologie 15: 58–64.

Ruthmann, A. and U. Terwelp. 1979. Disaggregation and reaggregation of cells of the primitive metazoan *Trichoplax adhaerens*. Differentiation 13: 185–198.

Ruthmann, A. and H. Wenderoth. 1975. Der DNA-Gehalt der Zellen bei dem primitiven Metazoon *Trichoplax adhaerens* F. E. Schulze. Cytobiologie 10: 421–431.

Schuchert, P. 1993. *Trichoplax adhaerens* (phylum Placozoa) has cells that react with the neuropeptide Rf amide. Acta Zoologica 74: 115–117.

Signorovitch, A., S. L. Dellaporta and L. W. Buss. 2006. Caribbean placozoan phylogeography. Biol. Bull. 211: 159–156.

Srivastava, M. and 21 others. 2008. The *Trichoplax* genome and the nature of placozoans. Nature 21: 995–961.

Voight, O., A. G. Collins, V. B. Pearse, J. S. Pearse, A. Endler, H. Hadrys and B. Schierwater. 2004. Placozoa—no longer a phylum of one. Curr. Biol. 14: R944–R945.

Porifera, General References

Antcliffe, J. B. 2013. Questioning the evidence of organic compounds called sponge biomarkers. Palaeontology 56: 917–925.

Becerro, M., M.-J. Uriz, M. Maldonado and X. Turon (eds.). 2012. *Advances in Marine Biology, Vol. 61*. Academic Press.

Bell, J. J., K. D. Simon, J. Timothy, W. M. Taylor and N. W. Webster. 2013. Could some coral reefs become sponge reefs as our climate changes? Global Change Biology. doi: 10.1111/gcb.12212

Bond, C. 1992. Continuous cell movements rearrange anatomical structures in intact sponges. J. Exper. Zool. 263: 284–302.

Boury-Esnault, N., L. deVos, C. Donadey and J. Vacelet. 1990. Ultrastructure of choanosome and sponge classification. Pp. 237–244 in K. Rützler (ed.), *New Perspectives in Sponge Biology*. Smithsonian Institution Press, Washington D.C.

Boury-Esnault, N. and B. G. M. Jamieson. 1999. Porifera. Pp. 1–20 in B. G. M. Jamieson (ed.), *Reproductive Biology of Invertebrates*. Oxford and IBH Publishing, New Delhi, India.

Boury-Esnault, N. and K. Rützler. 1997. Thesaurus of sponge morphology. Smithsonian Contrb. Zoology 596: 1–55. [An essential reference for entry into the world of spongology; also see De Vos et al. 1991.]

Boute, N. and 7 others. 1996. Type IV collagen in sponges, the missing link in basement membrane ubiquity. Biol. Cell. 88: 37–44.

Bowerbank, J. S. 1861, 1862. On the anatomy and physiology of the Spongiidae. Part 1: On the spicula. Philos. Trans. R. Soc. Lond. 148: 279–332. Part 2: Proc. R. Soc. Lond. 11: 372–375. [This and other work by J. S. Bowerbank stand as benchmark papers in nineteenth century poriferology, and for the serious invertebrate zoologist they are worthy of examination.]

Brien, P., C. Lévi, M. Sara, O. Tuzet and J. Vacelet. 1973. Spongiaires. In P. Grassé (ed.), *Traité de Zoologie*. Masson et Cie, Paris.

Brunton, F. R. and O. A. Dixon. 1994. Siliceous sponge-microbe biotic associations and their recurrence through the Phanerozoic as reef mound constructors. Palaois 9: 370–387.

Calcinai, B., C. Cerrano, C. Totti, T. Romagnoli and G. Bavestrello. 2006. Symbiosis of *Mycale* (*Mycale*) *vansoesti* sp. nov. (Porifera, Demospongiae) with a coralline alga from North Sulawesi (Indonesia). Invert. Biol. 125(3): 195–204.

Cárdenas, P., T. Pérez and N. Boury-Esnault. 2012. Sponge systematics facing new challenges. Pp. 80–209 in M. A. Becerro, M. J. Uriz, M. Maldonado and X. Turon (eds.), *Advances in Sponge Science: Physiology, Chemical and Microbial Diversity, Biotechnology*. Elsevier, Vol. 61. [An outstanding review of sponge systematics.]

Chanas, B. and J. R. Pawlik. 1996. Does the skeleton of a sponge provide a defense against predatory reef fish? Oecologia 107: 225–231.

Chaves, Fonnegra, A. and S. Zea. 2007. Observations on reef coral undermining by the Caribbean excavating sponge *Cliona delitrix* (Demospongiae, Hadromerida). Pp. 247–254 in M. R. Custódio, G. Lôbo-Hajdu, E. Hajdu and G. Muricy (eds.), *Porifera Research: Biodiversity, Innovation and Sustainability*. Série Livros 28, Museu Nacional, Rio de Janeiro.

Coutinho, C. C. and G. Maia. 2007. Mesenchymal cells in ancestral spongiomorph urmetazoa could be the mesodermal precursor before gastrulation origin. Pp. 281–295 in *Porifera Research: Biodiversity, Innovation and Sustainability*. Série Livros 28, Museu Nacional, Rio de Janeiro.

Cowart, J. D., T. P. Henkel, S. E. McMurray and J. R. Pawlik. 2006. Sponge orange band (SOB): a pathogenic-like condition of the giant barrel sponge, *Xestospongia muta*. Coral Reefs 25: 513.

Cox, G. and A. W. D. Larkum. 1983. A diatom apparently living in symbiosis with a sponge. Bull. Mar. Sci. 33(4): 943–945.

Custódio M. R., G. Lôbo-Hajdu and G. Hajdu E, Muricy (eds.). 2007. *Porifera Research: Biodiversity, Innovation and Sustainability*. Série Livros 28, Museu Nacional, Rio de Janeiro.

Dayton, P. K., G. A. Robilliard, R. T. Paine and L. B. Dayton. 1974. Biological accommodation in the benthic community at McMurdo Sound, Antarctica. Ecol. Monogr. 44: 105–128.

Dayton, P. K. 1979. Observations of growth, dispersal and population dynamics of some sponges in McMurdo Sound, Antarctica. Pp. 271–282 in C. Lévi and N. Boury-Esnault (eds.), *Biologie des Spongiaires*. Centre Nat. Recherche Scient, Paris.

Degnan, B. M., S. P. Leys and C. Larroux. 2005. Sponge development and antiquity of animal pattern formation. Integr. Comp. Biol. 45: 335–341.

De Vos, L., K. Rützler, N. Boury-Esnault, C. Donadey and J. Vacelet. 1991. *Atlas of Sponge Morphology*. Smithsonian Institution Press, Washington, D.C. [An essential reference for entry into the world of spongology; also see Boury-Esnault and Rützler 1997.]

Diaz, C. and B. B. Ward. 1999. Perspectives on sponge cyanobacterial symbioses. Mem. Queensland Mus. 44: 154.

Elliott, G. R. D. and S. P. Leys. 2007. Coordinated contractions effectively expel water from the aquiferous system of a freshwater sponge. J. Experimental Biol. 210: 3736–3748.

Ellwanger, K. and M. Mickel. 2006. Neuroactive substances specifically modulate rhythmic body contractions in the nerveless metazoan *Tethya wilhelma* (Demospongiae, Porifera). Frontiers Zool. 3: 7.

Ereskovsky, A. V. 2002. Polyaxial cleavage in sponges (Porifera): a new pattern of metazoan cleavage. Dokl. Biol. Sci. 386: 472–474.

Ereskovsky, A. V. 2010. *The Comparative Embryology of Sponges*. Springer, London. [A detailed and comprehensive review of the subject by an outstanding Russian spongologist.]

Ereskovsky, A. V., H. Keupp and R. Kohring (eds.). 1997. *Modern Problems of Poriferan Biology*. Berliner Geowissenschaftliche Abhandlung, Freie Universitat, Berlin.

Fell, P. E. 1995. Deep diapause and the influence of low temperature on the hatching of the gemmules of *Spongilla lacustris* (L.) and *Eunapius fragilis* (Leidy). Invert. Biol. 114(1): 3–8.

Fiore, C. L. and P. Cox Jutte. 2010. Characterization of macrofaunal assemblages associated with sponges and tunicates collected off the southeastern United States. Invert. Biol. 129(2): 105–120.

Fry, W. G. (ed.). 1970. *The Biology of the Porifera*. Academic Press, New York.

Gaino, E., M. Sciscioli, E. Lepore, M. Rebora and G. Corriero. 2006. Association of the sponge *Tethya orphei* (Porifera, Demospongiae) with filamentous cyanobacteria. Invert. Biol. 125(4): 281–287.

Garrone, R. and J. Pottu. 1973. Collagen biosynthesis in sponges: elaboration of spongin by spongocytes. J. Submicrosc. Cytol. 5: 199–218.

Goodwin, T. W. 1968. Pigments of Porifera. Pp. 53–64 in M. Florkin and B. T. Scheer (eds.), *Chemical Zoology, Vol. 2*. Academic Press, New York.

Goude, L., M. Norman and J. Finn. 2013. *Sponges. A Museum Victoria Field Guide*. Museum Victoria, Sydney, Australia. [A guide to the sponges of southern Australia.]

Haeckel, E. 1872. *Die Kalkschwämme, Eine Monographie*. Verlag von Georg Reimer, Berlin.

Haeckel, E. 1874. Die Gastraea-Theorie, die phylogenetische classification des thierreichs und die homologie der Keimblätter. Z. Natwiss. 8: 1–55.

Harrison, F. W. and R. R. Cowden (eds.). 1976. *Aspects of Sponge Biology*. Academic Press, New York.

Harrison, F. W. and L. De Vos. 1990. Porifera. In F. W. Harrison and J. A. Westfall (eds.), *Microscopic Anatomy of Invertebrates, Vol. 2*. Alan R. Liss, New York.

Hill, D. 1972. Archaeocyatha. In R. C. Moore (ed.), *Treatise on Invertebrate Paleontology 1*. Geological Soc. America and Univ. Kansas Press, Lawrence.

Hill, D. and E. C. Strumm. 1956. Tabulata. In R. C. Moore (ed.), *Treatise on Invertebrate Paleontology, F*. Geological Soc. America and Univ. Kansas Press, Lawrence.

Hooper, J. N. A. and R. W. M. Van Soest (eds.). 2002. *Systema Porifera. A Guide to the Classification of Sponges*. Kluwer Academic/Plenum, New York. [The outstanding outcome of six years of collaboration by 45 spongologist around the world. The classification is now beginning to be a bit outdated. Includes keys to subfamilies, genera, and subgenera. Also see the continuously-updated World Porifera Database, part of the World Register of Marine Species, or WoRMS: http://www.marinespecies.org/porifera/]

Hooper, J. N. A., R. W. M. van Soest and A. Pisera. 2011. Phylum Porifera Grant, 1826. In Zhang, Z.-Q. (ed.) Animal biodiversity: an outline of higher –level classification and survey of taxonomic richness. Zootaxa 3148: 13–18.

Hyman, L. H. 1940. *The Invertebrates, Vol. 1, Protozoa through Ctenophora*. McGraw-Hill, New York, pp. 284–364.

Jackson, D. J., L. Macis, J. Reitner, B. M. Degnan and G. Wörheide. 2007. Sponge paleogenomics reveals an ancient role for carbonic anhydrase in skeletogenesis. Science 316: 1893–1895.

Kaesler, R. L. (ed.). 2003. Porifera. In *Treatise on Invertebrate Paleontology, Part E (revised)*. Geological Soc. America and Univ. Kansas, Lawrence.

Kázmierczak, J. and S. Kempe. 1990. Modern cyanobacterial analogs of Paleozoic stromatoporoids. Science 250: 1244–1248.

Lang, J. D., W. D. Hartman and L. S. Land. 1975. Sclerosponges: primary framework constructors on the Jamaican deep forereef. J. Mar. Res. 33: 223–231.

Lecompte, M. 1956. Stromatoporoidea. In R. C. Moore (ed.), *Treatise on Invertebrate Paleontology, F*: F107–F114. Geological Soc. America and Univ. Kansas Press, Lawrence.

Lévi, C. and N. Boury-Esnault (eds.). 1979. *Sponge Biology*. Colloques Internationaux du Centre National de la Recherche Scientifique. Ed. Cen. Nat. Resch. Sci. No. 291.

Leys, S. P. 2003. Comparative study of spiculogenesis in demosponge and hexactinellid larvae. Micr. Res. Tech. 62: 300–311.

Leys, S. P. 2004. Gastrulation in sponges. Pp. 23–31 in C. Stern (ed.), *Gastrulation: From Cells to Embryos*. Cold Spring Harbor Press, Cold Spring Harbor.

Leys, S. P. 2007. Sponge coordination, tissues, and the evolution of gastrulation. Pp. 53–59 in M. R. Custódio, G. Lôbo-Hajdu, E. Hajdu and G. Muricy (eds.) *Porifera Research: Biodiversity, Innovation and Sustainability*. Série Livros 28, Museu Nacional, Rio de Janeiro.

Leys, S. P. and B. M. Degnan. 2001. Cytological basis of photoresponsive behavior in a sponge larva. Biol. Bull. 201: 323.

Leys, S. P. and A. V. Ereskovsky. 2006. Embryogenesis and larval differentiation in sponges. Can. J. Zool. 84(2): 262–287.

Leys, S. P. and G. O. Mackie. 1999. Propagated electrical impulses in a sponge. Mem. Queensland Mus. 4430: 342.

Leys, S. P., G. O. Mackie and H. M. Reiswig. 2007. *The Biology of Glass Sponges*. Advances in Marine Biology, Vol. 52. 145 pp.

Leys, S. P. and R. W. Meech. 2006. Physiology of coordination in sponges. Canadian J. Zool. 84(2): 288–306.

Leys, S. P., D. S. Rokhsar and B. M. Degnan. 2005. Quick guide – sponges. Cur. Biol. 15: R114-R115.

Li, C.-W., J.-Y. Chen and T.-E. Hua. 1998. Precambrian sponges with cellular structures. Science 279: 879–882.

López, S., B. Song, S. E. McMurray and J. R. Pawlik. 2008. Bleaching and stress in coral reef ecosystems: hsp70 expression by the giant barrel sponge *Xestospongia muta*. Mol. Ecol. 17: 1840–1849.

Love, G. D. and numerous others. 2009. Fossil steroids record the appearance of Demospongiae during the Cryogenian Period. Nature 457: 718–721.

Mah, J. L., K. K. Christensen-Dalsgaard and S. P. Leys. 2014. Choanoflagellate and choanocyte collar-flagellar systems and the assumption of homology. Evol. Dev. 16: 25–37.

Maldonado, M. 2004. Choanoflagellates, choanocytes, and animal multicellularity. Invert. Biol. 123: 1–22.

Maldonado, M. 2006. The ecology of the sponge larva. Canad. J. Zool. 84: 175–194.

Maldonado, M., X. Turon, M. A. Becerro and M. J. Uriz (eds.). 2010. Ancient animals, new challenges. Developments in sponge research. Hydrobiologia 687: 1–351

McMurray, S. E., J. E. Blum and J. R. Pawlik. 2008. Redwood of the reef: growth and age of the giant barrel sponge *Xestospongia muta* in the Florida Keys. Mar. Biol. 155: 159–171.

Meech, R. W. 2008. Non-neuronal reflexes: sponges and the origins of behaviour. Current Biol. 18: R70-R72.

Miller, G. 2009. On the origin of the nervous system. Science 325: 24–26.

Müller, W. E. G. (ed.). 2003. Sponges (Porifera). *Marine Molecular Biotechnology*. Springer-Verlag, Berlin.

Müller, W. E. G. and I. M. Müller. 2007. Porifera: an enigmatic taxon disclosed by molecular biology/cell biology. Pp. 89–106 in M. R. Custódio, G. Lôbo-Hajdu, E. Hajdu and G. Muricy (eds.) *Porifera Research: Biodiversity, Innovation and Sustainability*. Série Livros 28, Museu Nacional, Rio de Janeiro.

Nickel, M. 2004. Kinetics and rhythm of body contractions in the sponge *Tethya wilhelma* (Porifera: Demospongiae). J. Exper. Biol. 207: 4515–4524.

Norrevang, A. and K. G. Wingstrand. 1970. On the occurrence and structure of choanocyte-like cells in some echinoderms. Acta Zool. 51: 249–270.

Pansini, M., R. Pronzato, G. Bavestrello and R. Manconi (eds.). 2004. Sponge science in the new millennium. Boll. Mus. 1st Biol. Univ. Genova 68: 301–318.

Pomponi, S. A. 1980. Cytological mechanisms of calcium carbonate excavation by boring sponges. Int. Rev. Cytol. 65: 301–319.

Randall, J. E. and W. D. Hartman. 1968. Sponge-feeding fishes of the West Indies. Mar. Biol. 1: 216–225.

Reitner, J. and H. Keupp (eds.). 1991. *Fossil and Recent Sponges*. Springer-Verlag, Berlin.

Richards, G. S., E. Simionato, M. Perron, M. Adamska, M. Vervoort and B. M. Degnan. 2008. Sponge genes provide new insight into the evolutionary origin of the neurogenic circuit. Current Biology 18: 1156–1161.

Richter, C., M. Wunsch, M. Rasheed, I. Kötter and M. I. Badran. 2001. Endoscopic exploration of Red Sea coral reefs reveals dense populations of cavity-dwelling sponges. Nature 413: 726–730.

Ristau, D. A. 1978. Six new species of shallow-water marine demosponges from California. Proc. Biol. Soc. Wash. 91: 203–216.

Rützler, K. (ed.). 1990. *New Perspectives in Sponge Biology*. Smithsonian Institution Press, Washington, D.C.

Rützler, K. 2004. Sponges on coral reefs: a community shaped by competitive cooperation. Bol. Mus. Inst. Biol. Univ. Genova 68: 85–148.

Saleuddin, A. S. M. and M. B. Fenton (eds.). 2006. Biology of neglected groups: Porifera (Sponges). Canadian J. Zool. 82:2. [13 papers in a special issue]

Schmidt, I. 1970. Phagocytose et pinocytose chez les Spongillidae. Z. Vgl. Physiol. 66: 398-420.

Shore, R. E. 1972. Axial filament of siliceous sponge spicules, its organic components and synthesis. Biol. Bull. 143: 689–698.

Simpson, T. L. 1984. *The Cell Biology of Sponges*. Springer-Verlag, New York.

Turon, X., M. Codina, L. Tarjuelo, M.-J. Uriz and M. A. Becerro. 2000. Mass recruitment of *Ophiothrix fragilis* (Ophiuroidea) on sponges: settlement patterns and post-settlement dynamics. Mar. Ecol. Prog. Ser. 200: 201–212.

Vacelet, J. 2007. Diversity and evolution of deep-sea carnivorous sponges. Pp. 107–115 in M. R. Custódio, G. Lôbo-Hajdu, E. Hajdu and G. Muricy (eds.) *Porifera Research: Biodiversity, Innovation and Sustainability*. Série Livros 28, Museu Nacional, Rio de Janeiro.

Vacelet, J. and N. Boury-Esnault. 1995. Carnivorous sponges. Nature 373: 333–335.

Vacelet, J., N. Boury-Esnault, A. Flala-Medioni and C. R. Fisher. 1998. A methanotrophic carnivorous sponge. Nature 377: 296.

Van de Vyver, G. 1975. Phenomena of cellular recognition in sponges. Pp. 123–140 in A. Moscona and A. Monroy (eds.), *Current Topics in Developmental Biology, Vol. 4*. Academic Press, New York, pp. 123–140.

van Duyl, F. C. and 7 others. 2011. Coral cavity sponges depend on reef-derived food resources: stable isotope and fatty acid constraints. Marine Biology 158 (7): 1653–1666.

Van Soest, R. W. M. and 9 others. 2012. Global diversity of sponges (Porifera). PLoSONE 7 (4): e35105, 23 pp.

Van Soest, R. W. M., T. M. G. van Kempen and J. C. Braekman (eds.). 1994. *Sponges in Time and Space: Biology, Chemistry, Paleontology*. Balkema, Rotterdam.

Wiens, M. et al. 2003. The molecular basis for the evolution of the metazoan body plan: extracellular matrix-mediated morphogenesis in marine demosponges. J. Mol. Evol. 57: S60-S75.

Wilkinson, C. R. 1983. Net primary productivity in coral reef sponges. Science 219: 410–412.

Wilson, H. V. 1891. Notes on the development of some sponges. J. Morphol. 5: 511–519.

Wolfrath, B. and D. Barthel. 1989. Production of fecal pellets by the marine sponge *Halichondria panicea* Pallas, 1766. J. Exp. Mar. Biol. Ecol. 129: 81–94.

Wood, R. 1990. Reef-building sponges. Am. Sci. 78: 224–235.

Yin, Z., M. Zhu, E. H. Davidson, D. J. Bottjer, F. Zhao and P. Tafforeau. 2015. Sponge grade body fossil with cellular resolution dating 60 Myr before the Cambrian. PNAS PLUS, doi/10.1073/PNAS.1414577112

Porifera, Calcarea

Amano, S. and I. Hori. 2001. Metamorphosis of coeloblastula performed by multipotential larval flagellated cells in the calcareous sponge *Leucosolenia laxa*. Biol. Bull. 190: 161–172.

Borojevic, R., N. Boury-Esnault and J. Vacelet. 1990. A revision of the supraspecific classification of the subclass Calcinea (Porifera, Class Calcarea). Bull. Mus. Natn. Hist. Nat. 12: 243–276.

Eerkes-Medrano, D. I. and S. P. Leys. 2006. Ultrastructure and embryonic development of a syconoid calcareous sponge. Invert. Biol. 125(3): 177–194.

Ereskovsky A.V. and P. Willenz. 2008. Larval development in *Guancha arnesenae* (Porifera, Calcispongiae, Calcinea). Zoomorphology 127: 175–187.

Ledger, P. W. and W. C. Jones. 1978. Spicule formation in the calcareous sponge *Sycon ciliatum*. Cell Tiss. Res. 181: 553–567.

Leys, S. P. and D. Eerkes-Medrano. 2005. Gastrulation in calcareous sponges: in search of Haeckel's Gastraea. Integr. Comp. Biol. 45: 342–351.

Leys, S. P. and D. I Eerkes-Medrano. 2006. Feeding in a calcareous sponge: particle uptake by pseudopodia. Biol. Bull. 211: 157–171.

Porifera, Demospongiae

Amano, S. and I. Hori. 1996. Transdifferentiation of larval flagellated cells to choanocytes in the metamorphosis of the demosponge *Haliclona permollis*. Biol. Bull. 190: 161–172.

Ayling, A. L. 1983. Growth and regeneration rates in thinly encrusting Demospongiae from temperate waters. Biol. Bull. 165: 343–352.

Bautista-Guerrero, E., J. L. Carballo and M. Maldonado. 2010. Reproductive cycle of the coral-excavating sponge *Thoosa mismalolli* (Clionaidae) from Mexican Pacific coral reefs. Invertebrate Biology 129(4): 285–296.

Bonasoro, F, I. C. Wilkie, G. Bavestrello, C. Cerrano and M. D. C. Carnevali. 2001. Dynamic structure of the mesohyl in the sponge *Chondrosia reniformis* (Porifera: Demospongiae). Zoomorphology 121: 109–121.

Boury-Esnault, N. 2006. Systematics and evolution of Demospongiae. Can. J. Zool. 84: 205–224.

Bryan, P. G. 1973. Growth rate, toxicity and distribution of the encrusting sponge *Terpios* sp. (Hadromerida: Suberitidae) in Guam, Mariana Islands. Micronesica 9: 237–242.

Carballo, J. L., J. E. Sanchez-Moyano and J. C. Garcia-Gomez. 1994. Taxonomic and ecological remarks on boring sponges (Clionidae) from the Strait of Gibraltar: tentative bioindicators? Zool. J. Linn. Soc. 112: 407–424.

Carballo, J. L., J. A. Cruz-Barraza and P. Gómez. 2004. Taxonomy and description of clionaid sponges (Hadromerida, Clionaidae) from the Pacific Ocean of Mexico. Zoological Journal Linnean Society 141: 353–397.

Elvin, D. W. 1976. Seasonal growth and reproduction of an intertidal sponge, *Haliclona permollis* (Bowerbank). Biol. Bull. 151: 108–125.

Ereskovsky, A. V. 1999. Development of sponges of the order Haplosclerida. Russian J. Mar. Biol. 25(5): 36–371.

Ereskovsky, A. V. 2007. Sponge embryology: the past, the present, and the future. Pp. 41–52 in M. R. Custódio, G. Lôbo-Hajdu, E. Hajdu and G. Muricy (eds.) *Porifera Research: Biodiversity, Innovation and Sustainability*. Série Livros 28, Museu Nacional, Rio de Janeiro.

Ereskovsky A.V. and E. L. Gonobobleva. 2000. New data on embryonic development of *Halisarca dujardini* Johnston, 1842 (Demospongiae: Halisarcida). Zoosystema 22: 355–368.

Fell, P. E. 1969. The involvement of nurse cells in oogenesis and embryonic development in the marine sponge *Haliclona ecobasis*. J. Morphol. 127: 133–149.
Frost, T. M. 1991. Porifera. Pp. 95–124 in J. H. Thorp and A. P. Covich (eds.), *Ecology and Classification of North American Freshwater Invertebrates*. Academic Press, New York.
Frost, T. M. and C. E. Williamson. 1980. In situ determination of the effect of symbiotic algae on the growth of the freshwater sponge *Spongia lacustris*. Ecology 61: 1361–1370.
Gerrodette, T. and A. O. Fleschig. 1979. Sediment-induced reduction in the pumping rate of the tropical sponge *Verongia lacunosa*. Mar. Biol. 55: 103–110.
Gregson, P. R., B. A. Baldo, P. G. Thomas, R. J. Quinn, P. R. Bergquist, J. F. Stephens and A. R. Horne. 1979. Fluorine is a major constituent of the marine sponge *Halichondria moorei*. Science 206: 1108–1109.
Hatch, W. I. 1980. The implication of carbonic anhydrase in the physiological mechanism of penetration of carbonate substrata by the marine burrowing sponge *Cliona celata* (Demospongiae). Biol. Bull. 159: 135–147.
Leys, S. P. and H. M. Reiswig. 1998. Nutrient transport pathways in the neotropical sponge *Aplysina*. Biol. Bull. 195: 30–42.
Leys, S. P. and B. M. Degnan. 2002. Embryogensis and metamorphosis in a haplosclerid demosponge: gastrulation and differentiation of larval ciliated cells to choanocytes. Invert. Biol. 121: 171–189.
Maldonado, M., M. Durfort, D. A. McCarthy and C. M. Younge. 2003. The cellular basis of photobehavior in the tufted parenchymella larva of demosponges. Mar. Biol. 143: 427–441.
Manconi, R. and R. Pronzato. 2007. Gemmules as a key structure for the adaptive radiation of freshwater sponges: a morphofunctional and biogeographical study. Pp. 61–77 in M. R. Custódio, G. Lôbo-Hajdu, E. Hajdu and G. Muricy (eds.) *Porifera Research: Biodiversity, Innovation and Sustainability*. Série Livros 28, Museu Nacional, Rio de Janeiro.
Morrow, C. and P. Cárdenas. 2015. Proposal for a revised classification of the Demospongiae (Porifera). Front. Zool. doi: 10.1186/s12983-015-0099-8
Müller, W. E. G., M. Rothenberger, A. Bordiko, W. Tremel, A. Reiber and H. C. Schröder. 2005. Formation of siliceous spicules in the marine demosponge *Suberites domuncula*. Cell Tissue Res. 321: 285–297.
Pond, D. 1992. Protective-commensal mutualism between the queen scallop *Chlamys opercularis* (Linnaeus) and the encrusting sponge *Suberites*. J. Moll. Studies 58: 127–134.
Rasmont, R. 1962. The physiology of gemmulation in fresh water sponges. Pp. 1–25 in D. Rudnick (ed.), *Regeneration, 20th Growth Symposium*. Ronald Press, New York.
Reiswig, H. M. 1970. Porifera: sudden sperm release by tropical Demospongiae. Science 170: 538–539.
Reiswig, H. M. 1971. Particle feeding in natural populations of three marine Demosponges. Biol. Bull. 141: 568–591.
Reiswig, H. M. 1974. Water transport, respiration, and energetics of three tropical sponges. J. Exp. Mar. Biol. Ecol. 14: 231–249.
Reiswig, H. M. 1975. The aquiferous systems of three marine Demospongiae. J. Morphol. 145(4): 493–502.
Rützler, K. 1975. The role of burrowing sponges in bioerosion. Oecologia 19: 203–216.
Rützler, K. and G. Reiger. 1973. Sponge burrowing: fine structure of *Cliona lampa* penetrating calcareous substrata. Mar. Biol. 21: 144–162.
Simpson, T. L. and J. J. Gilbert. 1973. Gemmulation, gemmule hatching and sexual reproduction in freshwater sponges. I. The life cycle of *Spongilla lacustris* and *Tubella pennsylvanica*. Trans. Amer. Microsc. Soc. 92: 422–433.
Woollacott, R. M. and M. G. Hadfield. 1996. Induction of metamorphosis in larvae of a sponge. Invert. Biol. 115(4): 257–262.

Porifera, Hexactinellida

Aizenberg, J., J. C. Weaver, M. S. Thanawala, V. C. Sundar, D. E. Morse and P. Fratzl. 2005. Skeleton of *Euplectella* sp.: structural hierarchy from the nanoscale to the macroscale. Science 309: 275–278.
Boury-Esnault, N., S. Efremova, C. Bézac and J. Vacelet. 1999. Reproduction of a hexactinellid sponge: first description of gastrulation by cellular delamination in the Porifera. Invert. Reprod. Develop. 35: 187–201.
Conway, K. W., M. Drautter, J. V. Barrie and M. Neuweiler. 2001. Hexactinellid sponge reefs on the Canadian continental shelf: a unique "living fossil." Geosci. Can. 28(2): 71–78.
Leys, S. P. 1995. Cytoskeletal architecture and organelle transport in giant syncytia formed by fusion of hexactinellid sponge tissues. Biol. Bull. 188: 241–254.
Leys, S. P. 1999. The choanosome of hexactinellid sponges. Invert. Biol. 118: 221–235.
Leys, S. P. 2003. The significance of syncytial tissues for the position of the Hexactinellida in the Metazoa. Integr. Comp. Biol. 43: 19–27.
Leys, S. P., G. O. Mackie and R. W. Meech. 1999. Impulse conduction in a sponge. J. Exp. Biol. 202: 1139–1150.
Leys, S. P., G. O. Mackie and H. M. Reiswig. 2007. The biology of glass sponges. Adv. Mar. Biol. 52: 1–145. [An outstanding review of these unique animals.]
Reiswig, H. M. 1979. Histology of Hexactinellida (Porifera). Pp. 173–180 in C. Lévi and N. Boury-Esnault (eds.), *Sponge Biology*. Colloques Internat. C.N.R.S. 291.
Rosengarten, R. D., E. A. Sperling, M. A. Moreno, S. P. Leys and S. L. Dellaporta. 2008. The mitochondrial genome of the hexactinellid sponge *Aphrocallistes vastus*: evidence for programmed translational frameshifting. BMC Genomics 9: 33–42.
Yahel, G., F. Whitney, H. M Reiswig, D. I. Eerkes-Medrano and S. P. Leys. 2007. In situ feeding and metabolism of glass sponges (Hexactinellida, Porifera) studied in a deep temperate fjord with a remotely operated submersible. Limnol. Oceanogr. 52(1): 428–440.

Porifera, Homoscleromorpha

Boury-Esnault, N., A. Ereskovsky, C. Bézac and D. Tokina. 2003. Larval development in the Homoscleromorpha (Porifera, Demospongiae). Invert. Biol. 122: 187–202.
Boury-Esnault, N., D. V. Lavrov, C. A. Ruiz and T. Pérez. 2013 The integrative taxonomic approach applied to Porifera: a case study of the Homoscleromorpha. Integr. Comp. Biol. 53: 416–427
Ereskovsky, A.V. and D. B. Tokina. 2007. Asexual reproduction in homoscleromorph sponges (Porifera: Homoscleromorpha). Mar. Biol. 151: 425–434.
Ereskovsky, A.V., D. B. Tokina, C. Bézac and N. Boury-Esnault. 2007. Metamorphosis of cinctoblastula larvae (Homoscleromorpha, Porifera). J. Morphol. 268: 518–528.
Ivanišević, J., O. P. Thomas, C. Lejeusne, P. Chevaldonné and T. Pérez. 2011. Metabolic fingerprinting as an indicator of biodiversity: towards understanding interspecific relationships among Homoscleromorpha sponges. Metabolomics 7: 289–304.
Maldonado, M. and A. Riesgo. 2007. Intra-epithelial spicules in a homosclerophorid sponge. Cell Tissue Res. 328(3): 639–650.

Porifera, Phylogeny and Evolution

Baccetti, B., E. Gaino and M. Sará. 1986. A sponge with an acrosome: *Oscarella lobularis*. J. Ultrastruct. Mol. Struct. Res. 94: 195–198.

Borchiellini, C. M., C. Chombard, M. Manuel, E. Alivon, J. Vacelet and N. Boury-Esnault. 2004. Molecular orogeny of Demospongiae: implications for classification and scenarios of character evolution. Mol. Phylogenet. Evol. 32: 823–837.

Borchiellini, C. M., E. Manuel, N. Alivon, N. Boury-Esnault, J. Vacelet and Y. Le Parco. 2001. Sponge paraphyly and the origin of Metazoa. J. Evol. Biol. 14: 171–179.

Cavalier-Smith, T., E. E. Chao, N. Boury-Esnault and J. Vacelet. 1996. Sponge phylogeny, animal monophyly, and the origin of the nervous system: 18SrRNA evidence. Can. J. Zool.-Rev. Can. Zool. 74: 2031–2045.

Chombard, C. N. Boury-Esnault, S. Tillier and J. Vacelet. 1997. Polyphyly of "sclerosponges" (Porifera, Demospongiae) supported by 28S ribosomal sequences. Biol. Bull. 193: 359–367.

Chombard, C. N. Boury-Esnault and S. Tillier. 1998. Reassessment of homology of morphological characters in tetractinellid sponges based on molecular data. Syst. Biol. 47: 351–366.

Coutinho, C. C., R. N. Fonseca, J. J. C. Mansure and R. Borojevic. 2003. Early steps in the evolution of multicellularity: deep structural and functional homologies among homeobox genes in sponges and higher metazoans. Mech. Dev. 120: 429–440.

Degnan, B. M., S. P. Leys and C. Larroux. 2005. Sponge development and antiquity of animal pattern formation. Integr. Comp. Biol. 45: 335–341.

Dohrmann, M., D. Janussen, J. Reitner, A. G. Collins and G. Wörheide. 2008. Phylogeny and evolution of glass sponges (Porifera, Hexactinellida). Systematic Biol. 57: 388–405.

Ereskovsky, A. V. 2004. Comparative embryology of sponges and its application for poriferan phylogeny. Boll. Mus. Inst. Biol., Univ. Genova 68: 301–318.

Erpenbeck, D. and 7 others. 2012. Horny sponges and their affairs: on the phylogenetic relationships of keratose sponges. Mol. Phylogenet. Evol. 63: 809–816.

Fortunato, S. A. V., M. Adamski, O. Mendivil Ramos, S. Leininger, J. Liu, D. E. K. Ferrier and M. Adamska. 2014. Calcisponges have a ParaHox gene and dynamic expression of dispersed NK homeobox genes. Nature 514: 620–623.

Gazave, E., P. Lapébie, A. V. Ereskovsky, J. Vacelet, E. Renard, P. Cárdenas and C. Borchiellini. 2012. No longer Demospongiae: Homoscleromorpha formal nomination as a fourth class of Porifera. Hydrobiologia 68: 3–10.

Gazave, E. and 7 others. 2010. Molecular phylogeny restores the supra-generic subdivision of homoscleromorph sponges (Porifera, Homoscleromorpha). PLoS ONE 5(12): 1–15.

King, N. and S. B. Carroll. 2001. A receptor tyrosine kinase from choanoflagellates: molecular insights into early animal evolution. Proc. Natl. Acad. Sci. 98: 15032–15037.

Klautau, M., F. Azevedo, B. Cóndor-Luján, H. T. Rapp, A. Collins and C. A. M. Russo. 2013. A molecular phylogeny for the order Clathrinida rekindles and refines Haeckel's taxonomic proposal for calcareous sponges. Int. Comp. Biol. 53: 447–461

Larroux, C. and 9 others. 2006. Developmental expression of transcription factor genes in a demosponge: insights into the origin of metazoan multicellularity. Evol. Dev. 8(2): 150–173.

Lavrov, D. V. and B. F. Lang. 2005. Poriferan mtDNA and animal phylogeny based on mitochondrial gene arrangements. Systematic Biol. 54(4) 651–659.

Lavrov, D.V., W. Pett, O. Voight, G. Wörheide, L. Forget, B. F. Lang and E. Kayal. 2013. Mitochondrial DNA of *Clathrina clathrus* (Calcarea, Calcinea): six linear chromosomes, fragmented rRNAs, tRNA editing, and a novel genetic code. Mol. Biol. Evol. 30: 865–880.

Mah, J. L., K. K. Christensen-Dalsgaard and S. P. Leys. 2014. Choanoflagellate and choanocyte collar-flagellar systems and the assumption of homology. Evol. Develop. 16(1): 25–37.

Maldonado, M. 2004. Choanoflagellates, choanocytes, and animal multicellularity. Invert. Biol. 123: 1–22.

Manuel, M. 2006. Phylogeny and evolution of calcareous sponges. Can. J. Zool. 84: 225–241.

Manuel, M., C. Borchiellini, E. Alivon, Y. Le Parco, J. Vacelet and N. Boury-Esnault. 2003. Phylogeny and evolution of calcareous sponges: monophyly of Calcinea and Calcaronea, high levels of morphological homoplasy, and the primitive nature of axial symmetry. Syst. Biol. 52: 311–333.

Mehl, D. and H. M. Reiswig. 1991. The presence of flagellar vanes in choanomeres of Porifera and their possible phylogenetic implications. Z. Zool. Syst. Evolut.-forsch. 29: 312–319.

Mendivil Ramos, O., D. Barker and D. E K. Ferrier. 2012. Ghost loci imply Hox and ParaHox existence in the last common ancestor of animals. Curr. Biol. 22: 1951–1956.

Müller, W. E. G. 2001. How was the metazoan threshold crossed? The hypothetical Urmetazoa. Comp. Biochem. Physiol. A 129: 433–460.

Nosenko, T. and 12 others. 2013. Deep metazoan phylogeny: when different genes tell different stories. Mol. Phyl. Evol. 67: 223–233

Philippe, H., and 19 others. 2009. Phylogenomics revives traditional views on deep animal relationships. Curr. Biol. 19: 1–7

Redmond, N. E. and 8 others. 2011. Phylogenetic relationships of the marine Haplosclerida (phylum Porifera) employing ribosomal (28S rRNA) and mitochondrial (cox1, nad1) gene sequence data. PLoSONE 6 (9): e24344. doi: 10.1371/journal.pone.0024344

Redmond, N. E. and 11 others. 2013. Phylogeny and systematics of Demospongiae in light of new small-subunit ribosomal DNA (18S) sequences. Integrat. Comp. Biol. 53: 388–415

Thacker, R. W. and A. Collins (eds.). 2013. Assembling the poriferan tree of life. Int. Comp. Biol. 53: 373–530

Wang, X. and D. V. Lavrov. 2007. Mitochondrial genome of the homoscleromorph *Oscarella carmela* (Porifera, Demospongiae) reveals unexpected complexity in the common ancestor of sponges and other animals. Mol. Biol. Evol. 24(2): 363–373.

Whelan, N. V., K. M. Kocot, L. L. Moroz and K. M. Halanych. 2015. Error, signal, and the placement of Ctenophora sister to all other animals. PNAS 112(18): 5773–5778.

Woollacott, R. M. and R. L. Pinto. 1995. Flagellar basal apparatus and its utility in phylogenetic analyses of the Porifera. J. Morphol. 226: 247–265.

Wörheide, G. and 7 others. 2012. Deep phylogeny and evolution of sponges (phylum Porifera). Adv. Mar. Biol. 61: 2–78

Ziegler, B. and S. Rietschel. 1970. Phylogenetic relationships of fossil calcisponges. Symp. Zool. Soc. Lond. 25: 23–40.

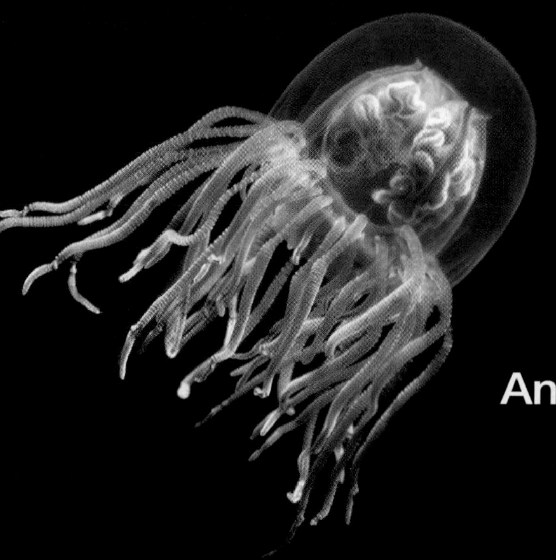

CHAPTER 7

Phylum Cnidaria

Anemones, Corals, Jellyfish, and Their Kin

The phylum Cnidaria is a highly diverse assemblage that includes jellyfish, sea anemones, corals, and the common laboratory *Hydra*, as well as many less familiar forms such as hydroids, sea fans, siphonophores, zoanthids, and myxozoans (Figure 7.1). There are about 13,400 living described species of cnidarians. Much of the striking diversity seen in this phylum results from three fundamental aspects of their natural history. First is the possession of **cnidae**—unique, tubular structures contained within cellular capsules that aid in prey capture, defense, locomotion, and attachment. No other animal group produces cnidae, and all cnidarians produce them. Second is the tendency to form colonies or assemblages of individuals by asexual reproduction; the colony can achieve dimensions and forms unattainable by single, nonmodular individuals. Third, many species of cnidarians exhibit a dimorphic life cycle that can include two entirely different morphologies: a **polypoid** form and an adult **medusoid** form. The dimorphic life cycle has major evolutionary implications touching on nearly every aspect of cnidarian biology.[1] The unique polyp form of Cnidaria, their **planula** larvae, and stinging or adhesive cnidae are three of the key synapomorphies that define the phylum.

Cnidarians are widely thought to be **diploblastic** metazoans at the tissue grade of construction. They possess radial or biradial symmetry, tentacles, cnidae, an endodermally derived incomplete **gastrovascular cavity** as their only "body cavity," and a middle layer (called mesenchyme, or mesoglea) derived primarily from ectoderm.[2] They lack cephalization, a centralized nervous system, and discrete respiratory, circulatory, and excretory organs (Box 7A). This basic body plan is retained in both the polypoid and medusoid forms (Figure 7.2). The primitive nature of the cnidarian body plan is exemplified by the fact that they have fewer cell types than any other animals except the sponges and mesozoans.

Classification of The Animal Kingdom (Metazoa)

Non-Bilateria*
(a.k.a. the diploblasts)
- PHYLUM PORIFERA
- PHYLUM PLACOZOA
- **PHYLUM CNIDARIA**
- PHYLUM CTENOPHORA

Bilateria
(a.k.a. the triploblasts)
- PHYLUM XENACOELOMORPHA

Protostomia
- PHYLUM CHAETOGNATHA

SPIRALIA
- PHYLUM PLATYHELMINTHES
- PHYLUM GASTROTRICHA
- PHYLUM RHOMBOZOA
- PHYLUM ORTHONECTIDA
- PHYLUM NEMERTEA
- PHYLUM MOLLUSCA
- PHYLUM ANNELIDA
- PHYLUM ENTOPROCTA
- PHYLUM CYCLIOPHORA

Gnathifera
- PHYLUM GNATHOSTOMULIDA
- PHYLUM MICROGNATHOZOA
- PHYLUM ROTIFERA

Lophophorata
- PHYLUM PHORONIDA
- PHYLUM BRYOZOA
- PHYLUM BRACHIOPODA

ECDYSOZOA

Nematoida
- PHYLUM NEMATODA
- PHYLUM NEMATOMORPHA

Scalidophora
- PHYLUM KINORHYNCHA
- PHYLUM PRIAPULA
- PHYLUM LORICIFERA

Panarthropoda
- PHYLUM TARDIGRADA
- PHYLUM ONYCHOPHORA
- PHYLUM ARTHROPODA
  - SUBPHYLUM CRUSTACEA*
  - SUBPHYLUM HEXAPODA
  - SUBPHYLUM MYRIAPODA
  - SUBPHYLUM CHELICERATA

Deuterostomia
- PHYLUM ECHINODERMATA
- PHYLUM HEMICHORDATA
- PHYLUM CHORDATA

*Paraphyletic group

---

[1]When both phases are present in a species' life cycle, it is said to undergo an **alternation of generations**, or as it is sometimes called, "**metagenesis**."

[2]There exists a suite of terms in zoological literature that apply to apparently diploblastic species that is frequently confused, misused, and generally messy. These terms include mesenchyme, mesoglea, collenchyme, parenchyme, and coenenchyme. In this book, these terms are used in the following ways. **Mesenchyme** (Greek, literally "middle juices") refers to a primitive connective tissue derived wholly or in part from ectoderm and located between the epidermis and the gastrodermis (endodermis). Mesenchyme generally consists of two components: a noncellular, jelly-like matrix called **mesoglea**, and various cells and cell products (e.g., fibers). When no (or very little) cellular material is present, this layer is simply called mesoglea. Mesenchyme is the typical middle layer of sponges (where it is called the **mesohyl**), and of members of the phyla Cnidaria and Ctenophora. In these diploblastic groups, where no true (endo-) mesoderm exists, the mesenchyme is fully ectodermally derived. When cellular material is sparse or densely packed, mesenchyme may sometimes be designated as **collenchyme** or **parenchyme**, respectively. The term parenchyme is also sometimes used for the mesenchymal layer of triploblastic acoelomate animals (such as platyhelminths and xenacoelomorphans), in which the dense layer includes tissues derived from both ecto- and endomesoderm.

In some colonial cnidarians, particularly anthozoan polyps, the individuals are embedded in and arise from a mass of mesenchyme perforated with gastrovascular channels that are continuous among the members of the colony. The term **coenenchyme** refers to this entire matrix of common basal material, which is itself covered by a layer of epidermis.

Adding to the potential confusion, the term mesenchyme is used in a second, very different way by some biologists. Vertebrate embryologists sometimes use the term to refer to that part of true (endo-) mesoderm from which all connective tissues, blood vessels, blood cells, the lymphatic system, and the heart are derived. Thus, to a vertebrate embryologist, the term "mesenchymal cell" often denotes any undifferentiated cell found in the embryonic mesoderm that is capable of differentiating into such tissues. Occasionally, one also sees the term "mesenchyme" used in this way by echinoderm embryologists and non-vertebrate chordate specialists. In these latter cases, researchers are generally referring to cells that are destined to become mesodermal and thus a better term might be "proto-" or "incipient mesodermal cells." Because of this confusion, some authors prefer to use the term mesoglea in lieu of mesenchyme when referring to the middle layers of sponges and diploblastic Metazoa. However, we adhere to the former definition of mesenchyme and hope that this note will lessen rather than add to the muddle.

A word of caution regarding spelling: the meanings of some of these terms can be altered by changing the terminal "e" to an "a." The termination "-chyme" is preferred for animals, "-chyma" for plants. Thus, mesenchyma refers to tissue lying between the xylem and phloem in plant roots; collenchyma refers to certain primordial leaf tissues. Parenchyma is a very general botanical term used in reference to various supportive tissues. Unfortunately, the same spelling is occasionally (improperly) used by zoologists.

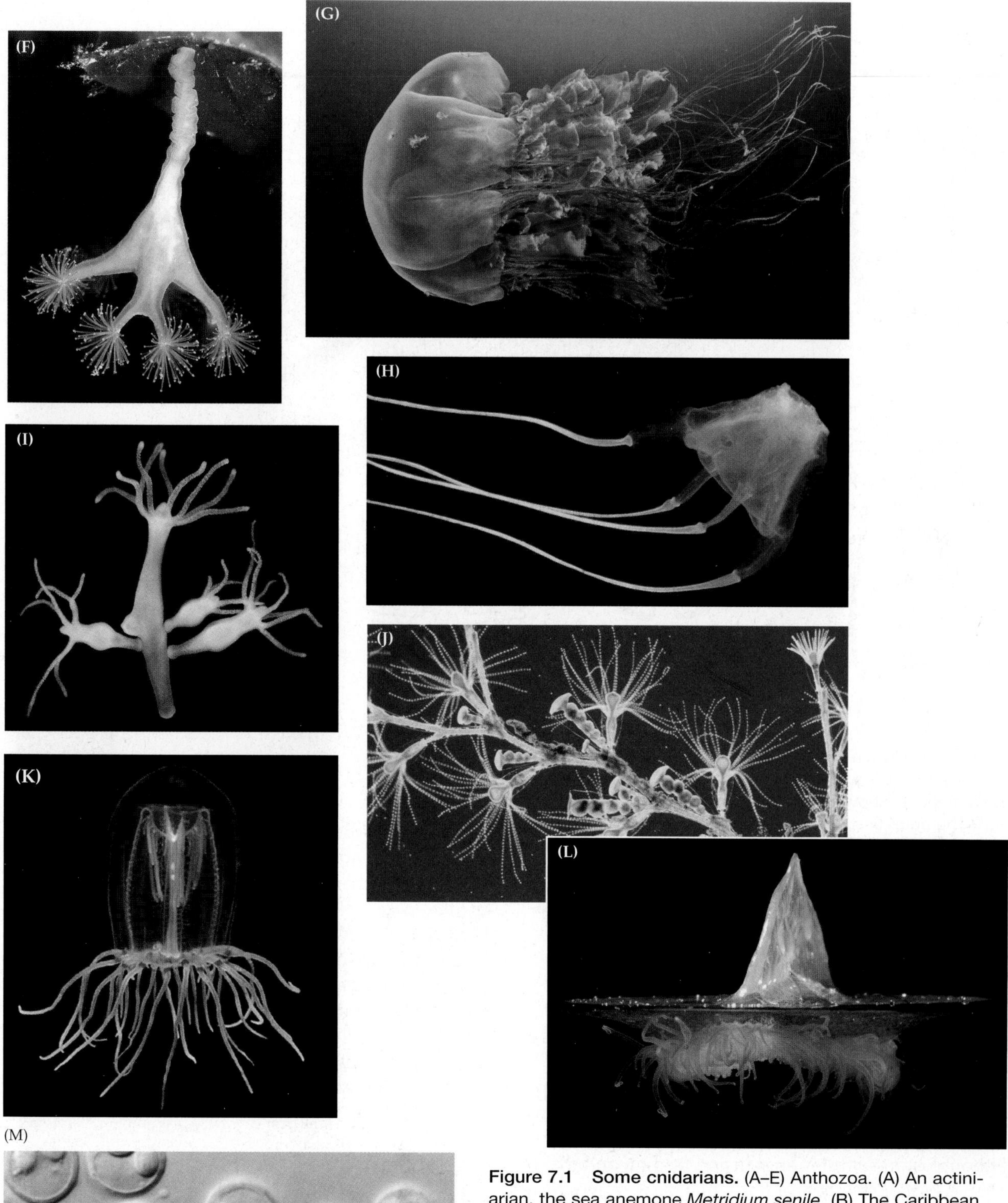

Figure 7.1 Some cnidarians. (A–E) Anthozoa. (A) An actiniarian, the sea anemone *Metridium senile*. (B) The Caribbean elkhorn coral, *Acropora palmata*. (C) The sea pen, *Ptilosarcus gurneyi* (Pennatulacea). (D) The yellow gorgonian *Eunicella cavolini* (Mediterranean Sea, Italy). (E) *Renilla*, the sea pansy (Pennatulacea). (F–M) Medusozoa. (F) *Lucernaria*, a staurozoan (White Sea, Russia). (G) A lion's mane jellyfish, *Cyanea capillata*, a semaeostoman scyphozoan from the northeast Pacific Ocean. (H) *Carybdea marsupialis*, a cubozoan; (I–L) Hydrozoa. (I) *Hydra*, an aberrant freshwater anthomedusan (shown here budding). (J) A colony of the leptomedusan *Gonothyrea*. (K) The medusa of *Polyorchis*, an anthomedusan. (L) A colonial hydrozoan, *Velella* ("by-the-wind sailor"). (M) *Myxobolus cerebralis*, myxozoan spores from an infected trout.

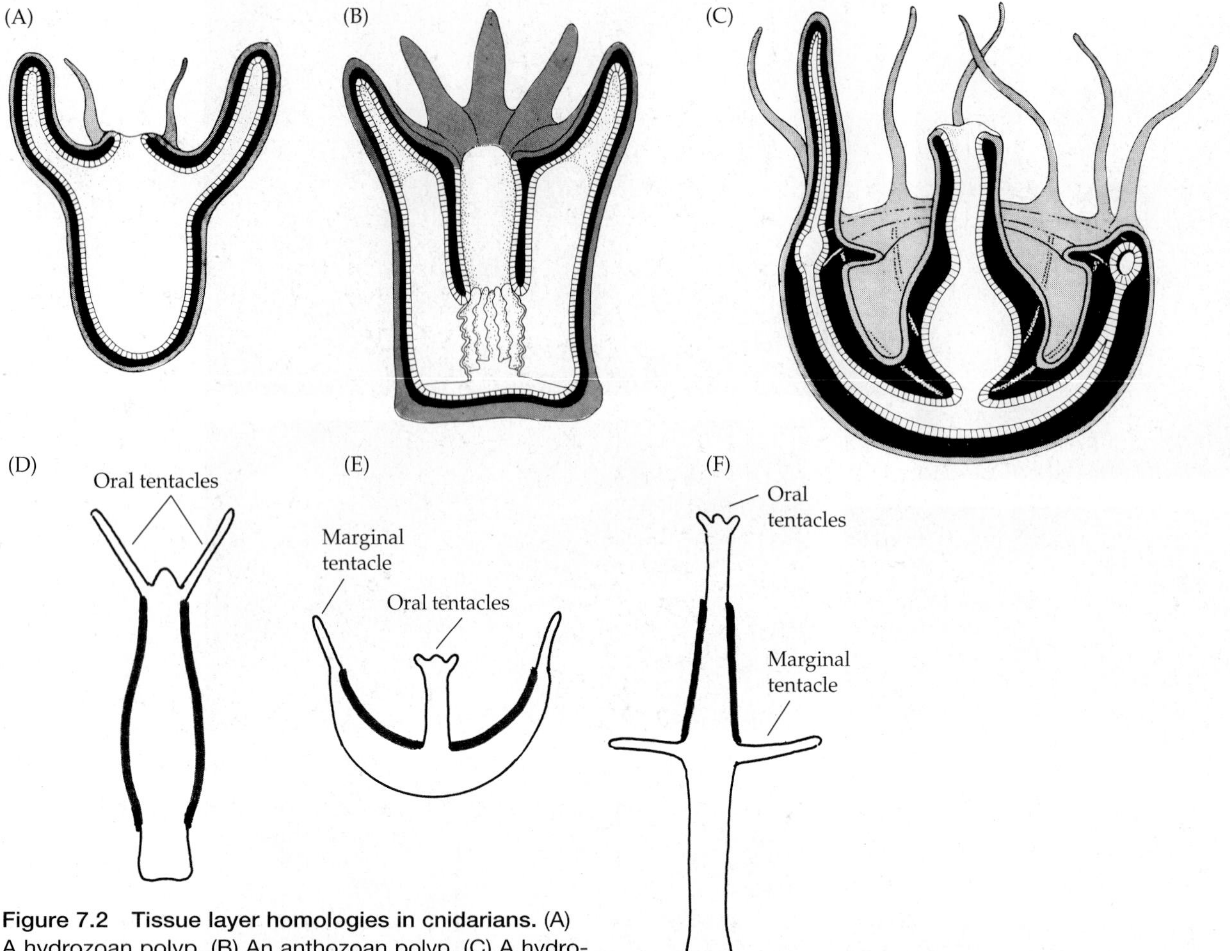

Figure 7.2 Tissue layer homologies in cnidarians. (A) A hydrozoan polyp. (B) An anthozoan polyp. (C) A hydrozoan medusa, shown "upside down" for similar orientation. The outer tissue layer is ectodermal (= epidermis); the inner tissue layer is endodermal (= gastrodermis); and the middle layer is the mesenchyme/mesoglea. Within the Hydrozoa, ectodermal tissue homologues may lie on the outer body wall as in *Hydra* (D) or within the marginal tentacles of hydromedusa and hydropolyps (E,F). OT = oral tentacles, MT = marginal tentacles.

In fact, cnidarians contain fewer cell types than does a single organ in some other Metazoa.

Cnidarians are mostly marine; only a few groups have successfully invaded fresh waters. Most cnidarians are sessile (polyps) or planktonic (medusae) carnivores, although some employ suspension feeding and many species harbor symbiotic intracellular algae from which they may secondarily derive energy. Cnidarians range in size from nearly microscopic polyps and medusae to individual jellyfish 2 m wide and with tentacles 25 m long. Colonies, such as the corals, may be many meters across. The phylum dates from the Precambrian, and its members have played important roles in various ecological settings throughout their long history, just as they do on modern coral reefs today.

Cnidarians play numerous roles in folklore around the world. In Samoa, the corallimorpharian anemone *Rhodactis howesii*, known as *mata-malu*, is served boiled as a festive holiday dish. Eaten raw, however, *mata malu* causes death and is a traditional device in Samoan suicide. Hawaiians refer to the zoanthid *Palythoa toxica* as *limu-make-o-Hana* ("the deadly seaweed of Hana"). Hawaiians used to smear their spear tips with this cnidarian, the toxin from which is called **palytoxin**. Interestingly, palytoxin may be produced by an unidentified symbiotic bacterium, not by the cnidarian itself. It is one of the most powerful toxins known, being more deadly than that of poison arrow frogs (batrachotoxin) and paralytic shellfish toxin (saxitoxin).

Taxonomic History and Classification

As is the case with sponges, the nature of cnidarians was long debated. In reference to their stinging tentacles, Aristotle called the medusae *Acalephae* (*akalephe*) and the polyps *Cnidae* (*knide*), both names derived from terms meaning "nettle." Renaissance scholars considered them plants, and it was not until the eighteenth century that the animal nature of cnidarians was widely recognized. Eighteenth-century naturalists classified them along with the sponges and a few other groups under Linnaeus's Zoophytes, a category for organisms

BOX 7A Characteristics of the Phylum Cnidaria

1. Diploblastic Metazoa with ectoderm and endoderm separated by an (primarily) ectodermally derived acellular mesoglea or partly cellular mesenchyme
2. Possess primary radial symmetry, often modified as biradial, quadriradial, or other variations on the radial theme; the primary body axis is oral–aboral
3. Possess unique stinging or adhesive structures called cnidae; each cnida resides in and is produced by one cell, the cnidocyte. The most common cnidae are called nematocysts.
4. Musculature formed largely of myoepithelial cells (= epitheliomuscular cells), derived from ectoderm and endoderm (adult epidermis and gastrodermis)
5. Exhibit alternation of asexual polypoid and sexual medusoid generations; but there are many variations on this basic theme (e.g., a medusoid stage is absent in the class Anthozoa)
6. The endodermally derived gastrovascular cavity (coelenteron) is the only "body cavity." The coelenteron is saclike, partitioned, or branched, but has only a single opening, which serves as both mouth and anus.
7. Without a head, centralized nervous system, or discrete gas exchange, excretory, or circulatory structures
8. Nervous system is a simple nerve net(s), composed of naked and largely nonpolar neurons.
9. Typically have planula larvae (ciliated, motile, gastrula larvae)

deemed somewhere between plants and animals. Lamarck instituted the group Radiata (or "Radiaires") for medusoid cnidarians, ctenophores, and echinoderms.

In the early nineteenth century the great naturalist Michael Sars demonstrated that medusae and polyps were merely different forms of the same group of organisms. Sars also demonstrated that the genera *Scyphistoma*, *Strobila*, and *Ephyra* actually represented stages in the life history of certain jellyfishes (scyphozoans). The names have been retained and are now used to identify these stages in the life cycle. Rudolph Leuckart eventually recognized the fundamental differences between the two great "radiate" groups, the Porifera/Cnidaria/Ctenophora and the Echinodermata, and in 1847 created the name Coelenterata (Greek *koilos*, "cavity"; *enteron*, "intestine") for the former group in his recognition of the "intestine" as the sole body cavity. In 1888 Berthold Hatschek split Leuckart's Coelenterata into the three phyla recognized today: Porifera, Cnidaria, and Ctenophora. Although some workers have been inclined to retain the cnidarians and ctenophores together within the Coelenterata, these two groups are universally recognized as distinct phyla, a view upheld by recent molecular analyses. Recently, some molecular analyses have suggested that Cnidaria and Ctenophora might be sister-groups, encouraging the resurrection of the name "Coelenterata," but other studies have failed to produce this sister relationship.

An important recent major advance in cnidarian systematics is the realization that the myxozoans, formerly classified as protists, are in fact, highly derived, parasitic cnidarians. Myxozoans had been considered protists until a century ago when Antonín Štolc (1899) proposed they should be considered reduced cnidarians because of the nematocyst-like architecture of their spores. Over time, the multicellular nature of Myxozoa, and the structural and developmental similarity of their **polar capsules** to cnidarian nematocysts, came to be recognized. Since the early 2000s, molecular data have supported the classification of Myxozoa as cnidarians. In 2002, the enigmatic metazoan *Buddenbrockia plumatellae,* first discovered in 1850, was reinterpreted as a myxozoan. *Buddenbrockia* is a motile vermiform creature that inhabits the body cavities of freshwater ectoprocts. It possesses four bands of longitudinal muscle cords, a cuticle-like body surface, no digestive tract, a cnidarian-like complement of polar capsules, and myxozoan-like infective spores (including polar capsules) that are formed inside the body cavity. Shortly after its rediscovery, it became clear that *Buddenbrockia plumatellae* is the same creature as *Tetracapsula bryozoides*, a known malacosporean (also described as *Myxosporidium bryozoides*), and furthermore, that another myxozoan, *Tetracapsula bryosalmonae* (later *Tetracapsuloides bryosalmonae*) was responsible for proliferative kidney disease (PKD) in salmonid fish. Clearly, malacosporeans are, in fact, Myxozoa.

While some molecular and developmental evidence based on the structure of the striated muscle has suggested that the diploblastic condition shared by Cnidaria and Ctenophora could in fact be derived from triploblastic states, most researchers agree that cnidarians are basal to triploblastic taxa. The Cnidaria appear to comprise three major lineages: Anthozoa, a polypoid lineage lacking medusae; Medusozoa, a diverse clade including species exhibiting primarily motile medusoid forms as well as forms with medusoid and polypoid life stages; and Myxozoa, a clade whose relationship to the other two lineages remains uncertain.

Within the Anthozoa, two main lineages appear to exist—the subclasses Octocorallia and Hexacorallia—but within these groups considerable revision has occurred, mainly involving consolidation of taxa previously considered distinct. Near the base of the medusozoan clade, the Staurozoa have been identified as a new cnidarian class, and the classes Scyphozoa and Cubozoa are identified as sister taxa. A possible sister taxon to the Hydrozoa is the class Polypoidozoa, an intracellular parasite of acipenceriform fish so far known from a single species. Hydrozoans themselves appear to include two major lineages, the subclasses Trachylina and Hydroidolina. As within the Anthozoa, considerable taxonomic revision within each hydrozoan subclass seems likely for some time to come.

PHYLUM CNIDARIA

SUBPHYLUM ANTHOZOA Class Anthozoa. Anemones, sea pens, soft corals, gorgonians, organ-pipe corals, and stony corals (Figure 7.1A–E).[3] Exclusively marine; solitary or colonial; without a medusoid stage (individuals all polypoid); cnidae epidermal and gastrodermal, composed of spirocysts and ptychocysts; coelenteron divided by longitudinal (oral–aboral) mesenteries, the free edges of which form thick, cordlike mesenterial filaments; mesenchyme thick; tentacles usually number 8 (Octocorallia) or occur in multiples of 6 (Hexacorallia), and contain extensions of the coelenteron; stomodeal pharynx (= actinopharynx) extends from the mouth into the coelenteron and bears one or more ciliated grooves (siphonoglyphs); polyps may reproduce both sexually and asexually; gametes arise from gastrodermis. About 6,225 described living species divided into two subclasses.

SUBCLASS HEXACORALLIA (= ZOANTHARIA) Anemones and stony corals. Solitary or colonial; naked, or with calcareous skeleton or chitinous cuticle, but never with isolated sclerites; mesenteries usually paired and in multiples of six; mesenteries bear longitudinal retractor muscles arranged so that those of each pair either face toward each other or away from each other; mesenterial filaments typically trilobed, with two ciliated bands flanking a central one bearing cnidocytes and gland cells; one to several circles of hollow tentacles arise from endocoels (the spaces between the members of each mesentery pair) and exocoels (the spaces between two mesenteries of different pairs); pharynx may have 0, 1, 2, or many siphonoglyphs; cnidae very diverse; endodermal zooxanthellae may be profuse.

ORDER CERIANTHARIA Cerianthids or tube anemones. Mesenteries complete, but with feeble musculature; mesenteries complete, with six primary mesenteries. Large, solitary, elongate polyps living in vertical tubes in soft sediments; tube constructed of interwoven specialized cnidae (ptychocysts) and mucus; aboral end lacks a pedal disc and possesses a terminal pore; long thin tentacles arise from margin of oral disc, fewer shorter labial tentacles encircle mouth; mesenteries complete; gonads occur only on alternate mesenteries; protandric hermaphrodites; some species possess long-lived pelagic larval stages (e.g., *Arachnanthus*, *Botruanthus*, *Ceriantheomorphe*, *Ceriantheopsis*, *Cerianthus*, *Pachycerianthus*).

[3] We use the term "anemone" in a general sense for any solitary anthozoan polyps, but restrict the term "sea anemone" to the anemones of the order Actiniaria.

ORDER ZOANTHIDEA Zoanthids. Colonial polyps arise from a basal mat or stolon containing gastrodermal solenia or canals; new polyps bud from gastrodermal solenia of stolons; pharynx flattened, with one siphonoglyph; mesenteries numerous, but with weak musculature; without intrinsic skeleton, but many species incorporate sand, sponge spicules, or other debris into the thick body wall; most with a thick cuticle; zooxanthellae abundant in some species; many species epizootic (e.g., *Epizoanthus*, *Isaurus*, *Isozoanthus*, *Palythoa*, *Parazoanthus*, *Thoracactus*, *Zoanthus*).

ORDER ACTINIARIA The sea anemones. Solitary or clonal; calcareous skeleton lacking, although some species secrete a chitinous cuticle; some harbor zooxanthellae; column often with specialized structures, such as warts or verrucae, acrorhagi, pseudotentacles, or vesicles; oral tentacles conical, digitiform, or branched, usually hexamerously arrayed in one or more circles; typically with two siphonoglyphs. A highly successful group with at least 1,000 living species in 41 families, the largest being the Actiniidae (e.g., the order includes *Actinia*, *Adamsia*, *Aiptasia*, *Alicia*, *Anthopleura*, *Anthothoe*, *Bartholomea*, *Bunodactis*, *Calliactis*, *Diadumene*, *Edwardsia*, *Epiactis*, *Halcampa*, *Haliplanella*, *Heteractis* (= *Radianthus*), *Liponema*, *Metridium*, *Nematostella*, *Peachia*, *Phyllodiscus*, *Ptychodactis*, *Stichodactyla*, *Stomphia*, *Triactis*).

ORDER ANTIPATHARIA Black or thorny corals. Upright, branched or whip-like colonies up to 6 m tall; calcareous axial skeleton, usually brown or black and covered by a thin coenosarc bearing small polyps, usually with 6 (but up to 24) nonretractable tentacles; with feeble mesenteries; skeleton produces thorns on its surface (e.g., *Antipathes*, *Bathypathes*, *Leiopathes*).

ORDER CORALLIMORPHARIA Solitary polyps, without a skeleton; lacking siphonoglyphs, ciliated bands on mesenterial filaments, and muscles in the pedal disk (e.g., *Amplexidiscus*, *Corynactis*, *Rhodactis*, *Ricordea*).

ORDER SCLERACTINIA (= MADREPORARIA) Stony corals. Colonial or solitary; polyp morphology almost identical to that of Actiniaria, except corals lack siphonoglyphs and ciliated lobes on the mesenterial filaments; zooxanthellae present in about half the known species; colony forms delicate to massive calcareous (aragonite) exoskeleton, with platelike skeletal extensions (septa). Over 1,300 extant species, in 24 families[4] (e.g., *Acropora*, *Agaricia*, *Astrangia*, *Balanophyllia*, *Dendrogyra*, *Flabellum*, *Fungia*,

[4] The taxonomy of the Scleractinia is in a state of confusion; many higher taxa appear to be nonmonophyletic. About 700 coral species live in deep water and do not form symbiotic relationships with zooxanthellae, and about two-dozen of these build reef structures.

Goniopora, *Letepsammia*, *Meandrina*, *Montipora*, *Oculina*, *Pachyseris*, *Porites*, *Psammocora*, *Siderastraea*, *Stylophora*).

SUBCLASS OCTOCORALLIA (= ALCYONARIA) Octocorals. Polyps with 8 hollow, marginal, pinnate tentacles, and 8 complete (perfect) mesenteries, each with retractor muscle on the sulcal side, facing the single siphonoglyph; most species with free or fused calcareous sclerites embedded in mesenchyme, although sclerites lacking in some species; stolons or coenenchyme connect the polyps. Three orders are currently recognized but considerable revision of subordinal groups is likely, particularly within the Alcyonacea.

ORDER HELIOPORACEA (= COENOTHECALIA) Helioporaceans. Colonies produce massive, rigid calcareous skeletons of aragonite crystals (not fused sclerites) similar to those of milleporid hydrozoans and stony corals; polyps monomorphic. Two families (Lithotelestidae and Helioporidae) and two genera (*Epiphaxum* and *Heliopora*, "blue coral").

ORDER ALCYONACEA Soft gorgonian and organ-pipe corals. Colonies encrusting or erect, often massive; usually fleshy and flexible, although the coenenchyme is sclerite-filled; polyps monomorphic or dimorphic; fleshy distal portions of polyps usually retractable into more compact basal portion; skeletal elements comprised of calcite and gorgonin. Twenty-nine recognized families, classified in 6 heterogeneous suborders that may be little more than morphological grades.

SUBORDER PROTOALCYONARIA Protoalcyonarians are deep-water octocorals with solitary polyps that reproduce exclusively by sexual means, (e.g., *Taiaroa* and possibly *Haimea*, *Hartea*, *Monoxenia*, *Psuchastes*). The status of this suborder is uncertain because phylogenetic studies suggest that it is nested deeply within the Octocorallia (its name is therefore misleading).

SUBORDER STOLONIFERA Stoloniferans. Simple polyps arise separately from ribbon-like stolons that forms an encrusting sheet or network; oral disc and tentacles retractable into a calyx (stiff proximal portion of polyp); mesenchyme with or without sclerites; in some (e.g., Cornulariidae) a thin, horny external skeleton may cover polyps and stolons (e.g., *Bathytelesto*, *Clavularia*, *Coelogorgia*, *Cornularia*, *Pseudogorgia, Sarcodictyon*, *Telesto*, *Tubipora* [organ-pipe or pipe-organ "coral," a monotypic genus, *T. Musica*, in which sclerites fuse to from a calcareous skeleton]).

SUBORDER ALCYONIINA Soft corals. Polyps united within a mass of fleshy coenenchyme; coenenchyme usually with sclerites, but lacking an axis (e.g., *Alcyonium*, *Nidalia*, *Sarcophyton*, *Studeriotes*, *Umbellulifera*, *Xenia*).

SUBORDER SCLERAXONIA Encrusting or upright, branched colonies with an inner axis (or axial-like layer) consisting mainly of fused sclerites. Includes *Corallium*, one of the precious corals used to make jewelry (also *Brianeum*, *Melithaea*, *Paragorgia*, *Parisis*).

SUBORDER HOLAXONIA Gorgonians. Upright, branched colonies with an inner axis consisting of scleroproteinous gorgonin without free spicules, with hollow cross-chambered central core, often with small amounts of embedded nonscleritic $CaCO_3$ (e.g., *Acanthogorgia*, *Eugorgia*, *Eunicea*, *Gorgonia*, *Ideogorgia*, *Muricea*, *Pseudopterogorgia, Swiftia*).

SUBORDER CALCAXONIA Gorgonians. Upright, branched or unbranched colonies with an inner axis consisting of scleroproteinous gorgonin without hollow cross-chambered central core, with large amounts of nonscleritic $CaCO_3$ as internodes or embedded in the gorgonin. Mostly found in the deep sea (e.g., *Australogorgia*, *Chrysogorgia*, *Dendrobrachia, Fanellia*, *Helicogorgia*, *Isis, Plumarella*, *Plumigorgia*, *Primnoa*, *Verrucella*).

ORDER PENNATULACEA Sea pens and sea pansies. Colonies complex and polymorphic; adapted for life on soft benthic substrata; often bioluminescent; elongate primary axial polyp extends length of the colony (to 1 m) and consists of a basal bulb or peduncle for anchorage and a distal stalk, the latter giving rise to dimorphic secondary polyps; coelenteron of axial polyp with skeletal axis of calcified horny material in canals (e.g., *Anthoptilum*, *Balticina*, *Cavernularia*, *Funiculina*, *Halipteris*, *Pennatula*, *Ptilosarcus*, *Renilla*, *Sarcoptilus, Stylatula*, *Umbellula*, *Virgularia*).

SUBPHYLUM MEDUSOZOA Medusozoans. Free-swimming and sessile forms; medusae produced through lateral budding and growth of the entocodon; epidermal gonads; loss of intramesogleal muscles associated with 4 peristomial pits; loss of primary polyp tentacles transformed into hollow structures; gastric filaments and coronal muscle absent. About 4,775 described living species in 5 classes.

CLASS STAUROZOA Stalked jellyfish. Small, sessile individuals that develop from benthic, non-ciliated planula larvae and have complex but poorly-understood life cycles; the planula develops into a sessile stauropolyp, and this into a stauromedusa with a stalked adhesive disc by which individuals attach to substratum (a highly motile medusa stage does not exist); stauromedusae have eight tentacle-bearing "arms"; sexual reproduction may be dominant in most species although asexual reproduction is known in at least one (*Haliclystus antarcticus*); ovaries with follicle cells. Occur mainly in shallow water at high latitudes (e.g., *Haliclystus*, *Lucernaria*; although *Kishinouyea* spp. are tropical). Molecular phylogenetic analyses indicate that despite considerable morphological convergence with trachyline hydrozoans, the minute Antarctic species, *Microhydrula limopsicola*, formerly identified within the Limnomedusae, is in fact a life stage within the life cycle of *H. antarctica*.

CLASS CUBOZOA Sea wasps and box jellyfish (Figures 7.1H; 7.15A). Medusae 1–30 cm, largely colorless; polyps each produce a single medusa by complete metamorphosis (strobilation does not occur), medusa bell nearly square in cross section; rhopalia with complex visual structures; hollow interradial tentacle(s) hang from bladelike pedalia, one at each corner of umbrella; unfrilled bell margin drawn inward to form a velum-like structure (the velarium) into which diverticula of the gut extend. Cnidae of the nematocyst type only; the sting is very toxic, in some cases fatal to humans, hence the name "sea wasps." Cubozoans are strong swimmers that occur in all tropical seas (and a few temperate regions, e.g., the west coast of South Africa and Namibia) but are especially abundant in the Indo-West Pacific region.

ORDER CHIRODROPIDA Pedalia branch into hand-like structure with each branch giving rise to a tentacle; externally fertilizing species including known deadly genera, although toxicity is variable within this order due to differences in tentacle surface area and means of venom delivery (e.g., *Chironex*, *Chiropsalmus*, *Chiropsella*).

ORDER CARYBDEIDA Ovoviviparous species; venom can induce Irukandji syndrome, a usually non-life threatening but physically and psychologically uncomfortable set of symptoms following envenomation. Five families, including the pelagic Alatinidae (*Alatina, Keesingia*), whose rhopaliar openings are t-shaped; two families that lack rhopaliar horns, and have rhopaliar openings that are frown-shaped, the Carukiidae (*Carukia*, *Gerongia*, *Malo, Morbakka*), and the Tamoyidae (e.g., *Tamoya*) which also lack gastric filaments; the Carybdeidae, *Carybdea*), whose ropaliar openings are heart-shaped; and the Tripedaliidae (*Tripedalia*, *Copula*), which engage in courtship and "copulatory" behavior.

CLASS SCYPHOZOA Jellyfish (Figure 7.1G). Medusoid stage predominates; polypoid individuals (scyphistomae) are small and inconspicuous but often long-lived; polyps lacking in some species; polyps produce medusae by asexual budding (strobilation); coelenteron divided by four longitudinal (oral–aboral) mesenteries; medusae acraspedote (without a velum), typically with a thick mesogleal (or collenchymal) layer, distinct pigmentation, filiform or capitate tentacles, and marginal notches producing lappets; sense organs occur in notches and alternate with tentacles; gametes arise from gastrodermis; cnidae present in epidermis and gastrodermis, nematocysts only; mouth may or may not be on a manubrium; usually without a ring canal. Scyphozoans are exclusively marine; planktonic, demersal, or attached. About 200 species are divided into three orders.

ORDER CORONATAE High bell divided into upper and lower regions by a coronal groove encircling exumbrella; margin of bell deeply scalloped by gelatinous thickenings termed pedalia, which give rise to tentacles, rhopalia, and marginal lappets; gonads present on the four gastrovascular septa. Small to moderate in size; primarily bathypelagic; some contain zooxanthellae (e.g., *Atolla*, *Linuche*, *Nausithoe*, *Periphylla*, *Stephanoscyphus*).

ORDER SEMAEOSTOMEAE Corners of mouth drawn out into four broad, gelatinous, frilly lobes; stomach with gastric filaments; hollow marginal tentacles contain extensions of radial canals; without coronal furrow or pedalia; gonads on folds of gastrodermis. This order contains most of the typical jellyfish of temperate and tropical seas; moderate to very large forms (some with bells reaching several meters in diameter) (e.g., *Aurelia*, *Chrysaora*, *Cyanea*, *Pelagia*, *Phacellophora*, *Sanderia*, *Stygiomedusa*).

ORDER RHIZOSTOMEAE Lacking a central mouth; frilled edges of the four oral lobes are fused over the mouth so that many suctorial "mouths" (ostioles) open from a complicated canal system on eight branching armlike appendages; bell without marginal tentacles or pedalia; stomach without gastric filaments; gonads on folds of gastrodermis. Small to large jellyfish that swim vigorously using a well-developed subumbrellar musculature; primarily occur in low latitudes (e.g., *Cassiopea*, *Cephea*, *Eupilema*, *Mastigias*, *Rhizostoma*, *Stomolophus*).

CLASS POLYPOIDOZOA Intracellular parasites of the oocytes of acipenceriform fish (e.g., sturgeons); binucleate cells within fish oocytes develop into inverted planulaform larvae which evert during host spawning into tentaclate stolons that possess cnidae and fragment when released into fresh water, dispersing as asexually reproducing medusoids whose gametes infect host fish. A single species, *Polypodium hydriforme*, currently defines this class.

CLASS HYDROZOA Hydroids and hydromedusae (Figure 7.1I–L and chapter opener photo). Alternation of generations occurs in most genera (typically asexual benthic polyps alternate with sexual planktonic medusae), although one or the other generation may be suppressed or lacking; medusae produced through lateral budding of the endocodon; medusoids often retained on the polyp; polyps usually colonial, with interconnected coelenterons; often polymorphic, individual polyps modified for various functions (e.g., gastrozooids feed, gonozooids are reproductive, dactylozooids are for defense and prey capture); exoskeleton when present usually of chitin or occasionally calcium carbonate (hydrocorals); coelenteron of polyps and medusae lacks a pharynx and mesenteries; mesoglea acellular; tentacles solid or hollow; cnidae occur only in epidermis; gametes arise from epidermal cells; medusae mostly small and transparent, nearly always craspedote (with a velum) and with a ring canal; mouth typically borne on pendant manubrium; medusae lack rhopalia. About 3,500 described species are contained in 6 orders, including some freshwater groups. Although the taxonomy of Hydrozoa is currently in a state of revision (and debate), there is considerable support for the existence of two monophyletic subclasses, the Trachylina and the Hydroidolina. The Siphonophora has previously been regarded as a subclass, but it now appears likely to be part of subclass Hydroidolina.

SUBCLASS TRACHYLINA Trachyline medusae. Polypoid generation minute or absent; medusae produce planula larvae that usually develop directly into actinula larvae, which metamorphose into adult medusae; medusae craspedote, with tentacles often arising from exumbrellar surface, well above bell margin; medusae mostly gonochoristic; currently including three suborders and over 150 known species.

ORDER LIMNOMEDUSAE (Figure 7.13A-C). Freshwater and marine hydromedusae with ecto-endodermal statocysts, hollow tentacles, and 4 (rarely 6) radial canals. The close relationships among freshwater genera (e.g., *Astrohydra, Craspedacusta*, *Limnocnida*) suggests that freshwater species share a common ancestor; these species are the only trachylines with true polyps, tiny structures that bud sexual medusae or asexual frustules which creep away to generate more polyps; families now considered within the Limnomedusae include the Olindiasidae (e.g., *Gonionemus*, *Olindias*, and the above freshwater genera), Monobrachiidae (e.g., *Monobrachium*), and Armohydridae, as well as the Geryoniidae (e.g, *Geryonia, Liriope*) which now appear to be Limnomedusae rather than Trachymedusae.

ORDER TRACHYMEDUSA Exclusively marine hydromedusae, usually with 8 radial canals and solid tentacles; includes 4 families—Halicreatidae, Petasidae, Ptychogastriidae, and Rhopalonematidae. The last family may include the actinulidans: free-living, solitary, minute (to 1.5 mm), motile, interstitial, apparently asexual polypoid hydrozoans with no medusa stage; (e.g., *Halammohydra*, *Otohydra*). This order may be polyphyletic.

ORDER NARCOMEDUSAE Pelagic or bathypelagic hydromedusae lacking radial canals, with lobed umbrella margins, broad, pouched stomachs, solid tentacles; includes 4 families: Aeginidae (e.g., *Aegina*), Cuninidae (e.g., *Cunia*), Solmarisidae (e.g., *Pegantha*), and Tetraplatiidae (e.g., the worm-like *Tetraplatia*, formerly considered a coronate scyphozoan).

SUBCLASS HYDROIDOLINA (Figure 7.13D-F). Hydroids and their medusa, and siphonophores. Polypoid generation often predominant; polyps may have a chitinous exoskeleton; oral tentacles filiform or capitate, rarely branched or absent; colonies often polymorphic; many do not release free medusae but release gametes from sporosacs or sessile attached medusoids (= medusoid buds, or gonophores) on colony; colonies gonochoristic. A large group, with over 75 described families and over 3,200 species. Hydroids occur at all depths; the polypoid forms are very common in the littoral zone.

ORDER THECATA (= LEPTOMEDUSAE, CALYPTOBLASTEA, OR LEPTOTHECATA) A diverse group of hydroids with polyps always colonial; hydranths and gonozooids encased in exoskeleton; free medusae usually absent, but when present flattened and with statocysts; medusae form gametes on subumbrella beneath radial canals; gonozooids (= gonangia) with blastostyle that produces medusae buds. Monophyly of the Leptothecata seems likely but relationships within this order await further study. These are some of the most common species of hydrozoans in the marine littoral zone (e.g., *Abietinaria*, *Aequorea*, *Aglaophenia*, *Bonneviella*, *Campanularia*, *Cuvieria*, *Gonothyrea*, *Lovenella*, *Obelia*, *Plumularia*, *Sertularia*).

ORDER ATHECATA (= ANTHOMEDUSAE, GYMNOBLASTEA, OR ANTHOATHECATA) Polyps solitary or colonial; hydranths and gonozooids lack exoskeleton; gonozooids produce free or sessile medusae; some groups produce gametes in transient sporosacs; free medusae tall and bell-shaped, without statocysts, with or without ocelli; medusae form gametes on subumbrella or manubrium. Current research suggests this order is polyphyletic.

SUBORDER CAPITATA A diverse group of hydroids possessing stenotele nematocysts and tentacles with rounded (capitate) tips at some stage in the life cycle; many have medusae with complex, cup-shaped ocelli. Current research suggests there are two clades, Zancleida and Corynida. Zancleida includes well-known polypoid forms (e.g., *Pennaria*, the Christmas tree hydroids) and the milleporid "fire corals" (known for their potent stinging nematocysts). Milleporids are distinguished by massive or encrusting calcareous coral-like skeletons, the calcareous matrix covered by thin epidermal layer; gastrozooids with short capitate tentacles; and each gastrozooid surrounded by 4–8 dactylozooid-like tentacles, each tentacle in a separate skeletal cup; gonophores housed in pits (ampullae) in skeleton; and, the small free medusae lack mouth, tentacles, and velum. Like the stony corals, milleporids rely on a commensal relationship with zooxanthellae and are thus restricted to the photic zone. Zancleida also includes well-known medusoid forms, formerly known as "chondrophorans," whose colonies may consist of gastrozooids, gonozooids, and dactylozooids, or as a solitary but highly specialized polypoid individual. Chondrophoran "zooids" are attached to a chitinous, multichambered, disclike float, that may or may not have an oblique sail (e.g., *Porpita*, *Vellela*), their "gonozooids" bear medusiform gonophores that are released and shed gametes, and most are richly supplied with zooxanthellae. The Corynida includes well-known medusoid forms such as *Coryne, Dipurena, Polyorchis*, and *Sarsia*. The status of the Corynida is presently unclear due to evidence that generic relationships may need revision.

SUBORDER APLANULATA Members of this taxon lack a ciliated planula stage, hence their name; the group is well supported by molecular phylogenetic analyses. This group includes *Corymorpha*, *Tubularia*, *Hydra*, and *Candelabrum*. *Hydra*, while often used to exemplify the Cnidaria, is actually quite unusual in life cycle and morphology.

SUBORDER FILIFERA This heterogeneous group of hydrozoans is currently divided into 4 major groups. The first group (Filifera I) is typified by the family Eudentridae, which is distinguished from other filiferans by the absence of desmoneme nematocysts, the presence of a styloid-shaped gonophore, and a trumpet-shaped hypostome. The second group (Filifera II) is less distinct but shares a solid ring canal and macrobasic eurytele cnidae, as well as reduction in hydranth tentacles. Many species in this group live on other invertebrates (e.g., *Hydrichthella* on octocorals; *Brinckmannia* within hexactinellid sponges; *Proboscidatyla* on sabellid polychaete tubes). Filifera III is typified by the families Hydractiniidae (e.g., *Clava*, *Hydractina*) and Stylasteridae (e.g., *Allopora*, *Stylaster*). Members of these families have polymorphic polyps and perisarc- or skeleton-covered stolons that can be colorful. The stylasterine skeleton is secreted within the epidermis and covered by a thick epidermal layer; a calcareous style often rises from base of polyp cup, hence the name "stylasterine"; polyps may have tentacles; free medusae are not produced, but sessile medusoid gonophores are retained in shallow chambers (ampullae) of the colony; several dactylozooids surround each gastrozooid, although polyp pits are joined. Filifera IV includes the families Bouganvilliidae (e.g., *Bouganvillia*, *Dicoryne*), Oceaniidae (e.g., *Cordylophora*), Pandeidae (e.g., *Pandea*) and Rathkeidae (e.g., *Rathkea*), which all bear gonophores on structures other than hydranths.

ORDER SIPHONOPHORA Siphonophores. Polymorphic swimming or floating colonies, with a number of distinct types of polyps and attached modified medusae; most have a gas-filled flotation zooid. Siphonophores are major oceanic predators, some reaching tens of meters in length; a majority of the species is bioluminescent. Historically divided into three groups based on body structure, molecular phylogenetic research now suggests the existence of two major clades—Cystonecta and Codonophora.

SUBORDER CYSTONECTA Possessing an anterior gas-filled float or pneumatophore, with a posterior, zooid-bearing siphosome (e.g., *Physalia*, *Rhizophysa*).

SUBORDER CODONOPHORA Structurally the more complex siphonophores, this suborder includes the monophyletic Calycophora and the paraphyletic Physonecta. The former possess an anterior nectosome, comprised of nectophores, swimming elements that propel the colony, and a posterior zooid-bearing siphosome but no pneumatophore (e.g., *Diphyes*, *Hippopodius*, *Sphaeronectes*). The Physonecta possess an anterior pneumatophore, a more posterior nectosome and a posterior siphosome (e.g., *Agalma*, *Apolemia*, *Bargmannia*, *Physophora*).

SUBPHYLUM MYXOZOA Intracellular parasites of poikilotherm vertebrates, annelids and bryozoans; possessing myxospores, cnida-like structures with two or more shell valves and polar capsules containing nematocyst-like filaments. This taxon contains two clades (Myxosporea and Malacosporea), well supported by molecular phylogenetic analyses. About 2,200 described species.

CLASS MYXOSPOREA Marine and freshwater species, usually with hard valves; life cycles include a myxosporean phase in vertebrates and an actinosporean phase within polychaete and sipunculan annelids. About 1,200 described species.

ORDER BIVALVULIDA Possessing two valves per myxospore; gut and tissue parasites (e.g., *Ceratomyxa*, *Henneguya*, *Myxidium*, *Myxobolus*, *Sphaerospora*; the genus *Myxidium* appears to be polyphyletic).

ORDER MULTIVALVULIDA Possessing more than two valves per myxospore (e.g., *Hexacapsula*, *Kudoa*, *Trilospora*).

CLASS MALACOSPOREA With a single order, Malacovalvulida, and a single family, Saccosporidae. Freshwater species characterized by soft-walled spores, bryozoans as invertebrate hosts, and spore formation within a saclike body form (e.g., *Buddenbrockia*, *Tetracapsuloides*).

The Cnidarian Body Plan

Although showing marked advances over Porifera and Placozoa, cnidarians still appear to possess only two embryonic germ layers—the ectoderm and the endoderm—which become the adult epidermis and gastrodermis, respectively. The middle mesoglea or mesenchyme in adults is derived largely from ectoderm and never produces the complex organs seen in triploblastic Metazoa (i.e., the Bilateria).

Whether or not a true basement membrane (= basal lamina) exists in cnidarians is debatable. We define a **basement membrane** as a thin sheet of extracellular matrix upon which an epithelial layer may rest; it contains collagen and other proteins, and it helps hold the epithelial cells in place. By this definition, placozoans lack a basement membrane, cnidarians and ctenophores possess one by way of the mesenchyme, and

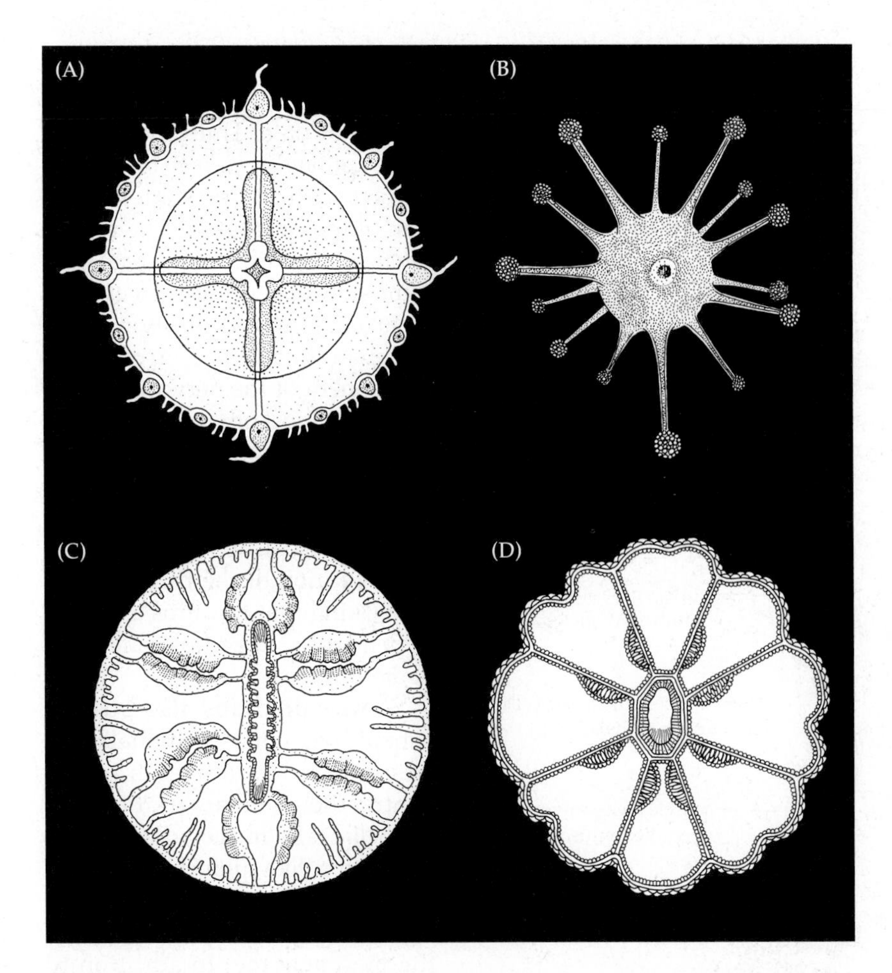

Figure 7.3 Cnidarian radial symmetries. (A) Quadriradial symmetry of a hydromedusa. (B) Radial symmetry of a hydrozoan polyp. (C) Biradial symmetry of an actiniarian polyp (a sea anemone). (D) Biradial symmetry of an octocoral polyp (Anthozoa).

In spite of the limitations of a diploblastic, radially symmetrical body plan, cnidarians are a highly successful and diverse group. Much of their success has resulted from their possession of cnidae and the diversity of different dimorphic life histories. Although polyps and medusae are very different in appearance, they are really variations on the basic cnidarian body plan (Figure 7.2). And yet, the two stages are vastly different ecologically, and their presence in a single life history allows an individual species to exploit different environments and resources, leading a "double life."

the Bilateria possess a well-developed, highly proteinaceous basement membrane. And, as we've seen, basement membranes also occur in some sponges.

The essence of the cnidarian body plan is radial symmetry (Figure 7.3). As discussed in Chapter 4, radial symmetry is associated with various architectural and strategic constraints. Cnidarians are either sessile, sedentary, or pelagic, and they do not engage in the active unidirectional movement seen in bilateral, cephalized creatures. Radial symmetry demands certain anatomical arrangements, particularly of those parts that interact directly with the environment, such as feeding structures and sensory receptors. Thus, we typically find a ring of tentacles encircling the body that can collect food from any direction, and a diffuse, noncentralized nerve net with radially distributed sense organs. These and other implications of radial symmetry are explored further throughout this chapter. Although polypoid and medusoid forms are often represented as inverted forms of one another (Figure 7.2A–C), studies of gene expression in cnidarians show that the oral tentacles of polyps and the marginal tentacles of medusae may not always be homologous structures (Figure 7.2D). As we will see, differences between polyps and medusae go well beyond how animals appear to be oriented along their oral-aboral axis.

The Body Wall

Cnidarian epithelia—the outer epidermis and the inner gastrodermis—include **myoepithelial cells** (Figures 7.4 and 7.19), which are primitive muscle cells. These columnar cells bear flattened, contractile, basal extensions called **myonemes** (Figure 7.19). In the epidermis, these cells are referred to as **epitheliomuscular cells**, and in the gastrodermis they are called **nutritive-muscular cells**. The myonemes rest against the middle mesoglea or mesenchyme, and the opposite ends of the cells form the outer body and gut surfaces. The myonemes run parallel to free surfaces and contain contractile myofibrils. Myonemes of neighboring cells are interconnected, often forming longitudinal and circular sheets capable of contracting like muscle layers of more advanced phyla. Primitive contractile cells similar to myoepithelial cells occur in poriferans—the contractile myocytes. Somewhat similar cells are even known among mammals, where they are found in association with certain secretory tissues.

Molecular studies of germ layer formation in anthozoans show that virtually all of the genes involved in endomesoderm formation in bilaterian embryos, including the core genes identified as components of an evolutionarily conserved endomesoderm ("kernel") gene regulatory network, are expressed in cnidarian

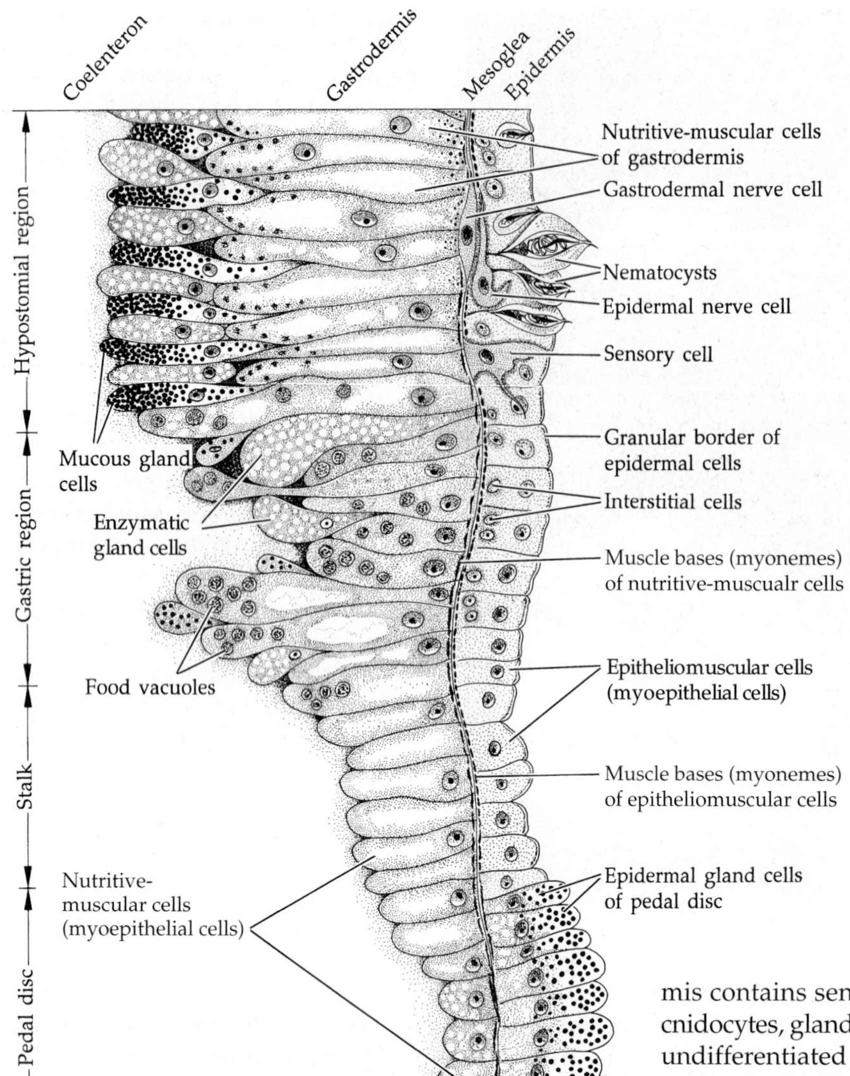

Figure 7.4 A hydrozoan polyp column wall (cross section) illustrates the basic cnidarian cell and tissue types.

epithelial tissue that lines the gastric cavity or the pharynx. This conservation in components of the endomesodermal gene regulatory network provides compelling evidence that the endodermal and pharyngeal tissue of cnidarians (and probably ctenophores) is homologous with that of the gut and oral ectoderm of bilaterians, and that both endoderm and mesoderm of bilaterians evolved from an ancestral endomesodermal layer. Fate mapping experiments have shown that the definitive endoderm is generated from the oral pole/animal hemisphere in both anthozoan cnidarians and ctenophore embryos. They also show that the single opening to the cnidarian and ctenophoran gut arises from the same region of embryo (the animal hemisphere) that forms the mouth in all other bilaterians. This suggests that the oral pole of adult ctenophores and cnidarians is homologous to the anterior pole of bilaterians, and that the single opening is homologous to the mouth of other bilaterians. This relationship is also supported by molecular data. Thus, the mouths of all animals appear to be homologous, with the possible exception of the chordates. (The mouth of chordates does not express the same suite of genes as seen in other metazoans, and its position forms independently of a circumoral component of the nervous system shared by most other metazoans. If this interpretation is correct, it suggests that the single opening to the cnidarian and ctenophoran (and perhaps also acoel) gut preceded the evolution of the through gut, suggesting that the anus arose independently in protostomes and deuterostomes.

Some cnidarians also possess subepidermal mesenchymal muscles, apparently derived from the contractile elements of the myoepithelial cells. In anemones, for example, cordlike sphincters are sunk below the epithelium and reside as distinct muscles wholly within the mesenchyme. In addition to epitheliomuscular cells, the epidermis contains sensory cells, cnida-bearing cells called cnidocytes, gland cells, and interstitial cells. The last are undifferentiated and capable of developing into other types of cells. The gastrodermis is histologically somewhat similar to the epidermis (Figure 7.4). Along with the nutritive-muscular cells it also contains cnidocytes (except in the Hydrozoa) and gland cells.

In hydrozoans, the middle layer is a rather simple, gel-like, largely acellular mesoglea. Scyphomedusae have a very thick mesogleal layer with scattered cells. In stauromedusans, cubozoans, and hydrozoans, mesoglea is non-cellular. In anthozoans the middle layer is often a thick and richly cellular mesenchyme (see footnote 2, earlier in the chapter). Myxozoans are primarily small and cellular without clearly defined body walls, with a few exceptions. The actinosporan larva of many myxosporeans has rays comprised of single cells that are used to attach to the mucous membranes of vertebrate hosts. The worm-like *Buddenbrockia* has four blocks of longitudinal musculature that lie beneath the epidermis and surround a hollow, gut-free body cavity.

The polypoid form Polyps are much more diverse than are medusae, largely as a result of their capacities for asexual reproduction and colony formation (Figures 7.5–7.12). This diversity of form has led to an extensive lexicon of terms for the parts of polyps. This expansive

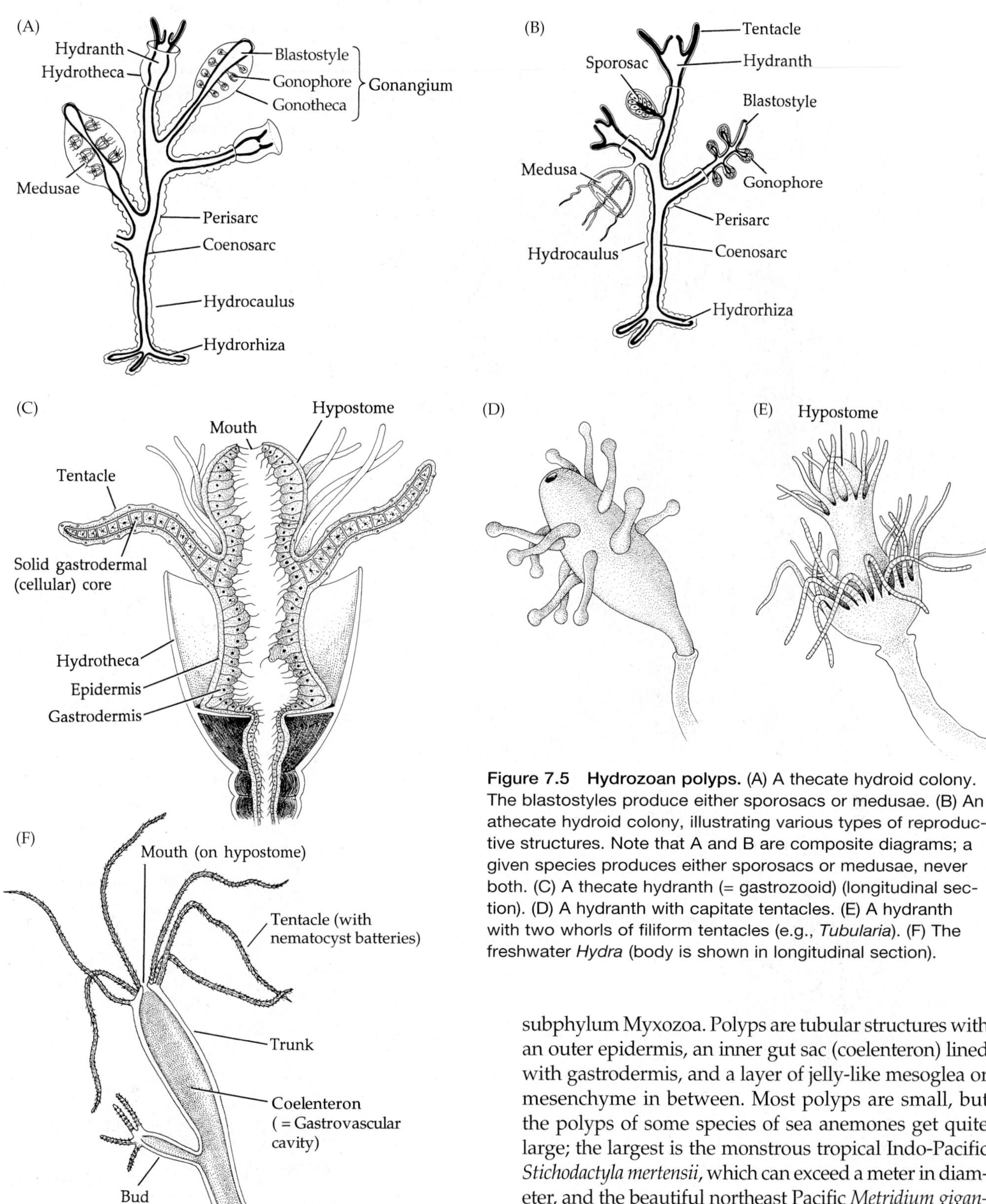

Figure 7.5 Hydrozoan polyps. (A) A thecate hydroid colony. The blastostyles produce either sporosacs or medusae. (B) An athecate hydroid colony, illustrating various types of reproductive structures. Note that A and B are composite diagrams; a given species produces either sporosacs or medusae, never both. (C) A thecate hydranth (= gastrozooid) (longitudinal section). (D) A hydranth with capitate tentacles. (E) A hydranth with two whorls of filiform tentacles (e.g., *Tubularia*). (F) The freshwater *Hydra* (body is shown in longitudinal section).

terminology may at first seem a bit overwhelming, but bear with us—we won't give you anything unnecessary here. The polypoid stage occurs in the subphyla Anthozoa and Medusozoa, although it is greatly modified in the Scyphozoa, Cubozoa, and Staurozoa. A clearly defined polypoid stage appears to be lacking in the subphylum Myxozoa. Polyps are tubular structures with an outer epidermis, an inner gut sac (coelenteron) lined with gastrodermis, and a layer of jelly-like mesoglea or mesenchyme in between. Most polyps are small, but the polyps of some species of sea anemones get quite large; the largest is the monstrous tropical Indo-Pacific *Stichodactyla mertensii*, which can exceed a meter in diameter, and the beautiful northeast Pacific *Metridium giganteum*, which can extend its column to a meter in height.

The fundamental polypoid symmetry is radial, although as a result of subtle modifications most species possess a biradial or quadriradial symmetry. The main body axis runs longitudinally through the mouth (oral end) to the base (aboral end) of the polyp. The aboral end may form a **pedal disc** for attaching to hard substrata (as in most common sea anemones); it may be a rounded structure—called a **physa**—adapted for digging and anchoring in soft substrata (as in burrowing anemones); or it may arise from a common mat, stalk, or stolon in colonial forms.

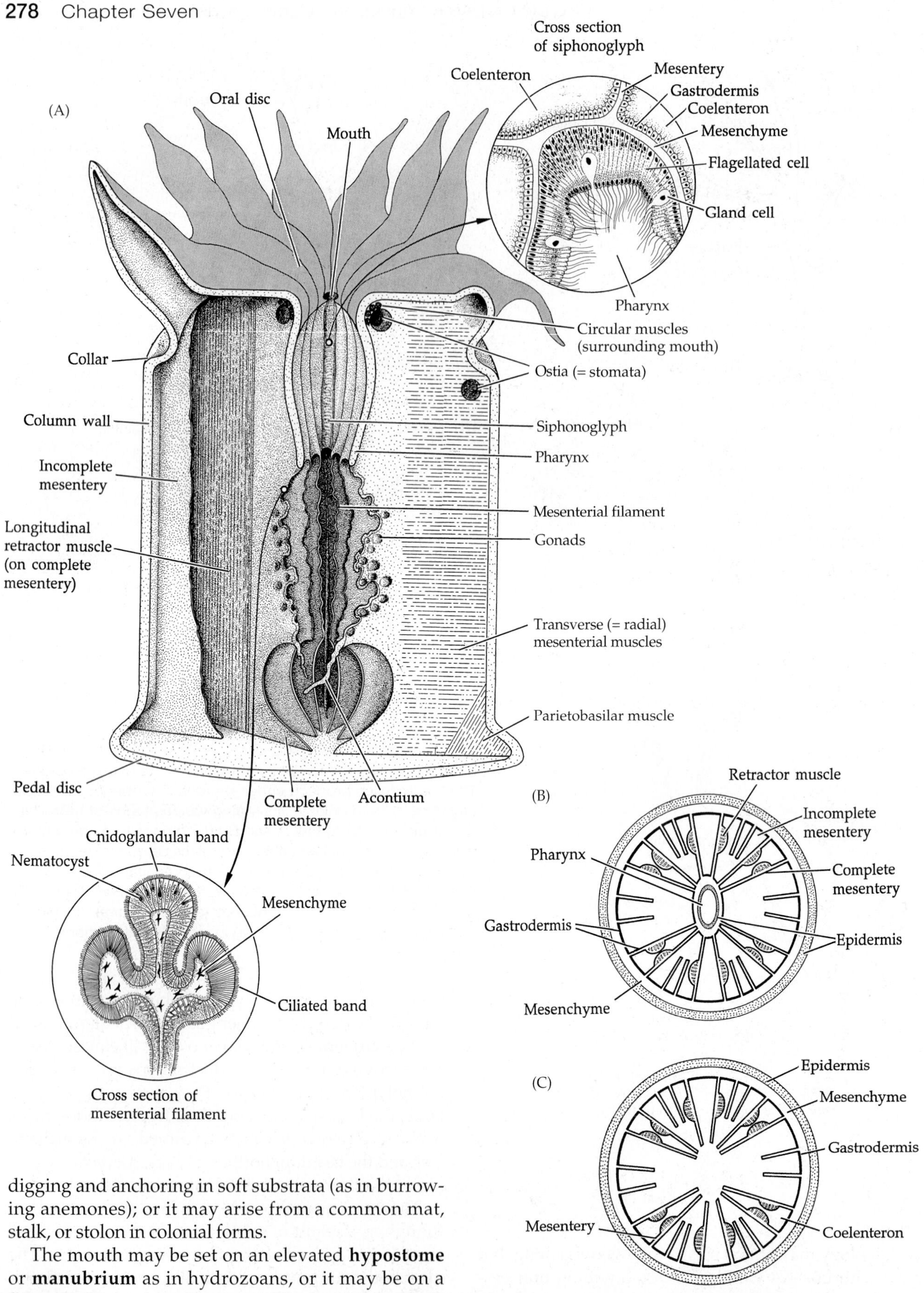

The mouth may be set on an elevated **hypostome** or **manubrium** as in hydrozoans, or it may be on a flat **oral disc** as in anthozoans (Figures 7.5 and 7.6). In anthozoans the mouth is usually slitlike and leads to a muscular, ectodermally derived pharynx that extends into the coelenteron (the gut cavity of a polyp).

Figure 7.6 An anthozoan polyp. (A) A sea anemone (longitudinal section). (B) Cross section taken at the level of the pharynx. (C) Cross section taken below level of the pharynx.

The pharynx usually bears from one to several ciliated grooves called **siphonoglyphs**, which drive water into the coelenteron (Figure 7.6). It is in part the presence of siphonoglyphs that gives these polyps a secondary biradial or quadraradial symmetry. The side of an anthozoan polyp that bears a single siphonoglyph is called the **sulcal side**, and the opposite side is called the **asulcal side**.

The **coelenteron**, or gastrovascular cavity, serves for circulation as well as digestion and distribution of food. In hydrozoan polyps, the coelenteron is a single, uncompartmentalized tube. In scyphozoan (scyphistomae) and staurozoan polyps, it is partially subdivided by four longitudinal, ridgelike mesenteries, and in some staurozoans and cubozoans additionally by a transverse **claustrum** (a partition running parallel to the umbrellar margin and dividing each gastric pouch into inner and outer pockets); in anthozoan polyps, it is extensively compartmentalized by mesenteries. Anthozoan mesenteries are projections of the inner body wall and thus are lined with gastrodermis and filled with mesenchyme. They extend from the inner body wall toward the pharynx, some or all of them fusing with it as complete mesenteries. Those that do not connect to the pharynx are called incomplete mesenteries. In anthozoan polyps, the free inner edge of each mesentery below the pharynx has a thickened, cordlike margin armed with cnidae, cilia, and gland cells and is called the **mesenterial filament** (Figures 7.6 and 7.20). In some sea anemones these filaments give rise to long threads, called **acontia**, that hang free in the gastrovascular cavity. Acontia function in defense and feeding (see the section on Feeding and Digestion section below). In most colonial anthozoans the cellular mesenchyme unites individual zooids (Figure 7.12). In some, such as the soft corals, gastrovascular cavities are connected to one another by canals called **solenia**.

The tentacles that surround the mouth contain hollow extensions of the coelenteron in anthozoans, whereas they house a solid core of packed gastrodermal cells in most hydrozoans. Tentacles may taper to a point (**filiform** tentacles) or may terminate in a conspicuous knob of cnidae (**capitate** tentacles). In some polyps the tentacles are branched, often as pinnately arranged pinnules (e.g., in the octocorals).

Branched hydrozoan colonies grow in two patterns (Figure 7.7). In **monopodial growth**, the first polyp elongates continuously from a growth zone at the distal end of the simple or branched stem of the colony, the **hydrocaulus.** This primary (axial) polyp may even lose its hydranth and persist merely as a stalk. The primary hydrocaulus gives rise to secondary polyps by lateral budding. These secondary polyps grow and may give rise to lateral tertiary polyps in the same fashion. In hydrozoan colonies developing by **sympodial growth**,

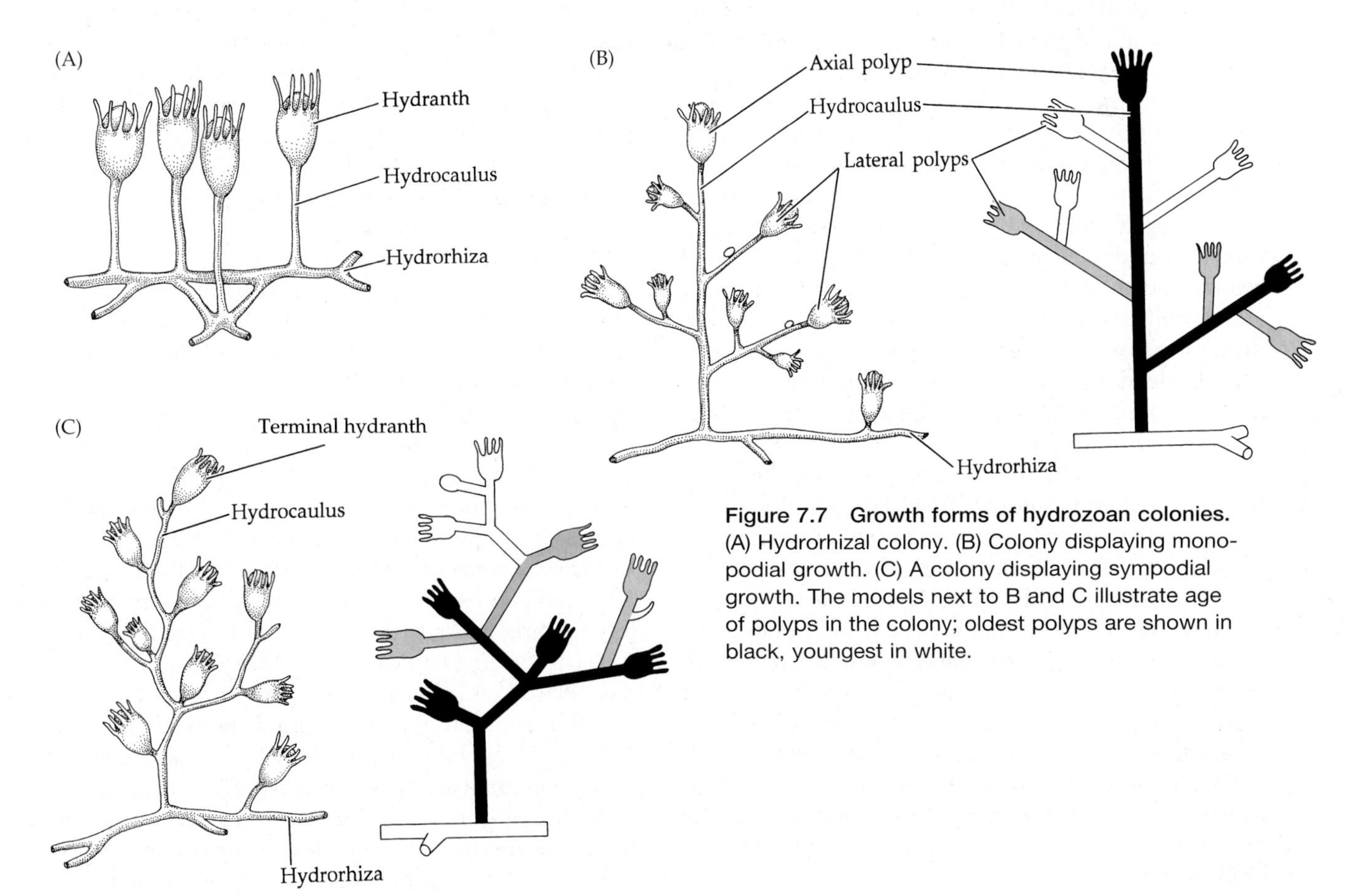

Figure 7.7 Growth forms of hydrozoan colonies. (A) Hydrorhizal colony. (B) Colony displaying monopodial growth. (C) A colony displaying sympodial growth. The models next to B and C illustrate age of polyps in the colony; oldest polyps are shown in black, youngest in white.

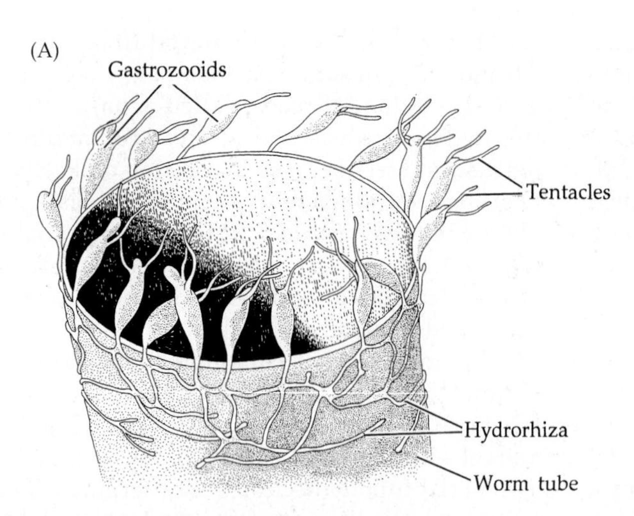

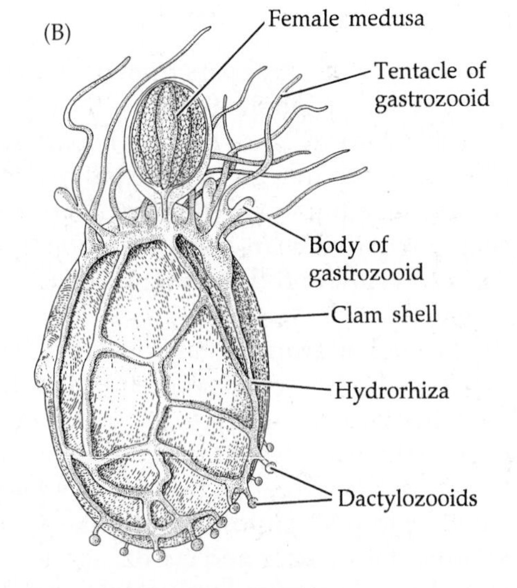

Figure 7.8 Diversity of form among the colonial Hydrozoa. (A) *Proboscidactyla*, a two-tentacled hydroid that lives around the open end of polychaete worm tubes. (B) *Monobrachium*, a one-tentacled hydroid that lives on clam shells. (C) *Hydractinia*, a colonial hydroid commensal on shells inhabited by hermit crabs. (D) The chondrophoran *Porpita* (aboral view). (E) A colony of the calcareous milleporid hydrocoral *Millepora* (fire corals). (F) A siphonophore, *Physalia* ("man-of-war"). (G) A colony of the calcareous stylasterine hydrocoral *Allopora*. (H) *Nectocarmen antonioi*, a colonial calycophoran siphonophore from California. (I) Another siphonophore.

the primary polyp does not continue to elongate but produces one or more lateral polyps by budding and then stops growing. The new polyps extend the colony upward some distance then stop growing and give rise to more new polyps by budding. In these colonies the main stem or axis actually represents the combined hydrocauli of many polyps and the age of the polyps decreases from base to tip along each branch.

Most marine hydroids are surrounded, at least in part, by a nonliving protein–chitin exoskeleton secreted by the epidermis and called the **perisarc** (Figure 7.5). Interestingly, this outer covering is absent in freshwater hydroids. The living tissue inside the perisarc is termed the **coenosarc**. The perisarc may extend around each hydranth and gonozooid as a **hydrotheca** and **gonotheca**, respectively. When this occurs, the hydroids are said to be **thecate**; hydroids whose perisarcs do not extend around the zooids are **athecate**.

A complex terminology has been developed to describe hydrozoan polyps, or "hydroids" as they are commonly called (Figures 7.5 and 7.7). A big reason for this special nomenclature is that hydroid colonies are usually polymorphic, containing more than one kind of polyp, or zooid. The term **hydranth** or **gastrozooid** refers to feeding zooids, which typically bear tentacles and a mouth. Other commonly occurring polyp types include defensive polyps (**dactylozooids**) and reproductive polyps (**gonozooids** or **gonangia**). Each zooid typically arises from a stalk, called a **hydrocaulus** (pl. hydrocauli). In most colonial hydrozoans, the individual polyps are anchored in a rootlike stolon called a **hydrorhiza**, which grows over the substratum. From the hydrorhiza arise the hydrocauli, bearing polyps singly or in clusters.

Gastrozooids capture and ingest prey and provide energy and nutrients to the rest of the colony. Dactylozooids, which occur in a variety of sizes and shapes, are heavily armed with cnidae. Often several dactylozooids surround each gastrozooid and serve for both defense and food capture. Gonozooids produce medusa buds called **gonophores** that are either released or retained on the colony. Whether released as free medusae or retained as attached gonophores, they produce gametes for the sexual phase of the hydrozoan life cycle. The living tissue (coenosarc) of the gonozooid is called the **blastostyle**; the gonophores arise from this tissue. When a **gonotheca** surrounds the blastostyle, the zooid is called a **gonangium**.

The most dramatic examples of polymorphism among polyps are seen in the hydrozoan order Siphonophora and the anthozoan order Pennatulacea. Siphonophores (Figures 7.8F,H,I and 7.9) are hydrozoan colonies composed of both polypoid and medusoid individuals, with as many as a thousand zooids in a single colony. This large order includes a great variety of unusual and poorly understood species, including the famous Portuguese man-of-war, *Physalia* (Figure 7.8F). The gastrozooids of siphonophores are actually highly modified polyps with a large mouth and one long, hollow feeding tentacle that bears many cnidae (Figure 7.9). This feeding tentacle reaches lengths of 13 m in the Atlantic species *Physalia physalis*. The nonfeeding dactylozooids also bear one long (unbranched) tentacle. The gonozooids are usually branched; they produce sessile gonophores that are never released as free medusae.

Siphonophores use one or more swimming bells (**nectophores**) or a gas-filled float (a **pneumatophore**), or both, to help maintain their position in the water.

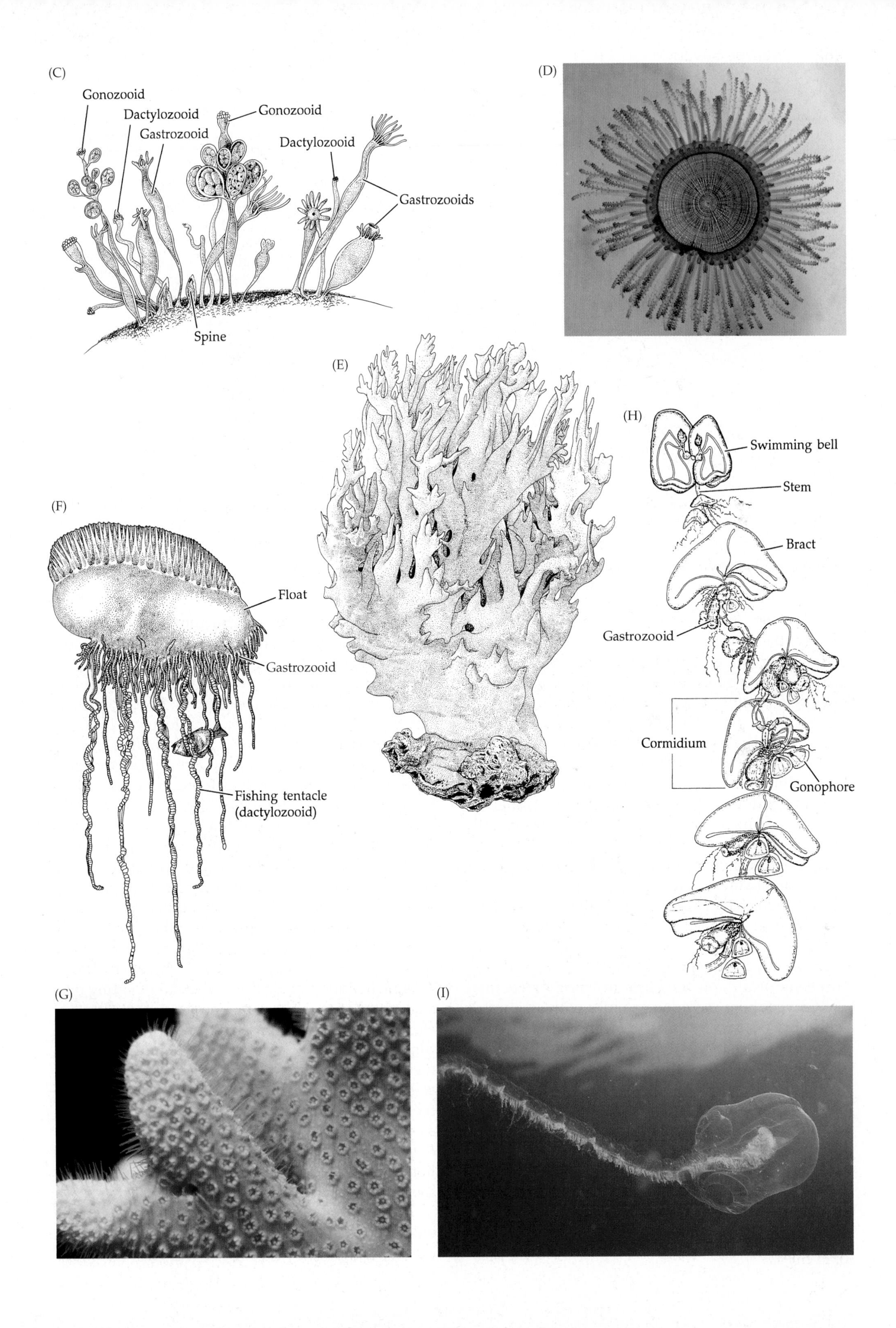
(C)
Gonozooid
Dactylozooid
Gastrozooid
Gonozooid
Dactylozooid
Gastrozooids
Spine
(D)
(E)
(F)
Float
Gastrozooid
Fishing tentacle
(dactylozooid)
(H)
Swimming bell
Stem
Bract
Gastrozooid
Cormidium
Gonophore
(G)
(I)

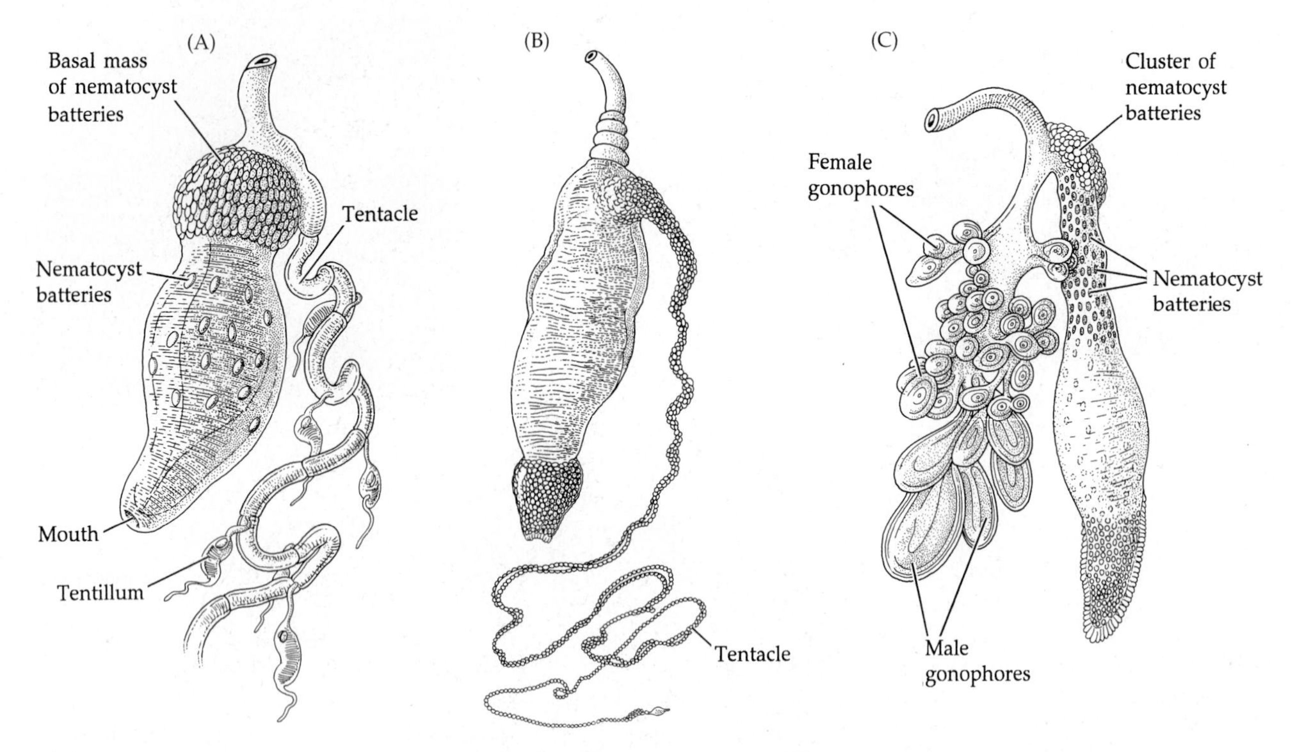

Figure 7.9 Siphonophore zooids. (A) A gastrozooid. (B) A dactylozooid. (C) A gonozooid.

Nectophores are true medusoid individuals with many of the structures common to free-swimming medusae, although each nectophore has lost its mouth, tentacles, and sense organs. The pneumatophore, once also thought to be a modified medusa, is now known to be derived directly from the larval stage and probably represents a highly modified polyp. Pneumatophores are double-walled chambers lined with chitin. Each float houses a gas gland, which consists of a mitochondria-laden glandular epithelium lining a chamber. The gland secretes a gas usually similar to air in composition, although in *Physalia* it apparently includes a surprisingly high proportion of carbon monoxide. Many siphonophores have mechanisms by which they regulate gas in their floats to keep the colony at a particular depth, much like the swim bladders of fishes.

Siphonophores were formerly classified into three suborders on the basis of colony structure: the Calycophora included colonies with swimming bells but no float; Physonecta included those with a small float and a long train of swimming bells; and Cystonecta had a large float and no bell. However, recent molecular evidence suggests that the Cystonecta are basal to other siphonophores, with the monophyletic Calyphora and the paraphyletic Physonecta grouped for now within the suborder Codonophora. Calycophorans have a long tubular stem extending from the swimming bell, from which various types of zooids bud in groups called **cormidia** (Figure 7.8H). Each cormidium acts as a colony-within-a-colony, and is usually composed of a shieldlike bract, a gastrozooid, and one or more gonophores that may function as swimming bells. The cormidia commonly break loose from the parent colony to live an independent existence, at which time they are termed **eudoxids**. The physonectans have an apical float with a long stem bearing a series of nectophores followed by a long train of cormidia. The cystonectans, including *Physalia*, usually have a large pneumatophore with a prominent budding zone at its base, which produces the various polyps and medusoids (Figure 7.8F).

Within the hydrozoan suborder Capitata, a group formerly known as the "chondrophora" is composed of colorful oceanic organisms that drift about on the sea surface in enormous flotillas, occasionally washing ashore to coat the beach with their bluish-purple bodies (Figures 7.1D and 7.10). Although they superficially resemble some siphonophores, current opinion holds these animals to be large, solitary, athecate hydranth polyps, floating upside down instead of sitting on a stalk attached to the bottom. The aberrant medusae of chondrophorans are short-lived and do not possess a functional mouth or gut, probably relying instead on their symbiotic zooxanthellae for nutrition. The aboral sail in *Velella* (the "by-the-wind-sailor") has no counterpart in sessile hydroids. In its ability to sail at an angle to the wind, *Velella* resembles the siphonophore *Physalia*, a similarity attributed to convergent evolution. Figure 7.10 compares a chondrophoran and a sessile hydroid, such as *Tubularia* or *Corymorpha*.

The apparent polyps of *Polypodium hydriforme* (the only species in the bizarre class Polypoidozoa) exist intracellularly within the ova of their fish hosts as an

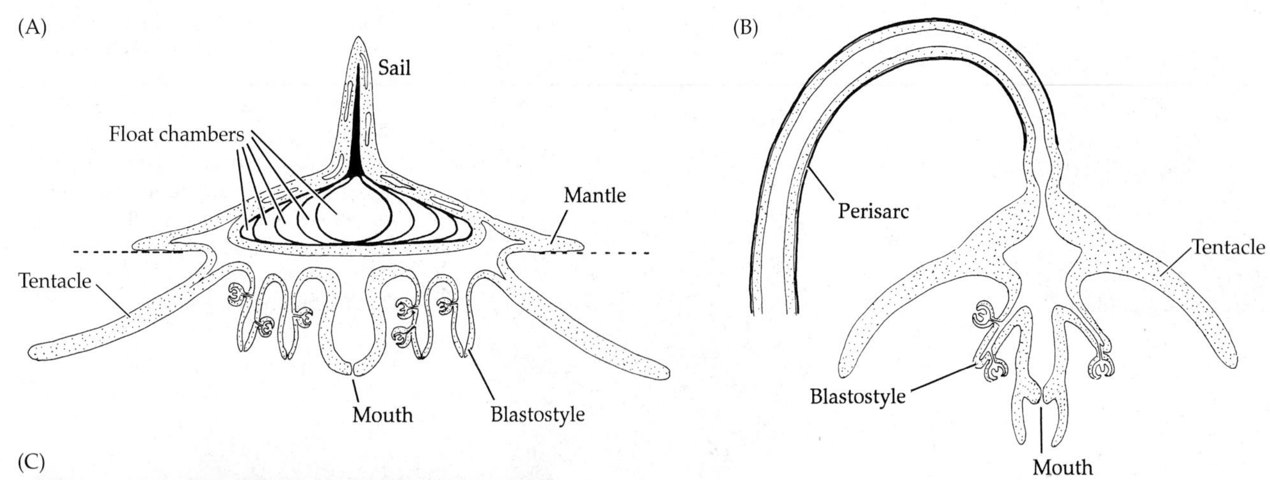

(C)

Figure 7.10 (A) A capitate hydrozoan such as Velella compared with (B) a sessile tubularian hydranth, in support of the view that "chondrophorans" are highly specialized, solitary tubularian zooids. (C) *Vellela* skeletons in beach wrack, coast of Washington.

inverted stolon whose digestive surface faces outward. Prior to host spawning when ova are provisioned with yolk, *Polypodium* everts to assume normal cell layer position, filling its gastric cavity with host yolk and revealing tentacles distributed along the stolon's length.

The pennatulaceans are the sea pens and sea pansies, and these are the most complex and polymorphic members of the class Anthozoa (Figure 7.1C,E and 7.11D,F). The colony is built around a main supportive stem, which is actually the primary polyp and buds lateral polyps in a regular fashion. The base of the primary polyp (the **peduncle**) is anchored in sediment, but the upper, exposed portion (the **rachis**) produces polyps in whorls or rows, or sometimes united in crescent-shaped "leaves." Often these polyps are of two distinct types. **Autozooids** bear tentacles and function in feeding; **siphonozooids** are small, have reduced tentacles, and serve to create water currents through the colony. In sea pens the rachis is elongated and cylindrical; in sea pansies it is flattened and shaped like a large leaf (Figure 7.1E). In the odd deep-sea genus *Umbellula*, the secondary polyps radiate outward to give the colony the appearance of a pinwheel set on the end of a tall narrow stalk. The first deep benthic photos of *Umbellula* had biologists scratching their heads for years, wondering to which phylum this preposterous creature might belong.

Gorgonians are also colonial anthozoans (Figures 7.1D and 7.12). Some grow in bushy shapes, whereas others are planar; size and shape of the colony are often mediated by the hydrodynamics of prevailing surge and currents. Where prevailing currents are more or less in one plane (although they may move in two directions back and forth), the branches of the colony tend to grow largely in one plane also—perpendicular to the flow. In regions of mixed currents, the same species tends to grow in two planes.

The medusoid form Free medusae occur only in the subphylum Medusozoa. Although variation in form exists, medusae are far less diverse than polyps and it is much easier to generalize about their anatomy. (After reading about the amazing variation in polypoid colonies, you're probably happy to hear this!) The relative uniformity of medusae is largely a result of their usually similar lifestyles in open water, and of their inability to form colonies by asexual reproduction. They are participants in colonial life only insofar as some remain attached to hydrozoan colonies to function as sessile gonophores. Sessile benthic medusae exist, but are rare (Figure 7.1F). Despite their simplicity relative to polypoid forms, medusae do vary considerably in size, ranging from 2.0 mm diameter hydromedusae to 2.0 m scyphomedusae, and depending on their habitat can exhibit particular morphological specializations associated with movement in a fluid medium. In many medusae each radial canal opens near the ring canal onto the subumbrellar surface by a pore. These are thought to be "excretory pores" for the ejection of wastes and indigestible material that have been seen passing through them. While polyps often represent the vegetative stage of development among cnidarians, most medusae are sexual forms.

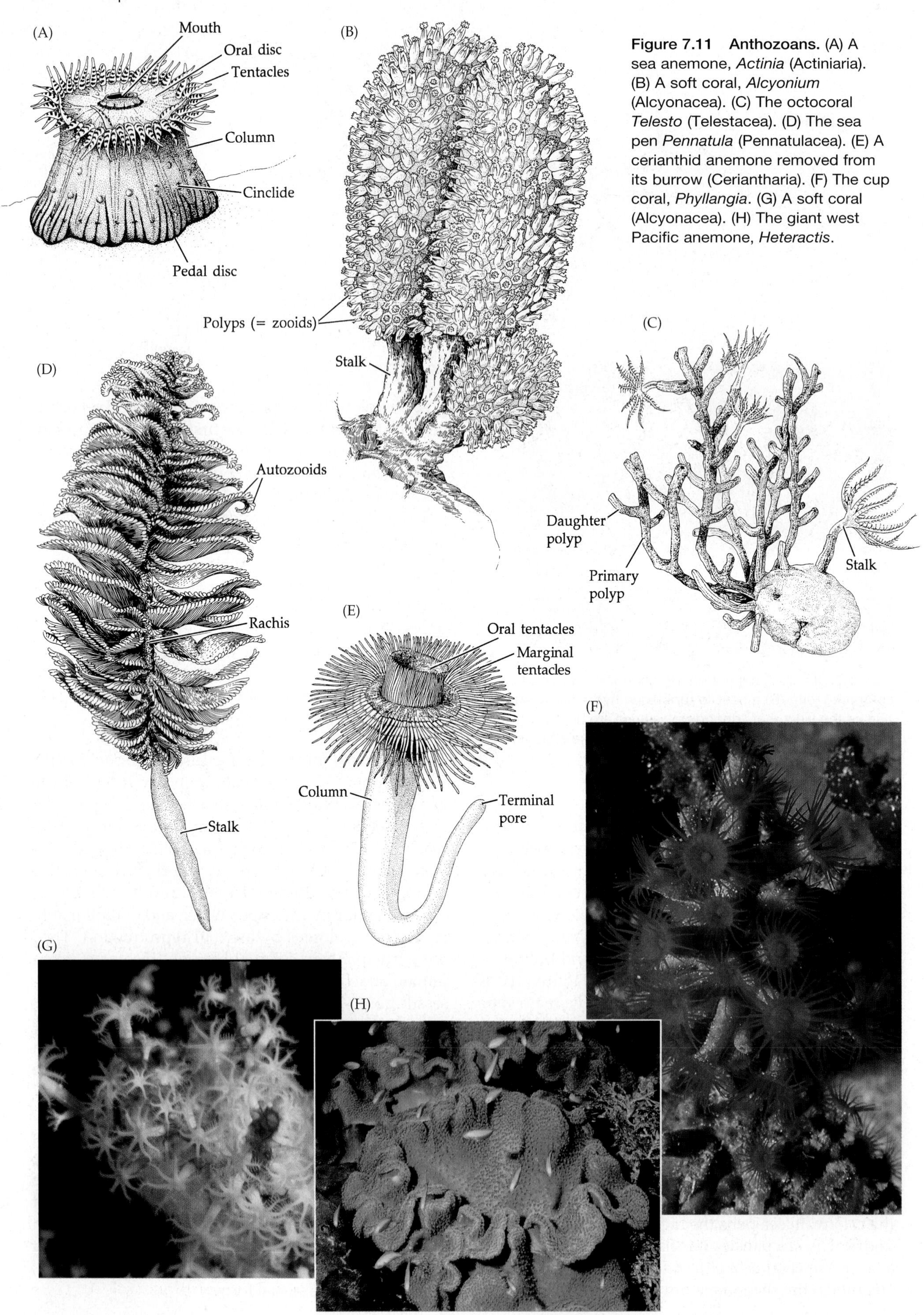

Figure 7.11 Anthozoans. (A) A sea anemone, *Actinia* (Actiniaria). (B) A soft coral, *Alcyonium* (Alcyonacea). (C) The octocoral *Telesto* (Telestacea). (D) The sea pen *Pennatula* (Pennatulacea). (E) A cerianthid anemone removed from its burrow (Ceriantharia). (F) The cup coral, *Phyllangia*. (G) A soft coral (Alcyonacea). (H) The giant west Pacific anemone, *Heteractis*.

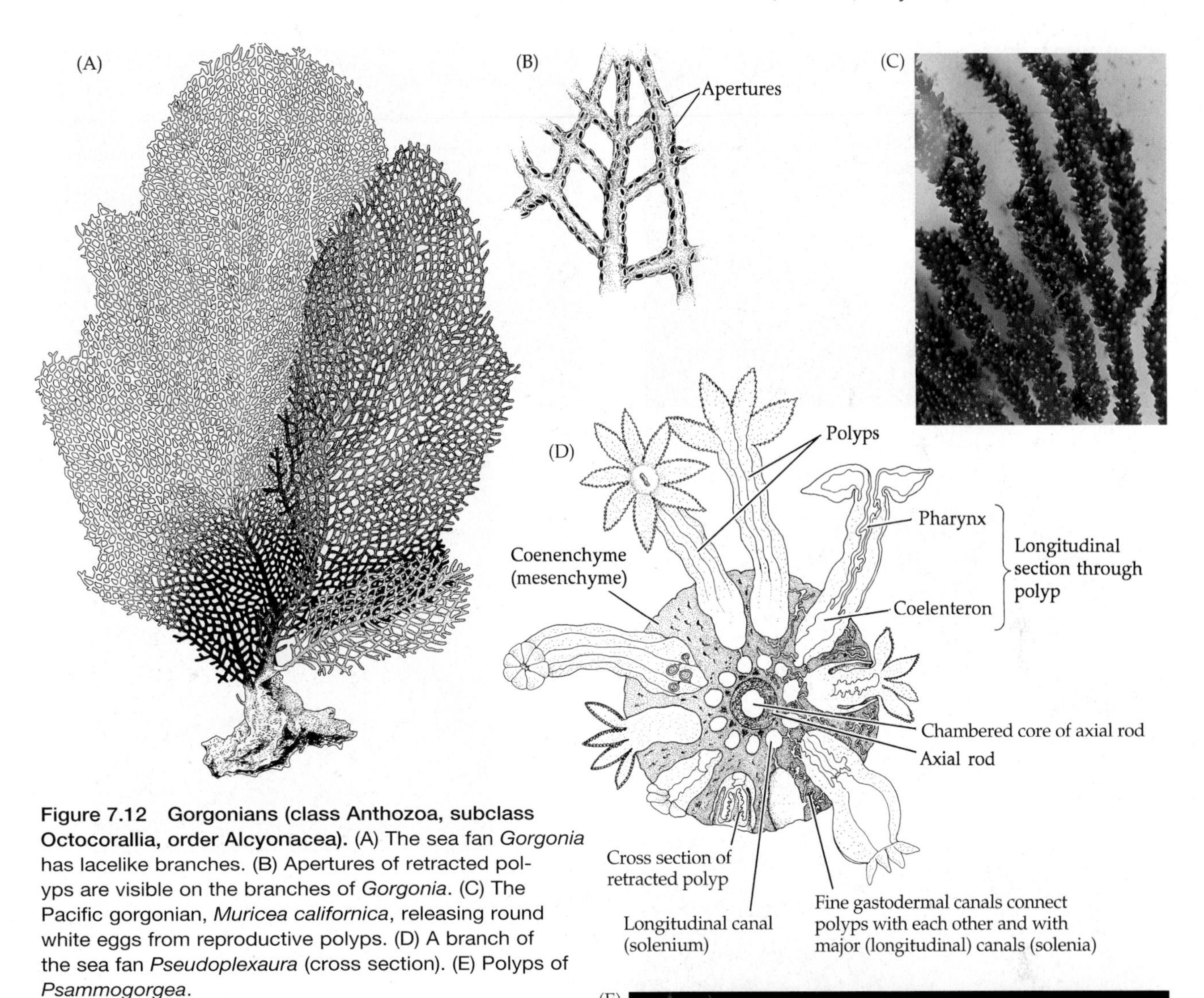

Figure 7.12 Gorgonians (class Anthozoa, subclass Octocorallia, order Alcyonacea). (A) The sea fan *Gorgonia* has lacelike branches. (B) Apertures of retracted polyps are visible on the branches of *Gorgonia*. (C) The Pacific gorgonian, *Muricea californica*, releasing round white eggs from reproductive polyps. (D) A branch of the sea fan *Pseudoplexaura* (cross section). (E) Polyps of *Psammogorgea*.

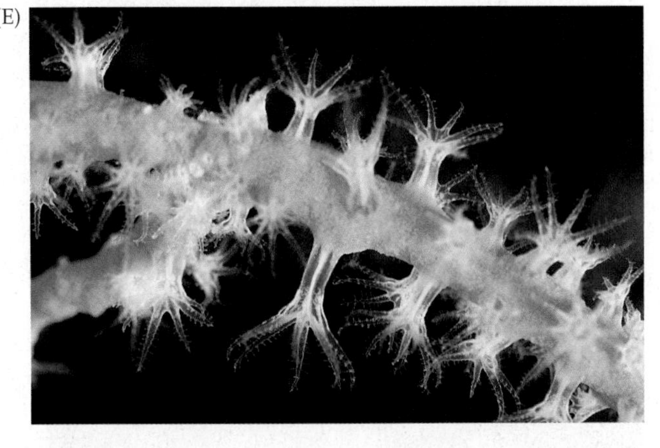

Even though the body walls of medusae and polyps are similar and both adhere to the general cnidarian body plan outlined earlier, their gross morphologies are adapted to their very different lifestyles. Medusae are bell-, dish-, or umbrella-shaped, and usually imbued with a thick, jellylike mesogleal layer (hence the name jellyfish, or simply "jellies"). The convex upper (aboral) surface is called the **exumbrella**; the concave lower (oral) surface is the **subumbrella**. The mouth is located in the center of the subumbrella, often suspended on a pendant, tubular extension called the **manubrium**, which is almost always present on hydromedusae (Figure 7.13), but usually reduced or absent in scyphomedusae (Figure 7.14).

The coelenteron or gastrovascular cavity occupies the central region of the umbrella and extends radially in the body via **radial canals**. In most hydromedusae (medusae of the class Hydrozoa), a marginal ring canal within the rim of the bell connects the ends of the radial canals. The presence of four radial canals and of tentacles in multiples of four (in hydromedusae) and the division of the stomach by mesenteries into four **gastric pouches** (in scyphomedusae) give most jellyfish a quadriradial (= tetramerous) symmetry (Figure 7.3A). Most hydromedusae have a thin circular flap of tissue, the **velum**, within the margin of the bell (Figure 7.13). Such medusae are termed **craspedote**. Those lacking a velum, such as scyphomedusae, are said to be **acraspedote** (Figures 7.14 and 7.15). The medusae of cubozoans possess an independently evolved structure, the **velarium**, which is structurally different but

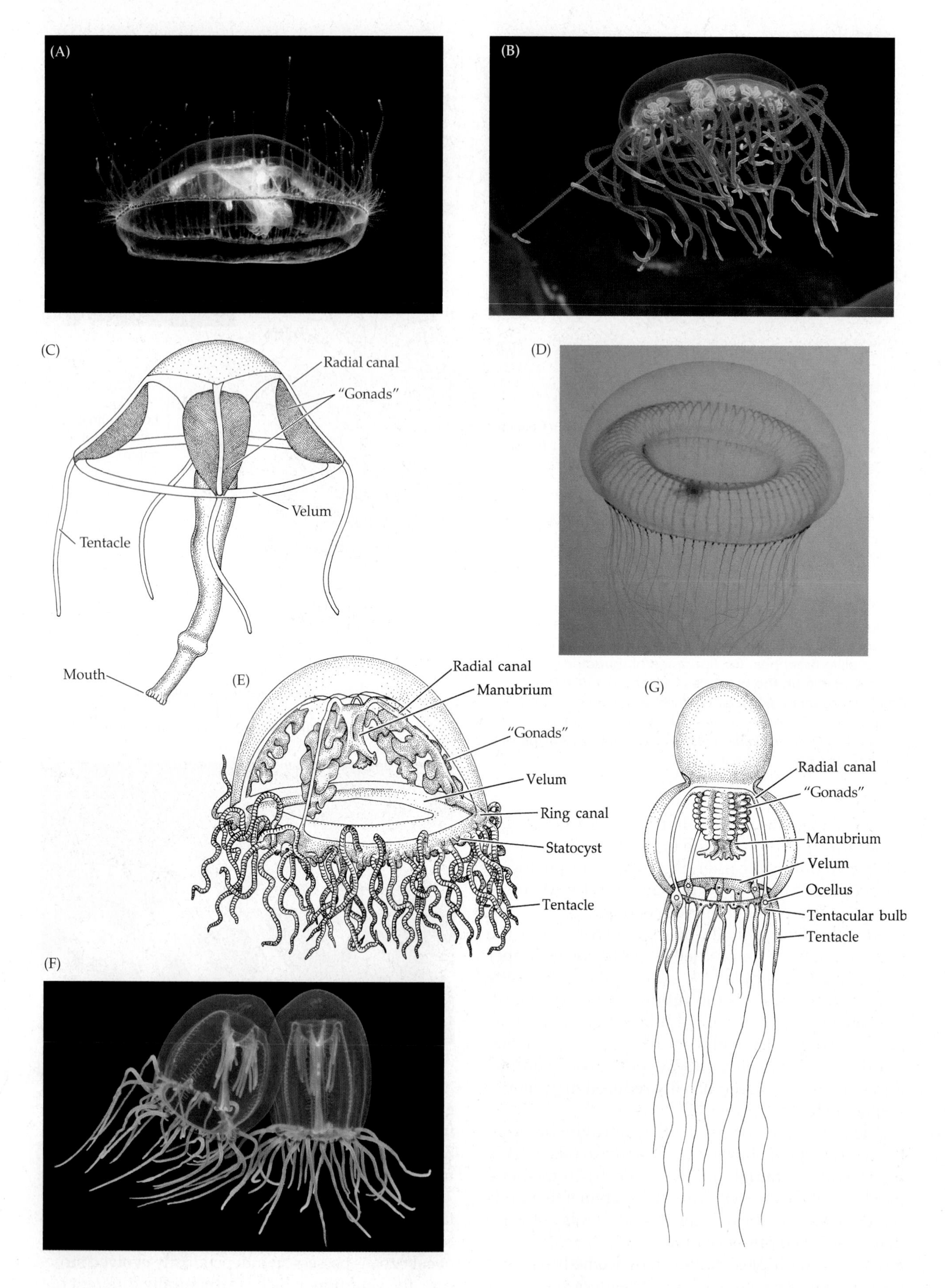
(A)
(B)
(C)
Radial canal
"Gonads"
Velum
Tentacle
Mouth
(D)
(E)
Radial canal
Manubrium
"Gonads"
Velum
Ring canal
Statocyst
Tentacle
(G)
Radial canal
"Gonads"
Manubrium
Velum
Ocellus
Tentacular bulb
Tentacle
(F)

◀ **Figure 7.13 Hydrozoan medusae.** (A–C) Limnomedusae (subclass Trachylina): (A) a freshwater limnomedusa, *Craspedacusta sowerbyi*. (B) A temperate coastal species, *Gonionemus vertens*. (C) *Liriope tetraphylla*. (D,E) Leptomedusae (subclass Hydroidolina): (D) *Aequoria victoria*. (E) Anatomy of a typical leptomedusan. (F,G) Anthomedusae (subclass Hydroidolina): (F) Two individuals of *Polyorchis* sp. (G) Anatomy of a typical anthomedusan.

functionally similar to the hydromedusan velum. As in polyps, the external surfaces of medusae are covered with epidermis, and the internal surfaces (coelenteron and canals) are lined with gastrodermis. The bulky, gelatinous middle layer is either a largely acellular mesoglea or a partly cellular mesenchyme.

Although morphologically similar, diversity within scyphozoan genera may be much greater than previously suspected. Molecular evidence suggests that *Aurelia aurita*, a cosmopolitan semaeostome species often used in invertebrate zoology laboratories to illustrate medusoid structure, may consist of 7 or more distinct species. While morphologically indistinguishable (so far), these genetically distinct populations may have diverged as early as the late Cretaceous (> 65 Ma).

Support

Cnidarians employ a wide range of support mechanisms. Polypoid forms rely substantially on the hydrostatic qualities of the water-filled coelenteron, which is constrained by circular and longitudinal muscles of the body wall. In addition, the mesenchyme may be stiffened with fibers, particularly in the anthozoans. Colonial anthozoans may incorporate bits of sediment and shell fragments onto the column wall for further support. Many colonial hydrozoans produce a flexible, horny perisarc, composed largely of chitin secreted by the epidermis. In medusae, the principal support mechanism is the middle layer, which ranges from a fairly thin and flexible mesoglea to an extremely thick and stiffened fibrous mesenchyme, which may be almost cartilaginous in consistency.

In addition to these soft or flexible support structures, there is an impressive array of hard skeletal structures of three fundamental types: horny or wood-like axial skeletal structures, calcareous sclerites, and massive calcareous frameworks. Horny axial skeletons occur in several groups of colonial anthozoans such as

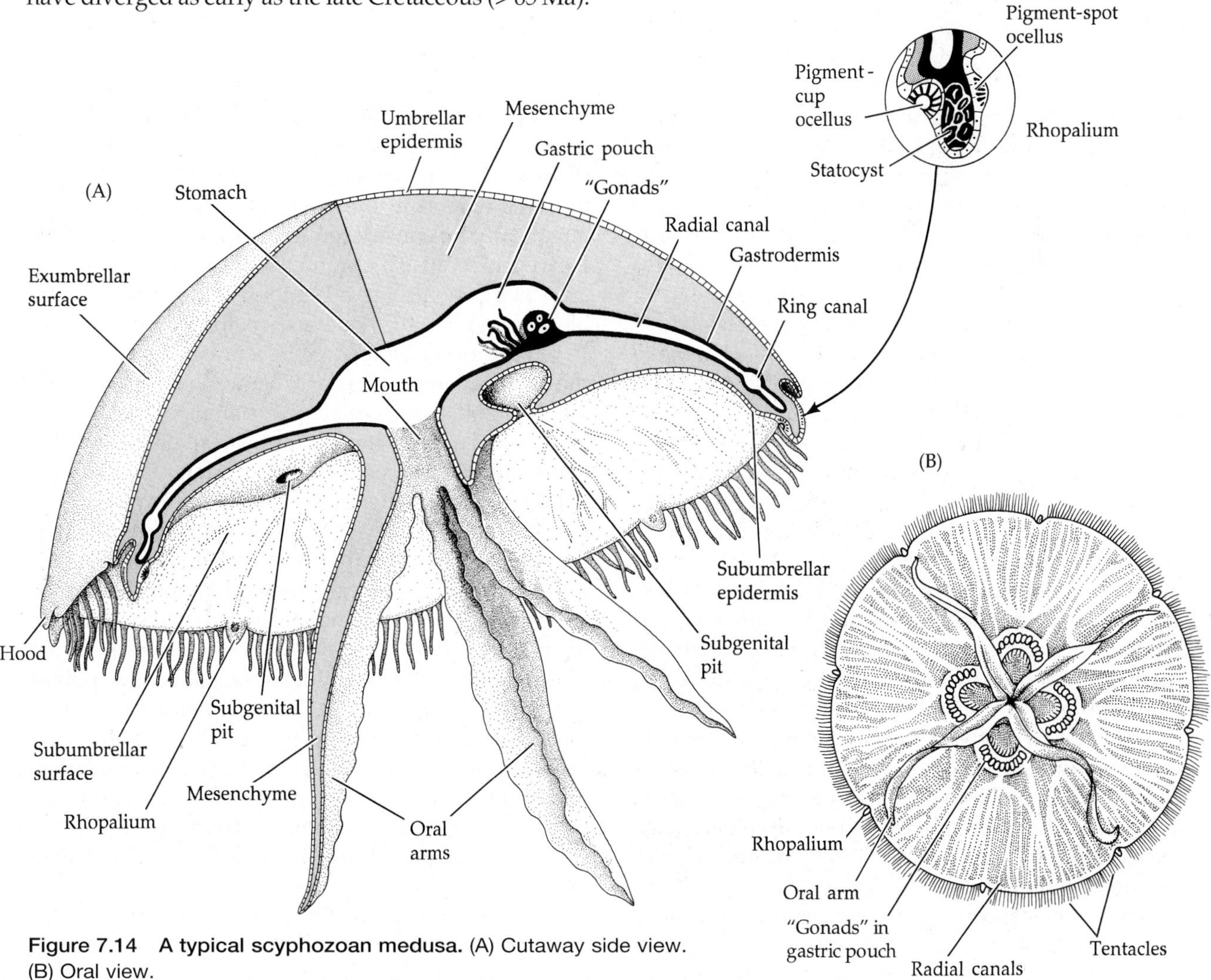

Figure 7.14 A typical scyphozoan medusa. (A) Cutaway side view. (B) Oral view.

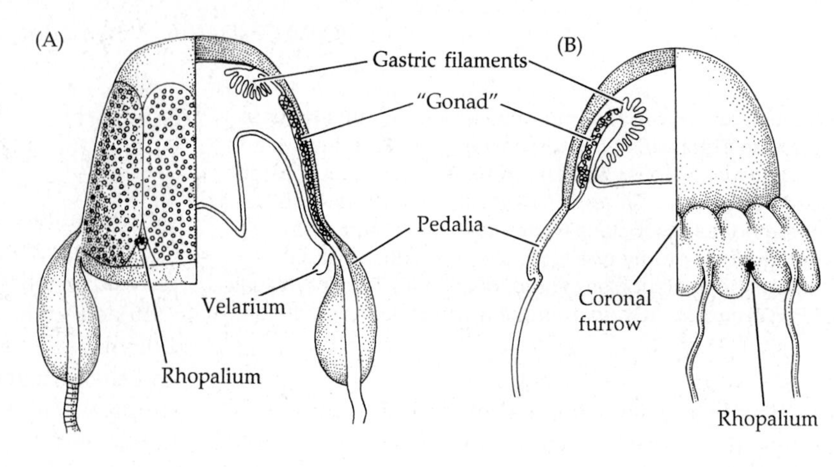

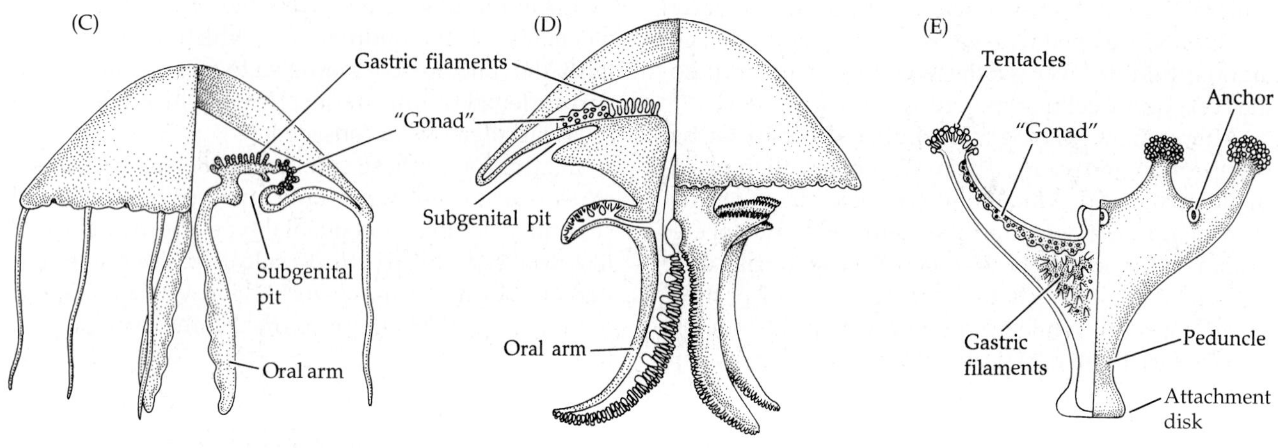

Figure 7.15 Anatomy of cubomedusae and scyphomedusae. (A) A cubomedusa. (B) A coronate scyphomedusa (order Coronatae). (C) A semaeostome scyphomedusa (order Semaeostomae). (D) A rhizostome scyphomedusa (order Rhizostomae). (E) A sessile staurozoan (class Staurozoa).

gorgonians, sea pens, and antipatharian corals (Figures 7.11 and 7.12). Amebocytes in the coenenchyme secrete a flexible or stiff internal **axial rod** as a supportive base embedded in the coenenchymal mass. Axial rods are protein–mucopolysaccharide complexes (called **gorgonin** for "gorgonian" corals in the octocoral order Alcyonacea), but little is known of their chemistry. In the antipatharians (black corals), the axial skeleton is so hard and dense that it is ground and polished to make jewelry (leading to a serious over-harvesting of these animals around the world).

In most octocorals, mesenchymal cells called **scleroblasts** secrete calcareous sclerites of various shapes and colors (Figure 7.16). It is usually these sclerites that give soft corals and gorgonians their characteristic color and texture. In many species, the sclerites become quite dense and may even fuse to form a more-or-less solid calcareous framework. The precious red coral *Corallium* is actually a gorgonian with fused red coenenchymal sclerites. In the stoloniferan organ-pipe corals (*Tubipora*), the sclerites of the body walls of the individual polyps are fused into rigid tubes. Invertebrate calcium carbonate skeletons do not usually have collagen incorporated into their framework, as occurs in vertebrates. However, in at least some gorgonians (e.g., *Leptogorgia*) the calcareous spicules do include a collagen component.

Massive calcareous skeletons are found in only certain groups of Anthozoa and Hydrozoa. The best known are the stony anthozoan corals (order Scleractinia), in which epidermal cells on the lower half of the column secrete a calcium carbonate skeleton (Figure 7.17). The skeleton is covered by the thin layer of living epidermis that secretes it, and thus it might technically be considered to be an internal skeleton. However, because the stony coral colony generally sits atop a large nonliving calcareous framework, most biologists speak of the skeleton as being external.

The entire skeleton of a scleractinian coral is termed the **corallum**, regardless of whether the animal is solitary or colonial; the skeleton of a single polyp, however, is called a **corallite**. The outer wall of the corallite is the **theca**; the floor is the **basal plate** (Figure 7.17). Rising from the center of the basal plate is often a supportive skeletal process called the **columella**. The basal plate and inner thecal walls give rise to numerous radially arranged calcareous partitions, the **septa**, which project inward and support the mesenteries of the polyp. Polyps occupy only the uppermost surface of the corallum. Skeletal thickness increases as polyps grow, and the bottoms of the corallites are sealed off by transverse calcareous partitions called **tabulae**, each of which becomes the basal support of a new polyp. The corallum can assume a great variety of shapes and sizes, from simple cup-shaped structures in solitary corals to large branching or encrusting forms in colonial species.

Members of the hydrozoan families Milleporidae (milleporids) and Stylasteridae (stylasterines) also produce calcareous exoskeletons, and they are often

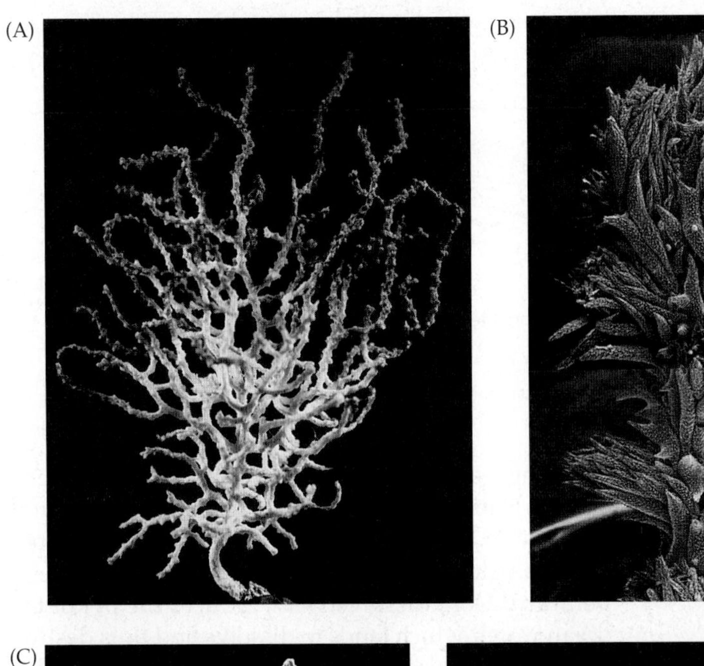

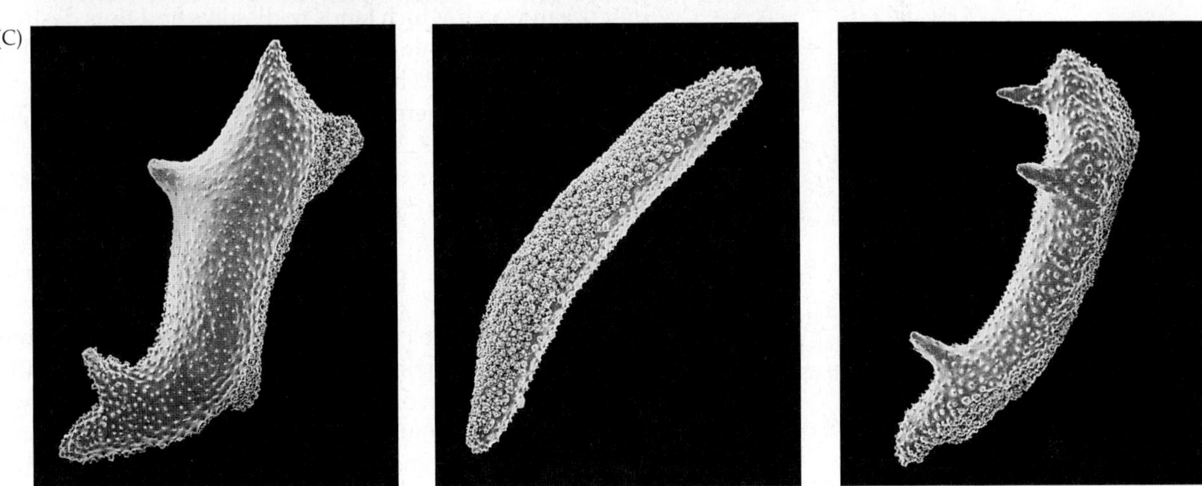

Figure 7.16 The skeleton of gorgonians, illustrated by SEMs at successively greater magnification of the gorgonian *Muricea fruticosa*. (A) A complete colony. (B) Colony branches bear whorls of polyps. (C) Sclerites from the tissues of a single polyp.

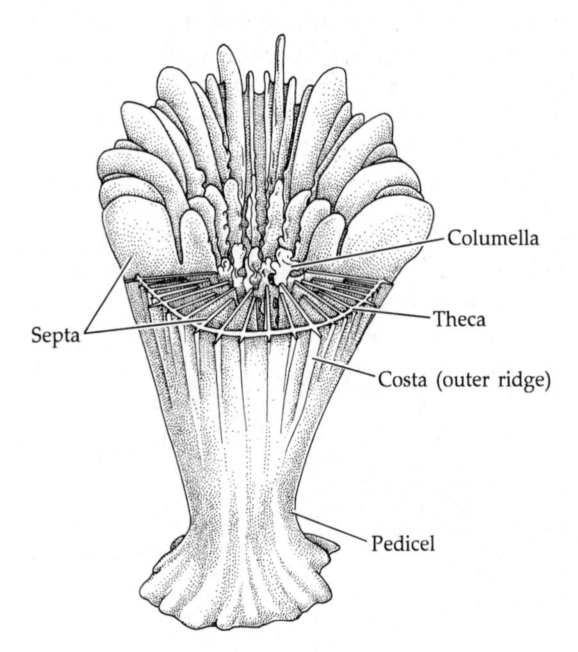

Figure 7.17 The corallite of a solitary scleractinian coral illustrating morphological features.

referred to as the hydrocorals. Like stony corals (Scleractinia), milleporid colonies (fire corals) may assume a variety of shapes, from erect branching forms to encrustations. The milleporid exoskeleton, termed a **coenosteum**, is perforated by pores of two sizes that accommodate two kinds of polyps (Figure 7.18). The gastrozooids live in large holes, or **gastropores**, and are surrounded by a circle of smaller **dactylopores**, which house the dactylozooids. Canals lead downward from the pores into the coenosteum and are closed off below by transverse calcareous tabulae. As growth proceeds and the colony thickens, new tabulae are formed, keeping the polyp pores at a more or less fixed depth. Hydrocoral colonies thus differ from scleractinian colonies in having the skeleton penetrated by living tissue. The stylasterine skeleton is similar to the milleporid skeleton, but the margins of the gastropores often bear notches that serve as dactylopores, and the gastrozooids and dactylozooids are supported by calcareous, spine-like **gastrostyles** and low ridges called **dactylostyles**, respectively. Stylasterine gonophores arise in

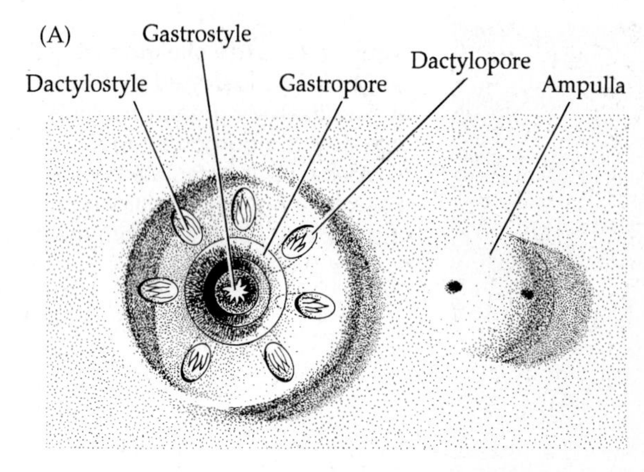

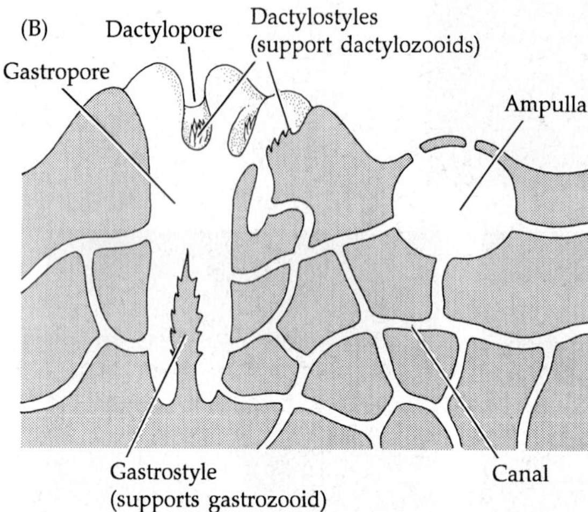

Figure 7.18 Hydrozoan skeletons. The stylasterine hydrocoral *Allopora* has a calcareous skeleton. Plane view, from above (A) and cross section through the skeleton (B).

chambers called **ampullae**, which connect to the feeding zooids through the coenosteum. In hydrocorals such as *Millepora*, ampullae open briefly to release large numbers of tiny medusae, which for each **coralla** (colony) contain either eggs or sperm, as milleporids are gonochoristic.

The calcium carbonate skeletons of cnidarians make them particularly vulnerable to a seldom appreciated outcome of continued carbon emissions and global

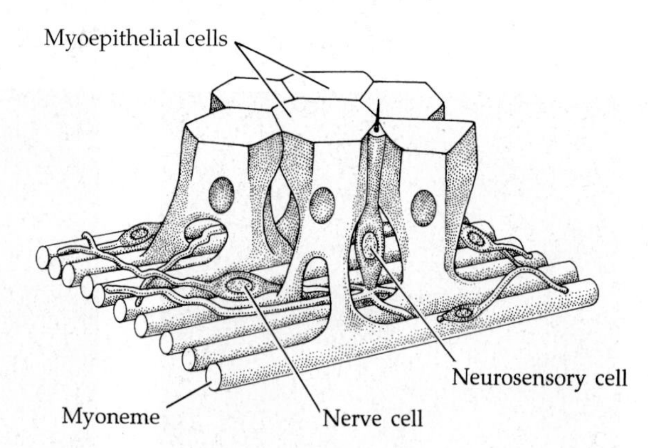

Figure 7.19 Myoepithelial cells and the nerve net of cnidarian epithelium.

warming—ocean acidification. Increased atmospheric carbon leads to more CO_2 dissolving into seawater (now estimated to occur at a rate of 1 million tons per hour), a reaction that accelerates with increasing temperature. More dissolved CO_2 reduces the pH of the ocean water, which tends to dissolve and thus destroy the primary building material of coral skeletons, calcium carbonate. Whereas calcification rates on the Great Barrier Reef increased 5.4% between 1900 and 1970, they dropped 14.2% between 1990 and 2005. The shells of echinoderms, molluscs, crustaceans, certain protists, and many other marine species are at similar and alarming risk.

Movement

The contractile elements of cnidarians are derived from their myoepithelial cells (Figure 7.19). In spite of the epithelial origin of these elements, for convenience we use the terms "muscles" and "musculature" for the sets of longitudinal and circular fibrils. In polyps, these two muscle systems work in conjunction with the gastrovascular cavity as an efficient hydrostatic skeleton, as well as providing a means of movement. However, unlike the fixed-volume hydrostatic skeletons of many animals (e.g., many worms), water can enter and leave the coelenteron of cnidarians, adding to its versatility as a support device. Polyp body musculature is most highly specialized and well developed in the anthozo-

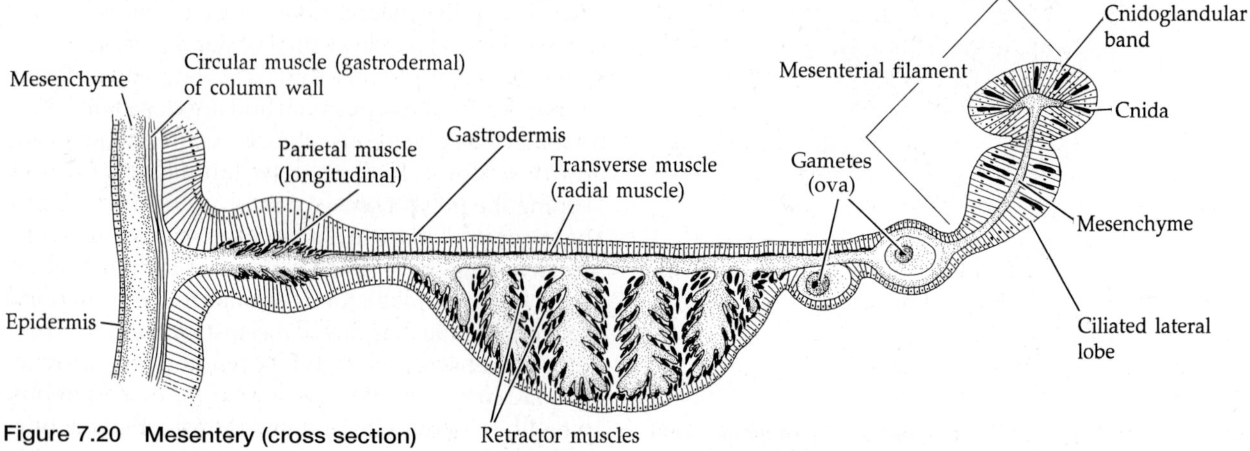

Figure 7.20 Mesentery (cross section) of a sea anemone (Actiniaria).

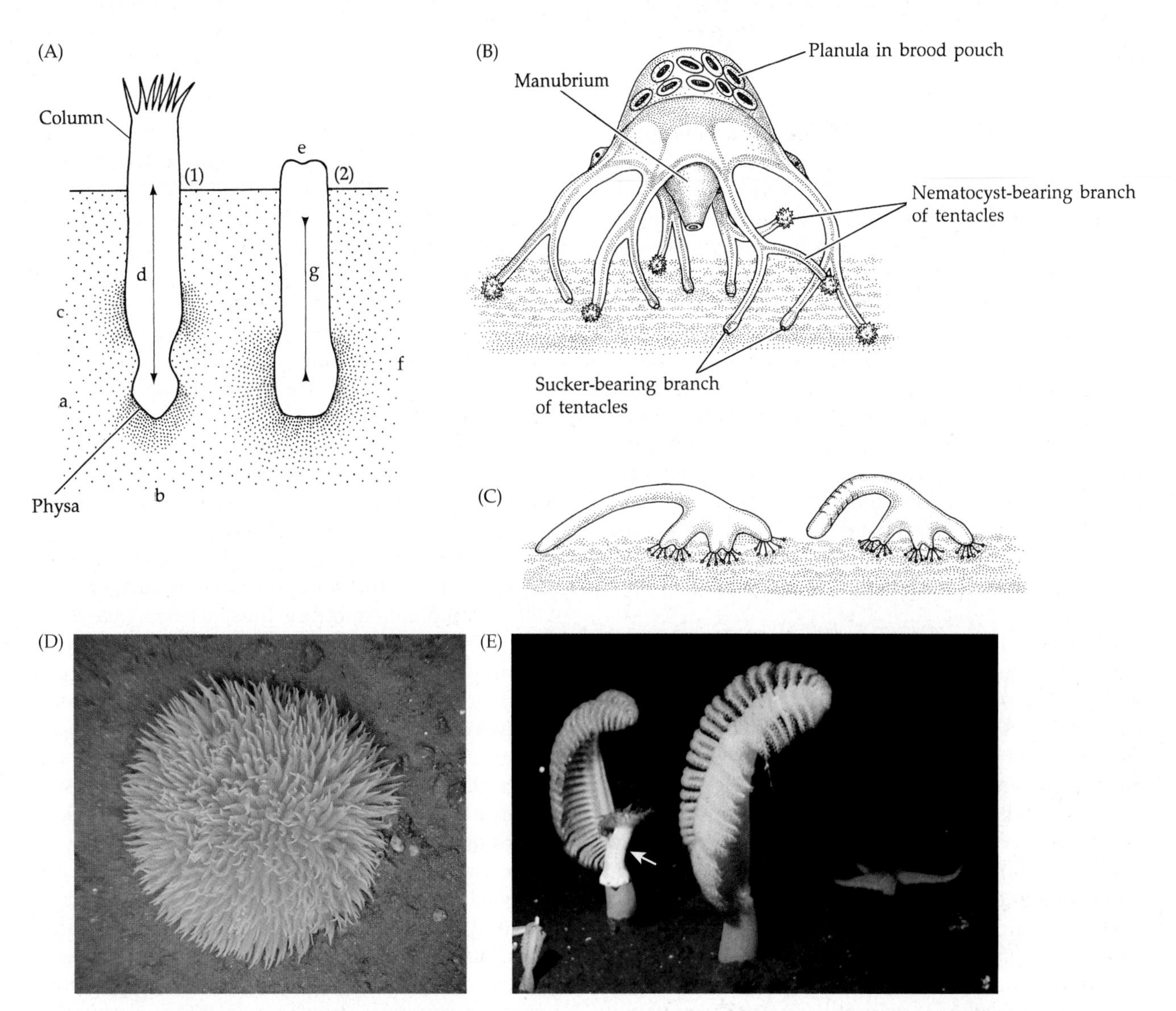

Figure 7.21 Benthic locomotion in some cnidarians. (A) A sea anemone burrowing: (1) eversion of the physa with displacement of sand (a) and further penetration (b) into substratum; the anemone is held by a column anchor (c) as extension (d) follows retraction in (2); with the tentacles folded inward (e), the physa is swollen to form an anchor (f), which allows retractor muscles (g) to pull the anemone into the sand. (B) The hydromedusan *Eleutheria*, which creeps about on its tentacles. (C) The staurozoan *Lucernaria*, which also creeps about on its tentacles. (D) *Liponema brevicornis*, a sea anemone that folds itself into a "ball" and rolls about on the sea floor with the bottom currents. (E) The sea anemone *Stomphia* (white arrow) swimming off the substratum by undulatory back-and-forth contractions of the column—an escape response to the predatory sea star *Gephyreaster swifti*, visible in this photo (Puget Sound, Washington).

ans, particularly the sea anemones, and many muscles lie in the mesenchyme. In anemones, the muscles of the column wall are largely gastrodermal, although epitheliomuscular cells occur in the tentacles and oral disc. Bundles of longitudinal fibers lie along the sides of the mesenteries and act as retractor muscles for shortening the column (Figure 7.20). Circular muscles derived from the gastrodermis of the column wall are also well developed. In most anemones, the circular muscles form a distinct sphincter at the junction of the column and the oral disc. Circular fibers also occur in the tentacles and the oral disc, and circular muscles surrounding the mouth can close it completely. When an anemone contracts, the upper rim of the column is pulled over to cover the oral disc. In many sea anemones, a circular fold—the **collar**, or **parapet**—occurs near the sphincter to further cover and protect the delicate oral surface upon contraction.

Most polyps are sedentary or sessile. Their movements consist mainly of food-capturing actions and the withdrawal of the upper portion of the polyp during body contractions. These activities are accomplished primarily by the epidermal muscles of the tentacles and oral disc, and by the strong gastrodermal muscles of the column. Circular muscles work in conjunction with the hydrostatic skeleton to distend the tentacles and body.

A variety of locomotor methods have evolved among polyps (Figure 7.21). Most can creep about slowly by using their pedal disc musculature. In some solitary hydrozoan polyps (e.g., *Hydra*), the column can bend far enough to allow the tentacles to contact and temporarily adhere to the substratum, whereupon the pedal disc releases its hold and the animal somersaults or moves like an inchworm. Simple polyps like the hydrozoan *Hydra* transfer fluid within their gastrovascular cavity using contractions of the peduncle; these contractions are biochemically mediated by RFamides, chemicals that induce cardiac contraction in higher metazoans, suggesting that muscular contractions in these distantly related taxa might share neurological similarities. A few sea anemones can detach from the substratum and actually swim away by "rapid" flexing or bending of the column (e.g., *Actinostola, Stomphia*); others swim by thrashing the tentacles (e.g., *Boloceroides*). These swimming activities are temporary behaviors, generally elicited by the approach or contact of a predator. In a few species of sea anemones, the basal disc may detach and secrete a gas bubble, permitting the polyp to float away to a new location.

Many species of small anthozoans can float hanging upside-down on the sea surface by using water surface-tension forces (e.g., *Epiactis, Diadumene*). Sea anemones of one family (Minyadidae) are wholly pelagic and float upside down in the sea by means of a gas bubble enclosed within the folded pedal disc. *Hydra* also is known to float upside down by means of a mucus-coated gas bubble on the bottom of its pedal disc. One of the oddest forms of polyp locomotion is that of the sea anemone *Liponema brevicornis* of the Bering Sea, which is capable of drawing itself into a tight ball that can be rolled around the sea floor by the bottom currents (Figure 7.21D). Even colonial sea pansies (Pennatulacea) are motile, in that they can use their muscular peduncle to move to different depths on the sea floor.

Most cerianthid anemones are burrowing, tube-building organisms (Figure 7.11E). They differ from the sea anemones (Actiniaria) in several important ways. They have no sphincter muscle, and their weak longitudinal gastrodermal muscles do not form distinct retractors in the mesenteries. As a result, cerianthids cannot retract the oral disc and tentacles as they withdraw into their tubes. In contrast to other anemones, however, they possess a complete layer of longitudinal epidermal muscles in the column, which allows a very rapid withdrawal response. The mere shadow of a passing hand will cause a cerianthid to rapidly pull itself deep into its long, buried tube.

In medusae, epidermal and subepidermal musculatures predominate, and the gastrodermal muscles that are so important in polyps are reduced or lacking. The epidermal musculature is best developed around the bell margin and over the subumbrellar surface. Here the muscle fibers usually form circular sheets called **coronal muscles** that are partly embedded in the mesenchyme or mesoglea. Contractions of the coronal muscles produce rhythmic pulsations of the bell, driving water out from beneath the subumbrella and moving the animal by jet propulsion. The restriction of striated myofibrils to epithelial cells appears to constrain the force with which bell musculature may contract, favoring either small solitary or groups of **prolate** (streamlined) bells that move by **jet propulsion** (e.g., Anthoathecata, Trachymedusae, Siphonophora, Cubozoa) or larger, **oblate** (flattened) bells that move by more gentle contractions of the bell margin, called **rowing** (e.g., Leptothecata, Narcomedusae, scyphozoan medusae).

The stiffened cellular collenchyme of scyphomedusae and cubomedusae includes elastic fibers that provide the antagonistic force to restore the bell shape between contractions. Many medusae also possess radial muscles that aid in opening the bell between pulses. In craspedote forms, the velum serves to reduce the size of the subumbrellar aperture, thus increasing the force of the water jet (Figure 7.13). The velarium of the fast-swimming cubomedusae has the same effect (Figure 7.15A), and the evolutionary forces that produced these two convergent features were probably similar.

Most medusae spend their time swimming upward in the water column, then sinking slowly down to capture prey by chance encounter, thereafter to pulsate upward once again. Some medusae have the ability to change direction as they swim, however, and many are strongly attracted to light (especially those harboring symbiotic zooxanthellae). Medusoid form also appears to correlate with feeding mode. Jet propulsion is associated with ambush foraging by medusa who lie motionless, waiting for motile prey to swim into their tentacles before rapidly consuming the ensnared prey, whereas rowing propulsion is associated with cruising foraging by medusa which swim continuously with tentacles extended to capture slow moving or floating prey. At least some medusae house their zooxanthellae in small pockets that remain contracted at night but expand during the day, exposing the algae to light.

Medusae can be abundant in certain localities. Some, such as the moon jelly *Aurelia* (Figure 7.22), are known to aggregate at temperature or salinity discontinuity layers in the sea where they feed on small zooplankters, which also concentrate at these boundaries. Large flotillas of scyphomedusae are sometimes seen at sea (e.g., *Phacellophora* in the eastern Pacific). A few unusual groups of medusae are benthic. Some hydromedusae (e.g., *Eleutheria, Gonionemus*) crawl about on algae or sea grasses by adhesive discs on their tentacles (Figure 7.21B). Members of the Class Staurozoa (e.g., *Haliclystus*) develop directly from the stauropolyp stage and affix to algae and other substrata by an aboral adhesive disc (Figure 7.1F). Aggregations are common in scypho- and cubomedusae, possibly to enhance feeding or defense. Almost all cubomedusae are tropical to subtropical in their range, but a large temperate

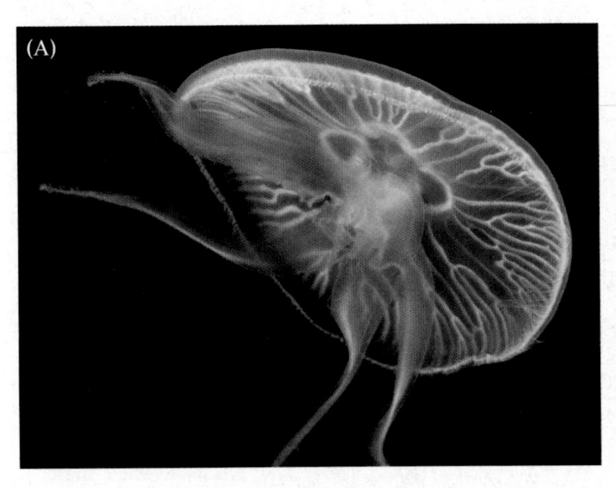

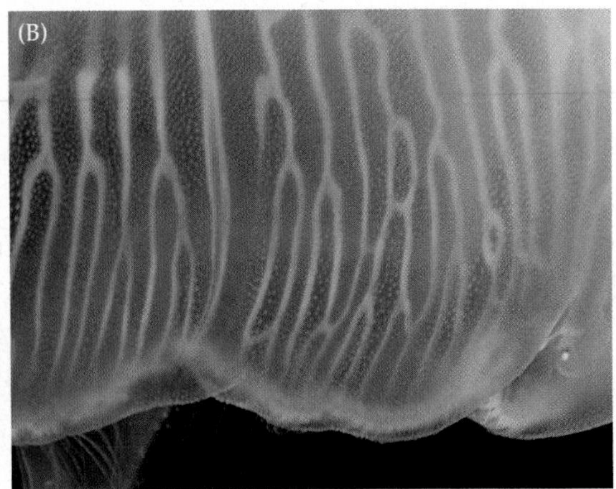

Figure 7.22 The semaeostoman medusa *Aurelia* (moon jellies) often form large swarms. (A) *Aurelia aurita*; notice elongated oral arms. (B) *Aurelia* radial canals and rhopalium.

species (*Carybdea branchi*) occurs on the Skeleton Coast of southwestern Africa where it can occur in dense "clouds" of an acre or more across.

Cnidae

Before considering feeding and other aspects of cnidarian biology, it is necessary to present some information on the structure and function of cnidae. **Cnidae** (sing. cnida), often referred to collectively as "nematocysts" in older works, are unique to the phylum Cnidaria. They have a variety of functions, including prey capture, defense, locomotion, and attachment. They are produced inside cells called **cnidoblasts**, which develop from interstitial cells in the epidermis and, in many groups, also in the gastrodermis. Once the cnida is fully formed, the cell is properly called a **cnidocyte**. During formation of a cnida, the cnidoblast produces a large internal vacuole in which a complex but poorly understood intracellular reorganization takes place. Cnidae may be complex secretory products of the Golgi apparatus of the cnidoblast. There is also some evidence that cnidae might have originated symbiogenetically from some ancient protist(s), and cnidalike structures have been reported from such diverse groups as dinoflagellates, "sporozoans," and microsporans.

Cnidae are among the largest and most complex intracellular structures known. When fully formed, they are cigar- or flask-shaped capsules, 5–100 μm or more long, with thin walls composed of a collagen-like protein. One end of the capsule is turned inward as a long, hollow, coiled, eversible tubule (Figure 7.23). The outer capsule wall consists of globular proteins of unknown function. The inner wall is composed of bundles of collagen-like fibrils having a spacing of 50–100 nm, with cross-striations every 32 nm (in the nematocysts of *Hydra*). The distinct pattern of mini-collagen fibers provides the tensile strength necessary to withstand the high pressure in the capsule. The entire structure is anchored to adjacent epithelial cells (supporting cells) or to the underlying mesenchyme.

When sufficiently stimulated, the tube everts from the cell. In members of the classes Hydrozoa, Scyphozoa, and perhaps Cubozoa, the capsule is covered by a hinged lid, or operculum, which is thrown open when the cnida discharges. In members of these three classes, each cnida bears a long cilium-like bristle called a **cnidocil**, a mechanoreceptor that elicits discharge when stimulated. The cnidocil responds to specific water-borne vibration frequencies. Anthozoan cnidae lack a cnidocil and have a tripartite apical flap instead of an operculum (anthozoan cnidae are only spirocysts and ptychocysts). Cnidocytes are most abundant in the epidermis of the oral region and the tentacles, where they often occur in clusters of wartlike structures called **nematocyst batteries**.

About 30 kinds of cnidae have been described (Figures 7.24 and 7.25). Combinations of cnida types, called **cnidomes**, occur in recognizable taxonomic patterns within Cnidaria, and these have had limited usefulness in analyzing phylogenetic patterns in the phylum. However, cnidae more or less sort out as three basic types. True **nematocysts** have double-walled capsules containing a toxic mixture of phenols and proteins. The tubule of most types is armed with spines or barbs that aid in penetration of and anchorage in the victim's flesh. The toxin is injected into the victim through a terminal pore in the thread or is carried into the wound on the tubule surface. **Spirocysts** have single-walled capsules containing mucoprotein or glycoprotein. Their adhesive tubules wrap around and stick to the victim rather than penetrating it. The capsule tubules of spirocysts never have an apical pore. Nematocysts occur in members of all cnidarian classes except within the Myxozoa (although polar capsules appear to be homologous structures; see following paragraph); spirocysts occur only in the Hexacorallia.

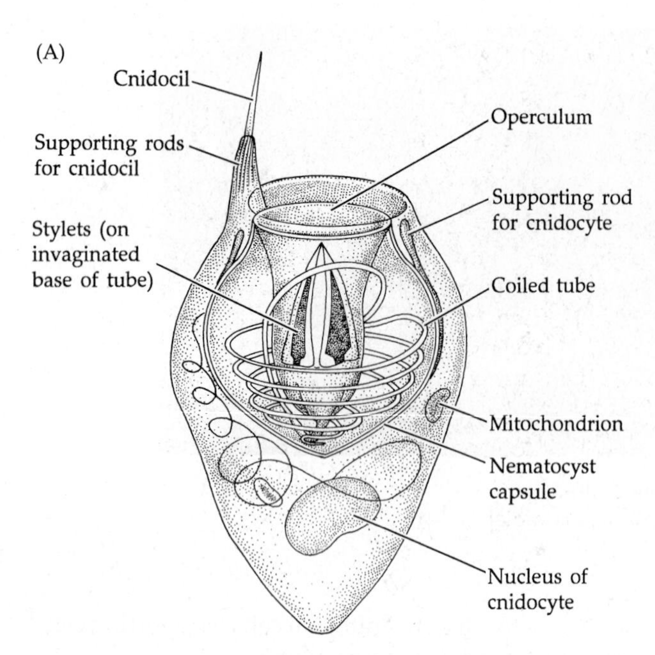

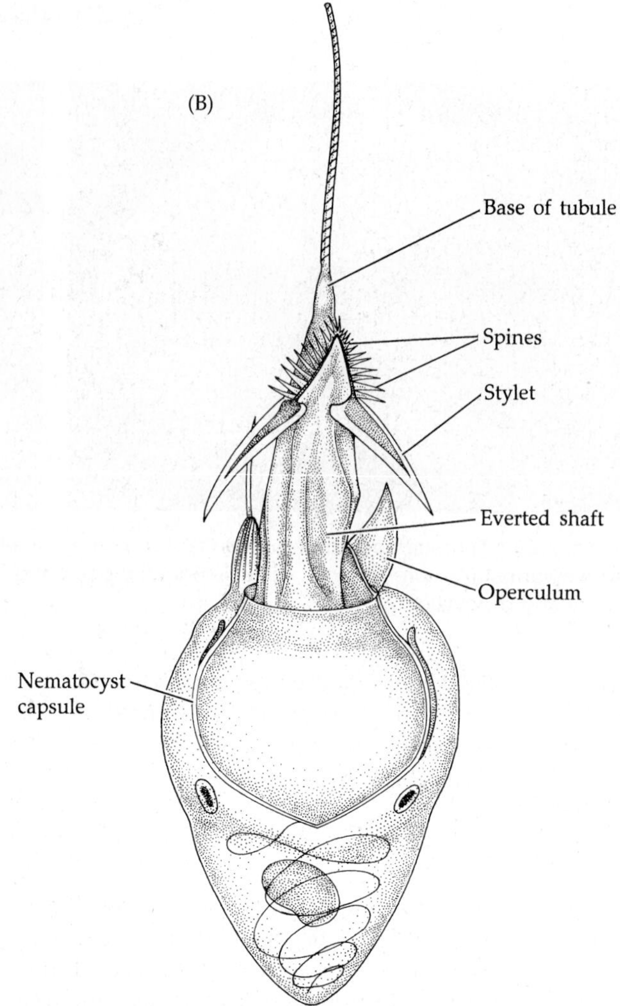

Figure 7.23 Nematocyst. (A) Before discharge. (B) After discharge.

The third kind of cnidae, the **ptychocyst**, differs morphologically and functionally from both nematocysts and spirocysts. The capsule tubule of a ptychocyst lacks spines and an apical pore and is strictly adhesive in nature. In addition, the tubule is folded into pleats rather than coiled within the capsule. Ptychocysts occur only in the cerianthids and function in forming the unique tube in which these animals reside.

The **polar capsules** of myxozoans are now widely viewed as homologous to other cnidarian cnidae, although they are simpler in form. Polar capsules are found in both myxosporean and actinosporean life stages of myxozoans. The capsules consist of a thick capsular wall; an eversible hollow filament that is spiraled along its length, may vary in length (up to 10× the length of the capsule) and is contiguous with the capsule; and a stopper-like structure that covers the inverted filament at its base. A variety of substances have been explored as possible inducers of polar filament extrusion and, like cnidae, polar capsules appear to be sensitive to pressure, extreme pH and K^+ concentrations. However, no consistent physical or chemical cue responsible for extrusion appears to exist across most taxa.

Cnidae have usually been viewed as independent effectors, and, indeed, they often discharge upon direct stimulation. However, experimental evidence suggests that the animals do have at least some control of the action of their cnidae. For example, starved anemones seem to have a lower firing threshold than satiated animals. It has also been demonstrated that stimulating discharge of cnidae in one area of the body results in discharge in surrounding areas. Still, either chemical and/or mechanical stimuli, initially perceived by the cnidocil or a similar structure, cause most cnidae to fire. Cnidarians are known to discharge their cnidae in the presence of various sugars and low-molecular-weight amino compounds.

The rapid protrusion of the tubule from a cnida is called **exocytosis**, and an individual cnida can be fired only once. Three hypotheses have been proposed to explain the mechanism of firing: (1) the discharge is the result of increased hydrostatic pressure caused by a rapid influx of water (the osmotic hypothesis); (2) intrinsic tension forces generated during cnidogenesis are released at discharge (the tension hypothesis); and (3) contractile units enveloping the cnida cause the discharge by "squeezing" the capsule (the contractile hypothesis). Because of the small size of cnidae and the extreme speed of the exocytosis process, these hypotheses have been difficult to test. Recent work using ultra high-speed microcinematography suggests that both the osmotic and tension models may be at work, and that capsules have very high internal pressures. The coiled capsular tubule is forcibly everted and thrown out of the bursting cell to penetrate or wrap around a portion of the unwary victim. It takes only a few milliseconds for the cnida to fire, and the everting tubule may reach a velocity of 2 m/sec—an acceleration force of about 40,000 g—making it one of the fastest cellular processes in nature. The firing mechanism of hydrozoan cnidae is thwarted by

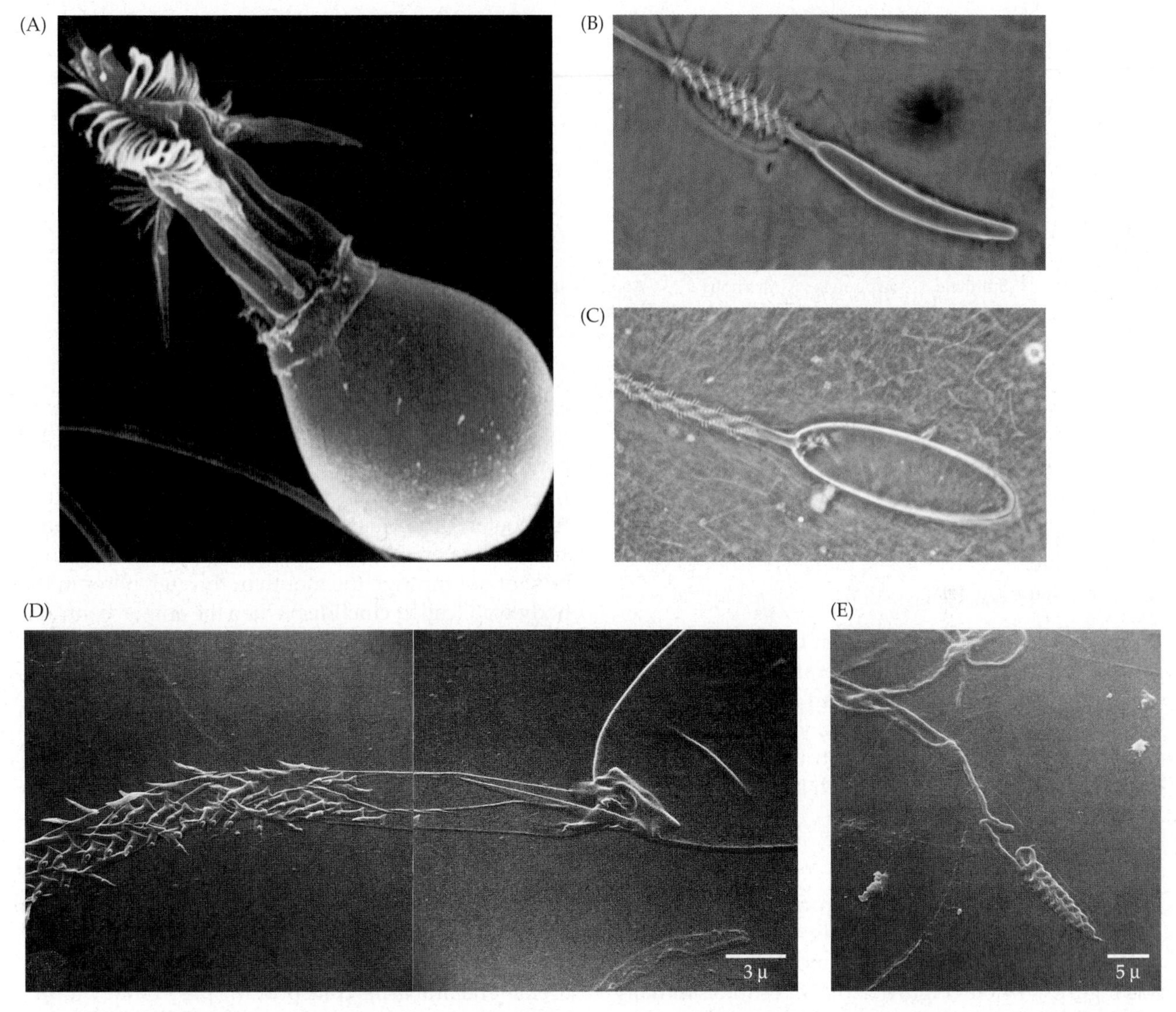

Figure 7.24 Discharged nematocysts. (A) The base of a discharged nematocyst from the hydrozoan *Hydra* (SEM). (B) A nematocyst of the anthozoan *Corynactis californica* (Corallimorpharia). The nematocyst has been "stopped" when partially everted; the everting tubule can be seen passing up tubule the already external region (light micrograph). (C) A fully everted nematocyst of *C. californica* (light micrograph). (D) A fully everted nematocyst of *C. californica* (SEM of the base of the everted thread and the tip of the capsule). (E) Everting nematocyst of the anthozoan coral *Balanophyllia elegans*.

certain nudibranch gastropods that, in order to feed upon and capture intact cnidae from their prey, release copious amounts of mucus to entangle and envelop hydroid tentacles containing still-intact cnidae.

Most nematocysts contain several different toxins that vary in activity and strength, but as a class of chemicals they are all potent biological poisons capable of subduing large, active prey, including fish. Most appear to be neurotoxins. The toxins of some cnidarians are powerful enough to affect humans (e.g., those of cubozoans; some jellyfish; certain colonial hydroids, such as *Lytocarpus*; many hydrocorals, such as *Millepora*; and many siphonophores, such as *Physalia*). The toxins of most scyphozoans are not strong enough to create problems for most people, unless they have an allergic reaction; even the effects of *Physalia* stings disappear within a few hours. However, toxins of most cubomedusans (box jellies) are another story altogether, and are estimated to be more potent than cobra venom. In tropical Australia, twice as many people die annually from box jellies as from sharks. Stings by *Chironex* (the "sea wasp") usually result in severe pain at least, and fatal respiratory or cardiac failure at worst. Both acidic and alkaline environments suppress nematocyst firing. Thus, if you come out of the surf with a jellyfish tentacle stuck to you, dousing it with urine (acidic) or baking soda (alkaline) may reduce the impact. Meat tenderizer or vinegar works somewhat, perhaps by denaturing the toxins or desensitizing the nematocysts, and in our experience following this with hot compresses seem to alleviate the pain. This is consistent with experiments showing that the lethality

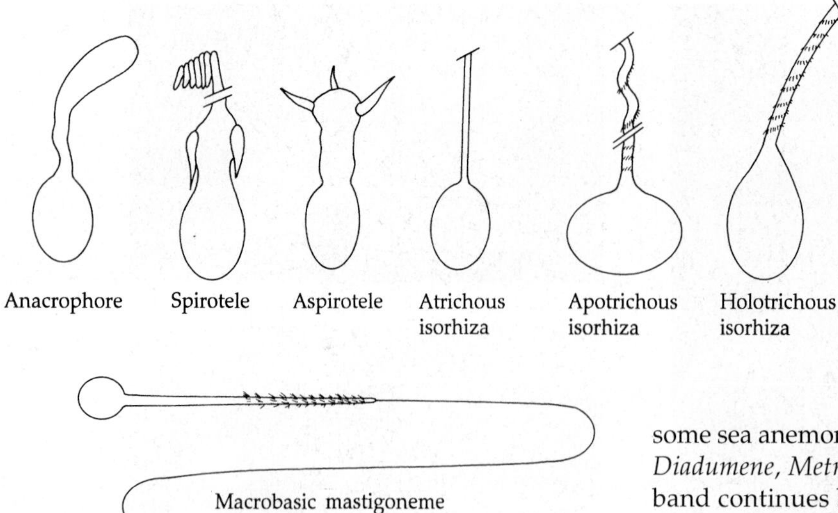

Figure 7.25 Some types of cnidae and their specialized nomenclature.

of box jellyfish (*Chironex*) venom to crayfish cardiac muscle was eliminated at temperatures approaching 60° C. If you want to swim in an area known to be frequented by dangerous jellies, you can always do what lifeguards in northern Australia do—don a pair of pantyhose (not fishnet!), which seem to offer some protection.

Feeding and Digestion

All cnidarians are carnivores (or parasites). Typically, nematocyst-laden feeding tentacles capture animal prey and carry it to the mouth region where it is ingested whole (Figure 7.26). Digestion is initially extracellular in the coelenteron. The gastrodermis is abundantly supplied with enzyme-producing cells that facilitate digestion (Figure 7.4). In many groups, gastrodermal cilia (or flagella) aid in mixing of the gut contents. In the absence of a true circulatory system, the gastrovascular cavity distributes the partially digested material. The larger the cnidarian, the more extensively branched or partitioned is its coelenteron. The product of this preliminary breakdown is a soupy broth, from which polypeptides, fats, and carbohydrates are taken into the nutritive-muscular cells by phagocytosis and pinocytosis. Digestion is completed intracellularly within food vacuoles. Undigested wastes in the coelenteron are expelled through the mouth. In cubozoans, and at least some hydrozoans, ciliary movement of material in the coelenteron is enhanced by peristaltic movements.

In anthozoans, the free edges of most of the gastrovascular mesenteries are thickened to form three-lobed mesenterial filaments (Figures 7.6 and 7.20). The lateral lobes are ciliated and aid in circulating the digestive juices in the coelenteron. The middle lobe, called the **cnidoglandular band**, bears cnidae and gland cells. In some sea anemones (e.g., *Aiptasia, Anthothoe, Calliactis, Diadumene, Metridium, Sagartia*), the cnidoglandular band continues beyond the base of the pharynx as a free thread called an acontium, which floats about in the coelenteron. The cnida-bearing acontia not only subdue live prey within the coelenteron, but they may be shot out through the mouth or through pores in the body wall (called **cinclides**) when the animal contracts violently; when this occurs, the acontia presumably play a defensive role.

Prey type appears to influence medusoid form. Most oblate medusae feed on small ciliated or soft-bodied prey, following their quarry by continuous rowing (using their large flat bell), and then using nematocysts on the tentacles, the oral arms, or both to capture their victims. *Pelagia noctiluca*, an open-sea diurnal migrator, follows the other migrating macrozooplankton upon which it feeds. *Pelagia* uses its marginal tentacles to paralyze and capture moving prey, then transports it to the oral arms dangling from the center of the subumbrella. The oral arms transport the prey to the mouth. Motionless prey may also be captured by the oral arms directly through chance contact. Most prolate medusa feed on actively swimming prey. These species often retract their tentacles while swimming to reduce drag. Cubomedusae such as *Chironex* spp. feed actively on fish, and can swim in bursts of up to five feet per second. Because of their higher metabolic demands, unlike other Medusozoa, they move semidigested food from their central gastrovascular cavity to canals lining the interior walls of each tentacle for absorption. Some cubomedusae catch and ingest fish and prawns matching their own diameter, and a large *Chironex* can eat fish 20–50 cm in length! Some cubomedusae are less active hunter-predators and feed primarily by passive hunting (e.g., *Tripedalia cystophora*, and perhaps also *Chironex fleckeri* and *Carybdea rastonii*), and in these cases the eyes are probably used to position themselves in the right food-rich habitat.

Several groups of cnidarians have adopted feeding methods other than the direct use of nematocyst-laden tentacles. One group of large tropical anemones in the order Corallimorpharia (e.g., *Amplexidiscus*) lacks nematocysts on the external surfaces of most tentacles.

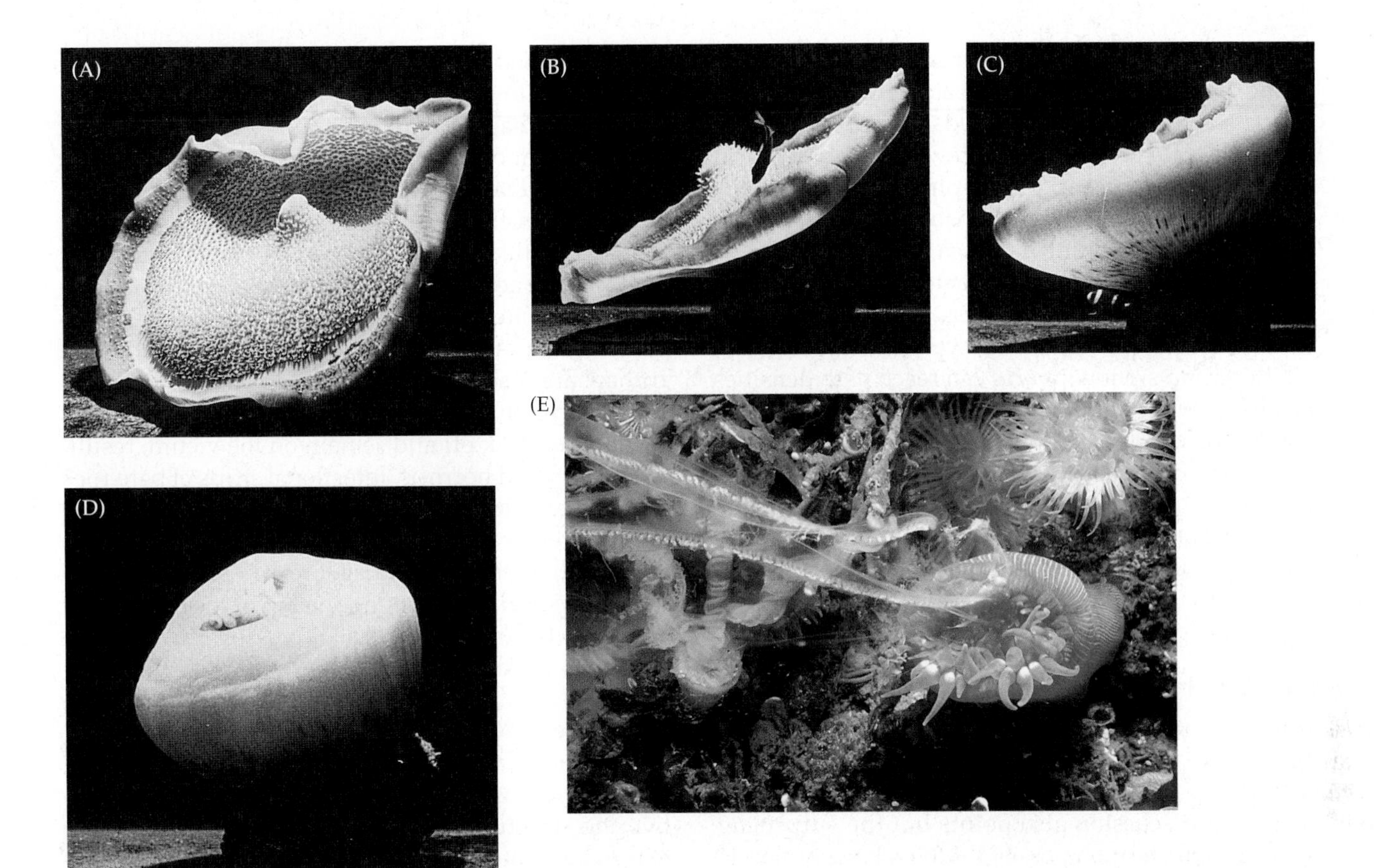

Figure 7.26 (A–D) Feeding sequence in the tropical sea anemone *Amplexidiscus senestrafer*. (A) An expanded oral disc has a tentacle-free area near the periphery, and an oral cone. (B) An expanded disc (side view). (C) Closure one-third complete, 1 second after stimulation of the oral disc. (D) Complete closure, 3 seconds after stimulation. (E) The temperate sea anemone *Epiactis prolifora* capturing a jellyfish (*Aequorea*?).

These remarkable anemones capture prey directly with the oral disc, which can envelop crustaceans and small fishes, rather like a fisherman's cast net (Figures 7.26A–D).

In addition to tentacular feeding on small plankters, many corals are capable of mucous-net suspension feeding, which is accomplished by spreading thin mucous strands or sheets over the colony surface and collecting organic particulate matter that rains down from the water. The food-laden mucus is driven by cilia to the mouth. In a few corals (e.g., members of the family Agariciidae), the tentacles are greatly reduced or absent, and all direct feeding is by the mucous-net suspension method. The amount of mucus produced by corals is so great that it is an important food source for certain fishes and other reef organisms, which feed directly off the coral or recover mucus sloughed into the surrounding sea water. Coral mucus released into the sea contains a variable mixture of macromolecular components (glycoproteins, lipids, and mucopolysaccharides) or a mucous lipoglycoprotein of specific character for a given species. These loose mucous webs, or flocs, are usually enriched by bacterial colonies and entrapped detrital materials, further enhancing their nutritional value.

The role of cnidarians as potentially significant members of food webs depends largely on location and circumstance. Stony corals obviously hold critical trophic positions in tropical reef environments, as do zoanthids and octocorals in many tropical and subtropical habitats. In many warm and temperate areas sea pens and sea pansies dominate benthic sandy habitats. Large scyphomedusae (e.g., *Aurelia, Cyanea, Pelagia, Phacellophora*) often occur in great swarms and may consume high numbers of larvae of commercially important fishes, as well as competing with other fishes for food. Swarms of jellyfish may be so dense that they clog and damage fishing nets and power plant intake systems. We once witnessed a swarm of *Phacellophora* in the Gulf of California that ran like a broad river from Loreto to La Paz, a distance of about 200 kilometers. Certain scyphozoans (*Chrysaora*) have undergone population blooms in their native habitats, perhaps due to climate shifts, while other species (*Phyllorhiza*) have become invasive after transport by ships or in shifting currents. In large numbers jellies significantly influence local fish and plankton populations.

Hydromedusae too are major components of temperate pelagic food webs. Members of several hydrozoan genera also occur in huge congregations in tropical seas, where they are important carnivores in the neustonic food web. Best known among these are the chondrophorans *Porpita* (which actively feeds on motile crustaceans, such as copepods) and *Velella* (which feeds on relatively passive prey, such as fish eggs and crustacean larvae), and the siphonophore *Physalia* (which actively catches and consumes fish). Other siphonophores inhabiting deep-sea environments (e.g., *Erenna* sp.) possess bioluminescent and red-fluorescent lures that may be important for capturing fish sensitive to long-wavelength light. Like scyphomedusae, hydromedusae and siphonophores can reach high densities in surface waters and have significant effects on zooplankton and human populations. The Chinese freshwater limnomedusa, *Craspedacusta sowerbyi*, is now established throughout the United States and Europe where it can undergo "blooms" and impact fisheries. The Portuguese man-of-war, *Physalia* sp., is a known menace for swimmers during summer months in coastal areas throughout the world.

Defense, Interactions, and Symbiosis

There are so many interesting aspects of cnidarian biology that do not fall neatly into our usual coverage of each group that we present this special section. The following discussion also points out the surprising level of sophistication possible at the relatively simple diploblastic, radiate grade of complexity.

In most cnidarians, defense and feeding are intimately related. The tentacles of most anemones and jellyfish usually serve both purposes, and the defense polyps (dactylozooids) of hydroid colonies often aid in feeding. In some cases, however, the two functions are performed by distinctly separate structures (as in many siphonophores).

Some species of acontiate sea anemones (e.g., *Metridium*) bear separate and distinct feeding tentacles and defense tentacles. Whereas the former usually move in concert to capture and handle prey, the defense tentacles move singly, in a so-called searching behavior, in which they extend to three or four times their resting length, gently touch the substratum, retract, and extend once more. Defense tentacles are used in aggressive interactions with other sea anemones, either those of a different species or nonclonemates of the same species. The aggressive behavior consists of an initial contact with the opponent followed by autonomous separation of the defense tentacle tip, leaving the tip behind attached to the other sea anemone. Severe necrosis develops at the site of the attached tentacle tip, occasionally leading to death. Defense tentacles develop from feeding tentacles and tend to increase under crowded conditions. The development involves loss of typical feeding tentacle cnidae (largely spirocysts) and acquisition of true nematocysts and gland cells, which dominate in defense tentacles. Similarly, elongated "sweeper tentacles" in many species of corals are used for defense and competition for space, by direct contact or release of toxic exudates.

The **acrorhagi** (= marginal tubercles or bumps) that ring the collar of some sea anemones (e.g., *Anthopleura*) also have a defensive function. These normally inconspicuous vesicles at the base of the tentacles bear nematocysts and usually spirocysts. In *A. elegantissima*, contact of an acrorhagi-bearing sea anemone with nonclonemates or other species causes the acrorhagi in the area of contact to swell and elongate. The expanded acrorhagi are placed on the victim and withdrawn, and the application may be repeated. Pieces of acrorhagial epidermis break off and remain on the victim, resulting in localized necrosis. Interclonal strips of bare rock are maintained by this aggressive behavior, and may help prevent overcrowding (Figure 7.27A). In addition to this behavior, the acrorhagi are exposed as a ring of nematocyst batteries around the top of the constricted column whenever an acrorhagi-bearing sea anemone contracts in response to violent stimulation. Other competitive interactions are known among stony corals (Figure 7.27C).

Octocorals, which lack toxic stinging nematocysts, have been shown to be a rich source of biologically active and structurally unusual compounds that appear to provide protection against predators and may allow them to colonize new habitats by causing tissue necrosis in potential competitors. These compounds include prostaglandins, diterpenoids, and nausea-inducing furanocembranolides including the descriptively named 11β, 12β-epoxypukalide. In contrast to many coastal and coral reef dwelling species, octocorals are remarkably free from predation except by the few species that are specialized to use them as food. While sclerites have also been proposed to provide antipredatory benefits, there is little clear evidence that sclerites decrease the nutritional value of octocorals sufficiently to deter predation. Thus, like sponges, octocorals appear to use secondary metabolic compounds as their primary anti-predator defense. Chemical defenses may have evolved to compensate for poor regenerative abilities in these slow growing cnidarians, or because their sessile habit makes them especially conspicuous to visual predators.

There are many examples of associations between cnidarians and other organisms, some of which are truly symbiotic, others of which are less intimate. With the exception of myxozoans, few groups of cnidarians are truly parasitic, although several species of hydroids infest marine fishes. The polyps of some of these hydroids lack feeding tentacles and occasionally even lack cnidae. The basal portion of the polyp erodes the fish's epidermis and underlying tissues, and nutrients are absorbed directly from the host. One species invades the ovaries of Russian sturgeons (a caviar feeder!).

(A)

(B)

(C)

Figure 7.27 (A) Defensive acrorhagi (white-tipped tentacles) on two sea anemones (*Anthopleura elegantissima*), engaging in territorial chemical combat. (B) Close-up showing acrorhagi of *Anthopleura elegantissima*. (C) Competition between true corals (Scleractinia) in the Virgin Islands. The coral *Isophyllia sinuosa* is seen extruding its mesenterial filaments and externally digesting the edge of a colony of *Porites astereoides*.

However, the Myxozoa, by any standard, are indeed parasitic. This group consists of about 1,200 species of tiny parasites, previously classified among the protists as the phylum Myxozoa. Morphological data, DNA sequence data, and the presence of metazoan Hox genes, all provide evidence that these strange creatures are allied with the cnidarians, possibly as a sister group to the Medusozoa (Figure 7.46). The coiled polar filaments housed within polar capsules of myxozoans are now viewed as modified nematocysts (Figure 7.28).

Myxozoan cnidarians infect annelids and various poikilothermic vertebrates, especially fishes (Figure 7.29). The life cycle begins when **actinospore** larvae (which may vary in form) are released from spores, contact the mucous membranes of an appropriate vertebrate host (either by ingestion or contact; Figure 7.29, part 1), and extrude their **polar filaments** to release sporoplasm into the host cells. Presporogonic development occurs within these cells, producing infective cell-doublets, which rupture host cells and disperse to infect other cells (Figure 7.29, part 2). Infection can spread to other tissues, particularly neural tissue, and can cause disruption of host tissue structure (e.g., a blackened tail in infected trout) or behavior (see below). Sporulation often occurs in particular tissues (e.g., cartilage in *M. cerebralis*), where a multinucleated **plasmodium** develops and produces sporoblasts with variable numbers of internal spores depending on the species (Figure 7.29, part 3).

Within plasmodia, valvogenic cells produce **spore valves**, which enclose capsulogenic cells that become polar capsules as well as sporoplasm. When complete, this process generates **myxospores** that are released by the vertebrate host and are infective to annelids (Figure

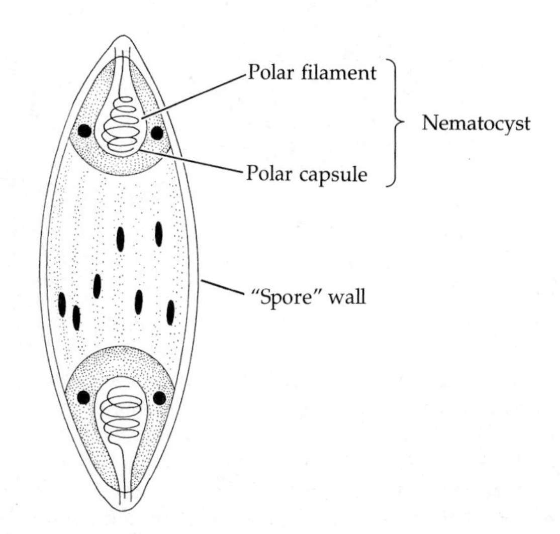

Figure 7.28 Previously considered to be protists, myxozoans are now viewed as highly specialized, parasitic cnidarians.

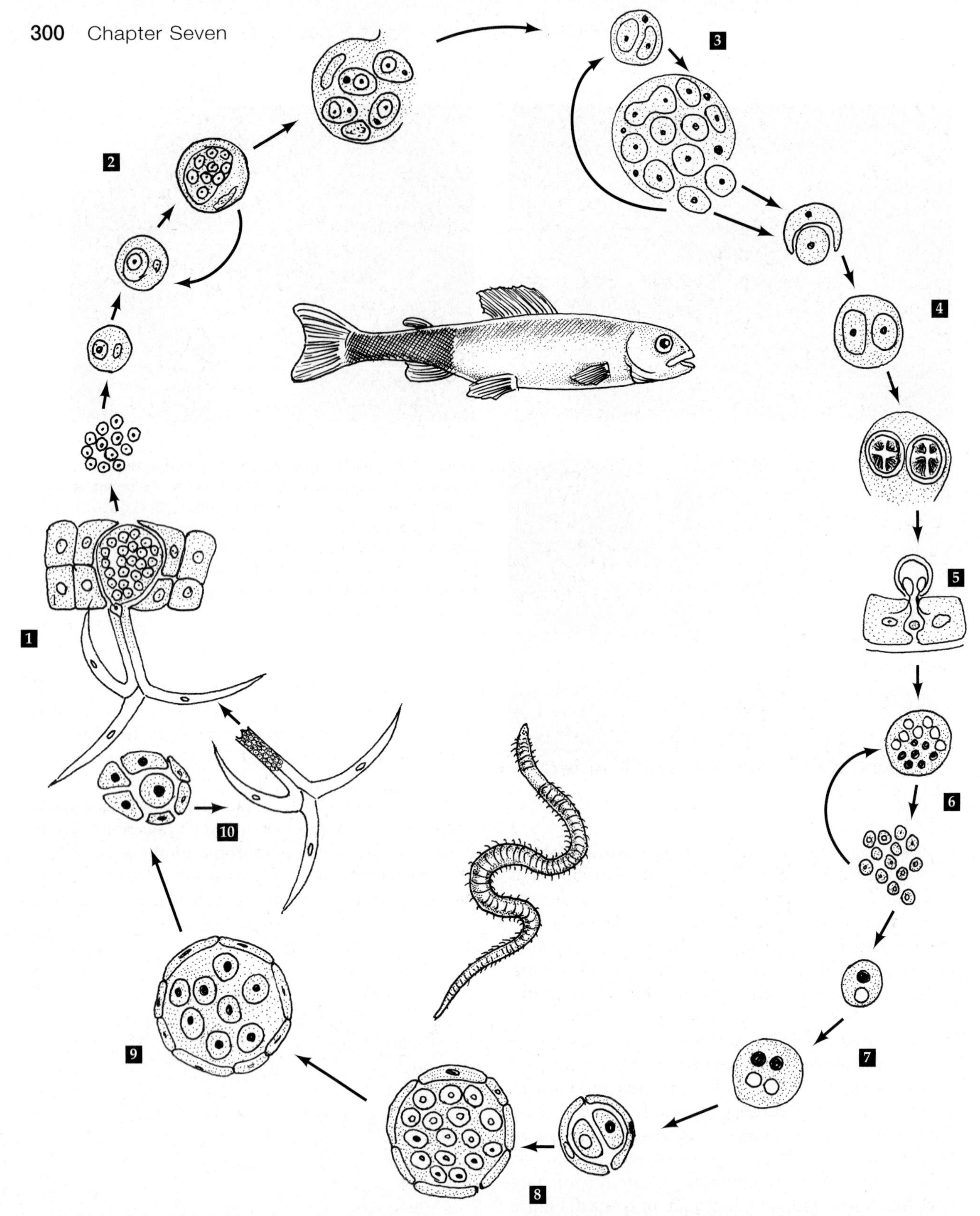

Figure 7.29 The life cycle of *Myxobolus cerebralis*: (1) Actinospore larvae attach to fish mucous membranes, extrude polar filaments and release sporoplasm into host cells. (2) Presporogonic development occurs within host cells producing infective cell-doublets, which rupture and infect other host cells; spreading infection disrupts host tissue causing a blackened tail in trout; (3) sporulation and multinucleated plasmodia develop in particular tissues (e.g., cartilage in *M. cerebralis*), further spreading infection; (4) within plasmodia, sporoblasts form internal myxospores that are (5) released by the vertebrate host and ingested by annelids, polar filaments facilitate penetration of gut cells; (6) multinucleate cells form, that infect other cells and generate plasmodia or (7) fuse with other cells to become binucleate cells, then differentiate into multinucleate cells with α or β nuclei and (8) become complimentary gametes; (9) gametes fuse to produce pansporocysts containing 8 zygotes; (10) zygotes differentiate into infective actinospores that are released in worm feces or remain within the worm's body; other fish are infected by contact with worm feces or by eating spore-bearing worms.

(A)

(B)

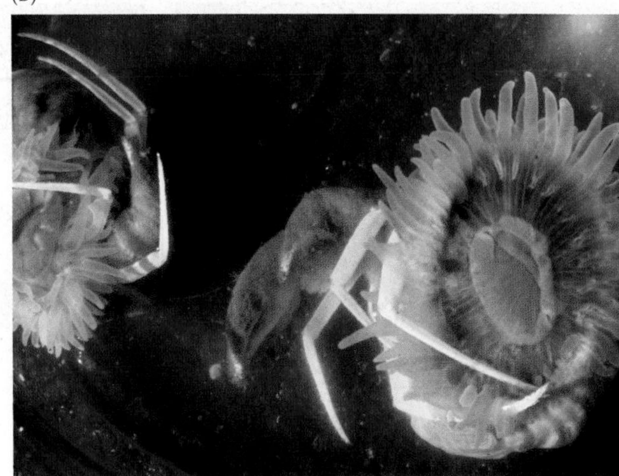

Figure 7.30 The golden "cloak anemone" (Anthozoa, Actiniaria) *Stylobates aenus.* (A,B) The anemone is forming a "shell," or carcinoecium, around the hermit crab *Parapagurus dofleini*. (C) The empty carcinoecium of *S. aenus*.

(C)

7.29, part 4). Polar filaments facilitate cell penetration of gut cells where multinucleate cells form by a process often called **schizogony** (although this is probably not the same process seen in sporozoan protists; Figure 7.29, parts 5 and 6). These cells then produce numerous uninucleate cells that may either generate other plasmodia or fuse with other cells to become binucleate within the worm gut. Binucleate cells differentiate into multinucleate cells with α or β nuclei that develop into complimentary gametes by the end of gametogony, which fuse to produce **pansporocysts** containing zygotes (Figure 7.29, parts 7–9). Zygotes differentiate into infective actinosporean spores that are released in worm feces or remain within the worm's body. Infection of the vertebrate host occurs when spores containing worm feces contact mucous membranes or when spore-bearing worms are ingested (Figure 7.29, part 10).

Most myxozoan species appear to include both a vertebrate and an invertebrate host in their life cycle. *Myxobolus cerebralis*, a parasite of freshwater fishes (especially trout, Figure 7.1M), devours the host's cartilage, leaving the fish deformed. Inflammation resulting from the infection puts pressure on nerves and disrupts balance, causing the fish to swim in circles—a condition known as whirling disease. When an infected fish dies, *M. cerebralis* spores are released from the decaying carcass and may survive for up to 30 years in sediment. Eventually, the spores are consumed by *Tubifex* worms (oligochaete annelids). They reside in this intermediate host until eaten by a new host fish.

Mutualism is common among cnidarians. Many species of hydroids live on the shells of various molluscs, hermit crabs, and other crustaceans. The hydroid gets a free ride and the host perhaps gains some protection and camouflage. Many members of the leptomedusan family Eirenidae (e.g., *Eugymnanthea*) occupy the mantle cavities of bivalves, where they protect their hosts against trematode parasites by consuming the infective sporocysts. Hydroids of the genus *Zanclea* are epifaunal on bryozoans, where they sting and discourage smaller predators and adjacent competitors, helping the bryozoan to survive and overgrow competing species. The bryozoan lends protection to the hydroid with its coarse skeleton, and the mutualism seems to allow both taxa to cover a larger area than either could individually. The bizarre, aberrant hydroid *Proboscidactyla* lives on the rim of polychaete worm tubes (Figure 7.8A) and dines on food particles dislodged by the host's activities. Another filiferan, *Brinckmannia hexactinellidophila*, lives within the tissues of Arctic glass sponges.

Some sea anemones attach to snail shells inhabited by hermit crabs. These partnerships are mutualistic; the sea anemone gains motility and food scraps while protecting the hermit crab from predators. The most extreme case of this mutualism might be that of the cloak anemones (e.g., *Adamsia*, *Stylobates*), which wrap themselves around the hermit crab's gastropod shell and grow as the crab does (Figure 7.30). Initially, the anemone's pedal disc secretes a chitinous cuticle over the small gastropod shell occupied by the hermit.

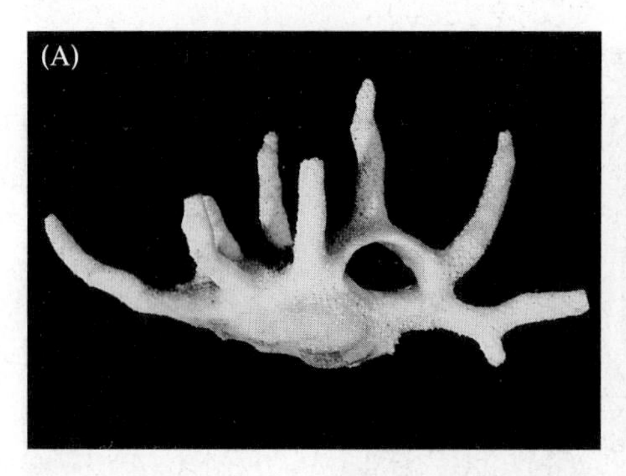

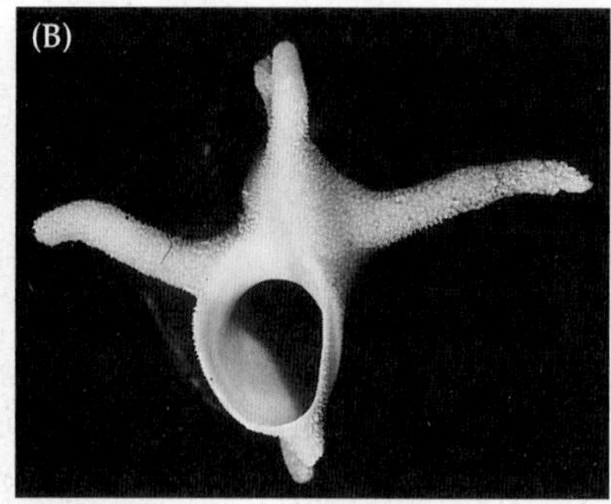

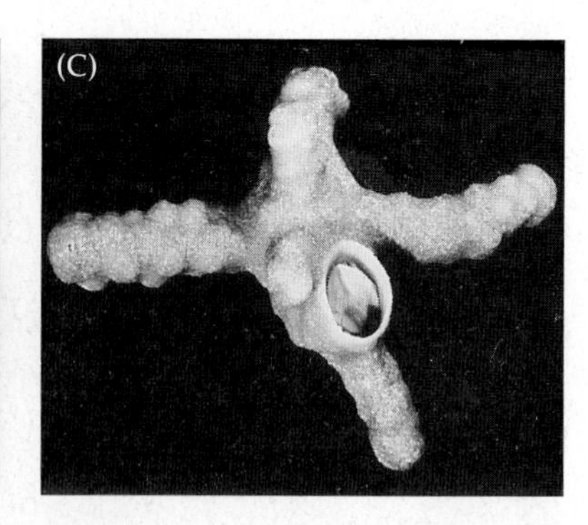

Figure 7.31 A case of remarkable evolutionary convergence. (A,B) The hydrozoan colony *Janaria mirabilis* (Athecata) forms a shell-like corallum inhabited by hermit crabs. (C) The ectoproct *Hippoporida calcarea*, which forms a similar structure, is also inhabited by hermit crabs.

Such fortunate crabs need not seek new, larger shells as they grow, for the cloak anemone simply grows and provides the hermit with a living protective cnidarian "shell," often dissolving the original gastropod shell over time. As if it were itself a gastropod, the sea anemone grows to produce a flexible coiled house called a **carcinoecium**. In fact, these odd anemone "shells" were initially described and classified as flexible gastropod shells. A similar relationship exists between some hermit crabs of the genus *Parapagurus* and certain species of *Epizoanthus*. The hydroid *Janaria mirabilis* secretes a "long-spined" shell-like casing that is inhabited by hermit crabs and, in an extraordinary case of evolutionary convergence, so does the bryozoan *Hippoporida calcarea* (Figure 7.31).

So effective are cnidae that many groups of animals have figured out ways to capture or otherwise utilize these structures for their own defense. Several aeolid sea slugs consume cnidarian prey, ingesting their unfired nematocysts and storing them in fingerlike processes on their dorsal surfaces. Once the nematocysts are in place, the sea slugs use them for their own defense. The ctenophore *Haeckelia rubra* feeds on certain hydromedusae and incorporates their nematocysts into its tentacles. The freshwater turbellarian flatworm *Microstoma caudatum* feeds on *Hydra*, risking being eaten itself, and then uses the stored nematocysts to capture its own prey. Several species of hermit crabs and brachyuran crabs carry sea anemones (e.g., *Calliactis*, *Sagartiomorpha*) on their shells or claws and use them as living weapons to deter would-be predators. The hermit crabs transfer their anemone partners to new shells, or the anemones move on their own, when the hermits take new shells. Some hermit crabs of the genus *Pagurus* often have their shell covered by a mat of symbiotic colonial hydroids (e.g., *Hydractinia*, *Podocoryne*). The presence of the hydroid coat deters more aggressive hermits (e.g., *Clibinarius*) from commandeering the pagurid's shell.

Several cases of fish–cnidarian symbiosis have been documented. The well-known association of anemone fishes (clown fishes) and their host sea anemones serves an obvious protective function for the fish. About a dozen species of sea anemones participate in this interesting relationship. The fish's ability to live among the sea anemone's tentacles is still not fully understood. However, the sea anemone does not voluntarily fail to spend its nematocysts on its fish partner; rather, the fish alters the chemical nature of its own mucous coating, perhaps by accumulating mucus from the sea anemone, thereby masking the normal chemical stimulus to which the anemone's cnidae would respond. *Neomus* are small fishes that live symbiotically among the tentacles of *Physalia* and appear to survive by simply avoiding direct contact with the beast. When stung accidentally, however, it shows a much higher survival rate than do other fishes of the same size. *Neomus* feeds on prey captured by its host.

A number of associations are known between cnidarians and crustaceans. Nearly all amphipods of the suborder Hyperiidea are symbionts on gelatinous zooplankters, including medusae. The nature of many of these associations is unclear, but various species of the amphipods are known to use their hosts as a nursery for the young and perhaps for dispersal. Some actually live among and eat the nematocyst-bearing parts of the host, such as the tentacles or oral arms. Many are commonly found inside the medusa's coelenteron, where they seem unaffected by the host's digestive enzymes. In a relationship similar to that of anemone fish, a few cases of anemone shrimp are known, at least one that is obligate for the shrimp (*Periclimenes brevicarpalis*).

One of the most noteworthy evolutionary achievements of cnidarians is their close relationship with unicellular photosynthetic partners. The relationship is widespread and occurs in many shallow-water cnidarians. The symbionts of freshwater hydrozoans (e.g., *Chlorohydra*) are single-celled species of green algae (Chlorophyta) called **zoochlorellae**. In marine cnidarians, the protists are unicellular cryptomonads and dinoflagellates called **zooxanthellae** (probably several genera including *Zooxanthella* [= *Symbiodinium*] and others) (Figure 7.32). These algae are capable of living

Figure 7.32 (A) An octocoral with zooxanthellae distributed throughout the gastrodermis (schematic section). (B) Cells of zooxanthellae in tissue of the giant green sea anemone *Anthopleura xanthogrammica*. (C) *Mastigias* sp., a rhizostoman medusa, harbors zooxanthellae in its cells.

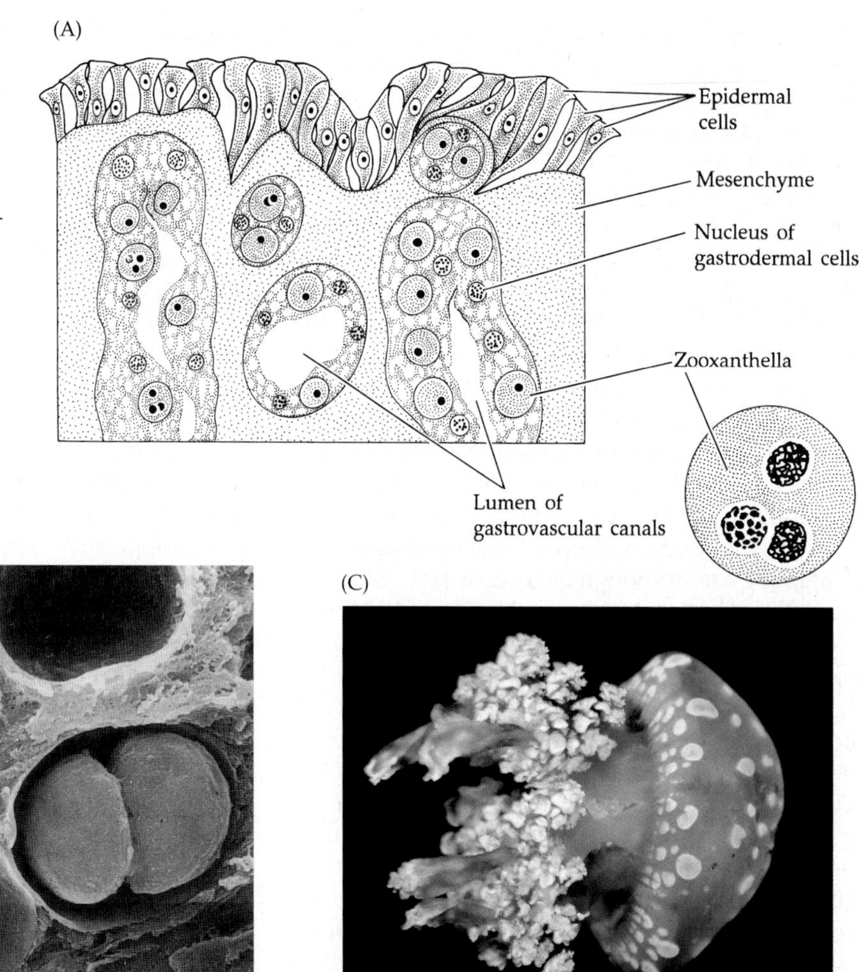

free from their hosts, and perhaps do so normally, but very little is known about their natural history. The algae typically reside in the host's gastrodermis or epidermis, although some cnidarians harbor extracellular zooxanthellae in the mesoglea. It is usually the algal symbionts that give cnidarians their green, blue-green, or brownish color. Corals that are reef-builders (i.e., **hermatypic corals**) typically harbor zooxanthellae (they are "zooxanthellate corals"). Resident populations of zooxanthellae in these corals may reach a density of 30,000 algal cells per cubic millimeter of host tissue (or from 1 to 2 × 10^6 cells per square centimeter of coral surface). Zooxanthellae also occur in many tropical octocorals, anemones, and zoanthids.

Surprisingly, both zoochlorellae and zooxanthellae occur within the tissues and cells of one group of sea anemones, *Anthopleura* of the northeast Pacific coast (*A. elegantissima* and *A. xanthogrammica*). Data suggest that zoochlorellae in these anemones photosynthesize more efficiently and grow faster at lower temperatures and light, whereas zooxanthellae do so under higher temperatures or light regimes. The two anemones are the most abundant rocky intertidal anemones in their range, from Alaska to Baja California, and the distributions of their two symbionts have been shown to be (predictably) related to latitude and intertidal position.

Even some scyphozoans harbor large colonies of zooxanthellae in their bodies, and it is now known that these protist colonies produce much of the energy required by their host jellyfish (e.g., *Cassiopea*, *Linuche*, *Mastigias*). Some of this information comes from studies on the scyphomedusa *Mastigias* (Figure 7.32C), which live in marine lakes on the islands of Palau, where they may occur in densities exceeding 1,000 per m^3. In these lakes, *Mastigias* makes daily vertical migrations between the oxygenated, nutrient-poor upper layers and the anoxic, nutrient-rich lower layers, as well as horizontal migrations to track the movement of the sun across the lake. This behavior appears to be related to the light and nutrient requirements of its symbiotic zooxanthellae. Unlike the zooxanthellae in benthic cnidarians, which tend to reproduce more-or-less evenly over a 24-hour period, the zooxanthellae of *Mastigias* show a distinct reproductive peak during the hours when their host occupies a position in the deeper nitrogen-rich layers of the lakes. This reproductive peak may be a result of the alga's use of free ammonia as a nutrient source.

Many cnidarians seem to derive only modest nutritional benefit from their algal symbionts, but in many others a significant amount of the hosts' nutritional needs appears to be provided by the algae. In such cases, a large portion of the organic compounds produced by photosynthesis of the symbiont may be passed on to the cnidarian host, probably as glycerol but also as glucose and the amino acid alanine. In return, metabolic wastes produced by the cnidarian provide the symbiotic alga with needed nitrogen and phosphorus. In corals, the symbiosis is thought to be important for rapid growth and for efficient deposition of the calcareous skeleton, and many corals can only form reefs when they maintain a viable dinoflagellate population in their tissues. Different coral species serve as hosts to genetically distinct algal symbiont taxa, which each appear to be adapted to their host as well as that host's particular ambient light regime. Although the precise physiological-nutritional link between corals and their zooxanthellae has been elusive, the algae clearly seem to increase the rate of calcium carbonate production. Corals and other cnidarians can be deprived of their algal symbionts by experimentally placing the hosts in dark environments. In such cases the algae may simply die, they may be expelled from the host, or they may (to a limited extent) actually be consumed directly by the host. Because they are dependent on light, zooxanthellate corals can live to depths of only 90 m or so. Most zooxanthellate corals also require warm waters and thus occur almost exclusively in shallow tropical seas (although zooxanthellae occur in some high-altitude anemones). Deep water and cold water corals also exist but tend to be entirely carnivorous. They grow at extremely slow rates and thus tend to produce reefs that have existed for thousands and even millions of years, providing a detailed record of changes in sea temperature.

Under stress, such as unusually high temperatures, corals may lose their zooxanthellae—a process known as **coral bleaching**. The long-term impact of coral bleaching, now accelerating throughout the world's tropics, perhaps due to a combination of warming seas and changes in oceanic acid–base balance due to increased atmospheric CO_2, is unclear. Certainly it seems detrimental in the short run, and often leads to death of entire coral colonies. In addition, anthropogenic pollution such as increases in phosphates, nitrates, and ammonia in the sea are enhancing growth of algae and bacteria that compete with coral. Caribbean reefs have been devastated over the last two decades, having lost about 80% of their coral cover. Interestingly, a few recent studies have suggested that bleaching might be an adaptive mechanism providing opportunity for acquiring new types of zooxanthellae better adapted to the changing environment. If true, it would remain to be seen if this symbiont switch could take place quickly enough to keep up with today's rapidly changing ocean chemistry. Evidence that selectivity between symbiotic partnerships may exist could mean slower symbiotic reassociations (i.e., certain combinations of hosts and algae are favored while others are impossible).

Loss of zooxanthellae by corals usually results in loss of the ability to secrete the calcium carbonate skeleton. The widespread disappearance of Caribbean corals is now considered responsible for a 32–72% decrease in reef fish populations, a potentially catastrophic change for coastal communities dependent on fishing. Coral reef biodiversity correlates with reef area, thus the long-term effects of reef loss are likely to be cumulative and difficult to reverse. However, one recent experiment found that colonies of some coral species that lost their calcium carbonate skeleton continued to exist as soft-bodied polyps. These recent discoveries have suggested a possible explanation for the geologically "sudden" appearance of the modern stony corals (Scleractinia) in the Middle Triassic, when geochemically perturbed oceans returned to "normal." Before this time, corals and reefs had disappeared from the fossil record for millions of years, but perhaps they continued to exist as "naked corals" (and thus not contributing to the fossil record).

Circulation, Gas Exchange, Excretion, and Osmoregulation

There is no independent circulatory system in cnidarians. The coelenteron serves this role to a limited extent by circulating partly digested nutrients through the interior of the body, absorbing metabolic wastes from the gastrodermis, and eventually expelling waste products of all types through the mouth. But large anemones and large medusae confront a serious surface area:volume dilemma. In such cases, the efficiency of the gastrovascular system as a transport device is enhanced by the presence of mesenteries in the anemones and the radially arranged canal system in the medusae. Cnidarians also lack special organs for gas exchange or excretion. The body wall of most polyps is either fairly thin or has a large internal surface area, and the thickness of many medusae is due largely to the gel-like mesoglea or mesenchyme. Thus, diffusion distances are kept to a minimum. Gas exchange occurs across the internal and external body surfaces. Facultative anaerobic respiration occurs in some species, such as anemones that are routinely buried in soft sediments. Nitrogenous wastes are in the form of ammonia, which diffuses through the general body surface to the exterior or into the coelenteron. In freshwater species there is a continual influx of water into the body. Osmotic stress in such cases is relieved by periodic expulsion of fluids from the gastrovascular cavity, which is kept hypoosmotic to the tissue fluids.

Nervous System and Sense Organs

Consistent with their radially symmetrical body plan, cnidarians generally have a diffuse, noncentralized

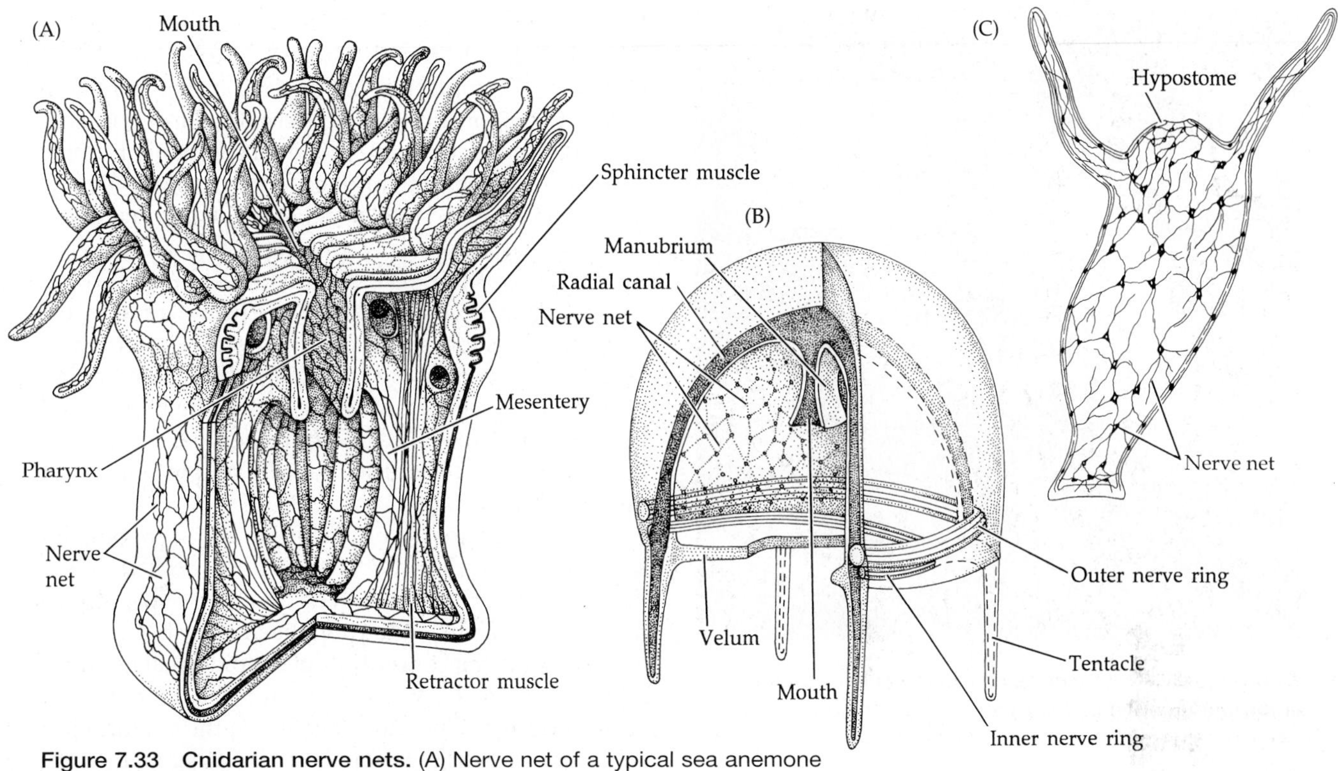

Figure 7.33 Cnidarian nerve nets. (A) Nerve net of a typical sea anemone (Anthozoa). (B) Nerve net in a hydromedusa (Hydrozoa). (C) Nerve net of *Hydra* (Hydrozoa).

nervous system, however great variation exists. Strong centralization is seen in many hydrozoans and cubozoans, and some workers feel those even qualify as central nervous systems. Nonetheless, the neurosensory cells of the system are the most primitive in the animal kingdom, being naked and largely nonpolar. Usually the neurons are arranged in two reticular arrays, called **nerve nets**, one between the epidermis and the mesenchyme and another between the gastrodermis and the mesenchyme (Figure 7.33). The presence of both ectodermal and endodermal nerve nets is unique to the Cnidaria. The subgastrodermal net is generally less well developed than the subepidermal net and is absent altogether in some species; in cubozoan polyps there is a nerve net *within* the gastrodermis. Some hydrozoan medusae possess one or two additional nerve nets, whereas in the polyps of hydrozoans and cubozoans there appears to be only a single epidermal nerve ring. Despite the seeming simplicity of cnidarian nervous systems, it has been shown that they possess at least some of the classic interneuronal and neuromuscular synapse neurotransmitters seen in Bilateria, including serotonin, suggesting that both catecholamine and indolamine neurotransmitters may be present (at least in sea anemones).

A few nerve cells and synapses are polarized and allow for transmission in only one direction, but most cnidarian neurons and synapses are nonpolar—that is, impulses can travel in either direction along the cell or across the synapse. Thus, sufficient stimulus sends an impulse spreading in every direction. In some cnidarians where both nerve nets are well developed, one net serves as a diffuse slow-conducting system of nonpolar neurons, and the other as a rapid through-conducting system of bipolar neurons.

Polyps generally have very few sensory structures. The general body surface has various minute hair-like structures developed from individual cells. These serve as mechanoreceptors, and perhaps as chemoreceptors, and are most abundant on the tentacles and other regions where cnidae are concentrated. They are involved in behavior such as tentacle movement toward a prey or predator and in general body movements. Some appear to be associated specifically with discharged cnidae, such as the ciliary cone apparatus of anthozoan polyps, which is believed to function like the cnidocil in hydrozoan and scyphozoan nematocysts (Figure 7.34). Oddly, these structures do not appear to be connected directly to the nerve nets. In addition, most polyps show a general sensitivity to light, not mediated by any known receptor but presumably associated with neurons concentrated in or just beneath the translucent surface of epidermal cells.

As might be expected, motile medusae have more sophisticated nervous systems and sense organs than do the sessile polyps (Figure 7.35). In many groups, especially the hydromedusae, the epidermal nerve net of the bell is condensed into two nerve rings near the bell

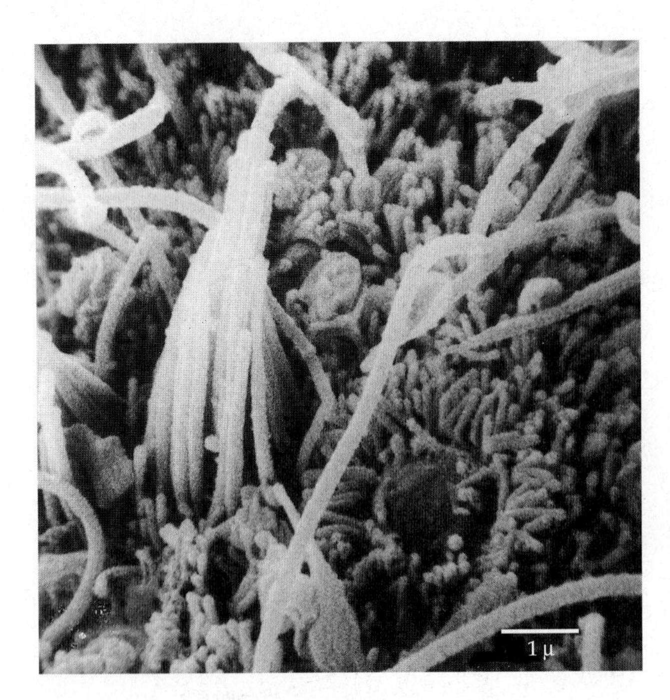

Figure 7.34 A ciliary cone on the tentacle of the corallimorpharian anemone *Corynactis californica* lies adjacent to cnidocyte (the circle of microvilli).

margin. These nerve rings connect with fibers enervating the tentacles, muscles, and sense organs. The inner ring stimulates rhythmic pulsations of the bell. This ring is also connected to statocysts, when present, on the bell margin, which is supplied with general sensory cells and with radially distributed ocelli and (probably) chemoreceptors. The general sensory cells are neurons whose receptor processes are exposed at the epidermal surface. The ocelli are usually patches of pigment and photoreceptor cells organized as a disc or a pit. Statocysts may be in the form of pits or closed vesicles, the latter housing a calcareous statolith adjacent to a sensory cilium. When one side of the bell tips upward, the statocysts on that side are stimulated. Statocyst stimulation inhibits adjacent muscular contraction and the medusa contracts muscles on the opposite side. Many medusae maintain themselves in a particular photoregime by directed swimming behaviors. This action is seen especially in those medusae harboring large populations of zooxanthellae, such as the medusa *Cassiopea*, which lies upside down on the shallow sea floor, exposing to light the dense zooxanthellae population residing in tissues of its tentacles and oral arms.

There is great variation among hydrozoan medusae visual systems. Some have structural eyes, but others have only a generally photosensitive neuronal system. The structural eyes are usually well-defined ocelli that very in complexity from a simple ectodermal layer with sensory and pigment cells, through pigment cup eyes, to small camera-type eyes with pigmented retina and lens- or cornea-like structures. The axons of the photosensory cells may join in bundles that collect to form an "optic nerve," that runs to the outer nerve ring in the bell. In addition, hydrozoans have been shown to possess a cytoplasmic conducting system similar in nature to that of sponges. Epidermal cells and muscle elements appear to be the principal components of the system. Although the impulse seems to move slowly, it is initiated by nerve cells and relies on gap junctions.

In the codonophoran siphonophores, a linear condensation of the nerve net produces longitudinal "giant axons" in the stem and the nerve tracts in the tentacles. This longitudinal "giant axon" is actually a neuronal syncytium that originates by fusion of neurons from the nerve net of the stem. The high-speed impulses in these large diameter nerve tracts enable codonophorans to contract rapidly and initiate a fast escape reaction.

The bell margins of cubomedusae and scyphomedusae usually bear club-shaped structures—called **rhopalia**—that are situated between a pair of flaps, or **lappets** (Figure 7.35A). The rhopalia are sensory centers, each containing a concentration of epidermal neurons, a pair of chemosensory pits, a statocyst, and eyes of various design. One pit is located on the exumbrellar side of the hood of the rhopalium, the other on the subumbrellar side.

Cubomedusae are strong swimmers that are able to make rapid directional changes in response to visual stimuli. They have been shown to be attracted to light, to avoid dark objects, and even to navigate around obstacles. They are active predators that "sleep" at night. Although cubomedusae have the basic cnidarian nerve net and subumbrellar nerve ring close to the bell margin (sometimes called the ring nerve), they also have the most elaborate visual system in Cnidaria, located on four rhopalia (sensory complexes). Each rhopalium has three sets of eyes: paired pit-shaped pigment cup eyes, paired slit-shaped pigment cup eyes, and two complex camera-type eyes with cornea, lens, and retina (one small upper lensed eye, and one large lower lensed eye). Although the retina of the camera eye is only one cell thick, it is multilayered, containing a sensory layer, a pigmented layer, a nuclear layer, and a region of nerve fibers. There number of sensory cells, or photoreceptors, in each of these remarkable eyes varies from about 300 to 1,000, depending on the species. Each rhopalium also has a crystalline concretion, the statolith, which is often called a statocyst. Neuronal signals from the rhopalia are presumably transmitted to swimming pacemaker neurons to direct visually-guided swimming movements. Giant neurons have been identified on either side of the cubomedusan rhopalium.

Despite the structural simplicity of the cubozoan planula larva, being composed of only six cell types, they possess rhabdomeric photoreceptors of the pigment-cup ocelli type. These comprise 10–15 ocelli arranged as individual photoreceptor cells. Each contains

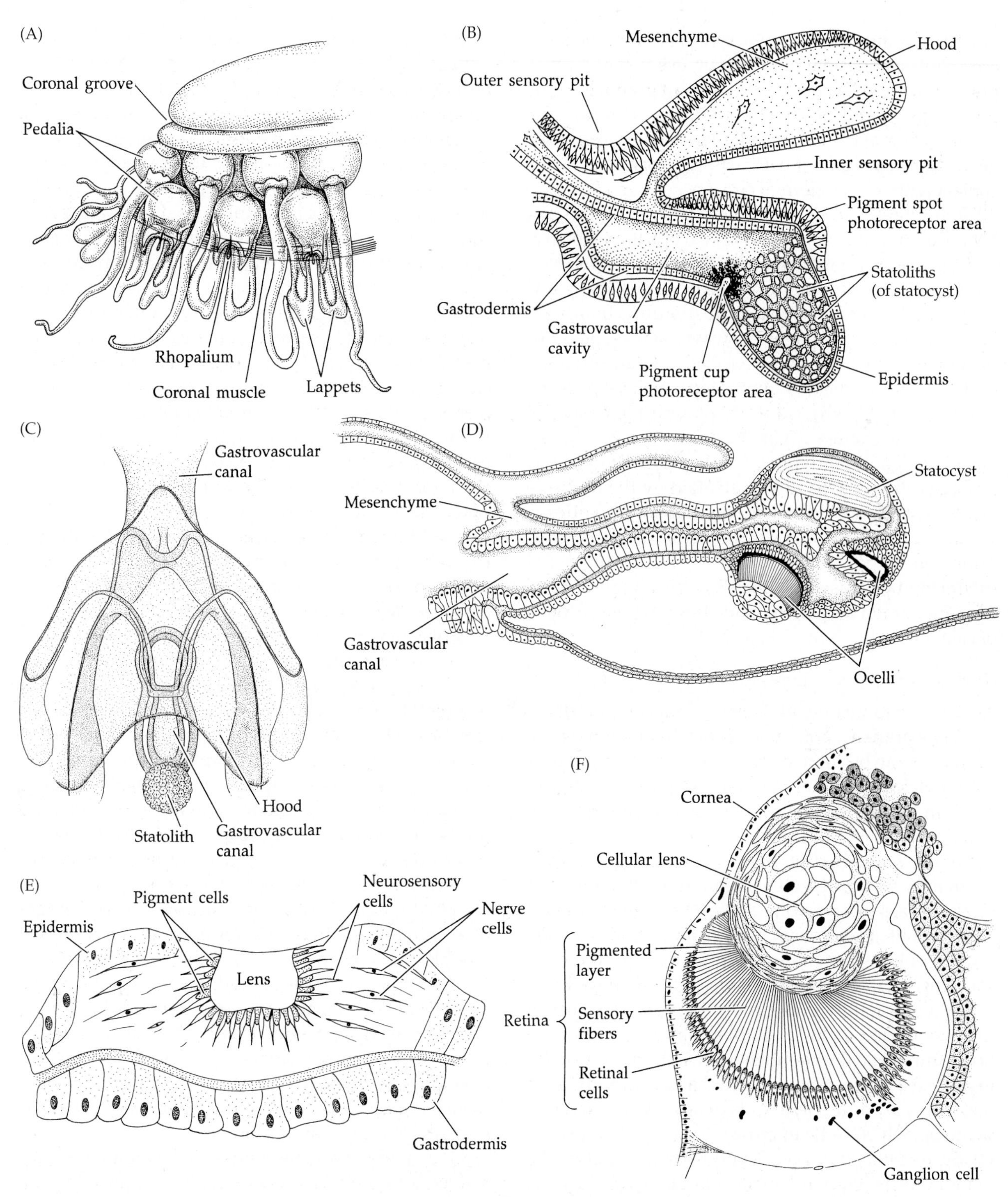

Figure 7.35 Sensory structures in medusae. (A) The rhopalia of the scyphomedusa *Atolla* are situated between the marginal lappets. (B) A rhopalium (section) has various sensory regions. (C) A rhopalium of *Aurelia* (diagrammatic). A portion of the gastrovascular canal has been cut away. (D) A cubozoan rhopalium (note the lower eye is not shown in this oblique section). (E) A pigment-cup ocellus (cross section) of a hydrozoan medusa. (F) The eye of a cubozoan (*Carybdea*) (cross section).

screening pigment in a cavity filled with microvilli (the sites of photoreception) and a single cilium. The cilium is the typical 9+2 structure and is likely motile, rather than sensory. There is no synaptic or electrical (gap junction) connection between the ocelli and any other cell in the larva. The ocelli of cubozoan planulae appear to represent one of the simplest visual systems in the animal kingdom.

Scyphozoan medusae have structurally simple visual eyespots located in the rhopalia, which also contain gravity-sensing structures, the statocysts. The eyespots are simple ectodermal pigment cup eyes, consisting of photosensory cells containing pigment granules, or of photosensory cells alternating with pigment cells, or pigment cup eyes with ectodermal sensory cells and gastrodermal pigment cells. The sensory cells make contact with the ectodermal nerve net.

Bioluminescence is common in cnidarians and has been documented in all classes except the Cubozoa and the poorly known Staurozoa. In some forms (e.g., many hydromedusae), luminescence consists of single flashes in response to a local stimulus. In others, bursts of flashes propagate as waves across the body or colony surface (e.g., sea pens and sea pansies). One of the most complicated luminescent behaviors occur in hydropolyps, where a series of multiple flashes is propagated. The sea pansy *Renilla* (an octocoral) also has very elaborate luminescent displays. Propagated luminescence is probably controlled by the nervous system, although this phenomenon is not well understood. In at least one hydromedusa (*Aequorea*), luminescence appears not to be the result of the usual luciferin–luciferase reaction. Rather, a high-energy protein, named **aequorin**, emits light in the presence of calcium.

Reproduction and Development

Asexual reproduction takes many forms in cnidarians, and regeneration after major injuries is common. Many anemones can be cut in half, and the two halves regenerate flawlessly. Sometimes injuries to the oral region result in the production of two or more mouths, each with its own set of feeding tentacles.

Sexual reproductive processes in cnidarians are intimately tied to the alternation of generations that characterizes this phylum. As you have already read, cnidarian life cycles often involve an asexually reproducing polyp stage, alternating with a sexual medusoid stage that produces a characteristic planula larva. Thus, we generally find a complex indirect or mixed life history that includes phases of asexual reproduction. Anthozoan development is least complex, usually involving a motile planula larva that settles and grows into a sessile adult polyp. Cubozoans also have motile planulae, which settle to grow into polyps that in turn produce sexual medusae. Scyphozoans too follow the planula-to-sessile-polyp mode, but the polypoid stage generates multiple juvenile medusa, called **ephyrae**, by transverse fission or **strobilation** at the polyp's oral end; ephyrae later mature as the sexual medusoid stage. Planulae, polyps, and medusa all appear in many hydrozoan species' life cycles. Medusae, when present, develop from a laterally budding tissue mass called the **entocodon**, but polyp or medusa stages may be entirely missing from certain life cycles. Because of these many variations in life cycle, we will discuss the cnidarian classes separately.

Anthozoan reproduction Members of this class are exclusively polypoid, and the variety of ways that new individuals can be produced asexually nearly defies imagination. Asexual reproduction is common in sea anemones, and longitudinal fission of polyps can result in two separate individuals, or the daughter anemones can remain close together to produce large groups (clones) of genetically identical individuals (e.g., seen in some species of *Anthopleura*, *Diadumene*, and *Metridium*). During longitudinal fission, the body column stretches to the point of ripping itself apart, each half then regenerating its missing parts. The less common process of pedal laceration (e.g., seen in some acontiate sea anemones: *Diadumene*, *Haliplanella*, *Metridium*) can also lead to clonal populations. During pedal laceration, the pedal disc spreads and the anemone simply glides away, leaving behind a circle of small fragments of the disc, each of which develops into a young sea anemone. This behavior is easily observed in aquaria, where anemones often affix themselves to the glass walls.

In addition to these two common modes of asexual reproduction, at least a few species of sea anemones undergo transverse fission (e.g., *Edwardsiella lineata*, *Nematostella vectensis*). Transverse fission is usually by way of a constriction and then separation in the lower part of the column that results in the formation of a small aboral compartment and a larger oral region, each of which then grows its missing region. Transverse fission has also been reported by a process called "polarity reversal," in which the aboral end spontaneously sprouts new tentacles and a new mouth; eventually a new physa forms in the midsection of the anemone and the two individuals separate (e.g., *Gonactinia*). Certain sagartiid sea anemones engage in intratentacular budding, wherein multiple mouth openings result from repeated longitudinal fission through the pharynges of existing individuals. This process produces band-like like colonies that resemble the elongated polyps of certain meandroid corals. Also, certain populations of *Anthopleura* engage in mesenterial budding of tiny polyps, which are brooded within the gastrovascular cavity before release from the parent anemone. In these anemones, no evidence of gonadal or gametic development has been found, suggesting that these populations are primarily if not entirely asexual.

One family of sea anemones (Boloceroidae) swims actively and produces new individuals from longitudinal fission, pedal laceration, and a bizarre process called tentacular shedding (or tentacle budding) in which tentacle fragments are pinched off at basal sphincters and brooded internally before dispersal. Additionally, certain anemones and at least one scleractinian coral (*Pocillopora damicornis*) are known to produce planula larvae parthenogenetically and brood them until release. Most surprising is the recent discovery that some sea anemones internally brood young that are asexually produced, by a mechanism not yet understood.

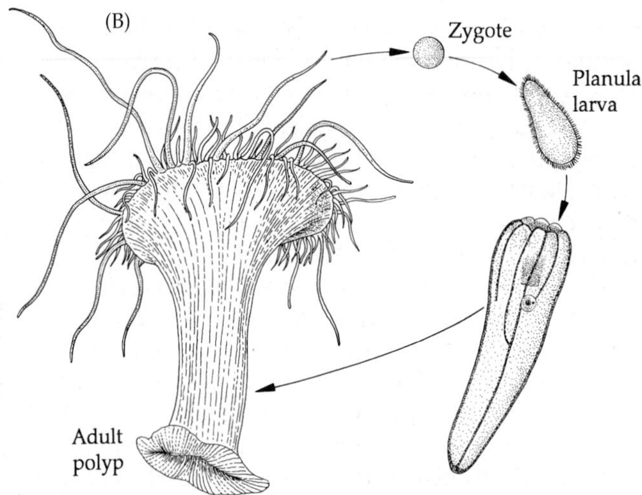

Figure 7.36 Reproduction in Anthozoa. (A) Asexual reproduction by longitudinal fission in the aggregating anemone *Anthopleura elegantissima*. (B) A typical anthozoan sexual life cycle: the adult polyp releases gametes which fuse externally, or fertilized eggs are released, and zygotes develop into a planula larvae; the larvae settle and transform directly into young polyps.

Anthozoans are typically gonochoristic, although most scleractinian corals are hermaphroditic. Fertilization can take place internally, but in most species it occurs externally, in the open sea. Eggs are free, or occasionally pooled into a gelatinous egg mass (even through fertilization). Sperm is usually equipped with flagellar and mitochondrial structures suited for propelling sperm to eggs, although structural differences among anthozoan sperm are remarkable indeed. Cleavage is typically radial and holoblastic, resulting in a hollow spherical, uniformly ciliated, coeloblastula. Gastrulation occurs by way of ingression or invagination, to produce distinct ectodermal and endodermal germ layers, and thence a ciliated planula larva. When invagination occurs, the blastopore remains open and sinks inward, drawing with it a tube of ectoderm that becomes the adult pharynx. Because the mouth forms at the site of gastrulation (at the animal pole), anemones/cnidarians are, by definition, true protostomes, suggesting that protostomy predated the cnidarian–bilaterian divergence. The planula larva may develop one or a few pairs of tentacles at the oral end, as well as a rudimentary pharynx and mesenteries, before settling. Some anthozoan planulae are planktotrophic, although very yolky ones do not feed. The ability of some larvae to feed allows them a potentially longer larval life, enhancing dispersal and selection of an appropriate settlement site. Anthozoan planulae (*Anthopleura*) also appear to obtain zooxanthellae by ingestion. In some species, the planulae develop up to eight complete mesenteries before settling, and this is the so-called **edwardsia stage**, named after the octamesenterial genus *Edwardsia*. The larva eventually settles on its aboral end and tentacles grow around the upwardly directed mouth and oral disc. A typical anthozoan life cycle is shown in Figure 7.36.

Octocorals are usually gonochoristic and often spawn synchronously, although the timing of gonadal development appears to be highly variable among temperate and tropical species, evidently due to variability in water temperature or resource availability, respectively. Although little is known about the reproductive biology of most cerianthárians, antipatharians, anemones, and stony corals may be gonochoristic or hermaphroditic. In some species, colonial forms can contain males, females, and hermaphrodites. The gametes arise from patches of tissue on the gastrodermis of all or only some mesenteries. Eggs are fertilized either in the coelenteron, followed by early development in the gut chambers, or more commonly outside the body, in the sea. A number of anemones brood their developing embryos internally or on the external body surface. The northeast Pacific sea anemone *Aulactinia incubans* releases its brooded young through a pore at the tip of each tentacle! Some corals undergo internal fertilization, brooding, and then release of planula larvae. The solitary coral *Balanophyllia elegans* builds skeletal chambers within which oocytes and embryos may be carried apart from the main digestive cavity, a structural arrangement that allows continued brooding through advanced developmental stages. These calcareous structures are preserved in the fossil record, perhaps providing clues to how parental care in these species evolved. Some octocorals (e.g., *Briareum*, *Alcyonium*) brood their embryos in a mucous coat on the body surface; then the planula larvae escape. Others shed their gametes and rely on external fertilization and planktonic development. *Heliopora coerulea* is a gonochoristic, hermatypic octocoral that broods its planulae at the surface of female colonies before their annual release.

Some coral planula larvae are long-lived, spending several weeks or months in the plankton, an obvious means of dispersal. Other corals release benthic planulae that crawl away from the parent and settle nearby. Many coral populations undergo synchronous spawning over large areas on reefs, a process mediated by moonlight sensitive molecules called **cryptochromes**, which have also been associated with control of circadian activity in vertebrates and insects. In some cases this synchrony is restricted to colonies of a single species, or is only loosely correlated with lunar cycles, but widespread synchronous spawning events involving over 100 different coral species have been reported (on the Great Barrier Reef, Australia), perhaps to satiate predators. Such events create a pulse of nutrients into the surrounding ecosystem, and may lead to hybridization within and among scleractinian corals. Verified cases of hybridization between members of different coral genera exist, and this may explain the great range of polymorphism seen in many coral "species." Because hybrid individuals can become secondarily clonal they may persist within populations for considerable durations, but the degree to which interbreeding by hybrids or their introgression into parental populations occurs remains unclear.

Sagartia troglodytes is the only sea anemone known to copulate. The coupling starts when a female glides up to a receptive male, whereupon their pedal discs are pressed together in such a way as to create a chamber into which the gametes are shed and fertilization occurs. The copulatory position, which forms a temporary marsupium, is maintained for several days, presumably until planula larvae have developed. This behavior may be an adaptation to areas of great water movement that might otherwise scatter gametes and reduce the probability of successful fertilization.

Recent work on the sea anemone *Nematostella vectensis* has revealed cnidarians to possess some (but not all) of the genes involved in dorsal–ventral patterning in the Bilateria. Although these homologs are expressed in somewhat haphazard ways during development, their expression has suggested to some workers the possibility that the oral–aboral polarity of cnidarians might be equivalent to the anterior–posterior polarity of bilaterians. In fact, homologues of 5 Hox genes known to regulate anterior–posterior axis patterning in Bilateria are also known from *N. vectensis*. These genes show a staggered domain expression in the oral–aboral axis patterning of the anemone. This suggests that the earliest stages of what was to become bilaterality had its roots at least as deep in the evolution of the Metazoa as the Cnidaria. Thus, what we recognize as "biradiality" in anemones (e.g., the right–left arrangement of feeding tentacles, pharyngeal siphonoglyph, coelenteron mesenteries) might, in fact, be a rudimentary form of bilaterality.

Scyphozoan reproduction The life cycles of most scyphozoans are poorly known because their benthic stages occur in yet unknown locations. However, known species share several common features. The asexual form of scyphozoan cnidarians is a small polyp called the **scyphistoma** (= scyphopolyp; Figure 7.37A). It may produce new scyphistomae by budding from the column wall or from stolons. At certain times of the year, generally in the spring, medusae are produced by repeated transverse fission of the scyphistoma, a process called **strobilation** (Figure 7.37B). During this process the polyp is known as a **strobila**. Medusae may be produced one at a time (**monodisc strobilation**), or numerous immature medusae may stack up like soup bowls and then be released one after the other as they mature (**polydisc strobilation**). Immature and newly released medusae are called **ephyrae** (sing. ephyra). An individual scyphistoma may survive only one strobilation event, or it may persist for several years, asexually giving rise to more scyphistomae and releasing ephyrae annually.

Ephyrae are very small larval animals with characteristically incised bell margins (Figure 7.37C). The ephyral arms, or primary tentacles, mark the position of what becomes the adult lappets and rhopalia. In some genera (e.g., *Aurelia*) the number of ephyral arms is quite variable (Figure 7.37D). Maturation involves growth between these arms to complete the bell. Development into sexually mature adult scyphomedusae takes a few months to a few years, depending on the species.

The gamete-forming tissue in adult scyphomedusae is always derived from the gastrodermis, usually on the floor of the gastric pouches, and gametes are generally released through the mouth. Most species are gonochoristic. Fertilization takes place in the open sea or in the gastric pouches of the female. Cleavage and blastula formation are similar to the processes in hydrozoans. Gastrulation is by ingression or invagination, and results in a mouthless, double-layered planula larva; when invagination occurs, the blastopore closes. The planula larva eventually settles and grows into a new scyphistoma.

The medusa phase clearly dominates the life cycles of most scyphozoans. The small polyp stage is often significantly suppressed or absent altogether. For example, many pelagic scyphomedusae have eliminated the scyphistoma, and the planula larva transforms directly into a young medusa (e.g., *Atolla*, *Pelagia*, *Periphylla*). In others, the larvae are brooded, developing in cysts on the parent medusa's body (e.g., *Chrysaora*, *Cyanea*). A few genera have branching colonial scyphistomae with a supportive skeletal tube and an abbreviated medusoid stage (e.g., *Nausithoe*, *Stephanoscyphus*). None, however, has lost the medusoid stage altogether. Some scyphozoan life cycles are shown in Figure 7.38.

Cubozoan reproduction The biology of cubozoans is not yet well known and the polyps of only a few

Figure 7.37 (A) Scyphozoan (Aurelia) scyphistoma (and one strobila), and (B) strobila. (C) A "typical" 8-armed ephyra. (D) A 12-armed ephyra.

species have been described. Apparently, each polyp metamorphoses directly into a single medusa, rather than undergoing "standard" strobilation as scyphozoan polyps do. Some cubozoan medusae engage in a form of copulation, in which sperm are transferred directly from the male to an adjacent female in the water column. In *Copula sivickisi* mature adults engage in courtship during which males transfer spermatophores to females who then insert them into their manubria. Females accept multiple spermatophores from multiple males, yet may only produce one embryo strand (a packet of fertilized ova that attaches to algae). During courtship, mature females, with bell margins greater than 5 mm, exhibit conspicuous velar spots that may provide a visual signal to courting males. Planula larvae of some cubozoans have been shown to possess many single-celled eyes, but no nervous system!

Staurozoan reproduction Reproduction in staurozoans has been observed in only a few species. In mature medusae, oocytes develop within follicular cells, a unique feature within the Cnidaria. Fertilization appears to occur in situ with zygotes shed into the water in summer. The zygotes then settle, and develop into creeping, nonciliated planulae, each with a fixed number of cells (n = 16). Planulae develop into a nonfeeding "microhydrula" stage which may asexually produce creeping frustrules, that later develop into stauropolyps, or the microhydrula may develop into stauropolyps themselves, which later develop into stauromedusae.

Hydrozoan reproduction Hydrozoan polyps reproduce asexually by budding. This is a rather simple process wherein the body wall evaginates as a bud, incorporating an extension of the gastrovascular cavity with it. A mouth and tentacles arise at the distal end, and eventually the bud either detaches from the parent and becomes an independent polyp or, in the case of colonial forms, remains attached. Asexual reproduction in this fashion creates larger and more complex polypoid colonies that have greater reproductive capacity and perhaps greater resistance to rapid or turbulent water flow.

Medusa buds, or gonophores, are also produced by polyps in a similar fashion, although the process is sometimes quite complex. A rather special kind of budding occurs in the siphonophores, in which the floating colonies produce chains of individuals called **cormidia**, which may break free to begin a new colony.

Certain hydromedusae also undergo asexual reproduction, either by the direct budding of young medusae (Figure 7.39) or by longitudinal fission. The latter process often involves the formation of multiple gastric pouches (**polygastry**) followed by longitudinal splitting, which produces two daughter medusae. In some species (e.g., *Aequorea macrodactyla*), direct fission

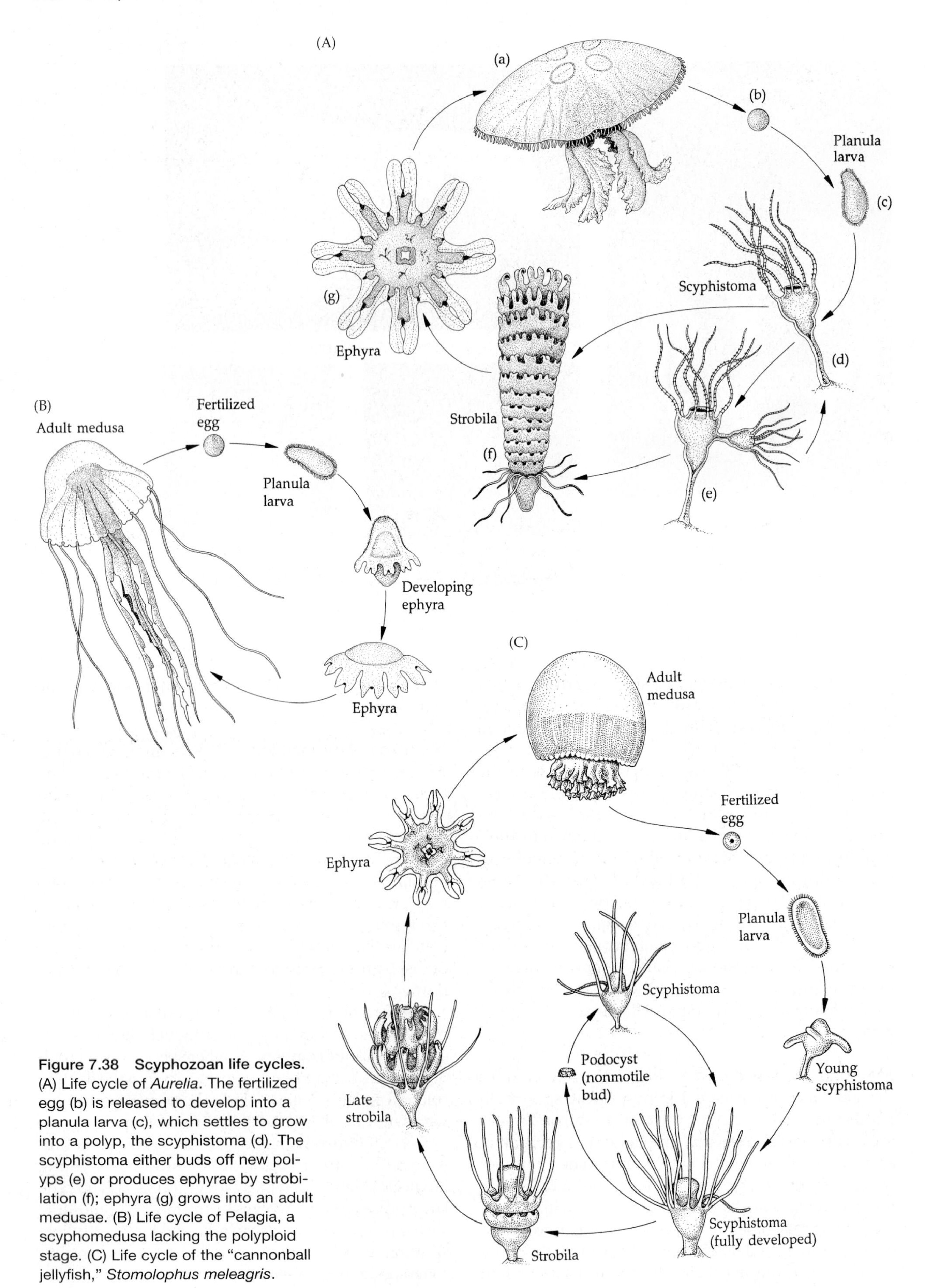

Figure 7.38 Scyphozoan life cycles. (A) Life cycle of *Aurelia*. The fertilized egg (b) is released to develop into a planula larva (c), which settles to grow into a polyp, the scyphistoma (d). The scyphistoma either buds off new polyps (e) or produces ephyrae by strobilation (f); ephyra (g) grows into an adult medusae. (B) Life cycle of Pelagia, a scyphomedusa lacking the polyploid stage. (C) Life cycle of the "cannonball jellyfish," *Stomolophus meleagris*.

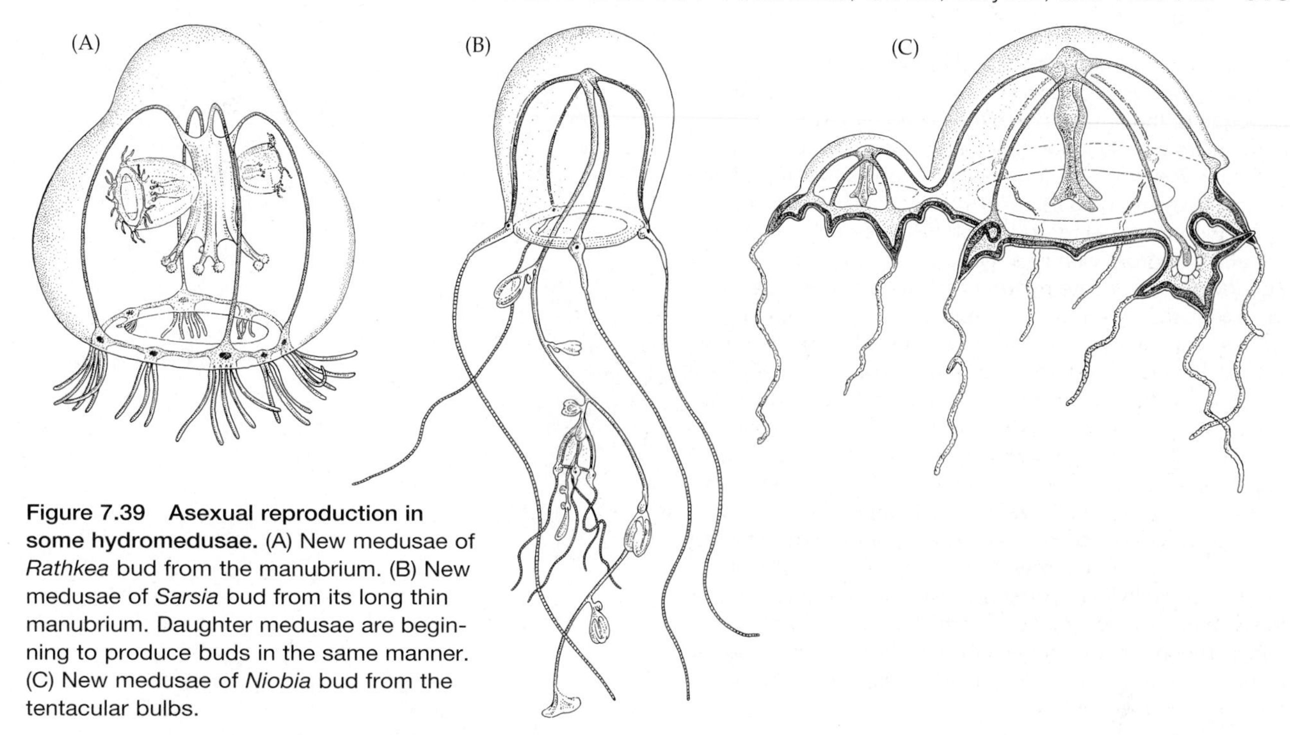

Figure 7.39 Asexual reproduction in some hydromedusae. (A) New medusae of *Rathkea* bud from the manubrium. (B) New medusae of *Sarsia* bud from its long thin manubrium. Daughter medusae are beginning to produce buds in the same manner. (C) New medusae of *Niobia* bud from the tentacular bulbs.

may take place. Polygastry does not occur during this process; instead, the entire bell folds in half, severing the stomach, ring canal, and velum (Figure 7.40). Eventually the whole medusa splits in half and each part regenerates the missing portions.

Cnidarians in general have a great capacity for regeneration, as exemplified by experiments on *Hydra*. The eighteenth-century naturalist Abraham Trembley had the clever idea of turning a *Hydra* inside out—and he did. To his delight, the animal survived quite well,

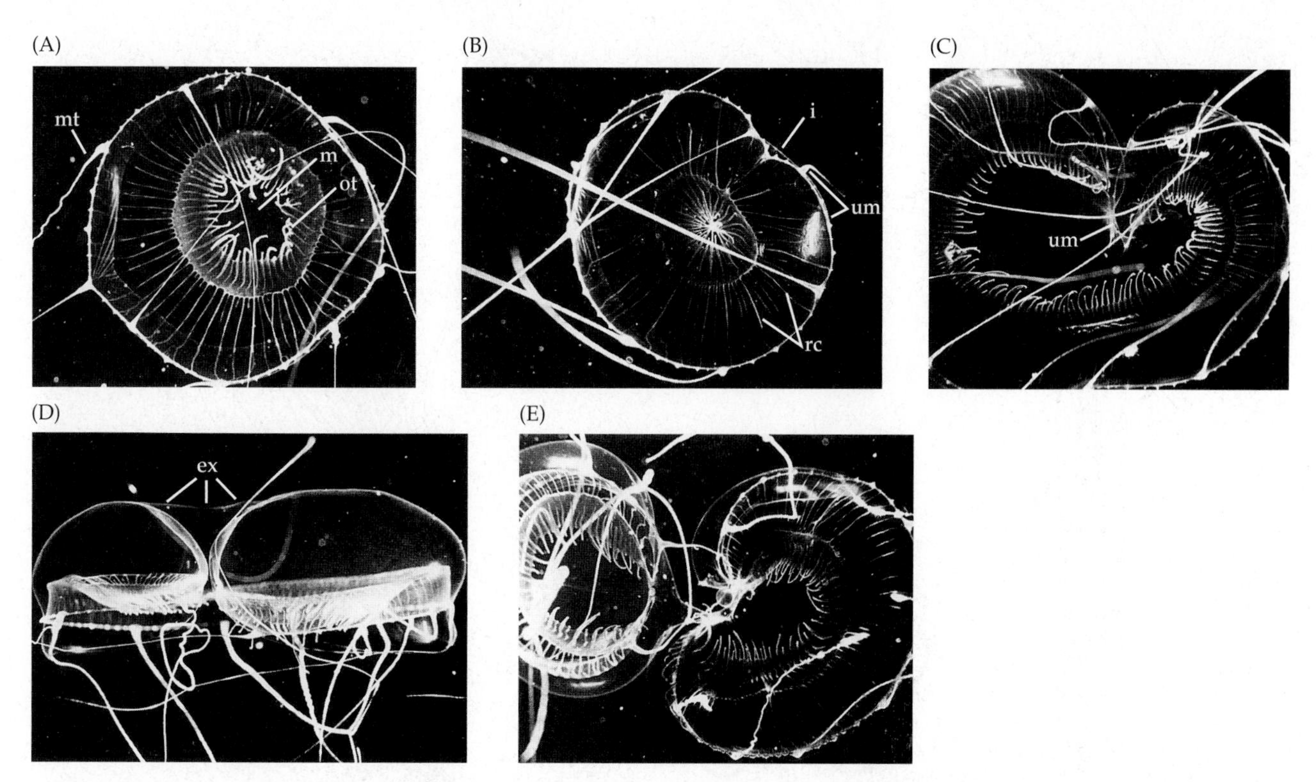

Figure 7.40 Asexual reproduction in the hydromedusa *Aequorea*. The sequence of photographs shows the direct fission of *A. macrodactyla*. (A) This oral view shows a nondividing medusa with its marginal fishing tentacles (mt) deployed. (B) Initiation of invagination (i). (C–E) A progression of the direct fission process. The oral (C) and marginal (D) views illustrate the severing of the umbrellar margin (um) and the separation of exumbrellar halves; (E) shows the exumbrellar surface (ex) beginning to pull apart, producing free-swimming daughter medusae; healing is nearly complete in the smaller daughter medusa on the left. ot: oral tentacles; m: mouth; rc: radial canals.

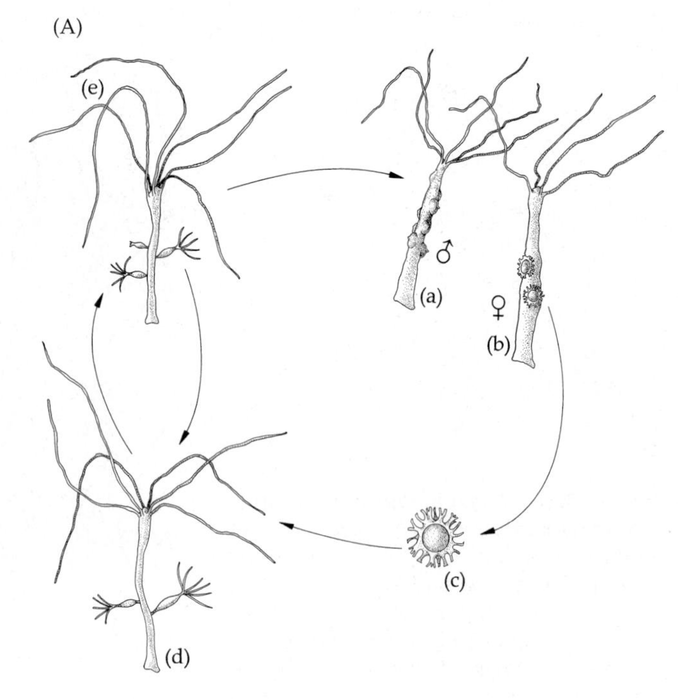

Figure 7.41 Some hydrozoan life cycles. (A) Life cycle of *Hydra*. Sperm produced by the male polyp (a) fertilizes the eggs of the female polyp (b). During cleavage, the eggs secrete a chitinous theca about themselves. After hatching, the embryos (c) grow into polyps that reproduce asexually by budding (d), until environmental conditions again trigger sexual reproduction. (B) Life cycle of *Obelia*, a thecate hydroid with free medusae. (C) Life cycle of *Tubularia*, an athecate hydroid that does not release free medusae. The polyp (a) bears many gonophores, whose eggs develop in situ into planulae (b) and then into actinula larvae (c) before release (d); the liberated actinula larvae (d) settle and transform directly into new polyps (e), which each proliferate to form a new colony (f). (D) Life cycle of a trachyline hydrozoan medusa without a polypoid stage (*Aglaura*). After fertilization, a gonochoristic adult (a) releases a planula larva (b), which adds a mouth and tentacles (c) to become an actinula larva (d). Subsequently the actinula larva becomes a young medusa (e). (E) Life cycle of a trachyline hydrozoan with a polypoid stage, the freshwater *Limnocnida*. Gonochoristic medusae (a) release fertilized eggs (b) that grow into planula larvae (c). Planula larvae settle to form small hydroid colonies (d), which bud off new medusae (e).

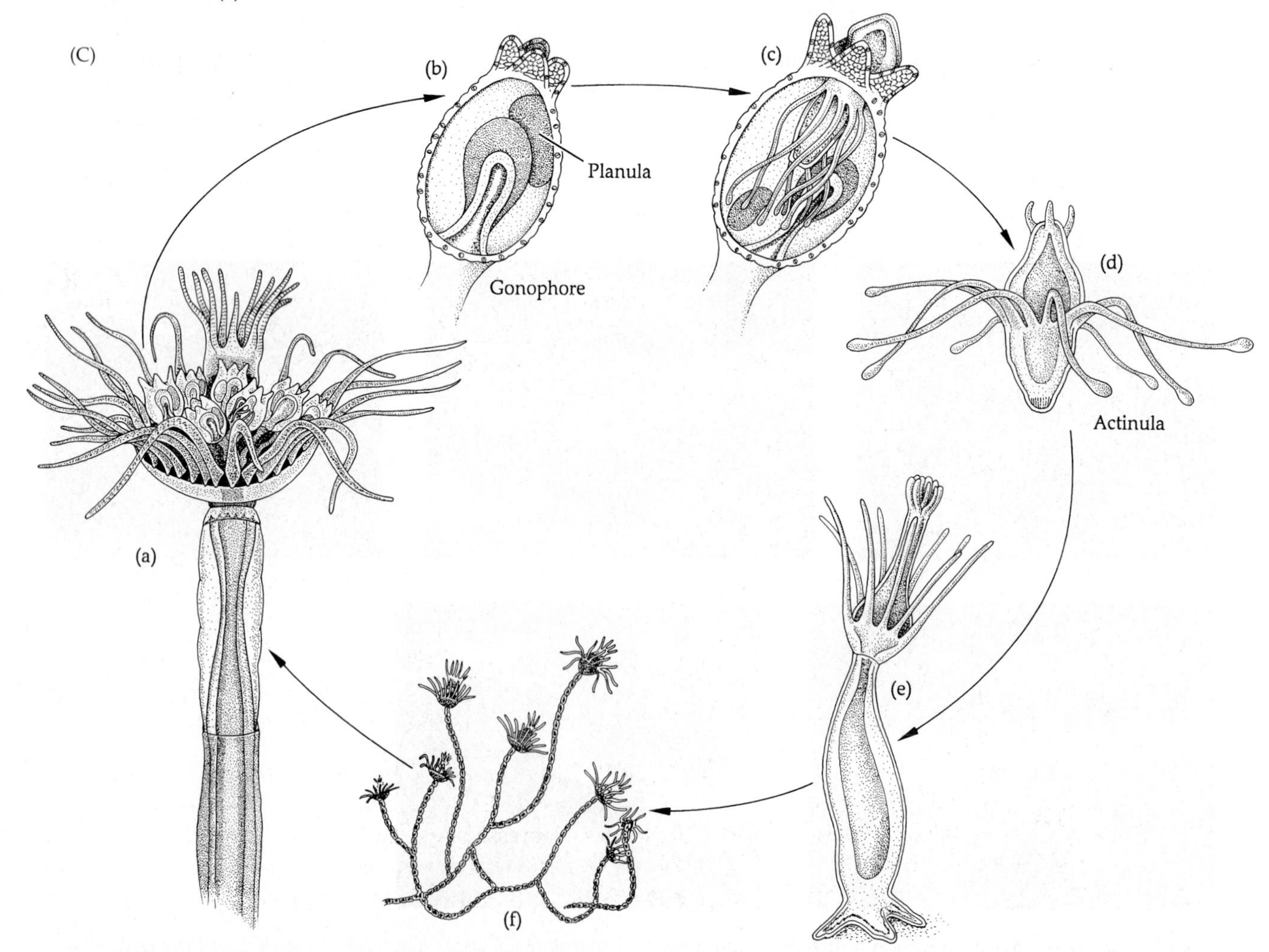

with the gastrodermal cells functioning as the "new epidermis" and vice versa. Cells removed from the body of a *Hydra* also have a modest degree of reaggregative ability, like that seen so dramatically in sponges. In some cases, entire animals will reconstruct from cells taken only from the gastrodermis or only from the epidermis. Although *Hydra* is an unusual and atypical cnidarian, this great capacity for cellular reorganization is

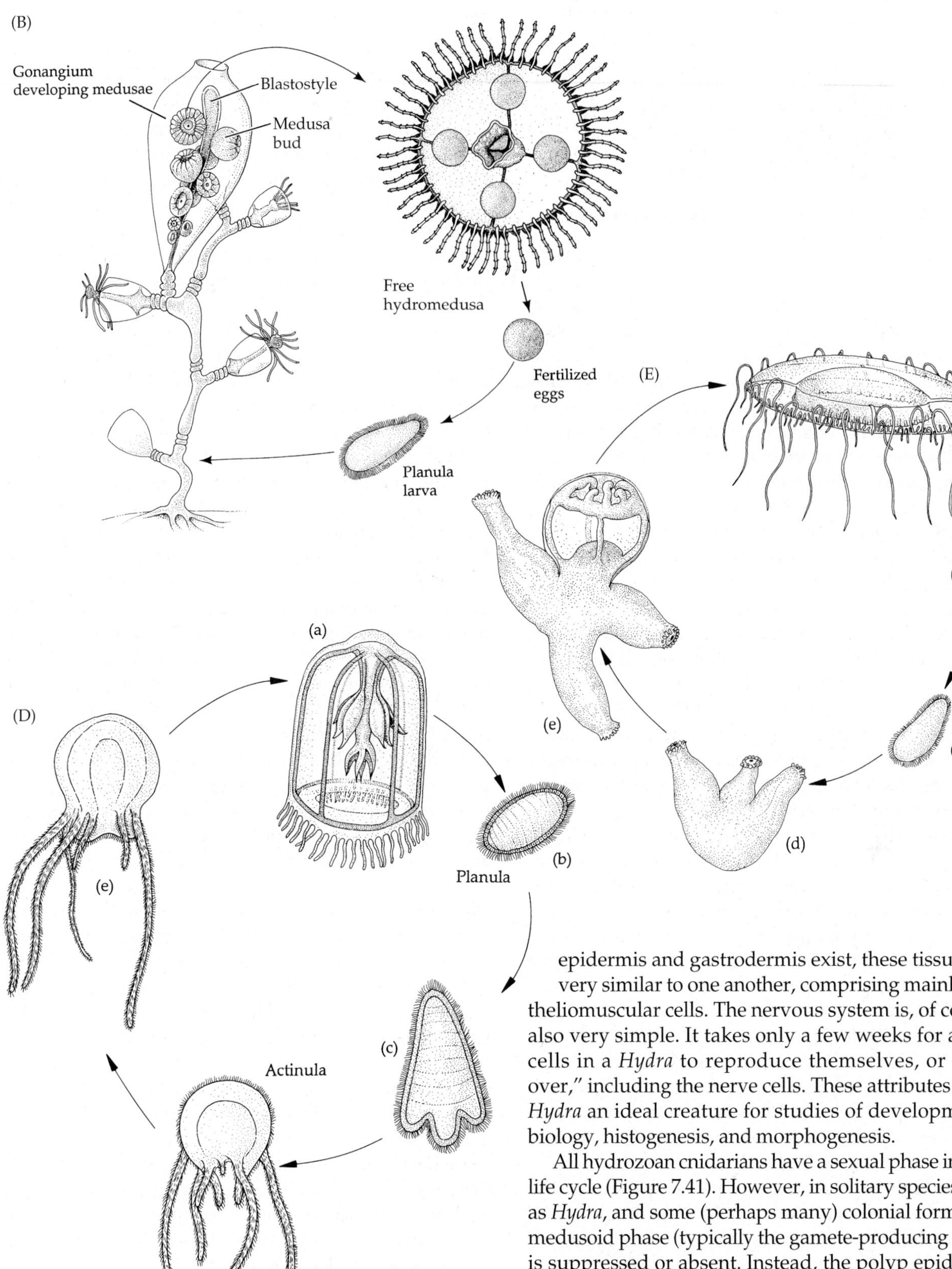

a reflection of the primitive state of tissue development in the animals belonging to this phylum.

A typical *Hydra* consists of only about 100,000 cells of roughly a dozen different types. Although distinct epidermis and gastrodermis exist, these tissues are very similar to one another, comprising mainly epitheliomuscular cells. The nervous system is, of course, also very simple. It takes only a few weeks for all the cells in a *Hydra* to reproduce themselves, or "turn over," including the nerve cells. These attributes make *Hydra* an ideal creature for studies of developmental biology, histogenesis, and morphogenesis.

All hydrozoan cnidarians have a sexual phase in their life cycle (Figure 7.41). However, in solitary species such as *Hydra*, and some (perhaps many) colonial forms, the medusoid phase (typically the gamete-producing stage) is suppressed or absent. Instead, the polyp epidermis develops simple, transient gamete-producing structures called **sporosacs** (Figures 7.5B). Most colonial hydroids produce medusa buds (gonophores) either from the walls of the hydranths or from separate gonozooids. The gonophores may grow into medusae that are released as free-living sexually reproducing individuals, or they may remain attached to the polyps as incipient medusae that produce gametes in place.

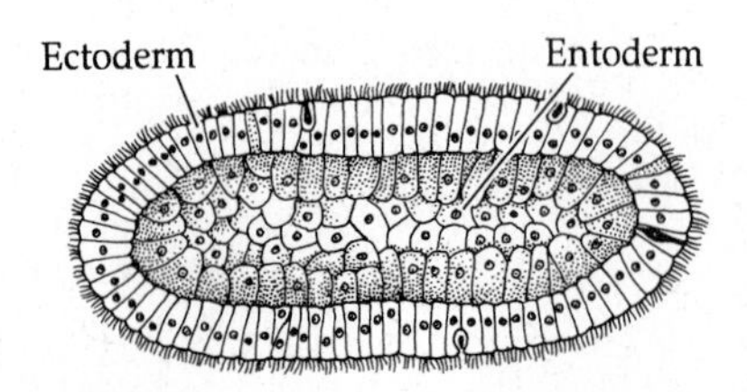

Figure 7.42 A typical solid hydrozoan planula larva resulting from ingression.

Milleporid hydrozoans produce short-lived sexual medusae that are released over several days during particular seasons. Temporal segregation of spawning seems to occur among colonies of different species. Colonies within species are gonochoristic and so release exclusively male medusae, bearing a sperm sac located at the vestigial manubrium, or female medusae bearing 3 to 5 ova in the subumbrella cavity. Male medusae appear to be released into the water column by their corallae a few hours before female medusae are released. Gametes are released by both medusa types, and fertilization occurs externally.

In free-living hydromedusae, germinal cells arise from interstitial epidermal cells that migrate to specific sites on the bell surface, where they consolidate into a temporary gonadal mass. Subsequently, gametogenic tissue appears on the surface of the manubrium, beneath the radial canals, or on the general subumbrellar surface. Hydromedusae are typically gonochoristic, with either sperm or eggs usually being released directly into the water, where fertilization occurs. In some, only sperm are released and fertilization occurs on or in the female medusa's body. Free hydromedusae are especially common in temperate waters, where they may be very abundant seasonally and easily collected in plankton nets. Siphonophores appear to be ancestrally gonochoristic but hermaphroditic forms exist among the Physonecta and exclusively among the Calycophorae (Codonophora). Planktonic dispersal of sexual forms would seem to mitigate against genetic differentiation of coastal populations, but considerable genetic population structure has been demonstrated among morphologically similar populations of *Obelia geniculata*, with estimated divergence times among reciprocally monophyletic lineages exceeding several million years.

Although several cleavage patterns occur in the Hydrozoa, the overall motif is fundamentally radial and holoblastic. A coeloblastula forms, which gastrulates by uni- or multipolar ingression to a stereogastrula. The interior cell mass is endoderm; the exterior cell layer is ectoderm (Figure 7.42). The stereogastrula elongates to form a unique, solid or hollow, nonfeeding, free-swimming **planula larva** (Figure 7.43). The planula larva is radially symmetrical, but it swims with a distinct "anterior–posterior" orientation. The ectodermal cells are monociliated and destined to become the adult epidermis; the endoderm is destined to become the adult gastrodermis. The trailing end of the larva (of all cnidarians) becomes the oral end of the adult, and even in the larval stage a mouth sometimes develops at this end. Hydrozoan planulae swim about for a few hours, a few days, or a few weeks before settling by attaching at the leading end. If the larva is still solid, then the endoderm hollows to form the coelenteron. The mouth opens at the unattached oral end and tentacles develop as the larva metamorphoses into a young solitary polyp. Studies have shown that a dramatic reorganization of the hydrozoan nervous system takes place during metamorphosis of the planula into the primary polyp (in *Pennaria*). The larval neurons degenerate and new neurons differentiate to reform a nerve net, and the overall distribution pattern of the nervous system changes dramatically.

This overview of the hydrozoan reproductive cycle covers most species (Figure 7.41), but in fact even more variety actually exists than we have space to discuss. For example, in some trachylines, the polypoid stage is apparently lost altogether. The medusae produce planula larvae that develop into actinula larvae, which metamorphose into adult medusae, bypassing any sessile polypoid phase. Some trachylines and some siphonophores undergo direct development, bypassing the larval stage altogether. Within the Trachylina, members of the family Rhopalonematidae (formerly within the order Actinulida) exhibit minute interstitial polyps that lack a medusoid stage (or possibly these forms represent creeping medusae) and have suppressed the larval phase. The adult polyp is ciliated and resembles an actinula larva (hence the name).

Myxozoan reproduction Myxozoan cnidarians have such aberrant life cycles that is it impossible to identify either polypoid or medusa stages within them. No planula stage is known to exist. Such marked differences from other cnidarian taxa no doubt contributed to the uncertain status of myxozoans within the phylum to date. Nevertheless, an alternation of sexual and asexual generations within vertebrate and annelid hosts does occur, albeit as parasitic forms as described above. Asexual reproduction occurs within the vertebrate host, producing infective myxospores as already described (Figure 7.29). Sexual reproduction occurs within annelids and other coelomate worms, generating infective actinosporean larvae. The separation of the sexual and asexual stages of the life cycle of these parasites is similar to that which occurs in other parasitic species and could suggest that that some developmental genetic homologies exist between the asexual and sexual stages of myxozoans and the polyp and medusa stages of other cnidarians. However, the reproductive modes of most myxozoans remain poorly known.

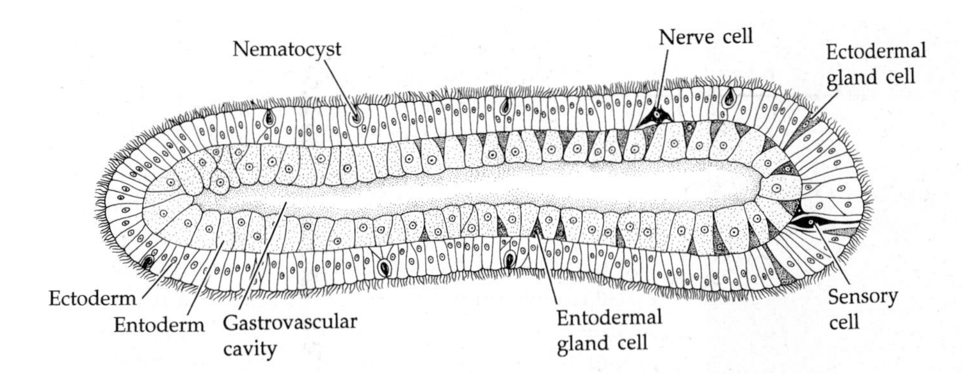

Figure 7.43 The hollow planula larva of the hydroid *Gonothyraea* (longitudinal section).

Cnidarian Phylogeny

Ediacaran Cnidaria?

Cnidarians have one of the longest fossil histories among the Metazoa. The first apparent cnidarian fossils date from the Ediacaran period, from the famous Ediacara Hills of South Australia, which contain possible medusae and polypoid colonies (e.g., sea pen-like creatures) that lived nearly 600 million years ago. However, these early circular and frondlike impressions cannot be unequivocally assigned to any modern Cnidarian taxa. Some researchers doubt that they represent true cnidarians, even though some of the circular forms have marginal tentacles such as seen in most medusae. Other possible neoproterozoic polypoids include impressions that may be the floats of the pelagic "chondrophorans," and fossils with triradiate symmetry (e.g., *Tribrachidium*), as well as pentiradiate and octoradiate symmetry, thus differing fundamentally from any living cnidarians. Some workers consider these fossils to be parts of other organisms, such as isolated holdfast structures; others have placed them into an extinct class of Cnidaria, the Trilobozoa. Still others view them as primitive triradiate echinoderms. It helps to have a good imagination if you're a paleontologist.

In the 1980s, Adolf Seilacher, frustrated by attempts to assign these Neoproterozoic era fossils to modern phyla, created an entirely new phylum, the Vendobionta (or Vendozoa) suggesting that it might be a sister-group to the Eumetazoa (metazoans above the poriferan grade). The Vendozoa, Seilacher noted, seemed to have a quiltlike construction of thin, lightly sclerotized but flexible, outer skin separated into compartments by more rigid internal struts (which in fossil impressions appear to be sutures). Later workers placed these fossils into an extinct cnidarian order, the Rangeomorpha, and called other, more bilaterally symmetrical forms Erinettomorpha. One of the most famous of these latter forms is the oval-shaped *Dickinsonia*, which may have measured over a meter in length but been only a few millimeters thickness. Some workers consider *Dickinsonia* to be medusa-like, but it has also been assigned to the Platyhelminthes, the Annelida (due to hints of possible body segmentation), to an extinct phylum (Proarticulata), and to an entirely new kingdom!

Attempts to place the Ediacaran fossils within the Cnidaria (or other extant phyla) have met with limited success until 2014. For example, differences in apparent growth patterns and frond structure between extant cnidarians (e.g., pennatulaceans) and frondlike Ediacaran species have argued against a direct cnidarian relationship. Then, in 2014, a 560-million-year-old likely-cnidarian fossil (named *Haootia quadriformis*) was reported from Newfoundland, with quadraradial symmetry and clearly preserved bundled muscular fibers. *Haootia* appears to be a nearly 6 cm long polyp, or perhaps an attached medusa—it greatly resembles modern species of Staurozoa.

Early trace fossils have also been assigned to the phylum Cnidaria. Burrows assigned to sea anemones occur in the early Cambrian, and some strata have meandering tracks that appear to be traces left from mucous trails of creeping anemones (some modern anemones move in this fashion, using ciliary activity). Remarkable embryos of probable scyphozoan affinity have also been reported form the Lower Cambrian of China.

A recent discovery of two new species of enigmatic diploblastic animals from southern Australian seabeds could provide an example of Ediacaran Cnidaria-like animals that survived to modern times. *Dendrogramma enigmatica* and *D. discoides* (Figures 7.44 and 7.45) possess a polyp-like structure, an asymmetrical branching digestive system with no apparent anus, a thick gelatinous mesoglea between the epidermis and gastrodermis, and other structural similarities shared with both ctenophores and cnidarians. But they lack the adhesive or stinging structures that characterize those taxa. *Dendrogramma* has an intriguing similarity to the fossil medusoid Ediacaran creatures (e.g., *Albumares*, *Anfesta*, *Margaritiflabellum*, and *Rugoconites*) that possessed a disk containing forked and radiating channels and a central stalk, possibly also with a terminal mouth. Initial preservation of *Dendrogramma* specimens in formalin has so far hindered molecular classification,

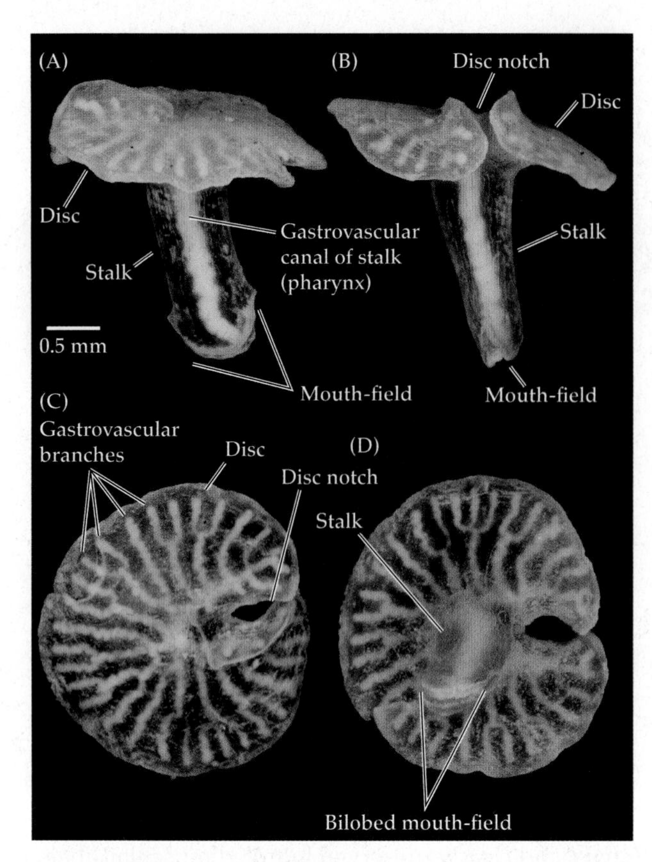

Figure 7.44 This new animal, *Dendrogramma enigmatica* was collected in deep waters off southern Australia in 1986. (A,B) Lateral views, (C) aboral view, (D) oral view.

leaving unresolved the possibility that Ediacaran medusoids still persist 540 million years after their presumed extinction. For now, the two living species have been placed in their own family, Dendrogrammatidae, as Metazoa *incertae sedis*.

Cnidarian Origins

The origin of the cnidarians is closely tied to the origin of the Metazoa themselves. Like Porifera, cnidarian body plans appear to have remained essentially unchanged since the Cambrian. The prevailing theory of metazoan origin sometimes called the **colonial theory**, depicts a colonial flagellated protist (e.g., a choanoflagellate) that gave rise to a hollow metazoan ancestor, termed a **blastea**, which in turn gave rise to a diploblastic planuloid animal called a **gastrea**. It is from these hypothetical ancestral forms that the Porifera and Cnidaria are said to have evolved. On the other hand, another view suggests that the ancestors of cnidarians were triploblastic, acoelomate organisms, perhaps something like rhabdocoel turbellarians, that underwent "degenerative evolution" to produce what we recognize today as the cnidarians. This view, called the **triploblastic theory** (or the **turbellarian theory**), usually holds the Anthozoa to be the most primitive cnidarian class and cites the "remnants" of bilateral symmetry in that class as evidence of a bilateral ancestry. As noted earlier in this chapter, certain molecular and developmental evidence supports this idea, but the weight of evidence is against this interpretation, and despite *Buddenbrockia*'s putative vestigial musculature, no living cnidarians show remnants of what might have been a triploblastic ancestry, either in their anatomy or their development. The triploblastic hypothesis remains controversial because many contemporary zoologists find it weak on several counts. "Degenerative evolution" (a poor choice of words) is a phenomenon primarily associated with the evolution of parasites or the exploitation of smallness (e.g., interstitial forms), wherein it may result in the reduction of certain systems and the specialized adaptive development of others. The idea of a free-living, triploblastic, bilateral, motile creature (such as a flatworm) taking up a sessile existence and transforming into a radially symmetrical, diploblastic, anthozoan polyp appears to be a highly unlikely evolutionary scenario. The adoption of radiality (or at least "functional" radiality) of bilaterally symmetrical animals is well documented in some taxa (e.g., Echinodermata), but does

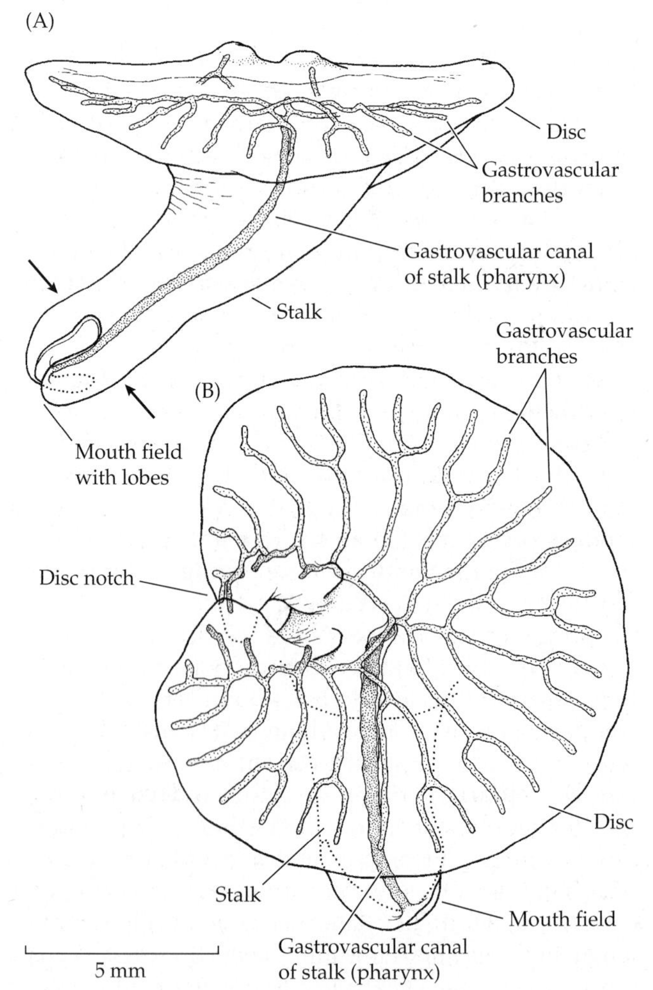

Figure 7.45 Diagrammatic rendering of *Dendrogramma*. (A) Lateral view, (B) aboral view.

not involve the kinds of "degeneration" required by the turbellarian theory. The transformation suggested by the triploblastic or turbellarian theory is complicated; it involves the loss or drastic simplification of many complex systems (notably the urogenital system) and major changes in fundamental body design. Both larvae and adults of extant cnidarians maintain a fundamental radial symmetry. The so-called remnant bilaterality of anthozoan polyps is not true bilaterality at all, but biradiality about an oral–aboral axis, which develops late in the ontogeny of these animals. The triploblastic/turbellarian theory is also weak on embryological grounds, such as differences in cleavage patterns and germ layer formation. And, perhaps most importantly, the most recent multi-gene phylogenetic analyses clearly position Cnidaria below the origin of bilaterality and likely as the sister group to the Bilateria.

In Chapter 5 we reviewed the important embryological differences between the coelomate metazoan clades known as the protostomes and deuterostomes. To the extent that these traits occur in noncoelomate Metazoa, it is of phylogenetic importance to note them. For example, radial cleavage is characteristic of the deuterostomes, but probably arose very early in metazoan evolution; it occurs in cnidarians and, in a slightly different form, in most sponges. Thus it appears to be the plesiomorphic type of cleavage among animals. Spiral cleavage, on the other hand, defines an entire clade of protostomes (the Spiralia), suggesting it is a more derived cleavage pattern in Metazoa. In fact, as more is learned about cnidarian genomes, the evidence seems to suggest that their genes share more similarities to deuterostome genes than to those of protostomes.

Relationships within Cnidaria

Anatomical evidence has been equivocal on the question of which living class of Cnidaria is the most ancestral. Some workers have favored a view that the first cnidarian was a medusa, based largely the conjecture that the sexual stage must have appeared first (the **medusoid hypothesis**). Other workers have favored the view that the polypoid form is primitive within Cnidaria, because polyps occur in all classes of the phylum, except the parasitic Myxozoa, and thus they represent part of the primitive cnidarian body plan (the **polypoid hypothesis**). Recent molecular work has clarified many relationships within Cnidaria, yet whether anthozoans or medusozoans represent the "basal" cnidarian group still remains debatable. Of course, the phylogenetic hypothesis that one adopts has great implications for interpreting character evolution within this group—and the weight of the evidence currently favors anthozoans as ancestral. Nevertheless, more work is needed to substantiate this view. Next, we present the implications of each hypothesis, knowing that further work will eventually substantiate or refute the following frameworks.

The polypoid hypothesis suggests that the exclusively polypoid class Anthozoa is closest to the ancestral cnidarian. This hypothesis has received increasing support and is the scheme presented in this chapter (Figure 7.46). Molecular and genetic data supporting this view continue to accumulate. For example, 18S rDNA sequence analyses place anthozoans as a sister group to all other cnidarians. In addition, we now know that, among cnidarians, only the anthozoans possess circular mitochondrial DNA, a trait they share with other Metazoa (including the Placozoa). Members of the Medusozoa clade all have linear mtDNA, viewed as a derived condition within cnidarians.

In the polypoid hypothesis, the ancestral cnidarian would possess some or all of the traits that define modern-day Anthozoa, whereas the clade comprising Staurozoa, Hydrozoa, Cubozoa, and Scyphozoa (i.e., the Medusozoa) is defined by the evolution of the cnidocil and linear mtDNA. Most workers also place the origin of the medusa at the base of the Medusozoa line, suggesting that all cnidarian medusae are homologous, whereas others have suggested that hydromedusae and scyphomedusae (and cubomedusae) arose independently. In any case, in the polypoid hypothesis, the medusa represents a derived dispersal stage. The position of the Myxozoa is still unclear.

In contrast, *the medusoid hypothesis* suggests that the Cnidaria arose from a swimming or creeping, ciliated, planuloid-like ancestor that, with the development of tentacles resulted an animal resembling an actinula larva. The transition from planula to actinula in the modern medusa form can be seen today in the life cycle of certain hydrozoans. Asexual reproduction, such as budding, by a benthic actinula larva could have secondarily led to the establishment of a distinct polypoid stage. If so, the polyp can be viewed as an extended larval form specialized for asexual reproduction and benthic existence. In this scenario, once the polypoid form became established some cnidarians began to suppress the medusoid phase of their life cycles, various degrees of which can be seen among modern hydrozoans. The epitome of this trend would be the class Anthozoa, whose members have no medusa stage at all. The sister-group of Cubozoa + Scyphozoa (known as the Acraspeda) would be defined by the same synapomorphies in both the polypoid and medusoid hypotheses—loss of gut mesenteries, loss of gastrodermal nematocysts, reduction of cells in the mesoglea, movement of the gamete-forming tissue to the epidermis, and the evolution of acraspedote medusae.

Phylogeny within each of the cnidarian classes is equally interesting but largely beyond the scope of this text. However, a few generalizations can be made about some important events. Coloniality has been a common and important evolutionary theme within Cnidaria. Coloniality in the hydrozoans probably arose by retention of young polyps during

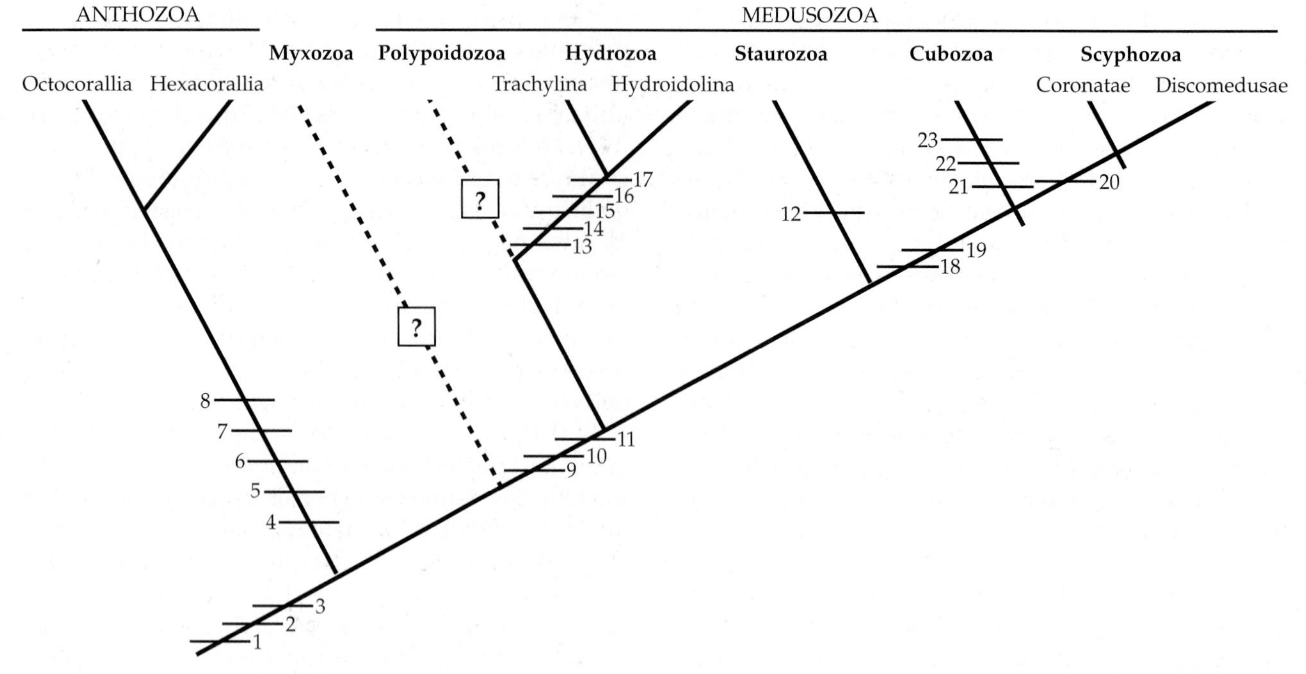

Figure 7.46 A molecular-based phylogeny of the Cnidaria, overlaid with key synapomorphies of the major lineages.
Synapomorphies of Cnidaria: (1) Cnidae; (2) Epitheliomuscular cells; (3) Planula larva.
Synapomorphies of Anthozoa: (4) Hexaradial/Octaradial symmetry; (5) Actinopharynx; (6) Siphonoglyph; (7) Mesenterial filaments in coelenteron; (8) Tripartite series of flaps on cnidae.
Synapomorphies of Medusozoa: (9) Medusae and alternation of generations; (10) Linear DNA; (11) Operculate cnidae (with cnidocil).
Synapomorphies of Staurozoa: (12) Evolution of unique life cycle, with stauropolyps and sessile stauromedusae.
Synapomorphies of Hydrozoa: (13) Relocation of gamete-forming tissue to epidermis; (14) Loss of gut mesenteries; (15) Simplification of middle layer to an acellular mesoglea; (16) Evolution of craspedote medusa form; (17) Loss of gastrodermal nematocysts.
Synapomorphies of Cubozoa-Scyphozoa: (18) Reduction or loss of polyp; (19) Rhopalia.
Synapomorphies of Scyphozoa: (20) Strobilation.
Synapomorphies of Cubozoa: (21) Boxlike medusae form; (22) Lensed rhopaliar eyes; (23) Velarium.

asexual reproduction, and this development ultimately led to the highly specialized colonial groups such as the Siphonophora, Milleporidae, and Stylasteridae. In the class Scyphozoa, evolution has clearly favored increasing specialization of the pelagic medusoid form and diminishing importance of the polypoid stage in their life cycle. Scyphomedusae and cubomedusae have evolved large size, special musculature, a cellular or fibrous mesenchyme, a complex gastrovascular system, and a fairly sophisticated sensory system.

Anthozoans are characterized by several unique synapomorphies: hexaradial or octaradial symmetry (transformed into biradial symmetry in most); the actinopharynx, siphonoglyphs, and unique mesenterial filaments in the coelenteron; absence of a cnidal operculum and cnidocil; tripartite flaps on the cnidae; and special ciliary cones associated with the cnidocytes. Anthozoans also possess a more complex gastrovascular and nervous system than seen among the Medusozoa, as well as a greater degree of cellularity of the mesenchyme.

Among members of the class Anthozoa, evolution has produced a grand series of experiments in colonial polypoid living, resulting in such "superorganisms" as stony corals, octocorals, pennatulaceans, and zoanthidians. Within the Hexacorallia, stony corals (scleractinians) first appear in the fossil record in the Middle Triassic period, about 237 million years ago, although the first scleractinians were not reef builders. Scleractinian origins (and radiations) are still not well understood. Suggested extinct ancestors of Scleractinia include three Paleozoic coral groups: the Rugosa, the Heterocorallia, and the Tabulata. The Rugosa (sometimes called horn corals) and Heterocorallia had their polyps divided by septa into 4 cycles (rather than 6 as in Scleractinia) and the septa arose in a pinnate fashion (rather than cyclically as in Scleractinia). Tabulate corals were nonseptate. The connection between rugose corals, which formed calcite skeletons, and modern scleractinians, whose skeletons are aragonite, is tightened by the discovery of Cretaceous scleractinians with calcite skeletons.

Within the Scleractinia, molecular evidence appears to confirm the existence of two clades, "robust" species with solid, heavily calcified skeletons forming massive or plate-like structures, and "complex" species with lighter, more porous and more complex skeleton structures. Within these clades, only the Poritiina and Dendrophiliina appear to be monophyletic, suggesting that relationships based on morphology alone are likely to be misleading in this group. Possibly contributing to this diversity, recent evidence suggests that coloniality and symbiosis with zooxanthellae have been repeatedly acquired and lost throughout the history of stony corals, a tendency that may have allowed scleractinians to diversify within reef and non-reef communities, as well as recover from repeated local extinctions over evolutionary time. An increase in polyp size within the Anthozoa seems to have occurred over time, along with the evolution of complex structural components of the mesenchyme and increasingly efficient musculature. Anthozoans, of course, have greatly exploited a commensal relationship with zooxanthellae—more so than members of the other classes. Convergent evolution has occurred frequently throughout the Cnidaria, as witnessed by such features as colonies, calcareous skeletons, the velum-velarium structures, and various means of suppressing the medusoid or polypoid stage in the life cycle.

The two recently recognized cnidarian lineages, Myxozoa and Polypoidozoa, represent experiments in parasitic life histories, with the first more successful than the second. Myxozoans now appear to be comprised of nearly 2,200 species that clearly fall within the Cnidaria, but whose relationships to one another are still uncertain. The morphology of the Polypoidozoa suggests that they arose from within the Medusozoa. However, the precise phylogenetic position of both taxa is uncertain at this time. Additional phylogenetic analyses of the Cnidaria in the coming years are certain to produce exciting new results.

Selected References

General References

Ball, E. E., D. C. Hayward, R. Saint and D. J. Miller. 2004. A simple plan–cnidarians and the origins of developmental mechanisms. Nat. Rev. Genet. 5: 567–577.

Bavestrello, G., C. Sommer and M. Sarà. 2002. Bi-directional conversion in *Turritopsis nutricula* (Hydrozoa). Sci. Mar. 56(2–3): 137–140.

Bayer, F. M. and H. B. Owre. 1968. *The Free-Living Lower Invertebrates*. Macmillan, New York.

Boardman, R. S., A. H. Cheetham and W. A. Oliver (eds.). 1973. *Animal Colonies*. Dowden, Hutchinson and Ross, Stroudsburg, Pennsylvania.

Bridge, D., C. W. Cunningham, R. DeSalle and L.W. Buss. 1995. Class-level relationships in the phylum Cnidaria: molecular and morphological evidence. Mol. Biol. Evol. 12(4): 679–689.

Byrum, C. A. and M. Q. Martindale. 2004. Gastrulation in the Cnidaria and Ctenophora. Pp. 33–50 in C. D. Stern (ed.), *Gastrulation: From Cells to Embryos*. Cold Spring Harbor Laboratory Press, Cold Spring Harbor, NY.

Cairns, S. D. and 11 other authors. 1991. *Common and scientific names of aquatic invertebrates form the United States and Canada: Cnidaria and Ctenophora*. Spec. Publ. No. 22, Amer. Fisheries Soc., MD.

Cairns, S. D. and I. G. Macintyre. 1992. Phylogenetic implications of calcium carbonate mineralogy in the Stylasteridae (Cnidaria: Hydrozoa). Palaios 7: 96–107.

Campbell, R. D. 1974. Cnidaria. Pp. 133–200 in A. C. Giese and J. S. Pearse (eds.), *Reproduction of Marine Invertebrates, Vol. 1*. Academic Press, New York.

Chang, E. S., M. Neuhof, N. D. Rubinstein, A. Diamant, H. Philippe, D. Huchon and P. Cartwright. 2015. Genomic insights into the evolutionary origin of Myxozoa within Cnidaria. PNAS Early Edition www.pnas.org/cgi/doi/10.1073/pnas.1511468112.

Cheng, L. 1975. Marine pleuston—animals at the sea–air interface. Oceanogr. Mar. Biol. Annu. Rev. 13: 181–212.

Collins, A. G. 2002. Phylogeny of Medusozoa and the evolution of cnidarian life cycles. J. Evol. Biol. 15: 418–432.

Collins, A. G. 2009. Recent insights into cnidarian phylogeny. Smithsonian Contrib. Mar. Sci. 38: 139–149.

Conklin, E. J. and R. N. Mariscal. 1977. Feeding behavior, ceras structure, and nematocyst storage in the aeolid nudibranch, *Spurilla neapolitana* (Mollusca). Bull. Mar. Sci. 27(4): 658–667.

Connor, J. L. and N. L. Deans. 2002. *Jellies. Living Art*. Monterey Bay Aquarium, Monterey, CA. [Beautiful photographs of rarely seen jellies; great accompanying text.]

Daly, M and 11 others. The phylum Cnidaria: A review of phylogenetic patterns and diversity 300 years after Linnaeus, Zootaxa 1668: 127–182.

Fautin, D. G. 2002. Reproduction in Cnidaria. Can. J. Zool. 80: 1635–1754.

Finnerty, J. R. and M. Q. Martindale. 1999. Ancient origins of axial patterning genes: Hox genes and ParaHox genes in the Cnidaria. Evol. Dev. 1: 16–23.

Finnerty, J. R., K. Pang, P. Burton, D. Paulson and M. Q. Martindale. 2004. Origins of bilateral symmetry: *Hox* and *Dpp* expression in a sea anemone. Science 304: 1335–1337.

Grassé, P.-P. 1994. *Traité de Zoologie, Tome 3, Fascicule 2: Cnidaires, Cténaires*. Masson et Cie, Paris.

Harrison, F. W. and J. Westfall (eds.). 1991. *Microscopic Anatomy of Invertebrates, Vol. 2. Placozoa, Porifera, Cnidaria, and Ctenophora*. Alan Liss, New York.

Holstein, T. and P. Tardent. 1983. An ultrahigh-speed analysis of exocytosis: nematocyst discharge. Science 223: 830–833.

Holstein, T. W., M. Benoit, G. v. Herder, G. Wanner, C. N. David and H. E. Gaub. 1994. Fibrous mini-collagens in *Hydra* nematocysts. Science 265: 402–404.

Hyman, L. H. 1940. *The Invertebrates, Vol. 1, Protozoa through Ctenophora*. McGraw-Hill, New York.

Just, J. J., R. M. Kristensen and J. Olesen. 2014. *Dendrogramma*, new genus, with two new non-bilaterian species from the marine bathyal of southeastern Australia (Animalia, Metazoa *incertae sedis*)—with similarities to some medusoids from the Precambrian Ediacara. PLoS ONE 9(9): 1–11 (e102976).

Liu, A. G., J. J. Matthews, L. R. Menon, D. McIlroy and M. D. Brasier. 2014. *Haootia quadriformis* n. gen., n. sp., interpreted as a muscular cnidarian impression from the Late Ediacaran Period (approx. 560 Ma). Proc. Roy. Soc. B 281: 20141202.

Lotan, A., L. Fishman, Y. Loya and E. Zlotkin. 1995. Delivery of a nematocyst toxin. Nature 375: 456.

Ma, H. and Y. Yang. 2010. *Turritopsis nutricula*. Nat. Sci. 8(2): 15.

Mackie, G. O. (ed.). 1976. *Coelenterate Ecology and Behavior*. Plenum Press, New York. [Though becoming dated, still one of the best treatments of the subject.]

Mackie, G. O. 2004. Central neuronal circuitry in the jellyfish *Aglantha*: a model "simple nervous system." Neuro-Signals 13: 5–19.

Mackie, G. O. 2004. Epithelial conduction: recent findings, old questions, and where do we go from here? Hydrobiologia 530/531: 73–80.

Mariscal, R. N. 1984. Cnidaria: Cnidae. Pp. 57–68 in J. Bereiter-Hahn, A. G. Batoltsy and K. Sylvia Richards (eds.), *Biology of the Integument: I. Invertebrates*. Springer Verlag, Berlin.

Mariscal, R. N. E. J. Conklin and C. H. Bigger. 1977. The ptychocyst, a major new category of cnida used in tube construction by a cerianthid anemone. Biol. Bull. 152: 392–405.

Mariscal, R. N., R. B. McLean and C. Hand. 1977. The form and function of cnidarian spirocysts. 3. Ultrastructure of the thread and function of spirocysts. Cell Tissue Res. 178: 427–433.

Marshall, A. T. 1996. Calcification in hermatypic and ahermatypic corals. Science 271: 637–639.

Martin, V. J. 1997. Cnidarians. The jellyfish and hydras. Pp. 57–86 in S. F. Gilbert and A. M. Raunio. *Embryology: Constructing the Organism*. Sinauer, Sunderland, MA.

Martin, V. J. 2000. Reorganization of the nervous system during metamorphosis of a hydrozoan planula. Invert. Biol. 119(3): 243–253.

Martindale, M. Q., K. Pang and J. R. Finnerty. 2004. Investigating the origins of triploblasty: 'mesodermal' gene expression in a diploblastic animal, the sea anemone *Nematostella vectensis* (phylum Cnidaria: class Anthozoa). Development 131: 2463–2474.

Matus, D. Q., K. Pang, H. Marlow, C. W. Dunn, G. H. Thomsen and M Q. Martindale. 2006. Molecular evidence for deep evolutionary roots of bilaterality in animal development. PNAS 103 (30): 11195–11200.

Miller, R. L. and C. R. Wyttenbach (eds.). 1974. The developmental biology of the Cnidaria. Am. Zool. 14: 440–866.

Muscatine, L. and H. M. Lenhoff (eds.). 1974. *Coelenterate Biology. Reviews and New Perspectives*. Academic Press, New York. [Though dated, still a highly useful volume with excellent reviews of histology, skeletal systems, cnidae, development, symbiosis, and bioluminescence.]

Philippe, H. and 19 others. 2009. Phylogenomics revives traditional views on deep animal relationships. Curr. Biol. 19: 1–7.

Pick, K. S. and 10 others. 2010. Improved phylogenomic taxon sampling noticeably affects nonbilaterian relationships. Mol. Biol. Evol. 27(9): 1983–1987.

Ryan, J. F. and numerous others. 2013. The genome of the ctenophore *Mnemiopsis leidyi* and its implications for cell type evolution. Science 342: 1242592.

Siepel, K. and V. Schmid. 2005. Evolution of striated muscle: jellyfish and the origin of triploblasty. Dev. Biol. 282: 14–26,

Tardent, P. and R. Tardent (eds.). 1980. *Developmental and Cellular Biology of Coelenterates*. Elsvierorth-Holland Biomedical Press.

Valentine, J. W. 1992. *Dickinsonia* as a polypoid organism. Paleobiology 18: 378–382.

Weill, R. 1934. Contribution a l'étude des Cnidaires et de leurs nématocystes. I. Recherches sur les nématocystes (morphologie-physiologie-dévelopment). II. Valeur taxonomique du cnidome. Travaux Station Zoologique de Wimereux, Volumes X-XI. [In this paper, Weill created a classification scheme for cnidocytes that largely remains in use today.]

Westfall, J. A., S. R. Elliott, P. S. MohanKumar and R. W. Carlin. 2000. Invert. Biol. 119(4): 370–378.

Williams, R. B., P. F. S. Cornelius, R. G. Hughes and E. A. Robson (eds.). 1991. *Coelenterate Biology: Recent Research on Cnidaria and Ctenophora*. Kluewer Academic Publishers.

Wyttenbach, C. R. (ed.). 1974. The developmental biology of the Cnidaria. Am. Zool. 14(2): 540–866. [Papers presented at a 1972 symposium.]

Zapata, F. and 14 others. 2015. Phylogenomic analyses support traditional relationships within Cnidaria. PLoS ONE. doi: 10:137/journal.pone.0139068

Zrzavy, J. 2001. The interrelationshps of metazoan parasites: a review of phylum- and higher-level hypotheses from recent morphological and molecular phylgenetic analyses. Folia Parasit. 48: 81–103.

Anthozoa

Adey, W. H. 1978. Coral reef morphogenesis: a multidimensional model. Science 202: 831–837.

Anderson, P. A. V. and J. F. Case. 1975. Electrical activity associated with luminescence and other colonial behavior in the pennatulid *Renilla kollikeri*. Biol. Bull. 149: 80–95.

Arai, M. N. 1965. The ceriantharian nervous system. Am. Zool. 5: 424–429.

Babcock, R., G. Bull, P. Harrison, A. Heyward, J. Oliver, C. Wallace and B. Willis. 1986. Synchronous spawnings of 105 scleractinian coral species on the Great Barrier Reef. Mar. Biol. 90: 379–394.

Batham, E. J. 1960. The fine structure of epithelium and mesoglea in a sea anemone. Quart. J. Microscop. Sci. 101: 481–485.

Benayahu, Y. and Y. Loya. 1983. Surface brooding in the Red Sea soft coral *Parerythropodium fulvum fulvum* (Forskäl, 1775). Biol. Bull. 165: 353–369.

Bigger, C. H. 1982. The cellular basis of the aggressive acrorhagial response of sea anemones. J. Morph. 173 (3): 259–278.

Birkeland, C. 1974. Interactions between a sea pen and seven of its predators. Ecol. Monogr. 44(2): 211–232.

Brugler, M. R. S. C. France. 2007. The complete mitochondrial genome of the black coral *Chrysopathes formosa* (Cnidaria: Anthozoa: Antipatharia) supports classification of antipatharians within the subclass Hexacorallia. Mol. Phylogen. and Evol. 42: 776–788.

Cook, C. B., C. F. D'Elia and G. Muller-Parker. 1988. Host feeding and nutrient sufficiency for zooxanthellae in the sea anemone *Aptasia pallida*. Mar. Biol. 98: 253–262.

Darwin, C. 1842. *The Structure and Distribution of Coral Reefs*. Reprinted in 1984 by the University of Arizona Press, Tucson.

De'Ath, G., J. M. Lough and K. E. Fabricius. 2009. Declining coral calcification on the Great Barrier Reef. Science 323: 116–119.

Ducklow, H. W. and R. Mitchell. 1979. Composition of mucus released by coral reef coelenterates. Limnol. Oceanogr. 24(4): 706–714.

Dunn, D. F. 1977. Dynamics of external brooding in the sea anemone *Epiactis prolifera*. Mar. Biol. 39: 41–49.

Dunn, D. F., D. M. Devaney and B. Roth. 1980. *Stylobates*: a shell-forming sea anemone (Coelenterata, Anthozoa, Actiniidae). Pac. Sci. 34: 379–388.

Epifanio, R. De A., D. L. Martins, R. Villaça and R. Gabriel. 1999. Chemical defenses against fish predation in three Brazilian octocorals: 11β,12β-Epoxypukalide as a feeding deterrent in *Phyllogorgia dilatata*. J. Chem. Ecol. 25: 2255–2265.

Fautin, D. G. 1991. The anemone fish symbiosis: what is known and what is not. Symbiosis. 10: 23–46.

Fautin, D. G., C.-C. Guo and J.-S. Hwang. 1995. Costs and benefits of the symbiosis between the anemone shrimp *Periclimenes brevicarpalis* and its host *Entacmaea quadricolor*. Mar. Ecol. Prog. Ser. 129: 77–84.

Fenical, W., R. K. Okuda, M. M. Bandurraga, P. Culver and R. S. Jacobs. 1981. Lophotoxin: a novel neuromuscular toxin from Pacific sea whips of the genus *Lophogorgia*. Science 212: 1512–1514.

Fricke, H. and L. Hottinger. 1983. Coral biotherms below the euphotic zone in the Red Sea. Mar. Ecol. Prog. Ser. 11: 113–117.

Gaino, E. M. B, M. Boyer and F. Scoccia. 2008. Sperm morphology in the black coral, *Cirrhipathes* sp. (Anthozoa, Antipatharia). Invert. Biol. 127(3) 249–258.

Gladfelter, E. H. 1983. Circulation of fluids in the gastrovascular system of the reef coral *Acropora cervicornia*. Biol. Bull. 165: 619–636.

Glynn, P. W. 1980. Defense by symbiotic Crustacea of host corals elicited by chemical cues from predators. Oecologia 47: 287–290.

Glynn, P. W. 1993. Coral reef bleaching: ecological perspectives. Coral Reefs 12: 1–17.

Glynn, P. W., G. M. Wellington and C. Birkeland. 1978. Coral reef growth in the Galápagos: limitation by sea urchins. Science 203: 47–49.

Godknechy, A. and P. Tardent. 1988. Discharge and mode of action of the tentacular nematocysts of *Anemonia sulcata*. Mar. Biol. 100: 83–92.

Goreau, T. F. 1963. Calcium carbonate deposition by coralline algae and corals in relation to their roles as reef builders. Ann. N.Y. Acad. Sci. 109: 127–167.

Grimstone, A. V., R. W. Horne, C. F. A. Pantin and E. A. Robson. 1958. The fine structure of the mesenteries of the sea-anemone *Metridium senile*. Quart. J. Microscop. Sci. 99: 523–540.

Hamner, W. M. and D. F. Dunn. 1980. Tropical Corallimorpharia (Coelenterata: Anthozoa): feeding by envelopment. Micronesica 16: 37–41.

Hand, C. and K. R. Uhlinger. 1995. Asexual reproduction by transverse fission and some anomalies in the sea anemone *Nematostella vectensis*. Invert. Biol. 114: 9–18.

Harrison, P. L., R. C. Babcock, G. D. Bull, J. K. Oliver, C. C. Wallace and B. L. Willis. 1984. Mass spawning in tropical reef corals. Science 232: 1186–1189.

Haussermann, V. and G. Forsterra. 2003. First evidence for coloniality in sea anemones. Mar. Ecol. Prog. Ser. 257: 2910294.

Howe, N. R. and Y. M. Sheikh. 1975. Anthopleurine: a sea anemone alarm pheromone. Science 189: 386–388.

Iglesias-Prieto, R., V. H. Beltrán, T. C. LaJeunesse, H. Reyes-Bonilla and P. E. Thomé. 2004. Different algal symbionts explain the vertical distribution of dominant reef corals in the eastern Pacific. Proc. Royal Soc. Lond. B 271: 1757–1763.

Isomura, N., K. Hamada and M. Nishihira. 2003. Internal brooding of clonal propagules by a sea anemone, *Anthopleura* sp. Invert. Biol. 122: 293–298.

Jennison, B. L. 1979. Gametogenesis and reproductive cycles in the sea anemone *Anthopleura elegantissima*. Can. J. Zool. 57: 403–411.

Jones, O. A. and R. Endean (eds.). 1973–1976. *Biology and Geology of Coral Reefs. Vol. I–III*. Academic Press, New York.

Josephson, R. K. and S. C. March. 1966. The swimming performance of the sea-anemone *Boloceroides*. J. Exp. Biol. 44: 493–506.

Kingsley, R. J., M. Tsuzaki, N. Watabe and G. L. Mechanic. 1990. Collagen in the spicule organic matrix of the gorgonian *Leptogorgia virgulata*. Biol. Bull. 179: 207–213.

Lang, J. 1973. Interspecific aggression by scleractinian corals: why the race is not only to the swift. Bull. Mar. Sci. 23: 269–279.

Le Goff-Vitry, M. C. A.D. Rogers and D. Baglowa. 2004, A deep-sea slant on the molecular phylogeny of the Scleractinia. Mol. Phylogen. Evol. 30:167–177.

Lewis, C. L. and M. A. Coffroth. 2004. The acquisition of exogenous algal symbionts by an octocoral after bleaching. Science 304: 1490–1492.

Levy, O., L. Appelbaum, W. Leggat, Y. Gothlif, D. C. Hayward, D. J. Miller and O. Hoegh-Guldberg. 2007. Light-responsive cryptochromes from a simple multicellular animal, the coral *Acropora millepora*. Science 318: 467–470.

Little, A. F., M. J. H. van Oppen and B. L. Willis. 2004. Flexibility in algal endosymbiosis shapes growth in reef corals. Science 304: 1492–1494.

Mariscal, R. N. 1974. Scanning electron microscopy of the sensory epithelia and nematocysts of corals and a corallimorpharian sea anemone. Proc. Second Int. Coral Reef Symp. 1: 519–532.

Marshall, A. T. 1996. Calcification in hermatypic and ahermatypic corals. Science 271: 637–639.

McFadden, C. S., S. C. France, J. A. Sánchez and P. Alderslade. 2006. A molecular phylogenetic analysis of the Octocorallia (Cnidaria: Anthozoa) based on mitochondrial protein-coding sequences. Mol. Phylog. Evol. 41: 513–527.

Meyer, J. L., E. T. Schultz and G. S. Helfman. 1983. Fish schools: an asset to corals. Science 220: 1047–1049.

Muscatine, L. and J. W. Porter. 1977. Reef corals: mutualistic symbioses adapted to nutrient-poor environments. Biol. Sci. 27: 454–459.

Patton, J. S., S. Abraham and A. A. Benson. 1977. Lipogenesis in the intact coral *Pocillopora capitata* and its isolated zooxanthellae: evidence for a light-driven carbon cycle between symbiont and host. Mar. Biol. 44: 235–247.

Pearse, V. B. 2001. Prodigies of propagation: the many modes of clonal replication in boloceroidid sea anemones (Cnidaria, Anthozoa, Actiniaria). Invert. Reprod. Dev. 41: 201–313.

Proceedings of the International Coral Reef Symposiums, 1969–present. [These edited volumes, 12 as of 2012, are major sources of information on corals and coral reefs. They can be individually searched at the ICRS Proceedings website: www.reefbase.org/resource_center/publication/icrs.aspx]

Purcell, J. E. and C. L. Kitting. 1982. Intraspecific aggression and population distributions of the sea anemone *Metridium senile*. Biol. Bull. 162: 345–359.

Roberts, J. M., A. J. Wheeler and A. Freiwald. 2006. Reefs of the deep: the biology and geology of cold-water coral ecosystems. Science 312: 543–547.

Reese, E. S. 1981. Predation on corals by fishes of the family Chaetodontidae: implications for conservation and management of coral reef ecosystems. Bull. Mar. Sci. 31(3): 594–604.

Reimer, A. A. 1971. Feeding behavior in the Hawaiian zoanthids *Palythoa* and *Zoanthus*. Pac. Sci. 25(4): 257–260.

Ribes, M., R. Coma, S. Rossi and M. Micheli. 2007. Cycle of gonadal development in *Eunicella singularis* (Cnidaria: Octocorallia): trends in sexual reproduction in gorgonians. Invertebr. Biol. 126: 307–317.

Richmond, R. H. 1985. Reversible metamorphosis in coral planula larvae. Mar. Ecol. Prog. Ser. 22: 181–185.

Rinkevich, B. and Y. Lola. 1983. Short-term fate of photosynthetic products in a hermatypic coral. J. Exp. Mar. Biol. Ecol. 73: 175–184.

Roberts, S. and M. Hirshfield. 2004. Deep-sea corals: out of sight, but no longer out of mind. Front. Ecol. Environ. 2(3): 123–130.

Schuhmacher, H. and H. Zibrowius. 1985. What is hermatypic? A redefinition of ecological groups in corals and other organisms. Coral Reefs 4: 1–9.

Schwarz, J. A., V. M. Weis and D. C. Potts. 2002. Feeding behavior and acquisition of zooxanthellae by planula larvae of the sea anemone *Anthopleura elegantissima*. Mar. Biol. 140: 471–478.

Sebens, K. P. 1981. Recruitment in a sea anemone population: juvenile substrate becomes adult prey. Science 213: 785–787.

Sebens, K. P. 1981. Reproductive ecology of the intertidal sea anemones *Anthopleura xanthogrammica* (Brandt) and *A. elegantissima* (Brandt): body size, habitat and sexual reproduction. J. Exp. Mar. Biol. Ecol. 54: 225–250.

Sebens, K. P. 1984. Agonistic behavior in the intertidal sea anemone *Anthopleura xanthogrammica*. Biol. Bull. 166: 457–472.

Sebens, K. P. and K. DeRiemer. 1977. Diel cycles of expansion and contraction in coral reef anthozoans. Mar. Biol. 43: 247–256.

Secord, D. and L. Augustine. 2000. Biogeography and microhabitat variation in temperate algal-invertebrate symbioses: zooxanthellae and zoochlorellae in two Pacific intertidal sea anemones, *Anthopleura elegantissima* and *A. xanthogrammica*. Invert. Biol. 119(2): 139–146

Schick, J. M. 1991. *A Functional Biology of Sea Anemones*. Chapman and Hall, London.

Spaulding, J. G. 1974. Embryonic and larval development in sea anemones (Anthozoa: Actiniaria). Am. Zool. 14: 511–520.

Stoddart, D. R. 1969. Ecology and morphology of recent coral reefs. Biol. Rev. 44: 433–498.

Stoddart, J. A. 1983. Asexual production of planulae in the coral *Pocillopora damicornis*. Mar. Biol. 76: 279–284.

Stolarski, J. A. Meibom, R. Przenioslo and M. Mazur. 2007. A Cretaceous scleractinian coral with a calcitic skeleton. Science 318: 92–94.

Tchernov, D., M. Y. Gorbunov, C. de Vargas, S. N. Yadav, A. J. Milligan, M. Häggblom and P. G. Falkowski. 2004. Membrane lipids of symbiotic algae are diagnostic of sensitivity to thermal bleaching in corals. PNAS 101(37): 13531–13535

Veron, J. 2000. *Corals of the World* [3 vols.]. Australian Institute of Marine Science, Townsville, Australia.

Watson, G. M. and D. A. Hessinger. 1989. Cnidocyte mechanoreceptors are tuned to the movements of swimming prey by chemoreceptors. Science 243: 1589–1591.

Watson, G. M. and R. N. Mariscal. 1983. Comparative ultrastructure of catch tentacles and feeding tentacles in the sea anemone *Haliplanella*. Tissue Cell 15(9): 939–953.

Watson, G. M. and R. N. Mariscal. 1983. The development of a sea anemone tentacle specialized for aggression: morphogenesis and regression of the catch tentacle of *Haliplanella luciae* (Cnidaria, Anthozoa). Biol. Bull. 164: 506–517.

Westfall, J. A. 1965. Nematocysts of the sea anemone *Metridium*. Am. Zool. 5: 377–393.

Williams, G. C. 1995. Living genera of sea pens (Coelenterata: Octocorallia: Pennatulacea): illustrated key and synopses. Zool. J. Linn. Soc.-Lon. 113: 93–140.

Williams, G. C. 1997. Preliminary assessment of the phylogenetics of pennatulacean octocorals, with a reevaluation of Ediacaran frond-like fossils, and a synthesis of the history of evolutionary thought regarding the sea pens. *Proceedings of the Sixth International Conference of Coelenterate Biology*: 497–509.

Williams, G. C. 2000. First record of a bioluminescent soft coral: description of a disjunct population of *Eleutherobia grayi* (Thomson and Dean, 1921) from the Solomon Islands, with a review of bioluminescence in the Octocorallia. Proc. Calif. Acad. Sci. 52 (17): 209–255.

Williams, R. B. 1975. Catch-tentacles in sea anemones: occurrence in *Haliplanella luciae* (Verrill) and a review of current knowledge. J. Nat. Hist. 9: 241–248.

Yonge, C. M. 1973. The nature of reef-building (hermatypic) corals. Bull. Mar. Sci. 23(1): 1–15. [Dated, but a great read.]

Cubozoa

Buskey, E. J. 2003. Behavioral adaptations of the cubozoan medusa *Tripedalia cystophora* for feeding on copepod (*Dioithona oculata*) swarms. Mar. Biol. 142: 225–232.

Coates, M. M., A. Garm, J. C. Theobald, S. H. Thompson and D.-E. Nilsson. 2006. The spectral sensitivity of the lens eyes of a box jellyfish, *Tripedalia cystophora* (Conant). J. Exper. Biol. 209: 3758–3765.

Chapman, D. M. 1978. Microanatomy of the cubopolyp, *Tripedalia cystophora* (class Cubozoa). Helgoländer Wiss Merresuntersuch 31: 128–168.

Garm, A., P. Ekström and D.-E. Nilsson. 2006. Rhopallia are integrated parts of the central nervous system in box jellyfish. Cell Tissue Res. 325: 333–343.

Gershwin, L-A and Gibbons, M. J. 2009. *Carybdea branchi*, sp. nov., a new box jellyfish (Cnidaria: Cubozoa) from South Africa. Zootaxa 2088: 41–50.

Hamner, W. M., M. S. Jones and P. P. Hamner. 1995. Swimming, feeding, circulation and vision in the Australian box jellyfish, *Chironex fleckeri* (Cnidaria, Cubozoa). Mar. Freshwater Res. 46: 985–990.

Laska, G. and M. Hündgen. 1984. Die ultrastruktur des neuromuskuläre systems der medusen von *Tripedalia cystopora* und *Carybdea marsupialis* (Coelenterata, Cubozoa). Zoomorphology104: 163–170.

Lewis, C and T. A. F. Long. 2005. Courtship and reproduction in *Carybdea sivickisi* (Cnidaria: Cubozoa) Mar. Biol. 147: 477–483.

Matsumoto, G. I. 1995. Observations on the anatomy and behaviour of the cubozoan *Carybdea rastonii* Haacke. Mar. Freshwater Behav. Phisiol. 26: 139–148.

Nilsson, D. E., M. M. Coates, I. Gislén, C. Skogh and A. Garm. 2005. Advanced optics in a jellyfish eye. Nature 435: 201–205.

Nordström, K., R. Wallén, J. Seymour and D. Nilsson. 2003. A simple visual system without neurons in jellyfish larvae. Proc. Roy. Soc. Lond. B 270: 2349–2354.

Parkefelt, L., C. Skogh, D.-E. Nilsson and P. Ekström. 2005. Bilateral symmetric organization of neural elements in the visual system of a coelenterate, *Tripedalia cystophora* (Cubozoa). J. Comp. Neurol. 492: 251–262.

Pearse, J. S. and V. B. Pearse. 1978. Vision in cubomedusan jellyfishes. Science 199: 458.

Seymour, J. 2002. One touch of venom. Nat. Hist. 2002(9): 72–75.

Skogh, C., A. Garm, D.-E. Nilsson and P. Ekström. 2006. Bilaterally symmetrical rhopalial nervous system of the box jellyfish *Tripedalia cystophora*. J. Morph. 267: 1391–1405.

Stewart, S. E. 1996. Field behavior of *Tripedalia cystephora* (class Cubozoa). Mar. Freshwater Behav. Phy. 27: 175–188.

Straehler-Pohl, I. and G. Jarms. 2005. Life cycle of *Carybdea marsupialis* Linnaeus, 1758 (Cubozoa, Carybdeidae) reveals metamorphosis to be a modified strobilation. Mar. Biol. 147: 1271–1277.

Toshino, S. and 8 others. 2013. Development and polyp formation of the giant box jellyfish *Morbakka virulenta* (Kishinouye, 1910) (Cnidaria: Cubozoa) collected from the Seto Inland Sea, western Japan. Plankton Benthos Res. 8(1): 1–8

Werner, B. 1971. Life cycle of *Tripedalia cystophora* Conant (Cubomedusae). Nature 232: 582–583.

Hydrozoa

Alvariño, A. 1983. *Nectocarmen antonioi*, a new Prayinae, Calycophorae, Siphonophorae, from California. Proc. Biol. Soc. Washington 96: 339–348.

Bellamy, N. and M. J. Risk. 1982. Coral gas: oxygen production in *Millepora* on the Great Barrier Reef. Science 215: 1618–1619.

Bieri, R. 1970. The food of *Porpita* and niche separation in three neuston coelenterates. Publ. Seto Mar. Biol. Lab. 27: 305–307.

Biggs, D. C. 1977. Field studies of fishing, feeding and digestion in siphonophores. Mar. Behav. Physiol. 4: 261–274.

Bode, H. R. 2003. Head regeneration in *Hydra*. Dev. Dyn. 22: 225–236.

Bode, P. M. and H. R. Bode. 1984. Patterning in *Hydra*. Pp. 213–241 in G. M. Malacinski and S. V. Bryant (eds.), *Pattern Formation: A Primer in Developmental Biology*. Macmillan, New York.

Burnett, A. L. (ed.). 1973. *Biology of Hydra*. Academic Press, New York.

Cairns, S. D. and J. L. Barnard. 1984. Redescription of *Janaria mirabilis*, a calcified hydroid from the eastern Pacific. Bull. South. Calif. Acad. Sci. 83: 1–11.

Cartwright, P., N. M. Evans, C. W. Dunn, A. M. Marques, M. P. Miglietta, P. Schuchert and A. G. Collins. 2008. Phylogenetics of Hydroidolina (Hydrozoa: Cnidaria). J. Mar. Biol. Assoc. U.K. 88(8): 1663–1672.

Collins, A. G. and 8 others. 2008. Phylogenetics of *Trachylina* (Cnidaria: Hydrozoa) with new insights on the evolution of some problematical taxa. J. Mar. Biol. Assoc. U.K. 88(8): 1673–1685.

Dunn, C. W., P. R. Pugh and S. H. D. Haddock. 2005. Molecular phylogenetics of the Siphonophora (Cnidaria), with implications for the evolution of functional specialization. Syst. Biol. 54: 916–935.

Eakin, R. M. and J. A. Westfall. 1962. Fine structure of photoreceptors in the hydromedusan, *Polyorchis penicillatus*. PNAS 48: 826–833.

Edwards, C. 1966. *Velella velella* (L.): The distribution of its dimorphic forms in the Atlantic Ocean and the Mediterranean, with comments on its nature and affinities. Pp. 283–296 in H. Barnes (ed.), *Some Contemporary Studies in Marine Science*. Allen and Unwin, London.

Evans, N. E., A. Lindner, E. V. Raikova, A. G. Collins and P. Cartwright. 2008. Phylogenetic placement of the enigmatic parasite, *Polypodium hydriforme*, within the phylum Cnidaria, BMC Evol. Biol. 8: 139–151.

Fields, W. G. and G. O. Mackie. 1971. Evolution of the Chondrophora: evidence from behavioural studies on *Velella*. J. Fish. Res. Bd. Can. 28: 1595–1602.

Francis, L. 1985. Design of a small cantilevered sheet: the sail of *Velella velella*. Pac. Sci. 39(1): 1–15.

Freeman, G. 1983. Experimental studies on embryogenesis in hydrozoans (Trachylina and Siphonophora) with direct development. Biol. Bull. 165: 591–618.

Fröbius, A. C., G. Genikhovich, U. Kurn, E. Anton-Erxleben and T. C. Bosch. 2003. Expression of developmental genes during early embryogenesis of *Hydra*. Dev. Genes Evol. 213: 445–455.

Govindarajan, A. F., K. M. Halanych and C. W. Cunningham. 2005. Mitochondrial evolution and phylogeography in the hydrozoan *Obelia geniculata* (Cnidaria). Mar. Biol. 146: 213–222.

Griffith, K. A. and A. T. Newberry. 2008. Effect of flow regime on the morphology of a colonial cnidarian. Invert. Biol. 127: 259–264.

Haddock, S. H. D., C. W. Dunn, P. R. Pugh and C. E. Schnitzler. 2005. Bioluminescent and red-fluorescent lures in a deep-sea siphonophore. Science 309: 263.

Koizumi, O., M. Itazawa, H. Mizomoto, S. Minobe, L. C. Javois, C. J. P. Grimmelikhuijzen and H. R. Bode. 1992. Nerve ring in the hypostome in *Hydra*. I. Its structure, development, and maintenance. J. Comp. Neurol. 326: 7–21.

Lane, C. E. 1960. The Portuguese man-of-war. Sci. Am. 202: 158–168.

Lenhoff, H. M. and W. F. Loomis (eds.). 1961. *The Biology of Hydra and of Some Other Coelenterates: 1961*. University of Miami Press, Coral Gables, Florida.

Lentz, T. L. 1966. *The Cell Biology of Hydra*. Wiley, New York.

Mackie, G. O. 1959. The evolution of the Chondrophora (Siphonophora: Disconanthae): New evidence from behavioral studies. Trans. R. Soc. Can. 53: 7–20.

Mackie, G. O. 1960. The structure of the nervous system in *Velella*. Q. J. Microsc. Sci. 101: 119–133.

Martin, R. and P. Walther. 2003. Protective mechanisms against the action of nematocysts in the epidermis of *Cratena peregrine* and *Flabellina affinis* (Gastropoda, Nudibranchia). Zoomorphology 122: 24–35.

Martin, W. E. 1975. *Hydrichthys pietschi*, new species (Coelenterata) parasitic on the fish, *Ceratias holboelli*. Bull. So. Calif. Acad. Sci. 74: 1–6. [A hydroid parasitic on a fish in California waters.]

Nawrocki, A. M., P. Schuchert and P. Cartwright. 2010. Phylogenetics and evolution of Capitata (Cnidaria: Hydrozoa), and the systematics of Corynidae. Zool. Scr. 39: 290–304.

Petersen, K. W. 1990. Evolution and taxonomy in capitate hydroids and medusae (Cnidaria: Hydrozoa). Zool. J. Linn. Soc.-Lond. 100: 101–231.

Purcell, J. E. 1980. Influence of siphonophore behavior upon their natural diets: Evidence for aggressive mimicry. Science 209: 1045–1047.

Purcell, J. E. 1984. The functions of nematocysts in prey capture by epipelagic siphonophores (Coelenterata, Hydrozoa). Biol. Bull. 166: 310–327.

Satterlie, R. A. 1985. Putative extraocellar photoreceptors in the outer nerve ring of *Polyorchis penicillatus*. J. Exper. Zool. 233: 133–137.

Schuchert, P. and H. M. Reiswig. 2006. *Brinckmannia hexactinellidophila*, n. gen., n. sp., a hydroid living in tissues of glass sponges of the reefs, fjords, and seamounts of Pacific Canada and Alaska. Can. J. Zool. 84: 564–572.

Shimizu, H., O. Koizumi and T. Fujisawa. 2004. Three digestive movements in Hydra regulated by the diffuse nerve net in the body column. J. Comp. Physiol. A 190: 623–630.

Shimizu, H. and H. Namikawab. 2009. The body plan of the cnidarian medusa: distinct differences in positional origins of polyp tentacles and medusa tentacles. Evol. Dev. 11: 619–621.

Shimomura, O. 1995. A short story of aequorin. Biol. Bull. 189: 1–5.

Singla, C. L. 1975. Statocysts of hydromedusae. Cell Tissue Res. 158: 391–407.

Soong, K. and L. C. Cho. 1998. Synchronized release of medusae from three species of hydrozoan fire corals. Coral Reefs. 17: 145–154.

Stretch, J. J. and J. M. King. 1980. Direct fission: An undescribed reproductive method in hydromedusae. Bull. Mar. Sci. 30: 522–526.

Totton, A. K. 1960. Studies on *Physalia physalis* (L.). Part 1, natural history and morphology. Discovery Rpt. 30: 301–367.

Wahle, C. M. 1980. Detection, pursuit, and overgrowth of tropical gorgonians by milleporid hydrocorals: *Perseus* and *Medusa* revisited. Science 209: 689–691.

West, D. A. 1978. The epithelio-muscular cell of hydra: its fine structure, three-dimensional architecture and relationship to morphogenesis. Tissue Cell 10: 629–646.

Myxozoa

Anderson, C. L., E. U. Canning and B. Okamura. 1998. A triploblast origin for Myxozoa? Nature 392: 346.

Bartošová, P. I. Fiala and V. Hypša. 2009. Concatenated SSU and LSU rDNA data confirm the main evolutionary trends within myxosporeans (Myxozoa: Myxosporea) and provide an effective tool for their molecular phylogenetics. Mol. Phylogenet. Evol. 53: 81–93.

Cannon, Q. and E. Wagner. 2003. Comparison of discharge mechanisms of cnidarian cnidae and myxozoan polar capsules. Rev. Fish. Sci. 11(3): 185–219.

Jimenez-Guri, E., H. Philippe, B. Okamura and P. W. H. Holland. 2007. *Buddenbrockia* is a cnidarian worm. Science 317: 116–118.

Kent, M. L. and 15 others. 2001. Recent advances in our knowledge of the Myxozoa. J. Eukaryot. Microbiol. 48(4): 395–413.

Monteiro, A. S., B. Okamura and P. W. H. Holland. 2002. Orphan worm finds a home: *Buddenbrockia* is a myxozoan. Mol. Biol. and Evol. 19: 968–971.

Okamura, B., A. Gruhl and J. L. Bartholomew (eds.). 2015. *Myxozoan Evolution, Ecology and Development*. Springer International Publishing, Switzerland.

Okamura, B., A. Curry, T. S. Wood and E. U. Canning. 2002. Ultrastructure of *Buddenbrockia* identifies it as a myxozoans and verifies the bilaterian origin of the Myxozoa. Parasitology 124: 215–223.

Scyphozoa

Alexander, R. M. 1964. Visco-elastic properties of the mesoglea of jellyfish. J. Exp. Biol. 41: 363–369.

Anderson, P. A. V. and G. O. Mackie. 1977. Electrically coupled photosensitive neurons control swimming in jellyfish. Science 197: 186–188.

Arai, M. N. 1997. *A Functional Biology of Scyphozoa*. Chapman and Hall, London.

Calder, D. R. 1971. Nematocysts of *Aurelia*, *Chrysaora* and *Cyanea* and their utility in identification. Trans. Am. Microscop. Soc. 90: 269–274.

Calder, D. R. 1982. Life history of the cannonball jellyfish, *Stomolophus meleagris* L. Agassiz, 1860 (Scyphozoa, Rhizostomida). Biol. Bull. 162: 149–162.

Costello, J. H., S. P. Colin and J. O. Dabri. 2008. Medusan morphospace: phylogenetic constraints, biomechanical solutions and ecological consequences. Invert. Biol. 127(3): 265–290.

Fancett, M. S. 1988. Diet and prey selectivity of scyphomedusae from Port Phillip Bay, Australia. Mar. Biol. 98: 503–509.

Fancett, M. S. and G. P. Jenkins. 1988. Predatory impact of scyphomedusae on ichthyoplankton and other zooplankton in Port Phillip Bay. J. Exp. Mar. Biol. Ecol. 116: 63–77.

Hamner, W. M., P. P. Hamner and S. W. Strand. 1994. Sun-compass migration by *Aurelia aurita* (Scyphozoa): population retention and reproduction in Saanich Inlet, British Columbia. Mar. Biol. 119: 347–356.

Hamner, W. M. and I. R. Hauri. 1981. Long-distance horizontal migrations of zooplankton (Scyphomedusae: *Mastigias*). Limnol. Oceanogr. 26: 414–423.

Larson, R. J. 1987. Trophic ecology of planktonic gelatinous predators in Saanich Inlet, British Columbia: diets and prey selection. J. Plankton Res. 9: 811–820.

Möller, H. 1984. Reduction of a larval herring population by jellyfish predator. Science 224: 621–622.

Purcell, J. E. 1985. Predation on fish eggs and larvae by pelagic cnidarians and ctenophores. Bull. Mar. Sci. 37: 739–755.

Rottini Sandrini, L. and M. Avian. 1989. Feeding mechanism of *Pelagia noctiluca* (Scyphozoa: Semaeostomeae); laboratory and open sea observations. Mar. Biol. 102: 49–55.

Russell, F. S. 1954, 1970. *Medusae of the British Isles, Vols. 1 and 2*. Cambridge University Press, London.

Shushkina, E. A. and E. I. Musayeva. 1983. The role of jellyfish in the energy system of Black Sea plankton communities. Oceanology 23: 92–96.

Widmer, C. L. 2006. Life cycle of *Phacellophora camtschatica* (Cnidaria: Scyphozoa). Invertebrate Biology 125: 80–90.

Staurozoa

Collins, A. G. and M. Daly. 2005. A new deepwater species of Stauromedusae, *Lucernaria janetae* (Cnidaria, Staurozoa, Lucernariidae), and a preliminary investigation of stauromedusan phylogeny based on nuclear and mitochondrial rDNA data. Biol. Bull. 208(3): 221–230.

Miranda, L. S., A. C. Morandini and A. C. Marques. 2009. Taxonomic review of *Haliclystus antarcticus* Pfever, 1889 (Stauromedusae, Staurozoa, Cnidaria), with remarks on the genus *Haliclystus* Clark, 1863. Polar Biol. 321: 1507–1519.

Miranda, L. S., A. G. Collins and A. C. Marques. 2010. Molecules clarify a cnidarian life cycle—The "Hydrozoan" *Microhydrula limopsicola* is an early life stage of the staurozoan, *Haliclystus antarcticus*. PLoS ONE 5: 1–9.

CHAPTER 8

Phylum Ctenophora

The Comb Jellies

Ctenophores (Greek *cten*, "comb"; *phero*, "to bear")—commonly called comb jellies, sea gooseberries, or sea walnuts—are transparent, gelatinous animals. Most of them are planktonic, living from surface waters to depths of at least 3,000 meters, and a few species are epibenthic. Their transparency and fragile nature make them difficult to capture or observe by traditional sampling methods such as towing or trawling with nets, and until the recent advent of manned submersibles and "blue water" SCUBA techniques they were thought to be only modestly abundant. However, they are now known to form a major portion of the planktonic biomass in many areas of the world, and they may periodically be the predominant zooplankters in some areas. About 100 species have been described, but there are probably many deep-sea forms yet to be discovered.

The ctenophores are radially (biradially) symmetrical, probably diploblastic animals, resembling cnidarians in many respects. This similarity is immediately obvious, for example, in features such as symmetry, a gelatinous mesenchyme or collenchyme, the absence of a body cavity between the gut and the body wall, and a relatively simple, netlike nervous system. Some zoologists, however, have viewed these similarities as convergent features resulting from adaptations to pelagic lifestyles. Ctenophores are significantly different from cnidarians in their more extensively organized digestive system, their wholly mesenchymal musculature, and certain other features (Box 8A). There remains uncertainty as to the origin of the ctenophoran mesenchyme; some evidence suggests it is primarily ectodermally derived, yet other evidence favors an endodermal origin. The latter possibility might at first suggest developmental homology with the true mesoderm of bilaterians. However, sequencing of the genome of *Pleurobrachia bachei* has shown that the genes involved in mesoderm development in bilaterians have no homologous counterpart in the mesenchymal development of Ctenophora.

Ctenophores also differ fundamentally from cnidarians in that they are monomorphic throughout their life histories, are never colonial, and lack

Classification of The Animal Kingdom (Metazoa)

Non-Bilateria*
(a.k.a. the diploblasts)
- PHYLUM PORIFERA
- PHYLUM PLACOZOA
- PHYLUM CNIDARIA
- **PHYLUM CTENOPHORA**

Bilateria
(a.k.a. the triploblasts)
- PHYLUM XENACOELOMORPHA

Protostomia
- PHYLUM CHAETOGNATHA

SPIRALIA
- PHYLUM PLATYHELMINTHES
- PHYLUM GASTROTRICHA
- PHYLUM RHOMBOZOA
- PHYLUM ORTHONECTIDA
- PHYLUM NEMERTEA
- PHYLUM MOLLUSCA
- PHYLUM ANNELIDA
- PHYLUM ENTOPROCTA
- PHYLUM CYCLIOPHORA

Gnathifera
- PHYLUM GNATHOSTOMULIDA
- PHYLUM MICROGNATHOZOA
- PHYLUM ROTIFERA

Lophophorata
- PHYLUM PHORONIDA
- PHYLUM BRYOZOA
- PHYLUM BRACHIOPODA

ECDYSOZOA

Nematoida
- PHYLUM NEMATODA
- PHYLUM NEMATOMORPHA

Scalidophora
- PHYLUM KINORHYNCHA
- PHYLUM PRIAPULA
- PHYLUM LORICIFERA

Panarthropoda
- PHYLUM TARDIGRADA
- PHYLUM ONYCHOPHORA
- PHYLUM ARTHROPODA
  - SUBPHYLUM CRUSTACEA*
  - SUBPHYLUM HEXAPODA
  - SUBPHYLUM MYRIAPODA
  - SUBPHYLUM CHELICERATA

Deuterostomia
- PHYLUM ECHINODERMATA
- PHYLUM HEMICHORDATA
- PHYLUM CHORDATA

*Paraphyletic group

any trace of an attached sessile stage. Ctenophores lack a hard skeleton, an excretory system, and a respiratory system. Most are simultaneous hermaphrodites capable of self-fertilization—an unusual quality among the Metazoa. A distinctive larval stage, the **cydippid larva**, is usually produced. Ctenophores are exclusively marine. They display a wonderful variety of shapes, and they range in size from less than 1 cm in height to ribbon-shaped forms 2 m long (Figure 8.1). Some have evolved rather bizarre body forms, and a few have taken up a benthic creeping existence. They occur in all the world's oceans and at all latitudes. Desiccated specimens of the genus *Pleurobrachia* are often found washed ashore after storms. However, in their planktonic environment, ctenophores are some of the most elegant and graceful creatures in the sea, and observing them in their natural habitat is a memorable experience.

BOX 8A Characteristics of the Phylum Ctenophora

1. Diploblastic (or possibly triploblastic) Metazoa, with ectoderm and endoderm separated by a cellular mesenchyme
2. Biradial symmetry; the body axis is oral–aboral
3. With adhesive exocytotic structures called colloblasts
4. Gastrovascular cavity (gut) is the only "body cavity"; gut with stomodeum and canals that branch complexly throughout body; gut ends in two small anal pores
5. Without discrete respiratory, excretory, or circulatory systems (other than the gut)
6. Nervous system in the form of a nerve net or plexus, but more specialized than that of cnidarians
7. Musculature always formed of true mesenchymal cells
8. Monomorphic, without alternation of generations and without any kind of an attached sessile life stage
9. With eight rows of ciliary plates (combs or ctenes) at some stage in their life history; comb rows controlled by unique apical sense organ
10. Some adults and most juveniles with a pair of long tentacles, often retractable into sheaths
11. Most are hermaphroditic; typically with a characteristic cydippid larval stage

Taxonomic History and Classification

Perhaps because many well-known ctenophores are brilliantly luminescent and are commonly seen from ships, the group has been known since ancient times. The first recognizable figures of ctenophores were drawn by a ship's doctor and naturalist in 1671. Linnaeus placed them in his group Zoophyta, along with various other "primitive" invertebrates. Cuvier classified them with medusae and anemones in Zoophytes. In the early nineteenth century, Johann Friedrich von Eschscholtz designed the first rational classification of pelagic medusae and ctenophores by creating the orders Ctenophorae (for comb jellies), Discophorae (for all the solitary cnidarian medusae), and Siphonophorae (for the colonial siphonophorans and chondrophorans). Eschscholtz viewed these orders as subdivisions of the class Acalepha, regarding them as intermediate between Zoophytes and Echinodermata (on the basis of the common presence of radial symmetry). Recall that it was Leuckart who, in 1847, first separated the Coelenterata from the echinoderms, although his Coelenterata also included sponges and ctenophores. Raspailia Vosmaer (in 1877) was responsible for removing the sponges and Hatschek (in 1889) for removing the ctenophores as separate groups.

Until recently, two classes of ctenophores were recognized: Nuda, for species lacking tentacles (the single order Beroida); and Tentaculata, for those species with tentacles (all other orders). However, the monophyly of these two groups had been questioned. Recent molecular analyses indicate that while the Ctenophora overall are monophyletic, the presence or absence of tentacles may provide little phylogenetic information.

Until more robust phylogenetic trees are achieved, we follow Rich Harbison and Larry Madin (1983) in simply dividing the ctenophores into 7 orders (containing 19 families). Figure 8.2 illustrates the general anatomy of the major groups. The position of the Ctenophora among the basal Metazoa is still under debate. Some molecular phylogenetic studies have suggested they might be basal animals, but most have found Porifera to be the basalmost Metazoa, with Ctenophora being close to Cnideria in the tree of life.

Figure 8.1 Representative ctenophores. (A) *Pleurobrachia* (order Cydippida). (B) *Beroe ovata* (order Beroida). (C) *Beroe*, showing iridescence in the comb rows. (D) An unidentified lobate ctenophore from 780 m depth in the Caribbean Sea. (E) Venus belt comb jelly, *Cestum veneris* (order Cestida), Socorro, Revillagigedo Islands, Mexico. (F) *Leucothea* (order Lobata). (G) Warty comb jelly or sea walnut, *Mnemiopsis leidyi* (order Lobata), Crimea, Ukraine. (H) An Antarctic cydippid ctenophore with two krill in its gut, and a third one being captured. (I) *Lyrocteis imperatoris*, an Indo-West Pacific platyctenid ctenophore. ▶

(A)
(B)
(C)
(D)
(E)
(F)
(G)
(I)
(H)

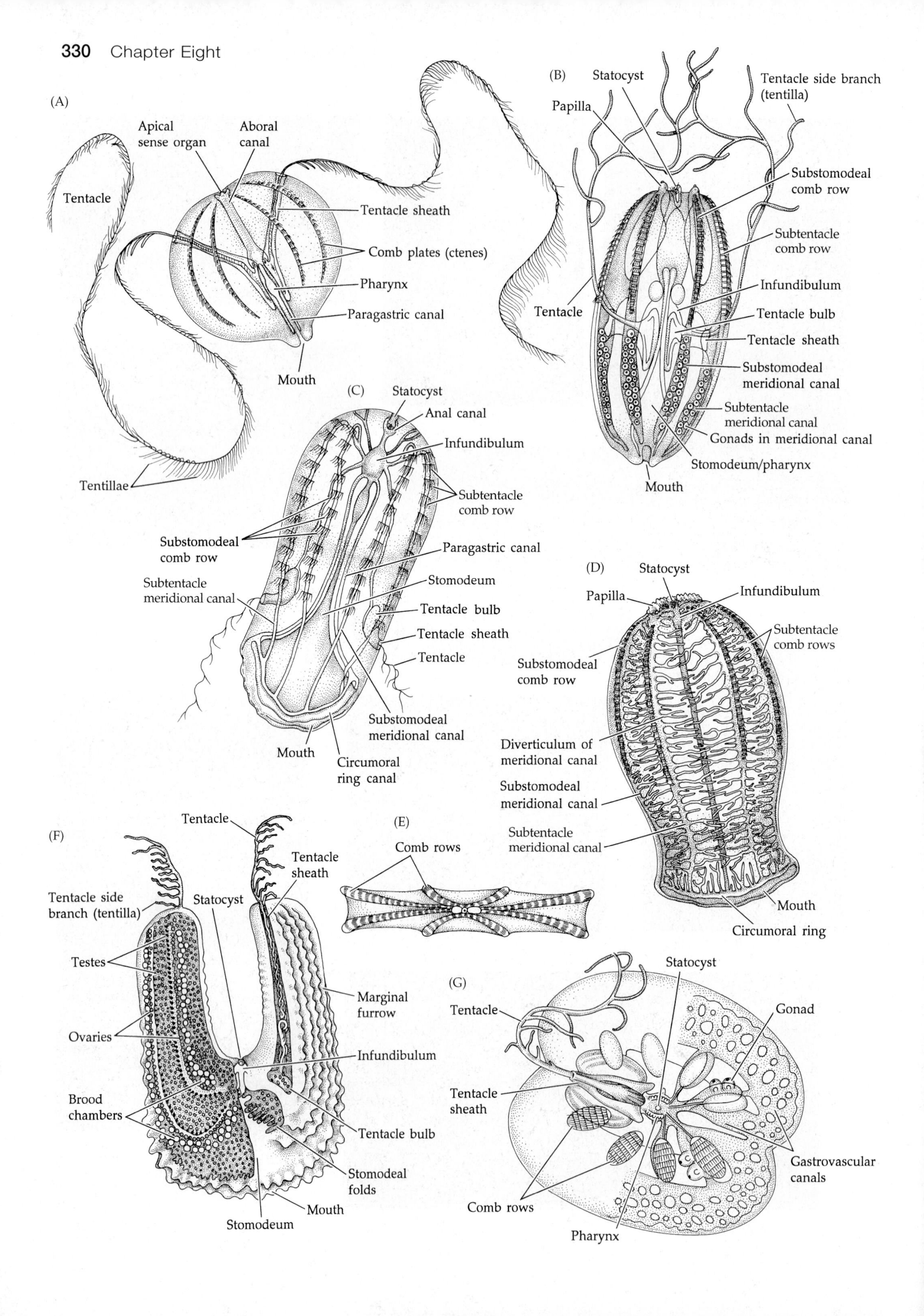
(A)
Apical sense organ
Aboral canal
Tentacle
Tentacle sheath
Comb plates (ctenes)
Pharynx
Paragastric canal
Mouth
Tentillae
(B)
Statocyst
Tentacle side branch (tentilla)
Papilla
Substomodeal comb row
Subtentacle comb row
Infundibulum
Tentacle
Tentacle bulb
Tentacle sheath
Substomodeal meridional canal
Subtentacle meridional canal
Gonads in meridional canal
Stomodeum/pharynx
Mouth
(C)
Statocyst
Anal canal
Infundibulum
Subtentacle comb row
Substomodeal comb row
Paragastric canal
Subtentacle meridional canal
Stomodeum
Tentacle bulb
Tentacle sheath
Tentacle
Substomodeal meridional canal
Mouth
Circumoral ring canal
(D)
Statocyst
Papilla
Infundibulum
Subtentacle comb rows
Substomodeal comb row
Diverticulum of meridional canal
Substomodeal meridional canal
Subtentacle meridional canal
Mouth
Circumoral ring
(E)
Comb rows
(F)
Tentacle
Tentacle sheath
Tentacle side branch (tentilla)
Statocyst
Testes
Marginal furrow
Ovaries
Infundibulum
Brood chambers
Tentacle bulb
Stomodeal folds
Mouth
Stomodeum
(G)
Statocyst
Tentacle
Gonad
Tentacle sheath
Gastrovascular canals
Comb rows
Pharynx

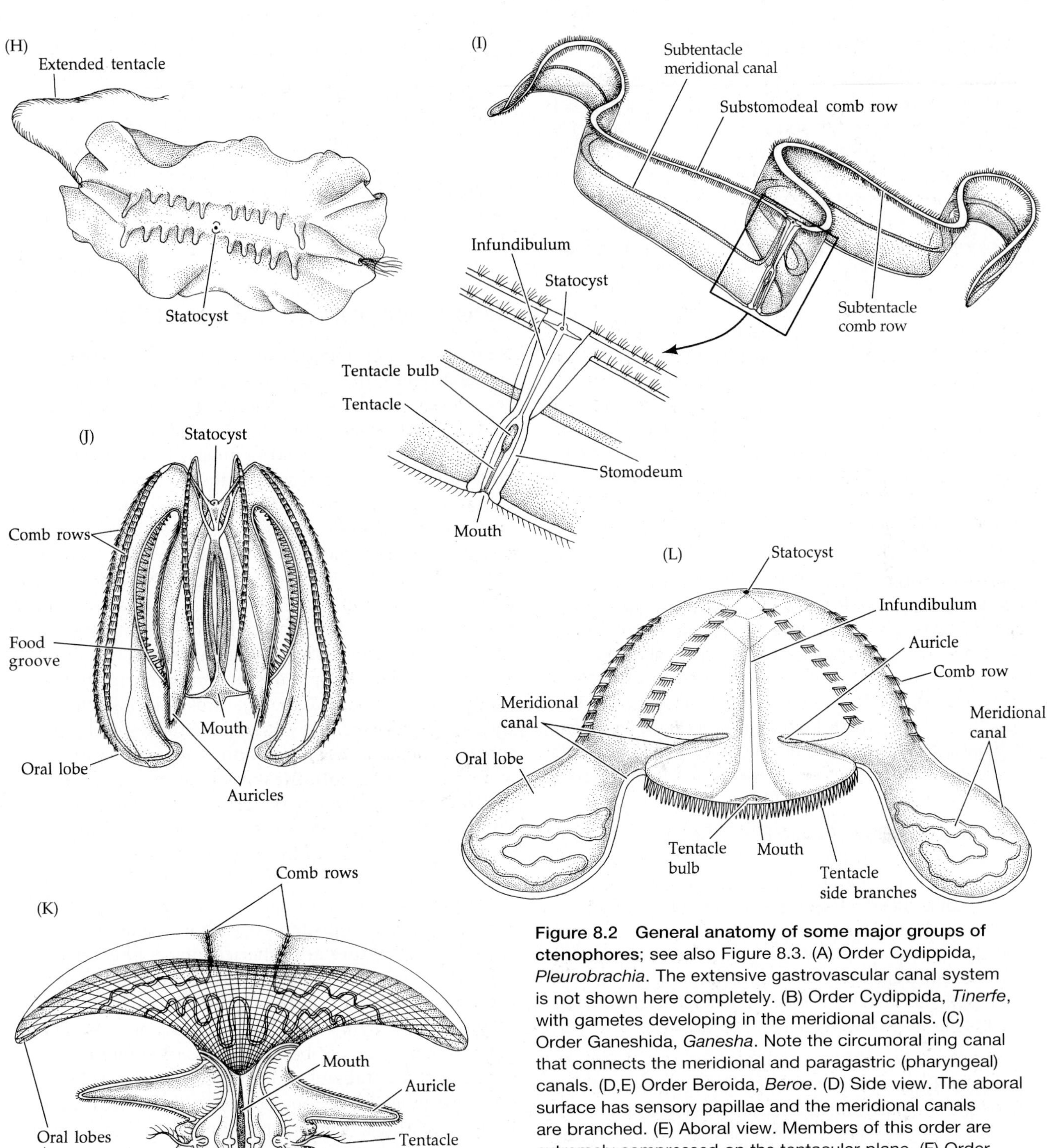

Figure 8.2 General anatomy of some major groups of ctenophores; see also Figure 8.3. (A) Order Cydippida, *Pleurobrachia*. The extensive gastrovascular canal system is not shown here completely. (B) Order Cydippida, *Tinerfe*, with gametes developing in the meridional canals. (C) Order Ganeshida, *Ganesha*. Note the circumoral ring canal that connects the meridional and paragastric (pharyngeal) canals. (D,E) Order Beroida, *Beroe*. (D) Side view. The aboral surface has sensory papillae and the meridional canals are branched. (E) Aboral view. Members of this order are extremely compressed on the tentacular plane. (F) Order Platyctenida, the odd-shaped *Lyrocteis*, shown here in layered cutaway view exposing various internal structures. (G) Order Platyctenida, *Ctenoplana* (aboral view). Only one tentacle is shown. (H) Order Platyctenida, *Coeloplana*. This ctenophore is a benthic form. (I) Order Cestida, *Cestum*. This ctenophore exhibits an extreme modification of body form. (J,K) Order Lobata, *Mnemiopsis*. (J) Side view. *Mnemiopsis* has oral lobes and auricles. (K) Oral view. Note the greatly expanded oral lobes with their distinctive pattern of muscle fibers. (L) Order Lobata, *Deiopea*.

PHYLUM CTENOPHORA

ORDER CYDIPPIDA (Figures 8.1A and 8.2A,B). Pelagic; with well developed comb rows; tentacles long and retractable into sheaths; body globular or ovoid, occasionally flattened in the stomodeal plane; meridional canals end blindly, paragastric canals (when present) end blindly at mouth. This order is widely viewed as polyphyletic and in need of revision. (e.g., *Aulococtena, Bathyctena, Callianira, Dryodora, Euplokamis, Haeckelia, Hormiphora, Lampea, Mertensia, Pleurobrachia, Tinerfe*)

ORDER PLATYCTENIDA (Figure 8.2F–H). Planktonic or benthic; most species greatly flattened, with part of stomodeum everted as a creeping sole; often with tentacle sheaths; tentacle canals bifid; gastrovascular system complex and anastomosing; most species possess anal pores; many are ectocommensals on other organisms (e.g., corals). Unlike most ctenophores, fertilization is often internal, and many platyctenids brood their embryos to the larval stage; asexual reproduction is common. (e.g., *Coeloplana, Ctenoplana, Gastra, Lyrocteis, Planoctena, Savangia, Tjalfiella, Vallicula*)

ORDER CESTIDA (Figures 8.1E and 8.2I). Pelagic; body extremely compressed in tentacular plane, and greatly elongated in stomodeal plane, producing a ribbon-like form up to 1 m long in some species; substomodeal comb rows elongated, extending along entire aboral edge; subtentacular meridional canals arise under subtentacular comb rows (*Cestum*) or equatorially from interradial canals (*Velamen*); paragastric canals extend along oral edge and fuse with meridional canals; tentacles and tentacle sheaths present. Two genera: *Cestum* and *Velamen.*

ORDER GANESHIDA (Figure 8.2C). Pelagic; body form somewhat intermediate between Cydippida and Lobata, compressed in tentacular plane; tentacles branched and with sheaths; interradial canals arise from infundibulum and divide into adradial canals, which join the aboral ends of the meridional canals; meridional canals and paragastric canals join and form a circumoral canal (as in Beroida); mouth large and expanded in tentacular plane; without auricles or oral lobes. One genus, *Ganesha*, with two known species.

ORDER LOBATA (Figures 8.1D,F–H and 8.2J–L). Pelagic; body compressed in tentacular plane; with a pair of characteristic oral lobes and four flaplike auricles; a ciliated auricular groove extends to base of auricles from each side of each tentacle base; paragastric and subtentacular meridional canals unite orally. (e.g., *Bolinopsis, Deiopea, Leucothea, Mnemiopsis, Ocyropsis*)

ORDER THALASSOCALYCIDA Pelagic; body extremely fragile, expanded orally into medusa-like bell, to 15 cm along tentacular axis; body slightly compressed in stomodeal plane; tentacle sheaths absent; tentacles arise near mouth and bear lateral filaments; comb rows short; mouth and pharynx borne on central conical peduncle; meridional canals long, describing complex patterns in bell; all meridional canals end blindly aborally. Monotypic: *Thalassocalyce inconstans.*

ORDER BEROIDA (Figures 8.1B–C and 8.2D,E). Pelagic; body cylindrical or thimble-shaped and strongly flattened in tentacular plane; tentacles and sheaths absent; aboral end rounded (*Beroe*) or with two prominent keels (*Neis*); stomodeum greatly enlarged; aboral sense organ well developed; comb rows present; meridional canals with numerous side branches; paragastric canals simple or with side branches. Without a cydippid larva phase. Two genera: *Beroe* and *Neis.*

The Ctenophoran Body Plan

Although ctenophores are among the most ancient of living animals, they do possess true tissues. Between the epidermis and the gastrodermis is a well-developed middle layer, which is always a cellular mesenchyme. Within this mesenchyme true muscle cells develop, a condition that also characterizes the triploblastic Metazoa although by different developmental pathways.

As we noted in the preceding chapter, a critical essence of the cnidarian and ctenophoran body plans is radiality (or biradiality); we have explained some of the structural constraints and advantages that derive from this symmetry. Thus, predictably, the nervous system of ctenophores is in the form of a simple, noncentralized nerve net, and the locomotor structures are arranged radially about the body. Other features that characterize the Ctenophora include: retractile **tentacles** and often **tentacle sheaths; anal pores**; adhesive prey-capturing structures called **colloblasts**; locomotor structures called **ctenes** or comb plates, arranged in **comb rows**; and an **apical sense organ** containing a statolith that regulates the activity of the comb rows. The sheathed tentacles, colloblasts, comb plates, and nature of the apical sense organ are unique features of ctenophores. Figures 8.2 and 8.3 illustrate ctenophoran anatomy.

Most ctenophores are spherical or ovoid in shape, although some species have evolved flattened shapes through compression and elongation in one of the two planes of body symmetry (Figures 8.1E and 8.2I). The general body plan can best be understood by first examining a generalized cydippid ctenophore (Figure 8.3). Specialists have long considered the cydippids to be ancestral within the phylum, although recent evidence suggests that the order Cydippida is probably polyphyletic. As in cnidarians, the principal axis is oral–aboral. The mouth is at the oral pole; the aboral pole bears the apical sense organ. On the surface of the body are eight equally spaced **meridional rows** of comb plates. Each comb plate, or ctene, is composed of a transverse band of long, fused (= compound) cilia. On each side of the body of many species is a deep, ciliated epidermal pouch (the tentacle sheath) from whose inner wall a tentacle arises. The tentacles are typically very long and contractile, and bear lateral branches called filaments, or **tentillae**. The epidermis of both the

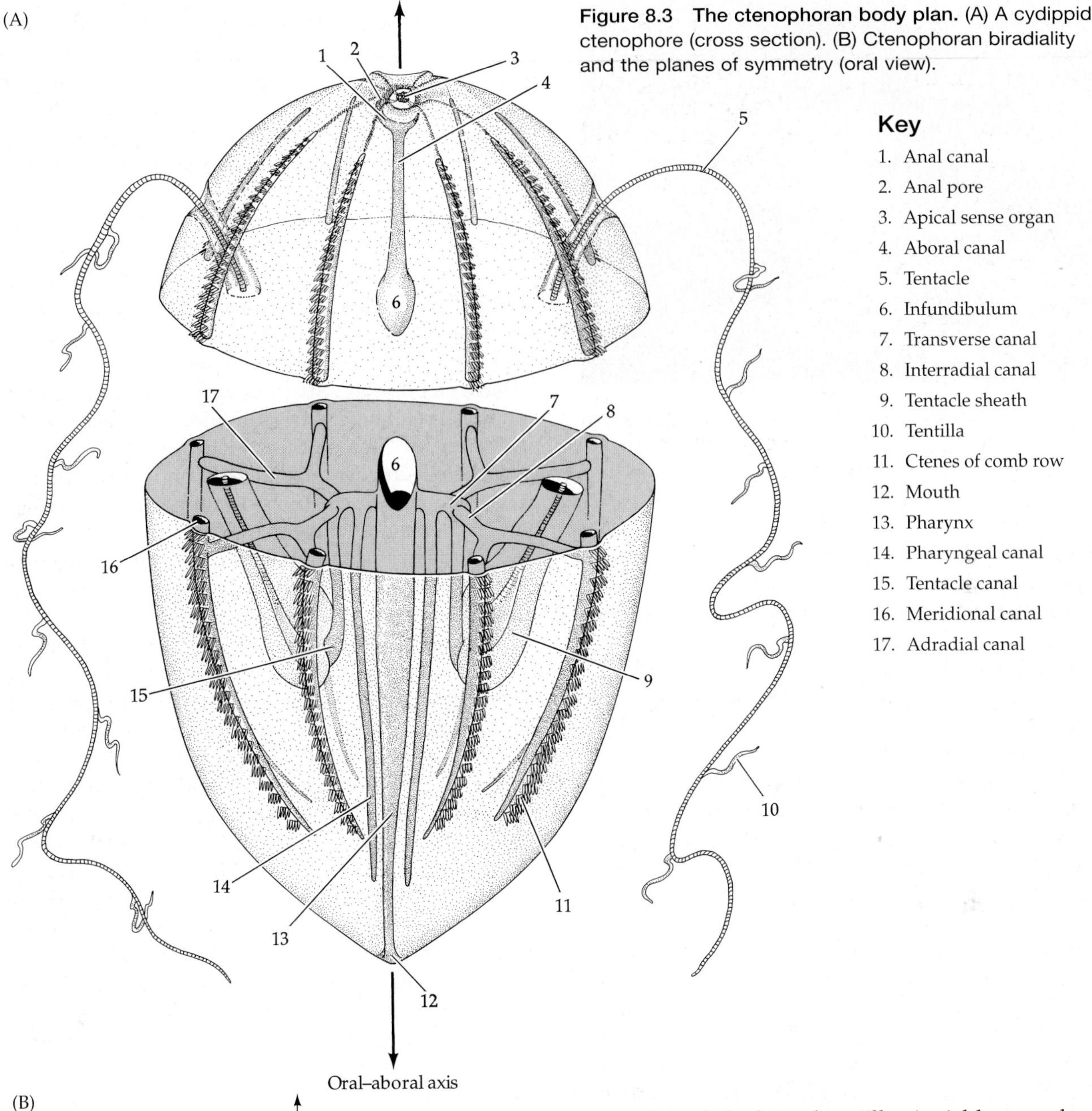

Figure 8.3 The ctenophoran body plan. (A) A cydippid ctenophore (cross section). (B) Ctenophoran biradiality and the planes of symmetry (oral view).

tentacle and the lateral tentillae is richly armed with colloblasts. Most species can retract the tentacles into the sheaths with muscles. It is the tentacles and certain aspects of the internal anatomy that give ctenophores a biradial symmetry (although some workers have described a more nuanced version of this in ctenophores, called "rotational symmetry"). The elongate **stomodeum** lies on the oral–aboral axis of the body. It is distinctly flattened in one plane of body symmetry, the **stomodeal plane** (Figure 8.3B). Bisecting the animal along the stomodeal plane separates the two tentacular halves of the body. The second plane of body symmetry, called the **tentacular plane**, is defined by the position of the tentacle sheaths.

Some variations of the basic ctenophoran body plan are illustrated in Figures 8.1 and 8.2. In members of the unique order Lobata (Figure 8.2J–L and chapter opener

Figure 8.4 *Thalassocalyce inconstans*, **in the monotypic order Thalassocalycida, with its expanded oral lobes that form a medusa-like bell.**

photo), the body is compressed in the tentacular plane and the oral end is expanded on each side into rounded, contractile **oral lobes**. The mouth sits on an elongate region, the base of which bears four long flaps called **auricles**. The tentacles are reduced and lack sheaths. From either side of each tentacle base, a ciliated **auricular groove** arises and extends to the auricles.

Members of the order Cestida (Figures 8.1E and 8.2I) are also compressed in the tentacular plane and extremely elongated in the stomodeal plane, giving these ctenophores a striking snake- or ribbon-like appearance. The sheathed tentacles are reduced and shifted alongside the mouth. Beroida are thimble-shaped and also flattened in the tentacular plane (Figure 8.1B,C and 8.2D,E). They lack tentacles and sheaths. In the single species of Thalassocalycida (*Thalassocalyce inconstans*), the body is expanded around the mouth to form a medusa-like bell (Figure 8.4).

The oddest ctenophores are members of the order Platyctenida (Figures 8.1I and 8.2G–H). Platyctenids are benthic and small, often less than 1 cm in length; in contrast to most pelagic ctenophores, they are pigmented rather than transparent. The body is oval and markedly flattened. Despite these unusual features, early naturalists recognized them as ctenophores by the presence of an apical sense organ, comb rows, and a pair of tentacles. Detailed studies have shown that the flattened oral surface is actually an everted portion of the pharynx! The platyctenid pharynx was, in a sense, preadapted to serve a dual purpose as a creeping foot or sole by its intrinsic musculature. Most of these animals crawl about on the sea bottom, but some are ectocommensals on alcyonarian cnidarians, echinoderms, or pelagic salps.

Support and Locomotion

Ctenophores rely primarily on their elastic **mesenchyme** for structural support. The watery gelatinous mesenchyme makes up most of the body mass; ctenophore dry weights are only about 4 percent of their live wet weights. The mesenchyme contains both elastic supportive cells and muscle cells, the general tonus of the latter being primarily responsible for maintaining body shape. Figure 8.5 shows a highly stylized cutaway section of a cydippid ctenophore and illustrates the arrangement of the supportive mesenchymal muscle fibers. Tension in the looped muscles tends to maintain the spherical geometry. Action of the radial muscles diminishes the radius and hence the circumference, and also serves to open the pharynx. These two muscle sets work antagonistically to one another.

Most ctenophores are pelagic. The gelatinous body and low specific gravity maintain relatively neutral buoyancy, allowing these creatures to float about with the ocean currents. Neutral buoyancy appears to be maintained by passive osmotic accommodation. Because buoyancy adjustments take time, ctenophores may temporarily accumulate at discontinuity layers in the sea, where a water mass of one density overlies a water mass of a slightly different density.

The beating of the ctenes provides most of the modest locomotor power that allows ctenophores to move up and down in the water column and to locate richer feeding sites or preferred environmental conditions. Each comb row comprises many ctenes. Each ctene consists of a transverse band of hundreds of very long, partly fused cilia (to 3.5 mm in length) that beat together as a unit. Ctenophores are the largest animals known to use cilia for locomotion. Each cilium has a typical 9+2 microtubule structure, but each also possesses a unique set of lamellae at the 3 and 8 doublets; these lamellae protrude to link together the adjacent cilia.

Ctenophores are beautiful animals to observe in life because their beating comb rows appear iridescent over a wide range of light intensities. This feature of

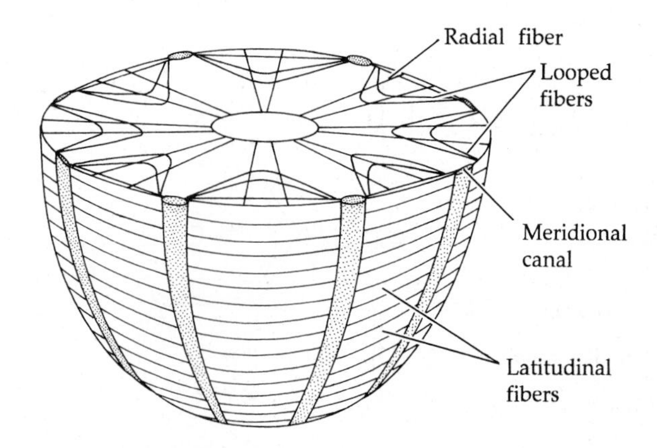

Figure 8.5 Stereogram of the arrangement of muscle fibers in ***Pleurobrachia*****, a cydippid ctenophore.** The diagram depicts a transverse section through the region of the pharynx; the gastrovascular system and tentacle sheaths have been omitted for clarity.

ctenophoran movement appears to be caused by the dense and highly regular packing of ciliary elements at the base of each comb row. While iridescence may serve to discourage predators or attract prey, it seems unlikely to function in social communication among ctenophores because of the simplicity of these animals' visual systems. Because of their size, ctenophore ctene plates move at low Reynolds numbers, where fluid flow is smoother but where water viscosity can impede ciliary movement. In contrast, the bodies of ctenophores move at higher Reynolds numbers, where viscosity is less important but water turbulence can impede or enhance animal movement through the medium.

As noted above, the mesenchymal musculature is used to maintain body shape and assist in feeding; it is involved in behaviors such as prey swallowing, pharyngeal contractions, and tentacle movements. Usually both longitudinal and circular muscles are present just beneath the epidermis. In the benthic and epifaunal platyctenids, stomodeal musculature facilitates a creeping locomotion. In the snakelike cestids, body muscles may generate graceful swimming undulations. Swimming in the lobate ctenophores is assisted by muscular flapping of their two oral lobes, and perhaps also by use of the four paddle-like auricles. The lobate species *Leucothea* (Figure 8.1F) can swim either by typical slow ctene propulsion or by rapid ctene propulsion; the latter is accomplished by an increased ciliary beat that produces a vortex wake, resulting in jet propulsion. Giant smooth muscle fibers—the first to be discovered in ctenophores—have been found in *Beroe*.

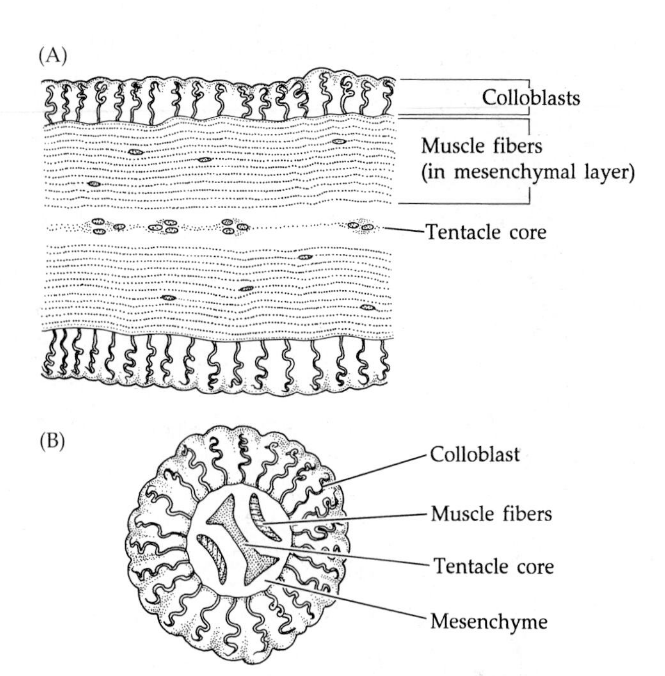

Figure 8.6 Ctenophore tentacle structure. (A) Longitudinal section of tentacle. (B) Cross section of a lateral filament (tentilla) of a tentacle.

Feeding and Digestion

Comb jellies, so far as is known, are mostly predatory in their habits. The long tentacles of cydippids (and of the larvae of most other forms) have a muscular core with a colloblast-laden epidermal covering (Figure 8.6). The tentacles trail passively or are "fished" by various swirling movements of the body. Upon contact with zooplankton prey, the colloblasts (sometimes called **lasso cells**) burst and discharge a strong adhesive material. Each colloblast develops from a single cell and consists of a hemispherical mass of secretory granules attached to the muscular core of the tentacle by a spiral filament coiled around a straight filament (Figure 8.7). The straight filament is actually the highly modified nucleus of the colloblast cell. The spiral filament, which uncoils upon discharge, adheres to the prey by the sticky material produced in the secretory granules. As the tentacles accumulate prey, they are periodically wiped across the mouth by muscular contractions, occasionally combined with a coordinated somersaulting action of the animal that brings the mouth to the trailing tentacle. In members of the orders Lobata and Cestida, which bear very short tentacles, small zooplankton are trapped in mucus on the body surface and then carried to the mouth by ciliary currents (along the ciliated **auricular grooves** in lobate forms and ciliated **oral grooves** in cestids). Most of the benthic platyctenids also feed by capturing zooplankton in a somewhat similar fashion.

In some areas of the world's seas, ctenophores may be the dominant macrozooplankters and planktonic predators (e.g., *Mertensia ovum* in the Arctic region). Early life stages of *Mnemiopsis leidyi* also consume significant quantities of microphytoplankton and microzooplankton, indicating a mixed diet. Stable isotope studies of the lobate ctenophore, *Bolitopsis infundibulum* indicate that these animals rely primarily on surface photosynthetic products for food, suggesting that some species may lie quite low in the food chain. These same animals also appear to supplement their diet with copepods, which may sink to waters layers near the sea floor during diapause, and where *B. infundibulum* may form large aggregations to feed on this abundant source of carbon. The tendency of ctenophores to form aggregations near food sources or at the water's surface may contribute to their patchy distribution in nature.

Some ctenophores prey on larger animals, especially gelatinous forms. The cydippid *Lampea* (formerly *Gastrodes*), for example, lives embedded in the body of pelagic tunicates of the genus *Salpa*, on which it feeds. Figure 8.8 is a series of remarkable photographs showing the cydippid ctenophore *Haeckelia* eating the tentacles of the trachyline hydromedusa, *Aegina*. After

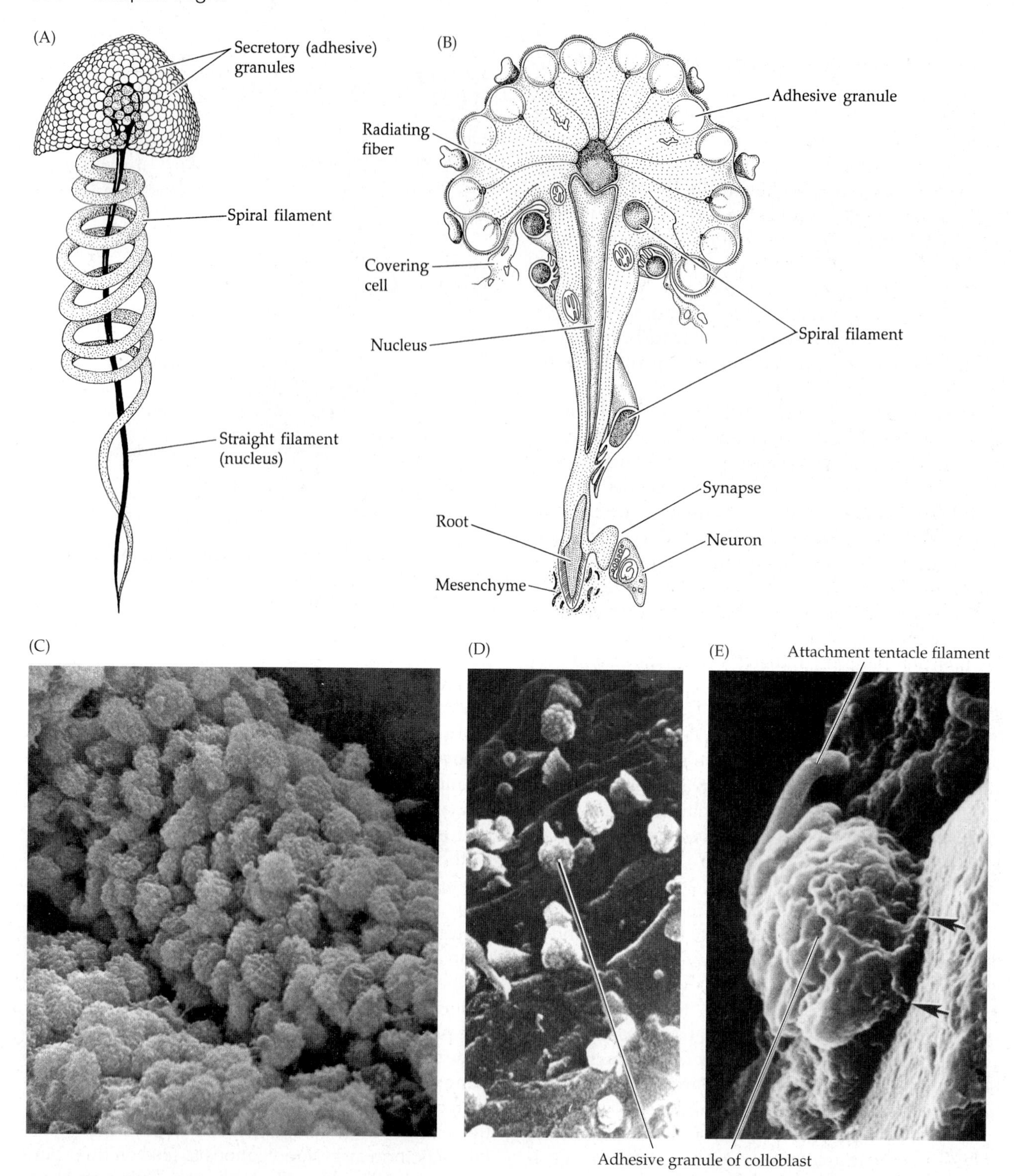

Figure 8.7 Colloblasts. (A) The functional parts of a colloblast. (B) Longitudinal section. (C) Colloblasts on the lateral tentacle filaments (tentillae) of *Pleurobrachia* (SEM). (D) Fired colloblasts of *Pleurobrachia*, showing adhesive granules attached to fragments of a copepod (small crustacean). (E) Fired colloblasts are still attached to the tentacle filament. The adhesive ends of the coiled filaments are stuck (arrows) to a bit of copepod.

consuming the tentacles one by one, *Haeckelia* retains the prey's unfired nematocysts, incorporates them into its epidermis, and uses them for its own defense. This phenomenon, known as **kleptocnidae**, occurs in several unrelated groups who prey on cnidarians.

Ctenophores were center stage in an ecological drama that played out in the Black Sea not long ago. In the 1980s, the predatory northwest Atlantic ctenophore *Mnemiopsis leidyi* was accidentally introduced into the Black Sea by way of ship ballast

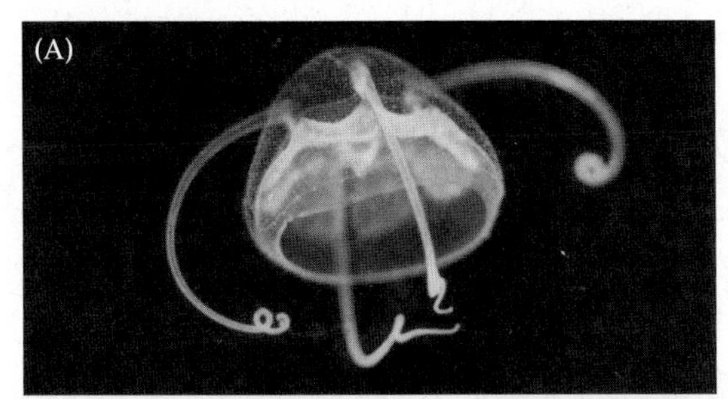

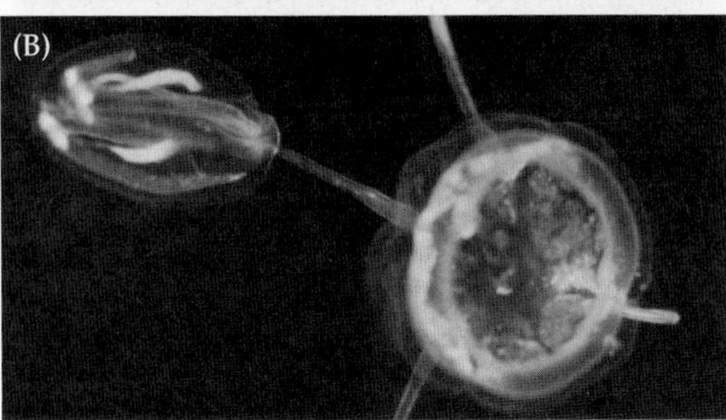

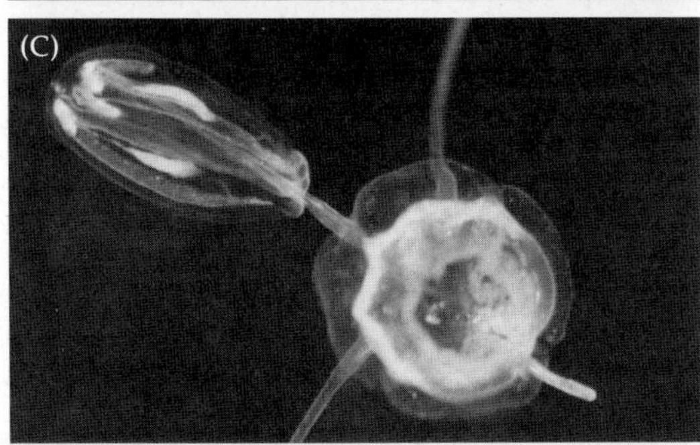

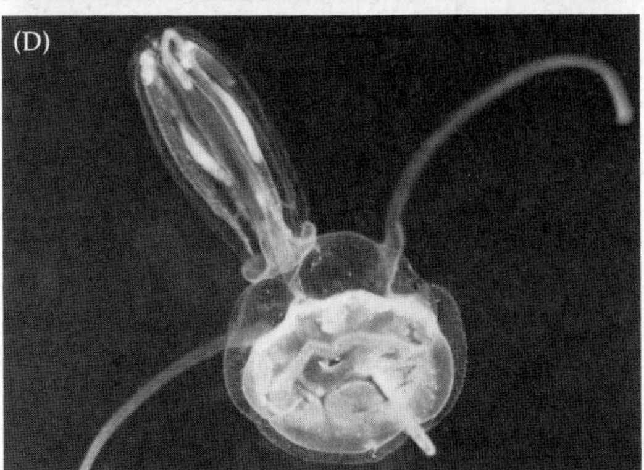

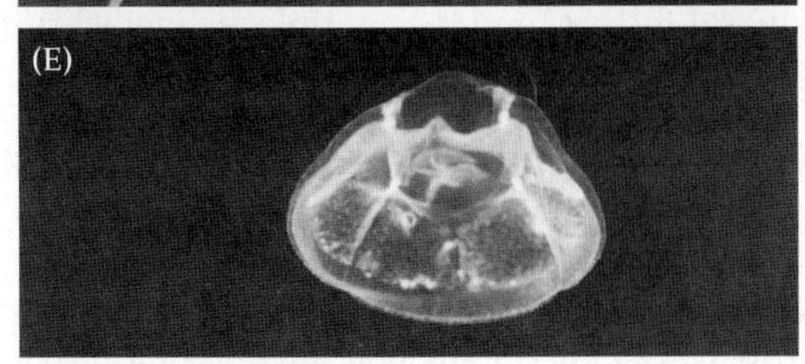

Figure 8.8 The cydippid ctenophore *Haeckelia rubra* (= *Euchlora rubra*) feeding on the trachyline hydromedusa *Aegina citrea*. (A) Intact specimen of *Aegina*, with all four tentacles present. (B) *Haeckelia* begins to consume one of *Aegina*'s tentacles. (C) Most of the first tentacle of the medusa has been ingested. (D) Same animals, 2 minutes after feeding began. (E) *Aegina* has lost all four of its tentacles to the hungry *Haeckelia*.

water. The few invaders quickly underwent explosive population growth, reaching biomass levels in excess of one kilogram per cubic meter by 1989, devastating the food web of the entire Black Sea Basin and causing a collapse of the anchovy fishery (one of the favored prey of *M. leidyi*). Then, in 1997, another ctenophore, *Beroe ovata*, was accidentally introduced to the Black Sea, probably also from ballast water. *Beroe ovata* feeds almost exclusively on *Mnemiopsis*, and its introduction resulted in a precipitous decline (perhaps extirpation) of *M. leidyi* in the sea, followed at last by the disappearance of *Beroe* itself. Introduced ctenophores continue to present a threat to local fisheries in many parts of the world and their numbers are now monitored closely, particularly in the Mediterranean Sea. Most recently, *M. leidyi* has invaded the Caspian and Baltic Seas.

The ctenophoran mouth opens into an elongate, highly folded, flattened, muscular, stomodeal pharynx. The epithelium of the pharynx is richly endowed with gland cells that produce the digestive juices. Large food items are tumbled within the pharynx by ciliary action. Digestion takes place extracellularly, mostly in the pharynx. The largely digested food passes via a small chamber (the **infundibulum**, funnel, or stomach) from the pharynx into a complex system of radiating gastrovascular canals (Figures 8.2 and 8.3). The details of the arrangement of the canals vary among different groups; the following description applies to the arrangement in a cydippid.

Two **paragastric** or **pharyngeal canals** recurve and lie parallel to the pharynx. Two **transverse canals** depart at right angles to the stomodeal plane and divide into three more branches. The middle branch of each triplet, **the tentacle canal**, leads to the base of the tentacle sheath. Each of the other two branches (the **interradial canals**) bifurcates to form a total of four **adradial canals** on each side of the animal. These in turn connect to the eight **meridional canals**, one beneath each comb row. Finally, an **aboral canal** passes from the infundibulum to the aboral pole, where it divides beneath the apical sense organ into four short canals, two ending blindly and two (the anal canals) opening to the outside via small **anal pores**. The anal pores serve as a rudimentary anus, assisting the mouth in the voiding of indigestible wastes. They may also serve as an exit for metabolic wastes.

Within this very complicated gastrovascular canal system, digestion is completed, nutrients are distributed through the body, and absorption takes place. Minute pores lead from the various canals into the mesenchyme (Figure 8.9). Surrounding these pores are circlets of ciliated gastrodermal cells called **cell rosettes**, which appear to regulate the flow of the digestive soup and perhaps also play a role in excretion. Except for the stomodeal pharynx, the gastrovascular system is lined by a simple epithelium of endodermal origin.

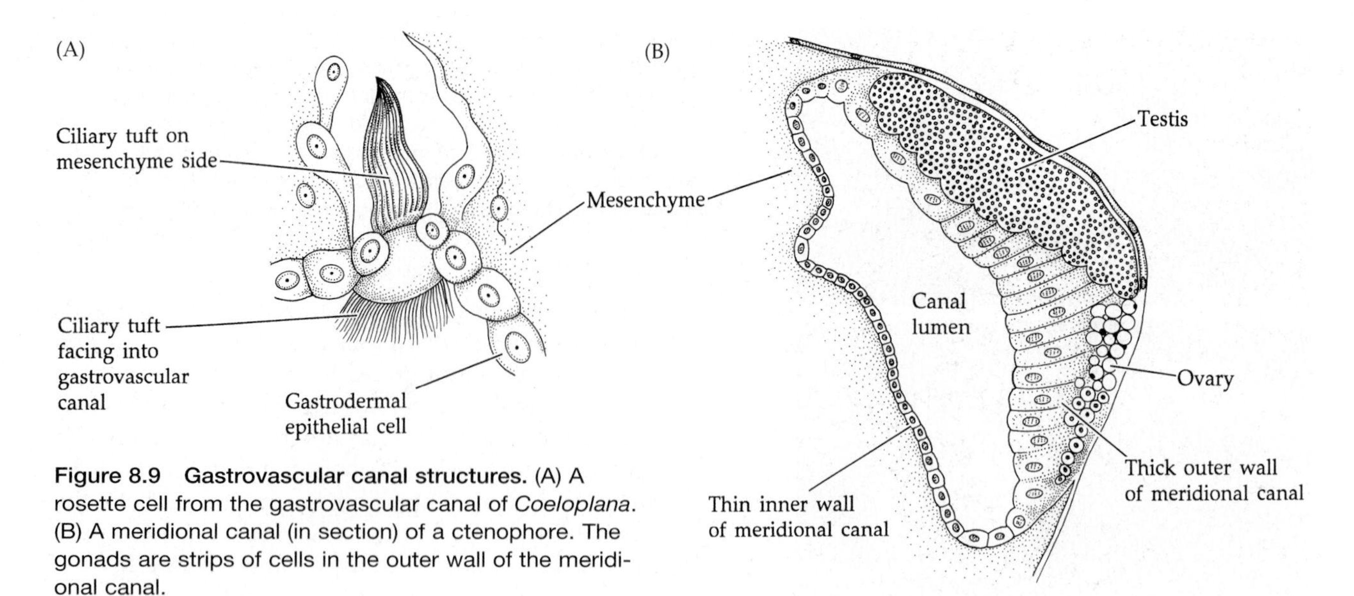

Figure 8.9 Gastrovascular canal structures. (A) A rosette cell from the gastrovascular canal of *Coeloplana*. (B) A meridional canal (in section) of a ctenophore. The gonads are strips of cells in the outer wall of the meridional canal.

Circulation, Excretion, Gas Exchange, and Osmoregulation

There is no independent circulatory system in ctenophores; as in cnidarians, the gastrovascular canal system serves in this role by distributing nutrients to most parts of the body. The gastrovascular system probably also picks up metabolic wastes from the mesenchyme for eventual expulsion out of the mouth or anal pores. The cell rosettes may also transport wastes to the gut. Gas exchange occurs across the general body surface and across the walls of the gastrovascular system. All of these activities are augmented by diffusion through the gelatinous mesenchyme. Movement of water over the body surface is enhanced by the beating of the comb plates. Thus, the extensive canal system and the ciliary bands help to overcome the problem of long diffusion distances.

Nervous System and Sense Organs

Although the nervous systems of both ctenophores and cnidarians are noncentralized nerve nets, there are certain important differences. In a ctenophore, nonpolar neurons form a diffuse **subepidermal plexus**. Beneath the comb rows, the neurons form **elongate plexes** or meshes such that they produce nervelike strands. The bases of the ctenes are thus in contact with a rich array of nerve cells. A similar concentrated plexus surrounds the mouth. However, as in cnidarians, no true ganglia occur, a condition that contrasts markedly with the presence of a centralized nervous system in bilateral Metazoa (the Bilateria).

The nerve nets of ctenophores consist of polygonal nerve cords spread under the ectodermal epithelium; these nerve nets show high levels of regional specialization and concentrations associated with the apical sensory organ/polar fields and tentacle bulbs, structures without clear homologs in any other animal group.

The apical sense organ is a statolith that functions in balance and orientation. The calcareous statolith is supported by four long tufts of cilia called **balancers** (Figure 8.10). The whole structure is enclosed in a transparent **dome** that apparently is also derived from cilia. From each balancer arises a pair of **ciliated furrows** (= **ciliated grooves**), each of which connects with one comb row. Thus, each balancer innervates the two comb rows of its particular quadrant. Tilting the animal causes the statolith to press more heavily on the downside balancers, and the resulting stimulus elicits a vigorous beating of the corresponding comb rows to right the body.

The two comb rows in each quadrant innervated by a single ciliated furrow beat synchronously. If a ciliated furrow is cut, the beating of the two corresponding comb rows becomes asynchronous. The normal direction of ciliary power strokes is toward the aboral pole, so that the animal is driven forward oral end first. The beat in each row, however, begins at the aboral end of the comb row and proceeds in metachronal waves toward the oral end (i.e., antiplectic metachrony). Stimulation of the oral end reverses the direction of both the wave and the power stroke. Removal of the apical sense organ or statolith results in an overall lack of coordination of the comb rows, and the injured ctenophore loses its ability to maintain a vertical position. The comb rows are very sensitive to contact; when a comb row is touched, many species retract it into a groove formed in the jelly-like body.

In cydippids and beroids, the stimulation for any given ctene to beat is triggered mechanically, by hydrodynamic forces arising from the movements of the preceding plate. However, in the lobate ctenophores, the ctenes are not coordinated in this mechanical fashion. In these animals a narrow tract of shorter cilia—the **interplate ciliated groove**—runs between successive ctenes and is responsible for coordinating their activity.

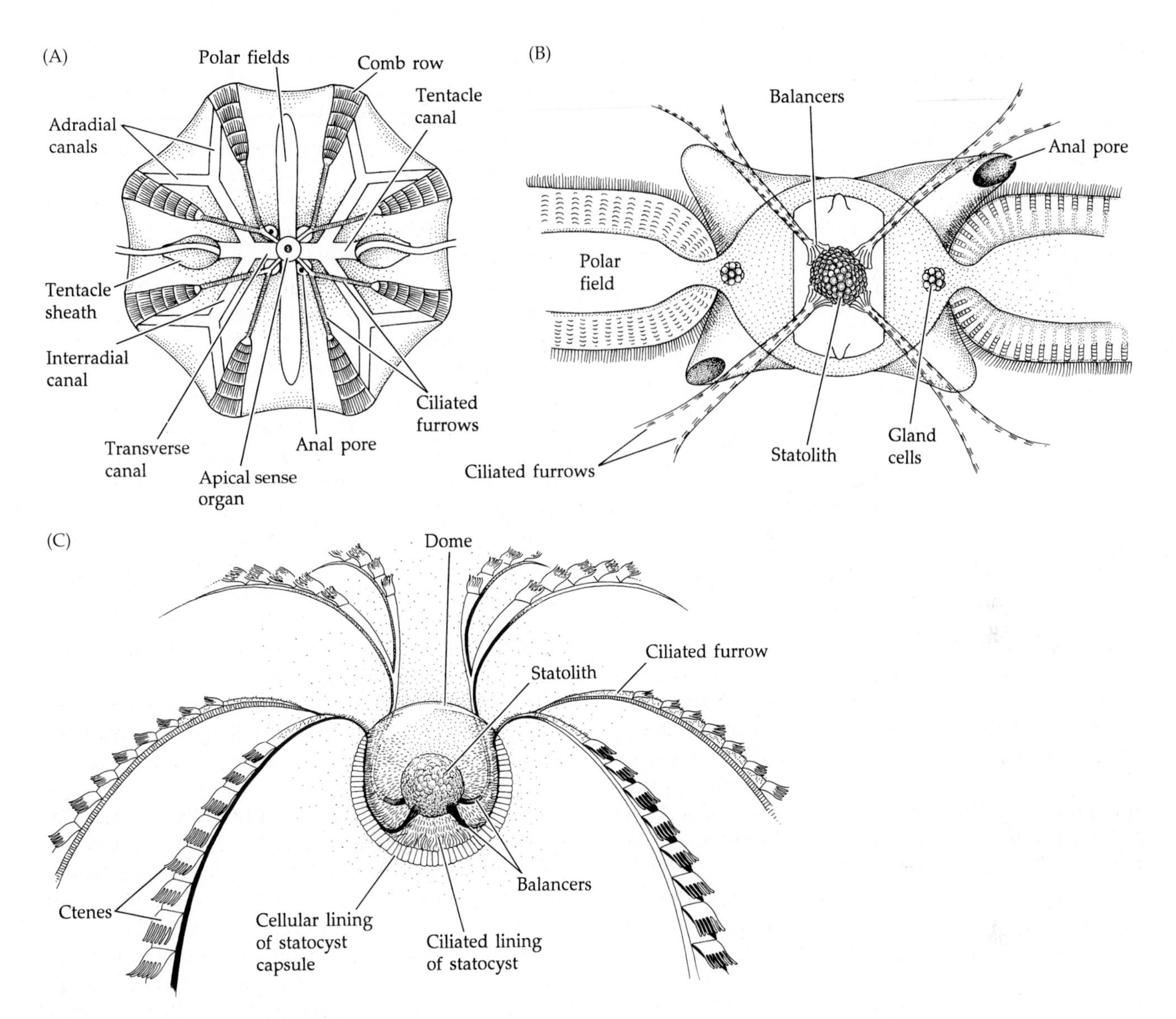

Figure 8.10 The ctenophore apical sense organ. (A) *Pleurobrachia* (aboral view). Note the relationship of the statocyst to the comb rows, the ciliated furrows, and the gastrovascular canals. (B) Apical sense organ of the cydippid *Hormiphora*. (C) Apical sense organ and its relationship to the eight comb rows.

It is not known how the cilia of the groove are coordinated or how the grooves stimulate the appropriate comb row, so these actions may also be mechanical. The interplate ciliated grooves develop only as the lobate ctenophores mature to adulthood; the free-swimming larvae resemble cydippids and lack the grooves.

In some ctenophores, two oval tracts of cilia called polar fields lie on the stomodeal plane of the aboral surface (Figure 8.10B). These structures are presumed to be sensory in function.

Reproduction and Development

Asexual reproduction and regeneration Ctenophores can regenerate virtually any lost part, including the apical sense organ. Entire quadrants and even whole halves will regenerate. Speculation that ctenophores may reproduce by fission or budding is still under investigation. Platyctenids reproduce asexually by a process that resembles pedal laceration in sea anemones; small fragments break free as the animal crawls, and each piece can regenerate into a complete adult.

Sexual reproduction and development Most ctenophores are hermaphroditic, but a few gonochoristic species are known (e.g., members of the genus *Ocyropsis*). The gonads arise on the walls of the meridional canals (Figure 8.9B). Pelagic ctenophores generally shed their gametes via the mouth into the surrounding seawater, where either self-fertilization or cross-fertilization takes place. Fertilization appears to take place within the meridional canals in *Pleurobrachia*. Special sperm ducts occur in at least some platyctenid species. The eggs are centrolecithal and formed in association with nurse cell complexes. Polyspermy is known to occur in some ctenophores. Those that free-spawn typically produce embryos that grow quickly to planktotrophic

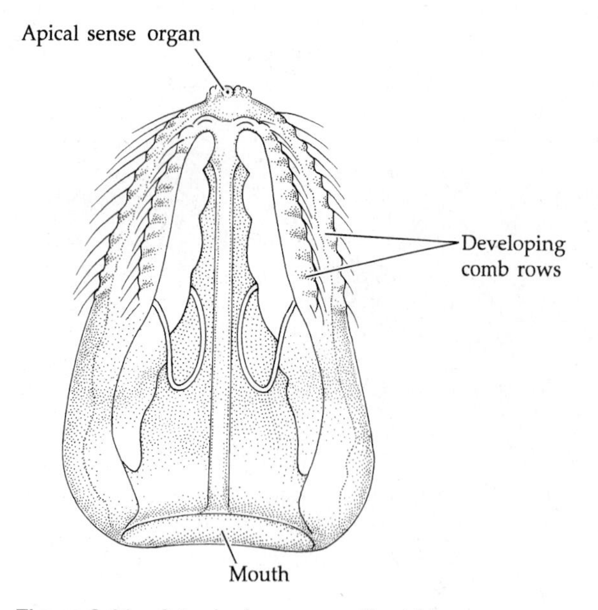

Figure 8.11 A typical young cydippid larva.

cydippid larvae (Figure 8.11), although species in the order Beroida lack this larval phase. Development is thus usually indirect, although growth to the adult is gradual rather than metamorphic. Some ctenophores are known to reproduce sexually before they have completed their larval development, a condition known as **dissogeny**. Precocious reproduction is known in at least two orders (Lobata: *Mnemiopsis leidyi*; Cydippida: *Pleurobrachia bachei*), and may reflect an evolutionary history favoring early reproduction, possibly because food supplies are only intermittently available, or because predation rates are high. In the benthic *Coeloplana* and *Tjalfiella*, fertilization is internal and embryos are brooded until a cydippid larva is formed and released. This mixed life history provides a means of dispersal for these benthic, sedentary animals.

Ctenophoran cleavage cannot easily be classified as either spiral or radial. During early cleavage, the first four blastomeres arise by the usual two meridional cleavages, which mark the adult planes of symmetry. The third division is also nearly vertical and results in a curved plate of eight cells (macromeres). The next division is latitudinal and unequal, giving rise to micromeres on the concave side of the macromere plate. The micromeres continue to divide and spread by epiboly over the aboral pole and eventually over the macromeres. The latter also invaginate, so the gastrula arises through a combination of epiboly and invagination. Thus, the micromeres become ectoderm and the macromeres become endoderm. Just prior to gastrulation, the macromeres divide and produce additional micromeres on the oral side of the embryo. Whereas the aboral micromeres become ectoderm, these oral micromeres are incorporated into the endoderm and, in at least some species, give rise to photoreceptor cells. There is some question about the fate of all of these oral micromeres. Metschnikoff (1885) suggested that these cells may contribute to the mesenchyme and may thus be viewed as true endomesoderm. Harbison (1985) also made a case for a triploblastic condition in ctenophores. However, as mentioned above, the recently sequenced genome of *Pleurobrachia bachei* shows no indication of developmental homology between ctenophoran mesenchyme and the mesoderm of true triploblast bilaterians.

As the micromeres cover the embryo to form the epidermis, four interradial bands of small, rapidly dividing cells become apparent. Eventually, each of these thickened ectodermal bands differentiates into two of the comb rows. The aboral ectoderm differentiates into the apical sense organ and its related parts; the oral ectoderm invaginates to form the stomodeum. The gastrovascular system develops from endodermal outgrowths and the tentacle sheaths arise as ectodermal invaginations from the points where the tentacles sprout. The embryo eventually develops into a free-swimming cydippid larva (Figure 8.11) that closely resembles adult ctenophores of the order Cydippida. Some authors have taken this as evidence that Cydippida houses the most primitive lineage of extant ctenophores.

The development of ctenophores differs markedly from that of cnidarians. In the latter group, early cleavage results in an irregular mass of cells whose fates are not clearly predictable until later development, and the mesenchyme is strictly ectodermal in origin. In the ctenophores, on the other hand, development is determinate and a very precise cleavage pattern unfolds, in which the ultimate morphology is definitely mapped. For example, if the two blastomeres of a 2-cell embryo are experimentally separated, the "half-embryos" develop into adults with exactly half the normal set of adult structures. Such results indicate that blastomere fate is highly determined, but they also suggest that inductive processes among embryonic cells and tissues are important in ctenophore embryogenesis, as is true in many bilaterian Metazoa. Additional data suggest that such processes can lead to more variation in cell fate than was previously suspected. Ctenophores lack the planula larva that characterizes cnidarians; instead, they produce a cydippid larval type having no obvious counterpart among the cnidarians. The presence of a brief "planula" stage has been reported in the development of the parasitic cydippid *Gastrodes parasiticum* (Komai 1922, 1963) that was said to burrow into the test of host salps, where it then developed into a free-swimming cydippid. This has not been confirmed by any subsequent research, and the nature of Komai's "planula" remains unresolved.

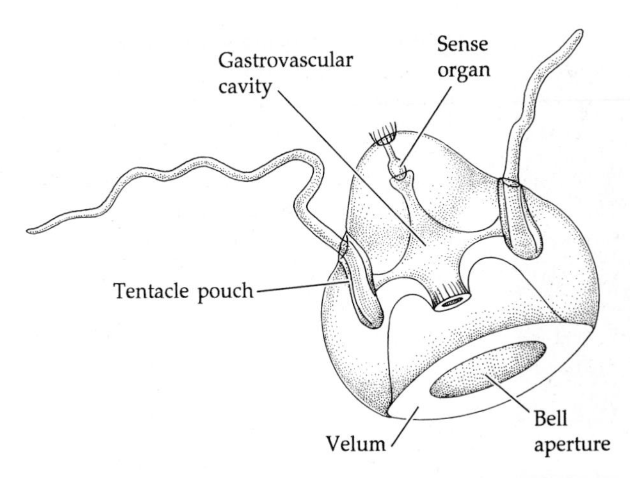

Figure 8.12 The aberrant cnidarian trachyline medusa *Hydroctena*, which superficially resembles a ctenophore in its possession of an apical sensory structure and tentacle pouches.

Ctenophoran Phylogeny

Even though the ctenophores and cnidarians are widely regarded as belonging to the same general grade of construction, it is difficult to link ctenophores to any specific cnidarian group. Although the phylogenetic relationships of the basal Metazoa (Porifera, Placozoa, Cnidaria, Ctenophora) remain enigmatic, some recent molecular phylogenies have united Cnidaria and Ctenophora as sister groups, reviving the old concept of "Coelenterata." Some zoologists have suggested that ctenophores arose from the hydrozoans, by way of an intermediate medusa possessing an aboral statocyst and two tentacle sheaths, such as is seen today in the aberrant trachyline medusa *Hydroctena* (Figure 8.12). In this medusa, the number of tentacles has been reduced to two, and these are set high on the bell, like the tentacles of trachylines in general. The tentacles also arise from deep epidermal pockets that resemble the tentacle sheaths of ctenophores. This argument is weakened by fossil evidence that Cambrian ctenophores probably lacked tentacles entirely. *Hydroctena* also has a single apical sense organ, although its construction differs from that of ctenophores. Several other trachyline medusae also have solitary aboral sense organs.

Although these similarities may suggest a relationship between ctenophores and trachyline cnidarians, and while ctenophores also show certain similarities to the scyphozoans and anthozoans, such as the stomodeum and the highly cellular mesenchyme, and the four-lobed gastrovascular cavity of the cydippid larva, as we have seen, ctenophores are really quite different from cnidarians in many fundamental ways. These differences are evident both in adult morphology and in patterns of development. Many of the similarities between ctenophores and cnidarian medusae may well be convergences reflecting adaptations to their similar lifestyles and, in fact, many gelatinous zooplankters show superficial similarities in body form and construction.

Some molecular phylogenies have placed the Ctenophora basal in the animal tree (below Porifera) as the earliest-diverging animal phylum. Perhaps supporting this hypothesis are next-generation sequencing studies that have shown the genomes of ctenophores and sponges to lack several homeobox-containing genes that are present within the genomes of placozoans, cnidarians, and bilaterians. Further, ctenophore genomes are unique in many respects. For example, they lack microRNA and microRNA-processing machinery, and their mitochondrial genomes are reduced.

The presence of mesenchymal muscle cells and gonoducts in some species, along with certain features of early cleavage, have led some zoologists to suggest a relationship between the ctenophores and the flatworms (Platyhelminthes; Chapter 10). Some researchers have viewed the ctenophores as ancestral to the flatworms; but a reverse scenario has also been suggested. The presence of benthic, crawling ctenophores (e.g., *Ctenoplana* and *Coeloplana*) has been used as evidence that ecological and anatomical intermediates between the two groups are plausible. However, the current consensus is that there is little evidence to link the ctenophores to the flatworms, and molecular phylogenetic studies do not link the two groups.

Phylogenetic relationships within the Ctenophora have been challenged by limited gene data and, in the case of the 18S rRNA gene, highly unequal substitution rates among the lineages. Within the Ctenophora, disagreement exists about whether the tentaculate or atentaculate condition is primitive, and whether the atentaculate lineage (the order Beroida) arose from somewhere among the tentaculate groups. The "atentaculate primitive" hypothesis is supported by the lack of tentacles among Cambrian forms. The modern number of 8-comb rows appears to have stabilized in fossil forms by about 400 million years ago. The enigmatic octoradiate Ediacaran *Eoandromeda* now appears likely to have been a ctenophore, whose comb rows spiraled down the length of its cone-shaped body (Figure 8.13). Limited molecular phylogenetic research to date has provided little resolution of internal ctenophoran relationships, though it suggests that the cydippid family Mertensiidae might be the sister group to all other ctenophores, and the orders Cydippida and Beroida might not be monophyletic. These studies have also supported the monophyly of the orders Lobata, Cestida, and Platyctenida.

(A)

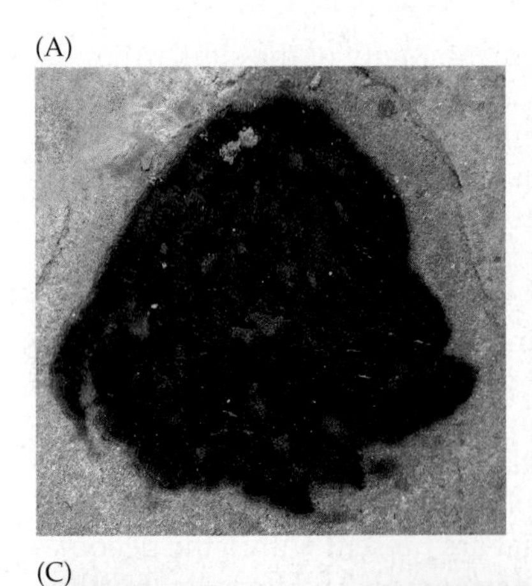

(B)

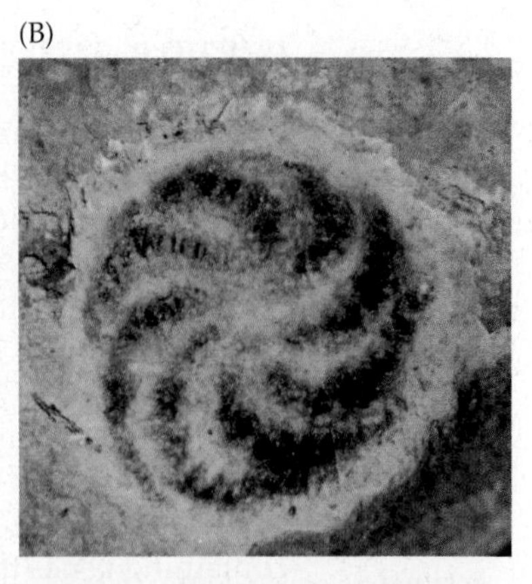

(C)

Figure 8.13 The Ediacaran fossil *Eoandromeda octobrachiata* from the Doushantuo Formation (580–551 Ma) in southern China. (A,B) carbonaceous impressions from the Institute of Geology, Chinese Academy of Geological Sciences, Bejing; (C) artist's rendering of live *Eoandromeda*.

Selected References

Abbott, J. F. 1907. Morphology of *Coeloplana*. Zool. Jahrb. Abt. Anat. Ontog. Tiere 24.

Arai, M. N. 1976. Behavior of planktonic coelenterates in temperature and salinity discontinuity layers. Pp. 211–218 in G. O. Mackie (ed.), *Coelenterate Ecology and Behavior*. Plenum, New York.

Boero, F. M. Putti, E. Trainito, E. Prontera, S. Piraino and T. A. Shiganova. 2009. First records of *Mnemiopsis leidyi* (Ctenophora) from the Ligurian, Thyrrhenian and Ionian Seas (Western Mediterranean) and first record of *Phyllorhiza punctata* (Cnidaria) from the Western Mediterranean. Aquatic Invasions 4(4): 675–680.

Carré, C. and D. Carré. 1980. Les cnidocysts du ctenophore *Euchlora rubra* (Kolliker 1853). Cah. Biol. Mar. 21: 221–226.

Carré, D., C. Rouvière and C. Sardet. 1991. In vitro fertilization in ctenophores: sperm entry, mitosis, and the establishment of bilateral symmetry in *Beröe ovata*. Dev. Biol. 147: 381–391.

Conway Morris, S. and D. H. Collins. 1996. Middle Cambrian ctenophores from the Stephen Formation, British Columbia, Canada. Phil. Trans. Biol. Sci. 351 (1337): 279–308.

Coonfield, B. R. 1936. Regeneration in *Mnemiopsis*. Biol. Bull. 71.

Costello, J. H. and R. Coverdale. 1998. Planktonic feeding and evolutionary significance of the lobate body plan with the Ctenophora. Biol. Bull. 195: 247–248.

Dawydoff, C. 1963. Morphologie et biologie des Ctenoplana. Arch. Zool. Exp. Gen. 75.

Dunn, C. W. and 17 others. 2008. Broad phylogenomic sampling improves resolution of the animal tree of life. Nature 452: 745–749.

Dunn, C. W., G. Giribet, G. D. Edgecombe and A. Hejnol. 2014. Animal phylogeny and its evolutionary implications. Ann. Rev. Ecol. Evol. Syst. 45: 371–395.

Dunn, C. W., S. P. Leys and S. H. D. Haddock. 2015. The hidden biology of sponges and ctenophores. Trends Ecol. Evol. 30(5): 282–291.

Farfaglio, G. 1963. Experiments on the formation of the ciliated plates in ctenophores. Acta Embryol. Morphol. Exp. 6: 191–203.

Franc, J.-M. 1978. Organization and function of ctenophore colloblasts: an ultrastructural study. Biol. Bull. 155: 527–541.

Freeman, G. 1976. The effects of altering the position of cleavage planes on the process of localization of developmental potential in ctenophores. Dev. Biol. 51: 332–337.

Freeman, G. 1977. The establishment of the oral–aboral axis in the ctenophore embryo. J. Embry. Exp. Morphol. 42: 237–260.

Harbison, G. R. and L. P. Madin. 1979. A new view of plankton biology. Oceanus 22(2): 18–27.

Harbison, G. R. and L. P. Madin. 1983. Ctenophora. Pp. 707–715 in S. P. Parker (ed.), *Synopsis and Classification of Living Organisms, Vol. 1*. McGraw-Hill, New York.

Harbison, G. R. 1985. On the classification and evolution of the Ctenophora. Pp. 78–100 in Morris et al. (eds.), *The Origins and Relationships of Lower Invertebrates*. Syst. Assoc. Spec. Vol. No. 28.

Harbison, G. R., L. P. Madin and N. R. Swanberg. 1984. On the natural history and distribution of oceanic ctenophores. Deep-Sea Res. 25: 233–256.

Harbison, G. R. and R. L. Miller. 1986. Not all ctenophores are hermaphrodites. Studies on the systematics, distribution, sexuality and development of two species of *Ocyropsis*. Mar. Biol. 90: 413–424.

Hernandez, M.-L. 1991. Ctenophora. Pp. 359-418 in F. Harrison and J. Westfall (eds.), *Microscopic Anatomy of Invertebrates, Vol. 2, Placozoa, Porifera, Cnidaria and Ctenophora*. Wiley-Liss, New York.

Hejnol, A. and 16 others. 2009. Assessing the root of bilaterians animals with scalable phylogenetic methods. Proc. R. Soc. B 276: 4261–4270.

Horridge, G. A. 1965. Macrocilia with numerous shafts from the lips of the ctenophore *Beröe*. Proc. Roy. Soc. London B 162: 351–364.

Horridge, G. A. 1965. Relations between nerves and cilia in ctenophores. Am. Zool. 5: 357–375.

Horridge, G. A. 1974. Recent studies on the Ctenophora. Pp. 439–468 in L. Muscatine and H. M. Lenhoff (eds.), *Coelenterate Biology*. Academic Press, New York.

Hyman, L. H. 1940. *The Invertebrates, Vol. 1, Protozoa through Ctenophora*. McGraw-Hill, New York, pp. 662–696.

Jékely, G., J. Paps and C. Nielsen. 2015. The phylogenetic position of ctenophores and the origin(s) of nervous systems. EvoDevo 6: 1–8.

Kideys, A. E. 2002. Fall and rise of the Black Sea ecosystem. Science 297: 1482–1484.

Komai, T. 1922. Studies on two aberrant ctenophores—*Coeloplana* and *Gastrodes*. Kyoto. [Published by the author.]

Komai, T. 1934. On the structure of *Ctenoplana*. Kyoto Univ. Col. Sci. Mem. (Ser. B) 9.

Komai, T. 1936. Nervous system, *Coeloplana*. Kyoto Univ. Col. Sci. Mem. (Ser. B) 11.

Komai, T. 1963. A note on the phylogeny of the Ctenophora. Pp. 181-188 in E. C. Dougherty (ed.), *The Lower Metazoa: Comparative Biology and Phylogeny*. University of California Press, Berkeley.

Komai, T. and T. Tokioka. 1940. *Kiyohimea aurita* n. gen., n. sp., type of a new family of lobate Ctenophora. Annot. Zool. Japan. 19: 43–46.

Komai, T. and T. Tokioka. 1942. Three remarkable ctenophores from the Japanese seas. Annot. Zool. Japan. 21: 144–151.

Kremer, P. 1977. Respiration and excretion by the ctenophore *Mnemiopsis leidyi*. Mar. Biol. 44: 43–50.

Kremer, P., M. F. Canino and R. W. Gilmer. 1986. Metabolism of epipelagic tropical ctenophores. Mar. Biol. 90: 403–412.

Kremer, P., M. R. Reeve and M. A. Syms. 1986. The nutritional ecology of the ctenophore *Bolinopsis vitrea*: comparisons with *Mnemiopsis mccradyi* from the same region. J. Plankton Res. 8: 1197–1208.

Link, J. S. and M. D. Ford. 2006. Widespread and persistent increase of Ctenophora in the continental shelf ecosystem off NE USA. Mar. Ecol. Prog. Ser. 320: 153–159.

Mackie, G. O., C. E. Mills and C. L. Singla. 1988. Structure and function of the prehensile tentilla of *Euplokamis* (Ctenophora, Cydippida). Zoomorphology 107: 319–337.

Madin, L. P. and G. R. Harbison. 1978a. *Bathocyroe fosteri* gen. et sp. nov., a mesopelagic ctenophore observed and collected from a submersible. J. Mar. Biol. Assoc. U.K. 58: 559–564.

Madin, L. P. and G. R. Harbison. 1978b. *Thalassocalyce inconstans*, new genus and species, an enigmatic ctenophore representing a new family and order. Bull. Mar. Sci. 28(4): 680–687.

Main, R. J. 1928. Observations on the feeding mechanism of a ctenophore, *Mnemiopsis leidyi*. Biol. Bull. 55: 69–78.

Martindale, M. Q. 1987. Larval reproduction in the ctenophore *Mnemiopsis mccradyi* (order Lobata). Mar. Biol. 94: 409–414.

Martindale, M. Q. and J. Q. Henry. 1995. Diagonal development: establishment of the anal axis in the ctenophore *Mnemiopsis leidyi*. Biol. Bull. 189: 190–192.

Martindale, M. Q. and J. Q. Henry. 1996. Development and regeneration of comb plates in the ctenophore *Mnemiopsis leidyi*. Biol. Bull. 191: 290–292.

Martindale, M. Q. and J. Q. Henry. 1997. Ctenophorans, the comb jellies. Pp. 87–111 in S. F. Gilbert and A. M. Raunio (eds.), *Embryology: Constructing the Organism*. Sinauer Associates, Sunderland, MA.

Martindale, M. Q. and J. Q. Henry. 1997. Reassessing embryogenesis in the Ctenophora: the inductive role of micromeres in organizing ctene row formation in the 'mosaic' embryo, *Mnemiopsis leidyi*. Development 124: 1999–2006.

Martindale, M. Q. and J. Q. Henry. 1999. Intracellular fate mapping in a basal metazoan, the ctenophore *Mnemiopsis leidyi*, reveals the origins of mesoderm and the existence of indeterminate cell lineages. Dev. Biol. 214: 243–257.

Matsumoto, G. I. 1991. Functional morphology and locomotion of the Arctic ctenophore *Mertensia ovum* (Fabricius) (Tentaculata: Cydippida). Sarsia 76: 177–185.

Matsumoto, G. I. and W. M. Hamner. 1988. Modes of water manipulation by the lobate ctenophore *Leucothea* sp. Mar. Biol. 97: 551–558.

Metschnikoff, E. 1885. Gastrulation und mesodermbildung der Ctenophoren. Z. Wiss. Zool. 42.

Mills, C. E. 1984. Density is altered in hydromedusae and ctenophores in response to changes in salinity. Biol. Bull. 166: 206–215.

Mills, C. 2012. Phylum Ctenophora: List of all valid species names. http://faculty.washington.edu/cemills/Ctenolist.html.

Mills, C. E. and R. L. Miller. 1984. Ingestion of a medusa (*Aegina citrea*) by the nematocyst-containing ctenophore *Haeckelia rubra* (formerly *Euchlora rubra*): phylogenetic implication. Mar. Biol. 78: 215–221.

Mills, C. E. and R. G. Vogt. 1984. Evidence that ion regulation in hydromedusae and ctenophores does not facilitate vertical migration. Biol. Bull. 166: 216–227.

Moroz, L. L. and 35 others. 2014. The ctenophore genome and the evolutionary origins of neural systems. Nature. doi: 10.1038/nature13400

Nielsen, C. 1987. *Haeckelia* (= *Euchlora*) and *Hydroctena* and the phylogenetic interrelationships of the Cnidaria and Ctenophora. Z. Zool. Syst. Evolutionsforsch. 25: 9–12.

Nosenko, T. and 12 others. 2013. Deep metazoan phylogeney: when different genes tell different stories. Mol. Phyl. Evol. 67: 223–233.

Ortolani, G. 1989. The ctenophores: a review. Acta. Embryol. Morphol. Exp. 10: 13–31.

Philippe, H. and 19 others. 2009. Phylogenomics revives traditional views on deep animal relationships. Curr. Biol. 19: 1–7.

Pianka, H. D. 1974. Ctenophora. Pp. 201–265 in A. C. Giese and J. S. Pearse (eds.), *Reproduction of Marine Invertebrates, Vol. 1*. Academic Press, New York.

Picard, J. 1955. Les nematocystes du ctenaire *Euchlora rubra* (Kolliker, 1953). Recl. Trav. Stn. Mar. Endoume-Marseille Fasc. Hors. Ser. Suppl. 15: 99–103.

Pick, K. S. and 10 others. 2010. Improved phylogenomic taxon sampling noticeably affects nonbilaterian relationships. Mol. Biol. Evol. 27: 1983–1987.

Podar, M. S., H. D. Haddock, M. L. Sogin and G. R. Harbison. 2001. Molecular phylogenetic framework for the phylum Ctenophora using 18S rRNA genes. Mol. Phylog. Evol. 21: 218–230.

Purcell, J. E. 2012. Jellyfish and ctenophore blooms coincide with human proliferations and environmental perturbations. Ann. Rev. Mar. Sci. 4: 209–235.

Rankin, J. J. 1956. The structure and biology of *Vallicula multiformis* gen. et sp. nov. a platyctenid ctenophore. Zool. J. Linn. Soc. 43: 55–71.

Reeve, M. R. and M. A. Walter. 1978. Nutritional ecology of ctenophores: a review of recent research. Pp. 249–289 in F. S. Russell and M. Yonge (eds.), *Advances in Marine Ecology, Vol. 15*. Academic Press, New York.

Robilliard, G. A. and P. K. Dayton. 1972. A new species of platyctenean ctenophore, *Lyrocteis flavopallidus* sp. nov., from McMurdo Sound, Antarctica. Can. J. Zool. 50: 47–52.

Ryan, J. F., K. Pang, NISC Comparative Sequencing Program, J. C. Mullikin, M. Q. Martindale and A. D. Baxevanis. 2010. The homeodomain complement of the ctenophore *Mnemiopsis leidyi* suggests that Ctenophora and Porifera diverged prior to the ParaHoxozoa. EvoDevo 1(9): 1–18.

Ryan, J. F., and 17 others. 2013. The genome of the ctenophore *Mnemiopsis leidyi* and its implications for cell type evolution. Science 342: 1336.

Shelton, G. A. B. (ed.). 1982. *Electrical Conduction and Behavior in "Simple" Invertebrates.* Oxford University Press, New York.

Simion, P., N. Bekkouche, M. Jager, E. Quéinnec and M. Manuel. 2014. Exploring the potential of small RNA subunit and ITS sequences for resolving phylogenetic relationships within the phylum Ctenophora. Zoology. doi: 10.1016/j.zool.2014.06.004

Stanlaw, K. A., M. R. Reeve and M. A. Walter. 1981. Growth rates, growth variability, daily rations, food size selection and vulnerability to damage by copepods of the early life history stages of the ctenophore *Mnemiopsis mccradyi*. Limnol. Oceanogr. 26: 224–234.

Stanley, G. D., Jr. and A. Sturmer. 1983. The first fossil ctenophore from the lower Devonian of West Germany. Nature 303: 518–520.

Sullivan, L. J. and D. J. Gifford. 2004. Diet of the larval ctenophore *Mnemiopsis leidyi* A. Agassiz (Ctenophora, Lobata). J. Plankton Res. 26: 417–431.

Tamm, S. L. 1973. Mechanisms of ciliary coordination in ctenophores. J. Exp. Biol. 59: 231–245.

Tamm, S. L. and S. Tamm. 1985. Visualization of changes in ciliary tip configuration caused by sliding displacement of microtubules in macrocilia of the ctenophore *Beröe*. J. Cell Sci. 79: 161–179.

Tang, F. S. Bengston, Y. Wang, X-L.Wang and C-Y. Yin. 2011. *Eoandromeda* and the origin of Ctenophora. Evol. & Develop. 13:5: 408–414.

Totton, A. 1954. Egg-laying in Ctenophora. Nature 174: 360.

Wallberg. A., A. Thollesson, J. S. Farris and U. Jondelius. 2004. The phylogenetic position of the comb jellies (Ctenophora) and the importance of taxonomic sampling. Cladistics 20: 558–578.

Welch, V., J. P Vigneron, V. Lousse and A. Parker. 2006. Optical properties of the iridescent organ of the comb-jellyfish *Beröe cucumis* (Ctenophora). Physical Review E 73: 041916.

Whelan, N. V., K. M. Kocot, L. L. Moroz and K. M. Halanych. 2015. Error, signal, and the placement of Ctenophora sister to all other animals. PNAS 112(18): 5773–5778.

CHAPTER 9

Introduction to the Bilateria and the Phylum Xenacoelomorpha

Triploblasty and Bilateral Symmetry Provide New Avenues for Animal Radiation

Along the evolutionary path from prokaryotes to modern animals, three key innovations led to greatly expanded biological diversification: (1) the evolution of the eukaryote condition, (2) the emergence of the Metazoa, and (3) the evolution of a third germ layer (triploblasty) and, perhaps simultaneously, bilateral symmetry. We have already discussed the origins of the Eukaryota and the Metazoa, in Chapters 1 and 6, and elsewhere. The invention of a third (middle) germ layer, the true **mesoderm**, and evolution of a bilateral body plan, opened up vast new avenues for evolutionary expansion among animals. We discussed the embryological nature of true mesoderm in Chapter 5, where we learned that the evolution of this inner body layer facilitated greater specialization in tissue formation, including highly specialized organ systems and condensed nervous systems (e.g., central nervous systems). In addition to derivatives of ectoderm (skin and nervous system) and endoderm (gut and its derivatives), triploblastic animals have mesodermal derivatives—which include musculature, the circulatory system, the excretory system, and the somatic portions of the gonads. Bilateral symmetry gives these animals two axes of polarity (anteroposterior and dorsoventral) along a single body plane that divides the body into two symmetrically opposed parts—the left and right sides. The evolution of bilaterality also resulted in **cephalization**, the concentration of sensory and feeding structures at a head end. Bilaterians further evolved a **complete gut** (or **through gut**), with a mouth and an anus and excretory organs in the form of protonephridia and metanephridia (except in the phylum Xenacoelomorpha, likely the most primitive living bilaterians).

As noted in Chapter 1, the oldest fossils thought to be bilaterians are from the Ediacaran period—embryos found in the Doushantuo deposits of China, dating 600–580 million years ago. The most recent molecular dating studies also suggest that the origin of the Bilateria was probably in the Ediacaran, about 630–600 million years ago, although some dated trees have estimated the origin even earlier.

Classification of The Animal Kingdom (Metazoa)

Non-Bilateria*
(a.k.a. the diploblasts)
- PHYLUM PORIFERA
- PHYLUM PLACOZOA
- PHYLUM CNIDARIA
- PHYLUM CTENOPHORA

Bilateria
(a.k.a. the triploblasts)
- **PHYLUM XENACOELOMORPHA**

Protostomia
- PHYLUM CHAETOGNATHA

SPIRALIA
- PHYLUM PLATYHELMINTHES
- PHYLUM GASTROTRICHA
- PHYLUM RHOMBOZOA
- PHYLUM ORTHONECTIDA
- PHYLUM NEMERTEA
- PHYLUM MOLLUSCA
- PHYLUM ANNELIDA
- PHYLUM ENTOPROCTA
- PHYLUM CYCLIOPHORA

Gnathifera
- PHYLUM GNATHOSTOMULIDA
- PHYLUM MICROGNATHOZOA
- PHYLUM ROTIFERA

Lophophorata
- PHYLUM PHORONIDA
- PHYLUM BRYOZOA
- PHYLUM BRACHIOPODA

ECDYSOZOA

Nematoida
- PHYLUM NEMATODA
- PHYLUM NEMATOMORPHA

Scalidophora
- PHYLUM KINORHYNCHA
- PHYLUM PRIAPULA
- PHYLUM LORICIFERA

Panarthropoda
- PHYLUM TARDIGRADA
- PHYLUM ONYCHOPHORA
- PHYLUM ARTHROPODA
  - SUBPHYLUM CRUSTACEA*
  - SUBPHYLUM HEXAPODA
  - SUBPHYLUM MYRIAPODA
  - SUBPHYLUM CHELICERATA

Deuterostomia
- PHYLUM ECHINODERMATA
- PHYLUM HEMICHORDATA
- PHYLUM CHORDATA

*Paraphyletic group

BOX 9A Characteristics of the Phylum Xenacoelomorpha

1. Soft-bodied, dorsoventrally flattened, acoelomate; almost exclusively marine worms
2. Epidermis with unique pulsatile bodies, found in no other Metazoan phylum; cilia of epidermis with distinctive arrangement of microfilaments, with the standard 9+2 arrangement extending for most of the shaft, but microfilament doublets 4–6 fail to reach the end of the cilium (the "xenacoelomorphan cilia")
3. Midventral mouth and incomplete gut (i.e., lacking an anus)
4. Largely lacking discrete organs (e.g., no discrete circulatory system, protonephridia or nephridia, or organized gonads)
5. Cerebral ganglion with a neuropil; with anterior statocysts and a diffuse intraepithelial nervous system
6. With circular and longitudinal muscles
7. Hox and ParaHox genes present (but fewer in number than in other metazoans)
8. With direct development (no larval forms)

Monophyly of the Bilateria is strongly supported by both molecular and morphological analyses. Anatomical synapomorphies of this clade include: presence of a third germ layer (mesoderm), bilateral symmetry, cephalization, and a body with both circular and longitudinal musculature—although there are prominent reversals or losses in all of these characters.

The field of molecular phylogenetics has recently provided us with better resolution of evolutionary relationships among animals. Molecular biology has also led to the expansion of a field called evolutionary developmental biology ("EvoDevo"), which attempts to understand the evolution of the molecular underpinnings of differences in the organization of animal body plans. In large part, understanding the evolution of animal body plans is about unraveling the transition from basal Metazoa (Porifera, Placozoa, Cnidaria, Ctenophora) to the Bilateria, and their subsequent radiation. However, genomic-level research has also shown that there is no simple relationship between genomic/molecular complexity and organismal/developmental complexity, so contrary to earlier assumptions the mere presence of members of conserved gene families (e.g., "segmentation genes") in an organism's genome reveals little about the evolutionary relationships of that organism to other animals. Thus, reconstructing the phylogenetic tree of the animal kingdom using conserved genes and morphological characters is a critical step. Ultimately, phylogenetics and EvoDevo together might allow us to reconstruct the nature of the first bilaterian animal—the so-called **Urbilaterian**.

The Basal Bilaterian

The notion that acoel flatworms are the most primitive living bilaterians has been around for many years. In the past, some workers have even speculated that acoels might be the most primitive living metazoans, having evolved from a ciliate protist ancestry (the "syncytial, or ciliate–acoel hypothesis"). But most biologists have favored some version of the "planuloid–acoeloid hypothesis," which postulated acoels to be the link between diploblasts (via a planula larva) and triploblasts. A third hypothesis, which gained traction for a short while, was that the simplicity of acoels was the result of *loss* of derived features from a more complex ancestor (the "archicoelomate hypothesis").

Recent molecular phylogenetic research strongly suggests that acoelomorph worms (Acoela and Nemertodermatida), perhaps along with worms of the genus *Xenoturbella*, are likely to be the oldest living bilaterian lineages, or perhaps basal deuterostomes—but either way, sharing a great many characteristics in common with an ancestral bilaterian, the so-called hypothetical "Urbilaterian." Acoelomorphs are small, direct-developing (no larval stage), unsegmented, ciliated worms. They have mesodermally-derived muscles (but no coelom, circulatory, or excretory system), multiple parallel longitudinal nerve cords and a centralized nervous system, and a single opening to the digestive cavity. Importantly, research suggests that their mouth—the single gut opening—does not derive from the blastopore (i.e., they are deuterostomous in their development). Thus, the origin of the Bilateria may have been accompanied by an embryological shift in the origin of the mouth from the blastopore (as seen in Cnidaria and Ctenophora) to elsewhere.

Many bilaterian phyla have a larval type that has been described as **primary larvae**. These are ciliated larvae with a characteristic **apical organ**—a true larval organ that disappears, in part or entirely, before or at metamorphosis. The apical organ is a putative sensory structure that develops from the most apical blastomeres during embryogenesis. It does not fit the narrow definition of a ganglion, because it seems to comprise only sensory cells. However, many spiralians develop lateral (cerebral) ganglia in close apposition to the apical organ, and this compound structure was called the "apical organ" in some of the older literature. It appears the apical organ is commonly used in larval settlement, and it is lost once a larva settles (the settlement process typically uses cells around the apical pole). In spiralian protostomes, the apical organ differentiates from the most apical cells (the $1a^1$-$1d^1$ cells in spiral-cleavage terminology). It is highly variable, and aspects of it are sometimes retained in the adult central nervous system. In the pilidium larva of nemerteans the apical organ is shed at metamorphosis together with the whole larval body, which in some cases is ingested by the emerging juvenile worm.

Deuterostomes are more difficult to interpret, and only in echinoderm and enteropneust (i.e., Ambulacraria) larvae are apical organs clearly present. Apical organs occur in cnidarian larvae, but not in Porifera, thus it has been proposed that the primary larva/apical organ could be a synapomorphy that defines a clade called Neuralia (i.e., Cnidara + Bilateria). In Cnidaria the apical organ consists of a group of monociliated nerve cells—upon settlement, the nervous system becomes reorganized and the larval nerve net is lost, with development of a new adult nerve net. Recent gene expression studies have demonstrated that the apical pole of cnidarians and the apical pole of bilaterians are probably homologous. Apical organs are apparently absent in Ecdysozoa and Chordata (except, perhaps, in the nonfeeding amphioxus larva).

Protostomes and Deuterostomes

Early in the evolution of bilaterians there was a split into two major lineages, which have long been called Protostomia and Deuterostomia. These groups were named over 100 years ago, and they were long defined on the basis of embryological principles. In protostomes the blastopore (the position in the embryo that typically gives rise to endodermal tissues) was said to give rise to the mouth ("protostome" = mouth first). Typically in deuterostomes, the blastopore gave rise to the adult anus, the mouth thus forming secondarily at a different location ("deuterostome" = secondary mouth). In both lineages, the blastopore sits at the vegetal pole of the embryo when gastrulation begins.

As molecular phylogenetic discoveries have reshuffled the animal phyla among the protostome and deuterostome lineages, a new view of embryological patterns has emerged. In the Deuterostomia (now defined as the phyla Echinodermata, Hemichordata, and Chordata), the blastopore *does* consistently give rise to the anus, and the mouth forms secondarily. But among the Protostomia, gastrulation is now known to be much more variable. In fact, we have learned that in protostomes, while the anus usually does form secondarily, the blastopore does not always give rise to the mouth, especially among animals in the large clade known as Spiralia (annelids, molluscs, nemerteans, and others). Even within the clade Ecdysozoa we now know deuterostomy can occur. For example, both nematomorphans and priapulans have deuterostomous development, with the blastopore giving rise to the anus (at the vegetal pole) and the mouth arising at the animal pole. Gene expression studies have shown that, in *Priapulus caudatus*, typical metazoan foregut and hindgut gene expressions accompany this development, and the hindgut/posterior markers *brachyury* (*bra*) and *caudal* (*cdx*) are expressed as the anus emerges from the blastopore. Continuing developmental work on crustaceans is revealing that most species probably also express a form of deuterostomy. And in Chaetognatha gastrulation occurs by invagination of the presumptive endoderm, leaving no blastocoel—the blastopore marks the eventual posterior end of the animal, and both mouth and anus form secondarily, thus also a deuterostome-like development. In fact, evidence is accumulating that mouth formation from oral ectoderm (in the animal hemisphere), typical of deuterostomy, may be ancestral in both protostomes and deuterostomes, and perhaps in Bilateria itself.

So we see that, although the names Protostomia and Deuterostomia are still used for the two main clades of Bilateria, the names themselves are no longer perfectly descriptive—they are **legacy names**. It has been suggested that new names should be coined for these two large clades, but as yet there has been no agreement on what these names might be. The largest animal phyla belong to Protostomia—Arthropoda (over a million described living species) and Mollusca (nearly 80,000 described living species)—as do the smallest animal phyla (Micrognathozoa and Placozoa, one described species each; Cycliophora, two described species), although several undescribed species are known to exist in these small phyla.

Today, the groups Protostomia and Deuterostomia constitute clades based mostly on molecular phylogenetic evidence, and morphological and developmental synapomorphies defining these two clades remain ambiguous. A probable synapomorphy of the Protostomia, as it is now constituted, is a central nervous system with a dorsal cerebral ganglion that usually has circumesophageal connectives to a pair of ventral nerve cords. Probable synapomorphies of the Deuterostomia are a trimeric body coelom condition and pharyngeal gill slits, at least primitively (trimery is lacking in the phylum Chordata, gill slits are absent in extant echinoderms but may have been present in some extinct, basal echinoderms). Although still somewhat controversial, the position of the phylum Xenacoelomorpha (acoels, nemertodermatids, and *Xenoturbella*) appears to be basal within Bilateria, this group not aligning strongly with either protostomes or deuterostomes (Box 9A).

The Protostomia contains 24 phyla, five of which still remain enigmatic in terms of their phylogenetic alignment: Chaetognatha, Platyhelminthes, Gastrotricha, Rhombozoa, and Orthonectida. Some molecular evidence suggests all of these but Chaetognatha probably belong in the clade known as Spiralia, and in fact Platyhelminthes and Rhombozoa do seem to show spiral cleavage (and some evidence suggests

Chaetognatha might also have spiral cleavage). Recent studies have suggested that Chaetognatha may be the sister group to Spiralia. Gastrotrichs have a unique, but non-radial embryogenesis, and the embryology of Orthonectida, Cycliophora, and Micrognathozoa is not yet known. Bryozoa and Brachiopoda clearly do not have spiral cleavage. Thus, we do not yet know if spiral cleavage is a synapomorphy of the clade that bears its name—i.e., "Spiralia" is another legacy name. It may eventually be shown that all of these phyla comprise a single clade and are descendants of a spirally-cleaving ancestor, making this cleavage pattern a valid synapomorphy for the group known as Spiralia. If spiral cleavage does prove to be a synapomorphy for Spiralia, its absence in some phyla would be viewed as the product of secondary modifications to the embryological process. Recall that in spiral cleavage, the 4d cell (also known as the mesentoblast) gives rise to most of the mesoderm, called endomesoderm. Most spiralians also generate some mesoderm from micromeres of the second or third quartet that are primarily responsible for ectoderm formation (thus it is called ectomesoderm); this commonly gives rise to larval musculature.

Some spirally-cleaving animals have a unique larval type, called the trochophore larva (e.g., Mollusca, Annelida, Nemertea, and possibly some others), and the clade name "Trochozoa" has been proposed for those phyla, although this clade gets very mixed support in molecular trees and might be paraphyletic. In addition, recent phylogenomic work suggests that these "trochophore phyla" may comprise a sister group to the Lophophorata (Phoronida, Bryozoa, Brachiopoda), and perhaps also including Entoprocta, as a larger grouping known as the Lophotrochozoa. Although DNA sequence data support the clade Spiralia, no unambiguous morphological synapomorphies that might define it have been identified. The phylogenetic relationships of the spiralian phyla remain to be sorted out, and so far their deep ancestry has defied clear resolution. However, two clades within Spiralia do seem to be well supported, and we treat these as chapters in this book; these are the clades Gnathifera (phyla Gnathostomulida, Micrognathozoa, Rotifera) and Lophophorata (phyla Phoronida, Bryozoa, Brachiopoda).

The other main protostome clade, Ecdysozoa, contains 8 phyla (and about 80% of animal species diversity) that all molt their cuticle at least once during their life history. The Ecdysozoa comprise three well-supported clades: Panarthropoda (Tardigrada, Onychophora, Arthropoda), Nematoida (Nematoda, Nematomorpha), and Scalidophora (Priapula, Kinorhyncha, Loricifera), the latter supported mostly by morphological data. The phylogenetic relationships of these three clades have not yet been determined, so they appear as an unresolved trichotomy in our tree of the Metazoa (Chapter 28). Morphological evidence suggests that Nematoida and Scalidophora are sister groups, and they share a number of morphological similarities (e.g., a circumoral collar-shaped brain composed of a ring neuropil with anterior and posterior concentrations of cell bodies). However, these morphological similarities might be plesiomorphic within Ecdysozoa, and most molecular analyses place Nematoida as a sister group to Panarthropoda. The internal relationships of the scalidophoran phyla also remain unclear. The most recent work on Panarthropoda suggests Onychophora may be the sister group to Arthropoda, and Tardigrada the sister group to those. Unlike the Spiralia, the Ecdysozoa can be defined by unambiguous morphological synapomorphies, including their three-layered cuticle that can be molted, a process regulated by ecdysteroid hormones in those groups where this is known. The cuticle consists of a proteinaceous exocuticle and an endocuticle with chitin or collagen, with the epicuticle forming from the apical zone of the epidermal microvilli. Ecdysozoans also lack external epithelial cilia, lack a primary larva or ciliated larva, and none of them has spiral cleavage. This clade was discovered in one of the first, pioneering studies using molecular sequence data (Aguinaldo et al. 1997).

The other great bilaterian clade, Deuterostomia, is quite small, comprising fewer than 100,000 living species in only three phyla: Echinodermata, Hemichordata, and Chordata. Although only a "side-branch" in the tree of life, we tend to give this clade exaggerated importance because, of course, it is the lineage to which we humans (and other vertebrates) belong. As noted above, this clade was originally defined largely on the basis of deuterostomous embryology. However, we now know that deuterostomous development occurs throughout its sister lineage, the Protostomia, leaving us with few definitive morphological or developmental features defining the Deuterostomia. However, as noted above, a trimeric body coelom and pharyngeal gill slits, at least primitively, may eventually be proven to be synapomorphies for the Deuterostomia. Deuterostomes also appear to possess a unique developmental gene, called *Nodal*.

Within the deuterostome clade, recent morphological and molecular work (and also Hox gene motifs) suggests that echinoderms and hemichordates are sister groups, constituting a clade called Ambulacraria, and this is the sister group to the phylum Chordata. If this assessment is correct, it means the features shared between chordates and hemichordates (long thought to comprise a sister group), such as gill slits, may have indeed been ancestral within Deuterostomia, but lost in the echinoderm line (and also in some hemichordate lineages), as suggested by the putative presence of gill slits in some extinct echinoderms. Gill slits in Deuterostomia have been shown to be homologous based on their gene expression patterns. Several deuterostome animals with gill slits are known from the fossil record, although it is not yet certain whether these belong to basal urochordates, to basal echinoderms, or to their own extinct lineages. Another feature shared between the Hemichordata and Chordata is the stomochord/notochord, long viewed as homologues. It is now thought that these structures might have had much earlier origins and may or may not be homologous, or that a group of vacuolated cells in ancestral Deuterostomia gave rise to these structures independently in hemichordates and chordates. Within Chordata, Urochordata is the sister group to Vertebrata (a clade known as Olfactores), and Cephalochordata is the sister group to those. There is some evidence that a fourth group, the genus *Xenoturbella* (or even the whole clade Xenacoelomorpha) might be near the base of the deuterostome line, but opposing evidence suggests *Xenoturbella* is more likely allied with the Acoelomorpha as an ancestral bilaterian clade, the view we follow in this book.

Deuterostomia is an ancient lineage, and dated phylogenetic trees (using fossils to date branching points) suggest the ancestral line existed well into the Precambrian. The oldest definitive fossil of Deuterostomia is a 530-million-year-old creature called *Yunnanozoon*, from the lower Cambrian Chengjiang biota of Yunnan Province, China, although the affinities of yunnanozoans are still uncertain.

The classification of Metazoa used in this book is shown in the box at the start of Chapters 6 through 27. You will notice that phyla are listed under clade names (most of which lack formal nomenclatural ranking). You will also notice that within these clades, there is often little phylogenetic structure indicated. This is because much of the branching pattern of the tree of life still remains to be discovered. Genomic data are still lacking for many groups, and in other cases data are available for only one or two species. Expanded taxonomic sampling, additional genomic data sets, and new analysis techniques should resolve the remaining questions of animal phylogeny over the coming decade.

Phylum Xenacoelomorpha

The acoels and the nemertodermatids have had a long journey. They were initially viewed as the most primitive living platyhelminths (true flatworms), due to their simple anatomy, and in fact, were thought by many to be the most primitive living Bilateria because most workers placed the phylum Platyhelminthes at the base of the bilaterian tree. As ultrastructural work revealed increasing complexity, opinion shifted, and from the 1960s to the turn of the century these worms, together known as the Acoelomorpha, were widely regarded not as primitive, but as secondarily reduced platyhelminths. However, as multigene phylogenetic analyses have begun to explore these small soft-bodied worms, it has become evident that they are indeed primitive bilaterians (perhaps diverging even before the protostome–deuterostome split), and not members of the phylum Platyhelminthes at all. Thus, the pendulum has swung 180 degrees, as is known to happen in phylogenetics. A growing knowledge base and new technologies can lead to major shifts in our understanding of life. In addition, molecular phylogenetics has shown a close relationship between the acoel and nemertodermatid worms, which is further supported by their unique ciliary rootlet system, perhaps the early cleavage pattern (i.e., the horizontal orientation of the second, asymmetric cleavage plane), and several other features described below.

Even more recently, another genus of small marine worms, *Xenoturbella*, was found to be allied closely with the Acoelomorpha, and a new phylum name was created to house these three worm groups—Xenacoelomorpha. The phylum currently contains about 400 species, two in the subphylum Xenoturbellida and 398 in the subphylum Acoelomorpha (mostly in the class Acoela). All described species are small, flattened, marine worms with an incomplete digestive system (i.e., lacking an anus) and lacking discrete excretory systems (however, there is an undescribed xenoturbellid species reported to be several centimeters in length).

DNA sequence analyses have suggested that Acoelomorpha are basal bilaterians and are likely the sister group of Xenoturbellida. Analyses have been divided on whether Xenoturbellida are deuterostomes or basal bilaterians, but the latter idea seems to have stronger support. However, the high evolutionary rate of analyzed genes in Acoelomorpha might be creating long-branch attraction problems and further studies are needed. Thus, although we recognize the phylum Xenacoelomorpha, and treat Acoelomorpha and Xenoturbellida as subphyla, it is possible that these two groups will eventually again be separated, with Acoelomorpha being placed at the base of the Bilateria, and Xenoturbellida within the Deuterostomia. We discuss each of the three curious worm groups (Acoela, Nemertodermatida, Xenoturbellida) separately below.

In addition to the molecular phylogenetic data that support an Acoela–Nemertodermatida sister group relationship, both groups have unique epidermal bodies that represent degenerating ciliated cells, the **pulsatile bodies** (and a type of pulsatile body also occurs in the xenoturbellids). These epidermal bodies are unknown from any other metazoan phylum. In Acoela, the cilia are retained in vacuoles prior to digestion, whereas in nemertodermatids the cilia appear to be lost before resorption begins. The musculature of acoels and nemertodermatids is also strikingly similar, yet different in some key aspects; acoels have a grid of orthogonal musculature with mainly ventral diagonal musculature, and a muscular posterior pharynx in what may be basal species. More derived acoels have more complex layers of diagonal muscles. Nemertodermatids seem to have an orthogonal grid and well-developed diagonal muscles throughout the body, but no evidence of a muscular pharynx. These anatomical features are described below.

In addition to pulsatile bodies, both Acoela and Nemertodermatida (and *Xenoturbella*) lack discrete excretory systems, the presence of which unites all other Bilateria, and their cerebral ganglion has a neuropil (i.e., it can be considered a true brain, but see below). Furthermore, they share a unique pattern of neurotransmitter activity, body-wall musculature, and mode of embryonic development. Hox and ParaHox genes are present in both groups, although these are not strictly similar. Both taxa appear to have the beginning of the extended central Hox set.

Although initially considered to be a turbellarian flatworm, the unusual anatomy of *Xenoturbella bocki* quickly distinguished it from platyhelminths, as well as from the Acoelomorpha. Phylogenetic (and even some morphological) studies initially linked *Xenoturbella* to deuterostomes. Sequences of Hox genes in *X. bocki* also suggested it could be a basal deuterostome with a reduced Hox gene complement. Additional work using the entire mitochondrial genome of *Xenoturbella* showed links with deuterostomes. However, the lack of typical deuterostome characteristics suggested that *Xenoturbella* might belong at the very base of the deuterostome tree. Other phylogenetic analyses, including nuclear genes from *X. bocki*, also suggested that *Xenoturbella* might be closely tied to the clade known as Ambulacraria (Echinodermata and Hemichordata). If these relationships are correct, developmental evidence of structures common to other Ambulacraria should exist, including gill slits, endostyle, and enterocoelic coelom formation. However, such evidence has not been found (although studies have been frustrated by the fact that *Xenoturbella* ova are very yolky, which obscures observation of early cleavage).

By 2009, large-scale molecular phylogenetic studies had begun to move *Xenoturbella* even further down the animal tree, suggesting it is sister to the Acoelomorpha (Acoela + Nemertodermatida), at the base of the Bilateria. The anatomical data seemed to agree with this linking, and it was eventually suggested that the three groups together warranted phylum status, the Xenacoelomorpha. Acoels have only three Hox genes (one each of the anterior, central, and posterior groups). Nemertodermatids have only two (a central and a posterior group). *Xenoturbella* has one anterior, two (or three) central, and one posterior gene. Platyhelminths, on the other hand, have an almost complete Hox cluster. The most recent phylogenetic studies on Acoelomorpha and *Xenoturbella* are still conflicting, plagued by long-branch attraction and small taxon sampling issues. Although we accept the phylum Xenacoelomorpha in this edition of *Invertebrates*, we recognize that the relationships of these three worm taxa are still subject to modification.

CLASSIFICATION OF PHYLUM XENACOELOMORPHA

Generally small, flattened or cylindrical, acoelomate marine worms with anterior statocysts, diffuse intraepithelial nervous system, midventral mouth, incomplete gut (i.e., lacking an anus), unique pulsatile bodies (unknown from any other Bilateria), and largely lacking discrete organs (e.g., without a discrete circulatory system, nephridia, or organized gonads). Cilia of epidermal cells with distinctive arrangement of microfilaments wherein the standard 9+2 arrangement extends for most of the ciliary shaft, but toward the end, microfilament doublets 4 through 7 end, leaving doublets 1–3 and 8–9, which continue to the end of the cilium. These **xenacoelomorphan cilia** are not known in any other animal phylum (although very similar cilia have been described from the pharynx of some enteropneust hemichordates). With both circular and longitudinal muscles. With direct development and no distinct larval forms. Two subphyla, Acoelomorpha and Xenoturbellida.

SUBPHYLUM ACOELOMORPHA The union of Acoela and Nemertodermatida as sister taxa is based on molecular phylogenetic evidence, as well as anatomical data. Both groups: lack discrete excretory systems (present in all other Bilateria), have cerebral ganglia with a neuropil, share a unique pattern of neurotransmitter activity and unique body-wall musculature, and go through a distinctive mode of embryonic development. Hox and ParaHox genes are present in both groups, although these are not strictly similar.

CLASS ACOELA Acoels lack a permanent digestive cavity. The pharynx, when present, is simple, leading to a solid syncytial or cellular endodermal mass. With a unique anterior statocyst containing one statolith, and biflagellate sperm with 2 flagella whose axonemes are incorporated into the sperm cell; endolecithal ova; without epithelial basal lamina, or discrete excretory or circulatory systems. Small (1–5 mm) worms, common in marine and brackish-water sediments; a few are planktonic or symbiotic. (e.g., *Actinoposthia, Amphiscolops, Antigonaria, Conaperta, Convoluta, Convolutriloba, Daku, Diopisthoporus, Eumecynostomum, Haplogonaria, Hofstenia, Isodiametra, Myopea, Oligochaerus* [with freshwater species], *Paratomella, Philactinoposthia, Polychoerus, Praesagittifera, Proporus, Solenofilomorpha, Symsagittifera, Waminoa*)

CLASS NEMERTODERMATIDA Interstitial or endosymbiotic marine worms possessing a ciliated, glandular epidermis and an anterior statocyst generally containing two statoliths; a proboscis with extensible filaments is present in some species; mouth may be present or absent; pharynx never present; gut cavity with small and relatively occluded, but with true epithelium and gland cells; uniflagellate sperm; with endolecithal ova; with limited basal lamina beneath the epidermis. One genus (*Meara*) contains species that are symbionts in sea cucumbers. (e.g., *Ascoparia, Flagellophora, Meara, Nemertinoides, Nemertoderma, Sterreria*)

SUBPHYLUM XENOTURBELLIDA Two described species, *Xenoturbellida bocki* and *X. westbladi* (but others are known to exist, and the species-level differences between the two described species have been questioned). Despite its simple body plan, *X. bocki* is a relatively large worm, reaching 4 cm in length, and some undescribed species may exceed that size. Xenoturbellids have a humplike structure in the anterior third of the body but lack other structural organs (other than a statocyst) and possess only a diffuse nervous system. These worms live in holes on sandy coastlines or deeper offshore muds and are specialized to eat molluscs.

Class Acoela

Acoels are mostly minute, marine or brackish-water, sediment- or surface-dwelling worms. They range in size from less than a millimeter to about a centimeter in length. Those inhabiting interstitial habitats are generally long and slender, whereas those inhabiting surfaces tend to be more disc shaped, broad, and flat. Swimming species are cylindrical with tapered ends, or occasionally enrolled sides. Epiphytic species are usually cone shaped with ventrally enrolled sides that may give the appearance of trailing "fins." A few species of acoels have also been found in the gut of echinoderms, in fresh water, and in hydrothermal vents. (Figures 9.1A–H)

Acoela lack a distinct internal body cavity—they are acoelomate (as are the other members of the phylum Xenacoelomorpha). Acoels also lack a structural gut, and this was actually the basis of the name Acoela. Instead, they possess a multinucleated mass (a syncytium) that phagocytizes ingested food particles (Figure 9.2). Larger species often supplement their nutritional requirements with endosymbiotic algae, which can contribute to the bright coloration seen in many (Figure 9.3A). Acoels living in the guts of other animals often have symbiotic bacteria inhabiting their epidermis.

Acoels possess both circular and longitudinal muscles. Their nervous system consists of an array of paired longitudinal nerve cords with a concentration of anterior sensory cells and a cerebral commissure (the "brain") (Figure 9.4). The anterior statocyst with a single statolith is distinctive in acoels and (along with simple, light sensitive eyes in a few species) appears to assist in maintaining the animal's orientation (Figure 9.1A–H). They lack sclerotized structures other than those associated with genitalia, although some species manufacture crystalline spicules in the parenchyma. They also possess aberrant, complex, biflagellate sperm that vary in the structure of the usual 9+2 arrangement of microtubules possessed by many metazoans. Acoels have direct development and exhibit no distinct larval forms.

Acoels were first described at the turn of the nineteenth century from northeast Atlantic coastlines. These and other early descriptions placed the Acoela within the turbellarian Platyhelminthes, and distinguished major subtaxa on the basis of the female reproductive system. Later revisions in the middle of the twentieth century established over 20 families, and most of the nearly 400 described species were based primarily on details of male copulatory structures. Similarities in internal anatomy, epidermal ciliation, and the appearance of epidermal "pulsatile bodies" led to combining Acoela with another turbellarian group, Nemertodermatida, as the Acoelomorpha.

The lack of hard anatomical features in these worms led workers to studies of microscopic ultrastructure using scanning and transmission electron microscopy, including investigations of muscle fiber orientation and structure (which distinguished several major lineages), sperm morphology, and spermatogenesis (which identified biflagellate sperm and unusual patterns of microtubules within sperm acrosomes), as well as neuroanatomy. Studies increased in number near the end of the twentieth century as the diversity of habitats investigated increased, including anoxic sulfide sands. 18S and 28S rRNA, mitochondrial DNA, and myosin heavy chain type II nucleotide sequences have all placed Acoelomorpha outside of the Platyhelminthes. Further systematic refinements within major acoel clades (notably the polyphyletic family Convolutidae), and developmental analyses, have corroborated genetic results that place acoels outside the Platyhelminthes. Much taxonomic revision is still underway, and about 9 to 20 families are currently recognized, depending on whose schema is followed.

Both molecular phylogenetics and EvoDevo research provide evidence that acoels likely lie at the base of the bilaterian tree. For example, the pattern of expression of *ClEvx* (a gene responsible for sensory specificity brain neurons) anterior and posterior to the statocyst in hatchling acoels is more similar to that found in cnidarians than it is to more derived bilaterians. Other studies indicate that *brachyury* (*bra*) and *goosecoid* (*gsc*), genes associated with the formation of the acoel mouth, are also expressed during mouth development in protostomes as well as deuterostomes, suggesting that acoel and bilaterian mouths are homologous. Studies of neural development and structure in the acoel *Symsagittifera* show that genes associated with brainlike structures are present, suggesting that such genetic machinery was in place in the Urbilaterian ancestor (if indeed acoels represent such an ancestor). The overall primitiveness of Acoela appears to also be reflected in their lack of a clearly differentiated gut or

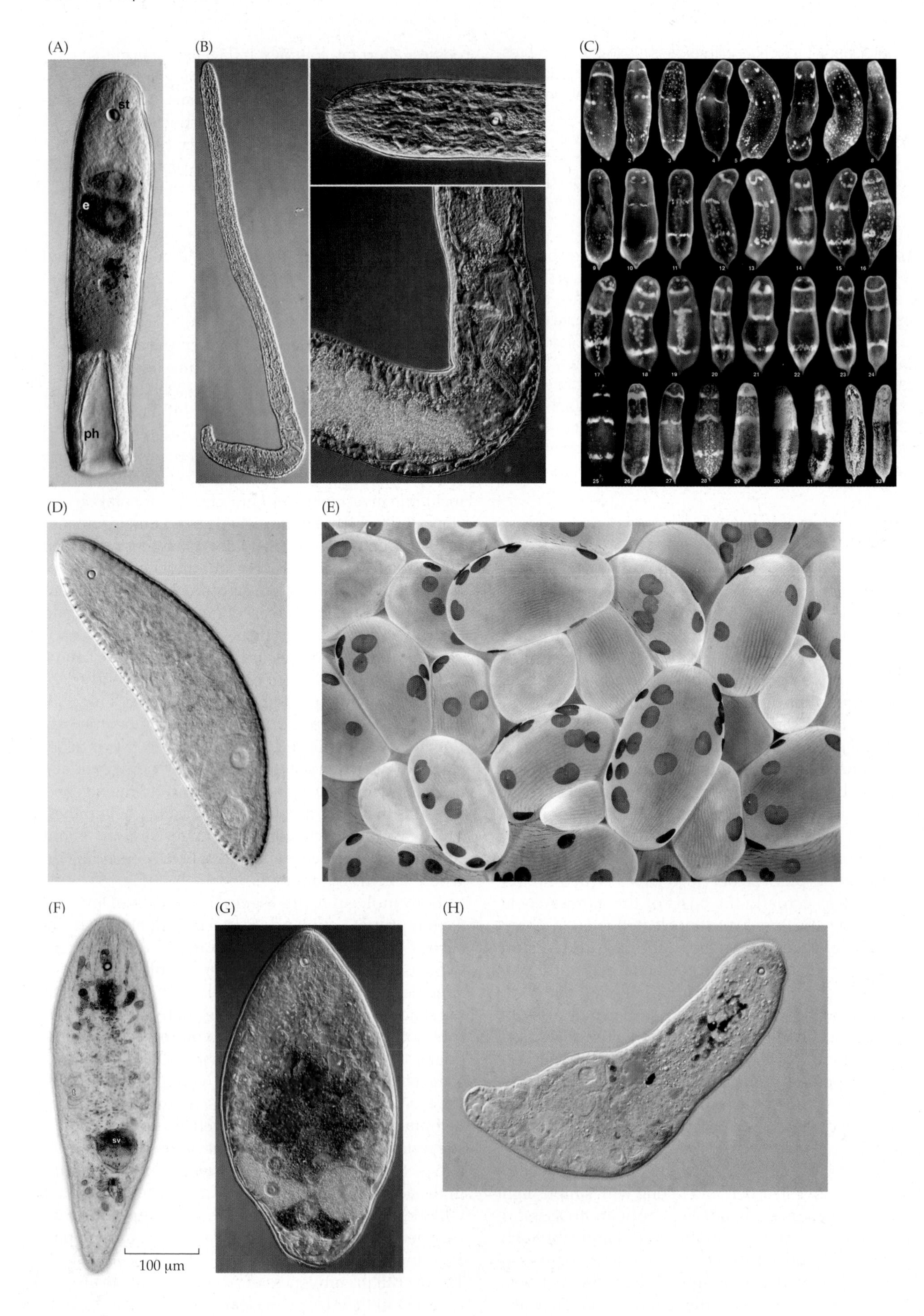
(A)
st
e
ph
(B)
(C)
(D)
(E)
(F)
sv
100 μm
(G)
(H)

◀ **Figure 9.1 Acoela.** (A) *Diopisthoporus lofolitus* (Diopisthoporidae). (B) *Paratomella rubra* (Paratomellidae). (C) Color variation in 33 specimens of *Hofstenia miamia* (Hofsteniidae) from the Caribbean. (D) *Philactinoposthia novaecaledoniae*, living specimen (Dakuidae). (E) *Waminoa* sp. (Convolutidae) on bubble coral (*Plerogyra sinuosa*). (F) *Daku riegeri* (Dakuidae). (G) *Eumecynostomum evelinae* (Mecynostomidae). (H) *Paramecynostomum diversicolor*.

excretory system, unencapsulated gonads, absence of ciliary or rhabdomeric eyes, lack of a basal lamina under the epidermis, and absence of a larval stage.

The Acoel Body Plan

Body Wall and External Appearance

Most acoels are tiny, just a few millimeters long. The smallest species tend to be interstitial, feeding on bacteria and organic particulates available on the surfaces or between the spaces of the sediments they inhabit. Infaunal species tend to be more elongate (Figure 9.1A,B,F,H). Larger species usually inhabit the surfaces of rocks, large algae, or cnidarians (Figure 9.1C–E,G). These more rapidly moving species are often predatory, gliding quickly on their ciliated surfaces and capturing prey with a raptorial "hood" that consists of lateral extensions of the body.

Large-bodied species, reaching lengths of 4–5 m, in some families (e.g., Convolutidae) have anterior ocelli (Figure 9.1G), whereas small-bodied species tend to lack these. Larger-bodied species also often have photosynthetic endosymbionts under their epidermis (Figures 9.1E and 9.3A–E). Endosymbiotic algae are contained within the bodies of many large species of acoels, and this association probably evolved more than once—both zoochlorellae and zooxanthellae have been identified among the various species. Algae are usually obtained during feeding by juvenile worms, but in some species can also be transmitted within oocytes by parents to their offspring (vertical transmission). In *Heterochaerus langerhansi*, the dinoflagellate *Amphidinium klebsii* resides below the epidermis and has been shown, using radioactively labeled carbon and nitrogen, to receive these substances from its host in the forms of CO_2 from respiration and excreted ammonia from protein metabolism. The rate of transfer is light dependent.

It has been suggested that acoel body pigmentation may have several contexts. One might be to provide protection from UV radiation for their symbiotic photosynthetic protists. A second context might be to provide cryptic coloration, either to make acoels inconspicuous to visual predators, or possibly to make them less visible to prey. *Hofstenia* species, also known as "panther worms," are highly polymorphic in dorsal pigmentation, with diverse patterns of dappling and striping of brown, yellow, and white colorations (Figure 9.1C).

The acoel body is completely covered with cilia, which may or may not also line the mouth and entrances to reproductive structures. The epithelium lacks a basal lamina (extracellular matrix, or ECM). Early researchers identified **pulsatile bodies** embedded within the epidermis of acoels (and nemertodermatids), which later proved to be clumps of ciliated cells in the process of being resorbed and replaced by the epidermis.

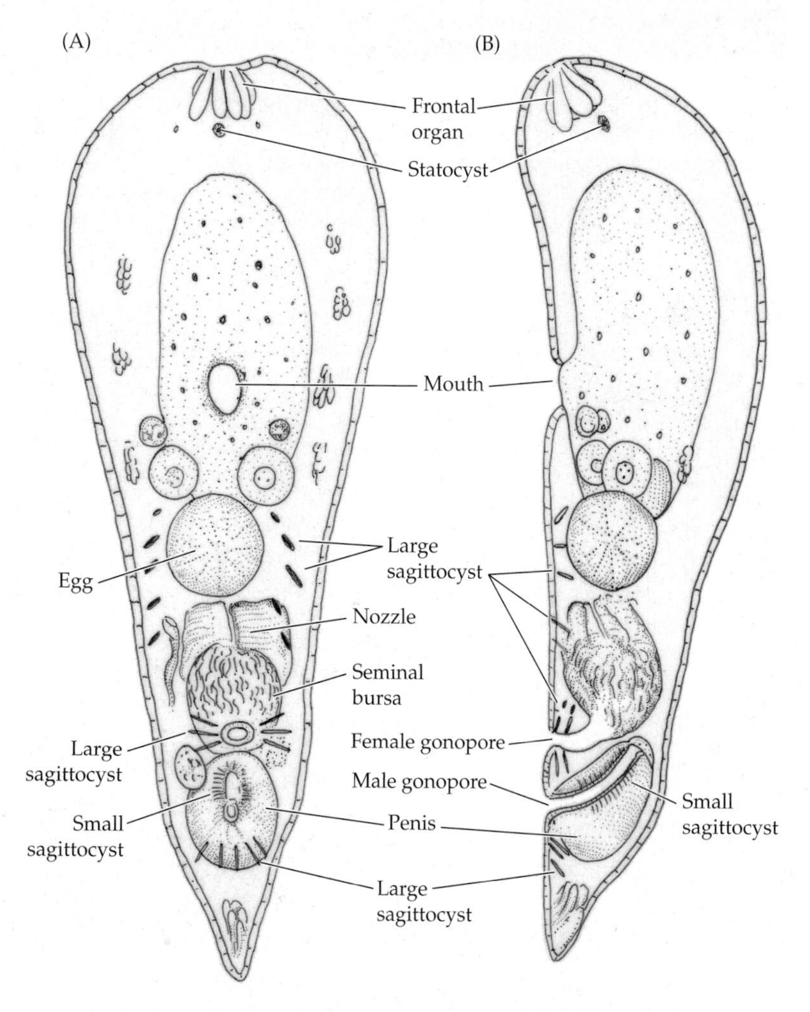

Figure 9.2 The anatomy of *Praesagittifera shikoki* (Acoela). (A) dorsal view, (B) lateral view.

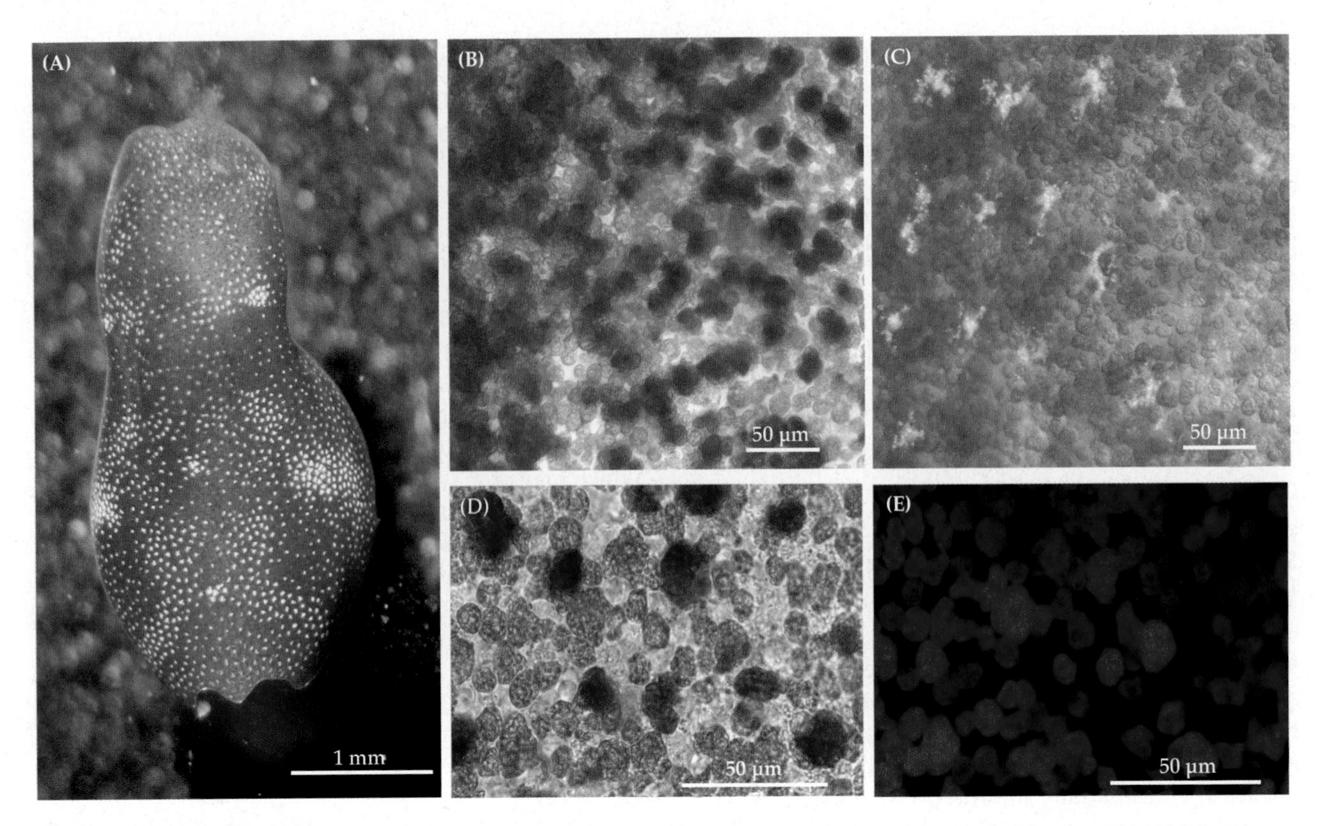

Figure 9.3 Bright coloration in *Convolutriloba longifissura* (Acoela). (A) Whole body (dorsal view). (B–E) Close-up views of the dorsal surface of *C. longifissura* showing endosymbiotic algae (B,D) transmission light; (C) incident light; (E) epifluorescent light (blue excitation). Note that (B,C) and (D,E) are paired images.

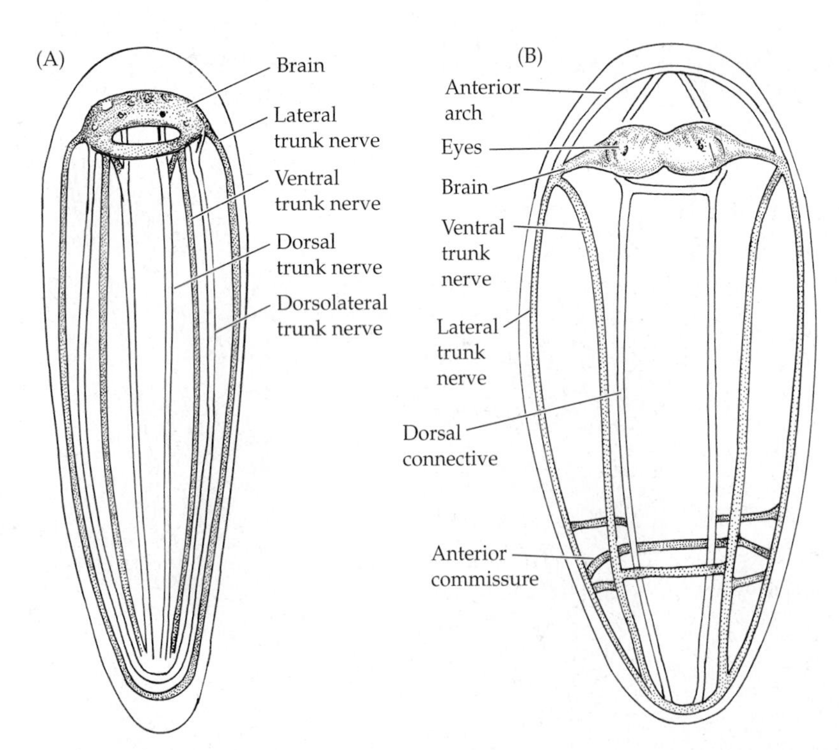

Figure 9.4 Comparison of the central nervous systems of (A) an acoel (*Actinoposthia beklemishevi*) and (B) a free-living flatworm (*Gievzstoria expedita*).

Mucus-producing **frontal organs**, that superficially resemble those of flatworms but are probably not homologous in structure, occur in most families. The ciliated epidermis of acoels also bears **rhabdoid glands** distributed over the body. The **rhabdoids** themselves are composed of mucopolysaccharides and are chemically as well as structurally distinct from the rhabdites of free-living flatworms, although their role in producing mucus to assist ciliary gliding appears to be similar. Most members of the family Sagittiferidae also have **sagittocysts** (Figure 9.5), complex needle-shaped secretory products (5–50 μm long) that are ejected with force in prey capture or for defense, and probably also to assist in sperm transfer during copulation (perhaps by perforating the partner's epidermis). Each sagittocyst arises from a **sagittocyte**, which is surrounded by tightly spiraled muscle filaments that expel the sagittocyst upon contraction (Figure 9.6).

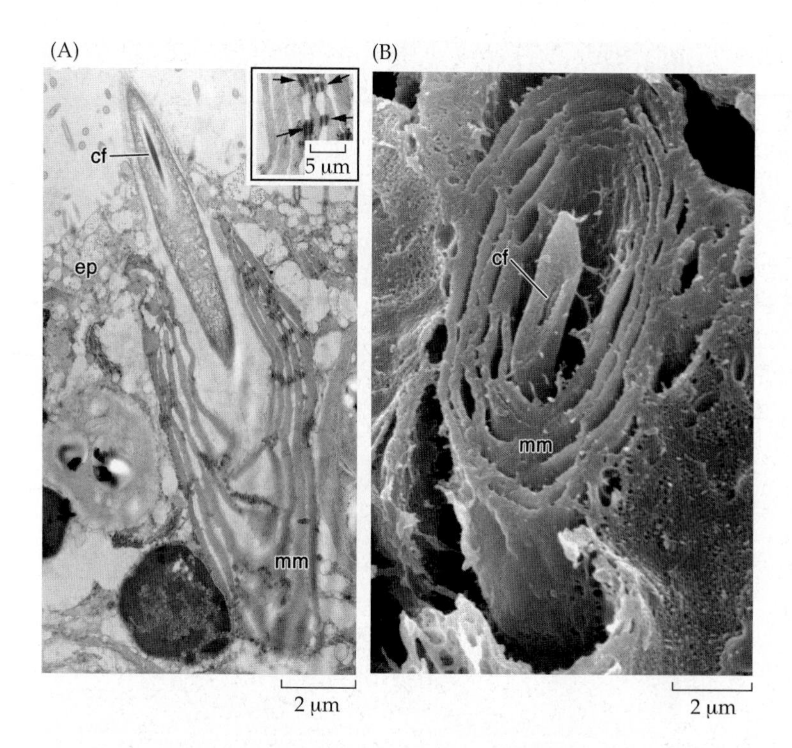

Figure 9.5 TEM images of a *Convolutriloba longifissura* (Acoela) sagittocysts. (A) A sagittocyst (cf = central filament of sagittocyst) during extrusion from the muscle mantle (mm) and penetration of the epidermis (ep). Inset shows higher magnification of the muscle mantle; arrows indicate the location of desmosomes linking the mantle layers. (B) Close-up of the cut surface of a sagittocyst within the muscle mantle.

The position of the mouth in acoels is highly variable. In families thought to be primitive, the mouth opens at the posterior end of the animal and leads to a distinct pharynx (Figure 9.1A). Other families have anteroterminal mouths, although most acoel mouths open midventrally (Figures 9.2 and 9.7B). Both a circulatory system and an excretory system, even in the form of protonephridia, are lacking in the Acoela. Male and female reproductive organs are visible through the body wall of smaller acoels (Figures 9.1A,F,H). In larger species they may protrude from the body surface (Figures 9.7D, 9.12B).

Body Musculature, Support, and Movement

The mesodermally derived musculature of acoels provides the primary means of support, whereas the body cilia (assisted by body muscles) provide for their gliding movement. The shape of the cilia is distinctive, having a marked shelf at the tip where doublets 4–7 terminate. The rootlet system that connects the cilia is also unique. Two lateral rootlets project from each cilium and connect to the tips of the adjacent cilia. From a caudal rootlet, two bundles of fibers project to join the kneelike bend of those same adjacent rootlets. Epidermal cilia of acoels beat in a coordinated fashion to create metachronal waves that move from anterior to posterior.

Abundant dorsoventral muscles serve to flatten the body, and muscles in the body wall generate bending, shortening, and lengthening movements (Figures 9.7A–E and 9.8A–G). The body wall musculature includes circular, diagonal, longitudinal, crossover,

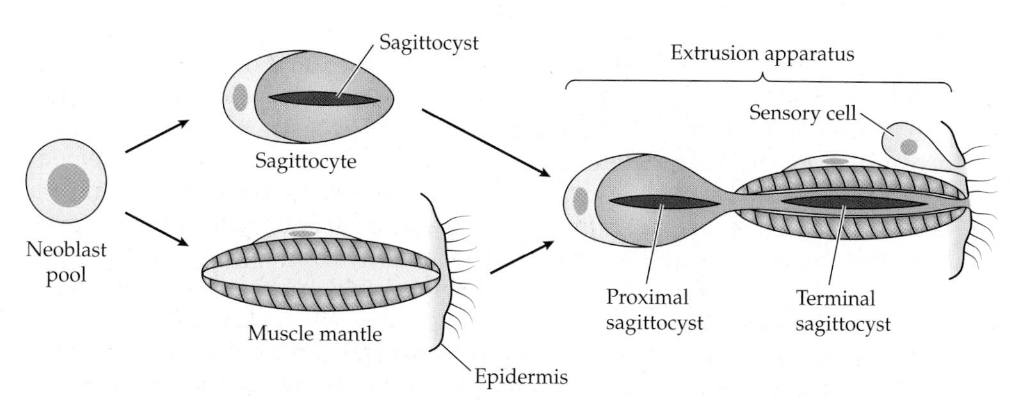

Figure 9.6 Formation and differentiation of sagittocytes and their muscle mantle from neoblast cells in Acoela. See text for description.

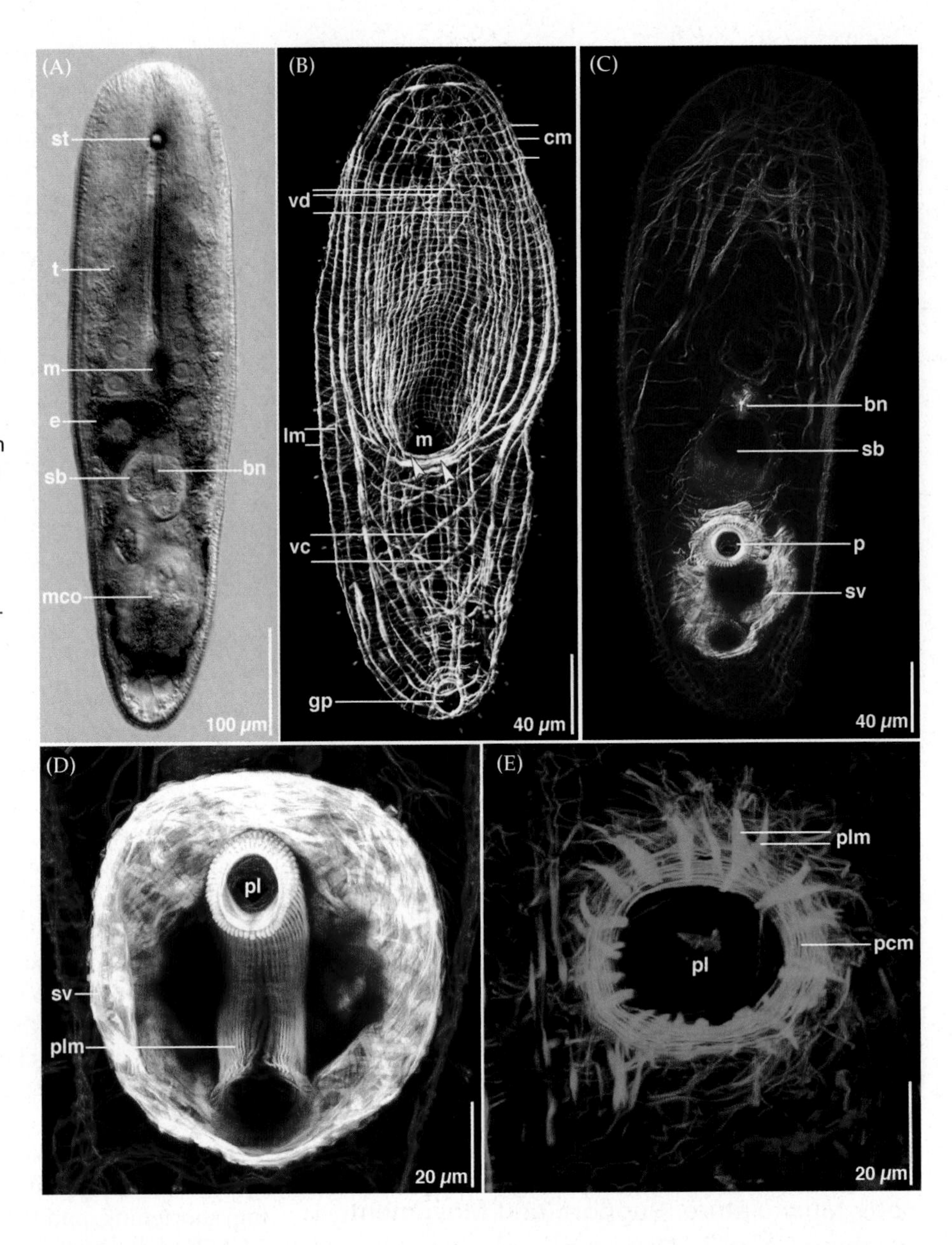

Figure 9.7 Musculature of acoels. (A) Whole mount of living specimen of *Isodiametra earnhardti*. (B) Ventral body-wall musculature of *Haplogonaria amarilla*. (C) Parenchymal musculature of *I. divae*, showing portions of copulatory organs. (D) Male copulatory organ of *I. divae*, showing musculature of seminal vesicle and invaginated penis. (E) Penis musculature of *Convoluta henseni*. Projections of musculature in whole-mount specimens of acoels stained with Alexa-488-labeled phalloidin and viewed using CLSM. bn: bursal nozzle; cm: circular muscle of body wall; e: egg; gp: gonopore; lm: longitudinal muscle of body wall; m: mouth; mco: male copulatory organ; p: penis; pcm: circular muscle of penis; pl: penis lumen; plm: longitudinal muscle of penis; sb: seminal bursa; st: statocyst; sv: seminal vesicle; t: testes; vc: ventral crossover muscle; vd: ventral diagonal muscle.

spiral, and even U-shaped muscles. Species lacking a pharynx appear to have specialized, complicated ventral musculature to compensate for the lack of a muscular food-moving structure and this allows body movements to force food through the mouth.

Nutrition, Excretion, and Gas Exchange

As juveniles, most acoels appear to feed on protists, including unicellular algae such as diatoms. Smaller species may continue this diet throughout their lives, whereas larger species (e.g., *Convoluta convoluta*) are often predaceous, hunting minute crustaceans but also feeding on larval molluscs and other worms. Smaller protists are captured as acoels glide over them with the syncytial gut extruded through the mouth such that it engulfs food with "amoeboid"-like movements. Larger prey are grasped with the anterior margin of the body and entrapped with mucus before being pressed toward the mouth. Swimming prey may also be rapidly captured and ingested, whereas dead material seems to be actively avoided.

Some acoels possess a pharynx, in some cases known as a **pharynx simplex** (Figure 9.1A), and this structure is variable among the families. In some cases the pharynx is a flexible, tube-shaped structure that can be everted from the mouth. The pharynx is anchored by musculature attaching to circular muscles within the body wall. In the larger, predaceous species, there is no oral sphincter but several layers of circular muscles interspersed with oblique and longitudinal muscles extend throughout the protrusible structure, which is attached to the body wall by densely packed muscles (Figure 9.7B,C). In cases where no distinct pharynx exists, muscle fibers encircle the mouth to form a sphincter.

Ingested prey is enclosed within vacuoles that drift within the digestive syncytium, and food is completely

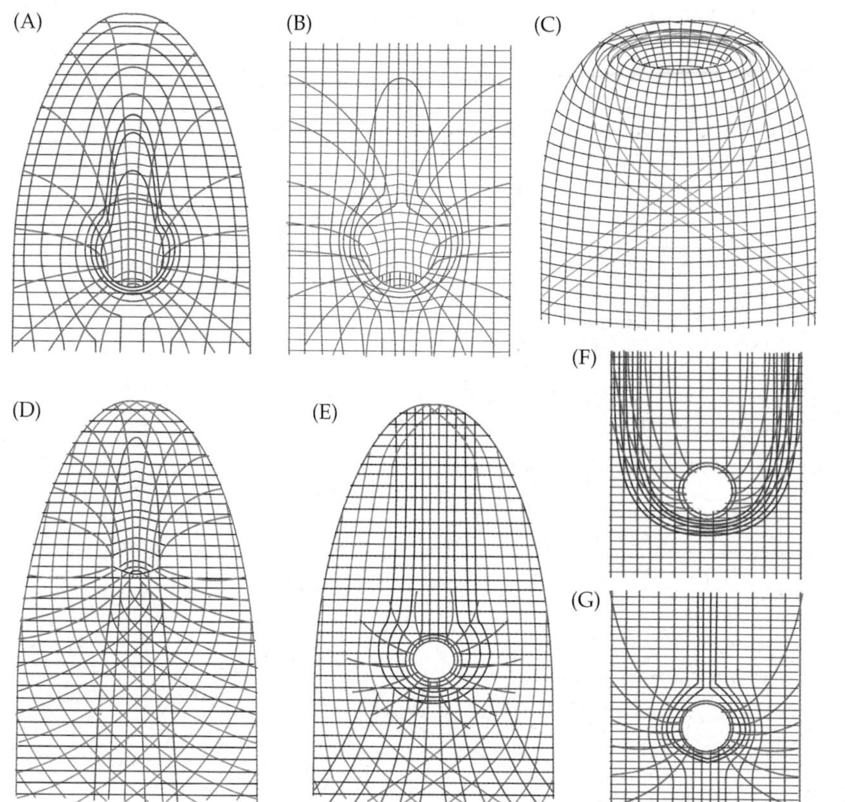

Figure 9.8 Schematic diagrams of patterns in acoel ventral body wall musculature. (A) *Myopea callaeum*. (B) *Solenofilomorpha crezeei*. (C) *Hofstenia miamia*. Notice that two layers of longitudinal musculature exist. (D) *Proporus bermudensis*. (E) *Conaperta thela*. Notice that pore muscles lie below longitudinal and spiral muscles. (F) *Paratomella unichaeta*. (G) *Sterreria psammicola*. Pharynx, if present, is not shown.

absorbed within 18 to 24 hours. The exoskeletons of hard-bodied prey such as crustaceans are voided through the mouth. Fat globules and occasional glycogen vacuoles stored within cells appear to be the primary forms of food reserve. A number of acoel species associate with corals (including *Waminoa* and several species of *Convolutriloba*). These associations appear to primarily benefit the acoels who likely feed on mucus produced by these cnidarians (Figure 9.1E). It has been suggested that the syncytial digestive system of acoels might be an extreme state of the condition seen in nemertodermatids, which have a small, relatively occluded gut lumen (and a remnant of a gut lumen is evident in the acoel *Paratomella rubra*).

The small size of acoels is sufficient to allow them to eliminate waste nitrogen and carbon dioxide, as well as obtain oxygen from the surrounding water, without a need for specific excretory or circulatory systems. Food vacuoles evidently serve to move materials from the digestive syncytium to other cells within the body.

Nervous Systems and Sense Organs

The central nervous system of acoels usually includes an anteriorly located cluster of large commissures and a few cell bodies that form a paired ganglia system with what some workers consider to be a minute neuropil (though it is quite rudimentary compared to other metazoans). Arising from this are three to five pairs of longitudinal nerve cords connected by an irregular network of transverse fibers (Figure 9.9). Typically there are single or paired dorsal nerve cords, and paired lateral and ventral cords. Peripheral neurons connect to epidermal sensory cells and to anterior light-sensitive cells that serve as simple eyes. There is no indication that the eyes have ciliary or rhabdomeric elements, and they are probably simple pigment cells with refractive inclusions and up to three nerve cells to relay the stimulus. This organization contrasts markedly with that of platyhelminths, where the brain consists of a comparatively dense ganglionic mass, the nervous system is primarily developed ventrally, and the nerve cords form an orthogonal nervous system composed of eight orthogons largely developed laterally and ventrally (Figure 9.4). Although organized as a bilobed structure, the acoel "brain" lacks the dense ganglionic cell mass (neuropil) seen in the Platyhelminthes.

The acoel statocyst is a fluid-filled, proteinaceous spherical capsule, 10–30 μm in diameter, surrounding a single retractile statolith (Figure 9.1A–H). The statolith appears to be a single spherical cell. The capsule enclosing the statolith comprises two unciliated cells. Behavioral observations indicate that acoels are capable of precise geotactic orientation, suggesting that movements of the statolith within the statocyst are detectable by the animal. Three pairs of muscle fibers insert into the membrane of the statocyst, evidently

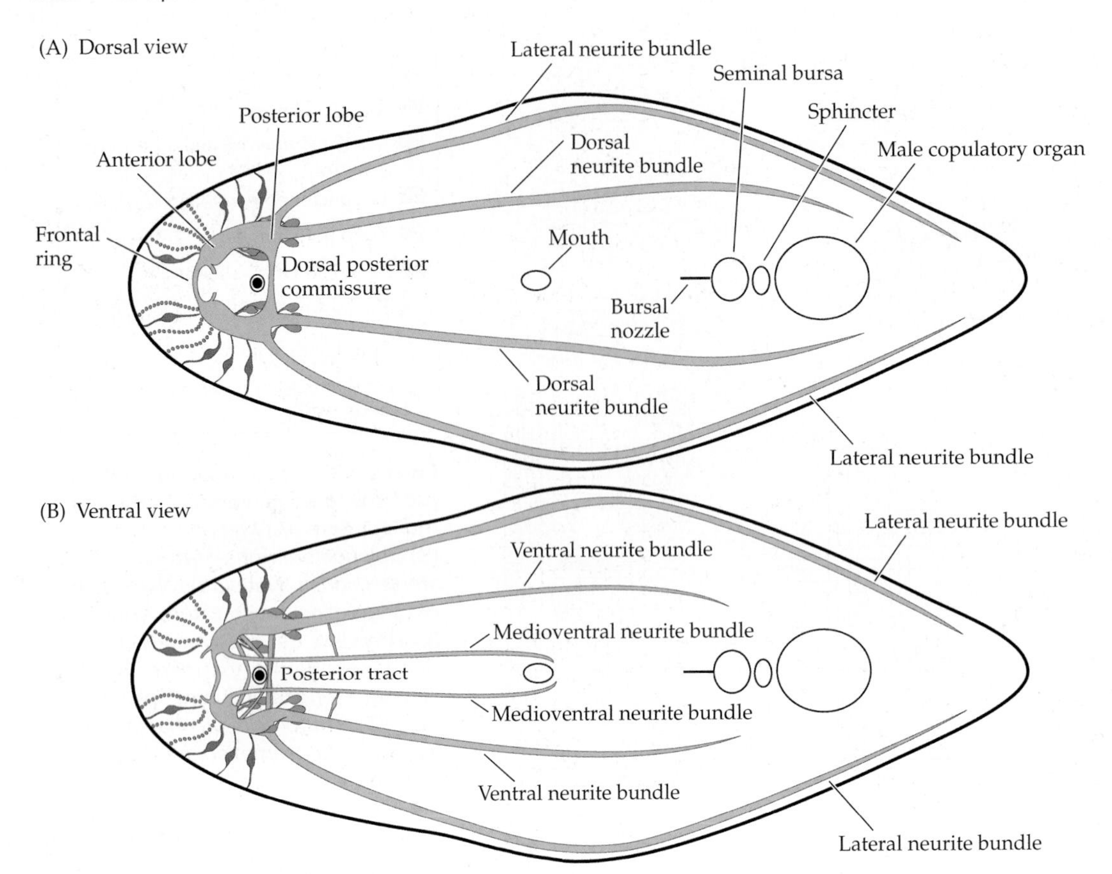

Figure 9.9 Diagram of the nervous system of *Isodiametra pulchra* (Acoela) revealed using nervous tissue-specific staining (green and magenta colors denote separate types of neural tissue in the bilobed acoel brain; cyan color is the central nervous system).

assisting in maintaining its position. While the cerebral commissure is closely associated with the statocyst, specific innervation of the structure is difficult to clearly identify, although a small nerve cushion created by two nerve bundles insert on the capsule and a cell body located at the ventral pole may be responsible for detecting deformation of statocyst fluid. Alternatively, positional information may be conveyed by the stretching of muscle fibers surrounding the statocyst. While statocysts appear in other metazoans, including cnidarians, ctenophores, platyhelminths, annelids, and others, statolith movement within the statocyst in these taxa is generally detected by cilia along the internal surface of the statocyst. The lack of these modifications within the Acoela appears to be unique.

Reproduction and Development

Acoels are capable of both sexual and asexual reproduction, and have considerable ability to regenerate cells through the actions of multipotent, mesodermally derived, neoblastlike cells. These structures were originally described in the Platyhelminthes, but analogous (or homologous) cells appear in the Acoela as well. These cells replace damaged or missing body components and appear to have few limitations in how they are able to repair or replace tissues, particularly epidermal cells.

Three distinct forms of asexual reproduction have been documented within the Acoela: transverse fission, longitudinal fission, and budding (Figure 9.10). Although

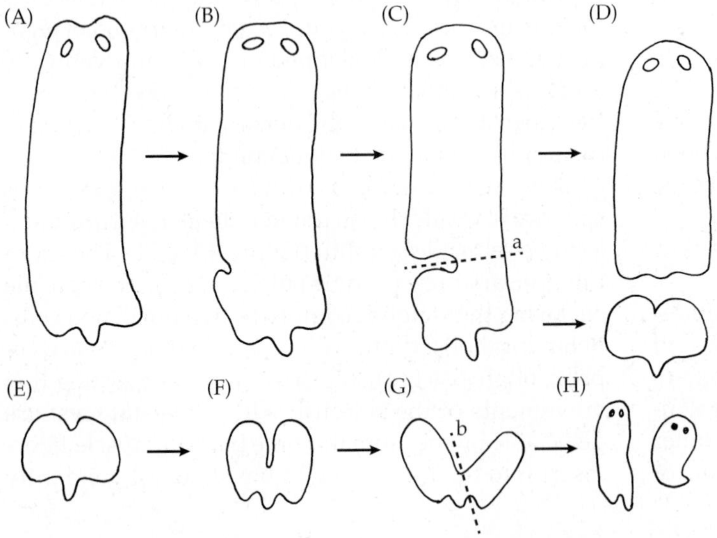

Figure 9.10 Modes of asexual reproduction in *Convoluta longifissura* (Acoela). (A) Intact animal. (B–D) Transverse fission; lower element of (D) shows "butterfly" stage preceding transversion fission (E–H).

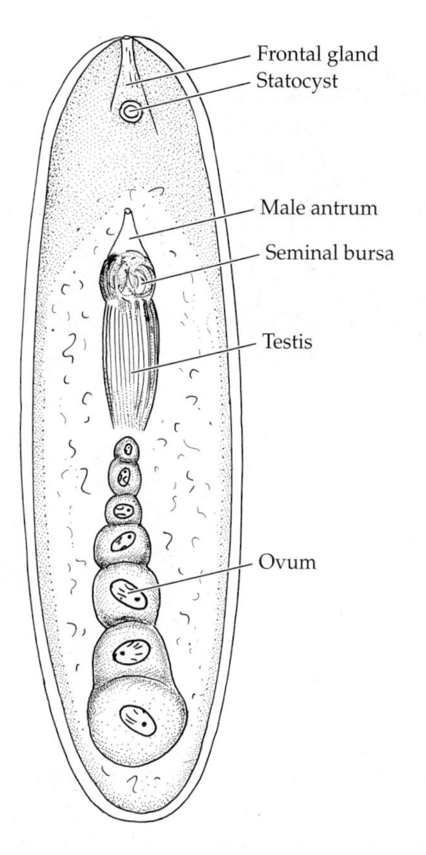

Figure 9.11 Internal organization of *Antigonaria arenaria* (Acoela).

capable of asexual reproduction, and while often found in large local abundances, acoels are not known to markedly increase their numbers asexually under natural conditions (as seen in many other asexual organisms), except perhaps in the family Paratomellidae.

Most acoels are simultaneous hermaphrodites (Figure 9.11), although some (e.g., all members of the family Solenofilomorphidae) are protandrous. Ovaries and testes may be paired or unpaired, with testes usually dorsal and ovaries more ventral (Figure 9.12A). In some species a single mixed gonad exists. In no cases are the gonads saccate—that is, the germ cells are not lined or discretely separated from the surrounding parenchyma.

Genitalia are usually visible near the posterior end of the animal. The penis is a muscular and glandular, or needle-like structure, often with multiple stylet-like elements (Figure 9.12B). Male intromittent organs, regardless of form, can be retracted into a seminal vesicle. During copulation the penis is everted through the gonopore that typically lies in a distinct antrum, or vestibule on the body surface. A separate female gonopore exists in some species. In others, the female pore connects directly to the male pore. In still others, no external female opening exists and insemination is hypodermic. Most, but not all acoels have the vagina positioned anterior to the penis.

A **seminal bursa** may exist that appears to receive sperm from mating partners either during copulation or after hypodermic insemination, and a sclerotized **bursal** or **vaginal nozzle** or sphincter regulates the

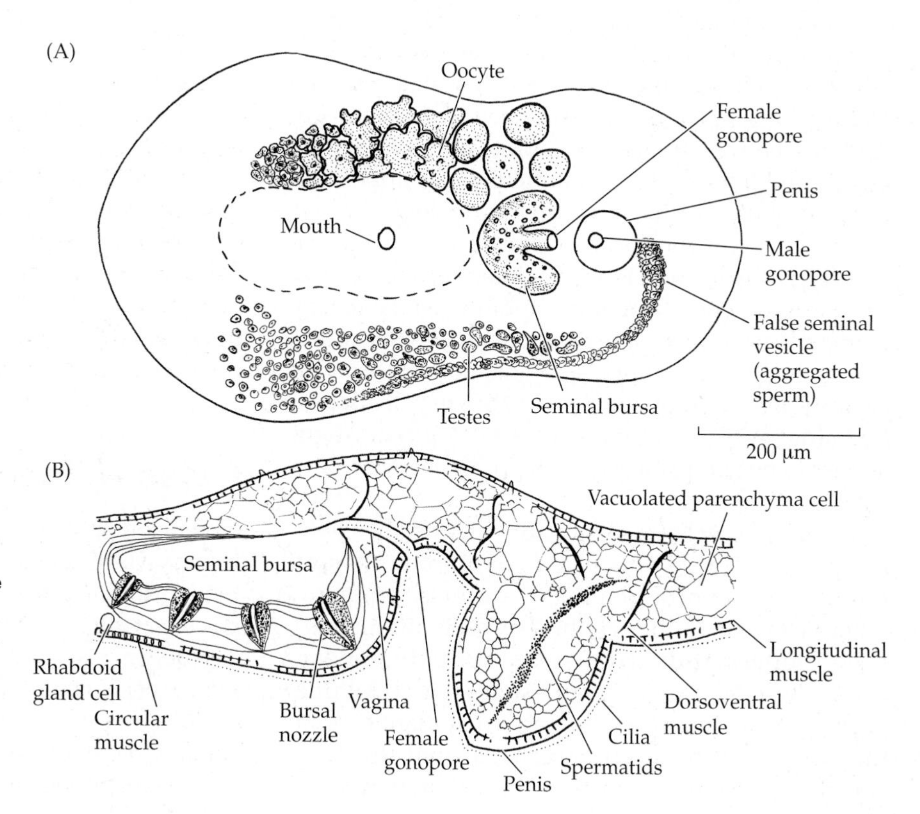

Figure 9.12 Acoela. Reproductive anatomy of *Polychoerus gordoni*. (A) Dorsal view; note in this species as in other Convolutidae, male and female gonads are paired, but are shown singly here. (B) Sagittal view of female and male reproductive anatomy.

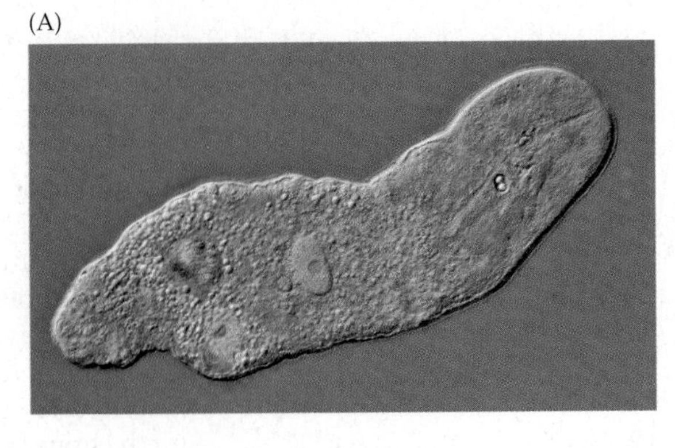

Figure 9.13 Nemertodermatids. (A) *Flagellophora apelti*. (B) *Sterreria* sp.

passage of sperm to eggs. These are among the few sclerotized structures in these soft-bodied worms, and they were considered important in early taxonomic treatments of Acoela. Like many persistent terms in invertebrate zoology, the name "nozzle" was coined by Libbie Hyman who thought these structures resembled the nozzle of a hose. In certain Convolutidae, multiple bursal nozzles may exist in the same individual (Figure 9.12B). While highly variable in form, seminal bursae and their associated structures appear to be homologous among all acoels.

Copulatory behavior was observed by Hyman to occur during daylight hours when high densities of individuals are maintained in aquaria or other confined spaces. Individuals approach one another and exchange quick touches or "nips" of the anterior ends. Larger individuals appear to initiate copulation, which proceeds after both individuals roll into a ball, and then unroll with their genitalia firmly engaged. Individuals mutually insert their penes into their partner's female gonopore and direct sperm and seminal fluid into the female bursa.

Acoel sperm are distinctively biflagellate, with the two axonemes of the flagella incorporated into the cell body (a condition also seen in Platyhelminthes). Several well-defined patterns exist in acoel sperm morphology, and these seem to be phylogenetically informative. Combined studies of 18S rRNA sequence data and sperm morphology have revealed remarkable concordance between these two sources of data. Seminal bursae and bursal nozzle complexity also appears to correlate with variation in sperm morphology.

Oocytes in the gonads of acoels give rise to endolecithal eggs. Fertilization is always internal, and zygotes are released either through the mouth, via the female gonopore, or through a rupture created in the epidermis by the growing embryos. Zygotes may be brooded or protected by encapsulation, but are usually deposited singly and undergo direct development. Eggs appear to be laid primarily at night, in flat gelatinous masses.

Embryonic development is direct, and the cleavage pattern of acoels appears to be a unique "**spiral duet**" cleavage program that is different from any other metazoan (although this has been questioned). Nemertodermatids, while exhibiting duet cleavage in the 4-cell stage, do not exhibit this spiral duet pattern, and their cleavage pattern is distinct from the Spiralia. It has been suggested that the acoel spiral duet-cleavage pattern might have evolved from an ancestral spiral quartet cleavage typical of the Spiralia. As in spiral quartet cleavage, the first horizontal cleavage is unequal and so produces micromeres, but it occurs at the two-cell stage instead of the four-cell stage, so the micromeres appear as duets instead of quartets. One is tempted to call this "bilateral cleavage," rather than spiral cleavage, and it is quite different from that of all other spiral-cleaving Metazoa. So far as is known, acoel embryos generate only endomesoderm, whereas most spiral-cleaving animals tend to also produce some ectomesoderm. Internal tissues arise either by delamination or by immigration of cells that form the ectoderm and mesoderm. By the time gastrulation is complete, the embryo has a layered appearance, with an outer epidermal primordium, and a middle layer of progenitor cells of muscles and neurons, while the innermost cells are those that will develop into the digestive syncytium. Endomesoderm forms from both of the third duet macromeres at the vegetal pole, whereas the mouth forms anteriorly as 1a micromere descendants expand around the posterior pole. An anus never forms.

Class Nemertodermatida

Nemertodermatida comprise a few dozen species of marine worms described from several locations around the world, including the Swedish coast (their original discovery site), the Mediterranean, Adriatic, North and Caribbean Seas, and the east coast of North America. Like acoels, nemertodermatids were once classified in the phylum Platyhelminthes. Nearly all known species are free-living, usually in fine sand, mud or gravel; however, species in the genus *Meara* are symbionts in the foreguts of holothurian echinoderms (sea cucumbers). Nemertodermatids range in length from a few millimeters to nearly a centimeter. They can be leaf shaped or narrow and elongated, and they may creep over the substratum or swim with serpentine movements. Their bodies are densely covered with locomotory cilia, and as a group they are easily recognized by an anterior statocyst containing two statoliths in separate chambers (Figure 9.13A,B)—although a few reports of one to four statoliths also exist in the literature. Some species possess an eversible proboscis associated with feeding (oddly, some of these species lack a distinct mouth) with numerous branches that extend anteriorly "like a witch's broom" (Figure 9.14).

The epidermal cells of nemertodermatids lack a true basal lamina but are connected to underlying muscle cells and to each other via a narrow extracellular matrix. Septate junctions between epidermal cells are lacking. As in the acoels, old or damaged ciliated epidermal cells are withdrawn into the body and resorbed, creating temporary structures called pulsatile bodies (Figure 9.15). Also as in acoels, species inhabiting the guts of other animals often have symbiotic bacteria inhabiting their epidermis (Figure 9.16). The form of the mouth and gut vary among species from temporary structures to a porelike opening and a narrow intestinal lumen, although a complete gut is lacking (there is no anus, nor is there a discrete pharynx). They lack discrete circulatory or excretory structures. Asexual reproduction has not been reported. Male sexual structures may consist of a simple, ciliated invagination of the epidermis or an eversible penis; seminal vesicles may be present. Where these structures are lacking, sperm appear to simply be ejected from the

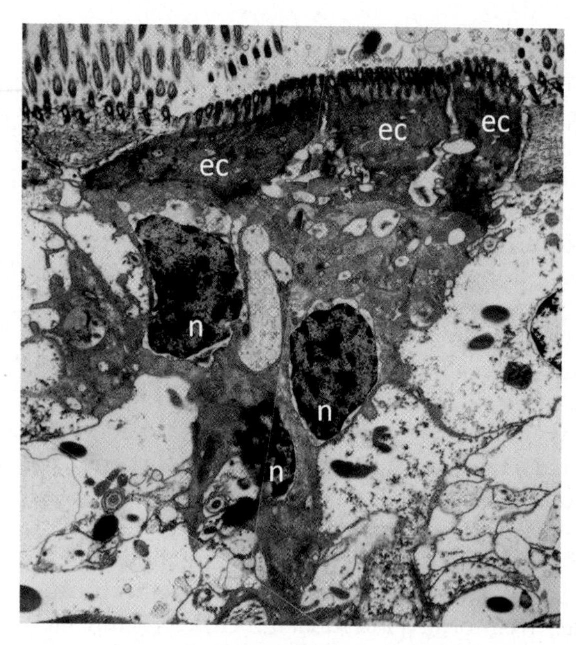

Figure 9.15 TEM micrograph showing a cross section the epidermis of the nemertodermatid, *Meara stichopi*. Three ciliated epidermal cells (ec), presumably worn or damaged and bearing only the dark stubs of locomotory cilia, are being compacted and withdrawn into the integument to be dissolved; the three dark structures are the epidermal cell nuclei (n).

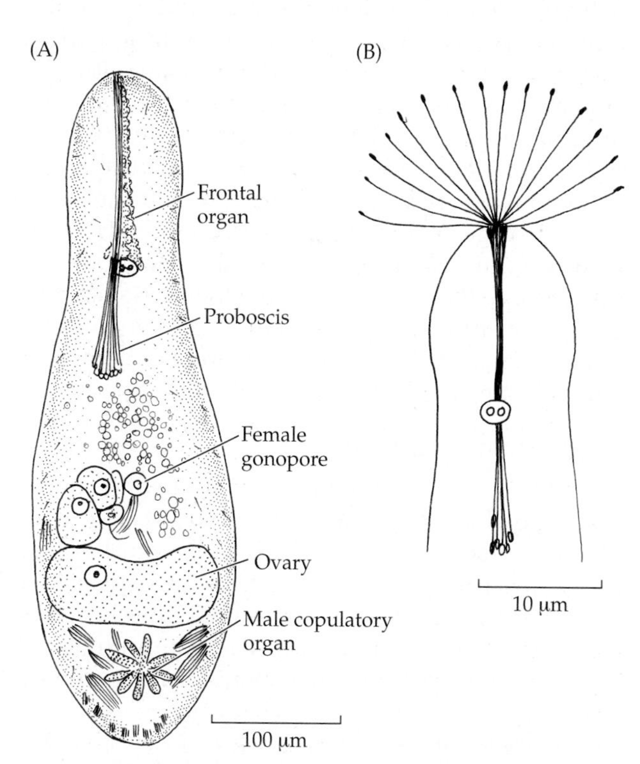

Figure 9.14 Nemertodermatids. *Flagellophora apelti*. (A) Dorsal view of mature specimen. (B) Protruded proboscis.

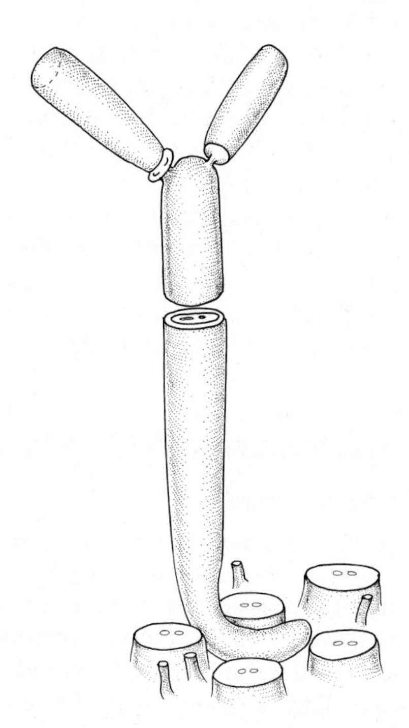

Figure 9.16 Nemertodermatida. Y-shaped elongated symbiotic bacteria associated with the epidermis of *Meara stichopi*.

male antrum. The female gonopore with an associated bursa is located dorsally in most species. Mature eggs are released through the mouth. Duet cleavage and direct development are similar to that observed in acoels (Figure 9.20).

The first described nemertodermatids was classified within the acoel Platyhelminthes by Otto Steinböck in 1930. Long before similar hyperbolic statements by now-deposed dictators, Steinböck, a colorful individual known to express himself in double-spaced capital letters with exclamation marks for emphasis, announced his discovery as "the mother of all turbellarians," possessing a "novel, two-stoned statocyst, an unusually thick and gland-rich epidermis, a peripheral nervous system, and a mixed, lacunar gonad without accessory organs." Going Steinböck one better, in 1940 Tor Karling removed the Nemertodermatida from the Acoela and other-than-turbellarians altogether because of their well-formed intestinal lumen, a structure lacking in acoels. The Acoela and Nemertodermatida were combined as sister taxa within the Platyhelminthes in 1985 with Ulrich Ehlers' recognition of the taxon Acoelomorpha, primarily based on ciliary structures. Additional work on Nemertodermatida has proceeded slowly because specimens are difficult to come by and because many characters can be highly variable within populations.

The relationship of *Meara stichopi* to its echinoderm hosts is poorly understood, but does not appear to be parasitic—hosts do not appear to be harmed by the presence of the worms. In fact, the relationship could be mutualistic, as nematodes have been found within the guts of endosymbiotic *Meara*. Symbiotic species of both *Meara* and *Nemertoderma* are known to possess elongated, y-shaped symbiotic bacteria (Figure 9.16). In *Meara*, these symbionts are found primarily on the ventral side of the host's body. The y-shape of the bacteria has been suggested to represent the mode of asexual reproduction because appendages are found only on certain bacterial cells. Ultrastructural studies indicate that bacteria occur only on the outside surface of their worm hosts, suggesting that the association between bacteria and host does not represent infection.

The Nemertodermatid Body Plan

General Body Structure

In general, nemertodermatids are small. The endosymbiotic *Meara stichopi* is usually less than 2 mm in length, free-living *Nemertoderma* average about 3 mm in length, and a few "giant" nemertodermatids grow to nearly one centimeter in length. Most individuals are colorless to yellow or red, but pigmentation can be variable within populations.

The epidermis of most species appears to contain numerous bottle-shaped mucous glands. Overall, the epidermis resembles that of nemertean worms, which led to the namesake "Nemertodermatid." In the genus *Nemertoderma*, these glands are more abundant at the apical pole, forming an anterior gland complex with separated, outwardly directed gland openings or necks. However, these openings are not grouped together in a regular way at an apical pore and thus do not form a "frontal organ" as has been described in Platyhelminthes. Nevertheless, the structures are sufficiently similar to that of turbellarians that earlier authors considered them to be homologous with the frontal organ structures of flatworms.

Cell and Tissue Organization

The epidermis of nemertodermatids is entirely ciliated. The cells are connected by an intracellular terminal web—a stratified structure composed of a closely woven inner layer of intensely staining fibrils overlain with more loosely packed fibrils, which bulges at the cell borders. Epidermal cells are joined apically by belt-like adherens junctions (belt-desmosomes) called **zonula adherens**. Interspersed among the cells are the necks of various glands and sensory receptors, particularly in the anterior region of the animal. The necks of glands appear to have associated muscular rings that may regulate the flow of gland contents (Figure 9.17E).

The ciliary rootlet structure is similar in Acoela and Nemertodermatida, one of the primary reasons workers grouped these two taxa together (as the Acoelomorpha). The rootlets of nemertodermatids include a rostrally-oriented rootlet and a caudally-oriented rootlet. In their original description, *Meara stichopi* was reported to possess "restitution cells," that appeared to contain ciliary structures in the process of being resorbed. Indeed, these cells represent structures similar to the pulsatile bodies reported in acoels, wherein worn cells are encapsulated and transported to the digestive tract for resorption (Figure 9.15). However, this feature is distinct in the nemertodermatids because the cilia detach from their basal apparatus before encapsulation, eliminating their ability to pulsate, causing some researchers to refer to them as "degenerating epidermal bodies."

Support and Movement

As in the acoels, body wall musculature of nemertodermatids has been extensively investigated using phalloidin-staining procedures.[1] Like most acoels, body musculature in nemertodermatids consists of outer circular and inner longitudinal muscle layers. Diagonal musculature typically is also present and varies among species, with fibers forming connec-

[1]Phalloidin is a naturally occurring toxin in the death cap mushroom (*Amanita phalloides*). Its toxicity is due to its ability to stabilize actin filaments within cells, and this attribute has led to it being widely used (fluorescently labeled) in research to visualize filamentous actin, such as muscle fibers.

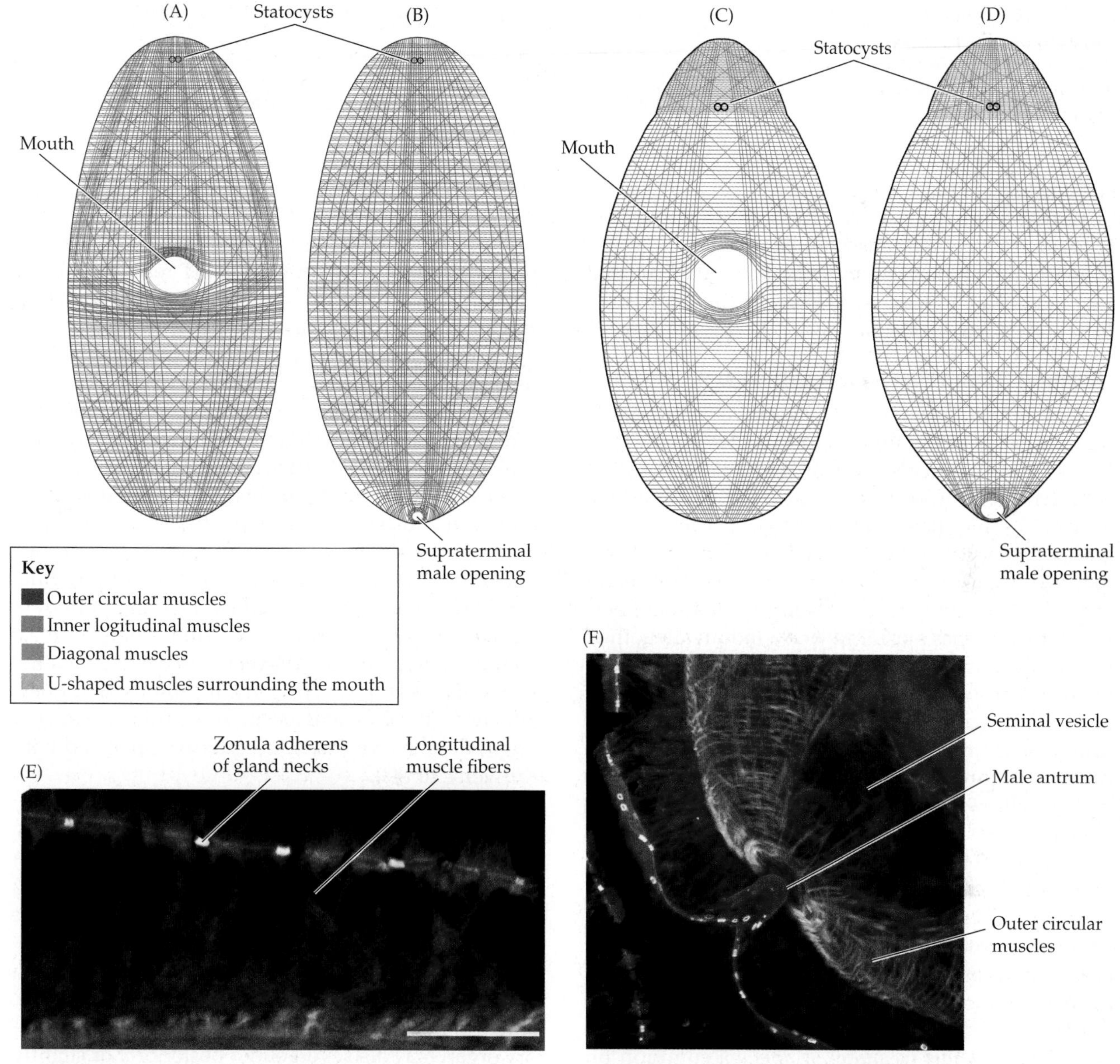

Figure 9.17 Schematic diagrams (A–D) and phalloidin-enhanced micrographs (E,F) of nemertodermatid musculature. (A) Ventral and (B) dorsal views of *Meara stichopi* (graphic showing muscle patterns). (C) Ventral and (D) dorsal views of *Nemertoderma westbladi* (graphic showing muscle patterns). Outer circular muscles (blue); inner longitudinal muscles (in red); diagonal muscles (in green); U-shaped muscles surrounding the mouth (in orange) on ventral side. (E) Lateral view of epidermis in *Nemertoderma westbladi*, showing longitudinal muscle fibers beneath circular ones in central space. Above this are two thin stained layers, the lower layer corresponding to the intracellular web, the upper layer corresponding to microvilli of the epidermal surface. The zonula adherens of the gland necks appear as brightly stained areas at this level. (F) Posterior body region of *Nemertoderma westbladi*, with invagination of body wall to form the male antrum; finer musculature of the seminal vesicle is visible in open space.

tions between layers in some (e.g., *M. stichopi*; Figure 9.17A,B) and forming distinct layers in others (e.g., *N. westbladi*; Figure 9.17C,D). Musculature surrounding the mouth also varies, being best developed in species with a permanent mouth. Musculature is also well developed around permanent genital openings (e.g., *M. stichopi*), but less so in species with transient genital orifices (e.g., *N. westbladi*).

The opening of the male gonopore and its associated antrum appear as an invagination of the entire body wall, and musculature associated with the seminal vesicle consists of a thin layer, present only in individuals with mature male organs (Figure 9.17F). Parenchymal muscles may also be present in individuals in all life stages, forming a three-dimensional network throughout parenchymal tissue. The statocyst is

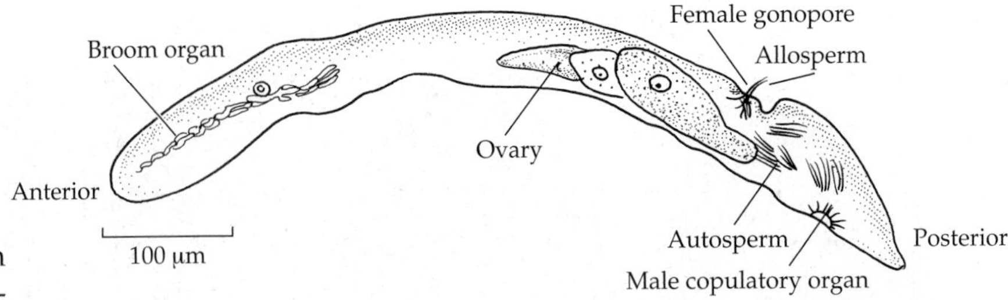

Figure 9.18 Nemertodermatida. Diagram of *Ascoparia* sp., showing location of dorsal female gonopore and subterminal male copulatory organ.

supported by muscles that attach dorsoposteriorly and anterolaterally to other body wall musculature. Nemertodermatids move by creeping on their ciliated surfaces or, in more elongated species, by undulating their bodies in a serpentine way.

Nutrition, Excretion, Gas Exchange

The gut of nemertodermatids has only a single opening, like that of cnidarians and other Xenacoelomorphs. However, unlike acoels, the nemertodermatid gut is not syncytial and instead contains a well-defined intestinal lumen. In some species, a cone of gut tissue has been reported to protrude and retract like a tongue to collect food particles. However, no known nemertodermatid possesses a structure recognizable as a muscular pharynx. Other species appear to lack a mouth altogether. In such species (e.g., *Flagellophora*), an anterior **broom organ** is reported although this structure does not seem to be directly connected to the gut. Instead it seems to consist of a bundle of up to 30 glands whose necks are protrusible through a canal at the anterior end of the body (Figure 9.14B). When opened, the broom organ appears to possess distal ends that are slightly swollen and possibly adhesive. Some researchers suggest that the lack of a mouth may represent an ancestral condition and that the mouth of nemertodermatids is a transient structure that appears during a limited part of postembryonic life, with the duration of persistence dependent upon the species.

Meara stichopi inhabits the foregut of the holothurian *Parastichopus tremulus*, a species common on Scandinavian coastlines, and appears to feed on detritus as well as upon nematodes within the guts of its host. Free-living species have been found with comparatively large turbellarians and nematodes within their guts.

As in acoels, the small body sizes of nemertodermatids allow them to eliminate waste nitrogen and carbon dioxide, as well as obtain oxygen from the surrounding water, without a need for discrete excretory or circulatory systems.

Nervous System

The nervous system of nemertodermatids is still poorly understood. Immunoreactivity studies to the neurotransmitter serotonin (5-hydroxytrypamine; 5-HT), and the regulatory neuropeptide FMRFamide, have shown considerable variation in responses in the species examined. In *Meara stichopi*, 5-HT reactivity reveals a subepidermal nerve net and two, loosely organized longitudinal nerve bundles along the length of the animal. In *Nemertoderma westbladi*, 5-HT reactivity shows a two-ringed, anterior commissure, with the rings converging near the statocyst, and connected by thin fibers. Two lateral fibers extend longitudinally from the commissure, as does a delicate curtain of evenly spaced finer longitudinal fibers that become indistinct caudally. FMRFamide immunoreactivity follows the same pattern as 5-HT reactivity in *M. stichopi* and *N. westbladi*. These results suggest that the nemertodermatid nervous system is distinct from the bi-lobed ganglionic brain and orthogon peripheral nervous systems of Platyhelminthes (i.e., paired longitudinal ventral nerve cords connected by a regular pattern of transverse commissures). Nemertodermatid central nervous systems are also distinct from the commissural brains of acoels (i.e., symmetrical commissural fibers with few cell bodies and 3–5 pairs of radially arranged longitudinal nerve cords, irregularly connected with transverse fibers).

Reproduction and Development

The reproductive anatomy and natural history of nemertodermatids is not well studied, and only a few species have been examined in this regard. The male gonopores in nemertodermatids appear to open dorsally (or supraterminally) and are associated with a muscular male antrum. In fully mature specimens, a muscular seminal vesicle and often a male copulatory organ may also evert either posteriorly or slightly dorsally (Figure 9.18). Female genitalia, if present, are located dorsally. *Flagellophora* seem to have a deep, well-defined invagination that may represent a female gonopore (Figure 9.14).

In *Meara stichopi*, follicular testes occupy most of the preoral part of the body. The ovary occupies the postoral part of the body and often contains one or more large ova within the posterior body region. The male intromittent organ opens terminally to slightly supraterminally in this species.

In general, Nemertodermatida have a 9+2 arrangement of microtubules in their uniflagellate sperm, a condition distinct from the variable microtubule arrangement and biflagellate condition of acoel sperm. Many field-collected nemertodermatids contain two types of sperm. **Autosperm** (sperm produced by the individual in which they are found) in *M. stichopi* are filiform, about 45–60 pm long, and under phase contrast microscopy show indistinct divisions of individual

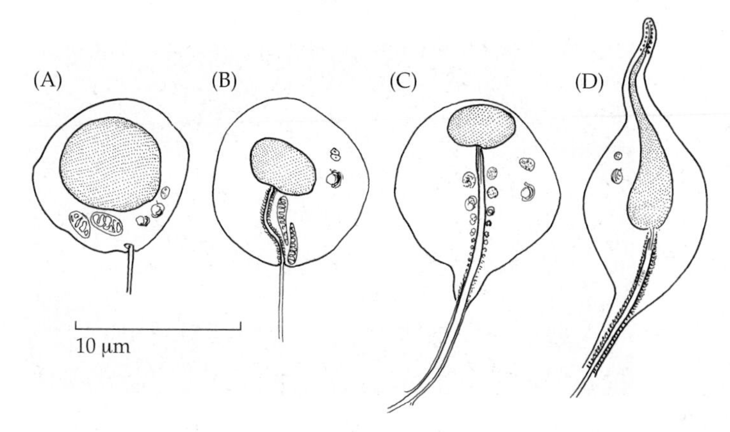

Figure 9.19 Nemertodermatida. Stages in spermatogenesis in *Meara stichopi*. (A) In early stages of spermatogenesis, the nucleus is large, with heterogeneous electron density; two mitochondria are visible. A single flagellum begins to form from a basal body situated near the cell membrane. (B) The nucleus shrinks and becomes homogenous in density. The mitochondria begin to elongate, and the basal body and associated fibers move into the cytoplasm toward a depression forming in the nucleus. Microtubules coil around the flagellar channel. (C) The mitochondria coil around the flagellar channel and a sheath grows from the cell to surround the proximal flagellum. (D) The cell and nucleus elongate to form the head of the spermatozoon. The mitochondria form tighter whorls and wander into the length of the flagellar sheath.

sperm into head, middle piece, and tail, as is typical of more derived spermatozoan forms within Metazoa (Figure 9.19). Some sperm appear to be coiled, corkscrew-fashion, over half of their length and are nonmotile within the animal producing them. **Allosperm** (sperm not produced by the individual in which they are found) are distinctive because they tend to be uncoiled and motile within the bodies of recipient individuals.

The pioneering investigator of acoels and nemertodermatids (what is now called Acoelomorpha), Einar Westblad, noticed that development of female reproductive structures seemed to precede that of male structures indicating that some nemertodermatids are protogynous. On the other hand, other authors have noted that "male maturity seems to precede female maturity," or they have specifically stated that individuals are protandrous. In *N. westbladi*, individuals were found to have matured as males, females, and hermaphrodites, with no clear evidence of either protandry or protogyny. Mature females contained a single large egg but also contained up to 10 ova, and allosperm were found in only a few individuals. In *Ascoparia neglecta*, an elongate species with no actual mouth in mature individuals, although male and female pores are visible, and individuals possess a male copulatory organ, allosperm appear to be contained in vacuoles near the vagina.

Many species appear to lack female genitalia, yet are found to contain autosperm that is clearly contained within male reproductive structures, as well as allosperm that appear to have been introduced to the individual housing it. Individuals bearing allosperm appear to include mature as well as immature individuals raising the possibility of sperm storage and sperm competition among individuals. Taken together, there appears to be great diversity in reproductive life history among nemertodermatids. Adults in some species appear to be smaller than juveniles suggesting that maturing individuals may cease to feed and complete their life history using stored food reserves or other resources.

Copulation has not been observed in any nemertodermatids. However, earlier workers suggested that worms may simply press their posterior ends together long enough for spermatids to get through the epithelium of the recipient worm. Since male structures are located subterminally, this behavior would have to occur with the recipient positioned somewhat dorsally, or an individual transferring sperm would have to undergo dorsiflexion to accomplish impregnation. In species lacking female structures, insemination appears to be hypodermic. Oviposition is accomplished by flexing the body into a dorsi-convex shape followed by protrusion of the circumoral area of the body to force individual eggs out of the mouth.

Development in nemertodermatids is similar to that of the Acoela. The first cleavage division is holoblastic (Figure 9.20). The second division results in the formation of micromeres and macromeres. In later 4-cell stages, the micromeres shift slightly clockwise (dexiotropic), resembling spiral cleavage, but this shift does not occur until well after the division has taken place. Nemertodermatids do not have the unique

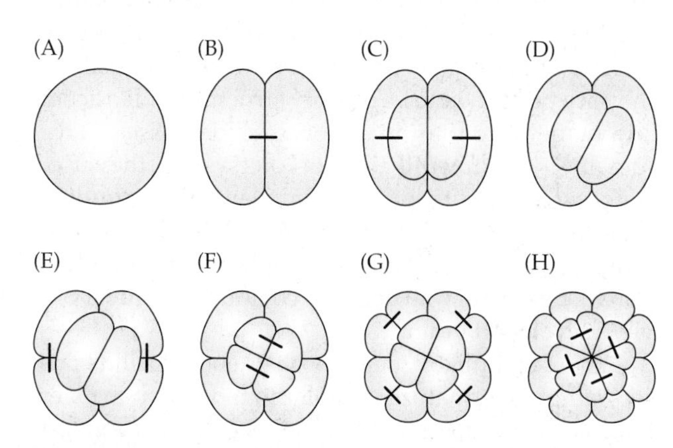

Figure 9.20 Diagram of duet cleavage in Nemertodermatida (*Nemertoderma westbladi*), viewed from the animal pole. Lines indicate cleaved relationships among cells. (A) Uncleaved zygote. (B) 2-cell stage. (C) Early 4-cell stage with micromeres oriented radially. (D) Late 4-cell stage with micromeres shifted. (E) 6-cell stage. (G) 12-cell stage with 8 macromeres and 4 micromeres. (H) 16-cell stage with 8 macromeres and 8 micromeres.

spiral duet-cleavage program seen in Acoela. Instead, cleavage starts out radial, but then takes place in a duet pattern, the micromeres shifting clockwise to produce a spiral-like pattern. The 4-cell divisions involve both macromeres, resulting a 6-cell embryo, followed by another division by the micromeres to yield an 8-celled embryo. This alternating pattern is followed until the 16-cell stage, similar to what is known in acoels as "duet cleavage," although in acoels, the first division involves a counterclockwise (levotropic) shift of the micromeres rather than the clockwise shift documented for *N. westbladi*. Nevertheless, the form of duet cleavage is similar in both taxa, suggesting to some researchers that this trait is ancestral in the Acoelomorpha.

Postembryonic development in *N. westbladi* appears to follow three life history phases. Hatchlings are nearly round; only slightly longer than wide (250 × 200 μm). These individuals grow into juveniles, which are bottle shaped and may reach nearly 1mm in length, but possess no discernable sexual structures. Mature specimens may be variable in size (averaging 450 μm in length), and possess a slightly more elongated shape as well as visible male copulatory organs and a small pointed tip at the posterior end formed by the male gonopore.

Subphylum Xenoturbellida

The free-living marine creatures known as xenoturbellids comprise just two described species, *Xenoturbella bocki* (Figure 9.21 and chapter opener photo) and *X. westbladi*, although several undescribed species are known to exist. They are delicate, ciliated worms with a very simple body plan, and are recognizable by their body furrows, the **horizontal furrows** (= side or anterolateral furrows), and a **ring furrow** (= equilateral furrow), the latter nearly crossing the animal's midline (Figure 9.22A). The furrows contain what has been interpreted as high concentrations of sensory cells, so they are presumed to be sensory structures. Like acoels and nemertodermatids, xenoturbellids possess a diffuse, basi- or intraepithelial nervous system, they use a statocyst for orientation, have circular and longitudinal muscles, and a mid-ventral mouth. Also like acoelomorphs, *Xenoturbella* lack a complete gut, organized gonads, excretory structures, coelomic cavities, or a well-developed brain.

However, unlike acoelomorphs, xenoturbellids possess simple spermatozoa, similar to those seen in externally fertilizing species. Also, muscle layers are connected by extensive interdigitations among the layers of cells as well as longitudinal muscles that are exceptionally robust. Their nervous system, while diffuse, is concentrated along its sensory furrows. Xenoturbellids also are generally larger in size, reaching up to 4 cm in length.

Sixten Bock (1884–1946), the great Swedish platyhelminth specialist, was collecting along the Swedish coast near The Sven Lovén Centre for Marine Sciences (then known as the Kristineberg Marine Station) in 1915 when he came across an odd-looking "flatworm." Bock never got around to identifying the creature, but Einar Westblad, another great platyhelminth specialist did, initially considered it an archoophoran turbellarian, along with similar specimens he had collected near Scotland and Norway. Eventually, in 1949, Westblad described the original specimens as *Xenoturbella bocki*, after its collector. The creatures caused immediate controversy because of their distinctive appearance. In 1999 the second species was described, and named after Westblad as *Xenoturbella westbladi*. By this time people were beginning to wonder about how unique these two "flatworms" were. The name *Xenoturbella* means "strange turbellarian" because, while they

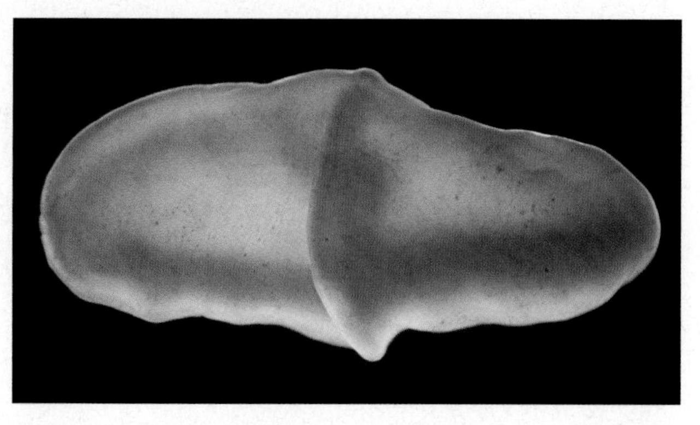

Figure 9.21 *Xenoturbella bocki*. Live specimen, from 80 m depth, off the west coast of Sweden.

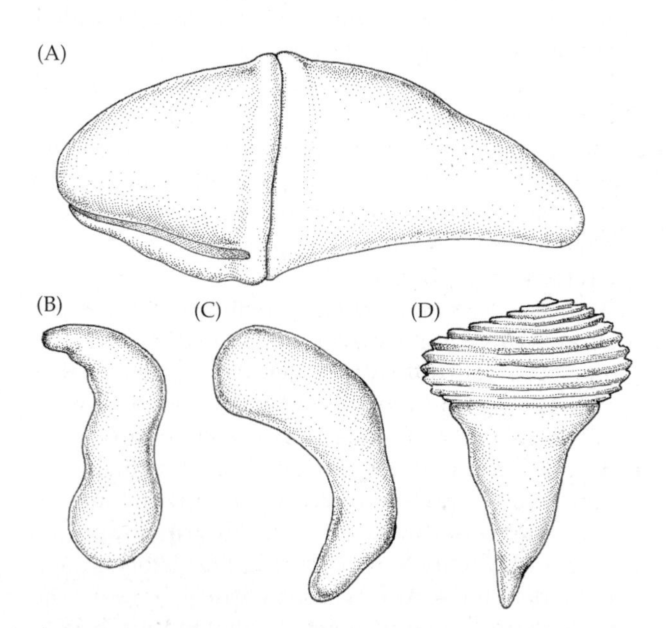

Figure 9.22 Line drawings of *Xenoturbella bocki* from live specimens. (A) Dorsolateral view showing the horizontal and ring furrows. (B,C) Lateral views of specimens moving by ciliary gliding (arrows indicate direction of movement). (D) Animal with contracted anterior end after exposure to $MgCl_2$.

resembled acoelomorphs overall, their epidermis was reminiscent of hemichordates, and their statocyst seemed similar to that of certain holothurians.

As morphological evidence accumulated on *Xenoturbella*, their relationship to flatworms began to be doubted, and by the late 1950s most researchers agreed that *Xenoturbella* was not a flatworm, but there was little consensus about what these animals actually were. Opinions on their identity ranged from considering them "among the coelenterates" to placing them as a sister taxon to the enteropneusts.

Then, in the late 1990s, analysis of ribosomal RNA on what appeared to be developing oocytes and embryos in some specimens led to the conclusion that *Xenoturbella* was in fact a highly degenerate mollusc, possibly some form of shell-less bivalve. However, subsequent investigations showed that these samples had been contaminated with gut contents containing mollusc DNA. Subsequent DNA studies suggested *Xenoturbella* might be a highly degenerate deuterostome, near the base of the deuterostome line or perhaps closely related to echinoderms and hemichordates (the clade known as Ambulacraria). Continued molecular phylogenetic studies have suggested that *Xenoturbella* is closely tied to acoels and nemertodermatids, and thus the new phylum name Xenacoelomorpha was created to house these three odd, primitive worms. While we accept this classification for this book, it is clear that the final resolution of *Xenoturbella* phylogenetic relationships is yet to be settled.

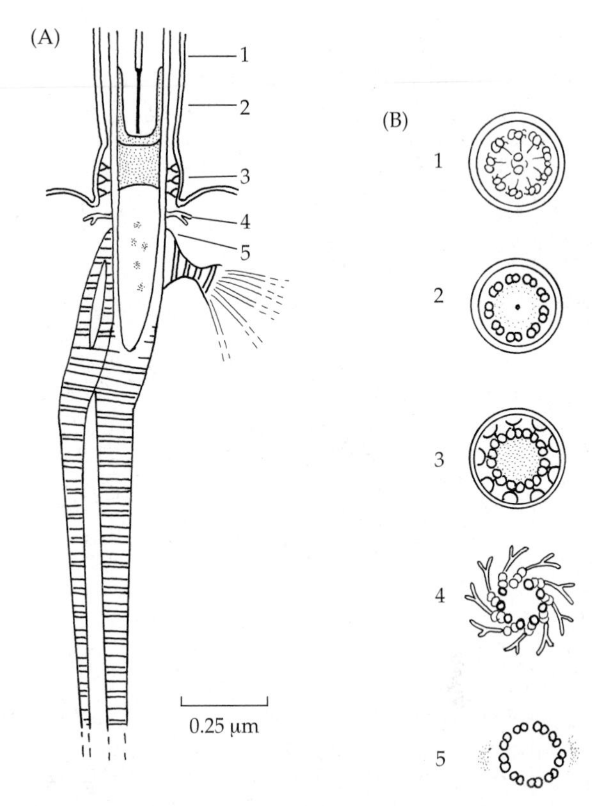

Figure 9.23 Diagram of the basal part of the cilium, basal body, and ciliary rootlets of *Xenoturbella bocki*. (A) Longitudinal median section of basal part of cilium. (B, 1–5) Cross sections of basal part of cilium and the basal body, showing the position of the microtubules at different levels. (1) Basal part of cilium. (2) Cup-shaped structure at the base of cilium. (3) Dense aggregation of granules and champagne-glass structures in the upper part of the basal body. (4) Centriolar triplet part of the basal body with winglike projections (the "alar sheets"). (5) Lower part of the basal body.

The Xenoturbellid Body Plan

General Body Structure

Most specimens of *Xenoturbella* are ovoid in shape, with a flattened ventrum and a length of 4 cm or less. These worms can be quite active and capable of considerable changes in shape (Figure 9.22B–D). The anterior region of most individuals is slightly lighter in color, and the horizontal furrows extend posteriorly, on either side, from the head end. Approximately midway down the body, these furrows nearly intersect with a ring furrow. The nervous system appears to be concentrated in these areas, suggesting a sensory function to these structures.

The epidermis of *X. bocki* consists of a layer of tall columnar cells with nuclei situated basally. These cells are densely ciliated, and are interspersed with unciliated or monociliated gland cells and ciliary receptors, the latter being most numerous in the horizontal furrows. The cilia themselves are attached to epidermal cells by several structures (Figure 9.23A). Each cilium ends in a basal body whose protruding **basal foot** has microtubules that extend into the epidermal cell. Two ciliary rootlets project from the basal body deeper into the epidermal cell; the thinner one, located on the same side of the cilium as the basal foot, projects straight into the cell, whereas the thicker rootlet has a knee-like bend. The cilia each have a distinctive arrangement of microfilaments in which the standard 9+2 arrangement extends for most of the ciliary shaft length, but near the end microfilament doublets 4–7 abruptly end, leaving only doublets 1–3 and 8–9 to continue on to the end of the cilium (Figure 9.23B). This "shelf" arrangement of microtubules is also present in Nemertodermatida and Acoela but is unknown in other known metazoan taxa (Figure 9.24).

The basal region of the epidermis houses the cell processes of the multiciliary cells, supporting cells, and a prominent intraepidermal nerve layer. The cell membrane of adjacent epidermal cells intermingle with each other, but tight couplings between the membranes of adjacent extensions do not appear to exist. However, where the cytoplasmic protrusions are shorter, they show a regular arrangement as if the two cells were held together by a zipper, but tight junctions,

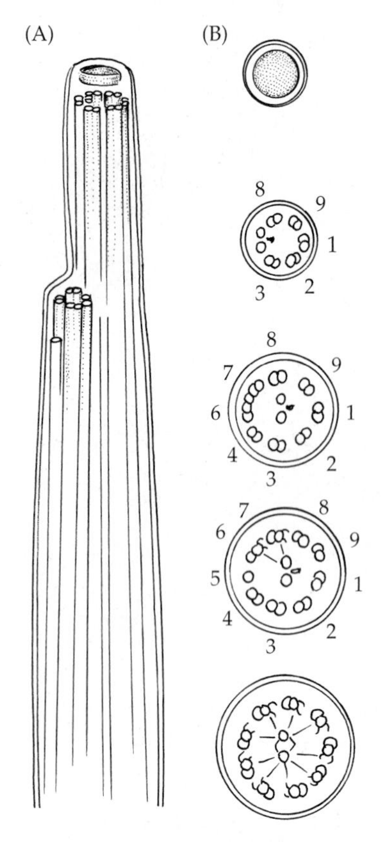

Figure 9.24 Diagram of the configuration of axonemal fibers within the distal shafts of epidermal cilia in *Xenoturbella bocki*. (A) Lateral view of the distal shaft showing the "shelf" located approximately 1.5 μm from the cilium tip. (B) Transverse sections of the cilium along its length; a 9+2 arrangement of axonemal fibers begins at the cilium base but microtubule doublets 4–7 end at the shelf.

desmosomes or gap junctions between cells have yet to be identified.

A number of workers have noted the similarities in both ciliary roots and ciliary tips in *Xenoturbella*, acoels, and nemertodermatids. Like these other groups, *Xenoturbella* is capable of withdrawing and resorbing worn out epithelial cells, yet there are differences in the character of this process. Whereas nemertodermatids do not withdraw still-motile ciliary cells, the withdrawn epidermal cells in *Xenoturbella* assume an orientation perpendicular to that of the other cells and retain some motility.

Support and Movement

Xenoturbella possess a highly muscular body wall (Figure 9.25A). An outer circular muscle layer surrounds a well-developed inner layer of longitudinal muscles, and with radial musculature extending from the gastrodermis to the outer circular layer of muscle cells (Figure 9.25B). The longitudinal layer of muscles is substantial, and individual muscle cells consists of a number of isolated fibers that when viewed in cross-section resemble a monolayered rosette.

No specialized parenchyma cells exist between the epidermis and the gastrodermis. However, all muscle cells tend to have numerous and well-defined cytoplasmic extensions with extensive mutual interdigitation. Tight attachment of adjacent cell membranes does not appear to exist, but connections resembling the zonula adherens in acoels and nemertodermatids are present, as is a fibrous subepidermal layer up to 5 μm thick. The extensive connections between muscle cells observed in *Xenoturbella* has been said to be reminiscent of hemichordates.

Xenoturbella inhabit marine mud bottoms at 20–120 m depth and move by ciliary gliding, without requiring modification of the body profile. The ventral surface is richly supplied with epidermal glands and moving animals leave behind a trail of mucus. While capable of considerable variation in body configuration due to powerful circular and longitudinal muscles, in most circumstances animals do not require such gymnastics in their basic activities.

Nutrition, Excretion, and Gas Exchange

Feeding by *X. bocki* occurs when individuals open their simple mouth and protrude their unciliated foregut. Extrusion of this structure appears to take place as a result of contractions of the surrounding body wall musculature, with relaxation of these muscles resulting in foregut retraction. The gut is cellular, but unciliated. Considerable attention has focused on the gut contents of *Xenoturbella*. Examination of mitochondrial DNA (cytochrome *c* oxidase subunit I sequence data) in the gut contents of *Xenoturbella* suggests that they feed primarily on bivalve prey, possibly in the form of eggs and benthic larvae. Such specificity suggests that these worms may be specialized predators, a hypothesis supported by the results of stable isotope studies indicating high ratios of N_{15} to N_{14} (3.42) characteristic of most predators. Two species of endosymbiotic bacteria have been described from the gut of *X. bocki*. Researchers have suggested that these bacteria might assist in nitrogen detoxification (given excretory organs are lacking) or might supply growth factors or chemical defenses to their hosts. Discrete excretory structures have not been described for *Xenoturbella*.

Nervous System and Sense Organs

The nervous system of *Xenoturbella* is a diffuse intraepithelial net without much of an anterior concentration and most researchers are reluctant to call this a brain. This arrangement is similar to that seen in some acoels and nemertodermatids, although the latter do have a small anterior concentration of neurons (larger than that of xenoturbellids). The sensory furrows of xenoturbellids appear to have slightly greater concentrations

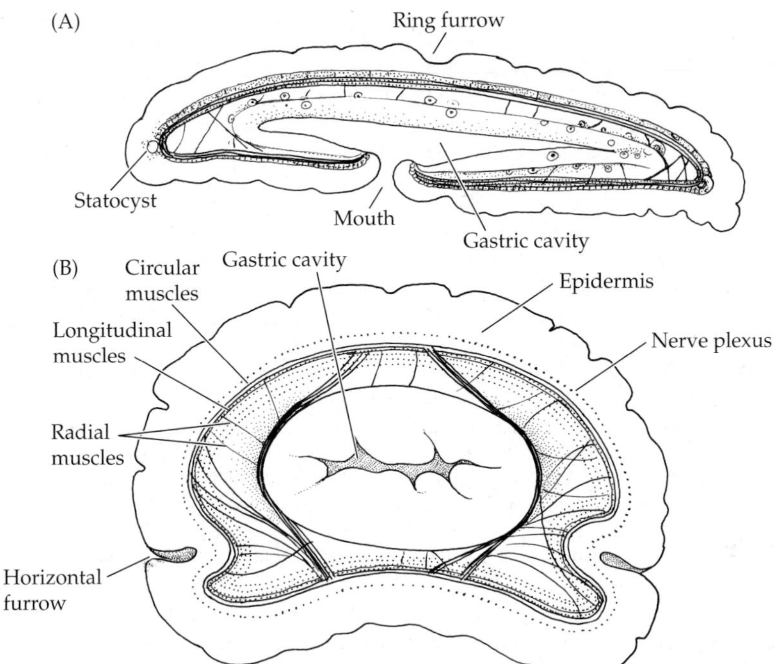

Figure 9.25 Internal morphology of *Xenoturbella bocki*. (A) Transverse section anterior to the mouth showing the gastric cavity. (B) Schematic diagram of longitudinal section anterior to mouth showing orientation of circular, longitudinal and radial muscles. Scale bars = 0.1 cm.

of neurons than other parts of their bodies. Like acoelomorphs, *Xenoturbella* have an anterior statocyst (Figure 9.25B), but the arrangement of muscles and neurons associated with this structure differs in that it appears to be embedded within the nerve net rather than specifically supplied with connecting commissures.

Reproduction and Development

Xenoturbellids are simultaneous hermaphrodites producing relatively large diameter, yolky eggs. Neither well-developed ovaries nor testes have been observed in adult individuals. In particular, male gonads appear to consist simply of a layer of male sex cells surrounding the gut. Sperm develop in clumps and appear to be of a "primitive" type, usually associated with external fertilization, wherein spermatids possess a small conical acrosome and a single flagellum. There are no copulatory organs and gametes appear to be spawned either through the gut or mouth opening. Although *Xenoturbella* has been said to have direct development, as in acoelomorphs, recent work has suggested the "hatching stage" might be called a larva; these are elongate/ovoid, swimming with a rotating motion with uniform ciliation, and have an apical tuft of cilia that are 20–30 μm in length. No mouth or blastopore has been seen in the larva.

Selected References

General References

Achatz, J. G., M. Chiodin, W. Salvenmoser, S. Tyler and P. Martinez. 2012. The Acoela: on their kind and kinships, especially with nemertodermatids and xenoturbellids (Bilateria *incertae sedis*). Org. Divers. Evol. doi: 10.1007/s13127-012-0112-4

Aguinaldo, A. M. A., J. M. Turbeville, L. S. Lindford, M. C. Rivera, J. R. Garey, R. A. Raff and J. A. Lake. 1997. Evidence for a clade of nematodes, arthropods and other moulting animals. Nature 387: 489–493.

Baguñà, J., P. Martínez, J. Paps and M. Riutort. 2008. Unraveling body plan and axial evolution in the Bilateria with molecular markers. Pp. 213–235 in A. Minelli and G. Fusco (eds.), *Evolving Pathways: Key Themes in Evolutionary Developmental Biology*. Cambridge Univ. Press, London.

Borner, J., P. Rehm, R. O. Schill, I. Ebersberger and T. Burmester. A transcriptome approach to ecdysozoan phylogeny. Mol. Phylog. Evol. 80: 79–87.

Bourlat, S. J. and A. Hejnol. 2009. Acoels. Cur. Biol. 19: R279–R280.

Brent, M. and 21 others. 2013. A comprehensive analysis of bilaterian mitochondrial genomes and phylogeny. Mol. Phylog. Evol. 69: 352–364.

Christiaen, L., Y. Jaszczyszyn, M. Kerfant, S. Kano, V. Thermes and J. S. Joly. 2007. Evolutionary modification of mouth position in deuterostomes. Semin. Cell. Dev. Biol. 18: 502–511.

Dunn, C. W. and 17 others. 2008. Broad taxon sampling improves resolution of the animal tree of life. Nature 452: 745–750.

Dunn, C. W., G. Giribet, G. D. Edgecombe and A. Hejnol. 2014. Animal phylogeny and its evolutionary implications. Ann. Rev. Ecol. Evol. Syst. 45: 371–395.

Edgecombe, G. D. and 8 others. 2011. Higher-level metazoan relationships: recent progress and remaining questions. Org. Divers. Evol. 11: 151–172.

Egger, B., D. and 24 others. 2009. To be or not to be a flatworm: the acoel controversy. PLoS ONE 4(5): e5502. doi: 10.1371/journal.pone.0005502

Ehlers, U. 1992. Frontal glandular and sensory structures in *Nemertoderma* (Nemertodermatida) and *Paratomella* (Acoela): ultrastructure and phylogenetic implications for the monophyly of the Euplathelminthes (Platyhelminthes). Zoomorphology 112: 227–236.

Fautin, D. and R. Mariscal. 1991. Placozoa, Porifera, Cnidaria and Ctenophora. *Vol. 2, Microscopic Anatomy of the Invertebrates.* Wiley-Liss, New York.

Fritzenwanker, J. H., J. Gerhart, R. M. Freeman Jr. and C. J. Lowe. 2014. The Fox/Forkhead transcription factor family of the hemichordate *Saccoglossus kowalevskii*. EvoDevo 5: 17.

Giese, A. C., J. S. Pearse and V. B. Pearse (eds.). 1991. *Reproduction of Marine Invertebrates, Vol. 6*. Boxwood Press, Pacific Grove, CA.

Gillis, J. A., J. H. Fritzenwanker and C. J. Lowe. 2012. A stem-deuterostome origin of the vertebrate pharyngeal transcriptional network. Proc. Royal Soc. B 279: 237–246.

Giribet, G. 2003. Molecules, development and fossils in the study of metazoan evolution: Articulata versus Ecdysozoa revisited. Zoology 106: 303–326.

Giribet, G. 2008. Assembling the lophotrochozoan (= Spiralian) tree of life. Phil. Trans. Roy. Soc. B, Biol. Sci. 363: 1513–1522.

Hejnol, A. and J. M. Martín-Durán. 2015. Getting to the bottom of anal evolution. Zoologischer Anzeiger. doi: 10.1016/j.jcz.2015.02.006

Hejnol, A. and M. Q. Martindale. 2008. Acoel development indicates the independent evolution of the bilaterian mouth and anus. Nature 456: 382–386.

Hejnol, A. and M. Q. Martindale. 2008. Acoel development supports a simple planula-like Urbilaterian. Philos. Trans. Roy. Soc. London. B, Biol. Sci. 363: 1493–1501.

Hejnol, A. and 16 others. 2009. Assessing the root of bilaterian animals with scalable phylogenomic methods. Proc. Roy. Soc. B, Biol. Sci. 276: 4261–4270.

Hinman, V. F. and E. H. Davidson. 2007. Evolutionary plasticity of developmental gene regulatory network architecture. Proc. Natl. Acad. Sci. 104: 19404–19409.

Holland, L. Z. and N. D. Holland. 2007. A revised fate map for amphioxus and the evolution of axial patterning in chordates. Integr. Comp. Biol. 47: 360–370.

Jager, M., R. Chiori, A. Allé, C. Dayraud, E. Quéinnec and M. Manuel. 2011. New insights on ctenophore neural anatomy: immunofluorescence study in *Pleurobrachia pileus* (Müller, 1776). J. Exp. Zool. B 316: 171–187.

Jägersten, G. 1972. *Evolution of the Metazoan Life Cycle*. Academic Press, London.

Laumer, C. E. and 10 others. 2015. Spiralian phylogeny informs the evolution of microscopic lineages. Curr. Biol. 25: 1–6.

Lowe, C. J. and A. M. Pani. 2011. Animal Evolution: a soap opera of unremarkable worms. Cur. Biol. 21(4): R151–R153.

Lundin, K. 1997. Comparative ultrastructure of the epidermal ciliary rootlets and associated structures in species of the Nemertodermatida and Acoela (Platyhelminthes). Zoomorphology 117: 81–92.

Marlétaz, F., and 11 others. 2006. Chaetognath phylogenomics: a protostome with deuterostome-like development. Cur. Biol. 16: R577–R578.

Marlow, H., M. A. Tosches, R. Tomer, P. R. Steinmetz, A. Lauri, T. Larsson and D. Arendt. 2014. Larval body patterning and apical organs are conserved in animal evolution. BMC Biol. 12(7): 1–17.

Martín-Durán, J. M., R. Janssen, S. Wennberg, G. E. Budd and A. Hejnol. 2012. Deuterostomic development in the protostome *Priapulus caudatus*. Curr. Biol. 22: 2161–2166.

Martindale, M. Q. and A. Hejnol. 2009. A developmental perspective: changes in the position of the blastopore during bilaterian evolution. Development Cell 17: 162–174.

Maxmen, A. 2011. A can of worms. Nature 470: 161–162.

Moreno, E., J. Pernmanyer and P. Martinez. 2011. The origin of patterning systems in Bilateria—insights from the Hox and ParaHox genes in Acoelomorpha. Genom. Proteom. Bioinform. 9(3): 65–76.

Nesnidal, M. P. and 9 others. 2013. New phylogenomic data support the monophyly of Lophophorata and an ectoproct-phoronid clade, and indicate that Polyzoa and Kryptrochozoa are caused by systematic bias. BMC Evol. Biol. 13: 253–272.

Nielsen, C. 2005. Trochophora larvae: cell-lineages, ciliary bands and body regions 2. Other groups and general discussion. J. Exp. Zool. B 304: 401–447.

Nielsen, C. 2012. How to make a protostome. Invertebr. Syst. 26: 25–40.

Nielsen, C. 2015. Larval nervous systems: true larval and precocious adult. J. Exper. Biol. 218: 629–636.

Nosenko, T., and 12 others. 2013. Deep metazoan phylogeny: when different genes tell different stories. Mol. Phylog. Evol. 67: 223–233.

Pahg, K. and M. Q. Martindale. 2008. Developmental expression of homeobox genes in the ctenophore *Mnemiopsis leidyi*. Dev. Genes Evol. 218: 307–319.

Pani, A. M., E. E. Mullarkey, J. Aronowicz, S. Assimacopoulos, E. A. Grove and C. J. Lowe. 2013. Ancient deuterostome origins of vertebrate brain signaling centres. Nature 483: 289–294.

Perea-Atienza, E., B. Gavilan, M. Chiodin, J. F. Abril, K. J. Hoff, A. J. Poustka and P. Martinez. 2015. The nervous system of Xenacoelomorpha: A genomic perspective. J. Exp. Biol. 218: 618–628. doi: 10.1242/jeb.110379

Phillippe, H. and 19 others. 2009. Phylogenomics revives traditional views on deep animal relationships. Curr. Biol. 19: 1–7.

Philippe, H. and 8 others. 2011. Acoelomorph flatworms are deuterostomes related to *Xenoturbella*. Nature 470: 255–258.

Pick, K. S. and 10 others. 2010. Improved phylogenomic taxon sampling noticeably affects nonbilaterian relationships. Mol. Biol. Evol. 27(9): 1983–1987.

Richter, S. and 20 others. 2010. Invertebrate neurophylogney: suggested terms and definitions for a neuroanatomical glossary. Front. Zool. 7: 29.

Ruiz-Trillo, I., J. Paps, M. Loukota, C. Ribera, U. Jondelius, J. Baguñà and M. Riutort. 2002. A phylogenetic analysis of myosin heavy chain type II sequences corroborates that Acoela and Nemertodermatida are basal bilaterians. Proc. Nat. Acad. Sci. 99: 11246–11251.

Ruiz-Trillo, I., M. Riutort, H. M. Fourcade, J. Baguñà and J. L. Boore. 2004. Mitochondrial genome data support the basal position of Acoelomorpha and the polyphyly of the Platyhelminthes. Mol. Phylog. Evol. 3: 321–332.

Ruiz-Trillo, I., M. Riutort, D. T. J. Littlewood, E. A. Herniou and J. Baguñà. 1999. Acoel flatworms: earliest extant bilaterian metazoans, not members of Platyhelminthes. Science 283: 1919–1923.

Stern, C. D. (ed.). 2004. *Gastrulation: From Cells to Embryos*. Cold Spring Harbor Laboratory Press, New York.

Stöger, I. and M. Schrödl. 2013. Mitogenomics does not resolve deep molluscan relationships (yet?). Mol. Phylog. Evol. 69: 376–392.

Struck, T. H. and 12 others. 2014. Platyzoan paraphyly based on phylogenomic data supports a noncoelomate ancestry of Spiralia. Mol. Biol. Evol. 31(7): 1833–1849.

Swalla, B. J. and A. B. Smith. 2008. Deciphering deuterostome phylogeny: molecular, morphological and palaeontological perspectives. Phil. Trans. Roy. Soc. B, Biol. Sci. 363: 1557–1568.

Telford, M. J. 2008. Xenoturbellida: the fourth deuterostome phylum and the diet of worms. Genesis 46: 580–586.

Telford, M. J. and D. T. J. Littlewood (eds.). 2009. *Animal Evolution: Genes, Genomes, Fossils and Trees*. Oxford University Press, Oxford.

Telford, M. J., A. E. Lockyer, C. Cartwright-Finch and D. T. J. Littlewood. 2003. Combined large and small subunit ribosomal RNA phylogenies support a basal position of the acoelomorph flatworms. Proc. Roy. Soc. London B, Biol. Sci. 270: 1077–1083.

Vannier, J., I. Calandra, C. Gaillard and A. Zylinska. 2010. Priapulid worms: pioneer horizontal burrowers at the Precambrian-Cambrian boundary. Geology 38: 711–714.

Vargas, P. and R. Zardoya (eds.). 2014. *The Tree of Life: Evolution and Classification of Living Organisms*. Sinauer Associates, Sunderland, MA.

Wägele, J. W. and T. Bartolomaeus (eds.). 2014. *Deep Metazoan Phylogeny: The Backbone of the Tree of Life. New Insights from Analyses of Molecules, Morphology, and Theory of Data Analysis*. De Gruyter, Berlin.

Wallberg, A., M. Curini-Galletti, A. Ahmadzadeh and U. Jondelius. 2007. Dismissal of Acoelomorpha: Acoela and Nemertodermatida are separate early bilaterian clades. Zool. Scripta 36: 509–523.

Webster, B. L., and 7 others. 2006. Mitogenomics and phylogenomics reveal priapulid worms as extant models of the ancestral ecdysozoan. Evol. Dev. 8: 502–510.

Whelan, N. V., K. M. Kocot, L. L. Moroz and K. M. Halanych. 2015. Error, signal, and the placement of Ctenophora sister to all other animals. PNAS 112(18): 5773–5778.

Zimmer, R. L. 1997. Phoronids, brachiopods, and bryozoans, the lophophorates. Pp. 279–305 in S. F. Gilbert and A. M. Raunio (eds.), *Embryology. Constructing the Organism*. Sinauer Associates, Sunderland.

Acoela

Achatz, J. G., M. Hooge, A. Wallberg, U. Jondelius and S. Tyler. 2010. Systematic revision of acoels with 9+0 sperm ultrastructure (Convolutida) and the influence of sexual conflict on morphology. J. Zool. Syst. Evol. Res. 48(1): 9–32.

Achatz, J. G. and P. Martinez. 2012. The nervous system of *Isodiametra pulchra* (Acoela) with a discussion on the neuroanatomy of the Xenacoelomorpha and its evolutionary implications. Front.Zool. 9: 27.

Barneah, O., I. Brickner, M. Hooge, V. M. Weis and Y. Benayahu. 2007. First evidence of maternal transmission of algal endosymbionts at an oocyte stage in a triploblastic host, with observations on reproduction in *Waminoa brickneri* (Acoelomorpha). Invert. Biol. 126(2): 113–119.

Boone, M., M. Willems, M. Claeys and T. Artois. 2011. Spermatogenesis and the structure of the testes in *Isodiametra pulchra* (Isodiametridae, Acoela). Acta Zoologica 92: 101–108.

Bourlat, S. J. and A. Hejnol. 2009. Acoels. Current Biology 19(7): 279–280.

Bush, L. F. 1981. Marine flora and fauna of the northeastern United States. Turbellaria: Acoela and Nemertodermatida. NOAA Technical Report NMFS Circular 440: 1–71.

Chiodin, M., A. Børve, E. Berezikov, P. Ladurner, P. Martinez and A. Hejnol. 2013. Mesodermal gene expression in the acoel *Isodiametra pulchra* indicates a low number of mesodermal cell types and the endomesodermal origin of the gonads. PLoS ONE 8(2): e55499.

Crezée, M. 1975. Monograph of the Solenofilomorphidae (Turbellaria: Acoela). Int. Rev. Ges. Hydrobiology 60: 769–845.

Ehlers, U. and B. Sopott-Ehlers. 1997. Ultrastructure of the subepidermal musculature of *Xenoturbella bocki*, the adelphotaxon of the Bilateria. Zoomorphology 117: 71–79.

Ferrero, E. 1973. A fine structural analysis of the statocyst in Turbellaria Acoela. Zool. Scripta 2: 5–16.

Gaerber, C. W., W. Salvenmoser, R. M. Rieger and R. Gschwentner. 2007. The nervous system of *Convolutriloba* (Acoela) and its patterning during regeneration after asexual reproduction. Zoomorphology 126: 73–87.

Gschwentner, R., S. Baric and R. Rieger. 2002. New model for the formation and function of sagittocysts: *Symsagittifera corsicae* n. sp. (Acoela). Invert. Biol. 212: 95–103.

Hanson, E. D. 1960. Asexual reproduction in acoelous turbellaria. Yale J. Biol. Med. 33: 107–11.

Haapkylä, J., A. S. Seymour, O. Barneah, I. Brickner, S. Hennige, D. Suggett and D. Smith. 2009. Association of *Waminoa* sp. (Acoela) with corals in the Wakatobi Marine Park, South-East Sulawesi, Indonesia. Mar. Biol. 156: 1021–1027.

Hejnol, A. and M. Q. Martindale. 2009. Coordinated spatial and temporal expression of Hox genes during embryogenesis in the acoel Convolutriloba longifissura. BMC Biol. 7:65.

Henry, J. Q., M. Q. Martindale and B. C. Boyer. 2000. The unique developmental program of the acoel flatworm, *Neochildia fusca*. Dev. Biol. 220: 285–295.

Hirose, E. and M. Hirose. 2007. Body colors and algal distribution in the acoel flatworm *Convolutriloba longifissura*: histology and ultrastructure. Zool. Sci. 24(12): 1241–1246.

Hooge, M. D. 2001. Evolution of body-wall musculature in the Platyhelminthes (Acoelomorpha, Catenulida, Rhabditophora). J. Morph. 249: 171–194.

Hooge, M. D. and S. Tyler. 2004. New tools for resolving phylogenies: a systematic revision of the Convolutidae (Acoelomorpha, Acoela). J. Zool. Sci. 43(2), 100–113.

Hooge, M. D. and S. Tyler. 2008. Concordance of molecular and morphological data: the example of the Acoela. Integr. Comp. Biol. 46 (2): 118–124.

Hooge, M., A. Wallberg, C. Todt, A. Maloy, U. Jondelius and S. Tyler. 2007. A revision of the systematics of panther worms (*Hofstenia* spp., Acoela), with notes on color variation and genetic variation within the genus. Hydrobiologia 592: 439–454.

Hyman, L. H. 1937. Reproductive system and copulation in *Amphiscolops langerhansi* (Turbellaria Acoela). Biol. Bull. 72: 319–326.

Jennings, J. B. 1957. Studies on feeding, digestion, and food storage in free-living flatworms (Platyhelminthes: Turbellaria). Biol. Bull. 112(1), 63–80.

Jondelius, U., A. Wallberg, M. Hooge and O. I. Raikova. 2011. How the worm got its pharynx: phylogeny, classification and Bayesian assessment of character evolution in Acoela. Syst. Biol. 60(6): 845–871, 2011

Kotikova, E. A. and O. I. Raikova. 2008. Architectonics of the central nervous system of Acoela, Platyhelminthes, and Rotifera. Zhurnal Evolyutsionnoi Biokhimii i Fiziologii 44(1): 83–93.

Nozawa, K., D. L. Taylor and L. Provasoli. 1972. Respiration and photosynthesis in *Convoluta roscoffensis* Graff, infected with various symbionts. Biol. Bull. 143: 420–430.

Petrov, A., M. Hooge and S. Tyler. 2004. Ultrastructure of sperm in Acoela (Acoelomorpha) and its concordance with molecular systematics. Invert. Biol. 123(3): 183–197.

Petrov, A., M. Hooge and S. Tyler. 2006. Comparative morphology of the bursal nozzles in Acoels (Acoela, Acoelomorpha). J. Morph. 267: 634–648.

Raikova, O. I., M. Reuter, M. K. S. Gustafsson, A. G. Maule, D. W. Halton and U. Jondelius. 2003. Evolution of the nervous system in *Paraphanostoma* (Acoela). Zool. Script., 33: 71–88.

Reuter M. and N. Kreshchenko. 2004. Flatworm asexual multiplication implicates stem cells and regeneration. Can. J. Zool. 82: 334–356.

Semmler, H., M. Chiodin, X. Bailly, P. Martinez and A. Wanninger. 2010. Steps towards a centralized nervous system in basal bilaterians: insights from neurogenesis of the acoel *Symsagittifera roscoffensis*. Develop. Growth Differ. 52: 701–713.

Smith, J. III, S. Tyler, M. B. Thomas and R. M. Rieger. 1982. The morphology of turbellarian rhabdites: phylogenetic implications. Trans. Amer. Microscop. Soc. 101(3): 209–228.

Smith, J. P. S. III and S. Tyler. 1986. Frontal organs in the Acoelomorpha (Turbellaria): ultrastructure and phylogenetic significance. Hydrobiology 132: 71–78.

Taylor, D. 1984. Translocation of carbon and nitrogen from the turbellarian host *Amphiscolops langerhansi* (Turbellaria: Acoela) to its algal endosymbiont *Amphidinium klebsii* (Dynophyceae). Zool. J. Linn. Soc. 80: 337–344.

Todt, C. 2009. Structure and evolution of the pharynx simplex in acoel flatworms (Acoela). J. Morph. 270: 271–290.

Yamazu, T. 1991. Fine structure and function of ocelli and sagittocysts of acoel flatworms. Hydrobiology 227: 273–282.

Nemertodermatida

Boone, M., W. Houthoofd, W. Bert and T. Artois. 2011. First record of Nemertodermatida from Belgian marine waters. Belg. J. Zool., 141 (1): 62–64.

Jiménez-Guri, E., J. Paps, J. García-Fernàndez and E. Saló. 2006. Hox and ParaHox genes in Nemertodermatida, a basal bilaterian clade. Int. J. Dev. Biol. 50: 675–679.

Jondelius, U., K. Larsson and O. Raikova. 2004. Cleavage in *Nemertoderma westbladi* (Nemertodermatida) and its phylogenetic significance. Zoomorphology 123: 221–225.

Lundin, K. 1998. Symbiotic bacteria on the epidermis of species of the Nemertodermatida (Platyhelminthes, Acoelomorpha). Acta Zool. (Stockholm) 79(3): 187–191.

Lundin, K. and J. Hendelberg. 1996. Degenerating epidermal bodies ("pulsatile bodies") in *Meara stichopi* (Platyhelminthes, Nemertodermatida). Zoomorphology 116: 1–5.

Lundin, K. and J. Hendelberg. 1998. Is the sperm type of the Nemertodermatida close to that of the ancestral Platyhelminthes? Hydrobiology 383: 197–205.

Meyer-Wachsmuth, I., M. C. Galletti and U. Jondelius. 2014. Hyper-Cryptic marine meiofauna: Species complexes in Nemertodermatida. PLoS ONE. doi: 10.1371/journal.pone.0107688

Meyer-Wachsmuth. I., O. I. Raikova and U. Jondelius. 2013. The muscular system of *Nemertoderma westbladi* and *Meara stichopi* (Nemertodermatida, Acoelomorpha). Zoomorphology. doi: 10.1007/s00435-013-0191-6

Raikova, O. I., M. Reuter, U. Jondelius and M. K. S. Gustafsson. 2000. The brain of the Nemertodermatida (Platyhelminthes) as revealed by anti-5HT and anti-FMRFamide immunostainings. Tissue Cell 32 (5) 358–365.

Sterrer. W. 1998. New and known Nemertodermatida (Platyhelminthes; Acoelomorpha)—a revision. Belg. J. Zool. 128: 55–92.

Xenoturbellida

Franzén, A. and B. A. Afzelius. 1987. The ciliated epidermis of *Xenoturbella bocki* (Platyhelminthes, Xenoturbellida) with some phylogenetic considerations. Zool. Scripta 16(1): 9–17

Fritsch, G. and 8 others. 2008. PCR survey of *Xenoturbella bocki* Hox genes. J. Exp. Zool. (Mol. Dev. Evol.) 310: 278–284.

Kjeldsen, K. U., M. Obst, H. Nakano, P. Funch and A. Schramm. 2010. Two types of endosymbiotic bacteria in the enigmatic marine worm Xenoturbella bocki. App. Environment. Microbiol. doi: 10.1128/AEM.01092-09.

Nakano, H. and 7 others. 2013. *Xenoturbella bocki* exhibits direct development with similarities to Acoelomorpha. Nat. Commun. 4: 1–6.

Nielsen, C. 2010. After all: *Xenoturbella* is an acoelomorph! Evol. Dev. 12(3): 241–243.

CHAPTER 10

Phylum Platyhelminthes

The Flatworms

The phylum Platyhelminthes (Greek *platy*, "flat"; *helminth*, "worm") includes about 26,500 extant species of free-living and parasitic worms; more than 100,000 undescribed species have been estimated to exist. Platyhelminths are triploblastic, nonsegmented, bilateral, acoelomate, soft-bodied, dorsoventrally flattened worms (Box 10A). They display a variety of body forms (Figure 10.1A–N) and inhabit a wide range of environments. About 75% of all described species are parasitic, mostly belonging to the infraclasses Monogenea and Trematoda (the flukes) and Cestoda (the tapeworms). Most of the free-living forms live in marine and freshwater benthic habitats and comprise a diverse group formerly known as "turbellarians"; a few are terrestrial and some are symbiotic in or on other invertebrates. Free-living marine flatworms are often some of the most colorful and graceful creatures found in shallow tropical waters and tide pools. As their name suggests, most flatworms are strikingly flattened dorsoventrally, although the body shape varies from broadly oval to elongate and ribbon-like; a few bear short tentacles at the anterior end or have other elaborations of the body surface. The free-living forms range from less than 1 mm to about 30 cm long; although most familiar species are 1–4 cm long, the vast majority of free-living flatworms are "microturbellarians." The largest of all platyhelminths are certain tapeworms that attain lengths of several meters (a tapeworm that infests blue whales grows to 10 m in length!).

The combined features of the platyhelminths represent a suite of attributes marking major advancements in the evolution of the Metazoa (Box 10A). Combined with a third germ layer (mesoderm), bilateral symmetry, and cephalization are some sophisticated organs and organ systems and a trend toward centralization of the nervous system. The solid (acoelomate) body plan usually includes a relatively dense mesenchyme (often called "parenchyme") between the gut and the body wall. The mesenchyme is not homogeneous, but comprises a multitude of differentiated cell types and small gaps or **lacunae**. Within the mesenchyme of most flatworms are discrete excretory–osmoregulatory structures, the protonephridia,

Classification of The Animal Kingdom (Metazoa)

Non-Bilateria*
(a.k.a. the diploblasts)
- PHYLUM PORIFERA
- PHYLUM PLACOZOA
- PHYLUM CNIDARIA
- PHYLUM CTENOPHORA

Bilateria
(a.k.a. the triploblasts)
- PHYLUM XENACOELOMORPHA

Protostomia
- PHYLUM CHAETOGNATHA

SPIRALIA
- **PHYLUM PLATYHELMINTHES**
- PHYLUM GASTROTRICHA
- PHYLUM RHOMBOZOA
- PHYLUM ORTHONECTIDA
- PHYLUM NEMERTEA
- PHYLUM MOLLUSCA
- PHYLUM ANNELIDA
- PHYLUM ENTOPROCTA
- PHYLUM CYCLIOPHORA

Gnathifera
- PHYLUM GNATHOSTOMULIDA
- PHYLUM MICROGNATHOZOA
- PHYLUM ROTIFERA

Lophophorata
- PHYLUM PHORONIDA
- PHYLUM BRYOZOA
- PHYLUM BRACHIOPODA

ECDYSOZOA

Nematoida
- PHYLUM NEMATODA
- PHYLUM NEMATOMORPHA

Scalidophora
- PHYLUM KINORHYNCHA
- PHYLUM PRIAPULA
- PHYLUM LORICIFERA

Panarthropoda
- PHYLUM TARDIGRADA
- PHYLUM ONYCHOPHORA
- PHYLUM ARTHROPODA
  - SUBPHYLUM CRUSTACEA*
  - SUBPHYLUM HEXAPODA
  - SUBPHYLUM MYRIAPODA
  - SUBPHYLUM CHELICERATA

Deuterostomia
- PHYLUM ECHINODERMATA
- PHYLUM HEMICHORDATA
- PHYLUM CHORDATA

*Paraphyletic group

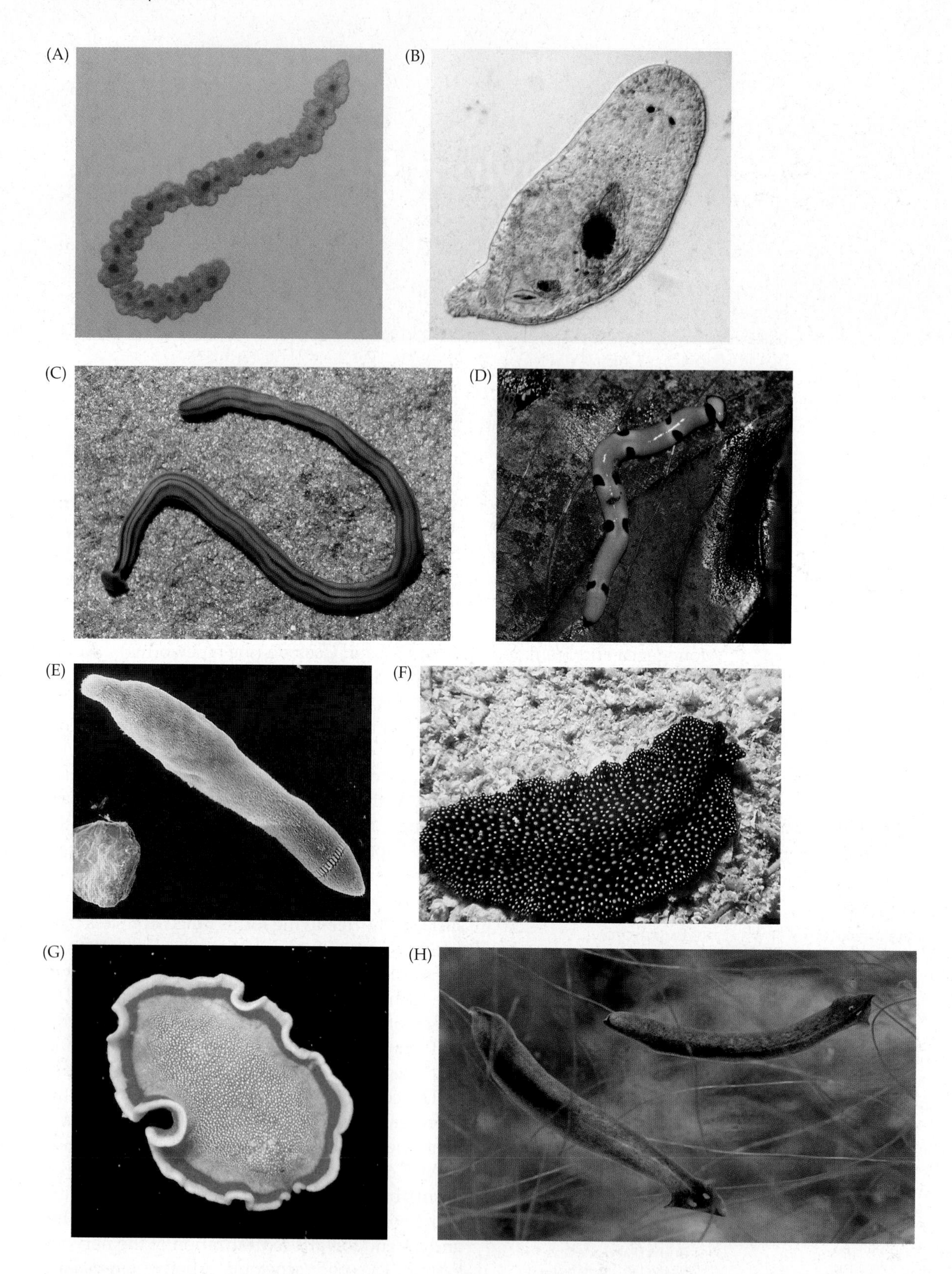

which are found in many invertebrate taxa, especially among protostomes. Most flatworms possess complex reproductive systems and an incomplete yet complex gut with a single opening serving for both ingestion and egestion. The mouth leads to a pharynx of varying complexity and then to a blind intestine. The gut is entirely lacking in tapeworms, as well as in a few other symbiotic species.

Taxonomic History and Classification

In his first edition of *Systema Naturae* (1735), Linnaeus established two phyla to encompass all of the known invertebrates. To one he assigned the insects and to the other the rest of the invertebrates. Linnaeus called this latter taxon Vermes (Greek, "worms"). By the thir-

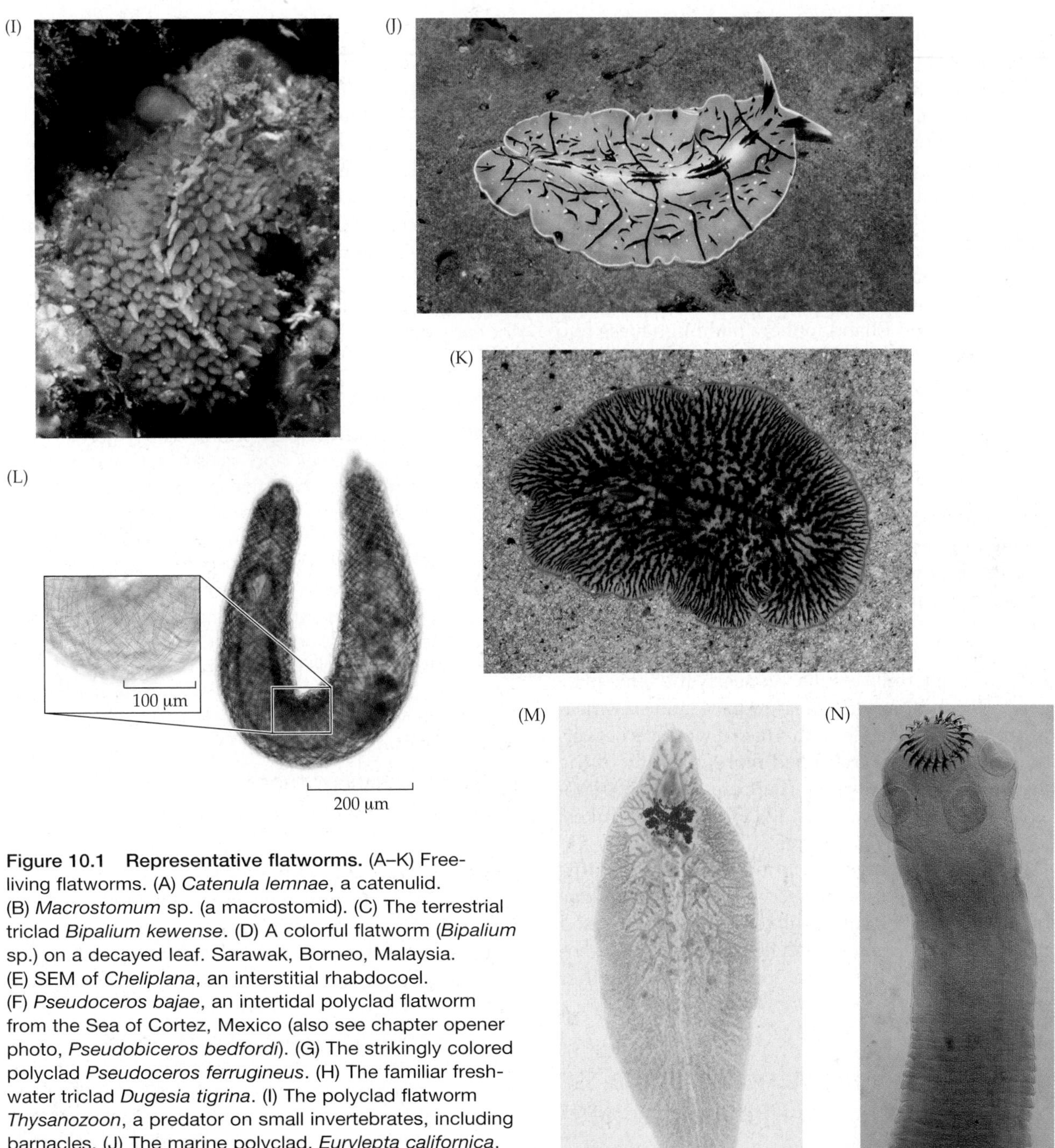

Figure 10.1 Representative flatworms. (A–K) Free-living flatworms. (A) *Catenula lemnae*, a catenulid. (B) *Macrostomum* sp. (a macrostomid). (C) The terrestrial triclad *Bipalium kewense*. (D) A colorful flatworm (*Bipalium* sp.) on a decayed leaf. Sarawak, Borneo, Malaysia. (E) SEM of *Cheliplana*, an interstitial rhabdocoel. (F) *Pseudoceros bajae*, an intertidal polyclad flatworm from the Sea of Cortez, Mexico (also see chapter opener photo, *Pseudobiceros bedfordi*). (G) The strikingly colored polyclad *Pseudoceros ferrugineus*. (H) The familiar freshwater triclad *Dugesia tigrina*. (I) The polyclad flatworm *Thysanozoon*, a predator on small invertebrates, including barnacles. (J) The marine polyclad, *Eurylepta californica*. (K) *Alloioplana californica*, a common polyclad on the Pacific Coast of North America. (L) *Acanthomacrostomum* sp. A meiofaunal flatworm that has a layer of calcareous spicules just below the epidermis. This skeletal framework may provide protection and support in aiding in locomotion. (M) The liver fluke *Fasciola hepatica* (Trematoda: Digenea). (N) Anterior end of the tapeworm *Taenia* (Cestoda: Cyclophyllidea).

teenth edition of *Systema Naturae* (1788), the various groups of flatworms were placed together in the order Intestina. During the early 1800s, several biologists, including Lamarck and Cuvier, questioned and rejected the concept of the phylum Vermes, although the taxon continued to surface from time to time and actually persisted into the twentieth century as a dumping ground for almost any creatures with wormlike bodies (and many that were not so wormlike).

During the nineteenth century, the flatworms were eventually separated from most other groups of worms and wormlike creatures. In 1851, C. Vogt isolated the flatworms and the nemerteans as a single taxon, which he called the Platyelmia, a name changed to Platyelminthes by Karl Gegenbaur in 1859. (Unfortunately, Gegenbaur also resurrected the phylum Vermes.) Gegenbaur's Platyelminthes (now Platyhelminthes) was eventually raised to the rank of phylum, comprising four classes:

Turbellaria, Nemertea, Trematoda, and Cestoda. In 1876 Charles Minot dropped the nemerteans from this assemblage, although many workers did not accept this change for several decades. Libbie Hyman (1951) asserted the monophyly of Platyhelminthes, a view shared by Tor Karling (1974), Peter Ax (1985) and Ulrich Ehlers (1985) who recognized three clades (Acoelomorpha, Catenulida, and Rhabditophora) that have since been supported by molecular phylogenetic studies. Later analyses of morphology and the advent of molecular gene sequence analyses have supported the monophyly of Catenulida and Rhabditophora (although these two major flatworm clades still lack unambiguous morphological synapomorphies), but suggested acoelomorphs were basal bilaterians rather than flatworms (see Chapter 9). With the removal of the Acoelomorpha, Platyhelminthes is now viewed as a monophyletic phylum.

The internal classification of flatworms has been subjected to frequent revisions. However, two recent studies by independent groups of researchers targeted platyhelminth phylogeny by analyzing different sets of thousands of genes (Egger et al. 2015; Laumer et al. 2015). Remarkably, these genomic studies, published at nearly the same time, resulted in nearly the same phylogeny. Our summary tree (Figure 10.34) and classification are based primarily on this recent work. However, students can expect continued reorganization of the flatworms in the near future. There are about 6,500 species of free-living flatworms, 12,000 species of flukes, and 8,000 species of tapeworms.

The free-living flatworms (formerly classified as "turbellarians," a category now known to be a paraphyletic group, although the name is still sometimes loosely used for the non-neodermatan platyhelminths) were previously grouped into two taxa on the basis of whether yolk is deposited within the cytoplasm of the ova (**endolecithal ova**) or separately, in special cells outside the ova (**ectolecithal ova**). Those with endolecithal ova were placed in the Archoophora and those with ectolecithal ova in the Neoophora. These names have been abandoned as formal taxa because they are now known to be nonmonophyletic. However, the names are still descriptive because the placement of yolk, as well as the uterine structures that mediate this process, provide additional characters for describing the various flatworm orders. The patterns also have important implications in the early development of these animals.

BOX 10A Characteristics of the Phylum Platyhelminthes

1. Parasitic or free-living, unsegmented worms (the Cestoda are strobilated)
2. Triploblastic, acoelomate, bilaterally symmetrical; dorsoventrally flattened
3. Spiral cleavage (in non-parasitic species) and 4d mesoderm
4. Complex, though incomplete, gut usually present; gut absent in some parasitic forms (Cestoda)
5. Cephalized, with a central nervous system comprising an anterior cerebral ganglion and (usually) longitudinal nerve cords connected by transverse commissures (ladderlike nervous system)
6. With protonephridia as excretory/osmoregulatory structures
7. Hermaphroditic, with complex reproductive system

PHYLUM PLATYHELMINTHES

SUBPHYLUM CATENULIDEA Catenulids (Figure 10.1A). Simple anterior pharynx and saclike gut; mesenchyme sometimes reduced to a fluid matrix (making them effectively blastocoelomate); sometimes with statocyst bearing 1 to 3 statoliths; brain lies at base or middle of preoral lobe; with a ciliated ventrolateral groove between base of preoral lobe and the rest of the body; tubules of excretory system mediodorsal; with a single biflagellate protonephridium; male genital opening dorsal and anterior; female ducts and accessory organs are lacking; with endolecithal ova and spiral cleavage. Catenulids are elongate freshwater and marine forms. A single order, Catenulida. (e.g., *Catenula*, *Paracatenula*, *Stenostomum*)

SUBPHYLUM RHABDITOPHORA With lamellated rhabdites, a duo-gland adhesive system and multiflagellate terminal cell in the protonephridia; most taxa with biflagellate sperm.

INFRAPHYLUM MACROSTOMORPHA Rhabditophora with duo-gland adhesive organ consisting of viscid gland and releasing gland necks emerging in a common collar of anchor cell microvilli; with a post-oral neural commissure, and a protrusible pharynx simplex; with aflagellate sperm, endolecithal ova, and spiral cleavage.

ORDER HAPLOPHARYNGIDA Minute worms (to 6 mm long) with a simple proboscis and pharynx; proboscis separate from pharynx and beneath the anterior tip of the body (reminiscent of nemerteans); anal pore weakly developed, but permanent; brain encapsulated by a unique membrane; oviduct posterior to male genital apparatus; male copulatory organ consists of a posterior prostatic vesicle and an anterior stylet apparatus. One genus (*Haplopharynx*) and at least three species

ORDER MACROSTOMIDA Macrostomids (Figure 10.1B). Simple pharynx; simple, saclike gut; with endolecithal ova and a common oviduct anterior to the male genital apparatus; small and predominately interstitial forms; marine and freshwater. (e.g., *Acanthomacrostomum*, *Macrostomum*, *Microstomum*)

INFRAPHYLUM TREPAXONEMATA Rhabditophorans with biflagellate sperm bearing a 9 × 2 + "1" pattern of microtubules.

SUPERCLASS AMPLIMATRICATA With a tendency to express an ample extracellular matrix; with spiral cleavage.

ORDER POLYCLADIDA Polyclads (Figure 10.1F,G,I–K). A diverse group of relatively large free-living flatworms with endolecithal ova; nearly all are marine; common in littoral zones throughout the world, especially in the tropics; predominately benthic and free-living. Some are so large and colorful as to be easily mistaken for sea slugs; some cases of mimicry are known. Many swim by graceful undulations of the body margins. A few are pelagic or symbiotic. (e.g., *Eurylepta*, *Hoploplana*, *Leptoplana*, *Notoplana*, *Planocera*, *Prostheceraeus*, *Pseudobiceros, Pseudoceros*, *Stylochus*, *Thysanozoon*)

ORDER PRORHYNCHIDA Freshwater and terrestrial flatworms, often bearing prominent anterior auricles and an anteriorly situated, complex pharynx; ovaries lecithoepitheliate, consisting of two kinds of cells, vitellocytes embracing the growing ovocytes, both types of cells produced by a common proximal germinal tissue. (e.g., *Geocentrophora, Prorhynchus*). Some classification schemes place the freshwater Prorhynchida and the marine Gnosonesimida within the Lecithoepitheliata, a taxon comprised of about 30 species united on the basis of an intermediate condition between endolecithal and ectolecithal ova, but recent research considers Lecithoepitheliata to be nonmonophyletic.

SUPERCLASS GNOSONESIMORA Marine forms with apparent lecithoepithelial development of ova but closer structural and molecular affinities to Euneoophora; with ectolecithal ova (cleavage not yet described).

ORDER GNOSONESIMIDA A single marine genus is known (*Gnosonesima*), bearing a coniform bulbous pharynx.

SUPERCLASS EUNEOOPHORA Rhabditophorans whose ovary is divided into germinal and vitelline cell producing parts; with ectolecithal ova.

CLASS RHABDOCOELA Rhabdocoels. Bulbous or sometimes plicate pharynx; simple saclike gut without diverticula; ectolecithal ova produced by ovaries that are usually fully separate from the yolk glands; spiral cleavage.

ORDER DALYTYPHLOPLANIDA A diverse group of free-living or ecto- or endosymbionts of marine and freshwater invertebrates. Molecular data suggests three main groups: neodalyellids, with an anterior mouth (e.g., *Anoplodium*, *Graffilla*, *Pterastericola*); thalassotyphloplanids, mostly marine species with the mouth not anterior (e.g., *Kytorhynchus*, *Mesostoma*, *Typhlorhynchus*); and limnotyphloplanids, mostly freshwater species (e.g., *Castrella*, *Dalyellia*, *Microdalyelliathis*). Limnotyphloplanids are small symbionts on freshwater decapod crustaceans, although a few live on other invertebrates or on turtles. (i.e., *Temnocephala*)

ORDER KALYPTORHYNCHIA Kalyptorhynchs (Figures 10.1E and 10.11F). Mouth not terminal; with a complex eversible proboscis at anterior end that is separate from the mouth and pharynx; free-living marine and freshwater species. (e.g., *Cheliplana*, *Cystiplex*, *Gnathorhynchus*, *Gyratrix*)

CLASS PROSERIATA Freshwater and marine free-living flatworms lacking lamellate rhabdites; bearing a cylindrical plicate pharynx; simple gut; spiral cleavage.

ORDER UNGUIPHORA Proseriates with pigment in the mantle cells of rhabdomeric receptors, and usually without a statocyst; molecular studies suggest that this order is in need of revision, and also that the enigmatic genus *Ciliopharyngiella* may belong here. (e.g., *Nema-toplana*, *Polystyliphora*)

ORDER LITHOPHORA Proseriates without pigment in the mantle cells and usually with a statocyst; recent molecular studies indicate that this order is monophyletic and suggests that families Coelogynoporidae and Calviriidae as well as Otoplanidae, Archimonocelididae and Monocelididae may belong in this group. (e.g., *Calviria*, *Coelogynopora*, *Otoplana, Otoplanella*, *Archimonocelis*, *Monocelis*)

CLASS ACENTROSOMATA Rhabditophorans lacking genes controlling formation of centrosomes, leading to loss of highly regulated spiral cleavage (and, in some lineages, blastomere anarchy) during early development.

SUBCLASS ADIAPHANIDA Although there are no clear synapomorphies for this clade, it is strongly supported by molecular analyses; the name is from the ancient Greek word for "opaque" referring to the fact that most species have opaque bodies.

ORDER TRICLADIDA Triclads. (Figure 10.1C,H) Freshwater, marine, terrestrial and parasitic forms; cylindrical plicate pharynx; gut three-branched with numerous diverticula; two germinaria located at anterior end of germo-vitelloducts. Most are free-living, including the familiar planarians. (e.g., *Bdelloura*, *Bipalium*, *Crenobia*, *Dugesia* [formerly *Planaria*], *Geoplana*, *Polycelis*, *Procotyla*)

ORDER PROLECITHOPHORA Prolecithopho rans. Pharynx plicate or bulbous; gut simple; sperm aflagellate, with extensive membranous folds. Reduction of duo-gland adhesive system complete; genitalia variable but male aperture often opens forward, female reproductive system often opens to a common pore; small, free-living, marine and fresh water. (e.g., *Plagiostomum*, *Urastoma*)

ORDER FECAMPIIDA Fecampiids. Endoparasites of various marine invertebrates and vertebrates; lacking a pharynx and intestine; with a ciliated epidermis but with vertical ciliary rootlets reduced. (e.g., *Fecampia*, *Glanduloderma*, *Kronborgia*, *Piscinquilinus*)

SUBCLASS BOTHRIONEODERMATA

INFRACLASS BOTHRIOPLANATA Freshwater flatworms bearing a three-branched, diverticulated gut, and medial-posteriorly directed plicate pharynx; multiple follicular vitellaria.

ORDER BOTHRIOPLANIDA Monospecific, *Bothrioplana semperi,* a freshwater scavenger and predator on small invertebrates.

INFRACLASS NEODERMATA Ciliated larval epidermis shed and replaced by a syncytial neodermis with subepithelial nuclei (the tegument); locomotory epidermal cilia bearing a single rootlet; epithelial sensory receptors with collars.

COHORT TREMATODA Digenean and aspidogastrean flukes (Figures 10.1I and 10.3A,B,E,F). With one or more suckers; lacking prohaptor and opisthaptor; male copulatory organ is a cirrus; 1 to 3 hosts during the life cycle, often including a mollusc; most are endoparasitic.

SUPERORDER ASPIDOGASTREA Aspidogastrean flukes.

ORDER ASPIDOGASTRIDA With a complex ventral sucker formed by lateral growth and subdivision of posterior part of sucker. Most with a single host (a mollusc) in life cycle; second host, when present, a turtle or a teleost fish; oral sucker absent. (e.g., *Aspidogaster*, *Cotylaspis*)

ORDER STICHOCOTYLIDA With complex ventral sucker formed by linear growth and subdivision of anterior part of sucker. Most with single host (a lobster) in life cycle; second host, when present, is a chondrichthyan fish. (e.g., *Stichocotyle*, *Rugogaster*)

SUPERORDER DIGENEA Digenean flukes.

ORDER DIPLOSTOMIDA Endoparasites of tetrapods; adult genital pore posterior to the ventral sucker; some with blood-dwelling adult stages. Usually with 3-host life cycles but occasionally with 2; first intermediate host is a mollusc. (e.g., *Schistosoma; Sanguinicola*)

ORDER PLAGIORCHIIDA An extraordinarily diverse order of vertebrate endoparasites with a wide range of life cycle characteristics and hosts. Usually with at least a 2-host life cycle; first intermediate host is a mollusc. (e.g., *Microphallus*, *Opisthorchis* [= *Clonorchis*], *Fasciola*, *Echinostoma*)

COHORT MONOGENEA Monogenean flukes (Figure 10.3C). Oral sucker (prohaptor) reduced or absent; posterior hook-bearing sucker (opisthaptor) present; ventral sucker (acetabulum) absent; gut bifurcate; 3 rows of cilia on oncomiracidium larva; life cycle involves only one host. Most are ectoparasitic, usually on fishes (some occur on turtles, frogs, hippos, copepods, or squids); a few are endoparasitic in ectothermic vertebrates. Although previously classified according to the relative complexity of the opisthaptor (Figure 10.8; simple, Monopisthocotylea; complex, Polyopisthocotylea), the evolutionary relationships within and among these taxa remain unclear. Nine orders are now recognized.

ORDER CAPSALIDEA Capsalids. Gill and skin ectoparasites of elasmobranch and teleost fishes; flattened, leaf like bodies with a simple aseptate or septate opisthaptor; with 3 pairs of median sclerites and 14 small hooklets at the periphery of the opisthaptor. (e.g., *Capsala*, *Benedeniella*, *Trochopus*)

ORDER CHIMAERICOLIDEA Gill parasites of holocephalan fishes. (e.g., *Chimaericola*)

ORDER DICLYBOTHRIIDEA Gill parasites of acipenseriform fishes. Oral sucker absent; mouth ventral; lateral sclerites on opisthaptor absent. (e.g., *Diclybothrium*, *Paradiclybothrium*)

ORDER DACTLOGYRIDEA Gill parasites of freshwater teleost fish; body with opisthaptor bearing 2-4 anchors with 14-16 marginal hooks, and 4 eyespots. (e.g., *Dactylogyrus*, *Ancyrocephalus*)

ORDER GYRODACTYLIDEA Skin and gill parasites of freshwater fish; body fusiform with terminal cephalic lobes; opisthaptor forms a half-oval and is armed with 16 marginal hooks and a pair of median anchors (hamuli) stabilized with median bars. (e.g., *Gyrodactylus*, *Paragyrodactyloides*, *Acanthoplacatus*)

ORDER MAZOCRAEIDEA Gill parasites of clupeid and scombrid fishes; oncomiracidium with one pair of fused eyes; opisthaptor with two pairs of lateral sclerites. (e.g., *Clupeocotyle*, *Mazocraes*, *Grubea*)

ORDER MONOCOTYLIDEA Gill and ectodermal tissue parasites of mostly tropical elasmobranchs; opisthaptor with a single central and multiple peripheral suckers (loculi), often armed with hamuli and marginal hooks. (e.g., *Monocotyle*, *Potamotrygonocotyle*)

ORDER MONTCHADSKYELLIDEA Gill parasites of tropical reef fishes. (e.g., *Montchadskyella*)

ORDER POLYSTOMATIDEA Skin, gill, and urogenital parasites of aquatic and semiaquatic tetrapods; opisthaptor well developed with 3 pairs of suckers or one sucker pair. (e.g., *Polystoma*, *Oculotrema*, *Metapolystoma*)

COHORT CESTODA Tapeworms and their relatives (Figures 10.1M and 10.4). Exclusively endoparasitic; in most, the body consists of an anterior scolex, followed by a short neck, and then a strobila composed of a series of "segments" or proglottids, although basal taxa are monozoic (not strobilated); digestive tract absent. Previous classifications recognized several subclasses; until taxonomic uncertainty is resolved, we recognize 16 orders.

ORDER AMPHILINIDEA Endoparasites in the guts or coelomic cavities of cartilaginous and certain primitive bony fishes, less commonly in turtles; leaflike bodies, lacking scolex and monozoic (not strobilated); 10 minute hooks may be present at posterior of body if retained from decacanth larvae that may develop in crustaceans. (e.g., *Amphilina*; *Austramphilina, Gyrometra*)

ORDER BOTHRIOCEPHALIDEA Gut parasites of teleost fishes and occasionally in acipenseriforms and amphibians; body strobilated with proglottids wider than long; scolex with a pair of bothria; occasionally with hooks; life cycle with 2–3 hosts, usually a crustacean first host and a teleost second host. (e.g., *Bothriocephalus*, *Triaenophorus*, *Polyonchobothrium*)

ORDER CARYOPHYLLIDEA Intestinal parasites of cypriniform and siluriform fishes; body monozoic (not strobilated); scolex often simple; with 2 host life cycles; oligochaetes as intermediate hosts. (e.g., *Archigetes*, *Paraglaridacris*)

ORDER CYCLOPHYLLIDEA Intestinal parasites of birds and mammals; body variable in size, scolex with 4 suckers, rostellum is present may be armed or not; most species are hermaphrodites although the family Dioecocestidae is gonochoristic); highly diverse and possibly the most derived order of cestodes; life cycles with 2–3 hosts with diverse invertebrate and vertebrate species as intermediate and paratenic hosts. (e.g., *Dipylidium*, *Hymenolepis*, *Moniezia*, *Taenia*)

ORDER DIPHYLLIDEA Intestinal parasites of elasmobranch fishes; body strobilated with midventral genital pores; scolex with 2 bothria and a cephalic peduncle; life cycles are poorly known but larval stages occur in marine crustaceans and molluscs. (e.g., *Echinobothrium*, *Ditrachybothridium*)

ORDER DIPHYLLOBOTHRIIDEA Intestinal parasites of piscivorous vertebrates, usually mammals; body strobilated but with variable external differentiation; scolex always unarmed, usually with paired attachment grooves (bothria); with 2–3 host life cycles; copepod crustaceans as first intermediate hosts, second intermediate hosts within vertebrates. (e.g., *Diphyllobothrium*, *Ligula*, *Spirometra*)

ORDER GYROCOTYLIDEA Intestinal parasites of holocephalan fishes although also reported in sharks; body stout, monozoic with a muscular anterior attachment organ; posterior body terminates in a rosette-like adhesive organ; lateral margins often ruffled; life cycles are unknown. (e.g., *Gyrocotyle, Gyrocotyloides*)

ORDER LECANICEPHALIDEA Small intestinal parasites of rays and occasionally sharks; although probably paraphyletic, species are characterized by a scolex with 4 suckers or bothria and an apical structure that may have tentacles, cones, or additional suckers; life cycles are poorly known by may include molluscs, crustaceans and teleosts as intermediate hosts. (e.g., *Polypocephalus*, *Quadcuspibothrium*, *Corrugatocephalum*)

ORDER LITOBOTHRIDEA Intestinal parasites of lamniform sharks; scolex with apical sucker and several muscular, cruciform pseudosegments; life cycles are unknown. (e.g., *Lithobothrium*)

ORDER ONCHOPROTEOCEPHALIDEA Small to medium sized intestinal parasites of elasmobranchs, as well as of freshwater fish, amphibians, reptiles, and occasionally mammals; strobila polyzoic, with proglottids, or with few anapolytic (non-detaching) proglottids; with lateral, irregularly alternating genital pores; scolex often with 4 muscular bothridia, unarmed or with one pair of hooks; occasionally with a rostellum-like apical structure; life cycles include 1 or 2 intermediate hosts (crustaceans or fish); this order has subsumed the former order Proteocephalidea and part of the Tetraphyllidea. (e.g., *Proteocephalus*, *Chambriella*, *Brachyplatysoma, Acanthobothrium*, *Platybothrium*)

ORDER PHYLLOBOTHRIIDEA Small to medium sized intestinal parasites of sharks, batoids and ratfish; strobili polyzoic and with proglottids, scolex often with 4 muscular bothridia. (e.g., *Calyptrobothrium*, *Chimaerocestos*, *Marsupiobothrium*)

ORDER RHINEBOTHRIIDEA Intestinal parasites of freshwater stingrays; scolex often with bothridial stalks. (e.g., *Spongiobothrium*)

ORDER SPATHEBOTHRIIDEA Intestinal parasites of Chondrichthyes and teleost fishes; body strobiliated without external differentiation of proglottids; male and female genital pores close together and alternating dorsally and ventrally the length of the body; with 2 host life cycles; crustaceans as intermediate hosts; progenesis (early maturation) of larvae is widespread. (e.g., *Spathobothrium, Bothrimonus*)

ORDER TETRABOTHRIIDEA Intestinal parasites of marine homeotherms inhabiting pelagic ecosystems; body with well-defined strobilation; scolex with 4 muscular bothridia of variable form; rostellum lacking; eggs with 3 membranes; life cycles are unknown but are likely to involve crustaceans, cephalopods and teleosts as intermediate hosts. (e.g., *Priapocephalus, Tetrabothrius, Trigonocotyle*)

ORDER TETRAPHYLLIDEA Intestinal parasites of elasmobranch fishes, rarely in holocephalan fishes; body strobilated with variable apolysis of proglottids; scolex often with 4 muscular bothridia that vary widely in form, often elongate, stalked and with hooks; life cycles are poorly known; but 3–5 hosts are likely and include molluscs, crustaceans, teleosts and marine mammals; a recent molecular analysis has shown this order to be nonmonophyletic, and genera are being redistributed to other orders (Onchoproteocephalidea, Phyllobothriidea). (e.g., *Rhoptrobothrium*, *Dinobothrium*)

ORDER TRYPANORHYCHA Intestinal and stomach parasites of elasmobranch fishes; body strobilated with lateral genital pores; scolex with 4 eversible tentacles each bearing a complex array of hooks; 2–3 host life cycles with crustaceans and teleosts and intermediate hosts. (e.g., *Dasyrhynchus*, *Halsiorhynchus*, *Otobothrium*)

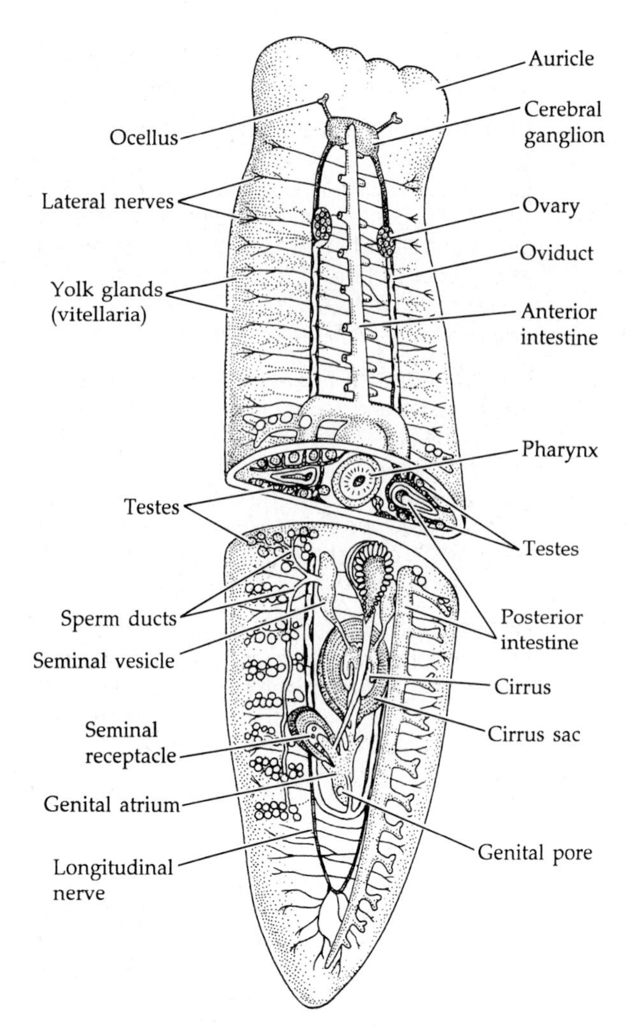

Figure 10.2 A generalized freshwater flatworm (order Tricladida).

The Platyhelminth Body Plan

Compared with phyla discussed in preceding chapters, flatworms display some of the most important advances found in the animal kingdom. In many ways, they are the prototypical acoelomate bilateria and, according to some hypotheses, they represent the basic body plan from which the protostomes were ultimately derived.

The evolution of the triploblastic condition and bilateral symmetry almost certainly occurred in concert with the evolution of sophisticated internal "plumbing" (organs and organ systems) and the tendency to centralize and cephalize the nervous system and to develop specialized units within the nervous system for sensory, integrative, and motor activities. It is likely that this first step toward bilaterality took place with the Xenacoelomorpha (Chapter 9), and progressed with the Platyhelminthes. With these features came unidirectional movement and a more active lifestyle than that of radially symmetrical animals. The primary evolutionary advantages of these coinciding changes derived chiefly from the ability of these "new" creatures to move around more or less freely and thus exploit previously unavailable survival strategies.

These strategies can be appreciated by examining the rather complex structural features displayed by the free-living flatworms (Figure 10.2). The presence of mesoderm allows the formation of a fibrous and muscular mesenchyme that provides structural support and allows patterns of locomotion not possible in diploblastic radiates. Elaborate reproductive systems evolved in the platyhelminths, providing for internal fertilization and enhancing the production of yolky and encapsulated eggs. Most flatworms have abandoned free-swimming larval stages for mixed and direct life histories. Osmoregulatory structures in the form of protonephridia were most likely instrumental in the invasion of fresh water.

This body plan is not without constraints, however. Higher energy demands accompany an active lifestyle. A major limiting factor for flatworms, functionally, is the absence of an efficient circulatory mechanism to move materials throughout the body. This problem is compounded by the lack of any special structures for gas exchange. These problems relate, of course, to the surface-to-volume dilemma discussed in Chapter 4. In the absence of circulatory and gas exchange structures, flatworms (particularly free-living ones) are constrained in terms of size and shape. They have remained relatively small and flat, with shapes that maintain short diffusion distances. The largest free-living flatworms have highly branched guts that assume much of the responsibility of internal transport.

Having a high surface-to-volume ratio and using the entire body surface for gas exchange create potential problems of ionic balance and osmoregulation in freshwater and terrestrial species, and of desiccation in intertidal and terrestrial habitats. The permeable body surface must be kept moist; thus, flatworms have invaded land rarely and mostly in very damp areas, although a few rhabdocoels, prorhynchids, and several parasitic species have developed desiccation-resistant life stages. Flatworms have, however, exploited a variety of marine and freshwater habitats, and are particularly successful as parasites and commensals, enjoying the benefits of living on or in their hosts.

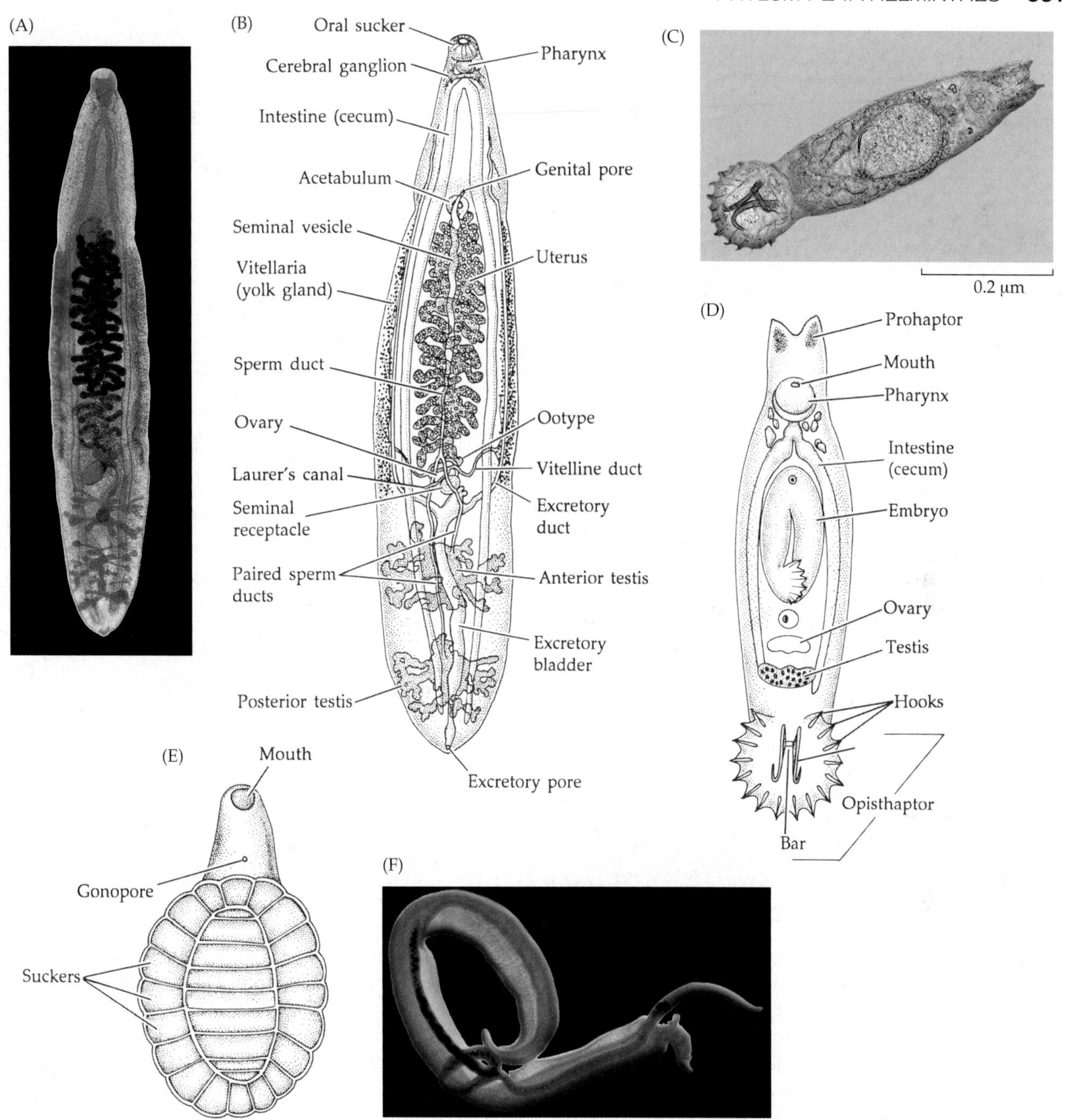

Figure 10.3 Representative flukes. (A,B) *Opisthorchis* (= *Clonorchis*) *sinensis*, a plagiorchiid trematode that inhabits human livers. (C,D) *Gyrodactylus* (Monogenea), an ectoparasite on fishes. (E) *Cotylaspis* (an aspidogastrid trematode). (F) The human blood fluke, *Schistosoma mansoni* (a copulating male and female), a diplostomid fluke and one of the few gonochoristic flatworms.

The ancestral flatworm is thought to have been a free-living form, from which the present-day catenulids and rhabditophorans evolved and diversified. The flukes and tapeworms evolved from within the rhabditophoran assemblage (discussed in more detail later in this chapter). Thus, in each of the following sections we first examine the basic features of the free-living flatworms and set the stage for understanding not only the diversity within that class but the derivation of the specialized parasitic taxa as well.

The anatomy of free-living flatworms, flukes, and tapeworms is shown in Figures 10.2 through 10.6. Free-living flatworm species vary in shape from broadly oval to ribbon-like, and are typically flattened dorsoventrally, although very small ones may be nearly cylindrical. The head is usually ill defined, except for the presence of sense organs. The mouth is often located ventrally, either near the middle of the body or more anteriorly, though again, there are exceptions (e.g., prorhynchids whose raptorial mouth is fully anterior; cestodes who have no mouth at all). Most flukes (Figure 10.3) are oval or leaf shaped and bear external attachment organs such as hooks and suckers. As their common name

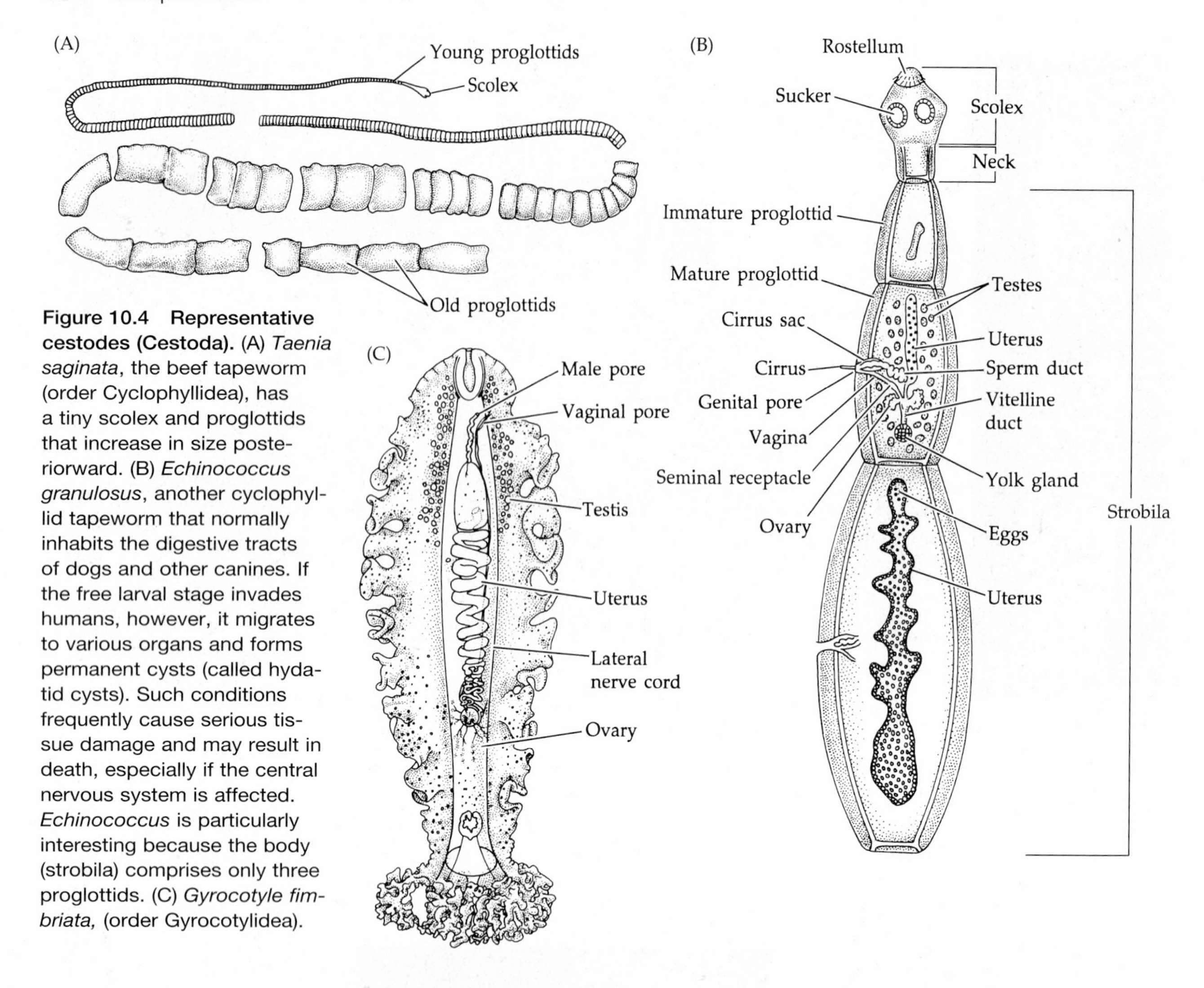

Figure 10.4 Representative cestodes (Cestoda). (A) *Taenia saginata*, the beef tapeworm (order Cyclophyllidea), has a tiny scolex and proglottids that increase in size posteriorward. (B) *Echinococcus granulosus*, another cyclophyllid tapeworm that normally inhabits the digestive tracts of dogs and other canines. If the free larval stage invades humans, however, it migrates to various organs and forms permanent cysts (called hydatid cysts). Such conditions frequently cause serious tissue damage and may result in death, especially if the central nervous system is affected. *Echinococcus* is particularly interesting because the body (strobila) comprises only three proglottids. (C) *Gyrocotyle fimbriata*, (order Gyrocotylidea).

suggests, the tapeworms are typically elongate and ribbon-like (Figure 10.4). Their anterior end forms a tiny **scolex**, modified for attachment within the host; the rest of the body is essentially a reproductive machine.

Tapeworms live in the guts of vertebrates. Most species belong to the Cestoda and possess three distinguishable regions of the body. The **scolex** serves for attachment and is usually armed with hooks and suckers. Immediately behind the scolex is a short region called the **neck**, followed by an elongated, segmented trunk, or **strobila**, consisting of individual **proglottids**. The proglottids bud (strobilate) from a germinal zone in the neck (or at the base of the scolex when a neck is absent). As new proglottids arise, older ones move posteriorly and mature, become inseminated, and fill with embryos. Strobilation in tapeworms is thus not by way of teloblastic growth (discussed in Chapters 14 and 20), and it is clearly not homologous to the true segmentation seen in annelids and arthropods.

Tapeworms in the orders Amphilinidea and Gyrocotylidea are somewhat flukelike in appearance. They lack a scolex, and the body is not divided into proglottids. They are placed within the Cestoda because of the absence of a digestive tract and because of certain features of the life cycle. They may represent the primitive, prestrobilation body plan of the Cestoda.

Body Wall

Free-living flatworms The body wall of these species is multilayered and complex (Figure 10.5). The epidermis is composed of a wholly or partially ciliated, syncytial or cellular epithelium, with gland cells and sensory nerve endings distributed in various patterns. Beneath the epidermis is a basement membrane, which is often thick enough to lend some structural support to the body. In Catenulida and Macrostomida, the basement membrane is apparently absent, but this condition is viewed as secondarily derived. Internal to the basement membrane are smooth muscle cells, frequently arranged in rather loosely organized outer circular, middle diagonal, and inner longitudinal layers. The area between the body wall and the internal organs is usually filled with a mesenchyme (often called a parenchyme) that includes a variety of loose and fixed cells, muscle fibers, and connective tissue. Many macrostomorphans appear to lack a cellular mesenchyme.

The gland cells of the body wall are generally derived from ectoderm. When mature, many of these

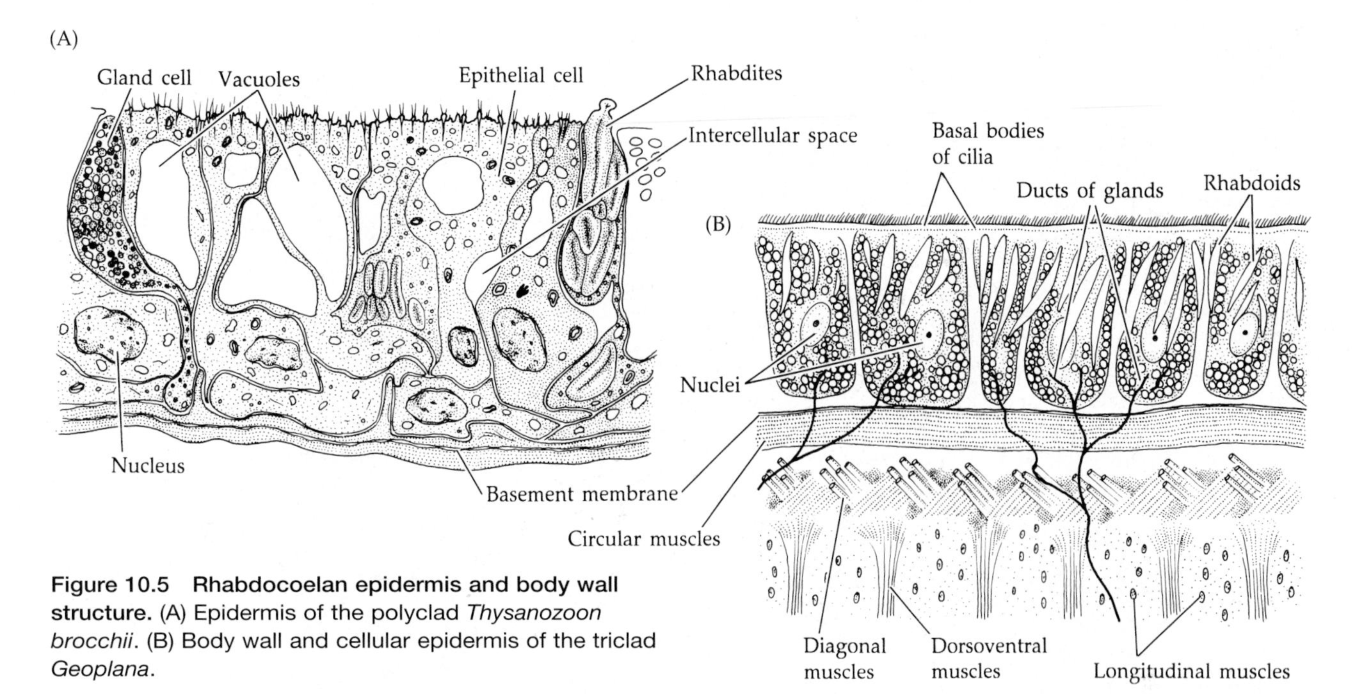

Figure 10.5 Rhabdocoelan epidermis and body wall structure. (A) Epidermis of the polyclad *Thysanozoon brocchii*. (B) Body wall and cellular epidermis of the triclad *Geoplana*.

cells lie in the mesenchyme with a "neck" extending between epidermal cells to the body surface. These cells produce mucous secretions that serve a number of functions. In semiterrestrial and intertidal flatworms, the mucus forms a moist covering that provides protection from desiccation and aids in gas exchange. Most benthic flatworms possess a ventral concentration of mucous gland cells that secrete a slime that aids in locomotion. Mucous secretion around the mouth aids in prey capture and swallowing. Other gland cells or complexes of cells provide granules containing adhesives for temporary attachment, as well as granules that break the attachment, occurring several times per second. Such **duo-gland adhesion** systems (often involving a third cell type that provides structural support) are widespread in free-living flatworms and other phyla with similar lifestyles. In some ectocommensal forms (e.g., *Bdelloura* triclads and various temnocephalid dalytyphloplanids; Figure 10.6) these adhesive glands are associated with special plates or suckers for attachment to the host.

In most free-living flatworms epidermal and subepidermal cells produce structures called **rhabdoids**

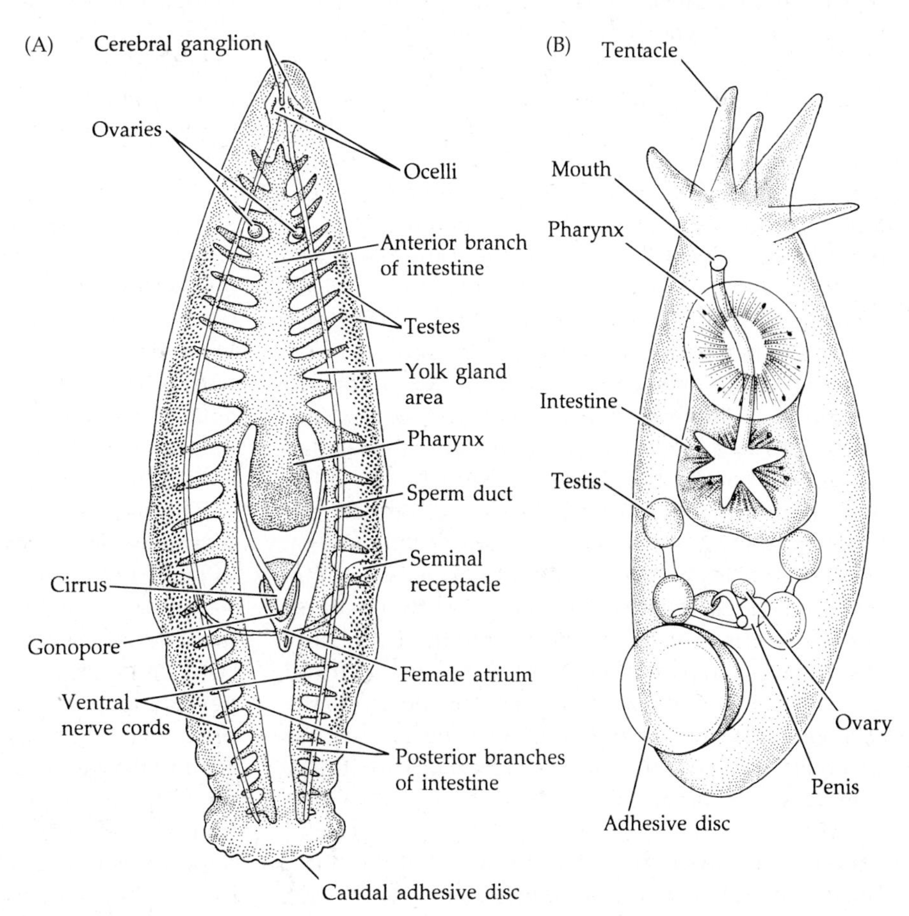

Figure 10.6 Two symbiotic euneoophoran flatworms with adhesive attachment organs. (A) *Bdelloura candida*, a triclad ectocommensal on horseshoe crabs (*Limulus*). (B) *Temnocephala caeca*, a rhabdocoel ectocommensal on *Phreatoicopis terricola* (a fresh-water isopod).

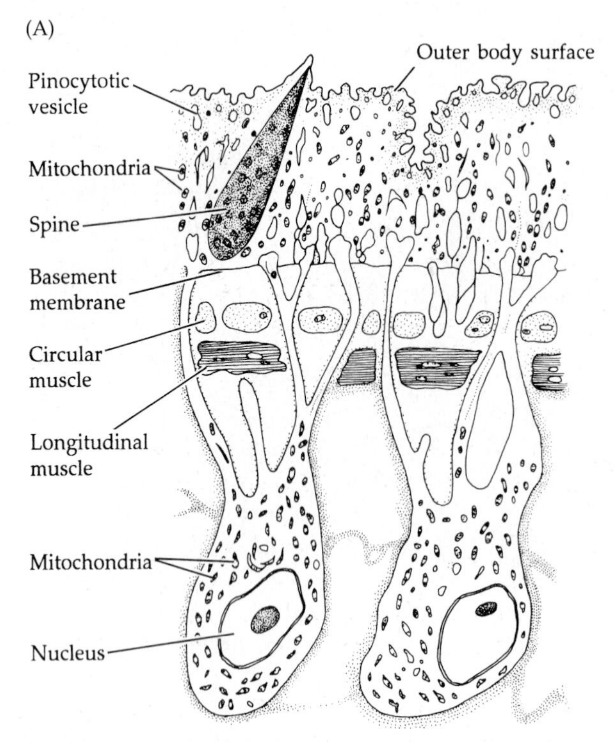

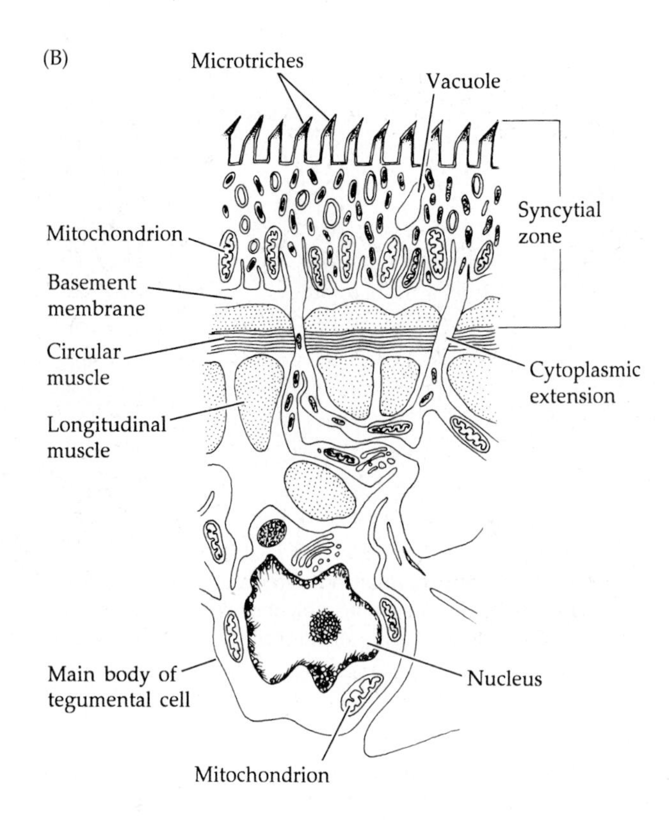

Figure 10.7 (A) The tegument and underlying body wall of a plagiorchiid fluke (*Fasciola hepatica*; longitudinal section). (B) The tegument and body wall of a cestode (cross section).

(Figure 10.5B). These rod-shaped cellular inclusions produce mucus when moved to the surface of the epithelium. Some other phyla (e.g., Xenacoelomorpha, Gastrotricha, Nemertea, Annelida) have similar, but probably convergently-evolved secretory cellular bodies. The mucus aids in ciliary gliding and also may help protect animals from desiccation and from predators. Rhabdoids that are produced by gland cells in the epidermis or mesenchyme are called **rhabdites**. These structures can reach the body surface through intercellular spaces in the epidermis (Figure 10.5A). They also contribute to mucus production and, in some species, release noxious chemicals for defense. Rhabdites are diagnostic of the huge flatworm clade (subphylum) Rhabditophora. Somewhat similar structures, called "false rhabdites," have been described from the other main flatworm clade (subphylum) Catenulidea, but they may not be homologous to those of Rhabditophora.

Some species of free-living flatworms (e.g., macrostomids, proseriates, polyclads) have prominent tubercles covering the dorsal surface and these probably also have a defensive role. In some species, unfired nematocysts from hydroid prey are transported to the tubercles. In others, such as species of *Thysanozoon*, the tubercles appear to release a powerful acid that may deter would-be predators.

Flukes and tapeworms Modifications of the outer body covering are common among parasites, and platyhelminths are no exception. Unlike the free-living flatworms, flukes and tapeworms possess an external covering called a **tegument**, formed of nonciliated cytoplasmic extensions of large cells whose cell bodies actually lie in the mesenchyme (Figure 10.7). The extensions fuse their margins such that the outer surface of the worm forms a functional syncytium. The tegument not only provides some protection but also is an important site of exchange between the body and the environment. Gases and nitrogenous wastes move across this surface by diffusion, and some nutrients, especially amino acids, are taken in by pinocytosis. In tapeworms, the uptake of nutrients occurs solely across the body wall, and the surface area of the tegument is greatly increased by many tiny folds called **microtriches** (Figure 10.7B). As one of nature's more remarkable adaptations, these folds may interdigitate with the intestinal microvilli of the host organism and aid in the absorption of nutrients.

The nature of the tegument in flukes and tapeworms is viewed by some zoologists as unique and of major phylogenetic importance. The larvae of these parasitic worms have a "normal" ciliated epidermis over at least part of their bodies. However, this epidermis is shed, and postlarval stages develop a new, syncytial body covering—the **neodermis**. This phenomenon appears to be a unique synapomorphy uniting the Monogenea, Trematoda, and Cestoda as a monophyletic taxon that Ehlers named Neodermata (in reference to the "new skin" of these animals), an hypothesis that is now substantiated by several molecular phylogenetic analyses.

The tegument/neodermis is underlain by a basement membrane, beneath which is the mesenchyme. Most flukes and tapeworms have circular and longitudinal muscles within the mesenchyme, and sometimes diagonal, transverse, and dorsoventral muscles as well. The mesenchyme varies from masses of densely packed cells to syncytial and fibrous networks with fluid-filled spaces. In some digenean flukes (Trematoda: orders Diplostomida and Plagiorchiida), spaces form vessels through the mesenchyme called **lymphatic channels**, which contain free cells that have been likened to lymphocytes. The mesenchyme also contains gland cells with connections to the surface of the body through the tegument. These gland cells are few in number compared with those of free-living flatworms, and they are primarily adhesive in nature and associated with certain organs of attachment.

One of the least explored yet most interesting attributes of tapeworms, and indeed of all intestinal parasites, is their ability to thrive in an environment of hydrolytic enzymes without being digested. In addition to constant replacement of the outer tegument by underlying cells, one popular hypothesis is that gut parasites produce enzyme inhibitors (sometimes called "antienzymes"). One study showed that *Hymenolepis diminuta* (a common tapeworm in rats and mice) releases proteins that appear to inhibit trypsin activity. This tapeworm can also regulate the pH of its immediate environment to about 5.0 by excreting organic acids; this acidic output may also inhibit the activity of trypsin.

Support, Locomotion, and Attachment

Only a very few flatworms possess any sort of special skeletal elements. In a few free-living species, tiny calcareous plates or spicules are embedded in the body wall (Figure 10.1L). Body support in all other flatworms is provided by the hydrostatic qualities of the mesenchyme, the elasticity of the body wall, and the general body musculature.

Most free-living, benthic species move on their ventral surface by cilia-powered gliding. Mucus provides lubrication as the animal moves and serves as a viscous medium against which the cilia act. Some of the larger or more elongate forms also use muscular contractions. The ventral surface of the body is thrown into a series of alternating transverse furrows and ridges that move as waves along the animal, propelling it forward. Muscular undulations of the lateral body margins allow some large polyclads to swim for brief periods of time. Muscular action allows the body to twist and turn, providing steerage. Some interstitial forms are highly elongate and use the body wall muscles to slither between sand grains. Many of these types of flatworms possess adhesive glands, the secretions of which provide temporary stickiness and enable the animals to gain purchase and leverage as they move.

Adult flukes lack external cilia, and their movement depends on their body wall muscles or on the body fluids of their host. Some move about slowly on or within their host by muscle action, and a few (e.g., blood flukes) are carried in the host's circulatory system. However, certain larval stages are highly motile and do swim using ciliary action. Once established within or on a host, it is usually advantageous for a fluke to stay more or less in one place. In that regard, nearly all of them are equipped with external organs for temporary or permanent attachment (Figures 10.3C and 10.8). Monogenean flukes typically have an anterior and a posterior adhesive organ called the **prohaptor** and the **opisthaptor**, respectively. The prohaptor consists of a pair of adhesive structures, one on each side of the mouth, bearing suckers or simple adhesive pads. The opisthaptor is usually the major organ of attachment, and includes one or more well developed suckers with hooks or claws.

The digenean flukes possess two hookless suckers. One, the **oral sucker**, surrounds the mouth, and the

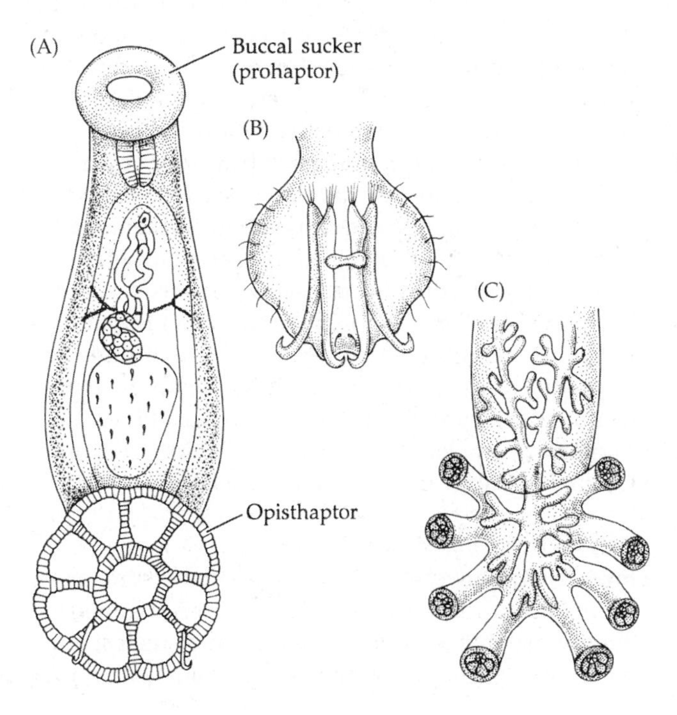

Figure 10.8 Some attachment organs of monogenean flukes. (A) *Anoplocotyloides papillata*. (B,C) Opisthaptors from monogenean flukes. (D) An unidentified fluke with suckered prohaptor and elaborate opisthaptor.

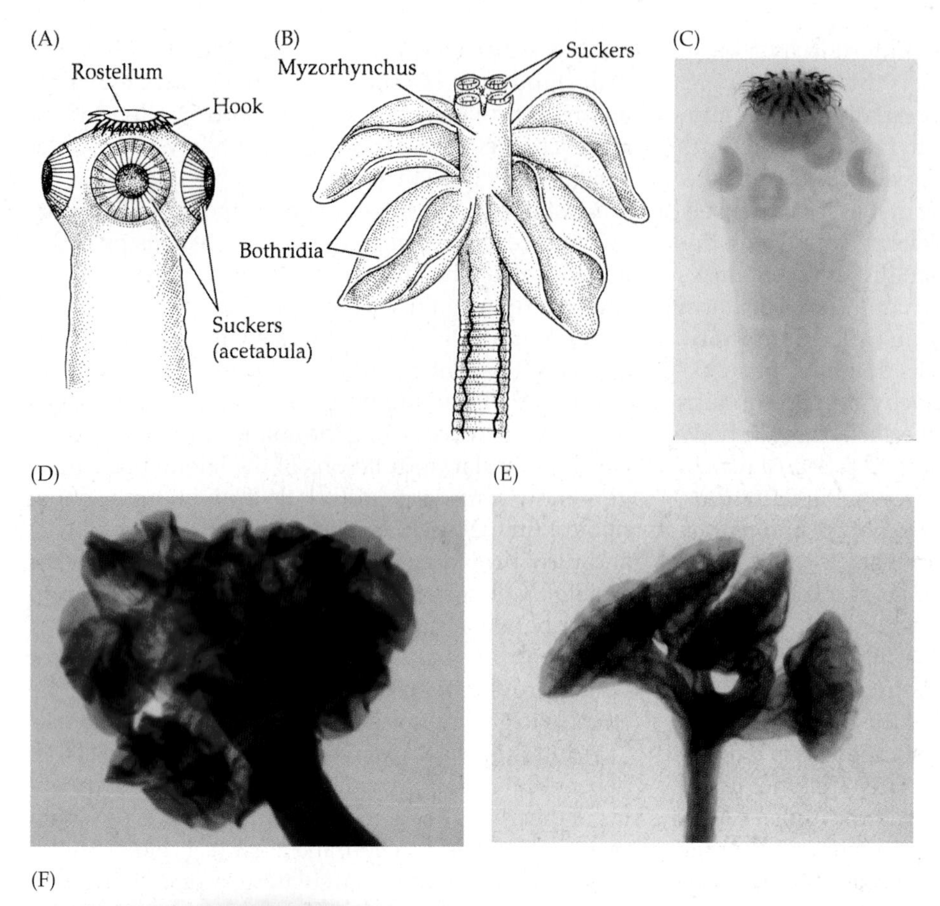

(F)

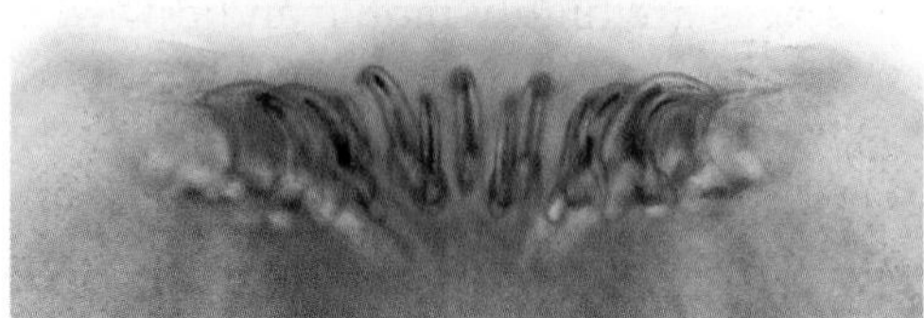

Figure 10.9 Scoleces of various cestodes. (A) Typical scolex with rostellum, hooks, and suckers (*Taenia solium*). (B) Complex scolex with suckered myzorhynchus and leaflike bothridia (*Myzophyllobothrium*). (C–F) Photos of four different scoleces.

other, the **acetabulum**, is located elsewhere on the ventral surface (Figure 10.3B). These suckers are usually supplied with adhesive gland cells, although the well-developed ones operate mainly on suction produced by muscle action. The aspidogastrean flukes lack an oral sucker but have a large, subdivided ventral sucker (Figure 10.3E).

Adult tapeworms (Cestoda) do not move around much, but they are capable of muscular undulations of the body. They remain fixed to the host's intestinal wall by the **scolex** (or, in the case of members of the cestode order Amphilinidea, by an anterior adhesive organ) and by the microtriches. The details of scolex anatomy (Figure 10.9) are extremely variable and have great importance in the taxonomy of tapeworms. The tip of the scolex in many cestodes (e.g., *Taenia*) is equipped with a movable hook-bearing **rostellum**, which is sometimes retractable into the scolex. In others (e.g., *Cephalobothrium*) the anterior end bears a protrusible sucker, or adhesive pad, called a **myzorhynchus**. The rest of the scolex bears various suckers or sucker-like structures and sometimes hooks or spines. There are three categories of adhesive suckers upon which ordinal and subordinal classification of cestodes is partially based. **Bothria** are elongate, longitudinal grooves on the scolex. They possess weak muscles but are capable of some sucking action. Bothria occur as a single pair and are typical of the orders Diphyllobothriidea (e.g., *Diphyllobothrium*) and Bothriocephalidea. Members of the nonmonophyletic order Tetraphyllidea (e.g., *Acanthobothrium*, *Phyllobothrium*) bear four symmetrically placed bothridia around the scolex. These foliose structures are often equipped with suckers at their anterior ends. The third and most familiar type of attachment structures on the scolex are true suckers, or **acetabula**. They are identical in structure and are probably homologous to the acetabula of digenean trematodes. There are usually four acetabula, placed symmetrically around the circumference of the scolex. They are characteristic of many members of the order Cyclophyllidea (e.g., *Dipylidium*, *Taenia*).

Feeding and Digestion

Free-living flatworms Most species are carnivorous predators or scavengers, feeding on nearly any available animal matter, or in the case of very small species, on bacteria or fungi in biofilms. A few are herbivorous on microalgae, and some species switch from herbivory to carnivory as they mature. Their prey includes almost any invertebrate small enough to be captured and ingested (e.g., protists, small crustaceans, worms, tiny gastropods). Some species graze on sponges, ectoprocts, and tunicates, and some consume the flesh of barnacles, leaving behind the empty shell. Most free-living flatworms locate food by chemoreception. Land planarians capture and consume earthworms (e.g., *Bipalium*), land snails (e.g., *Platydesmus*, *Endeavouria*), and insects (e.g., *Rhynchodemus*, *Microplana*). *Platydesmus manokwari*, the 6.5 cm long "New Guinea flatworm," is a highly invasive species that is a threat for endemic terrestrial molluscs; it has invaded throughout the Pacific, as well as in Europe, the Caribbean, and most recently Florida (USA).

More than 100 species of free-living flatworms are known to be symbiotic with other invertebrates. Some of these are simply commensals that derive some protection from their associations, showing only physical modifications for temporary attachment. Others, however, feed upon their hosts, causing various degrees of damage and displaying true physiological dependency in the relationship. While we can devote space to mentioning only a few examples of symbiotic free-living flatworms, recognition of these situations is of considerable importance. First, it emphasizes the evolutionary adaptability of the flatworm body plan; and second, it provides some essential foundation for our later discussion of the origins of the flukes and tapeworms. (For an excellent survey of the symbiotic free-living flatworms, see Jennings 1980).

Most of the symbiotic free-living flatworms belong to the infraclass Rhabdocoela. Species in the family Temnocephalidae (Figure 10.6B) are ectocommensals within the branchial chambers of freshwater decapod crustaceans, where they feed on microorganisms in the host's gas exchange currents. Temnocephalids also occur on aquatic insects, molluscs, turtles, and a few other kinds of hosts. Several families of Dalytyphloplanida include symbiotic members. For example, *Syndesmis* live within the gut and coelomic fluid of echinoids, where they feed on protists and bacteria, and some may devour cells of their hosts (Figure 10.10). The genera *Graffilla* and *Paravortex* include several species of parasites in the digestive tracts of gastropod and bivalve molluscs, where they derive nutrients from the host tissues. Members of the family Fecampiidae (*Fecampia*, *Kronborgia*, *Glanduloderma*) are parasites in marine crustaceans and certain polychaete worms, where they reside in the host's body fluids and absorb soluble organic nutrients. Non-rhabdocoelan symbiotic species include the triclad *Bdelloura* (Figure 10.6A), an ectocommensal on the gills of *Limulus*, the horseshoe crab. At least two polyclads live and feed on coral (*Prosthiostomum* on Hawaiian *Montipora*; *Amakusaplana acroporae* on Australian *Acropora*).

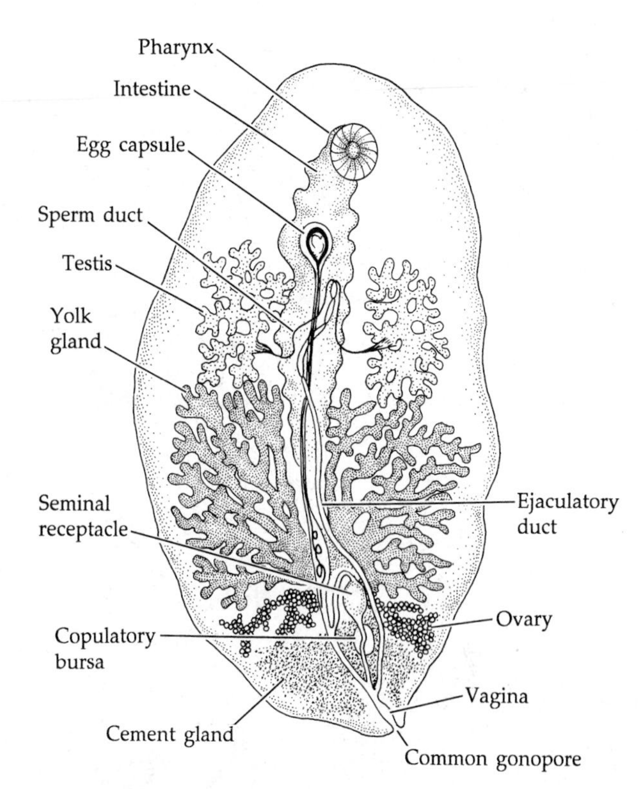

Figure 10.10 *Syndesmis*, a rhabdocoel from the gut of a sea urchin.

The digestive system of the free-living flatworms includes a mouth and a pharynx, which lead to an intestine, or **enteron**. Like that of cnidarians and Xenacoelomorpha, the free-living flatworm gut is incomplete, bearing a single opening, and thus may be called a **gastrovascular cavity**. The mouth varies in position from midventral to anterior. The pharynx is derived from embryonic ectoderm (i.e., it is stomodeal) and lined with epidermis. Epithelial **pharyngeal glands** are associated with the lumen of the pharynx; they produce mucus that aids in feeding and swallowing, and (in some species) proteolytic enzymes that initiate digestion outside the body.

The feeding methods of free-living flatworms vary with the size of the animal and the complexity of their food-getting apparatus, especially the pharynx. As noted in the classification scheme, the nature of the pharynx varies greatly among taxa. There are three basic pharynx types among the free-living flatworms: simple, bulbous, and plicate (Figures 10.11 and 10.12).

A simple pharynx (or **pharynx simplex**) is a short, ciliated tube connecting the mouth and intestine (Figure 10.12). This type of pharynx has been considered plesiomorphic within the phylum Platyhelminthes and is found in the orders Macrostomida and Catenulida. In all members of these orders, the pharynx leads to a simple saclike or elongate intestine generally lacking extensive diverticula. Free-living flatworms with a simple tubular pharynx are generally quite small, with the mouth located more or less midventrally. They usually

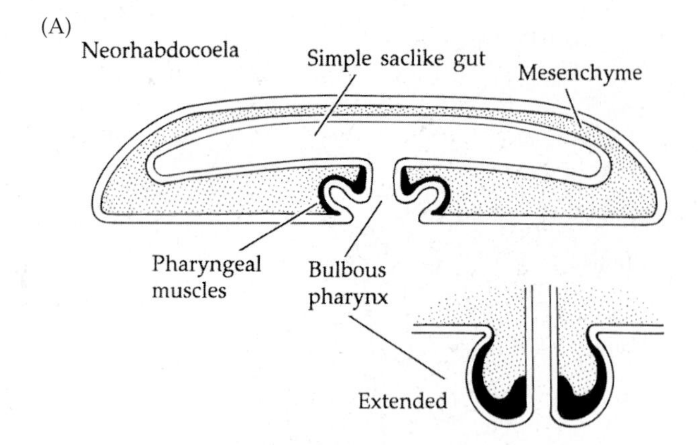

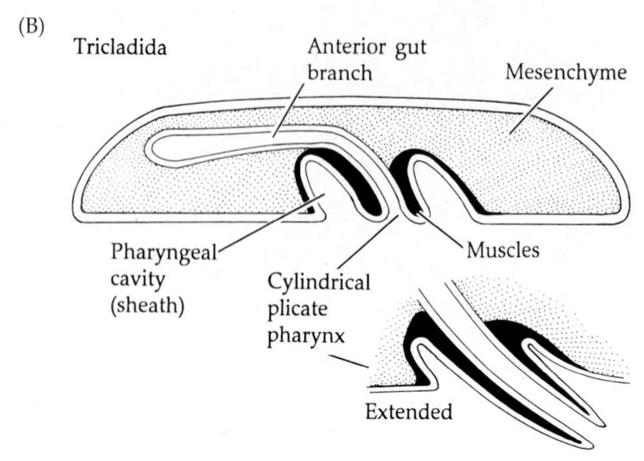

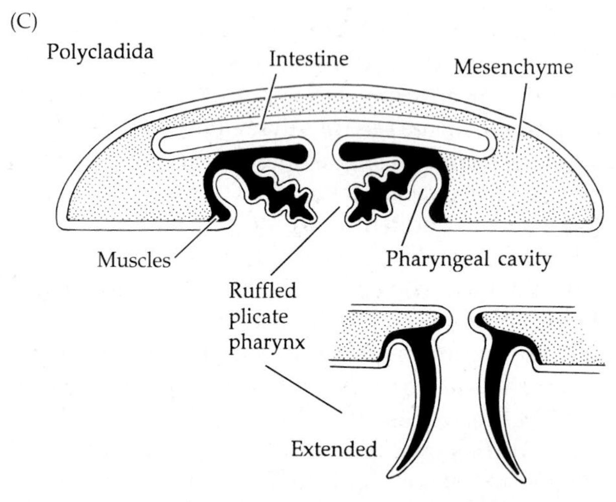

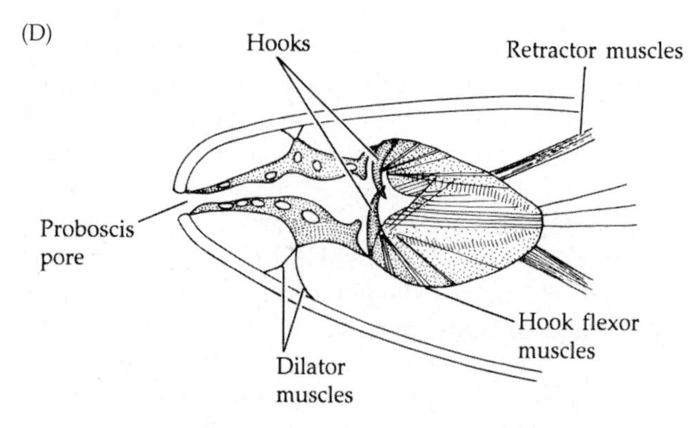

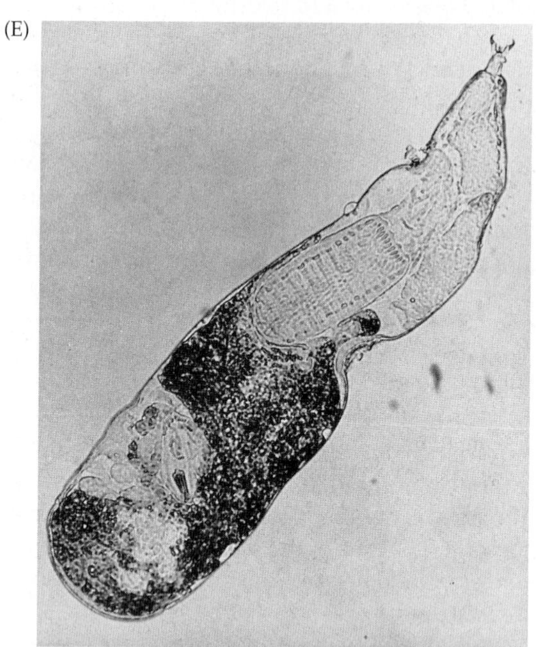

Figure 10.11 (A–C) The pharynges of three free-living flatworms (sagittal sections). (D) The proboscis apparatus (sagittal section) of the rhabdocoel *Gnathorhynchus* (order Kalyptorhynchia). (E) *Cheliplana*, another rhabdocoel (order Kalyptorhynchia), with jawed proboscis extended (top right).

feed by sweeping small organic particles and tiny prey into the pharynx by ciliary action. Those with an eversible pharynx usually fold their body around the prey or other food source and cover it with mucus from the epidermal glands. Then the pharynx is everted over or into the food item.

Some free-living flatworms, especially triclads, secrete digestive enzymes externally via special glands that empty through the pharyngeal lumen or from the tip of the pharynx. The food is partially digested and reduced to a soupy consistency prior to swallowing. Many other species swallow their food whole by the action of powerful pharyngeal muscles.

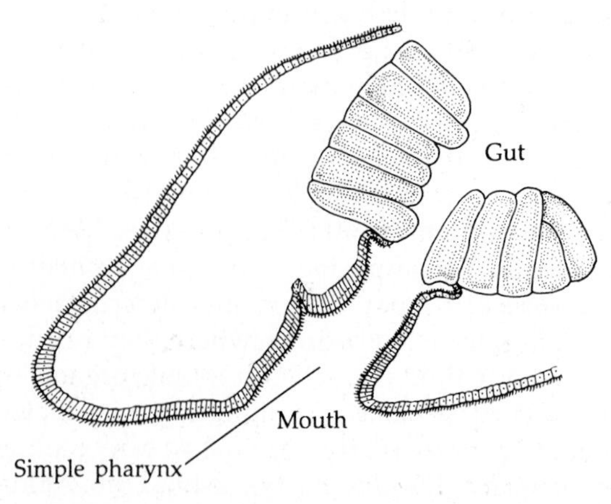

Figure 10.12 Sagittal section through anterior end of *Macrostomum* (Macrostomorpha), which has a simple tubular pharynx.

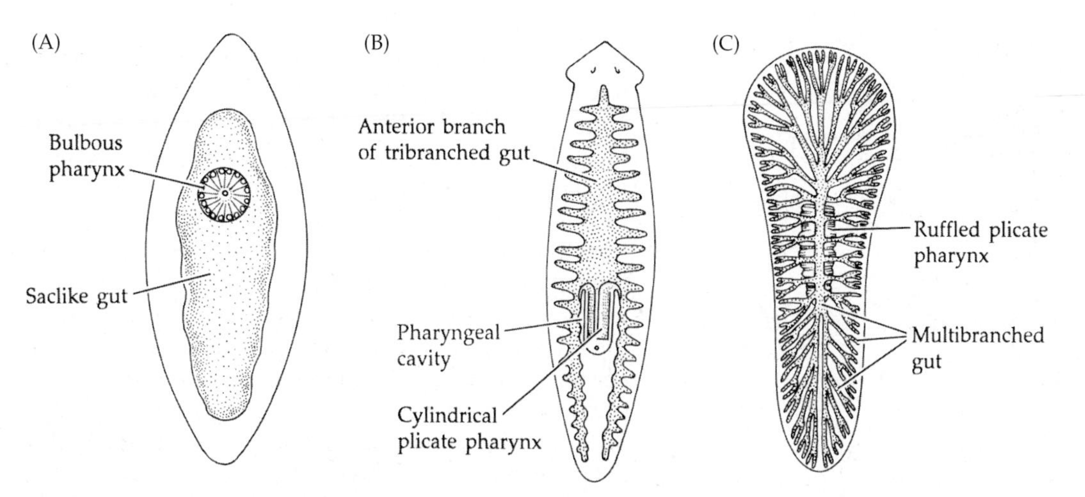

Figure 10.13 Pharynx type and gut shape combinations among free-living flatworms. (A) Rhabdocoela. (B) Tricladida. (C) Polycladida.

Rhabdocoels typically possess a slightly protractile, muscular, bulbous pharynx and a simple saclike gut (Figure 10.13A). Members of the Proseriata, Tricladida, and Polycladida have eversible plicate pharynges. The eversible portion of a plicate pharynx lies within a space called the pharyngeal cavity, which is produced by a muscular fold of the body wall (Figures 10.11 and 10.13).

Proseriates and triclads possess cylindrical plicate pharynges oriented along the body axis. Most polyclads have a ruffled, skirtlike plicate pharynx attached dorsally within the pharyngeal cavity. During feeding, a plicate pharynx is protruded by a squeezing action of extrinsic pharyngeal muscles. Once extended, the pharynx can be moved about by intrinsic muscles of its wall. Retractor muscles pull the pharynx back inside the cavity.

Active prey can be subdued in several ways. Some free-living flatworms produce mucus, which, in addition to entangling the prey, may contain narcotic or poisonous chemicals such as tetrodotoxin. A few flatworms use the sharp **stylet** of the copulatory organ to stab prey and even to introduce venom; one cannot help but concede the remarkable adaptive capacity of the flatworms. Members of the Kalyptorhynchia (Rhabdocoela) are unique among free-living flatworms in their possession of a muscular proboscis that is situated at the anterior end of the body and is separate from the mouth (Figure 10.11D,E); the proboscis, which in some species is armed with hooks, can be everted to grab prey.

Most of the free-living flatworms that possess a plicate pharynx are relatively large, especially the triclads and the polyclads. Associated with this large body size is an elaboration of the intestine. As the names imply, the triclad intestine comprises three main branches, one anterior and two posterior, each with numerous diverticula, whereas the intestine of polyclads is multibranched with diverticula (Figure 10.13B,C). These ramifications of the intestine provide not only an increased surface area for digestion and absorption, but also are a means of distributing the products of digestion through the relatively large body in the absence of a circulatory system. The lining of the intestine is a single cell layer of phagocytic nutritive cells and enzymatic gland cells (Figure 10.14). In some groups, the gastrodermis is ciliated.

In most free-living flatworms, initial digestion is extracellular, mediated by endopeptidases secreted from the pharyngeal glands or from enzymatic gland cells within the intestine. The partially digested material is distributed throughout the gut, then phagocytized by the intestinal cells, wherein final (intracellular) digestion occurs. There are, however, some notable exceptions to this sequence. Bowen (1980) described an interesting phagocytic process in the freshwater triclad *Polycelis tenuis*. Following the ingestion of tiny food

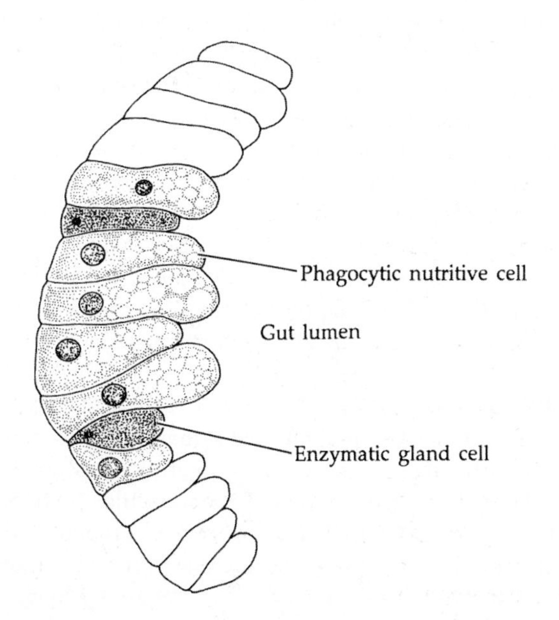

Figure 10.14 The gut lining (partial cross section) of a freshwater triclad contains enzymatic gland cells and phagocytic nutritive cells.

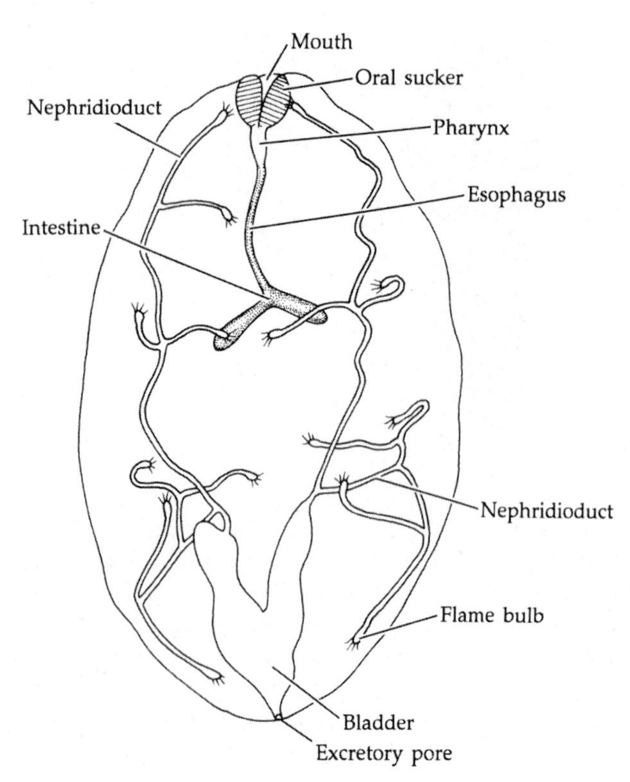

Figure 10.15 Gut and protonephridial system of *Microphallus* (order Plagiorchiida; see also Figure 10.3). In most monogenean flukes, the protonephridial ducts are separate and terminate anteriorly in separate pores.

particles or the preliminary extracellular digestion of larger food, the intestinal phagocytic cells extend processes into the gut lumen, nearly occluding the digestive cavity. These processes interdigitate to form a complex web, forcing food material into the phagocytes, where digestion is completed.

Since flatworms generally lack a through gut, any undigested material must be expelled through the mouth. As discussed in Chapter 4, the major limitation of single-opening guts is the restriction on regional specialization. However, an incipient anus occurs in several flatworms, suggesting that evolutionary "experimentation" with a complete gut began in this group. One macrostomorphan, *Haplopharynx rostratus*, possesses a minute anal pore, and some polyclads have pores at the ends of gut branches; some proseriatians (e.g., *Tabaota*) may form a temporary anus.

Flukes and flatworms Adult flukes feed on host tissues and fluids or, in some cases, material within the host's gut. Most of the food is taken in through the mouth by a pumping action of the muscular pharynx, but some organic molecules are picked up across the tegument by pinocytosis. The anterior part of the digestive system includes a mouth, a muscular pharynx, and a short esophagus. The esophagus leads to a pair of intestinal **ceca** (occasionally, a single cecum), which extend(s) posteriorly in the body (Figures 10.3 and 10.15). The lining of the ceca includes absorptive nutritive cells and enzymatic gland cells. Digestion is at least partly extracellular. Some flukes secrete enzymes from the gut out the mouth, or from the suckers, to partially digest host tissue prior to ingestion.

Cestodes lack any vestige of a mouth or digestive tract. All nutrients must be taken into the body across the tegument. Uptake probably occurs by pinocytosis and by diffusion across the increased surface area of the microtriches. Some work suggests that tapeworms are unable to take in large molecules and thus rely to a considerable extent on the digestive processes of their hosts and the secretion of enzymes outside their bodies to chemically reduce the size of potential nutrient material. It has also been proposed that the surface of the scolex may absorb host tissue fluids through the site of attachment to the gut wall.

Many parasites, in numerous phyla, have been shown to alter the behavior of their hosts in ways that are beneficial to themselves, usually by causing the host to position itself such that the odds of success in the next stage in the parasite's life cycle is improved. Parasitic platyhelminths are no exception. For example, larvae of the trematode, *Microphallus papillorobustus*, encyst in the nervous systems of two species of amphipods (*Gammarus* spp.). There, they induce photophilia in animals that are usually photophobic. Consequently, infected amphipods are more than twice as likely as uninfected animals to be eaten by seagulls (a potential final host). When the cestode *Eubothrium salvelini* reaches the developmental stage where it is infective to the final host (brook trout, *Salvelinus frontinalis*), its intermediate host—the copepod *Cyclops vernalis*—begins to swim more often than normal, making it more like to be eaten by the fish. The beetle, *Tribolium confusum*, is intermediate host to the tapeworm *Hymenolepis diminuta*. Evidence suggests the beetle might be more attracted to rat feces that contain the cestode's eggs than to uninfected feces. The same may be true of cockroaches (*Periplaneta americana*) confronted with the feces of rats parasitized by the acanthocephalan *Moniliformis moniliformis*.

Circulation and Gas Exchange

As mentioned earlier, except for the lymphatic channels in some flukes, flatworms lack special circulatory or gas exchange structures. This condition imposes restrictions on size and shape. The key to survival with such limitations and a generally solid mesenchyme is the maintenance of small diffusion distances. Thus, the flatness of their bodies facilitates gas exchange across the body wall, between the tissues and the environment; nutrients are distributed internally by the digestive system and by diffusion, which is aided by general body movements.

The endoparasitic flatworms are capable of surviving in areas of their host where oxygen is absent. In

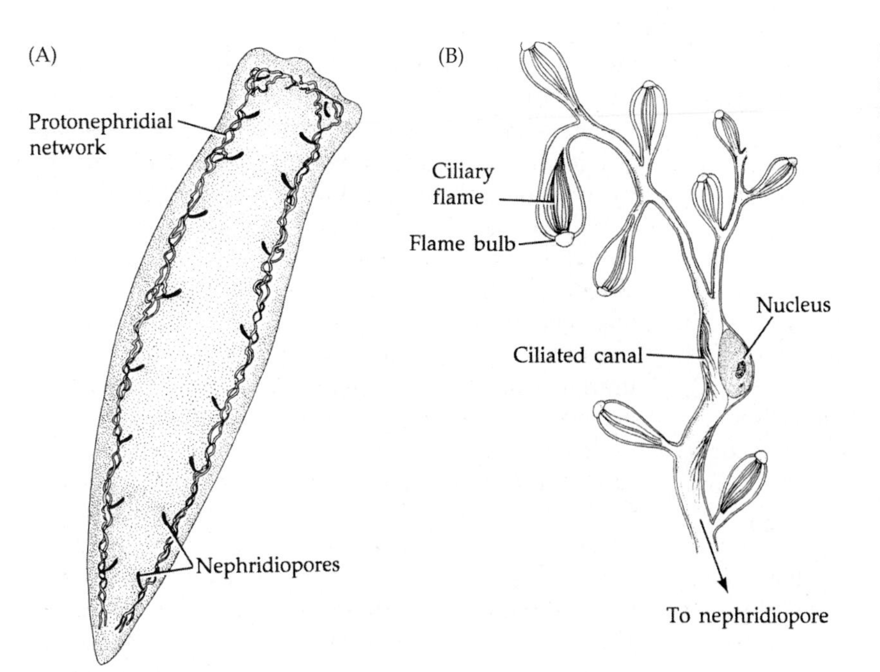

Figure 10.16 (A) The protonephridial system in a freshwater triclad. (B) The nephridial arrangement in a rhabdocoel that has anucleate flame bulbs attached to collecting tubules.

such cases, they rely on anaerobic metabolism, producing a variety of reduced end products (e.g., lactate, succinate, alanine, and long-chain fatty acids). Most of these adaptable animals also possess the appropriate enzymes for and are capable of aerobic respiration in the presence of oxygen.

Excretion and Osmoregulation

One of the major advances of flatworms over diploblastic animals is the development of protonephridia (see Chapter 4 for a review of nephridia among Metazoa). These structures occur in all free-living flatworms except some marine catenulids and consist of flame bulbs that may occur singly (as they do in some catenulids) or in pairs (from one to many pairs in different taxa). The protonephridia are connected to networks of collecting tubules that lead to one or more nephridiopores (Figure 10.16). Protonephridia in free-living flatworms function primarily as osmoregulatory structures. Freshwater species tend to have more protonephridia and more complex tubule systems than do their marine counterparts. Although a small amount of ammonia is released via the protonephridia, most metabolic wastes are lost by diffusion across the body wall.

Flukes also possess variable numbers of flame bulb protonephridia. Two nephridioducts drain the nephridia and lead to a storage area, or bladder, which in turn connects with a single posterior nephridiopore in the digenean flukes, or a pair of anterior pores in the monogenean types (Figure 10.15). Nitrogenous waste in the form of ammonia is excreted largely across the tegument. As in free-living flatworms, the protonephridia of flukes are primarily osmoregulatory.

Tapeworms possess numerous flame bulb protonephridia throughout the body. The flame bulbs drain to pairs of dorsolateral and ventrolateral nephridioducts that run the length of the body (Figures 10.15 and 10.17). Although some variation in plumbing occurs, the ventral ducts are typically connected to one another by transverse tubules near the posterior end of each proglottid. In relatively young worms that have not lost any proglottids (see the section on Reproduction and Development), the excretory ducts lead to a collecting bladder in the most posterior proglottid. Once this terminal proglottid is lost, the nephridioducts open separately to the outside on the posterior margin of the remaining hindmost proglottid.

There is still much to be learned about the protonephridia of cestodes. They probably function both in

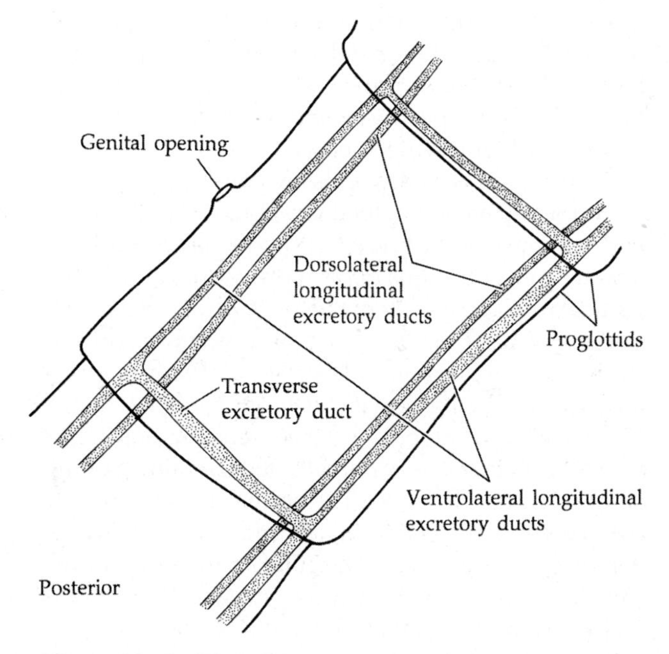

Figure 10.17 The arrangement of major protonephridial ducts in a cestode proglottid (ventral view).

excretion and osmoregulation. They may also serve to eliminate certain organic acid products of anaerobic cellular metabolism. Some experimental work indicates that tapeworms are capable of precipitating and storing some waste products within their proglottids.

Nervous System and Sense Organs

Free-living flatworms The nervous system of these species varies from a simple netlike nerve plexus with only a minor concentration of neurons in the head (similar to that seen in the phylum Xenacoelomorpha), to a distinctly bilateral arrangement with a well-developed cerebral ganglion and longitudinal nerve cords connected by transverse commissures (Figure 10.18). The more advanced condition is referred to as a ladder-like nervous system. Even many of those species that possess distinctly centralized nervous systems have a plexus formed by the repeated branching of nerve endings (e.g., polyclads). In general, larger flatworms show an increasing concentration of the peripheral nerves into fewer and fewer longitudinal cords and an accumulation of neurons in the head as an associative center or cerebral ganglion. Furthermore, they show a tendency to separate the elements of the nervous system into distinct sensory and motor pathways and to develop a circuitry that operates primarily on unidirectional impulse transmission.

The nervous system and sense organs of free-living flatworms appear to have been elaborated in association with bilateral symmetry and unidirectional movement. The result is a general concentration of sense organs at the anterior end of the body and an elaboration of those receptor types that are compatible with these animals' lifestyle. Tactile receptors are abundant over much of the body surface as sensory bristles projecting from the epidermis. These receptors tend to be concentrated at the anterior end and around the pharynx. Benthic free-living flatworms orient to the substratum by touch; they are positively thigmotactic ventrally and negatively thigmotactic dorsally.

Most free-living flatworms are equipped with chemoreceptors that aid in food location. Although sensitive over most of the body, free-living flatworms have distinct concentrations of chemoreceptors anteriorly, particularly on the sides of the head. Some forms, such as the familiar freshwater planarians, and prorhynchidans have the chemoreceptors located in flaplike processes called **auricles** on the head (Figures 10.2 and 10.18C), whereas others have these sense organs in ciliated pits, on tentacles, or distributed over much of the anterior end of the body. The epithelium bearing the chemoreceptors is often ciliated and frequently forms depressions or grooves. The cilia are the receptor organelles, but also circulate water, thus facilitating sensory input from the environment.

The utilization of chemoreception in locating food has been demonstrated in many free-living flatworms. Some are known to home in on concentrations of dissolved chemicals associated with potential food. Others, such as *Dugesia*, hunt by waving the head back and forth as they crawl forward, exposing the auricles to any chemical stimulus in their path. When exposed to diffuse chemical attractants, some free-living flatworms begin a trial-and-error behavior pattern. If unable to determine the direction of the attractant, the worm begins moving in a straight line. If the stimulus weakens, the animal makes apparently random turns until it encounters sufficient stimulus, then moves toward it. This behavior can eventually bring the animal near enough to the food source to home in on it directly. Some species orient to water movements by rheoreceptors located on the sides of the head (Figure 10.18D).

Statocysts are common in certain flatworms, notably in members of the Catenulida and Proseriata. These groups include mostly swimming and interstitial forms in which orientation to gravity cannot be accomplished by touch. When present, the statocyst is usually located on or near the cerebral ganglion. Ehlers (1991) presents details on the ultrastructure of some flatworm statocysts.

Most free-living flatworms possess photoreceptors in the form of inverted pigment-cup ocelli (Figure 10.18E). A few macrostomids possess simple pigment-spot ocelli, which are presumed to be primitive within the flatworms. Many species bear a single pair of ocelli on the head; but some, such as certain polyclads and terrestrial triclads, may have many pairs of eyes. In a few terrestrial forms (e.g., *Geoplana mexicana*) and many of the large tropical polyclads, numerous eyes extend along the edges of the body. Most free-living species are negatively phototactic. The dorsal placement of the eyes and the orientation of the pigment cups facilitate the detection of light direction as well as intensity.

Larvae of the polyclad flatworm *Pseudoceros canadensis* possess two dissimilar kinds of eyes. The right eye appears to be microvillar (i.e., rhabdomeric), but the left one has components of both microvillar and ciliary origin (Eakin and Brandenberger 1980). The histories of these two eye types were noted in Chapter 4. The discovery of both types of eyes in a flatworm larva suggests to some researchers the possibility that this animal stands at a major point of evolutionary divergence.

Neurosecretory cells have been known in free-living flatworms for more than three decades, and work continues on exploring their functions. These special cells are generally located in the cerebral ganglion, but they also occur along major nerve cords in at least some species. Neurosecretions play important roles in regeneration, asexual reproduction, and gonad maturation in free-living species that could serve as models for understanding similar functions in parasitic species.

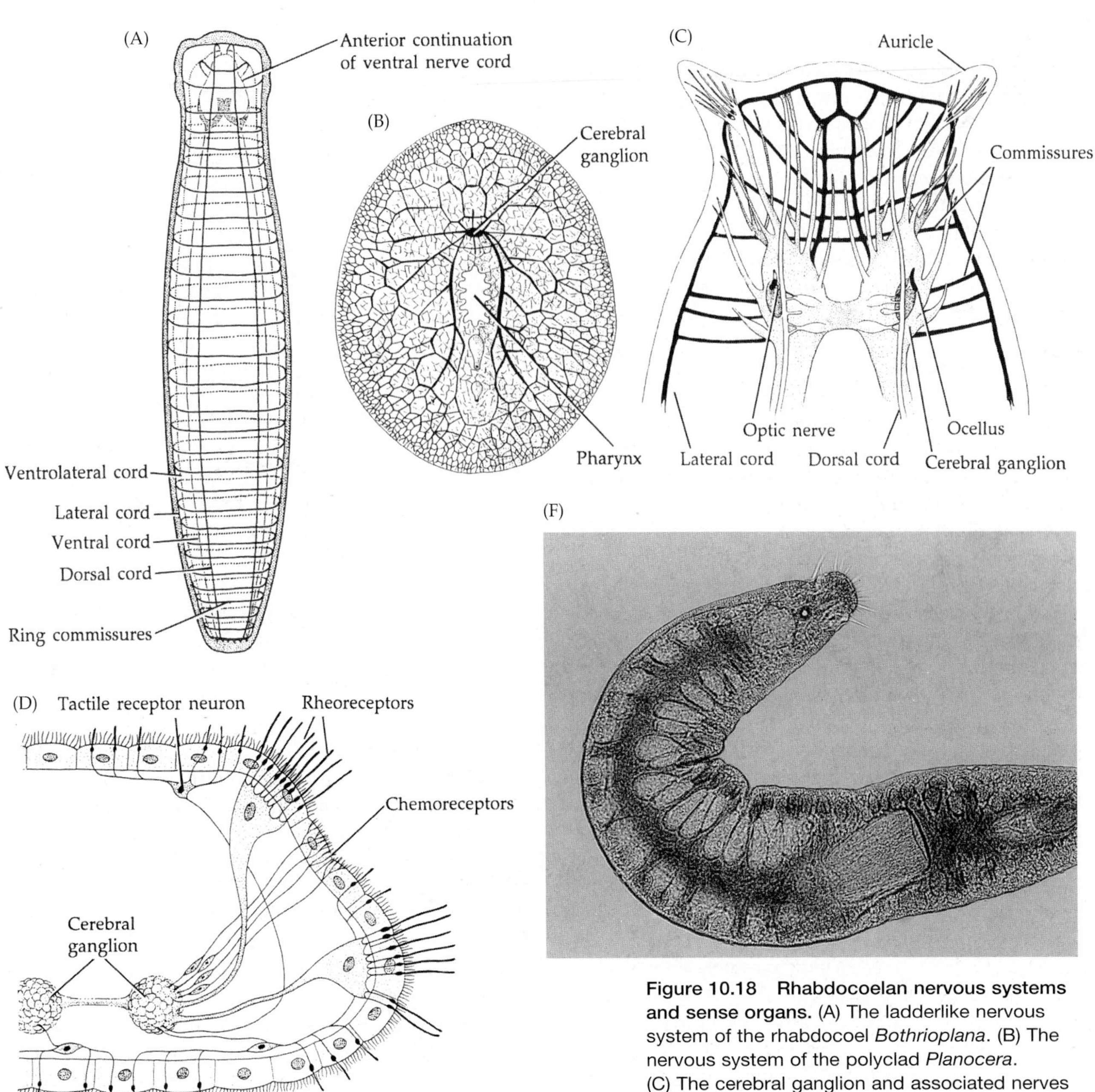

Figure 10.18 Rhabdocoelan nervous systems and sense organs. (A) The ladderlike nervous system of the rhabdocoel *Bothrioplana*. (B) The nervous system of the polyclad *Planocera*. (C) The cerebral ganglion and associated nerves in the triclad *Crenobia*. (D) The anterior end (cross section) of the rhabdocoel *Mesostoma*, showing tactile, chemo-, and rheoreceptors. (Rheoreceptors detect water movements over the surface of the animal.) (E) A typical free-living flatworm inverted pigment cup ocellus (section). (F) An interstitial free-living flatworm with a distinct statocyst and numerous anterior sensory bristles.

Flukes and tapeworms The nervous system of flukes is distinctly ladderlike and very similar to that in many free-living flatworms (Figure 10.19). The cerebral ganglion comprises two well-defined lobes connected by a dorsal transverse commissure. Nerves from the cerebral ganglion extend anteriorly to supply the area of the mouth, adhesive organs, and any cephalic sense organs. Extending posteriorly from the cerebral ganglion are up to three pairs of longitudinal nerve cords with transverse connectives. A pair of ventral cords is usually most well developed, and dorsal cords are present in the digenean flukes. Most flukes also have a pair of lateral nerve cords.

The suckers of flukes bear tactile receptors in the form of bristles and small spines. There is also some evidence of reduced chemoreceptors. Nearly all monogenean flukes possess a pair of rudimentary pigment-cup ocelli near the cerebral ganglion.

The cerebral ganglion of cestodes is usually a complex nerve ring located in the scolex (Figure 10.20). The ring bears ganglionic swellings and gives rise to a number of nerves. Anterior nerves, in the form of a ring or plexus, serve the rostellum (when present) and other attachment organs. Lateral cerebral ganglionic swellings give rise to a pair of major lateral longitudinal nerves, which extend the length of the animal. In each proglottid, these nerves bear additional ganglia from which transverse commissures arise and connect the two longitudinal cords. Additional longitudinal nerves are often present; the most typical pattern includes two pairs of accessory lateral cords—a pair of dorsal cords and a pair of ventral cords. As might be expected, sense organs are greatly reduced in cestodes and are limited to abundant tactile receptors in the scolex.

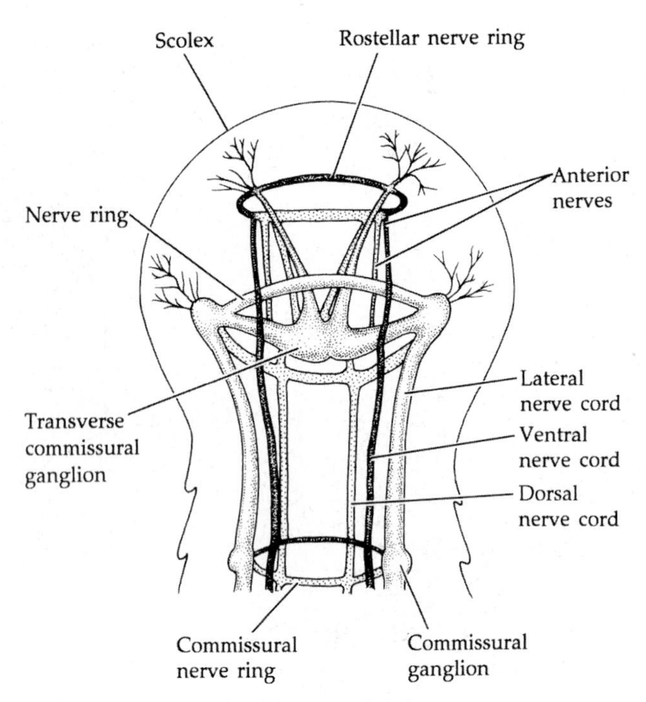

Figure 10.20 Cestode nervous system. The anterior end of *Moniezia*, a cyclophyllidean. The longitudinal cords extend the length of the animal.

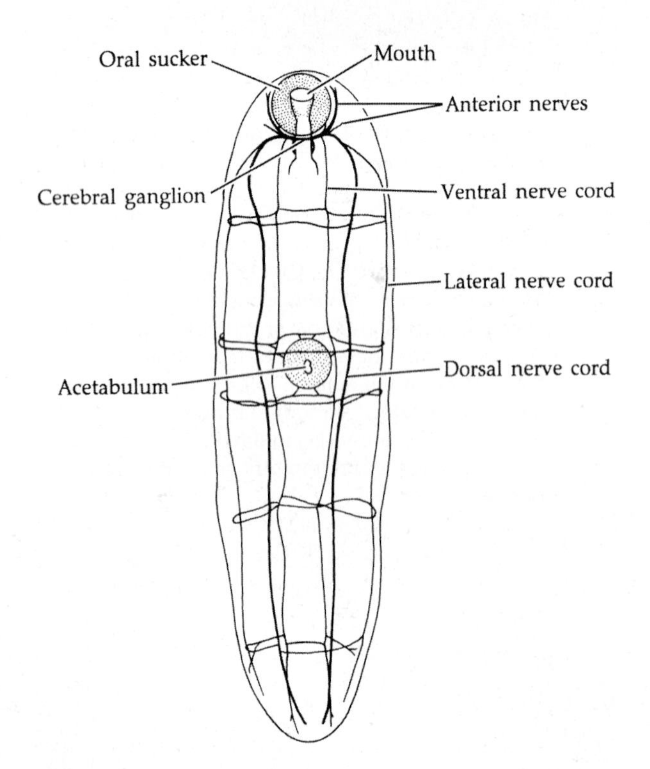

Figure 10.19 A generalized ladderlike nervous system of a trematode (ventral view).

Reproduction and Development

Asexual processes Asexual reproduction is common among freshwater and terrestrial free-living flatworms, and it generally occurs by transverse fission. Geographical variation exists in the tendency for individuals to undergo this process. In the catenulideans and macrostomorphans, an odd sort of multiple transverse fission occurs wherein the individuals thus produced remain attached to one another in a chain until they mature enough to survive alone (Figure 10.21A). Some freshwater triclads (e.g., *Dugesia*) split in half behind the pharynx, and each half goes its own way, eventually regenerating the lost parts. A few (e.g., *Phagocata*) reproduce by fragmentation, each part encysting until the new worm forms.

The remarkable regenerative abilities of free-living flatworms have been studied intensely for many years. Much of the experimental work has been conducted on the common triclad *Dugesia*, a familiar animal to beginning zoology students. Underlying all of the bizarre results of various surgeries performed on these animals (Figure 10.21B,C) is the fact that the cells of organisms like *Dugesia* are not totipotent; an anterior–posterior body polarity exists in terms of the regenerative capabilities of the cells. However, the cells in the midbody region are less fixed in their potential to produce other

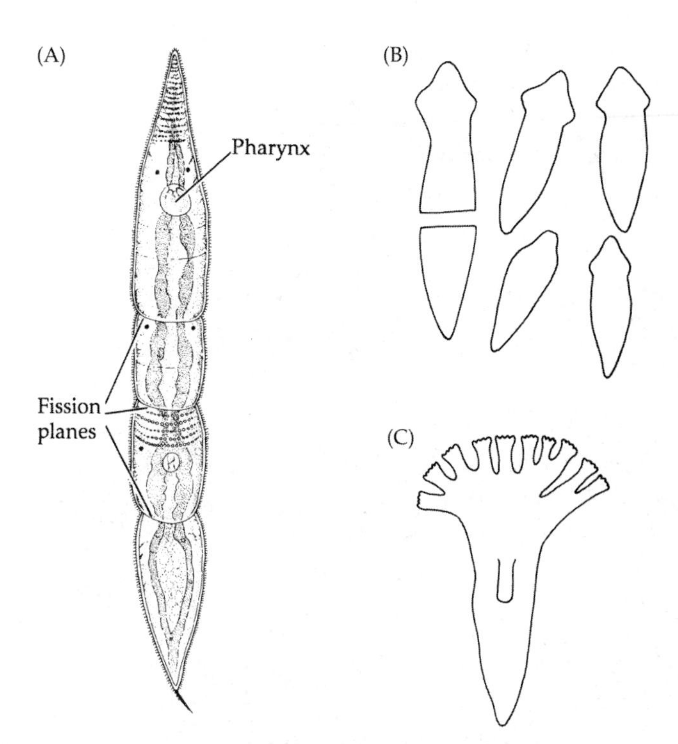

Figure 10.21 (A) Asexual reproduction by transverse fission in the catenulid flatworm, *Alaurina*. (B,C) Regeneration after experimental injuries in planarians.

parts of the body than are those toward the anterior or posterior ends. Thus, if the flatworm is cut through the middle of the body (as occurs in natural transverse fission), each half will regenerate the corresponding lost part. However, if the animal is cut transversely near one end—say, separating a small piece of the tail from the rest of the body—the larger piece will grow a new posterior end, but the piece of tail lacks the capability to produce an entire new anterior end. This gradient of cell potency has been of particular interest to cell biologists and medical researchers because of its relevance to healing and regeneration potential in higher animals.

Asexual reproduction is an important feature of the life cycle of flukes, where the ability to reproduce asexually helps ensure survival, particularly when potential mates may not be nearby.

Sexual reproduction: Free-living flatworms Most free-living flatworms are hermaphroditic and possess complex and highly diverse reproductive systems (Figure 10.22). The male system includes single (e.g., Macrostomorpha), paired (e.g., many Rhabdocoela), or multiple (e.g., Polycladida) testes. The testes are generally drained by collecting tubules that unite to form one or two sperm ducts, which often lead to a precopulatory storage area or seminal vesicle. Prostatic glands, which supply seminal fluid to the sperm, are often associated with and empty into the seminal vesicle. The seminal vesicle is typically part of a muscular chamber called the **male atrium**, which houses the copulatory organ. The actual organ of sperm transfer may be a papilla-like penis or an eversible cirrus, through which sperm are forced by muscular action of the atrium.

The female reproductive system is more variable than that of the male. Much of the variation is related to whether the flatworm in question produces **endolecithal** or **ectolecithal** ova—that is, whether the worm is described as **archoöphoran** or **neoöphoran**. The archoöphorans (e.g., Macrostomida, Polycladida) typically possess an organ that produces both eggs and yolk. The final product is endolecithal ova. Such an organ is called a **germovitellarium**, and may occur either singly or paired. In the neoöphorans (e.g., the order Prolecithophora and the entire superclass Euneoophora), the ovary (**germarium**) is separate from the yolk gland (**vitellarium**). Yolk-free eggs are produced by the ovary and then cellular yolk is transported through a vitelline duct and deposited alongside the ova inside the eggshell, a process resulting in ectolecithal ova.

In both cases, the eggs are typically moved via an oviduct toward the female atrium, which often bears special chambers for receipt and storage of sperm (i.e., copulatory bursa and seminal receptacle). Associated with this arrangement may be a variety of accessory glands, such as **cement glands**, for the production of shells and egg cases.

The male and female gonopores are often separate, the female opening usually located posterior to the male pore. In some species, however, the two systems share a common genital opening, and in a few the male atrium opens just inside the mouth. In the latter case, the mouth is referred to as an **orogenital pore**.

Mating is usually by mutual cross-fertilization. The two mates align themselves so that the male gonopore of each is pressed against the female gonopore of the mate (Figure 10.23A). The male copulatory organ, or **stylet** (or penis or cirrus) is everted by hydrostatic pressure caused by the muscles surrounding the atrium and is inserted into the mate's female atrium, where sperm are deposited. The mates then separate, each going its own way and carrying foreign sperm. Fertilization usually occurs as the eggs pass into the female atrium or within the oviduct itself. The zygotes are frequently stored for a period of time in special parts of the female system or in enlarged oviducts; any such storage area is called a **uterus**.

Some macrostomorph and polyclad flatworms exhibit hypodermic insemination, whereby the male stylet is thrust through the body wall of the mate and sperm are forcibly injected into the mesenchyme. In other polyclads, individuals receive spermatophores from partners on their dorsal surface. Whereas most flatworms mate in pairs, such dermal impregnation may occur in groups. By a method not yet understood,

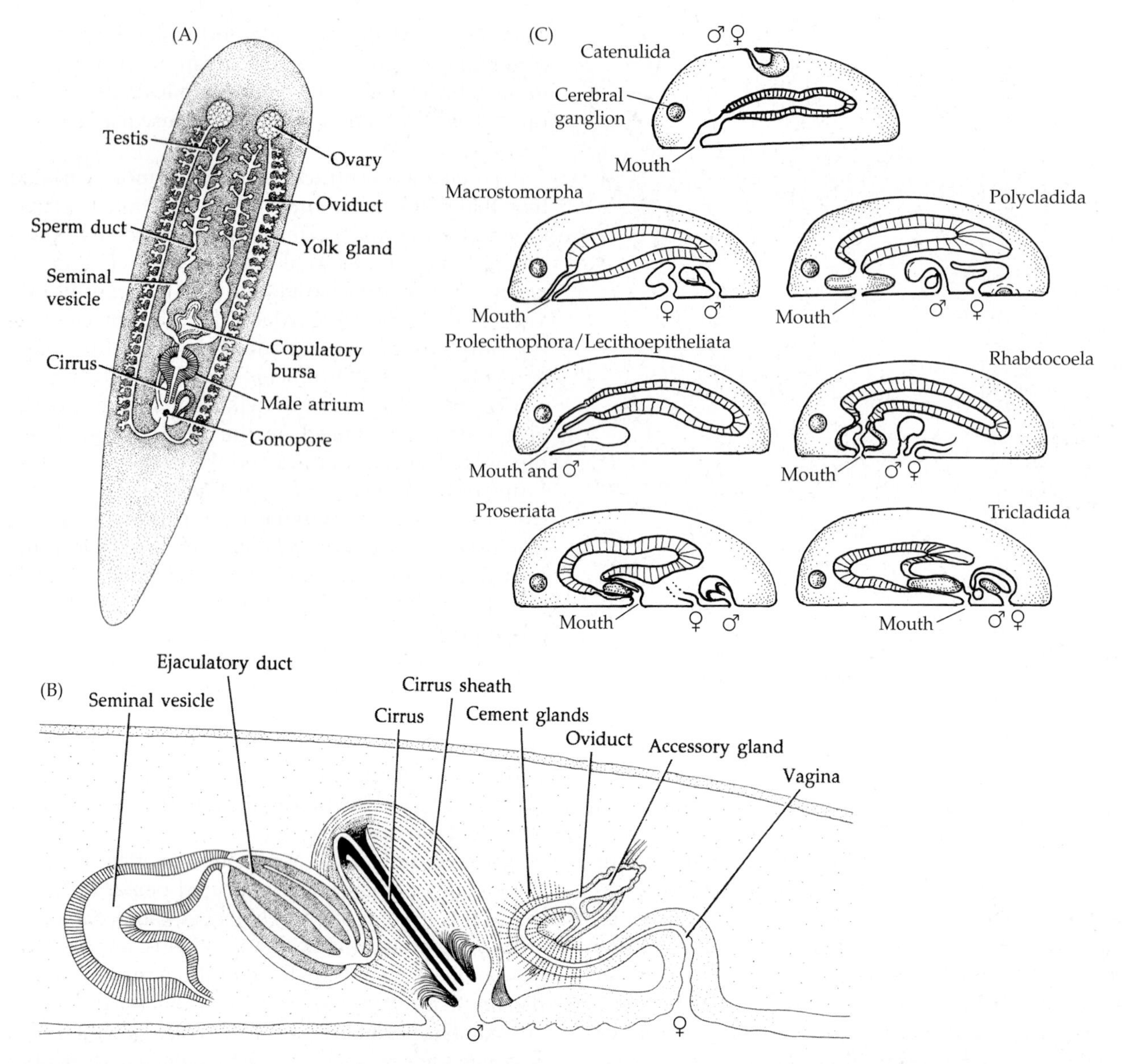

Figure 10.22 Reproductive systems of free-living flatworms. (A) Generalized triclad condition with separate ovaries and yolk glands (neoöphoran condition). (B) The copulatory structures of a triclad (sagittal section). (C) Generalized scheme of the relative position of male (♂) and female (♀) reproductive structures with respect to the location of the mouth in seven taxa.

after mating, the sperm find their way to the female system and fertilize the eggs.

In the genus *Macrostomum*, a close correspondence exists between the morphology of sperm, the morphology of the male stylet, and whether insemination is hypodermic or occurs by copulation (Figure 10.23C,D). Sperm in all *Macrostomum* have "feelers," anterior projections that embed themselves into the wall of the female reproductive tract in position to allow fertilization. However, depending on the species, sperm may also possess backwardly-directed bristles. Species in which the stylet is hooked engage in hypodermic insemination and sperm lack bristles. In contrast, species in which the stylet is not hooked tend to copulate, and sperm do possess bristles. Bristles appear to maintain sperm in position within the female reproductive tract after mutual insemination by mating pairs. After pairs separate, each individual applies its mouth to its own vagina and "sucks," evidently removing unattached sperm. The sperm that have imbedded their feelers into the female reproductive tract are in position for fertilization, and bristles appear to assist in this process. Unattached sperm are removed from the vagina by suction and are apparently eaten.

Ingestion of sperm from mating partners may in fact be widespread. In other free-living flatworms from several taxa, the presence of **Lang's vesicle** is associated with species that engage in copulation. Although this structure is thought to be a sperm storage organ its significance is still in doubt. The structure is usually

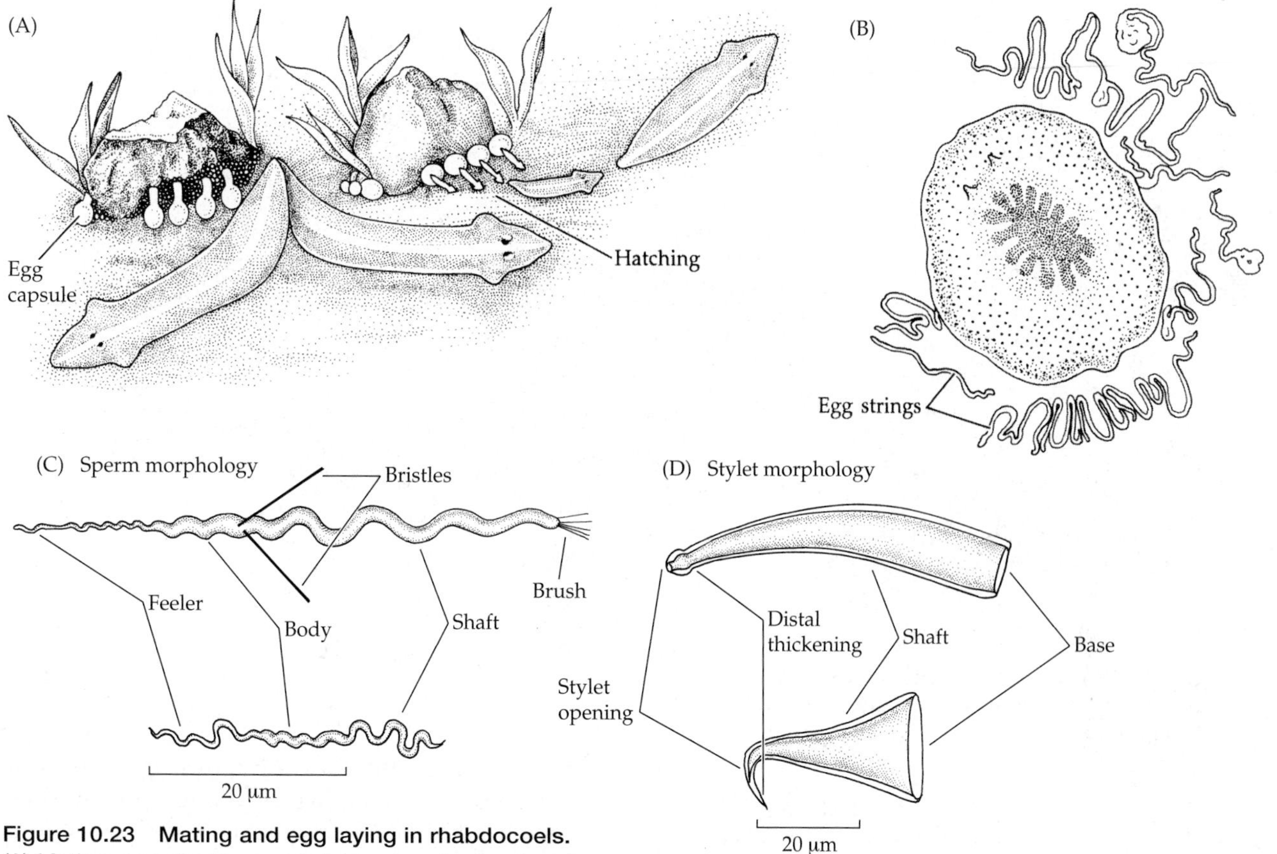

Figure 10.23 Mating and egg laying in rhabdocoels. (A) Mating, egg cluster, and hatching in a freshwater planarian. (B) Egg laying by the polyclad *Stylochus*. (C) Sperm and stylet morphology of *Macrostomum lignano*, a species in which individuals reciprocally copulate when mating. (D) Sperm and stylet morphology in *Macrostomum hystrix*, a species in which hypodermic insemination occurs.

attached to the digestive system suggesting that it may be involved in sperm digestion.

Once fertilization is accomplished, the zygotes are either retained by the parent within the uteri of the female reproductive tract or laid in various sorts of gelatinous or encapsulated egg masses (Figure 10.23B). Thus, most maternal free-living flatworms are obliged to contribute substantially toward the care of their embryos; they may be described as oviparous or ovoviviparous. Some freshwater triclads produce special overwintering zygotes, which are encapsulated and retained within the female reproductive tract until spring.

The general strategy of the vast majority of free-living flatworms is to produce relatively few zygotes, which are protected by brooding or encapsulation and undergo direct development. A few polyclads produce **Müller's larvae**, which swim about for a few days prior to settling and metamorphosing (Figures 10.24 and 10.25). This larva is equipped with eight ventrally directed ciliated lobes, by means of which it swims. A few species of parasitic polyclads of the genus *Stylochus* produce a **Götte's larva**, which bears four rather than eight lobes; and members of the freshwater catenulidean genus *Rhynchoscolex* pass through a vermiform stage in their development that has been referred to as a larval form.

Early embryogeny differs greatly between the archoöphoran and neoöphoran flatworms. The endolecithal ova of the archoöphorans undergo some form of spiral cleavage (although some authors question the details of this process). The pattern and cell fates in many of these archoöphoran embryos are distinctly

Figure 10.24 "Face-on" view of Müller's larva of a polyclad (*Planocera*).

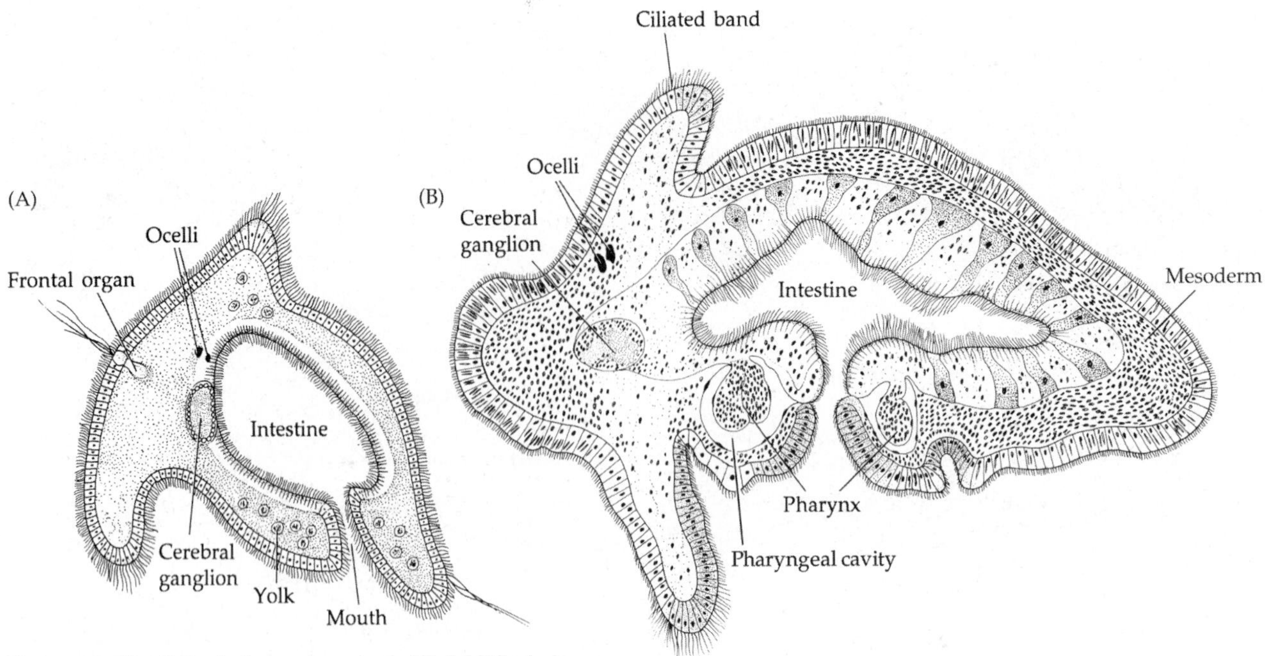

Figure 10.25 Polyclad development. (A) A Müller's larva (sagittal section). (B) A later larval stage, showing formation of the pharyngeal apparatus (sagittal section).

protostome-like, and this spiral cleavage has been well studied (especially in polyclads). Quartets of cells are produced, the fates of which may be described using Wilson's coding system (Chapter 5). By the end of spiral cleavage, the embryo is considered a stereoblastula, oriented with the derivatives of the first micromere quartet at the animal pole and the macromeres at the vegetal pole. The lq cells become the anterior ectoderm, cerebral ganglion, and most of the rest of the nervous system. The 2q derivatives contribute to ectoderm and ectomesoderm, particularly that of the pharyngeal apparatus and its associated musculature. The remainder of the ectoderm and probably some ectomesoderm are formed from the derivatives of the third micromere quartet. The 4d cell, normally associated solely with endomesoderm in typical protostomes, divides to produce a $4d_1$ and $4d_2$ cell in polyclads. The $4d_1$ gives rise to endoderm and thus to the intestine; the $4d_2$ produces the endomesoderm from which the body wall and mesenchymal muscles, much of the mesenchymal mass, and most of the reproductive system are derived. The remaining cells (4a, 4b, 4c, and the 4Q) include most of the yolk and are incorporated into the developing archenteron as embryonic food.

Gastrulation is by epiboly of the presumptive ectoderm derived from some of the cells of the first three micromere quartets. The ectoderm grows from the animal pole to the vegetal pole, surrounding the 4q and 4Q cells. At the vegetal pole the ectoderm turns inward as a stomodeal invagination, which later elaborates as the pharynx and connects with the developing intestine (Figure 10.25). As development proceeds, the embryo flattens, with the mouth directed ventrally, and hatches as a tiny flatworm. If development is indirect, the larva emerges about the time the intestine is hollowing.

Because of the deposition of extracellular yolk with ectolecithal ova, the development of neoöphoran species is highly modified from the plan described above. Certain species of rhabdocoels and triclads have been most extensively studied, and the two groups differ—especially in the early stages. In both cases, cleavage is so distorted that cell fates and germ layer formation cannot easily be compared with the typical spiralian pattern. In the rhabdocoels, early cleavage leads to the formation of three masses of cells positioned along the presumptive ventral surface of the embryo beneath the mass of yolk (Figure 10.26A). The cell masses then produce a layer of cells that extends around to enclose the yolk. This covering thickens to several cell layers, the innermost eventually becoming the intestinal lining (and enclosing the yolk), the outermost becoming the epidermis. The anterior cell mass produces the nervous system, the middle cell mass the pharynx and associated muscles. The posterior cell mass forms the rear portion of the worm and the reproductive system.

Early development in triclads differs from that of other neoöphorans. During early cleavage, the blastomeres are loose within a surrounding mass of fluid yolk, a condition described as "blastomere anarchy." A few of the blastomeres migrate away from the others and flatten to produce a thin membrane enclosing a packet of the yolk including the remaining blastomeres (Figure 10.26B,C). These migrating blastomeres are often called **hull cells**, but the yolk-enclosing **hull membrane** they create can be variable and may not be homologous among flatworm taxa. Additional yolk cells are produced as a syncytial mass around a group of developing embryos and encapsulated, as many as 40 per capsule. Through migration and differentiation of various blastomeres, each embryo forms a temporary intestine, pharynx, and mouth, through which it ingests the yolky syncytium. The embryonic mouth eventually closes and the wall of the embryo thickens to form anterior, middle, and posterior cell masses, whose fates are similar to those in rhabdocoels.

Sexual reproduction: Flukes Like the free-living flatworms, flukes are hermaphroditic and typically engage in mutual cross-fertilization. Self-fertilization occurs only in rare cases. There is a great deal of variation in the details of the reproductive systems among flukes, but most are built around a common plan similar to that in certain free-living flatworms (Figure 10.27). The male system includes a variable number of testes (usually many in the monogenean flukes and two in the digenean flukes), all of which drain to a common sperm duct that leads to a copulatory apparatus, usually an eversible cirrus. The lumen of the cirrus is continuous with that of the sperm duct, and their junction is frequently enlarged as a seminal vesicle. Prostatic glands are typically present, opening into the cirrus lumen near the seminal vesicle. All of these terminal structures are housed within a muscular **cirrus sac**, the contraction of which causes eversion of the cirrus as an intromittent organ (Figure 10.27C). The common genital pore opens ventrally near the anterior end of the animal and leads to a shallow atrium, usually shared by both the male and female systems. Many monogenean flukes have simpler male systems than that just described, often lacking much elaboration of the terminal structures and possessing a simple **penis papilla** rather than an eversible cirrus.

The female reproductive system (Figure 10.27B) usually bears a single ovary connected by a short oviduct to a region known as the **ootype**. The oviduct is joined by a yolk duct (= vitelline duct) formed by the union of paired ducts, which carry yolk from the multiple laterally placed yolk glands. A seminal receptacle is usually present as a blind pouch off the oviduct. Extending anteriorly to the genital atrium is a single uterus, which is sometimes modified as a **vagina** near the female gonopore.

Sperm are produced in the testes and stored prior to copulation in the seminal vesicle (Figure 10.27C). During mating, two flukes align themselves such that the cirrus of each can be inserted into the female orifice of the other. Sperm, along with semen from the prostatic glands, are ejaculated into the female system by muscular contractions. The sperm move to, and are stored within, the seminal receptacle, and the mates separate. As eggs pass through the oviduct to the ootype, they are fertilized by sperm released from the seminal receptacle into the oviduct.

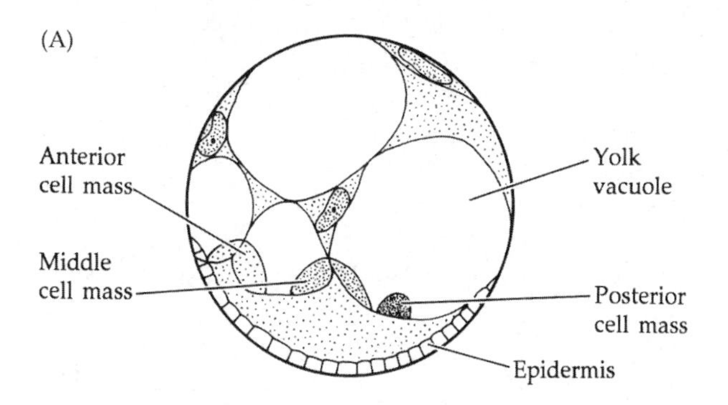

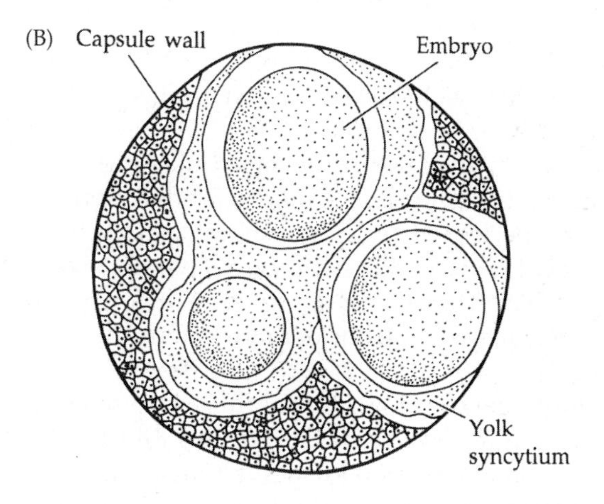

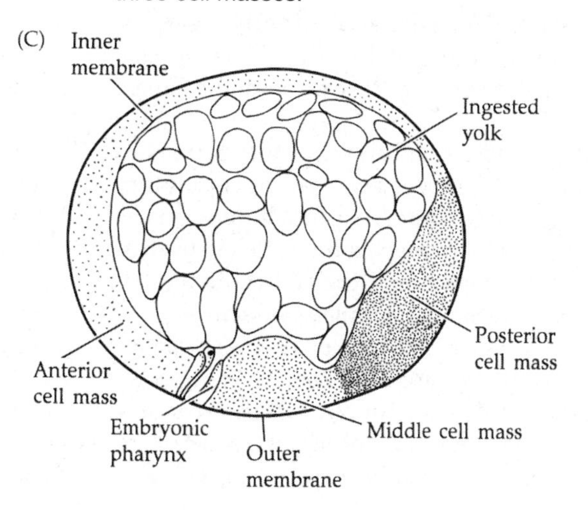

Figure 10.26 Neoöphoran development. (A) The embryo of a typical rhabdocoel has three cell masses with large, vacuolated external yolk cells. (B) A triclad egg capsule containing three embryos surrounded by yolk syncytium. (C) This single triclad embryo has ingested the yolk through the temporary embryonic pharynx, and shows the three cell masses.

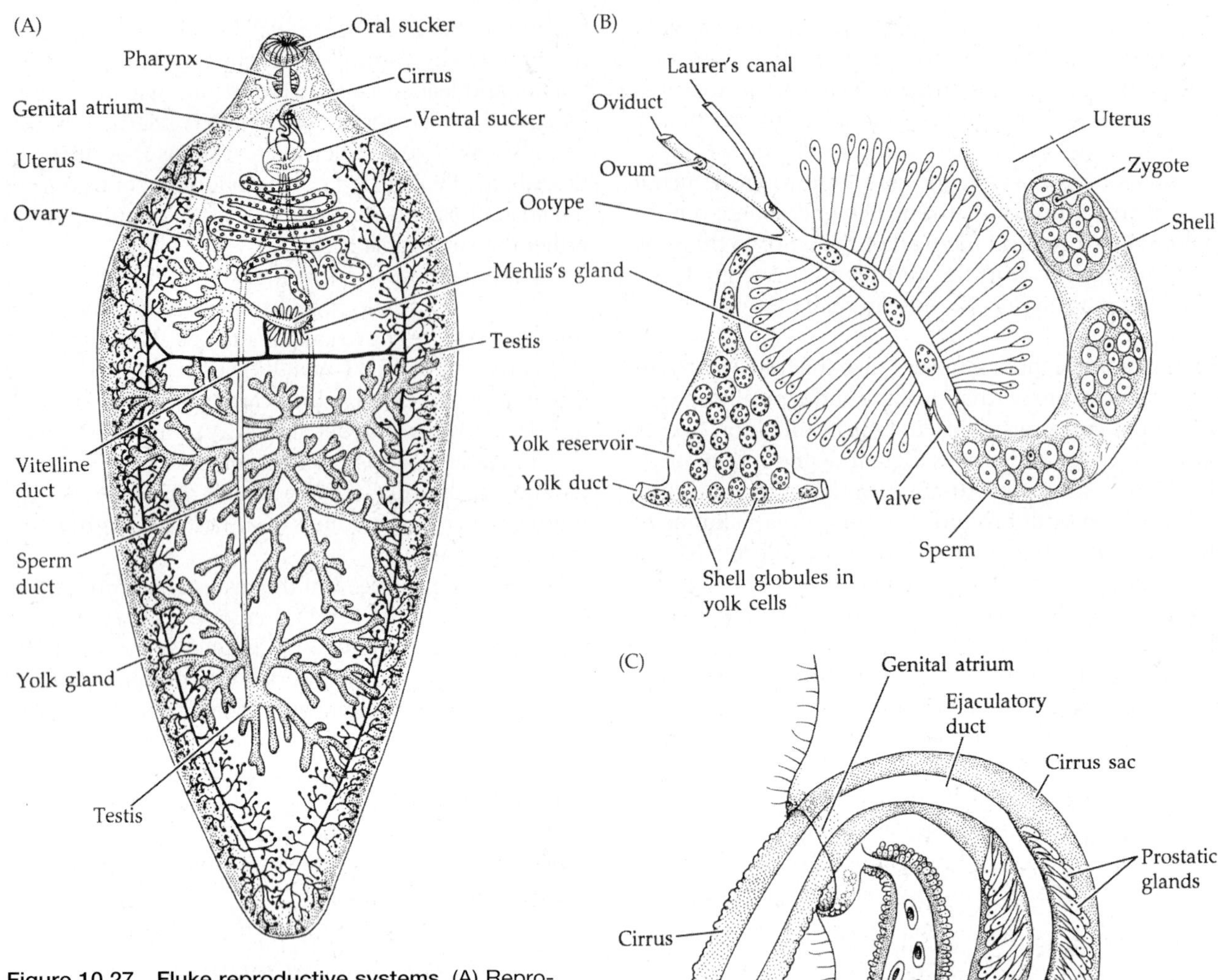

Figure 10.27 Fluke reproductive systems. (A) Reproductive structures of *Fasciola hepatica*. (B) The region of the oötype in *F. hepatica*. (C) Male copulatory apparatus with cirrus extended.

Flukes produce ectolecithal ova. The yolk glands produce yolk, which is deposited outside the eggs along with secretions that form a tough shell around the zygote. Thus encapsulated, the zygotes move from the ootype into the uterus, probably aided by secretions from clusters of unicellular **Mehlis's glands**. The zygotes may be stored within the uterus for various lengths of time prior to release through the female gonopore.

Some flukes possess an additional canal that arises from the oviduct and serves as a special copulatory duct. This duct, called **Laurer's canal**, opens on the dorsal body surface and receives the male cirrus during mating. A few polyclad and triclad free-living flatworms also possess a similar copulatory duct.

High fecundity is a general rule among parasites, and the flukes are no exception. The dangers of complex life cycles and host location result in extremely high mortality rates that must be offset by increased zygote production or asexual processes. Flukes may produce as many as 100,000 times as many eggs as free-living species.

The early stages of development in flukes are usually highly modified because of the ectolecithal nature of the ova. In species where little yolk is present, cleavage is holoblastic, and cell fates and germ layer formation have been traced accurately. Development is virtually always mixed, involving one or more independent larval stages.

The life cycles of monogenean flukes are relatively simple and involve only a single host. Most of the adults are ectoparasites on fishes, although some attach to turtles, various amphibians, and even some

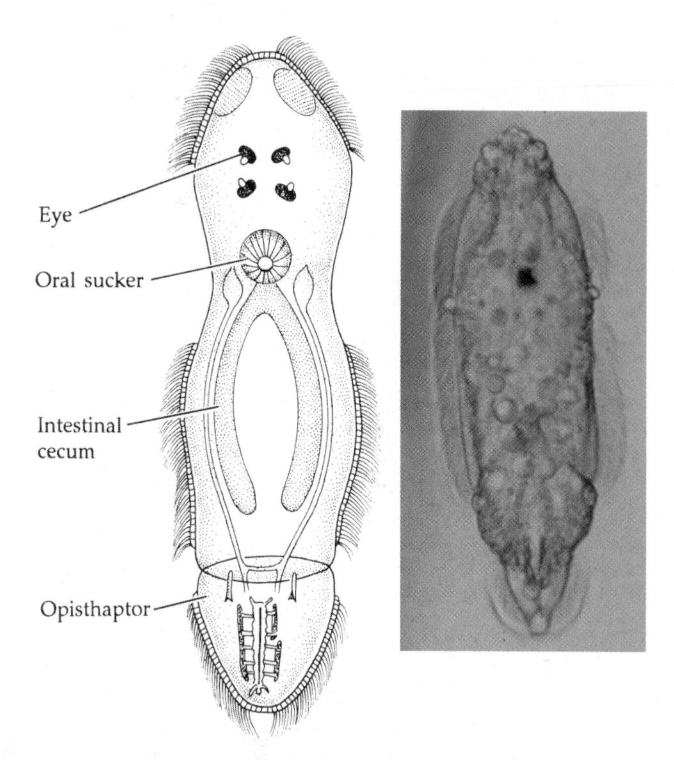

Figure 10.28 Photograph and drawing of an oncomiracidium larva of a monogenean fluke.

invertebrates. A few members of the Monogenea have taken up a mesoparasitic life style and reside in host body chambers that open to the environment (e.g., gill chambers, mouth, bladder, cloacal cavity). When the embryos are released from the uterus they often attach to the host tissue by means of special adhesive threads on the shell. Upon hatching, a larval stage called an **oncomiracidium** is released to the environment (Figure 10.28). The oncomiracidium is densely ciliated and swims about until it encounters another appropriate host. The cues larvae use to locate particular hosts may be tactile, chemical, or may vary depending on local densities of hosts and other larvae. The prohaptor and especially the opisthaptor develop during the larval stage and facilitate attachment to the new host, whereupon the larva metamorphoses to a juvenile fluke. It is about this time that the ciliated larval skin is shed and the tegument (= neodermis) forms. There are many variations on this basic life cycle among members of the class Monogenea; we present two in outline form in Figure 10.29.

Digenean flukes (orders Diplostomida and Plagiorchiida) include some of the most successful parasites known. A good deal of variation exists not only in adult morphology but also in life cycles (Figure 10.30). In general, eggs are produced by adult worms in their definitive host. After fertilization, the zygotes are eventually discharged via the host's feces, urine, or sputum. Upon reaching water they either are eaten by an intermediate host or hatch as free-swimming ciliated larvae called **miracidia**, which actively penetrate an intermediate host. Several asexual generations of larval forms occur in the intermediate host, eventually producing free-swimming forms called **cercaria**. The cercaria usually encyst within a second intermediate host, becoming **metacercaria**. Infection of the definitive host occurs when the metacercaria are eaten or, if there is no second intermediate host, when the cercaria penetrate directly. The larval skin is lost in the definitive host and the syncytial tegument develops.

In their adult stages, nearly all of the digenean flukes are endoparasites of vertebrates. They are known to inhabit nearly every organ of the body, and many are serious pathogens of humans and livestock. The intermediate hosts of all trematode flukes are gastropods, although some are known to use other invertebrates or even certain vertebrates. Most species of snails host one or more species of digenean flukes; the common California tidal flat gastropod *Cerithideoposis californica* serves as intermediate host to nearly two dozen species of digenean flukes, most of which ultimately infect shore birds. Almost all of the trematodes that infect *C. californica* castrate the snail, thus, theoretically, freeing up the host's energy reserves for their own purposes as they live out their life producing their own offspring. Space does not permit an account of more than a few of the life cycles of these worms. We begin below with a general case, using the Chinese liver fluke *Clonorchis sinensis* (order Plagiorchiida) as an example. This trematode is widespread in the Far East. It displays all the common stages found in the life cycles of most digenetic flukes. In northern Thailand a related fluke, *Opisthorchis viverrini,* has a similar life cycle and pathology.

The adult Chinese liver fluke usually lives within the branches of the bile duct in humans. This animal may reach several centimeters in length and in high numbers causes serious problems, including cholangiocarcinoma, a deadly cancer of the bile duct. While still in the uterus of the female reproductive tract, the zygotes develop to miracidia, each housed within its original egg case. Once released from the female system and passed out of the host with the feces, the miracidia are eaten by the first intermediate host, a snail of the genus *Parafossarulus*. The ciliated, swimming miracidium hatches from its egg case in the gut of the snail and migrates into the digestive gland. Here each miracidium becomes an asexually active form called a **sporocyst**, within which germinal cells become yet another larval form called a **redia**. Subsequently, germinal cells within the redia produce the cercaria larvae. This double sequence of rapid asexual reproduction results in perhaps 250,000 cercariae from each original miracidium!

The cercariae leave the snail, swim about, and enter the second intermediate host, the Chinese golden carp (*Macropodus opercularis*). The cercariae of *C. sinensis*

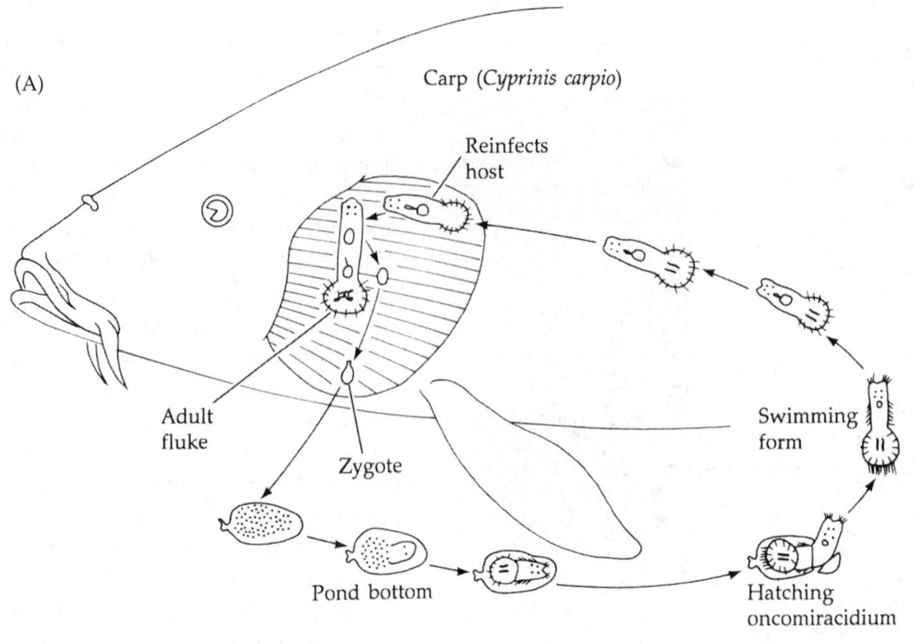
(A)
Carp (*Cyprinis carpio*)
Reinfects host
Adult fluke
Zygote
Pond bottom
Hatching oncomiracidium
Swimming form

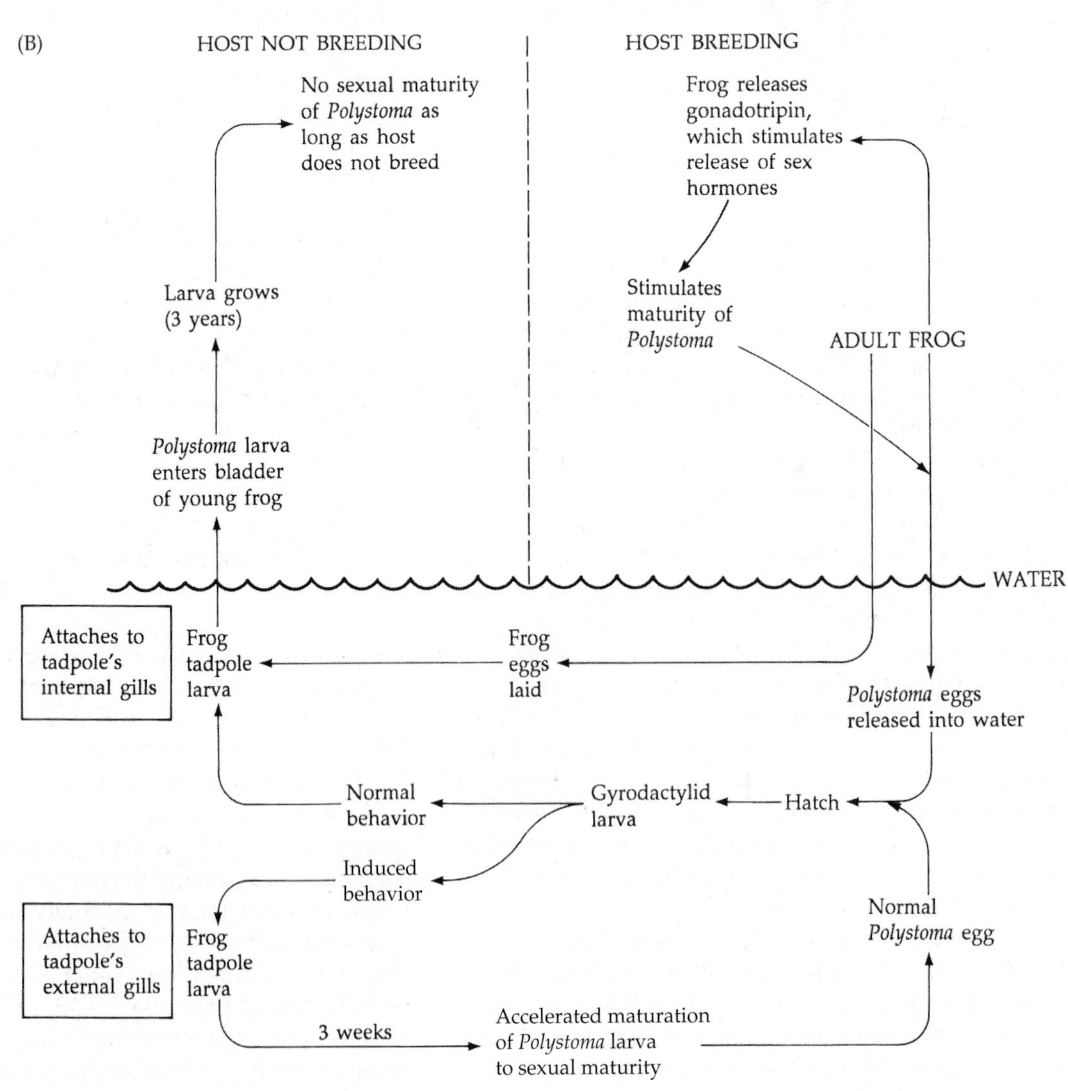
(B)
HOST NOT BREEDING
HOST BREEDING
No sexual maturity of *Polystoma* as long as host does not breed
Larva grows (3 years)
Polystoma larva enters bladder of young frog
Frog releases gonadotripin, which stimulates release of sex hormones
Stimulates maturity of *Polystoma*
ADULT FROG
WATER
Attaches to tadpole's internal gills
Frog tadpole larva
Frog eggs laid
Polystoma eggs released into water
Normal behavior
Gyrodactylid larva
Hatch
Induced behavior
Normal *Polystoma* egg
Attaches to tadpole's external gills
Frog tadpole larva
3 weeks
Accelerated maturation of *Polystoma* larva to sexual maturity

Figure 10.29 Life cycles of two monogenean trematodes. (A) The life cycle of *Dactylogyrus vastator*, a parasite of freshwater cyprinodont fishes. (B) The life cycle of *Polystoma integerrimum*, a parasite in the urinary bladders of frogs. This fascinating life history demonstrates the dramatic influences exerted by the developmental stage of the host on the development of the parasite. Under normal conditions, the adult fluke resides in the bladder of adult frogs. The fluke releases fertilized eggs into the water, where they hatch as oncomiracidia. These in turn become so-called gyrodactylid larvae, which attack the tadpole larval stages of the host. If the tadpole is very young, the fluke larvae attach to the external gills of the host and undergo precocious sexual maturation (progenesis) to produce more zygotes; these flukes die upon metamorphosis of the host. However, if the fluke larvae encounter more advanced tadpoles, they enter the branchial chambers and attach to the host's gills, where they reside until the host undergoes metamorphosis. At that time, the flukes leave the branchial chamber, migrate to the cloacal pore, and enter the host's bladder. Here the flukes live and grow, but they do not become sexually active until they are influenced by the host's sex hormones. Thus, sexual reproduction of the host and its parasites are synchronized—a pattern that guarantees availability of larval hosts for larval parasites!

burrow through the skin of the fish and encyst in the muscle tissue as metacercaria. If the fish is insufficiently cooked and then eaten by a human, the metacercaria survive and are released from their cysts by the action of the host's digestive enzymes. Once freed, they migrate into the bile duct, metamorphose into juvenile worms, mature, and complete the cycle. Adult flukes can inhabit their hosts for many years, causing irritation and blockages. With this general life cycle in mind, we refer you to Figure 10.30 for a brief overview of two additional examples.

A critical point here are the strategies for survival displayed by these parasites. There is, of course, no assurance of finding the proper hosts at the proper times, and mortalities are extremely high. The advantages of high specialization could be efficiency and reduced competition with other species once established in a proper host. However, multiple infections and within-host competition can be intense, favoring a wide range of competitive adaptations among species, including in at least one trematode species (*Himasthla*), soldier and reproductive castes within molluscan hosts, similar to those observed in social insects. Such adaptations imply the existence of strong selection, acting on larval as well as adult life stages. As we have emphasized earlier, the compensation for such high mortality is high fecundity coupled with asexual reproduction—and therein lies the expense.

Sexual reproduction: Tapeworms That the "business of animals is to reproduce themselves" is a lesson the cestodes demonstrate well. Most of their time, energy, and body mass are devoted to the production of more tapeworms. Like other flatworms, cestodes are hermaphroditic and practice mutual cross-fertilization when mates are available. However, some cestodes are known to self-fertilize. The orders Amphilinidea and Gyrocotylidea possess a single male and a single female reproductive system, whereas the more derived cestodes contain complete systems repeated within each proglottid. There is a good deal of variation in the details of these systems; the following is a generalized description of the male and female systems as they occur in a single proglottid of a cestode (Figure 10.31).

The testes are numerous. Some are scattered throughout the mesenchyme, but most are concentrated along the lateral margins. Collecting tubules lead from the testes to a single coiled sperm duct, which extends laterally (as a seminal vesicle) to a cirrus housed within a muscular cirrus sac. The male system empties into a common genital atrium.

The female system usually includes two ovaries from which an oviduct extends to an ootype surrounded by shell glands. The uterus is a branched blind sac extending from the ootype. A duct extends from the female gonopore in the genital atrium to the oviduct; its junction with the oviduct, near the tube, is swollen as a seminal receptacle. The portion of the duct near the genital atrium is called the vagina. A diffuse yolk gland empties via a vitelline duct into the oviduct.

During mutual cross-fertilization the cirrus of each mate is inserted into the vagina of the other. Many tapeworms double back on themselves so that two proglottids of the same worm cross-fertilize; in some species, self-fertilization is known to occur within a single proglottid. Sperm are injected into the vaginal duct and are stored in the seminal receptacle. Eggs are fertilized as they move through the oviduct from the ovaries to the ootype. Capsule material and yolk cells are deposited around each zygote, and the zygotes are moved into the uterus for temporary storage.

The reproductive systems mature and become functional with age as they are moved more posteriorly along the body by the production of new proglottids. If mating has occurred, the proglottids toward the posterior end become filled with an expanded uterus engorged with developing embryos (Figure 10.31B). These proglottids eventually break free from the body and are lost from the host with the feces, although in some cases the proglottids release embryos inside the host and the embryos pass out with the feces. Early development of tapeworm embryos is drastically modified from the pattern seen in free-living flatworms and varies somewhat among different groups. However, like all Neodermata, after fertilization the ectolecithal ova of cestodes show no vestiges of spiral cleavage, and even germ layer formation is often difficult or impossible to trace.

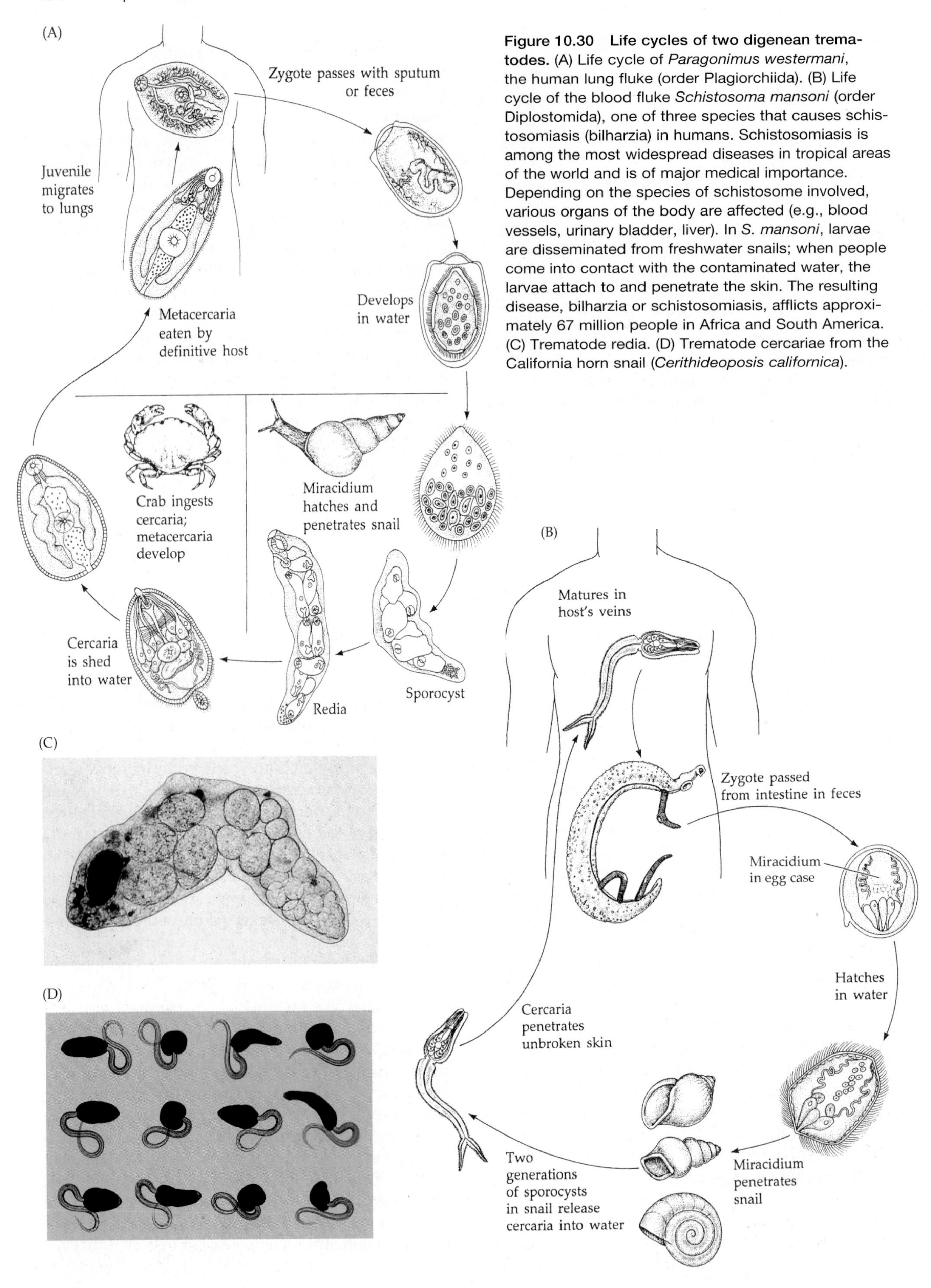

Figure 10.30 Life cycles of two digenean trematodes. (A) Life cycle of *Paragonimus westermani*, the human lung fluke (order Plagiorchiida). (B) Life cycle of the blood fluke *Schistosoma mansoni* (order Diplostomida), one of three species that causes schistosomiasis (bilharzia) in humans. Schistosomiasis is among the most widespread diseases in tropical areas of the world and is of major medical importance. Depending on the species of schistosome involved, various organs of the body are affected (e.g., blood vessels, urinary bladder, liver). In *S. mansoni*, larvae are disseminated from freshwater snails; when people come into contact with the contaminated water, the larvae attach to and penetrate the skin. The resulting disease, bilharzia or schistosomiasis, afflicts approximately 67 million people in Africa and South America. (C) Trematode redia. (D) Trematode cercariae from the California horn snail (*Cerithideoposis californica*).

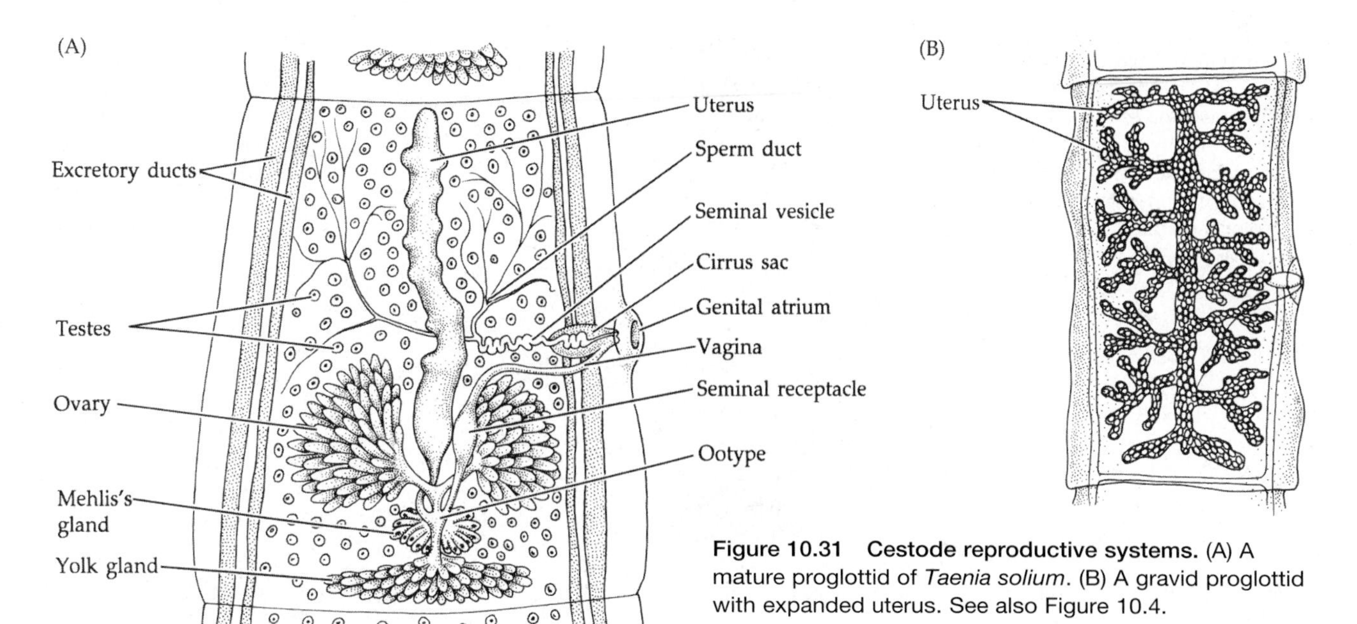

Figure 10.31 Cestode reproductive systems. (A) A mature proglottid of *Taenia solium*. (B) A gravid proglottid with expanded uterus. See also Figure 10.4.

Most adult tapeworms live in the digestive tracts of vertebrates and usually require one or more intermediate hosts to complete their life cycles. A few can complete their life cycle in a single host. Depending on the number of hosts and other factors, tapeworm life cycles are quite variable, and we describe only two examples.

Taenia saginata (order Cyclophyllidea) is known as the beef tapeworm, since cattle are the intermediate host. The adults, which may exceed 1 m in length, reside in human small intestines (Figure 10.32A). As proglottids mature, they are released in the host's feces. The fertilized eggs break free into the environment as the proglottids disintegrate. By this time each zygote has developed to a stage called an **oncosphere** surrounded by a resistant coat called an **embryophore**, which allows the embryo to remain in the environment for two or three months. Usually six tiny hooks are evident in the embryo; thus the oncosphere is sometimes called a **hexacanth larva**.

If grazing cattle ingest the oncosphere, it is released from its covering, the hexacanth larva penetrates the intestinal wall using its hooks, and is carried by the circulatory system to the cow's skeletal muscle. Here the hexacanth develops into a stage called the **cysticercus**, or bladder worm, which encysts in the connective tissue within the muscle of the intermediate host. Each cysticercus contains an invaginated developing scolex. If humans eat raw or poorly cooked infected beef, the scolex evaginates and attaches to the lining of the new host's small intestine, where the adult worm grows and matures. (Another tapeworm, *Taenia solium*, utilizes pigs as its intermediate host and follows a similar life cycle.)

The life cycles of some cestodes involve two or more hosts, such as that of *Diphyllobothrium latum*, the so-called broad fish tapeworm (order Diphyllobothriidea) (Figure 10.32B). Nearly any fish-eating mammal, including humans, can serve as the definitive host for this tapeworm. Encapsulated zygotes are released from mature proglottids and shed in the host's feces. After one or two weeks in water, the embryos develop to the oncosphere (hexacanth) stage. At this time, each oncosphere is encased in a ciliated embryophore and it hatches as a free-swimming larva called a **coracidium**. To successfully continue the life cycle, the coracidium must be eaten by the first intermediate host, a copepod (Crustacea). The cilia are shed and the released hexacanth larva bores through the gut wall into the host's body cavity, where it develops into a **procercoid** stage. Certain species of freshwater fish can serve as the second intermediate host. The fish eats the copepod, the procercoid bores through the gut and into the fish's muscle tissue, and there it grows into a segmented **plerocercoid** stage, complete with a tiny scolex. When a human consumes raw or undercooked infected fish, the plerocercoid attaches to the intestinal wall and matures.

Platyhelminth Phylogeny

Ideas about the origin of flatworms, their relationships to other phyla, and evolution within the group have been debated since early in the nineteenth century. The lack of a robust fossil record and their extreme anatomical simplicity have made morphology-based phylogenetic analysis difficult—there is a paucity of reliable characters that might identify and differentiate clades. In fact, unambiguous synapomorphies that define the phylum Platyhelminthes itself have yet to be identified. Platyhelminth fossils consist mainly of structures left on hosts by parasitic species, or occasional evidence

Figure 10.32 Two cestode life cycles. (A) The life cycle of the beef tapeworm, *Taenia saginata* (order Cyclophyllidea) (B) The life cycle of the broad fish tapeworm, *Diphyllobothrium latum* (order Diphyllobothriidea).

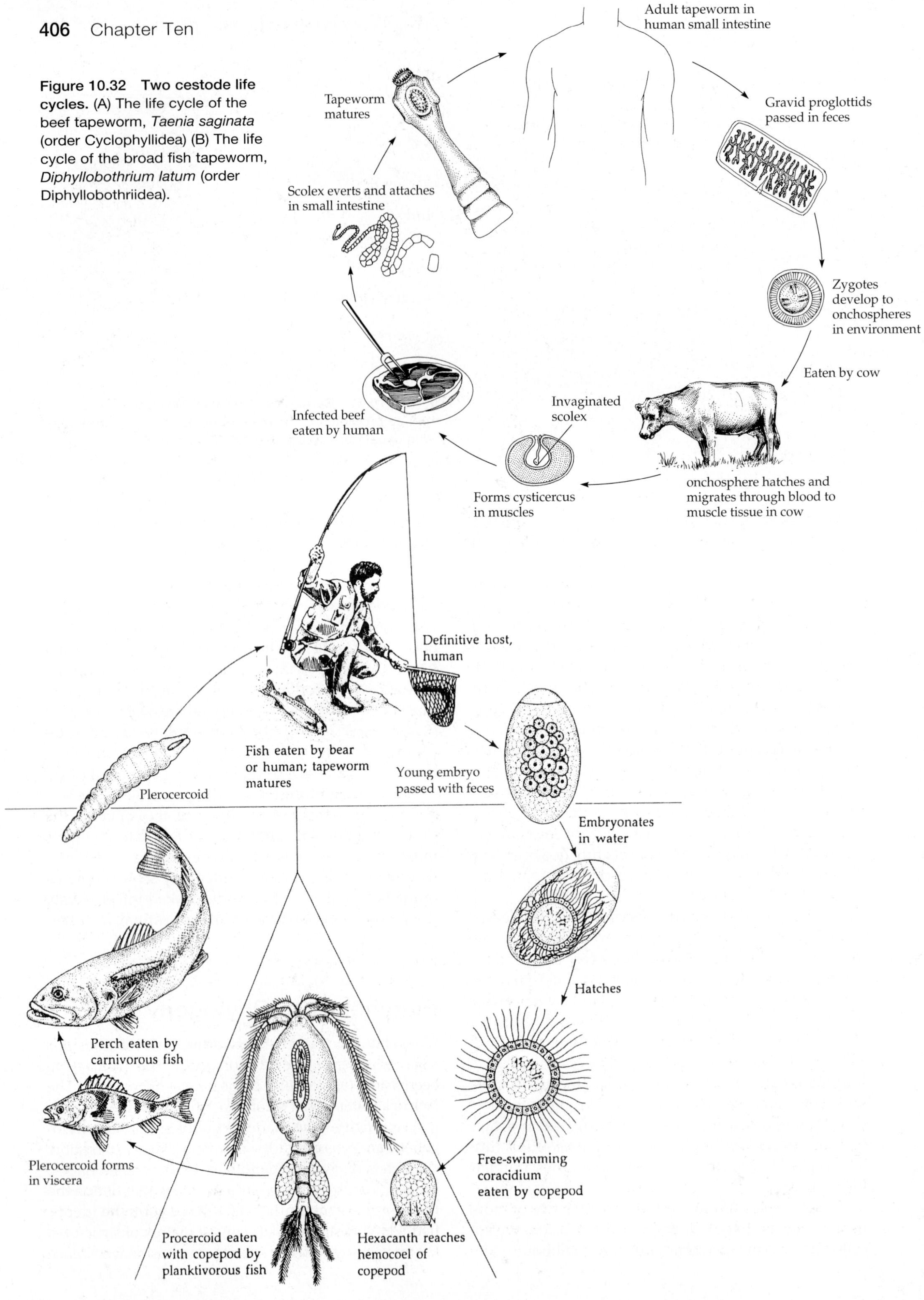

Adult tapeworm in human small intestine
Gravid proglottids passed in feces
Zygotes develop to onchospheres in environment
Eaten by cow
onchosphere hatches and migrates through blood to muscle tissue in cow
Invaginated scolex
Forms cysticercus in muscles
Infected beef eaten by human
Scolex everts and attaches in small intestine
Tapeworm matures

Definitive host, human
Fish eaten by bear or human; tapeworm matures
Plerocercoid
Young embryo passed with feces
Embryonates in water
Hatches
Free-swimming coracidium eaten by copepod
Hexacanth reaches hemocoel of copepod
Procercoid eaten with copepod by planktivorous fish
Plerocercoid forms in viscera
Perch eaten by carnivorous fish

from host coprolites, although some trace fossils from the early Paleozoic have been assigned to this phylum.

In the twentieth century, several popular hypotheses concerning the origin of flatworms (or acoels) appeared. The ciliate-to-acoel hypothesis (discussed in Chapter 5 as part of the syncytial theory of Hadzi and Hansen) has been abandoned by most modern zoologists. Another hypothesis was called the ctenophore–polyclad theory, which suggested the ctenophores gave rise to polyclad flatworms. This creative scenario envisioned a flattened ctenophore that assumed a benthic, crawling lifestyle, with the mouth directed against the substratum. By reducing the tentacles and moving them forward along with the apical sense organ, a bilateral condition was achieved. Couple these events with increased gut branching and the formation of a plicate pharynx, and a polyclad body plan is approximated—at least on paper. This hypothesis, too, no longer has much support.

We can safely assume that the original flatworm was free-living, although not necessarily assignable to any extant order. Peter Ax, Tor Karling, and Ulrich Ehlers have presented various versions of this "turbellarian archetype." These hypothetical ancestral flatworms are envisioned as having had a simple pharynx and a saclike gut without diverticula, such as seen in modern Catenulidea and Macrostomorpha (Figure 10.33A,B). There is also general agreement that the primitive flatworm was probably an archoöphoran with endolecithal ova, spiral cleavage, and a single-layered, completely ciliated epidermis.

In 1963, Ax suggested that flatworms might actually be more highly derived (in the protostome line) than previously thought, and that they might represent a series of reductions from a vermiform coelomate ancestor—a case of extreme reduction by paedomorphosis (the retention by adult individuals of traits seen in young), perhaps through neoteny, wherein the somatic development of adults is slowed or delayed. The idea had some support, especially with regards to the acoelomorphs (then included in Platyhelminthes). There was also an idea that platyhelminths arose through progenesis (a type of paedomorphosis in which sexual development of larvae or young is accelerated), from developmental stages of early protostomes prior to the embryonic appearance of the coelomic cavities—that is, rapid maturation of larval or other stages that had solid bodies or still contained the blastocoel. You are probably beginning to see how zoologists have struggled to understand what flatworms "really are"!

Fortunately, modern gene sequence analyses have begun to shed new light on the phylogeny of flatworms, including the removal of the Acoelomorpha (now in the phylum Xenacoelomorpha) and broad acceptance that platyhelminths comprise two major clades, Catenulida and Rhabditophora. Molecular phylogenies also support the monophyly of the

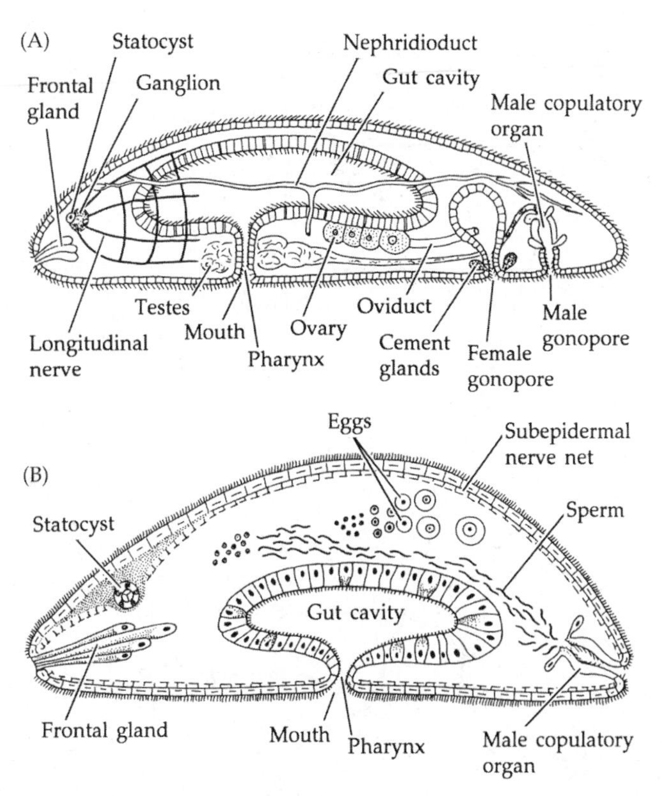

Figure 10.33 Two examples of hypothetical "turbellarian" archetypes. (A) Macrostomid-like archetype suggested by Peter Ax. (B) Archetype proposed by Tor Karling.

Neodermata and also of its three major clades—Trematoda, Monogenea, and Cestoda. Mitochondrial DNA analyses have suggested Monogenea is basal in the Neodermata clade, but 18S rRNA and nuclear "genomic" analyses (and morphology) studies suggest Trematoda is basal, and that Monogenea + Cestoda comprise a sister group. This uncertainty is shown by a trichotomy in our phylogenetic tree (Figure 10.34). Importantly, molecular phylogenies clearly place the flatworms within the protostome clade, and generally position them within the Spiralia. Therefore, while possibly somewhat removed from the base of the bilaterian tree, the morphological simplicity of platyhelminths likely represents an ancestral condition, and is less likely to be due a secondary loss or reduction of complex characters such as the coelom or segmentation.

The availability of genomic datasets for platyhelminths is uneven. High-quality reference genomes exist for some parasites such as *Schistosoma mansoni* and *Echinococcus multilocularis*, and detailed analyses of *S. haematobium* and *Taenia solium* as well as partial genomes of a couple dozen tapeworms, and about the same number for flukes, now exist. However, few if any free-living platyhelminths have been fully sequenced. A draft genome of *Macrostomum lignano* is currently underway, and a genomic compilation of the triclad *Schmidtea mediterranea* has been attempted but is not yet fully published. In the fluke *Schistosoma*, two

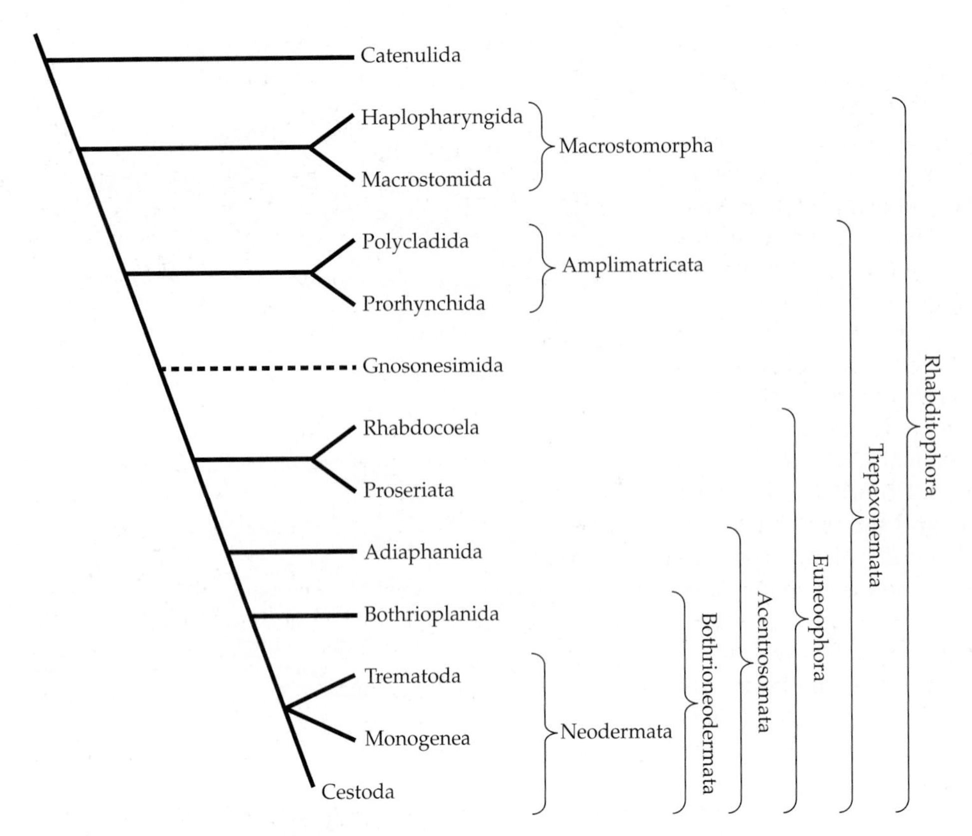

Figure 10.34 An evolutionary tree showing the relationships within Platyhelminthes, based on a consensus of molecular and morphological phylogenetics. Note that the order Proseriata, formerly classified among the Polycladidea, is now removed to the large unresolved clade classified as the polyphyletic subclass Eulecithophora. The relationships of the three Neodermata infraclasses also remain unresolved. See text for further details.

distinct mitochondrial gene orders have been found. However, all other parasitic species that have been examined have a similar gene order. The gene orders in free-living groups are often quite different from one another (and from the gene order of parasitic species). All this means resolution of platyhelminth phylogeny remains challenging.

The phylogeny of Platyhelminthes shown in Figure 10.34 is based primarily on two recent genomic studies (Laumer et al. 2015; Egger et al. 2015). This scheme is subject to revision, as many researchers are currently working to refine our understanding of platyhelminth phylogeny. Although many internal relationships remain largely enigmatic, the Catenulida are now widely viewed to be the most basal platyhelminths, and the sister group to Rhabditophora (the remaining platyhelminths). Catenulids are anatomically simple flatworms, with an unbranched pharynx and gut and a single biflagellate protonephridium. And, they retain the spiral cleavage and endolecithal ova that characterize most phyla in the clade Spiralia. Rhabditophora comprises two clades, Macrostomorpha and Trepaxonemata, the latter being home to the majority of species in the phylum and including some of the best known free-living groups as well as the flukes and tapeworms.

All of the earliest-branching clades of flatworms, Catenulida, Macrostomorpha, and Polycladida share spiral cleavage and endolecithal eggs with other spiralian protostomes. The terrestrial/freshwater group Prorhynchida, a proposed sister group to the mainly marine, diverse and well-known polyclad flatworms (Polycladida), retains spiral cleavage, but has shifted to ectolecithal eggs. Molecular evidence strongly supports the sister group relationship between the endolecithal polyclads and the ectolecithal Prorhynchida (a clade known as Amplimatricata), suggesting that prorhynchids independently developed ectolecithal ova. The discovery that Prorhynchida likely independently evolved ectolecithal ova led to abandonment of the old taxon "Neoophora," which had held *all* the ectolecithal taxa (i.e., Prorhynchida, Rhabdocoela, Proseriata, and Acentrosomata). The latter three groups now comprise the Euneoophora, the most highly derived flatworm lineage. These patterns confirm the basal nature of the archoöphoran plan. Gnosonesimora is likely the sister group to the Euneoophora. Recall that in the neoöphoran plan (seen in the order Prorhynchida and the superclass Euneoophora) there is an ovarian specialization into germinal and vitelline cell producing regions. It has been suggested that this separation of germ-line and yolk-line cells might allow these ectolecithal species to synthesize yolk at a higher rate than their endolecithal ancestors, thereby enhancing fecundity, and possibly allowing multiple embryos to develop within the same egg capsule, also enhancing fecundity by allowing conservation of resources for larval nutrition.

Phylogenetic relationships within the huge and complex Euneoophora lineage are still somewhat unsettled. The Rhabdocoela and Proseriata retain elements of spiral cleavage, suggesting their basal placement within this clade. Molecular analyses also suggest that rhabdocoels could be basal to the Proseriata, and that the latter might be the sister taxon to the Acentrosomata. Acentrosomata is notable for its loss of spiral cleavage. This is thought to be due to the evolutionary loss of three genes associated with centriole formation and function, which has been demonstrated in species of Adiaphanida, Bothrioplanida, and Neodermata. It is this feature that led to their grouping under the name Acentrosomata. Centrioles, of course, play a critical role in the development of spindle fibers during cell division. In some cases, early embryogenesis in acentrosomate species is so chaotic that the blastomeres undergo what has been called dispersive cleavage, wherein they temporarily lose physical contact with one another and "drift" in a matrix of yolk. This process has been given the colorful name blastomere anarchy, and it characterizes Adiaphanida (triclads, prolecithorans and fecampiids) and Bothrioplanida. Bothrioplanids appear to be the last free-living ancestor to the Neodermata. There is no trace of spiral cleavage in neodermatans either, although they are not generally characterized by blastomere anarchy. There seems to be no question about the monophyly of the Neodermata, all of which are ectolecithal, lack spiral cleavage, and have larvae that shed their cilia early in their life cycle (hence the name "neo-dermata"). However, there is still debate over relationships among the Monogenea, Cestoda, and Trematoda.

Selected References

General References

Auladell, C., J. Garcia-Valero and J. Baguña. 1993. Ultrastructural localization of RNA in the chromatoid bodies of undifferentiated cells (neoblasts) in planarians by the RNase-gold complex technique. J. Morph. 216: 319–326.

Ax, P. 1984. *Das phylogenetische System. Systematisierung der lebenden Natur aufgrund ihrer Phylogenese.* Gustav Fischer Verlag, Stuttgart.

Ax, P. 1985. The position of Gnathostomulida and Platyhelminthes in the phylogenetic system of the Bilateria. In S. Conway Morris, J. D. George, R. Gibson and H. M. Platt (eds.), *The Origins and Relationships of Lower Invertebrates.* Oxford University Press, Oxford.

Ax, P. 1996. *Multicellular Animals: A New Approach to the Phylogenetic Order in Nature.* Springer-Verlag, Berlin.

Baguña, J. and M. Riutort. 2004. Molecular phylogeny of the Platyhelminthes. Canadian Journal of Zoology 82: 168–193.

Boeger, W. A. and D. C. Kritsky. 1993. Phylogeny and revised classification of the Monogena Bychowsky, 1937 (Platyhelminthes). Syst. Parasitol. 26: 1–32.

Brooks, D. R. 1982. Higher classification of parasitic Platyhelminthes and fundamentals of cestode classification. In D. F. Mettrick and S. S. Dresser (eds.), *Parasites: Their World and Ours.* Elsevier Biomedical Press, Amsterdam.

Dailey, M. D. 1996. *Meyer, Olsen & Schmidt's Essentials of Parasitology,* 6th Ed. W. C. Brown, Dubuque, IA.

Ehlers, U. 1984. *Das phylogenetische System der Platyhelminthes.* Akademie der Wissenschaften und der Literatur, Mainz, and Gustav Fischer Verlag, Stuttgart.

Ehlers, U. 1986. Comments on a phylogenetic system of the Platyhelminthes. Hydrobiologia 132: 1–12.

Ehlers, U. 1995. The basic organization of the Plathelminthes. Hydrobiologica 305: 21–26.

Ellis, C.H. and A. Fausto-Sterling. 1997. Platyhelminthes: The flatworms. Pp. 115–130 in S. F. Gilbert and A. M. Raunio (eds.), *Embryology: Constructing the Organism.* Sinauer Associates, Sunderland, MA.

Florkin, M. and B. T. Scheer. 1968. *Chemical Zoology, Vol. 2.* Academic Press, New York.

Harrison, F. W. and B. J. Bogitsh (eds.). 1990. *Microscopic Anatomy of Invertebrates. Vol. III. Platyhelminthes and Nemertinea.* Wiley-Liss, New York.

Hyman, L. H. 1951. The Invertebrates, Vol. 2, Platyhelminthes and Rhynchocoela: The Acoelomate Bilateria. McGraw-Hill, New York.

Laumer, C. E., A. Hejnol and G. Giribet. 2015. Nuclear genomic signals of the "microturbellarian" roots of platyhelminth evolutionary innovation. eLife 2015; 4: e05503. doi: 10.7554/eLife.05503

Littlewood, D. T. J. and R. A. Bray. 2001. *Interrelationships of the Platyhelminthes.* Taylor & Francis, New York.

Llewellyn, J. 1965. The evolution of parasitic Platyhelminthes. Pp. 47–78 in A. E. R. Taylor (ed.), *Evolution of Parasites,* 3rd Ed. Symposium of the British Society for Parasitology. Blackwell, Oxford.

Llewellyn, J. 1986. Phylogenetic inference from platyhelminth life-cycle stages. Int. J. Parasitol. 17: 281–289.

Malakhov, V. V. and N. V. Trubitsina. 1998. Embryonic development of the polyclad turbellarian *Pseudoceros japonicus* from the Sea of Japan. Russian J. Mar. Biol. 24 (2): 106–113.

Martin, G. G. 1978. Ciliary gliding in lower invertebrates. Zoomorphologie 91: 249–262.

Marquardt, W. C., R. S. Demaree and R. B. Grieve. 2000. *Parasitology and Vector Biology.* 2nd Ed. Academic Press, San Diego, CA.

Maule, A. G. and N. J. Marks. 2006. *Parasitic Flatworms. Molecular Biology, Biochemistry, Immunology and Physiology.* CABI Publishing, Cambridge, MA.

Morris, S. C., J. D. George, R. Gibson and H. M. Platt (eds.). 1985. *The Origins and Relationships of Lower Invertebrates.* The Systematics Association, Spec. Vol. No. 28. Oxford. [Includes several papers addressing views on flatworm phylogeny; see especially papers by Ax, Ehlers, and Smith and Tyler.]

Newman, L. and L. Cannon. 2003. *Marine Flatworms. The World of Polyclads.* CSIRO Publishing, Australia.

Rieger, R. M. 1986. Über den Ursprung der Bilateria: Die Bedeutung der Ultrastrukturforschung für ein neues

Verstehen der Metazoenevolution. Verh. Dtsch. Zool. Ges. 79: 31–50.

Riutort León, M. 2014. Platyhelminthes. Pp. 270–280 in P. Vargas and R. Zardoya, *The Tree of Life*. Sinauer Associates, Sunderland, MA.

Roberts, L. S., J. Janovy, Jr., and S. Naylor. 2012. *Foundations of Parasitology*, 9th Ed. W.C. Brown, Dubuque, IA .

Ruppert, E. E. and P. R. Smith. 1988. The functional organization of filtration nephridia. Biol. Rev. 63:231–258.

Tyler, S., and M.S. Tyler. 1997. Origin of the epidermis in parasitic platyhelminths. Internat. J. Parasitol. 27: 715–738.

Tyler, S. and M. Hooge. 2004 Comparative morphology of the body wall of flatworms (Platyhelminthes). Can. J. Zool. 82: 194–210.

Free-living Flatworms

Baguña, J. and C. Boyer. 1990. Experimental embryology of the Turbellaria: Present knowledge, open questions, and future trends. In H. J. Marthy (ed.), *Experimental Embryology in Aquatic Plants and Animals*. Plenum, New York.

Baguña, J. and 7 others. 1994. Regeneration and pattern formation in planarians: Cells, Molecules, and Genes. Zool. Sci. 11: 781–795.

Bowen, I. D. 1980. Phagocytosis in *Polycelis tenuis*. Pp. 1–14 in D. C. Smith and Y. Tiffon (eds.), *Nutrition in the Lower Metazoa*. Pergamon Press, Oxford.

Boyer, B. C. 1989. The role of the first quartet micromeres in the development of the polyclad *Hoploplana inquilina*. Biol. Bull. 177: 338–343.

Boyer, B. C., J. Q. Henry and M. Q. Martindale. 1996. Dual origins of mesoderm in a basal spiralian: Cell lineage analysis in the polyclad turbellarian *Hoploplana inquilina*. Dev. Biol. 179: 329–338.

Collins, J. J. III and 7 others. 2010. Genome-wide analyses reveal a role for peptide hormones in planarian germline development. PLoS Biol. doi: 10.1371/journal.pbio.1000509

Crezee, M. 1982. Turbellaria. Pp. 718–740 in S. P. Parker (ed.), *Synopsis and Classification of Living Organisms, Vol. 1*. McGraw Hill, New York.

Eakin, R. M. and J. L. Brandenberger. 1980. Unique eye of probable evolutionary significance. Science 211: 1189–1190.

Egger, B., and 13 others. 2015. A transcriptomic-phylogenomic analysis of the evolutionary relationships of flatworms. Curr. Biol. 25, 1347–1353.

Ehlers, U. 1991. Comparative morphology of statocysts in the Platyhelminthes and Xenoturbellida. Hydrobiologia 227: 263–271.

Heitkamp, C. 1977. The reproductive biology of *Mesostoma ehrenbergii*. Hydrobiologia 55: 21–32.

Henley, C. 1974. Platyhelminthes (Turbellaria). In A. C. Giese and J. S. Pearse (eds.), *Reproduction of Marine Invertebrates, Vol. 1, Acoelomate and Pseudocoelomate Metazoans*. Academic Press, New York.

Hurley, A. C. 1976. The polyclad flatworm *Stylochus tripartitus* Hyman as a barnacle predator. Crustaceana 3(1): 110–111.

Hooge, M. D. and S. Tyler. 1999. Musculature of the faculative parasite *Urastoma cyprinae* (Platyhelminthes). J. Morphol. 241: 207–216.

Jennings, J. B. 1980. Nutrition in symbiotic Turbellaria. Pp. 45–56 in D. C. Smith and Y. Tiffon (eds.), *Nutrition in the Lower Metazoa*. Pergamon Press, Oxford.

Justine, J.-L. and 13 others. The invasive land planarian *Platydesmus manokwari* (Platyhelminthes, Geoplanidae): Records from six new localities, including the first in the USA. PeerJ3:e1037, doi 10.7717/peerj.1037

Karling, T. G. 1966. On nematocysts and similar structures in turbellarians. Acta Zool. Fenn. 116: 1–21.

Karling, T. G. and M. Meinander (eds.). 1978. The Alex Luther Centennial Symposium on Turbellaria. Acta Zool. Fenn. 154: 193–207.

Kato, K. 1940. On the development of some Japanese polyclads. Japan. J. Zool. 8: 537–573.

Laumer, C. E., and G. Giribet. 2014. Inclusive taxon sampling suggests a single, stepwise origin of ectolecithality in Platyhelminthes. Biol. J. Linn. Soc. Lond. 111: 570–588.

Lauer, D. M. and B. Fried. 1977. Observations on nutrition of *Bdelloura candida*, an ectocommensal of *Limulus polyphemus*. Am. Midl. Nat. 97(1): 240–247.

Martín-Durán, J. M. and B. Egger. 2012. Developmental diversity in free-living flatworms. EvoDevo 3:1–22. http://www.evodevojournal.com/content/3/1/7

McKanna, J. A. 1968. Fine structure of the protonephridial system in planaria. I. Female cells. Z. Zellforsch. Mikrosk. Anat. 92: 509–523.

Michiels, N.K., and L. J. Newman. 1998. Sex and violence in hermaphrodites. Nature 391: 647.

Moraczewski, J. 1977. Asexual reproduction and regeneration of *Catenula*. Zoomorphologie 88: 65–80.

Moraczewski, J., A. Czubaj and J. Bakowska. 1977. Organization and ultrastructure of the nervous system in Catenulida. Zoomorphologie. 87: 87–95.

Mueller, J. F. 1965. Helminth life cycles. Am. Zool. 5: 131–139.

Nentwig, M. R. 1978. Comparative morphological studies after decapitation and after fission in the planarian *Dugesia dorotocephala*. Trans. Am. Microsc. Soc. 97: 297–310.

Noreña, C., C. Damborenea and F. Brusa. 2015. Phylum Platyhelminthes. Pp. 181–203 in J. H. Thorp and D. C. Rogers (eds.), *Thorp and Covich's Freshwater Invertebrates*, 4th Ed. Academic Press, New York.

Ogren, R. E. 1995. Predation behavior of land planarians. Hydrobiologia 305: 105–111.

Prudhoe, S. 1985. *A Monograph on the Polyclad Turbellaria*. Oxford University Press, New York.

Rawlinson, K. A., D. M. Bolaños, M. K. Liana and M. K. Litvaitis. 2008. Reproduction, development and parental care in two direct-developing flatworms (Platyhelminthes: Polycladida: Acotylea). J. Nat. Hist. 42: 2173–2192.

Rawlinson, K, A. and J. S. Stella. 2012. Discovery of the corallivorous polyclad flatworm, *Amakusaplana acroporae*, on the Great Barrier Reef, Australia—the first report from the wild. PLoS ONE doi: 10.1371/journal.pone.0042240

Riser, N. W. and M. P. Morse (eds.). 1974. *Biology of the Turbellaria. Libbie H. Hyman Memorial Volume*. McGraw-Hill, New York.

Ruitort, M., M Alvarez-Presas, E. Lazaro, E. Sola and J. Paps. 2012. Evolutionary history of the Tricladida and the Platyhelminthes: an up-to-date phylogenetic and systematic account. Int. J. Dev. Biol. 56: 5–17.

Ruppert, E. E. 1978. A review of metamorphosis of turbellarian larvae. Pp. 65–81 in F. S. Chia and M. Rise (eds.), *Settlement and Metamorphosis of Marine Invertebrate Larvae*. Elsevier/North-Holland Biomedical Press, Amsterdam.

Salo, E., A. M. Muñoz-Marmol, J. R. Byascas-Ramírez, J. García-Fernández, A. Mirales, Z. Casali, M. Cormominas and J. Baguña. 1995. The freshwater planarian *Dugesia* (G.) *tigrina* contains a great diversity of homeobox genes. Hydrobiologia 305: 269–275.

Schärer, L., D. T. J. Littlewood, A. Waeschenbach, W. Yoshida, and D. B. Vizosoa. 2011. Mating behavior and the evolution of sperm design. Proc. Nat. Acad. Sci. 108: 1490–1495.

Smith, J., S. Tyler, M. B. Thomas and R. M. Rieger. 1982. The morphology of turbellarian rhabdites: Phylogenetic implications. Trans. Am. Microsc. Soc. 101: 209–228.

Sopott-Ehlers, B. 1991. Comparative morphology of photoreceptors in free-living platyhelminths: A survey. Hydrobiologia 227: 231–239.

Stocchino, G. A. and R. Manconi. 2013. Overview of life cycles in model species of the genus *Dugesia* (Platyhelminthes: Tricladida). Ital. J. Zool. 80(3): 319–328.

Van Steenkiste, N., B. Tessens, W. Willems, T. Backeljau, U. Jondelius, T. Artois. 2013. A comprehensive molecular phylogeny of Dalytyphloplanida (Platyhelminthes: Rhabdocoela) reveals multiple escapes from the marine environment and origins of symbiotic relationships. PLoS ONE 8(3): 1–13.

Vizoso, D. B., G. Reiger and L. Schärer. 2010. Goings-on inside a worm: Functional hypotheses derived from sexual conflict thinking. Biol. J. Linn. Soc. 99: 370–383.

Flukes (Trematoda and Monogenea)

Brooks, D. R., S. M. Bandoni, C. A. Macdonald and R. T. O'Grady. 1989. Aspects of the phylogeny of the Trematoda Rudolphi, 1808 (Platyhelminthes: Cercomeria). Can. J. Zool. 67: 2609–2624.

Bychowsky, B. E. 1957. *Monogenetic Trematodes: Their Systematics and Phylogeny*. American Institute of Biological Sciences, Washington, DC. [Translated from the Russian.]

Combes, C. and 16 others. 1980. *The World Atlas of Cercariae*. Museum National d'Histoire Naturelle, Paris.

Erasmus, D. A. 1972. *The Biology of Trematodes*. Crane, Russak, New York.

Hechinger, R. F., K. D. Lafferty, F. T. Mancini III, R. R. Warner and A. M. Kuris. 2009. How large is the hand in the puppet? Ecological and evolutionary factors affecting body mass of 15 trematode parasitic castrators in their snail host. Evol. Ecol. 23: 651–667.

Hechinger, R. F., A. C. Wood and A. M. Kuris. 2011. Social organization in a flatworm: trematode parasites form soldier and reproductive castes. Proc. Roy. Soc. B. 278: 656–665.

Kearn, G. C. 1986. Role of chemical substances from fish hosts in hatching and host-finding in monogeneans. J. Chem. Ecol. 12: 1651–1658.

Martin, W. E. 1972. An annotated key to the cercariae that develop in the snail *Cerithidea californica*. Bull. South. Calif. Acad. Sci. 71(1): 39–43.

Olson, P. D. and D. T. J. Littlewood. 2002. Phylogenetics of the Monogenea—evidence from a medley of molecules. Int. J. Parasitol. 32: 233–344.

Park, J.-K., K.-H. Kim, S. Kang, W. Kim, K. Eom and D. T. J. Littlewood. 2007. A common origin of complex life cycles in parasitic flatworms: Evidence from the complete mitochondrial genome of *Microcotyle sebastis* (Monogenea: Platyhelminthes). BMC Evolutionary Biology 7: 11. doi: 10.1186/1471-2148-7-11

Pearson, J. C. 1972. A phylogeny of life-cycle patterns of the Digenea. Adv. Parasitol. 10: 153–189.

Rohde, K. 1994. The minor groups of parasitic Platyhelminthes. Adv. Parasitol 33: 145–234.

Schell, S. C. 1982. Trematoda. Pp. 740–807 in S. P. Parker (ed.), *Synopsis and Classification of Living Organisms. Vol. 1*. McGraw-Hill, New York.

Smyth, J. D. and D. W. Halton. 1985. *The Physiology of Trematodes*, 2nd Ed. W. H. Freeman, San Francisco, CA.

Sproston, N. G. 1946. A synopsis of the monogenetic trematodes. Trans. Zool. Soc. Lond. 25: 185–600.

Yamaguti, S. 1963. *Systema Helminthum, Vol. 4, Monogenea and Aspidocatylea*. Interscience, New York.

Yamaguti, S. 1971. *Synopsis of Digenetic Trematodes of Vertebrates*, Vols. 1, 2. Keigaku, Tokyo.

Yamaguti, S. 1975. *A Synoptical Review of Life Histories of Digenetic Trematodes of Vertebrates*. Keigaku, Tokyo.

Cestoda

Aral, H. P. (ed.). 1980. *Biology of the Tapeworm* Hymenolepsis diminuta. Academic Press, New York.

Biserova, N. M., K. S. Margaretha, M. R. Gustafsson and N. B. Terenina. 1996. The nervous system of the pike-tapeworm *Triaenophorus nodulosus* (Cestoda: Pseudophyllidea): Ultrastructure and immunocytochemical mapping of aminergic and peptidergic elements. Invert. Biol. 115: 273–285.

Brooks, D. R., E. P. Hoberg and P. J. Weekes. 1991. Preliminary phylogenetic systematic analysis of the major lineages of the Eucestoda (Platyhelminthes: Cercomeria). Proc. Biol. Soc. Wash. 104(4): 651–668.

Kuchta, R., T. Scholz, J. Brabec and R. A. Bray. 2008. Suppression of the tapeworm order Pseudophyllidea (Platyhelminthes: Eucestoda) and the proposal of two new orders, Bothriocephalidea and Diphyllobothriidea. Inter. J. Parasitol. 38: 49–55.

Levron, C. J. Miquel, M. Oros and T. Scholz. 2010. Spermatozoa of tapeworms (Platyhelminthes, Eucestoda): Advances in ultrastructural and phylogenetic studies. Biol. Rev. 85: 325–345.

Schmidt, G. D. 1982. Cestoda. Pp. 807–822 in S. P. Parker (ed.), *Synopsis and Classification of Living Organisms, Vol. 1*. McGraw-Hill, New York.

Smyth, J. D. 1969. *The Physiology of Cestodes*. W. H. Freeman, San Francisco.

Uglem, G. L. and J. J. Just. 1983. Trypsin inhibition by tapeworms: Antienzyme secretion or pH adjustment. Science 220: 79–81.

Waeschenbach, A. B., L. Webster and D. T. J. Littlewood. 2012. Adding resolution to ordinal level relationships of tapeworms (Platyhelminthes: Cestoda) with large fragments of mtDNA. Mol. Phylog. Evol. 63: 834–847.

Wardle, R., J. McLeod and S. Radinovsky. 1974. *Advances in the Zoology of Tapeworms*. University of Minnesota Press, Minneapolis.

Yamaguti, S. 1959. *Systema Helminthum, Vol. 2, The Cestodes of Vertebrates*. Interscience, New York.

The rhombozoans and orthonectids were for many years treated as closely related groups, classes, or orders in a taxon of phylum rank: the Mesozoa. They were often allied with the flatworms (Platyhelminthes) because of their complex life cycle and superficially similar appearance to miracidia larvae of digenetic trematodes. At other times they were allied, surprisingly, with ciliate protists. Auguste Lameere (1922) even suggested that the orthonectids might have arisen from echiuran worms, apparently because some echiurans (e.g., *Bonellia*) show extreme sexual dimorphism, with tiny, reduced males. Lameere likened this feature to the dimorphic nature of some orthonectids. Neither the ciliate nor the echiuran hypothesis ever gained much favor. There are no proven fossils of these two groups. Recently, researchers have become convinced that Rhombozoa and Orthonectida are not at all closely related. We agreed with this view and abandoned the name Mesozoa as a formal taxon in the first edition (1990) of this textbook. Since then, molecular and ultrastructural work have supported the idea that these two groups each represent a distinct clade of highly modified parasitic protostomes.

Phylum Rhombozoa

Rhombozoans were described and named by A. Krohn in Germany in 1839. But it was not until 1876 that a careful study of these creatures was published by the Belgian zoologist Edouard van Beneden. He was convinced that these odd parasites represented a true link between the protists and the Metazoa, and it was van Beneden who coined the name "Mesozoa" (= middle animals) to emphasize this point of view. Stunkard (1982) considered Rhombozoa as a class of "Mesozoa" comprising the orders Dicyemida and Heterocyemida, and we retain those two subtaxa.[1] Rhombozoans have a simple, solid body construction—there are no body cavities or differentiated organs (e.g., chapter opener photo of a vermiform stage). An outer layer of somatic/nutritive cells surrounds an inner core of reproductive cells (or a single reproductive cell, depending on how one views it!). About 70 species have been described.

Recent molecular studies, including sequencing of 18S rRNA, Pax6, and other genes, as well as patterns of Hox (and other gene) expression, suggest that rhombozoans are probably highly reduced and specialized parasitic members of the protostome clade Spiralia. Dicyemids are more common and better understood than are the heterocyemids. Members of both groups are obligate symbionts in the nephridia of cephalopod molluscs.

The Dicyemida

Adult dicyemids are very small, just 0.5 to 3 mm long. There are two adult forms, **nematogens** and **rhombogens** that have nearly the same organization but that produce two distinct types of larvae: nematogens produce **vermiform larvae** asexually from an agamete in their body, whereas rhombogens produce **infusoriform larvae** from fertilized eggs.

The body of a nematogen consists of an outer sheath of ciliated somatic cells, the number of which has been constant for most, but not all, species that have been examined.[2] Within the covering of somatic cells lies a single, long **axial cell** (Figure 11.1A). Eight or nine somatic cells at the anterior end form a distinctive **polar cap** (or **calotte**). Immediately behind the polar cap are two parapolar cells. The rest of the 10 to 15 somatic cells are sometimes called trunk cells; the two posterior-most cells are the **uropolar cells**. Altogether, dicyemids have from about 8 to 40 body cells, depending on the species—one of the smallest cell counts among the Metazoa. As in other Bilateria, dicyemids have been shown to have three types of cell–cell junctions: septate, adherens, and gap junctions.

Young dicyemids are motile and swim about in the host cephalopod's urine by ciliary action. The adults, however, attach to the inner lining of the nephridia by their polar caps. There is no conclusive evidence that these animals cause damage to their hosts, but when present in very high numbers they may interfere with the normal flow of fluids through the nephridia. They attach by inserting the "head end"—the polar cap—into renal tubules or crypts of the host's kidney. Nematogens consume particulate and molecular nutrients from the host's urine by phagocytotic and pinocytotic action of their somatic cells. Once the adult has attached to the host, the somatic cilia might serve to keep fluids moving over the body, bringing nutrients in contact with the surface cells. Although in nature dicyemids appear to be obligatorily associated with cephalopods, they have been successfully maintained in experimental nutrient media.

What we know so far about dicyemid life history is rather bizarre, and the stages of the dicyemid life cycle that occur outside the host are still incompletely known. The host-dwelling portion of the life cycle includes both asexual and sexual processes, but without a regular alternation between them. Adult nematogens thus produce two kinds of offspring. One is the vermiform stage, in which the dicyemid exists as a vermiform larva formed asexually from an agamete cell; these grow to become new worms in the renal sac of the host. The other is the infusoriform larva, which develops from a fertilized egg produced around

[1]Some people refer to this phylum as simply "Dicyemida," but this is confusing and suggests that heterocyemids are not included; hence we retain the name Rhombozoa for the phylum.

[2]A constancy in the number of cells (in a given organ, or in the entire body of an animal) is called "eutely" and is a common feature of many microscopic and near-microscopic organisms.

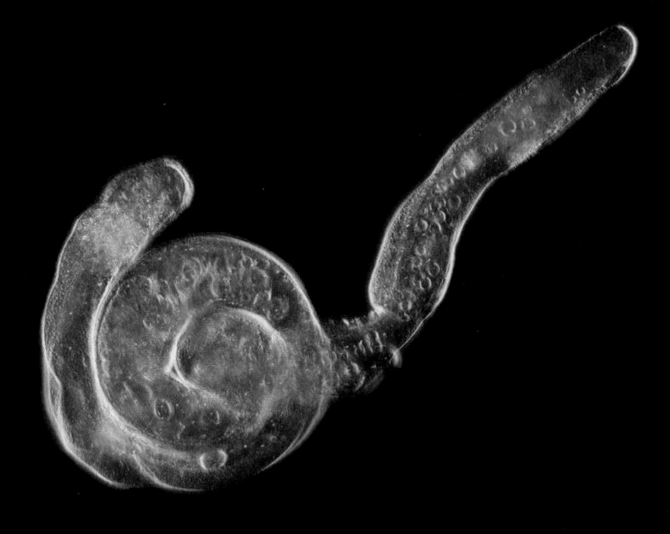

CHAPTER 11

Four Enigmatic Protostome Phyla

Rhombozoa, Orthonectida, Chaetognatha, Gastrotricha

In the 1979 book *The Medusa and the Snail,* Lewis Thomas wrote, "The only solid piece of scientific truth about which I feel totally confident is that we are profoundly ignorant about nature." Despite Dr. Thomas' pessimism, we have actually come a long way since 1979. But, alas, we are still uncertain about where four enigmatic animal phyla fit into the tree of life: Chaetognatha, Gastrotricha, Rhombozoa, and Orthonectida. Well, that's not quite true. Thanks mainly to molecular phylogenetic discoveries of the past decade, we do know that all four of these phyla belong to the Protostomia clade. And there is growing evidence that they could be spiralians, although this is still not certain. Taken together, these four phyla amount to only about 1,000 described species. We cover them together in this chapter not because they are closely related, but because they are, at this point in time, protostomes of uncertain affinity. Our classification in this book includes all but Chaetognatha in the clade Spiralia, even though their phylogenetic positions within Protostomia are still uncertain. Rhombozoans have a very spiralian-like embryogenesis, whereas gastrotrichs have a unique, nonradial and nonspiral embryogenesis; the embryology of orthonectids and chaetognaths is still poorly known. The most recent phylogenetic analyses suggest Chaetognatha may be the sister group to Spiralia, as shown in our tree of Metazoa (Chapter 28), or perhaps basal protostomes.

Each of these phyla is highly specialized for the niche they exploit: chaetognaths mainly inhabit the marine planktonic realm (although numerous species have adopted benthic lifestyles); gastrotrichs are minute species living meiofaunally or in surface detritus; rhombozoans are symbionts in the nephridia of cephalopod molluscs; and orthonectids are parasitic in a variety of invertebrates, including echinoderms, molluscs, nemerteans, free-living flatworms, and polychaete worms. All of these animals have been subjects of taxonomic and phylogenetic controversy since their discovery long ago.

Classification of The Animal Kingdom (Metazoa)

Non-Bilateria*
(a.k.a. the diploblasts)
- PHYLUM PORIFERA
- PHYLUM PLACOZOA
- PHYLUM CNIDARIA
- PHYLUM CTENOPHORA

Bilateria
(a.k.a. the triploblasts)
- PHYLUM XENACOELOMORPHA

Protostomia
- **PHYLUM CHAETOGNATHA**

SPIRALIA
- PHYLUM PLATYHELMINTHES
- **PHYLUM GASTROTRICHA**
- **PHYLUM RHOMBOZOA**
- **PHYLUM ORTHONECTIDA**
- PHYLUM NEMERTEA
- PHYLUM MOLLUSCA
- PHYLUM ANNELIDA
- PHYLUM ENTOPROCTA
- PHYLUM CYCLIOPHORA

Gnathifera
- PHYLUM GNATHOSTOMULIDA
- PHYLUM MICROGNATHOZOA
- PHYLUM ROTIFERA

Lophophorata
- PHYLUM PHORONIDA
- PHYLUM BRYOZOA
- PHYLUM BRACHIOPODA

ECDYSOZOA

Nematoida
- PHYLUM NEMATODA
- PHYLUM NEMATOMORPHA

Scalidophora
- PHYLUM KINORHYNCHA
- PHYLUM PRIAPULA
- PHYLUM LORICIFERA

Panarthropoda
- PHYLUM TARDIGRADA
- PHYLUM ONYCHOPHORA
- PHYLUM ARTHROPODA
  - SUBPHYLUM CRUSTACEA*
  - SUBPHYLUM HEXAPODA
  - SUBPHYLUM MYRIAPODA
  - SUBPHYLUM CHELICERATA

Deuterostomia
- PHYLUM ECHINODERMATA
- PHYLUM HEMICHORDATA
- PHYLUM CHORDATA

*Paraphyletic group

Phylum Chaetognatha has been revised by George Shinn.
Phylum Gastrotricha has been revised by Rick Hochberg.

hermaphroditic gonads called infusorigens, and these eventually escape the host to enter the sea. What stimulates production of infusoriform larvae, and how they find their new hosts are unknown. Notably, in sexual reproduction, both fertilization and embryonic development occur within the worm's body.

Asexual reproduction is curious indeed. The cytoplasm of the single axial cell of the nematogen contains numerous tiny cells called **axoblasts**. Immature vermiform organisms are produced asexually by a sort of embryogeny of individual axoblasts within the parent axial cell (Figure 11.2). The first division of an axoblast is unequal and produces a large presumptive axial cell and a small presumptive somatic cell. The presumptive somatic cell divides repeatedly, and its daughter cells move by an epiboly-like process (rapid growth of animal pole ectodermal cells to from a sheetlike covering to the vegetal cells) to enclose the presumptive axial cell, which has not yet divided. When this inner cell finally does divide, it does so unequally, and the smaller daughter cell is then engulfed by the larger one! The larger cell becomes the progeny's axial cell proper with its single nucleus, and the smaller engulfed cell becomes the progenitor of all future axoblasts within that axial cell. The "embryo," which now consists of its own central axial cell surrounded by somatic cells, elongates and the somatic cells develop cilia. The resulting structure is a miniature vermiform organism. The immature vermiform organism leaves the parent nematogen and swims about in the nephridial fluids. Eventually it attaches to the host and enters the adult stage of the life cycle.

The initiation of sexual reproduction in dicyemids may be a density-dependent phenomenon associated with high numbers of vermiform individuals within the host's nephridia. Some workers have suggested that the switch from asexual to sexual processes might be a response to some chemical factor that accumulates in the urine of the host. Other workers suggest that sexual reproduction in dicyemids is brought on by the sexual maturation of the host. In any event, once the vermiform adults become sexually "motivated" they are called rhombogens, and their somatic cells usually enlarge as they become filled with yolky material (Figure 11.3).[3] The axoblast of a rhombogen-stage individual develops into multicellular structures called **infusorigens**, each consisting of an outer layer of ova and an inner mass of sperm (Figure 11.3B). The infusorigens

[3]Because the reported differences between rhombogens and nematogens are not consistently found, some workers prefer to simply call them sexual and asexual vermiform adults, respectively.

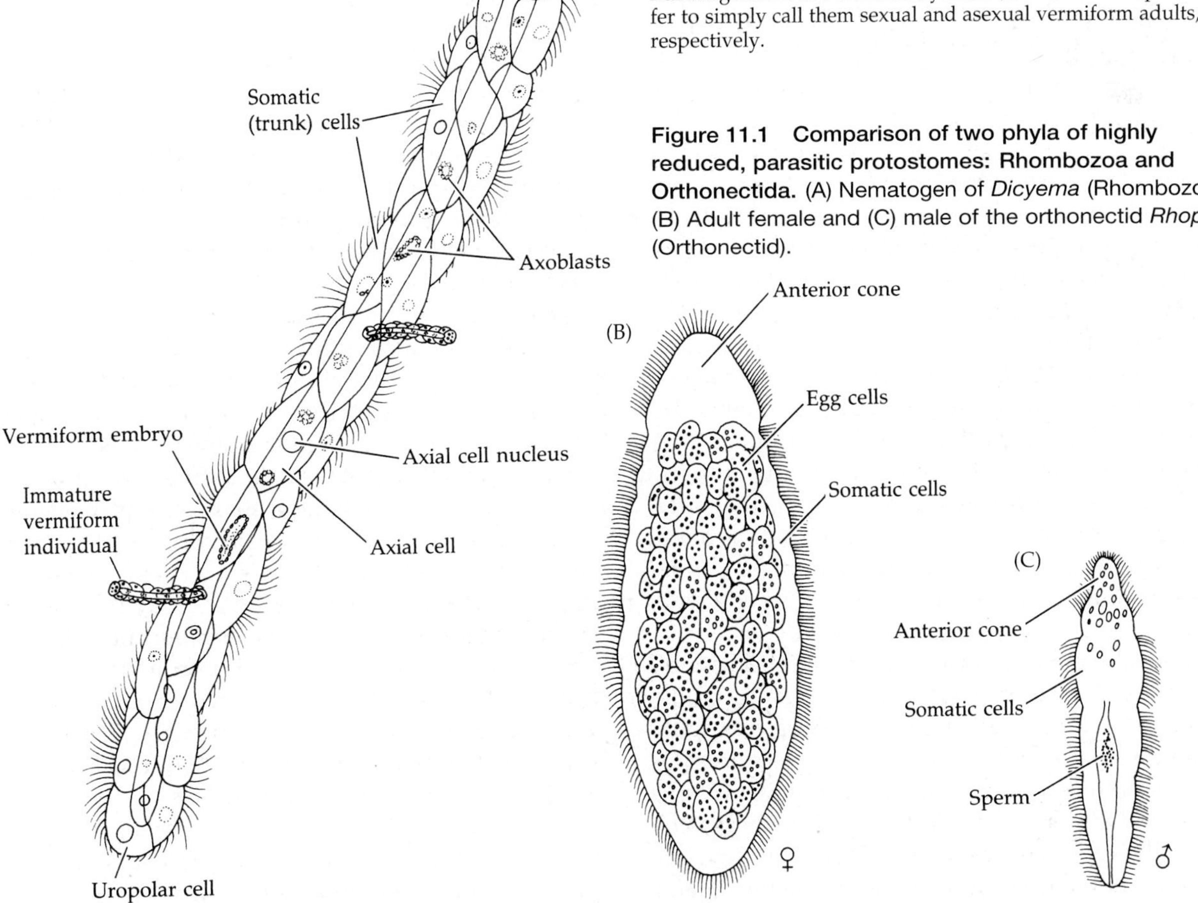

Figure 11.1 Comparison of two phyla of highly reduced, parasitic protostomes: Rhombozoa and Orthonectida. (A) Nematogen of *Dicyema* (Rhombozoa). (B) Adult female and (C) male of the orthonectid *Rhopalura* (Orthonectid).

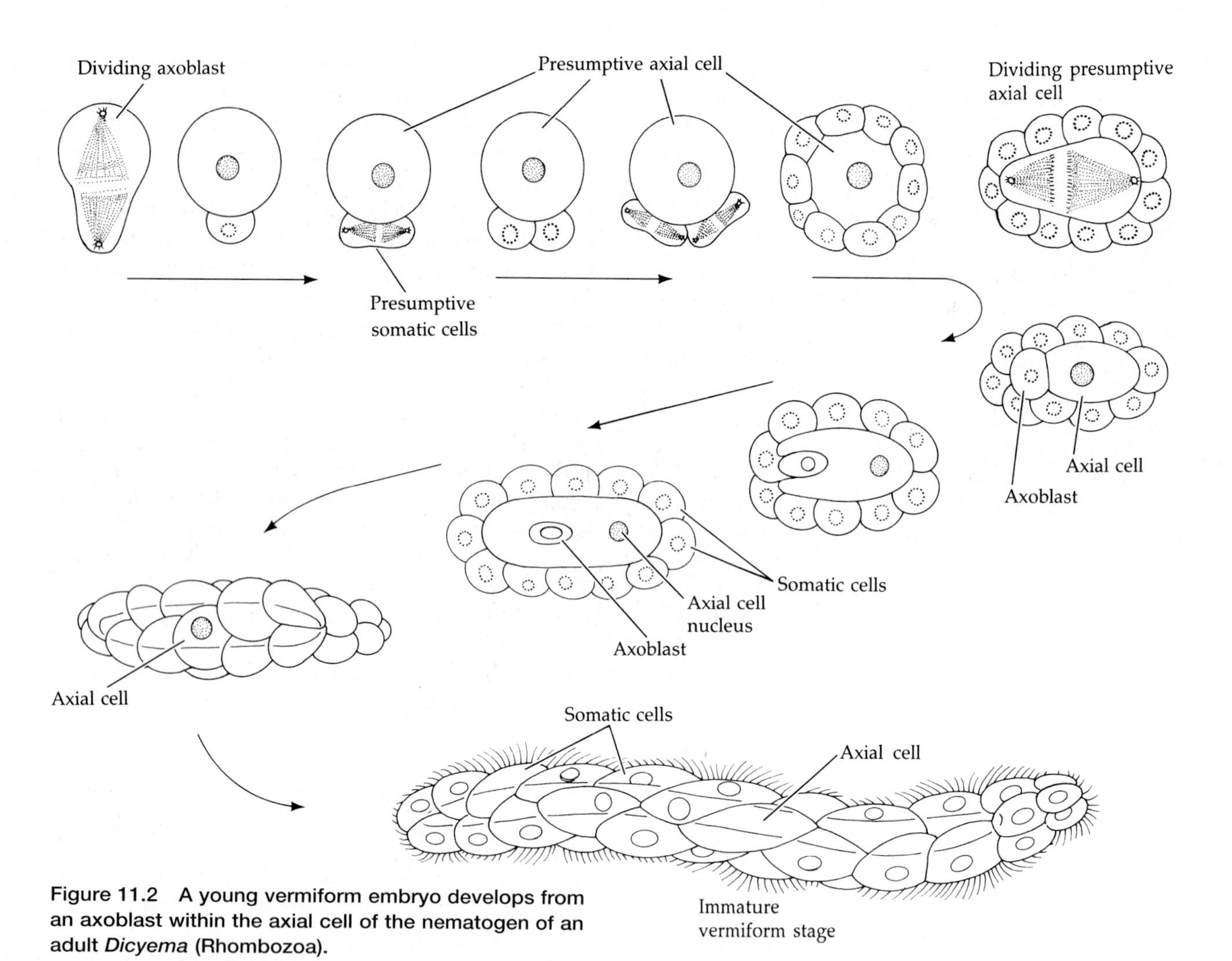

Figure 11.2 A young vermiform embryo develops from an axoblast within the axial cell of the nematogen of an adult *Dicyema* (Rhombozoa).

are retained within a vacuole of the parent's axial cell, and they are functionally hermaphroditic gonads. In each infusorigens, the centrally located sperm fertilize the peripherally arranged ova, each zygote then developing into a ciliated infusoriform larva (Figure 11.3C). This larva has a fixed number of cells; the two anteriormost cells—called apical cells—contain high-density substances within their cytoplasm. The rest of the surface cells are ciliated and form a sheath around a ring of capsule cells, which in turn enclose four central cells. The number of infusorigens, and the larvae they produce, seems to be related to the size of the adult rhombogen. The infusoriform larvae escape from the parent vermiform adult and pass out of the host's body with the urine.

The development of several species has been studied, and there is no clear evidence of germ-layer formation in the development of the infusoriform larva, with a rough distinction only as outer and inner cells. The outer cells occupy the dorsal and caudal surfaces of the embryo, and the inner cells are derived from the blastomeres of the vegetal hemisphere. The innermost germinal cells are derived from the cells that form the vegetal pole. The cleavage pattern of sexually-produced embryos in *Dicyema japonicum* has been described as holoblastic and spiral. A cavity (the urn cavity) appears among the ventral internal cells, but this small space is interpreted to appear secondarily and not to be a blastocoel.

The events of the dicyemid life cycle that occur outside the cephalopod host remain a mystery. Some workers have held to the view that the infusoriform larva enters an intermediate host (presumably some benthic invertebrate), but most of the evidence to date suggests that this is not the case. While much remains to be learned, the following scenario seems plausible. After leaving the host, the infusoriform larva sinks to the bottom of the ocean—the dense contents of the apical cells serving as ballast. The larva, or some persisting part of the larva (perhaps the innermost four cells), enters another cephalopod host. This infectious individual travels through the host, probably via the circulatory system, and enters the nephridia, where it becomes a so-called stem nematogen (Figure 11.3D). The stem nematogen is similar to the vermiform adult except that the former has three axial cells rather than one. Axoblasts within the axial cells of the stem nematogen give rise to more vermiform adults, just as the axoblasts

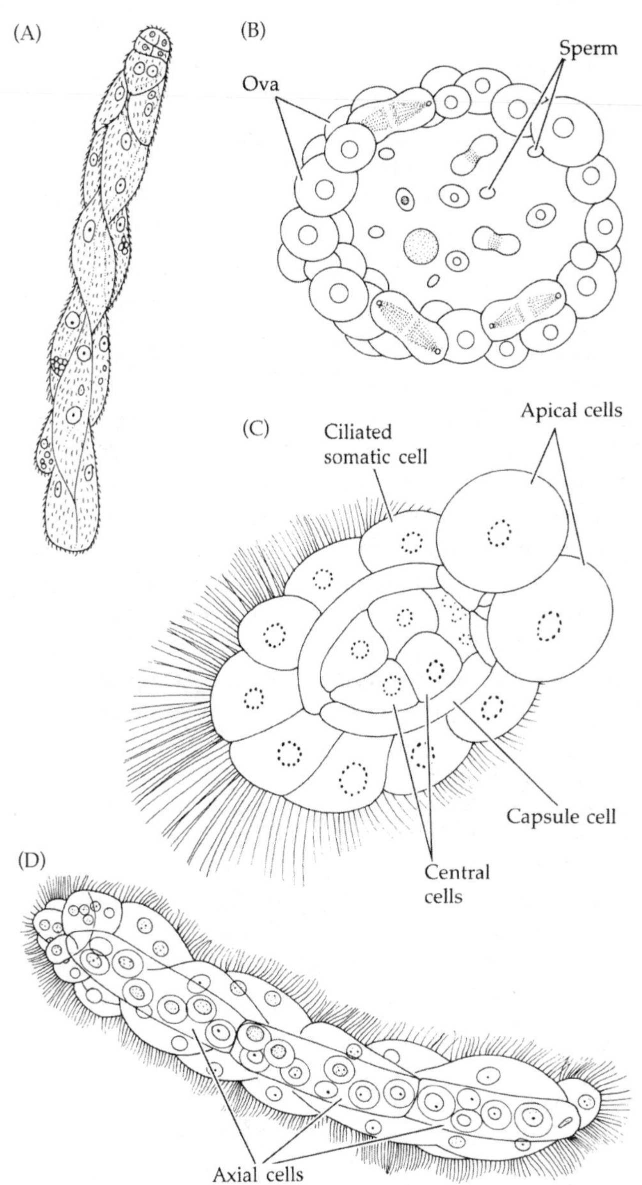

Figure 11.3 Sexual reproduction in dicyemid rhombozoans. (A) Adult, sexual (rhombogen) form. (B) Infusorigen of sperm and ova formed within the axial cell of the adult. (C) Infusoriform larva produced by fertilization. (D) Stem nematogen with three axial cells.

within the adults described earlier did. The vermiform adults produce more individuals like themselves until the onset of sexual reproduction is triggered again, presumably by the high population density. This putative life cycle is schematically represented in Figure 11.4.

The Heterocyemida

Only two species of heterocyemids have been described. *Conocyema polymorpha* lives in the nephridia of octopuses, and *Microcyema gracile* in cuttlefishes of the genus *Sepia*. These two heterocyemids differ from each other in certain respects.

The vermiform adult of *Conocyema* bears a polar cap of four enlarged cells and has a trunk of somatic cells around an inner axial cell; all the cells of the body lack cilia (Figure 11.5 A). The axial cell contains axoblasts, which give rise to ciliated "larvae" that escape from the parent, lose their cilia, and grow into more vermiform adults within the host. The individuals that produce the infusorigens lack a polar cap. They have only a very thin layer of somatic cells surrounding the axial cell, which contains the developing infusorigens (Figure 11.5B). The infusorigens produce infusoriform larvae similar to those of dicyemids. Details of the life cycle of *Conocyema* are lacking.

What little is known about *Microcyema* suggests a complex and distinctive life cycle. The vermiform adult consists of a single inner axial cell surrounded

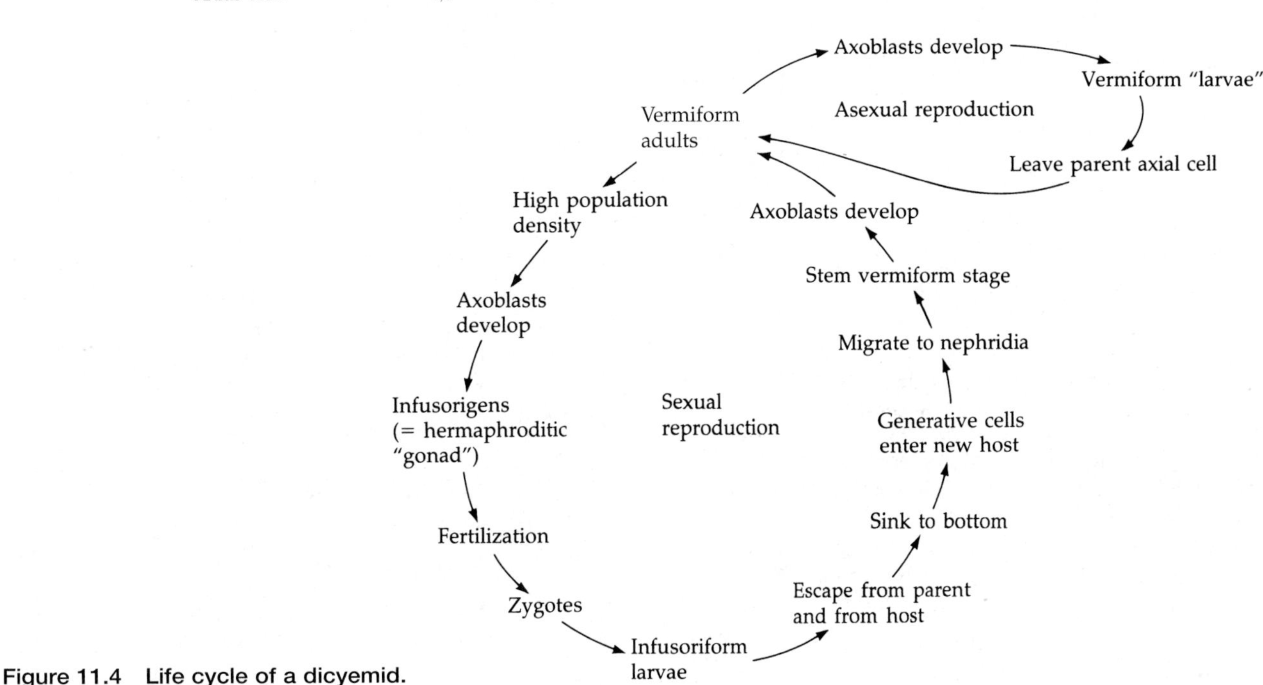

Figure 11.4 Life cycle of a dicyemid.

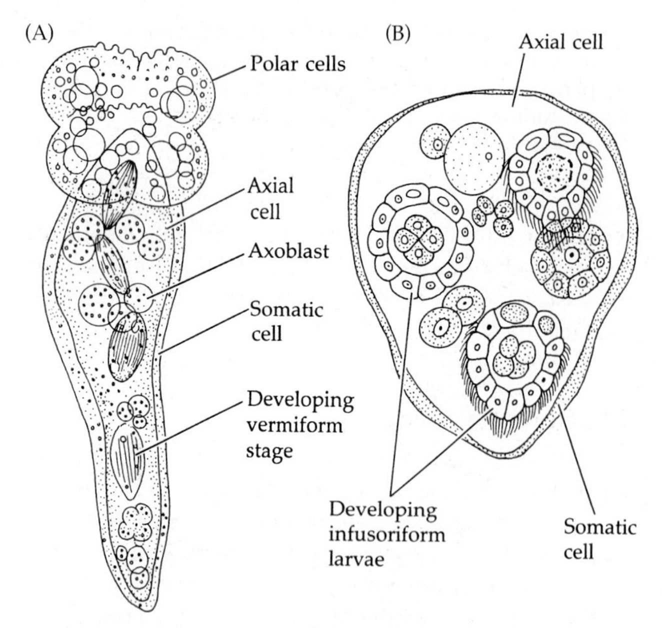

Figure 11.5 The heterocyemid rhombozoan, *Conocyema*. (A) Vermiform adult. (B) During the reproductive phase, infusoriform larvae are formed within the adult's axial cell (cross section).

by a somatic syncytium (Figure 11.6A). The axoblasts within the axial cell produce more vermiform adults by two very different methods. One sequence of events involves the formation of a multicellular "embryo" similar to that seen in dicyemids (Figure 11.6B). As the presumptive somatic cells surround the axial cell precursor, the cell boundaries of the somatic cells break down, resulting in an ameboid individual in which a syncytial mass surrounds the growing axial cell (Figure 11.6C,D). This individual apparently develops into a new vermiform adult. Another asexual process involves the formation of ciliated **Wagener's larvae** from the axoblasts (Figure 11.6E). These larvae leave the parent, swim about in the host's nephridial fluids, and eventually attach and metamorphose into more vermiform adults. *Microcyema* adults also produce infusorigens and infusoriform larvae much like those of the dicyemids. The infusoriform larvae apparently leave the host via the urine, but nothing is known about the stages of the life cycle outside the host. It is assumed that the infusoriform larva enters a host and matures into a ciliated nematogen, which has three axial cells. Such stem nematogens have been observed in host animals. Eventually, the cilia and the cell boundaries between adjacent somatic cells are lost (Figure 11.6F), and the animal develops into another vermiform adult.

Phylum Orthonectida

First described in 1877, orthonectids, like the rhombozoans, long defied placement in the tree of life. And like Rhombozoa, they have spent time allied with flatworms or viewed as some kind of "transition group" between the lower Metazoa and the Bilateria. Some

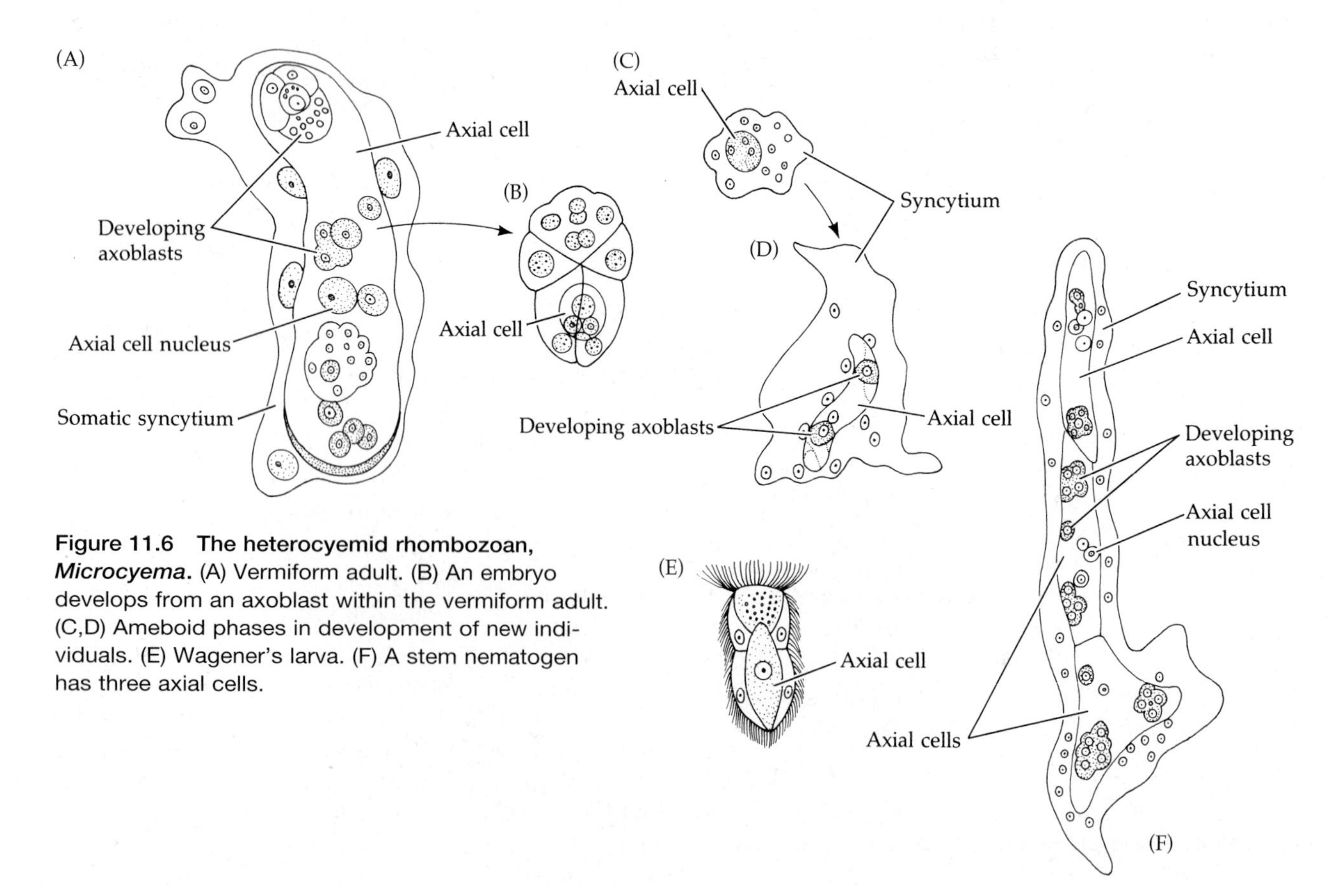

Figure 11.6 The heterocyemid rhombozoan, *Microcyema*. (A) Vermiform adult. (B) An embryo develops from an axoblast within the vermiform adult. (C,D) Ameboid phases in development of new individuals. (E) Wagener's larva. (F) A stem nematogen has three axial cells.

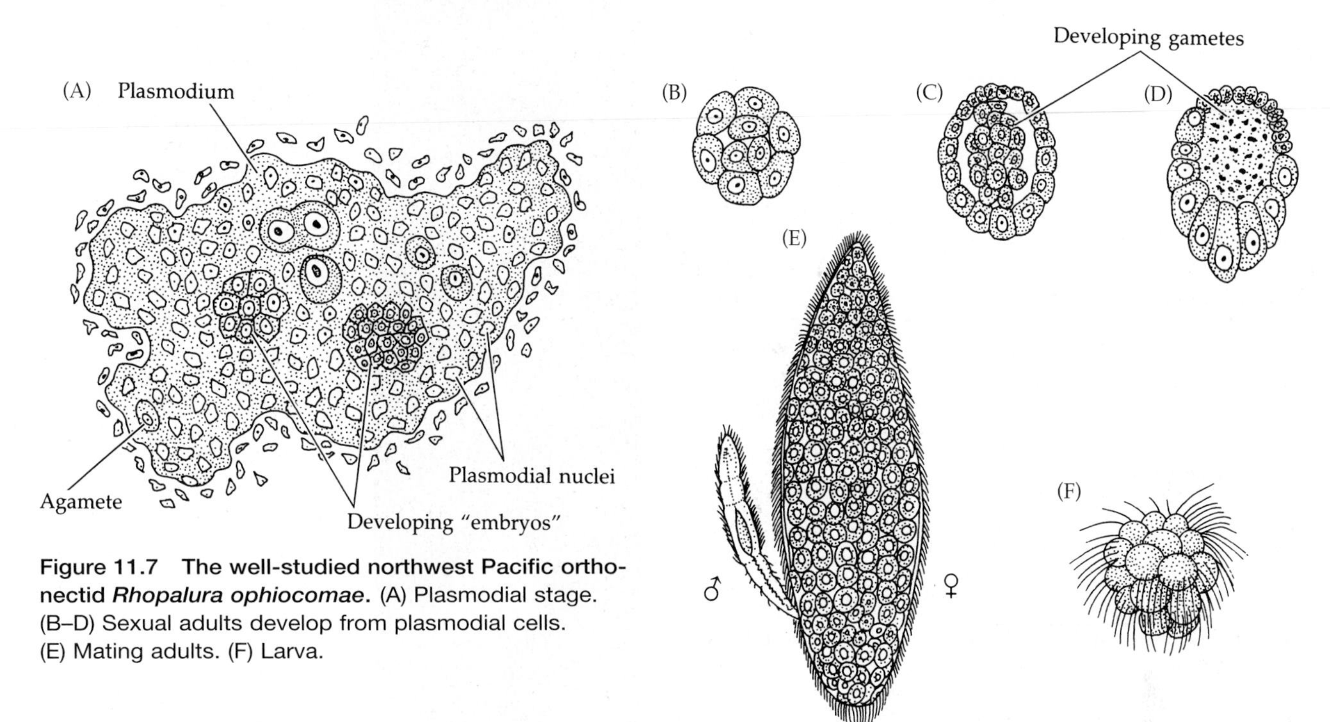

Figure 11.7 The well-studied northwest Pacific orthonectid *Rhopalura ophiocomae*. (A) Plasmodial stage. (B–D) Sexual adults develop from plasmodial cells. (E) Mating adults. (F) Larva.

workers even imagined them to be highly complex ciliate protists. But, preliminary molecular phylogenetic analyses (mainly 18S rDNA sequence data) have now placed them squarely within the Bilateria—not as sister to the Rhombozoa, but somewhere within the Protostomia. Thus they appear to be highly reduced, bilaterian parasites. Only about 21 species are known.

All orthonectids are morphologically simple and, like the Rhombozoa, they lack body cavities or discrete organs; there is no gut or nervous system. They are small, less than 1 mm in length, free swimming, and ciliated. A 2-layered cuticle has been described in one species (*Intoshia linei*). Most species are gonochoristic, but some are hermaphroditic (e.g., *Stoecharthrum*). The outer layer of the body consists of a defined number of cells, arranged more or less in rings. All are parasites of other invertebrates, and they have been reported from nemerteans, free-living platyhelminths, gastropods, bivalves, polychaete annelids, brittle stars, and ascidians.

The life cycles of some orthonectids are known and these differ markedly from those of the rhombozoans. Asexual individuals dominate the life cycle (Figure 11.7A). Some orthonectids cause severe damage to the host. For example, the well-known species *Rhopalura ophiocomae* (in the brittle star *Amphipholis squamata*) and *R. granosa* (in the bivalve mollusc *Heteranomia squamula*) destroy the gonads of their hosts.

The orthonectid *Rhopalura ophiocomae* is a well-studied parasite of the brittle star *Amphipholis squamata*, on the northeast Pacific shores. This orthonectid infects muscle cells underlying the peritoneum, adjacent to the genital bursae and gut of *Amphipholis*. The infected cells grow into large masses that bulge into the body coelom. The mass is enclosed by the coelomic peritoneum, giving it discrete form. These masses were long called "plasmodia" when it was thought that they were part of the orthonectid itself. Eventually, some of the orthonectid cells within the hypertrophied host cells develop into small (< 1 mm) ciliated adults that escape the host and swim free in the surrounding seawater. The organization of these free-living adults—with a central mass of developing gametes surrounded by a layer of ciliated cells—is reminiscent of rhombozoans. However, rhombozoans have a central axial cell within which germ cells (axoblasts) develop into both more worms and what appears to be hermaphroditic gonads, a process quite unlike anything seen in orthonectids. Also, rhombozoans lack free-living sexual forms.

Adult, free-living orthonectids are nonfeeding, ciliated, vermiform organisms that live for a short time in the sea. Females reach about 1 mm in length, males are generally smaller. They are filled with gametes (eggs, sperm, or both). Between the gametes and the outer somatic layer are what appear to be contractile cells, and in some species these have been described as circular and longitudinal muscles. Both males and females have genital pores, usually situated in the middle part of the body in a nonciliated ring of epidermal cells, and this is apparently where sperm is placed, although males lack a specialized copulatory organ. Copulation is very fast—30–40 seconds in *Intoshia variabili*. Sperm (*Rhopalura littoralis*) contain a single mitochondrion, lack an acrosome, and the tail is a simple flagellum with a 9+2 structure. The fertilized eggs develop into minute larvae that are the infective stage, and these larvae exit via the female's genital pore. When a new host is located, and the larvae have entered it, they shed their ciliated outer cells and, in a most remarkable process, the inner cells separate and scatter within the host's tissue spaces. These infective cells are called

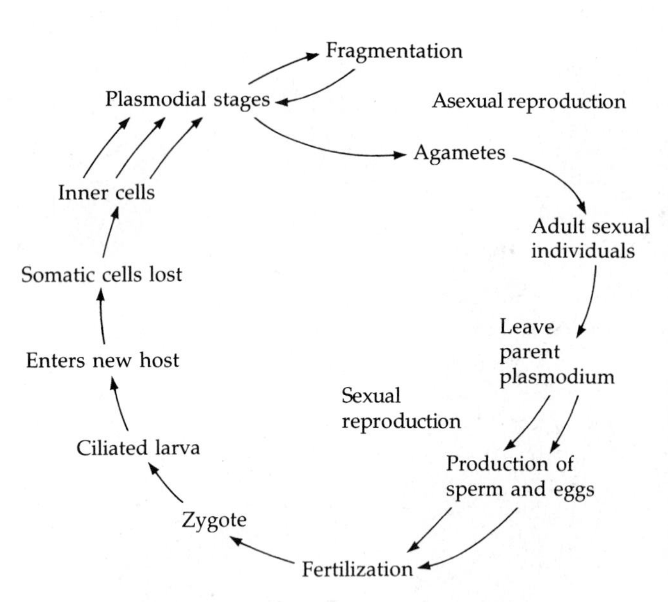

Figure 11.8 Generalized life cycle of an orthonectid.

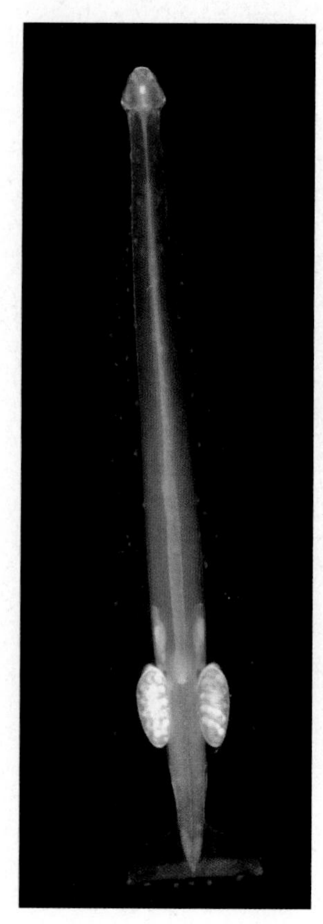

Figure 11.9 This stunning photograph by Eric Thuesen shows the deep-water chaetognath *Eukrohnia fowleri* (order Phragmophora) carrying its developing embryos in two temporary gelatinous pouches on either side of the body.

agametes (or "germline cells"). In *R. ophiocomae*, the larvae appear to enter the host brittle star through the genital bursae and gut. The agametes develop to become multicellular sexual forms, both males and females. When fully developed, the adults leave the host as tiny, nonfeeding worms that characteristically swim in straight lines (hence the phylum's name; *ortho*, "straight"; *necta*, "swimming"). Embryonic development has not yet been described for orthonectids. This life cycle is shown in Figure 11.8.

Phylum Chaetognatha

The chaetognaths, or arrow worms, comprise about 130 species of marine, mainly planktonic, voraciously predatory invertebrates (Figures 11.9 and 11.10). They are of moderate size, ranging from about 0.5 to 12 cm in

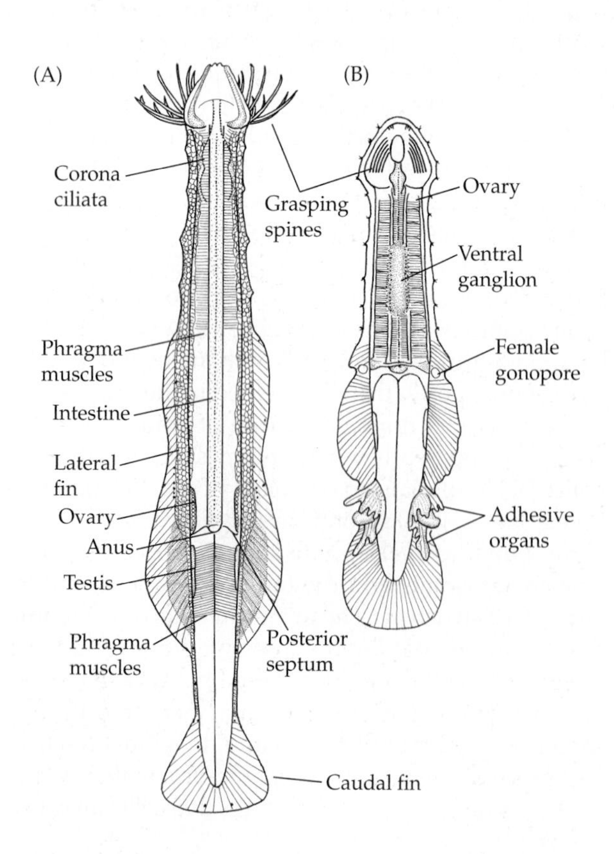

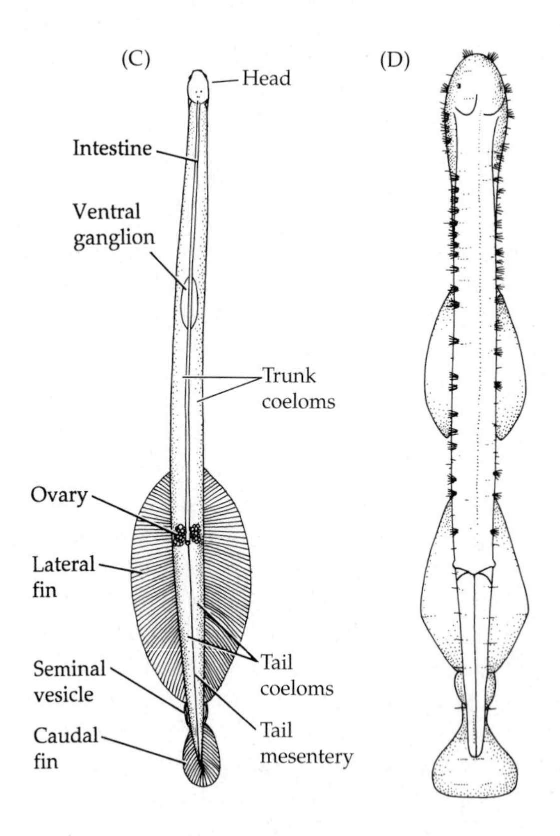

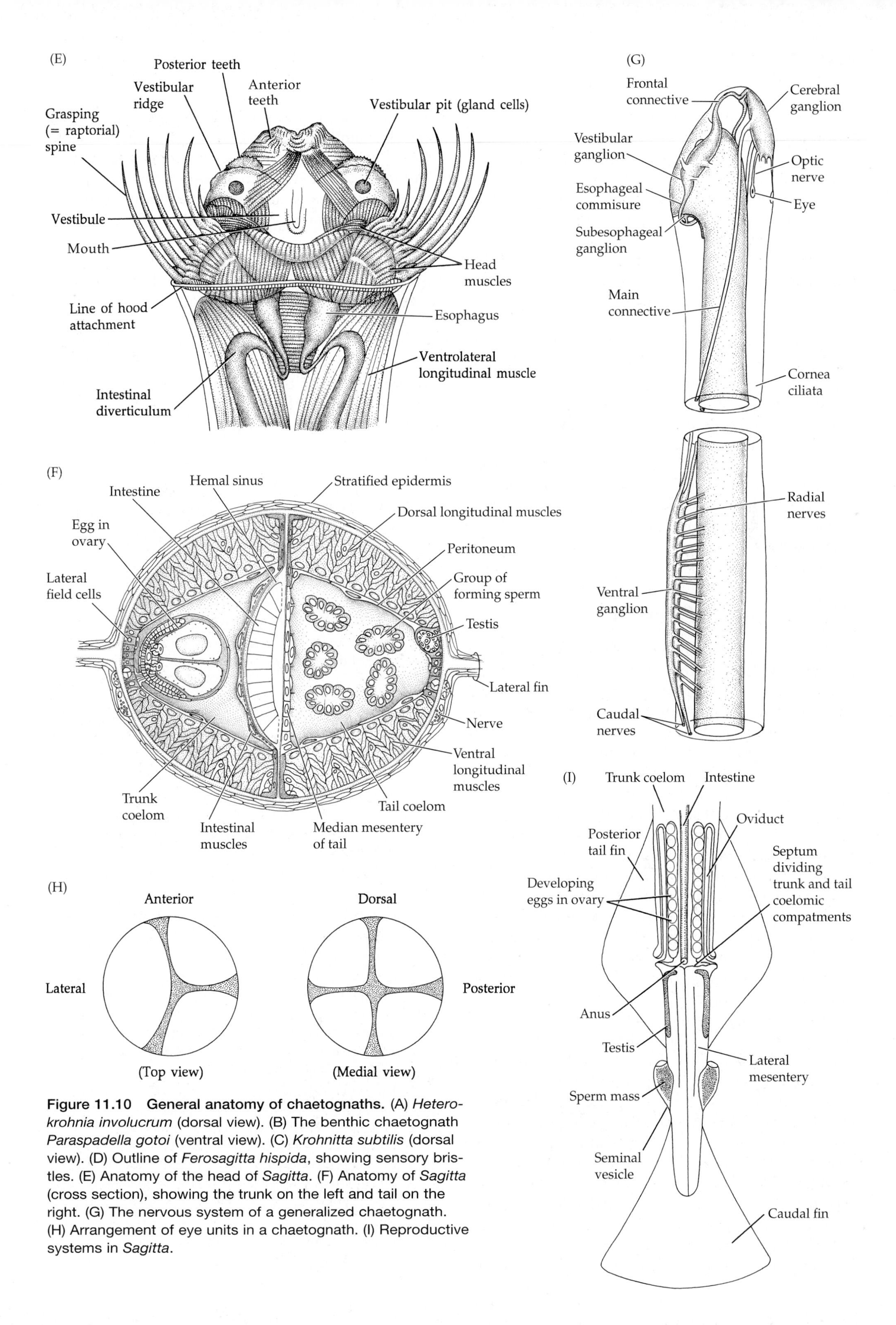

Figure 11.10 General anatomy of chaetognaths. (A) *Heterokrohnia involucrum* (dorsal view). (B) The benthic chaetognath *Paraspadella gotoi* (ventral view). (C) *Krohnitta subtilis* (dorsal view). (D) Outline of *Ferosagitta hispida*, showing sensory bristles. (E) Anatomy of the head of *Sagitta*. (F) Anatomy of *Sagitta* (cross section), showing the trunk on the left and tail on the right. (G) The nervous system of a generalized chaetognath. (H) Arrangement of eye units in a chaetognath. (I) Reproductive systems in *Sagitta*.

length. Arrow worms are distributed throughout the world's oceans and occur in some estuaries; they are ecologically important as consumers of copepods and other small zooplankton. Arrow worms often occur in very high numbers and sometimes dominate the biomass in mid-water plankton tows. Some species (e.g., *Spadella*) are epibenthic in shallow water. Special collection methods, such as use of submersibles, are turning up new species at great ocean depths and many of these live just off the deep ocean floor (e.g., *Heterokrohnia* and *Archeterokrohnia*). At least two deep-water species, *Caecosagitta macrocephala* and *Eukrohnia fowleri*, are bioluminescent, releasing luminous particles that form glowing clouds in the water (although their luciferin/luciferase-based luminescent organs are located in different parts of the body). Only two definitive fossil chaetognaths have been described, one from the Carboniferous (*Paucijaculum samamithion*) and one from the Lower Cambrian, about 520 million years ago (*Eognathacantha ercainella*), although it has been suggested that some of the Cambrian protoconodonts could be grasping spines of arrow worms.

Chaetognath Classification

The first record of a chaetognath was made by the Dutch naturalist Martinus Slabber in 1775, and the group name was first proposed by Leuckart in 1854. The systematic position of the group has been hotly debated ever since. Arrow worms have been at times allied with the catchall group Vermes, the molluscs, the arthropods, and certain blastocoelomates (particularly nematodes). For many years chaetognaths were considered deuterostomes, based on developmental features and their tripartite body coeloms. Although the question of chaetognath phylogenetic affinities is still somewhat unsettled, modern research is beginning to resolve some of the longstanding issues. For example, arrow worms are, without doubt, coelomate animals. This was revealed by early embryological studies, but it took electron microscopy to demonstrate that the adult body cavities are completely lined by mesodermally derived tissues.

Molecular phylogenetic methods consistently reveal that chaetognaths are neither deuterostomes nor a close sister group to the deuterostomes. Recent phylogenomic analyses have strongly suggested protostome affinities, branching near or at the base of that line. This is consistent with them having typical protostome ventral mid-body ganglia and circumesophageal fibers. The possibility of arrow worms being sister to all other protostomes is intriguing because it might explain their mix of protostome and deuterostome-like features. Thus, the enterocoelous formation of body cavities and the secondary appearance of the mouth, once considered to be their defining deuterostome attributes, may be ancestral (plesiomorphic) bilaterian features. Similarly, the enzyme GAMT (guanidinoacetate N-methyltransferase), which occurs in cnidarians, deuterostomes, and chaetognaths but not other protostomes, may represent a remnant ancestral feature of arrow worms that was lost in the rest of the protostome line.

Comparative morphologists have not yet identified any clear synapomorphies that unite chaetognaths with other protostome phyla. However, many unique features reveal the phylum status of arrow worms, including the moveable cuticular grasping spines on the head, a multilayered epidermis over most of the body, horizontal fins on the sides and posterior end, and closed seminal vesicles on the tail. Major characteristics of the group are listed in Box 11A.

BOX 11A Characteristics of the Phylum Chaetognatha

1. Bilateral protostomes, with streamlined, elongate, trimeric body comprising head, trunk, and tail region; single head coelom, and paired trunk and tail coeloms separated by transverse septa
2. Epidermis mostly stratified, with cuticle on ventral side of head; body with lateral and caudal fins, supported by "rays" consisting of elongate cytoskeleton-rich epidermal cells; nonmolting
3. Mouth surrounded by sets of long moveable grasping spines and short teeth used in prey capture; mouth set in ventral vestibule; anterolateral fold of body wall forms retractable hood that can enclose grasping spines
4. Longitudinal muscles of unusual type, arranged in quadrants; weak circular musculature consists of myoepithelial cells
5. No discrete gas exchange or excretory systems
6. Hemal system restricted to trunk, consisting of narrow peri-intestinal sinuses and larger sinuses in dorsal mesentery and posterior septum
7. Complete gut; anus ventral, at trunk–tail junction
8. Centralized nervous system with large dorsal (cerebral) and ventral (subenteric) ganglia connected by circumenteric connectives; ciliary fence receptors for detection of water-borne disturbances; anterior ciliary loop (= corona ciliata) of uncertain function. Inverted pigment-cup ocelli
9. Hermaphroditic, with internal fertilization and direct development. Cleavage equal, holoblastic, and perhaps modified spiral. Mesoderm and body cavities form by enterocoely. Although blastopore denotes posterior end of body, both mouth and anus form secondarily, subsequent to closure of the blastopore.
10. Strictly marine; raptorial carnivores; largely planktonic, but some benthic species are known

Morphological uniformity within the phylum has resulted in few features that can be used to establish genera and families, or to determine systematic relationships within the phylum. The features that are used may be subject to homoplasy. The following classification based on morphology is only partially corroborated by molecular systematic studies.

Order Phragmophora With ventral transverse muscles (phragma) in trunk or tail. Three families with about 60 species: Spadellidae (e.g., *Paraspadella*, *Spadella*), Eukrohniidae (*Eukrohnia*), Heterokrohniidae (*Archeterokrohnia, Heterokrohnia, Xenokrohnia*).

Order Aphragmophora Without ventral transverse muscles (phragma). Six families with about 70 species: Sagittidae (e.g., *Caecosagitta*, *Ferosagitta, Parasagitta*, *Sagitta*), Krohnittidae (*Krohnitta*), Pterosagittidae (*Pterosagitta*), Bathybelidae (*Bathybelos*), Krohnittellidae (*Krohnittella*), and Pterokrohniidae (*Pterokrohnia*).

The Chaetognath Body Plan

Externally, arrow worms are streamlined, with virtually perfect bilateral symmetry. Internally, transverse septa divide the coelomate body into a head, a trunk, and a tail. The head bears a ventrally placed mouth, set in a depression called the **vestibule**. The entire head region is elegantly adapted to a predatory lifestyle. Lateral to the mouth are long moveable **grasping spines**, or "hooks," and in front of the mouth are short **cuticular teeth**—both used in prey capture and ingestion (Figures 11.10E and 11.11). The dorsolateral margins of the head possess a muscularized fold of the body wall called the **hood**. Except during prey capture, this is drawn ventrally around the sides of the head, thereby enclosing the grasping spines and streamlining the head. A pair of small, pigmented photoreceptors lie dorsally on the head. Also dorsally is a distinctive ring of innervated ciliated cells, the **corona ciliata**, of unknown function. The trunk bears one or two pairs of **lateral fins** and the tail bears a single horizontal **tail fin**. Chaetognath fins are simple epidermal folds enclosing a thick sheet of supportive extracellular matrix; certain elongate epidermal cells form fin rays. The body surface bears many **ciliary fence receptors**, which are short rows of nonmotile cilia serving as mechanoreceptors for detecting disturbances in the water. The trunk contains the intestine, which terminates ventrally at the trunk–tail junction, and two ovaries. The female gonopores are located laterally at the posterior end of the trunk. The male reproductive system occupies the tail, which is commonly filled with masses of differentiating sperm. Paired seminal vesicles protrude laterally, between the lateral and caudal fins.

The chaetognath body plan couples structural simplicity with a high degree of specialization. We tend to view the specialized features of arrow worms as adaptations to a planktonic predatory life style but, with nowhere to hide, pelagic arrow worms must also be adept at avoiding their own predators. Most chaetognaths are as transparent as glass and spend much of their time suspended motionlessly in the water, both features that minimize detection by visual predators such as fish. A chaetognath's sensitivity to water-born disturbances is equally useful for detecting prey and approaching predators, and the ability to move quickly enables both prey capture and escape from potential predators.

Body wall, support, and movement

Over most of the body, the epidermis is a stratified epithelium. Flattened surface cells produce a thin layer of secretion over the body, much like in fish. Underlying epidermal cells are filled with cytoskeletal tonofilaments (supportive cytoskeletal microfilaments). The epidermis on the ventral and lateral parts of the head consist of a single layer of cuticularized cells. The cuticle is not molted. On all parts of the body, a thick basement membrane joins the epidermis to underlying tissues. Four large groups of striated longitudinal muscle dominate the body wall of the trunk and tail (Figure 11.10F). These are separated from the body cavities by a thin layer of squamous peritoneum (noncontractile peritoneal cells). In addition, a weak circular musculature exists in the form of myoepithelial "lateral field cells" and dorsomedial and ventromedial cells. In the Phragmophora, sheets of transverse muscle extend obliquely from the lateral body wall to the ventral midline. In all arrow worms, the head musculature is complex (Figure 11.10E).

The body cavities are true coeloms. The single head cavity is reduced by the elaborate cephalic musculature. In the trunk and tail, longitudinal mesenteries separate the coelomic space into left and right compartments and in the tail incomplete lateral mesenteries partially subdivide each tail compartment. The 1:2:2 tripartite arrangement of the coelom resembles the coelomic arrangement of deuterostomes and lophophorates, but this similarity may represent convergent evolution. It has even been suggested that the posterior-most transverse septum is not a embryological homologue to that seen in other deuterostomes, and that it forms later, in connection with the development of the gonads. The coelomic fluid is colorless in life, but it stains intensely, which suggests an abundance of dissolved organic molecules. Circulation of coelomic fluids in both trunk and tail is caused by ciliated peritoneal cells in the lateral body wall. Coelomocytes are lacking.

Body support in chaetognaths is provided by the hydrostatic quality of the coelom, crossed helical arrangement of collagen fibers in the basement membrane, and tonus of the body wall musculature. Locomotion of both pelagic and epibenthic species involves forward

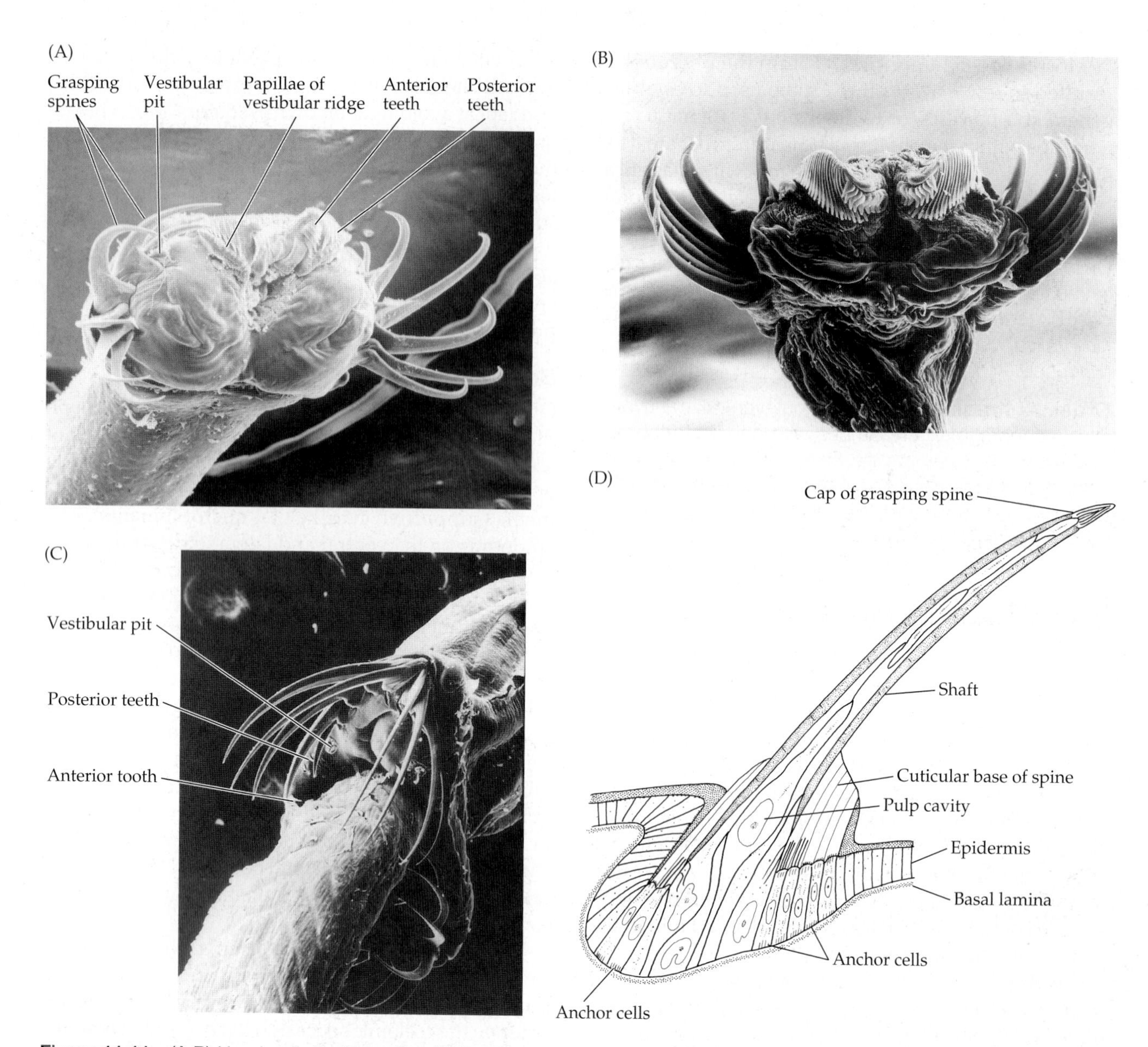

Figure 11.11 (A,B) Heads of chaetognaths. (A) *Zonosagitta pulchra*. Note the well-developed raptorial structures. (B) *Zonosagitta bedoti*, from the eastern Pacific. The hooks are clearly visible on either side of the head surrounding the exceptionally large number (17–20) of long, narrow, posterior teeth. The shorter anterior teeth lie just above the mouth. The vestibular ridge with its pores is partially visible behind the left set of posterior teeth. (C) The chaetognath in this photo, *Flaccisagitta hexaptera*, has partially swallowed a fish larva (probably an anchovy). A single anterior tooth projects down below the second hook on the left side of the photo, and two posterior teeth can be seen between the first and second hooks. The circular organ just below the first hook is the vestibular pit. (D) Diagram of a grasping spine of *Parasagitta elegans*.

darting motions caused by rapid alternating contractions of the dorsal- and ventral-longitudinal muscles. The fins are not used as propelling surfaces but are placed so that they slice through the water and serve as stabilizers. In pelagic species, brief swimming bursts alternate with quiescent periods, when the animals may slowly sink. This "hop-sink swimming," common among small planktonic invertebrates, probably constitutes a prey search behavior. Because swimming tends to be upward it helps the animals maintain their level in the water column. The fins increase resistance to sinking between swimming bursts. Species that make pronounced diurnal vertical migrations—up at night and down during the day—are presumably capable of extended bouts of swimming. Some species of arrow worms are neutrally buoyant due to hypertrophied intestinal cells or vacuolated epidermal cells that contain fluids less dense than seawater. The ventral transverse muscles in benthic chaetognaths (e.g., *Spadella*) might aid posturing or epibenthic movement.

Feeding and Digestion

Chaetognaths prey upon a variety of small pelagic animals, especially copepods. They can consume prey that are nearly as large as themselves, including small fish and other arrow worms! The complex musculature of the head operates the grasping spines and retraction of the hood during feeding. Both planktonic and benthic species are ambush predators, using ciliary receptors scattered over the body to detect movements of nearby prey. A chaetognath can determine both the direction and distance of potential prey at close range. Prey are ingested whole. The release of luminescent clouds in deep-sea species might serve to startle potential prey into movement, which can then be detected by the chaetognaths to facilitate predation. The luminescent cloud might also provide a means of escaping from predators.

Benthic forms, such as *Spadella*, feed while affixed to a substratum by adhesive secretions. As prey swim within reach, the anterior end is raised, and grasping spines flared. The prey is captured by a quick downward flex of the head while the rest of the body remains firmly attached. The grasping spines close around and manipulate the prey, orienting it for ingestion. Planktonic chaetognaths feed on prey approaching the body from all sides. Laterally positioned prey are captured by rapid flexure of the body and a short forward hop can be executed to grab prey located in front.

The grasping spines on left and right sides can be moved simultaneously or alternately, resulting in a surprising degree of dexterity during manipulation of prey. When rigid prey such as small crustaceans are captured, the chaetognath positions the victim longitudinally for swallowing. The spines and teeth contain α-chitin, hardened at their tips by silicon in the two species so far examined closely. While superficially resembling setae of arthropods or chaetae of annelids, each tooth and grasping spine is a complex structure produced by a group of specialized epidermal cells. The grasping spines of some species bear serrations. Teeth are commonly cuspidate, a shape that may aid in the penetration of prey, including the exoskeletons of small crustaceans. Some, if not most, species of arrow worms use a fast-acting neurotoxin, tetrodotoxin, to quell prey. Tetrodotoxin blocks sodium transport across cell membranes. Many marine bacteria synthesize tetrodotoxin, and in chaetognaths the toxin is probably produced by commensal bacteria (*Vibrio*) inhabiting the head or gut of the arrow worm. The toxin is presumably incorporated into the very sticky secretions produced by vestibular glands and, possibly, the esophageal glands.

The gut is a relatively simple straight tube extending from the mouth in the vestibule to the ventral anus at the trunk–tail junction (Figure 11.10A,C,I). The mouth leads to a short pharynx, which is equipped with mucus-secreting cells. Swallowing is accomplished by well-developed pharyngeal muscles aided by lubricants from the mucous cells. The gut narrows where it passes through the head–trunk septum, and it extends posteriorly as a long intestine. A short rectum joins the posterior intestine to the anus. Most digestion occurs extracellularly in the posterior intestine and can be extremely rapid. Orange carotenoid pigments derived from the prey are incorporated into the otherwise transparent tissues of some deep-water chaetognaths.

Circulation, Gas Exchange, and Excretion

A simple hemal system exists in chaetognaths, but it is easily overlooked because the blood is colorless and transparent. The hemal system consists of thin sinuses situated between the intestinal epithelium and surrounding myoepithelial peritoneum. Larger hemal sinuses exist in the dorsal mesentery, trunk/tail septum, ovaries, and body wall near the ventral ganglion. Transport through the hemal system is probably driven by the intestinal musculature. Even when the gut is empty, posteriorly directed peristalsis alternates with anteriorly directed peristalsis. The nutritive demands of various tissues in the tail are probably met by ultrafiltration across the tail side of the posterior septum, from the hemal sinus in the trunk/tail septum to the tail coelom. Gas exchange and excretion are apparently by diffusion through the body wall.

Nervous System and Sense Organs

Of paramount importance to the success of chaetognaths as active predators are features of the nervous system and associated sensory receptors. As we have seen in other groups, a body plan that emphasizes cephalization is usually an integral factor in adapting to a predatory lifestyle. The central nervous system of chaetognaths includes a large dorsal cerebral ganglion in the head. From this, paired nerves extend posteriorly to the eyes and corona ciliata, and two pairs of circumenteric connectives extend ventrally. Small anterior frontal connectives innervate the head muscles and gut. Larger main connectives extend posteroventrally to join a large ventral ganglion located in the trunk epidermis (Figure 11.10G). From the ventral ganglion, numerous peripheral nerves radiate to all parts of the trunk and tail, branching to form an elaborate intraepidermal nerve plexus. The ventral ganglion receives sensory input from the ciliary fence receptors on the body, and it controls swimming and other behaviors caused by the body wall musculature.

The ciliary fences are stereotypically arranged and are oriented either parallel or transversely relative to the long axis of the body, such that the entire body functions as an "antenna" for reception of nearby disturbances. Each ciliary fence contains 50–300 cells. Although chemoreceptors have yet to be positively identified in chaetognaths, they almost certainly exist. Candidates include the aforementioned corona ciliata, the pore-bearing vestibular ridges located just behind the teeth, and other tiny pores flanking the mouth.

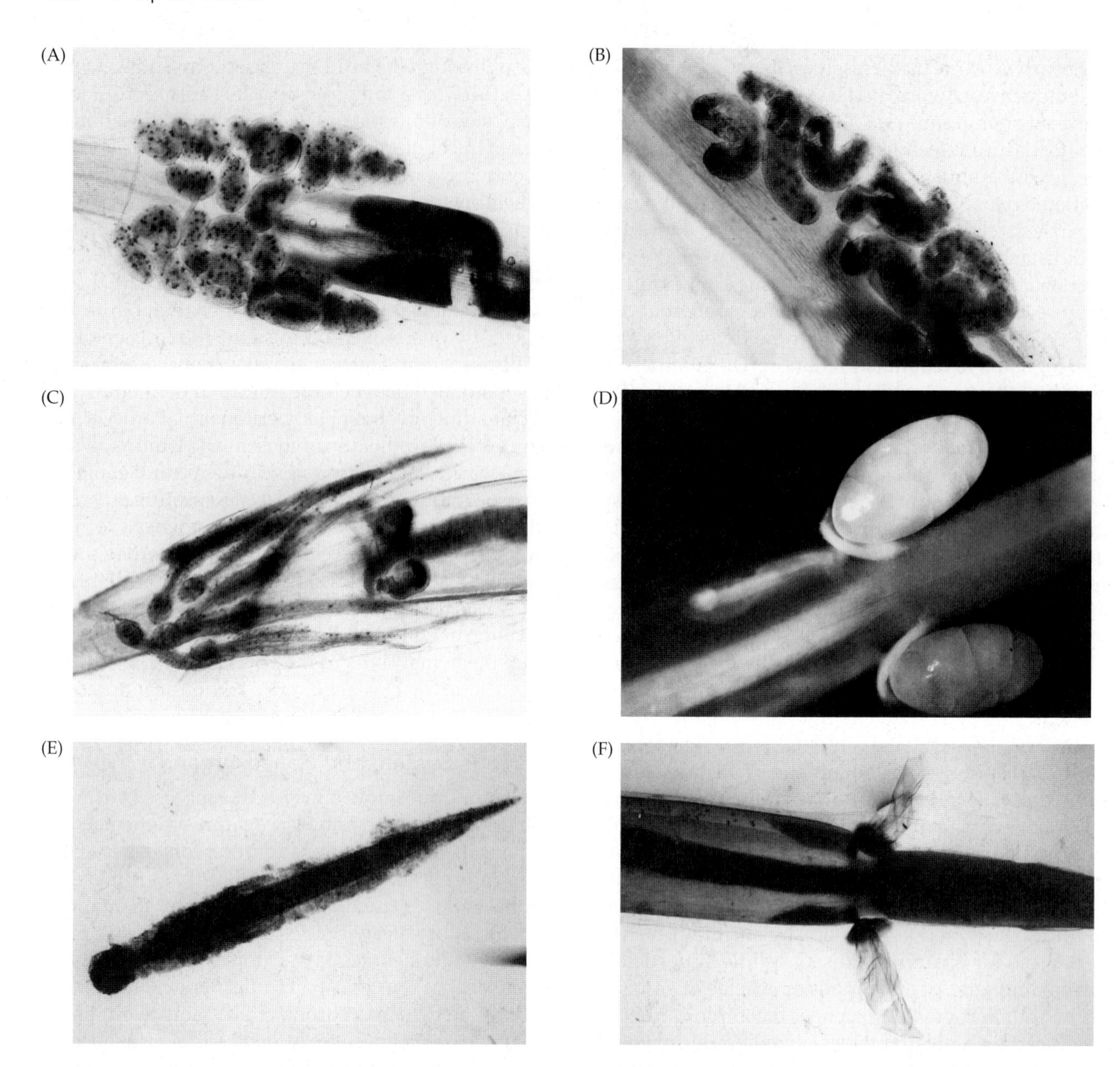

Figure 11.12 The chaetognath *Eukrohnia*, with temporary gelatinous marsupia housing the developing embryos. (A–C) *Eukrohnia bathypelagica* carrying developing embryos and young in the marsupium. (D) *Eukrohnia fowleri* carrying fertilized eggs in posterior marsupial sacs. (E) Young of *Eukrohnia fowleri* just after hatching. (F) *Eukrohnia fowleri* carrying the empty marsupial sacs from which the young have already escaped.

Most arrow worms possess a pair of eyes on the dorsal surface of the head. Typically, each eye has a large central pigment cell, indented to form seven cups containing the light sensitive parts of the ciliary receptor cells (Figure 11.10H). We can infer from their structure that chaetognaths have a nearly uninterrupted visual field enabling them to orient to light direction and intensity. The eyes of deep-water arrow worms are variously modified in ways that presumably increase sensitivity to light. For example, in certain species of *Eukrohnia*, the photoreceptor cells are directed outward, and each is capped by a transparent "lens." The eyes probably do not form images but are used for orientation during vertical migration.

Reproduction and Development

Arrow worms are hermaphroditic, with paired ovaries in the trunk and paired testes in the tail (Figure 11.10I). Groups of spermatogonia are released from the testes to the tail coeloms, where they circulate as the sperm mature. The filiform sperm are picked up by open ciliated funnels and conveyed down the sperm ducts to a pair of seminal vesicles, which bulge from the sides of the tail. In some species, gland cells secrete a spermatophore wall around the enclosed sperm. When filled with sperm, the seminal vesicles are conspicuously white. Release of sperm involves rupture of the seminal vesicles during mating. This can lead to self-fertilization in some species, but happens during mating in most species. Each ovary bears along its side an oviduct that leads to a genital pore just in front of the trunk–tail septum. Eggs develop within the ovaries where they are bathed in a nutritive fluid derived from the posterior hemal sinus. Fertilization occurs prior to ovulation as sperm received during mating pass from the oviduct through specialized accessory fertilization cells to the attached eggs. Immediately after fertilization, zygotes move into the oviduct for release to the environment.

Mating has been most extensively studied in some benthic spadellids (e.g., *Spadella celphaloptera, Paraspadella gotoi*) and the neritic *Ferosagitta hispida*. After a rather elaborate mating "dance," sperm from a seminal receptacle are deposited as a mass onto the mate's body. In *Paraspadella gotoi*, the sperm mass is precisely placed at the female gonopore, but in the others sperm masses are attached more anteriorly and columns of sperm stream over the epidermis of the recipient to enter the female gonopores. Benthic chaetognaths (e.g., *Spadella*) tend to deposit fertilized eggs on algae or other suitable substrata. Planktonic species typically shed floating embryos to the sea. *Pterosagitta draco* encloses the embryos in a large floating gelatinous mass and species of *Eukrohnia* carry developing embryos in gelatinous masses, one on either side of the body near the tail, until the young are ready to swim (Figure 11.12). *Ferosagitta hispida* descends and attaches developing eggs to stationary benthic objects. When food is abundant, new batches of sperm and eggs can be produced in daily succession.

Development is direct, lacking any larval stage or metamorphosis. The transparent eggs contain little yolk, and cleavage is holoblastic and equal. Both classical and modern studies have suggested a modified spiral pattern of cleavage, with unmistakable animal and vegetal cross-furrow cells, but this is easily overlooked because animal-pole cells ("micromeres") and vegetal-pole cells ("macromeres") are similar in size, and cleavage of arrow worms has been incorrectly described in the past as being radial. However, while a spiralian-like tetrahedral 4-cell embryo is derived via a levotropic blastomere displacement in the second cleavage, subsequent cleavage stages have not yet been documented and further research is needed to understand the nature of cleavage in these animals. The coeloblastula consists of pyramidal cells arranged around a small blastocoel (Figure 11.13).

Gastrulation (germ layer formation) occurs by invagination of the presumptive endoderm, leaving no blastocoel. The blastopore marks the eventual posterior end of the animal—both mouth and anus form secondarily,

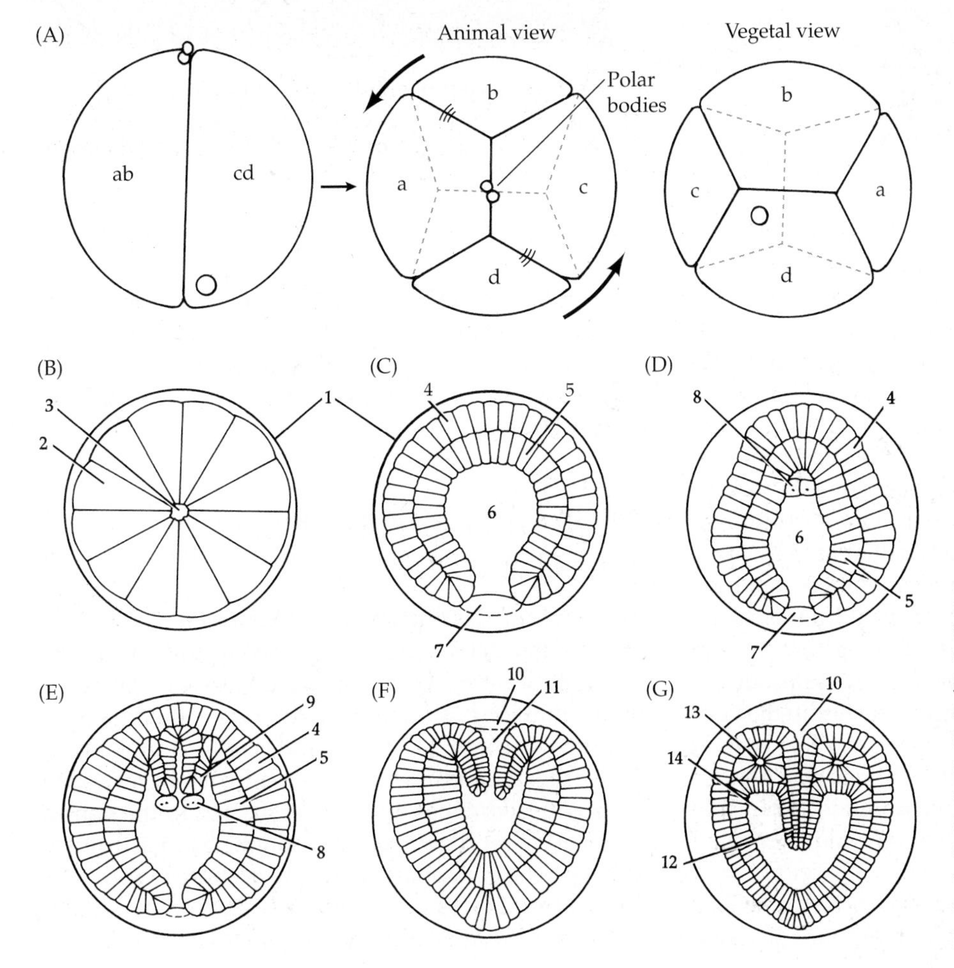

Figure 11.13 Early chaetognath development. (A) Four-celled embryo of *Paraspadella gotoi*, showing the spiral arrangement of blastomeres. (B) Early blastula. (C) Gastrula. (D) Later gastrula. (E) Production of mesodermal folds from archenteron. (F) Blastopore closure and secondary mouth opening with the formation of a stomodeum. (G) Formation of coelomic pouches.

(A)

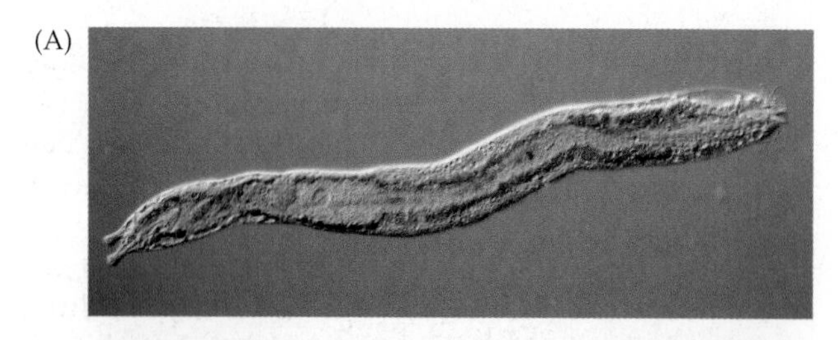

(B)

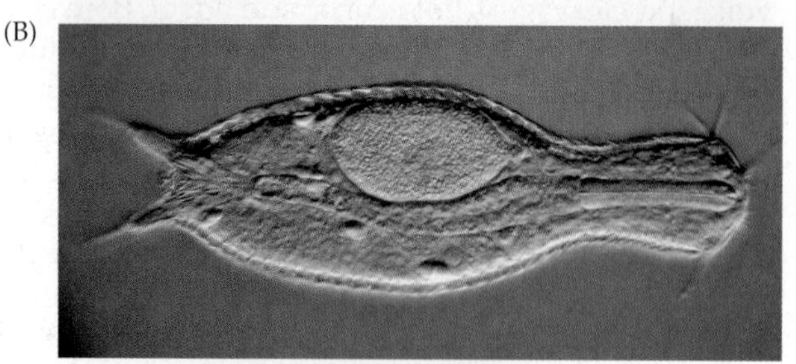

(C)

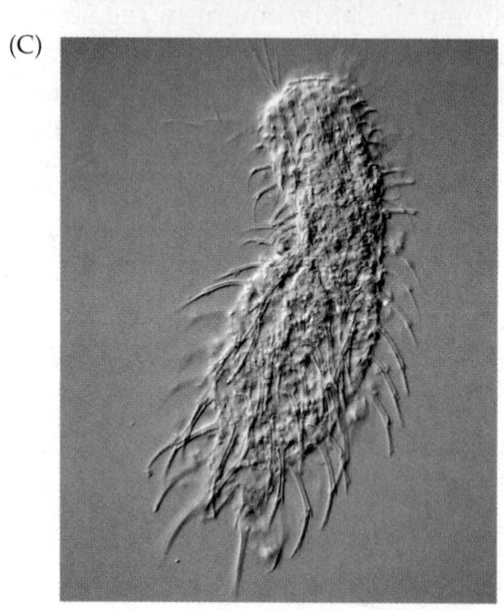

(D)

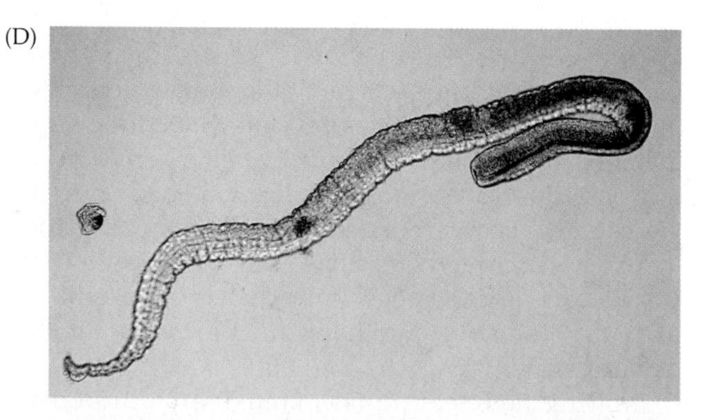

(E)

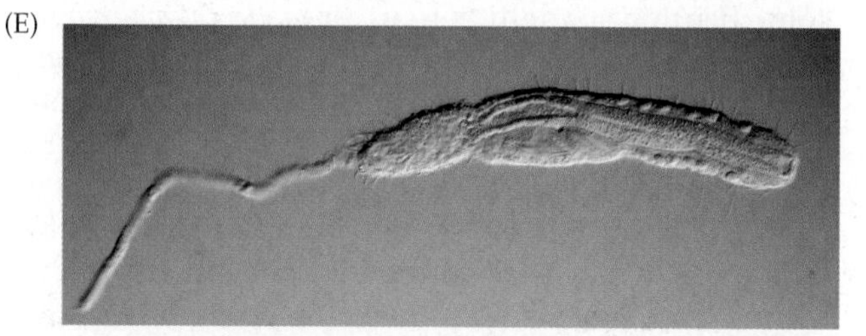

(F)

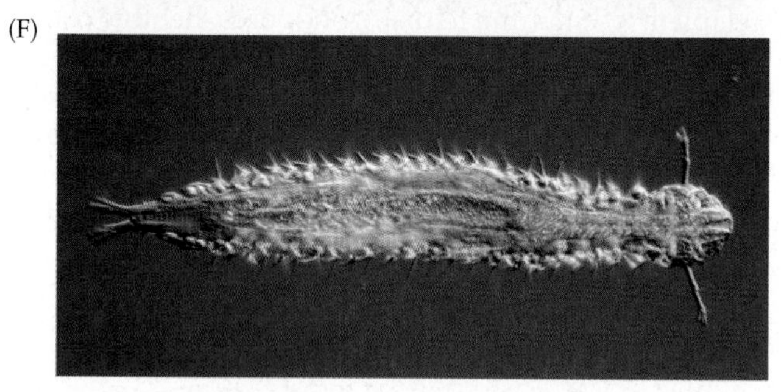

Figure 11.14 Representative gastrotrichs. (A–C) Light micrographs of order Chaetonotida. (A) *Neodasys*. (B) *Xenotrichula*. (C) *Chaetonotus*. (D–F) Light micrographs of order Macrodasyida. (D) *Megadasys*. (E) *Urodasys*. (F) *Xenodasys*.

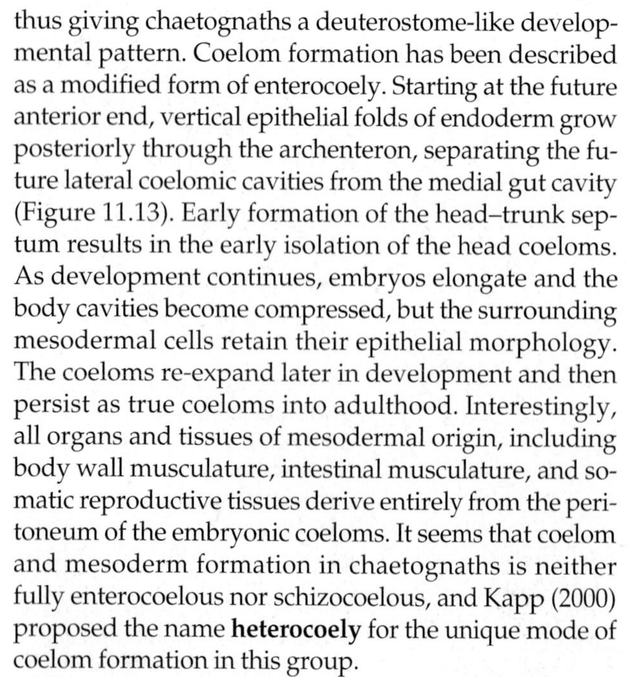

thus giving chaetognaths a deuterostome-like developmental pattern. Coelom formation has been described as a modified form of enterocoely. Starting at the future anterior end, vertical epithelial folds of endoderm grow posteriorly through the archenteron, separating the future lateral coelomic cavities from the medial gut cavity (Figure 11.13). Early formation of the head–trunk septum results in the early isolation of the head coeloms. As development continues, embryos elongate and the body cavities become compressed, but the surrounding mesodermal cells retain their epithelial morphology. The coeloms re-expand later in development and then persist as true coeloms into adulthood. Interestingly, all organs and tissues of mesodermal origin, including body wall musculature, intestinal musculature, and somatic reproductive tissues derive entirely from the peritoneum of the embryonic coeloms. It seems that coelom and mesoderm formation in chaetognaths is neither fully enterocoelous nor schizocoelous, and Kapp (2000) proposed the name **heterocoely** for the unique mode of coelom formation in this group.

Development from zygote release to hatching as a juvenile chaetognath is rapid, about 48 hours. Parental investment per embryo is small, the eggs contain little yolk, and they are abandoned soon after fertilization (except in brooding forms). The rapid development to a feeding juvenile is essential to the success of this life history strategy.

Phylum Gastrotricha: The Gastrotrichs

The phylum Gastrotricha (Greek *gasteros*, "stomach"; *trichos*, "hair") comprises about 800 species of marine, brackish, and freshwater Metazoa. Most species are less than 1 mm long, although a few reach 3 mm in length. Many gastrotrichs bear a superficial resemblance to flatworms and rotifers, but are easily distinguished by their adhesive tubes and cilia restricted to the ventral surface. Under high magnification the tiny gastrotrichs appear to be very charismatic invertebrates, with their "whiskered" faces and spiny or scaled, often bowling pin-shaped bodies. Figure 11.14 illustrates some of the

BOX 11B Characteristics of the Phylum Gastrotricha

1. Triploblastic, bilateral, unsegmented, acoelomate
2. Microscopic, body elongate or bowling pin-shaped
3. Bilayered cuticle, smooth or with scale or spine-like elaborations; with exocuticle covering entire body including all cilia; cuticle is not molted
4. Duo-gland adhesive tubes
5. Epidermis cellular or partly syncytial; epidermal cells monociliate or multiciliate; locomotory cilia on ventral surface only
6. Myoepithelial pharynx with triradiate lumen; complete gut
7. One pair or more of protonephridia, without special circulatory or gas exchange structures
8. Hermaphroditic, parthenogenetic, or hermaphroditic and parthenogenetic
9. Complex reproductive organs
10. Embryonic development poorly studied
11. Direct development

body forms within this phylum and some features of their external anatomy, and Box 11B lists the distinguishing features of this group. Marine gastrotrichs are meiofaunal and live in the interstitial spaces between sand grains from the intertidal zone to abyssal depths (> 2,500 m). Freshwater species are found in surface detritus or among the filaments of aquatic plants and plantlike protists; a few are semiplanktonic. They are sometimes abundant when found.

The gastrotrich body is typically divisible into a head and a trunk (Figures 11.14B,C and 11.15A), though many species lack such an obvious distinction. Externally, gastrotrichs possess a cuticle that surrounds all aspects of the body and is often smooth in appearance but may be elaborated into spines, bristles, scales, and/or plates (Figure 11.14). **Adhesive tubes**, a diagnostic character of most gastrotrichs, are also derived from the cuticle and generally contain a pair of glands that secrete adhesive and releaser substances (the **duo-gland adhesive system**). These tubes may be abundant (Macrodasyida) or sparse to absent (most Chaetonotida).

The gastrotrich head bears a terminal or subterminal mouth with stiff cilia or other cuticular elaborations. A ring of motile cilia often encompasses the head dorsally while other ciliary elaborations such as pits and tentacles are present on the sides (Figure 11.14F). A neck-like constriction behind the head is present in many but not all gastrotrichs. The gastrotrich trunk (the main part of the body) is generally vermiform (Macrodasyida, *Neodasys*) or bowling pin-shaped (Chaetonotida), and it contains most of the organ systems. The trunk terminates in a rounded, bilobed, tailed, or forked caudum that bears adhesive tubes.

The most recent molecular phylogenetic studies place gastrotrichs squarely within the protostome bilaterians and close to the base of the Spiralia, where they seem to arise near the Platyhelminthes or the Gnathifera (Gnathostomulida, Micrognathozoa, and Rotifera). Anatomically, however, gastrotrichs are more similar to Nematoida and Scalidophora clades because of their circumpharyngeal brain, bilayered cuticle, and myoepithelial pharynx.

Gastrotrich Classification

Order Chaetonotida (Figure 11.14A–C). Primarily freshwater, but also some marine, estuarine, and semiterrestrial (freshwater micohabitats) species; smooth or complex cuticle with a variable number of adhesive tubes (most with only two on caudal furca); pharynx with Y-shaped lumen, pharyngeal pores absent; one or more pair of protonephridia present; hermaphroditic, parthenogenetic, or a combination of hermaphroditic and parthenogenetic.

Suborder Multitubulatina Marine; strap-shaped body with smooth cuticle and anterior, posterior, and lateral adhesive tubes; hermaphroditic. Monogeneric: *Neodasys*, with two known species.

Suborder Paucitubulatina Mostly freshwater species but with many marine representatives; bowling pin-shaped body often covered by elaborate cuticle, usually with two terminal adhesive tubes on caudal furca; hermaphroditic, parthenogenetic, or both (e.g., *Aspidiophorus, Chaetonotus, Dasydytes, Lepidodermella*).

Order Macrodasyida (Figures 11.14D–F). Mostly marine and estuarine gastrotrichs with only a single freshwater genus (*Redudasys*); with smooth or elaborately sculptured cuticle, usually bearing numerous adhesive tubes along the head (ventral) and trunk (lateral, ventral, sometimes dorsal); pharynx with inverted Y-shaped lumen, pharyngeal pores present in all except species of *Lepidodasys*; usually with several pairs of protonephridia; hermaphroditic with complex reproductive organs; some may be parthenogenetic (e.g., *Dactylopodola, Macrodasys, Turbanella, Urodasys, Xenodasys*).

The Gastrotrich Body Plan

Body Wall

A gastrotrich's body is encased in a bilayered cuticle of varying thickness and complexity. It is never molted. The outer portion or exocuticle contains one or more lamellae that cover all external surfaces including locomotory and sensory cilia (Figure 11.15E); the inner portion or endocuticle is thick and fibrous and produces the spines, scales, and adhesive tubes (Figure 11.14). The epidermis may be cellular as in Macrodasyida or

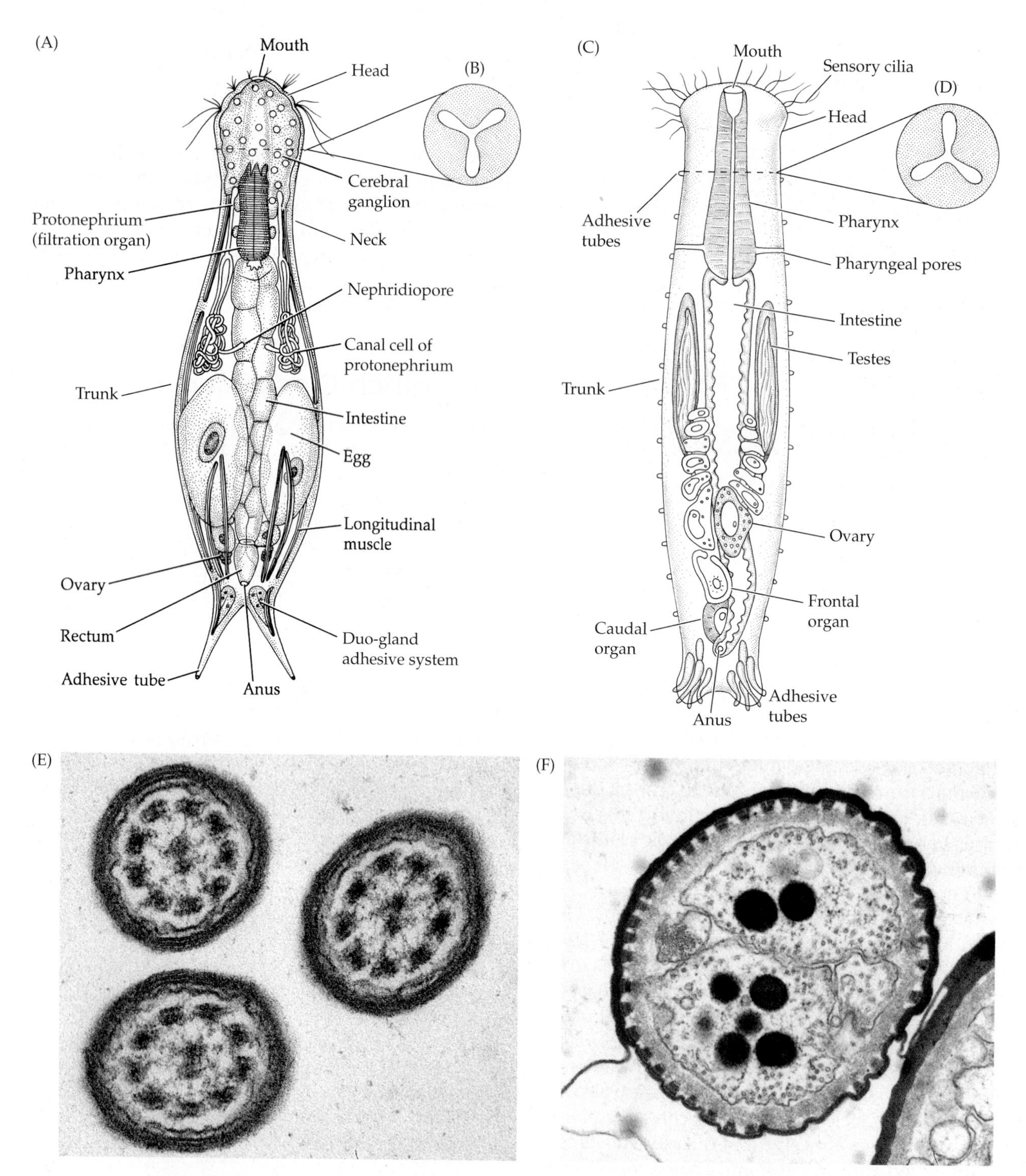

Figure 11.15 Anatomy of gastrotrichs. (A) Representative of Chaetonotida, *Paucitubulatina*. (B) Cross section through pharynx showing Y-shaped pharyngeal lumen. (C) Representative of Macrodasyida and (D) cross section through pharynx showing inverted Y-shaped pharyngeal lumen. (E–F) Transmission electron micrographs of (E) a cross section through locomotory cilia and (F) an adhesive tube.

partly syncytial as in most Chaetonotida; only ventral epidermal cells bear locomotory cilia (thus the derivation of the phylum name: "hairy belly"). Most species possess multiciliate epidermal cells, however some species possess monociliate cells. Duo-gland adhesive organs, consisting of at least one adhesive gland and one releaser gland (which are critical for rapid grip and release from sand grains in the interstitial environment), are present in all species and form cuticle covered neck-like tubes to the outside (Figure 11.15F).

Internal to the epidermis are connective tissues such as a basal lamina (absent in many species) or in some cases large vacuolar **Y-cells** (some Macrodasyida and *Neodasys*) that function in hydrostatic support; hemoglobin has been detected in the Y-cells of *Neodasys* and may be present in others. Muscles are present as circular, helicoidal, and longitudinal bands that insert on the cuticle via an intervening epithelial cell; dorsoventral and other muscle orientations may also be present.

Support and Locomotion

The cuticle and compact (acoelomate) body plan provide hydrostatic support while the complex array of muscles produce the bending and twisting movements characteristic of gastrotrich locomotion, and essential for life between sand grains (which are "boulder size" for millimeters-long gastrotrichs). Most animals are highly flexible and in some genera such as *Megadasys* (Figure 11.14D) can stretch their already elongate bodies to nearly twice gliding length or contract into a tight ball. The adhesive tubes play a significant role during these activities. Some semi-planktonic freshwater gastrotrichs such as species of Dasydytidae (Chaetonotida) have moveable spines that permit a form of saltatory swimming.

Feeding and Digestion

Most gastrotrichs feed by pumping small food items into the gut by action of the pharynx, a myoepithelial tube that is surrounded by complex layered muscles. The pharyngeal lumen is either Y-shaped (Chaetonotida) or inverted Y-shaped (Macrodasyida) (Figures 11.15B,D). In most macrodasyidans, a pair of pharyngeal pores connects the pharyngeal lumen to the outside, thereby permitting the release of excess water imbibed with the food (Figure 11.15C). Gastrotrichs feed on nearly any organic material, alive or dead, of appropriately small size (bacteria, detritus, protists, unicellular algae).

The mouth is terminal or subterminal and often surrounded by a ring of stiff cilia or cuticular spines that may aid in food capture. A short buccal cavity connects the mouth to the elongate muscular pharynx lined with cuticle (a stomodaeum) (Figure 11.14). In species of Chaetonotida, a pharyngeal "valve" may be present at the junction of the pharynx and intestine (midgut). The endodermally derived portion of the gut is a straight tube that narrows posteriorly and leads to a ventral anus. The anus is preceded by a cuticle-lined hindgut in Chaetonotida, while in most Macrodasyida a hindgut is absent and the anus is presumably permanent or temporary. Digestion and absorption occur in the intestine, which is comprised of cuboidal cells with a brush border; motile cilia may be present in some species and aid in food passage.

Circulation, Gas Exchange, Excretion, and Osmoregulation

Circulation and gas exchange occur by diffusion through the body wall. Some gastrotrichs are capable of anaerobic respiration when environmental conditions require it. Protonephridia are the primary excretory organs in gastrotrichs. Marine species such as most macrodasyidans and *Neodasys* possess multiple pairs of protonephridia that function solely in the removal of metabolic wastes. Their protonephridia are generally mono- or bicellular filtration organs each with a single flagellum; they connect to the outside by way of a single ciliary canal cell and nephridiopore cell. In species that live in a hypo-osmotic environment like most freshwater chaetonotidans, the single pair of protonephridia also function in osmoregulation. Their protonephridia are bicellular organs that lead to an elongate and convoluted canal cell (Figure 11.15A). The extra length of the canal cell in freshwater species is hypothesized to function for increased reabsorption of necessary salts and minerals from the primary urine before elimination.

Nervous System and Sense Organs

The central nervous system consists of a dorsal brain and one or more pairs of longitudinal nerve cords. The brain consists of numerous cells that sit atop a ring-like commissure around the anterior region of the pharynx (Figure 11.15A). Cerebral neurons innervate the head and pharynx. The nonganglionated nerve cords arise from the brain and are interconnected by thin commissures; they unite at the posterior end.

Sensory receptors are ciliary in structure and concentrated in the head region but may be distributed over the entire body. Ciliary pits and tentacles on the heads of some species are likely chemosensory, while stiff, bristlelike cilia on the head and trunk probably have a tactile function. Several species possess pigmented (red) ocelli that indicate probable photoreceptive capabilities; experiments have shown that species lacking ocelli still have spectral sensitivity. One species, *Pleurodasys helgolandicus*, possesses an unusual pair of drumstick-like organs that are hypothesized to function as gravity receptors.

Reproduction and Development

Most species of gastrotrichs are hermaphroditic or a combination of hermaphroditic and parthenogenetic. Strict parthenogenesis is rare and known from only a few species. The hermaphroditic reproductive system is best studied in Macrodasyida and often contains gametes and organs of sperm transfer and storage (Figure 11.15C). The male system comprises one or two testes with associated sperm ducts that lead to a single or paired ventral gonopores. A **caudal organ** that presumably functions in the transfer of sperm is present in most species, but only in Macrodasyidae and Thaumastodermatidae is it directly connected to the sperm ducts; some species bear a sclerotized stylet within the caudal organ.

The female system includes one or two ovaries in the posterior body region. Other reproductive organs may include a **frontal organ** (seminal receptacle) as in

Macrodasyida and *Neodasys*, or an **X-organ** as in most Chaetonotida. The ovaries are generally compact or somewhat diffuse. Oocytes mature anteriorly in most species and become positioned above the intestine and in proximity to the frontal organ when present; oviducts are rare but some species do have a distinct uterus. Yolk is produced within the oocyte itself. Mutual cross-fertilization leads to the direct exchange of sperm or spermatophores (via hypodermic impregnation or some other mechanism) in all species, but sperm is stored in the frontal organ of most macrodasyidans. The X-organ in chaetonotidans is a bilobed syncytial organ in the posterior trunk that develops during the sexual phase and is thought to play a role in reproduction, although its function has never been demonstrated. Fertilization is always internal and precedes the production of an eggshell by the egg itself. Oviposition is through the body wall of most species, and it might occur through a temporary or perhaps permanent pore, or in some cases via rupture of the body wall.

Historically, most freshwater chaetonotidans were assumed to be strict parthenogens until small packets of rodlike sperm were discovered next to their midgut. The sperm develop after the production of the parthenogenetic ova and in synchrony with the meiotic (sexual) ova, suggesting they are not vestigial male gametes; such a condition is referred to as **post-parthenogenetic hermaphroditism**. Cross-fertilization has never been observed in these species. Two forms of parthenogenetic eggs are produced: the first few eggs (tachyblastic) hatch quickly and are important for rapid population growth, while the last egg (opsiblastic) may overwinter before hatching. A third and more rare type of egg, called the plaque-bearing egg, is thought to be the result of sexual reproduction, but its development has never been followed.

Few embryological studies have been conducted on gastrotrichs and those that have detailed the main cleavage events are somewhat contradictory. In general, cleavage is holoblastic and adequal (i.e., the blastomeres are of approximately equal size), however, after the initial equatorial division, there is an apparent divergence in cleavage patterns between the two orders. The second cleavage plane of the posterior blastomere is meridional in macrodasyidans but equatorial in chaetonotidans, which leads to different fates for the developing tissues. In neither case is a spiral-like cleavage pattern evident. By the fifth division, a bilaterally symmetric embryo has developed. Gastrulation occurs by invagination followed by epiboly of the secondary ectoderm over presumptive mesoderm. A stomodaeum forms in both orders but a proctodeum has been noted only for chaetonotidans. The U-shaped embryo continues to grow and elongate in the eggshell for several hours (to several days) prior to hatching.

Development is direct and juveniles are small versions of adults, though without noticeable reproductive organs and with fewer adhesive tubes (in Macrodasyida). Maturation is rapid and the animals usually are sexually mature within a few days after hatching. Individuals of most species probably live only a few weeks.

Selected References

General References

Dunn, C. W., G. Giribet, G. D. Edgecombe and A. Hejnol. 2014. Animal phylogeny and its evolutionary implications. Annu. Rev. Ecol. Evol. S. 45: 371–395.

Lapan, E. and H. Morowitz. 1972. The Mesozoa. Sci. Am. 227(6): 94–101.

Shen, X. and 7 others. 2015. Phylomitogenomic analyses strongly support the sister relationship of the Chaetognatha and Protostomia. Zool. Scr. doi: 10.1111/zsc.12140

Stunkard, H. W. 1982. Mesozoa. Pp. 853–855 in S. Parker (ed.), *Synopsis and Classification of Living Organisms*. McGraw-Hill, New York.

Rhombozoa

Aruga, J., Y. S. Odaka, A. Kamiya and H. Furuya. 2007. *Dicyema* Pax6 and Zic: tool-kit genes in a highly simplified bilaterian. BMC Evol. Biol. 7: 201–216.

Furuya, H., F. G. Hochberg and K. Tsuneki. 2001. Developmental patterns and cell lineages of vermiform embryos in dicyemid mesozoans. Biol. Bull. 201: 405–416.

Furuya, H., F. G. Hochberg and K. Tsuneki. 2003. Reproductive traits in dicyemids. Mar. Biol. 142: 297–305.

Furuya, H., F. G. Hochberg and K. Tsuneki. 2004. Cell number and cellular composition in infusoriform larvae of dicyemid mesozoans (phylum Dicyemida). Zool. Sci. 21: 877–889.

Furuya, H., F. G. Hochberg and K. Tsuneki. 2007. Cell number and cellular composition in vermiform larvae of dicyemid mesozoans (phylum Dicyemida). J. Zool. 272: 608–618.

Furuya, H. and K. Tsuneki. 2003. Biology of dicyemid mesozoans. Zoolog. Sci. 20: 519–532.

Furuya, H. and K. Tsuneki. 2007. Developmental patterns of the hermaphroditic gonad in dicyemid mesozoans (phylum Dicyemida). Invert. Biol. 126(4): 295–306.

Furuya, H., K. Tsuneki and Y. Koshida. 1992. Development of the infusorium embryos of *Dicyema japonicum* (Mesozoa: Dicyemida). Biol. Bull. 183: 248–257.

Furuya, H., K. Tsuneki and Y. Koshida. 1997. Fine structure of a dicyemid mesozoans, *Dicyema acuticephalum*, with special reference to cell junction. J. Morphol. 231: 297–305.

Hochberg, F. G. 1983. The parasites of cephalopods: a review. Mem. Natl. Mus. Victoria 44: 109–145.

Horvath, P. 1997. Dicyemid mesozoans. Pp. 31–38 in S. Gilbert and A. Raunio (eds.), *Embryology: Constructing the Organism*. Sinauer Associates, Sunderland, MA.

Katayama, T., H. Wada, H. Furuya, N. Satoh and M. Yamamoto. 1995. Phylogenetic position of the dicyemid Mesozoa inferred form 18S rDNA sequences. Biol. Bull. 189: 81–90.

Kobayashi, M., H. Furuya and P. W. Holland. 1999. Dicyemids are higher animals. Nature 401: 762.

Kobayashi, M., H. Furuya and H. Wada. 2009. Molecular markers comparing the extremely simple body plan of dicyemids to that of lophotrochozoans: insight from the expression patterns of Hox, Otx, and brachyury. Evol. Dev. 11(5): 582–589.

Pearse, J. S. 1999. Rhombozoa. Pp. 276–278 in *Encyclopedia of Reproduction, Vol. 4*. Academic Press, New York.

Suzuki, T. G., K. Ogino, K. Tsuneki and H. Furuya. 2010. Phylogenetic analysis of dicyemid mesozoans (phylum Dicyemida) from innexin amino acid sequences: dicyemids are not related to Platyhelminthes. J. Parastiol. 96(3): 614–625.

Orthonectida

Atkins, D. 1933. *Rhopalura granosa* sp. nov. an orthonectid parasite of a lamellibranch *Heteranomia squamula* L. with a note on its swimming behavior. J. Mar. Biol. Assn. U.K. 19: 233–252.

Caullery, M. 1961. Classe des Orthonectides. Pp. 695–706 in P. Grassé (ed.), *Traité de Zoologie*. Masson et Cie, Paris.

Hanelt, B., D. Van Schyndel, C. M. Adema, L. A. Lewis and E. S. Loker. 1996. The phylogenetic position of *Rhopalura ophiocomae* (Orthonectida) based on 18S ribosomal DNA sequence analysis. Mol. Biol. Evol. 13(9): 1187–1191.

Kozloff, E. 1965. *Ciliocincta sabellariae* gen. and sp. n., an orthonectid mesozoan from the polychaete *Sabellaria cementarium* Moore. J. Parasitol. 51(1): 37–44.

Kozloff, E. 1969. Morphology of the orthonectid *Rhopalura ophiocomae*. J. Parasitol. 55(1): 171–195.

Kozloff, E. 1971. Morphology of the orthonectid *Ciliocincta sabellariae*. J. Parasitol. 57(3): 585–597.

Kozloff, E. 1992. The genera of the phylum Orthonectida. Cah. Biol. Mar. 33: 377–406.

Kozloff, E. 1994. The structure and origin of the plasmodium of *Rhopalura ophiocomae* (phylum Orthonectida). Acta Zool. 75: 191–199.

Pearse, J. S. 1999. Orthonectida. Pp. 532–533 in *Encyclopedia of Reproduction, Vol. 3*. Academic Press, New York.

Slyusarev, G. S. 2000. Fine structure and development of the cuticle of *Intoshia variabili* (Orthonectida). Acta Zool. 81 doi. 10.1046/j.1463-6395.2000.00030.x

Slyusarev, G. S. 2004. Fine structure and function of the genital pore of the female of *Intoshia variabili* (Orthonectida). Folia Parasitologica 51: 287–290.

Slyusarev, G. S. and M. Ferraguti. 2002. Sperm structure of *Rhopalura littoralis* (Orthonectida). Invert. Biol. 121(2): 91–94.

Chaetognatha

Alvariño, A. 1965. Chaetognaths. Annu. Rev. Oceanogr. Mar. Biol. 3: 115–194.

Bieri, R. 1966. The function of the "wings" of *Pterosagitta draco* and the so-called tangoreceptors in other species of Chaetognatha. Publ. Seto Mar. Biol. Lab. 14: 23–26.

Bieri, R. and E. V. Thuesen. 1990. The strange worm *Bathybelos*. Am. Sci. 78: 542–549.

Bone, Q., H. Kapp and A. C. Pierrot-Bults (eds.). 1991. *The Biology of Chaetognaths*. Oxford University Press, Oxford.

Burfield, S. 1927. *Sagitta*. Liverpool Mar. Biol. Comm., Mem. 28. Proc. Trans. Liverpool Biol. Soc. 41: 1–104.

Casanova, J.-P. 1994. Three new rare *Heterokrohnia* species (Chaetognatha) from deep sea benthic samples in the northeast Atlantic. Proc. Biol. Soc. Wash. 107(4): 743–750.

Chen, J.-Y. and D.-Y. Huang. 2002. A possible Lower Cambrian chaetognath (arrow worm). Science 298: 187.

Duvert, M. 1991. A very singular muscle: the secondary muscle of chaetognaths. Philos. Trans. R. Soc. Lond. B Biol. Sci. 332: 245–260.

Duvert, M. and C. Salat. 1990. Ultrastructural studies on the fins of chaetognaths. Tissue Cell 22: 853–863.

Eakin, R. M. and J. A. Westfall. 1964. Fine structure of the eye of a chaetognath. J. Cell Biol. 21: 115–132.

Feigenbaum, D. L. 1978. Hair-fan patterns in the Chaetognatha. Can. J. Zool. 56: 536–546.

Feigenbaum, D. L. and R. C. Maris. 1984. Feeding in the Chaetognatha. Annu. Rev. Oceanogr. Mar. Biol. 22: 343–392.

Ghirardelli, E. 1968. Some aspects of the biology of the chaetognaths. Adv. Mar. Biol. 6: 271–375.

Goto, T. and M. Yoshida. 1985. The mating sequence of the benthic arrow worm *Spadella schizoptera*. Biol. Bull. 169: 328–333.

Goto, T. M. Terazaki and M. Yoshido. 1989. Comparative morphology of the eyes of *Sagitta* (Chaetognatha) in relation to depth of habitat. Exp. Biol. 48: 95–105.

Goto, T., N. Takasu and M. Yoshida. 1984. A unique photoreceptive structure in the arrowworms *Sagitta crassa* and *Spadella schizoptera* (Chaetognatha). Cell Tissue Res. 235: 471–478.

Haddock, S. H. D. and J. F. Case. 1994. A bioluminescent chaetognath. Nature 367: 225–226.

Harzsch, S. and C. H. G. Müller. 2007. A new look at the ventral nerve centre of *Sagitta*: implications for the phylogenetic position of Chaetognatha (arrow worms) and the evolution of the bilaterian nervous system. Front. Zool. 4: 14.

Jordan, C. E. 1992. A model of rapid-start swimming at intermediate Reynolds number: undulatory locomotion in the chaetognath *Sagitta elegans*. J. Exp. Biol. 163: 119–137.

Kapp, H. 2000. The unique embryology of Chaetognatha. Zoologischer Anzeiger 239: 263–266.

Malakhov, V. V. and T. L. Berezinskaya. 2001. Structure of the circulatory system of arrow worms (Chaetognatha). Doklady Biol. Sci. 376: 78–80.

Matus, D. Q., K. M. Halanych and M. Q. Martindale. 2007. The Hox gene complement of a pelagic chaetognath, *Flaccisafitta enflata*. Integr. Comp. Biol. 47: 854–864.

Marletaz, F. and 8 others. Chaetognath transcriptome reveals ancestral and unique features among bilaterians. Genome Biol. 9: R94.

Moreno, I. (ed.). 1993. *Proceedings of the II International Workshop of Chaetognatha*. Universitat de les Illes Belears, Palma.

Marlétaz, F. and 11 others. 2006. Chaetognath phylogenomics: a protostome with deuterostome-like development. Curr. Biol. 16: R577–R578.

Muller, C. H., V. Rieger, Y. Perez, S. Harzsch. 2014. Immunohistochemical and ultrastructural studies on ciliary sense organs of arrow worms (Chaetognatha). Zoomorphology. doi: 10.1007/s00435-013-0211-6

Nagasawa, S. 1984. Laboratory feeding and egg production in the chaetognath *Sagitta crassa* Tokioka. J. Exp. Mar. Biol. Ecol. 76: 51–65.

Papillon, D. Y. Perez, X. Caubit and Y. L. Parco. 2006. Systematics of Chaetognatha under the light of molecular data, using the duplicated ribosomal 18S DNA sequences. Mol. Phylog. Evol. 38: 621–634.

Papillon, D., Y. Perez, L. Fasano, Y. L. Parco, X. Caubit. 2005. Restricted expression of a median Hox gene in the central nervous system of chaetognaths. Develop., Genes Evol. 215: 369–373.

Pierrot-Bults, A. C. and K. C. Chidgey. 1988. *Chaetognatha. Synopses of the British Fauna (New Series), No. 39*. E. J. Brill, New York.

Reeve, M. R. and M. A. Walter. 1972. Observations and experiments on methods of fertilization in the chaetognath *Sagitta hispida*. Biol. Bull. 143: 207–214

Rieger, V. Y. Perez, C. H. G. Müller, T. Lacalli, B. S. Hansson and S. Harzsch. 2011. Development of the nervous system in hatchlings of *Spadella cephaloptera* (Chaetognatha), and impilcations for nervous system evolution in Bilateria. Develop. Growth Differ. 53: 740–759.

Rieger, V. Y. Perez, C. H. G. Müller, E. Lipke, A. Sombke, B. S. Hansson and S. Harzsch. 2010. Immunohistochemical analysis and 3D reconstruction of the cephalic nervous system in

Chaetognatha: insights into the evolution of an early bilaterian brain? Invert. Biol. 129: 77–104.
Shimotori, T. and T. Goto. 2001. Developmental fates of the first four blastomeres of the chaetognath *Paraspadella gotoi*: relationship to protostomes. Develop. Growth Differ. 43: 371–382.
Shinn, G. L. 1993. The existence of a hemal system in chaetognaths. In I. Moreno (ed.), *Proceedings of the II International Workshop of Chaetognatha*. Universitat de les Illes Belears, Palma.
Shinn, G. L. 1994. Epithelial origin of mesodermal structures in arrow worms (phylum Chaetognatha). Am. Zool. 34: 523–532.
Shinn, G. L. 1994. Ultrastructural evidence that somatic "accessory cells" participate in chaetognath fertilization. Pp. 96–105 in Wilson, W.H., Stricker, S.A. and Shinn, G.L. (eds.), *Reproduction and Development of Marine Invertebrates*, Johns Hopkins University Press.
Shinn, G. L. 1997. Chaetognatha. Pp. 103–220 in F. W. Harrison and E. E. Ruppert, (eds.) *Microscopic Anatomy of Invertebrates, Vol. 15.* Wiley–Liss, New York.
Shinn, G. L. and M. E. Roberts. 1994. Ultrastructure of hatchling chaetognaths (*Ferosagitta hispida*): epithelial arrangement of the mesoderm and its phylogenotic implications. J. Morphol. 219: 143–163.
Telford, M. J. and P. W. H. Holland. 1997. Evolution of 28S ribosomal DNA in chaetognaths: duplicate genes and molecular phylogeny. J. Mol. Evol. 44: 135–144.
Terazaki, M. and C. B. Miller. 1982. Reproduction of meso- and bathypelagic chaetognaths in the genus *Eukrohnia*. Mar. Biol. 71: 193–196.
Terazaki, M., R. Marumo and Y. Fujita. 1977. Pigments of meso- and bathypelagic chaetognaths. Mar. Biol. 41: 119–125.
Thuesen, E. V. and R. Bieri. 1987. Tooth structure and buccal pores in the chaetognath *Flaccisagitta hexaptera* and their relation to the capture of fish larvae and copepods. Can. J. Zool. 65: 181–187.
Thuesen, E. V. and K. Kogure. 1989. Bacterial production of tetrodotoxin in four species of Chaetognatha. Biol. Bull. 176: 191–194.
Thuesen, E. V., F. A. Goetz and S. H. D. Haddock. 2010. Bioluminescent organs in two deep-sea arrow worms, *Eukrohnia fowleri* and *Caecosagitta macrocephala*, with further observations on bioluminescence in chaetognaths. Biol. Bull. 219: 100–111.
Takada, N., T. Goto and N. Satoh. 2002. Expression pattern of the *Brachyury* gene in the arrow worm *Paraspadella gotoi* (Chaetognatha) Genesis 32: 240–245.
Welsch, U. and V. Storch. 1982. Fine structure of the coelomic epithelium of *Sagitta elegans*. Zoomorphology 100: 217–222.

Gastrotricha

Balsamo, M., J. -L. d'Hondt, J. Kisielewski and L. Pierboni. 2008. Global diversity of gastrotrichs (Gastrotricha) in fresh waters. Hydrobiologia 595: 85–91.
Giere, O. 2009. *Meiobenthology*. 2nd Ed. Springer-Verlag, Berlin.
Hummon, M. R. and W. D. Hummon. 1983. Gastrotricha. Pp. 195–205 in K. G. Adiyodi and R. G. Adiyodi, (eds.), *Reproductive biology of invertebrates. Volume II: Spermatogenesis and sperm function.* John Wiley and Sons, London.
Hummon, W. D. and M. R. Hummon. 1983. Gastrotricha. Pp. 211–221 in K. G. Adiyodi and R. G. Adiyodi (eds.), *Reproductive Biology of Invertebrates. Volume I: Oogenesis, Oviposition, and Oosorption.* John Wiley and Sons, London.
Hummon, W. D. and M. R. Hummon. 1988. Gastrotricha. Pp. 81–85 in K. G. Adiyodi and R. G. Adiyodi (eds.), *Reproductive Biology of Invertebrates. Volume III: Accessory Sex Glands.* Oxford and IBH Publishing Company, New Delhi.
Hummon, W. D. and M. A. Todaro. 2010. Analytic taxonomy and notes on marine, brackish-water and estuarine Gastrotricha. Zootaxa 2392: 1–32.
Hochberg, R and Litvaitis, M. K. 2000. Phylogeny of Gastrotricha: a morphology-based framework of gastrotrich relationships. Biol. Bull. 198: 299–305.
Kieneke, A., P. Martínez Arbizu and W. H. Ahlrichs. 2007. Ultrastructure of the protonephridial system in *Neodasys chaetonotoideus* (Gastrotricha: Chaetonotida) and in the ground pattern of Gastrotricha. J. Morph. 268: 602–613.
Kieneke, A., W. H. Ahlrichs, P. Martínez Arbizu and T. Bartolomaeus. 2008. Ultrastructure of protonephridia in *Xenotrichula carolinensis syltensis* and *Chaetonotus maximus* (Gastrotricha: Chaetonotida): comparative evaluation of the gastrotrich excretory organs. Zoomorphology 127: 1–20.
Kieneke, A., P. Martínez Arbizu and W. H. Ahlrichs. 2008. Anatomy and ultrastructure of the reproductive organs in *Dactylopodola typhle* (Gastrotricha: Macrodasyida) and their possible functions in sperm transfer. Invert. Biol. 127(1): 12–32.
Kieneke, A., O. Riemann and W. H. Ahlrichs. 2008. Novel implications for the basal internal relationships of Gastrotricha revealed by an analysis of morphological characters. Zool. Scr. 37: 429–460.
Kieneke, A., W. H. Ahlrichs and P. Martinez Arbizu. 2009. Morphology and function of reproductive organs in *Neodasys chaetonotoideus* (Gastrotricha: *Neodasys*) with a phylogenetic assessment of the reproductive system in Gastrotricha. Zool. Scr. 38: 289–311.
Liesenjohann, T., Neuhaus, B. and A. Schmidt-Rhaesa. 2006. Head sensory organs of *Dactylopodola baltica* (Macrodasyida, Gastrotricha): a combination of transmission electron microscopical and immunocytochemical techniques. J. Morph. 267: 897–908.
Remane, A. 1929. Gastrotricha. Pp. 121–186 in W. Kükenthal and T. Krumbach (eds.), *Handbuch der Zoologie. Eine Naturgeschichte der Stämme des Tierreiches. 2. Band, 1. Hälfte, Vermes Amera.* Walter de Gruyter & Co., Berlin, Leipzig.
Ruppert, E. E. 1978. The reproductive system of gastrotrichs. II. Insemination in *Macrodasys*: a unique mode of sperm transfer in Metazoa. Zoomorphologie 89: 207–228.
Ruppert, E. E. 1988. Gastrotricha. Pp. 302–311 in R. P. Higgins and H. Thiel (eds.), *Introduction to the Study of Meiofauna.* Smithsonian Institution Press, Washington.
Ruppert, E. E. 1991. Gastrotricha. Pp. 41–109 in F. W. Harrison and E. E. Ruppert (eds.), *Microscopic Anatomy of Invertebrates. Vol. 4, Aschelminthes.* Wiley-Liss, New York.
Todaro, M. A., M. Dal Zotto, U. Jondelius, R. Hochberg, W. D. Hummon, T. Känneby and C. E. F. Rocha. 2012. Gastrotricha: a marine sister for a freshwater puzzle. PLoS ONE 7(2): e31740. doi: 10.1371/journal.pone.0031740
Weiss, M. J. 2001. Widespread hermaphroditism in freshwater gastrotrichs. Invert. Biol. 120: 308–341.

CHAPTER 12

Phylum Nemertea

The Ribbon Worms

Members of the phylum Nemertea (Greek, "a sea nymph") or Rhynchocoela (Greek *rhynchos*, "snout"; *coel*, "cavity") are commonly called ribbon worms. Figure 12.1 illustrates a variety of body forms within this taxon and the major features of their anatomy. These unsegmented vermiform animals are usually flattened dorsoventrally and are moderately cephalized; they possess highly extensible bodies. Many ribbon worm species are rather drab in appearance, but others are brightly colored or distinctively marked (e.g., the tropical eastern Pacific species *Baseodiscus punnetti*, above).

There are about 1,300 accepted, described species of nemerteans. They range in length from a few mm to several meters. Many can stretch easily to several times their contracted lengths (one specimen of *Lineus longissimus* reportedly measured 60 m in length). They are predominately benthic marine animals. A few, however, are planktonic, and some are symbiotic in molluscs, ascidians, or other marine invertebrates. A few freshwater and terrestrial species are known, the latter having some species of nearly cosmopolitan distribution.

Many features of the nemertean body plan (Box 12A) are similar to the conditions seen in flatworms (Platyhelminthes), and historically the two taxa were often grouped together as triploblastic acoelomate Bilateria. However, we now understand that nemerteans are truly coelomate, and that they are closely related to annelids, molluscs, and the lophophorate phyla. The similarities with the flatworms are reflected in the overall architecture of the nervous systems, the types of sense organs, and the protonephridial excretory structures. However, in other respects ribbon worms are more similar to the above-mentioned spiralian phyla. Nemerteans possess a through gut with an anus (a complete, one-way digestive tract), a closed circulatory system that is coelomic in nature, and an eversible proboscis surrounded by a hydrostatic cavity called the rhynchocoel. The circulatory system and the rhynchocoel are both coelomic cavities. The structure of

Classification of The Animal Kingdom (Metazoa)

Non-Bilateria*
(a.k.a. the diploblasts)
- PHYLUM PORIFERA
- PHYLUM PLACOZOA
- PHYLUM CNIDARIA
- PHYLUM CTENOPHORA

Bilateria
(a.k.a. the triploblasts)
- PHYLUM XENACOELOMORPHA

Protostomia
- PHYLUM CHAETOGNATHA

SPIRALIA
- PHYLUM PLATYHELMINTHES
- PHYLUM GASTROTRICHA
- PHYLUM RHOMBOZOA
- PHYLUM ORTHONECTIDA
- **PHYLUM NEMERTEA**
- PHYLUM MOLLUSCA
- PHYLUM ANNELIDA
- PHYLUM ENTOPROCTA
- PHYLUM CYCLIOPHORA

Gnathifera
- PHYLUM GNATHOSTOMULIDA
- PHYLUM MICROGNATHOZOA
- PHYLUM ROTIFERA

Lophophorata
- PHYLUM PHORONIDA
- PHYLUM BRYOZOA
- PHYLUM BRACHIOPODA

ECDYSOZOA

Nematoida
- PHYLUM NEMATODA
- PHYLUM NEMATOMORPHA

Scalidophora
- PHYLUM KINORHYNCHA
- PHYLUM PRIAPULA
- PHYLUM LORICIFERA

Panarthropoda
- PHYLUM TARDIGRADA
- PHYLUM ONYCHOPHORA
- PHYLUM ARTHROPODA
  - SUBPHYLUM CRUSTACEA*
  - SUBPHYLUM HEXAPODA
  - SUBPHYLUM MYRIAPODA
  - SUBPHYLUM CHELICERATA

Deuterostomia
- PHYLUM ECHINODERMATA
- PHYLUM HEMICHORDATA
- PHYLUM CHORDATA

*Paraphyletic group

This chapter has been revised by Gonzalo Giribet.

the proboscis apparatus is unique to nemerteans and represents a novel synapomorphy that distinguishes the Nemertea from all other invertebrate taxa.

Taxonomic History and Classification

The earliest report of a nemertean was that of William Borlase (1758), who described his specimen as "the sea long-worm" and categorized it "among the less perfect kind of sea animals." For nearly a century, most authors placed the ribbon worms with the turbellarian flatworms, although other writers suggested that they were allied to annelids (including sipunculans), nematodes, and even molluscs and insects. It was during this period that Georges Cuvier (1817) described a particular ribbon worm and called it *Nemertes*, from which the phylum name was eventually derived. However, it was not until 1851 that substantial evidence for the distinctive nature of the ribbon worms was published by Max Schultze, who described the functional morphology of the proboscis, established the presence of nephridia and an anus, and discussed many other features of these animals. Schultze even proposed the basis for classifying these worms, which is still largely employed today by most authorities. Interestingly, he persisted in considering them to be turbellarians, but he coined the names Nemertina and Rhynchocoela. Charles Minot separated the nemerteans from the flatworms in 1876, but it was not until the mid-twentieth century that the unique combination of characters displayed by the ribbon worms was fully accepted. Since that time they have been treated as a valid phylum. It was in the mid-1980s that James M. Turbeville, studying their ultrastructure, determined that nemerteans are more closely related to the coelomate protostomes than to platyhelminths, a result later corroborated by molecular sequence data and by the mode of development.

Classification

Since the publication of Schultze's classic accounts, the primary effort of nemertean taxonomists has been to refine the details of his scheme, which has generated surprisingly few controversies. The classification scheme used here is mostly based on the traditional one established by Wesley Coe in 1943 and refined with molecular phylogenetic analyses.

For most of the last 100 years, the traditional classification system of nemerteans has largely followed that of Gerarda Stiasny-Wijnhoff (1936), which accepted as classes Schultze's (1851) division of nemerteans into the classes Anopla and Enopla. Stiasny-Wijnhoff divided Anopla into Palaeonemertea and Heteronemertea, and Enopla into Hoplonemertea and Bdellonemertea. Hoplonemertea was further subdivided into Monostilifera and Polystilifera. Recent accounts of the systematics of Nemertea found a division between Palaeonemertea and Neonemertea, with Hubrechtidae as the sister group of Heteronemertea, constituting the clade Pilidiophora (see classification that follows). Furthermore, all molecular analyses have contradicted the ordinal status of Bdellonemertea, which is deeply nested within Hoplonemertea.

We therefore follow a system in which the phylum is subdivided into two main clades, Palaeonemertea and Neonemertea, the latter divided into Pilidiophora (including the traditional order Heteronemertea and the family Hubrechtidae) and Hoplonemertea (= Enopla of older classifications). The principal anatomical features used to distinguish between these clades include proboscis armature, mouth location relative to the position of the cerebral ganglion, gut shape, layering of the body wall muscles, position of the longitudinal nerve cords, and several larval features.

BOX 12A Characteristics of the Phylum Nemertea

1. Marine, freshwater, or terrestrial
2. Triploblastic, coelomate, bilaterally symmetrical unsegmented worms
3. Digestive tract complete, with an anus
4. With protonephridia (a few deep-sea pelagic species lack excretory systems)
5. With bilobed cerebral ganglion that surrounds proboscis apparatus (not the gut), and two or more longitudinal nerve cords connected by transverse commissures
6. With two or three layers of body wall muscles arranged in various ways
7. With a unique proboscis apparatus lying dorsal to the gut and surrounded by a coelomic hydrostatic chamber called the rhynchocoel
8. With a closed circulatory system; some species with hemoglobin
9. Most are gonochoristic; cleavage holoblastic; early development typically spiralian, and either direct or indirect
10. Asexual reproduction by fragmentation is not uncommon

PHYLUM NEMERTEA (= RHYNCHOCOELA)

CLASS PALAEONEMERTEA Unarmed nemerteans (Figure 12.1B). Proboscis not armed with stylets and not morphologically specialized into three regions. Mouth separate from proboscis pore and located directly below or somewhat posterior to cerebral ganglion. Two or three layers of body wall muscles, from external to internal either

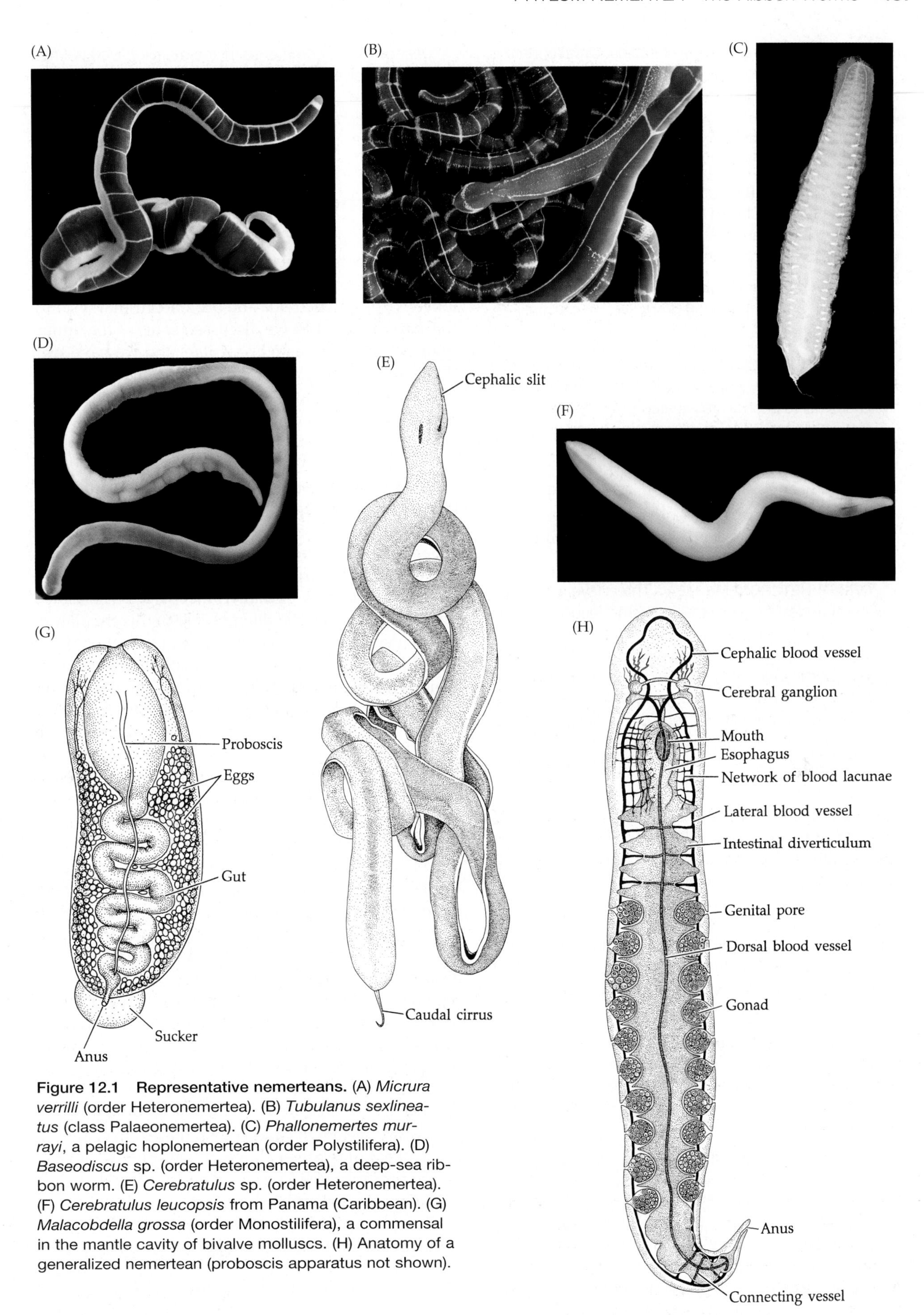

Figure 12.1 Representative nemerteans. (A) *Micrura verrilli* (order Heteronemertea). (B) *Tubulanus sexlineatus* (class Palaeonemertea). (C) *Phallonemertes murrayi*, a pelagic hoplonemertean (order Polystilifera). (D) *Baseodiscus* sp. (order Heteronemertea), a deep-sea ribbon worm. (E) *Cerebratulus* sp. (order Heteronemertea). (F) *Cerebratulus leucopsis* from Panama (Caribbean). (G) *Malacobdella grossa* (order Monostilifera), a commensal in the mantle cavity of bivalve molluscs. (H) Anatomy of a generalized nemertean (proboscis apparatus not shown).

circular-longitudinal or circular-longitudinal-circular; dermis thin and gelatinous, or absent; longitudinal nerve cords epidermal, dermal, or intramuscular within the longitudinal layer; cerebral organs and ocelli frequently lacking. Palaeonemerteans are marine, primarily littoral forms. (e.g., *Carinoma, Cephalothrix, Tubulanus*)

CLASS NEONEMERTEA Unarmed or armed nemerteans with separate or fused mouth and proboscis pores. Because Hubrechtidae was formerly placed in Palaeonemertea, most characters within Neonemertea are variable.

SUBCLASS PILIDIOPHORA Nemerteans with a pilidium larva (when present) and the proboscis musculature bilaterally arranged.

ORDER HUBRECHTIDA Unarmed nemerteans with separate mouth and proboscis pores. The brain and lateral nerve chords are situated below the epidermis in the connective tissue, whereas the cerebral sensory organ occupies a position relative to the brain and lateral blood vessel similar to that of heteronemerteans. Historically, hubrechtiids were regarded as a "transitional stage" between palaeonemerteans and basal heteronemerteans. All marine. (e.g., *Hubrechtia, Hubrechtella*)

ORDER HETERONEMERTEA Three layers of body wall muscles, from external to internal longitudinal-circular-longitudinal; dermis usually thick, partly fibrous; longitudinal nerve cords intramuscular, between outer longitudinal and middle circular layers; cerebral organs and ocelli usually present; development indirect. These nemerteans are primarily marine littoral forms, but a few freshwater species are known (Figure 12.1A,D–F). (e.g., *Baseodiscus, Cerebratulus, Lineus, Micrura, Nemertoscolex, Paralineus*)

SUBCLASS HOPLONEMERTEA (= ENOPLA) Typically armed nemerteans (Figure 12.1C,G). Proboscis usually armed with distinct stylets and morphologically specialized into three regions (except in *Bdellonemertea*); mouth and proboscis pore usually united into a common aperture; mouth located anterior to cerebral ganglion; longitudinal nerve cords within mesenchyme, internal to body wall muscles. With marine, freshwater and terrestrial species; many marine species symbiotic with or parasitic upon other invertebrates.

ORDER POLYSTILIFERA Stylet apparatus consists of many small stylets borne on a basal shield. All species are marine, either benthic or pelagic. (e.g., *Hubrechtonemertes, Nectonemertes, Pelagonemertes, Phallonemertes*)

ORDER MONOSTILIFERA Stylet apparatus consists of a single main stylet and two or more sacs housing accessory (replacement) stylets; proboscis apparatus is unarmed and opens into foregut in *Malacobdella*, which has a trunk with a large posterior sucker and a convoluted gut that lacks lateral diverticula. Most species are marine and benthic, but freshwater, terrestrial, and parasitic forms are known; *Malacobdella* species are commensal in the mantle cavities of marine bivalves and, in one species, a freshwater gastropod. (e.g., *Amphiporus, Annulonemertes, Carcinonemertes, Emplectonema, Geonemertes, Malacobdella, Ovicides, Paranemertes*)

The Nemertean Body Plan

The nemertean body plan is especially interesting, since it presents characteristics of both coelomates (the closed circulatory system and rhynchocoel) and acoelomates (the parenchyma and protonephridial system). In Chapters 4 and 10 we discussed some of the limitations of the acoelomate body plan, and we have seen the results of these constraints in our examination of the flatworms. It might be said that the ribbon worms have made the best of a rather difficult situation. Even though it is now clear that nemerteans have true coelomic cavities, these worms have relatively solid bodies. Thus they are at least functionally acoelomate. Recall that many of the problems inherent in acoelomate architecture are related to restricted internal transport capabilities. The presence of a circulatory system in nemerteans has largely eased this problem, and the functional anatomy of many other systems is related directly or indirectly to the presence of this circulatory mechanism. For example, nemertean protonephridia are usually intimately associated with the blood, from which wastes are drawn, rather than with the mesenchymal tissues as are flatworm protonephridia.

The increased capabilities for internal circulation and transport have allowed a number of developments that would otherwise be impossible. First, the circulatory system provides a solution to the surface-to-volume dilemma, and as a result nemerteans tend to be much larger and more robust than flatworms, having been largely relieved of the constraints of relying on diffusion for internal transport and exchange. Second, the digestive tract is complete and somewhat regionally specialized. With a one-way movement of food materials through the gut, and a circulatory system to absorb and distribute digested products, the anterior region of the gut has been freed for feeding and ingestion. Third, since the animal does not have to rely on diffusion for transport through a loosely organized mesenchyme, that general body area is freed for the development of other structures, notably the well developed layers of muscles. In summary, the presence of a circulatory system in concert with these other changes has resulted in relatively large, active animals, capable of more complex feeding and digestive activities than seen in most acoelomate metazoans.

This general body plan is enhanced by the presence of the unique proboscis apparatus (which usually functions in prey capture), the distinctly anterior location of the mouth, and well-developed cephalic sensory organs for prey location. Thus, while variation exists, the

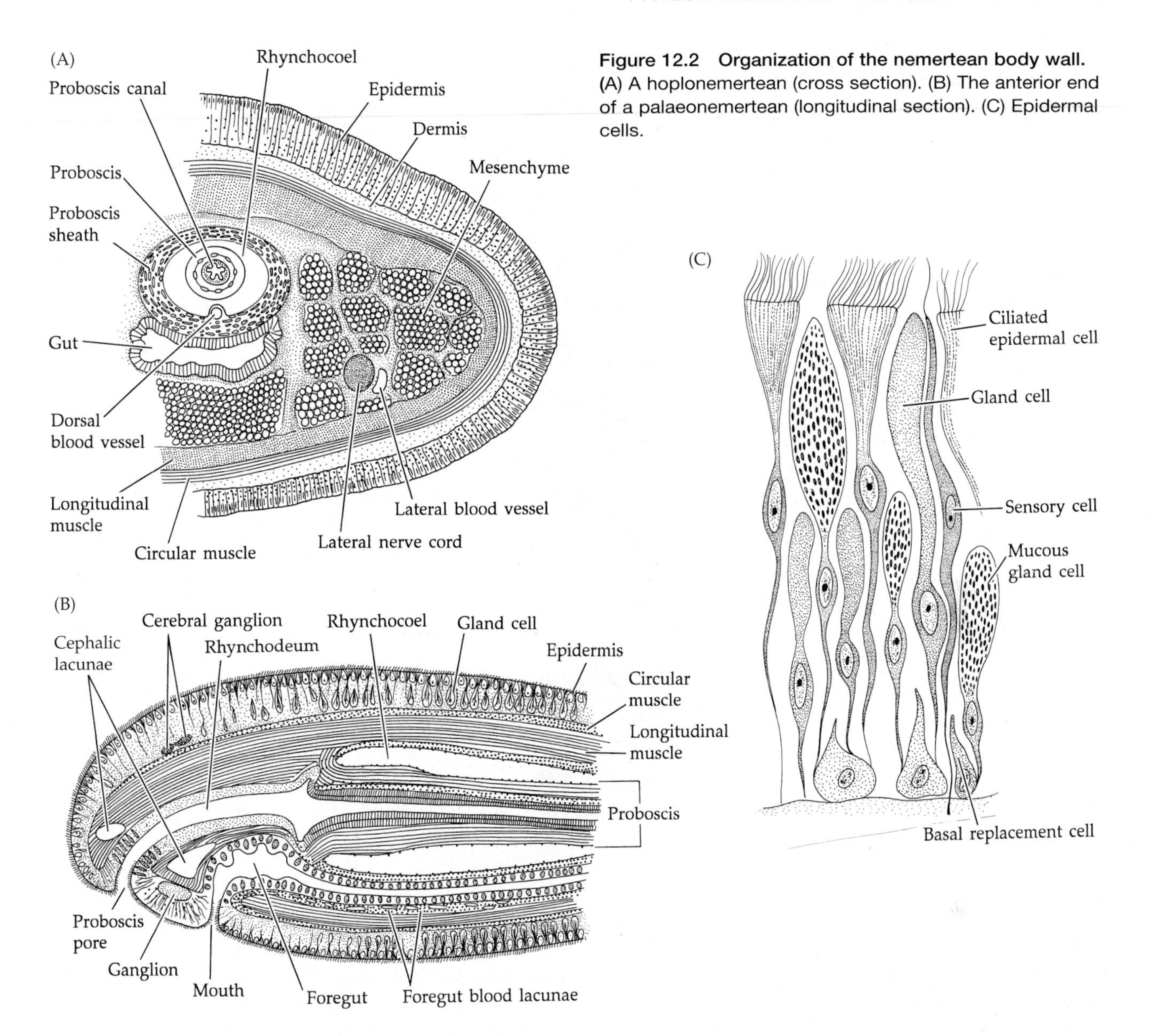

Figure 12.2 Organization of the nemertean body wall. (A) A hoplonemertean (cross section). (B) The anterior end of a palaeonemertean (longitudinal section). (C) Epidermal cells.

"typical" nemertean may be viewed as an active benthic hunter/tracker that moves among nooks and crannies preying on other invertebrates or even on some vertebrate prey and vertebrate remains.

Body Wall

The body wall of nemerteans comprises an epidermis, a dermis, relatively thick muscle layers surrounding the gut and other internal organs, and a mesenchyme of varying thickness (Figure 12.2). The epidermis is a ciliated columnar epithelium (Figure 12.2C). Mixed among the columnar cells are sensory cells (probably tactile), mucous gland cells, and basal replacement cells that may extend beneath the epidermis. Below the epidermis is the dermis, which varies greatly in thickness and composition. In some ribbon worms (e.g., the palaeonemerteans) the dermis is extremely thin or composed of only a homogeneous gel-like layer; in others (e.g., the heteronemerteans), it is typically quite thick and densely fibrous and usually includes a variety of gland cells. Beneath the dermis are well-developed layers of circular and longitudinal muscles. The organization of these muscles varies among taxa and may occur in either a two- or three-layered plan (Figure 12.3). The layering arrangement may also vary to some degree along the body length of individual animals. Internal to the muscle layers is a dense, more or less solid mesenchyme, although in some nemerteans the muscle layers are so thick that they nearly obliterate this inner mass. The mesenchyme includes a gel matrix and often a variety of loose cells, fibers, and dorsoventrally oriented muscles. Figure 12.3 depicts cross-sectional views of representatives of the two classes, showing mesenchyme thickness, muscles, placement of longitudinal nerve cords, major longitudinal blood vessels, and other features.

Support and Locomotion

In the absence of any rigid skeletal elements, the support system of nemerteans is provided by the muscles

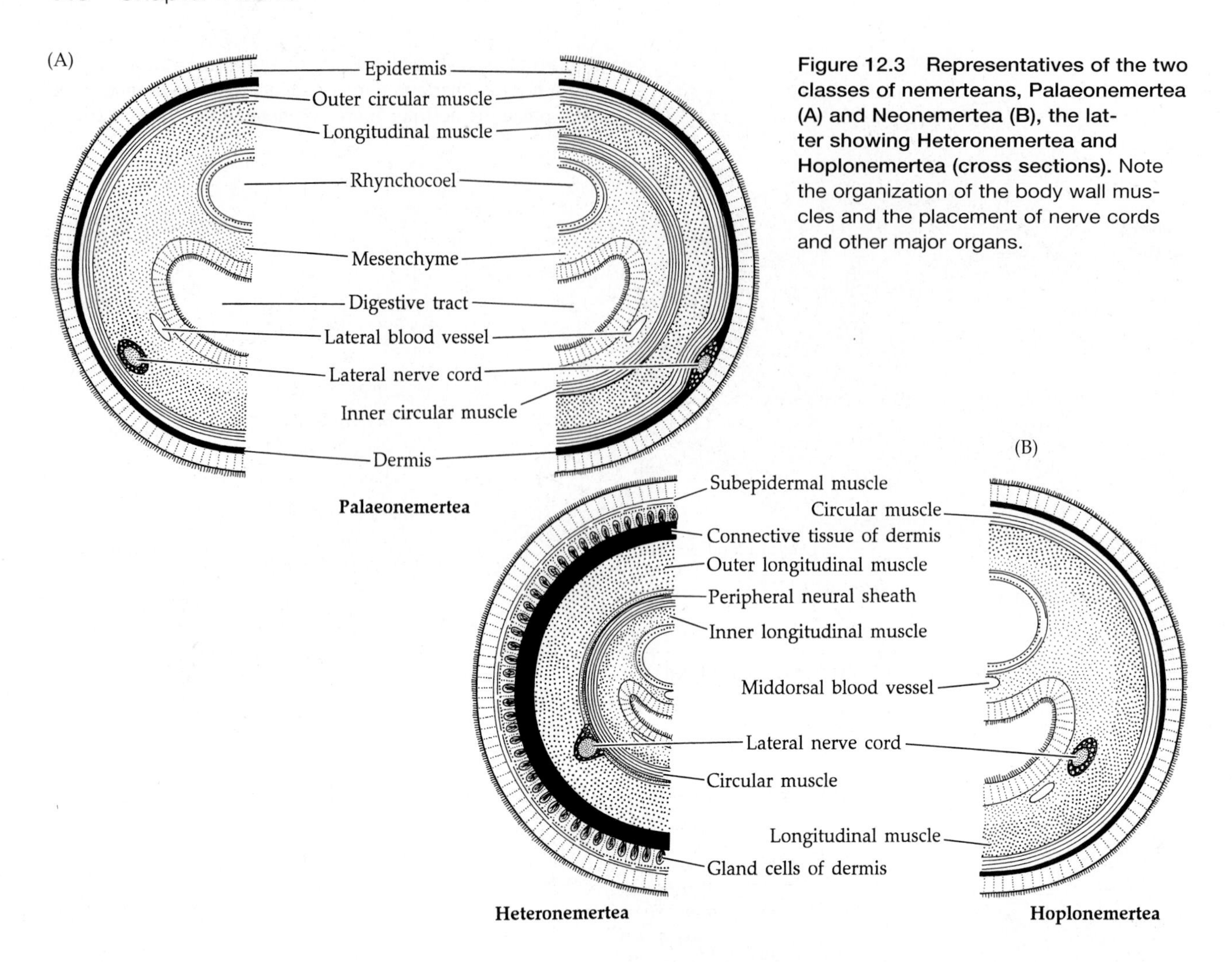

Figure 12.3 Representatives of the two classes of nemerteans, Palaeonemertea (A) and Neonemertea (B), the latter showing Heteronemertea and Hoplonemertea (cross sections). Note the organization of the body wall muscles and the placement of nerve cords and other major organs.

and other tissues of the body wall and by the hydrostatic qualities of the mesenchyme. These features permit dramatic changes in both length and cross-sectional shape and diameter, characteristics that are closely associated with locomotion and accommodation to cramped quarters. Most very small benthic ribbon worms are propelled by the action of their epidermal cilia. A slime trail is produced by the body wall mucous glands and provides a lubricated surface over which the worm slowly glides. Small nemerteans commonly live among the interstices of filamentous algae, under tidepool rocks, or in the spaces of other irregular surfaces such as those found in mussel beds and sand, mud, or pebble bottoms. Larger epibenthic ribbon worms and most of the burrowing forms employ peristaltic waves of the body wall muscles to propel them over moist surfaces or through soft substrata. Some of the larger forms (e.g., *Cerebratulus*) use undulatory swimming as a secondary means of locomotion, and perhaps as an escape reaction to benthic predators. Fully pelagic nemerteans (certain polystiliferan hoplonemerteans) generally drift or swim slowly. Some of the terrestrial forms produce a slime sheath through which they glide by ciliary action, and some use their proboscis for rapid escape responses.

Feeding and Digestion

Most ribbon worms are active predators on small invertebrates, but some are able to capture live fish, or even active invertebrates like cephalopods. Some are scavengers, feeding on all sorts of decaying animal matter, including large vertebrates, and yet others feed on plant material (at least under laboratory conditions). There is evidence to suggest that species of the commensal genus *Malacobdella,* which inhabit the mantle cavity of bivalve molluscs, feed largely on phytoplankton captured from their host's feeding and gas exchange currents. Field observations indicate that the diets of predatory forms may be either extremely varied or quite restricted, depending on the species. Some species are capable of tracking prey over long distances, whereas others must locate food by direct contact. Distance prey location and assessment of food acceptability are almost certainly chemotactic responses. Ribbon worms that actually hunt and track can recognize the trails left by potential prey, and they fire their proboscis along the trail ahead of them to capture food (Figure 12.4). Similar reactions are elicited when infaunal nemerteans encounter burrows in which potential prey might be located. Surface hunters that live in intertidal areas generally forage during high tides or

Figure 12.4 ***Paranemertes peregrina*** **(order Hoplonemertea) capturing a nereid polychaete.** The proboscis is coiled around the polychaete.

at night, and thus avoid the threats of desiccation and visual predators. However, members of some marine genera (e.g., *Tubulanus, Paranemertes, Amphiporus*) may frequently be seen during low tides on foggy mornings, gliding over the substratum in search of prey. The rapid expulsion of their proboscis and successful prey capture can be a memorable moment of high drama for tidepool enthusiasts.

The behavior involved in the capture and ingestion of live prey is significantly different from that associated with scavenging on dead material. In predation, the proboscis is employed both in capturing prey and in moving it to the mouth for ingestion. The proboscis is everted and wrapped around the victim (Figure 12.4). The prey is not only physically held by the proboscis but may be subdued or killed by its toxic secretions. In the Pacific species *Paranemertes peregrina,* which feeds primarily on nereid polychaetes, the glandular epithelium of the everted proboscis secretes a potent neurotoxin. Nemerteans with an armed proboscis (Hoplonemertea) actually use the stylets to pierce the prey's body (often numerous times) to introduce the toxin. Once captured, the prey is drawn to the mouth by retraction and manipulation of the proboscis; it is usually swallowed whole. The mouth is expanded and pressed against the food, and swallowing is accomplished by peristaltic action of the body wall muscles aided by ciliary currents in the anterior region of the gut. Scavenging, in contrast, usually does not involve the proboscis. The worm simply ingests the food directly by muscular action of body wall and foregut. In some predatory hoplonemerteans (those in which the lumen of the proboscis is connected with the anterior gut lumen), the foregut itself may be everted for feeding on animals too large to be swallowed whole. In such cases, fluids and soft tissues are generally sucked out of the prey's body. Polychaete worms and amphipod crustaceans seem to be favorite food items for many predatory nemerteans.

Species of the hoplonemertean genus *Carcinonemertes* are ectoparasites (egg predators) on brachyuran crabs. Different species inhabit different regions of the host's body, but all migrate to the egg masses on gravid female crabs and feed on the yolky eggs. In high numbers, these egg predators can kill all of the embryos in the host's clutch. Some studies have reported up to 99% infestation rates of *Carcinonemertes errans* on the commercially important Pacific Dungeness crab (*Cancer magister*), with up to 100,000 worms per host. This parasite has been implicated in past collapses of the central California Dungeness crab fishery. Several species of *Ovicides* (Carcinonemertidae) live on the abdomens of hydrothermal vent crabs in the Pacific, and *Nemertoscolex parasiticus* (Heteronemertea) lives in the coelomic fluid of the echiuran *Echiurus echiurus.*

The proboscis apparatus The proboscis apparatus is a complex arrangement of tubes, muscles, and hydraulic systems (Figure 12.5). The proboscis itself is an elongate, eversible, blind tube, and either is associated with the foregut or opens through a separate proboscis pore. The proboscis, which may be branching in some species, may be regionally specialized and bears stylets in various arrangements (Figures 12.5E–H). Nemertean stylets are nail-shaped structures that typically reach lengths of 50–200 μm. Each calcified stylet is composed of a central organic matrix surrounded by an inorganic cortex composed of calcium phosphate. The stylets are formed within large epithelial cells called styletocytes. Because growing ribbon worms must replace their stylets with new larger ones, and because they often lose the stylet during prey capture, new stylets are continuously produced in reserve stylet sacs and stored until needed, whereupon they are transported and affixed in their proper position.

The basic structure and action of the proboscis are most easily described where the apparatus is entirely separate from the gut. As shown in Figures 12.5A and B, the proboscis pore leads from the outside directly into the anterior proboscis lumen, called the rhynchodeum, the lining of which is continuous with the epidermis. Posterior to the rhynchodeum, the lumen continues as the proboscis canal that is surrounded by the muscular wall of the proboscis itself; these muscles are derived from the muscles of the body wall. The proboscis is surrounded by a closed, fluid-filled, coelomic space called the rhynchocoel, which in turn is surrounded by additional muscle layers. The inner blind end of the proboscis is connected to the posterior wall of the rhynchocoel by a proboscis retractor muscle. In a few taxa (e.g., *Gorgonorhynchus*), there is no retractor muscle, and eversion and retraction are accomplished hydrostatically.

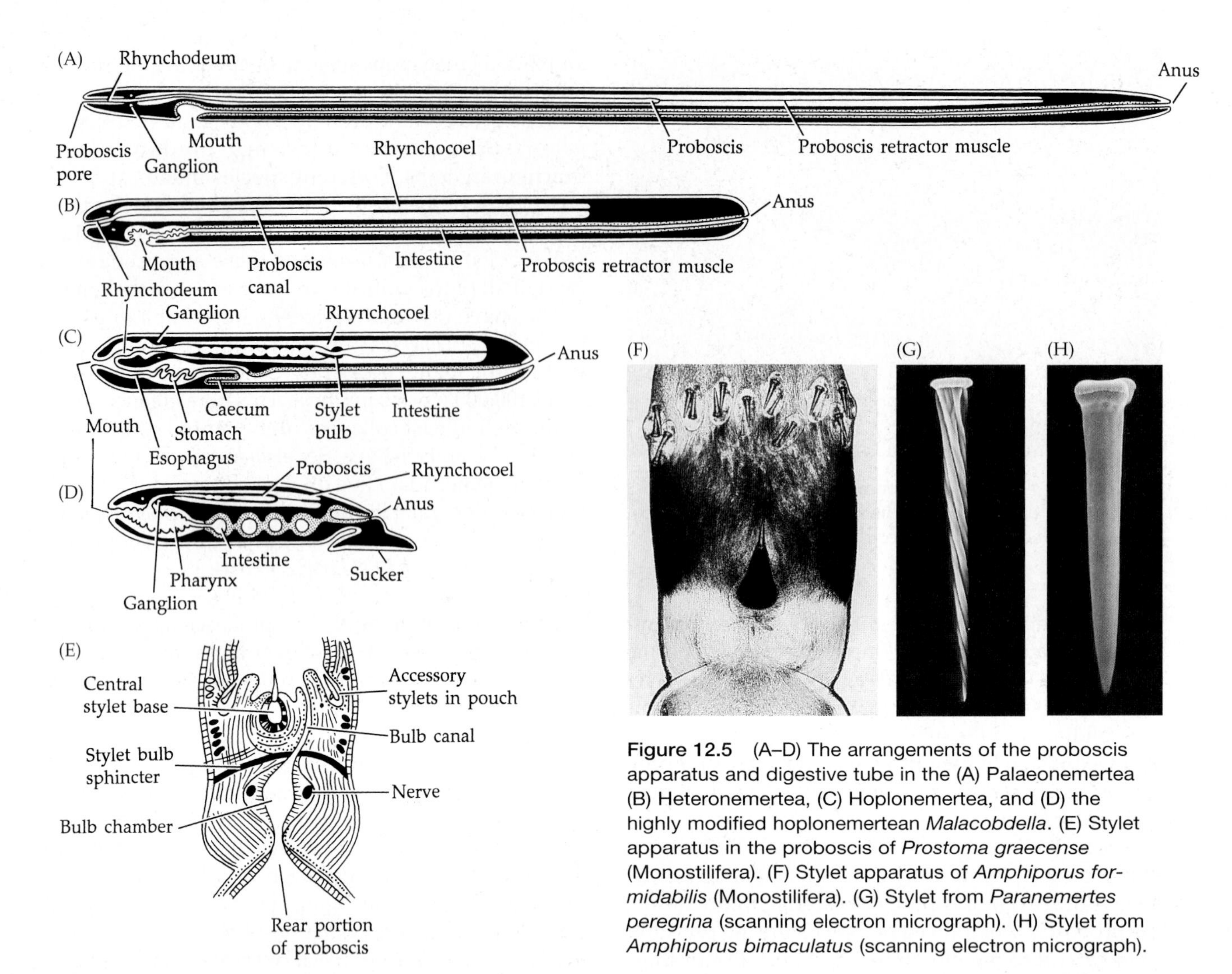

Figure 12.5 (A–D) The arrangements of the proboscis apparatus and digestive tube in the (A) Palaeonemertea (B) Heteronemertea, (C) Hoplonemertea, and (D) the highly modified hoplonemertean *Malacobdella*. (E) Stylet apparatus in the proboscis of *Prostoma graecense* (Monostilifera). (F) Stylet apparatus of *Amphiporus formidabilis* (Monostilifera). (G) Stylet from *Paranemertes peregrina* (scanning electron micrograph). (H) Stylet from *Amphiporus bimaculatus* (scanning electron micrograph).

Eversion of the proboscis (Figure 12.6) is accomplished by contraction of the muscles around the rhynchocoel; this increases the hydrostatic pressure within the rhynchocoel itself, squeezing on the proboscis and causing its eversion. The everted proboscis moves with the muscles in its wall; the proboscis is retracted back inside the body by the coincidental relaxation of the muscles around the rhynchocoel and contraction of the proboscis retractor muscle. The retracted proboscis may extend nearly to the posterior end of the worm, and usually only a portion of it is extended during eversion.

Digestive system Nemerteans have a complete through gut with an anus (Figures 12.6 and 12.7). Associated with the one-way movement of food from mouth to anus we find various degrees of regional specialization (both structural and functional) in the guts of ribbon worms. The mouth leads inward to an ectodermally derived foregut (stomodeum) consisting of a bulbous buccal cavity, sometimes a short esophagus, and a stomach. The stomach leads to an elongate intestine or midgut, which is more or less straight but usually bears numerous lateral diverticula. In *Malacobdella*, the intestine is loosely coiled and lacks diverticula; diverticula are also lacking in the strange "segmented" *Annulonemertes*. At the posterior end of the intestine is a short ectodermally derived hindgut (proctodeum) or rectum, which terminates in the anus. Elaborations on this basic plan are common in certain taxa and may include various ceca arising from the stomach or from the intestine at its junction with the foregut.

The entire digestive tube is ciliated, the foregut more densely than the midgut. The gut epithelium is basically columnar, mixed with gland cells. The foregut contains a variety of mucus-producing cells, sometimes multicellular mucous glands, and occasionally enzymatic gland cells in the stomach region. The midgut is lined with vacuolated ciliated columnar cells; these are phagocytic and bear microvilli, greatly increasing their surface area. Enzymatic gland cells are abundantly mixed with the ciliated cells of the midgut. The hindgut typically lacks gland cells. Food is moved

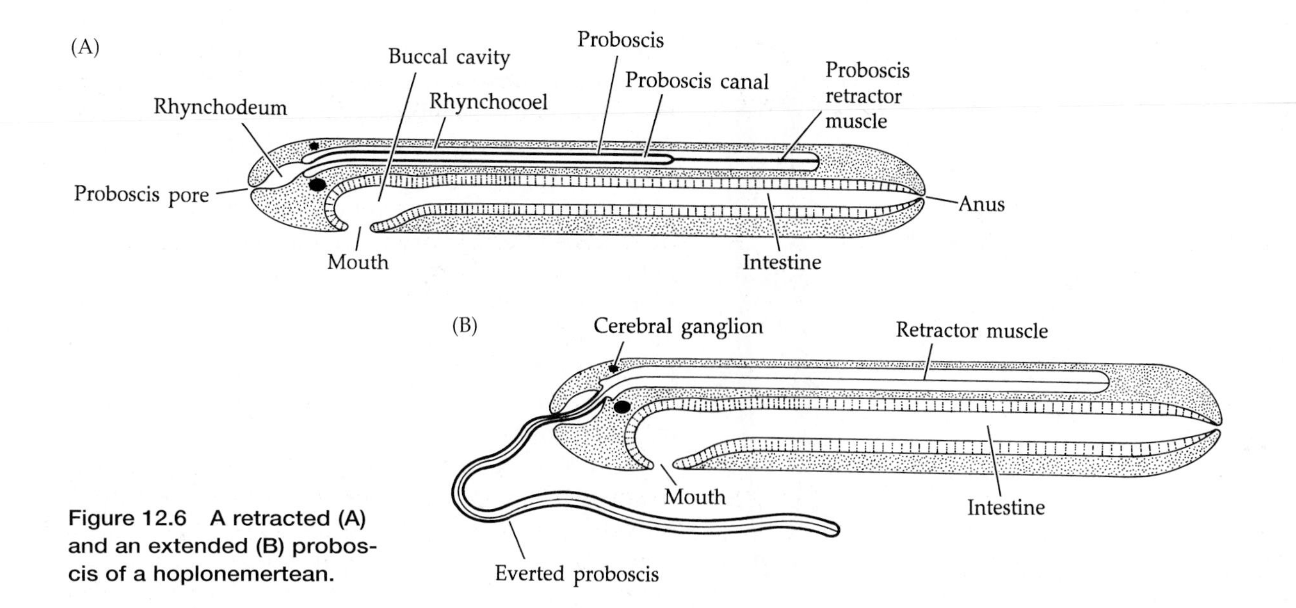

Figure 12.6 A retracted (A) and an extended (B) proboscis of a hoplonemertean.

through the digestive tract by cilia and by the action of the body wall muscles; there are usually no muscles in the gut wall itself, except in the foregut of some heteronemerteans.

The process of digestion in carnivorous nemerteans is a two-phase sequence of protein breakdown. The first step involves the action of endopeptidases released from gland cells into the gut lumen. This extracellular digestion is quite rapid and is followed by phagocytosis (and probably pinocytosis) of the partially digested material by the ciliated columnar cells of the midgut. Protein digestion is completed intracellularly by exopeptidases within the food vacuoles of the midgut epithelium. Lipases have been discovered in at least one species (*Lineus ruber*), and carbohydrases are known in the omnivorous commensal *Malacobdella*. Food is stored primarily in the form of fats, and to a much lesser extent as glycogen, in the wall of the midgut. Transportation of digested materials throughout the body is accomplished by the circulatory system, which absorbs these products from the cells lining the intestine. Indigestible materials are moved through the gut and out the anus.

Circulation and Gas Exchange

We have mentioned briefly the evolutionary and adaptive significance of the circulatory system in nemerteans and its general relationship to other systems and functions. This closed system consists of vessels and thin-walled spaces called lacunae (Figure 12.2B). There is a good deal of variation in the architecture of nemertean circulatory systems (Figure 12.8). The simplest arrangement occurs in certain palaeonemerteans in which a single pair of longitudinal vessels extends the length of the body, connecting anteriorly by a cephalic lacuna and posteriorly by an anal lacuna. Elaboration on this basic scheme may include transverse vessels

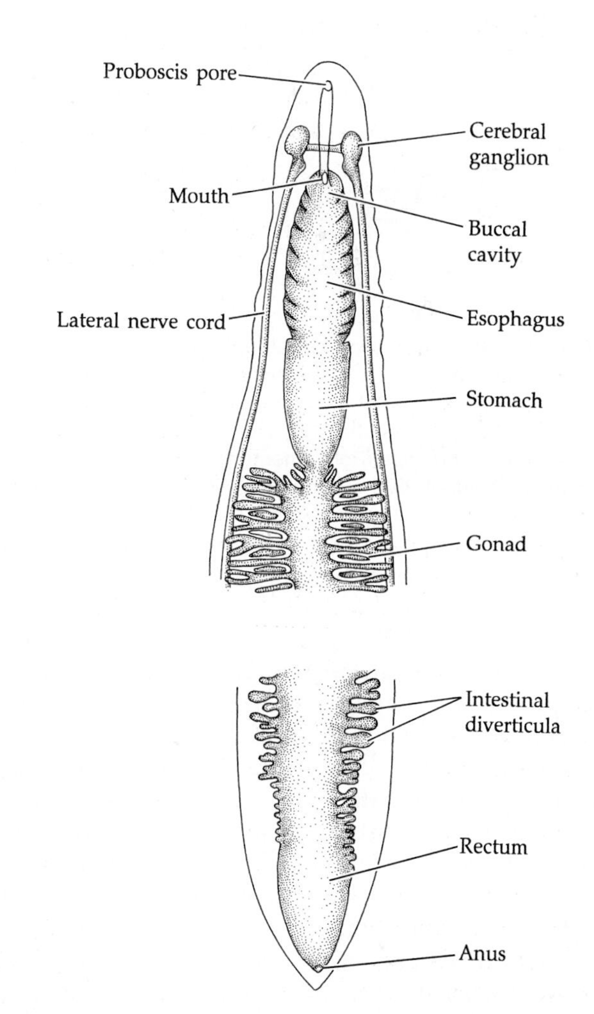

Figure 12.7 A nemertean digestive system. Anterior and posterior regions of the gut of *Carinoma* (ventral view).

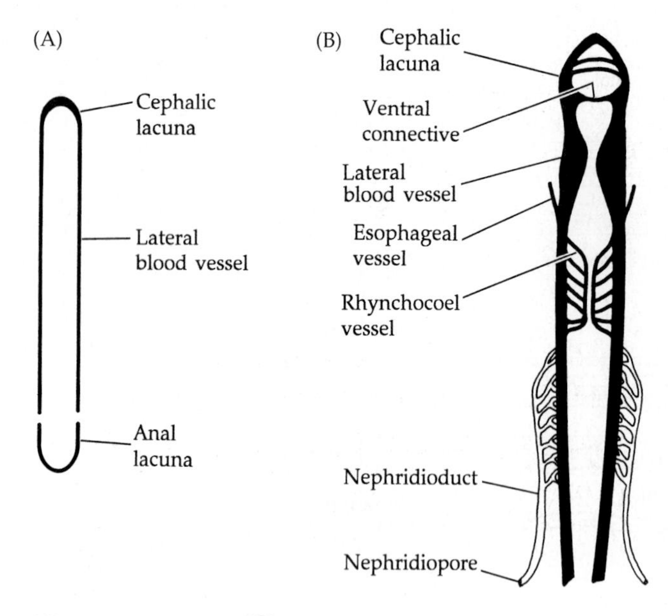

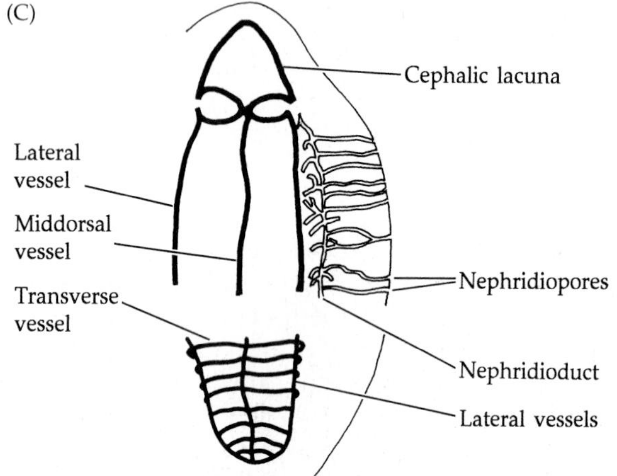

Figure 12.8 Nemertean circulatory systems. (A) The simple circulatory loop of *Cephalothrix* (Palaeonemertea) consists of a pair of lateral blood vessels connected by cephalic and anal lacunae. (B) The complex circulatory system of *Tubulanus* (Palaeonemertea). Note the intimate association of the nephridial system with the lateral blood vessels. (C) The circulatory system of *Amphiporus* (Hoplonemertea) includes a middorsal vessel and numerous transverse vessels.

between the longitudinal vessels, enlargement and compartmentalization of the lacunar spaces, and the addition of a middorsal vessel. The walls of the blood vessels are only slightly contractile, and general body movements generate most of the blood flow. There is no consistent pattern to the movement of blood through the system; it may flow either anteriorly or posteriorly in the longitudinal vessels, and currents often reverse directions.

The blood consists of a colorless fluid in which various cells are suspended. These cells can include pigmented corpuscles (yellow, orange, green, red), at least some of which contain hemoglobin, and a variety of so-called lymphocytes and leukocytes of uncertain function. The anatomical association of the circulatory system with other structures, as well as the composition of the blood, suggest several circulatory functions. Although conclusive evidence is lacking, the circulatory system appears to be involved with the transport of nutrients, gases, neurosecretions, and excretory products. Some intermediary metabolism probably occurs in the blood, as several appropriate enzymes have been identified in solution. The blood may also serve as an aid to body support through changes in hydrostatic pressure within the vessels and lacunar spaces. There is some evidence to support the idea that the blood may also function in osmoregulation.

Gas exchange in nemerteans is epidermal and does not involve any special structures. Oxygen and carbon dioxide diffuse readily across the moist body surface, which is usually covered with mucous secretions. Some robust forms (e.g., *Cerebratulus*) augment this passive exchange of gases across the skin with regular irrigation of the foregut, where there is an extensive system of blood vessels. In those species in which hemoglobin occurs, this pigment probably aids in oxygen transport or storage within the blood.

Excretion and Osmoregulation

The excretory system of most nemerteans consists of two to thousands of flame bulb protonephridia (Figures 12.8 and 12.9) similar to those found in free-living flatworms. However, apparently the deep-sea pelagic hoplonemerteans lack protonephridia altogether. The flame bulbs are usually intimately associated with the lateral blood vessels or less commonly with other parts of the circulatory system. The nephridial units are often pressed into the blood vessel walls, and in some instances the walls are actually broken down so that the nephridia are bathed directly in blood. In the simplest case, a single pair of flame bulbs leads to two nephridioducts, each with its own laterally placed nephridiopore. More complex conditions include rows of single flame bulbs or clusters of flame bulbs with multiple ducts. In some species the walls of the nephridioducts are syncytial and lead to hundreds or even thousands of pores on the epidermis. The most elaborate conditions occur in certain terrestrial nemerteans where approximately 70,000 clusters of flame bulbs (six to eight in each cluster) lead to as many surface pores. In some heteronemerteans (e.g., *Baseodiscus*), the excretory system discharges into the foregut.

The functioning of nemertean protonephridia in the excretion of metabolic wastes has not been well studied. The close association of the flame bulbs with the circulatory system suggests that nitrogenous wastes (probably ammonia), excess salts, and other metabolic products are removed from the blood as well as from the surrounding mesenchyme by the nephridia. If such is the case, it explains again the significance of the circulatory system in overcoming surface-to-volume

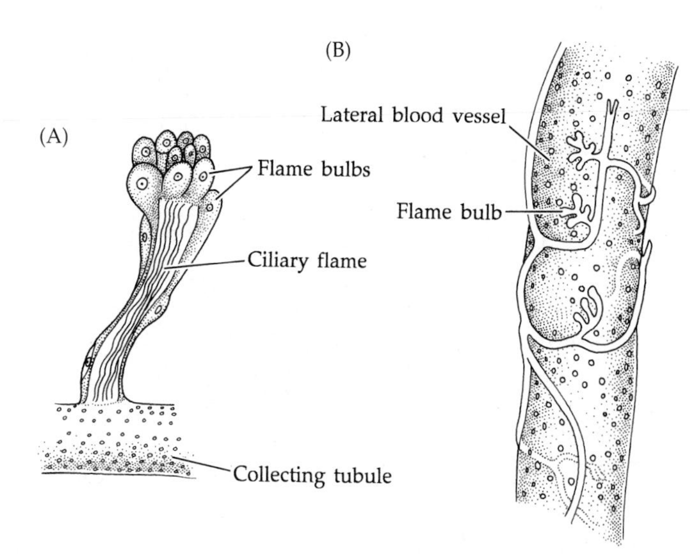

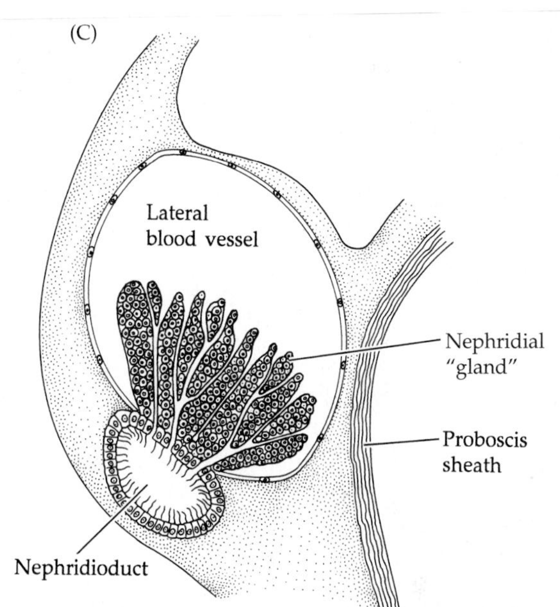

Figure 12.9 Nemertean excretory systems (see also Figure 12.8). (A) A protonephridial cluster of *Drepanophorus* (Hoplonemertea). (B) Nephridial ducts associated with a lateral blood vessel in *Amphiporus* (Hoplonemertea). (C) Excretory system of *Carinina* (Palaeonemertea) in which the secretory units (so-called nephridial gland) project into the lumen of the lateral blood vessel.

problems and the constraints of simple diffusion on body size. Relatively active animals produce large amounts of metabolic wastes. Dependence on diffusion alone would seriously limit any increase in body bulk, but the transport of these wastes from the tissues to the protonephridial system by circulatory vessels greatly eases this limitation. One of the most remarkable evolutionary achievements of the nemerteans has been their ability to grow to great size, particularly in length, without segmentation or the development of a large body cavity.

There is some morphological and experimental evidence that the protonephridia also play an important role in osmoregulation, especially in freshwater and terrestrial ribbon worms. It is in some of these forms, which are subjected to extreme osmotic stress, that the most elaborate excretory systems are found and these systems are probably associated with water balance. Furthermore, it appears that there may be a very complex interaction between the nervous system (neurosecretions), the circulatory system, and the nephridia to facilitate osmoregulatory mechanisms, but the details remain to be studied. Some members of Heteronemertea and Hoplonemertea have invaded fresh water and must combat water influx from their strongly hypotonic surroundings. Members of some genera (e.g., *Geonemertes*) are terrestrial, although restricted to moist shady habitats where they avoid serious problems of desiccation. In addition, they tend to cover their bodies with a mucous coat that reduces water loss. Those forms that inhabit marine subtidal or deep-water environments, or are endosymbiotic (one genus of Heteronemertea and several genera of Hoplonemertea), face little or no osmotic stress. But the many species found intertidally do face periods of exposure to air and to lowered (or elevated) salinities. Their soft bodies are largely unprotected, and they are relatively intolerant of fluctuations in environmental conditions. Intertidal nemerteans rely strongly on behavioral attributes to survive periods of potential osmotic stress and remain in moist areas during low tide periods. Burrowing in soft, water-soaked substrata, or living among algae or mussel beds, in cracks and crevices, or other areas that retain seawater at low tide are lifestyles illustrating how habitat preference and behavior prevent exposure to stress. In addition, most intertidal nemerteans are somewhat negatively phototactic, and many restrict their activities to night hours or to foggy or overcast mornings and evenings. A marine meiofaunal species from North Carolina lives in sediments at about 1 m depth above high tide level, probably relying on the water that fills the interstices of sand by capillarity.

Nervous System and Sense Organs

The basic organization of the nemertean nervous system reflects a relatively active lifestyle. Nemerteans are cephalized, especially in the anterior placement of the mouth and feeding structures, and we find related concentrations of sensory and other nervous elements in the head. The central nervous system of ribbon worms consists of a complex cerebral ganglion from which arises a pair of ganglionated, longitudinal (lateral) nerve cords (Figure 12.10A). The cerebral ganglion is formed of four attached lobes that encircle the proboscis apparatus (not the gut, as in many other

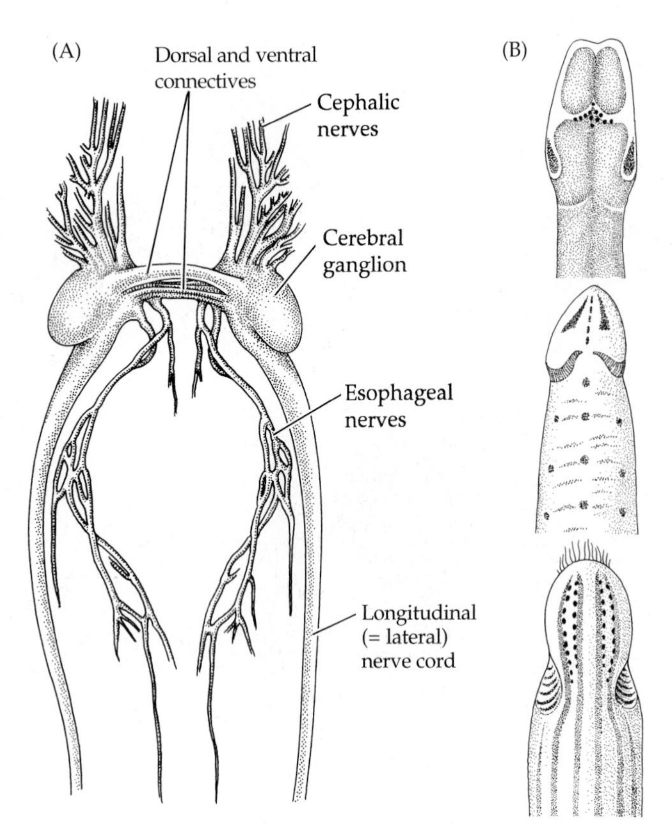

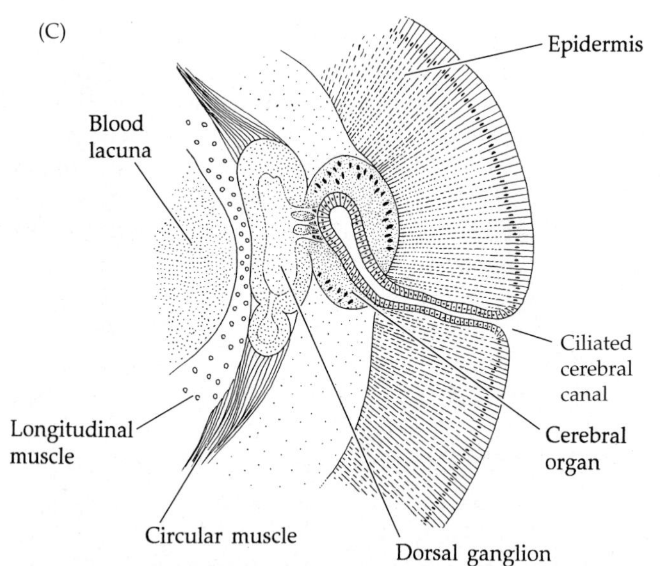

Figure 12.10 Nervous system and sense organs of nemerteans. (A) Anterior portion of the nervous system of *Tubulanus* (Palaeonemertea); see text for explanation of variations. (B) The cephalic slits and grooves and eye spots are visible on the heads of three nemerteans. (C) The cerebral organ of *Tubulanus* (cross section). Note the association of the organ with the cerebral canal, the nervous system, and the blood system. (D) Clusters of frontal glands occur in the anterior end of a hoplonemertean (longitudinal section).

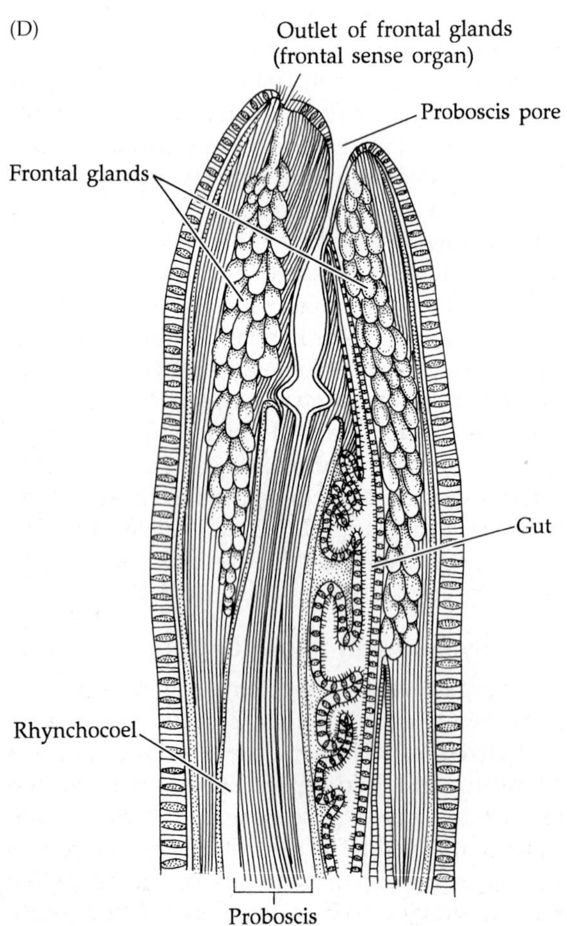

invertebrates). Each side of the cerebral ganglion includes a dorsal and a ventral lobe; the two sides are attached to one another by dorsal and ventral connectives. Several pairs of sensory nerves provide input directly to the cerebral ganglion from various cephalic sense organs. The main longitudinal nerve cords arise from the ventral lobes of the cerebral ganglion and pass posteriorly; they attach to each other at various points by branched transverse connectives and terminally by an anal commissure. The longitudinal nerves also give rise to peripheral sensory and motor nerves along the length of the body. Elaboration on this basic plan includes additional longitudinal nerve cords, frequently a middorsal one arising from the dorsal commissure of the cerebral ganglion, and a variety of connectives, nerve tracts, and plexus.

As noted in the classification scheme, the positions of the major longitudinal nerve cords vary among the nemertean orders (Figure 12.3). These changes in the position of the nerve cords from epidermal to mesenchymal correspond to general increases in body complexity and tendencies toward specialization. Most workers agree that these differences reflect a plesiomorphic (epidermal) to apomorphic (subepidermal) trend among these taxa.

Ribbon worms possess a variety of sensory receptors, many of which are concentrated at the anterior end and associated with an active, typically hunting lifestyle and with other aspects of their natural history. Nemerteans are very sensitive to touch. This tactile sensitivity plays a role in food handling, avoidance responses, locomotion over irregular surfaces, and mating behavior. Several types of modified ciliated epidermal cells are scattered over the body surface (especially abundant at the anterior and posterior ends) and are presumed to have a tactile function. The cells occur either singly or in clusters; some of the latter types are located in small depressions and can be thrust out from the body surface.

The eyes of ribbon worms are located anteriorly and number from two to several hundred; they can be arranged in various patterns (Figure 12.10B). Most of these ocelli are of the inverted pigment-cup type, similar to those seen in flatworms, although a few species possess lensed eyes. As discussed in Chapter 4, these types of eyes typically are sensitive to light intensity and light direction. They help the nemerteans avoid bright light and potential exposure to predators or environmental stresses.

Much of the sensory input important to nemerteans is chemosensory. These worms are very sensitive to dissolved chemicals in their environment and employ this sensitivity in food location, probably mate location, substratum testing, and general water analysis. Probably all nemerteans respond to contact with chemical stimuli, and many are capable of distance chemoreception of materials in solution. At least three different nemertean structures have been implicated (some through speculation) in the initiation of chemotactic responses: cephalic slits or grooves, cerebral organs, and frontal glands (= cephalic glands) (Figures 12.10B–D). **Cephalic slits** are furrows of variable depth that occur laterally on the heads of many ribbon worms (see also Figure 12.1E,F). These furrows are lined with a ciliated sensory epithelium supplied with nerves from the cerebral ganglion. Water is circulated through the cephalic slits and over this presumably chemosensory epithelial lining.

Most nemerteans possess a pair of the remarkably complex **cerebral organs** (Figure 12.10C). The core of each cerebral organ is a ciliated epidermal invagination (the cerebral canal), which is expanded at its inner end. These canals lead laterally to pores within the cephalic slits (when present) or else directly to the outside via separate pores on the head. The inner ends of the canals are surrounded by nervous tissue of the cerebral ganglion, and by glandular tissue, and they are often intimately associated with lacunar blood spaces. Cilia in the cerebral canal circulate water through the open portion of the organ; this activity intensifies in the presence of food. Nemerteans presumably use this mechanism when hunting and tracking prey or in other chemotactic responses. The association of the cerebral canals with glandular, nervous, and circulatory structures has led some workers to suggest an endocrine and/or neurosecretory function for the cerebral organs. Other suggestions have included auditory, gas exchange, excretory, and tactile activities. Cerebral organs are absent in several genera, including the symbiotic *Carcinonemertes* and *Malacobdella,* and the pelagic hoplonemerteans.

In the region anterior to the cerebral ganglion, large **frontal glands** open to the outside through a pitlike frontal sense organ (Figure 12.10D). These structures receive nerves from the cerebral ganglion and appear to be chemosensory, but solid evidence for this suggestion is lacking. Finally, statocysts have also been found in some nemerteans, including pelagic forms where geotaxis is an obvious advantage.

Reproduction and Development

Asexual processes Many nemerteans show remarkable powers of regeneration, and nearly all species can regenerate at least posterior portions of the body. Those with the greatest regenerative abilities are certain species of *Lineus,* which engage in a remarkable form of asexual reproduction on a regular basis by undergoing multiple transverse fission into numerous fragments. The fragments are often extremely small and the process is sometimes referred to simply as **fragmentation**. The small pieces often form mucous cysts within which the new worm regenerates; larger pieces grow into new animals without the protection of a cyst. In some nemerteans only anterior fragments can regenerate into new worms.

Sexual reproduction Nemerteans show remarkable variation in reproductive and developmental strategies. Most ribbon worms are gonochoristic, although protandric and even simultaneous hermaphrodites are known. The reproductive system of nemerteans has gonads that are simply specialized patches of mesenchymal tissue arranged serially along each side of the intestine and alternate with the midgut diverticula (Figure 12.11). In *Malacobdella* and a few others, the gonads are more or less packed within the mesenchyme (Figure 12.1H). In most nemerteans the development of gonads occurs along nearly the entire length of the body, but in a few species they are restricted to certain regions, usually toward the anterior end. The gonads begin to enlarge and hollow just prior to the onset of breeding activities. Specialized cells in the walls of the rudimentary ovaries and testes proliferate eggs and sperm into the lumina of the enlarging gonadal sacs. In females additional special cells are responsible for yolk production. There is evidence that maturation is under neurosecretory hormonal control, at least in some species. The secretions are probably from the cerebral organ complex.

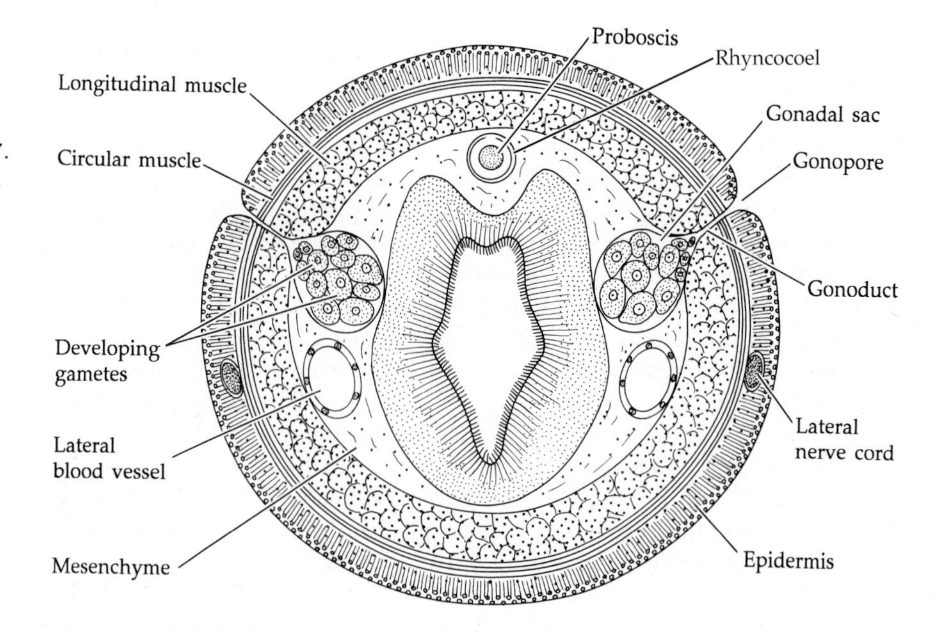

Figure 12.11 Arrangement of gonads in the palaeonemertean *Carinina* (cross section). Note the position of a pair of gonads in the mesenchyme. See also Figure 12.7.

With the proliferation of gametes, the gonadal sacs expand to almost fill the area between the gut and the body wall. When the animals are nearly ready to spawn, mating behavior is initiated and the worms become increasingly active. As mentioned earlier, mate location probably depends on chemotactic responses. The same is apparently true of spawning itself, at least for some species, because the presence of a ripe conspecific stimulates the release of gametes from other mature individuals. Experimental evidence indicates that physical contact is not necessary for such a spawning response; thus, some sort of pheromone is probably involved. In nature, however, spawning usually occurs in concert with actual physical contact; tactile responses evidently follow chemotactic mate location. During such mating activities, veritable knots of scores of worms may writhe in a mucus-covered mating mass. The coordinated release of ripe gametes under such conditions ensures successful fertilization. The gametes are extruded through temporary pores or through ruptures in the body wall. Rupture occurs by contraction of the body wall muscles or of special mesenchymal muscles surrounding the gonads.

Fertilization is often external, either free in the seawater or in a gelatinous mass of mucus produced by the mating worms. In the latter situation, actual egg cases are frequently formed, and part or all of the embryonic development occurs within them (Figure 12.12). Internal fertilization occurs in certain nemerteans. In some cases the sperm are released into the mucus surrounding the mating worms and then move into the ovaries of the female; once fertilized, the eggs are usually deposited in egg capsules, where they develop, although some Antarctic species brood with cocoons. Some terrestrial species are ovoviviparous; the embryos are retained within the body of the female and development is fully direct—an obvious advantage for surviving on land. Ovoviviparity is also known in a few other nemerteans, including deep-sea pelagic forms. Since the population densities of these pelagic worms are extremely low, they must presumably capitalize on the relatively infrequent encounters of males and females and ensure successful fertilization. In a few cases the males are equipped with suckers, which are used to clasp the female, or, rarely, with a protrusible penis, which is used to transfer sperm.

Regardless of the method of fertilization, development through the gastrula is similar among most of the nemerteans studied to date. Cleavage is holoblastic and spiral, producing either three (*Tubulanus*) or, more typically, four quartets of micromeres. A coeloblastula forms, and this often shows the rudiments of an apical ciliary tuft associated with a slight thickening of the blastula wall at or near the animal pole. Gastrulation is usually by invagination of the macromeres and the fourth micromere quartet to produce a coelogastrula. In at least one genus (*Prostoma*, a hermaphroditic freshwater form), gastrulation is by unipolar ingression of

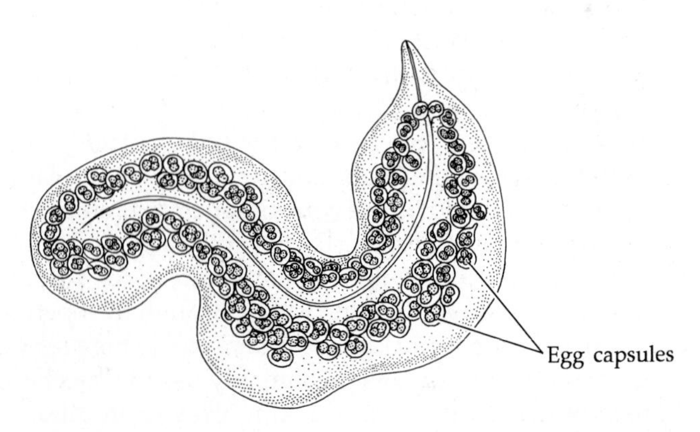

Figure 12.12 Egg capsules in the heteronemertean *Lineus ruber*.

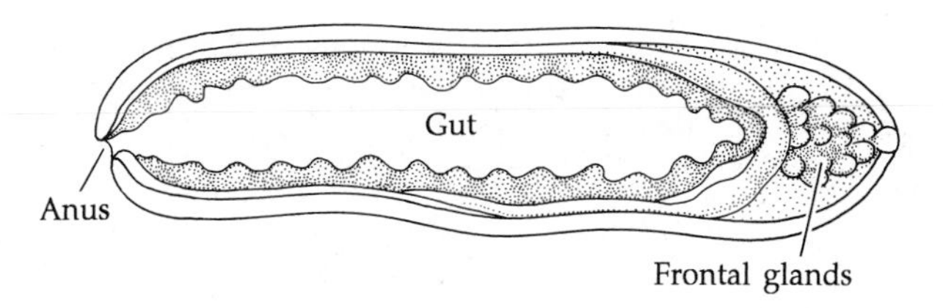

Figure 12.13 Hatching form produced by direct development in *Prosorhochmus*, a hoplonemertean.

the vegetal macromeres; this movement produces a stereogastrula, which later hollows. Mesoderm may originate in several ways, and in some cases the processes are poorly understood. In *Cerebratulus*, one of the best-studied nemerteans developmentally, ectomesoderm is derived from two blastomeres (3a and 3b), which give rise to the extensive array of the larval muscle cells. *Cerebratulus lacteus* also possesses a true mesentoblast (4d) which gives rise to a pair of small mesodermal bands, and scattered mesenchymal cells. This dual origin of the mesoderm, as both ectomesoderm and endomesoderm, appears to be a condition present in all spiralians. The gut is formed by all the fourth quartet micromeres as well as the vegetal macromeres (4A, 4B, 4C, 4D). Unraveling the embryogeny of nemerteans has led to the recent discovery that the rhynchocoel is formed by schizocoely, and it therefore represents a true coelomic cavity.

Developmental strategies are also varied among nemerteans. Members of Palaeonemertea and Hoplonemertea undergo direct development within egg cases. The embryos in these groups develop gradually to juvenile worms without any abrupt metamorphosis (Figure 12.13), but may have multiple invaginations and shedding of a transitory larval epidermis. These embryos are nourished by yolk until they hatch, whereupon they commence feeding. This is especially true of the palaeonemerteans, in which the proboscis apparatus is not fully formed at hatching (the proboscis of heteronemerteans and *Malacobdella* is functional at the time of hatching).

A recent study of embryogenesis of the palaeonemertean *Carinoma tremaphoros* has shown large squamous cells covering the entire larval surface except for the apical and posterior regions. Although apical and posterior cells continue to divide, in the large surface cells cleavage arrest and form a contorted preoral belt. Based on its position, cell lineage, and fate, it has been suggested that this belt corresponds to the prototroch of other trochozoans.

The Pilidiophora undergo a bizarre and fascinating pattern of indirect development. Most species in this subclass produce a free-swimming, planktotrophic larva called the **pilidium** (Figure 12.14). At this stage the gut is incomplete, consisting of a mouth located between a pair of flaplike ciliated lobes, a stomodeal foregut, and a blind intestine; the anus forms later as a proctodeal invagination. Interestingly, the intestinal diverticula of nemerteans do not form as evaginations of the gut wall but are produced by medial encroachments of mesenchyme, which press in the gut wall, thus creating the diverticula. As the pilidium swims and feeds, a series of invaginations in the larval ectoderm (Figure 12.14B) eventually pinch off internally to produce the presumptive adult ectoderm. Perhaps the most striking characteristic of the pilidium is the way the juvenile worm develops inside the larva from a series of isolated rudiments, called the imaginal discs. The paired cephalic discs, cerebral organ discs, and trunk discs originate as invaginations of larval epidermis and subsequently grow and fuse around the larval gut to form the juvenile (Figure 12.14C). The fully formed juvenile ruptures the larval body and, more often than not, devours the larva during catastrophic metamorphosis. In this way, the animal prepares for benthic life before it faces the rigors of

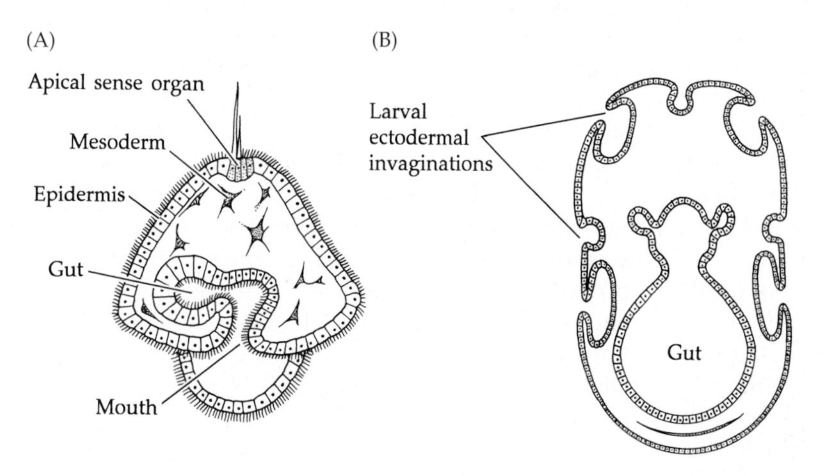

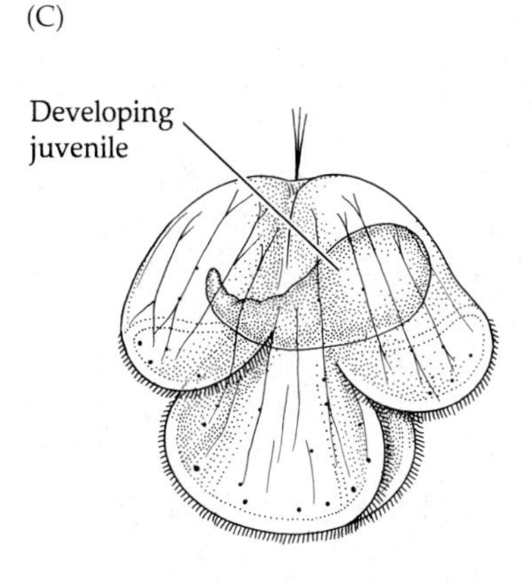

Figure 12.14 Development of a pilidium larva. (A) Pilidium larva (sagittal section). (B) A pilidium larva (transverse section) during invagination of larval ectoderm to form adult skin. (C) Late pilidium larva with juvenile formed within.

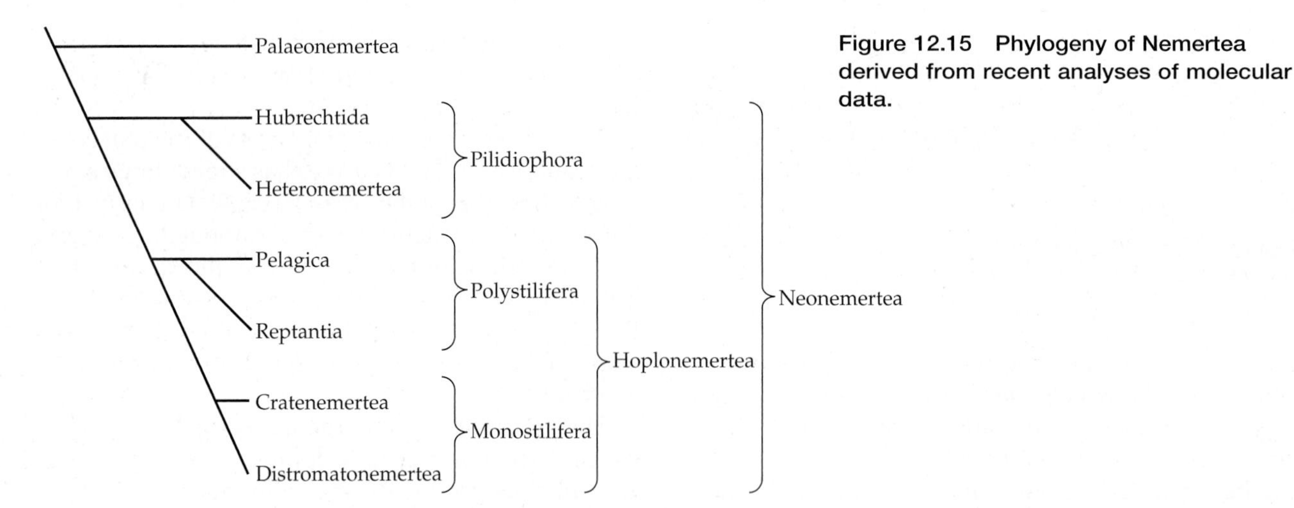

Figure 12.15 Phylogeny of Nemertea derived from recent analyses of molecular data.

settlement. When development is completed, the larval skin is shed and the juvenile assumes its life on the sea floor. Pilidia from different species vary in shape, size, and color. Modifications of pilidial development include **Desor's larva** in *Lineus viridis*, **Schmidt's larva** in *Lineus ruber*, and **Iwata's larva** in several *Micrura* species. Development of Desor's and Schmidt's larva is encapsulated, while the others have planktonic lecithotrophic development. In all of these cases, the pilidial lobes are lacking, the blastocoel is reduced, the larva does not feed but, similar to the canonical pilidium, the juvenile develops via imaginal discs.

Nemertean Phylogeny

The fossil record is, unfortunately, of little use in establishing the origin of nemerteans in geological time, but the group obviously diverged sometime after the origin of the spiralian bilateral condition. Although now viewed to be part of the spiralian clade, related to other spiralians with trochophore larvae, nemerteans had long been considered to be related to flatworms (Chapter 10). However, this idea is refuted by the discovery of the coelomic nature of the nemertean rhynchocoel and blood vessels, which place them closer to other phyla of coelomate spiralians. An older view is that nemerteans arose from an early archoöphoran "turbellarian" stock, perhaps sharing common ancestry with the macrostomid flatworms. The nemerteans and free-living platyhelminths display a number of similarities including protonephridia, types of ocelli, certain histological characteristics (especially of the epidermis), and the general organization of the nervous system. Furthermore, various ciliated slits and depressions among free-living flatworms resemble the cephalic slits and similar structures of the nemerteans. Some flatworms possess frontal (cephalic) glands long thought to be homologous to those of ribbon worms, but these could be symplesiomorphic for Spiralia or they could be convergences. Their mode or early egg cleavage places them as clear members of the clade Spiralia, and interpretations of the arrested cells of some embryos suggest a relationship to phyla with trochophore larvae—a relationship otherwise unsupported by any other anatomical characters. This is why their phylogenetic placement has been in flux for quite some time. However, molecular data strongly suggest a relationship to Mollusca, Annelida, and the lophophorate phyla (and probably also the Entoprocta).

The phylogenetic relationships among the various taxa of Nemertea have been examined in detail in the past few years and now include a phylogenomic analysis using hundreds of genes (Figure 12.15). These new studies are largely congruent with some of the traditional classifications, but they have found paraphyly of the former class Anopla, and include some important rearrangements, like the placement of Hubrechtida as sister group to Heteronemertea and not as a member of Palaeonemertea, or the placement of the former monogeneric order Bdellonemertea within Monostilifera. The former result is not really that surprising, as species of Hubrechtida have a mosaic of palaeonemertean and heteronemertean characters and some authors have considered them as a "transitional stage" between palaeonemerteans and basal heteronemerteans. Furthermore, the new evolutionary tree therefore implies that certain characters found in Anopla are actual symplesiomorphies for Nemertea, including the unarmed proboscis or the separation of the opening of mouth and proboscis. Within Hoplonemertea, Monostilifera appear divided into two clades, Cratenemertea and Distromatonemertea, whereas Polystilifera splits into the clades Pelagica and Reptantia.

One of the principal structural trends among the nemerteans is the internalization of the major longitudinal nerve cords. We assume that the earliest ribbon worms possessed epidermal nerve cords, as do some modern palaeonemerteans. Palaeonemerteans and pilidiophorans retain the plesiomorphic feature of the placement of the mouth posterior to the cerebral ganglion. The relatively simple and unarmed proboscis and the placement of the nerve cords external to the mesenchyme suggest further that these clades retain the plesiomophic character states among the ribbon worms. Pilidiophorans acquired indirect development, the unique formation of the double larval and adult ectoderm during metamorphosis, and the evolution of their unique arrangement of body wall muscles. The encapsulation, and thus functionally direct development, of those heteronemerteans with a Desor larva is almost certainly a secondary abandonment of free pilidial larval life.

The hoplonemerteans show some distinct changes from the members mentioned above. Most notable are the regional specialization and armature of the proboscis, the movement of the nerve cords to a mesenchymal position, and the movement of the mouth more anteriorly. The bdellonemerteans are a specialized offshoot of Monostilifera that displays significant modification for an endosymbiotic lifestyle, including simplification of the proboscis, coiling and increased relative length of the gut (probably associated with their herbivorous habits), a posterior body sucker, and decreased body length.

Selected References

Andrade, S. C. S. and 8 others. 2014. A transcriptomic approach to ribbon worm systematics (Nemertea): resolving the Pilidiophora problem. Mol. Biol. Evol. 31: 3206–3215.

Andrade, S. C. S. and 12 others. 2012. Disentangling ribbon worm relationships: multi-locus analysis supports traditional classification of the phylum Nemertea. Cladistics 28: 141–159.

Asakawa, M., K. Ito and H. Kajihara. 2013. Highly toxic ribbon worm *Cephalothrix simula* containing tetrodotoxin in Hiroshima Bay, Hiroshima Prefecture, Japan. Toxins (Basel) 5: 376–395.

Bayer, F. M. and H. B. Owre. 1968. *The Free-Living Lower Invertebrates.* Macmillan Company, New York.

Berg, G. 1985. *Annulonemertes* gen. nov., a new segmented hoplonemertean. Pp. 200–209 in C. Morris et al. (eds.), *The Origins and Relationships of Lower invertebrates*. Systematic Assocation, Special Volume No. 28. Oxford Press, London.

Berg, G. and R. Gibson. 1996. A redescription of *Nemertoscolex parasiticus* Greeff, 1879, an apparently endoparasitic heteronemertean from the coelomic fluid of the echiuroid *Echiurus echiurus* (Pallas). J. Nat. Hist. 30: 163–173.

Bianchi, S. 1969. On the neurosecretory system of *Cerebratulus marginatus* (Heteronemertini). Gen. Comp. Endocr. 12: 541–548.

Bierne, J. 1966. Localisation dans les ganglions cérébroides du centre regulateur de la maturation sexuelle chez la femalle de *Lineus ruber* Müller (Hétéronémertes). C. r. Hebd. Séanc. Acad. Sci. Paris 262: 1572–1575.

Coe, W. R. 1940. Revision of the nemertean fauna of the Pacific coasts of North, Central, and northern South America. Allan Hancock Pacific Expeditions 2(13): 247–323.

Coe, W. R. 1943. Biology of the nemerteans of the Atlantic coast of North America. Trans. Conn. Acad. Sci. 35: 129–328.

Gibson, R. 1972. *Nemerteans.* Hutchinson University Library, London.

Gibson, R. 1982. Nemertea. Pp. 823–846 in S. Parker (ed.), *Synopsis and Classification of Living Organisms. Vol. 1.* McGraw-Hill, New York.

Gibson, R. and J. Jennings. 1969. Observations on the diet, feeding mechanism, digestion and food reserves of the ectocommensal rhynchocoelan *Malacobdella grossa*. J. Mar. Biol. Assoc. U.K. 49: 17–32.

Gibson, R. and J. Moore. 1976. Freshwater nemertines. Zool. J. Linn. Soc. 58: 177–218.

Harrison, F. W. and B. J. Bogitsh (eds.). 1991. *Microscopic Anatomy of Invertebrates. Vol. 3. Platyhelmithes and Nemertina.* Wiley-Liss, New York.

Henry, J. Q. and M. Q. Martindale. 1998. Conservation of the spiralian developmental program: cell lineage of the nemertean, *Cerebratulus lacteus*. Dev. Biol. 201: 253–269.

Hiebert, L. S., G. Gavelis, G. von Dassow and S. A. Maslakova. 2010. Five invaginations and shedding of the larval epidermis during development of the hoplonemertean *Pantinonemertes californiensis* (Nemertea: Hoplonemertea). J. Nat. Hist. 44: 2331–2347.

Hyman, L. H. 1951. *The Invertebrates, Vol. 2. Platyhelminthes and Rhynchocoela: The Acoelomate Bilateria*. McGraw-Hill, New York.

Jespersen, A. and J. Lützen. 1988. The fine structure of the protonephridial system in the land nemertean *Pantinonemertes californiensis*. Zoomorphology 108: 69–75.

Jespersen, A. and J. Lützen. 1988. Ultrastructure and morphological interpretation of the circulatory system of nemerteans. Vidensk. Meddr. Dansk. Naturh. Foren. 147: 47–66.

Kvist, S., A. V. Chernyshev and G. Giribet. 2015. Phylogeny of Nemertea with special interest in the placement of diversity from Far East Russia and northeast Asia. Hydrobiologia. doi: 10.1007/s10750-015-2310-5

Kvist, S., C. E. Laumer, J. Junoy and G. Giribet. 2014. New insights into the phylogeny, systematics and DNA barcoding of Nemertea. Invertebr. Syst. 28: 287–308.

Laumer, C. E. and 10 others. 2015. Spiralian phylogeny informs the evolution of microscopic lineages. Curr. Biol. 25: 2000–2006.

Maslakova, S. A. 2010a. Development to metamorphosis of the nemertean pilidium larva. Front. Zool. 7: 30.

Maslakova, S. A. 2010b. The invention of the pilidium larva in an otherwise perfectly good spiralian phylum Nemertea. Integr. Comp. Biol. 50: 734.

Maslakova, S. A., M. Q. Martindale and J. L. Norenburg. 2004. Vestigial prototroch in a basal nemertean, *Carinoma tremaphoros* (Nemertea; Palaeonemertea). Evol. Dev. 6: 219–226.

McDermott, J. J. and P. Roe. 1985. Food, feeding behavior, and feeding ecology of nemerteans. Am. Zool. 25: 113–125.

Moore, J. and R. Gibson. 1981. The *Geonemertes* problem (Nemertea). J. Zool. Lond. 194: 175–201. [A revision of the world's terrestrial nemerteans.]

Norenburg, J. 1985. Structure of the nemertine integument with consideration of its ecological and phylogenetic significance. Am. Zool. 25: 37–51.

Norenburg, J. L. and S. A. Stricker. 2002. Phylum Nemertea. Pp. 163–177 in *Atlas of Marine Invertebrate Larvae.* Academic Press, San Diego.

Riser, H. W. 1974. Nemertinea. Pp. 359–389 in A. Giese and J. Pearse (eds.), *Reproduction of Marine Invertebrates, Vol. 1.* Academic Press, New York.

Riser, H. W. 1985. Epilogue: Nemertinea, a successful phylum. Am. Zool. 25: 145–151.

Shields, J. D. and M. Segonzac. 2007. New nemertean worms (Carcinonemertidae) on Bythograeid crabs (Decapoda: Brachyura) from Pacific hydrothermal vent sites. J. Crustacean Biol. 27: 681–692.

Strand, M., A. Herrera-Bachiller, A. Nygren and T. Kånneby. 2013. A new nemertean species: what are the useful characters for ribbon worm descriptions? J. Mar. Biol. Assoc. UK: 1–14.

Strand, M. and P. Sundberg. 2011. A DNA-based description of a new nemertean (phylum Nemertea) species. Mar.Biol. Res. 7: 63–70.

Sundberg, P. and M. Strand. 2010. Nemertean taxonomy — time to change lane? J. Zool. Syst. Evol. Res. 48: 283–284.

Stricker, S. A. 1985. The stylet apparatus of monostyliferous hoplonemerteans. Am. Zool. 25: 87–97.

Stricker, S. A., M. J. Cavey and R. A. Cloney. 1985. Tetracycline labeling studies of calcification in nemertean worms. T. Am. Microsc. Soc. 104: 232–241.

Sundberg, P., J. M. Turbeville and S. Lindh. 2001. Phylogenetic relationships among higher nemertean (Nemertea) taxa inferred from 18S rDNA sequences. Mol. Phylogenet. Evol. 20: 327–334.

Taboada, S., J. Junoy, S. C. S. Andrade, G. Giribet, J. Cristobo and C. Avila. 2013. On the identity of two Antarctic brooding nemerteans: redescription of *Antarctonemertes valida* (Bürger, 1893) and description of a new species in the genus *Antarctonemertes* Friedrich, 1955 (Nemertea, Hoplonemertea). Polar Biol. 36: 1415–1430.

Thollesson, M. and J. L. Norenburg. 2003. Ribbon worm relationships: a phylogeny of the phylum Nemertea. Proc. Biol. Sci. 270: 407–415.

Turbeville, J. M. 1986. An ultrastructural analysis of coelomogenesis in the hoplonemertine *Prosorhochmus americanus* and the polychaete *Magelona* sp. J. Morphol. 187: 51–56.

Turbeville, J. M., K. G. Field and R. A. Raff. 1992. Phylogenetic position of phylum Nemertini, inferred from 18S rRNA sequences: molecular data as a test of morphological character homology. Mol. Biol. Evol. 9: 235–249.

Turbeville, J. M. and J. E. Ruppert. 1985. Comparative ultrastructure and the evolution of nemertines. Am. Zool. 25: 53–71.

von Döhren, J. 2011. The fate of the larval epidermis in the Desor-larva of *Lineus viridis* (Pilidiophora, Nemertea) displays a historically constrained functional shift from planktotrophy to lecithotrophy. Zoomorphology 130: 189–196.

Wickham, D. E. 1979. Predation by the nemertean *Carcinonemertes errans* on eggs of the Dungeness crab, *Cancer magister*. Mar. Biol. 55: 45–53.

Wickham, D. E. 1980. Aspects of the life history of *Carcinonemertes errans* (Nemertea: Carcinonemertidae), an egg predator of the crab *Cancer magister*. Biol. Bull. 159: 247–257.

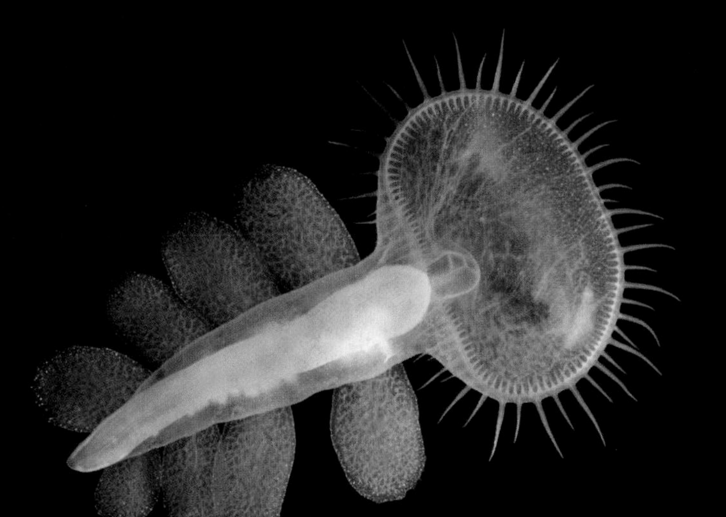

CHAPTER 13

Phylum Mollusca

Molluscs include some of the best-known invertebrates; almost everyone is familiar with snails, clams, slugs, squids, and octopuses. Molluscan shells have been popular since ancient times, and some cultures still use them as tools, containers, musical devices, money, fetishes, religious symbols, ornaments, and decorations and art objects. Evidence of historical use and knowledge of molluscs is seen in ancient texts and hieroglyphics, on coins, in tribal customs, and in archaeological sites and aboriginal kitchen middens or shell mounds. Royal or Tyrian purple of ancient Greece and Rome, and even Biblical blue (Num. 15:38), were molluscan pigments extracted from certain marine snails.[1] Many aboriginal groups have for millenia relied on molluscs for a substantial portion of their diet and for use as tools. Today, coastal nations annually harvest millions of tons of molluscs commercially for food.

There are approximately 80,000 described, living mollusc species and about the same number of described fossil species. However, many species still await names and descriptions, especially those from poorly studied regions and time periods, and it has been estimated that only about half of the living molluscs have so far been described. In addition to three familiar molluscan classes comprising the clams (Bivalvia), snails and slugs (Gastropoda), and squids and octopuses (Cephalopoda), five other extant classes exist: chitons (Polyplacophora), tusk shells (Scaphopoda), *Neopilina* and its kin (Monoplacophora), and the vermiform sclerite-bearing aplacophoran classes—Caudofoveata (or Chaetodermomorpha) and Solenogastres (or Neomeniomorpha). Although members of these eight classes differ enormously in superficial appearance, there is a suite of characters that diagnose their fundamental body plan (Box 13A).

Classification of The Animal Kingdom (Metazoa)

Non-Bilateria*
(a.k.a. the diploblasts)
- PHYLUM PORIFERA
- PHYLUM PLACOZOA
- PHYLUM CNIDARIA
- PHYLUM CTENOPHORA

Bilateria
(a.k.a. the triploblasts)
- PHYLUM XENACOELOMORPHA

Protostomia
- PHYLUM CHAETOGNATHA

SPIRALIA
- PHYLUM PLATYHELMINTHES
- PHYLUM GASTROTRICHA
- PHYLUM RHOMBOZOA
- PHYLUM ORTHONECTIDA
- PHYLUM NEMERTEA
- **PHYLUM MOLLUSCA**
- PHYLUM ANNELIDA
- PHYLUM ENTOPROCTA
- PHYLUM CYCLIOPHORA

Gnathifera
- PHYLUM GNATHOSTOMULIDA
- PHYLUM MICROGNATHOZOA
- PHYLUM ROTIFERA

Lophophorata
- PHYLUM PHORONIDA
- PHYLUM BRYOZOA
- PHYLUM BRACHIOPODA

ECDYSOZOA

Nematoida
- PHYLUM NEMATODA
- PHYLUM NEMATOMORPHA

Scalidophora
- PHYLUM KINORHYNCHA
- PHYLUM PRIAPULA
- PHYLUM LORICIFERA

Panarthropoda
- PHYLUM TARDIGRADA
- PHYLUM ONYCHOPHORA
- PHYLUM ARTHROPODA
  - SUBPHYLUM CRUSTACEA*
  - SUBPHYLUM HEXAPODA
  - SUBPHYLUM MYRIAPODA
  - SUBPHYLUM CHELICERATA

Deuterostomia
- PHYLUM ECHINODERMATA
- PHYLUM HEMICHORDATA
- PHYLUM CHORDATA

*Paraphyletic group

This chapter has been revised by Richard C. Brusca, David R. Lindberg, and Winston F. Ponder

[1]Archaeological sites in Israel reveal the probable use of two muricid snails (*Murex brandaris* and *Truncularioptsis trunculus*) as sources of the Royal purple dye.

Taxonomic History and Classification

Molluscs carry the burden of a very long and convoluted taxonomic history, in which hundreds of names for various taxa have come and gone. Aristotle recognized molluscs, dividing them into Malachia (the cephalopods) and Ostrachodermata (the shelled forms), the latter being divided into univalves and bivalves. Joannes Jonston (or Jonstonus) created the name Mollusca[2] in 1650 for the cephalopods and barnacles, but this name was not accepted until it was resurrected and redefined by Linnaeus nearly a hundred years later. Linnaeus's Mollusca included cephalopods, slugs, and pteropods, as well as tunicates, anemones, medusae, echinoderms, and polychaetes—but included chitons, bivalves, univalves, nautiloids, barnacles, and the serpulid polychaetes (which secrete calcareous tubes) in another group, Testacea. In 1795 Georges Cuvier published a revised classification of the Mollusca that was the first to approximate modern views. Henri de Blainville (1825) altered the name Mollusca to Malacozoa, which won little favor but survives in the terms malacology, malacologist, etc.

Much of the nineteenth century passed before the phylum was purged of all extraneous groups. In the 1830s, J. Thompson and C. Brumeister identified the larval stages of barnacles and revealed them to be crustaceans, and in 1866 Alexander Kowalevsky removed the tunicates from Mollusca. Separation of the brachiopods from the molluscs was long and controversial and not resolved until near the end of the nineteenth century.

The first sclerite-covered wormlike aplacophorans, members of what today we recognize as the class Caudofoveata, were discovered in 1841 by the Swedish naturalist Sven Lovén. He classified them with holothuroid echinoderms because of their vermiform bodies and the presence of calcareous sclerites in the body walls of both groups. In 1886, another Swede, Tycho Tullberg, described the first representative of the other aplacophoran group—the Solenogastres. Ludwig von Graff (1875) recognized both groups as molluscs and they were united in the Aplacophora in 1876 by Hermann von Ihering. The Aculifera hypothesis of Amélie Scheltema unites molluscs that possess calcareous sclerites by placing Polyplacophora as the sister taxon of the aplacophorans (Caudofoveata + Solengastres). Aculifera was sometimes also called Amphineura, although this latter term has also been used by some workers to refer only to chitons. **Sclerites** are spicules, scales, and so on that cover or are embedded in the epidermis of molluscs and are often calcified.

The history of classification of species in the class Gastropoda has been volatile, undergoing constant change since Cuvier's time. Most modern malacologists adhere more or less to the basic schemes of Henri Milne-Edwards (1848) and J. W. Spengel (1881). The former, basing his classification on the respiratory organs, recognized the groups Pulmonata, Opisthobranchia, and Prosobranchia. Spengel based his scheme on the nervous system and divided the gastropods into the Streptoneura and Euthyneura. In subsequent classifications, Streptoneura was equivalent to Prosobranchia; Euthyneura included Opisthobranchia and Pulmonata. The bivalves have been called Bivalvia, Pelecypoda, and Lamellibranchiata. More recently, anatomical, ultrastructural, and molecular studies have brought about considerable changes to molluscan classification, as outlined below. Many taxa have multiple names and the more commonly encountered ones are noted below.

Molluscan classification at the generic and species levels is also troublesome. Many species of gastropods

BOX 13A Characteristics of the Phylum Mollusca

1. Bilaterally symmetrical (or secondarily asymmetrical), unsegmented, coelomate protostomes
2. Coelom limited to small spaces in nephridia, heart and gonads
3. Principal body cavity is a hemocoel (open circulatory system)
4. Viscera concentrated dorsally as a "visceral mass"
5. Body covered by a cuticle-covered epidermal sheet of skin, the mantle
6. Mantle with shell glands that secrete calcareous epidermal sclerites, shell plates, or shells
7. Mantle overhangs and forms a cavity (the mantle cavity) in which are housed the ctenidia, osphradia, nephridiopores, gonopores, and anus
8. Heart situated in a pericardial chamber and composed of a single ventricle and one or more separate atria
9. Typically with large, well-defined muscular foot, often with a flattened creeping sole
10. Buccal region provided with a radula and muscular odontophore
11. Complete (through) gut, with marked regional specialization, including large digestive glands
12. With large, complex metanephridial "kidneys"
13. Cleavage spiral and embryogeny protostomous
14. With trochophore larva, and a veliger larva in two major groups

[2]The name of the phylum derives from the Latin *molluscus*, meaning "soft," in allusion to the similarity of clams and snails to the mollusca, a kind of Old World soft nut with a thin but hard shell. The vernacular for Mollusca is often spelled mollusks in the United States, whereas in most of the rest of the world it is typically spelled molluscs. In biology, a vernacular or diminutive name is generally derived from the proper Latin name; thus the custom of altering the spelling of Mollusca by changing the c to k seems to be an aberration (although it may have its historic roots in the German language, which does not have the free-standing c; e.g., Molluskenkunde). We prefer the more widely used spelling "molluscs," which seems to be the proper vernacularization and is in line with other accepted terms, such as molluscan, molluscoid, molluscivore, etc.

and bivalves are also burdened with numerous names (synonyms) that have been proposed for the same species. This tangle is partly the result of a long history of amateur shell collecting beginning with the natural history cabinets of seventeenth century Europe, which required documentation and promoted multiple taxonomies and names based only on shell characters. Today, species are recognized based on a combination of shell, anatomical, and, most recently molecular characters. However, because of the tremendous diversity of gastropods and bivalves many species still remain known only from their shells.

Only taxa with extant members are included in the following classification and not all families are listed in the taxonomic synopses. The classification is mostly ranked, but in a few cases unranked group names are used.[3] Examples of the major molluscan taxa appear in Figure 13.1.

ABBREVIATED CLASSIFICATION OF THE PHYLUM MOLLUSCA

CLASS CAUDOFOVEATA Caudofoveatan aplacophorans (spicule "worms")

CLASS SOLENOGASTRES Solenogaster aplacophorans (spicule "worms")

CLASS MONOPLACOPHORA Monoplacophorans. Deep sea, limpet-like

CLASS POLYPLACOPHORA Chitons, with eight shell valves

CLASS GASTROPODA Snails, slugs and limpet

SUBCLASS PATELLOGASTROPODA The true limpets

SUBCLASS VETIGASTROPODA "Primitive" marine top-shell snails, abalones and "limpets"

SUBCLASS NERITIMORPHA Marine, land and freshwater nerite snails and "limpets"

SUBCLASS CAENOGASTROPODA Marine, freshwater and land snails (creepers, periwinkles, conchs, whelks, cowries etc.) and some "limpets"

"ARCHITAENIOGLOSSA" Nonmarine basal caenogastropods (paraphyletic)

INFRACLASS SORBEOCONCHA All remaining caenogastropods

SUPERORDER CERITHIOMORPHA Creepers, turret shells, etc.

COHORT HYPSOGASTROPODA Higher caenogastropods

SUPERORDER LITTORINIMORPHA Periwinkles, cowries, triton shells, etc.

SUPERORDER NEOGASTROPODA Whelks, volutes, rock shells etc.

SUBCLASS HETEROBRANCHIA Marine, freshwater and land snails, most sea slugs, all land slugs, and some "false limpets"

"LOWER HETEROBRANCHIA" A few primitive heterobranch groups including sundial shells, valvatids, etc.

INFRACLASS EUTHYNEURA "Opisthobranchs" and "pulmonates"

COHORT NUDIPLEURA Side-gilled sea slugs and nudibranchs

COHORT EUOPISTHOBRANCHIA Bubble shells, sea hares, pteropods, etc.

COHORT PANPULMONATA "Pulmonates," pyramidellids, sacoglossan sea slugs, most land snails, all land slugs

CLASS BIVALVIA Clams and their kin (bivalves)

SUBCLASS PROTOBRANCHIA "Primitive" deposit-feeding bivalves

SUBCLASS AUTOBRANCHIA "Lamellibranch" suspension-feeding bivalves

COHORT PTERIOMORPHIA Mussels, oysters, scallops, and their kin

COHORT HETEROCONCHIA Marine and freshwater clams

MEGAORDER PALAEOHETERODONTA Freshwater clams (mussels), broch shells

MEGAORDER HETERODONTA Most marine clams

SUPERORDER ARCHIHETERODONTA A few families of primitive marine clams

SUPERORDER EUHETERODONTA The majority of marine and some freshwater clams

CLASS SCAPHOPODA Tusk shells

CLASS CEPHALOPODA Nautilus, squids, octopuses

SUBCLASS PALCEPHALOPODA

COHORT NAUTILIDIA Chambered nautilus

SUBCLASS NEOCEPHALOPODA

COHORT COLEOIDEA Octopuses, squids, cuttlefish

SUPERORDER OCTOPODIFORMES Octopuses, vampire squid

SUPERORDER DECAPODIFORMES Cuttlefish, squid

[3]Multitudes of extinct molluscs have been described. Perhaps the most well-known are some of the groups of cephalopods that had hard external shells, similar to those of living *Nautilus*. One of these groups was the ammonites. They differed from nautiloids in having shell septa that were highly fluted on the periphery, forming complex mazelike septal sutures. Ammonites also had the siphuncle lying against the outer wall of the shell, as opposed to the condition seen in many nautiloids where the siphuncle runs through the center of the shell.

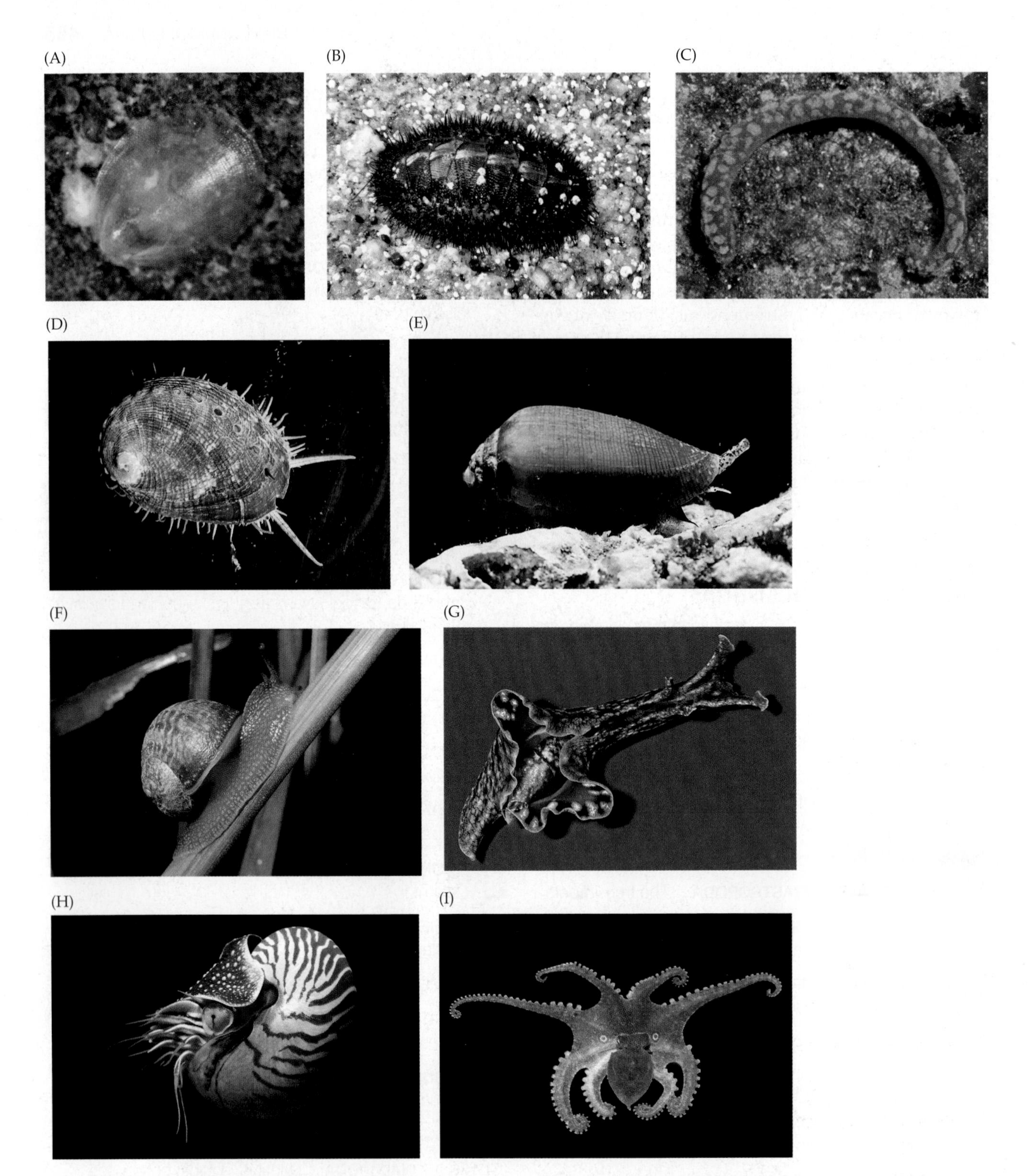

Figure 13.1 Morphological diversity among the molluscs. (A) *Laevipilina hyalina* (Monoplacophora). (B) *Mopalia muscosa*, the mossy chiton (Polyplacophora). (C) *Epimenia australis* (Solenogastres). (D) *Haliotis rufescens*, the red abalone, (Gastropoda); note the exhalant holes in the shell. (E) *Conus*, a predatory neogastropod; note anterior siphon extending beyond shell. (F) The common garden snail, *Cornu aspersum* (Gastropoda). (G) *Aplysia*, the sea hare (Gastropoda: Euopisthobranchia). (H) The chambered *Nautilus* (Cephalopoda). (I) *Octopus bimaculoides* (Cephalopoda). (J) *Sepioteuthis lessoniana*, the bigfin reef squid (Cephalopoda). (K) *Histioteuthis*, a pelagic squid (Cephalopoda). (L) *Fustiaria*, a tusk shell (Scaphopoda). (M) Scallops (Bivalvia: Pteriomorphia: Pectinidae), with a hermit crab in the foreground. (N) The giant clam *Tridacna maxima* (note zooxanthellate mantle), from the Marshall Islands, Northwest Pacific (Bivalvia: Heterodonta: Cardiida). (O) The European cockle *Acanthocardia tuberculata* (Bivalvia: Heterodonta: Cardiida). Note the partly extended foot. (P) *Lima*, a tropical clam that swims by clapping the valves together (Bivalvia). (Q) The highly modified bivalve *Brechites*. (Heterodonta: Poromyata). *Brechites* are known as watering pot shells. They begin their life as a typical small bivalve, but then secrete a large calcareous tube around themselves through which water is pumped for suspension feeding.

(J)
(K)
(L)
(M)
(N)
(O)
(P)
(Q)

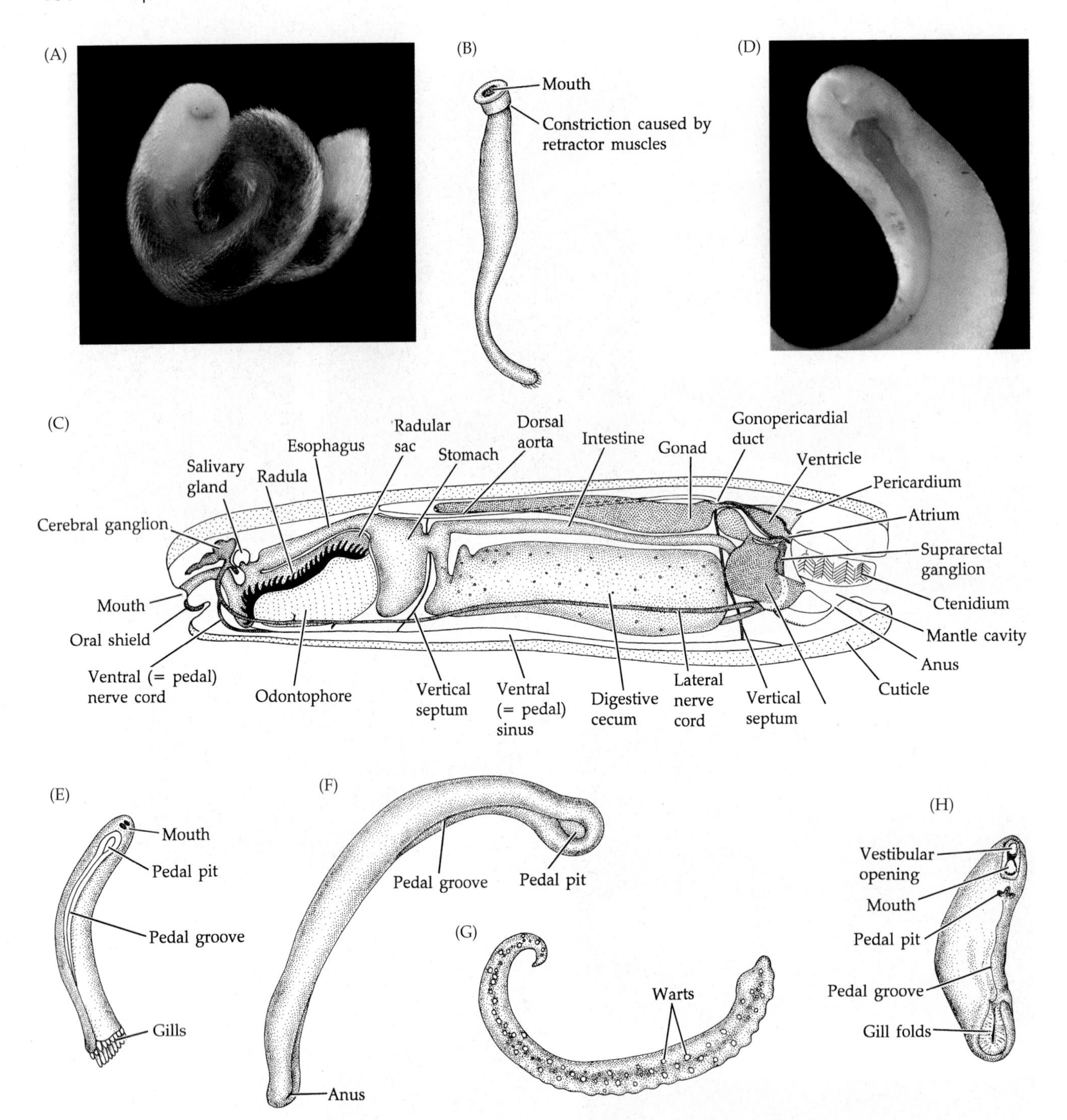

SYNOPSES OF MOLLUSCAN GROUPS

CLASS CAUDOFOVEATA (= CHAETODERMOMORPHA) (Figure 13.2A–C). Spicule "worms." Marine, benthic, burrowing; body vermiform, cylindrical, lacking any trace of a shell; body wall with a chitinous cuticle and imbricating scale-like aragonitic calcareous sclerites; mouth shield anterior to or surrounding the mouth; small posterior mantle cavity with a pair of bipectinate ctenidia; radula present; gonochoristic. Without foot, eyes, tentacles, statocysts, crystalline style, osphradia, or nephridia. About 120 species; burrow in muddy sediments and consume microorganisms such as foraminiferans. (e.g., *Chaetoderma, Chevroderma, Falcidens, Limifossor, Prochaetoderma, Psilodens, Scutopus*)

CLASS SOLENOGASTRES (= NEOMENIOMORPHA) (Figure 13.2D–K). Spicule "worms." Marine, benthic; body vermiform and nearly cylindrical; vestibulum (= atrium) with sensory papillae anterior to the mouth; small posterior mantle cavity lacking ctenidia but often with respiratory folds; body wall with a chitinous cuticle and imbued with calcareous sclerites (as spines or scales); with or without radula; hermaphroditic; pedal glands opening into a pre-pedal ciliary pit, foot weakly muscular, narrow, and can be retracted into a ventral furrow or "pedal groove." Without eyes, tentacles, statocysts, crystalline style, osphradia or nephridia. About 260 described species, but many undescribed species are thought to exist; epibenthic carnivores, often found on (and consuming) cnidarians and a few other types of

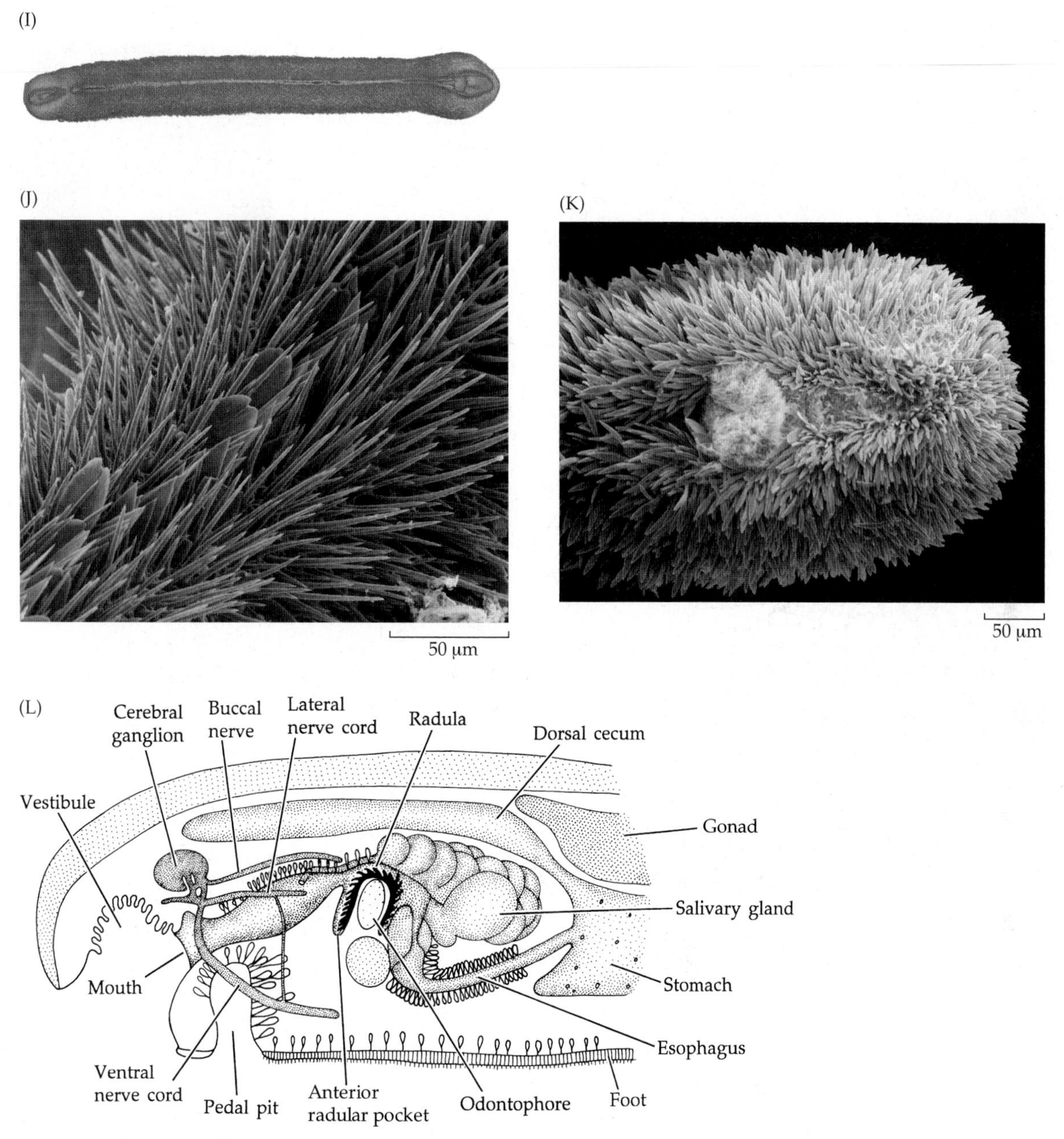

Figure 13.2 General anatomy of aplacophorans. (A–C) Caudofoveata. (A) *Chaetoderma productum*. (B) *Chaetoderma loveni*. (C) Internal anatomy of *Limifossor* (highly stylized sagittal section drawing). (D–L) Solenogastres. (D) *Kruppomenia minima*. (E) *Pruvotina impexa*, ventral view. (F) *Proneomenia antarctica*. (G) *Epimenia verrucosa*. The body is covered with warts. (H) *Neomenia carinata*, ventral view. (I) *Entonomenia tricarinata*, ventral view (X-ray micro-CT). (J) *Macellomenia morseae*. SEM of ventral surface showing two types of scale-like sclerites surrounding the foot and spiny sclerites covering the rest of the surface of the body. (K) *Macellomenia schanderi*. SEM of ventral surface of anterior end showing densely ciliated pedal pit and mouth. (L) Anterior region of *Spengelomenia bathybia* (highly stylized sagittal section drawing).

invertebrates. Solenogastres and Caudofoveata are probably sister groups and are sometimes regarded as subclasses within the class Aplacophora. (e.g., *Alexandromenia*, *Dondersia*, *Epimenia*, *Kruppomenia*, *Neomenia*, *Proneomenia*, *Pruvotina*, *Rhopalomenia*, *Spengelomenia, Wirenia*)

CLASS MONOPLACOPHORA Monoplacophorans. With a single, cap-like shell; foot forms weakly muscular ventral disc, with 8 pairs of retractor muscles; shallow mantle cavity around foot encloses 3–6 pairs ctenidia; 2 pairs gonads; 3–7 pairs nephridia; 2 pairs heart atria; a pair of statocysts; with radula and distinct but small head region; without eyes; short oral tentacles present around mouth; with posterior anus; without a crystalline style; gonochoristic or, rarely, hermaphroditic (Figures 13.1A and 13.3). Until the first living species (*Neopilina galatheae*) was discovered by the Danish Galathea Expedition in 1952, monoplacophorans were

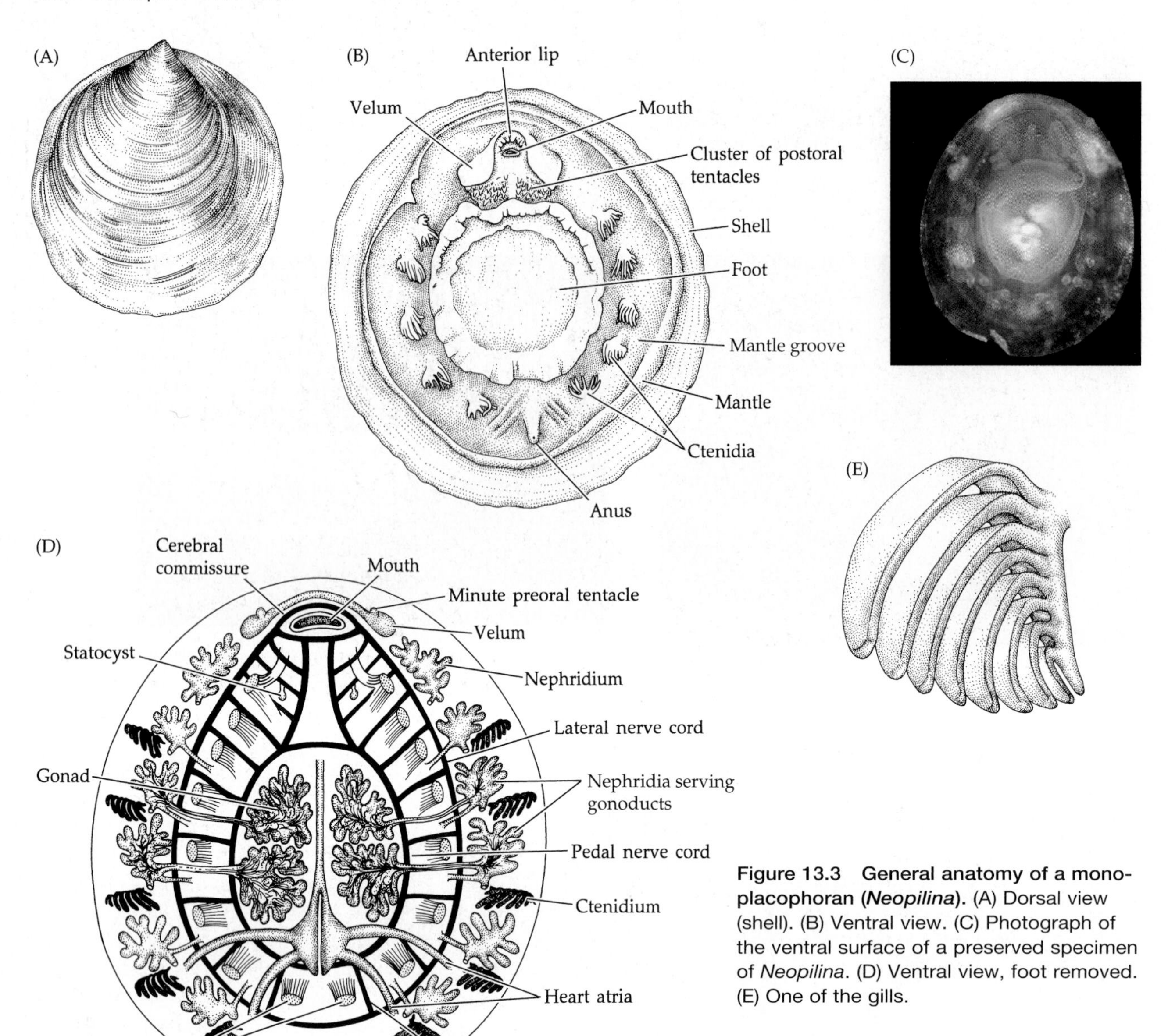

Figure 13.3 General anatomy of a monoplacophoran (*Neopilina*). (A) Dorsal view (shell). (B) Ventral view. (C) Photograph of the ventral surface of a preserved specimen of *Neopilina*. (D) Ventral view, foot removed. (E) One of the gills.

known only from lower Paleozoic fossils. Since then their unusual anatomy has been a source of much evolutionary speculation. Monoplacophorans are limpet-like in appearance, living species are less than 3 cm in length, and most live at considerable depths. About 30 described species, in 8 genera (*Adenopilina*, *Laevipilina*, *Monoplacophorus*, *Neopilina*, *Rokopella*, *Veleropilina*, *Vema*, *Micropilina*).

CLASS POLYPLACOPHORA Chitons (Figures 13.1B and 13.4). Flattened, elongated molluscs with a broad ventral foot and 8 dorsal shell plates (composed of aragonite); mantle forms thick girdle that borders and may partly or entirely cover shell plates; epidermis of girdle usually with calcareous spines, scales, or bristles; mantle cavity encircles foot and bears from 6 to more than 80 pairs of bipectinate ctenidia; 1 pair nephridia; head without eyes or tentacles; crystalline style, statocysts and osphradia absent; nervous system lacking discrete ganglia, except in buccal region; well-developed radula present. Shell canals (aesthetes) sometimes have shell eyes (Figure 13.43C,D). Marine, intertidal to deep sea. Chitons are unique in their possession of 8 separate shell plates, called valves, and a thick marginal girdle; about 850 described species in one living order.[4]

ORDER NEOLORICATA Shells with unique articulamentum layer, which forms insertion plates that interlock the valves.

SUBORDER LEPIDOPLEURIDA Chitons with outer edge of shell plates lacking attachment teeth; girdle not extending over plates; ctenidia limited to a few posterior pairs. (e.g., *Choriplax*, *Lepidochiton*, *Lepidopleurus*, *Oldroydia*)

SUBORDER CHITONIDA Outer edges of shell plates with attachment teeth; girdle not extending over plates, or extending partly over plates; ctenidia occupying most of mantle groove, except near anus. (e.g., *Callistochiton*, *Chaetopleura*, *Ischnochiton*,

[4]Uncommon, aberrant individuals have been found with only 7 valves.

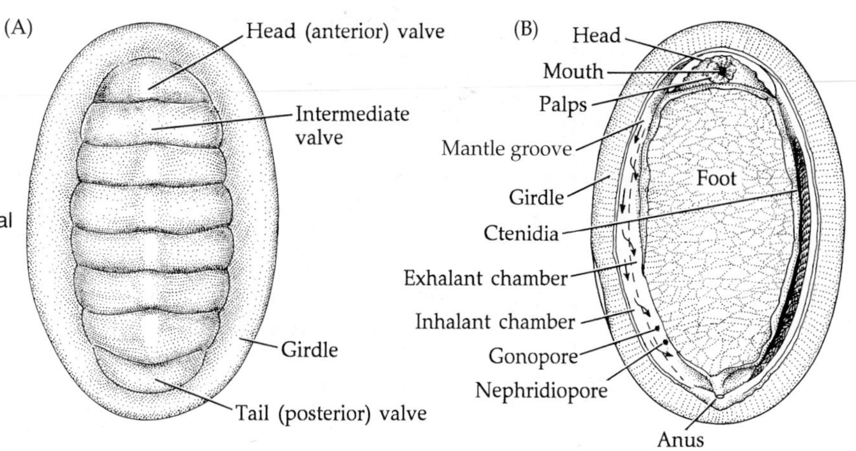

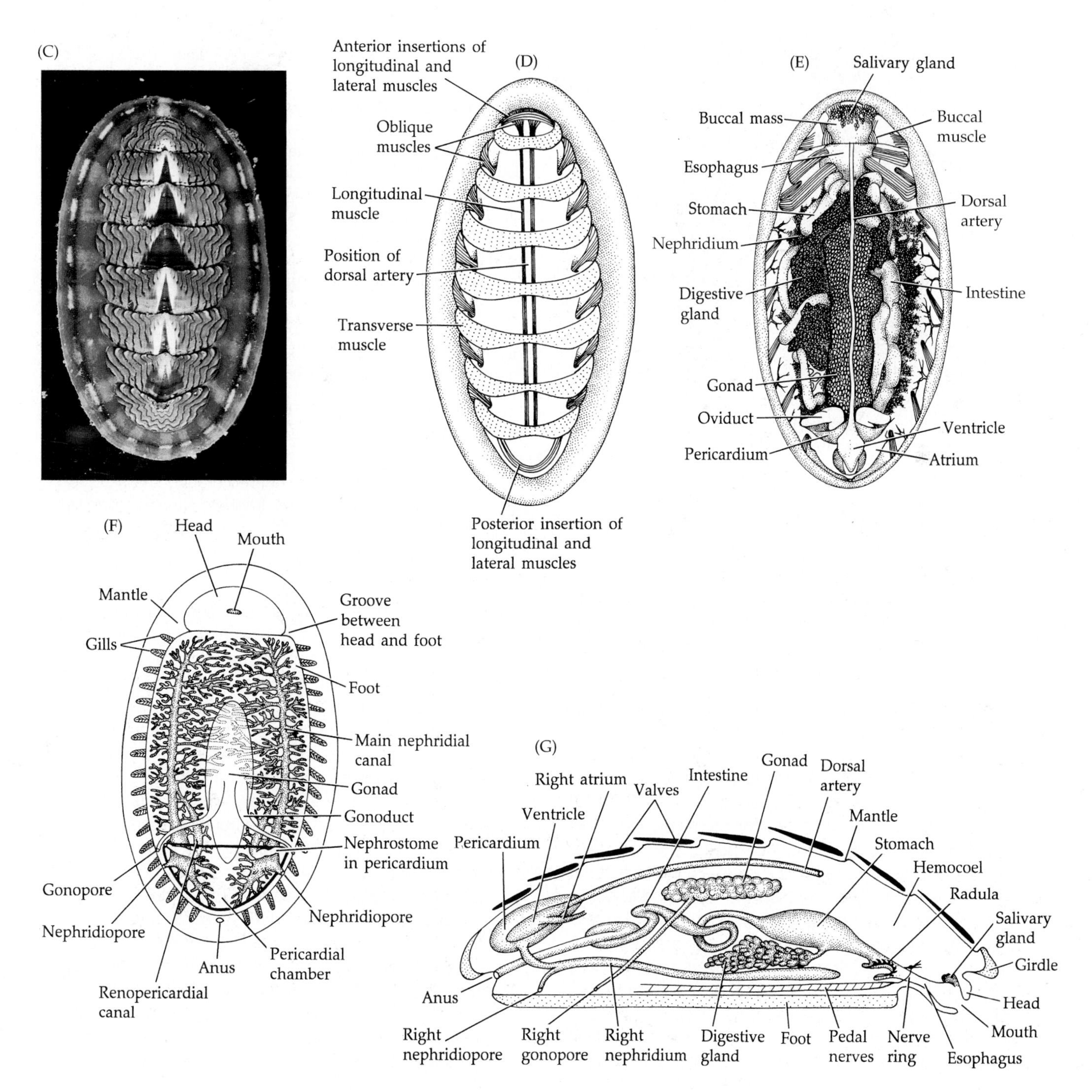

Figure 13.4 Generalized anatomy of chitons (Polyplacophora). (A,B) A typical chiton (dorsal and ventral views). (C) The Pacific lined chiton, *Tonicella lineata*. (D) Dorsal view of a chiton, shell plates (valves) removed. (E) Dorsal view of a chiton, dorsal musculature removed to reveal internal organs. (F) Dorsal view of a chiton, showing extensive nephridia. (G) The arrangement of internal organs in a chiton (lateral view).

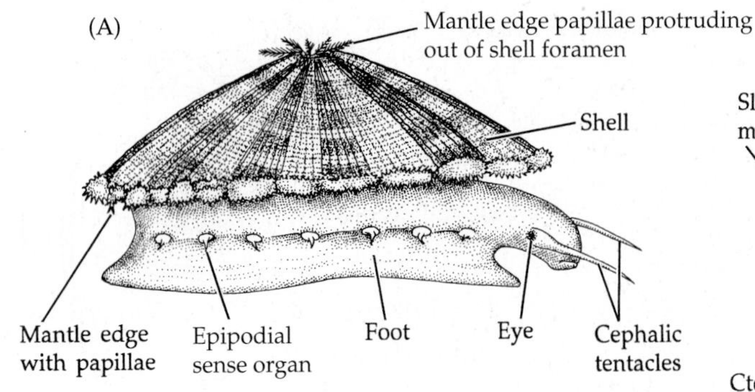

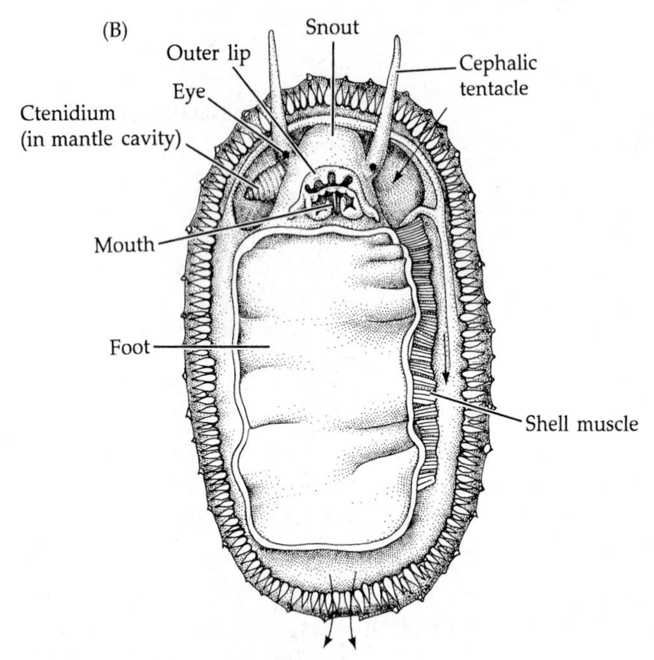

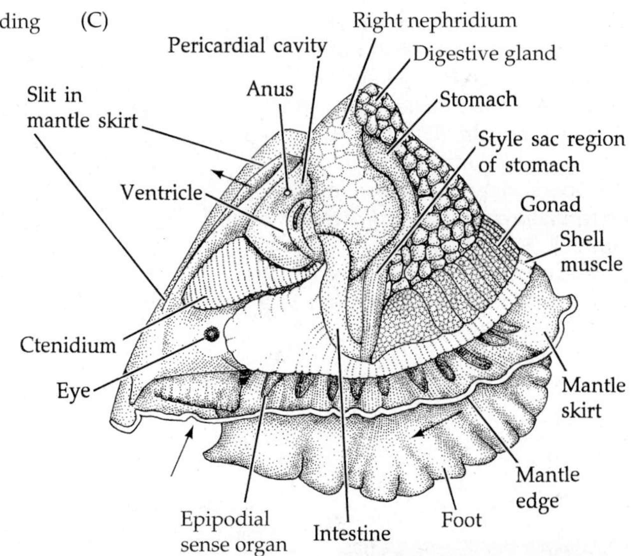

Figure 13.5 General anatomy of limpet-like gastropods. (A) The vetigastropod limpet *Fissurella* (Fissurellidae) (lateral view). (B) The patellogastropod limpet *Lottia* (Lottiidae) (ventral view). The arrows indicate the direction of water currents. (C) The vetigastropod limpet *Puncturella* (Fissurellidae), removed from shell and seen from the left. The arrows indicate the water currents. Certain structures are visualized through the mantle skirt: ctenidium, eye, anus, and epipodial sense organs.

Katharina, *Lepidozona*, *Mopalia*, *Nuttallina*, *Placiphorella*, *Schizoplax*, *Tonicella*)

SUBORDER ACANTHOCHITONIDA. Outer edge of shell plates with well-developed attachment teeth; shell valves partially or completely covered by girdle; ctenidia do not extend full length of foot. (e.g., *Acanthochitona*, *Cryptochiton*, *Cryptoplax*)

CLASS GASTROPODA Snails, limpets and slugs (Figures 13.1D–G, 13.5, 13.6, and 13.7). Asymmetrical molluscs with single, usually spirally coiled shell into which body can be withdrawn; shell lost or reduced in many groups; during development, visceral mass and mantle rotate 90–180° on foot (a process known as torsion), so mantle cavity lies anteriorly or on right side (rather than posteriorly as in other molluscs), and gut and nervous system are twisted; some taxa have partly or totally reversed the rotation (detorsion); with muscular creeping foot (modified in swimming and burrowing taxa); foot with operculum in larva and often in adult; head with eyes (often reduced or lost), and 1–2 pairs of tentacles, and a snout; most with radula and some with crystalline style, the latter being absent in most primitive and in many advanced groups; 1–2 nephridia; mantle (= pallium) usually forms anterior cavity housing ctenidia, osphradia, and hypobranchial glands; ctenidia sometimes lost and replaced with secondary gas exchange structures.

Gastropods comprise approximately 70,000 described living species of marine, terrestrial, and freshwater snails and slugs. The class was traditionally divided into three subclasses: prosobranchs (largely shelled marine snails), opisthobranchs (marine slugs), and pulmonates (terrestrial snails and slugs). However, recent anatomical and molecular studies have shown that classification to be incorrect, as reflected in the classification below.

SUBCLASS PATELLOGASTROPODA Cap-shaped (limpets) with porcelaneous, nonnacreous shell; operculum absent in adult; cephalic tentacles with eyes at outer bases; radula docoglossate, with iron impregnated teeth, rest of gut with large esophageal glands and simple stomach lacking a crystalline style; intestine long and looped; gill configuration variable, single bipectinate ctenidium sometimes present (Figure 13.5B), and/or with mantle groove secondary gills, or gills lacking; shell muscle divided into discrete bundles; mantle cavity without siphon or hypobranchial glands; 2 rudimentary osphradia; single atria; 2 nephridia; usually gonochoristic; nervous system weakly concentrated, pleural ganglia near pedal ganglia, pedal and lateral cords present. Primarily marine with a few estuarine species; herbivorous. The patellogastropods include 6 families: Patellidae (e.g., *Patella*, *Scutellastra*), Nacellidae (e.g., *Cellana*), Lottiidae (e.g., *Lottia*), Acmaeidae (e.g., *Acmaea*), Lepetidae (e.g., *Lepeta*), and Neolepetopsidae (e.g., *Neolepetopsis*). These are often regarded as the "true" limpets.

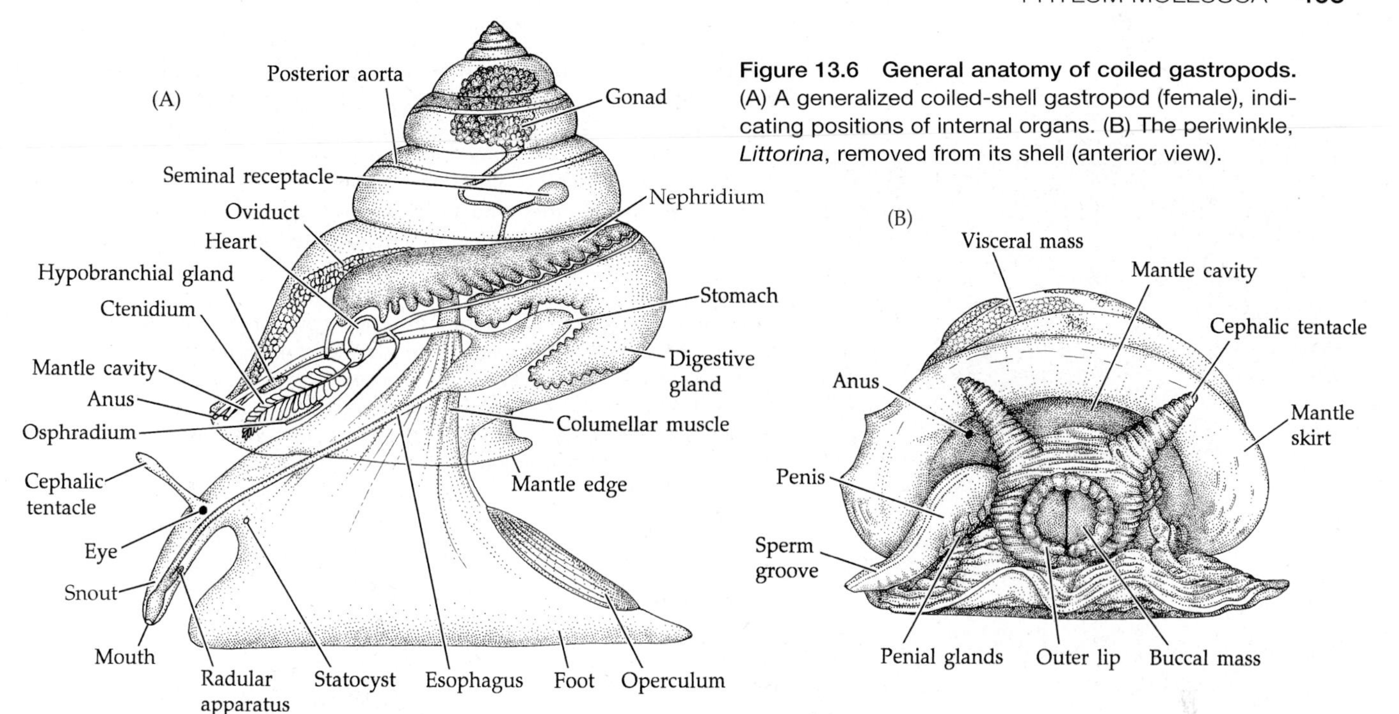

Figure 13.6 General anatomy of coiled gastropods. (A) A generalized coiled-shell gastropod (female), indicating positions of internal organs. (B) The periwinkle, *Littorina*, removed from its shell (anterior view).

SUBCLASS VETIGASTROPODA Shells both porcelaneous and nacreous; cephalic tentacles usually with eyes on short processes on outer bases; operculum usually circular, with a central nucleus and often many spirals, horny or calcareous; radula usually rhipidoglossate (with numerous transverse rows of teeth), rest of gut with esophagus having large glands, complex stomach with style sac but no crystalline style, looped intestine; 1–2 bipectinate ctenidia; shell muscles paired or single; mantle cavity with 2 hypobranchial glands, 2 atria, and 2 nephridia; usually gonochoristic; male generally without penis; nervous system weakly concentrated, ganglia poorly formed, pedal cords present; 1–2 osphradia, small, inconspicuous. All marine and benthic. Many species are microdetritivores or feed on films of bacteria or other organisms, or are microherbivores; some are macroherbivores, some grazing carnivores, and a few suspension feeders. Most gastropods found at hydrothermal vents, cold seeps, and on deep sea hard substrates are vetigastropods. Vetigastropods comprise about 30 families and while the internal classification remains unsettled, three main groups are often recognized, which we treat here as orders.

ORDER TROCHIDA Most of the vetigastropods, including the slit-shelled snails Pleurotomaridae (e.g., *Perotrochus*, *Pleurotomaria*), Scissurellidae (e.g., *Scissurella*) and Anatomidae (e.g., *Anatoma*), the abalones Haliotidae (e.g., *Haliotis*), the keyhole and slit limpets Fissurellidae (e.g., *Diodora*, *Fissurella*, *Lucapinella*, *Puncturella*), deep sea limpets comprising the Lepetellidae and related families (e.g., *Lepetella*, *Pseudococculina*), trochids Trochidae (e.g., *Trochus, Monodonta*) and related families such as Calliostomatidae (e.g., *Calliostoma*), Margaritidae (e.g., *Margarites*), Tegulidae (e.g., *Tegula*), and turbans Turbinidae (e.g., *Turbo*, *Astrea*).

ORDER NEOMPHALIDA Comprises many of the hot vent snails and limpets Neomphalidae (e.g., *Neomphalus*), Peltospiridae (e.g., *Peltaspira*), Lepetodrilidae (e.g., *Lepetodrilus*).

ORDER COCCULINIDA (= COCCULINIFORMES IN PART) The small, deep-sea wood and bone limpets Cocculinidae (e.g., *Cocculina*).

SUBCLASS NERITIMORPHA Shell coiled, limpet-like, or lost (Titiscaniidae). Shell porcelaneous, with interior whorls reabsorbed in many coiled groups; operculum typically present, of few spirals and with non-central nucleus, horny or calcified, usually with internal peg; shell muscle divided into discrete bundles; only left ctenidium present; hypobranchial glands often lost on left side; stomach highly modified; right nephridium incorporated into complex reproductive system with multiple openings into mantle cavity; radula rhipidoglossate; most species gonochoristic, with copulatory structures; nervous system with ganglia concentrated, pleural ganglia near pedal ganglia, pedal cords present. Globally distributed in marine, estuarine, freshwater, and terrestrial habitats. There are 9 families of neritimorphans, four of which, Helicinidae (e.g., *Alcadia, Helicinia*), Hydrocenidae (*Hydrocena, Georissa*), Proserpinellidae (e.g., *Proserpinella*), and Proserpinidae (*Proserpina*) are exclusively terrestrial; also Neritopsidae (neritopsids, *Neritopsis*), Titiscaniidae (titiscaniid, *Titiscania*), Neritidae (nerites, e.g., *Nerita*, *Theodoxus*), Neritiliidae (cave nerites, e.g., *Pisulina, Neritilia*) and Phenacolepadidae (*Phenacolepas*).

SUBCLASS CAENOGASTROPODA Shell mainly porcelaneous; operculum usually present and corneous, rarely calcified, with few spirals and usually with a non-central nucleus, mostly non-nacreous, rarely with internal peg(s); head with pair of cephalic tentacles, with eyes at outer bases; mantle cavity asymmetrical, with incurrent

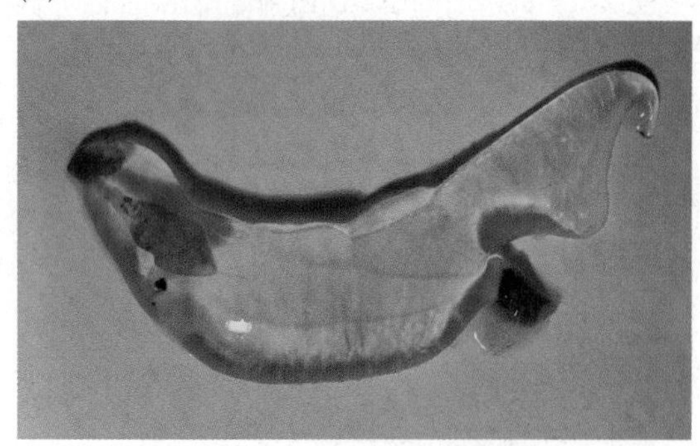

Figure 13.7 More gastropod anatomy; some caenogastropods (A–C) and heterobranchs (D–J). (A) The pelagic shelled heteropod *Carinaria* (Caenogastropoda). (B) Anatomy of *Carinaria*. (C) The shell-less heteropod *Pterotrachea* (Caenogastropoda). (D) The pelagic shelled pteropod *Clio* (Heterobranchia: Euopisthobranchia). The arrows indicate the direction of water flow; water enters all around the narrow neck and is forcibly expelled together with fecal, urinary, and genital products by contraction of the sheath. (E) A swimming pteropod, *Corolla* (Heterobranchia: Euopisthobranchia). (F–I) Various nudibranchs (Heterobranchia: Nudipleura). (F) A dorid nudibranch, *Diaulula*. (G) An aeolid nudibranch, *Phidiana*. (H) A dorid nudibranch, *Chromodoris geminus* (from the Red Sea). (I) The eastern Pacific "Spanish shawl" aeolid nudibranch, *Flabellina*. (J) The lettuce sea slug, *Tridachia crispata* (Heterobranchia: Panpulmonata), from the Caribbean.

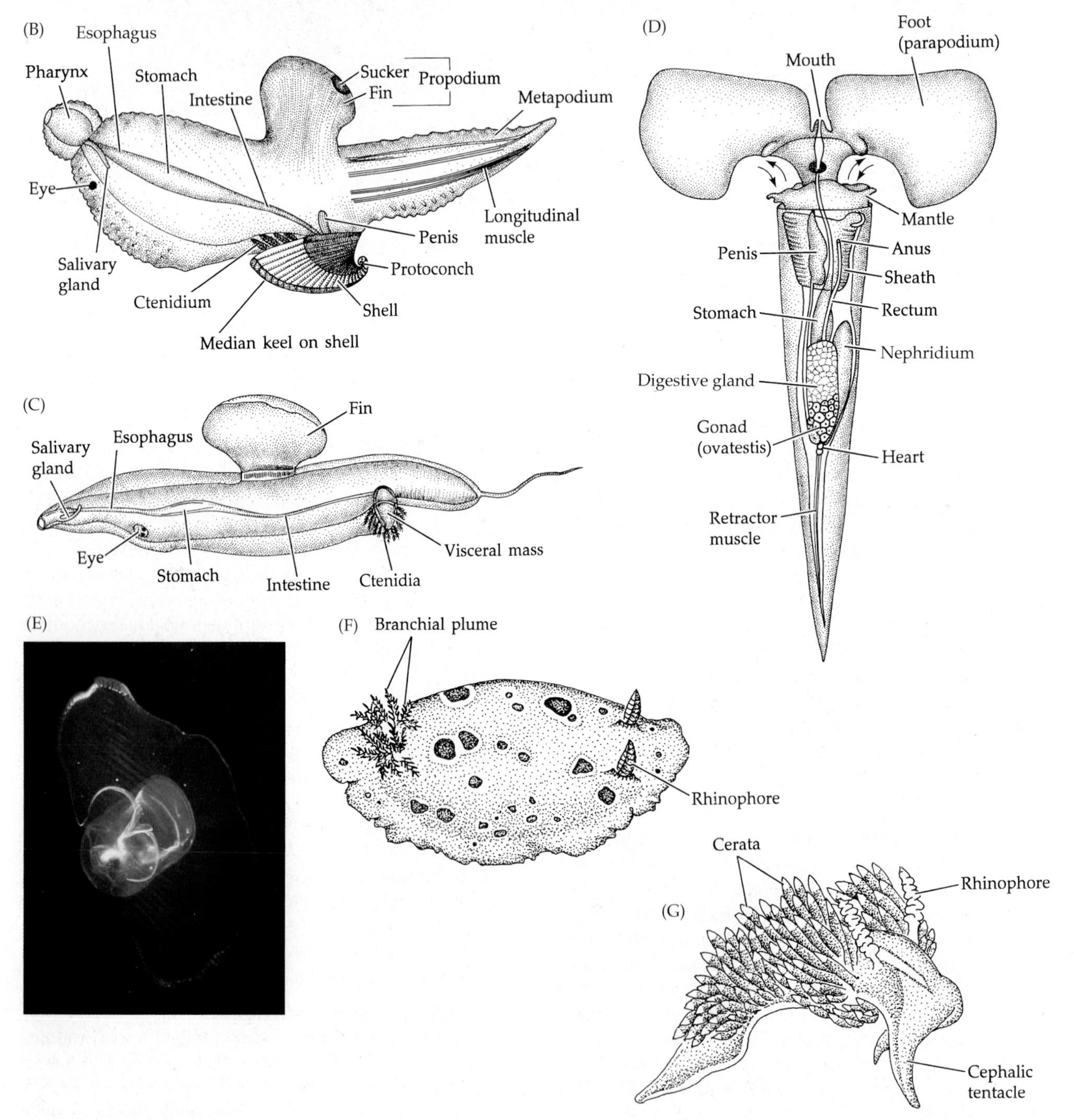

(H)
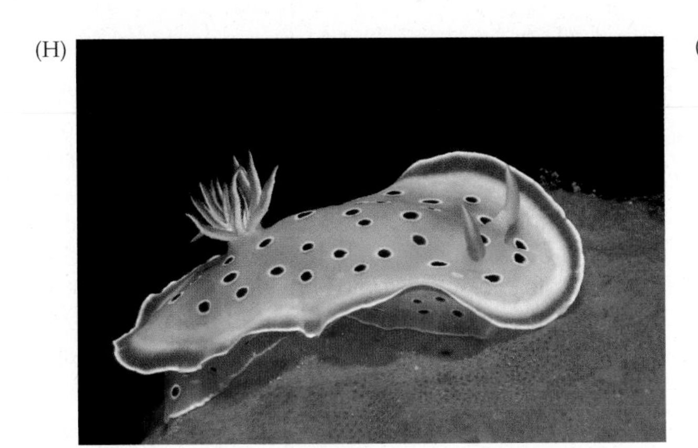

(I)

(J)

opening on anterior left, sometimes elaborated into an inhalant siphon; right ctenidium lost; left ctenidium monopectinate; left hypobranchial gland lost; right nephridium lost except for remnant incorporated into reproductive system; heart with only left atrium. Radula taenioglossate (7 rows of teeth), ptenoglossate (many rows of similar teeth), rachiglossate (1–3 rows of teeth), or toxoglossate (teeth modified as harpoons), or occasionally lost. Higher forms with concentrated ganglia, pleural ganglia usually near cerebral ganglia, pedal cords usually absent; osphradium conspicuous, often large, sometimes surface subdivided into lamellae. Most caenogastropods are gonochoristic. The caenogastropods, comprise the former "mesogastropods" and neogastropods, and they are often divided into two groups, as follows:

"ARCHITAENIOGLOSSA" Although this is not a monophyletic group, we retain it informally. Architaenioglossans differ from other caenogastropods in details of their nervous system and in the ultrastructure of their sperm and osphradia. They are divided among 10 families, including the freshwater Ampullariidae (apple snails, e.g., *Ampullaria, Pomacea, Pila*) and Viviparidae (river snails, e.g., *Viviparus*), and the terrestrial Cyclophoridae (e.g., *Cyclophorus*) and several related families such as Diplommatinidae (e.g., *Diplommatina* and *Opisthostoma*).

INFRACLASS SORBEOCONCHA This grouping contains all the rest of the caenogastropods. These are divided into two main groups, Cerithiomorpha and Hypsogastropoda.

SUPERORDER CERITHIOMORPHA Usually without a penis; eggs usually laid in jelly, often in strings, or are brooded. The anterior aperture may or may not have a notch, which houses a short siphon. Include marine, brackish, and freshwater species. About 19 families are recognized, including the marine Campanilidae (e.g., *Campanile*), Cerithiidae (horn shells, e.g., *Cerithidea, Cerithium*, *Liocerithium*), Siliquariidae (slit worm shells, e.g., *Siliquaria*), and Turritellidae (tower or turret shells, e.g., *Turritella*); and the freshwater Melanopsidae (e.g., *Melanopsis*), Thiaridae (e.g., *Thiara*), and Pleuroceridae (e.g., *Pleurocera*).

COHORT HYPSOGASTROPODA Comprises the remaining caenogastropods. The anterior mantle may be simple or can be enrolled forming an anterior siphon which emerges from an anterior notch in the aperture or, in some, is contained within an extension of the shell, the siphonal canal. Male with cephalic penis; eggs usually laid in capsules or sometimes brooded. Nervous system concentrated; operculum, if present, chitinous, rarely calcareous. This large group is divided into the Littorinimorpha and Neogastropoda.

SUPERORDER LITTORINIMORPHA Classification unsettled; includes the marine grazing snails Littorinidae (periwinkles, e.g., *Littorina*), a number of small-sized marine families including the Rissoidae (e.g., *Rissoa*, *Alvania*), and larger snails such as the Strombidae (conchs or strombids, e.g., *Strombus*), and the carrier shells Xenophoridae (e.g., *Xenophora*). Also includes the uncoiled suspension-feeding "worm" gastropods Vermetidae (e.g., *Serpulorbis*, *Dendropoma*) and the limpet-like Hipponicidae (e.g., *Hipponix*) which are deposit feeders, while Capulidae (e.g., *Capulus*) attach to other molluscs and mostly feed on their feces. The slipper shells Calyptraeidae (e.g., *Calyptraea*, *Crepidula*, *Crucibulum*) are suspension-feeders. The Carinariidae (one of several families of pelagic molluscs collectively called heteropods, e.g., *Carinaria*) also have a cap-shaped shell.[5] Cypraeidae (cowries, e.g., *Cypraea*) are herbivores or grazing carnivores, while several other littorinimorph snail-like families are strictly carnivorous, including Naticidae (moon snails, e.g., *Natica*, *Polinices*) that feed mostly on bivalves, the ascidian-feeding Eratoidae (coffee bean shells, e.g., *Erato*, *Trivia*), and the soft coral feeding Ovulidae (ovulids or egg shells, e.g., *Jenneria*, *Ovula*, *Simnia*). Tonnidae (tun shells, e.g., *Malea*) and related families such as Cassididae (helmet shells, e.g., *Cassis*) mainly feed on echinoderms, whereas Ficidae (fig shells, e.g., *Ficus*) are primarily polychaete feeders. Epitoniidae (wentletraps or epitoniids, e.g., *Epitonium*) feed on cnidarians, while the floating violet snails Janthinidae (e.g., *Janthina*) feed on siphonophores that drift on the surface of the ocean. Eulimidae are ectoparasites on echinoderms, and the sponge-feeding Triphoridae (e.g., *Triphora*) and Cerithiopsidae (e.g., *Cerithiopsis*) are highly diverse. There are some diverse families of small-sized freshwater snails such as the Hydrobiidae (e.g., *Hydrobia*) and several related families including the Pomatiopsidae (e.g., *Pomatiopsis*, *Tricula*), and there are also a few terrestrial taxa in families such as Pomatiasidae (e.g., *Pomatias*) and the otherwise mainly supralittoral Assimineidae.

SUPERORDER NEOGASTROPODA The most derived hypsogastropod clade. Radula rachiglossate or toxoglossate, with 1–5 teeth in each row; anterior siphon present; operculum, if present, chitinous; osphradium large and pectinate, lying near base of siphon. This highly diverse group comprises mostly carnivorous taxa.

The neogastropods comprise more than 30 living families of almost entirely marine snails, including: whelks such as Buccinidae (e.g., *Buccinum*, *Cantharus*, *Macron*, and the Asian freshwater genus *Clea*); Fasciolariidae (tulip shells and spindle shells, e.g., *Fasciolaria*, *Fusinus*, *Leucozonia*, *Troschelia*); Melongenidae (e.g., *Melongena*); Nassariidae (dog whelks and basket shells, e.g., *Nassarius*); dove shells Columbellidae (e.g., *Anachis*, *Columbella*, *Mitrella*, *Pyrene*, *Strombina*); harp shells Harpidae (e.g., *Harpa*); margin shells Marginellidae (e.g., *Marginella*, *Granula*); miter shells Mitridae (e.g., *Mitra*, *Subcancilla*) and Costellariidae (e.g., *Vexillum*, *Pusia*); rock shells and thaids Muricidae (e.g., *Hexaplex*, *Murex*, *Phyllonotus*, *Pterynotus*, *Acanthina*, *Morula*, *Neorapana*, *Nucella*, *Purpura*, *Thais*) and the related coral associated Coralliophilidae (e.g., *Coralliophila*, *Latiaxis*); Olividae (olive shells, e.g., *Agaronia*, *Oliva*); Olivellidae (e.g., *Olivella*); the volutes Volutidae (e.g., *Cymbium*, *Lyria*, *Voluta*) and nutmeg shells Cancellariidae (e.g., *Admete*, *Cancellaria*); cone shells Conidae (e.g., *Conus*) and the related Turridae (e.g., *Turris*) and several other allied families including the auger shells Terebridae (e.g., *Terebra*).

SUBCLASS HETEROBRANCHIA The heterobranchs were previously organized as two subclasses—Opisthobranchia (sea slugs and their kin) and Pulmonata (air breathing snails). Although this division was long accepted, recent morphological and molecular studies now divide the subclass into two main groups—an informal paraphyletic group often referred to as the "Lower Heterobranchia" (= Allogastropoda, Heterostropha) and the Euthyneura, which includes both pulmonates and the opisthobranchs.

The subclass Heterobranchia characterized by lacking a true ctenidium and, usually, a small to absent osphradium, a simple gut with the esophagus lacking glands, the stomach lacking a crystalline style in all but one group, and the intestine usually being short. The radula is highly variable ranging from rhipidoglossate to a single row of teeth or lost altogether. The shell may be well-developed, reduced, or absent; the operculum, if present, is horny; the larval shell is heterostrophic (i.e., coils in a different plane to the adult shell). The head bears one or two pairs of tentacles, with the eyes variously placed; all are hermaphroditic. The nervous system is streptoneurous or euthyneurous with various degrees of concentration of the ganglia; pleural ganglia near pedal or cerebral ganglia, pedal cords absent. Mostly benthic; with marine, freshwater and terrestrial species.

"LOWER HETEROBRANCHS" This informal group includes some snails long thought to be "Mesogastropoda," such as staircase or sundial shells Architectonicidae (e.g., *Architectonica, Philippia*) and some groups of small-sized marine snails including Rissoellidae (e.g., *Rissoella*), Omalogyridae (e.g., *Omalogyra*), and the freshwater Valvatidae (e.g., *Valvata*) and marine relatives such as Cornirostridae. These snails are superficially similar to caenogastropods, but often possess secondary gills and long cephalic tentacles with cephalic eyes set in the middle of their bases or on their inner sides. Another group included here are tiny interstitial slugs of the family Rhodopidae.

INFRACLASS EUTHYNEURA Include most of the former opisthobranchs and pulmonates. The euthyneuran body is characterized by: the shell being external or internal, or lost altogether; a heterostrophic larval shell; operculum horny, often absent in adult; body variously detorted; head usually with one or two pairs of tentacles, eyes on inner sides or on separate stalks; ctenidia and mantle cavity usually reduced or lost; hermaphroditic; euthyneurous with various degrees of nervous system concentration. Mostly benthic; with marine, freshwater, and terrestrial species.

The Euthyneura is divided into three major groups which we treat here as cohorts.

COHORT NUDIPLEURA Includes both the internal-shelled Pleurobranchidae (e.g., *Berthella*, *Pleurobranchus*) and the Nudibranchia (shell-less or "true"

[5]The term "heteropod" is an old taxonomic name now used informally for a group of planktonic, predatory caenogastropods that have a reduced shell or no shell at all.

nudibranchs) which includes many families, some examples being the doridoid nudibranchs such as Onchidorididae (e.g., *Acanthodoris, Corambe*), Polyceridae (e.g., *Polycera, Tambja*), Aegiretidae (e.g., *Aegires*), Chromodorididae (e.g., *Chromodoris*), Phyllidiidae (e.g., *Phyllidia*), Dendrodorididae (e.g., *Dendrodoris*), Discodorididae (e.g., *Discodoris, Diaulula, Rostanga*), Dorididae (e.g., *Doris*), Platydorididae (e.g., *Platydoris*), Hexibranchidae (e.g., *Hexabranchus*), Goniodorididae (e.g., *Okenia*), and the cladobranch nudibranchs including the Arminidae (*Armina*), Proctonotidae (e.g., *Janolus*), Embletoniidae (e.g., *Embletonia*), Scyllaeidae (e.g., *Scyllaea*), and Dendronotidae (e.g., *Dendronotus*). Also included are the cladobranch group collectively known as aeolididoids—including the Aeolidiidae (e.g., *Aeolidia*), Flabellinidae (e.g., *Coryphella*), Fionidae (e.g., *Fiona*), Facelinidae (e.g., *Hermissenda, Phidiana*), Tergipedidae (e.g., *Trinchesia*), Tethydidae (e.g., *Melibe*), and Glaucidae (e.g., *Glaucus*).

COHORT EUOPISTHOBRANCHIA Includes six main groups that could be treated at the ordinal level: (1) the basal acteonoideans including Acteonidae (barrel or bubble snails, e.g., *Acteon, Pupa, Rictaxis*); (2) several families grouped as Cephalaspidea, for example the slugs Aglajidae (e.g., *Aglaja, Chelidonura, Navanax*), Bullidae (bubble shells, e.g., *Bulla*), Haminoeidae (e.g., *Haminoea*), Retusidae (e.g., *Retusa*), and Scaphandridae (e.g., *Scaphander*); (3) the Runcinoidea, containing two families of tiny slugs, Ilbiidae (e.g., *Ilbia*) and Runcinidae (e.g., *Runcina*); (4) the Aplysiomorpha (= Anaspidea) or sea hares, including Aplysidae (e.g., *Aplysia, Dolabella, Stylocheilus*); (5) the pelagic pteropods, comprising two distant groups, the Thecosomata or shelled pteropods, which include the families Cavoliniidae (e.g., *Clio, Cavolinia*) and Limacinidae, (e.g., *Limacina*), and Gymnosomata or naked pteropods including Clionidae (e.g., *Clione*); and (6) the Umbrachulida, composed of the umbrella slugs, Umbraculidae (e.g., *Umbraculum*) and Tylodinidae (e.g., *Tylodina*).

COHORT PANPULMONATA This highly diverse group is characterized by: variable shell shape or loss of shell, minute or in moderate size; generally spirally coiled, planispiral, or limpet-shaped; usually without an operculum as adults; eyes at bases of sensory stalks; secondary gills present in some members (e.g., *Pyramidella, Siphonaria*); body detorted; nervous system highly concentrated (euthyneurous); mantle cavity-derived lung in the derived groups, with a contractile aperture in the Eupulmonata; marine (intertidal), brackish, freshwater and amphibious; includes freshwater limpets and the minute, meiofaunal Acochlidioidea (e.g., *Acochlidium, Unela*).

Other panpulmonate groups include: the sap-sucking sea slugs Sacoglossa (e.g., *Berthelinia, Elysia, Oxynoe, Tridachia,* and the "bivalve gastropods" Juliidae), which are shelled or shell-less; the small shell-less and sometimes spiculate acochlidioid slugs that are often interstitial and usually marine (although there are some freshwater species) and the ectoparasitic Pyramidellidae (e.g., *Odostomia, Pyramidella, Turbonilla, Amathina*), all of which were previously included in the Opisthobranchia. The remaining panpulmonates include all the members of the group previously known as Pulmonata, namely the mainly intertidal Siphonariidae (false limpets: e.g., *Siphonaria, Williamia*); two operculate families); the freshwater Glacidorbidae (e.g., *Glacidorbis*); the estuarine Amphibolidae (*Amphibola, Salinator*); the Hygrophila, including the mainly freshwater South American Chilinidae (e.g., *Chilina*), and freshwater Physidae (e.g., *Physa*), Planorbidae (e.g., *Bulinus, Planorbis, Ancylus*) and Lymnaeidae (e.g., *Lymnaea, Lanx*), these latter families being mostly snails, but some such as *Lanx* and *Ancylus* are limpets. The remaining "pulmonates" are contained within a superorder, Eupulmonata. The best known and largest group of eupulmonates is the order Stylommatophora, comprising the land snails and slugs. In some of the shelled forms, the shell is partly or completely enveloped by dorsal mantle. Their eyes are on the tips of long sensory stalks and there is an anterior pair of tentacles. Eupulmonates are all terrestrial and are an enormous group with over 26,000 described species in 104 families. Some of those included are the land snail families Helicidae (e.g., *Cornu* [= *Helix*], *Cepaea*), Achatinidae (e.g., *Achatina*), Bulimulidae (e.g., *Bulimulus*), Haplotrematidae (e.g., *Haplotrema*), Orthalicidae (e.g., *Liguus*), Cerionidae (e.g., *Cerion*), Oreohelicidae (e.g., *Oreohelix*), Pupillidae (e.g., *Pupilla*), Cerastidae (e.g., *Rhachis*), Succineidae (e.g., *Succinea*), and Vertiginidae (e.g., *Vertigo*), as well as terrestrial slug families such as Arionidae (e.g., *Arion*) and Limacidae (e.g., *Limax*).

The remaining Eupulmonata include the orders Systellommatophora and Ellobiacea. The former are sluglike, without internal or external shell; dorsal mantle integument forms a keeled or rounded notum; head usually with 2 pairs tentacles, upper ones forming contractile stalks bearing eyes. Included are the mainly marine family Onchidiidae (e.g., *Onchidella, Onchidium*) and the terrestrial Veronicellidae (e.g., *Veronicella*). The Ellobiacea includes the three superfamilies; the mainly supralittoral hollow-shelled ear snails Ellobioidea (, e.g., *Ellobium, Melampus, Carychium, Ovatella*), small intertidal snails or slugs of the Otinoidea (e.g., *Otina*), and the limpet-like intertidal Trimusculoidea (e.g., *Trimusculus*).

CLASS BIVALVIA (= PELECYPODA, = LAMELLIBRANCHIATA) Clams, oysters, mussels, scallops, etc. (Figures 13.1M–Q and 13.8). Laterally compressed; shell typically of two valves hinged together dorsally by elastic ligament and usually by shell-teeth; shells closed by adductor muscles derived from mantle muscles; head rudimentary, without eyes, tentacles or radula, but eyes may occur elsewhere on body; pair of large labial palps present composed of inner and outer parts that lie against one another; pair of statocysts present, associated with pedal ganglia, foot typically laterally compressed, often without a sole; 1 pair of large

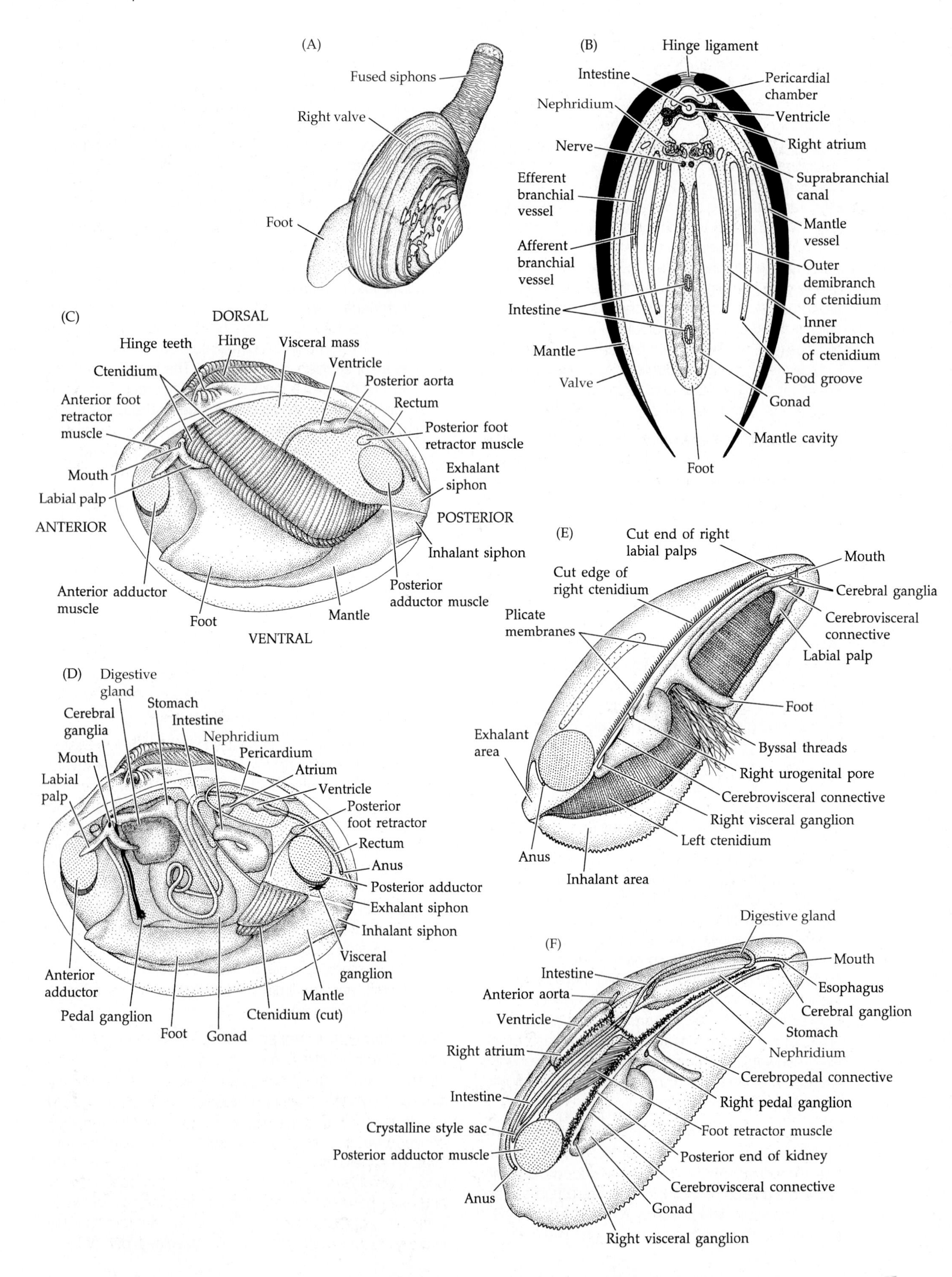

(A)
Fused siphons
Right valve
Foot
(B)
Hinge ligament
Intestine
Pericardial chamber
Nephridium
Ventricle
Right atrium
Nerve
Suprabranchial canal
Efferent branchial vessel
Mantle vessel
Afferent branchial vessel
Outer demibranch of ctenidium
Intestine
Inner demibranch of ctenidium
Mantle
Food groove
Valve
Gonad
Mantle cavity
Foot
(C)
DORSAL
Hinge teeth
Hinge
Visceral mass
Ctenidium
Ventricle
Posterior aorta
Anterior foot retractor muscle
Rectum
Posterior foot retractor muscle
Mouth
Exhalant siphon
Labial palp
ANTERIOR
POSTERIOR
Inhalant siphon
Anterior adductor muscle
Posterior adductor muscle
Foot
Mantle
VENTRAL
(D)
Digestive gland
Cerebral ganglia
Stomach
Intestine
Nephridium
Pericardium
Mouth
Atrium
Labial palp
Ventricle
Posterior foot retractor
Rectum
Anus
Posterior adductor
Exhalant siphon
Inhalant siphon
Visceral ganglion
Anterior adductor
Mantle
Pedal ganglion
Ctenidium (cut)
Foot
Gonad
(E)
Cut end of right labial palps
Mouth
Cut edge of right ctenidium
Cerebral ganglia
Cerebrovisceral connective
Plicate membranes
Labial palp
Foot
Exhalant area
Byssal threads
Right urogenital pore
Cerebrovisceral connective
Right visceral ganglion
Left ctenidium
Anus
Inhalant area
(F)
Digestive gland
Mouth
Intestine
Esophagus
Anterior aorta
Cerebral ganglion
Ventricle
Stomach
Nephridium
Right atrium
Cerebropedal connective
Intestine
Right pedal ganglion
Crystalline style sac
Foot retractor muscle
Posterior adductor muscle
Posterior end of kidney
Cerebrovisceral connective
Anus
Gonad
Right visceral ganglion

◀ **Figure 13.8 General anatomy of bivalves.** (A) *Tresus*, a deep-burrowing eulamellibranch (Mactridae), with a digging foot and long, fused siphons. (B) A typical eulamellibranch (cross section). (C) The eulamellibranch *Mercenaria* (Veneridae), with the left shell valve and mantle removed. (D) Internal anatomy of *Mercenaria*. The visceral mass is opened up, the foot is dissected, and most of the gills are cut away. (E) The common mussel, *Mytilus* (Mytilidae), seen from the right side after removal of the right shell valve and mantle. (F) *Mytilus*, with the visceral mass opened up, the foot dissected, and most of the gills cut away.

bipectinate ctenidia; large mantle cavity surrounds animal; mantle may be variously fused, sometimes forming extensions (siphons); 1 pair nephridia; nervous system simple, typically composed of cerebropleural, pedal and visceral ganglia

Bivalves are marine or freshwater molluscs, primarily microphagous or suspension feeders. The class includes about 9,200 living species represented at all depths and in all marine environments. Bivalve classification has been in a state of turmoil over the past 50 years and continues to be unsettled. Higher taxa have been delimited on the basis of shell characters (e.g., hinge anatomy, position of muscle scars), or, in other classifications, internal organ anatomy (e.g., ctenidia, stomach) have been used. However, beginning with the work of Giribet and Wheeler (2002), bivalve taxonomy has become more stable as both molecules and morphology have been combined with the fossil record to understand the relationships of the class.

SUBCLASS PROTOBRANCHIA Includes the former Palaeotaxodonta in part. Ctenidia are 2 pairs of simple, unfolded, bipectinate, plate-like leaflets suspended in the mantle cavity. The ctenidia are mainly respiratory structures, while the labial palps are the primary food collecting organs. These are the most primitive living bivalves, comprising two superorders.

SUPERORDER NUCULIFORMII (= OPPONO-BRANCHIA) Mantle open, with inhalant water entering anteriorly; shells with nacre and gill filaments along ctenidial axis arranged opposite one another; foot longitudinally grooved and with a plantar sole, without byssal gland; nervous system primitive, often with incomplete union of cerebral and pleural ganglia. Two orders.

ORDER NUCULIDA Shell aragonitic, interior nacreous or porcelaneous; periostracum smooth; shell valves equal and taxodont (i.e., the valves have a row of similar interlocking short teeth along the hinge margins); adductor muscles equal in size; with large labial palps extended as proboscides used for food collection; ctenidia small, for respiration; marine (particularly in the deep sea), mainly infaunal detritivores. (e.g., Nuculidae, *Nucula*).

ORDER SOLEMYIDA (= CRYPTODONTA) Shell valves thin, elongate, and equal in size; uncalcified along outer edges, without hinge teeth; anterior adductor muscle larger than posterior one; ctenidia large, used mainly for housing symbiotic bacteria. Gut reduced or absent. (e.g., Solemyidae, *Solemya*).

SUPERORDER NUCULANIFORMII Mantle fused posteriorly, with siphons, inhalant water enters posteriorly; shells without nacre and gill filaments along ctenidial axis alternate. Several mainly deep-sea families including Nuculanidae (e.g., *Nuculana*), Malletiidae (e.g., *Malletia*) and Sareptidae (e.g., *Yoldia*).

SUBCLASS AUTOBRANCHIA (= AUTOLAMELLIBRANCHIATA) Paired ctenidia, with very long filaments folded back on themselves so that each row of filaments forms two lamellae; adjacent filaments usually attached to one another by ciliary tufts (filibranch condition), or by tissue bridges (eulamellibranch condition). The greatly enlarged ctenidia are used in combination with the two pairs of labial palps in ciliary feeding; ctenidial surfaces capture water borne particles and transfer them to the labial palps where the capture debris is sorted and potential food particles routed to the mouth.

COHORT PTERIOMORPHIA (= FILIBRANCHIA) Ctenidia with outer fold not connected dorsally to visceral mass, with free filaments or with adjacent filaments attached by ciliary tufts (filibranch condition); shell aragonitic or calcitic, sometimes nacreous; mantle margin unfused, with weakly differentiated incurrent and excurrent apertures or siphons; foot well developed or extremely reduced; usually attached by byssal threads or cemented to substratum (or secondarily free). These primitive lamellibranchs include several divergent ancient lineages separated as orders.

ORDER MYTILIDA The true mussels, Mytilidae (e.g., *Adula, Brachidontes, Lithophaga, Modiolus, Mytilus*).

ORDER ARCIDA The ark shells Arcidae (e.g., *Anadara, Arca, Barbatia*), and dog cockles Glycymerididae (e.g., *Glycymeris*).

ORDER OSTREIDA The true oysters Ostreidae (true oysters, e.g., *Crassostrea, Ostrea*).

ORDER MALLDIDA (= PTERIOIDA, PTERIIDA) Pearl oysters and their relatives Pteriidae (e.g., *Pinctada, Pteria*), hammer oysters Malleidae (e.g., *Malleus*), and pen shells Pinnidae (e.g., *Atrina, Pinna*).

ORDER LIMOIDA File shells Limidae (e.g., *Lima*).

ORDER PECTINIDINA Scallops. Pectinidae (e.g., *Chlamys, Lyropecten, Pecten*), thorny oysters Spondylidae (e.g., *Spondylus*), jingle shells Anomiidae (e.g., *Anomia, Pododesmus*).

COHORT HETEROCONCHIA This clade encompasses the Paleoheterodonta and Heterodonta, previously treated as separate higher groups but shown to be sister groups in recent molecular phylogenies.

MEGAORDER PALAEOHETERODONTA Shell aragonitic, pearly internally; periostracum usually well developed; valves usually equal, with few hinge teeth; elongate lateral teeth (when present) are not separated from the large cardinal teeth; usually dimyarian; mantle opens broadly ventrally, mostly unfused posteriorly but with exhalant and inhalant apertures. About 1,200 species of marine and freshwater clams. Includes two very distinct groups classified as orders.

ORDER TRIGONIIDA The relictual marine broach shells (Trigoniidae), with only a few living species of *Neotrigonia* in Australia.

ORDER UNIONIDA Entirely freshwater, including the freshwater clams (or mussels), e.g., Unionidae (e.g., *Anodonta, Unio*), Margaritiferidae (e.g., *Margaritifera*), and Hyriidae (e.g., *Hyridella*).

MEGAORDER HETERODONTA Two main groups, ranked as superorders, are recognized—the Archiheterodonta (with a single living order) and the Euheterodonta (with four living orders).

SUPERORDER ARCHIHETERODONTA

ORDER CARDITIDA This group of primitive heterodonts is represented by the families Crassatellidae (e.g., *Crassatella*), Carditidae (e.g., *Cardita*) and Astartidae (e.g., *Astarte*).

SUPERORDER EUHETERODONTA

ORDER LUCINIDA Includes the families Luncinidae (e.g., *Lucina*, *Codakia*), a group with symbiotic bacteria in their gills and an anterior water current, and Thysiridae.

ORDER CARDIIDA (= VENERIDA) Usually thick-valved, equi-valved, and isomyarian, with posterior siphons. Includes: cockles and their kin, Cardiidae (e.g., *Cardium*, *Clinocardium*, *Laevicardium*, *Trachycardium,* and the giant clams e.g., *Tridacna*); surf clams, Mactridae (e.g., *Mactra*); solens, Solenidae (e.g., *Ensis*, *Solen*); tellins, Tellinidae (e.g., *Florimetis*, *Macoma*, *Tellina*); semelids, Semelidae (e.g., *Leptomya*, *Semele*); wedge shells, Donacidae (e.g., *Donax*); Venus clams, Veneridae (e.g., *Chione*, *Dosinia*, *Pitar*, *Protothaca*, *Tivela*); freshwater pea clams, Sphaeriidae (e.g., *Sphaerium, Pisidium*); the estuarine-to-freshwater Cyrenidae (e.g., *Corbicula, Batissa*); and zebra mussels, Dreissenidae (e.g., *Dreissena*). The latter two families contain important invasive species.

ORDER PHOLADIDA (= MYIDA) Thin-shelled burrowing forms with well-developed siphons. Includes: soft-shell clams, Myidae (e.g., *Mya*); rockborers or piddocks, Pholadidae (e.g., *Barnea*, *Martesia*, *Pholas*); shipworms, Teredinidae (e.g., *Bankia*, *Teredo*); and basket clams, Corbulidae (e.g., *Corbula*). The monophyly of this order is uncertain.

ORDER POROMYATA (= ANOMALODESMATA) Shells equivalved, aragonitic, of 2–3 layers, innermost consisting of sheet nacre; periostracum often incorporates granulations; with 0–1 hinge teeth; generally isomyarian, rarely amyarian; posterior siphons usually well developed; mantle usually fused ventrally, with anteroventral pedal gape, and posteriorly with ventral incurrent and dorsal excurrent apertures or siphons; ctenidia eulamellibranchiate or septibranchiate (modified as a muscular pumping horizontal septum). This ancient and very diverse group of marine bivalves includes about 20 living families, including the rare deep water Pholadomyidae (e.g., *Pholadomya*) and the aberrant watering pot shells Clavagellidae (e.g., *Brechites*), as well as Pandoridae, Poromyidae, Cuspidariidae, Laternulidae, Thraciidae, Cleidothaeridae, Myochamidae, and Periplomatidae.

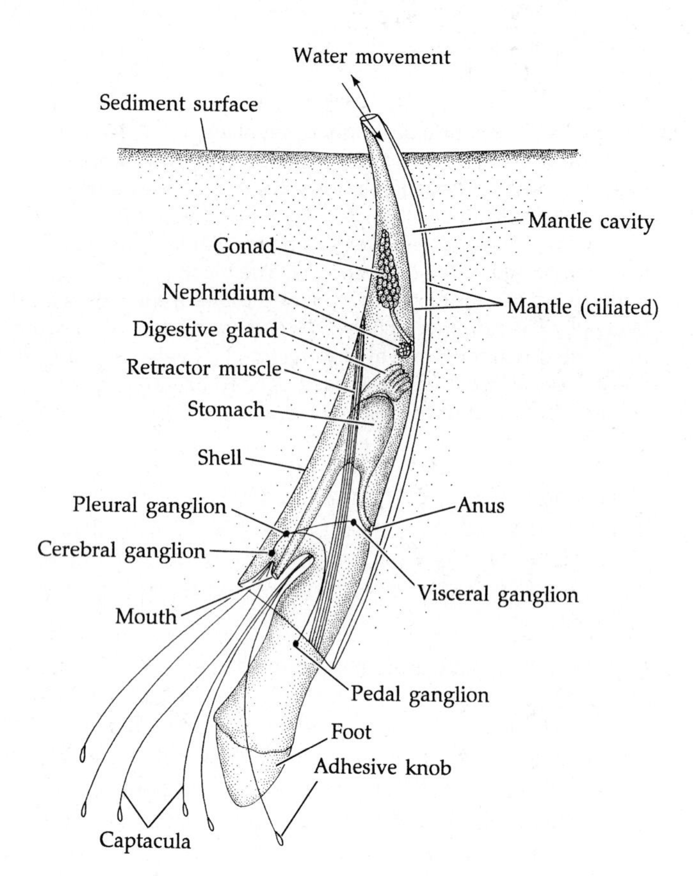

Figure 13.9 General anatomy of a scaphopod.

CLASS SCAPHOPODA Tusk shells (Figures 13.1L and 13.9). Shell of one piece, tubular, usually tapering, open at both ends; head rudimentary, projecting from larger aperture; mantle cavity large, extending along entire ventral surface; without ctenidia or eyes; with radula, long, snout-like "proboscis," and paired clusters of long, thin contractile tentacles with clubbed ends (captacula) that serve to capture and manipulate minute prey; heart absent; foot somewhat cylindrical, with epipodium-like fringe. Over 500 species of marine, benthic molluscs in fourteen families and two orders.

ORDER DENTALIIDA Shell regularly tapering. Paired digestive gland with a muscular foot terminating in epipodial lobes and a cone-shaped process. Several families including Dentaliidae (e.g., *Dentalium*, *Fustiaria*) and Laevidentaliidae (e.g., *Laevidentalium*).

ORDER GADILIDA Shell regularly tapering or bulbous with maximum shell diameter approximately nearer

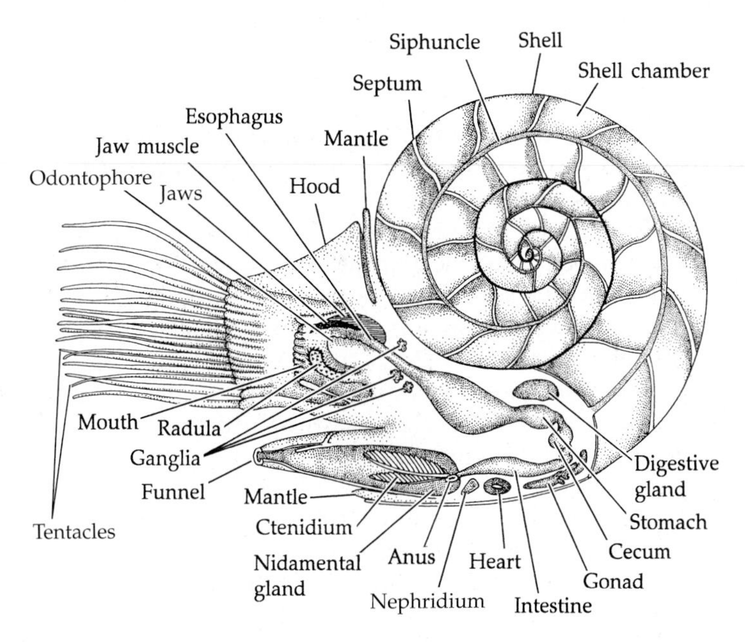

Figure 13.10 The anatomy of *Nautilus* (sagittal

center of shell. Single digestive gland and foot with a terminal disk surrounded by epipodial papilla. Families include Pulsellidae (e.g., *Pulsellum*, *Annulipulsellum*) and Gadilidae (e.g., *Cadulus*, *Gadila*).

CLASS CEPHALOPODA Nautilus, squids, cuttlefish, and octopuses (Figures 13.1H–K, 13.10, 13.11, 13.12, 13.17, and 13.22). With linearly chambered shell, usually reduced or lost in living taxa; if external shell present (*Nautilus*), animal inhabits last (youngest) chamber, with a filament of living tissue (the siphuncle) extending through older chambers; circulatory system largely closed; head with large, complex eyes and circle of prehensile arms or tentacles around mouth; with radula and beak; 1–2 pairs ctenidia, and 1–2 pairs complex nephridia; mantle forms large ventral mantle cavity containing 1–2 pairs of ctenidia; with muscular funnel (the siphon) through which water is forced, providing jet propulsion; some tentacles of male modified for copulation; benthic or pelagic, marine; about 700 living species.

SUBCLASS PALCEPHALOPODA Includes many fossil taxa, all with external shells, as well as the living pearly nautilus.

COHORT NAUTILIDIA (= TETRABRANCHIATA) The pearly nautilus. Shell external, many-chambered, coiled in one plane, exterior porcelaneous, interior nacreous (pearly); head with many (80–90) suckerless tentacles (4 modified as spadix in male for copulation and protected by a fleshy hood); 13 tooth rows in radula; beak of chitin and calcium carbonate; funnel of 2 separate folds; 2 pairs ctenidia ("tetrabranchiate"); 2 pairs nephridia; eyes like a pinhole camera, without cornea or lens; nervous system with anterior elements concentrated into a brain, optic lobes large; statocyst simple; without chromatophores or ink sac. Fossil record rich, but represented today by a single order (Nautilida) and single genus, the chambered or pearly nautilus (*Nautilus*), with five or six Indo-Pacific species (although a second, controversial genus, *Allonautilus*, has also been proposed).

SUBCLASS NEOCEPHALOPODA Includes one fossil group in addition to the coleoids. The shell is reduced and internal in most (and in all living taxa).

COHORT COLEOIDIA (= DIBRANCHIATA) Octopuses, squids, and their kin. Shell reduced and internal or absent; head and foot united into a common anterior structure bearing 8 or 10 prehensile appendages (arms and tentacles) bearing suckers and, often, cirri, one arm usually modified in male for copulation; 7 tooth rows in radula; with chitinous beak; funnel a single closed tube; 1 pair ctenidia ("dibranchiate"); 1 pair nephridia; eyes complex, with lens and often with cornea; nervous system well developed and concentrated into a brain; with a complex statocyst; with chromatophores and ink sac.

SUPERORDER OCTOPODIFORMES Members of this group, which includes the octopuses and vampire squid, do not have the head distinctly separated from the rest of the body; have 8 arms, with 2 additional retractile filaments in the vampire squids; lateral fins on the body are present or absent.

ORDER OCTOPODA Octopuses. Body short, round, usually without fins; internal shell vestigial or absent; 8 similar arms joined by web of skin (interbrachial web); suckers with narrow stalks, most are benthic. About 200 species, in two groups: the Incirrata, including the benthic octopuses and some pelagic taxa that lack fins and cirri with examples being Octopodidae (e.g., *Octopus*) and Argonautidae (*Argonauta*), the paper nautilus; and Cirrata, which are mainly pelagic deep sea cephalopods with fins and cirri, such as Cirroteuthidae (e.g., *Cirroteuthis*), Opisthoteuthidae (e.g., *Opisthoteuthis*), and Stauroteuthidae (e.g., *Stauroteuthis*).

ORDER VAMPYROMORPHA The vampire squid. Body plump, with 1 pair fins; shell reduced to thin, leaf-shaped, uncalcified, transparent vestige; 4 pairs equal-sized arms, each with one row of unstalked distal suckers and two rows of cirri; arms joined by extensive web of skin (interbrachial membrane); fifth pair of arms represented by 2 tendril-like, retractable filaments; hectocotylus lacking; radula well developed; ink sac degenerate; mostly deep water. One living species, *Vampyroteuthis infernalis*, that lives in the oxygen minimum zone of the deep sea.

SUPERORDER DECAPODIFORMES (= DECAPODA) Members of this group, which includes the squid and cuttlefish, have the head distinctly separated from the rest of the body; with 8 arms and 2 retractile (into pits) tentacles with suckers only on expanded tips; suckers with wide bases, sometimes with spines or hooks; lateral fins on body. The internal shell is large (as in the cuttlefish or in *Spirula*), reduced to an uncalcified gladius, or lost.

ORDER SPIRULIDA The only living species is the ram's horn, *Spirula spirula* (Spirulidae), a small deep sea squid with a coiled, internal, chambered shell.

ORDER SEPIIDA Cuttlefish. Body short, dorsoventrally flattened, with lateral fins; shell internal, calcareous, straight or slightly curved, chambered; or shell horny, or absent; 8 short arms, and 2 long tentacles; suckers lack hooks. Includes the shell-less Sepiolidae (e.g., *Rossia*, *Sepiola*), the Sepiidae (e.g., *Sepia*) with an internal calcareous shell, the cuttlebone, and Idiosepiidae (e.g., *Idiosepis*), being tiny squids that live in seagrass to which they attach with a special sucker. Their shell is reduced to a horny gladius.

ORDER MYOPSIDA Squids with the eye covered with a cornea and having a well developed gladius. Body elongate, tubular, with lateral fins. Loliginidae (e.g., *Loligo, Doryteuthis*).

ORDER OEGOPSIDA Includes the majority of squids (and the former Teuthoida in part); the eye lacks a cornea and the shell is a gladius. Body elongate, tubular, with lateral fins; suckers often with hooks. Some of the many families in this group include Architeuthidae (*Architeuthis*), Bathyteuthidae (e.g., *Bathyteuthis*; sometimes treated as a separate order), Chiroteuthidae (e.g., *Chiroteuthis*), Ommastrephidae (e.g., *Ommastrephes, Dosidiscus, Illex*), Gonatidae (e.g., *Gonatus*), Histoteuthidae (e.g., *Histioteuthis*), Lycoteuthidae (e.g., *Lycoteuthis*), and Octopoteuthidae (e.g., *Octopoteuthis*).

The Molluscan Body Plan

Mollusca is one of the most morphologically diverse phyla in the animal kingdom. Molluscs range in size from microscopic solenogasters, bivalves, snails, and slugs, to whelks attaining 70 cm in length, giant clams (Cardiidae) over a 1 m in length, and giant squids (*Architeuthis*) reaching at least 13 m in overall length (body plus tentacles). The giant Pacific octopus (*Octopus dofleini*) commonly attains an arm span of 3–5 m and a weight of over 40 kg. It is the largest living octopus, and one particularly large specimen was estimated to have an arm span of nearly 10 m and a weight of over 250 kg! Despite their differences, giant squids, cowries, garden slugs, eight-plated chitons, and wormlike aplacophorans are all closely related and share a common body plan (Box 13A). In fact, the myriad ways in which evolution has shaped the basic molluscan body plan provide some of the best lessons in homology and adaptive radiation in the animal kingdom.

Molluscs are bilaterally symmetrical, coelomate protostomes, but the coelom generally exists only as small vestiges around the heart (the **pericardial chamber**), the gonads, and parts of the nephridia (kidneys). The principal body cavity is a hemocoel composed of several large sinuses of the open circulatory system, except in some cephalopods that have a largely closed system. In general, the body comprises three distinguishable regions: a head, foot, and centrally concentrated visceral mass, but the configuration differs in different classes (Figure 13.13). The head may bear various sensory structures, most notably eyes and tentacles; statocysts may be located in the foot region and chemosensory structures can also be present.

The visceral mass is covered by a thick epidermal sheet of skin called the **mantle** (also known as the pallium), which is sometimes covered in cuticle and plays a critical role in the organization of the body. It secretes the hard calcareous skeleton, either as minute sclerites, or plates, that are embedded in the body wall or as a solid internal or external shell. Ventrally the body usually bears a large, muscular **foot**, which typically has a creeping **sole**.

Surrounding or posterior to the visceral mass is a cavity—a space between the visceral mass and folds of the mantle itself. This **mantle cavity** (also known as the pallial cavity) often houses the gills (the original molluscan gills are known as **ctenidia**), along with the openings of the gut, nephridial, and reproductive systems, and, in addition, special patches of chemosensory epithelium in many groups, notably the osphradia. In aquatic forms, water is circulated through this cavity, passing over the ctenidia, excretory pores, anus, and other structures.

Molluscs have a complete, or through-gut that is regionally specialized. The buccal region of the foregut typically bears a uniquely molluscan structure, the **radula**, which is a toothed, rasping, tongue-like strap used in feeding. It is located on a muscular **odontophore** that moves the radula through its feeding motions. The circulatory system usually includes a heart in a pericardial cavity and a few large vessels that empty into or drain hemocoelic spaces. The excretory system consists of one or more pairs of metanephridial kidneys (here simply referred to as nephridia), with openings (nephrostomes) to the pericardium via renopericardial canals and to the mantle cavity via the nephridiopore. The nervous system typically includes a dorsal cerebral ganglion, a nerve ring encircling the buccal area or esophagus, and two pairs of longitudinal nerve cords arise from paired pleural ganglia and connecting with the visceral ganglia more posteriorly in the body. Other anterior paired ganglia (buccal and labial) may be present. Pedal ganglia lie in the foot and may give off pedal nerve cords.

Gametes are produced by the gonad in the visceral mass and fertilization may be external or internal. Development is typically protostomous, with spiral cleavage and a trochophore larval stage. There is also a secondary larval form unique to gastropod and bivalve molluscs called the **veliger**.

Although this general summary describes the basic body plan of most molluscs, notable modifications occur and are discussed throughout this chapter. The eight classes are characterized above (see classification) and are briefly summarized below.

Some of the most bizarre molluscs are the "aplacophorans"—Solenogastres and Caudofoveata (Figure 13.1C and 13.2). Members of these groups are wormlike

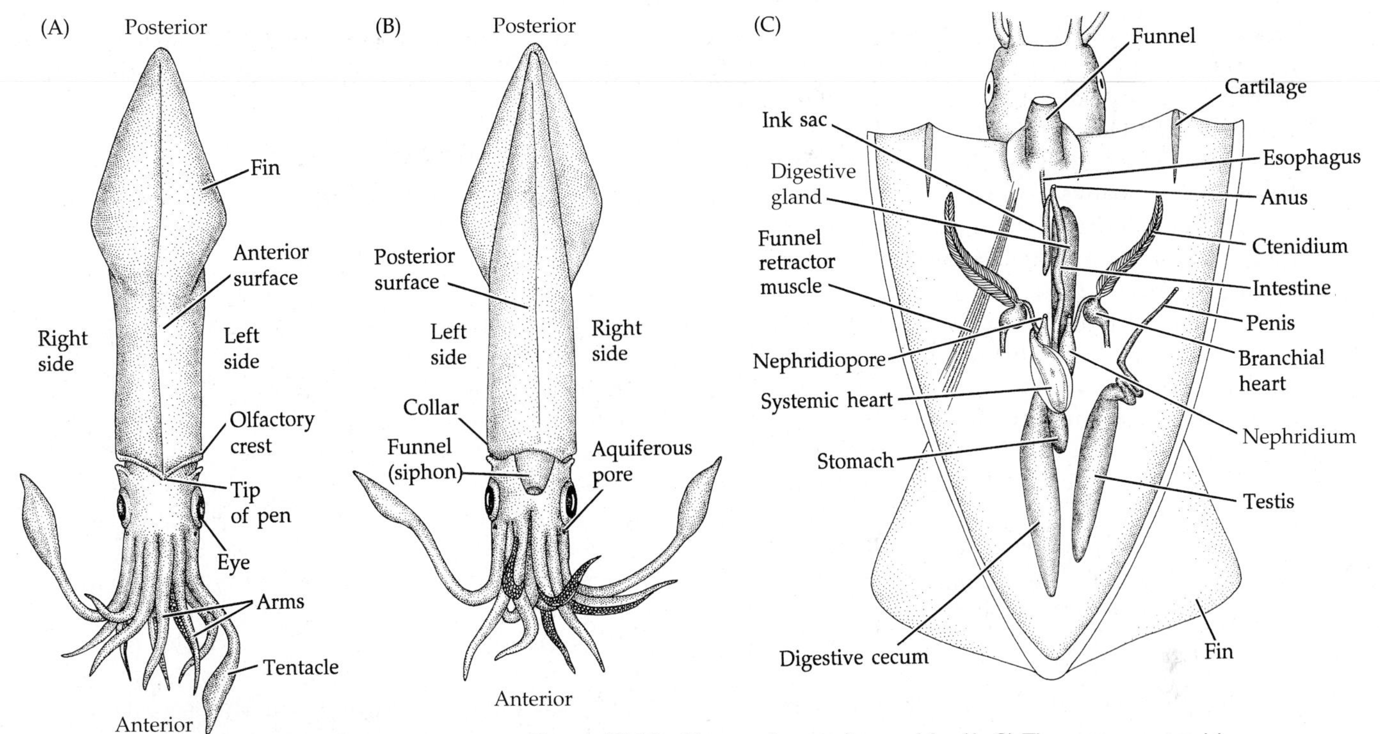

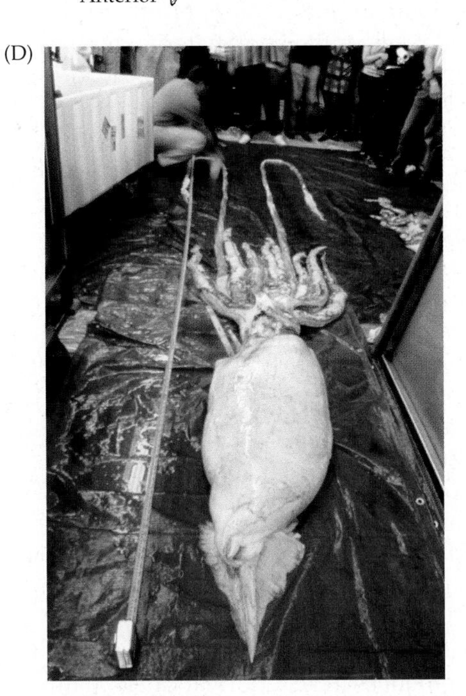

Figure 13.11 The anatomy of a squids. (A–C) The common squid, *Loligo*.(A) External morphology (anterior view). (B) External morphology (posterior view). (C) Internal anatomy of a male. The mantle is dissected open and pulled aside. (D) The giant squid (*Architeuthis dux*) netted off the coast of New Zealand in 1997.

and typically small, and either burrow in sediment (Caudofoveata) or may spend their entire lives on the branches of various cnidarians (Solenogastres) such as gorgonians upon which they feed. Caudofoveata lack a foot, but a reduced one is present in the Solenogastres, and neither group has a solid shell. Aplacophorans also have no distinct head, eyes, or tentacles. They were traditionally considered primitive molluscs that evolved before the appearance of solid shells, but some molecular and developmental data suggest they may actually be highly derived forms that have lost the shell and have secondarily acquired a simple body morphology.

Polyplacophorans, or chitons, are oval molluscs that bear eight (seven in Paleoloricata) separate articulating shell plates on their backs (Figures 13.1B and 13.4). They range in length from about 7 mm to over 35 cm. These marine animals are inhabitants of deep sea to intertidal regions around the world, at all latitudes.

Monoplacophorans are limpet-like molluscs with a single cap-shaped shell ranging from about 1 mm to about 4 cm in length (Figures 13.1A and 13.3). Most live in the deep sea, some at great depths (> 2,000 m). Their most notable feature is the repetitive arrangement of gills, gonads, and nephridia, a condition that has led some biologists to speculate that they must represent a link to some ancient segmented ancestor of the Mollusca (an idea no longer deemed reasonable).

Gastropods are by far the largest group of molluscs and include some of the best-studied species (Figures 13.1D–G, 13.5, 13.6, and 13.7). This class includes the common snails and slugs in all marine and many freshwater habitats, and they are the only molluscan class to have successfully invaded terrestrial environments. They are the only molluscs that undergo torsion during early development, a process involving a 90–180° rotation of the visceral mass relative to the foot (for details see section on torsion below).

Bivalves include the clams, oysters, mussels, and their kin (Figures 13.1M–Q and 13.8). They possess two separate shells, called **valves**. The smallest bivalves are members of the marine family Condylocardiidae some of which are about 1 mm in length; the largest are giant tropical clams (*Tridacna*), one species of which (*T. gigas*) may weigh over 400 kg! Bivalves inhabit all marine environments and many freshwater habitats.

Scaphopods, the tusk shells, live in marine surface sediments at various depths. Their distinctive single, tubular uncoiled shells are open at both ends and range from a few millimeters to about 15 cm in length (Figures 13.1L and 13.9).

The cephalopods are among the most highly modified molluscs and include the pearly nautilus, squids, cuttlefish, octopuses, and a host of extinct forms, including the ammonites (Figures 13.1I–K, 13.10, 13.11, 13.12, 13.17, and 13.22). This group includes the largest of all living invertebrates, the giant squid, with body and tentacle lengths around 13 m. Among living cephalopods, only the nautilus has retained an external shell. The cephalopods differ markedly from other molluscs in several ways. For example, they have a spacious body cavity that includes the pericardium, gonadal cavity, nephriopericardial connections, and gonoducts, all of which form an interconnected system representing a highly modified but true coelom. In addition, unlike all other molluscs, many coleoid cephalopods have a functionally closed circulatory system. The nervous system of cephalopods is the most sophisticated of all invertebrates, with unparalleled learning and memory abilities. Most of these modifications are associated with the adoption of an active predatory lifestyle by these remarkable creatures.

The Body Wall

The body wall of molluscs typically comprises three main layers: the cuticle (when present), epidermis, and muscles (Figure 13.15A). The cuticle is composed largely of various amino acids and sclerotized proteins (called conchin), but it apparently does not contain chitin (except in the aplacophorans). The epidermis is usually a single layer of cuboidal to columnar cells, which are ciliated on much of the body. Many of the epidermal cells participate in secretion of the cuticle. Other kinds of secretory gland cells can also be present, some of which secrete mucus and these can be very abundant on external surfaces such as the sole of the foot. Other specialized epidermal cells occur on the dorsal body wall, or mantle. Many of these cells constitute the molluscan **shell glands**, which produce the calcareous sclerites or shells characteristic of this phylum. Still other epidermal cells are sensory receptors. The epidermis and outermost muscle layer are often separated by a basement membrane and occasionally a dermis-like layer.

The body wall usually includes three distinct layers of smooth muscle fibers: an outer circular layer, a middle diagonal layer, and an inner longitudinal layer. The diagonal muscles are often in two groups with fibers running at right angles to each other. The degree of development of each of these muscle layers differs among the classes (e.g., in solenogasters the diagonal layers are frequently absent).

The Mantle and Mantle Cavity

The significance of the mantle cavity and its importance in the evolutionary success of molluscs has already been alluded to. Here we offer a brief summary of the nature of the mantle cavity, and its disposition in each of the major groups of molluscs.

The mantle, as the name implies, is a sheet-like organ that forms the dorsal body wall, and in most molluscs it grows during development to envelop the molluscan body and at its edge there are one or two folds that contain muscle layers and hemocoelic channels (Figure 13.15C). The outward growth creates a space lying between the mantle fold(s) and the body proper. This space, the mantle cavity, may be in the form of a groove surrounding the foot or a primitively posterior chamber through which water is passed by ciliary or, in more derived taxa, by muscular action. Generally, the mantle cavity houses the respiratory surface (usually the ctenidia or other gill-like structures), and receives the fecal material discharged from the anus and excretory waste from the kidney. Gametes are also primitively discharged into the mantle cavity. Incoming water provides a source of oxygen for respiration, a means of flushing waste and, in some instances, also carries food for suspension feeding.

The mantle cavity of chitons is a groove surrounding the foot (Figures 13.4A and 13.13A,B). Water enters the groove from the front and sides, passing medially over the ctenidia and then posteriorly between the ctenidia and the foot. After passing over the gonopores and nephridiopores, water exits the back end of the groove and carries away fecal material from the posteriorly located anus.

The aplacophorans have a small mantle cavity, with either a pair of ctenidia (Caudofoveata) or lamella-like folds or papillae on the mantle cavity wall (Solenogastres). The paired coelomoducts and the anus also open into the mantle cavity

The single mantle cavity of gastropods originates during development as a posteriorly located chamber. As development proceeds, however, most gastropods undergo a 180° rotation of the visceral mass and shell to bring the mantle cavity forward, over the head (Figures 13.5, 13.6, and 13.13C) (see section on torsion that follows). The different orientation does not affect the water flow, which still passes through this chamber through the ctenidia, and then past the anus, gonopores, and nephridiopores. A great many secondary modifications on this plan have evolved in the Gastropoda, including rerouting of current patterns; loss or modification of associated structures such as the gills, hypobranchial glands and sensory organs; and even "detorsion," as discussed in later sections of this chapter.

Figure 13.12. The anatomy of *Octopus*. (A) General external anatomy. (B) Right-side view of the internal anatomy. (C) Arm and sucker (cross section). (D) Tip of the hectocotylus arm. (E) The diminutive Eastern Pacific *Paroctopus digueti* well camouflaged on a sand bottom. (F) The tropical Pacific *Octopus chierchiae*. (G) The remarkable Indo-West Pacific *Abdopus horridus*.

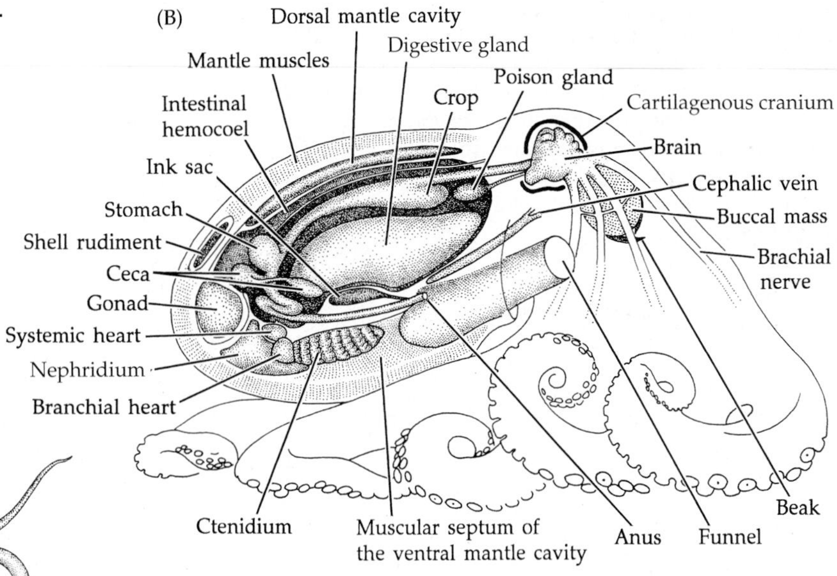

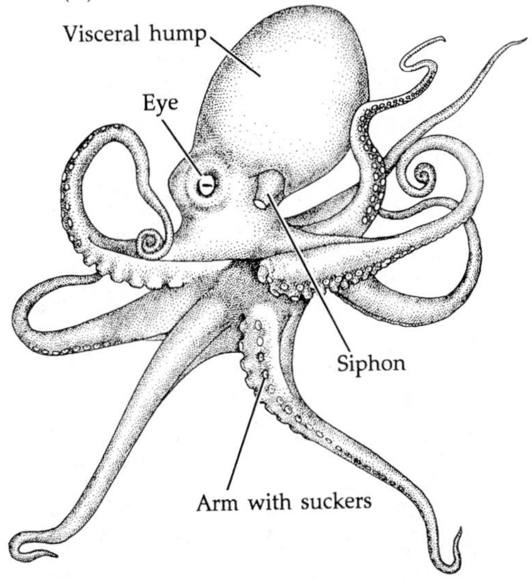

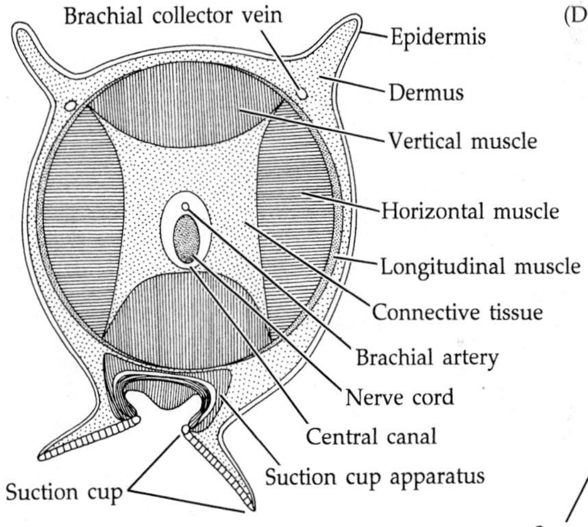

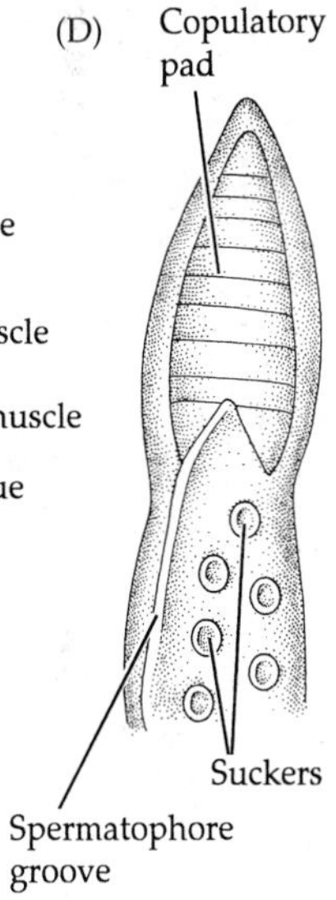

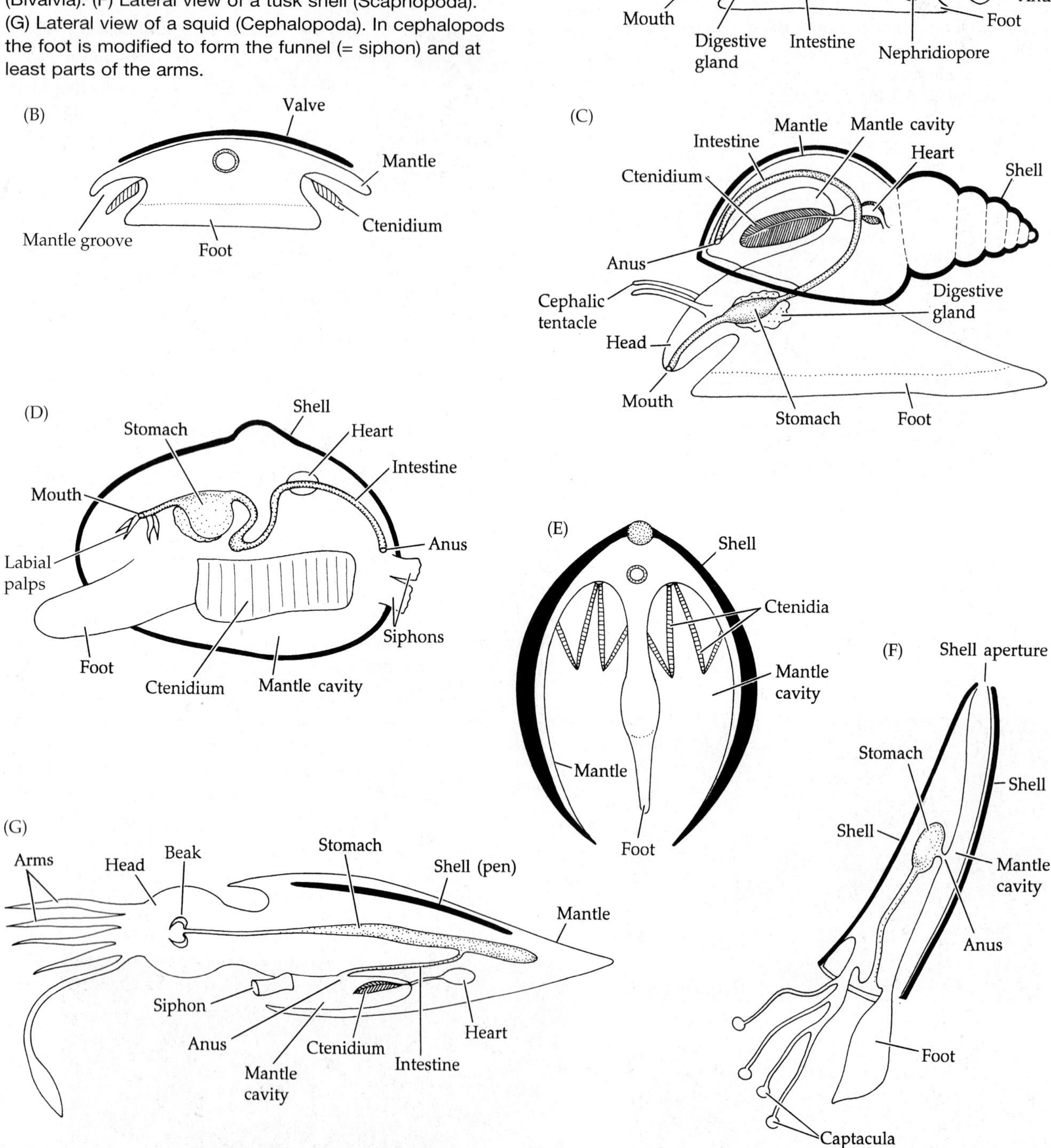

Figure 13.13 Modifications of the shell, foot, gut, ctenidia, and mantle cavity in five classes of molluscs. (A,B) Lateral and cross sections of a chiton (Polyplacophora). (C) Side view of a snail (Gastropoda). (D,E) Cutaway side view and cross section of a clam (Bivalvia). (F) Lateral view of a tusk shell (Scaphopoda). (G) Lateral view of a squid (Cephalopoda). In cephalopods the foot is modified to form the funnel (= siphon) and at least parts of the arms.

Bivalves possess a greatly enlarged mantle cavity that surrounds both sides of the foot and visceral mass (Figures 13.8 and 13.13D,E). The mantle lines the laterally placed shells, and the folds making up the mantle edges are often fused in various ways posteriorly to form inhalant and exhalant siphons, through which water enters and leaves the mantle cavity. The water passes over and through the ctenidia that, in autobranch bivalves, extract suspended food material as well as accomplishing gas exchange. The water flow then sweeps across the gonopores and nephridiopores, and lastly past the anus as it exits through the exhalant siphon.

Scaphopods have a tapered, tubular shell (Figures 13.9 and 13.13F). Water enters and leaves the elongate mantle cavity through the small opening in the top of the shell and flushes over the mantle surface, which, in the absence of ctenidia, is the site of gas exchange. The anus, nephridiopores, and gonopores also empty into the mantle cavity.

While no detailed studies of the functioning of the monoplacophoran mantle cavity have been made, observations of the first living specimens in 1977 revealed that the gills vibrated, apparently circulating water through the mantle groove. It was also noted that shell movement was accompanied by an acceleration of gill beating. Vibrating gills are unknown in other molluscs where ciliary action, sometimes assisted by muscular contractions, moves water through the mantle cavity. The anus, nephridiopores, and gonopores also open into the mantle cavity.

With the exception of the monoplacophorans, in all of the above cases, water is moved through the mantle cavity by the action of long lateral cilia on the ctenidia. But in cephalopods the ctenidial gills are not ciliated. Instead, in *Nautilus* a ventilatory current is passed through the mantle cavity by the unduatory movements of two muscular flaps associated with the funnel lobes. In the coleoid cephalopods, however, well-developed, highly innervated mantle muscles perform this function through the regular pulsation of the mantle wall. The exposed, fleshy body surface of squids and octopuses is, in fact, the mantle itself (Figures 13.11, 13.12, and 13.13G). Unconstrained by an external shell, the mantle of these molluscs expands and contracts to draw water into the mantle cavity and then forces it out through the narrow muscular **funnel** (= **siphon**). The forceful expelling of this jet of exhalant water can also provide a means of rapid locomotion for most cephalopods. In the mantle cavity the water passes through the ctenidia, and then past the anus, reproductive pores, and excretory openings.

The remarkable adaptive qualities of the molluscan body plan are manifested in these variations in the position and function of the mantle cavity and its associated structures. In fact, the nature of many other structures is also influenced by mantle cavity arrangement, as shown schematically in Figure 13.14. That molluscs have been able to successfully exploit an extremely broad range of habitats and lifestyles can be explained in part by these variations, which are central to the story of molluscan evolution. We will have more to say about these matters throughout this chapter.

The Molluscan Shell

Except for the two aplacophoran classes, all molluscs have solid calcareous shells (either aragonite or calcite) produced by shell glands in the mantle. In the Caudofoveata and Solenogastres, aragonite sclerites or scales are formed extracellularly in the mantle epidermis and are embedded in the cuticle. In the other classes molluscan shells vary greatly in shape and size, but they all adhere to the basic construction plan of calcium carbonate produced extracellularly, laid down on a protein matrix in layers, and often covered by a thin organic surface coating called a **periostracum** (called the **hypostracum** in chitons) (Figure 13.15). The periostracum is composed of a type of conchin (largely quinone-tanned proteins) similar to that found in the epidermal cuticle. The calcium layers have four crystal types: prismatic, spherulitic, laminar, and crossed structures. All incorporate conchin onto which the calcareous crystals precipitate. The majority of living molluscs have an outer prismatic layer and an inner porcelaneous, crossed layer. In monoplacophorans, cephalopods and in some gastropods and bivalves, an iridescent, nacre (laminar) layer replaces the layer of crossed crystals. Shells are often made up of multiple layers of different crystal types.

Molluscs are noted for their wonderfully intricate and often flamboyant shell color patterns and sculpturing (Figure 13.16), but very little is known about the evolutionary origins and functions of these features. Some molluscan pigments are metabolic by-products, and thus shell colors might largely represent strategically deposited food residues, while others appear to have no relationship to diet. Molluscan shell pigments include such compounds as pyrroles and porphyrins. Melanins are common in the integument (cuticle and epidermis), the eyes, and internal organs, but they are rare in shells.

Some shell sculpture patterns are correlated with specific behaviors or habitats. For example, shells with low spires are more stable in areas of heavy wave shock or on vertical rock surfaces. Similarly, the low, cap-shaped shells of limpets (Figures 13.5A and 13.16H,I) are presumably adapted for withstanding exposure to strong waves. Heavy ribbing, thick or inflated shells, and a narrow gape in bivalves are all possible adaptations to provide protection from predators. In some gastropods, fluted shell ribs help them land upright when they are dislodged from rocks. Several groups of soft-bottom benthic gastropods and bivalves have long spines on the shell that may help stabilize the animals in loose sediments as well as provide some protection from predators. Many molluscs, particularly clams, have shells covered with living epizootic organisms such as sponges, annelid tube worms, ectoprocts, and hydroids. Some studies suggest that predators have a difficult time recognizing such camouflaged molluscs as potential prey.

Molluscs may have one shell, two shells, eight shells, or no shell. In the latter case the outer body wall may contain calcareous sclerites of various sorts. In the aplacophorans, for example, the cuticular sclerites vary in shape and range in length from microscopic to about 4 mm. These sclerites are essentially crystals composed almost entirely of calcium carbonate as aragonite. Caudofoveates produce platelike cuticular sclerites that give their body surface a scaly texture and appearance. The sclerites in both taxa appear to be secreted by a

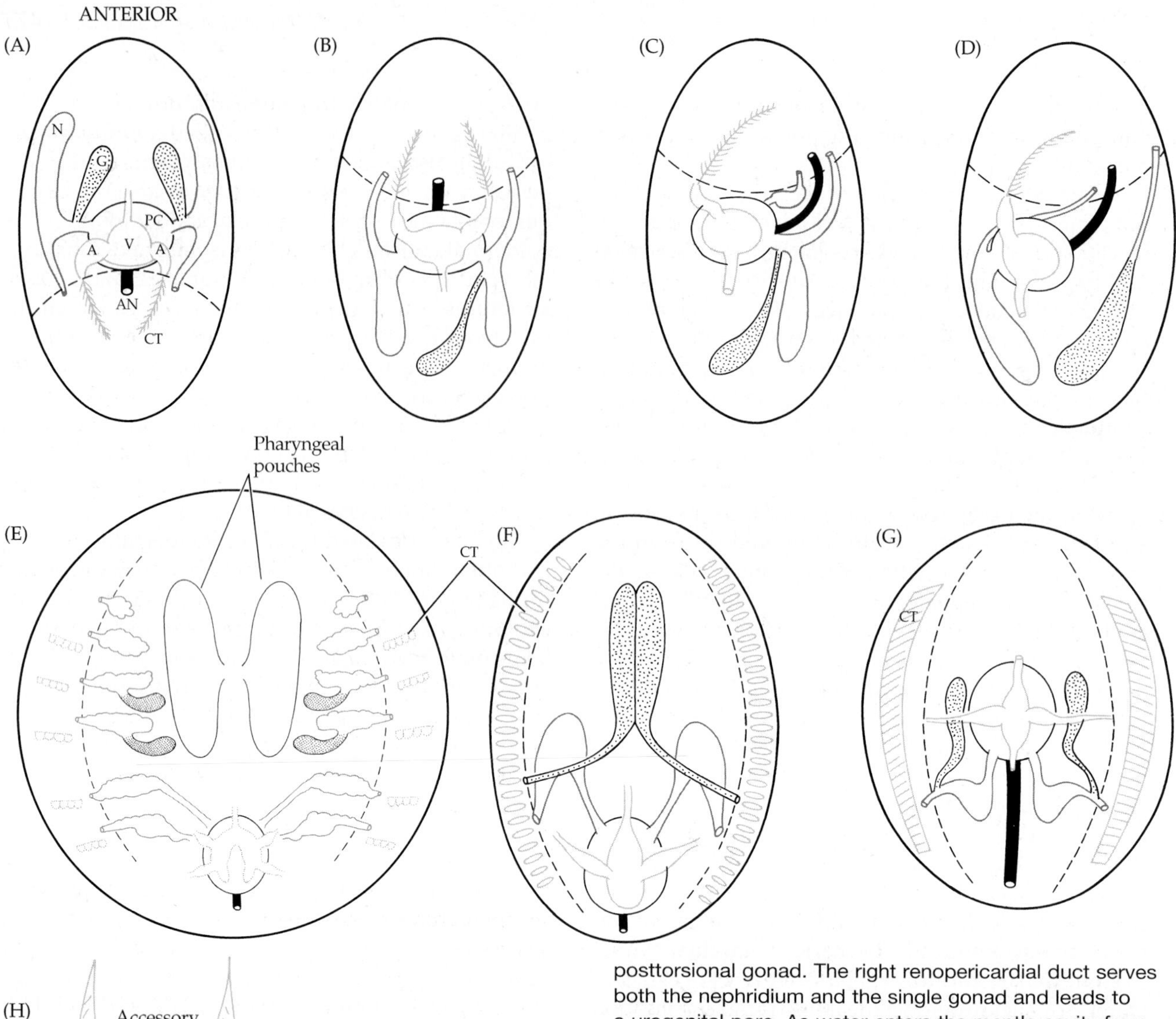

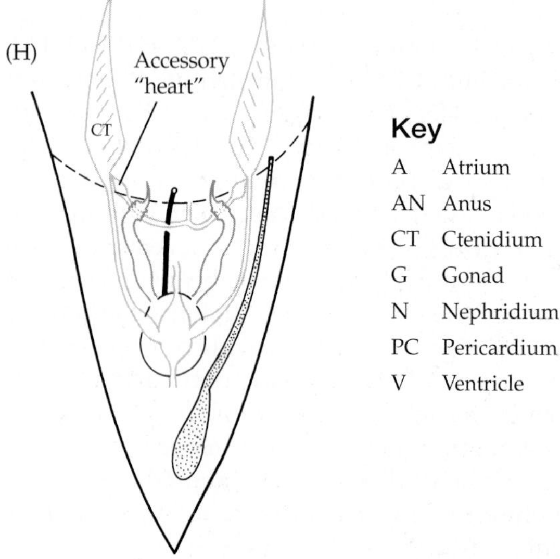

Key

A	Atrium
AN	Anus
CT	Ctenidium
G	Gonad
N	Nephridium
PC	Pericardium
V	Ventricle

Figure 13.14 Variations in the mantle cavity, circulatory system, ctenidia, nephridia, reproductive system, and position of the anus in molluscs (dorsal views). Although schematic, these drawings give some idea of the evolutionary changes in arrangement of these structures and systems in the phylum Mollusca. (A) A hypothetical, untorted, gastropod-like mollusc with a posterior mantle cavity and symmetrically paired atria, ctenidia, nephridia, and gonads. (B) A posttorsional vetigastropod (e.g., *Fissurella*) wherein all paired organs are retained except the left posttorsional gonad. The right renopericardial duct serves both the nephridium and the single gonad and leads to a urogenital pore. As water enters the mantle cavity from the front, it passes first over the two bipectinate ctenidia and then over the anus, nephridiopore, and urogenital pore before exiting through dorsal shell openings (e.g., holes or slits). (C) The patellogastropod (limpet), *Lottia*. Here the posttorsional right ctenidium and right atrium are lost, and the nephridiopore, anus, and urogenital pore are shifted to the right side of the mantle cavity, thus allowing a one-way, left-to-right water flow. A similar configuration and is also found in vetigastropods with single gills (e.g., *Tricolia*). (D) Most caenogastropods have a single, posttorsional left, monopectinate ctenidium, suspended from the roof of the mantle cavity. The right renopericardial duct has typically lost its association with the pericardium and is co-opted into the genital tract. Such isolation of the tract and gonad from the excretory plumbing has allowed the evolution of elaborate reproductive systems among "higher" gastropods (e.g., neritomorphs, caenogastropods, and heterobranchs) and probably important in the story of gastropod success. (E) The condition in monoplacophorans includes the serial repetition of several organs. (F) In polyplacophorans, the gonoducts and nephridioducts open separately into the exhalant regions of the lateral pallial grooves. (G) A generalized bivalve condition. The gonads and nephridia may share common pores, as shown here, or else open separately into the lateral mantle chambers. (H) The condition in a generalized cephalopod with a single, isolated reproductive system and an effectively closed circulatory system.

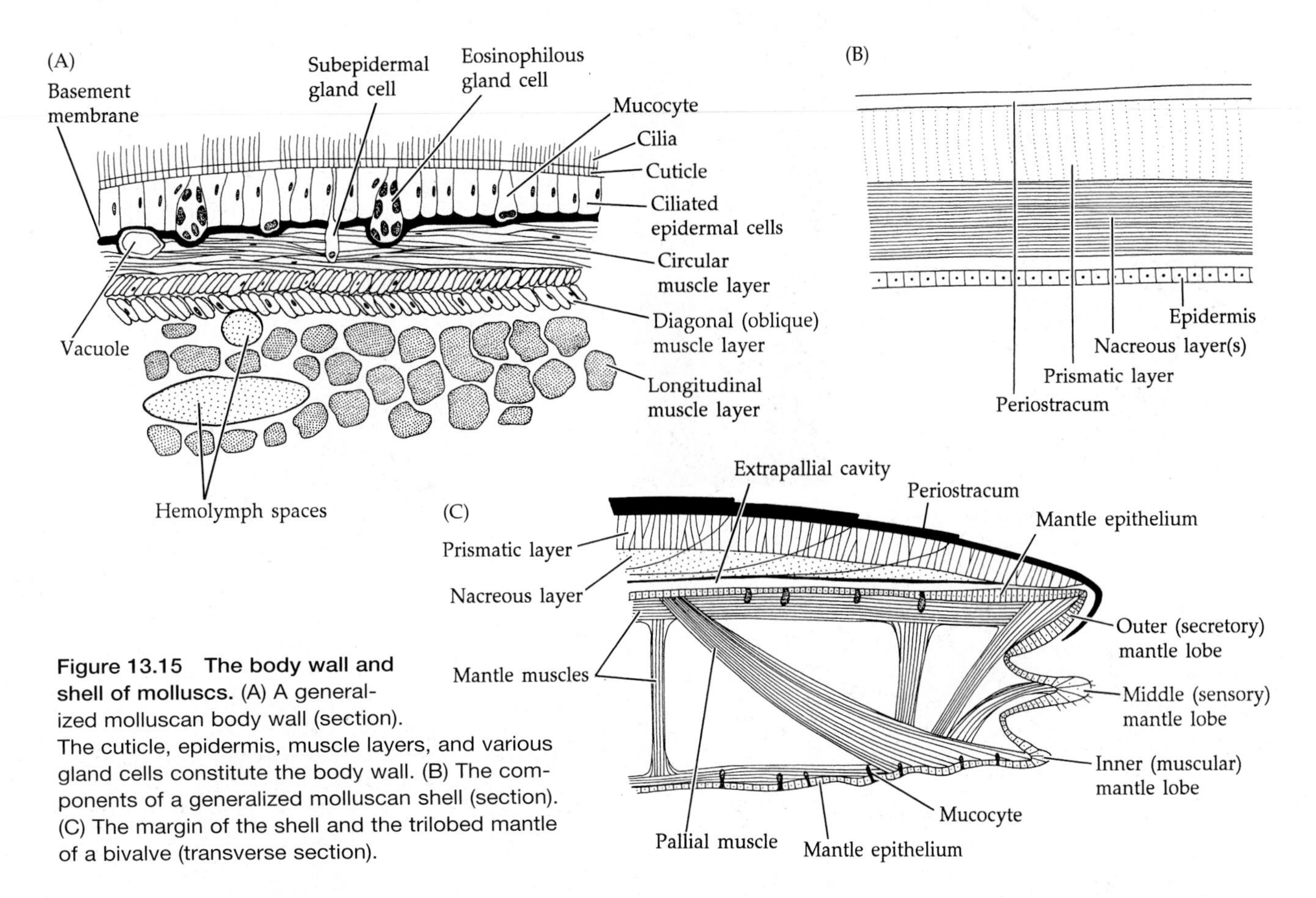

Figure 13.15 The body wall and shell of molluscs. (A) A generalized molluscan body wall (section). The cuticle, epidermis, muscle layers, and various gland cells constitute the body wall. (B) The components of a generalized molluscan shell (section). (C) The margin of the shell and the trilobed mantle of a bivalve (transverse section).

diffuse network of specialized groups of cells and different shapes are found in different regions of the body.

The eight transverse plates, or valves (Figures 13.4 and 13.16A–F), of polyplacophorans are encircled by and embedded in a thickened region of the mantle called the **girdle**. The size of the girdle varies from narrow to broad and may cover much of the valves. In the giant Pacific "gumboot" chiton, *Cryptochiton stelleri*, the girdle completely covers the valves. The girdle is thick, heavily cuticularized, and usually beset with calcareous sclerites, spines, scales, or noncalcareous bristles secreted by specialized epidermal cells. These sclerites are probably homologous with those in the body wall of aplacophorans.

The anterior and posterior valves of chitons are referred to as the end valves, or cephalic (= anterior) and anal (= posterior) plates; the six other valves are called the intermediate valves. Some details of chiton valves are shown in Figure 13.16A–F. The shells of chitons are three-layered, with an outer periostracum, a colored **tegmentum**, and an inner calcareous layer, or **articulamentum**. The periostracum is a very thin, delicate organic membrane and is not easily seen. The tegmentum is composed of organic material (probably a form of conchin) and calcium carbonate suffused with various pigments. It is penetrated by vertical canals that lead to minute pores in the surface of the valves. The pores are of two sizes: the larger ones (megapores) housing the **megaesthetes** and the smaller ones (micropores) the **microaesthetes**. In some species, megaesthetes may be modified as **shell eyes**, with compound lenses made of large crystals of araganite. The vertical aesthete canals arise from a layer of horizontal canals in the lower part of the tegmentum and the articulamentum (Figure 13.43C) and some pass through the articulamentum to join with nerves in the mantle at the lower edge of the shell valve. The articulamentum is a thick, calcareous, porcelaneous layer that differs in certain ways from the shell layers of other molluscs.

Monoplacophorans have a single, limpet-like shell with the apex situated far forward (Figures 13.1A and 13.3). The shell has a distinctive outer prismatic layer and an inner nacreous layer. As in chitons, the mantle encircles the body and foot as a circular fold, forming lateral mantle grooves.

The bivalves possess two shells, or **valves**, that are connected dorsally by an elastic, proteinaceous **ligament**, and enclose the body and spacious mantle cavity (Figures 13.1M–P, 13.8, and 13.16J,K). Shells of bivalves typically have a thin periostracum, covering two to four calcareous layers that vary in composition and structure. The calcareous layers are often aragonite or an aragonite/calcite mixture, and they usually have a substantial organic framework. The periostracum and organic matrix may account for over 70% of the shell's dry weight in some thin-shelled taxa. Each valve has a dorsal protuberance called the umbo, which is the oldest part of the shell. Concentric growth lines radiate

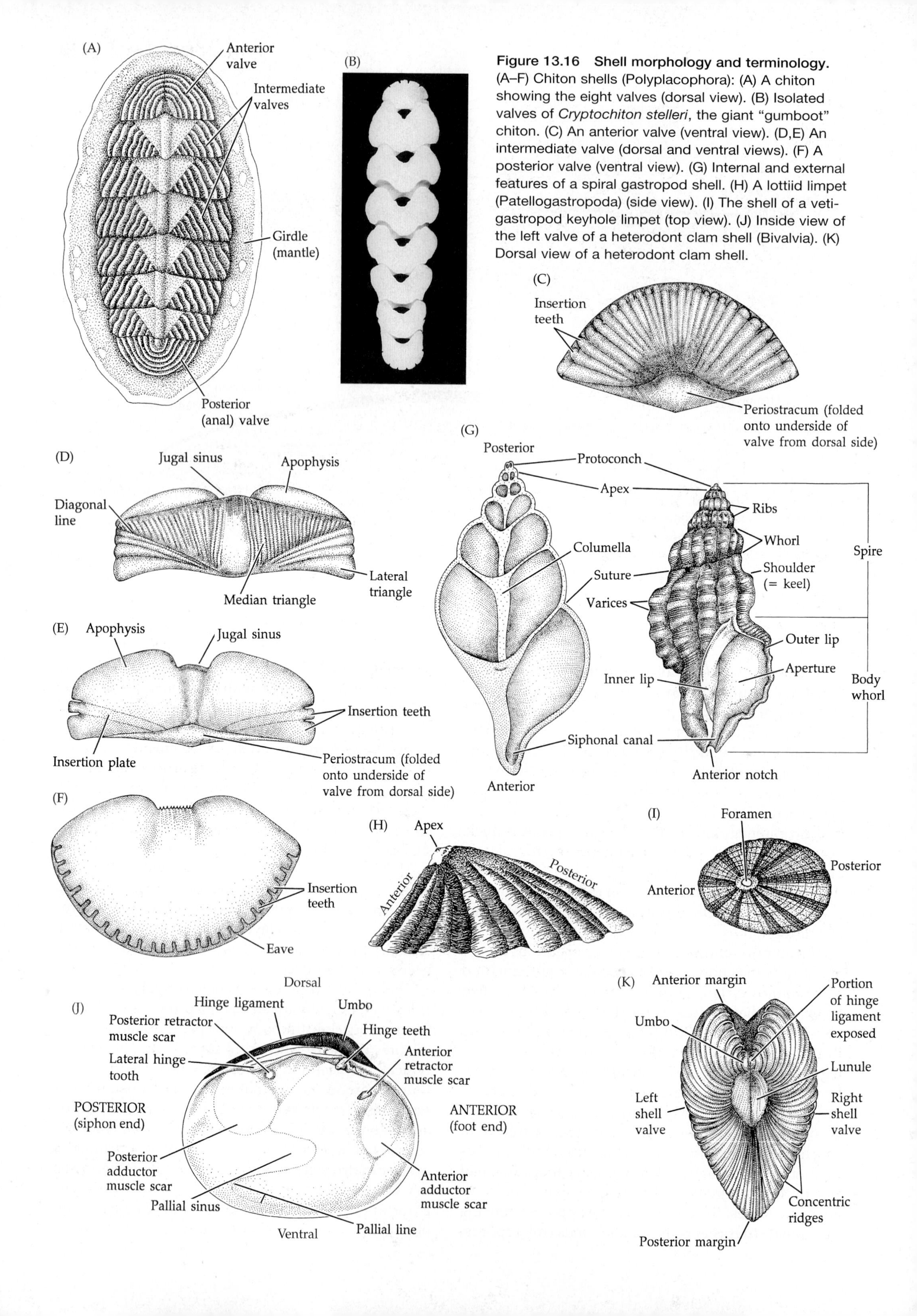

Figure 13.16 Shell morphology and terminology. (A–F) Chiton shells (Polyplacophora): (A) A chiton showing the eight valves (dorsal view). (B) Isolated valves of *Cryptochiton stelleri*, the giant "gumboot" chiton. (C) An anterior valve (ventral view). (D,E) An intermediate valve (dorsal and ventral views). (F) A posterior valve (ventral view). (G) Internal and external features of a spiral gastropod shell. (H) A lottiid limpet (Patellogastropoda) (side view). (I) The shell of a vetigastropod keyhole limpet (top view). (J) Inside view of the left valve of a heterodont clam shell (Bivalvia). (K) Dorsal view of a heterodont clam shell.

Figure 13.17 Two very different kinds of cephalopod shells. (A) The chambered shell of *Nautilus*, cut in longitudinal section. (B) The egg case "shell" of the paper nautilus, *Argonauta.*

outward from the umbo. When the valves are closed by contraction of the adductor muscles, the outer part of the ligament is stretched and the inner part is compressed. Thus, when the adductor muscles relax, the resilient ligament causes the valves to open. The hinge apparatus comprises various sockets and tooth-like pegs or flanges (hinge teeth) that align the valves and prevent lateral movement. In most bivalves, the adductor muscles contain both striated and smooth fibers, facilitating both rapid and sustained closure of the valves. This division of labor is apparent in some bivalves as for example in oysters where the large single adductor muscle is clearly composed of two parts, a dark striated region that functions as a rapid closure muscle, and a white smoother region that functions to hold the shell tightly closed for long periods of time.

The thin mantle lines the inner valve surfaces in bivalves and separates the visceral mass from the shell. The edge of a bivalve mantle bears three longitudinal ridges or folds—the inner, middle, and outer folds (Figure 13.15C). The innermost fold is the largest and contains radial and circular muscles, some of which attach the mantle to the shell. The line of mantle attachment appears on the inner surface of each valve as a scar called the **pallial line** (Figure 13.16J), and this scar is often a useful diagnostic character. The middle mantle fold is sensory in function, and the outer fold is responsible for secreting the shell. The cells of the outer lobe are specialized: the medial cells lay down the periostracum, and the lateral cells secrete the first calcareous layer. The entire mantle surface is then responsible for secreting the remaining innermost calcareous portion of the shell. A thin extrapallial space lies between the mantle and the shell, and it is into this space that materials for shell formation are secreted and mixed. Should a foreign object, such as a sand grain, lodge between the mantle and the shell, it may become the nucleus around which are deposited concentric layers of smooth nacreous or porcelaneous shell. The result is a pearl, either free in the extrapallial space or partly embedded in the growing shell.[6]

Scaphopod shells resemble miniature, hollow elephant tusks, hence the vernacular names "tusk shell" and "tooth shell" (Figures 13.1L and 13.9). The scaphopod shell is open at both ends, with the smaller opening at the dorsal end of the body. Most tusk shells are slightly curved, the concave side being the equivalent to the anterior of other molluscs. The mantle is large and lines the entire posterior surface of the shell. The dorsal aperture serves for both inhalant and exhalant water currents.

Most extant cephalopods have a reduced shell or are shell-less. A completely developed external shell is found only in fossil forms and the living species of *Nautilus*. In squids and cuttlefish the shell is reduced and internal, and in octopuses it is entirely lacking or present only as a small rudiment.

The shell of *Nautilus* is coiled in a planispiral fashion (whorls lie in a single plane) and has a thin periostracum (Figures 13.10, 13.17A, and 13.22B). *Nautilus* shells (and all cephalopod shells) are divided into internal chambers by transverse septa, and only the last chamber is occupied by the body of the living animal. As the animal grows, it periodically moves forward, and the posterior part of the mantle secretes a new septum behind it. Each septum is interconnected by a tube through which extends a cord of tissue called the **siphuncle**. The siphuncle helps to regulate buoyancy of the animal by varying the amounts of gas and fluid in the shell chambers. The shell is composed of an inner nacreous layer and an outer porcelaneous layer containing prisms of calcium carbonate and an organic matrix. The outer surface may be pigmented or pearly white. The junctions between septa and the shell wall are called **sutures**, and are simple and straight, or slightly waved (as in *Nautilus*), or were highly convoluted (as in the extinct

[6]Pearls are also found in some gastropods with nacreous inner shell layers, such as abalone.

ammonoids). In cuttlefish (order Sepiida), the shell is reduced and internal, with chambers that are very narrow spaces separated by thin septa. Like *Nautilus*, a cuttlefish can regulate the relative amounts of fluid and gas in its shell chambers. The small, coiled, septate, gas-filled shells of the deep-water squid *Spirula* are occasionally found washed up on beaches.

Fossil data suggest that the first cephalopod shells were probably small curved cones. From these ancestors both straight and coiled shells evolved, although secondary uncoiling probably occurred in several groups. Some straight-shelled cephalopods from the Ordovician period exceeded 5 m in length, and some Cretaceous coiled species had shell diameters of 3 m.

Gastropod shells are extremely diverse in size and shape (Figure 13.1D,G). The smallest are microscopic (less than 1 mm) and the largest may reach 70 cm in length. The "typical" shape is the familiar conical spiral wound around a central axis or columella (Figure 13.16G). The turns of the spire form whorls, demarcated by lines called sutures. The largest whorl is the last (or body) whorl, which bears the aperture through which the foot and head protrude. The traditional view of a coiled gastropod shell with the spire uppermost, is actually "upside down," since the lower edge of the aperture is anterior and the apex of the shell spire is posterior. The first few, very small, whorls at the apex are the larval shell, or **protoconch** (or its remnant), which usually differs in sculpturing and color from the rest of the shell. The last whorl and aperture may be notched and drawn out into an anterior **siphonal canal**, to house a siphon when present. A smaller posterior canal may also be present on the rear edge of the aperture that houses a siphon-like fold of the mantle where waste and water are expelled.

Every imaginable variation on the basic spiraled shell occurs among the gastropods (and some unimaginable): the shell may be long and slender (e.g., auger shells) or short and plump (e.g., trochids), or the shell may be flattened, with all whorls more-or-less in one plane (e.g., sundials). In some the spire may be more or less incorporated into the last whorl and eventually disappear from view (as in cowries). In some with a much larger last whorl, the aperture may be reduced to an elongated slit (Figure 13.1E (e.g., cowries, olives, and cones). In a few groups the shell may coil so loosely as to form a meandering wormlike tube (e.g., the so-called "tube snails," vermetids and siliquariids; Figure 13.19E). In a number of gastropod groups the shell may be reduced and overgrown by the mantle, or it may disappear entirely resulting in a slug (see below). Most gastropods spiral clockwise; that is, they show right-handed, or dextral, coiling. Some are sinistral (left-handed), and some normally dextral species may occasionally produce sinistral individuals. In limpets the shell is cap-shaped, with a low conical shape with no or little visible coiling (Figure 13.16H,I). The limpet shell form has been derived from coiled ancestors on numerous occasions during gastropod evolution.

Gastropod shells consist of an outer thin organic periostracum and two or three calcareous layers: an outer prismatic (or palisade) layer, and middle and inner lamellar or crossed layers. In many vetigastropods the inner layer is nacreous. In some patellogastropods up to six calcareous layers are distinguishable but in the great majority of living gastropods the shell structure is primarily one layer composed of crossed crystals (crossed lamellar shell structure). Gastropods in which the shell is habitually covered by mantle lobes lack a periostracum (e.g., olives and cowries), but in some other groups the periostracum is very thick and sometimes it is produced into lamellae or hairs. The prismatic and lamellate layers consist largely of calcium carbonate, either as calcite or aragonite. These two forms of calcium are chemically identical, but they crystallize differently and can be identified by microscopic examination of sections of the shell. Small amounts of other inorganic constituents are incorporated into the calcium carbonate framework, including chemicals such as phosphate, calcium sulfate, magnesium carbonate, and salts of aluminum, iron, copper, strontium, barium, silicon, manganese, iodine, and fluorine.

An intriguing aspect of gastropod evolution is shell loss and the achievement of the "slug" form. Despite the fact that evolution of the coiled shell led to great success for the gastropods—75% of all living molluscs are snails—secondary loss of the shell occurred many times in this class but mostly in various groups of euthyneurans such as the sea slugs and land slugs. In forms such as the land and sea slugs, the shell may persist as a small vestige covered by the dorsal mantle (e.g., in the euthyneuran sea slugs Aplysiinae and Pleurobranchidae, and the caenogastropod family Velutinidae), or as a small external rudiment as in the carnivorous land slug *Testacella*, or it may be lost altogether (e.g., the nudibranchs, the systellommatophorans and some terrestrial stylommatophoran slugs, and in the neritimorph slug *Titiscania*). In the nudibranchs (Nudibranchia) the larval shell is first covered, then resorbed, by the mantle during ontogeny. Shell loss occurred numerous times in gastropods, particularly among the sea slugs ("opisthobranchs") and stylommatophoran pulmonates. Shells are energetically expensive to produce and require a reliable source of calcium in the environment, so it might be advantageous to eliminate them if compensatory mechanisms exist. For example, most, if not all, sea slugs secrete chemicals that make them distasteful to predators. In addition, the bright coloration of many nudibranchs may serve a defensive function. In some species, the color matches the animal's background such as the small red nudibranch, *Rostanga pulchra*, which matches almost perfectly the red sponge on which it feeds. Many nudibranchs, are however,

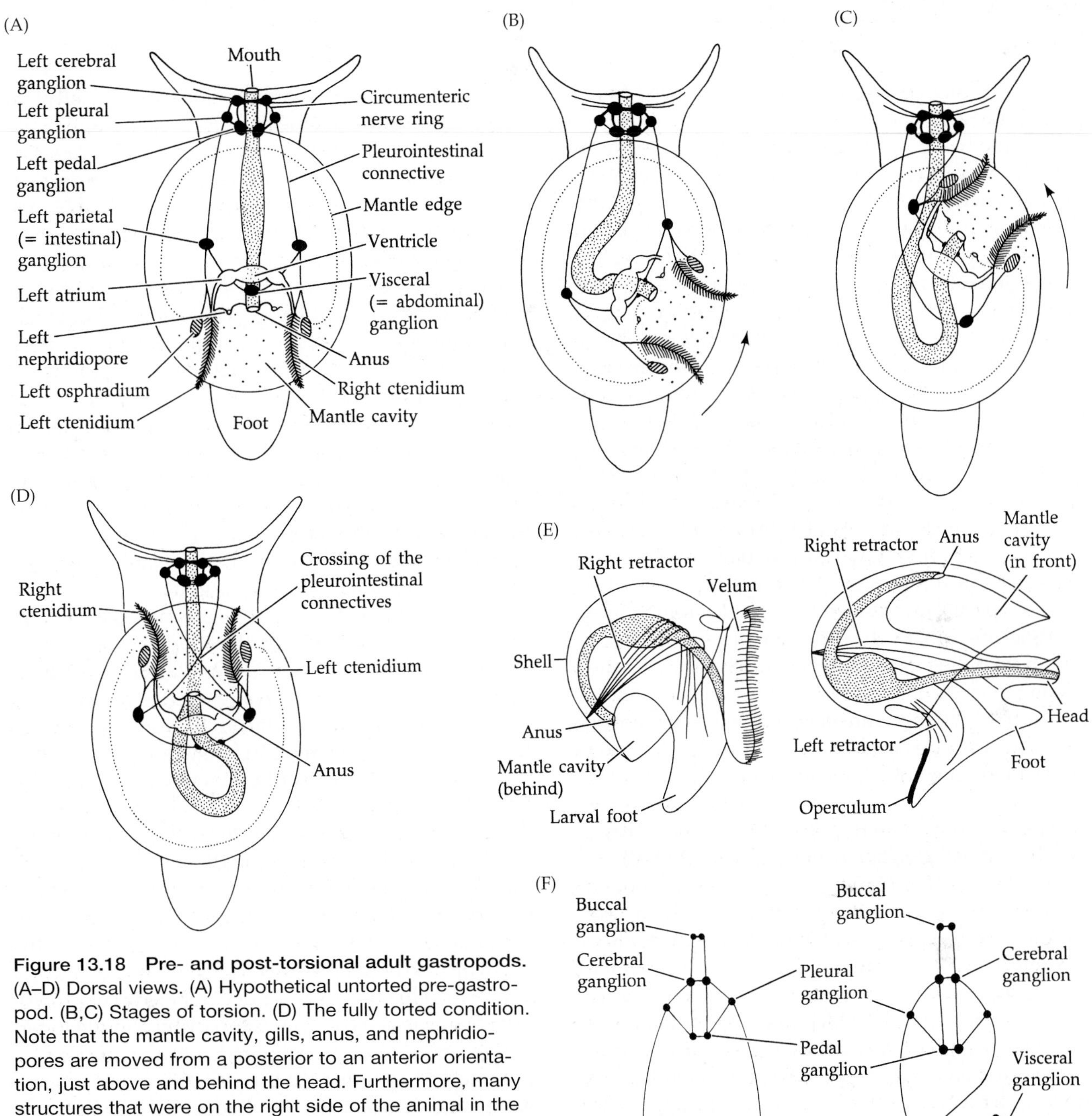

Figure 13.18 Pre- and post-torsional adult gastropods. (A–D) Dorsal views. (A) Hypothetical untorted pre-gastropod. (B,C) Stages of torsion. (D) The fully torted condition. Note that the mantle cavity, gills, anus, and nephridiopores are moved from a posterior to an anterior orientation, just above and behind the head. Furthermore, many structures that were on the right side of the animal in the pretorsional condition (e.g., the right gill, osphradium, heart atrium, and nephridiopore) are located on the left side after torsion has taken place (and the pretorsional left gill, osphradium, atrium, and nephridiopore subsequently occur on the right side). (E) A gastropod veliger larva, before and after torsion (lateral view). Note that after torsion the head can be withdrawn into the anterior mantle cavity. (F) Configuration of the principal ganglia and connectives of a hypothetical untorted and a torted adult gastropod.

conspicuous in nature. In these cases, the color may serve to warn predators of the noxious taste of the slug or, as suggested by Rudman (1991), predators may simply ignore such bright "novelties" in their environment.

Torsion, or "How the Gastropod Got Its Twist"

One of the most remarkable and dramatic steps taken during the course of molluscan evolution was the advent of torsion, a unique synapomorphy of gastropods, and it is quite unlike anything else in the animal kingdom. Torsion takes place during development in all gastropods, usually during the late veliger larval stage. It is a rotation of the visceral mass and its overlying mantle and shell as much as 180° with respect to the head and foot (Figures 13.18, and 13.53). The twisting is always in a counterclockwise direction (viewing the animal from above), and it is completely different from the phenomenon of coiling (= spiraling). During torsion, the mantle cavity and anus are moved from a posterior to an anterior position, somewhat above and behind the head. Visceral structures and incipient organs that were on the right side of the larval animal end up on the left side of the adult. The gut is twisted into a U-shape, and when the longitudinal nerve cords connecting the pleural to the visceral ganglia develop, they are crossed rather like a figure eight. Most veligers have nephridia, which reverse sides, but the adult gills and gonads are not fully developed when torsion occurs.

Torsion is usually a two-step process. During larval development, an asymmetrical velar or foot retractor muscle develops. It extends from the shell on the right, dorsally over the gut, and attaches on the left side of the head and foot. At a certain stage in the veliger's development, contraction of this muscle causes the shell and enclosed viscera to twist about 90° in a counterclockwise direction. This first 90° twist is usually rapid, taking place in a few minutes to a few hours. The second 90° twist is typically much slower and results from differential tissue growth. By the end of the process, the viscera have been pulled from above toward the left, ultimately leading to the figure-eight arrangement of the adult visceral nerves. But the figure-eight arrangement is not perfect as the left esophageal ganglion usually comes to lie dorsal to the gut and is thus called the supraesophageal (= supraintestinal) ganglion; however, the right esophageal ganglion lies ventral to the gut, as a subesophageal (= subintestinal) ganglion (Figures 13.18 and 13.40).

Gastropods that retain torsion into adulthood are said to be **torted**; those that have secondarily reverted back to a partially or fully untorted state in adulthood are **detorted**. The torted, figure-eight configuration of the nervous system is referred to as **streptoneury**. The detorted condition, in which the visceral nerves are secondarily untwisted, is referred to as **euthyneury**.

Detorted gastropods, such as many heterobranchs, undergo a postveliger series of changes through which the original torsion is reversed to various degrees. The process shifts the mantle cavity and at least some of the associated organs about 90° back to the right (as in many "pulmonates" and some sea slugs), or in some cases all the way back to the rear of the animal (the detorsion seen in some nudibranchs).

After torsion the anus lies in front; this means that the first gastropods could no longer grow in length easily. Subsequent increase in body size thus occurred by the development of loops or bulges in the middle portion of the gut region, thereby producing the characteristic coiled visceral hump. The first signs of torsion and coiling occur at about the same time during gastropod development. The earliest coiled gastropod shells in the fossil record include both planispiral and conispiral forms, and it is possible that coiling predated the appearance of torsion in gastropods. Once both features were established, they coevolved in various ways to produce what we see today in living gastropods.

The evolution of asymmetrically coiled shells had the effect of restricting the right side of the mantle cavity, a restriction that led to reduction or loss of the structures it contained on the adult right side (the original left ctenidium, atrium, and osphradium). At the same time, these structures on the adult left side (the original right ctenidium, atrium, and osphradium) tended to enlarge. Possibly correlated with torsion and coiling was the loss of the left post-torsional gonad. The single remaining gonad opens on the right side via the post-torsional right nephridial duct and nephridiopore. Patellogastropods and most vetigastropods retain two functional nephridia, although the post-torsional left one is often reduced. In other gastropods the post-torsional right nephridium is lost, but its duct and pore remain associated with the reproductive tract in neritimorphs and caenogastropods.

Such profound changes in spatial relations between major body regions as those brought about by torsion and spiral coiling in gastropods are rare among other animals. Several theories on the adaptive significance of torsion have been proposed and are still being argued. The great zoologist Walter Garstang suggested that torsion was an adaptation of the veliger larva that served to protect the soft head and larval ciliated velum from predators (see the section on development later in this chapter). When disturbed, the immediate reaction of a veliger is to withdraw the head and foot into the larval shell, whereupon the larva begins to sink rapidly. This theory may seem reasonable for evasion of very small planktonic predators, but it seems illogical as a means of escape from larger predators in the sea, which no doubt consume veligers whole—and any adaptive value to adults is not explained. Two zoologists finally tested Garstang's theory by offering torted and untorted abalone veligers to various planktonic predators; they found that, in general, torted veligers were not consumed any less frequently than untorted ones (Pennington and Chia 1985). Garstang first presented his theory in verse, in 1928, as he was often taken to do with his zoological ideas.

The Ballad of the Veliger, or
How the Gastropod Got Its Twist

The Veliger's a lively tar, the liveliest afloat,
A whirling wheel on either side propels his little boat;
But when the danger signal warns his bustling submarine,
He stops the engine, shuts the port, and drops below unseen.

He's witnessed several changes in pelagic motorcraft;
The first he sailed was just a tub, with a tiny cabin aft.
An Archi-mollusk fashioned it, according to his kind,
He'd always stowed his gills and things in a mantle-sac behind.

Young Archi-mollusks went to sea with nothing but a velum—
A sort of autocycling hoop, instead of pram—to wheel 'em;

And, spinning round, they one by one acquired parental features,
A shell above, a foot below—the queerest little creatures.
But when by chance they brushed against their neighbors in the briny,
Coelenterates with stinging threads and Arthropods so spiny,
By one weak spot betrayed, alas, they fell an easy prey—
Their soft preoral lobes in front could not be tucked away!
Their feet, you see, amidships, next the cuddly-hole abaft,
Drew in at once, and left their heads exposed to every shaft.
So Archi-mollusks dwindled, and the race was sinking fast,
When by the merest accident salvation came at last.
A fleet of fry turned out one day, eventful in the sequel,
Whose left and right retractors on the two sides were unequal:
Their starboard halliards fixed astern alone supplied the head,
While those set aport were spread abeam and served the back instead.
Predaceous foes, still drifting by in numbers unabated,
Were baffled now by tactics which their dining plans frustrated.
Their prey upon alarm collapsed, but promptly turned about,
With the tender morsel safe within and the horny foot without!
This manoeuvre (vide Lamarck) speeded up with repetition,
Until the parts affected gained a rhythmical condition,
And torsion, needing now no more a stimulating stab,
Will take its predetermined course in a watchglass in the lab.
In this way, then, the Veliger, triumphantly askew,
Acquired his cabin for'ard, holding all his sailing crew—
A Trochosphere in armour cased, with a foot to work the hatch,
And double screws to drive ahead with smartness and dispatch.
But when the first new Veligers came home again to shore,
And settled down as Gastropods with mantle-sac afore,
The Archi-mollusk sought a cleft, his shame and grief to hide,
Crunched horribly his horny teeth, gave up the ghost, and died.

Other workers have hypothesized that torsion was an adult adaptation that might have created more space for retraction of the head into the shell (perhaps also for protection from predators), or for directing the mantle cavity with its gills and water-sensing osphradia anteriorly. Still another theory asserts that torsion evolved in concert with the evolution of a coiled shell—as a mechanism to align the tall spiraling shells from a position in which they stuck out to one side (and were presumably poorly balanced and growth limiting), to a position more in alignment with the longitudinal (head–foot) axis of the body. The latter position would theoretically allow for greater growth and elongation of the shell while reducing the tendency of the animal to topple over sideways.

No matter what the evolutionary forces were that led to torsion in the earliest gastropods, the results were to move the adult anus, nephridiopores, and gonopores to a more anterior position, corresponding to the new position of the mantle cavity. It should be noted however, that the actual position and arrangement of the mantle cavity and its associated structures show great variation; in many gastropods these structures, while pointing forward, may actually be positioned further towards the posterior region of the animal's body. Torsion is not a perfectly symmetrical process.

Most of the stories of gastropod evolution focus on changes in the mantle cavity and its associated structures, and many of these changes seem to have been driven by some negative impacts of torsion. Many anatomical modifications of gastropods appear to be adaptations to avoid fouling, for without changing the original flow of water through the mantle cavity in a primitive gastropod with two ctenidia, waste from the centrally positioned anus (and perhaps the nephridia) would be dumped on top of the head and potentially pollute the mouth and ctenidia. Hence, it has long been hypothesized that the first step, subsequent to the evolution of torsion, was the development of slits or holes in the shell, thus altering water flow so that a one-way current passed first over the ctenidia, then over the anus and nephridiopore, and finally out the slit or shell holes. This arrangement is seen in some vetigastropods, such as the slit shells (Pleurotomarioidea) and abalone and keyhole limpets (Figures 13.1D, 13.16I, and 13.36). As reasonable as it sounds, there has been surprisingly little empirical evidence in support of this hypothesis. In addition, the adaptive significance of shell holes was examined by Voltzow and Collin (1995), who found that blocking the holes in keyhole limpets did not result in damage to the organs of the mantle cavity. Thus, the adaptive significance of torsion in gastropod evolution remains an open question.

Once evolutionary reduction or loss of the gill and osphradium on the right side had taken place, water flow through the mantle cavity was from left to right, passing through the left gill and osphradium first, then across the nephridiopore and anus, and on out the right side. This strategy also had the effect of allowing structures on the left side to enlarge and eventually to develop more control over water flow into and out of the mantle cavity, including the evolution of long siphons. While most gastropods have retained full or partial torsion, many of the heterobranch gastropods, all of which lost the original ctenidium, have undergone various degrees of detorsion, and a host of other modifications, perhaps in response to the absence of constraints originally brought on by torsion.

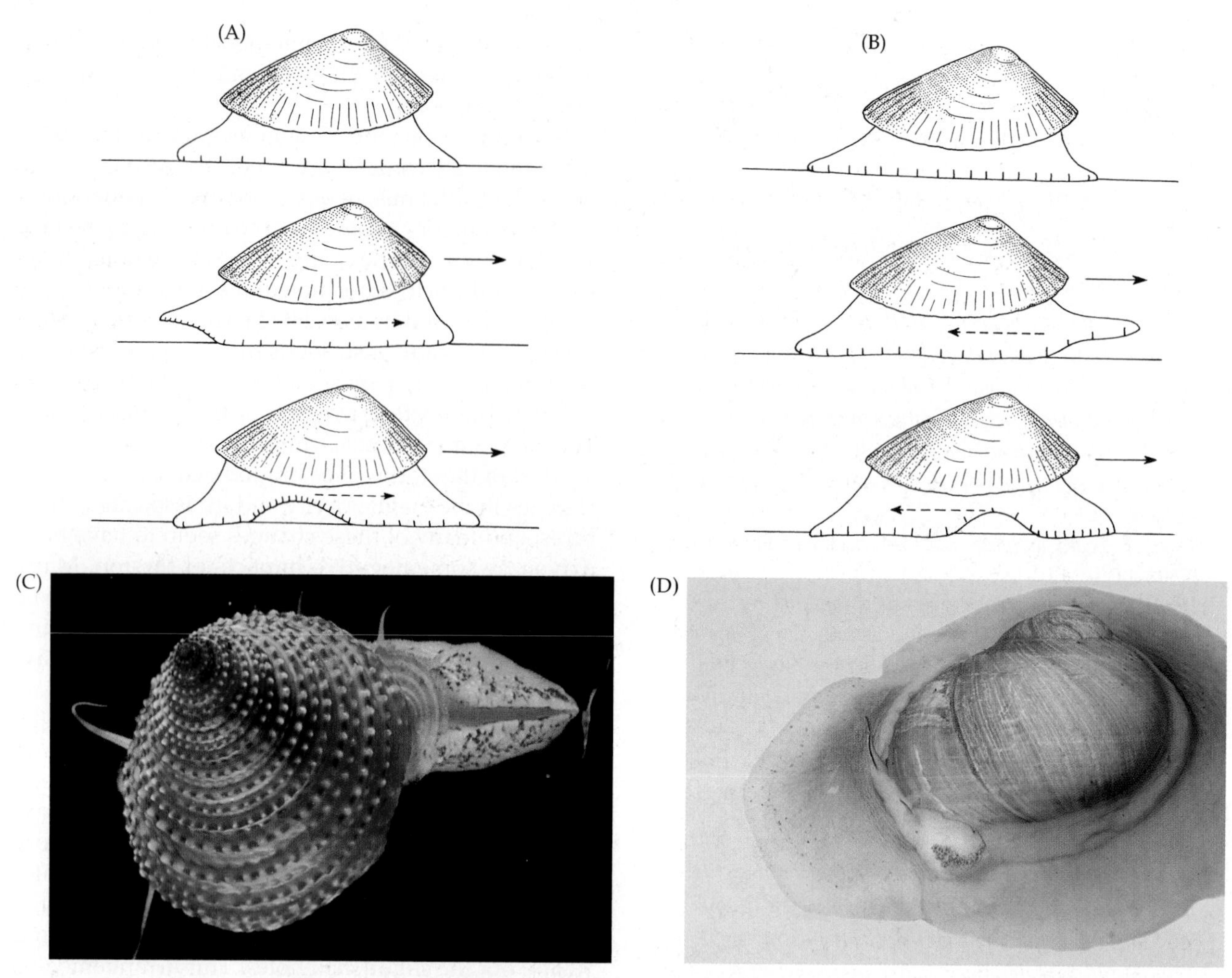

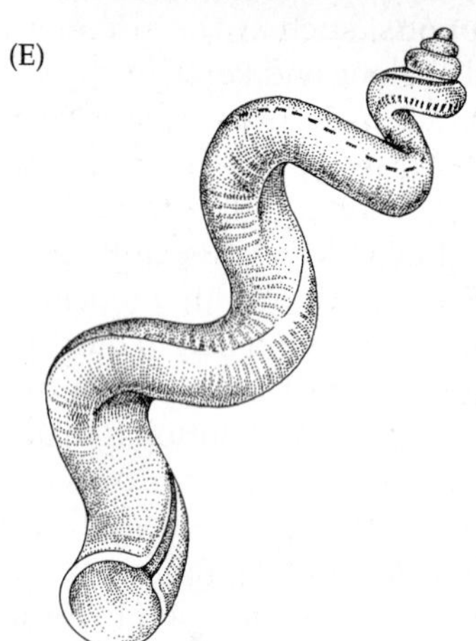

Figure 13.19 (A,B) Locomotion in a benthic gastropod moving to the right by waves of contraction of the pedal and foot muscles (solid arrow indicates direction of animal movement; dashed arrow indicates direction of muscle wave). In (A) the waves of contraction are moving in the same direction as the animal, from back to front (direct waves). Muscles at the rear of the animal contract to lift the foot off the substratum; the foot shortens in the contracted region and then elongates as it is placed back down on the substratum after the wave passes. In this way, successive sections of the foot are "pushed" forward. In (B) the animal moves forward as the contraction waves pass in the opposite direction, from front to back (retrograde waves). In this case, the pedal muscles lift the anterior part of the foot off the substratum, the foot elongates, is placed back on the substratum, then contracts to "pull" the animal forward, rather like "stepping." (C) *Calliostoma*, a vetigastropod (Calliostomatidae) adapted to crawling on hard substrata. Note the line separating the right and left sides of the trailing foot; the line denotes a separation of muscle masses that allows a somewhat "bipedal-like" motion as the animal moves (see text for further details). (D) The moon snail, *Polinices* (Naticidae), has a huge foot that can be inflated by incorporating water into a network of channels in its tissue, thus allowing the animal to plow through the surface layer of soft sediments. (E) *Tenagodus* (Siliquariidae), a sessile siliquariid worm snail.

Locomotion

The foot in aplacophorans is either rudimentary or lost (Figure 13.2). Caudofoveata are mostly infaunal burrowers and move by peristaltic movements of the body wall, using the anterior mouth shield as a burrowing device and anchor. The foot of solenogasters is only weakly muscular, and locomotion is primarily by slow ciliary gliding movements through or upon the substratum. Caudofoveata are mostly infaunal burrowers, and Solenogastres are largely symbiotic on various cnidarians. With the exception of these two groups, most other molluscs possess a distinct and obvious foot, with the exception of the cephalopods where it is very highly modified. In chitons, monoplacophorans and most gastropods the foot often forms a flat, ventral, creeping sole (Figures 13.3B, 13.4B, 13.5B, and 13.19). The sole is ciliated and imbued with numerous gland cells that produce a mucous trail over which the animal glides. In gastropods, enlarged pedal glands supply substantial amounts of mucus (slime), this being especially important in terrestrial species that must glide on relatively dry surfaces. In most gastropods, there is an anterior mucus gland, which opens in a slit on or just behind the anterior edge of the foot. This anterior lobe is called the **propodium**, the rest of the foot the **metapodium**. In some caenogastropods, an enlarged metapodial mucus gland opens into the middle of the sole. Small molluscs may move largely by ciliary propulsion but most move primarily by waves of muscular contractions that move along the foot.

The gastropod foot possesses sets of pedal retractor muscles, which attach to the shell and dorsal mantle at various angles. These and smaller muscles in the foot act in concert to raise and lower the sole or to shorten it in either a longitudinal or a transverse direction. Contraction waves may move from back to front (direct waves), or from front to back (retrograde waves) (Figure 13.19A,B). Direct waves depend on contraction of longitudinal and dorsoventral muscles beginning at the posterior end of the foot; successive sections of the foot are thus "pushed" forward. Retrograde waves involve contraction of transverse muscles interacting with hemocoelic pressure to extend the anterior part of the foot forward, followed by contraction of longitudinal muscles. The result is that successive areas of the foot are "pulled" forward (Figure 13.19A,B). In some gastropods the muscles of the foot are separated by a midventral line, so the two sides of the sole operate somewhat independently of each other. The right and left sides of the foot alternate in their forward motion, almost in a stepping fashion, resulting in a sort of "bipedal" locomotion.

Modifications of this general benthic locomotory scheme occur in many groups. Some gastropods, such as moon snails (Figure 13.19D), plow through the sediment, and some even burrow beneath the sediment surface. Such gastropods often possess an enlarged, shield-like propodium that acts like a plough and some naticids and cephalaspideans possess a dorsal flaplike fold of the foot that covers the head as a protective shield. Other burrowers, such as augers, dig by thrusting the foot into the substratum, anchoring it by engorgement with hemolymph, and then pulling the body forward by contraction of longitudinal muscles. In the conch *Strombus*, the operculum forms a large "claw" that digs into the substratum and is used as a pivot point as the animal thrusts itself forward like a pole-vaulter using its muscular, highly modified foot. In some heterobranchs, notably the sea hares (Aplysidae), lateral flaps of the foot expand dorsally as parapodia and these fuse dorsally in some species.

Some molluscs that inhabit high-energy littoral habitats, such as chitons and limpets, have a very broad foot that can adhere tightly to hard substrata. Chitons also use their broad girdle for additional adhesion to the substratum by clamping down tightly and raising the inner margin to create a slight vacuum. Some snails, such as the Vermetidae and Siliquariidae are entirely sessile, the former attached to hard substrata, the latter (Figure 13.19E) living in sponges. These gastropods have typical larval and juvenile shells; but after they settle and start to grow, the shell whorls become increasingly separated from one another, resulting in a corkscrew or twisted shape. Other gastropods, such as slipper shells, are sedentary. They tend to remain in one location and feed on organic particles in the surrounding water. The sole of the hipponicid limpets secretes a calcareous plate and the adults are thus oyster-like and deposit feed using their long snout.

Some limpets and a few chitons exhibit homing behaviors. These activities are usually associated with feeding excursions stimulated by changing tide levels or darkness, after which the animals return to their homesites which is seen as a scar or even a depression on the rock surface. Homing behaviors are also seen in some land snails and slugs.

Most bivalves live in soft benthic habitats, where they burrow to various depths in the substratum (Figure 13.20E–I). In these infaunal species the foot is usually blade-like and laterally compressed (the word pelecypod means "hatchet foot"), as is the body in general. The pedal retractor muscles in bivalves are somewhat different from those of gastropods, but they still run from the foot to the shell (Figure 13.8D). The foot is directed anteriorly and used primarily in burrowing and anchoring. It operates through a combination of muscle action and hydraulic pressure (Figure 13.20A–D). Extension of the foot is accomplished by engorgement with hemolymph, coupled with the action of a pair of pedal protractor muscles. With the foot extended, the valves are pulled together by the shell adductor muscles. More hemolymph is forced from the visceral mass hemocoel into the foot hemocoel, causing the foot to expand and anchor in the substratum. Once the foot is anchored, the anterior and posterior pairs of pedal retractor muscles contract and pull the shell downward. Withdrawal of the foot into the shell is accomplished by contraction of the pedal retractors coupled with relaxation of the shell adductor muscles. Many infaunal bivalves burrow upward in this same manner, but others back out by using hydraulic pressure to push against the anchored end of the foot. Most motile bivalves possess well-developed anterior and posterior adductor muscles (the dimyarian condition).

There are several groups of bivalves that have epifaunal lifestyles and are permanently attach to the substratum either by cementing one valve to a hard surface as in the true oysters such as the rock oysters (Ostreidae) and rock scallops (Spondylidae). Others use special anchoring threads (**byssal threads**), as

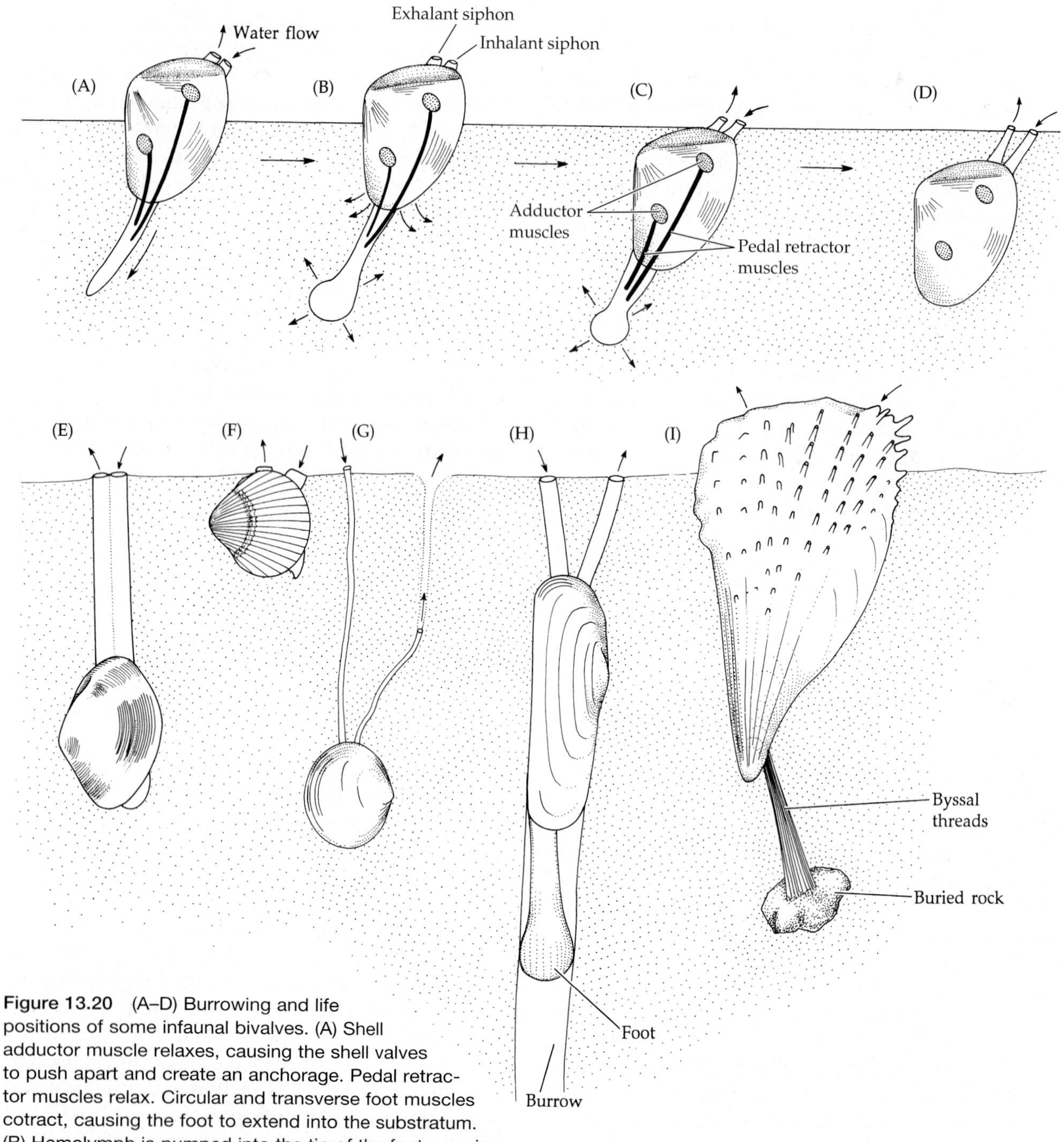

Figure 13.20 (A–D) Burrowing and life positions of some infaunal bivalves. (A) Shell adductor muscle relaxes, causing the shell valves to push apart and create an anchorage. Pedal retractor muscles relax. Circular and transverse foot muscles cotract, causing the foot to extend into the substratum. (B) Hemolymph is pumped into the tip of the foot, causing it to expand and form an anchorage. Siphons close and withdraw as the shell adductor muscles contract, closing the shell and forcing water out between the valves and around the foot. (C) Anterior and posterior pedal retractor muscles contract, pulling the clam deeper into substratum. (D) The shell adductor muscle relaxes to allow shell valves to push apart and create an anchorage in the new position. The foot is withdrawn. (E–I) Five bivalves in soft sediments; arrows indicate direction of water flow. (E) A deep burrower with long, fused siphons (*Tresus*). (F) A shallow burrower with very short siphons (*Clinocardium*). (G) A deep burrower with long, separate siphons (*Scrobicularia*). (H) The razor clam (*Tagelus*) lives in unstable sands and maintains a burrow into which it can rapidly escape. (I) The pen shell, *Atrina*, attaches its byssal threads to solid objects buried in soft sediments.

in mussels (Mytilidae) (Figure 13.21A,B), ark shells, and a number of other families including winged or pearl oysters (Pteriidae), and numerous other pteriomorphian bivalves including the Pinnidae and many Arcidae and Pectinidae. While the juveniles of many heterodont bivalves produce one or a few temporary byssal threads, a few species, such as the zebra mussel (*Dreissena*), remain byssally attached as adults.

The true oysters (Ostreidae) (including the edible American and European oysters) initially anchor as a settling veliger larva (called a **spat** by oyster farmers) by secreting a drop of adhesive from the byssus gland. Adults, however, have one valve permanently cemented to the substratum, with the cement being produced by the mantle.

Byssal threads are secreted as a liquid by the **byssus gland** in the foot. The liquid flows along a groove in the foot to the substratum, where each thread becomes tightly affixed. The threads are emplaced by the foot; once attached they quickly harden by a tanning process, whereupon the foot is withdrawn. A byssal thread retractor muscle may assist the animal in pulling against its anchorage. Mussels have a small, finger-like foot whose principal function is generation and placement of the byssal threads. Giant clams (Cardiidae) initially attach by byssal threads, but usually lose these as they mature and become heavy enough not to be cast about by currents (Figure 13.1N). In jingle shells (Anomiidae), the byssal threads run from the upper valve through a hole in the lower valve to attach to the substratum after which they become secondarily calcified. Byssal threads probably represent a primitive and persisting larval feature in those groups that retain them into adulthood, and many bivalves lacking byssal threads as adults utilize them for initial attachment during settlement.

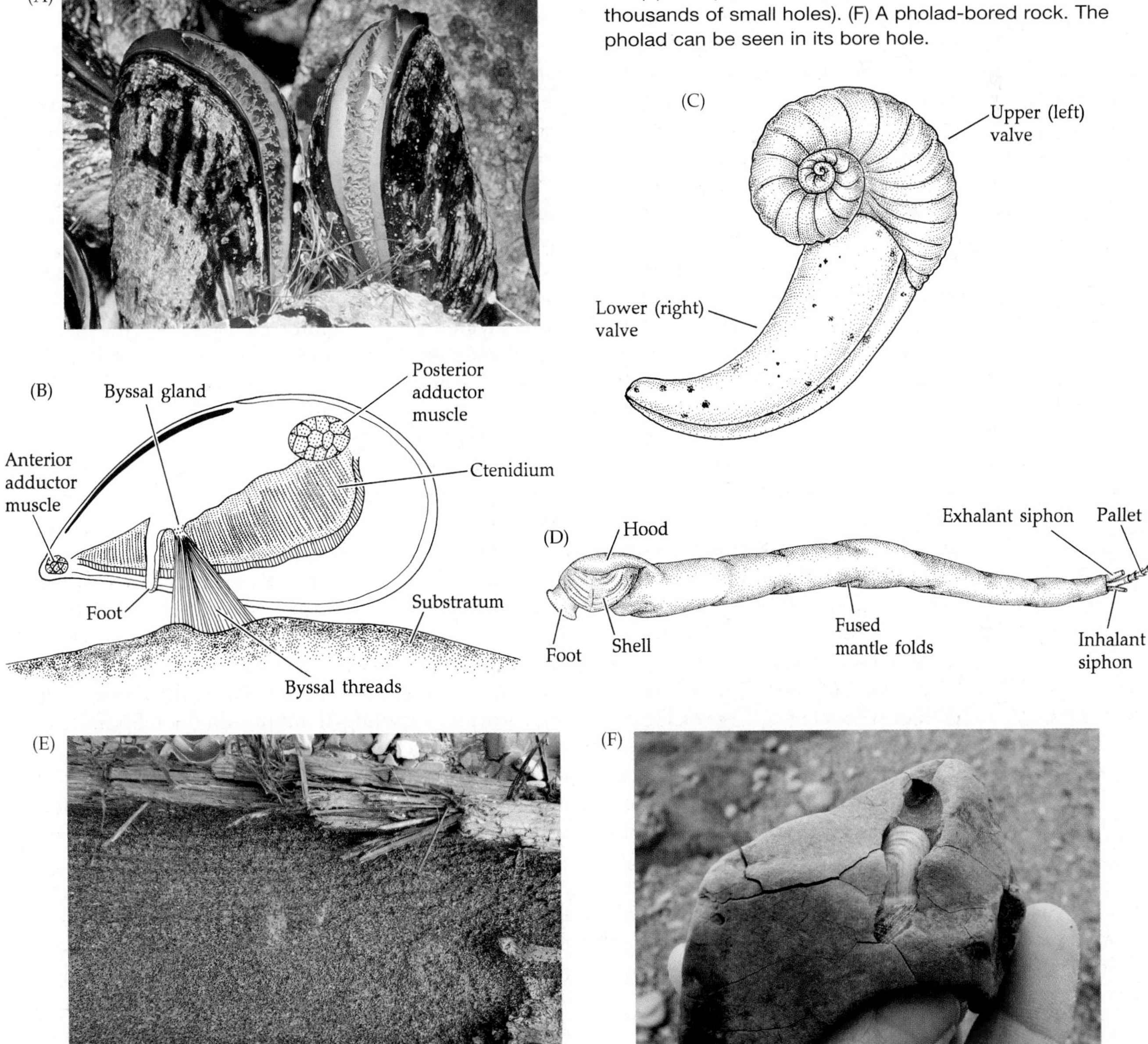

Figure 13.21 More bivalves. (A) A "bed" of mussels (*Mytilus californianus*) attached by byssal threads (close up of two mussels). (B) A mussel (lateral view, with left valve removed). (C) Shell of the Mesozoic rudist clam *Coralliochama*. (D) The wood-boring bivalve (shipworm) *Teredo*. The pallets (only one is shown) are a pair of shelly plates that close over the siphons when they are retracted. (E) A shipworm-bored piece of driftwood (notice the thousands of small holes). (F) A pholad-bored rock. The pholad can be seen in its bore hole.

In many families of attached bivalves, such as mussels and true oysters, the foot and anterior end are reduced. This often leads to a reduction of the anterior adductor muscle (**anisomyarian condition**) or its complete loss (**monomyarian condition**).

Great variation occurs in shell shape and size among attached bivalves. Some of the most remarkable were the Mesozoic rudists, in which the lower valve was horn-like and often curved, and the upper valve formed a much smaller hemispherical or curved lid (Figure 13.21C). Rudists were large, heavy creatures that often formed massive reef-like aggregations, either by somehow attaching to the substratum or by simply accumulating in large numbers on the seabed, in "log jams." These accumulations of fossil shells provide the spaces in which oil deposits formed in sediments in many parts of the Middle East and Caribbean.

Some originally attached bivalves have evolved to live freely upon the sea floor (e.g., some Pectinidae and Limidae) (Figure 13.1M). Some are capable of short bursts of "jet-propelled" swimming, which is accomplished by quickly clapping the valves together.

The habit of boring into hard substrata has evolved in several different bivalve lines. In all cases, excavation begins soon after larval settlement. As the animal bores deeper, it grows in size and soon becomes permanently trapped, with only the siphons protruding out of the original small opening. Boring is usually by a mechanical process; the animal uses serrations on the anterior region of the shells to abrade or scrape away the substratum. Some species also secrete an acidic mucus that partially dissolves or weakens hard calcareous substrata (limestone, coral, large dead shells). Some species bore into wood, such as *Martesia* (Phaladidae), *Xylophaga* (Xylophagidae), and nearly all species in the family Teredinidae (*Bankia*, *Teredo*). Teredinids, with their long wormlike bodies, are known as shipworms because of the destruction that they can cause to the wooden hulls of ships (or to wood pier pilings). In the teredinids the shells are reduced to small anterior bulb-like valves that serve as the drilling apparatus (Figure 13.21D,E). Some pholads bore into soft stone (e.g., *Pholas*), or into other substrata (e.g., *Barnea*; Figure 13.21E). Some species in the family Mytilidae also are borers, such as *Lithophaga*, which bores by mechanical and possibly chemical means into calcareous rocks, shells of various other molluscs (including chitons), and corals, and the genus *Adula*, which bores into soft rocks.

Scaphopods are adapted to infaunal habitats, burrowing vertically by the same basic mechanism used by many bivalves (Figures 13.1L and 13.9). The elongate foot is projected downward into soft substrata, whereupon a rim in the distal part of the foot is expanded to serve as an anchoring device; contraction of the pedal retractor muscles pulls the animal downward.

Perhaps the most remarkable locomotory adaptation of molluscs is swimming, which has evolved in several different taxa in several different ways, including by valve flapping in scallops. In most others of these groups, the foot is modified as the swimming structure. In the unique caenogastropod group known as heteropods, the body is laterally compressed, the shell is greatly reduced, the foot forms a fin, and the animal swims upside down (Figure 13.7A–C). Swimming has evolved several times in the heterobranchs, including the pteropods (sea butterflies), where the parapodial extensions of the foot form two long lateral fins that are used like oars (Figure 13.7D,E). Some nudibranchs also swim by graceful undulations of flaplike parapodial folds along the body margin or by vigorous undulations of the body. Although not technically swimming, violet shells (*Janthina*) float about the ocean's surface on a raft of bubbles secreted by the foot, and some planktonic nudibranchs (e.g., *Glaucus*, *Glaucilla*) stay afloat by use of an air bubble held in the stomach!

The champion swimmers are, of course, the cephalopods (Figures 13.1J,K and 13.22). These animals have abandoned the generally sedentary habits of other molluscs and have become highly effective swimming predators. Virtually all aspects of their biology have evolved to exploit this lifestyle. Most cephalopods swim by rapidly expelling water from the mantle cavity. In the coleoid cephalopods the mantle has both radial and circular muscle layers. Contraction of the radial muscles and relaxation of the circular muscles draws water into the mantle cavity while reversal of this muscular action forces water out of the mantle cavity. The mantle edge is clamped tightly around the head to channel the escaping water through a ventral tubular **funnel**, or **siphon** (Figure 13.11B,C). The funnel is highly mobile and can be manipulated to point in nearly any direction, thus allowing the animal to turn and steer. Squids attain the greatest swimming speeds of any aquatic invertebrates, and several species can even leave the water and propel themselves many feet into the air. Most octopuses are benthic and lack the fins and streamlined bodies characteristic of squids. Although octopuses still use water-powered jet propulsion, they more commonly rely on their long suckered arms for crawling about the sea floor. Some octopuses have even been observed moving about upright on only two tentacles—bipedal locomotion! Cuttlefish are slower than squids, and often use their fins for forward swimming as well as stabilization and to assist in steering and propulsion.

Nautilus move up and down in the water column on a diurnal cycle, often traveling hundreds of meters in each direction. They can actively regulate their buoyancy by secretion and reabsorption of shell chamber gases (chiefly nitrogen) by the cells of the siphuncle. The unoccupied chambers of these shells are filled partly with gas and partly with a liquid called the **cameral fluid**. The septa act as braces, giving the shells enough strength to withstand pressures at depth. As discussed earlier, each septum in nautiloid shells is perforated by a small hole, through which runs the siphuncle, which originates in the viscera and is enclosed in a porous calcareous tube. Various ions dissolved in the cameral fluid can be pumped through the porous outer layers into the cells of the siphuncular epithelium. When the cellular concentration of ions is high enough, the diffusion gradient thus created draws fluid from the shell chambers into the cells of the siphuncle while the fluid is replaced with gas. The result is an increase in buoyancy. By regulating this process, *Nautilus* may be able to remain neutrally buoyant at whatever depth they are. It was once thought that this gas–fluid "pump" mechanism allowed buoyancy changes sufficient to explain all the large-scale vertical movements of *Nautilus*, but density changes may not be the sole source of power for moving great distances up and down in the water column. *Nautilus* moves using jet propulsion by rapidly contracting its head, not by mantle muscle contraction.

Feeding

Two basic and fundamentally different types of feeding occur among molluscs: the first encompasses the feeding modes of most molluscs and includes micro- to macrophagy involving browsing and scraping, herbivory, carnivorous grazing and predation, while the second is suspension feeding (suspension microphagy). The basic mechanics of these two feeding modes are examined in Chapter 4. Here we briefly summarize the ways in which these feeding behaviors are employed by molluscs. In this section we also discuss a uniquely molluscan structure, the radula, which is used in microphagy, herbivory, and predation, and has become modified in a variety of unusual and interesting ways.

The buccal cavity may contain a pair of lateral jaws (or a single dorsal jaw), muscularized regions with chitinous plates that can be solid or composed of multiple small units. Molluscan jaws are highly variable. For example, in some heterobranchs the jaws can be quite complex, with distinct "teeth," in some carnivorous caenogastropods the jaws can be quite large, in cephalopods the jaws are modified to form the beak, and some lineages have no jaws at all including bivalves, which lack both jaws and radula.

The radula is usually a ribbon of recurved chitinous teeth (Figures 13.23–13.26). The teeth may be simple, serrate, pectinate, or otherwise modified. The radula often functions as a scraper to remove food particles for ingestion, although in many groups it has become adapted for other actions. A radula is present in the majority of the most primitive living molluscs and is therefore assumed to have originated in the earliest stages of molluscan evolution. In the aplacophoran groups the teeth, when present, may not be borne on a ribbon per se but on a relatively thin cuticle covering the foregut epithelium—perhaps the evolutionary forerunner of the ribbon-like radula. In some aplacophorans, the teeth form simple plates embedded in either side of the lateral foregut wall, while in others they form a transverse row, or up to 50 rows, with as many as 24 teeth per row.

(A)

(B)

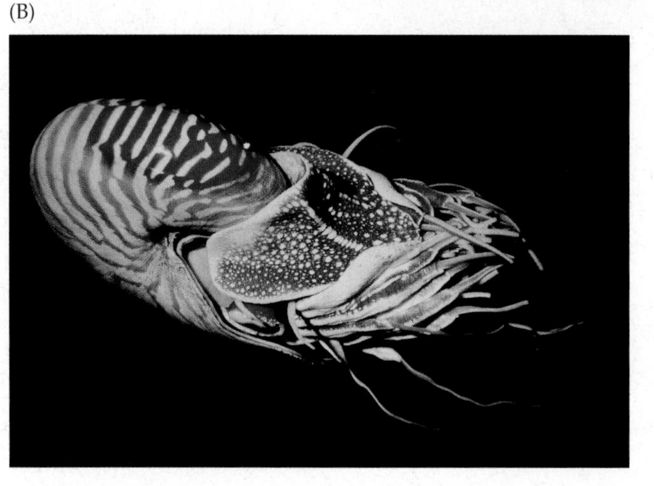

(C)

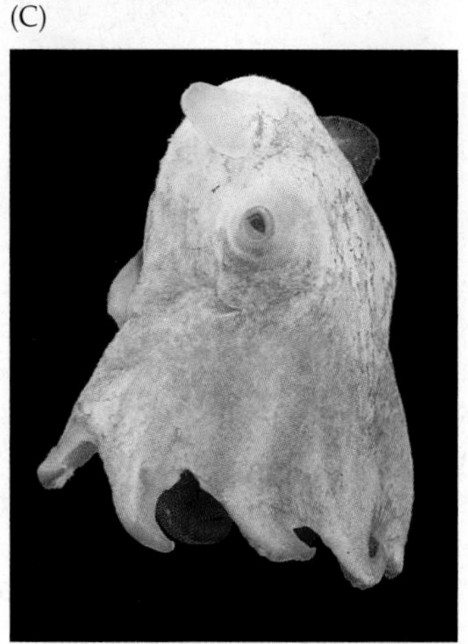

Figure 13.22 Swimming cephalopods. (A) *Sepia*, the cuttlefish. (B) *Nautilus*. (C) *Vampyroteuthis*, a "vampire" squid, viewed from the side.

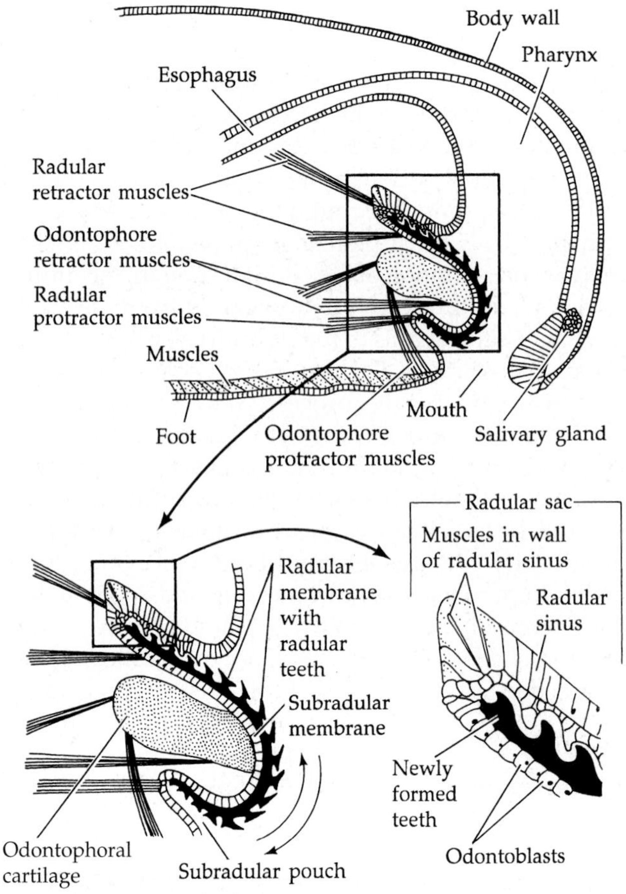

Figure 13.23 A generalized molluscan radula and associated buccal structures, at three "magnifications" (longitudinal section).

In gastropods and other molluscs (except bivalves) an **odontophore** projects from the floor of the pharynx or buccal cavity. It is a muscular structure bearing the complex tooth-bearing radular ribbon (Figure 13.23). The ribbon, called a **radular membrane**, is moved back and forth by sets of radular protractor and retractor muscles over cartilages encased in the odontophore (Figure 13.23). These cartilages are absent in many heterobranch gastropods. The radula originates in a **radular sac**, in which the radular membrane and new teeth are continually being produced by special cells called **odontoblasts**, to replace those lost by erosion during feeding. Measurements of radular growth indicate that up to five rows of new teeth may be added daily in some species. The odontophore itself is moved in and out of the buccal cavity during feeding by sets of odontophore protractor and retractor muscles, which also assist in applying the radula firmly against the substratum (Figures 13.23 and 13.24A,B).

The number of radular teeth ranges from a few to thousands and serves as an important taxonomic character in many groups. In some molluscs, the radular teeth are hardened with iron compounds, such as magnetite (in chitons) and goethite (in patellogastropods). Just as in many vertebrates, the radular teeth show adaptations to the type of food eaten. In vetigastropods (e.g., keyhole limpets, abalones, top shells), the **rhipidoglossate radulae** bear large numbers of fine marginal teeth in each row (Figures 13.25A and 13.26A). As the radula is pulled over the bending plane of the odontophore, these teeth act like stiff brushes, sweeping small particles to the midline where they are caught on the recurved parts of the central teeth, which draw the particles into the buccal cavity. Most vetigastropods are intertidal foragers

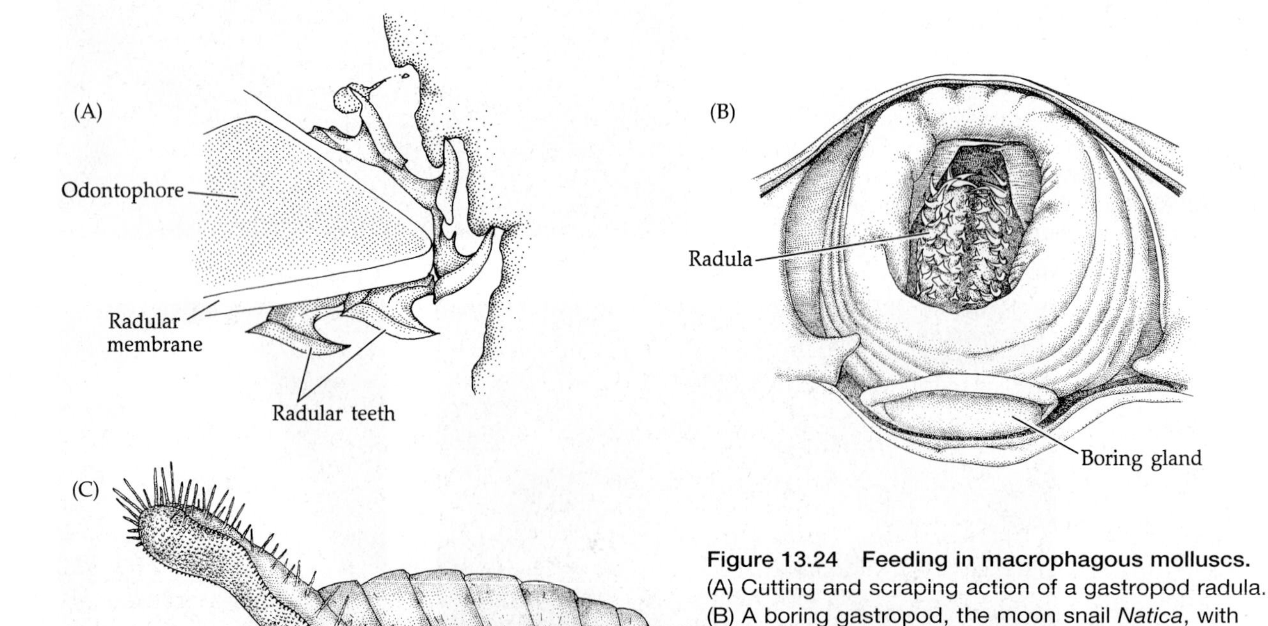

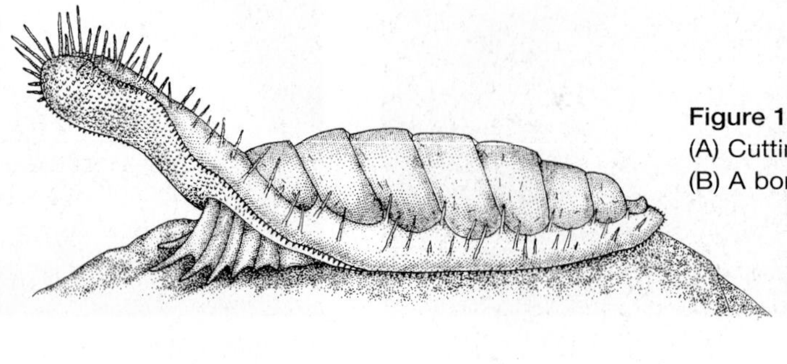

Figure 13.24 Feeding in macrophagous molluscs. (A) Cutting and scraping action of a gastropod radula. (B) A boring gastropod, the moon snail *Natica*, with radula visible in the mouth and the boring gland exposed (oral view). (C) The Pacific chiton *Placiphorella velata* in feeding position, with raised head flap ready to capture small prey.

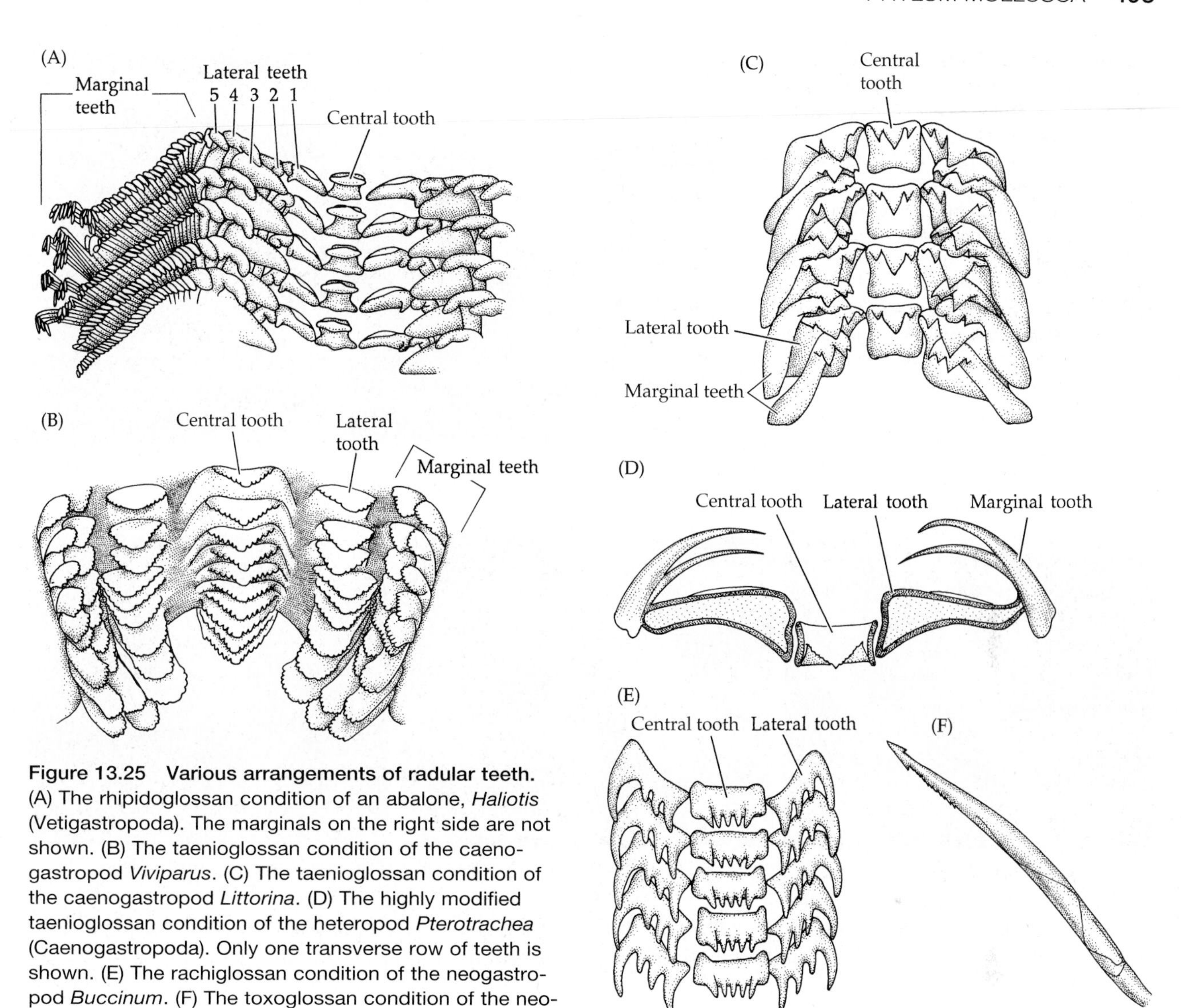

Figure 13.25 Various arrangements of radular teeth. (A) The rhipidoglossan condition of an abalone, *Haliotis* (Vetigastropoda). The marginals on the right side are not shown. (B) The taenioglossan condition of the caenogastropod *Viviparus*. (C) The taenioglossan condition of the caenogastropod *Littorina*. (D) The highly modified taenioglossan condition of the heteropod *Pterotrachea* (Caenogastropoda). Only one transverse row of teeth is shown. (E) The rachiglossan condition of the neogastropod *Buccinum*. (F) The toxoglossan condition of the neogastropod *Mangelia* (a single tooth).

that live on diatoms and other algae and microbes on the substratum. In contrast, patellogastropods (e.g., lottiid and patellid limpets) possess a **docoglossate radula**, which is impregnated with iron and bear relatively few teeth in each transverse row. Lottiid radulae, for example, have only one, two, or no marginal teeth, and only three pairs of lateral teeth per row (Figure 13.26B). The mucous trails left by some limpets (e.g., homing species such as the Pacific *Lottia gigantea* and *Collisella scabra*) actually serve as adhesive traps for the microalgae that are their primary food resource).

The radula of many caenogastropods is the **taenioglossate type**, in which there are only two marginal teeth in each row, along with three other teeth (laterals and central) (Figure 13.25B–D). In conjunction with the elaboration of jaws, taenioglossate radulae are capable of powerful rasping, which enables some littorid snails to feed by directly scraping off the surface cell layers of algae.

The most derived caenogastropods (Neogastropoda) usually have **rachiglossate radulae**, which lack marginal teeth altogether (Figures 13.25E and 13.26C,D). They use the remaining (one to three) teeth for rasping, tearing, or pulling. These snails are usually carnivores or carrion feeders although some members of one family, the Columbellidae, are herbivores. Caenogastropods of the families Muricidae and Naticidae eat other molluscs by boring through the prey's calcareous shell to obtain the underlying flesh. This ability to bore has evolved entirely independently in the two groups. It is mainly mechanical; the predator boring with its radula while holding the prey with the foot. The boring activity is complemented by the secretion of an acidic chemical from a **boring gland** (also called the "accessory boring organ"); the chemical is periodically applied to the drill hole to weaken the calcareous matrix. The boring gland of the neogastropod muricids is located on the foot while that of the littorinimorph naticids is located on the anterior end of the proboscis (Figure 13.24B). Boring gastropods such as the American drill (*Urosalpinx*) and the Japanese drill (*Rapana*) cause a loss of millions of dollars annually for oyster farms.

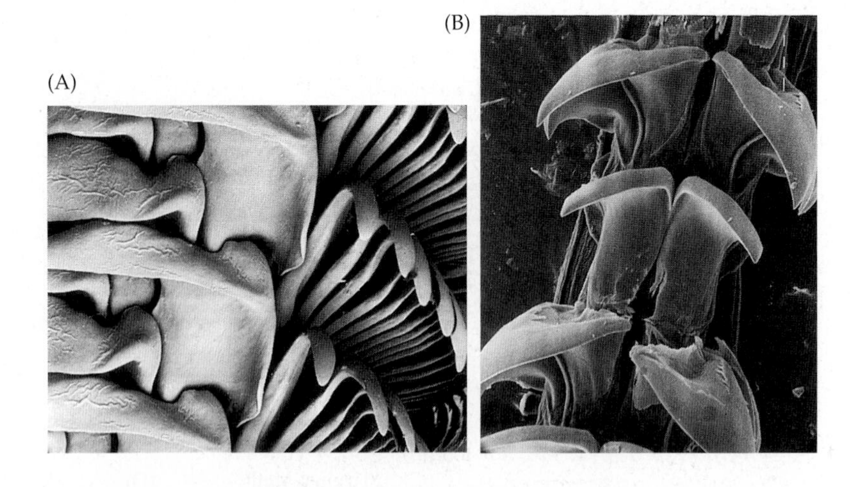

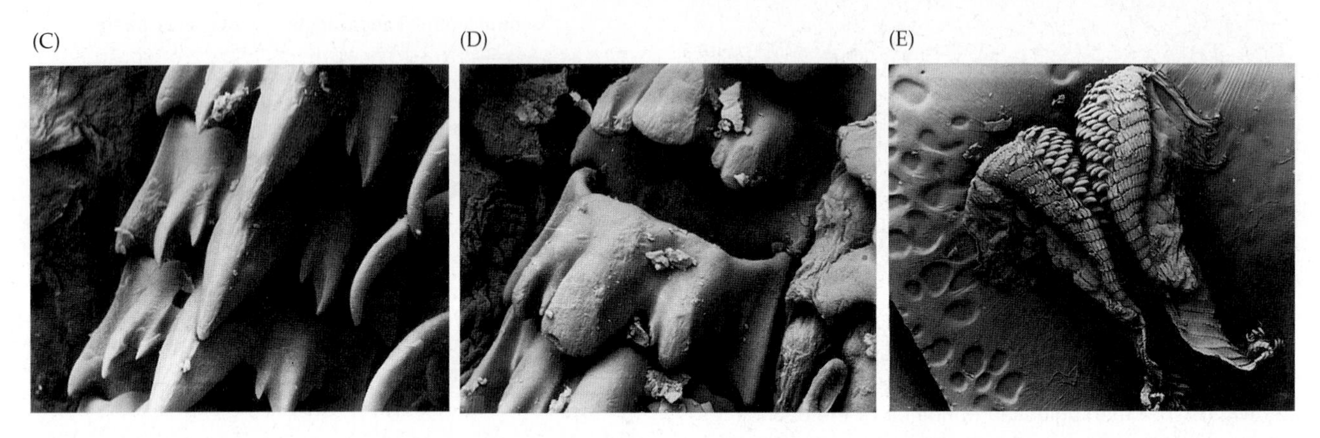

Figure 13.26 Gastropod radulae. (A) A closeup view of the rhipidoglossate radula of the abalone *Haliotis rufescens* (Vetigastropoda). Note the many hook-like marginal teeth. (B) The docoglossate radula of a lottiid limpet. (C) The serrated central teeth of a rachiglossate radula from the muricid whelk *Nucella emarginata*, a neogastropod (Caenogastropoda) that preys on small mussels and barnacles. (D) The worn radular teeth of *Nucella*. (E) The radula of the polycerid nudibranch *Triopha*, seen here in dorsal view as it rests in the animal.

Some carnivorous gastropods (e.g., *Janthina*) do not gnaw or rasp their prey, but swallow it whole. In these gastropods a **ptenoglossate radula** forms a covering of strongly curved spines over the buccal mass. The prey is seized by the quickly extruded buccal mass and simply pulled whole into the gut. A somewhat similar feeding method is seen in the carnivorous slug *Testacella*, where the hooked radula catches earthworms that the slug consumes whole. The nudibranch *Melibe* (chapter opener photo) uses its large hood to sweep the water for copepods, amphipods, and other small planktonic prey.

A few gastropods have lost the radula altogether and feed by sucking body fluids from their prey, a habit seen, for example, in some nudibranchs. Pyramidellids do this with the aid of a hypodermic stylet (a modified jaw) on the tip of an elongate proboscis.

One of the most specialized feeding modes in gastropods is seen in the cone snails (*Conus*), and relatives. Their toxoglossate radula is formed from a few harpoon-like, venom injecting teeth that are probably modified marginal teeth. The teeth (Figure 13.25F) are discharged from the end of a long proboscis that can be extended very rapidly to capture prey, usually a fish, a worm, or another gastropod, which is then pulled into the gut (Figure 13.27). The venom is injected through the hollow, curved radular teeth by contraction of a venom gland. A few Indo-West Pacific cones produce a potent neuromuscular toxin that has caused human deaths.

Among the most unusual gastropod feeding strategies are those that involve parasitism on fishes. For example, the neogastropod *Cancellaria cooperi* attaches to the Pacific electric ray and makes small cuts in the skin through which the proboscis is inserted to feed on the ray's blood and cellular fluids. Several other neogastropods parasitize "sleeping" reef fishes by inserting their proboscides into the host and sucking out fluids. Some other gastropods are known to parasitize various invertebrate hosts, notably the pyramidellids (on a variety of invertebrates, including other molluscs) and the Eulimidae on echinoderms, with the latter group including some internal parasites that have lost their shell and become wormlike.

Certain euthyneurans also show various radular modifications. Groups of "opisthobranchs" that feed on cnidarians, ectoprocts, and sponges, and those that scrape algae (e.g., aplysiids) usually have typical rasping radulae. In sacoglossans, however, the radula is modified as a single row of lance-like teeth that can pierce the cellulose wall of filamentous algae, allowing the gastropod to suck out the cell contents. A similar type of feeding strategy is also seen in the microscopic lower heterobranch *Omalogyra*.

Figure 13.27 Sequence of photographs of a cone shell (*Conus*) capturing and swallowing a small fish. The proboscis is extended and swept back and forth above the substratum in search of prey; when a fish is encountered, a poison-charged tooth of the toxoglossan radula is fired like a harpoon; and the prey is quickly paralyzed and ingested.

Aeolid nudibranchs (Figure 13.7G) have a well-deserved reputation for their particular mode of feeding, in which portions of their cnidarian prey are held by the jaws while the radula rasps off pieces for ingestion. A similar mode of feeding has also been observed in the caenogastropod family Epitoniidae.

Many aeolid nudibranchs engage in a remarkable phenomenon called **kleptocnidae**. Some of the prey's nematocysts are ingested unfired, passed through the nudibranch's gut, and eventually transported to lobes of the digestive gland in dorsal finger-like extensions called **cerata** (singular, ceras) (Figure 13.32D,E). How the nematocysts undergo this transport without firing is still a mystery. Popular hypotheses are that mucous secretions by the nudibranch limit the discharge, or that a form of acclimation occurs (like that suspected to occur between anemone fishes and their host anemones), or perhaps that only immature nematocysts survive, to later undergo maturation in the dorsal cerata. It may also be that, once the cnidocytes are digested, the nematocysts' firing threshold is raised, thereby preventing discharge. In any case, once in the cerata the nematocysts are stored in structures called cnidosacs and presumably help the nudibranch to fend off attackers. Discharge might even be under the control of the host nudibranch, perhaps by means of pressure exerted by circular muscle fibers around each cnidosac.

Some dorid nudibranchs also utilize their prey in remarkable ways. Many dorids secrete complex toxic compounds incorporated into mucus released from the mantle surface. These noxious chemicals act to deter potential predators. While some of these chemicals may be manufactured in some dorids, in most cases it appears that they are obtained from the sponges or ectoprocts on which they feed. Some species, such as the "Spanish dancer" nudibranch (*Hexabranchus sanguineus*), not only use a chemical from their prey (in this case a sponge) for its own defense, but deposit some of the noxious chemical on the egg mass, helping to protect the embryos until they hatch.

In polyplacophorans there are generally 17 teeth in each transverse row of the radula (a central tooth flanked by eight on each side). Most chitons are herbivorous grazers. Notable exceptions are certain members of the order Ischnochitonida (family Mopaliidae, e.g., *Mopalia*, *Placiphorella*), which are known to feed on both algae and small invertebrates. *Mopalia* consumes sessile invertebrates, such as barnacles, ectoprocts, and hydroids. *Placiphorella* captures live microinvertebrates (particularly crustaceans) by trapping them beneath its head-flap, a large anterior extension of the girdle (Figure 13.24C).

In monoplacophorans the radula consists of a ribbon-like membrane bearing a succession of transverse rows of 11 teeth each (a slender central tooth flanked on each side by five broader lateral teeth). Monoplacophorans are probably generalized deposit feeders that graze on minute organisms coating the substratum on which they live.

Cephalopods are predatory carnivores. Squids are some of the most voracious creatures in the sea, successfully competing with fishes. Octopuses are also active carnivores, preying primarily on crabs, bivalves and gastropods. Some species of *Octopus* bore through the shells of molluscan prey in a fashion similar to that of gastropod drills. Some even drill and prey upon their close relatives, the chambered nautiluses. They do not use the radula to drill, but instead use a rasp-like projection formed from the salivary papilla.

Using their impressive locomotor skills, most cephalopods hunt and catch active prey. Some octopuses, however, hunt "blindly," by tasting beneath stones with their highly sensitive suckers that are both

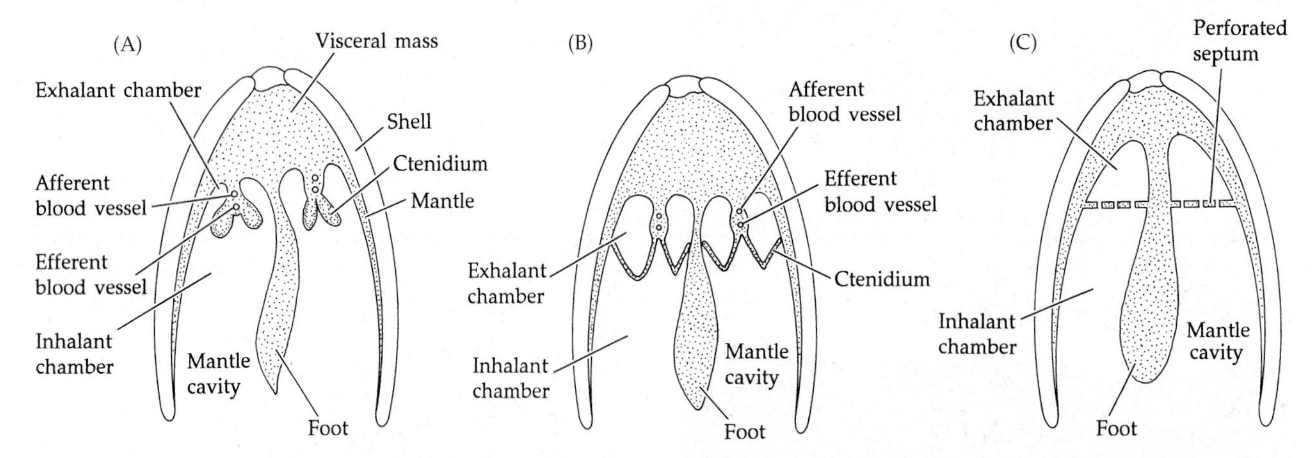

Figure 13.28 Arrangement of ctenidia in some bivalves (transverse sections) showing the following conditions: (A) protobranch, (B) lamellibranch, (C) septibranch.

mechano- and chemosensitive. In any event, once prey is captured and held by the arms, the cephalopod bites it with its horny beak (modified jaws) and injects a neurotoxin from modified salivary glands. This ability to quickly immobilize prey also helps prevent the soft-bodied cephalopod from being engaged in a potentially dangerous struggle.

Suspension feeding evolved in autobranch bivalves, and also several times in gastropods, and in most of these cases it involves modifications of the ctenidia that enabled the animal to trap particulate matter carried into the mantle cavity by incoming respiratory water current. The lamellar nature of molluscan gills exapted them for extracting suspended food particles. Increasing the size of the gills and the degree of folding also increased the surface area available for trapping particulate material. In suspension feeders, at least some of the gill and mantle cilia, which otherwise serve to remove potentially clogging sediment from the mantle cavity (as pseudofeces), are preadapted to transport particulate matter from the gills to the mouth region.

While collecting food on the gills has been adopted by autobranch bivalves and some groups of gastropods, other methods have also been employed. In the gill-less planktonic sea butterflies (pteropods), the ciliated "wings" or parapodia used for swimming (Figure 13.7D,E) also function as food-collecting surfaces or may cooperate with the mantle to produce large mucous sheets that capture small zooplankton. From the foot, ciliary currents carry mucus and food to the mouth. In some pteropods, the mucous sheet may be as large as 2 m across. Mucous sheet feeding is also employed by a few other gastropods including the intertidal limpet-like Trimusculidae and the vermetids (see below). However, the majority of suspension feeding gastropods, including a few vetigastropods such as the sand-beach trochids (*Umbonium, Bankivia*), the hot vent neomphalid (*Neomphalus*), and some marine caenogastropods (Calyptraeidae, e.g., *Crepidula*; Vermetidae, e.g., *Vermetus*; Turritellidae, e.g., *Turritella*) and the freshwater Viviparidae (e.g., *Viviparus*). The ctenidial filaments in these filter feeders are much elongated and the mantle waste rejection cilia have evolved into a food collecting groove that runs to the mouth. The radula in suspension-feeding gastropods is somewhat reduced, serving mainly to pull mucus-bound food into the mouth. Some feed entirely by suspension feeding, others use browsing to supplement this method.

The wormlike shell of the vermetids is permanently affixed to the substratum and while some adopt ciliary food collecting, others combine this with mucous net collecting or use the latter method exclusively. A special pedal gland in the reduced foot produces copious amounts of mucus that spreads into the water column as a sticky plankton trap. Periodically the net is hauled in by the foot and pedal tentacles, and a new one is quickly secreted. *Thylacodes arenarius*, a large Mediterranean species, casts out individual threads up to 30 cm long, whereas the gregarious California species *Thylacodes squamigerus* forms a communal net shared by many individuals.

The radula apparently disappeared early in the course of bivalve evolution, and in living species there is no trace of this structure or even the buccal cavity that contained it. Most autobranch bivalves use their large ctenidia for suspension feeding but the more primitive bivalves in the subclass Protobranchia are not suspension feeders but engage in a type of deposit-feeding microphagy. Protobranchs live in soft marine sediments and maintain contact with the overlying water either directly (e.g., *Nucula*) or by means of siphons (e.g., *Nuculana, Yoldia*). The two ctenidia are small, conforming to the primitive molluscan bipectinate plan of an elongated axis carrying a double row of lamellae (Figures 13.28A and 13.29). Protobranchs feed by means of two pairs of large labial palps flanking the mouth. The two innermost palps are the short **labial palps**, and the two outermost palps are formed into tentacular processes called **proboscides** (each being called a **palp proboscis**), which can be extended beyond the shell (Figure 13.29). During feeding the

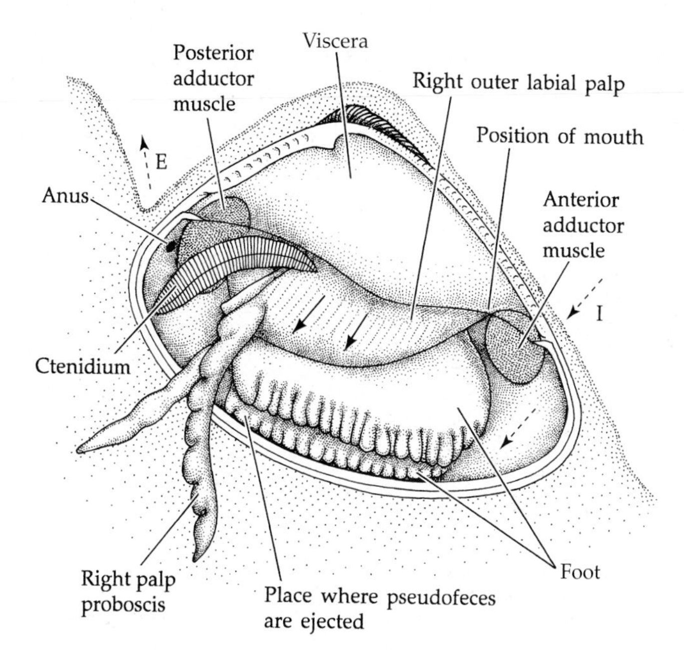

Figure 13.29 Feeding in the primitive bivalve *Nucula* (Protobranchia). The clam is seen from the right side, in its natural position in the substratum (right valve and right mantle skirt removed). Arrows show direction of ciliary currents in the mantle cavity and on the palps. Water currents are also shown in the (I) inhalant region and (E) exhalant region.

proboscides are extended into the bottom sediments. Detrital material adheres to the mucus-covered surface of the proboscides and is then transported by cilia to the labial palps, which function as sorting devices. Low-density particles are carried to the mouth; heavy particles are carried to the palp margins and ejected into the mantle cavity.

In the suspension feeding subclass Autobranchia, lateral cilia on the ctenidia generate a water current from which suspended particles are gleaned. Increased efficiency is achieved by various ctenidial modifications. The primary modification, seen in all living autobranch bivalves, has been the conversion of the original, small, triangular plates into V-shaped filaments with extensions on either side (Figures 13.28B and 13.30B). The arm of this V-shaped filament that is attached to the central axis of the ctenidium is called the descending arm; the arm forming the other half of the V is the ascending arm. The ascending arm is usually anchored distally by ciliary contacts or tissue junctions to the roof of the mantle, or to the visceral mass. Taken together, the two V-shaped filaments, with their double row of leaflets, form a W-shaped structure when seen in cross section.

Some pteriomorphian autobranch bivalves have **filibranch** ctenidia (e.g., mussels) wherein adjacent filaments are interlocked to one another by periodic clumps of specialized cilia, leaving long narrow slits in between (interfilament spaces) (Figure 13.30C,D). The spaces between the arms of the W's are exhalant suprabranchial chambers, which merge with the exhalant area in the posterior mantle cavity to be discharged; the spaces ventral to the W's are inhalant and communicate with the inhalant area of the mantle edge. Many other bivalves have **eulamellibranch** ctenidia, which are similar to the filibranch design but neighboring filaments are fused to one another by actual tissue junctions at numerous points along their length. This arrangement results in interfilament pores that are rows of ostia rather than the long narrow slits of filibranchs (Figure 13.30B,E,F). In addition, the ascending and descending halves of some filaments may be joined by tissue bridges that provide firmness and strength to the gill.

Both filibranch and eulamellibranch ctenidia are used to capture food. Water is driven from the inhalant to the exhalant parts of the mantle cavity by lateral cilia all along the sides of filaments in filibranchs, or by special lateral ostial cilia in eulamellibranchs (Figure 13.30E–F). As the water passes through the interfilament spaces it flows through rows of frontolateral cilia, which flick particles from the water onto the surface of the filament facing into the current. These feeding cilia are called **compound cirri**; they have a pinnate structure that probably increases their catching power. Mucus presumably plays some part in trapping the particles and keeping them close to the gill surface, although its precise role is uncertain. Bivalve ctenidia are not covered with a continuous sheet of mucus, as occurs in many other suspension-feeding invertebrates (e.g., gastropods, tunicates, amphioxus). Once on the filament surface, particles are moved by frontal cilia toward a food groove on the free edges of the ctenidium, and then anteriorly to the labial palps. The palps sort the material by size and perhaps also by quality before passing the food to the mouth. Rejected particles fall off the gill or palp edges into the mantle cavity as **pseudofeces**. This "filtration" of water by bivalves is quite efficient. The American oyster (*Crassostrea virginica*), for example, can process up to 37 liters of water per hour (at 24°C), and can capture particles as small as 1 μm in size. Studies on the common mussels *Mytilus edulis* and *M. californianus* suggest that these bivalves maintain pumping rates of about 1 liter per hour per gram of (wet) body weight.

Members of the superfamily Tellinoidea (including Tellinidae and Semelidae) are deposit feeders, sucking up surface detritus with their long, mobile inhalant siphon (Figure 13.20G) and using the large labial palps to pre-sort the particles before ingesting them.

Some members of the order Poromyata (Anomalodesmata) are known as septibranchs and are sessile predators and, unlike other autobranch bivalves, their gills are not used for feeding. Instead the ctenidia are very reduced and modified as a perforated but muscular septum that divides the mantle cavity into dorsal and ventral chambers (Figures 13.28C and 13.31A). The muscles are attached to the shell such that the septum can be raised or lowered within the mantle cavity. Raising the septum causes water to be sucked into the

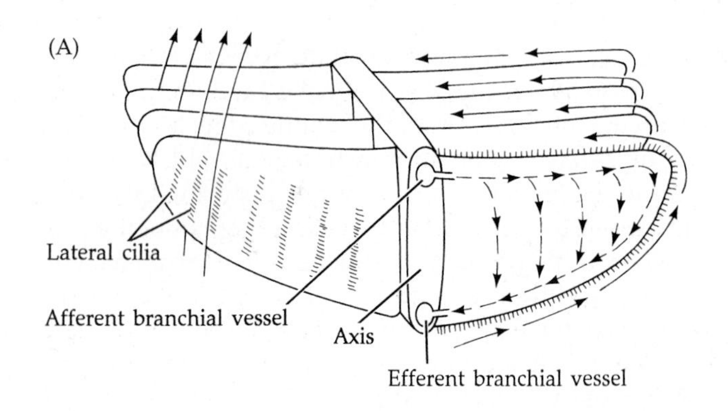

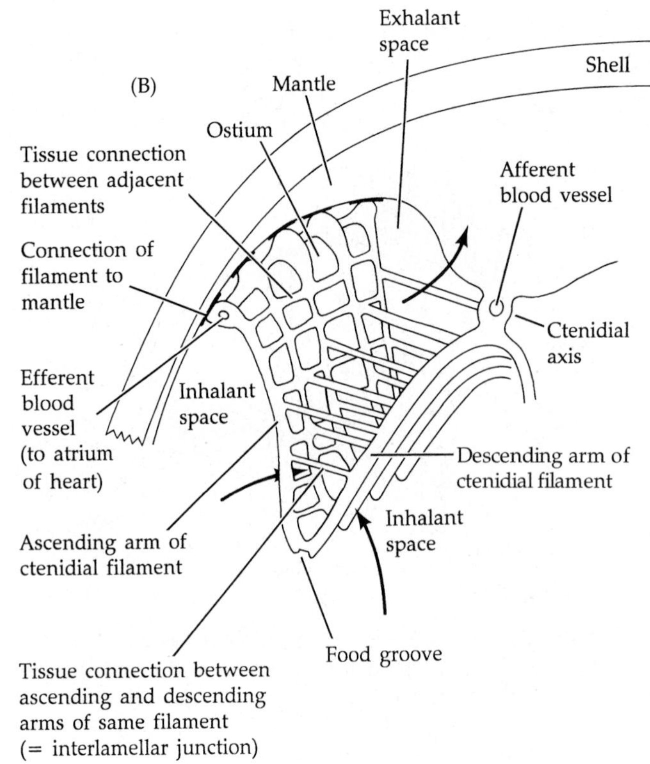

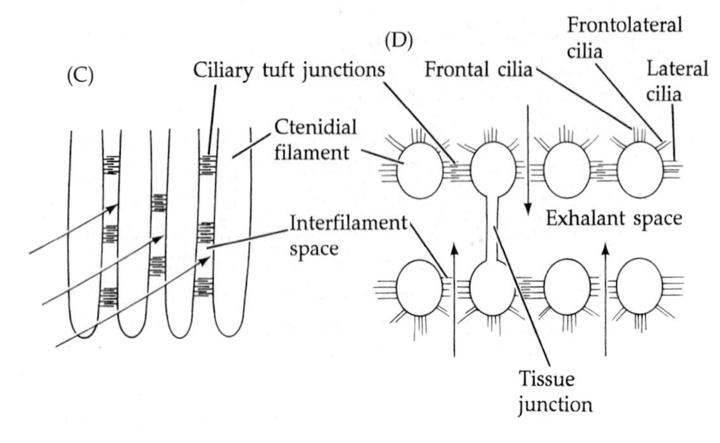

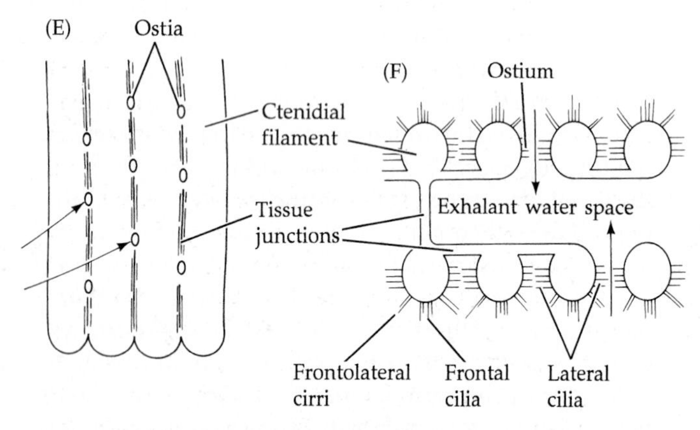

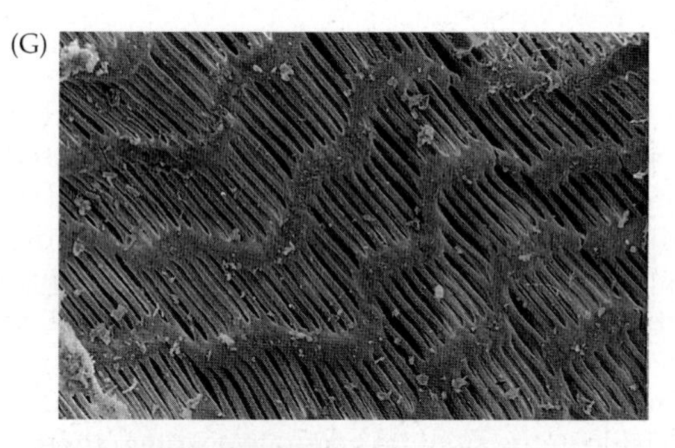

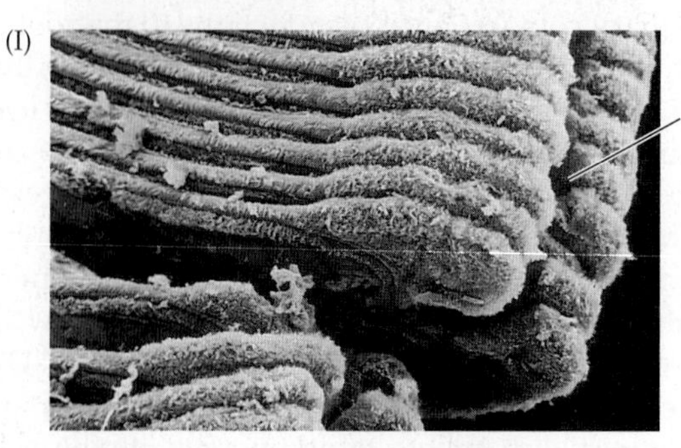

Figure 13.30 Ctenidial structure in bivalve molluscs. In all drawings, solid arrows indicate the direction of water flow (from inhalant space, between ctenidial filaments, to exhalant space). (A) Section through part of the gill axis in a nuculanid protobranch, showing four alternating filaments (leaflets) on each side. Dashed arrows indicate direction of hemolymph flow in the filament. (B) Highly schematic cutaway view showing four ctenidial filaments, and their interconnections, on one side of the body of a eulamellibranch. (C) Lateral view of four ctenidial filaments of a filibranch. (D) Cross section through ascending and descending arms of four filibranch ctenidial filaments. (E) Lateral view of four filaments of a eulamellibranch. (F) Cross section through ascending and descending arms of four eulamellibranch ctenidial filaments. (G) Ctenidial filaments of the mussel *Mytilus californianus* showing ciliary junctions and interfilament spaces. (H) Frontal ciliary tracts on ctenidial filaments of *Mytilus*. (I) Ventral gill edge of *Mytilus* showing food groove.

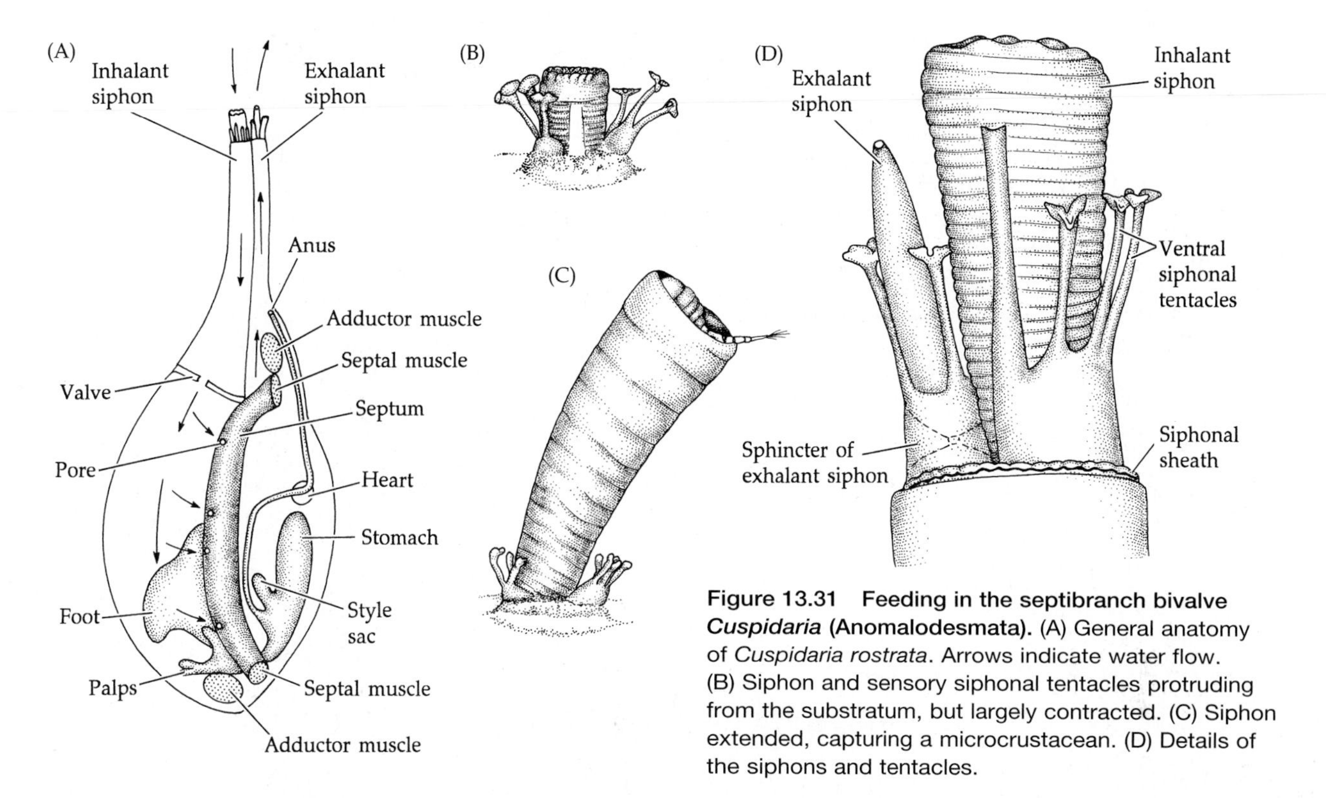

Figure 13.31 Feeding in the septibranch bivalve *Cuspidaria* (Anomalodesmata). (A) General anatomy of *Cuspidaria rostrata*. Arrows indicate water flow. (B) Siphon and sensory siphonal tentacles protruding from the substratum, but largely contracted. (C) Siphon extended, capturing a microcrustacean. (D) Details of the siphons and tentacles.

mantle cavity by way of the inhalant siphon; lowering the septum causes water to pass dorsally through the pores into the exhalant chamber. These movements also force hemolymph from mantle sinuses into the siphonal sinuses, thereby causing a rapid protrusion of the inhalant siphon, which can be directed toward potential prey (Figure 13.31B–D). In this fashion, small animals such as microcrustaceans are sucked into the mantle cavity, where they are grasped by muscular labial palps and thrust into the mouth; at the same time, the mantle tissue serves as the gas exchange surface.

While most pteriomorphians are restricted to epibenthic life, lacking siphons (Figure 13.21A, B), many heterodont bivalves live buried in soft sediments, where long siphons are utilized to maintain contact with the overlying water (Figures 13.8A and 13.20D–H).

Scaphopods consume foraminifera and other meiofauna taxa, diatoms, zooplankton, and interstitial detritus. Two lobes flank the head, each bearing numerous (up to several hundred) long, slender tentacles called **captacula** (Figures 13.9 and 13.13F). The captacula are extended into the substratum by metachronal beating of cilia on the small terminal bulb. Within the sediment organic particles and microorganisms adhere to the sticky terminal bulb; small food particles are transported to the mouth by way of ciliary tracts along the tentacles, while larger food items are transported directly to the mouth by muscular contraction of the captacula. A well-developed, large radula pulls the food into the mouth, perhaps partially macerating it in the process.

Several forms of symbiotic relationships have evolved within molluscs that are intimately tied to the host's nutritional biology. One of the most interesting of these relationships exists between many molluscs and sulfur bacteria. These molluscs appear to derive a portion of their nutritional needs from symbiotic, carbon-fixing sulfur bacteria, which usually reside on the host mollusc's gills. In some monoplacophorans (*Laevipilina antartica*) and gastropods (*Lurifax vitreus*, *Hirtopelta*) the bacteria are housed in special cavities called bacteriocytes in the mantle cavity. This mollusc–bacteria symbiosis has been recently documented from a variety of sulfide-rich anoxic habitats, including deep-sea hydrothermal vents, where geothermally produced sulfide is present, and from other reduced sediments, where microbial degradation of organic matter leads to the reduction of sulfate to sulfide (e.g., anoxic marine basins, seagrass bed and mangrove swamp sediments, pulp mill effluent sites, sewage outfall areas).

Members of some bivalve families, in particular the Solemyidae and Lucinidae, harbor sulfur bacteria in their enlarged gills which have the ability to directly oxidize sulfide. They do this by means of a special sulfide oxidase enzyme in the mitochondria. These bivalves inhabit reduced sediments where free sulfides are abundant. The ability to oxidize sulfide not only provides the bivalve with a source of energy to drive ATP synthesis, it also enables them to rid their body of toxic sulfide molecules that accumulate in such habitats. The nutrients obtained by this symbiosis are sufficient for the bivalve so, in solemyids, the gut is reduced or, in a few species, absent.

Another notable partnership exists between giant clams (*Tridacna*) and their symbiotic zooxanthellae (the dinoflagellate *Symbiodinium*). These clams live with

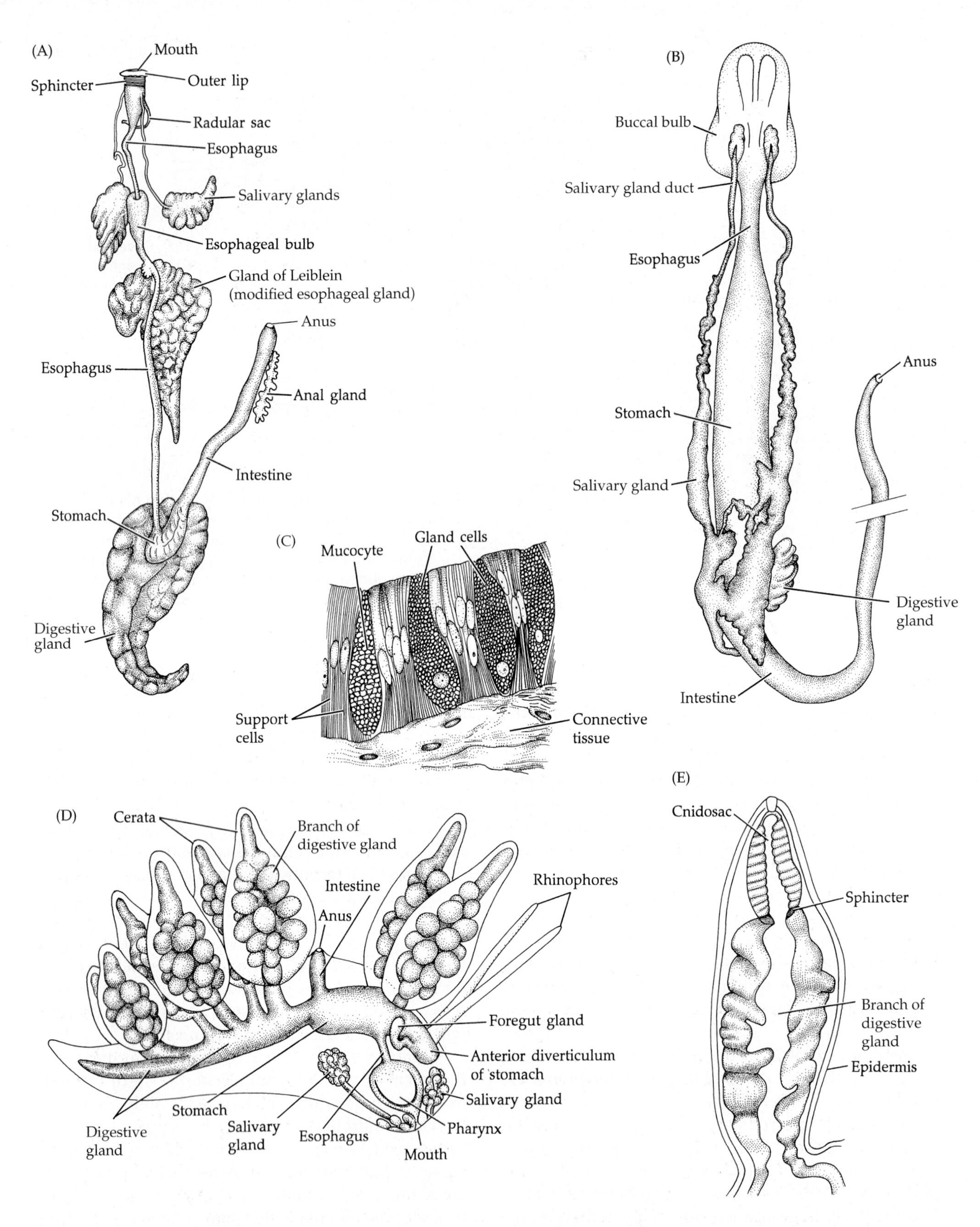

their dorsal side against the substratum, and they expose their fleshy mantle to sunlight through the large shell gape. The mantle tissues harbor the zooxanthellae. Many species have special lens-like structures that focus light on zooxanthellae living in the deeper tissues. A few other bivalves and certain sea slugs also maintain a symbiotic relationship with *Symbiodinium*. Several species of *Melibe, Pteraeolidia*, and *Berghia* harbor colonies of these dinoflagellates in "carrier cells" associated with their digestive glands. Experiments indicate that when sufficient light is available, host nudibranchs utilize photosynthetically fixed organic molecules produced by the alga to supplement their usual diet of prey. The dinoflagellates are probably not transmitted with the zygotes of the nudibranchs, each new generation thus requiring reinfection from the environment. A number

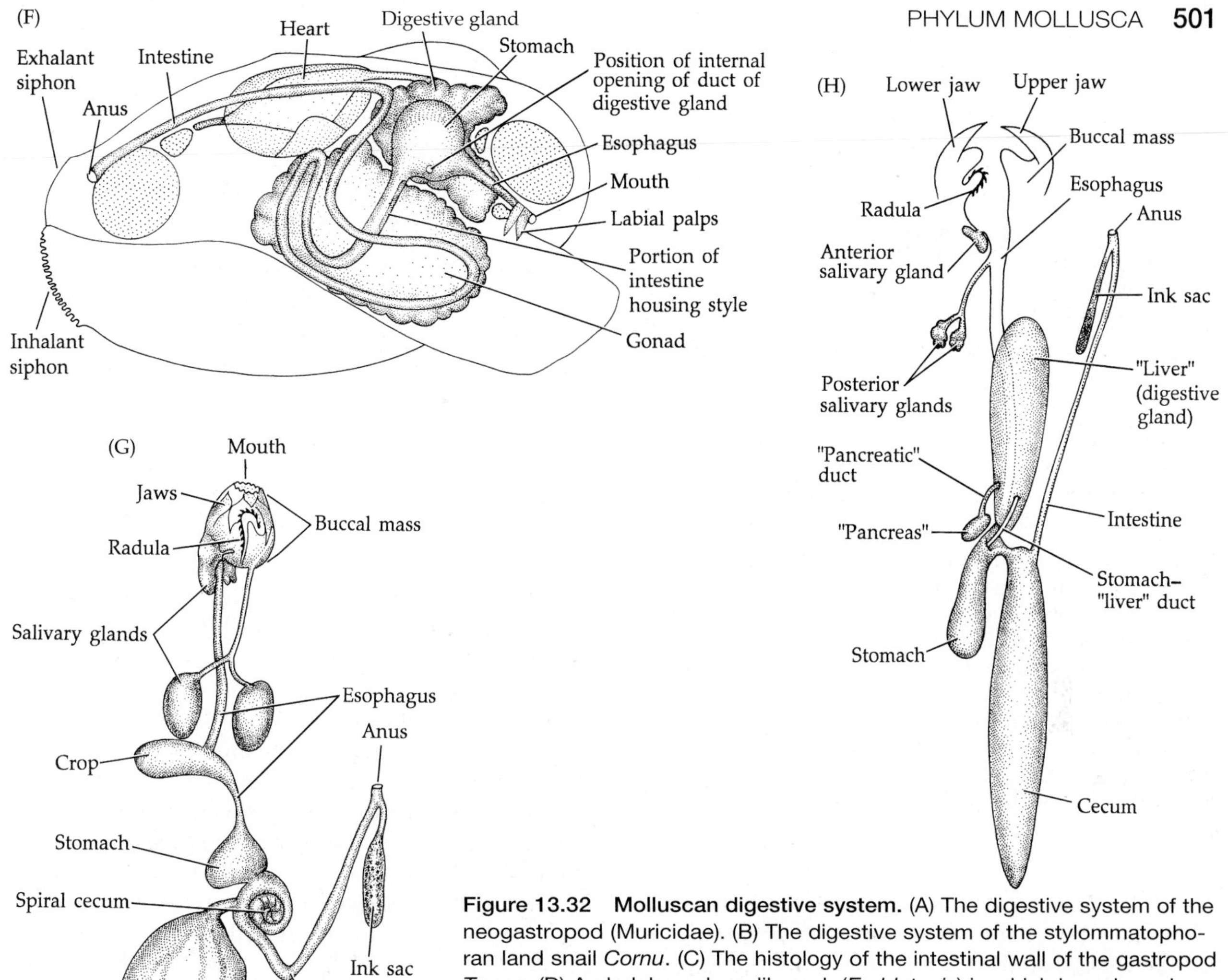

Figure 13.32 Molluscan digestive system. (A) The digestive system of the neogastropod (Muricidae). (B) The digestive system of the stylommatophoran land snail *Cornu*. (C) The histology of the intestinal wall of the gastropod *Tonna*. (D) A cladobranch nudibranch (*Embletonia*) in which large branches of the digestive gland fill the dorsal cerata. (E) A longitudinal section of the ceras of the aeolidid nudibranch *Trinchesia* showing the cnidosac where nematocysts (not shown) from this animal's cnidarian prey are stored. (F) Diagrammatic lateral view of the digestive tract and nearby organs of the unionoid clam *Anodonta*. (G) The digestive system of the cuttlefish *Eledone*. (H) The digestive system of the squid *Loligo*.

of aeolid nudibranchs accumulate zooxanthellae from their cnidarian prey. Some of the dinoflagellates end up inside cells of the nudibranch's digestive gland, but many others are released in the slug's feces, from where they may reinfect cnidarians. An even more remarkable phenomenon occurs in some members of another group of sea slugs, the Sacoglossa (e.g., *Placobranchus*). These sea slugs obtain functional chloroplasts from the green algae upon which they feed and incorporate them into their own tissues; the chloroplasts remain active for a period of time and produce photosynthetically fixed carbon molecules that are utilized by the hosts.

Still another unusual symbiosis occurs between an aerobic bacterium and the wood-boring marine shipworm bivalves (Teredinidae) (Figure 13.21D). Shipworms are capable of living on a diet of wood alone by harboring this cellulose-decomposing, nitrogen-fixing bacterium. The bivalve cultures the bacterium in a special organ associated with ctenidial blood vessels called the **gland of Deshayes**. The bacterium breaks down cellulose and makes its products available to its host. Nitrogen-fixing bacteria occur as part of the gut flora in many animals whose diet is rich in carbon but deficient in nitrogen (e.g., termites). However, shipworms are the only animals known to harbor a nitrogen fixer as a pure (single species) culture in a specialized organ (similar to the host nodule–*Rhizobium* symbiosis of leguminous plants).

In addition to the above and myriad other feeding strategies of molluscs, some species (notably some bivalves and sea slugs) probably obtain a significant portion of their nutritional needs by direct uptake of dissolved organic material from seawater, such as amino acids.

Digestion

Molluscs possess complete, or through guts, a few of which are illustrated in Figure 13.32. The mouth leads inward to a buccal cavity, within which the radular apparatus and jaws (when present) are located (Fig-

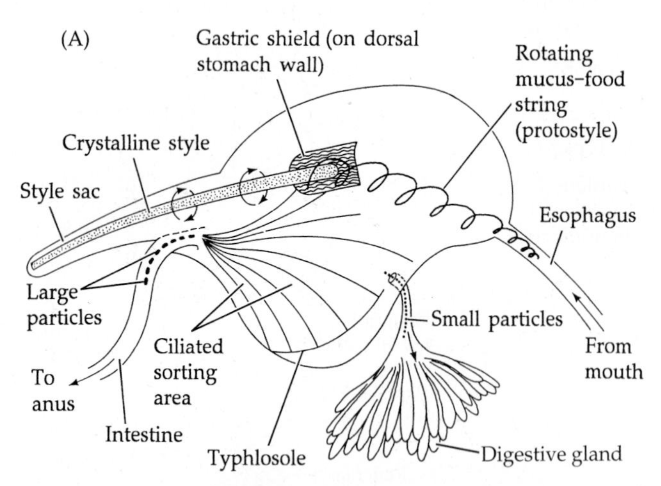

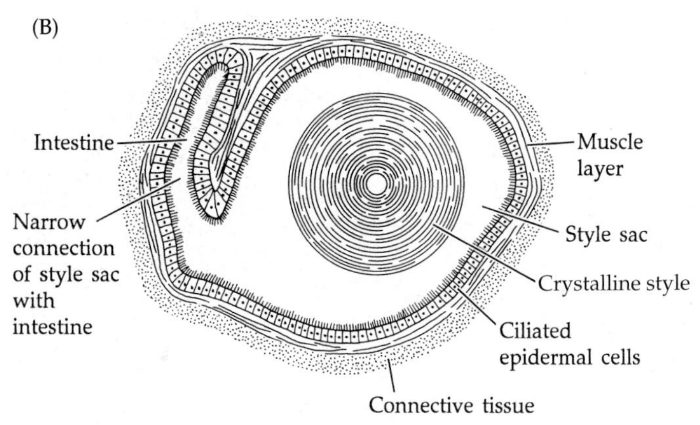

Figure 13.33 The molluscan stomach and style sac. (A) A generalized stomach and style apparatus of an autobranch bivalve. The crystalline style rotates against the gastric shield, releasing digestive enzymes and winding up the mucus–food string to assist in pulling it from the esophagus. Food particles are sorted in the ciliated, grooved sorting area: small particles are carried (in part by the typhlosole) to the digestive glands for digestion; large particles are carried to the intestine for eventual elimination. (B) A cross section of a style sac.

ure 13.23). The esophagus is generally a straight tube connecting the foregut to the stomach. Various glands are often associated with this anterior gut region, including some that produce enzymes and others that secrete a lubricant over the radula that are usually called salivary glands. In many herbivorous species (e.g., certain eupulmonates, anaspideans [*Aplysia*], and some cephalaspideans), a muscular **gizzard** (unrelated to the jaws) may be present for grinding up tough vegetable matter. The gizzard may have chitinous, or even calcareous plates or teeth. The stomach usually bears one or more ducts that lead to the large glandular digestive gland (variously called the digestive diverticula, digestive caeca, midgut glands, liver, or other similar terms). Several sets of digestive glands may be present. The intestine leaves the stomach and terminates as the anus, which is typically located in the mantle cavity in or near the exhalant water flow.

Once food has entered the buccal cavity of most molluscs, it is carried in mucous strings into the esophagus and then to the stomach. In cephalopods and some predatory gastropods, chunks of food or whole prey are swallowed by muscular action of the esophagus. The food is stored in the stomach or in an expanded region of the esophagus called the "crop," as in octopuses and *Nautilus*, and many gastropods. In many bivalves and gastropods, the stomach wall bears a chitinous **gastric shield** and a ciliated, ridged **sorting area** (Figure 13.33). The posterior stomach region (anterior in gastropods) is a **style sac**, which is lined with cilia and in autobranch bivalves and in some gastropods contains a **crystalline style** (Figure 13.33). This structure, which functions to aid in digestion, is a rodlike matrix of proteins and enzymes (often amylase) that are slowly released as the projecting end of the style rotates and grinds against the gastric shield that protects the otherwise delicate stomach wall. The gastric cilia and rotating style wind up the mucus and food into a string and draw it along the esophagus to the stomach. The style is produced by special cells of the style sac. The style of some bivalves is enormous, one-third to one-half the length of the clam itself. Particulate matter is swept against the stomach's anterior sorting region, which sorts mainly by size. Small particles are carried into the digestive glands, which arise from the stomach wall. Larger particles are passed along ciliated grooves of the stomach to the intestine. In more primitive bivalves (Protobranchia) and in many gastropods a crystalline style is absent but a style sac is often present, which contains a rotating mass of mucus mixed with particles that is termed a **protostyle**.

Extracellular digestion takes place in the stomach and lumina of the digestive glands, while absorption and intracellular digestion occur in the digestive gland cells and the intestinal walls. Extracellular digestion is accomplished by enzymes produced in foregut (e.g., salivary glands, esophageal pouches or glands, pharyngeal glands—sometimes called "sugar glands" because they produce amylase), the stomach, and the digestive glands. In primitive groups, intracellular digestion tends to predominate. In Solenogastres all digestive functions are accomplished in a uniform midgut lined by voluminous digestive and secretory cells. In most molluscs, ciliated tracts line the digestive glands and carry food particles to minute diverticula, where they are engulfed by phagocytic digestive cells of the duct wall. The same cells dump digestive wastes back into the ducts, to be carried by other ciliary tracts back to the stomach, from there to be passed out of the gut via the intestine and anus as fecal material. In most highly-derived groups (e.g., cephalopods and many gastropods), extracellular digestion predominates. Enzymes secreted primarily by the digestive glands

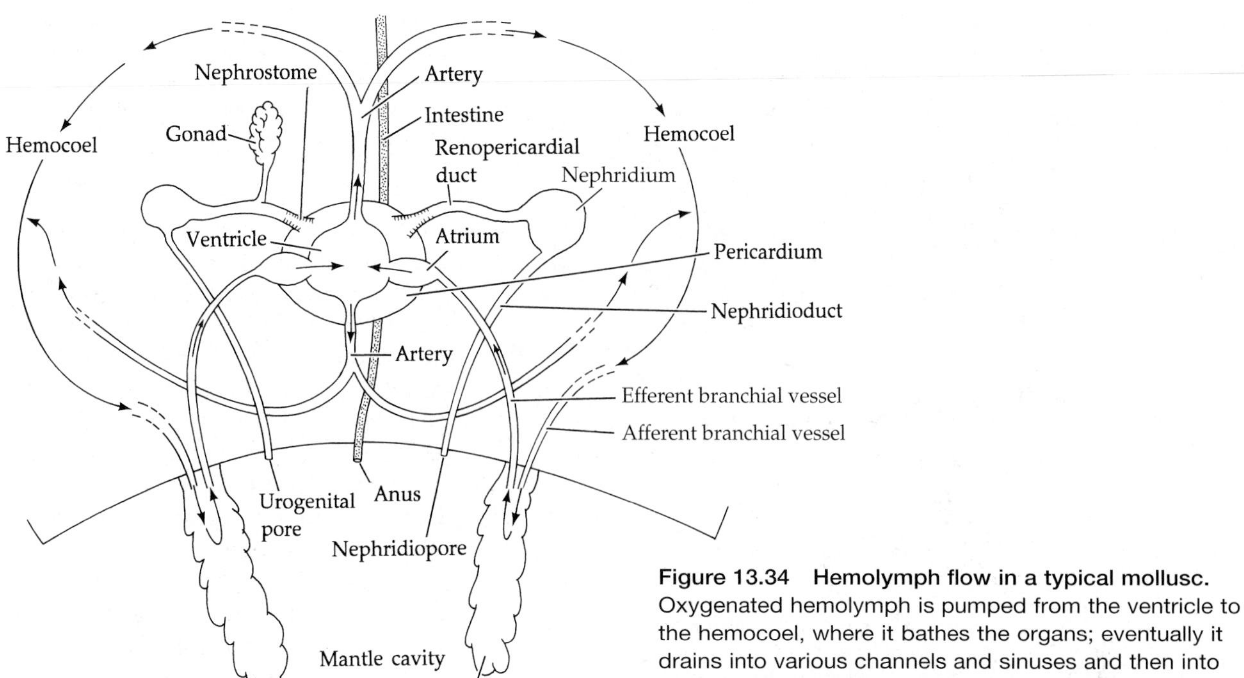

Figure 13.34 Hemolymph flow in a typical mollusc. Oxygenated hemolymph is pumped from the ventricle to the hemocoel, where it bathes the organs; eventually it drains into various channels and sinuses and then into the afferent branchial vessels, which enter the ctenidia. Oxygen is picked up in the ctenidia, and the hemolymph is then transported by the efferent branchial vessels to the left and right atria, through the ventricle, and then back to the hemocoel. Additional auxiliary pumping vessels occur in several taxa, particularly in active groups such as cephalopods.

and stomach digest the food, and absorption occurs in the stomach, digestive glands, and intestine.

Circulation and Gas Exchange

Although molluscs are coelomate protostomes, the coelom is greatly reduced. The main body cavity is, in most species, an open circulatory space or hemocoel, which comprises several separate sinuses, and a network of vessels in the gills, where gas exchange takes place. The blood of molluscs contains various cells, including amebocytes, and is referred to as hemolymph. It is responsible for picking up the products of digestion from the sites of absorption and for delivering these nutrients throughout the body. It usually carries in solution the copper-containing respiratory pigment hemocyanin. Some molluscs use hemoglobin to bind oxygen and many have myoglobin in the active muscle tissues, in particular those of the odontophore.

The heart lies dorsally within the **pericardial chamber**, and includes a pair of atria (often called auricles) and a single ventricle. In monoplacophorans and in *Nautilus* there are two pairs of atria, and in many gastropods there is only one (the left) that corresponds to the single gill. The atria receive the efferent branchial vessels, drawing oxygenated hemolymph from each ctenidium and passing it into the muscular ventricle, which pumps it anteriorly through a large anterior artery (the anterior or cephalic aorta). The anterior artery branches and eventually opens into various sinuses within which the tissues are bathed in oxygenated hemolymph. Return drainage through the sinuses eventually funnels the hemolymph back into the afferent branchial vessels. This basic pattern of molluscan circulation is shown diagrammatically in Figure 13.34, although it is modified to various degrees in different classes (Figure 13.35). In some cephalopods the circulatory system is secondarily closed (Figure 13.35C).

Most molluscs have ctenidia. But many have lost the ctenidia and rely either on secondarily derived gills or on gas exchange across the mantle or general body surface. In the primitive condition the ctenidium is built around a long, flattened axis projecting from the wall of the mantle cavity (Figure 13.30A). To each side of the axis are attached triangular or wedge-shaped filaments that alternate in position with filaments on the opposite side of the axis (except in nucuid protobranchs, where they are opposite). This arrangement, in which filaments project on both sides of the central axis, is called the **bipectinate condition**. There is one gill on each side of the mantle cavity, sometimes held in position by membranes that divide the mantle cavity into upper and lower chambers (Figure 13.28A,B). Lateral cilia on the gill draw water into the inhalant (ventral) chamber, from which it passes upward between the gill filaments to the exhalant (dorsal) chamber and then out of the mantle cavity (Figure 13.30A).

Two vessels run through each gill axis. The afferent vessel carries oxygen-depleted hemolymph into the gill, and the efferent vessel drains freshly oxygenated hemolymph from the gill to the atria of the heart, as noted above. Hemolymph flows through the filaments

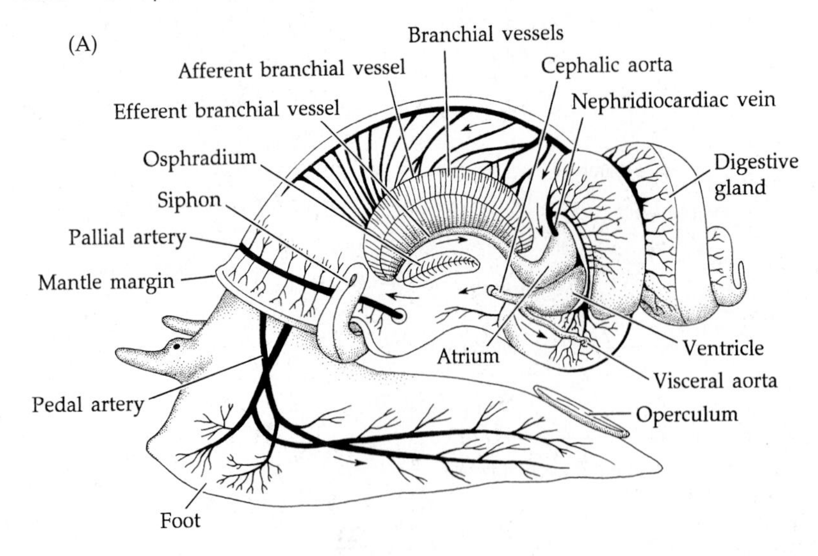

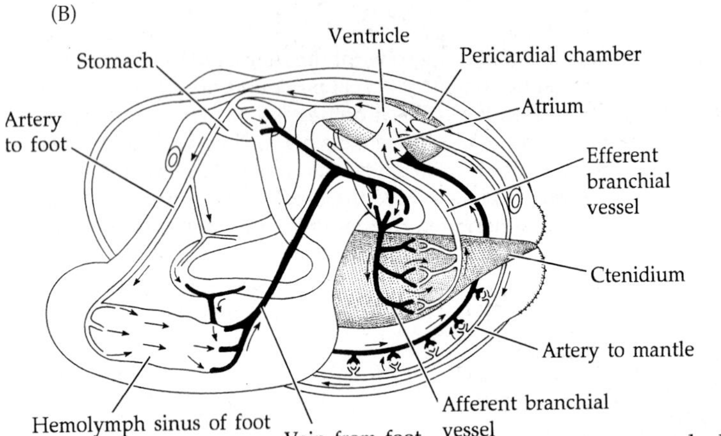

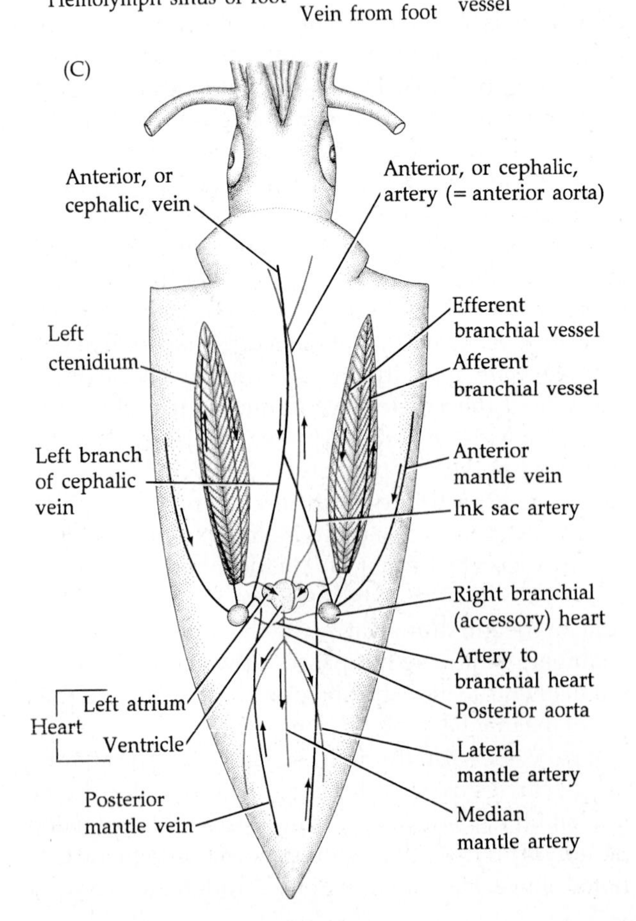

Figure 13.35 The circulatory systems of three molluscs. (A) A neogastropod *Buccinum* (hemolymph sinuses not shown). (B) A eulamellibranch unionid bivalve. (C) A squid, *Loligo*.

from the afferent to the efferent vessel. Ctenidial cilia move water over the gill filaments in a direction opposite to that of the flow of the underlying hemolymph in the branchial vessels. This countercurrent phenomenon enhances gas exchange between the hemolymph and water by maximizing the diffusion gradients of O_2 and CO_2 (Figure 13.30A). These presumed primitive bipectinate ctenidial gill conditions are expressed in several living groups, for example, in Caudofoveata, chitons, protobranch bivalves, and some gastropods.

As a result of torsion, gastropods evolved novel ways to circulate water over the gills before it comes into contact with gut or nephridial discharges. Some vetigastropods with two bipectinate ctenidia may accomplish this by circulating water in across the gills, then past the anus and nephridiopore, and away from the body via slits or holes in the shell. This circulation pattern is used by the slit shells (Pleurotomariidae) and the minute Scissurellidae and Anatomidae (Figure 13.36), abalones (Haliotidae) (Figure 13.1D), and volcano (or keyhole) limpets (Fissurellidae) (Figures 13.16H,I and 13.25A). Some specialists regard the Pleurotomariidae as "living fossils" that reflect an early gastropod character state since gastropods bearing slits are found amongst early gastropod fossils. Most other gastropods have lost the right ctenidium and with it the right atrium; inhalant water enters on the left side of the head and then passes through the mantle cavity and straight out the right side, where the anus and nephridiopore open. Other gastropods have lost both ctenidia and utilize secondary respiratory regions, either the mantle surface itself, expanded nephridial surfaces, or secondarily derived gills of one kind or another. Limpets of the genus *Patella* have rows of secondary gills in the mantle groove along each side of the body, superficially similar to the condition seen in chitons where multiple ctenidia are found.

In many gastropods one ctenidium is lost, e.g., patellogastropods, some vetigastropods, all neritimorphs and caenogastropods. In caenogastropods, the dorsal and ventral suspensory membranes seen in vetigastropod ctenidia are absent and the gill is attached directly to the mantle wall by the gill axis. The gill filaments on the attached side have been lost, while

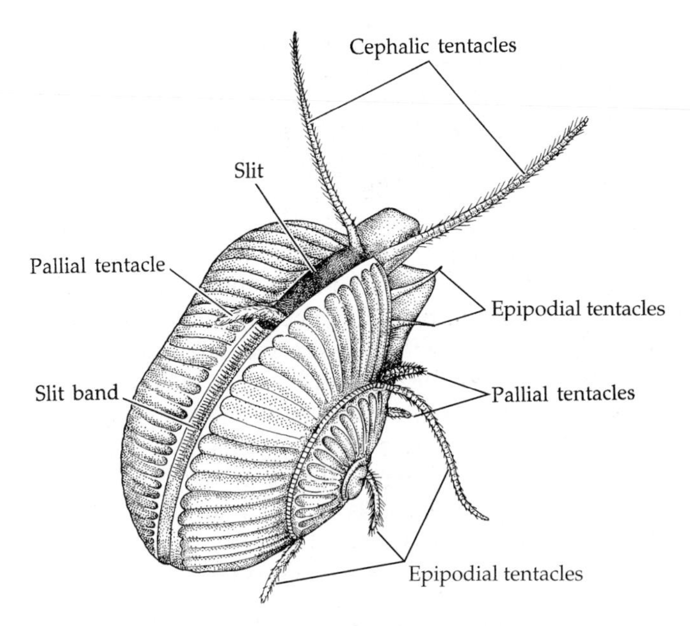

Figure 13.36 The slit-bearing vetigastropod *Anatoma*.

In chitons the mantle cavity is a groove extending along the ventral body margin and encircling the foot (Figure 13.4B). A large number of small bipectinate ctenidial gills lie laterally in this groove. The mantle is held tight against the substratum, largely enclosing

(A)

(B)

Figure 13.37 Land snails (Heterobranchia: Eupulmonata). (A) A terrestrial slug (*Arion lusitanicus*), showing the pneumostome that opens to the "lung" (Stylommatophora: Limacidae). (B) Terrestrial helicid snails (Stylommatophora: Helicidae) during summer dormancy in Sicily.

those of the opposite side project freely into the mantle cavity. This arrangement of filaments on only one side of the central axis is referred to as the **monopectinate (or pectinobranch) condition** (Figure 13.14D). Some caenogastropods have evolved inhalant siphons by extension and rolling of the anterior mantle margin (Figures 13.1E and 13.40A). In these cases the margin of the shell may be notched, or drawn out as a canal to house the siphon. The siphon provides access to surface water in burrowing species, and may also function as a mobile, directional organ used in conjunction with the chemosensory osphradium.

All heterobranchs have lost the typical ctenidia but some have a plicate, or folded, gill that has been considered by some to be a reduced ctenidium, but is now considered to be a secondary structure that has reformed in much the same location as the original ctenidial gill. Trends toward detorsion, loss of the shell, and reduction of the mantle cavity occur in many heterobranchs, and the process has apparently occurred several times within this group. Some nudibranchs have evolved secondary dorsal gas exchange structures called **cerata** or, in some nudibranchs, secondary gills that surround the anus (Figures 13.7F–J).

Wholly terrestrial gastropods lack gills and exchange gases directly across a vascularized region of the mantle, usually within the mantle cavity, the latter arrangement usually referred to as a **lung**. In marine, freshwater and terrestrial eupulmonates, the edges of the mantle cavity have become sealed to the back of the animal except for a small opening on the right side called a **pneumostome** (Figure 13.37A) that is controlled by a sphincter muscle (except in siphonariid limpets). Instead of having gills, the roof of the mantle cavity is highly vascularized. By arching and flattening the mantle cavity floor, air is moved into and out of the lung.

this groove except on either side at the anterior end to form incurrent channels, and in one or two places at the posterior end to form excurrent areas. Water enters the inhalant region of the mantle groove lateral to the gills, then passes between the gills into the exhalant region along the sides of the foot. Moving posteriorly, the current passes over the gonopores, nephridiopores, and anus before exiting (Figure 13.4B).

In bivalves the capacious mantle cavity allows the ctenidia to develop a greatly enlarged surface area, serving in most autobranch species for both gas exchange and feeding. Many of the morphological modifications of bivalve gills are described above in discussion of suspension feeding. In addition to the folded, W-shaped ctenidial filaments seen in many bivalves (Figure 13.28B), some forms (e.g., oysters) have plicate ctenidia. A plicate ctenidium has vertical ridges or folds, each ridge consisting of several ordinary ctenidial filaments. So-called "principal filaments" lie in the grooves between these ridges and their cilia are important in sorting particles from the ventilation and feeding currents. The plicate condition gives the ctenidium a corrugated appearance and further increases the surface area for feeding and gas exchange.

In spite of these modifications, the basic system of circulation and gas exchange in bivalves is similar to that seen in gastropods (Figure 13.35B). In most bivalves, the ventricle of the heart folds around the gut, so the pericardial cavity encloses not only the heart but also a short section of the digestive tract. The large mantle lines the interior of the valves and provides an additional surface area for gas exchange, which in some groups may be as important as the gills in this regard. For example, in lucinid bivalves where the gills are full of symbiotic bacteria, folds on the mantle act as a secondary gill, and in septibranchs, which have very reduced gills, the mantle surface is the principal area of gas exchange.

Most autobranch bivalves lack respiratory pigments in the hemolymph, although hemoglobin occurs in a few families and hemocyanin is found in protobranchs.

Scaphopods have lost the ctenidia, heart, and virtually all vessels. The circulatory system is reduced to simple hemolymph sinuses, and gas exchange takes place mainly across the mantle and body surface. A few ciliated ridges occur in the mantle cavity that may assist in maintaining water flow. A few tiny gastropods and at least one small monoplacophoran species lack a heart altogether.

No doubt associated with their large size and active lifestyle, cephalopods have a more developed circulatory system than other molluscs, and in the highly active decapodiforms (squid and cuttlefish) it is effectively closed, with many discrete vessels, secondary pumping structures, and capillaries (Figures 13.11C, 13.12B, and 13.35C). The result is increased pressure and efficiency of hemolymph flow and delivery. In most cephalopods, the pumping of blood into the ctenidia is assisted by muscular accessory branchial hearts, which boost the low venous pressure as the hemolymph enters the gills. The gills are not ciliated and their surface is highly folded, increasing their surface area for greater gas exchange necessary to meet the demands of their high metabolic rate.

In the Solenogastres, gills are absent but the mantle cavity surface may be folded or form respiratory papillae. Caudofoveates have a single pair of bipectinate ctenidia in the mantle cavity. Monoplacophoran gills are well-developed but weakly muscular and ciliated, and only have lamellae on one side of the gill axis; they occur as three to six pairs, aligned bilaterally within the mantle groove. The gills of monoplacophorans are thought to be modified ctenidia that vibrate and ventilate the groove where gas exchange occurs.

Excretion and Osmoregulation

The basic excretory structures of molluscs are paired tubular nephridia (often called kidneys) that are primitively similar to those of annelids. Typical nephridia are absent in the aplacophoran groups. Three, six, or seven pairs of nephridia occur in monoplacophorans, two pairs in the nautiloids, and a single pair in all other molluscs (except where one is lost in higher gastropods) (Figure 13.14). The nephrostome typically opens into the pericardial coelom via a renopericardial duct, and the nephridiopore discharges into the mantle cavity, often near the anus (Figures 13.14 and 13.34). In molluscs, the pericardial fluids (primary urine) pass through the nephrostome and into the nephridium, where selective resorption occurs along the tubule wall until the final urine is ready to pass out the nephridiopore. The pericardial sac and heart wall act as selective barriers between the open nephrostome and the hemolymph in the surrounding hemocoel and in the heart. Mollusc nephridia are rather large and saclike, and their walls are often greatly folded. In many species, afferent and efferent nephridial vessels carry hemolymph to and from the nephridial tissues (Figure 13.38). Sometimes a bladder is present just before the nephridiopore and sometimes a ureter forms a duct to carry urine well beyond the nephridiopore.

In many molluscs urine formation involves pressure filtration, active secretion, and active resorption. Aquatic molluscs excrete mostly ammonia, and most marine species are osmoconformers. In freshwater species the nephridia are capable of excreting a hyposomotic urine by resorbing salts and by passing large quantities of water. Terrestrial gastropods conserve water by converting ammonia to uric acid. Land snails are capable of surviving a considerable loss of body water, which is brought on in large part by evaporation and the production of the metabolically expensive slime trail. They often absorb water from the urine in the ureter. In many gastropods (e.g., neritimorphs,

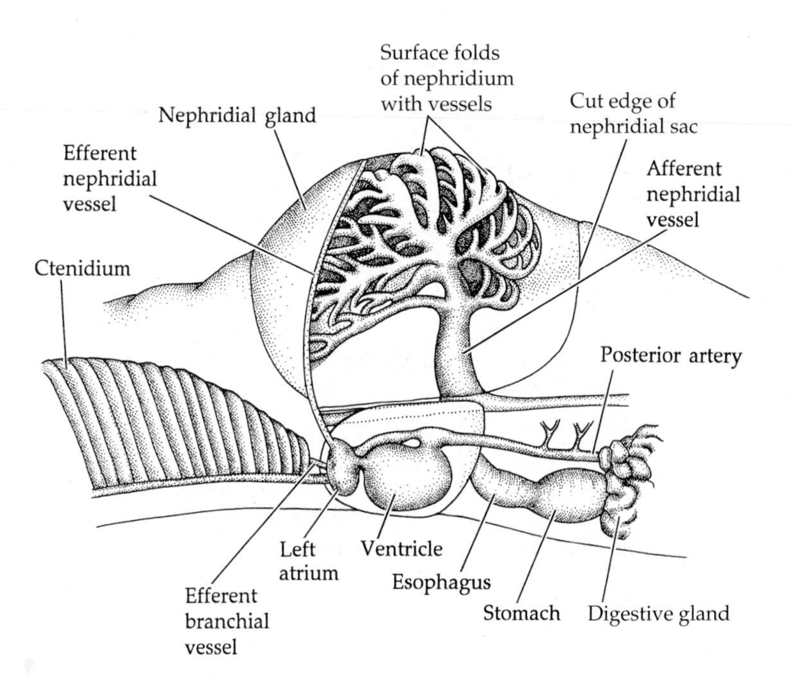

Figure 13.38 The nephridium and nearby organs of *Littorina* (cutaway view). The nephridial sac has been slit open.

caenogastropods, and heterobranchs), torsion is accompanied by loss of the adult right nephridium; in neritimorphs and caenogastropods, a small remnant contributes to part of the gonoduct. Some gastropods have lost the direct connection of the nephrostome to the pericardial coelom. In such cases the nephridium is often very glandular and served by afferent and efferent hemolymph vessels, and wastes are removed largely from the circulatory fluid.

In bivalves, the two nephridia are located beneath the pericardial cavity and are folded in a long U-shape. In autobranch bivalves, one arm of the U is glandular and opens into the pericardial cavity; the other arm often forms a bladder and opens through a nephridiopore in the suprabranchial cavity. In protobranchs, the unfolded walls of the tube are glandular throughout. The nephridiopores may be separate from or joined with the ducts of the reproductive system. In the latter case, the openings are urogenital pores.

In patellogastropods and vetigastropods and some other molluscs, the gonoduct fuses with the renopericardial canal, and the nephridiopore functions as a urogenital pore and discharges both excretory wastes and gametes. In some cases, as in one monoplacophoran, a few bivalves and in some vetigastropods, the urogenital pore may become glandular. In many bivalves and chitons the nephridium and gonad have separate ducts.

In monoplacophorans and chitons, the nephridia open into the exhalant regions of the mantle grooves; in scaphopods, the paired nephridia open near the anus. In most gastropods the nephridiopores open directly into the mantle cavity but in some, such as in stylommatophoran pulmonates, there is an elongate ureter that opens outside the enclosed lung (mantle cavity).

Cephalopods retain the basic nephridial plan, in which the nephridia drain the pericardial coelom by way of renopericardial canals and empty via nephridiopores into the mantle cavity. However, the nephridia bear enlarged regions called **renal sacs**. Before reaching the branchial heart, a large vein passes through the renal sac, wherein numerous thin-walled evaginations, called **renal appendages**, project off the vein. As the branchial heart beats, hemolymph is drawn through the renal appendages, and wastes are filtered across their thin walls into the nephridia. The overall result is an increase in excretory efficiency over the simpler arrangement present in other molluscs.

The fluid-filled nephridia of cephalopods are inhabited by a variety of commensals and parasites. The epithelium of the convoluted renal appendages provides an excellent surface for attachment, and the renal pores provide a simple exit to the exterior. Symbionts identified from cephalopod nephridia include viruses, fungi, ciliate protists, rhombozoans, trematodes, larval cestodes, and juvenile nematodes.

Nervous System

The molluscan nervous system is derived from the basic protostome plan of an anterior circumenteric arrangement of ganglia and paired ventral nerve cords. In molluscs, the more ventral and medial of the two pairs of nerve cords are called the **pedal cords** (or ventral cords); they innervate the muscles of the foot. The more lateral pair of nerves are the **visceral cords** (or lateral cords); they serve the mantle and viscera. Transverse commissures interconnect these longitudinal nerve cord pairs, creating a ladderlike nervous system. This basic plan is seen in the aplacophorans and polyplacophorans (Figure 13.39). The molluscan nervous system lacks the segmentally arranged ganglia of annelids and arthropods.

In the "simplest" molluscs—such as aplacophorans, monoplacophorans, and polyplacophorans—ganglia are poorly developed (Figure 13.39). A simple nerve ring surrounds the anterior gut, often with small cerebral ganglia on either side. Each cerebral ganglion, or the nerve ring itself, issues small nerves to the buccal region and gives rise to the pedal and the visceral nerve cords. Most other molluscs have more well-defined ganglia. Their nervous systems are built around

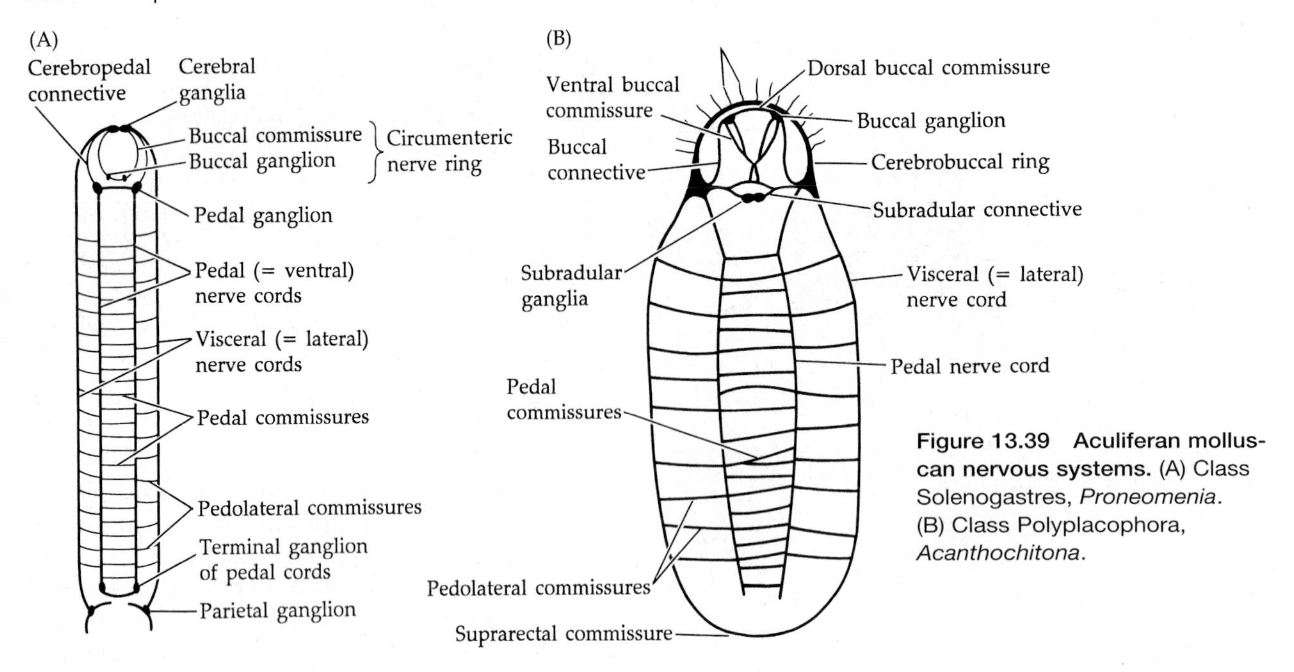

Figure 13.39 Aculiferan molluscan nervous systems. (A) Class Solenogastres, *Proneomenia*. (B) Class Polyplacophora, *Acanthochitona*.

three pairs of large ganglia that interconnect to form a partial or complete nerve ring around the gut (Figures 13.40 and 13.41). Two pairs, the cerebral and pleural ganglia, lie dorsal or lateral to the esophagus, and one pair, the pedal ganglia, lies ventral to the gut, in the anterior part of the foot. In cephalopods, bivalves, and advanced gastropods, the cerebral and pleural ganglia are often fused. From the cerebral ganglia, peripheral nerves innervate the tentacles, eyes, statocysts, and general head surface, as well as buccal ganglia, with special centers of control for the buccal region, radular apparatus, and esophagus. The pleural ganglia give rise to the visceral cords, which extend posteriorly, supplying peripheral nerves to the viscera and mantle. The visceral cords eventually join a pair of esophageal (= intestinal, = pallial) ganglia and from there pass on to terminate in paired visceral ganglia. The esophageal ganglia or associated nerves innervate the gills and osphradium, and the visceral ganglia serve organs in the visceral mass. The pedal ganglia also give rise to a pair of pedal nerve cords that extend posteriorly and provide nerves to muscles of the foot.

As described above, due to torsion, the posterior portion of the gastropod nervous system is twisted into a figure eight, a condition known as streptoneury (Figure 13.40A,B). In addition to twisting the nervous system, torsion brings the posterior ganglia forward. In many advanced gastropods this anterior concentration of the nervous system is accompanied by a shortening of some nerve cords and fusion of ganglia. In most detorted gastropods the nervous system displays a secondarily derived bilateral symmetry and more or less untwisted visceral nerve cords—a condition known as euthyneury (Figure 13.40C).

In bivalves, the nervous system is clearly bilateral, and fusion has usually reduced it to three large, distinct ganglia. Anterior cerebropleural ganglia give rise to two pairs of nerve cords, one extending posterodorsally to the visceral ganglia, the other leading ventrally to the pedal ganglia (Figure 13.41). The two cerebropleural ganglia are joined by a dorsal commissure over the esophagus. The cerebropleural ganglia send nerves to the palps, anterior adductor muscle, and mantle. The visceral ganglia issue nerves to the gut, heart, gills, mantle, siphon, and posterior adductor muscle.

The degree of nervous system development within the Cephalopoda is unequaled among invertebrates. The paired ganglia seen in other molluscs are not recognizable in cephalopods, where extreme cephalization has concentrated ganglia into lobes of a large brain encircling the anterior gut (Figure 13.42A). In addition to the usual head nerves originating from the dorsal part of the brain (more or less equivalent to the cerebral ganglia), a large optic nerve extends to each eye via a massive optic lobe. In most cephalopods, much of the brain is enclosed in a cartilaginous cranium. The pedal lobes supply nerves to the funnel, and anterior divisions of the pedal ganglia (called brachial lobes) send nerves to each of the arms and tentacles, an arrangement suggesting that the funnel and tentacles are derived from the molluscan foot. Octopuses may be the "smartest" invertebrates, for they can be quickly taught some rather complex memory-dependent tasks.

Squid and cuttlefish (Decapodiformes) have a rapid escape behavior that depends on a system of giant motor fibers that control powerful and synchronous contractions of the mantle muscles. The command center of this system is a pair of very large first-order giant neurons in the lobe of the fused visceral ganglia. Here, connections are made to second-order giant neurons that extend to a pair of large stellate ganglia. At the stellate ganglia, connections are made with third-order giant neurons that innervate the circular muscle fibers of the mantle (Figure 13.42D). Other nerves extend

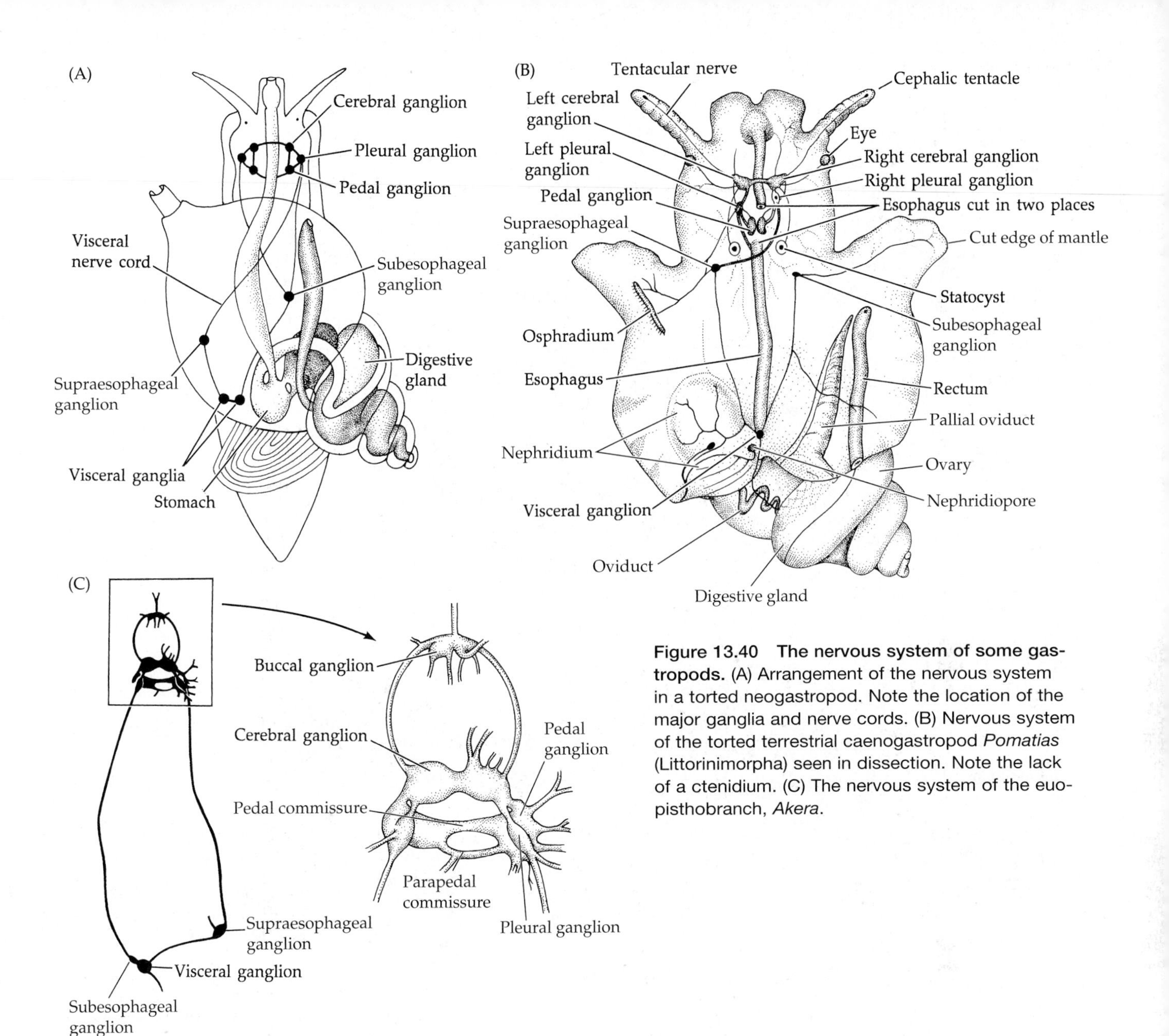

Figure 13.40 The nervous system of some gastropods. (A) Arrangement of the nervous system in a torted neogastropod. Note the location of the major ganglia and nerve cords. (B) Nervous system of the torted terrestrial caenogastropod *Pomatias* (Littorinimorpha) seen in dissection. Note the lack of a ctenidium. (C) The nervous system of the euopisthobranch, *Akera*.

posteriorly from the brain and terminate in various ganglia that innervate the viscera and structures in the mantle cavity.

For several decades neurobiologists have utilized the giant axons of *Loligo* as an experimental system for the study of nerve physiology and mechanics, and much of our fundamental knowledge of how nerve cells work is based on squid neurology. The sea hare *Aplysia*, and some eupulmonate snails have also been used in the same fashion and, although they lack giant axons, they possess exceptionally large neurons and ganglia that can be easily impaled with microelectrodes to discover the physiological secrets of such systems.

Sense Organs

With the exception of the aplacophorans, molluscs possess various combinations of sensory tentacles, photoreceptors, statocysts, and osphradia.

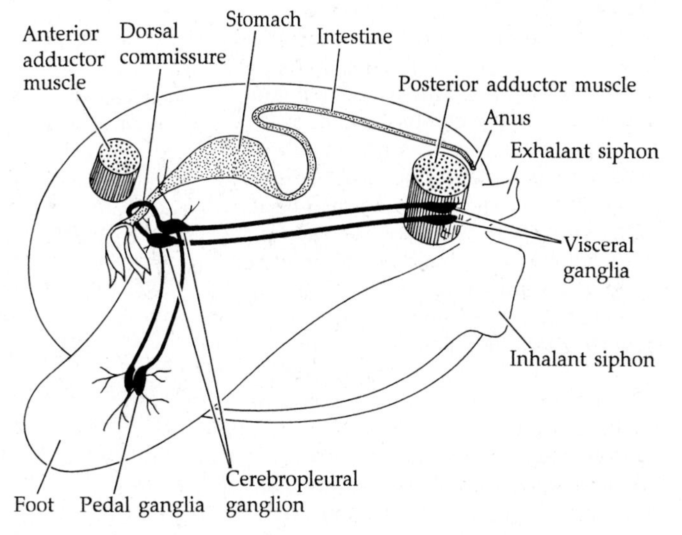

Figure 13.41 The reduced and concentrated nervous system of a typical autobranch bivalve.

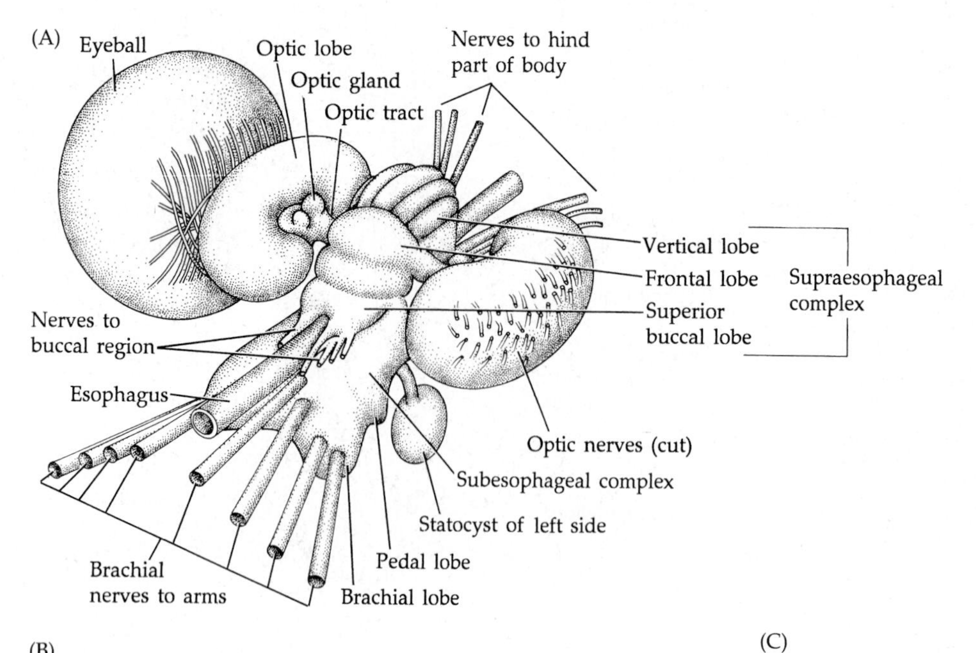

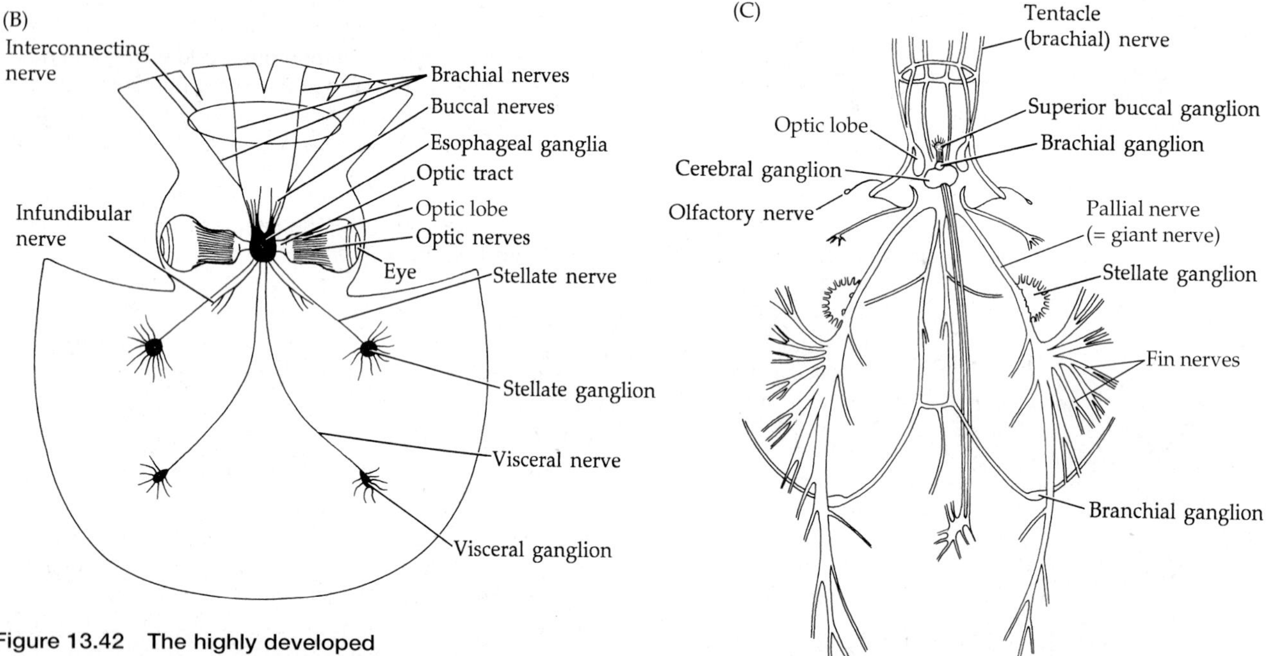

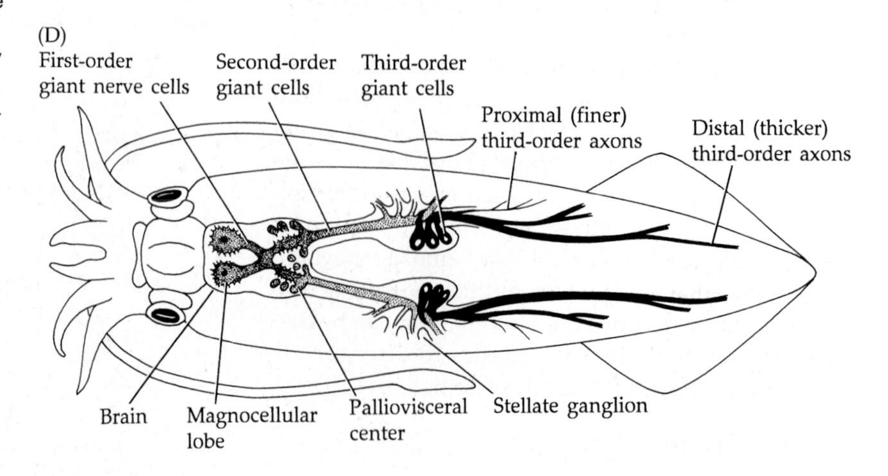

Figure 13.42 The highly developed nervous system of cephalopods. (A) The brain of an octopus. The lobes of the supraesophageal complex approximately correspond to the cerebral and buccal ganglia of other molluscs while the subesophageal complex comprises the fused pedal and pleurovisceral ganglia. About 15 structurally and functionally distinct pairs of lobes have been identified in the brain of octopuses. (B) Nervous system of an octopus. (C) Nervous system of a squid (*Loligo*). (D) Giant fiber system of a squid. Note that the first-order giant neurons possess an unusual cross connection, and that the third-order giant neurons are arranged so that motor impulses can reach all parts of the mantle-wall musculature simultaneously (as a result of the fact that impulses travel faster in thicker axons).

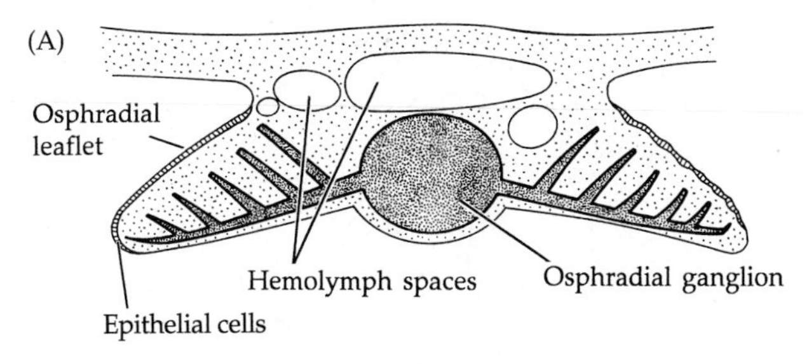

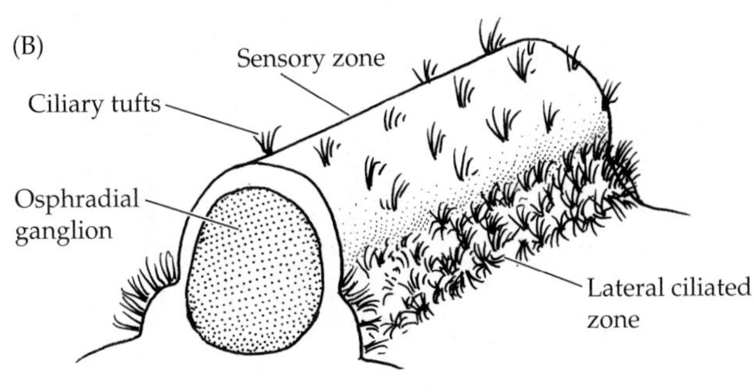

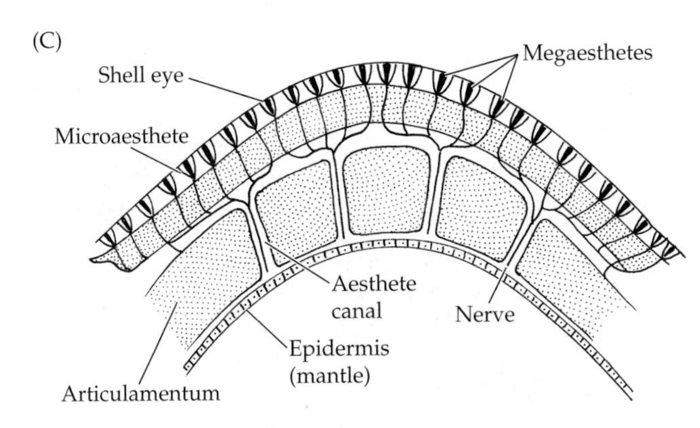

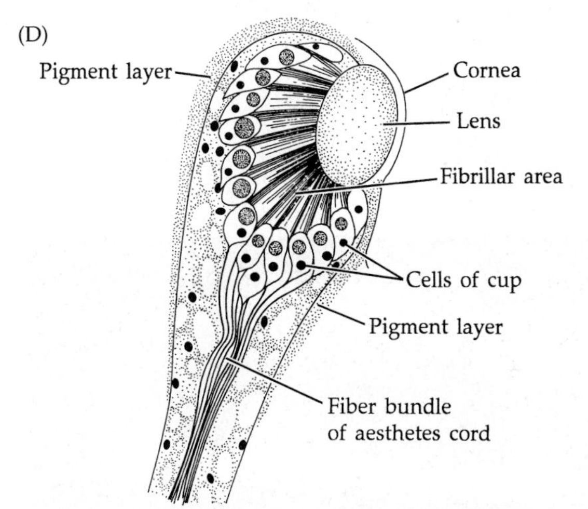

Figure 13.43 Two sensory organs of molluscs: osphradia and aesthetes. (A) Cross section of a bipectinate osphradium of the caenogastropod *Ranella* showing two leaflets. (B) Part of the osphradium of a littorinimorph caenogastropod such as *Littorina*. (C) One valve of a polyplacophoran (*Tonicia*). The aesthetes extend to the shell surface through megalopores and micropores. (D) Eye-bearing aesthetes (longitudinal section) in a megalopore of a chiton (*Acanthopleura*).

Osphradia are patches of sensory epithelium, located on or near the gills, or on the mantle wall (Figures 13.40B and 13.43A,B). They are chemoreceptors, and their cilia can also assist in mantle cavity ventilation in some caenogastropods. Little is known about the biology of osphradia, and their morphology and histology differs markedly within the phylum and even within some classes such as the gastropods.

In vetigastropods, a small osphradium is present on each gill; in those gastropods that possess one gill, there is only one osphradium, and it lies on the mantle cavity wall anterior and ventral to the attachment of the gill itself. Osphradia are reduced or absent in gastropods that have lost both gills, that possess a highly reduced mantle cavity, or that have taken up a strictly pelagic existence. Osphradia are best developed in benthic predators and scavengers, such as neogastropods and some other caenogastropods.

Most gastropods have one pair of sensory cephalic tentacles, but eupulmonates and many sea slugs possess two pairs. Many vetigastropods also have epipodial tentacles on the margin of the foot or mantle and there are also epipodial sense organs present (Figure 13.5A,C). The cephalic tentacles may bear eyes as well as tactile and chemoreceptor cells. Many nudibranchs have a pair of branching or folded anterior dorsal chemoreceptors called **rhinophores** (Figure 13.7F,G).

The primitive patellogastropods have simple pigment-cup eyes, while the more advanced gastropods have more complex eyes with a lens and often a cornea (Figure 13.44A,B,D). Most gastropods have a small eye at the base of each cephalic tentacle, but in some, such as the conch *Strombus* and some neogastropods, the eyes are enlarged and elevated on long stalks. The stylommatophoran and systellommatophoran pulmonates also have eyes placed on the tips of special optic tentacles and, in stylommatophorans, these tentacles have become olfactory organs.

Gastropods typically produce a mucopolysaccharide slime trail as they crawl. In many species the trail contains chemical messengers that other members of the species "read" by means of their excellent chemoreception. These chemical messengers may be simple trail markers, so one animal can follow or locate another, or they may be alarm substances that serve to warn others of possible danger on the path ahead. For example, when the carnivorous cephalaspidean sea slug *Navanax* is attacked by a predator, it quickly releases a yellow chemical mixture on its trail that causes other members of the species to abort their trail-following activity. Laboratory experiments have shown that at least one nudibranch (*Tritonia diomedea*) possesses geomagnetic orientation to the Earth's magnetic field. Motile gastropods usually possess a pair of closed statocysts near the pedal ganglia in the anterior region of the foot that contain either a single large statolith or several **statoconia** (much smaller particles).

Scaphopods lack eyes, tentacles, and osphradia typical of the epibenthic and motile molluscan groups. The captacula may function as tactile (as well as feeding) structures. Sense organs are found in the mantle edge surrounding the ventral aperture and at the dorsal water intake opening.

Bivalves have most of their sensory organs along the middle lobe of the mantle edge where they are in contact with the external environment (Figure 13.15C). These receptors may include mantle tentacles, which can contain both tactile and chemoreceptor cells. Such tentacles are commonly restricted to the siphonal areas, but in some swimming clams (e.g., *Lima, Pecten*) they may line the entire mantle margin. Paired statocysts usually occur in the foot near the pedal ganglia, and are

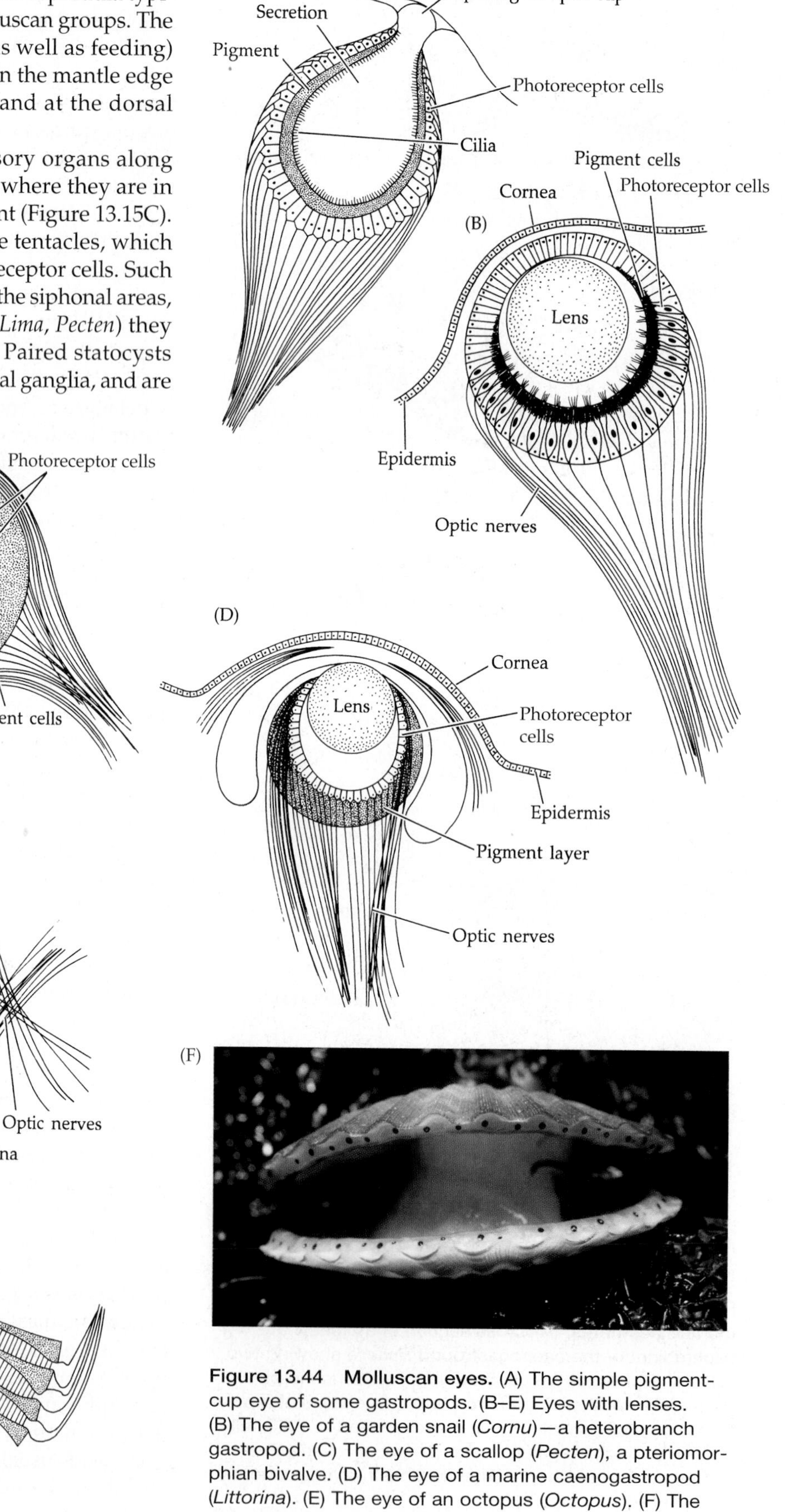

Figure 13.44 Molluscan eyes. (A) The simple pigment-cup eye of some gastropods. (B–E) Eyes with lenses. (B) The eye of a garden snail (*Cornu*)—a heterobranch gastropod. (C) The eye of a scallop (*Pecten*), a pteriomorphian bivalve. (D) The eye of a marine caenogastropod (*Littorina*). (E) The eye of an octopus (*Octopus*). (F) The queen scallop *Aequipecten opercularis*, showing its dark eyes along the mantle edges.

of particular importance in georeception by burrowing bivalves. Mantle eyes may also be present along the mantle edge or on the siphons and have evolved independently in a number of bivalve groups. In the spiny oyster *Spondylus* and the swimming scallop *Pecten*, these eyes are "mirror eyes" with a reflective layer (the **tapeum**) behind paired retinas. This layer reflects light back into the eye giving these bivalves a separate focal image on each retina–one from the lens and the other from the mirror. (Figure 13.44C–E). The bivalve osphradium lies in the exhalant chamber, beneath the posterior adductor muscle.

Chitons lack statocysts, cephalic eyes, and tentacles. Instead, they rely largely on two special sensory structures. These are the **adanal sensory structures** in the posterior portion of the mantle cavity and the **aesthetes**, which are a specialized system of photoreceptors unique to the class Polyplacophora. Aesthetes occur in high numbers across the dorsal surface of the shell plates. They are mantle cells that extend into the minute vertical canals (megalopores and micropores) in the upper tegmentum of the shell (Figure 13.43C,D). The canals and sensory endings terminate beneath a cap on the shell surface. Little is known about the functioning of aesthetes, but they apparently mediate light-regulated behavior. In at least one family (Chitonidae), some of them are modified as simple lensed eyes. The outer mantle surface of the girdle of many chitons is liberally supplied with tactile and photoreceptor cells (Figure 13.43D).

Like the rest of their nervous system, the sense organs of cephalopods are highly developed. The eyes are superficially similar to those of vertebrates (Figure 13.44E), and these two types of eyes are often cited as a classic example of convergent evolution. The eye of a coleoid cephalopod such as *Octopus* sits in a socket associated with the cranium. The cornea, iris, and lens arrangement is much like that of vertebrate eyes. Also as in vertebrates, the lens is suspended by ciliary muscles but has a fixed shape and focal length. An iris diaphragm controls the amount of light entering the eye, and the pupil is a horizontal slit. The retina comprises closely packed, long, rodlike photoreceptors whose sensory ends point toward the front of the eye; hence the cephalopod retina is the direct type rather than the indirect type seen in vertebrates. The rods connect to retinal cells that supply fibers to the large optic ganglia at the distal ends of the optic nerves. Unlike the eyes of vertebrates, the coleoid cornea probably contributes little to focusing because there is almost no light refraction at the corneal surface (as there is at an air–cornea interface). The coleoid eye accommodates to varying light conditions by changes in the size of the pupil and by migration of the retinal pigment. Coleoid eyes form distinct images (although octopuses are probably quite nearsighted) and experimental work suggests that they do not see colors other than as different shades of grey, although they can detect polarized light. In addition, coleoids can discriminate among objects by size, shape, and vertical versus horizontal orientation. The eyes of *Nautilus* are rather primitive relative to the eyes of coleoids. They lack a lens, and are open to the water through the pupil. They are thought to function in the same way that a pinhole camera does.

Coleoids have complex statocysts that provide information on static body position and on body motion. Those of *Nautilus* are relatively simple. In addition, the arms of coleoids are liberally supplied with chemosensory and tactile cells, especially on the suckers of benthic octopuses, which have extremely good chemical and textural discrimination capabilities. *Nautilus* is the only cephalopod with osphradia.

Cephalopod Coloration and Ink

Coleoid cephalopods are noted for their striking pigmentation and dramatic color displays. The integument contains many pigment cells, or chromatophores, most of which are under nervous control. Such chromatophores can be individually rapidly expanded or contracted by means of tiny muscles attached to the periphery of each cell. Contraction of these muscles pulls out the cell and its internal pigment into a flat plate, thereby displaying the color; relaxation of the muscles causes the cell and pigment to concentrate into a tiny, inconspicuous dot. Because these chromatophores are displayed or concealed by muscle action, their activity is extremely rapid and coleoid cephalopods can change color (and pattern) almost instantaneously. Chromatophore pigments are of several colors—black, yellow, orange, red, and blue. The chromatophore color may be enhanced by deeper layers of iridocytes that both reflect and refract light in a prismatic fashion. Some species, such as the cuttlefish *Sepia* and many octopuses, are capable of closely mimicking their background coloration (Figure 13.12E) as well as producing vivid contrasting colors (Figure 13.12F,G). Many epipelagic squids show a dark-above, light-below countershading similar to that seen in pelagic fishes. Most coleoids also undergo color changes in relation to behavioral rituals, such as courtship and aggression. In octopuses, many color changes are accompanied by modifications in the surface texture of the body, mediated by muscles beneath the skin—something like elaborate, controlled "gooseflesh."

In addition to the color patterns formed by chromatophores, some coleoids are bioluminescent. When present, the light organs, or photophores, are arranged in various patterns on the body, and in some cases even occur on the eyeball. The luminescence is sometimes due to symbiotic bacteria, but in other cases it is intrinsic. The photophores of some species have a complex reflector and focusing-lens arrangement, and some even have an overlying color filter or chromatophore shutter to control the color or flashing pattern. Most luminescent species are deep-sea forms, and little is known about the role of light production in their lives. Some appear to use the photophores to create a countershading effect, so as to appear less visible to predators (and prey) from below and above. Others living below the photic zone may use their glowing or flashing patterns as a means of communication, the signals serving to keep animals together in schools or to attract prey. The flashing may also play a role in mate attraction. The fire squid, *Lycoteuthis*, can produce several colors of light: white, blue, yellow, and pink. At least one genus of squid, *Heteroteuthis*, secretes a luminescent ink. The light comes from luminescent bacteria cultured in a small gland near the ink sac, from which ink and bacteria are ejected simultaneously.

In most coleoid cephalopods, a large ink sac is located near the intestine (Figure 13.32H). An ink-producing gland lies in the wall of the sac, and a duct runs from the sac to a pore into the rectum. The gland secretes a brown or black fluid that contains a high concentration of melanin pigment and mucus; the fluid is stored in the ink sac. When alarmed, the animal releases the ink through the anus and mantle cavity and out into the surrounding water. The cloud of inky material hangs together in the water, forming a "dummy" image that serves to confuse predators. The alkaloid nature of the ink may also act to deter predators, particularly fishes, and may interfere with their chemoreception.

Like virtually all other aspects of coleoid biology, the ability to change color and to defend against predators are part and parcel of their active hunting lifestyles. In the course of their evolution, coleoid cephalopods abandoned the protection of an external shell, becoming more efficient swimmers but also exposing their fleshy bodies to predators. The evolution of camouflage and ink production, coupled with high mobility and complex behavior, played a major role in the success of these animals in their radical modification of the basic molluscan body plan.

Reproduction

Primitively, molluscs are mostly gonochoristic, with a pair of gonads that discharge their gametes to the outside, either through the nephridial plumbing or through separate ducts. In species that free-spawn, fertilization is external and development is indirect. Many molluscs with separate gonoducts that store and transport the gametes also have various means of internal fertilization. In these forms, direct and mixed life history patterns have evolved.

Caudofoveata are gonochoristic with paired gonads, while Solenogastres are hermaphroditic with a pair of gonads (Figure 13.45). In both aplacophoran groups the gonads discharge gametes by way of short gonopericardial ducts into the pericardial chamber, from which they pass through gametoducts to the mantle cavity. In the Solengastres fertilization is internal and the young are sometimes brooded, while in the Caudofoveata the gametes are discharged into the surrounding seawater

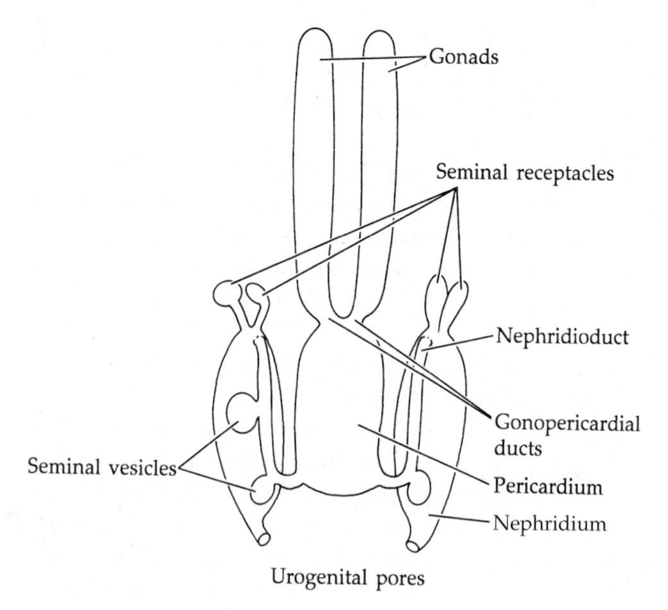

Figure 13.45 A solenogaster urogenital system.

where fertilization occurs. Monoplacophorans possess two pairs of gonads, each with a gonoduct connected to one of the pairs of nephridia (Figures 13.3D and 13.14E). One tiny monoplacophoran species, *Micropilina arntzi*, is a hermaphrodite and broods its embryos in its mantle cavity.

Most chitons are gonochoristic, although a few hermaphroditic species are known. In chitons, the two gonads are fused and situated medially in front of the pericardial cavity (Figure 13.4F). Gametes are transported directly to the outside by two separate gonoducts. The gonopores are located in the exhalant region of the mantle groove, one in front of each nephridiopore. Fertilization is external but can occur in the mantle cavity of the female. The eggs are enclosed within a spiny, buoyant membrane and are released into the sea individually or in strings. A few chitons brood their embryos in the mantle groove, and in one species (*Callistochiton viviparous*) development takes place entirely within the ovary.

In living gastropods, one gonad has been lost and the remaining one is usually located with the digestive gland in the visceral mass. The gonoduct is developed in association with the right nephridium in patellogastropods and vetigastropods (Figure 13.46A) while in neritimorphs and caenogastropods a vestige of the right nephridium is incorporated in the oviduct. In cases where the right nephridium is still functional in transporting excretory products, as in the patellogastropods and vetigastropods, the gonoduct is properly called a urogenital duct, because it discharges both gametes and urine.

Gastropods may be gonochoristic or hermaphroditic, but even in the latter case usually only a single gonad (an ovotestis) exists, although a few heterobranchs have separate male and female gonads (e.g.,

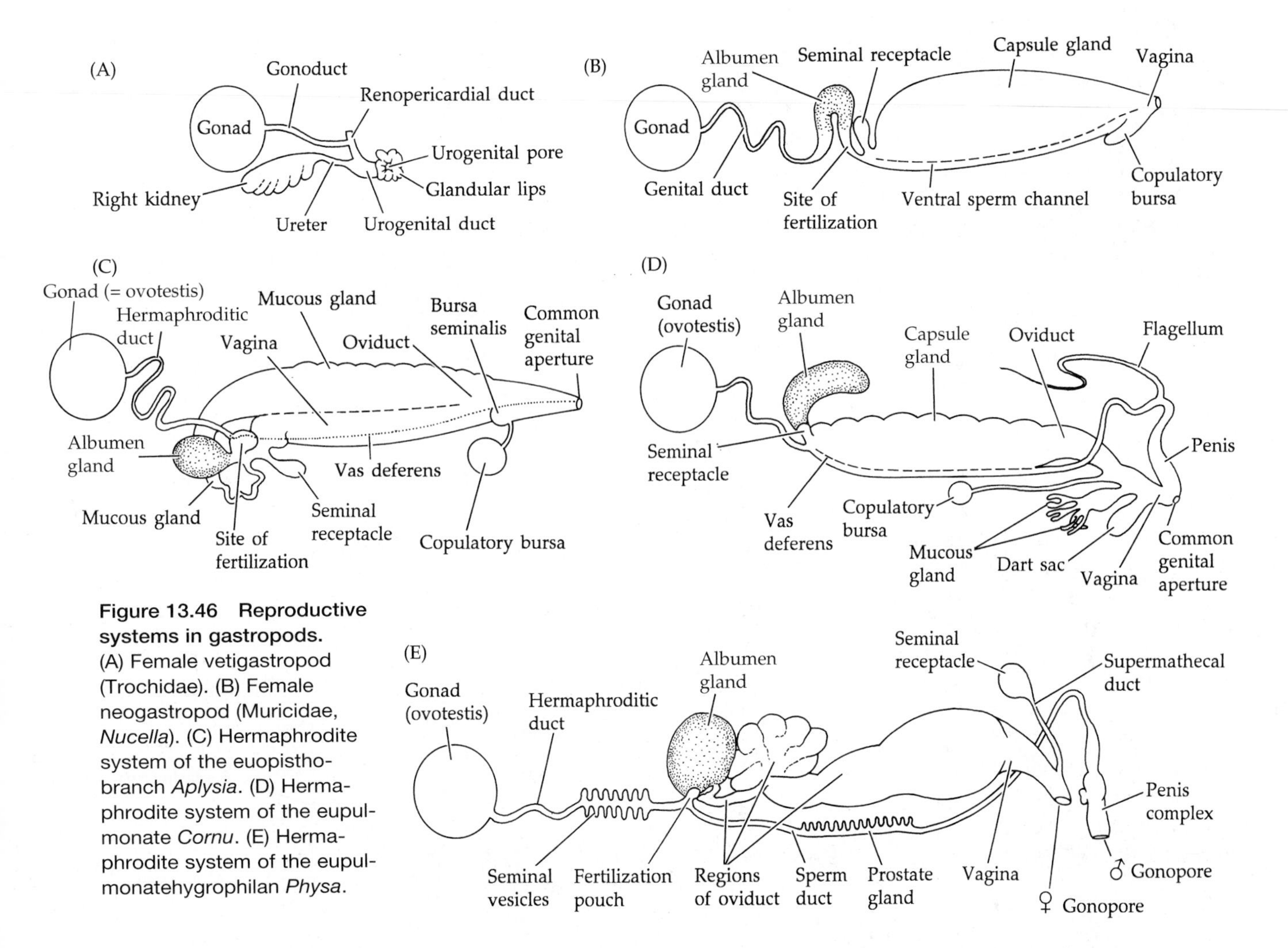

Figure 13.46 Reproductive systems in gastropods. (A) Female vetigastropod (Trochidae). (B) Female neogastropod (Muricidae, *Nucella*). (C) Hermaphrodite system of the euopisthobranch *Aplysia*. (D) Hermaphrodite system of the eupulmonate *Cornu*. (E) Hermaphrodite system of the eupulmonatehygrophilan *Physa*.

Omalogyra and in the mathildid *Gegania valkyrie*), while others are protandric. The commitment of the right nephridial plumbing entirely to serving the reproductive system was a major step in gastropod evolution. The isolation of the reproductive tract allowed its independent evolution, without which the great variety of reproductive and developmental patterns in gastropods may never have been realized.

In many gastropods with isolated reproductive tracts, the female system bears a ciliated fold or tube that forms a vagina and oviduct (or pallial oviduct). The tube develops inwardly from the mantle wall and connects with the genital duct. The oviduct may bear specialized structures for sperm storage or egg case secretion. An organ for storing received sperm, the seminal receptacle often lies near the ovary at the proximal end of the oviduct. Eggs are fertilized at or near this location prior to entering the long secretory portion of the oviduct. Many female systems also have a copulatory bursa, usually at the distal end of the oviduct, where sperm are received during mating. In such cases the sperm are later transported along a ciliated groove in the oviduct to the seminal receptacle, near where fertilization takes place. The secretory section of the oviduct may be modified as an albumin gland and a mucous or capsule gland. Many heterobranchs lay fertilized eggs in jelly-like mucopolysaccharide masses or strings produced by these glands. Most terrestrial pulmonates produce a small number of large, individual, yolky eggs, which are often provided with calcareous shells. Other pulmonates brood their embryos internally and give birth to juveniles. Many caenogastropods produce egg capsules in the form of leathery or hard cases that are attached to objects in the environment, thereby protecting the developing embryos. A ciliated groove is often present to conduct the soft egg capsules from the female gonopore down to a gland in the foot, where they are molded and attached to the substratum.

The male genital duct, or vas deferens, may include a prostate gland for production of seminal secretions. In many gastropods the proximal region of the vas deferens functions as a sperm storage area, or seminal vesicle. In many caenogastropods, neritimorphs, and lower heterobranchs the males have an external penis to facilitate transfer of sperm (Figures 13.6B and 13.47), and internal fertilization takes place prior to formation of the egg case. The penis is a long extension of the body wall usually arising behind the right cephalic tentacle. In these groups with a cephalic penis, most

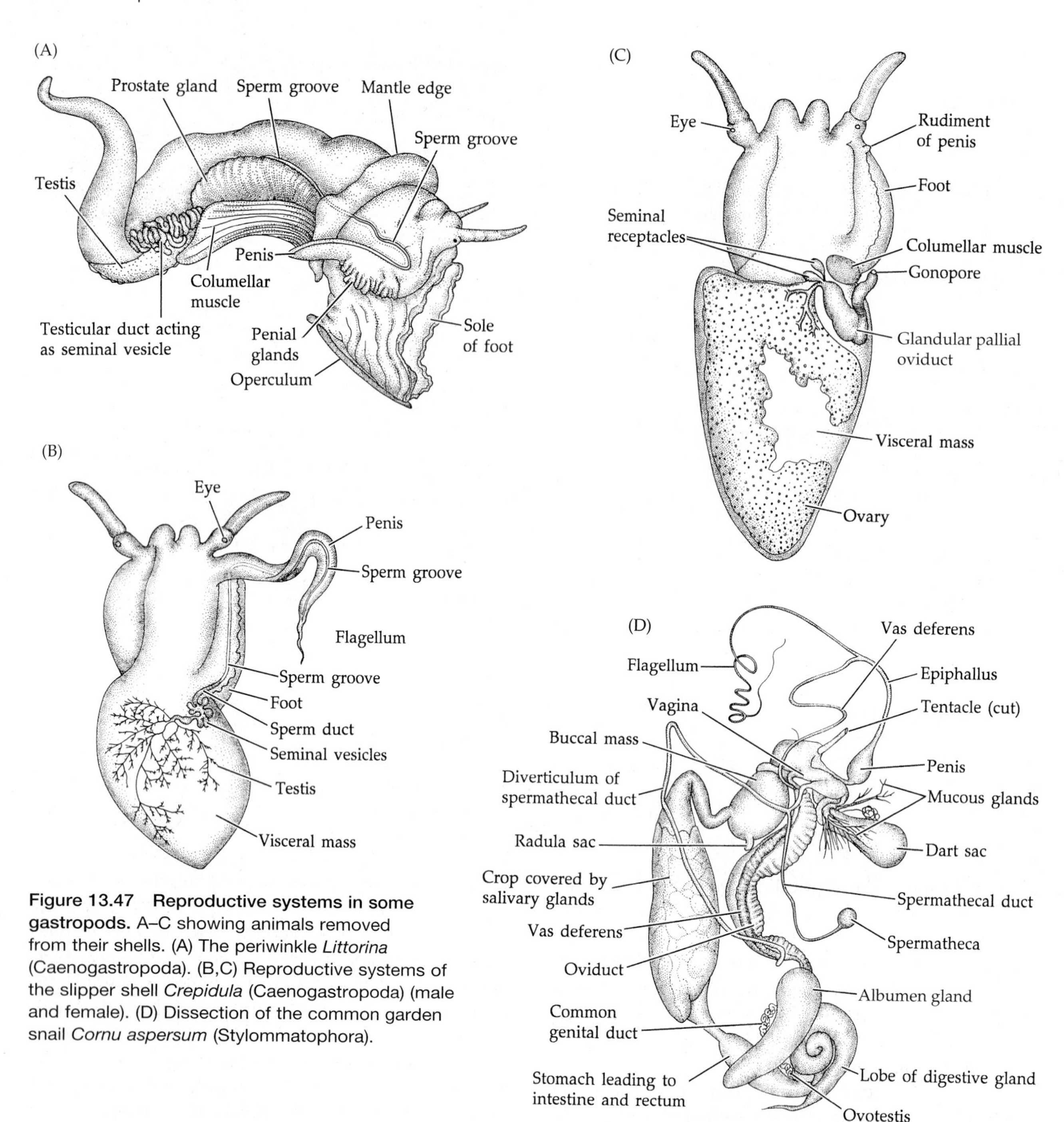

Figure 13.47 Reproductive systems in some gastropods. A–C showing animals removed from their shells. (A) The periwinkle *Littorina* (Caenogastropoda). (B,C) Reproductive systems of the slipper shell *Crepidula* (Caenogastropoda) (male and female). (D) Dissection of the common garden snail *Cornu aspersum* (Stylommatophora).

of the glandular parts of the reproductive system lie within the mantle cavity or may extend back alongside the nephridium. In most euthyneurans these parts of the reproductive system have migrated into the body cavity and the penis has become a retractile, internal structure. Sperm transfer in some gastropods involves the use of spermatophores either involving a penis or without one in the case of the cerithiomorph groups, and some others. In some, large parasperm are used to transport the normal sperm.

With both simultaneous and sequential hermaphrodite gastropods, copulation is the rule—either with one individual acting as the male and the other as the female, or with a mutual exchange of sperm between the two. Sedentary species, such as territorial limpets and slipper shells, are often protandric hermaphrodites. In slipper shells (*Crepidula*), individuals may stack one atop the other (Figure 13.48), with the more recently settled individuals being males on top of the stack, females on the bottom. Each male (Figure 13.47B) uses its long penis to inseminate the females (Figure 13.47C) below. Males that are in association with females tend to remain male for a relatively long period of time. Eventually, or if isolated from a female, the male develops into a female. Female slipper shells cannot switch back to males, because the masculine reproductive system degenerates during the sex change.

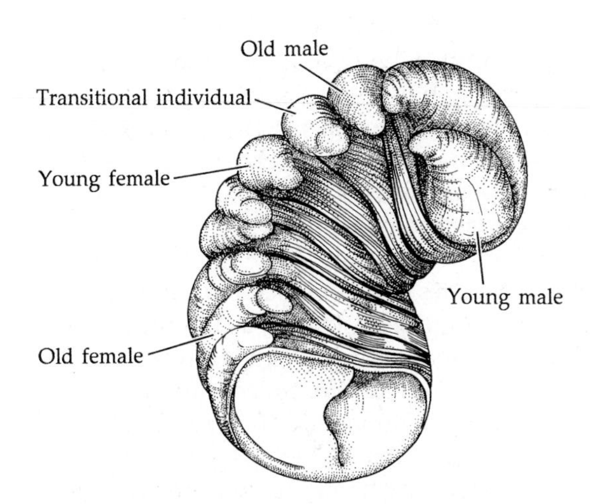

Figure 13.48 A stack of *Crepidula fornicata*, a slipper shell (Caenogastropoda) displaying sequential hermaphroditism.

Most eupulmonates are simultaneous hermaphrodites, although protandric hermaphrodites sometimes occur. In most simultaneous hermaphrodite euthyneurans a single complex gonad, the ovotestis, produces both eggs and sperm (Figures 13.46C–E and 13.47D) with the mature gametes leaving the ovotestis via the hermaphroditic duct. Euthyneuran reproductive systems are amazingly complex and varied in their plumbing and structure, and sometimes have separate male and female gonopores, or only a single common gonopore (Figure 13.46D,E).

Distinct precopulatory behaviors occur in a few groups of gastropods. These primitive courtship routines are best documented in land pulmonates and include behaviors such as oral and tentacular stroking, and intertwining of the bodies. In some pulmonates (e.g., the common garden snail, *Cornu*, formerly *Helix*) the vagina contains a dart sac, which secretes a calcareous harpoon. As courtship reaches its crescendo, and a pair of snails is intertwined, one will drive its dart into the body wall of the other, perhaps as a means of sexually arousing its partner.

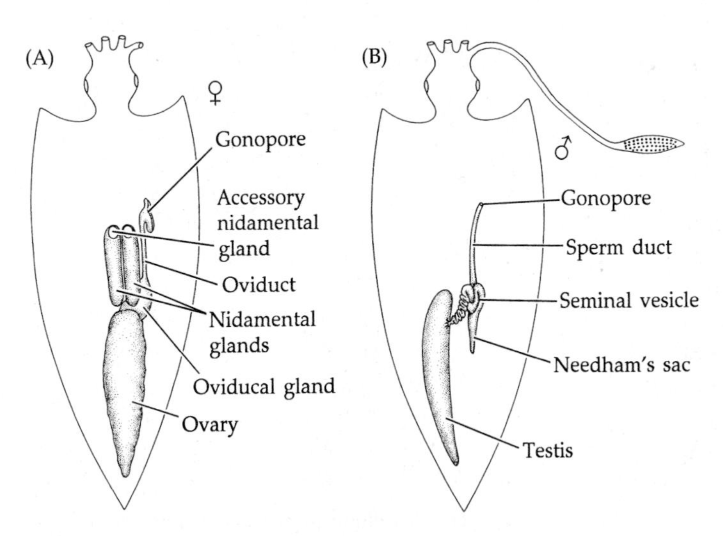

Figure 13.49 Reproductive systems in a coleoid cephalopod, the squid *Loligo*. (A) Female and (B) male.

Most bivalves are gonochoristic and retain the primitively paired gonads. However, the gonads are large and closely invested with the viscera and with each other, so an apparently single gonadal mass results. The gonoducts are simple tubes, and fertilization is usually external, although some marine and most freshwater species brood their embryos for a time. In primitive bivalves, the gonoducts join the nephridia and gametes are released through urogenital pores. In many advanced bivalves, the gonoducts open into the mantle cavity separately from the nephridiopores. Hermaphroditism occurs in some bivalves, including shipworms and some species of cockles, oysters, scallops, and others. Oysters of the genus *Ostrea* are sequential hermaphrodites, and most are capable of switching sex in either direction.

Cephalopods are almost all gonochoristic, with a single gonad in the posterior region of the visceral mass (Figures 13.11C, 13.12B, and 13.49). The testis releases sperm to a coiled vas deferens, which leads anteriorly to a seminal vesicle. Here various glands assist in packaging the sperm into elaborate spermatophores, which are stored in a large reservoir called **Needham's sac**. From there the spermatophores are released into the mantle cavity via a sperm duct. In females the oviduct terminates in one oviducal gland in squids, and two in octopuses. This gland secretes a protective membrane around each egg.

The highly developed nervous system of cephalopods has facilitated the evolution of some very sophisticated precopulatory behaviors, which culminate in the transfer of spermatophores from the male to the female. Because the oviducal opening of females is deep within the mantle chamber, male coleoids use one of their arms as an intromittent organ to transfer the spermatophores. These modified arms are called **hectocotyli** (Figures 13.12D and 13.49B). In squids and cuttlefish the right or left fourth arm is used; in octopuses it is the right third arm. In *Nautilus* four small arms form a conical organ, the **spadix**, that functions in sperm transfer. Hectocotylus arms have special suckers, spoonlike depressions, or superficial chambers for holding spermatophores during the transfer, which may be a brief or a very lengthy process.

Each spermatophore comprises an elongate sperm mass, a cement body, a coiled, "spring-loaded" ejaculatory organ, and a cap. The cap is pulled off as the spermatophore is removed from the Needham's sac in squids or by uptake of seawater in octopuses. Once the cap is removed, the ejaculatory organ everts, pulling the sperm mass out with it. The sperm mass adheres by means of the cement body to the seminal receptacle or mantle wall of the female, where it begins to disintegrate and liberate sperm for up to two days.

Precopulatory rituals in coleoid cephalopods usually involve striking changes in coloration, as the male tries to attract the female (and discourage other males in the area). Male squids often seize their female partner with the tentacles, and the two swim head-to-head through the water. Eventually the male hectocotylus grabs a spermatophore and inserts it into the mantle chamber of his partner, near or in the oviducal opening. Mating in octopuses can be a savage affair. The exuberance of the copulatory embrace may result in the couple tearing at each other with their sharp beaks, or even strangulation of one partner by the other as the former's arms wrap around the mantle cavity of the latter, cutting off ventilation. In many octopuses (e.g., *Argonauta, Philonexis*) the tip of the hectocotylus arm may break off and remain in the female's mantle chamber.[7]

As the eggs pass through the oviduct, they are covered with a capsule-like membrane produced by the oviducal gland. Once in the mantle cavity, various kinds of **nidamental glands** may provide additional layers or coatings on the eggs. In the squid *Loligo*, which migrates to shallow water to breed, the nidamental glands coat the eggs within an oblong gelatinous mass, each containing about 100 eggs. The female holds these egg cases in her arms and fertilizes them with sperm ejected from her seminal receptacle. The egg masses harden as they react with seawater and are then attached to the substratum. The adults die after mating and egg laying. Cuttlefish deposit single eggs and attach them to seaweed or other substrata. Many open-ocean pelagic coleoids have floating eggs, and the young develop entirely in the plankton. Octopuses usually lay grapelike egg clusters in dens in rocky areas, and many species care for the developing embryos by protecting them, and aerating and cleaning them by flushing the egg mass with jets of water. Octopuses and squids tend to grow quickly to maturity, reproduce, and then die, usually within a year or two. The pearly nautilus, however, is long-lived (perhaps to 25–30 years), slow growing, and able to reproduce for many years after maturity.

One of the most astonishing reproductive behaviors among invertebrates occurs in members of the pelagic octopod genus *Argonauta*, known as the paper nautiluses. Female argonauts use two specialized arms to secrete and sculpt a beautiful, coiled, calcareous shell into which eggs are deposited (Figure 13.17B). The thin-walled, delicate shell is carried by the female and serves as her temporary home and as a brood chamber for the embryos. The much smaller male often cohabits the shell with the female.

[7]The detached arm was mistakenly first described as a parasitic worm and given the genus name *Hectocotylus* (hence the origin of the term).

Development

Development in molluscs is similar in many fundamental ways to that of the other spiralian protostomes. Most molluscs undergo typical spiral cleavage, with the mouth and stomodeum developing from the blastopore, and the anus forming as a new opening on the gastrula wall (protostomous). Cell fates are also typically spiralian, including a 4d mesentoblast.

By the end of the 64-cell stage, the distinctive molluscan cross is formed by a group of apical micromeres ($1a^{12}$–$1d^{12}$ cells and their descendants, with cells $1a^{112}$–$1d^{112}$ forming the angle between the arms of the cross) (Figure 13.50). This configuration of blastomeres appears to be unique to the Mollusca. Beyond these generalities, a great deal of variation occurs in molluscan cleavage. As detailed studies are conducted on more and more species, the phylogenetic implications of these variations are being evaluated.

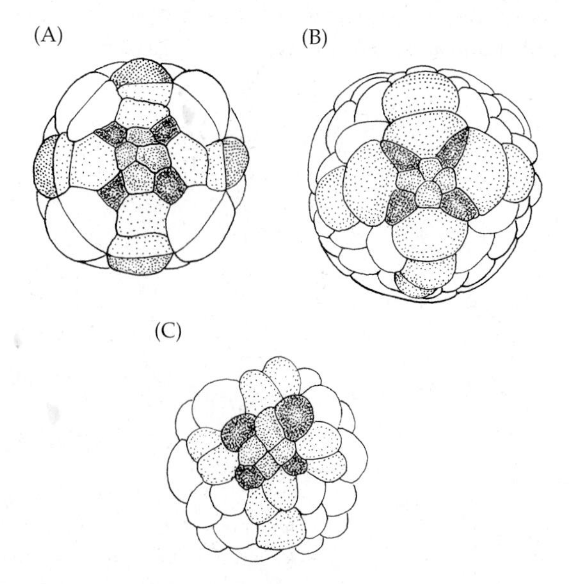

Figure 13.50 The "molluscan cross" of developing embryos. (A) Gastropoda (*Lymnaea*). (B) Polyplacophora (*Stenoplax*). (C) Aplacophora (*Epimenia*).

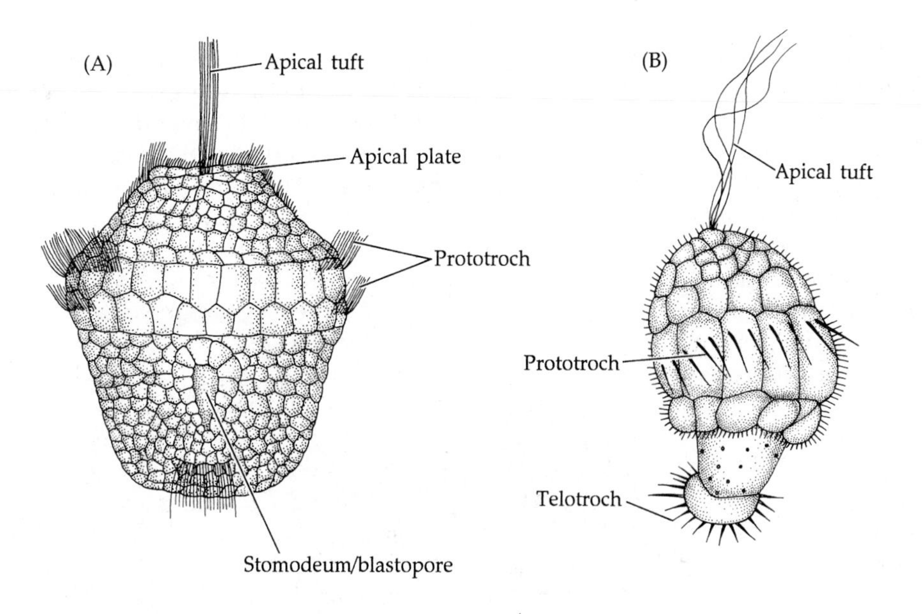

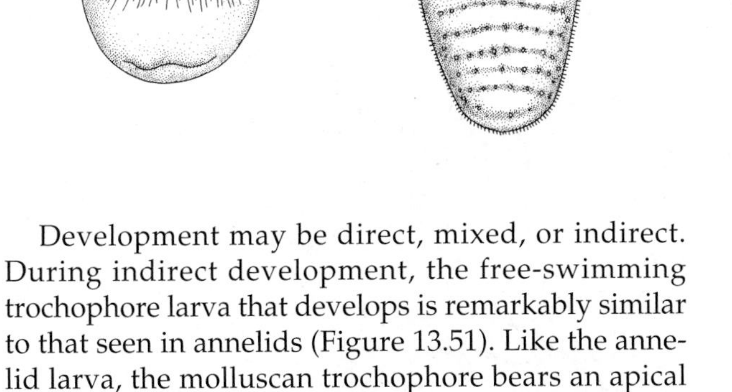

Figure 13.51 Molluscan trochophore larvae. (A) Generalized molluscan trochophore larva. (B) Trochophore of a solengaster aplacophoran. (C) Metamorphosis of a polyplacophoran from trochophore to juvenile.

Development may be direct, mixed, or indirect. During indirect development, the free-swimming trochophore larva that develops is remarkably similar to that seen in annelids (Figure 13.51). Like the annelid larva, the molluscan trochophore bears an apical sensory plate with a tuft of cilia and a girdle of ciliated cells—the prototroch—just anterior to the mouth.

In some free-spawning molluscs (e.g., chitons and Caudofoveata), the trochophore is the only larval stage, and it metamorphoses directly into the juvenile (Figure 13.51C). Solenogasters usually have a so-called "test cell larva," where a bell-shaped larval test encloses parts of the developing animal. But in other groups (e.g., gastropods and bivalves), the trochophore is followed by a uniquely molluscan larval stage called a veliger (Figure 13.52). The **veliger larva** may possess a foot, shell, operculum, and other adult-like structures. The most characteristic feature of the veliger larva is the swimming organ, or **velum**, which consists of two large ciliated lobes developed from the trochophore's prototroch. In some species the velum is also a feeding organ and is subdivided into four, five, or even six separate lobes (Figure 13.52C). Feeding (planktotrophic) veligers capture particulate food between opposed prototrochal and metatrochal bands of cilia on the edge of the velum, others are non-feeding (lecithotrophic) and live on yolk reserves. Eventually eyes and tentacles appear, and the veliger transforms into a juvenile, settles to the bottom, and assumes an adult existence.

Like gastropods, some bivalves have long-lived planktotrophic veligers, whereas others have short-lived lecithotrophic veligers. Many widely distributed species have very long larval lives that allow dispersal over great distances. A few bivalves have mixed development and brood the developing embryos in the suprabranchial cavity through the trochophore period; then the embryos are released as veliger larvae. Some marine and freshwater clams have direct development, as for example in the freshwater family Sphaeriidae where embryos are brooded between the gill lamellae and juveniles shed into the water when development is completed. Several unrelated marine groups have independently evolved a similar brooding behavior (e.g., *Arca vivipara*, some Carditidae, etc.).

In the freshwater mussels (Unionida), the embryos are also brooded between the gill lamellae, where they develop into veligers highly modified for a parasitic life on fishes, thereby facilitating dispersal. These parasitic larvae are called **glochidia** (Figure 13.52E). They attach to the skin or gills of the host fish by a sticky mucus, hooks, or other attachment devices. Most glochidia lack a gut and absorb nutrients from the host by means of special phagocytic mantle cells. The host tissue often forms a cyst around the glochidium. Eventually the larva matures, breaks out of the cyst, drops to the bottom, and assumes its adult life.

Among the gastropods, only the patellogastropods and vetigastropods that rely on external fertilization have retained a free-swimming trochophore larva. All other gastropods suppress the trochophore or pass through it quickly before hatching. In many groups embryos hatch as veligers (e.g., many neritimorphs, caenogastropods and heterobranchs). As with bivalves, some of these gastropods have planktotrophic veligers that may have brief or extended (to several months) free-swimming lives. Others have lecithotrophic veligers that remain planktonic only for short periods

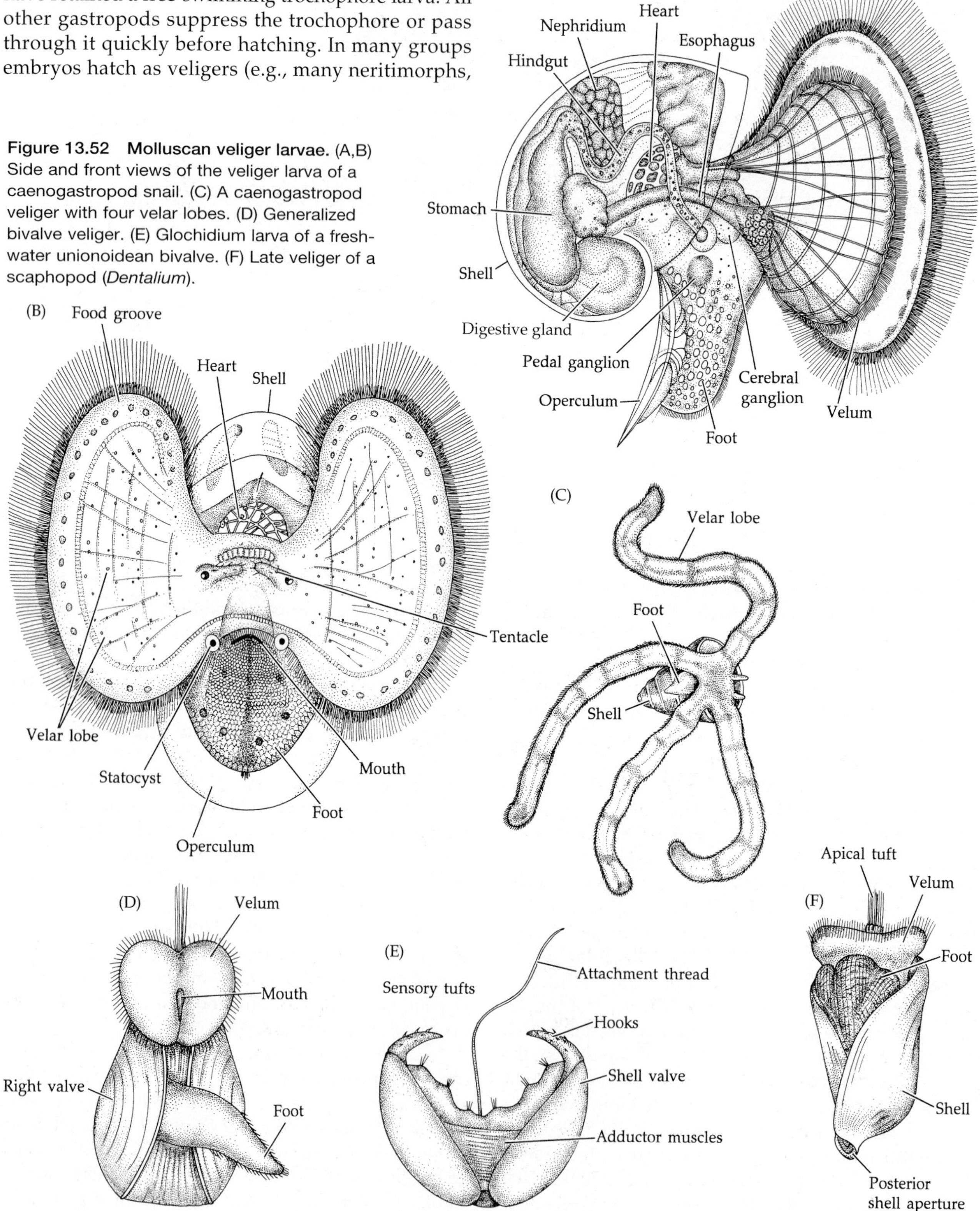

Figure 13.52 Molluscan veliger larvae. (A,B) Side and front views of the veliger larva of a caenogastropod snail. (C) A caenogastropod veliger with four velar lobes. (D) Generalized bivalve veliger. (E) Glochidium larva of a freshwater unionoidean bivalve. (F) Late veliger of a scaphopod (*Dentalium*).

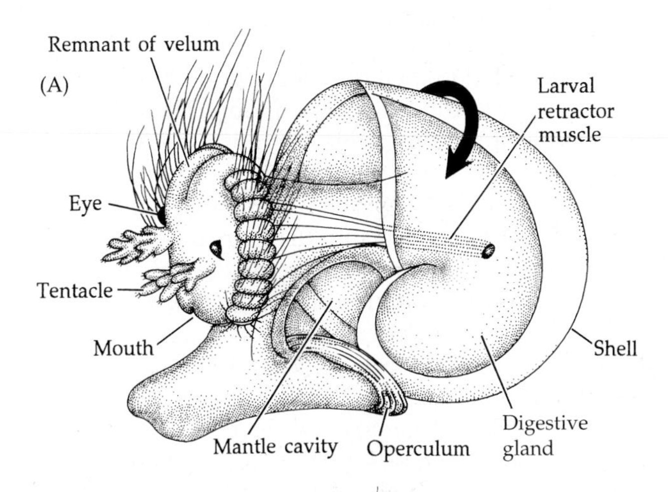

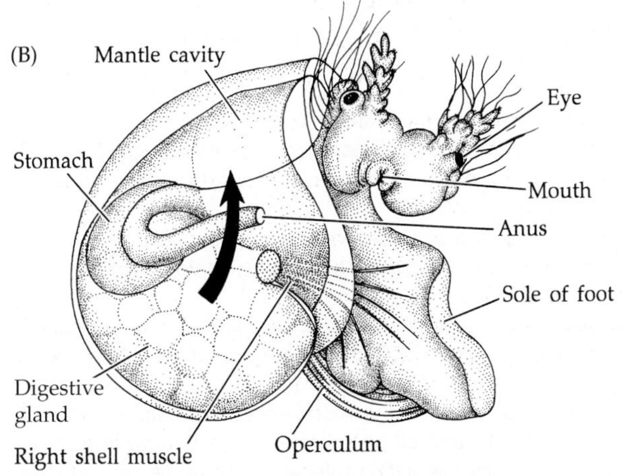

Figure 13.53 Settled larva of the abalone (*Haliotis*) undergoing torsion. (A) Left-side view after about 90° of torsion, with mantle cavity on the right side. (B) Torsion continues as the mantle cavity and its associated structures twist forward over the head.

(sometime less than a week). Planktotrophic veligers feed by use of the velar cilia, whose beating drives the animal forward and draws minute planktonic food particles into contact with the shorter cilia of a food groove. Once in the food groove, the particles are trapped in mucus and carried along ciliary tracts to the mouth.

Almost all pulmonates and many caenogastropods have direct development, and the veliger stage is passed in the egg case, or capsule. Upon hatching, tiny snails crawl out of the capsule into their adult habitat. In some neogastropods (e.g., certain species of *Nucella*), the encapsulated embryos cannibalize on their siblings, a phenomenon called **adelphophagy**; consequently, only one or two juveniles eventually emerge from each capsule.

It is usually during the veliger stage that gastropods undergo torsion (see previous discussion of torsion), when the shell and visceral mass twist relative to the head and foot (Figures 13.18 and 13.53). As we have seen, this phenomenon is still not fully understood, but it has played a major role in gastropod evolution.

Cephalopods produce large, yolky, telolecithal eggs. Development is always direct, the larval stages having been lost entirely during evolution of the yolk-laden embryo that develops within the egg case. Early cleavage is meroblastic and eventually produces a cap of cells (a discoblastula) at the animal pole. The embryo grows in such a way that the mouth opens to the yolk sac, and the yolk is directly "consumed" by the developing animal (Figure 13.54).

Molluscan Evolution and Phylogeny

The phylogenetic details of molluscan evolution have yet to be thoroughly elucidated. The phylum is highly diverse, and many named taxa below the class level are known to be polyphyletic or paraphyletic. The existence of a good fossil record (primarily of shells) has been both a blessing and a curse as efforts to trace the evolutionary history of molluscs have often been frustrated by the limited and sometimes confusing dataset provided by molluscan shells.

Until fairly recently, the idea of a "hypothetical ancestral mollusc" (affectionately known as HAM) was popular, the nature of which derived largely from early work of the eminent British biologist and "Darwin's Bulldog" T. H. Huxley. Detailed and sometimes highly imaginative descriptions of this hypothetical ancestral mollusc were proposed by various workers, even including speculations on its physiology, ecology and behavior (see Lindberg and Ghiselin 2003). The usefulness of HAM in molluscan evolutionary studies was questioned as zoology moved into an era of explicit phylogenetic analysis (i.e., cladistics). Thus, most workers now avoid the pitfalls of *a priori* construction of a

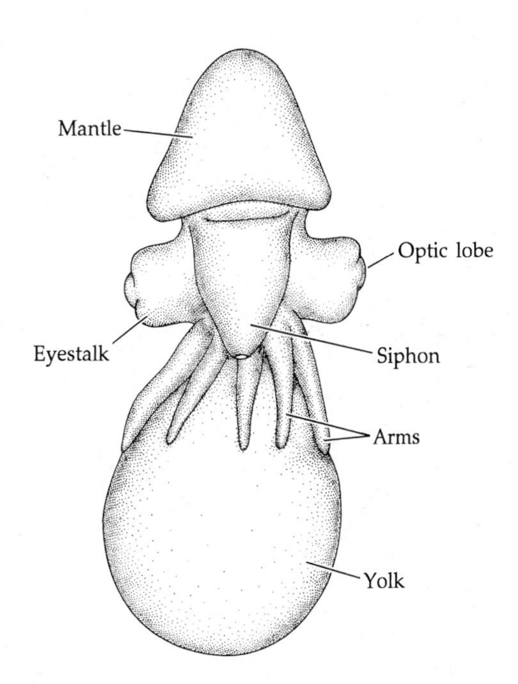

Figure 13.54 Juvenile coleoid cephalopod attached to and consuming its sac of yolk.

hypothetical ancestor, and instead analyze the evolutionary history of molluscs by phylogenetic inference. Although morphological analyses of molluscan relationships have differed in some details, the phylogenetic relationships resulting from this work have been similar. In contrast, more recent molecular analyses of molluscan relationships have produced several alternative trees depending on the molecular data type and analytical methods. Based on these recent phylogenetic studies, the probable molluscan common ancestor was small (~5 mm long), with a dorsal shell or cuticle, and a flattened ventral surface on which the animal moved by ciliary gliding. Our phylogeny (Figure 13.55) summarizes some current thinking on molluscan evolution. The characters used to construct the cladogram are enumerated in the figure legend and briefly summarized in the following discussion. The nodes on the cladogram have been lettered to facilitate the discussion.

Exactly where the molluscs arose within the spiralian clade, and their kinship to other spiralian phyla, are still matters of much debate. While some workers treat them as descendant from a segmented ancestor, most do not. We support the idea that molluscs arose from a schizocoelomate, nonsegmented precursor. Recent molecular studies (see References: Molluscan Evolution and Phylogeny) have suggested different sister-group relationships for the molluscs, though commonly the annelids are nested within these putative sister clades. However, as with many of the spiralian

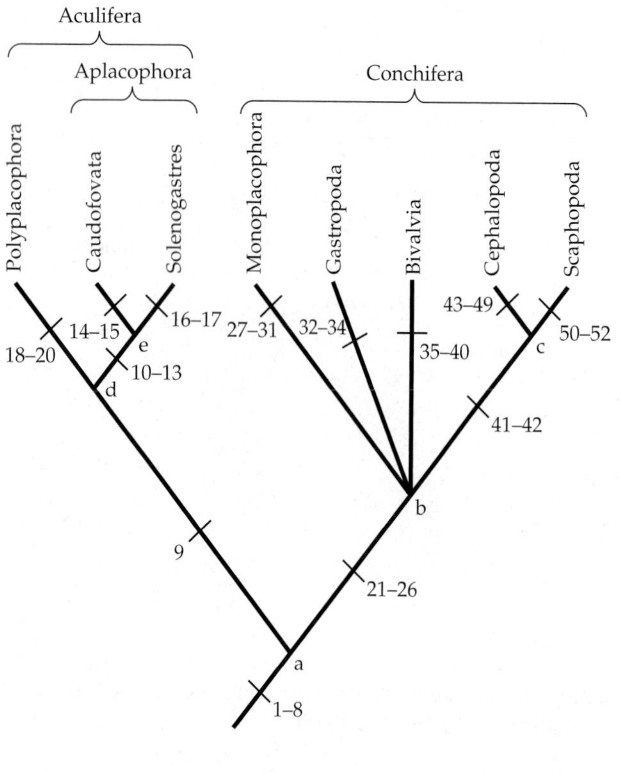

Figure 13.55 A cladogram depicting a conservative view of the phylogeny of the Mollusca based on current hypotheses (see Sigwart and Lindberg 2015 for alternative molluscan phylogenies). The numbers on the cladogram indicate suites of synapomorphies defining each hypothesized line or clade.
Synapomorphies of the phylum Mollusca defining node *a*: (1) reduction of the coelom and development of an open hemocoelic circulatory system; (2) dorsal body wall forms a mantle; (3) extracellular production of calcareous sclerites (and/or shell) by mantle shell glands; (4) ventral body wall muscles develop as muscular foot (or foot precursor); (5) radula; (6) chambered heart with separate atria and ventricle; (7) increase in gut complexity, with large digestive glands; (8) ctenidia.
Synapomorphies of the Aculifera (Aplacophora + Polyplacophora) defining node *d*: (9) sclerites.
Synapomorphies of the Aplacophora (Caudofoveata + Solenogastres) defining node *e*: (10) vermiform body; (11) foot reduced; (12) gonads empty into pericardial cavity, exiting to mantle cavity via U-shaped gametoducts; (13) without nephridia.
Synapomorphies of Caudofoveata: (14) calcareous sclerites of the body wall form imbricating scales; (15) complete loss of foot.
Synapomorphies of Solenogastres: (16) posterior end of reproductive system with copulatory spicules; (17) loss of ctenidia.
Synapomorphies of Polyplacophora: (18) shell with 8 plates (and with 8 shell gland regions), articulamentum layer, and aesthetes; (19) multiple ctenidia; (20) expanded and highly cuticularized mantle girdle that "fuses" with shell plates.
Synapomorphies of the Conchifera defining node *b*: (21) presence of a well-defined single shell gland region and larval shell (protoconch); (22) shell univalve (of a single piece; note: the bivalve shell is derived from the univalve condition); (23) shell of basically three-layers (periostracum, prismatic layer, lamellar or crossed layer); (24) mantle margin of three parallel folds, each specialized for specific functions; (25) statocysts; (26) viscera concentrated dorsally.
Synapomorphies of Monoplacophora: (27) 3–6 pairs ctenidia; (28) 3–7 pairs nephridia; (29) 8 pairs pedal retractor muscles; (30) 2 pairs gonads; (31) 2 pairs heart atria.
Synapomorphies of Gastropoda: (32) torsion; (33) cephalic tentacles; (34) operculum.
Synapomorphies of Bivalvia: (35) bivalve shell and its associated mantle and (in autobranch bivalves) ctenidial modifications; (36) loss of radula; (37) byssus (autobranchs); (38) lateral compression of body; (39) adductor muscles; (40) ligament.
Synapomorphies of the cephalopod–scaphopod line defining node c: (41) ano-pedal flexure; (42) new neuroanatomical features, including cerebral ganglia fusion and position.
Synapomorphies of Cephalopoda: (43) expansion of the coelom and closure of the circulatory system; (44) septate shell; (45) ink sac (in coleoids); (46) siphuncle; (47) beak-like jaws; (48) foot modified as prehensile arms/tentacles and funnel (=siphon); (49) development of large brain.
Synapomorphies of Scaphopoda: (50) tusk-shaped, shell open at both ends; (51) loss of heart and ctenidia; (52) captacula.

phyla, identification of the mollusc sister group remains a work in progress. Molluscs are clearly allied with the other spiralian protostomes (Platyhelminthes, Nemertea, Annelida), which are characterized by developmental features such as spiral cleavage, 4d mesentoblast, and trochophore-like larvae. But precisely where in the spiralian lineage they arose remains problematic.

The major steps in the evolution of what we generally think of as a "typical" mollusc—that is, a shelled mollusc, also remains controversial. Previous scenarios have often argued that this step took place after the origin of the aplacophorans, perhaps as molluscs adapted to active epibenthic lifestyles. These steps centered largely on the elaboration of the mantle and mantle cavity, the refinement of the ventral surface as a well-developed muscular foot, and the evolution of a consolidated dorsal shell gland and solid shell(s) in place of independent calcareous sclerites.

The description of a solenogaster larva by Pruvot in 1890, in which the dorsal surface was said to bear seven transverse bands of sclerites (described as "composite plates," reminiscent of chitons), led some workers to postulate that aplacophorans and polyplacophorans might be sister-groups; a relationships confirmed by several recent phylogenomic studies (e.g., Kocot et al. 2011). However, the discovery of a possible aplacophoran fossil having seven dorsal shell plates from Silurian deposits in England (*Acaenoplax*), as well as "footless" chitons (*Kulindroplax* and *Phthipodochiton*), have further confused the polarity of the aplacophoran-chiton character transformation. Adding to the confusion, there are fundamental differences between the shells of polyplacophorans and those of all other molluscs, an observation suggesting that the chitons and aplacophorans may stand alone as a unique radiation off the early molluscan line. Three hypotheses have been offered to explain this "shell problem" in molluscan evolution: (1) The multiplate shell may have been ancestral, the single-shell condition having evolved by coalescence of plates. (2) The single shell may have been ancestral, and the multiplate forms arose by subdivision of the single shell. (3) The single-shell and multishell designs arose independently from a shell-less ancestor, perhaps by way of sclerite consolidation. The presence of eight pairs of pedal retractor muscles in both polyplacophorans and monoplacophorans has been taken as evidence in favor of the first explanation. Acceptance of the first hypothesis suggests that the ancestor at node *a* in the cladogram in Figure 13.55 was a multivalved chiton-like creature. Acceptance of the second hypothesis implies that the ancestor at node *a* was a univalved, monoplacophoran-like ancestor. The third hypothesis postulates that the ancestor at node *a* lacked a solid shell altogether.

The primitive mantle and foot arrangement was probably somewhat similar to that in living polyplacophorans or monoplacophorans—that is, a large flattened sole was surrounded by a mantle groove. Because of their small size, specialized respiratory structures were probably not required in the first molluscs and gas exchange was through the dorsal epidermis. However, with the origin of the cuticle-covered mantle or dorsal shell covering this surface, posterior, specialized respiratory structures (ctenidia) originated and became associated with excretory and reproductive pores in a posterior mantle cavity. This arrangement would have been modified at least twice; in both the polyplacophorans and monoplacophorans the mantle cavity was lost as it became continuous with the expanded mantle groove alongside the foot and the ctenidia multiplied and extended anteriorly in the mantle groove. Secondary modifications of the shape of the foot and other features in bivalves and scaphopods allowed most of these animals to exploit infaunal life in soft sediments, and both of these taxa are highly adapted to sediment burrowing. However, these modifications are clearly convergent and scaphopods share other characters, including ano-pedal flexure, with cephalopods. Gastropods also undergo ano-pedal flexure, but this could be convergent according to some molecular studies. Scaphopods are also the last class of molluscs to appear in the fossil record (about 450 Ma, Late Ordovician).

Monoplacophorans share the character of a single (univalve) shell with other molluscs (other than bivalves and chitons). They also share a similar shell structure and a host of other features. The only synapomorphies defining the monoplacophorans seem to be their repetitive organs (multiple gills, nephridia, pedal muscles, gonads, and heart atria). The question of whether this multiplicity arose uniquely in the monoplacophorans or represents a symplesiomorphic retention of ancestral features from some unknown metameric ancestor (below node *a* on the cladogram) has not been resolved (see discussion below) and will likely require developmental studies on monoplacophorans to finally settle the question.

The bivalve line in the cladogram is defined by the presence of 2 shell valves, adductor muscles, reduction of the head region, decentralization of the nervous system and associated reduction or loss of certain sensory structures, and expansion and deepening of the mantle cavity.

Cephalopods are highly specialized molluscs and possess a number of complex synapomorphies. Primitive shelled cephalopods are represented today by only six species of *Nautilus*, although thousands of fossil species of shelled nautiloid cephalopods have been described. This highly successful molluscan class probably arose about 450 million years ago. The nautiloids underwent a series of radiations during the Paleozoic, but were largely replaced by the ammonoids after the Devonian period (325 million years ago). The ammonoids, in turn, became extinct around the Cretaceous–Tertiary boundary (65 million years ago). The origin of the coleoid cephalopods (octopuses, squids, and cuttlefish) is obscure, possibly dating back to the Devonian. They diversified mainly in the Mesozoic and became a highly successful group by exploiting a very new lifestyle, as we have seen.

The issue of ancestral metamerism in molluscs has been debated since the discovery of the first living monoplacophoran (*Neopilina galatheae*) in 1952. However, monoplacophorans are not the only molluscs to express serial replication or to have repeated organs reminiscent of metamerism (or "pseudometamerism," as some prefer to call it). Polyplacophorans have many serially repeated gills in the mantle groove and also typically possess eight pairs of pedal retractor muscles and eight shell plates. The two pairs of heart atria, nephridia, and ctenidia in *Nautilus* (and two pairs of retractor muscles in some fossil forms) have also been regarded by some workers as primitive metameric features.

The question is whether or not organ repetition in these molluscs represents vestiges of a true, or fundamental, metamerism in the phylum. If so, they represent remnants of an ancestral metameric body plan and may indicate a close relationship to annelids. On the other hand, organ repetition in certain molluscan groups may be the result of independent convergent evolution and not an ancestral molluscan attribute at all. And, nothing like the teloblastic metameric development of annelids is seen in molluscs. The genetic/evolutionary potential for serial repetition of organs is not uncommon and occurs in other non-annelid bilaterian phyla as well, e.g., Platyhelminthes, Nemertea, and Chordata.

The origin of molluscs themselves remains enigmatic. The excellent fossil record of this phylum extends back some 500 million years and suggests that the origin of the Mollusca probably lies in the Precambrian. Indeed, the late Precambrian fossil *Kimberella quadrata*, once thought to be a cnidarian, has been argued to have molluscan features, including perhaps a shell and muscular foot. However, recent examination of hundreds of specimens now suggests *Kimberella* more likely belongs to an extinct spiralian group.

The various ideas on the origin of the molluscs fall into three categories: molluscs were derived from (1) a free-living flatworm (Platyhelminthes) ancestor, (2) a nonsegmented coelomate protostome ancestor, or (3) a segmented ancestor, perhaps even a common ancestor with the annelids. The first hypothesis, known as the "turbellarian theory," was originally based upon the supposed homology and similarity in mode of locomotion between molluscs and flatworms by means of a "ventral mucociliary gliding surface." It suggests that either the molluscs were the first coelomate protostomes, or that they share a common ancestor with the first coelomates. However, most contemporary workers argue that the large pericardial spaces present in primitive molluscs (e.g., aplacophorans, monoplacophorans, polyplacophorans) point to a coelomate rather than an acoelomate (platyhelminth) ancestry, and the turbellarian theory enjoys little favor today.

The second theory advanced by Scheltema in the 1990s suggested that sipunculans (now placed within the Annelida) and molluscs might be sister groups, sharing, among other things, the unique "molluscan cross" during development. However, the idea that sipunculan embryogeny includes a molluscan cross blastomere configuration is no longer strongly supported. Scheltema also suggested that certain features of the sipunculan pelagosphera larva may be homologous to some molluscan structures. Indeed, molluscs share most of their typical spiralian features with the sipunculans, as well as the echiurids and other annelids (e.g., spiral cleavage, schizocoely, trochophore larvae). This leads to the third hypothesis, that molluscs and annelids are closely related and that molluscs might have arose from a segmented coelomate ancestor. Perhaps the three most striking synapomorphies distinguishing modern molluscs from annelids and most other spiralians are: the reduction of the coelom and the concomitant conversion of the closed circulatory system to an open hemocoelic one; the elaboration of the body wall into a mantle capable of secreting calcareous sclerites or shell(s); and, the unique molluscan radula. Identifying the sister group to the Mollusca remains a work in progress.

Selected References

The field of malacology is so large, has had such a long history, and has so embraced the mixed blessings of contributions from amateur shell collectors, that dealing with the literature is a daunting task. Many molluscs are of commercial importance (e.g., *Haliotis*, *Mytilus*, *Loligo*) and for these groups hundreds of studies appear annually; others are important laboratory/experimental organisms (e.g., *Loligo*, *Octopus*, *Aplysia*) and many papers are also published on these groups. New taxonomic monographs on various groups or geographical regions also appear each year, as do countless shell guides and coffee-table books. Distilling all of this into a small set of key references useful for entry into the professional literature is difficult; the list below is our attempt to do so.

General References

Beesley, P. L., G. J. B. Ross and A. Wells (eds.). 1998. Mollusca: *The Southern Synthesis. Fauna of Australia. Vol. 5.* CSIRO Publishing, Melbourne, Australia. [Perhaps the best general review of molluscan biology and systematics available. An extraordinary 2-volume text with chapters by leading specialists.]

Cheng, T. C. 1967. Marine mollusks as hosts for symbioses. Adv. Mar. Biol. 5: 1–424. [An exhaustive summary.]

Falini, G., S. Albeck, S. Weiner and A. Addadi. 1996. Control of aragonite or calcite polymorphism by mollusc shell macromolecules. Science 271: 67–69.

Giese, A. C. and J. S. Pearse (eds.). 1977. *Reproduction of Marine Invertebrates, Vol. 4, Molluscs: Gastropods and Cephalopods.* Academic Press, New York.

Giese, A. C. and J. S. Pearse (eds.). 1979. *Reproduction of Marine Invertebrates, Vol. 5, Molluscs: Pelecypods and Lesser Classes.* Academic Press, New York.

Harrison, F. W. and A. J. Kohn (eds). 1994. *Microscopic Anatomy of Invertebrates. Vols. 5 and 6. Mollusca.* Wiley-Liss, New York.

Heller, J. 1990. Longevity in molluscs. Malacologia 31(2): 259–295.

Hochachka, P. W. (ed.). 1983. *The Mollusca, Vol. 1, Metabolic Biochemistry and Molecular Biomechanics; Vol. 2, Environmental Biochemistry and Physiology.* Academic Press, New York.

Hyman, L. H. 1967. *The Invertebrates, Vol. 6, Mollusca I. Aplacophora, Polyplacophora, Monoplacophora, Gastropoda.* McGraw-Hill, New York. [Dated, but still one of the best general surveys of molluscan anatomy in English.]

Keen, A. M. and E. Coan. 1974. *Marine Molluscan Genera of Western North America: An Illustrated Key.* 2nd Ed. Stanford University Press, Stanford, California. [Technical keys.]

Kniprath, E. 1981. Ontogeny of the molluscan shell field. Zool. Scripta 10: 61–79.

Lydeard, C. and D. R. Lindberg (eds.). (2003). *Molecular systematics and phylogeography of mollusks.* Smithsonian Series in Comparative Evolutionary Biology.

Moore, R. C. (ed.). 1957–71. *Treatise on Invertebrate Paleontology. Mollusca, Parts 1–6 (Vols. I–N).* University of Kansas Press and Geological Society of America, Lawrence, KA.

Raven, C. P. 1958. *Morphogenesis: The Analysis of Molluscan Development.* Pergamon Press, New York.

Spanier, E. (ed.). 1987. *The Royal Purple and Biblical Blue. Argaman and Tekhelet. The Study of Chief Rabbi Dr. Isaac Herzog on the Dye Industries in Ancient Israel and Recent Scientific Contributions.* Keter Publishing House, Jerusalem.

Wilbur, K. M. (gen. ed.). 1983–1988. *The Mollusca. Vols. 1–12.* Academic Press, New York.

Caudofoveata and Solenogastres (the Aplacophorans)

García-Álvarez, O. and L. v. Salvini-Plawen. 2007. Species and diagnosis of the families and genera of Solenogastres (Mollusca). Iberus 25: 73–143.

Okusu, A. 2002. Embryogenesis and development of *Epimenia babai* (Mollusca: Neomeniomorpha). Biol. Bull. 203: 87–103.

Salvini-Plawen, L., v. (1980). A reconsideration of systematics in the Mollusca (phylogeny and higher classification). *Malacologia* 19: 249–278.

Scheltema, A. H. 1981. Comparative morphology of the radulae and the alimentary tracts in the Aplacophora. Malacologia 20: 361–383.

Scheltema, A. H. and C. Schander. 2000. Discrimination and phylogeny of solenogaster species through the morphology of hard parts (Mollusca, Aplacophora, Neomeniomorpha). Biol. Bull. 198: 121–151.

Scheltema, A. H., M. Tscherkassky and A. M. Kuzirian. 1994. Aplacophora. Pp. 13–54 in F. W. Harrison and E. E. Ruppert (eds.), *Microscopic Anatomy of Invertebrates. Vol. 5, Mollusca I.* Wiley-Liss, New York.

Scherholz, M., E. Redl, T. Wollesen, C. Todt and A. Wanninger. 2015. From complex to simple: Myogenesis in an aplacophoran mollusk reveals key traits in aculiferan evolution. BMC Evol. Biol. 15: 201.

Todt, C. 2013. Aplacophoran mollusks—still obscure and difficult? Am. Malacol. Bull. 31: 181–187.

Todt, C. and L. v. Salvini-Plawen. 2005. The digestive tract of *Helicoradomenia* (Solenogastres, Mollusca), aplacophoran molluscs from the hydrothermal vents of the East Pacific Rise. Inver. Biol. 124: 230–253.

Todt, C. and A. Wanninger. 2010. Of tests, trochs, shells, and spicules: Development of the basal mollusk *Wirenia argentea* (Neomeniomorpha) and its bearing on the evolution of trochozoan larval key features. Front. Zool. 7: 6.

Monoplacophora

Clarke, A. H. and R. J. Menzies. 1959. *Neopilina* (*Vema*) *ewingi*, a second living species of the Paleozoic class Monoplacophora. Science 129: 1026–1027.

Haszprunar, G. and Schaefer, K. (1997). Monoplacophora. Pp. 415–457 in F. W. Harrison and A. Kohn, *Mollusca 2. Microscopic anatomy of invertebrates, 6B.* Wiley-Liss, New York.

Lemche, H. 1957. A new living deep-sea mollusc of the Cambro-Devonian class Monoplacophora. Nature 179: 413–416. [Report of the first discovery of living monoplacophorans.]

Lemche, H. and K. G. Wingstrand. 1959. The anatomy of *Neopilina galatheae* Lemche, 1957. Galathea Rpt. 3: 9–71.

Lindberg, D. R. (2009). Monoplacophorans and the origin and relationships of mollusks. Evol. Educ. Outreach 2: 191–203.

Menzies, R. J. and W. Layton. 1962. A new species of monoplacophoran mollusc, *Neopilina* (*Neopilina*) *veleronis* from the slope of the Cedros Trench, Mexico. Ann. Mag. Nat. Hist. (13)5: 401–406.

Warén, A. 1988. *Neopilina goesi*, a new Caribbean monoplacophoran mollusk dredged in 1869. Proc. Biol. Soc. Wash. 101: 676–681.

Wilson, N. G., D. Huang, M. C. Goldstein, H. Cha, G. Giribet and G. W. Rouse. 2009. Field collection of *Laevipilina hyalina* McLean, 1979 from southern California, the most accessible living monoplacophoran. J. Mollus. Stud. 75: 195–197.

Wingstrand, K. G. 1985. On the anatomy and relationships of recent Monoplacophora. Galathea Rpt. 16: 7–94.

Polyplacophora

Boyle, P. R. 1977. The physiology and behavior of chitons. Ann. Rev. Oceanogr. Mar. Biol. 15: 461–509.

Eernisse, D. J. 1988. Reproductive patterns in six species of *Lepidochitona* (Mollusca: Polyplacophora) from the Pacific Coast of North America. Biol. Bull. 174(3): 287–302.

Fisher, v.-F. P. 1978. Photoreceptor cells in chiton esthetes. Spixiana 1: 209–213.

Kass, P. and R. A. VanBelle. 1998. *Catalogue of Living Chitons (Mollusca; Polyplacophora).* 2nd Ed. W. Backhuys, Rotterdam.

Nesson, M. H. and H. A. Lowenstam. 1985. Biomineralization processes of the radula teeth of chitons. Pp. 333–363 in J. L. Kirschvink et al. (eds.), *Magnetite Biomineralization and Magnetoreception in Organisms.* Plenum, New York.

Sigwart, J. D. (2008). Gross anatomy and positional homology of gills, gonopores, and nephridiopores in "basal" living chitons (Polyplacophora: Lepidopleurina). Am. Malacol. Bull. 25: 43–49.

Slieker, F. J. A. 2000. *Chitons of the World. An Illustrated Synopsis of Recent Polyplacophora.* L'Informatore Piceno, Ancona, Italy.

Smith, A. G. 1966. The larval development of chitons (Amphineura). Proc. Calif. Acad. Sci. 32: 433–446.

Sutton, M. D. and J. D. Sigwart, J. D. 2012. A chiton without a foot. Palaeontology 55: 401–411.

Gastropoda

Barker, G. M. (ed.). 2001. *The biology of terrestrial molluscs.* CABI Publishing, Wallingford, UK.

Barnhart, M. C. 1986. Respiratory gas tensions and gas exchange in active and dormant land snails, *Otala lactea*. Physiol. Zool. 59: 733–745.

Bouchet, P. 1989. A marginellid gastropod parasitizes sleeping fishes. Bull. Mar. Sci. 45(1): 76–84.

Brace, R. C. 1977. Anatomical changes in nervous and vascular systems during the transition from prosobranch to opisthobranch organization. Trans. Zool. Soc. London 34: 1–26.

Branch, G. M. 1981. The biology of limpets: Physical factors, energy flow and ecological interactions. Ann. Rev. Oceanogr. Mar. Biol. 19: 235 380.

Brunkhorst, D. J. 1991. Do phyllidiid nudibranchs demonstrate behaviour consistent with their apparent warning colorations? Some field observations. J. Moll. Stud. 57: 481–489.

Carlton, J. T., G. J. Vermeij, D. R. Lindberg, D. A. Carlton and E. C. Dudley. 1991. The first historical extinction of a marine invertebrate in an ocean basin: The demise of the eelgrass limpet *Lottia alveus*. Biol. Bull. 180: 72–80.

Chase, R. 2002. *Behavior and its neural control in gastropod molluscs*. Oxford University Press, Oxford.

Collier, J. R. 1997. Gastropods. The snails. Pp. 189–218 in S. F. Gilbert and A. M. Raunio, *Embryology. Constructing the Organism*. Sinauer Associates, Sunderland, MA.

Conklin, E. J. and R. N. Mariscal. 1977. Feeding behavior, ceras structure, and nematocyst storage in the aeolid nudibranch, *Spurilla neapolitana*. Bull. Mar. Sci. 27: 658–667.

Connor, V. M. 1986. The use of mucous trails by intertidal limpets to enhance food resources. Biol. Bull. 171: 548–564.

Cook, S. B. 1971. A study in homing behavior in the limpet *Siphonaria alternata*. Biol. Bull. 141: 449–457.

Croll, R. P. 1983. Gastropod chemoreception. Biol. Rev. 58: 293–319.

Fretter, V. 1967. The prosobranch veliger. Proc. Malac. Soc. London 37: 357–366.

Fretter, V. and M. A. Graham. 1962. *British Prosobranch Molluscs. Their Functional Anatomy and Ecology*. Ray Society, London.

Fretter, V. and J. Peake (eds.). 1975, 1978. *Pulmonates, Vol. 1, Functional Anatomy and Physiology; Vol. 2A, Systematics, Evolution and Ecology*. Academic Press, New York.

Fursich, F. T. and D. Jablonski. 1984. Late Triassic naticid drill holes: Carnivorous gastropods gain a major adaptation but fail to radiate. Science 224: 78–80.

Gaffney, P. M. and B. McGee. 1992. Multiple paternity in *Crepidula fornicata* (Linnaeus). Veliger 35(1): 12–15.

Garstang, W. (1928). Origin and evolution of larval forms. *Nature* 122: 366.

Gilmer, R. W. and G. R. Harbison. 1986. Morphology and field behavior of pteropod molluscs: Feeding methods in the families Cavoliniidae, Limacinidae and Peraclididae (Gastropoda: Thecosomata). Mar. Biol. 91: 47–57.

Gosliner, T. M. 1994. Gastropoda: Opisthobranchia. Pp. 253–355 in F. E. Harrison and A. J. Kohn (eds.), *Microscopic Anatomy of Invertebrates*. Wiley-Liss, New York.

Gould, S. J. 1985. The consequences of being different: Sinistral coiling in *Cerion*. Evolution 39: 1364–1379.

Greenwood, P. G. and R. N. Mariscal. 1984. Immature nematocyst incorporation by the aeolid nudibranch *Spurilla neapolitana*. Mar. Biol. 80: 35–38.

Haszprunar, G. 1985. The fine morphology of the osphradial sense organs of the Mollusca. I. Gastropoda, Prosobranchia. Phil. Trans. Roy. Soc. Lond. B 307: 457–496.

Haszprunar, G. 1985. The Heterobranchia: New concept of the phylogeny of the higher Gastropoda. Z. Zool. Syst. Evolutionsforsch. 23: 15–37.

Haszprunar, G. 1988. On the origin and evolution of major gastropod groups, with special reference to the Streptoneura. J. Moll. Stud. 54: 367–441.

Haszprunar, G. 1989. Die Torsion der Gastropoda — ein biomechanischer Prozess. Z. Zool. Syst. Evolutionsforsch. 27: 1–7.

Havenhand, J. N. 1991. On the behaviour of opisthobranch larvae. J. Moll. Stud. 57: 119–131.

Hickman, C. S. 1983. Radular patterns, systematics, diversity and ecology of deep-sea limpets. Veliger 26: 7–92.

Hickman, C. S. 1984. Implications of radular tooth-row functional integration for archaeogastropod systematics. Malacologia 25(1): 143–160.

Hickman, C. S. 1992. Reproduction and development of trochean gastropods. Veliger 35: 245–272.

Judge, J. and G. Haszprunar. 2014. The anatomy of *Lepetella sierrai* (Vetigastropoda, Lepetelloidea): implications for reproduction, feeding, and symbiosis in lepetellid limpets. Invert. Biol. 133: 324–339.

Kempf, S. C. 1984. Symbiosis between the zooxanthella *Symbiodinium* (= *Gymnodinium*) *microadriaticum* (Freudenthal) and four species of nudibranchs. Biol. Bull. 166: 110–126.

Kempf, S. C. 1991. A "primitive" symbiosis between the aeolid nudibranch *Berghia verrucicornis* (A. Costa, 1867) and a zooxanthella. J. Mol. Stud. 57: 75–85.

Lindberg, D. R. and R. P. Guralnick. 2003. Phyletic patterns of early development in gastropod molluscs. Evol. Dev. 5: 494–507.

Lindberg, D. R. and W. F. Ponder. 2001. The influence of classification on the evolutionary interpretation of structure: a re-evaluation of the evolution of the pallial cavity of gastropod molluscs. Org. Divers. Evol. 1: 273–299.

Linsley, R. M. 1978. Shell formation and the evolution of gastropods. Am. Sci. 66: 432–441.

Lohmann, K. J. and A. O. D. Willows. 1987. Lunar-modulated geomagnetic orientation by a marine mollusk. Science 235: 331–334.

Marcus, E. and E. Marcus. 1967. *American Opisthobranch Mollusks. Studies in Tropical Oceanography Series, No. 6.* University of Miami Press, Coral Gables, Florida. [Although badly out of date, this monograph remains a benchmark work for the American fauna.]

Marín, A and J. Ros. 1991. Presence of intracellular zooxanthellae in Mediterranean nudibranchs. J. Moll. Stud. 57: 87–101.

McDonald, G. and J. Nybakken. 1991. A preliminary report on a worldwide review of the food of nudibranchs. J. Moll. Stud. 57: 61–63.

Miller, S. L. 1974. Adaptive design of locomotion and foot form in prosobranch gastropods. J. Exp. Mar. Biol. Ecol 14: 99–156.

Miller, S. L. 1974. The classification, taxonomic distribution, and evolution of locomotor types among prosobranch gastropods. Proc. Malacol. Soc. London 41: 233–272.

Milne-Edwards H. 1848. Note sur la classification naturelle chez Mollusques Gasteropodes. Annales des Sciences Naturalles, Series 3, 9: 102–112.

Norton, S. F. 1988. Role of the gastropod shell and operculum in inhibiting predation by fishes. Science 241: 92–94.

Olivera, B. M. et al. 1985. Peptide neurotoxins from fish-hunting cone snails. Science 230: 1338–1343.

O'Sullivan, J. B., R. R. McConnaughey and M. E. Huber. 1987. A blood-sucking snail: The cooper's nutmeg, *Cancellaria cooperi* Gabb, parasitizes the California electric ray, *Torpedo californica* Ayres. Biol. Bull. 172: 362–366.

Pawlik, J. R., M. R. Kernan, T. F. Molinski, M. K. Harper and D. J. Faulkner. 1988. Defensive chemicals of the Spanish dancer nudibranch *Hexabranchus sanguineus* and its egg ribbons: Macrolides derived from a sponge diet. J. Exp. Mar. Biol. Ecol. 119: 99–109.

Pennington, J. T. and F. Chia. 1985. Gastropod torsion: A test of Garstang's hypothesis. Biol. Bull. 169: 391–396.

Perry, D. M. 1985. Function of the shell spine in the predaceous rocky intertidal snail *Acanthina spirata* (Prosobranchia: Muricacea). Mar. Biol. 88: 51–58.

Ponder, W. F. 1973. The origin and evolution of the Neogastropoda. Malacologia 12: 295–338.

Ponder, W. F. (ed.). 1988. Prosobranch phylogeny. Malacol. Rev., Supp. 4.

Ponder, W. F. and D. R. Lindberg. 1997. Towards a phylogeny of gastropod molluscs: an analysis using morphological characters. Zool. J. Linn. Soc. 119: 83–265.

Potts, G. W. 1981. The anatomy of respiratory structures in the dorid nudibranchs, *Onchidoris bilamellata* and *Archidoris pseudoargus*, with details of the epidermal glands. J. Mar. Biol. Assoc. U.K. 61: 959–982.

Rudman, W. B. 1991. Purpose in pattern: The evolution of colour in chromodorid nudibranchs. J. Moll. Stud. 57: 5–21.

Runham, N. W. and P. J. Hunter. 1970. *Terrestrial Slugs*. Hutchinson University Library, London. [A lovely review of slug biology.]

Salvini-Plawen, L. v. and G. Haszprunar. 1986. The Vetigastropoda and the systematics of streptoneurous Gastropoda. J. Zool. 211: 747–770.

Scheltema, R. S. 1989. Planktonic and non-planktonic development among prosobranch gastropods and its relationship to the geographic range of species. Pp. 183–188 in J. S. Ryland and P. A. Tyler (eds.), *Reproduction, Genetics and Distribution of Marine Organisms*. Olsen and Olsen, Denmark.

Seapy, R. and R. E. Young. 1986. Concealment in epipelagic pterotracheid heteropods (Gastropoda) and cranchiid squids (Cephalopoda). J. Zool. 210: 137–147.

Sleeper, H. L., V. J. Paul and W. Fenical. 1980. Alarm pheromones from the marine opisthobranch *Navanax inermis*. J. Chem. Ecol. 6: 57–70.

Spengel, J. W. (1881). Die Geruchsorgane und das Nervensystem der Mollusken. Z. Wiss. Zool. Abt. A. 35: 333–383.

Stanley, S. M. 1982. Gastropod torsion: Predation and the opercular imperative. Neues. Jahrb. Geol. Palaeontol. Abh. 164: 95–107. [Succinct review of ideas on why and how torsion evolved.]

Tardy, J. 1991. Types of opisthobranch veligers: Their notum formation and torsion. J. Moll. Stud. 57: 103–112.

Taylor, J. (ed.). 1996. *Origin and Evolutionary Radiation of the Mollusca*. Oxford Univ. Press, Oxford.

Taylor, J. D., N. J. Morris and C. N. Taylor. 1980. Food specialization and the evolution of predatory prosobranch gastropods. Paleontology 23: 375–410.

Thiriot-Quievreux, C. 1973. Heteropoda. Ann. Rev. Oceanogr. Mar. Biol. 11: 237–261.

Thompson, T. E. 1976. *Biology of Opisthobranch Molluscs, Vol. 1*. Ray Society, London.

Thompson, T. E. and G. H. Brown. 1976. *British Opisthobranch Molluscs*. Academic Press, London.

Thompson, T. E. and G. H. Brown. 1984. *Biology of Opisthobranch Molluscs, Vol. 2*. Ray Society, London.

van den Biggelaar, J. A. M. and G. Haszprunar. 1996. Cleavage patterns and mesentoblast formation in the Gastropoda: An evolutionary perspective. Evolution 50: 1520–1540.

Voltzow, J. and R. Collin. 1995. Flow through mantle cavities revisited: Was sanitation the key to fissurellid evolution? Invert. Biol. 114(2): 145–150.

Bivalvia

Ansell, A. D. and N. B. Nair. 1969. A comparison of bivalve boring mechanisms by mechanical means. Am. Zool. 9: 857–868.

Bieler, R. 2010. Classification of bivalve families. Malacologia 52: 114–133.

Bieler, R. and 19 others. 2014. Investigating the Bivalve Tree of Life—an exemplar-based approach combining molecular and novel morphological characters. Invertebr. Syst. 28: 32–115.

Beninger, P. G., S. St-Jean, Y. Poussart and J. E. Ward. 1993. Gill function and mucocyte distribution in *Placopecten magellanicus* and *Mytilus edulis* (Mollusca: Bivalvia): The role of mucus in particle transport. Mar. Ecol. Prog. Ser. 98: 275–282.

Bouchet, P., J.-P. Rocroi, R. Bieler, J. G. Carter and E. V. Coan. 2010. Nomenclator of bivalve families with a classification of bivalve families. Malacologia 52: 1–184.

Boulding, E. G. 1984. Crab-resistant features of shells of burrowing bivalves: Decreasing vulnerability by increasing handling time. J. Exp. Mar. Biol. Ecol. 76: 201–223.

Childress J. J., C. R. Fisher, J. M. Brooks, M. C. Kennicutt II, R. R. Bidigare and A. E. Anderson. 1986. A methanotrophic marine molluscan (Bivalvia, Mytilidae) symbiosis: Mussels fueled by gas. Science 233: 1306–1308.

Ellis, A. E. 1978. *British Freshwater Bivalve Mollusks*. Academic Press, London.

Goreau, T. F., N. I. Goreau and C. M. Yonge. 1973. On the utilization of photosynthetic products from zooxanthellae and of a dissolved amino acid in *Tridacna maxima*. J. Zool. 169: 417–454.

Harper, E. M., J. D. Taylor and J. A. Crame, J. A. (eds.) 2000. *The evolutionary biology of the Bivalvia*. Geological Society, London, Special Publications.

Jørgensen, C. B. 1974. On gill function in the mussel *Mytilus edulis*. Ophelia 13: 187–232.

Judd, W. 1979. The secretions and fine structure of bivalve crystalline style sacs. Ophelia 18: 205–234.

Kennedy, W. J., J. D. Taylor and A. Hall. 1969. Environmental and biological controls on bivalve shell mineralogy. Biol. Rev. 44: 499–530.

Manahan, D. T., S. H. Wright, G. C. Stephens and M. A. Rice. 1982. Transport of dissolved amino acids by the mussel, *Mytilus edulis*: Demonstration of net uptake from natural seawater. Science 215: 1253–1255.

Marincovich, L., Jr. 1975. Morphology and mode of life of the Late Cretaceous rudist, *Coralliochama orcutti* White (Mollusca: Bivalvia). J. Paleontol. 49(1): 212–223. [Describes the famous Punta Banda deposits of Baja California, Mexico.]

Meyhöfer, E. and M. P. Morse. 1996. Characterization of the bivalve ultrafiltration system in *Mytilus edulis*, *Chlamys hastata*, and *Mercenaria mercenaria*. Invert. Biol. 115(1): 20–29.

Morse, M. P., E. Meyhöfer, J. J. Otto and A. M. Kuzirian. 1986. Hemocyanin respiratory pigments in bivalve mollusks. Science 231: 1302–1304.

Morton, B. 1978. The diurnal rhythm and the processes of feeding and digestion in *Tridacna crocea*. J. Zool. 185: 371–387.

Morton, B. 1978. Feeding and digestion in shipworms. Ann. Rev. Oceanogr. Mar. Biol. 16: 107–144.

Owen, G. 1974. Feeding and digestion in the Bivalvia. Adv. Comp. Physiol. Biochem. 5: 1–35.

Pojeta, J., Jr. and B. Runnegar. 1974. *Fordilla troyensis* and the early history of pelecypod mollusks. Am. Sci. 62: 706–711.

Powell, M. A. and G. N. Somero. 1986. Hydrogen sulfide oxidation is coupled to oxidative phosphorylation in mitochondria of *Solemya reidi*. Science 233: 563–566.

Reid, R. G. B. and F. R. Bernard. 1980. Gutless bivalves. Science 208: 609–610.

Reid, R. G. B. and A. M. Reid. 1974. The carnivorous habit of members of the septibranch genus *Cuspidaria* (Mollusca: Bivalvia). Sarsia 56: 47–56.

Stanley, S. M. 1970. Relation of shell form to life habits of the Bivalvia (Mollusca). Geol. Soc. Am. Mem. 125: 1–296.

Stanley, S. M. 1975. Why clams have the shape they have: An experimental analysis of burrowing. Paleobiology 1: 48.

Taylor, J. D. 1973. The structural evolution of the bivalve shell. Paleontology 16: 519–534.

Trueman, E. R. 1966. Bivalve mollusks: Fluid dynamics of burrowing. Science 152: 523–525.

Vetter, R. D. 1985. Elemental sulfur in the gills of three species of clams containing chemoautotrophic symbiotic bacteria: A possible inorganic energy storage compound. Mar. Biol. 88: 33–42.

Waterbury, J. B., C. B. Calloway and R. D. Turner. 1983. A cellulolytic nitrogen-fixing bacterium cultured from the gland of Deshayes in shipworms (Bivalvia: Teredinidae). Science 221: 1401–1403.

Wilkens, L. A. 1986. The visual system of the giant clam *Tridacna*: Behavioral adaptations. Biol. Bull. 170: 393–408.

Yonge, C. M. 1953. The monomyarian condition in the Lamellibranchia. Trans. R. Soc. Edinburgh 62 (p. II): 443–478.

Yonge, C. M. 1973. Giant clams. Sci. Am. 232: 96–105.

Scaphopoda

Bilyard, G. R. 1974. The feeding habits and ecology of *Dentalium stimpsoni*. Veliger 17: 126–138.

Gainey, L. F. 1972. The use of the foot and captacula in the feeding of *Dentalium*. Veliger 15: 29–34.

Reynolds, P. D. 2002. The Scaphopoda. Adv. Mar. Biol. 42: 137–236.

Steiner, G. 1992. Phylogeny and classification of Scaphopoda. J. Mollus. Stud. 58: 385–400.

Trueman, E. R. 1968. The burrowing process of *Dentalium*. J. Zool. 154: 19–27.

Cephalopoda

Aronson, R. B. 1991. Ecology, paleobiology and evolutionary constraint in the octopus. Bull. Mar. Sci. 49(1–2): 245–255.

Barber, V. C. and F. Grazialdei. 1967. The fine structure of cephalopod blood vessels. Z. Zellforsch. Mikrosk. Anat. 77: 162–174. [Also see earlier papers by these authors in the same journal.]

Boyle, P. R. (ed.). 1983. *Cephalopod Life Cycles, Vols. 1–2*. Academic Press, New York.

Boyle, P. R. and S. v. Boletzky. 1996. Cephalopod populations: Definition and dynamics. Phil. Trans. Roy. Soc. B, Biol. Sci. 351: 985–1002.

Boyle, P. and P. Rodhouse. 2005. *Cephalopods: Ecology and Fisheries*. Blackwell Science Ltd., Oxford.

Clarke, M. A. 1966. A review of the systematics and ecology of oceanic squids. Adv. Mar. Biol. 4: 91–300.

Cloney, R. A. and S. L. Brocco. 1983. Chromatophore organs, reflector cells, irridocytes and leucophores in cephalopods. Am. Zool. 23: 581 592.

Denton, E. J. and J. B. Gilpin-Brown. 1973. Flotation mechanisms in modern and fossil cephalopods. Adv. Mar. Biol. 11: 197–264.

Fields, W. G. 1965. The structure, development, food relations, reproduction, and life history of the squid *Loligo opalescens* Berry. Calif. Dept. Fish Game Bull. 131: 1–108.

Fiorito, G. and P. Scotto. 1992. Observational learning in *Octopus vulgaris*. Science 256: 545–547.

Hanlon, R. T. and J. B. Messenger. 1996. *Cephalopod Behavior*. Cambridge Univ. Press.

Hochberg, F. G. 1983. The parasites of cephalopods: A review. Mem. Nat. Mus. Victoria Melbourne 44: 109–145.

House, M. R. and J. R. Senior. 1981. *The Ammonoidea: The Evolution, Classification, Mode of Life and Geological Usefulness of a Major Fossil Group*. Academic Press, New York.

Kier, W. M. 1991. Squid cross-striated muscle: The evolution of a specialized muscle fiber type. Bull. Mar. Sci. 49(1–2): 389–403.

Kier, W. M. and A. M. Smith. 1990. The morphology and mechanics of octopus suckers. Biol. Bull. 178: 126–136.

Lehmann, U. 1981. *The Ammonites: Their Life and Their World*. Cambridge University Press, New York.

McFell-Ngai, M. and M. K. Montgomery. 1990. The anatomy and morphology of the adult bacterial light organ of *Euprymna scolopes* Berry (Cephalopoda: Sepiolidae). Biol. Bull. 179: 332–339.

Mutvei, H. 1964. On the shells of *Nautilus* and *Spirula* with notes on the shell secretion in non-cephalopod molluscs. Ark. Zool. 16(14): 223–278.

Nixon, M. and J. B. Messenger (eds.). 1977. *The Biology of Cephalopods*. Academic Press, New York.

Packard, A. 1972. Cephalopods and fish: The limits of convergence. Biol. Rev. 47: 241–307.

Roper, C. F. E. and K. J. Boss. 1982. The giant squid. Sci. Am. 246: 96–104.

Saunders, W. B. 1983. Natural rates of growth and longevity of *Nautilus belauensis*. Paleobiology 9: 280–288.

Saunders, W. B., R. L. Knight and P. N. Bond. 1991. Octopus predation on *Nautilus*: Evidence from Papua New Guinea. Bull. Mar. Sci. 49(1–2): 280–287.

Saunders, W. B. and N. H. Landman. 2009. *Nautilus: The Biology and Paleobiology of a Living Fossil, Reprint with Additions*. Springer Science & Business Media.

Sweeney, M. J., C. F. E. Roper, K. A. M. Mangold, M. R. Clarke and S. V. Boletzky (eds.). 1992. "Larval" and juvenile cephalopods: a manual for their identification. Smithsonian Contrb. Zool. 513: 1–282.

Ward, P. D. 1987. *The Natural History of* Nautilus. Allen & Unwin, Boston.

Ward, P. D. and W. B. Saunders. 1997. *Allonautilus*, a new genus of living nautiloid cephalopod and its bearing on phylogeny of the Nautilida. J. Paleontol. 71: 1054–1064.

Ward, R., L. Greenwald, and O. E. Greenwald. 1980. The buoyancy of the chambered nautilus. Sci. Am. 243: 190–204.

Wells, M. J. 1978. *Octopus: Physiology and Behaviour of an Advanced Invertebrate*. Chapman and Hall, New York.

Wells, M. J. and R. K. O'Dor. 1991. Jet propulsion and the evolution of cephalopods. Bull. Mar. Sci. 49(1–2): 419–432.

Young, J. Z. 1972. *The Anatomy of the Nervous System of* Octopus vulgaris. Oxford University Press, New York.

Young, R. E. and F. M. Mencher. 1980. Bioluminescence in mesopelagic squid: Diel color change during counterillumination. Science 208: 1286–1288.

Molluscan Evolution and Phylogeny

Batten, R. L., H. B. Rollins and S. J. Gould. 1967. Comments on "The adaptive significance of gastropod torsion." Evolution 21: 405–406.

Bieler, R. 1992. Gastropod phylogeny and systematics. Ann. Rev. Ecol. Syst. 23: 311–338.

Bieler, R. and 19 others. 2014. Investigating the Bivalve Tree of Life—an exemplar-based approach combining molecular and novel morphological characters. Invertebr. Syst. 28: 32–115.

Dunn, C. W. and 17 others. 2008. Broad phylogenomic sampling improves resolution of the animal tree of life. Nature 452: 745.

Dunn, C. W., G. Giribet, G. D. Edgecombe and A. Hejnol. 2014. Animal phylogeny and its evolutionary implications. Ann. Rev. Eco. Evol. S. 45: 371–395.

Edgecombe, G. D. and 8 others. 2011. Higher-level metazoan relationships: recent progress and remaining questions. Org. Divers. Evol. 11: 151–172.

Eldredge, N. and S. M. Stanley (eds.). 1984. *Living Fossils*. Springer-Verlag, New York. [Includes chapters on Monoplacophora, Pleurotomaria, and *Nautilus*.]

Fedonkin, M. A. and B. M. Waggoner. 1997. The Late Precambrian fossil *Kimberella* is a mollusc-like bilaterian organism. Nature 388: 868–871.

Garstang, W. [Introduction by Sir A. Hardy]. 1951. *Larval Forms, and Other Zoological Verses*. Basil Blackwell, Oxford. [Reprinted in 1985 by the University of Chicago Press.]

Ghiselin, M. T. 1966. The adaptive significance of gastropod torsion. Evolution 20: 337–348.

Giribet, G. and W. Wheeler. 2002. On bivalve phylogeny: a high-level analysis of the Bivalvia (Mollusca) based on combined morphology and DNA sequence data. Invertebr. Biol. 121: 271–324.

Giribet, G., A. Okusu, A. R. Lindgren, S. W. Huff, M. Schrodl and M. K. Nishiguchi. 2006. Evidence for a clade composed of molluscs with serially repeated structures: monoplacophorans are related to chitons. PNAS 103: 7723–7728.

Giribet, G. 2014. On Aculifera: a review of hypotheses in tribute to Christoffer Schander. J. Nat. Hist. 48: 2739–2749.

Götting, K. 1980. Arguments concerning the descendence of Mollusca from metameric ancestors. Zool. Jahrb. Abt. Anat. 103: 211–218.

Götting, K. 1980. Origin and relationships of the Mollusca. Z. Zool. Syst. Evolutionsforsch. 18: 24–27.

Graham, A. 1979. Gastropoda. Pp. 359–365 in M. R. House (ed.), *The Origin of Major Invertebrate Groups.* Academic Press, New York.

Gutmann, W. F. 1974. Die Evolution der Mollusken-Konstruction: Ein phylogenetisches Modell. Aufsätze Red. Senckenb. Naturf. Ges. 25: 1–24.

Haas, W. 1981. Evolution of calcareous hardparts in primitive molluscs. Malacologia 21: 403–418.

Haszprunar, G. 1992. The first molluscs—small animals. Boll. Zool. 59: 1–16.

Haszprunar, G. 2000. Is the Aplacophora monophyletic? A cladistic point of view. Am. Malacol. Bull. 15(2): 115–130.

Hejnol, A. 2010. A twist in time—The evolution of spiral cleavage in the light of animal phylogeny. Integr. Comp. Biol. 50: 695-706.

Hickman, C. S. 1988. Archaeogastropod evolution, phylogeny and systematics: A re-evaluation. Malacol. Rev., Suppl. 4: 17–34.

Holland, C. H. 1979. Early Cephalopoda. Pp. 367–379 in M. R. House (ed.), *The Origin of Major Invertebrate Groups.* Academic Press, New York.

Ivantsov, A. 2012. Paleontological data on the possibility of Precambrian existence of mollusks. Pp. 153–179 in A. Fyodorov and H. Yakovlev (eds.), *Mollusks: Morphology, Behavior, and Ecology.* Nova Science Publishing, New York.

Johnson, C. C. 2002. The rise and fall of rudistid reefs. Amer. Sci. 90: 148–153.

Jörger, K. M., I. Stöger, Y. Kano, H. Fukuda, T. Knebelsberger and M. Schrödl. 2010. On the origin of Acochlidia and other enigmatic euthyneuran gastropods, with implications for the systematics of Heterobranchia. BMC Evol. Biol. 10: 323.

Kocot, K. M. and 10 others. 2011. Phylogenomics reveals deep molluscan relationships. Nature 477: 452–456.

Kocot, K. M. 2013. Recent advances and unanswered questions in deep molluscan phylogenetics. Am. Malacol. Bull. 31: 195–208.

Lauterbach, K.-E. von. 1983. Erörterungen zur Stammesgeschichte der Mollusca, insbesondere der Conchifera. Z. Zool. Syst. Evolutionsforsch. 21: 201–216.

Lindberg, D. R. and M. T. Ghiselin. 2003. Fact, theory and tradition in the study of molluscan origins. PCAS 54: 663–686.

Lindgren, A. R., G. Giribet and M. K. Nishiguchi. 2004. A combined approach to the phylogeny of Cephalopoda (Mollusca). Cladistics 20: 454–486.

Linsley, R. M. 1978. Shell form and the evolution of gastropods. Am. Sci. 66: 432–441.

Nesnidal, M. P. and 9 others. 2013. New phylogenomic data support the monophyly of Lophophorata and an Ectoproct-Phoronid clade and indicate that Polyzoa and Kryptrochozoa are caused by systematic bias. BMC Evol. Biol. 13: 253.

Okusu, A., E. Schwabe, D. J. Eernisse and G. Giribet. 2003. Towards a phylogeny of chitons (Mollusca, Polyplacophora) based on combined analysis of five molecular loci. Org. Divers. Evol. 3: 281–302.

Ponder, W. F. and D. R. Lindberg. 2008. *Phylogeny and Evolution of the Mollusca.* University of California Press, Berkeley, CA.

Runnegar, B. and J. Pojeta. 1974. Molluscan phylogeny: The paleontological viewpoint. Science 186: 311–317.

Ruppert, E. E. and J. Carle. 1983. Morphology of metazoan circulatory systems. Zoomorph. 103: 193–208.

Salvini-Plawen, L. v. 1977. On the evolution of photoreceptors and eyes. Evol. Biol. 10: 207–263.

Salvini-Plawen, L., v. 1990. Origin, phylogeny and classification of the phylum Mollusca. Iberus 9: 1– 33.

Scheltema, A. H. 1978. Position of the class Aplacophora in the phylum Mollusca. Malacologia 17: 99–109.

Scheltema, A. H. 1988. Ancestors and descendants: Relationships of the Aplacophora and Polyplacophora. Am. Malacol. Bull. 6: 57–68.

Scheltema, A. H. 1993. Aplacophora as progenetic aculiferans and the coelomate origin of mollusks as the sister taxon of Sipuncula. Biol. Bull. 184: 57–78.

Scherholz, M., E. Redl, T. Wollesen, C. Todt and A. Wanninger. 2013. Aplacophoran mollusks evolved from ancestors with polyplacophoran-like features. Curr. Biol. 23: 1–5.

Scherholz, M., E. Redl, T. Wollesen, C. Todt and A. Wanninger. 2015. From complex to simple: myogenesis in an aplacophoran mollusk reveals key traits in aculiferan evolution. BMC Evol. Biol. 15(1): 201.

Schrödl, M. and I. Stöger. 2014. A review on deep molluscan phylogeny: old markers, integrative approaches, persistent problems. J. Nat. Hist. 48: 1–32.

Sharma, P. P. and 12 others. 2012. Phylogenetic analysis of four nuclear protein-encoding genes largely corroborates the traditional classification of Bivalvia (Mollusca). Mol. Phylogenet. Evol. 65: 64–74.

Sigwart, J. D., I. Stoeger, T. Knebelsberger and E. Schwabe. 2013. Chiton phylogeny (Mollusca: Polyplacophora) and the placement of the enigmatic species *Choriplax grayi* (H. Adams and Angas). Invertebr. Syst. 27: 603–621.

Sigwart, J. D., C. Todt and A. H. Scheltema. 2014. Who are the 'Aculifera'? J. Nat. Hist. 48: 2733–2737.

Sigwart, J. D. and D. R. Lindberg. 2015. Consensus and confusion in molluscan trees: evaluating morphological and molecular phylogenies. Syst. Biol. 64: 384–395.

Smith, S. A. and 7 others. 2011. Resolving the evolutionary relationships of molluscs with phylogenomic tools. Nature 480: 364–367.

Stöger, I. and M. Schrödl. 2013. Mitogenomics does not resolve deep molluscan relationships (yet?). Mol. Phylogenet. Evol. 69: 376–392.

Struck, T. H. and 12 others. 2014. Platyzoan paraphyly based on phylogenomic data supports a noncoelomate ancestry of Spiralia. Mol. Biol. Evol. 31: 1833–1849.

Sutton, M. D. and J. D. Sigwart. 2012. A chiton without a foot. Palaeontology 55: 401–411.

Taylor, J. D. (ed.). 1996. *Origin and Evolutionary Radiation of the Mollusca.* New York, Oxford University Press.

Trueman, E. R. and M. R. Clarke. 1985. *Evolution. The Mollusca, 10.* New York, Academic Press.

Steiner, G. and M. Müller. 1996. What can 18S rDNA do for bivalve phylogeny? J. Mol. Evol. 43: 58–70.

Vagvolgyi, J. 1967. On the origin of molluscs, the coelom, and coelomic segmentation. Syst. Zool. 16: 153–168.

Wade, C. M., P. B. Mordan and B. Clarke. 2001. A phylogeny of the land snails (Gastropoda: Pulmonata). Proc. Royal Soc. London B. 268: 413–422.

Wagner, P. J. 2001. Gastropod phylogenetics: Progress, problems and implications. J. Paleontol. 75(6): 1128–1140.

Wilson, N. G., G. W. Rouse and G. Giribet. 2010. Assessing the molluscan hypothesis Serialia (Monoplacophora +

Polyplacophora) using novel molecular data. Mol. Phylogene. Evol. 54: 187–193.

Wingstrand, K. G. 1985. On the anatomy and relationships of recent Monoplacophora. Galathea Rpt. 16: 7–94.

Yonge, C. M. 1957. *Neopilina*: Survival from the Paleozoic. Discovery (London), June 1957: 255–256.

Yonge, C. M. 1957. Reflections on the monoplacophoran *Neopilina galatheae* Lemche. Nature 179: 672–673.

Zapata, F. and 8 others. 2014. Phylogenomic analyses of deep gastropod relationships reject Orthogastropoda. Proc. Royal Soc. London B. 281. doi: 10.1098/rspb.2014.1739

CHAPTER 14

Phylum Annelida

The Segmented (and Some Unsegmented) Worms

This chapter treats the segmented worms, or Annelida (Greek, *anellus*, "ringed"), which comprise about 20,000 described species. Annelids include the familiar earthworms and leeches, as well as various marine "sand worms," "tube worms," and an array of other descriptors (Figures 14.1–14.3). Some are tiny animals of the meiofauna; others, such as certain southern hemisphere earthworms and some marine species, can exceed 3 m in length. And, recently, phylogenetic studies have shown that several former phyla—including Sipuncula and Echiura—are also annelids (see the following section on Taxonomic History and Classification).

Annelids have successfully occupied virtually all habitats where sufficient water is available. They are particularly ubiquitous in the sea, but also abound in fresh water, and many live in damp terrestrial environments. There are also parasitic, mutualistic, and commensal species. Their success is no doubt due in part to the evolutionary plasticity of their segmented bodies and use of a wide variety of life histories and feeding strategies. Most annelids are characterized by having a head followed by a segmented body in which most internal and external parts are repeated with each segment, a condition referred to as **serial homology**. Serial homology refers to body structures with the same genetic and developmental origins that arise repeatedly during the ontogeny of an organism. In annelids this repetition of homologous body structures results in **metamerism**—body segmentation that arises by way of **teloblastic development** (the proliferation of paired, segmental mesodermal bands from teloblast cells at a posterior growth zone in the embryo). There are some dramatic exceptions to this for several annelid groups that have lost their metameric tendencies, or transformed from having an obviously segmented body (e.g., Echiuridae and Sipuncula). Annelids are triploblastic coelomate worms with a complete gut (with few exceptions), a closed circulatory system (again with

Classification of The Animal Kingdom (Metazoa)

Non-Bilateria*
(a.k.a. the diploblasts)
- PHYLUM PORIFERA
- PHYLUM PLACOZOA
- PHYLUM CNIDARIA
- PHYLUM CTENOPHORA

Bilateria
(a.k.a. the triploblasts)
- PHYLUM XENACOELOMORPHA

Protostomia
- PHYLUM CHAETOGNATHA

SPIRALIA
- PHYLUM PLATYHELMINTHES
- PHYLUM GASTROTRICHA
- PHYLUM RHOMBOZOA
- PHYLUM ORTHONECTIDA
- PHYLUM NEMERTEA
- PHYLUM MOLLUSCA
- **PHYLUM ANNELIDA**
- PHYLUM ENTOPROCTA
- PHYLUM CYCLIOPHORA

Gnathifera
- PHYLUM GNATHOSTOMULIDA
- PHYLUM MICROGNATHOZOA
- PHYLUM ROTIFERA

Lophophorata
- PHYLUM PHORONIDA
- PHYLUM BRYOZOA
- PHYLUM BRACHIOPODA

ECDYSOZOA

Nematoida
- PHYLUM NEMATODA
- PHYLUM NEMATOMORPHA

Scalidophora
- PHYLUM KINORHYNCHA
- PHYLUM PRIAPULA
- PHYLUM LORICIFERA

Panarthropoda
- PHYLUM TARDIGRADA
- PHYLUM ONYCHOPHORA
- PHYLUM ARTHROPODA
  - SUBPHYLUM CRUSTACEA*
  - SUBPHYLUM HEXAPODA
  - SUBPHYLUM MYRIAPODA
  - SUBPHYLUM CHELICERATA

Deuterostomia
- PHYLUM ECHINODERMATA
- PHYLUM HEMICHORDATA
- PHYLUM CHORDATA

*Paraphyletic group

This chapter has been revised by Greg Rouse except for the Sipuncula section, which has been revised by Gonzalo Giribet.

BOX 14A Characteristics of the Phylum Annelida

1. Bilaterally symmetrical, vermiform, protostomes. Terrestrial, fresh water, and marine
2. Segmented (segmentation secondarily lost in some groups, e.g., Sipuncula, Echiuridae); segments arise by teloblastic growth
3. Development typically protostomous, with holoblastic, spiral, 4d, schizocoelous embryogeny
4. Digestive tract complete, usually with regional specialization (secondarily lost in some groups)
5. With a closed circulatory system (secondarily lost in some groups); respiratory pigments may include hemoglobin, chlorocruorin and hemerythrin
6. Most adults possess metanephridia or, less commonly, protonephridia.
7. Nervous system well developed, with a dorsal cerebral ganglion, circumenteric connectives, and ventral ganglionated nerve cord(s)
8. With lateral, segmentally arranged epidermal chaetae (secondarily lost in some, e.g., Sipuncula)
9. In most, head composed of presegmental prostomium and peristomium (often with body segments fused to it; peristomium may be reduced). Echiuridae and Sipuncula without obvious prostomium/peristomium
10. Gonochoristic or hermaphroditic; development indirect or direct; often with trochophore larva

some exceptions), a well-developed nervous system, and excretory structures in the form of protonephridia or, more commonly, metanephridia (Box 14A). Marine annelids produce trochophore larvae, a feature shared with several other protostome taxa (e.g., Mollusca, Nemertea, Entoprocta). The story of annelid diversity and success is one of variation on this basic theme.

Taxonomic History and Classification

As mentioned in earlier chapters, the roots of modern animal classification can be traced to Linnaeus (1758), who placed all invertebrates except insects in the taxon Vermes. In 1802, Lamarck established the taxon Annelida; he had a reasonably good idea of their unity and of their differences from other groups of worms. He and many other workers recognized the affinity among most annelids, but Hirudinoidea (leeches and allies) were often erroneously allied with trematode platyhelminths.

Recently, some groups previously regarded as separate phyla have been accepted as annelids. These include Echiura, Sipuncula, and a group comprised of the former phyla Pogonophora and Vestimentifera (= Siboglinidae). Although now known to be highly modified annelids, these groups are so distinct that they are given detailed treatment at the end of this chapter. The overall classification within Annelida is also currently undergoing significant revisions and has yet to stabilize. Phylogenomic and other molecular phylogenetic analyses have shown that taxonomic groupings based on morphology are, in many cases, invalid. Annelida was traditionally divided into three classes. Polychaeta was the largest and most diverse group, the other two being Oligoghaeta (earthworms and their relatives) and the leeches (Hirudinoidea or Hirudinea). It is now clear that the closest relative of Hirudinoidea is the freshwater group Lumbriculidae, a taxon that is inside the Oligochaeta, and these two former classes (Hirudinoidea and Oligochaeta) are now referred to as Clitellata, a taxon that actually dates back to 1919. Furthermore, Clitellata has been shown to have arisen deep within Polychaeta, making "Polychaeta" synonymous with Annelida. Hence, all three of the former classes are no longer ranked as such, the Oligochaeta and Polychaeta are recognized as paraphyletic groups and no longer hold systematic ranking, and "oligochaete" and "polychaete" are informal names only. Owing to their novel morphology and lifestyle, Hirudinoidea are also given some detailed treatment at the end of this chapter.

SYNOPSES OF MAJOR GROUPS

As noted above, the taxonomy for Annelida has yet to stabilize, and our understanding of relationships within this phylum has undergone momentous changes in the last two decades. For a current assessment of the overall relationships among the major annelid groups see Figure 14.41, which is based on a synthesis of the phylogenomic studies of Weigert et al. (2014) and Andrade et al. (2015). No Linnaean ranks are used here with regard to the names, though those ending with "-idae" have traditionally carried the rank of family.

OWENIIDAE Fewer than 50 species (Figures 14.1A and 14.12B). Generally small-bodied (some reaching 10 cm), tube-dwelling annelids with numerous fine parapodial hooks. The prostomium may be lobed or folded into a ciliated crown (e.g., *Owenia*). The group is best known for its unusual and beautiful "mitraria" larvae (Figure 14.20H,I).

MAGELONIDAE Around 70 species, most in the genus *Magelona* (Figure 14.2A). Relatively similar in appearance, with shovel-shaped anterior ends and a pair of papillated palps. Thin, cylindrical worms, less than 1 mm wide, but reaching 15 cm in length. Live as burrowers in sands and muds; do not seem to form permanent tubes, though they do line their burrows with mucus.

CHAETOPTERIDAE Around 70 species, placed mostly in the genera *Chaetopterus*, *Mesochaetopterus*, *Phyllochaetopterus* or *Spiochaetopterus* (Figures 14.1B and 14.10C–E). Adults range in size from less than 1 cm to more than 40 cm, though there are fewer than 60 segments in

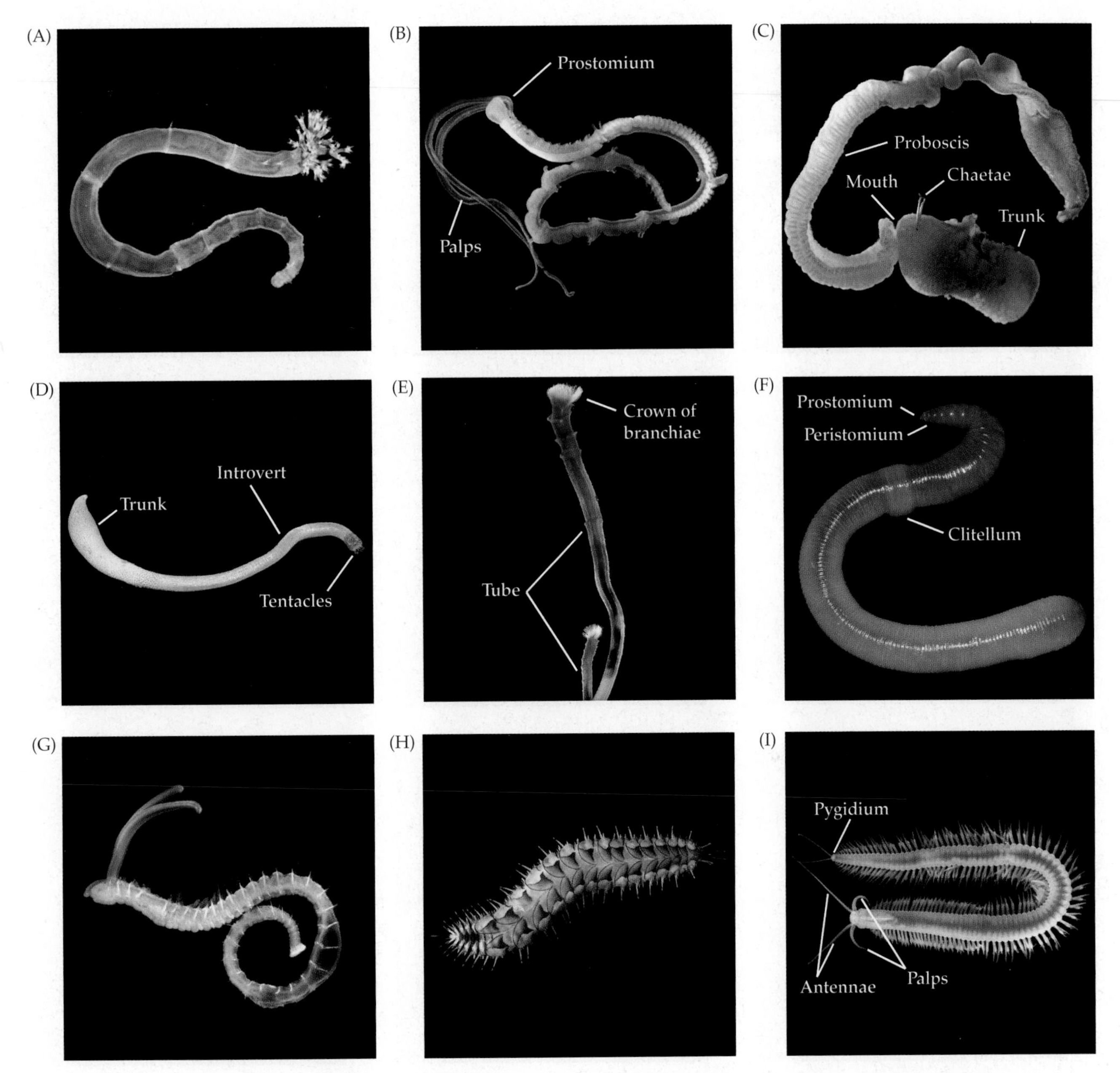

Figure 14.1 Representative annelids spanning the phylogenetic breadth of the group. (A) *Owenia* sp. (Oweniidae). (B) *Spiochaetopterus* sp. (Chaetopteridae). (C) *Protobonellia* sp. (Echiuridae). (D) *Phascolion* sp. (Sipuncula). (E) *Arcovestia ivanovi* (Siboglinidae). (F) *Lumbricus terrestris* (Lumbricidae). (G) *Boccardia proboscidea* (Spionidae). (H) *Harmothoe* sp. (Polynoidae). (I) *Dorvillea* sp. (Dorvilleidae).

most taxa. Body distinctly heteronomous, divided into two or three functional regions with varying parapodia. Chaetopterids live in straight or U-shaped tubes and most are mucous-net filter feeders, eating plankton and detritus pumped through the tube in water currents they generate. The group is most well known for the extraordinary filter-feeding mechanism employed by *Chaetopterus*, which is also renowned for the blue luminescence that it produces.

AMPHINOMIDA Includes Amphinomidae and Euphrosinidae; over 200 species (Figure 14.2B). The more motile forms, amphinomids, are often quite large (e.g., *Chloeia*, *Eurythoe, Hermodice*) and commonly referred to as fire worms. This name comes from a feature unique to Amphinomida, namely brittle calcareous chaetae that break off when touched and can be intensely irritating in the skin. Amphinomids are more common in shallow warm seas, whereas the more sessile Euphrosinidae (e.g., *Euphrosine*) are found in deeper colder waters.

SIPUNCULA Peanut worms. About 150 species, all marine (Figures 14.22–14.27). Formerly with the rank of phylum, phylogenetic studies have shown this group belongs within Annelida; with six families. Sipunculans range in length from less than 1 cm to about 50 cm. They are found from the intertidal zone to depths of over 5,000 meters. The body is sausage-shaped and has a retractable introvert that can withdraw into a much thicker trunk. The anterior end of the

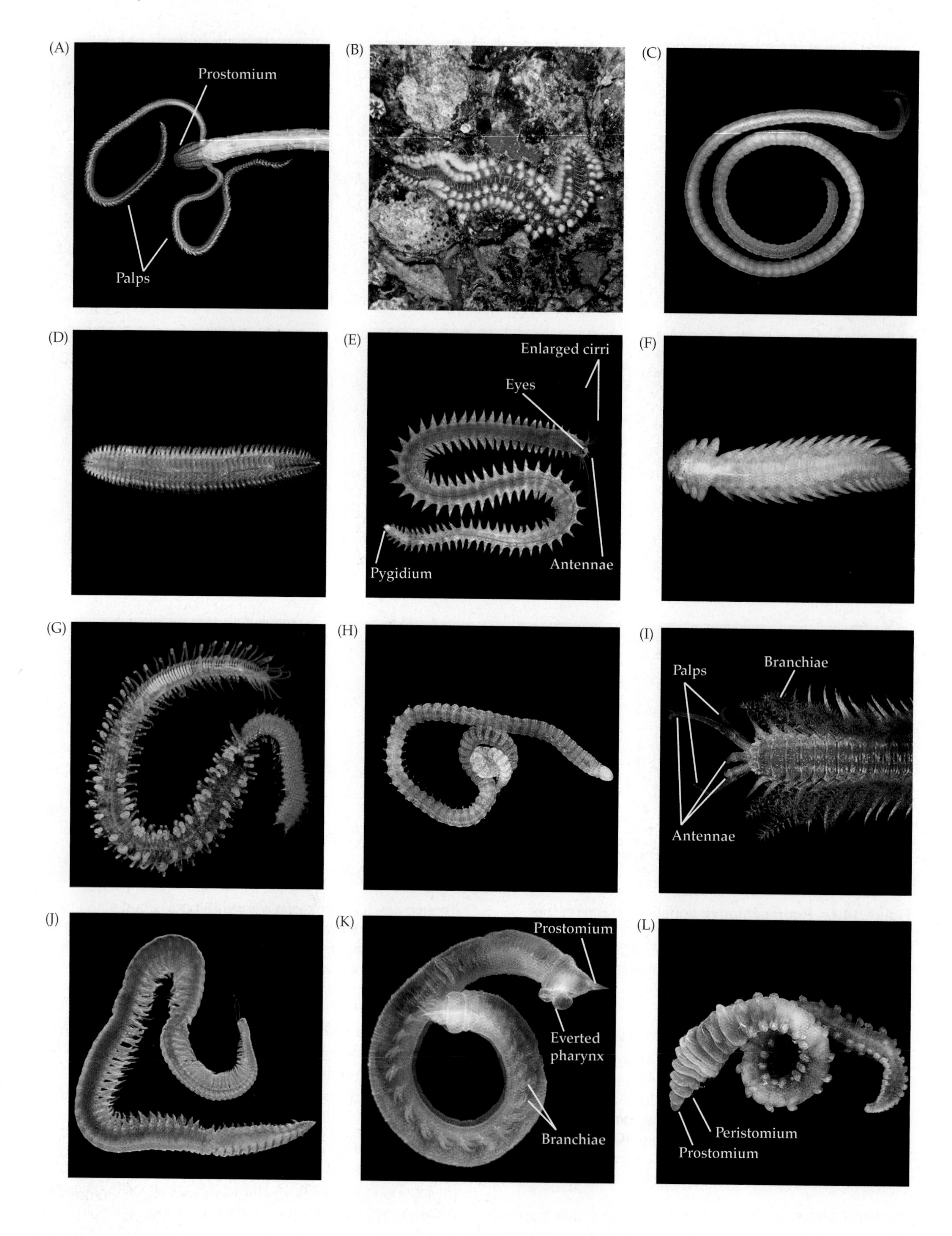
(A)
Prostomium
Palps
(B)
(C)
(D)
(E)
Enlarged cirri
Eyes
Pygidium
Antennae
(F)
(G)
(H)
(I)
Palps
Branchiae
Antennae
(J)
(K)
Prostomium
Everted pharynx
Branchiae
(L)
Peristomium
Prostomium

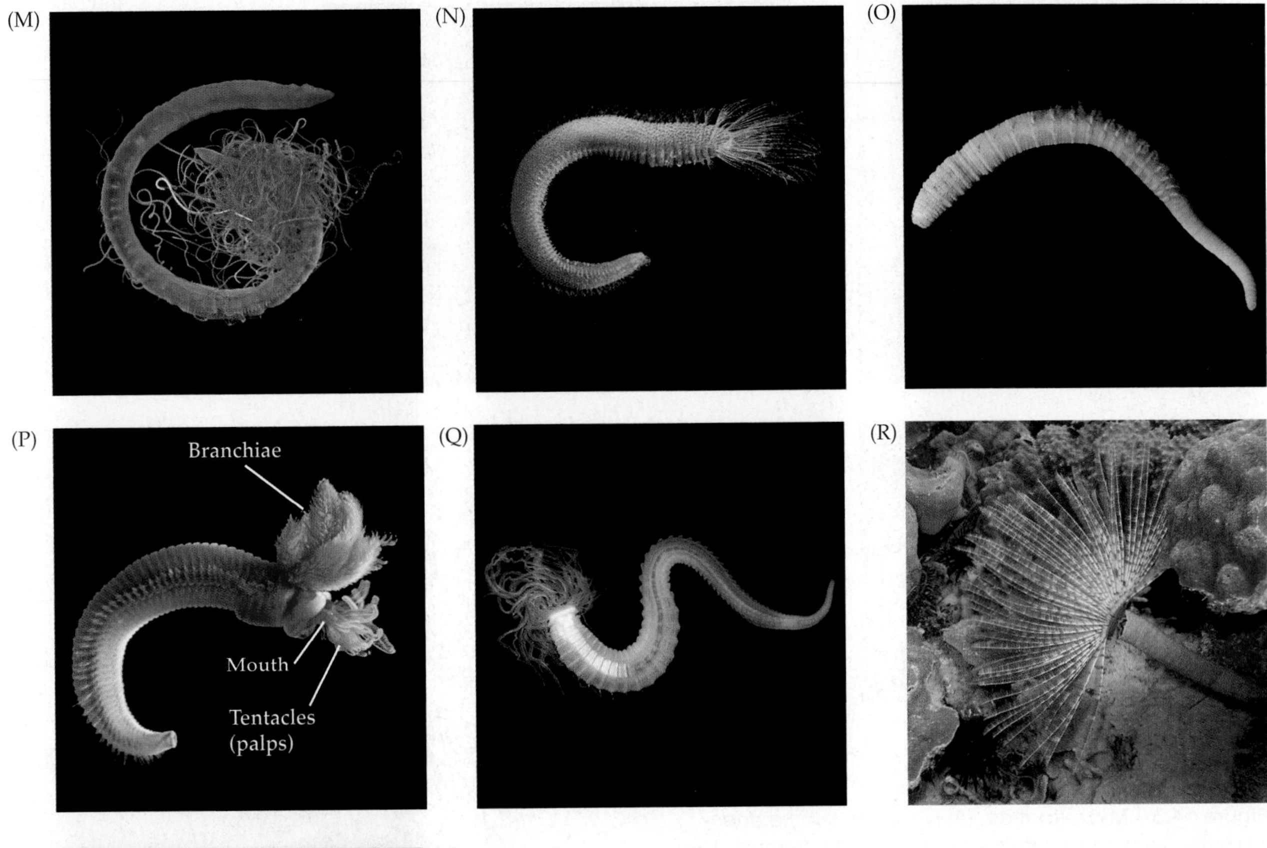

Figure 14.2 Further examples of annelid diversity. (A) *Magelona pitelkai* (Magelonidae). (B) *Hermodice carunculata* (Amphinomidae). (C) *Saccocirrus* sp. (Saccocirridae). (D) *Chrysopetalum* sp. (Chrysopetalidae). (E) *Platynereis dumerilii* (Nereididae). (F) *Lopadorhynchus* sp. (Lopadorhynchidae). (G) *Trypanosyllis californiensis* (Syllidae). (H) *Lumbrinereis* sp. (Lumbrineridae). (I) *Diopatra cuprea* (Onuphidae). (J) *Scoloplos armiger* (Orbiniidae). (K) *Thoracophelia mucronata* (Opheliidae). (L) *Capitella* sp. (Capitellidae). (M) *Cirriformia* sp. (Cirratulidae). (N) *Pherusa* sp. (Flabelligeridae). (O) *Abarenicola pacifica* (Arenicolidae). (P) *Paralvinella fijiensis* (Ampharetidae). (Q) *Amphitrite kerguelensis* (Terebellidae). (R) *Sabellastarte magnifica* (Sabellidae). (S) *Neosabellaria cementarium* (Sabellariidae).

introvert bears the mouth and feeding tentacles. It is when the introvert is retracted and the body is turgid that some species resemble a peanut (e.g., *Sipunculus*). Sipunculans are also notable among annelids in lacking a closed circulatory system, for their spacious coelom, and for the lack of obvious segmentation. The gut is U-shaped and coiled, with the anus located dorsally on the body near the introvert–trunk junction. The body surface is usually beset with minute bumps, warts, tubercles, or spines.

ERRANTIA This is an old taxonomic grouping that was recently resurrected by Torsten Struck and colleagues. It contains more than a quarter of all described annelid species diversity, in three major groups: Protodrilida, Eunicida, Phyllodocida (the latter two comprising a sister group called Aciculata).

PROTODRILIDA Over 60 species, mainly living interstitially in sediments (Figure 14.2C). Includes Protodrilidae, Protodriloididae, and Saccocirridae, best known for being members of the now defunct taxon Archiannelida. Taxa such as *Protodrilus* and *Saccocirrus* are found in medium to coarse sediments in shallow waters. Adult Protodrilida range in length from 2 to 30 mm and can have up to 200 segments. Protodrilidae lack chaetae, though these are found in Protodriloididae and Saccocirridae. All have a pair of palps emerging from the prostomium that form highly mobile sensory structures.

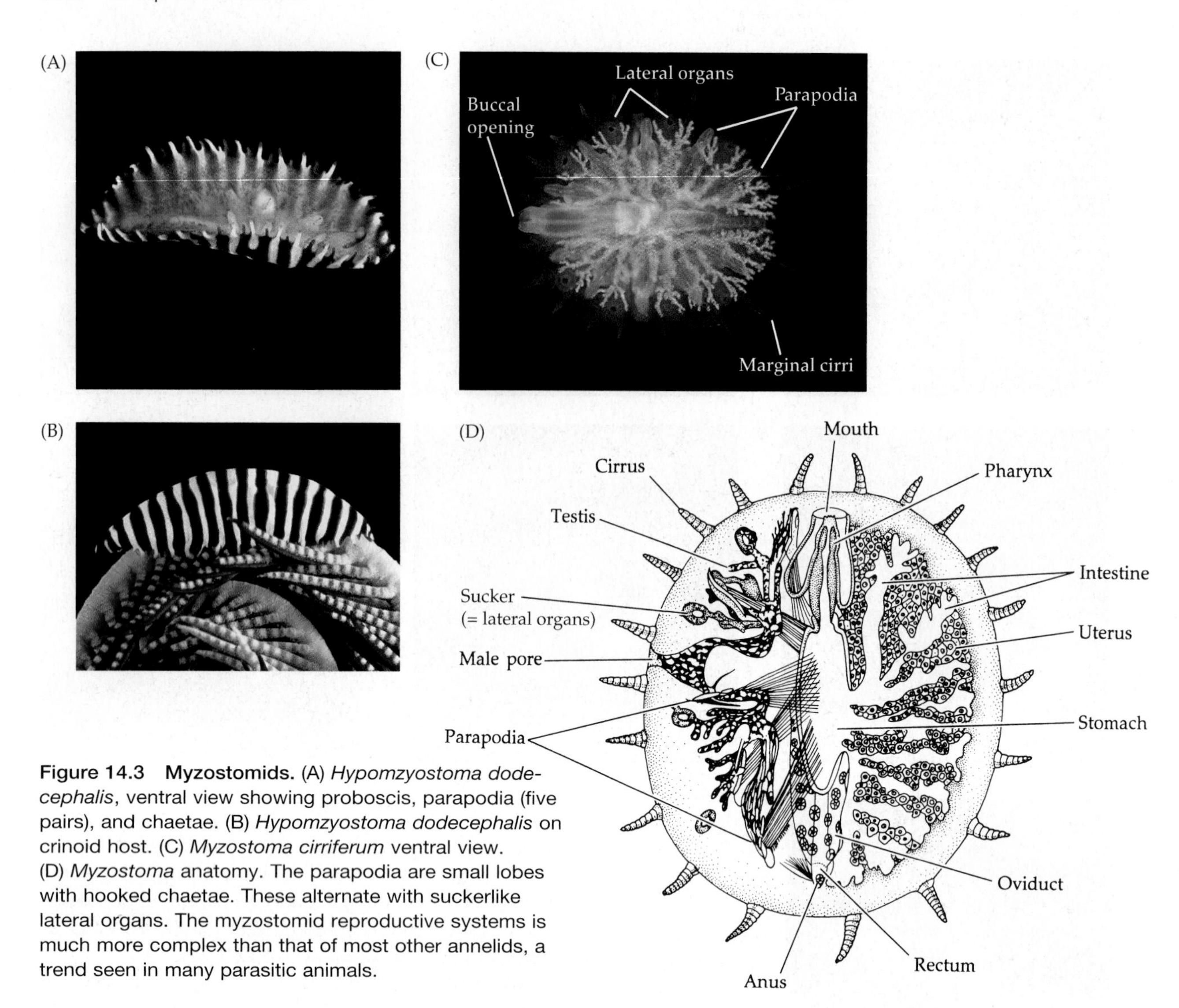

Figure 14.3 Myzostomids. (A) *Hypomzyostoma dodecephalis*, ventral view showing proboscis, parapodia (five pairs), and chaetae. (B) *Hypomzyostoma dodecephalis* on crinoid host. (C) *Myzostoma cirriferum* ventral view. (D) *Myzostoma* anatomy. The parapodia are small lobes with hooked chaetae. These alternate with suckerlike lateral organs. The myzostomid reproductive systems is much more complex than that of most other annelids, a trend seen in many parasitic animals.

EUNICIDA Eunicida is a well-delineated group of over 1,000 species that have a ventral muscularized pharynx with complex jaws that include elements such as ventral mandibles and dorsal maxillae. Their parapodia are supported by rodlike chaetae called aciculae and often bear compound chaetae. They generally have three antennae and a pair of sensory palps on the head. Eunicida include some of the smallest and largest known annelids. Currently with seven family-ranked taxa (four of which are described below).

DORVILLEIDAE 178 species (Figures 14.1I and 14.8F). Includes some of the smallest annelids (*Neotenotrocha*), while others may reach several centimeters and have a large number of segments (e.g., *Dorvillea*). The head often has three antennae and there can be a pair of palps that may either resemble or differ from the shape of the antennae. *Ophryotrocha* is a well-known "model annelid" as they are also easily kept in culture on spinach or similar foods. Dorvilleids are also very commonly found in extreme habitats such as hydrothermal vents, methane seeps, and whale falls, where they are bacteriovores.

EUNICIDAE 362 species (Figure 14.17H). A group that contains arguably the largest of annelids, some exceeding 3 m in length, with only a megascolecid earthworm species coming near this size. Usually with three antennae and a pair of palps that resemble antennae. Motile, though also living in burrows or temporary tubes in mucous or parchment-like tubes. Carnivores, omnivores or herbivores. Famous examples include Palolo worms and the Bobbitt worms. (e.g., *Eunice*, *Marphysa*, *Palola*)

LUMBRINERIDAE 275 species (Figure 14.2H). Thin and elongate and, except for *Lysarete* and *Kuwaita*; lacking head appendages. Most crawl about in algal mats and holdfasts, and small cracks in hard substrata; some burrow in sand or mud. Carnivores, scavengers, detritivores, and deposit feeders. (e.g., *Lumbrinerides*, *Lysarete*, *Ninoe*)

OENONIDAE 87 species. Elongate, with small parapodia; lacking head appendages, or with three small antennae. Resembling Lumbrineridae, though their jaws and chaetae differ. Often found in soft substrata where they burrow aided by secretion of copious amounts of mucus. Predatory carnivores; many have an endoparasitic juvenile stage living inside other annelids. (e.g., *Arabella*, *Drilonereis*)

ONUPHIDAE 272 species (Figure 14.2I). Closest relatives are Eunicidae, though onuphids generally have prominent gills on anterior segments. Most live in tubes, but others roam through sediments. Most tube dwellers are sessile (e.g. *Diopatra*), while others carry their tubes with them. Generally scavengers or predators. Interesting forms include quill worms (*Hyalinoecia*) and the Australian beach worms (*Australonuphis,* etc.).

PHYLLODOCIDA

Phyllodocida, with more than 4,600 species, is distinguished by its members having an axial muscular proboscis (often armed with two or more jaws). They tend to show fusion of some anterior segments with the head, often only identifiable by the retention of anterior enlarged cirri, and the head usually has two or three antennae and a pair of sensory palps. Like Eunicida, their parapodia are also often supported by aciculae and bear compound chaetae. Currently with around 20 family-ranked taxa; the more species-rich ones are introduced here.

APHRODITIFORMIA Scale worms comprise a diverse group of over 1,000 species in seven (or eight) family-ranked taxa (Figures 14.1H, 14.5E, 14.11A,B): Acoetidae, Aphroditidae, Eulepethidae, Polynoidae (contains most species), Iphionidae, Sigalionidae, and possibly Pholoidae. Aphroditiformia also includes the scaleless scale worms of the former family Pisionidae. Most are relatively short and somewhat flattened dorsoventrally; one highly unusual Antarctic species, *Eulagisca gigantea*, reaches a length of nearly 30 cm and a width of about 15 cm. Most also have relatively few segments at adulthood, some sigalionids being exceptions. Most of the dorsal surface is normally covered by transformed flattened cirri (called elytra, or scales), hence the common name. The cover photo of this book shows the scale worm *Arctonoe pulchra* on the warty sea cucumber (*Apostichopus parvimensis*); this species is a common commensal on echinoderms and molluscs (in mantle cavities) and also lives freely in tide pools of the temperate northeast Pacific. The eversible pharynx has one pair of jaws that close dorsoventrally somewhat like a parrot's beak. Most scale worms are motile but usually cryptic (under stones, etc.). Many are predators while others are bacteriovores. Many scale worms are commensals, living on the bodies or in the dwellings of other animals. (e.g., *Arctonoe*, *Gorgoniapolynoe*, *Halosydna*, *Harmothoe*, *Hesperonoe*, *Polynoa*)

CHRYSOPETALIDAE 135 species (Figure 14.2D). The name (Latin for "golden petals") refers to the shape and color of the golden, flattened notochaetae (called paleae) that cover the dorsal surface in many species (e.g., *Chrysopetalum*). These are small to moderately sized worms, with adults varying in length from 1 to 50 mm, and with as few as 10 to over 300 segments. Those with the dorsum covered by flattened paleae tend to be yellowish brown to golden, sometimes with transverse stripes. One clade, Calamyzinae, contains parasitic forms that live in bivalve molluscs (these were formerly in their own family, Nautiliniellidae, as they show little resemblance to other chrysopetalids; e.g., *Shinkai*).

GLYCERIDAE 89 species (Figures 14.5B, 14.8D and 14.12C). Cylindrical, tapered, homonomous body, usually red or pink, reaching 30 cm in length. Enormous eversible pharynx that can be retracted into 1/3 of the body length, armed with four hook-like jaws used in prey capture; each jaw has a venom gland. The pharynx is also used in burrowing. Most are infaunal burrowers in soft substrata. (e.g., *Glycera*, *Glycerella*, *Hemipodus*)

HESIONIDAE 172 species (Figure 14.11D). Generally beautiful worms, adults measuring from a few millimeters in length to more than 10 cm. Subtidal, especially on rocky and mixed bottoms, and increasingly known from hydrothermal vents and methane seeps. A famous hesionid is the ice worm, *Sirsoe methanicola,* which lives on methane hydrates in the Gulf of Mexico. Many Hesionidae have striking pigmentation patterns and some are commensals, particularly with echinoderms. The number of segments in adults may be fixed at 21 (e.g., *Hesione*, *Leocrates*), or varying up to about 50–60.

NEPHTYIDAE 142 species (Figure 14.5C). Usually long and slender, with well-developed parapodia but simple heads. They can be very abundant in shallow-water sediments, and there are often a number of different taxa in a single sediment sample. While they are easily identified to this family level, species identification is difficult. Adults range from a few millimeters to 30–40 cm and up to 150 segments. Eversible jawed pharynx used in prey capture and burrowing. (e.g., *Aglaophamus*, *Micronephtyes*, *Nephtys*)

NEREIDIDAE 691 species (Figures 14.2E, 14.8A–C, 14.13A and 14.17D). Among the best-known marine annelids and widely used in teaching, laboratories, and as fishing bait. Most nereidids are found in shallow waters, though some are found at hydrothermal vents. Small to very large (over 100 cm) with homonomous segments. Mostly errant predators or scavengers with well-developed eyes and parapodia. Immediately recognized by their pair of large, curved pharyngeal jaws (e.g., *Cheilonereis*, *Dendronereis*, *Neanthes*, *Nereis*, *Platynereis*). One group, Namanereidinae, is semiterrestrial or lives in freshwater. (e.g., *Lycastella*, *Namanereis*)

PHYLLODOCIDAE 417 species (Figures 14.5F, 14.8E and 14.16D). Thin, often elongate (over 50 cm) bodies of up to 700 homonomous segments; commonly active epibenthic predators on solid substrata; a few burrow in mud (e.g., *Eteone*, *Eulalia*, *Notophyllum*, *Phyllodoce*). One clade, Alciopinae, comprise holopelagic forms, in which the body is transparent except for pigment spots and a large pair of lensed eyes. (e.g., *Alciopa*, *Alciopina*, *Torrea*, *Vanadis*)

SPHAERODORIDAE 171 species. Easily recognized by conspicuous tubercles and/or papillae all over the body, generally arranged in transverse rows. Adults range from a few millimeters to several centimeters. Bodies can be short and grub-like with up to about 30 segments (e.g., *Sphaerodoridium, Sphaerodoropsis*), or elongate and slender with a larger number of segments. (e.g., *Ephesiella*, *Sphaerodorum*)

SYLLIDAE 906 species (Figures 14.2G, 14.11C 14.17C,G). Mostly small, homonomous worms found on various substrata. Best known for their diversity in reproductive biology, including various forms of epitoky. Adult sizes from 1 to 150 mm; with only a few segments or with many segments. Syllids show some of the most striking coloration patterns among annelids. Mainly predators on small invertebrates. The pharynx has a distinct barrel-shaped region that may be armed with a single tooth or a ring of small teeth for grasping prey. (e.g., *Autolytus*, *Brania*, *Odontosyllis*, *Syllis*, *Trypanosyllis*)

PELAGIC PHYLLODOCIDA A polyphyletic grouping of about 150 pelagic species. In addition to Alciopinae, several other groups of Phyllodocida have independently evolved into holopelagic forms, though the number of evolutionary events has yet to be fully resolved. Holopelagic forms are included in the families Iospilidae, Lopadorhynchidae (Figure 14.2F), Pontodoridae, Tomopteridae, Typhloscolecidae. Most are likely to be predators. Tomopterids are spectacular forms with transparent flattened bodies, finlike parapodia, and only a few chaetae.

SEDENTARIA Another previously defunct taxonomic group resurrected by Torsten Struck and colleagues and containing over 13,000 species. It comprises a series of major groups, though the relationships among them are not fully resolved. The former phyla Echiura and Pogonophora (and Vestimentifera) are placed here, as well as a variety of tube-dwelling and burrowing forms.

ORBINIIDAE 175 extant species (Figures 14.2J and 14.16G). Prostomium can be rounded or pointed, without appendages. Adults from 3 to 300 mm long and large forms can have several hundred segments; usually with an anterior "muscular" thoracic region and a more fragile abdomen. Usually with complex parapodia and a wide range of chaetae. Generally burrowing forms, they are usually found in the sediments of shallow bays and estuaries (e.g. *Orbinia*, *Scoloplos*), but also have been recorded from hydrothermal vents and methane seeps. (e.g. *Methanoaricia*)

CIRRATULIFORMIA

ACROCIRRIDAE Around 50 species. Often found intertidally, under rocks or in shallow sediments and muds. The head is simple with a rounded prostomium and a pair of grooved peristomial palps used for feeding. The first four segments each bear a pair of simple unbranched branchiae, which are easily lost. Adults reach 5 to 150 mm and have from as few as ten segments, to more than 200. Live acrocirrids tend to be yellowish to greenish-brown in color (e.g., *Acrocirrus*, *Macrochaeta*). An extraordinary new group of holopelagic Acrocirridae was recently described in which the anterior branchiae have transformed into bioluminescent structures that glow when shed, presumably to distract predators. (e.g., *Swima*)

CIRRATULIDAE Around 250 species (Figures 14.2M and 14.17B). Elongate, relatively homonomous, with up to 350 segments, often each with a pair of thread-like branchial filaments (e.g., *Cirratulus*, *Cirriformia*). In others there may be only four pairs of branchiae anteriorly (e.g., *Dodecaceria*), or none at all in smaller forms (e.g., *Ctenodrilus*). Cirratulids are mostly shallow-water burrowers lying just beneath the surface of the sediment, from where they extend their branchiae into the overlying water. Most are selective deposit feeders, extracting organic detritus from surface sediments using grooved palps that may be a simple pair (e.g., *Chaetozone*, *Dodecaceria*) or transformed into two clusters. (e.g., *Cirriformia, Timarete*)

FLABELLIGERIDAE 264 species (Figure 14.2N). Sometimes called bristle-cage worms. With papillae covering the body; papillae often sticky, hence some have a sediment coating. Others have a thick, transparent gelatinous sheath, allowing the body contents and green circulatory system to be seen. The head, which has a pair of grooved palps, and presumably some achaetous anterior segments bearing branchiae, is often retractable into the following anterior segments that bear a "cage" of protective chaetae. Mostly benthic, from the intertidal under stones to deep-sea muds, though there are two holopelagic groups, *Poeobius* and *Flota*, that were in their own families until recently. Adults are 5 mm to more than 10 cm in length. (e.g., *Brada*, *Pherusa*, *Spio*, *Spiophanes*)

OPHELIIDAE 153 species (Figure 14.2K). Homonomous, usually less than 3 cm long, with up to 60 segments. Body shape varies from short and thick to elongate and somewhat tapered. Most opheliids burrow in soft substrata, but some can swim by undulatory body movements. The eversible pharynx is unarmed. Most are direct deposit feeders. (e.g., *Armandia*, *Euzonus*, *Ophelia*, *Polyophthalmus*)

SPIONIDA

SPIONIDAE 527 species (Figures 14.1G and 14.9F). Body thin, elongate, homonomous. The head is simple with a pair of grooved peristomial palps. Most burrow, or form delicate sand or mud tubes. A few bore into calcareous substrata, including rocks and mollusc shells; most use the grooved peristomial palps to selectively extract food from the sediment surface. (e.g., *Polydora*, *Scolelepis*, *Spio*, *Spiophanes*)

SABELLARIIDAE 124 species (Figures 14.2S and 14.7H,I). Sabellariidae are easily distinguished from most other annelids by having an operculum that is developed from anterior segments and a robust tube built from coarse sand grains. The operculum comprises two fleshy lobes that are fused (e.g., *Sabellaria*), or free (e.g., *Lygdamis*), and it includes 1–3 rows of large golden or black stout chaetae (paleae). Generally found in intertidal to slightly subtidal areas, though deep-water forms are known. Sabellariids can form extensive biogenic reefs and can reach densities of up to 6000 individuals/m^2 (e.g., *Phragmatopoma*, *Sabellaria*), while others are solitary. (e.g., *Lygdamis*)

SABELLIDA

SABELLIDAE 460 species (Figures 14.2R, 14.7A and 14.16E). Commonly called fan worms or feather-duster worms. Sabellids live in sediment and mucous tubes and are common at all depths. The body is heteronomous, divided into a thorax and an abdomen, and can be 3 to 300 mm long. The thorax bears long dorsal "capillary" or hooded chaetae and ventral hooks, while the abdomen shows the inverse (chaetal inversion). The prostomium is a crown of branched, feathery palps (radioles) that projects from the tube and functions in gas exchange and ciliary suspension feeding (e.g., *Bispira*, *Eudistylia*, *Myxicola*, *Sabella*, *Schizobranchia*) and may bear simple (e.g., *Demonax*) or compound eyes (e.g., *Megalomma*). Some species bore into calcareous substrates (e.g., *Pseudopotamilla*). Sabellidae also once included Fabriciidae, all small-bodied feather-duster worms, but these have been shown to be more closely related to Serpulidae and were placed in their own family.

SERPULIDAE 397 species (Figures 14.7E,F,J and 14.20A–C). Similar to sabellids, except the secreted tube is calcareous and usually attached to rocks. The body is heteronomous, divided into thorax and abdomen, ranging in size from 3 to 200 mm. Anterior end bears radiolar crown as in sabellids; many species have one radiole transformed into an operculum that plugs the end of the tube when the worm withdraws (e.g., *Hydroides*, *Serpula*, *Spirobranchus*), though many lack this (e.g., *Filograna, Protula*). The most well-known serpulid group is *Spirobranchus*, the Christmas tree worms, that bore into coral substrates and have colorful spiral radiolar crowns (Figure 14.7J). One speciose clade, Spirorbinae, contains small-bodied forms that secrete coiled calcareous tubes. (e.g., *Circeis*, *Paralaeospira*, *Spirorbis*)

SIBOGLINIDAE 164 species (Figures 14.1E, 14.31–14.33). Previously known as the phyla Pogonophora and Vestimentifera, subsequent morphological and molecular analyses placed the group well inside Annelida; the original family name, Siboglinidae, has since been generally adopted. Siboglinidae all appear to live via symbiotic bacteria in their bodies and are generally found at greater than 1,000 m depth, even down to nearly 10,000 m, though exceptionally they are found in depths of less than 100 m. Though most are known from deep-sea muds (e.g., *Siboglinum*), mud volcanoes, or on sunken plant material (e.g., *Sclerolinum*), vestimentiferan siboglinids (e.g., *Riftia*) are spectacular members of hydrothermal vent communities or methane seeps (e.g., *Lamellibrachia*). Siboglinidae vary greatly in size, with adults of some *Siboglinum* reaching 5 cm in length (and only 0.1 mm in width), while *Riftia pachyptila*, are more than 150 cm in length and live in tubes more 2.5 m long. The most recently discovered group of Siboglinidae is the genus *Osedax*, a group that devours the bones of marine vertebrates by dissolving through them with tissue that resemble plant roots.

MALDANOMORPHA

ARENICOLIDAE 20 species (Figures 14.2O and 14.5H). Arenicolids or lug worms have simple heads and a thick, fleshy, heteronomous body divided into two or three distinguishable regions; the pharynx is unarmed but eversible and aids burrowing and feeding. Most arenicolids live in J-shaped burrows in intertidal and subtidal sands and muds, where they are direct deposit feeders. (e.g., *Abarenicola*, *Arenicola*)

MALDANIDAE 283 species (Figure 14.7C,D). Known as bamboo worms because of the long cylindrical segments with ridge-like parapodia that resemble stalks of bamboo. Like arenicolids, the head has no appendages; maldanids are 0.3 cm to more than 20 cm in length. They usually have 20–30 segments and the body does not taper posteriorly as in most other annelids.

TEREBELLIFORMIA

AMPHARETIDAE 282 species (Figures 14.2P and 14.11D). Ampharetids are tubicolous and easily distinguished from the similar Terebellidae in that their multiple grooved palps, usually called tentacles, can be retracted into the mouth. Generally four pairs of branchiae. Relatively uncommon in intertidal and shallow waters; in recent years, most new taxa have been described from deeper sediments. Includes the unusual hydrothermal vent group from the Pacific Ocean, the Pompeii worms, once in their own family, Alvinellidae (e.g., *Alvinella*, *Paralvinella*). Ampharetidae are 0.5 cm to 6 cm long as adults. The blood is often greenish owing to the presence of chlorocruorin, but is red from hemoglobin in some species. The body consists of two distinct regions in addition to the head, a thoracic region that generally has biramous parapodia and an abdomen that has neuropodia only. (e.g., *Ampharete, Amphicteis*, *Melinna*)

PECTINARIIDAE 53 species (Figures 14.7G and 14.9G). The ice-cream cone worms. Body short and conical with stout golden chaetae projecting from the head. Also easily recognizable from their elegant cone-shaped tubes, constructed from sand, small shells or other small particles, open at both ends. Body length

1 to 10 cm, with no more than 20 segments bearing chaetae. Multiple grooved palps ("tentacles") used to feed on detritus extracted from sediment. (e.g., *Pectinaria, Petta*)

TEREBELLIDAE 535 species (Figures 14.2Q and 14.9B–E). Tube dwelling, though many are burrowers or "creepers" and some are even capable of swimming. The grooved palps are often seen extending out over the sediment in shallow marine waters. When disturbed, the tentacles, often brightly colored, are retracted back toward the worm, though they are not retracted into the mouth. Usually three pairs of branchiae. The body usually consists of two distinct regions in addition to the head, a thoracic region that generally has biramous parapodia and a long tapering abdomen that has neuropodia only. (e.g., *Amphitrite, Pista, Polycirrus, Terebella, Thelepus*)

CAPITELLIDA

CAPITELLIDAE 173 species (Figure 14.2L). Easily recognized by the division of the body into an anterior region with capillary chaetae only and a posterior region with long-handled hooks. The body is a simple cylindrical shape, resembling Clitellata. The head has no appendages and is a simple conical structure in most. Body less than 1 cm to more than 20 cm in length; usually bright red. Extensions of the body wall, often erroneously called "branchiae" (erroneous, since capitellids lack a circulatory system), are present in abdominal segments of some taxa. Some, such as *Capitella,* are well known as pollution indicators since their numbers explode in nutrient-rich conditions that exclude many other annelid species.

ECHIURIDAE Around 200 species (Figures 14.1C and 14.28–14.30). Spoon worms, anchor worms. Unusual annelids in that they have a muscular extensible preoral proboscis at the anterior end of an apparently unsegmented trunk. The proboscis cannot be withdrawn into the mouth, but can be extended for several meters in some species. The trunk ranges from 1 to 40 cm in length and bears a single pair of chaetae anteriorly, or both anteriorly and with rings of chaetae posteriorly. Echiuridae were generally considered as annelids, but W. W. Newby proposed a separate phylum, Echiura, based on a detailed embryological study of *Urechis caupo.* His proposal was generally accepted until Damhnait McHugh provided molecular evidence that they are, in fact, annelids. Subsequent morphological studies supported her conclusion and recent molecular analyses have shown the group to be closely related to Capitellidae. Echiurids appear to be unsegmented, but the echiurid nervous system proceeds from anterior to posterior during embryogenesis, indicating the occurrence of a posterior growth zone (teloblasty). The taxonomy of the group has yet to be fully reconciled with its former status as a phylum. In line with the logic presented for Pogonophora and Vestimentifera becoming Siboglinidae, McHugh suggested they should be treated as a family Echiuridae, which is adopted here.

CLITELLATA Around 6,000 species. Earthworms, leeches, and related forms. Without parapodia; chaetae usually greatly reduced or absent. Hermaphroditic, with complex reproductive systems. A distinct ring, the clitellum, functions in cocoon formation; development is direct. With few or no chaetae; cephalic sensory structures reduced; body externally homonomous except for clitellum. Mostly terrestrial or freshwater annelids, although there are also many marine species.

CAPILLOVENTRIDAE Five species. The morphology and arrangement of their chaetae resemble those of some errantiate annelids, but the presence of a clitellum and other features reveals they are clitellates. Interestingly, both morphological and molecular evidence suggests they are the sister group to all other clitellates and lends support to the hypothesis that Clitellata has an aquatic origin. Two of the species are marine, two live in freshwater, and one in brackish water. A single genus, *Capilloventer*.

NAIDIDAE (formerly known as Tubificidae). 700 species. Range in length from a few mm to several cm. The best known of the freshwater clitellates, though some live in marine or brackish water. Some build tubes, others are burrowers. Usually the body is homonomous throughout with a simple head, though several species bear an elongate prostomial proboscis; some have branchiae. Many reproduce asexually, but most have gonads at some stage of development. Some species are very common in areas of high pollution. Some small, gutless species that rely on symbiotic bacteria for nutrition are known from tropical regions (e.g., *Inanidrilus*, *Olavius*). Other genera include *Branchiodrilus*, *Dero*, *Ripistes*, *Slavina*, *Stylaria, Branchiura*, *Clitellio*, *Limnodrilus*, and *Tubifex*.

CRASSICLITELLATA Earthworms and their allies (Figures 14.1F, 14.12F, 14.15F, and 14.18). Approximately 3,000 valid species that are mainly terrestrial, but some taxa (e.g., Biwadrilidae and Almidae) live in aquatic or semi-aquatic environments. The most species-rich families are Megascolecidae, Lumbricidae, and Glossoscolecidae. Crassiclitellata include the common terrestrial earthworms, including *Lumbricus terrestris*, which has been spread to soils across the world from its native Europe. Crassiclitellates are relatively large, with well-developed and complex reproductive systems. They are direct deposit feeders, living in soil, feeding on live and dead organic matter. The largest of all clitellates belong to Megascolecidae, for example the giant Gippsland earthworm, *Megascolides australis* (an endangered species native to Australia), can stretch to about 3 m in length, though their normal length is said to be around 1 m long and 2 cm in diameter.

ENCHYTRAEIDIDAE 700 species. Found mainly in soil, but also in a wide range of freshwater and marine/

brackish water habitats. Marine enchytraeids are common in intertidal sands, but they are also known from deep-sea sediments. Those of the genus *Mesenchytraeus* are known as ice worms as they thrive in glacial ice.

LUMBRICULIDAE Over 150 species (Figure 14.18H). Worms of moderate size, found in marshes, streams, and lakes. The group shows a great deal of endemism, mainly in Siberia (notably Lake Baikal) and the western parts of North America (e.g., *Lamprodilus*, *Rhynchelmis*, *Stylodrilus*, *Styloscolex*, *Trichodrilus*). This group is the closest relative to Hirudinoidea.

HIRUDINOIDEA Leeches and their relatives (Figures 14.34–14.40). Body with fixed number of segments, each with superficial annuli; chaetae generally absent; heteronomous, with clitellum and a posterior and usually an anterior sucker; most live in freshwater or marine habitats, a few are semiterrestrial; ectoparasitic, predaceous, or scavenging. There are three major clades within Hirudinoidea.

ACANTHOBDELLIDA Two species, *Acanthobdella peledina* and *Paracanthobdella livanowi*, both from northern hemisphere freshwater lakes (Figure 14.34A). Part of the animal's life is spent as an ectoparasite on freshwater fishes, especially salmonids and thymallids and presumably the rest of the time is spent in vegetation. Body with 30 segments, reaching 3 cm in length; with posterior sucker only in *Acanthobdella peledina*; chaetae on anterior segments; coelom partially reduced, but obvious and with intersegmental septa.

BRANCHIOBDELLIDA Around 150 species (Figure 14.34B). Usually less than 1 cm long; ectocommensal or ectoparasitic on freshwater crayfish; body with 15 segments; with anterior and posterior suckers; chaetae absent; coelom partially reduced, but spacious throughout most of the body. A single family, Branchiobdellidae. (e.g., *Branchiobdella*, *Cambarincola*, *Stephanodrilus*)

HIRUDINIDA Over 700 species (Figure 14.34C,D, 14.37, and 14.38). The "true" leeches. Around 100 species are marine, 90 are terrestrial, and the remainder freshwater. Many are ectoparasitic bloodsuckers, others are free-living predators or scavengers; some parasitic forms serve as vectors for pathogenic protozoa, nematodes, and cestodes. Body always with 34 segments; with anterior and posterior suckers; no chaetae; coelom reduced to a complex series of channels (lacunae). About 12 families, in two main groups. Rhynchobdellida (proboscis leeches) is a paraphyletic group of several families that contains the marine species as well as many freshwater forms (e.g., *Glossiphonia*, *Piscicola*). Arhynchobdellida is a clade that lacks a proboscis, although many have jaws; all are either freshwater or terrestrial. (e.g., *Erpobdella*, *Haemopsis*, *Hirudo*)

INCERTAE SEDIS

MYZOSTOMIDA 156 species (Figure 14.3). During the twentieth century, these animals were treated as annelids with a rank of family or order, or even as a separate class. Eeckhaut et al. (2000) suggested that Myzostomida are more closely related to Platyhelminthes than to any annelids. However, the pendulum swung back in favor of myzostomids being annelids starting with a molecular analysis by Bleidorn et al. (2007). Recent phylogenomic analyses have not been able to stabilize the phylogenetic position of the group, and while they clearly appear to be annelids, their closest relative is still not known, though there are morphological similarities with Phyllodocida. Myzostomids include several groups of flattened, oval, or elongate forms, always with five chaetae-bearing segments. They are mainly ectosymbionts, endosymbionts, or parasites of crinoid echinoderms, though a few are parasitic on Anthozoa.

The Annelid Body Plan

The annelid body is comprised of four regions: a presegmental region derived from the larval episphere, the prototroch region around the mouth, the serially repeated body segments, and the posterior pygidium. The episphere (see the section on Reproduction and Development) becomes the presegmental **prostomium**, and the prototroch and buccal region give rise to the **peristomium**, the region surrounding the mouth. The body segmentation of annelids is referred to as metamerism. The extreme posterior end of the body is the **pygidium** and it bears the anus and often some cirri. As with the prostomium, the pygidium is non-segmental and may well contain remnants of the larval body. From this basic scheme, the tremendous diversity of body forms seen in annelids is built. The majority of annelids show metamerism and a greatly elongate cylindrical body. They are triploblastic, have a through (complete) gut, and a spacious coelom. The exceptions with regard to metamerism are Sipuncula and Echiuridae, where this has been lost. In other groups, like Hirudinoidea and Myzostomida, the coelom is reduced. The gut (lost in some clitellates and in all Siboglinidae) is separated from the body wall by the coelom, except in those where the body cavity has been secondarily reduced.

The simplest annelid head is composed of a prostomium and a peristomium. The segmentation of the main body is usually visible externally as rings, or annuli, and is reflected internally by the serial arrangement of coelomic compartments separated from one another by intersegmental septa (Figure 14.4). This basic arrangement has been modified to various degrees among annelids, particularly by reduction in the size of the coelom or by loss of septa; the latter modification leads to fewer but larger internal compartments. The bodies of many annelids are **homonomous**, bearing segments that are very much alike.

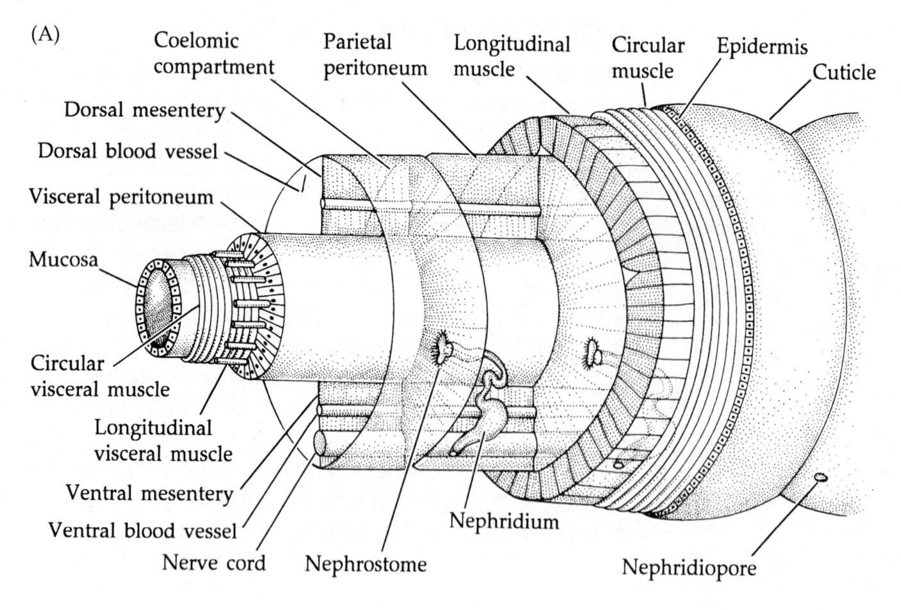

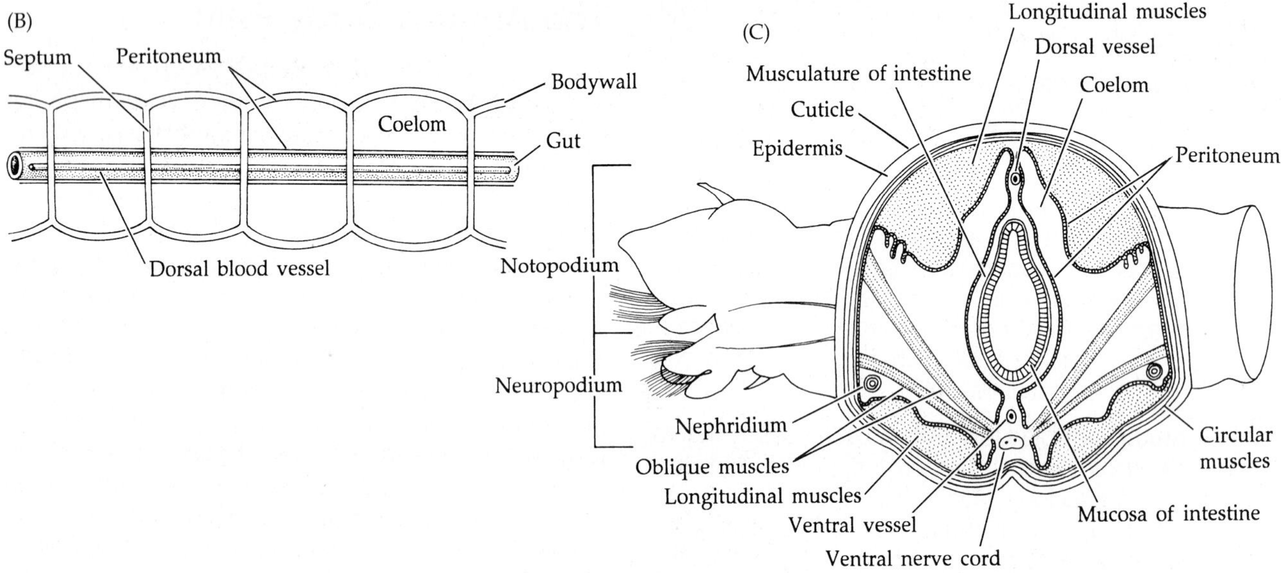

The current view of annelid phylogeny (Figure 14.41) hypothesizes Oweniidae, Magelonidae, Chaetopteridae, and Amphinomida forming a basal grade of four major groups, and in general these are homonomous annelids, suggesting that is the primitive condition for the phylum. Many others, particularly tubicolous forms, have groups of segments specialized for different functions and are thus **heteronomous**. Specialization of segment groups (**heteronomy**) has contributed greatly to morphological diversification among annelids. The hydraulic properties of different coelomic arrangements have allowed corresponding modifications in patterns of locomotion, which are responsible in part for the success of the annelids in a variety of habitats.

Body Forms

Most annelids are marine, living in habitats ranging from the intertidal zone to extreme depths. But quite a few inhabit brackish or fresh water, and the majority of Clitellata (thousands of species) live on land. Annelids range in length from less than 0.5 mm as adults for some interstitial species to over 3 m for some giant eunicids and megascolecid clitellates (earthworms). The myriad variations in body form among annelids can best be described relative to the basic annelid regions of a head, segmented trunk, and pygidium. Note though, that some groups such as Echiuridae, Hirudinea, Siboglinidae, and Sipuncula are so transformed from this condition, each in their own interesting ways, that they are treated as special sections near the end of this chapter. However, in general, the annelid head is composed of the prostomium and peristomium, which take various forms, and may also have one or more body segments fused with it, in which cases cirri and lateral chaetae may be present (Figure 14.8, 14.12A and 14.15). The prostomium and peristomium often bear appendages in the form of antennae and/or palps, or may be naked, as in many infaunal burrowers such as the clitellates. The nature of these head appendages

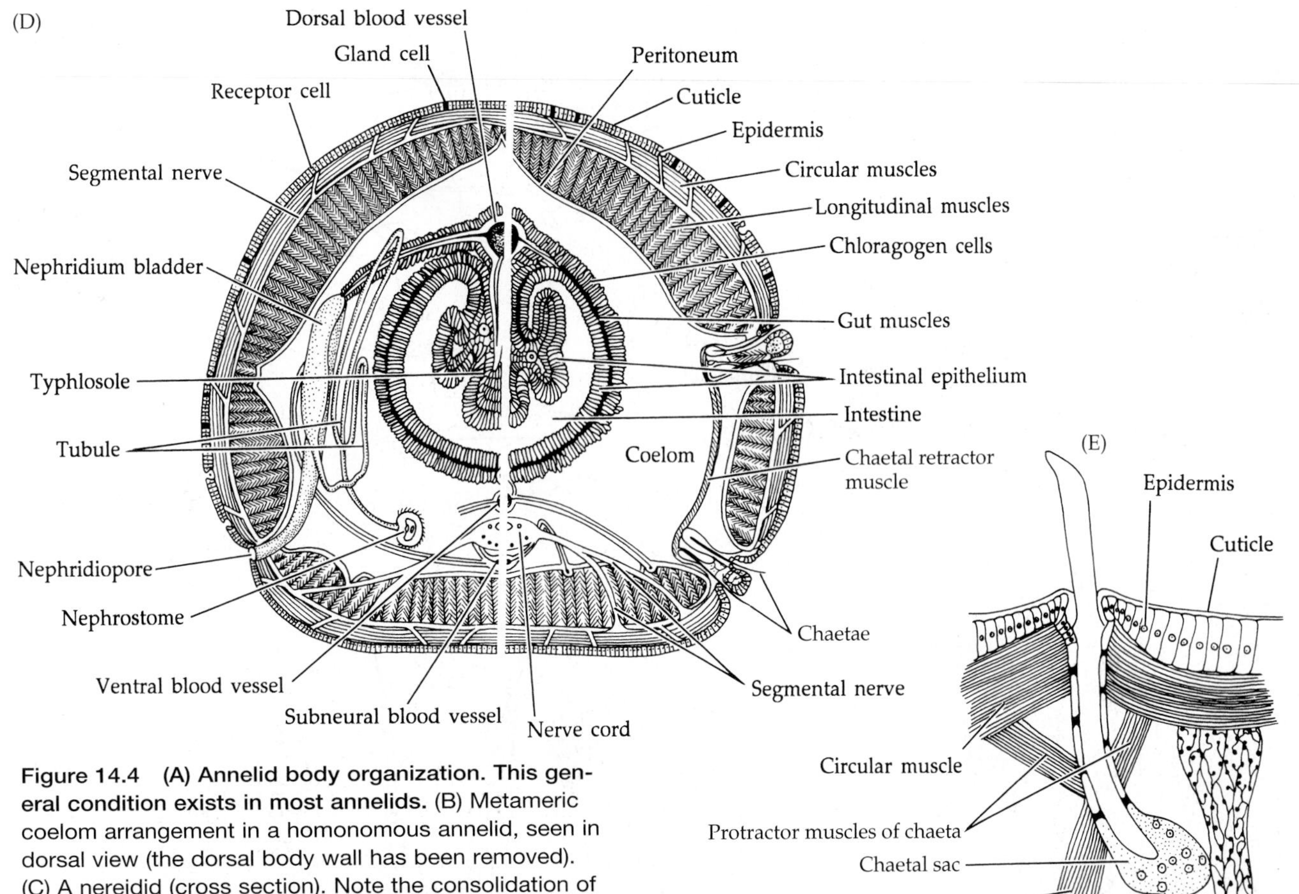

Figure 14.4 (A) Annelid body organization. This general condition exists in most annelids. (B) Metameric coelom arrangement in a homonomous annelid, seen in dorsal view (the dorsal body wall has been removed). (C) A nereidid (cross section). Note the consolidation of longitudinal muscles into nearly separate bands. (D) An earthworm (cross section). The left side of the illustration depicts a single nephridium and therefore the drawing is a composite of two segments; the right side of the illustration shows chaetae. (E) A chaeta and its associated musculature.

varies greatly and often reveals clues as to the worms' habits. The trunk may be homonomous or variably heteronomous as noted above, and each segment often bears a pair of unjointed appendages, called parapodia, and bundles of chaetae (Figure 14.4C and 14.5).

Chaetae are a unique feature of annelids, although very similar structures are found in some Brachiopoda. They come in a huge range of shapes and sizes and each is derived from the microvillar border of an invaginated epidermal cell and they are essentially bundles of parallel longitudinal canals, the walls of which are the sclerotized chitin (Figure 14.5). Chaetae have been lost in several annelid groups, most notably in Sipuncula and Hirudinea.

Parapodia, when present, are generally biramous, with a dorsal **notopodium** and a ventral **neuropodium**, each lobe with its own cluster of chaetae (Figures 14.4 and 14.5). However, annelids have evolved a huge diversity of parapodia that serve a variety of functions (locomotion, gas exchange, protection, anchorage, creation of water currents). In heteronomous annelids, the morphology of the parapodia may vary greatly in different body regions. Often, for example, parapodia in one region are modified as gills, in another region as locomotory structures, and elsewhere to assist in food gathering. In some cases, particularly in burrowing forms, as well as Sipuncula, Echiuridae and Clitellata, parapodia have been lost altogether.

Most annelids are gonochoristic and proliferate the gametes from the peritoneum into the coelom where they develop. Many annelids have broadcast spawning and produce free-swimming larvae, but there is a great variety of life history strategies in this phylum, with a wide range of kinds of parental care also present. Notably, many annelids are hermaphrodites, including all members of Clitellata, which typically exchange sperm and produce brooded or encapsulated embryos that develop directly to juveniles.

Body Wall and Coelomic Arrangement

The annelid body is covered by a thin cuticle, which is composed of scleroprotein and mucopolysaccharide fibers deposited by epidermal cellular microvilli. The epidermis is a columnar epithelium that is often ciliated on certain parts of the body. Beneath the epidermis lies a layer of connective tissue, circular muscles (sometimes absent) and thick longitudinal muscles,

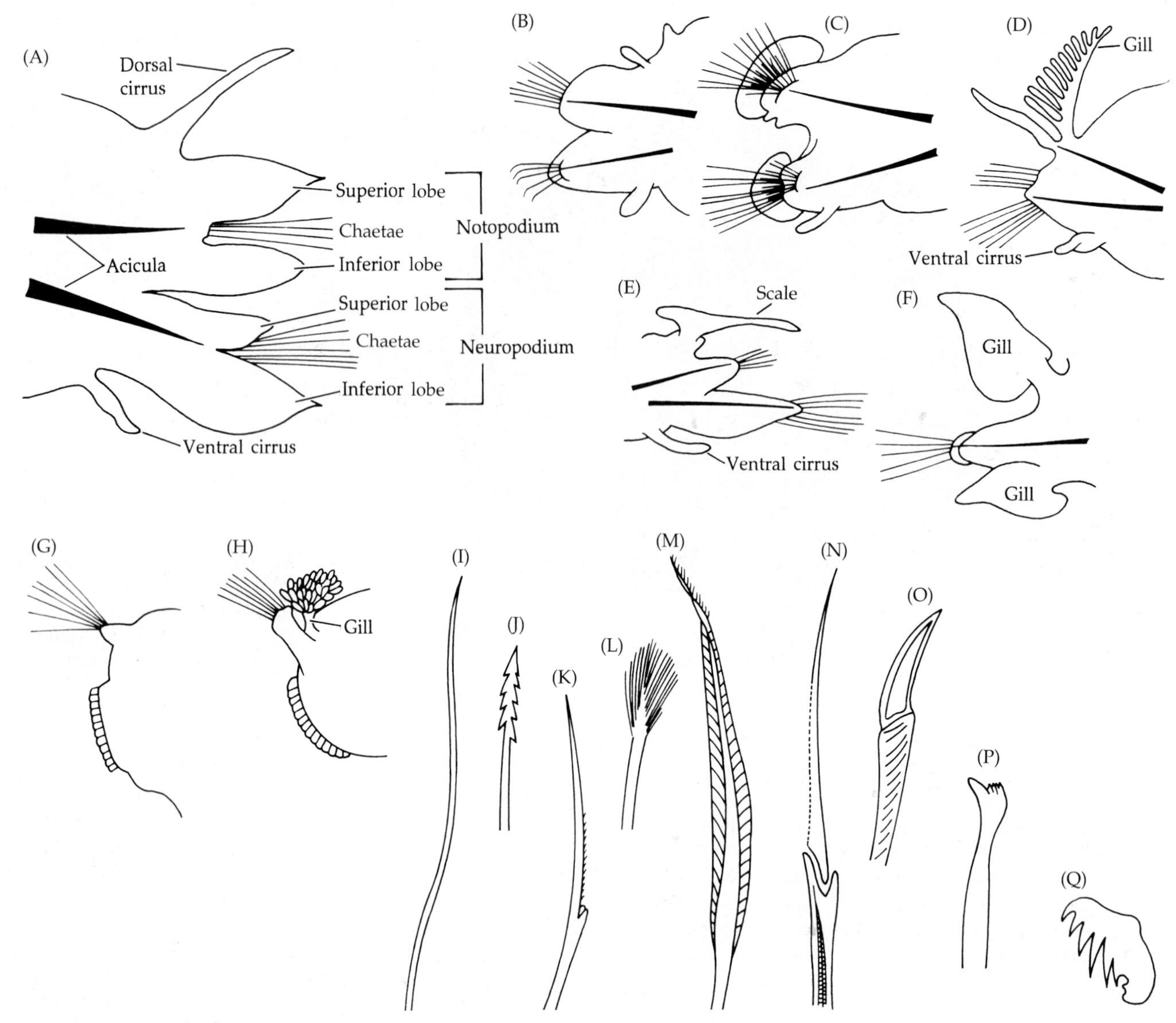

Figure 14.5 Parapodial and chaetal types among annelids. (A) A stylized parapodium. (B) The parapodium of a glycerid, with reduced lobes. (C) The parapodium of a nephtyid. (D) The parapodium of a eunicid with its modified notopodium; note the dorsal filamentous gill. (E) The parapodium of a polynoid (a scale worm) has the dorsal cirrus modified as a scale, or elytron. (F) The parapodium of a phyllodocid; the noto- and neuropodia are modified as gill blades. (G) The reduced parapodium of a tube-dwelling sabellid. (H) The parapodium of an arenicolid. (I–Q) Chaetae from various annelids. The classification of chaetae types rivals that of sponge spicules in complexity and terminology. A few general types are distinguished here as simple chaetae (I–M), compound chaetae (N–O), hooks (P), and uncini (Q).

the latter often arranged as four bands (Figure 14.4). The circular muscles do not form a continuous sheath, but are interrupted at least at the positions of the parapodia. The inner lining of the body wall is the peritoneum, which surrounds the coelomic spaces and lines the surfaces of internal organs. The coelom of annelids with homonomous bodies is generally arranged as laterally paired (i.e., right and left) spaces, serially (segmentally) arranged within the trunk. Dorsal and ventral mesenteries separate the members of each pair of coeloms, and muscular intersegmental septa isolate each pair from the next along the length of the body. In several annelid lineages, the intersegmental septa have been secondarily lost or are perforated, so in these animals the coelomic fluid and their contents, such as gametes, are continuous among segments. This is of course the case for Echiuridae and Sipuncula, where there are no segments.

In addition to the main body wall and septal muscles, other muscles function to retract protrusible or eversible body parts (e.g., branchiae, pharynx) and to operate the parapodia (Figure 14.4C). Each parapodium is an evagination of the body wall and contains a variety of muscles. Movable parapodia are operated primarily by sets of diagonal (oblique) muscles, which have their origin near the ventral body midline. These muscles branch and insert at various points inside the parapodium. In the clades Aciculata and Amphinomida, and some Orbiniidae, the large parapodia may contain stout internal chaetae called **aciculae** (Figure 14.5), on which some muscles insert and operate. In general, chaetae are also maneuvered by muscles and can usually be retracted and extended (quite unlike the setae of arthropods).

Support and Locomotion

Annelids provide a great example of the employment of coelomic spaces as a hydrostatic skeleton for body support. Coupled with the well-developed musculature, the metameric body, and the parapodia, this hydrostatic quality provides the basis for understanding locomotion in these worms. The genera *Nereis* and *Platynereis* (Phyllodocida, Nereididae) are errant, homonomous annelids that show a variety of locomotory patterns that are worth describing (Figure 14.2E and 14.6A–D). In such annelids the intersegmental septa are functionally complete, and thus the coelomic spaces in each segment can be effectively isolated hydraulically from each other. Modifications on this fundamental arrangement are discussed later.

In addition to burrowing (see below), *Nereis* can engage in three basic epibenthic locomotory patterns: slow crawling, rapid crawling, and swimming (Figure 14.6A–D). All of these methods of movement depend primarily on the bands of longitudinal muscles, especially the larger dorsolateral bands, and on the parapodial muscles. The circular muscles are relatively thin and serve primarily to maintain adequate hydrostatic pressure within the coelomic compartments. Each method of locomotion in *Nereis* (and similar forms) involves the antagonistic action of the longitudinal muscles on opposite sides of the body in each segment.

During movement, the longitudinal muscles on one side of any given segment alternately contract and relax (and are stretched) in opposing synchrony with the action of the muscles on the other side of the segment. Thus, the body is thrown into undulations that move in metachronal waves from posterior to anterior. Variations in the length and amplitude of these waves combine with parapodial movements to produce the different patterns of locomotion. The parapodia and their chaetae are extended maximally in a power stroke as they pass along the crest of each metachronal wave. Conversely, the parapodia and chaetae retract in the wave troughs during their recovery stroke. Thus, the parapodia on opposite sides of any given segment are exactly out of phase with one another.

When *Nereis* is crawling slowly, the body is thrown into a high number of metachronal undulations of short wavelength and low amplitude (Figure 14.6B). The extended parapodial chaetae on the wave crests are pushed against the substratum and serve as pivot points as the parapodia engage in its power stroke. As each parapodium moves past the crest, it is retracted and lifted from the substratum as it is brought forward during its recovery stroke. The main pushing force in this sort of movement is provided by the oblique parapodial muscles (Figure 14.4C). During rapid crawling, much of the driving force is provided by the longitudinal body wall muscles in association with the longer wavelength and greater amplitude of the body undulations (Figure 14.6C), which accentuate the power strokes of the parapodia.

Nereis can also leave the substratum to swim (Figure 14.6D). In swimming, the metachronal wavelength and amplitude are even greater than they are in rapid crawling. When watching a nereidid swim, however, one gets the impression that the "harder it tries" the less progress it makes, and there is some truth to this. The problem is that, even though the parapodia act as paddles pushing the animal forward on their power strokes, the large metachronal waves continue to move from posterior to anterior and actually create a water current in that same direction; this current tends to push the animal into reverse. The result is that *Nereis* is able to lift itself off the substratum, but then largely thrashes about in the water. This behavior is used primarily as a short-term mechanism to escape benthic predators rather than as a means to get from one place to another. There are, however, some annelid groups, such as Alciopinae (Phyllodocidae) and *Tomopteris* (Phyllodocida), whose members live their whole lives as pelagic swimming animals.

With these basic patterns and mechanisms in mind, we consider a few other methods of locomotion in annelids. *Nephtys* (Nephtyidae) superficially resembles *Nereis*, but its methods of movement are significantly different. Although *Nephtys* is less efficient than *Nereis* at slow walking, it is a much better swimmer, and it is also capable of effective burrowing in soft substrata. The large, fleshy parapodia serve as paddles and, when swimming, *Nephtys* does not produce long, deep metachronal waves. Rather, the faster it swims, the shorter and shallower the waves become, thus eliminating much of the counterproductive force described for *Nereis*. When initiating burrowing, *Nephtys* swims head first into the substratum, anchors the body by extending the chaetae laterally from the buried segments, and then extends the proboscis deeper into the sand. A swimming motion is then employed to burrow deeper into the substratum.

In contrast to the above descriptions, scale worms (Polynoidae; cover photo on book) have capitalized on the use of their muscular parapodia as efficient walking devices. The body undulates little if at all, and there is a corresponding reduction in the size of the longitudinal muscle bands and their importance in locomotion. These worms depend almost entirely on the action of

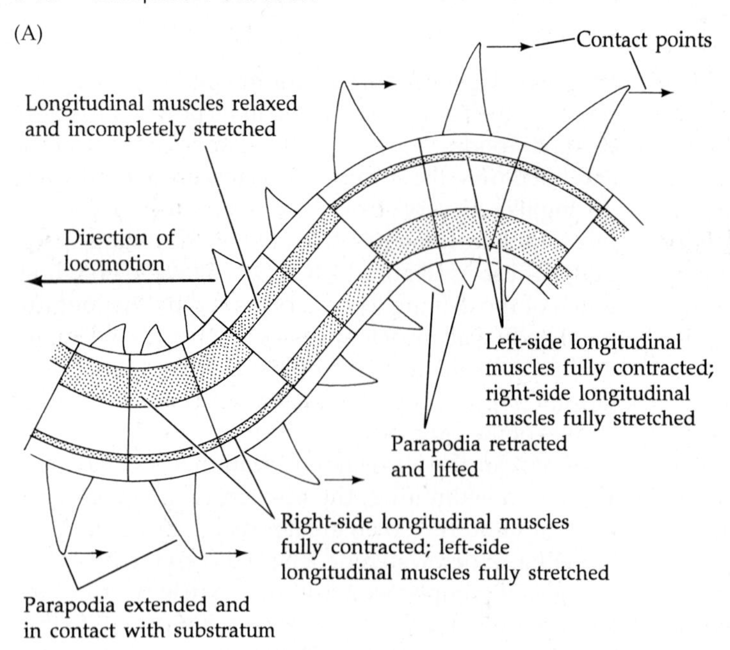

Figure 14.6 Patterns of locomotion in annelids. (A) Dorsal view of several segments of *Nereis* during crawling. Note the states of contraction of longitudinal muscles (stippled), the body curvature, and the retraction and extension of parapodia. (B–D) *Nereis* crawling and swimming. Note the changes in metachronal wavelength and amplitude. (E) Midsaggital section through an annelid. The perforated intersegmental septa allow peristaltic body contractions to cause volumetric changes in segments. (F) Burrowing movements in *Arenicola*. (G–J) An earthworm moving to the left. Every fourth segment is darkened for reference. The dotted line passes through a posteriorly moving point of contact with the substratum. (K) Several segments of an earthworm (sagittal section). Since each segment is a functionally isolated compartment, shortening and elongation accompany the contraction of longitudinal and circular muscles, respectively, while each segment essentially maintains a constant volume.

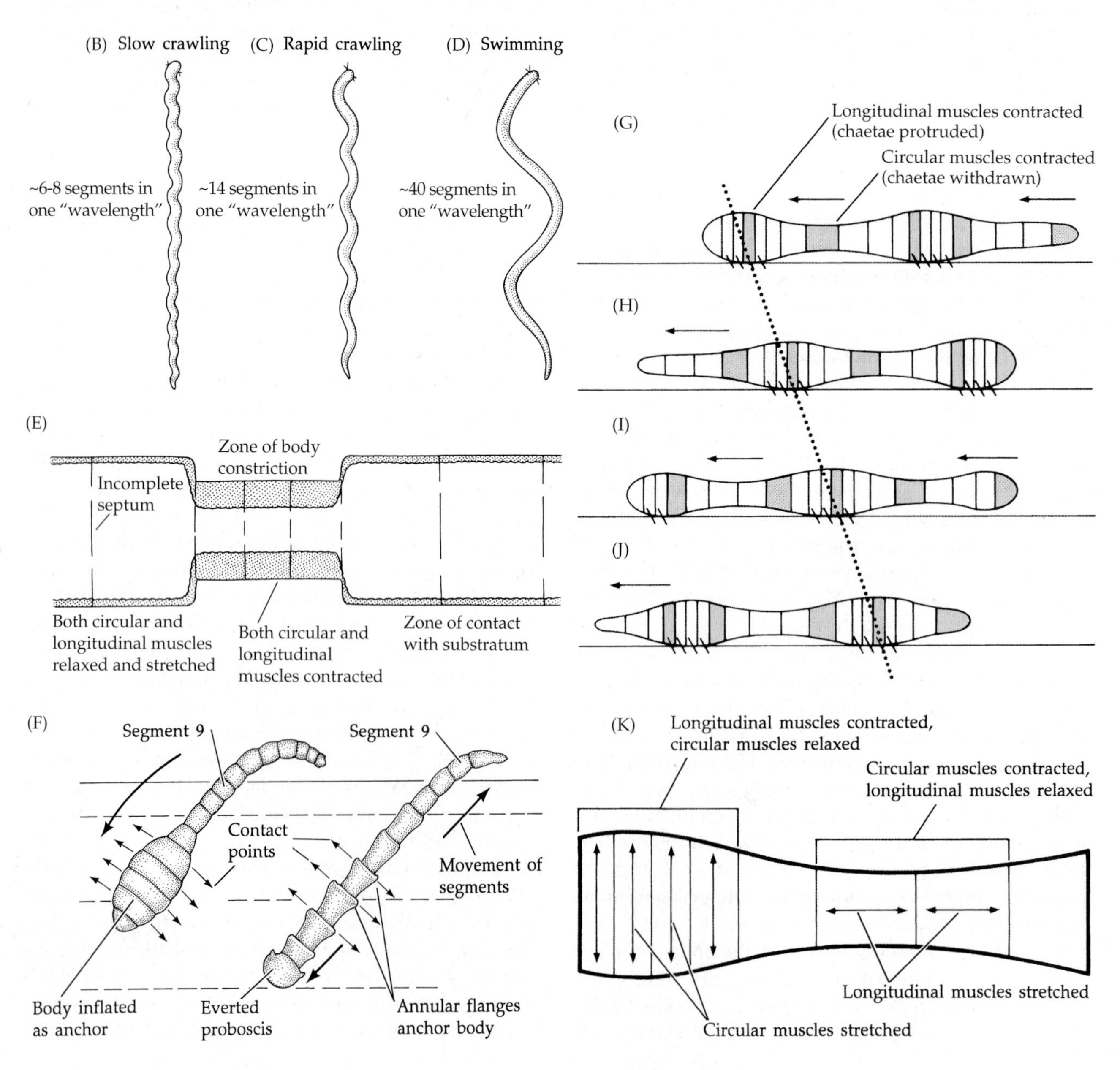

the parapodia for walking and most adult polynoids cannot swim, except for very short bursts.

Nereis is also a burrower and studies on *N. virens* showed that they can extend their burrows in muddy sediments by using mechanically efficient fracturing style of locomotion called "crack propagation." This is an efficient use of energy as the worm forces its way through the mud, and it involves everting the pharynx to apply dorsoventral forces to burrow walls that are amplified at the burrow tip making a crack in a forward direction. The worm then anchors itself by expanding the body laterally and pushing its narrow head into the crack, whereupon it repeats the process. Other annelid groups such as cirratulids, glycerids, and orbiniids are also known to use this form of burrowing. Burrowing in sand presents different challenges to burrowing in mud, since sand doesn't "crack," but instead needs to be fluidized. Sand burrowing has been described for *Arenicola* (Arenicolidae), which, like many burrowers, has lost most of the intersegmental septa, or have septa that are perforated. This means that segments are not of constant volume; in other words, a loss of coelomic fluid from one body region causes a corresponding gain in another. *Arenicola* also has reduced parapodia. The chaetae, or simply the surface of the expanded portions of the body, serve as anchor points, while the burrow wall provides an antagonistic force resisting the hydraulic pressure. *Arenicola* burrows in sand by first embedding and anchoring the anterior body region in the substratum. The anchoring is accomplished by contracting the circular muscles of the posterior portion of the body, thus forcing coelomic fluid anteriorly and causing the first few segments to swell. Then the posterior longitudinal muscles contract, thereby pulling the back of the worm forward. To continue the burrowing, a second phase of activity is undertaken. As the anterior circular muscles contract and the longitudinal bands relax, the posterior edges of each involved segment are protruded as anchor points to prevent backward movement; the proboscis is thrust forward, deepening the burrow. Then the proboscis is retracted, the front end of the body is engorged with fluid, and the entire process is repeated (Figure 14.6F).

Clitellates such as crassiclitellates (earthworms and their kin) face the problem of burrowing in soils, which may not crack or fluidize like marine sediments. It is argued that they solve this by eating their way through compacted soils, or pushing sediment aside in looser soils. In some cases they line the burrows with mucus to stop the burrows from collapsing. Movement in most clitellates involves the alternately contracting circular and longitudinal muscles within each segment, and this is seen in many other annelids with complete body septa. The shape of a segment changes from long and thin to short and thick with the respective muscle actions (Figure 14.6G–J). These shape changes move anteriorly along the body in a peristaltic wave generated by a sequence of impulses from the ventral nerve cord and associated motor neurons. So, at any moment during locomotion, the body of the worm appears as alternating thick and thin regions. Without some method of anchoring the body surface, this action would not produce any motion. The chaetae provide this anchorage as they protrude like barbs from the thicker portions of the body. When the longitudinal muscles relax and the circular muscles contract, the body diameter decreases and the chaetae are turned to point posteriorly and lie close to the body. As shown in Figure 14.6G–K, as the anterior end of the body is extended by circular-muscle contraction, the chaetae prevent backsliding, the head is pressed into the substratum, and the worm advances. The anterior end then swells by contraction of the longitudinal muscles, and the rest of the body is pulled along.

Most tube-dwelling annelids (Figure 14.7) are heteronomous and many, such as Terebelliformia, have rather soft bodies and relatively weak muscles. The parapodia are reduced, so the chaetae are used to position and anchor the animal in its tube. Movement within the tube is usually accomplished by slow peristaltic action of the body or by chaetae movements. When the anterior end is extended for feeding, it may be quickly withdrawn by special retractor muscles while the unexposed portion of the body is anchored in the tube. Annelid tubes provide protection as well as support for these soft-bodied worms, and also keep the animal oriented properly in relation to the substratum. Some annelids build tubes composed entirely of their own secretions. Most notable among these tube builders are the serpulids, which construct their tubes of calcium carbonate secreted by a pair of large glands near a fold of the peristomium called the collar. The crystals of calcium carbonate are added to an organic matrix; the mixture is molded to the top of the tube by the collar fold and held in place until it hardens.

Sabellids, close relatives of serpulids, produce parchment-like or membranous tubes of organic secretions molded by the collar. Some, such as *Sabella*, mix mucous secretions with size-selected particles extracted from feeding currents, then lay down the tube with this material (Figure 14.7A,B and 14.10A,B). Numerous other annelid groups form similar tubes of sediment particles collected in various ways and cemented together with mucus. Some of the most beautiful are made by pectinariids, the ice-cream cone worms, where the ornate tube is only one particle thick (Figure 14.7G).

A few annelids are able to excavate burrows by boring into calcareous substrata, such as rocks, coral skeletons, or mollusc shells (e.g., certain members of the families Cirratulidae, Eunicidae, Spionidae, Sabellidae). In extreme situations, the activity of the annelids may have deleterious effects on the "host." For example, a boring sabellid, *Terebrasabella*, which lives in abalone shells, can cause fatalities by making them

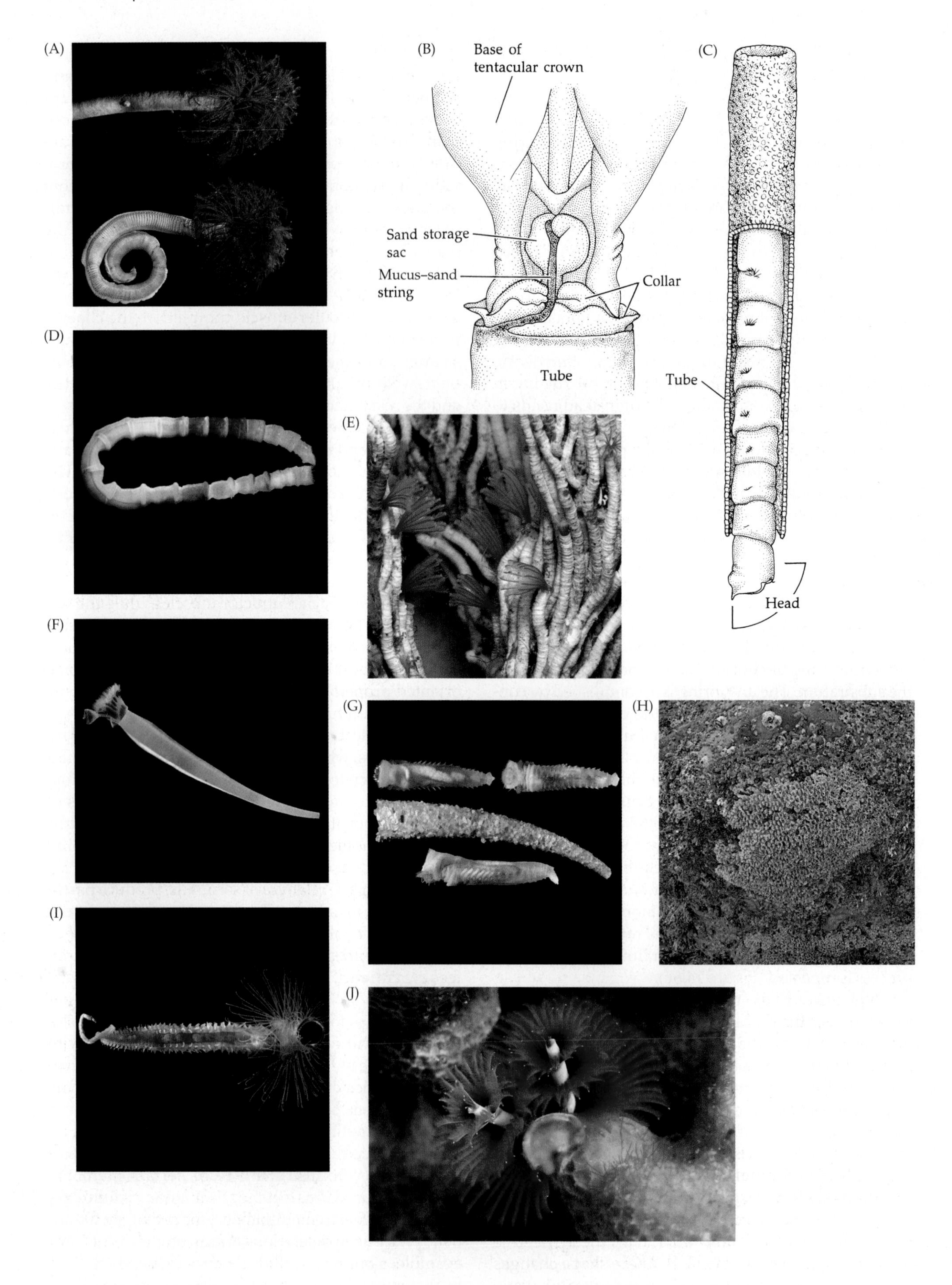
(A)
(B)
Base of
tentacular crown
Sand storage
sac
Mucus–sand
string
Collar
Tube
(C)
Tube
Head
(D)
(E)
(F)
(G)
(H)
(I)
(J)

◀ **Figure 14.7 Tube-dwelling annelids.** (A) *Eudistylia vancouveri* (Sabellidae in and out of its tube). (B) The ventral base of the radiolar crown of a sabellid. Note the addition of a mucus–sand mixture to the lip of the tube. (C) The bamboo worm, *Axiothella rubrocincta* (Maldanidae), oriented head down in its sand tube. (D) Bamboo worm *Clymenella* (Maldanidae) out of tube. (E) A cluster of serpulid tubes formed of calcium carbonate and cemented to the substratum. (F) *Ditrupa* (Serpulidae) in its calcareous tube. (G) The particulate tube of the ice-cream-cone worm, *Pectinaria* (Pectinariidae), and animals out of their tubes. (H) A colony of *Phragmatopoma californica* (Sabellariidae). (I) Specimen of *Phragmatopoma californica* (Sabellariidae) removed from its tube. (J) *Spirobranchus giganteus* (Serpulidae), with algae-encrusted operculum.

deform their shells. Species of *Polydora* (Spionidae) often excavate galleries in various calcareous substrata (e.g., shells) and have been responsible for killing oysters in commercially harvested areas of Europe, Australia, and North America (Figure 14.9F).

Feeding and Digestion

Feeding The great diversity of form and function among annelids has allowed them to exploit nearly all marine food resources in one way or another, and to be critical ecosystem components of most terrestrial soils. For convenience we have categorized annelids as raptorial, deposit, and suspension feeders (see Chapter 4). However, there are several feeding methods and dietary preferences within each of these basic designations. Following a discussion of selected examples of these feeding types, we mention a few of the symbiotic annelids.

The most familiar raptorial annelids are hunting predators belonging to the clade Errantia (e.g., many phyllodocids, syllids, nereidids and eunicids, all part of Aciculata). These animals tend toward homonomy and are capable of rapid movement across the substratum. For the most part they feed on small invertebrates. When prey is located by chemical or mechanical means, the worm everts its pharynx by quick contractions of the body wall muscles in the anterior segments, increasing the hydrostatic pressure in the coelomic spaces and causing the eversion. As a result of the design of the pharynx, the jaws (if present) gape at the anteriormost end when the pharynx is everted (Figure 14.8). Once the prey is positioned within the jaws, the coelomic pressure is released, the jaws collapse on the prey, and the proboscis and captured victim are pulled into the body by large retractor muscles. Many of these raptorial feeders can also ingest plant material and detritus. Some scavenge, feeding on almost any dead organic material they encounter. Raptorial feeding also occurs in some freshwater clitellates such as Lumbriculidae (e.g., *Phagodrilus*), which capture prey (often other clitellates!) with their muscular pharynx.

Some predatory annelids do not actively hunt. Many scale worms (Polynoidae) sit and wait for passing prey, then ambush it by sucking it into their mouth or grasping it with the pharyngeal jaws. In addition, not all raptorial annelids are surface dwellers. Some live in tubes (*Diopatra*) or in complex branched burrows (*Glycera*). Such annelids detect the presence of potential prey outside their tubes or burrows by chemosensory or vibration-sensory means and extend their everted proboscis to capture the prey. Some leave their residence to hunt for short periods of time (e.g., *Eunice aphraditois*; Figure 14.8G). Poison glands, associated with the jaws, occur in some genera (e.g., *Glycera*; Figure 14.8D).

A number of annelids are deposit feeders that are relatively unselective, simply ingesting the substratum and digesting the organic matter contained therein (e.g., members of Arenicolidae, Opheliidae, Maldanidae, and many clitellates). Lugworms, such as *Arenicola*, excavate an L-shaped burrow, which they irrigate with water drawn into the open end by peristaltic movements of the worm's body (Figure 14.9A). The water percolates upward through the overlying sediment and tends to liquefy the sand at the blind end of the L, near the worm's mouth. This sand is ingested by the muscular action of a bulbous proboscis. The water brought into the burrow also adds suspended organic material to the sand at the feeding site. The worm periodically moves to the open end of its tunnel and defecates the ingested sand outside the burrow in characteristic surface castings. Some maldanids live in straight vertical burrows, head down, and ingest the sand at the bottom (Figure 14.7C). They periodically move upward (backward) to defecate on the surface. A number of other direct deposit feeders (e.g., some opheliids) do not live in constructed burrows but simply move through the substratum ingesting sediments as they go. In high concentrations, populations of these annelids can pass thousands of tons of sediments through their guts each year—which has a significant impact on the nature of the deposits in which they live. Most terrestrial and many aquatic clitellates are at least in part direct deposit feeders. Earthworms burrow through the soil, ingesting the substratum as they move. As the soil is passed along the digestive tract, the organic material is digested and absorbed from the gut. The inorganic, indigestible material passes out the anus. Earthworms are said to "work" the soil in this manner, loosening and aerating it. Many of these terrestrial burrowers, including the common earthworm *Lumbricus*, are more selective and retrieve organic material from the surface. These worms can burrow to the surface of the soil and there use their sucker-like mouth to obtain relatively large pieces of food (e.g., partially decomposed leaves), which they carry back underground for ingestion.

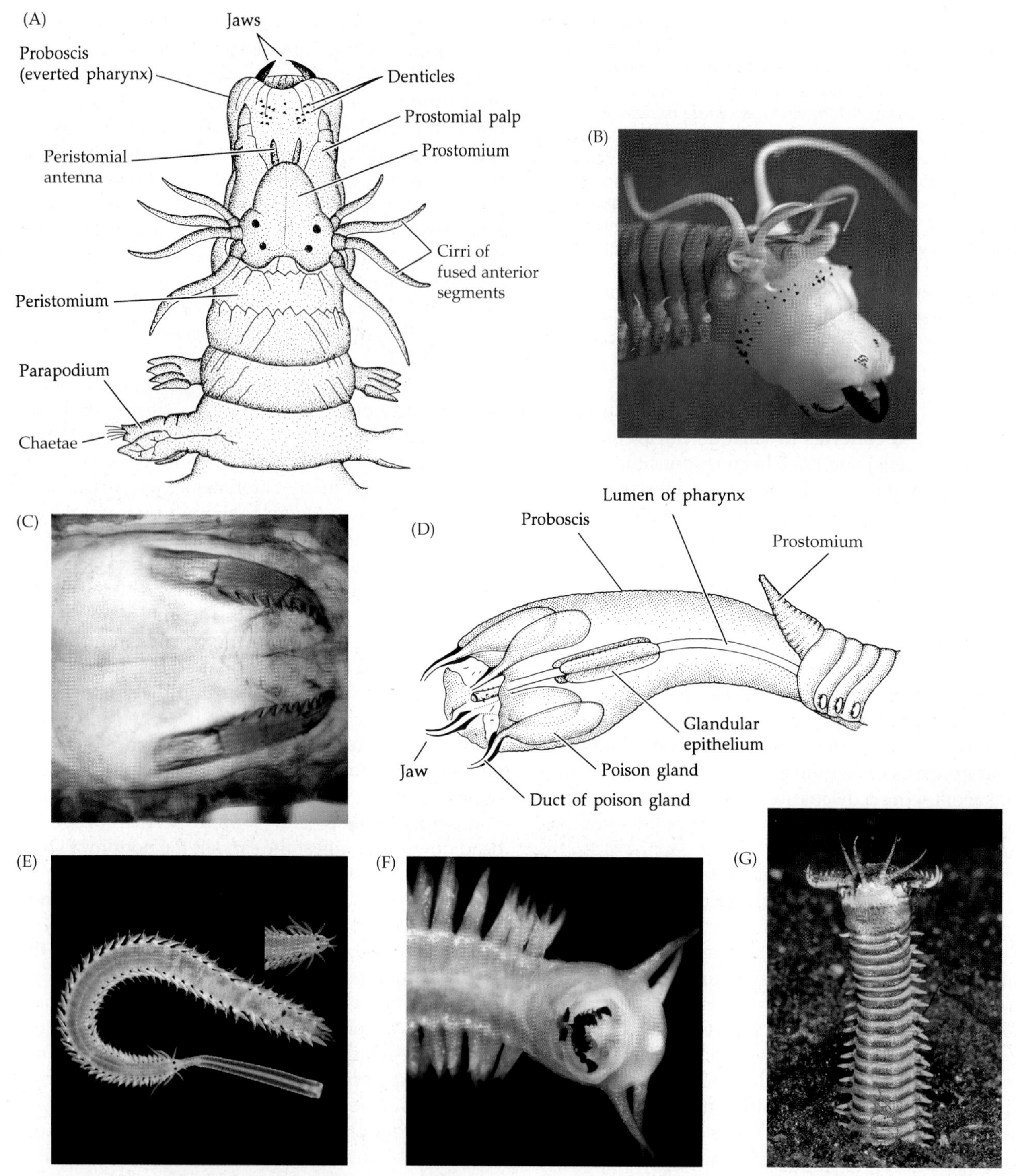

Figure 14.8 The eversible pharyngeal jaws of annelids. (A) *Nereis* (Nereididae) with jaws everted. (B) *Perenereis* (Nereididae) with jaws everted. (C) Nereidid with jaws retracted inside body. (D) *Glycera* (Glyceridae) with jaws everted. (E) *Eumida* (Phyllodocidae) with proboscis (lacking jaws but with numerous papillae) withdrawn and everted. (F) *Ophryotrocha* (Dorvilleidae). Complex jaws being everted. (G) The giant (1 to 3 m) tropical Bobbit worm, *Eunice aphroditois*, with its head extended from the sediment at night. The jaws are locked open while it awaits a passing victim, typically a fish. When the antennae or palps detect the prey, the jaws snap shut and it draws the victim into its burrow.

Selective deposit feeders are defined by their ability to effectively sort the organic material from the sediment prior to ingestion (e.g., many species of Terebellidae, Spionidae, and Pectinariidae). However, the methods used to sort food differs among these groups. Most terebellids (e.g., *Amphitrite*, *Pista*, *Terebella*) establish themselves in shallow burrows or permanent tubes (Figure 14.9B–E). The feeding

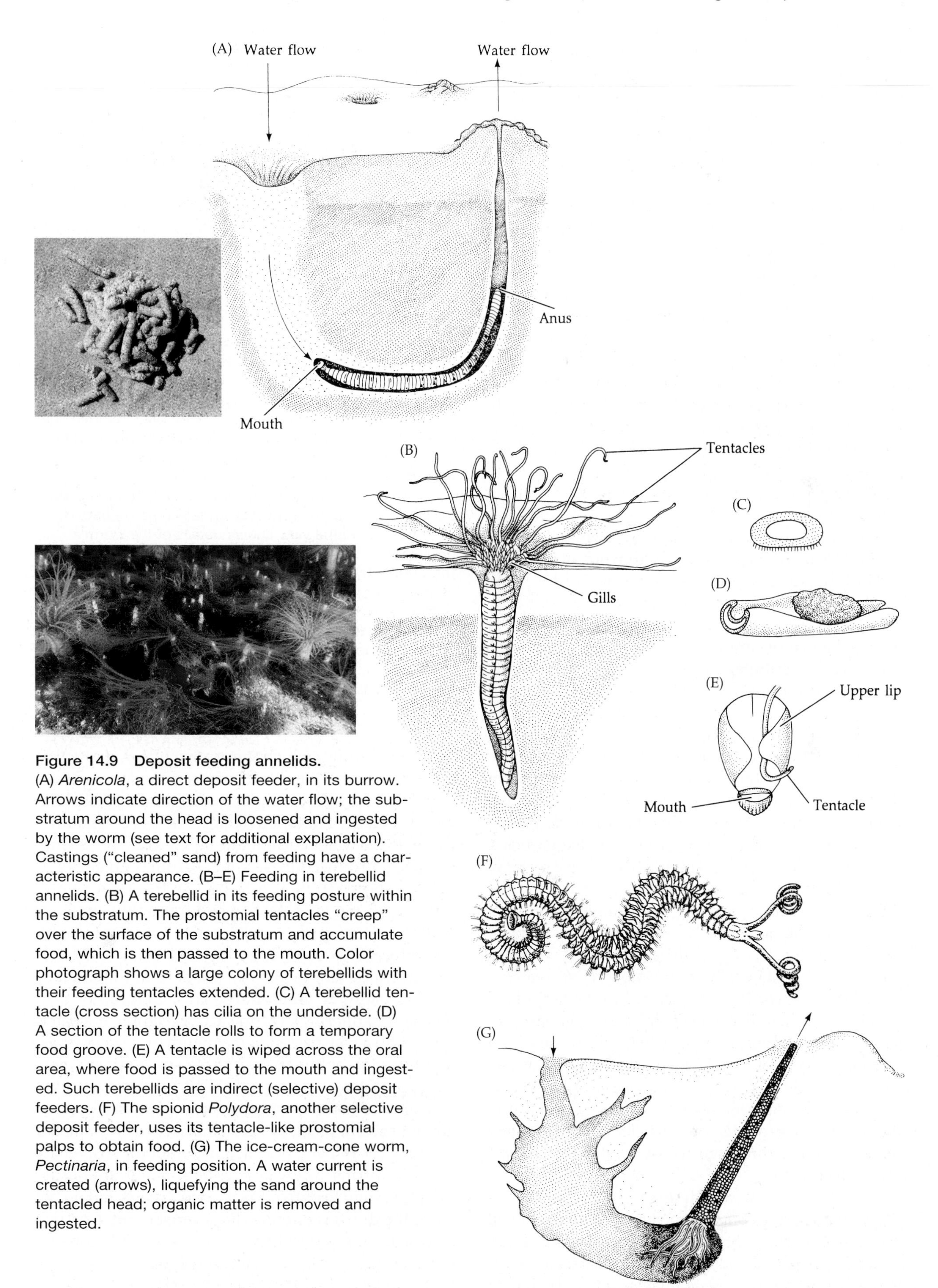

Figure 14.9 Deposit feeding annelids. (A) *Arenicola*, a direct deposit feeder, in its burrow. Arrows indicate direction of the water flow; the substratum around the head is loosened and ingested by the worm (see text for additional explanation). Castings ("cleaned" sand) from feeding have a characteristic appearance. (B–E) Feeding in terebellid annelids. (B) A terebellid in its feeding posture within the substratum. The prostomial tentacles "creep" over the surface of the substratum and accumulate food, which is then passed to the mouth. Color photograph shows a large colony of terebellids with their feeding tentacles extended. (C) A terebellid tentacle (cross section) has cilia on the underside. (D) A section of the tentacle rolls to form a temporary food groove. (E) A tentacle is wiped across the oral area, where food is passed to the mouth and ingested. Such terebellids are indirect (selective) deposit feeders. (F) The spionid *Polydora*, another selective deposit feeder, uses its tentacle-like prostomial palps to obtain food. (G) The ice-cream-cone worm, *Pectinaria*, in feeding position. A water current is created (arrows), liquefying the sand around the tentacled head; organic matter is removed and ingested.

tentacles are grooved prostomial palps that are extended over the substratum. These are extended by ciliary crawling, and can be retracted by muscles. Once extended, their epithelium secretes a mucous coat to which organic material, sorted from the sediment, adheres. The tentacle edges curl up to form a longitudinal groove along which food and mucus are carried by cilia to the mouth. Tube-dwelling spionids engage in a similar method of feeding. In these animals, the feeding structures are more muscular and derived from the peristomial palps (Figure 14.9F). They are swept through the water or brushed through the surface sediments, extracting food and moving it to the mouth.

Pectinaria, the ice-cream cone worms, live in a tube constructed of sand grains and shell fragments. The tube is open at both ends. The animal orients itself head down, with the posterior end of the tube projecting to the sediment surface (Figure 14.9G). Head appendages partially sort the sediment, and a relatively high percentage of organic matter is ingested. A number of other annelids employ these and other methods of selective deposit feeding.

Various forms of suspension feeding are accomplished by many tube-dwelling annelids (e.g., members of Serpulidae and Sabellidae), and by some that live in relatively permanent burrows (e.g., Chaetopteridae). The feeding structures of *Sabella* and many related types are a crown of branching prostomial palps called **radioles**. Some of these worms generate their own feeding currents, whereas others "fish" their tentacles in moving water. As food-laden water passes over the tentacles, the water is driven by cilia upward between the pinnules (branches) of the radioles (Figure 14.10A,B). Eddies form on the medial side (inside) of the tentacular crown and between the pinnules, slowing the flow of water, decreasing its carrying capacity, and thus facilitating extraction of suspended particles. The particles are carried, with mucus, along a series of small ciliary tracts on the pinnules to a groove along the main axis of each radiole. This groove is widest at its opening and decreases in width in a stepwise fashion to a narrow slot deep in the groove. By this means, particles are mechanically sorted into three size categories as they are carried into the groove. Typically, the smallest particles are carried to the mouth and ingested, the largest particles are rejected, and the medium-sized ones are stored for use in tube building. In the freshwater naidid clitellate genus *Ripistes*, long chaetae located on the anterior segments are waved about in the water and small detrital particles adhere to them; food material is then ingested by wiping the chaetae across the mouth.

Some members of Chaetopteridae, such as *Chaetopterus*, are among the most heteronomous of all annelids, and the body is distinctly regionally specialized (Figures 14.1B and 14.10C–E). *Chaetopterus* filters water for food. These animals reside in U-shaped burrows through which they move water, extracting suspended materials. Each body region plays a particular role in this feeding process. Segments 14–16 bear greatly enlarged notopodial fans that serve as paddles to create the water current through the burrow. A mucous bag, produced by secretions from segment 12, is held as shown in Figure 14.10C, so that water flows into the open end of the bag and through its mucous wall. Particles as small as 1 μm in diameter are captured by this structure, and there is some evidence that even protein molecules are held in the mucous net (probably by ionic charge attraction rather than mechanical filtering). During active feeding, the bag is rolled into a ball, passed to the mouth by a ciliary tract, and ingested every 15–30 minutes or so; then a new bag is produced.

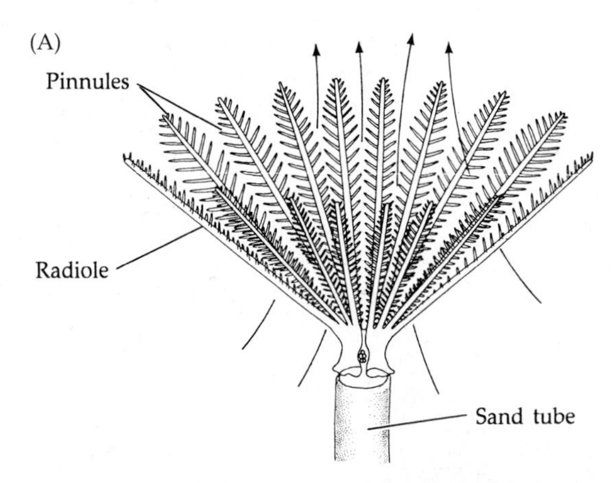

Figure 14.10 Two strategies of suspension feeding in annelids. (A,B) Suspension feeding by a sabellid. (A) Tentacular crown extended from tube and water currents (arrows) passing between tentacles. (B) A portion of a tentacle (radiole) in section. Various ciliary tracts remove particulate matter and direct it to the longitudinal groove on the radiole axis. Here, sorting by size occurs. Most of the largest particles are rejected, the smallest ones are ingested, and the medium-size particles are used in tube building. (C) *Chaetopterus* (Chaetopteridae) in its U-shaped burrow. The ventral view shows details of the worm's anterior end. A water current (arrows) is produced through the burrow by fan-shaped parapodia. Food is removed as the water passes through a secreted mucous bag. The bag is eventually passed to the mouth and ingested, food and all. See text for additional details. (D) The two ends of the U-shaped tube of a *Chaetopterus* emerging from the sediment. (E) *Chaetopterus* (Chaetopteridae) removed from its tube.

Symbiotic relationships with other animals occur among several groups of annelids. There are some interesting cases that reflect, again, the adaptive diversity of these worms. Many symbiotic annelids are hardly modified from their free-living counterparts and do not show the drastic adaptive characteristics often associated with this sort of life. For many, the relationship with their host is a loose one, the annelid often using the host merely as a protective refuge. We have already

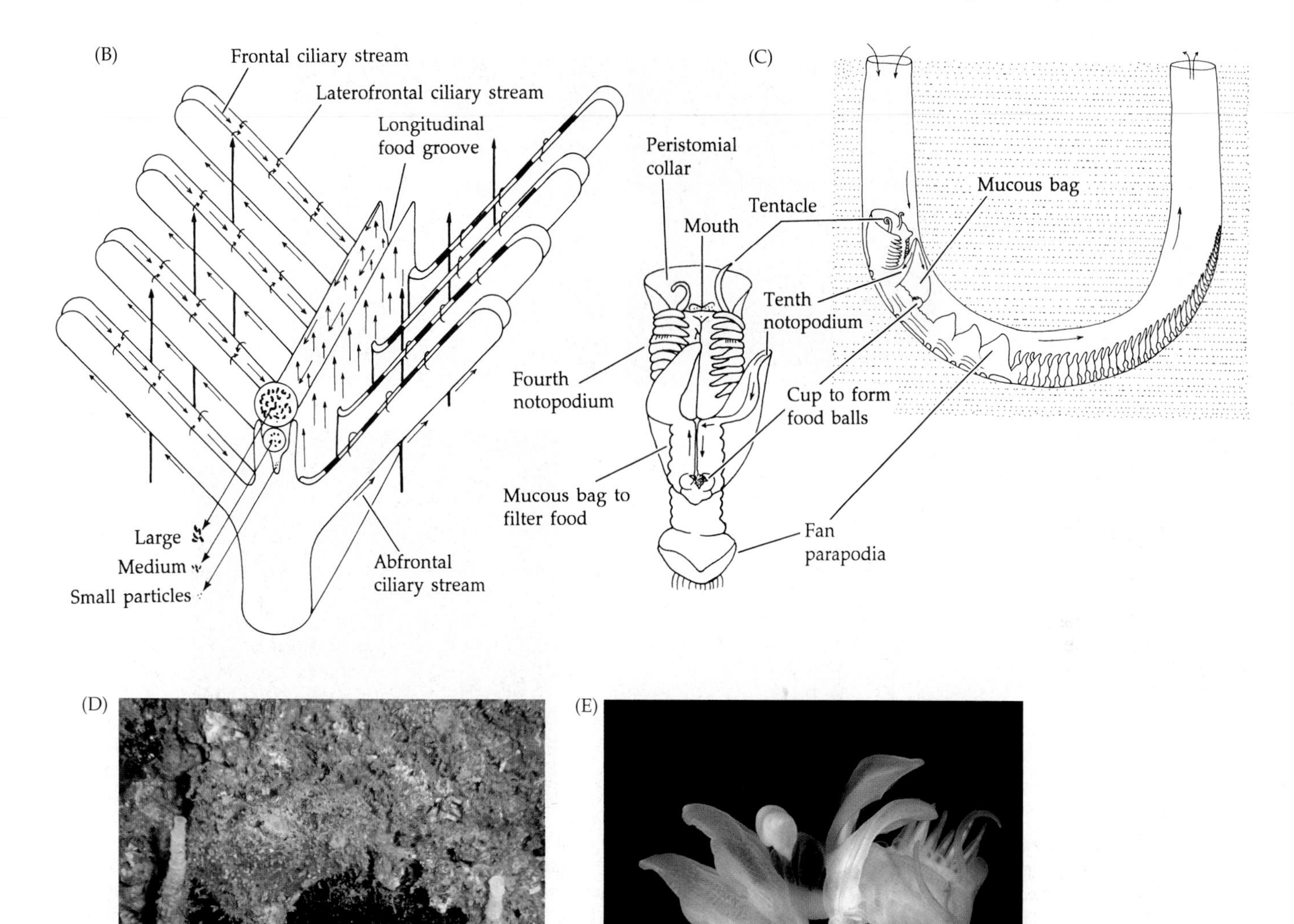

mentioned annelids that burrow into the shells of other invertebrates that are quite similar to their free-living relatives. Among the most commonly found commensal annelids are polynoid scale worms, especially members of the genera *Arctonoe, Halosydna,* and *Mallicephala,* which live on the bodies of various molluscs, echinoderms, and cnidarians (Figure 14.11A,B and this book's cover). A polynoid has even been discovered living as a commensal in the mantle cavity of giant deep-sea mussels residing near thermal vents on the East Pacific Rise. One scale worm, *Hesperonoe adventor*, inhabits the burrows of the Pacific innkeeper worm, *Urechis caupo* (Echiuridae; Figure 14.28G). One group of annelids, Myzostomida, is nearly always symbiotic with echinoderms, chiefly crinoids, but also asteroids and ophiuroids. Myzostomids show variety of adult body shapes and lifestyles, with most described species living freely on the exterior of their hosts as adults, while others live in galls, cysts, or in the host's mouth, digestive system, coelom, or even in the gonad. Myzostomid lifestyles range from stealing incoming food from the host's food grooves to consuming the host's tissue directly (Figure 14.3).

There are many examples of these rather informal associations: certain syllids that live and feed on hydroids, a nereidid (*Nereis fucata*) that resides in the shells of hermit crabs, and so on. Most of these animals do not feed upon their hosts, but prey upon tiny organisms that happen into their immediate environment. Others consume detritus or scraps from their host's meals.

A number of other odd associations are known among the annelids. Most oenonids live part of their

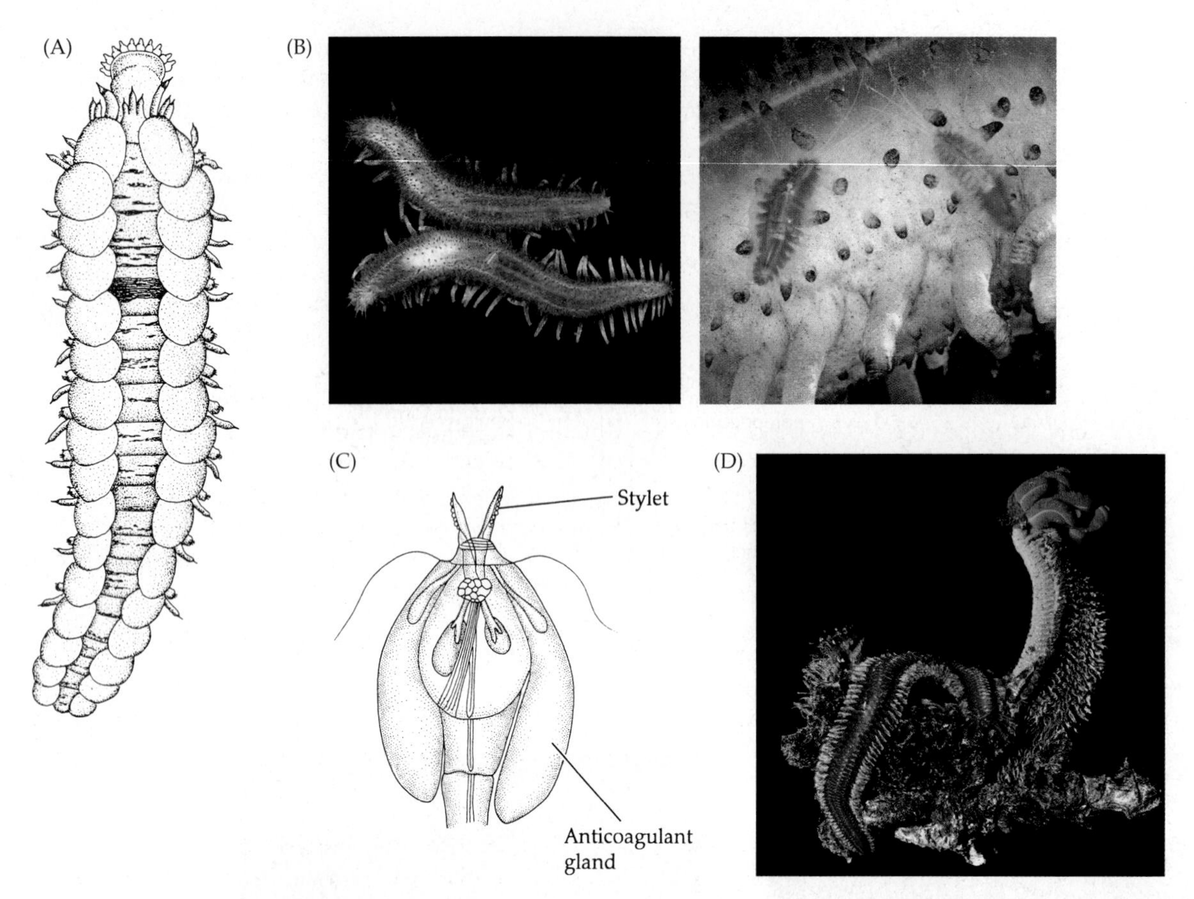

Figure 14.11 Symbiotic annelids. (A) *Arctonoe*, a polynoid that lives in the ambulacral grooves of starfish and the mantle chambers of certain molluscs (with proboscis extended). (B) Two specimens of *Macellicephala* sp. (Polynoidae) on *Pannychia* sp. (Holothuroidea). (C) The anterior end of *Ichthyotomous sanguinarius*, a parasite on fishes. The stylets anchor the worm to its host and the large glands secrete an anticoagulant. (D) Photo of *Alvinella pompejana* (Ampharetidae) partly removed from its tube, showing symbiotic bacteria all over its back, and also a specimen of the hesionid *Hesiolyra bergi* that is often seen in the tubes of *A. pompejana*.

lives as parasites in the bodies of echiurids and other annelids. Again, these endosymbionts show little structural modification associated with their lifestyles, other than a tendency for small body size and reduction in the pharyngeal jaws. An example of a fully parasitic annelid is *Ichthyotomus sanguinarius*. These small (1 cm long) worms attach to eels by a pair of stylets or jaws. The stylets are arranged so that when their associated muscles contract, the stylets fit together like the closed blades of a scissors. The stylets are thrust into the host, and when the muscles relax they open and anchor the parasite to the fish (Figure 14.11C). The Pacific hydrocoral *Allopora californica* typically harbors colonies of the spionid *Polydora alloporis*, whose paired burrow openings are often mistaken for the hydrozoan's polyp cups.

A most unusual symbiotic relationship exists between the strange ampharetid Pompeii worm (*Alvinella pompejana*; named after the deep-sea submersible Alvin) and a variety of marine chemoautotrophic sulfur bacteria. *Alvinella* is a notable member of deep hydrothermal vent communities of the East Pacific Rise (Figure 14.11D). It lives closer to the hot water extrusions than any other animal in the vent community, often near perforated structures called "snowballs" or "beehives" formed by the thermal plumes. Temperatures inside the tubes of *Alvinella* can reach an astonishing 80°C. The bodies of Pompeii worms are covered with unique vent bacteria. The worms are somehow protected from the hot temperatures, and they feed on these symbiotic bacteria. A hesionid worm, *Hesiolyra bergi*, is often found living in the tubes of *Alvinella pompejana* and may also be eating the bacteria or possibly preying on the worm itself (Figure 14.11D).

Over 80 species of gutless marine clitellates, all in the family Naididae, have been described in shallow coral-sand habitats and in anaerobic, sulfide-rich subsurface sediments. These worms typically harbor a range of coexisting subcuticular symbiotic bacteria (up to five species), whose precise role in the host's nutritional regimen is not yet fully understood. The endosymbiotic bacteria are clearly important to the worms; they are passed to the fertilized eggs during oviposition from storage areas next to the female's gonopore.

Digestion The gut of annelids is constructed on a basic plan of foregut, midgut, and hindgut; some examples are shown in Figure 14.12. The foregut is a stomodeum and includes the buccal capsule or tube, the pharynx, and at least the anterior portion of the esophagus. It is lined with cuticle, and the teeth or jaws, when present, are derived from scleroprotein produced along this lining. The jaws are often hardened with calcium carbonate or metal compounds. When present, the eversible portion of this foregut (the proboscis) is derived from the buccal tube or the pharynx. Various glands are often associated with the foregut, including poison glands (glycerids), esophageal glands (nereidids and others), and mucus-producing glands in several groups. In earthworms the posterior esophagus often bears enlarged regions forming a crop, where food is stored, and one or more muscular gizzards lined with cuticle and used to mechanically grind ingested material. The esophagus of many clitellates also has thickened portions of the wall in which are located lamellar evaginations lined with glandular tissue. (Figure 14.12B,G). These **calciferous glands** remove calcium from ingested material. The excess calcium is precipitated by the glands as calcite and then released back into the gut lumen. Calcite is not absorbed by the intestinal wall and so passes out of the body via the anus. In addition, the calciferous glands apparently regulate the level of calcium ions and carbonate ions in the blood and coelomic fluids, thereby buffering the pH of those fluids.

The endodermally-derived midgut generally includes the posterior portion of the esophagus and a long, straight intestine, the anterior end of which may be modified as a storage area, or stomach. The midgut may be relatively smooth, or its surface area may be increased by folds, coils, or many large evaginations (or ceca). The midgut is often histologically differentiated along its length. Typically, the anterior midgut (stomach or anterior intestine) contains secretory cells that produce digestive enzymes. The secretory midgut grades to a more posterior absorptive region. In many terrestrial clitellate species, the surface area of the intestine is enlarged by a middorsal groove called the **typhlosole**. Associated with the midgut of many clitellates, and some other annelids as well, are masses of pigmented cells called **chloragogen cells**. These modified peritoneal cells contain greenish, yellowish, or brownish globules that impart the characteristic coloration to this **chloragogenous tissue**. This tissue lies within the coelom, but is pressed tightly against the visceral peritoneum of the intestinal wall and typhlosole. Chloragogenous tissue serves as a site of intermediary metabolism (e.g., synthesis and storage of glycogen and lipids, deamination of proteins). It also plays a major role in excretion, as discussed below.

Toward the posterior end of the gut, there may be additional secretory cells that produce mucus, which is added to the undigested material during the formation of fecal pellets. Food is moved along the midgut by cilia and by peristaltic action of gut muscles, usually comprising both circular and longitudinal layers. A short rectum connects the midgut to the anus, located on the pygidium. A variety of digestive enzymes are known from different species. Predators tend to produce proteases, herbivores largely carbohydrases. Some omnivorous forms (e.g., *Nereis virens*) produce a mixture of proteases, carbohydrases, lipases, and even cellulase. Digestion is predominantly extracellular in the midgut lumen, although intracellular digestion is known in some groups (e.g., *Arenicola*). Some annelids harbor symbiotic bacteria in their guts that aid in the breakdown of cellulose and perhaps other compounds.

Circulation and Gas Exchange

Given the relatively large size of many annelids, the compartmentalization of their coelomic chambers, and the fact that only certain portions of their gut absorb digested food products, it is essential that a circulatory mechanism be present for internal transport and distribution of nutrients. Furthermore, many annelids have their gas exchange structures limited to particular body regions; thus they depend on the circulatory system for internal transport of gases.

It is easiest to understand the circulatory system of annelids by considering it in concert with their gas exchange structures, which are remarkably varied. In many annelids that lack appendages, the entire body surface functions in gas exchange (e.g., lumbrinerids, oenonids). Some of the active epibenthic forms utilize highly vascularized portions of the parapodia as gills. Special gas exchange structures, or branchiae, are found in the form of trunk filaments (cirratulids, orbiniids), anterior gills (ampharetids, terebellids), and tentacular, or branchial, crowns on the head (sabellids, serpulids, and siboglinids). Since the blood generally carries respiratory pigments, the anatomy of the circulatory system has evolved along with the structure and location of these gas exchange structures.

We again begin our examination with a homonomous annelid, such as *Nereis*, in which the parapodia are more or less similar to one another and the notopodia function as gills. The major blood vessels include a middorsal longitudinal vessel, which carries blood anteriorly, and a midventral vessel, which carries blood posteriorly. Exchange of blood between these vessels occurs through posterior and anterior vascular networks and serially arranged segmental vessels (Figure 14.13A). Anterior vessel networks are especially well developed around the muscular pharynx and the region of the cerebral ganglion.

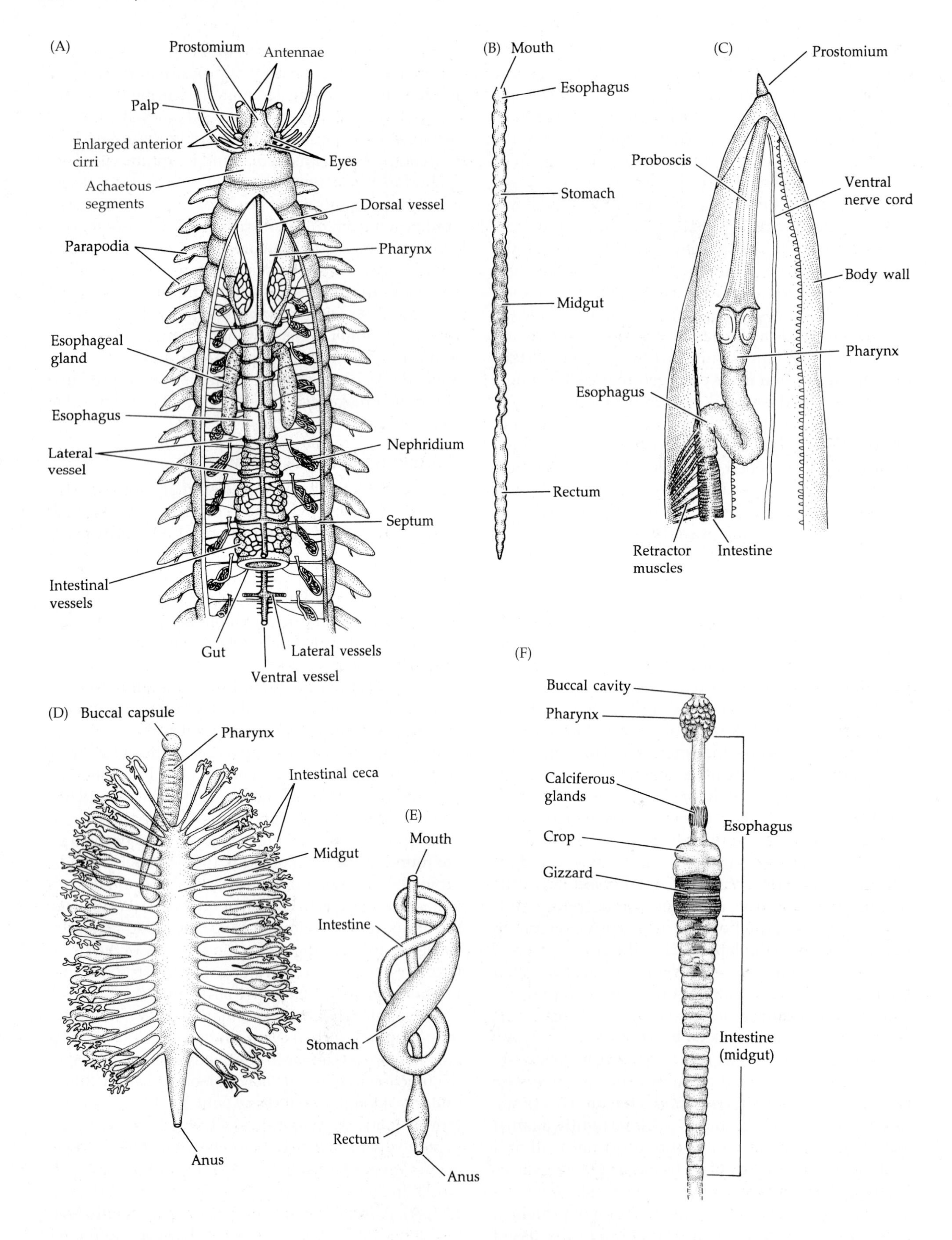
(A)
Prostomium
Antennae
Palp
Enlarged anterior cirri
Eyes
Achaetous segments
Dorsal vessel
Parapodia
Pharynx
Esophageal gland
Esophagus
Lateral vessel
Nephridium
Septum
Intestinal vessels
Gut
Lateral vessels
Ventral vessel
(B) Mouth
Esophagus
Stomach
Midgut
Rectum
(C)
Prostomium
Proboscis
Ventral nerve cord
Body wall
Pharynx
Esophagus
Retractor muscles
Intestine
(D) Buccal capsule
Pharynx
Intestinal ceca
Midgut
Anus
(E)
Mouth
Intestine
Stomach
Rectum
Anus
(F)
Buccal cavity
Pharynx
Calciferous glands
Esophagus
Crop
Gizzard
Intestine (midgut)

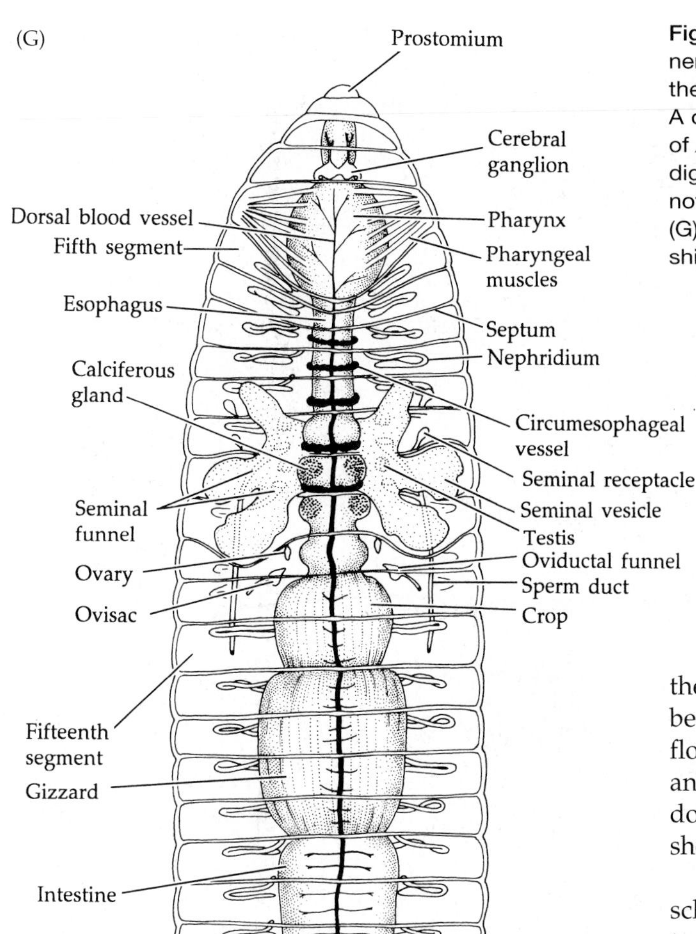

Figure 14.12 Annelid digestive systems. (A) A dissected nereidid (dorsal view). Note the regional specialization of the anterior gut. (B) The simple tubular gut of *Owenia*. (C) A dissected *Glycera* (dorsal view). (D) The multicecate gut of *Aphrodita*. (E) The coiled digestive tract of *Petta*. (F) The digestive tract of *Eisenia foetida* (Lumbricidae in dorsal view; note the marked regional specialization; hindgut not shown). (G) The foregut of *Lumbricus*. Note the positional relationship of the anterior gut regions to other organs.

The movement of blood in *Nereis* depends on the action of the body wall muscles and on intrinsic muscles in the walls of the blood vessels, especially the large dorsal vessel. There are no special hearts or other pumping organs. The blood passing through the various segmental vessels supplies the body wall muscles, gut, nephridia, and parapodia, as illustrated in Figure 14.13A. Note that the oxygenated blood is being returned to the dorsal vessel, thus maintaining a primary supply of oxygen to the anterior end of the animal, including the feeding apparatus and cerebral ganglion. In *Lumbricus* and many clitellates, three main longitudinal blood vessels extend most of the body length and are connected to one another in each segment by additional segmentally arranged vessels (Figure 14.13E,F). The largest longitudinal blood vessel is the dorsal vessel; the wall of this vessel is quite thick and muscular, and provides much of the pumping force for blood movement. Suspended in the mesentery beneath the gut is the longitudinal ventral vessel. The third longitudinal vessel lies ventral to the nerve cord and is called the subneural vessel. Exchanges between the longitudinal vessels occur in each segment through various routes supplying the body wall, gut, and nephridia (Figure 14.13F). Most of the exchanges between the blood and the tissues take place through capillary beds supplied by afferent and efferent vessels. Blood flows posteriorly in ventral and subneural vessels and anteriorly in the dorsal vessel; exchange between the dorsal and ventral vessels occurs in each segment, as shown in Figure 14.13E.

There are many variations on the basic circulatory schemes outlined above, and we mention only a few to illustrate the diversity within annelids. Drastic differences are present even among annelids of generally similar body forms. Among the homonomous forms, for example, the circulatory system may be reduced (e.g., Phyllodocidae), or lost (e.g., Capitellidae, Glyceridae, and Sipuncula). In some cases, reduction is probably associated with small size. This hypothesis, however, cannot be applied to capitellids and glycerids, many of which are large and quite active. In these worms and some others, the circulatory system is greatly reduced and has become fused with remnants of the coelom. The coelom of glycerids and capitellids contain red blood cells (with hemoglobin). Since glycerids have incomplete septa, the coelomic fluid can pass among segments, moved by body activities and ciliary tracts on the peritoneum. In their burrowing lifestyle, enlarged parapodial gills or delicate anterior gills would be disadvantageous; thus the general body surface has probably taken over the function of gas exchange and the coelom the function of circulation. A similar phenomenon has probably occurred in Sipuncula.

Compared with *Nereis* or *Lumbricus*, many annelids display additional blood vessels, modification of vessels, differences in blood flow patterns, and formation of large sinuses. As might be predicted, some striking differences are seen among certain heteronomous annelids with reduced parapodia and anteriorly located

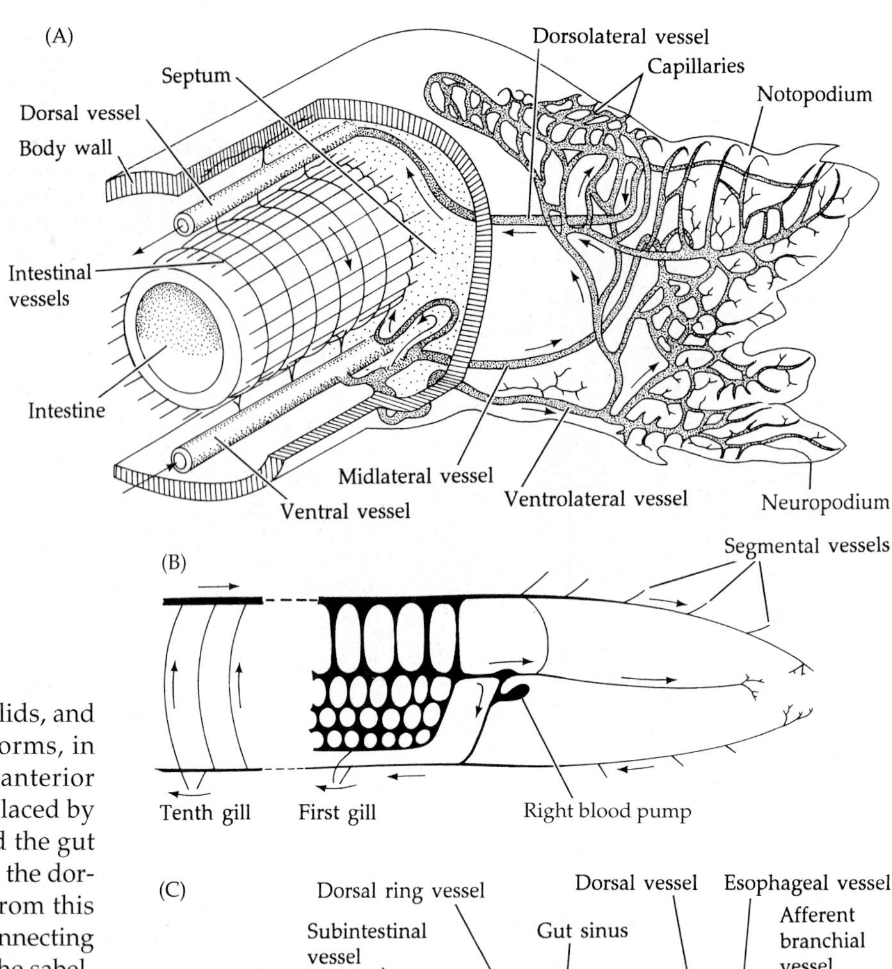

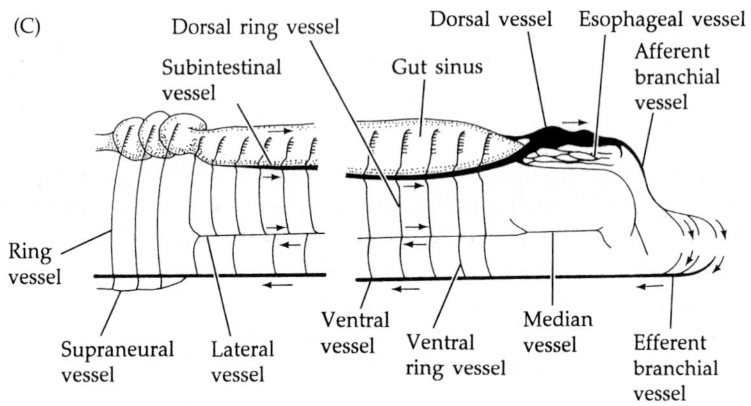

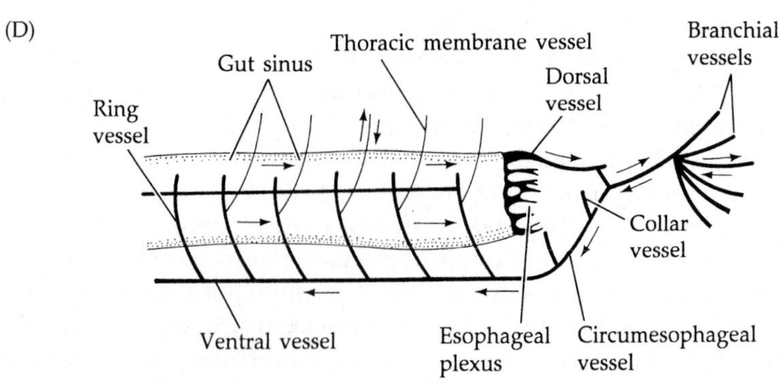

Figure 14.13 Annelid circulatory and gas exchange systems. Variations from the basic plan include additional vessels, sinuses associated with the foregut, and branchial vessels serving anterior gills. (A) A segment and parapodium (cutaway view) of a nereid. Note the major blood vessels and blood flow pattern (arrows). Blood flows anteriorly in the dorsal vessel and posteriorly in the ventral vessel. In such annelids the flattened parapodia serve as gills. (B–F) Circulatory patterns in an arenicolid (B), a terebellid (C), a serpulid (D), and *Lumbricus* (Clitellata). (E) Anterior blood vessels (lateral view) and (F) the circulatory pattern in one segment (cross section).

branchiae (e.g., terebellids, sabellids, and serpulids). In many of these worms, in the region of the stomach and anterior intestine, the dorsal vessel is replaced by a voluminous blood space called the gut sinus (Figure 14.13C,D). Usually, the dorsal vessel continues anteriorly from this sinus and it often forms a ring connecting with the main ventral vessel. In the sabellids and serpulids, a single, blind-ended vessel extends into each branchial tentacle. Blood flows in and out of these branchial vessels that, in some forms (e.g., serpulids), are equipped with valves that prevent backflow into the dorsal vessel. This two-way flow of blood within single vessels is quite different from the capillary exchange system in most closed vascular systems. A few clitellates possess extensions of the body wall that increase the surface area and function as simple gills (e.g., *Branchiura*, *Dero*), but most exchange gases across the general body surface.

Specialized pumping structures have evolved in a number of annelids. They are especially well developed in certain tube-dwelling forms where they compensate for the reduced effect of general body movements on circulation. These structures, sometimes called hearts, are often little more than an enlarged and muscularized portion of one of the usual vessels; the dorsal muscular vessel of chaetopterids is such a structure. Terebellids possess a "pumping station" at the base of the gills that functions to maintain blood pressure and flow within the branchial vessels (Figure 14.13B). A variety of similar structures are known among other annelids.

Most annelids contain respiratory pigment within their circulatory fluid, which is generally acellular. Those without any such pigment include some very

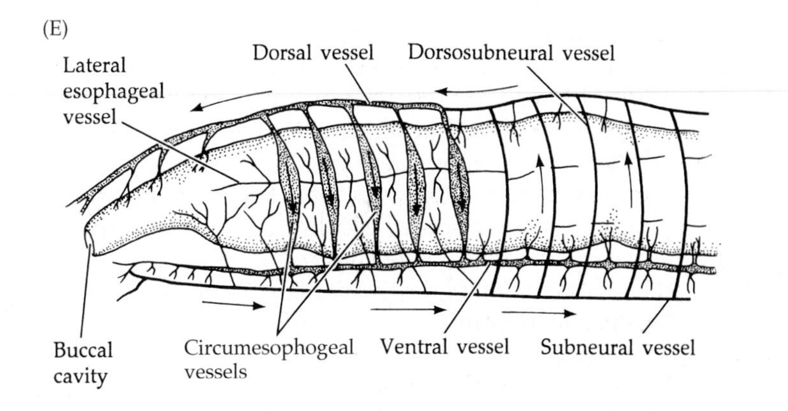

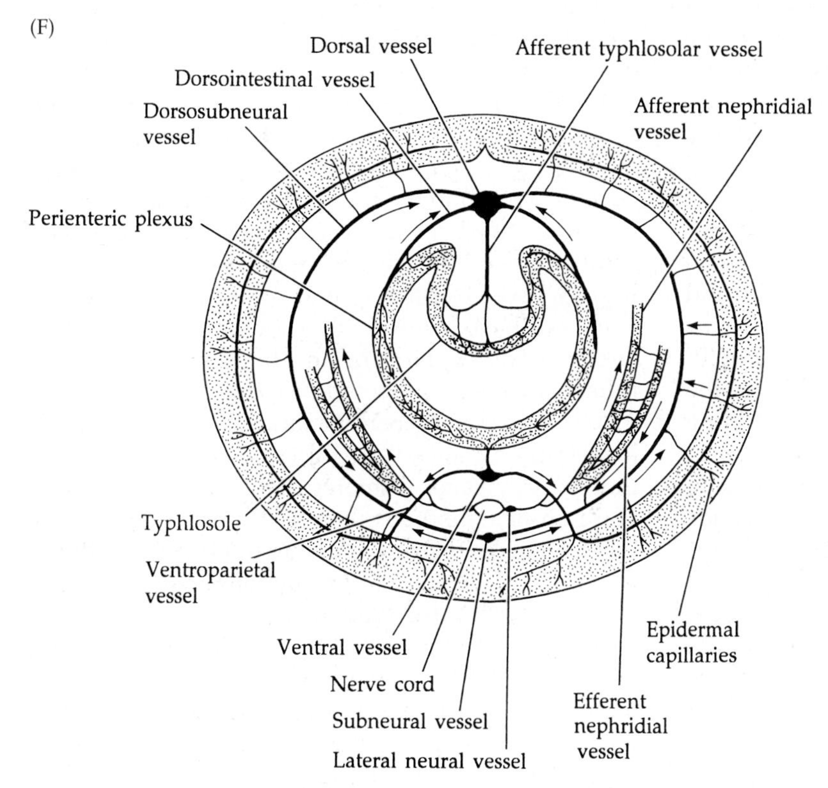

small forms and various syllids, phyllodocids, polynoids, aphroditids, *Chaetopterus*, and a few others. When a pigment is present, it is usually some type of hemoglobin, although chlorocruorin is common in some families (e.g., certain flabelligerids, sabellids, and serpulids), and hemerythrin occurs in magelonids, the latter now seen as one of the earliest branching annelid clades. Magelonidae are also interesting in that their blood contains cells (corpuscles). Some annelids have more than one type of pigment; for example, the blood of some serpulids contains both hemoglobin and chlorocruorin.

Annelid respiratory pigments may occur in the blood itself, the coelomic fluid, or both. With a few exceptions, blood pigments occur in solution and coelomic pigments are contained within corpuscles. The latter situation is generally associated with reduction or loss of the circulatory system (as in glycerids). The incorporation of coelomic pigments, usually hemoglobin, into cells is probably a mechanism to prevent the serious osmotic effects that would result from large numbers of free dissolved molecules in the body fluid. Corpuscular coelomic hemoglobins tend to be of much smaller molecular sizes than those dissolved in the blood plasma. The significance of this difference is not clear, but in the case of arenicolids it could be related to oxygen challenges as these worms often burrow in low-oxygen sediments.

The types of respiratory pigments and their disposition within the body of annelids are related at least in part to the lifestyles. As discussed in Chapter 4, different pigments—even different forms of the same pigment—have different oxygen loading and unloading characteristics. The nature of the pigments in a particular worm reflects its ability to store oxygen and then release it during periods of environmental oxygen depletion. A number of intertidal burrowing annelids take up and store oxygen during high tides and dissociate the stored oxygen during low tides. This sort of physiological cycle ameliorates the potential stress of oxygen depletion in the body during low tide periods. Some have more than one pigment type; for example, one form of hemoglobin for normal conditions and another form that stores oxygen and releases it during periods of stress. Some annelids (e.g., *Euzonus*) can actually convert to anaerobic metabolic pathways during extended periods of anoxic conditions. Many terrestrial clitellates are capable of sufficient gas exchange only when exposed to air; they will drown if submerged. (Remember, air contains far more oxygen than water.) We have all seen earthworms crawling about the surface following a heavy rain. One particular species of earthworm (*Alma emini*) has evolved a remarkable adaptation that allows it to survive the rainy season in its East African habitat. When rains cause its burrow to flood, the worm moves to the surface of the soil and forms a temporary opening. The worm then projects its posterior end out through the opening and rolls the sides of the body wall into a pair of folds, forming an open chamber that serves as a kind of "lung." The highly vascularized posterior epithelium enhances the exchange of gases. A number of aquatic clitellates can tolerate periods of low available oxygen and even anoxic conditions for short periods of time.

Excretion and Osmoregulation

In Chapter 4 we discussed nephridial organs in invertebrates. Thus far, we have seen various types of protonephridia, especially among the acoelomate and certain blastocoelomate Metazoa. In annelids, metanephridia are found and these structures commonly pass through a protonephridial stage during their development. Most annelids possess some type of metanephridia, often serially arranged as one pair per segment, with the pore in the segment posterior to the nephrostome. However, variations on this theme are many and protonephridia (secondarily derived in the sense that they do not fully develop into metanephridia and arrest at the protonephridial stage) are found in some adult annelids. The success of an animal with segmentally arranged isolated coelomic compartments depends on the physical and physiological maintenance of those separate segments. The removal of metabolic wastes (predominantly ammonia) and the regulation of osmotic and ionic balance must occur in each functionally isolated coelomic chamber. (See also the discussion and figures pertaining to excretion and osmoregulation in Chapter 4). While it was once assumed that annelids had separately derived excretory (nephridial) and gonoduct (coelomoduct) systems, the evidence now appears to suggest that the ducts exiting the body are for the most part nephridial only, but may serve both for excretion and spawning of gametes.

Protonephridia are found in certain adult annelids mostly in groups within Phyllodocida, and these appear to actually be metanephridia that have altered their development such that the funnel opening to the coelom connects to a nephridioduct (Figure 14.14A) that also leads to some flame cells. However, the vast majority of annelids possess paired metanephridia that each open to the coelom via a ciliated nephrostome. In a few annelids (e.g., capitellids), both metanephridia and separate coelomoducts occur, but in most there are metanephridia only. In certain annelid groups with incomplete or missing septa, the number of nephridia is reduced. In some, such as Terebelliformia, anterior nephridia are excretory only, whereas in the posterior region of the body, where the gametes are formed, nephridia are used for both spawning and presumably also excretion. The extreme case of this is found in groups such acrocirrids, cirratulids, serpulids, siboglinids, and sabellids where there is a single anterior pair of large excretory metanephridia. In sabellids and serpulids these lead to a single fused nephridioduct and common pore behind the head (Figure 14.14C). A typical clitellate nephridium is composed of a preseptal nephrostome (either open to the coelom or secondarily closed as a bulb), a short canal that penetrates the septum, and a postsegmental nephridioduct that is variably coiled and sometimes dilated as a bladder (Figure 14.14D). The nephridiopores are usually located ventrolaterally on body segments.

As described in Chapter 4, the open nephrostomes of metanephridia nonselectively pick up coelomic fluids. This action is followed by resorption of materials from the nephridium back into the body, either directly into the surrounding coelomic fluid, or into the blood in cases where extensive nephridial blood vessels are present (e.g., in some nereidids and in aphroditids). In either case the composition of the urine is quite different from that of the body fluids; the difference indicates a significant amount of physiological selectivity along the length of the nephridium.

Osmoregulation presents little problem for subtidal marine annelids living in relatively constant osmotic conditions. Intertidal and estuarine forms, however, must be able to withstand periods of stress associated with fluctuations in environmental salinities. Many species are osmoconformers (e.g., *Arenicola*), allowing the tonicity of their body fluids to fluctuate with changes in the environmental salinity. Most annelid osmoconformers have relatively simple metanephridia, with comparatively short nephridioducts and correspondingly weaker resorptive and regulatory capacities. Some also have relatively thin body wall musculature, and the body swells when in a hypotonic medium. It is likely that burrowers and tube dwellers face less osmotic stress than epibenthic forms, because the water in their tubes may be less subject to ionic variation than the overlying water. Osmoregulators, such as a number of estuarine nereidids, often have thicker body walls that tend to resist changes in shape and volume. When water enters the body from a hypotonic surrounding, the increased hydrostatic pressure generated within the coelom works against that osmotic gradient. In addition, regulators are able to maintain (within limits) a more or less constant internal fluid tonicity because of the greater selective capabilities of their more complex nephridia.

Ionic- and osmoregulation are major challenges for annelids in freshwater and terrestrial habitats, and few evolutionary lineages have solved these problems, the main example being the Clitellata. The moist, permeable surface necessary for gas exchange, and the severe osmotic gradients across the body wall, present potentially serious problems of water loss to terrestrial forms, and of water gain to freshwater forms. Both situations threaten the loss of precious diffusible salts. Passive diffusion of water and salts also occurs across the gut wall. The major organs of water and salt balance in freshwater annelids are, of course, the nephridia. Excess water is excreted and salts are retained by selective and active resorption along the nephridioduct. The problem in terrestrial species is more serious. Surprisingly, earthworms are not absolute osmoregulators; rather, they lose and gain water according to the amount of water in their environment. Various species can tolerate a loss of 20 to 75% of their body water and still recover. Under normal conditions, water

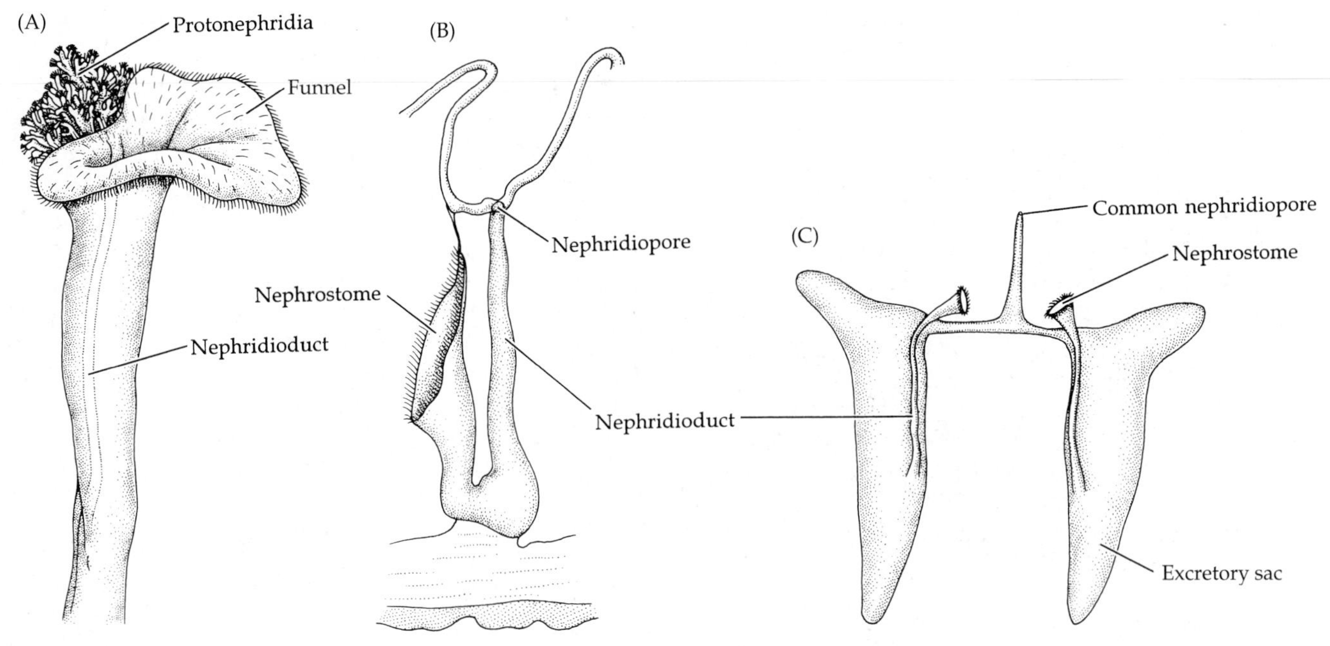

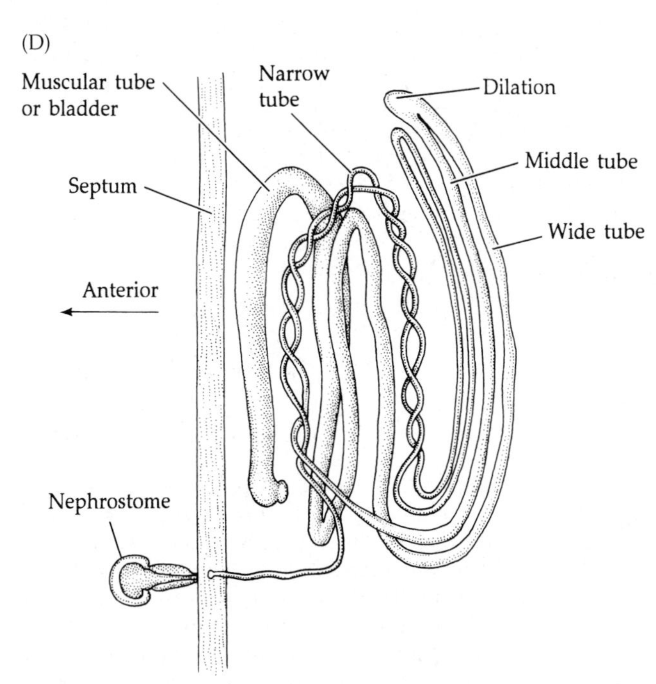

Figure 14.14 Annelid nephridia. (A) A protonephridium of a phyllodocid. Here a cluster of solenocytic protonephridia sits atop a nephridioduct that has a side funnel. (B) A metanephridium of a spionid. (C) A single pair of nephridia joined to a common duct in a serpulid. (D) A single nephridium and its relationship to a septum in *Lumbricus* (Clitellata). Evidence suggests that earthworm nephridia are highly selective excretory and osmoregulatory units. The nephridioduct is regionally specialized along its length. The narrow tube receives body fluids and various solutes, first from the coelom through the nephrostome and then from the blood via capillaries that lie adjacent to the tube. In addition to various forms of nitrogenous wastes (ammonia, urea, uric acid), certain coelomic proteins, water, and ions (Na^+, K^+, Cl^-) are also picked up. Apparently, the wide tube serves as a site of selective reabsorption (probably into the blood) of proteins, ions, and water, leaving the urine rich in nitrogenous wastes.

conservation by earthworms is probably accomplished in several ways. The production of urea allows the excretion of a relatively hypertonic urine compared with that of a strictly ammonotelic animal. There may also be active uptake of water and salts from food across the gut wall. Certainly there are behavioral adaptations for remaining in relatively moist environments, in addition to the physiological adaptations that allow these animals to tolerate temporary partial dehydration of their bodies.

Aquatic clitellates are ammonotelic, but most terrestrial forms are at least partially ureotelic. These wastes are transported to the nephridia via the circulatory system and by diffusion through the coelomic fluid. Uptake of materials into the nephridial lumen is partly nonselective from the coelom (in those worms with open nephrostomes), and partly selective across the walls of the nephridioduct from the afferent nephridial blood vessels. A significant amount of selective resorption occurs into the efferent blood flow along the distal portion of the nephridioduct, facilitating efficient excretion as well as ionic and osmoregulation.

Nervous System and Sense Organs

The central nervous system in annelids (as in protostomes in general) includes a dorsal cerebral ganglion, paired circumenteric connectives, and one or more ventral longitudinal nerve cords (Figure 14.15). The nervous system of Magelonidae, one of the groups near the base of the annelid tree, is unusual in that,

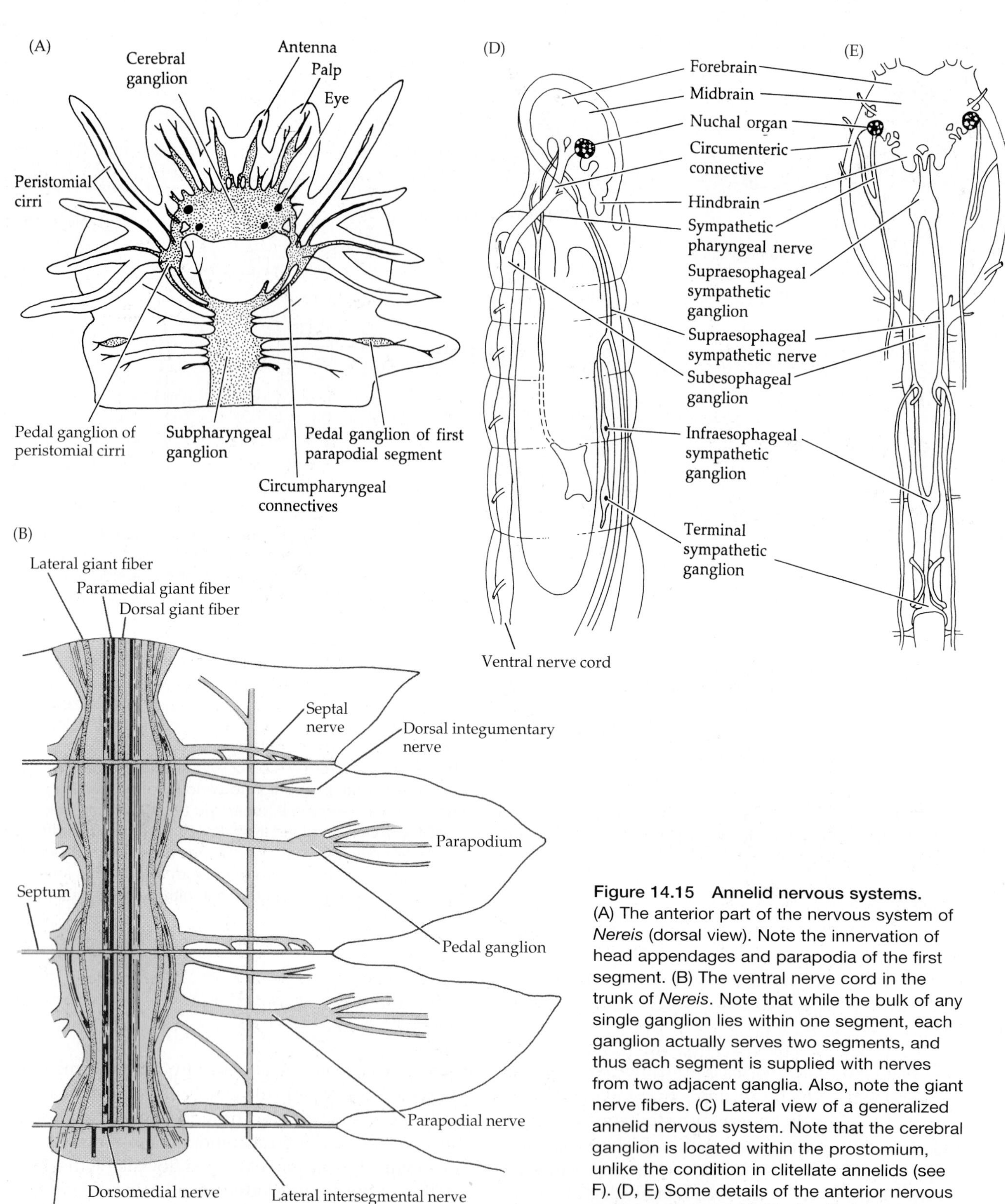

Figure 14.15 Annelid nervous systems. (A) The anterior part of the nervous system of *Nereis* (dorsal view). Note the innervation of head appendages and parapodia of the first segment. (B) The ventral nerve cord in the trunk of *Nereis*. Note that while the bulk of any single ganglion lies within one segment, each ganglion actually serves two segments, and thus each segment is supplied with nerves from two adjacent ganglia. Also, note the giant nerve fibers. (C) Lateral view of a generalized annelid nervous system. Note that the cerebral ganglion is located within the prostomium, unlike the condition in clitellate annelids (see F). (D, E) Some details of the anterior nervous systems of a eunicid, *Eunice* (D, *lateral view*; E, *dorsal view*). The cerebral ganglion is specialized into fore-, mid-, and hindbrain. (F) The nervous system of the clitellate *Lumbricus* in lateral view. Note that the cerebral ganglion is located behind the head, a development apparently associated with a reduction in the size of the prostomium.

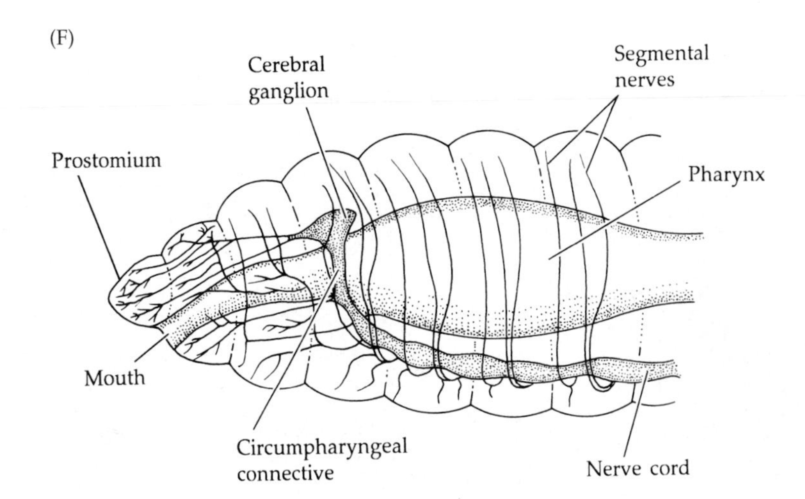

stomatogastric nerves associated with the foregut, especially with the operation of the proboscis or pharynx. The circumenteric connectives often bear ganglia from which nerves extend to the peristomial cirri, or else these appendages are innervated by nerves from the subenteric ganglion. The subenteric ganglion appears to exhibit excitatory control over the ventral nerve cord(s) and segmental ganglia.

The nerves that arise from the segmental ganglia innervate the body wall musculature and parapodia (via the pedal ganglia), and the digestive tract. The ventral nerve cord and sometimes the lateral nerves of most annelids contain some extremely long neurons, or giant fibers, of large diameter; these neurons facilitate rapid, "straight-through" impulse conduction, bypassing the ganglia (Figure 14.15B). Giant fibers are apparently lacking in some annelids (e.g., syllids), but are well developed in tube dwellers, such as sabellids and serpulids, permitting rapid contraction of the body and retraction into the tube. The central nervous system of clitellates consists of the usual annelid components: a supraenteric cerebral ganglion joined to a ganglionated ventral nerve cord by circumenteric connectives and a subenteric ganglion (Figure 14.15). With the reduction in head size, especially of the prostomium, the cerebral ganglion occupies a more posterior position than in the other annelids, often lying as far back as the third body segment. The paired ventral nerve cords are almost always fused as a single tract in clitellates, and it usually contains some giant fibers, especially in leeches.

unlike most annelids, it is largely intraepidermal and this might represent the primitive condition for the phylum. Another annelid group near the base of the tree, Oweniidae, has basioepithlial nerve cords, while almost all other annelids have subepidermal nerve cords.

The cerebral ganglion of annelids is usually bilobed and lies within the prostomium. One or two pairs of circumenteric connectives extend from the cerebral ganglion around the foregut and unite ventrally in the subenteric ganglion. Commonly, a pair of longitudinal nerve cords arises from the subenteric ganglion and extends the length of the body and these two separate trunks within the ventral cord has traditionally been viewed as the primitive condition. Ganglia are arranged along these nerve cords, one pair in each segment, and are connected by transverse commissures. Lateral nerves extend from each ganglion to the body wall and each bears a so-called **pedal ganglion**. This double nerve cord arrangement is common in certain groups of annelids, including sabellids and serpulids, though these are highly derived annelids. Interestingly, in amphinomids, which are found toward the base of the annelid tree, there are four longitudinal nerve cords, a medial pair and a lateral pair, the latter connecting the pedal ganglia. Similar, but perhaps nonhomologous, lateral longitudinal cords appear in some other annelid taxa that are considered to be relatively derived. In many other annelids there has been fusion of the medial nerve cords to form a single midventral longitudinal cord. The degree of fusion varies among taxa, and some retain separate nerve tracts within the single cord.

The cerebral ganglion is often specialized into three regions, typically called the forebrain, midbrain, and hindbrain. Generally, the forebrain innervates the prostomial palps, the midbrain the eyes and prostomial antennae or tentacles, and the hindbrain the chemosensory nuchal organs (Figure 14.15A,D,E). The circumenteric connectives arise from the fore- and midbrains. The midbrain also gives rise to a complex of motor

Annelids show an impressive array of sensory receptors. As would be expected, the kinds of sense organs present and the degree of their development vary greatly among annelids with different lifestyles. Certainly, the requirements for particular sorts of sensory information are not the same for a tube-dwelling sabellid as they are for an errant predatory nereidid or a burrowing arenicolid.

In general, annelids are highly touch-sensitive. Crawlers, tube dwellers, and burrowers depend on tactile reception for interaction with their immediate surroundings (locomotion, anchorage within their tube, and so on). Touch receptors are distributed over much of the body surface but are concentrated in such areas as the head appendages and parts of the parapodia. The chaetae are also typically associated with sensory neurons and serve as touch receptors. Some burrowers and tube dwellers have such a strong positive response to contact with the walls of their burrow or tube that the response dominates all other receptor input. Some of these annelids will remain in their burrow or tube regardless of other stimuli that would normally produce a negative response.

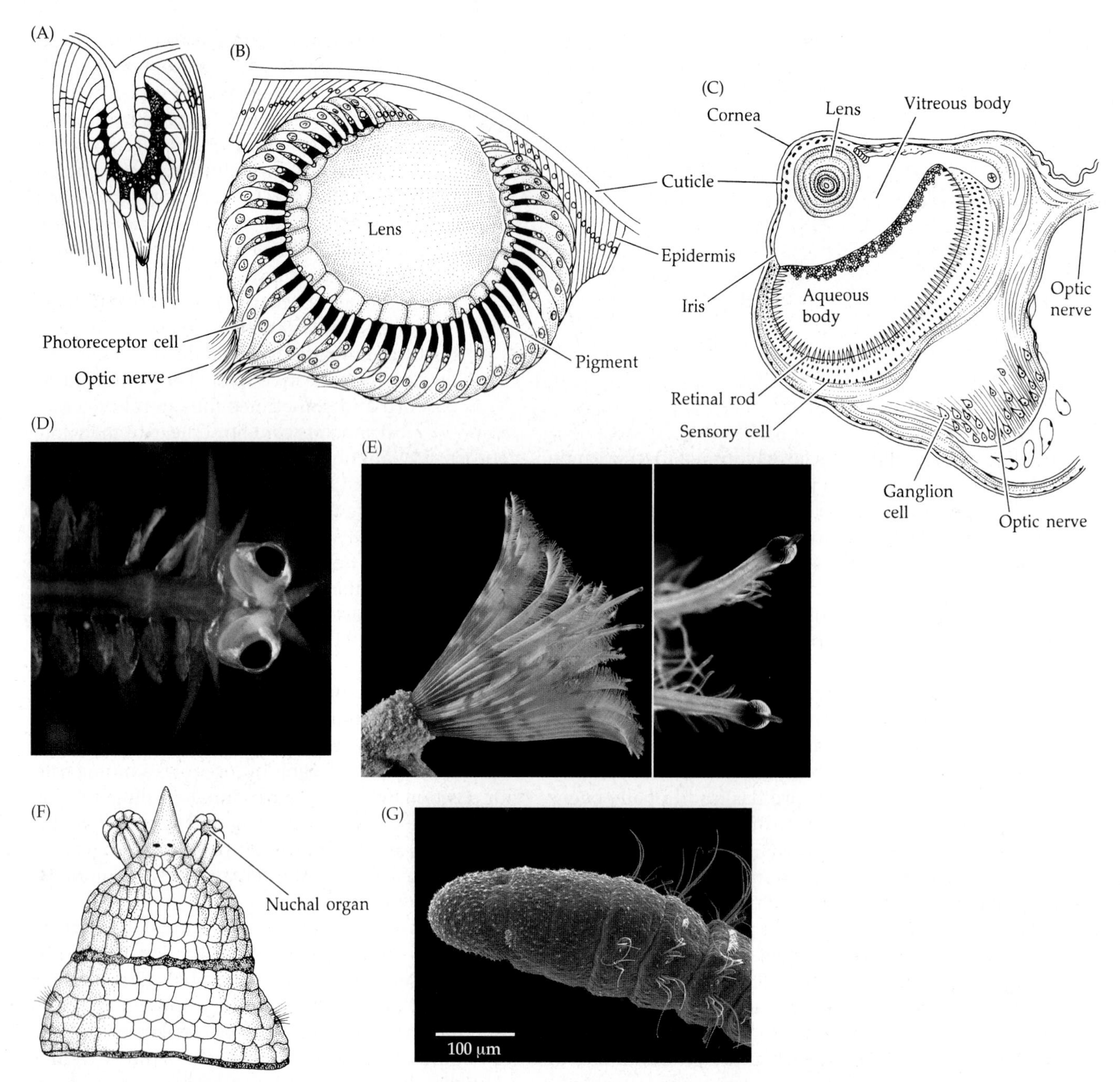

Figure 14.16 Annelid photoreceptors and nuchal organs. (A) Simple pigment cup eye of a chaetopterid. (B) Lensed pigment cup eye of a nereid. (C) A complex eye (section) of an alciopid (*Vanadis*). (D) Anterior end of an alciopin phyllodocid (ventral view) showing the large eye lobes. (E) *Megalomma* (Sabellidae) has compound eyes, similar to, but convergently evolved to those seen in arthropods. This pair of compound eyes allows the worm to detect potential predators and safely withdraw into the tube. (F) Nuchal organs of *Notomastus*. (G) SEM of *Proscoloplos* (Orbiniidae) showing nuchal organs on the prostomium.

Most annelids possess photoreceptors, although these structures are lacking in many burrowers (Figure 14.16). The best-developed annelid eyes occur in pairs on the dorsal surface of the prostomium. In some there is a single pair of eyes (e.g., most phyllodocids); in many there are two or more pairs (e.g., nereidids, polynoids, hesionids, many syllids). These prostomial eyes are direct pigment cups. They may be simple depressions in the body surface lined with retinular cells, or they may be quite complex, with a distinct refractive body or lens (Figure 14.16A–C). In nearly all cases, the eye units are covered by a modified section of the cuticle that functions as a cornea. The eyes of most annelids are capable of transmitting information on light direction and intensity, but in certain pelagic forms (e.g., alciopin phyllodocids) the eyes are huge and possess true lenses capable of accommodation and perhaps image perception (Figure 14.16C,D).

In addition to, or instead of, the prostomial eyes, some annelids bear photoreceptors on other parts of the body. A few species bear simple eyespots along the length of the body (e.g., the opheliid *Polyophthalmus*). Pygidial eyespots occur in newly settled sabellariids and some adult fabriciids (e.g., *Fabricia*) and sabellids (small ones such as *Amphiglena*). Interestingly, in these cases the animals crawl backward. Many sabellids and serpulids possess simple ocelli, or even compound eyes remarkably similar to those of arthropods, on the branchial crown tentacles, and they react to sudden decreases in light intensity by retracting into their tubes (Figure 14.16E.). This "shadow response" helps these sedentary worms avoid predators and can easily be demonstrated by passing one's hand to cast a shadow over a live worm.

Nearly all annelids are sensitive to dissolved chemicals in their environment. Most of the chemoreceptors are specialized cells that bear a receptor process extending through the cuticle. Sensory nerve fibers extend from the base of each receptor cell. Such simple chemoreceptors are often scattered over much of the worm's body, but they tend to be concentrated on the head and its appendages. Some annelids also possess ciliated pits or slits called **nuchal organs,** which are presumed to be chemosensory (Figure 14.16F,G). These structures are typically paired and lie posteriorly on the dorsal surface of the prostomium. In some forms (e.g., certain nereidids) the nuchal organs are simple depressions, whereas in others (e.g., opheliids) they are rather complex eversible structures equipped with special retractor muscles. In members of Amphinomidae, the nuchal organs are elaborate outgrowths of an extension of the prostomium called the **caruncle**.

Statocysts are common in some burrowing and tube-dwelling annelids (e.g., certain terebellids, arenicolids, and sabellids). A few forms possess several pairs of statocysts, but most have just a single pair, located near the head. These statocysts may be closed or open to the exterior, and the statolith may be a secreted structure or formed of extrinsic material, such as sand grains. It has been demonstrated experimentally that the statocysts of some annelids do serve as georeceptors and help maintain proper orientation when the bearer is burrowing or tube building.

A number of other structures of presumed sensory function occur in some annelids. These structures are often in the form of ciliated ridges or grooves occurring on various parts of the body and associated with sensory neurons. A variety of names have been applied to these structures, but in most cases their functions remain unclear. Annelids also possess organs or tissues of neurosecretory or endocrine functions. Most of the secretions appear to be associated with the regulation of reproductive activities, as discussed in the following section.

Reproduction and Development

Regeneration and asexual reproduction Annelids show various degrees of regenerative capabilities. Nearly all of them are capable of regenerating lost appendages such as palps, tentacles, cirri, and parapodia. Most of them can also regenerate posterior body segments if the trunk is severed. There are numerous exceptional cases of the regenerative powers of annelids. While regeneration of the posterior end is common, most cannot regenerate lost heads. However, sabellids, syllids, and some others can regrow the anterior end. The most dramatic regenerative powers among the annelids occur, oddly, in a few forms with highly specialized and heteronomous bodies. In *Chaetopterus*, for example, the anterior end will regenerate a normal posterior end as long as the regenerating part (the anterior end) includes not more than fourteen segments; if the animal is cut behind the fourteenth segment, regeneration does not occur. Furthermore, any single segment from among the first fourteen can regenerate anteriorly and posteriorly to produce a complete worm (Figure 14.17A). An even more dramatic example of regenerative power is known among certain species of *Dodecaceria* (Cirratulidae), which are capable of fragmenting their bodies into individual segments, each of which can regenerate a complete individual! A similar situation is seen in another cirratulid, *Ctenodrilus*, where segments transform into a series of heads that then proliferate further segments behind, resulting in a chain of individuals that eventually separates. A chain of six is shown in Figure 14.17B. Such clonal reproduction has been maintained by *Ctenodrilus* in culture for decades without any instances of sexual reproduction.

Regeneration appears to be controlled by neuroendocrine secretions released by the central nervous system at sites of regrowth. It is initiated by severing the elements of the nervous system. Initiation has been demonstrated experimentally by cutting the ventral nerve cord while leaving the body intact; the result is the formation of an extra part at the site of cutting (e.g., two "tails"). The actual mechanism of regeneration has been studied in a variety of annelids and, although the results are not entirely consistent, a general scenario can be outlined. Normal growth and addition of segments (in young worms) take place immediately anterior to the pygidium, in a region known as the growth zone. However, this growth zone is obviously not involved in regeneration. Rather, when the trunk is severed, the cut region heals over and then a patch of generative tissue, or blastema, forms. The blastema comprises an inner mass of cells originating from nearby tissues that were derived originally from mesoderm, and an outer covering of cells from ectodermally derived tissues such as the epidermis. These two cell masses act somewhat as a growth zone analogue, proliferating new body parts according to their tissue

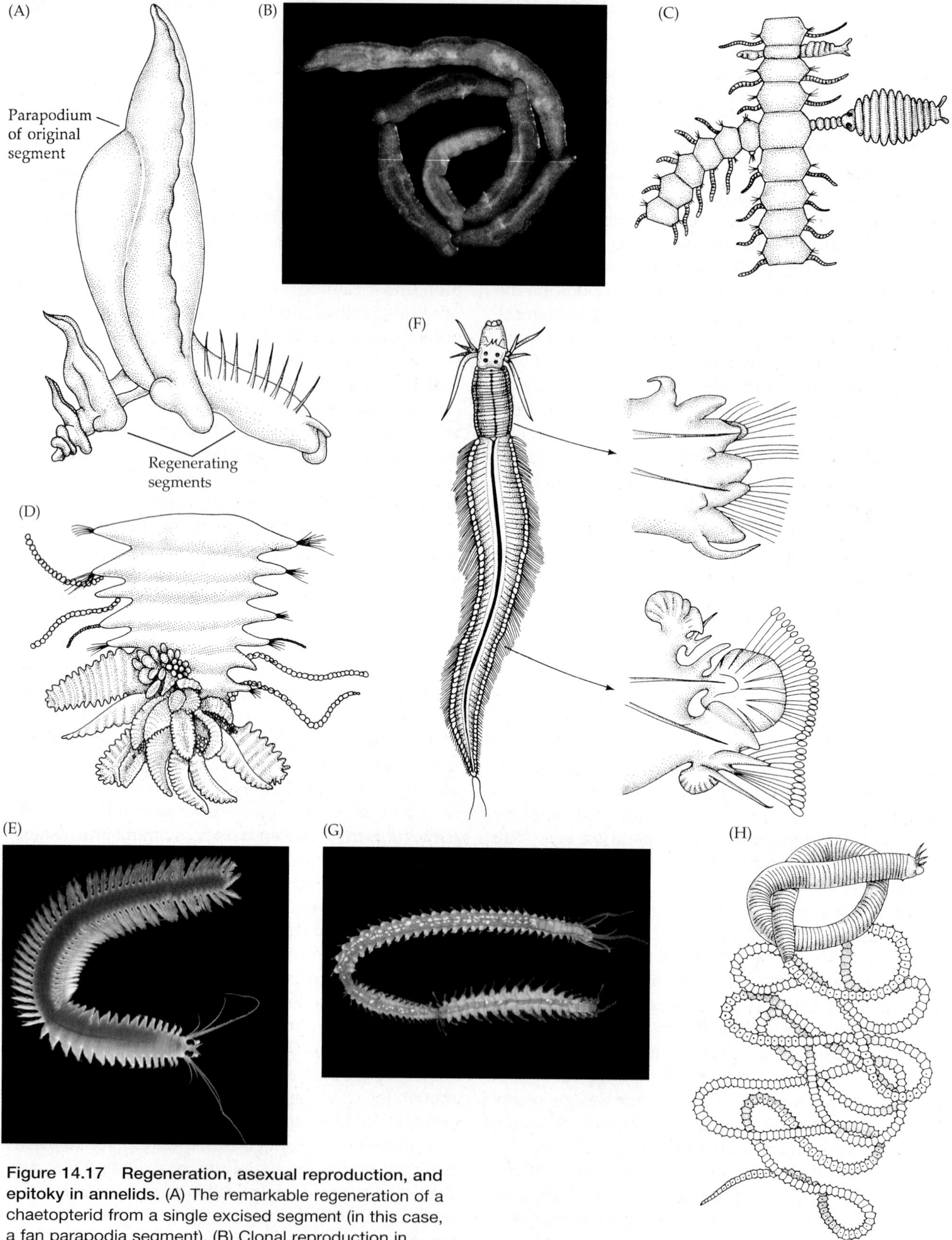

Figure 14.17 Regeneration, asexual reproduction, and epitoky in annelids. (A) The remarkable regeneration of a chaetopterid from a single excised segment (in this case, a fan parapodia segment). (B) Clonal reproduction in *Ctenodrilus* (Cirratulidae). Body segments transform into a series of heads that then proliferate further segments behind, resulting in a chain of individuals that eventually separates. (C) A portion of syllid (*Syllis ramosa*, but also occurs in *Ramisyllis*) in which reproductive individuals are budded (cloned) from the parent's parapodia. (D) The posterior end of *Typosyllis*, bearing a cluster of epitokes. (E) Another heteronereid. Note the enlarged eyes. (F) An epitokous nereidid, known as a heteronereid, *Nereis irrorata*. Note the dimorphic condition of the anterior and posterior parapodia. (G) Another syllid, *Myrianida gidholmi*, where there are two epitokes being produced in a series. The more mature epitoke (posterior) is full or eggs and has enlarged eyes. (H) The epitokous palolo worm, *Palola viridis* (Eunicidae).

origins. This process is coupled with the growth of the gut, which contributes parts of endodermal origin.

In addition, research has shown that relatively undifferentiated cells from mesenchyme-like layers of the body migrate to injured areas and contribute to various (and uncertain) degrees to the regenerative process. These so-called neoblast cells are ectomesodermal in origin, because they arise embryonically from presumptive ectoderm. During regeneration, they apparently contribute to tissues and structures normally associated with true mesoderm, and perhaps other germ layers as well. The implication here is that the germ layer of the precursor of a regenerated part may not correspond to the normal origin of that part. For example, regenerated coelomic spaces may be lined with tissue derived originally from ectoderm rather than from mesoderm.

A number of annelids use their regenerative powers for asexual reproduction. A few reproduce asexually by multiple fragmentation. We mentioned above the ability of *Dodecaceria* to regenerate complete individuals from isolated segments; this phenomenon occurs spontaneously and naturally in these animals as a highly effective reproductive strategy. Spontaneous transverse fragmentation of the body into two or several groups of segments also occurs in certain syllids, chaetopterids, cirratulids, and sabellids (Figure 14.17B). The point (or points) at which the body fragments is typically species-specific, and can be anticipated by an ingrowth of the epidermis that produces a partition across the body called a **macroseptum**. Asexual reproduction results in a variety of regeneration patterns, including chains of individuals, budlike outgrowths, or direct growth to new individuals from isolated fragments. Asexual reproduction in annelids may be under the same sort of neurosecretory control as that postulated for non-reproductive regeneration.

Sexual reproduction The majority of annelids are gonochoristic. Hermaphroditism is known in some sabellids, serpulids, certain freshwater nereidids, and isolated cases in other clades, but is notably found across all of Clitellata. The gametes arise by proliferation of cells from the peritoneum, these being released into the coelom as gametogonia or primary gametocytes. Formation of gametes may occur throughout the body or only in particular regions of the trunk. Within a reproductive segment, the production of gametes may occur all over the coelomic lining or only on specific areas.

The gametes generally mature within the coelom and are released to the outside by mechanisms such as metanephridia or a simple rupture of the parent body wall. Many species release eggs and sperm into the water, where external fertilization is followed by indirect development with planktotrophic trochophore larvae. Others display mixed life history patterns. In these forms, fertilization is internal, followed by brooding or by the production of floating or attached egg capsules. In most instances the embryos are released as free-swimming lecithotrophic trochophores. Some species brood their embryos on the body surface or in their tubes.

Many of the free-spawning annelids have evolved methods that ensure relatively high rates of fertilization. One of these methods is the fascinating phenomenon of **epitoky**, characteristic of many syllids, nereidids, and eunicids (Figure 14.17D–H). This phenomenon involves the production of a sexually reproductive worm called an **epitokous individual**. Epitokous forms may arise from nonreproductive (**atokous**) animals by a transformation of an individual worm, as in nereidids, or by the asexual production of new epitokous individuals, as in many syllids. So in nereidids, the whole body may transform into a sexual **epitokous** individual called a heteronereid. In these heteronereids the posterior body segments is swollen and filled with gametes, their associated parapodia become enlarged and natatory while the head develops large eyes and the gut atrophies (Figure 14.17E,F). In syllids, where the epitokous worm is clonally produced, the epitokes are formed as single clone, in a linear series (Figure 14.17G), or even as clusters of outgrowths from particular body regions (Figure 14.17D).

In any event, the epitokes are gamete-carrying bodies capable of swimming from the bottom upward into the water column, where the gametes are released. Epitoky is controlled by neurosecretory activity, and the upward migration of the epitokes is precisely timed to synchronize spawning within a population. The reproductive swarming of epitokes is linked with lunar periodicity. This activity not only ensures successful fertilization but establishes the developing embryos in a planktonic habitat suitable for the larvae. Perhaps the most famous of the epitokous worms are eunicids in the genus *Palola*, commonly known as palolo worms based on the Samoan name for them. Native Polynesian islanders have long been known to predict the swarming (typically to the day and hour) and collect the ripe epitokes, which are the released posterior end of the worm's body, under a full moon to feast on them (Figure 14.17H).

Clitellates are hermaphroditic and the various parts of the reproductive apparatus are restricted to particular segments, usually in the anterior portion of the worm (Figure 14.18). The arrangement of the reproductive system facilitates mutual cross-fertilization followed by encapsulation and deposition of the zygotes. The male system includes one or two pairs of testes located in one or two specific body segments. Sperm are released from the testes into the coelomic spaces, where they mature or are picked up by storage sacs (seminal vesicles) derived from pouches of the septal peritoneum (Figure 14.18B). There may be a single seminal

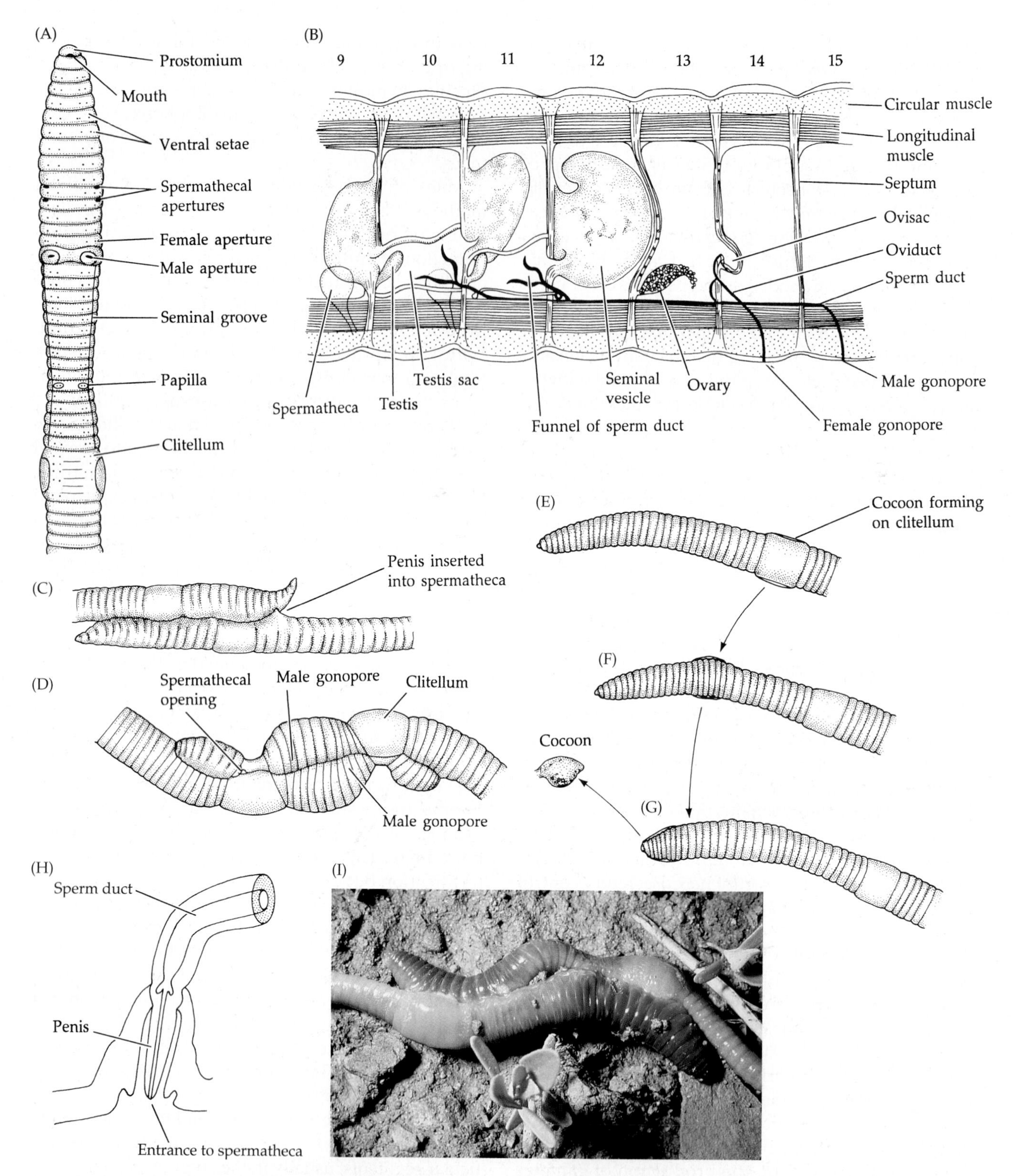

Figure 14.18 The reproductive system of *Lumbricus* (Clitellata) and mating in earthworms. (A) External structures associated with reproduction of *Lumbricus* (ventral view). (B) Segments 9–15 of *Lumbricus* (composite lateral view). (C,D) Copulating earthworms. (C) *Pheretima* transfers sperm directly from the male pore, through a penis, into the mate's spermatheca. (D) *Eisenia* uses indirect sperm transfer. As in *Lumbricus*, the sperm leave the male pores and travel along paired seminal grooves to the spermathecal openings of the mate. (E–G) An earthworm forming and releasing a cocoon. As the cocoon slides over the worm, it receives ova and sperm. (H) Engaged copulatory apparatus of *Rhynchelmis*, a lumbriculid with direct sperm transfer. (I) Copulating earthworms (*Lumbricus*).

vesicle or as many as three pairs in some earthworms. When mature, the sperm are released from the seminal vesicles, picked up by ciliated seminal (sperm) funnels, and carried by sperm ducts to paired gonopores. The female reproductive system consists of a single pair of ovaries located posterior to the male system (Figures 14.12F and 14.18B). Again, the ova are released into the adjacent coelomic space and sometimes stored until mature in shallow pouches in the septal wall called the ovisacs. Next to each ovisac is a ciliated funnel that carries the mature ova to an oviduct and eventually to the female gonopore. Most clitellates also possess one or two or more pairs of blind sacs called spermathecae (seminal receptacles) that open to the outside via separate pores (Figure 14.18A,B).

Of major importance to the overall reproductive strategy of clitellates is the unique region of glandular tissue called the **clitellum** (Latin for "saddle") (Figure 14.18A,C–G,I), a principal anatomical feature giving rise to the name Clitellata. The clitellum has the appearance of a thick sleeve that partially or completely encircles the worm's body. It is formed of secretory cells within the epidermis of particular segments. The exact position of the clitellum and the number of segments involved are consistent within any particular species. In freshwater forms the clitellum is located around the position of the gonopores, but in most earthworms it is posterior to the gonopores. There are three types of gland cells within the clitellum, each secreting a different substance important to reproduction: mucus that aids in copulation; the material forming the outer casing of the egg capsule (or cocoon); and albumin deposited with the zygotes inside the cocoon.

During copulation in most clitellates, the mating worms align themselves facing in opposite directions (Figure 14.18C–D,I) and mucous secretions from the clitellum hold them in this copulatory posture. Many clitellates position themselves so that the male gonopores of one are aligned with the spermathecal openings of the other. In such cases, special copulatory chaetae near the male pores or eversible penis-like structures aid in anchoring the mates together (Figure 14.18H). Some Crassicilitellata earthworms are not so accurate with their mating, and their copulatory position does not bring the male pores against the spermathecal openings. Instead, they develop external sperm grooves along which the male gametes must travel prior to entering the spermathecal pores. These grooves are actually formed temporarily by muscle contraction, and are covered by a sheet of mucus. Underlying muscles cause the grooves to undulate, and the sperm are transported along the body to their destination. Following the mutual exchange of sperm to the seminal receptacles of each mate, the worms separate, each functioning as an inseminated female.

From several hours to a few days following copulation between clitellate annelids, a sheet of mucus is produced around the clitellum and all the anterior segments. Then the clitellum produces the cocoon itself in the form of a leathery, proteinaceous sleeve. The cocoons of terrestrial species are especially tough and resistant to adverse conditions. Albumin is secreted between the cocoon and the clitellar surface. The amount of albumin deposited with the cocoon is much greater in terrestrial species than in aquatic forms. Thus formed, the cocoon and underlying albumin sheath are moved toward the anterior end of the worm by muscular waves and backward motion of the body. As it moves along the body, the cocoon first receives eggs from the female gonopores, and then sperm previously received from the mate and stored in the seminal receptacles. Fertilization occurs within the albumin matrix inside the cocoon (though not in leeches, see below). The open ends of the cocoon contract and seal as they pass off the anterior end of the body (Figure 14.18E–G). The closed cocoons are deposited in benthic debris by aquatic clitellates. Terrestrial forms deposit their cocoons in the soil at various depths, the particular depth depending on the moisture content of the substratum.

Development Early annelid development exemplifies a classic protostomous, spiralian pattern (Figures 14.19–14.21. The eggs are telolecithal with small to moderate quantities of yolk. Those with a period of encapsulation or brooding prior to larval release generally contain more yolk than those that free spawn. In any case, cleavage is holoblastic and clearly spiral. A coeloblastula or, in the cases of more yolky eggs, a stereoblastula develops and undergoes gastrulation by invagination, epiboly, or a combination of these two events. Gastrulation results in the internalization of the presumptive endoderm (the 4A, 4B, 4C, 4D and the 4a,

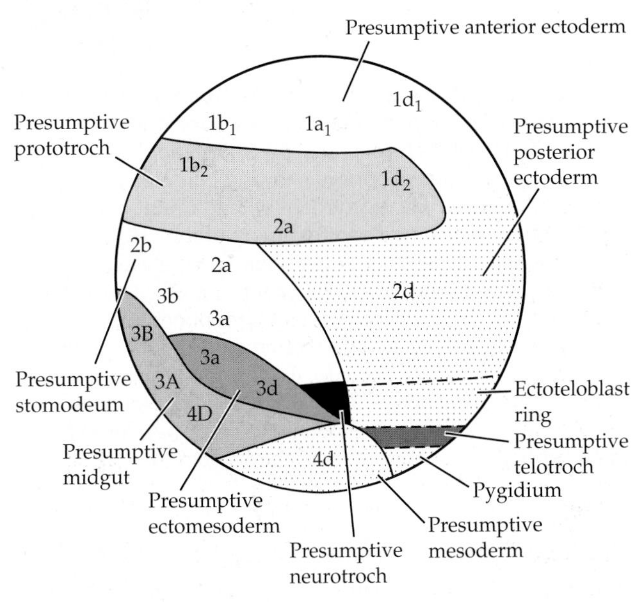

Figure 14.19 Fate map of a *Scoloplos* (Orbiniidae) blastula (viewed from the left side).

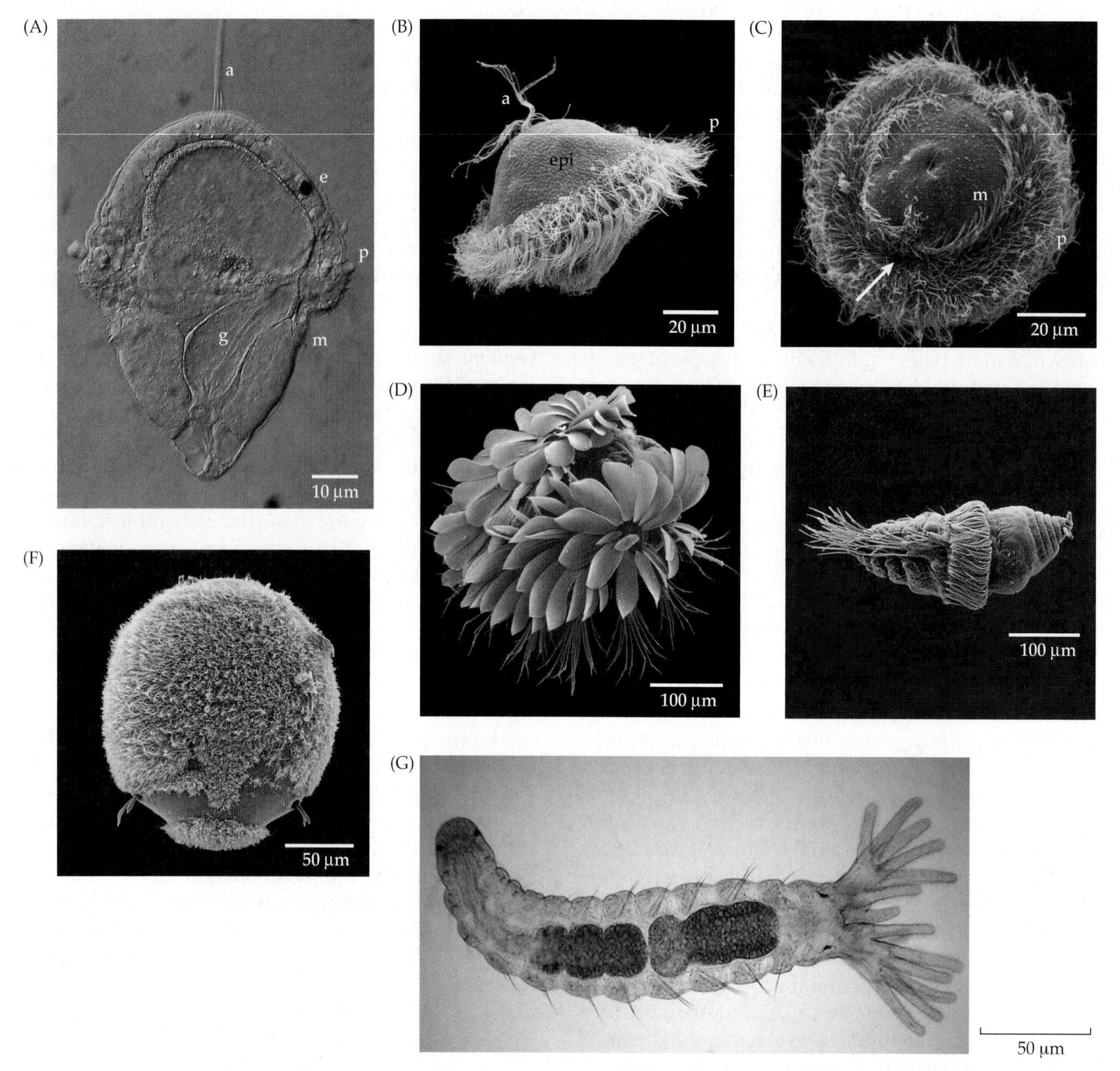

Figure 14.20 Some annelid larvae—trochophores and beyond. (A–C) Opposed-band planktotrophic trochophores of serpulid polychaetes. (A) *Spirobranchus giganteus* (Serpulidae) trochophore showing complete gut (g), apical tuft (a), eye (e), prototroch (p), and metatroch (m) (differential interference contrast micrograph). (B) SEM of *Spirobranchus giganteus* trochophore in side view showing apical tuft (a), episphere (epi), and prototroch (p). (C) SEM of *Spirobranchus giganteus* trochophore in posterior view showing prototroch (p), metatroch (m), and ciliated food groove (shown by arrow). (D) SEM of a planktonic 6-segmented nectochaete larva of a chrysopetalid. (E) SEM of a planktonic 5-segmented nectochaete larva of a glycerid. (F) Lecithotrophic trochophore of *Marphysa* (Eunicidae) with first segment developing and showing chaetae. (G) Some annelid larvae that show little, if any, metamorphosis. After 9 days in the tube a young juvenile of *Echinofabricia alata* (Fabriciidae), with a fully developed radiolar crown and gut, as well as anterior and posterior eyes, is ready to crawl away. (H–J) Others show dramatic metamorphosis, sometimes in just a few minutes as seen here in *Owenia* (Oweniidae). (H) Lateral view of an oweniid mitraria larva taken from plankton off Belize showing long larval chaetae and episphere with ciliated margin. (I) Close-up of mitraria episphere showing juvenile inside larval body. (J). Juvenile immediately after metamorphosis (taken only a few seconds after the previous micrograph). Larval chaetae have been shed, as has the episphere.

4b, 4c cells) and presumptive mesoderm (the 4d mesentoblast). The derivatives of the first three micromere quartets give rise to ectoderm and ectomesoderm, the latter producing various larval muscles between the body wall and the developing gut. As the endoderm hollows to produce the archenteron, a stomodeal invagination forms at the site of the blastopore and a proctodeal invagination produces the hindgut.

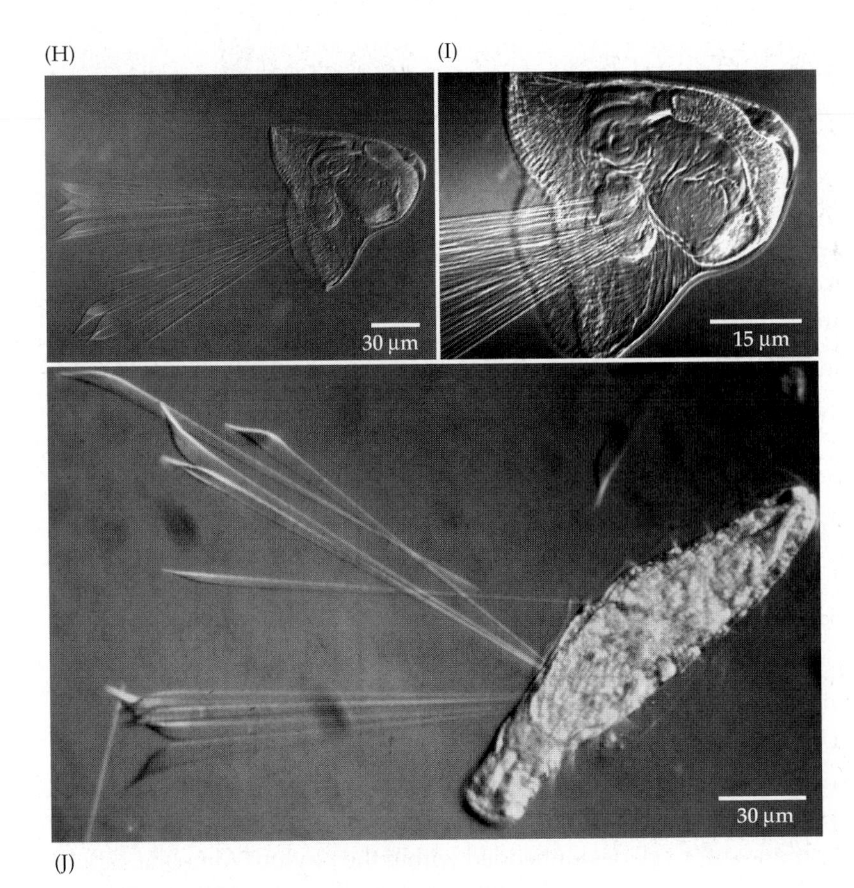

expand as paired coelomic spaces, which eventually obliterate the blastocoel. Thus the production of serially arranged coelomic compartments and the formation of segments are one and the same; the anterior and posterior walls of adjacent coelomic compartments form the intersegmental septa. Proliferation of segments by this process is called **teloblastic growth**. Externally, additional ciliary bands are added at each segment. These metatrochal bands aid in locomotion as the animal increases in size. Such segmented larvae are sometimes called **polytroch larvae**.

The fates of the various larval regions are now apparent (Figure 14.19, 14.21). The region anterior to the prototrochal ring (the **episphere**) becomes the prostomium, while the prototrochal area itself forms the peristomium. Note that these two parts are not involved in the proliferation of segments and are thus presegmental. However, in some annelids, one or more of the anterior trunk segments may be incorporated into the peristomium during growth. The segmental, metatrochal portion of the larva forms the trunk, and the growth zone and postsegmental pygidium remain as the corresponding adult body parts. The apical sense organ becomes the cerebral ganglion, which is eventually joined with the developing ventral nerve cord by the formation of circumenteric connectives. The body continues to elongate as more segments form, and the juvenile worm finally drops from the plankton and assumes the lifestyle of a young annelid. This whole affair was beautifully described in verse by the late Walter Garstang (1951), where he explained the development of *Phyllodoce* in the first part of his classic poem, "The Trochophores":

The trochophores are larval tops the Annelids set spinning

With just a ciliated ring—at least in the beginning—

They feed, and feel an urgent need to grow more like their mothers,

So sprout some segments on behind, first one, and then the others.

And since more weight demands more power, each segment has to bring

Its contribution in an extra locomotive ring:

With these the larva swims with ease, and, adding segments more,

Becomes a Polytrochula instead of Trochophore.

Then setose bundles sprout and grow, and the sequel can't be hid:

The larva fails to pull its weight, and sinks—an Annelid.

In most annelids, these early ontogenetic events result in a clearly recognizable trochophore larva characterized by a locomotory ciliary band just anterior to the region of the mouth (Figures 14.20 and 14.21). Trochophores are also seen in Mollusca and Nemertea (and possibly some other phyla). This ciliary band, the **prototroch**, arises from special cells, called trochoblasts, of the first and second quartets of micromeres. Most trochophores also bear an apical ciliary tuft associated with an apical sense organ derived from a plate of thickened ectoderm at the anterior end. In addition, there is often a perianal ciliary band called the telotroch and/or a midventral band called the neurotroch. Some trochophores may also have band called a metatroch. The mesentoblast divides to form a pair of cells called teloblasts, which in turn proliferate a pair of mesodermal bands, one on each side of the archenteron in the region of the hindgut, an area known as the growth zone (Figure 14.21B). Many trochophores bear larval sense organs such as ocelli, as well as a pair of larval protonephridia. Many trochophores develop bundles of mobile chaetae, which are known to serve as a defense against predators and to help retard sinking. Several annelid larvae are shown in Figure 14.20.

The larva grows and elongates by proliferation of tissue in the growth zone (Figure 14.21C), while segments are produced by the anterior proliferation of mesoderm from the teloblast derivatives on either side of the gut. These packets of mesoderm hollow (schizocoely) and

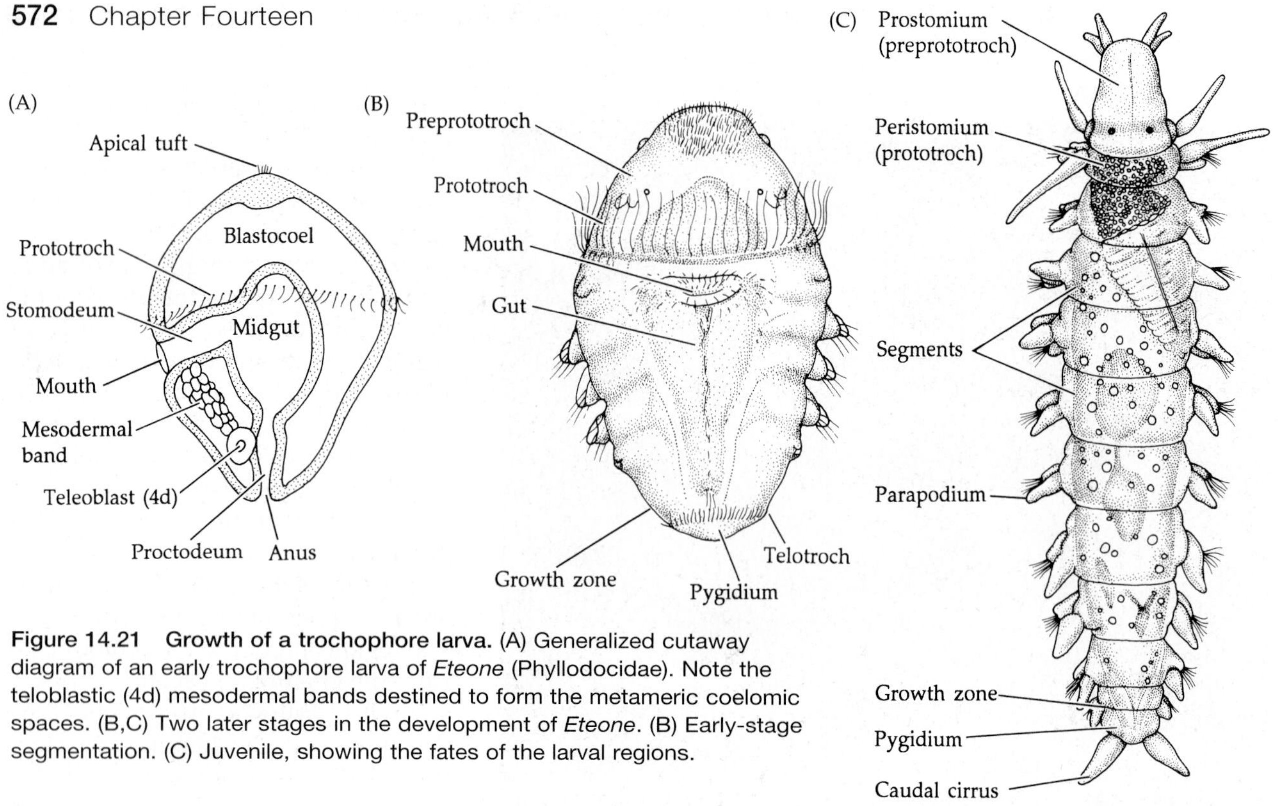

Figure 14.21 **Growth of a trochophore larva.** (A) Generalized cutaway diagram of an early trochophore larva of *Eteone* (Phyllodocidae). Note the teloblastic (4d) mesodermal bands destined to form the metameric coelomic spaces. (B,C) Two later stages in the development of *Eteone*. (B) Early-stage segmentation. (C) Juvenile, showing the fates of the larval regions.

Clitellates produce telolecithal ova, but the amount of yolk varies greatly and inversely with the amount of albumin secreted into the cocoon. The eggs of freshwater forms often contain relatively large amounts of yolk but are encased with only a small quantity of albumin. Conversely, the eggs of terrestrial species tend to have little yolk, but are supplied with large quantities of albumin on which the developing embryos depend for a source of nutrition. In any case, cleavage is holoblastic and unequal. And, although highly modified, evidence of the ancestral spiralian pattern is still apparent in cell placement and fates (e.g., an identifiable 4d mesentoblast homologue gives rise to the presumptive mesoderm). Development is direct, with no trace of a trochophore larval stage. However, the teloblastic production of coelomic spaces and segments is an obvious retained characteristic of the basic annelid developmental program. Development time varies from about one week to several months, depending on the species and environmental conditions. In climates where relatively severe conditions follow cocoon deposition, development time is usually long enough to ensure that the juveniles hatch in the spring. Under more stable conditions, development time is shorter and reproduction is less seasonal.

Sipuncula: The Peanut Worms

In the past, the coelomate worm phyla Sipuncula and Echiura were often dismissed in short fashion as "minor" or "lesser" groups. However, thanks to recent molecular phylogenetic research, we now know that these two groups are remarkably modified spiralian clades embedded within the phylum Annelida. Sipunculans and echiurans resemble one another in several respects, although they are not closely related, and they are often found in similar habitats. For these reasons, biologists who find interest in one of these groups often study both. Sipunculans are never as abundant or important ecologically as some other worms, especially the polychaetes and nematodes. Nonetheless, they display body plans that are different from any we have discussed so far and provide important lessons in functional morphology—and thus deserve special attention.

The clade Sipuncula (Greek *siphunculus*, "little tube") includes about 150 species in 16 genera and 6 families. Usually called "peanut worms," adult sipunculans show no evidence of segmentation or chaetae (two features viewed as characteristic of annelids). The body is sausage-shaped and divisible into a retractable introvert and a thicker trunk (Figure 14.22). It is when the introvert is retracted and the body is turgid that some species resemble a peanut. The anterior end of the introvert bears the mouth and feeding tentacles. The tentacles are derived from the regions around the mouth (**peripheral tentacles**) and around the nuchal organ (**nuchal tentacles**); differences in tentacular arrangements are of taxonomic importance. The gut is characteristically U-shaped and highly coiled, and the anus is located dorsally on the body near the introvert–trunk junction. The body surface is usually beset with minute bumps, warts, tubercles, or spines. Sipunculans range in length from less than 1 cm to about 50 cm, but most are 3–10 cm long. With the exception of the coiled gut, the body plan of sipunculans has remained largely

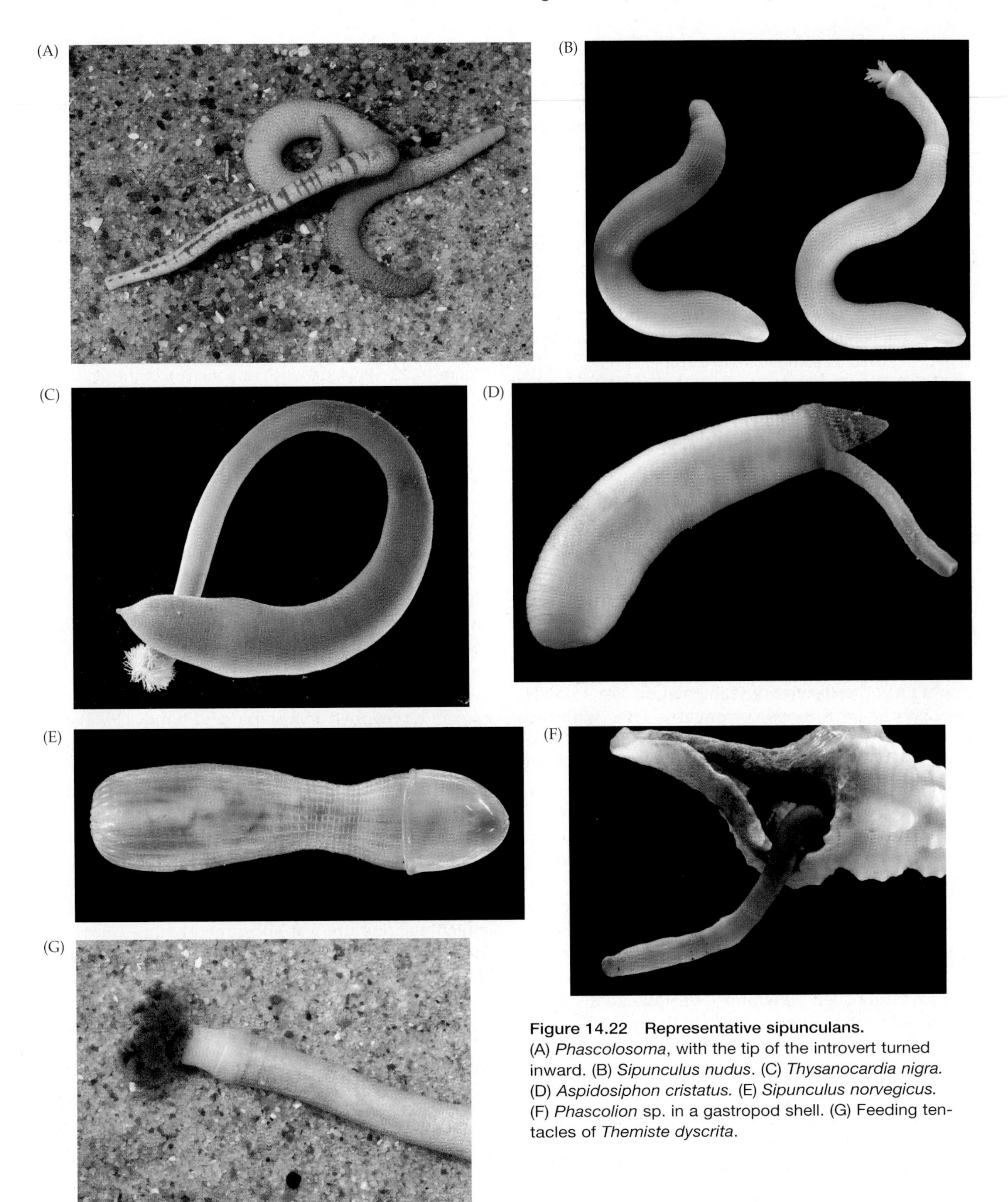

Figure 14.22 Representative sipunculans. (A) *Phascolosoma*, with the tip of the introvert turned inward. (B) *Sipunculus nudus*. (C) *Thysanocardia nigra*. (D) *Aspidosiphon cristatus*. (E) *Sipunculus norvegicus*. (F) *Phascolion* sp. in a gastropod shell. (G) Feeding tentacles of *Themiste dyscrita*.

unchanged since the Lower Cambrian, as judged by two recently discovered species from the Maotianshan Shales of Southwest China.

The coelom is well developed and unsegmented, forming a spacious body cavity. Metanephridia are present, with nephridiopores on the ventral body surface. There is no circulatory system, but the coelomic fluid includes cells containing a respiratory pigment. Most sipunculans are gonochoristic and reproduce by epidemic spawning. Development is usually indirect, typically protostomous, and includes a free-swimming larva.

Sipunculans are benthic and exclusively marine. They are usually reclusive, either burrowing into sediments or living beneath stones or in algal holdfasts. In tropical waters sipunculans are common inhabitants of coral and littoral communities where they often burrow into hard, calcareous substrata. Some inhabit abandoned gastropod shells, polychaete tubes, and other such structures. They are found from the intertidal zone to depths of over 5,000 meters, and some deep burrowing species may play an important role in influencing the ecology and geochemistry of the Nordic Seas region. In Southeast Asia (e.g., Vietnam), large sand-burrowing species are occasionally consumed as human food, and others are used as fishing bait in Europe.

The sipunculan body plan is founded on the qualities of the spacious body coelom. Uninterrupted by transverse septa, the coelomic fluid provides an ample circulatory medium for these sedentary worms. The coelom and associated musculature function as a hydrostatic skeleton and as a hydraulic system for locomotion, circulation of coelomic fluid, and introvert extension.

The embryonic and adult characteristics of sipunculans and echiurids place them solidly within the spiralian protostomes, and molecular data have provided strong evidence for their membership in the phylum Annelida, despite the lack of adult segmentation. One might view the absence of segmentation as a secondary loss of a partitioned coelom associated with the exploitation of a sedentary, burrowing lifestyle. The reduction in sensory receptors and simplification of the nervous system in general are explainable on this same basis, but echiurid neurogenesis shows clear traces of segmentation. However, many other species of annelids have evolved burrowing lifestyles while retaining their basic segmentation.

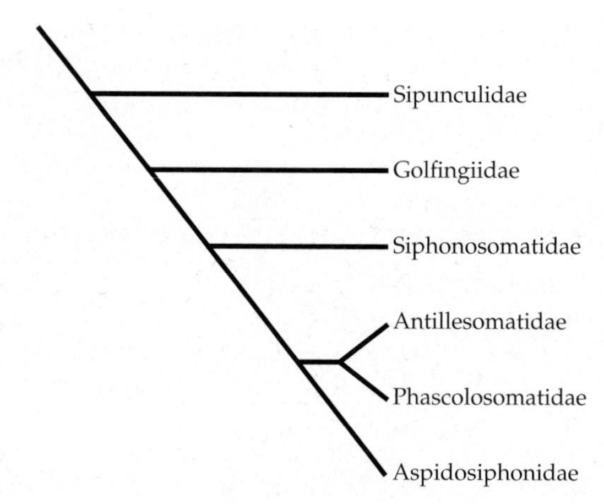

Figure 14.23 Phylogenetic relationships of the sipunculan families.

Sipunculan Classification

The first published illustrations of sipunculans were produced from woodcuts made in the mid-sixteenth century. Linnaeus included these animals in the twelfth edition of his *Systema Naturae* (1767) and placed them in the Vermes, along with so many other odds and ends. In the nineteenth century, Lamarck and Cuvier considered the sipunculans to be relatives of holothuroid echinoderms (sea cucumbers). No separate taxon was established for these worms until 1828, when Henri Marie Ducrotay de Blainville introduced the name Sipunculida and allied the group with certain parasitic helminths.

In 1847 Jean Louis Armand de Quatrefages invented the group Gephyrea to include sipunculans, echiurans, and priapulans. The Greek root *gephyra* means "bridge," as Quatrefages regarded these animals as intermediate between annelids and echinoderms. The gephyrean concept was founded on superficial characteristics, but it persisted well into the twentieth century even though many authors attempted to raise the constituent groups to individual phylum status. Finally, Libbie Hyman (1959), recognizing the polyphyletic nature of the Gephyrea, elevated the sipunculans to a separate phylum rank; her view was quickly accepted. At that time, however, no classes, orders, or families were recognized, and the phylum was divided into only genera and species. The Herculean effort by Alexander Stephen and Stanley Edmonds (1972) and subsequent modifications by other workers (e.g., Mary Rice 1982) led to a classification comprising 4 families and 16 genera. Later, Edward Cutler and Peter Gibbs (1985) applied modern phylogenetic methods to the sipunculans and produced a classification scheme of 2 classes, 4 orders, 6 families, and 17 genera, which has been used until recently. However, with the use of molecular phylogenetics, this classification was challenged, the phylum was subsumed within Annelida, and a new system, consisting of 6 families and 16 genera has been proposed (Figure 14.23).

PHYLUM SIPUNCULA

FAMILY SIPUNCULIDAE Large sipunculans, trunk to 45 cm in length. Introvert shorter than the trunk, covered with prominent papillae arranged irregularly. Hooks absent. Circular and longitudinal muscle layers divided into distinct bands. Body wall with coelomic extensions in the form of parallel longitudinal canals extending through most of the trunk length, or short diagonal canals limited in length to the width of one circular muscle band. Two protractor muscles may be present. With paired metanephridia. Development proceeds through a planktotrophic pelagosphera larva. Two genera: *Sipunculus*, *Xenosiphon*.

FAMILY GOLFINGIIDAE Heterogeneous group of small to medium-sized sipunculans (trunk no longer than 20 cm). Hooks may be deciduous, simple when present, not sharply curved and generally scattered, except in three species where hooks are arranged in rings. Body wall with a continuous muscle layer, except in *Phascolopsis*, where the longitudinal muscles are divided in anastomosing bands. With paired or single metanephridia. Seven genera: *Golfingia, Nephasoma, Onchnesoma, Phascolion, Phascolopsis, Themiste, Thysanocardia*.

FAMILY SIPHONOSOMATIDAE Large to medium-sized sipunculans (trunk to 50 cm in length). Introvert much shorter than trunk with prominent conical papillae and/or hooks arranged in rings. Body wall with small, irregular saclike coelomic extensions. Circular and longitudinal muscle layers gathered into anastomosing, sometimes indistinct bands. With paired metanephridia. Two genera: *Siphonomecus, Siphonosoma*.

FAMILY ANTILLESOMATIDAE Medium-sized sipunculans (trunk to 8 cm). Distal part of the introvert smooth and white, proximal portion bears dark papillae and is marked off by a distinctive collar. Hooks absent in adults, but a few hooks present in small individuals (<1 cm). Oral disc consisting of nuchal organ enclosed by numerous tentacles, which vary in number according to size (from 30 to 200 in adults). Body wall with longitudinal muscle layer gathered into anastomosing bands. Contractile vessel with many villi. Four introvert retractor muscles with the lateral pair often extensively fused. With paired metanephridia. Without an anal shield. Single genus: *Antillesoma*.

FAMILY PHASCOLOSOMATIDAE Small to medium-sized sipunculans (trunk to 12 cm in length). Hooks recurved, usually with internal structures, and closely packed in regularly spaced rings (absent in *Apionsoma trichocephalus*). External trunk wall rough with obvious papillae. In *Apionsoma*, papillae are concentrated at the posterior end of the trunk.

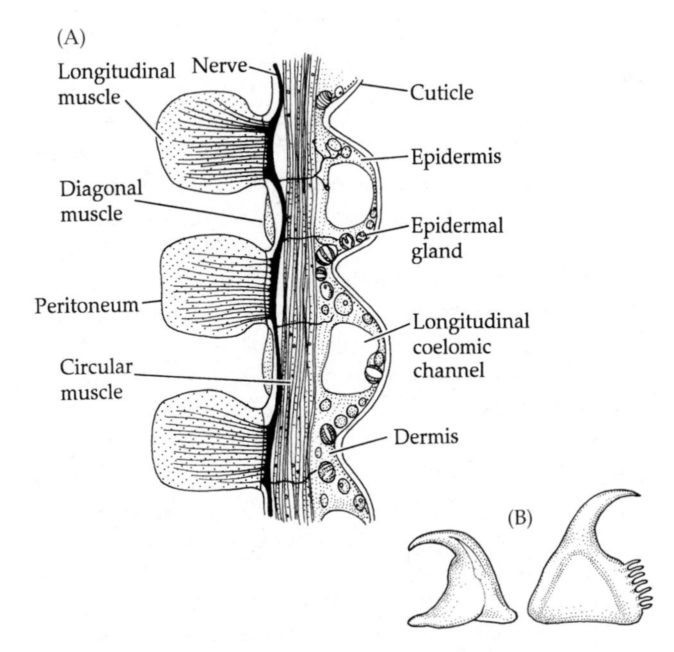

Figure 14.24 (A) The body wall of *Sipunculus nudus* (cross section). (B) Two types of cuticular hooks from sipunculans.

Internally, the longitudinal muscle is generally subdivided into anastomosing bands except in two subgenera–*Phascolosoma* (*Fisherana*) and *Apionsoma* (*Apionsoma*)–where this layer is thinner and continuous. Contractile vessel is smooth but may be large with bulbous pouches or swelling in a few *Phascolosoma*. With paired metanephridia. Two genera: *Apionsoma, Phascolosoma*.

FAMILY ASPIDOSIPHONIDAE Generally small (to 30 mm) sipunculans with smooth trunk and two retractor muscles. Introvert protruding at 45–90° angle ventral to main axis of trunk. Muscle layers generally smooth and continuous, or with longitudinal muscle layer separated in anastomosing bundles. Trunk with anterior, and sometimes posterior, anal shields derived from thickened cuticle or calcareous deposits. With paired metanephridia. Two genera: *Aspidosiphon, Cloeosiphon*.

The Sipunculan Body Plan

Body Wall, Coelom, Circulation, and Gas Exchange

The sipunculan body surface is covered by a well-developed cuticle that varies from thin on the tentacles to quite thick and layered over much of the trunk (Figure 14.22 and 14.24). The cuticle often bears papillae, warts, or spines of various shapes. Beneath the cuticle lies the epidermis, the cells of which are cuboidal over most of the body but grade to columnar and ciliated on the tentacles. The epidermis contains a variety of unicellular and multicellular glands called epidermal organs, some of which project into the cuticle and produce some of the surface papillae or knobs. Some of these glands are associated with sensory nerve endings; others are responsible for producing the cuticle or for mucus secretion. In both larval and adult sipunculans, epidermal organs are externally encapsulated by a cuticle and internally delimited by regular epidermal cells. If pocketed deeply into the body ball, epidermal organs are also delimited by subepidermal musculature.

Beneath the epidermis, especially where it is raised, is a connective tissue dermis of fibers and loose cells. From one to four large introvert retractor muscles extend from the body wall into the introvert where they insert on the gut just behind the mouth (Figure 14.25A,B). All species so far studied go through developmental stages with four retractor muscles, which are eventually reduced to a lower number in the adult. Species with crawling larvae have more strongly developed body wall musculature than those with swimming larvae.

In Sipunculidae, which are mostly large and include some warm-water species, the dermis also houses a system of coelomic extensions or channels (Figure 14.24). These **coelomic channels** may extend entirely through the muscle layers and into the epidermal

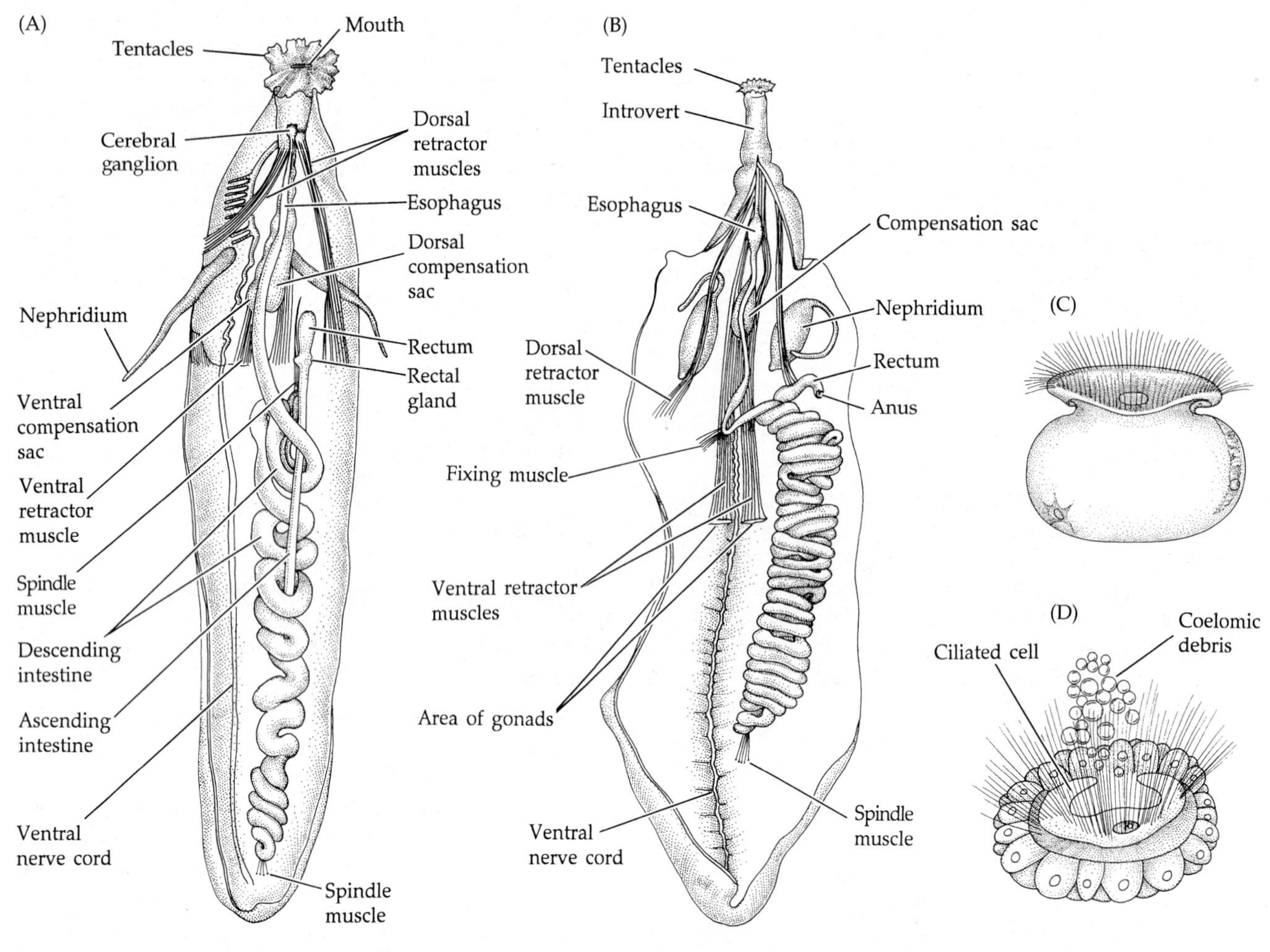

Figure 14.25 Sipunculan anatomy. (A) *Sipunculus nudus*. (B) *Golfingia vulgaris*. (C, D) Free coelomic urns from *Sipunculus* (C) and *Phascolosoma* (D). (E) The gut of *Golfingia* (partial section). Note the ciliated intestinal groove. (F) A nephridium of *Phascolosoma* (section).

layer, and they have a gas transport function in some species. They are connected to the trunk coelom by pores. The muscles of the body wall include outer circular and inner longitudinal layers and sometimes a thin middle layer of diagonal fibers. The longitudinal muscles form a continuous sheet in many sipunculans, but in some genera these muscles form distinct bundles or bands (e.g., *Phascolosoma*, *Phascolopsis*, *Siphonomecus*, *Siphonosoma*, *Sipunculus*, and *Xenosiphon*). This arrangement gives the internal surface of the body wall a ribbed appearance that is often visible through the animal's cuticle. While this character was once the basis for the family Sipunculidae, the family now has a more restricted definition and it has been shown that this feature is homoplastic.

A peritoneum lines the body wall and the internal organs. The coelom is a continuous space, but peritoneal mesenteries form incomplete partitions supporting the organs. In addition to the main body coelom and the coelomic channels in the body wall, a separate fluid-filled "coelom" called the compensation system is associated with the tentacles. The hollow tentacles contain lined spaces that are continuous with one or two sacs (the **compensation sacs**) that lie next to the esophagus (Figure 14.25A,B). Upon eversion of the introvert, circular body muscles apply pressure on these sacs and force the contained fluid into the tentacles, causing their erection.

The fluid of the body cavity contains a variety of cells and other inclusions. There are both granular and agranular amebocytes of uncertain function, and red blood cells containing hemerythrin. Also contained in the coelomic fluid are unique and fascinating multicellular structures called **urns**, some of which are fixed to the peritoneum and some of which swim free in the fluid (Figure 14.25C,D). The urns accumulate waste materials and dead cells by trapping them with cilia and mucus.

Gas exchange apparently varies among species. It has been suggested that rock borers and those that burrow in sediments of low oxygen content (e.g., *Themiste*) exchange gases largely across the tentacles, which extend into the overlying water. Other burrowers with long introverts may use the body surface of both the tentacles and the introvert for gas exchange, whereas others, such as the large-bodied species of the genus *Sipunculus*, may use the entire body surface. The coelomic fluid in the body cavity and in body wall channels provides a circulatory medium, aided by diffusion and body movements. The erythrocytes that circulate in the coelomic fluid of sipunculans contain intracellular polymeric hemerythrin for storing and transporting oxygen.

Support and Locomotion

Sipunculans are sedentary creatures. The general body shape is maintained by the muscles of the body wall and the hydrostatic skeleton established by the large coelom. The body is essentially a fluid-filled bag of constant volume, so any constriction at one point is accompanied by an expansion at another. Burrowing in soft substrata is accomplished by peristalsis, driven by the circular and longitudinal muscles of the body wall and by the action of the introvert. Movement through algal holdfasts and bottom rubble occurs in a similar manner. Some species with an anterior cuticular shield burrow into hard substrata and use the shield as a functional operculum to close the burrow entrance. Burrowing into hard substrata is probably accomplished by both mechanical and chemical means, the former using cuticular structures (such as spines and the posterior shield) as rasps, the latter facilitated by the secretions of epidermal glands.

The introvert is extended when the coelomic pressure is increased by contracting the circular muscles of the body wall. Withdrawal is accomplished by the retractor muscles, which pull from the mouth end, turning the introvert inward as the body wall muscles relax. When the introvert is fully extended, the tentacles are erected by increasing the pressure on the compensation sacs.

Sipunculans are highly tactile and strongly thigmotactic, requiring contact with their surroundings. Placed alone in a glass dish, they are rather inactive except for rolling the introvert in and out. However, if several sipunculans are placed together or with small stones or shell fragments, they soon respond by making contact with each other or surrounding objects.

Feeding and Digestion

There is surprisingly little information on the details of sipunculan feeding mechanisms. Indirect evidence from anatomy, gut contents, and general behavior suggests that these animals use different feeding methods in different habitats. Most of the sipunculans that can place their tentacles at a substratum–water interface are selective or nonselective detritivores (e.g., shallow burrowers, algal holdfast dwellers); they use the mucus and cilia on the tentacles to obtain food. Deeper burrowers in sand are direct deposit feeders. Some appear to be ciliary-mucus suspension feeders, using the tentacles to extract organic material from the water. Sipunculans that burrow in calcareous substrata use spines or hooks on their introverts to retrieve organic detritus within reach and ingest the material by retracting the introvert. Limited data suggest that at least some sipunculans take up dissolved organic compounds directly across the body wall. Some workers have speculated that up to 10% of these animals' nutritional requirements may be met in this fashion.

Because the anus is located anteriorly on the dorsal side of the body, the digestive tract is basically U-shaped, although highly coiled (Figure 14.25A,B)—the coiled gut is unique among annelids or any other invertebrate group. The mouth is terminal, located at the end of the introvert and is wholly or partially surrounded by the peripheral tentacles (e.g., Sipunculidae) or lies near the nuchal tentacles (e.g., Phascolosomatidae). The mouth leads inward to a short, muscular stomodeal pharynx, which is followed by an esophagus that extends through the introvert and into the trunk. The midgut consists of a long intestine composed of descending and ascending portions coiled together, although the coiling is not observed in the Cambrian species. Studies comparing the expression patterns of regulatory "gut genes" of *Themiste* and other annelids (with uncoiled guts) revealed differences in persistence and extension of endodermal expression of *FoxA*, which could be related to the U-shaped gut and coiling. The gut is usually supported by a threadlike spindle muscle that extends from the body wall near the anus through the coils to the end of the trunk and by several fixing muscles connecting the gut to the body wall. The ascending intestine leads to a short proctodeal rectum, terminating in the anus.

The intestine is ciliated and bears a distinct groove along its length (Figure 14.25E). This ciliated groove leads ultimately to a small pouch or diverticulum (the rectal gland) off the rectum. The function of this groove and diverticulum is unknown. The lumen epithelium of the descending intestine contains a variety of gland cells that are presumably the sources of digestive enzymes.

Excretion and Osmoregulation

Most sipunculans possess one pair of elongate, saclike, tubular metanephridia (nephromixia; Figure 14.25A,B,F) located ventrolaterally at the anterior end of the trunk. Two genera (*Onchnesoma* and *Phascolion*) have but a single nephridium. Species in these genera tend to be asymmetrically coiled. The nephridiopores are located ventrally on the anterior region of the trunk. The nephrostome lies close to the body wall, near the pore, and leads to a large nephridial sac that extends posteriorly in the trunk.

Sipunculans are ammonotelic. Nitrogenous wastes accumulate in the coelomic fluid and are excreted via the nephridia. The urns also play a major excretory role by picking up particulate waste material in the coelom. The fate of wastes accumulated by urns is unknown, but at least some is probably transported to the nephridia. Urns originate as fixed epithelial cell complexes in the peritoneum, where they trap and remove particulate debris. They are also known to secrete mucus in response to pathogens in the coelomic fluid. Fixed urns regularly detach and become free-swimming in the coelomic fluid. Not only do urns effectively cleanse the coelomic fluid, but they also participate in a clotting process when a sipunculan is injured. Free urns can be seen by preparing a wet slide of fresh coelomic fluid. They are usually obvious, moving about like little bumper cars, trailing strands of mucus and bits of particulate matter.

Sipunculans are basically osmoconformers and they are unknown from fresh and brackish-water habitats. Under normal conditions, the coelomic fluid is nearly isotonic to the surrounding seawater. However, when placed in hypotonic or hypertonic environments, the body volume increases or decreases, respectively. Interestingly, the rates of volume change differ when the animal is exposed to these opposing environments, suggesting that sipunculans are better at preventing water loss than at preventing water gain. This situation may be due to a differential permeability of the cuticle, or perhaps to some active mechanism of the nephridia. In any case, sipunculans rarely face severe osmotic problems in their usual environments, and even in laboratory experiments they are able to recover nicely from most conditions of osmotic stress.

Nervous System and Sense Organs

The general structure of the sipunculan nervous system is similar in many respects to that in other annelids. A bilobed cerebral ganglion lies dorsally in the introvert, just behind the mouth. Circumenteric connectives extend from the cerebral ganglion to a ventral nerve cord running along the body wall through the introvert and trunk (Figures 14.25A,B and 14.26). The adult ventral nerve cord is single, and there is no evidence of segmental ganglia. In *Phascolosoma agassizii*, a double ventral nerve cord forms initially, but later fuses into a single cord, and neurogenesis initially follows a segmental pattern similar to that of annelids. Starting out with paired FMRFamidergic and serotonergic axons, four pairs of associated serotonergic perikarya and interconnecting commissures form one after another in an anterior–posterior progression. In late-stage larvae, the two serotonergic axons of the VNCs fuse, the commissures disappear, and one additional pair of perikarya is formed. These cells (ten in total) migrate toward one another, eventually forming two clusters

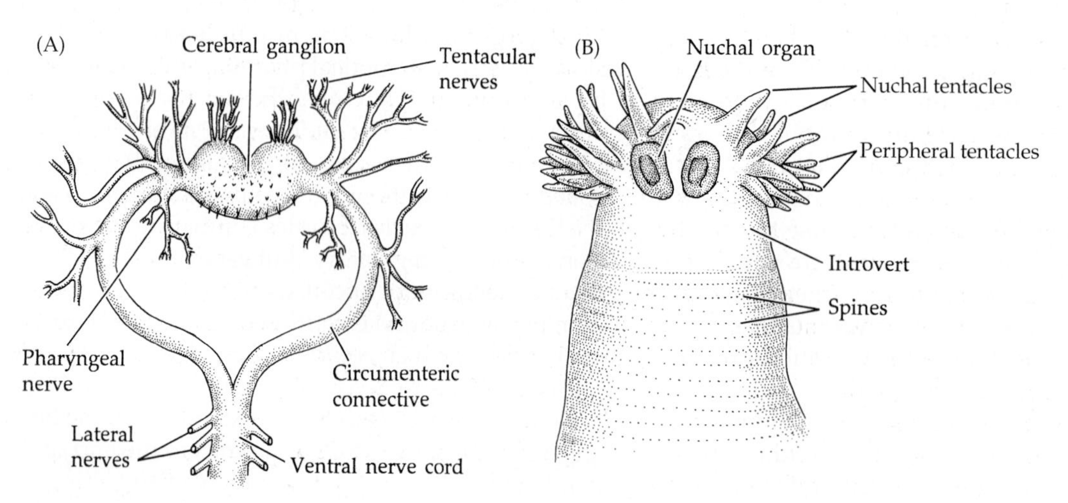

Figure 14.26 (A) Anterior portion of the central nervous system of *Golfingia*. (B) Anterior end of *Golfingia*. Note the nuchal organs.

of five cells each. These neural-remodeling processes result in the single nonmetameric central nervous system of the adult sipunculan. This ontogenetic example suggests that the ancestral sipunculan condition may have been a double ventral nerve cord like that found in some primitive annelids. Lateral nerves arise from the nerve cord and extend to the body wall muscles and sensory receptors in the epidermis.

Sensory receptors are widespread in sipunculans, but many are poorly understood. Tactile receptor cells are scattered over the body within the epidermis, as would be expected, and are especially abundant on and around the tentacles. Chemosensory **nuchal organs** are located on the dorsal side of the introvert in many species (Figure 14.26B). Many possess a pair of pigment-cup ocelli on the dorsal surface of the cerebral ganglion, and some have a so-called cerebral organ, consisting of a ciliated pit projecting inward to the cerebral ganglion. The cerebral organ may be involved in chemoreception or perhaps neurosecretion, as is similar in structure to that seen in nemerteans.

Reproduction and Development

Sipunculans possess reasonable powers of regeneration. Most species are able to regrow lost parts of the tentacles and even the introvert, and some can regenerate portions of the trunk and the digestive tract. It was long believed that sipunculans could not reproduce asexually. However, in the 1970s it was discovered that at least some species do possess this capacity. The process takes place by transverse fission of the body, whereby the worm divides into a small posterior fragment and a larger anterior portion. Both portions then regrow the missing parts. Regeneration from the small posterior part is quite remarkable, since most of the trunk, anterior gut, retractor muscles, nephridia, introvert, and so on must be regrown.

Except for *Golfingia minuta*, sipunculans are gonochoristic. (Facultative parthenogenesis has been reported in one species, *Themiste lageniformis*.) The gametes arise from the coelomic lining, often near the origins of the retractor muscles. Gametes are released into the coelom, where they mature. Ripe eggs and sperm are picked up selectively by the nephridia and stored in the sacs until released. The eggs are encased in a layered, porous covering. Males spawn first, probably in response to some environmental cue, and the presence of sperm in the water stimulates females to spawn.

Following external fertilization, the zygotes pass through typical spiralian development. In *Golfingia* cleavage is spiral and holoblastic, but the relative sizes of micromeres and macromeres differ among species, depending on the amount of yolk in the egg. Although traditionally sipunculan embryos were described as having a "molluscan cross," which suggested a close relationship with the Mollusca, this character is now known to have been misinterpreted, and sipunculans are now considered to be an early but highly derived offshoot of Annelida.

The cell fates are the same as those in most typical spiralians. The first three quartets of micromeres become ectoderm and ectomesoderm; the 4d cell produces endomesoderm; and 4a, 4b, 4c, and 4Q form the endoderm. The mouth opens at the site of the blastopore, and the surrounding ectodermal cells grow inward as a stomodeum. The anus breaks through secondarily on the dorsal surface (Figure 14.27). The 4d mesoderm proliferates as two bands, as it does in other annelids, but yields the major trunk coelom without segmentation.

Four different developmental sequences have been recognized among the sipunculans, including direct (e.g., *Golfingia minuta*, *Phascolion crypta*, *Themiste pyroides*) and indirect development. In direct development, the eggs are covered by an adhesive jelly and attached to the substratum after fertilization. The embryo develops directly to a vermiform individual that hatches as a minute juvenile sipunculan. The other three developmental patterns are indirect, involving various combinations of larval stages. In some species (e.g., *Phascolion strombi*), a free-living lecithotrophic trochophore larva develops and metamorphoses into a juvenile worm. The other two developmental patterns involve a second larval stage, the **pelagosphera larva**, that forms after a metamorphosis of the trochophore (Figure 14.27C). In some species, both the trochophore and the pelagosphera forms are lecithotrophic and relatively short-lived (e.g., some species of *Golfingia* and *Themiste*), while in others the pelagosphera larva is planktotrophic and may live for extended periods of time in the plankton (e.g., *Aspidosiphon parvulus*, *Sipunculus nudus*, members of the genus *Phascolosoma*).

The transformation of the trochophore to the pelagosphera larva involves a reduction or loss of the prototrochal ciliary band and the formation of a single metatrochal band for locomotion. The pelagosphera eventually elongates, settles, and becomes a juvenile sipunculan (Figure 14.27D).

Since the pelagosphera larva of some species are estimated to spend several months in the plankton and sipunculans display little morphological variation, a disproportionate number of species, when compared to any other marine worm groups, have been suggested to be cosmopolitan. However detailed developmental and molecular analyses of several supposed cosmopolitan species have shown that cosmopolitanism cannot be the norm, and that instead sipunculans have many cryptic or pseudocryptic species.

Echiuridae: The Spoon Worms

Echiuridae (Greek *echis*, "serpent-like") are secondarily unsegmented annelids. There are about 200 known

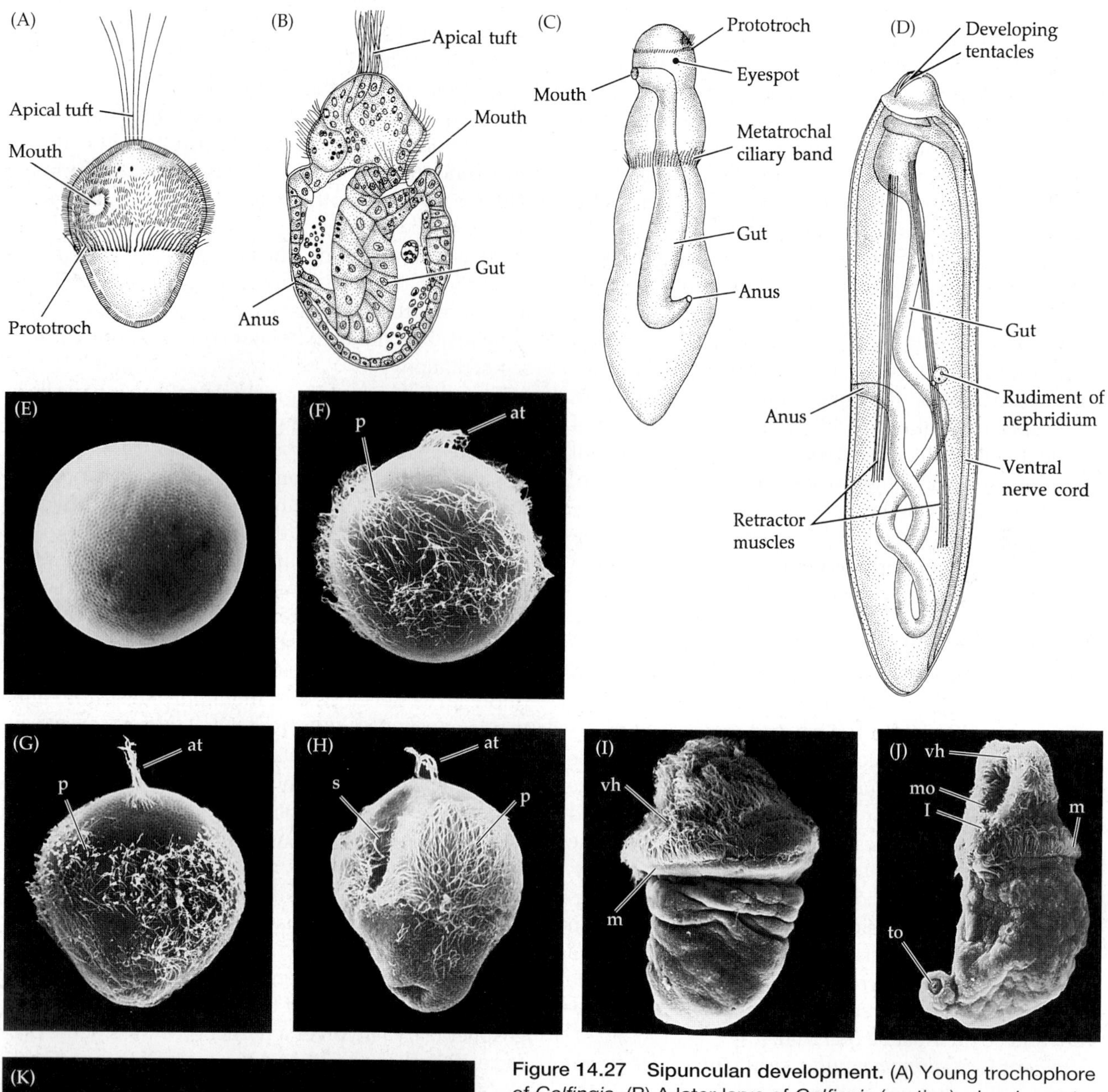

Figure 14.27 Sipunculan development. (A) Young trochophore of *Golfingia*. (B) A later larva of *Golfingia* (section), showing gut shape and placement of the anus. (C) The pelagosphera larva of *Phascolosoma* has an enlarged metatroch. (D) A juvenile sipunculan. (E–J) A series of scanning electron micrographs showing growth to the trochophore and metamorphosis to the pelagosphera in *Siphonosoma*. ap: apical tuft; m: metatroch; mo: mouth region; l: lower ciliated lip; p: prototroch; to: terminal attachment organ; vh: ventral ciliated head; s: stomodeum. (K) One-day-old juvenile of *Siphonosoma*. Note the coiled gut.

species. The vermiform body is divided into an anterior, preoral proboscis, and an enlarged trunk (Figure 14.28). The mouth is located at the anterior end of the trunk at the base of a proboscis groove, or gutter. The body surface may be smooth or somewhat warty, and often bears chaetae (e.g., *Protobonellia*). Most echiurids are quite large. The trunk may be from a few to as many as 40 cm long, but the proboscis may reach lengths of 1–2 meters (e.g., in *Ikeda*, Figure 14.29B). Some forms, such as the beautiful emerald green *Metabonellia*, show drastic sexual dimorphism, wherein "dwarf" males are less than 1 cm long (Figure 14.28E,F). *Bonellia*, and relatives such as *Metabonellia*, are also notable for females producing the compound bonellin, which appears to act as an antibiotic for the worm, but also as a sex hormone to create dwarf males. Echiurids are always ben-

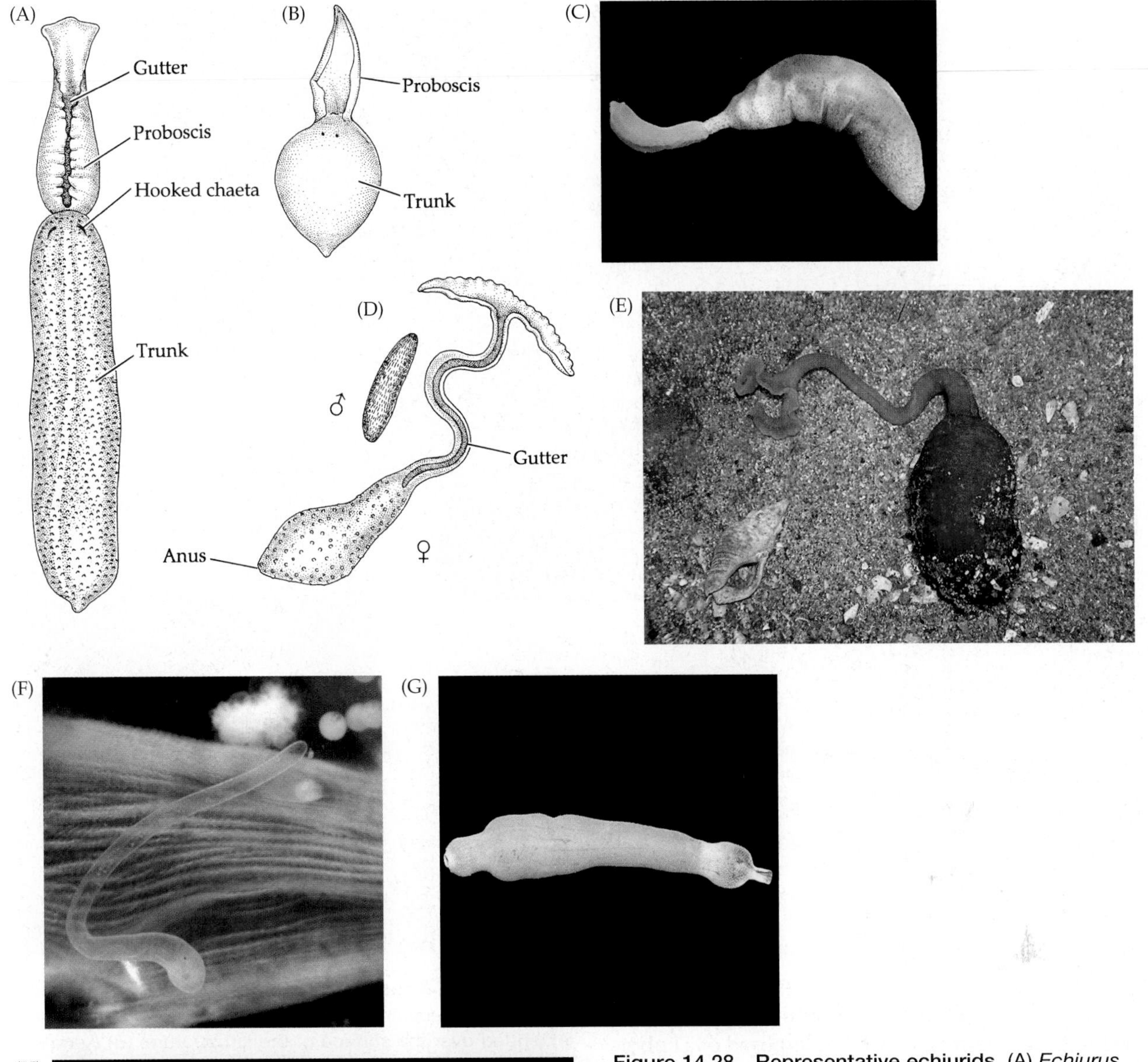

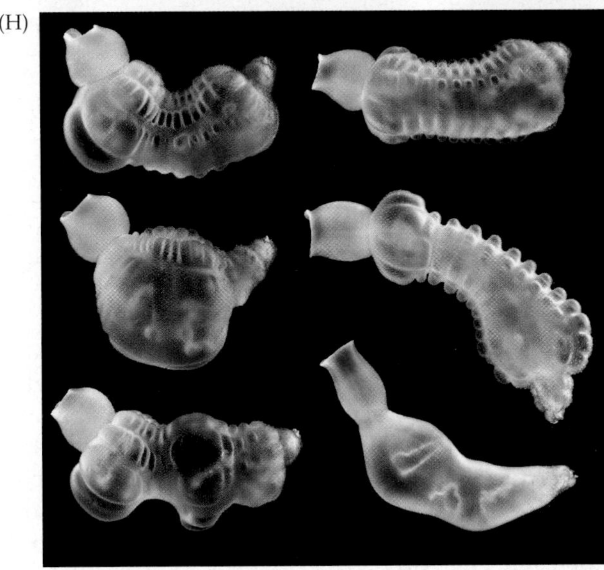

Figure 14.28 Representative echiurids. (A) *Echiurus*, anatomy. (B) *Listriolobus,* with the proboscis somewhat contracted. (C) *Anelassorhynchus porcellus* removed from under a rock, with proboscis somewhat contracted. (D) *Bonellia viridis.* Note the extreme sexual dimorphism between the large female (with her proboscis extended) and the tiny male. (E) Female *Metabonellia haswelli* removed from its burrow; with forked and somewhat contracted proboscis and 15 centimeter trunk. The green color comes from the pigment bonellin. (F) One centimeter long dwarf male of *Metabonellia haswelli*, removed from the nephridium of a female. (G) *Urechis caupo*, the "fat innkeeper." (H) *Ochetostoma* sp., from Papua New Guinea. Montage sequence of an individual with peristaltic waves along body.

thic and marine, though a few species are known from brackish-water habitats. Most burrow in sand or mud, or live in surface detritus or rubble. Some species typically inhabit rock galleries excavated by boring clams or other invertebrates. They are known from intertidal regions to a depth of 10,000 meters.

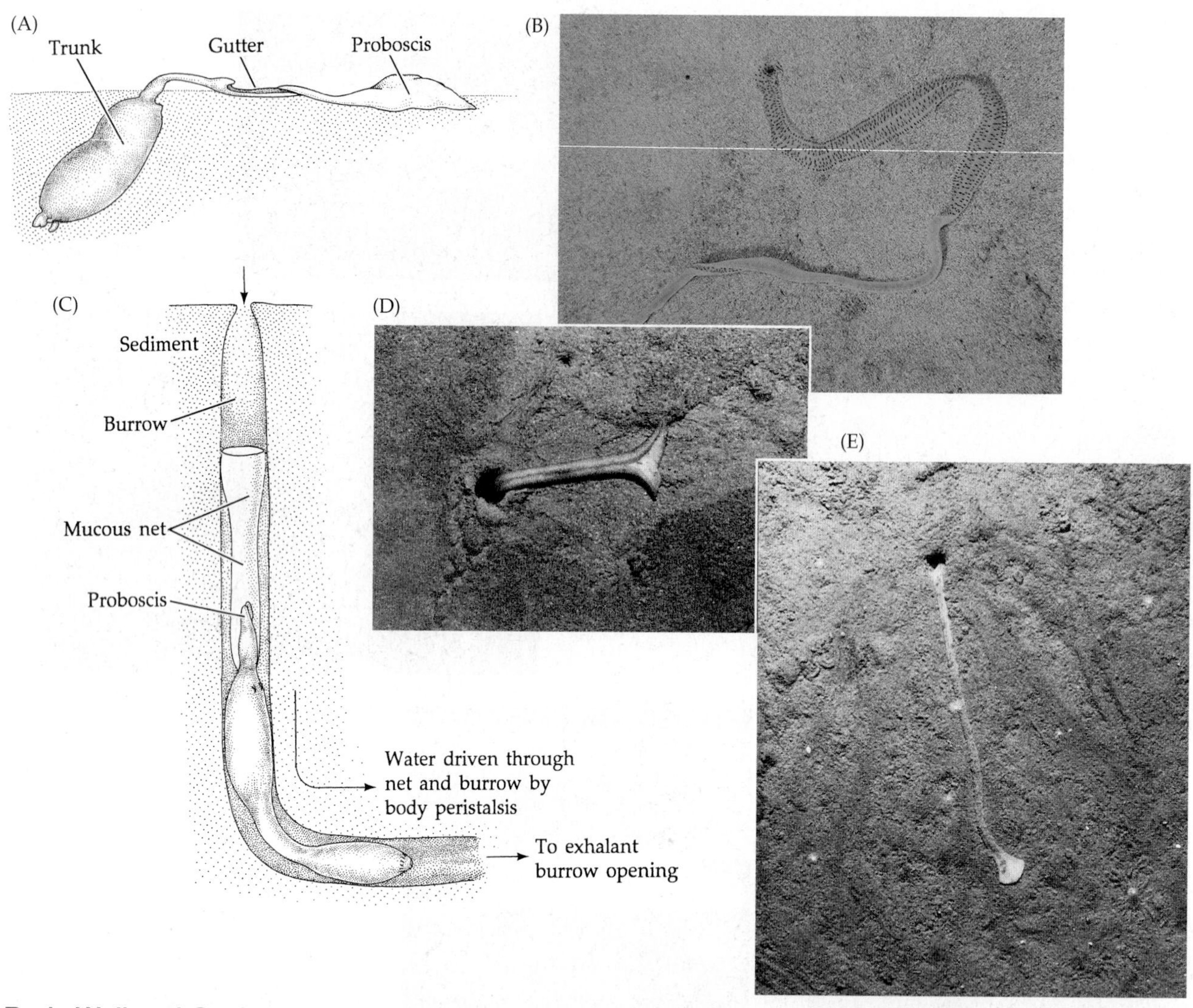

Figure 14.29 Feeding in echiurids. (A) *Tatjanellia grandis* in a "typical" echiurid feeding posture, with the proboscis extended over the surface of the substratum. (B) *Ikeda* sp., with its long proboscis extending from burrow in sediment. (C) A portion of the burrow of *Urechis caupo*. The worm is in its feeding position. (D,E) The proboscides of deep-sea echiurids living at depths of 2,635 m and 7,570 m, respectively.

Body Wall and Coelom

The body wall of echiurids has an outer thin cuticle covering the epidermis, which is composed of a cuboidal epithelium and contains a variety of gland cells. Epidermal chaetae occur in some species at either anterior trunk, posterior trunk, or both locations. Layers of circular, longitudinal, and oblique muscles form the bulk of the body wall, which is lined internally by the peritoneum. The epidermis is ciliated along the proboscis groove, or gutter. The coelomic cavity is spacious and occupies most of the trunk—it is interrupted only by partial mesenteries between the gut and the body wall. The coelomic fluid contains red blood cells, with hemoglobin in some species, and various types of amebocytes.

Support and Locomotion

The large trunk coelom provides a hydrostatic skeleton against which the body wall muscles operate. The non-septate coelom allows peristaltic movements as the animal burrows or moves through various sediments, gravel, or rubble. The proboscis is capable of shortening and lengthening, but it does not roll in and out as does the introvert of sipunculans. This characteristic, and the more conventional posterior position of the anus, allows one to easily distinguish the two groups.

Feeding and Digestion

Most echiurids feed on epibenthic detritus. Typically, the animal lies with the trunk more or less buried in the substratum, with the proboscis extended over the sediment (Figure 14.28C,E,G,H and 14.29). Densely packed gland cells of the proboscis epithelium secrete mucus, to which organic detrital particles adhere. The mucous coating and the food are moved along a ventral proboscis groove, often called the gutter, by ciliary action into the mouth.

An interesting exception to the above feeding method occurs in *Urechis*, worms that live in

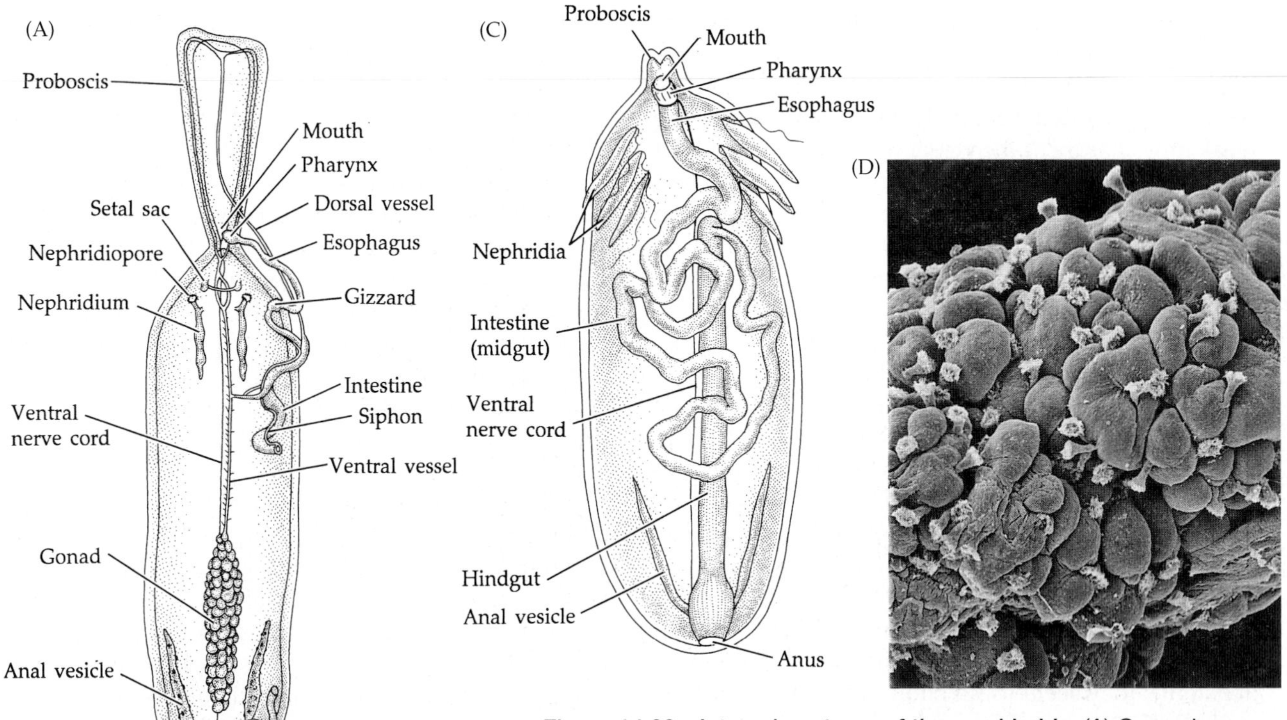

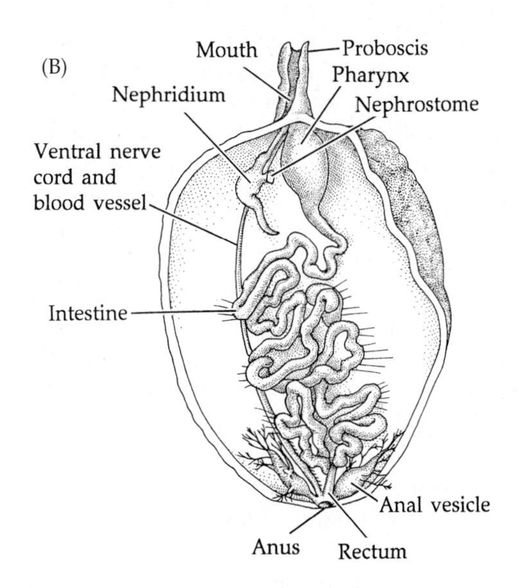

Figure 14.30 Internal anatomy of three echiurids. (A) General internal anatomy, as seen in *Echiurus*. (B) A dissected *Bonellia viridis* female (proboscis cut off). (C) Internal organs of *Urechis*. Note especially the enlarged midgut. (D) SEM of anal vesicles and excretory funnels of *Urechis* at 60×.

U-shaped burrows in soft substrata (Figure 14.28G). They have a short proboscis, unlike those of more typical echiurids, and engage in mucous-net filter feeding. A ring of glands located near the proboscis–trunk junction produces a funnel-shaped mucous net, which is attached to the burrow wall by the proboscis. Water is drawn through the burrow and the sheet of mucus by peristaltic movements of the body, and suspended food particles as small as 1 μm are caught in the fine-meshed net. Periodically the animal grasps the food-laden net with its proboscis and ingests it. The whole process is somewhat similar to the feeding behavior of another annelid, *Chaetopterus*.

The digestive tract is generally very long and coiled, leading from the mouth at the base of the proboscis to the posterior anus (Figure 14.30). The foregut may be regionally specialized as pharynx, esophagus, gizzard, and stomach, or it may be more or less uniform along its length. The midgut usually bears a longitudinal ciliated groove, the **siphon**, which probably aids the movement of materials through the gut. It may also shunt excess water from the main midgut lumen, thereby concentrating food and facilitating digestion. The hindgut, or cloaca, varies in structure among different species. In most echiurids, the cloaca bears a pair of large excretory diverticula called anal vesicles (see Excretion and Osmoregulation). In some species of *Urechis* the cloaca is enlarged and thin-walled. In such cases water is pumped in and out of the hindgut for gas exchange.

Not much is known about digestive physiology in echiurids. The epithelium of the midgut is rich in gland cells that presumably produce and secrete digestive enzymes. Digestion and absorption occur mainly in the midgut.

Circulation and Gas Exchange

Most echiurids possess a simple closed circulatory system, although it is absent from some forms (e.g., *Urechis*) in which case the coelomic fluid contains red blood cells with hemoglobin. The circulatory system generally includes dorsal and ventral longitudinal vessels in the trunk and median and lateral vessels in the proboscis (Figure 14.30A). There is no major pumping organ (except in *Ikeda*); the blood is transported by pressures generated from body movements and by the weak musculature of the vessel walls. The main site of gas exchange in *Urechis* is provided with oxygenated water by cloacal irrigation—water being pumped in and out of the anus by muscle action. In other echiurids it would appear that gas exchange also occurs across the surface of the proboscis.

Excretion and Osmoregulation

The excretory structures of echiurids include paired metanephridia and anal vesicles (Figure 14.30). The number of nephridia varies: one pair in *Bonellia* and *Metabonellia*; two pairs in *Echiurus*; three pairs in *Urechis*; and hundreds of pairs in *Ikeda*. When only one or a few pairs are present, the nephridia are located in the anterior region of the trunk and lead to nephridiopores on either side of the ventral midline. The degree to which these nephridia function in excretion is debatable; they seem to function primarily in picking up gametes from the coelom, having relinquished the major excretory responsibility to the anal vesicles. Echiurids are relatively poor osmoregulators providing an explanation as to why they are only generally in fully marine habitats.

The **anal vesicles** are hollow sacs arising as evaginations of the cloaca near the anus (Figure 14.30A–C). Each vesicle bears from about a dozen to as many as 300 ciliated funnels that open to the coelom. Few studies have been conducted on the function of these structures, but they apparently pick up wastes from the coelomic fluid and remove the material to the hindgut and anus.

Nervous System and Sense Organs

The nervous system of echiurids is simple, although constructed in a fashion generally similar to other annelids. An anteriorly located nerve ring extends around the gut and dorsally forward into the proboscis. Ventrally, in the trunk, the nerve ring meets at a subesophageal ganglion that connects with a seemingly single ventral nerve cord extending the length of the body (Figure 14.30A–C). There are no ganglia in this system and no obvious evidence of segmentation in adults. However, immunohistochemical methods and confocal laser-scanning microscopic studies have proposed a metameric organization of the nervous system in different larval stages of *Urechis caupo* and *Bonellia viridis*, which corresponds to the segmental arrangement of ganglia seen in other annelids. This suggests that the trunk of echiurids consists of a series of fused segments. Lateral nerves arise from the ventral nerve cord and extend to the body wall muscles.

Associated with the simple nervous system and infaunal sedentary lifestyle of echiurids is the absence of major sensory receptors. These animals are touch sensitive, especially on the proboscis, which can retract quickly upon detecting vibrations or water movements.

Reproduction and Development

Asexual reproduction is unknown in echiurids, and little work has been done on the powers of regeneration. At least a few display remarkable healing capabilities. For example, *Urechis caupo* (Figure 14.28G)—the "fat innkeeper"—is often found in bay muds that are subjected to heavy pressure from clam diggers. In some of these tidal flats, nearly every *Urechis* specimen bears scars, some nearly completely across the body—signs that the animal has survived cuts from the clammers' shovels.

Echiurid sexes are separate. The gametes are produced in special "gonadal" regions of the ventral peritoneum, and released into the coelom to mature. When ripe, the gametes accumulate in the nephridia (or nephridium, Figure 14.30C) until spawning occurs. The nephridia often swell enormously when packed with eggs or sperm. In most cases, epidemic spawning takes place and is followed by external fertilization.

Echiurid development is similar to that of other annelids, with spiral cleavage and lecithotrophic trochophore larvae (Figure 14.31C) that may drift in the plankton for up to three months as they gradually elongate to produce young worms (Figure 14.31D). The prototroch and the episphere regions develop into the proboscis, which is thus equivalent to the prostomium and peristomium in other annelids. The region behind the proboscis, apparently a series of fused segments, forms the trunk. As with other annelids the 4d mesentoblast proliferates the main trunk coelom.

Echiurids are well known for the sexual dimorphism and environmental sex determination found in a clade that includes genera such as *Bonellia*, *Ikeda*, and *Metabonellia*. In such species, the females are quite large, reaching lengths of up to 2 m, including the proboscis. The males, however, which can be from a few millimeters to a centimeter or so long, are simple-bodied, and often retain remnants of larval ciliation. They live on the female's body, or in her nephridia (Figures 14.28D–F and 14. 31A). Experiments have shown that the when exposed to females the larvae usually metamorphose into dwarf males, but normally differentiate into females when developing in the absence of females. This sexual determination is apparently caused by a masculinizing hormone produced by the skin of the female worm.

Siboglinidae: Vent Worms and Their Kin

Siboglinidae is an enigmatic family of tube-dwelling marine annelids containing four groups: Frenulata

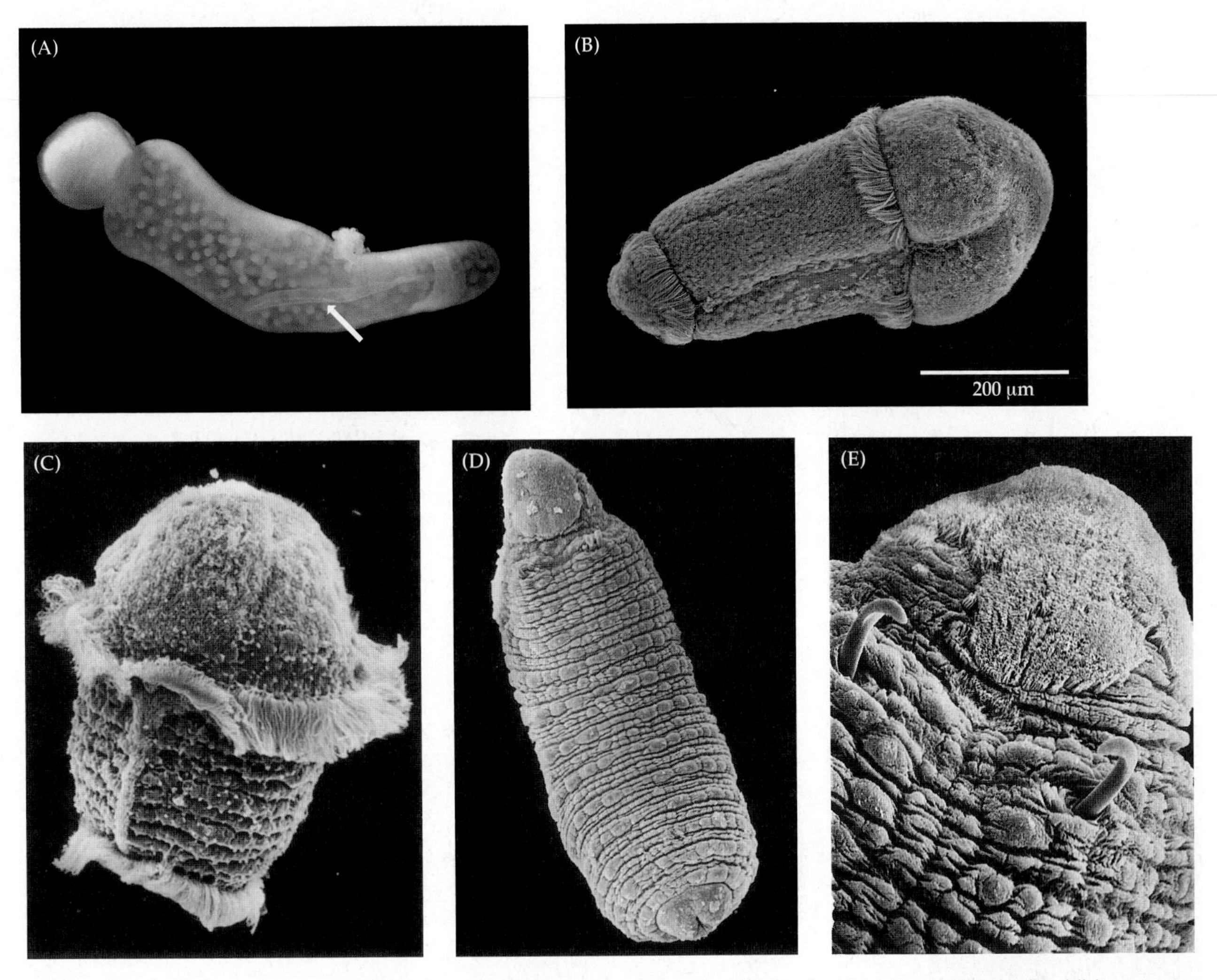

Figure 14.31 Reproduction in Echiurids. (A) A dwarf male (arrow) of *Metabonellia haswelli* inside the nephridium of a female, which is also packed with mature oocytes (yellow spheres). (B) *Metabonellia haswelli* trochophore larva in ventrolateral view, showing prototroch and telotroch and overall ciliation. This larva is sexually undifferentiated and will become a female if it settles on sediment, or a dwarf male if it settles on a female *Metabonellia.* (C) A *Urechis* trochophore larva (15 days) exhibits a prototroch, neurotroch, and telotroch. Note the segmented appearance of the trunk produced by distinct bands of epithelial cells. This has been dismissed previously as superficial but may be real. (D) A juvenile *Urechis*, after settling. (E) The anterior end of a juvenile (ventral view). Note the pair of hooked chaetae.

(Latin *frenulum,* "little bridle"), Vestimentifera (Latin *vestimentum,* "garment"; *ferre,* "to bear"), the genus *Sclerolinum* (Latin, *sclera,* "hard"; *linum,* "thread"), and the genus *Osedax* (Latin *Os,* "bone"; *edax,* "to devour"). Together these four worm groups comprise around 180 species of strange creatures that have been the subject of taxonomic and phylogenetic debate since their discovery some one hundred years ago (Figures 14.1E, 14.32 and 14.33). We recognize the family name Siboglinidae, although for a while these worms were classified in two distinct phyla, Pogonophora and Vestimentifera. Frenulata (e.g., *Lamellisabella, Siboglinum, Polybrachia*), which contains most of the described species, live in thin tubes buried in sediments at ocean depths from 100 to 10,000 m. Most frenulates are less than 1 mm in diameter, but can be 10 to 75 cm in length (Figure 14.32A–F). The tubes may be three or four times as long as the worm's body. The other main group, Vestimentifera (e.g., *Riftia, Sclerolinum, Tevina*) is generally found at methane seeps or hydrothermal vents (Figure 14.32G–I). These can be very large worms, with *Riftia pachyptila* reaching well over 1 meter in length and several cm in diameter. The anterior trunk of the body has a collar and two lateral extensions, the vestimentum, from which the group gets its name. One genus that is close to Vestimentifera, *Sclerolinum,* resides in wood fragments or other plant material or at seafloor mud volcanoes. A recently discovered group of siboglinids belongs to the genus *Osedax* (Figure 14.32J–L), and all of the 20+ known species consume the bones of vertebrates that have come to rest on the sea floor.

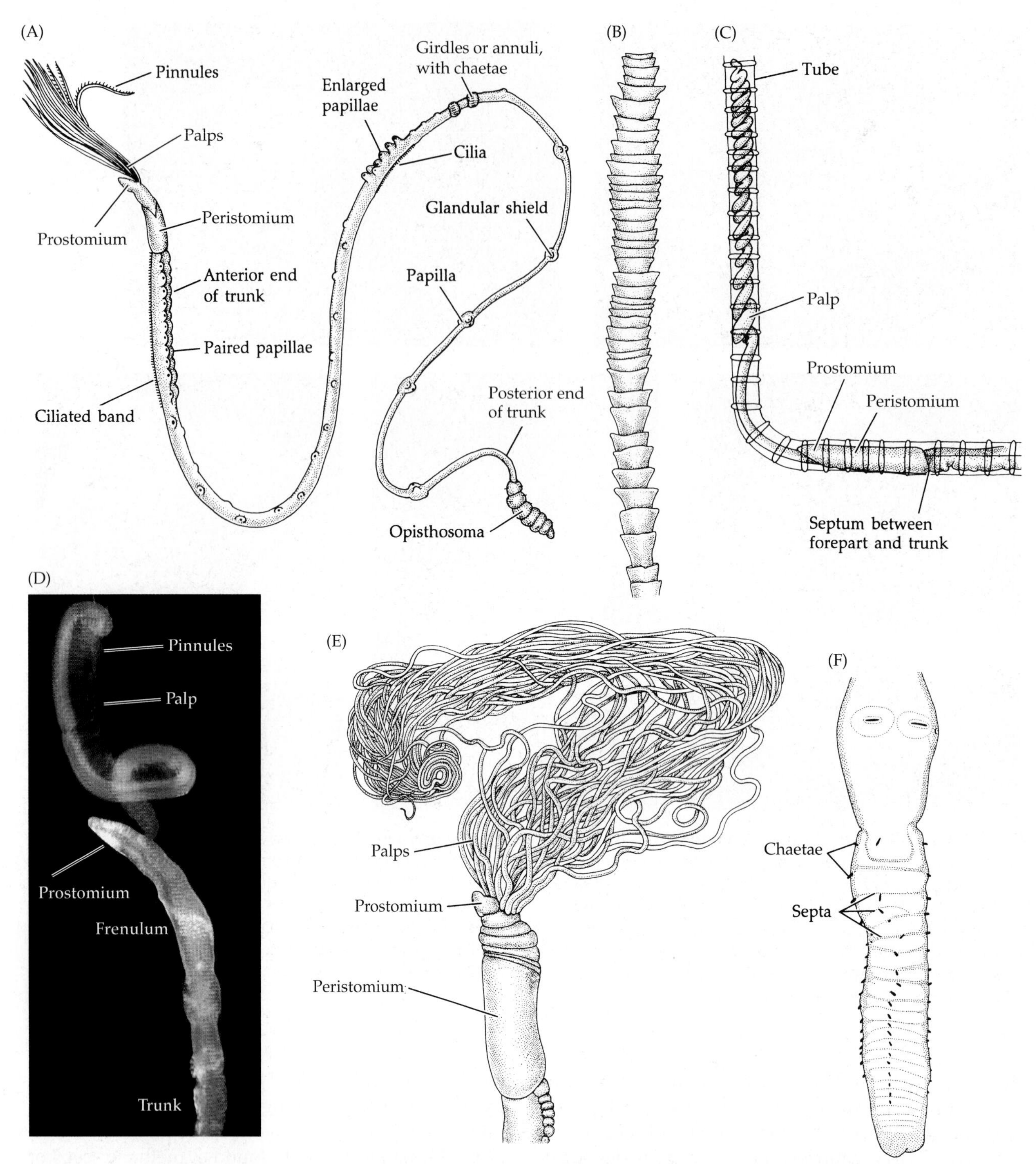

Figure 14.32 Siboglinidae. (A) A generalized frenulate siboglinid. (B) The tube of the frenulate *Lamellisabella*. (C) Anterior end of the monotentaculate frenulate *Siboglinum* in its tube. (D) Anterior end of living specimen of the frenulate *Siboglinum veleronis.* (E,F) Anterior and posterior ends of the frenulate *Polybrachia*. (E) The peristomium gives rise to multiple palps in *Polybrachia*. (F) Enlarged view of the opisthosoma of *Polybrachia*. (G) Living specimens of the deep-sea, hydrothermal vent vestimentiferan *Riftia pachyptila*. (H) *Riftia pachyptila* dissected from tube showing body regions. (I) The vestimentiferan *Tevnia jerichonana.* Anterior end emerging from the white chitinous tube. Whole specimen removed from tube showing vestimentum, long trunk, and segmented opisthosoma. (J) Dozens of Female *Osedax* emerging from a phalange (finger bone) of a gray whale. Each female has four palps. Note the large red blood vessels running along the trunk into the palps. (K) Female *Osedax frankpressi* with bone dissected away to reveal the ovisac and roots. (L) Diagram of *Osedax rubiplumus* with bone cutaway showing the major body regions. Note the placement of the harem of males in the transparent tube, where they lie near the female's oviduct.

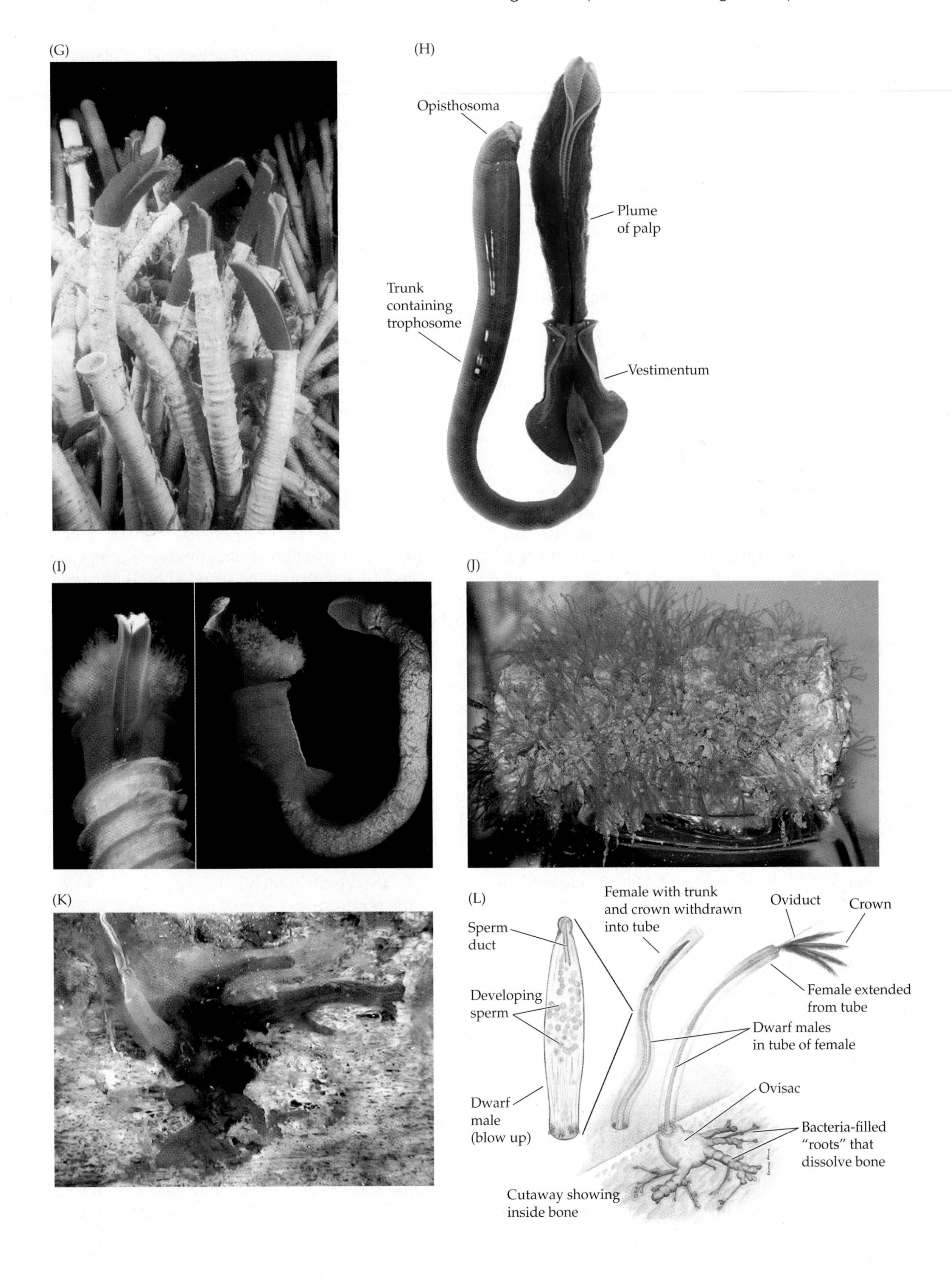
(G)
(H)
Opisthosoma
Plume of palp
Trunk containing trophosome
Vestimentum
(I)
(J)
(K)
(L)
Sperm duct
Developing sperm
Dwarf male (blow up)
Female with trunk and crown withdrawn into tube
Oviduct
Crown
Female extended from tube
Dwarf males in tube of female
Ovisac
Bacteria-filled "roots" that dissolve bone
Cutaway showing inside bone

The body of most Siboglinidae species is divided into four regions (Figure 14.32A). The head (prostomium, as in other annelids) bears from 1 to over 200 separate thin palps, or an elaborate crown of fused palps. These bear tiny side branches called **pinnules**. The **prostomium** (obscured in Vestimentifera) is followed by a short **peristomium** (possibly a segment) and then an elongate **trunk**, which is a single very lengthy segment often bearing various annuli, papillae, and a ring of chaetae. Behind the trunk is a series of segments with more chaetae called the **opisthosoma**.

Internally, siboglinids possess a closed circulatory system and well-developed nervous system. There is an anterior pair of large nephridia, as seen in close relatives such as Cirratuliformia and Sabellida. There is no digestive tract in the adults. There has been some confusion about the overall body orientation of these animals. For many years it was unclear whether the main longitudinal nerve cord was dorsal or ventral. Those who argued it was dorsal used this to justify the status of phylum for Pogonophora. Furthermore, developmental studies demonstrated a transitory gut in at least one species. However, it now seems clear that the nerve cord is ventral, as in other annelids and other protostomes.

Siboglinid Taxonomic History

Siboglinids have a complicated, though interesting, history. They were first studied in the early 1900s, following the expedition of the Dutch research vessel *Siboga* in Indonesia. The dredged, partial specimens (lacking the segmented opisthosoma) were given to the eminent French zoologist Maurice Caullery, who studied them for over 40 years and published several papers describing these strange worms. In 1914, Caullery named the originally collected specimens *Siboglinum weberi* and created the family name Siboglinidae for them, although he was unable to assign them to any known phylum. He suggested they were close to deuterostomes such as hemichordates, as they had what he interpreted to be a dorsal nerve cord. The next species to be described, *Lamellisabella zachsi* some 20 years later by a Russian annelid taxonomist (P. V. Uschakov), was placed as a new subfamily of Sabellidae. However, a Swedish annelid worker (K. E. Johansson) suggested that Uschakov had the worms upside down and said they were not annelids and coined the name Pogonophora for them as a new class of animals (again, without a phylum assignment). The problem in interpreting dorsal and ventral was related to the fact that adult animals lacked a mouth and a gut, features that normally helped with orientation. In 1944, Pogonophora was made a phylum and part of the Deuterostomia. A Russian specialist, A. V. Ivanov, described many new species of frenulate siboglinids in the following decades and developed an elaborate taxonomy of genera and families, strongly emphasizing that Pogonophora were deuterostomes. However, in 1964, the first whole specimens were recovered, with the segmented opisthosoma, and an annelid affinity for the group was again proposed by some workers. The discovery of the very large Vestimentifera in the late 1960s and 1970s, associated with seeps and vents, further complicated the controversies surrounding these animals as they were interpreted by one important worker (M. L. Jones) as being closer to annelids than to Pogonophora, leading to the two separate phyla, Pogonophora and Vestimentifera. By the 1990s, both morphology and DNA data began to make it clear that these worms should all be placed within Annelida. To clear up any lingering confusion the original name for the group, designated by Caullery, Siboglinidae is now used for the group as a whole.

The Siboglinidae Body Plan

The Tube, Body Wall, and Body Cavity

The elongate tubes of most siboglinids are composed of chitin and scleroproteins secreted by the epidermis. The tubes are often fringed, flared, or otherwise distinctively shaped and are frequently banded with yellow or brown pigment rings (Figures 14.1E and 14.32B,G). The upper end of the tube projects above the substratum so the worms' palps can extend into the water. The palps of Siboglinidae can be single as seen in most species (*Siboglinum*, Figure 14.32C,D), or a pair, or four as in *Osedax* (Figure 14.32J–L), and they also range in number from 14–40 (e.g., *Polybrachia*, Figure 14.32E) to hundreds (e.g. vestimentiferans, Figure 14.32G–I). The palps can be free from each other (e.g. *Osedax*, *Polybrachia*), or partially fused together by the cuticle in frenulates such as *Lamellisabella* and in all vestimentiferans. Vestimentiferans have a paired structure called the obturaculum as part of the crown, and its origin has yet to be resolved.

In frenulates the region immediately behind the prostomium is what appears to be a segment that bears a cuticular structure called the **frenulum**, while in vestimentiferans the equivalent region is called the **vestimentum**. These regions are likely involved in tube secretion. The bulk of the body, the trunk, is a single very elongate segment. In frenulates there is a girdle of chaetae halfway along the trunk, which lends support to it being a single segment. In Vestimentifera there are no chaetae on the trunk. The chaetae in front of the first opisthosomal septum in these groups are at the end of the trunk, and so correspond to the girdle of uncini-type chaete on the trunk of other siboglinids (Figure 14.5Q). In most siboglinids (except for *Osedax*) the remainder of the body is a short multi-segmented region called the opisthosoma that has peglike or hooked

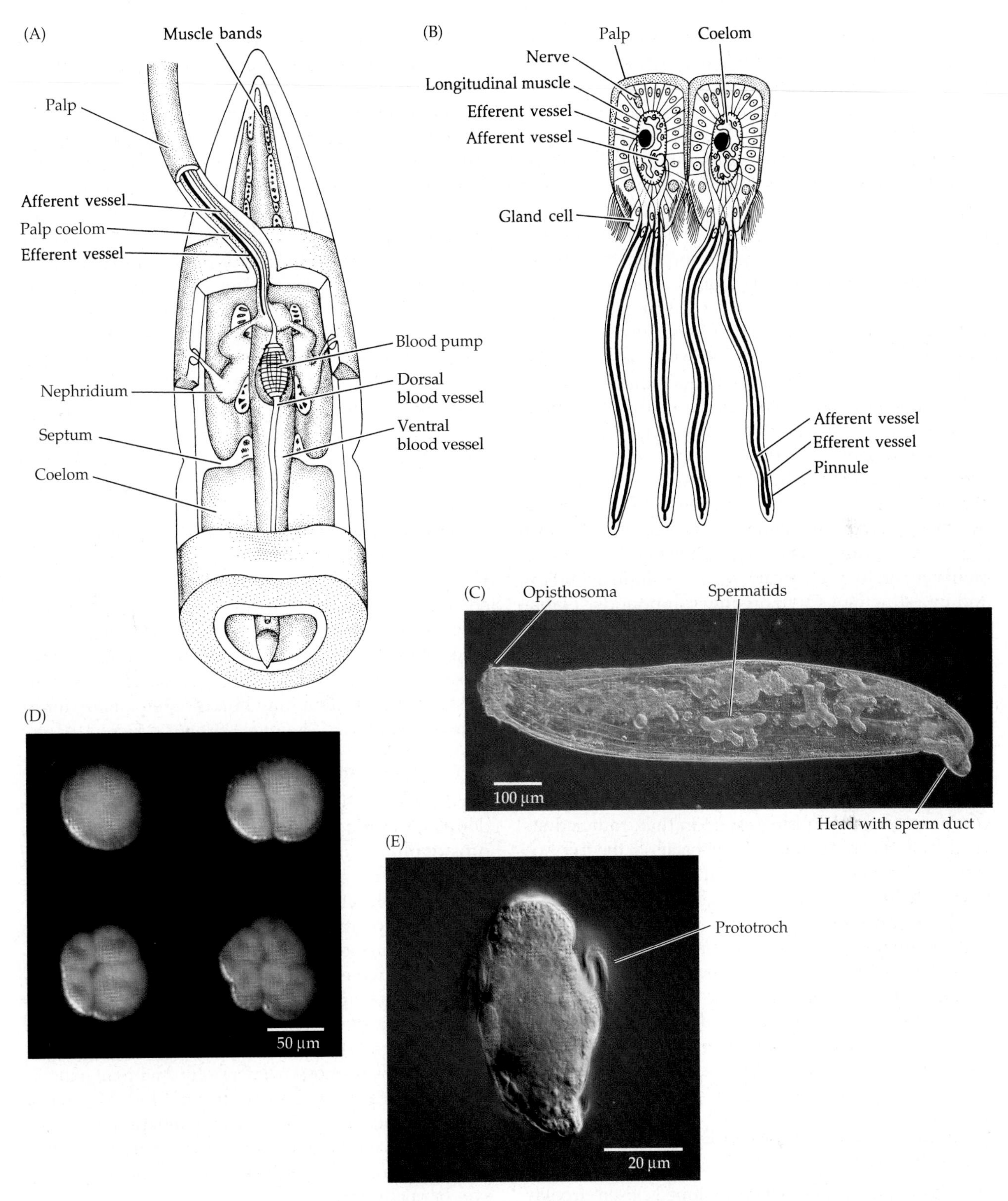

Figure 14.33 (A) The anterior end of the frenulate *Siboglinum* (cutaway dorsal view). (B) Two palps of the frenulate *Lamellisabella* (cross sections), showing elements of the circulatory system. (C) Dwarf male of *Osedax rubiplumus*. The male is basically an arrested larva; stage with a two-segmented opisthosoma. The body cavity contains developing sperm. (D) Fertilized egg and first three cleavage stages of *Osedax*. The 8-cell stage clearly shows normal spiral cleavage occurs in Siboglinidae. (E) Trochophore larva of *Osedax* (prior to formation of the opisthosoma).

chaetae. The body surface is covered by a typical annelid cuticle. There can be small papillae in various parts of the anterior and trunk-bearing thickenings of the cuticle referred to as **cuticular plaques**. The epidermis is mostly a cuboidal to columnar epithelium and includes various gland cells, papillae, and ciliary tracts. Beneath the epidermis is a thin layer of circular muscle and a thick layer of longitudinal muscle, the latter developed as bands or bundles in some parts of the body. There is a spacious coelom in the trunk and extensions extend into the head region and palps.

Nutrition

Siboglinidae lack a functional digestive tract, and for many years the method of nutrition of these relatively large worms was puzzling. Early suggestions were that the worms subsist on dissolved organic matter from the seawater flowing across the palps and from the muddy sediments in which the animals are buried. Other suggestions were that suspended particulate organic matter and/or bacteria were being filtered from the water column. These ideas have now been supplanted by the realization that all siboglinids have symbiotic bacteria in their bodies from which they derive their nutrition. This was first discovered in *Riftia pachyptila* by then graduate student Colleen Cavanaugh and colleagues. *Riftia* and its relatives inhabit the sulfide-rich waters near hydrothermal vents or methane seeps, and symbiotic chemoautotrophic gamma-Proteobacteria inhabit a region in the trunk called the **trophosome**. These symbionts generate ATP by carrying out sulfide oxidation and by reducing CO_2 to organic compounds. Transfer of organic matter from symbionts to host occurs through translocation of amino acids and other compounds released from the bacteria and direct digestion of symbiont cells. It appears that free-living stages of the bacteria are taken in by a transitory digestive tract or through the skin by early juvenile stages of the worms. A trophosome with symbiotic bacteria is now known for all siboglinids that have been examined. *Osedax* is unusual in this regard in that, rather than chemoautotrophic gamma-Proteobacteria, the trophosome houses heterotrophic gamma-Proteobacteria (belonging to the Oceanospirillales clade). This is because *Osedax* is relying on the organic material in the bone such as collagen, and possibly fats. The Oceanospiralleles bacteria can exploit these and then are in turn digested by the worms. In *Osedax* there are "roots" of tissue (Figure 14.32J–L) that ramify through the bone, and these roots contain the bacteria. The epidermis of the roots secretes acid to dissolve hard bone allowing access to the proteins and fats.

Circulation, Gas Exchange, Excretion, and Osmoregulation

Siboglinids all possess a well-developed closed circulatory system (Figure 14.33A). The major blood vessels are dorsal and ventral longitudinal vessels that extend nearly the entire length of the body. The dorsal vessel is swollen as a muscular pump in the anterior of the body. As in other annelids, blood flows anteriorly in the dorsal vessel and posteriorly in the ventral vessel. Anteriorly, the vessels branch extensively. Some of these vessels supply parts of the head region and lead to afferent and efferent vessels in the palps (Figure 14.33B). Exchange also occurs in the posterior part of the body through a series of connecting blood rings. In *Osedax* the blood vessels branch into the roots.

The blood contains hemoglobin in solution and a variety of cells and cell-like inclusions. Gas exchange probably takes place across the thin walls of the palps. In the species that live near hydrothermal vents, problems of oxygen supply and sulfur toxicity are especially critical. These worms, such as *Riftia pachyptila*, live where hot, anoxic, sulfurous vent water mixes with the surrounding cold, oxygenated water. The animals are thus exposed to potentially dramatic fluctuations in ambient temperatures and oxygen and sulfide availability. These worms possess very high concentrations of hemoglobin in the fluid of their body cavities as well as in their blood. This hemoglobin appears to retain its high affinity for oxygen across a wide temperature range. Furthermore, a unique sulfide-binding protein occurs in their blood, which serves to concentrate the sulfur and avoid sulfide toxicity, and to transport the sulfur to the chemoautotrophic bacteria. Interestingly, similar adaptations are known in the burrowing annelid *Arenicola*, which typically lives in anoxic muds in shallow marine environments.

A pair of large anterior metanephridia lies in close association with the circulatory system, as seen in the closely related Cirratuliformia and Sabellida. In some Vestimentifera there may be a single fused exterior pore or two pores.

Nervous System and Sense Organs

The nervous system of siboglinids is largely intraepidermal. A well-developed nerve ring just behind in the prostomium bears a large ventral ganglion. A single ventral nerve cord arises from this ganglion and extends through all body regions. In many cases there are ganglia (or at least nerve "bulges") at the junctions of the various body regions, and there is an enlargement of the nerve cord in the anterior part of the trunk. In frenulates that have been studied, the opisthosoma contains three distinct nerve cords that apparently bear segmental ganglia. The vestimentiferans, however, have a single opisthosomal nerve cord without ganglia.

Sense organs are poorly understood. The palps probably contain tactile receptors as the worms are sensitive to vibrations and can rapidly retract into their tubes if disturbed. No siboglinids are known to have eyes or nuchal organs.

Reproduction and Development

Nothing is known about asexual reproduction or regeneration in these animals. While there is simultaneous hermaphroditism known from one species (*Siboglinum fiordicum*), the sexes in Siboglinidae are otherwise separate and worms possess a pair of gonads in the trunk. In males of frenulates and vestimentiferans, paired sperm ducts extend from the testes to gonopores located near the anterior end of the trunk. As sperm move along these ducts, they are packaged. In frenulates these are as spermatophores, with an envelope around the sperm, while in vestimentiferans the sperm are in clusters called spermatozeugmata. *Osedax* is unusual in that the males are generally dwarfs (Figures 14.32L and 14.33C), which can be more than 100,000 times smaller than the female—developmentally arrested larvae that develop sperm. These live in the tubes of the females and use their limited yolk supply to make sperm before they die. Females can have up to 600 males in a "harem."

The female system of frenulates and vestimentiferans includes a pair of ovaries from which arise oviducts leading to gonopores on the sides of the trunk. In *Osedax*, the females have a large ovisac that lies in the "host" bone and can be up to one-third of the body mass. This leads to a single oviduct that runs dorsally along the trunk and ends in the plume of palps. The eggs of all siboglinids are spheroidal to elongate and development is lecithotrophic.

Fertilization is internal in Vestimentifera and *Osedax*, and this is likely also the case for Frenulata. In frenulates and vestimentiferans, males release spermatozeugmata or spermatophores, which drift to the plumes and tubes of nearby females. The sperm end up in the oviduct, which can have sperm storage areas. In *Osedax* the sperm somehow end up in the ovary itself. Brooding of larvae in the tube occurs in some frenulates, while in all vestimentiferans and *Osedax* the fertilized eggs are apparently released into the sea. Individual *Osedax* females can release 400–800 eggs daily and the larvae can live for several weeks in the plankton. Vestimentifera also produce numerous eggs and these also live for a month or more as larvae. Such high fecundity is necessary as the habitats for these worms is very restricted and most larvae are likely unsuccessful in finding a suitable place to settle.

Cleavage is spiral in all siboglinids and early reports of radial cleavage now appear to be erroneous (Figure 14.33D). Trochophore larvae develop (Figure 14.33E) and add some segments that will later become the opisthosoma. In the dwarf male of *Osedax*, two segments are formed than have hooked chaetae. A transitory digestive tract occurs during development and in juveniles of at least some vestimentiferans and frenulates. Inner cells near the presumptive posterior end of the embryo proliferate spaces within the opisthosomal segments as they develop, a process strikingly similar to the teloblastic growth and schizocoely seen in other annelids.

Hirudinoidea: Leeches and Their Relatives

Hirudinoids are clitellate annelids whose closest relatives are Lumbriculidae (Figure 14.34–14.41). They have a fixed number of segments, which are traditionally numbered with Roman numerals (Figure 14.34C). In an unusual tradition with respect to other annelids, the prostomium and peristomium (when present) are counted as segments, and so the bodies of branchiobdellids are counted as having 15 segments (though they lack a prostomium), acanthobdellids with 29 or 30 segments (lacking both a prostomium and peristomium), and hirudinids (leeches) have 34 segments. These segments are generally obscured by superficial annulations, giving the impression of many more segments (Figure 14.34). Externally, hirudinoids generally feature anterior and posterior suckers and clitellum, while lacking parapodia and chaetae (though Acanthobdellida have the latter in a few anterior segments). Internally, acanthobdellids have a coelom and segmental septa typical of annelids, while in branchiobdellids there are at least a few segments with coelomic spaces. However, the coelom of leeches is reduced to a series of interconnected channels and spaces, without serially arranged septa.

We usually think of leeches as the bloodsuckers popularized in adventure stories and films and this is arguably the ancestral state for Hirudinoidea. However, many hirudinids are free-living predators, and some are scavengers. Most range from less than 0.5 to about 2 cm in length, although one species from the Amazon basin, *Haementeria ghilianii*, may reach 45 cm in length. Acanthobdellids and branchiobdellids are exclusively freshwater groups, while leeches occur in both fresh and salt water, and many live in moist terrestrial environments as well. Those that are full- or part-time parasites feed on the body fluids of a variety of vertebrate and invertebrate hosts. Some of these leeches serve as the intermediate hosts and vectors for certain protist, nematode, and cestode parasites.

Many hirudinoids are cylindrical while others are flattened dorsoventrally. The body is usually divisible into five regions, although the points of division are somewhat arbitrary. The anteriormost head region is composed of the much-reduced (or absent) prostomium and peristomium and the anterior body segments. The anterior region usually bears a number of eyes and a ventral mouth surrounded by the oral or anterior sucker (note the sucker is lacking in *Acanthobdella peledina*). In leeches segments V–VIII form the preclitellar region, followed by the clitellar region (segments IX–XI). The clitellum is only apparent during periods of reproductive activity. The postclitellar or midbody region comprises segments XII–XXVII. The posterior region of the body includes the ventrally directed posterior sucker formed from seven fused segments (XXVIII–XXXIV). In branchiobdellids and acanthobdellids the counts are different owing to their smaller number of segments.

Other than the suckers, gonopores, and nephridiopores, hirudinoids have few distinctive external

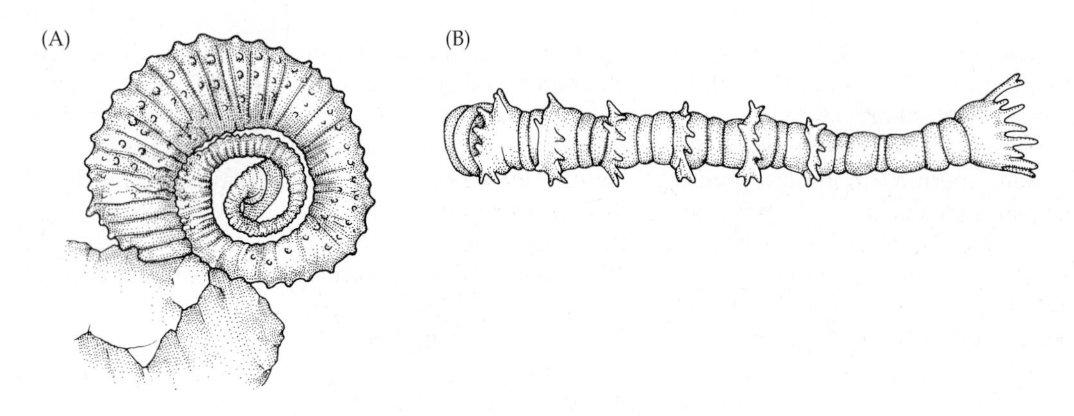

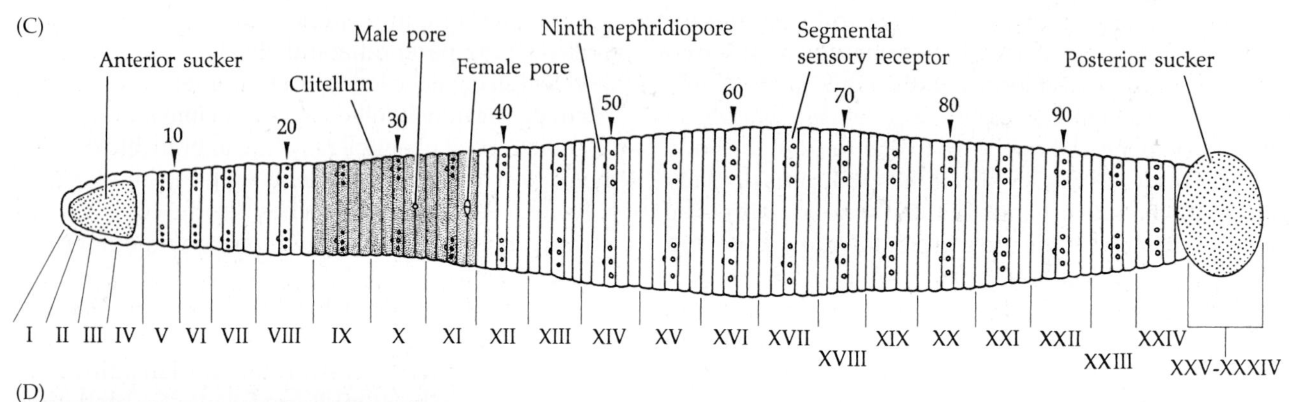

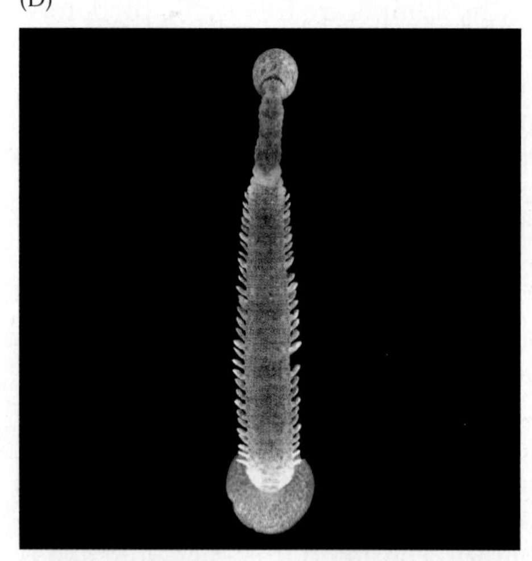

Figure 14.34 Representative hirudinoids. (A) *Acanthobdella peledina* (order Acanthobdellida), a part-time parasite of freshwater fishes. (B) *Stephanodrilus* (order Branchiobdellida). These small (less than 1 cm) worms live on freshwater crustaceans, especially crayfish. (C) Ventral view of *Hirudo medicinalis* (Arhynchobdellida). Roman numerals indicate true segments; Arabic numbers indicate superficial annuli. (D) *Branchellion parkeri*, Piscicolidae (Rhyncobdellida).

features. In a few forms the body surface bears tubercles, and some such as the piscicolid *Branchellion* have lateral gills (Figure 14.34D). Chaetae are present on the first few segments of acanthobdellids, but they do not occur in any other hirudinoids. Most is known about the anatomy and physiology of leeches (Hirudinida) and this is reflected in the summary below.

The Hirudinoidean Body Plan

Body Wall and Coelom

A cross-sectional view of a leech is dramatically different from that of other annelids, in large part because of a thick dermal connective tissue layer beneath the epidermis and the reduction of the coelom (Figure14.35). The typical circular and longitudinal muscles are present, as well as bands of dorsoventral and diagonal (oblique) muscles between the circular and longitudinal layers. The dense dermis fills the areas between the muscle bands. Septate coelomic compartments are present only in acanthobdellids (in the first five segments) and in the midbody region of branchiobdellids. In all Hirudinea, the coelom is limited to various channels and small spaces. These augment the circulatory system in rhynchobdellids (proboscis leeches) and completely replace it in arhynchobdellids (jawed leeches).

Support and Locomotion

With the exception of acanthobdellids, hirudinoids show a more or less solid body construction and the absence of a spacious coelom precludes kinds of locomotion seen in other annelids. They do not burrow, but are mostly surface dwellers, and the suckers in branchiobdellids and hirudinids serve as the points of contact with the substratum against which the muscle action can operate (Figure 14.36). Beginning with the posterior sucker attached, the circular muscles are contracted, making the entire body elongate forward, and then the anterior sucker is attached. Then the longitu-

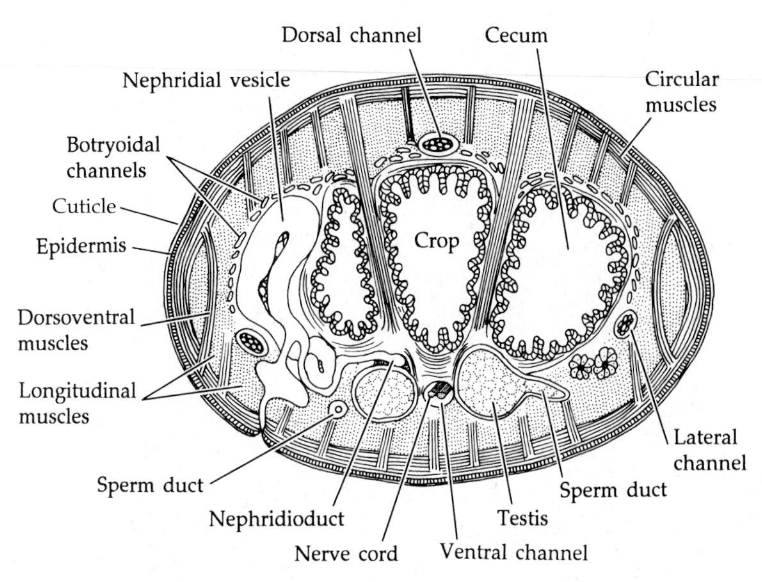

Figure 14.35 Body wall and general internal organization of *Hirudo* (cross section), in which the original circulatory system has been lost and replaced by coelomic channels.

dinal muscles contract, shortening the body and draw the posterior end forward. Some leeches are also capable of swimming by dorsoventral body undulations; this behavior is an important mechanism for locating and contacting nonbenthic hosts.

Feeding and Digestion

Like that of other annelids, the leech digestive tract includes a foregut, midgut, and hindgut (Figure 14.37). The foregut, as mentioned earlier, includes a mouth, buccal cavity, jaws or proboscis, pharynx, and esophagus. There are also masses of unicellular salivary glands that secrete hirudin in the jawed bloodsuckers and may produce enzymes to aid penetration of the proboscis in those parasitic forms that lack jaws. Posterior to the esophagus is the enlarged midgut, usually called the stomach or crop. This region bears large ceca in most leeches, providing a large storage capacity as well as a high surface area. In some kinds of leeches, the posterior midgut is structurally differentiated from the anterior portion. A short rectum connects the midgut to the anus, located dorsally near the junction of the body and the posterior sucker. Little is known about digestion in hirudinoids, except for some fragmentary information on blood-sucking leeches. Midgut enzymes apparently are limited to exopeptidases, which probably accounts for the extremely slow rate of digestion in these animals (a medicinal leech may take several months to digest the contents of a full blood meal).

Assessment of the distribution of blood-feeding and other feeding modes across the phylogeny of Hirudinoidea has led to the conclusion that blood-feeding is the ancestral condition for the group, with several losses of this feeding mode across the group. The acanthobdellids *Acanthobdella peledina* and *Paracanthobdella livanowi* live on the skin, usually fin rays, of freshwater fishes and suck the blood of their hosts, though they have been shown to also eat insect larvae. *Acanthobdella peledina* is found across northern Europe and Russia with *P. livanowi* limited to the Kamchatka Peninsula. There is record also from Alaska though the species identification needs further assessment. Branchiobdellids are tiny worms that live on freshwater crayfish. The anterior end of the pharynx bears a pair of toothed jaws (Figure 14.37). These animals eat other epizoites living on the host, but they also feed on the host's eggs and body fluids.

The majority of Hirudinida lineages ranked as families are ectoparasites that feed by sucking the blood or other body fluids from their hosts. The rest are predators on small invertebrates or scavenge on dead animal matter. Feeding involves either a protrusible pharyngeal proboscis or cutting structures in the form of slicing jaws or stylets. Members of the paraphyletic Rhynchobdellida possess a pharyngeal proboscis but lack jaws, whereas members of the Arhynchobdellida lack a proboscis and all but a few possess jaws (Figure 14.34). Because of their medical importance, much work has been done on the parasitic leeches, especially those that affect livestock, game animals, or humans. Blood-sucking leeches are not especially host-specific, and most do not remain attached to a single host for

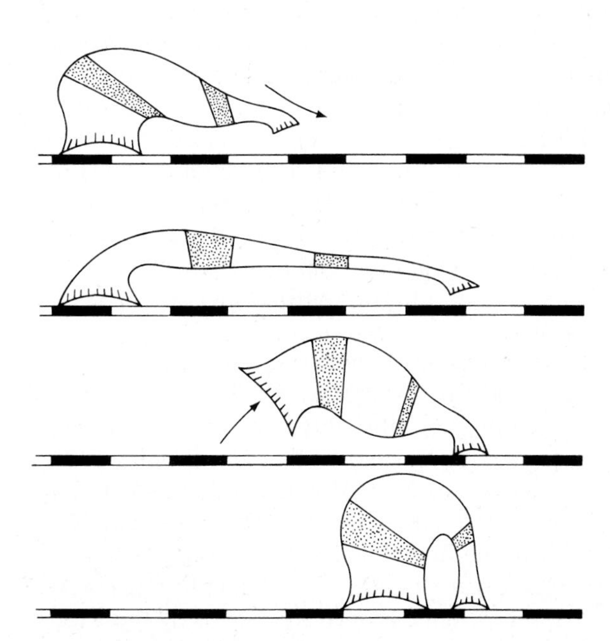

Figure 14.36 Locomotion in a leech, moving left to right, using the anterior and posterior suckers to progress in "inch worm" fashion.

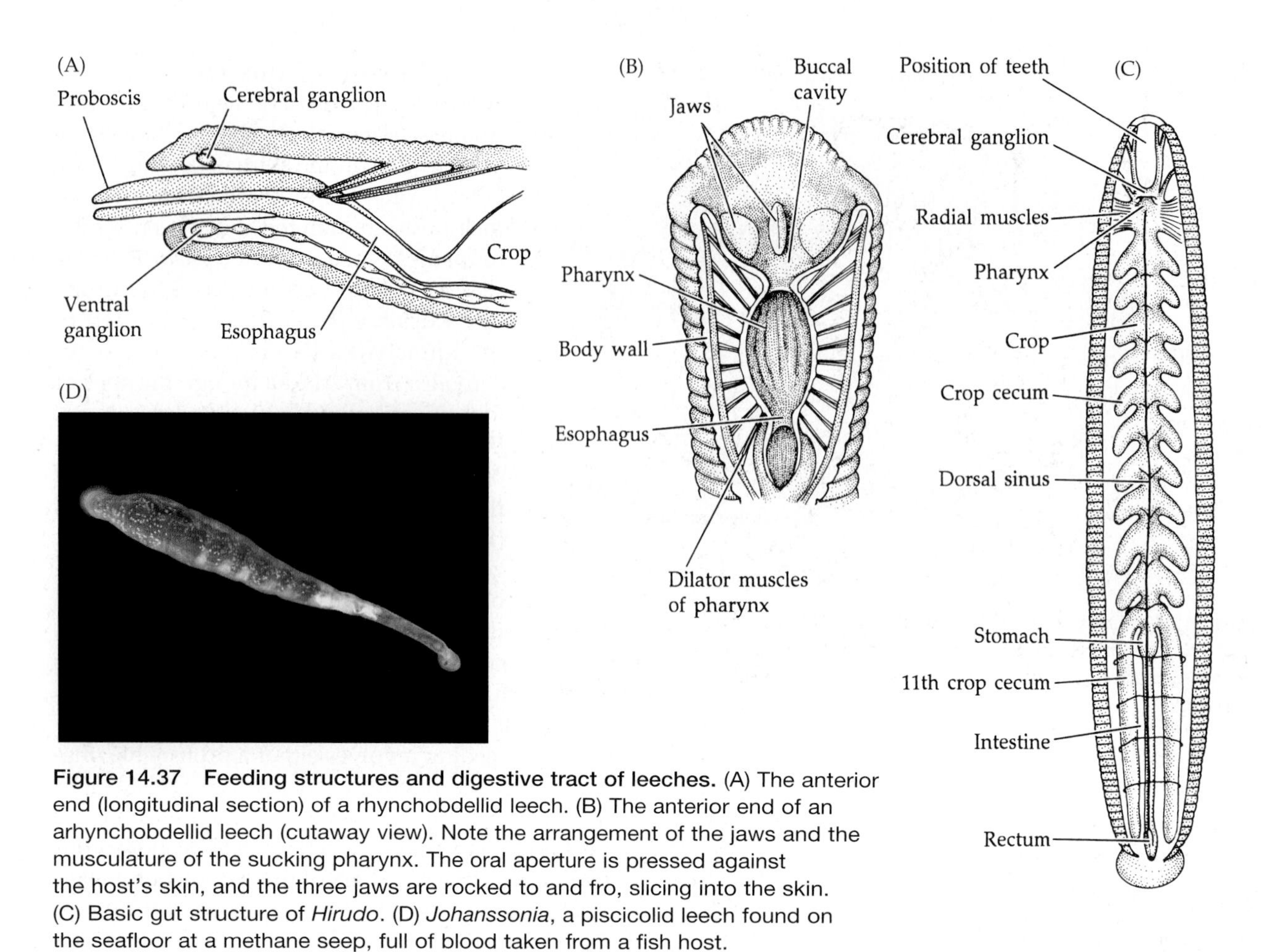

Figure 14.37 Feeding structures and digestive tract of leeches. (A) The anterior end (longitudinal section) of a rhynchobdellid leech. (B) The anterior end of an arhynchobdellid leech (cutaway view). Note the arrangement of the jaws and the musculature of the sucking pharynx. The oral aperture is pressed against the host's skin, and the three jaws are rocked to and fro, slicing into the skin. (C) Basic gut structure of *Hirudo*. (D) *Johanssonia*, a piscicolid leech found on the seafloor at a methane seep, full of blood taken from a fish host.

long periods of time; many of these leeches may feed by other means when not attached to a suitable host.

A few species of leeches feed exclusively on invertebrate hosts, including annelids (even other leeches), gastropods, and crustaceans, but the majority of them parasitize vertebrates. Some leeches are parasitic on members of particular groups of vertebrates. For example, most Piscicolidae (Rhynchobdellida) feed on the blood of fishes (including some deep-sea and hydrothermal vent fishes (Figure 14.37D), whereas Ozobranchidae (another family of Rhynchobdellida) seem to prefer aquatic reptiles such as turtles and crocodilians.

The European medicinal leeches such as *Hirudo medicinalis* and *Hirudo verbana,* which normally feed on amphibians, will obligingly consume human blood. Over-collecting has meant that *Hirudo medicinalis* is now classified as Near Threatened by the IUCN and it is protected across much of its range. The North American medicinal leech is *Macrobdella decora* and other species are used for medicinal purposes in Asia. The common name is derived from using these leeches to draw blood from people afflicted with a wide range of maladies for which such "bleeding" was once thought to be curative or restorative. Today, medicinal leeches are used to reduce hematomas in areas of the body that are difficult to treat surgically, and to avoid leaving scars. The anticoagulant pumped into the wound by the leech (hirudin) allows blood to continue to drain, even after the leech as been removed and so promotes healing. In addition to the anticoagulant, some leeches produce a number of other chemicals, including anesthetics and vasodilators. When feeding, the leech anchors to the host by the suckers and presses the mouth against the surface of the host's body. Jawed leeches have three bladelike jaws, each shaped like a half circle. The jaws are set at roughly 120-degree angles to one another so that the cutting edges form a Y-shaped incision. Muscles rock the jaws to and fro, making slices in the host's skin. The leech releases an anesthetic as it makes its incisions, then secretes hirudin into the wound, and blood is sucked from the host by the muscular pharynx. The anesthetic desensitizes the victim's skin, so that the worms can go unnoticed while they take their blood meal. While the predatory leeches eat frequently, the bloodsuckers feed at widely spaced, very irregular intervals, depending on the availability of hosts. When they do feed, they gorge themselves with several times their own weight in blood (Figure 14.37D).

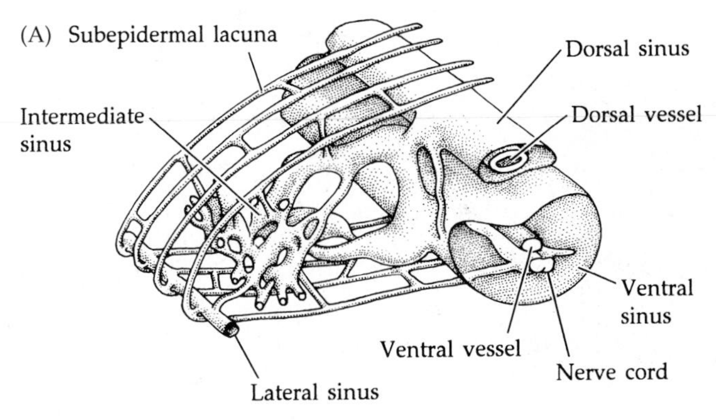

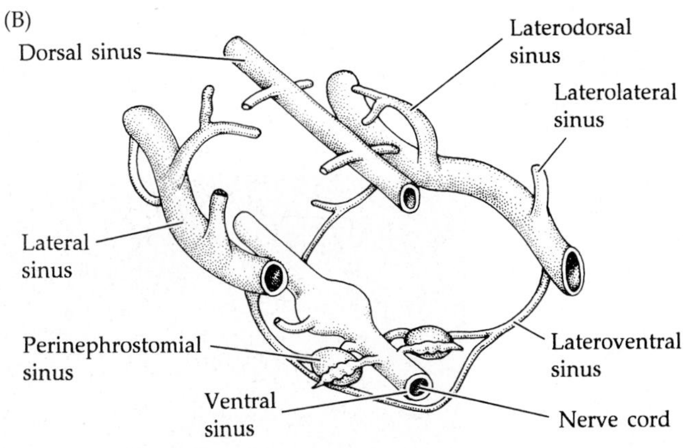

Figure 14.38 Coelomic and circulatory systems of two leeches. (A) A portion of the circulatory and coelomic systems of *Placobdella*, a rhynchobdellan leech in which the circulatory system persists and is associated with the coelomic channels. (B) A portion of the coelomic system in *Hirudo*, an arhynchobdellid leech. Here the circulatory system has been completely replaced by coelomic channels.

Circulation and Gas Exchange

Many leeches are relatively large and quite active. The drastic reduction of the coelom has led to a flattening of the body and the formation of extensive gut ceca or diverticula in some lineages. Both of these reduce internal diffusion distances. In Rhynchobdellida this system is a combination of a true closed circulatory system and the reduced coelomic spaces; in the Arhynchobdellida, the original circulatory system is completely replaced by one derived entirely from the reduced coelom. In both of these arrangements the circulatory fluid is moved through the system by the action of contractile vessels and by general body movements (Figure 14.38). Gas exchange is accomplished by diffusion across the body wall; gills are present only in some ozobranchids and piscocolids. Some leeches possess hemoglobin in solution in the circulatory fluid, thought to account for about 50 percent of their oxygen-carrying capacity.

Excretion and Osmoregulation

The excretory structures of hirudinoids are paired and segmentally arranged metanephridia that are usually absent from several anterior and posterior segments. The nephrostomes are ciliated funnels associated with coelomic circulatory vessels (where present). Ammonia is the main nitrogenous waste product eliminated via the nephridia. Apparently, particulate waste materials are engulfed by phagocytes, both in the coelomic fluid and in the "mesenchyme," but the eventual disposition of this material is not known.

Nervous System and Sense Organs

The nervous system of leeches has, and continues, to receive a great deal of attention from biologists. The leech nervous system—even that of large leeches—is composed of very few neurons, and these individual nerve cells are sufficiently large that their circuitry has been traced in great detail.

The hirudinoid nervous system includes a cerebral ganglion that is usually set back from the anterior end of the body at about the level of the pharynx (Figure 14.39). The cerebral ganglion, circumenteric connectives, and subenteric ganglion together form a nerve ring around the foregut. Two longitudinal nerve cords arise from the ventral portion of this ring

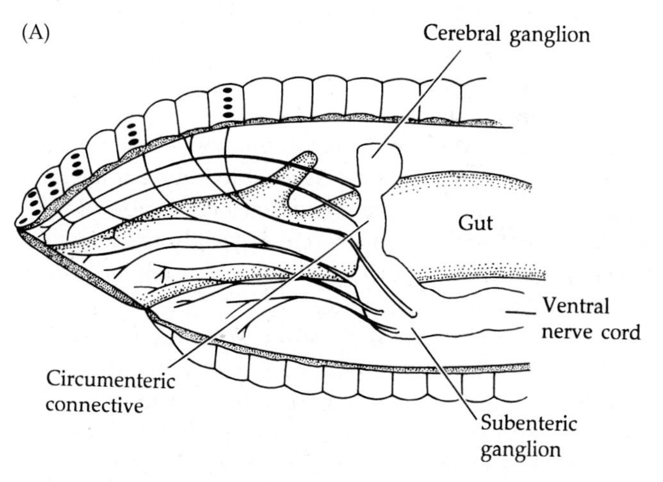

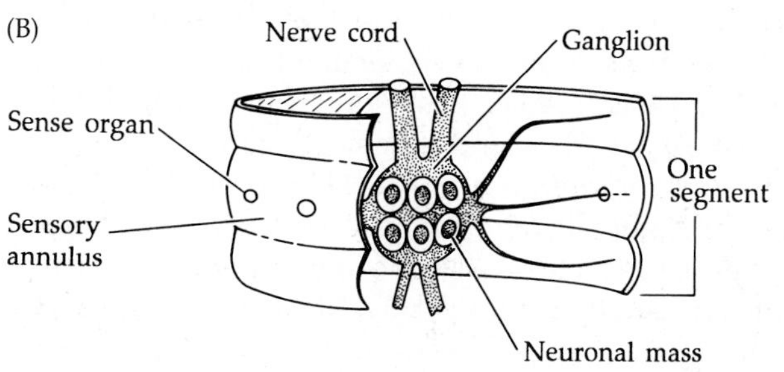

Figure 14.39 Leech nervous system. (A) The anterior nervous system (lateral view). (B) A generalized leech segment comprising three annuli, cut away to show segmental ganglion and innervation of epithelial sense organs.

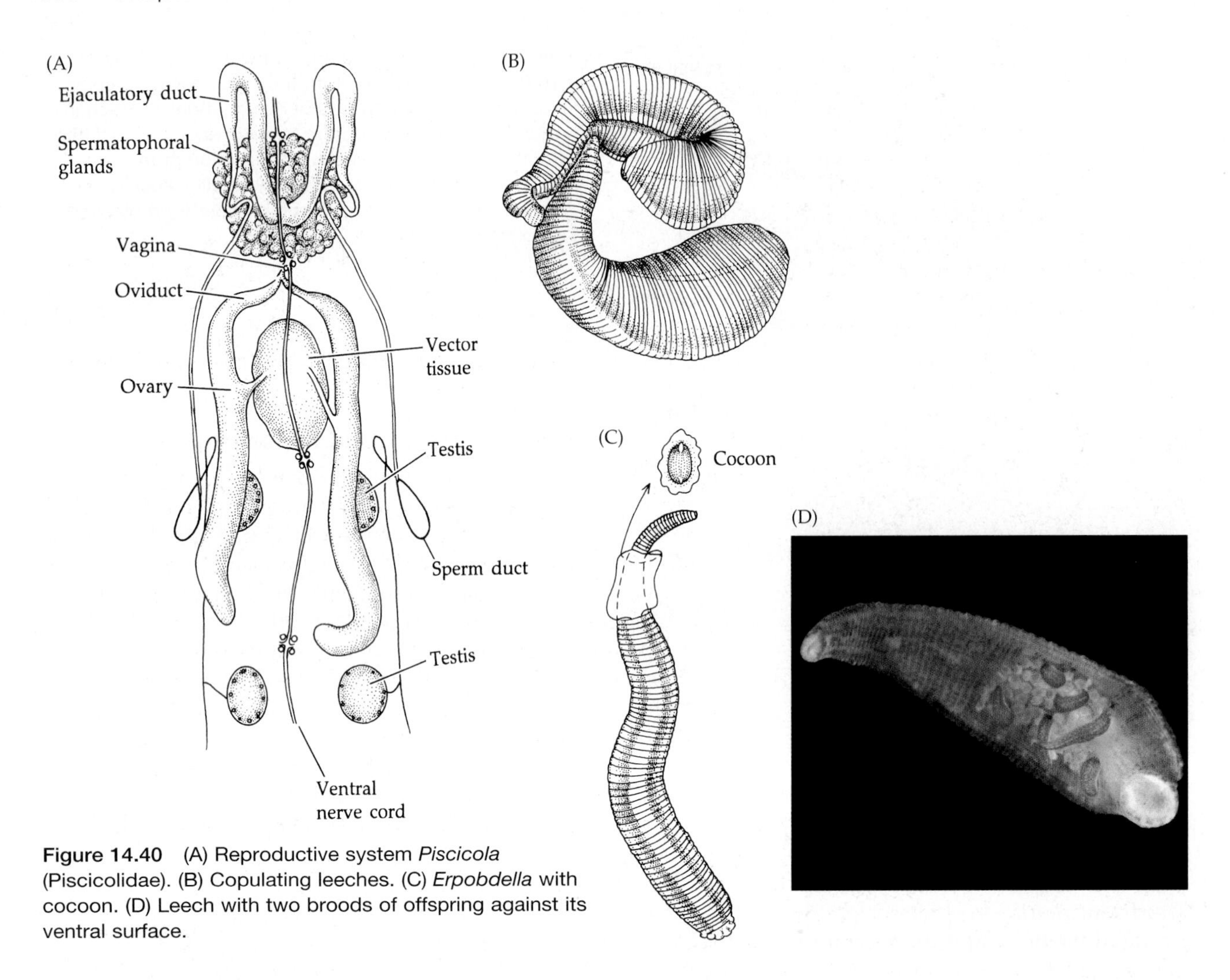

Figure 14.40 (A) Reproductive system *Piscicola* (Piscicolidae). (B) Copulating leeches. (C) *Erpobdella* with cocoon. (D) Leech with two broods of offspring against its ventral surface.

and extend posteriorly through the body. The nerve cords are separate in some areas, but the segmental ganglia are fused. Peripheral nerves include abundant sensory neurons from the cerebral ganglion and segmentally arranged motor and sensory neurons from the ventral nerve cord ganglia.

Leeches are extremely sensitive to certain environmental stimuli, although their sensory receptors are relatively simple. They have from two to ten dorsal eyes of varying complexity, and special sensory papillae that bear bristles extending from the body surface. Except for the eyes, the functions of various leech sense organs are not well understood, and most of the information is based on behavioral responses to different stimuli. Leeches tend to be negatively phototactic. However, some of the blood-sucking species react positively to light when preparing to feed. This behavioral change likely causes the leech to move into areas where a host encounter is more likely. Most leeches can also detect movement in their surroundings, as evidenced by their responses to shadows passing over them. This reaction has been noted particularly in leeches that attack fishes. Leeches also respond to mechanical stimulation in the forms of direct touch and vibrations in their environments. They are also chemosensitive and attracted to the secretions of potential hosts. Some aquatic and even terrestrial leeches that prefer warm-blooded mammalian hosts are apparently attracted to points of relatively high temperatures in their surroundings, thus aiding in food location. Standing in a leech-inhabited pond is a great way to observe first hand this rapid response, as these animals will detect your presence and begin swimming or crawling toward you within seconds.

Reproduction and Development

Hirudinoids, like all Clitellata, are hermaphroditic annelids and have complex reproductive organs (Figure

14.40) and direct development The male reproductive system includes a variable number of paired testes, usually from 5 to 10 pairs in leeches, arranged serially beginning in segment XI or XII (Figure 14.34C). The testes are drained by a pair of sperm ducts that lead to a copulatory apparatus and a single gonopore located midventrally on segment X. There is a single pair of ovaries in leeches, which may extend through several segments. Oviducts extend anteriorly from the ovaries and unite as a common vagina, which leads to the female gonopore on the midventral surface of segment XI, just behind the male pore.

The copulatory apparatus of leeches is often complex and varies in structure among species. Each sperm duct is coiled distally and enlarges as an ejaculatory duct. The two ducts join at a common glandular, muscular atrium. In arhynchobdellids, the atrium is modified as an eversible penis. The rhychobdellids lack a penis and the atrium functions as a chamber in which spermatophores are produced. In arhynchobdellids, mating worms align themselves so that the male pore of each rests over the female pore of the other. The penis is everted and inserted into the mate's vagina, where sperm are deposited. Fertilization takes place within the female reproductive system. Rhynchobdellids leeches use hypodermic impregnation. These animals grasp one another by their anterior suckers and bring their male pores in alignment with a particular region of the mate's body. The spermatophores are released onto the clitellar region of the mate, which then penetrate the body surface of the recipient and the sperm emerge beneath the epidermis. The sperm migrate to the ovaries by way of the coelomic channels and sinuses. In some of the piscicolid leeches there is a special "target" area, beneath which is a mass of tissue called vector tissue that connects with the ovaries via short ducts.

Cocoon formation in leeches is similar to that in other clitellates, with the clitellum producing a cocoon wall and albumin (Figure 14.40C). However, as the cocoon slides anteriorly past the female gonopore, it receives the zygotes or young embryos rather than separate eggs and sperm. The cocoons are deposited in damp soil by terrestrial species and even by a few aquatic forms that migrate to land for this process (e.g., *Hirudo*). Most aquatic forms deposit their cocoons by attaching them to the bottom or to algae; a few attach them to their hosts (e.g., some piscicolids). A few freshwater leeches display some degree of parental care for their cocoons. Some of these bury their cocoons and remain over them, generating ventilatory currents. Others attach the cocoons to their own bodies and brood them externally (Fig 14.40D). The development of leeches is similar to that described for other clitellates. Except for a few species, the amount of yolk is relatively small and development time quite short.

Annelid Phylogeny

There are undoubted annelid fossils that date back to the Cambrian period, with well-known examples being *Canadia* and *Burgessochaeta*. These show clear evidence of segmentation and chaetae, though it is a subject of debate as to whether they could be classified very closely with any extant lineage of annelids. But these were quite complex annelids and the early origins of the group are obscure. As outlined in the Introduction, the traditionally recognized classes were Polychaeta, Oligochaeta, and Hirudinea. The latter two are now viewed as the taxon Clitellata, since recognizing Hirudinea with class rank renders Oligochaeta paraphyletic (Figure 14.41). Furthermore, we now know that Polychaeta contains Clitellata, making it synonymous with Annelida and so the name "Polychaeta" is no longer used as a formal taxon.

In addition to the recognition that the traditional classification had to change, the past few decades has seen a series of major changes in annelid membership as well, mainly due to molecular-based phylogenetic studies. Echiura, Sipuncula, Pogonophora, and Vestimentifera, all of which were once considered separate phyla, are now clearly seen as part of Annelida. Hence, Annelida now encompasses many animals that do not conform to the traditional unique features of the group such as possession of body segmentation and chaetae. Even more problematic has been resolving the relationships among the groups that have long been regarded as part of the annelid radiation, a work that must still be regarded as very much in progress.

While a deep understanding of the origin and evolution of annelids still eludes us, considerable progress has been recently made with the use of Next Generation sequencing technology, which holds great promise. This means we will likely see a stable annelid phylogenetic tree in the coming years. Our current knowledge of annelid relationships, as used in this chapter, is based on the recent phylogenomic studies of Sónia Andrade and Anne Weigert and their colleagues, and these were used to generate the summary tree and classification used in this chapter (Figure 14.41). As future phylogenomic analyses are integrated with increasing knowledge of fossil annelids, we will be able to make better inferences as to the origin and evolution of this amazingly diverse group of animals.

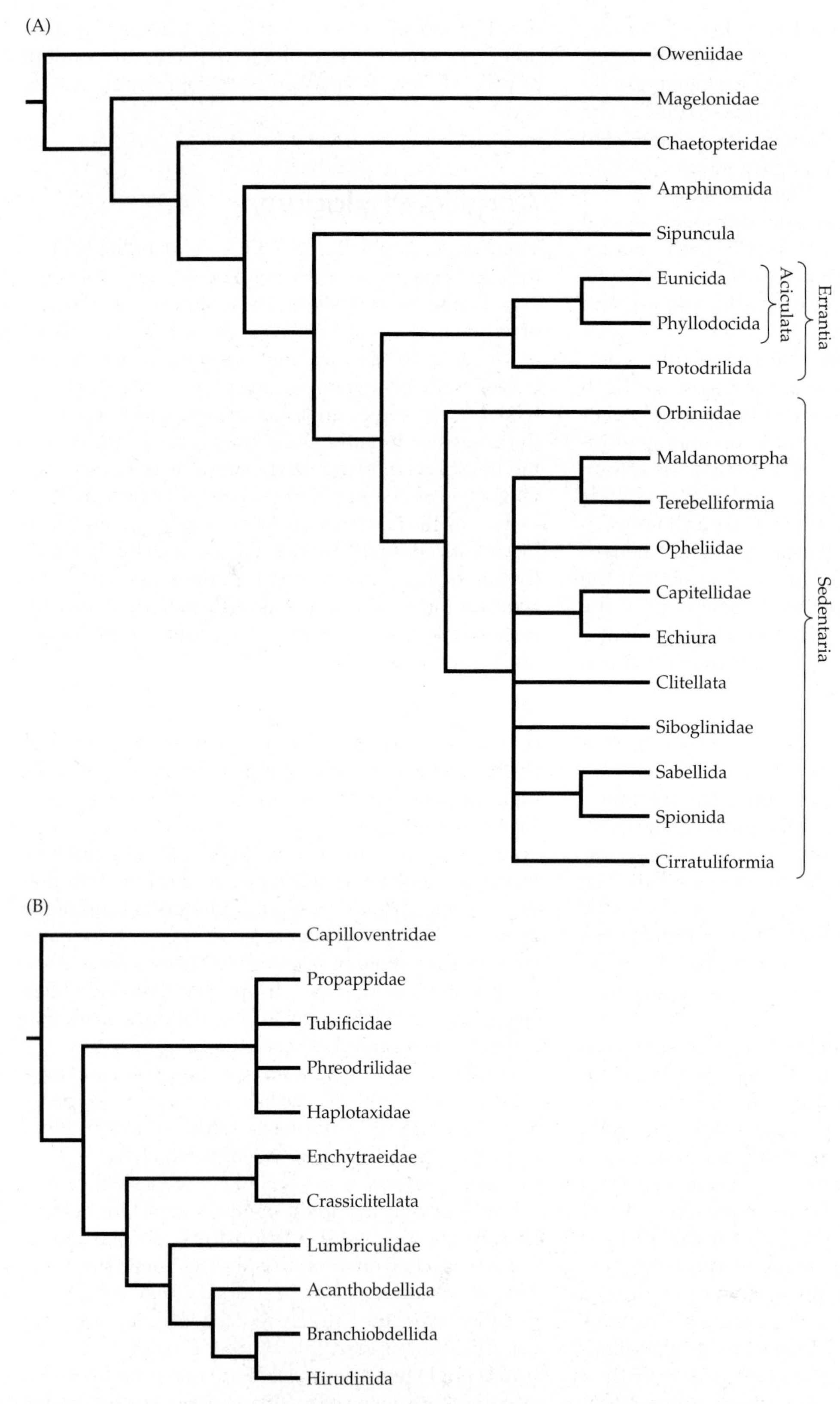

Figure 14.41 Phylogenetic relationships of annelids. (A) Annelida, based on areas of agreement in the phylogenomic analyses of Andrade et al. (2015) and Struck et al. (2015). (B) Clitellata based on topologies in Siddall et al. (2001), Rousset et al. (2008), and Marotta et al. (2008).

Selected References

General References

Andrade, S. C., M. Novo, G. Y. Kawauchi, K. Worsaae, F. Pleijel, G. Giribet and G. W. Rouse. 2015. Articulating "archiannelids": Phylogenomics and annelid relationships, with emphasis on meiofaunal taxa. Mol. Biol. Evol. doi: 10.1093/molbev/msv157

Bartolomaeus, T. 1994. On the ultrastructure of the coelomic lining in the Annelida, Sipuncula and Echiura. Microfauna Mar. 9: 171–220.

Bartolomaeus, T. 1997. Chaetogenesis in polychaetous Annelida—Significance for annelid systematics and the position of the Pogonophora. Zool-Anal. Complex. Sy. 100: 348–364.

Bartolomaeus, T. 1999. Structure, function and development of segmental organs in the Annelida. Hydrobiologia 402: 21–37.

Bartolomaeus, T. and B. Quast. 2005. Structure and development of nephridia in Annelida and related taxa. Hydrobiologia 535/536: 139–165.

Bely, A.E. 2006. Distribution of segment regeneration ability in the Annelida. Integr. Comp. Biol. 46: 508–518.

Brinkhurst, R. O. and D. G. Cook (eds.). 1980. *Aquatic Oligochaete Biology*. Plenum, New York.

Brown, S. C. 1977. Biomechanics of water-pumping by *Chaetopterus variapedatus* Renier: Kinetics and hydrodynamics. Biol. Bull. 153: 121–132.

Caspers, H. 1984. Spawning periodicity and habit of the palolo worm *Eunice viridis* in the Samoan Islands. Mar. Biol. 79: 229–236.

Clark, R. B. and D. J. Tritton. 1970. Swimming mechanisms of nereidiform polychaetes. J. Zool. London 161: 257–271.

Dales, R. P. 1977. The polychaete stomodeum and phylogeny. Pp. 525–546 in *Essays on polychaetous annelids in memory of Dr. Olga Hartman*. Allan Hancock Foundation, University of Southern California, Los Angeles, CA.

Dales, R. P. and G. Peter. 1972. A synopsis of the pelagic Polychaeta. J. Nat. Hist. 6: 55–92.

Desbruyeres, D., F. Gaill, L. Laubier, D. Prieur and G. Rau. 1983. Unusual nutrition of the Pompeii worm *Alvinella pompejana* (polychaetous annelid) from a hydrothermal vent environment: SEM, TEM, ^{13}C, and ^{15}N evidence. Mar. Biol. 75: 201–205.

Dordel, J., F. Fisse, G. Purschke and T. H. Struck. 2010. Phylogenetic position of Sipuncula derived from multi-gene and phylogenomic data and its implication for the evolution of segmentation. J. Zool. Syst. Evol. Res. 48: 197–207.

Dorgan, K. M. 2015. The biomechanics of burrowing and boring. J. Exp. Biol. 218: 176–183.

Dorgan, K. M., P. A. Jumars, B. Johnson, B. P. Boudreau and E. Landis 2005. Burrow extension by crack propagation: A worm minimizes its energy expenditure as it forges a path through mud sediment. Nature 253: 475.

Dorgan, K. M., C. J. Law and G. W. Rouse 2013. Meandering worms: mechanics of undulatory burrowing in muds. Proc. Roy Soc. B 280: 20122948.

Dunn C. W. and 17 others. 2008. Broad phylogenomic sampling improves resolution of the animal tree of life. Nature 452: 745–749.

Erséus, C. 2005. Phylogeny of oligochaetous Clitellata. Hydrobiologia 535: 357–372.

Fauchald, K. 1977. The polychaete worms: Definitions and keys to the orders, families, and genera. Nat. Hist. Mus. Los Angeles Co. Ser. 28: 1–190.

Fauchald, K. and P. A. Jumars. 1979. The diet of worms: A study of polychaete feeding guilds. Oceanogr. Mar. Biol. Annu. Rev. 17: 193–284.

Fauchald, K. and G. W. Rouse 1997. Polychaete systematics: Past and present. Zool. Scr. 26: 71–138.

Fischer, A. and U. Fischer. 1995. On the life-style and life-cycle of the luminescent polychaete *Odontosyllis enopla* (Annelida: Polychaeta). Invert. Biol. 114: 236–247.

Fischer, A. and H.-D. Pfannenstiel (eds.). 1984. *Polychaete Reproduction. Progress in Comparative Reproductive Biology*. Gustav Fischer Verlag, Stuttgart and New York.

Glasby, C. T. and Timm. 2008. Global diversity of polychaetes (Polychaeta; Annelida) in freshwater. Hydrobiologia 595: 107–115.

Gray, J. and H. W. Lissmann. 1938. Studies in animal locomotion. VII. Locomotory reflexes in the earthworm. J. Exp. Biol. 15: 506–517.

Gray, J. 1939. Studies in animal locomotion. VIII. The kinetics of locomotion of *Nereis diversicolor*. J. Exp. Biol. 16: 9–17.

Hartman, O. 1968. *Atlas of the Errantiate Polychaetous Annelids from California*. Allan Hancock Foundation, University of Southern California, Los Angeles, CA.

Hartman, O. 1969. *Atlas of the Sedentariate Polychaetous Annelids from California*. Allan Hancock Foundation, University of Southern California, Los Angeles, CA.

Hausen H. 2005. Comparative structure of the epidermis in polychaetes (Annelida). Hydrobiologia 535/536: 25–35.

Hausen H. 2005. Chaetae and chaetogenesis in polychaetes (Annelida). Hydrobiologia 535/536: 37–52.

Hessling, R. and W. Westheide. 2002. Are Echiura derived from a segmented ancestor? Immunohistochemical analysis of the nervous system in developmental stages of *Bonellia viridis*. J. Morphol. 252: 100–113.

Hermans, C. and R.M. Eakin. 1974. Fine structure of the eyes of an alciopid polychaete *Vanadis tagensis*. Z. Morphol. Tiere 79: 245–267.

Jamieson, B. G. M. 1981. *The Ultrastructure of the Oligochaeta*. Academic Press, London.

Jamieson, B. G. M. 2006. Non-leech Clitellata. Pp. 235–392 in G. W. Rouse and F. Pleijel (eds.), *Reproductive Biology and Phylogeny of Annelida*. Science Publishers, Inc., Enfield, New Hampshire.

Jumars, P. A., K. M. Dorgan and S. M. Lindsay. 2015. Diet of worms emended: An update of polychaete feeding guilds. Ann. Rev. Mar. Sci. 7: 497–520.

Kvist, S. and M. E. Siddall. 2013. Phylogenomics of Annelida revisited: a cladistic approach using genome-wide expressed sequence tag data mining and examining the effects of missing data. Cladistics 29: 435–448.

Law, C., K. M. Dorgan and G. W. Rouse 2014. Relating divergence in musculature within Opheliidae (Annelida) with different burrowing behaviors. J. Morph. 275: 548–571.

Lindsay S. M. 2009. Ecology and biology of chemoreception in polychaetes. Zoosymposia. 2:339–367.

Mangum, C. 1976. The oxygenation of hemoglobin in lugworms. Physiol. Zool. 49: 85–99.

Marotta, R., M. Ferraguti, C. Erseus and L. M. Gustavsson. 2008. Combined-data phylogenetics and character evolution of Clitellata (Annelida) using 18S rDNA and morphology. Zool. J. Linn. Soc. 154: 1–26.

Martin, D. and T. A. Britayev 1998. Symbiotic polychaetes: Review of known species. Oceanog. Mar. Biol. 36: 217–340.

Martin, P., E. Martinez-Ansemil, A. Pinder, T. Timm and M. J. Wetzel. 2007. A global assessment of the oligochaetous clitellate diversity in freshwater. Hydrobiologia 595: 117–127.

Mettam, C. 1967. Segmental musculature and parapodial movement of *Nereis diversicolor* and *Nephtys hombergi* (Annelida: Polychaeta). J. Zool. Lond. 153: 245–275.

Meyer N. P., M. J. Boyle, M. Q. Martindale and E. C. Seaver. 2010. A comprehensive fate map by intracellular injection of identified blastomeres in the marine polychaete *Capitella teleta*. EvoDevo. 1: 8–27.

Meyer N. P., A. Carrillo-Baltodano. R. E. Moore and E. C. Seaver. 2015. Nervous system development in lecithotrophic larval and juvenile stages of the annelid *Capitella teleta*. Front. Zool. doi: 10.1186/s12983-015-0108-y

Nicol, E. A. T. 1931. The feeding mechanism, formation of the tube, and physiology of digestion in *Sabella pavonia*. Trans. R. Soc. Edinburgh 56: 537–598.

Orrhage L. and M. Muller. 2005. Morphology of the nervous system of Polychaeta (Annelida). Hydrobiologia. 535: 79–111.

Osborn, K. J., S. H. D. Haddock, F. Pleijel, L. P. Madin and G. W. Rouse 2009. Deep-sea, swimming worms with luminescent "bombs." Science 325: 964.

Penry, D. L. and P. A. Jumars. 1990. Gut architecture, digestive constraints and feeding ecology of deposit-feeding and carnivorous polychaetes. Oecologia 82: 1–11.

Purschke G. 2005. Sense organs in polychaetes (Annelida) Hydrobiologia 535/536: 53–78.

Purschke G., D. Arendt, H. Hausen, M. C. M. Muller. 2006. Photoreceptor cells and eyes in Annelida. Arthropod. Struct. Dev. 35: 211–230.

Purschke G. and A. B. Tzetlin. 1996. Dorsolateral ciliary folds in the polychaete foregut: Structure, prevalence and phylogenetic significance. Acta Zool. 77: 33–49.

Rouse, G. W. 1999. Trochophore concepts: Ciliary bands and the evolution of larvae in spiralian Metazoa. Biol. J. Linn. Soc. 66: 411–464.

Rouse, G. W. 2000. Polychaetes have evolved feeding larvae several times. Bull. Mar. Sci. 67: 391–409.

Rouse, G. W. and K. Fitzhugh 1994. Broadcasting fables: Is external fertilization really primitive? Sex, size and larvae in sabellid polychaetes. Zool. Scr. 23: 271–312.

Rouse, G. W. and F. Pleijel. 2001. *Polychaetes*. Oxford Univ. Press, Oxford.

Rouse, G. W. and F. Pleijel (eds.) 2006. *Reproductive Biology and Phylogeny of Annelida*. Enfield, New Hampshire, Science Publishers, Inc.

Rousset, V., F. Pleijel, G. W. Rouse, C. Erseus and M. E. Siddall. 2007. A molecular phylogeny of annelids. Cladistics 23: 41–63.

Rousset, V., L. Plaisance, C. Erseus, M. E. Siddall and G. W. Rouse. 2008. Evolution of habitat preference in Clitellata (Annelida). Biol. J. Linn. Soc. 95: 447–464.

Ruby, E. G. and D. L. Fox. 1976. Anaerobic respiration in the polychaete *Euzonus* (*Thoracophelia*) *mucronata*. Mar. Biol. 35: 149–153.

Seymour, M. K. 1969. Locomotion and coelomic pressure in *Lumbricus*. J. Exp. Biol. 51: 47.

Stephensen, J. 1930. *The Oligochaeta*. Oxford University Press, New York.

Schroeder, P. C. and C. O. Hermans. 1975. Annelida: Polychaeta. Pp. 1–205 in A. C. Giese and J. S. Pearse (eds.), *Reproduction of Marine Invertebrates, Vol. 3*. Academic Press, New York.

Schulze, A. 2006. Phylogeny and genetic diversity of Palolo worms (*Palola*, Eunicidae) from the tropical North Pacific and the Caribbean. Biol. Bull. 210: 25–37.

Schulze, A. and L. E. Timm. 2011. *Palolo* and *un*: distinct clades in the genus *Palola* (Eunicidae, Polychaeta). Mar. Biodivers. 42:161–171.

Seaver E. C. 2014. Variation in spiralian development: insights from polychaetes. Int. J. Dev. Biol. 58: 457–467.

Seaver E. C., K. Thamm and S. D. Hill. 2005. Growth patterns during segmentation in the two polychaete annelids, *Capitella* sp. I and *Hydroides elegans*: comparisons at distinct life history stages. Evol. Dev. 7: 312–326.

Shain D. H. and 7 others. 2000. Morphologic characterization of the ice worm *Mesenchytraeus solifugus*. J. Morphol. 246:192–197.

Smetzer, B. 1969. Night of the *palolo*. Nat. Hist. 78: 64–71.

Struck, T. H., N. Schult, T. Kusen, E. Hickman, C. Bleidorn, D. McHugh and K. M. Halanych. 2007. Annelid phylogeny and the status of Sipuncula and Echiura. BMC Evol. Biol. 7: 1–34.

Struck, T. H. and 7 others. 2015. The evolution of annelids reveals two adaptive routes to the interstitial realm. Curr. Biol. 25: 1993–1999.

Summers M. M. and G. W. Rouse. 2014. Phylogeny of Myzostomida (Annelida) and their relationships with echinoderm hosts. BMC Evol. Biol. 14: 170.

Tzetlin, A. B. and G. Purschke G. 2005. Pharynx and intestine. Hydrobiologia 535/536: 199–225.

Van Praagh, B. D., A. L. Yen and P.K. Lillywhite. 1989. Further information on the giant Gippsland earthworm *Megascolides australis* (McCoy 1878). Victorian Nat. 106: 197–201.

Wallwork, J. A. 1983. *Earthworm Biology*. Edward Arnold and University Park Press, Baltimore.

Weigert, A and 10 others. 2014. Illuminating the base of the annelid tree using transcriptomics. Mol. Biol. Evol. 31: 1391–1401.

Wells, G. P. 1950. Spontaneous activity cycles in polychaete worms. Symp. Soc. Exp. Biol. 4: 127–142.

Worsaae K. and R. E. Kristensen. 2005. Evolution of interstitial Polychaeta (Annelida). Hydrobiologia 535/536: 319–340.

Hirudinoidea

Bielecki, A., J. M. Cichocka, I. Jeleń, P. Świątek, B. J. Płachno and D. Pikuła. 2014. New data about the functional morphology of the chaetiferous leech-like annelids *Acanthobdella peledina* (Grube, 1851) and *Paracanthobdella livanowi* (Epshtein, 1966) (Clitellata, Acanthobdellida). J. Morph. 275: 528–539.

Borda, E. and M. E. Siddall. 2004. Review of the evolution of life history strategies and phylogeny of the Hirudinida (Annelida: Oligochaeta). Lauterbornia 52: 5–25.

Brown, B. L., R. P. Creed and W. E. Dobson. 2002. Branchiobdellid annelids and their crayfish hosts: are they engaged in a cleaning symbiosis? Oecologia 132: 250–255.

Christensen, B. and H. Glenner. 2010. Molecular phylogeny of Enchytraeidae (Oligochaeta) indicates separate invasions of the terrestrial environment. J. Zool. Syst. Evol. Res. 48: 208–212.

Elliott, J. M. and U. Kutschera. 2011. Medicinal leeches: historical use, ecology, genetics and conservation. Freshwater Rev. 4: 21–41.

Gelder, S. R. and M. E. Siddall. 2001. Phylogenetic assessment of the Branchiobdellidae (Annelida, Clitellata) using 18S rDNA, mitochondrial cytochrome c oxidase subunit I and morphological characters. Zool. Scr. 30: 215–222.

Gray, J., H. W. Lissmann and R. J. Pumphrey. 1938. The mechanism of locomotion in the leech. J. Exp. Biol. 15: 408–430.

Holt, T. C. 1968. The Branchiobdellida: Epizootic annelids. Biologist 1: 79–94.

Kutschera, U. and P. Wirtz. 2001. The evolution of parental care in freshwater leeches. Theor. Biosci. 120: 115–137.

Mann, K. 1962. *Leeches (Hirudinea). Their Structure, Physiology, Ecology and Embryology*. Pergaman Press, London.

Martin, P. 2001. On the origin of the Hirudinea and the demise of the Oligochaeta. Proc. Roy. Soc. London B 268: 1089–1098.

Sawyer, R. T. 1986. *Leech Biology and Behaviour*. Oxford, Clarendon Press.

Sawyer, R. 1990. In search of the giant Amazon leech. Nat. Hist. 12: 66–67.

Siddall, M. E. and 7 others. 2001. Validating Livanow: molecular data agree that leeches, branchiobdellidans, and *Acanthobdella peledina* form a monophyletic group of oligochaetes. Mol. Phyl. Evol. 21: 346–351.

Skelton, J. and 8 others. 2013. Servants, scoundrels, and hitchhikers: current understanding of the complex interactions between crayfish and their ectosymbiotic worms (Branchiobdellida). Freshwater Sci. 32: 1345–1357.

Sket, B. and P. Trontelj 2007. Global diversity of leeches (Hirudinea) in freshwater. Hydrobiologia 595: 129–137.

Świątek, P. and 7 others. 2012. Ovary architecture of two branchiobdellid species and *Acanthobdella peledina* (Annelida, Clitellata). Zool. Anz. 251: 71–82.

Williams, B. W., S. R. Gelder, H. C. Proctor and D. W. Coltman. 2013. Molecular phylogeny of North American Branchiobdellida (Annelida: Clitellata). Mol. Phyl. Evol. 6: 30–42.

Siboglinidae

Arp, A. J. and J. J. Childress. 1983. Sulfide bonding by the blood of the hydrothermal vent tube worm *Riftia pachyptila*. Science 219: 295–297.

Bakke, T. 1977. Development of *Siboglinum fiordicum* Webb (Pogonophora) after metamorphosis. Sarsia 63: 65–73.

Bakke, T. 1980. Embryonic and post-embryonic development in the Pogonophora. Zool. Jb. Anat. 103: 276–284.

Caullery, M. 1944. Siboglinum Caullery, 1914. Type nouveau d'invertebres, d'affinites a preciser. Siboga-Expeditie 25: 1–26.

Cavanaugh, C. M., S. L. Gardiner, M. L. Jones, H. W. Janasch and J. B. Waterburg. 1981. Prokaryotic cells in the hydrothermal vent tube worm *Riftia pachyptila* Jones: Possible chemoautotrophic symbionts. Science 213: 340–342.

Felbeck, H. 1981. Chemoautotrophic potential of the hydrothermal vent tube worm, *Riftia pachyptila* Jones (Vestimentifera). Science 213: 366–338.

Gardiner, S. L. and M. L. Jones. 1994. On the significance of larval and juvenile morphology for suggesting phylogenetic relationships of the Vestimentifera. Amer. Zool. 34: 513–522.

George, J. D. and E. C. Southward. 1973. A comparative study of the setae of Pogonophora and polychaetous Annelida. J. Mar. Biol. Assoc. UK 53: 403–424.

Ivanov, A. V. 1963. *Pogonophora*. Academic Press, London. [English translation by D. B. Carlilse.]

Jones, M. L. 1981. *Riftia pachyptila* Jones: Observations on the vestimentiferan worm from the Galapagos Rift. Science 213: 333–336.

Jones, M. L. 1985. On the Vestimentifera, new phylum: Six new species and other taxa, from, hydrothermal vents and elsewhere. Bull. Biol. Soc. Wash. 6: 117–158.

Jones, M. L. and S. L. Gardiner. 1988. Evidence for a transitory digestive tract in Vestimentifera. Proc. Biol. Soc. Wash. 11: 423–433.

Land, J. van der and A. Nørrevang. 1977. Structure and relationships of *Lamellibrachia* (Annelida, Vestimentifera). Kongel. Dans. Vidensk. Selsk. Biol. Skr. 21: 1–102.

Marsh, A. G., L. S. Mullineaux, C. M. Young and D. T. Manahan 2001. Larval dispersal potential of the tubeworm *Riftia pachyptila* at deep-sea hydrothermal vents. Nature 411: 77–80.

Nørrevang, A. 1970. On the embryology of *Siboglinum* and its implications for the systematic position of the Pogonophora. Sarsia 42: 7016.

Pleijel, F., T. Dahlgren and G. W. Rouse 2009. Progress in science: from Siboglinidae to Pogonophora and Vestimentifera and back to Siboglinidae. CR Biol. 332: 140–148.

Rouse, G. W. 2001. A cladistic analysis of Siboglinidae Caullery, 1914 (Polychaeta, Annelida): Formerly the phyla Pogonophora and Vestimentifera. Zool. J. Linnean Soc. 132: 55–80.

Rouse, G. W., S. K. Goffredi and R. C. Vrijenhoek 2004. *Osedax*: Bone-eating marine worms with dwarf males. Science 305: 668–671.

Rouse, G. W. and 7 others. 2009. Spawning and development in *Osedax* boneworms. Mar. Biol. 156: 395–405.

Rouse, G. W., N. G. Wilson, K. Worsaae and R. C. Vrijenhoek 2015. A dwarf male reversal in bone-eating worms. Curr. Biol. 25: 236–241.

Rouse, G. W., K. Worsaae, S. B. Johnson, W. J. Jones and R. C. Vrijenhoek 2008. Acquisition of dwarf male "harems" by recently settled females of *Osedax roseus* n. sp. (Siboglinidae: Annelida). Biol. Bull. 214: 67–82.

Southward, E. C. 1978. Description of a new species of *Oligobrachia* (Pogonophora) from the North Atlantic, with a survey of the Oligobrachiidae. J. Mar. Biol. Assoc. U. K. 58: 357–365.

Southward, E. C. 1982. Bacterial symbionts in Pogonophora. J. Mar. Biol. Assoc. U. K. 62: 889–906.

Southward, E. C. 1988. Development of the gut and segmentation of newly settled stages of *Ridgeia* (Vestimentifera): implications for relationship between Vestimentifera and Pogonophora. J. Mar. Biol. Assoc. U.K. 68: 465–487.

Southward, E. C. and K. A. Coates 1989. Sperm masses and sperm transfer in a Vestimentiferan, *Ridgeia piscesae* Jones 1985 (Pogonophora Obturata). Can. J. Zool. 67: 2776–2781.

Southward, E. C. 1993. Pogonophora. Pp. 327–369 in F. W. Harrison and M. E. Rice (eds.), *Microscopic Anatomy of Invertebrates, Volume 12: Onychophora, Chilopoda, and Lesser Protostomata*. Wiley-Liss, New York..

Southward, E. C. 1999. Development of Perviata and Vestimentifera (Pogonophora). Hydrobiologia 402: 185–202.

Southward, E. C., A. Schulze and S. L. Gardiner. 2005. Pogonophora (Annelida): form and function. Hydrobiologia 535/536: 227–251.

Vrijenhoek, R. C., S. B. Johnson and G. W. Rouse 2008. Bone-eating *Osedax* females and their "harems" of dwarf males are recruited from a common larval pool. Mol. Ecol. 17: 4535–4544.

Vrijenhoek, R. C., S. B. Johnson and G. W. Rouse 2009. A remarkable diversity of bone-eating worms (*Osedax*: Siboglinidae: Annelida). BMC Biol. 7: 74.

Webb, M. 1964. The posterior extremity of *Siboglinum fiordicum* (Pogonophora). Sarsia 15: 33–36.

Webb, M. 1964. Additional notes on *Sclerolinum brattstromi* (Pogonophora) and the establishment of a new family, Sclerolinidae. Sarsia 16: 47–58.

Webb, M. 1969. *Lamellibrachia barhami*, gen. nov., sp. nov. (Pogonophora), from the northeast Pacific. Bull. Mar. Sci. 19: 18–47.

Worsaae, K. and G. W. Rouse 2010. The simplicity of males: progenetic dwarf males of four species of *Osedax* (Annelida) investigated by confocal laser scanning microscopy. J. Morphol. 271: 127–142.

Young, C. M., E. Vázquez, A. Metaxas and P. A. Tyler. 1996. Embryology of vestimentiferan tube worms from deep-sea methane/sulphide seeps. Nature 381: 514–516.

Sipuncula

Adrianov, A. V. and A. S. Maiorova. 2002. Microscopic anatomy and ultrastructure of a Polian vessel in the sipunculan *Thysanocardia nigra* Ikeda, 1904 from the Sea of Japan. Russ. J. Mar. Biol. 28: 100–106.

Boyle, M. J. and M. E. Rice. 2014. Sipuncula: an emerging model of spiralian development and evolution. Int. J. Dev. Biol. 58: 485–499.

Cutler, E. B. 1995. *The Sipuncula: Their Systematics, Biology, and Evolution*. Cornell University Press, New York.

Hansen, M. D. 1978. Food and feeding behavior of sediment feeders as exemplified by sipunculids and holothurians. Helgol. Wiss. Meeresunters. 31: 191–221.

Huang, D.-Y., J.-Y. Chen, J. Vannier and J. I. Saiz Salinas. 2004. Early Cambrian sipunculan worms from southwest China. Proc. Biol. Sci. 271: 1671–1676.

Kawauchi, G. Y., P. P. Sharma and G. G. Giribet. 2012. Sipunculan phylogeny based on six genes, with a new classification and the descriptions of two new families. Zool. Scr. 41: 186–210.

Kawauchi, G. Y. and G. Giribet. 2014. *Sipunculus nudus* Linnaeus, 1766 (Sipuncula): cosmopolitan or a group of pseudo-cryptic species? An integrated molecular and morphological approach. Mar. Ecol. 35:478–491.

Kristof, A., T. Wollesen and A. Wanninger. 2008. Segmental mode of neural patterning in Sipuncula. Cur. Biol. 18: 1129–1132.

Lemer S., G. Y. Kawauchi, S. C. S. Andrade, V. L. González, M. J. Boyle and G. Giribet. 2015. Re-evaluating the phylogeny of Sipuncula through transcriptomics. Mol. Phylog. Evol. 83: 174–83.

Mangum, C. P. and M. Kondon. 1975. The role of coelomic hemerythrin in the sipunculid worm *Phascolopsis gouldi*. Comp. Biochem. Physiol. 50A: 777–785.

Maxmen, A. B., B. F. King, E. B. Cutler and G. Giribet. 2003. Evolutionary relationships within the protostome phylum Sipuncula: a molecular analysis of ribosomal genes and histone H3 sequence data. Mol. Phylogenet. Evol. 27: 489–503.

Pilger, J. F. 1982. Ultrastructure of the tentacles of *Themiste lageniformis* (Sipuncula). Zoomorphologie 100: 143–156.

Pilger, J. F. 1987. Reproductive biology and development of *Themiste lageniformis*, a parthenogenetic sipunculan. Bull. Mar. Sci. 41: 59–67.

Pilger, J. F. 1997. Sipunculans and Echiurans. Pp. 167–168 in S. F. Gilbert and A. M. Raunio (eds.), *Embryology. Constructing the Organism*. Sinauer Associates, Sunderland, MA.

Purschke, G., F. Wolfrath and W. Westheide. 1997. Ultrastructure of the nuchal organ and cerebral organ in *Onchnesoma squamatum* (Sipuncula, Phascolionidae). Zoomorphology 117: 23–31.

Rice, M. E. 1978. Morphological and behavioral changes at metamorphosis in the Sipuncula. Pp. 83–102 in F. S. Chia and M. E. Rice (eds.), *Settlement and Metamorphosis of Marine Invertebrate Larvae*. Elsevier, New York.

Rice, M. E. 1967. A comparative study of the development of *Phascolosoma agassizii*, *Golfingia pugettensis*, and *Themiste pyroides* with a discussion of developmental patterns in the Sipuncula. Ophelia 4: 143–171.

Ruppert, E. E. and M. E. Rice. 1995. Functional organization of dermal coelomic canals in *Sipunculus nudus* (Sipuncula) with a discussion of respiratory designs in sipunculans. Invert. Biol. 114: 51–63.

Schulze, A., E. B. Cutler and G. Giribet. 2007. Phylogeny of sipunculan worms: A combined analysis of four gene regions and morphology. Mol. Phylogenet. Evol. 42: 171–192.

Schulze A and M. Rice. 2009. Musculature in sipunculan worms: ontogeny and ancestral states. Evol. Dev. 11: 97–108.

Stephen, A. C. and S. J. Edmonds. 1972. *The phyla Sipuncula and Echiura*. London: British Museum (Natural History).

Wanninger A, D. Koop, L. Bromham, E. Noonan and B. M. Degnan. 2005. Nervous and muscle system development in *Phascolion strombus* (Sipuncula). Dev. Genes Evol. 215: 509–18.

Wanninger A, A. Kristof and N. Brinkmann. 2009. Sipunculans and segmentation. CIB 2: 56–59.

Echiuridae

Arp, A. J., B. M. Hansen and D. Julian. 1992. Burrow environment and coelomic fluid characteristics of the echiuran worm *Urechis caupo* from populations at three sites in northern California. Mar. Biol. 113: 613–623.

Berec, L., J. Schembri and S. Boukal. 2005. Sex determination in *Bonellia viridis* (Echiura: Bonelliidae): population dynamics and evolution. Oikos 108: 473–484.

Biseswar, R. 2010. Zoogeography of the echiuran fauna of the Indo-West Pacific Ocean (Phylum: Echiura). Zootaxa 2727: 21–33.

Goto, R., T. Okamoto, H. Ishikawa, Y. Hamamura and M. Kato. 2013. Molecular phylogeny of echiuran worms (phylum Annelida) reveals evolutionary pattern of feeding mode and sexual dimorphism. PLoS ONE 8: e56809.

Gould-Somero, M. C. 1975. Echiura. Pp. 277–311 in A. C. Giese and J. S. Pearse (eds.), *Reproduction of Marine Invertebrates, Vol. 3*. Academic Press, New York.

Hessling, R. 2002. Metameric organisation of the nervous system in developmental stages of *Urechis caupo* (Echiura) and its phylogenetic implications. Zoomorphology 121: 221–234.

Hughes, D. J., S. J. Marrs, C. J. Smith and R. J. A. Atkinson. 1999. Observations of the echiuran worm *Bonellia viridis* in the deep basin of the northern Evoikos Gulf, Greece. J. Mar. Biol. Ass. U. K. 79: 361–363.

Jaccarini, V., P. J. Schembri and M. Rizzo. 1983. Sex determination and larval sexual interaction in *Bonellia viridis* Rolando (Echiura: Bonelliidae). J. Exp. Mar. Biol. Ecol. 66: 25–40.

Kato, C., J. Lehrke and B. Quast. 2011. Ultrastructure and phylogenetic significance of the head kidneys in *Thalassema thalassemum* (Thalassematinae, Echiura). Zoomorphology 130: 97–106.

Lehrke, J. and T. Bartolomaeus. 2009. Comparative morphology of spermatozoa in Echiura. Zool. Anz. J. Comp. Zool. 248: 35–45.

Lehrke, J., and T. Bartolomaeus, T. 2011. Ultrastructure of the anal sacs in *Thalassema thalassemum* (Annelida, Echiura). Zoomorphology 130: 39–49.

Newby, W. W. 1940. The embryology of the echiuroid worm *Urechis caupo*. Mem. Am. Phil. Soc. 16: 1–219.

Ohta, S. 1984. Star-shaped feeding traces produced by echiuran worms on the deep-sea floor of the Bay of Bengal. Deep-Sea Res. 31: 1415–1432.

Pilger, J. F. 1978. Settlement and metamorphosis in the Echiura: A review. Pp. 103–111 in F. S. Chia and M. E. Rice (eds.), *Settlement and Metamorphosis of Marine Invertebrate Larvae*. Elsevier, New York.

Pilger, J. F. 1993. Echiura. Pp. 185–236 in F. W. Harrison and M. E. Rice (eds.), *Microscopic Anatomy of Invertebrates Vol. 12*. Wiley-Liss, New York.

Pritchard, A. and F. N. White. 1981. Metabolism and oxygen transport in the innkeeper worm *Urechis caupo*. Physiol. Zool. 54: 44–54.

Schuchert, P. 1990. The nephridium of the *Bonellia viridis* Male (Echiura). Acta Zool. 71: 1–4.

Schuchert, P., and Rieger, R.M. 1990. Ultrastructure observations on the dwarf male *Bonellia viridis* (Echiura). Acta Zool. 71: 5–16.

Tanaka, M., T. Kon and T. Nishikawa. 2014. Unraveling a 70-year-old taxonomic puzzle: Redefining the genus *Ikedosoma* (Annelida: Echiura) on the basis of morphological and molecular analyses. Zool. Sci. 31: 849–861.

Tilic, E., J. Lehrke and T. Bartolomaeus. 2015. Homology and evolution of the chaetae in Echiura (Annelida). PLoS ONE 10: e0120002.

CHAPTER 15

Two Enigmatic Spiralian Phyla

Entoprocta and Cycliophora

Among the spiralian protostomes are two phyla that have been difficult to position in the tree of life, Entoprocta and Cycliophora. Entoprocts have been known since the nineteenth century, but cycliophorans were not discovered and described until 1995. Both are small marine phyla, although two freshwater entoproct species are known. Entoprocts resemble minute cnidarian polyps and are often colonial, whereas cycliophorans are solitary microscopic symbionts on the mouthparts of certain crustaceans. Both groups are acoelomate, and neither is encountered frequently enough to have acquired a generally accepted vernacular name. Although species in these two phyla do not resemble one another superficially, you will recognize some fundamental anatomical similarities as you read about them here. Some morphological work (and early molecular phylogenies) supported a grouping of Entoprocta + Cycliophora + Bryozoa (the "Polyzoa"), but phylogenetic studies over the past decade have removed the bryozoans from this group, leaving just the entoprocts and cycliophorans as a possible sister group buried somewhere deep in the Spiralia lineage (although some phylogenetic studies have also paired Entoprocta and Bryozoa as sister groups).

Classification of The Animal Kingdom (Metazoa)

Non-Bilateria*
(a.k.a. the diploblasts)
PHYLUM PORIFERA
PHYLUM PLACOZOA
PHYLUM CNIDARIA
PHYLUM CTENOPHORA

Bilateria
(a.k.a. the triploblasts)
PHYLUM XENACOELOMORPHA

Protostomia
PHYLUM CHAETOGNATHA

SPIRALIA
PHYLUM PLATYHELMINTHES
PHYLUM GASTROTRICHA
PHYLUM RHOMBOZOA
PHYLUM ORTHONECTIDA
PHYLUM NEMERTEA
PHYLUM MOLLUSCA
PHYLUM ANNELIDA
PHYLUM ENTOPROCTA
PHYLUM CYCLIOPHORA

Gnathifera
PHYLUM GNATHOSTOMULIDA
PHYLUM MICROGNATHOZOA
PHYLUM ROTIFERA

Lophophorata
PHYLUM PHORONIDA
PHYLUM BRYOZOA
PHYLUM BRACHIOPODA

ECDYSOZOA
Nematoida
PHYLUM NEMATODA
PHYLUM NEMATOMORPHA

Scalidophora
PHYLUM KINORHYNCHA
PHYLUM PRIAPULA
PHYLUM LORICIFERA

Panarthropoda
PHYLUM TARDIGRADA
PHYLUM ONYCHOPHORA
PHYLUM ARTHROPODA
SUBPHYLUM CRUSTACEA*
SUBPHYLUM HEXAPODA
SUBPHYLUM MYRIAPODA
SUBPHYLUM CHELICERATA

Deuterostomia
PHYLUM ECHINODERMATA
PHYLUM HEMICHORDATA
PHYLUM CHORDATA

*Paraphyletic group

Phylum Entoprocta: The Entoprocts

The phylum Entoprocta (Greek *entos*, "inside"; *proktos*, "anus"), or Kamptozoa (Greek *kamptos*, "bent"), includes about 200 described species of small, sessile, solitary or colonial creatures that superficially resemble cnidarian hydroids (Figure 15.1; Box 15A). All but two species are marine (the probably monotypic freshwater colonial *Urnatella* and one species of *Loxosomatoides*). Colonial forms live attached to various substrata, including algae, shells, and rock

---

Claus Nielsen has revised the section on Entoprocta, and Ricardo Cardoso Neves has revised the section on Cycliophora.

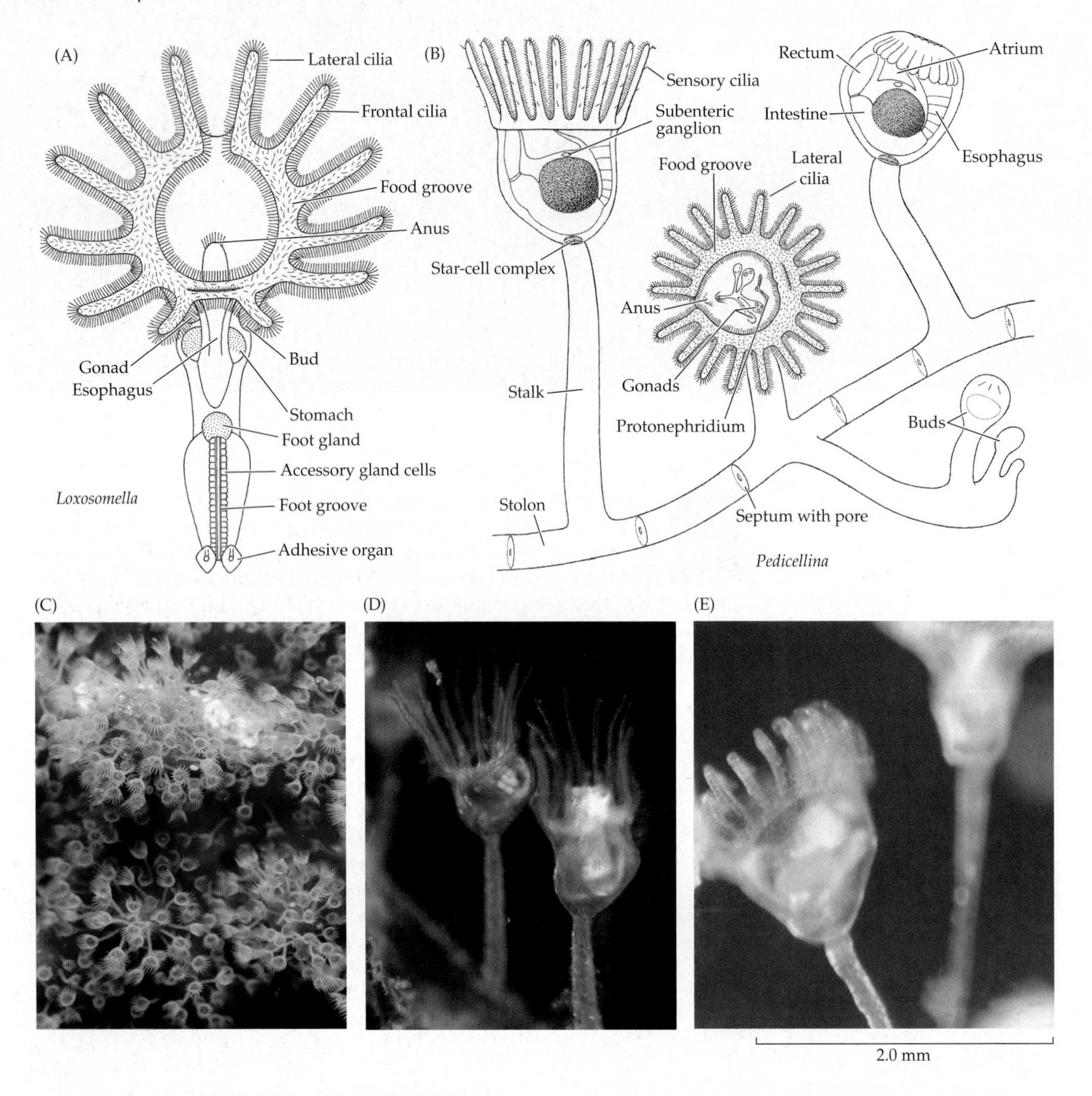

surfaces. Many solitary species are commensal on a variety of hosts, especially sponges (Figure 15.1C), polychaetes, and sipunculans, and are typically associated with just one or a few host species. Entoprocts are not uncommon in shallow water, and some species are known from depths as great as 500 meters. They are often mistaken for cnidarian polyps or otherwise overlooked by the casual tidepool observer because of their small size, but a quick examination through a hand lens reveals both their beauty and their true nature. Certain benthic flatworms and molluscs are known to feed on entoprocts.

The individual zooids of entoprocts are clearly bilateral, but the tentacle crown of most of the colonial species is almost circular. Each zooid consists of a cuplike body, the **calyx**, from which arises an almost closed horseshoe of ciliated tentacles (Figure 15.1). The tentacles encircle a concavity, called the **atrium** or vestibule, in which both the protonephridia and the gonoducts open. The mouth is situated on the rim carrying the tentacle crown, and the anus opens on a small anal cone in the atrium. The calyx is carried on a stalk, which in solitary forms attaches to the substratum, either directly or via a complicated foot. In colonial forms the stalk attaches to branched stolons or an enlarged basal plate.

The phylogenetic position of the entoprocts has been much discussed. Originally, they were treated as bryozoans, but when the larval protonephridium was discovered in 1877 (by Berthold Hatschek) they were placed in the Scolecida, with affinities to the Rotifera. In the 1920s, R. B. Clark discovered that they lack a coelom and they were then placed somewhere in the large group Spiralia.

(F)

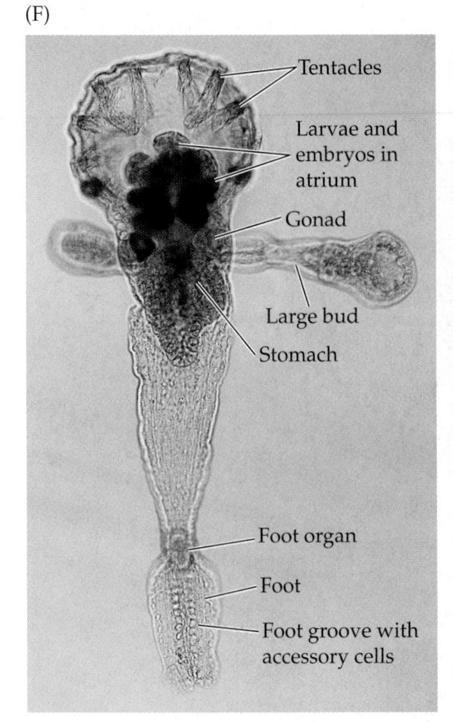

(G)

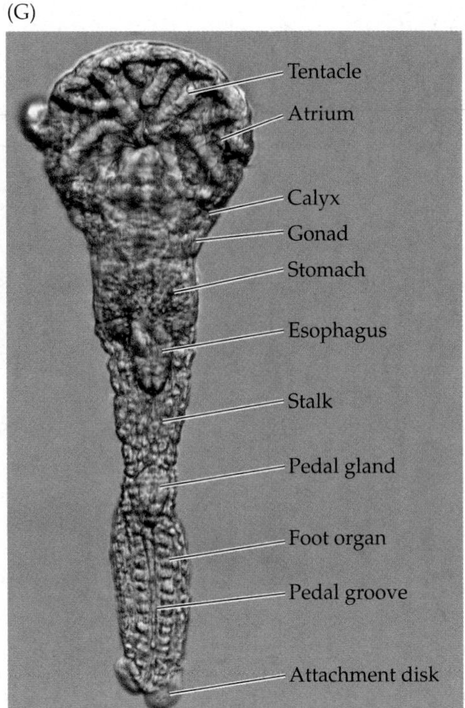

Figure 15.1 Entoprocts. (A, B) Diagrams of a solitary (*Loxosomella*) and a colonial (*Pedicellina*) entoproct. (C) Dense growth of an unidentified *Loxosomella* on a sponge; many calyces can be seen rising from their stalks. (D) Two zooids of *Pedicellina cernua*, each showing its calyx and stalk. (E) Zooids of an undescribed Australian species of *Barentsia*. (F) An undescribed species of *Loxosomella* from San Juan Island, Washington, with embryos, larvae, and buds. (G) Labeled micrograph of the meiofaunal species *Loxosomella vancouverensis*, from British Columbia (Canada). Described in 2012, this is the first free-living entoproct (and only the third entoproct species) reported from the west coast of North America. The other two known west coast species are both symbionts on the mantis shrimp *Pseudosquilla bigelowi*. *L. vancouverensis* is just 440 μm long, and has 14 tentacles and two attachment discs at the base of the foot organ. The foot adheres to the substrate (sand grains and shell bits), but can detach and reattach.

Their position among the Spiralia remains enigmatic, but recent work suggests they are closely allied to the Cycliophora, and possibly the Bryozoa. The phylum is commonly divided into four families as follows. Some specialists recognize orders on the basis of the presence or absence of a septum between the stalk and calyx, or on solitary versus colonial habits.

ENTOPROCT CLASSIFICATION

FAMILY LOXOSOMATIDAE Solitary; without a septum between stalk and calyx; usually commensal on other invertebrates; some are capable of limited movement on a suckered base or a complicated foot; muscles continuous from stalk to calyx. (e.g., *Loxosoma, Loxosomella*; see Figure 15.1A,C)

FAMILY LOXOKALYPODIDAE Colonial; without a septum between stalk and calyx; a few zooids attached to a basal plate; muscles continuous from stalk to calyx; ectocommensal on the polychaete *Glycera nana* in the northeastern Pacific. Monotypic: *Loxokalypus socialis*

FAMILY PEDICELLINIDAE Colonial; with incomplete stalk–calyx septum with star-cell complex; muscles extend length of stalk, but are not continuous with those of calyx; stalk

BOX 15A Characteristics of the Phylum Entoprocta

1. Triploblastic, bilateral, unsegmented, functionally acoelomate
2. Sessile, solitary or colonial (with stalks borne on stolons)
3. Cup-shaped body houses a U-shaped gut
4. Body carries a horseshoe of ciliated tentacles with mouth situated at base of the frontal tentacles and anus at the opposite side of the concavity
5. Body carried by a stalk situated on the convex side of the body (calyx)
6. With protonephridia
7. Hermaphroditic or gonochoristic
8. Cleavage spiral
9. With trochophore larvae
10. Nearly all species are shallow-water marine animals (two known freshwater species) inhabiting stones or algae; many of the solitary species are commensals of polychaetes, sponges, and other water-current producing invertebrates.

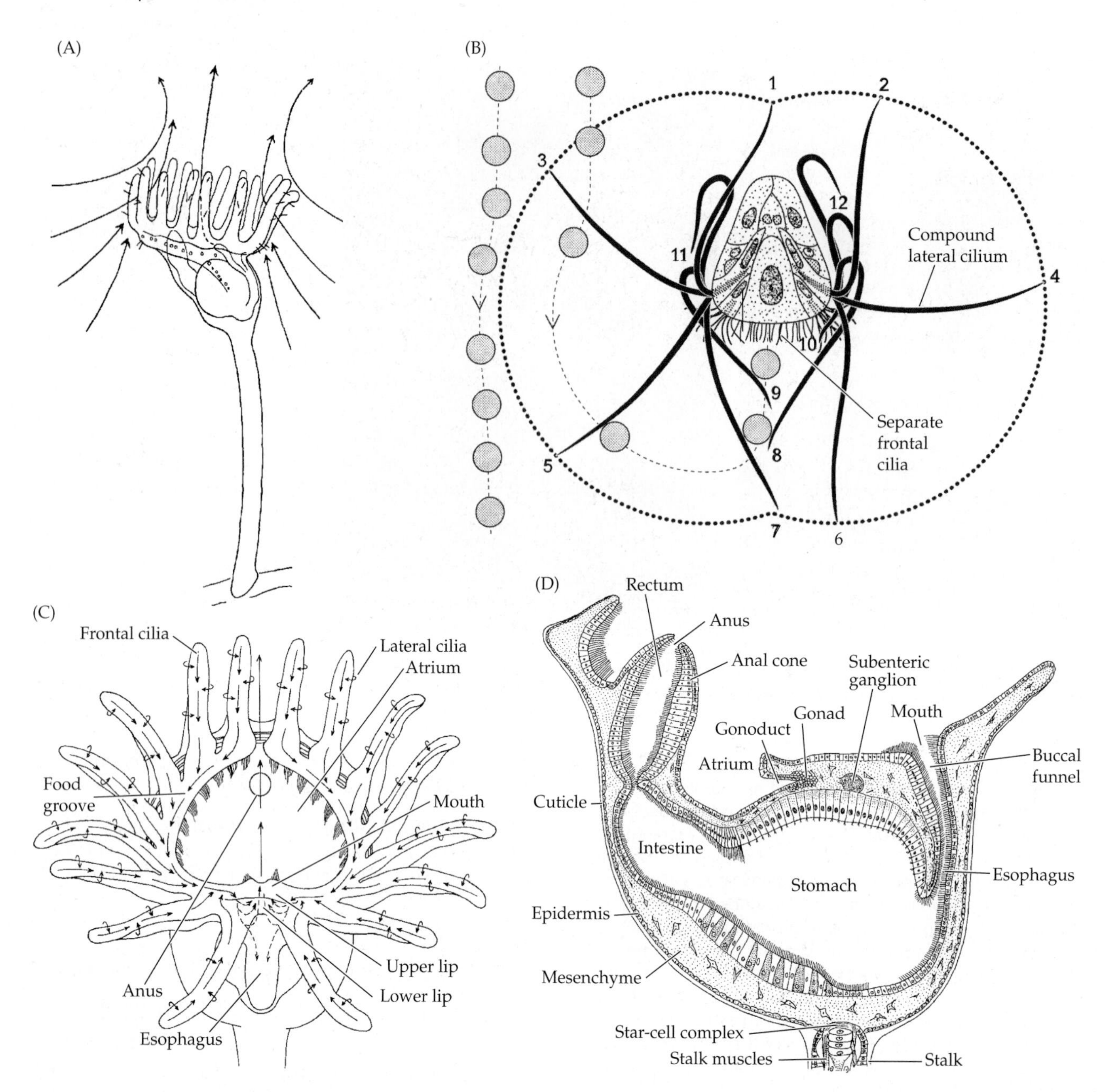

undifferentiated. (e.g., *Loxosomatoides, Myosoma, Pedicellina*; see Figure 15.1B,D)

FAMILY BARENTSIIDAE Colonial; with incomplete stalk–calyx septum; with star-cell complex; stalk differentiated into wide, muscular nodes and narrow, non-muscular rods (see Figure 15.1E). (e.g., *Barentsia, Urnatella*)

The Entoproct Body Plan

Body Wall, Support, and Movement

The calyx and stalk are covered by a thin cuticle that does not extend over the ciliated portion of the tentacles or the atrium. The stiff parts of the stalk of the barentsiids have a thick cuticle, and cuticular dorsal "shields" are characteristic of some of the colonial species. The epidermis is cellular, and the epidermal cells are cuboidal to flattened. Mesodermal and myoepithelial cells move the tentacles and curl them up when the tentacle crown is retracted; a ring muscle just below the tentacle base constricts the atrium over the retracted tentacles. Other mesodermal muscle bands compress the body (calyx) to extend the tentacles. Contractions of longitudinal and oblique muscles in the basal part of the body and in the stalk make characteristic nodding movements of the zooid possible, but also enable more complicated movements of the stalk. As mentioned above, there is no persistent body cavity, and the area between the gut and body wall is filled with mesenchyme (Figure 15.2D).

◀ **Figure 15.2 Entoproct feeding and digestion.** (A) Water currents of the tentacle crown of *Loxosomella leptoclini* set up by the lateral cilia. (B) Downstream-collecting by the catch-up principle. A cross section through a tentacle with the movements of the large compound lateral cilia illustrated through successive steps (numbered 1–12) of a cilium on each side. The compound cilium catches up with the grey particle and pushes it to the frontal band of small separate cilia. (C) Ciliary feeding currents of the tentacle crown of a *Loxosomella*. Particles caught by the lateral cilia are transported along the frontal side of the tentacle to the food groove and further to the mouth. (D) Diagram of a median section of the body of a *Barentsia*.

Feeding and Digestion

Entoprocts are ciliary suspension feeders. They collect food particles, mostly phytoplankton, from currents produced by the lateral cilia on the tentacles (Figure 15.2 A,C). The water currents pass between the tentacles and away from the atrium. Food particles are collected on the downstream-side of the ciliary band by the "catch-up method" in which the compound lateral cilia—several cilia working together as one unit—cut through the water to contact the particle and push it to the frontal band of separate cilia on the frontal side of the tentacle (Figure 15.2B). The particles are then transported along the tentacle to the horseshoe-shaped ciliary band along the food groove on the atrial ridge and to the mouth (Figure 15.2C).

Food particles are moved into the gut by cilia lining the buccal tube and by muscular contractions of the esophagus (Figure 15.2D). The esophagus leads to a spacious stomach, from which a short intestine extends to the rectum located within the anal cone. Food is moved through the gut by cilia. The stomach lining secretes digestive enzymes and mucus, which are mixed with the food by a tumbling action caused by the ciliary currents. Digestion and absorption probably occur within the stomach and intestine, where food is held for a time by an intestinal–rectal sphincter muscle.

Circulation, Gas Exchange, and Excretion

The gut also apparently serves as an excretory passage. Cells in the ventral stomach wall accumulate brownish spheres and release them into the stomach lumen, from which they are discharged through the anus. Adult entoprocts also possess a pair of flame bulb protonephridia located between the stomach and the atrium epithelium (Figure 15.1B). The freshwater species *Urnatella gracilis* has additional protonephridia both in the body and in the joints of the stalk. The protonephridia drain to a short common nephridioduct that leads to a pore on the surface of the atrium. In most species, protonephridia appear to be present also in the larvae.

Internal transport is largely through the expansive gut; diffusion distances through the mesenchyme are small between its lumen and the body wall. Colonial entoprocts have a so-called **star-cell organ** located near the stalk–calyx junction (Figure 15.2D). This structure functions as a heart by pulsating and pumping fluid from the calyx to the stalk. Gas exchange probably occurs over much of the body surface, particularly at the cuticle-free tentacles and atrium.

Nervous System

As is often the case in small sessile invertebrates, the nervous system is greatly reduced. A single ganglionic mass lies between the stomach and atrial surface, and is called the subenteric ganglion (Figures 15.1B and 15.2D). The subenteric ganglion gives rise to several pairs of nerves to the tentacles, calyx wall, and stalk. Unicellular tactile receptors are concentrated on the tentacles and scattered over much of the body surface. Ciliated papillae form lateral sense organs in some loxosomatids.

Reproduction and Development

Colony growth occurs by budding from the tips of the stolons (Figure 15.1A,B,F) or in some barentsiids from joints of the stalk. Solitary forms produce buds on the calyx in the region of the esophagus; when in an advanced state, they separate from the parent (Figure 15.3C).

Most, perhaps all, loxostomatids are hermaphroditic and many are protandric. Those that are thought to have separate sexes may also be protandric, but with a long time between the male and female phases. Colonial forms may have hermaphroditic or gonochoristic zooids, and colonies may contain one or both sexes. One or two pairs of gonads lie just beneath the surface of the atrium. Short gonoducts lead from the gonads to a common pore opening to the atrium (which serves as a brood chamber, somewhat like a kangaroo's pouch) (Figure 15.2D).

Sperm apparently are released into the water and then enter the female reproductive tract, with fertilization occurring in the ovaries or oviducts. As the zygote moves along the oviduct, cement glands secrete a tough surrounding membrane with a stalk by which the embryos are attached to the wall of the brood chamber. In a few species, the embryos are nourished by cells of the maternal atrium.

Cleavage in entoprocts is holoblastic and spiral. Nonsynchronous divisions produce five "quartets" of micromeres at about the 56-cell stage. Cell fates are similar to those in typical protostome development, including the derivation of mesoderm from the 4d mesentoblast. A coeloblastula forms and gastrulates by invagination. A larva develops that, in some species, resembles a typical, planktotrophic trochophore with proto- and metatroch used in swimming and downstream-collecting of food particles, a basic larval type among protostomes (Figure 15.3B). A few loxosomatids produce small eggs, and the

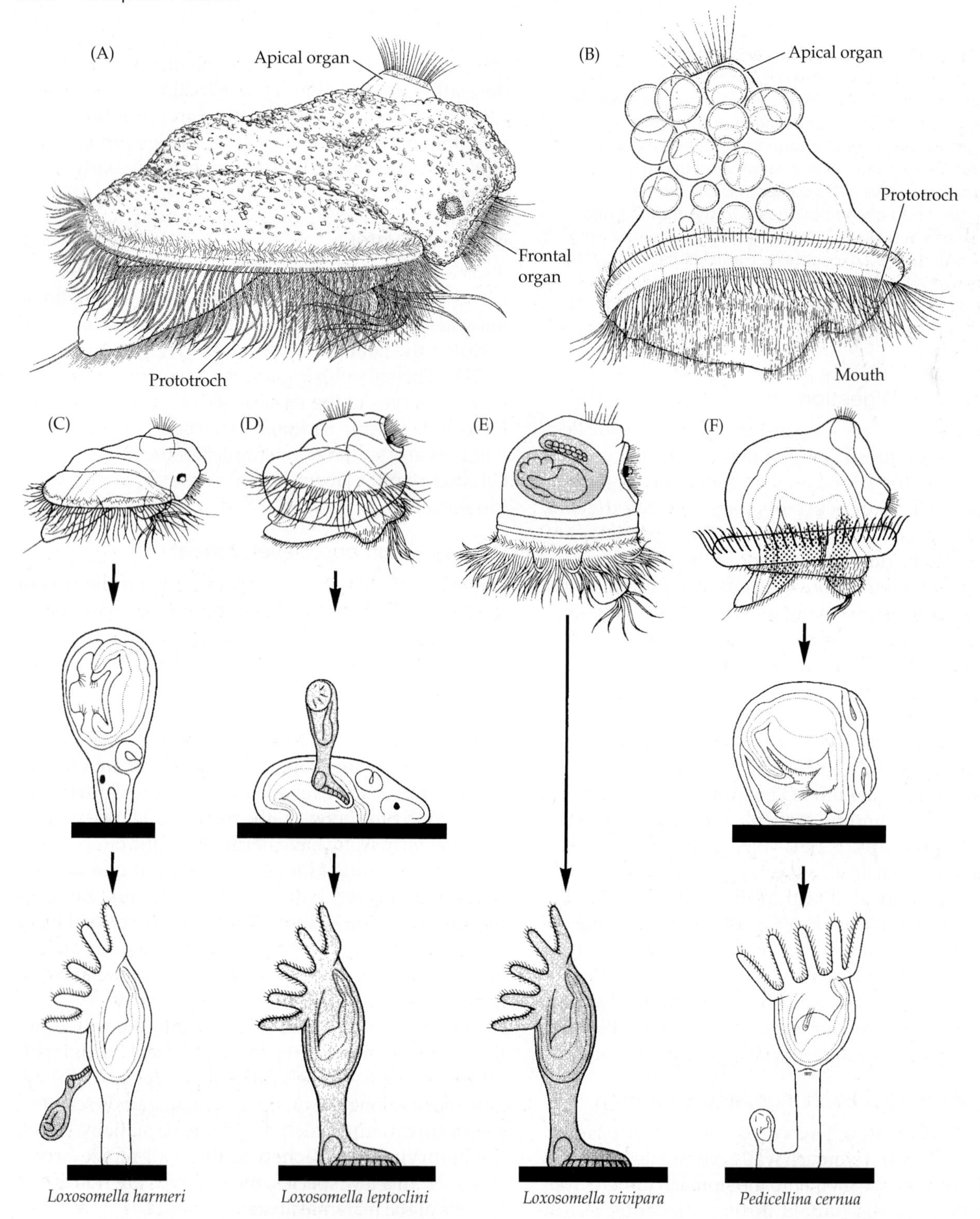

Figure 15.3 Entoproct larvae and metamorphoses. (A) Lecithotrophic larva of *Loxosomella harmeri*. (B) Planktotrophic larva of *Loxosoma pectinaricola*. (C–F) settling and metamorphosis. (C) The larva settles with the frontal organ and becomes a juvenile (*Loxosomella harmeri*). (D) The larva settles and develops one or two buds that detach as juveniles; the larval body disintegrates (*Loxosomella leptoclini*). (E) The large internal bud disrupts the larval body and settles as a juvenile (*Loxosomella vivipara*). (F) The larva settles with the area just above the prototroch and the gut rotates 180° with the mouth in front; the atrium reopens in the juvenile (*Pedicellina cernua*).

larvae apparently spend a considerable period feeding in the plankton. Most of the loxosomatids and all the colonial species produce larger eggs; the fully differentiated larvae break the egg envelope and begin feeding. After release from the mother, these larvae have a short free period, and some of them do not feed (Figure 15.3A). Settling and metamorphosis shows much variation. Some species of *Loxosomella* settle with a frontal

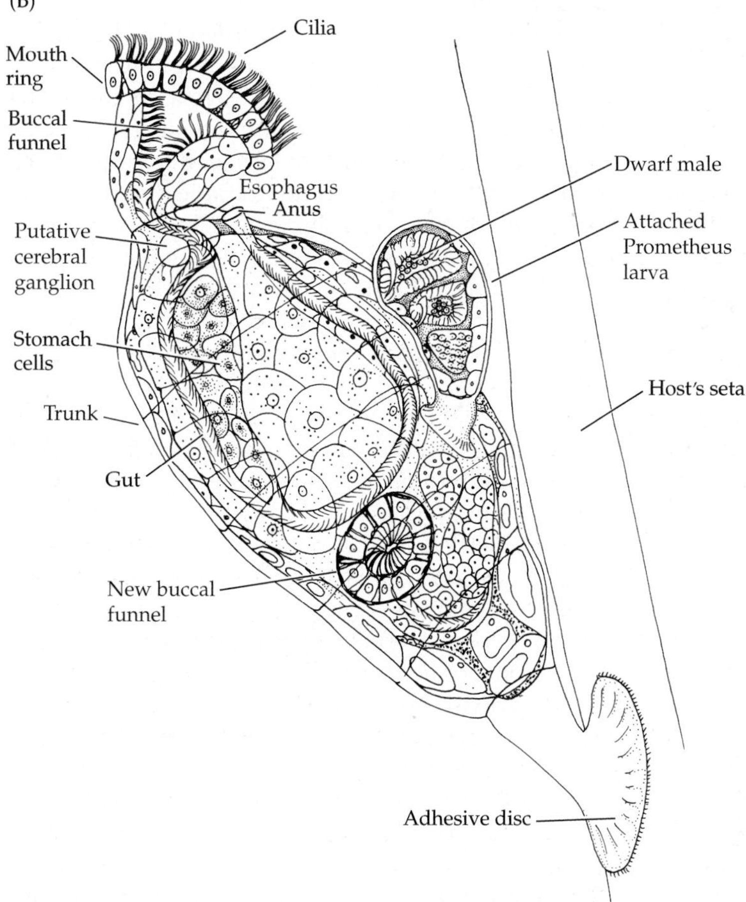

Figure 15.4 *Symbion pandora*. (A) Several individuals of *S. pandora* attached to the mouthparts of a Norwegian lobster (*Nephrops norvegicus*). (B) A feeding individual *Symbion* with a Prometheus larva attached to its trunk.

organ and the larval gut is retained in the adult (Figure 15.3C). Other loxosomatids develop external (Figure 15.3D) or internal (Figure 15.3E) buds from the episphere (the region above the prototroch) of the larva and the larval body disintegrates. The larvae of the colonial species settle by cells just above the retracted prototroch and undergo a remarkable unequal growth of the body mass to rotate the gut so that the ventral, atrial surface points away from the substratum (Figure 15.3F).

Phylum Cycliophora: The Cycliophorans

Symbion pandora is a microscopic marine animal first discovered in the 1960s living commensally on the mouthparts of Norway lobsters (Figure 15.4; Box 15B). Not described until 1995, this apparently acoelomate animal was recognized as a separate phylum, Cycliophora. Since then, a second species, *Symbion americanus*, has been described from the American lobster. And investigations of the mouthparts of the European lobster suggest that a third species might also exist. Additionally, the existence of one or more additional cryptic species on the American lobster has been suggested by molecular studies. Cycliophorans are high-

BOX 15B Characteristics of the Phylum Cycliophora

1. Triploblastic, bilateral, unsegmented, functionally acoelomate
2. Sessile, solitary symbionts of lobsters. Two described species and several undescribed species are known; so far all from Atlantic and Mediterranean lobsters
3. With layered cuticle, adhesive disc for attachment to host, and U-shaped gut
4. Suspension feeding, utilizing dense cilia situated on ring of multiciliate epidermal cells encircling opening of buccal funnel
5. Circulation and gas exchange by diffusion
6. With pair of protonephridia in the chordoid larva
7. With complex life cycle involving sexual and asexual stages that alternate through a succession of events
8. Cleavage apparently holoblastic, though pattern is unique among the Spiralia
9. Maturation of the male involving a marked reduction of internal body volume, mainly by massive nuclei loss
10. With unique Pandora, Prometheus, and chordoid larvae

ly host specific, though recent observations provided evidence for a symbiotic relationship between them and harpacticoid copepods. Although the role of these small crustaceans in the cycliophoran life cycle is unknown, two specimens of copepods carrying cycliophoran life history stages were collected from the mouthparts of a European lobster. Why these enigmatic little creatures only occur on lobsters is not known, and their odd ecology is surpassed only by their bizarre anatomy and life history.

The life cycle of a cycliophoran is complex and involves various sexual and asexual stages alternating through a succession of stages. The most conspicuous in the cycle is an approximately 350 μm sessile individual with a body divided into an anterior buccal funnel, an oval trunk, and a posterior adhesive disc by which the animal attaches to its host's setae (Figure 15.4). A layered cuticle covers the trunk and adhesive disc, the latter apparently composed entirely of cuticular material. Equipped with a U-shaped gut, this is the only cycliophoran stage able to feed. The region between the gut and body wall is packed with large mesenchymal cells and no evidence of a body cavity has been observed. Suspension feeding is enhanced by creating water currents with dense cilia that are situated on a ring of epidermal cells encircling the open end of the buccal funnel. These multiciliate epidermal cells alternate with contractile myoepithelial cells, which form a pair of sphincters involved in the closure of the mouth opening. The U-shaped gut is ciliated along its entire length. The buccal funnel leads to a curved esophagus and then to a stomach consisting of large gland cells penetrated by a narrow lumen. An intestine extends anteriorly to a short rectum and anus, located dorsally near the base of the buccal funnel. A complex sphincter is located proximally to the anus. The area between the gut and other organs is packed with a cellular mesenchyme. A pair of muscles span longitudinally along the buccal funnel, while several longitudinal muscle fibers (six in *S. americanus* and two to eight in *S. pandora*) span from the most distal region of the trunk until about two-thirds of its length.

Circulation and gas exchange are presumably accomplished by simple diffusion in these tiny animals. Two ganglia were described in the feeding individual: one ganglion located at the base of the buccal funnel, and the other partially surrounding the esophagus. However, further observations based on ultrastructural and immunohistochemical investigations have not confirmed the existence of these ganglia. Excretory organs have not been identified in the feeding individual, although a pair of protonephridia is present in one of the larval stages, the chordoid larva, which is described below.

Feeding stage individuals are able to reproduce asexually by a budding process that occurs inside the trunk and generates free-swimming stages, one at a time—these are the **Pandora larvae**, the **Prometheus larva**, and the female. In order to rapidly increase the population density on a single host, the feeding individual generates Pandora larvae (170 μm long), which possess a buccal funnel inside their bodies. Once released, the Pandora larvae settle close to the maternal individual on the lobster host and develop into new feeding stages.

In the sexual part of the life cycle, a free swimming Prometheus larva (120 μm long) settles on the trunk of a feeding individual and develops one to three dwarf males (40 μm long) inside its own body. Females that have also settled on the maternal individual carry a single oocyte, and this is impregnated by a dwarf male in a process that is not yet well understood. A number of morphological structures might be involved in impregnation. For instance, a small (~ 3 μm in diameter), circular structure surrounded by cilia is located medially on the ventral side of the female body. This structure was described as a putative gonopore, though a canal for the conductance of sperm is not evident. In addition, the external morphology of the male is characterized by a "penis" located ventroposteriorly, although its actual function as a true copulatory organ or as a mere anchoring and piercing device is uncertain. A recent investigation at the ultrastructural level revealed the presence of a sperm cell inside the penis of the dwarf male, which provides support for the former view. The female migrates from the body of the

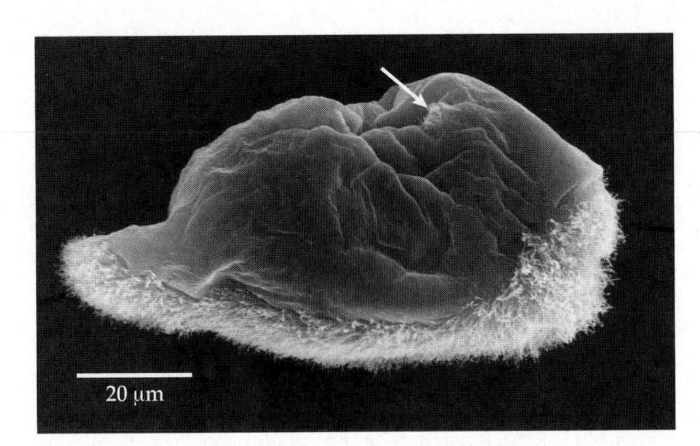

Figure 15.5 ***Symbion pandora*, chordoid larva.** Note the dense ciliated fields covering the ventral region of the body, from posterior (*left*) to anterior (*right*). The arrow points at the dorsal ciliated organ, which appears to have a sensory role.

maternal cycliophoran individual to the lobster host, where it settles into a sheltered area of the mouthparts, encysts, and the embryo develops into the so-called **chordoid larva**.

The chordoid larva (Figure 15.5) hatches from the cyst and settles on a new host, where it eventually develops into a new feeding stage. This is probably the avenue by which dispersal to new host individuals is achieved. The anatomy of the chordoid larva was first described by Funch (1996), who suggested that it is a modified trochophore larva, homologous to those of many spiralians (e.g., annelids, molluscs, sipunculans, echiurans). However, the overall neuroanatomy of this larval stage (below) much more closely resembles the condition of adult rather than larval stages of spiralian taxa. The chordoid larva is also characterized by a ventral chordoid organ that spans along the entire length of the body. The function of the chordoid organ is not known, although it might be supportive and locomotory; it is mainly muscular.

Knowledge of cycliophoran development is quite limited. Details on early embryogenesis derive from observations on settled females bearing an internal embryo. The eight-cell stage embryo is composed of a group of four macromeres and four micromeres. Cleavage appears to be holoblastic, though the arrangement of the blastomeres does not reveal any clear pattern similar to that seen in other spiralian taxa. As for asexual development, a recent study using serial block face-scanning electron microscopy provided new insights on the development of the dwarf male inside the attached Prometheus larva. During the maturation of a young male into a mature dwarf male, the body undergoes a reduction of about one third in the internal body volume, mainly by extensive loss of nuclei in the majority of its somatic cells (but especially in cells of the muscles and epidermis). Organs seen in mature dwarf males—muscles, brain, testis, glands, etc.—are already formed in the young male. The young male stage possesses about 200 nucleated cells, whereas the mature male stage comprises only around 50 nucleated cells; muscle and epidermal cells of the mature male lack nuclei. Thus, contrary to typical development of metazoans, the cycliophoran dwarf male does not grow after organogenesis is concluded.

The external morphology of the Prometheus larva, the Pandora larva, and the female is similar in many respects. For example, an anteroventral ciliated field and a posterior ciliated tuft, as well as four bundles of long cilia—which compose the sensory organ, or sensilla—are present in all these life cycle stages. In addition, all of these free-swimming stages possess a well-differentiated cuticle characterized by a polygonal sculptured surface. The external morphology of the chordoid larva differs from these stages in that it possesses anterior, ventral, and posterior ciliated fields, as well as a paired dorsal organ that appears to have a sensory role (Figure 15.5). The dwarf male is characterized by dense ventral and frontal ciliated fields, and sensilla situated laterally and frontally. The myoanatomy and nervous system of all cycliophoran free-swimming stages has been investigated by transmission electron and confocal laser scanning microscopy. In general, the musculature of the free stages is very complex and includes longitudinal muscles that span the body dorsally and ventrally, as well as dorsoventral muscles. All free-swimming stages possess a dorsal brain composed of a pair of lateral clusters of perikarya interconnected by a commissural neuropil. Moreover, the chordoid larva has four distinct ventral longitudinal neurites, while all the other free-swimming stages have only two. The typical spiralian apical organ is absent in all cycliophoran larval stages.

Selected References

Entoprocta

Franke, M. 1993. Ultrastructure of the protonephridia of *Loxosomella fauveli, Barentsia matsushimana* and *Pedicellina cernua*. Implications for the protonephridia in the ground pattern of the Entoprocta (Kamptozoa). Microfauna Mar. 8: 7–38.

Hausdorf, B. and 8 others. 2007. Spiralian phylogenomics supports the resurrection of Bryozoa comprising Ectoprocta and Entoprocta. Mol. Biol. Evol. 24(12): 2723–2729.

Nielsen, C. 1971. Entoproct life cycles and the entoproct/ectoproct relationship. Ophelia 9(2): 209–341.

Nielsen, C. 1989. Entoprocts. Synopses British Fauna, N.S. 41: 1–131. Brill, Leiden.

Nielsen, C. 2002. The phylogenetic position of Entoprocta, Ectoprocta, Phoronida and Brachiopoda. Integrative and Comparative Biology 42(3): 685–691.

Nielsen, C. 2010. A review of the taxa of solitary entoprocts (Loxosomellatidae). Zootaxa 2395: 45–56. [There is a beautiful figure with many line drawings of weird looking entoprocts, along with a species list and lengthy literature cited. This could be a guide for a future entoproctologist.]

Nielsen, C. and Å Jespersen. 1997. Entoprocta. Pp. 13–43 in F. W. Harrison (ed.), *Microscopic Anatomy of Invertebrates*, Vol. 13. Wiley-Liss, New York.

Nielsen, C. and J. Rostgaard. 1976. Structure and function of an entoproct tentacle with discussion of ciliary feeding types. Ophelia 15: 115–140.

Riisgård, H. U., C. Nielsen and P. S. Larsen. 2000. Downstream collecting in ciliary suspension feeders: the catch-up principle. Mar. Ecol. Prog. Ser. 207: 33–51.

Rundell, R. J. and B. S. Leander. 2012. Description and phylogenetic position of the first sand-dwelling entoproct from the western coast of North America: *Loxosomella vancouverensis* sp. nov. Marine Biology Research 8: 284–291.

Todd, J. A. and P. D. Taylor. 1992. The first fossil entoproct. Naturwiss. 79: 311–314.

Wasson, K. 1997a. Sexual modes in the colonial kamptozoan genus *Barentsia*. Biol. Bull. 193: 163–170.

Wasson, K. 1997b. Systematic revision of colonial kamptozoans (entoprocts) of the Pacific coast of North America. Zool. J. Linn. Soc. 121: 1–63.

Cycliophora

Baker, J. M. and G. Giribet. 2007. A molecular phylogenetic approach to the phylum Cycliophora provides further evidence for cryptic speciation in *Symbion americanus*. Zool. Scripta 36: 353–359.

Funch, P. 1996. The chordoid larva of *Symbion pandora* (Cycliophora) is a modified trochophore. J. Morphol. 230: 231–263.

Funch, P. and R. M. Kristensen. 1995. Cycliophora is a new phylum with affinities to Entoprocta and Ectoprocta. Nature 378: 661–662.

Funch, P. and R. M. Kristensen. 1997. Cycliophora. Pp. 409–474, in F. W. Harrison and E. E. Ruppert (eds.), *Microscopic Anatomy of Invertebrates, Vol. 13, Lophophorates, Entoprocta, and Cycliophora*. Wiley-Liss, New York.

Neves, R. C., R. M. Kristensen and A. Wanninger. 2009. Three-dimensional reconstruction of the musculature of various life cycle stages of the cycliophoran *Symbion americanus*. J. Morphol. 270: 257–270.

Neves, R. C., K. J. K. Sørensen, R. M. Kristensen and A. Wanninger. 2009. Cycliophoran dwarf males break the rule: high complexity with low cell-numbers. Biol. Bull. 217: 2–5.

Neves, R. C., M. R. Cunha, P. Funch, A. Wanninger and R. M. Kristensen. 2010. External morphology of the cycliophoran dwarf male: a comparative study of *Symbion pandora* and *S. americanus*. Helgoland Mar. Res. 64: 257–262.

Neves, R. C., M. R. Cunha, R. M. Kristensen and A. Wanninger. 2010. Comparative myoanatomy of cycliophoran life cycle stages. J. Morphol. 271: 596–611.

Neves, R. C., R. M. Kristensen and A. Wanninger. 2010. Serotonin immunoreactivity in the nervous system of the Pandora larva, the Prometheus larva, and the dwarf male of *Symbion americanus* (Cycliophora). Zool. Anzeiger 249: 1–12.

Neves, R. C., R. M. Kristensen and P. Funch. 2012. Ultrastructure and morphology of the cycliophoran female. J. Morphol. 273: 850–869.

Neves, R. C., C. Bailly and H. Reichert. 2014. Are copepods secondary hosts of Cycliophora? Org. Divers. Evol. 14: 363–367.

Neves, R. C. and H. Reichert. 2015. Microanatomy and development of the dwarf male of *Symbion pandora* (Phylum Cycliophora): new insights from ultrastructural investigation based on serial section electron. PLoS ONE. doi: 10.1371/journal.pone.0122364

Obst, M. and P. Funch. 2003. Dwarf Male of *Symbion pandora* (Cycliophora). J. Morphol. 255: 261–278.

Obst, M., P. Funch and G. Giribet. 2005. Hidden diversity and host specificity in cycliophorans: a phylogeographic analysis along the North Atlantic and Mediterranean Sea. Mol. Ecol. 14: 4427–4440.

Obst, M., P. Funch, R. M. Kristensen. 2006. A new species of Cycliophora from the mouthparts of the American lobster, *Homarus americanus* (Nephropidae, Decapoda). Org. Divers. Evol. 6: 83–97.

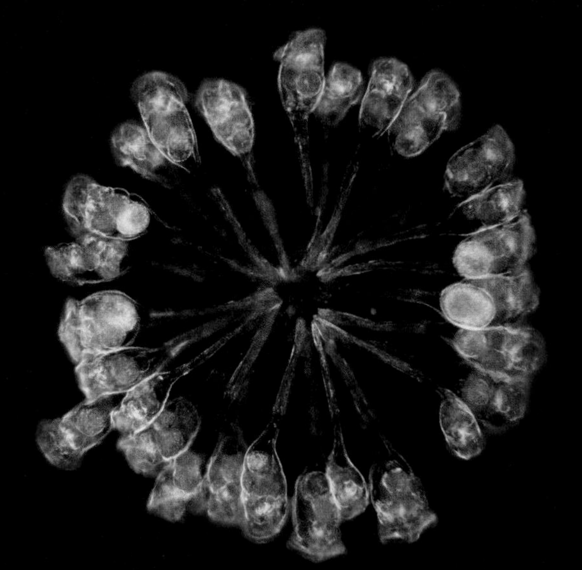

CHAPTER 16

The Gnathifera

Phyla Gnathostomulida, Rotifera (including Acanthocephala), and Micrognathozoa

The clade Gnathifera includes three phyla: Gnathostomulida, Micrognathozoa, and Rotifera, with the latter including the parasitic acanthocephalan worms (formerly a separate phylum). The name is derived from the Greek *gnathos*, "jaw" and the Latin *fera*, "to bear," and refers to the presence of pharyngeal hard parts, i.e., jaws that are either present or secondarily lost in all gnathiferan taxa. Despite their small size, gnathiferans show a remarkable complexity of anatomy, especially in their jaw structures (e.g., the mastax and trophi) and the organization of their muscular and nervous systems.

Until the mid-1990s, the Gnathostomulida and Rotifera were pooled together with other microscopic taxa in questionable groups such as "Aschelminthes" or "Nemathelminthes"—catchall groupings that were more or less solely characterized by hosting microscopic taxa with uncertain phylogenetic positions. At that time, the Acanthocephala was treated as a distinct phylum, but despite their macroscopic size and endoparasitic biology, they were already considered to be closely related to the Rotifera, based on ultrastructural similarities in their integuments. During the 1990s, a number of researchers began to investigate the phylogenetic positions of the aschelminth phyla. In 1995, two important papers (by W. H. Ahlrichs, R. M. Rieger and S. Tyler) suggested a sister-group relationship between Gnathostomulida and Rotifera, based on a proposed homology between the jaws in the two groups. The homology was supported by ultrastructural data from transmission electron microscopy, which demonstrated that jaws in both taxa are made up by rodlike elements that in cross section appear as translucent areas with a central, electron-dense core. Ahlrichs (1995) referred to this group as Gnathifera, and he further suggested that the

Classification of The Animal Kingdom (Metazoa)

Non-Bilateria*
(a.k.a. the diploblasts)
- PHYLUM PORIFERA
- PHYLUM PLACOZOA
- PHYLUM CNIDARIA
- PHYLUM CTENOPHORA

Bilateria
(a.k.a. the triploblasts)
- PHYLUM XENACOELOMORPHA

Protostomia
- PHYLUM CHAETOGNATHA

SPIRALIA
- PHYLUM PLATYHELMINTHES
- PHYLUM GASTROTRICHA
- PHYLUM RHOMBOZOA
- PHYLUM ORTHONECTIDA
- PHYLUM NEMERTEA
- PHYLUM MOLLUSCA
- PHYLUM ANNELIDA
- PHYLUM ENTOPROCTA
- PHYLUM CYCLIOPHORA

Gnathifera
- **PHYLUM GNATHOSTOMULIDA**
- **PHYLUM MICROGNATHOZOA**
- **PHYLUM ROTIFERA**

Lophophorata
- PHYLUM PHORONIDA
- PHYLUM BRYOZOA
- PHYLUM BRACHIOPODA

ECDYSOZOA

Nematoida
- PHYLUM NEMATODA
- PHYLUM NEMATOMORPHA

Scalidophora
- PHYLUM KINORHYNCHA
- PHYLUM PRIAPULA
- PHYLUM LORICIFERA

Panarthropoda
- PHYLUM TARDIGRADA
- PHYLUM ONYCHOPHORA
- PHYLUM ARTHROPODA
  - SUBPHYLUM CRUSTACEA*
  - SUBPHYLUM HEXAPODA
  - SUBPHYLUM MYRIAPODA
  - SUBPHYLUM CHELICERATA

Deuterostomia
- PHYLUM ECHINODERMATA
- PHYLUM HEMICHORDATA
- PHYLUM CHORDATA

*Paraphyletic group

Martin V. Sørensen wrote the introduction and revised sections on phyla Gnathostomulida and Rotifera. Katrine Worsaae and Reinhardt Møbjerg Kristensen wrote the section on phylum Micrognathozoa.

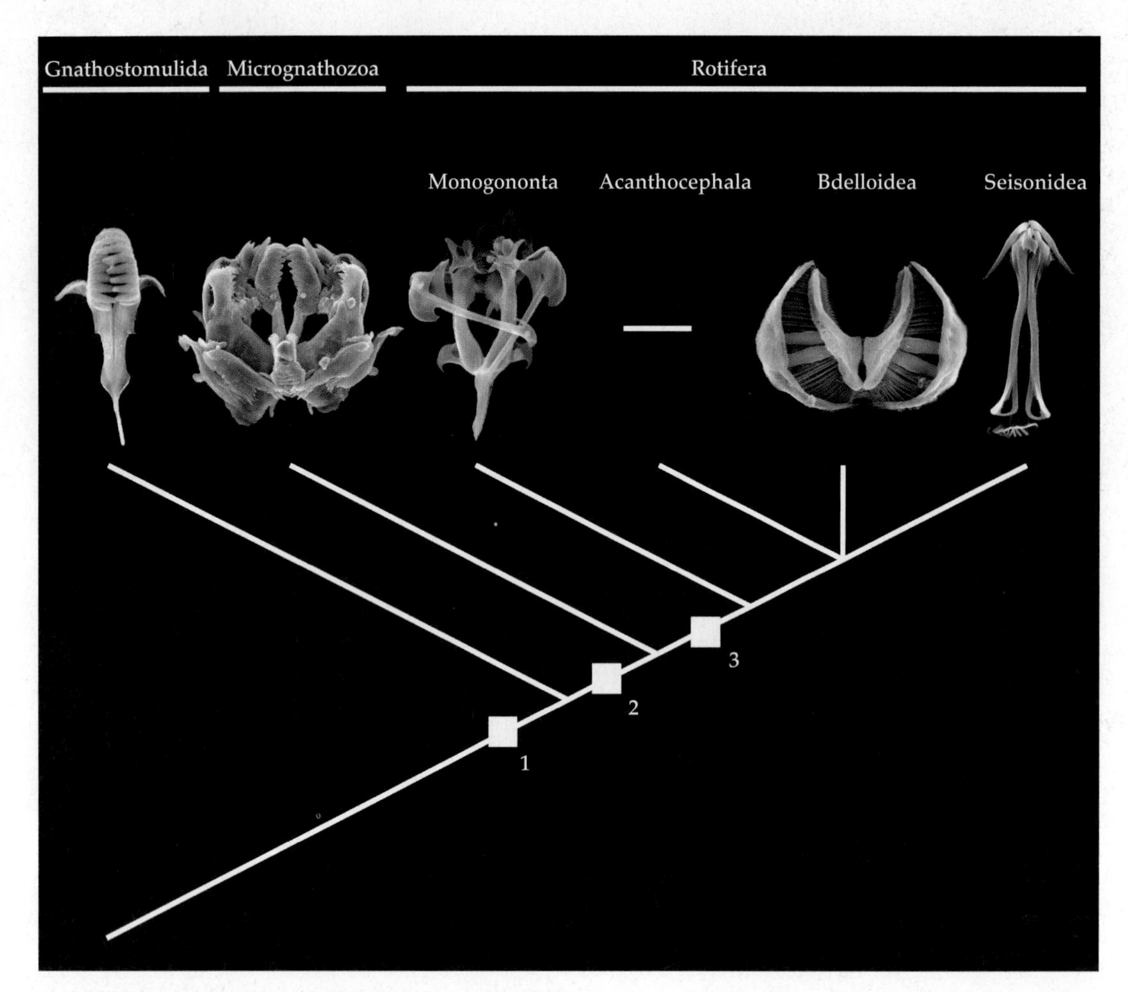

Figure 16.1 Phylogenetic tree showing jaw structure and the relationships within Gnathifera. Synapomorphies for the major clades are marked on the tree: (1) pharyngeal hard parts/jaws composed of rods made of a translucent material with an electron-dense core, (2) cellular epidermis with intracellular protein lamina, (3) syncytial epidermis.

endoparasitic acanthocephalans were actually no more than highly modified rotifers.

At the same time Gnathifera was taking its shape, another new animal was discovered in mosses from a cold spring in Greenland. It was a tiny, microscopic invertebrate that in some respects resembled a rotifer, and in other ways a gnathostomulid, but also possessed several characteristics that were not found in any other groups. The animal had jaws that were even more complex and numerous in elements than those found in the two other gnathiferan phyla. Transmission electron microscopy showed that the ultrastructure of the jaws was nearly identical with that of Rotifera and Gnathostomulida. Six years after its discovery, R. M. Kristensen and P. Funch (2000) named the animal *Limnognathia maerski*, and assigned it to a new animal group Micrognathozoa, which four years later was recognized as a third gnathiferan phylum.

Morphologically, Gnathifera appears to be a well-supported monophyletic group that is characterized by the presence of homologous pharyngeal hard parts, and at least one phylogenomic study has also found support for the clade Gnathifera. Morphology-based phylogenies indicate that Gnathostomulida branches off as the most basal gnathiferan, whereas Micrognathozoa and Rotifera (including Acanthocephala) appear as sister-groups supported by similarities in the ultrastructure of their integuments (Figure 16.1). Both free-living Rotifera and Acanthocephala (see below) have a syncytial epidermis where the outer cuticle has been replaced by an intracellular protein lamina in the epidermal cells. Micrognathozoa has a regular non-syncytial epidermis, but a similar intracellular protein lamina is found in its dorsal epidermal plates, and the presence of this lamina is considered synapomorphic for Micrognathozoa and Rotifera-Acanthocephala.

Several molecular studies initially suggested a close relationship between Gnathostomulida and Rotifera–Acanthocephala, but then in 2015, using larger datasets, the Micrognathozoa began to appear as a sister group to the Rotifera–Acanthocephala.

Based on ultrastructural similarities in their integuments, the microscopic rotifers and the macroscopic acanthocephalans have long been considered as likely sister taxa. However, more recently an increasing

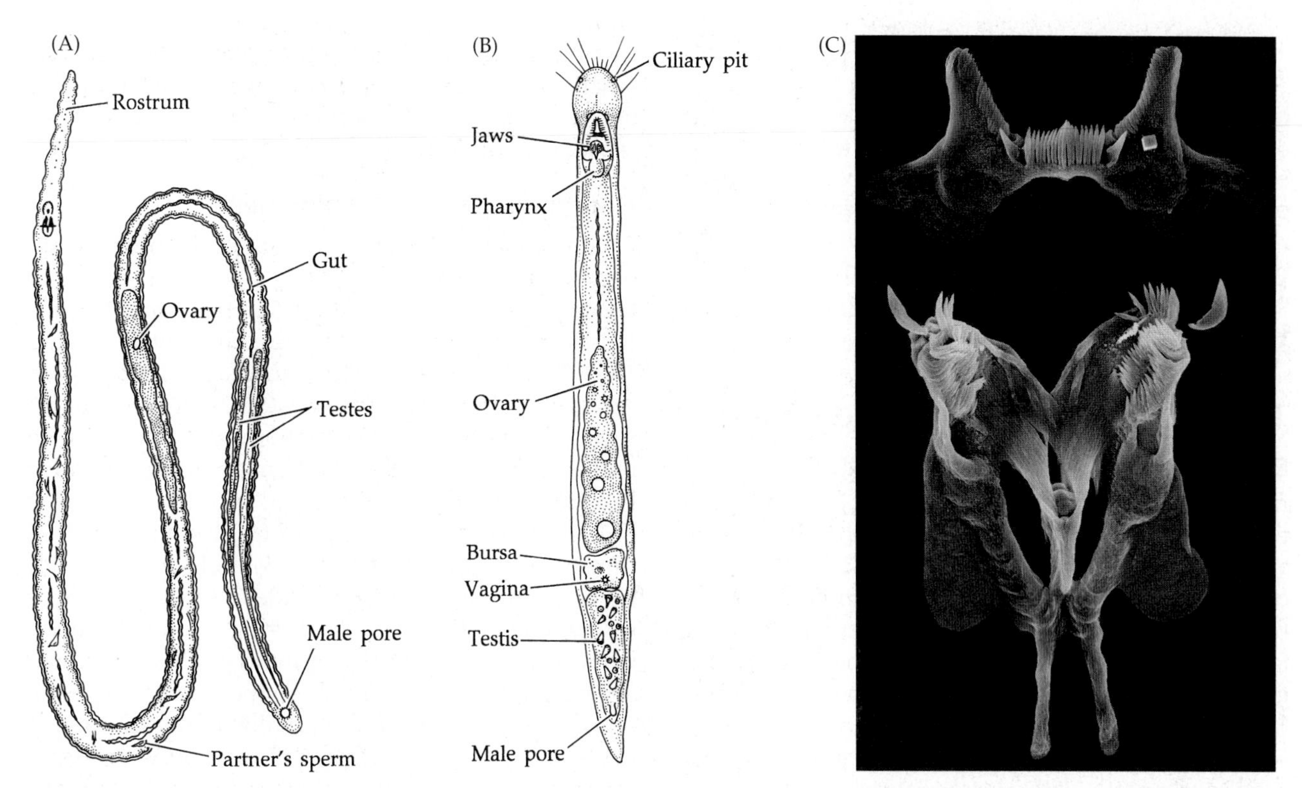

Figure 16.2 Representative gnathostomulids. (A) *Haplognathia simplex*. (B) *Austrognatharia kirsteueri*. (C) Basal plate and jaws of *Gnathostomula armata*.

amount of evidence from molecular studies have unambiguously shown that Acanthocephala is not the sister of Rotifera, but actually evolved from within the Rotifera—presumably as a clade that became obligate endoparasitic, and subsequently went through a series of dramatic morphological modifications and changes. Modern acanthocephalans are so modified and adapted to their endoparasitic lifestyle that it is hard to find comparative morphological characters that would place them inside Rotifera, but the molecular support is strong and includes studies based on selected target loci and expressed sequence tags, as well as the complete mitochondrial genomes.

Phylum Gnathostomulida: The Gnathostomulids

The phylum Gnathostomulida (Greek, *gnathos*, "jaw"; *stoma*, "mouth") includes about 100 species of minute vermiform hermaphroditic animals (Figure 16.2). These meiofaunal creatures were first described by Peter Ax in 1956 as turbellarians, but given their own phylum rank by Rupert Riedl in 1969. Gnathostomulids are found worldwide, interstitially in marine sands mixed with detritus, from the intertidal zone to depths of hundreds of meters. The tiny, elongate body (less than 2 mm long) is usually divisible into head, trunk, and in some species a narrow tail region. Distinguishing features of this phylum include a unique jawed pharyngeal apparatus and monociliated epidermal cells (Box 16A). The currently described 100 species and 26 genera are divided between two orders.

GNATHOSTOMULID CLASSIFICATION

ORDER FILOSPERMOIDEA Body usually very elongate, with slender rostrum; jaws relatively simple; male parts without injectory penis; sperm filiform (with one 9+2 flagellum);

BOX 16A Characteristics of the Phylum Gnathostomulida

1. Triploblastic, bilateral, unsegmented, vermiform acoelomates
2. Epidermis monolayered; all epithelial cells are monociliated
3. Gut incomplete (anus rudimentary, vestigial, or absent)
4. Pharynx with unique, complex jaw apparatus
5. Without circulatory system or special gas exchange structures
6. Excretion through protonephridia with monociliated terminal cells
7. Hermaphroditic
8. Cleavage spiral and development direct
9. Inhabit marine, interstitial environments

female parts without vagina and bursa. (3 genera: *Cosmognathia*, *Haplognathia*, and *Pterognathia*)

ORDER BURSOVAGINOIDEA Body usually not extremely elongate relative to width; head with shorter rostrum and often a constriction in the neck area; jaws complex; male parts with penis, with or without a stylet; sperm cells aflagellate, either dwarf cells or giant conuli; female parts with bursa and usually a vagina. (23 genera, including *Austrognatharia*, *Gnathostomula*, and *Onychognathia*)

The Gnathostomulid Body Plan

Body Wall, Support, and Locomotion

Each outer epithelial cell bears a single cilium by which the animal moves in a gliding motion. Movement is aided by body contortions produced by the contraction of thin strands of subepidermal (cross-striated) muscle fibers. These actions, plus reversible ciliary beating, facilitate twisting, turning, and crawling among sand grains, and allow limited swimming in some species. Mucous gland cells occur in the epidermis of at least some species. The body is supported by its more or less solid construction, with a loose mesenchyme filling the area between the internal organs.

Nutrition, Circulation, Excretion, and Gas Exchange

The mouth is located on the ventral surface at the "head–trunk" junction and leads inward to a complex muscular pharynx armed with pincerlike jaws and in some species an unpaired anterior basal plate (Figure 16.2). Curiously, the two known species of the genus *Agnathiella* have no jaws at all. Gnathostomulids ingest bacteria and fungi by snapping actions made by the jaws or scraping with the basal plate. The pharynx connects with a simple, elongate, saclike gut. A permanent, functional anus is not present, but in a few gnathostomulids a tissue connection between the posterior end of the gut and the overlying epidermis has been observed. This enigmatic feature has been variously interpreted as either a temporary anal connection to the exterior, as the remnant of an anus that has been evolutionarily lost, or as an incipient anus that has yet to fully develop.

These animals depend largely on diffusion for circulation and gas exchange. The excretory system is composed of serially arranged protonephridia that stretch from the pharyngeal region to the terminal end of the body. Like the epithelial cells, the protonephridial terminal cells are monociliated.

Nervous System

The nervous system is intimately associated with the epidermis and as yet is incompletely described. Various sensory organs, such as sensory ciliary pits and stiff sensoria formed by groups of joined cilia from monociliated cells are concentrated in the head region. Gnathostomulid specialists have attached a formidable array of names to these structures, which are of major taxonomic significance.

Reproduction and Development

Gnathostomulids are hermaphrodites. The male reproductive system includes one or two testes generally located in the posterior part of the trunk and tail; the female system consists of a single large ovary (Figure 16.2). Members of the order Bursovaginoidea possess a vaginal orifice and a sperm-storage bursa, both associated with the female gonopore, and a penis in the male system; members of the order Filospermoidea lack these structures.

Mating has been only superficially studied in gnathostomulids. Although the method of sperm transfer is not certain, suggestions include filiform sperm of filospermoid gnathostomulids boring through the body wall. Among some bursovaginoid gnathostomulids, sperm is transferred directly to the mating partner's bursa by hypodermic impregnation of the sclerotized penis stylet. In any case, these animals appear to be gregarious, to rely on internal fertilization, and to deposit zygotes singly in their habitat. Cleavage is reported as spiral and development is direct, but details on embryonic and juvenile development are generally scarce.

Phylum Rotifera: The Free-Living Rotifers

The phylum Rotifera (Latin *rota*, "wheel"; *fera*, "to bear") includes more than 2,000 described species of microscopic (about 100 to 1000 μm long), generally free-living animals. Furthermore, the parasitic, macroscopic acanthocephalan worms actually represent a rotifer in-group as well, but due to their considerable differences in biology and morphology, they will be discussed separately. Thus, in this section the name Rotifera refers to the microscopic, free-living rotifers only. The name "Syndermata" was proposed some time ago for a clade of Rotifera + Acanthocephala, but with the current understanding that these are not sister groups (the latter arose as a clade from within the former), that name is no longer useful.

Rotifers were discovered by the early microscopists, such as Antony van Leeuwenhoek in the late seventeenth century; at that time they were lumped with the protists as "animalcules" (mainly because of their small size). Besides the 2,050 or so known, morphologically recognizable species, complexes of cryptic speciation have been demonstrated for several morphospecies. For example, the species *Brachionus plicatilis* has been subject to intensive studies, and at least 22 cryptic species have been identified within this species group.

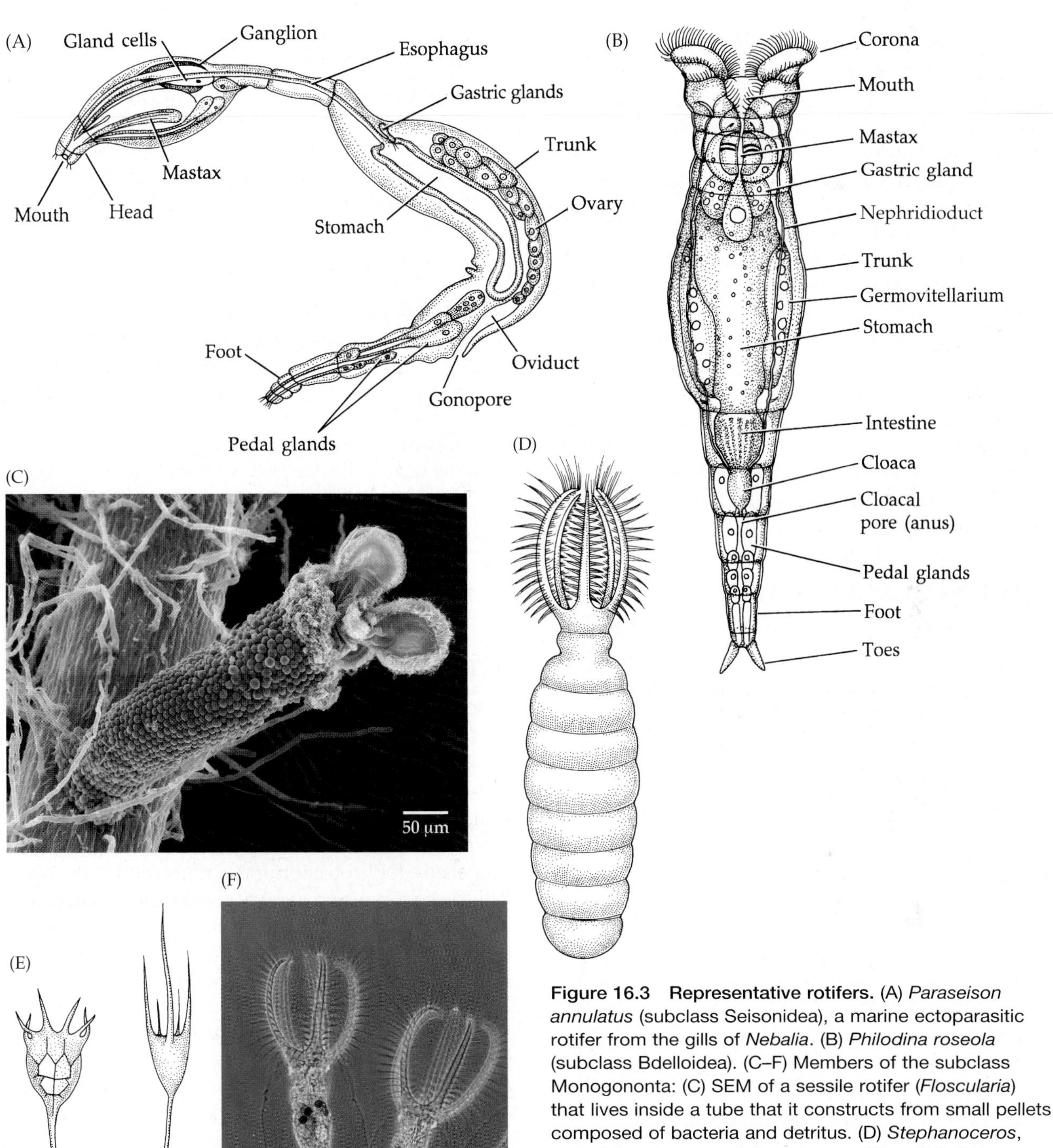

Figure 16.3 Representative rotifers. (A) *Paraseison annulatus* (subclass Seisonidea), a marine ectoparasitic rotifer from the gills of *Nebalia*. (B) *Philodina roseola* (subclass Bdelloidea). (C–F) Members of the subclass Monogononta: (C) SEM of a sessile rotifer (*Floscularia*) that lives inside a tube that it constructs from small pellets composed of bacteria and detritus. (D) *Stephanoceros*, one of the strange collothecacean rotifers with the corona modified as a trap. (E) The loricae of two loricate rotifers. (F) Live specimens of *Stephanoceros*.

Despite their small size, rotifers are actually quite complex and display a variety of body forms (Figure 16.3). Most are solitary, but some sessile forms are colonial, a few of which secrete gelatinous casings into which the individuals can retract (e.g., chapter opener photo of *Conochilus*). They are most common in fresh water, but many marine species are also known, and others live in damp soil or in the water film on mosses. They often comprise an important component of the plankton of fresh and brackish waters.

The body comprises three general regions—the head, trunk, and foot. The head bears a ciliary organ called the **corona**. When active, the coronal cilia often give the impression of a pair of rotating wheels, hence the derivation of the phylum name; in fact, rotifers were historically called "wheel animalcules." Members of this phylum are further characterized by being blastocoelomate, having an integument without an outer cuticle, but instead with a supportive intracellular protein lamina. They have a complete gut (usually), protonephridia, show a tendency to eutely, and often have syncytial tissues or organs (Box 16B). The pharynx is modified as a **mastax** comprising sets of internal jaws called **trophi**. The morphology of the trophi is of great systematic importance and often the main character to identify species and genera.

A highly surprising discovery about rotifers was made in 2008, when it was found that bdelloid rotifers have incorporated large numbers of genes from diverse foreign sources into their genomes, including bacteria, fungi, and plants. These foreign genes have accumulated mainly in the telomeric regions at the ends of chromosomes, and at least most of them seem to retain their functional integrity.

ROTIFER CLASSIFICATION

CLASS HEMIROTATORIA Endoparasites, ectoparasites, or free-living; this group is recognized only by molecular data.

SUBCLASS ACANTHOCEPHALA Macroscopic endoparasites; see chapter section below.

SUBCLASS BDELLOIDEA (Figure 16.3B) Found in freshwater, moist soils, and foliage (also marine, and terrestrial); corona typically well developed; trophi ramate (grinding). (20 genera, e.g., *Adineta, Embata, Habrotrocha, Philodina, Rotaria*)

SUBCLASS SEISONIDEA (Figure 16.3A) Epizoic on the marine leptostracan crustacean *Nebalia*; corona reduced to bristles; trophi fulcrate (piercing); males fully developed and considered to have diploid chromosome numbers; sexual females produce only mictic ova. (2 genera: *Paraseison* and *Seison*)

CLASS EUROTATORIA

SUBCLASS MONOGONONTA (Figure 16.3C–F) Predominately freshwater, some are marine; swimmers, creepers, or sessile; corona and trophi variable; males typically short lived, haploid, and reduced in size and complexity; sexual reproduction probably occurs at some point in the life history of all species; mictic and amictic ova produced in many species; single germovitellarium. (121 genera, e.g., *Asplanchna, Brachionus, Collotheca, Dicranophorus, Encentrum, Epiphanes, Euchlanis, Floscularia, Lecane, Notommata, Proales, Synchaeta, Testudinella*)

BOX 16B Characteristics of the Phylum Rotifera

1. Triploblastic, bilateral, unsegmented blastocoelomates
2. Gut complete and regionally specialized
3. Pharynx modified as a mastax, containing jawlike elements called "trophi"
4. Anterior end bears variable ciliated fields as a corona
5. Posterior end often bears toes and adhesive glands
6. Epidermis syncytial, with fixed number of nuclei; secretes extracellular glycocalyx and intracellular skeletal lamina (the latter forming a lorica in some species)
7. With protonephridia, but no special circulatory or gas exchange structures
8. With unique retrocerebral organ
9. Males generally reduced or absent; parthenogenesis common
10. With modified spiral cleavage
11. Inhabit marine, freshwater, or semiterrestrial environments; sessile or free-swimming

The Rotifer Body Plan

Body Wall, General External Anatomy, and Details of the Corona

Most rotifers possess a soft, gelatinous glycocalyx outside their epidermis, but unlike many other invertebrates, they have no external cuticle. Instead they have an intracellular protein lamina located inside the epidermis, for protection and stabilization of the body. This protein lamina may vary considerably in thickness and flexibility among the genera and families. Species with a very thin protein lamina are called "illoricate rotifers," and they often appear as very flexible and hyaline animals that contract completely when disturbed. In other species, the intracellular protein lamina is much thicker and forms a body-armor, called a **lorica**, and these species are referred to as "loricate rotifers." Another special condition of the rotifer epidermis regards the absence of walls between the epidermal cells, meaning that the epidermis is a syncytium with about 900 to 1,000 nuclei.

The body surface of many illoricate rotifers is annulated, allowing flexibility. The surface of loricate species often bears spines, tubercles, or other sculpturing (Figure 16.3E). Many rotifers bear single dorsal and paired lateral sensory antennae arising from various regions of the body. A foot is not present in all species, but when present it is often elongate, with cuticular annuli that permit a telescoping action. The distal portion of the foot often bears spines, or a pair of "toes" through which the ducts from pedal glands pass. The secretion from the pedal glands enables the rotifer to attach temporarily to the substratum. The foot is absent

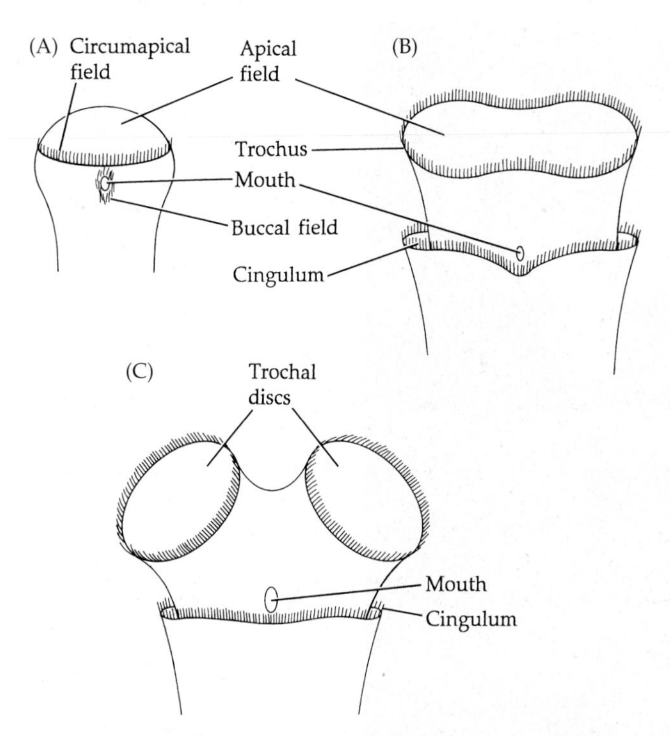

Figure 16.4 Modifications of the corona among selected rotifer types. (A) The presumed plesiomorphic condition has buccal and circumapical fields. (B) The circumapical field is separated into trochus and cingulum. The trochus is lobed, like that of *Floscularia*. (C) The trochus is separated into two trochal discs, as found in many bdelloid rotifers.

from some swimming forms (e.g., *Asplanchna*) and is modified for permanent attachment in sessile types (e.g., *Floscularia*).

The **corona** is the most characteristic external feature of rotifers. Its morphology varies greatly and, in some groups, the corona is an important taxonomic character. The presumed primitive condition is shown in Figure 16.4A. A well-developed patch of cilia surrounds the anteroventral mouth. This patch is the buccal field, or circumoral field, and it extends dorsally around the head as a ciliary ring called the circumapical field. The extreme anterior part of the head bordered by this ciliary ring is the apical field. The corona has evolved to a variety of modified forms in different rotifer taxa. In some species, the buccal field is quite reduced, and the circumapical field is separated into two ciliary rings, one slightly anterior to the other (Figure 16.4B). The anteriormost ring is called the trochus, the other the cingulum. In many bdelloid rotifers the trochus is a pair of well-defined anterolateral rings of cilia called trochal discs (Figure 16.4C), which may be retracted or extended for locomotion and feeding. It is the metachronal ciliary waves along these trochal discs that impart the impression of rotating wheels.

Many organs and tissues of rotifers display eutely: cell or nuclear number constancy. This condition is established during development, and there are no mitotic cell divisions in the body following ontogeny.

Body Cavity, Support, and Locomotion

Beneath the epidermis are various circular and longitudinal muscle bands (Figure 16.5); there are no sheets or layers of body wall muscles. The internal organs lie within a typically spacious, fluid-filled blastocoelom.

In the absence of a thick, muscular body wall, body support and shape are maintained by the intraepidermal skeletal lamina and the hydrostatic skeleton provided by the body cavity. In loricate species the integument is only flexible enough to allow slight changes in shape, so increases in hydrostatic pressure within the body cavity can be used to protrude body parts (e.g., foot, corona). These parts are protracted and retracted by various muscles (Figure 16.5), each consisting of only one or two cells.

Although a few rotifers are sessile, most are motile and quite active, moving about by swimming or creeping like an inchworm. Some are exclusively either swimmers or crawlers, but many are capable of both methods of locomotion. Swimming is accomplished by beating the coronal cilia, forcing water posteriorly along the body, and driving the animal forward, sometimes in a spiral path. When creeping, a rotifer attaches its foot with secretions from the pedal glands, then elongates its body and extends forward. It attaches the extended anterior end to the substratum,

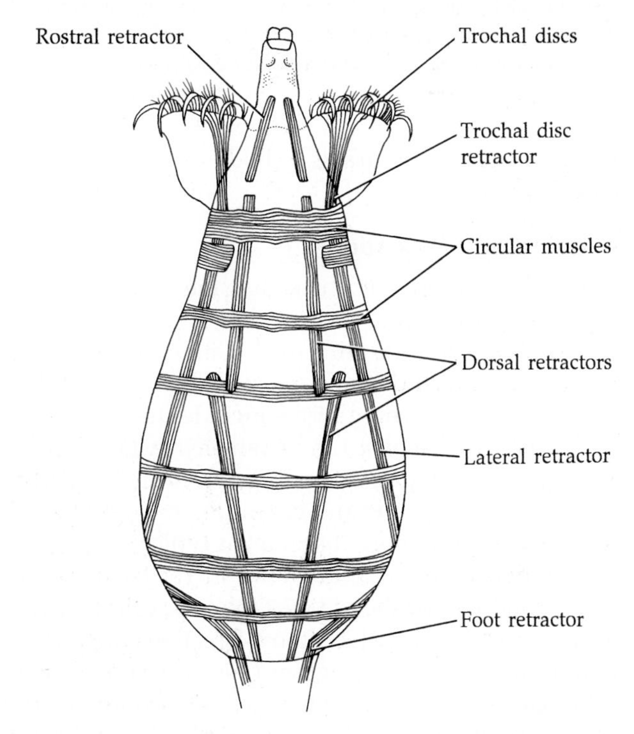

Figure 16.5 Major muscle bands of the bdelloid rotifer, *Rotaria* (dorsal view).

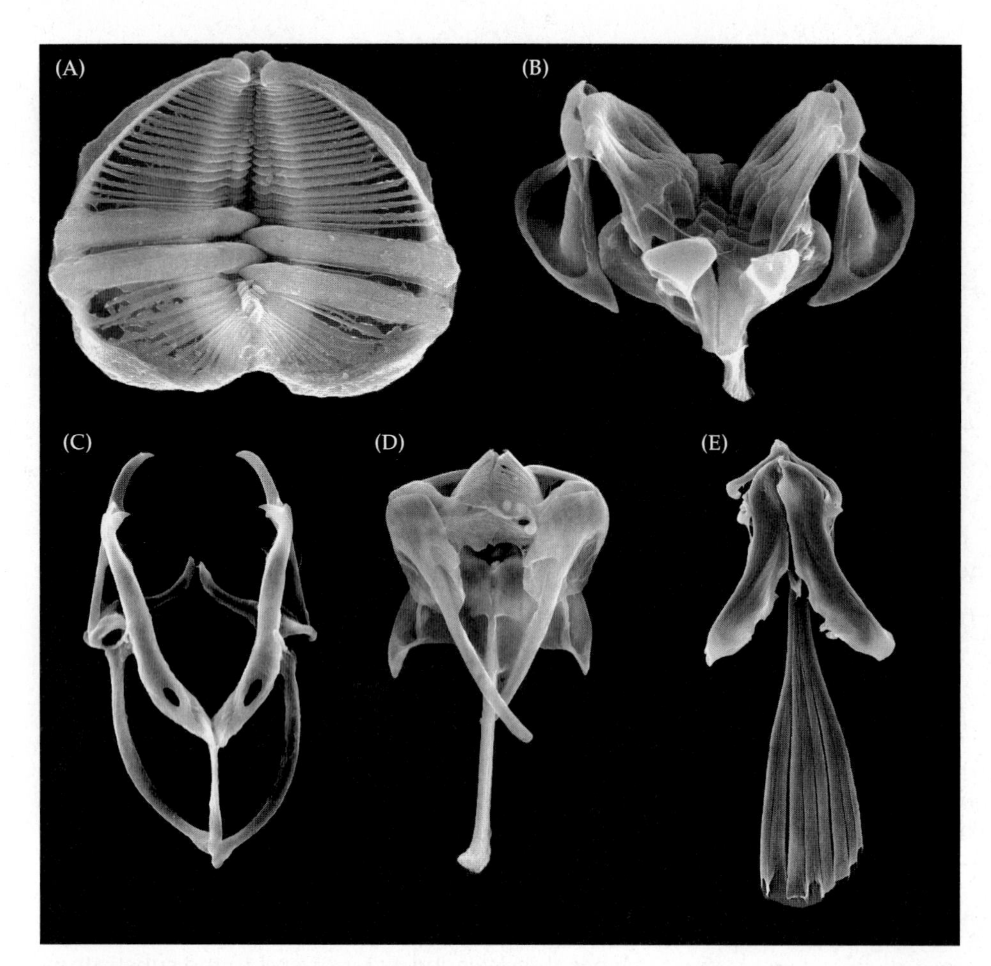

Figure 16.6 SEMs showing different rotifer trophi types. (A) *Dissotrocha aculeata* with the ramate trophus type, found in all bdelloids. (B) *Brachionus calyciflorus* with the malleate trophus type that characterizes several monogonont families. (C) *Encentrum astridae* with forcipate trophi, found in the monogonont family Dicranophoridae. (D) *Resticula nyssa* with its virgate trophi, typical for Notommatidae and several other monogonont families. (E) *Paraseison kisfaludyi*, fulcrate trophi of the ectoparasitic Seisonidea.

releases its foot, and draws its body forward by muscular contraction.

Feeding and Digestion

Rotifers display a variety of feeding methods, depending upon the structure of the corona (Figure 16.4) and the mastax trophi (Figure 16.6). Ciliary suspension feeders have well-developed coronal ciliation and a grinding mastax. These forms include the bdelloids, which have trochal discs and a ramate mastax (Figure 16.6A), and a number of monogonont rotifers, which have separate trochus and cingulum and a malleate mastax (Figure 16.6B). These forms typically feed on organic detritus or minute organisms. The feeding current is produced by the action of the cilia of the trochus (or trochal discs), which beat in a direction opposite to that of the cilia of the cingulum. Particles are drawn into a ciliated food groove that lies between these opposing ciliary bands and are carried to the buccal field and mouth.

Raptorial feeding is common in many species of Monogononta. Coronal ciliation in these rotifers is often reduced or used exclusively for locomotion. Raptorial feeders obtain food by grasping it with protrusible, pincerlike mastax jaws; most possess either a forcipate mastax (non-rotating) (Figure 16.6C) or an incudate mastax (rotating 90–180° during protrusion). Raptorial rotifers feed mainly on small animals but are known to ingest plant material as well. They may ingest their prey whole and subsequently grind it to smaller particles within the mastax, or they may pierce the body of the plant or animal with the tips of the mastax jaws and suck fluid from the prey (Figure 16.6D).

Some monogonont rotifers have adopted a trapping method of predation. In such cases the corona usually bears spines or setae arranged as a funnel-shaped trap (Figure 16.3D,F). The mouth in these trappers is located more or less in the middle of the ring of spines (rather than in the more typical anteroventral position); thus, captured prey is drawn to it by contraction of the trap elements. The mastax in trapping rotifers is often reduced.

A few rotifers have adopted symbiotic lifestyles. As noted in the classification scheme, seisonids live on

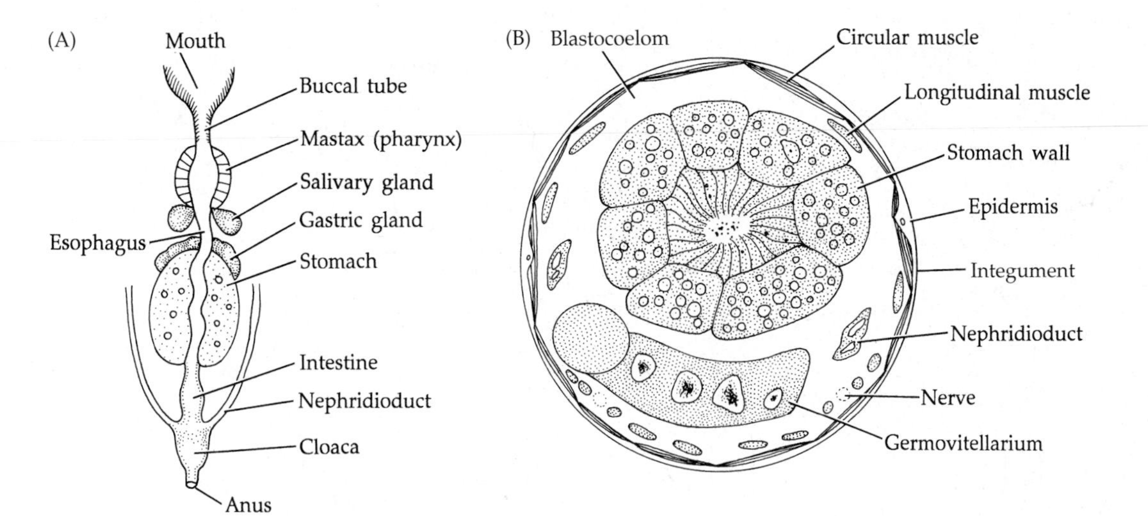

Figure 16.7 (A) Digestive system of a rotifer. (B) Cross section through the trunk.

marine leptostracan crustaceans of the genus *Nebalia*. These rotifers (*Seison* and *Paraseison*) crawl around the base of the legs and gills of their host, feeding on detritus and on the host's brooded eggs. It has been suggested that species of the predatory *Paraseison* may use the anterior tip of the fulcrum of its fulcrate trophi (Figure 16.6E) to pinch the cuticle of its leptostracan host and feed on its haemolymph. Some bdelloids (e.g., *Embata*) also live on the gills of crustaceans, particularly amphipods and decapods. There are isolated examples of endoparasitic rotifers inhabiting hosts such as *Volvox* (a colonial protist), freshwater algae, snail egg cases, and the body cavities of certain annelids and terrestrial slugs. Little is known about nutrition in most of these species.

The digestive tract of most rotifers is complete and more or less straight (Figure 16.7A). (The anus has been secondarily lost in a few species, and some have a moderately coiled gut.) The mouth leads inward to the pharynx (mastax) either directly or via a short, ciliated buccal tube. Depending on the feeding method and food sources, swallowing is accomplished by various means, including ciliary action of the buccal field and buccal tube, or a pistonlike pumping action of certain elements of the mastax apparatus. The mastax is ectodermal in origin. Opening into the gut lumen just posterior to the mastax are ducts of the salivary glands. There are usually two to seven such glands; they are presumed to secrete digestive enzymes and perhaps lubricants aiding the movement of the mastax trophi.

A short esophagus connects the mastax and stomach. A pair of gastric glands opens into the posterior end of the esophagus; these glands apparently secrete digestive enzymes. The walls of the esophagus and gastric glands are often syncytial. The stomach is generally thick walled and may be cellular or syncytial, usually comprising a specific number of cells or nuclei in each species (Figure 16.7B). The intestine is short and leads to the anus, which is located dorsally near the posterior end of the trunk. Except for *Asplanchna*, which lacks a hindgut, an expanded cloaca connects the intestine and anus. The oviduct and usually the nephridioducts also empty into this cloaca.

Digestion probably begins in the lumen of the mastax and is completed extracellularly in the stomach, where absorption occurs. In one large and enigmatic group of bdelloids the stomach lacks a lumen. Although much remains to be learned about the digestive physiology of rotifers, some experimental work indicates that diet has multiple and important effects on various aspects of their biology, including the size and shape of individuals as well as some life cycle activities (see Gilbert 1980).

Circulation, Gas Exchange, Excretion, and Osmoregulation

Rotifers have no special organs for internal transport or for the exchange of gases between tissues and the environment. The blastocoelomic fluid provides a medium for circulation within the body, which is aided by general movement and muscular activities. Small body size reduces diffusion distances and facilitates the transport and exchange of gases, nutrients, and wastes. These activities are further enhanced by the absence of linings and partitions within the body cavity, so the exchanges occur directly between the organ tissues and the body fluid. Gas exchange probably occurs over the general body surface wherever the integument is sufficiently thin.

Most rotifers possess one or several pairs of flame bulb protonephridia, located far forward in the body. A nephridioduct leads from each flame bulb to a collecting bladder, which in turn empties into the cloaca via a ventral pore. In some forms, especially the bdelloids, the ducts open directly into the cloaca, which is enlarged to act as a bladder (Figure 16.7A). The protonephridial system of rotifers is primarily osmoregulatory in function, and is most active in freshwater forms.

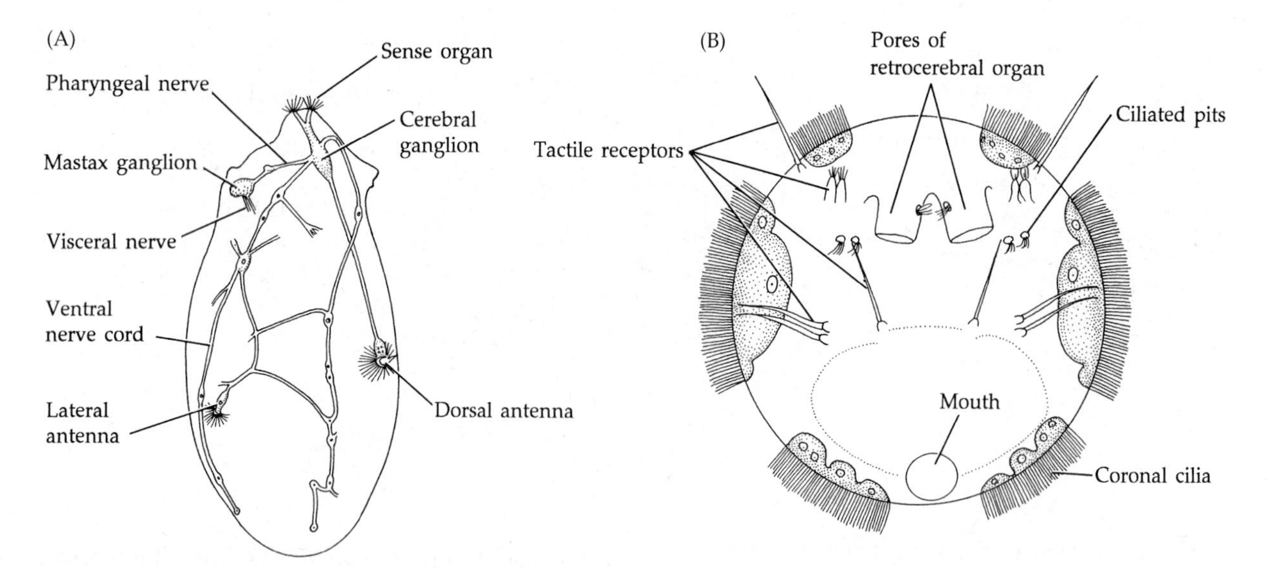

Figure 16.8 (A) The nervous system of *Asplanchna*. (B) The coronal area of *Euchlanis* (apical view). Note the various sense organs.

Excess water from the body cavity and probably from digestion is also pumped out via the anus by muscular contractions of the bladder. This "urine" is significantly hypotonic relative to the body fluids. It is likely that the protonephridia also remove nitrogenous excretory products from the body. This form of waste removal is probably supplemented by simple diffusion of wastes across permeable body wall surfaces.

Some rotifers (especially the freshwater and semiterrestrial bdelloids) are able to withstand extreme environmental stresses by entering a state of metabolic dormancy. They have been experimentally desiccated and kept in a dormant condition for as long as four years—reviving upon the addition of water. Some have survived freezing in liquid helium at –272°C and other severe stresses dreamed up by biologists.

Nervous System and Sense Organs

The cerebral ganglion of rotifers is located dorsal to the mastax, in the neck region of the body. Several nerve tracts arise from the cerebral ganglion, some of which bear additional small ganglionic swellings (Figure 16.8A). There are usually two major longitudinal nerves positioned either both ventrolaterally or one dorsally and one ventrally.

The coronal area generally bears a variety of touch-sensitive bristles or spines and often a pair of ciliated pits thought to be chemoreceptors (Figure 16.8B). The dorsal and lateral antennae are probably tactile. Some rotifers bear sensory organs, which are arranged as a cluster of micropapillae encircling a pore. These organs may be tactile or chemosensory. Most of the errant rotifers possess at least one simple ocellus embedded in the cerebral ganglion. In some, this cerebral ocellus is accompanied by one or two pairs of lateral ocelli on the coronal surface, and sometimes by a pair of apical ocelli in the apical field. The lateral and apical ocelli are multicellular epidermal patches of photosensitive cells. Pierre Clément (1977) described possible baro- or chemoreceptors in the body cavity that may help regulate internal pressure or fluid composition.

Associated with the cerebral ganglion is the so-called **retrocerebral organ**. This curious glandular structure gives rise to ducts that lead to the body surface in the apical field (Figure 16.8B). Once thought to be sensory in function, more recent work suggests that it may secrete mucus to aid in crawling.

Reproduction and Development

Parthenogenesis is probably the most common method of reproduction among rotifers. Other forms of asexual reproduction are unknown, and most groups show only very weak powers of regeneration. Most rotifers are gonochoristic; however, other than the Seisonidea, males are either reduced in abundance, size, and complexity, and with haploid chromosome numbers (Monogononta), or are still unknown (Bdelloidea). If you find a rotifer, the chances are good that it is a female.

The male reproductive system (Figure 16.9A) includes a single testis (paired in Seisonidea), a sperm duct, and a posterior gonopore whose wall is usually folded to produce a copulatory organ. Prostatic glands are sometimes present in the wall of the sperm duct. The males are short lived and possess a reduced gut unconnected to the reproductive tract.

The female system includes paired (Bdelloidea) or single (Monogononta) syncytial germovitellaria (Figure 16.9B). Eggs are produced in the ovary and receive yolk directly from the vitellarium before passing along the oviduct to the cloaca; in those forms that have lost the intestinal portion of the gut (e.g., *Asplanchna*), the oviduct passes directly to the outside via a gonopore. In the Seisonidea, there are no yolk glands.

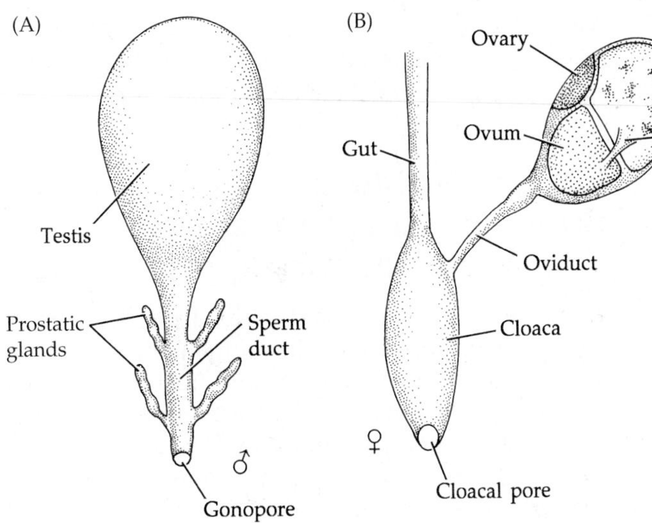

Figure 16.9 Male and female reproductive systems from a generalized monogonont rotifer.

In rotifers with a male form, copulation occurs either by insertion of the male copulatory organ into the cloacal area of the female or by hypodermic impregnation. In the latter case, males attach to females at various points on the body and apparently inject sperm directly into the blastocoelom (through the body wall). The sperm somehow find their way to the female reproductive tract, where fertilization takes place. The number of eggs produced by an individual female is determined by the original, fixed number of ovarian nuclei—usually 20 or fewer, depending on the species. Once fertilized, the ova produce a series of encapsulating membranes and are then either attached to the substratum or carried externally or internally by the brooding female.

Parthenogenesis is generally the rule among the bdelloids, but it is also a common and usually seasonal occurrence in the monogononts, where it tends to alternate with sexual reproduction. This cycle (Figure 16.10A) is an adaptation to freshwater habitats that are subject to severe seasonal changes. During favorable conditions, females reproduce parthenogenetically through the production of mitotically derived diploid ova (amictic ova). These eggs develop into more females without fertilization. However, when ova from amictic females are subjected to particular environmental conditions (so-called mixis stimuli), they develop into mictic females, which then produce mictic (haploid) ova by meiosis. The exact stimulus apparently varies among different species and may include such factors as changes in day length, temperature, food resources, or increases in population density. Although these cycles are commonly termed "summer" and "autumn cycles," this is a bit misleading because mixis can also occur during warm weather and many populations have several periods of mixis each year. Mictic ova require fertilization by male gametes to develop a new female individual, but if no males are present, the unfertilized mictic ova will instead develop into haploid males, which produce sperm by mitosis. These sperm fertilize other mictic ova, producing diploid, thick-walled, resting zygotes. The resting zygotic form is extremely resistant to low temperatures, desiccation, and other adverse environmental conditions. When favorable conditions return, the zygotes develop and hatch as amictic females (Figure 16.10B), completing the cycle.

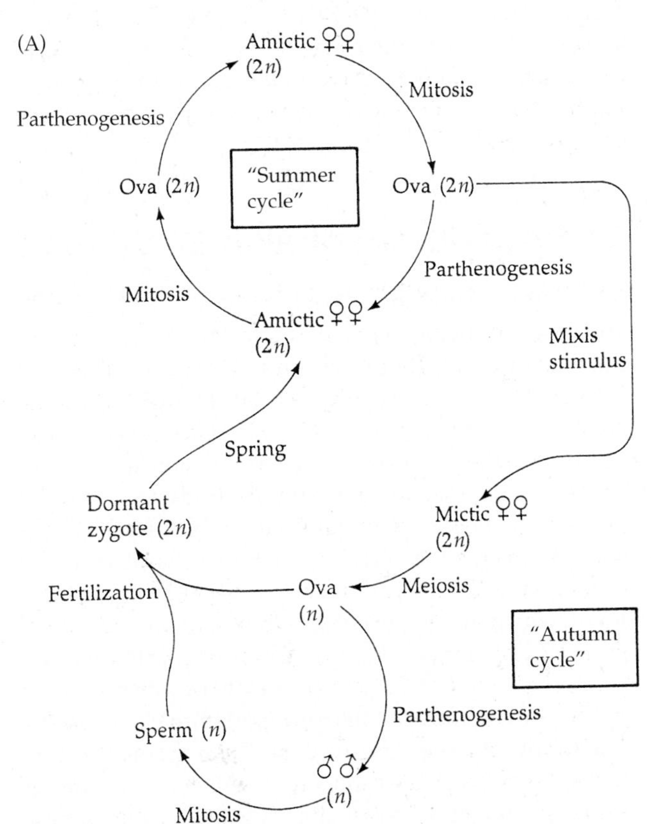

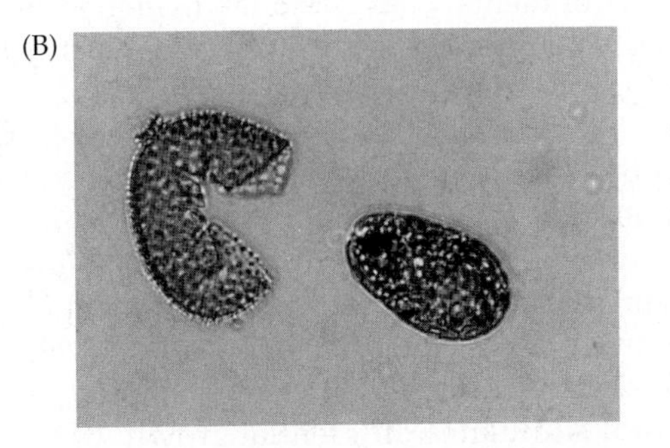

Figure 16.10 (A) Mictic/amictic alternation in the life cycle of a monogonont rotifer. (B) Micrograph of an amictic female hatching from an overwintering phase.

Bdelloids are prone to infection by an aggressive fungus, *Rotiferophthora angustispora*, that eats them from the inside out. Experiments have recently shown that the longer the rotifers remain dry and in a state of dormancy, the more likely they are to avoid infection by *R. angustispora*, suggesting that their adaptation for quiescence may also be an adaptation to avoid fungal predation.

Only a few studies have been conducted on the embryogeny of rotifers (see especially Pray 1965). In spite of the paucity of data, and some conflicting interpretations in the literature, it is generally thought that rotifers have modified spiral cleavage. However, detailed analyses of cell lineages are still needed to determine if the typical spiral pattern persists past the first couple of cell divisions in rotifers, especially with regard to the origin of the mesoderm. The isolecithal ova undergo unequal holoblastic early cleavage to produce a stereoblastula. Gastrulation is by epiboly of the presumptive ectoderm and involution of the endoderm and mesoderm; the gastrula gradually hollows to produce the blastocoel, which persists as the adult body cavity. The mouth forms in the area of the blastopore. Definitive nuclear numbers are reached early in development for those organs and tissues displaying eutely.

Errant rotifers undergo direct development, hatching as mature or nearly mature individuals. Sessile forms pass through a short dispersal phase, sometimes called a larva, which resembles a typical swimming rotifer. The "larva" eventually settles and attaches to the substratum. In all cases, there is a total absence of cell division during postembryonic life (i.e., they are eutelic).

Many rotifers exhibit developmental polymorphism, a phenomenon also seen in some protists, insects, and primitive crustaceans. It is the expression of alternative morphotypes under different ecological conditions, by organisms of a given genetic constitution (the differentiation of certain castes in social insects is one of the most remarkable examples of developmental polymorphism). In all such animals studied to date, the alternative adult morphotypes appear to be products of flexible developmental pathways, triggered by environmental cues and often mediated by internal mechanisms such as hormonal activities. In one well-studied genus of rotifers (*Asplanchna*), the environmental stimulus regulating which of several adult morphologies is produced is the presence of a specific molecular form of vitamin E—α-tocopherol. *Asplanchna* obtains tocopherol from its diet of algae or other plant material, or when it preys on other herbivores (animals do not synthesize tocopherol). The chemical acts directly on the rotifer's developing tissues, where it stimulates differential growth of the syncytial hypodermis after cell division has ceased. Predator-induced morphologies also occur among rotifers. *Keratella slacki* eggs, in the presence of the predator *Asplanchna* (both are rotifers), are stimulated to develop into larger-bodied adults with an extra long anterior spine, thus rendering them more difficult to eat.

Phylum Rotifera, Subclass Acanthocephala: The Acanthocephalans

As adults, the 1,200 or so described species of acanthocephalans are obligate intestinal parasites in vertebrates, particularly in birds and freshwater fishes. Larval development takes place in intermediate arthropod hosts. The name Acanthocephala (Greek *acanthias*, "prickly"; *cephalo*, "head") derives from the presence of recurved hooks located on an eversible proboscis at the anterior end. The rest of the body forms a cylindrical or flattened trunk, often bearing rings of small spines. Most acanthocephalans are less than 20 cm long, although a few species exceed 60 cm in length; females are generally larger than males. The digestive tract has been completely lost, and, except for the reproductive organs, there is significant structural and functional reduction of most other systems, a condition related to the parasitic lifestyles of these worms (Box 16C). The persisting organs lie within an open blastocoelom, partially partitioned by mesentery-like ligaments.

The acanthocephalans are usually divided into three groups based upon the arrangement of proboscis hooks, the nature of the epidermal nuclei, spination patterns on the trunk, and nature of the reproductive organs: Palaeacanthocephala (e.g., *Polymorphus, Corynosoma, Plagiorhynchus, Acanthocephalus*), Archiacanthocephala (e.g., *Moniliformis*), and Eoacanthocephala (e.g., *Neoechinorhynchus, Octospiniferoides*) (see Figure 16.11).

The Acanthocephalan Body Plan

Body Wall, Support, Attachment, and Nutrition

Adult acanthocephalans attach to their host's intestinal wall by their proboscis hooks, which are retractable into pockets, like the claws of a cat (Figure 16.11). The chemical nature of the hooks is not yet known. In nearly all species, the proboscis itself is retractable into a deep proboscis receptacle, enabling the body to be pulled close to the host's intestinal mucosa. Nutrients are absorbed through the body wall, and a gut is absent. The outer body wall is a multilayered, syncytial, living tegument, which overlies sheets of circular and longitudinal muscles. The tegument includes layers of dense fibers as well as what appear to be sheets of plasma membrane, and an intracellular protein lamina, such as the one found in free-living rotifers. The tegument is perforated by numerous canals that connect to a complex set of unique circulatory channels called the **lacunar**

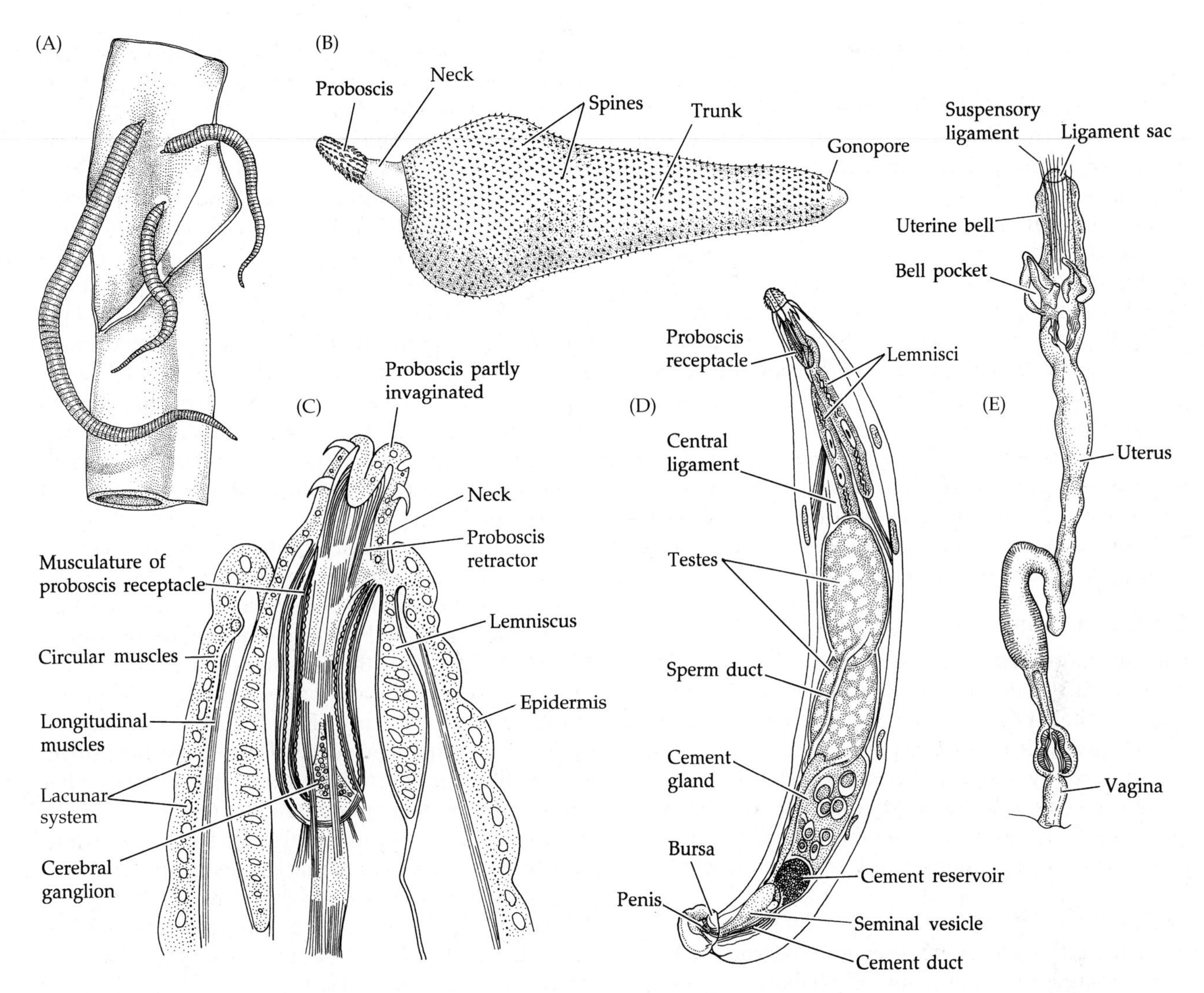

Figure 16.11 Representative acanthocephalans. (A) *Macracanthorynchus hirudinaceus*, an archiacanthocephalan, attached to the intestinal wall of a pig. (B) *Corynosoma*, a palaeacanthocephalan found in aquatic birds and seals. (C) Longitudinal section through the anterior end of *Acanthocephalus* (class Palaeacanthocephala). (D) An adult male eoacanthocephalan (*Pallisentis fractus*). (E) The isolated female reproductive system of *Bolbosoma*.

BOX 16C Characteristics of the Subclass Acanthocephala (Phylum Rotifera)

1. Triploblastic, bilateral, unsegmented blastocoelomates
2. Gut absent
3. Anterior end with hook-bearing proboscis
4. Tegument and muscles contain a unique system of channels called the lacunar system
5. Protonephridia absent except in a few species
6. With unique system of ligaments and ligament sacs partially partitioning the body cavity
7. With unique hydraulic structures called lemnisci that facilitate extension of proboscis
8. Gonochoristic
9. With acanthor larva
10. With modified spiral cleavage
11. All are obligate parasites in guts of vertebrates; many have complex life cycles.

system (Figure 16.11C). The tegumental channels near the body surface may facilitate pinocytosis of nutrients from the host. The body wall organization is such that each species has a distinct external appearance; some even appear to be segmented, but they are not.

At the junction of the proboscis and trunk, the epidermis extends inward as a pair of hydraulic sacs (**lemnisci**) that facilitate extension of the proboscis, as in free-living rotifers; the proboscis is withdrawn by retractor muscles. The lemnisci are continuous with each other and with a ring-shaped canal near the anterior end of the body, whereas their distal ends float free in the blastocoelom. This arrangement may help to circulate nutrients and oxygen from the body to the proboscis, although the actual function of the lemnisci is not known.

One or two large sacs lined with connective tissue arise from the rear wall of the proboscis receptacle and extend posteriorly in the body. These structures support the reproductive organs and divide the body into dorsal and ventral **ligament sacs** in the archiacanthocephalans and eoacanthocephalans, or produce a single ligament sac down the center of the body cavity in the palaeacanthocephalans (Figure 16.11D,E). Within the walls of these sacs are strands of fibrous tissue—the ligaments—that may represent remnants of the gut. The space between these internal organs is presumably a blastocoelom.

The body is supported by the fibrous tegument and the hydrostatic qualities of the blastocoelom and lacunar system. The muscles and ligament sacs add some structural integrity to this support system and canals of the lacunar system penetrate most of the muscles.

Circulation, Gas Exchange, and Excretion

Exchanges of nutrients, gases, and waste products occur by diffusion across the body wall (some Archiacanthocephala possess a pair of protonephridia and a small bladder). Internal transport is by diffusion within the body cavity and by the lacunar system, the latter functioning as a unique sort of circulatory system, which permeates most body tissues. The lacunar fluid is moved about by action of the body wall muscles.

Nervous System

As in many obligate endoparasites, the nervous system and the sense organs of acanthocephalans are greatly reduced. A cerebral ganglion lies within the proboscis receptacle (Figure 16.11C) and gives rise to nerves to the body wall muscles, the proboscis, and the genital regions. Males possess a pair of genital ganglia. The proboscis bears several structures that are presumed to be tactile receptors, and small sensory pores occur at the tip and base of the proboscis. Males have what appear to be sense organs in the genital area, especially on the penis.

Reproduction and Development

Acanthocephalans are gonochoristic and females are generally somewhat larger than males. In both sexes, the reproductive systems are associated with the ligament sacs (Figure 16.11E). In males, paired testes (usually arranged in tandem) lie within a ligament sac and are drained by sperm ducts to a common seminal vesicle. Entering the seminal vesicle or the sperm ducts are six or eight cement glands, whose secretions serve to plug the female genital pore following copulation. When nephridia are present, they also drain into this system. The seminal vesicle leads to an eversible penis, which lies within a genital bursa connected to the gonopore. This gonopore is often called a cloacal pore, because the bursa appears to be a remnant of the hindgut.

In females, a single mass of ovarian tissue forms within a ligament sac. Clumps of immature ova are released from this transient ovary and enter the body cavity, where they mature and are eventually fertilized. The female reproductive system comprises a gonopore, a vagina, and an elongate uterus that terminates internally in a complex open funnel called the **uterine bell** (Figure 16.11E). During mating the male everts the copulatory bursa and attaches it to the female gonopore. The penis is inserted into the vagina, sperm are transferred, and the vagina neatly capped with cement. Sperm then travel up the female system, enter the body cavity through the uterine bell, and fertilize the eggs.

Much of the early development of acanthocephalans takes place within the body cavity of the female. Cleavage is holoblastic, unequal, and likened to a highly modified spiral pattern. A stereoblastula is produced, at which time the cell membranes break down to yield a syncytial condition. Eventually, a shelled **acanthor larva** is formed (Figure 16.12). The embryo leaves the mother's body at this (or an earlier) stage. Remarkably, the uterine bell "sorts" through the developing embryos by manipulating them with its muscular funnel; it accepts only the appropriate embryos into the uterus. Embryos in earlier stages are rejected and pushed back into the body cavity, where they continue development. The selected embryos pass through the uterus and out the genital pore and are eventually released with the host's feces.

Once outside the definitive host, the developing acanthocephalan must be ingested by an arthropod intermediate host—usually an insect or a crustacean—to continue its life cycle. The acanthor larva penetrates the gut wall of the intermediate host and enters the body cavity, where it develops into an **acanthella** and then into an encapsulated form called a **cystacanth** (Figure 16.12). When the intermediate host is eaten by an appropriate definitive host, the cystacanth attaches to the intestinal wall of the host and matures into an adult.

Phylum Micrognathozoa: The Micrognathozoans

A new microscopic animal, *Limnognathia maerski,* was described in 2000 by Reinhardt Kristensen and Peter Funch from a cold spring at Disko Island, West Greenland. Due to the numerous unique features of this new microscopic animal, a new monotypic class, Micrognathozoa (Greek, *micro,* "small," *gnathos,* "jaw"; *zoa,* "animal") was erected. Though *L. maerski* shows

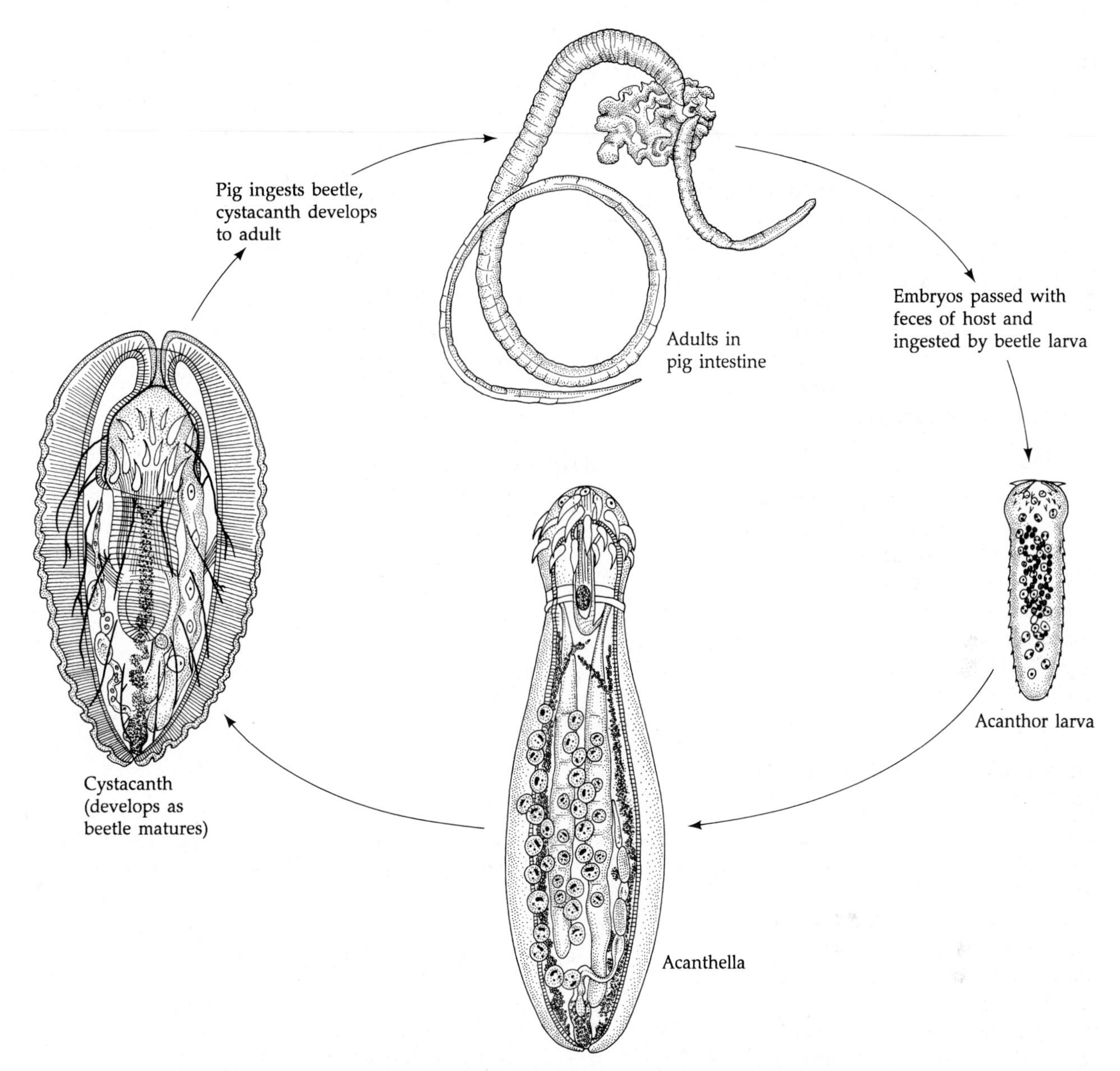

Figure 16.12 Life cycle of *Macracanthorhynchus hirudinaceus*, an intestinal parasite in pigs. The adults reside in the intestine of the definitive host and embryos are released with the host's feces. The encapsulated embryos are ingested by the secondary host, in this case, beetle larvae. Within the secondary host, the embryo passes through the acanthor and acanthella stages while the beetle grows, eventually becoming a cystacanth. When the beetle is ingested by a pig, the juvenile matures into an adult, thereby completing the cycle.

a superficial resemblance to microscopic annelids, its affiliation with Gnathostomulida and Rotifera was quickly established based on ultrastructural similarities of the epidermis and jaws. Jawlike structures are also found in other protostome taxa, such as the proboscises of kalyptorhynch turbellarians, in dorvilleid annelids, and aplacophoran molluscs, but studies of their ultrastructure show that none of these jaws are homologous with those of *L. maerski*. Early molecular phylogenetic studies showed that Micrognathozoa did not nest within either the two other gnathiferan phyla (Gnathostomulida and Rotifera), but next to them. For this reason, Micrognathozoa was given phylum status (Giribet et al. 2004). A later phylogenetic analysis based on transcriptomic data placed Micrognathozoa as the sister group to Rotifera (including Acanthocephala), with these two comprising the sister clade to Gnathostomulida. Micrognathozoa still includes only the single described species from Greenland, but two later records of morphologically similar micrognathozoans from geographically widely separated freshwater creeks in Antarctica and Great Britain will most likely prove to be distinct, cryptic species when DNA analyses have been completed. In the southern Indian ocean on Ile de la Possession (Crozet Islands) micrognathozoans were found to be numerous in

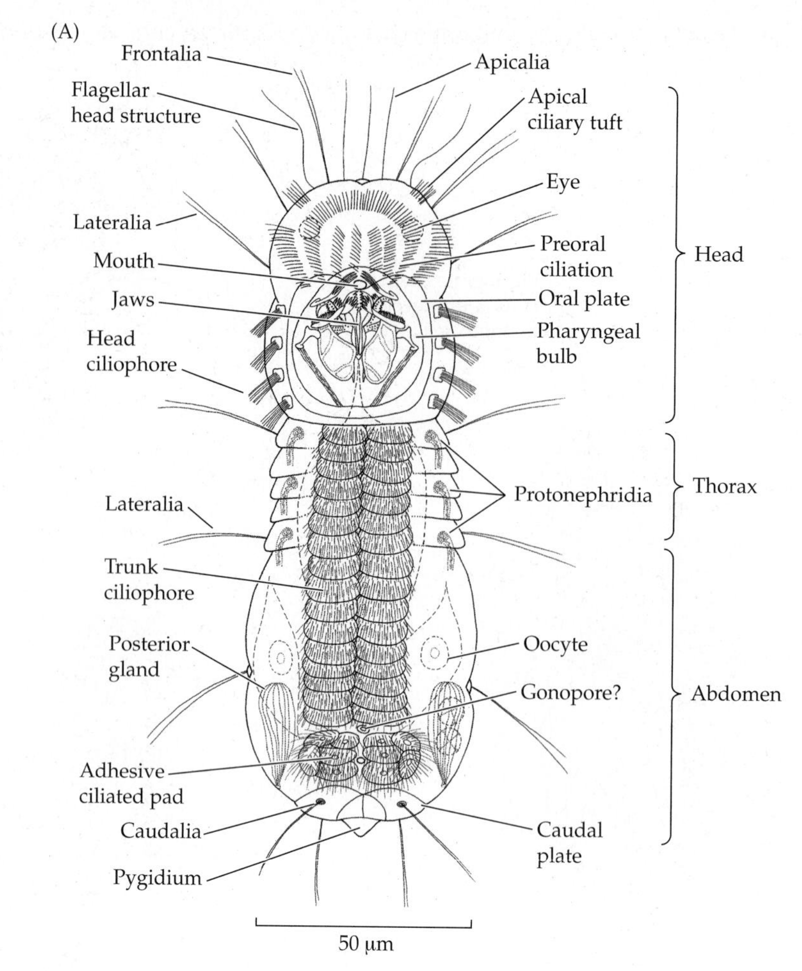

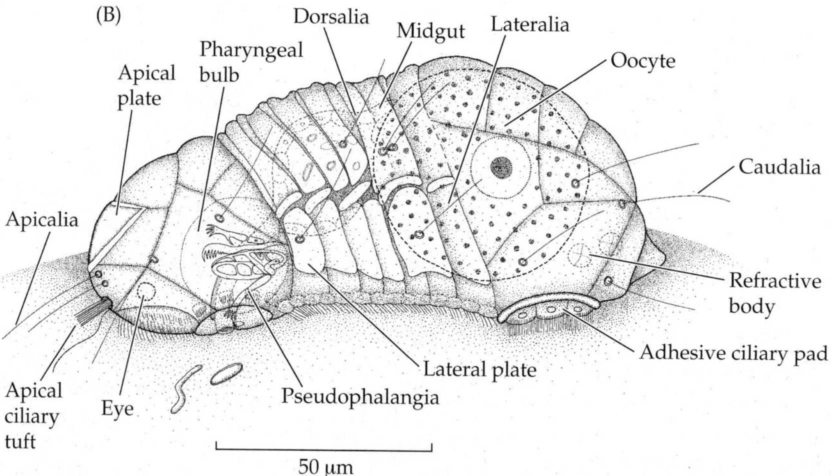

Figure 16.13 Micrognathozoa: ***Limnognathia maerski.*** (A) Ventral view. (B) Lateral view.

lakes and rivers, whereas in the United Kingdom only a few animals have been found from a stream in southern Wales (and only in winter), as well as an animal found on a single sand grain of river sediment in Lambourn Parish, Berkshire.

The Micrognathozoan Body Plan

Limnognathia maerski is an acoelomate animal ranging from 101 μm to 152 μm in adult length (juveniles measuring 85–107 μm in length). The adult body can be divided into three main regions: a head, an accordion-like thorax, and an abdomen (Figures 16.13 and 16.14); the head contains the prominent jaw apparatus (Box 16D).

Epidermis, Ciliation, and Body Wall Musculature

Despite their small size, micrognathozoans have a complex support system and body musculature. *Limnognathia maerski* has dorsal and lateral **epidermal plates** formed by an intracellular matrix as in rotifers and acanthocephalans (Figures 16.13 and 16.14). Ventral plates are lacking, but the "naked" epidermis has a thin extracellular glycocalyx layer and a true cuticular oral plate (described below). The animal lacks syncytia, a key character of Rotifera (and Acanthocephala); however, it possesses a unique form of gap junctions showing transverse electron dense bands in a zipper-like pattern (we call these **zip-junctions**) between the dorsal epidermal cells.

The ventral ciliation consists of an arched preoral ciliary field, four pairs of head **ciliophores** (synchronously beating multiciliated cells) surrounding the pharyngeal bulb, 18 pairs of ventral ciliophores located at the thorax and abdomen, and a posterior adhesive ciliary pad (Figure 16.13). The ventral paired ciliophores form the locomotory organ and are characterized by very long ciliary roots, originally mistaken for cross-striated muscles. These cells are highly similar to ciliophores found in the interstitial microscopic annelids *Diurodrilus* and *Neotenotrocha*. An adhesive ciliary pad consisting of five pairs of multiciliated cells is located posteriorly on the ventral side. A midventral pore exists between the clusters of ciliated cells, which may represent the female gonopore of the paired oviducts seemingly having a midventral common opening. The ciliary pad is very different from the adhesive toes of rotifers, gastrotrichs, and annelids, and the structure may be a unique synapomorphy for Micrognathozoa.

As in many marine interstitial animals (e.g., gnathostomulids, gastrotrichs, microscopic annelids), special forms of tactile bristles or sensoria are found on the body. The tactile bristle may consist of a single sensory

BOX 16D Characteristics of the Phylum Micrognathozoa

1. Triploblastic, bilateral, unsegmented, acoelomate
2. Epidermis with supporting dorsal and lateral plates (intracellular matrix)
3. Without a syncytial epidermis
4. Ventral ciliation consisting of preoral ciliary field and paired ciliophores (synchronously beating multiciliated cells) around mouth and along midline of thorax and abdomen
5. Sensory organs in the form of stiff monociliated cells supported by microvilli (collar receptors) and nonciliated internal eyes (phaosomes)
6. Posterior end with ciliated pad and one pair of glands
7. Mouth opening ventral, gut incomplete (the dorsal anus being temporary)
8. Pharyngeal apparatus containing complex jaw apparatus with four sets of jaw-like elements and several sets of striated muscles largely related to the fibularium and the main jaws
9. Three pairs of protonephridia with monociliated terminal cells
10. Without circulatory system or special gas exchange structures
11. Males unknown; probably parthenogenetic
12. Two female gonads in close contact with the midgut

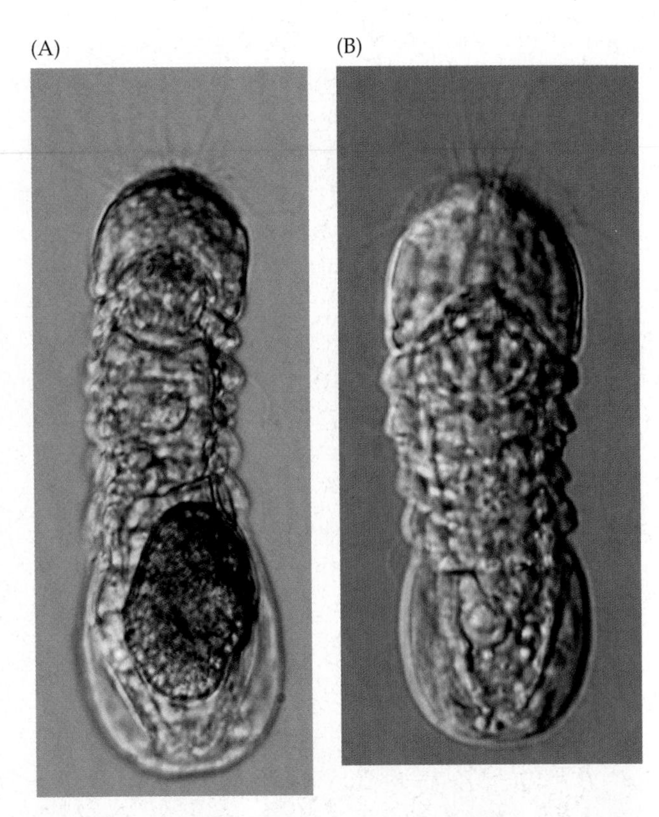

Figure 16.14 Micrognathozoa: *Limnognathia maerski*, light micrographs. (A) Adult female with mature egg (length 0.14 mm). (B) Juvenile with relatively large thorax/smaller abdomen and immature oocyte (length 0.09 mm).

cell, the **collar receptor**, with a single cilium in the middle surrounded by 8 or 9 microvilli. Two large, posterior glands were recently revealed by immunostaining (Figures 16.13 and 16.15), which by their simple configuration and homogenous content resemble mucus-secreting glands. They might have an adhesive function, together with the ciliary pad, but they do not resemble the more complex adhesive duo gland system found in the posterior end of gastrotrichs and the interstitial annelid *Diurodrilus*. Otherwise, no epidermal glands are known from micrognathozoans.

Limnognathia maerski has an elaborate body wall musculature, comprising seven main pairs of longitudinal muscles extending from head to abdomen, and 13 pairs of oblique dorsoventral muscles localized in the thoracic and the abdominal regions (Figure 16.16). The musculature further comprises several minor posterior muscles and fine anterior forehead muscle, as well as the prominent pharyngeal muscular apparatus (Figure 16.17B). Cross-striated muscles are found in both the body wall and the jaw musculature. The three main ventral longitudinal pairs and one dorsal pair of muscles (green and turquoise muscles, Figure 16.16) span the entire length of the body, and some fibers even branch off to continue anteriorly into the head and posteriorly into the abdomen, forming a fine muscular diversification. These muscles seemingly aid longitudinal contraction and ventral bending of the body. The 13 oblique dorsoventral muscles may function together with the longitudinal muscles as supporting semicircular body wall musculature. Their close approximation to the gut further suggests they may act as gut musculature, thus possibly compensating for the lack of outer or inner circular musculature in Micrognathozoa.

Locomotion

Micrognathozoans swim in a characteristic slow spiral motion when moving freely in the water column. It is a slow movement, very different from the rotifers. From video recordings, it seems that the trunk ciliophores are used both in swimming and in epibenthic crawling or gliding motions on the substrate. Gliding is accomplished by the rows of motile ciliophores, each with multiple cilia beating in unison, in the same way as seen in the annelid *Diurodrilus*. However, the preoral ciliary field does not seem to be involved in either swimming or gliding. *Limnognathia maerski* has never been observed moving backwards (as is common among gnathostomulids), not even when they reverse

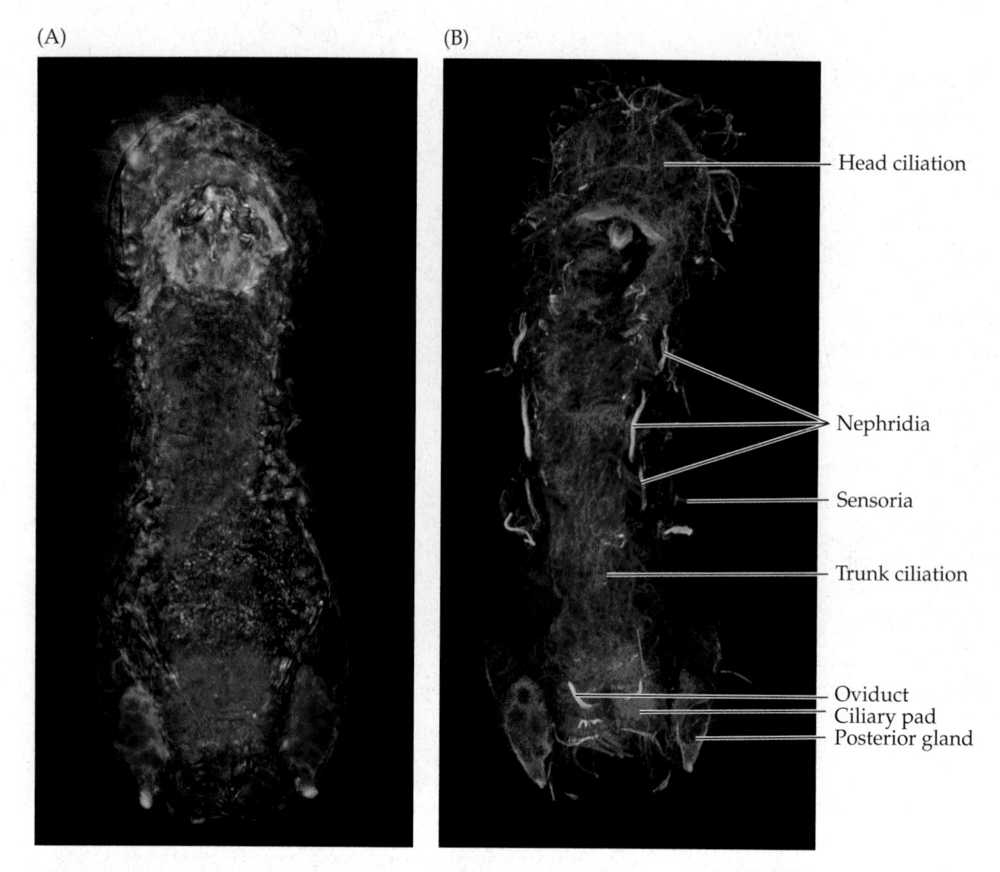

Figure 16.15 Micrognathozoa: confocal laser scanning microscopy images of *Limnognathia maerski*, maximum intensity projection of Z-stacks. (A) Antibody staining showing ventral ciliation in blue, pharyngeal musculature in green, posterior glands in red. (B) Depth coded projection of anti-acetylated α-tubulin immunoreactivity showing the ventral ciliation (red) and the ciliated three anterior pairs of nephridial ducts and one pair of posterior oviducts beneath the ciliary field (yellow).

the beating of their long cilia. In addition, an escape motion has been observed where contraction of trunk muscles creates rapid jerky movements.

Pharyngeal Apparatus, Feeding, and Digestion

The mouth opens ventrally on the anterior margin of the cuticular, nonciliated oral plate and leads into the pharyngeal cavity, followed by a short esophagus dorsal to the paired jaw apparatus, then continuing into the undifferentiated, nonciliated gut. The temporary anus is located dorsally and opens only periodically, as also seen in all gnathostomulids and in some gastrotrichs.

The less than 30 μm wide pharyngeal apparatus shows a complexity unseen in any other microscopic taxon, comprising numerous hard jaw parts and intricate musculature (Figure 16.17), as well as a buccal ganglion. The jaw parts comprise four main sets of sclerotonized, denticulated, hard elements (sclerites): the large paired **fibularium**, the **main jaws**, the **ventral jaws**, and the **dorsal jaws**. The largest sclerite in each jaw is the fibularium and it plays a central role in supporting the pharynx. Several subparts of the main sclerites have been described, including the anterior region of the ventral jaws called the **pseudophalangia**. So far, little is known about the functionality of this complex apparatus or the possible independent movement of all these parts, and only the pseudophalangia has been observed protruding from the mouth in fast snapping movements, possibly grasping food. The pharyngeal musculature is similarly complex and includes a major

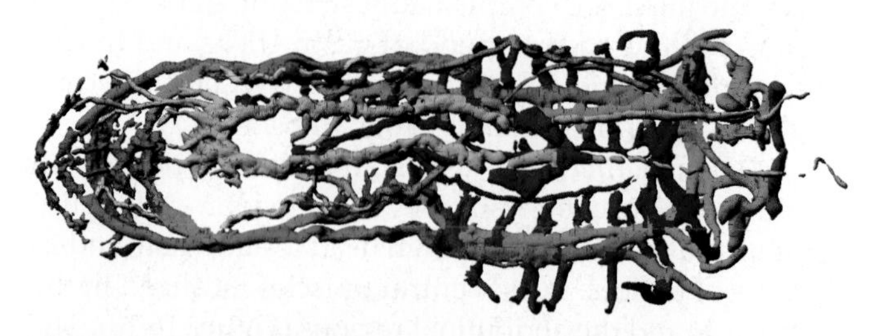

Figure 16.16 Micrognathozoa: *Limnognathia maerski* isosurface reconstruction of body wall musculature from confocal microscopy of phalloidin staining. Reconstruction showing 13 oblique dorsoventral pairs of muscles (red) and seven main pairs of longitudinal muscles: three ventral (green), two lateral (yellow and orange), and two dorsal pairs (blue) as well as additional minor muscles.

(A)

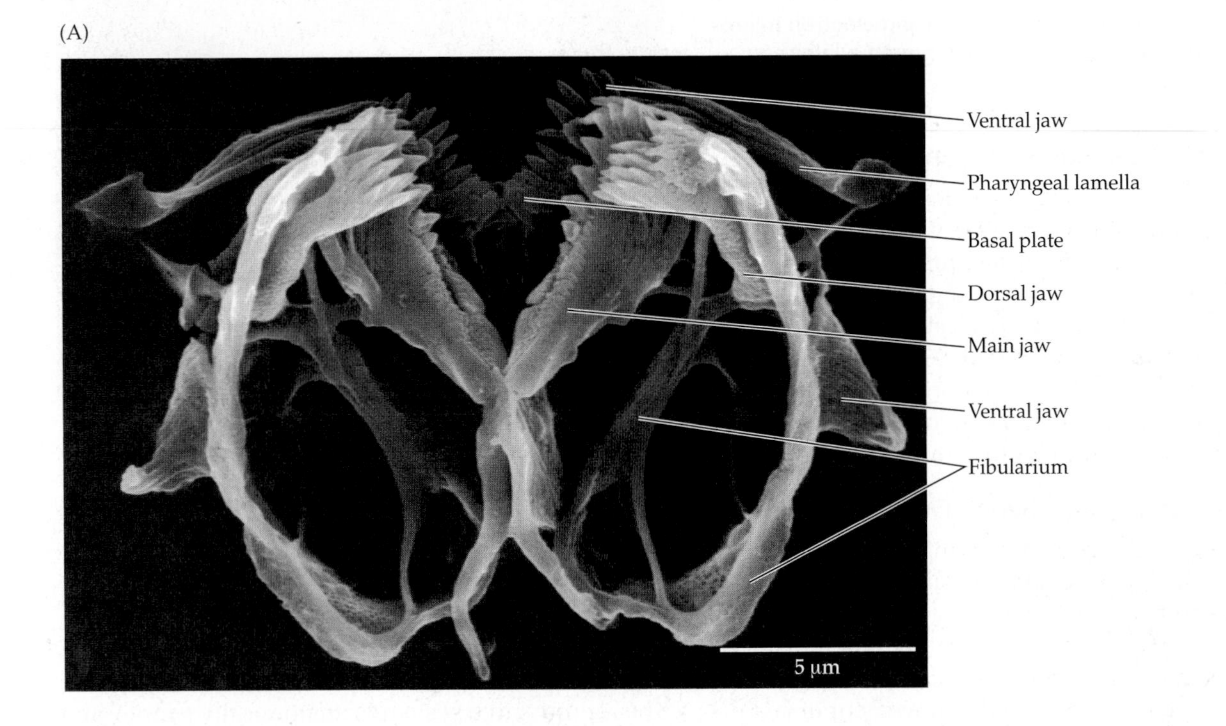

(B)

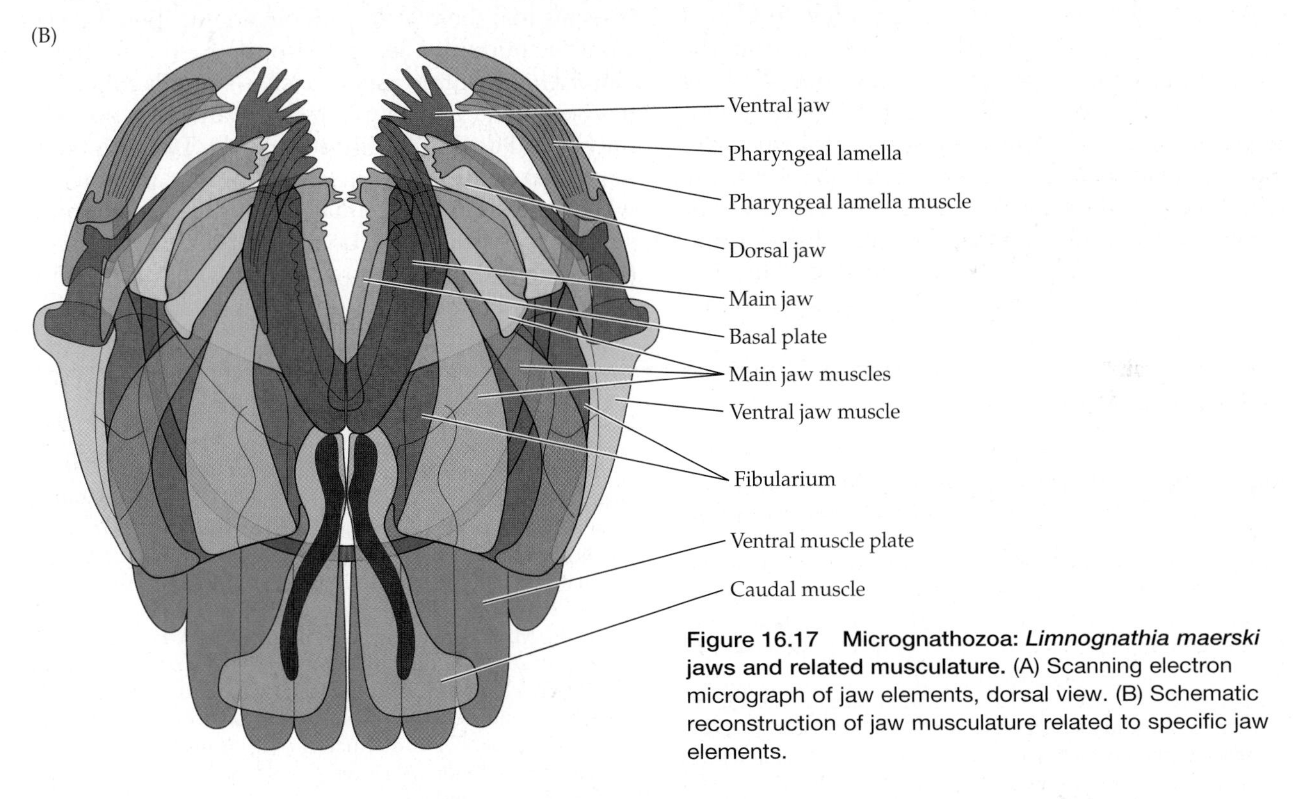

Figure 16.17 Micrognathozoa: *Limnognathia maerski* jaws and related musculature. (A) Scanning electron micrograph of jaw elements, dorsal view. (B) Schematic reconstruction of jaw musculature related to specific jaw elements.

ventral muscle plate supporting (and moving) the entire jaw apparatus, as well as several other paired and unpaired striated muscles (Figure 16.17B). The ventral muscle plate is formed by 8–10 longitudinal cross striated muscle fibers (purple muscles, Figure 16.17B) underlying the fibularium and enveloping the jaws laterally and caudally. This large muscle is unique to the Micrognathozoa, being absent in other gnathiferan phyla. The many paired and unpaired muscles seem mainly related to the fibularium and the main jaws, moving the jaws as well as supplying some of the minor jaw elements such as the accessory sclerites and the pharyngeal lamellae and allowing for the extrusion of the ventral jaws. The feeding biology of micrognathozoans is not well known. The animals are found on mosses or in the sediments, and video recordings have shown the animal eating bacteria on the surfaces of mosses and sand grains.

Figure 16.18 Micrognathozoa: scanning electron micrograph of *Limnognathia maerski* sculptured winter egg.

Circulation, Gas Exchange, and Excretion

Micrognathozoans are acoelomates with no circulatory system, and gas exchange takes place by diffusion across the epidermis. There are three pairs of protonephridia, two pairs in the thorax and one pair extending into the abdomen. The terminal cells are monociliated in contrast to the multiciliated terminal cells of Rotifera, but similar to those found in Gnathostomulida. It has been suggested that the monociliated condition is plesiomorphic within Protostomia.

Nervous System and Sense Organs

Micrognathozoa possess a seemingly simple nervous system consisting of an anterior, slightly bilobed, dorsal brain and two ventral nerve cords, extending from each of the lobes to the posterior abdomen. A large buccal ganglion is found within the pharyngeal apparatus, which may control the movement of jaw elements, and is possibly followed by a few indistinct posterior ganglia. Peripheral nerves extend from the cords, connecting to the sensory cilia. Some of these cilia are clearly monociliated collar receptors (one cilium surrounded by 8–9 microvilli), whereas others are more complex with several sensory cells involved. The terminology of the sensory structures is, from anterior to posterior: apicalia, frontalia, lateralia, dorsalia, and caudalia (Figure 16.13). In the anterior end of the animal, a pair of lateral hyaline vesicles is present. They may be unpigmented, inner eyes of the annelid type, the so-called phaosomes, and like these contain a dense layer of microvilli, but no ciliary structures.

Reproduction and Development

Only the female reproductive system has been found, suggesting the *Limnognathia maerski* is parthenogenetic. The reproductive system is anatomically simple, and it seems that the two ovaries obtain nutrition directly from the midgut, a feature also reported from freshwater chaetonotoid gastrotrichs. Though collecting has been done year round in Greenland, the species is only found during the short summer. Two egg types have been found, as in limnic gastrotrichs and rotifers, where the smooth egg may be a quick-developing summer egg and the strongly sculptured winter egg (Figure 16.18) may be a resting egg, not developing during the ten-month-long Arctic winter.

Selected References

Gnathifera

Ahlrichs, W. H. 1995. *Ultrastruktur und Phylogenie von* Seison nebaliae *(Grube 1859) und* Seison annulatus *(Claus 1876)*. Göttingen: Cuvillier Verlag.

Gazi, M., T. Sultana, G. Min, Y .C. Park, M. García-Varela, S. A. Nadler and J. Park. 2012. The complete mitochondrial genome sequence of *Oncicola luehei* (Acanthocephala: Archiacanthocephala) and its phylogenetic position within Syndermata. Parasitol. Int. 61: 307–316.

Giribet, G., M. V. Sørensen, P. Funch, R. M. Kristensen and W. Sterrer. 2004. Investigations into the phylogenetic position of Micrognathozoa using four molecular loci. Cladistics 20: 1–13.

Hyman, L. H. 1951. *The Invertebrates: Acanthocephala, Aschelminthes, and Entoprocta*. McGraw-Hill Publications, New York.

Kristensen, R. M. and P. Funch 2000. Micrognathozoa: a new class with complicated jaws like those of Rotifera and Gnathostomulida. J. Morphol. 246: 1–49.

Laumer, C. E. and 10 others. 2015. Spiralian phylogeny informs the evolution of microscopic lineages. Curr. Biol. 25: 2000–2006.

Min, G. and J. Park. 2009. Eurotatorian paraphyly: Revisiting phylogenomic relationships based on the complete mitochondrial genome sequence of *Rotaria rotatoria* (Bdelloidea: Rotifera: Syndermata). BMC Genomics 10: 533.

Near, T. J. 2002. Acanthocephalan phylogeny and the evolution of parasitism. Integ. Comp. Bio. 42: 668–677.

Rieger, R. M. and S. Tyler. 1995. Sister-group relationship of Gnathostomulida and Rotifera-Acanthocephala. Invertebr. Biol. 114: 186–188.

Sørensen, M. V., P. Funch,, E. Willerslev, A. J. Hansen and J. Olesen. 2000. On the phylogeny of the Metazoa in the light of Cycliophora and Micrognathozoa. Zool. Anz. 239: 297–318.

Sterrer, W. and M. V. Sørensen. 2015. Gnathostomulida. Pp. 135-196 in A. Schmidt-Rhaesa (ed.), *Handbook of Zoology. Gastrotricha, Cycloneuralia and Gnathifera*. Walter De Gruyter, GmbH, Berlin and Boston.

Witek, A., H. Herlyn, I. Ebersberger, D. B. Mark Welch and T. Hankeln. 2008. Support for monophyletic origin of Gnathifera from phylogenomics. Mol. Phyl. Evolut. 53: 1037–1041.

Witek, A., H. Herlyn, A. Meyer, L. Boell, G. Bucher and T. Hankeln. 2008. EST based phylogenomics of Syndermata questions monophyly of Eurotatoria. BMS Evolutionary Biology 8: 345.

Gnathostomulida

Ax, P. 1956. Die Gnathostomulida, eine rätselhafte Wurmgruppe aus dem Meeressand. Abh. Akad. Wiss. Lit. Mainz Math. Naturwiss. Kl. 8: 1–32.

Ax, P. 1965. Zur Morphologie und Systematik der Gnathostomulida. Untersuchungen an *Gnathostomula paradoxa* Ax. Z. Zool. Syst. Evolutionsforsch. 3: 259–296.

Herlyn, H. and U. Ehlers. 1997. Ultrastructure and function of the pharynx of *Gnathostomula paradoxa* (Gnathostomulida). Zoomorphol. 117: 135–145.

Jenner, R. A. 2004. Towards a phylogeny of the Metazoa: evaluating alternative phylogenetic positions of Platyhelminthes, Nermertea, and Gnathostomulida, with a critical reappraisal of cladistic characters. Contrib. Zool. 73: 3–163.

Knauss, E. B. 1979. Indication of an anal pore in Gnathostomulida. Zool. Scr. 8: 181–186.

Knauss, E. B. 1979. Fine structure of the male reproductive system in two species of *Haplognathia* Sterrer (Gnathostomulida, Filospermoidea). Zoomorphol. 94: 33–48.

Kristensen, R. M. and A. Nørrevang. 1977. On the fine structure of *Rastrognathia macrostoma* gen. et sp. n. placed in Rastrognathiidae fam. n. (Gnathostomulida). Zool. Scr. 6: 27–41.

Kristensen, R. M. and A. Nørrevang. 1978. On the fine structure of *Valvognathia pogonostoma* gen. et sp. n. (Gnathostomulida, Onychognathiidae) with special reference to the jaw apparatus. Zool. Scr. 7: 179–186.

Lammert, V. 1984. The fine structure of the spiral ciliary receptors in Gnathostomulida. Zoomorphol. 104: 360–364.

Lammert, V. 1989. Fine structure of the epidermis in the Gnathostomulida. Zoomorphol. 107: 14–28.

Lammert, V. 1991. Gnathostomulida. Pp. 20–39 in F. W. Harrison and E. E. Ruppert (eds.), *Microscopic Anatomy of Invertebrates, Vol. 4, Aschelminthes*. Wiley-Liss, New York.

Mainitz, M. 1979. The fine structure of gnathostomulid reproductive organs I. New characters in the male copulatory organ of Scleroperalia. Zoomorphologie 92: 241–272.

Müller, M. C. M. and W. Sterrer. 2004. Musculature and nervous of *Gnathostomula peregrina* (Gnathostomulida) shown by phalloidin labeling, immunohistochemistry, and cLSM, and their phylogenetic significance. Zoomorphol. 123: 169–177.

Riedl, R. J. 1969. Gnathostomulida from America. Science 163: 445–452.

Riedl, R. J. 1971. On the genus *Gnathostomula* (Gnathostomulida). Int. Rev. Ges. Hydrobiol. 56: 385–496.

Riedl, R. J. and R. Rieger. 1972. New characters observed on isolated jaws and basal plates of the family Gnathostomulidae (Gnathostomulida). Zool. Morphol. Tiere 72: 131–172.

Rieger, R. M. and M. Mainitz. 1977. Comparative fine structure of the body wall in Gnathostomulida and their phylogenetic position between Platyhelminthes and Aschelminthes. Z. Zool. Syst. Evolutionsforsch. 15: 9–35.

Sørensen, M. V. 2000. An SEM study of the jaws of *Haplognathia rosea* and *Rastrognathia macrostoma* (Gnathostomulida), with a preliminary comparison with the rotiferan trophi. Acta Zool. 81: 9–16.

Sørensen, M. V. 2002. Phylogeny and jaw evolution in Gnathostomulida, with a cladistic analysis of the genera. Zool. Scr. 31: 461–480.

Sørensen, M. V. and W. Sterrer 2002. New characters in the gnathostomulid mouth parts revealed by scanning electron microscopy. J. Morphol. 253: 310–334.

Sørensen, M. V., W. Sterrer and G. Giribet. 2006. Gnathostomulid phylogeny inferred from a combined approach of four molecular loci and morphology. Cladistics 22: 32–58.

Sørensen, M.V., S. Tyler, M. D. Hooge and P. Funch. 2003. Organization of pharyngeal hard parts and musculature in *Gnathostomula armata* (Gnathostomulida: Gnathostomulidae). Can. J. Zool. 81: 1463–1470.

Sterrer, W. 1971a. *Agnathiella beckeri* nov. gen. nov. spec. from southern Florida. The first gnathostomulid without jaws. Hydrobiologia 56: 215–225.

Sterrer, W. 1971b. On the biology of Gnathostomulida. Vie Milieu Suppl. 22: 493–508.

Sterrer, W. 1972. Systematics and evolution within the Gnathostomulida. Syst. Zool. 21: 151–173.

Sterrer, W. 1982. Gnathostomulida. Pp. 847–851 in S. Parker (ed.), *Synopsis and Classification of Living Organisms*. McGraw-Hill, New York.

Sterrer, W. 2011. Two species (one new) of Gnathostomulida (Bursovaginoidea: Conophoralia) from Barbados. Proc. Biol. Soc. Washington 124: 141–146.

Sterrer, W., M. Mainitz and R. M. Rieger. 1985. Gnathostomulida: Enigmatic as ever. Pp. 181–199 in S. C. Morris et al. (eds.), *The Origins and Relationships of Lower Invertebrates*. Syst. Assoc. Spec. Vol. No. 28, Oxford.

Tyler, S. and M. D. Hooge. 2001. Musculature of *Gnathostomula armata* Riedl 1971 and its ecological significance. Mar. Ecol. 22: 71–83.

Rotifera

Ahlrichs, W. H. 1995. *Ultrastruktur und Phylogenie von* Seison nebaliae *(Grube 1859) und* Seison annulatus *(Claus 1876)*. Cuvillier Verlag, Göttingen.

Ahlrichs, W. H. 1997. Epidermal ultrastructure of *Seison nebaliae* and *Seison annulatus*, and a comparison of epidermal structures within Gnathifera. Zoomorphology 117: 41–48.

Aloia, R. and R. Moretti. 1973. Mating behavior and ultrastructural aspects of copulation in the rotifer Asplanchna brightwelli. Trans. Am. Microsc. Soc. 92: 371–380.

Birky, W. 1971. Parthenogenesis in rotifers: The control of sexual and asexual reproduction. Am. Zool. 11: 245–266.

Birky, W. and B. Field. 1966. Nuclear number in the rotifer *Asplancha*: intraclonal variation and environmental control. Science 151: 585–587.

Clément, P. 1977. Ultrastructure research on rotifers. Arch. Hydrobiol. Beih. Ergebn. Limnol. 8: 270–297.

Clément, P. 1985. The relationships of rotifers. Pp. 224–247 in S. C. Morris et al. (eds.), *The Origins and Relationships of Lower Invertebrates*. Syst. Assoc. Spec. Vol. No. 28, Oxford.

Clément, P. 1987. Movements in rotifers: Correlations of ultrastructure and behavior. Hydrobiologia 147: 339–359.

Clément, P. and E. Wurdak. 1991. Rotifera. Pp. 219–297 in F. W. Harrison and E. E. Ruppert (eds.), *Microscopic Anatomy of Invertebrates, Vol. 4, Aschelminthes*. Wiley-Liss, New York.

Donner, J. 1966. *Rotifers*. Frederick Warne & Co., New York.

Felix, A., M. E. Stevens and R. L. Wallace. 1995. Unpalatability of a colonial rotifer, *Sinantherina socialis*, to small zooplanktiverous fishes. Invert. Biol. 114: 139–144.

Flot, J.-F. and 39 others. 2013. Genomic evidence for ameiotic evolution in the bdelloid rotifer *Adineta vaga*. Nature 500: 453–457.

Fontaneto, D., M. Kaya, E. A. Herniou and T. G. Barraclough. 2009. Extreme levels of hidden diversity in microscopic animals (Rotifera) revealed by DNA taxonomy. Mol. Phyl. Evolut. 53: 182–189.

Fontaneto, D., G. Melone and R. L. Wallace. 2003. Morphology of Floscularia *ringens* (Rotifera, Monogononta) from egg to adult. Invert. Biol. 122(3): 231–240.

García-Varela, M. and S. A. Nadler. 2006. Phylogenetic relationships among Syndermata inferred from nuclear and mitochondrial gene sequences. Mol. Phyl. Evolut. 40: 61–72.

Gilbert, J. J. 1980. Developmental polymorphism in the rotifer *Asplanchna sieboldi*. Am. Sci. 68: 636–646.

Gilbert, J. J. 1983. Rotifera. Pp. 181–209 in K. G. and R. G. Adiyodi, *Reproductive Biology of Invertebrates, Vol. 1*. John Wiley, London.

Gladyshev, E. A., M. Meselson and I. R. Arkhipova. 2008. Massive horizontal gene transfer in bdelloid rotifers. Science 320: 1210–1213.

Gómez, A., M. Serra, G. R. Carvalho and D. H. Lunt. 2002. Speciation in ancient cryptic species complexes: Evidence from the molecular phylogeny of *Brachionus plicatilis* (Rotifera). Evolution 56: 1431–1444.

Hejnol, A. 2010. A twist in time—the evolution of spiral cleavage in the light of animal phylogeny. Integr. Comp. Biol. 50(5): 695–706.

Hochberg, R. and M. K. Litvaitis. 2000. Functional morphology of the muscles in *Philodina* sp. (Rotifera: Bdelloidea). Hydrobiologia 432: 57–64.

Hochberg, R., S. O'Brien and A. Puleo. 2010. Behavior, metamorphosis, and muscular organization of the predatory rotifer *Acyclus inquietus* (Rotifera: Monogononta). Invertebr. Biol. 129: 210–219.

King, C. E. 1977. Genetics of reproduction, variation and adaptation in rotifers. Arch. Hydrobiol. Beih. 8: 187–201.

King, C. E. and M. R. Miracle. 1980. A perspective on aging in rotifers. Hydrobiologia 73: 13–19.

Koste, W. 1978. *Rotatoria. Die Rädertiere Mitteleuropas Begründer von Max Voight, Vols. I, II.* Borntraeger, Berlin. [The best compilation of rotifer species available as of 1978; cf. Segers 2007.]

Mark Welch, D. B. and M. Meselson, 2000. Evidence for the evolution of bdelloid rotifers without sexual reproduction or genetic exchange. Science: 288: 1211–1215.

Melone, G. and C. Ricci. 1995. Rotatory apparatus in bdelloids. Hydrobiologia 313/314: 91–98.

Pray, F. A. 1965. Studies on the early development of the rotifer *Monostyla cornuta* Muller. Trans. Am. Microsc. Soc. 84: 210–216.

Ricci, C., G. Melone and C. Sotgia. 1993. Old and new data on Seisonidea (Rotifera). Hydrobiologia 255/256: 495–511.

Segers, H. 2007. Annotated checklist for the rotifers (Phylum Rotifera), with notes on nomenclature, taxonomy and distribution. Zootaxa 1564: 1–104.

Segers, H. and G. Melone. 1998. A comparative study of trophi morphology in Seisonidea (Rotifera). J. Zool., London 244: 201–207.

Signorovitch, A., J. Hur, E. Gladyshev and M. Meselson. 2015. Allele sharing and evidence for sexuality in a mitochondrial clade of bdelloid rotifers. Genetics 200: 581–590.

Sørensen, M. V. 2002. Phylogeny and jaw evolution in Gnathostomulida, with a cladistic analysis of the genera. Zool. Scr. 31: 461–480

Sørensen, M. V. and G. Giribet. 2006. A modern approach to rotiferan phylogeny: Combining morphological and molecular data. Mol. Phyl. Evolut. 40: 585–608.

Sørensen, M. V., H. Segers and P. Funch. 2005. On a new *Seison* Grube, 1861 from coastal waters of Kenya, with a reappraisal of the classification of the Seisonida (Rotifera). Zool. Studies 44: 34–43.

Wallace, R. L. and R. A. Colburn. 1989. Phylogenetic relationships within phylum Rotifera: Orders and genus *Notholca*. Hydrobiologia 186/187: 311–318.

Wallace, R. L., T. W. Snell, C. Ricci and T. Nogrady, T. 2006. *Rotifera: Volume 1: Biology, Ecology and Systematics*. Kenobi Productions, Ghent and Backhuys Publishers, Leiden.

Wilson, C. G. And P. W. Sherman. 2010. Anciently asexual bdelloid rotifers escape lethal fungal parasites by drying up and blowing away. Science 327. doi: 10.1126/science.1179252

Acanthocephala

Abele, L. G. and S. Gilchrist. 1977. Homosexual rape and sexual selection in acanthocephalan worms. Science 197: 81–83.

Amin, O. M. 1982. Acanthocephala. Pp. 933–940 in S. Parker (ed.), *Synopsis and Classification of Living Organisms*. McGraw-Hill, New York.

Baer, J. C. 1961. Acanthocéphales. Pp. 733–782 in P. Grassé (ed.), *Traité de Zoologie, Vol. 4, Pt. 1.* Masson et Cie, Paris.

Crompton, D. W. T. and B. B. Nickol. 1985. *Biology of the Acanthocephala*. Cambridge University Press, New York.

Dunagan, T. T. and D. M. Miller. 1991. Acanthocephala. Pp. 299–332 in F. W. Harrison and E. E. Ruppert (eds.), *Microscopic Anatomy of Invertebrates, Vol. 4, Aschelminthes*. Wiley-Liss, New York.

Herlyn, H. 2001. First description of an apical epidermis cone in *Paratenuisentis ambiguus* (Acanthocephala: Eoacanthocephala) and its phylogenetic implications. Parasitol. Res. 87: 306–310.

Medoc, V. J. N. Beisel. 2008. An acanthocephalan parasite boosts the escape performance of its intermediate host facing non-host predators. Parasitology 135: 977–984.

Miller, D. M. and T. T. Dunagan. 1985. New aspects of acanthocephalan lacunar system as revealed in anatomical modeling by corrosion cast method. Proc. Helm. Soc. Wash. 52: 221–226.

Nicholas, W. L. 1973. The biology of Acanthocephala. Adv. Parasitol. 11: 671–706.

Weber, M. and 7 others. 2013. Phylogenetic analyses of endoparasitic Acanthocephala based on mitochondrial genomes suggest secondary loss of sensory organs. Mol. Phyl. Evol. 66: 182–89.

Wey-Fabrizius, A. R.and 7 others. 2014. Transcriptome data reveal Syndermatan relationships and suggest the evolution of endoparasitism in Acanthocephala via an epizoic stage. PLoS ONE 9(2): e88618.

Whitfield, P. J. 1971. Phylogenetic affinities of Acanthocephala: An assessment of ultrastructural evidence. Parasitology 63: 49–58.

Yamaguti, S. 1963. *Systema Helminthum, Vol. 5, Acanthocephala*. Interscience, New York.

Micrognathozoa

Bekkouche, N, R. M. Kristensen, A. Hejnol, M. V. Sørensen and K. Worsaae. 2014. Detailed reconstruction of the musculature in *Limnognathia maerski* (Micrognathozoa) and comparison with other Gnathifera. Front. Zool. 11: 71.

De Smet, W. H., 2002. A new record of *Limnognathia maerski* Kristensen & Funch, 2000 (Micrognathozoa) from the subantarctic Crozet Islands, with description of the trophi. J. Zool., London 258: 381–393.

Funch, P. and R. M. Kristensen, R.M. 2002. Coda: The Micrognathozoa—a new class or phylum of freshwater meiofauna. Pp. 337–348 in, S. D. rundle, A. L. Robertson and J. M. Schmid-Araya, *Freshwater Meiofauna: Biology and Ecology*. Blackhuys Publishers, Leiden.

Kristensen, R. M. and P. Funch. 2000. Micrognathozoa: A new class with complicated jaws like those of Rotifera and Gnathostomulida. J. Morphol. 246: 1–49.

Kristensen, R. M. 2002. An introduction to Loricifera, Cycliophora, and Micrognathozoa. Integ. and Comp. Biol. 42: 641–651.

Sørensen, M. V. 2003. Further structures in the Jaw apparatus of *Limnognathia maerski* (Micrognathozoa), with notes on the phylogeny of the Gnathifera. J. Morphol. 255: 131–145.

Sørensen, M. V. and R. M. Kristensen. 2015. Micrognathozoa. Pp. 197–216 in A. Schmidt-Rhaesa (ed.), *Handbook of Zoology, Gastrotricha, Cycloneuralia and Gnathifera*. Walter De Gruyter GmbH, Berlin and Boston.

Worsaae, K. and G. W. Rouse. 2008. Is *Diurodrilus* an annelid? J. Morphol. 269: 1426–1455.

CHAPTER 17

The Lophophorates

Phyla Phoronida, Bryozoa, and Brachiopoda

Since the late nineteenth century, three phyla—Phoronida, Bryozoa (= Ectoprocta), and Brachiopoda—all of which posses a unique feeding structure called a lophophore, had been regarded as a possible clade within the Deuterostomia (see box on this page). The primary unifying feature of the Lophophorata was the **lophophore** itself, a ciliated tentacular feeding structure, and the associated **epistome** (a muscular flap that functions to move captured food particles from the lophophore to the mouth). Their alliance with the deuterostomes had been based on a suite of developmental features, including: radial and indeterminate cleavage, enterocoely, and development of the mouth secondarily rather than from the blastopore (i.e., "deuterostomy"). Further, all three phyla develop an adult body that is organized into three distinct parts, each with its own coelomic cavity (or paired cavities). This trait was originally termed **archicoely**, in reference to the "archicoelomate" concept, the idea that three regionalized body coeloms is a fundamental feature of Deuterostomia. This archicoelomate condition is also seen in the deuterostome phyla Echinodermata, Hemichordata, and Chordata. The developmental features were viewed as differentiating them from protostomes, in which the blastopore develops into the mouth and radial cleavage typically does not occur. In deuterostomes, the mouth is formed as a secondary opening (i.e., *deutero*, "second"; *stome*, "mouth") on the anterior end of the embryo.

However, the advent of molecular phylogenetics threw us a curve ball, and it now seems certain that the three lophophorate phyla are allied with spiralian protostomes, not deuterostomes, even though their development seems to have a mix of protostomous and deuterostomous characteristics. It has long been suspected that something akin to radial cleavage might be the primitive cleavage pattern within Bilateria (most demosponges also have radial cleavage), and the discovery that the lophophorate phyla belong to the Spiralia suggests that radial cleavage

Classification of The Animal Kingdom (Metazoa)

Non-Bilateria*
(a.k.a. the diploblasts)
PHYLUM PORIFERA
PHYLUM PLACOZOA
PHYLUM CNIDARIA
PHYLUM CTENOPHORA

Bilateria
(a.k.a. the triploblasts)
PHYLUM XENACOELOMORPHA

Protostomia
PHYLUM CHAETOGNATHA

SPIRALIA
PHYLUM PLATYHELMINTHES
PHYLUM GASTROTRICHA
PHYLUM RHOMBOZOA
PHYLUM ORTHONECTIDA
PHYLUM NEMERTEA
PHYLUM MOLLUSCA
PHYLUM ANNELIDA
PHYLUM ENTOPROCTA
PHYLUM CYCLIOPHORA

Gnathifera
PHYLUM GNATHOSTOMULIDA
PHYLUM MICROGNATHOZOA
PHYLUM ROTIFERA

Lophophorata
PHYLUM PHORONIDA
PHYLUM BRYOZOA
PHYLUM BRACHIOPODA

ECDYSOZOA
Nematoida
PHYLUM NEMATODA
PHYLUM NEMATOMORPHA

Scalidophora
PHYLUM KINORHYNCHA
PHYLUM PRIAPULA
PHYLUM LORICIFERA

Panarthropoda
PHYLUM TARDIGRADA
PHYLUM ONYCHOPHORA
PHYLUM ARTHROPODA
SUBPHYLUM CRUSTACEA*
SUBPHYLUM HEXAPODA
SUBPHYLUM MYRIAPODA
SUBPHYLUM CHELICERATA

Deuterostomia
PHYLUM ECHINODERMATA
PHYLUM HEMICHORDATA
PHYLUM CHORDATA

*Paraphyletic group

Scott Santagata revised the section on phylum Phoronida, Claus Nielsen revised the section on phylum Bryozoa, and Carsten Lüter revised the section on phylum Brachiopoda.

and deuterostomy are retained (or reacquired) plesiomorphic developmental features of the lophophorate phyla. In fact, the protostome phylum Priapula is now also thought to have radial cleavage. And, in the case of the Phoronida, the mouth does develop from a portion of the blastopore (protostomously), and species with radial and spiral cleavage have both been documented (the cleavage pattern may depend on the size of the egg; see the section on phoronid reproduction and development). Brachiopods, on the other hand, develop the mouth (and the anus) secondarily and hence show deuterostomous development. Molecular phylogenetics also places the former deuterostome phylum Chaetognatha within the protostome lineage (chaetognaths might also have spiral cleavage, although this is not certain). In other words, developmental programs appear to be more flexible (and homoplasious) than once thought, and even phyla outside the classic Deuterostomia clade can exhibit a deuterostome type of development.

The archicoelomate concept recognizes three distinct body regions (an anterior **prosome**, middle **mesosome**, and posterior **metasome**), each with its own coelomic cavities (**protocoel**, **mesocoel**, and **metacoel**). This condition is also referred to as a tripartite or **trimeric body plan**. However, developmental research since 2000 has shown that the protocoel may not comprise a true coelom in all lophophorate species, or might be absent altogether in many, and we discuss this below for each of the phyla. The great zoologist Libbie Hyman noted in the 1950s that a correspondence between the adult coeloms of lophophorates and the trimeric body of deuterostomes was poorly supported by both embryology and by adult morphology. In fact, Hyman was dismayed with the mix of characteristics found in the Lophophorata. She was actually inclined to classify them with protostomes, despite the enterocoelous coelom formation in brachiopods and the lack of spiral cleavage in bryozoans and brachiopods. She eventually found comfort in concluding that "the lophophorates constitute a connecting link between the Protostomia and the Deuterostomia, but the details of this connection cannot be stated." Hardly a more glib phylogenetic statement could be made.

With molecular phylogenetics confirming that lophophorates are protostomes, use of the terms "prosome," "mesosome," and "metasome" to describe their body plan is perhaps ambiguous, and the homology of these regions to those of deuterostomes is now called into question. The same holds for the Entoprocta, another protostome group whose body regions are often referred to as "prosome," "mesosome," and "metasome." The tentacular crown of pterobranchs (phylum Hemichordata; Chapter 26) also possesses mesosomal tentacles with a coelomic lumen. However, the homology between pterobranch tentacles and a true lophophore is invalidated because in the former case the tentacles do not fully surround the mouth.

Even though the lophophorate phyla are no longer considered deuterostomes, recent molecular phylogenetic work does suggest they comprise a monophyletic clade, the Lophophorata. However, the precise placement of this clade within Spiralia remains enigmatic. And whether Phoronida is sister to Bryozoa or Brachiopoda is still being debated, but the latter grouping seems to be favored by both comparative anatomy and recent phylogenomic research.

At first glance, the three lophophorate phyla may not appear closely related to one another. However, in addition to molecular phylogenetic analyses, they are united by their common possession of a lophophore and epistome; no other animal phylum has these structures. The lophophore is a ciliated, tentacular outgrowth arising from the mesosome and containing extensions of the mesocoel. It surrounds the mouth but not the anus. All lophophorate animals are sessile, have a U-shaped gut, and typically very simple, often transient reproductive systems. Nearly all secrete outer casings in the form of tubes, shells, or compartmented exoskeletons.

With the exception of a few freshwater bryozoans, the lophophorates are exclusively marine. All are benthic, living either in tubes (phoronids) or in secreted shells, or casings. The phoronids are a small group, comprising only two genera and 11 known species of solitary or gregarious worms. Bryozoans, on the other hand, make up a diverse taxon of about 6,000 species of colonial forms. The brachiopods, or lamp shells, include about 400 extant species. However, the brachiopods have left a record of thousands of fossil species (over 15,000 extinct species) as evidence of a greater past, and they were well established by the early Cambrian. Brachiopods flourished in both abundance and diversity from the Ordovician through the Carboniferous, but they have declined in numbers and kinds ever since.

Taxonomic History of the Lophophorates

The lophophorates have had a long and torturous taxonomic history. The earliest records of lophophorates are of various bryozoans reported in the sixteenth century. With few exceptions, the early zoologists treated them as plantlike and included them in the taxon Zoophyta, a misconception that persisted into the 1700s. In 1729, Jean-André Peyssonal finally established the animal nature of bryozoans, and in 1742 Bernard de Jussieu noted the compartmentalized condition of the colonies and coined the term "polyps" to refer to the individual animals. Still, most well known workers of the day (e.g., Linnaeus and Cuvier) continued to insist on allying them with the cnidarians in the group Zoophyta.

Eventually (in 1820), H. M. de Blainville noted the complete gut of the bryozoans and "raised" them above the cnidarians. By 1831, two names had been coined for these animals—Bryozoa (by the German zoologist Christian Gottfried Ehrenberg) and Polyzoa (by the Englishman J. Vaughan Thompson). Just about the time the bryozoans were being recognized as a separate group (under one name or the other), concurrent events confused the issue further. The entoprocts (= Kamptozoa) were described, and most workers included them with the Bryozoa, while others recognized a relationship between bryozoans and other lophophorate phyla. All of this became terribly entangled in 1843 with Henri Milne-Edwards's concept of a taxon Molluscoides, which he established to include bryozoans and compound ascidians (chordates). Hatschek raised Bryozoa and Entoprocta to distinct phylum ranking in 1891. The brachiopods were known, at least from fossils, in the 1600s and were allied with the molluscs. This mistaken view was held until late in the nineteenth century.

Phoronids were first described from their larvae by Johannes Müller in 1846, who thought they were adults and named them *Actinotrocha brachiata*. In 1854, Karl Gegenbaur recognized these animals as larval stages. The adults were found in 1856 and described by T. S. Wright, who named them *Phoronis*. Finally, in 1867 the renowned embryologist Alexander Kowalevsky studied the metamorphosis of "Actinotrocha" and established the relationship between the two stages. To this day, the name "actinotroch" persists as a general term for the phoronid larva.

In 1857, Albany Hancock recognized the relationship between brachiopods and bryozoans, but the whole matter was mixed up with confusion over the entoprocts and Milne-Edwards's Molluscoides. Hinrich Nitsche made a valiant attempt in 1869 to separate the entoprocts and bryozoans, and in the 1880s Otis Caldwell and Thomas Masterman both put forth the idea of a close relationship between the three lophophorate phyla. This view was supported by Hatschek in 1888, when he suggested the establishment of a phylum Tentaculata to include the classes Phoronida, Bryozoa, and Brachiopoda (but excluding the entoprocts), thus establishing the grouping we recognize today as Lophophorata. Since that time several attempts to unite some or all of these animals into a single taxon have been made, including Anton Schneider's Lophophorata in 1902 (originally only for phoronids and bryozoans). Hyman (1959) retained separate phylum status for the three groups, an arrangement that has remained popular ever since.

In the recent past, some workers revived the idea of a possible entoproct–ectoproct affinity and argued that the two groups are closely allied with Cycliophora in a grouping called Polyzoa. However, this is not supported by the most recent large-scale phylogenetic work. An alliance between Bryozoa and Entoprocta can also be rejected on the basis of direct comparative grounds. Entoprocts do not possess the lophophore–epistome complex. Furthermore, they lack any vestiges of a coelom or trimeric body plan. The feeding currents are virtually opposite in the two groups, and the methods of food capture and transport are entirely different. Entoprocts possess ducted gonads and bryozoans do not. Cleavage in entoprocts is spiral, whereas it is radial in bryozoans. Larval forms and particularly metamorphosis are clearly different in the two groups. More importantly, if the three phyla comprising Polyzoa formed a monophyletic clade, then they should be defined by synapomorphies, but none have been found. Similarities such as budding and U-shaped guts are superficial and common to many colonial sessile animals. Thus, we conclude that Bryozoa and Entoprocta are distantly related protostomes, a conclusion supported by the most recent molecular phylogenetic studies. The name Ectoprocta is often used in lieu of Bryozoa, acknowledging the fact that the anus lies outside the lophophore ring, whereas in the Entoprocta it lies within the lophophore ring (and the mouth is on the rim carrying the tentacle crown).

The Lophophorate Body Plan

All lophophorates are built for benthic life and suspension feeding, the latter being a primary function of the lophophore itself. The anterior body region, or prosome, is reduced to a small flaplike structure called the epistome (lost in some lineages), which is associated with an overall reduction of the head, the elaboration of the mesosome as the lophophore, and a sessile lifestyle. The metasome houses the bulk of the viscera. The U-shaped gut is clearly advantageous for living encased in tubes, compartments, and shells; such animals do not "foul their nests," as it were. We have seen similar adaptations in some other groups with comparable habits, such as the recurved gut of sipunculans and the ciliated fecal-removal grooves of some tube-dwelling polychaetes. In lophophorates, this condition not only prevents fecal accumulation in the encasement, but generally brings the anus close to rejection currents produced by the cilia on the lophophore. In recent years, it has been shown that an epistome coelom (the protocoel) is absent in some phoronids and brachiopods, and all gymnolaemate bryozoans, whereas in phylactolaemate bryozoans there is an open connection between the epistomal cavity and the lophophoral coelom.

The phoronids most clearly display the above traits, and retain the vermiform shape of the probable ancestral form. Evolutionarily, bryozoans have exploited asexual reproduction, colonialism, and small size. Relieved of long-distance internal transport problems, the bryozoans have lost the circulatory and excretory systems. The brachiopods evolved a pair of valves or shells that encase and protect the body, including the lophophore. Thus, instead of exposing the lophophore in the water (as phoronids and bryozoans do), the brachiopods draw water into their mantle cavity for suspension feeding, an action analogous to that in bivalve molluscs. Sessile animals with soft parts (e.g., the lophophore), living on benthic surfaces, are exposed to potentially high levels of predation—thus the selective advantage of the brachiopod shell is clear. The phoronids are motile to the extent that their bodies can be retracted within their tubes, protecting the soft parts. Bryozoans are entirely anchored in their casings, but the lophophore itself is uniquely retractable into the body, as the result of a unique arrangement of muscles and hydraulic mechanisms.

BOX 17A Characteristics of the Phylum Phoronida

1. Trimeric, vermiform lophophorates
2. Adult mesoderm derived from both embryonic ectoderm and endoderm
3. Body divided into flap-like epistome (prosome), lophophore-bearing collar (mesosome), and elongate trunk (metasome)
4. Gut U-shaped, anus close to mouth
5. One pair of metanephridia in metasome (trunk)
6. Closed circulatory system
7. Gonochoristic or hermaphroditic, with transient peritoneal gonads
8. With indirect life histories; usually with a unique actinotroch larva
9. Radial and spiral-like, indeterminate cleavage; coeloblastulae gastrulate by invagination; blastopore becomes mouth (protostomous)
10. Marine benthic tube dwellers

Phylum Phoronida: The Phoronids

All phoronids occupy a cylindrical tube of their own secretion in which they can move freely. The animals typically occur in aggregations that, in some species, result from asexual propagation. Their chitinous tubes are usually cemented to hard substrata or buried vertically in soft sediments (Figure 17.1; Box 17A). They are assigned to only two genera— *Phoronis* and *Phoronopsis*—and 11 widely distributed species. The phylum name was apparently derived from the Latin term *Phoronis*, the surname of Io (who, according to mythology, was changed into a cow and roamed the Earth, eventually to be returned to her former body). Recall that these worms were first described from larval stages drifting in the sea, and only much later were the adults recognized as part of the same life cycle. Adult phoronids are found in intertidal sands or mud flats, and to depths of about 400 meters.

The Phoronid Body Plan

Most adult phoronids range from about 1 to 25 cm in length. The vermiform body shows little regional specialization, except for the distinct lophophore and a modest inflation of the end bulb, which houses the stomach and also aids in anchoring the animals in their tubes. The slitlike mouth is located between the tentacle-bearing lophophoral ridges and is overlaid by a flap, the epistome. The lateral aspects of the ridges are distinctly coiled and flank the dorsal anus and paired nephridiopores (Figures 17.2 and 17.3). Using the terms anterior and posterior when referring to phoronids can be misleading. During metamorphosis, the true anterior (mouth-bearing) and posterior (anus-bearing) ends are brought very close together by rapid growth and enlargement of the ventral surface (Figure 17.4). The dorsal surface is reduced to the small area between the adult mouth and anus. Because of these conditions, we refer to the "ends" of the adult worm as the lophophoral end and the stomachic (= ampullar) end (Figure 17.1A).

Body Wall, Body Cavity, and Support

The phoronid body wall includes an epidermis of columnar cells overlaid by a very thin cuticle. Within the epidermis are sensory neurons and various gland cells (Figure 17.2A), the latter being responsible for the production of mucus and chitin. The epidermis of the lophophore is densely ciliated. Internal to the epidermis and its basement membrane is a thin layer of circular muscle, some limited diagonal muscle, and a thick layer of longitudinal muscle. A peritoneum lines the longitudinal muscles and forms the outer boundary of the coelom.

Although the coelom was long described as being tripartite, there is structural variation among species in the epithelial lining of the most anterior cavity, the protocoel. Ultrastructural investigations have shown that the myoepithelial lining of the epistome lacks adherent cell junctions (a key component of true epithelial tissues) in at least two *Phoronis* species. However, adherent junctions are present in the myoepithelial lining of the epistome of *Phoronopsis harmeri*. True epithelial layers comprise the coelomic linings of the collar (mesocoel) and trunk (metacoel) body regions in both *Phoronis* and *Phoronopsis*. This leads to the question of whether the ancestral state of phoronid coelomic cavities was bipartite (and subsequently evolved to become tripartite in *Phoronopsis*) or tripartite (and was subsequently lost in *Phoronis*). Regardless of the validity of

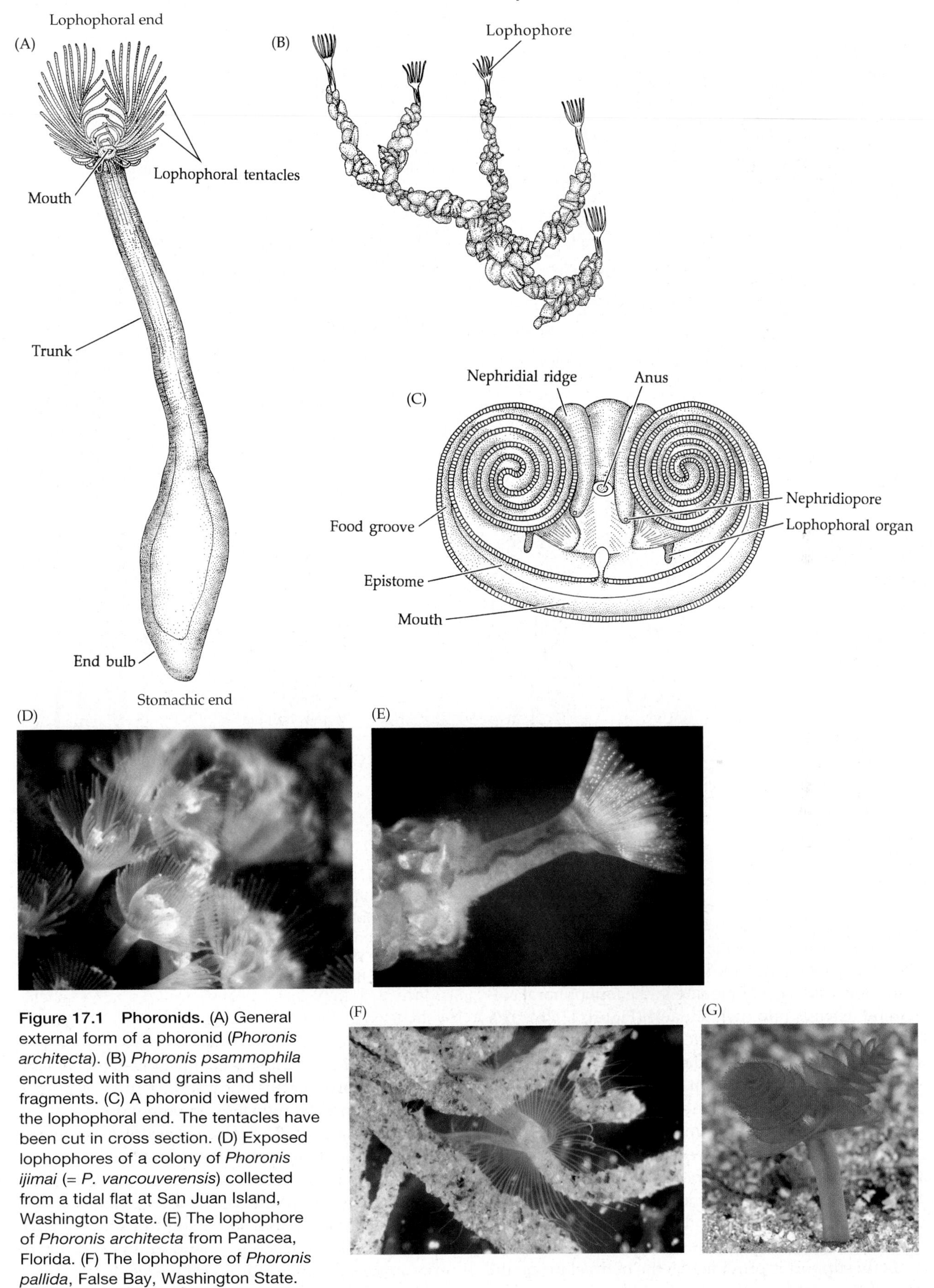

Figure 17.1 Phoronids. (A) General external form of a phoronid (*Phoronis architecta*). (B) *Phoronis psammophila* encrusted with sand grains and shell fragments. (C) A phoronid viewed from the lophophoral end. The tentacles have been cut in cross section. (D) Exposed lophophores of a colony of *Phoronis ijimai* (= *P. vancouverensis*) collected from a tidal flat at San Juan Island, Washington State. (E) The lophophore of *Phoronis architecta* from Panacea, Florida. (F) The lophophore of *Phoronis pallida*, False Bay, Washington State. (G) The phoronid *Phoronopsis californica*, with its spiral lophophore.

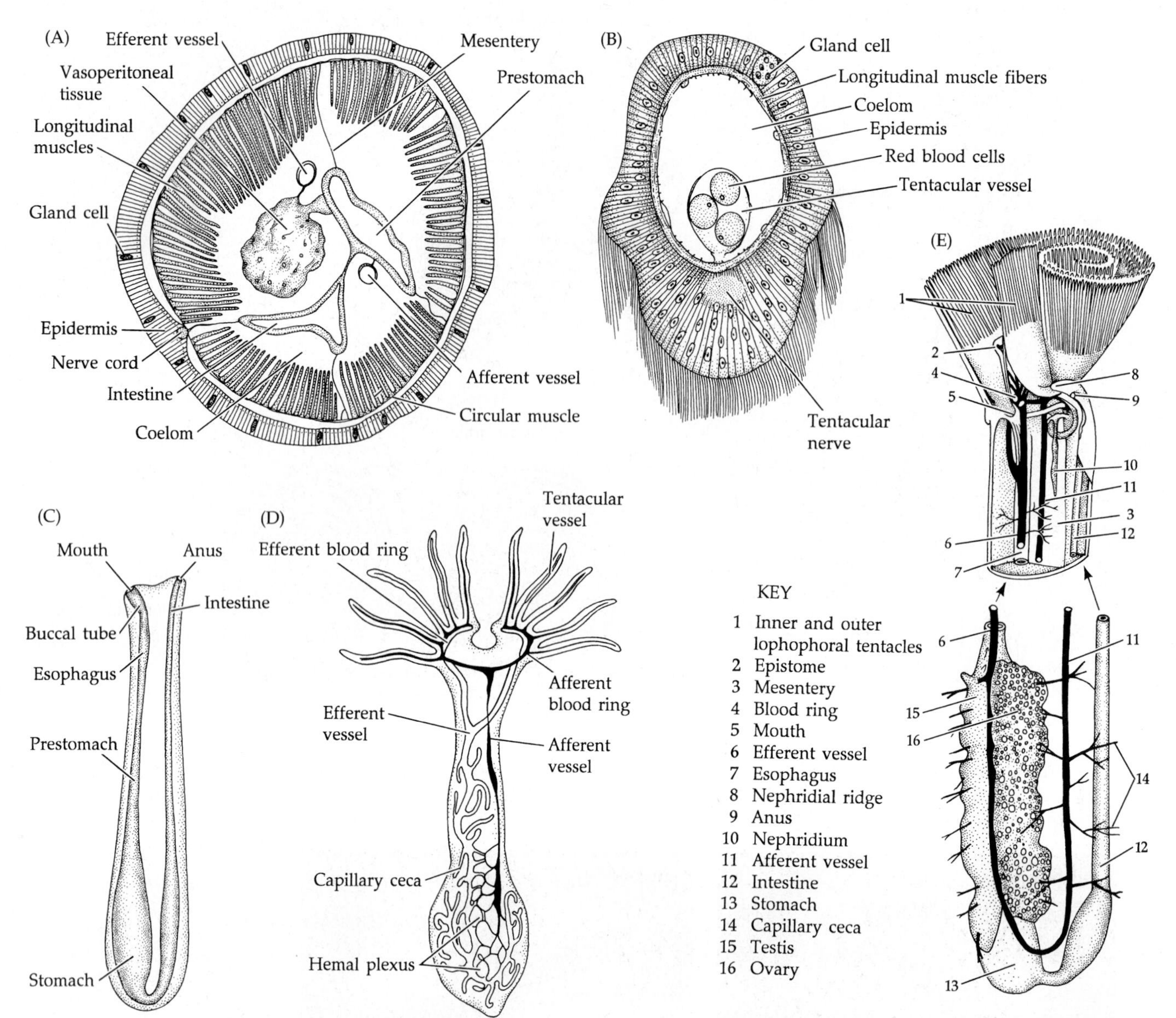

Figure 17.2 Phoronid internal anatomy. (A) The trunk of a phoronid in cross section. Note the body wall layers and coelomic partitioning. (B) A tentacle of the lophophore in cross section. (C) The digestive tract. (D) The circulatory system. (E) The major internal organs in the lophophoral and stomachic ends.

the "protocoel" as a true coelomic cavity, as an anatomical structure it is limited to a single small cavity within the epistome of only some species. The unpaired mesocoel comprises a coelomic ring in the lophophoral collar and extends into each tentacle (Figure 17.2B). The protocoel and mesocoel are connected to one another along the lateral regions of the epistome. The metacoel forms the main trunk coelom, which is separated from the mesocoel by a transverse septum. Ontogenetically, the metacoel is an uninterrupted cavity with only one midventral mesentery. However, secondary mesenteries form later in development, yielding four longitudinal spaces (Figure 17.2A). The coelomic fluid contains several kinds of freely wandering cells, or coelomocytes, including phagocytic amebocytes.

Body support is provided by the hydrostatic qualities of the coelomic chambers and by the tube. The muscles of the body wall are rather weak, particularly those in the circular layer, and once removed from their tubes phoronids are capable of only limited movement. Normally, however, the body wall of the end bulb is pressed against the tube, holding the worm in place. When disturbed, the animal simply contracts into the tube; the lophophore itself is not retractable.

The tube is secreted by epidermal gland cells. When first produced, the chitinous secretion is sticky, but upon contact with water it solidifies to a flexible parchment-like consistency. Sand grains and bits of other material adhere to the tube during the sticky phase of tube formation in those phoronids that inhabit soft substrata (e.g., *Phoronopsis harmeri*). In some species the tubes intertwine with one another, with the

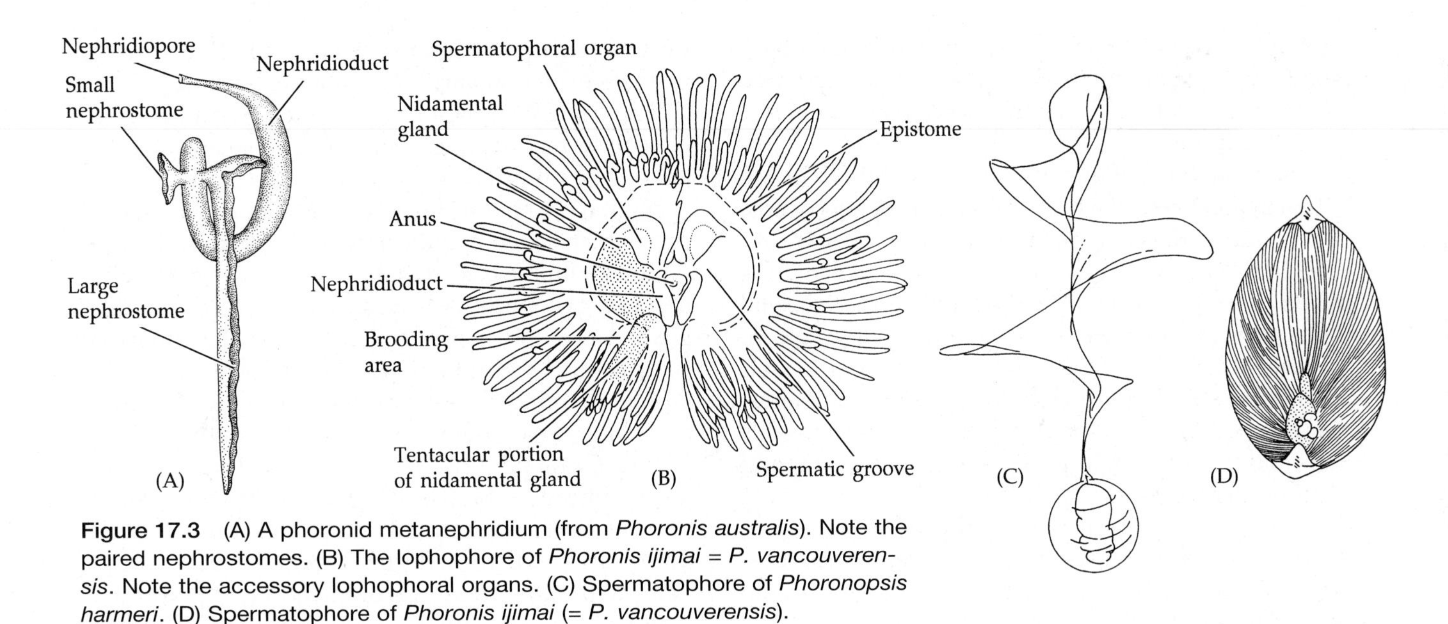

Figure 17.3 (A) A phoronid metanephridium (from *Phoronis australis*). Note the paired nephrostomes. (B) The lophophore of *Phoronis ijimai* = *P. vancouverensis*. Note the accessory lophophoral organs. (C) Spermatophore of *Phoronopsis harmeri*. (D) Spermatophore of *Phoronis ijimai* (= *P. vancouverensis*).

whole tangled aggregation attached to a substratum or actually embedded in calcareous stone or shells (e.g., *Phoronis hippocrepia*).

The Lophophore, Feeding, and Digestion

The tentacles of the **lophophore** are hollow, ciliated outgrowths of the mesosome, and each contains a blind-ended blood vessel and a coelomic extension (Figure 17.2B). The tentacles are in a double row, arising from two ridges, in a circle or a spiral as is *Phoronopsis californica* (Figures 17.1–17.3). The ridges lie close to one another and form a narrow food groove in which the slitlike mouth is located (Figure 17.1C). Because the sides of the lophophoral ridges are coiled, many tentacles are compacted into a small area.

Phoronids are ciliary-mucus suspension feeders. The lophophoral cilia generate a water current that passes down between the two rows of tentacles and then out between the tentacles. Food particles are trapped in mucus lining the food groove and then passed along the groove by cilia to the mouth. As the current passes between the tentacles and out of the area of the food groove, some water flow is directed over the anus and nephridiopores away from the animal (Figure 17.1C).

The digestive tract is U-shaped, but rather simple and not coiled (Figure 17.2C). The mouth is overlaid by the epistomal flap and leads inward to a short buccal tube, which is followed by an esophagus and a narrow prestomach. Within the end bulb the gut expands into a stomach, from which emerges the intestine. The intestine bends up toward the lophophore and leads to a short rectum and the anus. The gut is supported by peritoneal mesenteries (Figure 17.2A).

The majority of the digestive tract is apparently derived from endoderm, except for the anus and the initial formation of the larval intestine that are formed from an ectodermal invagination. Some parts are muscular, but only weakly so, and much of the movement of food is by ciliary action. A middorsal strip of densely ciliated cells arises in the prestomach and extends into the stomach, and it is probably responsible for directing food along that portion of the gut. Gland cells occur in the esophagus but their function remains uncertain. Transitory syncytial bulges in the stomach walls are the sites of intracellular digestion in that organ.

Circulation, Gas Exchange, and Excretion

Phoronids contain an extensive circulatory system comprised of two major longitudinal vessels between which blood is exchanged in the lophophoral and stomachic ends of the body (Figure 17.2D,E). Various names have been applied to these vessels depending on their positions in the body. These terms are often confusing because the positions of vessels vary along the length of the trunk, and they are not clearly dorsal, medial, lateral, or ventral as the names imply. Therefore, preference is given here to the terms afferent and efferent vessels, which refer to the direction of blood flow relative to the lophophore.

The afferent vessel extends unbranched from the region of the end bulb to the base of the lophophore. For most of its length it lies more or less between the descending and ascending portions of the gut. In the mesosome the afferent vessel forks, forming an afferent "ring" vessel (U-shaped) at the base of the lophophoral tentacles. A series of lophophoral vessels, one in each tentacle, arises from the afferent ring. Each of these vessels joins with an efferent "ring" vessel (also U-shaped), which drains blood from the lophophore. Thus the afferent and efferent blood rings lie against one another, and generally share openings into the lophophoral tentacles within which blood moves back and forth, as there is just a single vessel in each tentacle. One-way flap valves largely prevent backflow into the afferent ring.

The arms of the efferent ring unite to form the main efferent blood vessel, which extends through the trunk. This vessel connects with numerous branches or simple blind diverticula called capillary ceca, which bring blood close to the gut wall and other organs. In the end bulb, surrounding the stomach and first part of the intestine, blood flows from efferent to afferent vessels through spaces composing the **hemal (stomachic) plexus** (Figure 17.2D,E). Blood actually leaves the vessels here and flows through spaces between the organs and their bordering layers of peritoneum. Thus, technically speaking, the system is open at this point; however, blood flow is directed within the confines of these passages. Blood is moved through the circulatory system largely by muscular action of the blood vessel walls.

The intimate association of blood and the stomach wall suggests that nutrients are picked up from the stomach by the circulatory fluid and transported throughout the body. The tentacles of the lophophore are also probably the most important sites of gas exchange. Oxygenated blood flows from the lophophore into the efferent vessel, and is then distributed to all parts of the trunk. The blood contains nucleated red corpuscles, with hemoglobin as the respiratory pigment.

A pair of metanephridia lies in the trunk, and each bears two nephrostomes opening to the metacoel (Figure 17.2E and 17.3A). In each nephridium, the nephrostomes—one large and one small—join a curved nephridioduct, which leads to a nephridiopore adjacent to the anus. Although virtually nothing is known about excretory physiology in phoronids, particulate crystalline matter has been observed exiting the nephridiopores and probably represents precipitated nitrogenous waste products. The nephridia also function as pathways for the release of gametes. Osmoregulatory problems are presumably insignificant in subtidally occurring phoronid species. However, little is known about the osmotic challenges faced by those species inhabiting intertidal and/or estuarine habitats.

Nervous System

The nervous system of phoronids is rather diffuse and lacks a distinct cerebral ganglion. This condition is no doubt related to the sedentary lifestyle and overall reduction in cephalization in these worms. Most of the nervous system is intimately associated with the body wall, being either intraepidermal or immediately subepidermal. Throughout the body a layer of nerve fibers is present between the epidermis and the circular muscle layer. Simple sensory neurons project out from this layer, either singly or in bundles, and extend to the body surface as the only receptor structures. Motor neurons extend inward to the muscle layers.

The central nervous system includes a group of basiepidermal neuronal cell bodies concentrated between the mouth and the anus (the so called dorsal "ganglion" or neural plexus) connected to a collar nerve ring, which lies at the base of the lophophore and is continuous with the basiepidermal nerve layer. The collar nerve ring connects with frontal and abfrontal nerves of the tentacles and also with motor nerves that innervate some of the longitudinal muscles in the metasome. In addition, a bundle of sensory neurons extends from the nerve ring to each of the lophophoral organs.

Phoronids possess one or two longitudinal giant motor fibers in the trunk (absent in the very small *Phoronis ovalis*). When only one longitudinal fiber is present, it lies on the left side. This fiber actually originates within the right side of the nerve ring, through which it passes to emerge on the left side. This nerve fiber is intraepidermal except where it extends inward along the left nephridium. In those species where two longitudinal fibers are present, the right one originates on the left side of the nerve ring and extends to the opposite side of the body.

Reproduction and Development

Asexual reproduction by transverse fission or by a form of budding has been documented in a few species. Phoronids are also capable of regenerating lost body parts, and various parts of the lophophoral end are capable of autotomy.

Both gonochoristic and hermaphroditic species of phoronids are known, and in the latter case some are simultaneous hermaphrodites (e.g., *Phoronis ijimai* = *P. vancouverensis*). The gonads are transient and form as thickened areas of the peritoneum around the hemal plexus (Figure 17.2D). The resulting mass of gamete-forming tissue and blood sinuses is sometimes called vasoperitoneal tissue. Gametes are proliferated into the metacoel and typically are carried to the outside via the nephridia. Fertilization is usually internal, as found in *Phoronopsis harmeri*. In this species, the male lophophoral organs produce elaborately shaped spermatophores (Figure 17.3B–D) that are transferred to the metanephridial opening or tentacles of the females. In these females the lophophoral organs are called **nidamental glands** and serve as brooding areas. The complicated internal fertilization process in *Phoronopsis harmeri* (= *P. viridis*) was elucidated by Russel Zimmer (1972 and 1990). Spermatophores contacting the nephridiopore release ameboid masses of V-shaped sperm, which travel down the metanephridial funnels into the base of the trunk coelom near the ovary. If a spermatophore contacts a female's tentacle instead, the sperm enter the lophophoral coelom by lysis of the tentacular wall and then proceed to digest their way through the septum separating the mesocoel and metacoel into the trunk

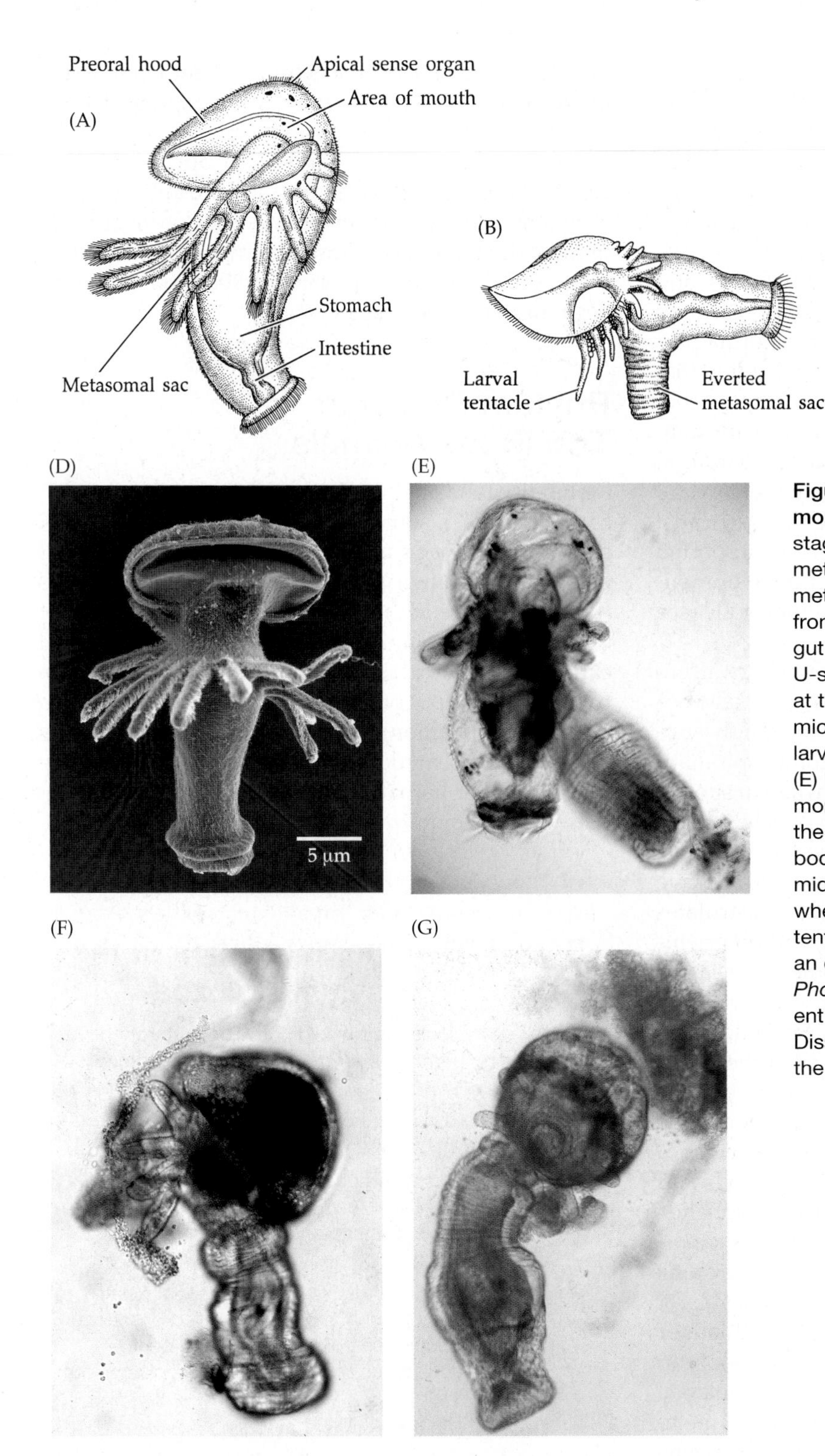

Figure 17.4 Phoronid larvae and metamorphosis. (A) Diagram of a typical mid-stage actinotroch larva. (B,C) Stages in the metamorphosis of an actinotroch larva. The metasomal (or juvenile trunk) sac everts from the larval body, drawing out the larval gut at its midpoint. The juvenile gut is now U-shaped, leaving the mouth and anus at the anterior end. (D) Scanning electron micrograph of an unidentified actinotroch larva collected from Tampa Bay, Florida. (E) Photograph of the initial stage of metamorphosis of *Phoronis hippocrepia* when the juvenile trunk sac everts from the larval body. (F) Photograph of a lateral view of a mid-metamorphic stage of *Phoronis pallida* when the gut is U-shaped and the larval tentacles are remodeled. (G) Photograph of an oral view of a late metamorphic stage of *Phoronis pallida* when the efferent and afferent vessels are filled with red blood cells. Dissociated cells at top are produced from the histolysis of hood tissues.

coelom, where fertilization occurs. Although fertilization has not actually been observed, Zimmer's experimental data, coupled with the fact that fertilized ova occur internally, suggest that this scenario is the likeliest explanation.

Developmental strategy differs among species, the particular pattern depending in part on the size of the egg and whether or not the embryo is brooded. The ova of broadcast-spawning species contain little yolk and develop quickly to planktotrophic **actinotroch larvae** (Figure 17.4). In species that possess nidamental glands, fertilization is followed by brooding until release at the actinotroch stage. The eggs of these species are moderately rich in yolk, providing nutrients for the embryos during the brooding period. *Phoronis ovalis* lacks nidamental glands, but the yolky eggs are shed into the maternal tube where they are brooded. Development in *P. ovalis* does not include a typical actinotroch; instead, the embryos emerge through the autotomized lophophoral end as ciliated, sluglike larvae that have a short, planktonic life.

There are aspects of actinotroch anatomy that link phoronid larvae to both trochophore and dipleurula-like larval forms (nervous system, ciliated bands, and cellular composition). However, considering recent interpretations of both trochophore and dipleurula concepts (e.g., Rouse 1999; Nielsen 2009), concluding that the actinotroch form is derived strictly from a trochophore or dipleurula-like larval form may be too constraining. It might be that plesiomorphic larval traits at the base of the bilaterian tree were mixed and extant larval forms exhibit a derived subset of these traits. Although this argument is speculative, it is supported by the broad general conservation of developmental programs that exist on a metazoan-wide scale (see Hejnol and Martin-Duran 2015).

Since their first description, the evolutionary affinities of phoronids have been debated, as some developmental and morphological characters seemingly support both a deuterostome and protostome (spiralian) origin. Early reports of spiral-like cleavage in species that broadcast spawn smaller (~60 μm) eggs have been corroborated by confocal and 4-D time-lapse microscopy. Species with larger (~100 μm) eggs typically have radial cleavage. A coeloblastula forms and gastrulates largely by invagination. Cell labeling and other embryological experiments show that some of the larval mesoderm is derived from ectodermal cells, especially where the ectoderm is in contact with the endoderm; the remaining mesoderm is derived from the endoderm. Furthermore, a portion of the blastopore forms the mouth. Along with these developmental and morphological features, molecular phylogenetic evidence supports phoronids being related to brachiopods and ectoprocts as well as spiralian animals such as annelids and molluscs.

With the exception of *Phoronis ovalis*, all phoronids produce distinctive **actinotroch larvae** (Figure 17.4A,D). The fully formed actinotroch bears a preoral hood, or lobe, over the mouth. As the actinotroch develops, an inpocketing (called the metasomal or juvenile trunk sac) forms on the ventral surface. At settlement and metamorphosis this sac everts, extending the ventral surface such that the anus and mouth remain close to one another as the gut is drawn out into the characteristic U-shape. As discussed previously the structure of the hood "coelom" of some *Phoronis* species varies and may collapse late in larval development. However, the hood coelom of *Phoronopsis harmeri* apparently has a true epithelial lining, at least a portion of which may be carried over during metamorphosis to form the coelomic lining of the juvenile epistome. Metamorphic remodeling of the larva into the juvenile form varies somewhat among species, but, in general, the neural and muscular tissues of the hood undergo histolysis. In some species the larval tentacles are remodeled into the juvenile lophophore during metamorphosis (Figure 17.4), but in other species the larval tentacles are shed, and the initial juvenile lophophore forms from separate tentacle rudiments. All of these dramatic tissue rearrangements can occur in as little as 15–20 minutes, after which the juvenile worms begin to secrete their tubes.

Phylum Bryozoa: The Moss Animals

Members of the phylum Bryozoa (Greek *bryon*, "moss"; *zoon*, "animal"), sometimes called Ectoprocta (Greek *ecto*, "outside"; *procta*, "anus"), are sessile colonies of zooids living in marine and freshwater environments (Box 17B, Figure 17.5). Most colonies are the product of budding from a single, sexually produced individual called an **ancestrula**, but colonies may also develop from various types of resting bodies. The colony form differs greatly among species, and the bushy species were earlier treated as plants, so specimens can be found in older herbariums. Each zooid typically consists of a polypide, with a tentacle crown and the gut.

BOX 17B Characteristics of the Phylum Bryozoa

1. Always colonial, with zooids developing through budding from a metamorphosed larva or from existing zooids
2. Each zooid consists of a polypide, which can be retracted into a cystid.
3. The polypide is composed of a ciliated tentacle crown used in filter feeding, a U-shaped gut, a simple nervous system, and various muscles associated with extension and retraction of the polypide.
4. The cystid wall consists of an extracellular gelatinous or chitinous cuticle, with a calcified basal layer in some groups, the epidermis, and the peritoneum; various types of muscles are associated with the body wall.
5. There are no circulatory or excretory organs.
6. Zooids are hermaphroditic in many species, but some species have male and female zooids in the same colony; gametes differentiate from transient patches of the body wall peritoneum or the funiculus.
7. Radial holoblastic cleavage; indirect development with planktotrophic cyphonautes larvae or with lecithotrophic coronate larvae; all internal larval organs degenerate after settling
8. Sessile in marine or freshwater habitats

Figure 17.5 Bryozoan diversity. (A–C) Ctenostomata. (A) *Triticella* growing on the antenna of a crustacean. (B) *Anguinella*. (C) *Alcyonidium* washed up on the shore. (D–F) Cheilostomata. (D) *Scruparia* growing on a coralline alga. (E) A small colony of *Membranipora* on a piece of kelp. (F) *Pentapora.* (G–I) Cyclostomata. (G) *Crisia,* note the white gonozooids on the two middle branches. (H) *Tubulipora.* (I) *Heteropora.* (J–L) Phylactolaemata. (J) *Plumatella*, part of a colony with many statoblasts in the body cavity. (K) *Cristatella.* (L) *Pectinatella*, large floating colonies (*ojassie*).

These can be retracted into a cystid, which is the body wall of the individual zooid or the wall that is common to several zooids (Figure 17.6). Marine bryozoans are known from all depths and latitudes, mostly growing on or attached to solid substrata. A few types form free colonies that creep on the bottom (Figure 17.8E). An Antarctic species forms gelatinous colonies on floating pieces of ice. A few species occur in fresh and brackish water. Large gelatinous colonies of *Pectinatella magnifica* are frequently encountered in streams and ponds east of the Mississippi River. Meter-large floating colonies of the same species occur in Lake Shoji in Japan, where they are called "ojassie." Littoral regions in most parts of the world harbor luxuriant growths of bryozoans that often cover large areas of rock surfaces and algal fronds. Some species have coral-like growth forms that can create miniature "reefs" in shallow-water habitats. Others form dense bushlike colonies or gelatinous spaghetti-like masses. Many encrusting forms grow on the shells or exoskeletons of other invertebrates and some bore into calcareous substrata. Most members of the gymnolaemate family Hippoporidridae grow on mollusc shells inhabited by hermit crabs, where evidence suggests a mutualistic relationship between the bryozoans and the crustaceans. Two classes are recognized here, but other classification schemes are in use. The fossil record indicates that the cheilostomes evolved from stem-group ctenostomes, and this probably holds for the cyclostomes as well. About 6,000 living species are recognized, and there is a rich fossil record going back to the Early Ordovician.

CLASSIFICATION: PHYLUM BRYOZOA

CLASS GYMNOLAEMATA Mainly marine bryozoans. Colonies with chitinous or calcified cystids; in calcified species, the cystid walls are primarily calcium carbonate, often with aragonite on the outer surface of the frontal walls; zooids rather small, monomorphic in some species, but especially the cheilostomes show several types of heterozooids; tentacle crowns circular or slightly oblique.

SUBCLASS EURYSTOMATA A highly diverse group of almost exclusively marine bryozoans. Colony form is extremely variable, soft or calcified, encrusting to arborescent; body wall lacks an entirely continuous layer of muscles; zooids variably modified from basic cylindrical form; zooids joined by pores through which cords of tissue extend and join with their neighbor's funiculus.

ORDER CTENOSTOMATA Colonies vary in shape from forms with hydroid-like individual zooids developing from free or creeping stolons to compact colonies; skeleton leathery, chitinous, or gelatinous, not calcified; openings through which tentacle crowns protrude lack operculum; zooids generally monomorphic. (e.g., *Aethozoon, Alcyonidium, Amathia, Bowerbankia, Flustrellidra, Nolella, Victorella*)

ORDER CHEILOSTOMATA Colony form varies, but generally of box-shaped zooids with calcareous walls; openings with operculum; zooids often polymorphic; embryos usually develop in various types of brooding structures. (e.g., *Bugula, Callopora, Carbasea, Cellaria, Conopeum, Cornucopina, Cribrilaria, Cryptosula, Cupuladria, Electra, Eurystomella, Flustra, Hippothoa, Primavelans* (formerly *Hippodiplosia*), *Membranipora, Metrarabdotos, Microporella, Pentapora, Porella, Pyripora, Rhamphostomella, Schizoporella, Selenella, Thalamoporella, Tricellaria*)

SUBCLASS STENOLAEMATA (with the only surviving subgroup Cyclostomata). Exclusively marine. Zooids housed in tubular, calcified skeletal compartments; zooids cylindrical or trumpet shaped, rarely polymorphic; each polypide surrounded by a coelomic sac, where the inner wall covers the gut whereas the outer wall is free, forming the unique, so-called membranous sac, which consists of the coelomic epithelium, a series of tiny ring-shaped muscles, and a basal membrane; the body cavity outside the membranous sac, the exosaccal cavity, is continuous throughout the colony, either through small interzooidal pores or through larger common cavities; funiculus inside the coelomic cavity, not connecting neighboring zooids; reproduction involves specialized female zooids in which the fertilized egg develops through polyembryony, so that the embryos in a gonozooid are a clone. (e.g., *Actinopora, Crisia, Diaperoecia, Disporella, Hornera, Idmodronea, Tubulipora*)

CLASS PHYLACTOLAEMATA Freshwater bryozoans. Colonies with chitinous or gelatinous cystids; zooids cylindrical, large, and monomorphic; tentacle crown large and usually horseshoe-shaped; body wall muscles well developed; body cavity of the zooids in open connection throughout the colonies; a cord of tissue, the funiculus, extends from the gut to the body wall, but not between zooids; most produce asexual resting bodies called statoblasts. (e.g., *Cristatella, Hyalinella, Lophophus, Lophopodella, Pectinatella, Plumatella*)

The Bryozoan Body Plan

Bryozoan specialists have developed a complicated terminology, especially concerning the morphology of the cystids, and some terms have, quite confusingly, been given various definitions. Early workers mistakenly thought that bryozoan zooids were actually composed of two organisms, the skeletal outer wall and the internal soft parts, which they named the cystid and polypide, respectively. These terms were redefined by Hyman (1959) and now have good meaning relative to the functional morphology of bryozoans. The **cystid** comprises the outer body wall—that is, the nonliving and living housing of each zooid. The **polypide** includes the **tentacle crown** (the lophophore) and soft

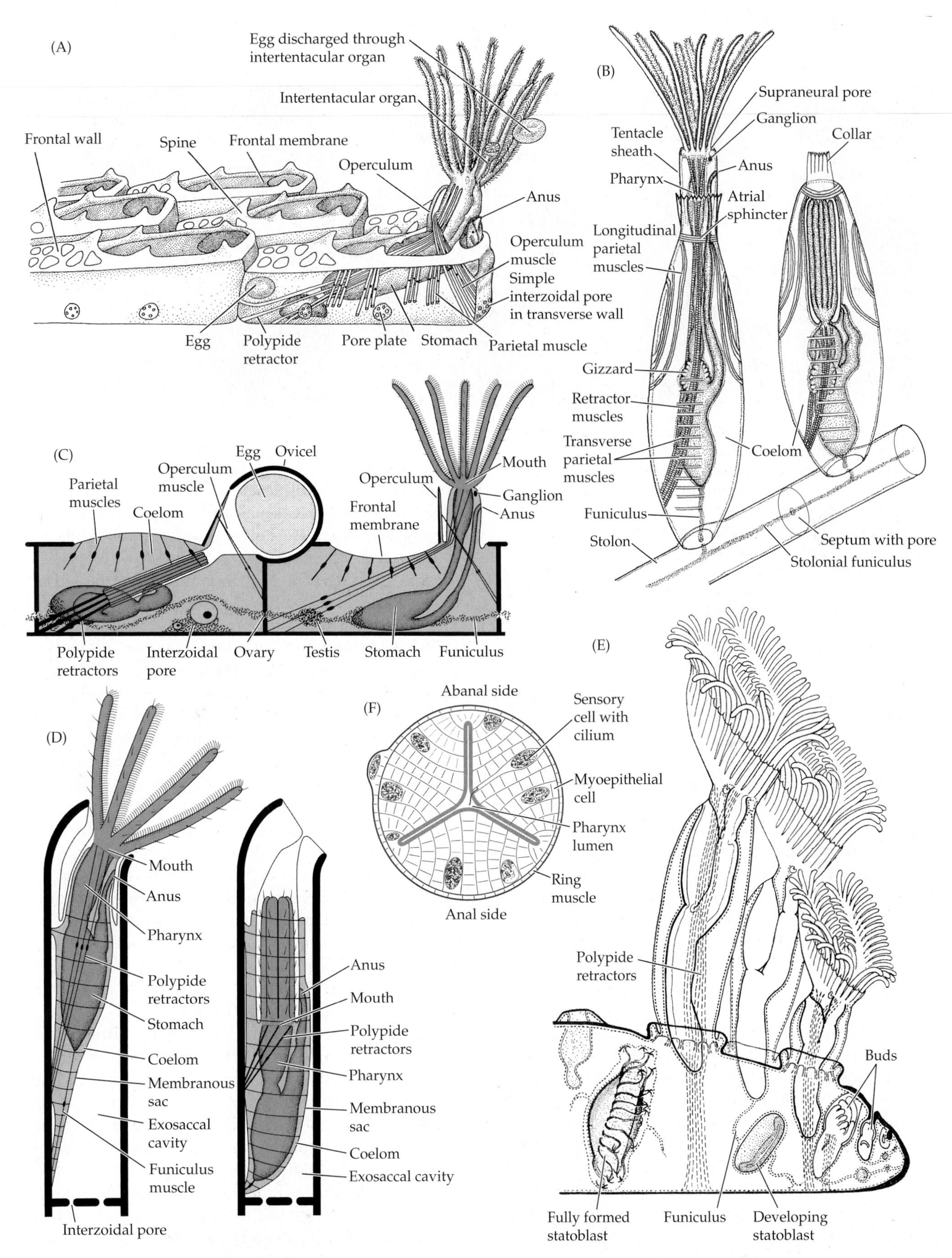

Figure 17.6 Bryozoan morphology. (A) General bryozoan structure, based on *Electra.* (B) Ctenostome anatomy, based on *Bowerbankia.* (C) Generalized cheilostome, note constant volume of body cavity plus polypides in extended and retracted states. (D) Cyclostome, based on *Crisia*. Note constant volume of the exosaccal cavity and of body cavity plus polypide in extended and retracted states. (E) Phylactolaemate anatomy, based on *Cristatella*. (F) Diagram of transverse section of the myoepithelial pharynx of *Crisia*.

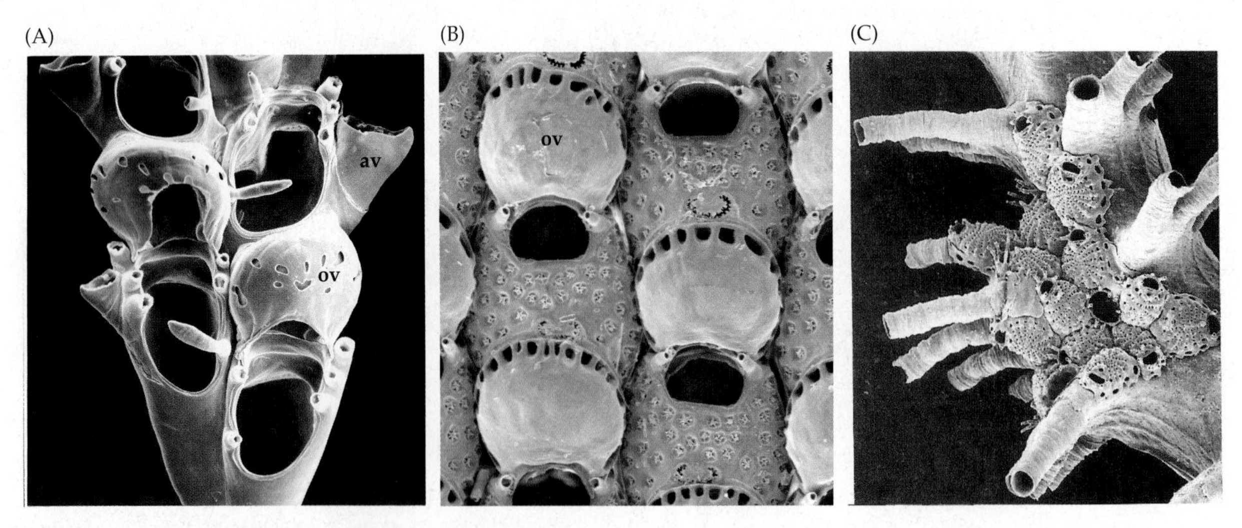

Figure 17.7 Zoecia of cheilostomes and cyclostomes (SEMs). (A) The cheilostome *Tricellaria occidentalis*; two zooids with ovicells (ov) and a lateral avicularium (av). (B) The cheilostome *Fenestrulina malusii*; rows of zooids with ovicells (ov). (C) Part of a colony of the cyclostome *Idmidronea* (the large tubular structures) with the small cheilostome *Cribrilaria* growing between the large tubes. (Note the small ancestrula in the lower middle of the colony, of a completely different structure with a large frontal membrane and small spines along the edge.)

viscera that can be protruded and retracted from the cystid (Figure 17.6). The opening in the cystid through which the tentacle crown extends is termed the **orifice** and bears a flaplike, chitinous covering, or operculum, in the cheilostomes.

The nature of the body wall differs among bryozoans, as does the form of the colony. It is entirely gelatinous or chitinous in the Ctenostomata and Phylactolaemata, but has an inner calcified layer in the Cheilostomata and Cyclostomata, called the **zoecium** (Figure 17.7). Many studies use colonies that have been cleaned of all organic material, so that the specific characters of the often highly ornamented zoecia can be studied. The chapter opener shows a gelatinous colony of *Alcyonidium* (Ctenostomata) from costal California waters.

The Gymnolaemata include a bewildering array of colony forms (Figure 17.5). Some of the ctenostomes have colonies in which the zooids arise from runners or stolons on the substratum (e.g., *Triticella, Bowerbankia*). This apparently represents the ancestral gymnolaemate pattern as demonstrated by the fossil record. In the more advanced forms, the zooids are more or less closely packed, with pores between neighboring zooids. These colonies may be sheet shaped, encrusting stones or algae; erect leaf shaped or arborescent; or in few cases discoidal and free-living.

In addition to the variation in overall form of the colonies, zooids of many cheilostomes and some cyclostomes are polymorphic (within a colony). Typical tentacle crown-bearing, feeding zooids are called **autozooids**. The widespread *Membranipora* normally consists only of autozooids. Most other types have both the normal autozooids and various types of nonfeeding zooids, called **heterozooids. Kenozooids** are reduced individuals modified for example for attachment to a substratum; various types of attachment discs, "holdfasts," and stolons are in this category. Many gymnolaemates possess the type of heterozooids called **avicularia**, each of which bears an operculum modified as a movable mandible (or jaw), which articulates against a rigid extension of the body wall called the rostrum (Figure 17.8). Avicularia may be large and take up a place in the general layer of zooids, others are large and situated at the side of the branched colonies (Figure 17.8D), but a much more common type is smaller and situated on the frontal wall or on the ovicells. A special type of avicularium is the pedunculate or bird's-head avicularium, which looks like the head of a parrot; it is attached to the colony through a stalk and able to perform nodding movements (Figure 17.8B,C). Avicularia, which were first described by Charles Darwin in 1845 based on observations of *Bugula* made aboard the *Beagle*, probably function primarily to defend the colony against small organisms and keep the surface clean of debris.

Another type of heterozooid is the **vibraculum** (Figure 17.8A,E), which is thought to be a modified avicularium, with a flagellum-like operculum that sweeps over the colony surface. They may help remove sediment particles and other material, but convincing evidence for this function is wanting. In the free-living, discoidal cheilostomes, such as *Selenella* and *Cupuladria*, the vibracula are used in locomotion (Figure 17.8E). Some of the arborescent cyclostomes have long anchoring zooids, but all cyclostomes have special female zooids, called gonozooids (see the section on reproduction).

Most members of the Phylactolaemata display either plumatellid or lophopodid colony forms (Figure 17.5K,L). **Plumatellid** colonies are usually erect

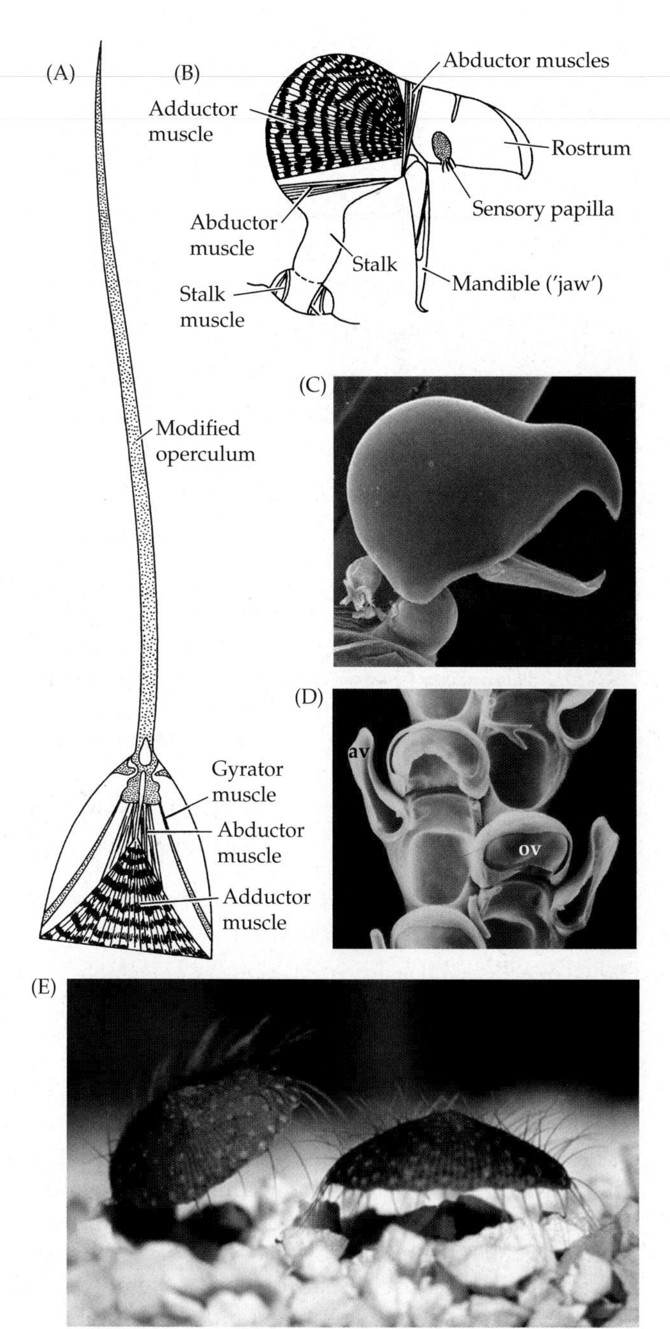

Figure 17.8 Heterozooids. (A) Diagram of a vibraculum; abm: abductor muscle; adm: adductor muscle; gyr: gyrator muscle. (B) Diagram of an avicularium. (C) SEM of an avicularium of *Bugula*. (D) Zoecia of *Tricellaria*; av: avicularium (without the organic mandible); ov: ovicell. (E) Two colonies of *Selenella* walking with vibracula.

or prostrate, and often are highly branched, like *Plumatella*. In **lophopodid** colonies, the gelatinous body wall forms the surface of irregular clumps from which the tentacle crowns protrude, as seen in *Pectinatella* and *Lophophus*. The colonies of the remarkable *Cristatella* are gelatinous, sluglike, and capable of locomotion, gliding over the substratum powered by the action of the cilia of the tentacle crown.

The Body Wall, Coelom, Muscles, and Movement

The body wall comprises the outer secreted body wall, which may be uncalcified or have a strong calcified inner layer (zoecium), the underlying epidermis, and the peritoneum. Sheets of circular and longitudinal muscles are present between the epidermis and peritoneum in phylactolaemates, but these muscle sheets are specialized in the other groups to form various groups of separate muscles.

Many bryozoans possess ornate and species-specific surface sculpturing including spines, pits, and protuberances. Experiments indicate that the cheilostomate *Membranipora* undergoes a rapid growth of new protective surface spines after being grazed by predatory nudibranchs. Some bryozoans also produce chemicals used as defense against would-be predators. These chemicals are probably all synthesized by symbiotic bacteria.

The mechanisms of tentacle crown retraction and extension (protraction) differ among bryozoan species. The specific mechanism depends largely upon the arrangement of muscles, the degree of rigidity of the cystid, and the morphology of the coelomic compartments, which function as a hydrostatic skeleton (see Figure 17.6).

In all bryozoans, the main coelom provides a fluid-filled space with muscles acting directly or indirectly to increase hydraulic pressure for extension of the tentacle crown. Thus, when the coelom is compressed the polypide is partially forced out of the cystid, thereby extending the tentacle crown. The protruded tentacles are straight or slightly curved, their shape probably being determined by the shape of their thickened basal membrane. The retractor muscles, which originate at the basal part of the cystid wall and insert on the thickened basal membrane surrounding the mouth, serve to pull the polypide back within the cystid. Generally, the methods of tentacle crown extension are common to all bryozoans, but the morphology of the muscle systems involved differs considerably. Below we describe a few examples of how these movements are accomplished and at the same time illustrate variations on the basic bryozoan body plan.

The phylactolaemates show the simplest mechanism. They extend their tentacle crowns by contraction of the muscles of the flexible body wall around the common body cavity. This action imposes pressure directly on the coelomic fluid and is similar to the mechanisms seen in many other coelomate animals. These bryozoans possess a ring-shaped, muscular diaphragm just internal to the orifice through which the tentacle crown protrudes. The diaphragm relaxes as the tentacle crown is protracted and serves to partially close off the orifice when the tentacle crown has been withdrawn by retractor muscles (Figure 17.6E).

Ctenostomes possess an uncalcified, flexible body wall composed of gelatinous, chitinous, or leathery material. Contraction of transverse parietal muscles pulls the cystid walls inward, thereby causing an increase in coelomic pressure that then extends the tentacle crown (Figure 17.6B). Retraction of the tentacle crown is accomplished by the usual retractor muscle, which is aided by longitudinal parietal muscles that pull in the atrial chamber. When the tentacle crown is fully retracted, a sphincter contracts to close the orifice and, in some species, folding a pleated collar over the end of the zooid.

Gymnolaemates show many intricate variations in the method of polypide extension and retraction, especially among the cheilostomes and the cyclostomes, but they all rely on the constant volume of the coelom, which functions as a hydrostatic skeleton.

Cheilostomes have cystids that are calcified to various degrees. A completely stiff body wall would make polypide movements impossible, so various areas of the cystids are uncalcified and can be moved in or out to by groups of special muscles of the body wall. In most species, the zooids are more or less boxlike (rather than erect or tubular). The outer surface of the box that bears the orifice is called the frontal surface. It has a partly uncalcified and flexible part called the frontal membrane, and contraction of parietal protractor muscles pulls the frontal membrane inward, thus increasing coelomic pressure and pushing out the polypide (Fig. 17.6C). There are, of course, variations on this general theme. An operculum closes over the orifice when the polypide is retracted, but the exposed frontal membrane presents a weakness in the defense against predation, and many species have evolved more or less complicated protective structures (Figure 17.9). Some forms, known as the **cribrimorphs**, bear hard spines that project over the membrane and in some cases actually meet and fuse to form a cage above the vulnerable area (Figure 17.7C). In others a calcified partition, called the frontal shield or **cryptocyst**, lies beneath the frontal membrane, separating it from the soft parts within. The cryptocyst bears pores through which the protractor muscles extend (Figure 17.9). The most drastic modifications are seen in the so-called ascophoran type (Figure 17.9), in which the frontal wall is heavily calcified except for a small pore below the orifice. This opening, the ascopore, leads into the invaginated sac-shaped frontal membrane, called the **ascus**. The parietal muscles insert on the ascus. Cheilostomes are the most diverse and species-rich group of bryozoans.

Cyclostomes have erect, tubular zooids surrounded by heavily calcified zoecia (Figure 17.6D). The inflexibility of the body wall and the absence of well-developed sheets of muscles preclude use of direct compression, so they have evolved a mechanism of polypide protraction unique among the bryozoans. The structural features associated with this mechanism comprise a synapomorphy on which this group was established

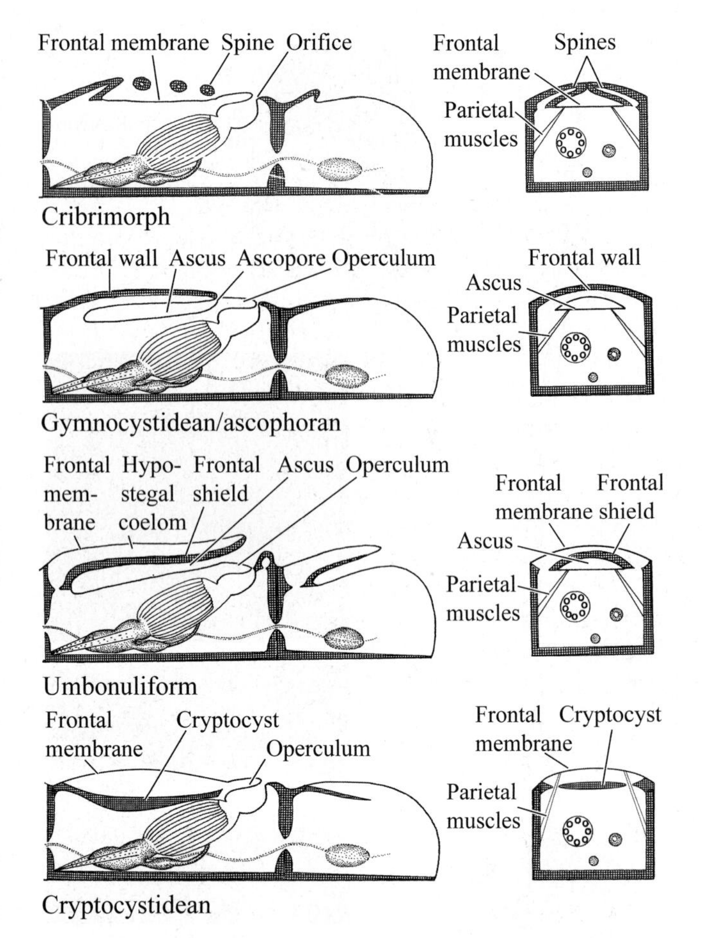

Figure 17.9 Frontal walls. Diagrams of four different types.

as a separate subclass. The key structure is the membranous sac, which is the detached peritoneum of the cystid wall plus its basal membrane and a series of tiny ring muscles. It is attached by ligaments to the body wall and separates the coelom from the outer exosaccal cavity. Distally, the exosaccal space harbors a series of fine atrial dilator muscles. Protraction of the polypide involves relaxation of the retractor and atrial sphincter muscles and contraction of the atrial dilators and the ring muscles of the membranous sac. With the constant volumes of both the coelom and the exosaccal cavity, the polypide is squeezed out. When a zooid is retracted by the large retractor muscles, the atrial dilators are relaxed and a special atrial sphincter effectively closes the inner end of the atrium.

Zooid Interconnections

As discussed in Chapter 4, clear definitions of the term colony are sometimes elusive. This difficulty arises because it is not always easy to tell where one individual ends and another begins, or because the degree of structural and functional communication among individuals is uncertain or variable. Bryozoan zooids, at least autozooids, are clearly demarcated by the elements of the polypide (tentacle crown, gut, ganglion, and so on), however the way in which the zooids are interconnected differs among groups.

In phylactolaemates the coelom is continuous among zooids, uninterrupted by septa (Figure 17.6E). Each zooid bears a tubular tissue cord called a **funiculus** that extends from the inner end of the curved gut to the body wall. All other bryozoans lack extensive coelomic connections, and the zooids are separated by various sorts of structural components.

Stoloniferous gymnolaemates (e.g., *Bowerbankia*; Figure 17.6B) have septa spaced along the stolons between the zooids. A cord of tissue passes along the stolons and through pores in each septum. This cord, called a **stolonal funiculus**, connects with the funiculus of each zooid arising from the stolon. In most nonstoloniferous gymnolaemates the cystid walls of adjacent zooids are penetrated by pores with tissue plugs, which, again, usually connect with the funiculus of associated zooids. In the cheilostomes the pores are covered by pore plates with a number of small openings

The walls of adjacent zooids of cyclostomes bear interzooidal pores or common distal cavities that allow communication of exosaccal coelomic fluid (Figure 17.6D). The funiculus is contained within the coelom with the rest of the viscera and attaches the gut to the body wall. The exchange of nutrients between zooids, for example, between the feeding autozooids and the nonfeeding gonozooids, remains a mystery.

It is clear that bryozoan zooids are interconnected structurally, either by direct sharing of coelomic spaces or by funicular tissue, except for the enigmatic cyclostomes. Functionally, these connections provide a means of distributing materials through the colony, and perhaps other communal activities as well. Other special functions of the funiculus are discussed later in the chapter.

The Tentacle Crown, Feeding, and Digestion

The tentacle crown (which is a lophophore) is horseshoe shaped in the phylactolaemates (except for the supposedly primitive *Fredericella*) and circular in the gymnolaemates. The tentacles are ciliated with the cilia arranged in characteristic bands related to the feeding mechanism. Bryozoans are suspension feeders, although supplemental methods occur. They feed largely on protists, but some species are able to capture and ingest larger organisms.

Upon extension, the tentacles of the circular tentacle crowns of gymnolaemates are erected in a funnel or bell-shaped arrangement around the mouth. Each tentacle bears four or five ciliary bands along its length, one frontal tract (lacking in the cyclostomes), and a pair of laterofrontal and lateral bands (Figure 17.10). During normal suspension feeding, the lateral cilia create a current that enters the open end of the funnel, flows toward the mouth, and then out between the tentacles. Some food particles are carried directly to the area of the mouth by the central flow of water. Other potential food, however, moves peripherally with the current toward the intertentacular spaces. When a particle contacts the rather stiff laterofrontal cilia, a tentacle flick is initiated moving the particle towards the central current (Figure 17.10E). Additional flicks may occur if particles have not reached the central current. A localized reversal of power stroke direction is initiated in the lateral cilia, and the particle may be tossed onto the frontal edge of the tentacle. The particle is repeatedly bounced in this fashion and is moved toward the mouth under the influence of a current generated by the frontal cilia.

Many bryozoans supplement suspension feeding by various means that allow them to capture relatively large food particles, including live zooplankton. Some species, such as *Bugula neritina*, are capable of trapping zooplankton or large algae by folding their tentacles over the prey and pulling it to the mouth. A number of bryozoans rock or rotate the entire tentacle crown, apparently sampling reachable water for food.

All the polypides in a colony transport water toward the surface of the colonies. This is no problem for branching, bushy types, but in colonies forming larger sheets the water must flow away again, and this has led to the formation of various types of **excurrent chimneys**. Polypides may simply bend away from a small area, which then creates a chimney. Large colonies of *Membranipora* show regularly spaced areas without polypides, which function as excurrent chimneys (Figure 17.11A). Other chimneys may be formed by small groups of nonfeeding male zooids or pregnant zooids with degenerated polypides. In leaf-shaped colonies (e.g., *Pentapora*), the edge of the leaves may function as excurrent areas (Figure 17.11B). The generation of strong excurrent water flow away from the colony surface helps to push nonfood material and feces far enough to reduce the possibility of recycling. Such currents, which move larger amounts of water over the tentacle crowns than could be moved by individual zooids, may be especially important to colonies inhabiting quiet water.

The digestive tract is U-shaped (Figures 17.6B,C and 17.12). The mouth lies within the tentacle ring, and in the Phylactolaemata it is overlain by the epistome. Cilia of the peristomial field keep the captured particles rotating in front of the mouth, and from time to time the mouth is opened and in the gymnolaemates the particles are sucked into the triradiate pharynx by contraction of the radiating epitheliomuscular cells (Figure 17.6F). A valve separates the lower end of the pharynx from the descending portion of the stomach, which is called the **cardia** and in some species is modified as a grinding gizzard. The cardia leads to a central stomach

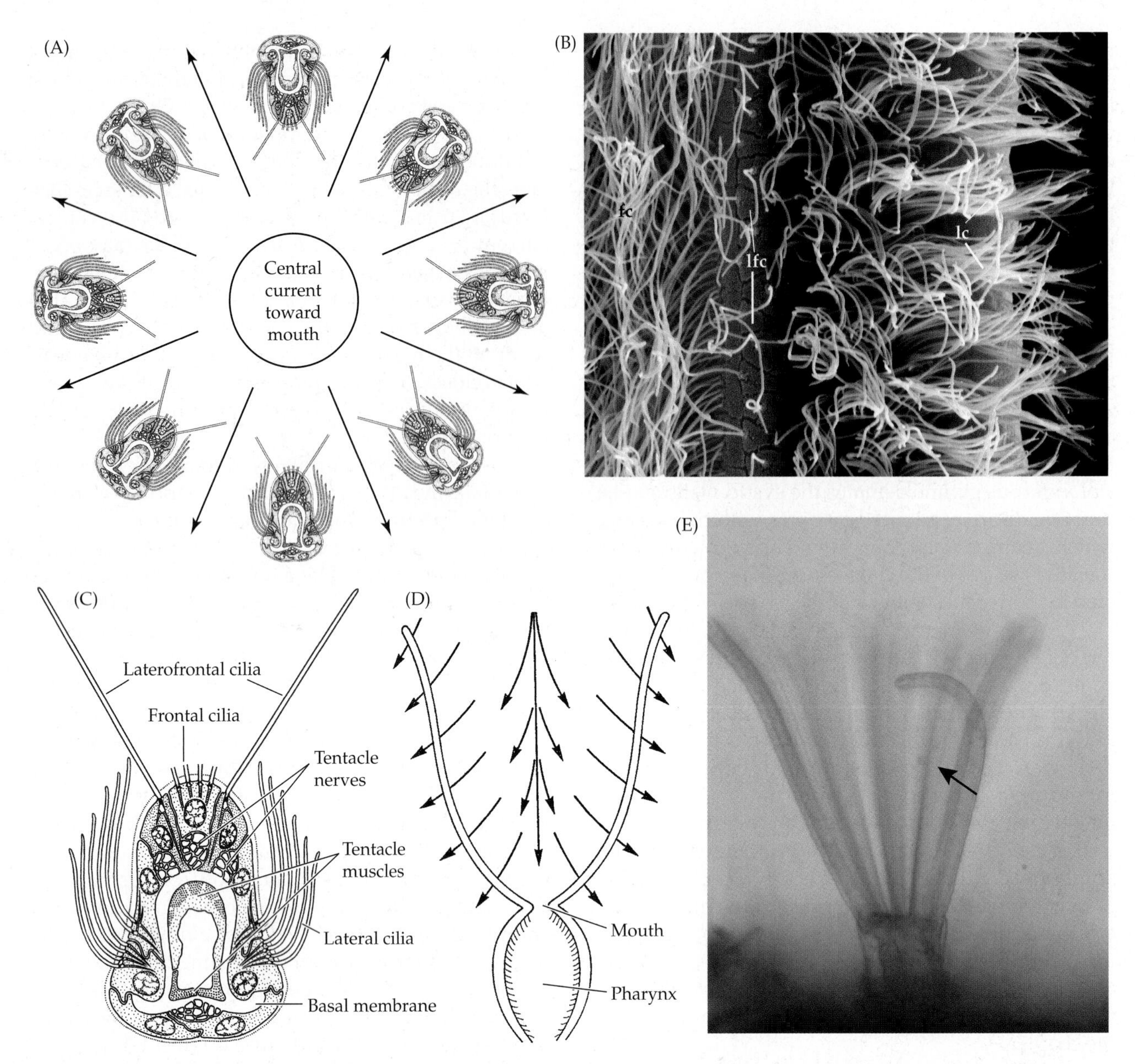

Figure 17.10 Tentacles and filter feeding. (A) Diagram of transverse section of a tentacle crown; the directions of the water currents set up by the lateral cilia are indicated. (B) SEM of one side of a tentacle of *Flustrellidra* with the three ciliary bands: fc: frontal cilia; lc: lateral cilia (note the metachronal waves); lfc: laterofrontal cilia (unnaturally curved due to fixation). (C) Transverse section of a tentacle of *Electra* (based on TEM). (D) Diagram of ciliary currents set up by the lateral cilia. (E) Tentacle flick. Note the small alga (arrow) that has just been pushed into the central current.

from which arises a large cecum; the funiculus attaches to the bottom of the cecum. The ascending portion of the stomach, or **pylorus**, arises from the central stomach and leads to a proctodeal rectum and the anus, which lies outside the tentacle crown ring. A sphincter controls the flow of material from the pylorus to the rectum. The phylactolaemates have a flat, funnel-shaped, ciliated pharynx without radiating muscles, and the hindgut is elongated as an intestine.

Digestion begins extracellularly in the cardia and central stomach, and is completed intracellularly in all parts of the stomach. Food is moved through the gut by peristalsis and cilia. Undigested material is rotated and formed into a spindle-shaped mass by the cilia of the pylorus and then passed to the rectum for expulsion.

Circulation, Gas Exchange, and Excretion

There is no structural circulatory system in bryozoans, so movement of metabolites within single zooids is by diffusion. Given the small size of these animals, intrazooid diffusion distances are small, and the coelomic fluid provides a medium for passive transport. Interzooid circulation is facilitated by the common coelom in phylactolaemates, the cystid pores in cyclostomes,

(A)

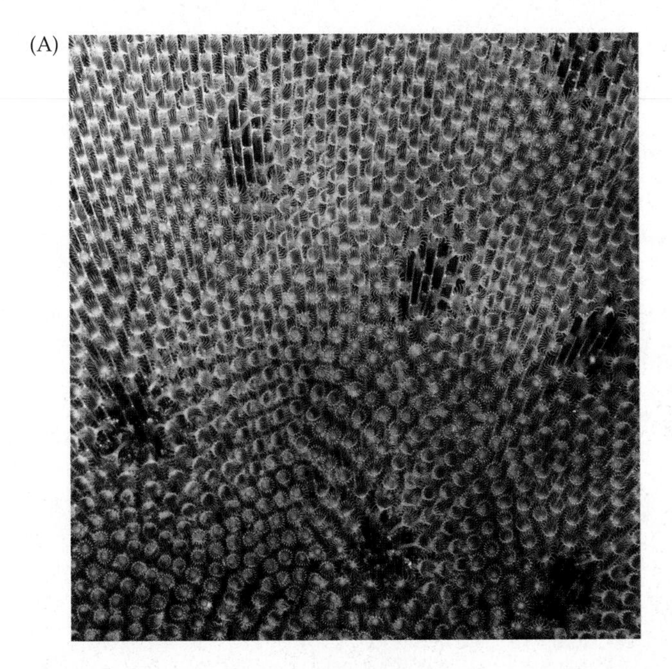

(B)

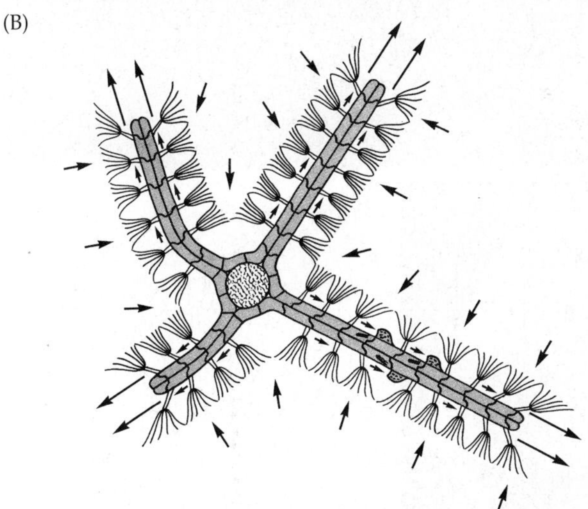

Figure 17.11 Excurrent chimneys. (A) *Membranipora*, the dark areas of empty cystids form the excurrent chimneys. (B) Diagram of a cross section of four leaves of a colony of *Primavelans* (formerly *Hippodiplosia*) showing the incoming currents set up by the tentacles (medium arrows), the currents along the colony surface below the tentacles (short arrows), and the excurrents (long arrows) at the edge of the leaves.

and the funicular cords of most eurystomates. Gas exchange occurs across the walls of the extended parts of the polypide, particularly the tentacle crown. Bryozoans contain no respiratory pigments, and gases are carried in solution.

Nondigested particles and waste products accumulate in the cells of the stomach. The elimination of these wastes is not fully understood, but apparently it occurs at least in part by the cyclic degeneration of the polypides that form structures called **brown bodies**. A new polypide is then formed through the usual budding process from the area of the orifice. In some eurystomates the brown body is left within the basal part of the cystid, and the number of regenerated polypides can be assessed by the number of brown bodies. In other species, the brown body is taken up into the gut of the developing polypide and expelled as the first fecal pellet when the new polypide first protrudes from the cystid. In most stoloniferous ctenostomes, the old cystid with its brown body drops from the colony, and an entire new zooid regenerates from the stolon. In all cases, it is presumed that metabolic wastes are precipitated and concentrated in the brown bodies and thus eliminated or at least rendered inert.

Nervous System and Sense Organs

In concert with their sessile lifestyle and the lack of a discrete head, the bryozoan nervous system and sense organs are predictably reduced. A neuronal mass, or cerebral ganglion, lies on the anal side of the pharynx (Figure 17.6B,C). Arising from this structure is a circumoral nerve ring. Nerves extend from the ring and ganglion to the viscera, and motor and sensory nerves extend into each tentacle. Interzooidal nerve fibers have been reported in some species, but their function remains unclear. The only known receptors are tactile cells on the tentacles (the laterofrontal ciliary cells [Figure 17.10B,C] and single cells on the abfrontal side) and on the sensory papilla of the avicularia (Figure 17.8B).

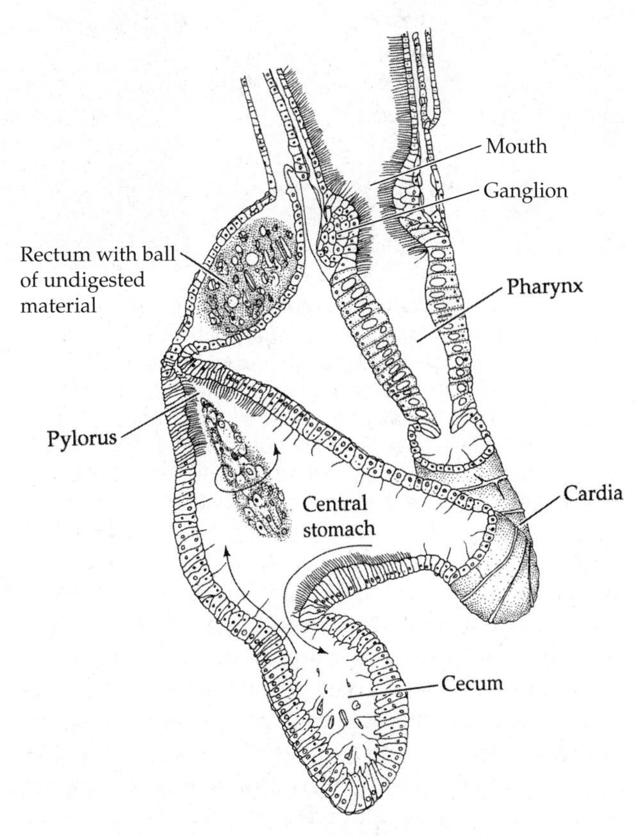

Figure 17.12 The gut of *Cryptosula*. The ingested particles rotate in the stomach and the rectum.

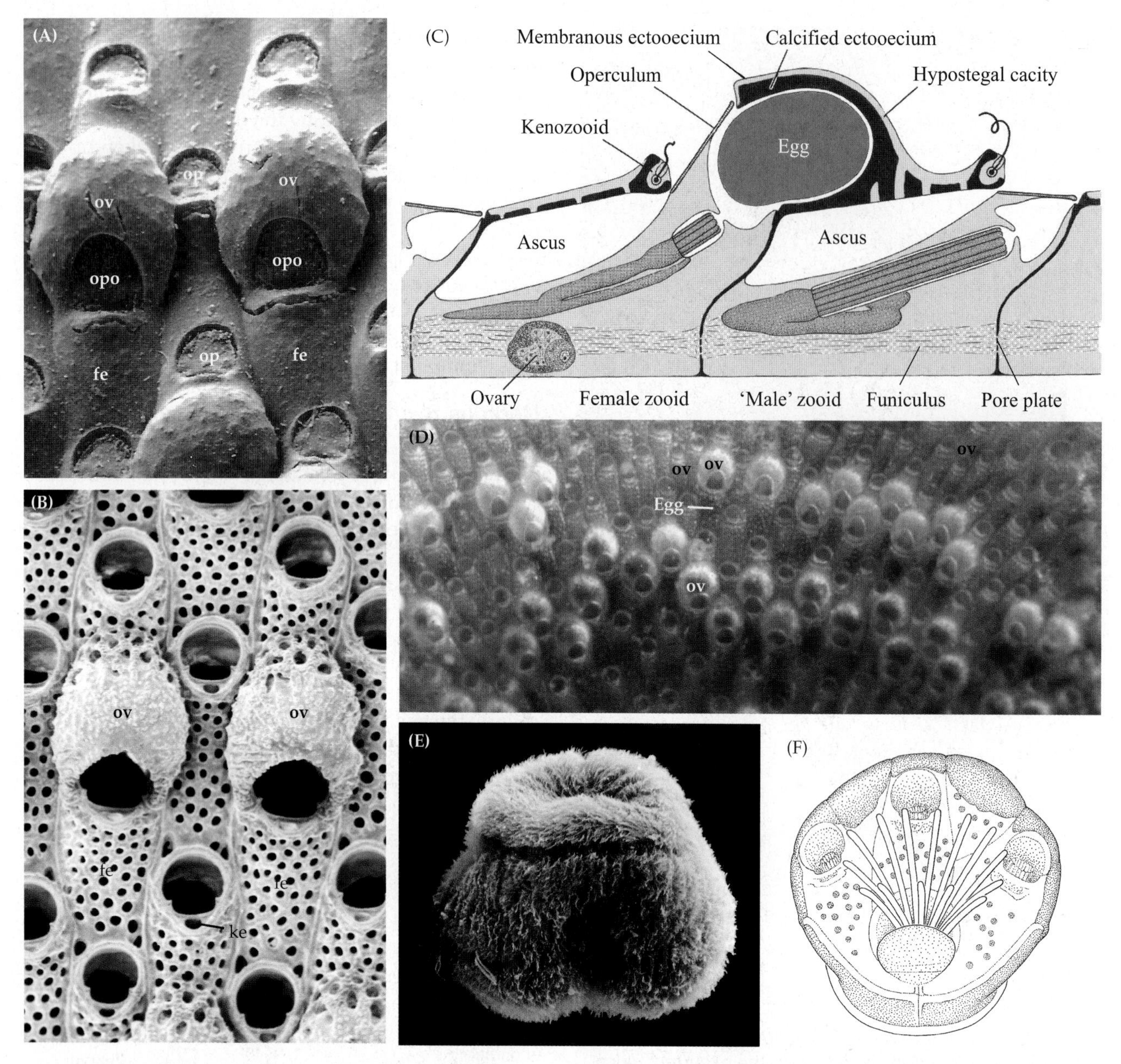

Figure 17.13 Reproduction of *Primavelans insculpta* (formerly *Hippodiplosia*). (A and B) SEM of two female zooids (fe) which have induced their distal neighbors to form an ovicell (ov); op: operculum of "male" zooid; opo: operculum of the ovicell (the cracks at the kenozooids are artifacts). (A) Uncleaned with the outer periostracum and the opercula retained. (B) Cleaned zoecium of a similar part of the colony; ke: kenozooid. (C) Diagram of a median section of a female and a "male" zooid. (D) Part of a colony with rows of female and "male" zooids; the upper labeled ovicell is newly formed and still empty, and large cherry-red eggs can be distinguished in the body cavity of the females; the lower ovicell contains a newly deposited egg. (E) SEM of newly released larva (note heavy ciliation). (F) The composite ancestrula with one zooid with protruded tentacles and three developing zooids.

The larvae have an apical organ, which supposedly is sensory, and a ciliated **pyriform organ** that is used in testing the substratum before settling. At least some bryozoans exhibit a marked negative geotaxis prior to settling. Experiments suggest that this geotaxis is a direct response to gravity, but the mechanism mediating this phenomenon is unknown. Many of the coronate larvae have well developed ocelli and are positively phototactic while free swimming. Settlement is often accompanied by a shift to a negative phototaxis.

Reproduction and Development

Sexual reproduction Bryozoan colonies are hermaphroditic, and some species may produce sperm and eggs in the same zooid (e.g., *Membranipora*). Other species (e.g., some cyclostomes) have colonies with separate male and female zooids of very different shape (see below). Rather diffuse ovaries usually arise from transient patches of the peritoneum at the cystid wall. Testes usually develop on the funiculus (Figure 17.6C). Sperm differentiates in the body cavity, and the ripe sperm moves to the coelom at the base of the tentacles and are eventually shed through small temporal pores at the tips of the abfrontal tentacles. The sperm enters the body cavity of a zooid with a ripe egg, probably through the supraneural pore located between the bases of two abfrontal tentacles, or the intertentacular organ, and fertilize the eggs. The few cheilostomes and ctenostomes that spawn fertilized eggs into the water (e.g., *Electra*, *Membranipora*, *Hislopia*, and some species of *Alcyonidium*) have the supraneural pore elevated on a pedestal called the **intertentacular organ** (Figure 17.6A). However, most species brood the eggs in more or less elaborate structures.

Phylactolaemates brood their embryos in embryo sacs produced by invaginations of the body wall. The eggs may leave the supraneural pore and become implanted into the embryo sacs in a method like that observed in ctenostomes (e.g., *Victorella*), but the process has not been observed.

Some ctenostomes (e.g., *Triticella*) attach the eggs to the atrial wall so that they become freely exposed when the polypides protrude. Others (e.g., *Victorella*) deposit the fertilized egg into a brood pouch formed by the atrial wall, and a further specialization is that the fertilized eggs are retained in the retracted atrium where the whole development takes place while the polypide degenerates (e.g., *Bowerbankia* and species of *Alcyonidium*).

A variety of brooding methods occurs among cheilostomes, usually involving the formation of an external brooding structure called an **ovicell** (Figures 17.6C, 17.7A,B, 17.13A–D, and 17.14A). In many genera (e.g., *Fenestrulina*, *Primavelans* [formerly *Hippodiplosia*], and *Tegella*), the developing female zooids induce the developing distal zooid (male or female) to produce an ovicell by a specialization of the proximal part of the frontal wall (Figure 17.13A–D). The usually very large egg is deposited into the ovicell where the development to the free-swimming lecithotrophic larval stage takes place. The opening may be covered by the operculum of the maternal zooid or by an extension of the distal wall of the maternal polypide. This extension is called the **oecial vesicle** and functions as a placenta in species with small eggs (e.g., *Bugula*) (Figure 17.14A).

Bryozoans undergo radial, holoblastic, nearly equal cleavage to form a coeloblastula or a stereoblastula. Subsequent development differs greatly among groups, but in all cases it involves a free-swimming dispersal form. Very little solid information exists on the derivation and fates of germ layers in bryozoans. This is especially true for mesoderm and coelomic linings.

In phylactolaemates, the fertilized egg enters a brood sac formed from the cystid wall, and later stages receive nourishment from the maternal zooid through various types of placentas. The coeloblastula develops into a cystid-like stage lacking endoderm and then generates one or more polypides through a normal budding process (see below). This "larva" is ciliated and escapes the embryo sac for a short swimming life before settling and metamorphosing.

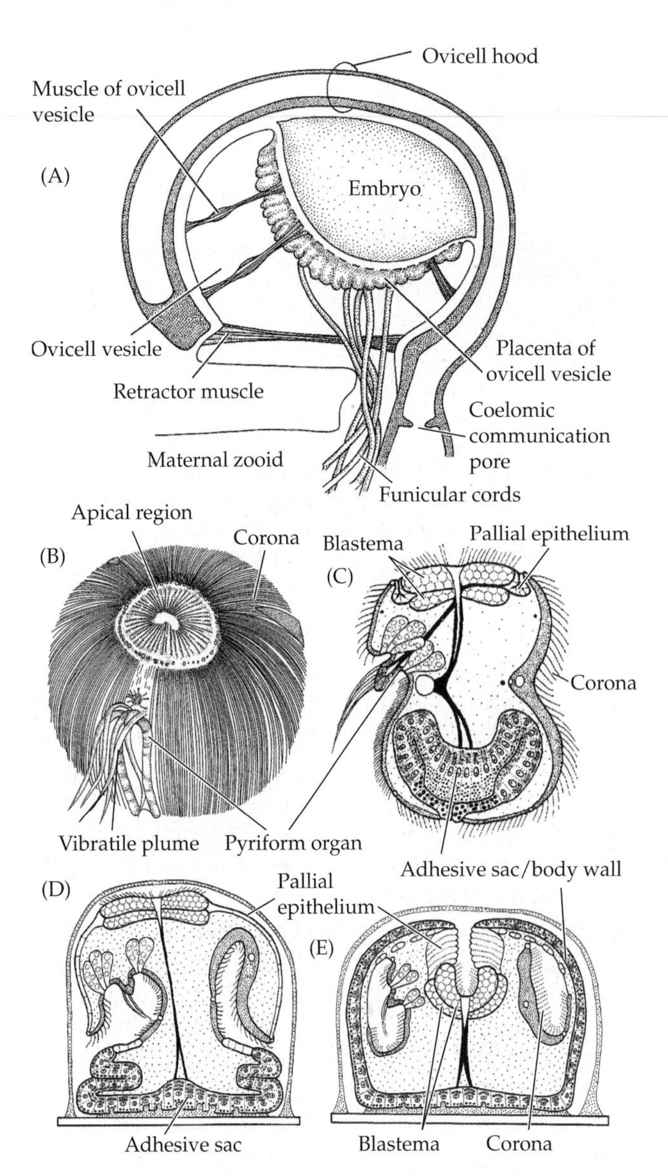

Figure 17.14 Brooding and larval metamorphosis of *Bugula neritina*. (A) Ovicell with an embryo that has grown large through nourishment from the placenta. (B) Drawing of a larva. (C–E) Larva, settling and metamorphosis. (C) Median section of the larva. (D) The adhesive sac is everted in contact with the substratum, and a fine pellicle has been secreted; the everted pallial epithelium covers the upper side of the ancestrula and the corona has become internalized. (E) The pallial epithelium has retracted again and the epithelium of the adhesive sac now covers the whole ancestrula, the polypide will differentiate from the blastema.

The coeloblastulae of eurystomates undergo a gastrulation where four cells divide so that one of each pair of daughter cells is shunted to the blastocoel as presumptive endoderm and mesoderm. Free-swimming larvae are eventually produced. Most of the species that free spawn have a characteristic laterally

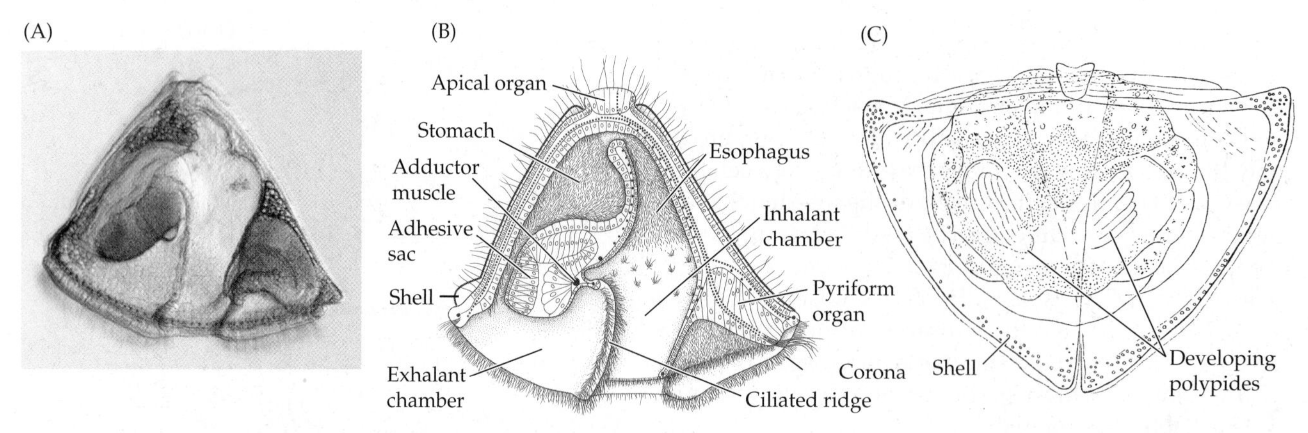

Figure 17.15 Larvae of *Membranipora*. (A) Cyphonautes larva. (B). Diagram of the morphology of the cyphonautes larva. (C) Newly settled larva, the shells are still covering the young ancestrula, which is developing a pair of polypides.

compressed, triangular larva called a **cyphonautes**, which swims by means of a ring of cilia at the "lower" edge of the triangular shells (Figure 17.15A,B). This larval type is found both in basal ctenostomes and cheilostomes. These larvae feed by means of the U-shaped ciliated ridge, which resembles and functions as one side of a tentacle. They have a complete gut and may remain in the plankton for months, whereas the larvae of brooding species lack a digestive tract and lead very short, pelagic lives (Figures 17.13E and 17.14B). These more or less globular larvae are called **coronate larvae**, because the ciliation that covers almost the whole body is believed to be the extended ciliary ring of the cyphonautes larva. Despite these differences, eurystomate larvae have some fundamental similarities. For example, they possess a complicated apical organ, a sensory pyriform organ complex, and a pouchlike **adhesive sac**, which are important in settling and metamorphosis (Figures 17.13E, 17.14B, and 17.15B).

The fertilized cyclostome egg cleaves to form a ball of cells, which then undergoes a budding process, forming secondary embryos, which in turn bud tertiary embryos. Dozens of small, solid, asexually produced embryos may result from a single primary ball of cells. This phenomenon, called **polyembryony**, is very unusual in the animal kingdom. Each embryo develops cilia and escapes as a completely ciliated larva with an invagination at both poles, one representing the apical organ/pole and the other is the adhesive sac. The larva settles after a short free period and undergoes a metamorphosis similar to that described below for the coronate eurystomate larvae, with the frontal wall covered by the everted chitinous covering of the apical invagination.

Bryozoan larvae are at first positively phototactic, and many possess pigment spots with a group of modified cilia that are thought to be light sensitive. Following a planktonic phase, the larvae usually become negatively phototactic and swim toward the bottom as they become "competent" (i.e., preparing to settle onto the substrate). Once in contact with the substratum, the pyriform organ complex is apparently used to test for chemical and tactile cues reflecting the suitability of the substratum for settling. Once a proper surface has been selected, the adhesive sac everts and secretes sticky material for attachment. After attachment, there is a thorough reorganization of tissue positions accompanied by histolysis of larval structures.

In gymnolaemates, the first zooid, the ancestrula, usually has one polypide, but two or more polypides are found in several species (Figures 17.13F, 17.15C). The settling cyphonautes larva everts the adhesive sac, which spreads out along the substratum and secretes the basal wall of the ancestrula. The shells open widely and become detached, while the underlying pallial epithelium secretes the frontal wall. The corona becomes enclosed in an inner ring-shaped cavity. Metamorphoses of the various types of coronate larvae show considerable variation on this theme. In one type (e.g., *Bugula*; Figure 17.14C–E), the periphery of the everted adhesive sac extends along the sides of the settled larva finally to cover the whole outer wall of the larval body; thus, the whole cystid wall originates from the adhesive sac. In other types (e. g., *Bowerbankia*), the adhesive sac retracts again, and the pallial epithelium expands to cover the whole surface of the ancestrula.

Asexual reproduction As in all colonial animals, asexual reproduction is an integral part of the life history of bryozoans and is responsible for colony growth and regeneration of zooids. Each colony begins from the tissues of a metamorphosed larva, which develops one or a few zooids from the body wall, or in some cheilostomates from a mass of undifferentiated cells (called a blastema) connected with the apical organ. As in sexual reproduction, this first zooid or group of zooids that develop before feeding begins is called the ancestrula (Figures 17.7C, 17.13F). It undergoes budding to produce daughter zooids, which subsequently form more buds, and so on. The initial group of daughter zooids may arise in a chainlike series, a plate, or a disc; the

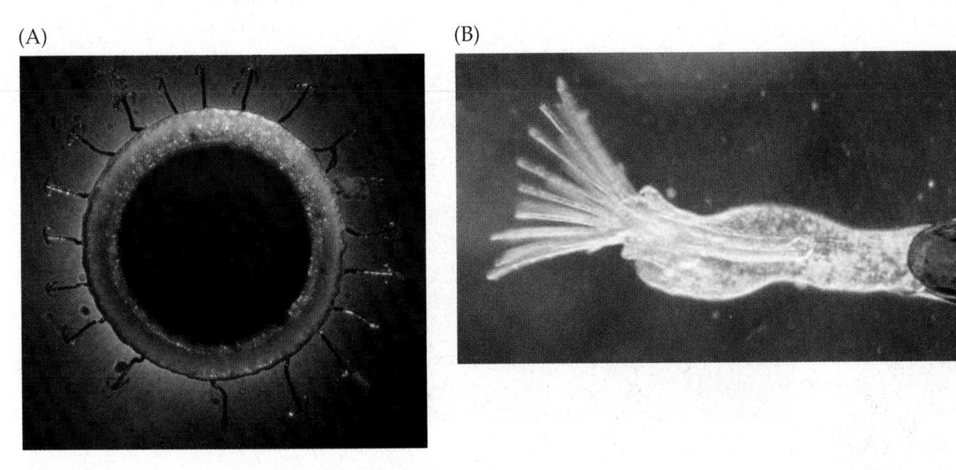

Figure 17.16 Statoblasts of phylactolaemates. (A) *Cristatella*. (B) Germinating statoblast of *Fredericella*.

budding pattern determines the growth form of the colony and is highly variable among species. The ancestrula is usually of a shape different from that of the budded zooids (Figure 17.7C) and, in a few species, changes in zooid morphology along the growing colony has been observed.

Budding involves only elements of the body wall. In most eurystomates, a partition forms that isolates a small chamber, the developing zooid, from the parent zooid. The bud initially includes only components of the cystid and an internal coelomic compartment. A new polypide is then generated from the living tissues of the bud (the epidermis and the peritoneum). The epidermis and peritoneum invaginate, the former producing the tentacle crown and the gut. The peritoneum produces all of the new coelomic linings and the funiculus. Budding in phylactolaemates is similar, except that the polypide develops first and only becomes isolated from the parent zooid in some species. Budding among cyclostomes is in need of further investigation.

In addition to budding, freshwater bryozoans (Phylactolaemata) reproduce asexually by the formation of **statoblasts** (Figures 17.6E and 17.16). These structures are extremely resistant to drying and freezing, and are often produced in huge numbers during adverse environmental conditions. Statoblasts generally form on the funiculus and include peritoneal and epidermal cells plus a store of nutrient material. Each cellular mass secretes a pair of chitinous protective valves, differing among species in shape and ornamentation. The parent colony usually degenerates, freeing the statoblasts. Some statoblasts sink to the bottom, but others float by means of enclosed gas spaces. Some bear surface hooks or spines and are dispersed by passive attachment to aquatic animals or vegetation (Figure 17.16A). With the return of favorable conditions, the cell mass generates a new zooid, which sheds its outer casing and attaches as a functional individual (Figure 17.16B).

Phylum Brachiopoda: The Lamp Shells

Members of the phylum Brachiopoda (Greek *brachium*, "arm"; *poda*, "feet") are called lamp shells because the shape of their exoskeleton resembles historical oil lamps (Box 17C and Figure 17.17). They have been known since at least the early Middle Ages and their images were published in books in the late sixteenth

BOX 17C Characteristics of the Phylum Brachiopoda

1. Enterocoelic, coelomate lophophorates
2. Epistome present, with or without coelomic lumen
3. Body enclosed between two shells (valves), one dorsal and one ventral
4. Usually attached to the substratum by a stalk, the pedicle
5. Valves lined (and produced) by mantle lobes formed by outgrowths of the body wall and creating a water-filled mantle cavity
6. Lophophores are circular to variably coiled, with or without internal skeletal support.
7. Gut U-shaped; anus present (Linguliformea, Craniiformea) or absent (Rhynchonelliformea)
8. One or two (Rhynchonellida) pairs of metanephridia
9. Circulatory system rudimentary and open
10. Most are gonochoristic and undergo mixed or indirect life histories.
11. Indirect developers with lecithotrophic larvae
12. Gametes develop from transient gonadal tissue on peritoneum of metacoel.
13. Cleavage holoblastic, radial, and nearly equal; coeloblastulae usually gastrulate by invagination; blastopore closes and mouth (and anus) form secondarily (deuterostomous development)
14. Solitary, benthic, marine

Figure 17.17 Representative brachiopods. (A) *Magellania venosa* (Rhynchonelliformea, Terebratelloidea). Reaching lengths of 9 cm, this is the largest living brachiopod; from Chilean fiords. (B) Close up of *M. venosa*, showing the lophophore inside the shell and branched egg-carrying mantle canals shining through the dorsal valve. (C) At less than 1 mm in length, *Gwynia capsula* (Rhynchonelliformea, Gwynioidea) is the smallest living brachiopod; the gaping valves expose the lophophore, forming a small circle of single tentacles. (D) *Thecidellina meyeri* (Rhynchonelliformea, Thecideoidea), with fully opened shell. The orange lophophoral tentacles form an effective filter for microplankton. (E) The limpet-like *Novocrania lecointei* (Craniiformea), which directly attaches to hard substrates with its ventral valve. (F) The organophosphatic shelled *Discinisca lamellosa* (Linguliformea, Discinoidea) with its fringing row of setae protruding from the mantle margin. (G) *Lingula* sp. (Linguliformea, Linguloidea), removed from its burrow. (H) *Lingula* sp. (Linguliformea, Linguloidea) in feeding posture. The arrows indicate the direction of water flow. (I) *Marginifera* sp., a spinose Permian brachiopod. (J) *Lingula* sp. for sale in a market in Southeast Asia, where it is locally consumed. (K) The articulate (rhynchonelliform) brachiopod *Frenulina sanguinolenta* from the tropical Pacific, showing the position of the lophophore. Attached to the shell is a purple bryozoan, *Disporella* sp.

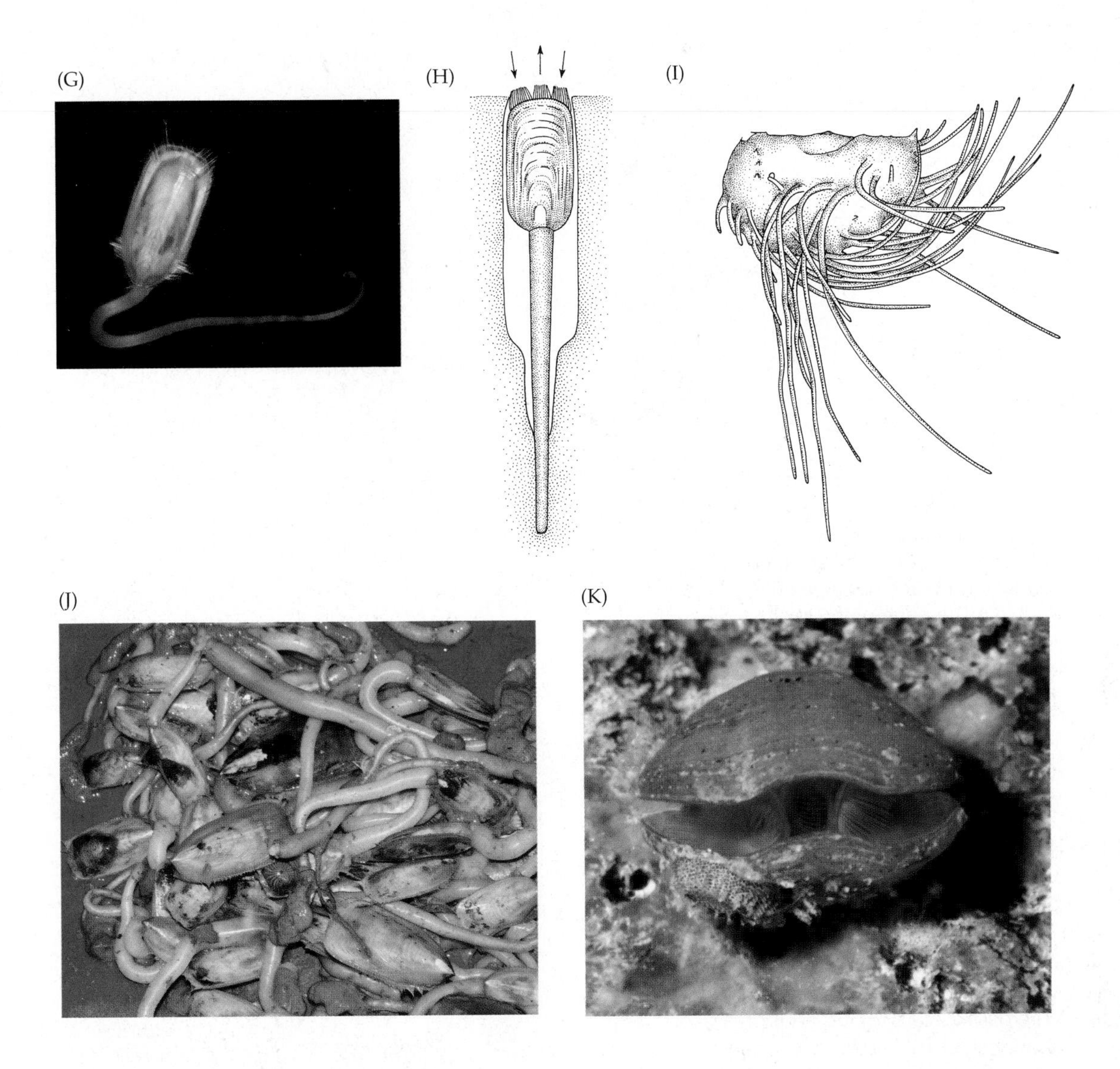

century. All are solitary, marine, benthic creatures. The body, including the lophophore, is enclosed between a pair of dorsoventrally oriented valves. Most brachiopods are attached to the substratum by a fleshy **pedicle** (Figure 17.18A,B). Some species lack a pedicle (e.g., *Novocrania*), and these usually cement themselves directly to a hard substratum. On the other hand, some species that possess a pedicle do not form permanent attachments, such as *Anakinetica cumingi*, which lies free, and *Lingula* spp., which anchor in loose sand (Figures 17.17G and 17.18B). A few species possess both unattached and attached populations (e.g., *Neothyris lenticularis* and *Terebratella sanguinea*).

The valves of the brachiopod shell are unequal, and are attached to one another either posteriorly by a tooth-and-socket hinge (Rhynchonelliformea) or simply by muscles (Linguliformea, Craniiformea). Brachiopods normally "sit" ventral side up, the pedicle usually arising from the ventral valve through a shell opening called the **foramen**.

Most extant brachiopods measure 2 to 4 cm along the greatest shell dimension, but range from below 1 mm to over 9 cm in extreme cases. Although they are known from nearly all ocean depths, they are most abundant on the continental shelf. The approximately 400 living species represent a small surviving fraction of the more than 15,000 extinct species that have been described. Their rich fossil record dates back at least 550 million years (Ediacaran period). Brachiopods, especially rhynchonelliforms, were among the most abundant animals of the Paleozoic, but they declined in numbers and diversity after that time. Charles Thayer (1985) presented experimental evidence that competition with epibenthic bivalve molluscs was at least partly responsible for the reduction in brachiopod diversity following their Paleozoic success.

CLASSIFICATION OF THE BRACHIOPODA

SUBPHYLUM LINGULIFORMEA Valves not hinged, attached by muscles only; valves of organophosphatic composition, including apatite (calcium phosphate), chitin, collagen, and proteins; pedicle usually with intrinsic muscles and a coelomic lumen; lophophore without internal skeletal support; anus present. Two extant superfamilies, Linguloidea

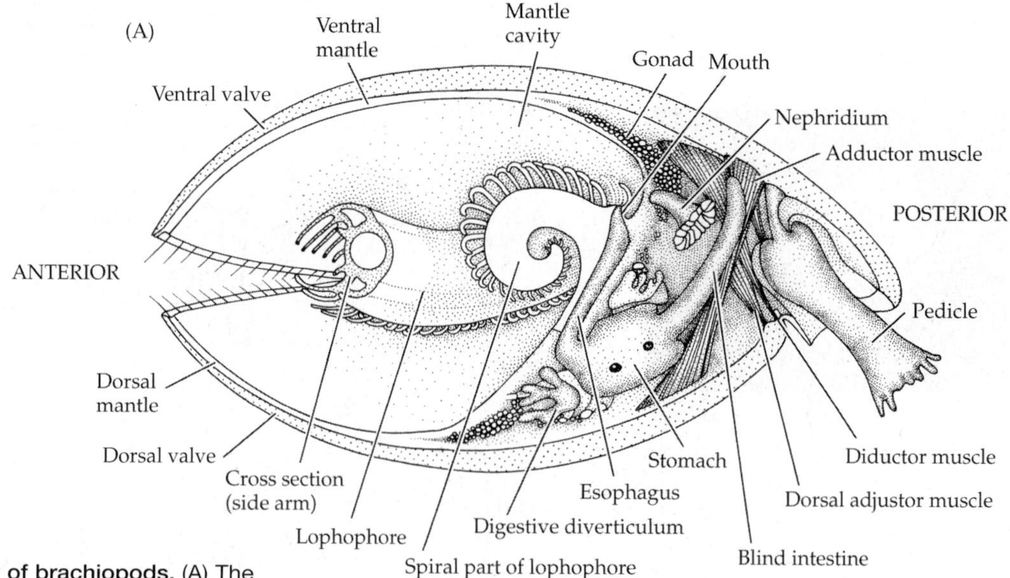

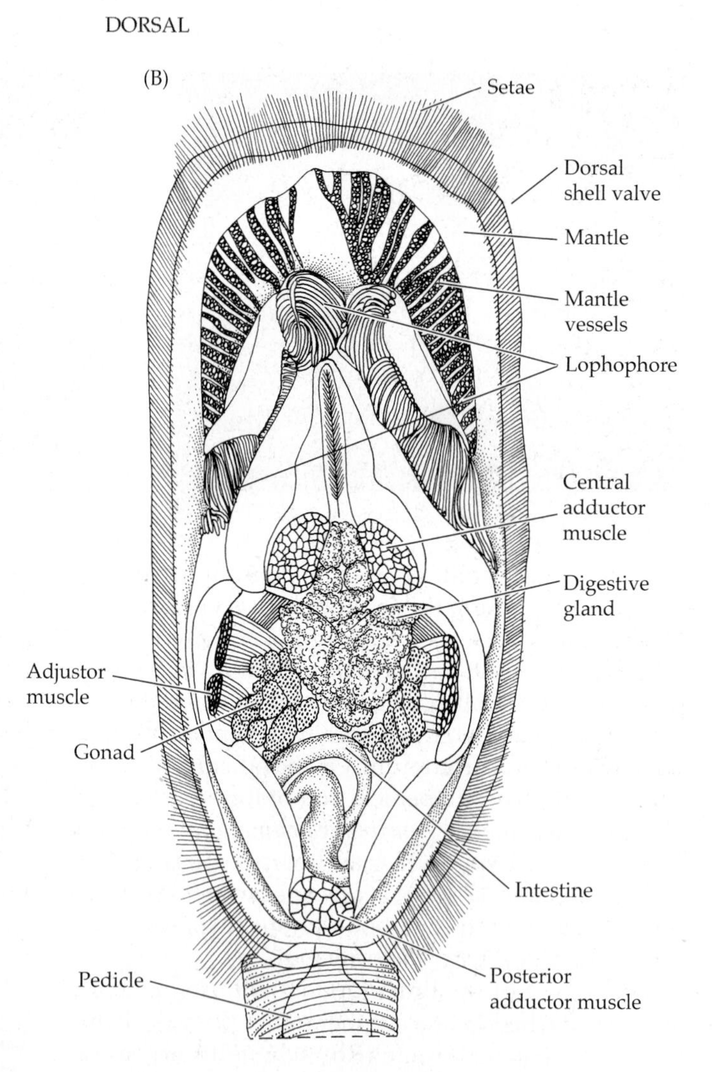

Figure 17.18 Anatomy of brachiopods. (A) The rhynchonelliform brachiopod *Terebratulina* sp. (cutaway view). (B) The linguliform brachiopod *Lingula* (ventral valve removed). (C) The edge of the shell and mantle of a rhynchonelliform (longitudinal section). (D) The mantle edge of *Notosaria nigricans* (Rhynchonelliformea, Hemithiridoidea) (inside view).

and Discinoidea, comprising about 25 extant species. (e.g., *Discinisca*, *Glottidia*, *Lingula*, *Pelagodiscus*)

SUBPHYLUM CRANIIFORMEA Valves not hinged, attached by muscles only; valves made of proteins and calcite (calcium carbonate); no pedicle; ventral valve directly cemented to hard substratum, variable from thin calcite lamella (e.g., *Novocrania anomala*) to massive cone-shaped valve (e.g., *Neoancistrocrania norfolki*); lophophore without internal skeletal support; anus present. One extant superfamily, Cranioidea, comprising about 20 extant species. (e.g., *Novocrania, Neoancistrocrania, Valdiviathyris*)

SUBPHYLUM RHYNCHONELLIFORMEA Valves articulated by tooth-and-socket hinge; valves composed of proteins and calcite (calcium carbonate); pedicle usually present, but lacking muscles and coelomic lumen; lophophore generally with internal supportive elements; gut ends blindly, anus lacking. Three extant orders: Rhynchonellida, Terebratulida, and Thecideida, with just over 350 species. (e.g., *Argyrotheca*, *Dallina*, *Frenulina*, *Gryphus*, *Hemithiris*, *Lacazella*, *Laqueus*, *Liothyrella*, *Magellania*, *Thecidellina*, *Terebratalia*, *Terebratella*, *Terebratulina*, *Tichosina*)

The Brachiopod Body Plan

The Body Wall, Coelom, and Support

The shells of brachiopods comprise an outer organic **periostracum** and an inner structural layer or layers composed variably of calcium carbonate (calcite), calcium phosphate (apatite), proteins, chitin, and collagen. Extant discinids may have siliceous nanoparticles covering their periostracum. Various spines were also present in some fossil species as outgrowths of the shell and served to anchor the animals in place (Figure 17.17I). In a fashion similar to that of molluscs, brachiopod shells are secreted by a **mantle**, which is divided into a dorsal and a ventral mantle formed as an out-

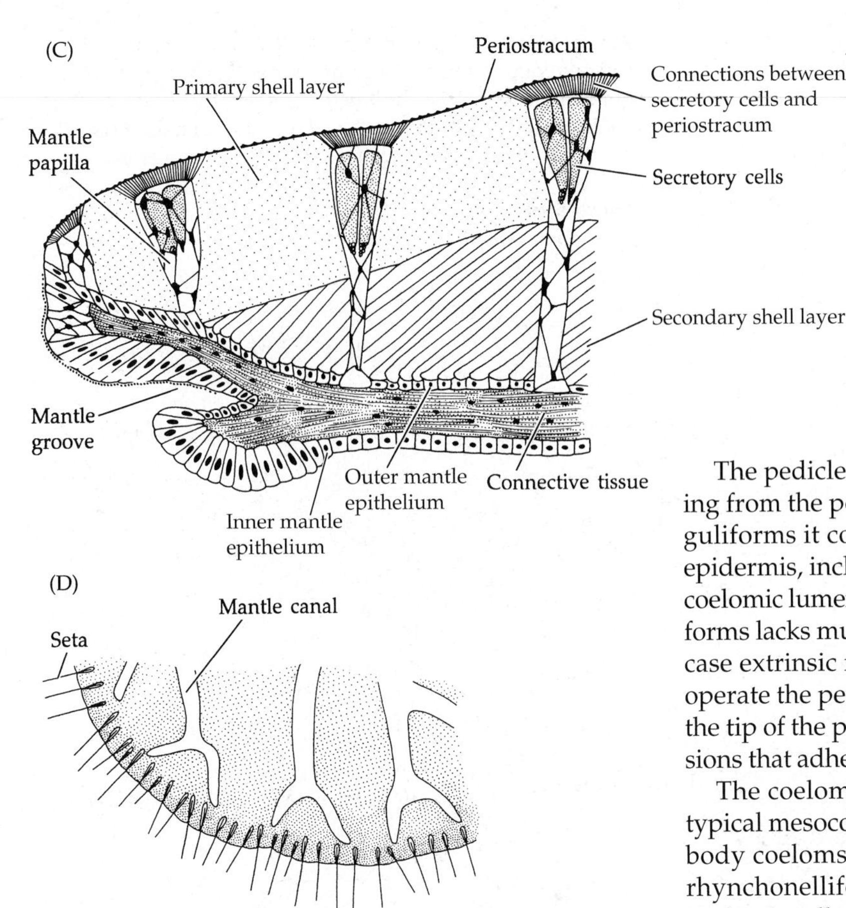

growth of the body wall (Figure 17.18). The periostracum is secreted by the mantle edges in a periostracal slot, and the inner shell layer by the general mantle surface, the outer mantle epithelium. Shells of many brachiopods bear perforations, or punctae, extending from their inner surfaces nearly to the periostracum and containing tiny tissue extensions of the mantle (Figure 17.18). The function of these mantle papillae is unknown, but some workers have suggested that they might serve as areas for food storage and gas exchange, or in some way deter the activities of borers. Shells that lack perforations are termed impunctate.

The soft mantle lines and is attached to the dorsal and ventral valves of the shell and forms the water-filled mantle cavity, which houses the lophophore. The mantle edges often bear chitinous setae, which may protect the fleshy tissue and perhaps serve to prevent the entrance of large particles into the mantle cavity.

The epidermal cells of the mantle and general body surface vary from cuboidal to columnar and are monociliated. Beneath the epidermis is a connective tissue layer of varying thickness, which in some species may house mesenchymal cells and can produce calcite spicules. The inner surface of the body wall is lined by both peritoneal and myoepithelial cells; these form the outer boundary of the coelom. Being folds of the body wall, the dorsal and ventral mantle contain extensions of the coelom, called **mantle canals** (Figure 17.18D).

The pedicle is an outgrowth of the body wall, arising from the posterior area of the ventral valve. In linguliforms it contains all the usual layers beneath the epidermis, including connective tissue, muscles, and a coelomic lumen. However, the pedicle of rhynchonelliforms lacks muscles and a coelomic cavity. In the latter case extrinsic muscle bands from the body wall itself operate the pedicle. In brachiopods that attach firmly, the tip of the pedicle bears papillae or fingerlike extensions that adhere tightly to the substratum.

The coelomic system of brachiopods includes the typical mesocoel and metacoel as the lophophoral and body coeloms, respectively. The epistome is solid in rhynchonelliforms, but in linguliforms may contain coelomic cells that are continuous with the lophophoral mesocoel. The coelomic fluid includes various coelomocytes, some of which contain hemerythrin.

The Lophophore, Feeding, and Digestion

Like that of phoronids and bryozoans, the lophophore of brachiopods comprises a ring of tentacles surrounding the mouth. In brachiopods however, the lophophore is produced as a pair of tentacle-bearing arms that extend anteriorly into the mantle cavity. The overall shape of the lophophore varies among taxa from a simple circular or U-shape to those with highly coiled arms. The brachiopod lophophore also differs in that it is always contained within the protection of the valves and is essentially immovable. In most brachiopods, the lophophore and tentacles are held in position by coelomic pressure, whereas in some rhynchonelliforms the dorsal valve produces a lophophore-supporting structure, the brachidium. In some groups (e.g., Thecideida) the dorsal valve has ridges and grooves instead of an erect lophophore skeleton that help support and position the lophophore.

In order to pass a water current through the mantle cavity, the two valves must be opened slightly. The mechanisms of valve operation differ among members of the three subphyla. In hinge-bearing rhynchonelliforms two pairs of muscles are responsible for valve movements—whereas a pair of diductor muscles opens the valves (Figures 17.18A,B and 17.19A), the adductor muscle pair is responsible for shell closure.

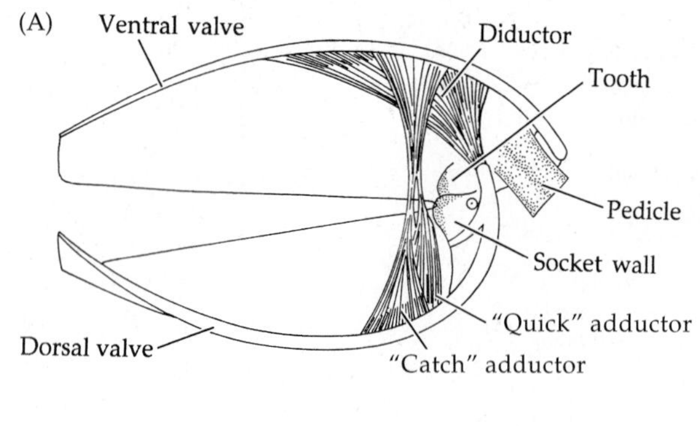

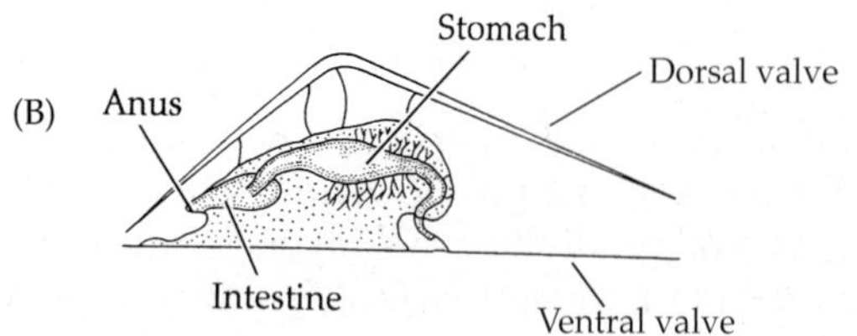

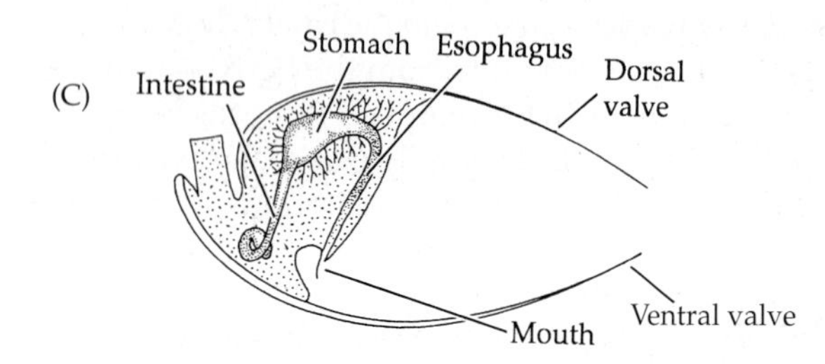

Figure 17.19 (A) The rhynchonelliform brachiopod *Calloria inconspicua* (Terebratelloidea) (ventral side up; cutaway view). Note the major muscles that operate the valves. (B) The complete gut of a craniiform. (C) The blind gut of a rhynchonelliform. (D) The nervous system of *Magellania flavescens* (Rhynchonelliformea, Terebratelloidea). Note the dorsal and ventral aspects on the left and right sides of the drawing, respectively.

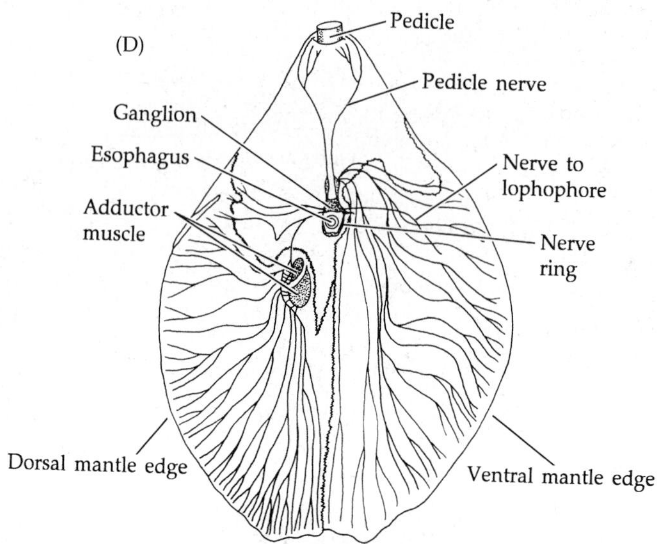

Linguliforms and craniiforms lack a hinge and do not possess diductor muscles. Instead, the gape is produced by contraction of the posterior adductor or the umbonal muscle and relaxation of different anterior muscle groups. Central adductor muscles are mainly used to close the valves. As both shell opening and closure depend on either rapid or long lasting active muscle contraction, diductor and adductor muscles contain striated as well as smooth fibers.

Feeding currents are generated by the lophophoral cilia. Specific incurrent and excurrent flow patterns occur, varying with shell morphology and the shape and orientation of the lophophore. In any case, water is directed over and between the tentacles before passing out of the mantle cavity (Figure 17.20A). Each tentacle bears lateral and frontal ciliary tracts (Figure 17.20B). The lateral cilia of adjacent tentacles overlap and redirect food particles from the water to the frontal cilia by beat reversal. The frontal cilia beat toward the base of the tentacles, helping to direct trapped food. The lophophoral ridge, or brachial axis, bears a brachial food groove within which food material is moved to the mouth (Figure 17.20C). Brachiopods feed on nearly any appropriately small organic particles, especially phytoplankton.

The digestive system is U-shaped (Figures 17.18A and 17.19B,C). The mouth is followed by a short esophagus, which extends dorsally and then posteriorly to the stomach. A digestive gland covers most of the stomach and connects to it via paired ducts. The intestine extends posteriorly, where it ends blindly in rhynchonelliforms or recurves as a rectum terminating in an anal opening in inarticulated brachiopods. In the latter case the anus opens either medially (craniiforms) or on the right side of the animal (linguliforms). The absence of an anus is almost certainly a secondary loss in rhynchonelliforms, and can be traced back to differences in the fate of the blastopore during early embryonic development in comparison to craniiforms. Little is known about digestion in brachiopods, but some work on *Lingula* indicates that it occurs intracellularly in the digestive gland.

Circulation, Gas Exchange, and Excretion

The brachiopod circulatory system is open, much reduced, and largely unstudied. A contractile heart lies in the dorsal mesentery just above the gut (*Novocrania* possesses several "hearts"). Leading anteriorly and posteriorly from the heart, the blood vessels form channels within the connective tissue of the mesenteries, thus no true vessels are present. These channels branch to various parts of the body, but the pattern of circulation is not fully understood. It appears that the blood is separate from the coelomic fluid, although both contain certain similar cells. The function of the circulatory system is thought to be largely restricted to nutrient distribution.

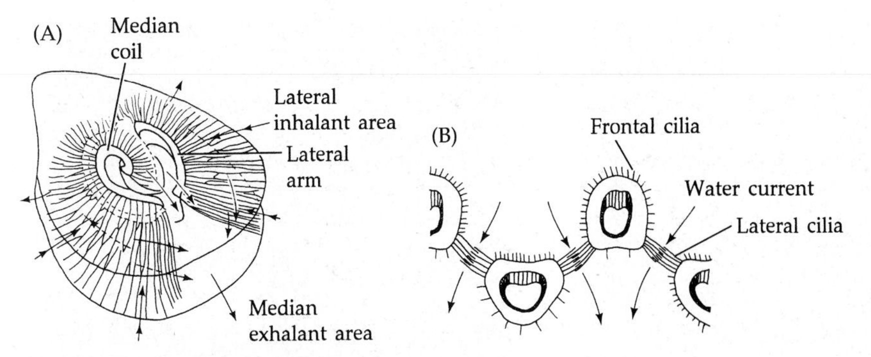

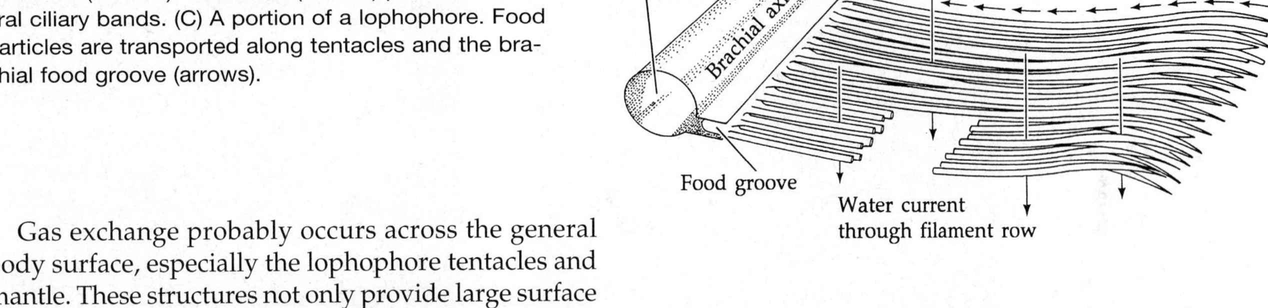

Figure 17.20 Feeding currents in brachiopods. (A) Feeding currents (arrows) of *Calloria inconspicua* (Rhynchonelliformea, Terebratelloidea). (B) Lophophoral tentacles (section). The water (arrows) passes over lateral ciliary bands. (C) A portion of a lophophore. Food particles are transported along tentacles and the brachial food groove (arrows).

Gas exchange probably occurs across the general body surface, especially the lophophore tentacles and mantle. These structures not only provide large surface areas but are also sites over which water moves and is brought close to underlying coelomic fluid. This general arrangement and the presence of hemerythrin in certain coelomocytes suggest that the coelomic fluid, not the blood, is the medium for oxygen transport.

Brachiopods possess one or two pairs of metanephridia, with the nephrostomes opening to the metacoel. The nephridioducts exit through pores into the mantle cavity. The nephridia function as gonoducts as well as discharging phagocytic coelomocytes that have accumulated metabolic wastes. In the distal part of the metanephridia (nephridial canal), columnar cells absorb large quantities of coelomic fluid by endocytosis suggesting secondary urine formation as in other metanephridial systems. However, primary urine formation through a podocyte filter system from the blood vessel into the coelom has only been assumed for the perioesophageal coelom in the rhynchonellid *Hemithiris psittacea*.

Nervous System and Sense Organs

The nervous system of brachiopods is somewhat reduced. A dorsal ganglion and a ventral ganglion lie against the esophagus and are connected by a circumenteric nerve ring. Nerves emerge from the ganglia and nerve ring and extend to various parts of the body, especially the muscles, mantle, and lophophore (Figure 17.19D). As usual, the array of sense organs in these animals is compatible with their lifestyle. The mantle edges and setae are richly supplied with sensory neurons, probably tactile receptors. There is also evidence that brachiopods are sensitive to dissolved chemicals, perhaps through surface receptors on the tentacles or mantle edge. Members of at least one burrowing species of *Lingula* possess a pair of statocysts that are associated with orientation in the substratum. Larvae of some rhynchonelliform groups possess eyes composed of two photoreceptor cells almost certainly mediating their photonegative behavior prior to settlement and metamorphosis. Comparable to the vertebrate eye, brachiopod larval eyes contain a ciliary opsin as the photosensitive pigment. The larval nervous system in brachiopods has been shown to contain both serotonergic and histaminergic neurons.

Reproduction and Development

Asexual reproduction does not occur in brachiopods. Most species are gonochoristic, with gametes developing from patches of transient gonadal tissue derived from the coelomic epithelium in the metacoel. Gametes are released into the metacoel and escape through the metanephridia. In most cases both eggs and sperm are shed freely, and fertilization is external. A few species, however, brood their embryos until the larval stage is reached. In these cases sperm are picked up in the water currents of females, and the eggs are retained in a brooding area where they are fertilized. *Argyrotheca*, for example, broods its embryos in special caverns formed by the mantle epithelium. Species of Thecideida form brood pouches either in the dorsal or ventral valves as derivatives of the lophophore epithelium. Others retain their embryos on the arms of the lophophore, in special regions of the mantle cavity, or in modified depressions in a valve.

(A)

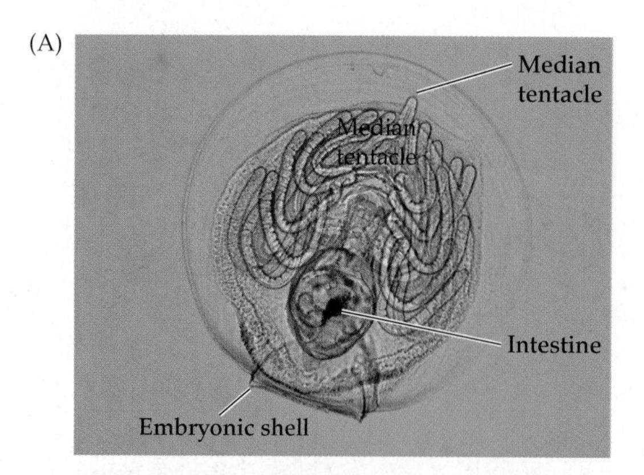
Median
tentacle
Median
tentacle
Intestine
Embryonic shell

(B)

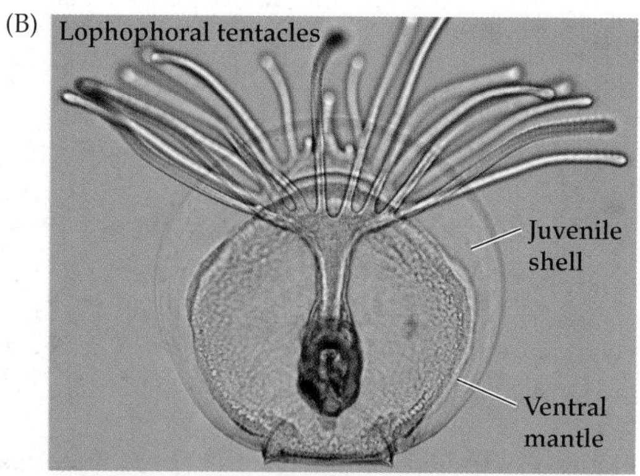
Lophophoral tentacles
Juvenile
shell
Ventral
mantle

(C)

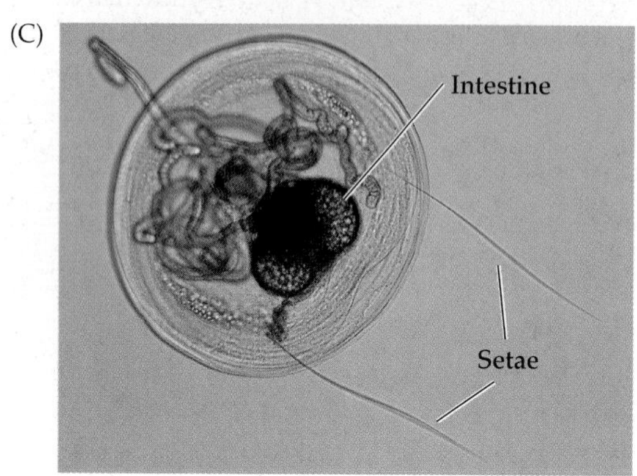
Intestine
Setae

(D)

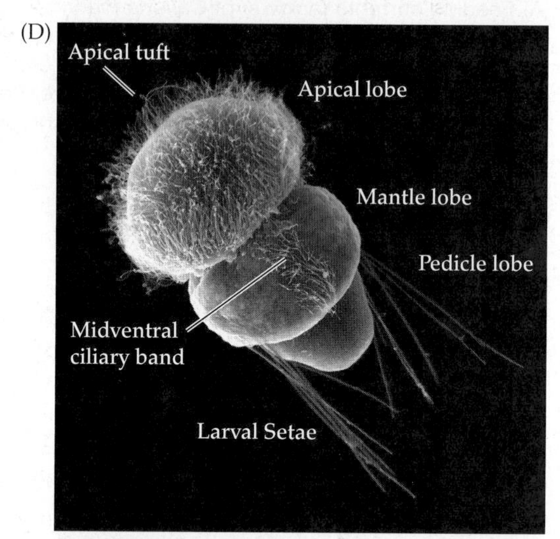
Apical tuft
Apical lobe
Mantle lobe
Pedicle lobe
Midventral
ciliary band
Larval Setae

(E)

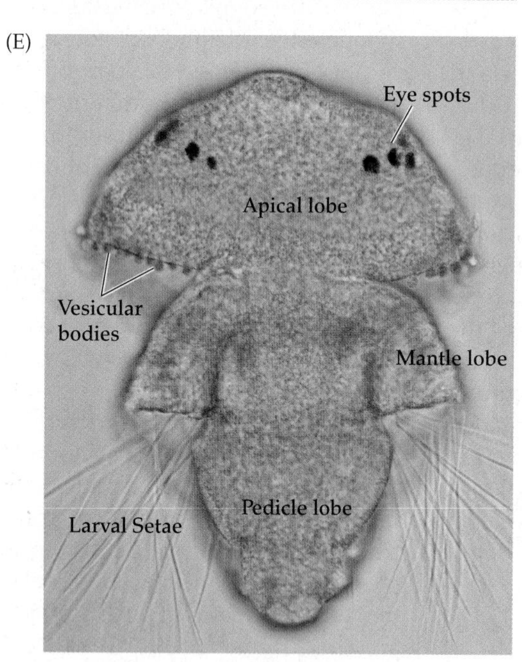
Eye spots
Apical lobe
Vesicular
bodies
Mantle lobe
Larval Setae
Pedicle lobe

(F)

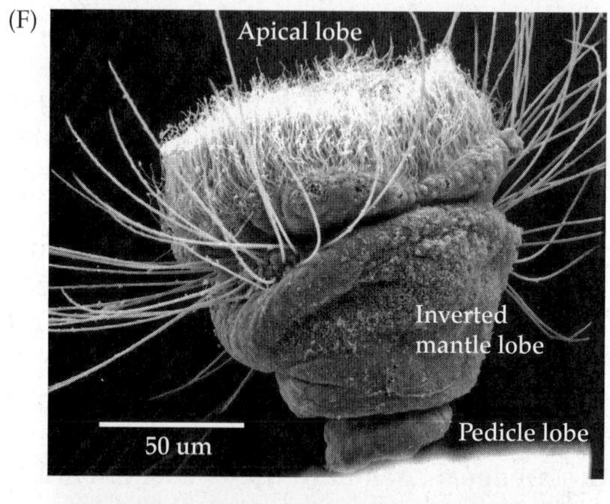
Apical lobe
Inverted
mantle lobe
50 um
Pedicle lobe

(G)

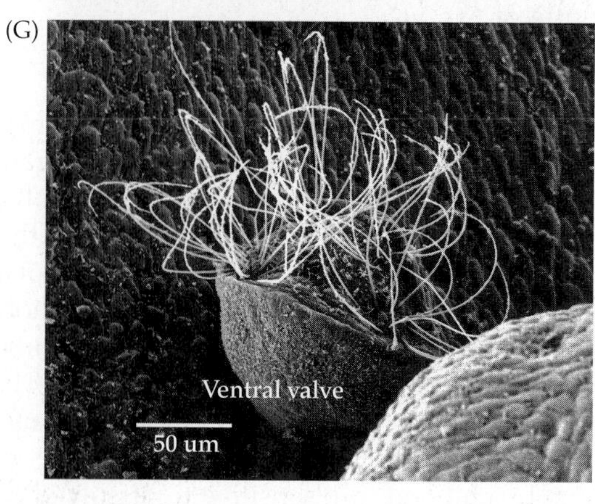
Ventral valve
50 um

(H)

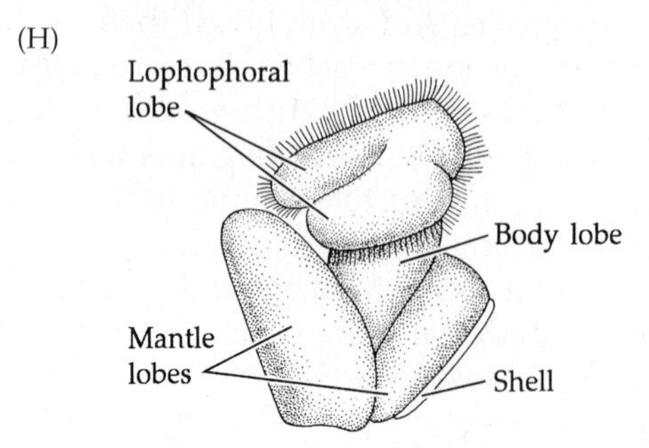
Lophophoral
lobe
Body lobe
Mantle
lobes
Shell

Figure 17.21 Brachiopod lobate larvae and metamorphosis. (A) Planktonic juvenile of *Glottidia* sp. (Linguliformea, Linguloidea) with contracted lophophore. (B) The same *Glottidia* juvenile with expanded lophophore during active swimming. (C) The shelled planktonic juvenile of a discinid (Linguliformea, Discinoidea). Note the pair of long posterior setae. (D) SEM photo of a three-lobed stage larva of *Macandrevia cranium* (Rhynchonelliformea, Zeillerioidea). Cilia are restricted to the apical lobe and a midventral band. (E) Three-lobed stage of *Terebratalia transversa* (Rhynchonelliformea, Laqueoidea) with two dorsolateral rows of red eye spots on the apical lobe. (F, G) Two subsequent metamorphic stages of *Laqueus erythraeus* (Rhynchonelliformea, Laqueoidea). Mantle lobe inversion starts immediately after attachment of the larva to the substrate. (H) Hatchling of *Lingula anatina* (Linguliformea, Linguloidea) already equipped with two shelled mantle lobes and a ciliated lophophore rudiment.

Cleavage is holoblastic, radial, and nearly equal; it leads to the formation of a coeloblastula. Gastrulation is by invagination, except in the brooding form *Lacazella*, where it apparently occurs by delamination. The blastopore closes and the mouth forms secondarily (i.e., deuterostomy). The anus, when present, breaks through late as the gut grows and approaches the body wall. Mesoderm formation starts by proliferation of archenteral cells and subsequent separation from the larval gut anlage. Only during metamorphosis the mesoderm cells diverge, thereby opening a lumen, the coelom. The origin of the prospective coelomic epithelium from archenteral cells can be interpreted as enterocoely.

Whether the developmental pattern is mixed or fully indirect, all brachiopods eventually enter a free-swimming stage (Figure 17.21). Although these planktonic forms differ morphologically they have been compared to trochophore larvae, in which case the ciliary band around the apical lobe could be interpreted as homologue to the prototroch of a traditional trochophore. Whereas lingulid brachiopods hatch as planktotrophic stages ("free-swimming juveniles") with a thorough and a first-formed ("embryonic") shell, all other groups hatch as nonfeeding stages. In discinids the hatchling has typical larval characters (e.g., larval setae) but quickly develops tentacles, a functional digestive system, and a first-formed shell while still in the plankton stage. Craniiforms and rhynchonelliforms have a lecithotrophic larval stage that only develops tentacles and a shell, and starts feeding upon settlement and metamorphosis.

In linguliforms, such as *Lingula*, the lophophore of the planktonic stage (Figure 17.21A,B) can be protruded out from between the mantle lobes and function to propel and feed the larva. As the mantle lobes and shell bear some weight, the developmental stage immediately sinks when the lophophoral tentacles are retracted into the mantle cavity. Extension of the growing pedicle while on the seafloor marks the start of the juvenile's benthic life. Thus, there is no drastic metamorphosis at the time of settling (Figure 17.21A,B).

The free-swimming larvae of rhynchonelliforms are regionalized into an anterior (apical) lobe, a mantle lobe, and a pedicle lobe (Figure 17.21C–F). As craniiforms lack a pedicle throughout their life, their larval body has only an apical and a posterior lobe, the latter bearing three pairs of larval setal bundles. After a short larval life of a few days to weeks (depending on the ambient temperatures), the nonfeeding brachiopod larvae settle and metamorphose. As the pedicle lobe attaches to the substratum (only rhynchonelliforms), the mantle lobes flex forward over the anterior lobe. The now exterior surfaces of the mantle lobes commence secretion of the valves, while the anterior lobe differentiates into the body and the lophophore.

Selected References

General References

Gee, H. 1995. Lophophorates prove likewise variable. Nature 374: 493.

Giese, A. C., J. S. Pearse and V. B. Pearse (eds.). 1991. *Reproduction of Marine Invertebrates. Vol. VI, Echinoderms and Lophophorates.* The Boxwood Press, Pacific Grove, California.

Halanych, K. M., J. D. Bacheller, A. M. A. Aguinaldo, S. M. Liva, D. M. Hillis and J. A. Lake. 1995. Evidence from 18S ribosomal DNA that the lophophorates are protostome animals. Science 267: 1641–1642.

Hejnol, A. and J. M. Martín-Durán. 2015. Getting to the bottom of anal evolution. Zool. Anz. 1–13. doi: 10.1016/j.jcz.2015.02.006

Helmkampf, M., I. Bruchhaus and B. Hausdorf. 2008. Phylogenomic analyses of lophophorates (brachiopods, phoronids and bryozoans) confirm the Lophotrochozoa concept. Proc. R. Soc. Lond. B 275: 1927–1933.

Hyman, L. H. 1959. *The Invertebrates, Vol. 5, Smaller Coelomate Groups.* McGraw-Hill, New York.

Jefferies, R. P. S. 1986. *The Ancestry of the Invertebrates.* British Mus. (Natural History), London.

Larwood, G. and B. R. Rosen. 1970. *Biology and Systematics of Colonial Organisms.* Academic Press, NY.

Lüter, C. 2000. The origin of the coelom in Branchiopoda and its phylogenetic significance. Zoomorphology 120: 15–28.

Mackey, L. Y., B. Winnepenninckx, R. DeWachter, T. Backeljau, P. Emschermann and J. R. Garey. 1996. 18S rRNA suggests that Entoprocta are protostomes, unrelated to Ectoprocta. J. Mol. Evol. 42: 552–559.

McCammon, H. M. and W. A. Reynolds (organizers). 1977. Symposium: Biology of Lophophorates. Am. Zool. 17: 3–150.

Moore, R. C. (ed.). 1965. *Treatise on Invertebrate Paleontology. Pts. G and H (Vols. 1 and 2).* Geological Society of America, Inc. and The University of Kansas, Lawrence, Kansas.

Morris, S. C. 1995. Nailing the lophophorates. Nature 375: 365–366.

Nielsen, C. 2009. How did indirect development with planktotrophic larvae evolve? Biol. Bull. 216: 203–215.

Zimmer, R. L. 1973. Morphological and developmental affinities of the lophophorates. Pp. 593–600 in G. P. Larwood (ed.), *Living and Fossil Bryozoa*. Academic Press, London.

Zimmer, R. L. 1997. Phoronids, brachiopods and bryozoans: the lophophorates. Pp. 279–308 in S. F. Gilbert and A. M. Raunio, *Embryology, Constructing the Organism*. Sinauer Associates, Sunderland, MA.

Brachiopoda

Altenburger, A. and A. Wanninger. 2010. Neuromuscular development in *Novocrania anomala*: evidence for the presence of serotonin and a spiralian-like apical organ in lecithotrophic brachiopod larvae. Evol. Dev. 12: 16–24.

Cohen, B. L. and A. Weydmann. 2005. Molecular evidence the phoronids are a subtaxon of brachiopods (Brachiopoda: Phoronata) and that genereic divergence of metazoan phyla began long before the early Cambrian. Org. Divers. Evol. 5: 253–273.

Gutman, W. F., K. Vogel and H. Zorn. 1978. Brachiopods: biochemical interdependencies governing their origin and phylogeny. Science 199: 890–893.

James, M. A. 1997. Brachiopoda: internal anatomy, embryology, and development. Pp. 297–407 in F. W. Harrison and R. M. Woollacott (eds.), *Microscopic Anatomy of Invertebrates, Vol. 13, Lophophorates, Entoprocta, and Cycliophora*. Wiley-Liss, New York.

Kaulfuss, A., R. Seidel and C. Lüter. 2013. Linking micromorphism, brooding and hermaphroditism in articulate brachiopods: insights from Caribbean *Argyrotheca* (Brachiopoda). J. Morph. 274: 361–376.

Lüter, C. 2000. The origin of the coelom in Brachiopoda and its phylogenetic significance. Zoomorphology 120: 15–28.

Lüter, C. 2001. Brachiopod larval setae: a key to the phylum's ancestral life cycle? Pp. 46–55 in C. H. C. Brunton, L. R. M. Cocks and S. L. Long (eds.), *Brachiopods: Past and Present*. Taylor and Francis, London.

Lüter, C. 2004. How brachiopods get covered with nanometric silicon chips. Proc. R. Soc. Lond. B Suppl. 6 (Biology Letters): S465–S467.

Lüter, C. 2007. Anatomy. Pp. 2321–2355 in P. A. Selden (ed.), *Treatise on Invertebrate Paleontology, part H: Brachiopoda, revised, Vol. 6*. The Geological Society of America and University of Kansas Press, Boulder and Lawrence.

MacKay, S. and R. A. Hewitt. 1978. Ultrastructure studies on the brachiopod pedicle. Lethaia 11: 331–339.

Nielsen, C. 1991. The development of the brachiopod *Crania* (*Neocrania*) *anomala* (O. F. Müller) and its phylogenetic significance. Acta. Zool. 72 (1): 7–28.

Passamaneck, Y., N. Furchheim, A. Hejnol, M. Q. Martindale and C. Lüter. 2011. Ciliary photoreceptors in the cerebral eyes of a protostome larva. EvoDevo 2: 6.

Richardson, J. R. 1981. Brachiopods in mud: Resolution of a dilemma. Science 211: 1161–1163.

Rudwick, M. J. S. 1970. *Living and Fossil Brachiopods*. Hutchinson University Library, London.

Santagata, S. 2011. Evaluating Neurophylogenetic Patterns in the Larval Nervous Systems of Brachiopods and Their Evolutionary Significance to Other Bilaterian Phyla. J. Morphol. 272: 1153–1169.

Steele-Petrovic, H. M. 1976. Brachiopod food and feeding processes. Paleontology 19(3): 417–436.

Thayer, C. W. 1985. Brachiopods versus mussels: competition, predation, and palatability. Science 228(4707): 1527–1528.

Watabe, N. and C.-M. Pan. 1984. Phosphatic shell formation in atremate brachiopods. Am. Zool. 24: 977–985.

Williams, A. 1997. Brachiopoda: introduction and integumentary system. Pp. 237–296 in F. W. Harrison and R. M. Woollacott (eds.), *Microscopic Anatomy of Invertebrates, Vol. 13, Lophophorates, Entoprocta, and Cycliophora*. Wiley-Liss, New York.

Williams, A., M. A. James, C. C. Emig, S. Mackay and M. C. Rhodes. 1997. Anatomy. Pp. 7–188 in Kaesler R. L. (ed.), *Treatise on Invertebrate Paleontology, Pt. H: Brachiopoda, Revised, Vol.1*. The Geological Society of America and The University of Kansas, Boulder and Lawrence.

Williams, A., C. Lüter and M. Cusack. 2001. The nature of siliceous mosaics forming the first shell of the brachiopod *Discinisca*. J. Struct. Biol. 134: 25–34.

Williams, A., S. Mackay and M. Cusack. 1992. Structure of the organo-phosphatic shell of the brachiopod *Discinisca*. Phil. Trans. Roy. Soc. Lond. B 337: 83–104.

Bryozoa

Carle, K. J. and E. E. Ruppert. 1983. Comparative ultrastructure of the bryozoan funiculus: a blood vessel homologue. Z. Zool. Syst. Evol. 21: 181–193.

Fuchs, J., M. Obst and P. Sundberg. 2009. The first comprehensive molecular phylogeny of Bryozoa (Ectoprocta) based on combined analyses of nuclear and mitochondrial genes. Mol. Phylogen. Evol. 52: 225–233.

Gruhl, A. 2010. Ultrastructure of mesoderm formation and development in *Membranipora membranacea* (Bryozoa: Gymnolaemata). Zoomorphology 129: 45–60.

Hayward, P. J. 1985. *Ctenostome Bryozoans*. Synopses of the British Fauna, New Series, Vol. 33. Brill/Backhuys, London.

Hayward, P. J. and J. S. Ryland 1985. *Cyclostome Bryozoans*. Synopses of the British Fauna, New Series, Vol. 34. Brill/Backhuys, London.

Hayward, P. J. and J. S. Ryland 1998. *Cheilostomatous Bryozoa Part I. Aeteoidea - Cribrilinoidea*. Synopses of the British Fauna, New Series, Vol. 10 (2nd Ed.), The Linnean Society of London.

Hayward, P. J. and J. S. Ryland 1998. *Cheilostomatous Bryozoa Part 2. Hippothooidea - Celleporelloidea*. Synopses of the British Fauna, New Series, Vol. 14 (2nd Ed.), The Linnean Society of London.

Mukai, H., K. Terakado and C. G. Reed 1997. Bryozoa. Pp. 45–206 in F. W. Harrison (ed.), *Microscopic Anatomy of Invertebrates*, Vol. 13. Wiley-Liss, New York.

Nielsen, C. 1970. On metamorphosis and ancestrula formation in cyclostomatous bryozoans. Ophelia 7: 217–256.

Nielsen, C. 1981. On morphology and reproduction of *"Hippodiplosia" insculpta* and *Fenestrulina malusii* (Bryozoa, Cheilostomata). Ophelia 20: 91–125.

Nielsen, C. and K. Worsaae 2010. Structure and occurrence of cyphonautes larvae (Bryozoa, Ectoprocta). J. Morphol. 271: 1094–1109.

Ostrovsky, A. N., D. P. Gordon and S. Lidgard. 2009. Independent evolution of matrotrophy in the major classes of Bryozoa: transitions among reproductive patterns and their ecolgidal background. Mar. Ecol. Prog. Ser. 378: 113–124.

Ostrovsky, A., C. Nielsen, N. Vávra and E. Yagunova. 2009. Diversity of brood chambers in calloporid bryozoans (Gymnolaemata, Cheilostomata): comparative anatomy and evolutionary trends. Zoomorphology 128: 13–35.

Reed, C. G. 1991. Bryozoa. Pp. 85–245 in A. C. Giese, J. S. Pearse and V. B. Pearse (eds.), *Reproduction of Marine Invertebrates, Vol. 6*. Boxwood Press, Pacific Grove, CA.

Reed, C. G. and R. M. Woollacott. 1982. Mechanisms of rapid morphogenetic movements in the metamorphosis of the bryozoan *Bugula neritina* (Cheilostomata, Cellularioidea). I. Attachment to the substratum. J. Morphol. 172: 335–348.

Reed, C. G. and R. M. Woollacott. 1983. Mechanisms of rapid morphogenetic movements in the metamorphosis of the bryozoan *Bugula neritina* (Cheilostomata, Cellularioidea): II. The role of dynamic assemblages of microfilaments in the pallial epithelium. J. Morphol. 177: 127–143.

Riisgård, H. U. and P. Manríquez. 1997. Filter-feeding in fifteen marine ectoprocts (Bryozoa): particle capture and water pumping. Mar. Ecol. Prog. Ser. 154: 223–239.

Riisgård, H. U., B. Okamura and P. Funch. 2010. Particle capture in ciliary filter-feeding gymnolaemate and phylactolaemate bryozoans – a comparative study. Acta Zool. (Stockh.) 91: 416–425.

Ryland, J. S. 1970. *Bryozoans*. Hutchington, London.

Santagata, S. and W. C. Banta. 1996. Origin of brooding and ovicells in cheilostome bryozoans: interpretive morphology of *Scrupocellaria ferox*. Trans. Amer. Microsp. Soc. 115(2): 170–180.

Soule, D. F., J. D. Soule and H. W. Chaney. 1995. *The Bryozoa*. In Taxonomic Atlas of the Santa Maria Basin and Westen Santa Barbara Channel. Santa Barbara Museum of Natural History, Santa Barbara, CA.

Taylor, P. D. 1985. Carboniferous and Permian species of the cyclostome bryozoan *Corynotrypa* Bassler, 1911 and their clonal propagation. Bull. British Mus. (Nat. Hist.), Geol. 38: 359–372.

Taylor, P. D., A. B. Kudryavtsev and J. W. Schopf. 2008. Calcite and aragonite distributions in the skeletons of bimineralic bryozoans as revealed by Raman spectroscopy. Invert. Biol. 127(1): 87–97.

Winston, J. E. 1978. Polypide morphology and feeding behavior in marine ectoprocts. Bull. Mar. Sci. 28: 1–31.

Winsten, J. E. 1984. Why bryozoans have avicularia—a review of the evidence. Amer. Mus. Novitates, No. 2789, 26 pp.

Woollacott, R. M. and R. L. Zimmer. 1972. Origin and structure of the brood chamber in *Bugula neritina* (Bryozoa). Mar. Biol. 16: 165–170.

Xia, F.-S., S.-G. Zhang and Z.-Z. Wang. 2007. The oldest bryozoans: new evidence from the Late Tremadocian (Early Ordovician) of East Yangtse Gorges in China. J. Paleont. 81: 1308–1326.

Phoronida

Bartolomaeus, T. 2001. Ultrastructure and formation of the body cavity lining in *Phoronis muelleri* (Phoronida, Lophophorata). Zoomorphology 120: 135–148.

Cohen, B. and A. Weydmann. 2005. Molecular evidence that phoronids are a subtaxon of brachiopods (Brachiopoda: Phoronata) and that genetic divergence of metazoan phyla began long before the early Cambrian. Org. Divers. Evol. 5: 253–273.

Cohen, B. L. 2012. Rerooting the rDNA gene tree reveals phoronids to be "brachiopods without shells"; dangers of wide taxon samples in metazoan phylogenetics (Phoronida; Brachiopoda). Zool. J. Linn. Soc. 167: 82–92.

Chernyshev, A. V. and E. N. Temereva. 2010. First report of diagonal musculature in phoronids (Lophophorata: Phoronida). Dokl. Biol. Sci. 433: 264–267.

Emig, C. C. 1974. The systematics and evolution of the phylum Phoronida. Z. Zool. Syst. Evol. 12(2): 128–151.

Emig, C. C. 1977. The embryology of Phoronida. Am. Zool. 17: 21–38.

Emig, C. C. 1982. Phoronida. In S. P. Parker (ed.), *Synopsis and Classification of Living Organisms*. McGraw-Hill, New York.

Freeman, G. 1991. The bases for and timing of regional specification during larval development in *Phoronis*. Develop. Biol. 147: 157–173.

Freeman, G., Martindale MQ. 2002. The origin of mesoderm in phoronids. Develop. Biol. 252: 301–311.

Garlick, R. L., Williams B. J., Riggs, A. F. 1979. The hemoglobins of *Phoronopsis viridis*, of the primitive invertebrate phylum Phoronida: characterization and subunit structure. Arch. Biochem. Biophys. 194: 13–23.

Gruhl, A., P. Grobe and T. Bartolomaeus. 2005. Fine structure of the epistome in *Phoronis ovalis*: significance for the coelomic organization in Phoronida. Invert. Biol. 124: 332–343.

Hermann, K. 1997. Phoronida. Pp. 207–235 in F. W. Harrison and R. M. Woollacott (eds.), *Microscopic Anatomy of Invertebrates, Vol. 13, Lophophorates, Entoprocta, and Cycliophora*. Wiley-Liss, New York.

Hirose, M., R. Fukiage, T. Katoh and H. Kajihara. 2014. Description and molecular phylogeny of a new species of *Phoronis* (Phoronida) from Japan, with a redescription of topotypes of *P. ijimai* Oka, 1897. Zookeys 398: 1-31.

Nesnidal, M. P. and 9 others. 2013. New phylogenomic data support the monophyly of Lophophorata and an ectoproct-phoronid clade and indicate that Polyzoa and Kryptrochozoa are caused by systematic bias. BMC Evol. Biol. 13: 253.

Pennerstorfer, M. and G. Scholtz. 2012. Early cleavage in *Phoronis muelleri* (Phoronida) displays spiral features. Evol. Dev. 14: 484–500.

Santagata, S. 2002. Structure and metamorphic remodeling of the larval nervous system and musculature of *Phoronis pallida* (Phoronida). Evol. Dev. 4: 28–42.

Santagata, S. 2004. A waterborne behavioral cue for the actinotroch larva of *Phoronis pallida* (Phoronida) produced by *Upogebia pugettensis* (Decapoda: Thalassinidea). Biol. Bull. 207: 103–115.

Santagata, S. 2004. Larval development of *Phoronis pallida* (Phoronida): implications for morphological convergence and divergence among larval body plans. J. Morphol. 259: 347–358.

Santagata, S. 2011. Evaluating neurophylogenetic patterns in the larval nervous systems of brachiopods and their evolutionary significance to other bilaterian phyla. J. Morphol. 272: 1153–1169.

Santagata, S. and R. L. Zimmer. 2002. Comparison of the neuromuscular systems among actinotroch larvae: systematic and evolutionary implications. Evol. & Develop. 4: 43–54.

Santagata, S. and B. L. Cohen. 2009. Phoronid phylogenetics (Brachiopoda; Phoronata): evidence from morphological cladistics, small and large subunit rDNA sequences, and mitochondrial *cox1*. Zool. J. Linn. Soc. 157: 34–50.

Silén, L. 1954. Developmental biology of the Phoronidea of the Gullmar Fjord area of the west coast of Sweden. Acta Zool. 35: 215–257.

Temereva, E. N. and V. V. Malakhov. 2006. Trimeric coelom organization in the larvae of *Phoronopsis harmeri* Pixell, 1912 (Phoronida, Lophophorata). Dokl. Biol. Sci. 410: 396–399.

Temereva, E. N. and V. V. Malakhov. 2011. Organization of the epistome in *Phoronopsis harmeri* (Phoronida) and consideration of the coelomic organization in Phoronida. Zoomorphology 130: 121–134.

Temereva, E and A. Wanninger. 2012. Development of the nervous system in *Phoronopsis harmeri* (Lophotrochozoa, Phoronida) reveals both deuterostome- and trochozoan-like features. BMC Evol. Biol. 12: 121.

Temereva, E. N. and E. B. Tsitrin. 2013. Development, organization, and remodeling of phoronid muscles from embryo to metamorphosis (Lophotrochozoa: Phoronida). BMC Dev. Biol. 13: 1–24.

Zimmer, R. L. 1967. The morphology and function of accessory reproductive glands in the lophophores of *Phoronis vancouverensis* and *Phoronopsis harmeri*. J. Morphol. 121(2): 159–178.

Zimmer, R. L. 1972. Structure and transfer of spermatozoa in *Phoronopsis viridis*. In C. J. Arceneaux (ed.), 30th Annual Proceedings of the Electron Microscopical Society of America.
Zimmer, R. L. 1978. The comparative structure of the preoral hood coelom. Pp. 23–40 in F. S. Chia and M. E. Rice (eds.), *Settlement and metamorphosis of marine invertebrate larva*. New York, Elsevier.
Zimmer, R. L. 1980. Mesoderm proliferation and function of the protocoel and metacoel in early embryos of *Phoronis vancouverensis* (Phoronida). Zool. Jb., Anat. 103: 219–233.
Zimmer, R. L. 1991. Phoronida. Pp. 1–45 in J. S. Pearse, V. B. Pearse and A. C. Giese (eds.), *Reproduction of Marine Invertebrates, volume VI Echinoderms and Lophophorates*. Boxwood Press, Pacific Grove, CA.

CHAPTER 18

The Nematoida

Phyla Nematoda and Nematomorpha

This chapter is the first of seven that describe the animals belonging to a clade known as Ecdysozoa. Ecdysozoa is one of the two main protostome clades (the other being Spiralia) and it contains eight phyla and about 83% of animal species diversity (see Chapter 9), much of which is contained within the arthropods and nematodes. All ecdysozoans molt their cuticle at least once during their life history. The group comprises three well-supported subclades: Nematoida (phyla Nematoda, Nematomorpha), Scalidophora (phyla Kinorhyncha, Priapula, Loricifera), and Panarthropoda (phyla Onychophora, Tardigrada, Arthropoda). The relationships of these three subclades have not been firmly determined, and they appear as an unresolved trichotomy in our tree of the Metazoa (see Chapter 28). However, some evidence suggests that Nematoida and Scalidophora comprise a sister group, and they do share a number of morphological similarities (e.g., a circumoral collar- or ring-shaped brain composed of a ring neuropil, or network of nervous tissue, with anterior and posterior somata). This putative clade has been given the name Cycloneuralia. However, recent molecular phylogenetic analyses have failed to find strong support for this clade. Unlike the Spiralia, Ecdysozoa can be defined by morphological synapomorphies, including their three-layered cuticle that is molted, a process regulated by ecdysteroids hormones. The cuticle consists of a proteinaceous exocuticle and an endocuticle with chitin or collagen, with the epicuticle forming from the apical zone of the epidermal microvilli. Ecdysozoans also lack external epithelial cilia, lack a primary (ciliated) larva with an apical organ, and unlike most other protostomes, ecdysozoans do not undergo spiral cleavage.

The phyla Nematoda (roundworms) and Nematomorpha (horsehair worms) comprise the clade Nematoida, supported by both morphological and molecular analyses. As in the other ecdysozoan

Classification of The Animal Kingdom (Metazoa)

Non-Bilateria*
(a.k.a. the diploblasts)
- PHYLUM PORIFERA
- PHYLUM PLACOZOA
- PHYLUM CNIDARIA
- PHYLUM CTENOPHORA

Bilateria
(a.k.a. the triploblasts)
- PHYLUM XENACOELOMORPHA

Protostomia
- PHYLUM CHAETOGNATHA

SPIRALIA
- PHYLUM PLATYHELMINTHES
- PHYLUM GASTROTRICHA
- PHYLUM RHOMBOZOA
- PHYLUM ORTHONECTIDA
- PHYLUM NEMERTEA
- PHYLUM MOLLUSCA
- PHYLUM ANNELIDA
- PHYLUM ENTOPROCTA
- PHYLUM CYCLIOPHORA

Gnathifera
- PHYLUM GNATHOSTOMULIDA
- PHYLUM MICROGNATHOZOA
- PHYLUM ROTIFERA

Lophophorata
- PHYLUM PHORONIDA
- PHYLUM BRYOZOA
- PHYLUM BRACHIOPODA

ECDYSOZOA

Nematoida
- **PHYLUM NEMATODA**
- **PHYLUM NEMATOMORPHA**

Scalidophora
- PHYLUM KINORHYNCHA
- PHYLUM PRIAPULA
- PHYLUM LORICIFERA

Panarthropoda
- PHYLUM TARDIGRADA
- PHYLUM ONYCHOPHORA
- PHYLUM ARTHROPODA
  - SUBPHYLUM CRUSTACEA*
  - SUBPHYLUM HEXAPODA
  - SUBPHYLUM MYRIAPODA
  - SUBPHYLUM CHELICERATA

Deuterostomia
- PHYLUM ECHINODERMATA
- PHYLUM HEMICHORDATA
- PHYLUM CHORDATA

*Paraphyletic group

The section on phylum Nematoda has been revised by S. Patricia Stock. Phylum Nematomorpha (and the introductory text) has been revised by Andreas Schmidt-Rhaesa.

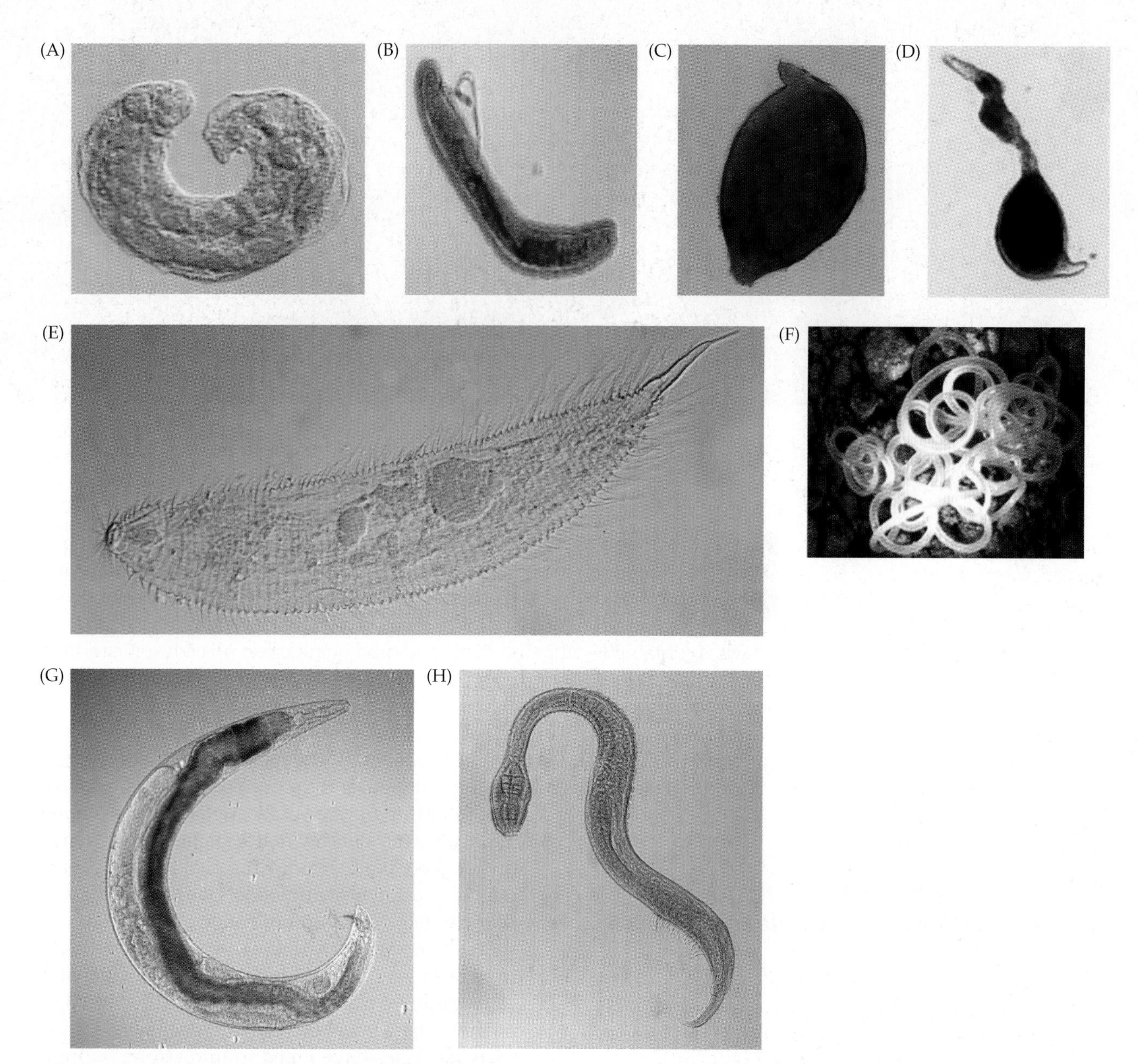

Figure 18.1 Examples of the phylum Nematoda. (A) *Allantonema* sp. (class Chromadorea), an insect parasite. (B) *Sphaerulariopsis* sp. (class Chromadorea), an insect parasite. (C) *Heterordera glycines* (class Chromadorea), a plant parasite. (D) *Tylenchulus semipenetrans* (class Chromadorea), a plant parasite. (E) *Greeffiella* sp. (class Enoplea), a free-living marine species. (F) *Hexamermis* sp. (class Enoplea), a parasite of various invertebrates. (G) *Heterorhabditis bacteriophora* (class Chromadorea), an insect pathogen. (H) *Draconema* sp. (class Enoplea), a free-living marine species.

taxa, nematoids lack locomotory cilia and have a cuticle that is molted during growth. The cuticle contains chitin in all ecdysozoans, at least in some stages of the life cycle or in particular regions of their bodies. In the case of nematoids, chitin has been shown only in the pharyngeal cuticle of nematodes and in the juvenile cuticle of nematomorphans.

Nematodes and nematomorphans are both slender worms. Specimens of the nematode order Mermithida (Figure 18.1F) can be easily confused with nematomorphans, because both are insect parasites with a free-living phase and large body size (usually in the range of several tens of centimeters). However, they are differentiated by external differences in the surface structure of the cuticle and the shape of the ends of their bodies. Nematomorphans emerge from their hosts as adults and possess diagnostic characters for identification to species; mermithid nematodes emerge as postparasitic juveniles and must undergo a final molt to reach the adult stage.

Nematodes and nematomorphans share a number of other features, for example the absence of circular musculature in the body wall and the absence of protonephridia. The epidermis has epidermal cords, and it is likely that two such cords, dorsal and ventral, are ancestral for the clade Nematoida. The epidermal cords include longitudinal nerve strands. Males in both phyla possess a cloaca, the exception being the marine nematomorphans (genus *Nectonema*) in which the posterior part of the intestine is reduced and therefore cannot form a cloaca. In female nematodes the genital opening is separate from the anus and often far away from it, in the middle region of the body. One further corresponding character is the possession of sperm that lack cilia, although both taxa have highly modified spermatozoa that are not comparable with each other in their structure.

Nematodes molt four times during their development; the pre-adult stages resemble the adults in general and are therefore often called juveniles (although the term "larvae" is often applied as well). In nematomorphans, only one molt has been observed. This takes place at the end of a growth phase of the parasitic juvenile; its thin cuticle is therefore capable of enormous growth. Nematomorphans possess a true larva, which is microscopic (< 1 mm) and morphologically completely different than the adult.

While Nematoda is a species-rich taxon with a great variety of life styles that has adapted to nearly every habitat on earth, Nematomorpha are all very similar in their life cycle, which includes a parasitic phase and a free-living phase for reproduction.

Phylum Nematoda: Roundworms and Threadworms

Nematoda is one of the most diverse groups of metazoans, with estimates ranging from 100,000 to 100 million species, although only about 25,000 have so far been named and described. An enormous amount of literature exists on nematodes (Greek *nema*, "thread"; *odes* "resembling")—roundworms and threadworms—much of it dealing with the parasitic species of economic or medical importance. Many of the large parasitic forms, such as the trichina worm (*Trichinella spiralis*), have been known since ancient times. However, the small free-living types were not discovered until after the invention of the microscope. Some authorities on the group prefer the shortened phylum name Nemata, although Nematoda is more commonly used.

Roundworms have been characterized as a "tube within a tube," referring to the linearity of the body and the alimentary tract and other organs. However, not all species have the typical thread-like appearance (Figure 18.1). Their size may vary from a few microns to meters in length. For example, one of the smallest species known is *Greeffiella minutum*, a coral reef species that is only 80 μm long. At the opposite end of the size spectrum is *Placentonema gigantisima*, the sperm whale nematode—the largest roundworm known, reaching over 8 m in length.

Nematodes are nearly ubiquitous, and they exist in almost every habitat and ecosystem on Earth. Ecologically, they can be divided into free-living and parasitic species. At the beginning of the 1900s, Nathan Cobb, known as the father of nematology in the United States, described the diversity of nematodes like this: "*If all the matter in the universe except nematodes were swept away, would our world still be recognizable? … if we could then investigate it, we should find its mountains, hills, valleys, rivers, lakes, and oceans represented by a film of nematodes.*" Indeed, nematodes can be found in aquatic (marine and fresh water) and terrestrial ecosystems ranging from the tropics to the poles, and at all elevations. In 2011, a 0.5 mm-long nematode was even found living in ancient waters several miles below the Earth's surface in South Africa. Named *Halicephalobus mesphisto*, after Mephistopheles (in reference to the devil in the Faust legend), this is the deepest-living metazoan yet discovered. It is thought to inhabit water-filled rock fractures and feed on subterranean bacteria.

Some species are generalists, but many have very specific habitats. For example, the sour paste nematode, *Panagrellus redivivus*, described by Linnaeus in 1776, was isolated from bookbinding glue. Another species, *Steinernema scapterisci*, is a parasite of crickets and mole crickets, and the species *Dioctophyma renale* parasitizes only the right kidney of monkeys.

Marine roundworms are considered the most diverse and widespread group of nematodes, occurring from shores to the abyss. However, in spite of their abundance, most marine nematodes are poorly known and their importance in benthic systems is little appreciated, though their relative abundance is sometimes used in biomonitoring. Some soil environments yield as many as three million nematodes per square meter. Many free-living soil species are used as indicators in biodiversity assessments and biomonitoring.

An exception to this relative obscurity is the free-living soil nematode *Caenorhabditis elegans*, which is considered a model organism in many fields including neurobiology, developmental biology, toxicology, and genetics. Many scientists around the world focus their research on *C. elegans* with the goal of fully understanding every aspect of its biology and the developmental fate of every embryonic cell. *Caenorhabditis elegans* has a number of features that make it not just relevant but quite powerful as a model for biological research. For example, it is easy and inexpensive to maintain in laboratory conditions, it has a short life cycle (about 3 days), it produces a large number (300+) of offspring, and it has a transparent body that allows easy observation of all the cells. *C. elegans* was the first multicellular organism to have its entire genome sequenced, with the surprising finding that 40% of its genes have human matches!

Nematodes have developed a multitude of parasitic life styles, which is why they can affect almost

BOX 18A Characteristics of the Phylum Nematoda

1. Triploblastic, bilateral, vermiform, unsegmented blastocoelomates
2. Body round in cross section and covered by a layered cuticle; growth in juveniles (four stages) usually accompanied by cuticular molting
3. With unique cephalic sense organs called amphids; some have caudal sense organs called phasmids
4. Complete digestive system; various mouth structures arranged in radially symmetrical pattern
5. Most with unique excretory system, composed of one or two renette cells or a set of collecting tubules
6. Without special circulatory or respiratory systems
7. Body wall has only longitudinal muscles (no circular muscles)
8. Epidermis cellular or syncytial, forming longitudinal cords housing nerve cords
9. Gonochoristic (males commonly with 'hooked' posterior end); wide range of reproductive modes
10. With unique cleavage pattern; not unambiguously radial or spiral
11. Inhabit marine, fresh water and terrestrial environments, free living and parasitic

every animal and plant on Earth. In fact, it has been estimated that over one-third of all humans, mainly in the developing world, carry a nematode infection. Some cause serious damage to crops and livestock, and some are pathogenic to humans. Most pet owners eventually encounter parasitic nematodes, as they are commonly seen in the feces and vomit of dogs and cats. The filarid worm, *Onchocerca volvulus*, causes an eye disease in humans called "river blindness" and is thought to infect nearly 20 million people in Latin America and Africa.

Nematodes are vermiform blastocoelomates with thin unsegmented bodies that are usually distinctly round in cross section (Box 18A). They possess digestive, nervous, excretory, and reproductive systems, but lack a discrete circulatory or respiratory system. To the untrained and unaided eye, most nematodes look very much alike, but there are significant variations in external body form (Figure 18.1).

Nematode Classification

Molecular phylogenetics, bioinformatics, and digital communication technologies have substantially impacted nematode systematics over the past two decades. Blaxter et al. (1998) created the first molecular framework for the classification of Nematoda, using small subunit (SSU) rDNA sequences from 53 nematode species. Since then, SSU rDNA sequences available in public databases have increased enormously. The molecular data published to date have resulted in a new classification system that is relatively stable. This molecular framework has confirmed the presence of three early nematode lineages, named Chromadoria, Enoplia, and Dorylaimia (Figure 18.2). However, the exact order of appearance of these three lineages is not yet resolved. The phylum comprises two classes, in both the classical and modern classification: Chromadorea and Enoplea. It is beyond the scope of this text to present much of the exhaustive classification scheme of the nematodes; for a more detailed classification see De Ley and Blaxter 2002, Holterman et al. 2006, and Meldal et al. 2007.

CLASS CHROMADOREA With porelike or slitlike amphids, which vary from labial pores or slits to postlabial elaborate coils and spirals; cuticle usually annulated, sometimes ornamented with projections and setae; phasmids present or absent, generally posterior; esophagus usually divided into bulbs, with 3 to 5 esophageal glands; excretory system glandular or tubular, female with one or two ovaries; caudal alae present or absent. This class contains a single subclass, Chromadoria, and numerous orders, including the following:

ORDER AREOLAIMIDA (e.g., *Aphanolaimus*)

ORDER CHROMADORIDA (e.g., *Achromadora, Atrochromadora*)

ORDER DESMODORIDA (e.g., *Draconema, Ethmoliamus, Heterordera*)

ORDER DESMOCOLECIDA (e.g., *Greeffiella*)

ORDER MONHYSTERIDA (e.g., *Parastomonema*)

ORDER OXYURIDA (e.g., *Oxyuris*)

ORDER PLECTIDA (e.g., *Anaplectus, Aphanolaimus*)

ORDER RHABDITIDA

SUBORDER CEPHALOBINA (e.g., *Cephalobus, Plectonchus*)

SUBORDER DIPLOGASTERINA (e.g., *Diplogaster*)

SUBORDER MYOLAIMINA (e.g., *Myolaimus*)

SUBORDER RHABDITINA (e.g., *Caenorhabditis, Chronogaster, Oesophagostomum, Rhabditis*)

SUBORDER SPIRURINA (e.g., *Ascaris, Camallanus, Onchocerca, Parascaris, Placentonema, Rhigonema, Spironoura, Wuchereria*)

SUBORDER TYLENCHINA (e.g., *Allantonema, Aphelenchus, Bursaphelenchus, Criconema, Helioctylenchus, Globodera, Heterodera, Heterorhabditis, Meloidogyne, Steinernema, Tylenchulus, Sphaerulariopsis*)

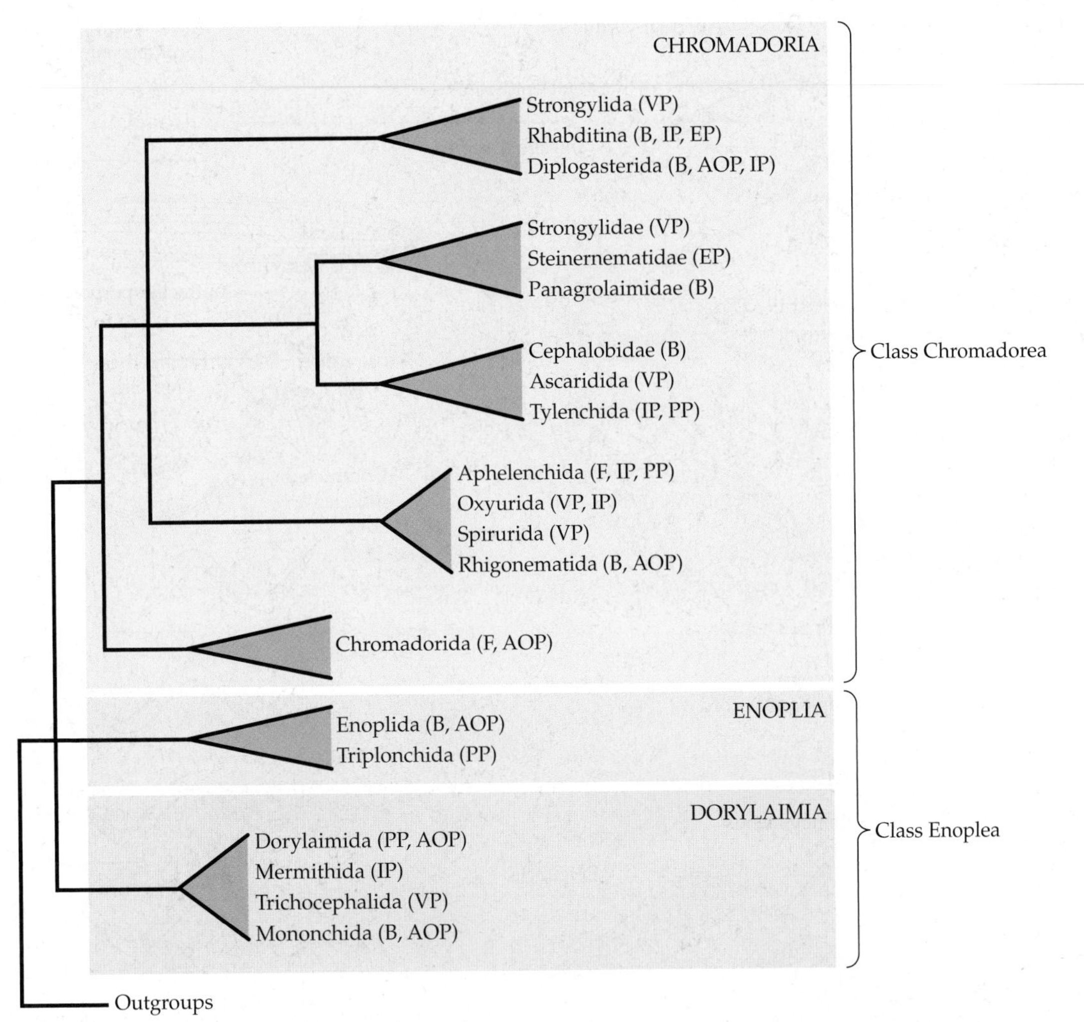

Figure 18.2 Nematoda molecular phylogenetic framework. VP: vertebrate parasites; IP: invertebrate parasites; EP: entomopathogens; B: bacterivores; AOP: algivores-omnivores-predators; PP: plant parasites; F: fungivores. (Based on Blaxter et al. 1998. Note that not all orders or families are included.)

CLASS ENOPLEA With pocketlike amphids, not spiral, usually postlabial; cuticle smooth or finely striated; phasmids present or absent; esophagus cylindrical or bottle-shaped with 3 to 5 esophageal glands, a stichosome and trophosome may be present (e.g., Mermithida); simple nontubular excretory system, usually a single cell; female generally with two ovaries; male generally with two testes; caudal alae rare. This class contains two subclasses and numerous orders, including the following:

SUBCLASS ENOPLIA

ORDER ENOPLIDA

ORDER TRIPLONCHIDA

ORDER TRICHURIDA (e.g., *Trichuris*)

SUBCLASS DORYLAIMIA

ORDER DIOCTOPHYMATIDA

ORDER DORYLAIMIDA (e.g., *Dorylaimus*, *Xiphinema*)

ORDER ISOLAIMIDA

ORDER MARIMERMITHIDA

ORDER MERMITHIDA (e.g., *Hexamermis*)

ORDER MONONCHIDA (e.g., *Mononchus*)

ORDER MUSPICEIDA

ORDER TRICHOCEPHALIDA (e.g., *Trichinella*)

The Nematode Body Plan

Body Wall, Support, and Locomotion

The nematode body is covered by a well-developed and complexly layered cuticle secreted by the epidermis (Figure 18.3). The cuticle is mainly composed of lipids and proteins associated with mucopolysaccharides. Collagen, a structural protein, is the major component of the cuticle (> 80%). Nematodes lack chitin in their cuticle, although it is present in the egg shell. The cuticle forms a flexible exoskeleton that invaginates at the mouth, cloaca, and rectum as well at the amphids, phasmid secretory–excretory pore, and vulva. The cuticle is responsible in part for allowing

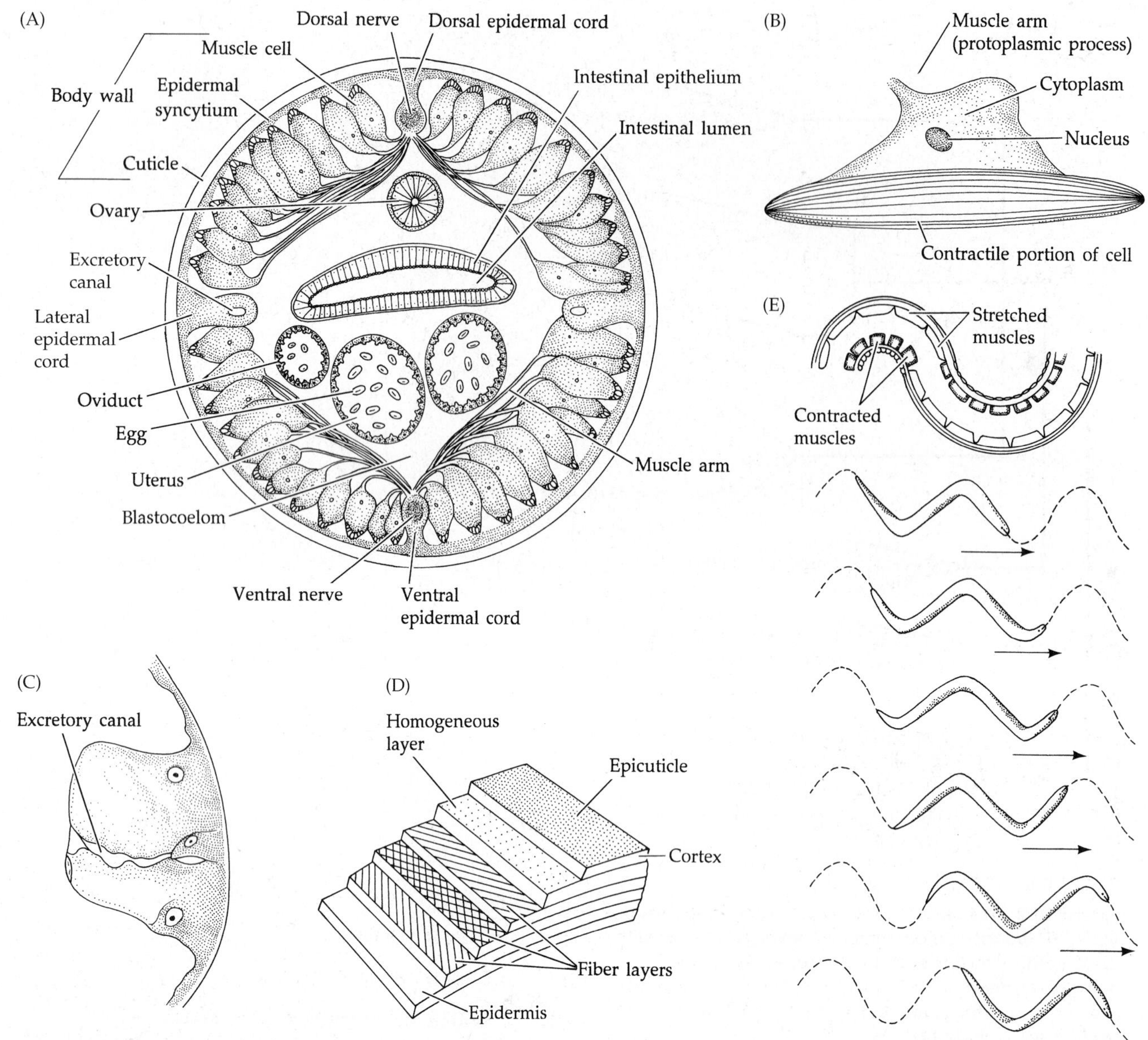

Figure 18.3 (A) Stylized section through a female nematode such as *Ascaris*. (B) A single longitudinal muscle cell, illustrating the origin of the muscle arm. (C) The lateral epidermal cord of *Cucullanus* (order Rhabditida). (D) The layers of the cuticle. (E) Undulatory locomotion in a free-living nematode results from the action of the longitudinal muscle fibers. The concave areas along the body represent positions of muscle contractions; the convex areas are regions of the muscle stretching. Leverage is gained against surrounding objects of the substratum in the environment.

nematodes to live in hostile environments, such as dry terrestrial soils and the digestive tracts of host animals, for it drastically reduces the permeability of the body wall. Predominantly terrestrial or parasitic nematodes usually have a dense, fibrous inner layer of the cuticle, whereas most of the free-living marine and freshwater forms lack this inner layer. The cuticle is a complex structure and it is highly variable among nematodes. It may be relatively smooth, or covered with sensory setae and wartlike bumps. The cuticle in many roundworms has rings or **annules**, or it is marked with longitudinal ridges and grooves (Figure 18.4). In many marine forms, the cuticle contains radially arranged rods, punctuations, or other inclusions of various shapes. As a nematode grows, it sheds its cuticle and grows a new one through a series of four molts during its lifetime.

The epidermis varies among the different taxa from cellular to syncytial, and it is often thickened as dorsal, ventral, and lateral longitudinal cords (Figure 18.3A,C). The dorsal and ventral thickenings house longitudinal nerve cords; the lateral thickenings contain excretory canals (when present, as they are in *Ascaris*) and neurons. Internal to the epidermis is a relatively thick layer of obliquely striated longitudinal muscle arranged in four quadrants. The muscles are connected to the

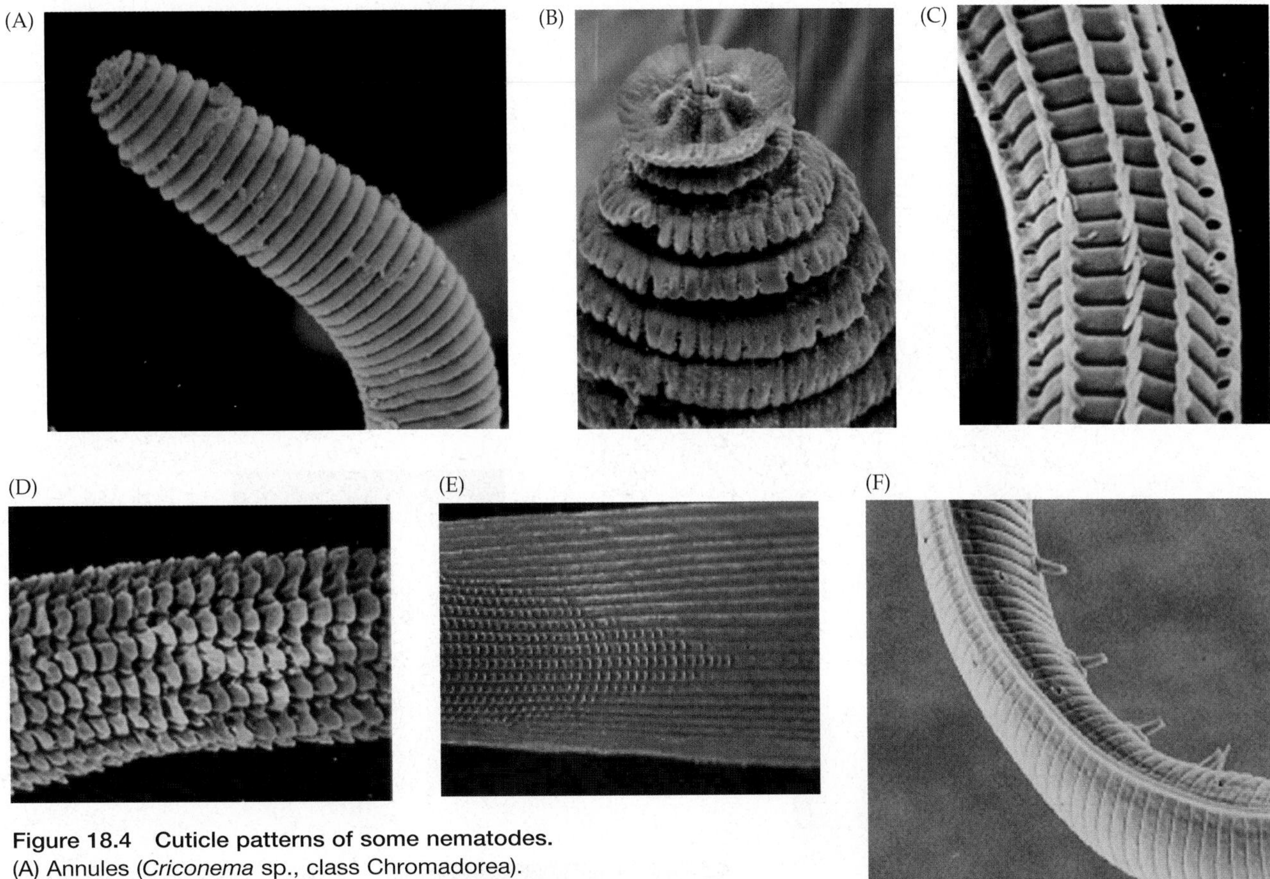

Figure 18.4 Cuticle patterns of some nematodes. (A) Annules (*Criconema* sp., class Chromadorea). (B) Close-up of annules showing crenate pattern (*Criconema* sp., class Chromadorea). (C) *Chronogaster* sp., class Chromadorea). (D) Scale-like ornamentation of annules (criconematid nematode, class Chromadorea). (E) Tessellate pattern and longitudinal striae (*Heterorhabditis* sp., infective juvenile stage, class Chromadorea). (F) Transverse striae and tubular supplement (plectid nematode, class Enoplea).

dorsal and ventral nerve cords by unique extensions called **muscle arms** (Figure 18.3A,B). This arrangement is different from the usual neuromuscular junctions in most other animals; in nematodes the connections are made by extensions of the muscle cells rather than of the neurons. Oddly, a similar condition apparently exists in the cephalochordate *Branchiostoma* (Chapter 27), presumably a case of convergent evolution. Also, in nematomorphans (horsehair worms) and gastrotrichs, the longitudinal muscles bear extensions suggested to be a possible homologue of the nematode muscle arms. There is no circular muscle layer in nematodes, or in nematomorphans, a condition viewed as homologous by some workers.

The fluid-filled blastocoelom is not spacious. The apparently large body cavity seen in many laboratory specimens is an artifact caused by shrinkage of tissues in alcohol. Modern microscopy techniques reveal that the organs of most nematodes occupy nearly all of the internal space. The cuticle provides most of the body support in nematodes. In the absence of circular body wall muscles, some types of locomotion, such as peristaltic burrowing, are impossible. Nematode movements depend on waves of muscle contractions, which travel backward along the body. These movements drive the worm forward, whether it is swimming through water or crawling through soil particles. The typical pattern of nematode locomotion involves contractions of longitudinal muscles, producing a whiplike undulatory motion (Figure 18.3E). Among the free-living nematodes this movement pattern relies on contact with environmental substrata, against which the body pushes. The muscles act against the hydrostatic skeleton and the cuticle, which serve as antagonistic forces to the muscle contractions. The portion of the collagenous fibers of the cuticle that are crossed are nonelastic, but their overlapping arrangement allows shape changes as the body undulates. When placed in a fluid environment and deprived of contact with solid objects, benthic nematodes thrash about rather inefficiently. Some actually do swim (but not very well), and some are able to crawl along using various cuticular spines, grooves, ridges, and glands to gain purchase on the substratum (Figures 18.1E,H and 18.4). For nematodes, the energy cost of swimming is trivial. Contrastingly, the energy cost of locomotion for many larger animals corresponds to a considerable fraction of their metabolic rate.

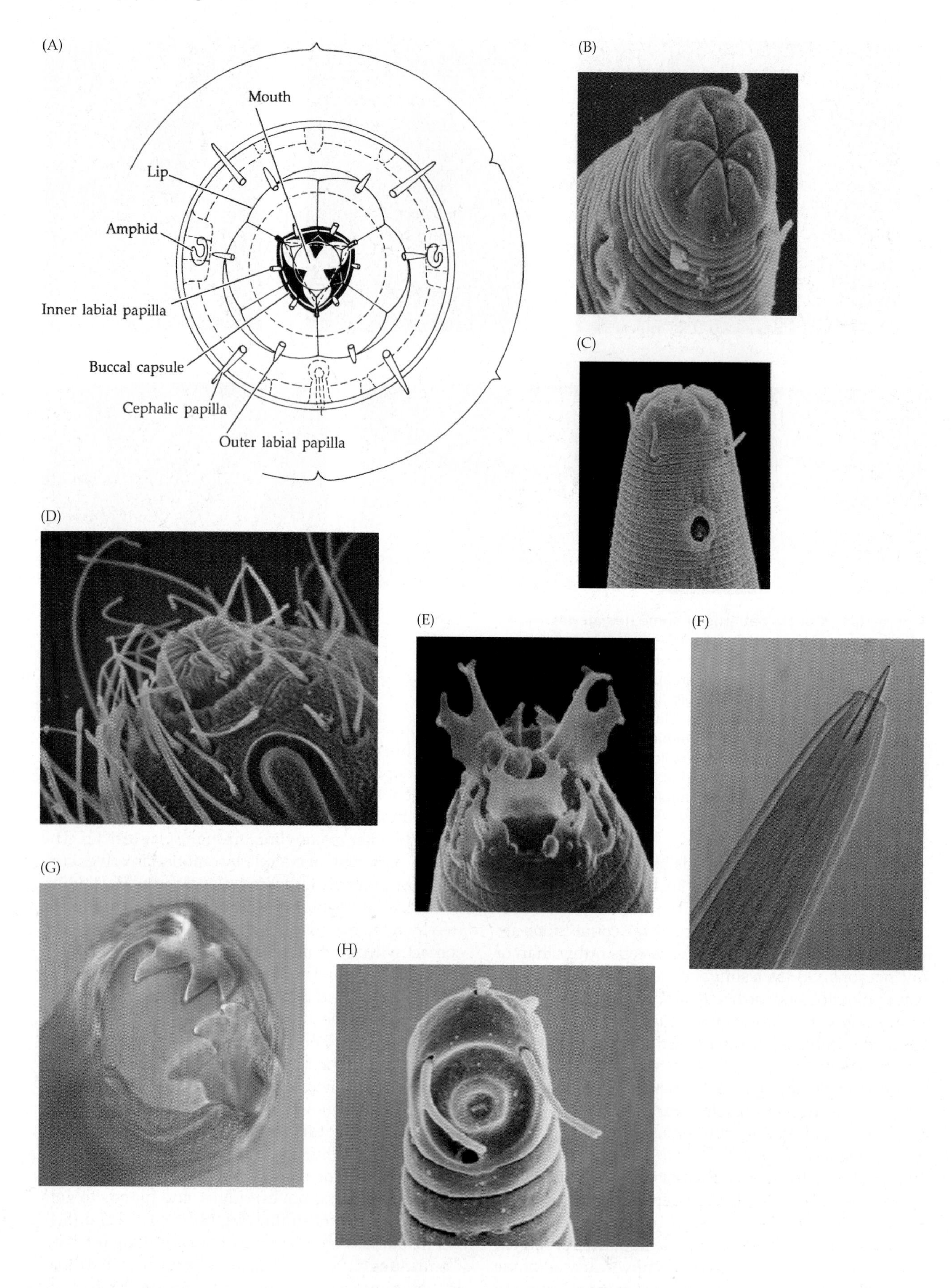
(A)
Mouth
Lip
Amphid
Inner labial papilla
Buccal capsule
Cephalic papilla
Outer labial papilla
(B)
(C)
(D)
(E)
(F)
(G)
(H)

Figure 18.5 Modifications of anterior end among selected nematodes. (A) The front end (in face view) of a generalized nematode. Note basic radial symmetry of the parts. (B) Anterior end (face view) of a typical free-living nematode showing four cephalic papillae and six symmetrical lips each bearing a labial papillae and lateral amphids. (C) Anterior end of *Anaplectus* sp. (class Chromadorea), a free-living freshwater nematode, showing a circular amphid and cephalic papillae. (D) Anterior end of *Draconema* sp. (class Chormadorea), a free-living marine nematode showing an amphid and setae. (E) Anterior end of *Cervidellus* sp. (class Chromadorea), a free-living soil nematode, showing probolae (used for sorting of food). (F) Anterior end of *Dorylaimus* sp. (class Enoplea), a free-living soil nematode, showing the protruded stylet used to puncture prey. (G) Anterior end of *Ancylostoma caninum*, a vertebrate parasite, showing cuticular teeth that are used like a lathe on the host's intestine. (H) Anterior end of *Aphanolaimus* sp. (class Chromadorea), a free-living freshwater nematode with spiral amphids and long cephalic papillae.

Feeding and Digestion

The vast range of environments inhabited by nematodes is paralleled by the diversity of feeding habits and associated anatomical and behavioral adaptations. Many infaunal nematodes are direct deposit feeders. Others are detritivores or microscavengers, living in or on dead organisms or fecal material. Many of these species apparently do not feed directly on the carcasses they inhabit, but on the microorganisms (fungi and bacteria) that grow in the decomposing organic matter. Many free-living nematodes are predatory carnivores, feeding on a variety of other small animals including other nematodes. Others feed on diatoms, algae, and bacteria. Plant parasitic nematodes use a specialized oral structure, the **stylet**, to pierce individual root cells and suck out the contents (Figure 18.5).

Bacterial symbioses occur in some nematodes. For example, species of *Artomonema* and *Parastomonema* that inhabit sulfur-rich sediments harbor chemoautotrophic bacteria in their highly reduced guts. These worms meet their nutritional requirements by absorbing some of the metabolic products of the bacteria. Some other nematodes (family Stilbonematidae) inhabit sulfide-rich oxygen-poor sediments, where symbiotic bacteria coat their cuticle. These nematodes are able to twist themselves in such a way that they can feed on the bacteria "farmed" on their body surface.

Insect pathogenic nematodes (Steinernematidae and Heterorhabditidae) carry gram-negative bacteria in their intestines that help them kill their insect hosts. The bacteria, once introduced into the host, grow and reproduce to become food for the nematode, which also grows and reproduces in the insect cadaver.

Nematode parasites have been described from nearly every group of plants and animals, and they have evolved a multitude of life history patterns. In invertebrates and vertebrates (including humans), nematodes subsist on a variety of body fluids and organs, where they can cause extreme tissue damage.

Nematode digestive tracts vary greatly in complexity and regional specialization. The anteriorly located mouth is usually surrounded by **lips** (one dorsal and two ventral parts, some or all of which may be fused together and may form elaborate prolongations), with each lip typically bearing a papilla. **Labial probolae** are flaplike cuticular processes surrounding the mouth opening in many cephalobid nematodes—there are three flaps, one dorsal and two subventral. **Cephalic probolae** are simple or complex cuticular processes arising from the lips in many cephalobid nematodes; there are six in total—two lateral, two subdorsal, and two subventral. Lips, labial and cephalic probolae, spines, teeth, jaws, and other armature are arranged in radially symmetrical patterns (Figure 18.5). The mouth leads to a buccal cavity or stoma, the shape of which is variable depending on the feeding habits of the nematodes. (The stoma has been the subject of detailed developmental and histological studies.) The stoma connects to the pharynx (Figure 18.6).

The basic structural features of the pharynx are a tri-radiated lumen, muscles, one or more valves, and dorsal and ventral glands and neurons. The pharynx is elongate and may be subdivided into distinct muscular and glandular regions, the details of which are of considerable taxonomic importance (Figure 18.6A–G). The muscles of the pharynx pump food material from the buccal cavity into the intestine. In most nematodes the intestine is a simple tube composed of a single layer of cells ensheathed by a basal lamina. In some cases it is possible to discern an anterior, middle, and posterior region of the intestine. The posterior portion of the intestine leads to a short proctodeal (i.e., ectodermally-derived) rectum and a subterminal anus on the ventral surface of the body (Figure 18.6L). In males the rectum opens up into a cloaca, which also receives products of the reproductive system.

The esophageal glands and the midgut lining secrete digestive enzymes into the intestinal lumen. Initial digestion is extracellular; final intracellular digestion occurs in the intestine, following absorption across the surfaces of the microvilli of the intestinal cells (Figure 18.6K). In mermithid nematodes (most of which are parasites of insects) the pharynx is modified as a **stichosome**, a region of large glandular cells (stichocytes) important in protein synthesis. Also in mermithids, modification of the intestine creates a **trophosome**, a region of enlarged, solid cells that form a closed food storage organ.

Circulation, Gas Exchange, Excretion, and Osmoregulation

Nematodes have no specialized circulatory or gas exchange structures. As in many other very small invertebrates, particularly blastocoelomates, diffusion

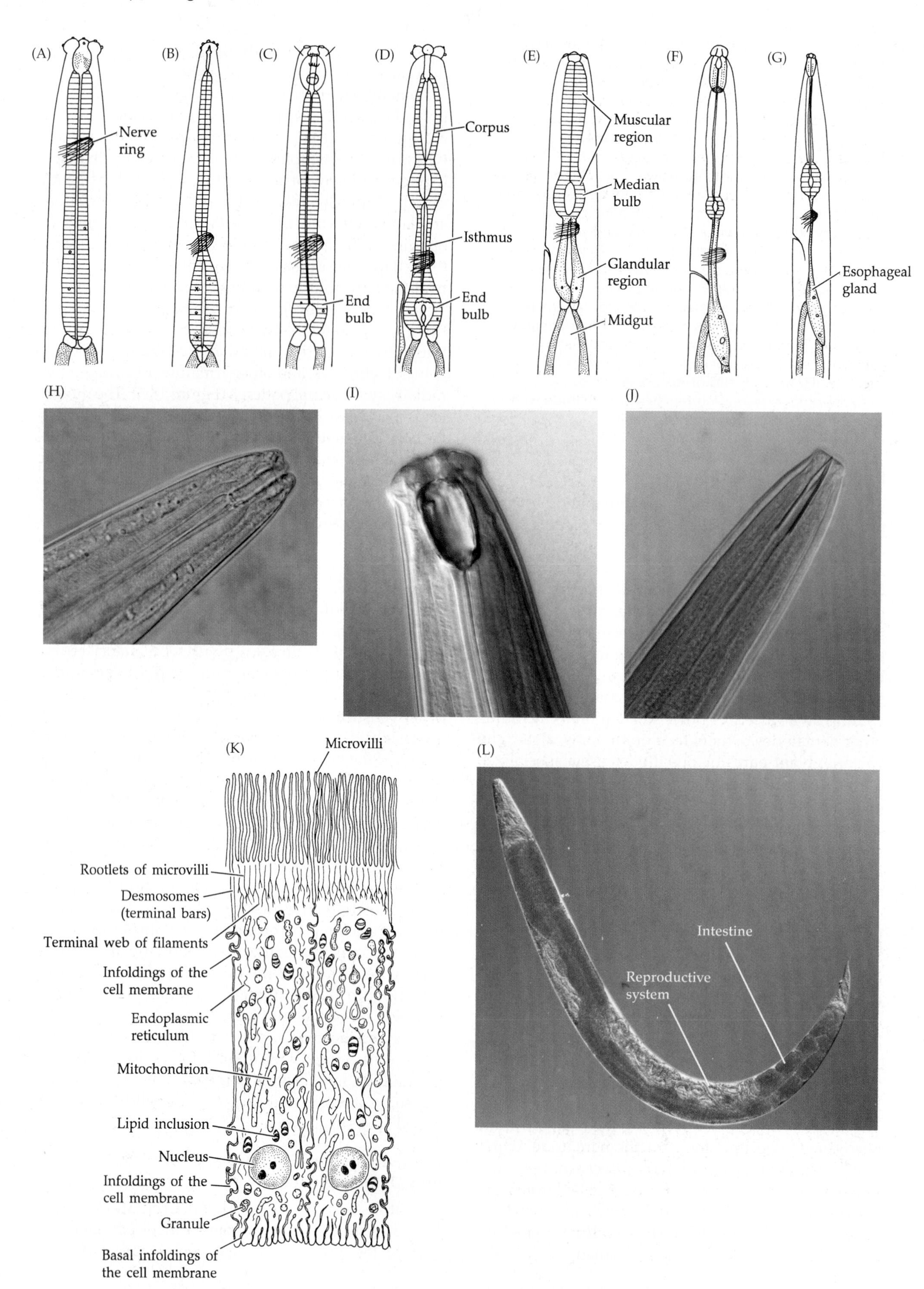
(A)
(B)
(C)
(D)
(E)
(F)
(G)
Nerve ring
Corpus
Isthmus
End bulb
End bulb
Muscular region
Median bulb
Glandular region
Midgut
Esophageal gland
(H)
(I)
(J)
(K)
Microvilli
Rootlets of microvilli
Desmosomes (terminal bars)
Terminal web of filaments
Infoldings of the cell membrane
Endoplasmic reticulum
Mitochondrion
Lipid inclusion
Nucleus
Infoldings of the cell membrane
Granule
Basal infoldings of the cell membrane
(L)
Intestine
Reproductive system

◀ **Figure 18.6** (A–G) Variation in pharynx structure gut anatomy among different nematodes. Note the different degrees of regional specialization. (A) Cylindrical pharynx (*Mononchus*, class Enoplia). (B) Dorylaimoid pharynx (*Dorylaimus*, class Enoplea). (C) Bulboid pharynx (*Ethmoliamus*, class Chromadorea). (D) Rhabditoid pharynx (*Rhabditis*, class Chromadorea). (E) Diplogasteroid pharynx (*Diplogaster*, class Chromadorea). (F) Tylenchoid pharynx (*Helicotylenchus*, class Chromadoria). (G) Aphelenchoid pharynx (*Aphelenchus*, class Chromadorea). (H) Anterior end of *Rhabditis* (class Chromadoria), a free-living soil bacterivore showing tubular stoma (lateral view). (I) Anterior end of *Mononchus* (class Enoplia), free-living predatory soil nematode showing cylindrical stoma (lateral view). (J) Anterior end of *Dorylaimus*, showing stylet (lateral view). (K) Intestinal epithelium of *Ascaris* (class Chromadorea). (L) Digestive tract and reproductive system of *Steinernema* female (class Chromadorea).

and movement of body cavity fluids accomplish these functions. Some parasitic nematodes possess a form of hemoglobin in these fluids that presumably transports and stores oxygen. Both aerobic and anaerobic metabolic pathways are found among the nematode groups, and many of these worms are able to shift from one mechanism to the other according to environmental oxygen concentrations. Facultative anaerobiosis is surely significant in parasitic nematodes and those that live in other anoxic environments. The blastocoelom serves as a circulatory system, transporting molecules such as CO_2 from the tissues of origin (mostly muscle and reproductive tissue) to the epidermis and intestine for excretion into the environment.

Nematodes have no recognizable kidney, and soluble waste products are apparently concentrated prior to their elimination. For this purpose they have unique excretory structures that are apparently not homologous to any of the protonephridial types found in other Metazoa. In fact, there exists a rather clear evolutionary sequence of different excretory structures among nematodes (Figure 18.7A). The presumed ancestral condition occurs in certain free-living taxa and has been modified among other groups, especially within specialized parasitic forms. In many free-living threadworms, the system comprises one or two glandular **renette cells** that connect directly to a midventral excretory pore (Figure 18.7E,F), and sometimes a third

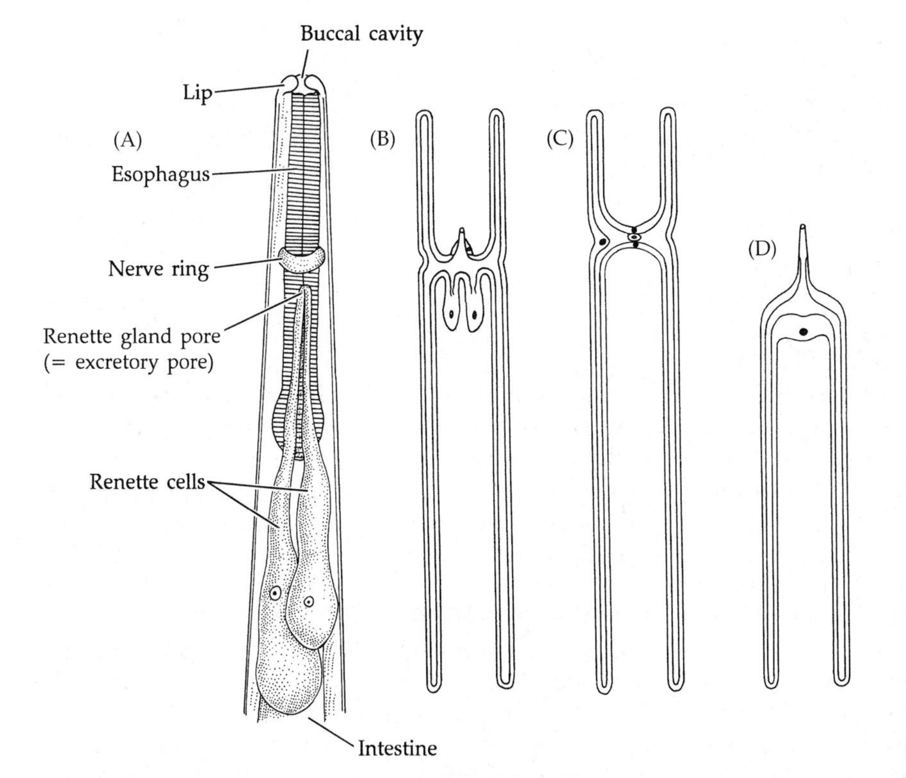

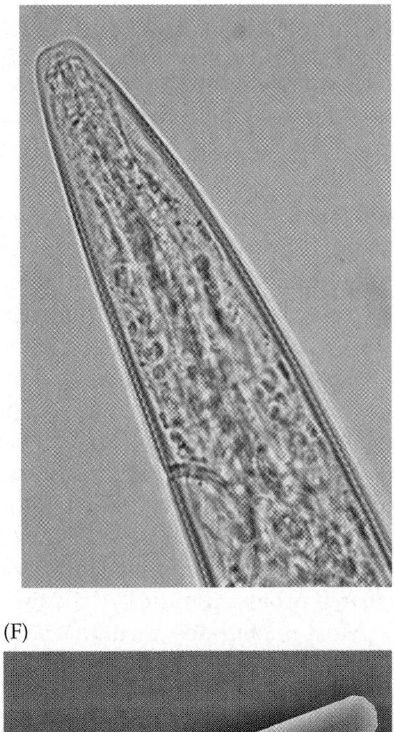

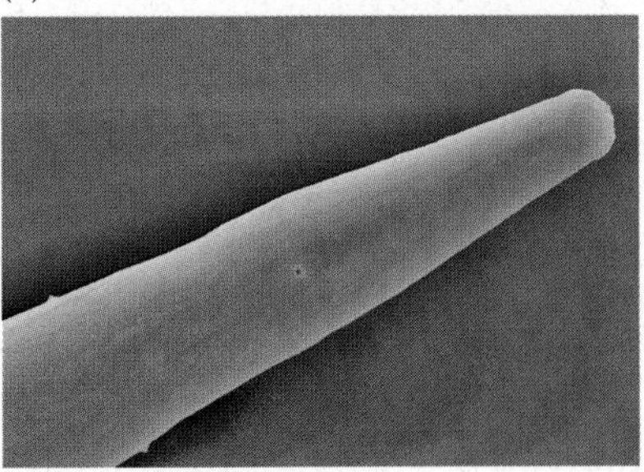

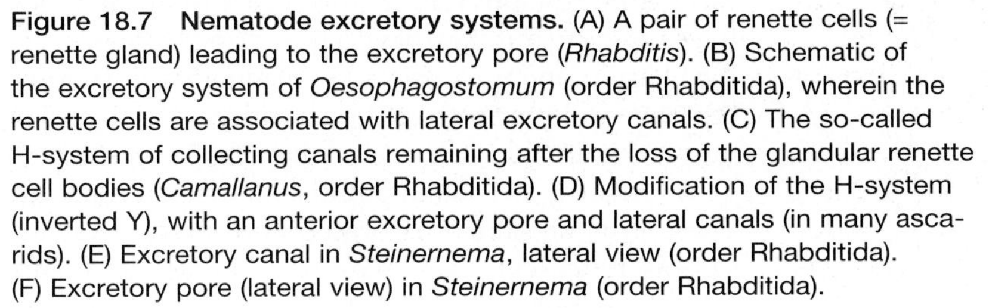

Figure 18.7 Nematode excretory systems. (A) A pair of renette cells (= renette gland) leading to the excretory pore (*Rhabditis*). (B) Schematic of the excretory system of *Oesophagostomum* (order Rhabditida), wherein the renette cells are associated with lateral excretory canals. (C) The so-called H-system of collecting canals remaining after the loss of the glandular renette cell bodies (*Camallanus*, order Rhabditida). (D) Modification of the H-system (inverted Y), with an anterior excretory pore and lateral canals (in many ascarids). (E) Excretory canal in *Steinernema*, lateral view (order Rhabditida). (F) Excretory pore (lateral view) in *Steinernema* (order Rhabditida).

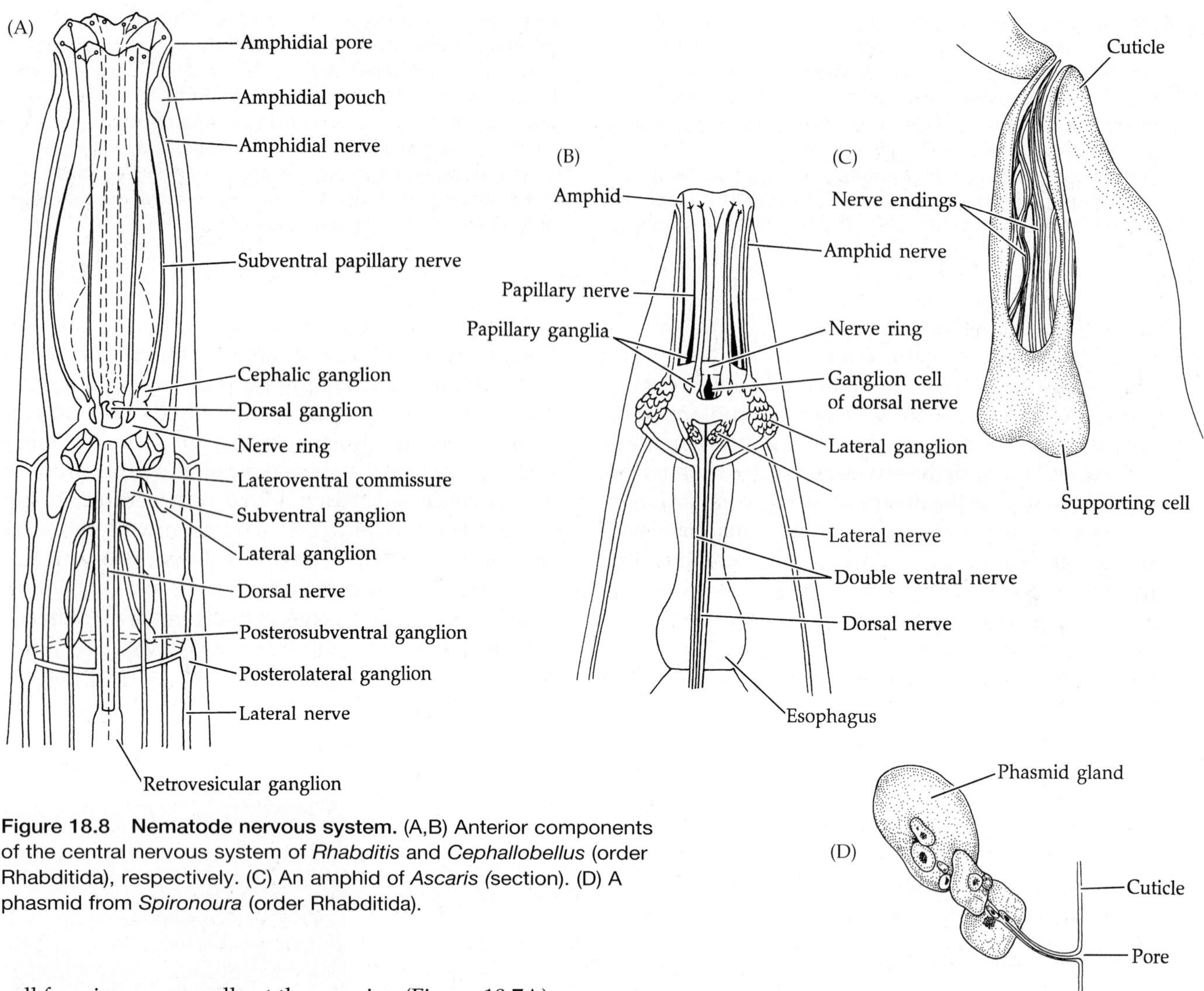

Figure 18.8 Nematode nervous system. (A,B) Anterior components of the central nervous system of *Rhabditis* and *Cephallobellus* (order Rhabditida), respectively. (C) An amphid of *Ascaris (*section). (D) A phasmid from *Spironoura* (order Rhabditida).

cell forming an ampulla at the opening (Figure 18.7A). Modifications to this system often include various arrangements of intracellular collecting ducts within the cytoplasm of extensions of the renette cells (Figure 18.7B). In many parasitic species the renette cell bodies are lost completely, leaving only the system of tubules in an H or inverted Y pattern (Figure 18.7C,D). Many members of the subclass Enoplia lack renette cells altogether. Instead they have numerous unicellular units distributed along the entire length of the body. Each cell opens to the outside via a duct and a pore. If these cells are excretory in function, they may represent nonciliated protonephridia.

Most nematodes are ammonotelic, although some excrete increased amounts of urea when in a hypertonic environment. Apparently much of the loss of nitrogenous wastes is across the wall of the midgut, and the renette cells are primarily osmoregulatory. Water balance is also aided by the activities of other tissues, organs, and structures. In some roundworms, the cuticle is differentially permeable to water in that it allows water to enter but not to leave the body. This condition is advantageous under conditions of potential desiccation, but it presents problems in hypotonic environments where excess water must be eliminated; such elimination is apparently accomplished by the renette cells (when present), by the gut lining, and by the epidermis. Marine species do not osmoregulate well and desiccate rapidly when exposed to air.

Nervous System and Sense Organs

The neuroanatomy of nematodes was studied first by R. Hesse (1892) and Richard Goldschmidt (1908, 1909) who showed that nematodes have 250 to 302 neurons and their position and structure is roughly the same throughout the phylum (Figure 18.8A,B). Nematodes have no brain, but they have a nerve ring that encircles the pharynx and is associated with longitudinal nerves that extend anteriorly and posteriorly. The anterior nerves (six in total) connect with cephalic sensory organs (sensilla) and labial papillae in the anterior end. These structures are chemoreceptors or have a chemosensory function in these worms' highly touch-oriented world (e.g., interstitial, parasitic, and soil habitats). **Amphids** are paired organs located laterally on the head. They consist of an external pore leading inward to a short duct and amphidial pouch. The pouch is associated with a unicellular gland and an amphidial nerve from the cerebral nerve ring (Figure 18.8A,C), although there is some variation in structural details among species. The receptor sites of amphids are derived from modified cilia, but recall that motile cilia do not occur in nematodes. Specialists think that the amphids are chemosensory in function.

Via a series of associated ganglia, longitudinal nerves also extend posteriorly through the epidermal cords (Figure 18.8A). The major nerve trunk is ventral and includes both motor and sensory fibers. It is formed from the union of paired nerve tracts that arise ventrally on the nerve ring and fuse posteriorly, where the main trunk bears ganglia. The dorsal nerve cord is motor, and the less well-developed lateral nerve tracts are predominantly sensory. Lateral commissures connecting some or all of the longitudinal nerves occur in many nematodes.

Most members of the class Chromadorea (parasitic forms) possess a posteriorly located pair of glandular structures called **phasmids** (Figure 18.8D). These structures are also considered to be chemoreceptors. Some freshwater and marine free-living nematodes (class Enoplea) possess a pair of anterior pigment-cup ocelli as well, and at least some nematodes contain proprioceptor cells in the lateral epidermal cords. These sensory cells contain a cilium and appear to monitor bending of the body during locomotion.

Reproduction, Development, and Life Cycles

Most nematodes are gonochoristic and show some degree of sexual dimorphism (Figure 18.9A,B). The female reproductive system (Figure 18.9A) usually consists of one or two elongate ovaries that gradually hollow as oviducts and then enlarge as uteri (Figures 18.9C–E). The uteri converge to form a short vagina connected to the single gonopore, exiting via the vulva. The uteri converge to form a short vagina connected to the single gonopore. The female gonopore is completely separate from the anus, opening on the ventral surface near the middle of the body or sometimes right above the anus.

Males tend to be smaller than females and are often sharply curved posteriorly. The male reproductive system (Figure 18.9B,F,G) typically includes one (or two) threadlike tubular testes, each of which is regionally differentiated into a distal germinal zone, a middle growth zone, and a proximal maturation zone near the junction with the sperm duct. The sperm duct extends posteriorly, where it enlarges as a seminal vesicle leading to a muscular ejaculatory duct that joins the hindgut near the anus. Some species have prostatic glands that secrete seminal fluid into the ejaculatory duct. Most male nematodes possess a copulatory apparatus, including one or two cuticular **spicules** that transfer sperm (Figure 18.9E–G). Another structure called the **gubernaculum** may be present (Figure 18.9G). The gubernaculum is a sclerotized region of the dorsal wall of the cloaca that serves to anchor and guide the spicules during copulation.

Prior to copulation the males produce sperm (round or elongate depending on the species) and store them in the seminal vesicle, while the females produce eggs that are moved into the hollow uteri. Potential mates make contact (females of some species are known to produce male-attracting pheromones), and the male usually wraps his curved posterior end around the body of the female near her gonopore (Figure 18.9H). Thus positioned, the copulatory structures are inserted into the vagina, and sperm are transferred by contractions of the ejaculatory duct. Fertilization usually occurs within the uteri. A relatively thick double-layered shell forms around each zygote; the inner layer is derived from the fertilization membrane and the outer layer is produced by the uterine wall. The zygotes are usually deposited in the environment where development takes place. In some female nematodes the zygotes hatch inside, a process known as **endotokia matricida**, which causes the female's death. The juveniles that hatch within the mother remain there obtaining nutrients from the mother until its body ruptures liberating the progeny into the environment.

In addition to the general description given above, two relatively uncommon reproductive processes occur in nematodes. In the few known hermaphroditic species, sperm and egg production take place within the same gonad (an ovitestis). Sperm formation precedes egg production, so the animals are technically protandric; but they do not engage in cross-fertilization as occurs in most sequential hermaphrodites. Rather, the sperm are stored until ova are produced, and self-fertilization occurs. Parthenogenesis also occurs in a few species of nematodes. Sperm and eggs are produced by separate males and females that then engage in typical copulation. However, the sperm do not fuse with the egg nuclei, but apparently serve only to stimulate cleavage.

As stated earlier, development among free-living nematodes is typically direct, although the term "larva" is often used for juvenile stages. Cleavage is holoblastic and subequal, but the pattern appears to be unique among the Metazoa. The orientation of blastomeres during early cleavage is fairly consistent among those nematodes that have been studied, but it cannot be readily assigned to a clearly radial or spiral pattern. Figure 18.10 illustrates this cleavage pattern and some details of cell fates. A stereoblastula or slightly hollow coeloblastula forms and undergoes gastrulation by epiboly of the presumptive ectoderm combined with an inward movement of presumptive endoderm and mesoderm. After a specific point in development, few nuclear divisions occur, and most subsequent growth,

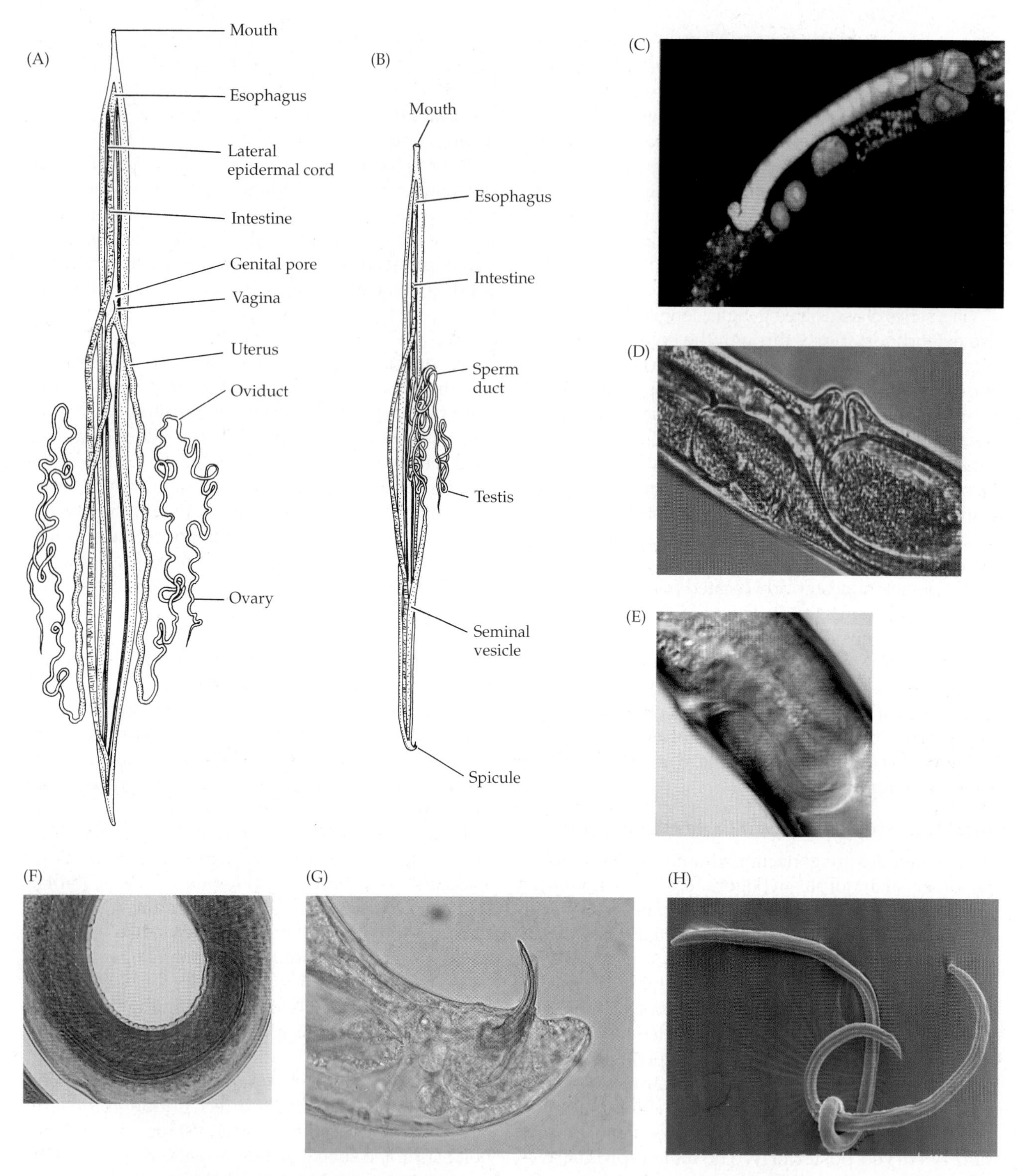

Figure 18.9 Nematode reproductive systems. (A,B) Female and male reproductive system of *Ascaris* (dissected). (C) Confocal microscopy image of female ovary of *Heterorhabditis bacteriophora* (an insect pathogen). (D) Liplike vulva and vagina of *Plectonchus* sp., a soil microbivore. (E) The insect parasite, *Hexamermis* sp., has an S-shaped vagina, with the gonopore surrounded by a slightly-raised vulva. (F) Spicule of a mermithid nematode, an insect parasite. (G) Spicules and gubernaculum of *Steinernema* sp., an insect pathogen. (H) Male and female *Plectonchus* nematodes mating (scanning electron micrograph).

even after hatching, is via the enlargement of existing cells. Four sequential cuticular molts during juvenile life usually accompany growth.

Life Cycles of Some Parasitic Nematodes

The study of parasitic nematode life cycles is a field unto itself, and we present only a few illustrative

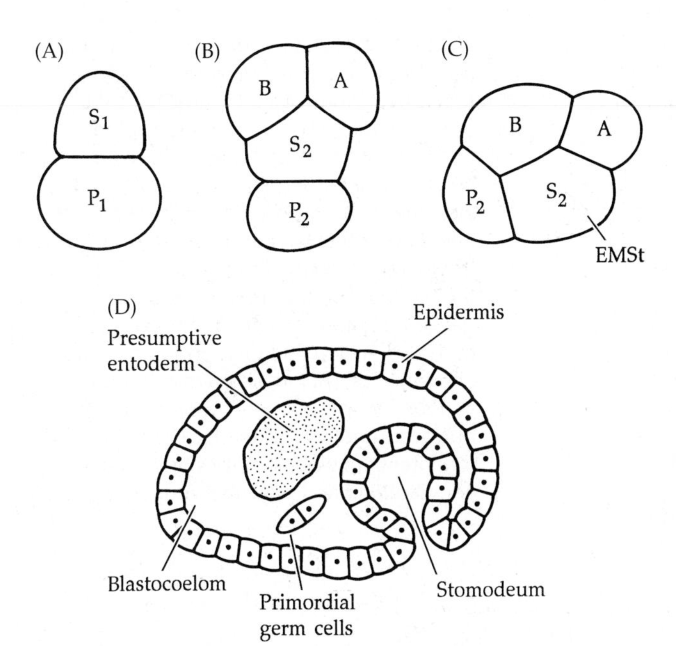

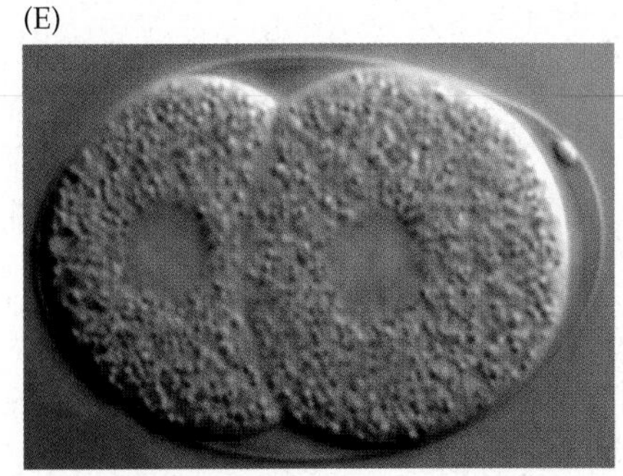

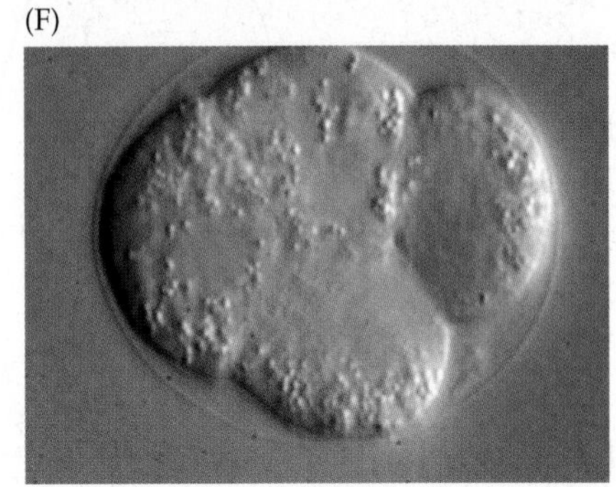

Figure 18.10 Early embryogeny in *Parascaris equorum* (class Chromadorea). (A) Two-cell stage, coded S_1 and P_1 cells. (B,C) Four-cell stage. (B) The S_1 cell divides to produce cells A and B; the P_1 cell divides to produce cells S_2 and P_2. (C) The P_2 cell migrates to the presumptive posterior end of the embryo. (D) A later stage following gastrulation, during the formation of the stomodeum. The S_2 (EMSt, endodermal/mesodermal/stomodeal) cell has divided to produce and E and an St cell. The E cell forms the endoderm upon gastrulation, the M cell becomes mesoderm and the St cell forms the region of the stomodeum; the A and B cells produce most of the ectoderm, the P_2 cell becomes some of the posterior ectoderm and the primordial germ cells, and the mesoderm. (E,F) *Steinernema carpocapsae* embryo with two (E) and four (F) blastomeres.

life history cycles here (Figures 18.11 and 18.12). An example of a simple parasitic life cycle is that of the whipworm *Trichuris trichiura* (class Enoplea) (Figure 18.11A). These relatively large (3–5 cm long) nematodes reside and mate in the human gut, and the fertilized eggs are passed with the host's feces. Reinfection occurs when another host ingests the embryos. A more complicated life cycle is that of the trichina worm *Trichinella spiralis* (Figure 18.11B). A mammalian host acquires *Trichinella* by ingesting raw or poorly cooked meat containing encysted "larvae."

Most parasites of vertebrates belong to the class Chromadorea, which includes such notables as hookworms, pinworms, ascarid worms, and filarid worms. One of the most dramatic nematode infections in humans is filariasis, caused by any of a number of secernentean nematodes called the filarids. These parasites require an intermediate host, typically a blood-sucking insect (e.g., fleas, biting flies, mosquitoes). One such filarid is *Wuchereria bancrofti*, whose vector is a mosquito. When present in humans in high numbers, masses of adult *Wuchereria* block lymphatic vessels and cause fluid accumulation (edema) and severe swelling, a condition resulting in grotesque enlargement of body parts and known as lymphatic filariasis, or **elephantiasis**. Such infections often affect the legs and arms, or the scrotum in males and breasts in females. Some 120 million people in 73 countries are infected with lymphatic filariasis.

Plant parasitic nematodes can be found in both nematode classes. They cause billions of dollars annually in crop losses worldwide. One of the most damaging and widely distributed plant parasites is root-knot (*Meloidogyne*) and cyst nematodes (e.g., *Heterodera, Globodera*). Nematodes usually cause damage to the roots of plants, including the formation of visible galls (e.g., by root-knot nematodes). Figure 18.12A shows the life cycle of a root-knot nematode that attaches to the roots of a host tomato plant. However other species can cause damage to the aerial parts of the plants including leaves, stem, flowers, and seeds. Some species transmit plant pathogens (viruses and bacteria) through their feeding activity on roots. For example the *Xiphinema index* vectors the grapevine fan-leaf virus, an important disease of grapes. Some species, such as the pine wood nematode *Bursaphelenchus xylophilus*, are important forestry parasites in Asia and America and Europe.

Onchocerca volvulus, a filarid nematode, is the causative agent of river blindness, transmitted by bites from certain black fly species (Simulidae: *Simulium*). The disease (and the parasite) comes out of Africa, where it is endemic in 27 countries, and it also occurs in six Latin American countries as a result of the slave trade. If left to run its course, river blindness is a disease for life. Adult worms can live 12 to 15 years and be sexually

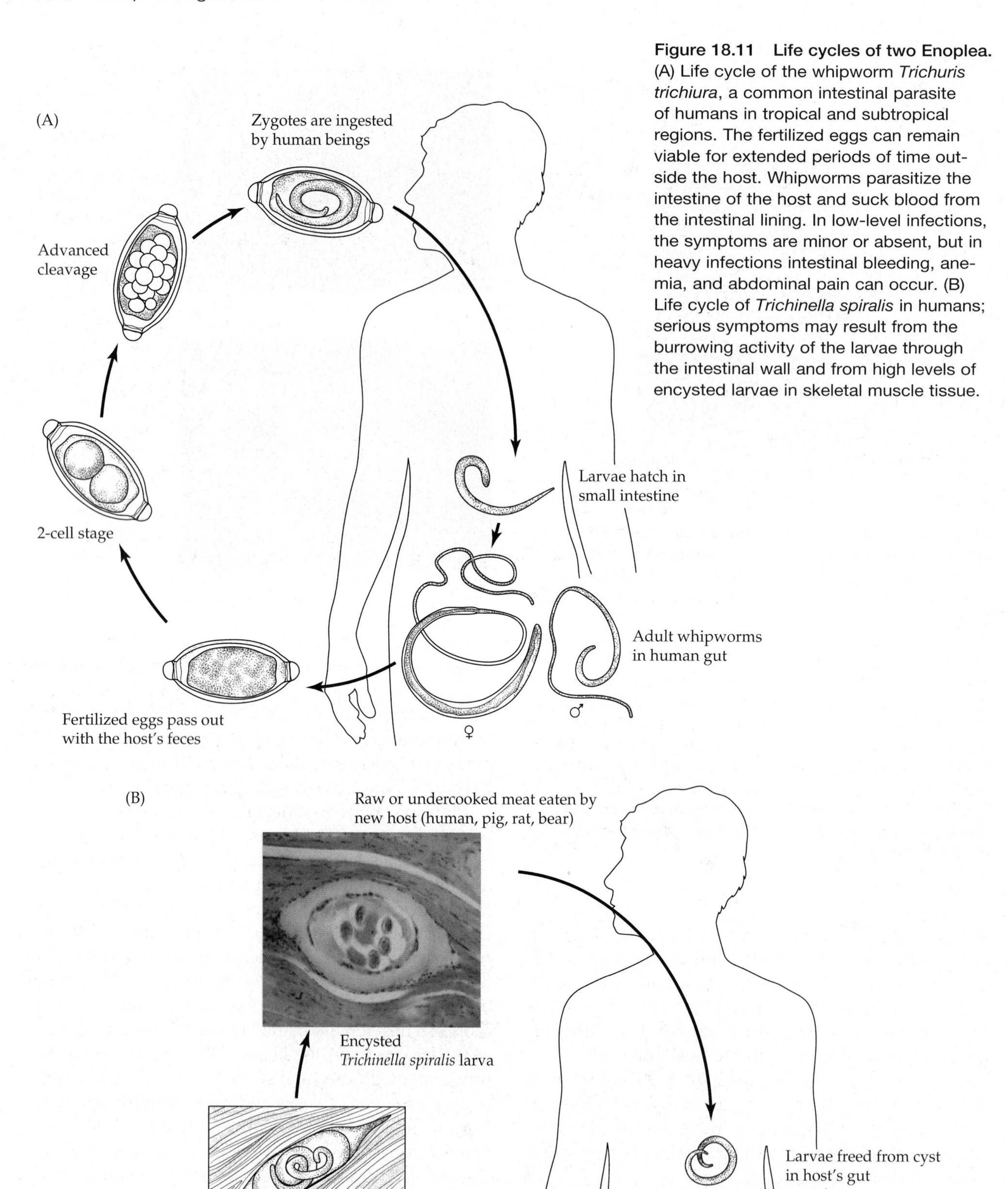

Figure 18.11 Life cycles of two Enoplea. (A) Life cycle of the whipworm *Trichuris trichiura*, a common intestinal parasite of humans in tropical and subtropical regions. The fertilized eggs can remain viable for extended periods of time outside the host. Whipworms parasitize the intestine of the host and suck blood from the intestinal lining. In low-level infections, the symptoms are minor or absent, but in heavy infections intestinal bleeding, anemia, and abdominal pain can occur. (B) Life cycle of *Trichinella spiralis* in humans; serious symptoms may result from the burrowing activity of the larvae through the intestinal wall and from high levels of encysted larvae in skeletal muscle tissue.

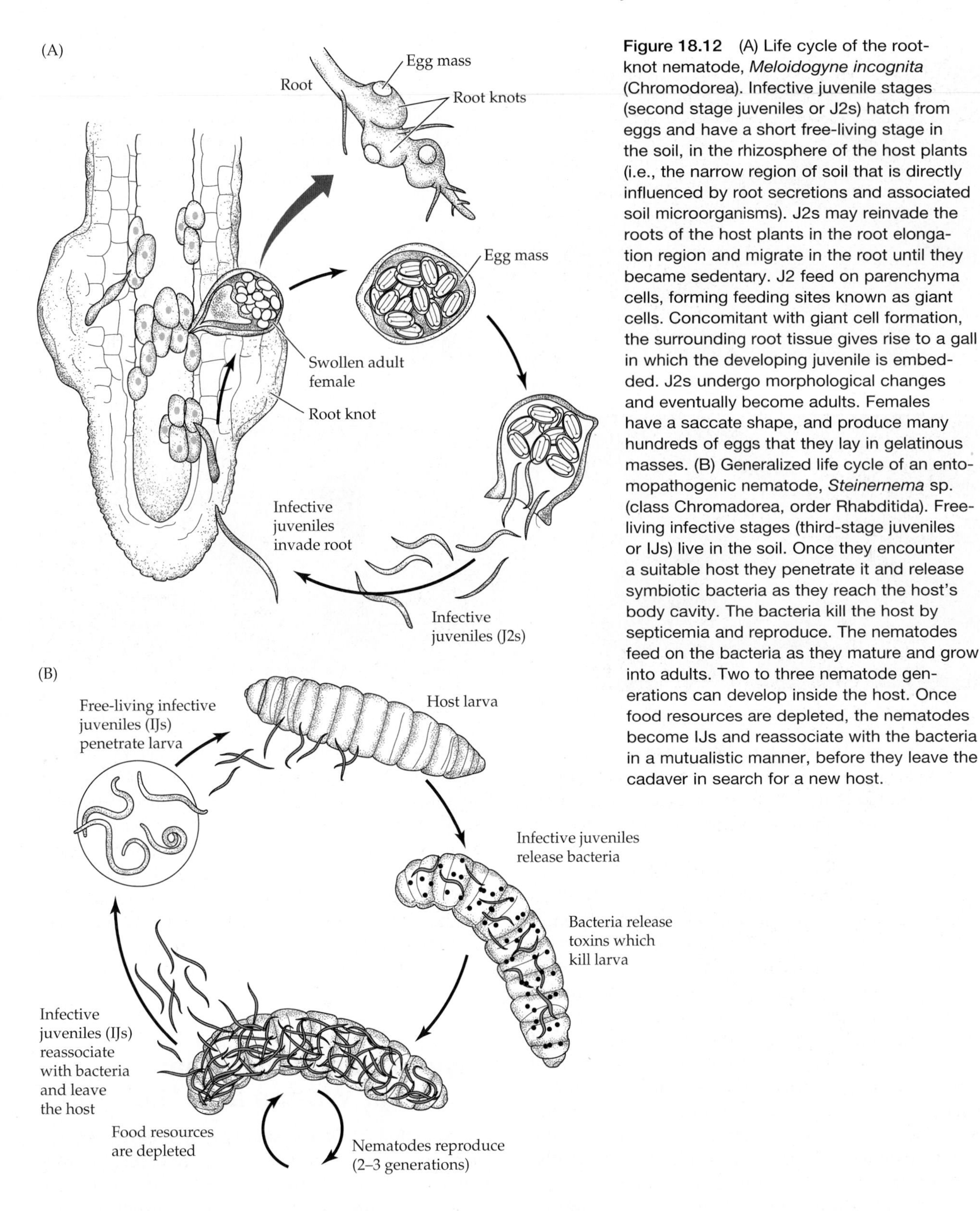

Figure 18.12 (A) Life cycle of the root-knot nematode, *Meloidogyne incognita* (Chromodorea). Infective juvenile stages (second stage juveniles or J2s) hatch from eggs and have a short free-living stage in the soil, in the rhizosphere of the host plants (i.e., the narrow region of soil that is directly influenced by root secretions and associated soil microorganisms). J2s may reinvade the roots of the host plants in the root elongation region and migrate in the root until they became sedentary. J2 feed on parenchyma cells, forming feeding sites known as giant cells. Concomitant with giant cell formation, the surrounding root tissue gives rise to a gall in which the developing juvenile is embedded. J2s undergo morphological changes and eventually become adults. Females have a saccate shape, and produce many hundreds of eggs that they lay in gelatinous masses. (B) Generalized life cycle of an entomopathogenic nematode, *Steinernema* sp. (class Chromadorea, order Rhabditida). Free-living infective stages (third-stage juveniles or IJs) live in the soil. Once they encounter a suitable host they penetrate it and release symbiotic bacteria as they reach the host's body cavity. The bacteria kill the host by septicemia and reproduce. The nematodes feed on the bacteria as they mature and grow into adults. Two to three nematode generations can develop inside the host. Once food resources are depleted, the nematodes become IJs and reassociate with the bacteria in a mutualistic manner, before they leave the cadaver in search for a new host.

active for 9 to 11 years. Adults live in large nodules under the skin—a telltale diagnostic feature. Each nodule harbors 2 to 3 female individuals measuring up to 50 cm in length (but only about 0.5 mm in diameter), and the much smaller males (to 4 cm long) migrate to these sites to inseminate the females. Once fertilized, females release a staggering 1,300 to 1,900 microfilariae worms (third "larval stage" or L3) per day throughout their reproductive lives. The microfilariae are also the infective stages in this disease. It is the microfilariae stage that is transmitted to humans by the biting black flies, which serve as intermediate hosts to the worms. Once in the blood stream, these tiny (250–300 μm long) microfilariae circulate throughout the host's body, some eventually arriving in the eyes where they invest the vitreous chamber, retina, and optic nerve. Here,

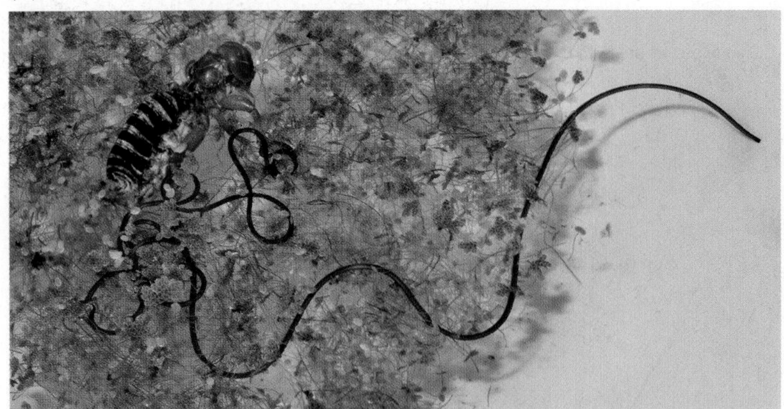

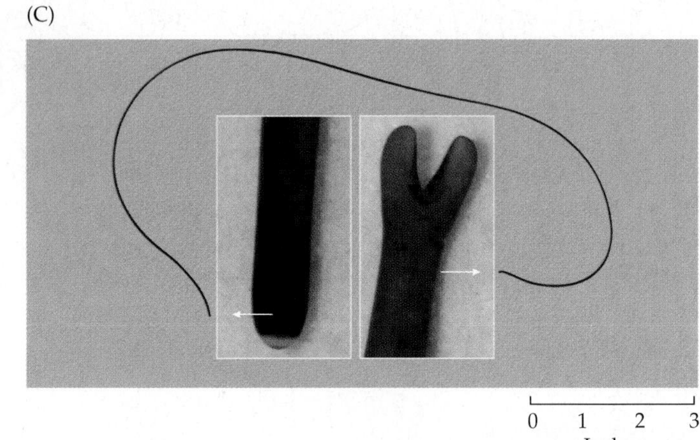

Figure 18.13 ***Gordius robustus*****, a common North American horsehair worm.** (A) The juvenile worm emerging from its insect host (a Jerusalem cricket). (B) Close up of the worm. (C) Anterior (left) and posterior (right) ends of adult male worm.

when the microfilariae die, they cause scarification, leading to visual impairment and eventually blindness. It is estimated that 37 million people in Africa have onchocerciasis, hundreds of thousands of them being blind.

In the 1970s, the drug ivermectin was developed from a bacteria (*Streptomyces avermectinius*) found on a golf course in Japan. The drug kills or permanently sterilizes the adult worm, with few side effects to the host animal. At first this drug was broadly developed for use on husbandry animals, cats, and dogs, but in 1996 the FDA approved its use in humans. Recent studies have shown that a large part of the pathogenicity of filarial worms is due to immune response of the host toward the nematode's naturally-occurring *Wolbachia* bacterial symbionts. Also, it has been demonstrated that elimination of *Wolbachia* from the worms generally results in either their death or sterility. Consequently, current strategies for control of filarial nematode diseases include elimination of *Wolbachia* via administration of antibiotics such as doxycycline, which are more effective and less harmful than anti-nematode medications.

Many nematodes parasitize invertebrates in both aquatic and terrestrial ecosystems. One of the most interesting groups is the so-called insect-pathogenic (or entomopathogenic) nematodes. These nematodes associate (mutualistic symbiosis) with certain gram-negative bacteria, which they vector in their intestines. Once the nematodes enter a suitable host they release the bacteria, which kill the insect host by a massive septicemia in 24 to 48 hours. The bacteria then become the nematodes' food allowing them to mature and reproduce in the insect cadaver. Figure 18.12B shows a generalized the life cycle. These nematodes have been successfully used in the control of many agricultural insect pests. The worms with their symbiotic bacteria are raised and sold worldwide as biological control agents.

Phylum Nematomorpha: Horsehair Worms and Their Kin

The Nematomorpha is a small group of large worms. There are about 360 described species of nematomorphans (Greek, *nema*, "thread"; *morph*, "shape"), commonly called the hair, horsehair, or gordian worms. The phylum name (and the common name of hair and horsehair worms) derives from the threadlike or hairlike shape of these animals (Figure 18.13A,B), and from the belief held for some time after their discovery in the fourteenth century that they actually arose from the hairs of horses' tails. They are generally from 1 to 3 mm in diameter and up to 1 m in length. Many of the highly elongate forms tend to twist and turn upon themselves in such a way as to give the appearance of complicated knots, and thus the name gordian worms. In the American midwest large clumps of intertwined groups of *Gordius difficilis* are commonly found in cold, spring-fed habitats. Most species occur in fresh water, but five marine species have also been described. Characteristics of this phylum are listed in Box 18B.

Nematomorphans were once thought to have a persistent blastocoelom, but recent work suggests this is not the case. Apparently the embryological blastocoel (the coeloblastula) is fully obliterated by the invasion of mesenchyme and organs during development, and neither larvae nor adults have a significant body cavity. In mature animals, the gonads become very spacious and occupy a large volume of the body. The digestive tract is largely nonfunctional, and nematomorphans appear to lack any structural excretory mechanisms. Like nematodes, nematomorphans possess only longitudinal (and no circular) musculature—in contrast, the closely related ecdysozoan phyla Kinorhyncha, Priapula, and Loricifera have both longitudinal and circular musculature in the body wall. There are no functional cilia in these animals.

The larvae of nematomorphans are parasitic in arthropods. Most adults live in fresh water, among litter and algal mats near the edges of ponds and streams. Members of the enigmatic genus *Nectonema* are pelagic in coastal marine environments.

BOX 18B Characteristics of the Phylum Nematomorpha

1. Triploblastic, bilaterally symmetrical, unsegmented, vermiform; body long and very thin
2. Body cavity largely obliterated by mesenchyme and organs
3. Cuticle well developed; a single cuticular molt has been reported in some species
4. Body without functional cilia or flagella
5. Body wall has only longitudinal muscles (no circular muscles)
6. Gut reduced to various degrees
7. With unique juvenile stage that is parasitic in arthropods
8. Epidermis forms cords housing longitudinal nerves
9. Without special excretory, circulatory, or gas exchange organs
10. Gonochoristic
11. With unique cleavage pattern; not unambiguously radial or spiral
12. Mostly inhabit freshwater and wet terrestrial habitats. Almost all are parasites during the juvenile phase. A few are marine planktonic species.

NEMATOMORPHAN CLASSIFICATION

ORDER NECTONEMATOIDEA Marine, planktonic; with a double row of natatory setae along each side of body; with dorsal and ventral longitudinal epidermal cords and integrated dorsal and ventral nerve cords; small body spaces may be evident, though fluid filled; gonads single; larvae parasitize decapod crustaceans. Monogeneric: *Nectonema* (5 known species)

ORDER GORDIOIDEA Fresh water; lack lateral rows of setae; with a single, ventral epidermal cord in a submuscular position, including the single nerve cord; body cavity largely filled with mesenchyme; older individuals develop spacious gonads; gonads paired; larvae parasitize aquatic and terrestrial insects, such as grasshoppers and crickets. (e.g., *Chordodes, Gordius, Paragordius*)

The Nematomorphan Body Plan

Body Wall, Support, and Locomotion

The general organization of the body wall of nematomorphans is similar in many aspects to that of the nematodes (Figure 18.14A–C). The cuticle of adult specimens, secreted by the epidermis, is very thick (especially in the gordioids) and comprises an outer homogeneous layer and an inner, lamellate, fibrous layer. The homogeneous layer often forms bumps, warts, or papillae (collectively called **areoles**; Figure 18.14D), some of which bear apical spines or pores. The function of the areoles is unknown, but spined ones may be touch sensitive and pore-bearing ones may produce a lubricant. Some species have two or three caudal lobes at the posterior end.

The unciliated epidermis covers the entire body and rests upon a thin basal lamina. The epidermis is produced into ventral (Gordioidea) or dorsal and ventral (Nectonematoidea) epidermal cords containing longitudinal nerve tracts. In Gordioidea the ventral cord is in a submuscular position and connected to the epidermis by a thin lamella. Beneath the epidermis is a single layer of longitudinal muscle cells; as in nematodes, there is no layer of circular muscle in the body wall.

The body cavity of nematomorphans is very small and filled with fluid (e.g., *Nectonema*), or altogether absent and filled with mesenchyme (see Figure 18.14A,B). Body support comes mainly from the well-developed cuticle. Locomotion in the planktonic *Nectonema* is by undulatory swimming using the body wall muscles and the natatory bristles (Figure 18.15F) or by passive flotation in nearshore currents, aided by the bristles, which provide resistance to sinking. Freshwater nematomorphans (order Gordioidea) use their longitudinal muscles to move by undulations or by coiling and uncoiling movements.

Feeding and Digestion

Nematomorphans might feed only during the parasitic juvenile stage, when they absorb nutrients across their

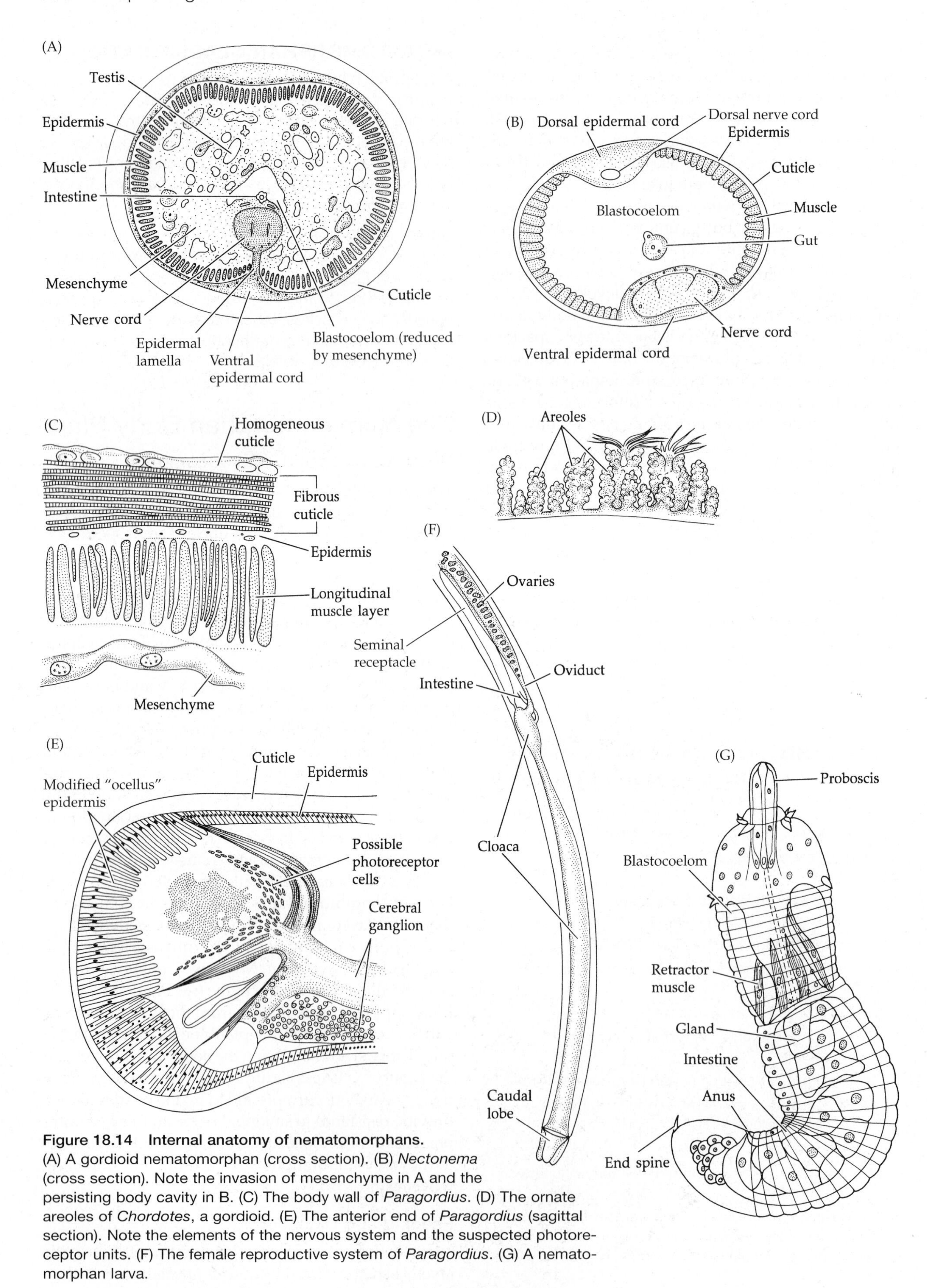

Figure 18.14 Internal anatomy of nematomorphans. (A) A gordioid nematomorphan (cross section). (B) *Nectonema* (cross section). Note the invasion of mesenchyme in A and the persisting body cavity in B. (C) The body wall of *Paragordius*. (D) The ornate areoles of *Chordotes*, a gordioid. (E) The anterior end of *Paragordius* (sagittal section). Note the elements of the nervous system and the suspected photoreceptor units. (F) The female reproductive system of *Paragordius*. (G) A nematomorphan larva.

thin body wall from the host's tissue and fluids, leaving adults to rely entirely on nutrients stored during the juvenile parasitic life. There is some evidence that adults of a few species might take in nutrients by absorbing small organic molecules across the body wall, but this seems a remote possibility due to the thickness of the adult cuticle and the reduced state of the intestine. There is growing evidence that parasitism by nematomorphans in insects can alter the host's behavior in ways that cause the insect to position itself near or in standing water, ensuring that the adults emerge in an appropriate habitat.

The digestive tract of nematomorphans is a simple elongate tube running the length of the body (Figure 18.14E). In juveniles, the gut is actively involved with uptake and storage of nutrients, transported across the body wall (although possible nutrient uptake through the mouth cannot be excluded). The intestine (midgut) is a thin-walled tube and may serve an excretory function as well as a digestive one. The hindgut is proctodeal (ectodermally-derived), functions as a cloaca, and receives the reproductive ducts. In *Nectonema* a tiny mouth leads to a midgut that deteriorates posteriorly and is not connected to the genital opening.

Circulation, Gas Exchange, Excretion, and Osmoregulation

Very little is known about the physiology of nematomorphans. Internal transport is undoubtedly by diffusion through body fluids and mesenchyme, and is probably aided by body movement. The free-living gordian worms are presumably obligate aerobes as adults and are restricted to moist environments with ample available oxygen. The threadlike body results in short diffusion distances between the environment and the body organs and tissues.

Excretory and osmoregulatory functions probably operate on a strictly cellular level, as there are no protonephridia or other known special structures for these functions. Some workers, however, have speculated that the cells of the midgut may function in the excretion of metabolic wastes, and that they may have a structure similar to the Malpighian tubules of insects.

Nervous System and Sense Organs

Like the nervous systems of nematodes and some other small Metazoa, the nervous system of nematomorphans is closely associated with the epidermis. The cerebral ganglion is a circumesophageal mass of nervous tissue, the majority of it ventral to the esophagus, located in an enlarged region of the head called the **calotte**, in reference to the skull caps worn by ecclesiastics (Figures 18.14E and 18.15B). In gordioids a single midventral nerve cord arises from the cerebral ganglion and extends the length of the body. It is attached to the epidermis by a tissue connection called the epidermal lamella (Figure 18.14A). *Nectonema* possesses an additional dorsal, intraepidermal nerve cord.

All nematomorphans are touch-sensitive, and some are apparently chemosensitive. Adult males are able to detect and track mature females from a distance. However, the structures associated with these sensory functions are a matter of speculation. Presumably, some of the cuticular areoles are tactile, and perhaps others are chemoreceptive. Members of the genus *Paragordius* possess modified epidermal cells in the calotte that contain pigment and may be photoreceptors (Figure 18.14E), although this function has not been confirmed. Other possible sensory structures are four "giant cells" near the cerebral ganglion in *Nectonema*.

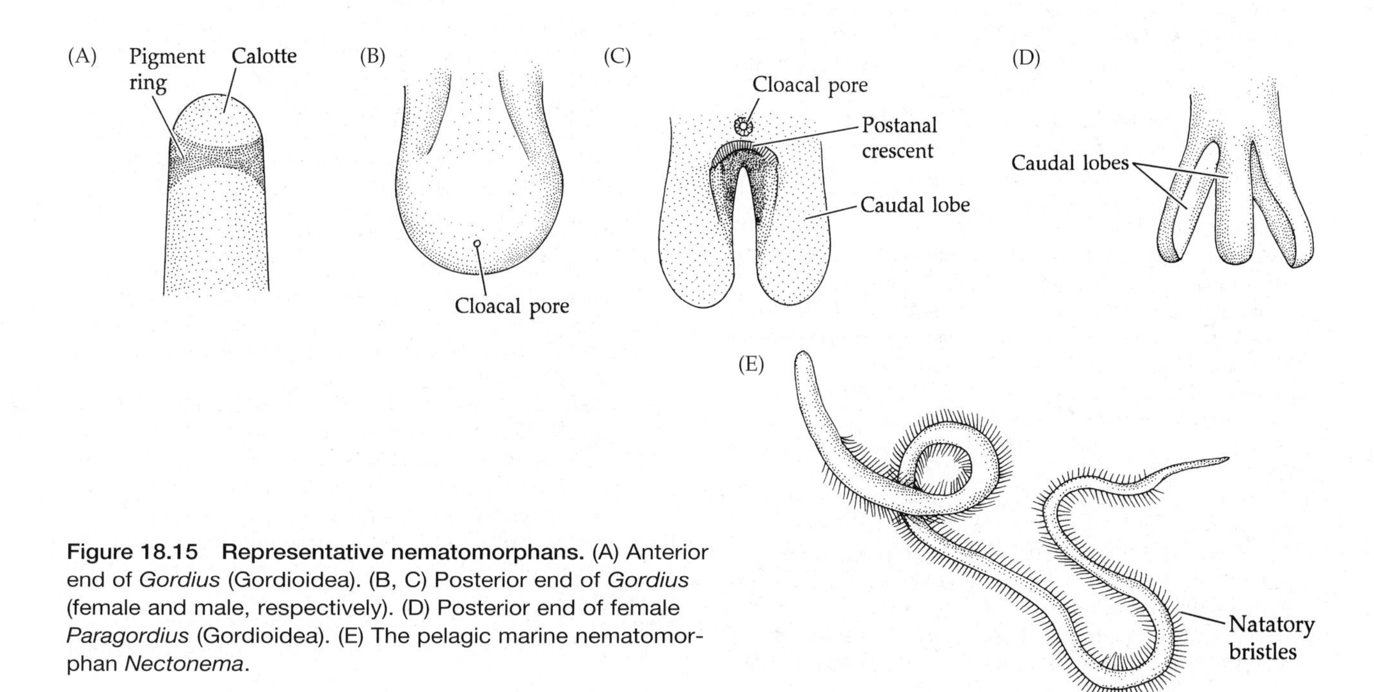

Figure 18.15 Representative nematomorphans. (A) Anterior end of *Gordius* (Gordioidea). (B, C) Posterior end of *Gordius* (female and male, respectively). (D) Posterior end of female *Paragordius* (Gordioidea). (E) The pelagic marine nematomorphan *Nectonema*.

Reproduction and Development

Nematomorphans are gonochoristic and display some sexual dimorphism (Figure 18.15C–E). The male reproductive system includes one (*Nectonema*) or two (gordioids) testes (Figure 18.14A). Each testis opens to the cloaca via a short sperm duct. The reproductive system of female gordioids is not completely understood. They possess a pair of elongate ovaries with repeated pouches. A seminal receptacle as an extension of the cloaca is present (Figure 18.14G). *Nectonema* contains no discrete ovary; instead, the germinal cells occur as scattered oocytes in the body cavity.

Mating has been studied in some gordioids. The females remain relatively inactive, but males become highly motile during the breeding season and respond to the presence of potential mates in their environment. Once a male locates a receptive female, he wraps his body around her and deposits a drop of sperm near her cloacal pore. The sperm find their way into the seminal receptacle and are stored while the ova mature. The eggs are fertilized internally and laid in gelatinous strings.

Early development has been studied in only a few species of gordioid nematomorphans. Cleavage is holoblastic, but not clearly spiral or radial. A coeloblastula forms, then gastrulates by invagination of the presumptive endoderm. Mesodermal cells proliferate into the blastocoel from the area around the blastopore. The anus and cloacal chamber also form from the area of the blastopore. A nematomorphan larva develops (Figure 18.14G) and emerges from the egg case. The following life cycle is not completely resolved and may differ among species. In most cases, the larvae infect intermediate (paratenic) hosts and encyst in their tissue without further development. Such development occurs only in an appropriate arthropod host, which it probably enters by being eaten together with the intermediate host. Intermediate hosts include a variety of aquatic animals, while final hosts are mainly terrestrial insects such as beetles, crickets, cockroaches, mantids, and even some flies, myriapods, and back swimmers (*Notonecta*). It is likely that metamorphosing insects with aquatic larvae play the most important role as intermediate hosts. Within the host's hemocoel, the larva grows into a juvenile nematomorphan, which emerges from the host when it is mature. A single cuticular molt has been reported in some species, taking place shortly before the young worm emerges. Juveniles are nearly full size when they leave their host and do not grow much as adults. For most of the two dozen or so horsehair worms of North America, the hosts are not known.

Selected References

Nematoda

Adamson, M. L. 1987. Phylogenetic analysis of the higher classification of the Nematoda. Can. J. Zool. 65: 1478–1482.

Allgen, C. A. 1947. West American marine nematodes. Vid. Medd. Dansk. Naturhist. Foren. (Copenhagen) 110: 65–219.

Baldi, C., S. Cho and R. E. Ellis. 2009. Mutations in two independent pathways are sufficient to create hermaphroditic nematodes. Science 326: 1002–1005.

Bauer-Nebelsick, M. M. Blumer, W. Urbanick and J. A. Ott. 1995. The glandular sensory organ of Desmodoridae (Nematoda): Ultrastructure and phylogenetic implications. Invert. Biol. 114: 211–219.

Bird, A. F. 1984. Growth and moulting in nematodes: moulting and development of *Rotylenchulus reniformis*. Parasitol. 89: 107–119.

Bird, A. F. and J. Bird. 1971. *The Structure of Nematodes*, 2nd Ed. Academic Press, New York.

Blaxter, M. L. and 11 others. 1998. A molecular evolutionary framework for the phylum Nematoda. Nature 392: 71–75.

Borner, J., P. Rehm, R. O. Schill, I. Ebersberger and T. Bumester. 2014. A transcriptome approach to ecdysozoan phylogeny. Molecular Phylogenetics and Evolution 80: 79–87.

Chitwood, B. G. and M. B. Chitwood. 1974. *Introduction to Nematology*. University Park Press, Baltimore.

Crofton, H. D. 1966. *Nematodes*. Hutchinson University Library, London.

Croll, N. A. and B. E. Matthews. 1977. *Biology of Nematodes*. Wiley, New York.

De Ley, P. and M. L. Blaxter. 2002. Systematic position and phylogeny. Pp. 1–30 in D. L. Lee (ed.), *The Biology of Nematodes*, Taylor and Francis, London.

De Ley, P. and M. Blaxter. 2004. A new system for Nematoda: combining morphological characters with molecular trees, and translating clades into ranks and taxa. Nematology Monographs and Perspectives 2: 633–653.

Deutsch, A. 1978. Gut ultrastructure and digestive physiology of two marine nematodes, *Chromadorina germanica* (Butschli, 1874) and *Diplolaimella* sp. Biol. Bull. 155: 317–355.

Ehlers, U. 1994. Absence of a pseudocoel or pseudocoelom in *Anoplostoma vivipara* (Nematoda). Microfauna Marina 9: 345–350.

Goldschmidt, R. 1908. Das nervesystem von *Ascaris lumbricoides* und *megalocephala*. Ein Versuch, in den Aufbau eines einfachen Nevernsystems einzudringam. Zweiter teil. Zeitschrift für wissenschaflichte Zoologie, 90: 73–136.

Goldschmidt, R. 1909. Das nervesystem von *Ascaris lumbricoides* und *megalocephala*. Ein Versuch, in den Aufbau eines einfachen Nevernsystems einzudringam, II. Zweiter teil. Zeitschrift für wissenschaflichte Zoologie, 92: 306–357.

Harris, J. E. and H. D. Crofton. 1957. Structure and function of nematodes. J. Exp. Biol. 34: 116–155.

Heip, C., M. Vinex and G. Vranken. 1985. The ecology of marine nematodes. Oceanogr. Mar. Biol. Annu. Rev. 23: 399–489.

Hesse, R. 1892. Über das Nervensystem von *Ascaris lumbricoides* und *Ascaris megalocephala*. Zeitschrift für wissenschaflichte Zoologie Abt. A 90: 73–136.

Holterman, M., A. van der Wurff, S. van den Elsen, H. van Megen, T. Bongers, O. Holovachov and J. Helder. 2006. Phylum-wide analysis of SSU rDNA reveals deep phylogenetic relationships among nematodes and accelerated evolution toward crown clades. Mol. Biol. Evol. 23: 1792–1800.

Hope, W. D. 1967. Free living marine nematodes from the west coast of North America. Trans. Amer. Micros. Soc. 86: 307–334.

Hope, W. D. and S. L. Gardiner. 1982. Fine structure of a proprioceptor in the body wall of the marine nematode

Donostoma californicum Steiner and Albin, 1933 (Enoplida = Leptostomatidae). Cell Tissue Res. 225: 1–10.

Hope, W. D. and D. G. Murphy. 1972. A taxonomic hierarchy and checklist of the genera and higher taxa of marine nematodes. Smithson. Contrib. Zool. 137. [Includes an outstanding bibliography of papers dealing with marine nematodes.]

Kampfer, S., C. Sturmbauer and J. Ott. 1998. Phylogenetic analysis of rDNA sequences from adenophorean nematodes and implications for the Adenophorea-Secernentea controversy. Invert. Biol. 117: 29–36.

Lee, D. L. and H. J. Atkinson. 1977. *Physiology of Nematodes*, 2nd Ed. Columbia University Press, New York.

Maas, A., D. Waloszek, J. T. Haug and K J. Müller. 2007. A possible larval roundworm from the Cambrian Orsten and its bearing on the phylogeny of Cycloneuralia. Mem. Assoc. Australasian Paleontologists 34: 499–519. [This is a fossil of uncertain affinity, but thought to be related to the Nematoida and Scalidophora.]

Maggenti, A. R. 1982. Nemata. Pp. 880–929 in S. Parker (ed.), *Synopsis and Classification of Living Organisms.* McGraw-Hill, New York.

Meldal, B. H. and 13 others. 2007. An improved molecular phylogeny of the Nematoda with special emphasis on marine taxa. Mol. Phylogenet. Evol. 42: 622–636.

Nicholas, W. L. 1975. *The Biology of Free-Living Nematodes.* Clarendon Press, Oxford.

Poinar, G. O., Jr. 1983. *The Natural History of Nematodes.* Prentice-Hall, Englewood Cliffs, New Jersey.

Roggen, D. R., D. J. Raski and N. O. Jones. 1966. Cilia in nematode sensory organs. Science 152: 515–516.

Rota-Stabelli, O. and 8 others. 2011. A congruent solution to arthropod phylogeny: phylogenomic, microRNAs and morphology support monophyletic Mandibulata. Proc. R. Soc. B, 278: 298–306.

Schaefer, C. 1971. Nematode radiation. Syst. Zool. 20: 77–78.

Somers, J. A., H. H. Shorey and L. K. Gastor. 1977. Sex pheromone communication in the nematode, *Rhabditis pellio*. J. Chem. Ecol. 3: 467–474.

Stock, S. P. 2005. Insect-parastic nematodes: from lab curiositeis to model organisms. J. Invert. Pathol. 89: 57–66.

Stock, S. P. 2009. Molecular approaches and the taxonomy of insect-parasitic and pathogenic nematodes. Pp. 71–100 in S. P. Stock, J. Vandenberg, I. Glazer and N. Boemare (eds.), *Insect Pathogens: Molecular Approaches and Techniques.* CABI Publishing–CABI, Wallingford, England, UK.

Stock, S. P. and H. Goodrich-Blair. 2008. Nematode-bacterium symbioses: crossing kingdom and disciplinary boundaries. Symbiosis 46: 61–64.

Stock, S. P. and D. J. Hunt. 2005. Morphology and systematics of nematodes used in biocontrol. Pp. 3–43 in P. S. Grewal, R. U. Ehlers and D. Shapiro-Ilan (eds.), *Nematodes as Biocontrol Agents.* CABI, New York.

Wright, K. A. 1991. Nematoda. Pp. 111–195 in F. W. Harrison and E. E. Ruppert (eds.), *Microscopic Anatomy of Invertebrates, Vol. 4, Aschelminthes.* Wiley-Liss, New York.

Yeats, G. W. 1971. Feeding types and feeding groups in plant and soil nematodes. Pedobiologia 11: 173–179.

Zuckerman, B. M., W. F. Mai and R. A. Rhode (eds.), 1971. *Plant Parasitic Nematodes. Vol. 1, Morphology, Anatomy, Taxonomy, and Ecology. Vol. 2, Cytogenetics, Host–Parasite Interactions, and Physiology.* Academic Press, New York.

Nematomorpha

Biron, D. G. and 9 others. 2006. "Suicide" of crickets harbouring hairworms: a proteomics investigation. Inst. Mol. Biol. 15: 731–742.

Bleidorn, C., A. Schmidt-Rhaesa and J. R. Garey. 2002. Systematic relationships of Nematomorpha based on molecular and morphological data. Invert. Biol. 121: 357–364.

Bresciani, J. 1991. Nematomorpha. Pp. 197–218 in F. W. Harrison and E. E. Ruppert (eds.), *Microscopic Anatomy of Invertebrates, Vol. 4, Aschelminthes*, Wiley-Liss, New York.

Hanelt, B. and J. Janovy. 1999. The life cycle of a horsehair worm, *Gordius robustus* (Nematomorpha: Gordioidea). J. Parasitol. 85: 139–141.

Hanelt, B. and J. Janovy. 2003. Spanning the gap: experimental determination of paratenic host specificity of horsehair worms (Nematomorpha: Gordiida). Invertebr. Biol. 122: 12–18.

Hanelt, B. and J. Janovy. 2004. Untying the Gordian knot: the domestication and laboratory maintenance of a gordian worm, *Paragordius varius* (Nematomorpha: Gordiida). J. Nat. Hist. 38: 939–950.

Hanelt, B., F. Thomas and A. Schmidt-Rhaesa. 2005. Biology of the phylum Nematomorpha. Adv. Parasitol. 59: 243–305.

Hanelt, B., M. Bolek and A. Schmidt-Rhaesa. 2012. Going solo: discovery of the first parthenogenetic gordiid (Nematomorpha: Gordiida). PlosONE 7(4): e34472. doi: 10.1371/journal.pone.0034472

Malakhov, V. V. and S. E. Spiridonov. 1984. The embryogenesis of *Gordius* sp. from Turmenia, with special reference to the position of the Nematomorpha in the animal kingdom. Zool. Z. 63: 1285–1296.

Poinar, G. O. and A. M. Brockerhoff. 2001. *Nectonema zealandica* n. sp. (Nematomorpha: Nectonematoidea) parasitizing the purple rock crab *Hemigrapsus edwardsi* (Brachyura: Decapoda) in New Zealand, with notes on the prevalence of infection and host defense reactions. Syst. Parasitol. 50: 149–157.

Sato, T., K. Watanabe, M. Kanaiwa, Y. Niizuma, Y. Harada and K. D. Laferty. 2011. Nematomorph parasites drive energy flow through a riparian ecosystem. Ecology 92.

Schmidt-Rhaesa, A. 2005. Morphogenesis of *Paragordius varius* (Nematomorpha) during the parasitic phase. Zoomorphology 124: 33–46.

Schmidt-Rhaesa, A. 2012. Nematomorpha. Pp. 29–145 in A. Schmidt-Rhaesa (ed.), *Handbook of Zoology. Gastrotricha, Cycloneuralia and Gnathifera. Volume 1: Nematomorpha, Priapulida, Kinorhyncha and Loricifera*. De Gruyter, Berlin.

Schmidt-Rhaesa, A., G. Pohle, J. Gaudette and V. Burdett-Coutts. 2012. Lobster (*Homarus americanus*), a new host for marine horsehair worms (*Nectonema agile*, Nematomorpha). J. Mar. Biol. Assoc. UK 10: 1017.

Schmidt-Rhaesa, A., D. G. Biron, C. Joly and F. Thomas. 2005. Host-parasite relations and seasonal occurrence of *Paragordius tricuspidatus* and *Spinochordodes tellinii* (Nematomorpha) in Southern France. Zool. Anz. 244: 51–57.

Skaling, B. and B. M. MacKinnon. 1988. The absorptive surfaces of *Nectonema* sp. (Nematomorpha: Nectonematoidea) from *Pandalus montagui*: Histology, ultrastructure, and absorptive capabilities of the body wall and intestine. Can. J. Zool. 66: 289–295.

Szmygiel, C., A. Schmidt-Rhaesa, B. Hanelt and M. G. Bolek. 2014. Comparative descriptions of non-adult stages of four genera of gordiids (phylum: Nematomorpha). Zootaxa 3768: 101–118

Thomas, F., A. Schmidt-Rhaesa, G. Martin, C. Manu, P. Durand and F. Renaud. 2002. Do hairworms (Nematomorpha) manipulate the water-seeking behaviour of their terrestrial hosts? J. Evol. Biol. 15: 356–361.

CHAPTER 19

The Scalidophora

Phyla Kinorhyncha, Priapula, and Loricifera

Scalidophora is an ecdysozoan clade that, together with Nematoida (Nematoda and Nematomorpha), is often considered to comprise a larger clade, Cycloneuralia. However, molecular phylogenetics is still divided on this view, and some work suggests Nematoida is a sister clade to the Panarthropoda. In any case, adult nematoidans and scalidophorans retain the embryonic blastocoel into adulthood (although recent studies suggest it might be obliterated in adult Nematomorpha and in some Kinorhyncha). The clade Scalidophora accommodates three phyla: Kinorhyncha, Priapula, and Loricifera, two of which (Kinorhyncha and Loricifera) are exclusively meiofaunal. As with all other ecdysozoans, the scalidophoran taxa lack cilia for locomotion and feeding, and they shed their cuticle during growth or periodic molts. Chitin is present in the cuticle of all three scalidophoran phyla, whereas collagen only has been demonstrated from macrofaunal priapulans where it occurs in the intestinal epithelium and in the connective tissue beneath the epidermis.

Classification of The Animal Kingdom (Metazoa)

Non-Bilateria*
(a.k.a. the diploblasts)
- PHYLUM PORIFERA
- PHYLUM PLACOZOA
- PHYLUM CNIDARIA
- PHYLUM CTENOPHORA

Bilateria
(a.k.a. the triploblasts)
- PHYLUM XENACOELOMORPHA

Protostomia
- PHYLUM CHAETOGNATHA

SPIRALIA
- PHYLUM PLATYHELMINTHES
- PHYLUM GASTROTRICHA
- PHYLUM RHOMBOZOA
- PHYLUM ORTHONECTIDA
- PHYLUM NEMERTEA
- PHYLUM MOLLUSCA
- PHYLUM ANNELIDA
- PHYLUM ENTOPROCTA
- PHYLUM CYCLIOPHORA

Gnathifera
- PHYLUM GNATHOSTOMULIDA
- PHYLUM MICROGNATHOZOA
- PHYLUM ROTIFERA

Lophophorata
- PHYLUM PHORONIDA
- PHYLUM BRYOZOA
- PHYLUM BRACHIOPODA

ECDYSOZOA

Nematoida
- PHYLUM NEMATODA
- PHYLUM NEMATOMORPHA

Scalidophora
- **PHYLUM KINORHYNCHA**
- **PHYLUM PRIAPULA**
- **PHYLUM LORICIFERA**

Panarthropoda
- PHYLUM TARDIGRADA
- PHYLUM ONYCHOPHORA
- PHYLUM ARTHROPODA
  - SUBPHYLUM CRUSTACEA*
  - SUBPHYLUM HEXAPODA
  - SUBPHYLUM MYRIAPODA
  - SUBPHYLUM CHELICERATA

Deuterostomia
- PHYLUM ECHINODERMATA
- PHYLUM HEMICHORDATA
- PHYLUM CHORDATA

*Paraphyletic group

The three scalidophoran phyla may not seem like they have much in common, given the variation in body plans: the segmented trunk of kinorhynchs, the worm-like body of priapulans, and the loricated trunk (i.e., covered by a hard protective shell) of loriciferans. However, the three groups share important similarities in their head morphology. In all three phyla the head has the form of a uniquely built eversible introvert, and is defined by a multilayered cuticle, a circumenteric "brain," and characteristic sensory or locomotory **scalids** whose structure is continuous with the epidermal cuticle.

When the head is withdrawn, the mouth cone is retracted like a tube, whereas the introvert is inverted so that the parts that are exterior and anterior in the protruded head become interior

---

The sections on phyla Kinorhyncha and Priapula, and the introduction have been revised by Martin Vinther Sørensen. The section on phylum Loricifera has been revised by Reinhardt Møbjerg Kristensen.

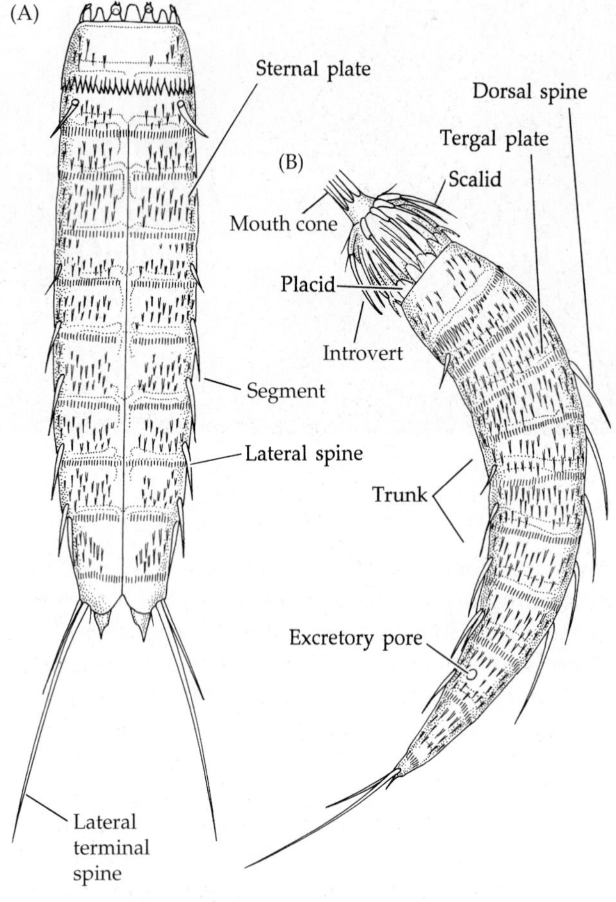

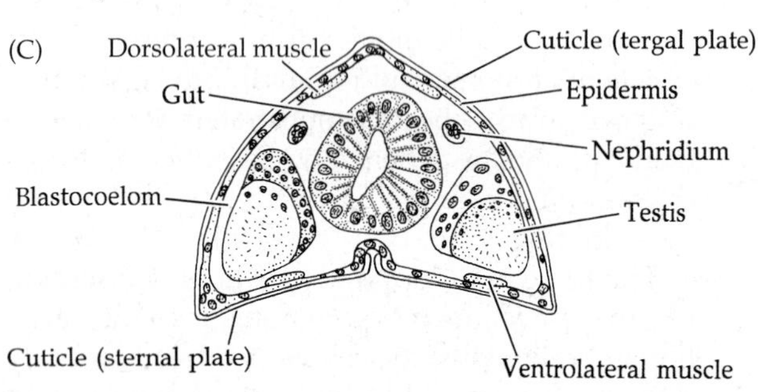

Figure 19.1 (A,B) External anatomy of the kinorhynch *Echinoderes*. (A) Ventral view with head retracted. (B) Lateral view with head extended. (C) A trunk segment (cross section) of a kinorhynch; note the arrangement of the body wall structures and the organs within the blastocoelom. (D) *Echinoderes*, a meiofaunal species, with head extended.

and posterior in the retracted head (Figure 19.1A,B). Among kinorhynchs and loriciferans, the mouth cone is usually large and well developed, with structures such as large oral styles or a mouth tube with oral ridges. However, among priapulans, a mouth cone is only found in some of the meiofaunal taxa, and even among these species, the mouth only constitutes a relatively small region between the pharynx and the introvert. The introvert is always well developed in scalidophorans, and always equipped with sensory or locomotory scalids. The priapulan scalids are small and hook shaped, whereas kinorhynchs and loriciferans have longer spino- or clavoscalids.

From a morphological perspective, scalidophoran monophyly is more or less undisputed. However, it has not yet been possible to confirm this monophyly with molecular sequence data. Comprehensive genomic data are available from priapulans, and to some extent from kinorhynchs as well, and their close relationship has been confirmed by several analyses. However, data from only a few selected loci are available for loriciferans, and only very few molecular analyses of metazoan or ecdysozoan relationships have included representatives of all three scalidophoran taxa. One analysis of ecdysozoan relationships, based on 18S rRNA sequences supplemented with Histone 3 data, included two species of loriciferans, and the results confirmed a close relationship between kinorhynchs and priapulans, but it also suggested a sister-group relationship between Loricifera and Nematomorpha, thus moving the loriciferans from Scalidophora to Nematoida. This result contradicts morphology, since the mouth cone and introvert are scalidophoran synapomorphies. However, the result might also point towards an alternative hypothesis, suggesting that possession of a mouth cone and an introvert was part of the original ecdysozoan or cycloneuralian body plan, and that these characters are therefore plesiomorphic for the scalidophoran taxa. To some extent this is supported by the fossil record that includes a broad variety of scalidophoran stem- and crown-group taxa. Many palaeoscolecidans, for example, have toothed pharynxes and scalid-bearing introverts.

The fossil record holds a diverse scalidophoran fauna. Some fossil taxa, such as *Priapulites*, unquestionably belong to the phylum Priapula, whereas others appear to be either priapulan (e.g., *Ottoia*, *Selkirkia*, *Xiaoheiqingella*, *Yunnanpriapulus*) or loriciferan (*Sirilorica*) stem- or crown-group taxa. Fossil kinorhynchs have not yet been found. A possible stem-group scalidophoran could be *Markuelia*—an annulated, introvert-bearing worm that is known from the lower Cambrian.

Phylum Kinorhyncha: The Kinorhynchs

Among the most intriguing of the "little animalcules" are members of the phylum Kinorhyncha (Greek *kineo*, "movable"; *rhynchos*, "snout"), sometimes called "mud dragons." Since their discovery in 1841 on the northern coast of France, about 200 species of kinorhynchs have been described from around the world, nearly all of which are less than 1 mm in length. Most live in marine sands or muds, from the intertidal zone to a depth of 7,800 meters. However, specimens have also been recorded from algal mats or kelp holdfasts, sandy beaches, and brackish estuaries; others live on hydroids, ectoprocts, sponges, or sediment aggregations in polychaete tubes or on shells of molluscs.

The body of a kinorhynch is composed of three regions: A head, a short neck region, and a segmented trunk (Figure 19.1). Families and genera are often diagnosed by the composition of the segments (which are sometimes called "zonites"), whereas species differences often relate to distribution of spines and setae. The kinorhynchs are one of three truly segmented metazoan clades (the other two being Annelida and the Panarthropoda), although the genetic developmental process of segment formation in kinorhynchs is not yet well known. The major characteristics of kinorhynchs are given in Box 19A.

The kinorhynch head is formed by a retractable mouth cone with nine outer **oral styles**, and an introvert with numerous appendages called **spinoscalids** (Figure 19.2). By using the ten anterior-most primary spinoscalids as markers, the remaining spinoscalids

BOX 19A Characteristics of the Phylum Kinorhyncha

1. Triploblastic, bilaterally segmented
2. Body blastocoelomate or, in some species, acoelomate
3. Body divided into head, neck, and 11-segmented trunk
4. Head divided into mouth cone and introvert with scalids
5. Segments formed by complete cuticular rings or composed of tergal and sternal plates
6. Epidermis cellular without locomotory cilia
7. Gut complete
8. With one pair of protonephridia in segment 8, but without special circulatory or gas exchange structures
9. Direct development through six juvenile stages; juveniles molt their cuticle at the transition between each stage
10. Gonochoristic
11. Embryonic development and cleavage patterns not well understood
12. Inhabit marine, benthic habitats, although a few brackish species, and species inhabiting epizoic or epiphytic habitats are known

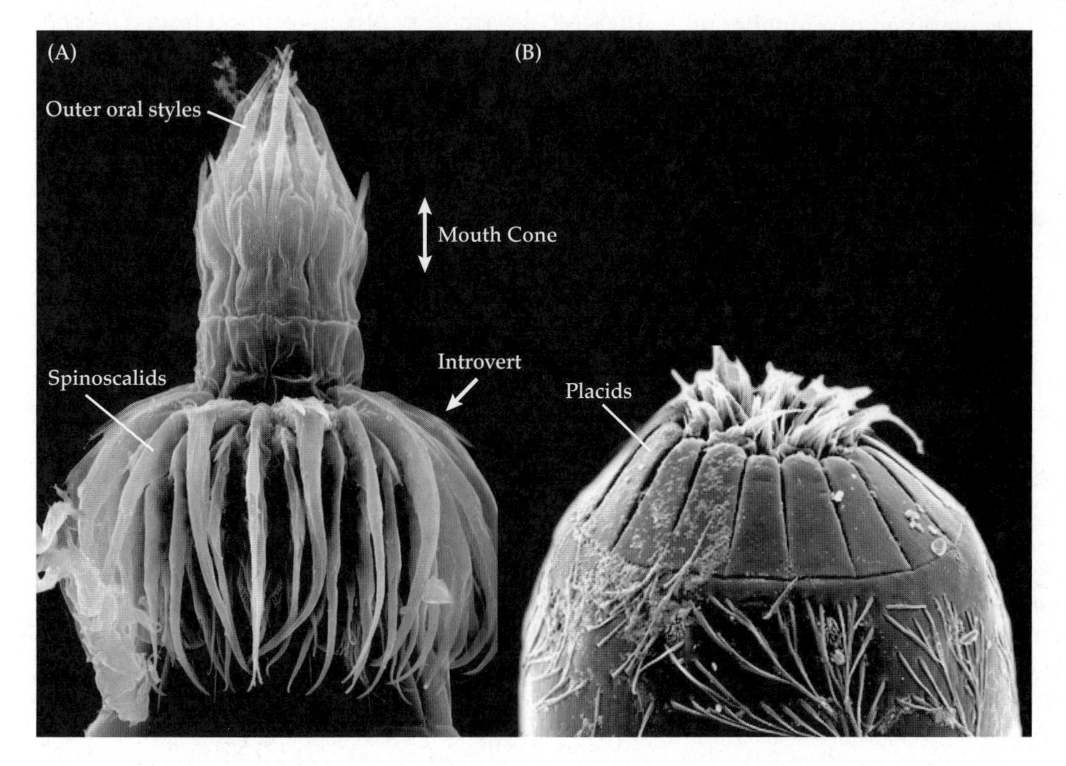

Figure 19.2 Scanning electron microscopy showing extended and retracted head of a kinorhynch. (A) Extended head, showing the external morphology of the mouth cone and introvert. (B) Retracted head, with the placids of the neck acting as closing apparatus.

Figure 19.3 Scanning electron microscopy of Kinorhyncha, showing (A) the allomalorhagid genus *Pycnophyes*, and the cyclorhagid genera (B) *Centroderes* and (C) *Meristoderes*. (D) Close-up of right sternal plate. Note the articulations with the left sternal plate, and the tergal plate.

on the introvert can be divided into ten sectors, where the spinoscalids follow either a bilateral or pentaradial symmetry pattern. Functionally, the two units can be distinguished by the way they move when the head is retracted into the trunk. The mouth cone is retracted like a tube, meaning that the proximal end of the mouth cone is the first to disappear when the mouth cone muscles drag the mouth cone into the introvert. When the head retraction continues into the trunk, the introvert is inverted, meaning that the anterior-most spinoscalids and the part of the introvert that joins the mouth cone are the first to retract.

The kinorhynch neck is composed of a number of plates, called **placids**, and when the head is retracted, the placids act as a closing apparatus (Figure 19.2B). In some genera the placids are hard structures that are easily recognized, whereas in other genera they are thinner and softer, and more difficult to see. Within the genus *Cateria*, the placids are extremely soft and almost impossible to identify, whereas they are completely fused into a neck ring in the genus *Franciscideres*. The number of placids differs among genera, and range from four to sixteen.

The kinorhynch trunk consists of eleven segments in all adults. Most trunk segments are made up of a dorsal (tergal) plate and a pair of ventral (sternal) plates (Figures 19.1A and 19.3). The anus is located on the last segment and is usually flanked by strong lateral terminal spines.

During juvenile development, and in some species during adult life as well, kinorhynchs shed their cuticle. This supports their position among the ecdysozoan protostomes, even though it is still uncertain whether molting is mediated by the hormone ecdysone.

KINORHYNCH CLASSIFICATION

CLASS CYCLORHAGIDA Neck usually with 14–16 placids that work as a closing apparatus, although in one group a clamshell-like first trunk segment may assist; trunk oval, round, or slightly triangular in cross section; numerous cuticular trunk spines and tubes present; trunk segments usually with dense covering of cuticular hairs or denticles. The order includes genera such as *Centroderes* (Figure 19.3B), *Echinoderes* (Figure 19.1D), *Meristoderes* (Figure 19.3C), and *Semnoderes*.

CLASS ALLOMALORHAGIDA Neck with 9 or fewer placids. Trunk either triangular or completely round in cross section. Midterminal spine never present in adults. The class includes genera such as *Paracentropyhyes*, *Neocentrophyes*, *Pycnophyes, Dracoderes*, and *Franciscideras* (Figure 19.3A) that previously were assigned to the order Homalorhagida (now considered paraphyletic).

The Kinorhynch Body Plan

Body Wall

Beneath the thick, well-developed, chitinous cuticle is a nonciliated epidermis (Figure 19.1C), which contains elements of the nervous system. Largely internal to the epidermis but still attached to the cuticle are bands of cross-striated, dorsolateral, and ventrolateral intersegmental muscles. Some of the anterior longitudinal muscle bands serve as head retractors. A series of metamerically arranged dorsoventral muscles creates the increased hydrostatic pressure that protracts the head and **oral cone** when the retractor muscles relax (Figure 19.4).

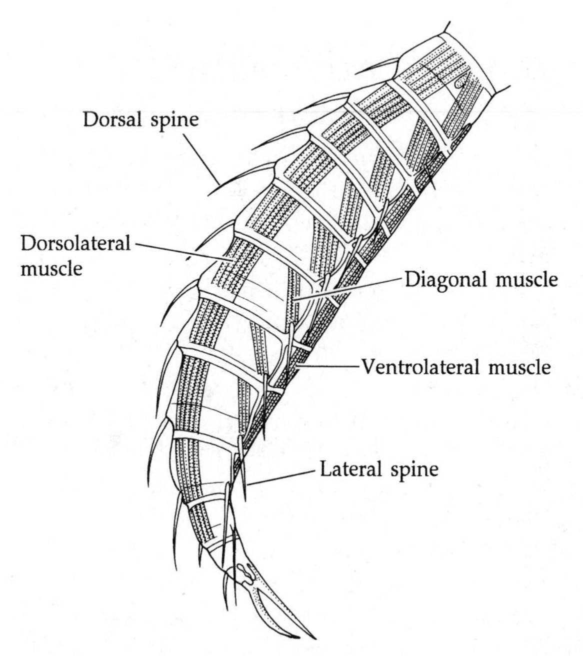

Figure 19.4 The trunk muscles of a kinorhynch (lateral view).

Support and Locomotion

Body shape in kinorhynchs is more or less fixed by the rigid plates of the supportive cuticular exoskeleton, but the animals are able to flex and even twist at the points of articulation between adjacent segments. Segmented trunk musculature facilitates movements of the articulated cuticular plates along the anterior–posterior axis. Intersegmental muscle fibers assist with dorsoventral and lateral trunk movements. In the absence of external cilia, burrowing is accomplished by extending the head into the substratum, anchoring the anterior spines, and then pulling the rest of the body forward. It has also been suggested that the numerous glandular cells that open along the trunk may be involved in locomotion, i.e., by creating a push in a forward direction when mucus is released from the glands.

Feeding and Digestion

Kinorhynchs are probably direct deposit feeders, ingesting the substratum and digesting the organic material or eating unicellular algae contained therein; they have been found with their guts full of benthic diatoms. However, the details of feeding and the exact nature of their food are not known.

The mouth leads into a buccal cavity located within the mouth cone and subsequently to a muscular pharynx (Figure 19.5). The buccal cavity, pharynx, and esophagus are lined with cuticle. Protractors, retractors, and circular muscles coordinate eversion and retraction of the introvert and mouth cone. Various paired glands are often associated with the esophagus, but their functions are uncertain. The esophagus connects with an elongate, straight midgut, which leads to a short, cuticle-lined proctodeal hindgut (rectum), and the anus on the last segment. So-called "digestive glands" often arise from the midgut.

Circulation, Gas Exchange, Excretion, and Osmoregulation

Gas exchange takes place by diffusion across the body wall, and circulation and internal transport is accomplished primarily by body movements. Kinorhynchs possess one pair of solenocyte protonephridia in segment 8, each with a short nephridioduct that opens laterally on segment 9. Excretory and osmoregulatory physiology is likely interesting, although remains largely unstudied; some estuarine and intertidal species can tolerate large fluctuations in salinities, and some live in salinities as low as 6 parts per thousand in the Gulf of Finland.

Nervous System and Sense Organs

The central nervous system of kinorhynchs is relatively simple and is intimately associated with the epidermis.

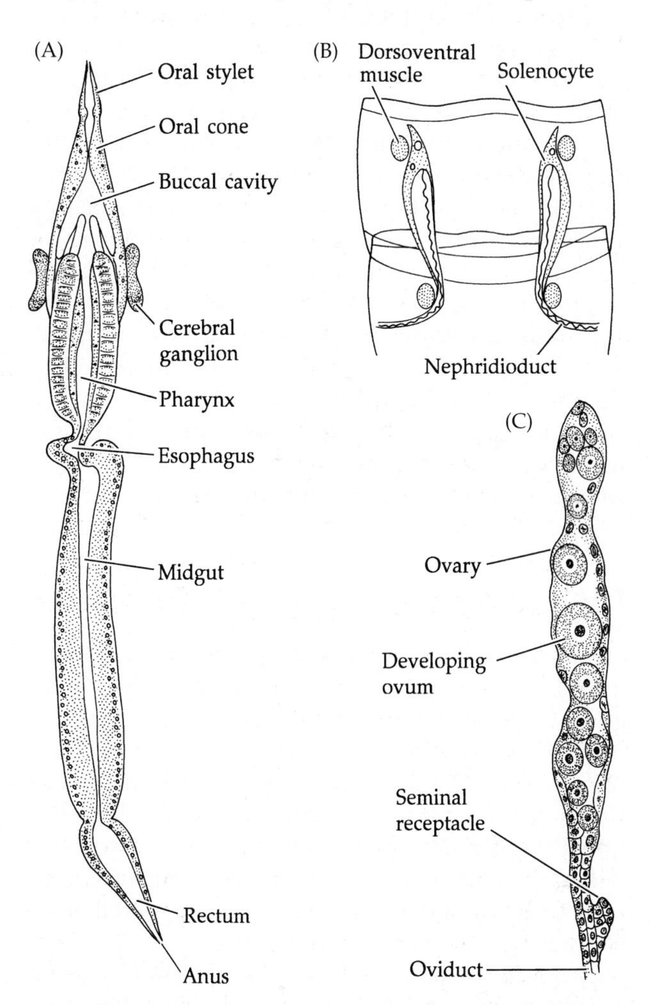

Figure 19.5 Internal anatomy of kinorhynchs. (A) The digestive tract of *Pycnophyes*. (B) The nephridial arrangement in two segments of *Pycnophyes*. (C) The simple female reproductive system of a kinorhynchs.

A series of ten connected ganglia is arranged in a ring around the pharynx. Each ganglion probably originally gave rise to one longitudinal nerve cord, eight of which are retained by most species. The two midventral nerve cords are most prominent, and they bear ganglia in each segment. Sensory receptors include tactile bristles, spines, and sensory spots on the trunk. Microvillar eyespots are present on the pharyngeal nerve ring in at least some species.

Reproduction and Development

Kinorhynchs are gonochoristic and possess relatively simple reproductive systems. External morphological differences between males and females are usually restricted to the most posterior segments. In both sexes, paired saclike gonads lead to short gonoducts that open separately on the terminal segment. In the males, the gonopore may be associated with two or three cuticular, hollow, penile spines that presumably aid in copulation, although their precise function is uncertain. The female gonads comprise both germ cells and nutritive (yolk-producing) cells. Each oviduct bears a diverticulum that forms as a seminal receptacle prior to ending at a gonopore between segments 10 and 11.

Mating has never been observed in kinorhynchs, and egg laying and early development (e.g., cleavage patterns) have not been adequately studied. Fertilized eggs are coated with a protective envelope of silt and detritus, and then deposited in sediment. Embryos develop directly and hatch as juveniles with nine segments. The juvenile will reach maturity after six molts, and gradually change from juvenile to adult morphology during these molts. New segments are added by elongation, and subsequently subdivision of the terminal segment. The mature condition with eleven trunk segments is often reached in juvenile stage 5. Molting among adults has been reported as well.

BOX 19B Characteristics of the Phylum Priapula

1. Triploblastic, bilaterally symmetrical, unsegmented (may be superficially annulated), and vermiform
2. Body cavity lining probably not peritoneal; presumably blastocoelomate
3. Introvert with hooked spines
4. Nervous system radially arranged and largely intraepidermal
5. Gut complete
6. With (often multicellular) protonephridia associated with the gonads as a urogenital system
7. Many with unique caudal appendage (in adults) that may serve for gas exchange
8. No circulatory system
9. Thin cuticle is periodically shed
10. With unique loricate larva, with cuticular lorica that is shed at metamorphosis to adult stage
11. Gonochoristic
12. Cleavage radial-like
13. Marine and benthic; most are burrowers

Phylum Priapula: The Priapulans

Twenty extant species (and several fossil species) are assigned to the phylum Priapula (from *Priapos*, the Greek god of reproduction, symbolized by his enormous penis). These odd creatures were recorded in Linnaeus's *Systema Naturae*, where he mentioned the species *Priapus humanus* (literally, "human penis"). Since that time priapulans (or priapulids) have been associated with several different invertebrate phyla, but today both morphological and molecular evidence support their alliance with the other scalidophoran ecdysozoans, perhaps most closely related to the kinorhynchs. The major characteristics of priapulans are given in Box 19B.

Three genera (*Tubiluchus*, *Meiopriapulus*, and *Maccabeus*) comprise meiofaunal species that measure 1.5 mm to about 5 mm, whereas the remaining four genera (*Acanthopriapulus*, *Halicryptus*, *Priapulopsis*, and *Priapulus*) are macrofaunal, with the largest species, the Alaskan *Halicryptus higginsi*, reaching a length of almost 40 cm. Priapulans are cylindrical vermiform creatures. The body comprises an introvert, sometimes a neck region, a trunk, and sometimes a "tail" or caudal appendage (Figure 19.6 and opening chapter photograph). The whole introvert can be inverted, and it is equipped with short, hook-shaped scalids that are considered homologous with the spinoscalids of loriciferans and kinorhynchs. When present, the caudal appendage varies in shape and function among species.

Large priapulans are active infaunal burrowers in relatively fine marine sediments and occur primarily in boreal and cold temperate seas. A few species construct tubes. Adults of one species, *Halicryptus spinulosus*, lives in anoxic, sulfide-rich sediments in the Baltic Sea and shows remarkable abilities to both tolerate and detoxify high sulfide levels in its body. In contrast, the small meiofaunal forms burrow or live interstitially among sediment particles, and they appear to be most abundant in subtropical and tropical waters.

Stem- and crown-group priapulans are rather common in the fossil record. They may have been one of the major predator groups in Cambrian seas, and they were almost certainly more abundant in Paleozoic times than they are today. The most abundant fossil species, *Ottoia prolifica*, resembles the extant *Halicryptus spinulosus*. Recent work confirms that *O. prolifica* (and the closely related *O. tricuspida*), which date from the middle Cambrian, were serious carnivores, their introvert being lined with rows of hooks, teeth, and spines for dealing with their prey.

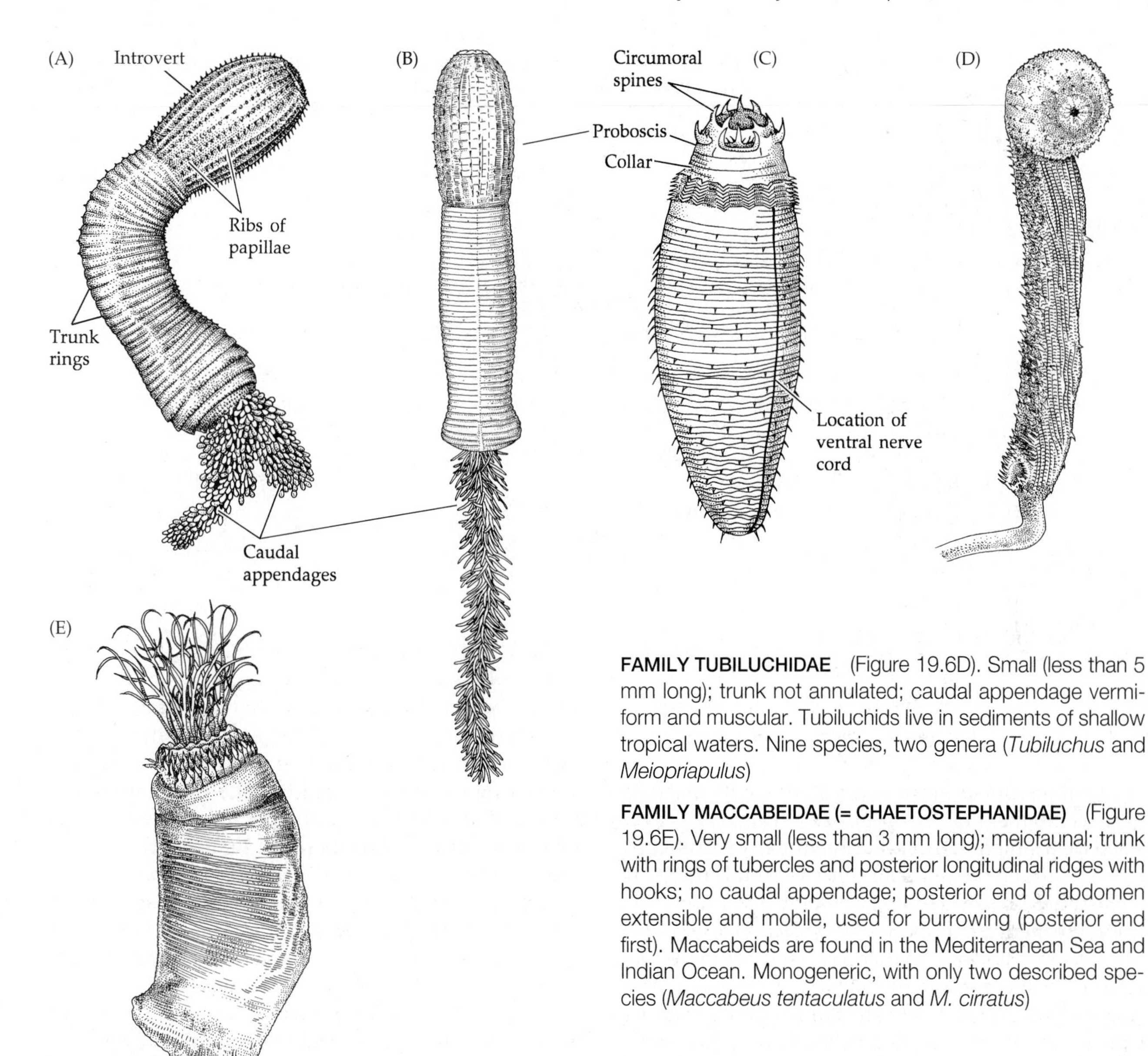

Figure 19.6 Representative priapulans. (A) *Priapulus bicaudatus* (Priapulidae). The North Atlantic species *Priapulus caudatus* is shown in the chapter opener photo. (B) *Priapulopsis australis* (Priapulidae). (C) *Halicryptus* (Priapulidae). (D) *Tubiluchus corallicola* (tubiluchidae). (E) *Maccabeus tentaculatus* (Maccabeidae).

PRIAPULAN CLASSIFICATION

Classes and orders are usually not recognized within the Priapula; the phylum is divided into the three extant families:

FAMILY PRIAPULIDAE (Figure 19.6A–C). Relatively large (4–40 cm); trunk with superficial annulations; caudal appendage (absent from *Halicryptus spinulosus*) either a grapelike cluster of fluid-filled sacs (called vesiculae) or a muscular extension with cuticular hooks. Nine species, four genera: *Acanthopriapulus, Halicryptus, Priapulopsis*, and *Priapulus*

FAMILY TUBILUCHIDAE (Figure 19.6D). Small (less than 5 mm long); trunk not annulated; caudal appendage vermiform and muscular. Tubiluchids live in sediments of shallow tropical waters. Nine species, two genera (*Tubiluchus* and *Meiopriapulus*)

FAMILY MACCABEIDAE (= CHAETOSTEPHANIDAE) (Figure 19.6E). Very small (less than 3 mm long); meiofaunal; trunk with rings of tubercles and posterior longitudinal ridges with hooks; no caudal appendage; posterior end of abdomen extensible and mobile, used for burrowing (posterior end first). Maccabeids are found in the Mediterranean Sea and Indian Ocean. Monogeneric, with only two described species (*Maccabeus tentaculatus* and *M. cirratus*)

Priapulan Body Plan

Body Wall, Support, and Locomotion

The priapulan body is covered by a thin, flexible cuticle that forms a variety of spines, warts, and tubercles (Figure 19.6). Hook-shaped scalids are often present around the mouth and on the introvert, and these are considered homologous with the spinoscalids of kinorhynchs and loriciferans. In all three groups the scalids function as sensory structures and assist in locomotion. Priapulans move through sediments by means of the introvert and peristaltic body muscle action. The cuticle contains chitin and is periodically shed as the animal grows. Beneath the cuticle lies an unciliated epidermis of thin, elongate cells with large fluid-filled intercellular spaces. Beneath the epidermis are well-developed layers of circular and longitudinal muscles. There are also complex muscle layers and bands associated with the pharynx, and a set of introvert retractor muscles (Figure 19.7A).

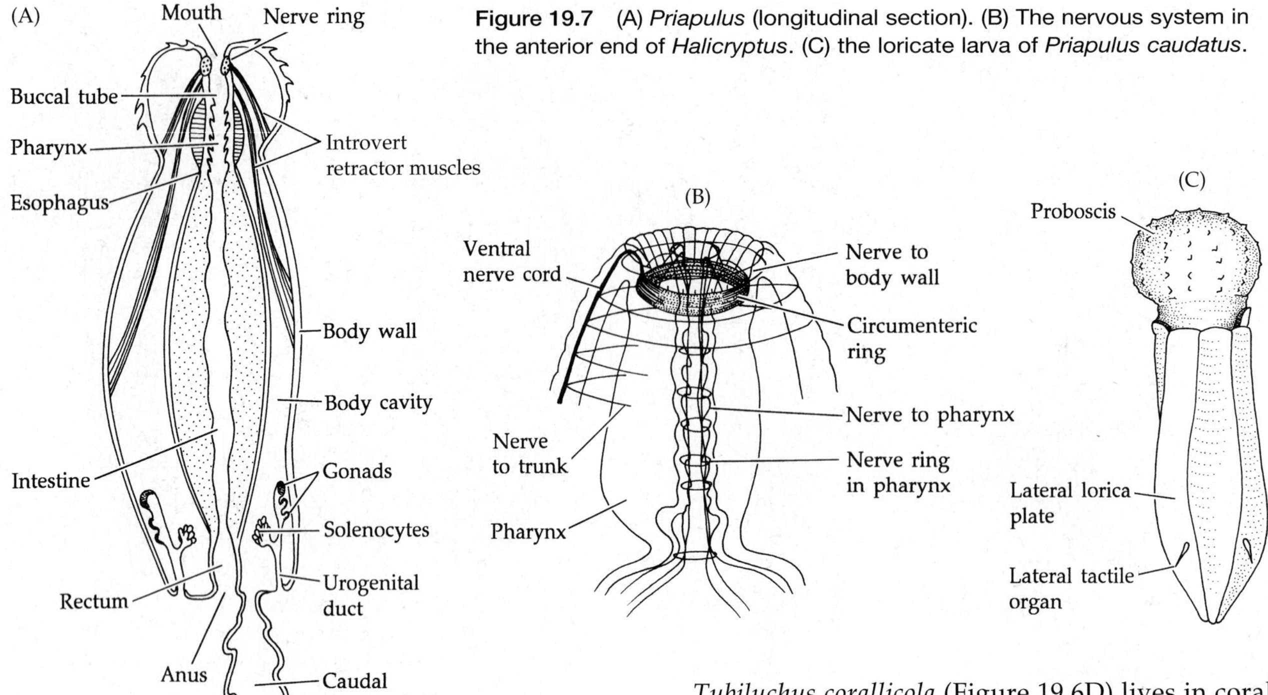

Figure 19.7 (A) *Priapulus* (longitudinal section). (B) The nervous system in the anterior end of *Halicryptus*. (C) the loricate larva of *Priapulus caudatus*.

Adult priapulans have a large body cavity that may extend into the caudal appendage. The body cavity is lined by a simple noncellular membrane secreted by surface cells on the retractor muscles, and is therefore a blastocoel (blastocoelom). The only exception from this is the meiofaunal species *Meiopriapulus fijiensis* that does possess a small epithelium-lined compartment around the foregut, and this cavity is apparently a true coelom (Storch et al. 1989). The fluid of the body cavity contains motile phagocytic amebocytes and free erythrocytes with hemerythrin.

Maintenance of body form and support are provided by the hydrostatic skeleton of the body cavity. The contraction of circular muscles around this cavity also facilitates protrusion of the introvert by increasing the internal pressure. Priapulans that move through the substratum do so largely by peristaltic burrowing, probably using the various hooks and other cuticular extensions to hold one part of the body in place while the rest is pushed or pulled along. *Maccabeus* is thought to use its ring of posterior cuticular spines for anchorage within its burrow (Figure 19.6E).

Feeding and Digestion

The majority of priapulans (i.e., members of the family Priapulidae) live in soft sediments and prey on soft-bodied invertebrates such as polychaetes. During feeding, a portion of the toothed, cuticle-lined pharynx is everted through the mouth at the end of the extended introvert. As the prey is grasped, the pharynx is inverted; the introvert then retracts and the prey is drawn into the gut.

Tubiluchus corallicola (Figure 19.6D) lives in coral sediments and feeds on organic detritus. The pharynx is lined with pectinate teeth, which the animal uses to sort food material from the coarse sediment particles. *Maccabeus tentaculatus*, a tube dweller, is a trapping carnivore. Eight short scalids, called trigger spines, arranged around the mouth opening are assumed to be touch-sensitive and these are in turn ringed by 25 branched scalids (Figure 19.6E). It is suspected that when a potential prey touches the trigger spines, the other scalids quickly close as a trap.

The digestive system is complete and either straight or slightly coiled (Figure 19.7A). The portion of the gut that lies roughly within the bounds of the introvert comprises the buccal tube, pharynx, and esophagus, all of which are lined with cuticle and together constitute a stomodeum. In members of the genus *Tubiluchus*, the stomodeum also includes a region behind the esophagus called a **polythridium**, which bears a circlet of two rows of plates and may be used to grind food. The midgut or intestine is the only endodermally derived section of the digestive tract and is followed by a short proctodeal rectum. The anus is located at the posterior end of the abdomen, either centrally or slightly to one side. Nothing is known about the digestive physiology of priapulans, although it is likely that digestion and nutrient absorption occur in the midgut.

Circulation, Gas Exchange, Excretion, and Osmoregulation

Internal transport takes place through diffusion and movement of the fluid in the body cavity. The presence of the respiratory pigment hemerythrin in the body fluid cells suggests oxygen transport or storage functions, and many priapulans are known to live in marginally anoxic muds. In those species with a vesiculate caudal appendage, the lumen of that structure is continuous with the main body cavity and such caudal appendages may function as gas exchange surfaces.

Clusters of solenocyte protonephridia lie in the posterior portion of the body cavity and are associated with the gonads as a urogenital system or complex (Figure 19.7A). Priapulan protonephridia are possibly unique in being composed of two or more terminal cells. A pair of urogenital pores opens near the anus. Priapulans inhabit both hyper- and hyposaline environments, so the protonephridia may function in both osmoregulation and excretion.

Nervous System and Sense Organs

The priapulan nervous system is intraepidermal and is constructed for the most part on a radial plan within the cylindrical body (Figure 19.7B). The brain is ring shaped and wrapped around the buccal tube in the anterior part of the introvert. The main ventral nerve cord arises from this ring and gives off a series of ring nerves and peripheral nerves along the body. In addition, longitudinal nerves extend from the main nerve ring along the inner pharyngeal lining and are connected by the ring commissures.

Priapulans have several types of sensory structures. All introvert scalids, except the locomotory scalids in *Meiopriapulus fijiensis*, probably have a sensorial function, and smaller sensory structures—the so-called **flosculi**—are distributed over the trunk region. A flosculus is a sensory structure that includes a central sensory cilium surrounded by a circle or collar of seven modified microvilli. Besides scalids and flosculi, different setae, tubuli, and anal hooks may have receptor cells and hence serve a sensory function.

Reproduction and Development

Priapulans are gonochoristic, although males are unknown in *Maccabeus tentaculatus*. The reproductive organs are similarly placed and connected in both sexes. The paired gonads are drained by genital ducts, which are joined by collecting tubules from the protonephridia to form a pair of urogenital ducts exiting posteriorly through urogenital pores (Figure 19.7A).

Priapulans free-spawn (first the males and then the females), and fertilization is external. Cleavage is holoblastic, appears to be radial, and results in a coeloblastula (some accounts differ) that undergoes invagination. Gene expression studies on developing embryos of *Priapulus caudatus* have led to the surprising discovery that the anus is formed directly from the blastopore, whereas the mouth opening is formed by a group of ectodermal cells at the animal hemisphere of the embryo, thus these animals have deuterostomous development. Gastrulation starts at the 64-cell stage with the invagination of the vegetal-most cells into a small blastocoel and the epibolic movement of the animal blastomeres toward the vegetal pole.

Direct development occurs only in *Meiopriapulus fijiensis*, in which the females brood the embryos. All other species develop through a series of loricate larvae (Figure 19.7C), which in some ways resemble loriciferans. The trunk of the larva is encased within a thick cuticular lorica into which the introvert can be withdrawn. The larvae live in benthic muds and are probably detritivores. The lorica is periodically shed as the larva grows; it is finally lost as the animal metamorphoses to a juvenile priapulan. At that time the caudal appendage forms in those species that possess this structure. The larvae of different genera vary somewhat in cuticular shape and ornamentation. Detailed larval development is best known for *Priapulus caudatus*. The newly hatched larva has a functional introvert, but mouth and anus are still closed, and no lorica is present yet. The lorica appears after the first molt, and additional scalids are added to the introvert. A mouth appears after the next molt, and during the following molts the trunk changes shape from being circular in cross section to have the flattened, bilateral symmetrical shape that typically are found in the *Priapulus* larvae (Wennberg et al. 2009). The exact number of larval stages is uncertain.

Phylum Loricifera: The Loriciferans

It may be apparent to you by now that interstitial habitats (the meiobenthic realm) are home for a host of bizarre and specialized creatures. Recent studies by two German zoologists, D. Waloszek and K. J. Müller, have revealed that a rich meiofaunal ecosystem was already in place as early as the upper Cambrian Era. Studies on modern meiofauna continue to reveal new animals, previously undescribed taxa, and myriad examples of convergent evolution associated with success in this environment. Among these recently described groups is the Loricifera, first named and described by Reinhardt Kristensen in 1983. The name Loricifera (Latin *lorica*, "corset"; *ferre*, "to bear") refers to the well-developed cuticular lorica encasing most of the body (Figure 19.8; Box 19C). The description of the phylum was initially based upon a single widespread species, *Nanaloricus mysticus*, but 34 other species have since been described. Most loriciferans have been found at depths of about 20–450 m in coarse marine sediments. One species, *Pliciloricus hadalis*, was collected in the western Pacific at a depth of over 8,000 m. Others, as yet undescribed, have been recorded from additional deep-sea muddy bottom locations. Most recently, a new type of larval form, called the Shira-larva, was described from more than 3,000 meters; it resembles Cambrian fossils from Australia. To date, all loriciferans have been placed in three families (Nanaloricidae, Pliciloricidae, and Urnaloricidae) in a single order, Nanaloricida. All

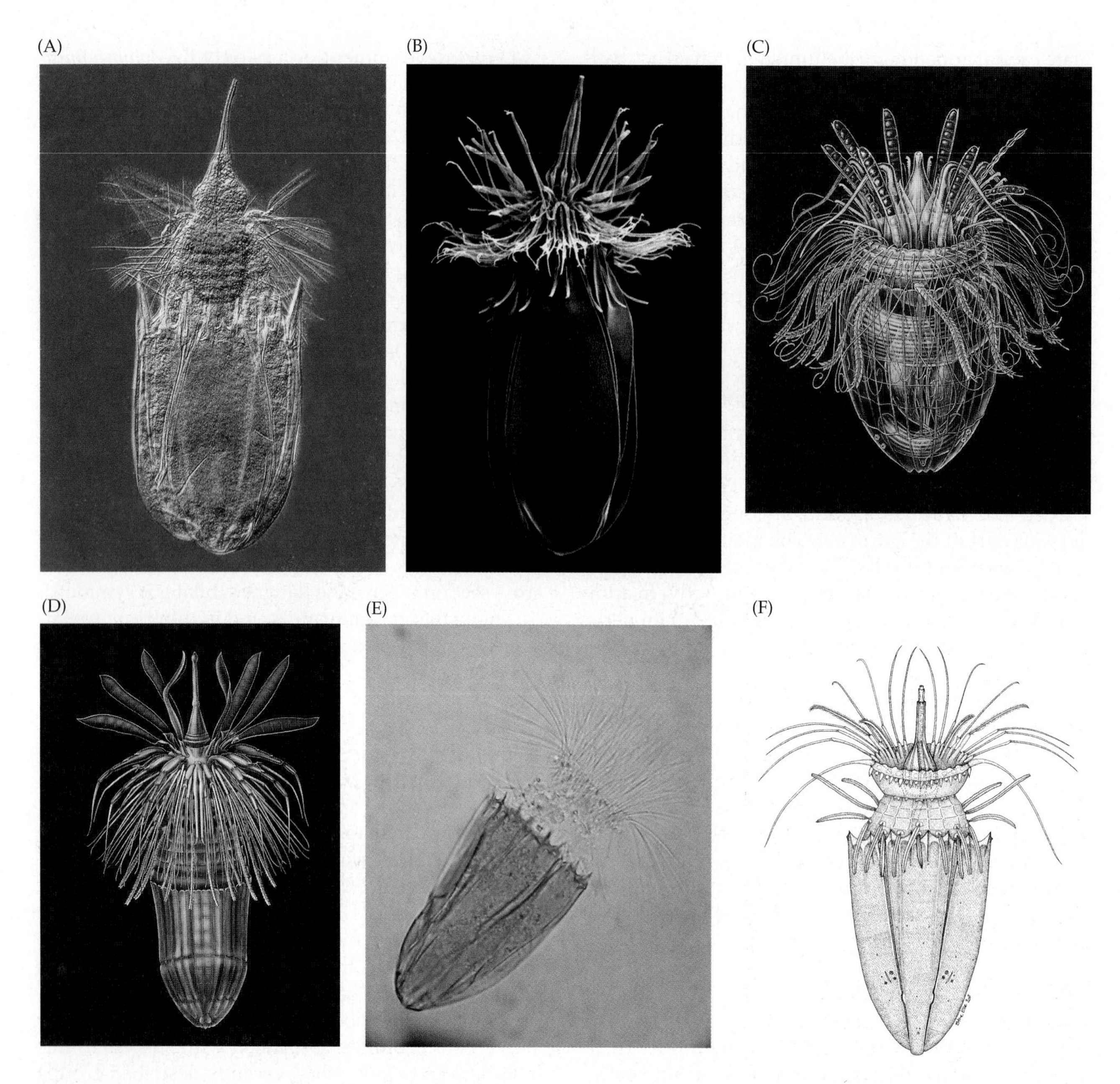

Figure 19.8 Representative loriciferans. (A) *Nanloricus mysticus*, the first described loriciferan (photograph of the male allotype specimen for the species, from Roscoff, France). (B) *Nanloricus* sp. (an undescribed species from northwest France). (C) *Pliciloricus enigmaticus* (male), from the U.S.A. (D) *Pliciloricus gracilis*, male, from the U.S.A. (E,F) *Spinoloricus cinziae*, dorsal views in micrograph and line drawing. This is the first known metazoan living in an anoxic environment; from a deep basin in the Mediterranean Sea.

are free living. Figures 19.8 and 19.9 provide overviews of loriciferan anatomy.

In 2010, three species of loriciferans (all new to science) were discovered living and reproducing in a fully anoxic, high sulfide, hypersaline basin in the deep Mediterranean Sea. Although prokaryotes and a few protists have long been known to spend their entire life in anoxic conditions, this was the first report of a metazoan proven to do so and to have obligate anaerobic metabolism. All three species lack mitochondria, but have hydrogenosomes in their cells, as seen in some ciliate protists. In this deep basin, these loriciferans occur in the highest abundance ever recorded for this phylum (up to 701 individuals per square meter). The new species *Spinoloricus cinziae* from the hypersaline basin L'Atalante was described in 2014.

The loriciferan body is minute (85–800 μm long; the giant is *Titaniloricus inexpectatovus*) but complex, containing over 10,000 cells. It is divided into a head (introvert and mouth cone), neck, thorax, and loricate abdomen. The head, neck, and most of the thorax can telescope into the lorica. The mouth is located at the end of an introvert, called the **mouth cone**, that projects from the head and contains 6 protrusible **oral stylets** in some species. Furthermore, the mouth cone also

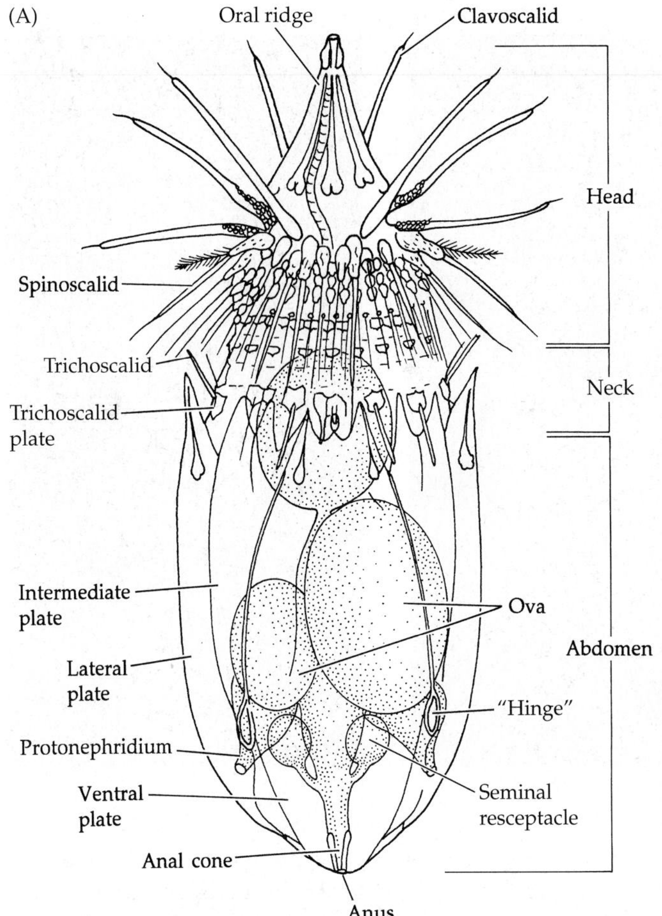

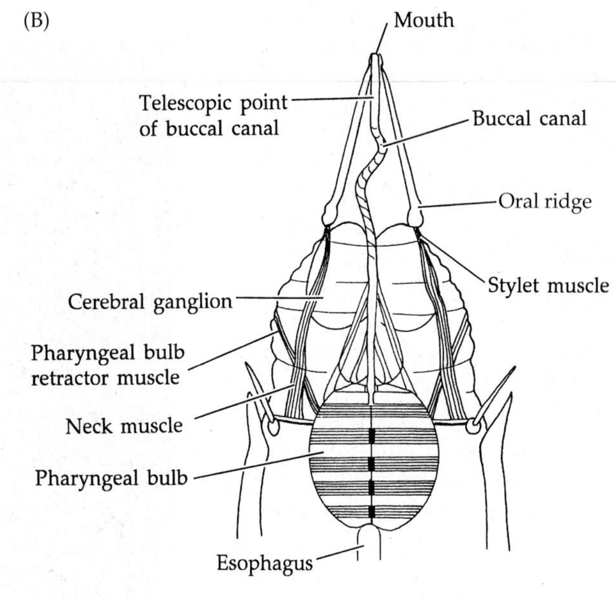

Figure 19.9 Anatomy of Loricifera. (A) An adult female loriciferan, *Nanaloricus mysticus* (ventral view). (B) The anterior end of *Nanaloricus mysticus*.

can have up to 8 **oral ridges**, each with a furca at the base where the 8 muscles attach. Nine rings of spine-like **scalids** (200–400 scalids in all) of various shapes also protrude from the spherical head, most apparently with intrinsic muscles. In the pliciloricids, some of these projections bear joints near their bases, reminiscent of articulated limbs. The first ring consists of 8 anteriorly directed **clavoscalids**; the remaining 8 rings of **spinoscalids** are directed posteriorly. In at least some species of Nanaloricidae, the number of clavoscalids differs between males and females. The males may have 6 of the clavoscalids each split into 3 branches, resulting in a total of up 20 clavoscalids. The clavascalids seems to be chemoreceptors, while the spinoscalids both are locomotory and mechanoreceptors. The 15 trichoscalids of the neck can be double or single. Recently it was discovered that the trichoscalids are attached to large cross-striated muscles, and in family Nanaloricidae the adults use them to "jump" more than three times their own length, whereas in the family Pliciloricidae the trichoscalids maybe used for swimming.

The lorica comprises 6 to 10 plates in the family Nanaloricidae, which bear anteriorly directed spines around the base of the neck. In the family Pliciloricidae there are no plates, but there are cuticular plicae. In the genus *Pliciloricus* there are about 22 plicae, but in the genera *Rugiloricus* and *Titaniloricus* there are 30 to 60 plicae. Beneath the cuticle and the body wall are several muscle bands, including those responsible for retraction of the anterior parts. The body cavity is presumably a blastocoelom (as is typical of the Scalidophora)

BOX 19C Characteristics of the Phylum Loricifera

1. Triploblastic, bilaterally symmetrical, unsegmented, never vermiform
2. Body cavity lining probably not peritoneal; presumably blastocoelomate
3. Body divided into a head (with a mouth cone and an introvert), neck, and thorax, and loricate abdomen; the head-neck-thorax telescopes into the lorica
4. Mouth located on mouth cone, beset with spines (stylets); head (introvert) and neck with 7–9 rings of scalids
5. Gut complete
6. No apparent circulatory or gas exchange systems
7. One pair of protonephridia, situated in gonads
8. Gonochoristic, parthenogenic (neotenous larvae, or hermaphroditic; development always includes a unique, toed larva
9. All inhabit marine interstitial environments or deep-sea muds

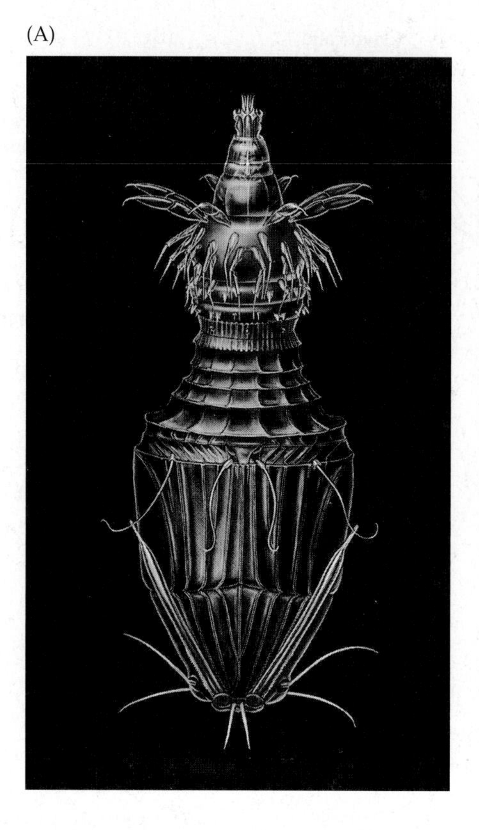

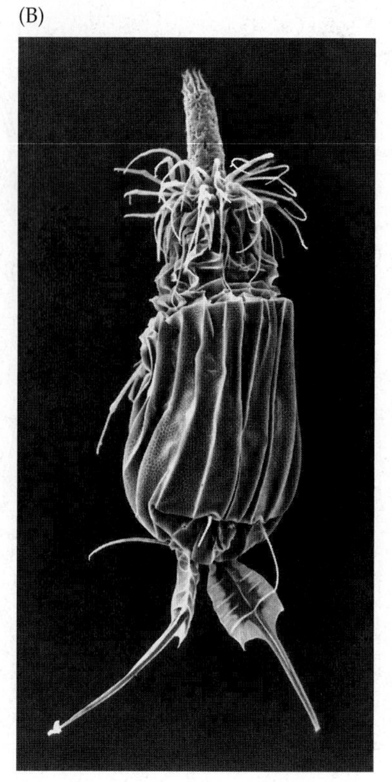

Figure 19.10 The Higgins larvae of two loriciferans. (A) Higgins larva of *P. gracilis* (ventral view; note the oral stylets and the tubular toes). (B) The Higgins larva of *Armorloricus elegans* (dorsal view; note the toes have mucrones for swimming).

and varies from rather spacious in pliciloricids to virtually absent in *Nanaloricus* (which would thus be acoelous in the adult). Two muscularized **toes** occur caudally in the larva, on the abdomen, and function in both locomotion and adhesion. In the family Nanaloricidae, the Higgins larva has leaflike cuticular structures called **mucrones**. These structures may be involved in the swimming behavior of the larva.

The digestive tract is complete. A long, tubular, buccal canal leads from the mouth to a muscular pharynx bulb and esophagus (Figure 19.9B). The pharynx has both circular and longitudinal muscles. The lumen of these anterior gut structures is lined with cuticle. Behind the esophagus is a long midgut leading to a short, cuticle-lined rectum and an anus located on an anal cone. One pair of salivary glands is associated with the buccal tube. Little is known about feeding in loriciferans, but some apparently eat bacteria.

The central nervous system includes a large circumpharyngeal ganglion and a number of smaller ganglia associated with the various body regions and parts. A large ventral nerve cord also bears ganglia. At least some of the scalids are probably sensory in function; however, videos have shown that at least the trichoscalids are involved in a jumping behavior.

Loriciferans are gonochoristic, although one recently described species (*Rugiloricus renaudae*) is hermaphroditic. Males and females differ externally in the form and number of certain scalids. The male reproductive system comprises two dorsal testes in the abdominal body cavity, probably a blastocoelom. The female system includes a pair of ovaries and probably in a few species of *Nanaloricus* also 2 seminal receptacles. Fertilization is suspected to be internal. Loriciferans have one or two pairs of monoflagellate protonephridia that are uniquely located within the gonads.

Little is yet known about early development in loriciferans. In most species, a feeding, benthic **Higgins larva** develops. This larva (Figure 19.10) is built along the same general body plan as the adult, but it possesses a pair of "toes" at the posterior end that are used for locomotion. These toes are thought to also have adhesive glands at their bases. Early stages apparently contain yolk reserves in the cells of the body cavity. The cuticle is periodically shed as the individuals grow in size. In some species the larva metamorphoses to a so-called "postlarva" (juvenile), which resembles an adult female but lacks ovaries. In others, larval metamorphosis involves a molt directly to an adult.

The original discovery of loriciferans was a fortunate event because the first species described, *Nanaloricus mysticus* (Nanaloricidae), proved to have the simplest life cycle in the phylum with the only life stages being the Higgins larva, free-living postlarva and adult. In the second described genus, *Rugiloricus*, a strongly modified postlarval stage was discovered (in *R. cauliculus*). Successive complexity was added to loriciferan life cycles with the discovery of parthenogenetic phases, and then with the newly discovered Urnaloricidae, which lack free adults. And we now know that deep-sea species hasten their life cycle by producing neotenous larvae, which produce from 2 to 4 additional larvae by parthenogenesis. The larvae of some loriciferans encyst for a period of time before metamorphosis, and this process might include a new type of larva—the "ghost larva." The newest discovered loops in life cycle are shown in Figure 19.11. The species *Rugiloricus manuelae* lacks free-living postlarvae, similar to what is seen in the genus *Pliciloricus*. The larvae of some loriciferans encyst for a period of time before metamorphosis. Figure 19.11 shows a typical life cycle for species in the family Pliciloricidae.

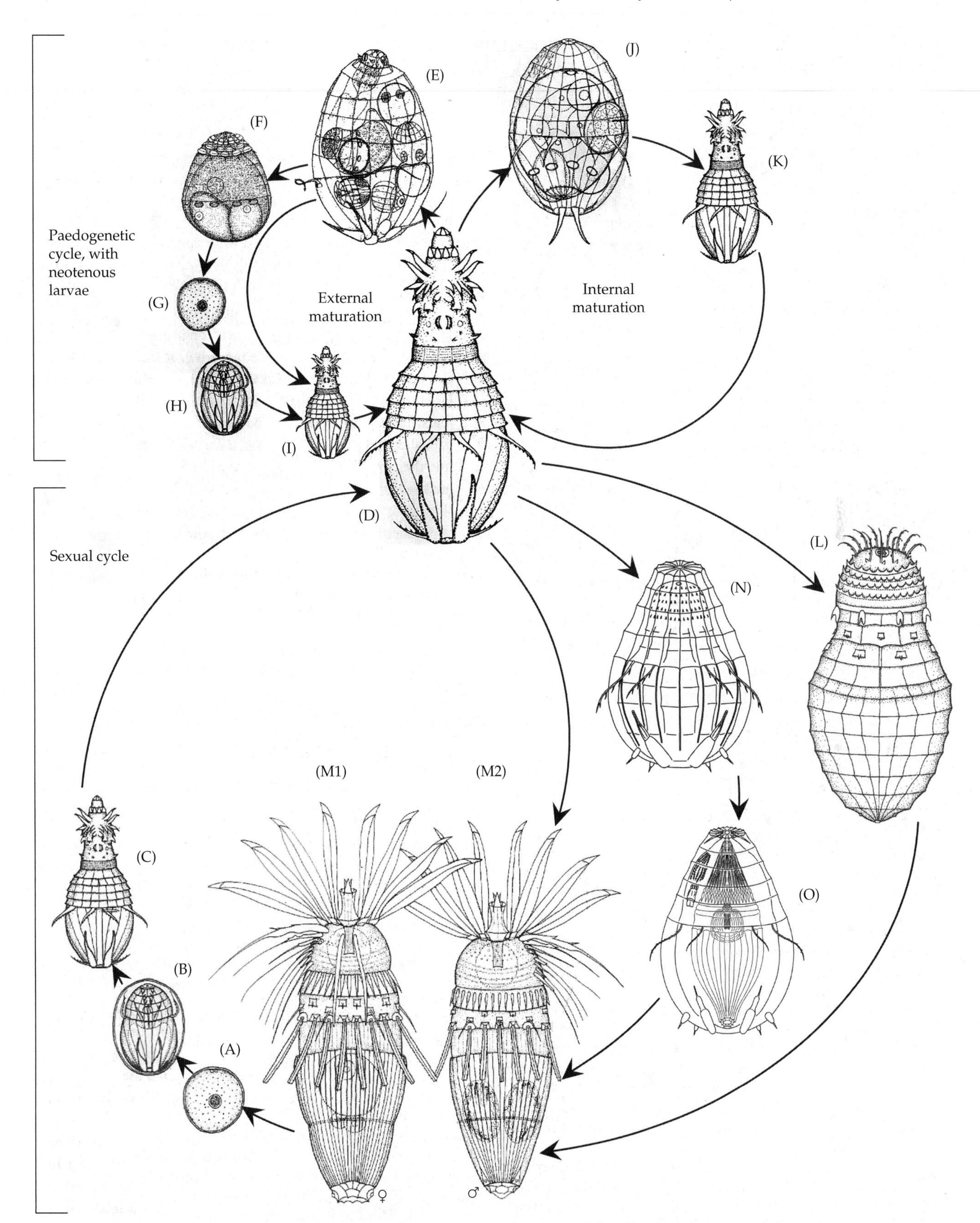

Figure 19.11 The life cycle of Pliciloricidae, with all the newly discovered "loops" in the life history. (A) Egg. (B) Embryo in the egg. (C) First instar Higgins larva. (D) Late instar Higgins larva. (E) Penultimate larva with ghost larva and embryos inside. (F) Free ghost larva. (G) Egg. (H) Embryo in egg. (I) First instar Higgins larva. (J) Penultimate larva with ghost larva and embryos inside. (K) First instar Higgins larva. (L) Free postlarva. (M1) Adult female. (M2) Adult male. (N) Last instar Higgins larva with postlarva inside. (O) Last instar Higgins larva with post-larva and young adult inside

Selected References

General References

Adrianov, A. V. and V. V. Malakhov. 1999. Cephalorhyncha of the World Ocean. KMK Scientific Press, Moscow.

Borner, J., P. Rehm, R. O. Schill, I. Ebersberger and T. Burmester. 2014. A transcriptome approach to ecdysozoan phylogeny. Molecular Phylogenetics and Evolution 80: 79–87.

Danovaro, R., A. Dell'Anno, A. Pusceddu, C. Gambi, I. Heiner and R. M. Kristensen. 2010. The first Metazoa living in permanently anoxic conditions. BMC Biology 8: 30.

Dong, X. and 10 others. 2010. The anatomy, taphonomy, taxonomy and systematic affinity of *Markuelia*: Early Cambrian to Early Ordovician scalidophorans. Palaeontology 53: 1291–1314.

Dunn, C. W., G. Giribet, G. D. Edgecombe and A. Hejnol. 2014. Animal phylogeny and its evolutionary implications. Annu. Rev. Ecol. Evol. Syst. 45: 371–395.

Edgecombe, G.D. and 8 others. 2011. Higher-level metazoan relationships: recent progress and remaining questions. Org. Divers. Evol. 11: 151–172.

Giese, A. C. and J. S. Pearse. 1974. *Reproduction of Marine Invertebrates, Vol. 1, Acoelomate and Pseudocoelomate Metazoans*. Academic Press, New York.

Higgins, R. P. and H. Thiel (eds.). 1988. *Introduction to the Study of Meiofauna*. Smithsonian Institution Press, Washington, DC.

Hyman, L. H. 1951. *The Invertebrates, Vol. 3, Acanthocephala, Aschelminthes, and Entoprocta*. McGraw-Hill, New York.

Kristensen, R. M. 1995. Are Aschelminthes pseudocoelomate or acoelomate? Pp. 41–43 in G. Lanzavecchia, R. Valvassori and M. D. Candia Carnevali (eds.), *Body Cavities: Function and Phylogeny*. Selected Symposia and Monographs U. Z. I.

Maas, A., D. Waloszek, J. T. Haug and K J. Müller. 2009. Loricate larvae (Scalidophora) from the Middle Cambrian of Australia. Mem. Assoc. Australasian Paleontol. 37: 281–302.

Müller, K. J., D. Walossek and A. Zakharov. 1995. "Orsten" type phosphatized soft-integument preservation and a new record from the Middle Cambrian Kuonamka Formation in Siberia. Neus Jahrb. Geol. Paläontol. Abh. 197: 101–118.

Roberts, L. S. and J. Janovy, Jr. 1996. *Foundations of Parasitology*, 5th Ed. Wm. C. Brown, Chicago.

Rota-Stabelli, O. and 8 others. 2011. A congruent solution to arthropod phylogeny: phylogenomic, microRNAs and morphology support monophyletic Mandibulata. Proceedings of the Royal Society B, 278: 298–306.

Ruppert, E. E. 1991. Introduction to the aschelminth phyla: A consideration of mesoderm, body cavities, and cuticle. Pp.1–17 in F. W. Harrison and E. E. Ruppert (eds.), *Microscopic Anatomy of Invertebrates, Vol. 4, Aschelminthes.* Wiley-Liss, New York.

Schmidt-Rhaesa, A. (ed.) 2013. *Handbook of Zoology. Gastrotricha, Cycloneuralia and Gnathifera. Vol. 1: Nematomorpha, Priapulida, Kinorhyncha, Loricifera.* De Gruyter, Berlin/Boston.

Schmidt-Rhaesa, A., T. Bartolomaeus, C. Lemburg, U. Ehlers and J. R. Garey. 1998. The position of the Arthropoda in the phylogenetic system. J. Morphol. 238: 263–285.

Sørensen, M.V., M. B. Hebsgaard, I. Heiner, H. Glenner, E. Willerslev and R. M. Kristensen. 2008. New data from an enigmatic phylum: evidence from molecular sequence data supports a sister-group relationship between Loricifera and Nematomorpha. J. Zool. Syst. Evol. Res. 46: 231–23.

Telford, M. J., S. J. Bourlat, A. Economou, D. Papillon and O. Rota-Stabelli. 2008. The evolution of the Ecdysozoa. Phil. Trans. Roy. Soc. B, Biol. Sci. 363: 1529–1537.

Wilson, R. A. and L. A. Webster. 1974. Protonephridia. Biol. Rev. 49: 127–160.

Kinorhyncha

Adrianov, A. V. and V. V. Malakhov. 1994. *Kinorhyncha: Structure, Development, Phylogeny and Taxonomy*. Nauka Publishing. [In Russian.]

Brown, R. 1988. Morphology and ultrastructure of the sensory appendages of a kinorhynch introvert. Zool. Scr. 18: 471–482.

Dal Zotto, M., M. Di Domenico, A. Garraffoni and M. V. Sørensen. 2013. *Franciscideres* gen. nov.—a new, highly aberrant kinorhynch genus from Brazil, with an analysis of its phylogenetic position. Syst. Biodiv. 11: 303–321.

Herranz, M., M. J. Boyle, F. Pardos and R. C. Neves. 2014. Comparative myoanatomy of *Echinoderes* (Kinorhyncha): a comprehensive investigation by CLSM and 3D reconstruction. Front. Zool. 11: 31.

Herranz, M., F. Pardos and M. J. Boyle. 2013. Comparative morphology of serotonergic-like immunoreactive elements in the central nervous system of kinorhynchs (Kinorhyncha, Cyclorhagida). J. Morphol. 274: 258–274.

Higgins, R. P. 1974. Kinorhynchs. Pp. 507–518 in A. G. Giese and J. S. Pearse (eds.), *Reproduction of Marine Invertebrates.* Academic Press, New York.

Kozloff, E. 2007. Stages of development, from first cleavage to hatching, of an *Echinoderes* (Phylum: Kinorhyncha: Class Cyclorhagida). Cah. Biol. Mar. 48: 199–206.

Kristensen, R. M. and R. P. Higgins. 1991. Kinorhyncha. Pp. 378–404 in F. W. Harrison and E. E. Ruppert (eds.), *Microscopic Anatomy of Invertebrates, Vol. 4, Aschelminthes.* Wiley-Liss, New York.

Müller, M. C. M., A. Schmidt-Rhaesa. 2003 Reconstruction of the muscle system in *Antygomonas* sp. (Kinorhyncha, Cyclorhagida) by means of phalloidin labeling and cLSM. J. Morphol. 256: 103–110.

Neuhaus, B. 1994. Ultrastructure of alimentary canal and body cavity, ground pattern, and phylogenetic relationships of the Kinorhyncha. Microfauna Mar. 9: 61–156.

Neuhaus, B. 1995. Postembryonic development of *Paracentrophyes praedictus* (Homalorhagida): Neoteny questionable among the Kinorhyncha. Zool. Scr. 24: 179–192.

Neuhaus, B. 2013. 5. Kinorhyncha (= Echinodera). Pp. 181–348 in Schmidt-Rhaesa, A. (ed.), *Handbook of Zoology. Gastrotricha, Cycloneuralia and Gnathifera. Vol. 1: Nematomorpha, Priapulida, Kinorhyncha, Loricifera.* De Gruyter, Berlin/Boston.

Sørensen, M. V. 2013. Phylum Kinorhyncha. Zootaxa 3703: 63–66.

Sørensen, M. V., G. Accogli and J. G. Hansen. 2010. Postembryonic development in *Antygomonas incomitata* (Kinorhyncha: Cyclorhagida). J. Morphol. 271: 863–882.

Sørensen, M. V., M. Dal Zotto, H. S. Rho, M. Herranz, N. Sánchez, F. Pardos and H. Yamasaki. 2015. Phylogeny of Kinorhyncha based on morphology and two molecular loci. PLoS ONE 10(7): e0133440.

Sørensen, M.V. and F. Pardos. 2008. Kinorhynch systematics and biology an introduction to the study of kinorhynchs, inclusive identification keys to the genera. Meiofauna Mar. 16, 21–73.

Yamasaki, H., S. F. Hiruta and H. Kajihara. 2013. Molecular phylogeny of kinorhynchs. Mol. Phyl. Evolut. 67: 303–310.

Priapula

Alberti, G. and V. Storch. 1988. Internal fertilization in a meiobenthic priapulid worm: *Tubiluchus philippinensis* (Tubiluchidae, Priapulida). Protoplasma 143: 193–196.

Calloway, C. B. 1975. Morphology of the introvert and associated structures of the priapulid *Tubiluchus corallicola* from Bermuda. Mar. Biol. 31: 161–174.

Calloway, C. B. 1988. Priapulida. Pp. 322–327 in P. Higgins and H. Thiel (eds.), *Introduction to the Study of Meiofauna*, Smithsonian Institution Press, Washington, DC.

Hammon, R. A. 1970. The burrowing of *Priapulus caudatus*. J. Zool. 162: 469–480.

Higgins, R. P. and V. Storch. 1989. Ultrastructural observations of the larva of *Tubiluchus corallicola* (Priapulida). Helgoländer Meeresuntersuchungen 43(1): 1–11.

Higgins, R. P. and V. Storch. 1991. Evidence for direct development in *Meiopriapulus fijiensis*. Trans. Am. Microsc. Soc. 110: 37–46.

Holger Roth, B. and A. Schmidt-Rhaesa. 2010. Structure of the nervous system in *Tubiluchus troglodytes* (Priapulida). Invert. Biol. 129: 39–58.

Huang, D. Y., J. Vannier and J. Y. Chen. 2004. Recent Priapulidae and their Early Cambrian ancestors: comparisons and evolutionary significance. Geobios 37: 217–228.

Joffe, B. I. and E. A. Kotikova. 1988. Nervous system of *Priapulus caudatus* and *Halicryptus spinulosus* (Priapulida). Proc. Zool. Inst. USSR Acad. Sci. 183: 52–77.

Lang, K. 1948. On the morphology of the larva of *Priapulus*. Arkiv. Zool. 41A, art. nos. 5 and 9.

Martín-Durán, J. M., R. Janssen, S. Wennberg, G. E. Budd and A. Hejnol. 2012. Deuterostome development in the protostome *Priapulus caudatus*. Curr. Biol. 22: 2161–2166.

Morris, S. C. 1977. Fossil priapulid worms. Palaeontol. Assoc. Lond. Spec. Pap. Palaeontol. 20: 1–95.

Morris, S. C. and R. A. Robison. 1985. Middle Cambrian priapulids and other soft-bodied fossils from Utah and Spain. Univ. Kansas Paleontol. Contr. 117: 1–22.

Oeschger, R. and K. B. Storey. 1990. Regulation of glycolytic enzymes in the marine invertebrate *Halicryptus spinulosus* (Priapulida) during environmental anoxia and exposure to hydrogen sulfide. Mar. Biol. 106(2): 261–266.

Oeschger, R. and R. D. Vetter. 1992. Sulfide detoxification and tolerance in *Halicryptus spinulosus* (Priapulida): A multiple strategy. Mar. Ecol. Prog. Ser. 86: 167–179.

Por, F. D. and H. J. Bromley. 1974. Morphology and anatomy of *Maccabeus tentaculatus* (Priapulida: Seticoronaria). J. Zool. 173: 173–197.

Rothe, B. H. and A. Schmidt-Rhaesa. 2010. Structure of the nervous system in *Tubiluchus troglodytes* (Priapulida). Invert. Biol. 129: 39–58.

Schmidt-Rhaesa, A. 2013. 4. Priapulida. Pp. 147–180 in Schmidt-Rhaesa, A. (ed.), *Handbook of Zoology. Gastrotricha, Cycloneuralia and Gnathifera. Vol. 1: Nematomorpha, Priapulida, Kinorhyncha, Loricifera.* De Gruyter, Berlin/Boston.

Schmidt-Rhaesa, A., B. H. Rothe and A. G. Martínez. 2013. *Tubiluchus lemburgi*, a new species of meiobenthic Priapulida. Zool. Anz. 253: 158–163.

Schreiber, A. and V. Storch. 1992. Free cells and blood proteins of *Priapulus caudatus* Lamarck (Priapulida). Sarsia 76: 261–266.

Shapeero, W. L. 1962. The epidermis and cuticle of *Priapulus caudatus* Lamarck. Trans. Am. Microsc. Soc. 81(4): 352–355.

Shirley, T. C. and V. Storch. 1999. *Halicryptus higginsi* n. sp. (Priapulida)—a giant new species from Barrow, Alaska. Invert. Biol. 118: 404–413.

Sørensen, M. V., H. S. Rho, W. Min and D. Kim. 2012. A new recording of the rare priapulid *Meiopriapulus fijiensis*, with comparative notes on juvenile and adult morphology. Zool. Anz. 251: 364–371.

Storch, V. 1991. Priapulida. Pp. 333–350 in F. W. Harrison and E. E. Ruppert (eds.), *Microscopic Anatomy of Invertebrates, Vol. 4, Aschelminthes*, Wiley-Liss, New York.

Storch, V., R. P. Higgins and M. P. Morse. 1989. Internal anatomy of *Meiopriapulus fijienses* (Priapulida). Trans. Am. Microsc. Soc. 108: 245–261.

Storch, V., R. P. Higgins and H. Rumohr. 1990. Ultrastructure of the introvert and pharynx of *Halicryptus spinulosus* (Priapulida). J. Morphol. 206: 163–171.

Storch, V., R. P. Higgins, P. Anderson and J. Svavarsson. 1995. Scanning and transmission electron microscopic analysis of the introvert of *Priapulopsis australis* and *Priapulopsis bicaudatus* (Pripulida). Invert. Biol. 114: 64–72.

van der Land, J. 1970. Systematics, zoogeography, and ecology of the Priapulida. Zool. Verh. Rijksmus. Nat. Hist. Leiden 112: 1–118.

van der Land, J. and A. Nørrevang. 1985. Affinities and intraphyletic relationships of the Priapulida. Pp. 261–273 in S. C. Morris et al. (eds.), *The Origins and Relationships of Lower Invertebrates.* Syst. Assoc. Spec. Vol. No. 28. Oxford Press.

Vannier, J., I. Calandra, C. Gaillard and A. Zylinska. 2010. Priapulid worms: pioneer horizontal burrowers at the Precambrian-Cambrian boundary. Geology 38: 711–714.

Webster, B. L., and 7 others. 2006. Mitogenomics and phylogenomics reveal priapulids worms as extant models of the ancestral ecdysozoan. Evol. Dev. 8: 502–510.

Webster, B. L., J. A. Mackenzie-Dodds, M. J. Telford and D. T. J. Littlewood. 2007. The mitochondrial genome of *Priapulus caudatus* Lamarck (Priapulida: Priapulidae). Gene 389: 96–105.

Welsch, U., R. Erlinger and V. Storch. 1992. Glycosaminoglycans and fibrillar collagen in Priapulida: a histo- and cytochemical study. Histochemistry 98: 389–397.

Wennberg, S. A., R. Janssen and G. E. Budd. 2008. Early embryonic development of the priapulids worm *Priapulus caudatus*. Evol. Dev. 10: 326–338.

Wennberg, S. A., R. Janssen and G. E. Budd. 2009. Hatching and earliest larval stages of the priapulid worm *Priapulus caudatus*. Invert. Biol. 128: 157–171.

Loricifera

Gad, G. 2004. A new genus of Nanaloricidae (Loricifera) from deepsea sediments of volcanic origin in the Kilinailau Trench north of Papua New Guinea. Helgoland Mar. Res. 58: 40–53.

Gad, G. 2005. Giant Higgins-larvae with paedogenetic reproduction from the deep sea of the Angola Basin: evidence for a new life cycle and for abyssal gigantism in Loricifera? Organisms, Diversity and Evolution 5, suppl. 1: 59–75.

Heiner, I. and R. M. Kristensen. 2008. *Urnaloricus gadi* nov. gen. et nov. sp. (Loricifera, Urnaloricidae nov. fam.), an aberrant Loricifera with a viviparous pedogenetic life cycle. J. Morphol. doi: 10.1002/jmor.10671

Higgins, R. P. and R. M. Kristensen. 1986. New Loricifera from southeastern United States coastal waters. Smithson. Contrib. Zool. 438: 1–70.

Higgins, R. P. and R. M. Kristensen. 1988. Loricifera. Pp. 319–321 in R. P. Higgins and H. Theil (eds.), *Introduction to the Study of Meiofauna.* Smithsonian Institution Press, Washington, DC.

Kristensen, R. M. 1983. Loricifera, a new phylum with Aschelminthes characters from the meiobenthos. Z. Zool. Syst. Evolutionsforsch. 21: 163–180.

Kristensen, R. M. 1991. Loricifera. Pp. 351–375 in F. W. Harrison and E. E. Ruppert (eds.), Microscopic Anatomy of Invertebrates, Vol. 4, Aschelminthes, Wiley-Liss, New York.

Kristensen, R. M. 1991. Loricifera: A general biological and phylogenetic overview. Vehr. Dtsch. Zool. Ges. 84: 231–246.

Kristensen, R. M., I. Heiner and R. P. Higgins. 2007. Morphology and life cycle of a new loriciferan from the Atlantic coast of Florida with an emended diagnosis and life cycle of Nanaloricidae (Loricifera). Invert. Biol. 126: 120–137

Kristensen, R. M., R. C. Neves and Gad, G. 2013. First report of Loricifera from the Indian Ocean: a new *Rugiloricus*-species represented by a hermaphrodite. Cah. Biol. Mar. 54: 161–171.

Kristensen, R. M. and Y. Shirayama. 1988. *Pliciloricus hadalis* (Pliciloricidae), a new loriciferan species collected from the Izu-Ogasawara Trench, Western Pacific. Zool. Sci. 5: 875–881.

Neves, R., X. Bailly, F. Leasi, H. Reichert, M. V. Sørensen and R. M. Kristensen. 2013. A complete three-dimensional reconstruction of the myoanatomy of Loricifera: comparative morphology of an adult and a Higgins larva stage. Front. Zool. 10: 19.

Neves, R. C., C. Gambi, R. Danovara and R. M. Kristensen. 2014. *Spinoloricus cinziae* (Phylum Loricifera) a new species from a hypersaline anoxic deep basin in the Mediteterranean Sea. Syst. Biodivers. 12 (4): 489–502.

Neves, R. C. and R. M. Kristensen. 2013. A new type of loriciferan larva (Shira larva) from the deep sea of Shatsky Rise, Pacific Ocean. Org. Divers. Evol. doi: 10.1007/s13127-013-0160-4

Neves, R.C., X. Bailly, F. Leasi, H. Reichert, M. V. Sørensen and R. M. Kristensen. 2013. A complete three-dimensional reconstruction of the myoanatomy of Loricifera: comparative morphology of an adult and a Higgins larva stage. Front. Zool. 10 (19): 1–21.

Pardos, F. and R. M. Kristensen. 2013. First record of Loricifera from the Iberian Peninsula, with the description of *Rugiloricus manuelae* sp. nov. (Loricifera, Pliciloricidae). Helgol. Mar. Res. doi: 10.1007/s10152-013-0349-0

Peel, J. S., M. Stein and R. M. Kristensen. 2013. Life cycle and morphology of a Cambrian stem-lineage loriciferan. PLoS ONE 8: e73583.

CHAPTER 20

The Emergence of the Arthropods

Tardigrades, Onychophorans, and the Arthropod Body Plan

Arthropods constitute 81.5% of all described living animal species. They are so abundant, so diverse, and play such vital roles in all of Earth's environments that we devote five chapters to them. The present chapter is divided into three parts, first introducing the close allies of the arthropods—the phyla Tardigrada (water bears) and Onychophora (velvet worms; *Peripatus* and their kin). Then we introduce the arthropods themselves, exploring the general body plan and basic unifying features of the phylum and how this combination of features has led to its preeminent success. Finally, we provide a brief overview of arthropod evolution. Detailed treatments of the living arthropod subphyla (Crustacea, Hexapoda, Myriapoda, Chelicerata) are provided in Chapters 21–24.

Classification of The Animal Kingdom (Metazoa)

Non-Bilateria*
(a.k.a. the diploblasts)
- PHYLUM PORIFERA
- PHYLUM PLACOZOA
- PHYLUM CNIDARIA
- PHYLUM CTENOPHORA

Bilateria
(a.k.a. the triploblasts)
- PHYLUM XENACOELOMORPHA

Protostomia
- PHYLUM CHAETOGNATHA

SPIRALIA
- PHYLUM PLATYHELMINTHES
- PHYLUM GASTROTRICHA
- PHYLUM RHOMBOZOA
- PHYLUM ORTHONECTIDA
- PHYLUM NEMERTEA
- PHYLUM MOLLUSCA
- PHYLUM ANNELIDA
- PHYLUM ENTOPROCTA
- PHYLUM CYCLIOPHORA

Gnathifera
- PHYLUM GNATHOSTOMULIDA
- PHYLUM MICROGNATHOZOA
- PHYLUM ROTIFERA

Lophophorata
- PHYLUM PHORONIDA
- PHYLUM BRYOZOA
- PHYLUM BRACHIOPODA

ECDYSOZOA

Nematoida
- PHYLUM NEMATODA
- PHYLUM NEMATOMORPHA

Scalidophora
- PHYLUM KINORHYNCHA
- PHYLUM PRIAPULA
- PHYLUM LORICIFERA

Panarthropoda
- **PHYLUM TARDIGRADA**
- **PHYLUM ONYCHOPHORA**
- **PHYLUM ARTHROPODA**
  - SUBPHYLUM CRUSTACEA*
  - SUBPHYLUM HEXAPODA
  - SUBPHYLUM MYRIAPODA
  - SUBPHYLUM CHELICERATA

Deuterostomia
- PHYLUM ECHINODERMATA
- PHYLUM HEMICHORDATA
- PHYLUM CHORDATA

*Paraphyletic group

The close relationship among the Onychophora, Tardigrada, and Arthropoda has rarely been questioned, and this clade is known as the **Panarthropoda**. Panarthropoda comprises one of the three great clades of Ecdysozoa, or the molting protostomes. The Panarthropoda arose in ancient Precambrian seas 550 to 600 million years ago, and the arthropods, in particular, have undergone a tremendous evolutionary radiation since then. There are over one million described living arthropod species, as compared with 1,200 tardigrades and 200 onychophorans (Table 20.1). For years it was thought that tardigrades were the sister group to the Arthropoda, but recent molecular phylogenies suggest that onychophorans are the sister group to Arthropoda, and that Tardigrada is the sister group to that pairing. However, the final word on this has probably not yet been written.

Today, arthropods occur in virtually every environment on Earth, exploiting every imaginable lifestyle (Figure 20.1C–J). Modern forms range in size from tiny mites and crustaceans less than 1 mm long to Japanese spider crabs (*Macrocheira kaempferi*) with leg spans that reach nearly 4 m. Our inadequate knowledge of Earth's biodiversity

---

The section on Onychophora has been revised by Gonzalo Giribet. The section on Tardigrada has been revised by Reinhardt Møbjerg Kristensen and Richard C. Brusca.

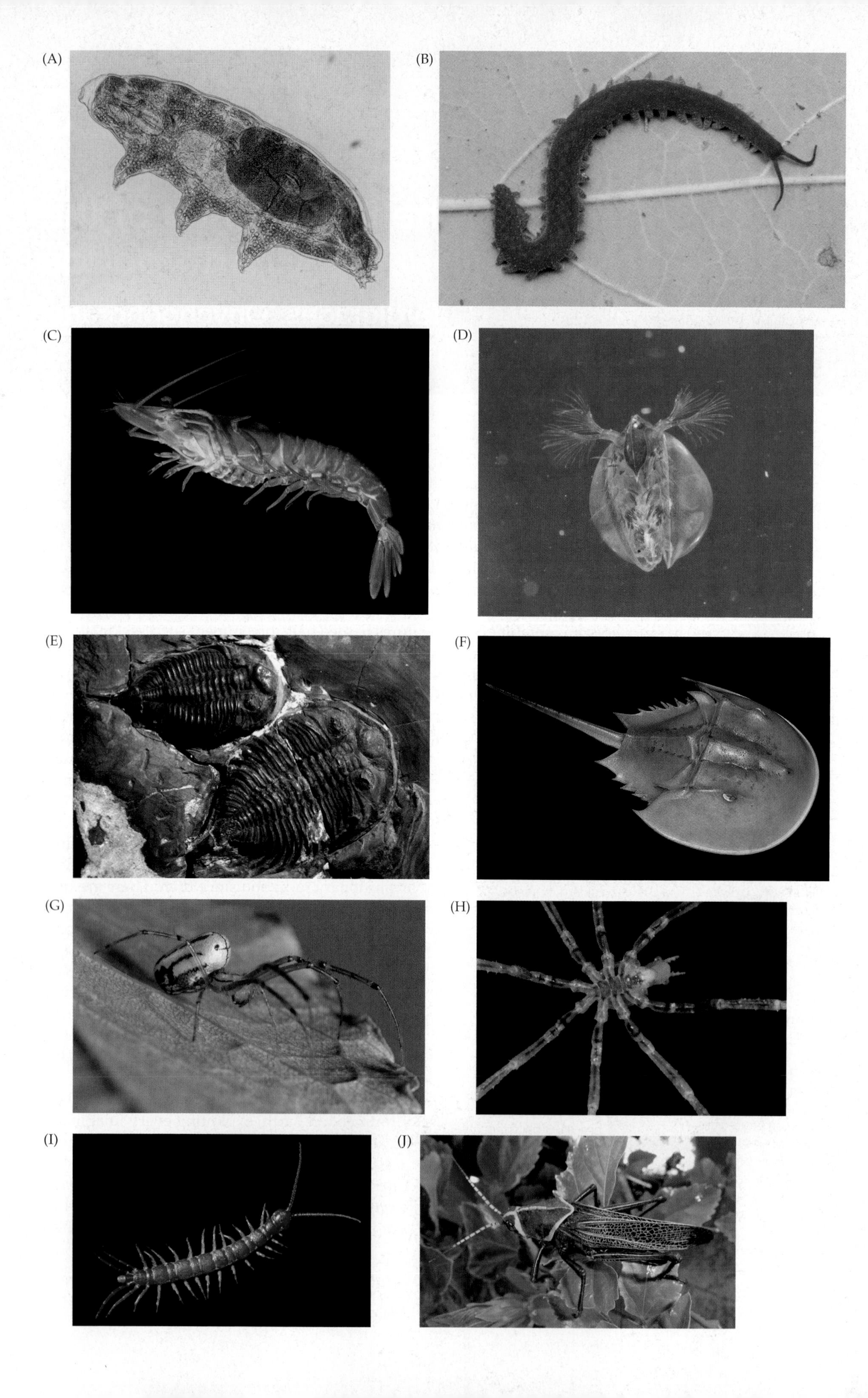

(A)
(B)
(C)
(D)
(E)
(F)
(G)
(H)
(I)
(J)

◀ **Figure 20.1 Examples of the Panarthropoda.** (A) Phylum Tardigrada; a water bear with developing oocytes. (B) Phylum Onychophora; a velvet worm. (C,D) Phylum Arthropoda, subphylum Crustacea. (C) The oceanic shrimp, *Sicyonia ingentis*. (D) A clam shrimp (order Diplostraca). (E) Phylum Arthropoda, subphylum Trilobita. (F,G,H) Phylum Arthropoda, subphylum Chelicerata. (F) A horseshoe crab, *Limulus* (Class Merostomata). (G) An orchard spider, *Leucauge* (a long-jawed orb-weaver; class Arachnida). (H) A pycnogonid (class Pycnogonida). (I) Phylum Arthropoda, subphylum Myriapoda. A lithobiomorphan centipede. (J) Phylum Arthropod, subphylum Hexapoda. A horse lubber grasshopper (*Taeniopoda eques*), from Arizona.

is apparent when we consider the estimates of numbers of *undescribed species* of arthropods, which range from 3 million to over 100 million species. Most of this undiscovered diversity resides among the insects and mites on land and the crustaceans in the sea. Whether we lean toward the conservative or the liberal estimates, we are struck by the reality that no other phylum approaches the magnitude of species richness seen in arthropods. One might say, the modern world belongs to the arthropods. Yet, despite their overwhelming diversity, the arthropods share a suite of fundamental similarities and a distinct unifying body plan that we describe in this chapter. But first, we begin our discussion with the less diverse, but equally charismatic relatives of the arthropods—the water bears and velvet worms.

BOX 20A Characteristics of the Phylum Tardigrada

1. With modified, or non-teloblastic segmentation; 4 pairs of legs
2. Malpighian tubules endodermally derived; without cuticle
3. Muscles are isolated bands.
4. With non-jointed, telescopic legs lacking intrinsic musculature
5. With a unique nerve connection between the first brain lobe and first ventral trunk ganglion, not seen in any other Metazoa
6. Body coelom greatly reduced and functioning as a hemocoel (defined coelomic compartments restricted to gonadal cavities)
7. With thin, uncalcified cuticle
8. Body wall without sheetlike circular muscle layer
9. Some with striated muscle
10. Mouth terminal or ventral
11. Without discrete blood vessels, gas exchange structures, and true metanephridia
12. Capable of pronounced anabiosis/cryptobiosis

Phylum Tardigrada

The first recorded observation of a tardigrade was by Eichhorn in 1767. Since then, about 1,200 species have been described. Fossil tardigrades are very rare, and until very recently the only described fossils were from North American Cretaceous amber—*Beorn leggi* from Canada and *Milnesium swolenskyi* from New Jersey (USA). But recently, rock fossils of tardigrades were discovered in the Lower Cambrian Chengjiang deposits of China and the mid-Cambrian Orsten deposits of Siberia, specimens from the latter having only three pairs of legs (Box 20A).

Recent phylogenomic analyses, as well as neuroanatomical studies, support the traditional view that tardigrades are part of the Panarthropoda clade. Although the phylogenetic position of tardigrades relative to Onychophora is still debated, the most recent evidence suggests Onychophora is the sister group to

TABLE 20.1 A Comparison of the Panarthropod Phyla

Phylum	Number of described species	Malpighian tubules	Muscles isolated as bands	Jointed legs	Compound eyes
Tardigrada	1,200	Present; endodermally derived	To some degree	No	No
Onychophora	200	Absent	Partly	No	No
Crustacea	70,000	Absent	Strongly	Yes	Yes
Hexapoda	About 926,990	Present; ectodermally derived	Strongly	Yes	Yes
Myriapoda	16,360	Present; ectodermally derived	Strongly	Yes	Yes
Chelicerata	113,335	Present; endodermally derived	Strongly	Yes	Yes
Trilobita (extinct)	>15,000	Unknown	Strongly	Yes	Yes

(A) (B)

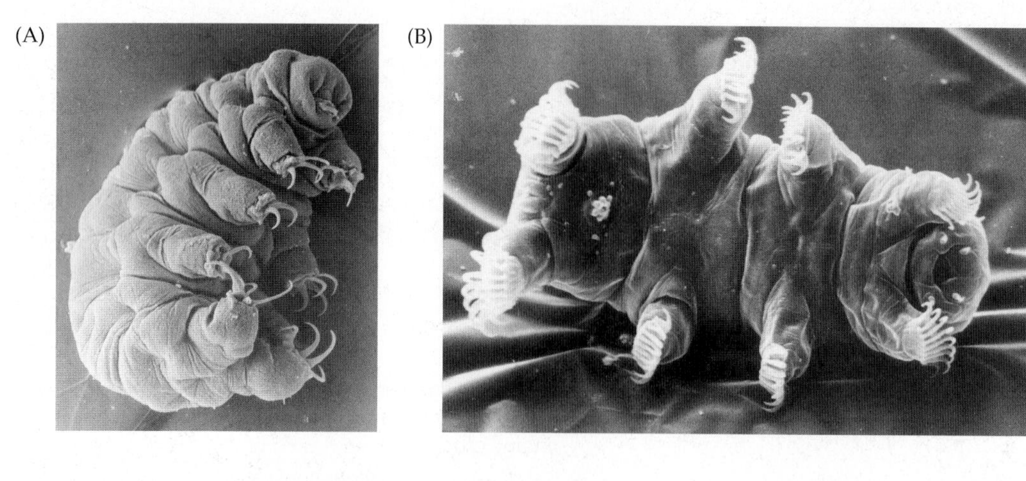

(C)

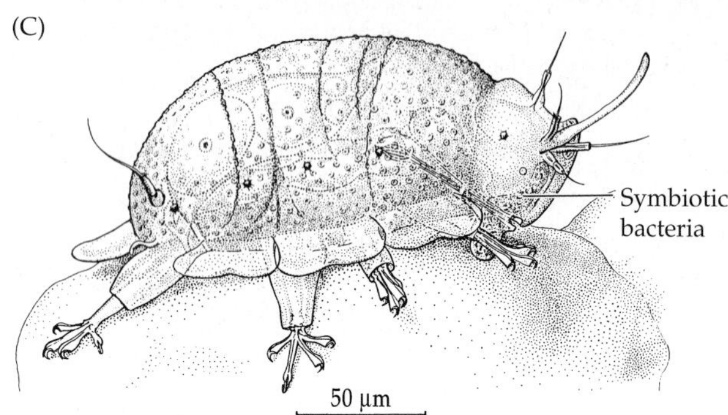

(D) (E)

(F)

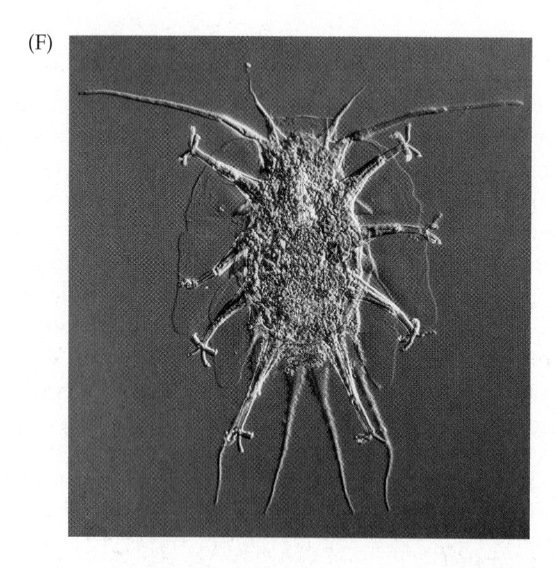

Figure 20.2 Phylum Tardigrada. Representative tardigrades. (A) *Halobiotus crispae*, a marine species common on brown algae in Greenland. This species undergoes a yearly cyclomorphosis involving a special hibernation stage (the pseudosimplex) during which it overwinters in the icy Greenland littoral zone. (B) *Echiniscoides sigismundi* (ventral view), a littoral species from Denmark. (C) *Wingstrandarctus corallinus*. (D,E) *Styraconyx qivitoq* (ventral and dorsal views), a tardigrade that lives on ectoprocts and has been collected only in Greenland. (F) A marine water bear, *Florarctus heimi* (Arthrotardigrada), from coal sand at Heron Island, Australia. A male with extremely long primary clavae (chemoreceptors).

Arthropoda, while Tardigrada is likely the sister group to that grouping.

Most living tardigrade species are found in semi-aquatic habitats such as the water films on mosses, lichens, liverworts, and certain angiosperms, or in soil and forest litter. Others live in various freshwater and marine benthic habitats, both deep and shallow, often interstitially or among shore algae. A few, very interesting species have been reported from hot springs. Some marine species are commensals on the pleopods of isopods or the gills of mussels; others are ectoparasites on the epidermis of holothurians or barnacles. Tardigrades occasionally occur in high densities, up to 300,000 per square meter in soil and more than 2,000,000 per square meter in moss. All are small, usually on the order of 0.1–0.5 mm in length, although some 1.3 mm giants have been reported.

Under the microscope, tardigrades resemble miniature eight-legged bears, especially as they move with a lumbering, ursine gait—hence the name Tardigrada (Latin *tardus*, "slow"; *gradus*, "step"). Their locomotion,

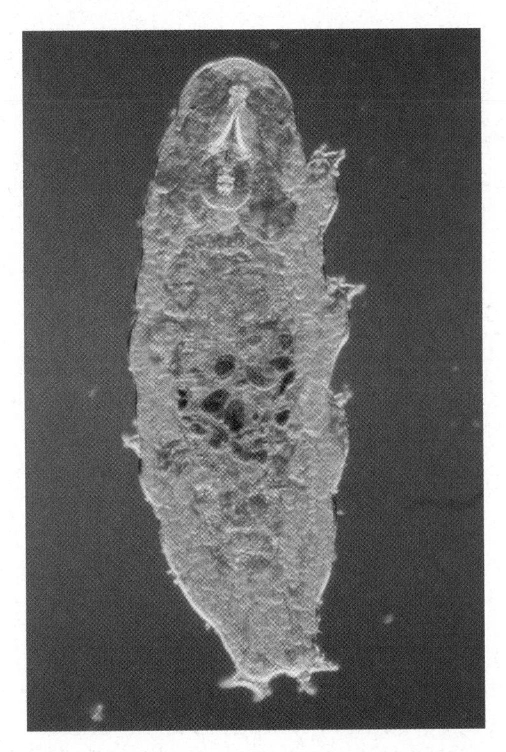

Figure 20.3 The giant eutardigrade *Richtersius coronifer*, from moss on the Swedish island of Öland. The cryptobiotic tun of this species survived experiments in outer space.

paunchy body, and clawed legs have earned them the nickname "water bears" (Figures 20.1A and 20.2–20.10).

Most terrestrial tardigrade species are widespread and many might be cosmopolitan. A major factor in their wide distribution is probably the fact that their eggs, cysts, and tuns (see below) are light enough and resistant enough to be carried great distances either by winds or on sand and mud clinging to the feet of insects, birds, and other animals. The minute sizes and precarious habitats of water bears have resulted in numerous traits also seen in some blastocoelomate groups that live in similar habitats.

Tardigrades are well known for their remarkable powers of **anabiosis** (a state of dormancy that involves greatly reduced metabolic activity during unfavorable environmental conditions) and **cryptobiosis** (an extreme state of anabiosis in which all external signs of metabolic activity are absent). Soil and fresh-water tardigrades, during dry periods when the ponds or vegetation they inhabit becomes desiccated, encyst by pulling in their legs, losing body water, and secreting a double-walled cuticular envelope around the shriveled body. Such anabiotic cysts maintain a very low basal metabolism. Further reorganization (or "deorganization") of the body can result in a single-walled **tun** stage, in which body metabolism is undetectable (a cryptobiotic state). The extreme expression cryptobiosis is the **anhydrobiosis** state, in which terrestrial tardigrades can survive as dry tuns for many years.

The resistant qualities of tardigrade tuns in anhydrobiosis have been demonstrated by experiments in which individuals have recovered after immersion in extremely toxic compounds such as brine, ether, absolute alcohol, and even liquid helium. They have survived temperatures ranging from +149°C to –272°C, on the brink of absolute zero. They have also survived high vacuums, intense ionizing radiation, and long periods with no environmental oxygen whatsoever. Following desiccation, when water is again available, the animals swell and become active within a few hours. Many rotifers, nematodes, mites, and a few insects are also known for their anabiotic powers, and these groups often occur together in the surface water films of plants such as mosses and lichens. Tardigrades are also found on the Greenland ice sheet, in the so-called cryoconite holes formed by stardust and dark terrestrial material such as soot and volcanic dust, where few other metazoans can survive. Experiments with tardigrades in anhydrobiosis have demonstrated they can even survive at least several months in the outer space. The BIOPAN 6 experiment in 2007 showed that tardigrades from mosses (Figure 20.3) could survive outside the spacecraft in anhydrobiosis for several months. In the most extreme case, embryos of *Milnesum tardigradum* kept for three months in the outer space hatch had 100% survival after rehydration.

One marine tardigrade genus (*Echiniscoides*) survives quite well with a life cycle that regularly alternates between active (Figure 20.2B) and tun stages, and it can even survive an experimentally induced cycle forcing it to undergo cryptobiosis every six hours! Evidence indicates that the tardigrade aging process largely ceases during cryptobiosis, and that by alternating active and cryptobiotic periods tardigrades may extend their life span to several decades. One rather sensational report described a dried Italian museum specimen of moss that yielded living tardigrades when moistened after 120 years on the shelf! However, this specimen died shortly later. Recent genomic analyses of tardigrades indicate this group has acquired a huge number of foreign genes into its own genome, by way of horizontal gene transfer, from various Eubacteria, Archaea, and even plants and fungi. These adopted genes, which may comprise as much as 17.5% of the tardigrades DNA, might play a role in the ability of water bears to withstand extreme stresses.

In certain areas of extreme environmental conditions, marine tardigrades may undergo an annual cycle of **cyclomorphosis** (rather than the cryptobiosis typical of terrestrial and freshwater forms). During cyclomorphosis, two distinct morphologies alternate. For example, *Halobiotus crispae*, a littoral species first found in Greenland, (Figures 20.2A and 20.4), has a summer morph and a winter morph. The latter is a special hibernation stage called the **pseudosimplex** that is resistant to freezing temperatures and perhaps

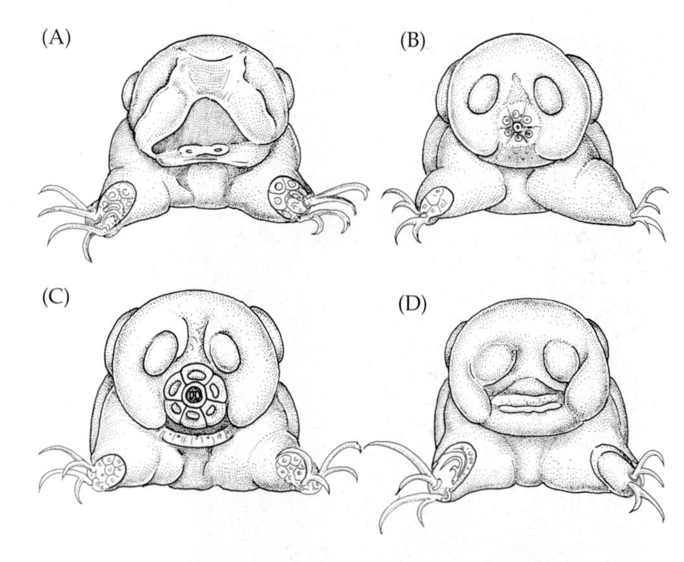

Figure 20.4 Cyclomorphosis in the marine tardigrade, *Halobiotus crispae* from Greenland; frontal views. (A) The hibernation stage, pseudosimplex 1 (winter form, with double cuticle). (B) Pseudosimplex 2 (spring form, with thin cuticle and small claws). (C) The sexually mature stage (summer form, with large claws). (D) The molting (simplex) stage, with closed mouth opening (molting animals can be found year round).

low salinities. In contrast to cryptobiotic tuns, the pseudosimplex is active and motile. Cyclomorphosis is coupled with gonadal development, and in Greenland only the summer morph is sexually mature.

The phylum Tardigrada comprises 22 families in three classes: Heterotardigrada, Mesotardigrada, and Eutardigrada. The classes are defined largely on the basis of the details of the head appendages, the nature of the leg claws, and the presence or absence of "Malpighian tubules." The class Mesotardigrada is interesting because the single species *Thermozodium esakii* is known only from the hot springs of Unzen Park, near Nagasaki, Japan (and it has not been found since

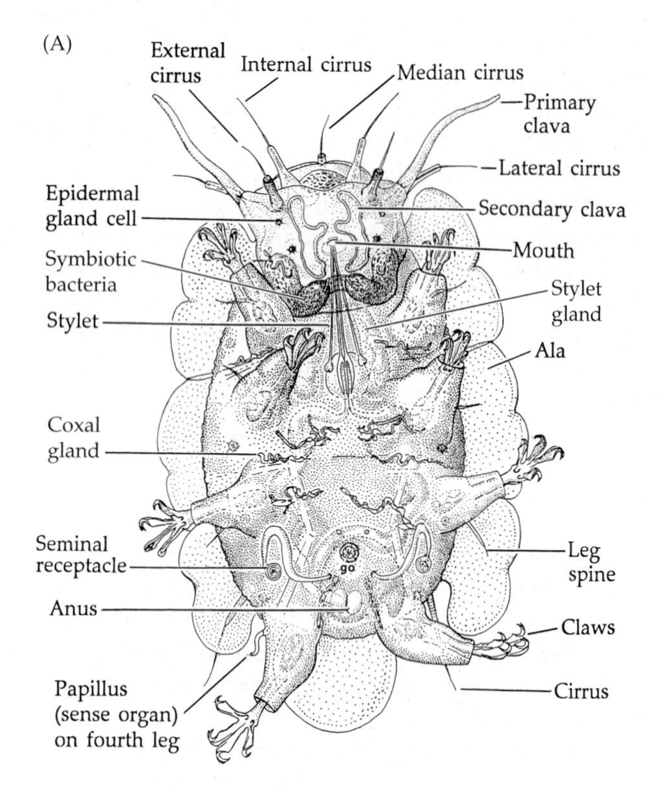

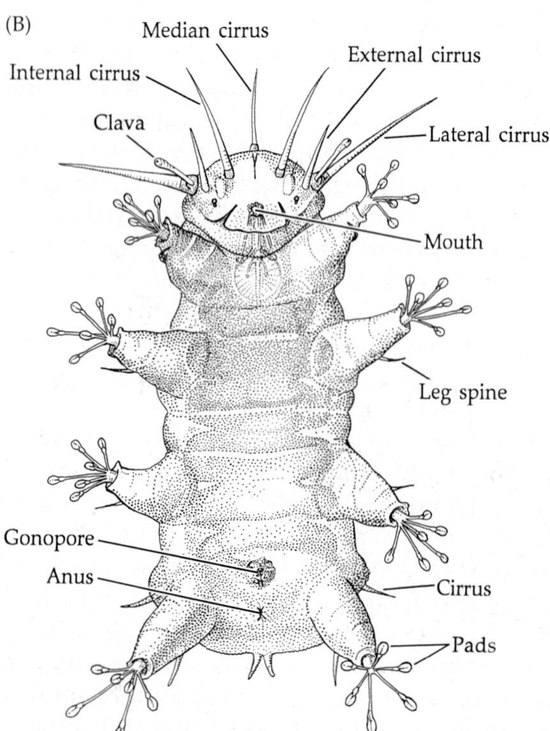

Figure 20.5 Tardigrade anatomy. (A) *Wingstrandarctus corallinus* (ventral view), an inhabitant of shallow, sandy marine habitats in Australia and Florida. (B) *Batillipes noerrevangi* (ventral view). (C) Generalized *Echiniscus* (dorsal view).

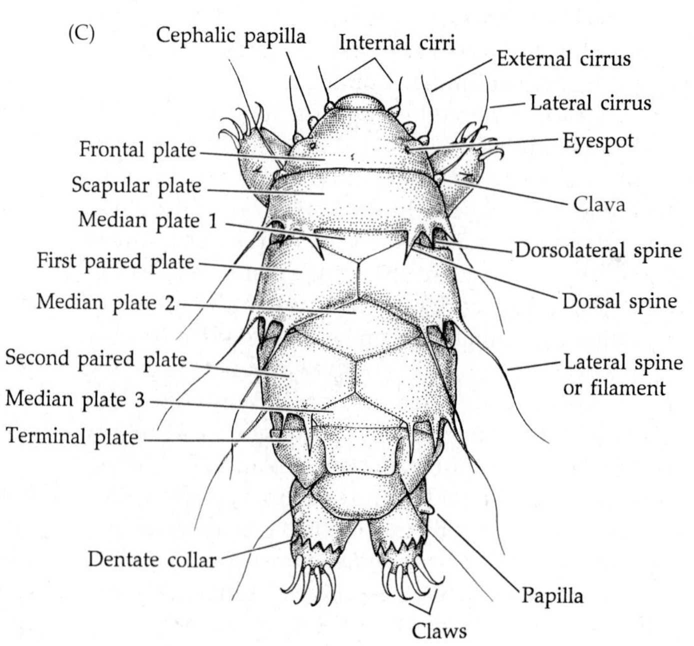

Figure 20.6 Body plan of a generalized tardigrade.

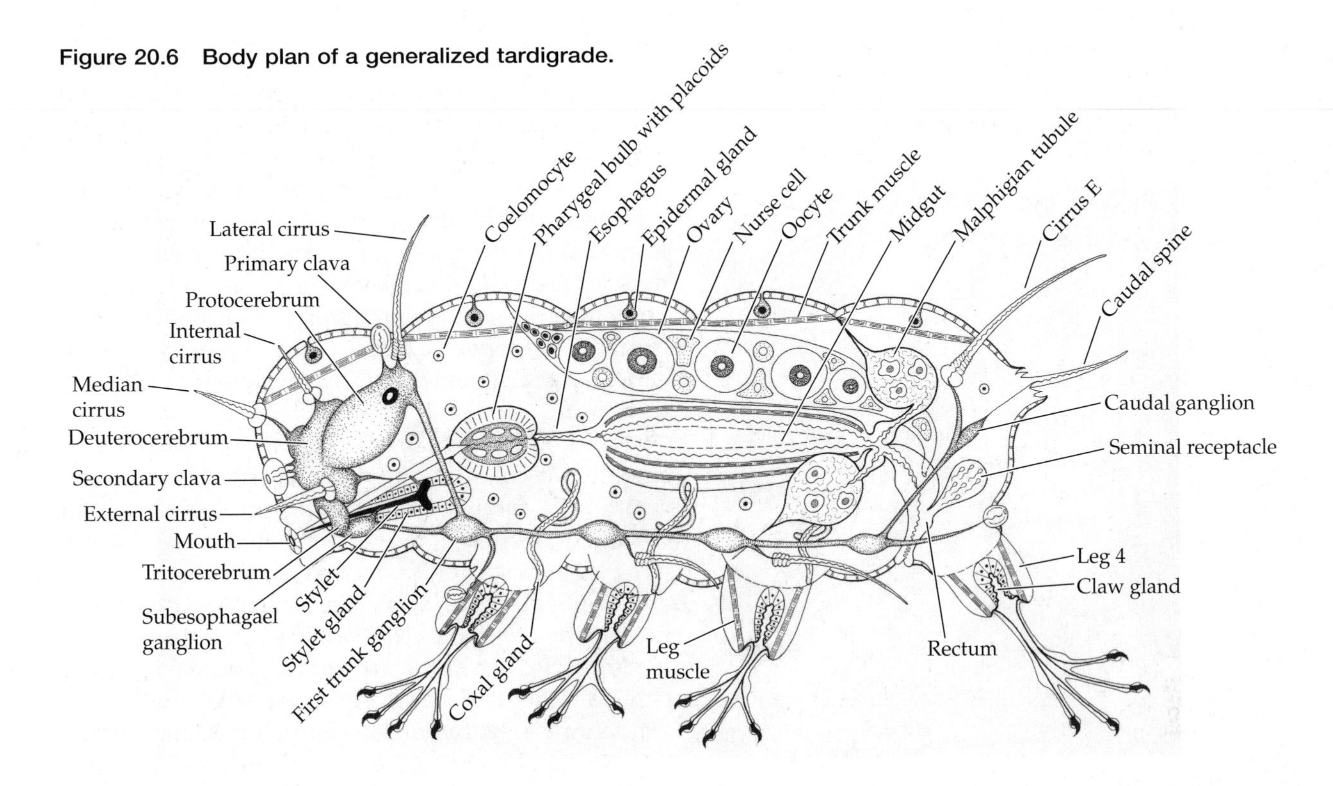

the end of the Second World War). Heterotardigrades include the orders Arthrotardigrada (nearly all marine) and Echiniscoidea (marine–limnic–terrestrial). Eutardigrades comprises two orders, Apochela (limnic–terrestrial) and Parachela (limnic–terrestrial, except for two secondarily marine genera).

The Tardigrade Body Plan

Tardigrades have four pairs of ventrolateral legs (Figures 20.2, 20.5, and 20.6). The legs of eutardigrades (Figure 20.2A) are short, hollow extensions of the body wall, essentially lobopodal in design, and similar to the lobopodal legs of onychophorans. Each leg terminates in one or as many as a dozen or so "claws" (Figures 20.2B and 20.7). The claws may be modified as adhesive pads or discs, or the claws may resemble those of onychophorans (Figure 20.7). In the marine arthrotardigrades (Figures 20.2C and 20.5A), the legs are strongly telescopic, and the leg consists of a so-called coxa, femur, tibia, and tarsus; however there is no evidence that these leg "segments" are homologous to those of arthropod legs.

As in onychophorans, the body is covered by a thin, uncalcified cuticle that is periodically molted. It is often ornamented and occasionally divided into symmetrically arranged dorsal and lateral (rarely ventral) plates (Figures 20.2 and 20.5C). These plates may be homologous with the sclerites of arthropods, but this is not certain. The cuticle shares some features with both onychophorans and arthropods, but it is also

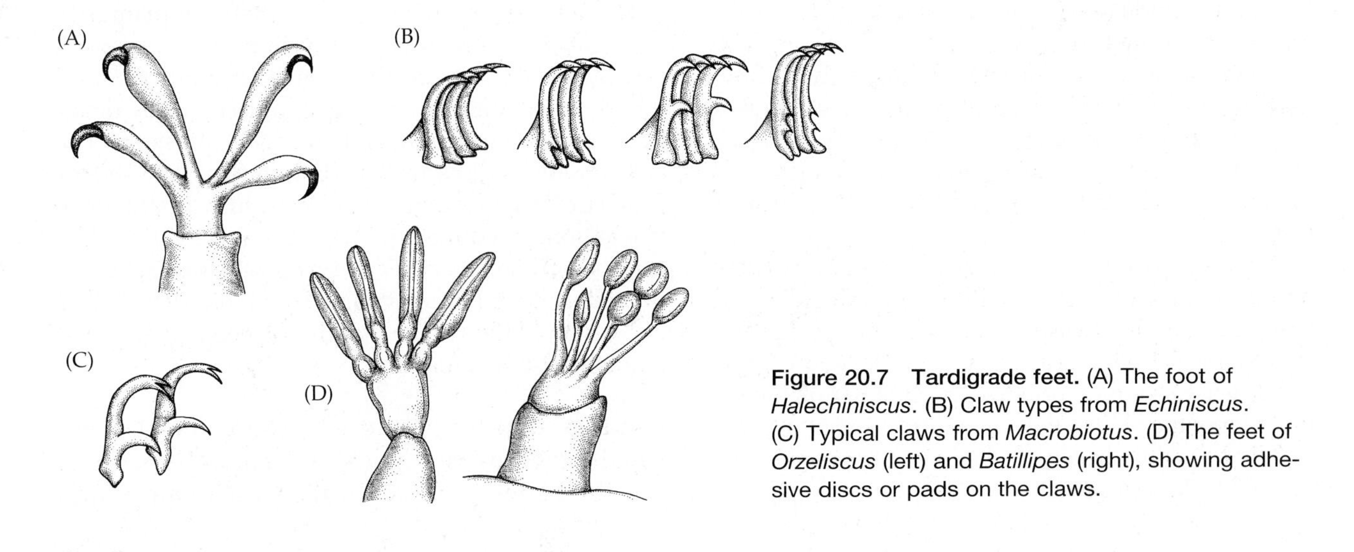

Figure 20.7 Tardigrade feet. (A) The foot of *Halechiniscus*. (B) Claw types from *Echiniscus*. (C) Typical claws from *Macrobiotus*. (D) The feet of *Orzeliscus* (left) and *Batillipes* (right), showing adhesive discs or pads on the claws.

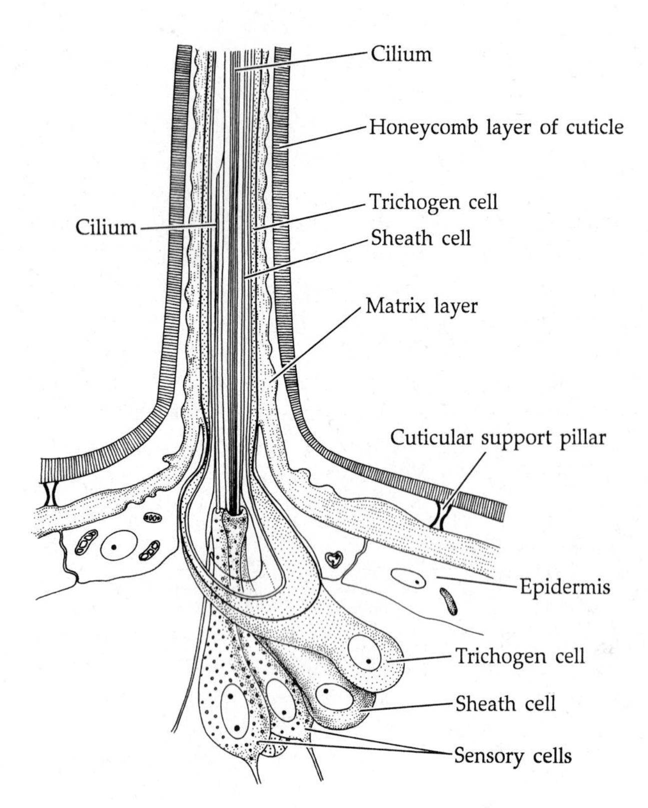

Figure 20.8 Base of an external bristle of *Batillipes noerrevangi* (longitudinal section), showing the relationship between the various cells and the cuticle.

unique in certain ways—for example, the epicuticle often has support pillars (Figure 20.8). Elsewhere the cuticle comprises up to seven distinguishable layers, contains various sclerotized ("tanned") proteins and always chitin in the procuticle, and it lines the foregut and rectum. The cuticle is secreted by an underlying epidermis, which is composed of a constant cell number in many (but not all) species. Such eutely is common in minute metazoans, and we have noted several other examples among the blastocoelomate phyla (e.g., rotifers). Growth in tardigrades proceeds by molts, as in onychophorans and arthropods, with sexual maturity being attained after three to six instars. Not only is the cuticle molted—the buccal canal, the two stylets, the two stylet supports, and the distal part of legs with claws or toes are also re-formed during ecdysis. Both the claws and the stylets are formed in special glands—called **stylet** and **claw glands** (Figures 20.5A and 20.6). During ecdysis the animal cannot eat, and the mouth opening is closed. This stage is called the **simplex stage**.

Although the body is quite short, it is nevertheless homonomous and rather weakly cephalized. Nonmarine tardigrades are often colorful animals, exhibiting shades of pink, purple, green, red, yellow, gray, and black. Color is determined by cuticular pigments, the color of the food in the gut, or the presence of granular bodies with carotene in the coelomocyte cells suspended in the hemocoel.

Like the coelom of arthropods and onychophorans, the adult coelom of tardigrades is greatly reduced and confined largely to the gonadal cavities. The main body cavity is thus a hemocoel, and the colorless body fluid directly bathes the internal organs and body musculature. The musculature of tardigrades is quite different from that of onychophorans, in which the body wall muscles are in sheetlike layers; in tardigrades there is no circular muscle layer in the body wall, and the muscles occur in separate bands extending between subcuticular attachment points, as they do in arthropods (Figure 20.6).

It was long thought that tardigrades possessed only smooth muscle, in contrast to the striated muscles of arthropods, and in the past this feature was used as an argument against a close relationship between these two phyla. However, we now know that both smooth and striated muscles occur in tardigrades, the latter predominantly in the class Arthrotardigrada. The striated muscles are of the arthropod type, being cross-striated rather than obliquely-striated like those of onychophorans. Numerous fine structural details of the muscle attachment regions are also shared between tardigrades and arthropods. R. M. Kristensen has suggested that a partial shift from arthropod-like striated muscle to smooth muscle in some tardigrades might have accompanied a transition from the marine to the terrestrial environment, and it might be functionally tied to the phenomenon of cryptobiosis. Furthermore, both slow and fast nerve fibers occur in tardigrades, the former predominating in the somatic musculature and the latter in the leg musculature. However, the leg musculature appears to be entirely extrinsic, like that of onychophorans, with one muscle attachment near the tip of the leg and the other within the body proper. Many of the muscle bands in tardigrades consist of only a single muscle cell or a few large muscle cells each.

Locomotion The concentration of muscles as discrete units and the thick cuticle in tardigrades demands a different locomotory strategy than the primarily hydrostatic system seen in onychophorans. Instead, tardigrades use a step-by-step gait controlled by independent antagonistic sets of muscles or by flexor muscles that work against hemocoelic pressure. The claws, pads, or discs at the ends of the legs are used for purchase and for clinging to objects, such as strands of vegetation or sediment particles (Figure 20.7). And, at least one marine species *Tholoarctus natans* is capable of limited jellyfish-like swimming by use of a bell-shaped expansion of the cuticle margin to keep it suspended just above the sediment.

Feeding, digestion, and excretion Tardigrades usually feed on the fluids inside plant or animal cells by piercing the cell walls with a pair of **oral**

stylets. Soil-dwelling species feed on bacteria, algae, and decaying plant matter or are predators on small invertebrates. Carnivorous and omnivorous tardigrades have a terminal mouth; herbivorous and detritivorous ones have a ventral mouth.

The mouth opens into a short stomodeal buccal tube, which leads to a bulbous, muscular pharynx (Figure 20.6). A large pair of salivary glands flank the esophagus and produce digestive secretions that empty into the mouth cavity; these glands also are responsible for the production of a new pair of oral stylets with each molt, hence they are often referred to as **stylet glands** (Figures 20.5A and 20.6). The muscular pharynx produces suction that attaches the mouth tightly to a prey item during feeding and pumps the cell fluids out of the prey and into the gut. In many species there is a characteristic arrangement of chitinous rods, or **placoids**, within an expanded region of the pharynx. These rods provide support for the musculature of that region and may contribute to masticating action. The pharynx empties into an esophagus, which in turn opens into a large intestine (midgut), where digestion and absorption take place. The short hindgut (the cloaca or rectum) leads to a terminal anus. In some species defecation accompanies molting, with the feces and cuticle being abandoned together. Several marine arthrotardigrades have symbiotic bacteria in cephalic vesicles (Figures 20.2C and 20.5A). In *Wingstrandarctus* these bacteria may be used when the tardigrades are starving. The vesicles can be empty when the tardigrades have a full, green gut content, indicating they have been feeding on plant material.

At the intestine–hindgut junction in freshwater and terrestrial species of eutardigrades there are three large glandular structures that are called Malpighian tubules (Figure 20.6), each consisting of only about nine cells. The precise nature of these organs is not well understood, but they are probably not homologous to the Malpighian tubules of arthropods. In at least one marine eutardigrade genus (*Halobiotus*), the Malpighian tubules are greatly enlarged and have an osmoregulatory function. It is probable that some excretory products are absorbed through the gut wall and eliminated with the feces; other waste products may be deposited in the old cuticle prior to molting. In terrestrial heterotardigrades, excretory and ion/osmoregulatory structures are found between the second and third pairs of legs. However, in several marine arthrotardigrades segmental trunk glands are open at the base of the legs (Figures 20.5A and 20.6), and a single pairs of these segmental glands is located in the head. These glands may be homologous with the coxal glands of arthropods.

Circulation and gas exchange Perhaps because of their small size and moist habitats, tardigrades have no traces of discrete blood vessels, gas exchange structures, or true metanephridia; consequently, they rely on diffusion through the body wall and the extensive body cavity. The body fluid contains numerous cells (sometimes called coelomocytes) credited with a storage function.

Nervous system and sense organs The nervous system of tardigrades is built on the arthropod plan and is distinctly metamerous. A large dorsal cerebral ganglion is connected to a subpharyngeal ganglion by a pair of commissures surrounding the buccal tube (Figure 20.6). From the subpharyngeal ganglion, a pair of ventral nerve cords extends posteriorly, connecting a chain of four pairs of ganglia that serve the four pairs of legs. As a unique apomorphy for all tardigrades, there exists a nerve connective between the first brain lobe and the first ventral trunk ganglion. Sensory bristles or spines occur on the body, particularly in the anterior and ventral region and on the legs. The structure of these bristles is essentially homologous to that of arthropod setae (Figure 20.8). A pair of sensory eyespots is often present inside the dorsal brain. Each eyespot consists of five cells, one of which is a pigmented light-sensitive cell with both a microvilli and modified ciliary structure. Recent research using neural markers has suggested tardigrades may have a one-segmented head, but subsequent work has shown that the brain and sense organs are tripartite, suggesting more work is needed.

The anterior end of many heterotardigrades bears long sensory **cirri**, and most of these species also have one to three pairs of hollow anterior sensory structures called **clavae** that are probably chemosensory in nature (Figures 20.2C and 20.5A,B,C). The clava appears structurally similar to the olfactory seta of many arthropods. Many males have longer clavae than females.

Reproduction and development Many tardigrades are gonochoristic, with both sexes possessing a single saclike gonad lying above the gut. In males, the gonad terminates as two sperm ducts, suggesting that the single gonad is derived from an ancestral paired condition. The ducts extend to a single gonopore, which opens just in front of the anus or into the rectum. In females a single oviduct (right or left) opens either through a gonopore anterior to the anus or into the rectum (which in this case is called a cloaca) (Figure 20.6). There are two complex seminal receptacles (e.g., Arthrotardigrada) that open separately (Figure 20.5A), or a single small seminal receptacle (e.g., some Eutardigrada) that opens into the rectum near the cloaca (Figure 20.6).

Males are unknown in some terrestrial genera, but most tardigrades that have been studied copulate and lay eggs, and copulation is amazingly diverse among these little water bears. Parthenogenesis may be common in some terrestrial species, notably those in which males are unknown. Hermaphroditism has also been reported in a few genera. Dwarf males have been recently discovered in several marine genera. In some tardigrades the male deposits sperm directly into the female's seminal receptacles (or cloaca) or into the body cavity by cuticular penetration. In the latter case fertilization takes place in the ovary. In other tardigrades, a wonderfully curious form of indirect fertilization takes place: the male deposits sperm beneath the cuticle of the female prior to her molt, and fertilization occurs when she later deposits eggs in the shed cuticular cast (Figure 20.9). Several studies have shown tardigrade sperm to be uniflagellate. In at least a few marine species, a primitive courtship behavior exists, wherein the male strokes the female with his cirri. Thus stimulated, the female deposits her eggs on a sand grain, upon which the male then spreads his sperm.

Females lay from 1 to 30 eggs at a time, depending on the species. In strictly aquatic species the fertilized eggs are either left in the shed cuticle or glued to a submerged object. The eggs of many terrestrial species bear thick, sculptured shells that resist drying (Figure 20.10). Some species alternate between thin-walled and thick-walled eggs, depending on environmental conditions.

There have been only a few studies on tardigrade embryology. Development appears to be direct and rapid, but indeterminate. Cleavage has been described as holoblastic and radial. A blastula develops, with a small blastocoel; eventually it proliferates an inner mass of endoderm that later hollows to form the archenteron. Stomodeal and proctodeal invaginations develop, completing the digestive tube. Subsequent to gut formation, five pairs of archenteric coelomic pouches are said to appear off the gut, reminiscent of the enterocoelous development of many deuterostomes. The first pair arises from the stomodeum (ectoderm) and the last four pairs from the midgut (endoderm). The two posterior pouches fuse to form the gonad; the others disappear as their cells disperse to form the body musculature.

Tardigrades were long thought to have typical teloblastic segmentation. Typically, this means there is a growth zone from which the new body segments arise. However, recent work has shown that tardigrades develop all four pairs of legs and the coeloms very early in embryogenesis, and the legs are present when they hatch. So teloblastic segmentation may be modified in living species. However, extinct Cambrian tardigrades from Siberian limestone may have had teloblastic segmentation—the smallest specimens have only three pairs of legs, and larger specimens have limb buds for the fourth pair of legs. Development in living species is typically completed in 14 days or less, whereupon the young use their stylets to break out of the shell (Figure 20.10C).

Juveniles lack adult coloration, have fewer lateral and dorsal spines and cirri, and may have reduced numbers of claws. At birth, the number of cells in the body is relatively fixed, and growth is primarily by increases in cell size rather than in cell number. In nature, these remarkable animals may live only a few months or may survive for a great many years.

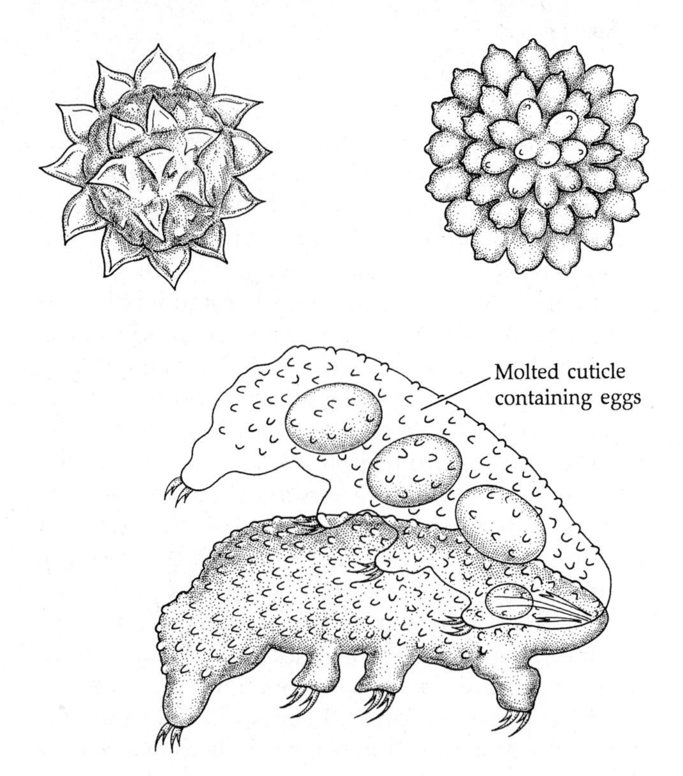

Figure 20.9 (A) Sculptured eggs of terrestrial tardigrades. (B) A female *Hypsibius annulatus* in the process of molting an egg-containing cuticle.

Phylum Onychophora

The first living onychophoran (Greek *onycho*, "talon"; *phora*, "bearer") was described by the Reverend Lansdown Guilding in 1826 as a leg-bearing slug (a mollusc). Since that initial discovery, 200 or so species of onychophorans have been described, 180 of which are considered valid, and probably at least that many more remain to be discovered (Figure 20.1A–G). All the living species are terrestrial, constituting the only animal phylum that is entirely land-bound, although the fossil record informs us that they originated from marine ancestors. And, we now know that onychophoran relatives known as lobopodians were part of the explosive marine diversification in the early Cambrian (see Chapter 1). Their fossils have been found in middle Cambrian marine faunas at several localities (e.g., *Aysheaia pedunculata* from the famous middle Cambrian Burgess Shale deposits of British Columbia, Canada; and *Aysheaia prolata* from a similar deposit in Utah), in the remarkable Chengjiang Lower Cambrian (520–530 Ma) deposits of China, and in the equally stunning Swedish Upper Cambrian Orsten fauna. Perhaps the most famous lobopodan fossil is the amazing Cambrian genus *Hallucigenia*, long a mystery because it was originally interpreted in an upside-down orientation, but later turned right side up and discovered to possess long dorsal spines (Figure 20.12A). The fossil

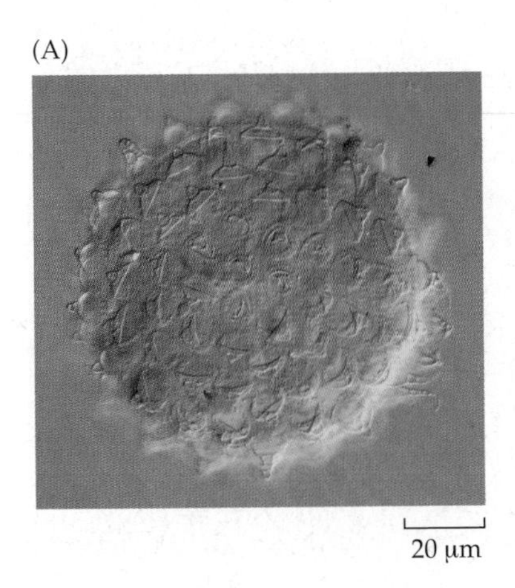

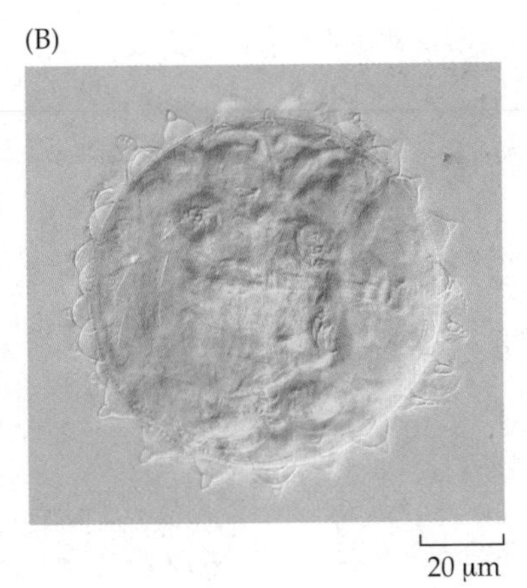

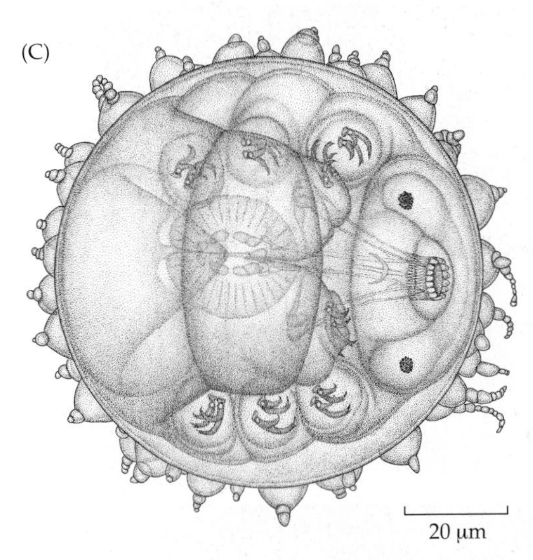

Figure 20.10 The egg and developing embryo of *Austeruseus faeroensis* (Eutardigrada). (A) Surface ornamentation of egg. (B) Developing embryo inside the egg shell. (C) Drawing of developing embryo inside the egg shell.

record of crown-group Onychophora is, however, restricted to two Carboniferous deposits, one from the Middle Pennsylvanian Mazon Creek (USA), and one from the Stephanian deposits of Montceau-les-Mines (France), in addition to a few amber deposits from the Cretaceous of Myanmar, and the early Tertiary of the Baltic Sea and Dominican Republic (Figure 20.11).

Onychophora constitute an old group that has changed very little over the past 310 million years, but at some point in its long history successfully invaded the terrestrial environment prior to the Carboniferous, when the first unambiguous crown-group onychophorans first appear in the fossil record. Like arthropods and tardigrades, onychophorans are segmented animals, and they have features that are somewhat intermediate between those of tardigrades and arthropods. The three phyla together comprise the clade called Panarthropoda. Due to their special type of segmentation, in the past, workers regarded onychophorans as "living fossils," or "missing links" between other segmented soft-bodied animals like annelids, and the arthropods. However, molecular phylogenetic studies now place the annelids in the Spiralia, quite distant from the Panarthropoda, which are ecdysozoans. Their close relationship to arthropods has triggered a great deal of research on onychophorans, aimed at understanding the early steps in the evolution of Earth's largest animal phylum. Indeed, research on velvet worms has undergone a revival in the twenty-first century with a plethora of studies focusing on their development, the nervous system, segmentation, as well as taxonomic and biogeographic treatments.

Living onychophorans comprise two families, Peripatidae and Peripatopsidae (Figure 20.11). The former is circumtropical in distribution (with multiple species in the Neotropics, one in West Africa, and a few in southeast Asia), whereas the latter is circumaustral (confined to the temperate Southern Hemisphere), with species in Chile, South Africa, New Guinea, Australia, and New Zealand. During dry periods, velvet worms retire to protective burrows or other retreats where they can preserve moisture and become inactive. During wet periods they can be found actively hunting at night, inside damp fallen tree logs, or in leaf litter in the regions where they live. Onychophorans live for several years, during which time periodic molting takes place, as often as every two weeks in some species (Box 20B).

ONYCHOPHORA

FAMILY PERIPATIDAE 19 to 43 leg pairs, genital opening between the penultimate pair; with a diastema on the inner blades of the jaws. Tropical distribution, 74 species in 10 genera: *Eoperipatus* (Southeast Asia), *Mesoperipatus* (West Africa), and a series of poorly characterized genera from the Neotropics, *Epiperipatus*, *Heteroperipatus*, *Macroperipatus*, *Oroperipatus*, *Peripatus*, *Plicatoperipatus*, *Speleoperipatus*, *Typhloperipatus*.

FAMILY PERIPATOPSIDAE 13 to 29 leg pairs, genital opening between the last pair; without a diastema on the inner blades of the jaws. Temperate to tropical in the southern hemisphere, 106 species in 39 genera, e.g., *Austroperipatus*, *Cephalofovea*, *Euperipatoides*, *Kumbadjena*, *Nodocapitus*, *Occiperipatoides*, *Ooperipatellus*, *Ooperipatus*, *Opisthopatus*, *Paraperipatus*, *Peripatoides*, *Peripatopsis*, *Phallocephale*, *Planipapillus*, *Tasmanipatus*.

Figure 20.11 Some onychophorans. (A) *Peripatoides* sp. from Waikato, North Island, New Zealand. (B) *Peripatoides aurorbis*, Kahurango National Park, North Island, New Zealand. (C) *Peripatopsis moseleyi*, Karkloof Nature Reserve East, Kwazulu-Natal, South Africa. (D) *Peripatopsis capensis*, Marloth Nature Reserve, South Africa. (E) Unidentified Peripatidae from the Central Amazon Conservation Complex, Roraima, Brazil. (F) *Peripatopsis alba*, Wynberg Cave, Table Mountain National Park, South Africa. (G) An undescribed fossil onychophoran from the Stephanian deposits of the Montceau-les-Mines Lagerstätte, late Carboniferous of France.

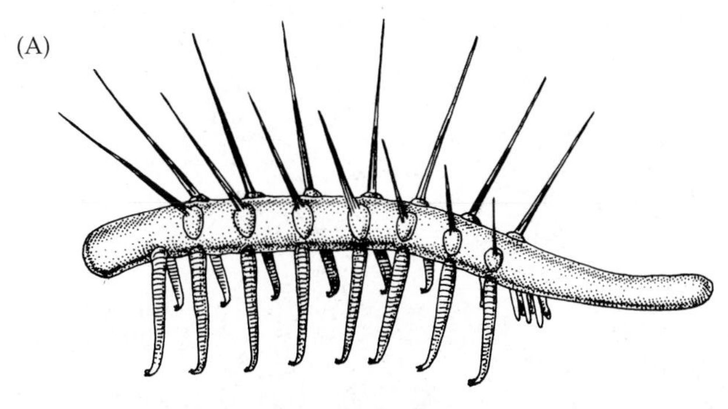

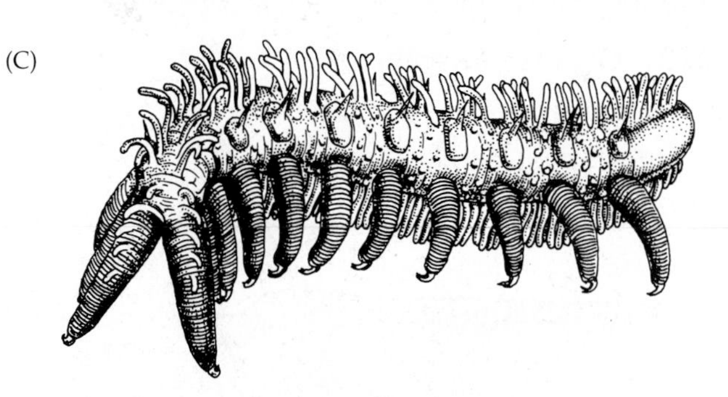

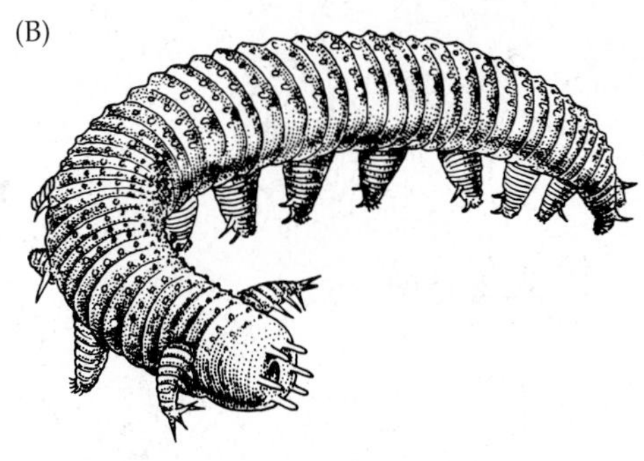

Figure 20.12 Reconstructions of Cambrian marine lobopodians. (A) The enigmatic *Hallucigenia sparsa*, with two rows of long dorsal spines. (B) *Onychodictyon ferox* from the Lower Cambrian Chengjiang deposits of China, with dorsal spines and papillae. (C) *Aysheaia pedunculata*, from middle Cambrian Burgess shale deposits. (D) In 2015 the remarkable lobopodan fossil *Collinsium ciliosum* was described from the Cambrian Xiaoshiba deposit in southern China (not far from the famous Chengjiang deposit). Like *Hallucigenia*, its dorsum is covered with long hard spines, but with many more than seen in *Hallucigenia*—up to 72 in fact. It also differs from *Hallucigenia* in having distinctly different anterior and posterior legs; the front legs being brushlike and probably functioning as feeding appendages, the rear legs being clawed and likely adapted for clinging to a sponge or cnidarian or other substratum. Being one of the earliest animals to develop armor, and reaching lengths of nearly 10 cm, this species would have been a striking and formidable Cambrian marine creature. It is further evidence that the ancestors to modern Onychophora were far more morphologically and ecologically diverse than today's velvet worms. A reconstruction of *C. ciliosum* is shown below the photograph of the actual fossil.

BOX 20B Characteristics of the Phylum Onychophora

1. Segmentation probably teloblastic; 13–43 pairs of legs
2. Muscles are isolated bands.
3. With superficially annulated, non-jointed, telescopic, lobopodal legs lacking intrinsic musculature; legs with terminal claw (or hooks)
4. Cerebral ganglion (brain) lies dorsal to pharynx, probably with only two pairs of ganglia. Protocerebrum innervates eyes and antennae; deutocerebrum innervates jaws.
5. With jaws, probably homologous to chelicerae of Chelicerata, or antennae of myriapods and hexapods (first antennae of crustaceans)
6. Paired ventral nerve cords differ from those of tardigrades and arthropods, with no relationship between arrangement of median or ring commissures and position of legs; cell bodies of serotonergic neurons not arranged in the segmentally repeated and bilaterally symmetrical pattern characteristic of arthropods (instead, the neurons are scattered in an apparently random fashion along the length of the nerve cord)
7. With slime papillae, possibly from modified nephridia (innervated by ventral nerve cords)
8. Embryonic coelomic cavities fuse with spaces of primary body cavity (blastocoel); adult body thus a mixocoel/hemocoel; with unique subcutaneous vascular channels (hemal channels).
9. Body wall with sheetlike muscle layer
10. Gas exchange by tracheae and spiracles (probably not homologous to arthropod trachea)
11. Almost all species gonochoristic

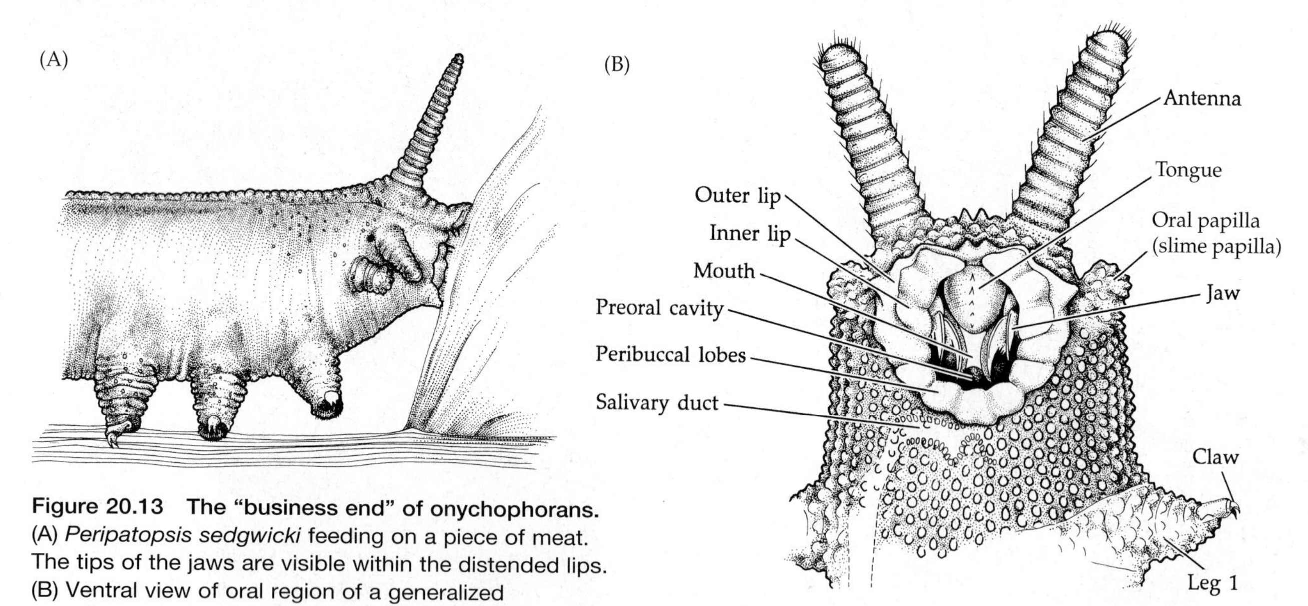

Figure 20.13 The "business end" of onychophorans. (A) *Peripatopsis sedgwicki* feeding on a piece of meat. The tips of the jaws are visible within the distended lips. (B) Ventral view of oral region of a generalized onychophoran.

The Onychophoran Body Plan

Modern onychophorans loosely resemble caterpillars, ranging from 5 mm to 15 cm in adult length. Within a given species, males are smaller than females and normally have fewer legs. Little cephalization is externally visible, and body segmentation is homonomous, where all legs differ little. Three paired appendages are found on the head: one pair of fleshy annulated antennae, a single pair of jaws, and a pair of fleshy **oral papillae** ("**slime papillae**"), resembling a small leg, lying posterior to the mouth (Figure 20.13). Circular lips surround the jaws. Beady eyes are located at the bases of the antennae in most species, but a few species are eyeless (Figure 20.14). The anterior head appendages are followed by 13–43 pairs of simple **lobopodal (saclike) walking legs**. A series of **ventral and preventral organs** is common among onychophorans and serve as attachment sites for segmental limb depressor muscles. The origin of these structures can be traced back in the embryo as lateroventral segmental, ectodermal thickenings not associated with the development of the nervous system. Although the head appendages and lobopodia are superficially annulated, they are not jointed or segmented, nor do they possess intrinsic (segmental) musculature.

The homology of onychophoran head structures with those of annelids and arthropods has long been a matter of debate, but the issue now seems resolved, in part thanks to the use of DNA labeling, immunocytochemistry, and neuronal tracing techniques in developing embryos. The eyes and antennae are innervated by the protocerebrum, unlike in any living arthropod where the protocerebrum innervates only the eyes. The jaws are homologous to the chelicerae of chelicerates, or to the antennae of myriapods and hexapods, or the first antennae of crustaceans, and are innervated by the deutocerebrum. Although the slime papillae align with the pedipalps of chelicerates and the second antennae of crustaceans, they are innervated by the ventral nerve cords, and not by the tritocerebrum as in arthropods, as the brain of velvet worms seems to comprise only two pairs of ganglia.

Similarly, the paired ventral nerve cords are very different from those of tardigrades and arthropods, as there is no relationship between the arrangement of median or ring commissures and the position of the legs. In addition, the cell bodies of serotonergic neurons are not arranged in the segmentally repeated and bilaterally symmetrical pattern characteristic of arthropods and annelids. Rather, these neurons are scattered in an apparently random fashion along the length of the onychophoran nerve cord. In summary, the overall arrangement of nerve pathways in the onychophoran nerve cord differs markedly from the rope-ladder arrangement in arthropods. In fact, it is more similar to the arrangement of orthogonally crossing nerve pathways found in the nervous systems of various worm-like protostomes, than to those of tardigrades or arthropods.

Externally, the unjointed, fleshy nature of the head appendages, the structure of the jaws, and the legs appear quite different to those of arthropods. As such, the serially arranged, clawed lobopodial appendages of onychophorans (including certain enigmatic fossil forms) have no clear counterpart in the animal kingdom.

The body is covered by a thin chitinous cuticle (containing α-chitin, as in arthropods) that is molted, as is in all other ecdysozoans. However, unlike that of arthropods, the cuticle of onychophorans is soft, thin, flexible, very permeable, and not divided into articulating plates or sclerites, but instead is annulated like that

(A)

(B)

Figure 20.14 Although a few species of onychophorans are blind, most have small beady eyes that are homologous to the simple eyes of arthropods. (A) *Epiperipatus* sp. from Reserva Ducke, Manaus, Amazonia, Brazil. (B) *Peri-patopsis moseleyi*, Karkloof Nature Reserve, Kwazulu-Natal, South Africa.

of priapulans, exhibiting multiple plicae per segment. Beneath the cuticle is a thin epidermis, which overlies a connective tissue dermis and layers of circular, diagonal, and longitudinal muscles (Figure 20.15). The body surface of onychophorans is covered with wartlike tubercles or papillae of multiple kinds, usually arranged in rings or bands around the trunk and appendages that are of taxonomic importance. The tubercles are covered with minute scales. Most onychophorans are distinctly colored blue, green, orange, or black, and the papillae, sometimes quite colorful, and scales give the body surface a velvety sheen—hence the common name "velvet worms."

The coelom formation in relation to that of arthropods has received considerable recent attention and is restricted almost entirely to the gonadal cavities in the adult onychophoran. In the Neotropical species *Epiperipatus biolleyi* the fate of the embryonic coelomic cavities has been studied in detail, providing evidence that embryonic coelomic cavities fuse with spaces of the primary body cavity (blastocoel). During embryogenesis, the somatic and splanchnic portions of the mesoderm separate and the former coelomic linings are transformed into mesenchymatic tissue. The resulting body cavity therefore represents a mixture of primary and secondary (coelomic) body cavities, i.e., the "mixocoel," but the homology of the segmental coeloms and nephridia in onychophorans and arthropods (and annelids) is not supported from the point of view of comparative anatomy. The hemocoel is also arthropod-like, being partitioned into sinuses, including a dorsal pericardial sinus.

Locomotion The segmentally paired walking legs of onychophorans are conical, unjointed, ventrolateral lobes with a multispined terminal claw (sometimes called hooks). When the animal is standing or walking, each leg rests on three to six distal transverse pads (Figure 20.15). These lobopodal legs are filled with hemocoelic fluid and contain only extrinsic muscle insertions. Walking is accomplished by leg mechanics combined with extension and contraction of the body by hydrostatic forces exerted via the hemocoel. Waves of contraction pass from anterior to posterior. When a segment is elongated, the legs are lifted from the ground and moved forward. When a segment contracts, a pulling force is exerted and the more anterior legs are held against the substratum. The overall effect is reminiscent of some types of polychaete locomotion, wherein the parapodia are used mainly for purchase rather than as legs or paddles.

The body muscles are a combination of smooth and obliquely striated fibers and are arranged similarly to those of annelids. The thin cuticle, soft body, and

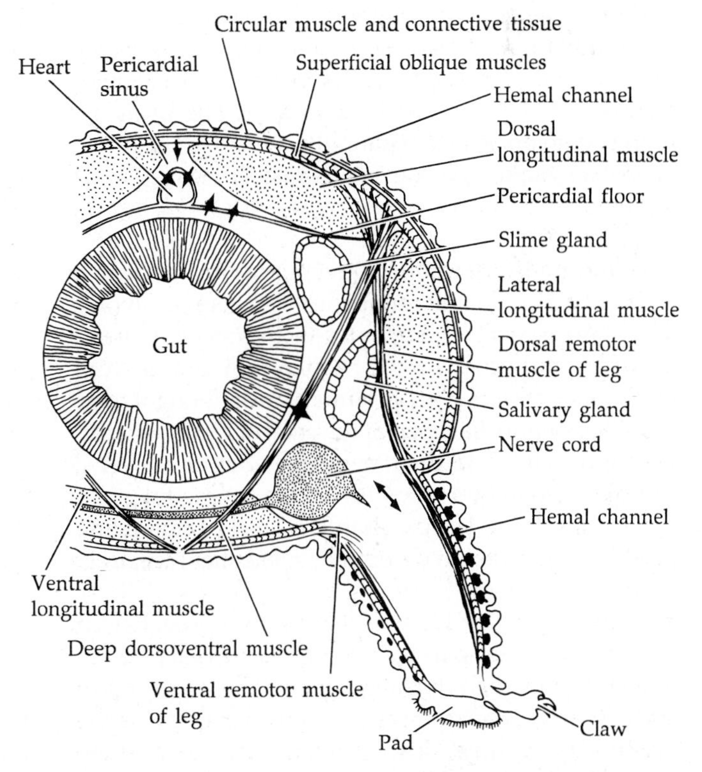

Figure 20.15 Body segment and leg of *Peripatopsis* (transverse section). The arrows indicate direction of blood flow.

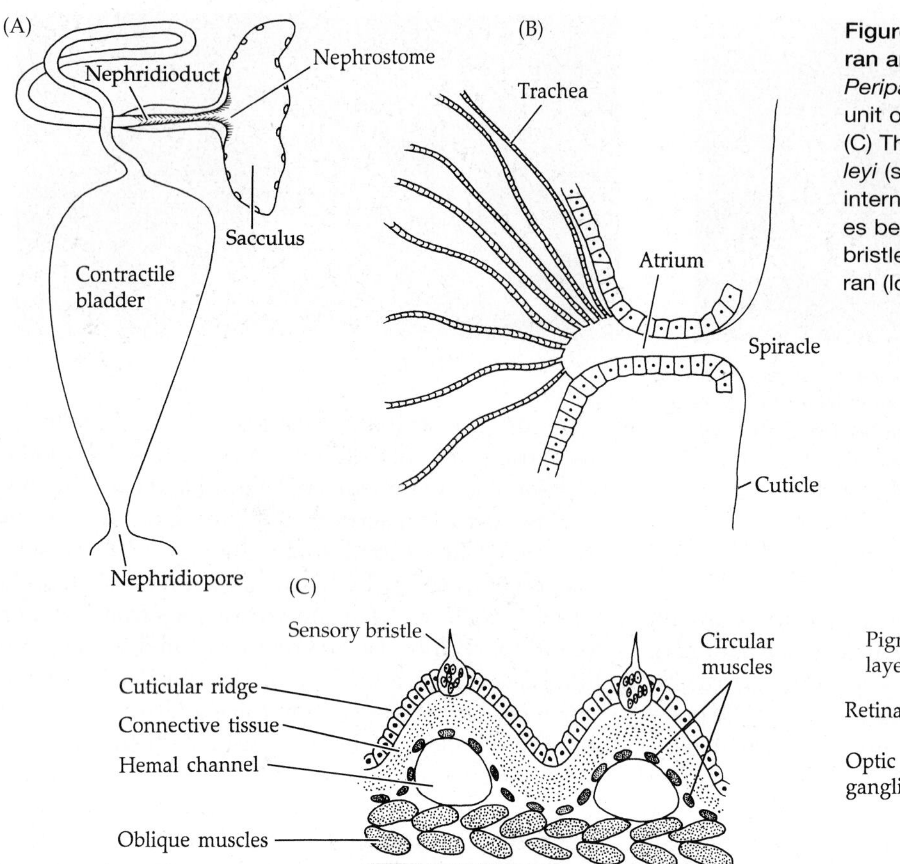

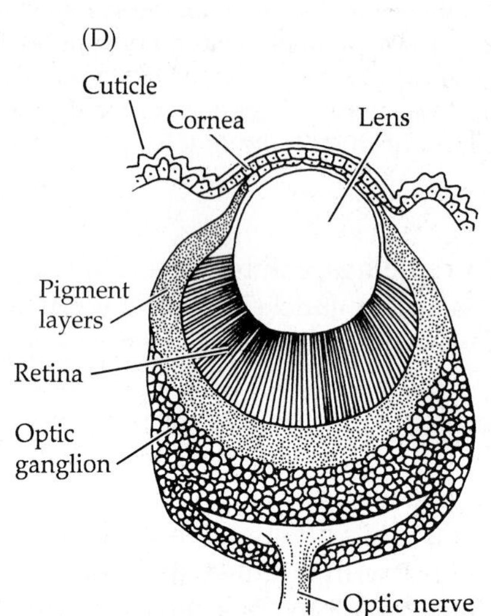

Figure 20.16 Some onychophoran anatomy. (A) Nephridium from *Peripatopsis capensis*. (B) A tracheal unit of *P. capensis* (cross section). (C) The body wall of *Peripatopsis moseleyi* (section). Note the hemal channels internal to each annular ridge. The ridges bear papillae surmounted by sensory bristles. (D) The eye of an onychophoran (longitudinal section).

hydrostatic body plan allow onychophorans to crawl and force their way through narrow passages in their environment. As we saw in the annelids, the efficiency of a hydrostatic skeleton is enhanced by internal longitudinal communication of body fluids. The ancestors of the onychophorans apparently expanded the blood vascular system to achieve a hemocoelic hydrostatic skeleton.

Feeding and digestion Onychophorans occupy a niche similar to that of centipedes. They are carnivores that prey on small invertebrates such as snails, worms, termites, and other insects, which they pursue into cracks and crevices. Special **slime glands**, thought to be modified nephridia, open at the ends of the oral papillae (Figure 20.17); through these openings an adhesive is discharged in two powerful streams, sometimes to a distance of 30 cm. The glue hardens quickly, entangling prey (or would-be predators) for subsequent leisurely dining.

The jaws are used to grasp and cut up prey. Paired salivary glands, also thought to be modified nephridia, open into a median dorsal groove on the jaws (Figure 20.13). Salivary secretions pass into the body of the prey and partly digest it; the semiliquid tissues are then sucked into the mouth, after they ingest their own glue. The mouth opens into a chitin-lined foregut, composed of a pharynx and esophagus. A large, straight intestine is the principal site of digestion and absorption. The hindgut (rectum) usually loops forward over the intestine before passing posteriorly to the anus, which is located ventrally or terminally on the last body segment.

Circulation and gas exchange The circulatory system of onychophorans is arthropod-like and linked to the hemocoelic body plan. A tubular heart is open at each end and bears a pair of lateral ostia in each segment. The heart lies within a pericardial sinus. Blood leaves the heart anteriorly and then flows posteriorly within the large hemocoel via body sinuses, eventually reentering the heart by way of the ostia. The blood is colorless, containing no oxygen-binding pigments. Onychophorans possess a unique system of subcutaneous vascular channels, called **hemal channels** (Figure 20.15 and 20.16C). These channels are situated beneath the transverse rings, or ridges, of the cuticle. A bulge in the layer of circular muscle forms the outer wall of each channel, and the oblique muscle layer forms the inner wall. The hemal channels may be important in the functioning of the hydrostatic skeleton. Thus the superficial annulations of the onychophoran body are external manifestations of the subcutaneous hemal channels, as it may also be in some Cambrian lobopodians.

Gas exchange is by tracheae that open to the outside

through the many small spiracles located between the bands of body tubercles. Each tracheal unit is small and supplies only the immediate tissue near its spiracle (Figure 20.16). Anatomical data suggest that the tracheal system is not homologous to those of insects, arachnids, or terrestrial isopods, but has been independently derived in each of these terrestrial groups, including in Onychophora.

Excretion and osmoregulation A pair of nephridia lies in each leg-bearing body segment, except the one possessing the genital opening (Figures 20.16A and 20.17). The nephridiopores are situated next to the base of each leg, except at the fourth and fifth legs where the nephridia open through distal nephridiopores on the transverse pads of the legs themselves. The nephridial anlagen develops by reorganization of the lateral portion of the embryonic coelomic wall that initially gives rise to a ciliated canal. All other structural components, including the **sacculus**, merge after the nephridial anlagen has been separated from the remaining mesodermal tissue. The nephridial sacculus thus does not represent a persisting coelomic cavity, as previously thought, since it arises de novo during embryogenesis. There is no evidence for "nephridioblast" cells participating in the nephridiogenesis of Onychophora, which is in contrast to the general mode of nephridial formation in Annelida. Each set of sacculus + nephridioduct together is called a **segmental gland**. The nephridioduct, or tubule, enlarges to form a contractile bladder just before opening to the outside via the nephridiopore. The nature of the excretory wastes is not known. The anterior nephridia are thought to be modified as the salivary glands and slime glands, in addition to the nephridial anlagen in the antennal segment, and the posterior ones form the gonoducts in females.

The legs of some onychophorans, such as *Peripatus*, bear thin-walled eversible sacs or vesicles that open to the exterior near the nephridiopores by way of minute pores or slits. These vesicles may function in taking up moisture, as do the coxal glands of many myriapods, hexapods, and arachnids. They are everted by hemocoelic pressure and pulled back into the body by retractor muscles.

Nervous system, sense organs, and behavior The nervous system of onychophorans is ladder-like in structure, but differs considerably from those of annelids and arthropods. A large bilobed cerebral ganglion ("brain") lies dorsal to the pharynx. A pair of ventral nerve cords, without segmental ganglia, is connected by non-segmental ring commissures, which are also connected to multiple longitudinal nerve tracts. The protocerebrum innervates the eyes and antennae; the deutocerebrum innervates the jaws. The paired leg nerves are the only segmental structures in the onychophoran nerve cords, and the somata of

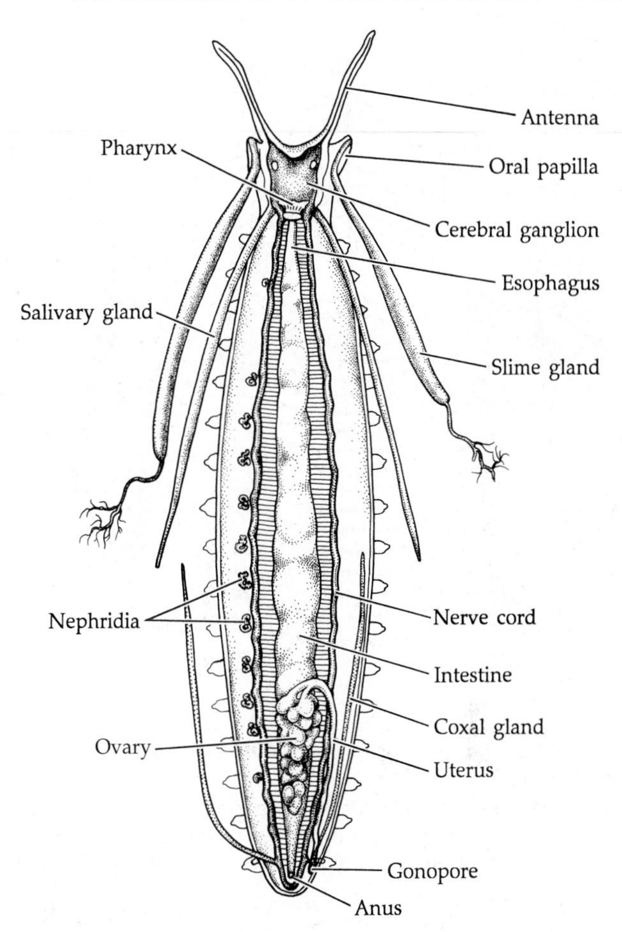

Figure 20.17 Internal anatomy of a generalized onychophoran.

serotonin-like immunoreactive neurons do not show any ordered arrangement but are instead scattered throughout the entire length of each nerve cord, showing neither a serially iterated nor a bilaterally symmetric pattern, in contrast to the strictly segmental arrangement of serotonergic neurons in arthropods (Figure 20.18). The general body surface, especially the larger tubercles, is supplied with sensillae that might be homologous to those of tardigrades and arthropods.

Onychophorans are nocturnal animals and photophobic. There is a small dorsolateral eye at the base of each antenna. The eyes are of the direct rhabdomeric type, with a large chitinous lens and a relatively well-developed retinal layer (Figure 20.16D).

The presence in onychophorans of only one optic neuropil, and eye development from an ectodermal groove, correspond with the median ocelli rather than compound eyes of arthropods. In addition, there are some parallels in the innervation pattern between onychophoran eyes and the median ocelli of arthropods, since both are associated with the central (rather than lateral) part of the brain. It has therefore been interpreted that there may be particular correspondences between the eyes of Onychophora, median ocelli of Chelicerata, and nauplius eyes of Malacostraca since in all these taxa there is a visual input to the central body.

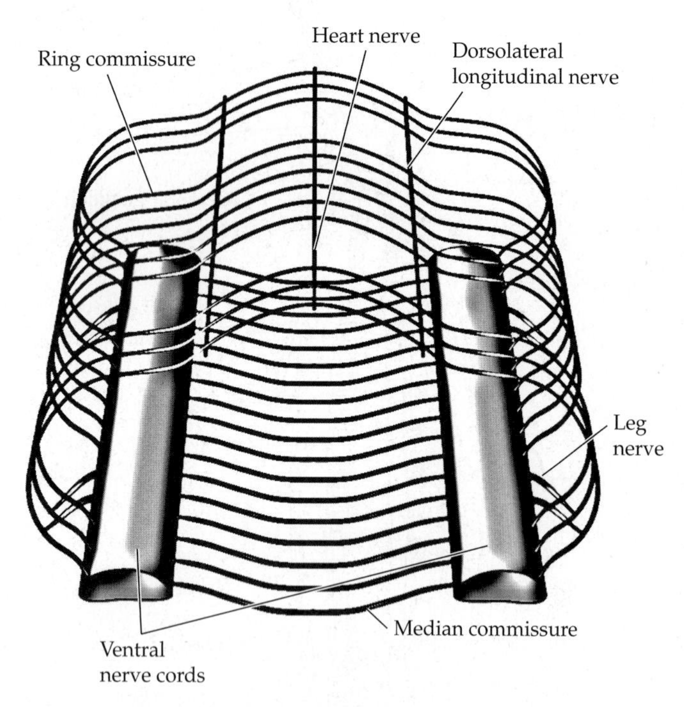

Figure 20.18 Schematic diagram of the arrangement of the main serotonin-like immunoreactive nerve tracts in the onychophoran trunk. Innervation of gut, various fiber networks, nephridial nerves, and ventral and lateral branches contributing to the subepidermal network are not shown.

Males of onychophorans have a special **crural gland**, a type of exocrine gland that opens at the base of the legs. This gland secretes a pheromone that attracts conspecific males.

Reproduction and development With the exception of one known parthenogenetic species from Trinidad (*Epiperipatus imthurni*), all onychophorans are gonochoristic. Most females have a pair of largely fused ovaries in the posterior region of the body (Figure 20.17). Each ovary connects to a gonoduct (oviduct), and each gonoduct to a uterus. The uteri open through a posteroventral gonopore. Males are smaller than females and have a pair of elongate, separate testes. Paired sperm ducts join to form a single tube in which sperm are packaged into spermatophores up to 1 mm in length. The male gonopore is also located posteroventrally.

Copulation has been observed in a few onychophorans. In the South African genus *Peripatopsis* the male deposits spermatophores seemingly at random on the general body surface of the female. The presence of the spermatophores stimulates special amebocytes in her blood to bring about a localized breakdown of the integument beneath the spermatophore. Sperm then pass from her body surface into her hemolymph, through which they eventually reach the ovaries, where fertilization takes place. In some onychophorans a portion of the uterus is expanded as a seminal receptacle, but sperm transfer in these species is not well understood. Insemination can also be vaginal or facultative between vaginal and dermal. Several Eastern Australian species present a diversity of male extra-genital sexual structures in the form of organs on the dorsal surface of the head, are highly elaborate in some species, and may be involved in the mechanics of sperm transfer.

Embryological work on onychophorans has revealed some unusual features. For example, onychophorans may be oviparous, viviparous, or ovoviviparous. Females of oviparous species (e.g., *Ooperipatus*) have an ovipositor and produce large, oval, yolky eggs with chitinous shells. Evidence suggests that this is the primitive onychophoran condition, even though living oviparous species are rare. The eggs of oviparous onychophorans contain so much yolk that early, superficial, intralecithal cleavage takes place, with the eventual formation of a germinal disc similar to that seen in many terrestrial arthropods. Most living onychophorans, however, are viviparous and have evolved a highly specialized mode of development associated with small, spherical, nonyolky eggs. Interestingly, most Old World viviparous species, although developing at the expense of maternal nutrients, lack a placenta, whereas all New World viviparous species have a placental attachment to the oviducal wall (Figure 20.19A). Placental development is viewed as the most advanced condition in onychophorans.

The yolky eggs of lecithotrophic species have a typical centrolecithal organization. Cleavage is by intralecithal nuclear divisions, similar to that seen in many groups of arthropods. Some of the nuclei migrate to the surface and form a small disc of blastomeres that eventually spreads to cover the embryo as a blastoderm, thus producing a periblastula. Simultaneously, the yolk mass divides into a number of anucleate "yolk spheres" (Figure 20.19B–D).

Nonyolky and yolk-poor eggs are initially spherical, but once within the oviduct, they swell to become ovate. As cleavage ensues, the cytoplasm breaks up into a number of spheres. The nucleate spheres are the blastomeres, and the anucleate ones are called **pseudoblastomeres** (Figure 20.19E–G). The blastomeres divide and form a saddle of cells on one side of the embryo (Figure 20.19G). The pseudoblastomeres disintegrate and are absorbed by the dividing blastomeres. The saddle expands to cover the embryo with a one-cell-thick blastoderm around a fluid-filled center.

Placental oviparous species have even smaller eggs than do nonplacental species, and the eggs do not swell after release from the ovary. Further, these eggs are not enclosed in membranes. Cleavage is total and equal, yielding a coeloblastula. The embryo then attaches to the oviducal wall and proliferates as a flat placental plate. As development proceeds, the embryo moves progressively down the oviduct and eventually attaches in the uterus. Gestation may be quite long, up to 15 months, and the oviduct/uterus often contains a

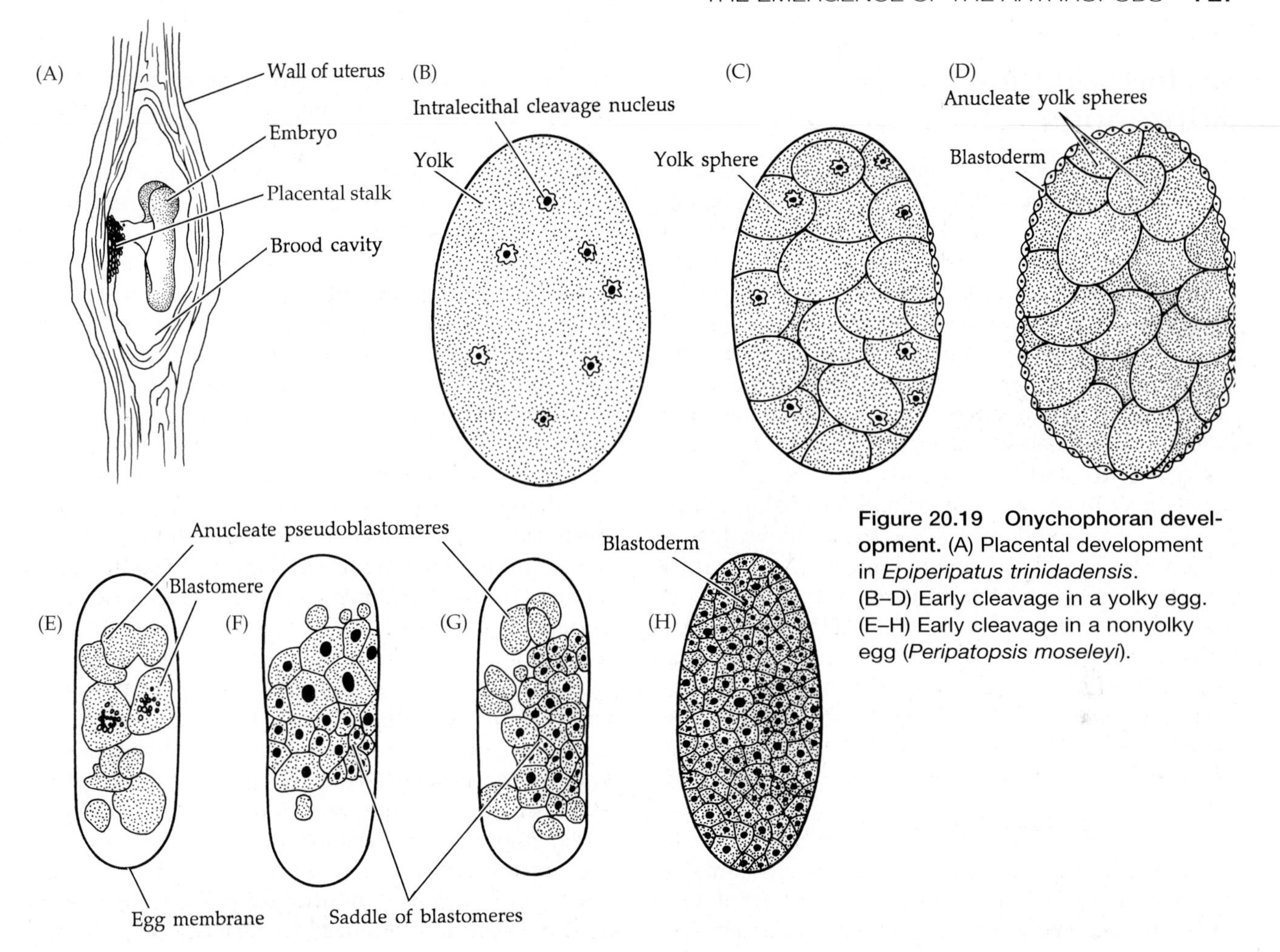

Figure 20.19 Onychophoran development. (A) Placental development in *Epiperipatus trinidadensis*. (B–D) Early cleavage in a yolky egg. (E–H) Early cleavage in a nonyolky egg (*Peripatopsis moseleyi*).

series of developing embryos of different ages. Sperm can remain motile in the seminal receptacle for up to 6 months.

Development after the formation of the blastula is remarkably similar among the few species of onychophorans that have been studied. Gastrulation in onychophorans involves very little actual cell migration. Cells of the presumptive areas undergo immediate organogenesis by direct proliferation. This process involves the proliferation of small cells into the interior of the embryo through and around the yolk mass or fluid-filled center and the production by surface cells of the germinal centers of limb buds and other external structures. All onychophorans have direct development. In all species that have been studied, the full complement of segments and adult organ systems is attained before they hatch or are born as juveniles.

Onychophorans are also unusual in that neither a presegmental acron nor a postsegmental pygidium or telson can be differentiated. In onychophorans, even though growth is teloblastic, the growth zone from which the trunk segments arise appears to be postanal. When the last mesoderm has been formed, the growth-zone ectoderm apparently develops directly into the anal somite with no postsegmental ectoderm remaining. Onychophorans thus have the last pair of legs in subterminal position, as opposed to tardigrades, with the fourth pair of legs being terminal—a character that seems to differentiate them from the Cambrian lobopodians.

Systematics and biogeography The phylogenetic relationships of onychophorans have received little attention until recently, when the use of DNA sequence data has been generalized and used in many evolutionary studies at multiple hierarchical levels, illustrating in several cases the existence of cryptic species. A recent analysis of multiple DNA markers has shown strong evidence for the division of Onychophora into Peripatidae and Peripatopsidae, with both families diverging around the Carboniferous period, before the breakup of Pangaea. Peripatidae, with their tropical distribution have a species-poor lineage in Southeast Asia, a monotypic genus in West Africa, and most of its diversity in the Neotropics, including many Caribbean islands, and the group has been diversifying since around the Permian period. Peripatopsidae, restricted to the former circum-Antarctic Gondwanan landmasses divides into a South African/Chilean clade and into an Australian/New Zealand clade with monophyletic groups in Western Australia, Eastern Australia and Tasmania/New Zealand, diversifying initially around the Jurassic period, long after its divergence from Peripatidae.

An Introduction to the Arthropods

With over a million described living species, and 3–100 times that many still remaining to be described, the phylum Arthropoda is unprecedented in its diversity. There are five clearly distinguished groups of arthropods, which are usually recognized as subphyla: Trilobita (trilobites and their kin, with a fossil record from the early Cambrian to the end of the Permian), Crustacea (crabs, shrimps, etc.), Hexapoda (insects and their kin), Myriapoda (centipedes, millipedes, and their kin), and Chelicerata (horseshoe crabs, eurypterids, arachnids, and pycnogonids). After 150 years of debate over the evolutionary relationships of arthropods, molecular systematics has finally allowed us to largely resolve the phylogeny among these groups (Figure 20.38). The basic features of the arthropod body plan are listed in Box 20C. Some of these features are unique to the phylum Arthropoda and thus represent defining synapomorphies; others also occur in closely related taxa, such as the onychophorans and tardigrades, and hence are symplesiomorphies within the panarthropod line.

Taxonomic History and Classification

As mentioned earlier, Linnaeus recognized six major groups of animals (Vermes, Insecta, Pisces, Amphibia, Aves, and Mammalia), placing all of the invertebrates except the insects in a single group—Vermes. In the early 1800s such famous zoologists as Lamarck and Cuvier presented substantial reorganizations of Linnaeus's earlier scheme, and it was during this period that the various arthropod taxa began to emerge. Lamarck recognized four basic arthropod groups: Cirripedia (barnacles), Crustacea, Arachnida, and Insecta. He placed the ostracods with the brachiopods and, of course, he did not realize the crustacean nature of the barnacles. Cuvier joined the arthropods and annelids in his Articulata (referring to the segmented nature of these animals), and Lankester also classified them together with rotifers in his Appendiculata. The great zoologists Hatschek, Haeckel, Beklemishev, Snodgrass, Tiegs, Sharov, and Remane all viewed the Articulata as a discrete phylum, including in it at various times the groups Echiura, Sipuncula, Onychophora, Tardigrada, and Pentastomida. It was Leuckart who, in 1848, separated out the arthropods as the distinct phylum we recognize today; Von Siebold coined the name Arthropoda in the same year, noting the jointed legs (Greek *arthro*, "jointed"; *pod*, "foot") as the group's principal distinguishing attribute. Haeckel published the first evolutionary tree of the arthropods in 1866. And, beginning in 1997 with the benchmark study by Anna Marie Aguinaldo and colleagues, molecular phylogenetics began to reveal that annelids and arthropods are not at all as closely related as long thought, the former now being placed in the clade Spiralia, with arthropods being in the clade Ecdysozoa (also see Giribet 2003). Brief diagnoses of the four living arthropod subphyla are provided below, and detailed treatments of these groups are presented in Chapters 21 through 24.

It is important to offer a word of caution about the use of terminology among the various groups of arthropods. Because the Arthropoda is such a vast and diverse lineage, specialists usually concentrate on only one or a few groups. Thus, over time, slightly different terminologies have evolved. Students sometimes feel overwhelmed by arthropod terminology—for example, the hindmost region of the body may be called an abdomen or pleon (as in insects and crustaceans), an opisthosoma (in chelicerates), or a pygidium (in trilobites). But there is a greater, though more subtle danger to this mixed terminology. Different terms for similar parts or regions in different taxa do not necessarily imply nonhomology; conversely, the same term applied to similar parts of different arthropods does not always imply homology. To deal with these problems in this text, we have made an effort to achieve consistency in terminology as much as possible, to simplify word use and spelling, and to indicate homologies (and nonhomologies) where known.

Although we do not treat them in this book, of all fossil invertebrates trilobites are perhaps the most symbolic of ancient and exotic faunas. The subphylum Trilobita, or Trilobitomorpha (Latin *trilobito*, "three-lobed"; Greek morph, "form"), includes over 15,000 species of arthropods known only from the fossil record (Figure 20.36 and 20.37). They were restricted to (and characteristic of) Paleozoic seas. Trilobites dominate the fossil record of the Cambrian and Ordovician periods (551 to 444 Ma) and continued to be important components of marine communities until the Permo-Triassic mass extinction that marked the end of the Paleozoic era. Because of their hard exoskeletons (made of chitin and calcium carbonate, as in modern crustaceans and horseshoe crabs), great abundances, and broad distributions, the trilobites left a rich fossil record, and more is known about them than about most other extinct taxa. Most of the present world's land areas were submerged during various parts of the Paleozoic, so trilobites are found in marine sedimentary rocks worldwide.

The trilobite body was divided into three tagmata: cephalon, thorax, and pygidium (abdomen). The segments of the cephalon and pygidium were fused, while those of thorax were free. The body was demarcated by two longitudinal grooves into a median and two lateral lobes ("tri-lobite"). The cephalon had one pair of preoral antennae; all other appendages were postoral and more or less similar to one another, with a robust locomotory telopod to which was attached at the base a long filamentous branch (thought to be a protopodal exite). Most seem to have had compound eyes.

BOX 20C Characteristics of the Phylum Arthropoda

1. Body segmented, both internally and externally; segments arise by teloblastic growth (showing *en* gene expression)
2. Minimally, body divided into head (cephalon) and trunk; commonly with further regional body specialization or tagmosis; typically with a head shield or carapace covering fused head segments
3. Head with labrum (or clypeolabrum) (showing *Distal-less* gene expression) and with nonsegmental acron.
4. Cuticle forms well developed exoskeleton, generally with thick sclerotized plates (sclerites) consisting of dorsal tergites, lateral pleurites, and ventral sternites; cuticle of exoskeleton consists of chitin and protein (including resilin), with varying degrees of calcification; without collagen
5. Each true body segment primitively with a pair of segmented (jointed), ventrally attached appendages, showing a great range of specialization among the various taxa; appendages composed of a proximal protopod and a distal telopod (both multiarticulate); protopodal articles may bear medal endites or lateral exites
6. Cephalon with a pair of lateral faceted (compound) eyes and one to several simple median ocelli; the compound eyes, ocelli, or both have been lost in several groups
7. Coelom reduced to portions of the reproductive and excretory systems; main body cavity is an open hemocoel (= mixocoel); circulatory system largely open; dorsal heart is a muscular pump with lateral ostia for blood return
8. Gut complex and highly regionalized, with well-developed stomodeum and proctodeum; digestive material (and often also the feces) encapsulated in a chitinous peritrophic membrane
9. Nervous system with dorsal (supraenteric) ganglia (= cerebral ganglia), circumenteric (circumesophageal) connectives, and paired, ganglionated ventral nerve cords, the latter often fused to some extent.
10. Growth by ecdysone-mediated molting (ecdysis); with cephalic ecdysial glands
11. Muscles metamerically arranged, striated, and grouped in isolated, intersegmental bands; dorsal and ventral longitudinal muscles present; intersegmental tendon system present; without circular somatic musculature
12. Most are gonochoristic, with direct, indirect, or mixed development; some species parthenogenetic

Although trilobites were exclusively marine, they exploited a variety of habitats and lifestyles. Most were benthic, either crawling about over the bottom or plowing through the top layer of sediment. Most benthic species were a few centimeters long, although some giants reached lengths of 60–70 centimeters. A few trilobites appear to have been planktonic; they were mostly small forms, less than 1 cm long and equipped with spines that presumably aided in flotation. Most of the benthic trilobites were probably scavengers or direct deposit feeders, although some species may have been predators that laid partially burrowed in soft sediments and grabbed passing prey. Some workers speculate that at least some trilobites may have suspension fed by using the filamentous parts of their appendages. One group of Olenidae may have had symbiotic relationships with sulfur bacteria; these late Cambrian–early Ordovician trilobites had vestigial mouthparts and large "gill filaments" (epipods) that might have been sites for bacterial cultivation.

SYNOPSES OF THE LIVING ARTHROPOD SUBPHYLA

SUBPHYLUM CRUSTACEA Crabs, lobsters, shrimps, beach hoppers, pillbugs, etc. About 70,000 described living species. Body usually divided into three tagmata: head (cephalon), thorax, and abdomen (the notable exception being the class Remipedia, which has only head + trunk); appendages uniramous or biramous; 5 pairs of cephalic appendages—the preoral first antennae (antennules) and 4 pairs of postoral appendages: second antennae (which migrate to a "preoral position" in adults), mandibles, first maxillae (maxillules), and second maxillae; cerebral ganglia tripartite (with deutocerebrum); with compound eyes usually having tetrapartite crystalline cone; gonopores located posteriorly on thorax or anteriorly on abdomen. Crustacea is now known to be a paraphyletic group, because the Hexapoda arose from within it (see Chapter 21).

SUBPHYLUM HEXAPODA Insects and their kin; monophyletic. Nearly a million described living species. Body divided into three tagmata: head (cephalon), thorax, abdomen; with 4 pairs cephalic appendages: antennae, mandibles, maxillae, and labium (fused second maxillae); 3-segmented thorax with uniramous legs; cerebral ganglia tripartite (with deutocerebrum); with compound eyes having tetrapartite crystalline cone; gas exchange by spiracles and tracheae; with ectodermally derived (proctodeal) Malpighian tubules; gonopores open on abdominal segment 7, 8, or 9 (see Chapter 22).

SUBPHYLUM MYRIAPODA Millipedes, centipedes, etc.; monophyletic. Over 16,350 described living species. Body divided into two tagmata, head (cephalon) and long, homonomous, many-segmented trunk; with 4 pairs cephalic appendages (antennae, mandibles, first maxillae, second maxillae); first maxillae free or coalesced; second maxillae absent or partly (or wholly) fused; all appendages uniramous; cerebral ganglia with deutocerebrum, but lacking tritocerebrum; living species mostly lack compound eyes (but they are present in Scutigeromorpha); with ectodermally derived (proctodeal) Malpighian tubules; gonopores on third or last trunk somite (see Chapter 23).

SUBPHYLUM CHELICERATA Horseshoe crabs, scorpions, spiders, mites, "sea spiders," etc.; monophyletic. About 113,335 described species. Body divided into two tagmata, anterior prosoma (cephalothorax) and posterior opisthosoma (abdomen); opisthosoma with up to 12 segments (plus telson); prosoma of 6 somites, each with a pair of uniramous appendages (chelicerae, pedipalps, 4 pairs of legs); gas exchange by gill books, book lungs, or tracheae; excretion by coxal glands and/or endodermally derived (midgut) Malpighian tubules; with simple medial eyes and lateral compound eyes; cerebral ganglia tripartite, the deutocerebrum innervating the cheliceres (see Chapter 24).

The Arthropod Body Plan and Arthropodization

If we are to grasp the "essence of arthropod," we must first understand the effects of one of the major synapomorphies of this phylum—the hard, jointed exoskeleton. Just imagine living your life encased in a rigid exoskeleton, a permanent suit of armor if you will. What kinds of structural and functional problems would have to be solved in order for such an animal to survive? The approach arthropods evolved to live in their hard casing comprises a suite of adaptations collectively known as arthropodization. Arthropodization has some of its roots in the Onychophora and Tardigrada, but came to full fruition in the phylum Arthropoda itself.

Being encased in an exoskeleton resulted in some obvious constraints on growth and locomotion. The fundamental problem of locomotion was solved by the evolution of body and appendage joints and highly regionalized muscles. Flexibility was provided by thin intersegmental areas (joints) in the otherwise rigid exoskeleton, imbued with a unique and highly elastic protein called resilin. As the muscles became concentrated into intersegmental bands associated with the individual body segments and appendage joints, the circular muscles were lost almost entirely.

With the loss of peristaltic capabilities resulting from body rigidity and the loss of circular muscles, the coelom became nearly useless as a hydrostatic skeleton. The ancestral body coelom was lost, and an open circulatory system evolved—the body cavity became a **hemocoel**, or blood chamber, in which the internal organs could be bathed directly in body fluids.[1] But the large bodies of these animals still required some way of moving the blood around through the hemocoel, hence a highly muscularized dorsal vessel was developed as a pumping structure—a heart. The excretory organs became closed internally, thereby preventing the blood from being drained from the body. Surface sense organs (the "arthropod setae") differentiated, becoming numerous and specialized and acquiring various devices for transmitting sensory impulses to the nervous system in spite of the hard exoskeleton. Gas exchange structures evolved in various ways that overcame the barrier of the exoskeleton.

For these animals, now encased in a rigid outer covering, growth was no longer a simple process of gradual increase in body size. Thus the complex process of **ecdysis**, a specific hormone-mediated form of molting or cuticular shedding, was "perfected." As we have already seen, some form of ecdysis occurs in all eight phyla belonging to the clade Ecdysozoa. In arthropods, it is through the process of ecdysis (molting) that the exoskeleton is periodically shed to allow for an increase in real body size. If we add to this suite of events the notion of arthropods invading terrestrial and freshwater environments, the evolutionary challenges become compounded by osmotic and ionic stresses, the necessity for aerial gas exchange, and the need for structural support and effective reproductive strategies.

While the origin of the exoskeleton demanded a host of coincidental changes to overcome the constraints it placed on arthropods, it clearly endowed these animals with great selective advantages, as evinced by their enormous success. One of the key advantages is the protection it provides. Arthropods are armored not only against predation and physical injury, but also against physiological stress. In many cases the cuticle provides an effective barrier against osmotic and ionic gradients, and as such is a major means of homeostatic control. It also provides the strength needed for segmental muscle attachment and for predation on other shelled invertebrates.

If we start with a generalized, rather homonomous arthropod prototype with a fairly high number of segments, and with paired appendages on each of those segments, we can set the stage for arthropod diversification. The diversity seen today has resulted largely from the differential specialization of various segments, regions, and appendages. The arthropod body has itself undergone various forms of regional specialization, or **tagmosis**, to produce segment groups specialized for different functions. These specialized body regions (e.g., the head, thorax, and abdomen) are called **tagmata**. Tagmosis is mediated by Hox genes and the other developmental genes they influence. Our emerging understanding of Hox genes tells

[1]The hemocoel is not a true coelom, either evolutionarily or ontogenetically, but it may be viewed as a persistent blastocoelic remnant. Thus, one might at first reason that the arthropods technically are blastocoelomates. However, the absence of a large body coelom in arthropods is a secondary condition resulting from a loss of the ancestral coelomic body cavity during the evolution of the arthropod body plan, not a primary condition like that seen in the true blastocoelomates, at least some of which probably never have had a true coelom in their ancestry. A similar secondary loss of the coelom has occurred, in a different way, in the molluscs (see Chapter 13).

us that the most fundamental aspects of animal design arise from spatially restricted expression of these "master developmental genes." However, tagmosis varies among the arthropod groups (see the classification above). The genetic and evolutionary plasticity of regional specialization, like limb variation, has been of paramount importance in establishing the diversity of the arthropods and their dominant position in the animal world.

One of the best examples of arthropod tagmosis is revealed by the expression pattern of the segment polarity gene *engrailed* (*en*) in the head, or cephalon, of Crustacea, Hexapoda, and Myriapoda, three arthropod subphyla comprising the clade known as Mandibulata. In each of these subphyla, the same seven head regions emerge during embryogenesis. The most anterior region is the presegmental acron. Following the acron is the first true segment (the protocerebral, or ocular segment) that lacks appendages (see Table 20.2). Next come the first and second antennal segments, the mandibular segment, and the first and second maxillary segments. This six-segmented (plus the acron) head pattern was long ago used in support of a grouping known as Mandibulata, which is currently supported by molecular phylogenetic data.

As we discuss the various aspects of the arthropod body plan below, and in subsequent chapters, do not lose sight of the "whole animal" and the "essence of arthropod" described in this section.

The Body Wall

A cross section through a body segment of an arthropod reveals a good deal about its overall architecture (Figure 20.20). As noted above, the body cavity is an open hemocoel, and the organs are bathed directly in the hemocoelic fluid, or blood (although some blood vessels do occur, notably in Crustacea). The body wall is composed of a complex, layered **cuticle** secreted by the underlying epidermis (Figure 20.21). The epidermis, often referred to in arthropods as the **hypodermis**, is typically a simple cuboidal epithelium. In general, each body segment (or **somite**) is "boxed" by skeletal plates called **sclerites**. Each somite typically has a large dorsal and ventral sclerite, the **tergite** and **sternite** respectively.[2] The side regions, or **pleura**, are flexible unsclerotized areas in which are embedded various minute, "floating" sclerites, the origins of which are hotly debated. The legs (and wings) of arthropods articulate in this pleural region. Numerous secondary deviations from this plan exist, such as fusion or loss of adjacent sclerites. Muscle bands are attached at points where the inner surfaces of sclerites project inward as ridges or tubercles, called **apodemes**.

The structure of the multilayered arthropod cuticle is similar to that of other ecdysozoans, although the layers can have different names in different phyla. Figure 20.21 illustrates the cuticles of an insect and a marine crustacean. The outermost layer is the **epicuticle**, which is itself multilayered. The external surface of the epicuticle is a protective lipoprotein layer—sometimes called the cement layer. Beneath this is a waxy layer that is especially well developed in terrestrial arachnids and insects. The waxes in this layer, which are long-chain hydrocarbons and the esters of fatty acids and alcohols, provide an effective barrier to water loss and, coupled with the outer lipoprotein layer, protection against bacterial invasion. These outermost two layers of the epicuticle largely isolate the arthropod's internal environment from the external environment.

[2]The terms tergum, sternum, and pleuron are often used interchangeably with tergite, sternite, and pleurite. However, the term tergum refers more precisely to the dorsal, or tergal region, and sternum to the ventral, or sternal region. Thus we restrict the use of the terms tergite (pl., tergites) and sternite (pl., sternites) to the specific skeletal plates, or sclerites. In some cases, the sternum is formed of multiple fused sternites.

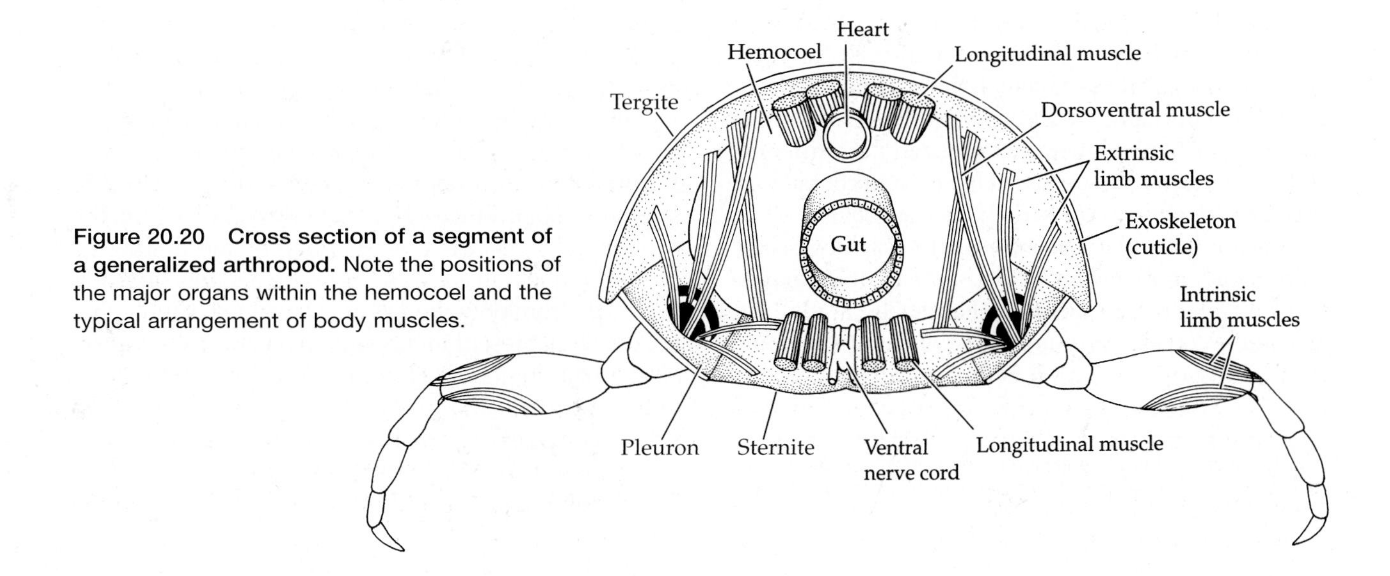

Figure 20.20 Cross section of a segment of a generalized arthropod. Note the positions of the major organs within the hemocoel and the typical arrangement of body muscles.

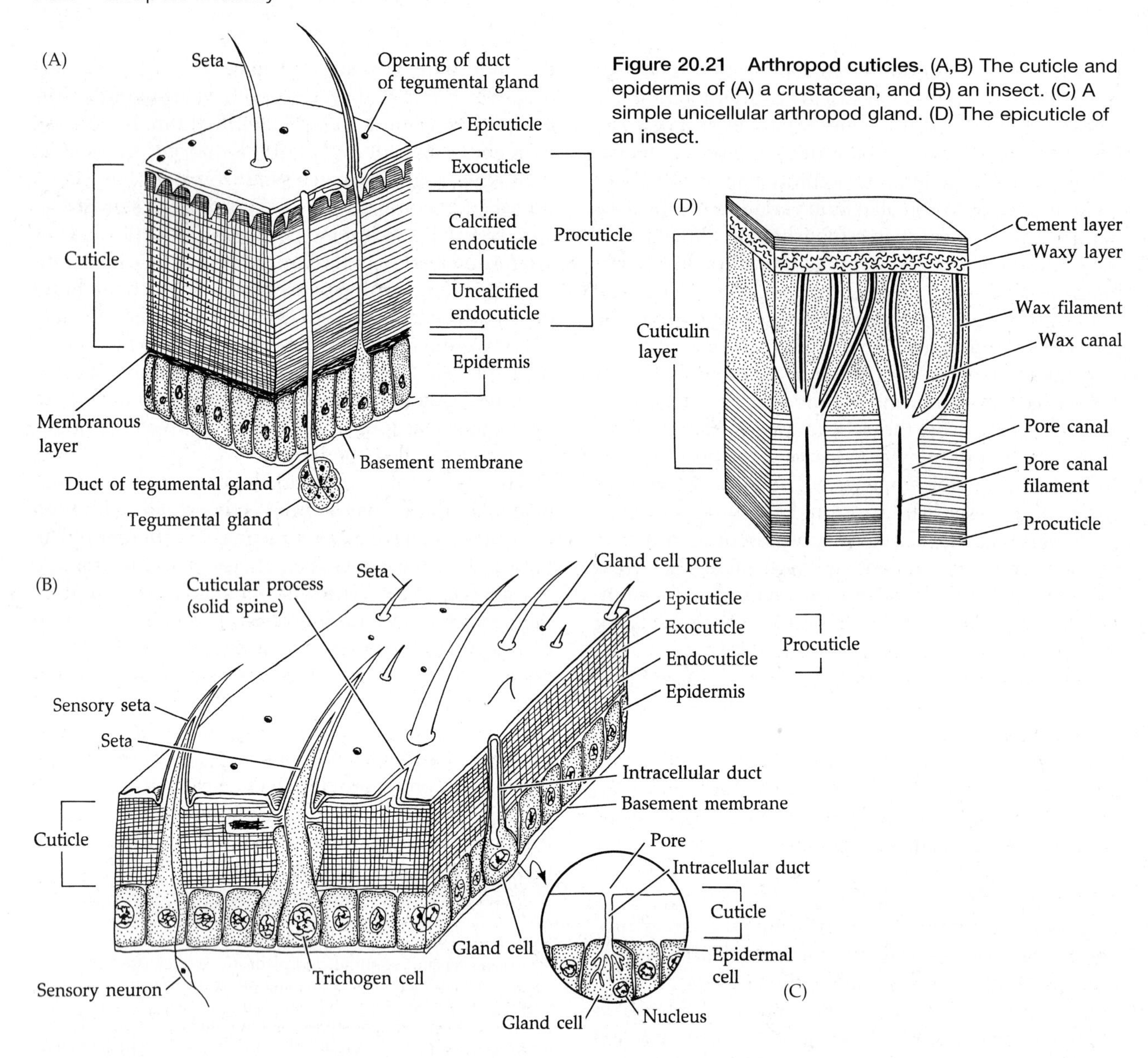

Figure 20.21 Arthropod cuticles. (A,B) The cuticle and epidermis of (A) a crustacean, and (B) an insect. (C) A simple unicellular arthropod gland. (D) The epicuticle of an insect.

No doubt the development of the epicuticle was critical to the invasion of land and fresh water by various arthropod lineages. The innermost layer of the epicuticle is a **cuticulin layer**, which consists primarily of proteins and is particularly well developed in insects. The cuticulin layer usually has two components: a thin but dense outer layer and a thicker, somewhat less dense inner layer. The cuticulin layer is involved in the hardening of the exoskeleton, as discussed below, and contains canals through which waxes reach the waxy layer.

Beneath the epicuticle is the relatively thick **procuticle**, which may be subdivided into an outer **exocuticle** and an inner **endocuticle** (Figure 20.21A,B).[3] The procuticle consists primarily of layers of protein and chitin (but no collagen). It is intrinsically tough, but flexible. In fact, certain arthropods possess rather soft and pliable exoskeletons (e.g., many insect larvae, parts of spiders, some small crustaceans). However, in most arthropods, the cuticle is hard and inflexible except at the joints, a condition brought about by one or both of two processes: sclerotization and mineralization.

Cuticular hardening by sclerotization (**tanning**) occurs to various degrees in all arthropods. The layered arrangement of untanned proteins yields a flexible structure. To produce a rigid sclerotized structure, the protein molecules are cross-bonded to one another by orthoquinone linkages. The bonding agent is typically produced from polyphenols and catalyzed by polyphenol oxidases present in the protein layers of the cuticle. Sclerotization generally begins in the cuticulin layer of the epicuticle and progresses into the procuticle to various degrees, where it is associated with a distinct darkening in color. The relationship between cuticular hardening, joints, and molting is discussed in the

[3]Caution: Some authors use the term "endocuticle" to refer to the entire procuticle of crustaceans, and use the two subdivision terms only when referring to insects.

section dealing with support, locomotion, and growth. Mineralization of the skeleton is largely a phenomenon of crustaceans, millipedes, and horseshoe crabs, and is accomplished by the deposition of calcium carbonate in the outer region of the procuticle.

The epidermis is responsible for the secretion of the cuticle, and as such contains various unicellular glands (Figure 20.21C), some of which bear ducts to the surface of the cuticle. Because the cuticle is secreted by the cells of epidermis, it often bears their impressions in the form of microscopic geometric patterns. The epidermis is underlain by a distinct basement membrane that forms the outer boundary of the body cavity or hemocoel.

Arthropod Appendages

Appendage anatomy In an evolutionary sense, one might be tempted to say "arthropods are all legs." Certainly, much of arthropod evolution has been about the appendages, modified in myriad ways over the 600-million-year history of this group. The unique combination of body segmentation and serially homologous appendages, in combination with the evolutionary potential of developmental genes, has allowed arthropods to develop modes of locomotion, feeding, and body region/appendage specialization that have been unavailable to the other metazoan phyla. The enormous variety of limb designs in arthropods has, unfortunately, also driven zoologists to create a plethora of terms to describe them. Read on, and we will try to walk you through this terminological jungle in the clearest fashion we can.

Primitively, every true body somite, or segment, probably bore a pair of appendages, or limbs. Arthropod appendages are articulated outgrowths of the body wall, equipped with sets of extrinsic muscles (connecting the limb to the body) and intrinsic muscles (wholly within the limb). The limbs of the other panarthropod phyla (Tardigrada and Onychophora) do not have intrinsic musculature and thus must rely strongly on hydraulics for movement. In the arthropods, muscles move the various limb segments or pieces, which are called **articles** (or podites).[4] The limb articles are organized into two groups, the basalmost group constituting the **protopod** (= sympod) and the distalmost group constituting the **telopod** (Figure 20.22). Whether the protopod is composed of one or more articles, the basalmost article is always called the **coxa** (in living arthropods). The telopod arises from the distalmost protopodite, or protopodal article. Sometimes the exoskeleton of the telopodites becomes annulated, forming a flagellum, as in the antennae of many arthropods, but these annuli should not be confused with true articles.

A great variety of additional structures can arise from articles of the protopod (the protopodites), either laterally (collectively called **exites**) or medially (collectively called **endites**) (Figure 20.22A). Evolutionary creativity among the protopodal exites has been exceptional among arthropods. In crustaceans and trilobites they form a diversity of structures such as gills, gill cleaners, and swimming paddles. Broad or elongate exites that function as gills or gill cleaners are often called **epipods**. Exites may become annulated, like the flagella of some antennae. Protopodal exites probably gave rise to the wings of insects. Protopodal endites, on the other hand, often form grinding surfaces, or "jaws," usually termed **gnathobases.** Figure 20.22 illustrates some arthropod appendage types and the terms applied to their parts.[5]

Appendages with large exites, such as gills, gill cleaners, or swimming paddles (the latter often developed in combination with a paddle-like telopod) are often called **biramous limbs** (or, sometimes, triramous or polyramous limbs). Biramous limbs occur only in crustaceans and trilobites, although their ancestral occurrence in chelicerates is suggested by the gills and other structures that may be derivatives of early limb exites. In crustaceans, the exite on the last protopodite can be as large as the telopod itself, and in these cases it is termed an **exopod**, the telopod then being called the **endopod** (Figure 20.22A). Biramous limbs are commonly associated with swimming arthropods, and in crustaceans in which they are greatly expanded

---

[4]Although some authors refer to the articles of the appendages as "segments," we attempt to restrict the use of the latter term to the true body segments, or somites, reserving the term "articles" for the separate "segments" of the limb.

[5]A basic morphological view of arthropod limb evolution is rooted in 30 years of detailed comparative morphology by Jarmila Kukalová-Peck, who studies both fossil and living arthropod limbs. In the Kukalová-Peck model, the ancestral arthropod appendage comprised a series of 11 articles (4 protopodites, 7 telopodites), each of which could theoretically bear an articulated endite or exite. The number of articles in her arthropod limb ground plan is not so important as her concept of a single series of articles, with endites and exites that specialized to become the diversity of structures seen in modern taxa. Over evolutionary time, so her theory goes, the basalmost protopodites fused with the pleural region of the body to form pleural sclerites in various taxa. On the thoracic segments of hexapods, the exite of the first protopodite (the epicoxa) migrated dorsally and gave rise to insect wings (see Chapter 22). On the other hand, many of the earliest known arthropod fossils (including trilobites) have protopods of a single article, as do many living arthropods, suggesting to some workers that multiarticulate protopods might be derived conditions. Kukalová-Peck's hypothesis, however, holds that such uniarticulate protopods represent cases in which protopodal articles have fused together (e.g., in trilobites) or migrated onto the pleural region of the body somites. The number of articles in the telopods of living arthropods varies greatly, reflecting, in Kukalová-Peck's view, various kinds of loss or fusion of articles. The elegance of Kukalová-Peck's theory is that it simply explains the origin of all arthropod limb structures. Viewing the arthropod limb ground plan as a series of articles from which endites and exites were modified in a variety of ways eliminates 100 years of confusion over the nature of uniramous, biramous, and polyramous limbs (these terms now having little phylogenetic significance).

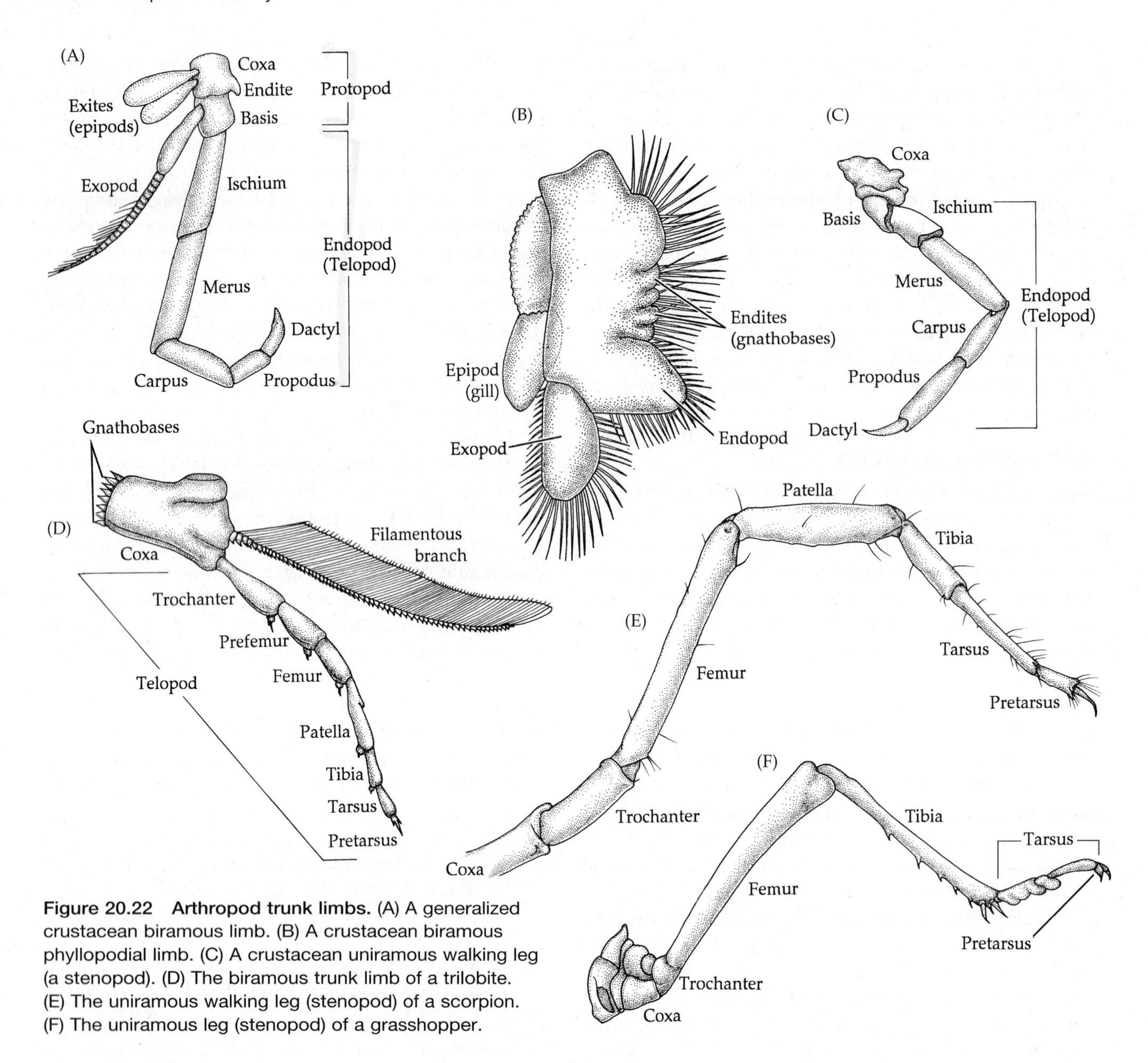

Figure 20.22 Arthropod trunk limbs. (A) A generalized crustacean biramous limb. (B) A crustacean biramous phyllopodial limb. (C) A crustacean uniramous walking leg (a stenopod). (D) The biramous trunk limb of a trilobite. (E) The uniramous walking leg (stenopod) of a scorpion. (F) The uniramous leg (stenopod) of a grasshopper.

and flattened (e.g., Cephalocarida, Branchiopoda, Phyllocarida), they may also be called foliacious limbs, or **phyllopodia** (Greek *phyllo*, "leaf-shaped"; *podia*, "feet") (Figure 20.22B).

Appendages without large exites are called **uniramous limbs** (or **stenopods**; Greek *steno*, "narrow"; *podia*, "feet") (Figure 20.22C). Uniramous limbs are characteristic of the chelicerates, hexapods, myriapods, and some crustaceans, although these appendages were probably secondarily derived from biramous limbs on more than one occasion. Uniramous legs are typically ambulatory (walking legs).

The combination of protopodal and telopodal articles, and their evolutionarily "plastic" endites and exites, has created in arthropods a veritable "Swiss army knife" of appendages. This diversity has no equal in the animal kingdom, and it has played a pivotal role in the evolutionary success of the phylum. As you peruse the following chapters, be sure to notice the phenomenal array of limb morphologies and adaptations among the arthropods.

Appendage evolution The amazing diversity seen in arthropod limbs has come about through the unique potential of homeobox (Hox) genes and other developmental genes, and the downstream genes they regulate, which are conserved and yet flexible in their expression. We are just beginning to understand how these genes work, and new information in this field is appearing so fast that we hesitate to go into great detail—our understanding of arthropod developmental biology is literally changing from one month to the next! We now know that the fates of arthropod appendages are largely under the ultimate control of Hox genes, which dictate where body appendages form and the general types of appendages that form. Hox genes can either suppress limb development or modify it to create alternative appendage morphologies. These unique genes have played major roles in the evolution of new body plans among arthropods (and other phyla).

A good example of the evolutionary potential of Hox genes is seen in the abdominal limbs of insects. Abdominal limbs (prolegs) occur on the larvae (but not the adults) of various insects in several orders, and they are ubiquitous in the order Lepidoptera (i.e., caterpillars). Abdominal limbs were almost certainly present in the crustacean ancestry of insects. Hence, prolegs may have reappeared in groups such as the Lepidoptera through something as simple as the derepression of an ancestral limb development program (i.e., they are a Hox gene–mediated atavism). We now know that proleg formation is initiated during embryogenesis by a change in the regulation and expression of the bithorax gene complex (which includes the Hox genes *Ubx*, *abdA*, and *AbdB*).

Molecular developmental biology has also begun to unravel the origins of arthropod appendages themselves. We now know that appendage development is orchestrated by a complex of developmental genes, in particular the genes *Distal-less* (*Dll*) and *Extradenticle* (*Exd*). Evidence suggests that *Exd* is necessary for the development of the proximal region of arthropod limbs (the protopod), whereas *Dll* is expressed in the distal region of developing appendages (the telopod). Thus the protopod and the telopod of arthropod appendages are somewhat distinct, each under its own genetic control and each, presumably, free to respond to the whims and processes of evolution. So, whether an arthropod mandible is a "telomeric," or "whole-limb," appendage (i.e., built of all, or most, of the full complement of articles) or a "gnathobasic" appendage (i.e., built of only the basalmost, or protopodal, articles) depends on whether or not (or how much) the gene *Dll* is expressed.

Dll is expressed throughout the development of the multiarticulate, telomeric chelicerae, and pedipalps of chelicerates, but only transiently in myriapod mandibles and in crustacean mandibles lacking palps. It is expressed throughout embryogeny in crustacean mandibles (in the mandibular palp), but not at all in the mandibles of hexapods. This expression pattern suggests that only the chelicerae and pedipalps of chelicerates are fully telomeric appendages, although the palp of the crustacean mandible represents the telopod of that limb. It also suggests that the development of a telopod is an evolutionarily flexible feature that can easily show homoplasy (i.e., parallelism). *Dll* is also expressed in the endites of arthropod limbs (e.g., in the phyllopodous limbs of Branchiopoda). In fact, *Dll* is an ancient gene that occurs in many animal phyla, where it is expressed at the tips of ectodermal body outgrowths in such different structures as the limbs of vertebrates, the parapodia and antennae of polychaete worms, the tube feet of echinoderms, and the siphons of tunicates.

So, we see that despite their considerable diversity, all arthropod limbs have a common ground plan and similar genetic mechanisms in their development. For example, developmental biology has now identified the appendage homologies of the head region for the major arthropod groups. A good example of developmental gene potential is revealed in Table 20.2, where we see the diversity of homologous head

TABLE 20.2 Homologies of Arthropod (and Onychophoran) Anterior/Head Appendages[a]

Segment	Onychophora	Pycnogonida	Chelicerata	Myriapoda	Crustacea	Hexapoda
1 (= protocerebral segment, ocular segment)	Antennae					
2 (= deutocerebral segment)	Jaws	Chelifores	Chelicerae	Antennae	First antennae (= antennules)	Antennae
3 (= tritocerebral segment)	Slime papillae (Note: Onychophorans probably do not have a tritocerebrum)	Palps	Pedipalps	–	Second antennae	–
4	Leg pair	Leg pair	Leg pair	Mandibles	Mandibles	Mandibles
5	Leg pair	Leg pair	Leg pair	First maxillae	First maxillae (= maxillules)	Maxillae
6	Leg pair	Leg pair	Leg pair	Second maxillae (or without appendages)	Second maxillae	Labium (= second maxillae)

[a]Gene expression studies have revealed the likelihood that the anteriormost appendages (the "protocerebral appendages") have been lost in all modern arthropods. A labrum (or "upper lip") occurs in all arthropod subphyla, and also in Onychophora. Its function is thought to be to prevent food particles from escaping ingestion. The evolutionary origin of the labrum has long been a mystery, and most workers have not regarded it as a true appendage of the head. However, recent gene expression studies suggest it might represent the remnant of a pair of ancient fused appendages. In insects and spiders, the labrum has been shown to arise from two primordia that fuse during embryogeny, controlled by the genes *decapentaplegic* (*dpp*) and *wingless* (*wg*), the same gene expressions as seen in the buds of regular appendages in arthropods.

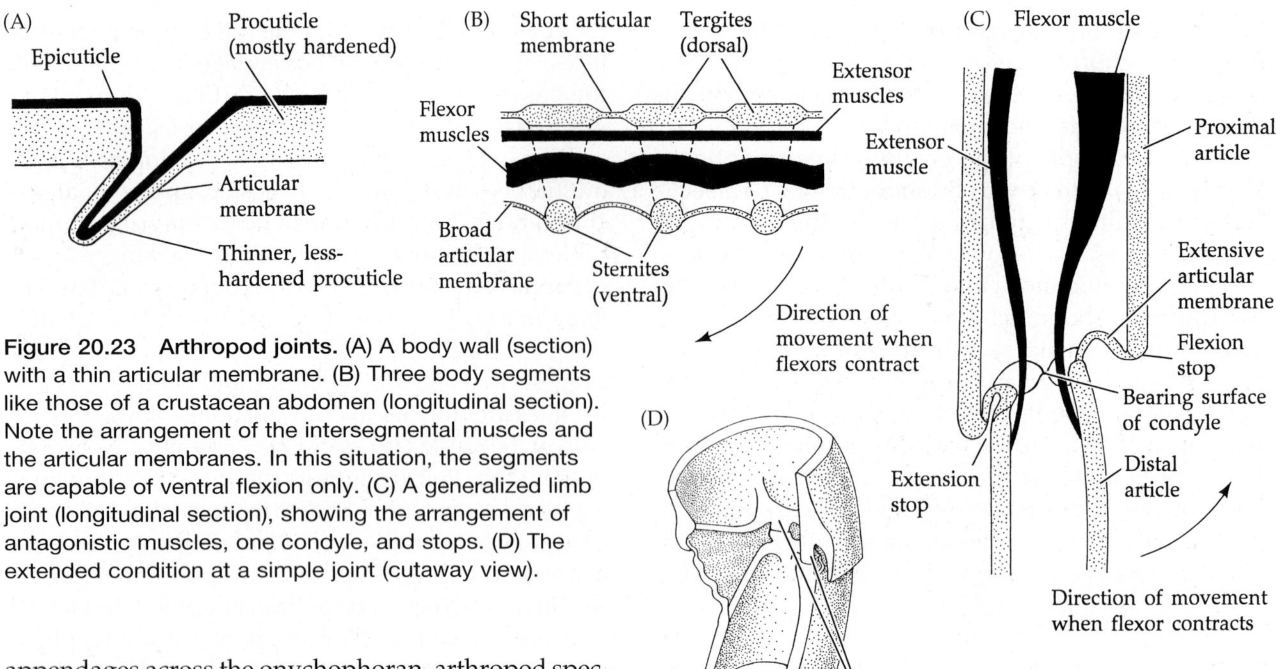

Figure 20.23 Arthropod joints. (A) A body wall (section) with a thin articular membrane. (B) Three body segments like those of a crustacean abdomen (longitudinal section). Note the arrangement of the intersegmental muscles and the articular membranes. In this situation, the segments are capable of ventral flexion only. (C) A generalized limb joint (longitudinal section), showing the arrangement of antagonistic muscles, one condyle, and stops. (D) The extended condition at a simple joint (cutaway view).

appendages across the onychophoran–arthropod spectrum. Imagine the jaws of velvet worms, chelicerae of scorpions, and antennae of insects all being derived from the same ancestral appendages! Evidence now suggests that the uniramous legs of hexapods (and perhaps also the myriapods) arose from the biramous (or uniramous) appendages of crustacean ancestors. Although morphologists have struggled to establish homologies among the specific articles, or podites, of arthropod legs, considerable debate still exists on that issue. But we are beginning to acquire the developmental genetic toolkit needed to resolve this and other issues across the arthropod subphyla.

Support and Locomotion

Arthropods rely on the exoskeleton for support and maintenance of body shape. Their muscles are arranged as short bands that extend from one body segment to the next, or across the joints of appendages and other regions of articulation. An understanding of the nature of these articulation points—the areas where the cuticle is notably thin and flexible—is crucial to an understanding of the action of the muscles and hence of locomotion. In contrast to most of the exoskeleton, the articulations (joints) between body and limb segments are bridged by areas of very thin, flexible cuticle in which the procuticle is much reduced and unhardened (Figure 20.23). These thin areas are called **arthrodial or articular membranes**. Generally, each articulation is bridged by one or more pairs of antagonistic muscles. One set of muscles, the flexors, acts to bend the body or appendage at the articulation point; the opposing set of muscles, the extensors, serves to straighten the body or appendage.

Joints that operate as described above generally articulate in only a single plane (much like your own knee or elbow joints). Such movement is limited not only by the placement of the antagonistic muscle sets, but also by the structure of the hard parts of the cuticle that border the articular membrane. In such cases the articular membrane may not form a complete ring of flexible material, but will be interrupted by points of contact between hard cuticle on either side of the joint. These contact points, or bearing surfaces, are called **condyles** and serve as the fulcrum for the lever system formed by the joint. A dicondylic joint allows movement in one plane, but not at angles to that plane. The motion at a joint is also usually limited by hard cuticular processes called locks or stops, which prevent overextension and overflexion (Figure 20.23C,D).

Some joints are constructed to allow movement in more than one plane, much like a ball-and-socket joint. For example, in most arthropods the joints between walking legs and the body (the coxal–pleural joints) lack large condyles, and the articular membranes form complete bands around the joints. In other cases, two adjacent dicondylic limb joints articulate at 90° to one another, forming a gimbal-like arrangement that facilitates movement in two opposing planes.

Arthropods have evolved a plethora of locomotor devices for movement in water, on land, and in the air. Only the vertebrates can boast a similar range of abilities, albeit utilizing a far less diverse set of mechanisms. Like so many other aspects of arthropod biology, their methods of movement reflect the extreme evolutionary plasticity and adaptive qualities associated with the segmented body and appendages.

Movement through water involves various patterns of swimming that include smooth paddling by

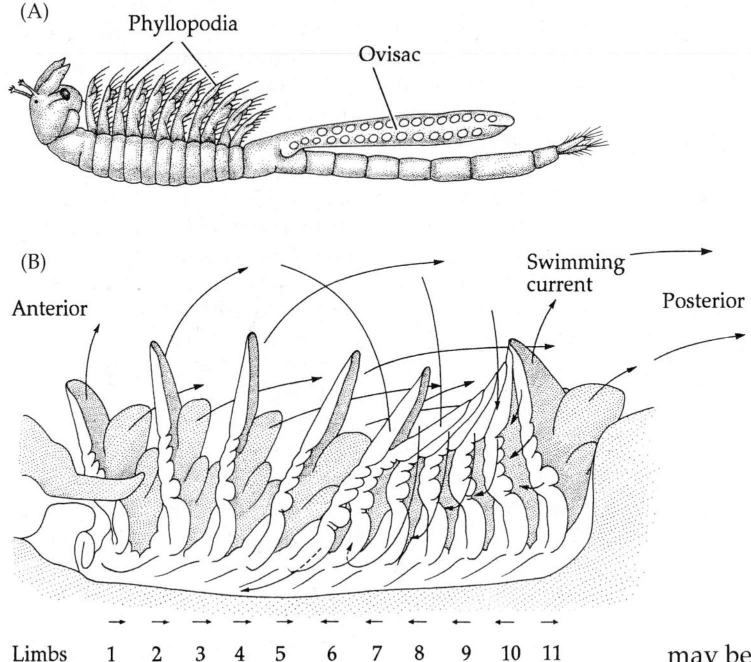

Figure 20.24 Swimming motions in a primitive crustacean. (A) A fairy shrimp (Anostraca) on its back in its normal swimming posture. (B) The appendages "in motion," producing a posteriorly directed flow of water that propels the animal forward. Arrows near the bases of the appendages indicate feeding currents. The small arrows below the drawing indicate the direction of movement of each numbered appendage at this moment in the anterior progression of the metachronal wave. Water is drawn into the interlimb spaces as adjacent appendages move away from one another, and water is pressed out of the spaces as adjacent limbs move together. The lateral articles of these phyllopodial appendages are hinged in such a way that they extend on the power stroke to present a large surface area and collapse on the recovery stroke, thereby producing less drag.

shrimps, jerky stroking by certain insects and small crustaceans, and startling backward propulsion by tail flexion in lobsters and crayfish. Aerial locomotion has been mastered by the pterygote (winged) insects, but is also practiced by certain spiders that drift on threads of silk. Many arthropods burrow or bore into various substrata (e.g., ants, bees, termites, burrowing crustaceans). Some terrestrial arthropods that are normally associated with the ground engage in short-term aerial movements that serve as escape responses. Some, like fleas, simply jump, whereas others jump and glide, giving us possible clues to the evolutionary origin of flight. Some crustaceans jump as well, such as the familiar beach hoppers (amphipods) that bound away over the sand when disturbed. Arthropods that move in contact with the surface of the substratum, under water or on land, by various forms of walking, creeping, crawling, or running are referred to as pedestrian or reptant.

All of the common forms of arthropod locomotion except flight depend on the use of typical appendages and thus are based on the principles of joint articulation described above, coupled with specialized architecture of the appendages. Below we discuss some aspects of two fundamental types of appendage-dependent locomotion in arthropods, swimming and pedestrian locomotion, exploring variations on these methods and others in subsequent chapters.

Many examples of swimming arthropods are found among the crustaceans. Most swimming crustaceans (e.g., anostracans and shrimps) and even those that swim only infrequently (e.g., isopods and amphipods) employ ventral, flaplike setose appendages as paddles (Figure 20.22B). The appendages used for swimming may be restricted to particular body regions (e.g., the abdominal swimmerets of shrimps, stomatopods, and isopods; the metasomal limbs of swimming copepods) or may occur along much of the trunk (e.g., the appendages of anostracans, remipedes, and cephalocarids) (Figure 20.24). These appendages engage in a backward power (propulsive) stroke and a forward recovery stroke. In all cases, the appendages are constructed in such a way that on the recovery stroke, they are flexed and the flaps and marginal setae passively "collapse" to reduce the coefficient of friction (drag). On the power stroke the limbs are held erect, with their largest surface facing the direction of limb movement, thus increasing thrust efficiency (by increasing the coefficient of friction and the distance through which the limb travels). These swimming appendages typically articulate with the body only on a plane parallel to the body axis. Less sophisticated swimming is accomplished in other arthropods by use of various other appendages, including the antennae of many minute crustaceans and larvae and the thoracic stenopods of many aquatic insects.

Pedestrian locomotion in arthropods is highly variable, both among different groups and even in individual animals. With the exception of a few strongly homonomous "vermiform" types (e.g., centipedes and millipedes), most arthropods are incapable of lateral body undulations. Thus, they cannot amplify the stride length of their appendages by body waves (as many polychaetes do, for example). Walking arthropods depend almost entirely on the mobility of specialized groups of appendages. The structure of these ambulatory legs is quite different from that of paddle-like swimming appendages, and their action is much more complex and variable.

Consider the general movement of an ambulatory leg as it passes through its power and recovery strokes

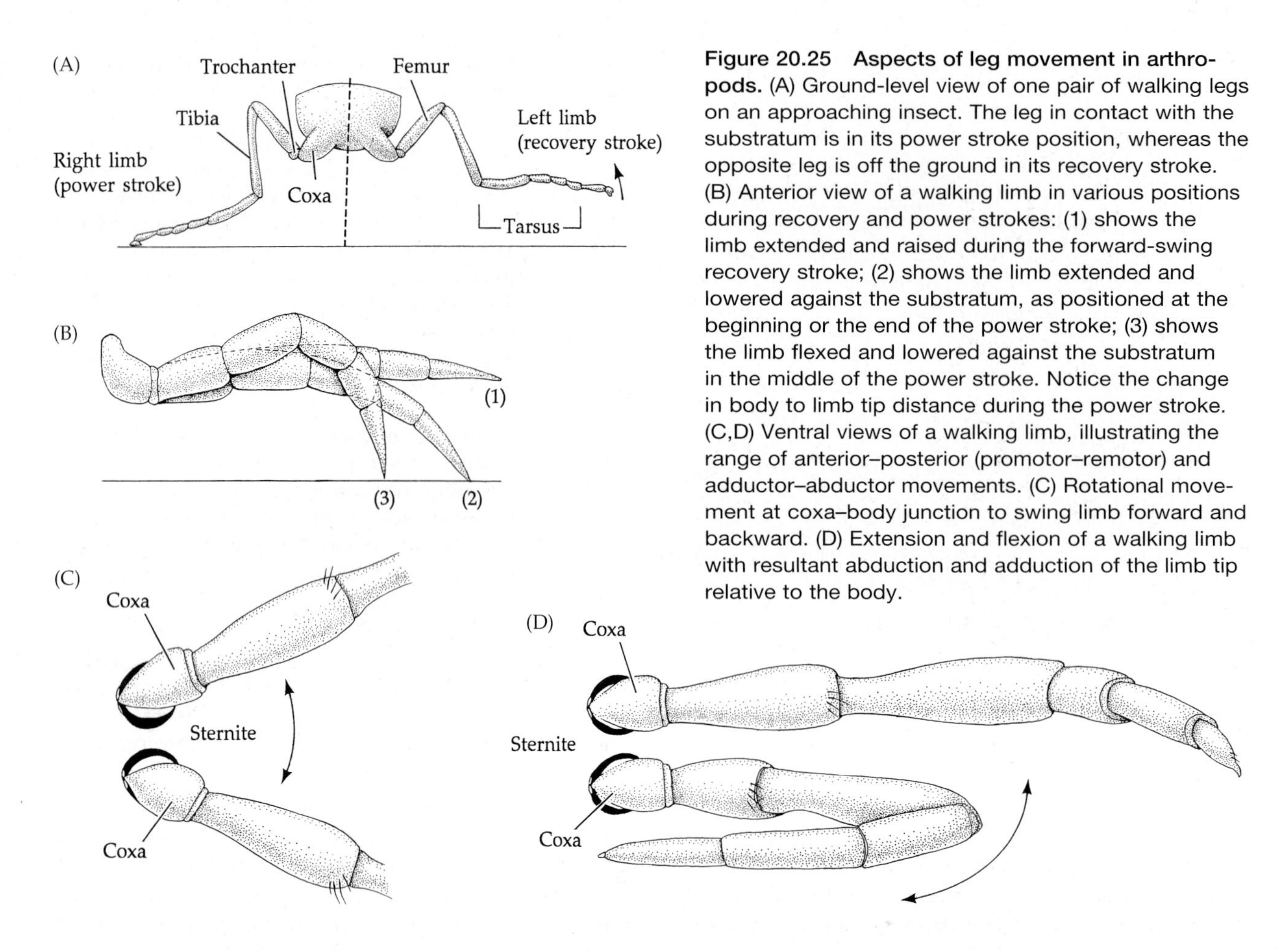

Figure 20.25 Aspects of leg movement in arthropods. (A) Ground-level view of one pair of walking legs on an approaching insect. The leg in contact with the substratum is in its power stroke position, whereas the opposite leg is off the ground in its recovery stroke. (B) Anterior view of a walking limb in various positions during recovery and power strokes: (1) shows the limb extended and raised during the forward-swing recovery stroke; (2) shows the limb extended and lowered against the substratum, as positioned at the beginning or the end of the power stroke; (3) shows the limb flexed and lowered against the substratum in the middle of the power stroke. Notice the change in body to limb tip distance during the power stroke. (C,D) Ventral views of a walking limb, illustrating the range of anterior–posterior (promotor–remotor) and adductor–abductor movements. (C) Rotational movement at coxa–body junction to swing limb forward and backward. (D) Extension and flexion of a walking limb with resultant abduction and adduction of the limb tip relative to the body.

(Figure 20.25). At the completion of the power stroke, the appendage is extended posteriorly and its tip is in contact with the substratum. The recovery stroke involves lifting the limb, swinging it forward, and placing it back down on the substratum; by then the limb is extended anterolaterally. The power stroke is accomplished by first flexing and then extending the leg while the tip is held in place against the substratum. Thus the body is first pulled and then pushed forward by each limb.

These complicated movements obviously would not be possible if all of the limb joints and limb–body joints were dicondylic articulations in the same plane, parallel to the body axis. The leg must be able to move up and down as well as forward and backward, and the action at each joint must be coordinated with the actions of all the others. In general, the distal limb joints are dicondylic, with articulation (and movement) planes parallel to the limb axis. They allow the appendage to flex and extend, that is, to move the tip closer to (adduction) or farther from (abduction) the point of limb origin. The actions of these joints typically involve the usual sets of antagonistic flexor and extensor muscles described earlier. In some arachnids and a few crustaceans, however, certain limb joints lack extensor muscles, and the limbs are extended by an increase in blood pressure. Raising and lowering of the limb are also accomplished by extensor and flexor muscles, which thus serve as levators and depressors, respectively; the muscles in the proximal leg joints usually serve these purposes.

Anterior–posterior limb movements are accomplished in two basic ways. First, the ball-and-socket type of joint at the point of limb–body articulation typically carries out these actions in most crustaceans, insects, and myriapods. Promotor and remotor muscles that are associated with these joints rotate the limb forward and backward, respectively. Second, many arachnids accomplish multidirectional limb movements by using only uniplanar dicondylic joints. In these arthropods, one or more of the proximal joints articulate perpendicular to the limb axis, and thus to the rest of the limb joints, providing forward and backward movement.

Understanding how a single limb moves does not, of course, describe the locomotion of the whole animal. The various patterns of pedestrian locomotion in arthropods, called gaits, are the result of many factors (e.g., leg number, leg movement sequences, stride lengths, speed). The number of patterns is great, but it is limited by certain biological and physical constraints. Speed is limited by rates of muscle contraction and the necessity for coordinating leg movements to avoid tangling. Furthermore, the animal must maintain an appropriate distribution of legs at all times in various phases of power and recovery strokes so that its weight is fully supported.

The gaits of insects have been more extensively studied than those of other arthropods. Studies on insects and myriapods led to an attempt to establish principles under which all pedestrian arthropod locomotion could be unified. The most frequently used descriptions of arthropod walking, crawling, and running are based on the "metachronal model." The basic idea of this model is that the legs on each side of the body move in metachronal (repeated) waves from back to front and that the waves overlap to various degrees, depending on the speed of movement. This model does work for some arthropods, some of the time, but things are not so simple, and attempts to over generalize have been misleading. A good deal of the work on crustaceans and arachnids (and even insects) indicates that leg movement sequences, stepping patterns, stride lengths, and other characteristics are extremely variable, even within individuals, and depend on a host of factors other than speed. The actions of the joints are coordinated by information supplied to the central nervous system by proprioceptors in the joints themselves.

Growth

The imposition of a rigid exoskeleton on the arthropods precludes growth by means of a gradual increase in external body size. Rather, an overall increase in body size takes place in staggered increments associated with the periodic loss of the old exoskeleton and the deposition of a new, larger one (Figure 20.26). The process of shedding the exoskeleton is called **molting**, and it is a phenomenon characteristic of arthropods and all other ecdysozoans. The molting process varies in detail even among the arthropods. It has been best studied in certain insects and crustaceans, and the description below is based primarily on those two groups. We first outline the basic steps in the arthropod molt cycle and then briefly discuss the hormonal control of those events. In all arthropods (and probably all ecdysozoans) molting is regulated by a hormone called **ecdysone**; thus, the entire molting process in these groups is referred to as **ecdysis.**

The stages between molts are called **intermolts** (or **instars**, in insects). In arthropods, it is during these intermolt stages that real tissue growth occurs, although with no increase in external size. When such tissue growth reaches the point at which the body "fills" its exoskeletal case, the animal usually enters a physiological state known as **premolt or proecdysis**. During this stage there is active preparation for the molt, including accelerated growth of any regenerating parts. Certain epidermal glands secrete enzymes that begin digesting the old endocuticle, thus separating the exoskeleton from the epidermis. In many crustaceans, some of the calcium is removed from the cuticle during this period

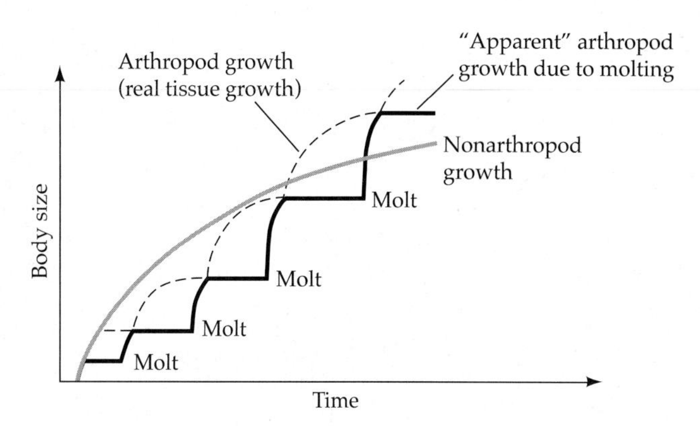

Figure 20.26 Arthropod versus non-arthropod growth. The heavy solid line indicates the incremental ("stair-step") growth pattern of an arthropod as measured by changes in external body size associated with molts. The dotted line depicts real tissue growth in the same arthropod. The gray line depicts the typical growth of a non-arthropod.

and stored within the body for later redeposition. As the old cuticle is loosened and thinned, the epidermis begins secreting a soft new cuticle. Figure 20.27 depicts some of these events.

Once the old cuticle has been substantially loosened and a new cuticle formed, actual molting occurs. The old cuticle splits in such a way that the animal can wriggle free and pull itself out. The lines along which the cuticle splits vary among the arthropods but are consistent within particular groups. It is important to remember that all cuticular linings are lost during ecdysis, including the linings of the foregut and hindgut, the eye surfaces, and the cuticle that lines every pit, groove, spine, and seta on the body surface. When you see a cast-off intact exoskeleton, or **exuvium**, of an arthropod, you are bound to be impressed with its wonderfully perfect detail. It is at first difficult to imagine how the animal could extricate itself from each and every tiny part of the old cuticle (Figure 20.27C,D). The ability to do so depends, of course, on the great flexibility of the body within its new and unhardened exoskeleton.

As soon as the arthropod emerges from its old cuticle, and while the new cuticle is still soft and pliable, its body swells rapidly by taking up air or water. Once the new cuticle is thus enlarged, the animal enters a postmolt period (**postecdysis**) during which the cuticle is hardened by sclerotization and/or the redeposition of calcium salts. The excess water (or air) is then actively pumped from the body, and real tissue growth occurs during the subsequent **intermolt** period. During the sclerotization process, the cuticle becomes drier, stiffer, and resistant to chemical and physical degradation through the molecular cross-linking process in the protein–chitin matrix described earlier.

We have stressed the adaptive significance of the arthropod exoskeleton in terms of its protective and

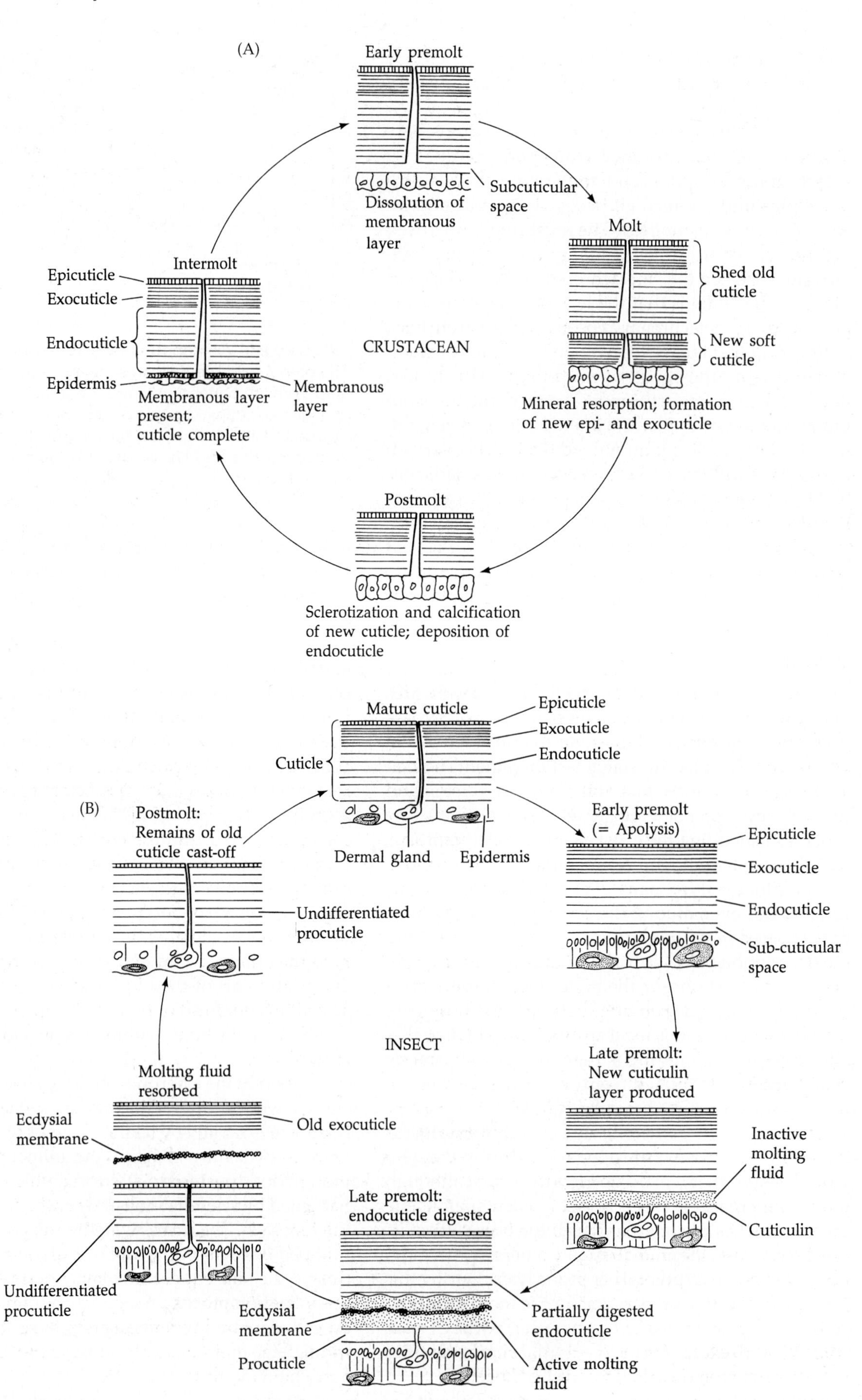
(A)
Early premolt
Subcuticular space
Dissolution of membranous layer
Molt
Shed old cuticle
New soft cuticle
Mineral resorption; formation of new epi- and exocuticle
Intermolt
Epicuticle
Exocuticle
Endocuticle
Epidermis
Membranous layer
Membranous layer present; cuticle complete
CRUSTACEAN
Postmolt
Sclerotization and calcification of new cuticle; deposition of endocuticle
(B)
Mature cuticle
Cuticle
Epicuticle
Exocuticle
Endocuticle
Dermal gland
Epidermis
Postmolt: Remains of old cuticle cast-off
Undifferentiated procuticle
Early premolt (= Apolysis)
Epicuticle
Exocuticle
Endocuticle
Sub-cuticular space
INSECT
Molting fluid resorbed
Ecdysial membrane
Old exocuticle
Undifferentiated procuticle
Late premolt: New cuticulin layer produced
Inactive molting fluid
Cuticulin
Late premolt: endocuticle digested
Ecdysial membrane
Procuticle
Partially digested endocuticle
Active molting fluid

(C)

(D)

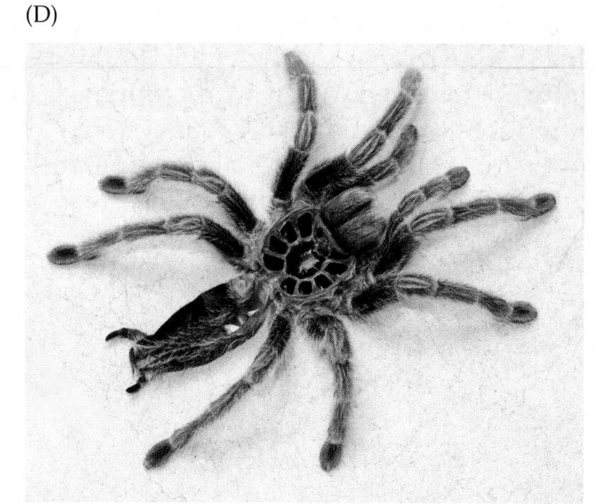

Figure 20.27 Arthropod molting. Schematic representations of some events in the molting of (A) a crustacean and (B) an insect. The separation of the old cuticle from the body is generally accomplished by dissolution of the membranous layer in large crustaceans and by digestion of the inner boundary of the cuticle in insects. (C) A swimming crab in the final stage of extracting itself from its old, molted exoskeleton; only the chelipeds remain to be pulled out of the exuvium. (D) The cast-off exoskeleton of a tarantula.

supportive qualities. However, during the **postmolt period**, before the new exoskeleton is hardened, the animal is quite vulnerable to injury, predation, and osmotic stress. Many arthropods become reclusive at this time, hiding in protective nooks and crannies and not even feeding when in this "soft-shell" condition. The time required for hardening of the new exoskeleton varies greatly among arthropods, generally being longer in larger animals. The well-known and delectable "soft-shell crabs" of the eastern United States are simply blue crabs (*Callinectes*) caught during their postmolt period.

Many genes and a complex hormonal system regulate the molt cycle (Figure 20.28). Several models have been proposed to explain the hormonal pathways involved in molting in insects and crustaceans, and the picture is still somewhat incomplete. The hormonal activities of the crustacean ecdysial cycle have been most extensively studied in decapods. In some (e.g., lobsters and crayfish), molting occurs periodically throughout the animal's life, but in many others (e.g., copepods and some crabs), molting, and therefore growth, ceases at some point, and a maximum size is attained. Animals that have engaged in their final molt are said to have entered a state of **anecdysis**, or permanent intermolt—they are in their final instar. Among insects, molting is largely associated with metamorphosis from one developmental stage to the next (e.g., pupa to adult), and, except for the most primitive hexapods, adults do not molt (i.e., they are in anecdysis).

In both crustaceans and hexapods (and probably all arthropods), the initiation of molting, beginning with the events of proecdysis, is brought about by the action of a molting hormone called ecdysone. Apparently, however, the pathways controlling the secretion of ecdysone are different in insects and crustaceans, as diagrammed in Figure 20.28. In crustaceans ecdysone is secreted by an endocrine gland called the Y-organ located at the base of the antennae or near the mouthparts. The action of the Y-organ is controlled by a complex neurosecretory apparatus located near the eyes or in the eyestalks. During the intermolt period, a **molt-inhibiting hormone (MIH)** is produced by neurosecretory cells of the X-organ, located in a region of the eyestalk nerve (or ganglion) called the medulla terminalis (Figure 20.28A). MIH is carried by axonal transport to a storage area called the sinus gland, which appears to control MIH release into the blood. As long as sufficient levels of MIH are present in the blood, the production of ecdysone by the Y-organ is inhibited.

The active premolt and subsequent molt phases are initiated by sensory input to the central nervous system. The stimulus is external for some crustaceans (e.g., day length or photoperiod for certain crayfishes) and internal for others (e.g., growth of soft tissues in certain crabs). External stimuli are transmitted via the central nervous system to the medulla terminalis and X-organ (Figure 20.28B). Appropriate stimuli inhibit the secretion of MIH, ultimately resulting in the production of ecdysone and the initiation of a new molt cycle.

The sequence of events in insects is somewhat different from that in crustaceans in that a molt inhibitor is apparently not involved. When an appropriate stimulus is introduced to the central nervous system, certain neurosecretory cells in the cerebral ganglia are activated. These cells, which are located in the pars intercerebralis, secrete **ecdysiotropin**. This hormone is carried by axonal transport to the corpora cardiaca, paired neural masses associated with the cerebral ganglia. Here, thoracotropic hormone is produced and carried to the prothoracic glands, stimulating them to produce and release ecdysone (Figure 20.28C).

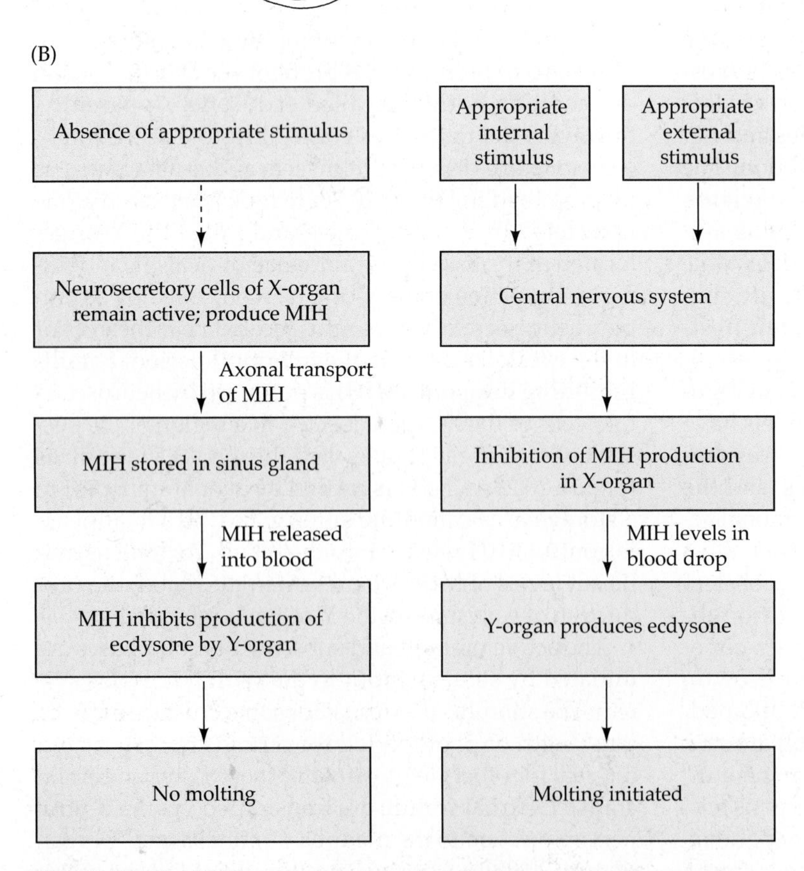

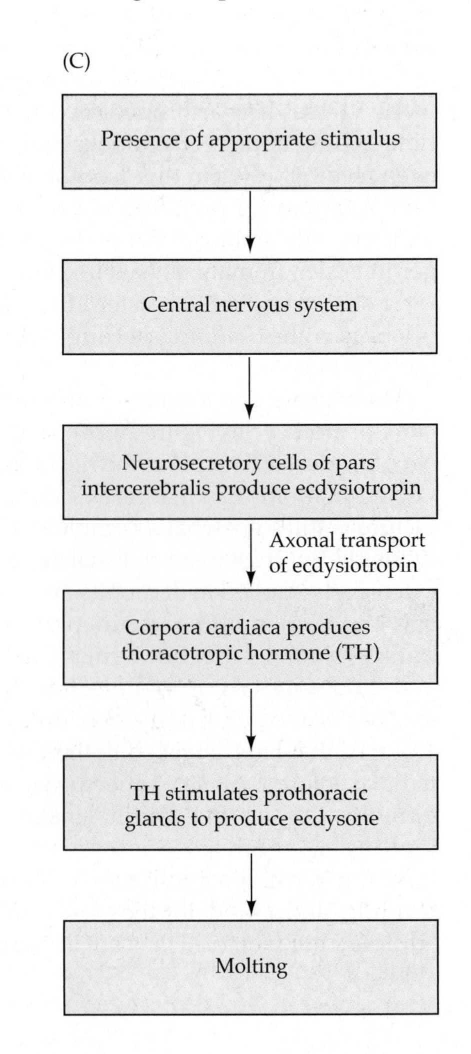

Figure 20.28 (A) The neurosecretory apparatus in a crustacean eyestalk. (B) Flow diagram of events inhibiting and initiating molting in crustaceans. (C) Flow diagram of events initiating molting in an insect.

The Digestive System

It will come as no surprise that the great diversity among arthropods is reflected in their display of nearly every feeding method imaginable. As a group, the only real constraint on arthropods in this regard is the absence of external, functional cilia. Evolutionarily, many arthropods have overcome even this limitation and suspension feed by other means. So varied are arthropod feeding strategies that we postpone discussion of them to the sections and chapters on particular taxa, and here attempt only to generalize about the basic structure and function of arthropod digestive systems.

The digestive tract of arthropods is complete (a through gut) and generally straight, extending from a ventral mouth on the head to a posterior anus. Various appendages (the mouthparts and other associated appendages) may be associated with processing food and moving it to the mouth. Regional specialization of the

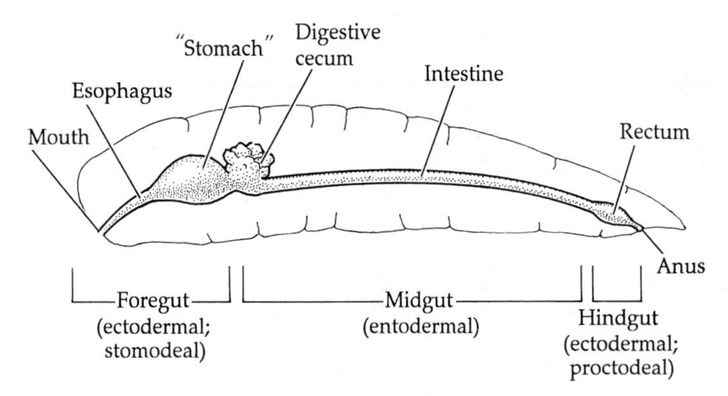

Figure 20.29 The major gut regions of arthropods. The myriad variations on this theme are discussed in subsequent chapters on particular taxa.

gut occurs in most taxa. In almost all cases there is a well-developed, cuticle-lined, stomodeal foregut and proctodeal hindgut, connected by an endodermally derived midgut (Figure 20.29). In general, the foregut serves for ingestion, transport, storage, and mechanical digestion of food; the midgut for enzyme production, chemical digestion, and absorption; and the hindgut for water absorption and the preparation of fecal material. The midgut typically bears one or more evaginations in the form of **digestive ceca** (often referred to as the "digestive gland," "liver," or "hepatopancreas"). The number of ceca and the arrangement of the other gut regions vary among the different taxa. A characteristic feature of arthropods (and tardigrades) is the enclosure of the material being digested in the hindgut within a permeable **peritrophic membrane**, which allows digestive fluids to flow in and water and nutrients to flow out. Arthropod feces are typically "packaged" in the remains of the peritrophic membrane.

The various terrestrial arthropods have convergently evolved many similar features as adaptations to life on land. Many of these convergent structures are associated with (although not necessarily derived from) the gut. For example, excretory structures called **Malpighian tubules** (see below) that develop from the midguts or hindguts of insects, arachnids, myriapods, and tardigrades appear to be convergences (i.e., nonhomologous structures). The excretory structures of onychophorans also used to be called Malpighian tubules, but they have recently been shown to be complex metanephridia with secondarily derived, closed end sacs. Many unrelated terrestrial taxa have special **repugnatorial glands**, which may or may not be associated with the gut and which produce noxious substances used to deter predators. Many different groups of terrestrial arthropods also have evolved the ability to produce silks or silk-like substances for use outside their bodies. These silk-like fibers are produced by nonhomologous structures among different arthropods. Although they vary greatly in chemical composition, all share a common molecular feature that gives them strength and elasticity; they are composed of regular assemblies of long-chain macromolecules (most being fibrous proteins) and many also incorporate collagens. Modified salivary glands are common silk-producing organs, but silks are also secreted by the digestive tract, Malpighian tubules, accessory reproductive glands, and assorted dermal glands. Silk production occurs in chelicerates (false scorpions, spiders, and mites), many insect orders (such as adult webspinners or Embioptera, and the larvae of commercial silkworm moths *Bombyx* and *Anaphe*), some myriapods, and some crustaceans (e.g., amphipods). Arthropod silks are used in the production of cocoons, egg cases, webs, larval "houses," flotation rafts, prey entrapment threads, draglines, spermatophore receptacles, intraspecific recognition devices, and other sundry items. The truly spectacular array of silk uses by spiders is discussed in Chapter 24. Silk production and use provide one of the more spectacular examples of evolutionary convergence seen in the arthropods.

Circulation and Gas Exchange

A major aspect of the arthropod body plan is reflected in the nature of the circulatory system. The largely open hemocoelic system is in part a result of the imposition of the rigid exoskeleton and the loss of an internally segmented and fluid-filled coelom. We have seen that isolated coelomic spaces (such as the segments of annelids) require a closed circulatory system to service them, but this requirement is not present in arthropods. Furthermore, without a muscular, flexible body wall to augment blood movement, a pumping mechanism becomes necessary, resulting in the elaboration of a muscular heart. The result is a system wherein the blood is driven from the heart chamber through short vessels and into the hemocoel, where it bathes the internal organs. The blood returns to the heart via a noncoelomic pericardial sinus and perforations in the heart wall called ostia (Figure 20.30). The blood flows back to the heart along a decreasing pressure gradient resulting from lowered pressure within the pericardial sinus as the heart contracts. The complexity of the circulatory system varies greatly among arthropods, the differences being dependent in large part on body size and shape. These differences include variations in the size and shape of the heart (Figure 20.30B–D), the number of ostia, the length and number of vessels, the arrangement of hemocoelic sinuses, and the circulatory structures associated with gas exchange.

Arthropod blood, or **hemolymph**, serves to transport nutrients, wastes, and usually gases. It includes a variety of types of amebocytes and, in some groups, clotting agents. The blood of many kinds of small arthropods is colorless, simply carrying gases in solution. Most of the larger forms, however, contain hemocyanin, and a few contain hemoglobin. Both pigments are always dissolved in the hemolymph rather than contained within

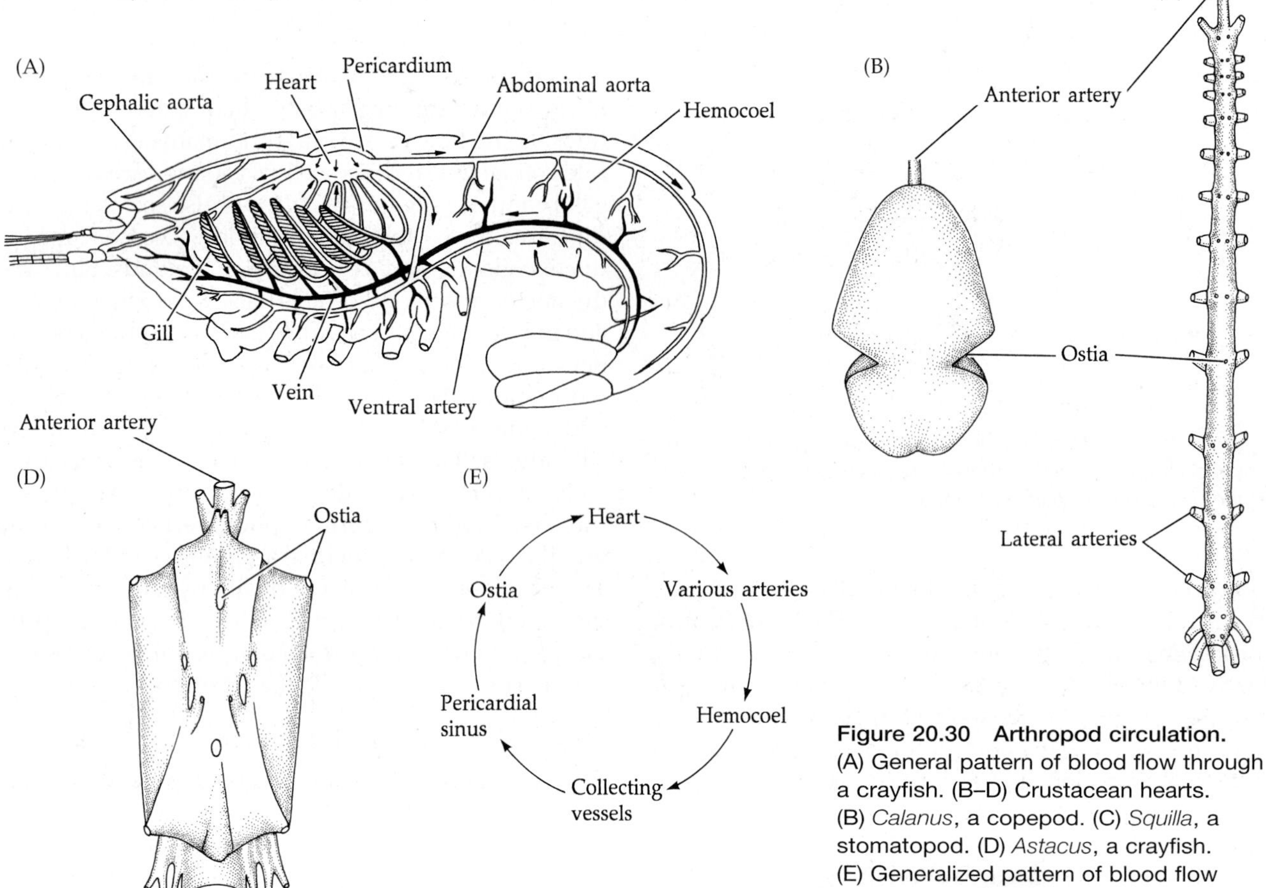

Figure 20.30 Arthropod circulation. (A) General pattern of blood flow through a crayfish. (B–D) Crustacean hearts. (B) *Calanus*, a copepod. (C) *Squilla*, a stomatopod. (D) *Astacus*, a crayfish. (E) Generalized pattern of blood flow through an arthropod.

cells. In most groups of arthropods, the circulatory route takes at least some of the blood past the gas exchange surfaces (e.g., gills) before returning to the heart.

One of the major evolutionary problems arising from the acquisition of a relatively impermeable exoskeleton involves gas exchange, particularly for terrestrial arthropods. On land, any increase in cuticular permeability to facilitate gas exchange also increases the threat of water loss. Remember that gas exchange surfaces not only must be permeable but also must be kept moist (see Chapter 4). Evolutionarily, the challenge for the arthropods becomes one of disrupting the integrity of the exoskeleton in such a way as to allow gas exchange without seriously jeopardizing the survival of the animal by abandoning the principal benefits of the exoskeleton.

The design of arthropod gas exchange structures has taken one form in aquatic groups and quite another in terrestrial taxa (Figure 20.31). The former is best exemplified by the crustaceans and the latter by the insects and terrestrial chelicerates. Some very tiny crustaceans (e.g., copepods) with a low surface area-to-volume ratio exchange gases cutaneously across the general body surface or at thin cuticular areas such as articulating membranes. However, most of the larger crustaceans have evolved various types of gills in the form of thin-walled, hemolymph-filled cuticular evaginations. Gills are commonly branched or folded, providing large surface areas (Figure 20.31A). The gills of some crustaceans (e.g., euphausids) are exposed, unprotected, to the surrounding medium, whereas in others (e.g., crabs and lobsters) the gills are carried beneath protective extensions of the exoskeleton.

The most successful terrestrial arthropods—the insects and arachnids—have evolved gas exchange structures in the form of invaginations of the cuticle, rather than the evaginations seen in aquatic crustaceans. Obviously, external gills would be unacceptable in dry conditions, but placed internally, these gas exchange structures remain moist and act as humidity chambers, allowing oxygen to enter solution for uptake. Many arachnids possess invaginations called **book lungs**, which are sacciform pockets with thin, highly folded walls called lamellae (Figure 20.31C). Hexapods, myriapods, and many arachnids possess inwardly directed branching tubules called **tracheae**, which open externally through pores called **spiracles** (Figure 20.31B). In insect tracheal systems the inner ends of the tubules lie in the hemocoel or are embedded in organ tissues, allowing direct gas exchange between the air and the blood and internal organs. The trachea of arachnids are probably not directly homologous to those of insects. Some terrestrial crustaceans (isopods) have **pseudotracheae** on the abdominal appendages; these structures are short, branching tubes that bring air close to the blood-filled spaces in these appendages (Figure 20.31D).

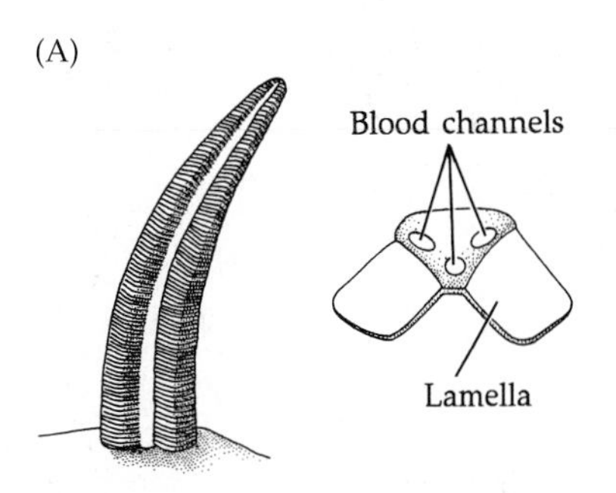

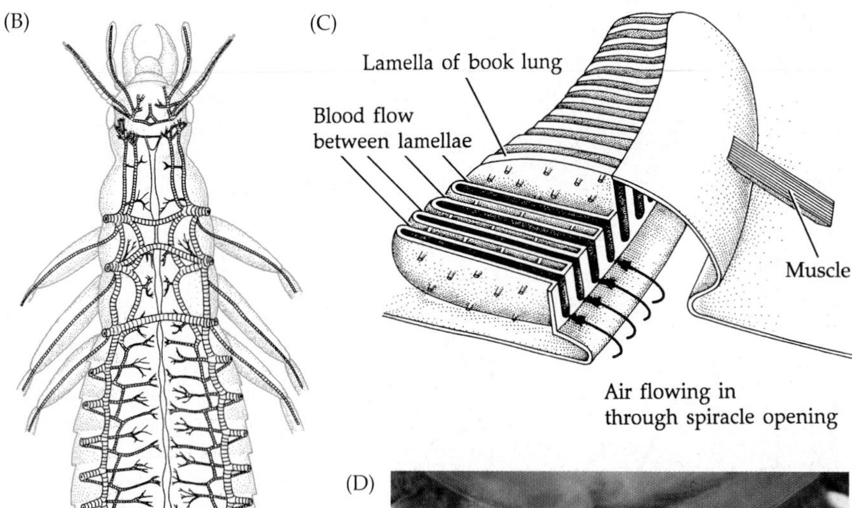

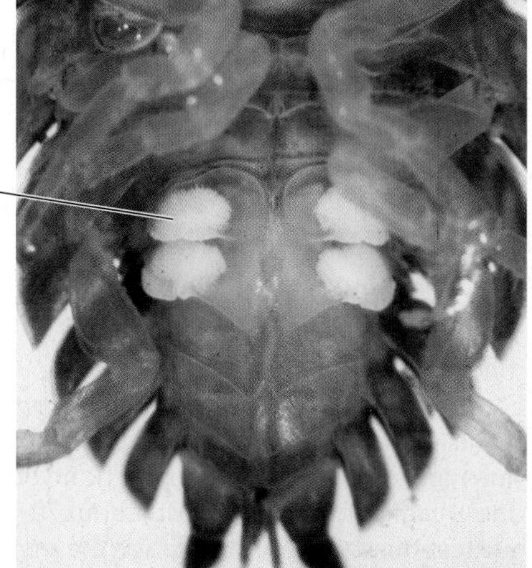

Figure 20.31 Arthropod gas exchange structures. (A) One type of crustacean gill (surface view and cross section). (B) Tracheal system in a beetle. (C) An arachnid book lung (cutaway view). (D) The abdomen of a terrestrial isopod or "pillbug," *Porcellio* (ventral view). Note the pseudotracheae ("white bodies") on the abdominal appendages, or pleopods.

Excretion and Osmoregulation

With the evolution of a hemocoelic circulatory system in arthropods, nephridia with open nephrostomes became functionally untenable. It simply would not do to drain the blood directly from an open hemocoel to the outside. Arthropods have evolved a variety of highly efficient excretory structures that share a common adaptive feature in that they are internally closed. In addition to this major difference between the nephridia of arthropods and those of other coelomate protostomes, there has been a reduction in the overall number of excretory units.

In many arthropods, portions of the excretory units are coelomic remnants and are formed in various body segments during development. In most adult crustaceans only a single pair of nephridia (nephromixia) persists, usually associated with particular segments of the head (i.e., as antennal glands or maxillary glands) (Figure 20.32A). In arachnids there may be as many as four pairs of nephridia (and in onychophorans, many more) opening at the bases of the walking legs (i.e., coxal glands).

A second type of excretory structure occurs in four terrestrial panarthropod taxa: arachnids, myriapods, insects, and tardigrades. These structures, known as **Malpighian tubules**, arise as blind tubules extending into the hemocoel from the gut wall (Figure 20.32C). However, anatomical and developmental evidence suggests that these Malpighian tubules may have evolved independently in each of these groups, representing yet another exemplary case of convergent evolution among the Arthropoda.

The excretory physiology of arthropods is a complex and extensively studied topic, and we present only a very general summary here. The various nephridial types of coelomic origin are functionally much more complex and efficient than open metanephridia. The uptake of materials from the hemocoel by the inner ends of these nephridia apparently involves passive movement in response to filtration pressure, as well as active transport. The fluid entering the nephridium is generally similar in composition to the hemolymph itself, but as it passes along the plumbing system of the nephridium, a good deal of selective reabsorption occurs, particularly of salts and nutrients such as glucose. Thus the urine exiting the nephridial pores is markedly different from the hemolymph and represents a concentration of nitrogenous waste products (Figure 20.32B).

Malpighian tubules accomplish the same process, but they must rely on assistance from the gut. Malpighian tubule uptake from the hemocoel is relatively nonselective, and the resulting "primary urine" (containing nutrients, water, salts, and so on) is emptied directly into the gut. Very little reabsorption of non-waste material occurs along the length of the tubule itself. The hindgut is mostly responsible for

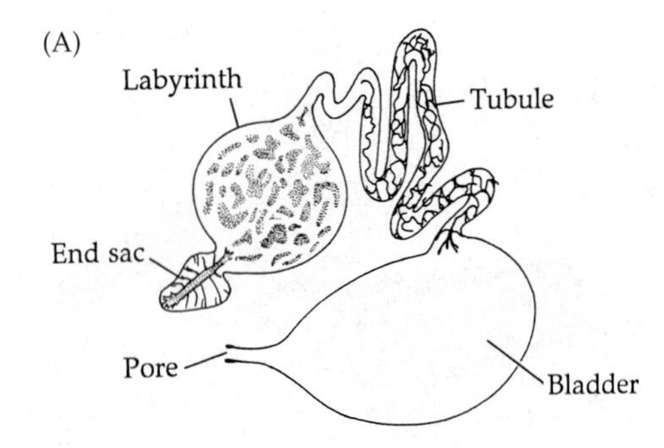

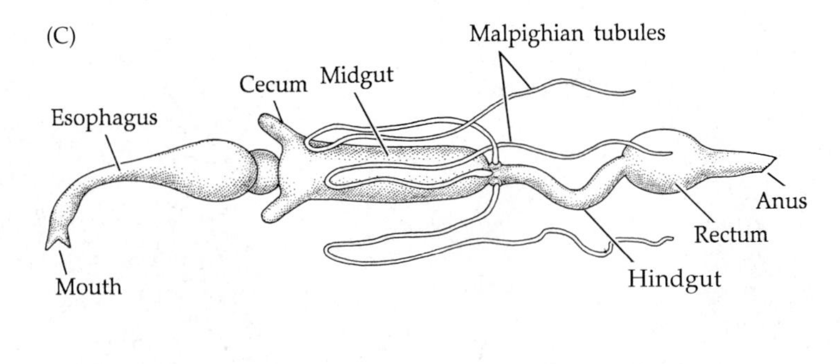

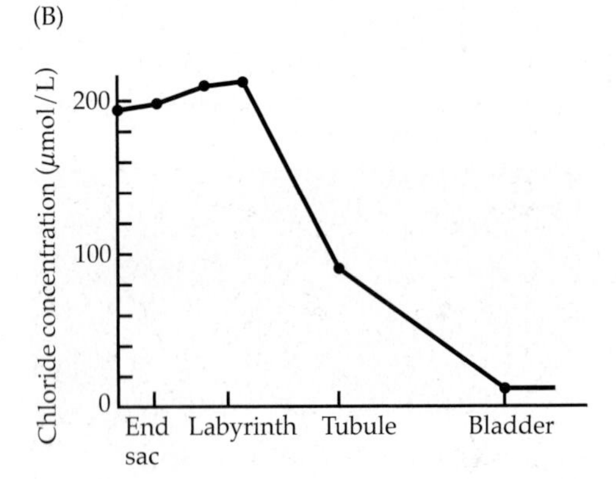

Figure 20.32 Arthropod excretory structures. (A) The antennal gland of a decapod crustacean (section). (B) Changes in chloride content of the excretory fluid in different regions of a decapod antennal gland. Note the active resorptive capabilities of the structure. (C) An insect gut. Note the attachments of Malpighian tubules.

concentrating the urine by reabsorbing the non-waste fractions. The ability of the gut to reabsorb water plays a critical role in osmoregulation in terrestrial and freshwater arthropods. Like most aquatic invertebrates, marine crustaceans excrete most (about 70–90%) of their nitrogenous wastes as ammonia; the remainder is excreted in the forms of urea, uric acid, amino acids, and some other compounds. Terrestrial arachnids, myriapods, and insects excrete predominantly uric acid (via the hindgut and anus). In Chapter 4 we reviewed some of the relationships between excretory products and osmoregulation in terms of adaptation to terrestrial habitats. The ability to produce large quantities of uric acid, and thus conserve water, has doubtless contributed significantly to the success of arachnids and insects on land. The crustaceans, on the other hand, have not been able to make a major shift from ammonotelism to uricotelism. Only the terrestrial crustaceans (i.e., isopods—woodlice and pillbugs) show a slight increase in uric acid excretion over that of their marine counterparts.

Nervous System and Sense Organs

The general plan of the arthropod nervous system is similar to that seen in many other protostomes, and many obvious homologies exist (Figure 20.33A). The arthropod brain (**cerebral ganglia**) comprises three regions, each of which originates from a pair of coalesced ganglia. The two pairs of anteriormost ganglia are the protocerebrum and deutocerebrum. The posterior-most ganglion, the tritocerebrum, usually forms circumenteric connectives around the esophagus to a ventral subesophageal (subenteric) ganglion. The latter is formed by the coalescence of several other head ganglia, usually those associated with the mandibles and maxillae. A double or single, ganglionated ventral nerve cord extends through some or all of the body segments. Each of these regions gives rise to a major pair of nerves to particular head appendages (Figure 20.33B,C). In extant arthropods, the protocerebrum innervates the eyes, whereas the deutocerebrum innervates the antennae (only the first antennae in Crustacea), the chelifores of pycnogonids, and the cheliceres of chelicerates. Recall that in Onychophorans the protocerebrum innervates both the eyes and the antennae, and the deutocerebrum innervates the jaws.[6]

The segmental ganglia of the ventral nerve cord show various degrees of linear fusion with one another in different groups of arthropods. Hence, just as tagmosis is reflected in the joining of body segments externally, it is also apparent in the union of groups of ganglia along the ventral nerve cord. These modifications of the central nervous system are examined more closely in the following chapters on the arthropod subphyla.

Although the presence of an exoskeleton has had little evolutionary effect on the structure of the cerebral ganglia and nerve cord, it has had a major effect on the nature of sensory receptors. Unmodified, the exoskeleton would impose an effective barrier between the environment and the epidermal sensory nerve endings. Hence, most of the external mechanoreceptors and chemoreceptors are actually cuticular processes (setae,

[6]The concentration of nervous tissue in the arthropod head has been called a brain, cerebrum, cerebral ganglion, and cerebral ganglionic mass. Some of these terms may seem anthropomorphic, or might suggest the presence of a single head ganglion. In fact, the cerebral ganglia comprise a cluster of associated ganglia (concentrations of nervous tissue composed primarily of neuronal cell bodies)—hence the term cerebral ganglia is probably the most accurate.

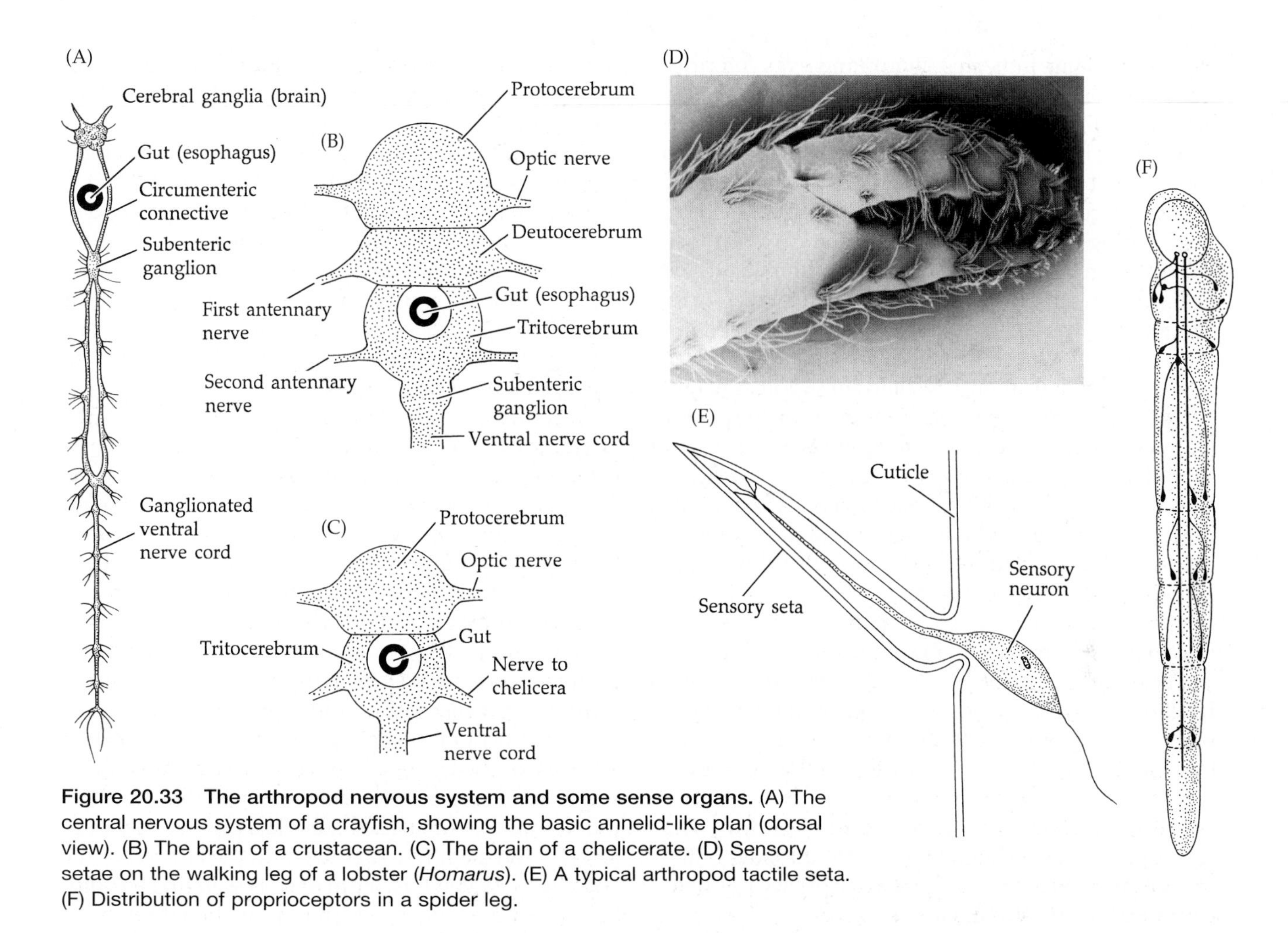

Figure 20.33 The arthropod nervous system and some sense organs. (A) The central nervous system of a crayfish, showing the basic annelid-like plan (dorsal view). (B) The brain of a crustacean. (C) The brain of a chelicerate. (D) Sensory setae on the walking leg of a lobster (*Homarus*). (E) A typical arthropod tactile seta. (F) Distribution of proprioceptors in a spider leg.

hairs, bristles), pores, or slits, collectively called sensilla.

Most arthropod tactile receptors (mechanoreceptors) are cuticular projections in the form of movable bristles or setae, the inner ends of which are associated with sensory neurons (Figure 20.33D,E). When the cuticular projections are touched, that movement is translated into a deformation of the nerve ending, thereby initiating a nerve impulse. Sensitivity to environmental vibrations is similar to tactile reception. Sensilla in the form of fine "hairs" or setae are mechanically moved by external vibrations and impart that movement to underlying sensory neurons. Some terrestrial arthropods bear thin, membranous cuticular windows overlying chambers lined with sensory nerves. When struck by airborne vibrations (e.g., sound), these windows vibrate in turn and impart the stimulus to the chamber and thence to the nerves below.

We have seen in soft-bodied invertebrates that chemoreception is usually associated with ciliated epithelial structures (e.g., nuchal organs, ciliated pits), across which dissolved chemicals diffuse to nerve endings. In arthropods, in the presence of a relatively impermeable cuticle and in the absence of free cilia, such arrangements are obviously not possible. Thus, many arthropods possess special thin or hollow setae, often associated with the head appendages, with permeable cuticular coverings or minute pores that bring the environment into contact with chemoreceptor neurons.

Proprioception is of particular importance to animals with jointed appendages, such as arthropods and vertebrates. The way in which these stretch receptors span the joints enables them to convey information to the central nervous system about the relative positions of appendage articles or body segments (Figure 20.33F). Through this system, an arthropod (or vertebrate) knows where its appendages are, even without seeing them. Arthropod versions of these "strain gauges" are called **campaniform sensilla** in hexapods, **slit sensilla** in arachnids, and **force-sensitive organs** in most crustaceans. Despite subtle differences in their anatomy, all are linked to the central nervous system in similar ways, and all record exoskeletal strain by means of neuronal stretching or deformation. Arthropods possess three basic kinds of photoreceptors, including simple ocelli, complex lensed ocelli, and faceted or compound eyes. Ocelli were described in Chapter 4; although they vary in anatomical detail, their basic structure and operation is consistent in all invertebrates. Compound eyes, though found in all arthropod subphyla, have been lost or modified in various groups throughout the phylum. Because of their unique structure and function, compound eyes are described here in some detail.

As their name indicates, **compound eyes** comprise from a few to many distinct photoreceptive units, called **ommatidia**. Each ommatidium is supplied with its own nerve tracts leading to the major optic nerve, and each has its own field of vision through square or hexagonal cuticular facets on the eye surface. The visual fields of neighboring ommatidia overlap to some extent, such that a shift in the position of an object within the visual field generates impulses from several ommatidia; hence compound eyes are especially suitable for detecting movement. However, visual acuity is affected by the degree of overlap among the fields of vision of neighboring ommatidia—the greater the overlap, the poorer the visual acuity. In general, compound eyes with many small facets probably produce higher-resolution images than eyes with fewer, larger facets. Note that the function of an ommatidium is to concentrate light from a reasonably narrow direction into a receptor area, and an individual ommatidium cannot "focus" in the sense of image formation. Image formation is the result of multiple signals from multiple ommatidia.

The following discussion describes the structure and function of compound eyes using the crustacean–hexapod model (Figure 20.34). Each ommatidium is covered by a modified portion of the cuticle called the **cornea** (= corneal lens); the special epidermal cells that produce the corneal elements are called **corneagen cells**. The corneagen cells may later withdraw to the sides of the ommatidium to form (usually two) **primary pigment cells** (= iris cells). When viewed externally, the facets on the surface of each cornea produce the characteristic mosaic pattern so frequently photographed by microscopists. The core of each ommatidium comprises a group of **crystalline cone cells** and the crystalline cone that they produce, sometimes a crystalline cone stalk, and a basal **retinula** (= retinular element). There are typically four (rarely three or five) crystalline cone cells; an ommatidium with a four-part crystalline cone is highly diagnostic of crustacean–hexapod eyes and is termed a **tetrapartite ommatidium**.[7] The crystalline cone is a hard, clear structure bordered laterally by the primary pigment cells. The retinula is a complex structure formed from several retinular cells, which are the actual photosensitive units that give rise to the sensory nerve tracts. These retinular cells, usually numbering 8 but ranging from 5 to 13 in various derived conditions, are arranged in a cylinder along the long axis of the ommatidium. The retinular cells are surrounded by secondary pigment cells, which isolate each ommatidium from its neighbors. The core of the cylinder is the **rhabdome**, which is made up of rhodopsin-containing microtubular folds (microvilli) of the cell membranes of the retinular cells. Each retinular cell's contribution to the rhabdome is called a **rhabdomere**. The microvilli of the rhabdomeres extend toward the central axis of the ommatidium at right angles to the long axis of the retinular cell.

The initiation of an impulse depends upon light striking the rhabdome portion of the retinular element. Light that enters through the facet of a particular ommatidium is directed to its rhabdome by the lens-like qualities of the cornea and the crystalline cone. The lens has a fixed focal length, so accommodation to objects at different distances is not possible. Light is shared among all the rhabdomeres of a given rhabdome, although not necessarily equally.[8]

In contrast to those of insects and crustaceans, the lateral eyes of most myriapods are probably not true compound eyes, but comprise clusters of simple ocelli. There is evidence, however, that the eyes of scutigeromorph centipedes (and of some fossil myriapods) may be true compound eyes. The only chelicerates with typical compound eyes, the xiphosurans, have ommatidia that differ in most details from the insect–crustacean design. There is some evidence that the lateral eye groups of terrestrial chelicerates may be derived from reduced and fused compound eye ommatidia, and Silurian scorpions had huge bulging compound eyes. The eyes of chelicerates are relatively larger than those of insects or crustaceans, and their ommatidia have an indeterminate number of cells. The pigment cells are arranged in a cuplike manner, and the bottom of the "cup" is occupied by a sheet of cells secreting a protuberance of the cuticle, which is a functional but not a morphological equivalent of a crystalline cone. A special "eccentric cell"—a large specialized photoreceptor—found in chelicerate ommatidia has no equivalent in the insect or crustacean retina.

Compound eyes evolved early in arthropod history and were present in trilobites and many stem groups, including anomalocaridids. The versatility of compound eyes is to a large degree a result of the distal and proximal screening pigments located in cells that wholly or partially surround the core of the ommatidium (Figure 20.34B–D). Distal screening pigments are located in the primary pigment cells (iris cells), and proximal pigments are often located in the retinular cells and secondary pigment cells. In many cases, these screening pigments are capable of migrating in response to varying light conditions and thus changing their positions somewhat along the length of the ommatidium. In bright light, the screening pigments may disperse so that nearly all of the light that strikes a particular rhabdomere must have entered through the facet of its ommatidium. In other words, the screening pigment prevents light that strikes the facets at an angle from passing through one ommatidium and into

[7]The crystalline cone cells are often called Semper cells (especially by entomologists).

[8]The compound eyes of one large group of crustaceans, the maxillopodans, differ considerably from those of other crustaceans; see Chapter 21.

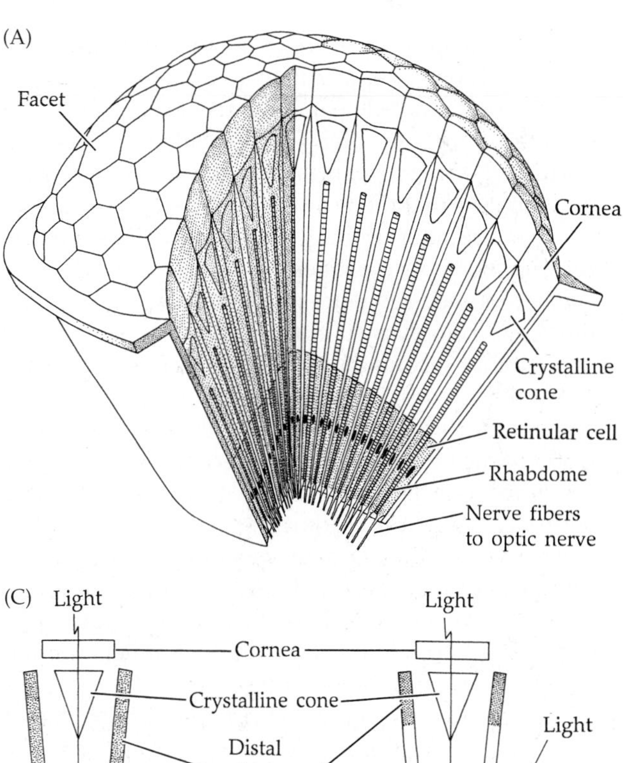

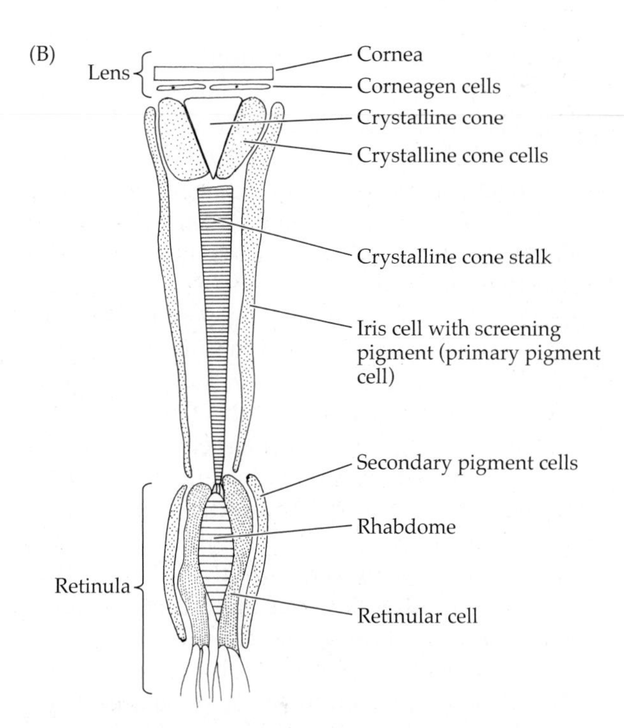

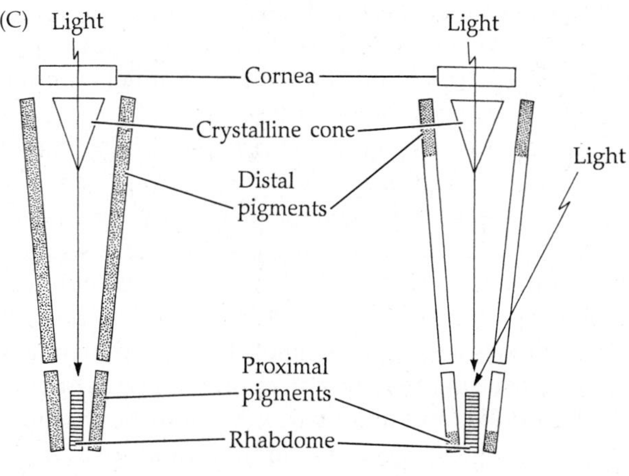

Figure 20.34 Arthropod compound eyes. (A) A compound eye (cutaway view). (B) A single ommatidium. (C,D) Major ommatidial elements in (C) an appositional, or light-adapted, eye and (D) a superpositional, or dark-adapted, eye.

another. Many crustacean eyes are fixed in this condition. Such **appositional eyes** (= light-adapted eyes) are thought to maximize resolution, in that the image from the visual field of each ommatidium is maintained as a discrete unit. Conversely, under conditions of dim light, screening pigments may concentrate, usually distally, thereby allowing light to pass through more than one ommatidium before striking rhabdomeres. The result is that the image formed by each ommatidium is superimposed on the images formed by neighboring ommatidia. This design has the advantage of producing enhanced irradiances on the retinula, but at the cost of reduced resolution. Many crustacean eyes are fixed in this condition also. Such **superpositional eyes** (= dark-adapted eyes) function as efficient light-gathering structures while sacrificing some visual acuity and image formation capabilities.

Some arthropod groups possess compound eyes that are always either appositional or superpositional; thus they lack the ability to switch back and forth with varying light conditions. For example, maxillopodans and branchiopods apparently all possess appositional eyes. However, within the two principal malacostracan clades, Eucarida and Peracarida, both types of eyes occur (e.g., isopods and amphipods have appositional eyes, but mysids have superpositional eyes). Furthermore, crustacean larvae that possess compound eyes almost always have the appositional type, which metamorphose into superpositional eyes in those groups that possess them in adulthood.

Among the arthropods, compound eyes elevated on stalks occur only in certain crustaceans (and perhaps a few Paleozoic trilobites and anomalocaridids). Biologists have long argued over the derivation of such stalked eyes, and the matter is still far from settled—are they the primitive condition in arthropods or in crustaceans, or have they been derived multiple times from sessile-eyed ancestors (the latter seems far more likely). The eyestalk is much more than a device to support and move the eye. Eyestalk movements are produced by up to a dozen or more muscles with complex motor innervation. In most malacostracans the eyestalks contain several optic ganglia separated by chiasmata, as well as important endocrine organs, usually including the sinus gland and X-organ. Thus, neither loss of eyestalks nor convergent recreation of these structures would have been a simple evolutionary feat. The loss of functional eyes is a common evolutionary pathway among the Crustacea (and other arthropods), especially in species that inhabit subterranean, deep-sea, or interstitial habitats. But among those clades with stalked eyes, the eyestalk remains even when the eye itself degenerates—testimony to importance of this complex bit of anatomy.

Reproduction and Development

The great diversity of adult form and habit among arthropods is also reflected in their reproductive and developmental strategies. The extreme evolutionary and ontogenetic plasticity of arthropods has led to a great deal of convergence and parallelism as different groups have developed similar structures under similar selective conditions or pressures.

Nearly all arthropods are gonochoristic, and most engage in some sort of formal mating. Fertilization is usually, although not always, internal and is often followed by brooding or some other form of parental care, at least during early development. Development is frequently mixed, with brooding and encapsulation followed by larval stages, although direct development occurs in many groups.[9]

Arthropod eggs are centrolecithal (Figure 20.35), but the amount of yolk varies greatly and results in different patterns of early cleavage. Cleavage is holoblastic in the relatively weakly-yolked eggs of xiphosurans, some scorpions, and various crustaceans (e.g., copepods and barnacles), and is meroblastic in the strongly-yolked eggs of most insects and many other crustaceans. A number of arthropods exhibit a unique form of meroblastic cleavage that begins with nuclear divisions within the yolky mass (Figure 20.29). These intralecithal nuclear divisions are followed by a migration of the daughter nuclei to the periphery of the cell and subsequent partitioning of the nuclei by cell membranes. These processes typically result in a periblastula that consists of a single layer of cells around an inner yolky mass. Holoblastic cleavage is usually more or less radial in appearance.

One of the most striking features of arthropods is their segmentation and the complex way in which it is embryologically derived. This developmental process is very similar to that seen in annelids (and in onychophorans and tardigrades), and until the advent of molecular phylogenetic techniques it was taken as evidence of a close annelid–arthropod shared ancestry. The process is called **teloblastic segmental growth** (or simply teloblasty) (Greek *telos*, "end"; *blasto*, "bud"). It is characterized by a progressive, anterior-to-posterior addition of segments from a distinct posterior growth zone situated near the anus (i.e., in front of the terminal, or anal, segment—often called the **telson** or **pygidium**). So programmed is the development of the segments that the growth zone is often composed of a fixed number of uniquely identifiable stem cells (**teloblasts**) whose progeny undergo a predictable and stereotyped sequence of segmentally iterated cell divisions. Teloblasty is further characterized by the formation of secondary body cavities (coelomic cavities) at the growth zone as segments are elaborated. In arthropods, the adult body cavity is derived from the fusion of the blood vascular system with these transitory embryonic coelomic cavities. Although the coelomic cavities are secondarily lost during arthropod embryogenesis, they form by way of schizocoely, originating from a pair of caudally situated mesodermal cell bands. These bands have long been thought to originate from a 4d (mesentoblast) cell, as in annelids, but in fact there is little evidence for this.

Not only do the segments form in a similar fashion in annelids and arthropods, but the two groups also appear to use a similar genetic mechanism (especially the products of the *engrailed* gene) to define segmental boundaries and polarity. The *engrailed* (*en*) gene belongs to a class of genes known as segment polarity genes. It plays a key role in determining and maintaining the posterior cell fates of segmental structures in all arthropods (and in annelids and onychophorans). Because this gene occurs in iterated transverse stripes in the posterior portion of each developing segment in the germ band of all annelids and arthropods, it can be used to clarify ambiguities in segment position and number. Embryonic *Hedgehog* signaling pathways also play similar roles in annelids and arthropods, in both cases playing a crucial role in the axial patterning of developing segments.

[9]Direct development in arthropods is sometimes called **amorphic development**, and indirect development is often called **anamorphic development.**

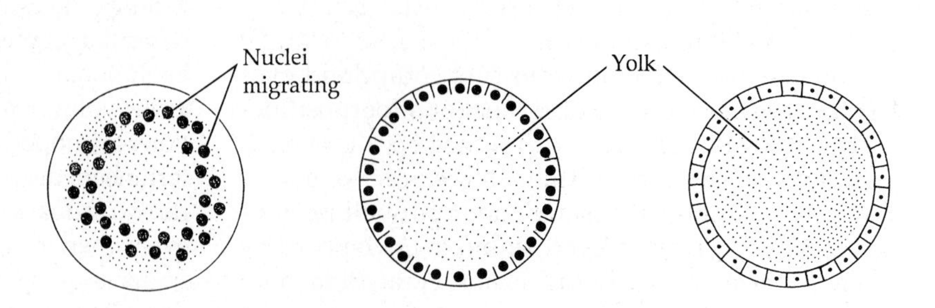

Figure 20.35 Superficial cleavage of a centrolecithal egg and the formation of a periblastula in arthropods. (A) Centrolecithal egg. (B) Intralecithal nuclear divisions following fertilization. (C) Migration of nuclei to the periphery of the cell. (D,E) The periblastula is produced by a partitioning of nuclei as cell membranes form.

In arthropods, the transitory embryonic coelomic cavities are lined with simple epithelium, but this gradually disappears as the embryo develops, and the large adult body cavity comes to be lined by an extracellular matrix. Hence, the body cavity of adult arthropods is derived by the fusion of a primary body cavity (the blood vascular system) and a secondary body cavity (the coelom) and is therefore termed a hemocoel (or mixocoel). Segmentally arranged pairs of transitory coelomic cavities, and their embryonic fusion, have been reported in the early developmental stages of virtually every onychophoran and arthropod that has been studied.

The Evolution of Arthropods

The Origin of Arthropods

In 1997, a ground-breaking analysis of animal phylogeny based on 18S rRNA sequence data led Aguinaldo and her colleagues to propose a new hypothesis of arthropod relationships—that they belong to a clade of animals that we now know also includes nematodes, nematomorphans, kinorhynchs, priapulans, and loriciferans (as well as tardigrades and onychophorans), but *not* annelids. The newly recognized clade was called **Ecdysozoa**. Today, scores of molecular phylogenetic studies support this view of animal relationships and it is now broadly accepted that the protostomes comprise two main clades: Ecdysozoa and Spiralia. One of the most revolutionary outcomes is compelling evidence that Annelida and Panarthropoda are not sister groups (a long held hypothesis based on morphological and developmental studies). This modern view of animal relationships stimulated new morphological and developmental investigations, and significant differences between the annelids and arthropods have begun to emerge, the most obvious being fundamental differences in the adult body cavity and cleavage patterns during embryogenesis. However, the relationships among the three main clades within Ecdysozoa is still uncertain and we depict the clade Panarthropoda as an unresolved trichotomy with the other two ecdysozoan clades—Nematoida and Scalidophora—on our phylogenetic tree of Metazoa (Chapter 28). Molecular clock estimates place the origin of Arthropoda at around 600 million years ago.

Evolution within the Arthropoda

With nearly 1.2 million described species, arthropods are arguably the most successful animal phylum on Earth. They encompass an unparalleled range of structural and taxonomic diversity, have a rich fossil record, and have become favored study animals of evolutionary developmental biology. Some arthropod "model systems" (e.g., *Drosophila*) have been studied intensively for many decades. Arthropods were among the earliest animals to evolve; even the Ediacaran fauna of the late Proterozoic included animals regarded by some as stem arthropods (including, perhaps, anomalocaridids among others, some of which had stalked eyes as in modern Crustacea). For all these reasons, we know a lot about arthropods.

Arthropod evolutionary history has been a favorite pastime of zoologists for centuries. Every imaginable phylogenetic tree has been proposed for the relationships among the crustaceans, hexapods, chelicerates and myriapods at one time or another. However, since the mid-1980s, a virtual explosion of new information on this phylum has appeared, much of it concerning phylogeny, paleontology, gene expression, and developmental biology. If we examine the evolution of arthropods in light of these recent discoveries, a new phylogeny of the Arthropoda begins to come into focus (Figure 20.38). We do not yet have enough information to fill in all of the details of this tree, but some clear relationships have emerged.

The most important ancient arthropod fossils are those in which even the soft parts of the animal were preserved—the so-called ancient *Lagerstätten*, such as the faunas of the Upper Cambrian Orsten deposits of Sweden, the middle Cambrian Burgess Shale of Canada (and elsewhere), and the Lower Cambrian Chengjiang (520 Ma) deposits of China. Recent discoveries from these well-preserved deposits have shown that the fossil record of Crustacea and Pycnogonida dates to at least the early Cambrian, and possibly to the late Precambrian. In 2015, Ma and colleagues described the brains of a lower Cambrian (517 Ma) shrimp-like crustaean, *Fuxianhuia protensa*, that were basically identical to those of modern Crustaceans. These extraordinary faunas are now informing us that crustaceans possibly predate the appearance of trilobites in the fossil record, and some workers consider the first arthropods to be crustaceans (or "protocrustaceans" or "crustaceamorphans").

The Chengjiang fauna includes at least a hundred species of animals, many without hard skeletons, including the first known members of many modern groups. However, it is the arthropods that dominate this fauna, including trilobites and bradoriid crustaceans (and also tardigrades and onychophorans). The largest of the Chengjiang animals are the anomalocaridids, stem arthropods that were common during the Cambrian and are also known from a single described species (*Schinderhannes bartelsi*) from the Lower Devonian Hunstrück Slate of Germany (see Figure 1.3C). Some anomalocaridids reached 2 m in length; most were predators, but some were probably suspension feeders. The Chengjiang fauna is very similar to that of the Burgess Shale, and it demonstrates that the arthropods were already far advanced by this early date. In fact, arthropods may have been the dominant animals in terms of species diversity since the

Cambrian, and they constitute over one-third of all species described from Lower Cambrian strata.

The spectacular research by Klaus Müller and Dieter Waloszek since the 1980s on microscopic arthropods from the Upper Cambrian Orsten deposits of Sweden have brought to light a rich fauna of crustaceans, many of which closely resemble modern groups such as onychophoran-like lobopodians, tardigrades, pentastomids, cephalocarids, mystacocarids, and branchiopods. Orsten-type preservation is a secondary phosphatization of the upper cuticle, apparently soon after death of the animal as no further destruction takes place. Such preservation can produce exquisite three-dimensional fossils with all the details of eyes, limbs, setae, cuticular pores and other structures less than a micrometer in size. It generally occurs with a post-mortem embedding of the animals in limestone (that later formed into nodules). Since the fossils themselves are phosphatized, they can be etched from their surrounding limestone rock with weak acids. These exceptional fossils, which are mostly microscopic arthropods, were first discovered, in southern Sweden, by the German paleontologist Klaus Müller in the mid-1960s. Orsten-type fossils are now known from several continents from the early Cambrian (520 Ma) to the Early Cretaceous (100 Ma). The recovery of these three-dimensionally preserved animals and the developmental series that have been found (with successive larval, juvenile, and adult stages) have provided us with information on the detailed anatomy of the body segments and appendages of many ancient stem arthropods, or crustaceamorphans. The Orsten fauna shows that Cambrian Crustacea had all the attributes of modern crustaceans, such as compound eyes, a head shield, naupliar larvae (with locomotory first antennae), and biramous appendages on the second and third head somites (the second antennae and mandibles).

A modern view of arthropod phylogeny thus places a panorama of crustacean-like early Cambrian lineages at its base—a diverse array of early arthropods that had typical crustacean, or crustaceamorph bodies, eyes, development, and naupliar larvae, though perhaps with a smaller number of fused head somites than seen in modern Crustacea. Early in the Cambrian, the trilobites may have emerged from this crustaceamorph stem line, radiating rapidly to become the most abundant arthropods of Paleozoic seas, but then abruptly disappearing in the Permian–Triassic extinction. Next to appear were probably the chelicerates, in the form of giant marine water scorpions nearly 3 m long (eurypterids) and their kin, which had appeared at least by the Ordovician; by the Silurian, eurypterids had probably become keystone predators in the marine realm. Also by the Silurian, the chelicerates had invaded land and begun to leave a fossil record of terrestrial arachnids. By the late Ordovician or early Silurian the first myriapods had evolved, perhaps marine creatures, and about 15 million years later terrestrial millipedes appear in the fossil record. The last major arthropod group to appear was the hexapods, making their appearance in the Devonian, or perhaps the Silurian, and radiating rapidly to dominate the terrestrial world, ultimately qualifying the Cenozoic to be called "the age of insects."

The relationship of trilobites to living arthropods is still being debated (Figures 20.36 and 20.37). The two main opposing views argue them close to either the Chelicerata or the Mandibulata. Our view, in contrast, positions them as one of many extinct lineages stemming from a crustaceamorph stem line ancestry (Figure 20.38).

This model of a Paleozoic crustaceamorph stem line spinning off the other arthropod subphyla one after another differs considerably from earlier ideas of arthropod evolution. Nonetheless, a great deal of information now supports this new view—one in which the Crustacea are recognized as an ancient paraphyletic assemblage that was the "mother of all modern Arthropoda."

Arthropods are the first land animals for which we have a paleontological record. The first land arthropods appeared in the late Ordovician or early Silurian (arachnids, millipedes, centipedes), and these fossils represent the first terrestrial invertebrates for which we have direct evidence. Indeed, animal life on land might not have been possible before the late Ordovician, when terrestrial plants first made their appearance. The first insects in the fossil record are 390-million-year-old Devonian springtails (Collembola) and bristletails (Archaeognatha). By the mid-Paleozoic all four living arthropod subphyla were in existence and had already undergone substantial radiation. By the Middle Devonian, centipedes, millipedes, mites, amblypygids, opilionids, scorpions, pseudoscorpions, and hexapods were all well established. Hence, terrestrial arthropods seem to have undergone major radiations in the Silurian. The presence of a wide variety of predatory terrestrial arthropods during the early Paleozoic suggests that complex terrestrial ecosystems were in place at least as early as the late Silurian. However, interestingly, molecular dating studies consistently suggest the origin of terrestrial arthropods much earlier, as early as 510 million years ago

Work by the great comparative biologist Robert Snodgrass in the 1930s established a benchmark in arthropod biodiversity research. The Snodgrass classification embraced three important hypotheses: (1) arthropods constituted a monophyletic taxon; (2) myriapods and hexapods were sister groups, together forming a taxon called Atelocerata (= Tracheata,

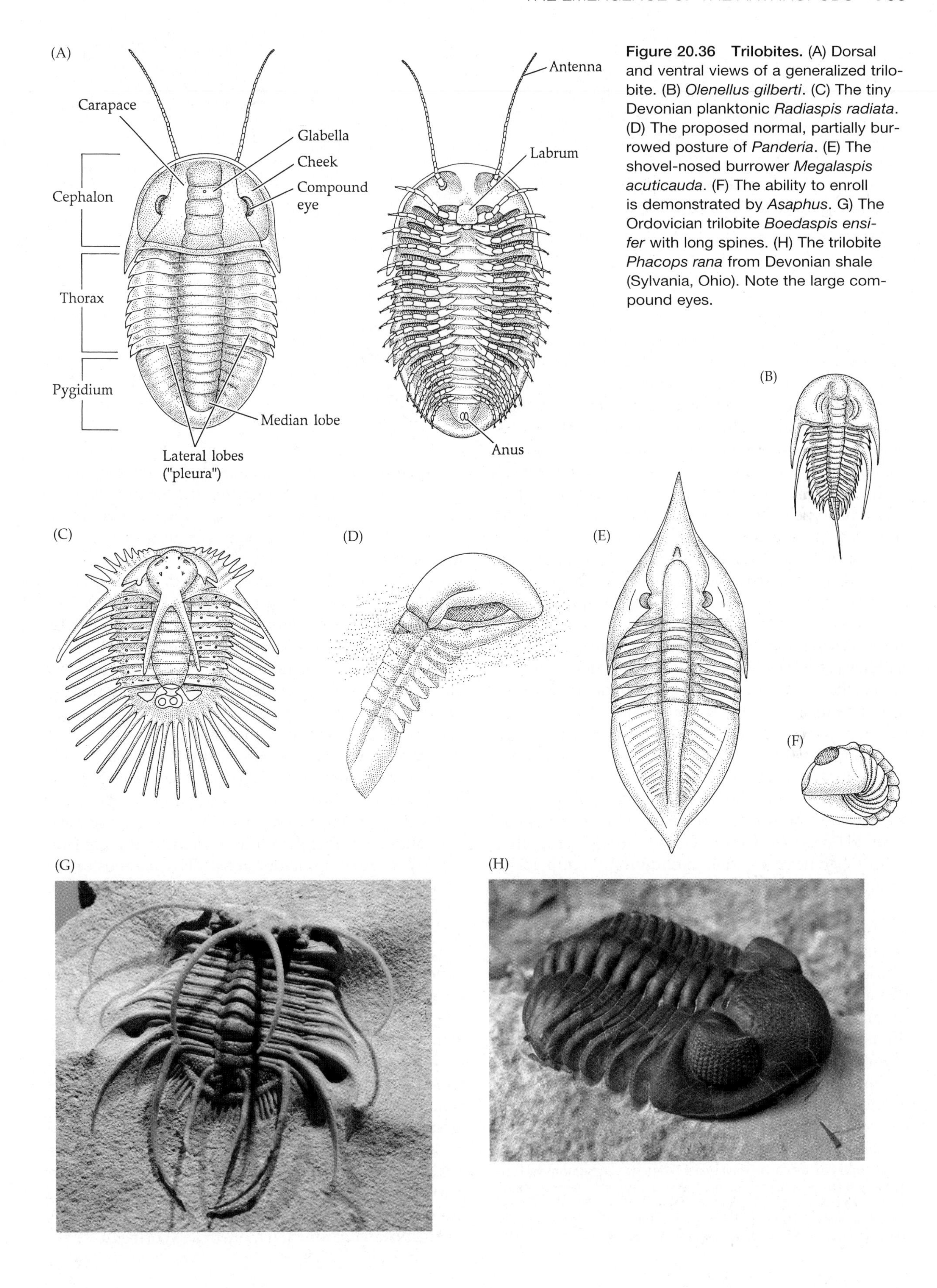

Figure 20.36 Trilobites. (A) Dorsal and ventral views of a generalized trilobite. (B) *Olenellus gilberti*. (C) The tiny Devonian planktonic *Radiaspis radiata*. (D) The proposed normal, partially burrowed posture of *Panderia*. (E) The shovel-nosed burrower *Megalaspis acuticauda*. (F) The ability to enroll is demonstrated by *Asaphus*. G) The Ordovician trilobite *Boedaspis ensifer* with long spines. (H) The trilobite *Phacops rana* from Devonian shale (Sylvania, Ohio). Note the large compound eyes.

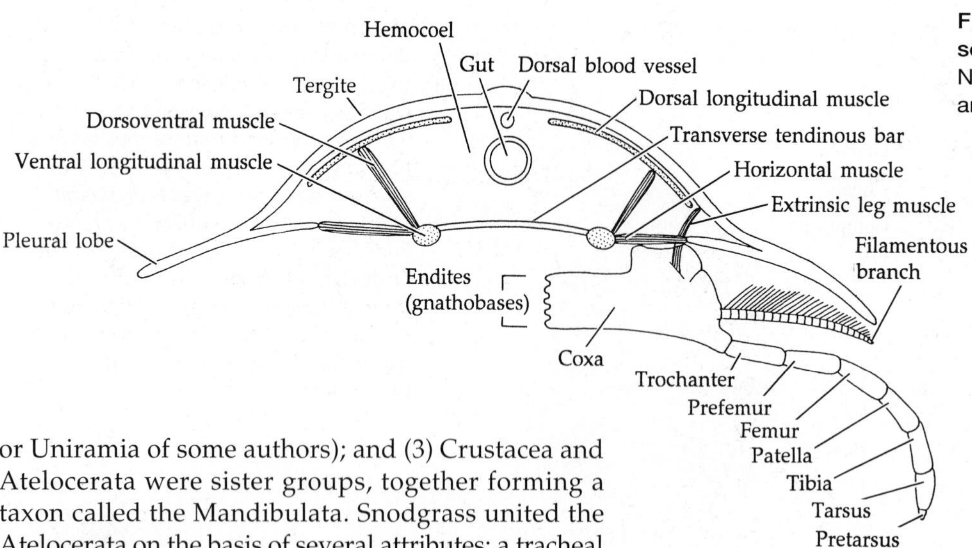

Figure 20.37 A trilobite segment (cross section). Note the extended pleura and the general leg structure.

or Uniramia of some authors); and (3) Crustacea and Atelocerata were sister groups, together forming a taxon called the Mandibulata. Snodgrass united the Atelocerata on the basis of several attributes: a tracheal respiratory system, uniramous legs, Malpighian tubules for excretion, and loss of the second antennae (as the name Atelocerata implies). The Mandibulata were united on the basis of the mandibles (which are almost certainly homologous in these taxa) and a similar head and head appendage structure.

It was not until the late 1980s that Snodgrass's longstanding view of arthropod relationships began to be seriously questioned. While the Mandibulata is still supported as a monophyletic group, it now appears almost certain that Hexapoda arose from somewhere within Crustacea, a clade called Pancrustacea, and this then is the sister group to the Myriapoda. It turns out that Snodgrass, and others, were misled by the similarities between myriapods and hexapods, which are now known to be convergences (or parallelisms) resulting from adaptations to terrestrial habitats. When fossils are included in morphological phylogenetic analyses, the Crustacea are seen to arise at the very base of the arthropod tree as a sequence of taxa from which the other subphyla emerge—much as shown in Figure 20.38. While the crustacean sister group to the Hexapoda remains enigmatic, several recent multigene analyses suggest it might be the Remipedia or a Cephalocarida + Remipedia clade, although these crustaceans are all restricted to the marine environment today and there is no paleontological evidence that they invaded land in the past. While we treat "Crustacea" as a subphylum in this book, remember that the discovery that insects are "flying crustaceans" makes the subphylum Crustacea a paraphyletic group (if the Hexapoda are not also included).[10]

Research in the field of developmental biology and gene expression has shown that the Arthropoda are rich with homoplasy, and it is now clear that much of the difficulty in reconciling earlier morphological trees and molecular trees was due to the high levels of parallel evolution seen within the Arthropoda. For example, many characters shared between hexapods and myriapods, which in the past had been assumed to be synapomorphies, are now interpreted as parallelisms (or convergences), including such things as the tracheal gas exchange system, uniramous legs, Malpighian tubules, organs of Tömösvary, loss of the second antennae, and loss of mandibular palps.

The rigid, compartmentalized bodies of arthropods have allowed for modes of body region specialization unavailable to most other metazoan phyla. We now know that the fates of segmental units and their appendages are under the ultimate orchestration of Hox and other developmental genes. These genes select the critical developmental pathways to be followed by groups of cells during morphogenesis. Hox genes determine such basic body architecture as the dorsoventral and anterior–posterior body axes and where body appendages form. Hox genes can either suppress limb development or modify it to create alternative appendage morphologies. These unique genes have played major roles in the evolution of new body plans among arthropods (and the Metazoa in general). The degree to which such developmental genes have been conserved is remarkable, and most of them probably date back at least to the Cambrian. For example, homologues of the developmental gene *Pax-6* seem to dictate where eyes will develop in all animal phyla. *Pax-6* is so similar in protostomes (e.g., insects) and deuterostomes (e.g., mammals) that their genes can be experimentally interchanged and still function more or less correctly.

[10]Although the idea of a Crustacea–Hexapoda sister-group relationship may seem new to many, the idea was actually proposed at the turn of the twentieth century, when William T. Calman presented detailed arguments from comparative anatomy that argued for an alliance between the Crustacea and the Hexapoda.

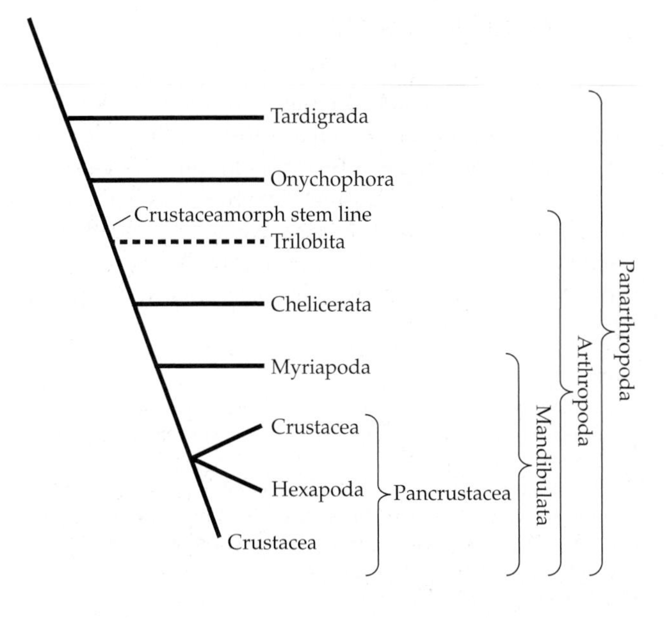

Figure 20.38 A phylogeny of the Panarthropoda. Based on a synthesis of paleontological and phylogenetic studies, the arthropods are seen to be rooted in a crustacean-like ancestral line, from which the living subphyla emerged. The subphylum Crustacea is paraphyletic (unless the Hexapoda are included). The position of the Trilobita is uncertain. See text for details.

It has been known since the late nineteenth century that the compound eyes of hexapods and crustaceans possess many complex homologous features, and that they differ markedly from the eyes of myriapods and chelicerates. Recall that in hexapods and crustaceans, each ommatidium consists of a cuticular corneal lens that is, at least partly, secreted by two cells—termed primary pigment cells in Hexapoda and corneagen cells in Crustacea. The crystalline cone, produced by four Semper cells, is fundamentally tetrapartite. A retinula is present, typically composed of eight retinular cells. This common anatomical plan is unique to the Hexapoda + Crustacea and constitutes strong evidence of a close relationship between the two groups. Dohle (2001) even proposed a name on this basis for the clade, the "Tetraconata," although the name "Pancrustacea" has had more popularity.[11]

Recent research on the anatomy and development of the arthropod central nervous system has identified many other neurological features that appear to be unique to the Hexapoda + Crustacea. In fact, Strausfeld (1998) developed a phylogeny of the Arthropoda based solely on 100 anatomical features of the cerebral ganglia. In particular, numerous anatomical similarities exist in the elements of the optic lobes and "midbrain." Developmental similarities suggest a fundamental distinction between embryonic development in the hexapod + crustacean central nervous system and in that of myriapods and chelicerates. In hexapods and crustaceans, development of the CNS begins with the delamination of enlarged cells, called neuroblasts, from the ectoderm (neuroectoderm) of each segment. The neuroblasts aggregate to form the segmental ganglia. The formation of neuroblasts is highly stereotyped and predictable in both Crustacea and Hexapoda. These large neuroblasts are of a special type, regarded as stem cells, and they divide unequally to generate a specific number of neurons. So far, 29–31 neuroblasts have been identified in each segment of hexapods and 25–30 in crustacean segments, many of which appear to be homologous between the two subphyla. Nothing resembling these stem cell neuroblasts has been seen in myriapods. In both hexapods and crustaceans, longitudinal connectives of the CNS originate in the segmental neurons, whereas in myriapods they derive from neurons in the cerebral ganglia that send their axons posteriorly to set up long parallel connectives. Thus, myriapod segmental ganglia receive contributions from more widely distributed neurons.

Their great size range, especially at the smaller end of the scale, adapts arthropods for a great variety of ecological niches. The Cambrian Orsten deposits inform us that a whole fauna of interstitial/meiofaunal arthropods already existed as early as the mid-Cambrian, and this habitat has continued to be rich in adaptive radiation and specialized species ever since. Similar small-body-size niches are filled by arthropods in a great many environments today. We find high diversities of minute arthropods in habitats such as marine sediments, coral reefs, among the fronds of algae, on mosses and other primitive plants, and on the bodies of every kind of animal imaginable. Small insects and mites have exploited virtually every terrestrial microhabitat available. The arthropods (insects) were presumably the first flying animals, and the ability to fly also led them into niches that other invertebrates simply could not penetrate.

[11]Although tetrapartite ommatidia appear to be restricted to the Hexapoda and Crustacea, they may not be a synapomorphy for this clade. We don't yet know when the tetrapartite condition first appeared within the Crustacea or the Arthropoda, but it probably was well before the insects emerged (perhaps in the early Cambrian crustacean stem line). Crustacean fossils from Cambrian Lagerstätten deposits have eyes that strongly resemble modern crustacean eyes, at least superficially. In any case, tetrapartite ommatidia are probably a symplesiomorphy retained in both Hexapoda and crown Crustacea (their transformation in myriapods perhaps being a synapomorphy defining that clade). The compound eyes of xiphosurans are very different from those of other Arthropoda, suggesting they might have evolved apart from the tetrapartite condition seen in Crustacea and Hexapoda (or that they are somehow derived from the crustacean tetrapartite condition).

Selected References

Here we provide only a small sampling of the vast literature on arthropods. We have selected works that are broad in coverage as well as those dealing with the onychophorans, tardigrades, and trilobites. Several of these references are collections of papers, and in this and the following arthropod chapters we have generally not cited the papers they contain separately. References specific to individual subphyla are provided in Chapters 21–24.

General References

Averof, M. and N. H. Patel. 1997. Crustacean appendage evolution associated with changes in Hox gene expression. Nature 388: 682–686.

Bartolomaeus T. and H. Ruhberg. 1999. Ultrastructure of the body cavity lining in embryos of *Epiperipatus biolleyi* (Onychophora, Peripatidae): A comparison with annelid larvae. Invert. Biol. 118(2): 165–174.

Bergström, J. 1987. The Cambrian *Opabinia* and *Anomalocaris*. Lethaia 20: 187–188.

Briggs, D. E. G. 1994. Giant predators from the Cambrian of China. Science 264: 1283–1284.

Briggs, D. E. G., D. H. Erwin and F. J. Collier. 1994. *The Fossils of the Burgess Shale*. Smithsonian Institution Press, Washington, D.C.

Budd, G. E. 2008. Head structure in upper stem-group euarthropods. Palaeontology 51: 561–573.

Cisne, J. L. 1979. The visual system of trilobites. Palaeontology 22: 1–22.

Cloudsley-Thompson, J. L. 1958. *Spiders, Scorpions, Centipedes and Mites: The Ecology and Natural History of Woodlice, Myriapods and Arachnids*. Pergamon Press, New York.

Cronin, T. W. 1986. Optical design and evolutionary adaptation in crustacean compound eyes. J. Crust. Biol. 6(1): 1–23.

Daley, A. C., G. E. Budd, J.-B. Caron, G. D. Edgecombe and D. Collins. 2009. The Burgess Shale anomalocaridid *Hurdia* and its significance for early eurarthropod evolution. Science 323: 1597-1600.

Dray, N. and 12 others. 2010. Hedgehog signaling regulates segment formation in the annelid *Platynereis*. Science 329: 339–342.

Eguchi, E. and Y. Tominaga. 1999. *Atlas of Arthropod Sensory Receptors*. Springer-Verlag, New York.

Ewing, A. W. 1991. *Arthropod Bioacoustics*. Cornell University Press, Ithaca, NY.

Fortey, R. A. 2001. Trilobite systematics: The last 75 years. J. Paleontol. 75: 1141–1151.

Gilbert, S. F. and A. M. Raunio. 1997. *Embryology: Constructing the Organism*. Sinauer Associates, Sunderland, MA.

Gould, S. J. 1989. *Wonderful Life: The Burgess Shale and the Nature of History*. W. W. Norton, New York. [Classic S. J. Gould.]

Gould, S. J. 1995. Of tongue worms, velvet worms, and water bears. Nat. Hist., January: 6–15.

Grenacher, H. 1879. *Untersuchungen über das Sehorgan der Arthropoden, insbesondere der Spinnen, Insecten und Crustaceen*. Vandenhoek and Ruprecht, Göttingen. [Probably the first detailed description of compound eyes in insects and crustaceans, noting their homology.]

Gupta, A. P. (ed.) 1983. *Neurohemal Organs of Arthropods: Their Development, Evolution, Structures, and Functions*. Charles C. Thomas, Springfield, IL.

Hallberg, E. and R. Elofsson. 1989. Construction of the pigment shield of the crustacean compound eye: A review. J. Crust. Biol. 9(3): 359–372.

Harrison, F. W. and M. E. Rice (eds.). 1998. *Microscopic Anatomy of Invertebrates. Vol. 12. Onychophora, Chilopoda, and Lesser Protostoma*. Wiley-Liss, New York.

Herreid, C. F. and C. R. Fourtner (eds.) 1981. *Locomotion and Energetics in Arthropods*. Plenum Press, New York.

Hou, X. and J. Bergström. 1997. *Arthropods of the Lower Cambrian Chengjiang Fauna of Southwest China. Fossils and Strata, No. 45.* Universitetsforlaget, Oslo.

Kaestner, A. 1968, 1970. *Invertebrate Zoology. Vols. 2 and 3*. Wiley, New York. [These volumes do an excellent job of describing arthropod morphology; adroitly translated from the German by H. W. and L. R. Levi.]

Kenchington, W. 1984. Biological and chemical aspects of silks and silk-like materials produced by arthropods. S. Pac. J. Nat. Sci. 5: 10–45.

Kimm, M. A. and N.-M. Prpic. 2006. Formation of the arthropod labrum by fusion of paired and rotated limb-bud-like primordia. Zoomorphology 125: 147–155.

Kühl, G., D. E. G. Briggs and J. Rust. 2009. A great-appendage arthropod with a radial mouth from the Lower Devonian Hunsrück Slate, Germany. Science 323: 771-773.

Kunze, P. 1979. Apposition and superposition eyes. Pp. 441–502 in H. Autrum (ed.), Comparative physiology and evolution of vision in invertebrates. Part A: Invertebrate photoreceptors. *Handbook of Sensory Physiology, Vol. 7, Part 6*. Springer-Verlag, Berlin.

Levi-Setti, R. 2014. *The Trilobite Book*. University of Chicago Press, Chicago. [Coffee-table paleontology at its best; a wonderful book.]

Linzen, B. et al.. 1985. The structure of arthropod hemocyanins. Science 229: 519–524. [A review of these large copper-based, oxygen-carrying proteins in arthropods.]

Ma, X., X. Hou, G. D. Edgecombe and N. J. Strausfeld. 2012. Complex brain and optic lobes in an early Cambrian arthropod. Nature 490: 258–262.

Manuel, M., M. Jager, J. Murienne, C. Clabaut and H. LeGuyade. 2006. Hox genes in sea spiders and the homology of arthropod head segments. Dev. Genes Evol. 216: 481–491.

Mass, A. and D. Waloszek. 2001. Cambrian derivatives of the early arthropod stem lineage, pentastomids, tardigrades and lobopodians—an "Orsten" perspective. Zool. Anz. 240: 451–459.

Mayer, G. 2006. Structure and development of anychophoran eyes: What is the ancestral visual organ in arthropods? Arthropod Struct. Dev. 35: 231–245.

Mayer, G., P. M. Whitington, P. Sunnucks, and H.-J. Pflüger. 2010. A revision of brain composition in Onychophora (velvet worms) suggests that the tritocerebrum evolved in arthropods. BMC Evol. Biol. 10: 10: 255.

Minelli, A., G. Boxshall and G. Fusco (eds.). 2013. *Arthropod Biology and Evolution: Molecules, Development, Morphology*. Springer, Heidelberg.

Müller, C. H. G., J. Rosenberg, S. Richter and V. B. Meyer-Rochow. 2003. The compound eye of *Scutigera coleoptrata* (Linnaeus, 1758) (Chilopoda: Notostigmophora): an ultrastructural reinvestigation that adds support to the Mandibulata concept. Zoomorphology. 122:191–209.

Müller, K. J. and D. Walossek. 1985. A remarkable arthropod fauna from the Upper Cambrian "Orsten" of Sweden. T. Roy. Soc. Edin: Earth. 76: 161–172.

Müller, K. J. and D. Walossek. 1986. Arthropod larvae from the Upper Cambrian of Sweden. T. Roy. Soc. Edin: Earth. 77: 157–179.

Müller, K. J. and D. Walossek. 1986. *Martinssonia* elongate gen. et sp. n., a crustacean-like euarthropod from the Upper Cambrian "Orsten" of Sweden. Zool. Scripta 15: 73–92.

Ou, Q., D. Shu and G. Mayer. 2012. Cambrian lobopodians and extant onychophorans provide new insights into early cephalization in Panarthropoda. Nat. Commun. 3: 1261–1272.

Snodgrass, R. E. 1951. *Comparative Studies on the Head of Mandibulate Arthropods*. Comstock Publishing Associates, Ithaca, NY.

Snodgrass, R. E. 1952. *A Textbook of Arthropod Anatomy*. Cornell University Press, Ithaca, NY. [Obviously dated, but still an excellent introduction to arthropod anatomy.]

Strausfeld, N. J. 2012. *Arthropod Brains: Evolution, Functional Elegance and Historical Significance*. Belknap Press, Cambridge.

Strausfeld, N. J., M. C. Strausfeld, S. Stowe, D. Rowell and R. Loesel. 2006. The organization and evolutionary implications of neuropil and their neurons in the brain of the onychophoran *Euperipatoides rowelli*. Arthropod Struct. Dev. 135: 169–196.

Tu, A. T. (ed.). 1984. *Handbook of Natural Toxins. Vol. 2, Insect Poisons, Allergens, and Other Invertebrate Venoms*. Marcel Dekker, New York. [Summarizes information on insects, chilopods, and arachnids.]

Waloszek, D., J. Chen, A. Mass and X. Wang. 2005. Early Cambrian arthropods—new insights into arthropod head and structural evolution. Arthropod Struct. Dev.34: 189–205.

Whittington, H. B. 1971. Redescription of *Marrella splendens* (Trilobitoidea) from the Burgess Shale, Middle Cambrian, British Columbia. Bull. Geol. Surv. Can. 209: 1–24.

Whitington, P. M., T. Meier and P. King. 1991. Segmentation, neurogenesis, and formation of early axonal pathways in the centipede *Ethmostigmus rubripes* (Brandt). Roux's Arch. Dev. Biol. 199: 349–363.

Zill, S. N. and E. Seyfarth. 1996. Exoskeletal sensors for walking. Sci. Am. 275(1): 86–90.

Tardigrada

Bertolani, R. and D. Grimaldi. 2000. A new eutardigrade (Tardigrada: Milnesiidae) in amber from Upper Cretaceous (Turonian) of New Jersey. Pp. 103–110 in *Studies on Fossils in Amber, with Particular Reference to the Cretaceous of New Jersey*. Backhuys Publishers, Leiden, The Netherlands.

Boothby, T. C. and 12 others. 2015. Evidence for extensive horizontal gene transfer from the draft genome of a tardigrade. PNAS doi: 10.1073/pnas.1510461112

Bussers, J. C. and C. Jeuniaux. 1973. Structure et composition de la cuticle de *Macrobiotus* et de *Milnesium tardigradum*. Ann. Soc. R. Zool. Belg. 103: 271–279.

Dewel, R. A. and W. C. Dewel. 1996. The brain of *Echiniscus viridissimus* Peterfi, 1956 (Heterotardigrada): a key to understanding the phylogenetic position of tardigrades and the evolution of the arthropod head. Zool. J. Linn. Soc. 116: 35–49.

Dewel, R. A., D. R. Nelson and W. C. Dewel. 1993. Chapter 5. Tardigrada. In F. W. Harrison and M. E. Rice (eds.), *Microscopic Anatomy of Invertebrates, Volume 12: Onychophora, Chilopoda and Lesser Protostomata*. Wiley-Liss, New York.

Eibye-Jacobsen, J. 1997. Development, ultrastructure and function of the pharynx of *Halobiotus crispae* Kristensen, 1982 (Eutardigrada). Acta Zool. 78(4): 329–347.

Hejnol, A. and R. Schnabel. 2005. The eutardigrade *Thulinia stephaniae* has an indeterminate development and the potential to regulate early blastomere ablations. Development 132: 1349–1361.

Jönsson, K. I., M. Harms-Ringdahl and J. Torudd. 2005. Radiation tolerance in the eutardigrade *Richtersius coronifer*. Int. J. Radiat. Biol. 81: 649–656.

Jørgensen, A., S. Faurby, J. G. Hansen, N. Møbjerg and R. M. Kristensen. 2010. Molecular phylogeny of Arthrotardigrada (Tardigrada). Mol. Phylogen. Evol. 54: 1006–1015.

Kinchin, I. M. 1994. *The Biology of Tardigrades*. Portland Press, London.

Kristensen, R. M. 1978. On the structure of *Batillipes noerrevangi* Kristensen, 1978. 2. The muscle attachments and the true cross-striated muscles. Zool. Anz. Jena 200 (3/4): 173–184.

Kristensen, R. M. 1981. Sense organs of two marine arthrotardigrades (Heterotardigrada, Tardigrada). Acta Zool. 62: 27–41.

Kristensen, R. M. 1982. The first record of cyclomorphosis in Tardigrada, based on a new genus and species from Arctic meiobenthos. Z. Zool. Syst. Evolutionsforsch. 20: 249–270.

Kristensen, R. M. 1984. On the biology of *Wingstrandarctus corallinus* nov. gen. et spec., with notes on the symbiontic bacteria in the subfamily Florarctinae (Arthrotardigrada). Vidensk. Medd. Dan. Naturhist. Foren. København 145: 201–218.

Kristensen, R. M. and R. P. Higgins. 1984. A new family of Arthrotardigrada (Tardigrada: Heterotardigrada) from the Atlantic Coast of Florida, U.S.A. Trans. Am. Microsc. Soc. 103: 295–311.

Kristensen, R. M. and R. P. Higgins. 1984. Revision of *Styraconyx* (Tardigrada: Halechiniscidae), with descriptions of two new species from Disko Bay, west Greenland. Smithsonian Contrb. Zool. 391. Smithsonian Institution Press, Washington, D.C.

Marcus, E. 1929. Tardigrada. In H. G. Bronn (ed.), *Klassen und Ordnungen des Tierreichs, Bd. 5, Abt. 4*. Akad. Verlagsgesellschaft, Frankfurt.

Marley, N. J., S. J. McInnes and C. J. Sands. 2011. Phylum Tardigrada: A re-evaluation of the Parachela. Zootaxa 2819: 51–64.

Mayer G., S. Kauschke, J. Rudiger and P. A. Stevenson. 2013. Neural markers reveal a one-segmented head in tardigrades (water bears). PLoS ONE. 8:e59090.

Møbjerg, N. and C. Dahl. 1996. Studies on the morphology and ultrastructure of the Malpighian tubules of *Halobiotus crispae* Kristensen, 1982 (Eutardigrada). Zool. J. Linn. Soc. 116: 85–99.

Møbjerg, N., A. Jørgensen, J. Eibye-Jacobsen, K. A. Halberg, D. Persson and R. M. Kristensen. 2007. New records on cyclomorphosis in the marine eutardigrade *Halobiotus crispae*. J. Limnol. 66, Suppl. 1: 132–140.

Møbjerg, N., K. A. Halberg, A. Jørgensen, D. Persson, M. Bjørn, H. Ramløv and R. M. Kristensen. 2011. Survival in extreme environments—on the current knowledge of adaptations in tardigrades. Acta Physiol. 202: 409–420.

Nelson, D. R. 1991. Tardigrada. Pp. 501–521 in J. H. Thorp and A. P. Covich (eds.), *Ecology and Classification of North American Freshwater Invertebrates*. Academic Press, New York.

Persson, D., K. A. Halberg, A. Jørgensen, C. Ricci, N. Møbjerg and R. M. Kristensen. 2011. Extreme stress tolerance in tardigrades: Surviving space conditions in low earth orbit. J. Zool. Syst. Evol. Res. 49: 90–97.

Persson D. K., K. A. Halberg, A. Jørgensen, N. Møbjerg and R. M. Kristensen. 2012. Neuroanatomy of *Halobiotus crispae* (Eutardigrada: Hypsibiidae): tardigrade brain structure supports the clade Panarthropoda. J. Morph. 273 (11): 1227–1245.

Persson, D., K. A. Halberg, A. Jørgensen, N. Møbjerg and R. M. Kristensen. 2014. Brain anatomy of the marine tardigrade *Actinarctus doryphorus* (Arthrotardigrada). J. Morph. 275(2): 173–190.

Rebecchi, L., T. Altiero T and R. Guidetti. 2007. Anhydrobiosis: the extreme limit of desiccation tolerance. Invert. Survival J. 4: 65–81.

Schmidt-Rhaesa, A and J. Kulessa. 2007. Muscular architecture of *Milnesium tardigradum* and *Hypsibius* sp. (Eutardigrada, Tardigrada) with some data on *Ramazzottius oberhauseri*. Zoomorphology 126: 265–281.

Schulze, C., R. C. Neves and A. Schmidt-Rhaesa. 2014. Comparative immunohistochemical investigation on the nervous system of two species of Arthrotardigrada (Heterotardigrada (Heterotardigrada, Tardigrada). Zool. Anz. 253:225–235.

Wiederhöft, H. and H. Greven. 1996. The cerebral ganglia of *Milnesium tardigradum* Doyère (Apochela, Tardigrada): three dimensional reconstruction and notes on their ultrastructure. Zool. J. Linn. Soc. 116: 71–84.

Wright, J. C. 2001. Cryptobiosis 300 years on from van Leuwenhoek: what have we learned about tardigrades? Zool. Anz. 240: 563–582.

Zantke, J., C. Wolff and G. Scholtz. 2008. Three-dimensional reconstruction of the central nervous system of *Macrobiotus hufelandi* (Eutardigrada, Parachela): implications for the phylogenetic position of Tardigrada. Zoomorphology 127: 21–36.

Onychophora

Daniels, S. R., D. E. McDonald and M. D. Picker. 2013. Evolutionary insight into the *Peripatopsis balfouri* sensu lato species complex (Onychophora: Peripatopsidae) reveals novel lineages and zoogeographic patterning. Zoologica Scripta 42: 656–674.

de Sena Oliveira, I., V. M. St. J. Read and G. Mayer. 2012. A world checklist of Onychophora (velvet worms), with notes on nomenclature and status of names. ZooKeys 211: 1–70.

de Sena Oliveira, I., N. N. Tait, I. Strübing and G. Mayer. 2013. The role of ventral and preventral organs as attachment sites for segmental limb muscles in Onychophora. Front. Zool. 10: 73.

Dzik, J. and G. Krumbiegel. 1989. The oldest "onychophoran" Xenusion: A link connecting phyla? Lethaia 22: 169–182.

Eakin, R. M. and J. A. Westfall. 1965. Fine structure of the eye of *Peripatus* (Onychophora). Z. Zellforsch. Mik. Ana. 68: 278–300.

Eliott, S., N. N. Tait and D. A. Briscoe. 1993. A pheromonal function for the crural glands of the onychophoran *Cephalofovea tomahmontis* (Onychophora, Peripatopsidae). J. Zool, 231: 1–9.

Eriksson, B. J., N. N. Tait, G. E. Budd, R. Janssen and M. Akam. 2010. Head patterning and Hox gene expression in an onychophoran and its implications for the arthropod head problem. Dev. Genes Evol. 220: 117–122.

Frase, T. and S. Richter. 2013. The fate of the onychophoran antenna. Dev. Genes Evol. doi: 10.1007/s00427-013-0435-x

Grimaldi, D., M. S. Engel and P. C. Nascimbene. 2002. Fossiliferous Cretaceous amber from Myanmar (Burma): Its rediscovery, biotic diversity, and paleontological significance. Amer. Mus. Novitates 3361: 74.

Hoyle, G. and M. Williams. 1980. The musculature of *Peripatus dominicae* and its innervation. Phil. Trans. R. Soc. Lond. Ser. B 288: 481–510.

Koch, M., B. Quast and T. Bartolomaeus. 2014. Coeloms and nephridia in annelids and arthropods. Pp. 173–284 in J. W. Wägele and T. Bartolamaeus (eds.), *Deep Metazoan Phylogeny: the Backbone of the Tree of Life*. De Gruyter, Berlin.

Manton, S. M. and N. Heatley. 1937. The feeding, digestion, excretion and food storage of *Peripatopsis*. Phil. Trans. R. Soc. Lond. Ser. B 227: 411–464.

Mayer, G. 2006. Origin and differentiation of nephridia in the Onychophora provide no support for the Articulata. Zoomorphology 125: 1–12.

Mayer, G. 2006. Structure and development of onychophoran eyes: What is the ancestral visual organ in arthropods? Arthropod Struct. Dev. 35:231–245.

Mayer, G. and S. Harzsch. 2007. Immunolocalization of serotonin in Onychophora argues against segmental ganglia being an ancestral feature of arthropods. BMC Evol. Biol.7: 118.

Mayer, G. and S. Harzsch. 2008. Distribution of serotonin in the trunk of *Metaperipatus blainvillei* (Onychophora, Peripatopsidae): Implications for the evolution of the nervous system in Arthropoda. J. Comp. Neurol. 507: 1196–1208.

Mayer, G. and M. Koch. 2005. Ultrastructure and fate of the nephridial anlagen in the antennal segment of *Epiperipatus biolleyi* (Onychophora, Peripatidae)—evidence for the onychophoran antennae being modified legs. Arthropod Struct. Dev. 34: 471–480.

Mayer, G. and 9 others. 2013. Selective neuronal staining in tardigrades and onychophorans provides insights into the evolution of segmental ganglia in panarthropods. BMC Evol. Biol. 13: 230.

Mayer, G., H. Ruhberg and T. Bartolomaeus. 2004. When an epithelium ceases to exist: an ultrastructural study on the fate of the embryonic coelom in *Epiperipatus biolleyi* (Onychophora, Peripatidae). Acta Zool. 85: 163–170.

Mayer, G., P. M. Whitington, P. Sunnucks and H.-J. Pflüger. 2010. A revision of brain composition in Onychophora (velvet worms) suggests that the tritocerebrum evolved in arthropods. BMC Evol. Biol. 10: 255.

Monje-Nájera, J. 1995. Phylogeny, biogeography and reproductive trends in the Onychophora. Zool. J. Linn. Soc. 114: 21–60.

Murienne, J., S. R. Daniels, T. R. Buckley, G. Mayer and G., Giribet. 2014. A living fossil tale of Pangaean biogeography. Proc. R. Soc. B Biol. Sci. 281: 20132648.

Nylund, A., H. Ruhberg, A. Tjoenneland and B. A. Meidell. 1988. Heart ultrastructure in four species of Onychophora, and phylogenetic implications. Zool. Beitr. n.f. 32: 17–30.

Oliveira, I. S. and 8 others. 2012. Unexplored character diversity in Onychophora (velvet worms): a comparative study of three peripatid species. PLoS ONE 7: e51220.

Pacaud, G., W. Rolfe, F. R. Schram, S. Secretan and D. Sotty. 1981. Quelques invertébrés noveaux du Stéphanien de Montceau-les-Mines. Null. S. H. N. Autun 97: 37–43.

Peck, S. B. 1975. A review of the New World Onychophora, with the description of a new cavernicolous genus and species from Jamaica. Psyche 82: 341–358.

Perrier, V. and S. Charbonnier. 2014. The Montceau-les-Mines Lagerstätte (Late Carboniferous, France). Compt. Rend. Palevol. 13: 353–367.

Poinar Jr., G. 1996. Fossil velvet worms in Baltic and Dominican amber: Onychophoran evolution and biogeography. Science 273: 1370–1371.

Ramsköld, L. 1992. Homologies in Cambrian Onychophora. Lethaia 25: 443–460.

Ramsköld, L. 1992. The second leg row of *Hallucigenia* discovered. Lethaia 25: 321–324.

Ramsköld, L. and X. Hou. 1991. New early Cambrian animal and onychophoran affinities for enigmatic metazoans. Nature 351: 225–228.

Ruhberg, H. and W. B. Nutting. 1980. Onychophora: Feeding, structure, function, behaviour and maintenance. Berh. Naturwiss. Ver. Hamburg (NF) 24: 79–87.

Smith, M. R. and J. Ortega-Hernández. 2014. *Hallucigenia's* onychophoran-like claws and the case for Tactopodia. Nature 514: 363–366.

Tait, N. N. and D. A. Briscoe. 1990. Sexual head structures in the Onychophora: unique modifications for sperm transfer. J. Nat. Hist. 24: 1517–1527.

Tait, N. N., and J. M. Norman. 2001. Novel mating behaviour in *Florelliceps stutchburyae* gen. nov., sp. nov. (Onychophora: Peripatopsidae) from Australia. J. Zool. (London) 253: 301–308.

Thompson, I., Jones, D. S. 1980. A possible onychophoran from the Middle Pennsylvanian Mazon Creek beds of Northern Illinois. J. Paleontol. 54: 588–596.

Walker, M. H. 1995. Relatively recent evolution of an unusual pattern of early embryonic development (long germ band?) in a South African onychophoran, *Opisthopatus cinctipes* Purcell (Onychophora: Peripatopsidae). Zool. J. Linn. Soc. 114: 61–75.

Walker, M. H., E. M. Roberts, T. Roberts, G. Spitteri, M. J. Streubig, J. L. Hartland and N. N. Tait. 2006. Observations on the structure and function of the seminal receptacles and

associated accessory pouches in ovoviviparous onychophorans from Australia (Peripatopsidae; Onychophora). J. Zool. (London) 270: 531–542.
Whitington, P. M. and G. Mayer. 2011. The origins of the arthropod nervous system: insights from the Onychophora. Arthropod Struct. Dev. 40: 193–209.

Arthropod Phylogeny and Evolution

Aguinaldo, A. M. A., J. M. Turbeville, L. S. Linford, M. C. Rivera, J. R. Garey, R. A. Raff and J. A. Lake. 1997. Evidence for a clade of nematodes, arthropods and other moulting animals. Nature 387: 489–493.
Akam, M. 2000. Arthropods: Developmental diversity within a (super) phylum. Proc. Natl. Acad. Sci. U.S.A. 97: 4438–4441.
Averof, M. and N. H. Patel. 1997. Crustacean appendage evolution associated with changes in Hox gene expression. Nature 388: 682–686.
Bergström, J. 1992. The oldest arthropods and the origin of the Crustacea. Acta Zool. 73(5): 287–292.
Borner, J., P. Rehm, R. O. Schill, I. Ebersberger and T. Burmester. 2014. A transcriptome approach to ecdysozoan phylogeny. Mol. Phylogenet. Evol. 80: 79–87
Briggs, D. E. G. and H. B. Whittington. 1987. The affinities of the Cambrian animals *Anomalocaris* and *Opabinia*. Lethaia 20: 185–186.
Brusca, R. C. 2000. Unraveling the history of arthropod biodiversification. Ann. Missouri Bot. Garden 87: 13–25.
Budd, G. E. 1996. The morphology of *Opabinia regalis* and the reconstruction of the arthropod stem-group. Lethaia 29: 1–14.
Calman, W. T. 1909. Crustacea. Part VII, Fasicle 3. In R. Lankester, A Treatise on Zoology. Adam and Charles Black (Londen. Reprinted by A. Asher and Co., Amsterdam, 1964). [A dated, but delightful book.]
Campbell, L. I. and 9 others. 2008. MicroRNAs and phylogenomics resolve the relationships of Tardigrada and suggest that velvet worms are the sister group of Arthropoda. Proc. Natl. Acad. Sci. 108: 15920–15924.
Chen, J.-Y., L. Ramsköld and G.-Q. Zhou. 1994. Evidence for monophyly and arthropod affinity of Cambrian giant predators. Science 264: 1304–1308.
Chen, J., D. Waloszek and A. Maas. 2004. A new "great-appendage" arthropod from the Lower Cambrian of China and homology of chelicerate chelicerae and raptorial antero-ventral appendages. Lethaia 37: 3–20.
Deuve, T. (ed.) 2001. Origin of the Hexapoda. Ann. Soc. Entomol. France 37 (1/2): 1–304. [Papers presented at conference held in Paris, January 1999.]
Edwards, J. S. and M. R. Meyer. 1990. Conservation of antigen 3G6: A crystalline cone constituent in the compound eye of arthropods. J. Neurobiol. 21: 441–452.
Eldredge, N. 1977. Trilobites and evolutionary patterns. Pp. 30–332 in A. Hallam (ed.), *Patterns of Evolution*. Elsevier, Amsterdam.
Eldredge, N. and S. M. Stanley (eds.). 1984. *Living Fossils*. Springer-Verlag, New York.
Ericksson, B. J. and G. E. Budd. 2000. Onychophoran cephalic nerves and their bearing on our understanding of head segmentation and stem-group evolution of Arthropoda. Arthropod Struct. Dev. 29: 197–209.
Giribet, G. 2003. Molecules, development and fossils in the study of metazoan evolution: Articulata versus Ecdysozoa revisited. Zoology 106: 303-326.
Giribet, G. and G. D. Edgecombe. 2012. Reevaluating the arthropod tree of life. Ann. Rev. Entomol. 57: 167–186.
Harzsch, S. and D. Walossek. 2001. Neurogenesis in the developing visual system of the branchiopod crustacean *Triops longicaudatus* (LeConte, 1846): Corresponding patterns of compound-eye formation in Crustacea and Insecta? Dev. Genes Evol. 211: 37–43.
Harzasch, S., M. Wildt, B. Battelle and D. Waloszek. 2005. Immunohistochemical localization of neurotransmitters in the nervous system of larval *Limulus polyphemus* (Chelicerata, Xiphosura): evidence for a conserved protocerebral architecture in Euarthropoda. Arthropod Struct. Dev. 34: 327–342.
Hou, X. and J. Bergström. 1995. Cambrian lobopodians—ancestors of extant onychophorans? Zool. J. Linn. Soc. 114: 3–19.
Hou, X. H. and S. Weiguo. 1988. Discovery of Chengjiang fauna at Meichucun, Jinning, Yunnan. Acta Palaeontol. Sinica 27(1): 1–12.
Hou, X., L. Ramsköld and J. Bergström. 1991. Composition and preservation of the Chengjiang fauna—A lower Cambrian soft-bodied biota. Zool. Scripta 20: 395–411.
Jager, M., J. Murienne, C. Clabaut, J. Deutsch, H. Le Guyader and M. Manuel. 2006. Homology of arthropod anterior appendages revealed by Hox gene expression in a sea spider. 441: 506–508.
Janssen, R., W. G. M. Damen and G. E. Budd. 2011. Expression of *collier* in the premandibular segment of myriapods: Support for the traditional Atelocerata concept of a case of convergence? BMC Evol. Biol. 11: 50.
Jeram, A. J., P. A. Selden and D. Edwards. 1990. Land animals in the Silurian: Arachnids and myriapods from Shropshire, England. Science 250: 658–661.
Legg, D. A., M. D. Sutton, G. D. Edgecombe and J.-B. Caron. 2012. Cambrian bivalved arthropod reveals origin of anrthropodization. Proc. R. Soc. B doi: 10.1098/rspb.2012.1958
Koenemann, S. and R. Jenner (eds.). 2005. *Crustacea and Arthropod Relationships*. Crustacean Issues 15, CRC Press/Balkema, Leiden.
Ma, X., G. D. Edgecombe, X. Hou, T. Goral and N. J. Strausfeld. 2015. Preservational pathways of corresponding brains of a Cambrian euarthropod. Curr. Biol. 25: 1–7.
Ma, X., X. Hou and J. Bergström. 2009. Morphology of *Luolishania longicruris* (Lower Cambrian, Chengjiang Lagestätte, SW China) and the phylogenetic relationships within lobopodians. Arthropod Struct. Dev. 38: 271–291.
Maas, A., G. Mayer, R. M. Kristensen and D. Waloszek. 2007. A Cambrian micro-lobopodian and the evolution of arthropod locomotion and reproduction. Chinese Sciences Bulletin 52(24): 3385–3392.
Maas, A. and D. Waloszek. 2001. Cambrian derivatives of the early arthropod stem lineage, pentastomids, tardigrades and lobopodians—an "Orsten" perspective. Zool. Anz. 240: 451–559.
Martinsson, A. (ed.) 1975. *Evolution and Morphology of the Trilobita, Trilobitoida and Merostomata. Fossils and Strata, No. 4.* Universitentsforlaget, Oslo.
Maxmen, A., W. E. Browne, M. Q. Martindale and G. Giribet. 2005. Neuroanatomy of sea spiders implies an appendicular origin of the protocerebral segment. Nature 437: 1144–1148.
McMenamin, M. A. S. 1986. The Garden of Ediacara. Palaios 1: 178–182.
Melzer, R. R., R. Diersch, D. Nicastro and U. Smola. 1997. Compound eye evolution: Highly conserved retinula and cone cell patterns indicate a common origin of the insect and crustacean ommatidium. Naturwissenschaften 84: 542–544.
Melzer, R. R., C. Michalke and U. Smola. 2000. Walking on insect paths? Early ommatidial development in the compound eye of the ancestral Crustacea, *Triops cancriformis*. Naturwissenschaften 87: 308–311.
Meusemann, K. and 15 others. 2010. A phylogenomic approach to resolve the arthropod tree of life. Mol. Biol. Evol. 27(11): 2451–2464.

Mikulic, D. G., D. E. G. Briggs and J. Kluessendorf. 1985. A new exceptionally preserved biota from the lower Silurian of Wisconsin, U.S.A. Phil. Trans. R. Soc. Lond. B 311: 75–85.

Minelli, A., G. Boxshall and G. Fusco (eds.). 2013. *Arthropod Biology and Evolution*. Springer, New York. [A timely and important summary.]

Misof, B. and 100 others. 2014. Phylogenomics resolves the timing and pattern of insect evolution. Science 346: 763–767.

Norstad, K. 1987. Cycads and the origin of insect pollination. Am. Sci. 75: 270–278.

Osorio, D., M. Averof and J. P. Bacon. 1995. Arthropod evolution: Great brains, beautiful bodies. Trends Ecol. Evol. 10(11): 449–454.

Paulus, H. F. 2000. Phylogeny of the Myriapoda-Crustacea-Insecta: A new attempt using photoreceptor structure. J. Zool. Syst. Evol. Res. 38: 189–208.

Persson, D. K., K. A. Halberg, A. Jørgensen, N. Møbjerg and R. M. Kristensen. 2012. Neuroanatomy of *Halobiotus crispae* (Eutardigrada: Hypsibiidae): Tardigrade brain structure supports the clade Panarthropoda. J. Morphol. 273: 1227–1245.

Popadic, A., D. Rusch, M. Peterson, B. T. Rogers and T. C. Kaufman. 1996. Origin of the arthropod mandible. Nature 380: 395.

Ramsköld, L. 1991. New early Cambrian animal and onychophoran affinities of enigmatic metazoans. Nature 351: 225–228.

Reiger, J. C., J. W. Shultz and R. E. Kambic. 2005. Pancrustacaean phylogeny: hexapods are terrestrial crustaceans and maxillio-pods are not monophyletic. Proc. R. Soc. B. 272: 395-401.

Regier, J. C. and 7 others. 2010. Arthropod relationships revealed by phylogenomic analysis of nuclear protein-coding sequences. Nature 463: 1079–1083.

Robison, R. A. 1985. Affinities of *Aysheaia* (Onychophora), with description of a new Cambrian species. J. Paleontol. 59: 226–235.

Rota-Stabelli, O., A. C. Daley and D. Pisani. 2013. Molecular timetrees reveal a Cambrian colonization of land and a new scenario for ecdysozoan evolution. Curr. Biol. 23: 392–398.

Scholtz, G. 2001. Evolution of developmental patterns in arthropods—the analysis of gene expression and its bearing on morphology and phylogenetics. Zoology 103: 99–111.

Scholtz, G. 2002. The Articulata hypothesis—or what is a segment. Org. Divers. Evol. 2: 197–215.

Scholtz, G. and G. D. Edgecombe. 2006. The evolution of arthropod heads: reconciling morphological, developmental and palaeontological evidence. Dev. Genes Evol. 216: 395–415.

Shear, W. A. and J. Kukalová-Peck. 1990. The ecology of Paleozoic terrestrial arthropods: The fossil evidence. Can. J. Zool. 68: 1807–1834.

Simpson, P. 2001. A review of early development of the nervous system in some arthropods: Comparison between insects, crustaceans and myriapods. Ann. Soc. Entomol. France 37(1/2): 71–84.

Snodgrass, R. E. 1938. Evolution of the Annelida, Onychophora and Arthropoda. Smithsonian Misc. Coll. 97: 1–159. [This paper codified the concepts of Mandibulata and Atelocerata.]

Stollewerk, A. and A. D. Chipman. 2006. Neurogenesis in Myriapods and chelicerates and its importance for understanding arthropod relationships. Integrative and Comparative Biology 46: 195–206.

Strausfeld, N. J. 1998. Crustacean-insect relationships: The use of brain characters to derive phylogeny amongst segmented invertebrates. Brain Behav. Evol. 52: 186–206.

Strausfeld, N. J., E. K. Bushbeck and R. S. Gomez. 1995. The arthropod mushroom body: Its roles, evolutionary enigmas and mistaken identities. Pp. 349–381 in O. Breidbach and W. Kutsch (eds.), *The Nervous Systems of Invertebrates: An Evolutionary and Comparative Approach*. Birkhäuser Verlag, Basel.

Strausfeld, N. J. and F. Hirth. 2013. Deep homology of arthropod central complex and vertebrae basal ganglia. Science 340: 157–161.

Strausfeld, N. J., C. M. Strausfeld, R. Loesel, D. Rowell and S. Stowe. 2006. Arthropod phylogeny: onychophoran brain organization suggests an archaic relationship with a chelicerate stem lineage. Proc. R. Soc. B 273: 1857–1866.

Van Roy, P., A. C. Daley and D. E. G. Briggs. 2015. Anomalocaridid trunk limb homology revealed by a giant filter-feeder with paired flaps. Nature 522: 77–80.

Von Reumont, B. M. and 12 others. 2012. Pancrustacean phylogeny in the light of new phylogenomic data: Support for Remipedia as the possible sister group of Hexapoda. Mol. Biol. Evol. 29: 1031–1045.

Waloszek, D. 1995. The Upper Cambrian *Rehbachiella*, its larval development, morphology and significance for the phylogeny of Branchiopoda and Crustacea. Hydrobiologia 298: 32: 1–13.

Waloszek, D. 2003. The "Orsten" window—a three-dimensionally preserved Upper Cambrian meiofauna and its contribution to our understanding of the evolution of Arthropoda. Paleontol. Res. 7: 71–88.

Waloszek, D. and A. Maas. 2005. The evolutionary history of crustacean segmentation: a fossil-based perspective. Evol. Dev.7: 515–527.

Waloszek, D., A. Maas, J. Chen and M. Stein. 2007. Evolution of cephalic feeding structures and the phylogeny of Arthropoda. Palaeo 254: 273–287.

Walossek, D. and K. J. Müller. 1990. Upper Cambrian stem-lineage crustaceans and their bearing upon the monophyletic origin of Crustacea and the position of *Agnostus*. Lethaia 23: 409–427.

Walossek, D. and K. J. Müller. 1992. The "Alum Shale Window"–contribution of "Orsten" arthropods to the phylogeny of Crustacea. Acta Zool. 73(5): 305–312.

Walossek, D. and K. J. Müller. 1997. Cambrian "Orsten"-type arthropods and the phylogeny of Crustacea. Pp. 139–153 in R. A. Fortey and R. H. Thomas (eds.), *Arthropod Relationships*. Chapman and Hall, New York.

Whittington, H. B. and D. E. G. Briggs. 1985. The largest Cambrian animal, *Anomalocaris*, Burgess Shale, British Columbia. Phil. Trans. R. Soc. Lond. B 309: 569–609.

Whittington, P. M., D. Leach and R. Sandeman. 1993. Evolutionary change in neural development within the arthropods: axonogensis in the embryos of two crustaceans. Development 118: 449–461.

Zhang, X.-G., D. J. Siveter, D. Waloszek and A. Maas. 2007. An epipodite-bearing crown-group crustacean from the Lower Cambrian. Nature 449: 595–598.

CHAPTER 21

Phylum Arthropoda

Crustacea: Crabs, Shrimps, and Their Kin

Crustaceans are one of the most popular invertebrate groups, even among nonbiologists, for they include some of the world's most delectable gourmet fare, such as lobsters, crabs, and shrimps (Figure 21.1). There are an estimated 70,000 described living species of Crustacea, and probably five or ten times that number waiting to be discovered and named. They exhibit an incredible diversity of form, habit, and size. The smallest known crustaceans are less than 100 μm in length and live on the antennules of copepods. The largest are Japanese spider crabs (*Macrocheira kaempferi*), with leg spans of 4 m, and giant Tasmanian crabs (*Pseudocarcinus gigas*) with carapace widths of 46 cm. The heaviest crustaceans are probably American lobsters (*Homarus americanus*), which, before the present era of overfishing, attained weights in excess of 20 kilograms. The world's largest land arthropod by weight (and possibly the largest land invertebrate) is the coconut crab (*Birgus latro*), weighing in at up to 4 kg. Crustaceans are found at all depths in every marine, brackish, and freshwater environment on Earth, including in pools at 6,000 m elevation (fairy shrimp and cladocerans in northern Chile). A few have become successful on land, the most notable being sowbugs and pillbugs (the terrestrial isopods).[1]

Crustaceans are commonly the dominant organisms in aquatic subterranean ecosystems, and new species of these stygobionts continue to be discovered as new caves are explored. They also dominate ephemeral pool habitats, where many undescribed species are known to occur.[2] And crustaceans are the most widespread, diverse, and

Classification of The Animal Kingdom (Metazoa)

Non-Bilateria*
(a.k.a. the diploblasts)
- PHYLUM PORIFERA
- PHYLUM PLACOZOA
- PHYLUM CNIDARIA
- PHYLUM CTENOPHORA

Bilateria
(a.k.a. the triploblasts)
- PHYLUM XENACOELOMORPHA

Protostomia
- PHYLUM CHAETOGNATHA

SPIRALIA
- PHYLUM PLATYHELMINTHES
- PHYLUM GASTROTRICHA
- PHYLUM RHOMBOZOA
- PHYLUM ORTHONECTIDA
- PHYLUM NEMERTEA
- PHYLUM MOLLUSCA
- PHYLUM ANNELIDA
- PHYLUM ENTOPROCTA
- PHYLUM CYCLIOPHORA

Gnathifera
- PHYLUM GNATHOSTOMULIDA
- PHYLUM MICROGNATHOZOA
- PHYLUM ROTIFERA

Lophophorata
- PHYLUM PHORONIDA
- PHYLUM BRYOZOA
- PHYLUM BRACHIOPODA

ECDYSOZOA

Nematoida
- PHYLUM NEMATODA
- PHYLUM NEMATOMORPHA

Scalidophora
- PHYLUM KINORHYNCHA
- PHYLUM PRIAPULA
- PHYLUM LORICIFERA

Panarthropoda
- PHYLUM TARDIGRADA
- PHYLUM ONYCHOPHORA
- **PHYLUM ARTHROPODA**
  - **SUBPHYLUM CRUSTACEA***
  - SUBPHYLUM HEXAPODA
  - SUBPHYLUM MYRIAPODA
  - SUBPHYLUM CHELICERATA

Deuterostomia
- PHYLUM ECHINODERMATA
- PHYLUM HEMICHORDATA
- PHYLUM CHORDATA

*Paraphyletic group

This chapter has been revised by Richard C. Brusca and Joel W. Martin

[1]When most people hear the word "shrimp" they think of edible shrimps, two crustacean groups nested within the order Decapoda (in the suborders Dendrobranchiata and Pleocyemata). However, the term "shrimp" is applied to a number of long-tailed crustaceans, many not closely related at all to decapods. So, in this general sense, there are fairy shrimps, tadpole shrimps, mantis shrimps, etc. In much of the English-speaking world, the word "prawn" is used for the edible shrimps, thus eliminating some of the confusion.

[2]One study of ephemeral pools in northern California discovered 30 probable undescribed and unnamed crustacean species (King et al. 1996).

(A)
(B)
(C)
(D)
(E)
(F)
(G)
(H)
(I)
(J)
(K)
(L)
(M)

Figure 21.1 Diversity among the Crustacea. (A–D) Classes Remipedia, Cephalocarida, and Branchiopoda. (A) *Speleonectes ondinae*; note the remarkably homonomous body of this swimming crustacean (Remipedia). (B) A cephalocarid. (C) A tadpole shrimp (Branchiopoda; Notostraca) from an ephemeral pool. (D) A clam shrimp (Branchiopoda: Diplostraca), carrying eggs. (E–Q) Class Malacostraca. (E) The fiddler crab, *Uca princeps* (Decapoda: Brachyura). (F) The giant Caribbean hermit crab, *Petrochirus diogenes* (Decapoda: Anomura). (G) A hermit crab (*Paragiopagurus fasciatus*) removed from its snail shell (Decapoda: Anomura). (H) The coconut crab, *Birgus latro* (Decapoda: Anomura), climbing a tree. (I) *Euceramus praelongus*, the New England olive pit porcelain crab (Decapoda: Anomura). (J) The pelagic lobsterette, *Pleuroncodes planipes* (Decapoda: Anomura). (K) The cleaner shrimp, *Lysmata californica* (Decapoda: Caridea). (L) The unusual rock-boring lobster shrimp, *Axius vivesi* (Decapoda: Axiidea). (M) The Hawaiian regal lobster, *Enoplometopus* (Decapoda: Achelata). (N,O) Two unusual amphipods (Peracarida: Amphipoda). (N) *Cystisoma*, a huge (some exceed 10 cm), transparent, pelagic hyperiid amphipod. (O) *Cyamus scammoni*, a parasitic caprellid amphipod that lives on the skin of gray whales. (P) A male and female cumacean; note the eggs in the marsupium of the female (Peracarida: Cumacea). (Q) *Ligia pacifica* (Peracarida: Isopoda), the rock louse; *Ligia* are inhabitants of the high spray zone on rocky shores worldwide. (R) The mysid *Siriella* sp. (Peracarida: Mysida) with a parasitic epicaridean isopod (Peracarida: Isopoda: Dajidae) attached to its carapace (Western Australia). (S,T) Class Thecostraca. (S) Acorn barnacles, *Semibalanus balanoides* (Cirripedia: Thoracica). (T) *Lepas anatifera* (Cirripedia: Thoracica), a pelagic barnacle hanging from a floating timber.

(Continued on next page)

(U)

(V)

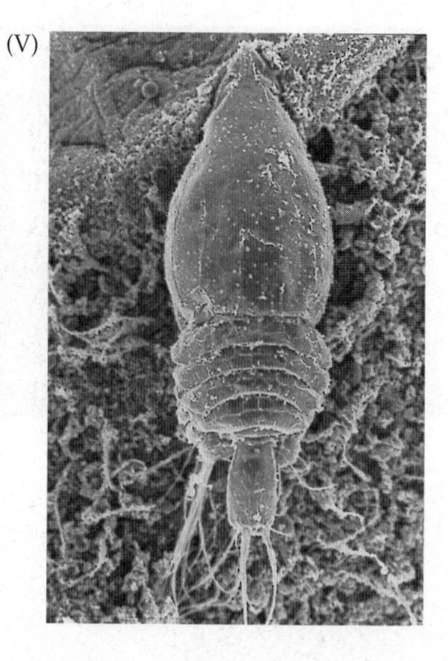

Figure 21.1 (continued) Diversity among the Crustacea. (U) Class Copepoda; the calanoid copepod *Gaussia*, a common planktonic genus. (V) Class Tantulocarida, *Deoterthron*, parasitic on other crustaceans.

BOX 21A Characteristics of the Subphylum Crustacea

1. Body composed of a 6-segmented head, or cephalon (plus the acron), and a long postcephalic trunk; trunk divided into two more or less distinct tagmata (e.g., thorax and abdomen) in all but the remipedes and ostracods (Figure 21.2)
2. Cephalon composed of (anterior to posterior): presegmental (indistinguishable) acron, protocerebral segment (lacking appendages), antennular segment, antennal segment, mandibular somite, maxillulary somite, and maxillary somite; one or more anterior thoracomeres may fuse with the head in members of some classes (e.g., Remipedia, and Malacostraca), their appendages forming maxillipeds
3. Cephalic shield or carapace present (highly reduced in anostracans, amphipods, and isopods)
4. Appendages multiarticulate, uniramous or biramous
5. Mandibles usually multiarticulate limbs that function as biting, piercing, or chewing/grinding jaws
6. Gas exchange by aqueous diffusion across specialized branchial surfaces, either gill-like structures or specialized regions of the body surface
7. Excretion by true nephridial structures (e.g., antennal glands, maxillary glands)
8. Both simple ocelli and compound eyes occur in most taxa (not Remipedia), at least at some stage of the life cycle; compound eyes often elevated on stalks
9. Gut with digestive ceca
10. With nauplius larva (unknown from any other arthropod subphylum); development mixed or direct

abundant animals inhabiting the world's oceans. The biomass of one species, the Antarctic krill (*Euphausia superba*), has been estimated at 500 million tons at any given time, probably surpassing the biomass of any other group of marine animals (and rivaling that of the world's ants). The range of morphological diversity among crustaceans far exceeds that of even the insects. Many species of Crustacea are threatened by environmental degradation, over 500 are listed on the IUCN Red List, and about two dozen are protected by the U.S. Environmental Protection Agency. Crustaceans are also the most common invasive invertebrates, and well over 100 species have established themselves in marine and estuarine waters of North America alone.

Because of their taxonomic diversity and numerical abundance, it is often said that crustaceans are the "insects of the sea." We prefer to think of insects as "crustaceans of the land." And indeed, there is now strong phylogenetic evidence that the insects arose from a branch within the Crustacea.

Despite the enormous morphological diversity seen among crustaceans (Figures 21.1 to 21.20), they display a suite of fundamental unifying features (Box 21A). In an effort to introduce both the diversity and the unity of this enormous group of arthropods, we first present a classification and synopses of the major taxa. We then discuss the biology of the group as a whole, drawing examples from its various members. As you read this chapter, we ask that you keep in mind the general account of the arthropods presented in Chapter 20.

Classification of The Crustacea

Crustaceans have been known to humans since ancient times and have provided us with sources of both food and legend. It is somewhat comforting to carcinologists (those who study crustaceans) to note that Can-

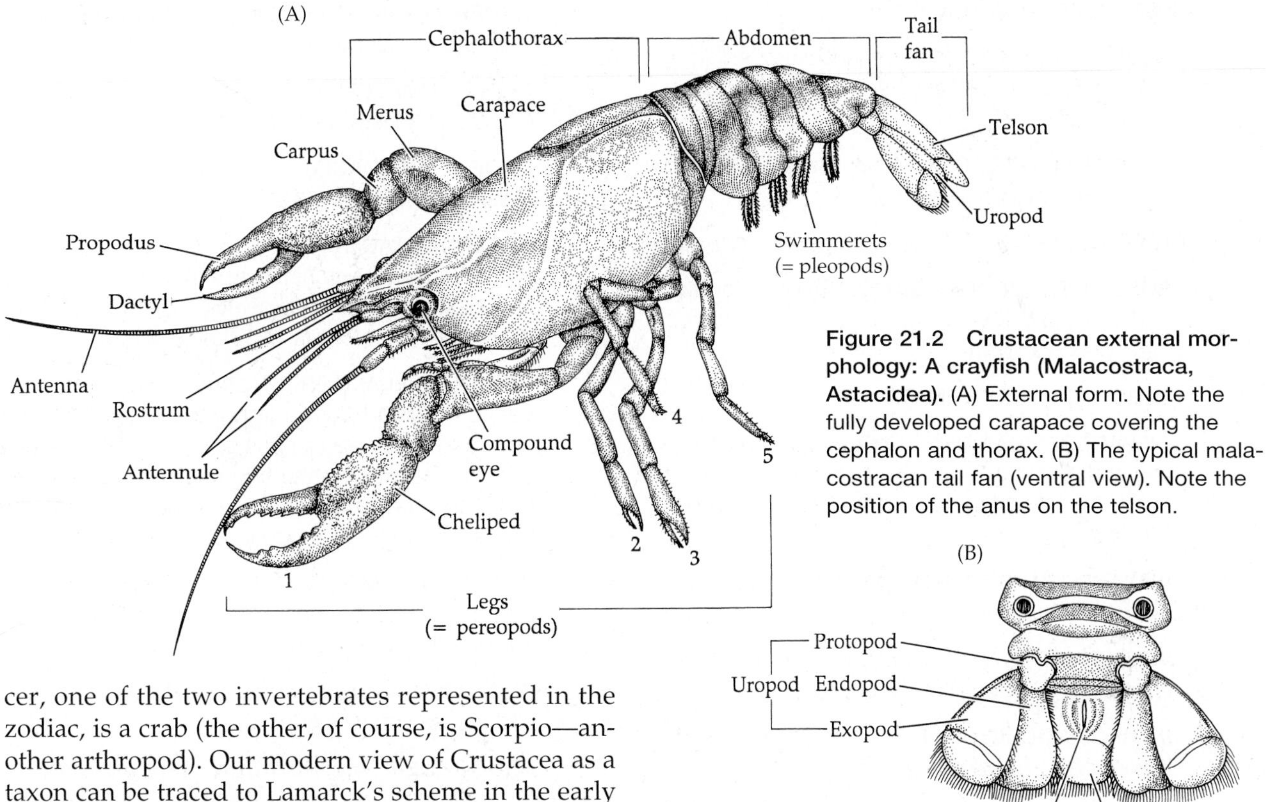

Figure 21.2 Crustacean external morphology: A crayfish (Malacostraca, Astacidea). (A) External form. Note the fully developed carapace covering the cephalon and thorax. (B) The typical malacostracan tail fan (ventral view). Note the position of the anus on the telson.

cer, one of the two invertebrates represented in the zodiac, is a crab (the other, of course, is Scorpio—another arthropod). Our modern view of Crustacea as a taxon can be traced to Lamarck's scheme in the early nineteenth century. He recognized most crustaceans as such, but placed the barnacles and a few others in separate groups. For many years barnacles were classified with molluscs because of their thick, calcareous outer shell. Crustacean classification as we know it today was more or less established during the second half of the nineteenth century, although internal revisions continue. Martin and Davis (2001) presented an overview of crustacean classification, and readers are referred to that publication for a window into the labyrinthine history of this subphylum. Our classification recognizes 11 classes, following Ahyong et al. (2011) and Martin et al. (2014).

CLASSIFICATION OF CRUSTACEA

CLASS REMIPEDIA Remipedes. One living order, Nectiopoda (e.g., *Cryptocorynectes*, *Godzillius*, *Lasionectes*, *Pleomothra*, *Speleonectes*)

CLASS CEPHALOCARIDA Cephalocarids (e.g., *Chiltoniella*, *Hampsonellus*, *Hutchinsoniella*, *Lightiella*, *Sandersiella*)

CLASS BRANCHIOPODA Branchiopods

ORDER ANOSTRACA Fairy shrimps (e.g., *Artemia*, *Branchinecta*, *Branchinella*, *Streptocephalus*)

ORDER NOTOSTRACA Tadpole shrimps (*Lepidurus*, *Triops*)

ORDER DIPLOSTRACA The "bivalved" branchiopods

SUBORDER LAEVICAUDATA Flat-tailed clam shrimps (e.g., *Lynceus*)

SUBORDER ONYCHOCAUDATA Clam shrimps, cladocerans (water fleas), and cyclestherians

INFRAORDER SPINICAUDATA Clam shrimps (e.g., *Cyzicus*, *Eulimnadia*, *Imnadia*, *Metalimnadia*)

INFRAORDER CLADOCEROMORPHA Water fleas and cyclestheriids

TRIBE CYCLESTHERIDA Monotypic: *Cyclestheria hislopi*

TRIBE CLADOCERA Water fleas (e.g., *Anchistropus*, *Daphnia*, *Leptodora*, *Moina*, *Polyphemus*)

CLASS MALACOSTRACA

SUBCLASS PHYLLOCARIDA

ORDER LEPTOSTRACA Leptostracans or nebaliaceans (e.g., *Dahlella*, *Levinebalia*, *Nebalia*, *Nebaliella*, *Nebaliopsis*, *Paranebalia*)

SUBCLASS HOPLOCARIDA

ORDER STOMATOPODA Mantis shrimps (e.g., *Echinosquilla*, *Gonodactylus*, *Hemisquilla*, *Squilla*)

SUBCLASS EUMALACOSTRACA

SUPERORDER SYNCARIDA Syncarids

ORDER BATHYNELLACEA (e.g., *Bathynella*)

ORDER ANASPIDACEA (e.g., *Anaspides*, *Psammaspides*)

SUPERORDER EUCARIDA

ORDER EUPHAUSIACEA Euphausids, or krill (e.g., *Bentheuphausia*, *Euphausia*, *Meganyctiphanes*, *Nyctiphanes*)

ORDER AMPHIONIDACEA Amphionids. Monotypic: *Amphionides reynaudii*

ORDER DECAPODA Crabs, shrimps, lobsters, etc.

SUBORDER DENDROBRANCHIATA Penaeid and sergestid shrimps (e.g., *Lucifer*, *Penaeus*, *Sergestes*, *Sicyonia*)

SUBORDER PLEOCYEMATA

INFRAORDER CARIDEA Caridean and procaridean shrimps (e.g., *Alpheus*, *Crangon*, *Hippolyte*, *Lysmata*, *Macrobrachium*, *Palaemon*, *Pandalus*, *Pasiphaea*)

INFRAORDER PROCARIDOIDA "Primitive" shrimp (*Procaris, Vetericaris*)

INFRAORDER STENOPODIDEA Stenopodidean shrimps (e.g., *Spongicola*, *Stenopus*)

INFRAORDER GLYPHEIDEA "Primitive" lobsters (*Neoglyphea, Laurentaeglyphea*)

INFRAORDER POLYCHELIDA Deep-sea slipper lobsters (e.g., *Polycheles, Stereomastis*)

INFRAORDER BRACHYURA "True" crabs (e.g., *Actaea*, *Callinectes*, *Cancer*, *Cardisoma*, *Carcinus*, *Grapsus*, *Hemigrapsus*, *Maja*, *Menippe*, *Ocypode*, *Pachygrapsus*, *Pinnotheres*, *Polydectus*, *Portunus*, *Scylla*, *Uca, Xantho*)

INFRAORDER ANOMURA Hermit crabs, galatheid crabs, sand crabs, porcelain crabs, etc. (e.g., *Birgus, Coenobita, Emerita, Galathea, Hippa, Kiwa, Lithodes, Lomis, Paguristes, Pagurus, Petrochirus, Petrolisthes, Pleuroncodes, Pylopagurus*)

INFRAORDER ASTACIDEA Crayfishes and clawed (chelate) lobsters (e.g., *Astacus, Cambarus, Homarus, Nephrops*)

INFRAORDER ACHELATA (formerly Palinura). Palinurid, spiny, and Spanish (slipper) lobsters (e.g., *Enoplometopus, Evibacus, Ibacus, Jassa, Jasus, Palinurus, Panulirus, Scyllarus*)

INFRAORDER AXIIDEA Lobster shrimps (e.g., *Axiopsis*, *Axius*, *Calocarides*, *Calocaris*, *Callianassa*, *Neaxius*)

INFRAORDER GEBIIDEA Mud and ghost shrimps (e.g., *Axianassa*, *Gebiacantha*, *Thalassina*, *Upogebia*)

SUPERORDER PERACARIDA

ORDER MYSIDA Mysids or opossum shrimps (e.g., *Acanthomysis, Hemimysis, Mysis, Neomysis*)

ORDER LOPHOGASTRIDA Lophogastrids (e.g., *Gnathophausia, Lophogaster*)

ORDER CUMACEA Cumaceans (e.g., *Campylaspis, Cumopsis, Diastylis, Diastylopsis*)

ORDER TANAIDACEA Tanaids (e.g., *Apseudes, Heterotanais, Paratanais, Tanais*)

ORDER MICTACEA Mictaceans (e.g., *Mictocaris*, etc.). Some researchers separate out the family Hirsutiidae as a distinct order (Bochusacea, containing *Hirsutia*, *Montucaris* and *Thetispelecaris*).

ORDER SPELAEOGRIPHACEA Spelaeogriphaceans. Four described living species (*Potiicoara braziliense*s, *Spelaeogriphus lepidops,* and two species of *Mangkurtu*), and two known fossil species (the Carboniferous *Acadiocaris novascotica* and the Upper Jurassic *Liaoningogriphus quadripartitus*)

ORDER THERMOSBAENACEA Thermosbaenaceans (e.g., *Halosbaena, Limnosbaena, Monodella, Theosbaena, Thermosbaena, Tulumella*)

ORDER ISOPODA Isopods (sea slaters, rock lice, pillbugs, sowbugs, roly-polies)

SUBORDER ANTHURIDEA (e.g., *Anthura*, *Colanthura*, *Cyathura*, *Mesanthura*)

SUBORDER ASELLOTA (e.g., *Asellus*, *Eurycope*, *Jaera*, *Janira*, *Microcerberus*, *Munna*)

SUBORDER CALABOZOIDEA (*Calabozoa*)

SUBORDER CYMOTHOIDA (e.g., *Aega*, *Bathynomus*, *Cymothoa*, *Cirolana*, *Rocinela*)

SUBORDER EPICARIDEA (e.g., *Bopyrus*, *Dajus*, *Hemiarthrus*, *Ione*, *Pseudione*)

SUBORDER GNATHIIDEA (e.g., *Gnathia*, *Paragnathia*)

SUBORDER LIMNORIDEA (e.g., *Limnoria*, *Keuphylia*, *Hadromastax*)

SUBORDER MICROCERBERIDEA (e.g, *Atlantasellus*, *Microcerberus*)

SUBORDER ONISCIDEA (e.g., *Armadillidium*, *Ligia*, *Oniscus*, *Porcellio*, *Trichoniscus*, *Tylos*, *Venezillo*)

SUBORDER PHREATOICIDEA (e.g., *Mesamphisopus*, *Phreatoicopis*, *Phreatoicus*)

SUBORDER PHORATOPIDEA (*Phoratopus*)

SUBORDER SPHAEROMATIDEA (e.g., *Ancinus*, *Bathycopea*, *Paracerceis, Paradella*, *Sphaeroma*, *Serolis*, *Tecticeps*)

SUBORDER TAINISOPIDEA (*Pygolabis*, *Tainisopus*)

SUBORDER VALVIFERA (e.g., *Arcturus*, *Idotea*, *Saduria*)

ORDER AMPHIPODA Amphipods—beach hoppers, sand fleas, scuds, skeleton shrimps, whale lice, etc.

SUBORDER GAMMARIDEA (e.g., *Ampithoe, Anisogammarus, Corophium, Eurythenes, Gammarus, Niphargus, Orchestia, Phoxocephalus, Talitrus*)

SUBORDER HYPERIIDEA (e.g., *Cystisoma, Hyperia, Phronima, Primno, Rhabdosoma, Scina, Streetsia, Vibilia*)

SUBORDER CAPRELLIDEA (e.g., *Caprella, Cyamus, Metacaprella, Phtisica, Syncyamus*)

SUBORDER INGOLFIELLIDEA (e.g., *Ingolfiella, Metaingolfiella*)

CLASS THECOSTRACA Barnacles and their kin

SUBCLASS FACETOTECTA Monogeneric (*Hansenocaris*): the mysterious "y-larvae," a group of marine nauplii and cyprids for which adults are unknown

SUBCLASS ASCOTHORACIDA Two orders (Laurida, Dendrogastrida) of parasitic thecostracans (e.g., *Ascothorax, Dendrogaster, Laura, Synagoga, Zoanthoecus*)

SUBCLASS CIRRIPEDIA Cirripedes, the barnacles, and their kin

SUPERORDER THORACICA True barnacles. Four orders, Lepadiformes (pedunculate or goose barnacles: e.g., *Lepas*), Ibliformes, Scalpelliformes (*Pollicipes, Scalpellum*), and Sessilia (sessile or acorn barnacles: e.g., *Balanus, Chthamalus, Conchoderma, Coronula, Tetraclita, Verruca*)

SUPERORDER ACROTHORACICA Burrowing "barnacles." Two orders, Cryptophialida and Lythoglyptida (e.g., *Cryptophialus, Trypetesa*)

SUPERORDER RHIZOCEPHALA Parasitic "barnacles." Two orders, Kentrogonida and Akentrogonida (e.g., *Heterosaccus, Lernaeodiscus, Mycetomorpha, Peltogaster, Sacculina, Sylon*)

CLASS TANTULOCARIDA Deep water, marine parasites (e.g., *Basipodella, Deoterthron, Microdajus*)

CLASS BRANCHIURA Fish lice, or argulids. A single family (Argulidae) (e.g., *Argulus, Chonopeltis, Dipteropeltis, Dolops*)

CLASS PENTASTOMIDA Tongueworms. Two orders, numerous families (e.g., *Cephalobaena, Linguatula, Pentastoma, Waddycephalus*)

CLASS MYSTACOCARIDA Mystacocarids, with a single family (Derocheilocarididae), and 13 species (e.g., *Ctenocheilocaris, Derocheilocaris*)

CLASS COPEPODA

SUBCLASS PROGYMNOPLEA

ORDER PLATYCOPIOIDA Platycopioids (e.g., *Antrisocopia, Platycopia*)

SUBCLASS NEOCOPEPODA

SUPERORDER GYMNOPLEA

ORDER CALANOIDA Calanoids (e.g., *Bathycalanus, Calanus, Diaptomus, Eucalanus, Euchaeta*)

SUPERORDER PODOPLEA

ORDER CYCLOPOIDA Cyclopoids (e.g., *Cyclopina, Cyclops, Eucyclops, Lernaea, Mesocyclops, Notodelphys*)

ORDER GELYELLOIDA Gelyelloids (e.g., *Gelyella*)

ORDER HARPACTICOIDA Harpacticoids (e.g., *Harpacticus, Longipedia, Peltidium, Porcellidium, Psammus, Sunaristes, Tisbe*)

ORDER MISOPHRIOIDA Misophriods (e.g., *Boxshallia, Misophria*)

ORDER MONSTRILLOIDA Monstrilloids (e.g., *Monstrilla, Stilloma*)

ORDER MORMONILLOIDA Mormonilloids. Monogeneric: *Mormonilla*

ORDER POECILOSTOMATOIDA Poecilostomatoids (e.g., *Chondracanthus, Erebonaster, Ergasilus, Pseudanthessius*)

ORDER SIPHONOSTOMATOIDA Siphonostomatoids (e.g., *Clavella, Nemesis, Penella, Pontoeciella, Trebius*)

CLASS OSTRACODA Ostracods

SUBCLASS MYODOCOPA

ORDER MYODOCOPIDA (e.g., *Cypridina, Euphilomedes, Eusarsiella, Gigantocypris, Skogsbergia, Vargula*)

ORDER HALOCYPRIDA (e.g., *Conchoecia, Polycope*)

SUBCLASS PODOCOPA

ORDER PODOCOPIDA (e.g., *Cypris, Candona, Celtia, Darwinula, Limnocythre*)

ORDER PLATYCOPIDA (e.g., *Cytherella, Sclerocypris*)

ORDER PALAEOCOPIDA (e.g., *Manawa*)

Synopses of Crustacean Taxa

The following descriptions of major crustacean taxa will give you an idea of the range of diversity within the group and the variety of ways in which these successful animals have exploited the basic crustacean body plan.

Subphylum Crustacea

Body composed of a 6-segmented cephalon (plus presegmental acron), or head, and multisegmented postcephalic trunk; trunk divided into thorax and abdomen (except in remipedes and ostracods); segments of cephalon bear first antennae (antennules), second antennae, mandibles, maxillules, and maxillae (see Table 20.2); one or more anterior thoracomeres may fuse with the head (e.g., Remipedia and Malacostraca), their appendages forming maxillipeds (secondarily modified for feeding); cephalic shield or carapace present (secondarily lost in some groups); with antennal glands or maxillary glands (excretory nephridia); both simple ocelli and compound eyes in most groups, at least at some stage of the life cycle; compound eyes stalked in many groups; with nauplius larval stage (suppressed or bypassed in some groups), and often a series of additional larval stages. There are an estimated 70,000 living species, in over 1,000 families.[3]

Class Remipedia

Body of two regions, a cephalon and an elongate homonomous trunk of up to 32 segments, each with a pair of flattened limbs. Cephalon with a pair of sensory preantennular frontal processes; first antennae biramous; trunk limbs laterally directed, biramous, paddle-like, but without large epipods; rami of trunk limbs (exopod and endopod) each of three or more articles; without a carapace, but with cephalic shield covering head; midgut with serially arranged digestive ceca; first trunk segment fused with head and bearing one pair of prehensile maxillipeds; labrum very large, forming a chamber (atrium oris) in which reside the "internalized" mandibles; maxillules function as hypodermic fangs; last trunk segment partly fused dorsally with telson; telson with caudal rami; segmental double ventral nerve cord; eyes absent in living species; male gonopore on trunk limb 15, female on 8; up to 45 mm in length. The above diagnosis is for the 24 known living remipedes (order Nectiopoda); the fossil record is currently based on a single poorly preserved specimen (order Enantiopoda) (Figures 21.1A 21.3D–F, 21.21D, 21.22F, and 21.31E,F).

The discovery of living remipedes in 1981 by Jill Yager, strange vermiform crustaceans first collected from a cavern in the Bahamas, gave the carcinological world a turn. The combination of features distinguishing these creatures is puzzling, for they possess characteristics that are certainly very primitive (e.g., long, homonomous trunk; double ventral nerve cord; segmental digestive ceca; cephalic shield) as well as some attributes traditionally recognized as advanced (e.g., maxillipeds; nonphyllopodous [though flattened], biramous limbs). They swim about on their backs as a result of metachronal beating of the trunk appendages, similar to anostracans. The remipedes are thus reminiscent of two other primitive classes, the branchiopods and cephalocarids. However, the laterally directed limbs are unlike those of any other crustacean, and the "internalized" mandibles and the poison-injecting hypodermic maxillules are unique (the complex venom contains neurotoxins, peptidases, and chitinases). The presence of the preantennular processes is also puzzling, although similar structures are known to occur in a few other crustaceans. Some phylogenetic analyses based on morphological data suggest that remipedes may be the most primitive living crustaceans, whereas molecular data remain ambiguous on the subject or, in some cases, ally them with cephalocarids and/or the Hexapoda.

All of the living remipedes discovered thus far are found in caves (usually with connections to the sea) in the Caribbean Basin, Indian Ocean, Canary Islands, and Australia. The water in these caves is often distinctly stratified, with a layer of fresh water overlying the denser salt water in which the remipedes swim. Remipedes hatch as lecithotrophic naupliar larvae, which is also unusual given their habitat (most cave crustaceans have direct development). Postnaupliar development is largely anamorphic; juveniles have fewer trunk segments than do the adults. Based on the three pairs of raptorial, prehensile cephalic limbs (and direct observations), it was long thought that remipedes were strictly predators. However, studies by Stefan Koenemann and his colleagues have suggested they might also be capable of suspension feeding.

Class Cephalocarida

Head followed by 8-segmented thorax with each segment bearing limbs, a 11-segmented limbless abdomen, and a telson with caudal rami; common gonopore on protopods of sixth thoracopods; carapace absent but head covered by cephalic shield; thoracopods 1–7 biramous and phyllopodous, with large flattened exopods and epipods (exites) and stenopodous endopods; thoracopods 8 reduced or absent; maxillae resemble thoracopods; no maxillipeds; eyes absent; nauplii with antennal glands, adults with maxillary glands and (vestigial) antennal glands (Figures 21.1T, 21.3A–C, and 21.21A).

Cephalocarids are tiny, elongate crustaceans ranging in length from 2 to 4 mm. There are 12 species in

[3]Segments of the thorax are called thoracomeres (regardless of whether or not any of these segments are fused to the head), whereas appendages of the thorax are called thoracopods. The term "pereon" refers to that portion of the thorax not fused to the head (when such fusion occurs), and the terms "pereonites" (= pereomeres) and "pereopods" are used for the segments and appendages, respectively, of the pereon. Hence, on a crustacean with the first thoracic segment (thoracomere 1) fused to the head, thoracomere 2 is typically called pereonite 1, the first pair of pereopods represents the second pair of thoracopods, and so on. Be assured that we are trying to simplify, not confuse, this issue. Also, we caution you that the homology of the thorax and abdomen among the major crustacean lineages is probably more reverie than reality; the segmental homologies of the thorax and abdomen have not yet been unraveled among the crustacean classes. For summaries of naupliar development across the group and larval features in all crustaceans, see Martin et al. (2014). In most crustacean species the number of body segments is fixed, but in at least two groups (notostracans and remipedes) the segment number can vary within a given species.

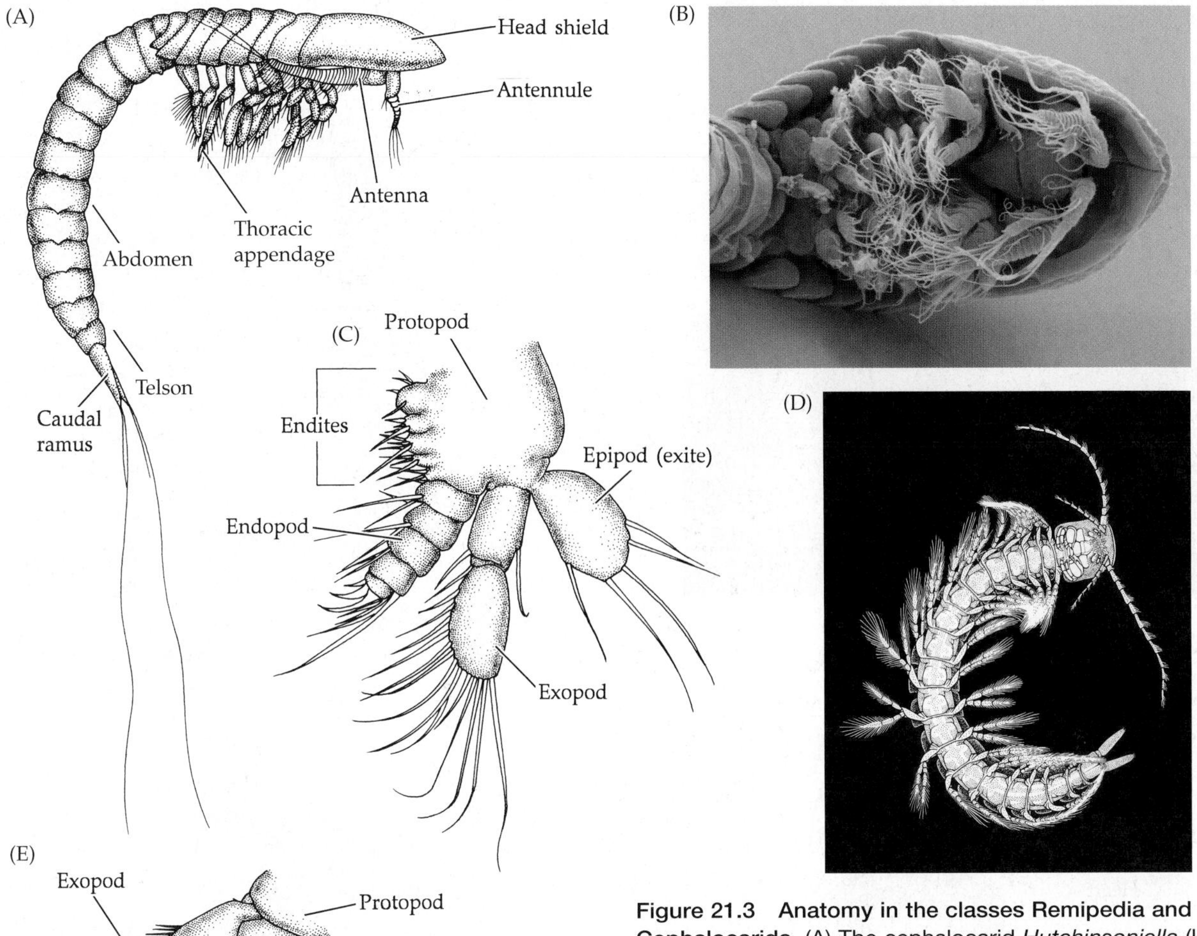

Figure 21.3 Anatomy in the classes Remipedia and Cephalocarida. (A) The cephalocarid *Hutchinsoniella* (lateral view). (B) SEM of head and thorax of a cephalocarid. (C) First trunk limb of the cephalocarid, *Lightiella*. (D) The remipede, *Speleonectes* (ventral view). (E) Tenth trunk limb of the remipede, *Lasionectes*. (F) The anterior end of a remipede (ventral view). In both the cephalocarids and the remipedes, the trunk is a long, homonomous series of somites with biramous swimming appendages. In cephalocarids the first trunk appendages are like all the others, which bear large swimming epipods (exites). In remipedes the first trunk somite is fused to the head, and its appendages are maxillipeds.

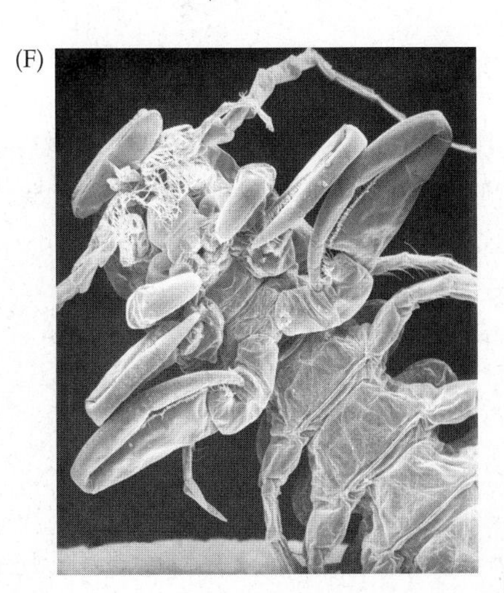

5 genera. All are benthic marine detritus feeders. Most are associated with sediments covered by a layer of flocculent organic detritus, although some have been found in clean sands. They occur from the intertidal zone to depths of over 1,500 m, from the North and South Pacific, to the North and South Atlantic, and the Mediterranean. Hatching is at a slightly advanced naupliar stage (called a metanauplius). Most researchers agree that cephalocarids are very primitive crustaceans, largely because of their relatively homonomous body form, undifferentiated maxillae, long and gradual (anamorphic) larval development, and shape of the trunk limbs (from which limbs of all other crustaceans seem to be easily derived).

Class Branchiopoda

Number of segments and appendages on thorax and abdomen vary, the latter usually lacking appendages; carapace present or absent; telson usually with caudal rami; body appendages generally phyllopodous; maxillules and maxillae reduced or absent; no maxillipeds (Figures 21.1C,D, 21.4, 21.21B, 21.31C, and 21.35B).

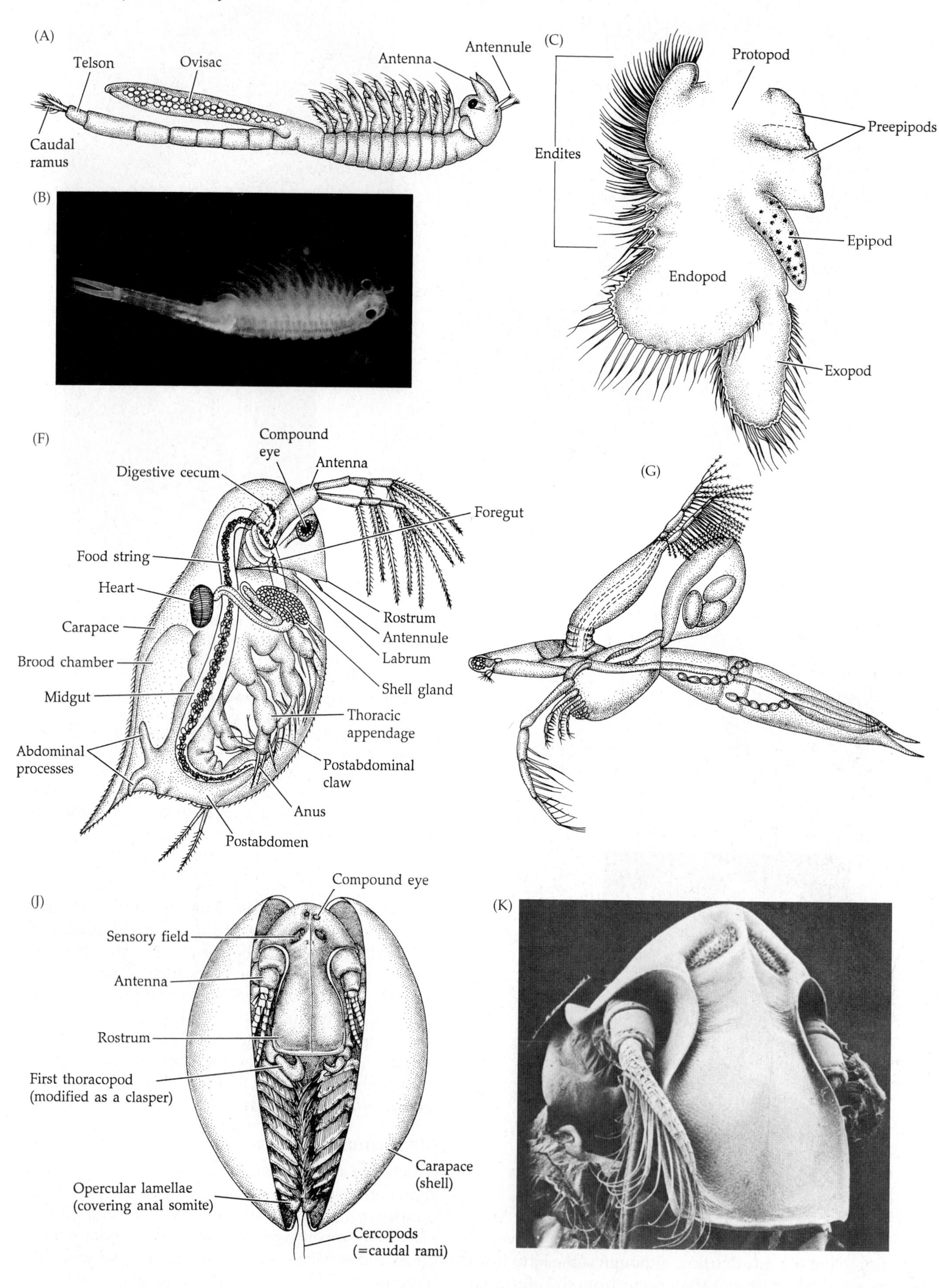
(A)
Telson
Ovisac
Antenna
Antennule
Caudal ramus
(B)
(C)
Protopod
Preepipods
Endites
Epipod
Endopod
Exopod
(F)
Compound eye
Antenna
Digestive cecum
Foregut
Food string
Heart
Carapace
Brood chamber
Midgut
Abdominal processes
Rostrum
Antennule
Labrum
Shell gland
Thoracic appendage
Postabdominal claw
Anus
Postabdomen
(G)
(J)
Compound eye
Sensory field
Antenna
Rostrum
First thoracopod (modified as a clasper)
Carapace (shell)
Opercular lamellae (covering anal somite)
Cercopods (=caudal rami)
(K)

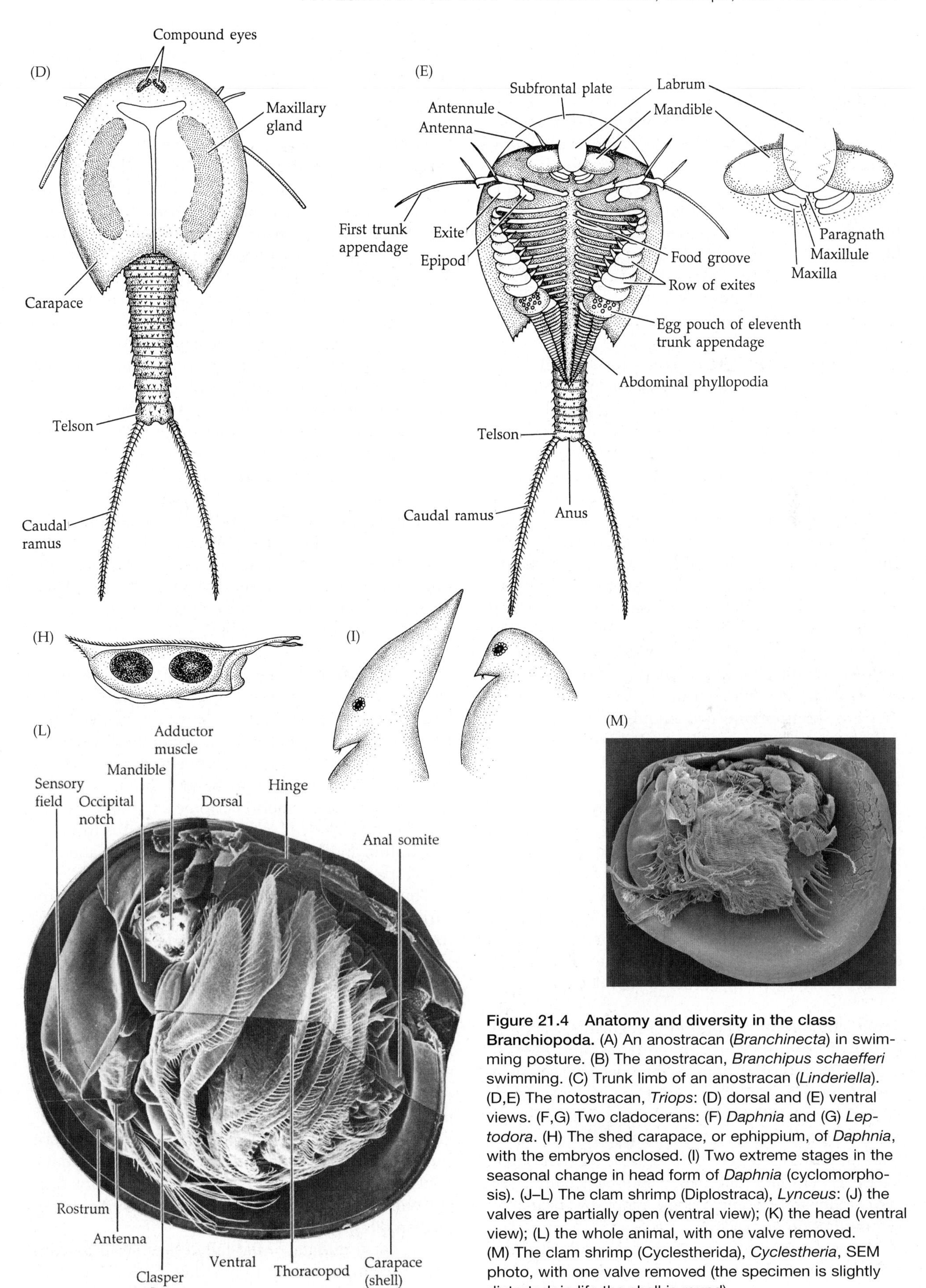

Figure 21.4 Anatomy and diversity in the class Branchiopoda. (A) An anostracan (*Branchinecta*) in swimming posture. (B) The anostracan, *Branchipus schaefferi* swimming. (C) Trunk limb of an anostracan (*Linderiella*). (D,E) The notostracan, *Triops*: (D) dorsal and (E) ventral views. (F,G) Two cladocerans: (F) *Daphnia* and (G) *Leptodora*. (H) The shed carapace, or ephippium, of *Daphnia*, with the embryos enclosed. (I) Two extreme stages in the seasonal change in head form of *Daphnia* (cyclomorphosis). (J–L) The clam shrimp (Diplostraca), *Lynceus*: (J) the valves are partially open (ventral view); (K) the head (ventral view); (L) the whole animal, with one valve removed. (M) The clam shrimp (Cyclestherida), *Cyclestheria*, SEM photo, with one valve removed (the specimen is slightly distorted; in life the shell is round).

years or decades until the next adequate rains appear. As diverse as the branchiopods might appear, both morphological and molecular analyses indicate that they comprise a monophyletic group. Development is remarkably uniform across the group, with naupliar larvae distinguishable by several unique features. Several taxonomic names have been proposed to cluster alleged sister-groups within the Branchiopoda, but there is as yet little consensus, and branchiopod relationships remain unsettled. Nearly 1,000 extant species have been described, and at least one fossil branchiopod (*Rehbachiella*) is known from the middle Cambrian, confirming that they are an ancient group.

Order Anostraca Postcephalic trunk divisible into appendage-bearing thorax of 11 segments (17 or 19 in members of the family Polyartemiidae) and abdomen of 8 segments plus telson with caudal rami; gonopores on genital region of abdomen; trunk limbs biramous and phyllopodous; small cephalic shield present; paired, large, stalked compound eyes and a single median simple (naupliar) eye.

The anostracans are commonly called fairy shrimps and brine shrimps. They differ from other branchiopods in lacking a carapace. There are just over 300 living species in the order, worldwide, most of which are less than 1 cm in length, although a few giants attain lengths of 10 cm. These animals inhabit ephemeral ponds (including snowmelt ponds and desert pools), hypersaline lakes, and marine lagoons. In many areas they are an important food resource for water birds. Anostracans are sometimes united with the extinct order Lipostraca as the subclass Sarsostraca. Their fossil record dates back to the Devonian.

Anostracans swim ventral side up by metachronal beating of the trunk appendages. Many use these limb movements for suspension feeding. Some other species scrape organic material from submerged surfaces, and at least two species (*Branchinecta gigas, B. raptor*) are specialized as predators on other fairy shrimps. Although most anostracans live in isolated ponds, their eggs might be transported during passage through the gut of predatory diving beetles (Dytiscidae).

Order Notostraca Thorax of 11 segments, each with a pair of phyllopodous appendages; abdomen of "rings," each formed of more than one true segment; each anterior ring with several pairs of appendages; posterior rings lack appendages; telson with long caudal rami; gonopores on last thoracomere; broad, shield-like carapace fused only with head, but extending to loosely cover thorax and part of abdomen; paired, sessile compound eyes and a single simple eye near anterior midline on carapace.

Notostracans are often called tadpole shrimps. There are only about a dozen living species, in two genera (*Triops* and *Lepidurus*) placed in a single family, Triopsidae. Most species are 2–10 cm long. The common name derives from the general body shape: the broad carapace and narrow "trunk" give the animals a superficial tadpole-like appearance.

Notostracans inhabit inland waters of all salinities, but none occur in the ocean. Of the two known genera, *Triops* (Figure 21.4D,E) lives only in temporary waters, and its eggs are capable of surviving extended dry periods. Most species of *Lepidurus* live in temporary ponds, but at least one species (*L. arcticus*) inhabits permanent ponds and lakes. However, all species are short-lived, and most complete their lifecycle in just 30 to 40 days. *Triops* is of some economic importance in that large populations often occur in rice paddies and destroy the crop by burrowing into the mud and dislodging young plants. Tadpole shrimps mostly crawl, but they are also capable of swimming for short periods by beating the thoracic limbs. They are omnivorous, feeding mostly on organic material stirred up from the sediments, although some scavenge or prey on other animals, including molluscs, other crustaceans, frog eggs, and even frog tadpoles and small fishes. Some species of notostracans are exclusively gonochoristic, but others may include hermaphroditic populations (often those populations living at high latitudes). Earlier reports of parthenogenetic populations have been questioned.

Order Diplostraca The clam shrimps and cladocerans (water fleas) comprise two groups of closely related branchiopods known as the Diplostraca (Figure 21.4J–L). They share the feature of a uniquely developed, large, "bivalved" carapace that covers all or most of the body. Most diplostracans are benthic, but many swim during reproductive periods. Some are direct suspension feeders, whereas others stir up detritus from the substratum and feed on suspended particles, and others scrape pieces of food from the sediment.

The species formerly lumped together as clam shrimps are now partitioned between the two diplostracan suborders, Laevicaudata and Onychocaudata. Diplostracans share several features: body divided into cephalon and trunk, the latter with 10–32 segments, all with appendages, and with no regionalization into thorax and abdomen; trunk limbs phyllopodous, decreasing in size posteriorly; males with trunk limbs 1, or 1–2, modified for grasping females during mating; trunk typically terminates in spinous anal somite or telson, usually with robust caudal rami (cercopods); gonopores on eleventh trunk segment; bivalved carapace completely encloses body; valves folded (Spinicaudata, Cyclestherida) or hinged (Laevicaudata) dorsally; usually with a pair of sessile compound eyes and a single, median, simple eye.

The Laevicaudata, or flat-tailed clam shrimps, contains the single family Lynceidae, with close to 40 species in three genera, characterized by a hinged, globular carapace that encloses the entire animal. The Onychocaudata contains the spiny-tailed or "true" clam shrimp in the infraorder Spinicaudata, and the cladocerans (commonly called water fleas) in the infraorder Cladoceromorpha. Spinicaudatans include the commonly encountered freshwater genera *Limnadia, Eulimnadia, Leptestheria*, and *Cyzicus*. The Cladoceromorpha contains the well-known cladoceran water fleas *Daphnia, Moina, Diaphonosoma*, and *Leptodora*, as well as the unique genus *Cyclestheria* (in the Cyclestherida). Confusingly, *Cyclestheria* is also commonly called a clam shrimp, though it is more closely related to the water fleas).

The common name "clam shrimp" derives from the clamlike appearance of the valves, which usually bear concentric growth lines reminiscent of bivalved molluscs. The approximately 200 species of clam shrimps (including laevicaudatans, spinicaudatans, and *Cyclestheria*) live primarily in ephemeral freshwater habitats worldwide. *Cyclestheria hislopi*, the only member of the Cyclestherida, inhabits permanent freshwater habitats throughout the world's tropics, and is one of the most widespread animals on Earth. *Cyclestheria* is also the only clam shrimp with direct development, the larval and juvenile stages being passed within the brood chamber, one of the features allying it with the cladocerans.

In cladocerans, the carapace is never hinged (only folded dorsally, like a taco) and never covers the entire body, and appendages do not occur on all the trunk somites. The body segmentation is generally reduced. The thorax and abdomen are fused as a "trunk" bearing 4–6 pairs of appendages anteriorly and terminating in a flexed "postabdomen" with clawlike caudal rami. Trunk appendages are usually phyllopodous. The carapace usually encloses the entire trunk, but not the cephalon, serving as a brood chamber (and greatly reduced to this function) in some species; a single median compound eye is always present.

The cladocerans, or water fleas, include about 400 species of predominantly freshwater crustaceans, although several American marine genera and species are known (e.g., *Evadne, Podon*). Although there are relatively few species, the group exhibits great morphological and ecological diversity. Most cladocerans are 0.5–3 mm long, but *Leptodora kindtii* reaches 18 mm in length. Except for the cephalon and large natatory antennae, the body is enclosed by a folded carapace, which is fused with at least some of the trunk region. The carapace is greatly reduced in members of the families Polyphemidae and Leptodoridae, in which it forms a brood chamber.

Cladocerans are distributed worldwide in nearly all inland waters. Most are benthic crawlers or burrowers; others are planktonic and swim by means of their large antennae. One genus (*Scapholeberis*) is typically found in the surface film of ponds, and another (*Anchistropus*) is ectoparasitic on *Hydra*. Most of the benthic forms feed by scraping organic material from sediment particles or other objects; the planktonic species are suspension feeders. Some (e.g., *Leptodora, Bythotrephes*) are predators on other cladocerans.

In sexual reproduction, fertilization generally occurs in a brood chamber between the dorsal surface of the trunk and the inside of the carapace. Most species have direct development. In the family Daphnidae the developing embryos are retained by a portion of the shed carapace, which functions as an egg case called an **ephippium** (Figure 21.4H), whereas in the Chydoridae the ephippium remains attached to the entire shed carapace. *Leptodora* exhibits a heterogenous life cycle, alternating between parthenogenesis and sexual reproduction, the latter of which results in free-living larvae (metanauplii hatch from the shed resting eggs).

Cladoceran life histories are often compared with those of animals such as rotifers and aphids. Dwarf males occur in many species in all three groups, and parthenogenesis is common. Members of two cladoceran families that undergo parthenogenesis (Moinidae and Polyphemidae) produce eggs with very little yolk. In these groups the floor of the brood chamber is lined with glandular tissue that secretes a fluid rich in nutrients, which is absorbed by the developing embryos. Periods of overcrowding, adverse temperatures, or food scarcity can induce parthenogenetic females to produce male offspring. Occasional periods of sexual reproduction have been shown to occur in most parthenogenetic species. Many planktonic cladocerans undergo seasonal changes in body form through succeeding generations of parthenogenetically produced individuals, a phenomenon known as cyclomorphism (Figure 21.4I).

Class Malacostraca

Body of 19–20 segments, including 6-segmented cephalon, 8-segmented thorax, and 6-segmented pleon (7-segmented in leptostracans), plus telson; with or without caudal rami; carapace covering part or all of thorax, or reduced, or absent; 0–3 pairs of maxillipeds; thoracopods primitively biramous, uniramous in some groups, phyllopodous only in members of the subclass Phyllocarida; antennules and antennae usually biramous; abdomen (pleon) usually with 5 pairs of biramous pleopods and 1 pair of biramous uropods; eyes usually present, compound, stalked or sessile; mainly gonochoristic; female gonopores on sixth, and male pores on eighth thoracomeres. When uropods are present, they are often broad and flat, lying alongside the broad telson to form a tail fan.

Most classification schemes divide the more than 40,200 species of malacostracans into three subclasses, Phyllocarida (leptostracans), Hoplocarida (stomatopods), and the megadiverse Eumalacostraca. The phyllocarids are typically viewed as representing the

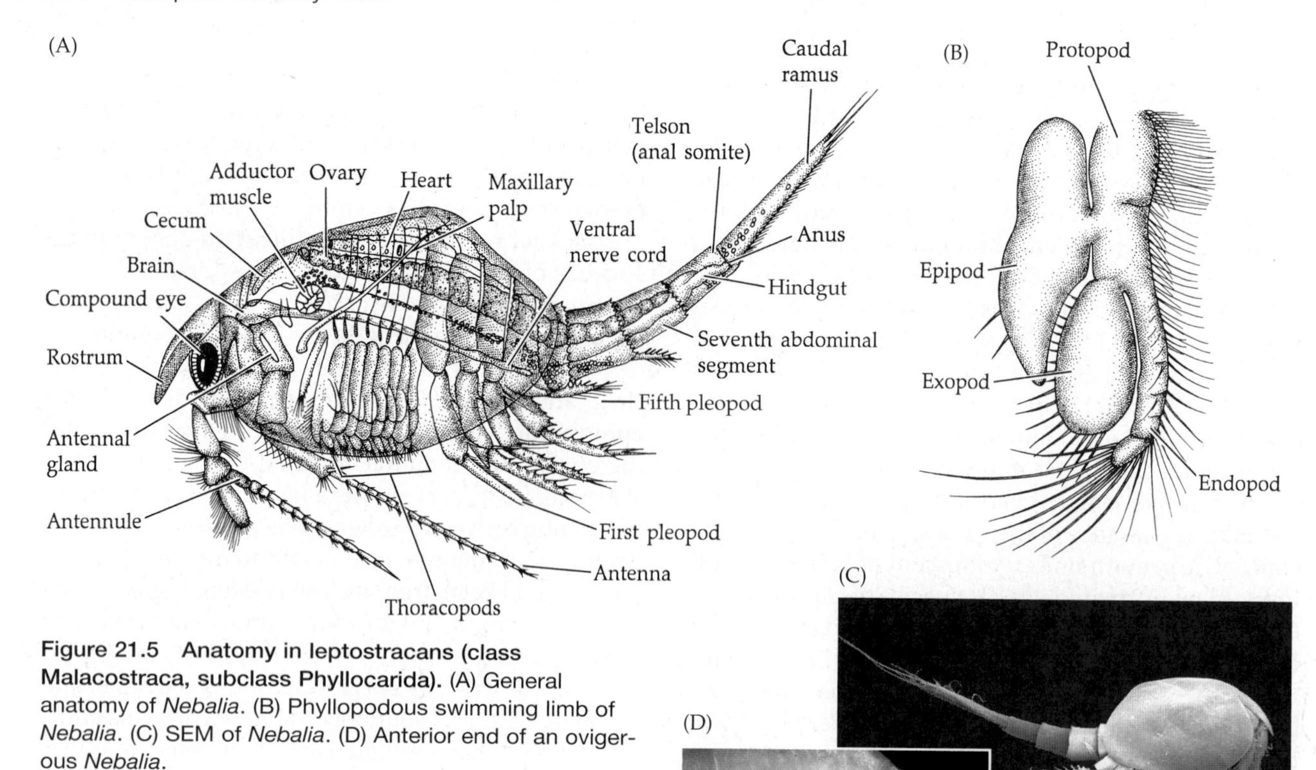

Figure 21.5 Anatomy in leptostracans (class Malacostraca, subclass Phyllocarida). (A) General anatomy of *Nebalia*. (B) Phyllopodous swimming limb of *Nebalia*. (C) SEM of *Nebalia*. (D) Anterior end of an ovigerous *Nebalia*.

primitive malacostracan condition (6-8-7 body segments plus telson; Figure 21.5). The basic eumalacostracan body plan, characterized by the 6-8-6 (plus telson) arrangement of body segments, was recognized in the early 1900s by W. T. Calman, who termed the defining features of the Eumalacostraca "caridoid facies" (Figure 21.6). Much work has been done since Calman's day, but the basic elements of his caridoid facies are still present in all members of the subclass Eumalacostraca.

Subclass Phyllocarida

Order Leptostraca With the typical malacostracan characteristics, except notable for presence of seven free pleomeres (plus telson) rather than six, generally taken to represent the primitive condition for the class. Also, with phyllopodous thoracopods (all similar to one another); no maxillipeds; large carapace covering thorax and compressed laterally so as to from an unhinged bivalved "shell," with an adductor muscle; cephalon with a movable, articulated rostrum; pleopods 1–4 similar and biramous, 5–6 uniramous; no uropods; paired stalked compound eyes; antennules biramous; antennae uniramous; adults with both antennal and maxillary glands (Figures 21.5 and 21.21C).

The subclass Phyllocarida includes about 40 species in 10 genera. Most are 5–15 mm long, but *Nebaliopsis typica* is a giant at nearly 5 cm in length. The leptostracan body form is distinctive, with its loose bivalved

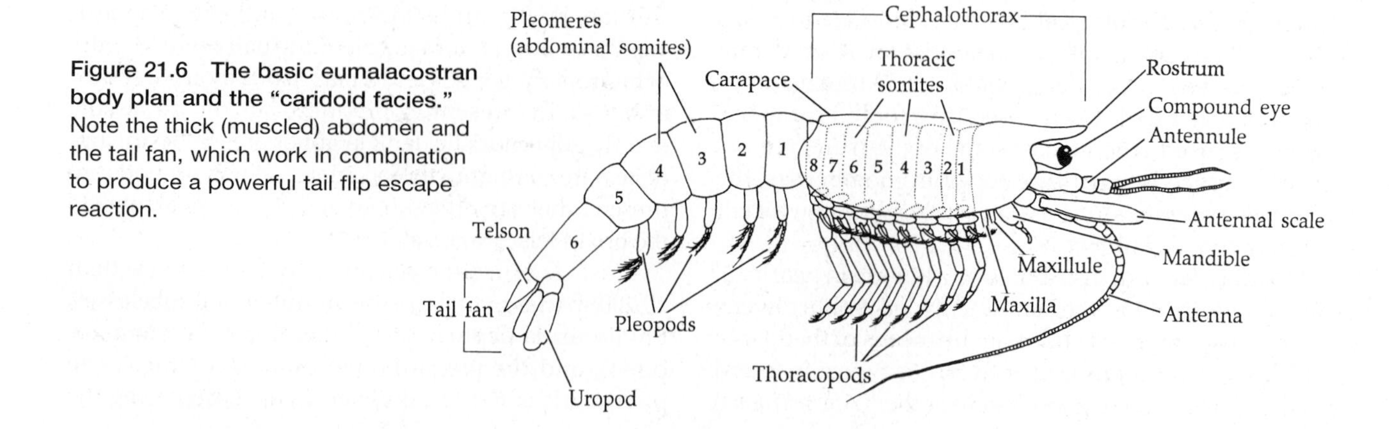

Figure 21.6 The basic eumalacostran body plan and the "caridoid facies." Note the thick (muscled) abdomen and the tail fan, which work in combination to produce a powerful tail flip escape reaction.

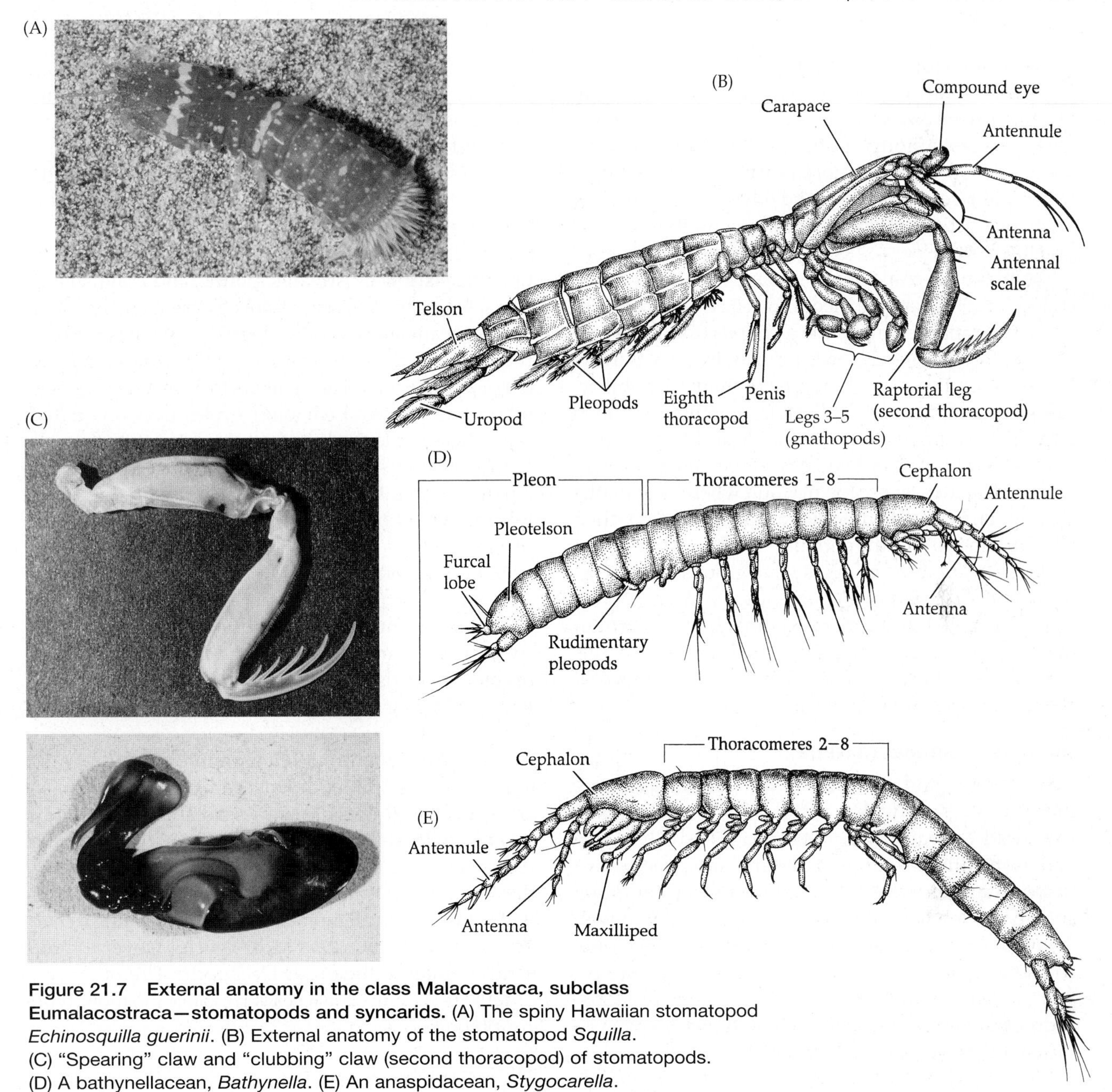

Figure 21.7 External anatomy in the class Malacostraca, subclass Eumalacostraca—stomatopods and syncarids. (A) The spiny Hawaiian stomatopod *Echinosquilla guerinii*. (B) External anatomy of the stomatopod *Squilla*. (C) "Spearing" claw and "clubbing" claw (second thoracopod) of stomatopods. (D) A bathynellacean, *Bathynella*. (E) An anaspidacean, *Stygocarella*.

carapace covering the thorax, a protruding rostrum, and an elongate abdomen. All leptostracans are marine, and most are epibenthic from the intertidal zone to a depth of 400 m; *Nebaliopsis typica* is bathypelagic. Most species seem to occur in low-oxygen environments. One species, *Dahlella caldariensis*, is associated with the hydrothermal vents of the Galapagos and the East Pacific Rise. *Speonebalia cannoni* is known only from marine caves.

Most leptostracans suspension feed by stirring up bottom sediments. They are also capable of grasping relatively large bits of food directly with the mandibles. Some are carnivorous scavengers, and some are known to aggregate in areas on the sea floor where large amounts of detritus accumulate. In many species the antennae or antennules of males are modified to hold females during copulation.

Subclass Hoplocarida

Order Stomatopoda Carapace covering portion of head and fused with thoracomeres 1–4; head with movable, articulated rostrum; thoracopods 1–5 uniramous and subchelate, second pair massive and raptorial (all five are sometimes called "maxillipeds" or gnathopods because they are involved in feeding); thoracopods 6–8 biramous, ambulatory; pleopods biramous, with dendrobranchiate-like gills on exopods; antennules triramous; antennae biramous, with large, paired, stalked compound eyes that are unique in the animal kingdom (Figures 21.7A–C, 21.27D, and 21.33K).

All 500 or so living hoplocarids are placed in the order Stomatopoda, known as mantis shrimps. They are relatively large crustaceans, ranging in length from 2 to 30 cm. Compared with that of most malacostracans, the muscle-filled abdomen is notably robust.

Most stomatopods are found in shallow tropical or subtropical marine environments. Nearly all of them live in burrows excavated in soft sediments or in cracks and crevices, among rubble, or in other protected spots. All species are raptorial carnivores, preying on fishes, molluscs, cnidarians, and other crustaceans. The large, distinctive subchelae of the second thoracopods act either as crushers or as spears (Figure 21.7C).

Stomatopods crawl about using the posterior thoracopods and the flaplike pleopods. They also can swim by metachronal beating of the pleopods (the swimmerets). For these relatively large animals, living in narrow burrows requires a high degree of maneuverability. The short carapace and the flexible, muscular abdomen allow these animals to twist double and turn around within their tunnels or in other cramped quarters. This ability facilitates an escape reaction whereby a mantis shrimp darts into its burrow rapidly head first, then turns around to face the entrance.

Stomatopods are one of only two groups of malacostracans that possess pleopodal gills. Only the isopods share this trait, but the pleopods are quite different in the two groups. The tubular, thin, highly branched gills of stomatopods provide a large surface area for gas exchange in these active animals.

Subclass Eumalacostraca

Head, thorax, and abdomen of 6-8-6 somites respectively (plus telson); with 0, 1, 2, or 3 thoracomeres fused with head, their respective appendages usually modified as maxillipeds; antennules and antennae primitively biramous (but often reduced to uniramous); antennae often with scalelike exopod; most with well developed carapace, secondarily reduced in syncarids and some peracarids; gills primitively as thoracic epipods; tail fan composed of telson plus paired uropods; abdomen long and muscular. Three superorders: Syncarida, Eucarida, and Peracarida.

Superorder Syncarida Without maxillipeds (Bathynellacea) or with one pair of maxillipeds (Anaspidacea); no carapace; pleon bears telson with or without furcal lobes; at least some thoracopods biramous, eighth often reduced; pleopods variable; compound eyes present (stalked or sessile) or absent (Figure 21.7D,E). There are about 285 described species of syncarids in two orders, Anaspidacea and Bathynellacea.[4] To many workers, the syncarids represent a key group in eumalacostracan evolution, and they may represent an ancient relictual taxon that now inhabits refugial habitats. Through studies of the fossil record and extant members of the order Anaspidacea (e.g., *Anaspides*), it has been suggested that syncarids may encompass the most primitive living eumalacostracan body plan. Bathynellaceans occur worldwide in interstitial or groundwater habitats, whereas the anaspididaceans are strictly Gondwanan in distribution. Many Anaspidacea are endemic to Tasmania, where they inhabit freshwater environments, such as open lake surfaces, streams, ponds, and crayfish burrows. No syncarids are marine. These reclusive eumalacostracans show various degrees of what some have regarded as paedomorphism, including small size (Anaspididae includes members to 5 cm, whereas most others are less than 1 cm long), eyelessness, and reduction or loss of pleopods and some posterior pereopods. Bathynellaceans are small (1–3 mm long), possess 6 or 7 pairs of long, thin swimming legs, and have a pleotelson formed by the fusion of the telson to the last pleonite.

Syncarids either crawl or swim. Little is known about the biology of most species, although some are considered omnivorous. Unlike most other crustaceans, which carry the eggs and developing early embryos, syncarids lay their eggs or shed them into the water following copulation.

[4]Until recently a third syncarid order was recognized, the Stygocaridacea, endemic to the Southern Hemisphere. Most workers now agree that the stygocarids should be reduced to the rank of family within the order Anaspidacea.

Superorder Eucarida Telson without caudal rami; 0, 1, or 3 pairs of maxillipeds; carapace present, covering and fused dorsally with head and entire thorax; usually with stalked compound eyes; gills thoracic. Although members of this group are highly diverse, they are united by the presence of a complete carapace that is fused with all thoracic segments, forming a characteristic cephalothorax. Most species (several thousand) belong to the order Decapoda. The other two orders are the Euphausiacea (krill), and the monotypic Amphionidacea.

Order Euphausiacea Euphausids are distinguished among the eucarids by the absence of maxillipeds, the exposure of the thoracic gills external to the carapace, and the possession of biramous pereopods (the last 1 or 2 pairs sometimes being reduced). They are shrimplike in appearance. Adults have antennal glands. Most of them have photophores on the eyestalks, the bases of the second and seventh thoracopods, and between the first 4 pairs of abdominal limbs.

The 90 or so species of euphausids are all pelagic and range in length from 4 to 15 cm. The pleopods function as swimmerets. Euphausids are known from all oceanic environments to depths of 5,000 m. Most species are distinctly gregarious, and species that occur in huge schools (**krill**) provide a major source of food for larger nektonic animals (baleen whales, squids, fishes) and even some marine birds. Krill densities, particularly for *Euphausia superba*, often exceed 1,000 animals/m^3 (614

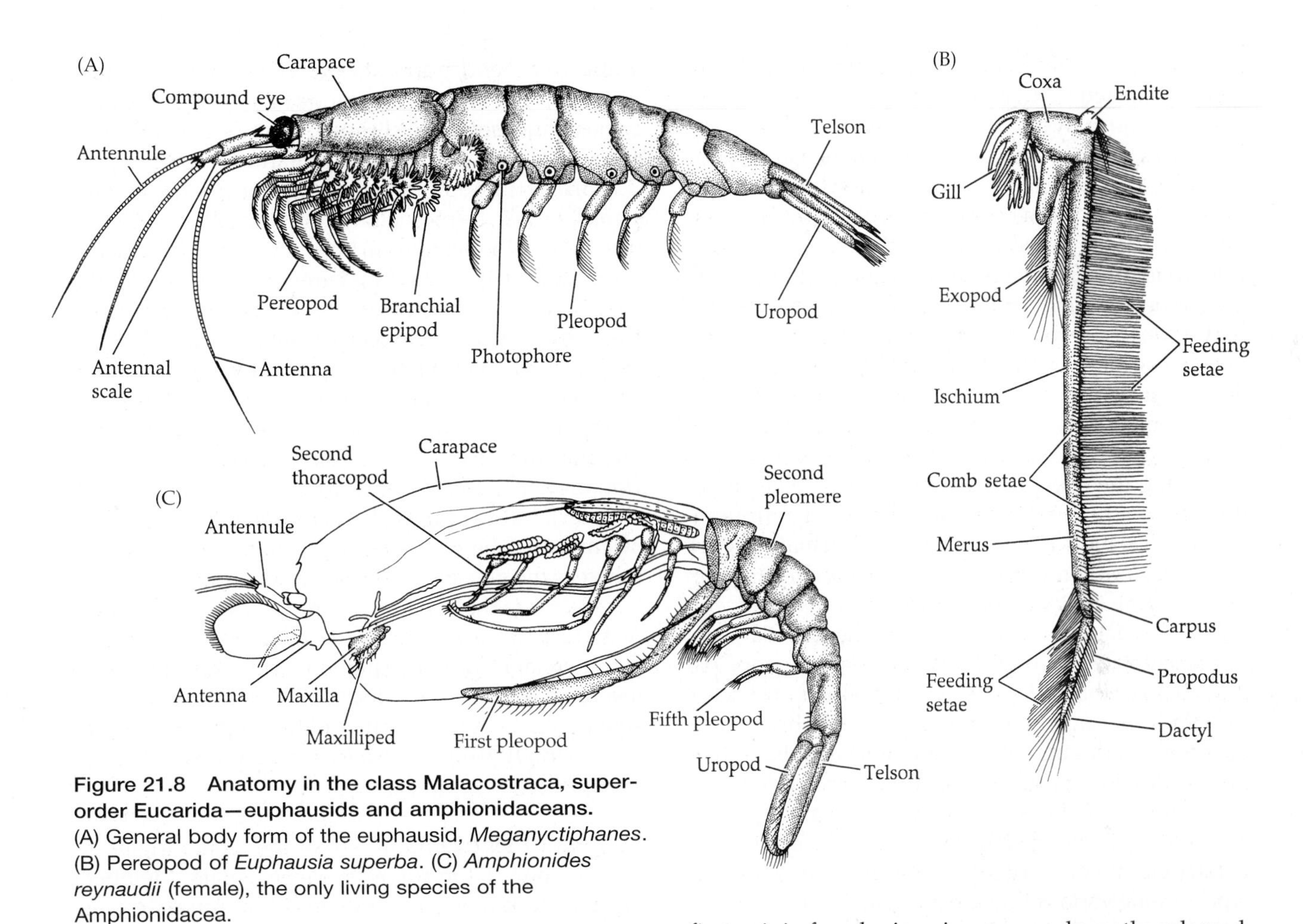

Figure 21.8 Anatomy in the class Malacostraca, superorder Eucarida—euphausids and amphionidaceans. (A) General body form of the euphausid, *Meganyctiphanes*. (B) Pereopod of *Euphausia superba*. (C) *Amphionides reynaudii* (female), the only living species of the Amphionidacea.

g wet weight/m^3).[5] Generally, euphausids are suspension feeders, although predation and detritivory also occur (Figures 21.8A,B and 21.21E).

Order Amphionidacea The single known species of the order Amphionidacea, *Amphionides reynaudii*, possesses an enlarged cephalothorax covered by a thin, almost membranous carapace that extends to enclose the thoracopods. The thoracopods are biramous with short exopods. The first pair is modified as maxillipeds, and the last pair is absent in females. Some of the mouthparts are highly reduced in females. The pleopods are biramous and natatory, except that the first pair in females is uniramous and greatly enlarged, perhaps functioning to form a brood pouch extending under the thorax. Females have a reduced gut and apparently do not feed. *Amphionides* is a worldwide member of marine oceanic plankton and occurs to a depth of 1,700 m (Figure 21.8C).

Order Decapoda The decapods are among the most familiar eumalacostracans. They possess a well developed carapace enclosing a branchial chamber, but they differ from other eucarid orders in always possessing 3 pairs of maxillipeds, leaving 5 pairs of functional uniramous or weakly biramous pereopods (hence the name, Decapoda); one (or more) pairs of anterior pereopods are usually clawed (chelate). Adults have antennal glands. Rearrangement of the subtaxa within this order is a popular carcinological pastime (see Martin and Davis 2001 for an entry into the vast literature on decapod classification). In vernacular terms, nearly every decapod may be recognized as some sort of shrimp, crab, lobster, or crayfish.

We do not want to belabor the issue of decapod gill nomenclature. However, the gills play a prominent role in the taxonomy of this group; thus, we provide brief descriptions of the basic types. All decapod gills arise as thoracic coxal exites (epipods), but their final placement varies. Those that remain attached to the coxae are

[5]Where krill densities exceed about 100 grams per cubic meter, they are often fished commercially. Krill schools can extend for tens of miles, contain millions of tons of krill, and stain the ocean red with their surface swarms in coastal waters. Large baleen whales can eat a ton of krill in one mouthful. Seals, fish, squid, and humans also eat krill. Krill fishing has been banned in most of North America, but it continues in Japan where tens of thousands of tons are landed annually and used mainly as feed for farmed fish. The largest krill fishery is in the ocean surrounding Antarctica, where they have been harvested commercially since the 1970s. In the 1980s, large fleets from the Soviet Union caught up to 400,000 tons of Antarctic krill annually, but the current catch is now down to about 120,000 tons (taken by Japan, Korea, Norway, Poland, the Ukraine, and the U.S.).

podobranchs (= "foot gills"), but others eventually become associated with the articular membrane between the coxae and body and are thus called arthrobranchs (= "joint gills"). Some actually end up on the lateral body wall, or side-surface of the thorax, as **pleurobranchs** (= "side gills"). The sequence by which some of these gills arise ontogenetically varies. For example, in the Dendrobranchiata and the Stenopodidea, arthrobranchs appear before pleurobranchs, whereas in members of the Caridea the reverse is true. In most of the other decapods the arthrobranchs and pleurobranchs tend to appear simultaneously. These developmental differences may be minor heterochronic dissimilarities and of less phylogenetic importance than actual gill anatomy.

Among the decapods, the gills can also be one of three basic structural types, described as dendrobranchiate, trichobranchiate, and phyllobranchiate (Figure 21.28B–D). All three of these gill types include a main axis carrying afferent and efferent blood vessels, but they differ markedly in the nature of the side filaments or branches. Dendrobranchiate gills bear two principal branches off the main axis, each of which is divided into multiple secondary branches. Trichobranchiate gills bear a series of radiating unbranched tubular filaments. Phyllobranchiate gills are characterized by a double series of platelike or leaflike branches from the axis. Within each gill type, there may be considerable variation. The occurrences of these three major gill types among various taxa are presented below.

Close inspection of the proximal parts of the pereopods usually reveals another decapod feature: in most forms, the basis and ischium are fused (as a basiischium), with the point of fusion often indicated by a suture line. Tegumental glands are also a ubiquitous feature among the Decapoda. These glands originate below the epidermal cells and produce a fluid that opens on the surface of the cuticle. They have been reported from gills, legs, pleopods, and uropods. The roles of tegmental glands are not well known, and they have been suspected to be involved in cuticular tanning, the production of mucus by the mouthparts, the production of cement substance involved with egg attachment, and possibly also grooming.

The 18,000 or so living species of decapods comprise a highly diverse group. They occur in all aquatic environments at all depths, and a few spend most of their lives on land. Many are pelagic, but others have adopted benthic sedentary, errant, or burrowing lifestyles. Decorating of the exoskeleton is frequently seen among the decapods, especially in spider crabs (Brachyura: Majoidea), which use Velcro-like hooked setae to attach dead or living plants and animals; decorating has been shown to reduce predation through camouflage and/or chemical deterrence. Decapod feeding strategies include suspension feeding, predation, herbivory, scavenging, and more. Most workers recognize two suborders: Dendrobranchiata and Pleocyemata.

Suborder Dendrobranchiata This group includes over 500 species of decapods, most of which are penaeid and sergestid shrimps. As the name indicates, these decapods possess dendrobranchiate gills (Figure 21.28B), a unique synapomorphy of the taxon. One genus, *Lucifer*, has secondarily lost the gills completely. The dendrobranchiate shrimps are further characterized by chelae on the first three pereopods, copulatory organs modified from the first pair of pleopods in males, and ventral expansions of the abdominal tergites (called pleural lobes). Generally, none of the chelipeds is greatly enlarged. In addition, females of this group do not brood their eggs. Fertilization is external, and the embryos hatch as nauplius larvae (see the section on Reproduction and Development later in this chapter). Many of these animals are quite large, over 30 cm long. The sergestids are pelagic and all marine, whereas the penaeids are pelagic or benthic, and some occur in brackish water. Some dendrobranchiates (e.g., *Penaeus, Sergestes, Acetes, Sicyonia*) are of major commercial importance in the world's shrimp fisheries, most of which are now being exploited beyond sustainable levels and often with fishing techniques that are highly habitat-destructive (Figures 21.9A and 21.33G).

Suborder Pleocyemata All of the remaining decapods belong to the suborder Pleocyemata. Members of this taxon never possess dendrobranchiate gills. The embryos are brooded on the female's pleopods and hatch at some stage later than the nauplius larva. Included in this suborder are several kinds of shrimps, the crabs, crayfish, lobsters, and a host of less familiar forms. Most current workers now recognize 11 infraorders within the Pleocyemata, as we have done below, but a number of other schemes have been proposed and persist in the literature. One older approach divided decapods into two large groups, called the Natantia and Reptantia—the swimming and walking decapods, respectively. Although these terms have largely been abandoned as formal taxa, they still serve a useful descriptive purpose, and one continues to see references to natant decapods and reptant decapods.

Infraorder Procarididea The procarids consist of a single family containing two genera of shrimp, *Procaris* and *Vetericaris*, that have been called "primitive shrimps." Like caridean shrimp (below) and most other pleocyemates, they bear flattened, platelike (nondendrobranch) gills, but they lack claws on any of their legs, have a very leglike (pediform) third maxilliped, and have epipods on all of the maxillipeds and pereopods, assumed to be an ancestral condition because most decapod groups no longer bear the full complement. They are known from anchialine habitats—inland pools with connections to the sea (Figure 21.9B).

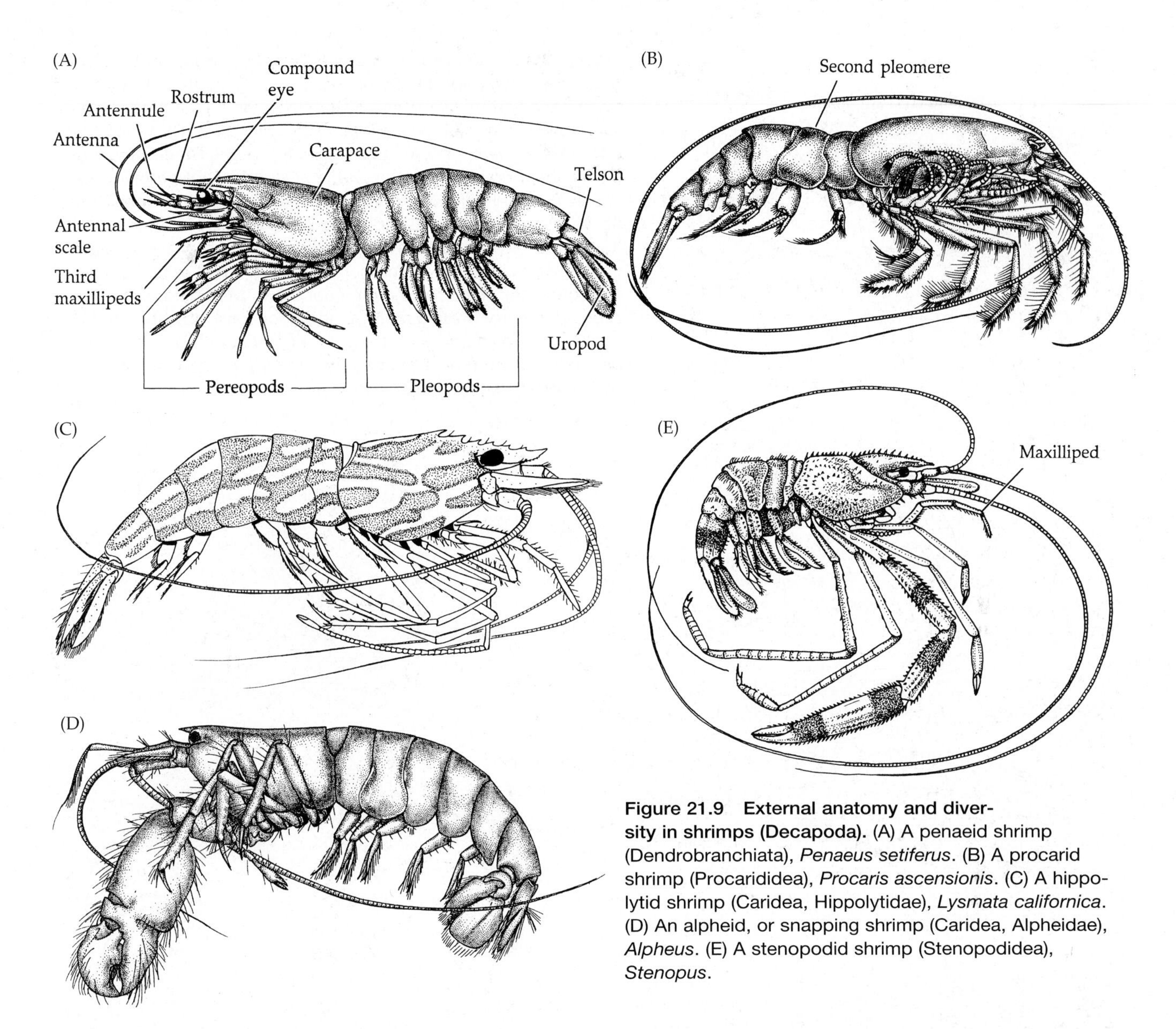

Figure 21.9 External anatomy and diversity in shrimps (Decapoda). (A) A penaeid shrimp (Dendrobranchiata), *Penaeus setiferus*. (B) A procarid shrimp (Procarididea), *Procaris ascensionis*. (C) A hippolytid shrimp (Caridea, Hippolytidae), *Lysmata californica*. (D) An alpheid, or snapping shrimp (Caridea, Alpheidae), *Alpheus*. (E) A stenopodid shrimp (Stenopodidea), *Stenopus*.

Infraorder Caridea The nearly 3,500 living species in this infraorder are generally referred to as the caridean shrimps. These swimming decapods have phyllobranchiate gills. The first 1 or 2 pairs of pereopods are chelate and variably enlarged. The second abdominal pleura (side walls) are distinctly enlarged to overlap both the first and third pleura. The first pleopods are generally somewhat reduced, but not much modified, in the males (Figures 21.1K, 21.9C,D, 21.24D, and 21.31D).

Infraorder Stenopodidea The 70 or so species in this infraorder belong to three families. The first 3 pairs of pereopods are chelate, and the third pair is significantly larger than the others. The gills are trichobranchiate. The first pleopods are uniramous in males and females, but are not strikingly modified. The second abdominal pleura are not expanded as they are in carideans (Figures 21.9E and 21.31B).

These colorful shrimps are usually only a few centimeters long (2–7 cm). Most species are tropical and associated with shallow benthic environments, especially with coral reefs; others are known from the deep sea. Many are commensal, and the group includes the cleaner shrimps (e.g., *Stenopus*) of tropical reefs, which are known to remove parasites from local fishes. Stenopodids often occur as male–female couples. Perhaps the most noted example of this bonding is associated with the glass sponge (*Euplectella*) shrimp, *Spongicola venusta*: A young male and female shrimp enter the atrium of a host sponge, eventually growing too large to escape and thus spending the rest of their days together.

Infraorder Brachyura These are the so-called "true crabs." The abdomen is symmetrical but highly reduced and flexed beneath the thorax, and uropods are usually absent. The body, hidden beneath a well-developed carapace, is distinctly flattened dorsoventrally and often expanded laterally. The gills are typically phyllobranchiate, but exceptions occur.

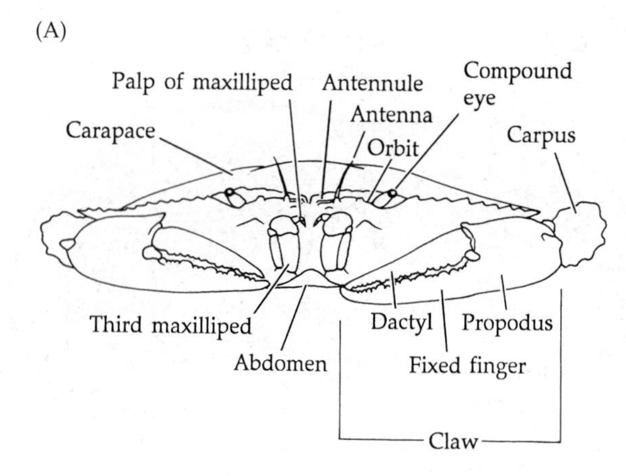

Figure 21.10 Anatomy and diversity of the "true," or brachyuran, crabs (Decapoda: Brachyura). (A,B) General crab anatomy: frontal and ventral views of a swimming crab (family Portunidae). (C) A spider crab (family Majidae), *Loxorhynchus*. (D) A kelp crab (Majidae), *Pugettia*. (E) An arrow crab (Majidae), *Stenorhynchus*. (F) A cancer crab (family Cancridae), *Cancer*. (G) A grapsid crab (family Grapsidae), *Pachygrapsus*. (H) A pinnotherid or pea crab (family Pinnotheridae), *Parapinnixa*. (I) A xanthid crab (family Xanthidae), *Trapezia*. Many members of this genus are obligate commensals in scleractinian corals. (J) A dromiid crab (family Dromiidae), *Hypoconcha* (anterior view). Members of the Dromiidae carry bivalve mollusc shells (or other objects) on their backs. (K) A calappid crab (family Calappidae), *Hepatus* (anterior view). (L) Ventral views of a female (upper photo) and (lower photo) male *Hemigrapsus sexdentatus*.

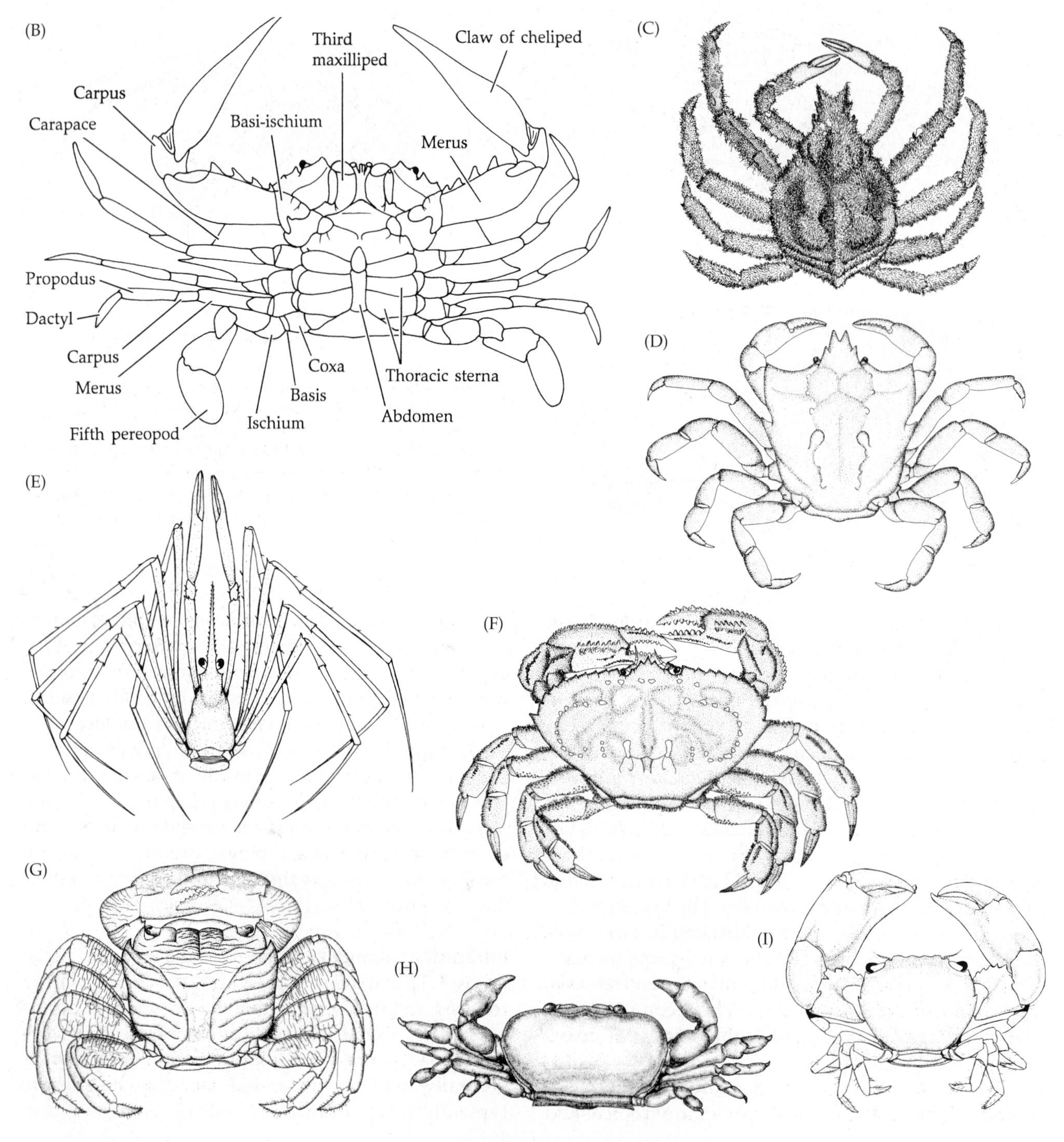

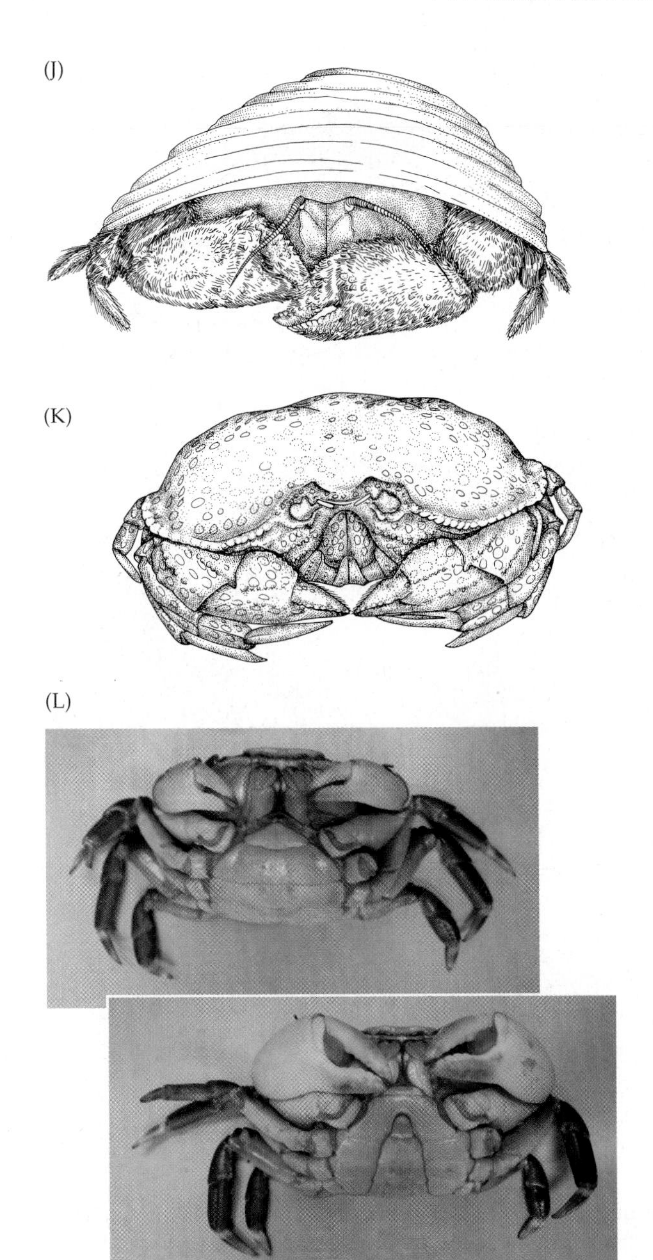

The first pereopods are chelate and usually enlarged. Pereopods 2 to 5 are typically simple, stenopodous walking legs, although in some groups, the fifth pereopods are also chelate. The eyes are positioned lateral to the antennae. Males lack pleopods 3 to 5. The distinctive larval stage is called a zoea; its carapace is spherical and bears a ventrally directed rostral spine (or no spine) (Figures 21.1E, 21.10, 21.27H, 21.28F,G, 21.29C, 21.32, and 21.33H,I).

Brachyuran crabs are mostly marine, but freshwater, semi-terrestrial, and moist terrestrial species occur in the tropics. The land crabs (certain species in the families Gecarcinidae, Ocypodidae, Grapsidae, etc.) are still dependent on the ocean for breeding and larval development. The surprisingly large number of freshwater crabs (about 3,000 species, classified into about a dozen families) all have direct development, incubate their embryos, and are independent of sea water. Some freshwater crabs are intermediate hosts of *Paragonimus*, a cosmotropical parasitic human lung fluke, and others are obligate phoretic hosts of larval black flies (*Simulium*), the vector for *Onchocerca volvulus* (the causative agent of river blindness). A number of species carry other invertebrates on their carapace (e.g., sponges, tunicates) or on their claws (e.g., anemones); these associations are generally thought to be mutualistic, providing camouflage or predator deterrence for the crab while their partner is moved about in the environment and may feed off debris from their host's feeding activities. The opener photo for this chapter shows the tropical Pacific teddy-bear crab, *Polydectus cupulifer*, which is densely covered with setae and frequently carries a sea anemone on each cheliped. In the northeast Pacific, megalopae and juveniles of the crab *Cancer gracilis* ride (and feed) on the bell of certain jellyfish; individuals of *Phacellophora camtschatica* (Scyphozoa) have been found with hundreds of *C. graciils* megalopae. There are about 7,000 described species of Brachyura.

Infraorder Anomura This group includes hermit crabs, galatheid crabs, king crabs, porcelain crabs, mole crabs, and sand crabs. The abdomen may be soft and asymmetrically twisted (as in hermit crabs) or symmetrical, short, and flexed beneath the thorax (as in porcelain crabs and others). Those with twisted abdomens typically inhabit gastropod shells or other empty "houses" not of their own making. King crabs (Lithodidae and Hapalogastridae) probably evolved out of hermit crab-like ancestors. Carapace shape and gill structure vary. The first pereopods are chelate; the third pereopods are never chelate. The second, fourth, and fifth pairs are usually simple, but occasionally they are chelate or subchelate. The fifth pereopods (and sometimes the fourth) are generally much reduced and do not function as walking limbs; the fifth pereopods function as gill cleaners and often are not visible externally. The pleopods are reduced or absent. The eyes are positioned medial to the antennae. The zoea larva is similar to that of the true crabs but is typically longer than broad, with the rostral spine directed anteriorly. Most anomurans are marine, but a few freshwater and semi-terrestrial species are known. The so-called "yeti crab" (*Kiwa hirsuta*) was discovered in 2005 from 2,200 m-deep hydrothermal vents south of Easter Island. It is remarkable for its "garden" of filamentous bacteria that grow on the long setae of the exoskeleton; the bacteria are heterotrophic, utilizing sulfides in the deep environment (the precise role of the bacteria, of several species, in the life history of the yeti crab is not yet well understood). A second species of *Kiwa* was described in 2011 (Figures 21.1F–J, 21.11C–G, 21.24A–C, 21.31A, and 21.33I).

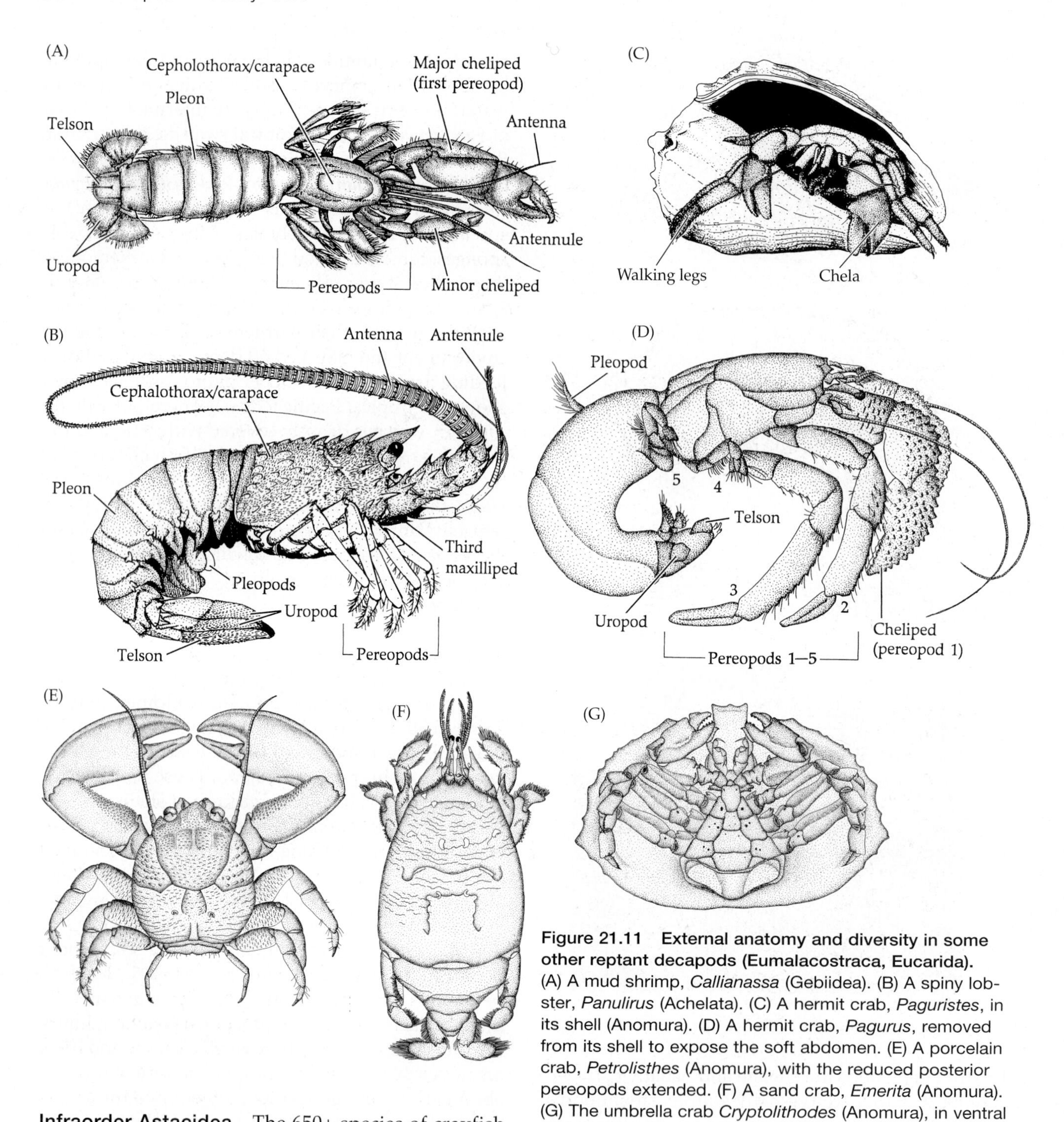

Figure 21.11 External anatomy and diversity in some other reptant decapods (Eumalacostraca, Eucarida). (A) A mud shrimp, *Callianassa* (Gebiidea). (B) A spiny lobster, *Panulirus* (Achelata). (C) A hermit crab, *Paguristes*, in its shell (Anomura). (D) A hermit crab, *Pagurus*, removed from its shell to expose the soft abdomen. (E) A porcelain crab, *Petrolisthes* (Anomura), with the reduced posterior pereopods extended. (F) A sand crab, *Emerita* (Anomura). (G) The umbrella crab *Cryptolithodes* (Anomura), in ventral view.

Infraorder Astacidea The 650+ species of crayfish and clawed lobsters comprise some of the most familiar decapods (Figure 21.2). As in most other decapods, the dorsoventrally flattened abdomen terminates in a strong tail fan. The gills are trichobranchiate. The first 3 pairs of pereopods are always chelate, and the first pair is greatly enlarged. *Homarus americanus*, the American or Maine lobster, is strictly marine and is the largest living crustacean by weight (the record weight being over 20 kilograms). Most crayfish live in fresh water, but a few species live in damp soil, where they may excavate extensive and complex burrow systems. The 600+ species of freshwater crayfish comprise a monophyletic group that is sister to clawed lobsters. Over 425 species of crayfish occur in North America alone, where they show high levels of endemicity to particular regions or river drainages. (Figures 21.27E,G and 21.29B).

Infraorder Achelata This group includes the coral and spiny lobsters (family Palinuridae) and slipper lobsters (family Scyllaridae). The name Achelata comes from the fact that they lack chelae on all pereopods as adults (except for a small grooming claw on pereopod 5 in some females). The flattened abdomen bears a tail fan; the carapace may be cylindrical or flattened dorsoventrally; the gills are trichobranchiate. The large, flattened larvae, called phyllosomas because of their leaf-like appearance, are unique and distinctive. All species are marine, and they are found in a variety of habitats throughout the tropics. Many species produce sounds by rubbing a process (the plectrum) at the base of the antennae against a "file" on the head (Figures 21.1M, 21.11B, 21.30A,C, and 21.33L).

Infraorders Gebiidea and Axiidea These two infraorders, traditionally referred to collectively as the Thalassinidea, have recently been recognized as distinct. The vernacular term "thalassinid" is still sometimes used to refer to them together. The mud and ghost shrimps are particularly difficult to place within the decapods. Sometimes they are depicted as related to the crayfish and chelate lobsters (Astacidea), and sometimes they are grouped with the hermit crabs and their relatives (Anomura). These decapods have a symmetrical abdomen that is flattened dorsoventrally and extends posteriorly as a well-developed tailfan. The carapace is somewhat compressed laterally, and the gills are trichobranchiate. The first 2 pairs of pereopods are chelate, and the first pair is generally much enlarged. Most of these animals are marine burrowers or live in coral rubble. They generally have a rather thin, lightly sclerotized cuticle, but some (e.g., members of the family Axiidae) have thicker skeletons and are more lobster-like in appearance. Gebiideans (particularly *Upogebia*, *Callianassa*, and related genera) often occur in huge colonies on tidal flats, where their burrow holes form characteristic patterns on the sediment surface (Figures 21.1L and 21.11A).

Infraorders Glypheidea and Polychelida The glypheids are something of a relict group, represented by two living genera (*Neoglyphea* and *Laurentaglyphea*), each with a single species, of a formerly diverse group known from the fossil record. Polychelids are a small group of blind deep-sea lobsters, notable for having chelae on all of their pereopods and unusual, large, globate larvae (called eryoneicus larvae) unique among the decapods.

Superorder Peracarida Telson without caudal rami; 1 (rarely 2–3) pair of maxillipeds; maxilliped basis typically produced into an anteriorly directed, bladelike endite; mandibles with articulated accessory processes in adults, between molar and incisor processes, called the **lacinia mobilis**; carapace, when present, not fused with posterior pereonites and usually reduced in size; gills thoracic or abdominal; with unique, thinly flattened thoracic coxal endites, called oostegites, that form a ventral brood pouch or marsupium in females all species except members of the order Thermosbaenacea (the latter using the carapace to brood embryos); young hatch as **mancas**, a prejuvenile stage lacking the last pair of thoracopods (no free-living larvae occur in this group) (Figures 21.12–21.15).

The roughly 25,000 species of peracarids are divided among nine orders. The peracarids are an extremely successful group of malacostracan crustaceans and are known from many habitats. Although most are marine, many also occur on land and in fresh water, and several species live in hot springs at temperatures of 30–50°C. Aquatic forms include planktonic as well as benthic species at all depths. The group includes the most successful terrestrial crustaceans—the pillbugs and sowbugs of the order Isopoda—and a few amphipods that have invaded land and live in damp forest leaf litter or gardens. Peracarids range in size from tiny interstitial forms only a few millimeters long to planktonic amphipods over 12 cm long (*Cystisoma*), deepsea necrophagous amphipods exceeding 34 cm (*Alicella gigantea*), and benthic isopods growing to 50 cm in length (*Bathynomus giganteus*). These animals exhibit all sorts of feeding strategies; a number of them, especially isopods and amphipods, are commensals or parasites.

Order Mysida Carapace well developed, covering most of thorax, but never fused with more than four anterior thoracic segments; maxillipeds (1–2 pairs) not associated with cephalic appendages; thoracomere 1 separated from head by internal skeletal bar; abdomen with well developed tail fan; pereopods biramous, except last pair, which are sometimes reduced; pleopods reduced or, in males, modified; compound eyes stalked, sometimes reduced; gills absent; usually with a statocyst in each uropodal endopod; adults with antennal glands (Figures 21.12A,B, 21.30B, and 21.33C).

There are more than 1,050 species of mysids, ranging in length from about 2 mm to 8 cm. Most swim by action of the thoracic exopods. Mysids are shrimplike crustaceans that are often confused with the superficially similar euphausids (which lack oostegites and uropodal statocysts). Mysids are pelagic or demersal and are known from all ocean depths; a few species occur in freshwater. Some species are intertidal and burrow in the sand during low tides. Most are omnivorous suspension feeders, eating algae, zooplankton, and suspended detritus. In the past, mysids were combined with lophogastrids and the extinct Pygocephalomorpha as the "Mysidacea."

Order Lophogastrida Similar to mysids, except for the following: maxillipeds (1 pair) are associated with the cephalic appendages; thoracomere 1 not separated from head by internal skeletal bar; pleopods well developed; gills present; adults with both antennal and maxillary glands; without statocysts; all 7 pairs of pereopods well developed and similar (except among members of the family Eucopiidae, in which their structure varies) (Figures 21.12C,D and 21.21G).

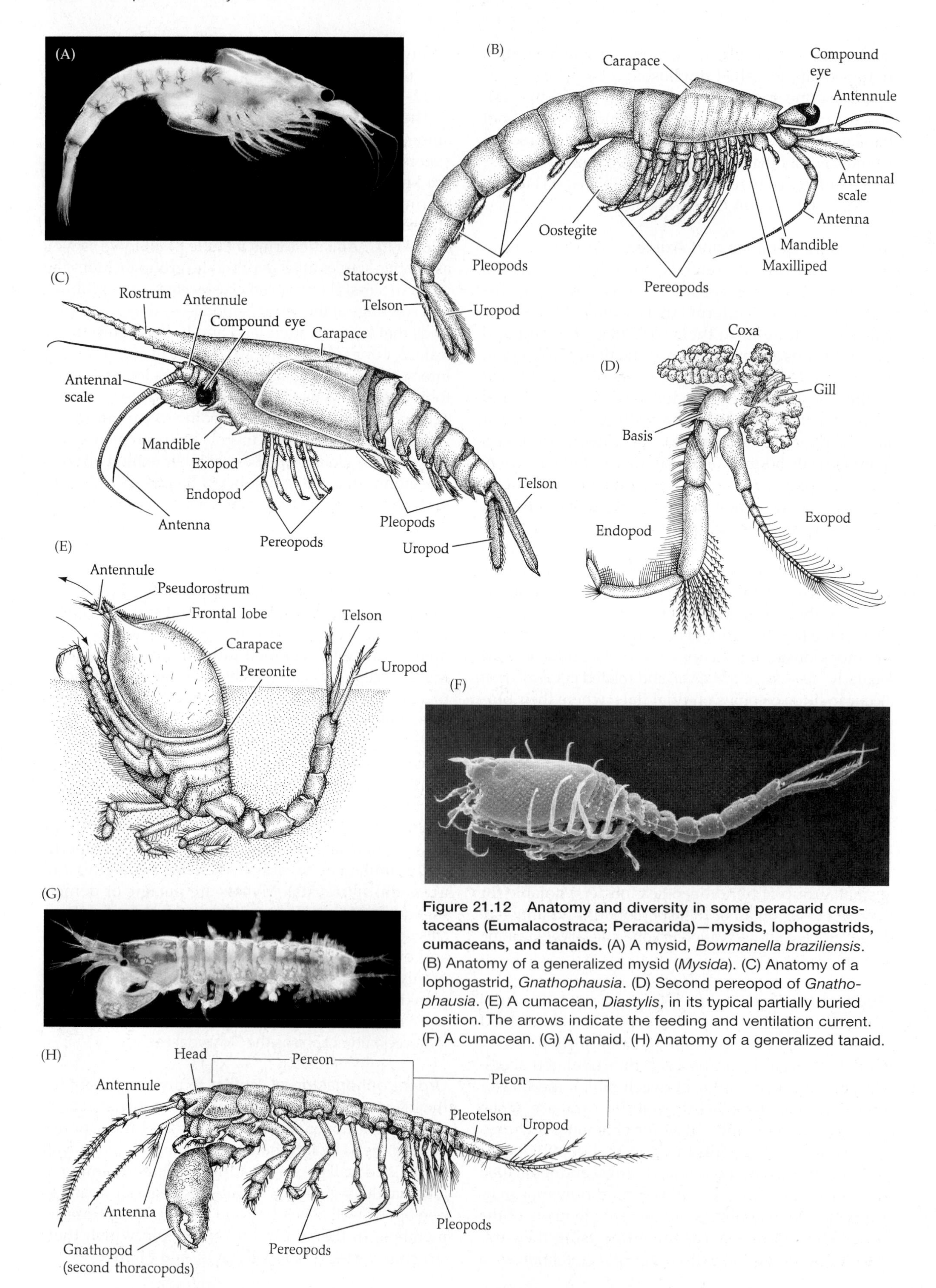

Figure 21.12 Anatomy and diversity in some peracarid crustaceans (Eumalacostraca; Peracarida)—mysids, lophogastrids, cumaceans, and tanaids. (A) A mysid, *Bowmanella braziliensis*. (B) Anatomy of a generalized mysid (*Mysida*). (C) Anatomy of a lophogastrid, *Gnathophausia*. (D) Second pereopod of *Gnathophausia*. (E) A cumacean, *Diastylis*, in its typical partially buried position. The arrows indicate the feeding and ventilation current. (F) A cumacean. (G) A tanaid. (H) Anatomy of a generalized tanaid.

There are about 60 known species of lophogastrids, most of which are 1–8 cm long, although the giant *Gnathophausia ingens* reaches 35 cm. All are pelagic swimmers, and the group has a cosmopolitan oceanic distribution. Lophogastrids are primarily predators on zooplankton.

Order Cumacea Carapace present, covering and fused to first three thoracic segments, whose appendages are modified as maxillipeds, the first with modified branchial apparatus associated with branchial cavity formed by carapace; pereopods 1–5 ambulatory, simple, 1–4 may be biramous; pleopods usually absent in females and present in males; telson sometimes fused with sixth pleonite, forming pleotelson; uropods styliform; compound eyes absent, or sessile and usually fused (Figures 21.1P and 21.12E,F).

Cumaceans are small, odd-looking crustaceans with a large, bulbous anterior end and a long, slender posterior—resembling horizontal commas! The great carcinologist Waldo Schmitt referred to them as "little wonders and queer blunders." They occur worldwide and include about 1,500 species, most of which are between 0.1 and 2 cm long, though some species in cold water reach 3 cm in length. Most are marine, although a few brackish-water species are known. They live in association with bottom sediments, but are capable of swimming and probably leave the bottom to breed. Most are deposit feeders or predators on the meiofauna, others eat the organic film on sand grains.

Order Tanaidacea Carapace present and fused with first two thoracic segments; thoracopods 1–2 are maxillipeds, the second being chelate; thoracopods 3–8 are simple, ambulatory pereopods; pleopods present or absent; uropods biramous or uniramous; telson and last one or two pleonites fused as pleotelson; adults with maxillary and (vestigial) antennal glands; compound eyes absent, or present and on cephalic lobes. Members of this order are known worldwide from benthic marine habitats; a few live in brackish or nearly fresh water. Most of the 1,500 or so species are small, ranging from 0.5 to 2 cm in length. They often live in burrows or tubes and are known from all ocean depths. Many are suspension feeders, others are detritivores, and still others are predators (Figure 21.12G,H).

Order Mictacea Without a carapace, but with a well developed head shield fused with first thoracomere and produced laterally over bases of mouthparts; 1 pair of maxillipeds; pereopods simple, 1–5 or 2–6 biramous, exopods natatory; gills absent; pleopods reduced, uniramous or biramous; uropods biramous, with 2–5 segmented rami; telson not fused with pleonites; stalked eyes present (*Mictocaris*) but lacking any evidence of visual elements, or absent (*Hirsutia*) (Figure 21.13D–E).

Mictacea is the most recently (1985) established peracaridan order. The order was erected to accommodate two species of unusual crustaceans: *Mictocaris halope* (from marine caves in Bermuda) and *Hirsutia bathyalis* (from a benthic sample 1,000 m deep in the Guyana Basin off northeastern South America). A third species of Mictacea was described in 1988 from Australia, and a fourth from the Bahamas in 1992; there are now 6 species known. Mictaceans are small, 2–3.5 mm in length. *Mictocaris halope* is the best known of these species because many specimens have been recovered and some have been studied alive. It is pelagic in cave waters and swims by using its pereopodal exopods. The status of the Mictacea as a monophyletic grouping and its relationships to other peracarid orders is a subject of ongoing debate. Some workers recognize the family Hirsutiidae (containing *Hirsutia, Montucaris,* and *Thetispelecaris*) as a separate order, the Bochusacea (male pleopods biramous).

Order Spelaeogriphacea Carapace short, fused with first thoracomere; 1 pair of maxillipeds; pereopods 1–7 simple, biramous, with shortened exopods; exopods on legs 1–3 modified for producing currents, on legs 4–7 as gills; pleopods 1–4 biramous, natatory; pleopod 5 reduced; tail fan well developed; compound eyes nonfunctional or absent, but eyestalks persist (Figures 21.13A and 21.21H). The order Spelaeogriphacea is currently known from only four living species. These rare, small (less than 1 cm) peracarids were long known only from a single species living in a freshwater stream in Bat Cave on Table Mountain, South Africa. A second species is known from a freshwater cave in Brazil, and a third and fourth species were described from an aquifer in Australia. Little is known about the biology of these animals, but they are suspected to be detritus feeders. Like thermosbaenaceans, spelaeogriphaceans are thought to be relicts of a more widespread shallow-water marine Tethyan fauna stranded in interstitial and ground-water environments during periods of marine regression.

Order Thermosbaenacea Carapace present, fused with first thoracomere and extending back over 2–3 additional segments; 1 pair of maxillipeds; pereopods biramous, simple, lacking epipods and oostegites; carapace forms dorsal brood pouch (unlike all other peracarids, which form the brood pouch from ventral oostegites); 2 pairs of uniramous pleopods; uropods biramous; telson free or forming pleotelson with last pleonite; eyes absent (Figure 21.13B,C). About 34 species of thermosbaenaceans are recognized in seven genera. *Thermosbaena mirabilis* is known from freshwater hot springs in North Africa, where it lives at temperatures in excess of 40°C. Several species in other genera occur in much cooler fresh waters, typically in groundwater or in caves. Other species are marine or inhabit underground anchialine pools. Limited data suggest that thermosbaenaceans feed on plant detritus.

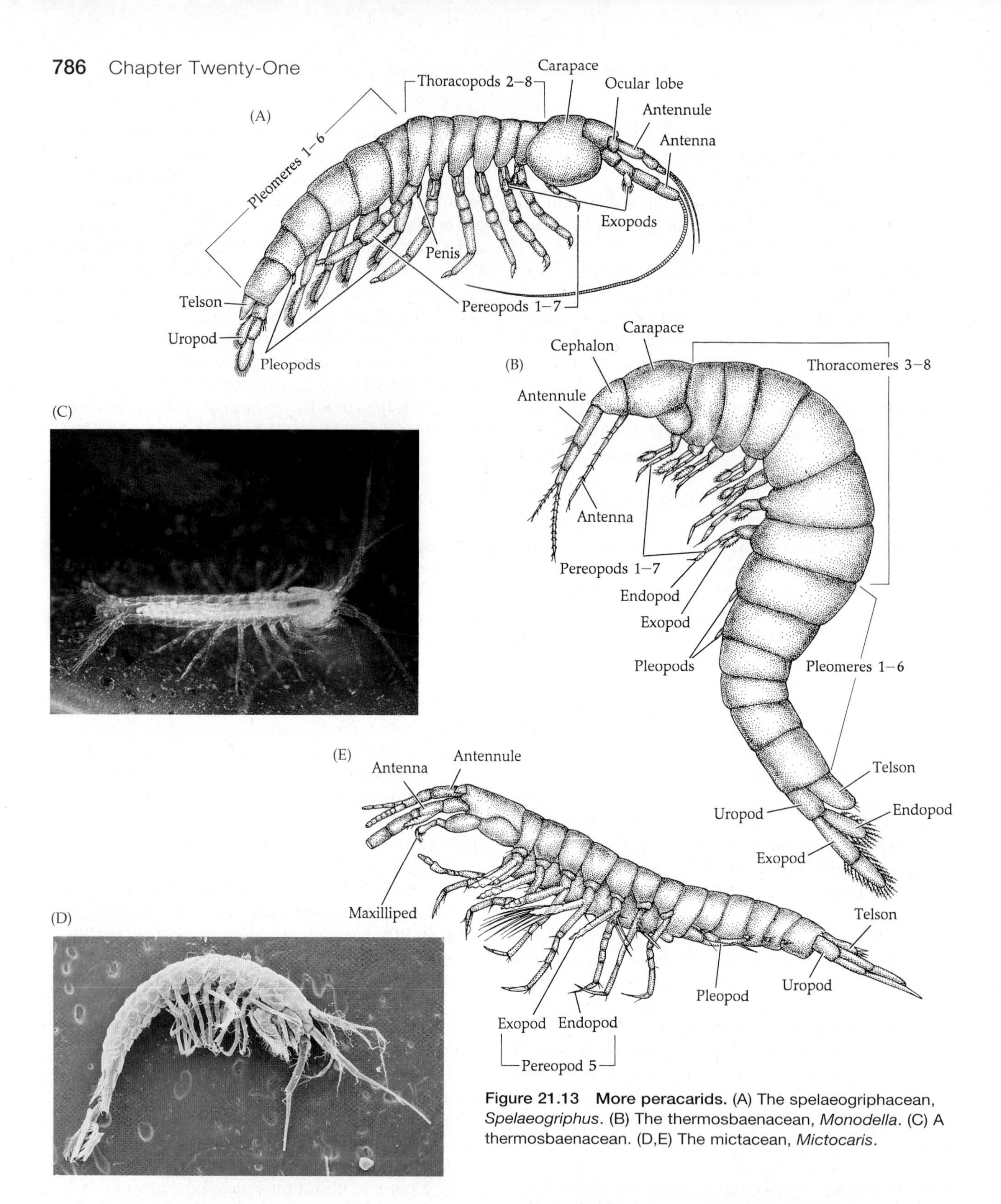

Figure 21.13 More peracarids. (A) The spelaeogriphacean, *Spelaeogriphus*. (B) The thermosbaenacean, *Monodella*. (C) A thermosbaenacean. (D,E) The mictacean, *Mictocaris*.

Order Isopoda Carapace absent; first thoracomere fused with head; 1 pair of maxillipeds; 7 pairs of uniramous pereopods, the first of which is sometimes subchelate, others usually simple (gnathiids have only five pairs of pereopods, as thoracopod 2 is a maxillipedal "pylopod" and thoracopod 8 is missing); pereopods variable, modified as ambulatory, prehensile, or swimming; in the more derived suborders pereopodal coxae are expanded as lateral side plates (coxal plates); pleopods biramous and well developed, natatory and for gas exchange (functioning as gills in aquatic taxa, and with air sacs called pseudotrachea in most terrestrial Oniscidea); adults with maxillary and (vestigial) antennal glands; telson fused with one to six pleonites, forming pleotelson; eyes usually sessile and compound, absent from some, pedunculate in most Gnathiidea; with biphasic molting (posterior region molts before anterior region) (Figures 21.1Q, 21.14, 21.21I, 21.28H,I, and 21.33M).

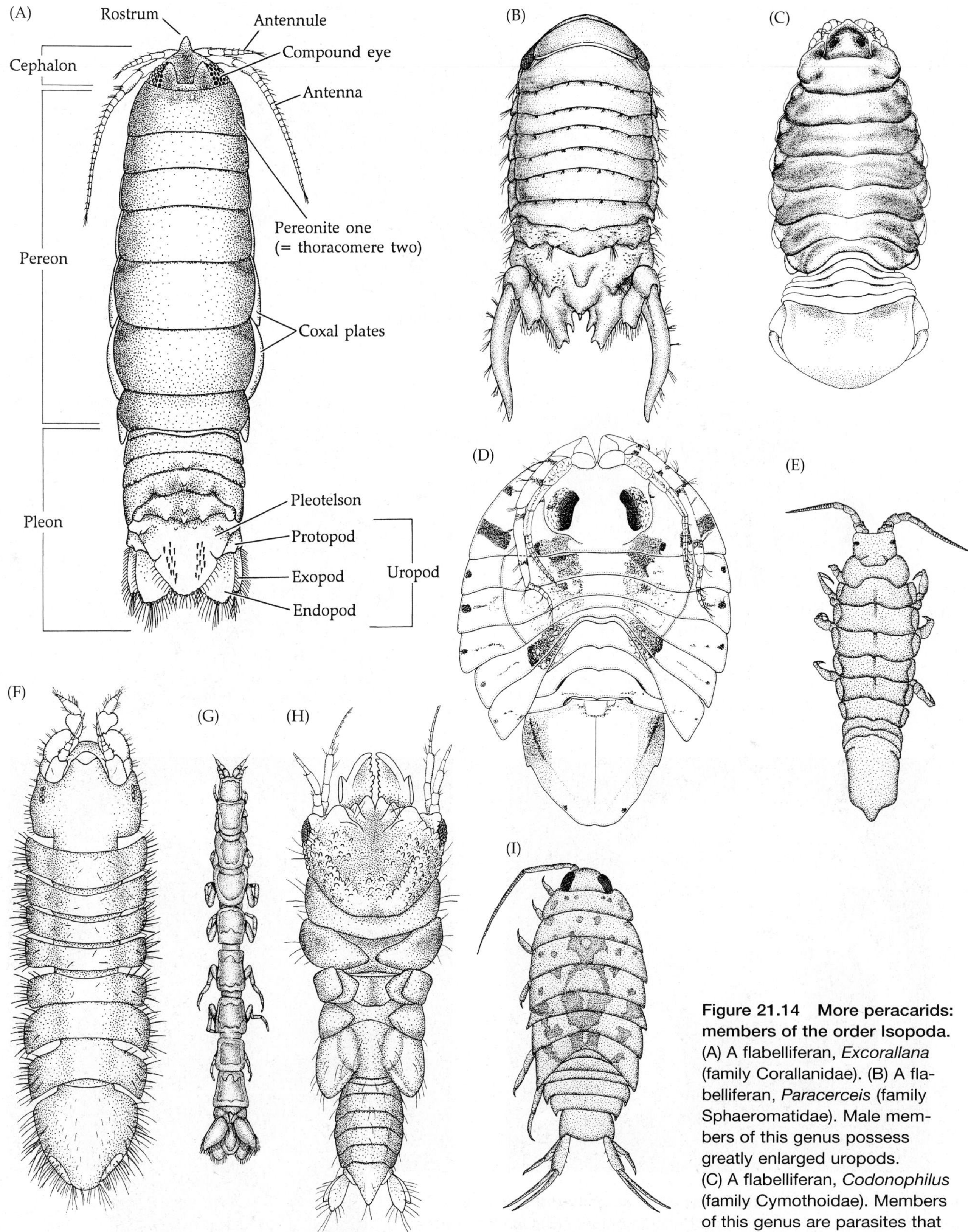

Figure 21.14 More peracarids: members of the order Isopoda. (A) A flabelliferan, *Excorallana* (family Corallanidae). (B) A flabelliferan, *Paracerceis* (family Sphaeromatidae). Male members of this genus possess greatly enlarged uropods. (C) A flabelliferan, *Codonophilus* (family Cymothoidae). Members of this genus are parasites that attach to the tongues of various marine fishes. (D) A flabelliferan, *Heteroserolis* (family Serolidae). (E) A valviferan, *Idotea* (family Idoteidae). (F) An asellote, *Joeropsis* (family Joeropsididae). (G) An anthurid, *Mesanthura* (family Anthuridae). (H) A gnathiidean, *Gnathia* (family Gnathiidae). Note the grossly enlarged mandibles characteristic of male Gnathiidea. (I) An oniscidean, *Ligia* (the common seashore "rock louse").

The isopods comprise about 10,500 marine, freshwater, and terrestrial species, ranging in length from 0.5 to 500 mm, the largest being species of the benthic genus *Bathynomus* (Cirolanidae). They are common inhabitants of nearly all environments, and some groups are exclusively (e.g., Bopyridae, Cymothoidae) or partly

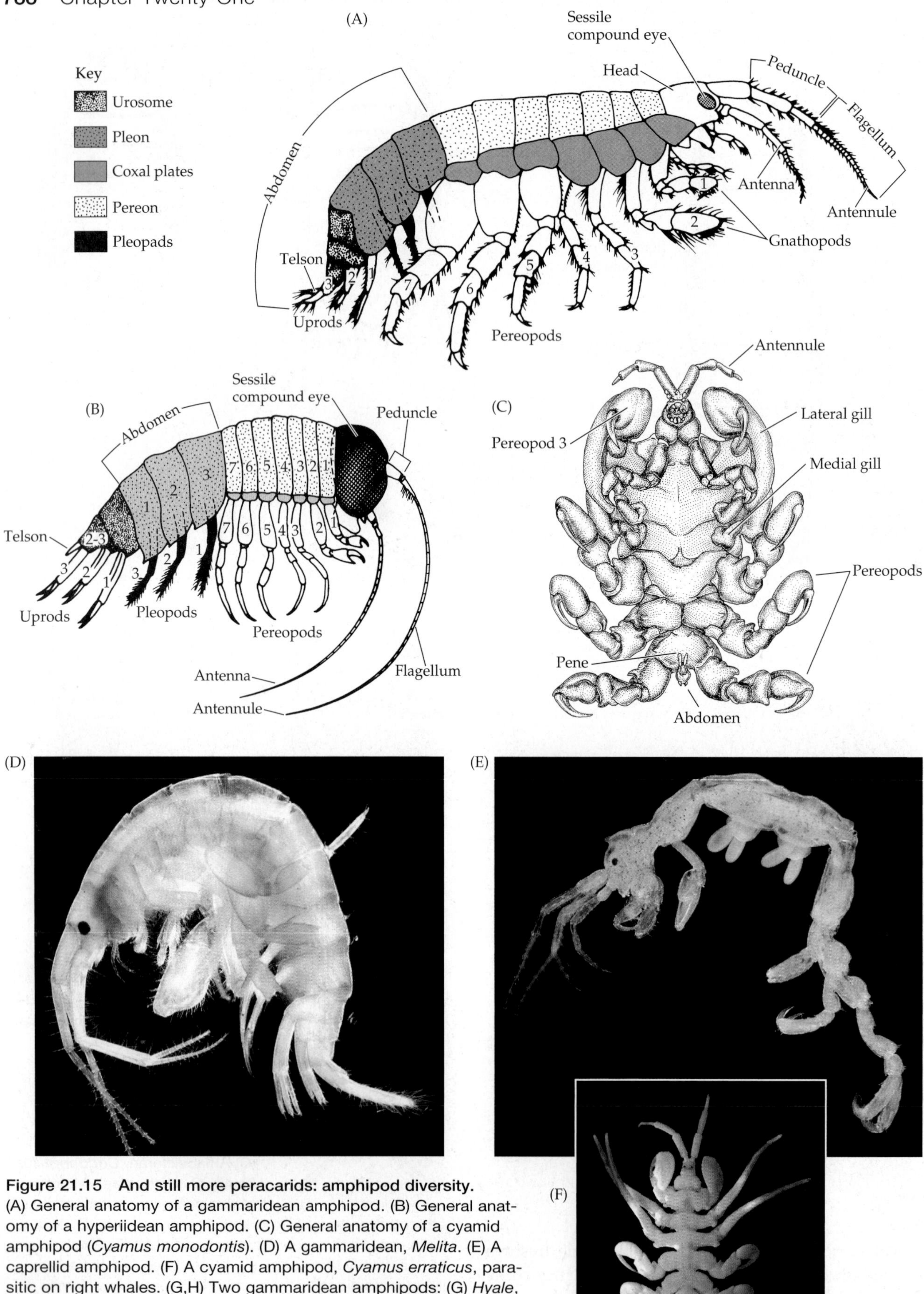

Figure 21.15 And still more peracarids: amphipod diversity. (A) General anatomy of a gammaridean amphipod. (B) General anatomy of a hyperiidean amphipod. (C) General anatomy of a cyamid amphipod (*Cyamus monodontis*). (D) A gammaridean, *Melita*. (E) A caprellid amphipod. (F) A cyamid amphipod, *Cyamus erraticus*, parasitic on right whales. (G,H) Two gammaridean amphipods: (G) *Hyale*, a beach hopper, and (H) *Heterophlias*, an unusual, dorsoventrally flattened species. (I–K) Three hyperiidean amphipods: (I) *Primno*, (J) *Leptocottis*, and (K) a hyperiid on its host medusa. (L) A free-living caprellid, *Caprella*. (M) An ingolfiellid amphipod, *Ingolfiella*.

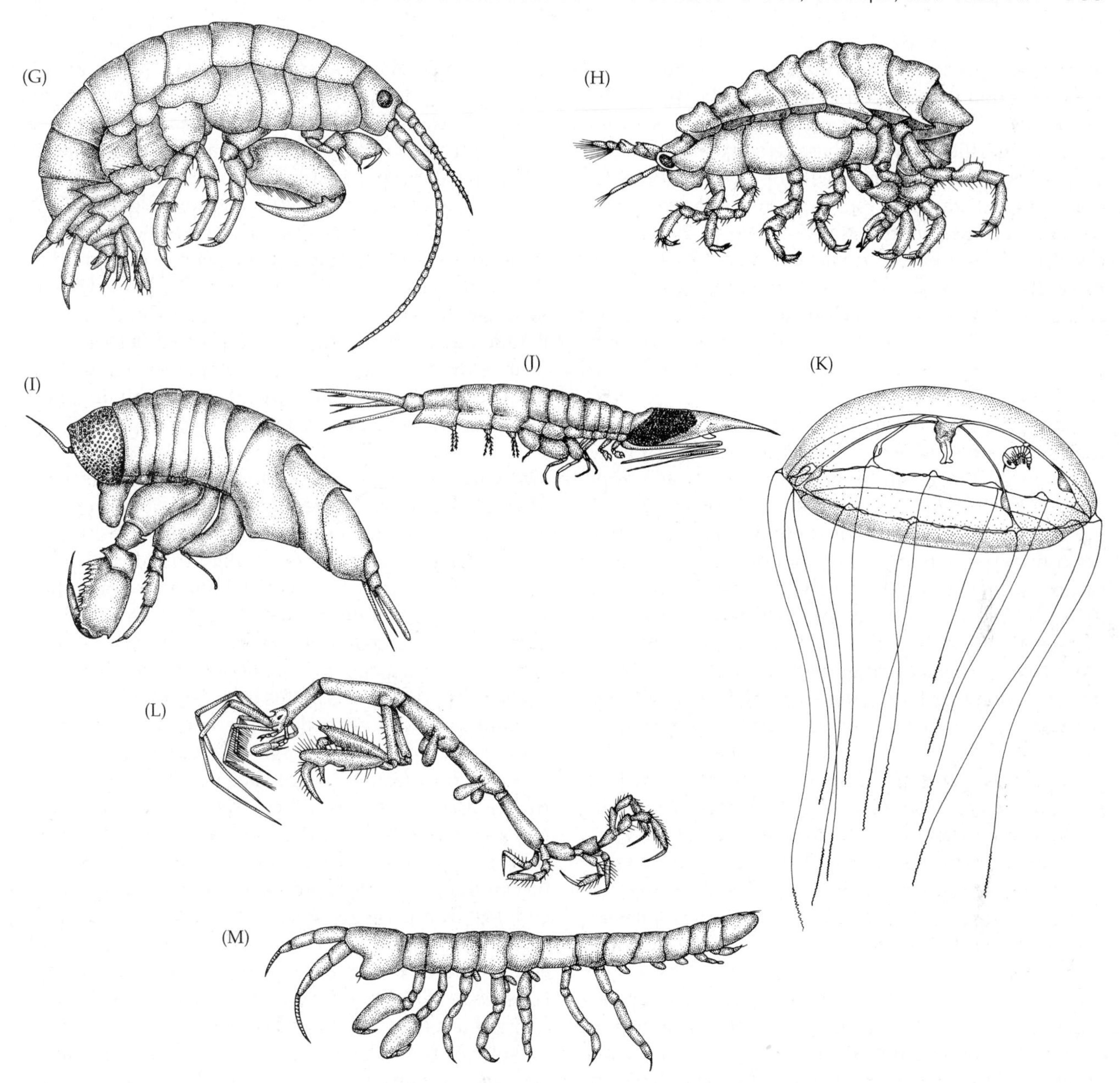

(e.g., Gnathiidae) parasitic. The suborder Oniscidea includes about 5,000 species that have invaded land (pillbugs and sowbugs); they are the most successful terrestrial crustaceans. Their direct development, flattened shape, osmoregulatory capabilities, thickened cuticle, and aerial gas exchange organs (pseudotrachea) allow most oniscideans to live completely divorced from aquatic environments. Fossils as old as 325 million years (Carboniferous) have been reported.

Isopod feeding habits are extremely diverse. Many are herbivorous or omnivorous scavengers, but direct plant feeders, detritivores, and predators are also common. Some are parasites (e.g., on fishes or on other crustaceans) that feed on the tissue fluids of their hosts. Overall, grinding mandibles and herbivory seem to represent the primitive state, with slicing or piercing mandibles and predation appearing later in the evolution of several isopod clades.

Order Amphipoda Carapace absent; first thoracomere fused to head; 1 pair of maxillipeds; 7 pairs of uniramous pereopods, with first, second, and sometimes others frequently modified as chelae or subchelae; pereopodal coxae expanded as lateral side plates (coxal plates); gills thoracic (medial pereopodal epipods); adults with antennal glands; abdomen "divided" into two regions of three segments each, an anterior "pleon" and posterior urosome, with anterior appendages as typical pleopods and urosomal appendages modified as uropods; telson free or fused with last urosomite; other urosomites sometimes fused; compound eyes sessile, absent in some, huge in many (but not all) members of the suborder Hyperiidea (Figures 21.1N,O, 21.15, 21.23, 21.27F, and 21.29D).

Isopods and amphipods share many features and are often said to be closely related. Earlier workers recognized these similarities (e.g., sessile compound eyes, loss of carapace, and presence of coxal plates) and classified them together as the "Edriopthalma" or "Acarida." However, recent work suggests that many similarities between these two taxa may be convergences or parallelisms. The roughly 11,000 species of amphipods range in length from tiny 1 mm forms to giant deep-sea benthic species reaching 29 cm, and a group of planktonic forms exceeds 10 cm. They have invaded most marine and freshwater habitats and often constitute a large portion of the biomass in many areas. The largest known amphipod is *Alicella gigantea*, a cosmopolitan marine species living at depths up to 7,000 m.

The principal suborder is Gammaridea. A few gammarideans are semiterrestrial in moist forest leaf litter or on supralittoral sandy beaches (e.g., beach hoppers); a few others live in moist gardens and greenhouses (e.g., *Talitrus sylvaticus* and *T. pacificus*). They are common in subterranean groundwater ecosystems of caves, the majority being stygobionts—obligatory groundwater species characterized by reduction or loss of eyes, pigmentation, and occasionally appendages. About 900 species of stygobiontic amphipods have been described, including the diverse genera *Niphargus* (in Europe) and *Stygobromus* (in North America), each with over 100 described species. However, most of the gammaridean amphipods are marine benthic species, and a few have adopted a pelagic lifestyle, usually in deep oceanic waters. There are many intertidal species, and a great many of these live in association with other invertebrates and with algae. Domicolous gammaridean amphipods in at least three families spin silk from their legs that is used for consolidating the walls of their tube or shelter.

The suborder Hyperiidea includes exclusively pelagic amphipods that have apparently escaped the confines of benthic life by becoming associated with other plankters, particularly gelatinous zooplankton such as medusae, ctenophores, and salps. The hyperiideans are usually characterized by huge eyes (and a few other inconsistent features), but several groups bear eyes no larger than those of most gammarideans. The Hyperiidea are almost certainly a polyphyletic group, and it is thought that several lineages are derived independently from various gammaridean ancestors, although a modern phylogenetic analysis has yet to be attempted. The precise nature of the relationships between hyperiideans and their zooplankton hosts remains controversial. Some appear to eat host tissue, others may kill the host to fashion a floating "home," and still others may utilize the host merely for transport or as a nursery for newly hatched young. Specimens of the scyphozoan *Phacellophora camtschtica* have been found with nearly 500 *Hyperia medusarum* riding (and feeding) on it.

There are two other small amphipod suborders: Ingolfiellidea and Caprellidea. The first suborder contains only about 40 species, most of which live in subterranean fresh and brackish waters, although a few are marine and interstitial. Little is known about their biology. The 300 or so species of caprellid amphipods ("skeleton shrimp") are highly modified for clinging to other organisms, including filamentous algae and hydroids. In most species the body and appendages are very narrow and elongated. In one family of caprellids, the Cyamidae (with 28 species), individuals are obligate symbionts on cetaceans (whales, dolphins, and porpoises) and have flattened bodies and prehensile legs.

In addition to parasitism, amphipods exhibit a vast array of feeding strategies, including scavenging, herbivory, carnivory, and suspension feeding.

"Maxillopodans" The following seven classes—Thecostraca, Tantulocarida, Branchiura, Pentastomida, Mystacocarida, Copepoda, and Ostracoda—historically were united in a group called the Maxillopoda that is now known to be artificial (nonmonophyletic). However, many of these "maxillopodan" groups share some basic characteristics. These include a body of five cephalic, six thoracic, and four abdominal somites, plus a telson, but reductions of this basic 5-6-4 body plan are common, and different specialists sometimes interpret the nature of these tagmata in different ways, leading to some confusion. Additional characters that most of these classes have in common are: thoracomeres variously fused with cephalon; usually with caudal rami; thoracic segments with biramous (sometimes uniramous) limbs, lacking epipods (except in many ostracods); abdominal segments lack typical appendages; carapace present or reduced; with both simple and compound eyes, the latter being unique, with three cups, each with tapetal cells (an arrangement still often referred to as the maxillopodan eye).

Most of these former "maxillopodans" are small crustaceans, barnacles being a notable exception. They are generally recognizable by their shortened bodies, especially the reduced abdomen, and by the absence of a full complement of legs. The reductions in body size and leg number, emphasis on the naupliar eye, minimal appendage specialization, and certain other features have led biologists to hypothesize that paedomorphosis (progenesis) played a role in the origin of some of these classes. That is, in many ways, they resemble early postlarval forms that evolved sexual maturity before attaining all the adult features. Over 26,000 species have been described within the seven classes.

Class Thecostraca

This group includes the barnacles, parasitic ascothoracids, and mysterious "y-larvae." The thecostracan clade is defined by several rather subtle synapomorphies of cuticular fine structure, including cephalic chemosensory structures known as lattice organs. The group is also supported by molecular phylogenetic analyses. All taxa have pelagic larvae, the terminal instar of which possesses prehensile antennules and is specialized for locating and attaching to the substratum of the sessile adult state.

Subclass Ascothoracida About 125 described species of parasites of anthozoans and echinoderms. Although greatly modified, they retain a bivalved carapace and the full complement of thoracic and abdominal segments (facts that suggest they might be the most primitive living thecostracans). Ascothoracids generally have mouthparts modified for piercing and sucking body fluids, but some live inside other animals and absorb the host's tissue fluids. In at least one species, *Synagoga mira*, males retain the ability to swim throughout their lives, attaching only temporarily while feeding on corals (Figure 21.16F).

Subclass Cirripedia Primitively with tagmata as in the class, but in most groups the adult body is modified for sessile or parasitic life; thorax of six segments with paired biramous appendages; abdomen without limbs; telson absent in most, although caudal rami persist on abdomen in some; nauplius larva with frontolateral horns; unique, "bivalved" cypris larva; adult carapace "bivalved" (folded) or forming fleshy mantle; first thoracomere often fused with cephalon and bearing maxilliped-like oral appendages; female gonopores near bases of first thoracic limbs, male gonopore on median penis on last thoracic or first abdominal segment; compound eyes lost in adults (Figures 21.1S,T, 21.16A–E, 21.25, 21.26, 21.27B,C, 21.32E, and 21.33F).

The 1,285 or so described cirripede species are mostly free-living barnacles, but this group also includes some strange parasitic "barnacles" rarely seen except by specialists. The common acorn and goose barnacles belong to the superorder Thoracica. The superorder Acrothoracica consists of minute animals that burrow into calcareous substrata, including corals and mollusc shells (Figure 21.16G). The superorder Rhizocephala are exceptionally modified parasites of other crustaceans, especially decapods (Figure 21.16H).

If the body plan of the cirripedes is derived from something similar to what is seen in other "maxillopodan" classes, then it has been so extensively modified that its basic features are nearly unrecognizable in adults of this subclass. The abdomen is greatly reduced in adults, and also in most cypris larvae. In cyprids (cypris larvae) the carapace is always present and "bivalved," the two sides being held by a transverse cypris adductor muscle; in adults the carapace may be lost (Rhizocephala) or modified as a membranous, saclike mantle (thoracicans and acrothoracicans). In the barnacles (Thoracica), it is this mantle that produces the familiar calcareous plates that enclose the body. Cyprids and adult acrothoracicans share a unique tripartite crystalline cone structure in the compound eye, a feature not known from any other crustacean group and perhaps a vestige of the ancestral thecostracan body plan. Most species of barnacles are hermaphrodites, whereas separate sexes are the rule in acrothoracicans and rhizocephalans, and some androgonochoristic species (males + hermaphrodites; e.g., *Scalpellum*) have also been reported.

Locomotion in barnacles is generally confined to the larval stages, although adults of a few species are specifically adapted to live attached to floating objects (e.g., seaweeds, pumice, logs) or nektonic marine animals (e.g., whales, sea turtles). Others are often found on the shells and exoskeletons of various errant invertebrates (e.g., crabs and gastropods), which inadvertently provide a means of transportation from one place to another. Thoracican and acrothoracican barnacles use their feathery thoracopods (cirri) to suspension feed. Barnacles in the family Coronulidae are suspension feeders that attach to whales and turtles (e.g., *Chelonibia, Platylepas, Stomatolepas, Coronula, Xenobalanus*). The 265 or so known species of rhizocephalans are all endoparasitic in other Crustacea, and are the most highly modified of all cirripedes. They mainly inhabit decapod crustaceans, but a few are known from isopods, cumaceans, stomatopods, and even thoracican barnacles. Some even parasitize freshwater and terrestrial crabs. The body consists of a reproductive part (the externa) positioned outside the host's body, and an internal, ramifying, nutrient-absorbing part (the interna).[6]

Subclass Facetotecta Monogeneric (*Hansenocaris*): The "y-larvae," a half-dozen small (250–620 mm) marine nauplii and cyprids (Figure 21.16I). Although known since Hansen's original description in 1899, the adult stage of these animals has still not been identified. However, a stage subsequent to the y-larva, the slug-like ypsigon stage, has been induced by treating y-larvae with molting hormones. The prehensile antennules and hooked labrum of the y-cyprids and the degenerative nature of the ypsigon suggest that the adults are parasitic in yet-to-be-identified hosts.

Class Tantulocarida

Bizarre parasites of deep-water crustaceans. Juveniles with cephalon, 6-segmented thorax, and abdomen of up to 7 segments; cephalon lacking appendages (other than antennules in one known stage only) but with an internal median stylet; thoracopods 1–5 biramous, 6

[6]Rhizocephalans of the family Sacculinidae infest only decapod crustaceans and have been suggested as biological control agents for invasive exotics such as the green crab (*Carcinus maenas*), which are upsetting coastal ecosystems worldwide. Sacculinds have the ability to take control over such major host functions as molting and reproduction, and also to compromise the host's immune system.

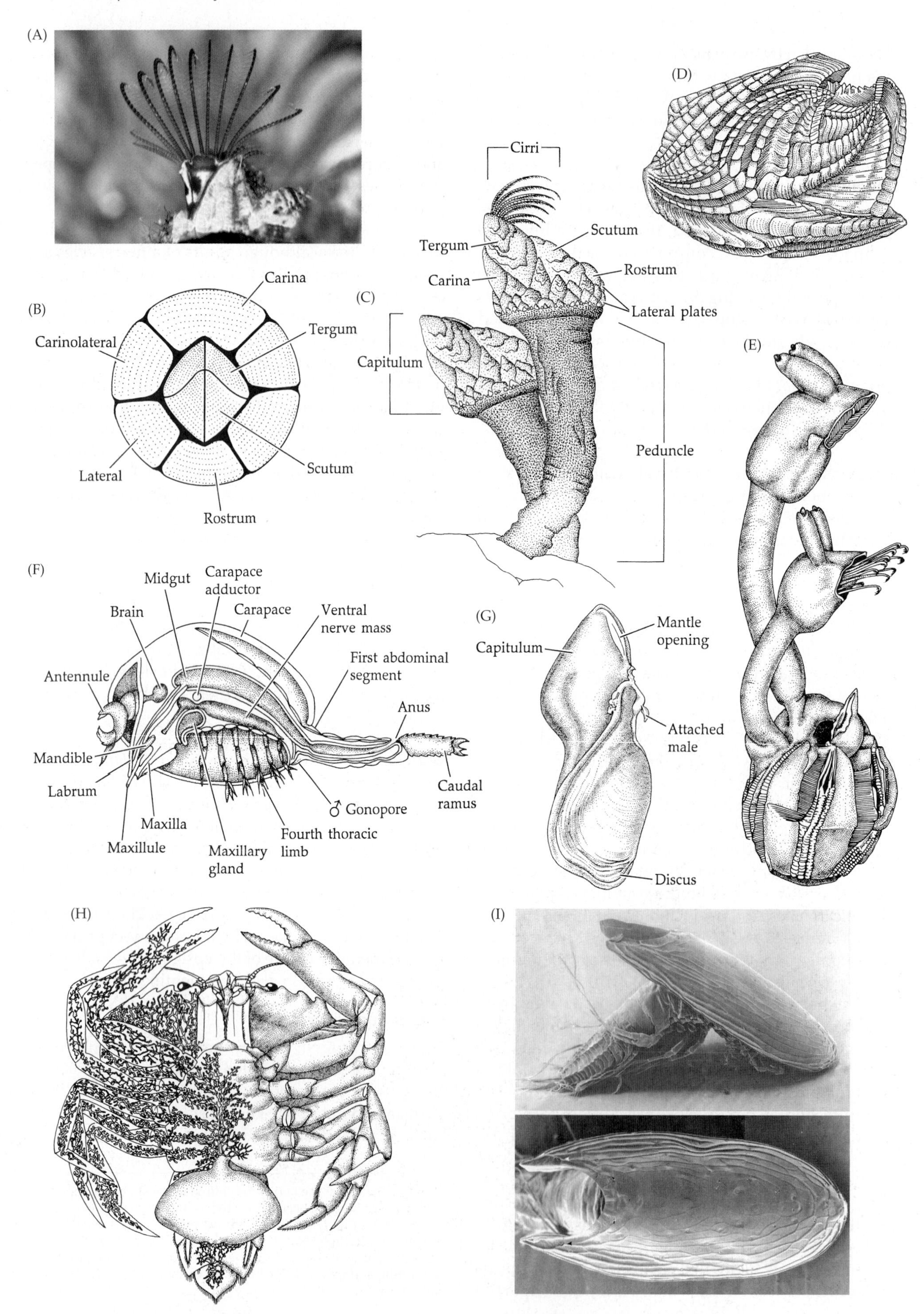
(A)
(B)
Carina
Tergum
Carinolateral
Lateral
Scutum
Rostrum
(C)
Cirri
Tergum
Scutum
Rostrum
Carina
Lateral plates
Capitulum
Peduncle
(D)
(E)
(F)
Midgut
Carapace adductor
Brain
Carapace
Ventral nerve mass
First abdominal segment
Antennule
Anus
Mandible
Labrum
Caudal ramus
♂ Gonopore
Maxilla
Fourth thoracic limb
Maxillule
Maxillary gland
(G)
Mantle opening
Capitulum
Attached male
Discus
(H)
(I)

◀ **Figure 21.16 Anatomy and diversity in the class Thecostraca—barnacles and their kin.** (A–E) Thoracican barnacles. (A) A sessile (acorn) barnacle with its cirri extended for feeding. (B) Plate terminology in a balanomorph (acorn) barnacle. (C) The lepadomorph (stalked) barnacle *Pollicipes polymerus*. (D) *Verruca*, the "wart" barnacle. (E) Two thoracican barnacles that live in association with each other and with whales. The stalked barnacle *Conchoderma* attaches to the sessile barnacle *Coronula*, which in turn attaches to the skin of certain whales. (F) The ascothoracican *Ascothorax ophiocentenis*, a parasite that feeds periodically on echinoderms (longitudinal section). (G) An acrothoracican, *Alcippe*. Note the highly modified female and the tiny attached male. This species bores into calcareous substrata such as coral skeletons. (H) A crab (*Carcinus*) infected with the rhizocephalan *Sacculina carcini*. The crab's right side is shown as transparent, exposing the ramifying body of the parasite. (I) A cypris y-larva, in lateral and dorsal views.

uniramous; abdomen without appendages but with caudal rami; adults highly modified, with "unsegmented" sacciform thorax and a reduced abdomen bearing a uniramous penis on the first segment; female gonopores on fifth thoracic segment.

The tiny tantulocarids are less than 0.5 mm long. They attach to their hosts by penetrating the body with a protruding cephalic stylet. The young bear natatory thoracopods. About three dozen species in 22 genera have been described (Figures 21.1V and 21.17). They are known from abyssal depths to the intertidal zone, from polar to tropical waters, and from anchialine pools and hydrothermal vents, always as parasites on other crustaceans. Until recently, members of this group had been assigned to various parasitic groups of Copepoda and Cirripedia. In 1983 Geoffrey Boxshall and Roger Lincoln proposed the new class Tantulocarida. Some early work on the group supported a view of these animals as maxillopodans, although the presence of six or seven abdominal segments in juveniles of some species is inconsistent with this view; they are now believed to be allied to the Thecostraca. The life cycle is unique and includes a larva called the tantulus.

Class Branchiura

Body compact and oval, head and most of trunk covered by broad carapace; antennules and antennae reduced, the latter sometimes absent; mouthparts modified for parasitism; no maxillipeds; thorax reduced

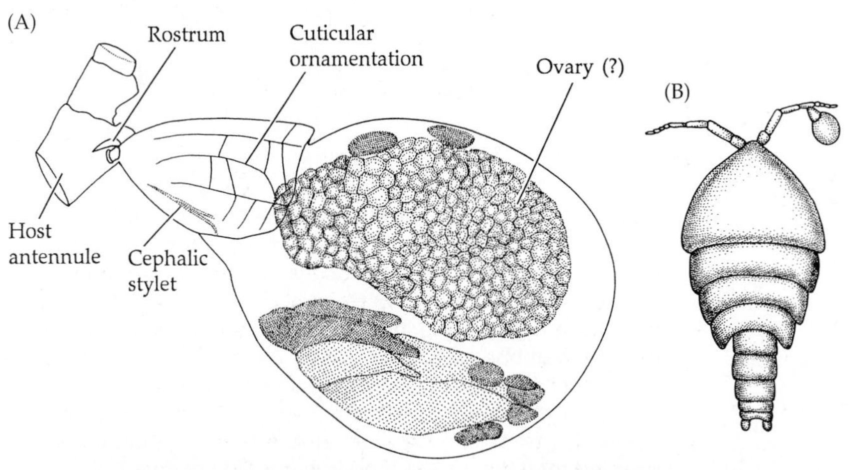

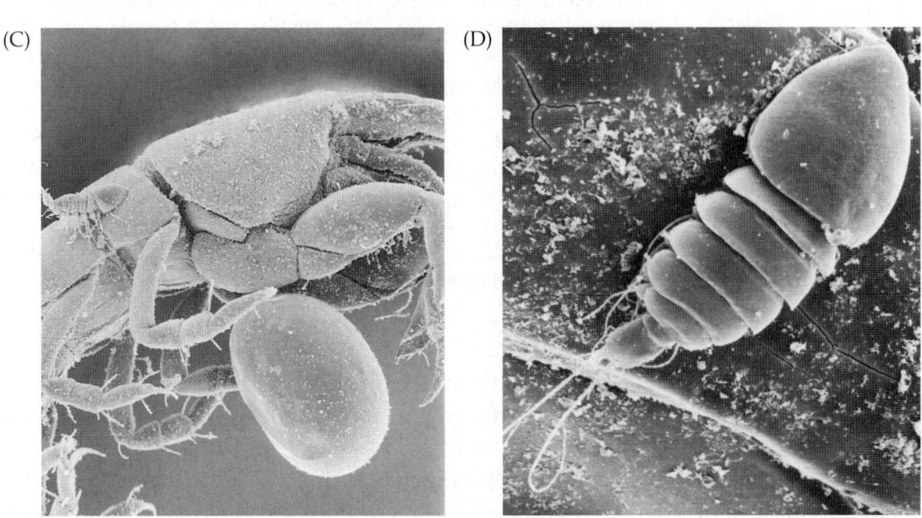

Figure 21.17 Anatomy in the class Tantulocarida. (A) An adult *Basipodella atlantica*. Note the absence of an abdomen and the modifications for parasitic life. (B) *Basipodella* attached to the antenna of a copepod host. (C,D) *Microdajus pectinatus* on a crustacean host, adult, and juvenile (SEM).

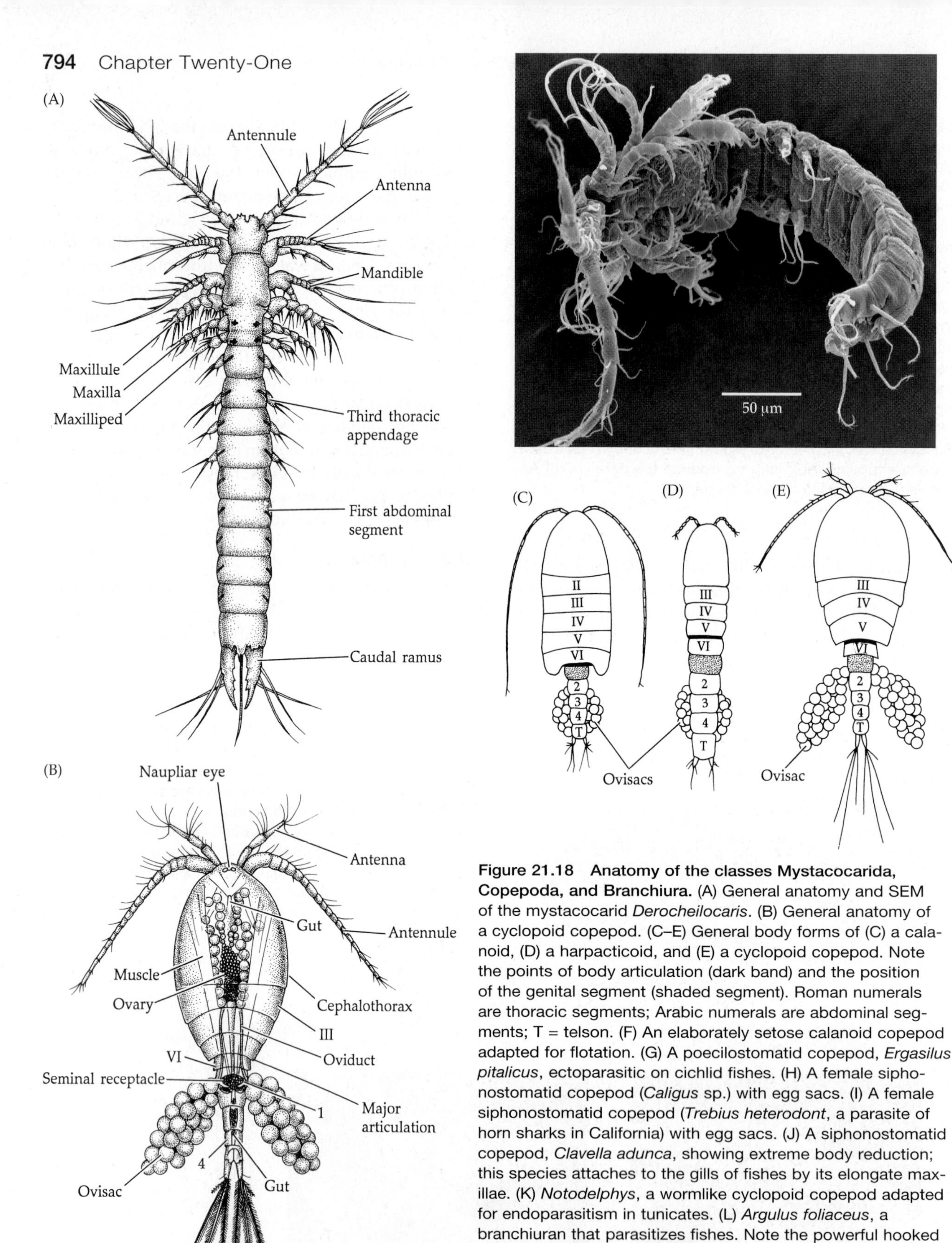

Figure 21.18 Anatomy of the classes Mystacocarida, Copepoda, and Branchiura. (A) General anatomy and SEM of the mystacocarid *Derocheilocaris*. (B) General anatomy of a cyclopoid copepod. (C–E) General body forms of (C) a calanoid, (D) a harpacticoid, and (E) a cyclopoid copepod. Note the points of body articulation (dark band) and the position of the genital segment (shaded segment). Roman numerals are thoracic segments; Arabic numerals are abdominal segments; T = telson. (F) An elaborately setose calanoid copepod adapted for flotation. (G) A poecilostomatid copepod, *Ergasilus pitalicus*, ectoparasitic on cichlid fishes. (H) A female siphonostomatid copepod (*Caligus* sp.) with egg sacs. (I) A female siphonostomatid copepod (*Trebius heterodont*, a parasite of horn sharks in California) with egg sacs. (J) A siphonostomatid copepod, *Clavella adunca*, showing extreme body reduction; this species attaches to the gills of fishes by its elongate maxillae. (K) *Notodelphys*, a wormlike cyclopoid copepod adapted for endoparasitism in tunicates. (L) *Argulus foliaceus*, a branchiuran that parasitizes fishes. Note the powerful hooked suckers (modified maxillules) on the ventral surface.

to four segments, with paired biramous appendages; abdomen unsegmented, bilobed, limbless, but with minute caudal rami; female gonopores at bases of fourth thoracic legs, male with single gonopore on midventral surface of last thoracic somite; paired, sessile compound eyes and one to three median simple eyes (Figure 21.18L).

The Branchiura comprise about 230 species of ectoparasites on marine and freshwater fishes. The

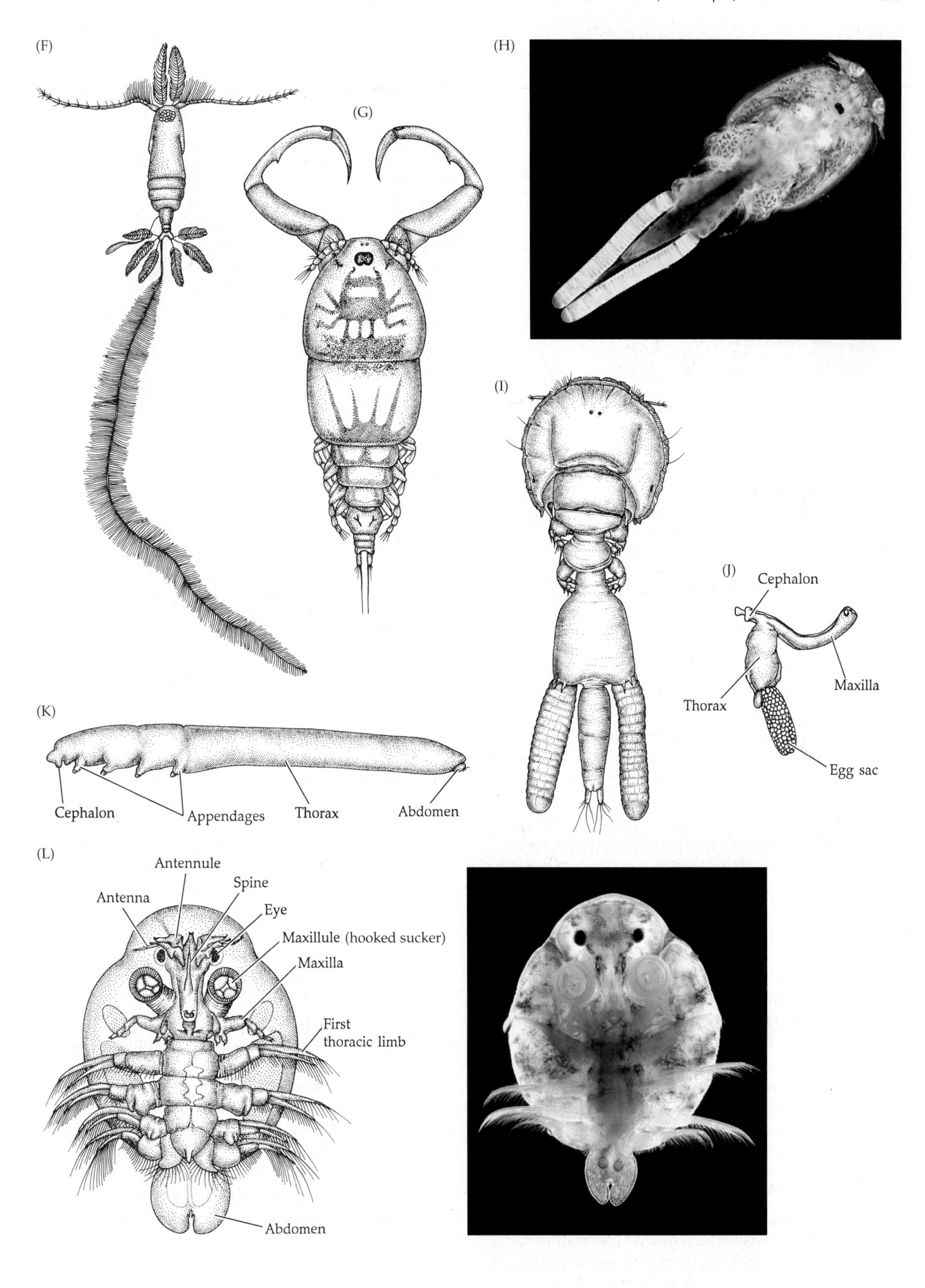
(F)
(G)
(H)
(I)
(J)
Cephalon
Maxilla
Thorax
Egg sac
(K)
Cephalon
Appendages
Thorax
Abdomen
(L)
Antennule
Spine
Antenna
Eye
Maxillule (hooked sucker)
Maxilla
First thoracic limb
Abdomen

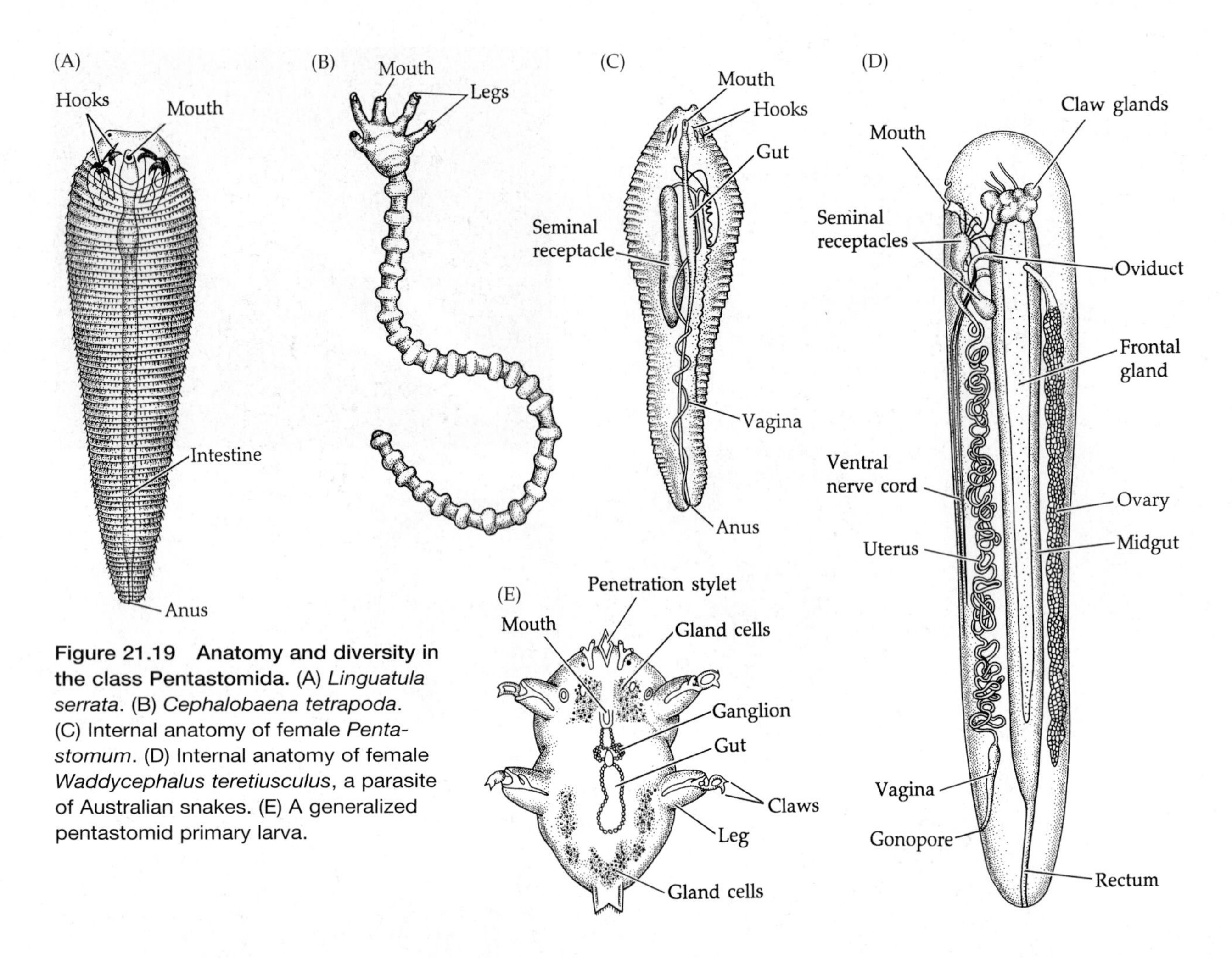

Figure 21.19 Anatomy and diversity in the class Pentastomida. (A) *Linguatula serrata*. (B) *Cephalobaena tetrapoda*. (C) Internal anatomy of female *Pentastomum*. (D) Internal anatomy of female *Waddycephalus teretiusculus*, a parasite of Australian snakes. (E) A generalized pentastomid primary larva.

antennules generally bear hooks or spines for attachment to their host fish. The mandibles are reduced in size and complexity, bear cutting edges, and are housed within a styliform "proboscis" apparatus. The maxillules are clawed in *Dolops*, but they are modified as stalked suckers in the other genera (*Argulus, Chonopeltis, Dipteropeltis*). The uniramous maxillae usually bear attachment hooks. The thoracopods are biramous and used for swimming when the animal is not attached to a host. Branchiurans feed by piercing the skin of their hosts and sucking blood or tissue fluids. Once they locate a host, they crawl toward the fish's head and anchor in a spot where water flow turbulence is low (e.g., behind a fin or gill operculum).

Members of the genus *Argulus* occur worldwide, and can pose a serious problem to aquaculture, but members of the other genera have restricted distributions. *Chonopeltis* is found only in Africa, *Dipteropeltis* in South America, and *Dolops* in South America, Africa, and Tasmania.

Class Pentastomida

Obligatory parasites of various amphibians, reptiles, birds, and mammals. Adults inhabit respiratory tracts (lungs, nasal passages, etc.) of their hosts. Body highly modified, wormlike, 2–13 cm in length. Adult appendages reduced to 2 pairs of head appendages, lobelike and with chitinous claws used to cling to host. Body cuticle nonchitinous and highly porous. Body muscles somewhat sheetlike, but clearly segmental and cross-striated. Mouth lacks jaws; often on end of snoutlike projection; connected to a muscular pumping pharynx used to suck blood from host. The combination of the snout and the 2 pairs of legs gives the appearance of there being five mouths, hence the name (Greek *penta*, "five"; *stomida*, "mouths"). In many species the appendages are reduced to no more than the terminal claws. No specific gas exchange, circulatory, or excretory organs. Gonochoristic; females larger than males. About 130 described species, including two cosmopolitan species that can occasionally infest humans (Figure 21.19).

For years it was believed that pentastomids were allied with the fossil lobopodians, onychophorans, and tardigrades as some kind of segmented, vermiform, proto-arthropod creature—and some workers still entertain this idea. However, independent molecular studies suggest the pentastomids are highly modified crustaceans, perhaps derived from the Branchiura. Corroboration has come from cladistic analyses of sperm and larval morphology, nervous system anatomy, and cuticular fine structure.

Work on the Swedish Orsten fauna indicates that pentastomid-like animals had appeared as early as the late Cambrian (500 Ma), long before the land vertebrates had evolved. What might the original hosts of these parasites have been? Conodont fossils are common in all the Cambrian localities that have yielded pentastomids, raising the possibility that conodonts (also long a mystery, but now widely regarded as parts of early fishlike vertebrates) may have been at least one of the original hosts of these early Pentastomida.

Class Mystacocarida

Body divided into cephalon and 10-segmented trunk; telson with clawlike caudal rami; cephalon characteristically cleft; all cephalic appendages nearly identical, antennae and mandibles biramous; antennules, maxillules, and maxillae uniramous; first trunk segment bears maxillipeds but is not fused with cephalon; no carapace; gonopores on fourth trunk segment; trunk segments 2–5 with short, single-segment appendages (Figure 21.18A).

There are only 13 described species of mystacocarids, eight in the genus *Derocheilocaris* and five in *Ctenocheilocaris*. Most are less than 0.5 mm long, although *D. ingens* reaches 1 mm. The head is marked by a transverse "cephalic constriction" between the origins of the first and second antennae, perhaps a remnant of primitive head segmentation. In addition, the lack of fusion of the cephalon and maxillipedal trunk segment, the simplicity of the mouth appendages, and other features have led some workers to propose that the mystacocarids are among the most primitive living crustaceans. These attributes may, however, simply be related to a neotenic origin and specialization for interstitial habitats.

Mystacocarids are marine, interstitial crustaceans that live in littoral and sublittoral sands throughout the world's temperate and subtropical seas. Their rather vermiform body and small size are clearly adaptations to life among sand grains. Mystacocarids are thought to feed by scraping organic material from the surfaces of sand grains with their setose mouthparts.

Class Copepoda

Without a carapace, but with a well developed cephalic shield; single, median, simple maxillopodan eye (sometimes lacking); one or more thoracomeres fused to head; thorax of six segments, the first always fused to the head and with maxillipeds; abdomen of five segments, including anal somite (= telson); well developed caudal rami; abdomen without appendages, except an occasional reduced pair on the first segment, associated with the gonopores; point of main body flexure varies among major groups; antennules uniramous, antennae uniramous or biramous; 4–5 pairs of natatory thoracopods, most locked together for swimming; posterior thoracopods always biramous (Figures 21.1U, 21.18B–K, 21.27A, 21.30D, and 21.33D).

There are more than 12,500 described species of copepods. They can be incredibly abundant in the world's seas, and also in some lakes—by one estimate, they outnumber all other multicellular forms of life on Earth. Most are small, 0.5–10 mm long, but some free-living forms exceed 1.5 cm in length, and certain highly modified parasites may reach 25 cm. The bodies of most copepods are distinctly divided into three tagmata, the names of which vary among authors. The first region includes the five fused head segments and one or two additional fused thoracic somites; it is called a cephalosome (= cephalothorax) and bears the usual head appendages and maxillipeds. All of the other limbs arise on the remaining thoracic segments, which together constitute the metasome. The abdomen, or urosome, bears no limbs. The appendage-bearing regions of the body (cephalosome and metasome) are frequently collectively called the prosome. The majority of the free-living copepods, and those most frequently encountered, belong to the orders Calanoida, Harpacticoida, and Cyclopoida, although even some of these are parasitic. We focus here on these three groups and then briefly discuss some of the other, smaller orders and their modifications for parasitism. The calanoids are characterized by a point of major body flexure between the metasome and the urosome, marked by a distinct narrowing of the body. They possess greatly elongate antennules. Most of the calanoids are planktonic, and as a group they are extremely important as primary consumers in freshwater and marine food webs. The point of body flexure in the orders Harpacticoida and Cyclopoida is between the last two (fifth and sixth) metasomal segments. (Note: Some authors define the urosome in harpacticoids and cyclopoids as that region of the body posterior to this point of flexure.) Harpacticoids are generally rather vermiform, with the posterior segments not much narrower than the anterior; cyclopoids generally narrow abruptly at the major body flexure. Both the antennules and the antennae are quite short in harpacticoids, but the latter are moderately long in cyclopoids (although never as long as the antennules of calanoids). The antennae are uniramous in cyclopoids but biramous in the other two groups. Most harpacticoids are benthic, and those that have adapted to a planktonic lifestyle show modified body shapes. Harpacticoids occur in all aquatic environments; encystment is known to occur in at least a few freshwater and marine species. Cyclopoids are known from fresh and salt water, and most are planktonic.

The nonparasitic copepods move by crawling or swimming, using some or all of the thoracic limbs. Many of the planktonic forms have very setose appendages, offering a high resistance to sinking. Calanoids are predominantly planktonic feeders. Benthic harpacticoids are often reported as detritus feeders, but many feed predominantly on microorganisms living on the surface of detritus or sediment particles (e.g., diatoms, bacteria, and protists).

Of the seven remaining orders, the Mormonilloida are planktonic; the Misophrioida are known from deep-sea epibenthic habitats as well as anchialine caves in both the Pacific and Atlantic; and the Monstrilloida are planktonic as adults, but the larval stages are endoparasites of certain gastropods, polychaetes, and occasionally echinoderms. Members of the orders Poecilostomatoida and Siphonostomatoida are exclusively parasitic and often have modified bodies. Siphonostomatoids are endo- or ectoparasites of various invertebrates as well as marine and freshwater fishes; they are often very tiny and show a reduction or loss of body segmentation. Poecilostomatoids parasitize invertebrates and marine fishes, and may also show a reduced number of body segments. The Platycopioida are benthic forms known primarily from marine caves; the Gelyelloida are known only from European groundwaters.

Class Ostracoda

Body segmentation reduced, trunk not clearly divided into thorax and abdomen, with 6 to 8 pairs of limbs (including the male copulatory limb); trunk with 1 to 3 pairs of limbs, variable in structure; caudal rami present; gonopores on lobe anterior to caudal rami; carapace bivalved, hinged dorsally and closed by a central adductor muscle, enclosing body and head; carapace highly variable in shape and ornamentation, smooth or with various pits, ridges, spines, etc.; most with one simple median naupliar eye (often called a "maxillopodan eye") and sometimes weakly stalked compound eyes (in Myodocopida); adults with maxillary and (in some) antennal glands; males with distinct copulatory limbs; caudal rami (furca) present (Figure 21.20).

Ostracods include about 30,000 described living species of small bivalved crustaceans, ranging in length from 0.1 to 2.0 mm, although some giants (e.g., *Gigantocypris*) reach 32 mm. They superficially resemble clam shrimps in having the entire body enclosed within the valves of the carapace. However, ostracod valves lack the concentric growth rings of clam shrimps, and there are major differences in the appendages. The shell is usually penetrated by pores, some bearing setae, and is shed with each molt. A good deal of confusion exists about the nature of ostracod limbs, and homologies with other crustacean taxa (and even within the Ostracoda) are unclear—this confusion is reflected in the variety of names applied by different authors. We have adopted terms here that allow the easiest comparison with other taxa.

Ostracods possess the fewest limbs of any crustacean class. The four or five head appendages are followed by 1–3 trunk appendages. Superficially, the (second) maxillae appear to be absent; however, the highly modified fifth limbs are in fact these appendages. The trunk seldom shows external evidence of segmentation, although all 11 postcephalic somites are discernable in some taxa. The trunk limbs vary in structure among taxa and on individuals. The third pair of trunk limbs bears the gonopores and constitutes the so-called copulatory organ.

Ostracods are one of the most successful groups of crustaceans. They also have the best fossil record of any arthropod group, dating from at least the Ordovician, and an estimated 65,000 fossil species have been described. Most are benthic crawlers or burrowers, but many have adopted a suspension-feeding planktonic lifestyle, and a few are terrestrial in moist habitats. One species is known to be parasitic on fish gills—*Sheina orri* (Myodocopida, Cypridinidae). Ostracods are abundant worldwide in all aquatic environments and are known to depths of 7,000 m in the sea. Some are commensal on echinoderms or other crustaceans. A few podocopans have invaded supralittoral sandy regions (members of the family Terrestricytheridae), and members of several families inhabit terrestrial mosses and humus. Two principal taxa (ranked as subclasses here) are recognized within the Ostracoda: Myodocopa and Podocopa.

Myodocopans are all marine. Most are benthic, but the group also includes all of the marine planktonic ostracods. The largest of all ostracods, the planktonic *Gigantocypris*, is a member of this group. Myodocopans include scavengers, detritus feeders, suspension feeders, and some predators. There are two orders: Myodocopida and Halocyprida.

Podocopans include predominantly benthic forms; although some are capable of temporary swimming, none are fully planktonic. Their feeding methods include suspension feeding, herbivory, and detritus feeding. The Podocopa are divided into three orders: the exclusively marine Platycopida, the ubiquitous Podocopida, and the Palaeocopida. The Palaeocopida were diverse and widespread in the Paleozoic, but are represented today only by the extremely rare Punciidae (known from a few living specimens, and from dead valves dredged in the South Pacific).

The Crustacean Body Plan

We realize that the above synopses are rather extensive, but the diversity of crustaceans demands emphasis before we attempt to generalize about their biology. The evolutionary success of crustaceans, like that of other arthropods, has been closely tied to modifications of the jointed exoskeleton and appendages, the latter having an extensive range of modifications for a great variety of functions.

The most basic crustacean body plan is a **head** (**cephalon**) followed by a long body (**trunk**) with many

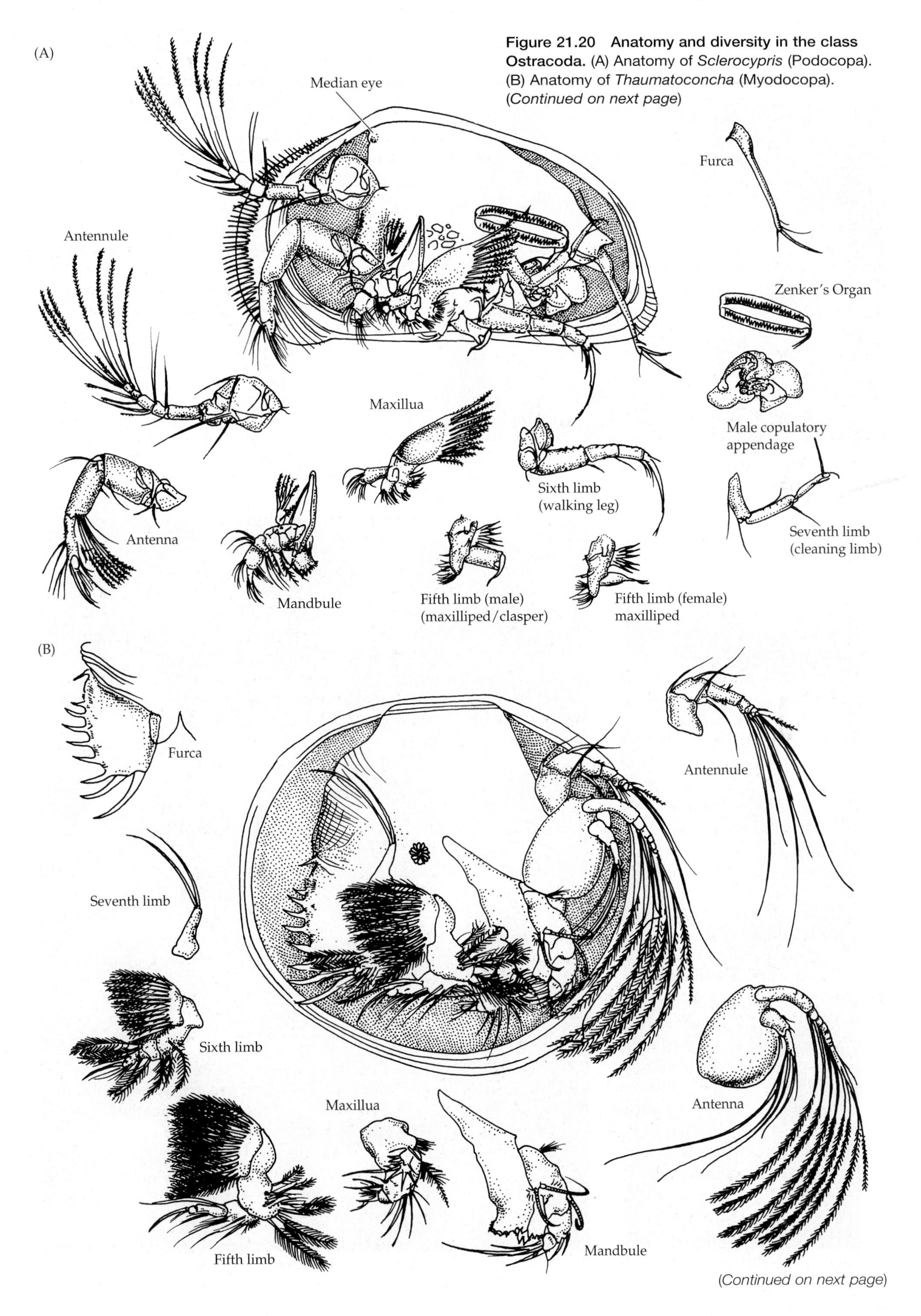

Figure 21.20 Anatomy and diversity in the class Ostracoda. (A) Anatomy of *Sclerocypris* (Podocopa). (B) Anatomy of *Thaumatoconcha* (Myodocopa). (*Continued on next page*)

(*Continued on next page*)

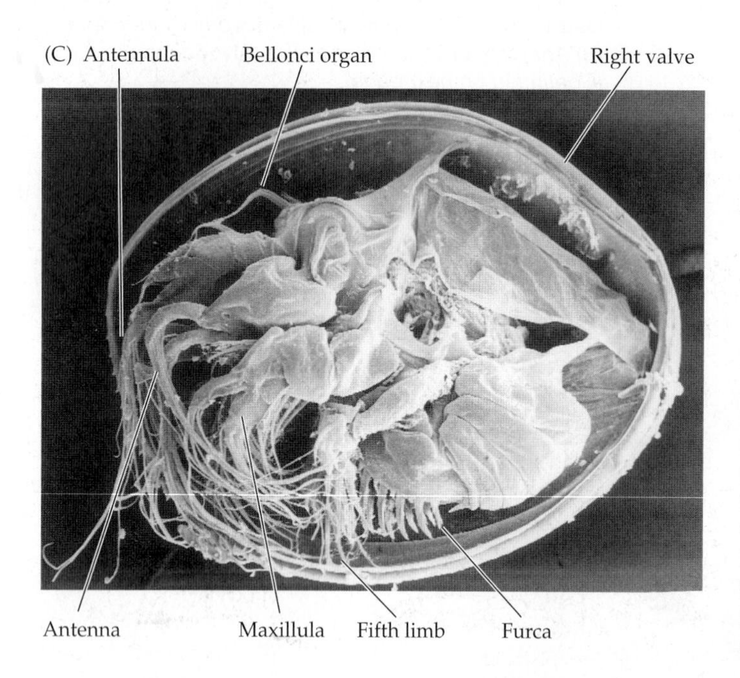

Figure 21.20 (continued) Anatomy and diversity in the class Ostracoda. (C) Internal view of *Metapolycope* (Myodocopa), left valve removed. (D) The highly ornate *Eusarsiella* (Myodocopa); side view and edge view, showing the ornate shell. (E) Examples of genera from the major ostracod groups (scale bar = 1.0 mm). A: *Vargula* (Myodocopa, Myodocopida). B: *Eusarsiella* (Myodocopa, Myodocopida). C: *Polycope* (Myodocopa, Halocyprida). D: *Cytherelloidea* (Podocopa, Platycopida). E: *Propontocypris* (Podocopa, Podocopida). F: *Macrocypris* (Podocopa, Podocopida). G: *Saipanetta* (Podocopa, Podocopida). H: *Neonesidea* (Podocopa, Podocopida). I: *Triebelina* (Podocopa, Podocopida). J: *Candona* (Podocopa, Podocopida). K: *Ilyocypris* (Podocopa, Podocopida). L: *Cyprinotus* (Podocopa, Podocopida). M: *Potamocypris* (Podocopa, Podocopida). N: *Hemicytherura* (Podocopa, Podocopida). O: *Acanthocythereis* (Podocopa, Podocopida). P: *Celtia* (Podocopa, Podocopida). Q: *Limnocythere* (Podocopa, Podocopida). R: *Sahnicythere* (Podocopa, Podocopida). S: *Darwinula* (Podocopa, Podocopida). All arrows point anteriorly.

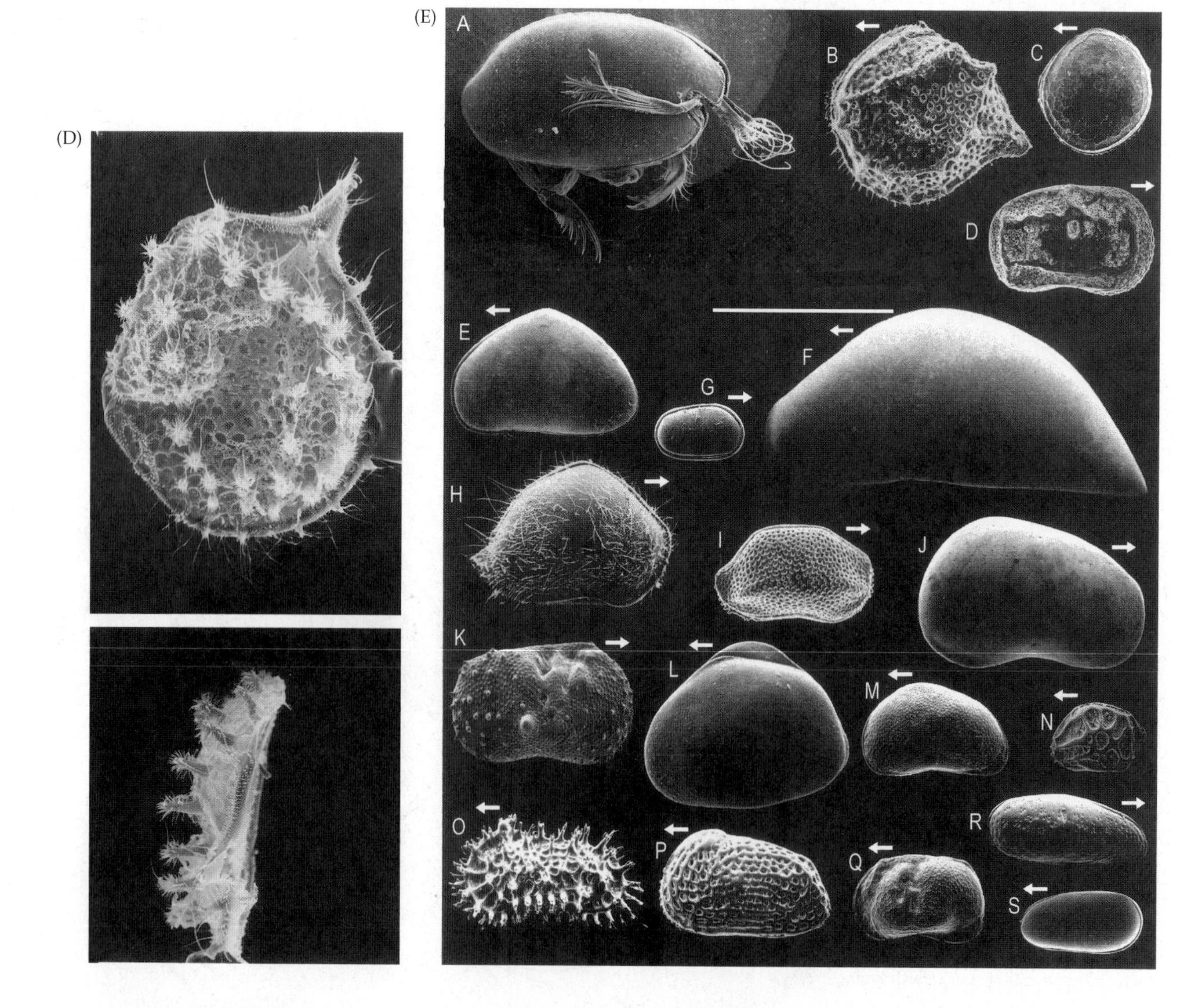

similar appendages, as seen in the class Remipedia (Figures 21.1A and 21.3D,E). In the other crustacean classes, however, various degrees of tagmosis occur, and the cephalon is typically followed by a trunk that is divided into two distinct regions, a **thorax** and an **abdomen**. All crustaceans possess, at least primitively, a **cephalic shield** (**head shield**) or a **carapace**. The cephalic shield results from the fusion of the dorsal head tergites to form a solid cuticular plate, often with ventrolateral folds (**pleural folds**) on the sides. Head shields are found in ancient Cambrian fossil crustaceans (e.g., from the Orsten fauna), and they are characteristic of the classes Remipedia and Cephalocarida; they also occur in some former maxillopodan groups and in malacostracans. The carapace is a more expansive structure, composed of the head shield and a large fold of the exoskeleton that probably arises (primitively) from the maxillary somite. The carapace may extend over the body dorsally and laterally as well as posteriorly, and it often fuses to one or more thoracic segments, thereby producing a **cephalothorax** (Figure 21.2A). Occasionally, the carapace may grow forward beyond the head as a narrow rostrum.

Most of the differences among the major groups of crustaceans, and the basis for much of their classification, arise from variations in the number of somites in the thorax and abdomen, the form of their appendages, and the size and shape of the carapace. A brief skimming of the synopses (above) and the corresponding figures will give you some idea of the range of variation in these characteristics.

Uniformity within the subphylum Crustacea is demonstrated particularly by the consistency of elements of the cephalon and the presence of a nauplius larva. Except for a few cases of secondary reduction, the head of all crustaceans has 5 pairs of appendages. From anterior to posterior, these are the **antennules** (first antennae), **antennae** (second antennae), **mandibles**, **maxillules** (first maxillae), and **maxillae** (second maxillae). The presence of 2 pairs of antennae is, among arthropods, unique to the Crustacea (as is the nauplius larva, although similar "head larvae" are known from other arthropod groups in the fossil record).[7] Although the eyes of some crustaceans are simple, most possess a pair of well developed compound eyes, either set directly on the head (sessile eyes) or borne on distinct movable stalks (stalked eyes).

In many crustaceans, from one to three anterior thoracic segments (**thoracomeres**) are fused with the cephalon. The appendages of these fused segments are typically incorporated into the head as additional mouthparts called **maxillipeds**. In the class Malacostraca, the remaining free thoracomeres are together termed the **pereon**. Each segment of the pereon is called a **pereonite** (= pereomere), and their appendages are called **pereopods**. The pereopods may be specialized for walking, swimming, gas exchange, feeding, and/or defense. Crustacean thoracic (and pleonal) appendages might be primitively biramous, although the uniramous condition is seen in a variety of taxa. The general crustacean limb is composed of a basal **protopod** (= sympod), from which may arise medial **endites** (e.g., gnathobases), lateral **exites** (e.g., **epipods**), and two rami, the endopod and exopod. Members of the classes Remipedia, Cephalocarida, Branchiopoda, and some ostracods possess appendages with uniarticulate (single-segment) protopods; the remaining classes usually have appendages with multiarticulate protopods (Table 21.1).[8]

The abdomen, called a **pleon** in malacostracans, is composed of several segments, or **pleonites** (= pleomeres), followed by a postsegmental plate or lobe, the **anal somite** or **telson**, bearing the anus (Figure 21.2B). In primitive crustaceans this anal somite bears a pair of appendage-like or spinelike processes conventionally called **caudal rami**. In the Eumalacostraca, the anal somite lacks caudal rami and is followed by a dorsal flattened cuticular flap; this flap is sometimes referred to as the telson.

In general, distinctive abdominal appendages (**pleopods**) occur only in the malacostracans. These appendages are almost always biramous, and often they are flaplike and used for swimming (e.g., Figures 21.9–21.15). The posteriorly directed last pair(s) of abdominal appendages is usually different from the other pleopods, and are called **uropods**. Together with the telson, the uropods form a distinct **tail fan** in many malacostracans (Figure 21.2B).

Crustaceans produce a characteristic, and unique, larval stage called the **nauplius** (Figures 21.25B,C and 21.33D), which bears a median simple (naupliar) eye and 3 pairs of setose, functional head appendages—destined to become the antennules, antennae, and mandibles. Behind the head segments is a growth-zone and the telson. In many groups (e.g., Peracarida, and most of Decapoda), however, the free-living nauplius larva is absent or suppressed. In such cases, development is either fully direct or mixed, with larval hatching taking place at some postnaupliar stage (Table 21.2). Often other larval stages follow the nauplius (or other hatching stage) as the individual passes through a series of molts, during which segments and appendages are gradually added. A recent compilation of all crustacean larval forms (Martin et al. 2014) includes

[7]Cambrian crustacean-like larvae in the fossil record termed "head larvae" may be precursors to the distinctive nauplius stage seen in many modern crustaceans. See Martin et al. (2014)

[8]The term "peduncle" is a general name often applied to the basal portion of certain appendages; it is occasionally (but not always) used in a way that is synonymous with protopod. As noted in Chapter 20, the exopod might be no more than a highly modified exite that evolved from an ancestral uniramous condition.

TABLE 21.1 Comparison of Distinguishing Features among the 11 Crustacean Classes (and in Orders of the Class Branchiopoda and Subclasses of the Class Eumalacostraca)

Taxon	Carapace or cephalic shield	Body tagmata and number of segments in each (excluding telson)	Thoracopods	Maxillipeds
Class Remipedia	Cephalic shield	Cephalon (6) Trunk (up to 32)	Not phyllopodous	1 pair
Class Cephalocarida	Cephalic shield	Cephalon (6) Thorax (8) Abdomen (11)	Phyllopodous	None
Class Branchiopoda, Order Anostraca	Cephalic shield	Cephalon (6) Thorax (usually 11) Abdomen (usually 8)	Phyllopodous	None
Class Branchiopoda, Order Notostraca	Carapace	Cephalon (6), Thorax (11), Abdomen (many segments)	Phyllopodous	None
Class Branchiopoda, Order Diplostraca	Carapace (bivalved, hinged or folded)	Cephalon 6, Trunk 10–32, or obscured)	Phyllopodous	None
Class Malacostraca, Subclass Phyllocarida	Large, folded carapace covers thorax	Cephalon (6), Thorax (8), Abdomen (7)	Phyllopodous	None
Class Malacostraca, Subclass Hoplocarida	Well developed carapace covers thorax	Cephalon (6), Thorax (8), Abdomen (6)	Not phyllopodous	Five paris of thoracopods referred to as maxillipeds
Class Malacostraca, Subclass Eumalacostraca	Carapace well developed or secondarily reduced or lost	Cephalon (6), Thorax (8), Abdomen (6)	Not phyllopodous; uniramous in many	0 to 3 pairs
Class Thecostraca	Carapace bivalved (at some stage), often modified as a mantle	Cephalon (6), Thorax (6), Abdomen (4)	Not phyllopodous, often reduced	None
Class Tantulocarida	Cephalic shield	Cephalon (6), Thorax (6), Abdomen (up to 7)	Not phyllopodous, greatly reduced	None
Class Branchiura	Carapace broad, covering head and trunk	Cephalon (6), Thorax (6), Abdomen (4?)	Not phyllopodous (but all natatory)	None
Class Pentastomida	None	Not distinguishable; body vermiform	None	None
Class Mystacocarida	Cephalic shield	Cephalon (6), Trunk (10)	Not phyllopodous	1 pair
Class Copepoda	Cephalic shield	Cephalon (6), Thorax (6), Abdomen (5)	Not phyllopodous; natatory, often reduced	Usually 1 pair
Class Ostracoda	Carapace, bivalved	Subdivisions not clear; 6–8 pairs of limbs	Not phyllopodous, reduced	None

keys to the distinctive naupliar larvae in all groups that hatch as a nauplius as well as synopses of larval development in all crustaceans.

Locomotion

Crustaceans move about primarily by use of their limbs (Figure 21.21), and lateral body undulations are unknown. They crawl or swim, or more rarely burrow, "hitchhike," or jump. Many of the ectoparasitic forms (e.g., branchiurans, certain isopods, and copepods) are largely sedentary on their hosts, and most cirripedes are fully sessile.

Swimming is usually accomplished by a rowing action of the limbs. Archetypical swimming is exemplified by crustaceans with relatively undifferentiated trunks and high numbers of similar biramous appendages (e.g., remipedes, anostracans, notostracans). In general, these animals swim by posterior to anterior metachronal beating of the trunk limbs (Figure 21.22 and Chapter 20). The appendages of such crustaceans are often broad and flattened, and they usually bear fringes of setae that increase the effectiveness of the power stroke. On the recovery stroke the limbs are flexed, and the setae may collapse, reducing resistance.

Antennules	Antennae	Compound eyes	Abdominal appendages	Gonopore location
Biramous	Biramous	Absent	All trunk appendages similar	♂: protopods of trunk segment 15; ♀: trunk segment 8
Uniramous	Biramous	Absent	None	Common pores on protopods of thoracopods 6
Uniramous	Uniramous	Present	None	♀: on segment 12/13 or 20/21
Uniramous	Vestigial	Present	Present (posteriorly reduced)	Thoracomere 11 (both sexes)
Uniramous	Biramous	Present	All trunk appendages similar, or posterior segments limbless	Variable, on trunk segment 9 or 11, or on apodous posterior region (some Cladocera)
Biramous	Uniramous	Present	Pleopods (posteriorly reduced)	♂: coxae of thoracopods 8; ♀: coxae of thoracopods 6
Triramous	Biramous	Present, well developed	Well developed pleopods with gills; 1 pair of uropods	♂: coxae of thoracopods 8; ♀: coxae of thoracopods 6
Uniramous or biramous	Uniramous or biramous	Present, well developed	Usually 5 paris of pleopods, 1 pair of uropods	♂: coxae of thoracopods 8 or sternum of thoracomere 8; ♀: coxae of thoracopods 6 or sternum of thoracomere 6
Uniramous	Biramous	Absent (in adults)	None	Variable; ♂ openings usually on trunk segment 4 or 7; ♀ on 1, 4, or 7
None (paired in one stage)	None	Absent	None	♀ openings on trunk segment 5
Uniramous, reduced	Biramous, reduced	Present	None	♂ single opening on thoracomere 6; ♀ gonopores at base of thoracopod 4
None	None	Absent	None	Common gonopore, location variable
Uniramous	Biramous	Absent	None	♂ and ♀ openings on trunk segment 4
Uniramous	Uniramous or biramous	Absent	None	Gonopores usually on urosomal segment 1
Uniramous	Biramous	Absent (typically)	None	Gonopores on lobe anterior to caudal rami

In members of some groups (e.g., Cephalocarida, Branchiopoda, Leptostraca), large exites or epipods arise from the base of the leg, producing broad, "leafy" limbs called **phyllopodia**. These flaplike structures aid in locomotion and may also serve as osmoregulatory (branchiopods) or gas exchange (cephalocardis and leptostracans) surfaces (Figure 21.21A–C). Although such epipods increase the surface area on the power stroke, they also are hinged so that they collapse on the recovery stroke, reducing resistance. Metachronal limb movements are retained in many of the "higher" swimming crustaceans, but they tend to be restricted to selected appendages (e.g., the pleopods of shrimps, stomatopods, amphipods, and isopods; the pereopods of euphausids and mysids). In swimming euphausids and mysids the thoracopods beat in a metachronal rowing fashion, with the exopod and setal fan extended on the power stroke and flexed on the recovery stroke. The movements and nervous–muscular coordination of crustacean limbs are deceivingly complex. In the common mysid *Gnathophausia ingens*, for example, twelve separate muscles power the thoracic exopod alone (three that are extrinsic to the exopod, five in the limb peduncle, and four in the exopodal flagellum).

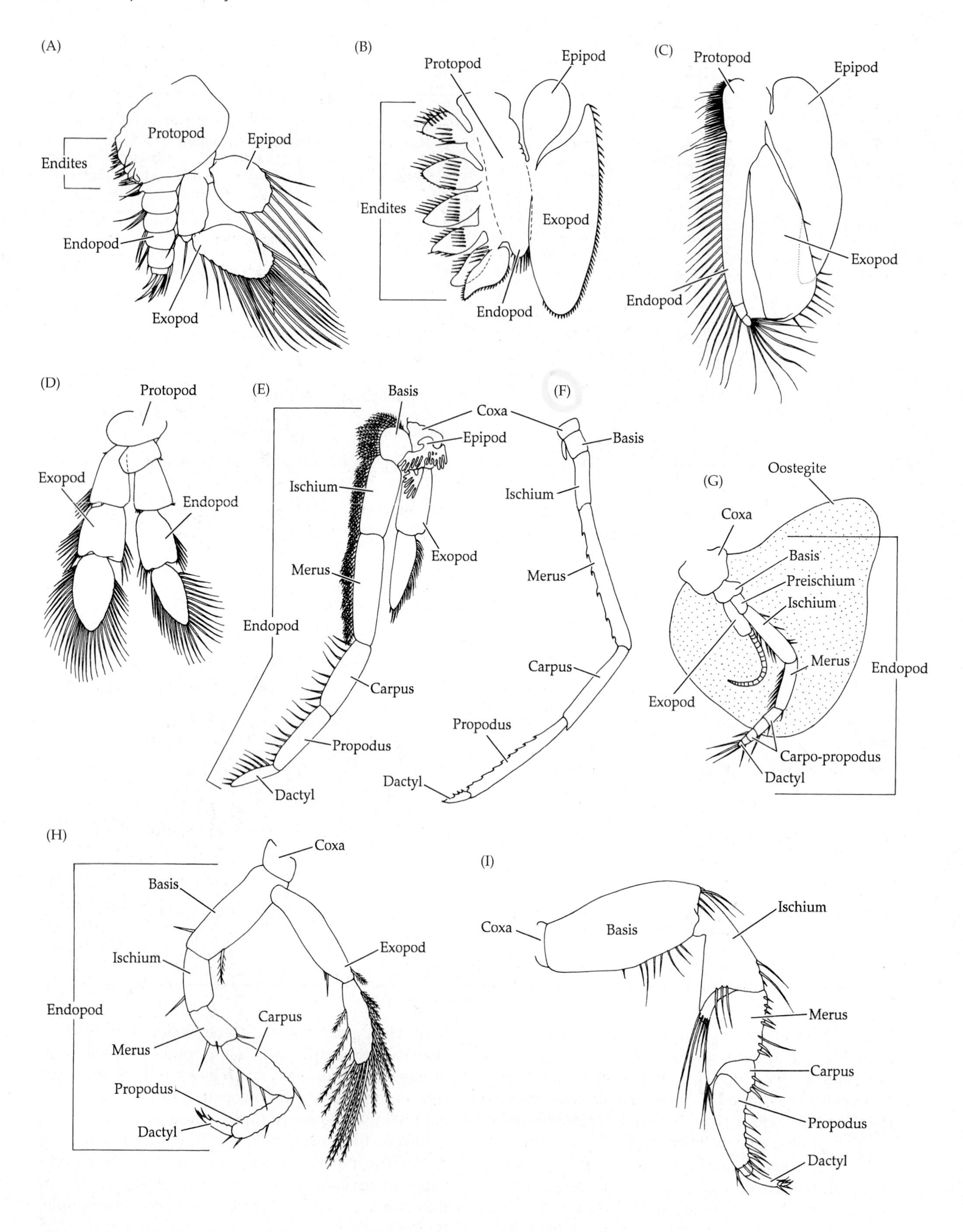
(A)
Endites
Protopod
Epipod
Endopod
Exopod
(B)
Protopod
Epipod
Endites
Exopod
Endopod
(C)
Protopod
Epipod
Exopod
Endopod
(D)
Protopod
Exopod
Endopod
(E)
Basis
Coxa
Epipod
Ischium
Exopod
Merus
Endopod
Carpus
Propodus
Dactyl
(F)
Basis
Ischium
Merus
Carpus
Propodus
Dactyl
(G)
Oostegite
Coxa
Basis
Preischium
Ischium
Merus
Endopod
Exopod
Carpo-propodus
Dactyl
(H)
Coxa
Basis
Ischium
Exopod
Endopod
Carpus
Merus
Propodus
Dactyl
(I)
Coxa
Basis
Ischium
Merus
Carpus
Propodus
Dactyl

◀ **Figure 21.21 Generalized thoracic appendages of various crustaceans.** (A–C) Biramous, phyllopodous thoracopods. (A) Cephalocarida. (B) Branchiopoda. Dashed lines indicate fold or "hinge" lines. (C) Leptostraca (Phyllocarida). (D) A biramous, flattened, but nonphyllopodous thoracopod: Remipedia. (E–I) Stenopodous thoracopods. (E) Euphausiacea. (F) Caridea (Decapoda). (G) Lophogastrida (Peracarida). (H) Spelaeogriphacea (Peracarida). (I) Isopoda (Peracarida). Because of the presence of large epipods on the legs of the cephalocarids, branchiopods, and phyllocarids, some authors refer to them as "triramous" appendages. However, smaller epipods also occur on many typical "biramous" legs, so this distinction seems unwarranted (and confusing). Note that in the four primitive groups of crustaceans (cephalocarids, branchiopods, phyllocarids, and remipedes) the protopod is composed of a single article. And in branchiopods and leptostracans, the articles of the endopod are not clearly separated from one another. In the higher crustaceans (Malacostraca and former "maxillopodans") the protopod comprises two or three separate articles, although in most former maxillopodans these may be reduced and not easily observed. In the lophogastrid (G), the large marsupial oostegite characteristic of female peracarids is shown arising from the coxa. In two groups (amphipods and isopods) all traces of the exopods have disappeared, and only the endopod remains as a long, powerful, uniramous walking leg.

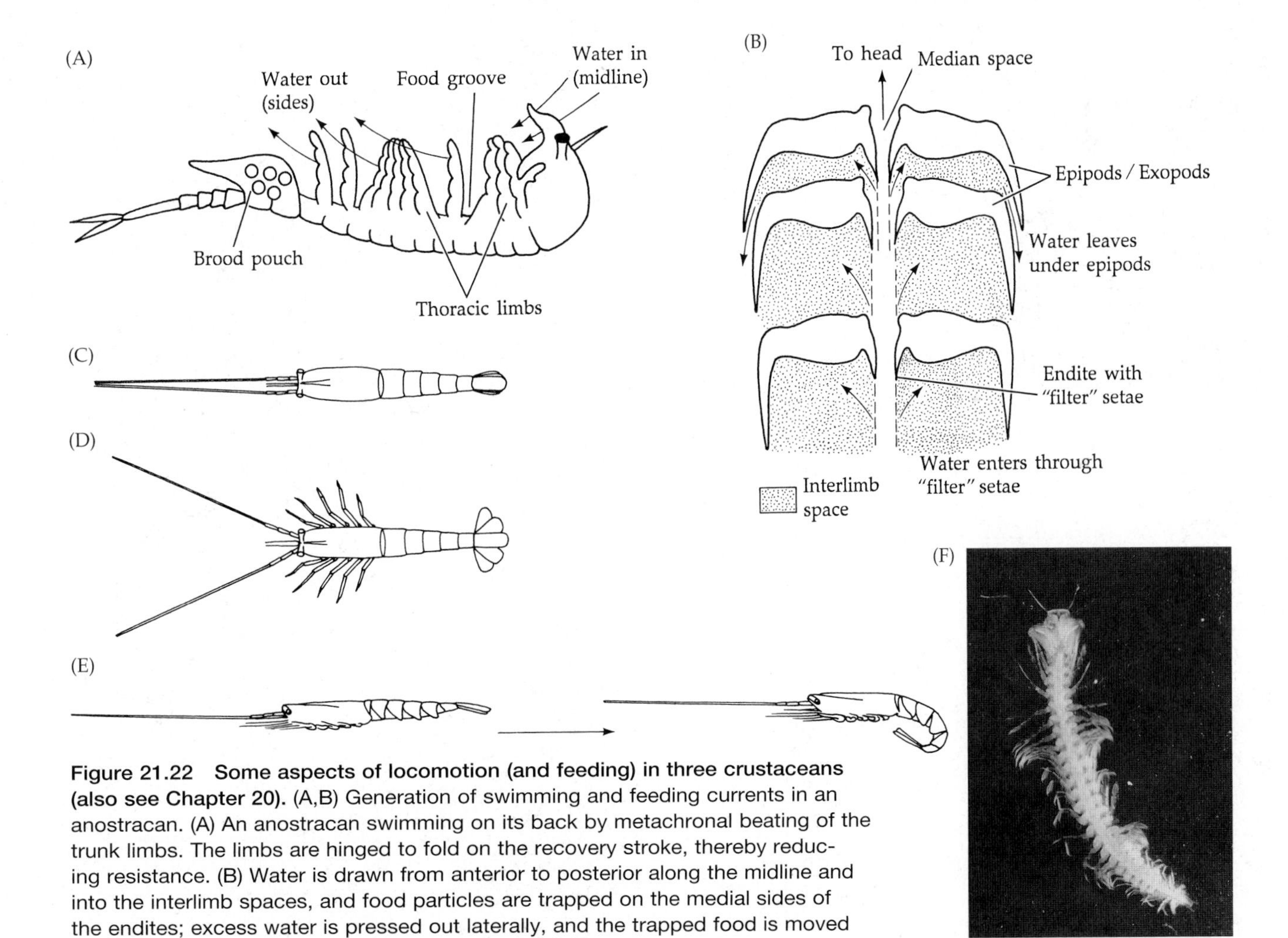

Figure 21.22 Some aspects of locomotion (and feeding) in three crustaceans (also see Chapter 20). (A,B) Generation of swimming and feeding currents in an anostracan. (A) An anostracan swimming on its back by metachronal beating of the trunk limbs. The limbs are hinged to fold on the recovery stroke, thereby reducing resistance. (B) Water is drawn from anterior to posterior along the midline and into the interlimb spaces, and food particles are trapped on the medial sides of the endites; excess water is pressed out laterally, and the trapped food is moved anteriorly to the mouth. (C–E) Locomotion in the postlarva of *Panulirus argus*. (C) Normal swimming posture when moving forward slowly. (D) Sinking posture with appendages flared to reduce sinking rate. (E) A quick retreat by rapid tail flexure (the "caridoid escape reaction"), a method commonly employed by crustaceans with well developed abdomens and tail fans. (F) A swimming remipede, *Lasionectes*. Note the metachronal waves of appendage movement.

TABLE 21.2 Summary of Crustacean Reproductive Features[a]

Taxon	Development type, or larval type at hatching	Hermaphroditic (at least some species)	Gonochoristic	Parthenogenesis (in at least some species)
Class Remipedia	Nauplius (anamorphic)	Yes	No	No
Class Cephalocarida	Metanauplius (anamorphic)	Yes	No	No
Class Branchiopoda, order Anostraca	Nauplius or metanauplius (anamorphic)	No	Yes	Yes
Class Branchiopoda, order Notostraca	Nauplius or metanauplius (anamorphic), or direct development	Yes	Yes	Yes
Class Branchiopoda, order Diplostraca	Nauplius (partly anamorphic)	No	Yes	Yes
Class Ostracoda	Direct, or with bivalved nauplius/ metanauplius with anamorphic development	No	Yes	Yes
Class Mystacocarida	Metanauplius (partly anamorphic)	No	Yes	No
Class Copepoda	Nauplius (metamorphic)	No	Yes	No
Class Branchiura	Metanauplius-like (partly anamorphic), or with direct development	No	Yes	No
Class Thecostraca	?	Yes	Yes	No
Class Tantulocarida	Nauplius? + tantulus (metamorphic)	No	Yes	?
Class Malacostraca, subclass Phyllocarida	Direct development	No	Yes	No
Class Malacostraca, subclass Hoplocarida	Zoea larva ("antizoea" or "pseudozoea") (metamorphic)	No	Yes	No
Class Malacostraca, subclass Eumalacostraca, superorder Syncarida	Direct development	No	Yes	No
Class Malacostraca, subclass Eumalacostraca, superorder Peracarida	Direct development	No	Yes	No
Class Malacostraca, subclass Eumalacostraca, superorder Eucarida, order Euphausiacea	Nauplius (metamorphic)	No	Yes	No
Class Malacostraca, subclass Eumalacostraca, superorder Eucarida, order Amphionidacea	Amphion (modified zoea) (weakly metamorphic)	No	Yes	No
Class Malacostraca, subclass Eumalacostraca, superorder Eucarida, order Decapoda	Pre-zoea or zoea, and with a nauplius in Dendrobranchiata (metamorphic), or direct development	Rare	Yes	No

[a] Also see Martin et al. 2014, Table 55.1.

Comments
Eight naupliar stages (2 orthonauplii and 6 metanauplii), followed by a "pre-juvenile" stage, have been reported.
One or two eggs at a time are fertilized and carried on genital processes of the first pleonites.
Embryos usually shed from ovisac early in development; resistant (cryptobiotic) fertilized eggs accommodate unfavorable conditions.
Eggs briefly brooded, then deposited on substrata; resistant (cryptobiotic) fertilized eggs accommodate unfavorable conditions.
Most cladocerans undergo direct development (*Leptodora* hatches as nauplii or metanauplii). Clam shrimps carry developing embryos on the thoracopods prior to releasing them as nauplii or metanauplii.
Embryos usually deposited directly on substrata; many myodocopans and some podocopans brood embryos between valves until hatching as a reduced adult; no metamorphosis; up to 8 preadult instars.
Little studied, but apparently the eggs are laid free and up to 7 preadult stages may occur.
Usually with 6 naupliar stages leading to a second series of 5 "larval" stages called copepodites.
Embryos are deposited; only *Argulus* is known to hatch as metanauplii, others have direct development and hatch as juveniles.
Six naupliar stages followed by a unique larval form called a cypris larva.
A benthic non-feeding stage, possibly a nauplius, has been reported but not formally described. Development entails complex metamorphosis with infective "tantulus larva."
Undergo direct development in the female brood pouch, emerging as a postlarval/prejuvenile stage called a "manca."
Eggs brooded or deposited in burrow; hatch late as clawed pseudozoea larvae, or earlier as unclawed antizoea larvae. Both molt into distinct erichthus (and some into alima) larvae before settling on bottom as post-larvae or juveniles.
All free larval stages have been lost; eggs deposited on substratum.
Embryos brooded in marsupium of female typically formed from ventral coxal plates called oostegites; usually released as mancas (subjuveniles with incompletely developed 8th thoracopods). Brood pouch (marsupium) in Thermosbaenacea formed by dorsal carapace chamber.
Embryos shed or briefly brooded; typically undergo nauplius-metanauplius-calyptopis-furcilia-juvenile developmental series.
Eggs apparently brooded in female brood pouch; hatching likely occurs as eggs float upward after release; hatching probably as an amphion larva (a zoea with only one pair of thoracopods modified as maxillipeds), with perhaps 20 stages that metamorphose gradually to a subadult.
Dendrobranchiata shed embryos to hatch in water as nauplii or pre-zoea; all others brood embryos (on pleopods), which do not hatch until a pre-zoeal or zoeal stage (or later).

Recall from our discussions in Chapters 4 and 20 that at the low Reynolds numbers at which small crustaceans (such as copepods or larvae) swim about, the netlike setal appendages act not as a filtering net, but as a paddle, pushing water in front of them and dragging the surrounding water along with them due to the thick boundary layer adhering to the limb. Only in larger organisms, with Reynolds numbers approaching 1, do setose appendages (e.g., the feeding cirri of barnacles) begin to act as filters, or rakes, as the surrounding water acts less viscous and the boundary layer thinner. Of course, the closer together the setae and setules are placed, the more likely it is that their individual boundary layers will overlap; thus densely setose appendages are more likely to act as paddles.

Not all swimming crustaceans move by typical metachronal waves of limb action. Certain planktonic copepods, for example, move haltingly and depend on their long antennules and dense setation for flotation between movements (Figure 21.18F). Watch living calanoid copepods and you will notice that they may move slowly by use of the antennae and other appendages, or in short jerky increments, often sinking slightly between these movements. The latter type of motion results from an extremely rapid and condensed metachronal wave of power strokes along the trunk limbs. Although the long antennae may appear to be acting as paddles, they actually collapse against the body an instant prior to the beating of the limbs, thus reducing resistance to forward motion. Some other planktonic copepods create swimming currents by rapid vibrations of cephalic appendages, by which the body moves smoothly through the water. "Rowing" does occur in the swimming crabs (family Portunidae) and some deep-sea asellote isopods (e.g., family Eurycopidae), both of which use paddle-shaped posterior thoracopods to scull about.

Most eumalacostracans with well-developed abdomens exhibit a form of temporary, or "burst," swimming that serves as an escape reaction (e.g., mysids, syncarids, euphausids, shrimps, lobsters, and crayfish). By rapidly contracting the ventral abdominal (flexor) muscles, such animals shoot quickly backward, the spread tail fan providing a large propulsive surface (Figure 21.22C–E). This behavior is sometimes called a **tail-flip**, or "**caridoid escape reaction**."

Surface crawling by crustaceans is accomplished by the same general sorts of leg movements described in the preceding chapters for insects and other arthropods: by flexion and extension of the limbs to pull or push the animal forward. Walking limbs are typically composed of relatively stout, more or less cylindrical articles (i.e., stenopodous limbs) as opposed to the broader, often phyllopodous limbs of swimmers (see Figure 21.21 for a comparison of crustacean limb types). Walking limbs are lifted from the substratum and moved forward during their recovery strokes; then they are placed against the substratum, which provides purchase as they move posteriorly through their power strokes, pulling and then pushing the animal forward. Like many other arthropods, crustaceans generally lack lateral flexibility at the body joints, so turning is accomplished by reducing the stride length or movement frequency on one side of the body, toward which the animal turns (like a tractor or tank slowing one tread). Many crustaceans migrate; perhaps the most famous is the Chinese mitten crab (*Eriocheir sinensis*), which spends most of its life in fresh water, but returns to the sea to breed. These crabs have been found over 1000 km upriver from the sea—testimony to their superb locomotory ability. Perhaps not unexpectedly, *E. sinensis* is also an important (and destructive) invasive species in North America and Europe. It has been accidently introduced into the Great Lakes several times, but has not yet been able to establish a permanent population.

Most walking crustaceans can also reverse the direction of leg action and move backward, and most brachyuran crabs can walk sideways. Brachyuran crabs are perhaps the most agile of all crustaceans. The extreme reduction of the abdomen in this group allows for very rapid movement because adjacent limbs can move in directions that avoid interference with one another (and much the same thing has happened, independently, in many anomurans with reduced abdomens). Brachyuran crab legs are hinged in such a way that most of their motion involves lateral extension (abduction) and medial flexion (adduction) rather than rotation frontward and backward. As a crab moves, its limbs move in various sequences, as in normal crawling, but those on the leading side exert their force by flexing and pulling the body toward the limb tips, while the opposite, trailing, legs exert propulsive force as they extend and push the body away from the tips. Still, this motion is simply a mechanical variation on the common arthropodan walking behavior. Many crustaceans move into mollusc shells or other objects, carrying these about as added protection. In most cases, the exoskeleton of the crustacean is reduced, especially the abdomen (e.g., hermit crabs). Of course, crustaceans grow in size as they go through their molts, so these mobile-home-carrying crustaceans must continually find larger shells to inhabit. Many hermit crabs assemble in congregations to exchange shells in a sort of group "passing of the shells" as they move into vacated larger abodes.

In addition to these two basic locomotor methods ("typical" walking and swimming by metachronal beating of limbs), many crustaceans move by other specialized means. Ostracods, cladocerans, and clam shrimp (Diplostraca), most of which are largely enclosed by their carapaces (Figures 21.4F,G,J,L,M and 21.20), swim by rowing with the antennae. Mystacocarids crawl in interstitial water using various head appendages. Most semi-terrestrial amphipods known as "beach hoppers" (e.g., *Orchestia* and *Orchestoidea*) execute dramatic jumps by rapidly extending the urosome and its appendages (uropods), reminiscent of the jumping of springtails described in Chapter 22. Most caprellid amphipods (Figure 21.15E) move about in inchworm fashion, using their subchelate appendages for clinging. There are also a number of crustacean burrowers, and even some that build their own tubes or "homes" from materials in their surroundings. Many benthic amphipods, for example, spin silk-lined mud burrows in which they reside. One species, *Pseudamphithoides incurvaria* (suborder Gammaridea), constructs and lives in an unusual "bivalved pod" cut from the thin blades of the same alga on which it feeds (Figure 21.23A). Another gammaridean amphipod, *Photis conchicola*, actually uses empty gastropod shells in a fashion similar to that of hermit crabs (Figure 21.23C). "Hitchhiking" (phoresis) occurs in various ectosymbiotic crustaceans, including isopods that parasitize fishes or shrimps and hyperiidean amphipods that ride on gelatinous drifting plankters.

In addition to simply getting from one place to another in their usual day-to-day activities, many crustaceans exhibit various migratory behaviors, employing their locomotor skills to avoid stressful situations or to remain where conditions are optimal. A number of planktonic crustaceans undertake daily vertical migrations, typically moving upward at night and to greater depths during the day. Such vertical migrators include various copepods, cladocerans, ostracods, and hyperiid amphipods (the latter may make their migrations by riding on their gelatinous hosts). Such movements place the animals in their near-surface feeding grounds during the dark hours, when there is probably less danger of being detected by visual predators. In the daytime, they move to deeper, perhaps safer, water. These crustaceans can form enormous shoals that contribute to the deep scattering layer seen on ship's sonar. Many intertidal crustaceans use their locomotor abilities to change their behaviors with the tides. Crab larvae in particular are known to migrate upward or downward according to daily rhythms, taking advantage of incoming or outgoing tides to move in and out of estuaries. Some anomuran and brachyuran crabs simply move in and out with the tide, or seek shelter beneath rocks when the tide is out, thus avoiding the problems of air exposure. One of the most interesting locomotor behaviors among crustaceans is the mass migration of the spiny lobster, *Panulirus argus*, in the Gulf of Mexico and northern Caribbean. Each autumn, lobsters queue up in single file and march in long lines across the seafloor for several days. They move from

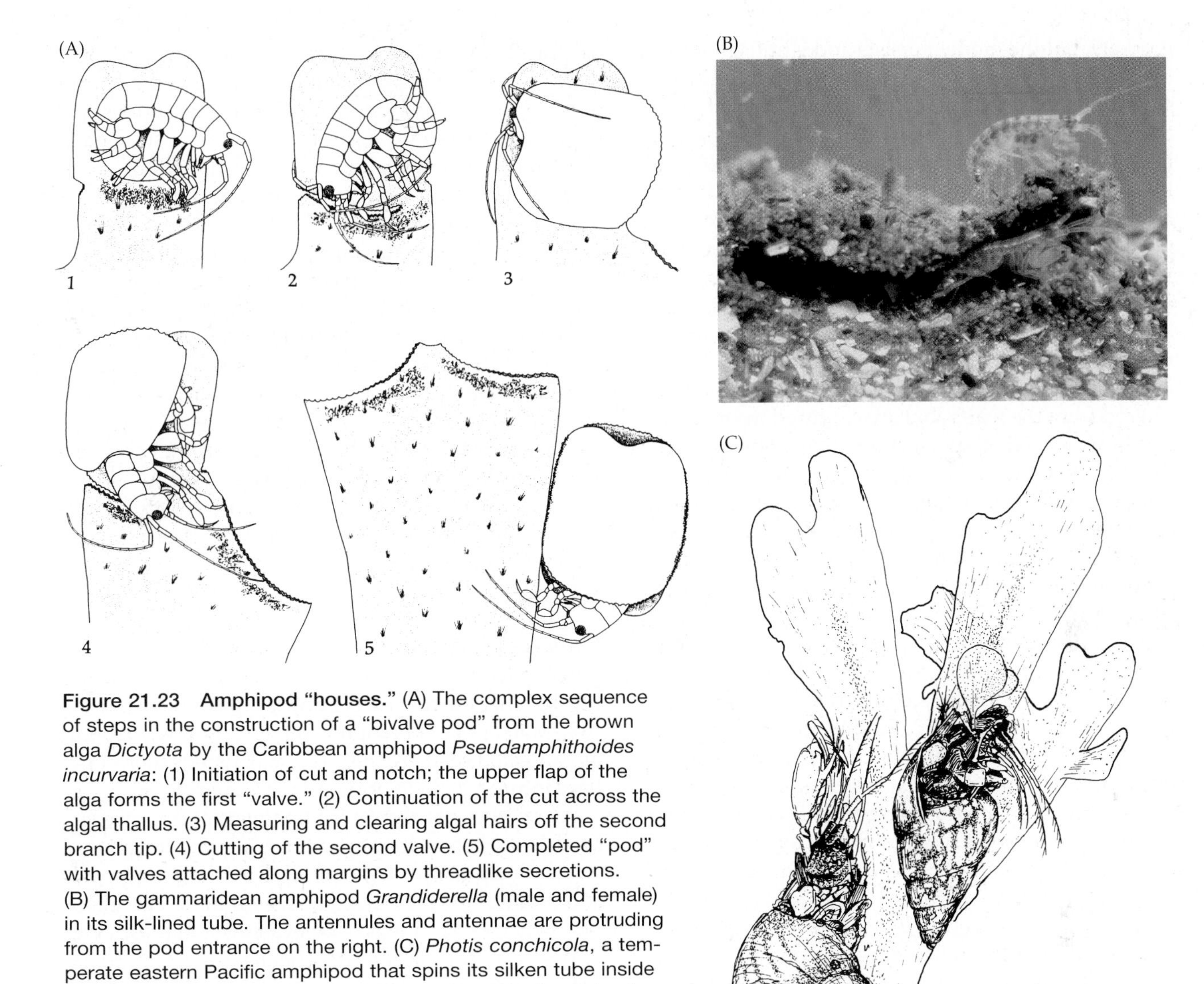

Figure 21.23 Amphipod "houses." (A) The complex sequence of steps in the construction of a "bivalve pod" from the brown alga *Dictyota* by the Caribbean amphipod *Pseudamphithoides incurvaria*: (1) Initiation of cut and notch; the upper flap of the alga forms the first "valve." (2) Continuation of the cut across the algal thallus. (3) Measuring and clearing algal hairs off the second branch tip. (4) Cutting of the second valve. (5) Completed "pod" with valves attached along margins by threadlike secretions. (B) The gammaridean amphipod *Grandiderella* (male and female) in its silk-lined tube. The antennules and antennae are protruding from the pod entrance on the right. (C) *Photis conchicola*, a temperate eastern Pacific amphipod that spins its silken tube inside a minute snail shell, which is then carried about in the style of hermit crabs.

shallow areas to the edges of deeper oceanic channels. This behavior is apparently triggered by winter storm fronts moving into the area, and it may be a means of avoiding rough water conditions in the shallows.

Feeding

With the exception of ciliary mechanisms, crustaceans have exploited virtually every feeding strategy imaginable (and some "unimaginable"). Even without cilia, many crustaceans generate water currents and engage in various types of suspension feeding. We have selected a few examples to demonstrate the range of feeding mechanisms that occur in this group.

In some crustaceans the action of the thoracic limbs simultaneously creates the swimming and suspension feeding currents. As the metachronal wave of appendage motion passes along the body, adjacent limb pairs are alternately moved apart and then pressed together, thus changing the size of each interlimb space (Figure 21.22A,B; see also Chapters 4 and 20). Surrounding water is drawn into an interlimb space as the adjacent appendages move away from one another, and waterborne particles are trapped by setae on the endites as the appendages then close. From here, the trapped particles are moved to a midventral food groove and then anteriorly, toward the head. This mechanism of forming a boxlike "filter press" with setose phyllopodous limbs is the typical suspension feeding strategy of cephalocarids, most branchiopods, and many malacostracans.

Planktonic copepods were long thought to "filter" feed by generating lateral feeding gyres or currents by movements of the antennae and mouth appendages. It was believed that these gyres swept in small particles that were directly filtered by the maxillae. This classic idea of maxillary filtration, built on work by H. G. Cannon in the 1920s, has been questioned by recent workers, but the model persists and is still commonly presented in general works. As mentioned in Chapter 4, we now know that copepods and other small planktonic crustaceans live in a world of low Reynolds numbers, a world dominated by viscosity rather than by inertia. Thus, the setose mouth appendages behave more like paddles than like sieves, with a water layer near the limb adhering to it and forming part of the "paddle." As the maxillae move apart, parcels of water containing food are drawn into the interlimb space. As the maxillae press together, the "parcel" is moved forward to the endites of the maxillules, which push it into the mouth. Thus, food particles are not actually filtered from the water, but are captured in small parcels of water. High-speed cinematography indicates that copepods are capable of capturing individual algal cells, one at a time, by this "hydraulic vacuum" method. In fact, copepods are probably fairly selective about what they consume, which includes everything from protistan microplankton (e.g., diatoms) to other small crustaceans.

Sessile thoracican barnacles feed by using their long, feathery, biramous thoracopods, called **cirri**, to filter feed on suspended material from the surrounding water (Figure 21.16A,C,E). Studies indicate that barnacles are capable of trapping food particles ranging from 2 μm to 1 mm, including detritus, bacteria, algae, and various zooplankters. Many barnacles are also capable of preying on larger planktonic animals by coiling a single cirrus around the prey, in tentacle fashion. In slow-moving or very quiet water, most barnacles feed actively by extending the last three pairs of cirri in a fanlike manner and sweeping them rhythmically through the water. The setae on adjacent limbs and limb rami overlap to form an effective filtering net. The first three pairs of cirri serve to remove trapped food from the posterior cirri and pass it to the mouthparts. In areas of high water movement, such as wave-swept rocky shores, barnacles often extend their cirri into the backwash of waves, allowing the moving water to simply run through the "filter," rather than moving the cirri through the water. In such areas you will often see clusters of barnacles in which all the individuals are oriented similarly, taking advantage of this labor-saving device.

Most krill (euphausids) feed in a fashion similar to barnacles, but while swimming. The thoracopods form a "feeding basket" that expands as the legs move outward, sucking food-laden water in from the front. Once inside the basket, particles are retained on the setae of the legs as the water is squeezed out laterally. Other setae comb the food particles out of the "trap" setae, while yet another set brushes them forward to the mouth region.

Sand crabs of the genus *Emerita* (Anomura) use their long, setose antennae in a fashion similar to that of barnacle cirri that "passively" strain wave backwash (Figure 21.24A; see also Chapter 4). *Emerita* are adapted to living on wave-swept sand beaches. Their compact oval shape and strong appendages facilitate burrowing in the unstable substratum. They burrow posterior end first in the area of shallow wave wash, with the anterior end facing upward. Following a breaking wave, as the water rushes seaward, *Emerita* unfurls its antennae into the moving water along the surface of the sand. The fine setose mesh traps protists and phytoplankton from the water, and the antennae then brush the collected food onto the mouthparts. Many porcelain and hermit crabs also engage in suspension feeding. By twirling their antennae in various patterns these anomurans create spiraling currents that bring food-laden water toward the mouth (Figure 21.24B,C). Food particles then become entangled on setae of the mouth appendages and are brushed into the mouth by the endopods of the third maxillipeds. Many of these animals also feed on detritus by simply picking up particles with their chelipeds.

Some mud, ghost, and lobster shrimps, such as *Callianassa* and *Upogebia* (Axiidea, Gebiidea), suspension feed within their burrows. They drive water through the burrow by beating the pleopods, and the first two pairs of pereopods remove food with medially directed setal brushes. The maxillipeds then comb the captured particles forward to the mouth.

Most other crustacean feeding mechanisms are less complicated than suspension feeding and usually involve direct manipulation of food by the mouthparts and sometimes the pereopods, especially chelate or subchelate anterior legs. However, even in these cases of direct manipulation, the sheer number of crustacean appendages adapted for feeding makes "simple" feeding a surprisingly complex affair, with the various mouthparts taking on tasks that include tasting (via sensory setae), holding, chewing, scraping, and macerating.

Many small crustaceans may be classified as microphagous selective deposit feeders, employing various methods of removing food from the sediments in which they live. Mystacocarids, many harpacticoid copepods, and some cumaceans and gammaridean amphipods are referred to as "sand grazers" or "sand lickers." By various methods these animals remove detritus, diatoms, and other microorganisms from the surfaces of sediment particles. Interstitial mystacocarids, for example, simply brush sand grains with their setose mouthparts. On the other hand, some cumaceans pick up an individual sand grain with their first pereopods and pass it to the maxillipeds, which in turn rotate and tumble the particle against the margins of the maxillules and mandibles. The maxillules brush and the mandibles scrape, removing organic material. Some sand-dwelling isopods may employ a similar feeding behavior.

Predatory crustaceans include stomatopods, remipedes, and most lophogastrids, as well as many species

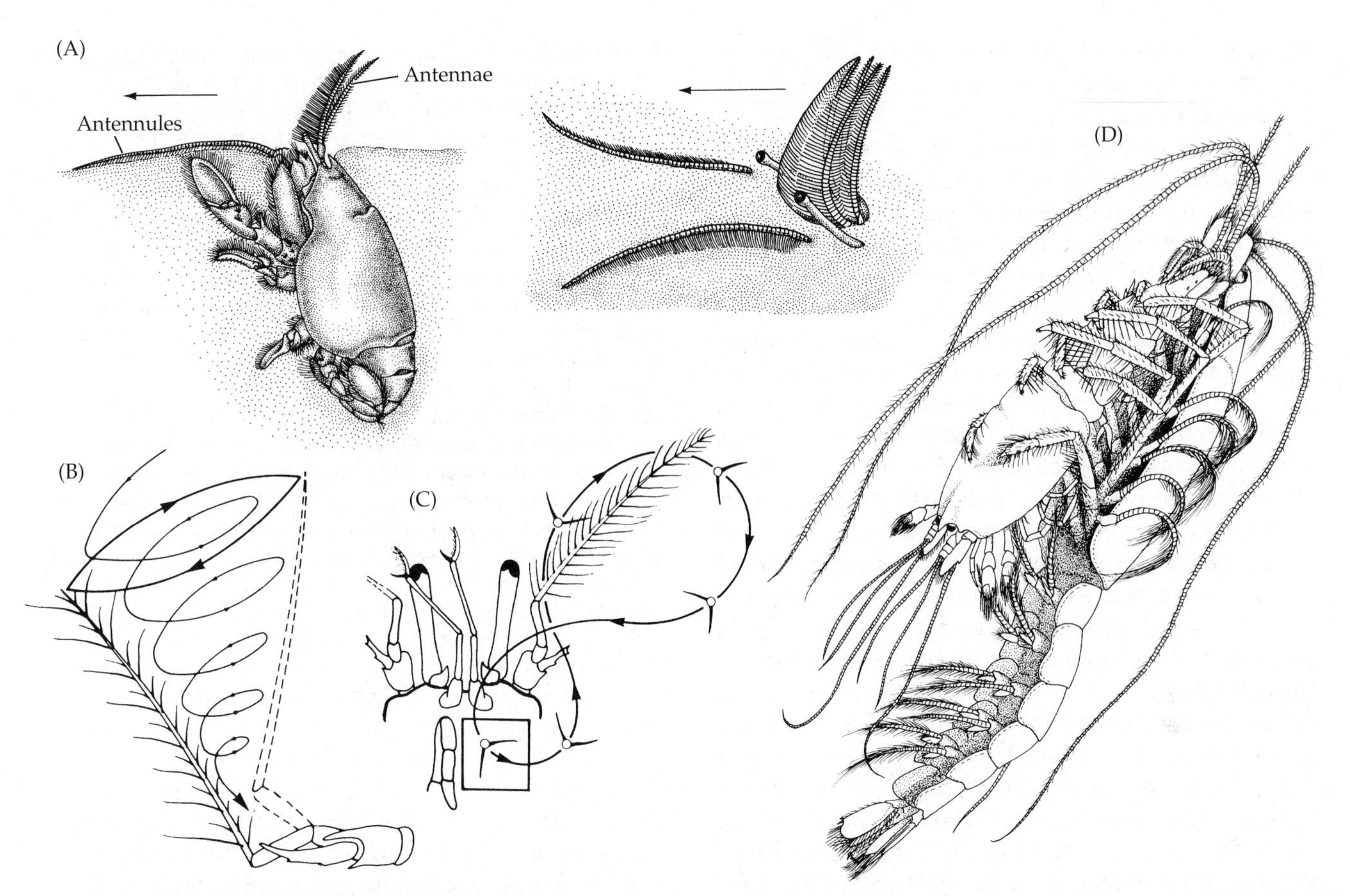

Figure 21.24 Some crustacean feeding mechanisms (see also Figure 21.22). (A) Suspension feeding in the sand crab *Emerita*. The arrows point seaward and indicate the direction of water movement as waves recede. The antennules direct water through branchial chambers. The antennae remove food particles from the water and then brush them onto the mouthparts. (B,C) The suspension feeding hermit crabs (B) *Australeremus cooki* and (C) *Paguristes pilosus* twirl their antennae, either in a circle or a figure eight, to create water currents that pull food particles to the mouth region. (D) The predatory shrimp *Procaris ascensionis* (Caridea) is shown here munching on another shrimp (*Typhlatya*) as it holds the prey in a "cage" formed by the pereopodal endopods.

of anostracans, cladocerans, copepods, ostracods, cirripedes, anaspidaceans, euphausids, decapods, tanaids, isopods, and amphipods. Predation typically involves grasping the prey with chelate or subchelate pereopods (or sometimes directly with the mouth appendages), or even with the antennae in the case of predatory anostracans, followed by tearing, grinding, or shearing with various mouthparts, particularly the mandibles. Perhaps the most highly adapted predatory specialists are the stomatopods (Figure 21.7), which possess greatly enlarged, raptorial subchelate limbs, which they use to stab or to club and smash prey. Some species search out prey, but many sit in ambush at their burrow entrance. The actual attack generally follows visual detection of a potential prey item, which may be another crustacean, a mollusc, or even a small fish. Once captured and stunned or killed by the raptorial claws, the prey is held against the mouthparts and shredded into ingestible pieces.

Although the cave-dwelling and often presumed "primitive" shrimp *Procaris* is omnivorous, its predatory behavior is particularly interesting. Its prey includes other crustaceans, particularly amphipods and shrimps. After *Procaris* locates a potential victim (probably by chemoreception), it moves quickly to the prey and grasps it within a "cage" formed by the pereopodal endopods (Figure 21.24D). Once captured, the prey is eaten while the shrimp swims about. Apparently the third maxillipeds press the prey against the mandibles, which bite off chunks and pass them to the mouth.

The remipedes capture prey with their raptorial mouth appendages (Figures 21.1A, 21.3D–F, and 21.22F), then immobilize the victim with an injection from the hypodermic maxillules. It is suspected that tissues are then sucked out of the prey by action of a mandibular mill and muscular foregut. They are probably also facultative suspension feeders and scavengers.

Another fascinating adaptation for predation can be seen in many species of Alpheidae (the snapping shrimps, e.g., *Alpheus, Synalpheus*) (Figure 21.9D). In such species, one of the chelipeds is much larger than the other, and the movable finger is hinged in such a way that it can be held open under muscle tension and then snapped quickly closed; this forceful closing produces a loud popping sound and a pressure or "shock" wave in the surrounding water. Some species appear to use this mechanism in ambushing prey (although most alpheids are probably omnivores and even include algae in their diet). When feeding in a predatory mode, the shrimp sits at its burrow entrance with the antennae extended. When a potential prey approaches (usually a small fish, crustacean, or annelid), the shrimp "pops" its cheliped, and the resulting pressure wave stuns the victim, which is then quickly pulled into the burrow and consumed. These shrimps typically live in male-female pairs within the burrow, and prey captured by one individual is shared with its partner. Two mechanisms have been proposed for the production of the "pop" and associated shock wave. In some species the pop seems to be created by mechanical impact of the dactylus hammering into the propodus. A second mechanism recently proposed is the collapse of cavitation bubbles, which are created by the rapid closure of the claw (in excess of 100 km/sec). Captivation occurs in liquids when bubbles form and then implode around an object (it is a well-known phenomenon damaging ship propellers). Both mechanisms create shock waves. Some boring (endolithic) alpheids even use the claw snap to break off pieces of the rock into which they are digging. In some areas of the world, populations of snapping shrimp are so large that their noise disturbs underwater communication. In the Caribbean, colonies of alpheids have been likened to those of social insects (e.g., *Synalpheus regalis* colonies in Belize inhabit sponges and contain up to 350 sibling males and females, and a single dominant breeding female).

Many crustaceans emerge from the benthos under cover of darkness to feed or mate in the water column. Many predatory isopods emerge at night to feed on invertebrates or fish, particularly weak or diseased fish (or fish caught in fishing nets).

Macrophagous herbivorous and scavenging crustaceans generally feed by simply hanging onto the food source and biting off bits with the mandibles (a feeding technique similar to that of grasshoppers and other insects). Notostracans, some ostracods, and many decapods, isopods, and amphipods are scavengers and herbivores. Certain isopods in the family Sphaeromatidae bore into the aerial roots of mangrove trees. Their activities often result in root breakage followed by new multiple root initiation, creating the stiltlike appearance characteristic of red mangroves (*Rhizophora*). A number of crustaceans are full-time or part-time detritivores; many scavenge directly on detritus, but others (e.g., cephalocarids) stir up the sediments in order to remove organic particles by suspension feeding.

Finally, several groups of crustaceans have adopted various degrees of parasitism. These animals range from ectoparasites with mouthparts modified for piercing or tearing and sucking body fluids (e.g., many copepods, branchiurans, tantulocarids, several isopod families, and at least one species of ostracod) to the highly modified and fully parasitic rhizocephalans, whose bodies ramify throughout the host tissue and absorb nutrients directly (Figures 21.25 and 21.26).

Figure 21.25 The remarkable life cycle of the rhizocephalan cirripede *Peltogaster paguri*, a kentrogonid parasite of hermit crabs. (A) The mature reproductive portion of the parasite (externa) produces numerous broods of male and female larvae, which are released as nauplii (B,C) and eventually metamorphose into cypris larvae (D). Female cyprids settle on the thorax and limbs of host crabs (E) and undergo a major internal metamorphosis into the kentrogon form (F), which is provided with a pair of antennules and an injection stylet. The kentrogon's viscera metamorphoses into an infective stage, the vermigon, which is transferred to the host through the hollow stylet. Inside the host, the vermigon grows with rootlets that ramify throughout much of the host's body; it is now called the interna (G). Eventually the female parasite emerges on the abdomen of the host as a virginal externa (H). When the externa acquires a mantle pore, or aperture, it becomes attractive to male cyprids (I). Male cyprids settle within the aperture, transform into a trichogon form, and implant part of their body contents in the female's receptacles (J). The deposit proceeds to differentiate into spermatozoa, which fertilize the eggs of the female. (K) The dissected externa, with its rootlets, of *Peltogaster*, removed from its host. Note the mantle aperture. ▶

Rhizocephalans, which are cirripeds that have been highly modified to become internal parasites of other crustaceans, are some of the most bizarre organisms in the animal kingdom. They may have typical cirripede nauplius and cypris larvae, but in this group the cyprid will settle only upon another crustacean, selected to be the unfortunate host.

The most complex rhizocephalan life cycle is that of the suborder Kentrogonida, which are obligate parasites of decapods. In this group, a settled female cypris larva undergoes an internal reorganization that rivals that of caterpillar pupae in scope, developing an infective stage, called the **kentrogon**, beneath the cyprid exoskeleton. Once fully developed, the kentrogon forms a hollow cuticular structure, the **stylet**, which injects a motile, multicellular, vermiform creature called the **vermigon** into the host. The vermigon is the active infection stage. It has a thin cuticle and epidermis, several types of cells, and the anlagen of an ovary. It invades the host's hemocoel by sending out long, branching, hollow rootlets that penetrate most of the host's body and draw nutrients directly from the hemocoel. So profound is the intrusion by the rootlets that the parasite takes over nearly complete control of the host's body, altering its morphology, physiology, and behavior. Once the parasite invades the host's gonads, parasitic castration may

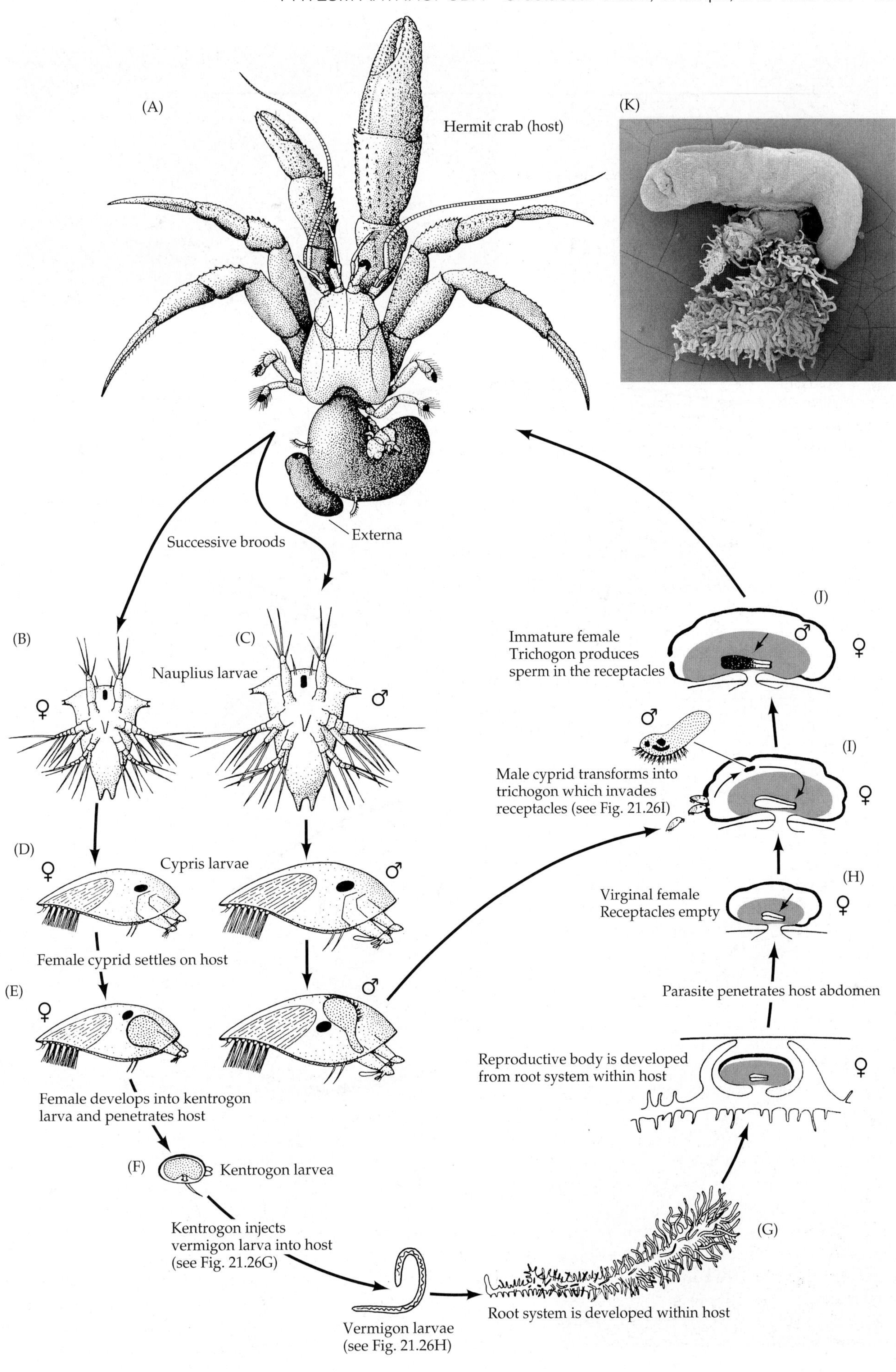
(A)
Hermit crab (host)
(K)
Externa
Successive broods
(B)
(C)
Nauplius larvae
♀
♂
(D)
Cypris larvae
♀
♂
Female cyprid settles on host
(E)
♀
♂
Female develops into kentrogon
larva and penetrates host
(F)
Kentrogon larvea
Kentrogon injects
vermigon larva into host
(see Fig. 21.26G)
Vermigon larvae
(see Fig. 21.26H)
Root system is developed within host
(G)
Reproductive body is developed
from root system within host
♀
Parasite penetrates host abdomen
Virginal female
Receptacles empty
(H)
♀
Male cyprid transforms into
trichogon which invades
receptacles (see Fig. 21.26I)
♂
(I)
♀
Immature female
Trichogon produces
sperm in the receptacles
(J)
♂
♀

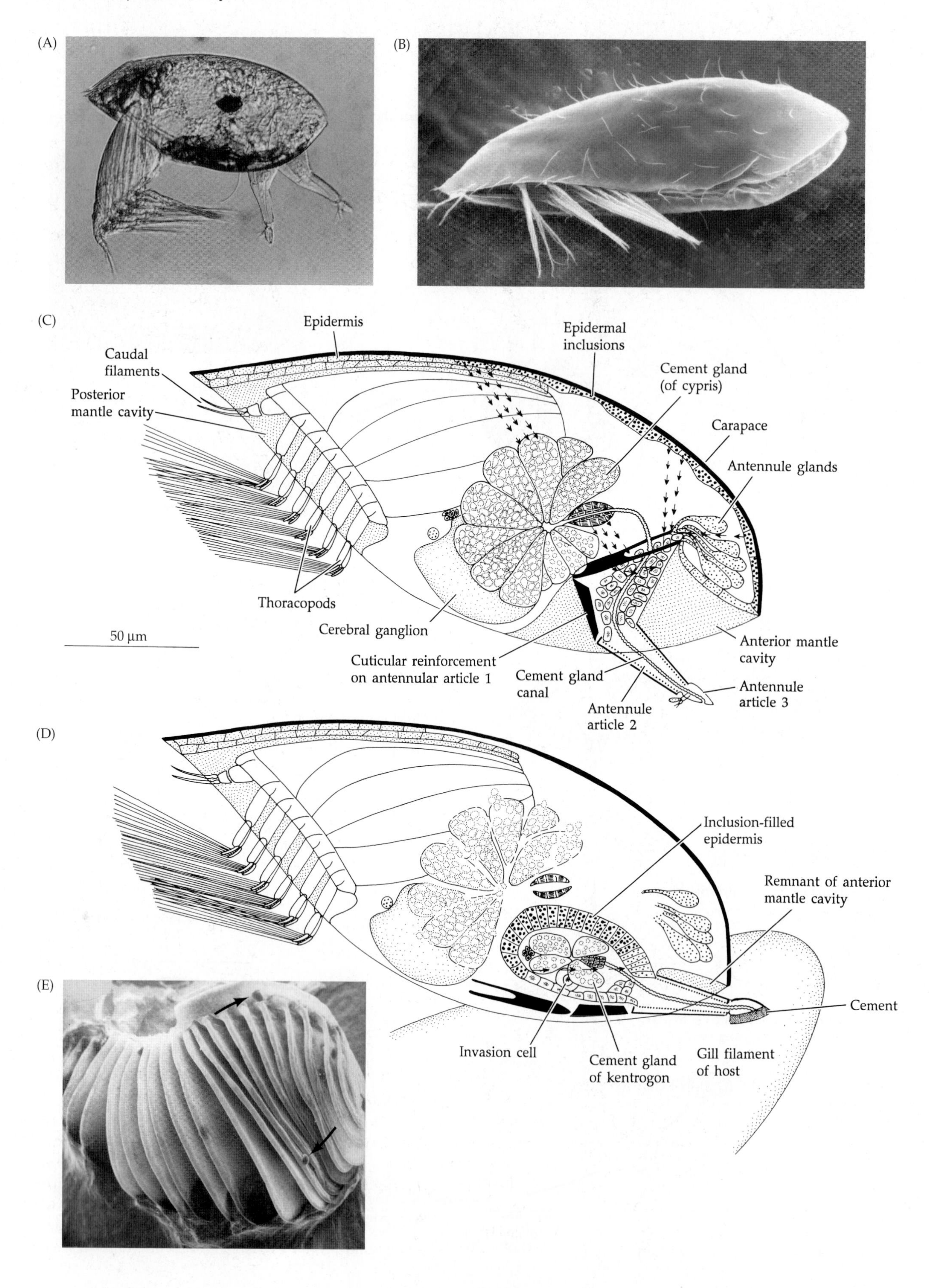
(A)
(B)
(C)
Epidermis
Epidermal inclusions
Caudal filaments
Cement gland (of cypris)
Posterior mantle cavity
Carapace
Antennule glands
Thoracopods
Cerebral ganglion
50 μm
Anterior mantle cavity
Cuticular reinforcement on antennular article 1
Cement gland canal
Antennule article 3
Antennule article 2
(D)
Inclusion-filled epidermis
Remnant of anterior mantle cavity
Cement
Invasion cell
Cement gland of kentrogon
Gill filament of host
(E)

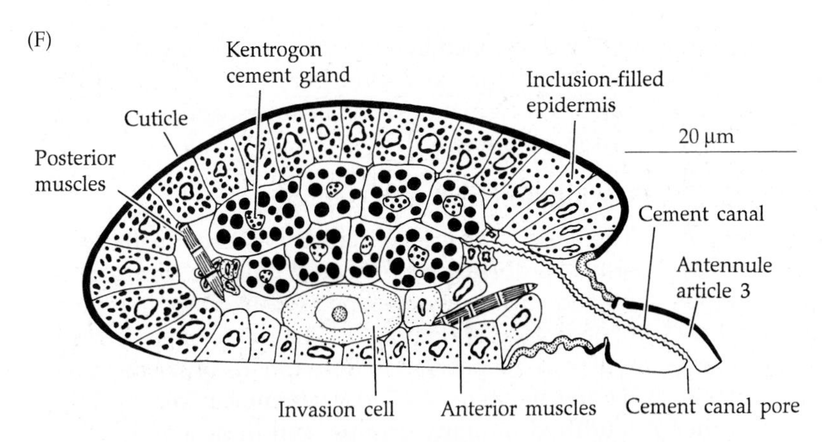

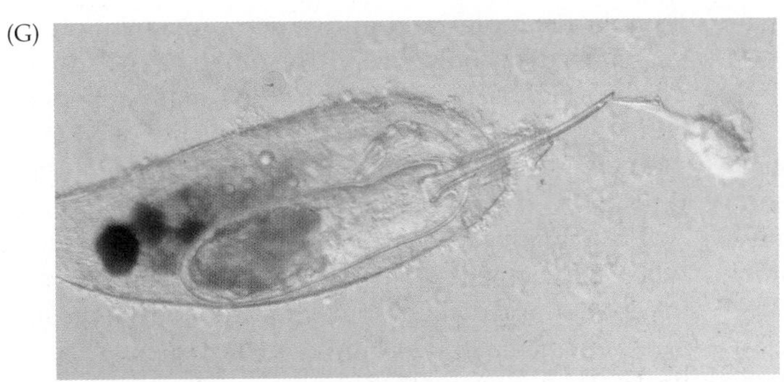

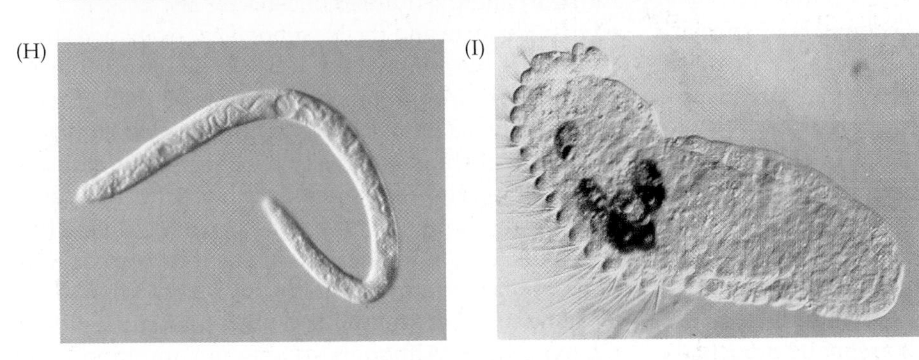

Figure 21.26 Various stages in the life cycle of parasitic rhizocephalans. (See Figure 21.25 for a full description of a rhizocephalan (kentrogonid) life cycle.) (A–D) Larvae of *Lernaeodiscus porcellanae*: (A) A live cyprid. (B) A cyprid (lateral view; SEM). (C,D) Diagrams of cyprid larvae before and after settlement (right side of carapace removed; naupliar eye omitted). The dotted line in the second antennular article indicates the primordial kentrogon cuticle, and the placement of muscle fibers in the cyprid are indicated by arrows; the muscles are hypothesized to effect formation of the kentrogon and separation of the old cyprid from the kentrogon. In D, kentrogon formation is complete. (E) A whole gill (SEM) of a host crab, *Petrolisthes cabriolli* (Anomura), with several attached kentrogons (arrows). (F) A 2-hour-old kentrogon (sagittal section). (G) A kentrogen injecting a vermigon via the stylet. (H) A vermigon. (I) A trichogon.

result (i.e., the gonads of parasitized crabs never produce mature gametes). Thus the host is transformed into a slave that serves the needs of its master. The internal root system, or **interna**, eventually develops an external reproductive body (the **externa**), where egg production occurs. A male cyprid settles on the externa, transforms into a minute sexually mature instar called a **trichogon**, and moves into the ovary-filled externa to take up residence, where its sole function is to produce sperm. It takes only one or two trichogons to stimulate the female ovaries to mature and begin releasing eggs into the chamber of the externa. Externa that fail to obtain male trichogons eventually die. The males are thus parasitic on the female (which is itself parasitic on the crustacean host), and kentrogonids are gonochoristic. A mature externa, usually arising from the host's abdomen, will produce a succession of larval broods, molting after each larval release (it is the only part of the rhizocephalan body that molts). The larvae are lecithotrophic and develop through several nauplius stages to the cyprid (Figures 21.25 and 21.26).

Members of the rhizocephalan order Akentrogonida parasitize a much wider range of crustacean hosts and do not have a kentrogon stage in their life cycle. Instead of injecting a vermigon, the female cyprid has long, slender antennules that it uses to attach to the abdomen of the host, one of which actually penetrates the host's cuticle, becomes hollow, and serves for the passage of embryonic cells from cyprid larva to host. Male cyprids somehow find infected hosts and penetrate them in the same fashion as the females, releasing their sperm in such a way that they actually enter the body of the female parasite.

The akentrogonid life cycle is similar to that of the Kentrogonida although, in some cases, more than one individual parasite might infect a single host, leading to multiple externas. And in at least one genus (*Thompsonia*), multiple externa can develop from a single infection. The anatomy of the externa is more variable than in kentrogonids, and the embryos develop directly into cypris larvae—there are no free-swimming nauplius stages. The extreme sexual size dimorphism of kentrogonids does not occur in Akentrogonida, and no migrating trichogon stage has been observed.

Isopods of the family Cymothoidae use modified mouthparts to suck the body fluids of their fish hosts. They attach to their host's skin or gills or, in some genera, to the tongue. One species, *Cymothoa exigua*, sucks so much blood out of its host's tongue (the spotted rose snapper, in the Gulf of California) that the tongue completely degenerates. But, remarkably, the fish doesn't die; the isopod remains attached with its clawed legs to the basal muscles of the missing tongue, and the fish use the crustacean as a "replacement tongue" to continue to feed. This is one of the few known cases of a parasite functionally replacing a host organ it destroys.

Digestive System

The digestive system of crustaceans includes the usual arthropod foregut, midgut, and hindgut. The foregut and hindgut are lined with a cuticle that is continuous with the exoskeleton and molted with it (i.e., it is ectodermally derived). The stomodeal foregut is modified in different groups, but usually includes a relatively short pharynx–esophagus region followed by a stomach. The stomach often has chambers or specialized regions for storage, grinding, and sorting; these structures are best developed in the malacostracans (Figure 21.27G). The midgut forms a short or long intestine—the length depending mainly on overall body shape and size—and bears variably placed digestive ceca. The ceca are serially arranged only in the remipedes. In some malacostracans, such as crabs, the ceca fuse to form a solid glandular mass, best called a **digestive gland** but sometimes called a midgut gland or hepatopancreas, within which are many branched, blind tubules. The digestive gland in some crustaceans has been shown to store lipids. The hindgut is usually short, and the anus is generally borne on the anal somite or telson, or on the last segment of the abdomen (when the anal somite or telson is reduced or lost).

Examples of some crustacean digestive tracts are shown in Figure 21.27. After ingestion, the food material is usually handled mechanically by the foregut. This may involve simply transporting the food to the midgut or, more commonly, processing the food in various ways prior to chemical digestion. For example, the complex foregut of decapods (Figure 21.27G) is divided into an anterior cardiac stomach and a posterior pyloric stomach. Food is stored in the enlarged portion of the cardiac stomach and then moved a bit at a time to a region containing a gastric mill, which usually bears heavily sclerotized teeth. Special muscles associated with the stomach wall move the teeth, grinding the food into smaller particles. The macerated material then moves into the back part of the pyloric stomach, where sets of filtering setae prevent large particles from entering the midgut. This type of foregut arrangement is best developed in macrophagous decapods (scavengers, predators, and some herbivores). Thus the food can be taken in quickly, in big bites, and mechanically processed afterward.

Circulation and Gas Exchange

The basic crustacean circulatory system usually consists of a dorsal ostiate heart within a pericardial cavity and variously developed vessels emptying into an open hemocoel (Figure 21.27). But the open hemocoel, which in the past led to descriptions of crustaceans as having an open circulatory system, has become highly modified in many groups, and in decapods at least there is a complex and intricate arrangement of true vessels that has only recently begun to be recognized. The heart is absent in most ostracods, many copepods, and many cirripedes. In some groups the heart is replaced or supplemented by accessory pumping structures derived from muscular vessels.

The primitive heart structure in crustaceans is a long tube with segmental ostia, a condition retained in part in cephalocarids and in some branchiopods, leptostracans, and stomatopods. However, the general shape of the heart and the number of ostia are also closely related to body form and the location of gas exchange structures. The heart may be relatively long and tubular and extend through much of the postcephalic region of the body, as it does in the remipedes, anostracans, and leptostracans, or it may tend toward a globular or box shape and be restricted to the thorax (e.g., as in cladocerans), where it may be associated with the thoracic gills (e.g., as in decapods). The intimate coevolution of the circulatory system with body form and gill placement is best exemplified when comparing closely related groups. Although isopods and amphipods, for instance, are both peracarids, their hearts are located largely in the pleon and in the pereon, respectively, corresponding to the pleopodal and pereopodal gill locations.

The number and length of blood vessels and the presence of accessory pumping organs are related to body size and to the extent of the heart itself. In most non-malacostracans, for example, there are no arterial vessels at all; the heart pumps blood directly into the hemocoel from both ends. These animals tend to have short bodies, long hearts, or both, an arrangement that facilitates circulation of the blood to all body parts. Sessile forms, such as most cirripedes, have lost the heart altogether, although it is replaced by a vessel pump in the thoracicans. Large malacostracans tend to have well developed vessel systems, thus ensuring that blood flows throughout the body and hemocoel and to the gas exchange structures (Figure 21.27D,E). Recent studies using corrosion casting techniques have shown just how complex these vessel systems can be and have even called into question the paradigm of "open" vs. "closed" circulatory systems in invertebrates. Large or

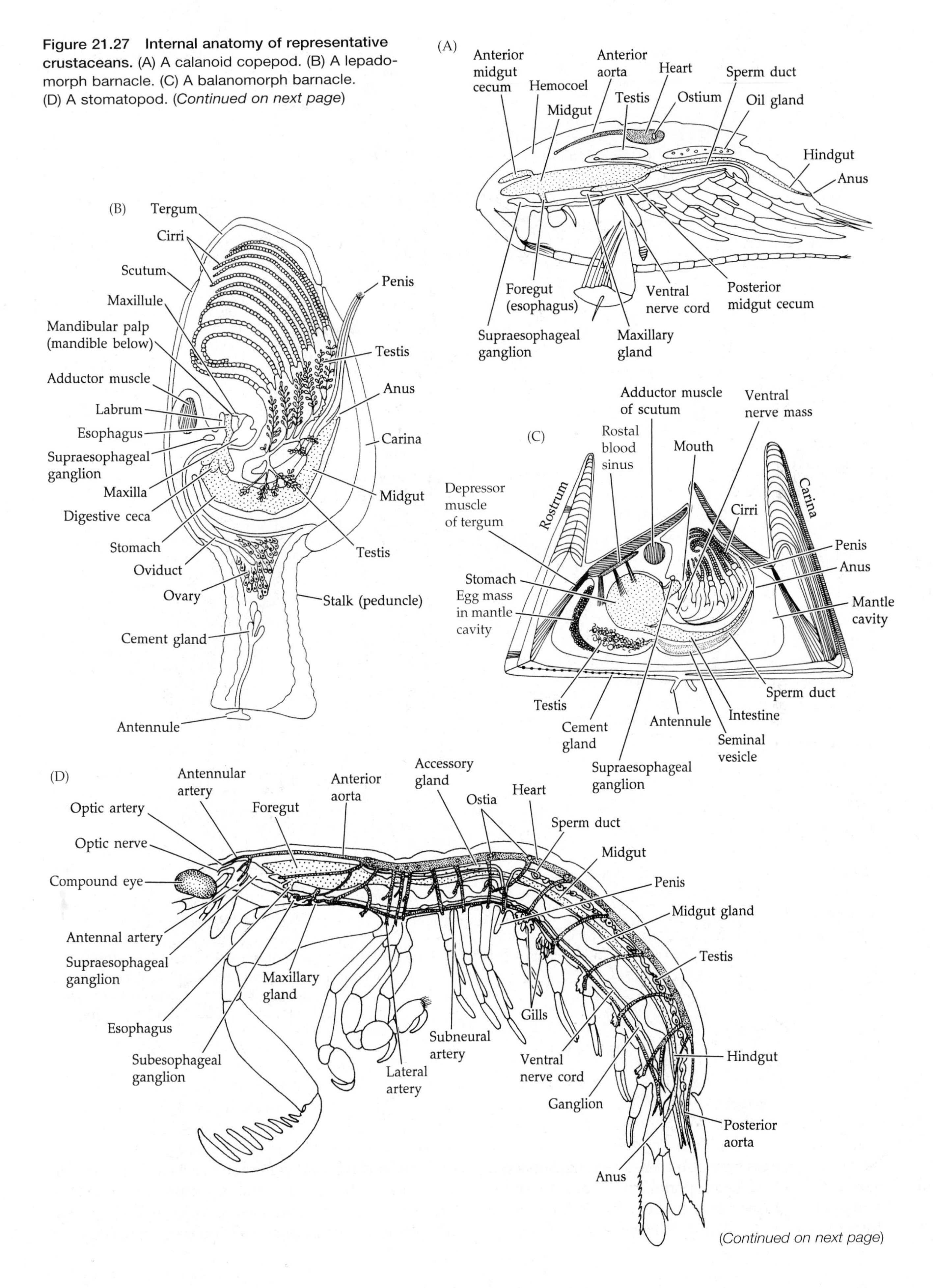

Figure 21.27 Internal anatomy of representative crustaceans. (A) A calanoid copepod. (B) A lepadomorph barnacle. (C) A balanomorph barnacle. (D) A stomatopod. *(Continued on next page)*

(Continued on next page)

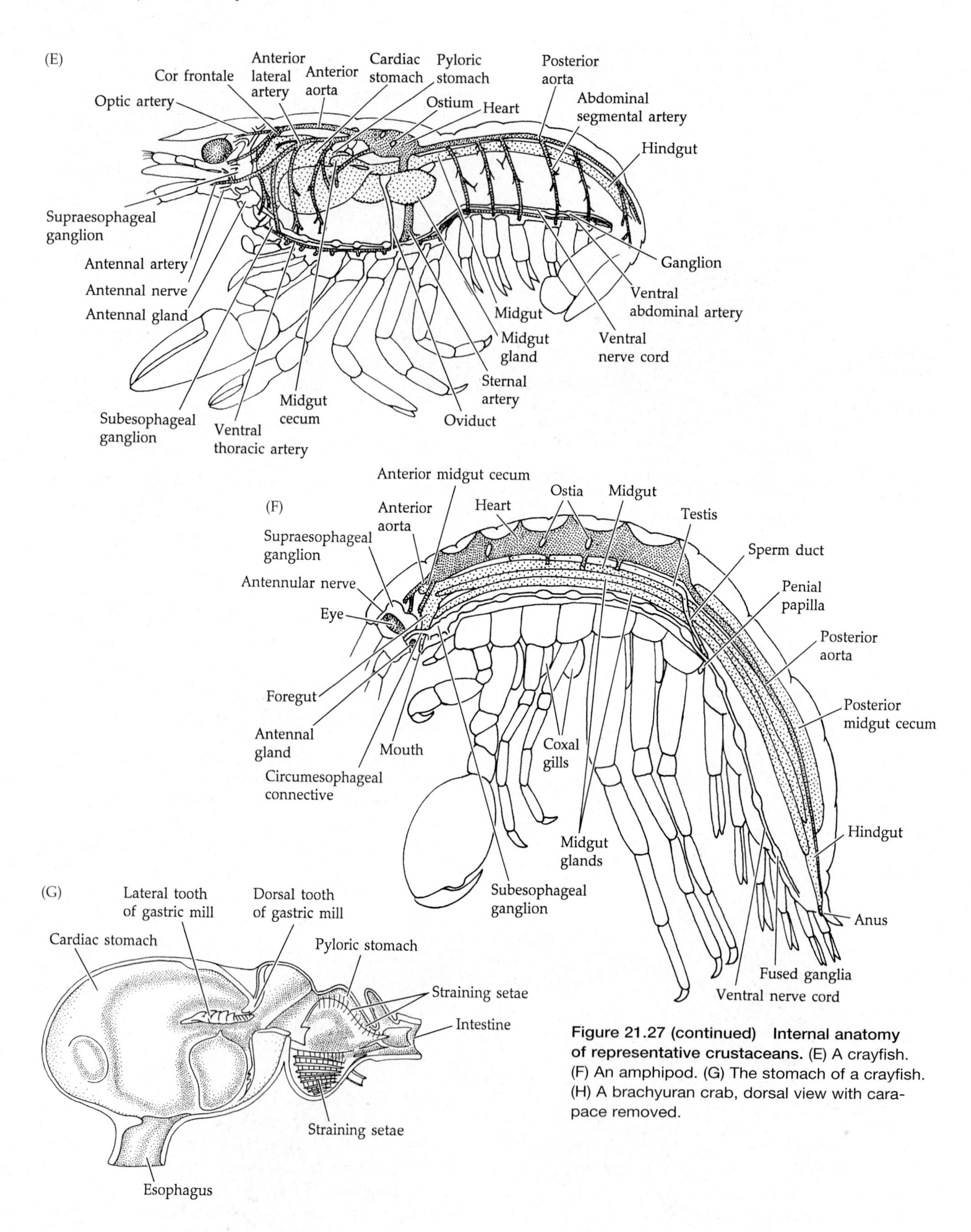

Figure 21.27 (continued) Internal anatomy of representative crustaceans. (E) A crayfish. (F) An amphipod. (G) The stomach of a crayfish. (H) A brachyuran crab, dorsal view with carapace removed.

active crustaceans may also possess an anterior accessory pump called the **cor frontale**, which helps maintain blood pressure, and often a venous system for returning blood to the pericardial chamber.

Crustacean blood contains a variety of cell types, including phagocytic and granular amebocytes and special wandering explosive cells that release a clotting agent at sites of injury or autotomy. In

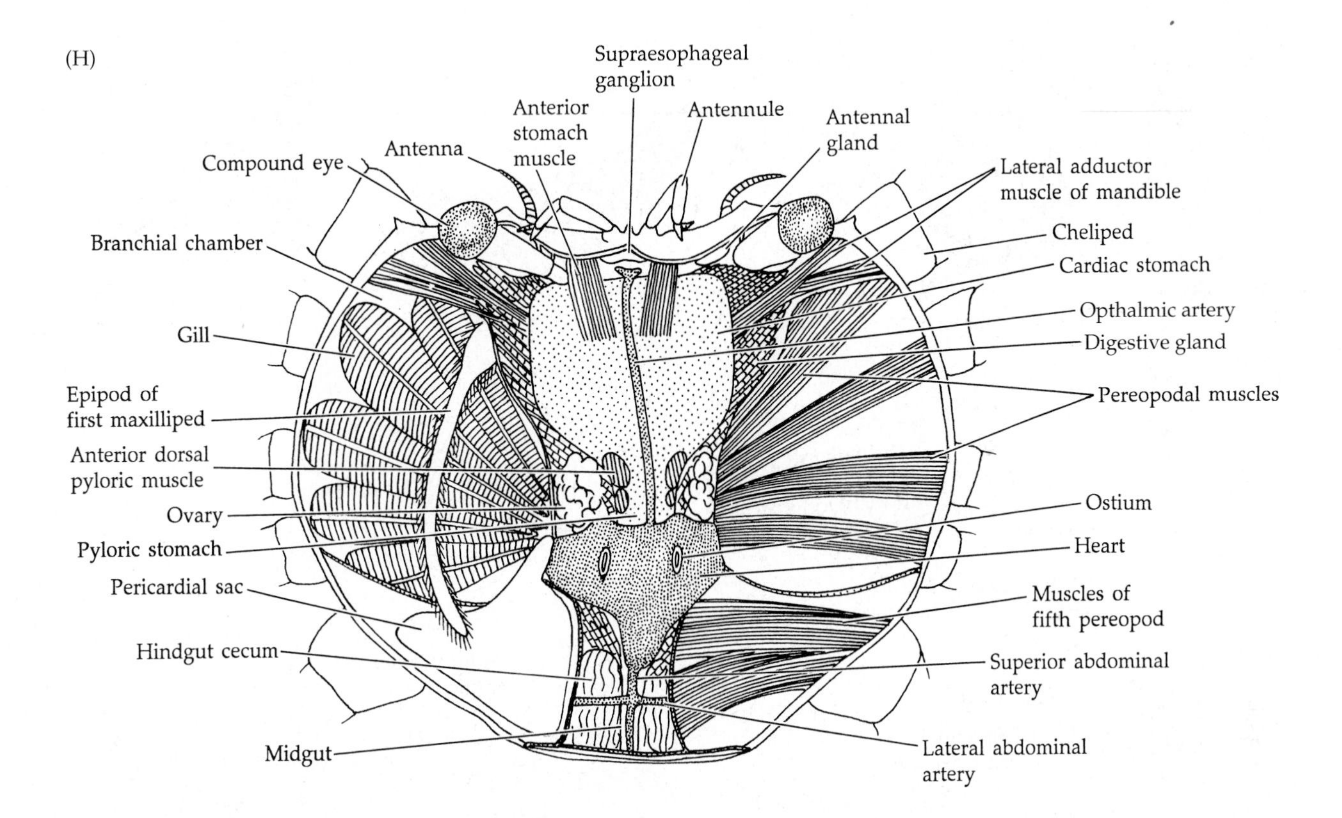

non-malacostracans, oxygen is either carried in solution or attached to dissolved hemoglobin. Most malacostracans possess hemocyanin in solution (although some contain hemoglobin within tissues). Hemoglobin uses iron as the oxygen-binding site, whereas hemocyanin uses copper. The latter can give a bluish color to the hemolymph; carotenoid pigments frequently give hemolymph a pale brown or orange color. Oxygen-binding pigments are never carried in corpuscles as they are in the vertebrates.

We have mentioned the form and position of gas exchange organs (gills) for some groups of crustaceans in the taxonomic synopses. Some small forms (e.g., copepods, some ostracods) lack distinct gills and rely on cutaneous exchange, which is facilitated by their relatively thin cuticles and high surface area-to-volume ratio. In the small forms of other groups a thin, membranous inner lining of the carapace serves this purpose (e.g., Cladocera, Cirripedia, Leptostraca, Cumacea, Mysida, clam shrimps, and even some members of the Decapoda).

Most crustaceans, however, possess distinct gills of some sort (Figure 21.28). These structures are commonly derived from epipods (exites) on the thoracic legs that have been modified in various ways to provide a large surface area. The inner hollow chambers of these gills are confluent with the hemocoel or their vessels. Although their structure varies considerably (recall the various decapod gills described earlier), they all operate on the basic principles of gas exchange organs addressed in Chapter 4 and throughout this text: the circulatory fluid is brought close to the oxygen source in an organ with a relatively high surface area. The gills provide a thin, moist, permeable surface between the internal and external environments. The gills of stomatopods and isopods (Figure 21.28H,I) are formed from the abdominal pleopods. In the first case they are branched processes off the base of the pleopods, but in the isopods the flattened pleopods themselves are vascularized and provide the necessary surface area for exchange. Stomatopods also have epipodal gills on the thoracopods, but these are highly reduced.

For gills to be efficient, a flow of water must be maintained across them. In stomatopods and aquatic isopods a current is generated by the beating of the pleopods. Similarly, the pereopodal gills of euphausids are constantly flushed by water as the animal swims. In many crustaceans, however, the gills are concealed to various degrees and require special mechanisms in order to produce the ventilating currents. In most decapods, for example, the gills are contained in branchial chambers formed between the carapace and the body wall (Figure 21.28). Thus the delicate gills, while still technically outside the body, appear to be (and are protected as though they were) internal organs. While such an arrangement provides protection from damage to the fragile gill filaments, the openings to the chambers are generally small, restricting the passive flow of water. Not surprisingly, the solution to this dilemma comes once again from the evolutionary plasticity of crustacean appendages. Most decapods have elongate exopods on the maxillae, called gill bailers or scaphognathites, that vibrate to create ventilating currents through the branchial

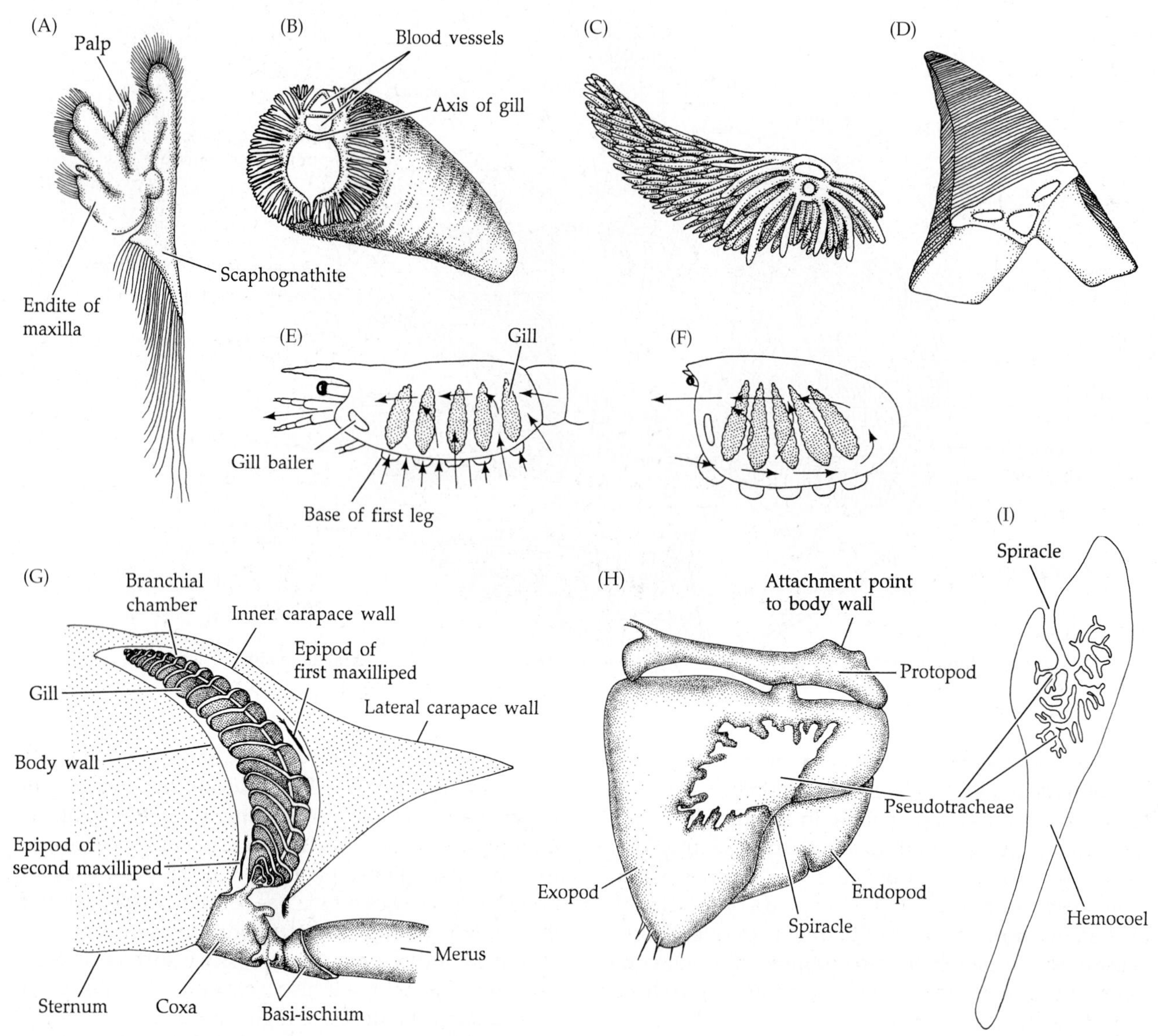

Figure 21.28 Gas exchange structures. (A) Maxilla of the shrimp *Pandalus*. Note the setose scaphagnathite used to generate the ventilating current. (B–D) Cross sections of types of decapod gills: (B) Dendrobranchiate. (C) Trichobranchiate. (D) Phyllobranchiate. (E,F) Paths of ventilating currents through the left branchial chambers of (E) a shrimp and (F) a brachyuran crab. (G) The branchial chamber (cross section) of a brachyuran crab, showing the position of a single phyllobranchiate podobranch. (H,I) A pleopod of the terrestrial isopod *Porcellio* (surface view and section). Note the pseudotracheae.

chambers (Figure 21.28A). These currents typically enter from the sides and rear through small openings around the coxae of the pereopods (called in crabs **Milne-Edwards openings**, after their discoverer), and exit anteriorly from under the carapace in the vicinity of the mouth field (and antennal glands). They can be easily seen by observing a crab or lobster in quiet water. The flow rate of the currents can be altered, depending on environmental factors, and can also be reversed, thus allowing certain decapods to burrow in sand or mud with only their front ends exposed to the water.

The positioning of the gills in branchial chambers protects them from desiccation during low tides and thus enables many crustaceans to live in littoral habitats; diffusion of respiratory gases commonly continues even during low tides. Some decapods have even invaded land, especially certain crayfish and the anomuran and brachyuran crabs known as land crabs (e.g., the hermit crab *Coenobita* and the coconut crab *Birgus*; Figure 21.1H). In these semi-terrestrial species the gills are typically reduced in size. In *Birgus* the original gills are very small, and the vascularized cuticular surface of the gill chamber is used for gas exchange. Although young *Birgus* may carry shells (or coconuts) to protect their soft abdomen, adults do not and the abdomen is hardened. Another striking decapod adaptation to life in air is displayed by the sand-bubbler crabs of the Indo-Pacific region (family Dotillidae: *Scopimera, Dotilla*). These crabs possess membranous discs on their legs or sternites that were once thought to be auditory organs (tympana), but are now thought to function as gas exchange surfaces.

The most successful crustaceans on land are not the decapods, however, but the familiar sowbugs and pillbugs. The success of these oniscidean isopods (e.g., *Porcellio*) is due in part to the presence of aerial gas exchange organs called **pseudotrachea** (Figure 21.28H,I). These organs are inwardly directed, moderately branched, thin-walled, blind sacs located in some of the pleopodal exopods, connected to the outside via small pores (similar to insect tracheae and spiracles). Air circulates through these sacs, and gases are exchanged with the blood in the pleopods. Thus, in these animals the original aquatic pleopodal gills have been refashioned for air breathing by moving the exchange surfaces inside, where they remain moist. The superficially similar tracheal systems of isopods, insects, and arachnids evolved independently, by convergence, in association with other adaptations to life on land.

Excretion and Osmoregulation

Like other fundamentally aquatic invertebrates, crustaceans are ammonotelic, whether in fresh water or seawater or on land. They release ammonia both through nephridia and by way of the gills. As discussed in Chapter 20, most crustaceans possess nephridial excretory organs in the form of either antennal glands or maxillary glands (Figures 21.5A and 21.27). These are serially homologous structures, constructed similarly but differing in the position of their associated pores (at the base of the second antennae or the second maxillae, respectively). The inner blind end is a coelomic remnant of the nephridium called the sacculus, which leads through a variably coiled duct to the pore. The duct may bear an enlarged bladder near the opening. Antennal glands are sometimes called "green glands."

Most crustaceans have only one pair of these nephridial organs, but lophogastrids and mysids have both antennal and maxillary glands, and a few others (cephalocarids and a few tanaids and isopods) have well developed maxillary and rudimentary antennal glands. Most non-malacostracans have maxillary glands, as do stomatopods, cumaceans, and most tanaids and isopods. Adult ostracods have maxillary glands, but antennal glands also occur in freshwater species. All of the other malacostracans have antennal glands.

Blood-filled channels of the hemocoel intermingle with branched extensions of the sacculus epithelium, creating a large surface area across which filtration occurs. The cells of the sacculus wall also actively take up and secrete material from the blood into the lumen of the excretory organ. These processes of filtration and secretion are to some degree selective, but most of the regulation of urine composition is accomplished by active exchange between the blood and the excretory tubule. These activities not only regulate the loss of metabolic wastes but are also extremely important in water and ion balance, particularly in freshwater and terrestrial crustaceans.

The excretion and osmoregulation carried out by antennal and maxillary gland activity are supplemented by other mechanisms. The cuticle itself acts as a barrier to exchange between the internal and external environments and, as we have mentioned, is especially important in preventing water loss on land or excessive uptake of water in fresh water. Moreover, thin areas of the cuticle, especially the gill surfaces, serve as sites of waste loss and ionic exchange. The epipods on the legs of Branchiopoda were long assumed to function in gas exchange (as "gills") but they are now known to serve primarily as sites of osmoregulation (hence, the taxonomic name Branchiopoda, meaning gill-footed, is a misnomer!). Phagocytic blood cells and certain regions of the midgut are also thought to accumulate wastes. In some terrestrial isopods, ammonia actually diffuses from the body in gaseous form.

Nervous System and Sense Organs

The central nervous system of crustaceans is constructed in concert with the segmented body structure, along the same lines as seen in other arthropods (Figure 21.29). In the more primitive condition it is ladderlike, the segmental ganglia being largely separate and linked by transverse commissures and longitudinal connectives (Figure 21.29A). The crustacean brain is composed of three fused ganglia, the two anterior being the dorsal (supraesophageal) protocerebrum and deutocerebrum, which are thought to be preoral in origin. From the protocerebrum, optic nerves innervate the eyes. From the deutocerebrum, antennulary nerves run to the antennules, while smaller nerves innervate the eyestalk musculature. The third ganglion of the brain is the posterior tritocerebrum, which presumably represents the first postoral somite ganglion. The tritocerebrum forms a pair of circumenteric connectives that extend around the esophagus to a subesophageal or subenteric ganglion and link the brain with the ventral nerve cord bearing the segmental body ganglia. From the tritocerebrum also arise the antennary nerves as well as certain sensory nerves from the anterior region of the head.

The nature of the ventral nerve cord often clearly reflects the influence of body tagmosis. In crustaceans with relatively homonomous bodies (e.g., remipedes, cephalocarids, and anostracan and notostracan branchiopods), the ganglia associated with each postantennary segment remain separate along the ventral nerve cord. In more heteronomous forms, however, a single large subenteric ganglionic mass is formed by the fusion of ganglia associated with the postoral cephalic segments (e.g., those of the mandibles, maxillules, maxillae, and, when present, maxillipeds). The ganglia

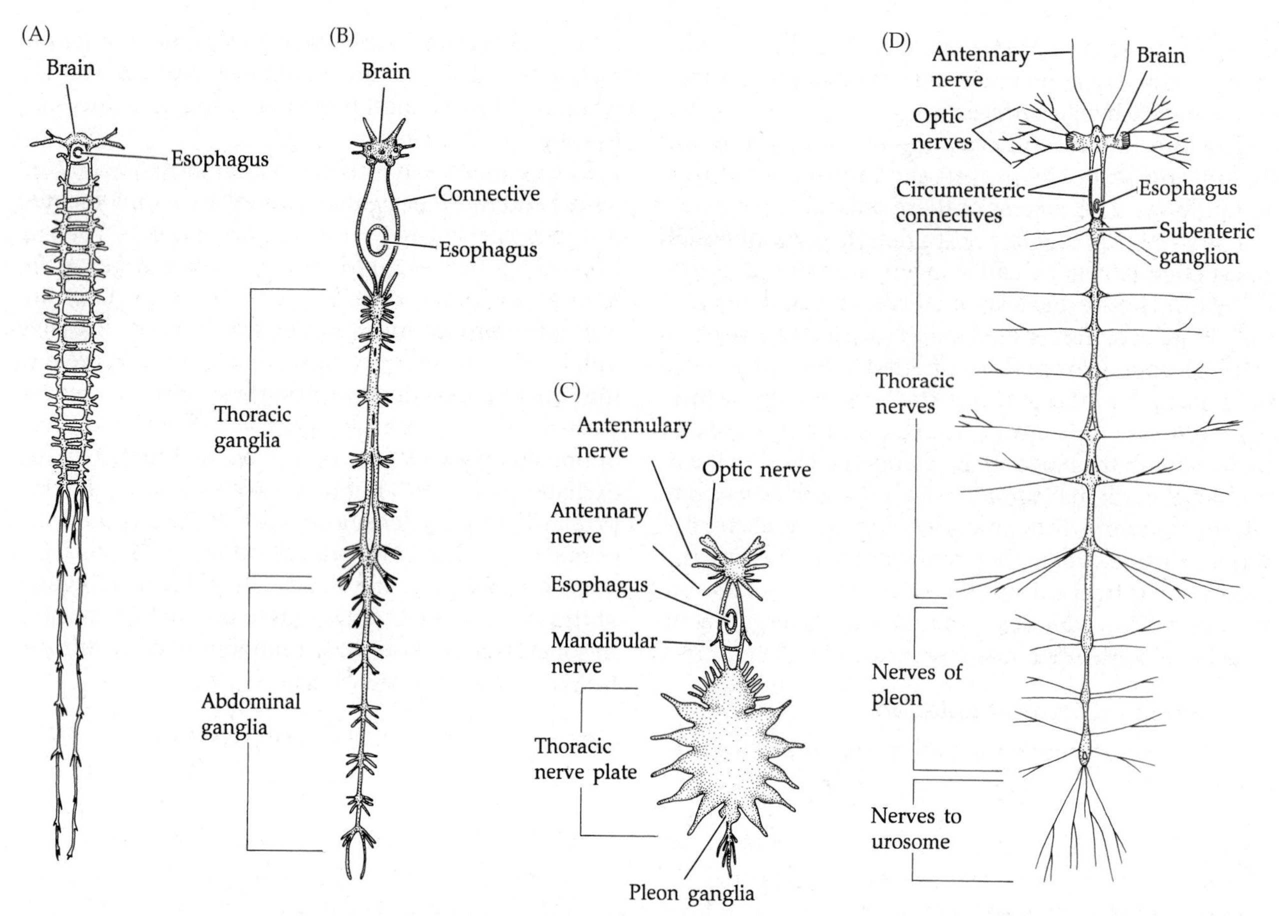

Figure 21.29 Central nervous systems of four crustaceans. (A) The ladderlike system of an anostracan. Note the absence of well developed ganglia in the posterior, apodous, portion of the trunk. (B) Elongate metameric system of a crayfish. (C) Highly compacted system of a brachyuran crab, wherein all thoracic ganglia have fused and the abdominal ganglia are reduced. (D) Nervous system of a hyperiid amphipod. Note the loss of the urosomal ganglia typical of all amphipods.

of the thorax and abdomen may also be variably fused, depending on segment fusion and body compaction. For example, in most long-bodied decapods (lobsters and crayfish), the thoracic and abdominal ganglia are largely fused across the body midline, but remain separate from one another longitudinally (Figure 21.29B). However, in short-bodied decapods (e.g., crabs), all of the thoracic segmental ganglia are fused to form a large ventral nerve plate, and the abdominal ganglia are much reduced (Figure 21.29C).

Most crustaceans have a variety of sensory receptors that transmit information to the central nervous system in spite of the imposition of the exoskeleton (as previously explained for arthropods in general) (Figure 21.30). Among the most obvious of these sensory structures are the many innervated setae or **sensilla** that cover various regions of the body and appendages (Figure 21.31). Studies of these structures suggest most function as both mechanoreceptors (sensing touch and currents) and chemoreceptors. Most crustaceans also possess special chemoreceptors in the form of clumps or rows of soft, tubular, cuticular processes called **aesthetascs** (Figure 21.30A) located on the first antennae. In decapods, hundreds of neurons can innervate each aesthetasc. Thermoreceptors probably occur in many crustaceans (those living near hydrothermal vents in the deep sea would seem likely candidates) but are not yet documented. However, behaviors related to thermal avoidance and temperature preferences have been shown, with thermosensitivities reported from 0.2 to 2.0° C. A pair of unique sensory structures whose function is unknown, called **frontal processes**, occurs on the head of remipedes. Many crustaceans have dorsal organs, poorly understood glandular–sensory structures on the head, which actually constitute several different types of sensory structures that may or may not be homologous.

Like all arthropods, crustaceans contain well-developed proprioceptors that provide information about body and appendage position and movement during locomotion. A few taxa within the class Malacostraca possess statocysts, which either are fully closed and contain a secreted statolith (e.g., mysids, some anthurid isopods) or open to the outside through a small pore and contain a statolith formed of sand grains (e.g., many decapods) (Figure 21.30B,C). In the latter case the statocyst not only serves as a georeceptor, but also detects the angular and linear acceleration of the body relative to the surrounding water as well as the movement of water past the animal (i.e., the statolith is rheotactic). And, in some shrimp, the statocyst is apparently also engaged in hearing.

There are two types of rhabdomeric photoreceptors among crustaceans, median simple eyes and lateral compound eyes; both are innervated by the protocerebrum. Many species possess both kinds of eyes, either simultaneously or at different stages of development. The compound eyes may be sessile or stalked. Stalked compound eyes occur in the Anostraca, many Malacostraca, and perhaps some Cumacea (and perhaps also some trilobites). These are the only examples of moveable stalked compound eyes in the animal kingdom.

The median eye generally first appears during the nauplius larval stage, and for that reason it is often called a **naupliar eye**. Like the nauplius larva itself, the median eye is thought to be an ancestral (defining) feature of the Crustacea; it is secondarily reduced or lost in many taxa in which the corresponding larval stage is suppressed. Median eyes are in a sense "compound" in that they are composed of more than one photoreceptor unit (Figure 21.30D). There are typically three such units in the median eyes of nauplii and up to seven in the eyes of adults in which they persist. Except for their basic rhabdomeric nature, however, the structure of median eye units is unlike that of the ommatidia of true compound eyes. The former are inverse pigment cups, each with relatively few retinular (photoreceptor) cells. Cuticular lenses are present over the median eyes of most ostracods and some copepods. Simple crustacean eyes probably function only to detect light direction and intensity. Such information is of particular value as a means of orientation in planktonic forms without compound eyes, such as nauplius larvae, many copepods, etc. In some branchiopods, a space above the median eye is connected to the external environment by a small pore, perhaps indicating an invagination of the eye at some point in the distant past, although the nature and function of the pore is not known.

The structure and function of compound eyes (ommatidia) were reviewed in Chapter 20. In terms of

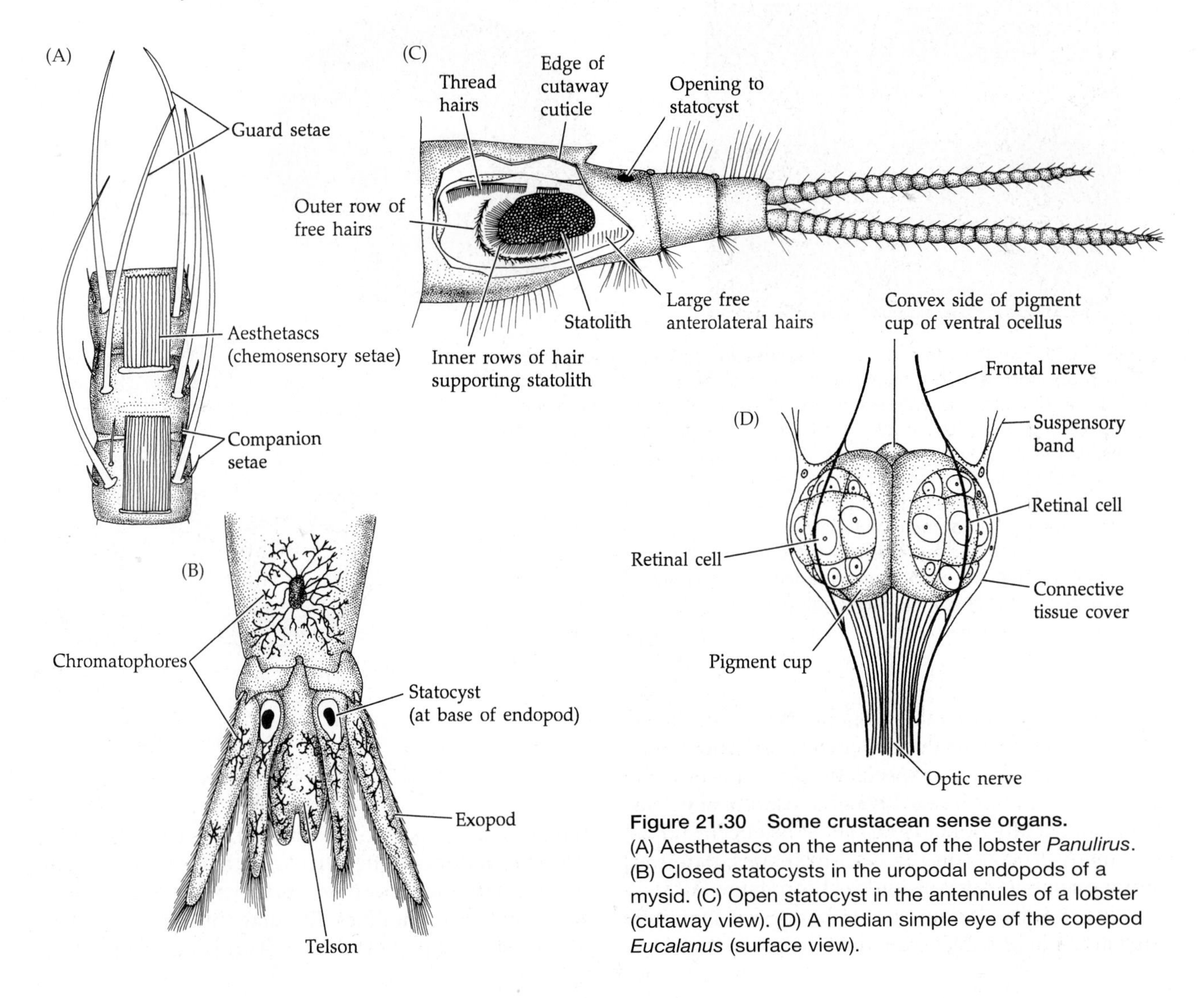

Figure 21.30 Some crustacean sense organs. (A) Aesthetascs on the antenna of the lobster *Panulirus*. (B) Closed statocysts in the uropodal endopods of a mysid. (C) Open statocyst in the antennules of a lobster (cutaway view). (D) A median simple eye of the copepod *Eucalanus* (surface view).

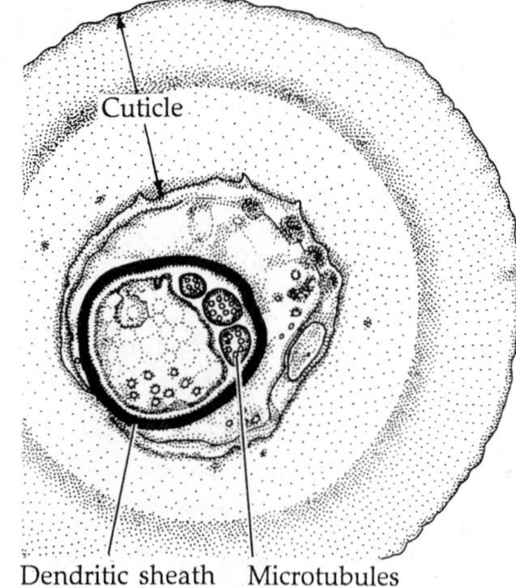

Figure 21.31 Sensilla of selected crustaceans. (A) Serrate seta (mechanoreceptor) of the second maxilliped of the anomuran *Petrolisthes armatus* (×428). (B) Sockets of serrate setae of the third maxilliped of the cleaning shrimp *Stenopus hispidus* (×228). (C) Current receptor from the anterior trunk limbs of the notostracan *Triops*. (D) Chemosensory seta from the first pereopod of the freshwater shrimp *Atya* (×5700); note the characteristic apical pore. (E) A dual receptor (mechanoreceptor and chemoreceptor) seta from the maxilla of the remipede *Speleonectes tulumensis* (×4560). (F) Cross section of a dual receptor (micrograph and interpretive drawing). Note the microtubules within the dendrites and the dendritic sheath that attaches to the cuticle of the setal shaft (×17,100).

visual capacity, much more work has been done on the eyes of insects than on those of crustaceans, and we are left with a good deal of speculation in terms of what crustaceans actually see. Although most probably lack the visual acuity of many insects, it has been shown that many crustaceans can discern shapes, patterns, and movement; color vision has been demonstrated in some species (various species have been shown to respond to light waves from the blue-green region to the ultraviolet and far-red spectra, at least to 470–570 nm). Stomatopods, in particular, have been shown to have extremely complex and sensitive eyes with a wide range of capabilities. Running through the center of each stomatopod eye is a midband of 6 ommatidial rows—which in most species contain 4 rows that make up a color vision system with 12 visual pigments—as well as 2 rows that analyze both linearly and circularly polarized light. The ommatidia that deal with color have a three-tier structure: the top tier contains an ultraviolet-sensitive pigment, and below this are 2 tiers with different pigments sensitive to wavelengths in the human visible spectrum. The upper and lower halves of each eye also have overlapping visual fields, so that each eye can act as a stereoscopic rangefinder. The 12 different photoreceptor types each sample a narrow set of wavelengths ranging from deep UV to far red (300 to 720 nm).

Although both insects and crustaceans have tetrapartite ommatidia, there are certain structural differences between the compound eyes of insects and those of crustaceans, probably as a result of adaptation to the requirements of aerial and aquatic vision. In water, light has a more restricted angular distribution, a lower intensity, and a narrower range of wavelengths than it does in air. Contrast is also somewhat reduced in water. All of these factors place a premium on enhancing the sensitivity and contrast perception of the eyes of aquatic creatures. Mounting the eyes on stalks is one dramatic way in which many crustaceans increase the amount of information available to the eyes, by increasing the field of view and binocular range. Eyestalks are complex structural features with a dozen or so muscles controlling their movement.

Typical tetrapartite compound eyes are lacking in the small crustaceans formerly combined as "maxillopodans" (copepods, barnacles, etc.), but various forms of "compound eyes" do occur among the Branchiura, Ostracoda (Cypridinacea), and Cirripedia. Eyes in the first two taxa most closely resemble those of other crustaceans in general structure and may be homologous with them. In the Cirripedia, the median eye and two lateral eyes are all derived from a single tripartite ocellar eye of the nauplius larva, which splits into its three components, each forming an adult photoreceptor following metamorphosis of the nauplius into a cypris larva. All three of these eyes thus appear to be composed of simple ocelli, although the lateral eyes have three photoreceptor cells and for this reason are often called "compound eyes." Rhizocephalan nauplii also have a tripartite nauplius eye, which persist into the cyprid larval stage.

Compound eyes are lacking altogether in many crustacean taxa (e.g., Copepoda, Mystacocarida, Cephalocarida, Tantulocarida, Pentastomida, Remipedia, and some Ostracoda). Members of some other groups possess compound eyes only in late larval stages and lose them at metamorphosis (e.g., some cirripedes). Reduction or loss of eyes is also common in many deep-sea species, burrowers, cave dwellers, and parasites.

Crustaceans have complex endocrine and neurosecretory systems, although our understanding of these systems is far from complete. In general, the phenomena of molting (see Chapter 20), chromatophore activity, and various aspects of reproduction are under hormonal and neurosecretory control. Interesting recent work indicates that juvenile hormone-like compounds, long thought to occur only in insects, may also occur in at least some crustaceans. (Juvenile hormones are a family of compounds that regulate adult metamorphosis and gametogenesis in insects.) Bioluminescence also occurs in several crustacean groups. It is common among pelagic decapods, and it has also been reported in certain myodocopan ostracods, hyperiid amphipods, and copepod larvae.

Reproduction and Development

Reproduction We have often mentioned the relationships between an animal's reproductive and developmental pattern and its lifestyle and overall survival strategy. With the exception of purely vegetative processes such as asexual budding, the crustaceans have managed to exploit virtually every life history scheme imaginable. The sexes are usually separate, although hermaphroditism is the rule in remipedes, cephalocarids, most cirripedes, and a few decapods. Sequential hermaphroditism is not uncommon and usually is expressed as protandry (individuals first mature as males, then later become females), although protogyny occurs in a few species (e.g., the marine isopod *Gnorimosphaeroma orgonense*). In addition, parthenogenesis is known in some branchiopods and certain ostracods. In one species of clam shrimp (*Eulimnadia texana*) a rare type of mixed mating system exists, called **androgonochorism** (i.e., androdioecy in plants), in which males coexist with hermaphrodites, but there are no true females. Androgonochorism is rare, but is also known in the nematode *Caenorhabditis elegans*, some thoracican barnacles (e.g., *Balanus galeatus*, *Scalpellum scalpellum*), and from several other branchiopod crustaceans.

The reproductive systems of crustaceans are generally quite simple (Figure 21.27). The gonads are derived from coelomic remnants and lie as paired elongate structures in various regions of the trunk. In many cirripedes, however, the gonads lie in the cephalic region. In some cases the paired gonads are partially or wholly fused into a single mass. A pair of gonoducts extends from the gonads to genital pores located on one of the trunk segments, either on a sternite, on the arthrodial membrane between the sternite and leg protopods, or on the protopods themselves. In many crustaceans the paired penes are fused into a single median penis (e.g., in tantulocarids, cirripedes, and some isopods). The female system sometimes includes seminal receptacles. The position of the gonopores varies among the classes (Table 21.1).

The curious phenomenon of **intersex**—having both male and female secondary sexual characteristics—is widespread among crustaceans. Intersexual development is associated directly with the presence of endoparasites such as bacteria and microsporidia, and it has also been correlated with the presence of endocrine-disrupting pollutants that seem to induce opposing secondary sexual characterisitcs.

Most crustaceans copulate, and many have evolved courtship behaviors, the most elaborate and well known of which occur among the decapods. Although many crustaceans are gregarious (e.g., certain planktonic species, barnacles, many isopods and amphipods), most decapods live singly except during the mating season. More or less permanent, or at least seasonal, pairing is known among many crustaceans (e.g., stenopodid shrimps; certain parasitic and commensal isopods; pinnotherid "pea" crabs, which often live as pairs in the mantle cavities of bivalve molluscs or in burrows of thalassinid shrimps).

Even the parasitic pentastomids copulate (within the host's respiratory system) and have internal fertilization, relying on a transfer of sperm to the female's vagina by way of the male's cirri (penes). Pentastomid early embryos metamorphose into a so-called primary larva with two pairs of double-clawed legs and one or more piercing stylets (Figure 21.19E). The larvae may be autoinfective in the primary host, or they may migrate to the host's gut and pass out with the feces. In the latter case, an intermediate host is required, which may be almost any kind of vertebrate. The larvae bore through the gut wall of the intermediate host, where they undergo further development to the infective stage. Once the intermediate host is consumed by a definitive host (usually a predator), the parasite makes its way from the new host's stomach up the esophagus, or bores through the intestinal wall, eventually settling in the respiratory system.

Mating in nonpaired crustaceans requires mechanisms that facilitate location and recognition of partners. Among decapods, and perhaps many other crustaceans, scattered individuals apparently find one another either by distance chemoreception (pheromones) or through synchronized migrations associated with lunar periodicity, tidal movements, or some other environmental cue. Contact sex pheromones are also utilized. Males of some marine myodocopan ostracods (some Halocyprididae, some Cypridinidae) produce complex bioluminescent displays, similar to those of fireflies, to attract females. Once prospective mates are near each other, recognition of conspecifics of the opposite sex may involve several mechanisms. Over 60 speceis of cypridinid ostracods engage in luminescent courtship behavior in the Caribbean alone. Apparently, most decapods employ chemotactic cues requiring actual contact. Vision is known to be important in the stenopodid shrimps (most of which live in pairs), in certain anomurans (e.g., the family Porcellanidae), and in brachyurans (many grapsids and ocypodids).

Figure 21.32 Reproduction in Crustacea. (A–D) Mating behaviors of the fiddler crab *Uca*. (A) Two males in ritualized combat for the favor of a female, while she watches (B). (C) A single male waving his enlarged cheliped to attract a female. (D) A male fiddler crab engaged in claw-waving behavior to attract a female. (E) A balanomorph barnacle, with cirri and groping penis extended, impregnating a neighbor. The advantage of a long penis in sessile animals is made obvious by this illustration. (F) Ventral views of a male and female brachyuran crab, *Cancer magister*, showing the modified pleopods (setose appendages to retain eggs in female; modified as gonopod in male). (G) A copulating pair of *Hemigrapsus sexdentatus*. (H) The planktonic copepod *Sapphirina*, with egg sacs. ▶

A good deal of work has been done on fiddler crabs of the genus *Uca* (family Ocypodidae). In these species, males engage in dramatic cheliped waving (of their greatly enlarged chela, or **major claw**, which can account for more than 50% of the male's total mass—more than the largest rack of a male elk!) to attract females and repel competing males (Figure 21.32A–D). In addition, males produce sounds by stridulation and substratum thumping, which are thought to attract potential mates. Mating generally takes place once the male has enticed the female into his burrow. Males of some fiddler crab species build sand structures at the entrance of their burrow, which have been shown to attract females.

Among many crustaceans, the external sexual characteristics are associated with the actual mating process. In some males, particular appendages, such as the antennae of anostracans and some cladocerans, ostracods, and copepods, are modified for grasping the female. Additionally, many males bear special sperm transfer structures, in the form of either modified appendages or special penes such as those of the thoracican barnacles (Figure 21.32E), anostracans, and ostracods. Examples of modified appendages include the last trunk limbs of copepods and the anterior pleopods of most male malacostracans (called gonopods in most malacostracans, or **petasma** in Dendrobranchiata) (Figure 21.32F). Sperm are transferred either loose in seminal fluid or (in many malacostracans and in copepods) packaged in spermatophores. Motile flagellated sperm occur only in some of the former maxillopodan groups; in other crustaceans the sperm are nonmotile. Crustacean sperm are highly variable in shape, even bizarre in many instances, often being large round or stellate cells that move by pseudopods or are, seemingly, nonmotile.[9] Sperm are deposited directly into the oviduct or into a seminal receptacle in or near the female reproductive system. In some crustaceans

[9]The sperm of freshwater ostracods are the longest in the animal kingdom, relative to body size (up to 10× body length). Even though their sperm are aflagellate, they are filiform and range from several hundred microns to millimeters in length.

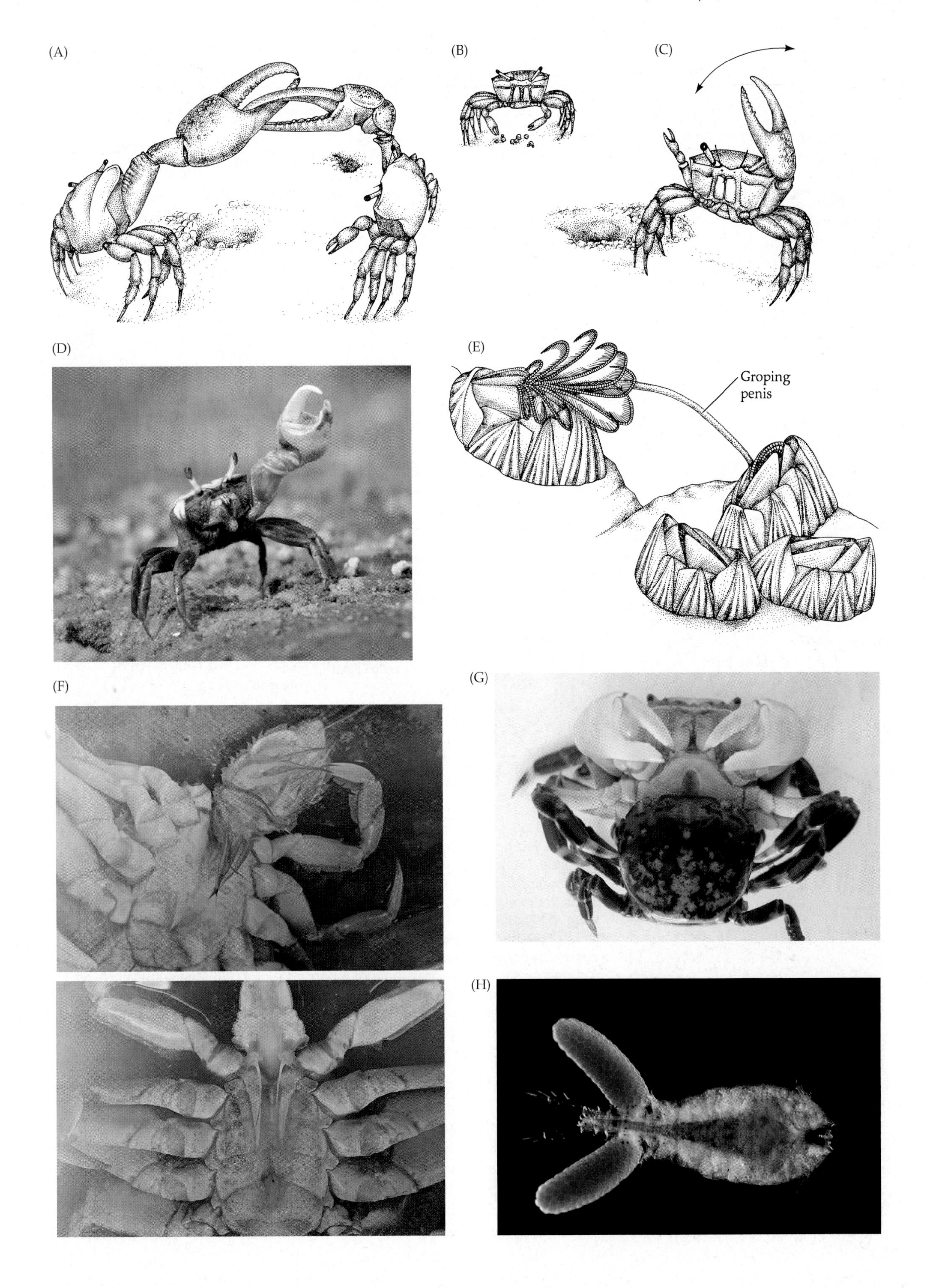
(A)
(B)
(C)
(D)
(E)
Groping
penis
(F)
(G)
(H)

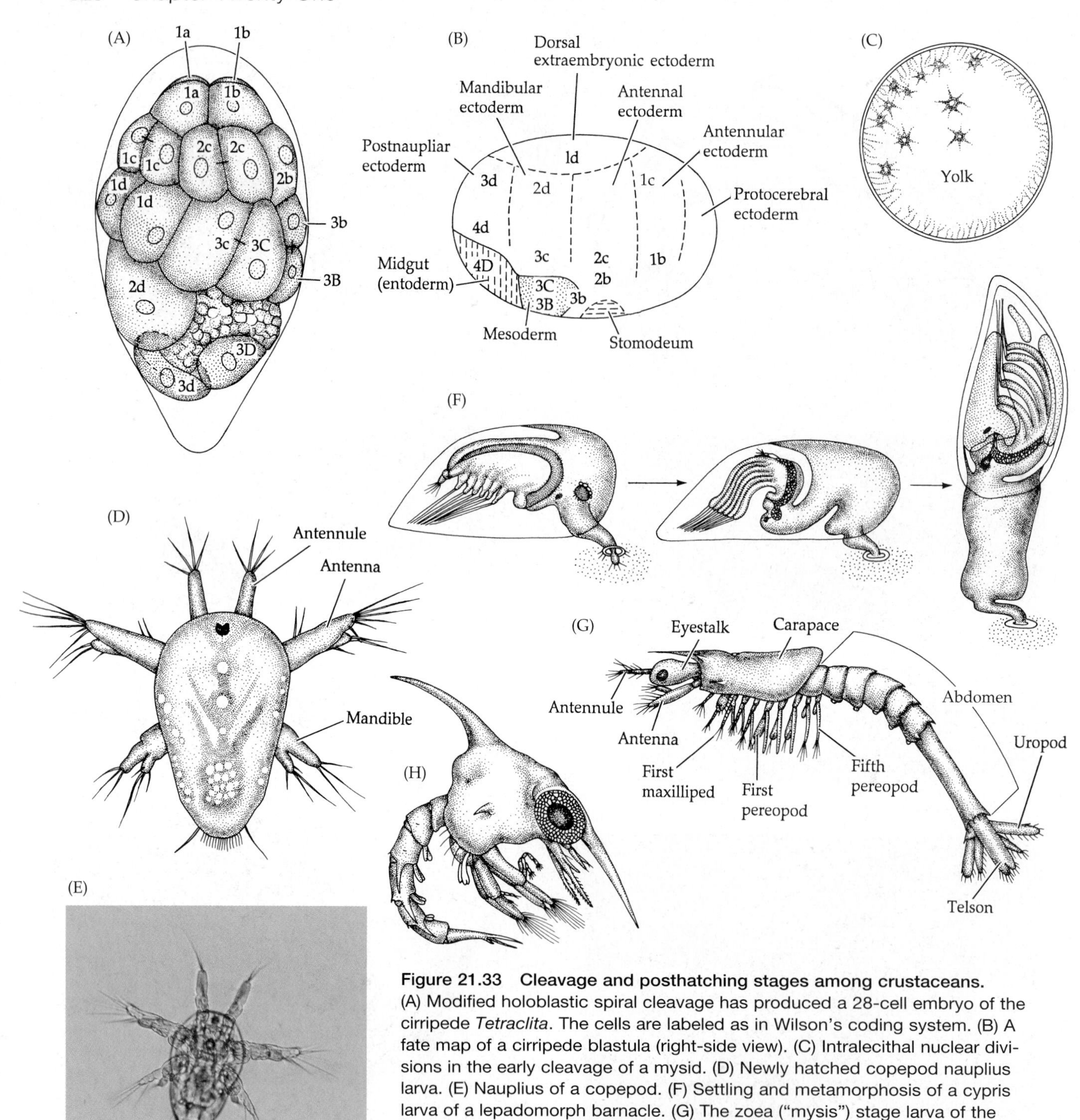

Figure 21.33 Cleavage and posthatching stages among crustaceans. (A) Modified holoblastic spiral cleavage has produced a 28-cell embryo of the cirripede *Tetraclita*. The cells are labeled as in Wilson's coding system. (B) A fate map of a cirripede blastula (right-side view). (C) Intralecithal nuclear divisions in the early cleavage of a mysid. (D) Newly hatched copepod nauplius larva. (E) Nauplius of a copepod. (F) Settling and metamorphosis of a cypris larva of a lepadomorph barnacle. (G) The zoea ("mysis") stage larva of the dendrobranchiate shrimp *Penaeus*. (H) Zoea larva of the brachyuran crab *Callinectes sapidus*. (I) Zoea larva of a porcelain crab. (J) Megalopa larvae of the xanthid crab *Menippe adina*. (K) The characteristic antizoea larva of a stomatopod. (L) The translucent, paper-thin phyllosoma larva of the lobster *Jassa*. (M) Cryptoniscus stage (not a true larva) of the epicaridean isopod *Probopyrus bithynis*. (N) Cyprid larva of a barnacle.

females can store sperm for long periods (e.g., several years in the lobster *Homarus*), thus facilitating multiple broods from single inseminations.

The great majority of crustaceans brood their eggs until hatching occurs, and a variety of brooding strategies has evolved. Peracarids brood the developing embryos in a marsupium, a ventral brood pouch formed from inwardly directed plates of the leg coxae called **oostegites** (thermosbaenaceans are an exception among the Peracarida and use the carapace as a brood chamber). Other crustaceans attach the embryos to endites on the bases of the legs or to the pleopods

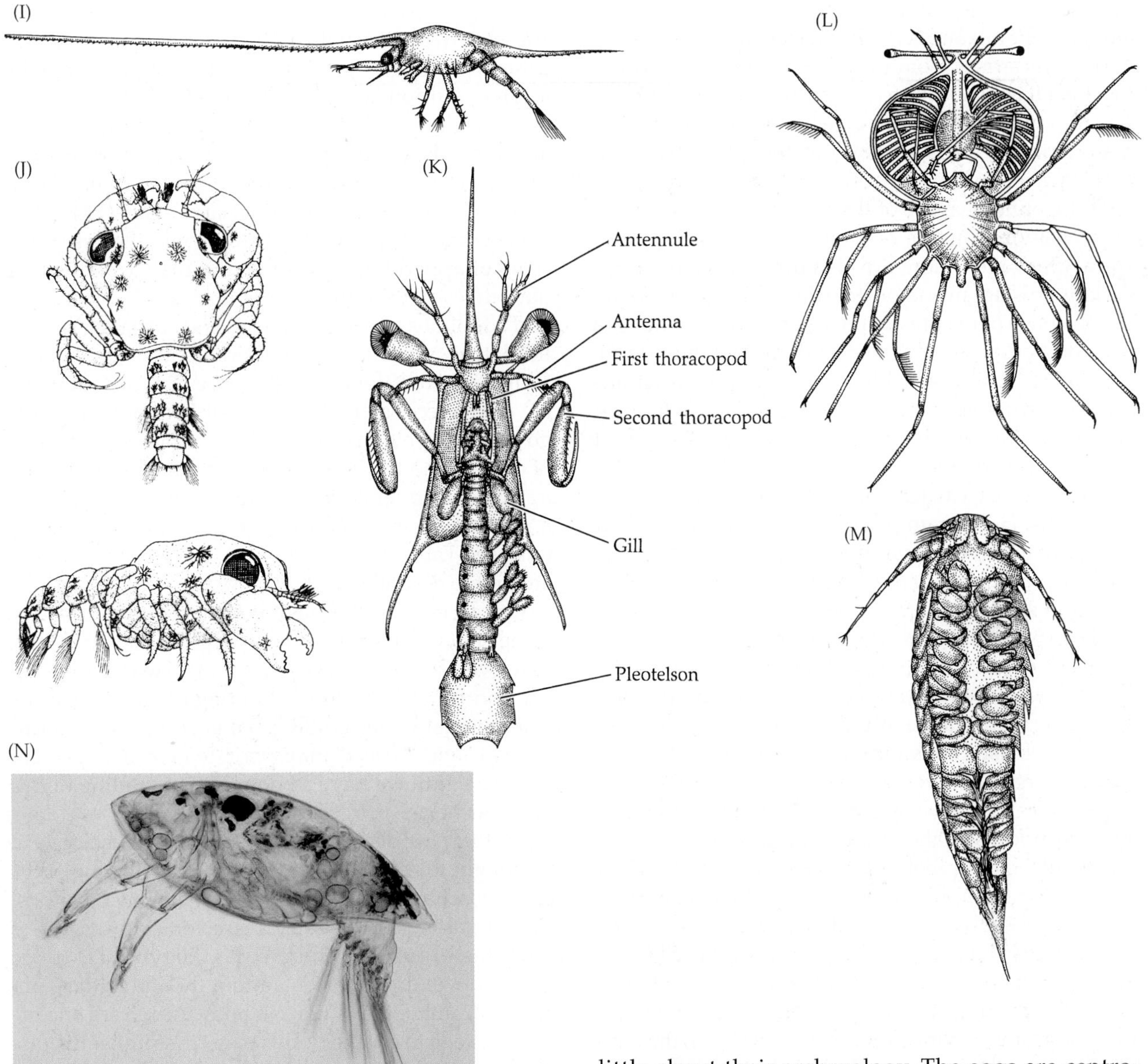

(Figure 21.32F), usually using mucus secreted by specialized glands. In some cladocerans, brooding takes place in a dorsal brood chamber formed by the carapace. However, the syncarids, almost all dendrobranchiate shrimps, and most euphausids shed the zygotes directly into the water. A few others deposit their fertilized eggs in the environment, usually attaching them to some object (e.g., branchiurans, some ostracods, many stomatopods). These deposited embryos may be abandoned or, as is the case in stomatopods, carefully tended by the female. Nonetheless, parental protection of the embryos until they hatch as larvae or juveniles is typical in crustaceans. Thus, crustaceans usually engage in mixed or direct life histories (Table 21.2).

Development Although crustaceans are the most widespread animals on Earth, we know surprisingly little about their embryology. The eggs are centrolecithal, with various amounts of yolk. The amount of yolk greatly influences the type of early cleavage and is often related to the time of hatching (Chapter 4). As far as is known, the zygotes of most non-malacostracans undergo some form of holoblastic cleavage, as do those of syncarids, euphausids, penaeids, amphipods, and parasitic isopods. However, cleavage patterns are extremely variable, ranging from equal to unequal and from radial-like to spiral-like. The occurrence of modified spiral cleavage (Figure 21.33A) reported from some crustaceans had generally viewed as evidence of close ties between the crustaceans and other spiralian groups such as the annelids. However, the generality of spiral-like cleavage among crustaceans has been called into question with evidence from molecular phylogenetics that indicates arthropods belong to the Ecdysozoa clade of Protostomia, not the Spriralia (where annelids are nested). And in some crustacean groups, the cell lineages and germ layer origins are quite different from those of typical spirally cleaving embryos. For example, in barnacles the mesodermal germ layer arises from the 3A, 3B, and 3C cells, and the 4d cell contributes to ectoderm (Figure 21.33B)—whereas typical spiral cleavage involves a 4d origin of mesoderm. Euphausids were long thought to have spiral cleavage, but recent work indicates they do not—only the oblique angle of the mitotic spindles during the transition from the 2- to the 4-cell stage resembles spiral cleavage. There are also differences between various crustacean and other arthropod taxa involving the positions of the presumptive germ layers relative to one another, especially the endoderm and mesoderm. We want to emphasize, however, that such variations are not surprising in such a diverse and ancient taxon and do not negate the fundamental similarities that unite the Arthropoda.

Meroblastic cleavage is the rule among many malacostracans. Here again, the exact pattern varies, but it generally involves intralecithal nuclear divisions followed by nuclear migration to the periphery of the embryo and subsequent partitioning of the nuclei into a cell layer around a central yolky mass (Figure 21.33C).

The form of the blastula and the method of gastrulation are dependent primarily on the preceding cleavage pattern and hence ultimately on the amount of yolk. Holoblastic cleavage may lead to a coeloblastula that undergoes invagination (as in syncarids) or ingression (as in many copepods and some cladocerans and anostracans). Other crustaceans form a stereoblastula followed by epibolic gastrulation (e.g., cirripedes). Most cases of meroblastic cleavage result in a periblastula and the subsequent formation of germinal centers.

Crustaceans share a characteristic larval stage known as the nauplius larva, denoted by the appearance of three pairs of appendage-bearing somites (Figure 21.33D).[10] In those groups having little yolk in their eggs, the nauplius is generally free living. In those species with yolky eggs, the nauplius stage is generally passed through as part of a longer period of embryonic development (or a long brood period), and it is sometimes referred to as an **egg nauplius**. Free-living nauplii are usually planktotrophic, and their release corresponds to the depletion of stored yolk. However, in a few groups of crustaceans (e.g., euphausids and dendrobranchiate shrimps), the nauplius exhibits lecithotrophy.

Crustacean development is either direct, with the embryos hatching as juveniles that resemble miniature adults, or mixed, with embryos brooded for a brief or prolonged period and then hatching as a distinct larval form. These larval forms may pass through several subsequent phases, with each phase containing slightly different morphological steps called stages, before the adult condition is achieved. Direct development occurs in some cladocerans and branchiurans, and in all ostracods, phyllocarids, syncarids, and peracarids. Ostracods are typically viewed as having direct development, and they lack a distinct larval stage. However, some ostracod species hatch with only the first three pairs of appendages present, and they are thus true nauplii, even though they are in a bivalved carapace and add limbs gradually (the juvenile instars resemble miniature adults). All other crustaceans have some form of mixed development. The larval stages that have been recognized in crustacean groups that undergo mixed development have been assigned a plethora of names, and the homologies among these forms are not always understood. The more commonly encountered developmental forms are summarized below (also see Table 21.2 and Figure 21.33), but we do not attempt to describe them all.

Crustacean development is sometimes described as being either epimorphic, metamorphic, or anamorphic. However, we caution readers that a clear evolutionary and functional understanding of crustacean developmental stages is still lacking, and thus the terms "mixed" and "direct" may be preferable, and less ambiguous, until we have a better understanding of this phenomenon.

Epimorphic development is direct; in crustaceans it is thought to result from a delay in the hatching of the embryo, which causes the nauplius (and any other possible larval stages) to be suppressed or absent.

Metamorphic development is the type of extreme mixed development seen among the Eucarida; it includes dramatic transitions in body form from one life history stage to another. (This pattern is similar to holometabolous development in insects—for example, the transformation of a caterpillar into a butterfly.) In general, up to five distinct preadult, or larval, phases may be recognized among crustaceans: **nauplius**, **metanauplius**, **protozoea**, **zoea**, and "**postlarva**." The zoeal phase shows the greatest diversity in form among the various taxa, and especially in decapods it has been given different names in different groups (e.g., acanthosoma, antizoea, mysis, phyllosoma, pseudozoea). It is common for the zoeal phase to contain a large number of stages, each differing only slightly from the one preceding it.[11] Regardless of name, zoea are characterized by the presence of natatory exopods on some or all of the thoracic appendages and by the pleopods being absent (or rudimentary). Use of the term "postlarva" is unfortunate, as these stages differ dramatically from the preceding larvae as well as from the adults; they

[10]It was not until J. V. Thompson discovered the nauplius larvae of barnacles in the nineteenth century that this group was finally classified as Crustacea, a discovery that also marked the first use of larval features in understanding the phylogeny of marine invertebrates. A recent atlas of crustacean larvae (Martin et al. 2014) includes historical notes on larval development in all major crustacean groups.

[11]The zoea larvae of panulirid lobsters (phyllosoma larvae) are large, bizarre-appearing creatures (Figure 21.33L), which can occur in such large numbers as to be a favorite food of tuna.

represent unique transitional stages (both morphologically and ecologically). Examples in the Decapoda include the **megalopa** of true crabs and the **puerulus** larvae of spiny lobsters. In this phase of development, the role of swimming has switched from the thoracopodal limbs to the appendages of the abdomen.

Anamorphic development is a less extreme type of indirect development in which the embryo hatches as a nauplius larva, but the adult form is achieved through a series of gradual changes in body morphology as new segments and appendages are added (it is similar in many ways to hemimetabolous development in insects). In other words, the postnaupliar stages gradually take on the adult form with succeeding molts; the classic example of anamorphic development is often said to be the Anostraca. Cephalocarida, Remipedia, many Branchiopoda, and Mystacocarida are anamorphic—the nauplius larva grows by a series of molts that add new segments and appendages gradually as the adult morphology appears. In many groups hatching is somewhat delayed, and the emergent nauplius larva is termed a metanauplius. The basic nauplius possesses only three body somites, while the metanauplius has a few more; however, both possess only three pairs of similar-appearing appendages (which become the adult antennules, antennae, and mandibles). The end of the naupliar/metanaupliar stage is defined by the appearance of the fourth pair of functional limbs, the maxillules. In copepods a postnaupliar stage called a **copepodite** (simply a small juvenile) is often recognized.

The most extreme forms of metamorphic, or mixed, development occur in the malacostracan superorder Eucarida. The most complex developmental sequences are seen among the dendrobranchiate shrimps, which hatch as a typical nauplius larva that eventually undergoes a metamorphic molt to become a protozoea larva, with sessile compound eyes and a full complement of head appendages. The protozoea, after several molts, becomes a zoea larva, with stalked eyes and three pairs of thoracopods (as maxillipeds). The zoea eventually yields a juvenile stage (the "postlarva," a better term for which is "decapodid") that resembles a miniature adult, but is not sexually mature. In some other eucarid groups (Caridea and Brachyura) the postlarva is called a **megalopa**, and in the Anomura it is often called a **glaucothoe**; in both cases there are setose natatory pleopods on some or all of the abdominal somites. In other eucarids, some (or all) of these stages are absent.

Various other terms have been coined for different (or similar) developmental stages. For example, the modified zoeal stages of some stomatopods are called **antizoea** and **pseudozoea** larvae, and the advanced zoeal stage of many other malacostracans is often called a **mysis larva**. In euphausids, the nauplius is followed by two stages, the **calyptopis** and the **furcilia**, which roughly correspond to protozoea and zoea stages, before the juvenile morphology is attained.

From this wealth of terms and diversity of developmental sequences, we can draw two important generalizations concerning the biology and evolution of the crustaceans. First, different developmental strategies reflect adaptations to different lifestyles. In spite of many exceptions, we can cite the early release of dispersal larvae by groups with limited adult mobility, such as thoracican barnacles, and by those whose resources may not permit production of huge quantities of yolk, such as the copepods. At the other end of this adaptive spectrum is the direct development of peracarids—a major factor allowing the invasion of land by certain isopod lineages. Between these extremes we see all degrees of mixed life histories, with larvae being released at various stages following brooding and care. Second, because developmental stages also evolve, an analysis of developmental sequences can sometimes provide information about the radiation of the principal crustacean lineages. For example, the evolution of oostegites and of direct development combine as a unique synapomorphy of the Peracarida. Similarly, the addition of a unique larval form, such as the **cypris larva** that follows the nauplius in the cirripedes, can be viewed as a unique specialization that demarcates that group (Cirripedia). The cyprid either hatches as the only free-living larva, or it is the final larval stage after a series of lecithotrophic or planktotrophic nauplius larval stages.

It should also be noted that the branchiopods and some freshwater ostracods have evolved specialized ways of coping with the harsh conditions of many freshwater environments. Parthenogenesis, for example, is common in freshwater ostracods. Other adaptations include production of special overwintering forms, usually eggs or zygotes that can survive extreme cold, lack of water, or anoxic conditions. Perhaps most remarkable in this respect are the large-bodied branchiopods (fairy shrimp, tadpole shrimp, and clam shrimp) whose encysted embryos are capable of an extreme state of anaerobic quiescence, or diapause. During these resistant stages, the metabolic rate of the embryos may drop to less than 10% of their normal rate.

Many crustaceans have **indeterminate growth**, that is, they continue to molt throughout their life. In contrast, other species have **determinate growth** and cease molting following puberty (this life history stage is sometimes referred to as the terminal molt, or terminal anecdysis). In some species, the terminal molt is sex specific; for example, in American blue crabs (*Callinectes sapidus*) only females have a terminal molt.

Crustacean Phylogeny

Countless phylogenetic studies have been published on the topic of crustacean evolution. General agreement has been reached in some areas, but despite a great deal of effort, many fundamental mysteries remain unsolved, including the broad structure of the Pancrustaca tree. The use of molecular gene sequence data in crustacean phylogenetics is only just beginning. While in some cases it has resolved longstanding issues or confirmed previous hypotheses, in other cases it has led to still more questions. There are several particularly problematic issues. What are the basalmost living crustaceans, what sort of body did they have, and what are the relationships among them? What are the relationships of, and among, the taxa that were once united as "maxillopodans" (the copepods, branchiurans, thecostracans, tantulocarids, mystacocarids, and pentastomids)? Where do the Ostracoda, Cephalocarida, and Remipedia fit into crustacean phylogeny? What are the relationships among the Peracarida, especially of the many orders and families of Isopoda and Amphipoda? What are the major decapod lineages and how are they related to one another? What group of crustaceans is represented by the mysterious "y-larvae" (the Facetotecta), beyond the subsequent sluglike ypsigon stage that is clearly still not an adult? And, of course, what is the crustacean sister group to the Hexapoda?

Debates on crustacean phylogeny often center on whether paddle-legged Crustacea are ancestral or derived. The paddle-legged ancestry hypothesis holds that the first crustaceans had leaflike (phyllopodous) thoracic legs that were used both for swimming and for suspension feeding, as seen in the living cephalocarids, leptostracans, and many branchiopods. Or that the first crustaceans had simple, paddle-like legs that were used for swimming, but not for feeding; instead, the tasks of feeding were undertaken by the cephalic appendages—a plan perhaps best represented among living crustaceans by the remipedes. However, the opposing view, that paddle-legged crustaceans are more derived, is supported by the recent multigene studies.

A large molecular study suggested a highly-derived sister group relationship between the two most many-segmented and "wormlike" groups, the cephalocarids and remipedes, both being paddle-legged groups, uniting them in a proposed clade (called Xenocarida) that would be the sister to the hexapods (Regier et al. 2010; Figure 21.34B). Another recent phylogenomic study concluded that Remipedia are the closest extant relatives of insects (Misof et al. 2014), and some comparative neurological work also supports this view. In contrast, a traditional, morphology-based phylogeny of the Crustacea hypothesizes a paddle-legged ancestry, but places the Remipedia at the base of the crustacean tree (Figure 21.34A). Comparative spermatological studies, on the other hand, seem to ally the remipedes with certain former maxillipodans. Clearly, we have a good way to go before we will deeply understand the phylogeny of Crustacea.

The Regier et al. (2010) study, summarized here in Figure 21.34B, strongly supported some traditionally recognized groupings, such as Branchiopoda, Thecostraca (barnacles and their kin), and Malacostraca, but challenged others, and proposed several new names and clades (e.g., Oligostraca and Vericrustacea) that have yet to withstand the test of time.

In the 1950s, Russian biologist W. N. Beklemischev and Swedish carcinologist E. Dahl independently proposed that the copepods and several related classes constitute a monophyletic clade. Dahl proposed the class Maxillopoda for these taxa, and that term has been employed often since then. Characters shared by these small taxa include the shortening of the thorax to six or fewer segments and of the abdomen to four or fewer segments, reduction of the carapace (or, in the case of ostracods and cirripedes, extreme modification of the carapace), loss of abdominal appendages, and other associated changes, all thought to be tied to early paedomorphic events during the larval (or postlarval) stage of this lineage as it began to radiate (an idea first proposed in 1942 by R. Gurney). However, molecular phylogenetic studies since then have shown Maxillopda to be a nonmonophyletic grouping, and the taxon has been abandoned by most modern workers. Yet two of those groups, the copepods and thecostracans, have since been reunited in one study (Oakley et al. 2013) that posits a clade (the Hexanauplia) based on the same number (6) of naupliar larval stages in these groups.

The monophyletic nature of the class Malacostraca has rarely been questioned. Within the Malacostraca are two clades: Leptostraca, which have phyllopodous limbs and seven abdominal somites, and Eumalacostraca, which lack phyllopodous limbs and have six abdominal segments. Whether the Hoplocarida are members of the Eumalacostraca or deserving of their own subclass of the Malacostraca is still a subject of debate, but we consider them a separate subclass here. Hoplocarids and eumalacostracans also have the sixth abdominal appendages modified as uropods (which work in conjunction with the telson as a tail fan). Relationships among the three main eumalacostracan lines (syncarids, peracarids, and eucarids), and even within the Eucarida, are far from settled and have provided zoologists with many generations of lively debate.

The class Branchiopoda is usually monophyletic in molecular analyses, but it is difficult to define on the basis of unique synapomorphies because it shows such great morphological variation. But larval (naupliar) characters, such as the reduced and tubular first antenna and uniramous mandibles, strongly support their monophyly, as do almost all studies employing molecular data. As is true with so many crustacean groups, our early classifications greatly underappreciated their diversity, and relationships among the constituent groups are far more complex than was first thought. Apparently some branchiopods have secondarily lost

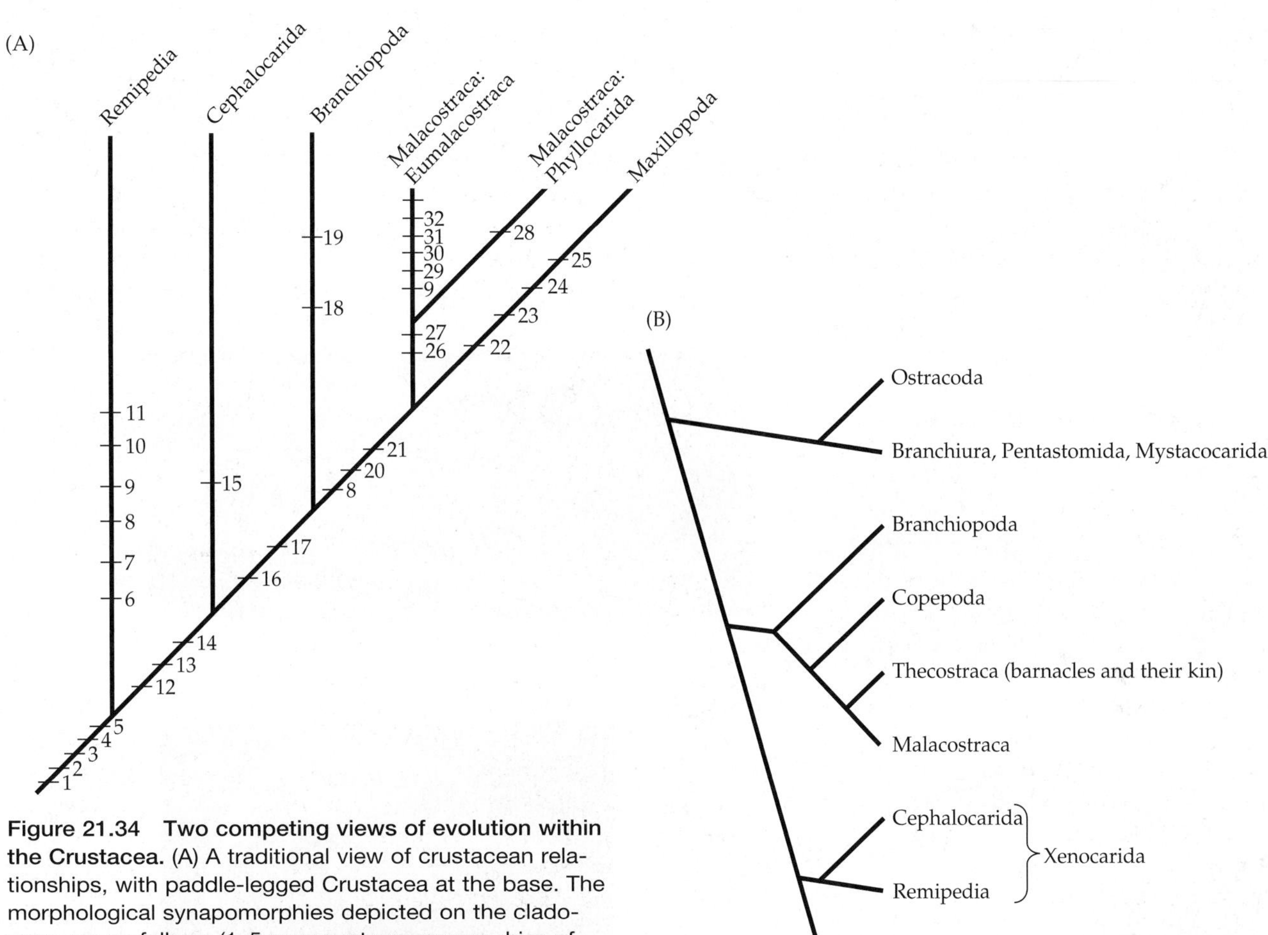

Figure 21.34 Two competing views of evolution within the Crustacea. (A) A traditional view of crustacean relationships, with paddle-legged Crustacea at the base. The morphological synapomorphies depicted on the cladogram are as follows (1–5 represent synapomorphies of the subphylum "Crustacea"): 1: head composed of 4 or 5 fused segments (plus acron) with 2 pairs of antennae and 2 or 3 pairs of mouth appendages; 2: biramous second antennae; 3: nauplius larva; 4: phyllopodous body limbs (with large epipods); 5: with head shield or small carapace; 6: raptorial mouth appendages; 7: mouth appendages situated in posteriorly directed atrium; 8: anterior thoracopods (one or more pairs) modified as maxillipeds (a highly variable trait that occurs in remipedes, malacostracans, and some former "maxillopodans"); 9: loss of phyllopodous condition on trunk appendages; 10: trunk appendages oriented laterally; 11: maxillules function as hypodermic fangs; 12: postcephalic trunk regionalized as thorax and abdomen; 13: loss of internal organ homonomy (e.g., segmental gut ceca); 14: reduction in number of body segments; 15: reduction of abdomen (to 11 segments); 16: fully developed carapace (reduced in several subsequent lineages); 17: reduction of abdomen to fewer than 9 segments; 18: reduction (or loss) of abdominal appendages; 19: first and second maxillae reduced or lost; 20: thorax shortened to fewer than 11 segments; 21: abdomen shortened to fewer than 8 segments; 22: with maxillopodan naupliar eye; 23: thorax of 6 or fewer segments; 24: abdomen of 4 or fewer segments; 25: genital appendages on the first abdominal somite (associated with male gonopores); 26: 8-segmented thorax and 7-segmented abdomen (plus telson); 27: male gonopores fixed on thoracomere 8/females on thoracomere 6; 28: carapace forms large "folded" structure enclosing most of body; 29: abdomen reduced to 6 segments (plus telson); 30: last abdominal appendages modified as uropods and forming tail fan with telson; 31: caridoid tail flip locomotion (escape reaction); 32: thoracopods with stenopodous endopods; 33: replacement of thoracic suspension feeding and phyllopodous thoracic limbs with cephalic feeding and nonphyllopodous thoracic limbs. Note that loss of the phyllopodous trunk limbs (character 9) has occurred several times, in the Remipedia, Eumalacostraca, some lineages of Branchiopoda, and most of the former "maxillopodan" groups. (B) Phylogeny of the Pancrustacea clade as derived by Regier et al. 2010.

the carapace, and others have secondarily lost most or all of the abdominal appendages.

That a group of crustaceans gave rise to the megadiverse group of arthropods called the Hexapoda (insects and their kin) is now almost universally agreed upon, based on a wide array of evidence that includes molecular sequence data and neuroanatomy. This realization renders the group called Crustacea paraphyletic. The monophyletic clade containing both crustaceans and insects is most often referred to as the Pancrustacea (see Chapter 20), but it is also sometimes called the Tetraconata, a name that recognizes the square shape of the ommatidia of many species. Earlier work suggested that Branchiopoda was the likely sister

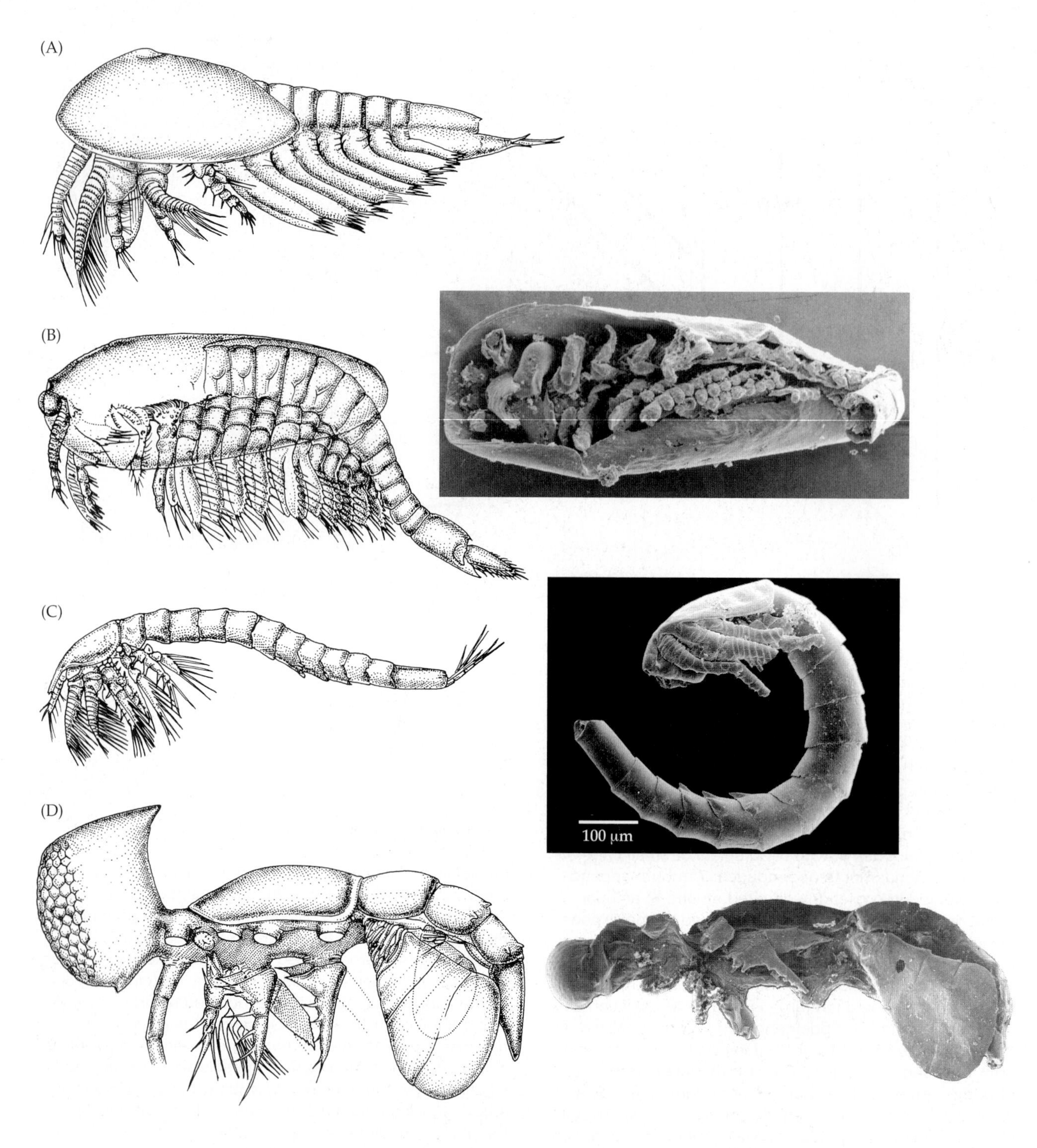

group to Hexapoda, but recent research supports the origin of the Hexapoda from either the Remipedia or a remipede-cephalocarid clade—the later grouping (Remipedia-Cephalocarida-Hexapoda) has been named Miracrustacea, meaning "surprising crustaceans." Like the arthropods in general, crustaceans exhibit high levels of evolutionary parallelism and convergence and many apparent reversals of character states. This genetic flexibility is no doubt due in part to the nature of the segmented body, the serially homologous appendages, and the flexibility of developmental genes, which, as we have stressed, provide enormous opportunity for evolutionary experimentation. Any conceivable cladogram of crustacean phylogeny will require the acceptance of considerable homoplasy.

Fossil data (including that of the Orsten fauna, Figure 21.35) seems to favor phyllopodous limbs as the primitive condition. However, developmental studies following the expression of *Distal-less* and other developmental genes suggest that the early embryogeny of limbs is very similar among crustaceans. For example, trunk limbs always emerge as ventral, subdivided limb buds. In phyllopodous limbs, the subdivisions of these limb buds grow to become the endites and

(E)

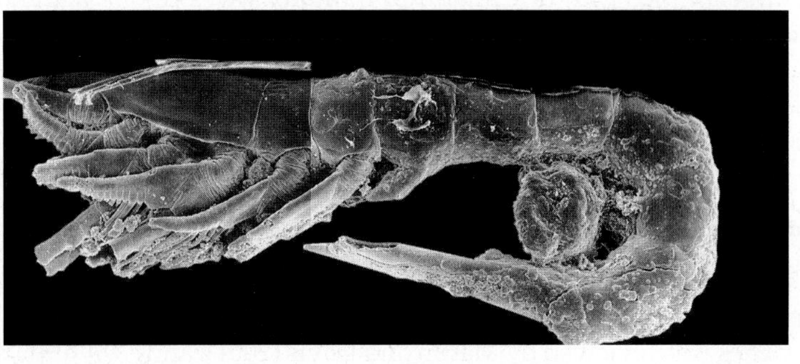

(F)

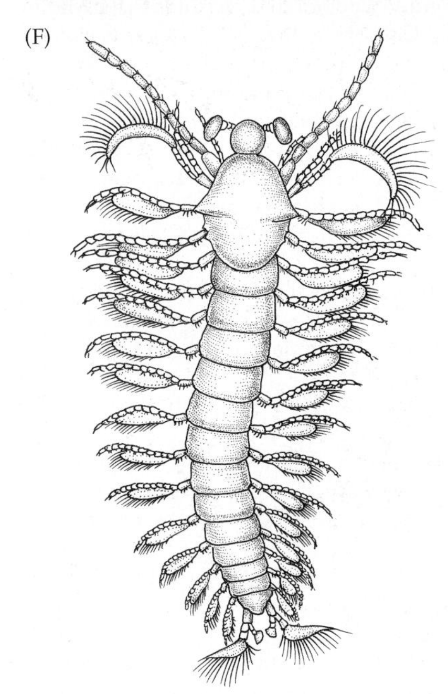

Figure 21.35 (A–E) Examples of Upper Cambrian (~510 Ma), probably meiofaunal crustacean fossils from the spectacular Swedish Orsten deposits. This ancient crustacean fauna possessed the key attributes of modern Crustacea, including compound eyes, head shields/carapaces, naupliar larvae (with locomotory first antennae), and biramous appendages on the second and third head segments (the second antennae and mandibles). (A) *Bredocaris*. (B) *Rehbachiella*, an early branchiopod, lateral (drawing) and ventral (SEM) views. (C) *Skara*, drawing and SEM. (D) *Cambropachycope clarksoni*, a bizarre species with an expanded head and two pairs of enlarged thoracopods, drawing and SEM. (E) *Martinssonia elongata*, SEMs of first larva and postlarval stage. (F) *Ercaia minuscula*, an early Cambrian (520 Ma) crustacean from South China.

the endopod of the natatory/filtratory adult limbs. In stenopodous limbs, the same limb bud subdivisions end up developing into the actual segments of the adult limb. Hence, the endites of phyllopodous limbs appear to be homologous to the segments of the stenopodous limbs. This discovery supports an emerging view of developmental plasticity in arthropod limbs, and it suggests that relatively simple genetic "switches" can account for major differences in adult morphologies. Thus, it is quite plausible that stenopodous limbs have evolved multiple times from phyllopodous ancestors, and this is the scenario depicted in the cladogram in Figure 21.34A.

Work by Klaus Müller and Dieter Waloszek on three-dimensionally preserved microscopic arthropods from the middle Cambrian Orsten (around 510 Ma) deposits of Sweden has documented a diverse fauna of minute crustaceans and their larvae. Among them, for example, is *Skara* (Figure 21.35C), a cephalocarid- or mystacocarid-like crustacean for which both naupliar larvae and adults have been recovered (the nauplius larvae are only a couple hundred microns long; adults are about 1 mm in length). *Skara* and many other Orsten Crustacea were probably meiofaunal animals not unlike modern marine meiofaunal crustaceans (e.g., the mystacocarids). Dozens of Orsten microcrustacea have so far been described (Figure 21.35). Recently, a beautifully preserved fossil crustacean, *Ercaia minuscula*, was described from the middle Cambrian (520 Ma) of South China. It has an untagmatized, 13-segmented trunk with serially repeated biramous appendages and a head with stalked eyes and five pairs of head appendages, including two pairs of antennae. *Ercaia minuscula* is only 2–4 mm long (hence the species name), and bears resemblances to both cephalocarids and maxillopodans (Figure 21.35F).

Studies on the Swedish Orsten fauna (510 Ma), the Middle Cambrian (520 Ma) Burgess Shale-like deposits from around the world, and the Lower Cambrian (530 Ma) Chengjiang fossils from China have shown that Cambrian Crustacea had all the attributes of modern crustaceans, such as compound eyes, an acron, distinct head and trunk tagmata, at least four head appendages, a carapace (or head shield), naupliar (or "head") larvae (with locomotory first antennae), and biramous appendages on the second and third head somites (the second antennae and mandibles). We now know that the crustaceans are an ancient group. Their fossil record dates back to the early Cambrian, or likely the Ediacaran period if some arthropod fossils from those strata are viewed as Crustacea. The earliest known crustacean larval stage is a fossil of a slightly advanced nauplius (called a metanauplius) from the early

Cambrian (525 Ma) of China that in many ways resembles the naupliar larvae of modern cirripedes. This is remarkable, as it confirms not only of the great age of the Crustacea as a whole but also the age of groups within the Crustacea, such as the cirripedes, that were already becoming distinct. Depending on one's definition of "Crustacea," it may even be that the first arthropods were themselves crustaceans.

Selected References

The amount of published literature on crustaceans is vast. Much of the key work on classification and phylogenetics was reviewed by Martin and Davis (2001), and on larval development by Martin et al. (2014). We refer readers to those works for an entrée into those fields.

General References

Ahyong, S. T. and 12 others. 2011. Subphylum Crustacea Brünnich, 1772. Pp. 164–191 in, A.-Q. Zhang (ed.), Animal biodiversity: an outline of higher-level classification and survey of taxonomic richness. Zootaxa 3148.

Anderson, D. T. 1994. *Barnacles: Structure, Function, Development and Evolution*. Chapman and Hall, London.

Arhat, A. and T. C. Kaufman. 1999. Novel regulation of the homeotic gene Scr associated with a crustacean leg-to-maxilliped appendage transformation. Development 126: 1121–1128.

Averof, M. and N. H. Patel. 1997. Crustacean appendage evolution associated with changes in Hox gene expression. Nature 388: 682–686.

Baker, A. de C., B. P. Boden and E. Brinton. 1990. *A Practical Guide to the Euphausiids of the World*. Natural History Museum Publications, London.

Barnard, J. L. 1991. The families and genera of marine gammaridean Amphipoda (except marine gammaroids). Rec. Aust. Mus., Suppl 13 (1/2): 1–866. [A benchmark compilation by one of the world's foremost, and most colorful, carcinologists.]

Barnard, J. L. and C. M. Barnard. 1983. *Freshwater Amphipoda of the World*. Hayfield Associates, Mt. Vernon, VA.

Bauer, R. T. 1981. Grooming behavior and morphology in the decapod Crustacea. J. Crust. Biol. 1: 153–173.

Bauer, R. T. 1987. Stomatopod grooming behavior: Functional morphology and amputation experiments in *Gonodactylus oerstedii*. J. Crust. Biol. 7: 414–432.

Bauer, R. T. 2004. *Remarkable Shrimps*. Univeristy of Oklahoma Press, Norman.

Bauer, R. T. and J. W. Martin (eds.). 1991. *Crustacean Sexual Biology*. Columbia University Press, New York.

Bliss, D. E. (gen. ed.). 1982-1990. *The Biology of Crustacea. Vols. 1–4*. Academic Press, New York.

Bowman, T. E. and H.-E. Gruner. 1973. The families and genera of Hyperiidea (Crustacea: Amphipoda). Smithson. Contrib. Zool. 146: 1–64.

Bowman, T. E., S. P. Garner, R. R. Hessler, T. M. Iliffe and H. L. Sanders. 1985. Mictacea, a new order of Crustacea Peracarida. J. Crust. Biol. 5: 74–78.

Boxshall, G. and S. Halsey. 2004. *An introduction to copepod diversity*. The Ray Society (London), Publ. No. 166. [A 2 volume work providing a family-by-family account of all copepods.]

Brtek, J. and G. Mura. 2000. Revised key to families and genera of the Anostraca with notes on their geographical distribution. Crustaceana 73: 1037–1088.

Brusca, G. J. 1981. Annotated keys to the Hyperiidea (Crustacea: Amphipoda) of North American coastal waters. Allan Hancock Found. Tech. Rep. 5: 1–76.

Brusca, G. J. 1981. On the anatomy of *Cystisoma* (Amphipoda: Hyperiidea). J. Crust. Biol. 1: 358–375.

Brusca, R. C. 1981. A monograph on the Isopoda Cymothoidae (Crustacea) of the Eastern Pacific. Zool. J. Linn. Soc. 73(2): 117–199.

Brusca, R. C. and M. Gilligan. 1983. Tongue replacement in a marine fish (*Lutjanus guttatus*) by a parasitic isopod (Crustacea: Isopoda). Copeia 3: 813–816. [The first known case of a parasite functionally replacing a host organ.]

Brusca, R. C., S. Taiti and V. Coelho. 2001. A Guide to the Marine Isopods of Coastal California. http://phylogeny.arizona.edu/tree/eukaryotes/animals/arthropoda /crustacea / isopoda/ isopod_lichen/bruscapeet.html. [Comprehensive coverage of the Northeastern Pacific isopod fauna.]

Brusca, R. C., R. Wetzer and S. France. 1995. Cirolanidae (Crustacea; Isopoda; Flabellifera) of the tropical eastern Pacific. Proc. San Diego Nat. Hist. Soc., No. 30.

Burukovskii, R. N. 1985. *Key to Shrimps and Lobsters*. A. A. Balkema, Rotterdam.

Calman, R. T. 1909. Crustacea. In R. Lankester (ed.), *A Treatise on Zoology, Pt. 7*. Adam and Charles Black, London. [Obviusly dated, but still a solid benchmark work on Crustacean anatomy.]

Cameron, J. N. 1985. Molting in the blue crab. Sci. Am. 252: 102–109.

Carpenter, J. H. 1999. Behavior and ecology of *Speleonectes epilimnius* (Remipedia, Speleonectidae) from surface water of an anchialine cave on San Salvador Island, Bahamas. Crustaceana 72: 979–991.

Carter, J. W. 1982. Natural history observations on the gastropod shell-using amphipod *Photis conchicola* Alderman, 1936. J. Crust. Biol. 2: 328–341.

Chace, F. A., Jr. 1972. The shrimps of the Smithsonian-Bredin Caribbean expeditions with a summary of the West Indies shallow-water species. Smithson. Contrib. Zool. 98: 1–180. [A useful entrée into the marine shrimps of the Caribbean Region.]

Chang, E. S. 1985. Hormonal control of molting in decapod Crustacea. Am. Zool. 25: 179–185.

Chapman, M. A. and M. H. Lewis. 1976. *An Introduction to the Freshwater Crustacea of New Zealand*. Collins, Auckland.

Cohen, A. C. and J. G. Morin. 1990. Patterns of reproduction in ostracodes: A review. J. Crust. Biol. 10: 84–211.

Cohen, A. C. and J. G. Morin. 2003. Sexual morphology, reproduction and the evolution of bioluminescence in Ostracoda. Paleontological Society Papers 9: 37–70.

Crane, J. 1975. *Fiddler Crabs of the World (Ocypodidae: Genus* Uca*)*. Princeton University Press, Princeton.

Cronin, T. W. 1986. Optical design and evolutionary adaptation in crustacean compound eyes. J. Crust. Biol. 6: 1–23.

Darwin, C. 1852, 1854. *A Monograph on the Subclass Cirripedia. Vols. 1–2*. Ray Society, London. [Still the starting place for barnacle taxonomy.]

De Grave, S., and 17 others. 2009. A classification of living and fossil genera of decapod crustaceans. Raffles Bull. Zool. Supp. 21: 1–109.

De Jong-Moreau, L. and J.-P. Casanova. 2001. The foreguts of the primitive families of the Mysida (Crustacea, Peracarida): A transitional link between those of the Lophogastrida

(Crustacea, Mysidacea) and the most evolved Mysida. Acta Zool. 82: 137–147.

Derby, C. D. 1982. Structure and function of cuticular sensilla of the lobster *Homarus americanus*. J. Crust. Biol. 2: 1–21.

Derby, C. D. and J. Atema. 1982. The function of chemo- and mechanoreceptors in lobster (*Homarus americanus*) feeding behaviour. J. Exp. Biol. 98: 317–327.

Duffy, E. J. and M. Thiel (eds.). 2007. Evolutionary Ecology of Social and Sexual Systems: Crustaceans as Model Organisms. Oxford University Press.

Efford, I. E. 1966. Feeding in the sand crab *Emerita analoga*. Crustaceana 10: 167–182.

Elofsson, R. 1965. The nauplius eye and frontal organs in Malacostraca. Sarsia 19: 1–54.

Elofsson, R. 1966. The nauplius eye and frontal organs of the non-Malacostraca. Sarsia 25: 1–28.

Factor, J. R. (ed.). 1995. *Biology of the Lobster* Homarus americanus. Academic Press, San Diego.

Fanenbruck, M., S. Harzsch and J. W. Wägele. 2004. The brain of the Remipedia (Crustacea) and an alternative hypothesis on their phylogenetic relationships. PNAS 101(11): 3868–3873.

Fitzpatrick, J. F., Jr. 1983. *How to Know the Freshwater Crustacea*. Wm. C. Brown, Dubuque, IA.

Forest, J. 1999a. *Traité de Zoologie. Anatomie, Systématique, Biologie. Tome VII, Fascicule II. Généralités (suite) et Systématique*. Crustacés. Masson, Paris.

Forest, J. (ed.) 1999b. *Traité de Zoologie. Anatomie, Systématique, Biologie. Tome VII, Fascicule IIIA*. Crustacés Péracarides. Masson, Paris.

Forest, J. (founding editor). 2004–present. *The Crustacea*. Brill, Leiden. [This series of 14 planned volumes, each with separate editors, is an update to the 1999 *Traité de Zoologie*]

Fryer, G. 1964. Studies on the functional morphology and feeding mechanism of *Monodella argentarii* Stella (Crustacea: Thermosbaenacea). Trans. R. Soc. Edinburgh 66(4): 49–90.

Garm, A. 2004. Revising the definition of the crustacean seta and setal classification systems based on examinations of the mouthpart setae of seven species of decapods. Zoological Journal of the Linnean society 142: 233–252.

Gerrish, G. A. and J. G. Morin. 2008. Life cycle of a bioluminescent marine ostracode, *Vargula annecohenae* (Myodocopida: Cypridinidae). J. Crustacean Biol. 28(4): 669–674.

Ghiradella, H. T., J. Case and J. Cronshaw. 1968. Structure of aesthetascs in selected marine and terrestrial decapods: Chemoreceptor morphology and environment. Am. Zool. 8: 603–621.

Gilchrist, S. and L. A. Abele. 1984. Effects of sampling parameters on the estimation of population parameters in hermit crabs. J. Crust. Biol. 4: 645–654. [Includes a good literature list on shell selection by hermit crabs.]

Glenner, H. 2001. Cypris metamorphosis, injection and earliest internal development of the rhizocephalan *Loxothylacus panopaei* (Gissler). Crustacea: Cirripedia: Rhizocephala: Sacculinidae. J. Morphol. 249: 43-75.

Glenner, H. and M. B. Hebsgaard. 2006. Phylogeny and evolution of life history strategies of the parasitic barnacles (Crustacea, Cirripedia, Rhizocephala). Mol. Phylog. Evol. 41: 528-538.

Glenner, H., J. T. Høeg, J. J. O'Brien and T. D. Sherman. 2000. Invasive vermigon stage in the parasitic barnacles *Loxothylacus texanus* and *L. panopaei* (Sacculinidae): Closing of the rhizocephalan life-cycle. Mar. Biol. 136: 249–257.

Goffredi, S. K., W. J. Jones, H. Erhlich, A. Springer and R. C. Vrijenhoek. 2008. Epibiotic bacteria associated with the recently discovered Yeti crab, *Kiwa hirsuta*. Environ. Microbiol. 10: 2623–2634.

Goldstein, J. S., H. Matsuda, T. Takenouchi and M. J. Butler IV. 2008. The complete development of larval Caribbean spiny lobster *Panulirus argus* (Latreille, 1804) in culture. J. Crustacean Biol. 28: 306-327.

Gordon, I. 1957. On *Spelaeogriphus*, a new cavernicolous crustacean from South Africa. Bull. Br. Mus. Nat. Hist. Zool. 5: 31–47.

Govind, C., M. Quigley and K. Mearow. 1986. The closure muscle in the dimorphic claws of male fiddler crabs. Biol. Bull. 170: 481–493.

Grey, D. L., W. Dall and A. Baker. 1983. *A Guide to the Australian Penaeid Prawns*. North Territory Govt. Printing Office, Australia.

Grindley, J. R. and R. R. Hessler. 1970. The respiratory mechanism of *Spelaeogriphus* and its phylogenetic significance. Crustaceana 20: 141–144.

Grygier, M. J. 1982. Sperm morphology in Ascothoracida (Crustacea: Maxillopoda): Confirmation of generalized nature and phylogenetic importance. Int. J. Invert. Reprod. 4: 323–332.

Grygier, M. J. 1987. New records, external and internal anatomy, and systematic position of Hansen's Y-larvae (Crustacea: Maxillopoda: Facetotecta). Sarsia 72: 261–278.

Guinot, D., D. Doumenc and C. C. Chintiroglou. 1995. A review of the carrying behaviour in Brachyuran crabs, with additional information on the symbioses with sea anemones. Raffles Bull. Zool. 43(2): 377–416.

Haig, J. 1960. The Porcellanidae (Crustacea: Anomura) of the eastern Pacific. Allan Hancock Pacific Expeditions 24: 1–440.

Hallberg, E. and R. Elofsson. 1983. The larval compound eye of barnacles. J. Crust. Biol. 3: 17–24.

Hamner, W. M. 1988. Biomechanics of filter feeding in the Antarctic krill *Euphausia superba*: Review of past work and new observations. J. Crust. Biol. 8: 149–163.

Harbison, G. R., D. C. Biggs and L. P. Madin. 1977. The associations of Amphipoda Hyperiidea with gelatinous zooplankton. II. Associations with Cnidaria, Ctenophora and Radiolaria. Deep-Sea Res. 24(5): 465–488.

Harrison, F. W. and A. G. Humes. 1992. *Microscopic Anatomy of Invertebrates. Vols. 9 and 10, Crustacea and Decapod Crustacea*. Wiley-Liss, New York. [Two outstanding volumes in this fine series.]

Harvey, A. H., J. W. Martin and R. Wetzer. 2002. Crustacea. Pp. 337–369 in C. Young, M. Sewell and M. Rice (eds.), *Atlas of Marine Invertebrate Larvae*. Academic Press, London.

Heart, R. W., W. W. Price, D. M. Knott, R. A. King and D. M. Allen. 2006. *A Taxonomic Guide to the Mysids of the South Atlantic Bight*. NOAA Professional Paper NMFS 4.

Hegna, T. A. and E. Lazo-Wasem. 2010. *Branchinecta brushi* n. sp. (Branchiopoda: Anostraca: Branchinectidae) from a volcanic crater in northern Chile (Antofagasta Province): a new altitude record for crustaceans. J. Crustacean Biol. 30: 445–464.

Herrnkind, W. F. 1985. Evolution and mechanisms of single-file migration in spiny lobster: Synopsis. Contrib. Mar. Sci. 27: 197–211.

Hessler, R. R. 1982. The structural morphology of walking mechanisms in eumalacostracan crustaceans. Phil. Trans. R. Soc. Lond. Ser. B 296: 245–298. [An outstanding review.]

Hessler, R. R. 1985. Swimming in Crustacea. Trans. R. Soc. Edinburgh 76: 115–122.

Hessler, R. R. and R. Elofsson. 1991. Excretory system of *Hutchinsoniella macracantha* (Cephalocarida). J. Crust. Biol. 11: 356–367.

Hessler, R. R. and J. Yager. 1998. Skeletomusculature of trunk segments and their limbs in *Speleonectes tulumensis* (Remipedia). J. Crustacean Biol. 18: 111–119.

Høeg, J. T. and G. A. Kolbasov. 2002. Lattice organs in y-cyprids of the Facetotecta and their significance in the phylogeny of the Crustacea Thecostraca. Acta Zool. 83: 67–79.

Holthuis, L. B. 1980. Shrimps and prawns of the world: An annotated catalogue of species of interest to fisheries. FAO Species Catalogue, Vol. 1/Fisheries Synopses 125: 1–261.

Holthuis, L. B. 1991. *Marine lobsters of the world: An annotated and illustrated catalogue of species of interest to fisheries known to date. FAO Species Catalogue, Vol. 13.* FAO, Rome.

Holthuis, L. B. 1993. *The recent genera of the caridean and stenopodidean shrimps (Crustacea, Decapoda) with an appendix on the order Amphionidacea.* Nat. Natuurhistorisch. Mus., Leiden.

Horch, K. W. and M. Salmon. 1969. Production, perception and reception of acoustic stimuli by semiterrestrial crabs. Forma Functio 1: 1–25.

Holmes, J. and A. Chivas (eds.). 2002. *The Ostracoda: Applications in Quaternary Research.* AGU Geophysical Monograph.

Huvard, A. L. 1990. The ultrastructure of the compound eye of two species of marine ostracods (Ostracoda: Cypridinidae). Acta Zool. 71: 217–224.

Huys, R., G. A. Boxshall and R. J. Lincoln. 1993. The tantulocarid life cycle: The circle closed? J. Crust. Biol. 13: 432–442.

Ingle, R. W. 1980. *British Crabs.* Oxford University Press, Oxford.

Ivanov, B. G. 1970. On the biology of the Antarctic krill *Euphausia superba.* Mar. Biol. 7: 340.

Jamieson, B. G. M. 1991. Ultrastructure and phylogeny of crustacean spermatozoa. Mem. Queensland Mus. 31: 109–142.

Jensen, G. C. 2014. *Crabs and Shrimps of the Pacific Coast. A Guide to Shallow-Water Decapods from Southeastern Alaska to the Mexican Border.* MolaMarine, Bremerton, WA. [A comprehensive natural history of northeastern Pacific decapods.]

Jones, D. and G. Morgan. 2002. *A Field Guide to Crustaceans of Australian Waters.* Reed New Holland, Sydney.

Jones, N. S. 1976. *British Cumaceans.* Academic Press, New York.

Kabata, Z. 1979. *Parasitic Copepoda of British Fishes.* The Ray Society, London.

Kaestner, A. 1970. *Invertebrate Zoology. Vol. 3, Crustacea.* Wiley, New York. [An excellent resource; translated from the 1967 German second edition by H. W. Levi and L. R. Levi.]

Kennedy, V. S. and L. E. Cronin (eds.). 2007. *The Blue Crab:* Callinectes sapidus. Maryland Sea Grant.

Kensley, B. and R. C. Brusca (eds.) 2001. *Isopod Systematics and Evolution.* Balkema, Rotterdam.

King, J. L., M. A. Simovich and R. C. Brusca. 1996. Endemism, species richness, and ecology of crustacean assemblages in northern California vernal pools. Hydrobiologia 328: 85–116.

Koehl, M. A. R. and J. R. Strickler. 1981. Copepod feeding currents: Food capture at low Reynolds numbers. Limnol. Oceanogr. 26: 1062–1073.

Koenemann, S., F. R. Schram, T. M. Iliffe, L. M. Hinderstein and A. Bloechl. 2007. Behavior of Remipedia in the laboratory, with supporting field observations. J. Crustacean Biol. 27(4): 534–542.

Land, M. F. 1981. Optics of the eyes of *Phronima* and other deep-sea amphipods. J. Comp. Physiol. 145: 209–226.

Land, M. F. 1984. Crustacea. Pp. 401–438 in M. A. Ali (ed.), *Photoreception and Vision in Invertebrates.* Plenum, New York.

Lang, K. 1948. *Monographie der Harpacticoiden.* Hakan Ohlssons, Lund. [1,682 pp.—whew!]

Laval, P. 1980. Hyperiid amphipods as crustacean parasitoids associated with gelatinous zooplankton. Oceanogr. Mar. Biol. Annu. Rev. 18: 11–56.

Maas, A. and D. Waloszek. 2001. Larval development of *Euphausia superba* Dana, 1852 and a phylogenetic analysis of the Euphausiacea. Hydrobiologia 448: 143-169.

MacPherson, E., W. Jones and M. Segonzac. 2005. A new squat lobster family of Galatheoidea (Crustacea, Decapoda, Anomura) from the hydrothermal vents of the Pacific-Antarctic Ridge. Zoosystema 27(4): 709–723.

Madin, L. P. and G. R. Harbison. 1977. The associations of Amphipoda Hyperiidea with gelatinous zooplankton. I. Associations with Salpidae. Deep-Sea Res. 24: 449–463.

Maitland, D. P. 1986. Crabs that breathe air with their legs—*Scopimera* and *Dotilla.* Nature 319: 493–495.

Manning, R. B. 1969. *Stomatopod Crustacea of the Western Atlantic.* University of Miami Press, Coral Gables, FL.

Manning, R. B. 1974. *Crustacea: Stomatopoda. Marine flora and fauna of the northeastern U.S.* NOAA Tech. Rpt., Nat. Mar. Fish. Serv. Circular 386.

Marshall, S. M. 1973. Respiration and feeding in copepods. Adv. Mar. Biol. 11: 57–120.

Martin, J. W. and D. Belk. 1988. Review of the clam shrimp family Lynceidae (Stebbing, 1902) (Branchiopoda: Conchostraca) in the Americas. J. Crust. Biol. 8: 451–482.

Martin, J. W. and G. E. Davis. 2001. *An updated classification of the Recent Crustacea.* Nat. Hist. Mus. Los Angeles Co., Sci. Ser. No. 39. [A synthesis of the literature.]

Martin, J. W., J. Olesen, and J. T. Høeg (editors). 2014. *Atlas of Crustacean Larvae.* Johns Hopkins University Press.

Mauchline, J. 1980. The biology of mysids and euphausiids. Adv. Mar. Biol. 18: 1–681.

McCain, J. C. 1968. The Caprellidae (Crustacea: Amphipoda) of the western North Atlantic. U.S. Nat. Mus. Bull. 278: 1–147.

McGaw, I. J. 2005. The decapod crustacean cardiovascular system: A case that is neither open nor closed. Microscopy and Microanalysis 11: 18–36.

McLaughlin, P. A. 1974. The hermit crabs of northwestern North America. Zool. Verh. Rijksmus. Nat. Hist. Leiden 130: 1–396.

McLay, C. L. 1988. *Crabs of New Zealand.* Leigh Lab. Bull. 22: 1–463.

Miller, D. C. 1961. The feeding mechanism of fiddler crabs with ecological considerations of feeding adaptations. Zoologica 46: 89–100.

Morin, J. G. and A. C. Cohen. 2010. It's all about sex: bioluminescent courtship displays, morphological variation and sexual selection in two new genera of Caribbean ostracodes. J. Crustacean Biol. 30: 6–67.

Müller, K. J. 1983. Crustaceans with preserved soft parts from the Upper Cambrian of Sweden. Lethaia 16: 93–109.

Müller, K. J. and D. Walossek. 1985. Skaracarida, a new order of Crustacea from the Upper Cambrian of Västergötland, Sweden. Fossils and Strata 17: 1–65.

Müller, K. J. and D. Walossek. 1986. *Martinssonia elongata* gen. et sp. n., a crustacean-like euarthropod from the Upper Cambrian "Orsten" of Sweden. Zoologica Scripta 15: 73–92.

Müller, K. J. 1986. Arthropod larvae from the Upper Cambrian of Sweden. Trans. R. Soc. Edinburgh, Earth Sci. 77: 157–179.

Newman, W. A. and R. R. Hessler. 1989. A new abyssal hydrothermal verrucomorphan (Cirripedia: Sessilia): the most primitive living sessile barnacle. Trans. San Diego Nat. Hist. Soc. 21: 259–273.

Newman, W. A. and A. Ross. 1976. Revision of the balanomorph barnacles; including a catalog of the species. San Diego Soc. Nat. Hist. Mem. 9: 1–108.

Ng, P.K.L., D. Guinot, and P.J.F. Davie. 2008. Systema Brachyurorum: Part I. An annotated checklist of extant brachyuran crabs of the world. Raffles Bull. Zool. Supplement 17: 1–286.

Nolan, B. A. and M. Salmon. 1970. The behavior and ecology of snapping shrimp (Crustacea: *Alpheus heterochelis* and *Alpheus normanni*). Forma Functio 2: 289–335.

Oeksnebjerg, B. 2000. The Rhizocephala of the Mediterranean and Black Seas: taxonomy, biogeography, and ecology. Israel J. Zool. 46 (1): 1–102.

Olesen, J. 1999. Larval and post-larval development of the branchiopod clam shrimp *Cyclestheria hislopi* (Baird, 1859)

(Crustacea, Branchiopoda, Conchostraca, Spinicaudata). Acta Zool. 80: 163–184.

Olesen, J. 2001. External morphology and larval development of *Derocheilocaris remanei* Delamare-Deboutteville & Chappuis, 1951 (Crustacea, Mystacocarida), with a comparison of crustacean segmentation and tagmosis patterns. Biologiske Skrifter 53: 1–59.

Olesen, J., J. W. Martin and E. W. Roessler. 1996. External morphology of the male of *Cyclestheria hislopi* (Baird, 1859) (Crustacea, Branchiopoda, Spinicaudata), with a comparison of male claspers among the Conchostraca and Cladocera and its bearing on phylogeny of the "bivalved" Branchiopoda. Zoologica Scripta 25: 291–316.

Pabst, T. and G. Scholz. 2009. The development of phyllopodous limbs in Leptostraca and Branchiopoda. J. Crustacean Biol. 29(1): 1–12.

Pennak, R. W. and D. J. Zinn. 1943. Mystacocarida, a new order of Crustacea from intertidal beaches in Massachusetts and Connecticut. Smithson. Misc. Coll. 103: 1–11.

Pérez Farfante, I. and B. F. Kensley. 1997. Penaeoid and sergestoid shrimps and prawns of the world. Keys and diagnoses for the families and genera. Mem. Mus. Nation. d'Hist. Natur. 175: 1–233.

Perry, D. M. and R. C. Brusca. 1989. Effects of the root-boring isopod *Sphaeroma peruvianum* on red mangrove forests. Mar. Ecol. Prog. Ser. 57: 287–292.

Persoone, G., P. Sorgeloos, O. Roels and E. Jaspers (eds.) 1980. *The Brine Shrimp Artemia*. Universa Press, Wetteren, Belgium.

Reiber, C. L., and I. J. McGaw. 2009. A review of the "open" and "closed" circulatory systems: new terminology for complex invertebrate circulatory systems in light of current findings. Internat. J. Zool. (2009), Article ID 301284. doi: 10.1155/2009/301284

Riley, J. 1986. The biology of pentastomids. Adv. Parasitol. 25: 45–128.

Roer, R. and R. Dillaman. 1984. The structure and calcification of the crustacean cuticle. Am. Zool. 24: 893–909.

Sanders, H. L. 1955. The Cephalocarida, a new subclass of Crustacea from Long Island Sound. Proc. Natl. Acad. Sci. U.S.A. 41: 61–66.

Sanders, H. L. 1963. The Cephalocarida: Functional morphology, larval development, comparative external anatomy. Mem. Conn. Acad. Arts Sci. 15: 1–80.

Schembri, P. J. 1982. Feeding behavior of 15 species of hermit crabs (Crustacea: Decapoda: Anomura) from the Otago region, southeastern New Zealand. J. Nat. Hist. 16: 859–878.

Schmitt, W. L. 1965. *Crustaceans.* University of Michigan Press, Ann Arbor. [A wonderful, timeless little volume.]

Scholtz, G. 1995. Head segmentation in Crustacea—an immunocytochemical study. Zoology 98: 104–114.

Scholtz, G. and W. Dohle. 1996. Cell lineage and cell fate in crustacean embryos: A comparative approach. Int. J. Dev. Biol. 40: 211–220.

Scholtz, G., N. H. Patel and W. Dohle. 1994. Serially homologous engrailed stripes are generated via different cell lineages in the germ band of amphipod crustaceans (Malacostraca, Peracarida). Int. J. Dev. Biol. 38: 471–478.

Schram, F. R. (gen. ed.). 1983–2014. *Crustacean Issues. Vols. 1–19*. A. A. Balkema (Rotterdam) and CRC Press (Boca Raton). [A series of topical symposium volumes, each edited by a specialist, e.g., phylogeny, biogeography, growth, barnacle biology, biology of isopods, history of carcinology.]

Schram, F. R. and J. C. von Vaupel Klein (eds.). 2010-2014. *Treatise on Zoology—Anatomy, Taxonomy, Biology. The Crustacea, 9.* [5 volumes] Brill, London.

Schram, F. R., J. Yager and M. J. Emerson. 1986. *The Remipedia. Pt. I, Systematics.* San Diego Soc. Nat. Hist. Mem. 15.

Scott, R. 2003. *Darwin and the Barnacle: The Story of One Tiny Creature and History's Most Spectacular Sceintific Breakthrough.* W.W. Norton & Co., New York.

Shuster, S. M. 2008. The expression of crustacean mating strategies. Pp. 224–250 in, R. Oliveira et al. (eds.), *Alternative Reproductive Tactics.* Cambridge Univ. Press.

Skinner, D. M. 1985. Interacting factors in the control of the crustacean molt cycle. Am. Zool. 25: 275–284.

Smirnov, N. N. and B. V. Timms. 1983. A revision of the Australian Cladocera (Crustacea). Rec. Aust. Mus. Suppl. 1: 1–132.

Smith, R. J. and K. Martens. 2000. The ontogeny of the cypridid ostracod *Eucypris virens* (Jurine, 1820) (Crustacea, Ostracoda). Hydrobiologia 419: 31–63.

Smit, N. J. and A. J. Davies. 2004. The curious life-style of the parasitic stages of gnathiid isopods. Advances in Parasitilogy 58: 290–391.

Snodgrass, R. E. 1956. Crustacean metamorphosis. Smithson. Misc. Contrib. 131(10): 1–78. [Dated, but still a good introduction to the subject.]

Stebbing, T. R. R. 1893. *A History of Crustacea.* D. Appleton and Co., London. [Still a great read.]

Steinsland, A. J. 1982. Heart ultrastructure of *Daphnia pulex* De Geer (Crustacea, Branchiopoda, Cladocera). J. Crust. Biol. 2: 54–58.

Stepien, C. A. and R. C. Brusca. 1985. Nocturnal attacks on nearshore fishes in southern California by crustacean zooplankton. Mar. Ecol. Prog. Ser. 25: 91–105.

Stock, J. 1976. A new genus and two new species of the crustacean order Thermosbaenacea from the West Indies. Bijdr. Dierkdl. 46: 47–70.

Strickler, R. 1982. Calanoid copepods, feeding currents and the role of gravity. Science 218: 158–160.

Sutton, S. L. 1972. *Woodlice.* Ginn and Co., London. [Most of what you always wanted to know about pillbugs and roly-polies.]

Sutton, S. L. and D. M. Holdich (eds.) 1984. *The Biology of Terrestrial Isopods.* Clarendon Press, Oxford. [The rest of what you always wanted to know about pillbugs and roly-polies.]

Takahashi, T. and J. Lützen. 1998. Asexual reproduction as part of the life cycle in *Sacculina polygenea* (Cirripedia: Rhizocephala: Sacculinidae). J. Crustacean Biol. 18: 321-331.

Thorp, J. H. and A. P. Covich (eds.). 2009. *Ecology and Classification of North American Freshwater Invertebrates.* 3rd Ed. Academic Press, New York. [Includes comprehensive overviews of American freshwater crustacea.]

Tomlinson, J. T. 1969. The burrowing barnacles (Cirripedia: Order Acrothoracica). U.S. Nat. Mus. Bull. 259: 1–162.

Tóth, E and R. T. Bauer. 2007. Gonopore sexing technique allows determination of sex ratios and helper composition in eusocial shrimps. Mar. Biol. 151: 1875–1886.

Van Name, W. G. 1936. The American land and freshwater isopod Crustacea. Bull. Am. Mus. Nat. Hist. 71: 1–535. [Badly in need of updating; no other keys are available to this poorly known fauna.]

Vinogradov, M. E., A. F. Volkov and T. N. Semenova. 1982 (1996). *Hyperiid Amphipods (Amphipoda, Hyperiidea) of the World Oceans.* Translated from the Russian by D. Siegel-Causey for the Smithsonian Institution Libraries, Washington, D.C.

Wagner, H. P. 1994. A monographic review of the Thermosbaenacea. Zoologische Verhandelingen 291: 1–338.

Walker, G. 2001. Introduction to the Rhizocephala (Crustacea: Cirripedia). J. Morphol. 249: 1–8.

Wallosek, D. 1993. The Upper Cambrian *Rehbachiella* and the phylogeny of Branchiopoda and Crustacea. Fossils and Strata 32: 1–202.

Wanninger, A. (ed.). 2015. *Evolutionary Developmental Biology of Invertebrates 4: Ecdysozoa II: Crustacea.* Springer-Verlag, Wien.

Warner, G. F. 1977. *The Biology of Crabs*. Van Nostrand Reinhold, New York.

Waterman, T. H. (ed.). 1960, 1961. *The Physiology of Crustacea. Vols. 1–2*. Academic Press, New York. [Dated, but still useful.]

Waterman, T. H. and A. S. Pooley. 1980. Crustacean eye fine structure seen with scanning electron microscopy. Science 209: 235–240.

Watling, L. and M. Thiel (eds.). 2013–2015. *The Natural History of Crustacea, Vols 1-4*. Oxford Univ. Press, Oxford.

Weeks, S. C. 1990. Life-history variation under varying degrees of intraspecific competition in the tadpole shrimp *Triops longicaudatus* (Le Conte). J. Crust. Biol. 10: 498–503.

Wenner, A. M. (ed.) 1985. *Crustacean Growth: Factors in Adult Growth* and *Larval Growth*. A. A. Balkema, The Netherlands.

Wiese, K. 2000. *The Crustacean Nervous System*. Springer-Verlag, New York.

Williams, A. B. 1984. *Shrimps, Lobsters, and Crabs of the Atlantic Coast of the Eastern United States, Maine to Florida*. Smithsonian Institution Press, Washington, D.C. [An outstanding reference by one of the grand gentleman of carcinology.]

Williams, A. B. 1988. *Lobsters of the World: An Illustrated Guide*. Osprey Books, New York.

Wingstrand, K. G. 1972. Comparative spermatology of a pentastomid *Raillietiella hemidactyli* and a branchiuran crustacean *Argulus foliaceus* with a discussion of pentastomid relationships. Biol. Skr. 19: 1–72.

Yagamuti, S. 1963. *Parasitic Copepoda and Branchiura of Fishes*. Wiley, New York.

Yager, J. 1981. Remipedia, a new class of Crustacea from a marine cave in the Bahamas. J. Crust. Biol. 1: 328–333.

Yager, J. 1991. The Remipedia (Crustacea): Recent investigation of their biology and phylogeny. Verhandlungen der Deutschen Zoologischen Gesellschaft, Stuttgart 84: 261–269.

Yager, J. and W. F. Humphreys. 1996. *Lasionectes esleyi*, sp. nov., the first remipede crustacean recorded from Australia and the Indian Ocean, with a key to the world species. Invert. Taxon. 10: 171–187.

Phylogeny and Evolution

See chapters 20 and 28 for references on general arthropod phylogeny.

Almeida, W. de O. and M. L. Christoffersen. 1999. A cladistic approach to relationships in Pentastomida. J. Parasitol. 85: 695–704.

Andrew, D. R. 2011. A new view of insect-crustacean relationships II. Inferences from expressed sequence tags and comparisons with neural cladistics. Arthropod Structure & Development 40: 289–302.

Boxshall, G. A. 1991. A review of the biology and phylogenetic relationships of the Tantulocarida, a subclass of Crustacea recognized in 1983. Verhandlungen der Deutschen Zoologischen Gesellschaft 84: 271–279.

Bracken-Grissom, H. D., and 16 others. 2014. Emergence of the lobsters: Phylogenetic relationships, morphological evolution and divergence time comparisons of a fossil rich group (Achelata, Astacidea, Glypheidea, Polychelida). Syst. Biol. 63(4): 457–479.

Bracken, H. D., De Grave, S., Toon, A., Felder, D. L. & Crandall, K. A. 2009. Phylogenetic position, systematic status, and divergence time of the Procarididea (Crustacea: Decapoda). Zool. Scripta 39: 198–212.

Brusca, R. C. and G. D. F. Wilson. 1991. A phylogenetic analysis of the Isopoda (Crustacea) with some classificatory recommendations. Mem. Queensland Mus. 31: 143–204.

Castellani, C., A. Maas, D. Waloszek and J. T. Haug. 2011. New pentastomids from the Late Cambrian of Sweden—deeper insight of the ontogeny of fossil tongue worms. Palaeontographica, Abt. A: Palaeozoology-Stratigraphy 293: 95–145.

Chen, Y.-U., J. Vannier and D.-Y. Huang. 2001. The origin of crustaceans: new evidence from the early Cambrian of China. Proc. Royal Soc. London 268: 2181–2187.

De Grave, S., and 17 others. 2009. A classification of living and fossil genera of decapod crustaceans. Raffles Bull. Zool. Suppl. 21: 1–109.

Edgecombe, G. 2010. Arthropod phylogeny: an overview from the perspectives of morphology, molecular data and the fossil record. Arthropod Structure and Development 39: 74–87.

Gale, A. S. 2014. Origin and phylogeny of verrucomorph barnacles (Crustacea, Cirripedia, Thoracica). J. Syst. Palaentol. doi: 10.1080/14772019.2014.954409

Gale, A. S. and A. M. Sørensen. 2014. Origin of the balanomorph barnacles (Crustacea, Cirripedia, Thoracica): new evidence from the Late Cretaceous (Campanian) of Sweden. J. Syst. Palaenotol. doi: 10.1080./14772019.2014.954824

Giribet, G., and G. D. Edgecomb. 2012. Reevaluating the Arthropod Tree of Life. Ann. Rev. Entomology 57: 167–186.

Ho, J. S. 1990. Phylogenetic analysis of copepod orders. J. Crust. Biol. 10: 528–536.

Huys, R. and G. A. Boxshall. 1991. *Copepod Evolution*. The Ray Society, London.

Jenner, R. A. 2010. Higher-level crustacean phylogeny: consensus and conflicting hypotheses. Arthropod Struct. Dev. 39: 143–153.

Koenemann, S. and R. A. Jenner. 2005. *Crustacea and Arthropod Relationships*. Taylor & Francis, New York.

Lavrov, D. V., W. M. Brown and J. L. Boore. 2004. Phylogenetic position of the Pentastomida and (pan)crustacean relationshps. Proceedings of the Royal Society of London B, 271: 537–544.

Lefébure, T., C. J. Douady, M. Gouy, and J. Gibert. 2006. Relationship between morphological taxonomy and molecular divergence within Crustacea: Proposal of a molecular threshold to help species delimitation. Mol. Phylog. Evol. 40: 435–447.

Luque, J. 2014. The oldest higher true crabs (Crustacea: Decapoda: Brachyura): insights from the Early Cretaceous of the Americas. Palaentology. doi: 10.1111/pala.12135

Martin, J. W. 2013. Arthropod Evolution and Phylogeny. Pp. 34–37 in *McGraw-Hill Yearbook of Science & Technology for 2013*. McGraw-Hill, New York.

Martin, J. W., K. A. Crandall and D. L. Felder (eds.). 2009. *Decapod Crustacean Phylogenetics. Crustacean Issues 18*. CRC Press, Taylor & Francis, Boca Raton, Florida.

McLaughlin, P. A. 1983. Hermit crabs—are they really polyphyletic? J. Crust. Biol. 3: 608–621.

Morrison, C. L. A. W. Harvey, S. Lavery, K. Tieu, Y. Huang and C. W. Cunningham. 2002. Mitochondrial gene rearrangements confirm the parallel evolution of the crab-like form. Proc. Royal Soc. London, Biol. Sci. 269: 345–350.

Meusemann, K., and 15 others. 2010. A phylogenomic approach to resolve the arthropod tree of life. Mol. Biol. Evol. 27: 2451–2464.

Misof, B., and 100 others. 2014. Phylogenomics resolves the timing and patern of insect evolution. Science 346(6210): 763–767.

Negrea, S., N. Botnariuc and H. J. Dumont. 1999. Phylogeny, evolution and classification of the Branchiopoda (Crustacea). Hydrobiologia 412: 191–212.

Oakley, T. H., J. M. Wolfe, A. R. Lindgren, and A. K. Zaharoff. 2013. Phylotranscriptomics to bring the understudied into the fold: monophyletic Ostracoda, fossil placement, and pancrustacean phylogeny. Mol. Biol. Evol. 30 (1): 215–233.

Olesen, J. 2000. An updated phylogeny of the Conchostraca—Cladocera clade (Branchiopoda, Diplostraca). Crustaceana 73: 869–886.

Olesen, J., and S. Richter. 2013. Onychocaudata (Branchiopoda: Diplostraca), a new high-level taxon in branchiopod systematics. J. Crust. Biol. 33: 62–65.

Pérez-Losada, M., J. T. Høeg, G. A. Kolbasov and K. A. Crandall. 2002. Reanalysis of the relationships among the Cirripedia and Ascothoracida, and the phylogenetic position of the Facetotecta using 18S rDNA sequences. J. Crust. Biol. 22: 661–669.

Pérez-Losada, M., J. T. Høeg, N. Simon-Blecher, Y. Achituv, D. Jones and K. A. Crandall. 2014. Molecular phylogeny, systematics and morphological evolution of the acorn barnacles (Thoracica: Sessilia: Balanomorpha). Mol. Phylog. Evol. 81: 147–158.

Regier, J. C., J. W. Shultz, R. E. Kambic. 2005. Pancrustacean phylogeny: hexapods are terrestrial crustaceans and maxillopods are not monophyletic. Proc. Royal Soc. B 272 (1561): 395–401

Regier, J. C. and 7 others. 2010. Arthropod relationships revealed by phylogenomic analysis of nuclear protein-coding sequences. Nature 463: 1079–1083.

Remigio, E. A. and P. D. Hebert. 2000. Affinities among anostracan (Branchiopoda) families inferred from phylogenetic analyses of multiple gene sequences. Mol. Phylogen. Evol. 17: 117–128.

Richter, S. and G. Scholtz. 2000. Phylogenetic analysis of the Malacostraca (Crustacea). J. Zool. Syst. Evol. Res. 39: 113–136.

Schnabel, K. E., S. T. Ahyong and E. W. Maas. 2011. Galatheoidea are not monophyletic: Molecular and morphological phylogeny of the squat lobsters (Decapoda: Anomura) with recognition of a new superfamily. Mol. Phylogen. Evol. 58: 157–168.

Spears, T. and L. G. Abele. 1999. The phylogenetic relationships of crustaceans with foliacious limbs: An 18S rDNA study of Branchiopoda, Cephalocarida, and Phyllocarida. J. Crust. Biol. 19: 825–843.

Spears, T. and L. G. Abele. 2000. Branchiopod monophyly and interordinal phylogeny inferred from 18S ribosomal DNA. J. Crust. Biol. 20: 1–24.

Spears, T., L. G. Abele and M. A. Applegate. 1994. Phylogenetic study of cirripedes and selected relatives (Thecostraca) based on 18S rDNA sequence analysis. J. Crust. Biol. 14: 641–656.

Stemme, T., T. M. Iliffe, B. M. von Reumont, S. Koenemann, S. Harzsch and G. Bicker. 2013. Serotonin-immunoreactive neurons in the ventral nerve cord of Remipedia (Crustacea): support for a sister group relationship of Remipedia and Hexapoda? BMC Evol. Biol. 13: 119.

Sternberg, R. V., N. Cumberlidge and G. Rodríguez. 1999. On the marine sister groups of the freshwater crabs (Crustacea: Decapoda). J. Zool. Sys. Evol. Res. 37: 19–38.

Storch, V. and B. G. M. Jamieson. 1992. Further spermatological evidence for including the Pentastomida (tongue worms) in the Crustacea. Int. J. Parasitol. 22: 95–108.

Tam, Y. K. and I. Kornfield. 1998. Phylogenetic relationships of clawed lobster genera (Decapoda: Nephropidae) based on mitochondrial 16S rRNA gene sequences. J. Crust. Biol. 18(1): 138–146.

Tsang, L. M., T.-Y. Chan, S. T. Ahyong and K. H. Chu. 2011. Hermit to king, or hermit to all: Multiple transitions to crab-like forms from hermit crab ancestors. Syst. Biol. doi: 10.1093/sysbio/syr063

Von Reumont, B. M. and 12 others. 2012. Pancrustacean phylogeny in light of new phylogenomic data: support for remipedia as the possible sister group of Hexapoda. Mol. Biol. Evol. 29: 1031–1045.

Walker-Smith, G. K. and G. C. B. Poore. 2001. A phylogeny of the Leptostraca (Crustacea) from Australia. Mem. Mus. Victoria 58: 137–148.

Waloszek, D. 2003. Cambrian "Orsten"-type preserved arthropods and the phylogeny of Crustacea. Pp. 69–87 in A. Legakis et al. (eds), *The New Panorama of Animal Evolution.* PENSOFT Publishers, Sofia, Moscow.

Waloszek, D., J. Chen, A. Maas and X. Wang. 2005. Early Cambrian arthropods – new insights into arthropod head and structural evolution. Arthropod. Struct. Dev. 34(2): 189–205.

Waloszek, D. and A. Maas. 2005. The evolutionary history of crustacean segmentation: a fossil-based perspective. Evol. Dev. 7: 515–527.

Waloszek, D. and K. J. Müller. 1990. Stem-lineage crustaceans from the Upper Cambrian of Sweden and their bearing upon the position of Agnostus. Lethaia 23: 409–427.

Waloszek, D. and K. J. Müller. 1997. Cambrian "Orsten"-type arthropods and the phylogeny of Crustacea. Pp. 139–153 in R. A. Fortey (ed.), *Arthropod Relationships*. Chapman and Hall, London.

Waloszek, D., J. E. Repetski and A. Maas. 2005. A new Late Cambrian pentastomid and a review of the relationships of this parasitic group. Transactions of the Royal Society of Edinburgh: Earth Sciences 96: 163–176.

Wilson, K., V. Cahill, E. Ballment and J. Benzie. 2000. The complete sequence of the mitochondrial genome of the crustacean *Penaeus monodon*: are malacostracan crustaceans more closely related to insects than to branchiopods. Mol. Biol. Evol. 17: 863–874.

Yan-bin, S., R. S. Taylor and F. R. Schram. 1998. New spelaeogriphaceans (Crustacea: Peracarida) from the Upper Jurassic of China. Contr. Zool. 83(4): 1–14.

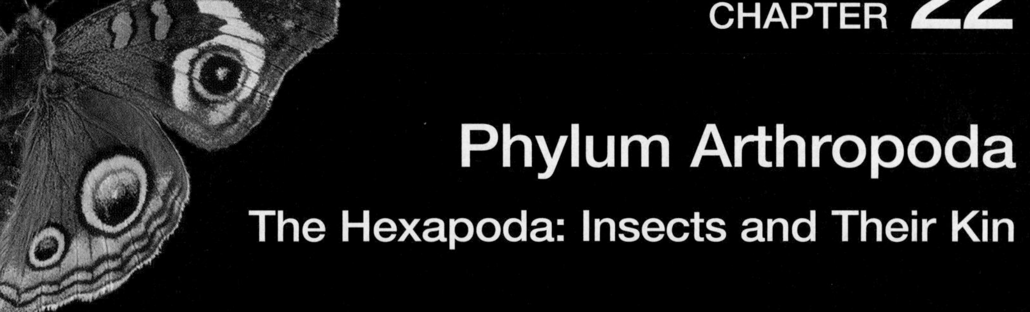

CHAPTER 22

Phylum Arthropoda

The Hexapoda: Insects and Their Kin

Hexapods stand out among all other invertebrates for being far and away the most diverse group of animals on Earth, the only invertebrates to fly, and the only terrestrial invertebrates to undergo indirect development or complete metamorphosis.

The arthropod subphylum Hexapoda comprises the class Insecta and three other small, closely related, wingless, insect-like groups: Collembola, Protura, and Diplura. The Hexapoda are united on the basis of a distinct body plan of a head, 3-segmented thorax, and 11-segmented abdomen, 3 pairs of thoracic legs, a single pair of antennae, 3 sets of "jaws" (mandibles, maxillae, and labium), an aerial gas exchange system composed of tracheae and spiracles, Malpighian tubules formed as proctodeal (ectodermal) evaginations, and, among the Pterygota, wings (Box 22A). The presence of a thorax fixed at 3 segments, each with a pair of walking legs, is a unique synapomorphy for the Hexapoda. Other synapomorphies include the presence of a large fat body (mainly concentrated in the abdomen), and fusion of the second maxillae to form a lower lip (the labium).

Hexapods evolved on land; groups inhabiting aquatic environments today have secondarily invaded those habitats through behavioral adaptations and modifications of their aerial gas exchange systems. The earliest undisputed fossils of hexapods are early Devonian (412 Ma). However, there are Silurian trace fossils that are very hexapod-like, and molecular clock data suggest an Early Ordovician origin for Hexapods at about 479 million years ago and an early Silurian origin about 441 million years ago for Insecta.

The most spectacular evolutionary radiation among the Hexapoda (in fact among all eukaryotic life) has, of course, been within the insects, which inhabit nearly every conceivable terrestrial and freshwater habitat and, less commonly, even the sea surface and the marine littoral region. Insects are also found in such unlikely places as oil swamps and seeps, sulfur springs, glacial streams, and brine ponds. They often live where few other

Classification of The Animal Kingdom (Metazoa)

Non-Bilateria*
(a.k.a. the diploblasts)
- PHYLUM PORIFERA
- PHYLUM PLACOZOA
- PHYLUM CNIDARIA
- PHYLUM CTENOPHORA

Bilateria
(a.k.a. the triploblasts)
- PHYLUM XENACOELOMORPHA

Protostomia
- PHYLUM CHAETOGNATHA

SPIRALIA
- PHYLUM PLATYHELMINTHES
- PHYLUM GASTROTRICHA
- PHYLUM RHOMBOZOA
- PHYLUM ORTHONECTIDA
- PHYLUM NEMERTEA
- PHYLUM MOLLUSCA
- PHYLUM ANNELIDA
- PHYLUM ENTOPROCTA
- PHYLUM CYCLIOPHORA

Gnathifera
- PHYLUM GNATHOSTOMULIDA
- PHYLUM MICROGNATHOZOA
- PHYLUM ROTIFERA

Lophophorata
- PHYLUM PHORONIDA
- PHYLUM BRYOZOA
- PHYLUM BRACHIOPODA

ECDYSOZOA

Nematoida
- PHYLUM NEMATODA
- PHYLUM NEMATOMORPHA

Scalidophora
- PHYLUM KINORHYNCHA
- PHYLUM PRIAPULA
- PHYLUM LORICIFERA

Panarthropoda
- PHYLUM TARDIGRADA
- PHYLUM ONYCHOPHORA
- **PHYLUM ARTHROPODA**
  - SUBPHYLUM CRUSTACEA*
  - **SUBPHYLUM HEXAPODA**
  - SUBPHYLUM MYRIAPODA
  - SUBPHYLUM CHELICERATA

Deuterostomia
- PHYLUM ECHINODERMATA
- PHYLUM HEMICHORDATA
- PHYLUM CHORDATA

*Paraphyletic group

This chapter has been revised by Wendy Moore.

BOX 22A Characteristics of the Subphylum Hexapoda

1. Body composed of 20 true somites (plus acron) organized as a head (6 somites), thorax (3 somites) and abdomen (11 somites). Due to fusion of somites, these body segments are not always externally obvious.
2. Head segments bear the following structures (from anterior to posterior): compound eyes and ocelli; antennae; clypeolabrum; mandibles; maxillae, labium (fused second maxillae). Ocelli (and compound eyes) are secondarily lost in some groups.
3. Legs uniramous; present on the three thoracic segments of adults; legs composed of 6 articles: coxa, trochanter, femur, tibia, tarsus, pretarsus; tarsus often subdivided; pretarsus typically clawed
4. Gas exchange by spiracles and tracheae
5. Gut with gastric (digestive) ceca
6. With large fat body (mainly concentrated in abdomen)
7. Fused exoskeleton of head forms unique internal tentorium
8. With ectodermally derived Malpighian tubules (proctodeal evaginations)
9. Gonopores open on the last abdominal segment, or on abdominal segment 7, 8, or 9
10. Gonochoristic; direct or indirect development

animals or plants can exist. It is no exaggeration to say that insects rule the land. Their diversity and abundance defy imagination (Figures 22.1–22.7).

We do not know how many species of insects there are, or even how many have been described. Published estimates of the number of described species range from 890,000 to well over a million (we calculate about 926,990). An average of about 3,500 new species have been described annually since the publication of Linnaeus's *Systema Naturae* in 1758, although in recent years the average has climbed to 7,000 new species annually. Estimates of the number of insect species remaining to be described range from 3 million to 100 million. The Coleoptera (beetles), with an estimated 380,000 described species, is far and away the largest insect order (more than a quarter of all animal species are beetles). The beetle family Curculionidae (the weevils) contains about 65,000 described species (nearly 5% of all described animal species). The rich diversity of insects seems to have come about through a combination of advantageous features, including the evolutionary exploitation of developmental genes working on segmented and compartmentalized bodies, coevolution with plants (particularly the flowering plants), miniaturization, and the invention of flight.

Insects are not only diverse, but also incredibly abundant. For every human alive, there are an estimated 200 million insects. Howard Ensign Evans estimated that an acre of ordinary English pasture supports an astonishing 248,375,000 springtails and 17,825,000 beetles. In tropical rain forests, insects can constitute 40% of the total animal biomass (dry weight), and the biomass of the ants can be far greater than that of the combined mammal fauna (up to 15% of the total animal biomass). A single colony of the African driver ant *Anomma wilverthi* may contain as many as 22 million workers. Based on his research in the tropics, biodiversity sleuth Terry Erwin has calculated that there are about 3.2×10^8 individual arthropods per hectare, representing more than 60,000 species, in the western Amazon. In Maryland, a single population of the mound-building ant *Formica exsectoides* comprised 73 nests covering an area of 10 acres and containing approximately 12 million workers. Termites have colonies of similar magnitudes. E. O. Wilson has calculated that, at any given time, 10^{15} (a million billion) ants are alive on Earth!

In most parts of the world, insects are among the principal predators of other invertebrates. Insects are also key items in the diets of many terrestrial vertebrates, and they play a major role as reducer-level organisms (detritovores and decomposers) in food webs. Due to their sheer numbers, they constitute much of the matrix of terrestrial food webs. Their biomass and energy consumption exceed those of vertebrates in most terrestrial habitats. In deserts and in the tropics, ants replace earthworms as the most abundant earth movers (ants are nearly as important as earthworms even in temperate regions). Termites are among the chief decomposers of dead wood and leaf litter around the world, and without dung beetles African savannahs would be buried under the excrement of the tens of thousands of large grazing mammals.

Without insects, life as we know it would cease to exist. In fact, E. O. Wilson has stated, "so important are insects and other land-dwelling arthropods that if all were to disappear, humanity probably could not last more than a few months." Eighty percent of the world's crop species, including food, medicine, and fiber crops, rely on animal pollinators, nearly all of which are insects. Insects also play key roles in pollinating wild, native plants. Beekeeping began long ago, at least by 600 BC in the Nile Valley and probably well before that. The first migratory beekeepers were Egyptians who floated hives up and down the Nile to provide pollination services to floodplain farmers while simultaneously producing a honey crop. Domestic honeybees (*Apis mellifera*), introduced to North America from Europe in the mid-1600s, are now the dominant pollinators of most food crops grown around the world, and they play some role

(A)

(B)

(C)

(D)

Figure 22.1 Representatives of the three orders of entognathous (noninsect) hexapods. (A) *Anurida granaria*, a springtail (order Collembola). (B) *Ptenothrix* sp., a springtail showing entognathous mouthparts. (C) A dipluran from New Zealand (order Diplura). (D) A proturan from British Columbia (order Protura).

in pollinating 80% of the crop varieties grown in the United States (they are estimated to be directly responsible for $10 to $20 billion in crops annually).[1]

Interactions between insects and flowering plants have been going on for a very long time, beginning over 100 million years ago with the origin of the angiosperms and accelerating with the ascendancy of these flowering plants during the early Cenozoic. Millions of years of plant–insect coevolution have resulted in flowers with anatomy and scents that are finely tuned to their insect partners. In exchange for pollination services, flowers provide insects with food (nectar, pollen), shelter, and chemicals used by the insects to produce such things as pheromones. In general, insect pollination is accomplished coincidentally, as the pollinators visit flowers for other reasons. But in a few cases, such as that of the yucca moths of the American Southwest (*Tegiticula* spp.), the insects actually gather up pollen and force it into the receptive stigma of the flower, initiating pollination. The moth's goal is to assure a supply of yucca seed for its larvae, which develop within the yucca's fruits. Some insects also play important roles as seed dispersers, especially ants. More than 3,000 plant species (in 60 families) are known to rely on ants for the dispersal of their seeds.

Like all other animals on Earth, insects are facing enormous threats of extinction. Certainly many thousands of species have become extinct over the past century as a result of rampant land use change and deforestation. With accelerating biodiversity losses worldwide, estimates of the number of insect species that have already gone extinct range into the millions. Further, widespread and often inappropriate use of

[1] The study of bees is called "melittophily." The study of honeybees (*Apis* spp.) and their management is called "apiculture." The management of bumblebees (*Bombus* spp.) is "bombiculture." The ritualized keeping of stingless bees is "meliponiculture." The Asian honeybee, *Apis dorsata*, not yet introduced to the U.S., is a giant reaching an inch in length. Its droppings are quite noticeable, and the mass defecations of these bees at sunset create a "golden shower" that provides significant nutrient enrichment to tropical soils. Their droppings were once confused with the dreaded "yellow rain," a deadly form of biochemical warfare that poisoned thousands of villagers during the Vietnam War.

pesticides has created a "pollination crisis" in many parts of the world, as pollinating insects are locally extirpated and native plant and domestic crop pollination plummets.

As valuable as insects are to human life, some species seem to conspire to give the whole group a bad name. Some insect species are pests and consume about a third of our potential annual harvest, and some other species transmit many major human diseases. Every year we spend billions of dollars on insect control. Malaria, transmitted by mosquitoes, kills 1 to 3 million people annually (mostly children), and each year nearly 500 million people contract the disease. It is the leading cause of death from infectious disease, and it has plagued humans for at least 3,000 years (Egyptian mummies have been found with malaria antigens in their blood). One of the most widespread and fastest spreading human viral diseases, dengue, is transmitted by mosquitoes of the genus *Aedes*. Dengue is essentially an urban disease, almost entirely associated with anthropogenic environments because its main vectors, *A. albopictus* and *A. aegypti*, breed primarily in artificial containers (e.g., flower vases, discarded tires, water tanks). In recent years, these species have spread throughout the tropics, often following the international trade in used tires. Although these mosquitoes have not penetrated very far into the temperate zones, they have become established in the southern United States. A variety of mosquitoes transmits the filarial nematode *Wuchereria bancrofti*, the causative agent of lymphatic filariasis ("elephantiasis") throughout the world's tropics. Chagas' disease (American trypanosomiasis) is transmitted by certain hemipteran bugs in the subfamily Triatominae (family Reduviidae), and causes chronic degenerative disease of the heart and intestine. The species of triatomine bugs that occur in the southwestern United States tend not to defecate when they feed, greatly reducing the possibility of humans contracting Chagas' disease in that area. However, global warming is now increasing the spread of *Aedes*, *Culex*, triatomines, and other tropical disease vectors.

The natural history writer David Quammen, speaking of the blood-sucking varieties of mosquitoes, notes that the average blood meal of a female (the only sex that feeds on blood) amounts to 2.5 times the original weight of the insect—the equivalent, Quammen notes, "of Audrey Hepburn sitting down to dinner and getting up from the table weighting 380 pounds, then, for that matter, flying away." But as Quammen also points out, mosquitoes have made tropical rain forests (the most diverse ecosystems on Earth) largely uninhabitable to humans, thus helping to preserve them!

The enormous diversification of Hexapoda is often attributed to the evolution of three key innovations: the ability to fly, the ability to fold back their wings, and the evolution of holometabolous development (= indirect development, = complete metamorphosis). The persistence of the main lineages of insects since the Devonian and their ecological and morphological versatility have undoubtedly contributed to making Hexapoda the dominant group in extant terrestrial ecosystems, with respect to species diversity, functional diversity, and overall biomass. Obviously, the subject of insect biology, or entomology, is a discipline in its own right, and a multitude of books and college courses on the subject exists. If we apportioned pages to animal groups on the basis of numbers of species, overall abundance, or economic importance, insect chapters could easily fill 90% of this textbook. The Selected References at the end of this chapter provide entry into some of the current literature on insects.

Because the Hexapoda comprises such a large and diverse assemblage of arthropods, we first present a brief classification of the 31 recognized orders, followed by more detailed synopses, brief diagnoses, and comments, on each order. These two sections serve as a preface to the body plan discussion that follows, and also provide a reference that the reader can turn to as needed.

CLASSIFICATION OF THE SUBPHYLUM HEXAPODA

Our classification scheme recognizes 31 orders of living hexapods. Three entognathous, or non-insect hexapod orders (the Entognatha), are basal to the monophyletic class Insecta. The Insecta comprise two monophyletic sister groups, the order Archaeognatha (with monocondylic mandibles) and all the others (with dicondylic mandibles). The subclass Pterygota, or flying insects, comprise two groups: the Palaeoptera (Ephemeroptera, Odonata), which may be a paraphyletic group, and the monophyletic infraclass Neoptera. We recognize three superorders within the modern winged insects, or Neoptera: the Polyneoptera (Plecoptera, Zoraptera, Blattodea, Mantodea, Dermaptera, Orthoptera, Phasmida, Grylloblattodea, Embioptera, Mantophasmatodea), the Acercaria (Psocodea, Thysanoptera, Hemiptera), and the Holometabola (all the remaining orders). Within the Holometabola we recognize four well supported, but unranked clades: Coleopetrida (Coleoptera, Strepsiptera), Neuropterida (Neuroptera, Megaloptera, Rhapidioptera), Antliophora (Mecoptera, Siphonaptera. Diptera), and Amphiesmenoptera (Trichoptera, Lepidoptera). With nearly a million named species of Hexapoda, we have opted to include representative families of only the most diverse and common orders in the taxonomic synopses.

SUBPHYLUM HEXAPODA

(Note: "Entognatha" and "Palaeoptera" are likely paraphyletic groups)

Entognatha

Order Collembola: Springtails
Order Protura: Proturans
Order Diplura: Diplurans

CLASS INSECTA (= ECTOGNATHA)

Order Archaeognatha: Jumping bristletails

Dicondylia

Order Thysanura (= Zygentoma): Silverfish

Subclass Pterygota: Winged insects

Palaeoptera: Ancient winged insects

Order Ephemeroptera: Mayflies
Order Odonata: Dragonflies and damselflies

Infraclass Neoptera: Modern, wing-folding insects

Superorder Polyneoptera

Order Plecoptera: Stoneflies
Order Blattodea: Cockroaches and termites
Order Mantodea: Mantises
Order Phasmida (= Phasmatodea): Stick and leaf insects
Order Grylloblattodea: Rock crawlers
Order Dermaptera: Earwigs
Order Orthoptera: Locusts, katydids, crickets, grasshoppers
Order Mantophasmatodea: Heel-walkers or gladiators
Order Embioptera (= Embiidina): Web-spinners
Order Zoraptera: Zorapterans

Superorder Acercaria (= Paraneoptera)

Order Thysanoptera: Thrips
Order Hemiptera: True bugs
Order Psocodea: Book lice, bark lice, true lice

Superorder Holometabola

Order Hymenoptera: Ants, bees, wasps

Coleopterida

Order Coleoptera: Beetles
Order Strepsiptera: Twisted-wing parasites

Neuropterida

Order Megaloptera: Alderflies, dobsonflies, fishflies
Order Raphidioptera: Snakeflies
Order Neuroptera: Lacewings, ant lions, mantisflies, owlflies

Antliophora

Order Mecoptera: Scorpionflies, hangingflies, snow scorpionflies
Order Siphonaptera: Fleas
Order Diptera: True flies, mosquitoes, gnats

Amphiesmenoptera

Order Trichoptera: Caddisflies
Order Lepidoptera: Butterflies, moths

Hexapod Classification

Subphylum Hexapoda

Body differentiated into head (acron + 6 segments), thorax (3 segments), and abdomen (11 or fewer segments); cephalon with one pair lateral compound eyes and often with a triad or pair of medial ocelli; with one pair of uniramous multiarticulate antennae, mandibles, and maxillae; second pair of maxillae fused to form a complex labium; each thoracic segment with one pair of uniramous legs; wings often present on second and third thoracic segments (in pterygote insects); abdomen without fully developed legs, but "prolegs" (presumably homologous to the ancestral arthropod abdominal appendages) occur in at least seven orders (in adults of some Diplura, Thysanura, and Archaeognatha; in larvae of some Diptera, Trichoptera, Lepidoptera, and Hymenoptera); abdomen with a large fat body; gonopores open on the last abdominal segment or on the seventh, eighth, or ninth abdominal segment; paired cerci often present; males commonly with intromittent and clasping structures; development direct, involving relatively slight changes in body form (ametabolous or hemimetabolous), or indirect with striking changes (holometabolous).

Entognatha

Mouthparts with bases hidden within the head capsule (ento-gnathous); mandibles with single articulation; most or all antennal articles with intrinsic musculature; wingless; without, or with poorly developed, Malpighian tubules; legs with one (undivided) tarsus. The three orders of entognathous hexapods do not form a monophyletic group. While the entognathous conditions of Collembola and Protura appear to be homologous (these two orders are often placed together in the class Ellipura), entognathy in the Diplura may be a product of convergent evolution. Recent data from paleontology, comparative anatomy, and molecular phylogenetics suggest that the Diplura are the sister group of the Insecta, and are therefore more closely related to the Insecta than they are to the other entognathous orders.

Order Collembola Approximately 6,000 described species (Figure 22.1A,B). Small (most less than 6 mm); biting–chewing mouthparts; with or without small compound eyes; ocelli vestigial; antennae 4-articulate, first 3 articles with intrinsic muscles; tarsus of legs indistinct (perhaps fused with tibia); pretarsus of legs with single claw; abdomen with a reduced number of segments (six); first abdominal segment with ventral tube (collophore) of unknown function; third abdominal segment with small process (retinaculum); a forked tail-like appendage on fourth or fifth abdominal segment (furcula); without cerci; with gonopores on last abdominal segment; without Malpighian tubules; often without spiracles or tracheae.

The earliest known hexapods in the fossil record are collembolans. *Rhyniella praecursor* and other species from the Lower Devonian closely resemble some modern collembolan families. Unlike other hexapods, which breathe using internal tubes called trachea, springtails breathe air directly through their cuticle and epidermis. Their cuticle repels water, allowing them to live in moist environments without suffocating. They also have a remarkable system for escaping predators. While at rest, the furcula is retracted under the abdomen and held in place by the retinaculum. When the furcula and retinaculum disassociate, the furcula swings downwards with such force that it hits the substrate, and quickly propels the collembolan high into the air. Many workers believe that springtails evolved via neoteny.

Order Protura Approximately 200 described species (Figure 22.1D). Minute (smaller than 2 mm); whitish; without eyes, abdominal spiracles, hypopharynx, or cerci; Malpighian tubules are small papillae; sucking mouthparts; stylet-like mandibles; vestigial antennae; first pair of legs carried in an elevated position and used as surrogate "antennae"; pretarsus of legs with single claw; abdomen 11-segmented, with a telson (perhaps reminiscent of their crustacean ancestors); the segmental nature of this telson or twelfth "segment" has not been confirmed; first 3 abdominal segments with small appendages; without external genitalia, but male gonopores on protrusible phallic complex; gonopores on last abdominal segment; with or without tracheae; simple development.

Proturans are the only hexapods with anamorphic development, a type of development in which a new abdominal segment is added with each instar (or molt). All other insects have epimorphic development, in which segmentation is complete before hatching. Proturans are rare, and live in leaf litter, moist soils, and rotting vegetation.

Order Diplura Approximately 800 described species (Figure 22.1C); fossils date to the Carboniferous. Small (less than 4 mm); whitish; without eyes, ocelli, external genitalia or Malpighian tubules; chewing mouthparts; abdomen 11-segmented, but embryonic segments 10 and 11 fuse before hatching; gonopores on ninth abdominal segment; 7 pairs of lateral abdominal leglets; 2 caudal cerci; with tracheae and up to 7 pairs of abdominal spiracles; antennae multiarticulate, each article with intrinsic musculature; simple development. Most species live in mesic habitats beneath rocks, rotting logs, leaf litter, humus, and soil.

Class Insecta

Mouth appendages ectognathous (exposed and projecting from the head capsule); mandibles with two points of articulation (except Archaeognatha); intrinsic musculature of antennal articles greatly reduced; antennal pedicel with a mechanoreceptor that perceives the movement of the flagellum, called the Johnston's organ; head with a tentorial bridge connecting the posterior tentorial arms; tarsi subdivided into tarsomeres; with well-developed Malpighian tubules; ovipositor formed from modifications of the appendages of abdominal segments 8 and 9. The Insecta comprise two clades, the order Archaeognatha (with monocondylic mandibles) and all other insects (the clade Dicondylia, with dicondylic mandibles). Dicondylia also comprise two clades, the order Thysanura and the subclass Pterygota (winged insects).

Order Archaeognatha Approximately 390 described species (Figure 22.2A,B). Small (to 15 mm), wingless (perhaps secondarily), resembling silverfish but body more cylindrical; ocelli present; compound eyes large and contiguous; body usually covered with scales; mandibles biting–chewing; with a single condyle (articulation point); maxillary palp large and leglike; tarsi 3-articulate; middle and hind coxae usually with exites ("styli"); abdomen 11-segmented, with 3 to 8 pairs of lateral leglets ("styli") and 3 caudal filaments; simple development. Jumping bristletails are usually found in grassy or wooded areas under leaves, bark, or stones.

Dicondylia

Includes Thysaura and the Pterygota. These insects have mandibles with a two condyles (articulation points).

Order Thysanura Approximately 450 described species (Figure 22.2C,D). Small, wingless, resembling Archaeognatha but with flattened body; with or without ocelli; compound eyes reduced, not contiguous; body usually covered with scales; mandibles biting–chewing; antennae multiarticulate, but only basal article with musculature; tarsi 3- to 5-articulate; abdomen 11-segmented, with lateral leglets (often called styli) on segments 2–9, 7–9, or 8–9; 3 caudal cerci; female gonopores on eighth abdominal segment, male gonopores on tenth; without copulatory organs; with tracheae; simple development. Silverfish occur in leaf litter or under bark or stones, or in buildings, where they may feed on wallpaper paste, bookbindings, and the starch sizing of some fabrics.

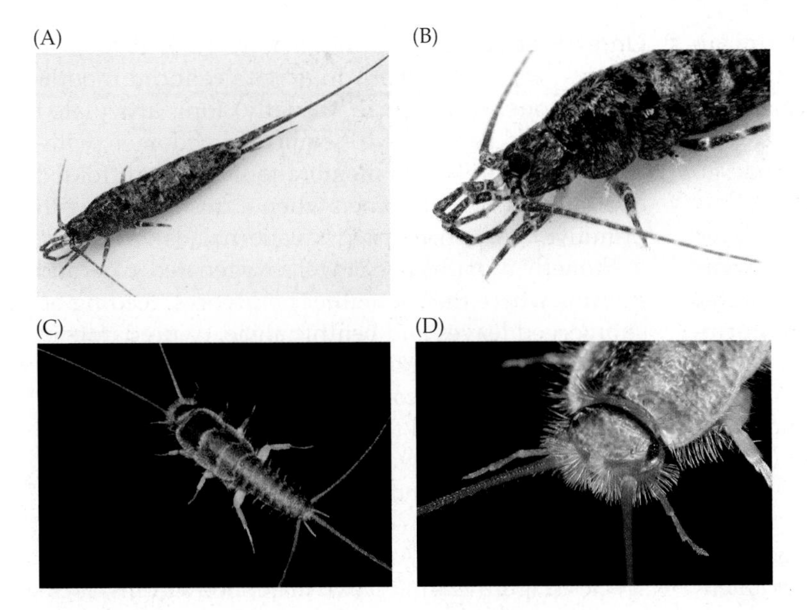

Figure 22.2 Representatives of the two orders of wingless Insecta. (A) Jumping bristletail (order Archaeognatha) and (B) Close-up of jumping bristletail head, note the eyes are contiguous. (C) A silverfish (order Thysanura). (D) Close-up of sliverfish head. Note that eyes are not contiguous, but set wide apart, and that the bodies of both of these groups are covered with scales.

Subclass Pterygota

The winged insects (with a pair of wings on the second and third thoracic segments), the forewings (front wings) and hindwings; wings may be secondarily lost in one or both sexes, or modified for functions other than flight; adults without abdominal leglets except on genital segments; female gonopores on eighth abdominal segment, male on tenth; female often with ovipositor; molting ceases at maturity.

Palaeoptera

Wings cannot be folded, and when at rest wings are either held straight out to the side or vertically above the abdomen (with dorsal surfaces pressed together); wings always membranous, with many longitudinal veins and cross veins; wings tend to be fluted, or accordion-like; antennae highly reduced or vestigial in adults; hemimetabolous development; larvae aquatic. Two extant orders; many extinct groups.

Order Ephemeroptera Approximately 2,500 described species (Figure 22.3A). Adults with vestigial mouthparts, minute antennae, and soft bodies; wings held vertically over body when at rest; forewings present; hindwings absent or present but much smaller than forewings; long, articulated cerci, usually with medial caudal filament; male with first pair of legs elongated for clasping female in flight; second and third legs of male, and all legs of female, may be vestigial or absent (Polymitarcyidae); abdomen 10-segmented; larvae aquatic; young (nymphs) with paired articulated lateral gills, caudal filaments, and well developed

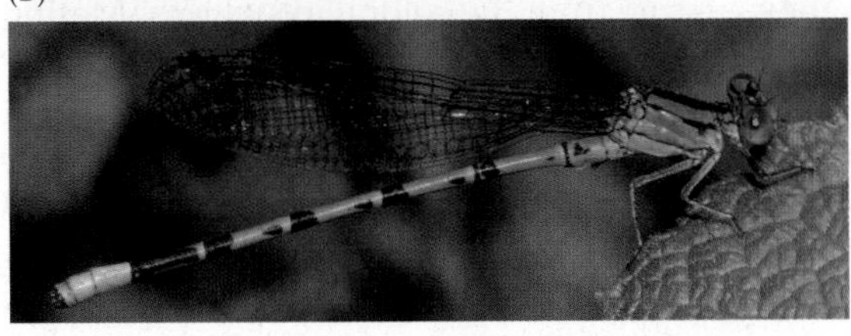

Figure 22.3 Representatives of the two orders of Palaeoptera. (A) A mayfly (order Ephemeroptera). (B) A damselfly (order Odonata). (C) A dragonfly (order Odonata). Note all of these insects hold their wings either out flat or straight up, but they cannot fold their wings over their back.

mouthparts; adults preceded by winged subimago stage. The mayfly subimago (subadult) is the only winged insect known to undergo an additional molt. It is the one exception to the rule, that an insect with wings is a mature adult insect and will never undergo another molt.

Mayflies are primitive winged insects in which the aquatic nymphal stage dominates the life cycle. Larvae hatch in fresh water and become long-lived nymphs, passing through many instars. Mayfly nymphs are important food for many stream and lake fishes. Adults eclose in synchrony and live only a few hours or days (hence the name, "ephemeral winged"), do not feed, and copulate in the air, sometimes in large nuptial swarms.

Order Odonata Approximately 6,000 described species (Figure 22.3B,C). Adults with small filiform antennae, large compound eyes, and chewing mouthparts with massive mandibles; larval labium modified into prehensile organ; two pairs of large wings, held outstretched (dragonflies) or straight up over body (damselflies) when at rest; abdomen slender and elongate, 10-segmented; male with accessory genitalia on second and third abdominal sternites; eggs and larvae aquatic, with caudal or rectal gills.

Dragonflies and damselflies are spectacular insects with broad public appeal, not only for their beauty but because they are fast flyers and consume large numbers of insect pests, including mosquitoes, on the wing. Larvae and adults are both highly active predators, with larvae consuming various invertebrates and adults capturing other flying insects. Many species are 7–8 cm long, while some extinct forms had a wingspan of over 70 cm.

Infraclass Neoptera

The modern, wing-folding insects. Modifications of the sclerites and an associated muscle at the base of the wings, allow neopterans to rotate their wing joint and fold back their wings when they are not flying. This is one of the most important evolutionary innovations in hexapods. Wing folding allows insects to protect their fragile wings, especially from abrasion, thereby allowing them to live in tight spaces such as crevices under bark, under rocks, in burrows, nests, and tunnels.

Superorder Polyneoptera

Polyneopterans are a morphologically diverse group of insects with biting–chewing mouthparts and hemimetabolous development. The phylogenetic relationships among the orders within Polyneoptera, and the monophyly of the group itself have long been controversial. However, recent phylogenomic work based upon 1478 single-copy nuclear genes, recovered strong support that the 10 polyneopteran orders form a monophyletic group (Misof et al. 2014).

Order Plecoptera Approximately 1,700 described species (Figure 22.4A). Adults with reduced mouthparts, elongate antennae, (usually) long articulated cerci, soft bodies, and a 10-segmented abdomen; without ovipositor; wings membranous, pleated, folded over and around abdomen when at rest; wings with primitive venation; nymphs aquatic (naiads), with gills.

Stonefly nymphs live in well-oxygenated lakes and streams where they are either herbivores, feeding on submerged leaves and benthic algae, or predators of other aquatic arthropods. They are an important food resource for fish. Plecopteran nymphs are intolerant of water pollution, and their presence is often used as a bioindicator of high water quality. Adults of most species are short-lived and die soon after mating.

Order Blattodea Approximately 7,000 described species (Figure 22.4B,C). Two major body forms: cockroaches and termites. Cockroaches have a dorsoventrally flattened body; large pronotum, with expanded margins extending over head; forewings (when present) leathery; hindwings expansive and fanlike; ovipositor reduced; cerci multiarticulate; legs adapted for running; eggs laid in cases (ootheca). Termites are small; soft-bodied; wings equal-sized, elongate, membranous, dehiscent (shed by breaking at basal line of weakness); antennae short, filamentous, with 11–33 articles; cerci small to minute; ovipositor reduced or absent; many with rudimentary or no external genitalia; marked polymorphism.

Of the 4,000 described species of cockroaches, fewer than 40 are domestic (household inhabitants). Some species are omnivores, while others are restricted in diet. Most species are tropical, but some live in temperate habitats, caves, deserts, and ant and bird nests. Some live in and feed on wood and have intestinal flora that aid in cellulose digestion (*Cryptocercus*). These wood-feeding cockroaches gave rise to the termites (infraorder Isoptera), a strictly social group of insects, usually with three distinct types of individuals, or castes, in a species: workers, soldiers, and alates (reproductive individuals). Workers are generally sterile, blind individuals with normal mandibles; they are responsible for foraging, nest construction, and caring for members of the other castes. Soldiers are blind, usually sterile, wingless forms with powerful enlarged mandibles used to defend the colony. Alates have wings and fully formed compound eyes. They are produced in large numbers at certain times of the year, whereupon they emerge from the colony in swarms. Mating occurs at this time, and individual pairs start new colonies. Wings are shed after copulation. Colonies form nests (termitaria) in wood that is in or on the ground. Workers harbor a variety of symbiotic cellulose-digesting flagellate protists in special chambers in the hindgut. Some families contain symbiotic bacteria that serve the same purpose. Termites often occur in enormous

Figure 22.4 Representatives of the ten orders of Polyneoptera. (A) A stonefly (order Plecoptera). (B) A cockroach (order Blattodea). (C) Termites (order Blattodea, infraorder Isoptera). (D) A mantid (order Mantodea). (E) A stick insect (order Phasmida). (F) A rock crawler (order Grylloblattodea). (G) Earwigs (order Dermaptera). (H) A horse lubber grasshopper (order Orthoptera). (I) A heel-walker (order Mantophasmatodea), inset SEM of the tarsus of a heelwalker showing enlarged arolium. (J) A webspinner (order Embioptera). (K) A zorapteran (order Zoraptera).

numbers; one spectacular estimate suggests that there are about three-quarters of a ton of termites for every person on Earth! There is strong morphological and molecular evidence that order Blattodea is the sister group of the order Mantodea. Together these two groups are referred to as the Dictyoptera.

Order Mantodea Approximately 2,400 described species (Figure 22.4D). First pair of legs large and raptorial; prothorax elongate; head highly mobile due to cervical sclerites that lend structural and musculature support, with very large compound eyes, not covered by the pronotum; forewings thickened, hindwings membranous; abdomen 11-segmented, 10 visible segments and the fragmented epiproct (which is made up of a median and two lateral components); reduced ovipositor made up of three valvular structures; male genitalia made up of three phallomeric lobes; one pair multiarticulate cerci; styli sometimes present in males.

Mantises are obligate predators, mostly on insects and spiders. While many species are highly cryptic in both color and structural morphology, some species feature brightly colored patches on the anteroventral surface of their forecoxae to use in threat displays or courtship displays. They have very good eyesight, which they use to locate and track prey before striking with their raptorial forelegs, using either an ambush or cursorial hunting strategy. Females lay many eggs together in an ootheca, a protective matrix of hardened foam, which is characteristic of this order. Mantises are distributed around the world, with their greatest diversity in the Indo-Malaysian, tropical African, and Neotropical regions.

Order Phasmida Approximately 3,000 described species (Figure 22.4E). Body cylindrical or markedly flattened dorsoventrally, usually elongate; biting–chewing mouthparts; prothorax short, with specialized glands for the excretion of noxious chemicals when disturbed; meso- and metathorax greatly elongate; forewings absent, or forming small to moderately elongate, leathery tegmina; hindwings absent, reduced to leathery tegmina, or fanlike; tarsomeres with ventral adhesive pads (euplantulae); short, unsegmented cerci; male with vomer—a specialized sclerite on venter of abdominal segment 10; ovipositor of female weak.

Stick and leaf insects ("walking sticks"), and other phasmids are large, tropical, predominantly nocturnal herbivores and are some of the most spectacular and oddest of all the insects. Although resembling orthopterans in basic form, they are clearly a distinct radiation. Their ability to mimic plant parts is legendary, and many have evolved as perfect mimics of twigs, leaves, bark, broken branches, moss or lichens. Stick and leaf insects are the only hexapods that can regenerate lost limbs (e.g., due to predation) like their crustacean ancestors; regeneration occurs during molts, with the new legs markedly smaller than original limbs but increasing in relative size with each successive molt (as in Crustacea). Phasmids are the only insect order with species-specific egg morphology across the entire order. The eggs of some species even mimic angiosperm seeds and are spread via ant-mediated dispersal. The order includes the longest extant insects on Earth, including the worlds longest at just over 56 cm in length, though some are less than 4 cm. Sexual dimorphism is so striking that males and females have often mistakenly been given different species names. Phasmids are facultatively parthenogenetic; in the absence of males, females can produce viable eggs that are genetic clones of their mother.

Order Grylloblattodea Approximately 30 species (Figure 22.4F). Slender, elongate, cylindrical, wingless insects, usually 15–30 mm long. Body usually pale or golden and finely pubescent; compound eyes small or absent; no ocelli; mouthparts mandibulate; antennae long and filiform, of 23–45 antennomeres, cerci long, 8-segmented; terminal sword-shaped ovipositor of similar length as cerci.

Rock crawlers were not discovered until 1914, and today 33 species are known, roughly half of which are from North America. They inhabit cold, rocky habitats, including snow fields below glaciers and ice caves. Most species cannot tolerate warm temperatures, but thrive at below freezing temperatures. Due to the cold temperature at which they live, growth and development are very slow. Rock crawlers may require up to seven years to complete a single generation. They are nocturnal scavengers on dead insects and other organic matter.

Order Dermaptera Approximately 1,800 described species (Figure 22.4G). Cerci usually form heavily sclerotized posterior forceps; forewings (when present) form short, leathery tegmina, without veins and serving as covers for the semicircular, membranous hindwings (when present); ovipositor reduced or absent.

Earwigs are common in urban environments. Most appear to be nocturnal scavenging omnivores. The forceps are used in predation, for defense, to hold a mate during courtship, for grooming the body, and for folding the hindwings under leathery forewings. Some species eject a foul-smelling liquid from abdominal glands when disturbed. Most species are tropical, although many also inhabit temperate regions.

Order Orthoptera Approximately 23,000 described species (Figure 22.4H). Pronotum unusually large, extending posteriorly over mesonotum; forewings with thickened and leathery region (tegmina), occasionally modified for stridulation or camouflage; hindwings membranous, fanlike; hindlegs often large, adapted for jumping; auditory tympana present on forelegs and abdomen; tarsomeres with ventral adhesive pads (euplantulae); ovipositor large; male genitalia complex; cerci distinct, short, and jointed.

Grasshoppers and their kin are common and abundant insects at all but the coldest latitudes. This order includes some of the largest living insects. Most are herbivores, but many are omnivorous, and some are predatory. Stridulation, which is common among males, is usually accomplished by rubbing the specially modified forewings (tegmina) together, or by rubbing a ridge on the inside of the hind femur against a special vein of the tegmen. No orthopterans stridulate by rubbing the hindlegs together, as is often thought. Common families include Acrididae (short-horned grasshoppers), Gryllotalpidae (mole crickets), Gryllacridadae (Jerusalem crickets), Tetrigidae (pygmy grasshoppers); Tridactylidae (pygmy mole crickets); Gryllacridadae (cave crickets); Gryllidae (crickets); Tettigoniidae (long-horned grasshoppers, katydids).

Order Mantophasmatodea Approximately 20 species (Figure 22.4I). Head hypognathous, with generalized mouthparts; antennae long, filiform; ocelli absent; wings entirely lacking; coxae elongate; tarsi with 5 tarsomeres, pretarsus of all legs with an unusually large arolium; cerci short, one-segmented.

The Mantophasmatodea is the most recently described order of insects (2002) and the only new insect order described since 1914. The order includes several living species (from Namibia, South Africa, Tanzania, and Malawi) and six fossil species (5 from Baltic amber and one fossil from China). They resemble a mix between praying mantises and phasmids, but molecular evidence indicates that they are most closely related to the Grylloblattodea.

Mantophasmatodeans have several distinct characteristics, including a hypognathous head and, most strikingly, when walking all species keep the fifth tarsomere and pretarsus (a greatly enlarged arolium plus two tarsal claws) of each leg turned upwards and off the substrate, giving them the appearance of "walking on their heels." Both sexes are wingless. They are highly flexible along their longitudinal body axis enabling them to clean their external genitalia with their mouthparts. During the day they hide in bushes, rock crevices or clumps of grass and prey on spiders and insects at night. Males and females produce percussive signals by tapping their abdomens on the substrate to locate mates.

Order Embioptera (Embiidina) Approximately 400 described species (Figure 22.4J). Males somewhat flattened; females and young cylindrical. Most about 10 mm long, however Southeast Asian species in the genus *Ptilocerembia* are approximately 20 mm long. Antennae filiform; ocelli lacking; chewing mouthparts; head prognathous; legs short and stout; tarsi 3-articulate; hind femora greatly enlarged. The basal article of the front tarsus is enlarged and contains glands that produce silk, which is spun from a dense field of hollow hairlike structures on the ventral surface. Males of most species are winged, but some are wingless; females and nymphs are always wingless. Abdomen 10-segmented, with rudiments of the eleventh segment, and a pair of short cerci.

Webspinners are small, slender, chiefly tropical insects. They live gregariously in silken galleries that they construct in leaf litter, under or on stones, in soil cracks, in bark crevices, and in epiphytic plants. Wings are made rigid for flight and flexible in galleries, by regulating the hemocoelic fluid pressure in the radial blood sinuses in both pairs of wings. They feed mostly on dead plant material and also graze on the outer bark of trees, and on mosses and lichens.

Order Zoraptera Approximately 40 species (Figure 22.4K). Minute (to 3 mm); termite-like; colonial; wingless or with wings; wings eventually shed; antennae moniliform, 9-articulate; abdomen short, oval, 10-segmented; chewing mouthparts; simple development. These uncommon insects are usually found in gregarious colonies in dead wood, but they do not have a division of labor or polymorphism (as in termites and ants). They feed chiefly on mites and other small arthropods. All extant species are classified in the single genus *Zorotypus*. The sister group relationship of this order remains controversial and elusive.

Superorder Acercaria

The superorder Acercaria is sometimes referred to as the Paraneoptera or the "hemipteroids." These insects are characterized by (usually) short antennae, enlarged cibarial (feeding) muscles, visible externally as an enlarged portion of the head; lacinia slender and elongate; sucking mouthparts, tarsi with three or fewer tarsomeres, absence of cerci, lack of true male gonopods, wings (when present) with reduced venation, and hemimetabolous development (although life cycles in several groups includes one or two inactive pupa-like stages).

Order Thysanoptera Approximately 5,000 described species (Figure 22.5D). Slender, minute (0.5–1.5 mm) insects with long, narrow wings (when present) bearing long marginal setal fringes; mouthparts form a conical, asymmetrical sucking beak; left mandible a stylet, right mandible vestigial; with compound eyes; antennae with 4–10 flagellomeres; abdomen 10-segmented; without cerci; tarsi 1–2 segmented, with an eversible, pretarsal eversible adhesive sac, or arolium. Thrips are mostly herbivores or predators, and many pollinate flowers. They are known to transmit plant viruses and fungal spores. "Thrips" is both singular and plural.

Order Hemiptera Approximately 85,000 species (Figure 22.5A,B). Piercing–sucking mouthparts form an articulated beak, mandibles and first maxillae stylet-like, lying in dorsally grooved labium; forewings either completely membranous, or hardened basally and membranous only distally; hindwings membranous; pronotum large.

Hemipterans occur worldwide and in virtually all habitats. Hemipterans are liquid feeders. Most feed on the xylem or phloem of plants, although many feed on the hemolymph of arthropods or blood of vertebrates and some are specialized ectoparasites. They are of considerable economic importance because many

(A)

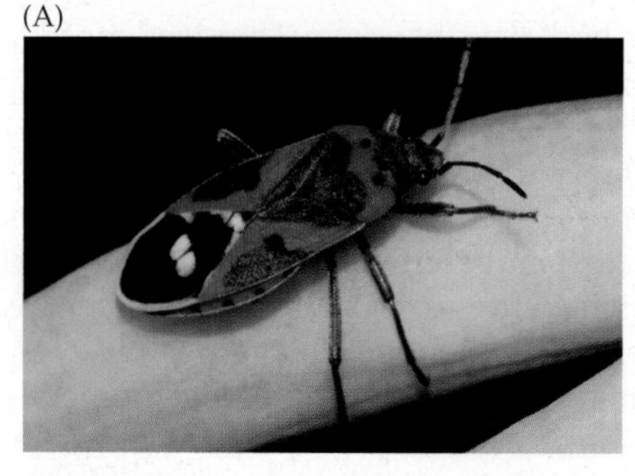

(B)

(C)

(D)

Figure 22.5 Representatives of the three orders of Acercaria. (A) A true bug (order Hemiptera). (B) A cicada (order Hemiptera). (C) Two bark lice (order Psocodea). (D) A thrips (order Thysanoptera). Note that "thrips" is both singular and plural.

are serious crop pests. Members of one subfamily of Reduviidae (Triatominae, the assassin or kissing bugs) transmit Chagas disease. Others have more positive economic importance to humans, such as the cochineal bugs (Dactylopiidae), from which a safe red dye (cochineal) is extracted for use in the food industry. Shellac is made from lac, a chemical produced by members of the family Kerriidae (lac insects). One of the most famous hemipterans are the 17-year cicadas (genus *Magiciada*), which have a very long life cycle and synchronized development and can reach plague-like levels of abundance (to 3.7 million individuals per hectare). Common predacious hemipteran families include Nepidae (water scorpions), Belostomatidae (giant water bugs or "toe biters"), Corixidae (water boatmen), Notonectidae (backswimmers), Gerridae (water striders), Saldidae (shore bugs), Cimicidae (bedbugs), and Reduviidae (assassin bugs).

Many others are plant feeders (hence the common name "plant bugs"). Heavy infestations of these insects on plants may cause wilting, stunting, or even death, and some are vectors of important plant diseases. Common herbivorous families include the Cicadidae (cicadas), Cicadellidae (leafhoppers), Fulgoridae (planthoppers), Membracidae (treehoppers), Cercopidae (spittle bugs and froghoppers), Aleyrodidae (whiteflies), and Aphidae (aphids), as well as members of the large superfamily Coccoidea (coccoids, scale insects, mealybugs, and many others).

Order Psocodea Approximately 8,500 described species (Figure 22.5C). Until recently, species in this order were classified in two separate orders: the Psocoptera (the book and bark lice) and the Phthiraptera (the true lice). Psocopterans are small (1–10 mm long); with long, filiform, multiarticulate antennae; short prothorax; meso- and metathorax often fused; chewing mouthparts; abdomen 9-segmented; without cerci. Phthirapterans are even smaller (less than 5 mm), wingless, blood-sucking, obligate ectoparasites of birds and mammals; thoracic segments completely fused; cuticle largely membranous and expandable to permit engorgement; compound eyes absent or of 1–2 ommatidia; ocelli absent; piercing–sucking mouthparts retractable into a buccal pouch; antennae short (5 or fewer flagellomeres), exposed or concealed in grooves beneath the head; with 1 pair dorsal thoracic spiracles and 6 or fewer abdominal spiracles; without cerci; females lack ovipositor.

Psocids—the book and bark lice—generally feed on algae and fungi, and occur in suitably moist areas (e.g., under bark, in leaf litter, under stones, in human habitations where humid climates prevail). They are often pests that get into various stored food products or consume insect and plant collections; some species live in books and eat the bindings. Commonly called sucking lice ("Anoplura") and biting lice ("Mallophaga"), phthirapterans spend their entire life on one host. Eggs (nits) are usually attached to the hair or feathers of the host, although the human body louse (a "sucking louse") may attach eggs to clothing. No biting lice are known to infest humans. Posthatching development comprises three nymphal instars. Some species that infest domestic birds and mammals are of economic significance.

Superorder Holometabola

Holometabola is a monophyletic group strongly supported by both morphological and molecular data. One of its key synapomorphies is indirect (holometabolous) development, with distinct egg, larval, pupal, and adult states. During the pupal stage, most tissues undergo a complete reorganization (e.g., larval eyes [stemmata] disintegrate and adult compound eyes and ocelli form *de novo*). Another key apomorphy is the presence of internal wing buds (imaginal discs) in the larval stage, these clusters of early embryonic cells arise from localized invaginations of the ectoderm in the early embryo and form adult structures during pupation.

Order Hymenoptera Approximately 115,000 described species (Figure 22.6A). Mouthparts often elongate and modified for ingesting floral nectar, although mandibles usually remain functional; labium often (bees) distally expanded as paired lobe-like structures called glossae and paraglossae; three ocelli; usually with two pairs of membranous wings; hindwings small, coupled to forewings by hooks (hamuli); wing venation highly reduced; antennae well developed, of various forms and with 3–70 flagellomeres, reduced metathorax usually fused to first abdominal segment; males with complex genitalia; females with ovipositor (in most), modified for sawing, piercing, or stinging.

The earliest fossil Hymenoptera date from the Triassic (220–207 Ma). Ants, bees, wasps, sawflies, and their relatives are all active insects with a tendency to form polymorphic social communities. Two suborders are generally recognized. Suborder Symphyta contains the primitive, wasplike, "thick-waisted" hymenopterans (sawflies, horntails, and their kin). They rarely show conspicuous sexual dimorphism and are always fully winged. The first and second abdominal segments are broadly joined. Larvae are mostly caterpillar-like, with a well-developed head capsule, true legs, and often also with abdominal prolegs. Suborder Apocrita contains the "narrow-waisted" hymenopterans (true wasps, bees, and ants), in which the first and second abdominal segments are joined by a distinct and often elongate constriction. Adults tend to be strongly social and display marked polymorphism. Social communities often include distinct castes of queens, haploid males, parthenogenetic females, and individuals with other sex-related specializations, as well as non-reproducing worker and soldier forms. Common families of Hymenoptera include the Apidae (bumblebees and honeybees), Formicidae (ants), Vespidae (yellow jackets, hornets, paper wasps, and potter wasps, Halictidae (sweat bees), Sphecidae (sand wasps, digger wasps, and mud-daubers), and three large groups of parasitic wasps (Ichneumonidae, Braconidae, Chalcidoidea).

Coleopterida

The Coleopterida includes two orders the Coleoptera (beetles) and the Strepsiptera (twisted wing parasites). This clade includes the most diverse order of insects (Coleoptera) and one with the most highly modified and morphologically and developmentally aberrant groups of parasites (Strepsiptera). Largely due to their bizarre combination of morphological characters, and the high rate of molecular evolution in many strepsipteran genes, the sister group relationship of the Strepsiptera eluded researchers until very recently. In the past, Strepsiptera were considered at various times to be close relatives of either Hymenoptera, Diptera, or Coleoptera. Recent genome-wide analyses and new morphological evidence strongly suggests that they are in fact the sister group to the Coleoptera.

Order Coleoptera Approximately 380,000 described species (Figure 22.6B). Body usually heavily sclerotized; forewings sclerotized and modified as rigid covers (elytra) over hindwings and body; membranous hindwings fold both transversely and longitudinally, and are often reduced or absent; biting-chewing mouthparts; antennae usually with 8–11 flagellomeres; prothorax large and mobile; mesothorax reduced; abdomen typically of 5 (or up to 8) segments; without ovipositor; male genitalia retractable.

Coleoptera is the largest order of insects. Many hypotheses have been proposed to explain the extraordinary diversity of beetles, including: (1) their age, the oldest fossils date to the early Permian but they most probably arose in the late Carboniferous (300 Ma); (2) their heavily sclerotized bodies, including their protective elytra, and general lack of exposed membranous surfaces facilitates their adaptation to a wide variety of tight narrow spaces and reduced their risk of predation; and (3) the their co-evolution with the great angiosperm radiation in the Cretaceous period. Today, beetles range from minute (0.35 mm, *Nanosella fungi*) to very large (20 cm, *Titanus giganteus*) and occur in all the world's environments (except the open sea). Some of the world's strongest animals are beetles: rhinoceros beetles can carry up to 100 times their own weight for short distances, and 30 times their weight indefinitely (equivalent to a 150-pound man walking with a Cadillac on his head—without tiring). Humans have had a long fascination with beetles, and beetle worship can be traced back to at least 2500 BC. (The venerated scarab of early Egyptians was actually a dung beetle.)

Some common coleopteran families include Carabidae (ground beetles), Dytiscidae (predaceous diving beetles), Gyrinidae (whirligig beetles), Hydrophilidae (water scavenger beetles), Staphylinidae (rove beetles), Cantharidae (soldier beetles), Lampyridae (fireflies and lightning bugs), Phengodidae (glowworms), Elateridae (click beetles), Buprestidae (metallic wood-boring beetles), Coccinellidae (ladybird beetles), Meloidae (blister beetles), Tenebrionidae (darkling beetles), Scarabaeidae (scarab beetles, dung beetles, June "bugs"), Cerambycidae (long-horned beetles), Chrysomelidae (leaf beetles), Curculionidae (weevils), Brentidae (primitive weevils), and Ptiliidae (featherwinged beetles, the smallest of all beetles, some with body lengths of just 0.35 mm).

Order Strepsiptera Approximately 600 described species (Figure 22.6C). Extreme sexual dimorphism; males free-living and winged; females wingless, usually parasitic. Females of free-living species with distinct head, simple antennae, chewing mouthparts, and compound eyes. Females of parasitic species neotenous, larviform, usually without eyes, antennae, and legs; with indistinct body segmentation. Male antennae often with elongate processes on flagellomeres; forewings reduced to club-like structures resembling halteres of Diptera; hindwings large and membranous, with reduced venation; raspberry-like eyes.

Most of these minute insects are parasitic on other insects. Adult females of parasitic species are larviform and most commonly live between the abdominal sclerites of flying insects that pollinate flowers, such as bees and wasps. Winged males find the females on the bee or wasp abdomen and mate with her. The fertilized eggs hatch into first instar larvae inside their mother's body. These larvae, called triungulins, have well developed eyes and legs and they actively crawl out of their mother to invade the soil and vegetation. The triungulins eventually locate a new host insect and enter it, wherein they molt into a legless wormlike larval stage that feeds in the host's body cavity. Pupation also takes place within the host's body, where the females remain for the rest of their lives and the free-living males emerge as fully formed adults.

Neuropterida

The three orders of Neuropterida have always been considered close relatives. Some workers subsume the Megaloptera and Raphidioptera within the Neuroptera, but all three groups are monophyletic making such a taxonomic decision arbitrary. These three orders have two pairs of membranous wings with many cross veins, 5-segmented tarsi, adults with mandibles, and decticous pupae (with articulated mandibles).

Order Megaloptera Approximately 300 described species (Figure 22.6D). Ocelli present or absent; larvae aquatic, with lateral abdominal gills. Megalopterans (alderflies, dobsonflies, fishflies) strongly resemble neuropterans (and are often regarded as a suborder), but their hindwings are broader at the base than the forewings, and the longitudinal veins do not have branches near the wing margin. Larvae of some megalopterans (hellgrammites) are commonly used as fish bait.

Order Raphidioptera Approximately 260 described species (Figure 22.6E). Snakeflies strongly resemble neuropterans (and are often regarded as a suborder), but are unique in having the prothorax elongate (as in the mantises), but the forelegs similar to the other legs. The head can be raised above the rest of the body, as in a snake preparing to strike. Adults and larvae are predators on small insect prey.

Order Neuroptera Approximately 6,000 described species (Figure 22.6F). Adults soft-bodied; with two pairs of similar, highly veined wings held tent-like over the abdomen when at rest; with biting–chewing mouthparts; abdomen 10-segmented; without cerci; larvae with mandibles and maxillae co-adapted to create a sucking tube; mouth closed off by modified labrum and labium; well developed legs; larval midgut is closed off posteriorly, larval waste accumulates until the adult emerges; Malpighian tubules secrete silk, via the anus, to construct the pupal cocoon.

The lacewings, ant lions, mantisflies, spongillaflies, and owlflies form a complex group, the adults of which are often important predators of insect pests (e.g., aphids). The larvae of many species have piercing–sucking mouthparts, and those of other species are predaceous and have biting mouthparts. The pupae are often unusual in possessing free appendages and functional mandibles used for defense; they may actively walk about prior to the adult molt, but do not feed. Common families include Chrysopidae (green lacewings), Myrmeleontidae (ant lions); Ascalaphidae (owlflies); Mantispidae (mantidflies).

Antliophora

The Antliophora include three orders, the Mecoptera, Siphonaptera, and Diptera. There is strong molecular and morphological support for the monophyly of this group. The males of all members of Antliophora have a sperm pump, a structure that aids in sperm transfer during copulation. Many other more characters also define this grouping, most of which have to do with relatively subtle aspects of the adult mouthparts.

Order Mecoptera Approximately 600 described species (Figure 22.6G). Two pairs of similar, narrow, membranous wings, held horizontally from sides of body when at rest; antennae long, slender, and of many flagellomeres (about half the body length); head with ventral rostrum and reduced biting mouthparts; long, slender legs; mesothorax, metathorax, and first abdominal tergum fused; abdomen 11-segmented; female with two cerci; male genitalia prominent and complex, at apex of attenuate abdomen and often resembling a scorpion's stinger. The larvae of some species are remarkable in having compound eyes, a condition unknown among larvae of other insects having complete metamorphosis.

Figure 22.6 Representatives of the eleven orders of Holometabola. (A) A paper wasp (order Hymenoptera). (B) A pleasing fungus beetle, *Gibbifer californicus* (order Coleoptera). (C) Female twisted wing parasites (order Strepsiptera) visible between the abdominal sclerites of a wasp (order Hymenoptera). (D) An alderfly (order Megaloptera). (E) A snakefly (order Raphidioptera). (F) A green lacewing (order Neuroptera). (G) A scorpionfly (order Mecoptera). (H) A adult male *Oropsylla montana* flea (order Siphonaptera). (I) A golden dung fly (*Scathophaga stercoraria)* (order Diptera). (J) A caddisfly (order Trichoptera). (K) A luna moth (order Lepidoptera).

Mecopterans are usually found in moist places, often in forests, where most are diurnal flyers. They are best represented in the Holarctic region. Some feed on nectar; others prey on insects or are scavengers. There are several families, including Panorpidae (scorpionflies), Bittacidae (hangingflies), and Boreidae (snow scorpionflies).

Order Siphonaptera Approximately 3,000 described species (Figure 22.6H). Small (less than 3 mm long); wingless; body laterally compressed and heavily sclerotized; short antennae lie in deep grooves on sides of head; mouthparts piercing–sucking; compound eyes often absent; legs modified for clinging and (especially hindlegs) jumping; abdomen 11-segmented; abdominal segment 10 with distinct dorsal pincushion-like sensillum, containing a number of sensory organs; without ovipositor; pupal stage passed in a cocoon.

Adult fleas are ectoparasites on mammals and birds, from which they take blood meals. They occur wherever suitable hosts are found, including the Arctic and Antarctic. Larvae usually feed on organic debris in the nest or dwelling place of the host. Host specificity is often weak, particularly among the parasites of mammals, and fleas regularly commute from one host species to another. Fleas act as intermediate hosts and vectors for organisms such as plague bacteria, dog and cat tapeworms, and various nematodes. Commonly encountered species include *Ctenocephalides felis* (cat flea), *C. canis* (dog flea), *Pulex irritans* (domestic flea), and *Diamus montanus* (western squirrel flea).

Order Diptera Approximately 135,000 described species (Figure 22.6I). Adults with one pair of membranous mesothoracic forewings and a metathoracic pair of club-like halteres (organs of balance); head large and mobile; compound eyes large; antennae primitively filiform, with 7 to 16 flagellomeres and often secondarily annulated (reduced to only a few articles in some groups); mouthparts adapted for sponging, sucking, or lapping; mandibles of blood-sucking females developed as piercing stylets; hypopharynx, laciniae, galeae, and mandibles variously modified as stylets in parasitic and predatory groups; labium forms a proboscis ("tongue"), consisting of distinct basal and distal portions, the latter in higher families forming a sponge-like pad (labellum) with absorptive canals; mesothorax greatly enlarged; abdomen primitively 11-segmented, but reduced or fused in many higher forms; male genitalia complex; females without true ovipositor, but many with secondary ovipositor composed of telescoping posterior abdominal segments; larvae lack true legs, although ambulatory structures (prolegs and "pseudopods") occur in many.

The true flies (which include mosquitoes and gnats) are a large and ecologically diverse group, notable for their excellent vision and aeronautic capabilities. The mouthparts and digestive system are modified for a fluid diet, and several groups feed on blood or plant juice. Dipterans are vastly important carriers of human diseases, such as sleeping sickness, yellow fever, African river blindness, and various enteric diseases. It has been said that mosquitoes (and the diseases they carry) prevented Genghis Khan from conquering Russia, killed Alexander the Great, and played pivotal roles in both world wars. Myiasis—the infestation of living tissue by dipteran larvae—is often a problem for livestock and occasionally for humans. Many dipterans are also beneficial to humans as parasites or predators of other insects and as pollinators of flowering plants. Dipterans occur worldwide and in virtually every major environment (except the open sea). Some breed in extreme environments, such as hot springs, saline desert lakes, oil seeps, tundra pools, and even shallow benthic marine habitats.

Some common dipteran families include Asilidae (robber flies), Bombyliidae (bee flies), Calliphoridae (blowflies, bluebottles, greenbottles, screwworm flies, etc.), Chironomidae (midges), Coelopidae (kelp flies), Culicidae (mosquitoes: *Culex*, *Anopheles*, etc.), Drosophilidae (pomace or vinegar flies; often also called "fruit flies"), Ephydridae (shore flies and brine flies), Glossinidae (tsetse flies), Halictidae (sweat bees), Muscidae (houseflies, stable flies, etc.), Otitidae (picture-winged flies), Sarcophagidae (flesh flies), Scatophagidae (dung flies), Simuliidae (blackflies and buffalo gnats), Syrphidae (hover flies and flower flies), Tabanidae (horseflies, deerflies, and clegs), Tachinidae (tachinid flies), Tephritidae (fruit flies), and Tipulidae (crane flies).

Amphiesmenoptera

The Amphiesmenoptera contains the orders Trichoptera and Lepidoptera. Some characteristics that unite these two groups include the presence of hairy wings (the hairs are modified into scales in Lepidoptera), the females are the heterogametic sex, and the larval labium and hypopharynx are fused into a composite lobe with the opening of an associated silk gland at its apex.

Order Trichoptera Approximately 12,000 described species (Figure 22.6J). Adults resemble small moths, but with body and wings covered with short hairs; two pairs of wings, tented in oblique vertical plane (rooflike) over abdomen when at rest; compound eyes present; mandibles minute or absent; antennae usually as long or longer than body, setaceous; legs long and slender; larvae and pupae mainly in fresh water, adults terrestrial; larvae with abdominal prolegs on terminal segment.

The freshwater larvae of caddisflies construct fixed or portable "houses" (cases) made of sand grains, wood fragments, or other material bound together by silk emitted through the labium. Larvae are primarily herbivorous scavengers; some use silk to produce food-filtering devices. Most larvae inhabit benthic habitats in temperate streams, ponds, and lakes. Adults are strictly terrestrial and have liquid diets.

Order Lepidoptera Approximately 120,000 described species (Figure 22.6K). Minute to large; sucking mouthparts; mandibles usually vestigial; maxillae coupled, forming a tubular sucking proboscis, coiled between labial palps when not in use; head, body, wings, and legs usually densely scaled; compound eyes well developed; usually with two pairs of large and colorfully scaled wings, coupled to one another by various mechanisms; protibia with epiphysis used

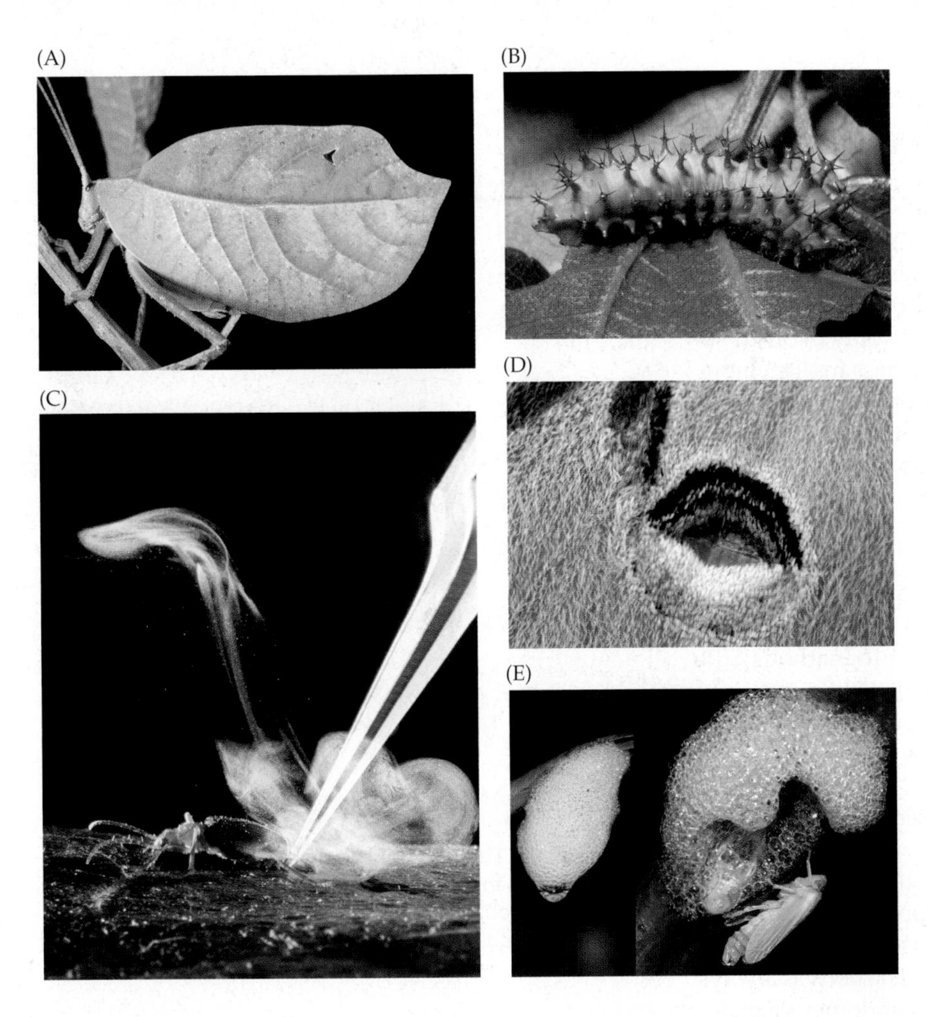

Figure 22.7 Defense, warning coloration, and camouflage in insects. (A) A leaf-mimicking katydid (Orthoptera: Tettigoniidae) from Ecuador is perfectly camouflaged. (B) The bright coloration of the silkmoth caterpillar (Lepidoptera: Saturniidae: *Hyalophora euryalus*), warns potential predators of the defensive hairs on its back. (C) A bombardier beetle (Coleoptera: Carabidae: *Brachinus*) sprays noxious benzoquinones in defense while being "attacked" by forceps. (D) Close-up of an "eye spot" on a luna moth (see Figure 22.6K) (Lepidoptera: Saturniidae). Eyespots attract predators away from the moth's head and body and encourage predators to focus on their wings. Since butterflies and moths can fly even with slight wing damage, eyespots are a commonly used defense mechanism. (E) A spittlebug (Hemiptera: Cercopoidea) produces a cover of frothed-up plant sap resembling saliva which hides the nymph from the view of predators and parasites.

for cleaning antennae; male genitalia complex; females with ovipositors.

Butterflies and moths are among the best known and most colorful of all the insects. The adults are primarily nectar feeders, and many are important pollinators, some of the best known being the large hawk, or sphinx, moths (Sphingidae). A few tropical species are known to feed on animal blood, and some even drink the tears of mammals. The larvae (caterpillars) feed on green plants. Caterpillars have three pairs of thoracic legs and a pair of soft prolegs on each of abdominal segments 3–6; the anal segment bears a pair of prolegs or claspers. Butterflies can be distinguished from moths by two features: their antennae are always long and slender, ending in a knob (moth antennae are never knobbed), and their wings are typically held together above the body at rest (moths never hold their wings in this position). Over 80% of the described Lepidoptera are moths.

Other common families include Psychidae (bag worms); Cossidae (carpenter worms); Pyralidae (snout moths); Saturniidae (silk moths); Sphingidae (hawk moths); Hesperiidae (skippers); Geometridae (inch worms); Arctiidae (tiger moths); Noctuidae (owlet moths); Papillionidae (swallowtails); Nymphalidae (brush-footed butterflies); Pieridae (whites and sulphurs); Lycaenidae (gossamer-winged butterflies); and Riodinidae (metalmarks).[2]

The Hexapod Body Plan

General Morphology

In Chapter 20 we briefly discussed the various advantages and constraints imposed by the phenomenon of arthropodization, including those associated with the establishment of a terrestrial lifestyle. Departure from the ancestral aquatic environment necessitated the evolution of stronger and more efficient support and locomotory appendages, special adaptations to withstand osmotic and ionic stress, and aerial gas exchange structures. The basic arthropod body plan included many preadaptations to life in a "dry" world. As we have seen, the arthropod exoskeleton inherently provides physical support and protection from predators, and

[2]One of the best known butterfly taxonomists was the great Russian novelist Vladimir Nabokov (*Lolita, Pale Fire, The Gift*), who left Saint Petersburg in 1917 to travel around Europe and eventually settled in the United States (first working at the American Museum of Natural History in New York, then at Cornell University). Nabokov was a specialist on the blue butterflies (Polyommatini) of the New World and a pioneer anatomist, coining such alliterative anatomical terms as "alula" and "bullula." Butterflies, real and imaginary, flit through 60 years of Nabokov's fiction, and many lepidopterists have named butterflies after characters in his life and writings (e.g., species epithets include *lolita, humbert, ada, zembla,* and *vokoban*—a reversal of Nabokov). Nabokov's descriptions of Lolita were patterned after his species descriptions of butterflies (e.g., "her fine downy limbs").

by incorporating waxes into the epicuticle, the insects, like the arachnids, acquired an effective barrier to water loss. Similarly, within the Hexapoda, the highly adaptable, serially arranged arthropod limbs evolved into a variety of specialized locomotory and food-capturing appendages. Reproductive behavior became increasingly complex, and in many cases highly evolved social systems developed. Within the class Insecta, many taxa underwent intimate coevolution with land plants, particularly angiosperms. The adaptive potential of insects is evident in the many species that have evolved striking camouflage, warning coloration, and chemical defense (Figure 22.7).

Non-insect hexapods (proturans, collembolans, and diplurans—the entognathous hexapods) differ from insects in several important ways. The mouthparts are not fully exposed (i.e., they are **entognathous**), the mandibles have a single point of articulation, development is always simple, the abdomen may have a reduced number of segments, and they never developed flight.

Insects are primitively composed of 20 somites (as in the Eumalacostraca; Chapter 21), although these are not always obvious. The consolidation and specialization of these body segments (i.e., tagmosis) has played a key role in hexapod evolution and has opened the way for further adaptive radiation. The body is always organized into a head, thorax, and abdomen (Figures 22.8 and 22.9), comprising 6, 3, and 11 segments respectively. In contrast to marine arthropods, a true carapace never develops in hexapods. In the head, all body sclerites are more or less fused as a solid head capsule. In the thorax and abdomen, the adult sclerites usually develop embryologically such that they overlap the primary segment articulations, forming secondary segments, and these are the "segments" we typically see when we examine an insect externally (e.g., the tergum and sternum of each adult abdominal secondary segment actually overlap its adjacent anterior primary segment) (Figure 22.10). The primitive (primary) body segmentation can be seen in unsclerotized larvae by the insertions of the segmental muscles and transverse grooves on the body surface.

Most insects are small, between 0.5 and 3.0 cm in length. The smallest are the thrips, feather-winged beetles, and certain parasitic wasps, which are all nearly microscopic. The largest are certain beetles, orthopterans, and stick insects, the latter attaining lengths greater than 56 cm. However, certain Paleozoic species grew to twice that size. To familiarize you with the hexapod body plan and its terminology, we briefly discuss each of the main body regions (tagmata) below.

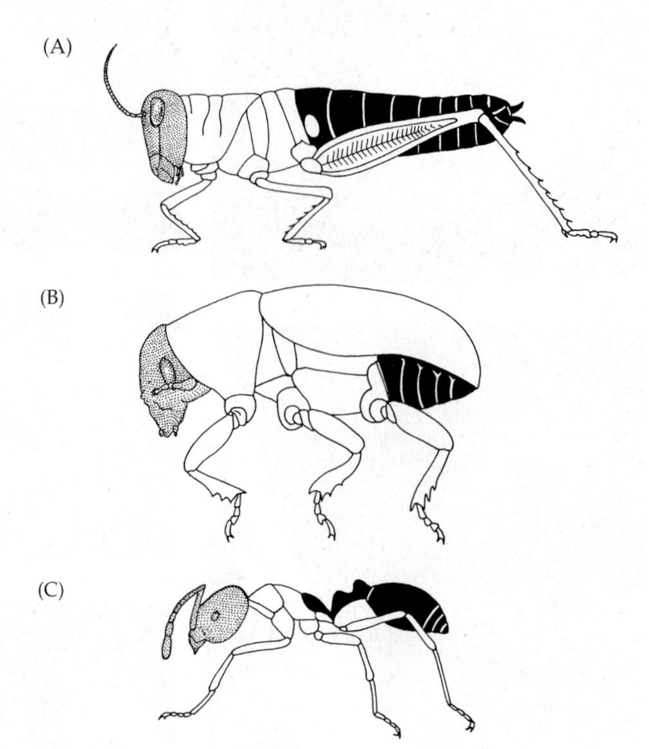

Figure 22.8 The principal body regions of hexapods, illustrated by three kinds of insects. (A) A grasshopper (wings removed). (B) A beetle. (C) An ant. In each case, the stippled region is the head; the white region is the thorax; the black region is the abdomen.

The hexapod head The hexapod head comprises an acron and six segments, bearing (from anterior to posterior) the eyes, antennae, clypeolabrum, and three pairs of mouth appendages (mandibles, maxillae, labium) (Figure 22.11). Compound eyes, as well as three **simple eyes** (**ocelli**) are typically present in adult hexapods. The median (anterior) ocellus is thought to have arisen through the fusion of two separate ocelli. The internal manifestation of the fused exoskeleton of the head forms a variety of apodemes, braces, and struts collectively called the **tentorium**. Externally, the head may also bear lines that may demarcate its original segmental divisions, and others that represent the dorsal (and ventral) ecdysial lines, where the head capsule splits in immature insects and which persist as unpigmented lines in some adults. Still other lines represent inflections of the surface associated with internal apodemes.

The antennae (Figure 22.12) are composed of three regions: the scape, pedicel, and multijointed sensory flagellum. The **scape** and **pedicel** constitute the **protopod**; the **flagellum** represents the telopod. Among the entognathous insects, muscles intrinsic to the scape, pedicel and flagellum are retained. But in the class Insecta, the intrinsic muscles of the antennae have been lost except for those in the scape. In addition, in many insects the joints, or **flagellomeres**, of the flagellum may have been secondarily subdivided (or annulated) to produce additional unmuscled joints, and increasing the length and flexibility of the antenna.

The mouth is bordered anteriorly by the clypeolabrum, posteriorly by the labium, and on the sides by

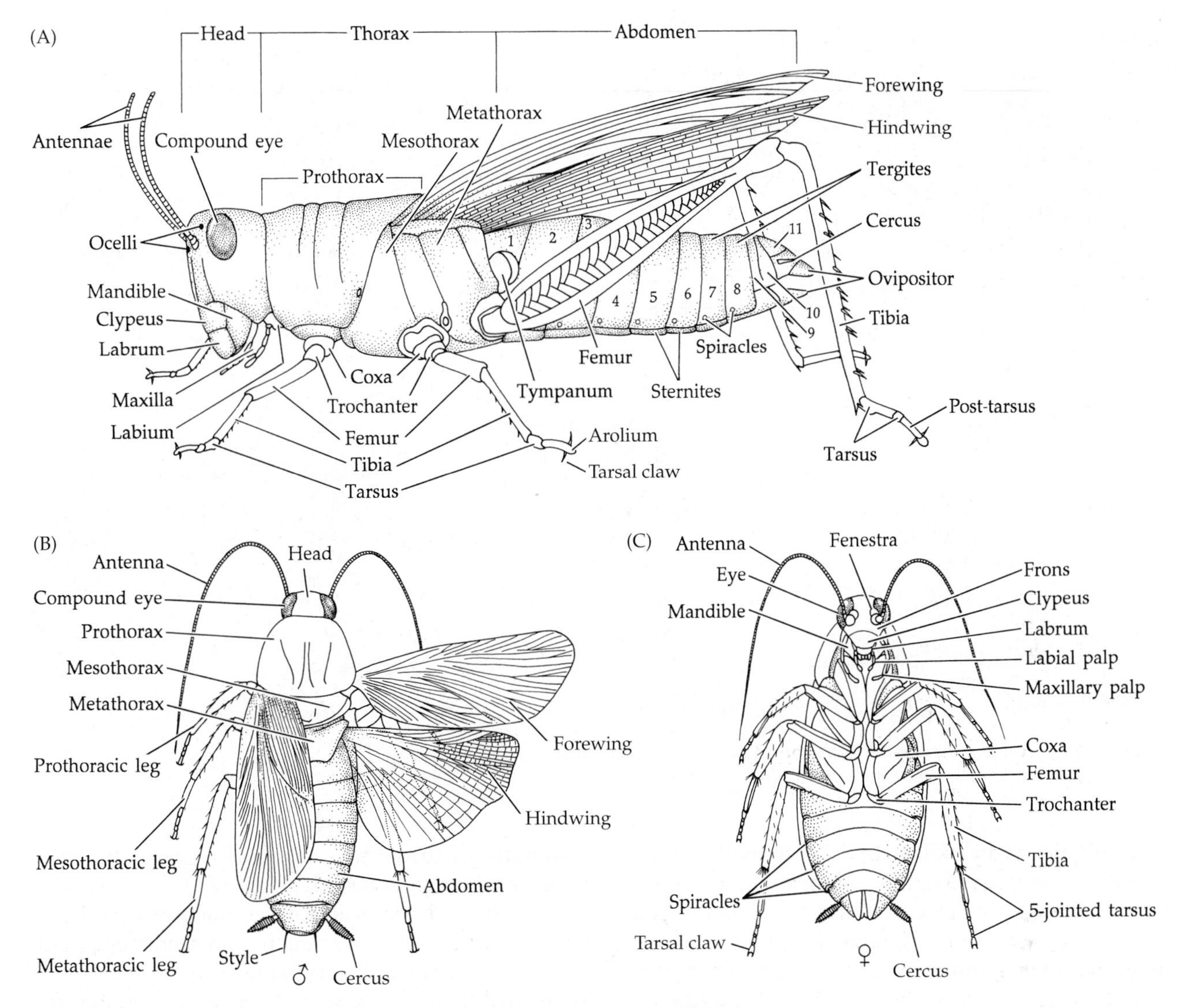

Figure 22.9 General body anatomy of insects. (A) A grasshopper (order Orthoptera). (B,C) Dorsal and ventral views of a cockroach (order Blattodea).

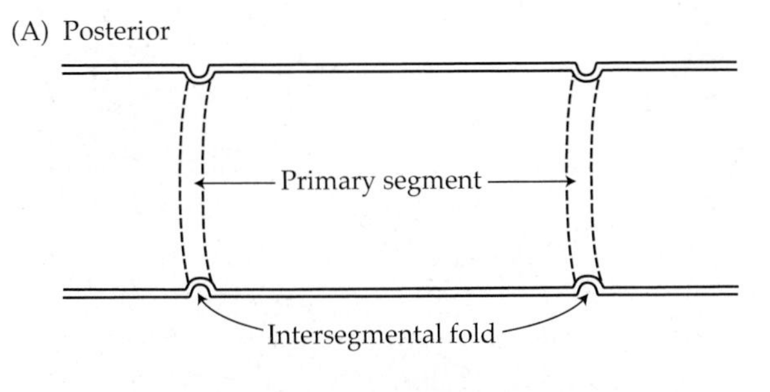

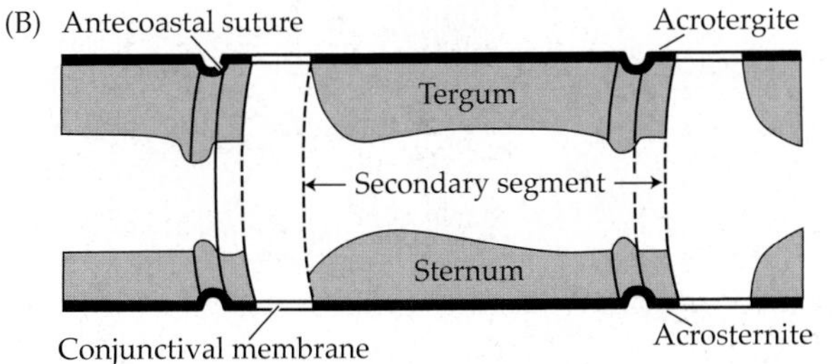

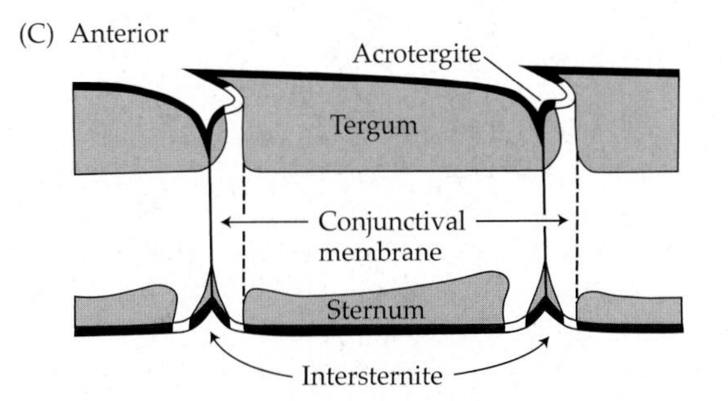

Figure 22.10 The ontogenesis of insect body segments. (A) Primary segmentation. (B) Simple secondary segmentation. (C) More advanced secondary segmentation.

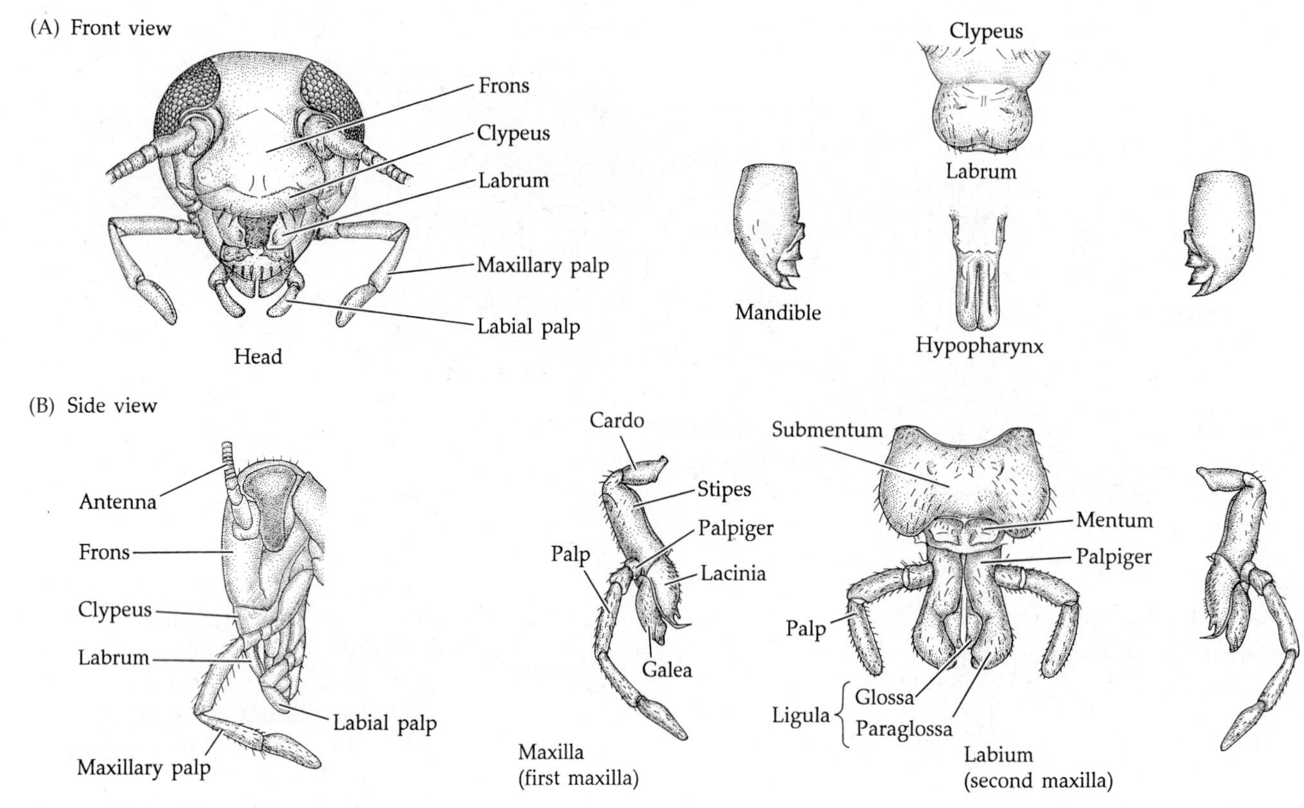

Figure 22.11 The mouth appendages of a typical biting–chewing insect: a grasshopper (order Orthoptera). (A) Front view. (B) Side view.

the mandibles and maxillae. In the entognathans, the mouthparts are sunk within the head capsule and largely hidden from view. In contrast, the mouthparts of insects are exposed (**ectognathous**) and ventrally projecting (**hypognathous**). However, in some insects, the orientation of the head has changed so that they are **prognathous** (projecting anteriorly) or **opisthognathous** (projecting posteriorly; Figure 22.13).

The labrum is a movable plate attached to the margin of the clypeus (a projecting frontal head piece), and together they form the **clypeolabrum**. Some workers regard the clypeolabrum to be an independently

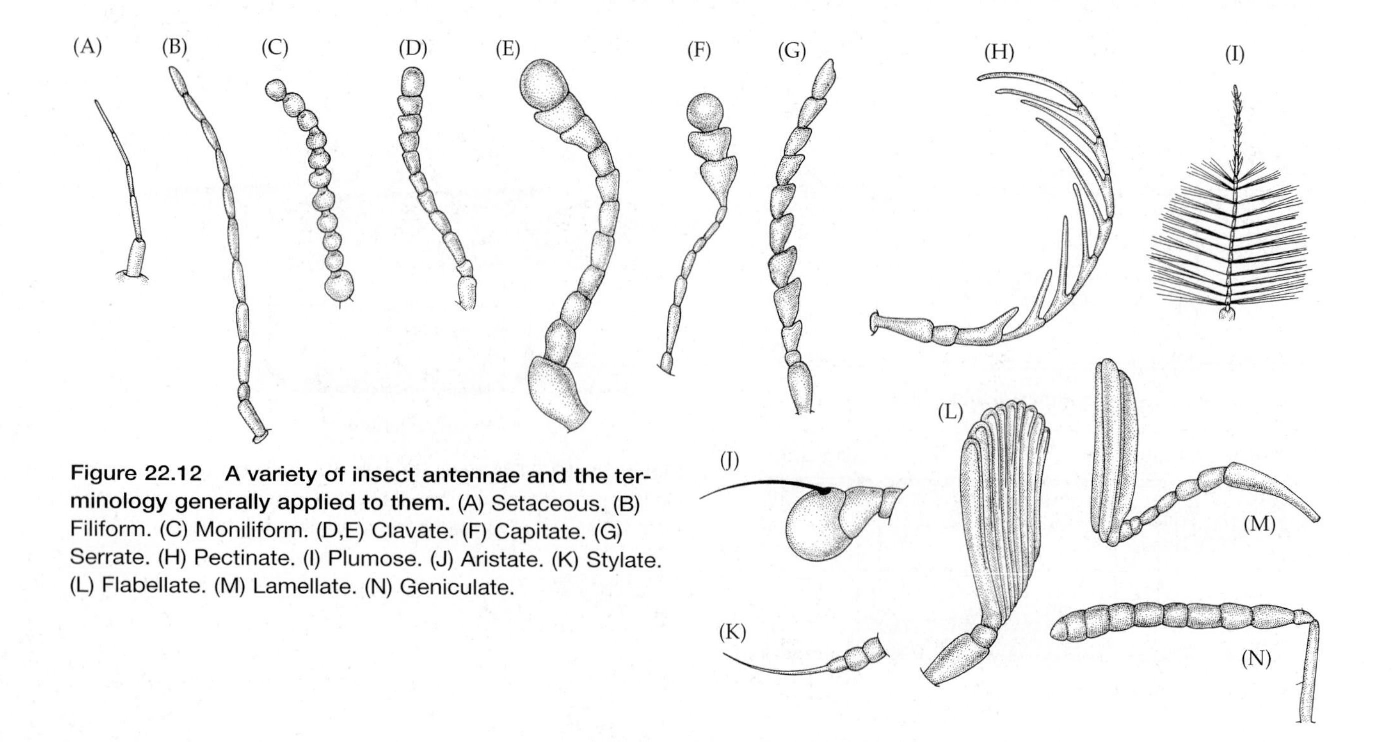

Figure 22.12 A variety of insect antennae and the terminology generally applied to them. (A) Setaceous. (B) Filiform. (C) Moniliform. (D,E) Clavate. (F) Capitate. (G) Serrate. (H) Pectinate. (I) Plumose. (J) Aristate. (K) Stylate. (L) Flabellate. (M) Lamellate. (N) Geniculate.

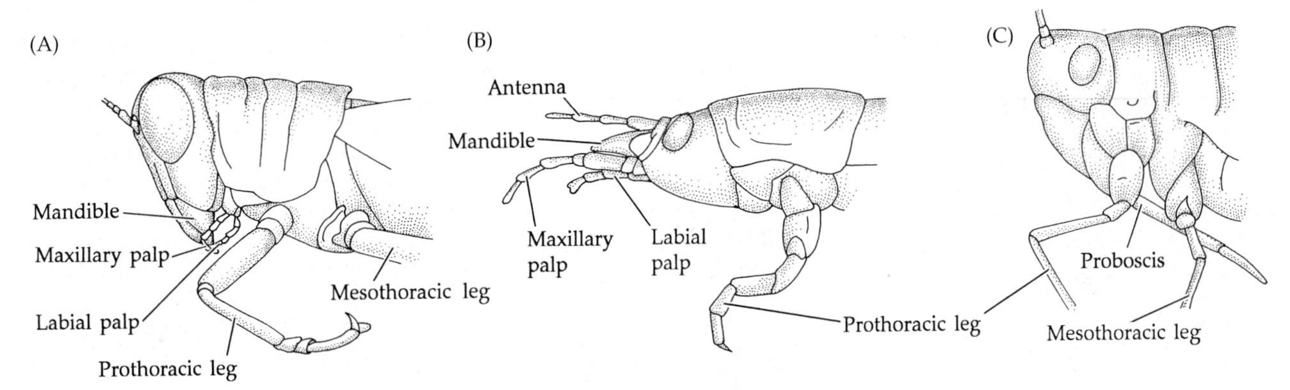

Figure 22.13 Different positions of head and mouthparts relative to the rest of the body. (A) Hypognathous condition (grasshopper). (B) Prognathous condition (beetle larva). (C) Opisthorhynchous condition (aphid).

derived structure of the exoskeleton; others believe it could be the fused appendages of the first or third head segment. The mandibles (Figure 22.14) are strongly sclerotized, usually toothed, and lack a palp. In most insects the mandible is of one article, but in some primitive groups (and fossil taxa) it is composed of several articles. However, the gene *Distal-less* (*Dll*) is apparently never expressed in the embryogeny of hexapodan mandibles, suggesting that they are fully gnathobasic (i.e., protopodal). The maxillae are generally multiarticulate and bear a palp of 1–7 articles. The labium comprises the fused second maxillae and typically bears two palps. In addition to these appendages, there is a median, unpaired, tonguelike organ called the hypopharynx that projects forward from the back of the preoral cavity. The salivary glands open through the hypopharynx.[3]

In the past variations in feeding appendages were often used to define the major insect clades, and more recent molecular phylogenetic work has largely corroborated those ideas although a number of cases of convergent evolution in mouthparts have also been revealed. In sucking insects both mandibles and maxillae may be transformed into spear-like structures (stylets), or the mandibles may be absent altogether (Figure 22.15 and 22.24). In most Lepidoptera, the maxillae form an elongate coiled sucking tube, the proboscis. In some insects (e.g., Hemiptera), the labium is drawn out into an elongate trough to hold the other mouthparts, and its palps may be absent; in others (e.g., Diptera), it is modified distally into a pair of fleshy porous lobes called labellae.

The hexapod thorax The three segments of the thorax are the **prothorax**, **mesothorax**, and **metathorax** (Figure 22.9). Their tergites carry the same prefixes: **pronotum**, **mesonotum**, **metanotum**. In the winged insects (Pterygota), the mesothorax and metathorax are enlarged and closely united to form a rigid **pterothorax**. Wings, when present, are borne on these two segments and articulate with processes on the tergite (notum) and pleura of the somites. The prothorax is sometimes greatly reduced, but in some insects it is greatly enlarged (e.g., beetles) or even expanded into a large shield (e.g., cockroaches). The lateral, pleural sclerites are complex and thought to be derived, at least in part, from subcoxal (protopodal) elements of the ancestral legs that became incorporated into the lateral body wall. The sternites may be simple, or may be divided into multiple sclerites on each segment.

Each of the three thoracic somites bears a pair of legs (Figure 22.9), composed of two parts, a proximal protopod and a distal telopod. The protopod (sometimes

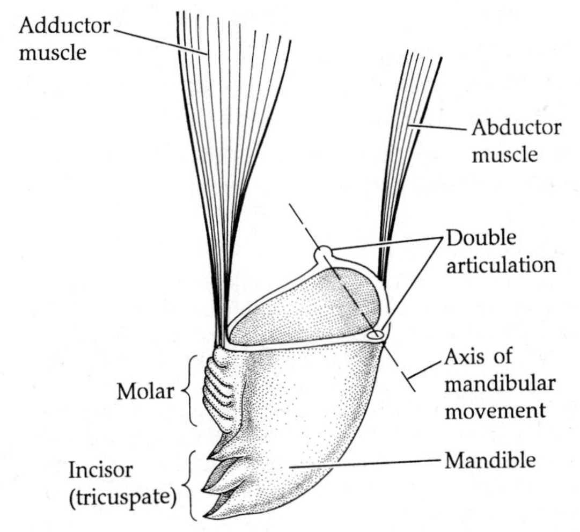

Figure 22.14 The musculature of an insect mandible.

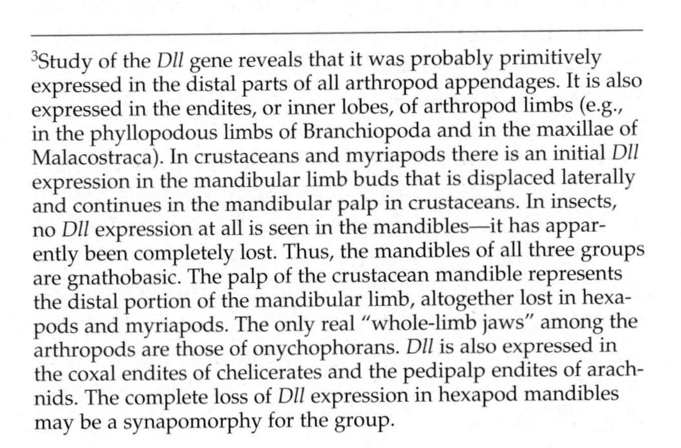
[3]Study of the *Dll* gene reveals that it was probably primitively expressed in the distal parts of all arthropod appendages. It is also expressed in the endites, or inner lobes, of arthropod limbs (e.g., in the phyllopodous limbs of Branchiopoda and in the maxillae of Malacostraca). In crustaceans and myriapods there is an initial *Dll* expression in the mandibular limb buds that is displaced laterally and continues in the mandibular palp in crustaceans. In insects, no *Dll* expression at all is seen in the mandibles—it has apparently been completely lost. Thus, the mandibles of all three groups are gnathobasic. The palp of the crustacean mandible represents the distal portion of the mandibular limb, altogether lost in hexapods and myriapods. The only real "whole-limb jaws" among the arthropods are those of onychophorans. *Dll* is also expressed in the coxal endites of chelicerates and the pedipalp endites of arachnids. The complete loss of *Dll* expression in hexapod mandibles may be a synapomorphy for the group.

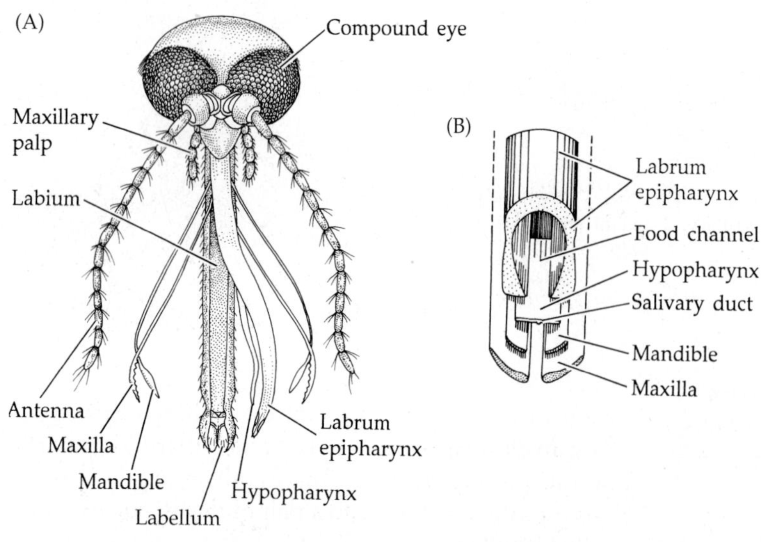

Figure 22.15 **A variety of insect mouth appendages, specialized for different types of feeding habits.** (A,B) Piercing–sucking mouthparts of a mosquito (Diptera). Note the complex stylet structure in (B). (C) Sucking mouthparts of a honeybee (Hymenoptera). (D) Sucking mouthparts of a butterfly (Lepidoptera). (E) Sponging mouthparts of a false blackfly (Diptera). (For an illustration of biting–chewing mouthparts, see Figure 22.11.)

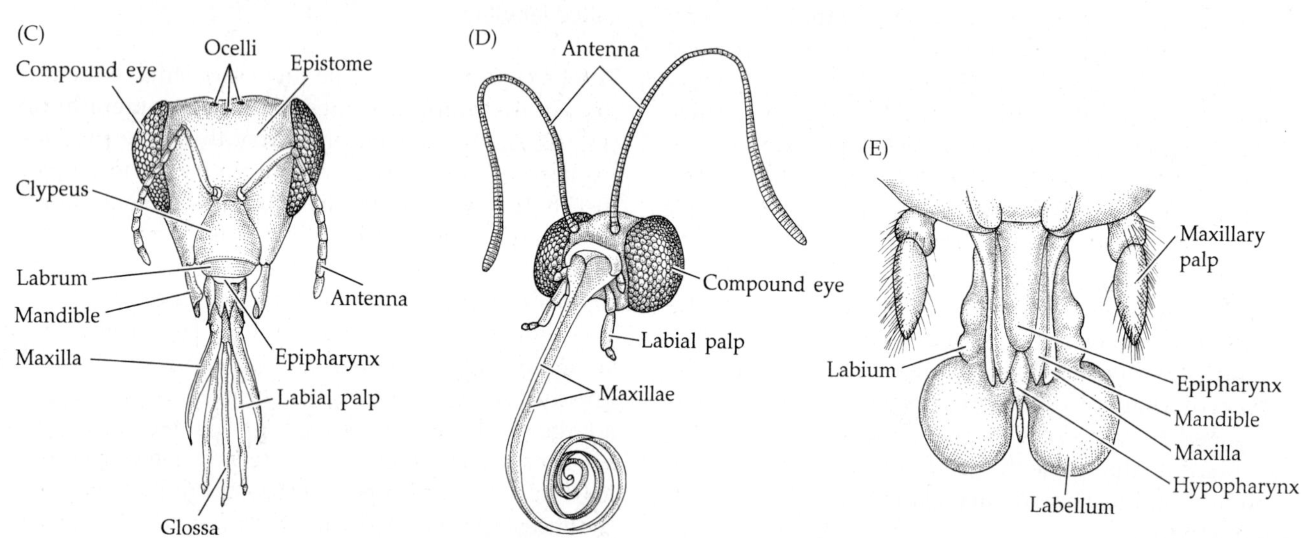

called a "coxopodite" by entomologists) is composed of two articles (coxa, trochanter), and the telopod is composed of four articles (**femur, tibia, tarsus, pretarsus**) (Figure 22.16). The tarsus is often subdivided into additional pseudoarticles called tarsomeres. The basal hexapods have a single tarsus (Protura and Diplura) or an indistinct tarsus (Collembola; probably fused with the tibia). In the Archaeognatha the tarsus is usually composed of three tarsomeres, and in the Pterygota it is composed of one, three, or five tarsomeres.[4] Whatever the number of tarsal articles, no intrinsic musculature occurs in them, and they are thus viewed as subdivisions of a single original article. The whole length of the tarsus is crossed by the tendon of the flexor muscle of the pretarsus, whose fibers usually arise on the tibia (Figure 22.17). The pretarsus is a minute article that usually bears a pair of lateral claws. The pretarsus in Collembola and Protura bears a single median claw. A single claw also occurs in many holometabolous larvae and some pterygote adults. But in most hexapods, the pretarsus bears a pair of lateral claws, and many also have a median **arolium** (which functions as an adhesive pad on smooth surfaces), **unguitractor plate**, or median claw.

In insects, many adult organs derive from clusters of early embryonic cells called **imaginal discs**, which arise from localized invaginations of the ectoderm in the early embryo. The embryonic thorax contains three pairs of leg discs, and as development proceeds, these discs develop a series of concentric rings, which are the presumptive leg articles. The center of the disc corresponds to the distalmost articles (tarsus and pretarsus) of the future leg, while the peripheral rings correspond

[4]It is unclear whether having one or several tarsomeres is the primitive condition for the Hexapoda. Homologization of the leg articles among the various arthropod groups is a popular and often rancorous pastime. The issue is further confused when the number of articles differs from the norm (e.g., some insects have two trochanters, the number of tarsi varies from one to five, etc.). The protopods of myriapods, chelicerates, and trilobites all seem to consist of a single basal article (usually called the coxa). In the latter case, Kukalová-Peck hypothesizes that three "missing" protopodal articles were lost by fusion among themselves and with the pleural region. See Chapter 20 for a discussion of ancestral arthropod legs.

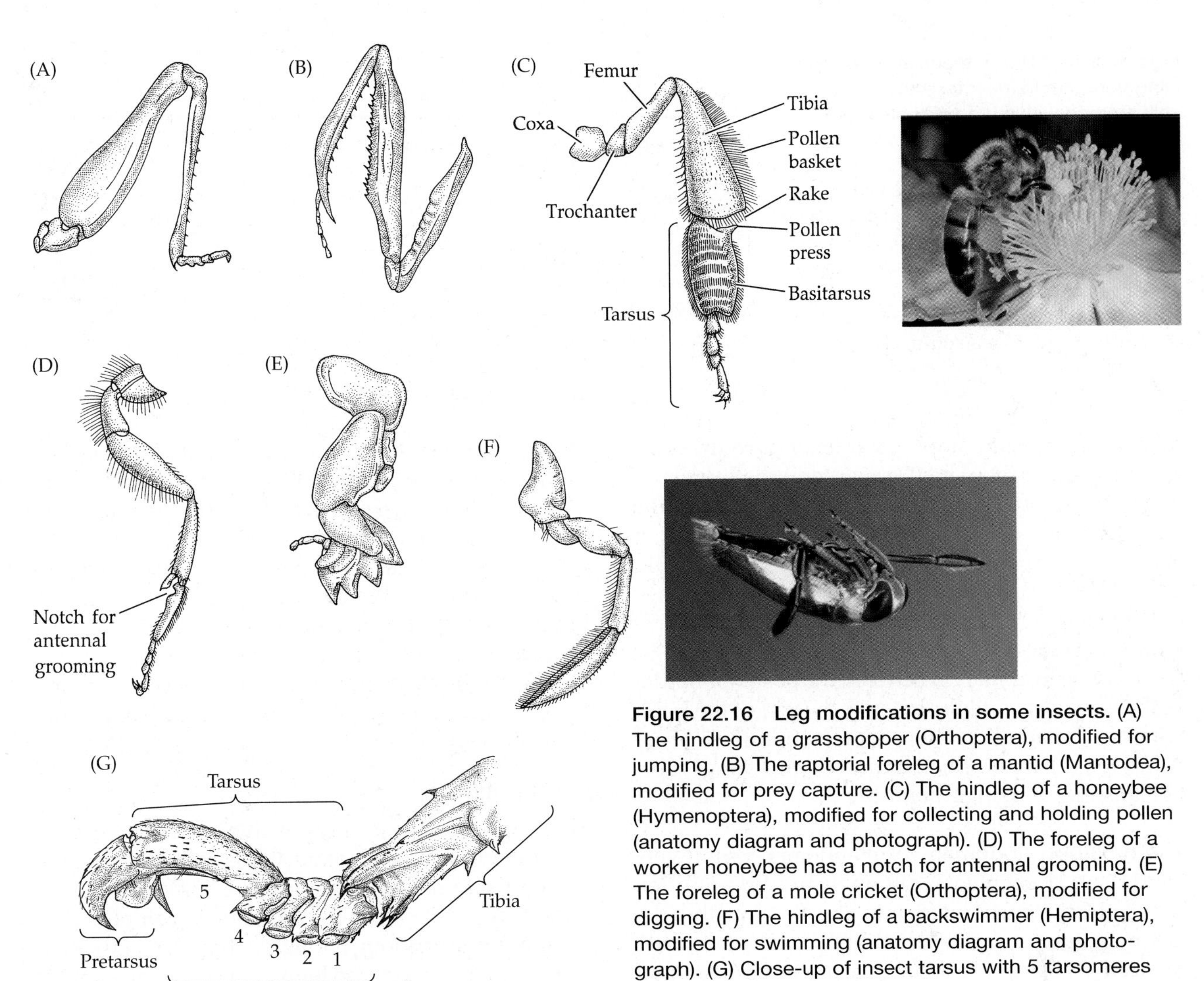

Figure 22.16 Leg modifications in some insects. (A) The hindleg of a grasshopper (Orthoptera), modified for jumping. (B) The raptorial foreleg of a mantid (Mantodea), modified for prey capture. (C) The hindleg of a honeybee (Hymenoptera), modified for collecting and holding pollen (anatomy diagram and photograph). (D) The foreleg of a worker honeybee has a notch for antennal grooming. (E) The foreleg of a mole cricket (Orthoptera), modified for digging. (F) The hindleg of a backswimmer (Hemiptera), modified for swimming (anatomy diagram and photograph). (G) Close-up of insect tarsus with 5 tarsomeres and a pretarsus with 2 lateral claws flanking an arolium.

to its proximal region (coxa, trochanter). During embryogenesis, the leg telescopes out as it subdivides into the component articles. The gene *Distal-less* (*Dll*) is expressed in the presumptive distal region of the limb, while the gene *Extradenticle* (*Exd*) is necessary for the development of the proximal portion of the limb. Thus the protopod and telopod of the legs are each under their own genetic control.

Within the Pterygota, most species also have a pair of wings on the second and third thoracic segments. Wing morphology has been more extensively used in insect classification than any other single structure. Wings are often the only remains of insects preserved in fossils. The wings of modern insects develop as evaginations of the integument, with thin cuticular membranes forming the upper and lower surfaces of each wing. Wing **veins**, which contain circulating hemolymph, anastomose and eventually open into the body. The arrangement of veins in insect wings provides important diagnostic characters at all taxonomic levels. The origin and homologization of wing venation has been heavily debated over the decades. Most workers use a consistent naming system that recognizes six major veins: costa (C), subcosta (SC), radius (R), media (M), cubitus (CU), and anal (A) (Figure 22.18). Areas in the wings that are enclosed by longitudinal and cross veins are called **cells**, and these too

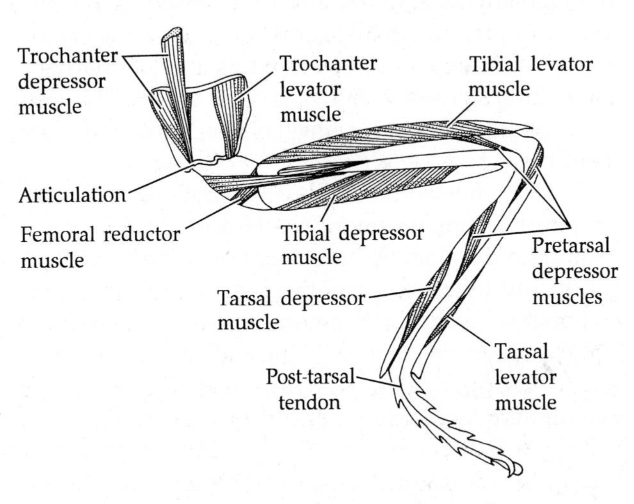

Figure 22.17 The musculature of an insect leg.

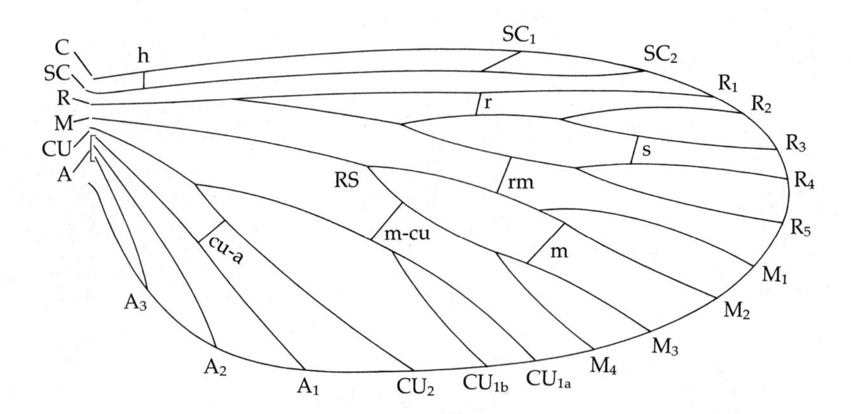

Figure 22.18 The nomenclature of basic wing venation in insects. Although the cells formed within the veins also have names, only the names of the veins are given here. Longitudinal veins are coded with capital letters, cross veins with lowercase letters. Longitudinal veins: costa (C); subcosta (SC); radius (R); radial sector (RS); media (M); cubitus (CU); anal (A). Cross veins: humeral (h); radial (r); sectorial (s); radiomedial (rm); medial (m); mediocubital (m-cu); cubitoanal (cu-a).

have a somewhat complex nomenclature. In some groups (e.g., Orthoptera, Dermaptera) the forewings develop heavily sclerotized regions called **tegmina** (sing. tegmin), used for protection, stridulation, or other purposes. In many sedentary, cryptic, parasitic, and insular lineages the wings have become shortened (brachypterous) or lost (apterous). Insects often couple their wings together for flight by means of hook-like devices along the margin between the posterior border of the forewings and the anterior margin of the hindwings (e.g., **hamuli** in Hymenoptera; **frenula** in many Lepidoptera). In these insects the coupled wings function together as a single unit.

The hexapod abdomen The abdomen primitively comprises 11 segments, although the first is often reduced or incorporated into the thorax, and the last may be vestigial. Abdominal pleura are greatly reduced or absent. The occurrence of true (though minute) abdominal leglets (sometimes called "prolegs" or "styli") on the pregenital segments is commonplace among the apterygotes and also occurs in the larvae of many pterygotes (e.g., the legs of caterpillars). In addition, transitory limb buds or rudiments appear fleetingly in the early embryos of some species, presumably harking back to the deep evolutionary past. Abdominal segments 8–9 (or 7–9) are typically modified as the **anogenital tagmata**, or **terminalia**, the exposed parts being the genitalia. The female median gonopore occurs behind sternum 7 in Ephemeroptera and Dermaptera, and behind sternum 8 or 9 in all other orders. The anus is always on segment 11 (which may be fused with segment 10).

There is enormous complexity in both clasping and intromittent organs among the Hexapoda, and a correspondingly sharp disagreement over the homologization and terminology of these structures. In general, females discriminate among males on the basis of sensory stimuli produced by the male genitalia; hence selection pressure has been a powerful force in the evolution of these structures (in both sexes). The most primitive male architecture can be seen in apterygotes and Ephemeroptera, in which the penes are paired and contain separate ejaculatory ducts. In most other insects, however, the intromittent organ develops late in embryogeny by fusion of the genital papillae to form a median, tubular, often eversible endophallus, with the joined ejaculatory ducts opening at a gonopore at its base. The external walls may be sclerotized or modified in a wide variety of ways, and the whole organ is known as the **aedeagus**. Some workers consider the aedeagus to be derived from segment 9; others regard it as belonging to segment 10. A pair of sensory cerci (sing. cercus) often project from the last abdominal segment.

Locomotion

Walking Hexapods rely on their sclerotized exoskeleton for support on land. Their limbs provide the physical support needed to lift the body clear of the ground during locomotion. In order to accomplish this, the limbs must be long enough to hold the body high off the ground, but not so high as to endanger stability. Most hexapods maintain stability by having the legs in positions that suspend the body in a sling-like fashion and keep the overall center of gravity low (Figure 20.19).

The basic structure of arthropod limbs was described in Chapter 20. In hexapods (and crustaceans) the anterior–posterior limb movements take place between the coxae and the body proper (in contrast to most arachnids, in which the coxae are immovably fixed to the body and limb movement occurs at more distal joints). Like the power controlled by the range of gears in an automobile, the power exerted by a limb is greatest at low speeds and least at higher speeds. At lower speeds the legs are in contact with the ground for longer periods of time, thus increasing the power, or force, that can be exerted during locomotion. In burrowing forms the legs are short, and the gait is slow and powerful as the animal forces its way through soil, rotting wood, or other material. Longer limbs reduce the force, but increase the speed of a running gait, as do limbs capable of swinging through a greater angle. Limbs long in length and stride are typical features of the fastest-running insects (e.g., tiger beetles, Carabidae).

One of the principal problems associated with increased limb length is that the field of movement of one

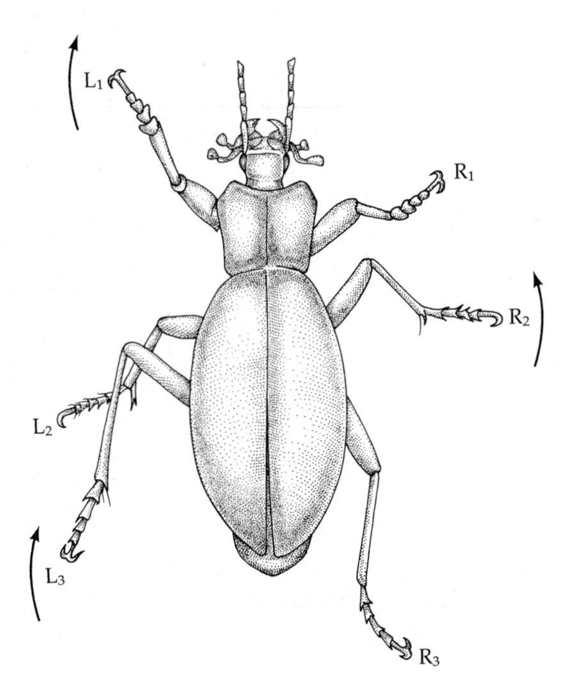

Figure 22.19 A beetle walking. The alternating tripod gait consists of alternate stepping with two sets of three legs; thus the body is always supported by a triad of legs. Here, three legs (L1, R2, and L3) are moving forward while the other three (R1, L2, and R3) are on the ground.

limb may overlap that of adjacent limbs. Interference is prevented by the placement of the tips of adjacent legs at different distances from the body (Figure 22.19). Thus fast-running insects usually have legs of slightly different lengths. Insects usually move their legs in an alternating tripod sequence. Balance is maintained by always having three legs in contact with the ground.

Like many spiders, some insects can walk on water, and they do so in much the same way—by balancing the pull of gravity on their featherweight bodies with the physical principles of buoyancy and surface tension. Insects (and spiders) that walk on water don't get wet because their exoskeletons are coated with waxes that repel water molecules. The water surface, held taut by surface tension, bends under each leg to create a depression, or dimple, that works to push the animal upward in support. Water walking occurs in many insect groups, notably the Hemiptera (e.g., water striders), Coleoptera (e.g., whirligig beetles), and Collembola (some springtails).[5]

Many insects are good jumpers (e.g., fleas, springtails, most orthopterans), but the click beetles (Elateridae) are probably the champions. It has been calculated that a typical click beetle (e.g., *Athous haemorrhoidalis*), when jackknifing into the air to escape a predator, generates 400 g of force, with a peak deceleration of 2,300 g.

Flight Among the many remarkable advances of insects, flight is perhaps the most impressive. Insects were the first flying animals, and throughout the history of life on Earth no other invertebrates have learned the art of true flight. The wingless insects belong either to groups that have secondarily lost the wings (e.g., fleas, lice, certain scale insects) or to primitive taxa (the apterygotes) that arose prior to the evolution of wings. In three orders, the wings are effectively reduced to a single pair. In beetles, the forewings are modified as a protective dorsal shield (**elytra**). In dipterans, the hindwings are modified as organs of balance (**halteres**). The halteres beat with the same frequency as the forewings, functioning as gyroscopes to assist in flight performance and stability—flies fly very well.

Compared with an insect, an airplane is a simple study in aerodynamics. Planes fly by moving air over a fixed wing surface, the leading edge of which is tilted upward, forcing the air to travel farther (thus faster) over the top of the wing than the bottom, resulting in a vortex that creates a lift. But conventional fixed-wing aerodynamic theory is insufficient to understand insect flight. Insect wings are anything but fixed. Insects, of course, fly by flapping their wings to create vortices, from which they gain lift, but these vortices slip off the wings with each beat, and new vortices are formed with each alternate stroke. Beating insect wings trace a figure eight pattern, and they also rotate at certain crucial moments. Thus, each cycle of flapping creates dynamic forces that fluctuate drastically. By complex actions of wing orientation, insects can hover, fly forward, backward, and sideways, negotiate highly sophisticated aerial maneuvers, and land in any position. To complicate matters even more, in the case of small insects (and most are small, the average size of all insects being just 3–4 mm), the complex mechanics of flight take place at very low Reynolds numbers (see Chapter 4), such that the insect is essentially "flying through molasses." As a result of these complex mechanics, insect flight is energetically costly, requiring metabolic rates as high as 100 times the resting rate.

Each wing articulates with the edge of the notum (thoracic tergite), but its proximal end rests on a dorsolateral pleural process that acts as a fulcrum (Figure 22.20). The wing hinge itself is composed in large part of resilin, a highly elastic protein that allows for rapid, sustained movement. The complex wing movements are made possible by the flexibility of the wing itself and by the action of a number of different muscle sets that run from the base of the wing to the inside walls of the thoracic segment on which it is borne. These direct flight muscles serve to raise and lower the wings and to tilt their plane at different angles (somewhat like altering the blade angles on a helicopter) (Figure

[5]On an undisturbed surface, water molecules are attracted to their neighbors beside and below, resulting in a flat skin of molecules that exerts only horizontal tensile forces; it is this "elasticity" of the surface that we call surface tension.

Figure 22.20 **A typical insect wing hinge arrangement.** This transverse section through the thoracic wall of a grasshopper shows the base of the wing and the wing hinge. (A) Entire hinge area. (B) Enlargement of hinge section.

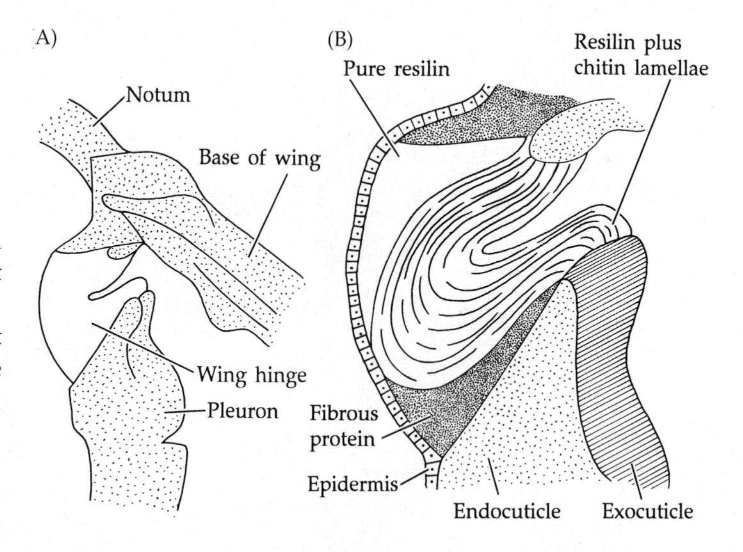

22.21). However, except in palaeopterans (Odonata and Ephemeroptera), the direct flight muscles are not the main source of power for insect wing movements. Most of the force comes from two sets of indirect flight muscles, which neither originate nor insert on the wings themselves (Figures 22.21 and 22.22).

Dorsal longitudinal muscles run between apodemes at the anterior and posterior ends of each winged segment. When these muscles contract, the segment is shortened, which results in a dorsal arching of the segment roof and a downstroke of the wings. Dorsoventral muscles, which extend from the notum to the sternum (or to basal leg joints) in each wing-bearing segment, are antagonistic to the longitudinal muscles. Contraction of the dorsoventral muscles lowers the roof of the segment. In doing so, it raises the wings. Thus, wing flapping in most insects is primarily generated by rapid changes in the walls and overall shape of the mesothorax and metathorax. Other, smaller thoracic muscle sets serve to make minor adjustments to this basic operation.

Insects with low wing-flapping rates (e.g., dragonflies, orthopterans, mayflies, and lepidopterans) are limited by the rate at which neurons can repeatedly fire and muscles can execute contractions. However, in insects with high wing-flapping rates (e.g., dipterans, hymenopterans, and some coleopterans), an entirely different regulatory mechanism has evolved. Once

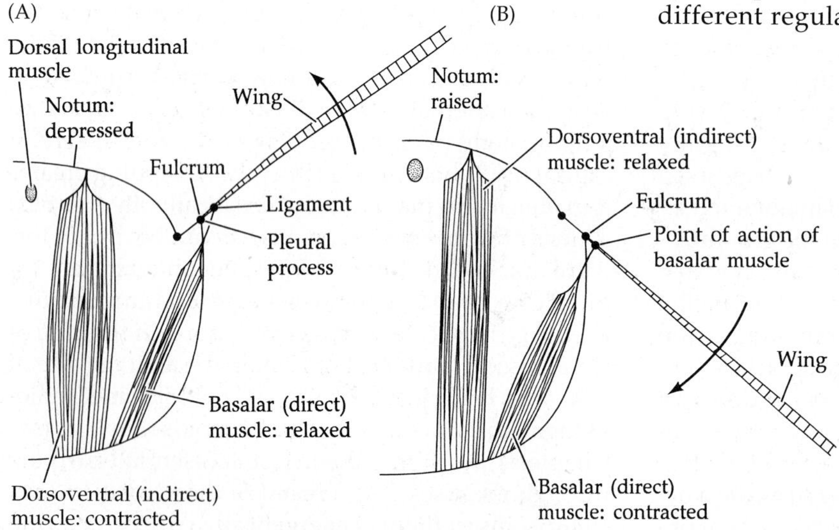

(C)

Figure 22.21 **Wing movements of a primitive insect such as a dragonfly, in which direct wing muscles cause depression of the wings.** Dots represent pivot points, and arrows indicate the direction of wing movement. (A) The dorsoventral muscles contract to depress the notum as the basalar muscles relax, a combination forcing the wings into an upstroke. (B) The dorsoventral muscles relax as the basalar muscles contract, a combination pulling the wings into a downstroke and relaxing (and raising) the notum. (C) Thorax of a dragonfly showing wing attachment to the notum of thoracic segments 2 and 3.

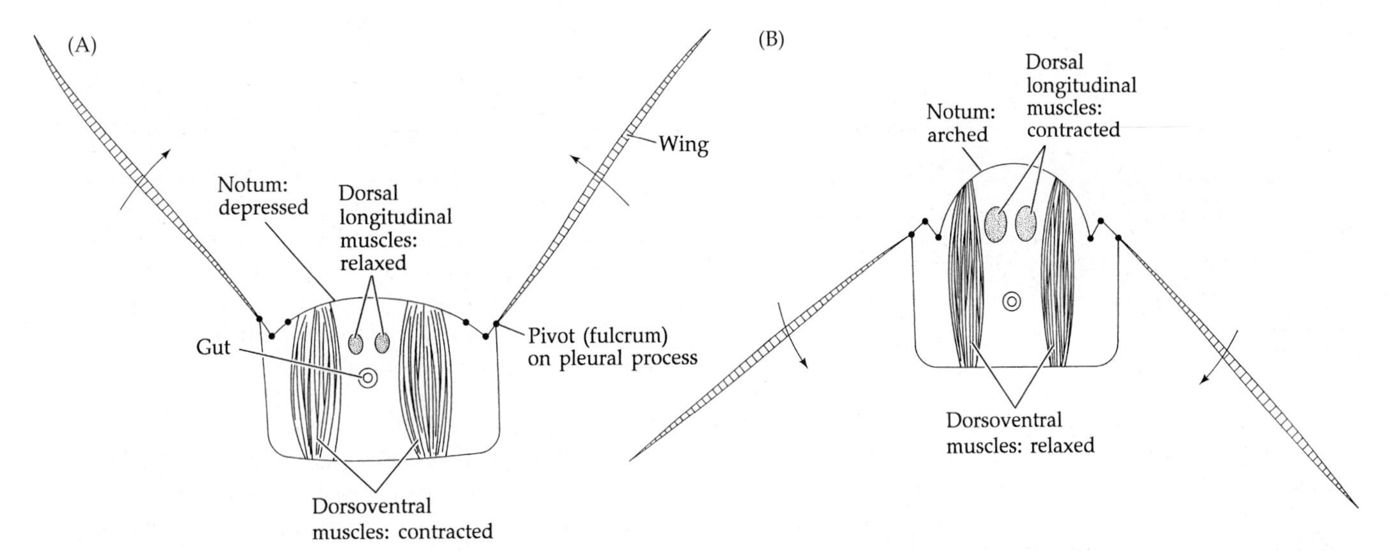

Figure 22.22 Wing movements of an insect such as a fly or hemipteran, in which both upward and downward movements of the wings are produced by indirect flight muscles. In these transverse sections of a thoracic segment, dots represent pivot points, and arrows indicate the direction of wing movement. Only two sets of muscles are shown. (A) The dorsoventral muscles contract, depressing the thoracic notum and forcing the wings into upstroke. (B) The dorsoventral muscles relax as the dorsal longitudinal muscles contract to "pop up" the notum, elevating it and forcing the wings into a downstroke.

flight has been initiated and a high wing-flapping rate attained (up to 100 beats/second), myogenic control takes over. This mechanism exploits the elastic–mechanical properties of the exoskeleton. When one set of indirect muscles contracts, the thorax is deformed. Upon relaxation of the muscles, there is an elastic rebound of the thoracic exoskeleton, which stretches the second set of indirect muscles and thus directly stimulates their contraction. This contraction establishes a second deformation, which in turn stretches and stimulates the first muscle group. Once initiated, this mechanism is nearly self-perpetuating, and the nonsynchronous firing of neurons serves only to keep it in action.

Not all insects utilize wings to travel through the air. Many small and immature insects are effectively dispersed by wind power alone. Some first-instar lepidopterans use silk threads for dispersal (as do spiders and mites). Tiny scale insects are commonly collected in aerial nets. In fact, studies have revealed the existence of "aerial plankton" consisting of insects and other minute arthropods, extending to altitudes as high as 14,000 feet. Most are minute winged forms, but wingless species are also common.

The Origin of Insect Flight

For many decades, two competing views of insect wing origin have dominated. In general, these views can be termed the paranotal lobe hypothesis and the appendage hypothesis. The former holds that wings evolved by way of a gradual expansion of lateral folds of the thoracic tergites (paranotal lobes), which eventually became articulated and muscled to form wings. The latter hypothesizes that wings evolved from pre-existing articulated structures on the thoracic appendages, such as gills or protopodal exites on the legs. There is also tantalizing evidence from the fossil record suggesting that the first pterygote insects possessed appendages on the prothorax, called "winglets," that may have been serially homologous to modern wings, implying that the loss of prothoracic proto-wings might have taken place in the early evolutionary history of the Hexapoda.

The paranotal lobe hypothesis was first proposed by Müller in 1873, saw a resurgence of popularity in the middle of the twentieth century, and has lost favor in recent years. It suggests that wings originated as lateral aerodynamic flaps of the thoracic nota that enabled insects to alight right side up when jumping or when blown about by the wind. These stabilizing paranotal lobes later evolved hinged structures and muscles at their bases. The occurrence of fixed paranotal lobes in certain ancient fossil insects has been cited in support of the paranotal lobe hypothesis (Figure 22.23). However, recent studies suggest that these primitive paranotal lobes might have been used for other purposes, such as covering the spiracular openings or gills in amphibious insects, protecting or concealing the insects from predators, courtship displays, or thermoregulation by absorption of solar radiation.

The appendage hypothesis (also known as the "gill or branchial theory," "exite theory," or "leg theory") also dates back to the nineteenth century, but was resurrected by the great entomologist V. B. Wigglesworth in the 1970s, and was championed by J. Kukalová-Peck since the 1980s. It is the more favored hypothesis of wing origin today, based on recent paleontological work, microscopic anatomy, and molecular developmental biology. It suggests that insect wings are derived from thoracic appendages—from protopodal exites, in Wigglesworth and Kukalová-Peck's view. These proto-wing appendages might have first

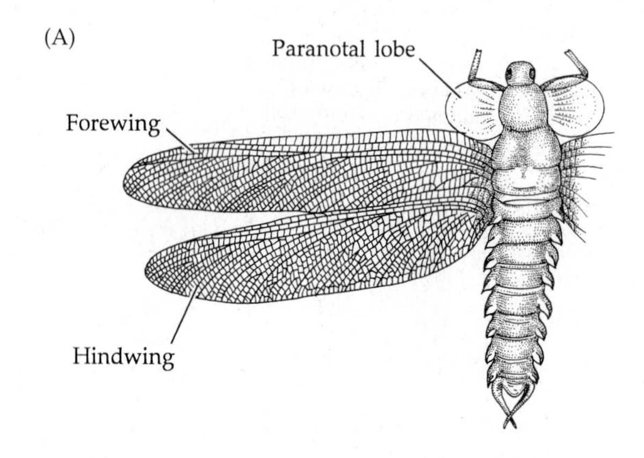

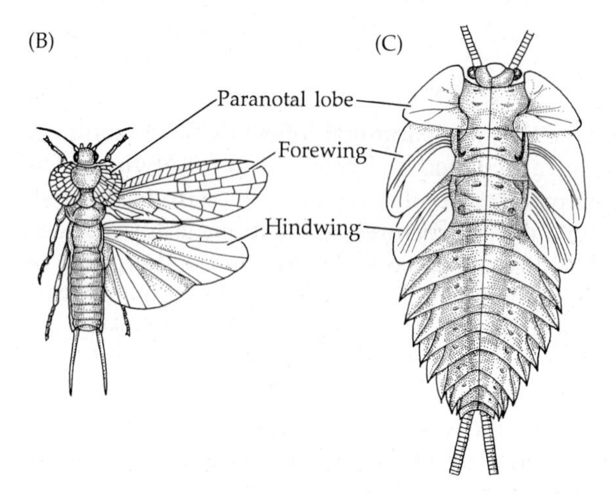

Figure 22.23 Fossil insects with paranotal lobes on the prothorax. (A) *Stenodictya lobata*. (B) *Lemmatophora typa*. (C) Nymphal stage of *Rochdalia parkeri*, a Paleozoic terrestrial palaeodictyopteran. In this species, all three thoracic segments appear to have had "articulated" thoracic lobes.

functioned as aquatic gills or paddles, or as terrestrial gliding structures. The paired abdominal gills of mayflies have been suggested as serial homologues of such "proto-wings." In Kukalová-Peck's version of this hypothesis, the first protopodal leg article (the epicoxa) fused with the thoracic pleural membrane early in the evolution of the Arthropoda, as did the second article (the precoxa) in the ancient hexapods, with both migrating dorsally off the leg and onto the body proper. In insects, the epicoxa eventually fused with the tergite, its exite enlarging to form the proto-wing, and eventually the true wing. The precoxa formed the pleural sclerite providing the ventral articulation of the wing. Wing veins might have evolved from cuticular ridges that served to strengthen these structures, and eventually to circulate blood through them.[6]

Kukalová-Peck's theory of wing evolution finds support in molecular developmental studies, which have shown that the cells that give rise to the wing primordium derive from the same cluster of cells that form the leg primordium, from which they segregate, migrating dorsally to a position below the tergum. Recent studies on gene expression also support the origin of wings from legs. The genes *pdm* and *apterous* are expressed in the wing (and leg) primordia of all insects. Expression of both genes appears to be necessary for normal wing formation. In malacostracan crustaceans (but not in branchiopod crustaceans) these same genes are expressed, in a similar manner, in the formation of the leg rami (the exopod and endopod).

[6]Unspecialized coxal exites can be seen on the legs of some living archaeognathans (bristletails) and in numerous extinct hexapods.

Feeding and Digestion

Feeding Every conceivable kind of diet is exploited by species within the Hexapoda, whose feeding strategies include herbivory, carnivory, and scavenging, as well as a magnificent array of commensalism and parasitism. This "nutritional radiation" has played a key role in the phenomenal evolution among the Insecta. A comprehensive survey of insect feeding biology alone could easily fill a book this size. Setting aside symbiotic relationships for a moment, in the most general sense insects can be classified as (1) biters–chewers, (2) suckers, or (3) spongers (Figure 22.24).

Biters–chewers, such as the grasshoppers, have the least modified mouthparts, so we describe them first. The maxillae and labium of these insects have well developed leglike palps (Figure 22.24A) that help them hold food in place, while powerful mandibles cut off and chew bite-sized pieces. The mandibles lack palps (in all insects) and typically bear small, sharp teeth that work in opposition as the appendages slide against each other in the transverse (side to side) fashion characteristic of most arthropod jaws. Biting–chewing insects may be carnivores, herbivores, or scavengers. In many plant eaters, the labrum bears a notch or cleft in which a leaf edge may be lodged while being eaten. Some of the best examples of this feeding strategy are seen among the Orthoptera (locusts, grasshoppers, crickets), and most people have witnessed the efficient fashion in which these insects consume their garden plants! Equally impressive are the famous leafcutter ants of the Neotropics, which can denude an entire tree in a few days. Leafcutter ants have a notable feeding adaptation: when cutting leaf fragments, they produce high-frequency vibrations with an abdominal stridulatory organ. This stridulation is synchronized with movements of the mandible, generating complex vibrations. The high vibrational acceleration of the mandible appears to stiffen the material being cut, just as soft material is stiffened with a vibratome for sectioning in a laboratory. Leafcutters don't eat the leaves they cut; instead, they carry them into an underground nest, where they use them to grow a fungus on which they feed. Several other insect groups have evolved associations with fungi, and in almost every case these relationships are obligate and mutualistic—neither partner can live without the other.

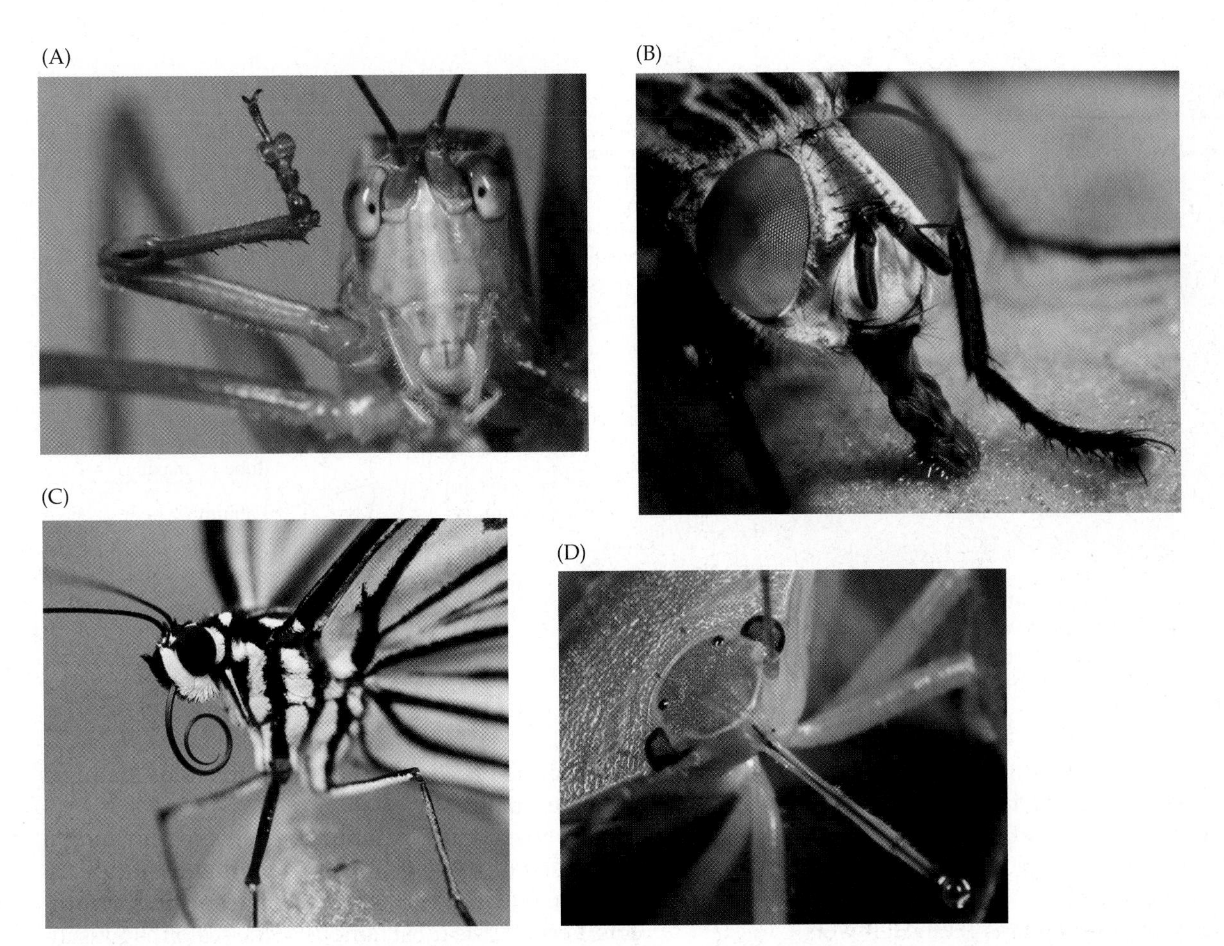

Figure 22.24 Mouthparts specialized for different modes of feeding. (A) Biting-chewing mouthparts of a grasshopper. Note the well developed leglike palps. (B) Sponging mouthparts of a fly. (C) Sucking mouthparts of a butterfly. (D) Piercing mouthparts of a true bug.

In sucking insects the mouthparts are markedly modified for the consumption of liquid foods, generally plant saps or nectars or animal blood or cell fluids (Figure 22.24D). Sucking mouthparts and liquid diets have clearly evolved many times in different insect lines—further testimony to the frequency of evolutionary convergence in arthropods and the developmental adaptability of their appendages. In some sucking insects, such as mosquitoes, feeding is initiated by piercing the victim's epidermal tissue; this mode of feeding is referred to as piercing–sucking. Other insects, such as butterflies and moths that feed on flower nectar, do not pierce anything and are merely suckers.

In all sucking insects the mouth itself is very small and hidden. The mouthparts, instead of being adapted for handling and chewing solid pieces of food, are elongated into a needle-like beak adapted for a liquid diet. Different combinations of mouth appendages constitute the beak in different taxa. True bugs (Hemiptera), which are piercer–suckers, have a beak composed of five elements: an outer troughlike element (the labium) and, lying in the trough, four very sharp stylets (the two mandibles and two maxillae). The stylets are often barbed to tear the prey's tissues and enlarge the wound. The labrum is in the form of a small flap covering the base of the grooved labium. When piercer–suckers feed, the labium remains stationary, and the stylets do the work of puncturing the plant (or animal) and drawing out the liquid meal.

Different variations of piercing–sucking mouthparts are found in other insect taxa. In mosquitoes, midges, and certain biting flies (e.g., horseflies) there are six long, slender stylets, which include the labrum–epipharynx and the hypopharynx as well as the mandibles and first maxillae (Figure 22.15A). Other biting flies, such as the stable fly, have mosquito-like mouthparts but lack mandibles and maxillae altogether. Fleas (Siphonaptera) have three stylets: the labrum–epipharynx and the two mandibles. Thrips have unusual mouthparts: the right mandible is greatly reduced, making the head somewhat asymmetrical, and the left mandible, first maxillae, and hypopharynx make up the stylets.

Lepidopterans are nonpiercing sucking insects in which the paired first maxillae are enormously elongated, coiled, and fused to form a tube through which flower nectar is sucked (Figure 22.24C); the mandibles are vestigial or absent (Figure 22.15D). The mouthparts of bees are similar: the first maxillae and labium are modified together to form a nectar-sucking tube, but the mandibles are retained and used for

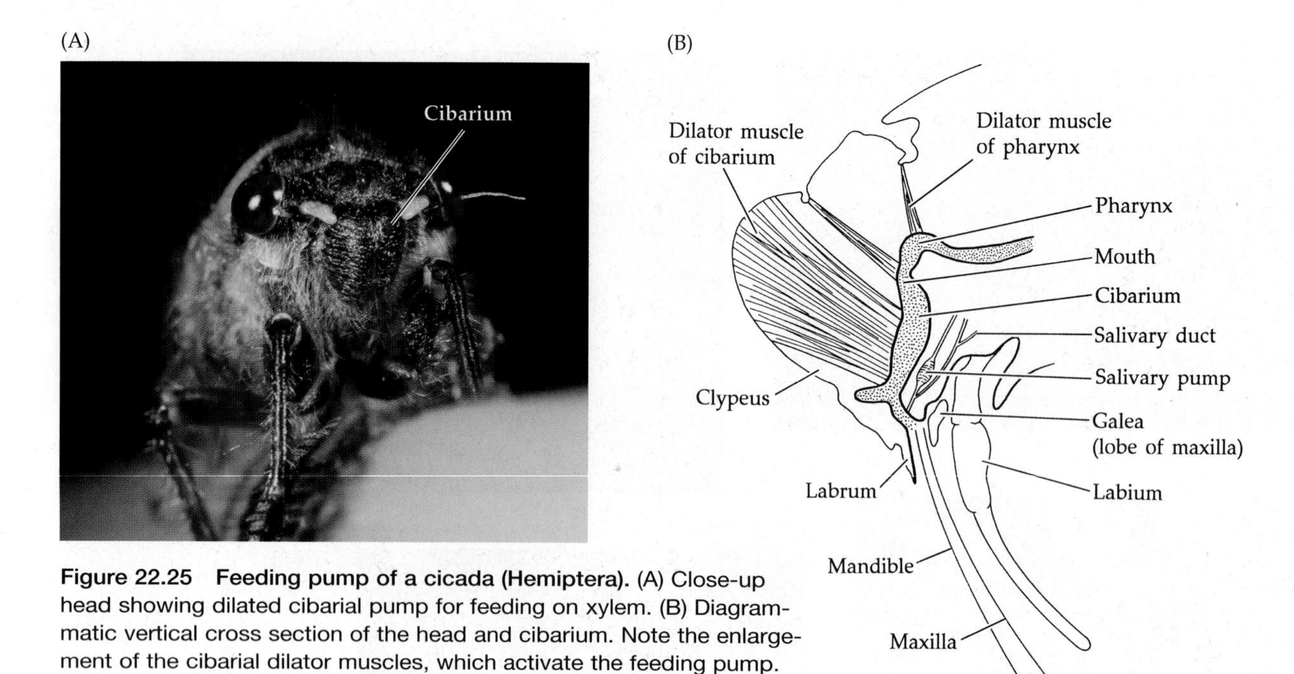

Figure 22.25 Feeding pump of a cicada (Hemiptera). (A) Close-up head showing dilated cibarial pump for feeding on xylem. (B) Diagrammatic vertical cross section of the head and cibarium. Note the enlargement of the cibarial dilator muscles, which activate the feeding pump.

wax manipulation during hive construction (Figure 22.15C). The collected nectar is stored in a special "sac" in the foregut and carried back to the hive where it is converted into honey, which is stored as a food reserve. Bees in an average hive consume about 500 pounds of honey per year—we humans get the leftovers.

Associated with sucking mouthparts are various mechanisms for drawing liquid food into the mouth. Most piercer–suckers rely largely on capillary action, but others have developed feeding "pumps." Often the pump is developed through elongation of the preoral cavity, or **cibarium**, which by extension of the cuticle around the mouth becomes a semi-closed chamber connecting with the alimentary canal (Figure 22.25). In these cases, cibarial muscles from the clypeus are enlarged to make a powerful pump. In lepidopterans, dipterans, and hymenopterans the cibarial pump is combined with a pharyngeal pump, which operates by means of muscles arising on the front of the head. Specialized salivary glands are also often associated with sucking mouthparts. In some hemipterans a salivary pump forces saliva through the feeding tube and into the prey, softening tissues and predigesting the liquid food. In mosquitoes, saliva carries blood thinners and anticoagulants (and often parasites such as *Plasmodium*, which causes malaria).

In spongers, such as most flies (order Diptera), the labium is typically expanded distally into a **labellum** (Figures 22.15E and 25.24B). Fluid nutrients are transported by capillary action along minute surface channels from the labellae to the mouth. In many spongers, such as houseflies, saliva is exuded onto the food to partly liquefy it. In strict spongers, the mandibles are absent. In biting spongers, such as horseflies, the mandibles serve to slice open a wound in the flesh, thus exposing the blood and cellular fluids to be sponged up by the labellae.

Many insects are scatophagous, feeding on animal feces. Most of these groups have biting mouthparts, but some (such as certain flies) have sucking mouthparts. Perhaps the most famous of the scatophagous insects are the dung beetles, or tumblebugs (certain beetles in the families Scarabaeidae and Histeridae). These remarkable insects harvest animal dung by biting or slicing off pieces with specialized head or leg structures and working them into a ball. They may roll the dung ball a considerable distance, and eventually bury it in the soil, whereupon females deposit eggs within it. Larvae are thus assured of a ready food supply. Dung balls may even be maneuvered by a pair of dung beetles pushing and pulling in a cooperative effort.

There are many symbiotic insects, and two orders are composed entirely of wingless parasites, most of which spend their entire lives on their host: some Psocodea (lice) and Siphonaptera (fleas). Bird lice are common, and lice are also found on dogs, cats, horses, cattle, and other mammals. Biting lice have broad heads and biting mouthparts used to chew epithelial cells and other structures on the host's skin. Sucking lice have narrow heads and piercing–sucking mouthparts, which they use to suck blood and tissue fluids from their host, always a mammal. Unlike most arthropod parasites, lice (of both types) spend their entire lives on the bodies of their hosts, and transmission to new hosts is by direct contact. For this reason most lice show a high degree of host specificity. Eggs, or nits, are attached by the female to the feathers or hair of the host, where they develop without a marked metamorphosis. Many lice, particularly those whose diet is chiefly keratin, possess symbiotic intracellular bacteria that appear to aid in the digestion of their food. These bacteria are passed to the offspring by way of the insects' eggs. Similar bacteria occur in ticks, mites, bedbugs, and some blood-sucking dipterans.

None of the biting lice are known to infest people or to transmit human disease microorganisms, although one species acts as an intermediate host for certain dog tapeworms. The sucking lice, on the other hand, include two genera that commonly infest humans (*Pediculus* and *Phthirus*). The latter genus includes the notorious *P. pubis*, the human pubic "crab" louse (which often occurs on other parts of the body as well). A number of sucking lice are vectors for human disease organisms. The most common reaction to infestation with lice—a condition known as **pediculosis**, or being lousy—is simple irritation and itching caused by the anticoagulant injected by the parasite during feeding. Chronic infestation with lice among certain footloose travelers is manifested by leathery, darkened skin—a condition known as **vagabond's disease**.

Fleas (order Siphonaptera) are perhaps the best known of all insect parasites. Nearly 1,500 species from birds and mammals have been described. Unlike lice, fleas are holometabolous, passing through egg, larval, pupal, and adult stages. Some species of fleas live their entire lives on their host, although eggs are generally deposited in the host's environment and larvae feed on local organic debris. Larvae of domestic fleas, including the rare human flea (*Pulex irritans*), feed on virtually any organic crumbs they find in the household furniture or carpet. Upon metamorphosis to the adult stage, fleas may undergo a quiescent period until an appropriate host appears. A number of serious disease organisms are transmitted by fleas, and at least 8 of the 60 or so species of fleas associated with household rodents are capable of acting as vectors for bubonic plague bacteria.

Other insect orders contain primarily free-living insects, but include various families of parasitic or micropredatory forms, or groups in which the larval stage is parasitic but the adults are free-living. Most of these "parasites" do not live continuously on a host and have feeding behaviors that fall into a gray zone between true obligate parasitism and predation. Such insects are sometimes classed as intermittent parasites, or micropredators. Bedbugs (Hemiptera, Cimicidae), for example, are minute flattened insects that feed on birds and mammals. However, most live in the nest or sleeping area of their host, emerging only periodically to feed. The common human bedbugs (*Cimex lectularius* and *C. hemipterus*) hide in bedding, in cracks, in thatched roofs, or under rugs by day and feed on their host's blood at night. They are piercer–suckers, much like the sucking lice. Bedbugs are not known to transmit any human diseases, although when present in large numbers they can be troublesome (in South America, as many as 8,500 bugs have been found in a single adobe house). Mosquitoes (family Culicidae), on the other hand, are vectors for a large number of disease-causing microorganisms, including *Plasmodium* (responsible for malaria; Figure 3.16), yellow fever, viral encephalitis, dengue, and lymphatic filariasis (with its gross symptom, elephantiasis, resulting from blockage of lymph ducts). Kissing bugs (Hemiptera, Reduviidae, Triatominae) also have a casual host relationship. They live in all kinds of environments, but often inhabit the burrows or nests of mammals, especially rodents and armadillos, as well as birds and lizards. They feed on the blood of these and other vertebrates, including dogs, cats, and people. Their host specificity is low. Several species are vectors of mammalian trypanosomiasis (*Trypanosoma cruzi*, the causative agent of Chagas' disease). The tendency of some species to bite on the face (where the skin is thin) has resulted in the common name.

In the dipteran family Calliphoridae, larvae are saprophagous, coprophagous, wound feeding, or parasitic. The parasitic species include earthworms, locust egg cases, termite colonies, and nestling birds among their hosts, and several parasitize humans and domestic stock (e.g., *Gochliomyia americana*, the tropical American screwworm).

Many insect parasites of plants cause an abnormal growth of plant tissues, called a gall. Some fungi and nematodes also produce plant galls, but most are caused by mites and insects (especially hymenopterans and dipterans). Parasitic adults may bore into the host plant or, more commonly, deposit eggs in plant tissues, where they undergo larval development. The presence of the insect or its larva stimulates the plant tissues to grow rapidly, forming a gall. The adaptive significance (for insects) of galls remains unclear, but one popular theory is that their production interferes with the production of defensive chemicals by the plant, thus rendering gall tissues more palatable. A somewhat similar strategy is used by leaf miners, specialized larvae from several orders (e.g., Coleoptera, Diptera, Hymenoptera) that live entirely within the tissues of leaves, burrowing through and consuming the most digestible tissues.

An interesting predatory strategy is that of New Zealand glowworms (*Arachnocampa luminosa*), which live in caves and in bushes along riverbeds. These larvae of small flies produce a bright bioluminescence in the distal ends of the Malpighian tubules, which lights up the posterior end of the body. (The light peaks at 485 nm wavelength.) Each larva constructs a horizontal web from which up to 30 vertical "fishing lines" descend, each with a regularly spaced series of sticky droplets. Small invertebrates (e.g., flies, spiders, small beetles, hymenopterans) attracted to the light are caught by the fishing lines, hauled up, and eaten. Harvestmen (Opiliones), the main predators of glowworms, use the light to locate their prey!

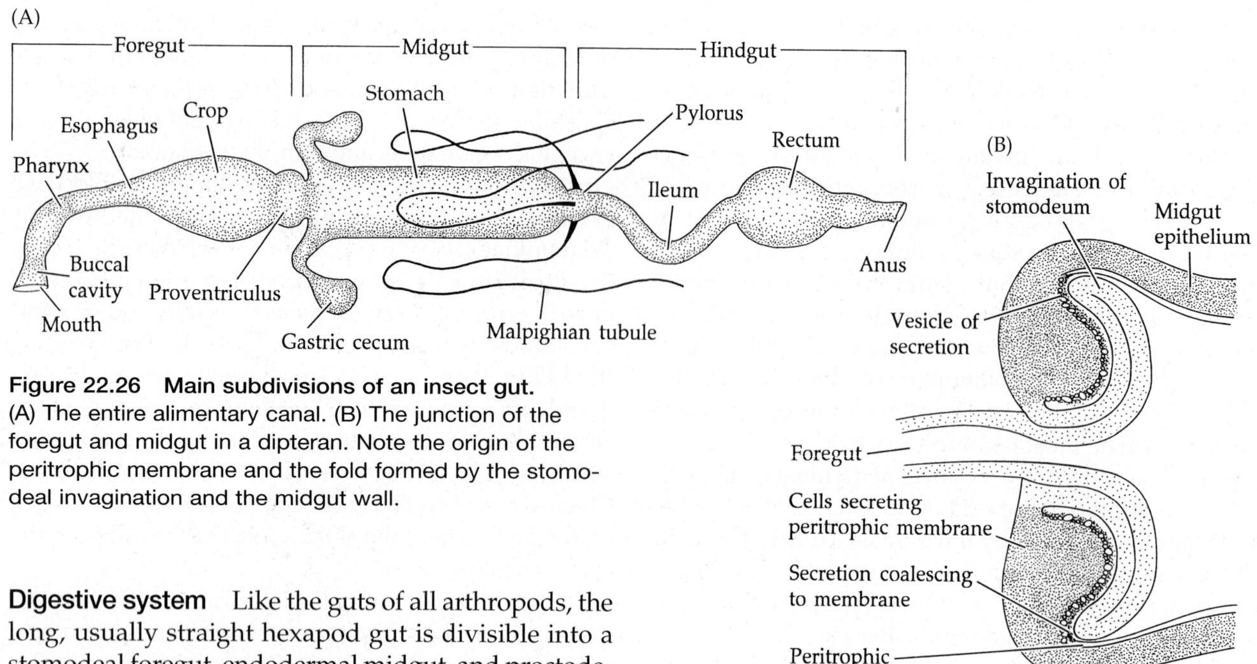

Figure 22.26 Main subdivisions of an insect gut. (A) The entire alimentary canal. (B) The junction of the foregut and midgut in a dipteran. Note the origin of the peritrophic membrane and the fold formed by the stomodeal invagination and the midgut wall.

Digestive system Like the guts of all arthropods, the long, usually straight hexapod gut is divisible into a stomodeal foregut, endodermal midgut, and proctodeal hindgut (Figure 22.26). Salivary glands are associated with one or several of the mouth appendages (Figure 22.27). The salivary secretions soften and lubricate solid food, and in some species contain enzymes that initiate chemical digestion. In larval moths (caterpillars), and in larval bees and wasps, the salivary glands secrete silk used to make pupal cells.

All hexapods, as well as most other arthropods that consume solid foods, produce a **peritrophic membrane** in the midgut (Figure 22.26B). This sheet of thin chitinous material may line the midgut or pull free to envelop and coat the food particles as they pass through the gut. The peritrophic membrane serves to protect the delicate midgut epithelium from abrasion. It is permeated by microscopic pores that allow passage of enzymes and digested nutrients. In many species, production of this membrane also takes place in the hindgut, where it encapsulates the feces as discrete pellets.

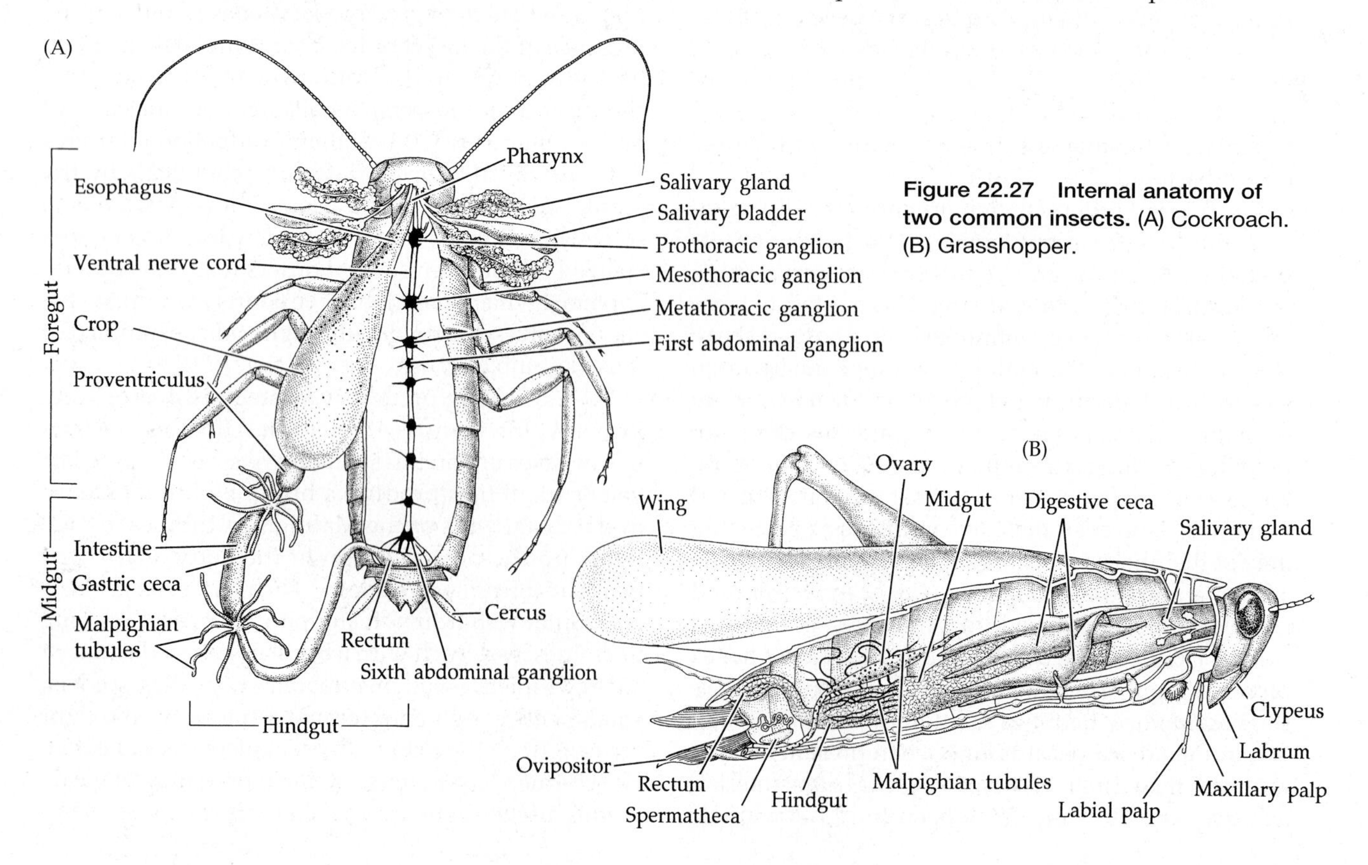

Figure 22.27 Internal anatomy of two common insects. (A) Cockroach. (B) Grasshopper.

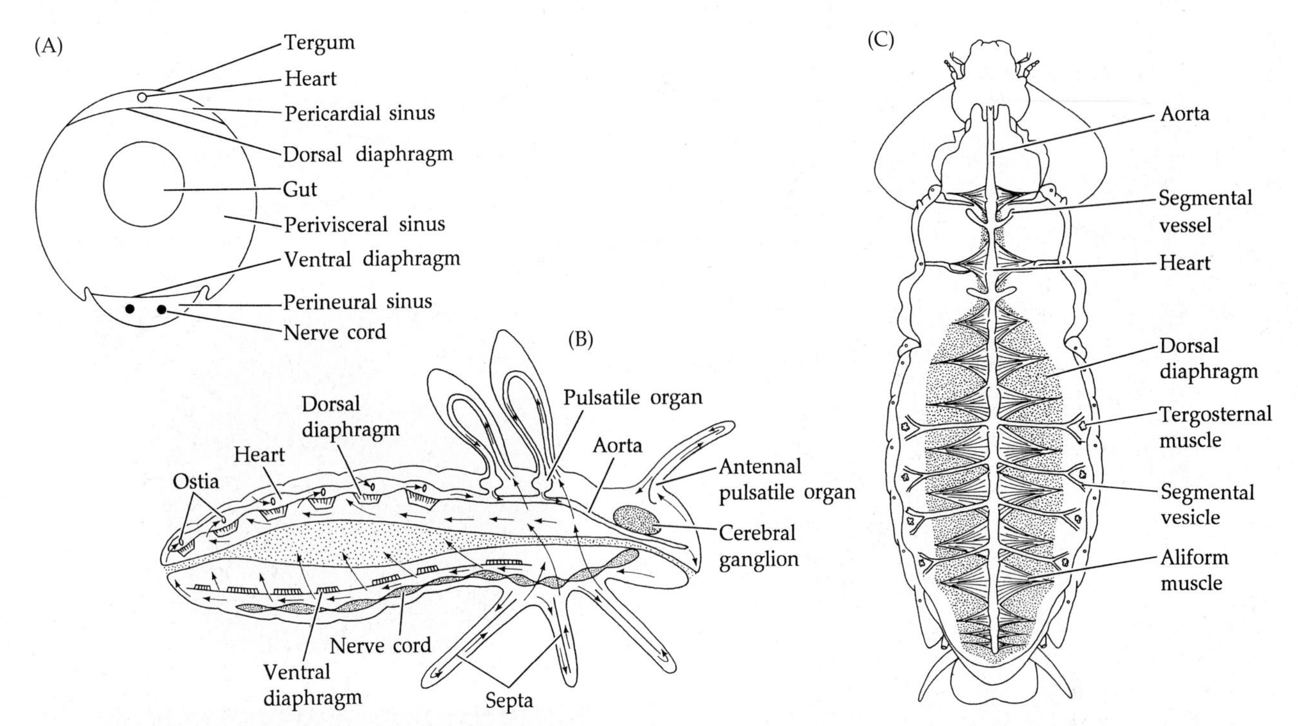

Figure 22.28 The circulatory system of insects. (A) An insect abdomen (cross section). Note the division of the hemocoel into three chambers (a dorsal pericardial sinus, a ventral perineural sinus, and a central perivisceral sinus). These chambers are separated by diaphragms lying on frontal planes. (B) Blood circulation in an insect with a fully developed circulatory system (longitudinal section). Arrows indicate the circulatory course. (C) A cockroach (ventral dissection). Note the dorsal and segmental vessels. The dorsal diaphragm and aliform (heart) muscles are continuous over the ventral wall of the heart and vessels, but they are omitted from the diagram for clarity.

Along with their vast range of feeding habits, insects have evolved a number of specialized digestive structures. The foregut is typically divided into a well defined pharynx, esophagus, crop, and proventriculus (Figures 22.26 and 22.27). The pharynx is muscular, particularly in the sucking insects, in which it commonly forms a **pharyngeal pump**. The crop is a storage center whose walls are highly extensible in species that consume large but infrequent meals. The **proventriculus** regulates food passage into the midgut, either as a simple valve that strains the semifluid foods of sucking insects or as a grinding organ, called a gizzard or gastric mill, that masticates the chunks ingested by biting insects. Well-developed gastric mills have strong cuticular teeth and grinding surfaces that are gnashed together by powerful proventricular muscles.

The midgut (= stomach) of most insects bears gastric ceca that lie near the midgut–foregut junction and resemble those of crustaceans. These evaginations serve to increase the surface area available for digestion and absorption. In some cases the ceca also house mutualistic microorganisms (bacteria and protists). The insect hindgut serves primarily to regulate the composition of the feces and perhaps to absorb some nutrients. Digestion of cellulase by termites and certain wood-eating roaches is made possible by enzymes produced by protists and bacteria that inhabit the hindgut.

Clusters of fat cells create a **fat body** in the hemocoel of many insects, which is most conspicuous in the abdomen but also extends into the thorax and head. The fat body is a unique organ to the insects and is often likened to the vertebrate liver and the chlorogogen tissue in annelids. The fat body not only stores lipids, proteins and carbohydrates, but also synthesizes proteins. Many insects do not feed during their adult life; instead, they rely on stored nutrients accumulated in the larval or juvenile stages and stored in the fat body.

Circulation and Gas Exchange

The hexapod circulatory system includes a dorsal tubular heart that pumps the hemocoelic fluid (blood) toward the head. The heart narrows anteriorly into a vessel-like aorta, from which blood enters the large hemocoelic chambers, through which it flows posteriorly, eventually returning to the pericardial sinus and then to the heart via paired lateral ostia (Figure 22.28). In most insects the heart extends through the first nine abdominal segments; the number of ostia is variable. Accessory pumping organs, or **pulsatile organs**, often occur at the bases of wings and of especially long appendages, such as the hindlegs of grasshoppers, to assist in circulation and maintenance of blood pressure.

The heart is a rather weak pumping organ, and blood is moved primarily by routine muscular activity of the body and appendages. Hence, circulation is slow and system pressure is relatively low. Like many

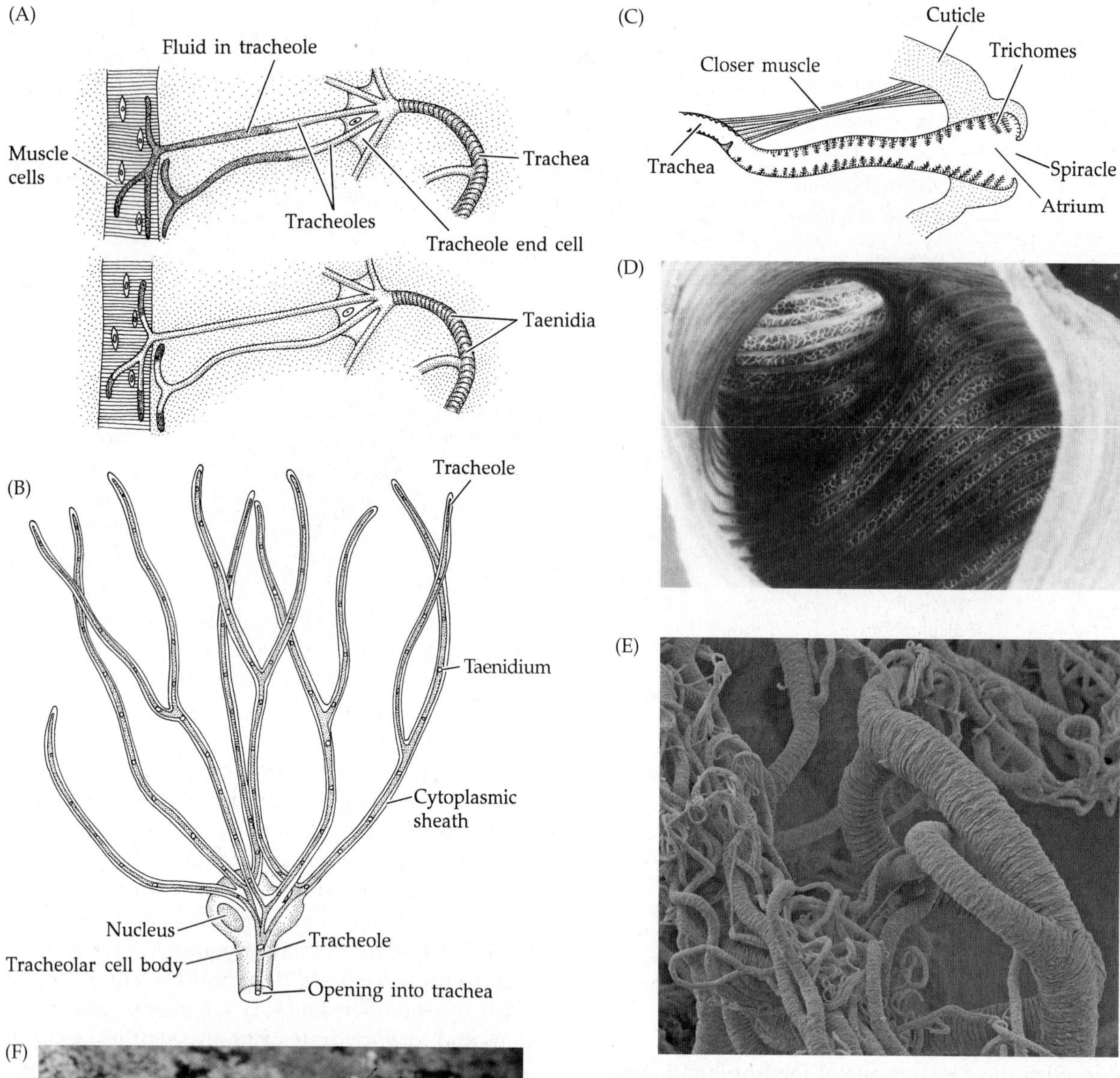

Figure 22.29 The tracheal system of insects. (A) Tracheoles and flight muscle cells. Note the region where the trachioles become functionally intracellular within the muscle fibers. The upper figure depicts a situation in which the muscle cells are well oxygenated, oxygen demand is low, and fluid accumulates in the trachioles. The lower figure depicts muscle cells that are oxygen-deficient. Decreased fluid volumes in the tracheoles allow the tissues increased access to oxygen. (B) A tracheole end cell. The taenidia are rings that serve to keep the lumen of the trachioles open. (C) A generalized insect spiracle (longitudinal section). Note the dust-catching spines (trichomes) within the atrium. (D) View inside a prothoracic trachea of the honeybee, *Apis mellifera*. (960×) (E) SEM of trachea and tracheoles of a carabid beetle. (F) Dragonfly nymph molting into an adult. Note the long white tracheae being pulled out of the exuvia by the emerging adult.

arachnids, some hexapods use the hydraulic pressure of the hemocoelic system in lieu of extensor muscles. In this way, for example, butterflies and moths unroll their maxillary feeding tubes.

Many types of hemocytes have been reported from the blood of insects. None functions in oxygen storage or transport, but several are apparently important in wound healing and clotting. Nutrients, wastes, and hormones can be efficiently carried by this system, but respiratory oxygen cannot (some CO_2 does diffuse into the blood). The active lifestyles of these terrestrial animals require special structures to carry out the tasks of respiratory gas exchange and excretion. These structures are the tracheal system and the Malpighian tubules, described below.

Desiccation is one of the principal dangers faced by terrestrial invertebrates. Adaptations to terrestrial life always involve some degree of compromise between water loss and gas exchange with the atmosphere. Even though the general body surface of insects may be largely waterproof, the gas exchange surfaces cannot be.

In some minute hexapods, such as Collembola, gas exchange occurs by direct diffusion across the body surface. However, the vast majority of hexapods rely on a tracheal system (Figure 22.29). As explained in Chapter 20, **tracheae** are extensive tubular invaginations of the body wall, opening through the cuticle by pores called spiracles. Up to ten pairs of spiracles can occur on the pleural walls of the thorax and abdomen. Since tracheae are epidermal in origin, their linings are shed with each molt. The cuticular wall of each trachea is sclerotized and usually strengthened by rings or spiral thickenings called **taenidia**, which keep the tube from collapsing but allow for changes in length that may accompany body movements. The tracheae originating at one spiracle commonly anastomose with others to form branching networks penetrating most of the body. In some insects it appears that air is taken into the body through the thoracic spiracles and released through the abdominal spiracles, thus creating a flow-through system.

Each spiracle is usually recessed in an atrium, whose walls are lined with setae or spines (**trichomes**) that prevent dust, debris, and parasites from entering the tracheal tubes. A muscular valve or other closing device is often present and is under control of internal partial pressures of O_2 and CO_2. In resting insects most of the spiracles are generally closed. Ventilation of the tracheal system is accomplished by simple diffusion gradients, as well as by pressure changes induced by the animal itself. Almost any movement of the body or gut causes air to move in and out of some tracheae. Telescopic elongation of the abdomen is used by some insects to move air in and out of the tracheal tubes. Many insects have expanded tracheal regions called tracheal pouches, which function as sacs for air storage.

Because the blood of hexapods does not transport oxygen, the tracheae must extend directly to each organ of the body, where their ends actually penetrate the tissues. Oxygen and CO_2 thus are exchanged directly between cells and the small ends of the tracheae, the tracheoles. In the case of flight muscles, where oxygen demand is high, the tracheal tubes invade the muscle fibers themselves. **Tracheoles**, the innermost parts of the tracheal system, are thin-walled, fluid-filled channels that end as a single cell, the tracheole end cell (= tracheolar cell) (Figure 22.29). The tracheoles penetrate every organ in the body, and gas exchange takes place directly between the body cells and the tracheoles. Unlike tracheae, tracheoles are not shed during ecdysis. The tracheoles are so minute (0.2–1.0 μm) that ventilation is impossible, and gas transport here relies on aqueous diffusion. This ultimate constraint on the rate of gas exchange may be the primary reason terrestrial arthropods never achieved extremely large sizes.

In aquatic insects the spiracles are usually nonfunctional, and gases simply diffuse across the body wall directly to the tracheae. A few species retain functional spiracles; they hold an air bubble over each opening, through which oxygen from the surrounding water diffuses. The air bubbles are held in place by secreted waxes and by patches of hydrophobic hairs in densities that may exceed 2 million per square millimeter. Most aquatic insects, particularly larval stages, have **gills**—external projections of the body wall that are covered by thin, unsclerotized cuticle and contain blood, tracheae, or air bubbles (Figure 22.30). The gills contain channels that lead to the main tracheal system. In some aquatic insects, such as dragonfly nymphs, the rectum bears tiny branched tubules called **rectal accessory gills**. By pumping water in and out of the anus, these insects exchange gases across the increased surface area of the thin gut wall. There are analogous examples of hindgut respiratory irrigation in other, unrelated invertebrate groups (e.g., echiurids, holothurians).

Excretion and Osmoregulation

The problem of water conservation and the nature of the circulatory and gas exchange systems in terrestrial arthropods necessitated the evolution of entirely new structures to remove metabolic wastes. Like the gas exchange surfaces, the excretory system is a site of potential water loss, because nitrogenous wastes initially occur in a dissolved state. These problems are

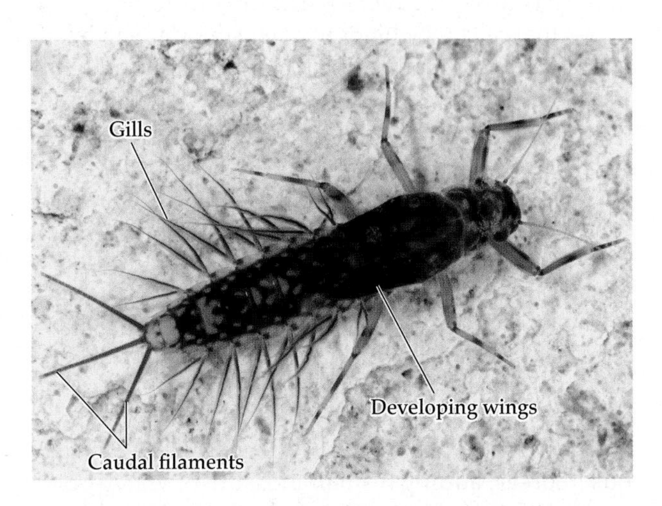

Figure 22.30 Aquatic nymph of a mayfly, *Paraleptophlebia* (Ephemeroptera), with lateral abdominal gills.

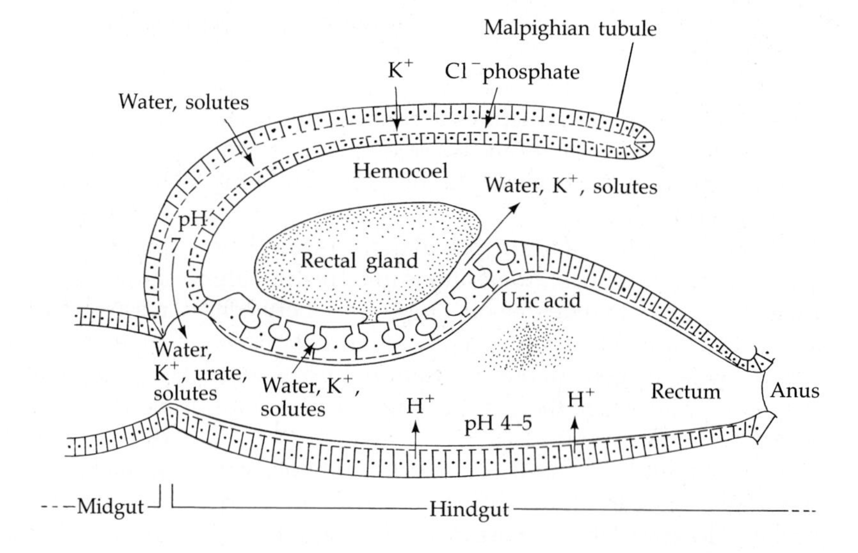

Figure 22.31 A single Malpighian tubule opening into the hindgut at its junction with the midgut. Arrows indicate the flow of materials.

compounded in small terrestrial organisms, such as many hexapods, because of their large surface area-to-volume ratios. And water loss problems are even more severe in flying insects, because flight is probably the most metabolically demanding of all locomotor activities.

In most terrestrial arthropods, the solution to these problems is Malpighian tubules. In the Hexapoda, these unbranched outgrowths of the gut arise near the junction of the midgut and hindgut (Figures 22.26, 22.27, and 22.31). Their blind distal ends extend into the hemocoel and lie among various organs and tissues. Up to several hundred Malpighian tubules may be present.

In the absence of sufficient blood pressure for typical excretory filtration, hexapods use osmotic pressure to achieve the same result. Various ions, especially potassium, are actively transported across the Malpighian tubule epithelium from the blood into the tubule lumen (Figure 22.31). The osmotic gradient maintained by this ion transport mechanism enables water and solutes to move from the body cavity into the tubules, and thence into the gut. Water and other metabolically valuable materials are selectively reabsorbed into the blood across the wall of the hindgut, while the Malpighian filtrate left behind is mixed with the other gut contents. Reabsorption of water, amino acids, salts, and other nutrients may be enhanced by the action of special cells in thickened regions called rectal glands. The soluble potassium urate from the Malpighian tubules has, at this point in the gut, been precipitated out as solid uric acid as a result of the low pH of the hindgut (pH 4–5). Uric acid crystals cannot be reabsorbed into the blood, hence they pass out the gut with the feces. Insects also possess special cells called nephrocytes or pericardial cells that move about in certain areas of the hemocoel, engulfing and digesting particulate or complex waste products.

The hexapod cuticle is sclerotized or tanned to various degrees, adding a small measure of waterproofing. But more importantly, a waxy layer occurs within the epicuticle, which greatly increases resistance to desiccation and frees insects to fully exploit dry environments. In many terrestrial arthropods (including primitive insects) an eversible coxal sac (not to be confused with the coxal glands of arachnids) projects from the body wall near the base of each leg. It is thought that the coxal sacs assist in maintaining body hydration by taking up water from the environment (e.g., dewdrops). Many insects collect environmental water by various other devices. Some desert beetles (Tenebrionidae) collect atmospheric water by "standing on their heads" and holding their bodies up to the moving air so that humidity can condense on the abdomen and be channeled to the mouth for consumption.

Insects that inhabit desert environments have a much greater tolerance of high temperatures and body water loss than do insects in mesic environments, and they are particularly good at water conservation and producing insoluble nitrogenous waste products. They also have behavioral traits, such as nocturnal activity cycles and dormancy periods that enhance water conservation. Upper lethal temperatures for desert species commonly range to 50°C. The spiracles are often covered by setae or depressed below the cuticular surface. Many xeric insects also undergo periods of dormancy (i.e., diapause or aestivation) during some stage of the life cycle, characterized by a lowering of the basal metabolic rate and cessation of movement, which allow them to withstand prolonged periods of temperature and moisture extremes. Some even utilize evaporative cooling to reduce body temperatures. The long-chain hydrocarbons that waterproof the epicuticle also are more abundant in xeric insects.

Nervous System and Sense Organs

The hexapod nervous system conforms to the basic arthropod plan described in Chapter 20 (Figures 22.32 and 22.33). The two ventral nerve cords, as well as the segmental ganglia, are often largely fused. In dipterans, for example, even the three thoracic ganglia are fused into a single mass. The largest number of free ganglia occurs in the primitive wingless insects, which have as many as eight unfused abdominal ganglia. Giant fibers have also been reported from several insect orders.

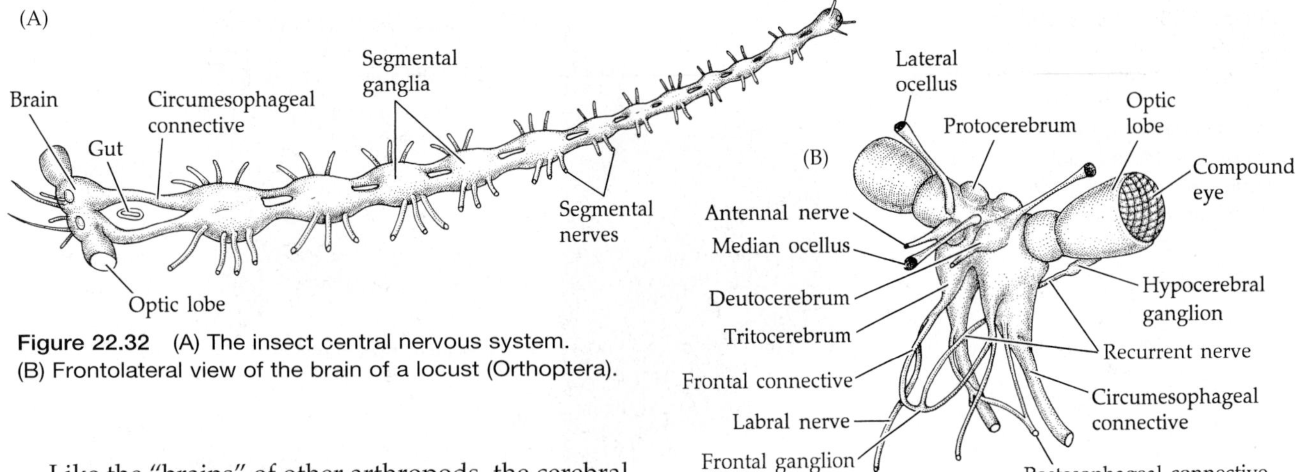

Figure 22.32 (A) The insect central nervous system. (B) Frontolateral view of the brain of a locust (Orthoptera).

Like the "brains" of other arthropods, the cerebral ganglia of insects comprise three distinct regions: the protocerebrum, the deutocerebrum, and the tritocerebrum. The subesophageal ganglion is composed of the fused ganglia of the fourth, fifth, and perhaps the sixth head segments and controls the mouthparts, salivary glands, and some other local musculature.

Insects possess a hypocerebral ganglion between the cerebral ganglion and the foregut. Associated with this ganglion are two pairs of glandular bodies called the **corpora cardiaca** and the **corpora allata** (Figure 22.33). These two organs work in concert with the prothoracic glands and certain neurosecretory cells in the protocerebrum. The whole complex is a major endocrine center that regulates growth, metamorphosis, and other functions (see Chapter 20).

Hexapods typically possess simple ocelli in the larval, juvenile, and often adult stages. When present in adults, they usually form a triad or a pair on the anterodorsal surface of the head. The compound eyes are well developed, resembling those of Crustacea, and are image-forming. Most adult insects have a pair of compound eyes (Figure 22.34), which bulge out to some extent, giving these animals a wide field of vision in almost all directions. Compound eyes are greatly reduced or absent in parasitic groups and in many cave-dwelling forms. The general anatomy of the arthropod compound eye was described in Chapter 20, but several distinct structural trends are found in hexapod eyes, as we describe below.

The number of ommatidia apparently determines the overall visual acuity of a compound eye; hence large eyes are typically found on active, predatory insects such as dragonflies and damselflies (order Odonata), which may have over 10,000 ommatidia in each eye. On the other hand, workers of some ant species have but a single ommatidium per eye (ants live in a world of chemical communication)! Similarly, larger facets capture more light and are typical of nocturnal insects. In all cases, a single ommatidium consists of two functional elements: an outer light-gathering part composed of a lens and a crystalline cone, and an inner sensory part composed of a rhabdome and sensory cells (Figure 22.34).

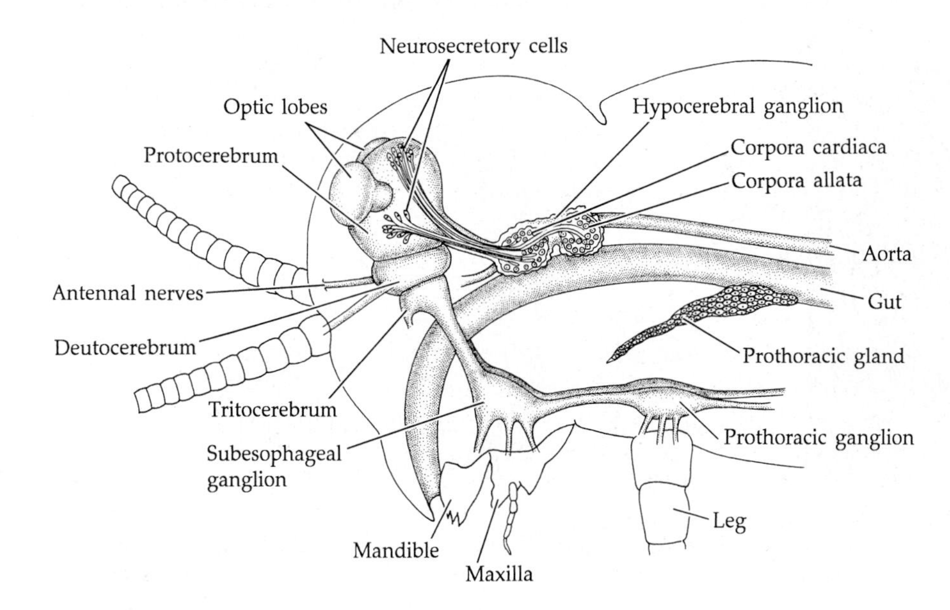

Figure 22.33 Endocrine organs and central nervous system in the head and thorax of a generalized insect. These organs all play roles in the control of molting and metamorphosis.

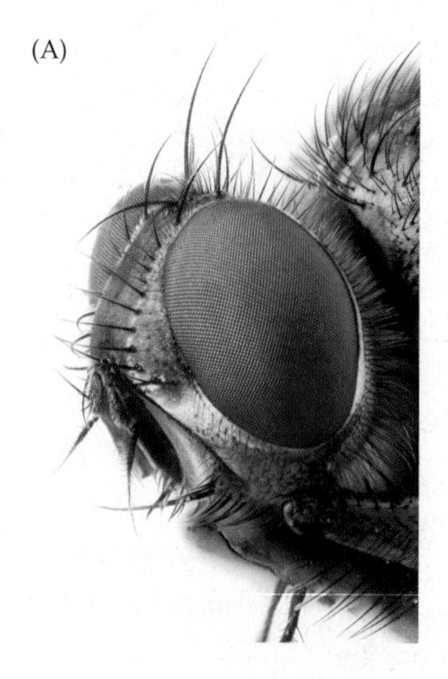

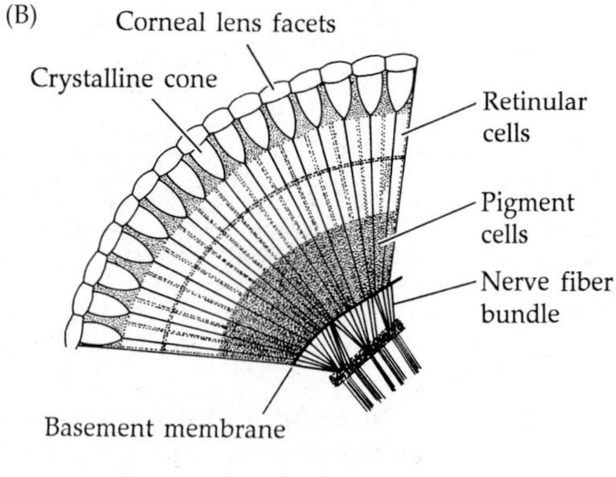

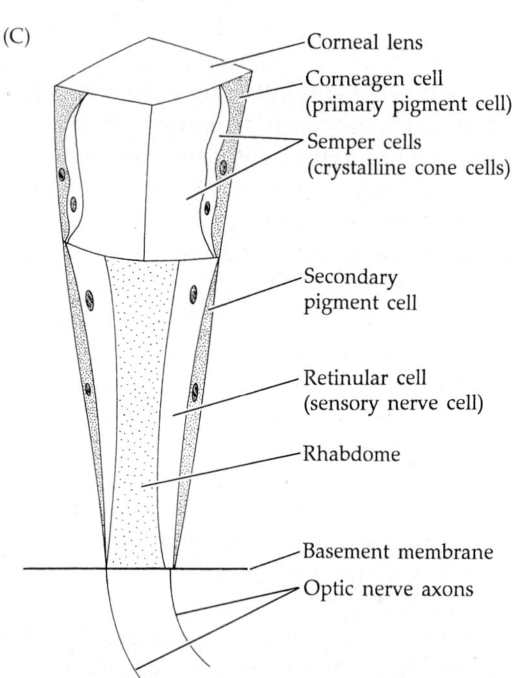

Figure 22.34 Compound eyes of insects. (A) Close-up of the compound eyes of a tachinid fly (Diptera) with many ommatidia. (B) A generalized insect compound eye (cross section). (C) A single ommatidium from a eucone compound eye.

In some insects the outer surface of the cornea (lens) is covered with minute conical tubercles about 0.2 μm tall and arranged in a hexagonal pattern. It is thought that these projections decrease reflection from the surface of the lens, thus increasing the proportion of light transmitted through the facet. Insect eyes in which the crystalline cone is present are called **eucone eyes** (Figure 22.34B). Immediately behind the crystalline cone (in eucone eyes) are the elongate sensory neurons, or retinular cells. Primitively, each ommatidium probably contained eight retinular cells arising from three successive divisions of a single cell. This number is found in some insects today, but in most it is reduced to six or seven, with the other one or two persisting as short basal cells in the proximal region of each ommatidium. Arising from each retinular cell is a neuronal axon that passes out through the basement membrane at the back of the eye into the optic lobe. There is no true optic nerve in insects; the eyes connect directly with the optic lobe of the brain. The rhabdomeres consist of tightly packed microvilli that are about 50 nm in diameter and hexagonal in cross section. The retinular cells are surrounded by 12 to 18 secondary pigment cells, which isolate each ommatidium from its neighbors.

The general body surface of hexapods, like that of other arthropods, bears a great variety of microscopic

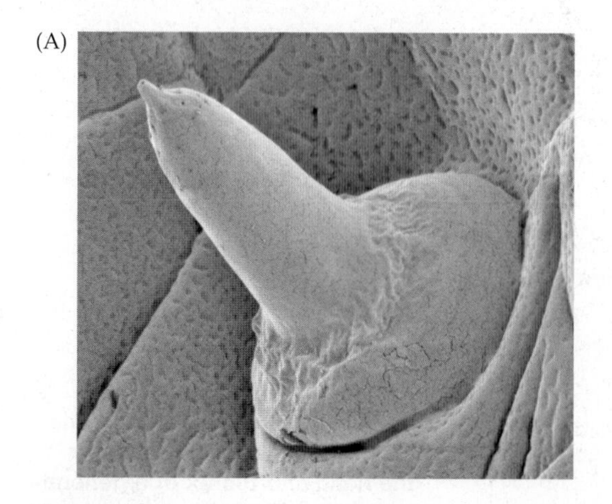

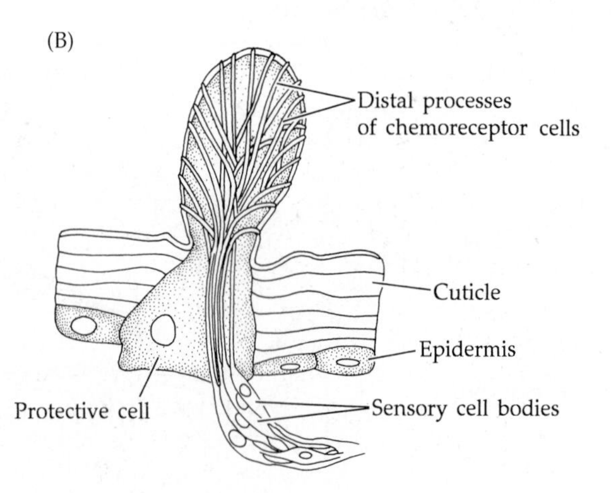

Figure 22.35 Chemosensory peg organs. (A) SEM of a peg organ from an antenna of a beetle. (B) Diagram of a peg organ in cross section from the antenna of a grasshopper.

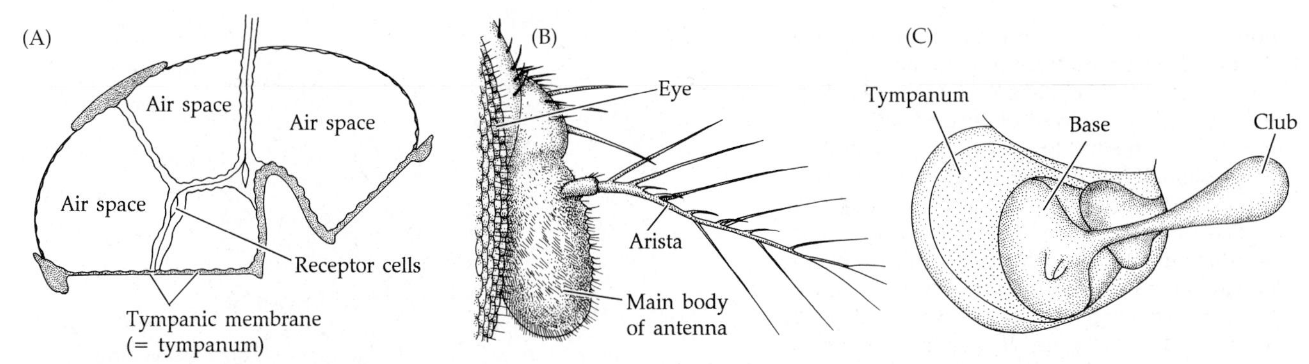

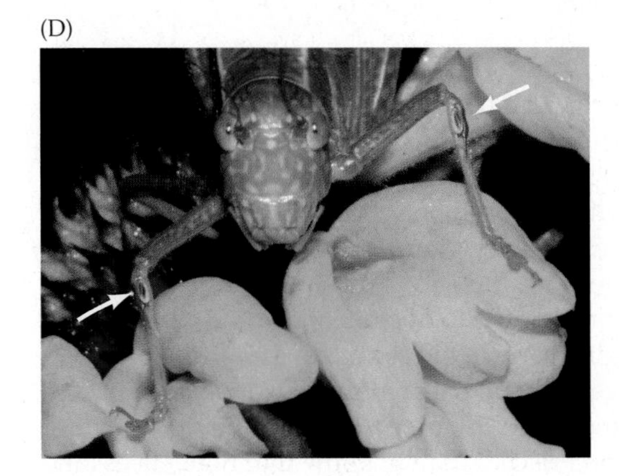

Figure 22.36 Insect "ears." Insect auditory organs (phonoreceptors) differ widely in their anatomy and location. (A) The "ear" of noctuid moths (Lepidoptera) is a pressure receiver used to detect the ultrasonic cries of hunting bats. It is similar to most insect "ears" in comprising a tympanic membrane backed by a tracheal air space. Two receptor cells attach to the tympanum. (B) In *Drosophila* (Diptera), a feathery seta called an arista arises on the third antennal segment. The arista detects air movements, thus responding to sound through interaction with vibrating air particles. It is used to detect the calling song of the species. (C) In the "ear" of a water boatman (Hemiptera), the tympanum is covered by the base of a club-shaped cuticular body that protrudes outside the body. The club performs rocking movements that allow some frequency analysis of the songs of other water boatmen. (D) Tympani (arrows) on the tibiae of katydid forelegs.

sensory hairs and setae, known collectively as **sensilla** (sing., sensillum). The incredible diversity of these cuticular surface structures has only begun to be explored, primarily by scanning electron microscopy. Sensilla are most heavily concentrated on the antennae, mouthparts, and legs. Most appear to be tactile or chemosensory. Club-shaped or peg-shaped chemosensory setae, usually called **peg organs** and resembling the aesthetascs of crustaceans, are particularly common on the antennae of hexapods (Figure 22.35).

Insects have internal proprioceptors called **chordotonal organs**. These structures stretch across joints and monitor the movement and position of various body parts. Phonoreceptors also occur in most insect orders. These structures may be simple modified body or appendage setae, or antennae, or complex structures called **tympanic organs** (Figure 22.36). Tympanic organs generally develop from the fusion of parts of a tracheal dilation and the body wall, which form a thin tympanic membrane (= tympanum). Receptor cells in an underlying air sac, or attached directly to the tympanic membrane, respond to vibrations in much the same fashion as they do in the cochlea of the human inner ear. Many insects can discriminate among different sound frequencies, but others may be tone-deaf. Tympanic organs may occur on the abdomen, the thorax, or the forelegs. Several insects that are prey to bats have the ability to hear the high frequencies of bat echolocation devices, and they have evolved flight behaviors to avoid these flying mammals. For example, some moths, when they hear a bat's echolocation (generally above the range of human ears), will fold their wings and suddenly drop groundward as an evasive maneuver. Praying mantises, whose sonar detection device is buried in a groove on the ventral side of the abdomen, throw out the raptorial forelimbs and elevate the abdomen. These movements cause the insect to "stall" and go into a steep roll, which it pulls out of at the last minute with a "power dive" that effectively avoids bat predators.

Sound communication in insects, like light communication in fireflies (and some ostracods), is a species-specific means of mate communication. Several insect groups (e.g., some orthopterans, coleopterans, dipterans, and hemipterans) possess sound-producing structures. Male flies of the genus *Drosophila* create species-specific mating songs by rapidly vibrating the wings or abdomen. These "love songs" attract conspecific females for copulation. It has been demonstrated that the rhythm of the male's song is encoded in genes inherited from his mother, on the X chromosome, whereas the song's "pulse interval" is controlled by genes on autosomal chromosomes.

Cicadas may possess the most complex sound-producing organs in the animal kingdom (Figure 22.37). The ventral metathoracic region of male cicadas bears two large plates, or opercula, that cover a complex system of vibratory membranes and resonating chambers. One membrane, the tymbal, is set vibrating by special muscles, and other membranes in the resonating chambers amplify its vibrations. The sound leaves the cicada's body through the metathoracic spiracle.

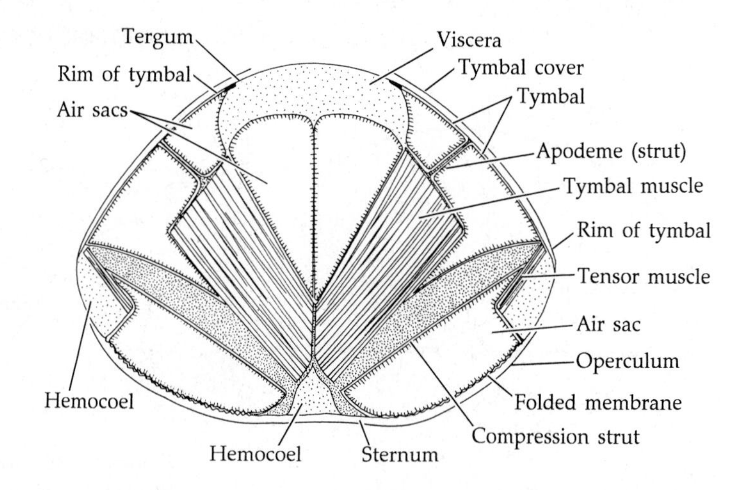

Figure 22.37 Sound production structure in cicadas (Hemiptera) from the first abdominal segment (section). Sound is produced by buckling of the tymbal, a thin disc of cuticle. The tymbal muscle is connected to the tymbal by a strut. Contraction of this muscle causes the tymbal to buckle inward, thereby producing a click that is amplified by resonance in the underlying air sacs. On relaxation, the elasticity of the muscle causes the tymbal to buckle out again. On the underside of this abdominal segment, a folded membrane can be stretched to tune the air sacs to the resonant frequency of the tymbal.

Numerous families of beetles and bugs utilize water surfaces as a substratum both for locomotion and for communication by waves or ripples. Such insects produce a signal with simultaneous vertical oscillations of one or more pairs of legs, and sometimes also with distinct vertical body motions. The wave patterns produced are species-specific. Potential prey trapped on a surface film may also be recognized in this fashion, just as spiders recognize prey by web vibrations. Limited data suggest that the receptor organs for ripple communication are either specialized sensilla on the legs or special proprioceptors between joints of the legs or antennae, perhaps similar to the tarsal organs of scorpions (Chapter 24).

A number of insects are bioluminescent, the most familiar being beetles of the family Lampyridae, known as lightning bugs or fireflies. In the tropics, where they are especially abundant, fireflies are sometimes kept in containers and used as natural flashlights, and women may wrap them in gauze bags worn as glowing hair ornaments. The light of luminescent insects ranges from green through red and orange, depending on the species and the precise chemical nature of the luciferin–luciferase system involved. Light-producing organs are typically composed of clusters of light-producing cells, or **photocytes**, backed by a layer of reflecting cells and covered with a thin, transparent epidermis. The photocytes are richly supplied with tracheae, oxygen being necessary for the chemical reaction. Each species of firefly, and of most other glowing insects, has a distinct flash pattern, or code, to facilitate mate recognition and communication.

One of the most sophisticated communication behaviors among insects may be the famous honeybee "waggle dance." Each day forager bees leave their colony to locate new food sources (e.g., fresh flower blooms). They fly meandering search forays until a good source is located. Then they return to the hive along a straight flight path (a "bee line"); while doing so, they are thought to imprint a navigational "map" from the colony to the food source. Most behaviorists believe that this information is communicated to hive-mates in a complex tail-wagging dance that allows other bees from the hive to fly directly to the new feeding ground. The forager bee also carries food odors (nectar samples), pollen, and various other odors clinging to the hairs on her body. She can also mark the food source with a pheromone produced in a special gland, called the **Nasanov gland**. All of these clues help her hive-mates find the new food source. Karl von Frisch was the first person to document all these attributes of bee foraging early in the twentieth century.

A large body of research on bee navigation has accumulated since the pioneering "dancing bee studies" of von Frisch. We now know that honeybees (and solitary bees) have outstanding vision. Much of the bee's daily activity, including navigation and flower recognition, relies strongly on ultraviolet vision. Bees appear to utilize a hierarchical series of flight orientation mechanisms; when the primary mechanism is blocked, a bee can switch to a secondary system. The primary navigation system utilizes the pattern of polarized ultraviolet sunlight in the sky. This pattern depends on the location of the sun as determined by two coordinates, the azimuth and the elevation. Bees and many other animals that orient to the sun have a built-in ability to compensate for both hourly changes (elevation) and seasonal changes (azimuth) in the sun's position with time. On cloudy days, when the sun's light is largely depolarized, bees cannot rely on their ultraviolet celestial navigation mechanism and thus may switch to their second-order navigational system: navigation by landmarks (foliage, rocks, and so on) that were imprinted during the most recent flight to the food source. Limited evidence suggests that some form of tertiary backup system may also exist.

Thus, if the honeybee dance model is correct,[7] honeybees must simultaneously process information concerning time, the direction of flight relative to the sun's azimuth, the movement of the sun, the distance flown, and local landmarks (not to mention complications due to other factors, such as crosswinds), and in doing so reconstruct a straight-line heading to inform

[7]The honeybee dance hypothesis is not without its detractors, and some workers doubt its existence altogether; see the General References section of this chapter for a glimpse at the history of the honeybee dance controversy.

their hive-mates. If recent evidence is correct, bees (like homing pigeons) may also detect Earth's magnetic fields with iron compounds (magnetite) located in their abdomens. Bands of cells in each abdominal segment of the honeybee contain iron-rich granules, and nerve branches from each segmental ganglion appear to innervate these tissues.[8]

In some insects the ocelli are the principal navigation receptors. Some locusts and dragonflies and at least one ant species utilize the ocelli to read compass information from the blue sky. As in bees, the pattern of polarized light in the sky seems to be the main compass cue. In some species, both ocelli and compound eyes may function in this fashion. Many (probably most) insects also see ultraviolet light.

Perhaps the most famous insect navigators are North America's monarch butterflies (*Danaus plexippus*). Each autumn, monarchs migrate up to 4,000 km from breeding grounds in the eastern United States and Canada to over-wintering sites in the mountains of Michoacán, in central Mexico. They make this remarkable journey by orienting with a Sun compass, using the Sun's changing azimuth (and knowledge of the relative time of day) to direct their movements. On cloudy days, when a precise solar azimuth is unobtainable, monarchs still manage to orient towards the south-southwest, suggesting that they also have a backup mechanism of orientation, such as a geomagnetic compass. Monarchs are one of a small group of animal species for which a sun compass orientation mechanism has been shown to exist.

Many insects release noxious quinone compounds to repel attacks. Perhaps best known in this regard are certain Tenebrionidae, many of which stand on their heads to do so. But the champions of this chemical warfare strategy are definitely the bombardier beetles, members of the carabid subfamilies Brachininae and Paussinae, which expel quinone compounds at temperatures reaching 100°C (Figure 22.7C).

Reproduction and Development

Reproduction Hexapods are dioecious, and most are oviparous. A few insects are ovoviviparous, and many can reproduce parthenogenetically. Most insects rely on direct copulation and insemination. Reproductively mature insects are termed adults or **imagos**. Female imagos have one pair of ovaries, formed of clusters of tubular ovarioles (Figure 22.38A). The oviducts unite as a common duct before entering a genital chamber. Seminal receptacles (spermathecae) and accessory glands also empty into the genital chamber. The genital chamber opens, via a short **copulatory bursa** (= vagina), on the sternum of the eighth, or occasionally the seventh or ninth, abdominal segment. The male reproductive system is similar, with a pair of testes, each formed by a number of sperm tubes (Figure 22.38B). Paired sperm ducts dilate into seminal vesicles (where sperm are stored) and then unite as a single ejaculatory duct. Near this duct, accessory glands discharge seminal fluids into the reproductive tract. The lower end of the ejaculatory duct is housed within a penis, which extends posteroventrally from the ninth abdominal sternite.

Courtship behaviors in insects are extremely diverse and often quite elaborate, and each species has its own recognition methods. Courtship may consist of simple chemical or visual attraction, but more typically it involves pheromone release, followed by a variety of displays, tactile stimulation, songs, flashing lights, or other rituals that may last for hours. The subject of insect courtship is a large and fascinating study of its own. Although the field of pheromone biology is still in its infancy, sexual attractant or aggregation pheromones have been identified from about 500 different insect species (about half of which are synthesized and sold commercially for pest control purposes).

Most insects transfer sperm directly as the male inserts either his **aedeagus** (Figure 22.38B,D) or a gonopod into the genital chamber of the female. Special abdominal claspers, or other articulated cuticular structures on the male, often augment his copulatory grip. Such morphological modifications are species specific and thus serve as valuable recognition characters, both for insect mates and insect taxonomists. Copulation often takes place in mid-flight. In some of the primitive wingless insects, and in the odonatans, sperm transfer is indirect. In these cases, a male may deposit his sperm on specialized regions of his body to be picked up by the female; or he may simply leave the sperm on the ground, where they are found and taken up by females. In bedbugs (order Hemiptera, family Cimicidae) males use the swollen penis to pierce a special region of the female's body wall; sperm are then deposited directly into an internal organ (**the organ of Berlese**). From there they migrate to the ovaries, where fertilization takes place as eggs are released.

Sperm may be suspended in an accessory gland secretion, or, more commonly, the secretion hardens around the sperm to produce a spermatophore. Females of many insect species store large quantities of sperm within the spermathecae. In some cases sperm from a single mating is sufficient to fertilize a female's eggs for her entire reproductive lifetime, which may last a few days to several years.

[8]Many animals possess magnetotactic capabilities, including some molluscs, hornets, salmon, tuna, turtles, salamanders, homing pigeons, cetaceans, and even bacteria and humans. Magnetotactic bacteria swim to the north in the Northern Hemisphere, to the south in the Southern Hemisphere, and in both directions at the geomagnetic equator. In all these cases, iron oxide crystals in the form of magnetite have been shown to underlie the primary detection devices. However, in honeybees, the iron-containing structures are trophocytes that contain paramagnetic magnetite. These magnetotactic trophocytes surround each abdominal segment and are innervated by the central nervous system.

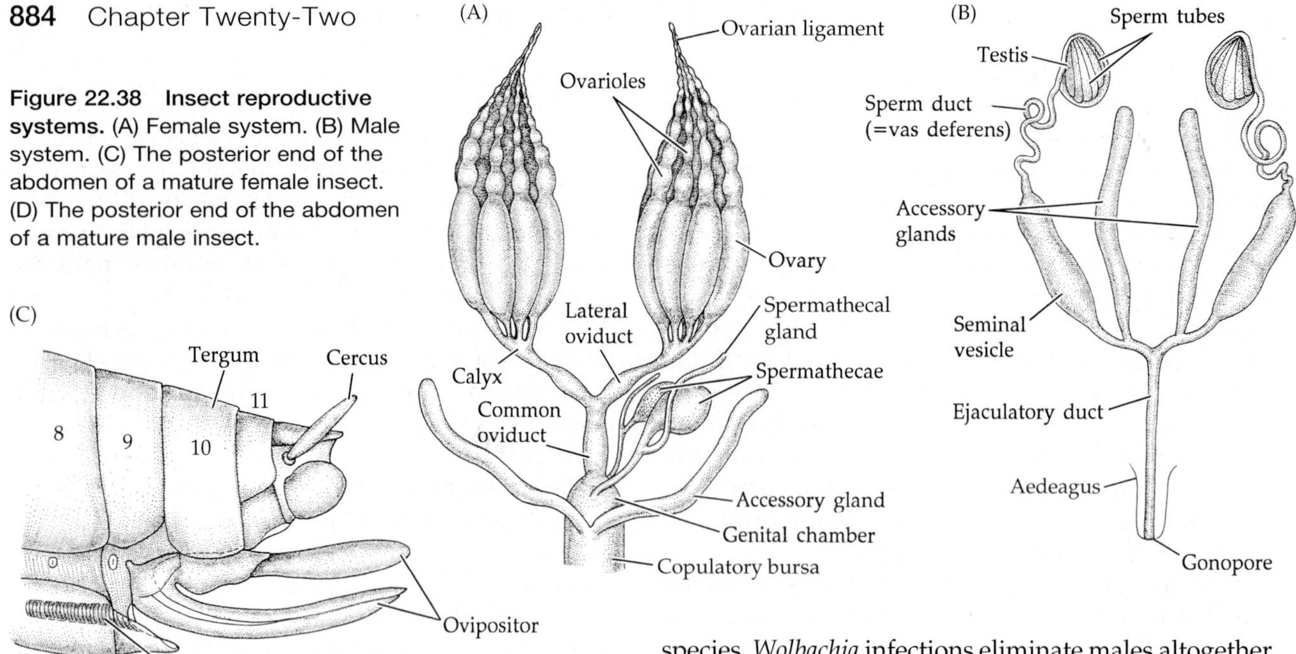

Figure 22.38 Insect reproductive systems. (A) Female system. (B) Male system. (C) The posterior end of the abdomen of a mature female insect. (D) The posterior end of the abdomen of a mature male insect.

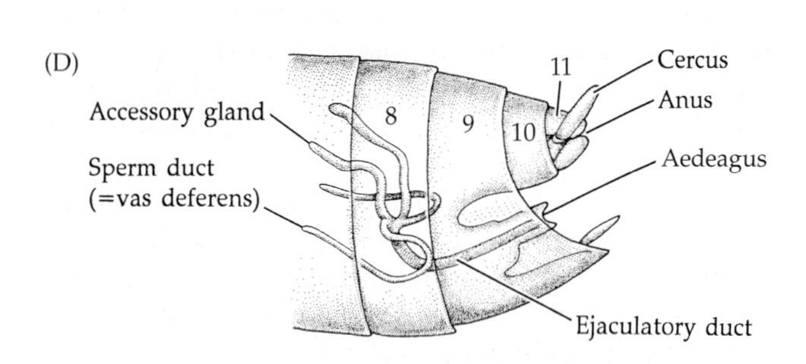

Insect eggs are protected by a thick membrane (the chorion) produced within the ovary. Fertilization occurs as the eggs pass through the oviduct to be deposited. Accessory glands contribute adhesives or secretions that harden over the zygotes. In many species, cuticular extensions around the gonopore of the female form an **ovipositor** (Figure 22.38C), with which she places the eggs in a brooding site that will afford suitable conditions for the young once they hatch (such as in a shallow underground chamber, in a plant stem, or within the body of a host insect). Although 50–100 eggs are usually laid at a time, as few as one and as many as several thousand are deposited by some species. Some insects, such as cockroaches, enclose several eggs at a time in a protective egg case.

Parthenogenesis is common in a variety of insect groups. It is used as an alternative form of reproduction seasonally by a number of insect taxa, particularly those living in unstable environments. In the Hymenoptera (bees, wasps, ants), it is also used as a mechanism for sex determination. In these cases, diploid fertilized eggs become females, and haploid unfertilized eggs develop into males. Infections by the bacterium *Wolbachia*, a frequent parasite of arthropod reproductive systems, are known to affect reproduction in many insect species. In some cases, infections result in infertility, whereas in others they transform males into functional females. But in some wasp species, *Wolbachia* infections eliminate males altogether by disrupting the first cell division of the egg, resulting in diploid eggs that can develop only as females—thus creating parthenogenetic strains of normally sexual wasps. Such asexual strains of wasps will revert back to dioecy if the *Wolbachia* dies out.[9]

Embryology As discussed in Chapter 20, the large centrolecithal eggs of arthropods are often very yolky, a condition resulting in modifications of the cleavage pattern. Although vestiges of what have been interpreted as holoblastic spiral cleavage are still discernible in some crustaceans, the hexapods show almost no trace of spiral cleavage at all. Instead, most undergo meroblastic cleavage by way of intralecithal nuclear divisions, followed by migration of the daughter nuclei to the peripheral cytoplasm (= periplasm). Cytokinesis does not occur during these early nuclear divisions (up to 13 cycles), which thus generate a syncytium, or plasmodial phase of embryogenesis. The nuclei continue to divide until the periplasm is dense with nuclei, whereupon a syncytial blastoderm exists. Eventually, cell membranes begin to form, partitioning uninucleate cells from one another. At this point the embryo is a periblastula, comprising a yolky sphere containing a few scattered nuclei and covered by a thin cellular layer (Figure 22.39).

Along one side of the blastula a patch of columnar cells forms a germinal disc, sharply marked off from the thin cuboidal cells of the remaining blastoderm (Figure 22.39A). From specific regions of this disc, presumptive endodermal and mesodermal cells begin to proliferate as germinal centers. These cells migrate inward during gastrulation to lie beneath their parental cells, which now form the ectoderm. The mesoderm proliferates inward as a longitudinal gastral groove

[9] *Wolbachia pipientis* are maternally transmitted, gram-negative, obligate intracellular bacteria found in filarial nematodes, crustaceans, arachnids, and at least 20% of all insect species. Many *Wolbachia* increase their prevalence in populations by manipulating host reproductive systems.

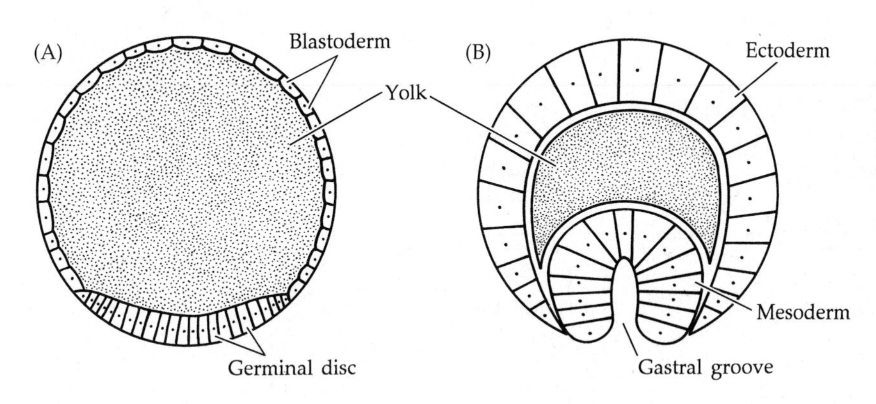

Figure 22.39 Early stages of insect development. (A) The blastoderm (blastula) of a generalized insect, subsequent to cytokinesis (cross section). Note the thickened germinal disc. (B) An early gastrula of a honeybee (cross section). Note the gastral groove and the proliferation of mesoderm.

(Figure 22.39B). The cells of the developing gut usually surround and gradually begin absorbing the central yolky mass of the embryo, and paired coelomic spaces appear in the mesoderm.

As segments begin to demarcate and proliferate, each receives one pair of mesodermal pouches and eventually develops appendage buds. As the mesoderm contributes to various organs and tissues, the paired coelomic spaces merge with the small blastocoel to produce the hemocoelic space. The mouth and anus arise by ingrowths of the ectoderm that form the proctodeal foregut and hindgut, which eventually establish contact with the developing endodermal midgut.

Polyembryony occurs in a number of insect taxa, particularly parasitic Hymenoptera. In this form of development the early embryo splits to give rise to more than one developing embryo. Thus, from two to thousands of larvae may result from a single fertilized egg, which is often deposited in the body of another (host) insect.

Post-embryonic development Within Hexapoda there are three main types of development: **ametabolous** (direct, or amorphic development), **hemimetabolous**, and **holometabolous** (indirect, or complete development). Figure 22.40 depicts these development types. Species in the most primitive wingless hexapod

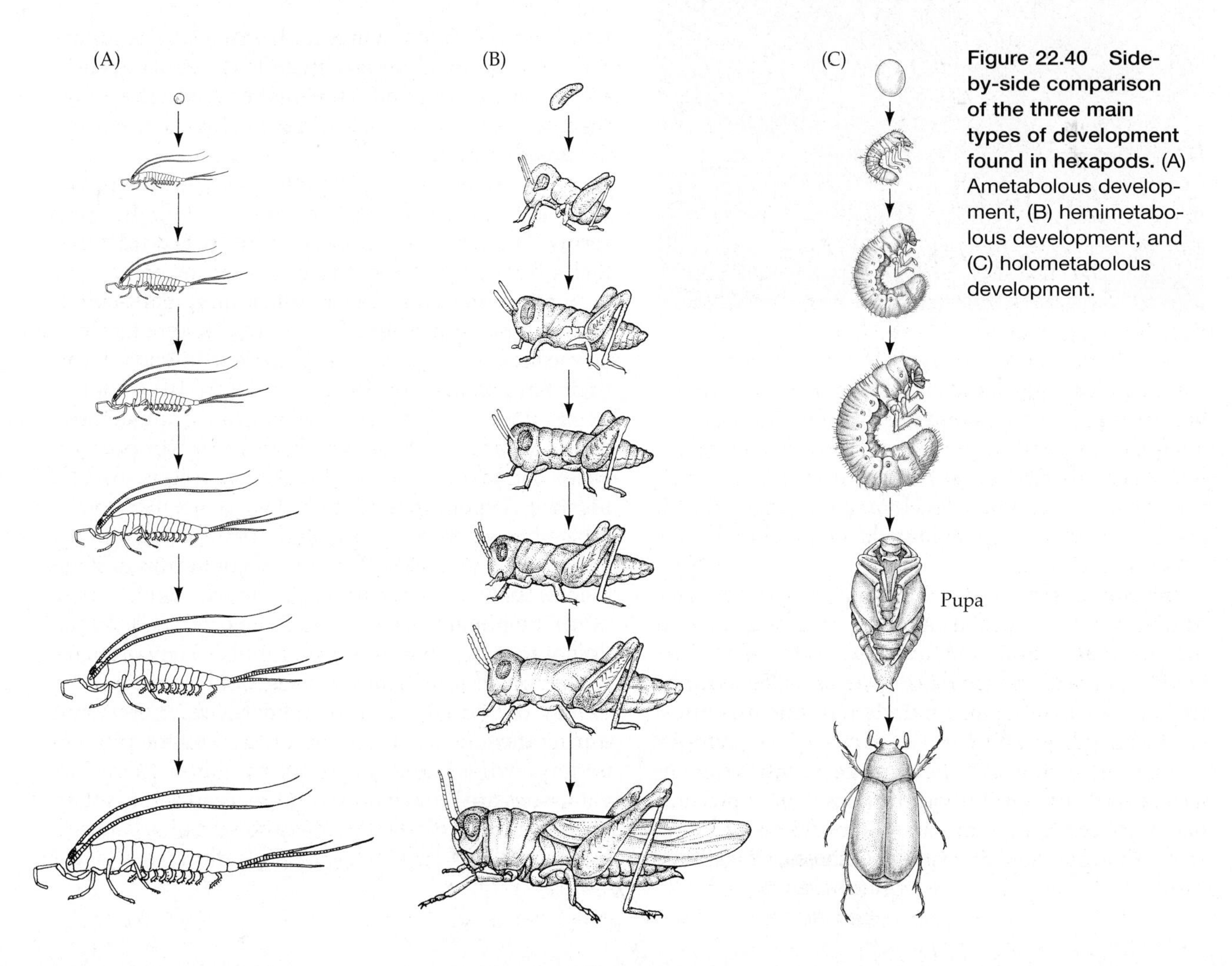

Figure 22.40 Side-by-side comparison of the three main types of development found in hexapods. (A) Ametabolous development, (B) hemimetabolous development, and (C) holometabolous development.

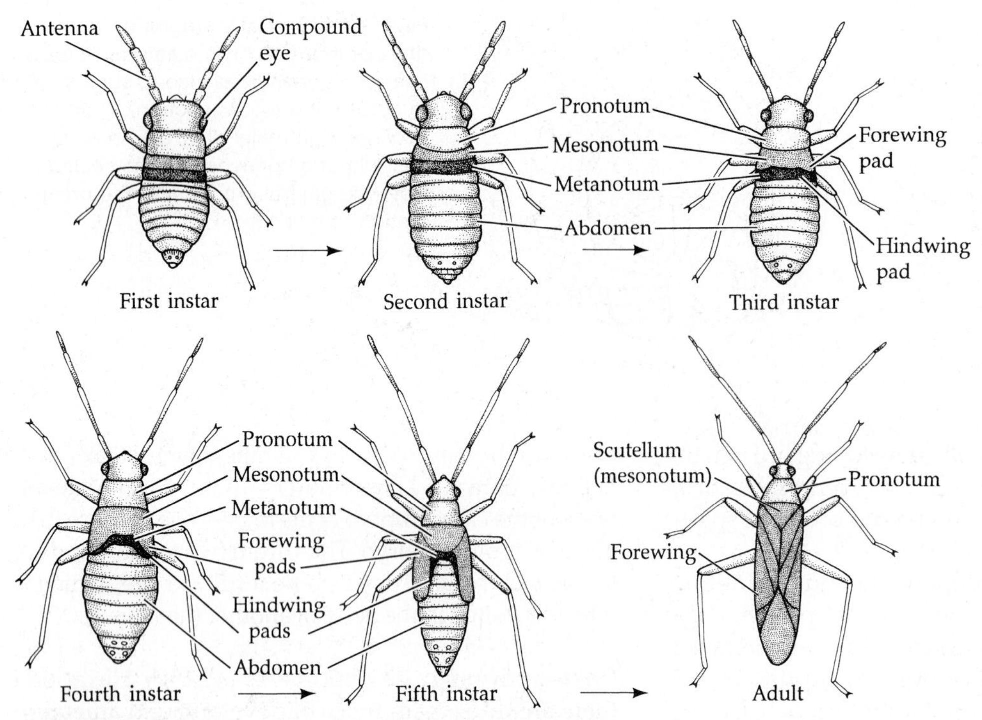

Figure 22.41 Illustrations and photo of the life stages of milkweed bugs (Hemiptera), illustrating the main life stages of hemimetabolous development. Note the wing pads in instars 3–5.

orders undergo ametabolous development. The young hatch out as juveniles closely resembling the adult, or imago, condition but the overall body size increases with each successive molt. Winged insects undergo either hemimetabolous development (Figures 22.40B and 22.41) or holometabolous development (Figures 22.40C and 22.42).

In hemimetabolous development, principal changes during growth are in body size and proportions and in the development of wings and sexual structures. The juveniles of hemimetabolous insects are called **nymphs** (terrestrial juveniles) or **naiads** (aquatic juveniles, such as mayflies, dragonflies, damselflies). Nymphs and adults often live in the same general habitat; naiads and their respective adults do not. Nymphs and naiads possess compound eyes, antennae, and feeding and walking appendages similar to those of the adults. However functional wings and sexual structures are always lacking, although juveniles have wing rudiments called wing pads or wing buds, and the wings themselves become exposed for the first time during the preadult molt.

Holometabolous insects hatch as vermiform larvae that bear no resemblance whatsoever to the adult forms. These larvae are so different from adults that they are often given separate vernacular names; for example, butterfly larvae are called caterpillars, fly larvae maggots, and beetle larvae grubs. Holometabolous larvae lack compound eyes (and often antennae), and their natural history differs markedly from that of adults. Their mouthparts may be wholly unlike those of adults, and external wing buds are never present. Often the greater part of a holometabolous insect's lifetime is spent in a series of larval instars. Larvae typically consume vast quantities of food and attain a larger size than adults. Termination of the larval stage is accompanied by **pupation**, during which (in a single molt) the pupal stage is entered (Figure 22.43). **Pupae** do not feed or move about very much. They often reside in protective niches in the ground, within plant tissues, or housed in a **cocoon**. Energy reserves stored during the long larval life are utilized by the pupa to undergo whole scale body transformation. Many larval tissues are broken down and reorganized to attain the adult form; external wings and sexual organs are formed. The remarkable transformation from larval

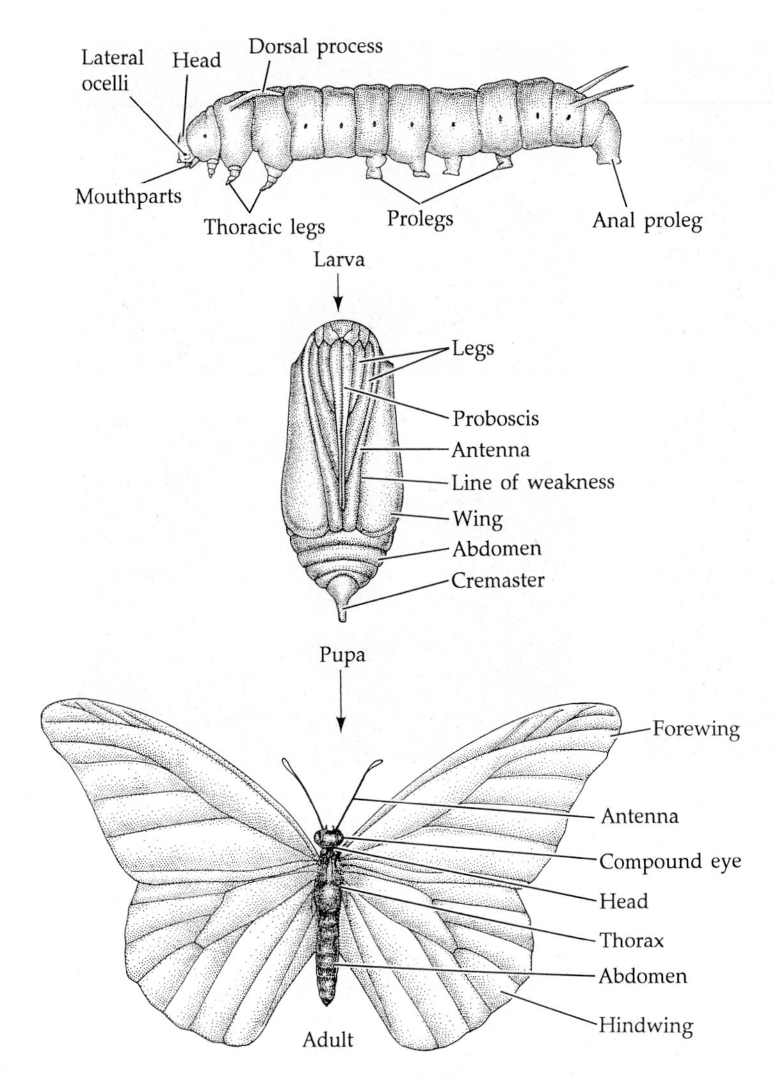

Figure 22.42 Life stages of a butterfly (Lepidoptera), illustrating the main life stages of holometabolous development.

stage to adult stage in holometabolous insects is one of the most impressive achievements of animal evolution (Figure 22.42), and it is on par with the transformation of crustaceans through a series of larval stages to adulthood (Figure 21.33).

The role of ecdysone in initiating molting is described in Chapter 20. This hormone works in conjunction with a second endocrine product in controlling the sequence of events in insect metamorphosis. This second product, juvenile hormone, is manufactured and released by the corpora allata, a pair of glandlike structures associated with the brain (Figure 22.33). When ecdysone initiates a molt in an early larval instar, the accompanying concentration of juvenile hormone in the hemolymph is high. A high concentration of juvenile hormone ensures a larva-to-larva molt. After the last larval instar is reached, the corpora allata ceases to secrete juvenile hormone. Low concentrations of juvenile hormone result in a larva-to-pupa molt. Finally, when the pupa is ready to molt, juvenile hormone is absent from the hemolymph altogether; this deficiency leads to a pupa-to-adult molt.

Hexapod Evolution

The hexapods were among the first animals to colonize and exploit terrestrial and freshwater ecosystems. The fossil record is good, with about 1,263 recognized families (by comparison, there are 825 recognized fossil families of tetrapod vertebrates). The oldest known fossil insects are from the Early Devonian, which has led to the hypothesis that hexapods originated in the late Silurian with the earliest terrestrial ecosystems. The remarkable diversification of insects is undoubtedly related to the evolution of wings, and insects are the only group of invertebrates with the ability to fly. Fossil winged insects exist from the Late Mississippian (~324Ma), which suggests a pre-Carboniferous origin of insect flight. However, the description of *Rhyniognatha* (~412Ma) from a mandible, potentially indicative of a winged insect, suggests a late Silurian to Early Devonian origin of winged insects. Divergence time estimates based on a molecular phylogenetic analysis of many nuclear, protein-coding genes (Misof et al. 2014) corroborate an origin of winged insect lineages during this time period, which implies that the ability to fly emerged after the establishment of complex terrestrial ecosystems. Since then, insects have shaped Earth's terrestrial ecosystems, coevolving with another hyperdiverse terrestrial group, the flowering plants, ultimately qualifying the Cenozoic to be called "the age of insects."

By the Carboniferous, various modern insect orders were flourishing, although many were quite unlike today's fauna. Some Carboniferous hexapods are notable for their gigantic size, such as silverfish (Thysanura) that reached 6 cm in length and dragonflies with wingspans of about 70 cm. In addition to the living orders of insects, at least ten other orders arose and radiated in late Paleozoic and early Mesozoic times, then went extinct.

The Permian saw an explosive radiation of holometabolous insects, although many groups went extinct in the great end-Permian extinction event (Chapter 1). In fact, relatively few groups of Paleozoic insects survived into the Mesozoic, and many recent families first appeared in the Jurassic. By the Cretaceous, most modern families were extant, insect sociality had evolved, and many insect families had begun their intimate relationships with angiosperms. Tertiary insects were essentially modern and included many genera indistinguishable from the Recent (Holocene) fauna.

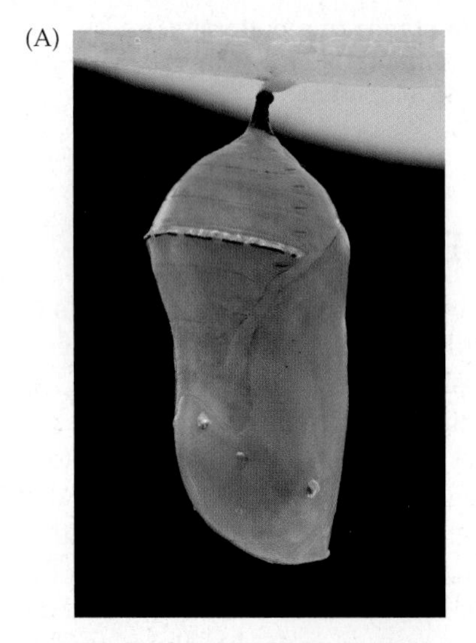

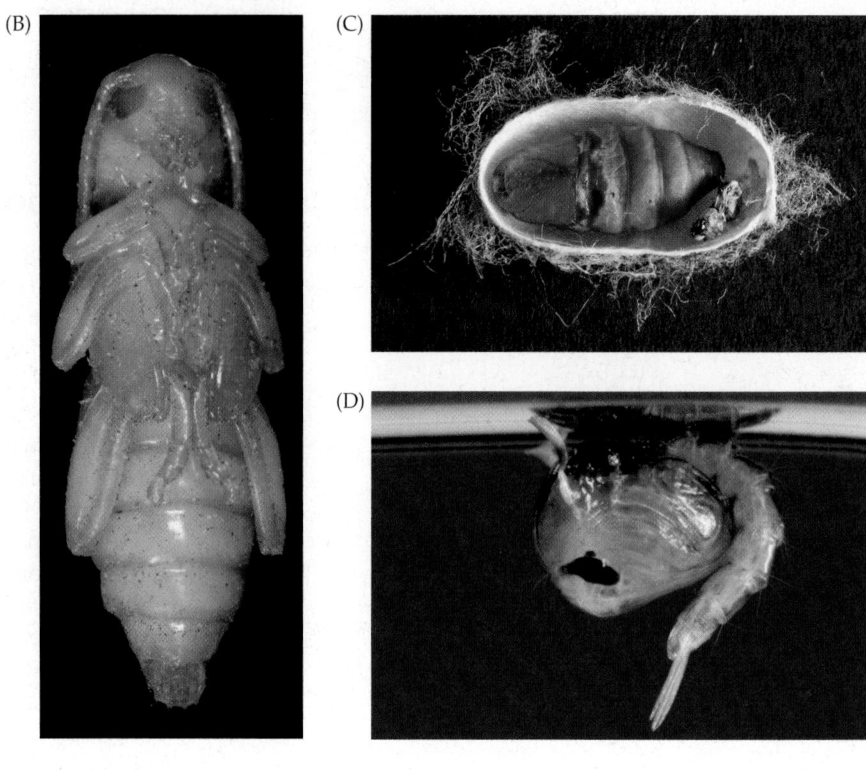

Figure 22.43 Four different types of insect pupae. (A) Chrysalis of a monarch butterfly (Lepidoptera: Nymphalidae: *Danaus plexippus*). (B) Pupa of a longhorn beetle (Coleoptera: Cerambycidae: *Xylotrechus nauticus*). (C) Cocoon of a silkworm moth (Lepidoptera: Saturniidae: *Bombyx mori*) sliced open to reveal the pupa inside. The shed skin (exuviae) of the final larval instar is visible to the right of the pupa inside the cocoon. (D) Pupa of a mosquito (Diptera: Culicidae).

Our current view of the evolutionary relationships of hexapod orders is presented in Figure 22.44. This tree is based on a recent publication (Misof et al. 2014) presenting a molecular phylogenetic analysis of 1,478 single-copy nuclear, protein-coding genes (see Chapter 2 for detailed tree with divergence time estimates.) This is by far the most data rich analysis of the hexapods conducted to date. The results of this analysis support many of the long held views of hexapod evolution but also provide novel support for some parts of the tree that had been difficult to resolve based on smaller molecular sets and by analyses of morphological characters alone.

The hexapods are divided between three entognathous orders and the Insecta. The three entognathous orders (the Collembola, Protura, and Diplura) all have internalized mouthparts. Most current workers regard the Collembola + Protura to be a monophyletic group (called the Ellipura), and this relationship is strongly supported by the most recent analyses. However, whether the Diplura are more closely related to the Collembola + Protura, or to the Insecta (a view persuasively argued by Kukalová-Peck) remains hotly debated and unresolved. Thus, the entognathous hexapods may or may not form a monophyletic group. Among the potential synapomorphies that may unite all three orders is entognathy itself (the overgrowth of the mouthparts by oral folds from the lateral cranial wall). In addition, the Malpighian tubules and compound eyes are reduced—compound eyes are degenerate in Collembola and absent in the extant Diplura and Protura. However, these reductions could be convergences resulting from small body size. Our evolutionary tree thus depicts an unresolved trichotomy at the base of the Hexapoda, and we treat the entognathous hexapods as a potentially paraphyletic group in our classification.

The monophyly of the Insecta (Archaeognatha, Thysanura, and Pterygota) (Figure 22.44B) is undisputed. The principal synapomorphies of this group include the structure of the antenna, with its lack of muscles beyond the first segment (scape); the presence of a group of special chordotonal organs (vibration sensors) in the second antennal segment (pedicel) called the Johnston's organ; a well developed posterior tentorium (forming a transverse bar); subdivision of the tarsus into tarsomeres; females with ovipositor formed by gonapophyses (limb-base endites) on segments 8 and 9; and long, annulated, posterior terminal filaments (cerci).

The Insecta have traditionally been divided into the wingless insects (Archaeognatha and Thysanura) and the winged insects (Pterygota). However, on the basis of molecular studies and the presence of a dicondylic mandible, the Thysanura (silverfish) are now thought to be the sister group of the Pterygota.

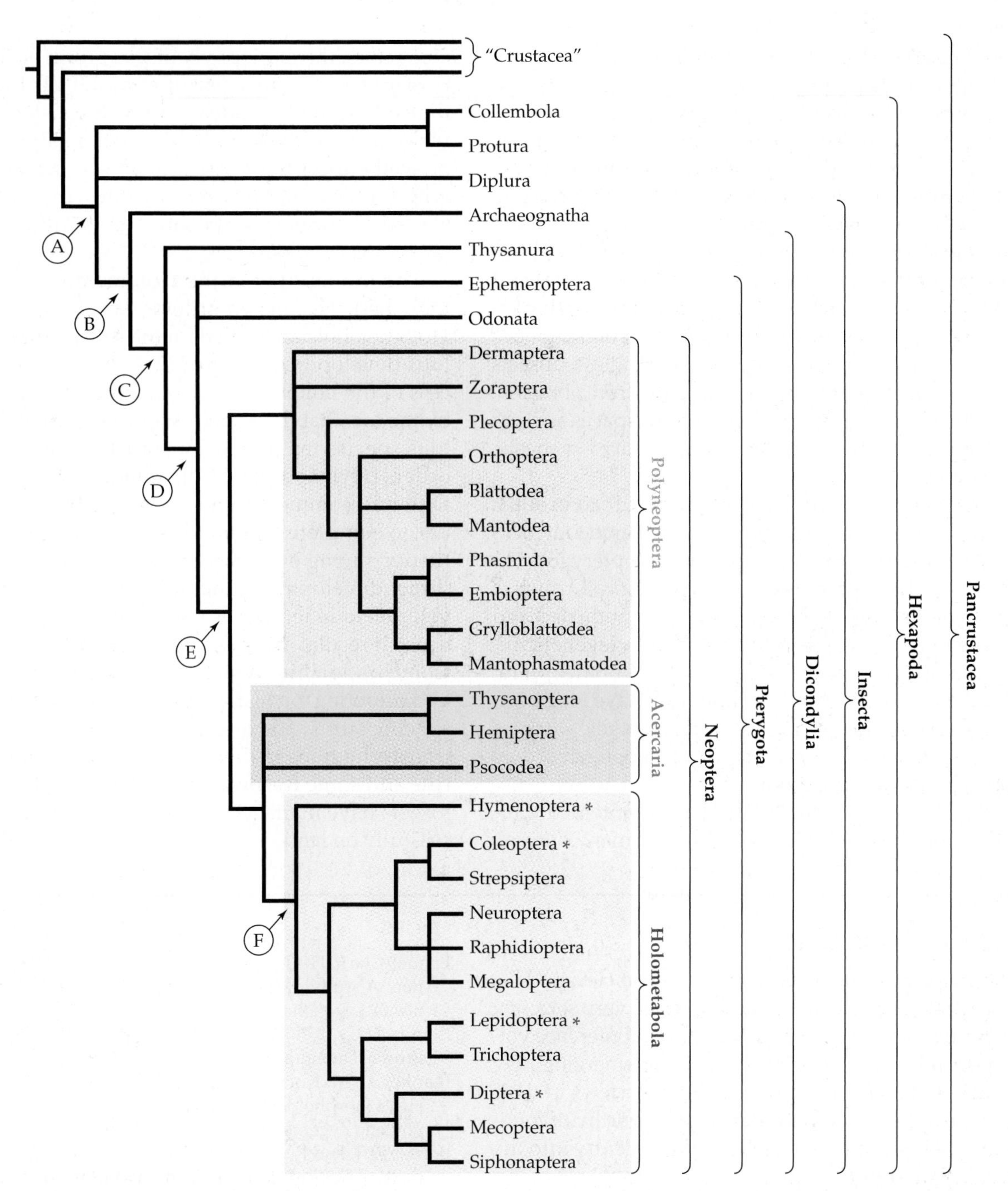

Figure 22.44 Phylogeny of the 31 orders of Hexapoda. Major clades are indicated to the right of the tree. The most recent common ancestor of extant taxa with traits that led to the success of the Hexapods are denoted with circled letters to the left of the tree. These key innovations include: (A) arthropod body subdivided into three tagmata: the head, 3-segmented thorax with one pair of uniramous legs on each segment, and an 11-segmented abdomen (synapomorphies of Hexapoda); (B) the evolution of external mouthparts (an apomorphy of the class Insecta); (C) the evolution of dicondylic mandibles (an apomorphy of the Dicondylia) (D) the evolution of wings (an apomorprhy of Pterygota); (E) the evolution of wing folding (an apomorphy of Neoptera); (F) the evolution of holometabolous development (= indirect development, or complete metamorphosis; an apomorphy of the Holometabola). The most diverse insect orders, commonly known as the "big four," are denoted with asterisks. Crustacea, a paraphyletic group due to its exclusion of the Hexapoda, is shown at the base of the tree.

They are classified together in a lineage known as the Dicondylia (see Figure 22.44C).

The Pterygota are, of course, distinguished by the presence of wings on the mesothorax and metathorax of adults. There is wide agreement that wings evolved only once in Hexapoda along the branch leading to the Pterygota (Figure 22.44D). However, the Pterygota comprise two groups, Palaeoptera and Neoptera, with fundamentally different wing types. The Palaeoptera ("ancient wings") include the Odonata (dragonflies and damselflies) and Ephemeroptera (mayflies), which are characterized by many-veined, netlike wings that cannot be folded over the back. It is still unclear whether or not the two palaeopterous orders, Ephemeroptera and Odonata, form a monophyletic group and they are therefore depicted in a trichotomy with Neoptera in our summary tree (Figure 22.44D). The Neoptera ("modern wings") have reduced wing venation, but more importantly, they can rotate their wing joint and fold back their wings when they are not flying (Figure 22.44E). This is one of the most important evolutionary innovations in hexapods. Wing folding allows insects to protect their fragile wings, especially from abrasion, thereby allowing them to live in tight spaces such as crevices under bark, under rocks, in burrows, nests, and tunnels.

The Neoptera is divided into three broad groups: Polyneoptera (10 orders); Acercaria, or Paraneoptera (3 orders); and the Holometabola, or Endopterygota (11 orders). Polyneopterans are a morphologically diverse group of insects with biting–chewing mouthparts and hemimetabolous development. The phylogenetic relationships among the orders within Polyneoptera, and the monophyly of the group itself have long been controversial. However, recent phylogenomic work recovered strong support that the polyneopteran orders form a monophyletic group.

The Acercaria includes the Thysanoptera, Hemiptera, and the Psocodea. Supported by many derived characters, the monophyly of this group is likely (but controversial), since Misof et al. (2014) found evidence that the Psocodea may be the sister group to Holometabola. Although this relationship appears to be well-supported by their analyses (see Figure 2.6) with high bootstrap support values, this measure of statistical support is of controversial utility for such large molecular data sets.

The monophyly of the Holometabola is well established. The eleven orders placed together in the Holometabola are united on the basis of holometabolous development (Figure 22.44F). The eminent success of the holometabolous lifestyle is demonstrated by the fact that their species outnumber hemimetabolous species ten to one. The most species rich insect orders (Hymenoptera, Coleoptera, Lepidoptera, and Diptera), commonly known as the "big four" all undergo complete metamorphosis. There is a popular theory among evolutionary biologists that views indirect development, including holometabolous development in insects, as selectively advantageous because it results in the ecological segregation of adults from young, thus avoiding intraspecific competition and allowing each stage to develop its own suite of specific survival strategies. We have seen that such transformations in development are common in marine and some freshwater invertebrates, but only the insects have managed to exploit this strategy so successfully on land.

Selected References

The amount of published information on Hexapoda is overwhelming. We have thus had to be very selective in our reference list, emphasizing texts, conference volumes, and journal articles that present emerging new fundamental knowledge, are solid overviews in their respective fields, or are older classics worthy of mention. These will provide readers with an entry into the primary literature.

General References

Arnett, R. H. 1985. *American Insects: A Handbook of the Insects of North America*. Van Nostrand Reinhold, New York.

Arnett, R. H., N. M. Downie and H. E. Jaques. 1980. *How to Know the Beetles*. Wm. C. Brown, Dubuque, IA.

Arnett, Jr., R. H., H. Ross and M. C. Thomas. 2001. *American Beetles, Vol. 1*. CRC Press, Boca Raton, FL.

Arnett, Jr. R. H., H. Ross, M. C. Thomas, P. E. Skelley and J. H. Frank. 2002. *American Beetles, Vol. 2*. CRC Press, Boca Raton, FL.

Barth, F. G. 1985. *Insects and Flowers: The Biology of a Partnership*. Translated by M. A. Bierderman-Thorson. Princeton University Press, Princeton, NJ.

Bate, M. and A. Martinez (eds.). 1993. *The Development of Drosophila melanogaster*. CSH Laboratory Press, New York.

Batra, S. W. T. and L. R. Batra. 1967. The fungus gardens of insects. Sci. Am. 217(5): 112–120.

Bennet-Clark, H. C. and E. C. A. Lucey. 1967. The jump of the flea: A study of the energetics and a model of the mechanisms. J. Exp. Biol. 47: 59–76.

Bland, R. G. 1978. *How to Know the Insects,* 3rd Ed.. Wm. C. Brown, Dubuque, IA.

Blanke, A., M. Koch, B. Wipfler, F. Wilde and B. Misof. 2014. Head morphology of *Tricholepidion gertschi* indicates monophyletic Zygentoma Front. Zool. 11: 16.

Borrer, D. J. and R. E. White. 1970. *A Field Guide to the Insects of America North of Mexico*. Houghton Mifflin, Boston.

Johnson, N. F. and C. A. Triplehorn. 2004. *Borror and Delong's Introduction to the Study of Insects,* 7th Ed. Saunders, New York. [A standard reference and identification guide for many generations of entomology students.]

Brown, M. 1994. Interactions between germ cells and somatic cells in *Drosophila melanogaster*. Semin. Dev. Biol. 5: 31–42.

Buchmann, S. L. and G. P. Nabhan. 1996. *The Forgotten Pollinators*. Island Press, Washington, DC.

Butler, C. G. 1967. Insect pheromones. Biol. Rev. 42: 42–87.

Carroll, S. B., S. D. Weatherbee and J. A. Langeland. 1995. Homeotic genes and the regulation of insect wing number. Nature 375: 58–61.

Chapela, I. H., S. R. Rehner, T. R. Schultz and U. G. Mueller. 1994. Evolutionary history of the symbiosis between fungus-growing ants and their fungi. Science 266: 1691–1697.

Chapman, R. F., S. J. Simpson, and A. E. Douglas. (eds.) 2013. *The Insects, Structure and Function,* 5th Ed. Cambridge University Press, Cambridge. [One of the best references on general insect anatomy and biology.]

Cheng, L. 1976. *Marine Insects*. North-Holland, Amsterdam/American Elsevier, New York.

Chu, H. F. 1949. *How to Know the Immature Insects*. Wm. C. Brown, Dubuque, IA. [Greatly in need of revision.]

Clements, A. N. 1992, 1999. *The Biology of Mosquitoes. Vols. 1 and 2*. Chapman and Hall, New York.

Cook, O. F. 1913. Web-spinning fly larvae in Guatemalan caves. J. Wash. Acad. Sci. 3(7): 190–193.

Crosland, M. W. J. and R. Crozier. 1986. *Myrmecia pilosula*, an ant with only one pair of chromosomes. Science 231: 1278–1284.

CSIRO. 1991. *The Insects of Australia: A Textbook for Students and Research Workers, Vols. 1 and 2*. 2nd Ed. Cornell University Press, Ithaca, NY. [An outstanding treatment of Australian biodiversity.]

Dethier, V. G. 1963. *The Physiology of Insect Senses*. Methuen, London.

Douglass, J. K. and N. J. Strausfeld. 1995. Visual motion detection circuits in flies: Peripheral motion computation by identified small field retinotopic neurons. J. Neurosci. 15: 5596–5611.

Douglass, J. K. and N. J. Strausfeld. 1996. Visual motion detection circuits in flies: Parallel direction- and non-direction-sensitive pathways between the medulla and lobula plate. J. Neurosci. 16: 4551–4562.

Doyle, J. A. 2012. Molecular and fossil evidence on the origin of Angiosperms. Annu. Rev. Earth Planet. Sci. 40: 301–326.

Dyer, F. C. and J. L. Gould. 1983. Honeybee navigation. Am. Sci. 71: 587–597.

Ellington, C. P. 1984. The aerodynamics of flapping animal flight. Am. Zool. 24: 95–105.

Engel, M. S. and D. A. Grimaldi. 2004. New light shed on the oldest insect. Nature 427: 627–630.

Erwin, T. L. 1982. Tropical forests: their richness in Coleoptera and other arthropod species. Coleopterists Bull. 36(1): 74–75.

Erwin, T. L. 1983. Tropical forest canopies, the last biotic frontier. Bull. Ent. Soc. Am. 29(1): 14–19.

Erwin, T. L. 1985. The taxon pulse: A general pattern of lineage radiation and extinction among carabid beetles. In G. E. Ball (ed.), *Taxonomy, Phylogeny and Zoogeography of Beetles and Ants: A Volume Dedicated to the Memory of Philip Jackson Darlington, Jr., 1904–1983*. W. Junk, Publ., The Hague.

Erwin, T. L. 1991. An evolutionary basis for conservation strategies. Science 253: 750–752.

Erwin, T. L. 1991. How many species are there? Revisited. Cons. Biol. 5(3): 330–333.

Evans, H. E. 1984. *Insect Biology: A Textbook of Entomology*. Addison-Wesley, Reading, MA.

Evans, H. E. 1993. *Life on a Little-Known Planet. A Biologist's View of Insects and Their World*. Lyons Press/Rowman & Littlefield, Lanham, MD.

Fent, K. and R. Wehner. 1985. Ocelli: A celestial compass in the desert ant *Cataglyphis*. Science 228: 192–194.

Fletcher, D. J. C. and M. Blum. 1983. Regulation of queen number by workers in colonies of social insects. Science 219: 312–314.

Glassberg, J. 2000. *Butterflies through Binoculars: The West*. Oxford University Press, Oxford. [There are several companion volumes, by the same publisher, for the East Coast, Africa, India, etc.]

Gotwald, W. H., Jr. 1996. *Army Ants: The Biology of Social Predation*. Comstock Books, Ithaca, NY.

Gould, J. L. 1976. The dance-language controversy. Q. Rev. Biol. 51: 211–243.

Gould, J. L. 1985. How bees remember flower shapes. Science 227: 1492–1494.

Gould, J. L. 1986. The locale map of honeybees: Do insects have cognitive maps? Science 232: 861–863.

Gwynne, D. T. 2001. *Katydids and Bush-Crickets: Reproductive Behavior and Evolution of the Tettigonidae*. Comstock Publishing Associates, Ithaca, NY.

Hardie, R. J. (ed.) 1999. *Pheromones of Non-Lepidopteran Insects Associated with Agricultural Plants*. Oxford University Press, Oxford.

Heinrich, B. 1993. *The Hot-Blooded Insects: Strategies and Mechanisms of Thermoregulation*. Harvard University Press, Cambridge, MA.

Hermann, H. R. (ed.). 1984. *Defense Mechanisms in Social Insects*. Praeger, New York.

Herreid, C. F. II and C. R. Fourtner. 1981. *Locomotion and Energetics in Arthropods*. Plenum, New York.

Hinton, H. E. 1981. *Biology of Insect Eggs, Vols. 1–3*. Pergamon Press, Elmsford, NY. [A benchmark study of insect egg morphology and biology; no embryology.]

Hodgson, C. J. 1994. *The Scale Insect Family Coccidae*. Oxford University Press, Oxford.

Hölldobler, B. 1971. Communication between ants and their guests. Sci. Am. 224(3): 86–93.

Hölldobler, B. and E. O. Wilson. 1990. *The Ants*. Belknap Press, Cambridge, MA.

Holt, V. M. 1973. *Why Not Eat Insects?* Reprinted from the original (1885) by E. W. Classey Ltd., 353 Hanworth Rd., Hampton, Middlesex, England. [Ninety-nine pages of fun and recipes.]

Huffaker, C. B. and R. L. Rabb (eds.). 1984. *Ecological Entomology*. Wiley-Interscience, New York.

Kerkut, G. A. and L. I. Gilbert. 1985. *Comprehensive Insect Physiology, Biochemistry and Pharmacology*. Pergamon Press, Elmsford, NY. [In 13 volumes.]

Klass, K.-D., O. Zompro, N. P. Kristensen and J. Adis. 2002. Mantophasmatodea: A new insect order with extant members in the Afrotropics. Science 296: 1456–1459.

Labandeira C. C. 2006. Silurian to Triassic plant and hexapod clades and their associations: New data, a review, and interpretations. Arthro Syst. Phylo. 64: 53–94.

Labandeira,C. C., S. L. Tremblay, K. E. Bartowski and L. VanAller Hernick. 2014. Middle Devonian liverwort herbivory and antiherbivore defence. New Phytol. 202: 247–258.

Lehmkuhl, D. M. 1979. *How to Know the Aquatic Insects*. Wm. C. Brown, Dubuque, IA.

Lewis, T. (ed.). 1984. *Insect Communication*. Academic Press, Orlando, FL. [A definitive review as of the early 1980s.]

Matheson, A., S. L. Buchmann, C. O'Toole, P. Westrich and I. H. Williams (eds.) 1996. *The Conservation of Bees*. Academic Press, Harcourt Brace, London.

Matsuda, R. 1965. *Morphology and evolution of the insect head. Memoirs of the American Entomological Institute, No. 4*. American Entomological Institute, Ann Arbor, MI.

Matsuda, R. 1970. *Morphology and evolution of the insect thorax.* Memoirs of the Entomological Society of Canada, No. 76. Entomological Society of Canada, Ottawa.

Matsuda, R. 1976. *Morphology and Evolution of the Insect Abdomen*. Pergamon Press, Oxford.

Merlin, C., R. J. Gegear and S. M. Reppert. 2009. Antennal circadian clocks coordinate sun compass orientation in migratory monarch butterflies. Science 325: 1700–1704.

Merritt, R. W., K. W. Cummins, and M. B. Berg (eds.). 2008. *An Introduction to the Aquatic Insects of North America*, 4th Ed. Kendall/Hunt, Dubuque, IA. [Excellent keys to the North American groups.]

Michelsen, A. 1979. Insect ears as mechanical systems. Am. Sci. 67: 696–706.

Michener, C. D. 1974. *The Social Behavior of the Bees: A Comparative Study*. Belknap Press/Harvard University Press, Cambridge, MA.

Nieh, J. C. 1999. Stingless-bee communication. Am. Sci. 87: 428–435.

Pearce, M. J. 1998. *Termites: Biology and Pest Management*. Oxford University Press, Oxford.

Perez, S. M., O. R. Taylor and R. Jander. 1997. A sun compass in monarch butterflies. Nature 387: 29.

Phelan, P. L. and T. C. Baker. 1987. Evolution of male pheromones in moths: Reproductive isolation through sexual selection? Science 235: 205–207.

Prestwick, G. D. 1987. Chemistry of pheromone and hormone metabolism in insects. Science 238: 999–1006.

Price, P. W., T. M. Lewinsohn, G. Wilson Fernandes and W. W. Benson (eds.) 1991. *Plant–Animal Interactions. Evolutionary Ecology in Temperate and Tropical Regions*. Wiley-Interscience, New York.

Prokop, J., A. Nel and I. Hoch. 2005. Discovery of the oldest known Pterygota in the lower Carboniferous of the Upper Silesian Basin in the Czech Republic (Insecta: Archaeorthoptera). Geobios 38: 383–387.

Robinson, G. E. 1985. The dance language of the honeybee: The controversy and its resolution. Am. Bee J. 126: 184–189.

Rockstein, M. (ed.). 1964–65. *The Physiology of Insecta, Vols. 1–3*. Academic Press, New York.

Rosin, R. 1984. Further analysis of the honeybee "Dance Language" controversy. I. Presumed proofs for the "Dance Language" hypothesis by Soviet scientists. J. Theor. Biol. 107: 417–442.

Rosin, R. 1988. Do honeybees still have a "dance language"? Am. Bee J. 128: 267–268.

Ross, K. G. and R. W. Matthews (eds.). 1991. *The Social Biology of Wasps*. Comstock Books, Ithaca, NY.

Rota-Stabelli, O., A.C. Daley and D. Pisani. 2013. Molecular timetrees reveal a Cambrian colonization of land and a new scenario for ecdysozoan evolution Curr. Biol. 23: 392–398.

Roth, S., J.Molina and R. Predel. Biodiversity, ecology, and behavior of the recently discovered insect order Mantophasmatodea, Front. Zool. 11:70.

Saunders, D. S. 1982. *Insect Clocks,* 2nd Ed. Pergamon Press, Elmsford, NY. [Good introduction to photoperiodism.]

Schmitt, J. B. 1962. The comparative anatomy of the insect nervous system. Annu. Rev. Entomol. 7: 137–156.

Schuh, R. T. and J. A. Slater. 1995. *True Bugs of the World (Hemiptera: Heteroptera)*. Comstock Books, Ithaca, NY.

Schwalm, F. E. 1988. *Insect Morphogenesis. Monographs in Developmental Biology, Vol. 20*. Karger, Basel.

Schwan, F. E. 1997. Arthropods: The insects. Pp. 259–278 in S. F. Gilbert and A. M. Raunio, *Embryology: Constructing the Organism*. Sinauer Associates, Sunderland, MA.

Scott, J. A. 1986. *The Butterflies of North America: A Natural History and Field Guide*. Stanford University Press, Stanford, CA.

Snodgrass, R. E. 1935. *Principles of Insect Morphology*. McGraw-Hill, New York. [An early classic; still useful.]

Snodgrass, R. E. 1944. *The feeding apparatus of biting and sucking insects affecting man and animals. Smithsonian Miscellaneous Collections, Vol. 104, No. 7*. Smithsonian Institution, Washington, DC.

Snodgrass, R. E. 1952. *A Textbook of Arthropod Anatomy*. Cornell University Press, Ithaca, NY. [Generations of subsequent books and reports have relied heavily on the information and figures contained in this work and Snodgrass's 1935 text.]

Snodgrass, R. E. 1960. *Facts and theories concerning the insect head. Smithsonian Miscellaneous Collections, Vol. 152, No. 1*. Smithsonian Institution, Washington, DC.

Somps, C. and M. Luttges. 1985. Dragonfly flight: Novel uses of unsteady separated flows. Science 228: 1326–1329.

Strausfeld, N. J. 1976. *Atlas of an Insect Brain*. Springer, Heidelberg.

Strausfeld, N. J. 1996. Oculomotor control in flies: From muscles to elementary motion detectors. Pp. 277–284 in P. S. G. Stein and D. Stuart (eds.), *Neurons, Networks, and Motor Behavior*. Oxford University Press, Oxford.

Strausfeld, N. J. 2012. *Arthropod Brains. Evolution, Functional Elegance, and Historical Significance*. Belknap Press, Cambridge, MA.

Strausfeld, N. J. and J.-K. Lee. 1991. Neuronal basis for parallel visual processing in the fly. Visual Neuroscience 7: 13–33.

Stubs, C. and F. Drummond (eds.). 2001. *Bees and Crop Pollination: Crisis, Crossroads, Conservation*. Entomological Society of America, Annapolis.

Tauber, M. J., C. A. Tauber and S. Masaki. 1986. *Seasonal Adaptations of Insects*. Oxford University Press, New York. [A comprehensive treatment of insect life cycles.]

Tilgner, E. 2002. Mantophasmatodea: A new insect order? Science 297: 731a.

Treherne, J. E. and J. W. L. Beament (eds.). 1965. *The Physiology of the Insect Central Nervous System*. Academic Press, New York.

Treherne, J. E., M. J. Berridge and V. B. Wigglesworth. 1963–1985. *Advances in Insect Physiology, Vols. 1–18*. Academic Press, New York.

Unarov, B. P. 1966. *Grasshoppers and Locusts*. Cambridge University Press, Cambridge.

Usinger, R. L. (ed.). 1968. *Aquatic Insects of California, with Keys to North American Genera and California Species*. University of California Press, Berkeley, CA.

Vane-Wright, R. I. and P. R. Ackery. 1989. *The Biology of Butterflies*. University of Chicago Press, Chicago, IL.

Veldink, C. 1989. The honey-bee language controversy. Interdiscip. Sci. Rev. 14(2): 170–175.

von Frisch, K. 1967. *The Dance Language and Orientation of Bees*. Translated by Leigh E. Chadwick. Belknap Press, Cambridge, MA.

Wenner, A. M. and P. H. Wells. 1987. The honeybee dance language controversy: The search for "truth" vs. the search for useful information. Am. Bee J. 127: 130–131.

Wigglesworth, V. B. 1954. *The Physiology of Insect Metamorphosis*. Cambridge University Press, Cambridge. [A long-standing classic; dated but still useful.]

Wigglesworth, V. B. 1984. *Insect Physiology,* 8th Ed. Chapman and Hall, London.

Williams, C. B. 1958. *Insect Migration*. Collins, London.

Wilson, E. O. 1971. *The Insect Societies*. Harvard University Press, Cambridge, MA.

Wilson, E. O. 1975. Slavery in ants. Sci. Am. 232(6): 32–36.

Winston, M. L. 1987. *The Biology of the Honeybee*. Harvard University Press, Cambridge, MA.

Winston, M. L. 1992. *Killer Bees*. Harvard University Press, Cambridge, MA.

Hexapod Evolution

Andersson, M. 1984. The evolution of eusociality. Annu. Rev. Ecol. Syst. 15: 165–189.

Beutel, R.G. and 9 others. 2011. Morphological and molecular evidence converge upon a robust phylogeny of the megadiverse Holometabola. Cladistics 27: 341–355.

Bitsch, C. and J. Bitsch. 2000. The phylogenetic interrelationships of the higher taxa of apterygote hexapods. Zool. Scripta 29: 131–156.

Dell'Ampio, E. and 15 more. 2014. Decisive data sets in phylogenomics: lessons from studies on the phylogenetic relationships of primarily wingless insects. Mol. Biol. Evol. 31: 239–249.

Deuve, T. (ed.) 2001. Origin of the Hexapoda. Ann. Soc. Entomol. France 37 (1/2): 1–304. [Papers presented at Conference held in Paris, 1999; see review in Brusca, R. C., 2001, J. Crustacean Biol. 21(4): 1084–1086.]

Diaz-Benjumea, F. J., B. Cohen and S. M. Cohen. 1994. Cell interaction between compartments established the proximal–distal axis of *Drosophila* legs. Nature 372: 175–179.

Douglas, M. M. 1980. Thermoregulatory significance of thoracic lobes in the evolution of insect wings. Science 211: 84–86.

Futuyma, D. J. and M. Slatkin (eds.). 1983. *Coevolution*. Sinauer Associates, Sunderland, MA.

Gaunt, M. W. and M. A. Miles. 2002. An insect molecular clock dates the origin of the insects and accords with paleontological and biogeographic landmarks. Mol. Biol. Evol. 19 (5): 748–761.

Goodchild, A. J. P. 1966. Evolution of the alimentary canal in the Hemiptera. Biol. Rev. 41: 97–140.

Hennig, W. 1981. *Insect Phylogeny*. Wiley, New York.

Hoy, R. R., A. Hoikkala and K. Kaneshiro. 1988. Hawaiian courtship songs: Evolutionary innovation in communication signals of *Drosophila*. Science 240: 217–220.

Ishiwata, K., G. Sasaki, J. Ogawa, T. Miyata and Z.-H. Su. 2011. Molecular phylogenetic analyses support the monophyly of Hexapoda and suggest the paraphyly of Entognatha. Mol. Phylogenet. Evol. 58: 169–180.

Kukalová-Peck, J. 1983. Origin of the insect wing and wing articulation from the insect leg. Can. J. Zool. 61: 1618–1669.

Kukalová-Peck, J. 1987. New Carboniferous Diplura, Monura, and Thysanura, the hexapod ground plan, and the role of thoracic side lobes in the origin of wings (Insecta). Can. J. Zool. 65: 2327–2345.

Kukalová-Peck, J. 1991. The "Uniramia" do not exist: The ground plan of the Pterygota as revealed by Permian Diaphanopterodea from Russia (Insecta: Paleodictyopteroidea). Can. J. Zool. 70: 236–255.

Kukalová-Peck, J. and C. Brauckmann. 1990. Wing folding in pterygote insects, and the oldest Diaphanopteroda from the early Late Carboniferous of West Germany. Can. J. Zool. 68: 1104–1111.

Labandeira, C., B. Beall and F. Hueber. 1988. Early insect diversification: Evidence from a lower Devonian bristletail from Quebec. Science 242: 913–916.

Letsch, H. and S. Simon. 2013. Insect phylogenomics: new insights on the relationships of lower neopteran orders (Polyneoptera). Syst. Entomol. 38: 783–793.

Li, H. and 6 others. 2015. Higher-level phylogeny of paraneopteran insects inferred from mitochondrial genome sequences. Scientific Reports 5 (8527). doi:10.1038/srep08527

Misof, B. and 100 others. 2014. Phylogenomics resolves the timing and pattern of insect evolution. Science 346: 763–767.

Nilsson, D.-E. and D. Osorio. 1997. Homology and parallelism in arthropod sensory processing. Pp. 333–347 in R. A. Fortey and R. H. Thomas, *Arthropod Relationships*. Chapman and Hall, London.

Osorio, D. and J. P. Bacon. 1994. A good eye for arthropod evolution. BioEssays 16: 419–424.

Panganiban, G., A. Sebring, L. Nagy and S. Carroll. 1995. The development of crustacean limbs and the evolution of arthropods. Science 270: 1363–1366.

Panganiban, G. et al. 1997. The origin and evolution of animal appendages. Proc. Natl. Acad. Sci. U.S.A. 94: 5162–5166.

Regier, J. C. and J. W. Shultz. 1997. Molecular phylogeny of the major arthropod groups indicates polyphyly of crustaceans and a new hypothesis of the origin of hexapods. Mol. Biol. Evol. 14: 902–913.

Robertson, R. M., K. G. Pearson and H. Reichert. 1981. Flight interneurons in the locust and the origin of insect wings. Science 217: 177–179.

Sander, K. 1994. The evolution of insect patterning mechanisms: A survey. Development (Suppl.) 1994: 187–191.

Strausfeld, N. J. 1998. Crustacean–insect relationships: The use of brain characters to derive phylogeny amongst segmented invertebrates. Brain Behav. Evol. 52: 186–206.

Strausfeld, N. J., E. K. Bushbeck and R. S. Gomez. 1995. The arthropod mushroom body: Its roles, evolutionary enigmas and mistaken identities. Pp. 349–381 in O. Breidbach and W. Kutsch (eds.), *The Nervous Systems of Invertebrates: An Evolutionary and Comparative Approach*. Birkhäuser Verlag, Basel.

Tautz, D., M. Friedrich and R. Schröder. 1994. Insect embryogenesis: What is ancestral and what is derived. Development (Suppl.). 1994: 193–199.

Therianos, S., S. Leuzinger, F. Hirth, C. S. Goodman and H. Reichert. 1995. Embryonic development of the *Drosophila* brain: Formation of commissural and descending pathways. Development 121: 3849–3860.

Thomas, J. B., M. J. Bastiani and C. S. Goodman. 1984. From grasshopper to *Drosophila*: A common plan for neuronal development. Nature 310: 203–207.

Thornhill, R. and J. Alcock. 1983. *The Evolution of Insect Mating Systems*. Harvard University Press, Cambridge, MA.

Whiting, M. F., J. C. Carpenter, Q. D. Wheeler and W. C. Wheeler. 1997. The Strepsiptera problem: Phylogeny of the holometabolous insect orders inferred from 18S and 28S ribosomal DNA sequences and morphology. Syst. Biol. 46(1): 1–68.

Wilson, E. O. 1985. The sociogenesis of insect colonies. Science 228: 1489–1425.

Wootton, R. J., J. Kukalová-Peck, D. J. S. Newman and J. Muzón. 1998. Smart engineering in the mid-Carboniferous: How well could Palaeozoic dragonflies fly? Science 282: 749–751.

Yoshizawa, K. 2011. Monophyletic Polyneoptera recovered by wing base structure. Syst. Entomol. 36: 377–394.

CHAPTER 23

Phylum Arthropoda

The Myriapods: Centipedes, Millipedes, and Their Kin

The arthropod subphylum Myriapoda includes four groups traditionally ranked as classes: Chilopoda (centipedes), Diplopoda (millipedes), Pauropoda (pauropods), and Symphyla (symphylans) (see classification on p. 897). All modern myriapods are terrestrial, but the lineage probably began its evolution in the aquatic realm. The first fossil records of millipedes are from the late Ordovician or early Silurian, and some of these are thought to represent marine species. Fossil evidence suggests that myriapods (millipedes) did not make their first appearance on land until the mid-Silurian. Figure 23.1 illustrates a variety of myriapod types. More than 16,000 living species have been described so far.

Centipedes and millipedes are familiar arthropods, easily distinguished from all other terrestrial invertebrates by a body divided into just two tagmata, the cephalon and the long, homonomous, and many-segmented trunk arranged with many pairs of articulated legs. As in the Hexapoda, the head is equipped with one pair of antennae, and the mouthparts are represented by the mandibles and the first maxillae; a second pair of maxillae is present in the centipedes and symphylans, but in millipedes and pauropods the corresponding segment bears no appendages.

Millipedes are slow-moving detritus feeders that generally spend their time burrowing through soil and litter, consuming plant remains and converting vegetable matter into humus. In tropical environments where earthworms are often scarce, millipedes may be the major soil-forming animals. Most of their trunk segmental units are actually **diplosegments** bearing two pairs of legs each (Figure 23.2F). When threatened, many millipedes roll up into a flat coil, and some can roll up into perfect balls (like many isopods). Millipedes typically have many trunk segments, but some have only 9 to 12. Despite their vernacular name (*millipedes*), which has equivalents in many languages, no millipede has a thousand legs; the record holder

Classification of The Animal Kingdom (Metazoa)

Non-Bilateria*
(a.k.a. the diploblasts)
- PHYLUM PORIFERA
- PHYLUM PLACOZOA
- PHYLUM CNIDARIA
- PHYLUM CTENOPHORA

Bilateria
(a.k.a. the triploblasts)
- PHYLUM XENACOELOMORPHA

Protostomia
- PHYLUM CHAETOGNATHA

SPIRALIA
- PHYLUM PLATYHELMINTHES
- PHYLUM GASTROTRICHA
- PHYLUM RHOMBOZOA
- PHYLUM ORTHONECTIDA
- PHYLUM NEMERTEA
- PHYLUM MOLLUSCA
- PHYLUM ANNELIDA
- PHYLUM ENTOPROCTA
- PHYLUM CYCLIOPHORA

Gnathifera
- PHYLUM GNATHOSTOMULIDA
- PHYLUM MICROGNATHOZOA
- PHYLUM ROTIFERA

Lophophorata
- PHYLUM PHORONIDA
- PHYLUM BRYOZOA
- PHYLUM BRACHIOPODA

ECDYSOZOA

Nematoida
- PHYLUM NEMATODA
- PHYLUM NEMATOMORPHA

Scalidophora
- PHYLUM KINORHYNCHA
- PHYLUM PRIAPULA
- PHYLUM LORICIFERA

Panarthropoda
- PHYLUM TARDIGRADA
- PHYLUM ONYCHOPHORA
- **PHYLUM ARTHROPODA**
  - SUBPHYLUM CRUSTACEA*
  - SUBPHYLUM HEXAPODA
  - **SUBPHYLUM MYRIAPODA**
  - SUBPHYLUM CHELICERATA

Deuterostomia
- PHYLUM ECHINODERMATA
- PHYLUM HEMICHORDATA
- PHYLUM CHORDATA

*Paraphyletic group

This chapter has been revised by Alessandro Minelli.

Figure 23.1 Representative myriapods. (A) The European centipede *Himantarium gabrielis*, whose leg count may reach 179 pairs. (B) *Scutigera coleoptrata*, the common house centipede. (C) The Southwestern desert centipede *Scolopendra heros*. (D) A millipede from East Africa. (E) The Southwestern desert millipede *Orthoporus ornatus*. (F) An unidentified pauropod from Tasmania (Australia). (G) An unidentified symphylan from Tasmania (Australia). (H) A bioluminescent millipede (*Motyxia*) from California.

(*Illacme plenipes*, a California species) reaches the impressive number of 375 pairs of legs.

Millipedes are incapable of strong biting and at first sight seem only to rely on their calcified cuticle and their rolling ability for defense, but the majority of these arthropods possess repugnatorial glands with lateral openings (**ozopores**) that secrete volatile toxic liquids (Figure 23.2F). Their chemical weaponry is surprisingly diverse: it includes the benzoquinones, hydroquinones, phenol, and acetates of long-chain carboxylic acids produced by juloids; and the benzoic acid, benzaldehyde and hydrogen cyanide produced by polydesmoids. A few tropical species have toxins powerful enough to cause blisters on human skin. Pill millipedes, such as the common European species of the genus *Glomeris*, lack lateral ozopores, but from along their dorsal midline they squeeze out drops of fluid containing quinazolinones, which belong to the same class of substances as the synthetic drug Quaalude (methaqualone), a powerful sedative. One common household species in North America (introduced long ago from Asia), the polydesmidan *Oxidus gracilis*, releases strong-smelling defensive chemicals when injured. Members of the subclass Penicillata, which are small, soft bodied, and lack repugnatorial glands, have evolved a mechanical defense strategy of throwing stiff bristles from their posterior end at ants and other predators. Given their range of defensive tactics, it is not surprising that many millipedes have aposematic (warning) coloration, usually bright reds, yellows, and oranges, and that some soil-dwelling vertebrates (lizards and worm snakes) have evolved look-alike coloration (Batesian mimicry).

Millipedes in the family Xystodesmidae (order Polydesmida) produce hydrogen cyanide as a defensive chemical, and use aposematic coloration to warn predators of their toxicity. Coloration patterns in these species include bright yellows, oranges, reds, and violet. One nocturnal genus in this family, *Motyxia*, known only from California, does not display conspicuous coloration. Remarkably, however, they produce a novel green bioluminescent glow at a dominant wavelength of 495 nm by way of a biochemical source of light produced in their exoskeleton. The source is a photoprotein with a chromophore containing a porphyrin as its functional group. This is the only known instance of light production via a photoprotein on land, all other examples being marine. The structure of the millipede's photoprotein is unknown and its homology to molecules of other animals is uncertain. Scientists have speculated that the emitted light could be a sexual signal to attract mates, or an aposematic warning glow to announce the presence of a cyanide-based chemical defense. So far as is known, millipedes in the order Polydesmida all are blind, suggesting that the predator-warning hypothesis might be more likely. Other than the genus *Motyxia*, there are no other confirmed accounts of bioluminescence among millipedes.

In contrast to millipedes, most centipedes are fast-running predators, with one pair of legs per segment, and with poison claws at the tip of the first legs, thus turned into a peculiar kind of maxillipeds called **forcipules** or **prehensors**. Centipedes never roll into a coil when threatened; instead, they usually strike out with the poison claws with which they can inflict a painful bite. A few *Scolopendra* species have warning coloration—black and red banding (Figure 23.1C). Some lithobiomorph centipedes bear large numbers of unicellular repugnatorial glands on the last four pairs of legs, which they kick in the direction of an enemy—a behavior that throws out sticky droplets of noxious secretions. Despite their vernacular name (*centi–pedes*), these creatures rarely have as many as a hundred legs and only in the order Geophilomorpha is this attained (some species have up to 191 pairs of legs).

CLASSIFICATION OF THE SUBPHYLUM MYRIAPODA

CLASS CHILOPODA (CENTIPEDES)

Subclass Notostigmophora

Order Scutigeromorpha

Subclass Pleurostigmophora

Order Craterostigmomorpha
Order Geophilomorpha
Order Lithobiomorpha
Order Scolopendromorpha

CLASS DIPLOPODA (MILLIPEDES)

Subclass Penicillata

Order Polyxenida

Subclass Chilognatha

INFRACLASS PENTAZONIA

Order Glomeridesmida
Order Glomerida
Order Sphaerotheriida

INFRACLASS HELMINTHOMORPHA

Superorder Colobognatha

Order Platydesmida
Order Polyzoniida
Order Siphonocryptida
Order Siphonophorida

Superorder Eugnatha

Order Julida
Order Spirobolida
Order Spirostreptida
Order Callipodida
Order Chordeumatida
Order Stemmiulida
Order Siphoniulida
Order Polydesmida

CLASS PAUROPODA (PAUROPODS)

CLASS SYMPHYLA (SYMPHYLANS)

Myriapod Classification

Subphylum Myriapoda

Terrestrial mandibulate and tracheate arthropods provided with one pair of antennae; mandibles with articulating endite; first maxillae free or fused together; second maxillae partly or wholly fused, or absent; postcephalic trunk of numerous, mostly similar segments (or diplosegments), specialized segments (or diplosegments) being confined to either end of the series or those bearing sexual openings or sexual appendages; appendages of head and trunk uniramous; most lineages with lateral eyes, perhaps representing the product of disintegrated compound eyes (compound eyes are largely retained only in scutigeromorph centipedes); exoskeleton without well-developed wax layer (except in some desert species); without endodermally derived digestive ceca; ectodermally derived (proctodeal) Malpighian tubules assist in excretion; often with organs of Tömösváry (see Nervous Systems and Sense Organs further in this chapter); copulation often indirect (see p. 907); development direct, but often with postembryonic increase in the number of segments and appendages. About 16,350 described, living species.

Class Chilopoda

With numerous freely articulating trunk segments, each with one pair of legs, the first pair modified as large poison claws (called prehensors or forcipules) held under the head like mouthparts; antennae simple, of fixed (14 in the Geophilomorpha, 17 in most Scolopendromorpha) or variable (in excess of 100 in some cave-dwelling Lithobiomorpha) number of articles (segments); both pairs of maxillae may be medially coalesced; cuticle stiff but uncalcified; gonopores on last true body segment; spiracles mostly lateral; with 15 to 191 leg-bearing body segments (number always odd in the adults); legs long, extending laterally such that the body is carried close to the ground; last pair of legs often sexually dimorphic, extending backwards and not used for locomotion.

Centipedes of the subclass Notostigmophora have middorsal spiracles and very short tracheae that end on the dorsal vessel; hemolymph containing hemocyanin; dome-shaped head, compound eyes; and 15 pairs of long, thin legs. The subclass Pleurostigmophora includes centipedes with lateral spiracles, a flattened head, a variable number of legs, and no respiratory pigments. The largest centipedes belong to the order Scolopendromorpha (either 21 or 23 pairs of legs; 39 or 43 only in one species, *Scolopendropsis duplicata*), although the highest leg count occurs in the long, thin geophilomorphs (27 to 191 pairs of legs). The globally-distributed order Scolopendromorpha contains the species most people readily recognize as centipedes. North America is the only inhabited continent that has no native species of scutigeromorph centipedes, although the long-legged, swift-moving European species, *Scutigera coleoptrata*, has become widely introduced and is often encountered in cool moist areas of houses. Approximately 3,300 living species of chilopods have been described.

Class Diplopoda

Trunk segments fused by pairs into 9 to 192 leg-bearing diplosegments; most of these with two pairs of legs, spiracles, ganglia, and heart ostia; each diplosegment with one tergite, two pleura, and one to three sternites; first trunk segment legless, modified as collum; diplosegments 2–4 (and also 5 in the order Spirobolida) with one pair of legs each; total number of leg pairs in adults ranging from 11 up to 375; antennae simple, generally 7-jointed; first maxillae fused into a gnathochilarium; second maxillae absent; gonopores opening on third trunk segment, on or near coxae of second pair of legs; cuticle usually calcified; many species capable of rolling into a tight coil; spiracles located ventrally, typically in front of leg coxae, and never valvular (i.e., they cannot be closed); legs ventrally positioned and generally short, such that the body is carried close to the ground. On some or most trunk segments many millipedes have lateral repugnatorial glands that secrete noxious chemicals that can irritate the skin and eyes. A few millipedes spray their secretions up to 40 cm. About 12,000 species of millipedes have been described.

The segmental nature of the trunk rings was recently challenged by insights obtained from gene expression studies demonstrating that dorsal and ventral halves of the body are decoupled in their early development. The body itself is usually elongate. Most common forms are subcylindrical and wormlike, short legged and capable of rolling into a tight spiral when disturbed; or are broadly flattened and equipped with lateral wing- or keel-like cuticular extensions of the body wall called paranota that may cover the legs. Filamentous outgrowths (spinnerets) bearing the openings of silk-producing glands are sometimes present at the caudal end of the body. Female gonopores open on eversible vulvae, sometimes transformed into ovipositors or into sclerotized appendages (**cyphopods**). Males gonopores lie on leg bases or on paired, rarely unpaired, penes. Sperm transfer is either direct or mediated by specialized copulatory legs called **gonopores**. In millipedes (Diplopoda) gonopores are present only in males, though not in all groups. (In centipedes gonopods may occur in either sex.) Helminthomorphan millipedes have one or two pairs of gonopods closely behind the body ring that bears the gonopores, whereas in Pentazonian millipedes the last pair of legs (the telopods) are modified as gonopods. Spiracles (tracheal openings) are situated close to the leg bases. In the subclass Penicillata, the exoskeleton is soft; in the Chilognatha, the cuticle is calcified, hard, and rigid.

The main subgroups of the Diplopoda are defined by the following apomorphies:

- **Penicillata** Cuticle soft, not calcified; 9–11 diplosegments covered by tufts of bristles; males without copulatory appendages
- **Chilognatha** Cuticle calcified; wide tracheal pouches; sperm actively transferred by male into the female genital opening
- **Pentazonia** Ventral plate (stigmatic plate, traditionally called sternite) of diplosegments divided along the midline; tracheae anastomosing; terminal leg pair of male transformed in genital appendages (telopods); intestine N-shaped, that is, bent twice. Glomeridesmids are flattened millipedes with 21–22 diplosegments, and cannot roll into a ball. Glomerids and sphaerotheriids are short, round, 12- or 13-segmented millipedes that can roll into a perfect ball.
- **Helminthomorpha** More than 21 body rings; stigmatic plates undivided; lateral ozopores (often limited to some diplosegments and sometimes lost); tracheae inserted in a cluster
- **Colobognatha** Gnathochilarium without palps; adults protecting egg clutch by curling around them; first postembryonic stage with four pairs of

legs; ninth and tenth leg pair of adult males transformed into sexual appendages (gonopods)

- **Eugnatha** Mandible with two articulated basal articles; pleurites fused to tergites; eighth (sometimes also ninth) leg pair of adult males transformed into sexual appendages (gonopods). This is by far the largest group of millipedes, including common cylindrical species (e.g., orders Julida, Spirobolida, and Spirostreptida) as well as the Polydesmida, often provided with lateral keel, or paranota, that confer them a somehow flattened aspect.

Class Pauropoda

Minuscule (0.3–2 mm) whitish-brownish myriapods; eyeless; with four- or six-segmented biramous antennae with three flagella and a unique candelabra-shaped or globular sense organ; trunk of 11 segments, tergites weakly sclerotized except in a few families; 8–11 pairs of legs (none on the first trunk segment); tergites of trunk segments usually covering more than one segment each; first maxillae fused into a gnathochilarium; second maxillae absent; most species without tracheal or circulatory systems; gonopores on third trunk segment. Although found in all parts of the world, pauropods are not common; they occur mainly in moist soils and woodland litter. About 850 species have been described.

Class Symphyla

Small (1–9 mm, but up to 30 mm in *Hanseniella magna*); eyeless; trunk with 14 segments, last fused to telson; first 12 trunk segments each with a pair of legs (the first one occasionally vestigial); penultimate segment with spinnerets and a pair of long sensory hairs; dorsal surface with 15–24 tergal plates; soft uncalcified cuticle; antennae long, threadlike, of up to 50 segments; first maxillae medially coalesced; second maxillae completely fused to form a complex labium; one pair of spiracles opening on the head; tracheae supply first 3 trunk segments; gonopores open on third trunk segment. Symphylans, generally uncommon, occur in soil and rotting vegetation. About 200 species have been described.

BOX 23A Characteristics of the Subphylum Myriapoda

1. Body of two tagmata: head and multisegmented trunk
2. All appendages multiarticulate and uniramous*
3. Head appendages, from anterior to posterior, are antennae, mandibles, first maxillae, and second maxillae; second maxillae may be fused into a single flaplike structure called a "labium," or they may be absent; first and second maxillae often bear palps.
4. Without a carapace
5. With an aerial gas exchange system composed of tracheae and spiracles (possibly evolved multiple times and convergent with those in the Hexapoda)
6. With one or two pairs of ectodermally derived (proctodeal) Malpighian tubules (probably convergent with those in the Hexapoda)
7. Most with lateral eye, possibly corresponding to modified (disintegrated) compound eyes
8. Gut simple, without digestive ceca
9. Gonochoristic; with direct development

*Pauropod antennae are branched, but whether this represents a vestige of a primitive crustacean-like condition or is secondarily derived is not known.

The Myriapod Body Plan

The myriapods differ from the hexapods and crustaceans in their retention of a largely homonomous trunk with its paired segmental appendages. The head appendages and legs are very similar to those of the insects. Metamerism is also evident internally in structures such as the segmental heart ostia, tracheae, and ganglia. The key features that distinguish the Myriapoda are listed in Box 23A.

The centipedes (chilopods) bear one pair of walking legs per segment (Figure 23.2A–D). Most species are 1–2 cm in length, although the smallest are just 3–4 mm long and some tropical giants attain lengths of nearly 30 cm.

In the millipedes (diplopods) the head is followed by a limbless segment called the **collum**, a heavily sclerotized collar between head and trunk (Figure 23.2E). Each of the next three segmental units bears a single pair of legs. The trunk is articulated into diplosegments, each formed by fusion of two somites and bearing a double complement of metameric organs (Figure 23.2F). Millipedes range in length from about 0.5 to 30 cm. The cuticle of millipedes is particularly robust, being well sclerotized and usually calcified.

Pauropods are minute, soft-bodied, eyeless, soil-inhabiting myriapods (Figure 23.1F). They are less than 2 mm in length and have 11 trunk segments. As in millipedes, some segmental pairing occurs, and usually only six tergites are visible dorsally.

Symphylans (Figure 23.1G) are also minute, eyeless myriapods, only one species exceeding 1 cm in length. The trunk has 12 (rarely 11) pairs of leg. Some tergites are divided, and 15–24 tergites are usually visible dorsally. The thirteenth body segment bears spinnerets, a pair of long sensory hairs, and a tiny postsegmental telson. Like pauropods, symphylans inhabit loose soil and humus.

Head and Mouth Appendages

In myriapods, the eyes and protocerebrum are located in a preantennal position. The antennary segment

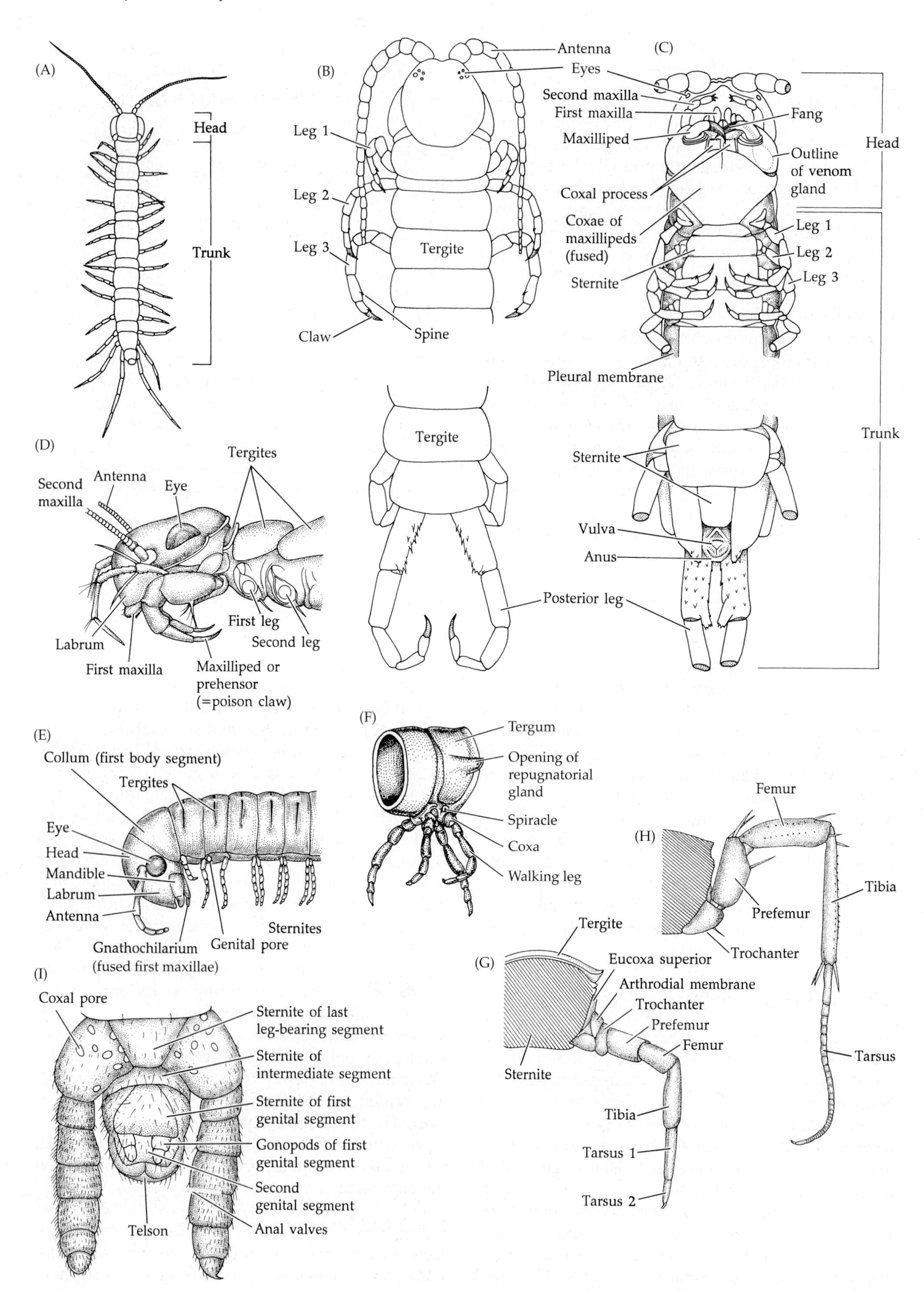
(A)
Head
Trunk
(B)
Antenna
Eyes
Leg 1
Leg 2
Leg 3
Tergite
Claw
Spine
(C)
Second maxilla
First maxilla
Fang
Maxilliped
Outline of venom gland
Head
Coxal process
Coxae of maxillipeds (fused)
Sternite
Leg 1
Leg 2
Leg 3
Pleural membrane
Trunk
Tergite
Sternite
Vulva
Anus
Posterior leg
(D)
Second maxilla
Antenna
Eye
Tergites
First leg
Second leg
Labrum
First maxilla
Maxilliped or prehensor (=poison claw)
(E)
Collum (first body segment)
Tergites
Eye
Head
Mandible
Labrum
Antenna
Gnathochilarium (fused first maxillae)
Genital pore
Sternites
(F)
Tergum
Opening of repugnatorial gland
Spiracle
Coxa
Walking leg
(H)
Femur
Tibia
Prefemur
Trochanter
Tarsus
(G)
Tergite
Eucoxa superior
Arthrodial membrane
Trochanter
Prefemur
Femur
Sternite
Tibia
Tarsus 1
Tarsus 2
(I)
Coxal pore
Sternite of last leg-bearing segment
Sternite of intermediate segment
Sternite of first genital segment
Gonopods of first genital segment
Second genital segment
Telson
Anal valves

◀ **Figure 23.2 General body anatomy of myriapods.** (A) The two body regions of myriapods; shown is a lithobiomorph centipede. (B,C) A female scolopendromorph centipede (dorsal and ventral views), with middle body segments omitted. (D) Head of a scutigeromorph centipede. (E) Anterior end of a millipede (Diplopoda). (F) A single diplosegment from the polydesmid millipede *Gigantowales chisholmi* (Diplopoda). (G,H) Trunk segments in two centipedes shown as transverse sections; note the variation in limb morphology. (G) *Lithobius*. (H) *Scutigera*. (I) Ventral view of the terminal segments of a male geophilomorph centipede (*Strigamia*).

bears the antennae and houses the deutocerebrum. The next segment (the premandibular segment) houses the tritocerebrum but lacks appendages. The three following segments carry, in turn, the mandibles, the first and the second pair of maxillae. The second maxillae are fused as a "labium" in symphylans, but have been lost altogether in millipedes and pauropods. In diplopods and pauropods the first maxillae fuse to form a flaplike gnathochilarium. The ganglia of the mandibular and maxillary segments are generally fused to form the subesophageal ganglion. The antennae of millipedes are primarily composed of seven distinct antennomeres and a short, retractile tip bearing four apical sensory cones. Like the Crustacea and Hexapoda, the mouth field of the Myriapoda is bordered by a labrum anteriorly, by a hypopharynx posteriorly. The latter arises as a flaplike (or tonguelike) outgrowth of the body wall through which open the salivary glands.

Locomotion

Myriapods rely on their well-sclerotized exoskeleton for body support. In centipedes, the legs are long but extend laterally, keeping the body close to the ground to maintain stability while at the same time providing for long strides and generally rapid locomotion. In millipedes, the legs arise ventrally and are short, again keeping the body close to the ground while providing for powerful, though slow, locomotion.

The basic design of arthropod limbs was described in Chapter 20. The power exerted by a limb is greatest at low speeds and least at high speeds. At lower speeds the legs are in contact with the ground for longer periods of time, and in the case of myriapods more legs are contacting the ground at any given moment. In burrowing forms, such as many millipedes, the legs are short, and the gait is slow and powerful as the animals bulldoze their way through soil or rotting wood. In centipedes, which mostly run at high speeds, less than half the limbs may touch the ground at any given moment, and for shorter periods of time (Figure 23.3). Longer limbs increase the speed of a running gait, and limbs long in length and stride are typical features of the fastest-running centipedes and symphylans.

Locomotor repertoires typically evolve in concert with the overall habits of animals, particularly feeding behaviors. Most centipedes are surface-dwelling predators that must move quickly to capture their prey. *Scutigera* is a centipede that qualifies as a world-class runner, reaching speeds up to 42 cm/second when in pursuit of prey. On the other hand, most millipedes are detritivores that burrow through soil, leaf litter, or rotten timber in search of food. Thus millipedes tend to have shorter legs and slower, but more powerful gaits.

One group of centipedes lacks the high-speed modifications of their relatives. Most geophilomorphs burrow, aided by dilation of the body and use of the trunk musculature in a fashion similar to that of earthworms. This peristalsis-like motion is rare in arthropods because of their rigid exoskeleton. However, geophilomorphs have enlarged areas of flexible cuticle on the sides of the body between the tergites and sternites; these enlarged areas (pleura) allow them to significantly alter their body diameter. Other centipedes have smaller, flexible pleural areas that allow some degree of lateral undulatory motion. Centipede legs are attached to these flexible regions. Geophilomorphs are the most legged of all centipedes, having 27 to 191 pairs.

In fast-running centipedes the limbs of each succeeding segment are slightly longer than the immediately anterior pair. The legs of most myriapods move in clear metachronal waves from posterior to anterior (Figure 23.3A–C). Unlike most arthropods, millipedes move the two pairs of legs on each diplosegment synchronously.

Thanks to a well-developed musculature and to the soft integument, pauropods are able to both elongate the body and shorten it by a high degree of contraction, allowing them to creep in narrow irregular spaces; adult pauropods also run swiftly in short rushes.

Feeding and Digestion

Most centipedes are active, aggressive predators on smaller invertebrates, particularly worms, snails, and other arthropods. Their first trunk appendages form large claws, called **prehensors** or **forcipules**. These raptorial limbs are located ventral to the mouth field (Figure 23.2D), and are used to stab prey and inject poison. The poison is produced in large venom glands, generally located in the basal articles of the forcipules. The poison is so effective that large centipedes, such as the tropical *Scolopendra*, can subdue small vertebrates. Some centipedes actually rear up on their hind legs and capture flying insects! The prehensors and second maxillae hold the prey, while the mandibles and first maxillae bite and chew. The bite of even the most dangerous centipede is normally not fatal to people, but the poison can cause a reaction similar to that accompanying a serious wasp or scorpion sting.

The feeding strategy of millipedes is quite different. Most millipedes are slow-moving detritivores with a preference for dead and decaying plant material, although some species (e.g., *Blaniulus guttulatus*) have

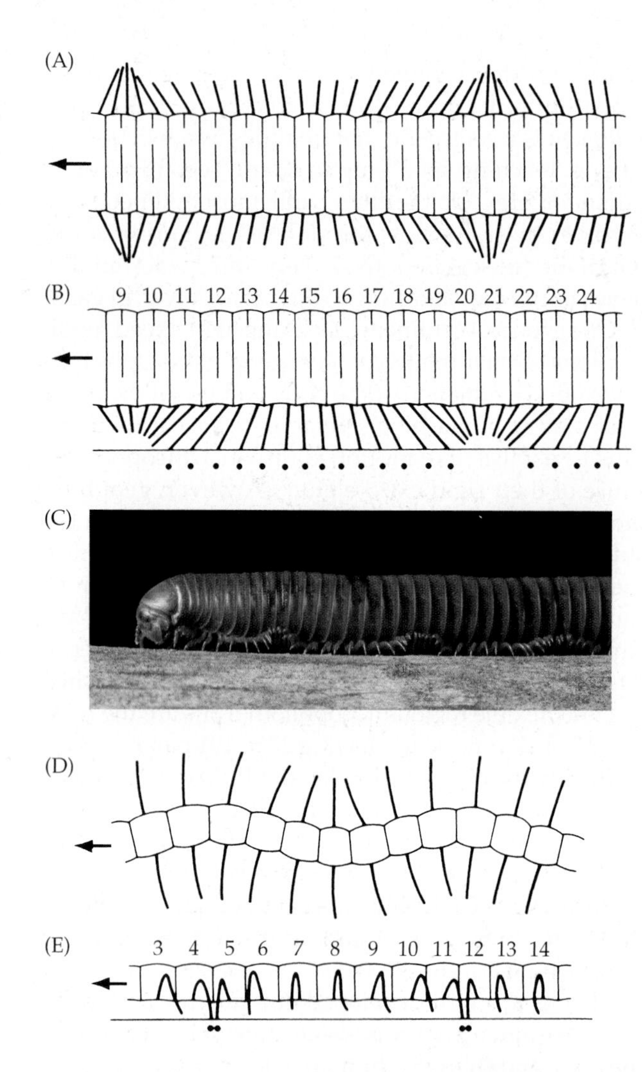

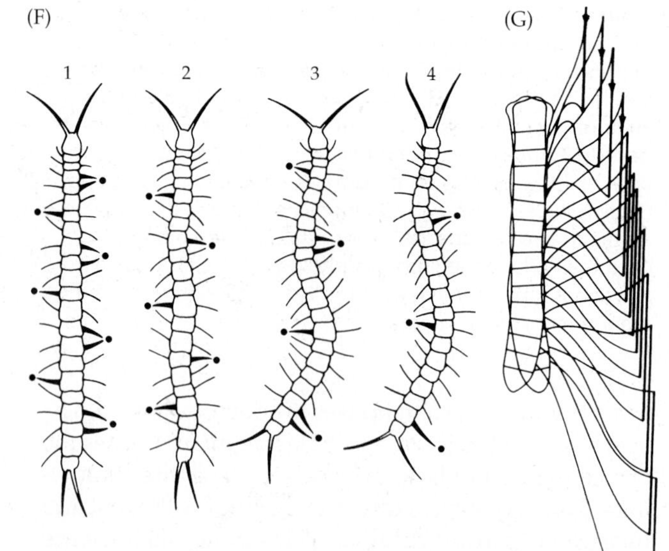

Figure 23.3 Locomotion in millipedes and centipedes. (A,B) A typical millipede (such as *Spirostreptus* or *Gymnostreptus*) in motion. Note the 16 diplosegments (with 32 pairs of legs). The dorsal view (A) shows left and right leg sets exactly in phase with each other. The lateral view (B) of the same animal shows that the majority of the limb tips are on the ground at once, an arrangement yielding a slow but powerful gait. (C) A millipede (*Narceus americanus*) in motion. Also see chapter opener photo of *N. americanus* (the giant North American millipede). (D,E) A typical centipede (such as *Scolopendra* or *Cryptops*) in motion. Note the 12 segments, each with one pair of legs. The dorsal view (D) shows the limb pairs in opposite phase and the undulations of the body that accentuate stride length. The lateral view (E) of the same animal illustrates that fewer than one-third of the limb tips are on the ground at any one time, an arrangement yielding the short, swift strokes typical of a rapid but weak gait. Arrows indicate the animal's direction of travel; dots are points of leg-tip contact with the substratum. (F) Locomotion in a scolopendrid centipede at various speeds. Parts 1–4 show the body waves and leg actions at increasing speeds. Limbs shown with heavy lines are in their power strokes, with the tips against the substratum (dots); limbs depicted by thin lines are in various stages of their recovery strokes. Notice that at maximum speed the animal is still supported by a tripod stance. (G) Field of leg movements in a running centipede, *Scutigera*. The heavy vertical lines trace the movement of the tips of each leg during the propulsive backstroke. Note the gradual increase in limb length posteriorly that allows for unencumbered overstepping even when the full swing of the limb is used.

been reported to cause damage to cultivated plants. Millipedes play an important role in the recycling of leaf litter in many parts of the world. Most bite off large pieces of plant tissue with their powerful mandibles, mix it with saliva as they chew it, and then swallow it. Some, such as the tropical Siphonophorida, are believed to feed on the juices of living plants and fungi. In these groups the labrum, gnathochilarium, and reduced mandibles are modified into a suctorial piercing beak. A few odd groups of millipedes have evolved predaceous lifestyles and feed as centipedes do.

The feeding biology of pauropods is not well understood, but most appear to feed on fungi and decaying plant and animal matter. *Millitauropus*, however, feeds on springtails and their eggs.

Symphylans are primarily herbivores, although a few have adopted a carnivorous, scavenging lifestyle. Many symphylans consume live plants. One species, *Scutigerella immaculata*, is a serious pest in plant nurseries and flower gardens, where it has been reported at abundances exceeding 90 million per acre.

Like the guts of all arthropods, the long, usually straight myriapod gut is divisible into an ectodermal foregut, endodermal midgut, and ectodermal hindgut (Figure 23.4). There are no digestive ceca branching off the gut. Salivary glands are associated with the mouth appendages. The salivary secretions soften and lubricate solid food, and in some species contain enzymes that initiate chemical digestion. The mouth leads inward to a long esophagus, which is sometimes expanded posteriorly as a storage area, or crop and gizzard (as in most centipedes). The gizzard often contains cuticular spines that help strain large particles from the food entering the midgut, where absorption occurs (Figure

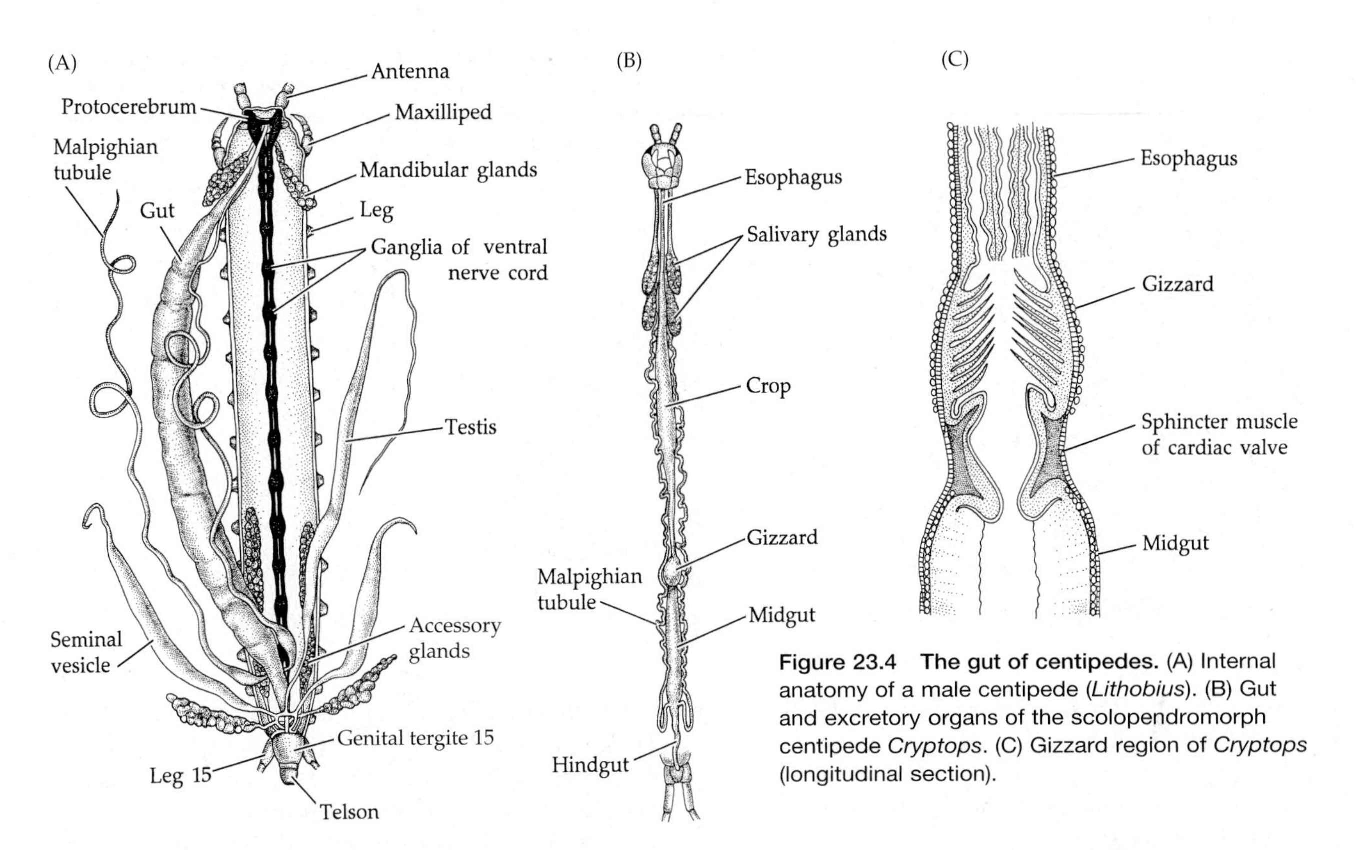

Figure 23.4 The gut of centipedes. (A) Internal anatomy of a male centipede (*Lithobius*). (B) Gut and excretory organs of the scolopendromorph centipede *Cryptops*. (C) Gizzard region of *Cryptops* (longitudinal section).

23.4C). As in other arthropods, the gut of myriapods produces a peritrophic membrane—a sheet of thin, porous, chitinous material that lines and protects the midgut, and may pull free to envelop and coat food particles as they pass through the gut. The midgut connects to a short, proctodeal hindgut that terminates at the anus.

Circulation and Gas Exchange

The myriapod circulatory system includes a dorsal tubular heart that pumps the hemocoelic fluid (blood) toward the head. The heart narrows anteriorly to a vessel-like aorta, from which blood flows posteriorly through large hemocoelic chambers before returning to the pericardial sinus and then back to the heart via paired lateral ostia. Circulation is slow, and system pressure is relatively low. In diplopods, the heart bears two pairs of ostia in each diplosegment; one pair of ostia occurs in each segment in the chilopods. The copper-containing respiratory pigment hemocyanin has been found in the blood of scutigeromorph centipedes and a few millipedes.

Most myriapods rely on a tracheal system for gas exchange (Figure 23.5). As explained in Chapter 20, tracheae are extensive tubular invaginations of the body wall opening through the cuticle by pores called spiracles. The tracheae originating at one spiracle often anastomose with others to form branching networks penetrating most of the body. Spiracles are placed segmentally, although not necessarily on all segments. In most chilopods the spiracles are located in the membranous pleural region just above and behind the base of the leg; in the lithobiomorphs and in most scolopendromorphs, spiracle-bearing segments alternate with segments without spiracles. In the scutigeromorph centipedes, the respiratory openings are dorsal and lead to thick bundles of very short tracheae that terminate onto the heart. Diplopods usually bear two pairs of spiracles per diplosegment, just anterior to the leg coxae. In the symphylans a single pair of spiracles opens on the sides of the head, and the tracheae supply only the first three trunk segments. Except for a few primitive species, pauropods lack a tracheal system.

Each spiracle is usually recessed in an atrium, the walls of which are lined with setae or spines (trichomes) that prevent dust, debris, and parasites from entering the tracheal tubes. The spiracles are often surrounded by a sclerotized rim or lip, the **peritrema**, which also aids in excluding foreign particles. A muscular valve or other closing device is present in many centipedes and under control of internal partial pressures of O_2 and CO_2. As in insects, ventilation of the tracheal system is accomplished by simple diffusion gradients, as well as by pressure changes induced by the animal's movements. The blood of myriapods appears to play a limited role in oxygen transport (except in the very active scutigeromorphs). Hemocyanin has recently been discovered in the blood of some large millipedes, but at present, we have limited information about its role.

The innermost parts of the tracheal system are the tracheoles, which are thin-walled, fluid-filled channels

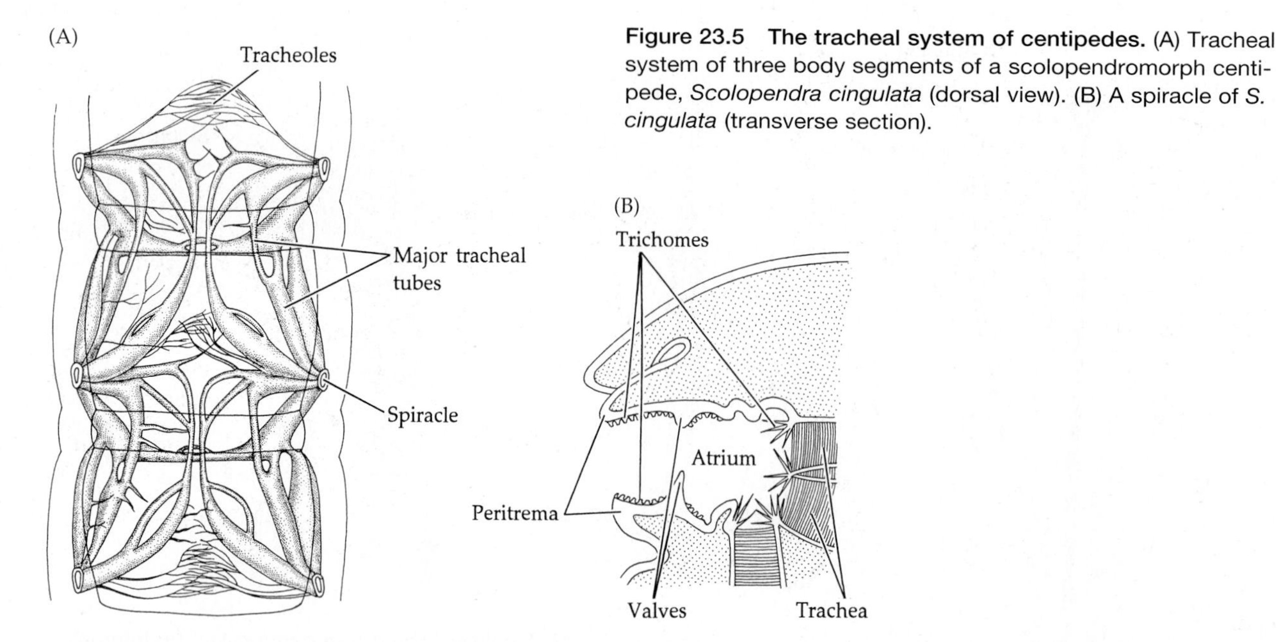

Figure 23.5 The tracheal system of centipedes. (A) Tracheal system of three body segments of a scolopendromorph centipede, *Scolopendra cingulata* (dorsal view). (B) A spiracle of *S. cingulata* (transverse section).

that end as a single cell, the tracheole end cell (= tracheolar cell). Unlike tracheae, tracheoles are not shed during ecdysis. The tracheoles are so minute (0.2–1.0 mm) that ventilation is impossible and gas transport here relies on aqueous diffusion.

Excretion and Osmoregulation

When myriapods invaded the land, the problem of water conservation necessitated the evolution of entirely new structures to remove metabolic wastes—Malpighian tubules (Figure 23.4). These excretory organs function much as those of insects (Chapter 22). Myriapods usually possess only one pair of Malpighian tubules, but there are three tubules in Craterostigmomorpha and four in Scutigeromorpha. In those species that are largely confined to moist habitats and nocturnal activity patterns, a significant portion of the excretory wastes may be ammonia rather than uric acid.

The Malpighian tubules of myriapods were long assumed to be homologous to those of hexapods. However, it is now accepted that these two groups do not share an immediate common ancestry; Hexapoda appears to have evolved from a crustacean ancestry. Therefore these excretory structures represent another striking case of convergent evolution.

The myriapod cuticle is sclerotized and calcified to various degrees, adding a measure of waterproofing, but aside from a few species, such as *Orthoporus ornatus* of the American semideserts, it lacks the waxy layer seen in insects. For this reason, myriapods rely to a considerable extent on behavioral strategies to avoid desiccation. Many live in humid or wet environments, or are active only during cool periods. Other myriapods stay hidden in cool or moist microhabitats, such as under rocks, during hot hours of the day or dry periods.

A few geophilomorph centipedes live in intertidal habitats, under rocks or in algal mats. A few millipedes have semiaquatic habits, for example, the littoral julid *Thalassisobates litoralis* that lives among the stranded decaying matters on the shores of the Mediterranean Sea. However, this species does not spend much time under water, unlike some subterranean species from caves in southern Europe, or the Amazonian *Myrmecodesmus adisi* whose juveniles spend up to 11 months under water. A few geophilomorph centipedes, also from periodically inundated Amazonian forests, tolerate month-long submersion. When under water, these myriapods apparently rely on plastron respiration, in other words, they use water-dissolved oxygen that reaches their tracheae via minute air bubbles trapped into the spiracles or retained by a specialized integument. Although the tracheal system of myriapods greatly resembles that of insects, there is evidence that it also evolved independently.

Nervous System and Sense Organs

The myriapod nervous system conforms to the basic arthropod plan described in Chapter 20. Very little secondary fusion of ganglia occurs, and the ventral nerve cord retains much of its primitive double nature, with a pair of fused ganglia in each segment. Millipedes possess two pairs of fused ganglia in each diplosegment.

As in other arthropods, the cerebral ganglion of myriapods comprises three distinct regions: the preantennal protocerebrum (associated with the eyes, if these are present), the deutocerebrum (associated with the antennae), and the tritocerebrum. A subesophageal ganglion controls the mouthparts, salivary glands, and some local musculature (Figure 23.6).

Myriapods typically possess eyes, although these are lacking in pauropods, symphylans, all geophilomorphs and many scolopendromorphs among the centipedes; and siphoniulids, siphonophorids, and polydesmidans among the millipedes. Many trogloditic

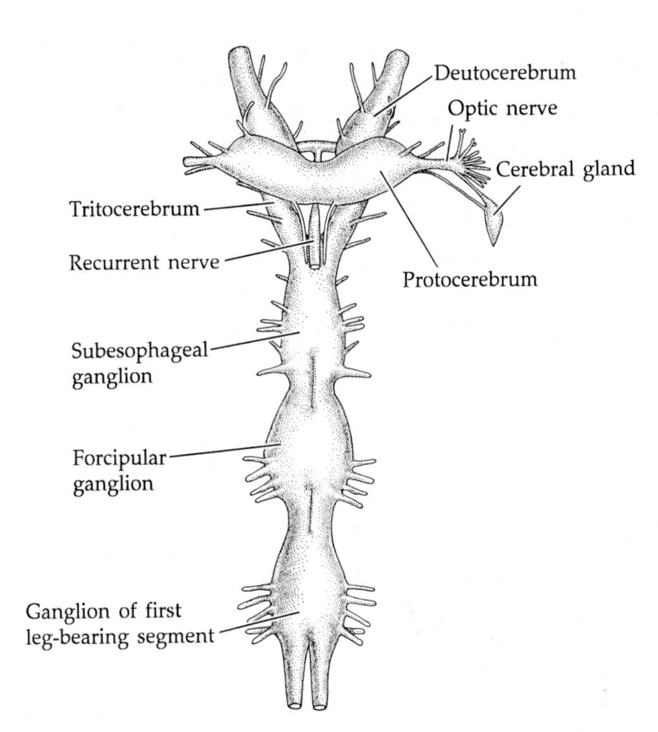

Figure 23.6 The brain and anterior ganglia of a centipede, *Lithobius forficatus* (dorsal view).

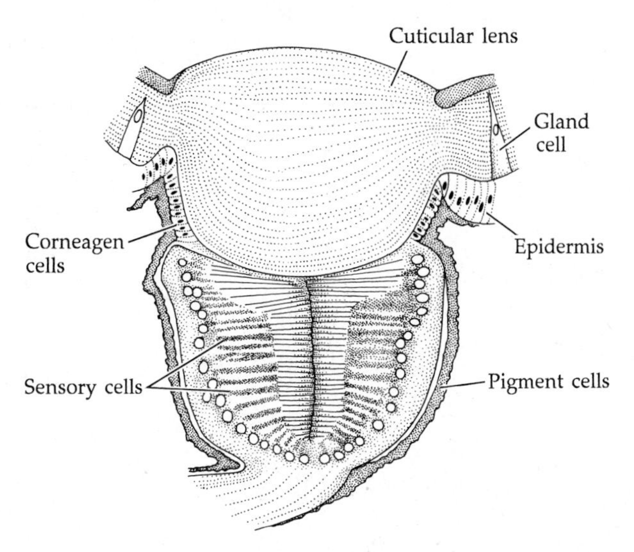

Figure 23.7 An eye of a centipede (section).

(cave dwelling) species belonging to other groups, especially to the lithobiomorph centipedes, are also blind, especially the lithobiomorph centipedes. Centipedes can possess a few or many independent eyes (Figure 23.7). In the scutigeromorph centipedes, up to 200 of these eyes cluster to form a sort of "compound eye." However, the eyes of living centipedes appear to function only in the detection of light and dark, and not in image formation. Many diplopods have from 2 to 80 eyes arranged variously on the head. Some diplopods also possess integumental photoreceptors, and many eyeless species exhibit negative phototaxis.

Centipedes and millipedes are noted for their highly sensitive antennae, which are richly supplied with tactile and chemosensory setae. In many species, an **organ of Tömösváry** is located at the base of each antenna (Figure 23.8). Each part of this paired organ consists of a disc with a central pore where the ends of sensory neurons converge. The exact role of this organ has yet to be clearly established, and speculation has run the gamut from chemosensation to pressure sensation to humidity detection to audition (sound or vibration detection). The last idea is probably the most popular today. Whether the organ of Tömösváry might detect aerial vibrations (auditory impulses) or only ground vibrations is also a debated question.

Reproduction and Development

Myriapods are gonochoristic and oviparous, although parthenogenesis occurs in several families of millipedes and in some centipedes, pauropods, and symphylans. Many myriapods, like most arachnids, rely on indirect copulation and insemination. Packets of sperm (spermatophores) are deposited in the environment, or are held by the male and picked up by the female. All myriapods have direct development, with the young hatching as "miniature adults," although often with fewer body segments. Beyond these generalizations, some specifics for each major group are discussed below.

Chilopoda Female centipedes possess a single elongate ovary located above the gut, whereas males have from 1 to 26 testes similarly placed (Figure 23.9). The oviduct joins with the openings from several accessory glands and a pair of seminal receptacles just internal to the gonopore, which is situated on the genital segment (the legless segment in front of the telson). The gonopore of females is usually flanked by a pair of small grasping appendages, or gonopods, that manipulate the males' sperm packets. In males the testes join the

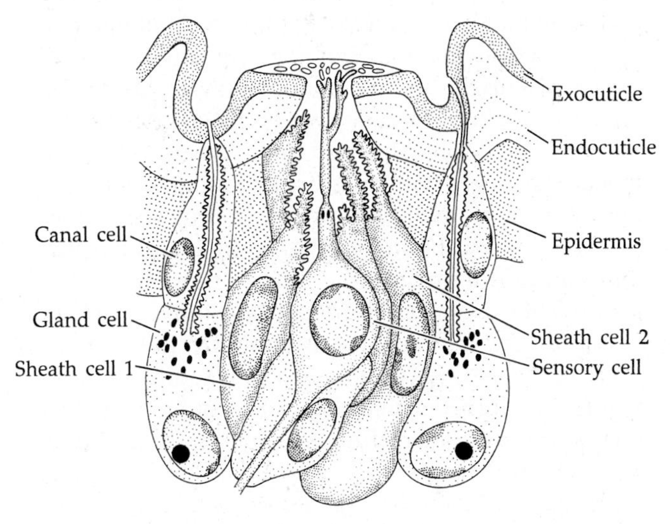

Figure 23.8 The organ of Tömösváry from the centipede *Lithobius forficatus*.

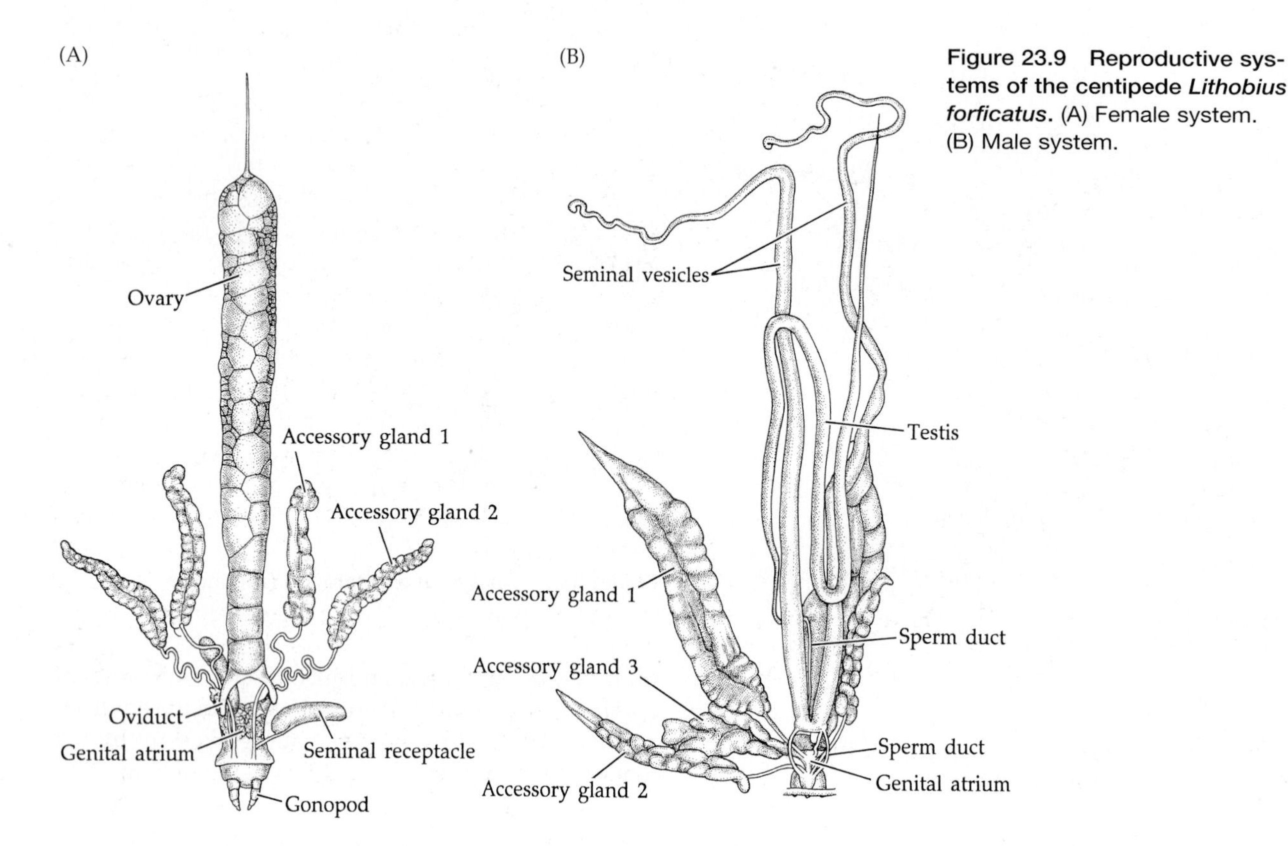

Figure 23.9 Reproductive systems of the centipede *Lithobius forficatus*. (A) Female system. (B) Male system.

ducts of several accessory glands and a pair of seminal vesicles near the gonopore, which opens on the ventral surface of the genital segment. The male gonopore also lies between a pair of small gonopods.

Sperm are packed into spermatophores, which are transferred to the female. In some cases (e.g., scutigeromorphs), males deposit spermatophores directly on the ground and the females simply pick them up. In most species, however, females produce a silken **nuptial net**, which is spun from modified genital glands, and males deposit a spermatophore on this net (Figure 23.10). Mating pairs of centipedes typically perform courtship behavior, stroking each other with their antennae and often moving the large (up to several millimeters) spermatophore into the nuptial net. Eventually the female picks up the spermatophore with her gonopods and inserts it into her gonopore. Fertilization occurs as the eggs pass through the gonoduct. Females often coat the fertilized eggs with moisture and fungicides before depositing them in the ground or in rotting vegetation.

In the Craterostigmomorpha, Scolopendromorpha, and Geophilomorpha, the female remains coiled around her brood until hatching (Figure 23.10D). Parental caretaking is almost unknown in the other myriapod class, except for a few polyzoniid and platydesmid millipedes that also exhibit brood care. In a few platydesmidan millipedes, embryo brooding is provided by the male.

Scutigeromorphs, lithobiomorphs, and craterostigmomorphs hatch with 4, 7, and 12 pairs of legs, respectively, and obtain the final number of 15 leg pairs following one (craterostigmomorphs) or a few molts but continue molting until they reach maturity and beyond (hemianamorphosis). Scolopendromorphs and geophilomorphs, however, develop by epimorphosis. That is, they reach the final number of body segments during the embryonic development; no segment or leg pair is added during the postembryonic life.

Diplopoda Millipedes possess a single pair of elongate gonads in both sexes. Unlike those of chilopods, millipede gonads lie between the gut and the ventral nerve cord. The gonopores open on the third (genital) trunk segment. In females each oviduct opens separately into a genital atrium (= vulva) near the coxae of the second pair of legs. A groove in the vulva leads into one or more seminal receptacles. Males possess a pair of penes, or **gonapophyses**, also on the coxae of the second pair of legs.

The mandibles are used to transfer sperm in the "pill millipedes," but more commonly one or both pairs of legs of the seventh or eighth trunk segment are modified as copulatory appendages, or gonopods, and serve this purpose. Males bend the anterior body segments back under the trunk until the penes and gonopods make contact, whereupon the latter pick up a spermatophore in preparation for mating.

In most millipedes, mating occurs by indirect copulation, in which the male and female genital openings never actually make contact, although the male and female may embrace with their legs and mandibles and may coil their trunks together. Sometimes a pair remains

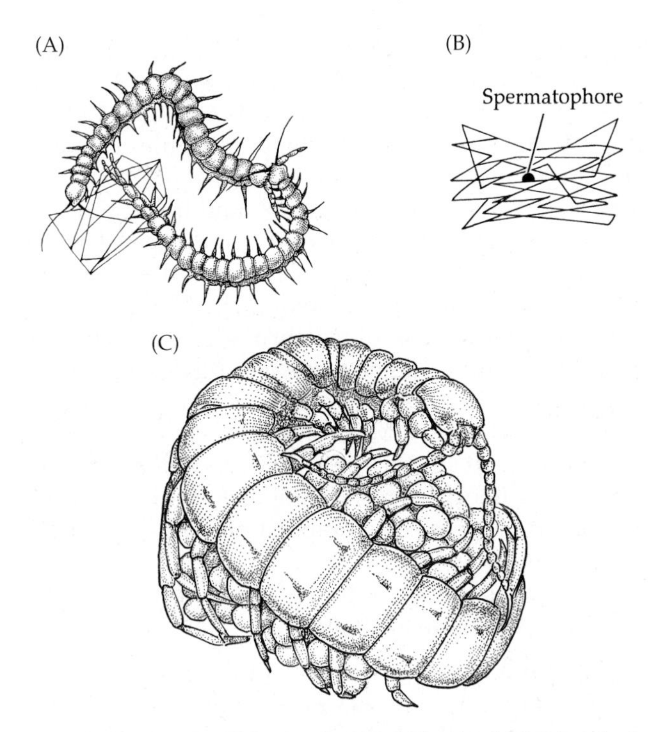

Figure 23.10 (A–C) Reproductive behavior in the tropical centipede *Scolopendra cingulata*. (A) A male and female in mating position over the nuptial net. (B) A nuptial net with a spermatophore. (C) A female and her eggs. (D) The giant North American desert centipede *Scolopendra heros* embracing her young.

in such an embrace for up to two days. Pheromones are known to play an important role in many species. Antennal tapping, head drumming, and stridulation may be included in the mating repertoire. Males of the family Sphaerotheriidae produce sounds during courtship by rubbing the telopods against the back edge of the body. The *Sphaerotherium* male usually stridulates when in contact with a female to initiate mating, probably to prevent the female from rolling into a ball or to stimulate her to uncoil if she has already volvated.

Males often have an array of secondary sexual characteristics, including strong modifications of the first pair of legs and head. In species of the genus *Chordeuma* in Europe, dorsal glands on trunk ring XVI of the male produce a secretion on the dorsal surface upon which females feed prior to mating.

Fertilization takes place as the eggs are laid. In pentazonian millipedes, the last pair of male legs is modified into strong claspers, used to hold the female in place. In these millipedes, males mold soil into a small cup into which they ejaculate sperm. The sperm-filled cup is then passed to the female, who is firmly held by the male's large claspers. Eggs are usually laid in the soil, although some species also produce silk bags to house the eggs. Some species fashion a nest of soil and humus reinforced with parental feces, or the entire nest may be composed of fecal material. One to 300 eggs are laid at a time.

Most newborn millipedes have only three pairs of legs; new segments and leg pairs are added at each molt, according to one of three possible developmental schedules. In Pentazonia (as in the pill millipede *Glomeris*) the final number of segments and leg pairs is obtained after some molts, but the millipede must undergo further molts without segmental increase before achieving adult size and sexual maturity (hemianamorphosis). In polydesmidans and other helminthomorphs, the achievement of the final number of segments and leg pairs coincides with obtaining the adult condition and no further molt occurs (teloanamorphosis). In the remaining helminthomorphs, new segments and leg pairs are added at every molt and the animal continues molting even after reaching sexual maturity (euanamorphosis).

The males of some European julids may undergo periodomorphosis: specific environmental conditions may induce sexually mature males to molt to intercalary males with regressed gonopods, thereby suspending reproductive activity. This is resumed after one or more additional molts, with newly differentiated gonopods.

Pauropoda In female pauropods a single ovary lies beneath the gut, but in the males the testes are located above the gut. As in millipedes, the gonopores (a median, unpaired opening in the female, two paired openings in the male) are on the third trunk segment. Females take up a spermatophore, often after the male has suspended it from one or a few silk threads spanning two stones or leaves. Fertilization is internal. The yolky eggs are laid in decaying wood. Embryonic development ends with an initial motionless pupoid stage with minute oral and anal openings and inarticulated traces of the antennae and of the first two pairs of legs. A molt changes it into the first active juvenile, with three pairs of legs. The development then proceeds by anamorphosis through four juveniles and one adult stage.

Symphyla Symphylans display one of the more unusual methods of fertilization in the animal kingdom. Paired gonads discharge through gonopores on the third trunk segment in both sexes. Males deposit spermatophores in the environment. At least some symphylans construct stalk-like structures, topped with spermatophores. When a female encounters one of these sperm packages, she bites off the spermatophores and stores the sperm in her preoral cavity. When her own eggs are ripe, the female removes them from her gonopore with her mouthparts and cements them to moss or some other substrata. Fertilization occurs during this process as she coats each egg with the stored sperm. Young symphylans hatch with only about half the adult number of trunk segments and appendages.

In the postembryonic development, first-order and second-order molts regularly alternate until maturity is reached. A new two-segment unit is added after each first-order molt, but only one more pair of legs is differentiated, the other pair differentiating after the subsequent second-order molt at which time no new segments are added. Mature symphylans continue to molt throughout life (up to 40 times in *Scutigerella immaculata*), without any further addition of segments or legs.

Embryonic development

The indirect copulation and external development typical of myriapods demands large amounts of stored nutrients (yolk) in the eggs. The yolky eggs of myriapods have shifted almost entirely to meroblastic cleavage. Most myriapods undergo early cleavage by intralecithal nuclear divisions, followed by a migration of the daughter nuclei to the peripheral cytoplasm (= periplasm). Here they continue to divide until the periplasm is dense with nuclei, whereupon cell membranes begin to form, partitioning uninucleate cells from one another. At this point the embryo is a periblastula, comprising a yolky sphere that has a few scattered nuclei and is covered by a thin cellular layer, or blastoderm.

Along one side of the blastula a patch of columnar cells forms a germinal disc, sharply marked off from the thin cuboidal cells of the remaining blastoderm. From specific regions of this disc, presumptive endodermal and mesodermal cells begin to proliferate as germinal centers. These cells migrate inward during gastrulation to lie beneath their parental cells, which now form the ectoderm. The mesoderm proliferates inward as a longitudinal gastral groove; the cells of the developing gut usually surround and gradually begin absorbing the large central yolky mass of the embryo.

Segments begin to demarcate and proliferate and eventually develop appendage buds. The mouth and anus arise by ingrowths of the ectoderm that form the fore- and hindguts, which eventually establish contact with the developing endodermal midgut.

Variations on this basic scheme occur in groups with secondary yolk reduction, such as in pauropods and symphylans. In these groups a largely holoblastic cleavage takes place, and a coeloblastula actually forms. In most chilopods, a type of cleavage occurs that is somewhat intermediate between total and superficial cleavage. After a few intralecithal nuclear divisions, the yolk breaks up into blocks called yolk pyramids. Gradually the yolk pyramids disappear, and development shifts to the superficial type.

Myriapod Phylogeny

Myriapoda and Hexapoda have long been considered sister groups, together forming a clade called Atelocerata, in turn regarded as sister to the Crustacea within the Mandibulata. The sister-group relationship between myriapods and hexapods was based upon several seemingly strong synapomorphies, including proctodeal Malpighian tubules, loss of the second pair of antennae, uniramous legs, and tracheal respiratory systems. However, recent work agrees that the hexapods arose from within Crustacea, and thus the putative synapomorphies linking Hexapoda to Myriapoda must have evolved through convergent or parallel evolution. Some molecular phylogenies suggest that the Myriapoda are sister to the Chelicerata (within a clade called Myriochelata or Paradoxopoda), but most recent studies converge in restoring the Myriapoda as a lineage of the Mandibulata, although sister to the Pancrustacea (= Hexapoda plus a paraphyletic Crustacea).

The monophyly of the four myriapod classes is supported by the following traits:

- **Chilopoda** Appendages of first trunk segment housing a poison gland; 15 pairs (secondarily more) of walking legs; alternation of long and short tergites, with reversal on segments 7–8, and spiracles associated with the segments bearing long tergites
- **Diplopoda** Trunk formed by diplosegments; antenna with eight articles, the distal article bearing apical sensory cones; spermatozoon without flagellum
- **Pauropoda** Antennae branching, with a special sensory organ (globulus); paired pseudoculi on lateral sides of head capsule
- **Symphyla** Single pair of tracheal stigmata on sides of head capsule; eyes absent; female spermathecae formed by paired lateral pockets in mouth cavity; twelve pairs of trunk legs; unpaired genital opening; paired terminal spinnerets

The Chilopoda are probably sister to the remaining myriapod classes, which are collectively classified as the Progoneata, meaning that their genital opening is anterior (in fact, in the region of the third body segment) rather than posterior as in the centipedes (Opisthogoneata). Within the Progoneata, Pauropoda and Diplopoda form a clade (Dignatha) defined by the lack of limbs on the post-maxillary segment, spiracles opening at the bases of the walking legs, a motionless pupoid stage after hatching, and the first free-living juvenile having three pairs of legs.

Myriapods with calcified exoskeletons are reasonably well-preserved in the fossil records, the oldest specimens being from the middle Silurian (425 Ma), although older trace fossils from the Late Ordovician (some 450 Ma) have also been attributed to them. Both modern clades of chilognathan millipedes

(Pentazonia and Helminthomorpha) are documented since the Upper Carboniferous. Extinct myriapod taxa are the Arthropleurida (which includes the giant *Arthropleura*, up to 2.6 m long), Eoarthropleurida and Microdecemplicida, all Paleozoic. From middle Silurian to late Carboniferous are documented the Archypolypoda, a group of millipedes with the trunk composed of alternating smaller and larger units with one leg pair each. Myriapods with uncalcified cuticle such as centipedes and penicillate millipedes are fairly common in amber and are extensively documented since the Cretaceous, but the oldest centipedes (*Devonobius*) are known from the Middle Devonian (380 Ma).

Selected References

General References

Adis, J. 1992. How to survive six months in a flooded soil: strategies in Chilopoda and Symphyla from Central Amazonian floodplains. Stud. Neotrop. Fauna Envir. 27: 117–129.

Akkari, N., H. Enghoff and A. Minelli. 2014. Segmentation of the millipede trunk as suggested by a homeotic mutant with six extra pairs of gonopods. Front. Zool. 11: 6.

Anderson, B. D., J. W. Shultz and B. C. Jayne. 1995. Axial kinematics and muscle activity during terrestrial locomotion of the centipede *Scolopendra heros*. J. Exptl. Biol. 198: 1185–1195.

Barber, A. D. 2009a. *Centipedes. Synopses of the British fauna (New series) 58*. Field Studies Council, Shrewsbury.

Barber, A. D. 2009b. Littoral myriapods: a review. Soil Org. 81: 735–760.

Bonato, L., S. Bevilacqua and A. Minelli. 2009. An outline of the geographical distribution of world Chilopoda. Contr. Nat. Hist., Bern 12: 489–503.

Bonato, L., A. Minelli, L. Drago and L. A. Pereira. 2015. The phylogenetic position of *Dinogeophilus* and a new evolutionary framework for the smallest epimorphic centipedes (Chilopoda: Epimorpha). Contrb. Zool. 84(3): 237–253.

Bush, S. P., B. O. King, R. L. Norris and S. A. Stockwell. 2001. Centipede envenomation. Wild. Environ. Med. 12: 93–99.

Chagas, A. Jr., G. D. Edgecombe and A. Minelli. 2008. Variability in trunk segmentation in the centipede order Scolopendromorpha: a remarkable new species of *Scolopendropsis* Brandt (Chilopoda: Scolopendridae) from Brazil. Zootaxa 1888: 36–46.

Causey, N. B., D. L. Tiemann, 1969. A revision of the bioluminescent millipedes of the genus *Motyxia* (Xystodesmidae: Polydesmida). Proc. Amer. Phil. Soc. 113, 14-33.

Chipman, A. D. and M. Akam. 2008. The segmentation cascade in the centipede *Strigamia maritima*: involvement of the Notch pathway and pair-rule gene homologues. Dev. Biol. 319: 160–169.

Damen, W. G. M., N.-M. Prpic and R. Janssen. 2009. Embryonic development and the understanding of the adult body plan in myriapods. Soil Org. 81: 337–346.

Damsgaard, C., A. Fago, S. Hagner-Holler, H. Malte, T. Burmester and R. E. Weber. 2013. Molecular and functional characterization of hemocyanin of the giant African millipede, *Archispirostreptus gigas*. J. Exp. Biol 216: 1616–1623.

Dove, H. and A. Stollewerk. 2004. Comparative analysis of neurogenesis in the myriapod *Glomeris marginata* (Diplopoda) suggests more similarities to chelicerates than to insects. Development 130: 2161–2171.

Drago, L., G. Fusco, E. Garollo and A. Minelli. 2011. Structural aspects of leg-to-gonopod metamorphosis in male helminthomorph millipedes (Diplopoda). Front. Zool. 8: 19.

Eisner, T., H. Eisner, J. Hurst, F. Kafatos and J. Meinwald. 1963. Cyanogenic glandular apparatus of a millipede. Science 139: 1218–1220.

Enghoff, H., W. Dohle and J. G. Blower. 1993. Anamorphosis in millipedes (Diplopoda): The present state of knowledge with some developmental and phylogenetic considerations. Zool. J. Linn. Soc. 109: 103–234.

Fusco, G. 2005. Trunk segment numbers and sequential segmentation in myriapods. Evol. Dev. 7: 608–617.

Fusco, G. and A. Minelli. 2013. Arthropod segmentation and tagmosis. Pp. 197–221 in A. Minelli, G. Boxshall and G. Fusco (eds.), *Arthropod Biology and Evolution. Molecules, Development, Morphology*. Springer, Heidelberg.

Gilbert, S. F. 1997. Arthropods: The crustaceans, spiders, and myriapods. Pp. 237-257 in S. F. Gilbert and A. M. Raunio (eds.), *Embryology: Constructing the Organism*. Sinauer Associates, Sunderland, MA.

Harzsch, S., R. R. Melzer and C. H. G. Müller. 2006. Mechanisms of eye development and evolution of the arthropod visual system: the lateral eyes of Myriapoda are not modified insect ommatidia. Org. Div. Evol. 7: 20–32.

Hoffman, R. L. 1999. Checklist of the millipeds of North and Middle America. Spec. Publ. 8, Virginia Mus. Nat. Hist.

Hoffman, R. L. et al. 1982. Chilopoda–Symphyla–Diplopoda–Pauropoda. Pp. 681–726 in S. P. Parker (ed.), *Synopsis and Classification of Living Organisms*. McGraw-Hill, New York.

Hoffman, R. L., S. I. Golovatch, J. Adis and J. W. de Morais. 1996. Practical keys to the orders and families of millipedes of the Neotropical region (Myriapoda: Diplopoda). Amazoniana 14: 1–35.

Hopkin, S. P. and H. J. Read. 1992. *The Biology of Millipedes*. Oxford University Press, New York.

Hughes, C. L. and T. C. Kaufman. 2002a. Exploring myriapod segmentation: The expression patterns of *even-skipped*, *engrailed*, and *wingless* in a centipede. Dev. Biol. 247: 47–61.

Hughes, C. L. and T. C. Kaufman. 2002b. Exploring the myriapod body plan: Expression patterns of the ten Hox genes in a centipede. Development 129: 1225–1238.

Janssen, R. 2011. Diplosegmentation in the pill millipede *Glomeris marginata* is the result of dorsal fusion. Evol. Dev. 13: 477–487.

Janssen, R., N.-M. Prpic and W. G. M. Damen. 2004. Gene expression suggests decoupled dorsal and ventral segmentation in the millipede *Glomeris marginata* (Myriapoda: Diplopoda). Dev. Biol. 268: 89–104.

Kettle, C., J. Johnstone, T. Jowett, H. Arthur and W. Arthur. 2003. The pattern of segment formation, as revealed by *engrailed* expression, in a centipede with a variable number of segments. Evol. Dev. 5: 198–207.

Kuse, M., A. Kanakubo, S. Suwan, K. Koga, M. Isobe and O. Shimomura. 2001. 7,8-dihydropterin-6-carboxylic acid as light emitter of luminous millipede, *Luminodesmus sequoiae*. Bioorg. Med. Chem. Lett. 11: 1037–1040.

Lesniewska, M., L. Bonato, A. Minelli and G. Fusco. 2009. Trunk anomalies in the centipede *Stigmatogaster subterranea* provide insight into late embryonic segmentation. Arthr. Struct. Dev. 38: 417–426.

Lewis, J. G. E. 1981. *The Biology of Centipedes*. Cambridge University Press, Cambridge-London-New York.

Marek, P. and W. Moore. 2015. Discovery of a glowing millipede in California and the gradual evolution of bioluminescence in Diplopoda. Proc. Natl. Acad. Sci. doi: 10.1073/pnas.1500014112

Marek, P., D. Papaj, J. Yeager, S. Molina and W. Moore. 2011 Bioluminescent aposematism in millipedes. Curr. Biol. 21(18): 680–681.

McElroy, W. D., M. DeLuca and J. Travis 1967. Molecular uniformity in biological catalyses. The enzymes concerned with firefly luciferin, amino acid, and fatty acid utilization are compared. Science 157: 150–160.

Minelli, A. 1993. Chilopoda. Pp. 57–114 in Harrison, F. W. and M. E. Rice (eds.) 1997. *Microscopic Anatomy of Invertebrates. Vol. 12, Onychophora, Chilopoda, and Lesser Protostomata*. Wiley-Liss, New York.

Minelli, A. (ed.) 2011. *Treatise on Zoology–Anatomy, Taxonomy, Biology. The Myriapoda. Volume 1*. Brill, Leiden.

Minelli, A. (ed.). 2015. *The Myriapoda. Volume 2. Treatise on Zoology-Anatomy, Taxonomy, Biology*. Brill, Leiden.

Minelli, A. and S. I. Golovatch. 2013. Myriapods. Pp. 421–432 in Levin, S. A. (ed.), *Encyclopedia of Biodiversity, 2nd Ed., Volume 5*. Academic Press, Waltham, MA.

Molinari, J., E. E. Gutierrez; A. A. De-Ascenção, J. M. Nassar, A. Arends and R. J. Marquez. 2005. Predation by giant centipedes, *Scolopendra gigantea*, on three species of bats in a Venezuelan cave. Caribb. J. Sci. 41: 340–346.

Müller, C. H. G., J. Rosenberg, S. Richter and V. B. Meyer-Rochow. 2003. The compound eye of *Scutigera coleoptrata* (Linnaeus, 1758) (Chilopoda: Notostigmophora): an ultrastructural reinvestigation that adds support to the Mandibulata concept. Zoomorphology 122: 191–209.

Shelley, R. M. 2002. A synopsis of the North American centipedes of the order Scolopendromorpha (Chilopoda). Virginia Mus. Nat. Hist. Mem. No. 5: 1–108.

Shelley, R. M. 2003. A revised, annotated, family-level classification of the Diplopoda. Arthropoda Selecta 11: 187–207.

Shelley, R. M. and S. I. Golovatch. 2011. Atlas of myriapod biogeography. I. Indigenous ordinal and supra-ordinal distributions in the Diplopoda: Perspectives on taxon origins and ages, and a hypothesis on the origin and early evolution of the class. Insecta Mundi 0158: 1–134.

Sierwald, P. and J. E. Bond. 2007. Current status of the myriapod class Diplopoda (millipedes): taxonomic diversity and phylogeny. Annu. Rev. Entomol. 52: 401–420.

Summers, G. 1979. An illustrated key to the chilopods of the north-central region of the United States. J. Kansas Entomol. Soc. 52: 690–700.

Undheim, E. A. and G. F. King. 2011. On the venom system of centipedes (Chilopoda), a neglected group of venomous animals. Toxicon 57: 512–524.

Wesener, T., J. Köhler, S. Fuchs and D. VandenSpiegel. 2011. How to uncoil your partner – "mating songs" in giant pill millipedes (Diplopoda: Sphaerotheriida). Naturwissenschaften 98: 967–975.

Wirkner, C. S. and G. Pass. 2002. The circulatory system in Chilopoda: functional morphology and phylogenetic aspects. Acta Zool. (Stockholm) 83: 193–202.

Myriapod Evolution

Almond, J. E. 1985. The Silurian–Devonian fossil record of the Myriapoda. Phil. Trans. R. Soc. Lond. B 309: 227–237.

Blanke, A. and T. Wesener. 2014. Revival of forgotten characters and modern imaging techniques help to produce a robust phylogeny of the Diplopoda (Arthropoda, Myriapoda). Arthr. Struct. Devel. 43: 63–75.

Bonato, L., L. Drago and J. Murienne 2013. Phylogeny of Geophilomorpha (Chilopoda) inferred from new morphological and molecular evidence. Cladistics 30: 485–507.

Bonato, L. and A. Minelli. 2002. Parental care in *Dicellophilus carniolensis* (C. L. Koch, 1847): New behavioural evidence with implications for the higher phylogeny of centipedes (Chilopoda). Zool. Anz. 241: 193–198.

Kime, R. D. and S. I. Golovatch. 2000. Trends in the ecological strategies and evolution of millipedes (Diplopoda). Biol. J. Linn. Soc. 69: 333–349.

Regier, J. C. and 7 others. 2010. Arthropod relationships revealed by phylogenomic analysis of nuclear protein-coding sequences. Nature 463: 1079–1083.

Brewer, M. S. and J. E. Bond. 2013. Ordinal-level phylogenomics of the arthropod class Diplopoda (millipedes) based on an analysis of 221 nuclear protein-coding loci generated using next-generation sequence analysis. PLoS ONE 8: e79935.

Burmester, T. 2002. Origin and evolution of arthropod hemocyanins and related proteins. J. Comp. Physiol. B 172: 95–107.

Chipman, A. D. and 105 others. 2014. The first myriapod genome sequence reveals conservative arthropod gene content and genome organisation in the centipede *Strigamia maritima*. PLoS Biol. 12(11): e1002005.

Edgecombe, G. D. 2010. Arthropod phylogeny: An overview from the perspectives of morphology, molecular data and the fossil record. Arthr. Struct. Dev. 39: 74–87.

Edgecombe, G. D., L. Bonato and A. Minelli. 2014. Geophilomorph centipedes from the Cretaceous amber of Burma. Palaeontology 57: 97–110.

Edgecombe, G. D., A. Minelli and L. Bonato. 2009. A geophilomorph centipede (Chilopoda) from La Buzinie amber (Late Cretaceous: Cenomanian), SW France. Geodiversitas 31: 29–39.

Kusche, K. and T. Burmester. 2001. Diplopod hemocyanin sequence and the phylogenetic position of the Myriapoda. Mol. Biol. Evol. 18: 1566–1573.

Kusche, K., A. Hembach, S. Hagner-Holler, W. Gebauer and T. Burmester. 2003. Complete subunit sequences, structure and evolution of the 6 × 6-mer hemocyanin from the common house centipede, *Scutigera coleoptrata*. Eur. J. Biochem. 270: 2860–2868.

Minelli, A., A. Chagas-Júnior and G. D. Edgecombe. 2009. Saltational evolution of trunk segment number in centipedes. Evol. Dev. 11: 318–322.

Murienne, J., G. D. Edgecombe and G. Giribet. 2010. Including secondary structure, fossils and molecular dating in the centipede tree of life. Mol. Phylog. Evol. 57: 301–313.

Pisani, D., L. L. Poling, M. Lyons-Weiler and S. B. Hedges. 2004. The colonization of land by animals: molecular phylogeny and divergence times among arthropods. BMC Biol. 2: 1.

Podsiadlowski, L., H. Kohlhagen and M. Koch. 2007. The complete mitochondrial genome of *Scutigerella causeyae* (Myriapoda: Symphyla) and the phylogenetic position of Symphyla. Mol. Phylog. Evol. 45: 251–260.

Regier, J. C. and 7 others. 2010. Arthropod relationships revealed by phylogenomic analysis of nuclear protein-coding sequences. Nature 463: 1079–1083.

Shear, W. A. and G. D. Edgecombe. 2010. The geological record and phylogeny of the Myriapoda. Arthr. Struct. Dev. 39: 174–190.

Wilson, H. M. 2001. First Mesozoic scutigeromorph centipede, from the Lower Cretaceous of Brazil. Palaeontology 44: 489–495.

CHAPTER 24

Phylum Arthropoda

The Chelicerata

The arthropod subphylum Chelicerata includes Xiphosura (horseshoe crabs), Arachnida (spiders, scorpions, mites, ticks, daddy longlegs, and several less familiar groups), and Pycnogonida (sea spiders); the first two taxa together are referred to as Euchelicerata. In addition to the 113,000 or so described living species there is an impressive array of fossil forms (ca. 2,000 species), such as the Paleozoic giant water scorpions (eurypterids), some of which were nearly three meters long. Chelicerata had their origin in ancient Cambrian seas. Pycnogonida and Xiphosura remain entirely marine, but Arachnida invaded land long ago and today are mostly terrestrial. Among the Metazoa, Chelicerata are second only to the insects in species diversity. A few kinds, such as some mites, have secondarily invaded various aquatic habitats. On land, arachnids have adapted to virtually every imaginable situation and lifestyle, with the mites having an enormous diversity of life cycles.

In addition to the basic characteristics common to all arthropods, chelicerates are distinguished by several unique features (Box 24A). The body is typically divided into two main regions—the prosoma and opisthosoma (Figures 24.1–24.3)—but this is different in mites and ticks (see below).[1]

In contrast to most other arthropods, it is not possible to delineate a discrete "head" in chelicerates. There are no antennae, but generally all six segments of the prosoma bear appendages. The first pair of appendages is the chelicerae, followed by the pedipalps, and then four pairs of walking legs. In the pycnogonids there may exist an additional pair of appendages, the ovigers, between the pedipalps and first walking legs, and the number of walking legs can be greater than four pairs. The chelicerae and pedipalps are specialized for an enormous variety of roles

Classification of The Animal Kingdom (Metazoa)

Non-Bilateria*
(a.k.a. the diploblasts)
- PHYLUM PORIFERA
- PHYLUM PLACOZOA
- PHYLUM CNIDARIA
- PHYLUM CTENOPHORA

Bilateria
(a.k.a. the triploblasts)
- PHYLUM XENACOELOMORPHA

Protostomia
- PHYLUM CHAETOGNATHA

SPIRALIA
- PHYLUM PLATYHELMINTHES
- PHYLUM GASTROTRICHA
- PHYLUM RHOMBOZOA
- PHYLUM ORTHONECTIDA
- PHYLUM NEMERTEA
- PHYLUM MOLLUSCA
- PHYLUM ANNELIDA
- PHYLUM ENTOPROCTA
- PHYLUM CYCLIOPHORA

Gnathifera
- PHYLUM GNATHOSTOMULIDA
- PHYLUM MICROGNATHOZOA
- PHYLUM ROTIFERA

Lophophorata
- PHYLUM PHORONIDA
- PHYLUM BRYOZOA
- PHYLUM BRACHIOPODA

ECDYSOZOA

Nematoida
- PHYLUM NEMATODA
- PHYLUM NEMATOMORPHA

Scalidophora
- PHYLUM KINORHYNCHA
- PHYLUM PRIAPULA
- PHYLUM LORICIFERA

Panarthropoda
- PHYLUM TARDIGRADA
- PHYLUM ONYCHOPHORA
- **PHYLUM ARTHROPODA**
  - SUBPHYLUM CRUSTACEA*
  - SUBPHYLUM HEXAPODA
  - SUBPHYLUM MYRIAPODA
  - **SUBPHYLUM CHELICERATA**

Deuterostomia
- PHYLUM ECHINODERMATA
- PHYLUM HEMICHORDATA
- PHYLUM CHORDATA

*Paraphyletic group

This chapter has been revised by Gonzalo Giribet and Gustavo Hormiga.

[1]These two body regions are sometimes also referred to as the cephalothorax and abdomen, but they are not homologous to those regions in other arthropods.

BOX 24A Characteristics of the Subphylum Chelicerata

1. Body composed of two tagmata: the prosoma and the opisthosoma. Prosoma composed of six somites, often covered by a carapace-like dorsal shield. Opisthosoma composed of up to 12 somites and a postsegmental telson; subdivided into two parts in some groups.
2. Appendages of prosoma are chelicerae, pedipalps, and four pairs of walking legs; antennae are absent. All appendages are multiarticulate.
3. Gas exchange by book gills, book lungs, tracheae, or through the cuticle
4. Excretion by coxal glands and/or Malpighian tubules. (Chelicerate Malpighian tubules are probably not homologous to those of insects, terrestrial isopods, or tardigrades.)
5. With simple median eyes and lateral (compound in xiphosurans) eyes
6. Gut with two to six pairs of digestive ceca
7. Mostly gonochoristic

in the various chelicerate groups, including sensation, locomotion, and copulation; some of the walking legs also may assume some of these functions, especially tactile (e.g., the first or second legs of some arachnid orders) or copulatory (e.g., the third leg of the male ricinuleids). The opisthosoma usually bears a terminal postsegmental telson. In some Xiphosura, and in some arachnid orders, the opisthosoma may be subdivided into two regions, a mesosoma and a metasoma (Figure 24.1). A detailed discussion of chelicerate anatomy is provided in the body plan section below.

Chelicerata is a large and diverse taxon comprising three rather different groups: Xiphosura, which includes the horseshoe crabs; Arachnida, including the terrestrial spiders, scorpions, and mites, among others; and Pycnogonida, the "sea spiders." To assist readers in gaining a grasp of this large and diverse subphylum, we treat the Euchelicerata and Pycnogonida separately in this chapter.

SUBPHYLUM CHELICERATA

CLASS PYCNOGONIDA The Sea Spiders

CLASS EUCHELICERATA The Euchelicerates

SUBCLASS MEROSTOMATA

ORDER EURYPTERIDA† Extinct giant water scorpions (Ordovician to Permian), 246 species

ORDER XIPHOSURA Horseshoe crabs *Limulus*, *Carcinoscorpius*, *Tachypleus*

ORDER CHASMATASPIDIDA† Extinct lineage (Ordovician to Devonian) or grade giving origin to Eurypterida

SUBCLASS ARACHNIDA Spiders, scorpions, mites, ticks, and their kin. Sixteen extant orders

ORDER AMBLYPYGI Whip spiders, tailless whip scorpions. (e.g., *Acanthophrynus*, *Damon*, *Heterophrynus*, *Stegophrynus*, *Tarantula*)

ORDER ARANEAE True spiders

SUBORDER MESOTHELAE "Segmented" spiders. One family, Liphistiidae. (e.g., *Heptathela*, *Liphistius*)

SUBORDER OPISTHOTHELAE "Modern" spiders

INFRAORDER MYGALOMORPHAE Tarantula-like spiders; orthognathous chelicerae. About 15 families, including the following:

FAMILY CTENIZIDAE Trapdoor spiders. (e.g., *Cyclocosmia*, *Ummidia*)

FAMILY EUCTENIZIDAE (e.g., *Eucteniza*, *Aptostichus*)

FAMILY ATYPIDAE Purse-web spiders. (e.g., *Atypus*, *Sphodros*)

FAMILY THERAPHOSIDAE Tarantulas and bird spiders. (e.g., *Acanthoscurria*, *Aphonopelma*, *Theraphosa*)

FAMILY DIPLURIDAE Diplurid funnel-web spiders. (e.g., *Australothele*, *Diplura*, *Euagrus*)

FAMILY HEXATHELIDAE Hexathelid funnel-web spiders. (e.g., *Atrax*, *Hadronyche*, *Macrothele*)

INFRAORDER ARANEOMORPHAE "Typical" spiders; labidognathous chelicerae. About 75 families including the following:

FAMILY HYPOCHILIDAE Lampshade weavers. (e.g., *Hypochilus*, *Ectatosticta*)

FAMILY GRADUNGULIDAE Long-claw spiders. (e.g., *Gradundula*, *Progradungula*)

FAMILY AUSTROCHILIDAE Missing-link spiders. (e.g., *Austrochilus*, *Hickmania*)

FAMILY FILISTATIDAE Crevice weavers. (e.g., *Filistata*, *Kukulcania*)

FAMILY OONOPIDAE Goblin spiders. (e.g., *Costarina*, *Oonops*, *Opopaea*, *Orchestina*)

FAMILY DYSDERIDAE Woodlouse hunters. (e.g., *Dysdera*, *Harpactocrates*, *Rhode*)

FAMILY PHOLCIDAE Cellar spiders, daddy longleg spiders. (e.g., *Modismus*, *Pholcus*, *Spermophora*)

FAMILY SCYTODIDAE Spitting spiders. (e.g., *Scytodes*)

FAMILY SICARIIDAE Brown recluse spiders. (e.g., *Loxosceles*, *Sicarius*)

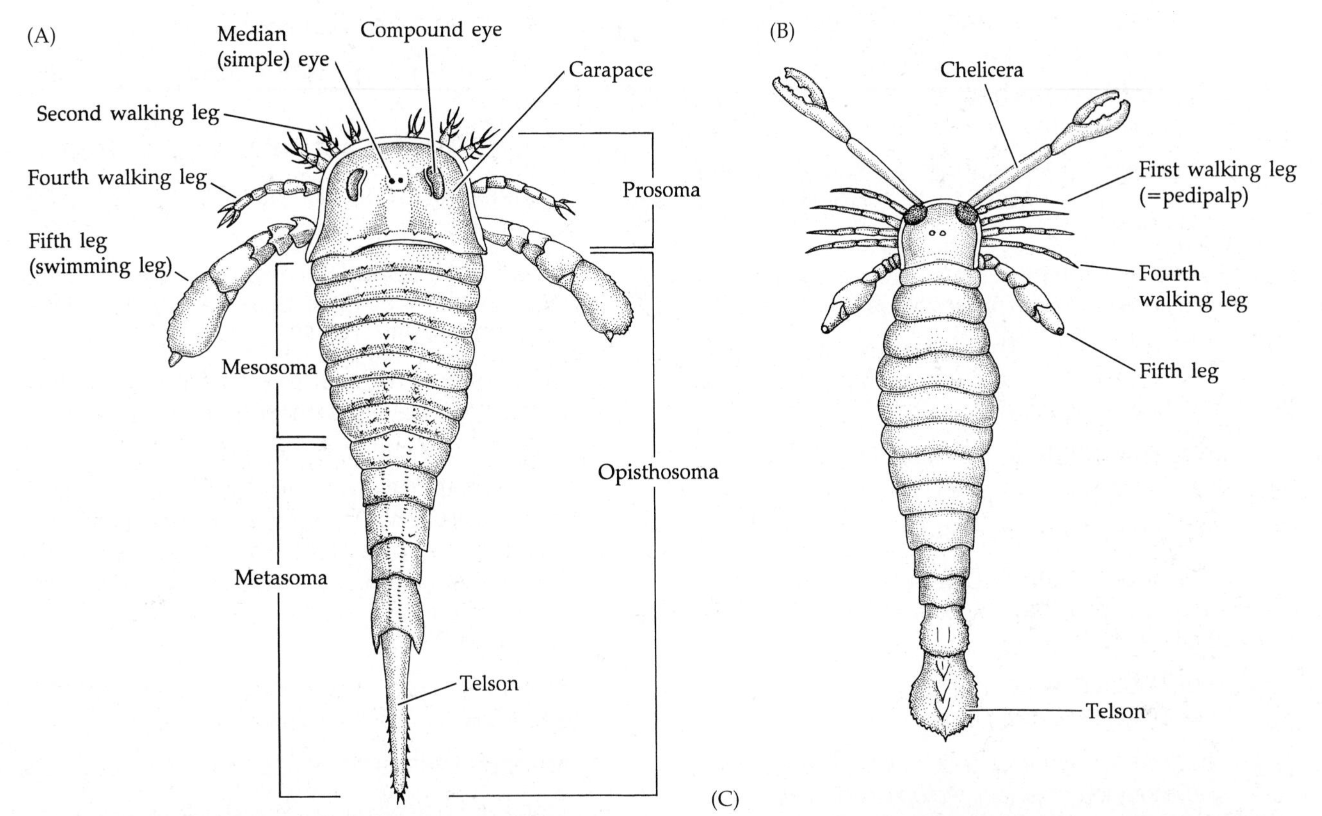

(C)

Figure 24.1 Water scorpions (subclass Merostomata, order Eurypterida), an extinct group of chelicerates. (A) *Eurypterus* (dorsal view). (B) *Pterygotus buffaloensis*, which reached lengths of nearly 3 meters. (C) *Eurypterus remipes*, a species from the Silurian period of the Paleozoic. Eurypterids flourished in Paleozoic seas; some probably invaded fresh water and perhaps even land.

FAMILY ARCHAEIDAE Assassin spiders. (e.g., *Austrarchaea, Eriauchenius*)

FAMILY ERESIDAE Velvet spiders. (e.g., *Eresus, Stegodyphus*)

FAMILY THERIDIIDAE Cobweb and widow spiders. (e.g., *Anelosimus*, *Argyrodes*, *Episinus*, *Latrodectus*, *Parasteatoda*, *Steatoda*)

FAMILY ULOBORIDAE Cribellate orb-weavers. (e.g., *Hyptiotes*, *Uloborus*)

FAMILY DEINOPIDAE Ogre-faced spiders. (e.g., *Deinopis, Menneus*)

FAMILY ARANEIDAE Ecribellate orb-weavers. (e.g., *Araneus*, *Argiope*, *Cyrtophora*, *Gasteracantha*, *Mastophora*, *Nephila*, *Neoscona*, *Zygiella*)

FAMILY TETRAGNATHIDAE Longjawed orb-weavers. (e.g., *Dolichognatha*, *Leucauge*, *Meta*, *Pachygnatha, Tetragnatha*)

FAMILY MIMETIDAE Pirate spiders. (e.g., *Australomimetus*, *Ero, Mimetus*)

FAMILY LINYPHIIDAE Sheet-web spinners, dwarf spiders. (e.g., *Agyneta*, *Dubiaranea*, *Erigone*, *Haplinis*, *Oedothorax*, *Linyphia, Neriene*)

FAMILY ANAPIDAE Ground orb-weavers. (e.g., *Acrobleps, Anapis, Comaroma, Micropholcomma*)

FAMILY CLUBIONIDAE Sack-spiders. (e.g., *Clubiona, Elaver*)

FAMILY AGELENIDAE Funnel-weavers. (e.g., *Agelena, Coelotes, Tegenaria*)

FAMILY LYCOSIDAE Wolf spiders. (e.g., *Lycosa, Pardosa, Pirata, Schizocosa*)

FAMILY CTENIDAE Wandering spiders. (e.g., *Anahita, Ctenus, Cupiennius, Phoneutria*)

FAMILY PISAURIDAE Nursery-web spiders. (e.g., *Architis, Dolomedes, Pisaura*)

FAMILY OXYOPIDAE Lynx spiders. (e.g., *Oxyopes, Peucetia*)

FAMILY THOMISIDAE Crab spiders. (e.g., *Diaea, Misumena, Thomisus, Xysticus*)

FAMILY SPARASSIDAE Huntsman spiders. (e.g., *Delena, Heteropoda, Micrommata, Olios, Pandercetes*)

FAMILY GNAPHOSIDAE Ground spiders. (e.g., *Drassodes, Eilica, Gnaphosa, Micaria, Zelotes*)

FAMILY SALTICIDAE Jumping spiders. (e.g., *Habronattus, Lyssomanes, Phidippus, Portia, Salticus, Spartaeus*)

ORDER HAPTOPODA† Extinct group (Carboniferous). (e.g., *Plesiosiro*)

ORDER OPILIONES Harvestmen, daddy longlegs

SUBORDER CYPHOPHTHALMI Mite harvestmen. (e.g., *Cyphophthalmus, Siro, Pettalus, Neogovea, Stylocellus, Troglosiro*)

SUBORDER EUPNOI Daddy longlegs. (e.g., *Caddo, Leiobunum, Neopilio, Phalangium*)

SUBORDER DYSPNOI Daddy longlegs, hard-bodied daddy longlegs. (e.g., *Acropsopilio, Ischyropsalis, Nemastoma, Ortholasma, Sabacon, Trogulus*)

SUBORDER LANIATORES Armored harvestmen. (e.g., Equitius, *Fumontana, Sitalcina, Sandokan, Stygnomma, Zalmoxis, Vonones*)

ORDER PALPIGRADI Palpigrades, micro-whip scorpions. (e.g., *Allokoenenia, Eukoenenia, Koenenia, Leptokoenenia, Prokoenenia*)

ORDER PHALANGIOTARBIDA† Extinct group (Devonian to Permian) abundant in the Upper Carboniferous coal measures of Europe and North America. (e.g., *Phalangiotarbus*)

ORDER PSEUDOSCORPIONES Pseudoscropions, false scorpions. (e.g., *Chelifer, Chitrella, Chthonius, Dinocheirus, Garypus, Menthus, Pseudogarypus*)

ORDER RICINULEI Ricinuleids or hooded-tick spiders. Three genera: *Cryptocellus, Pseudocellus, Ricinoides*

ORDER SCHIZOMIDA Schizomids. (e.g., *Agastoschizomus, Megaschizomus, Nyctalops, Protoschizomus, Schizomus*)

ORDER SCORPIONES Scorpions. (e.g., *Androctonus, Bothriurus, Buthus, Centruroides, Chactus, Chaerilus, Diplocentrus, Hadrurus, Hemiscorpion, Nebo, Parabuthus, Paruroctonus, Tityus, Vaejovis*)

SUPERFAMILY BUTHOIDEA (e.g., *Androctonus, Buthus, Centruroides*)

SUPERFAMILY CHAERILOIDEA (e.g., *Chaerilus*)

SUPERFAMILY PSEUDOCHACTOIDEA (e.g., *Troglokhammouanus*)

SUPERFAMILY IUROIDEA (e.g., *Iurus*)

SUPERFAMILY BOTHRIUROIDEA (e.g., *Bothriurus*)

SUPERFAMILY CHACTOIDEA (e.g., *Euscorpius*)

SUPERFAMILY SCORPIONOIDEA (e.g., *Diplocentrus*)

ORDER SOLIFUGAE Sun spiders, wind spiders, camel spiders. (e.g., *Biton*, *Branchia*, *Dinorhax*, *Galeodes*, *Solpuga*)

ORDER TRIGONOTARBIDA† Extinct group (Silurian to Permian) abundant in Carboniferous coal measures. (e.g., *Trigonotarbus*).

ORDER URARANEIDA† Extinct group (Devonian to Permian) established in 2008 for two fossils previously thought to be spiders. Uraraneids were spider-like animals with a segmented opisthosoma and a flagelliform telson and they could produce silk from spigots. (e.g., *Attercopus*)

ORDER UROPYGI Whip scorpions or vinegaroons. (e.g., *Albaliella*, *Chajnus*, *Mastigoproctus*)

SUPERORDER PARASITIFORMES Mites and ticks

ORDER HOLOTHYRIDA Ca. 25 species mainly feeding on the fluids of dead arthropods. (e.g., *Allothyrus, Neothyrus*)

ORDER IXODIDA Hard ticks and soft ticks. (e.g., *Amblyomma, Dermacentor, Ixodes, Rhipicephalus, Argas, Nuttalliella*)

ORDER MESOSTIGMATA Free-living predatory mites. (e.g., *Uropoda, Zeroseius*)

SUPERORDER OPILIOACARIFORMES

ORDER OPILOACARIDA Primitive mites. (e.g., *Opilioacarus, Neocarus*)

SUPERORDER ACARIFORMES Mites, dust mites, and "chiggers". (e.g., *Demodex, Dermatophagoides, Halotydeus, Penthaleus, Scirus, Tydeus*)

ORDER TROMBIDIFORMES A large, diverse order of mites, comprising more than 22,000 described species in 125 families. (e.g., *Trombidium, Scirus, Demodex, Tydeus*)

SUBORDER PROSTIGMATA (e.g., *Halotydeus, Penthaleus*)

SUBORDER SPHAEROLICHIDA

ORDER SARCOPTIFORMES Large order of mites, comprising around 230 families and 15,000 described species.

SUBORDER ORIBATIDA "Chewing" mites, moss mites, or beetle mites. (e.g., *Archegozetes, Conoppia*)

SUBORDER ASTIGMATA "Biting mites," dust mites, many parasites of vertebrates. (e.g., *Acarus, Dermatophagoides*)

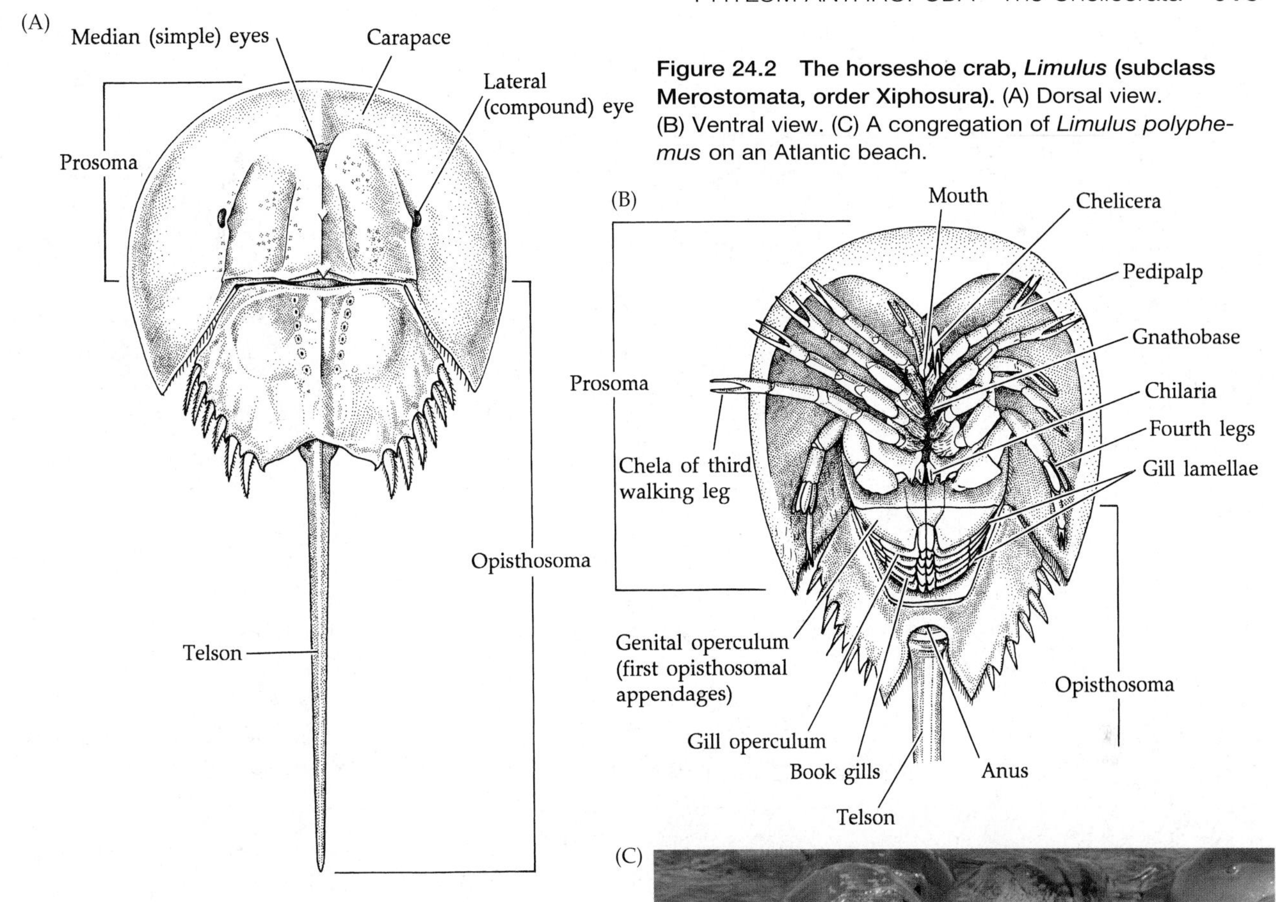

Figure 24.2 The horseshoe crab, *Limulus* (subclass Merostomata, order Xiphosura). (A) Dorsal view. (B) Ventral view. (C) A congregation of *Limulus polyphemus* on an Atlantic beach.

Chelicerate Classification

Class Euchelicerata

Body composed of two tagmata: prosoma and opisthosoma. Prosoma composed of presegmental acron plus 6 segments, often covered by a carapace-like dorsal shield; with simple median eyes and compound or simple lateral eyes. Opisthosoma of up to 12 segments (subdivided into two parts in some groups) and a postsegmental telson. Prosomal appendages are chelicerae, pedipalps, and four pairs of walking legs; antennae absent; all appendages uniramous and multiarticulated.

Subclass Merostomata

Prosoma covered by a large, hardened, carapace-like shield; pedipalps similar to walking legs; opisthosoma undivided or divided into mesosoma and metasoma; with flaplike appendages as book gills; telson long and spiked. The merostomates diversified during the great Cambrian invertebrate radiation. Fossil eurypterids and xiphosurans are known from the Ordovician, and they flourished in the Silurian and Devonian. Alas, only four species of merostomates have survived to modern times, all xiphosurans—the horseshoe crabs.

Order Eurypterida† The extinct giant water scorpions (Figure 24.1). Opisthosoma divided, with scalelike appendages on mesosoma; metasoma narrow (e.g., *Eurypterus*, *Pterygotus*); 246 extinct species. Eurypterids represented a zenith in arthropod body size, some being nearly 3 m long (e.g., *Pterygotus*). These giant chelicerates roamed ancient seas and freshwater environments well into the Permian and were very abundant during their heyday. There is evidence that some species even became amphibious or semiterrestrial. Eurypterids were probably capable of swimming as well as crawling. The last pair of prosomal limbs were greatly enlarged, flattened distally, and probably used as paddles. The chelicerae were extremely reduced in some species, but well developed and chelate in others—evidence that eurypterids radiated ecologically and exploited a variety of food resources and feeding strategies.

Order Xiphosura Certain extinct forms and the living horseshoe crabs (Figure 24.2). Opisthosoma unsegmented and undivided, but with six pairs of flaplike appendages, the first pair fused medially as a genital

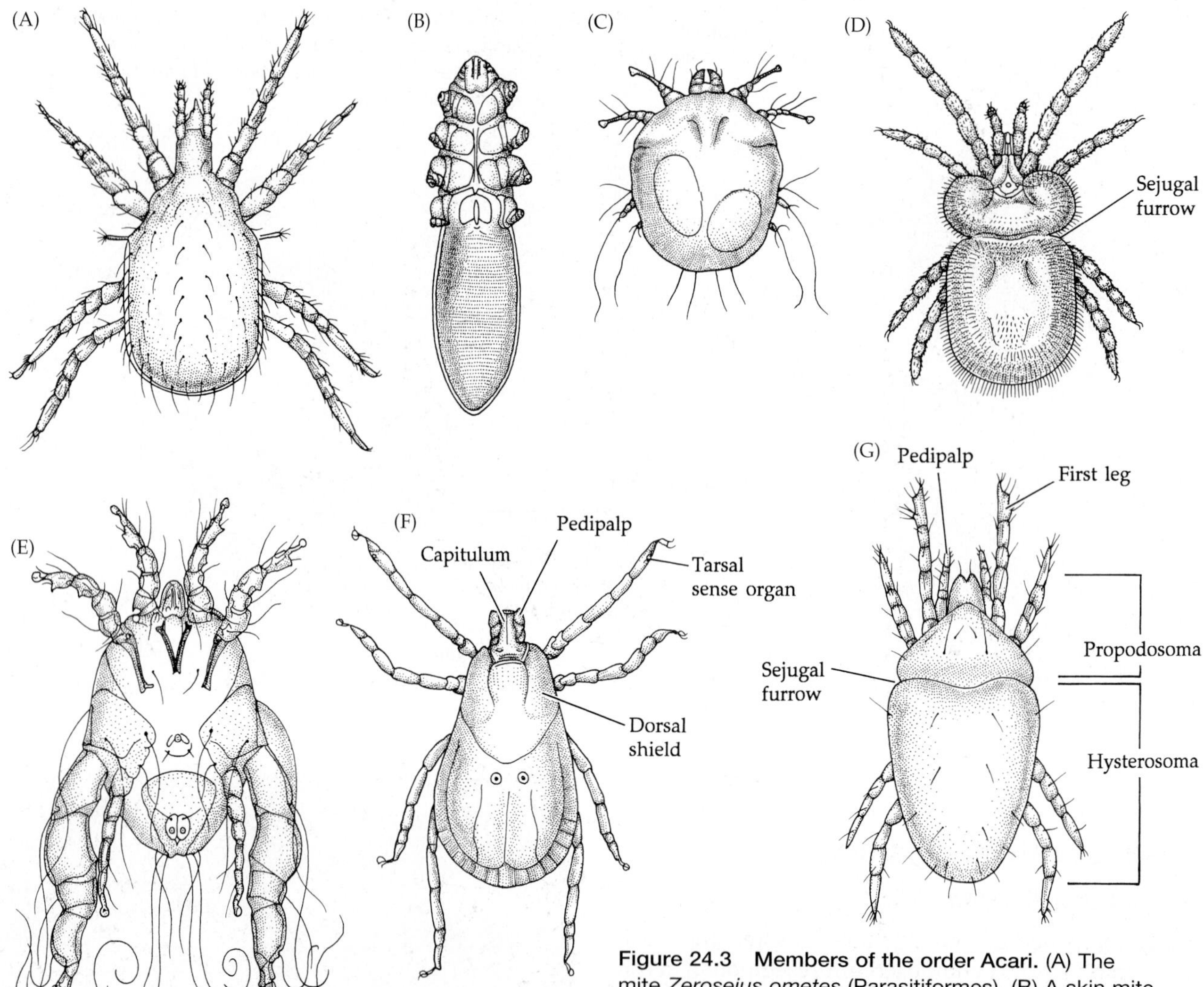

Figure 24.3 Members of the order Acari. (A) The mite *Zeroseius ometes* (Parasitiformes). (B) A skin mite, *Demodex* (Acariformes), with very short legs. (C) A mange mite, *Sarcoptes scabiei* (Acariformes). (D) An adult chigger mite, *Trombicula* (Acariformes). Note the sejugal furrow between the second and third legs. (E) One of the feather mites, *Analges* (Acariformes). (F) A tick, *Dermacentor* (Parasitiformes). (G) *Tydeus starri* (Acariformes), with a deep sejugal furrow. (H) An opilioacarid mite, *Neoacarus* sp. (Opilioacariformes) from South Africa. (I) Phoretic mites on the harvestman *Rhampsinitus transvaalicus*. (J) Phoretic mites on the harvestman *Sadocus polyacanthus*. (K) A Chyzeriidae mite.

operculum over the gonopores, the last five pairs modified as book gills. Pedipalps and walking legs chelate; last (fourth) pair of legs splayed distally for support on soft sediments (e.g., *Carcinoscorpius, Limulus, Tachypleus*).

Extant members of the order Xiphosura (the horseshoe crabs) are regarded as "living fossils" and are, of course, far better known than their extinct relatives. Especially well studied is *Limulus polyphemus* (Figure 24.2C), the common horseshoe crab of the Atlantic and Gulf coasts of North America and a favorite laboratory animal of physiologists. Living xiphosurans inhabit shallow marine waters, generally on clean sandy bottoms, where they crawl about or burrow just beneath the surface, preying on other animals or scavenging. The chelicerae are smaller than the other appendages, being composed of only three articles. Each walking leg is formed of seven articles (coxa, trochanter, femur, patella, tibia, tarsus, pretarsus), the last two of which form the chelae (Figure 24.2B). The coxal endites of the pedipalps and first three walking legs are modified as gnathobases for mastication. Arising from the coxae of the fourth walking legs are tiny appendage-like processes called flabella. In addition, just posterior and medial to the last walking legs is a pair of reduced appendages called chilaria. Their function is unknown, and there is some controversy about their evolutionary significance—some specialists believe that they may reflect an additional opisthosomal segment.

The four living species of Xiphosura belong to three geographically distinct genera in the family Limulidae. *Limulus* (*L. polyphemus*) is restricted to eastern North America, ranging from Nova Scotia to the Yucatán region of Mexico; *Tachypleus* (*T. gigas* and *T. tridentatus*) occurs in Southeast Asia; *Carcinoscorpius* (*C.

rotundicauda) has been collected only from Malaysia, Thailand, and the Philippines, and is capable of migrating upriver into freshwater. Another 98 extinct species have been described, the oldest being a species of *Lunataspis* from 445 million year old deposits in Manitoba (Canada).

Subclass Arachnida

Prosoma wholly or partly covered by a carapace-like shield; opisthosoma segmented or unsegmented, divided (in scorpions) or undivided; opisthosomal appendages absent or modified as spinnerets (spiders) or pectines (scorpions); penes absent (except in some Opiliones and some mites); gas exchange by tracheae, book lungs, or both, but also through the cuticle; nearly all terrestrial; over 110,000 species, in 16 orders.

"Acari" The term "Acari" has been traditionally used for several lineages of mites, ticks, and chiggers (Figure 24.3) that are now thought to be non-monophyletic and that are currently classified in at least six orders in two main clades. These two clades of uncertain phylogenetic position within the arachnid tree of life are: Parasitiformes (encompassing the orders Holothyrida, Ixodida, Mesostigmata) + Opilioacariformes (order Opilioacarida); and Acariformes (encompassing the orders Trombidiformes and Sarcoptiformes). These two lineages include a huge morphological disparity and a large diversity of ecological and reproductive strategies. For convenience the six acarine orders (in three superorders) are discussed together below.

In mites the prosoma and opisthosoma are fused, and a region of flexible cuticle (the **cirumcapitular furrow**) separates the chelicerae and pedipalps from the rest of the body. This anterior body region is called the **capitulum** or gnathosoma, and the remainder of the body is called the **idiosoma**, which is in turn divided into **podosoma** and **opisthosoma**. Another nomenclature divides the body into proterosoma and hysterosoma: the proterosoma includes the gnathosoma and the first two pedigerous segments, and the hysterosoma includes the last two pedigerous segments and the opisthosoma. The old junction between prosoma and opisthosoma is usually visible as the **disjugal furrow**, but more often these are indistinguishably fused; sometimes with a secondary sejugal furrow between the second and third walking legs in Acariformes; chelicerae pincer-like or styliform; pedipalpal coxae uniquely fused with head elements; opisthosoma unsegmented except in a few primitive forms; gas exchange cutaneous or by tracheae; eyes present or absent; males of some species with a penis.

Mites and ticks comprise the largest group of arachnids. There are approximately 54,600 described species, and some experts suggest that there may be as many as one million or more yet to be described! Because of the size and diversity of these orders, we give a somewhat extended account. Even so, it is not possible to do justice here to the vast range of forms and lifestyles represented by the mites and ticks.

Mites and ticks occur worldwide. Most are terrestrial, many are parasitic, and some have invaded aquatic environments, both freshwater and marine. Their tremendous evolutionary success, especially the Acariformes, is reflected in their species diversity and in their extremely varied lifestyles. This success is probably due, at least in part, to a compact body and reduced size. By coupling small size with the inherent arthropod quality of segment and appendage specialization, mites have exploited myriad microhabitats unavailable to larger animals. If the group is polyphyletic, then several ancestral arachnid lineages evolved convergently to capitalize on miniaturization.

Recent classification schemes divide the members of the old "Acari" into two or three groups. One of these is a group of omnivorous and predatory forms in the order Opilioacarida (Figure 24.3H). They are characterized by the retention of opisthosomal segmentation (at least ventrally) and the presence of a transverse groove (the disjugal furrow) separating the prosoma from the opisthosoma. These mites are found on tropical forest floors and in arid temperate habitats.

The remainder of the thousands of mites and ticks are placed in Parasitiformes and Acariformes, the latter housing most of the species. Opilioacarida form a clade with Parasitiformes, and in some classifications they are indeed considered members of this superorder. The body of parasitiform mites is undivided dorsally, there being no clear transverse groove (Figure 24.3A). Parasitiformes include free-living and symbiotic forms in all parts of the world. The free-living species inhabit various terrestrial habitats, including leaf litter, decaying wood and other organic debris, mosses, insect and small mammal nests, and soil. Most of these mites are predators on small invertebrates. Many species are full-time or part-time symbionts on other animals, either as immatures or as adults. Their hosts are frequently other arthropods, such as centipedes, millipedes, ants, and especially beetles. In some cases the relationship is truly parasitic, in others it is phoretic (Figure 24.3I,J), and in many instances the nature of the association is unknown.

The most familiar members of the Parasitiformes are the ticks (families Argasidae and Ixodidae; Figure 24.3F). Ticks are blood-sucking ectoparasites on vertebrates (one species, *Aponomma ecinctum*, lives on beetles). They are the largest-bodied members of the entire order, and some swell to 2 to 3 cm in length during a blood meal. The chelicerae are smooth and adapted for slicing skin, and together with the pedipalp coxae form a structure called the capitulum. The ixodids are called the hard ticks because of a sclerotized shield covering the entire dorsum. They parasitize reptiles, birds, and mammals. They generally remain attached to their hosts for days or even weeks, feeding on blood. Some are vectors of important diseases, including *Dermacentor andersoni* (the vector of Rocky Mountain spotted fever) and *Boophilus annulatus* (the vector of Texas cattle fever). Lyme disease (first described in Old Lyme, Connecticut, in 1975) is a bacterial disease that is harbored in deer and certain rodents and is transmitted by several tick species in North America, including the western black-legged tick *Ixodes pacificus*.

Soft ticks (Argasidae) lack the heavily sclerotized dorsal shield of hard ticks. They are usually rather transient parasites on avian and mammalian hosts (particularly bats), typically feeding for less than an hour at a time. When not attached to a host, these ticks remain hidden in cracks and crevices or buried in soil. Disease-carrying soft ticks include *Argas persicus* (the vector of fowl spirochetosis) and *Ornithonodorus moubata* (the vector of African relapsing fever, or "tick fever").

The vast and diverse Acariformes, once considered to be polyphyletic, are now recognized as a well-supported clade. These mites usually have bodies divided into two regions, but not as the normal prosoma and opisthosoma, which are fused. Rather, the disjugal furrow is lost, and a secondarily evolved sejugal furrow partly or wholly traverses the dorsum between the origins of the second and third pairs of walking legs (Figures 24.3D,G). In some acariform mites this division is secondarily lost.

Free-living acariform mites are found in virtually every conceivable situation: in soil, leaf litter, decaying organic matter, mosses, lichens, and fungi; under bark; on freshwater algae; in sands; on seaweed; at all altitudes and at most ocean depths. "Chiggers" is a name given to the larvae of acariform mites of several genera, notably *Trombicula*. They include both herbivores (some fungivores) and predators, and their feeding methods are diverse. Many ingest solid as well as liquid food, and a few aquatic forms actually suspension feed. One species, *Agauopsis auzendei*, is known from hydrothermal vent environments. Certain groups are serious pests that destroy stored grain crops and other food products. On the other hand, some predatory acariform mites have been used for biological pest control of other arthropods—even other mites!

Most of the symbiotic acariform mites are parasites on vertebrate and invertebrate hosts. Various species parasitize marine and freshwater crustaceans, freshwater insects, marine molluscs, terrestrial arthropods, the pulmonary chambers of terrestrial snails and slugs, the outer surfaces of all groups of terrestrial vertebrates, and the nasal passages of amphibians, birds, and mammals. In addition to direct parasitism, many mites are phoretic, using their hosts for dispersal. There are also many plant-eating mites that are considered parasitic.

A great many acariform mites cause economic or medical problems by direct predation and parasitism, by acting as disease vectors, or by feeding on stored food products. The family Penthaleidae includes the red-legged earth mite (*Halotydeus destructor*) and the winter grain mite (*Penthaleus major*), both of which are serious pests of many important crops. Members of the superfamily Eriophyoidea are vermiform mites adapted to feeding on various plants. This group includes the gall and leaf-curl mites, as well as a number of others that serve as vectors for certain disease-causing viruses (e.g., wheat and rye mosaic viruses). Another family of mites (Demodicidae) includes parasites of hair follicles and sebaceous glands of mammals. Two species, *Demodex folliculorum* and *D. brevis*, occur specifically in the hair follicles and sebaceous glands, respectively, of the human forehead. Another, *D. canis*, causes mange in dogs. Some other problems caused by acariform mites include subcutaneous tumorlike growths in humans, various sorts of skin irritations, mange in many domestic animals, reduced wool production in sheep, and feather loss in birds.

With a mixture of reluctance and relief, we leave the mites; interested readers should consult the references at the end of this chapter for sources of additional information.

Order Amblypygi Whip spiders (Figure 24.4A,B). Prosoma undivided, covered by a carapace-like shield and connected to opisthosoma by narrow pedicel; opisthosoma segmented but undivided; with two pairs of book lungs; telson absent; chelicerae biarticulated, modified as spider-like fangs; pedipalps raptorial; first pair of legs are greatly elongate as antenniform sensory appendages. With eight eyes, except in some cave-dwelling species. Molting occurs throughout life as the animals keep growing in size.

The 160 or so extant species of amblypygids (there are 9 described fossil species) are commonly known as whip spiders or tailless whip scorpions, reflecting their similarities to both spiders and uropygids (whip scorpions). Externally they resemble whip scorpions, they have some internal characters related to spiders, but they lack spinnerets and poison glands. Amblypygids are widely distributed in warm humid areas, where they are found under bark, logs, or in leaf litter, and in similar protected habitats; several species are cave dwellers. Some species are able to stay underwater hours at a time or live in beaches; others live in extremely dry environments, including deserts. One West African small blind species inhabits termite nests.

Most species are less than 5 cm in length, but the first pair of legs may be as long as 25 cm! These limbs are used as touch and chemoreceptors (like antennae in other, nonchelicerate arthropods) and are important in prey location because these animals hunt at night. Amblypygids walk sideways with these long "feelers" extended, sensing for potential prey. Once located, the prey is grasped by the enlarged spiny pedipalps and torn open by the chelicerae. The body fluids are then sucked from the victim and ingested. Many species of amblypygid have complex mating behaviors.

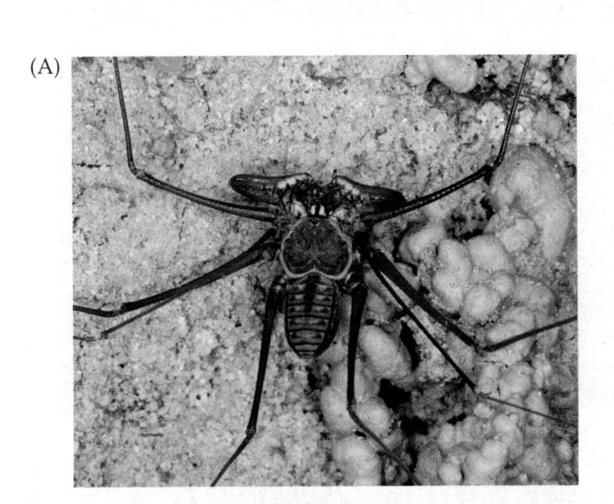

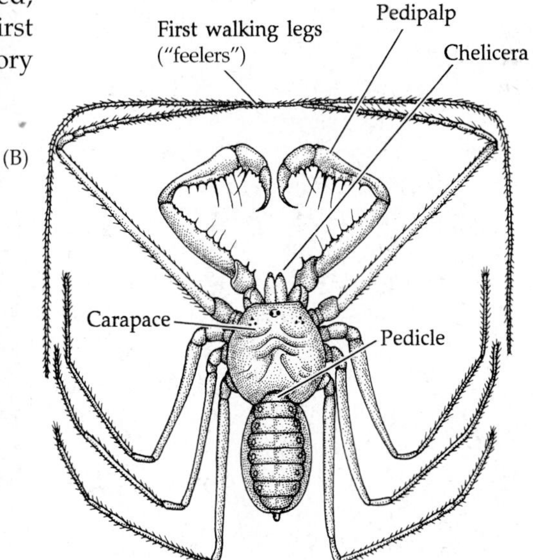

Figure 24.4 Some examples of arachnids other than spiders and scorpions. (A,B) Whip spiders (order Amblypygida): (A) *Heterophrynus batesii* (from the Amazon); (B) *Stegophrynus dammermani*. (C,D) Daddy longlegs, or harvestment (order Opiliones): (C) A daddy longlegs; (D) An armored harvestman (Laniatores), *Phareicranaus manauara*. (E) *Koenenia* (Palpigradi). (F,G,H) Pseudoscorpions (order Pseudoscorpiones): (F) *Chelifer cancroides*, a frequent hitchhiker on houseflies; (G) An unidentified female pseudoscorpion from New Zealand carrying its eggs; (H) A pseudoscorpion grabbing on to a harvestman's leg as a form of phoresy. (I,J) Ricinuleids (order Ricinulei): (I) *Ricinoides crassipalpe*, dorsal view; (J) A nymph of *Ricinoides atewa*. (K) A schizomid (order Schizomida), *Nyctalops crassicaudatus*. (L,M) Sun spiders (order Solifugae): (L) *Galeodes arabs*; (M) *Eremobates*. (N,O) Whip scorpions (order Uropygi): (N) *Mastigoproctus*, (O) *Thelyphonellus amazonicus*. (P) The cave palpigrade *Eukoenenia spelaea.* (Q) An unidentified schizomid from Panama.

(*Continued on next page*)

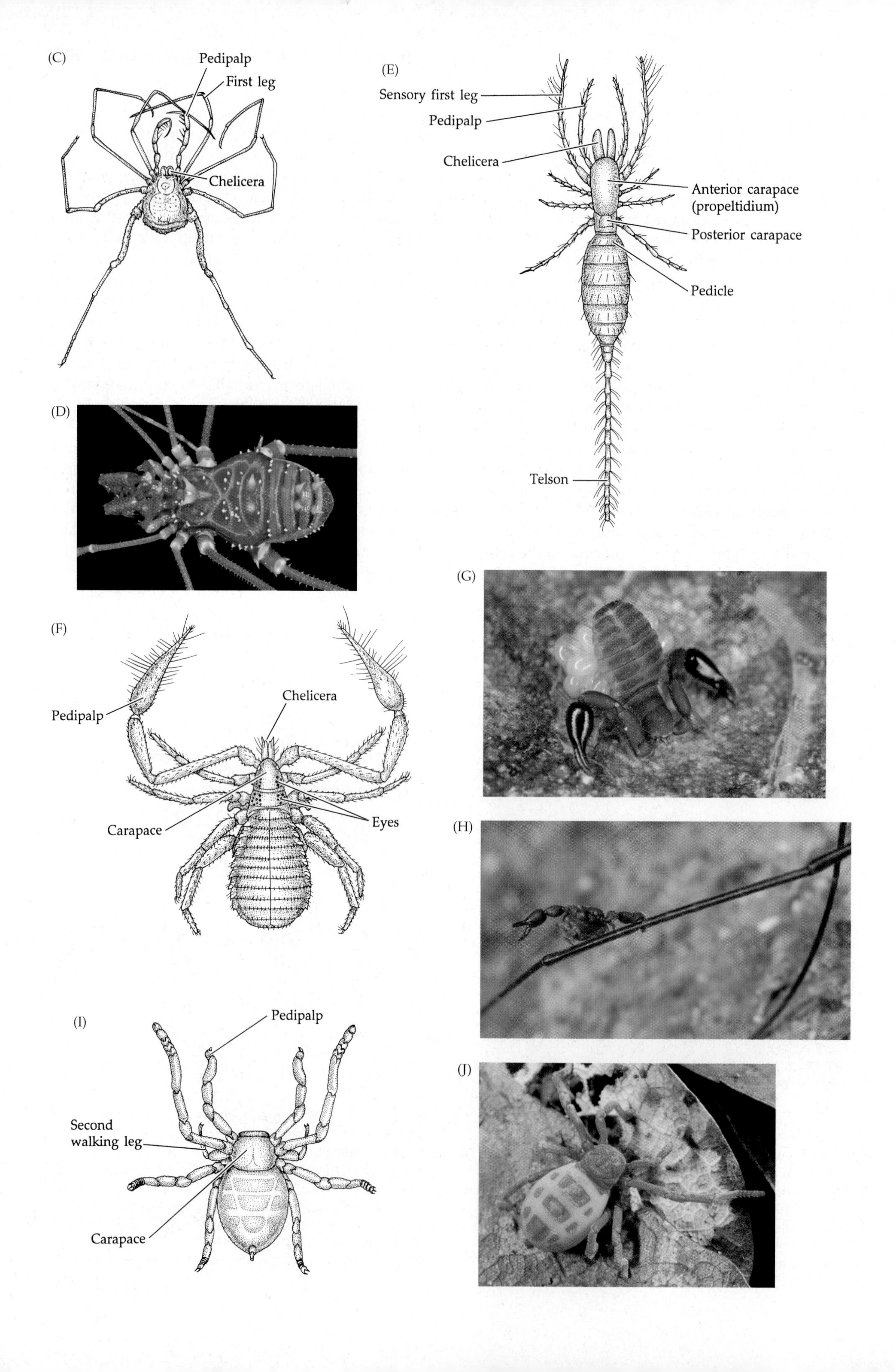
(C)
Pedipalp
First leg
Chelicera
(D)
(E)
Sensory first leg
Pedipalp
Chelicera
Anterior carapace
(propeltidium)
Posterior carapace
Pedicle
Telson
(F)
Pedipalp
Chelicera
Carapace
Eyes
(G)
(H)
(I)
Pedipalp
Second
walking leg
Carapace
(J)

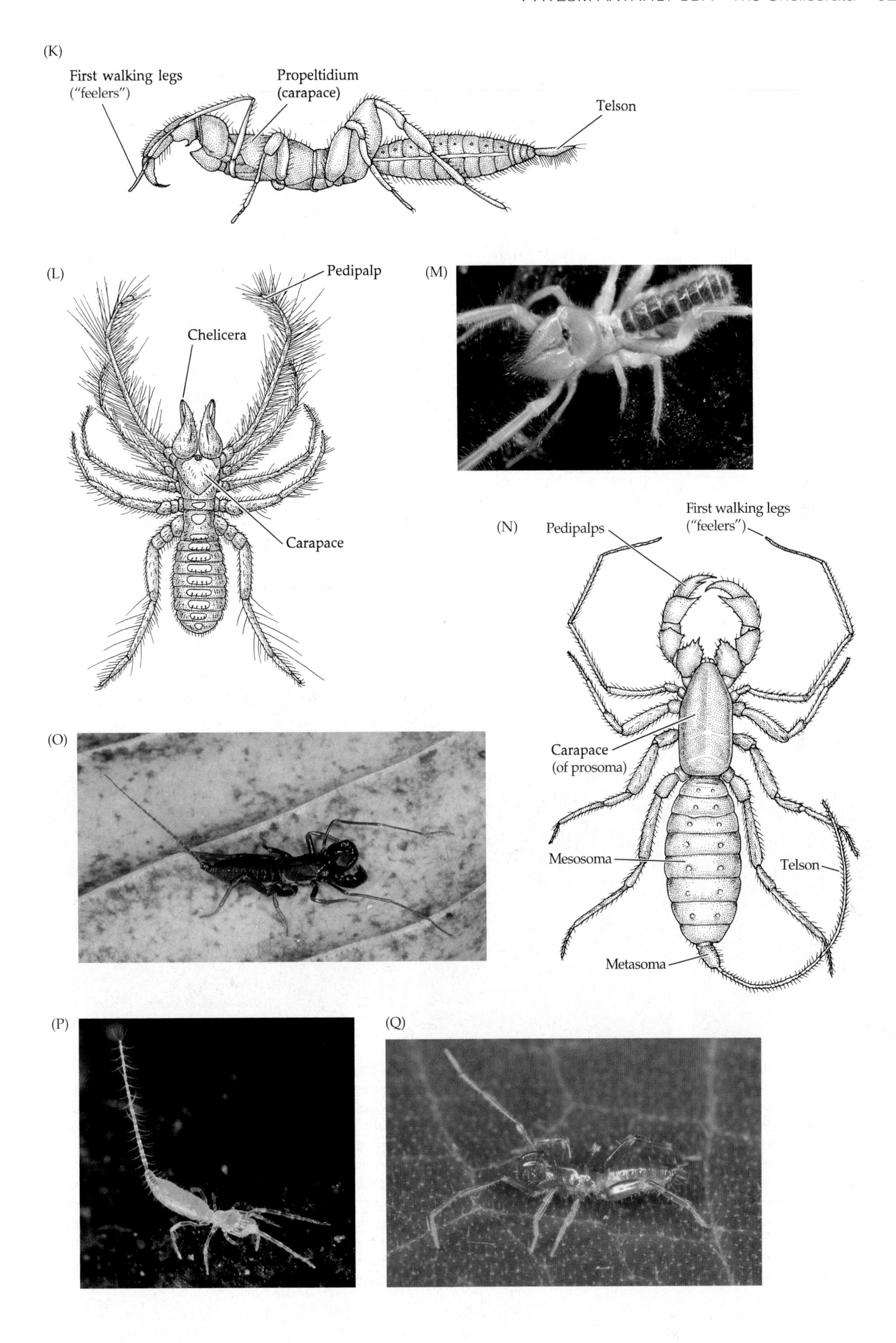
(K)
First walking legs
("feelers")
Propeltidium
(carapace)
Telson
(L)
Pedipalp
Chelicera
Carapace
(M)
(N)
Pedipalps
First walking legs
("feelers")
Carapace
(of prosoma)
Mesosoma
Telson
Metasoma
(O)
(P)
(Q)

Order Araneae Spiders (Figure 24.5). Spiders, of course, are the most familiar of all chelicerates, and they are one of the most abundant groups of land animals. In fact, they might be the most commonly encountered, macroscopic terrestrial invertebrates (along with isopods and flies) in the world! They have successfully exploited nearly every terrestrial environment, and many freshwater and intertidal habitats as well. Spiders display a truly staggering array of lifestyles—yet they all utilize a fairly uniform body plan and the vast majority are generalist predators. Prosoma undivided, covered by a carapace-like shield and attached to opisthosoma by narrow pedicel; opisthosoma undivided and unsegmented except in liphistiids and some mygalomorph families, which have discrete tergites (i.e., "segmented abdomens"); chelicerae modified as fangs, served by venom glands located in the prosoma (absent in the families Uloboridae and Holarchaeidae); pedipalps modified in the adult males as copulatory organs; opisthosoma bears book lungs and/or tracheae, silk-producing glands, and spinnerets, the last being highly modified appendages that

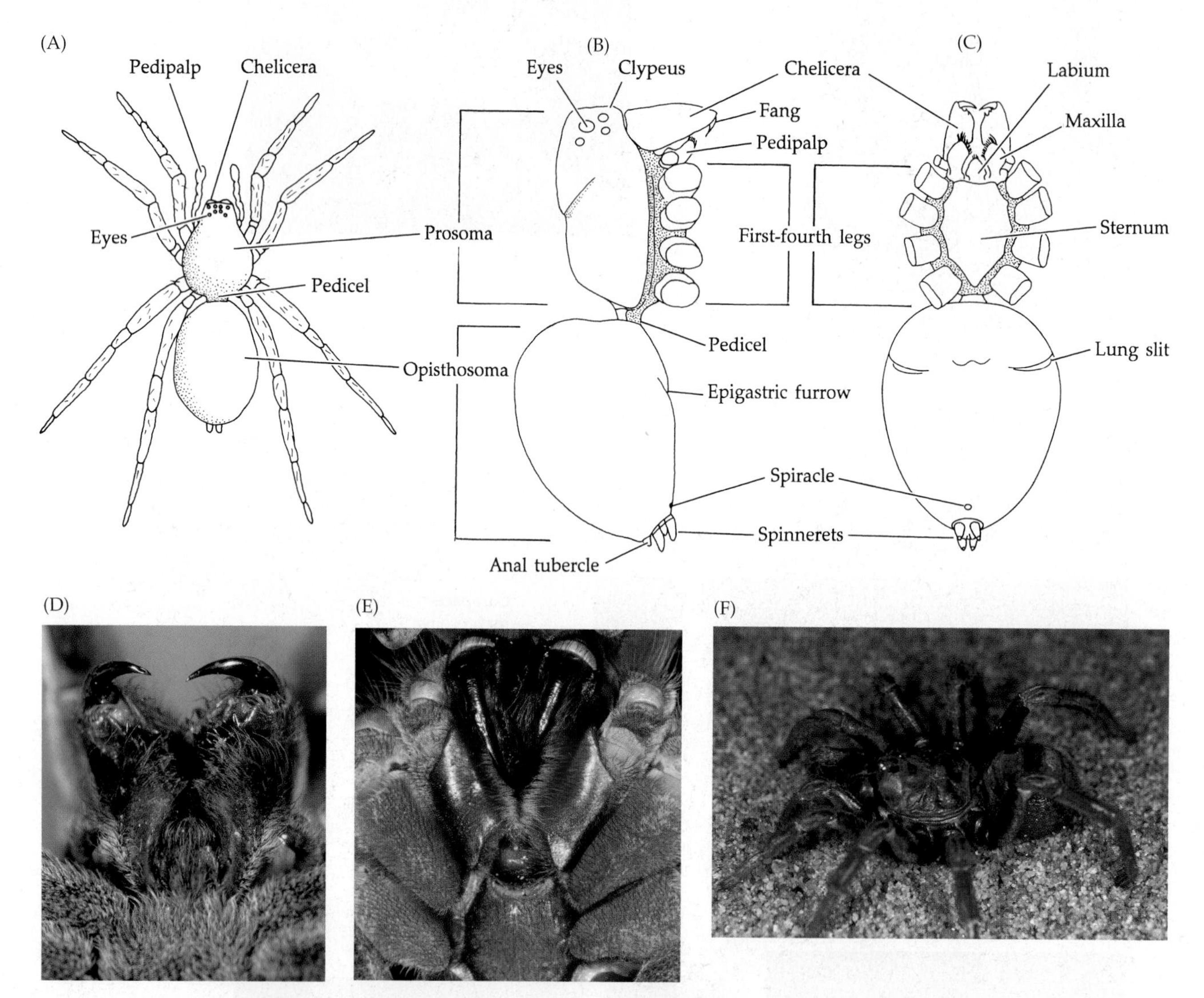

Figure 24.5 Spiders. (A–C) A generalized spider. (A) Dorsal view. In (B) (lateral view) and (C) (ventral view) the legs have been omitted, except the coxae. (D) The cheliceral fangs of *Phoneutria*, a ctenid spider (family Ctenidae) from Ecuador with a highly toxic venom (the poison pores are visible at the end of the fangs). (E) Fangs of *Theraphosa blondi* (family Theraphosidae) from Brazil. (F) A liphistiid (suborder Mesothelae) from Thailand. Note the evidence of opisthosomal segmentation. (G, H) The orientation and plane of motion of the chelicerae of an orthognathous (G; see also D) and a labidognathous (H; see also E) spider. (I) The notorious black widow spider, *Latrodectus* (family Theridiidae). (J) An orb-weaving spider from Madagascar, *Caerostris* sp. (family Araneidae). (K) *Theraphosa blondi* (family Theraphosidae), the Goliath tarantula, from the Brazil Amazon region; this is the largest extant spider. (L) *Cupiennius coccineus* (family Ctenidae), from the rainforests of Costa Rica, a favorite study subject of neurobiologists. (M) A sparassid spider from Brazil (family Sparassidae). (N) Liphistiid spider "trap door" from Thailand (closed and open). Note the silk triplines radiating from the entrance. (O) *Myrmecotypus* sp. (Corinnidae), from the rainforests of Brazil, a spider that mimics ants; ant mimicry has evolved multiple times among spiders.

(G)
(H)
(I)
(J)
(K)
(L)
(M)
(N)
(O)

serve to spin the silk produced by the silk glands; most with eight eyes. Most cease to molt once adulthood is achieved.

The body is divided into a fused prosoma and a fleshy, less sclerotized, opisthosoma. The first opisthosomal segment forms a narrow pedicel, which joins the two body regions. Each chelicera has two segments; the basal segment is short and conical and the distal segment is a hard, curved fang, usually bearing a pore from the duct of the poison gland (Figures 24.5D,E and 24.16A). A medial flap (the labium) projects ventrally over the mouth. Each pedipalp is composed of six segments, and in most spiders the proximal segments are enlarged as lobes called endites, which bear gnathobases that flank the mouth and are used to handle and grind food (sometime endites are also referred to as maxillae, although endites are not homologous to the maxillae of crustaceans and insects). Beyond the gnathobases, the pedipalps extend forward as tactile organs in females and juveniles, but they are highly modified as sperm transfer organs in mature males. In a few species of spiders (e.g., tarantulas) the pedipalps function as walking "legs." The ventral surface of the prosoma bears a cuticular plate, or sternum, around which arise the four pairs of walking legs, each composed of eight articles.

The opisthosoma bears openings for gas exchange, the gonads and associated structures, the spinnerets, and the anus. On the ventral surface posterior to the pedicel is a transverse **epigastric furrow** in which the gonopores are located. In most females, a slightly elevated plate, the **epigynum**, lies in front of the epigastric furrow and bears the openings to the seminal receptacles. Lateral to the furrow are spiracles that open to the book lungs, or anterior tracheae. The hind part of the opisthosoma bears the anus and the spinnerets (the latter are situated near the middle of the opisthosoma in the Mesothelae). The ability of spiders to produce silk and fashion it into a great variety of functional devices has been a major factor in their evolutionary success, as we will see below.

There are more than 45,000 described spider species, classified into about 112 extant families. The largest spiders by leg span are the giant huntsman spiders (Sparassidae), some of which reach 30 cm (12 in) in diameter. The largest spider by weight is probably the goliath birdeater tarantula (*Theraphosa blondi*), which can weigh over 170 g (6.0 oz). The smallest adult spiders are found in the mircro-orb-weaver genus *Patu* (Symphytognathidae), the males measuring under 0.4 mm. The order Araneae is divided into two suborders. The suborder Mesothelae includes the "segmented" spiders (family Liphistiidae, about 90 species), which are characterized by persisting segmentation of the opisthosoma (Figure 24.5F), with conspicuous tergites, and mid-abdominal location of the spinnerets. Most are 1–3 cm long and build burrows with a trapdoor over the entrance; some have silk triplines that converge radially onto the entrance (Figure 24.5N). All are predators, feeding on animals that pass within striking distance of the burrow opening.

The other suborder, Opisthothelae (with more than 45,000 species), comprises two large groups: the Infraorder Mygalomorphae (tarantula-like spiders) and the Infraorder Araneomorphae ("typical" spiders). Almost all members of both infraorders have an unsegmented opisthosoma and posteriorly located spinnerets, but they can be differentiated by the nature of the chelicerae (Figures 24.5G,H). Mygalomorph spiders have chelicerae that articulate in a manner that enables movement of the appendages parallel to the body axis (**orthognathous**), whereas araneomorph spiders have chelicerae that move at right angles to the body axis (**labidognathous**). An intermediate condition, known as **plagiognath** chelicerae, can be found in liphistiids, in some mygalomorphs (e.g., Hexathelidae), and some araneomorphs (e.g., Hypochilidae).

Order Opiliones Daddy longlegs, or harvestmen, including mite harvestmen (Figure 24.4C,D). Divided into four extant suborders (Cyphophthalmi, and the three Phalangida suborders: Eupnoi, Dyspnoi, and Laniatores), and an extinct one (Tetrophthalmi). With an important fossil record, including multiple Paleozoic species. Prosoma entire or divided (as it also is in schizomids, palpigrades, and solifuges), sometimes into a proterosoma and two free segments. The prosoma joins broadly to the often segmented opisthosoma. Chelicerae small, of three articles, pincer-like; pedipalps long and leglike or well developed into a prehensile appendage, especially in Laniatores. Blind or with one pair of median eyes; whether the eyes of some Cyphophthalmi are median or lateral, remains debated. One pair of tracheae on opisthosoma; male with spermatopositor organ (in Cyphophthalmi) or penis (in Phalangida); female with an ovipositor organ.

The order Opiliones is a large and diverse group of about 6,500 species, recently suggested to be closely related to Solifugae and Ricinulei (a trio of orders with a completely segmented opisthosoma and a tracheal tubule system for respiration). Opiliones are known from nearly all climatic regions of the world, including subarctic areas, but are most abundant in tropical and temperate regions of the southern Hemisphere. Most species have small bodies, less than 2 cm in length, but many Phalangida have very long legs (up to 10 cm).

Opiliones prefer damp, shaded areas and are commonly found in leaf litter, on trees and logs in dense forests, and in caves. They feed on various small invertebrates and also scavenge on dead animal and plant matter. Food is grasped by the pedipalps and passed to the chelicerae for chewing. These animals are among the few arachnids capable of ingesting solid particles instead of the usual liquefied food. Opiliones possess a pair of defensive repugnatorial glands, which produce noxious secretions containing quinones and phenols. As discussed later, most arachnids mate by indirectly passing sperm from the male to the female. Phalangida (and some mites) are the only arachnids in which males use a penis for direct copulation, while in Cyphophthalmi the male uses a "penis-like" spermatopositor organ to place a spermatophore near the female gonostome.

Order Palpigradi Palpigrades or micro-whip scorpions (Figure 24.4E). Minute arachnids with the prosoma divided into proterosoma, covered by the carapace-like propeltidium, followed by two free segments and joined to opisthosoma by a narrow pedicel; opisthosoma segmented and divided into broad mesosoma and short narrow metasoma, the latter bearing a long multiarticulated telson; eyes absent.

Palpigrades are delicate animals that walk by sensing the substrate with what seems a nervous behavior of the first pair of walking legs, and use their unmodified palps for walking (unlike all other arachnids). While moving, most palpigrades keep the first legs held upwards, moving it laterally. It is possible that the uplifted flagellum is associated with perception of the environment. These small, often highly translucent arachnids range in size from 0.65 mm in *Eukoenenia grassii* to 2.8 mm in the "giant" *E. draco* from caves on the island of Mallorca, Spain. These tiny arachnids have undergone a great deal of evolutionary reduction in association with their small size and cryptic habits. They are colorless, have very thin cuticles, and have lost the circulatory and gas exchange organs. Most are found under rocks or in caves, in very humid environments, and some live on sandy beaches. One species has recently been reported to feed on heterotrophic Cyanobacteria. The mode of sperm transfer in these arachnids remains unknown.

These rare animals have been recorded from widely separated parts of the world, and their biology and biogeography remain poorly known. The phylogeny of palpigrades has been recently investigated. The living members of the order are currently divided in two families: Eukoeneniidae, with four genera and 85 species, and Prokoeneniidae, with two genera and seven species. However, the position of Palpigradi among the arachnid orders is still being debated.

Order Pseudoscorpiones False scorpions (Figure 24.4F,G). Prosoma covered by a dorsal carapace-like shield, but clearly segmented ventrally. Opisthosoma undivided but with 11–12 segments and broadly joined to prosoma. Chelicerae chelate bearing silk glands; pedipalps large and scorpion-like, with venom glands in most species; eyes present or absent.

There are about 3,500 described species of false scorpions, with the body lengths of adults generally ranging from 0.7 to 5 mm, but the largest of which, *Garypus titanius* from Ascension Island, reaches 12 mm in length. The group is cosmopolitan and found in a great variety of habitats—under stones, in litter, in soil, under bark, and in animal nests. One genus, *Garypus*, is found on sandy and cobble marine beaches, and one species, *Chelifer cancroides*, typically cohabits with humans and is found in homes worldwide.

These strange little creatures resemble scorpions in general appearance, but they lack the elongation of the opisthosoma and telson and do not have a stinging apparatus. However, they may possess poison glands in the pedipalps with which they immobilize their prey, usually other tiny arthropods (e.g., mites). Once captured, the victim is torn open by the chelicerae and the body fluids are sucked out. Also, like spiders, they produce silk, but the silk glands open on the chelicerae, with the glands located in the prosoma. This silk is used to build silken chambers for hibernation, molting, and egg laying.

A number of false scorpions use larger arthropods as temporary "hosts" for purposes of dispersal. This phenomenon of "hitchhiking," known as phoresy, typically involves females that grab onto the larger host animal with their pedipalps. The host is often a flying insect or other arthropod (Figure 24.4H). The above-mentioned cosmopolitan species, *Chelifer cancroides*, for example, is frequently encountered as a phoretic "guest" on houseflies.

Order Ricinulei Ricinuleids or hooded-tick spiders (Figure 24.4I,J). Prosoma fully covered by a carapace-like shield and broadly joined with opisthosoma. Opisthosoma unsegmented, with paired tracheae; with a prosomal–opisthosomal locking mechanism. Chelicerae pincer-like, covered with a flaplike hinged plate named the **cucullus**; pedipalps small, with coxae fused medially. Eyes absent in extant species, but some have a lighter patch in the side of the prosoma that is thought to be light sensitive; two pairs of lateral eyes present in some fossil species. Third legs of males modified for sperm transfer. Cuticle very thick (Figure 24.4I,J).

There are only about 76 described species of ricinuleids; all are less than 11 mm long and live in caves and tropical forest leaf litter in West Africa (*Ricinoides*) and tropical America (*Cryptocellus* and *Pseudocellus*), but fossil species are known from Myanmar Cretaceous amber. All are slow-moving predators on other small invertebrates. Ricinuleids have a hexapod larva and three other developmental stages (protonymph, deutonymph, and tritonymph); they may take 1 to 2 years to reach maturity and live 5 to 10 years. In immatures, the tergites and sternites are more widely spaced, and there are fewer than in the adults, where they almost touch.

Order Schizomida Schizomids (Figure 24.4K). Prosoma divided; first four segments (the proterosoma) covered by a short carapace-like shield (the propeltidium), followed by two free segments, the mesopeltidium and metapeltidium. Opisthosoma segmented and divided; mesosoma with one pair of book lungs; metasoma with short, thin telson. Eyes present or absent.

There are about 260 species of small (less than 1 cm) arachnids in this order. Although some authors classify them as a suborder of Uropygi, they are distinguished from true whip scorpions by the divisions of the prosoma and the shorter telson. Schizomids live in leaf litter, under stones, in caves, and in burrows and are most common in tropical and subtropical areas of Asia, Africa, and the Americas. A few temperate species are known.

The first walking legs are sensorial and resemble those of the closely-related uropygids; together, the two comprise a clade. Like the uropygids, schizomids are predators on small invertebrates, and they also possess opisthosomal repugnatorial glands.

Order Scorpiones True scorpions (Figure 24.6). Body clearly divided into three regions: prosoma, mesosoma, and metasoma. Prosomal segments fused and covered by a carapace-like shield. Opisthosoma elongate, segmented, and divided into mesosoma and metasoma of seven and five segments, respectively; telson spinelike, with poison gland. Chelicerae of three articles; pedipalps large, chelate, of six articles. With a pair of median eyes and sometimes additional pairs of lateral eyes, but some species are blind. First mesosomal segment bears a gonopore covered by genital operculum; second mesosomal segment bears a pair of unique sensory appendages called pectines; third through sixth mesosomal segments each with a pair of book lungs. Metasoma without appendages.

Scorpions are among the most ancient terrestrial arthropods and some consider them the most primitive arachnids, although recent phylogenetic analyses suggest that they are related to the other arachnids with book lungs, or Tetrapulmonata (Araneae, Amblypygi, Uropygi, and Schizomida). Others believe that they evolved from aquatic ancestors, perhaps from the eurypterids or a common ancestor, and invaded land during the Carboniferous period. All of the roughly 2,068 known species are terrestrial predators. They inhabit a variety of environments, ranging from deserts to tropical rain forests, where some arboreal species occur, and even deep caves. They are notably absent from colder regions of the world. Scorpions include the largest living arachnids, with some reaching lengths of 18 cm.

The chelicerae are short and bear gnathobases for grinding food. The pedipalps are large, with the last two articles forming grasping chelae. The walking legs each comprise eight articles (coxa, trochanter, femur, patella, tibia, metatarsus, tarsus, pretarsus). The ventral surface of the mesosoma bears the genital pore, a pair of pectines, and four pairs of spiracular openings to the book lungs. The anus is located on the last true segment of the metasoma, and is followed by a stinging apparatus derived from the telson and bearing a sharp barb called the aculeus.

Scorpions are admired and feared in many cultures. They are well known for glowing in the dark under UV light; even a cast-off exuvium glows. They display complex mating behaviors and maternal care.

Order Solifugae Sun spiders, camel spiders, or wind spiders (Figure 24.4L,M). Prosoma divided into a proterosoma covered by a carapace-like shield and two free segments; opisthosoma undivided, but with eleven segments, bearing three pairs of tracheae; chelicerae huge and held forward, with two articles; pedipalps long and leglike; propeltidium with one pair of eyes. On the ventral side of the fourth pair of legs, solifuges have up to ten malleoli or racquet organs, a unique sensory organ that occurs in this order.

Most of the 1,116 or so species of solifuges (or solpugids) live in tropical and subtropical desert environments of America, Asia, and Africa. In contrast to many arachnids, they are often daytime hunters, hence the common name "sun spiders." They are also known as "wind spiders" because the males run at high speeds, "like the wind," or "camel spiders" because they are often found in deserts.

Some solifuges are only a few millimeters long, but many reach lengths of up to 7 cm. The feeding habits of many solifuges are unknown. Among those that have been studied, most are omnivorous, but they frequently show a preference for termites or other arthropods. Lacking poison, they rip their live prey apart with their strong chelicerae.

Order Uropygi Whip scorpions and vinegaroons (Figure 24.4N,O). Prosoma elongated and covered by a carapace-like shield. Opisthosoma segmented and divided; mesosoma broad with two pairs of book lungs; metasoma short with long, whiplike telson; first walking legs elongated and multiarticulated distally. With one pair of median and four or five pairs of lateral eyes.

Whip scorpions are moderately large arachnids reaching lengths of up to 8 cm. There are about 110 known species of living Uropygi (or Thelyphonida), most of which occur in Southeast Asia; a few are known from the southern United States, Mexico, and parts of Central and South America; and some species in Africa have probably been introduced. With the exception of a few desert species, whip scorpions live under rocks, in leaf litter, or in burrows in relatively humid tropical and subtropical habitats. The telson is sensitive to light, and most uropygids are negatively phototactic and active only at night.

The elongate first walking legs are held forward as "feelers," aiding the animals in their nocturnal hunting

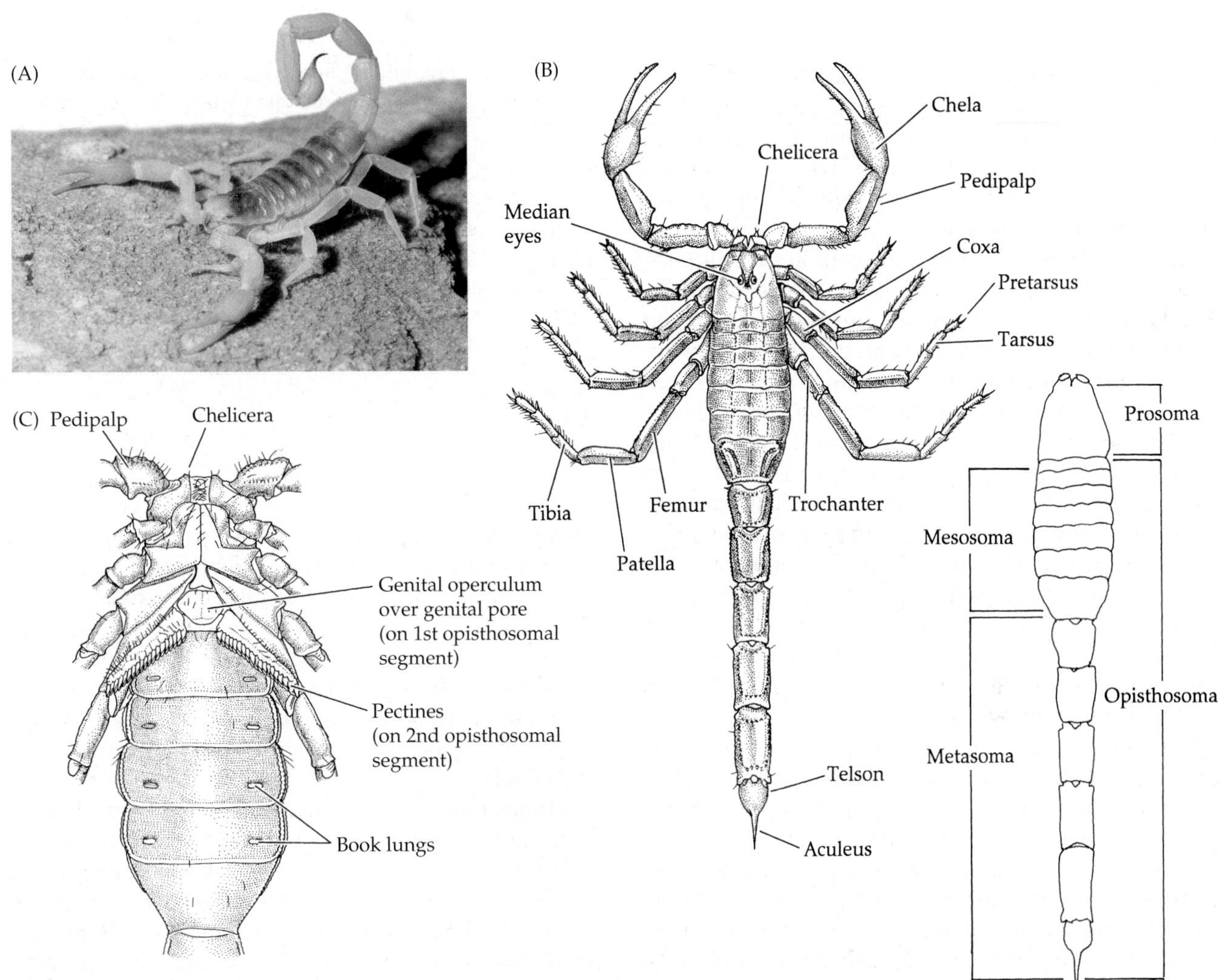

Figure 24.6 **Scorpions (subclass Arachnida, order Scorpiones).** (A) A desert scorpion, *Androctonus*. (B,C) *Buthus martensi* (dorsal and ventral views).

excursions. They feed on various small invertebrates by grasping prey with the pedipalps and grinding it with the chelicerae. The long telson is constantly moving about, probably serving as a sensory structure in case a predator moves in from behind.

Whip scorpions possess a pair of repugnatorial glands that open near the anus. When a uropygid is threatened by a potential predator, it raises the opisthosoma and sprays an acidic liquid from these glands on the would-be attacker. Some forms (e.g., *Mastigoproctus*) produce a secretion high in acetic acid, earning them the common name "vinegaroons." The spray nozzle of vinegaroons is like a turret, such that they can spray in almost any direction, even toward the front of their head. Most live in self-made burrows.

Uropygi are closely related to Schizomida. In fact, some authors use the name Thelyphonida for the whip scorpions and the term Uropygi to refer to both Schizomida + Thelyphonida. This clade is, in turn, closely related to Amblypygi and Araneae, constituting the group Tetrapulmonata, or the lung-bearing arachnids.

The Euchelicerate Body Plan

Looking over the taxonomic summaries above will give you a good sense of diversity within Euchelicerata. This section covers the general biology and structure of members of this large class, with an emphasis on the xiphosurans, spiders, and scorpions. We hope to convey an impression not only of diversity but also of unity within this group and to reinforce the concept of the evolutionary plasticity of the arthropod body plan in general.

As noted earlier, the body of chelicerates is typically divided into two main regions, the **prosoma** and the **opisthosoma** (Figures 24.1, 24.2, 24.5, and 24.6); a discrete head is not recognizable. The prosoma includes a presegmental acron and seven segments; the opisthosoma includes up to twelve segments and a postsegmental, postanal telson. As in other arthropod groups, these basic body regions have undergone various degrees of specialization and tagmosis. In most euchelicerates the entire prosoma is fused and covered by a carapace-like shield. However, in certain groups (e.g., schizomids, palpigrades, solifuges, and some harvestmen) the prosoma is divided into three parts: a **proterosoma**, comprising the first four segments, all fused and covered by a carapace-like shield (often called the propeltidium), and two free segments (often called the **mesopeltidium** and **metapeltidium**). The opisthosoma may be undivided, as it is in most spiders (except in the southeast Asian Liphistiomorpha, where the segments are visible), or divided into an anterior **mesosoma** and posterior **metasoma**, as it is in scorpions and in eurypterids.

The appendages further distinguish the euchelicerates from other arthropods. There are no antennae and the first segment bears the eyes, with all six following segments of the prosoma bearing appendages. The first pair of appendages is embryologically postoral, often pincer-like, **chelicerae**. During embryogeny, the chelicerae migrate to a position lateral to the mouth, or even preoral, as compared with adults of most groups; here they serve as fangs or grasping structures during feeding. The chelicerae can be formed of two or three articles and are followed by a pair of postoral **pedipalps**, which are usually elongate or, more rarely, chelate. The pedipalps are usually sensory in function, but in some groups (e.g., scorpions) they aid in feeding and defense. The remaining four segments of the prosoma typically bear four pairs of walking legs.

The number of segments and appendages on the opisthosoma varies. In general, appendages are absent or very reduced, although in the horseshoe crabs they persist as large platelike limbs, called **book gills**, that function in locomotion and gas exchange. In most chelicerates, the opisthosomal limbs are greatly reduced and persist only as specialized structures, such as the silk-producing spinnerets of spiders or the pectines of scorpions.

In summary, we may define the Euchelicerata as arthropods in which the body is divided into two regions (or two tagmata): a prosoma and an opisthosoma (with modifications of this body plan in mites). Additionally, the first two pairs of appendages are chelicerae and pedipalps, and the remaining four pairs of prosomal appendages are walking legs. Evolutionarily, this body plan has been a highly successful one.

As mentioned earlier, the great success of spiders seems to have been due in large part to the evolution of complex behaviors associated with silk and web production, particularly among araneomorph spiders, whose silk fibers are stronger than those of mygalomorphs and liphistiids. Because we are paying special attention to members of the order Araneae in this chapter, and because silk production is so important to nearly all facets of their lives, we present a special section on spider silk and its uses.

Spinnerets, Spider Silk, and Spider Webs

Spider silk is a complex fibrous protein composed mostly of the amino acids glycine, alanine, and serine.

Figure 24.7 Silk glands and spinnerets in spiders. ▶ (A) The silk glands and spinnerets of *Nephila*, the golden silk spider. Only one member of each pair of glands is shown. (B) The spinnerets of the orb-weaver *Araneus* (external view). (C) Cutaway view of the posterior end of the opisthosoma (*Tegenaria*). Note the spinneret muscles. (D) The cribellar spigots of *Hypochilus* (SEM, 1,600×). (E) The cribellate silk of *Uloborus*. (F) SEMs of the comb-like calamistrum on the metatarsus of the fourth walking legs of *Amaurobius similis*, used to brush the silk threads as they emerge from the cribellum. (G) An SEM (237×) of the spinning apparatus of *Amaurobius similis*. The plate-like structure anterior to the three pair of spinnerets is the cribellum.

Spider silk proteins are known as **spidroins**. The silk is produced in a liquid, water-soluble form that transforms into an insoluble thread after it is extruded from the body. This near instantaneous transformation involves as much as a tenfold increase in the molecular weight of the silk protein, which is enhanced by the formation of intermolecular bonds.

Spider silks are among the toughest energy absorbing materials known. Some types of silk require up to ten times more energy to fracture than an equivalent volume of the synthetic fiber Kevlar (Kevlar is five times stronger than steel in terms of tensile strength-to-weight ratio). Some spider silks can stretch as much as rubber. Spider silk is so strong that in some parts of the world (the Solomon Islands, for example) people actually use it as fishing nets by stacking numerous webs on top of one another.

The spider silk-producing organs are located in the opisthosoma and comprise several types of **silk glands**, depending on the taxonomic group. For example, an adult female orb-weaver (e.g., *Araneus*) has seven gland types, each producing silks with different physicochemical properties and functions. The liquid silk produced by these glands (also called "dope") is secreted into ducts that open to the outside through spigots located on the spinnerets (Figure 24.7). As the liquid silk travels along this duct, ionic exchanges occur and water is removed. A muscle-controlled valve toward the distal end of the duct is also involved in the process that causes a phase shift from liquid to a solid fiber. The apparatus spins the silk into threads of different thicknesses and properties. By spinning various kinds of silks in different diameters, spiders produce a variety of threads with unique properties for specific functions and at particular times. The **spinnerets** are highly modified opisthosomal appendages, and they retain some musculature that allows for their movement during spinning (Figure 24.7C). The spinnerets are located in body segments 11 and 12 (opisthosomal segments 4 and 5).

The numbers and kinds of glands and spinnerets vary. Ancestrally, spiders had four pairs of spinnerets

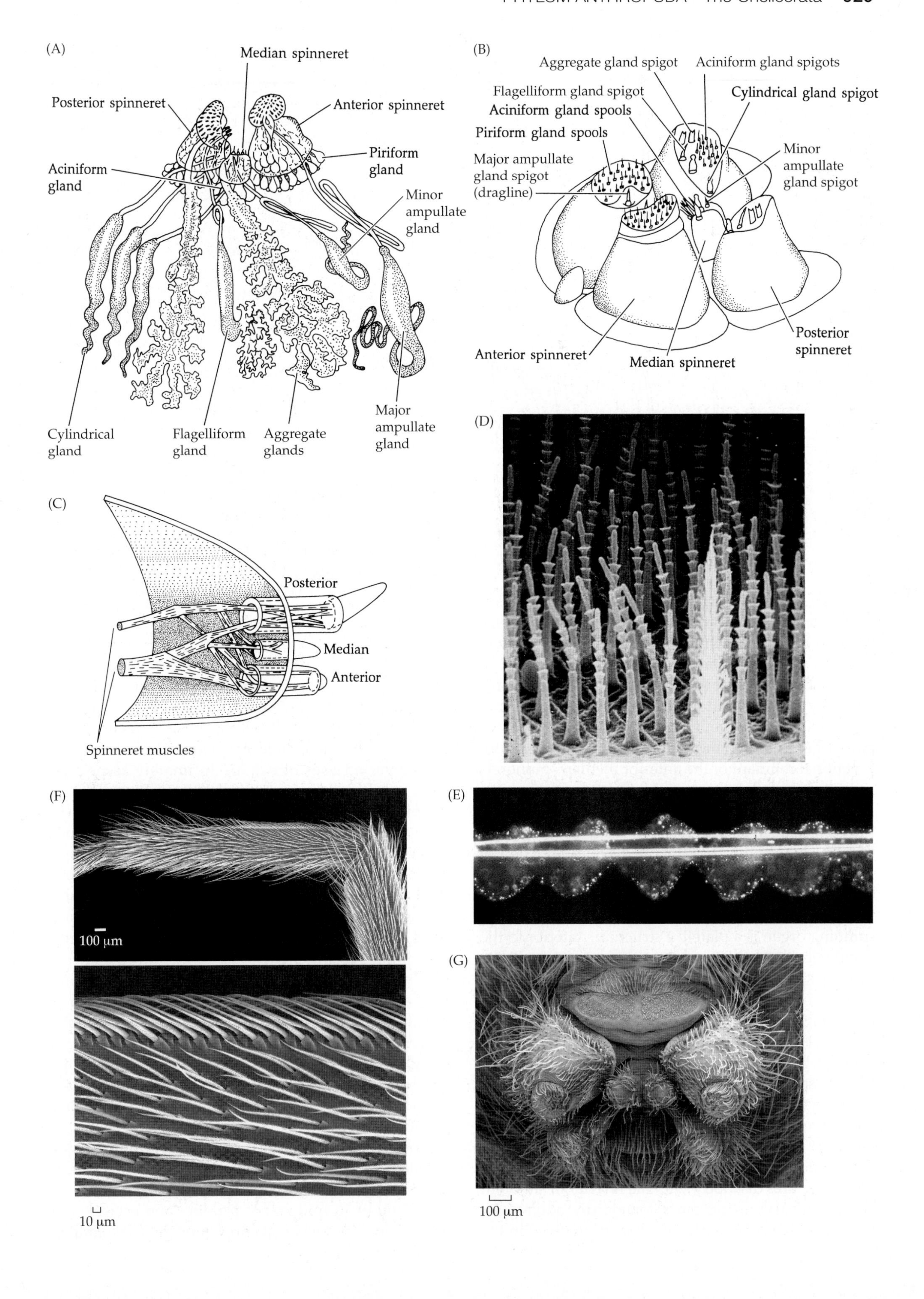
(A)
Median spinneret
Posterior spinneret
Anterior spinneret
Piriform gland
Aciniform gland
Minor ampullate gland
Cylindrical gland
Flagelliform gland
Aggregate glands
Major ampullate gland
(B)
Aggregate gland spigot
Aciniform gland spigots
Flagelliform gland spigot
Cylindrical gland spigot
Aciniform gland spools
Piriform gland spools
Major ampullate gland spigot (dragline)
Minor ampullate gland spigot
Anterior spinneret
Median spinneret
Posterior spinneret
(C)
Posterior
Median
Anterior
Spinneret muscles
(D)
(F)
100 μm
10 μm
(E)
(G)
100 μm

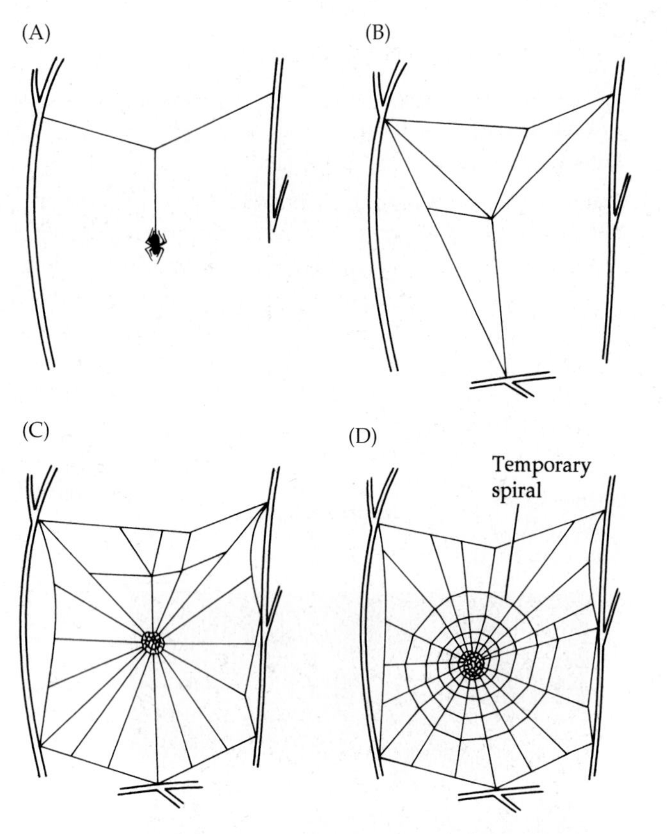

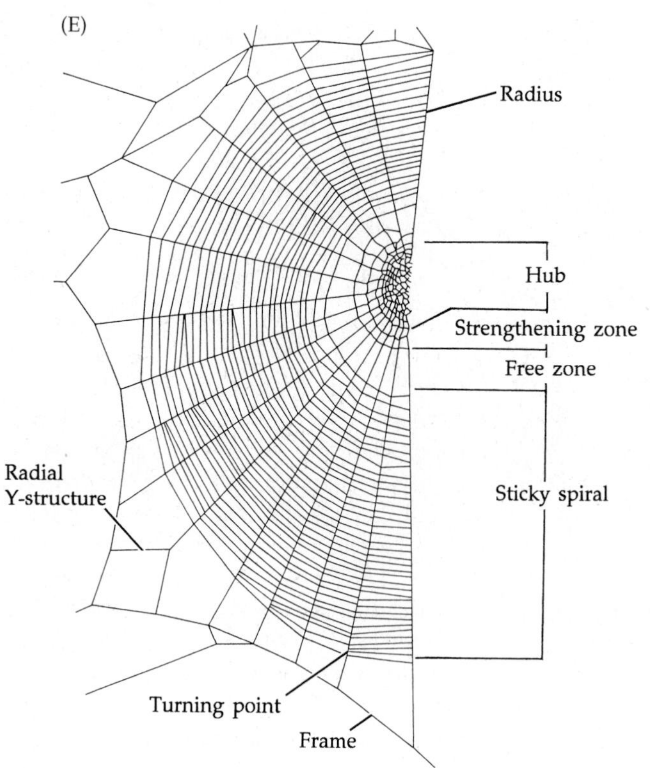

Figure 24.8 Construction of an orb web. (A) Formation of the Y-shaped frame. (B,C) Addition of radial threads. (D) Addition of the temporary spiral or working platform. (E) A portion of the completed orb web, with the temporary spiral replaced by the sticky spiral.

(two pairs per segment): anterior lateral, anterior median, posterior lateral, and posterior median. Mesothelae spiders retain the four pairs of spinnerets (although in some species the anterior median pair is reduced to nonfunctional knobs). Most extant spiders have three pairs (by the loss or modification of the anterior median spinnerets), although some have only one or two pairs of spinnerets. In many spiders there is a nonfunctional remnant of the anterior median spinnerets known as the **colulus**.

Another spinning organ, the **cribellum** (Figure 24.7D), occurs in several families of araneomorph spiders. Although the cribellum is a synapomorphy of Araneomorphae, it has been independently lost multiple times in this large clade. As the colulus, the cribellum is a homolog of the anterior median spinnerets. This spinning organ is a platelike structure, anterior to the spinnerets, that bears many small spigots (up to 40,000 in some species). Cribellate silk is emitted in many extremely fine threads and then combed into a delicate mesh (Figure 24.7E) over a pair of axial fibers from the spinnerets by a row of special macrosetae (the calamistrum) on the metatarsus of the fourth legs (Figure 24.7F). In some species the calamistrum consists of two rows of macrosetae. Cribellate orb-weavers (e.g., the Uloboridae) use the resulting mesh, known as a "hackle band," to build the sticky spiral of their orbs. The cribellum and the calamistrum appear after the third molt, and in males both structures disappear with the final molt, although nonfunctional remnants can be found in the adult males of some species. Similarly, the spigots involved in the spinning of the viscid sticky silk of ecribellate orb-weavers disappear in the final molt of males (there are exceptions, such as the adult males of many erigonine linyphiids, which retain this triplet of spigots).

The varied uses of silk are intimately associated with nearly every aspect of the lifestyle and habits of spiders, as explained more fully in following sections. Different types of silk are used as safety lines and climbing lines (drag lines), to anchor or cement drag lines to the substrate, for the construction of retreats and eggsacs, to wrap prey for brief storage, and to line burrows. In addition, several kinds of silks are used to build foraging webs. Adult males of many araneomorph spiders also use silk to build a sperm web, which is then used to deposit sperm (through the gonopore, in the abdomen) prior to charging their copulatory organs in the pedipalps (a process termed "sperm induction"). Sperm webs are built using special silk glands that open through a set of spigots along the anterior margin of the epigastric furrow (these male spigots are known as "epiandrous fusules"). The number and arrangement of spigots in the spinnerets and of epiandrous fusules varies taxonomically. Many newly hatched spiderlings spin long, thin, threads to ride the wind for aerial dispersal ("ballooning").

Although all spiders can produce and use silk, not all of them build webs for prey capture (for example, jumping spiders do not build capture webs but often make silk retreats). In those spiders that use foraging webs, the diversity of web types and architectures is extraordinary, ranging from the highly geometrical, densely woven, micro-orb webs of symphytognathids to the deceivingly simple single line snares of mastophorine araneids that mimic moth pheromones. Almost everything in between these extremes of web types seems to exist or be possible, and much of spider web architecture and natural history remains to be discovered and/or described, particularly in the tropical regions. The construction and use of prey-capture webs are discussed in more detail in the section on feeding. In general, capture webs serve both to trap and retain prey long enough to be subdued and, through vibrations, to signal the spider of its presence. Indeed, most spiders live in a world dominated by vibrations, a world where a potential meal, predator, or mate reveals itself with characteristic resonance patterns. We present here a brief description of the construction of a familiar orb web, as commonly built by members of the families Araneidae and Tetragnathidae (see chapter opener photo of *Eriophora* sp., an aracnid orb-weaver from the Amazon region of Brazil).

The spinning of an orb web takes place in three stereotypical phases, which are apparently genetically programmed (Figure 24.8). The first phase is the construction of a supportive Y-shaped framework and a series of radiating threads. The upper branches of the Y are initially laid down as a horizontal thread between two objects in the spider's environment. The spider sits in one spot and secretes the thread into the air; the loose end is then carried by air movements and flutters about until it contacts and sticks to an object. The spider then tacks down its end of the thread, moves to the center of the horizontal line, and drops itself on a vertical thread. The vertical thread is pulled taut and attached, thereby producing the Y-shaped frame. The intersection of the three branches of the Y becomes the hub of the final web, and it is from this point that the radii are extended. Radial threads are then attached to the frame threads. Once this initial phase is complete, the spider quickly lays down a nonsticky temporary spiral thread, starting from the hub, as the second phase of construction (the temporary spiral is left in the finished orb web of some species, such as those in the genera *Nephila* and *Phonognatha*). This spiral thread, along with the initial framework of threads, serves as a working platform during the third and final phase of web building—the production of the sticky spiral or prey trap. In ecribellate orb-weavers this final thread, the sticky spiral, is always coated with a gluey glycoprotein that automatically assumes a beaded distribution after it is deposited (Figure 24.9A). As the sticky spiral is laid down, the temporary spiral is removed or eaten.

Some orb-weavers (e.g., many species of *Argiope* and *Cyclosa*) produce a dense mesh of silk, called a **stabilimentum**, across the hub of their webs (Figure 24.9B). The form and materials used for stabilimenta vary widely taxonomically, and its phylogenetic distribution implies that stabilimenta have evolved independently multiple times. Consequently, its function also may vary across groups. Stabilimenta have been hypothesized to serve as a camouflaging device, to aid in thermoregulation, to have a defensive function, to warn large flying animals such as birds of the otherwise difficult to see web (and thus avoiding web damage) and as a visual guide to lure prey towards the web (by reflecting UV light and imitating floral patterns). Many orb-weavers

(A)

(B)

Figure 24.9 (A) A thread of the sticky spiral of an orb web with evenly distributed adhesive droplets. (B) An *Argiope* (Araneidae) from Madagascar on its orb web with a stabilimentum. (C) Funnel web of a funnel web spider (Dipluridae).

(C)

(e.g., *Araneus*) can produce an entire web in less than 30 minutes, and most build a new web every night, but some have long lasting webs (e.g., species of *Nephila* and *Cyrtophora*). When building new orb webs, spiders do not "waste" the silk of the old web, but eat it before or during the production of the new one. Radioactive labeling experiments show that the silk proteins from eaten webs appear at the spigots in new threads soon after ingestion, often within a few minutes!

Orb web building behavior is highly stereotypical and programmed, although there is some plasticity in the process as spiders can, to some extent, adapt their web building behavior to a specific situation. In general, web architecture is fairly constant within species, in the sense that under similar circumstances and physical conditions different individuals of the same ontogenetic stage and sex will build very similar webs. Among orb-weavers of the same species, the number of radii and spiral turns varies little. Web building appears to depend entirely on tactile input—normal webs are produced at night and by experimentally blinded spiders. Even gravity does not seem to be necessary, as illustrated by two famous orb-weaver spiders sent into space aboard Skylab.

Orb-weavers are an ancient group, with the oldest fossils being of Jurassic age and the oldest fossilized orb-web spiders from the Cretaceous. Two groups of spiders make similar orb webs using different types of silks but with presumably homologous stereotypical behaviors. These are the deinopoids (Deinopoidea, two families), which used cribellate silk for the capture spiral, and the araneoids (Araneoidea, about 18 families), which use a novel type of sticky, viscid, and gluey silk. During the last few decades the dominant paradigm has been that deinopoids and araneoids are reciprocally monophyletic and that together they form a clade (known as Orbiculariae). Recently, molecular studies (including phylogenomic approaches using thousands of genes) have rejected the long-accepted monophyly of Orbiculariae, by placing the cribellate orb-weavers (Deinopoidea) with other groups and not with the ecribellate orb-weavers (Araneoidea). This hypothesis implies independent origins for the two types of orb webs (cribellate and ecribellate) or a much more ancestral origin of the orb web with subsequent loss in the so-called RTA clade (a large lineage that includes cursorial spiders such as salticids and wolf spiders). Either alternative begs for a major reevaluation of our current understanding of the spider evolutionary chronicle.

For over a century specialists have debated about the origin and evolution of spider silk and webs. The silks and fibrous proteins found in many arthropod groups (e.g., in Lepidoptera and numerous other insect taxa, in pseudoscorpions, tanaid crustaceans, and others) seem to have evolved independently of the silks of spiders. Silk production is ancestral in spiders, and all spiders retain both the ancestral and less complex silk types and the more recently evolved silk glands. This implies that silk gland evolution is correlated with selection for additional, not alternative, silk types and functions. It has been suggested that spider silk glands are derived from epidermal invaginations. In males, the epiandrous glands are derived from dermal glands. Silk gland spigots may have evolved from setae (cuticular "hairs") or from sensory setae (such as morphologically similar chemoreceptor setae of mygalomorphs). It is unclear why spider silk first evolved, but many researchers favor the idea that its origin was linked to wrapping and protecting eggs.

Some spider species cooperate in web building, prey capture, and even spiderling rearing. Such social activity has been reported for some 20 species in at least six families. In spiders, group living takes two different forms: cooperative (nonterritorial permanent social or quasi-social) and colonial (communal territorial or territorial permanent-social). Cooperative social species (e.g., several species in the theridiid genus *Anelosimus* and in the eresid *Stegodyphus*) have communal nests and capture webs that they inhabit throughout the entire lifetime of the individual, with colony members cooperating in prey capture and raising young. Colonial species (e.g., the uloborid *Philoponella republicana*) form aggregations, but individuals in the colony generally forage and feed alone, and there is no maternal care beyond the egg stage.

The theridiid *Anelosimus eximius* (Figure 24.15A), and some other species in this genus of cobweb weavers, exhibit the highest level of sociality known to spiders. These social species can form very large colonies, up to thousands of individuals, sometimes covering entire trees. These individuals live in a communal web and cooperate in brood care, prey capture, feeding, and web building.

Locomotion

Euchelicerate locomotion follows the principles of arthropod joint articulation and leg movements discussed in Chapter 20. Except for the xiphosurans and a few aquatic arachnids, the legs must also be strong enough to support the body on land. Walking by terrestrial euchelicerates demands that the body be supported off the substratum and that the four pairs of legs move in sequences that maintain the animal's balance.

The xiphosurans are slow benthic crawlers and shallow burrowers, utilizing their stout prosomal limbs to push their heavy bodies over and through the sand. The legs are close together (Figure 24.2B), and thus coordination of movement sequences is essential. Horseshoe crabs are also able to swim, upside down, by beating the opisthosomal appendages.

The details of scorpion walking patterns are fairly well understood. During simple forward walking, each of the eight limbs moves through the usual power and recovery strokes. In scorpions (and many other

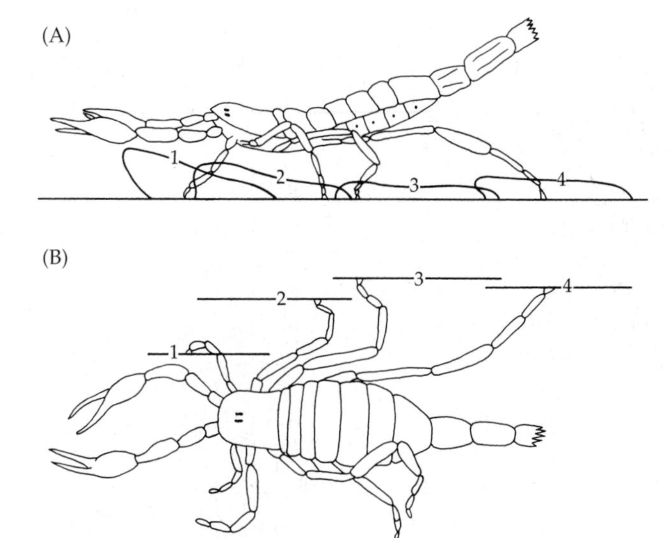

Figure 24.10 Walking in scorpions. Numbers denote legs 1–4. (A) The paths of the leg tips during their recovery strokes. Notice that the anterior legs are lifted higher off the ground than are the posterior legs. (B) Distance of each leg tip from the body.

arthropods) the joints between the coxae and the body are virtually immovable and do not contribute to the overall motion of the limbs. Not all of the walking legs move in the same pattern. The tips of the anterior legs are brought quite high off the ground during their recovery strokes; they may be used to feel ahead as the animal moves (Figure 24.10A). Also, the tips of each pair of legs extend different distances from the body, allowing for stride overlap without contact between legs (Figure 24.10B). Leg movements in scorpions do not follow the usual metachronal model. Rather, the typical sequence of movement along one side of the body is leg 4, then 2, then 3, and finally 1, with the legs on the opposite side being generally, but not precisely, out of phase. The somewhat staggered overlap of the movement sequences produces a smooth gait during forward walking, unlike the jerky motion of an insect moving by the alternating tripod pattern described in Chapter 22. Like many arthropods, scorpions are also able to change speeds, turn abruptly, walk backward, and burrow in loose sand.

Spiders have evolved a number of methods of locomotion, all involving the usual leg motions inherent in jointed arthropod limbs, which involve flexors and extensor muscles. However in spiders the femur–patella and the metatarsus–tarsus joints lack extensor muscles. In these later joints extension is accomplished by increasing hemolymph pressure through a hydraulic mechanism. During normal spider walking, the eight ambulatory limbs move in a so-called diagonal rhythm sequence (Figure 24.11). That is, legs 2 and 3 on one side of the body are moving simultaneously with legs 1 and 4 on the opposite side. This gait maintains a four-point stance while distributing the body weight more or less equally among those appendages in contact with the substratum. During very slow walking, however, posterior to anterior metachronal waves of limb motion are detectable. The stride lengths of the legs overlap somewhat and vary with speed and direction of movement. The placement of the legs on the body, and the arcs through which they swing, prevent contact between legs.

A number of spiders (e.g., members of the family Salticidae) are also capable of jumping (Figure 24.12A). Propulsion is achieved primarily by a rapid extension of the fourth pair of legs. Once the spider is airborne, the front legs are extended forward and used in landing. Salticids jump during normal locomotion, and also when capturing prey and escaping from predators.

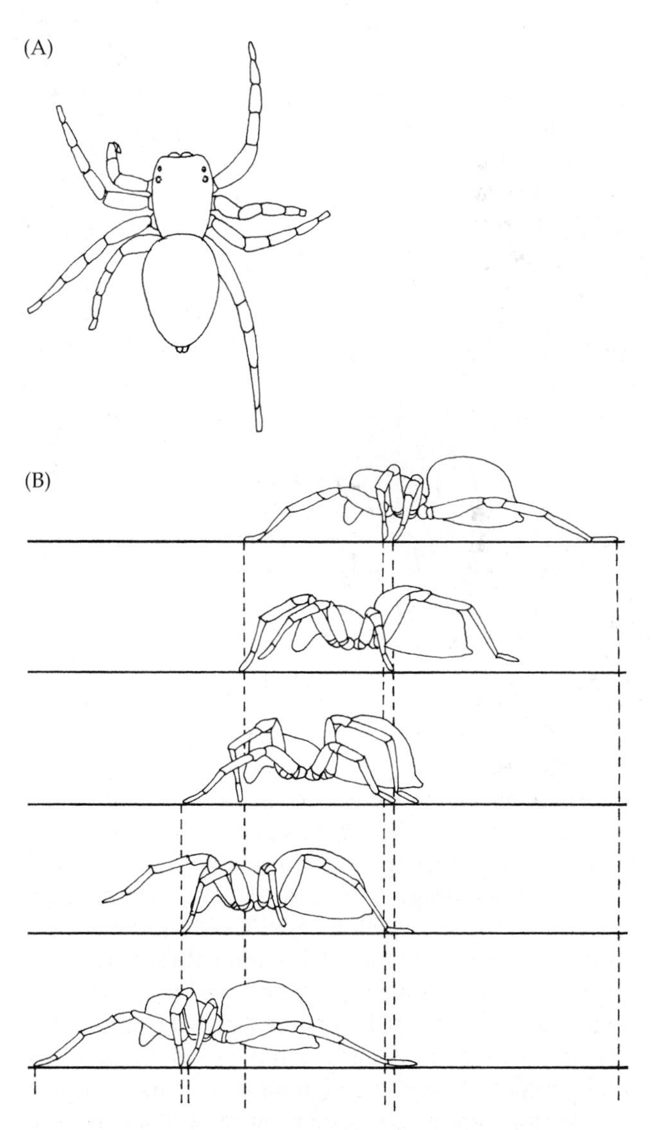

Figure 24.11 Walking in spiders. (A) A salticid with legs in walking positions (dorsal view). (B) A lycosid during slow walking (lateral view). Vertical dotted lines connect each leg through its power and recovery strokes relative to the forward progress of the body. Note the large degree of overlap in leg movements, particularly of the first two legs. Tangling is prevented in part by keeping the tips of adjacent legs at different distances from the body.

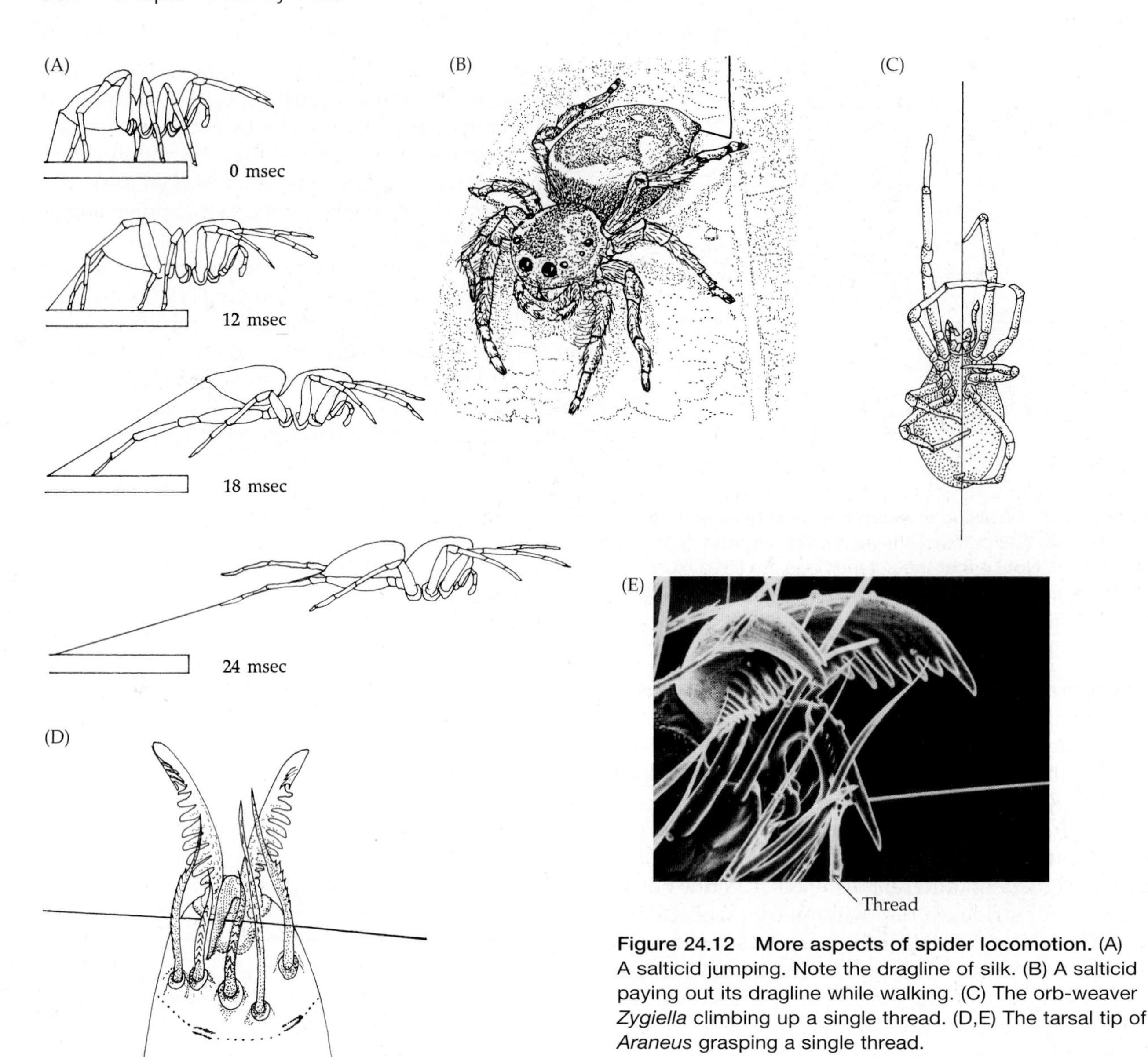

Figure 24.12 More aspects of spider locomotion. (A) A salticid jumping. Note the dragline of silk. (B) A salticid paying out its dragline while walking. (C) The orb-weaver *Zygiella* climbing up a single thread. (D,E) The tarsal tip of *Araneus* grasping a single thread.

Silk plays an important role in various methods of spider locomotion. When walking or jumping, most spiders continually produce a strong thread behind them, called a dragline (Figure 24.12B), produced by the major ampullate silk glands. The dragline is periodically cemented to the substratum with silk from the piriform glands, providing a safety line for the wandering spider. Thus, a spider brushed from a surface does not fall to the ground; instead, it pays out dragline silk and dangles like a tethered mountaineer who has lost his footing. Silk is also used to provide a substratum on which spiders move. Web spinners crawl over (e.g., diplurids and agelenids) or under (e.g., linyphiids and cyatholipids) their sheet webs with limb movements more or less like those used in normal walking, except that stride lengths must correspond to the distances between threads in the web. Many spiders are able to move about on a single thread (Figure 24.12C). This may involve dropping vertically as a thread is paid out from the spinnerets, climbing up a vertical thread, or moving while hanging upside down on a horizontal thread (they do not walk atop single threads, tightrope fashion). Most spiders that are capable of these sorts of activities possess intricately fashioned "thread clamps" on some or all of the leg tips (Figure 24.12D,E).

Many spiders also excavate burrows, and a few lycosoids (such as the fishing spiders of the genus *Dolomedes*) can walk on water (Figure 24.13). One species, *Argyroneta aquatica* (family Cybaeidae), actually lives under water. *Argyroneta* can walk on submerged substrates and builds a "diving bell" that allows gas exchange to oxygenate the air trapped in it; they spend most of their time in this bell when not wandering around in search of prey. There are also several intertidal species that can tolerate submergence at high tides, in some cases by utilizing the air bubbles trapped in empty barnacle shells or other such structures.

(A)

(B)

Figure 24.13 (A) A wolf spider walking on the surface of water. (B) A diving spider (*Argyroneta*) with its "diving bell."

Feeding and Digestion

Feeding The basic chelicerate feeding strategy is one of prey capture followed by extensive external digestion and then ingestion of liquefied food or, more rarely, of small particulate material. There are exceptions to this pattern, of course, most of which involve drastic modifications of the mouthparts. For example, we have already mentioned the varied feeding habits among the mites and ticks, many of which are herbivores or parasites with piercing mouthparts. Whereas many chelicerates are highly specialized in terms of their feeding behavior, others are generalists. Horseshoe crabs, for example, commonly feed on a variety of invertebrates, including worms, molluscs (especially bivalves), crustaceans, and other infaunal and epibenthic creatures, but they also scavenge on almost any organic material. Food is gathered by any of the chelate appendages and passed to the gnathobases along the ventral midline, where it is ground to small bits and then moved forward to the mouth, as in many non-chelicerate arthropods. This mode of feeding by ingesting solid food is also found in Opiliones, now thought to be one of the earliest-diverging arachnid orders.

Scorpions feed mostly on insects, although some large species occasionally eat snakes and lizards. Most are nocturnal and detect prey mainly by highly sensitive mechanoreceptors. A fascinating method of prey location by the Mojave Desert sand scorpion, *Paruroctonus mesaensis*, has been described by Phillip Brownell (see References). Brownell noticed that *Paruroctonus* ignores both airborne vibrations (e.g., wingbeats) and visual input, but responds immediately to nearby prey that are in contact with the sand. This scorpion is even able to detect buried prey, which it immediately uncovers and attacks. Apparently the scorpion senses subtle mechanical waves set up in the loose sand by movements of the prey. Special mechanoreceptors in the walking legs are stimulated as the waves pass beneath the scorpion's limbs. The information is processed to determine the direction and approximate distance to the prey source.

Once a scorpion locates a victim, the prey is grasped by the chelate pedipalps. The opisthosoma is arched over the prosoma, bringing the telson and stinging apparatus into position to inject venom (Figure 24.14). Muscular contractions force the venom through a pore in the aculeus and into the prey. The venom is a

(A)

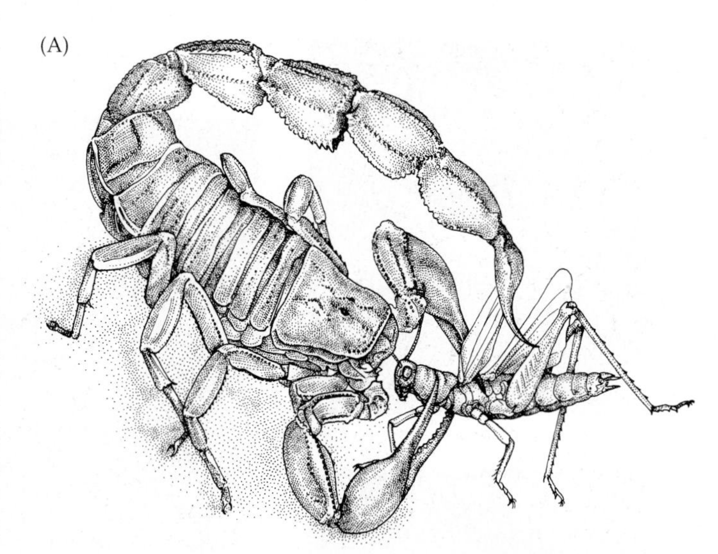

(B)

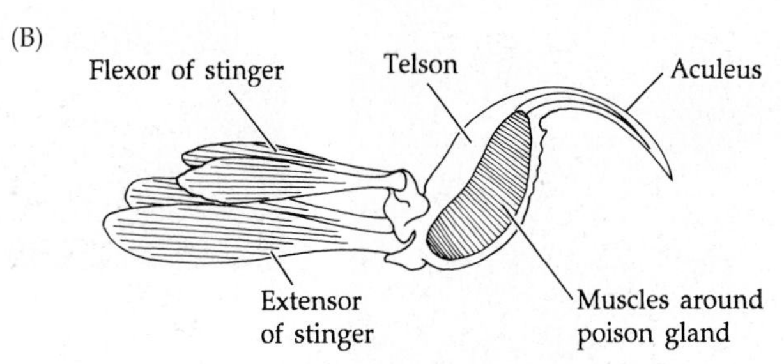

Figure 24.14 (A) The scorpion *Androctonus australis*, applying its stinging apparatus to a grasshopper while holding the prey with its pedipalps. (B) The scorpion's telson and stinging apparatus. The telson is normally held flexed; extension drives the aculeus into the prey.

neurotoxin that can quickly paralyze and kill most prey; in fact, some scorpion venoms can kill large animals, including humans. Perhaps the most familiar of these dangerous scorpions are two species that occur in the American Southwest: the bark scorpion (*Centruroides exilicauda*) and the stripe-tailed scorpion (*Vaejovis spinigerus*). A North African species, *Androctonus australis*, produces a venom considered to be as potent as that of cobras. Other genera of potentially dangerous scorpions include *Buthus* and *Parabuthus* (southern Europe, Africa, and the Middle East) and *Tityus* (South America).

The sequence of events in the ingestion of food by scorpions is typical of many arachnids. Once it has been captured and stung, the prey is passed to the chelicerae, which tear it into small bits. The gnathobases grind and mash the food as digestive juices are released through the mouth; this process reduces the food to a semiliquid form. As this organic "soup" is ingested, the hard parts are discarded, and more bits of food are moved between the gnathobases for processing.

Virtually all spiders are predatory carnivores, although young spiderlings in certain families may consume pollen and fungal spores caught on web silk. On some rare cases, some species exploit other food sources, such as nectar, but the vast majority are generalist carnivores. With the exception of a few families (e.g., Salticidae, Oxyopidae, Thomisidae, and Lycosidae), spiders hunt or feed mainly at night. Most spiders can be separated into two broad categories based on prey-capture strategies. The first is the more sedentary spiders that use some sort of silken web, trap, or net to catch prey. The second is the "wandering" spiders that actively hunt or ambush prey without the direct use of silk (although many wrap their victims after capture).

Earlier, we described the construction of the familiar orb web (Figure 24.8). When a potential prey item, such as a flying insect, strikes and adheres to a web, its movements send out vibrations that alert the spider to the presence of food. The spider then moves rapidly to the victim and bites it. The immense majority of spiders are solitary, but as we have mentioned few live and feed communally (Figure 24.15A).

Another type of silken trap is the horizontal sheet web produced by members of several families, such as the Linyphiidae and Agelenidae. Sheet webs are suspended by a network of supporting threads (Figure 24.15B). Insects become entangled in the sheet web, or in the supporting threads, until the prey drops to the sheet. Once on the sheet, the prey is captured.

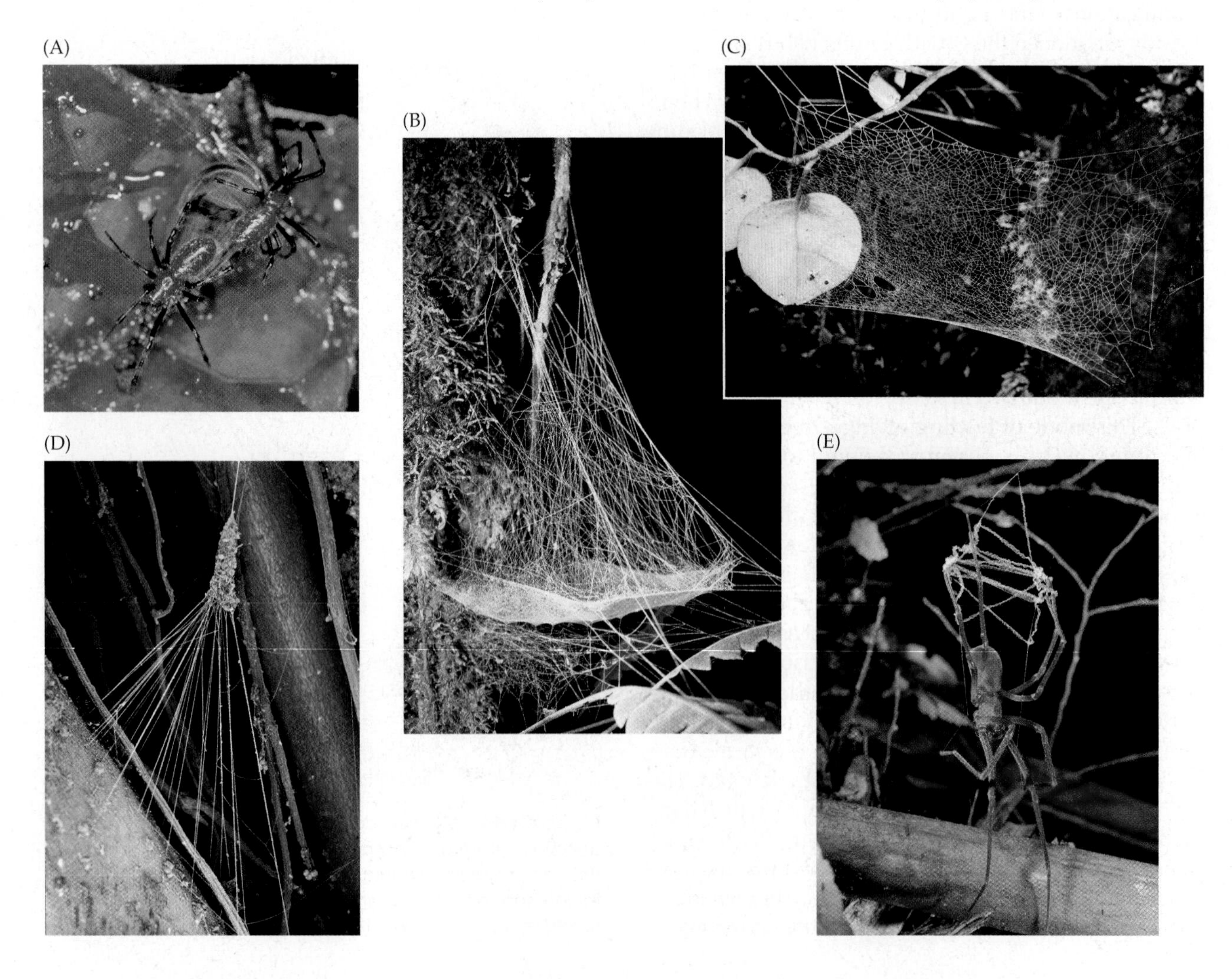

Many theridiid spiders build vertical frame webs (Figure 24.15C). Near their attachment to the substratum, the vertical trap threads are beaded with sticky liquid from silk glands (aggregate glands). Walking prey contact these sticky droplets and become trapped. Upon sensing movement in the web, the spider rushes to the prey, wraps it in silk, and bites it.

Members of the genus *Hyptiotes* (family Uloboridae) spin modified orb webs consisting of only three sectors (Figure 24.15D). The spider produces a tension thread from the point of convergence of the radii and a short attachment thread stuck to the substrate; the spider's body acts as a bridge between these two silk threads. When an insect strikes the web, the spider releases the tension thread, and the web, called a spring trap, collapses around the prey.

Although it is not known what the ancestral web architecture of spiders might have been, araneologists have traditionally looked at liphistiids (Mesothelae; Figure 24.5N) to conjecture about the biology of primitive spiders. Many liphistiids build simple subterranean silken tubes with a single opening from which threads radiate outward (Figure 24.5N). The spider resides in the tube, and the threads serve as "fishing lines" or trip

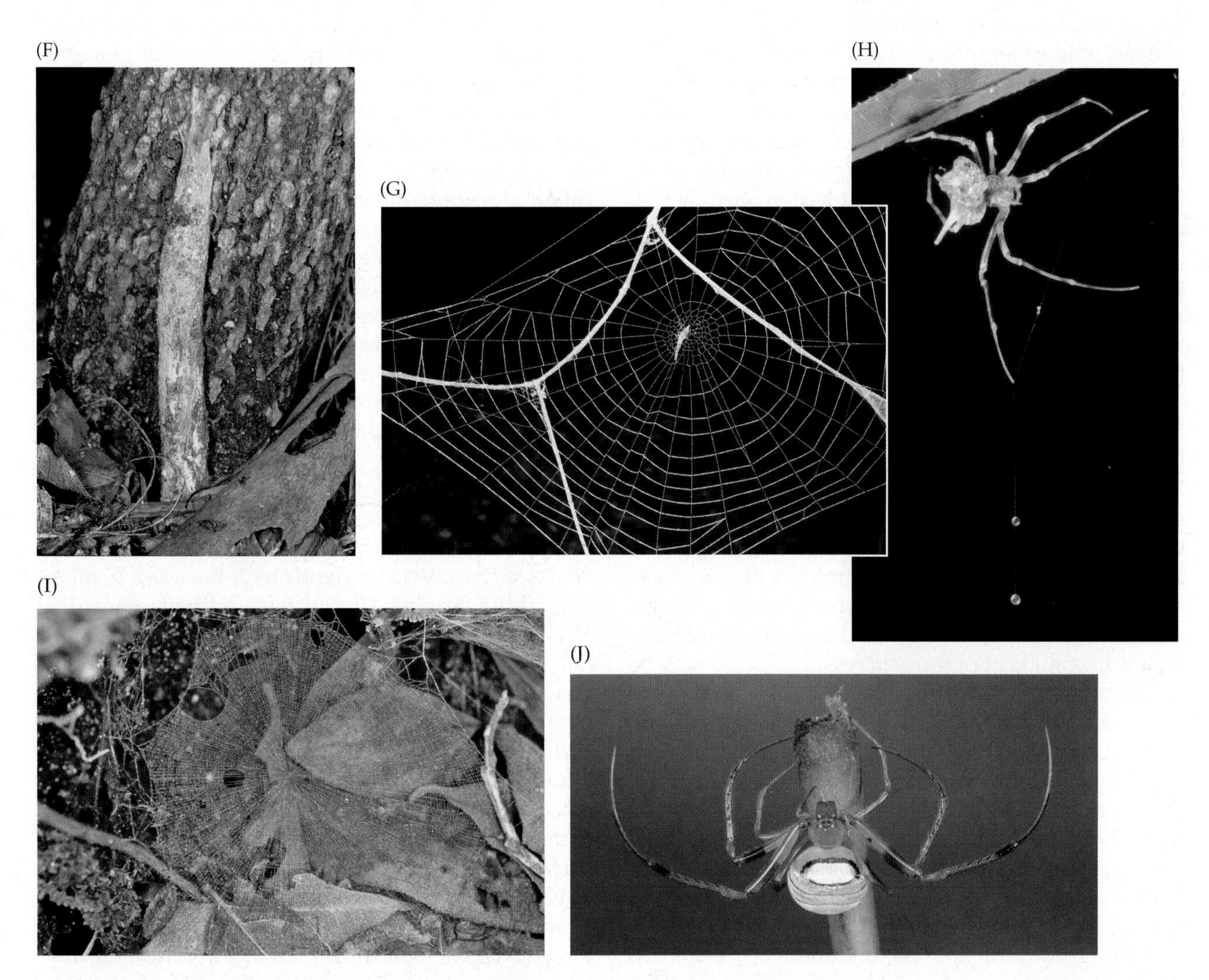

Figure 24.15 The use of spider silk for prey capture. (A) Several *Anelosimus eximius* (Theridiidae) and their prey in Ecuador, an unusual case of communal prey capture. (B) The horizontal sheet web of *Frontinella* (Linyphiidae) in the Dominican Republic. (C) The delicate sheet web of a New Zealand, Synotaxidae, *Runga*. (D) The web of an *Achaearanea* from Ecuador (Theridiidae). (E) The Australian gradungulid *Progradungula otwayensis* holding its cribellate silk catching ladder. (F) The purse web of the North American mygalomorph spider *Sphodros* (Atypidae). When the spider senses the presence of an insect on the tube, it stalks it through the silk with its formidable chelicerae. (G) The cribellate orb web of *Sybota*, an uloborid from Chile. (H) *Exechocentrus lancearius* (Araneidae), a bolas spider from Madagascar that presumably uses chemical mimicry to lure its prey. (I) The micro-orb web of *Patu* (Symphytognathidae) built on the forest leaf litter in the Dominican Republic. Symphytognathids include the smallest known adult spiders (less than half a millimeter). (J) The araneophagic pirate spider *Gelanor latus*, from Brazil (Mimetidae). Mimetids are members of the ecribellate orb-weaving clade and have abandoned building capture webs.

lines that allow the spider to detect passing prey, thus extending its sensory zone as far as these lines reach. An interesting modification of this system is seen in the purse web of some atypids (family Atypidae), such as *Atypus* and *Sphodros* (Figure 24.15F). Here the silken tube is mostly buried beneath the soil, with just a short portion of the blind end lying horizontally above the surface or vertical against a tree trunk or a rock. Insects crawling over the exposed tube are detected by the spider, and the orthognathous chelicerae make two parallel slices in the tube wall near the prey. Their large cheliceral fangs are extended through the incisions to grab the victim and pull it through the tube wall. After the prey is killed, the tear in the tube is repaired.

Most spiders simply set their "traps" and wait for prey, but others actually manipulate silken structures to catch insects. The ogre-faced spiders (family Deinopidae) produce a rectangular web of cribellate threads, which is a highly modified orb web, and hold it between their front walking legs. Deinopids forage at night and have extraordinary visual capabilities, especially *Deinopis*, with a pair of enormous posterior median eyes that are ultrasensitive to light (Figure 24.21A). When an insect is visually detected, the spider sweeps the net around it. The bolas spiders, such as *Mastophora* (family Araneidae), are among the most bizarre and specialized hunters. While hanging from a suspension thread, the bolas spider "throws" a catching thread tipped with sticky liquid to capture its prey (Figure 24.15H). Bolas spiders of the genus *Mastophora* hunt at night using chemical mimicry: they specialize in feeding on male moths of the genus *Spodoptera*. *Mastophora* releases an airborne chemical that mimics the sex pheromone of the female *Spodoptera*, thus attracting males within reach of the catching thread and greatly increasing the likelihood of capture success.

Whereas most sedentary spiders detect their prey by sensing web vibrations, the "wandering" spiders use a variety of methods. Some wolf spiders and most jumping spiders locate prey visually, while many others sense vibrations (e.g., wingbeats or walking movements) or rely simply on accidental contact. Some actually chase prey; others, such as certain trapdoor spiders, lie in ambush and wait for victims to come close enough to grab.

Among jumping spiders, members of the genus *Portia* require special mention. Unlike most spiders, whose behaviors appear to be genetically programmed, *Portia* species exhibit trial and error strategies and learning in their predation on other spiders. *Portia* really doesn't look much like a spider; rather, it resembles crumpled leaves or twigs. Some species utilize this camouflage in what has been described as aggressive mimicry. To do so, *Portia* climbs onto the web of another spider and begins to generate vibrations that mimic those of a trapped insect. When the resident spider responds, *Portia* attacks and kills it. More remarkable is that *Portia* often tries various vibration patterns on a web, and when a successful one is found, continues to use only that one on the webs of the same prey species in the future. Aggressive mimicry is also used by pirate spiders (Mimetidae) (Figure 24.15J), which feed mainly on other spiders.

Regardless of the method of prey location and capture, once contact is made, spiders pull their prey to the chelicerae and bite it, inserting the fangs and injecting venom from poison glands within the prosoma (Figure 24.16A). The prey is quickly immobilized or killed by the poison. An interesting exception to this grabbing-and-biting pattern is displayed by the spitting spiders (family Scytodidae), some of which are social. The poison glands of *Scytodes* include a posterior glue-producing portion along with the usual venom-secreting cells in the anterior region. A mixture of venom and glue is shot very fast from pores in the cheliceral fangs by muscular contraction of the glands; thus the prey is captured without direct contact (Figure 24.16B).

Many spiders wrap their prey in silk to some extent prior to feeding, even if silk is not used in its actual capture. Many hold the victim and wrap it prior to biting it (e.g., theridiids and araneids). Very active insects caught in orb webs are generally wrapped immediately, thereby preventing possible damage to the web or its owner (Figure 24.16D). Potentially dangerous prey, such as large stinging insects, are generally handled in this manner.

Nearly all spiders possess poison glands that produce proteinaceous neurotoxins, although venom composition is highly heterogeneous and varies widely across species. Venom glands have been lost in members of the families Uloboridae and Holarchaeidae. The toxicity of spider venom is quite variable, and only about two dozen species are considered dangerous to humans. Among these are the black widows (*Latrodectus* species), a Brazilian wolf spider (*Lycosa erythrognatha*), the brown recluse spiders (such as *Loxosceles reclusa*), the Australian funnel-web spiders (such as *Atrax robustus*), and some species belonging to the tropical wandering spider family Ctenidae (e.g., *Phoneutria fera*).

Uloborid spiders lack the poison glands typical of almost all other spiders and wrap their prey with up to hundreds of meters of silk that severely compresses and physically damages them. This compaction probably functions to facilitate these spiders' unusual method of feeding, which involves covering the entire surface of the prey with digestive fluid. Spiders, like most other chelicerates, ingest their food in a liquid or semiliquid form. The chelicerae of most spiders have dentate gnathobases with which the prey is mechanically pulverized, while at the same time the food is flooded with digestive juices. Except for the hard parts, the prey is thus reduced to a partially digested broth. Bristles bordering the mouth and thousands of

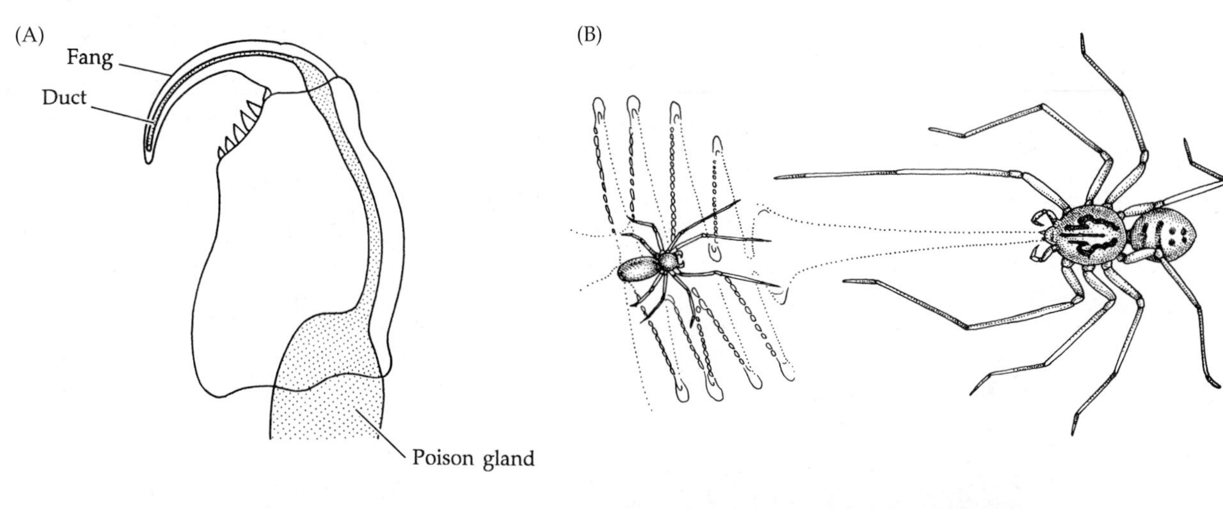

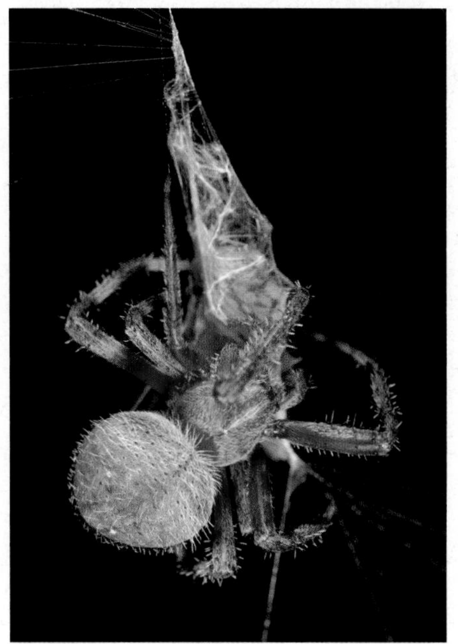

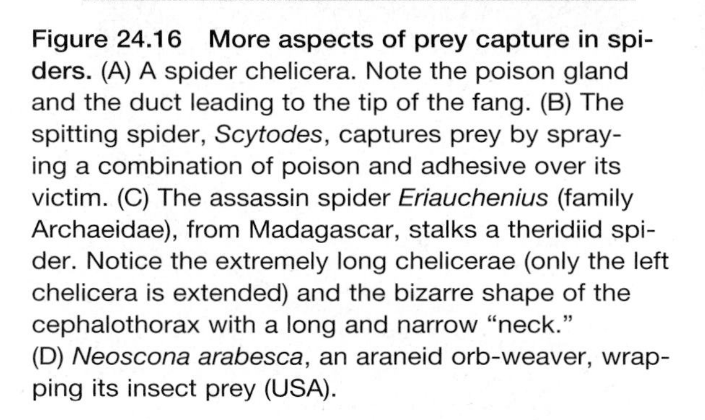

Figure 24.16 More aspects of prey capture in spiders. (A) A spider chelicera. Note the poison gland and the duct leading to the tip of the fang. (B) The spitting spider, *Scytodes*, captures prey by spraying a combination of poison and adhesive over its victim. (C) The assassin spider *Eriauchenius* (family Archaeidae), from Madagascar, stalks a theridiid spider. Notice the extremely long chelicerae (only the left chelicera is extended) and the bizarre shape of the cephalothorax with a long and narrow "neck." (D) *Neoscona arabesca*, an araneid orb-weaver, wrapping its insect prey (USA).

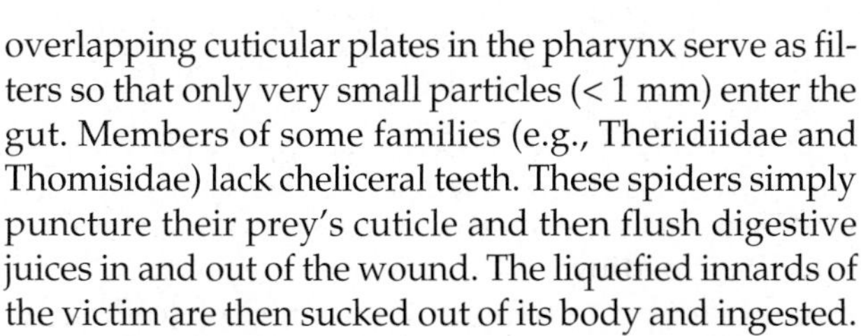

overlapping cuticular plates in the pharynx serve as filters so that only very small particles (< 1 mm) enter the gut. Members of some families (e.g., Theridiidae and Thomisidae) lack cheliceral teeth. These spiders simply puncture their prey's cuticle and then flush digestive juices in and out of the wound. The liquefied innards of the victim are then sucked out of its body and ingested.

The fascinating biology of two species of jumping spiders (Salticidae) that feed preferentially on mosquitoes was recently described; both are highly specialized mosquito assassins. One species, *Evarcha culicivora*, like an eight-legged vampire, is particularly drawn to female mosquitoes whose guts are filled with blood (especially species of *Anopheles*). Eating blood-filled mosquitoes also allows this spider to acquire a scent that is attractive to members of the opposite sex. Another species, *Paracyrba wanlessi*, preferentially feeds on mosquito larvae lurking in pools of water inside bamboo.

Digestion The digestive system of chelicerates follows the basic arthropod plan of foregut, midgut, and hindgut, the first and last parts being lined with cuticle (Figure 24.17). The foregut is often regionally specialized. In the xiphosurans (e.g., *Limulus*), the foregut loops anteriorly to form an esophagus, crop, and gizzard, the last of these bearing sclerotized ridges that grind ingested particles (Figure 24.17A). In many arachnids, portions of the foregut are modified as pumping organs for sucking in liquefied food. In scorpions this function is served by a muscular pharynx, and in spiders by an elaborate sucking stomach (Figure 24.17B). The pharynx of spiders may contain chemosensory cells that function as taste receptors.

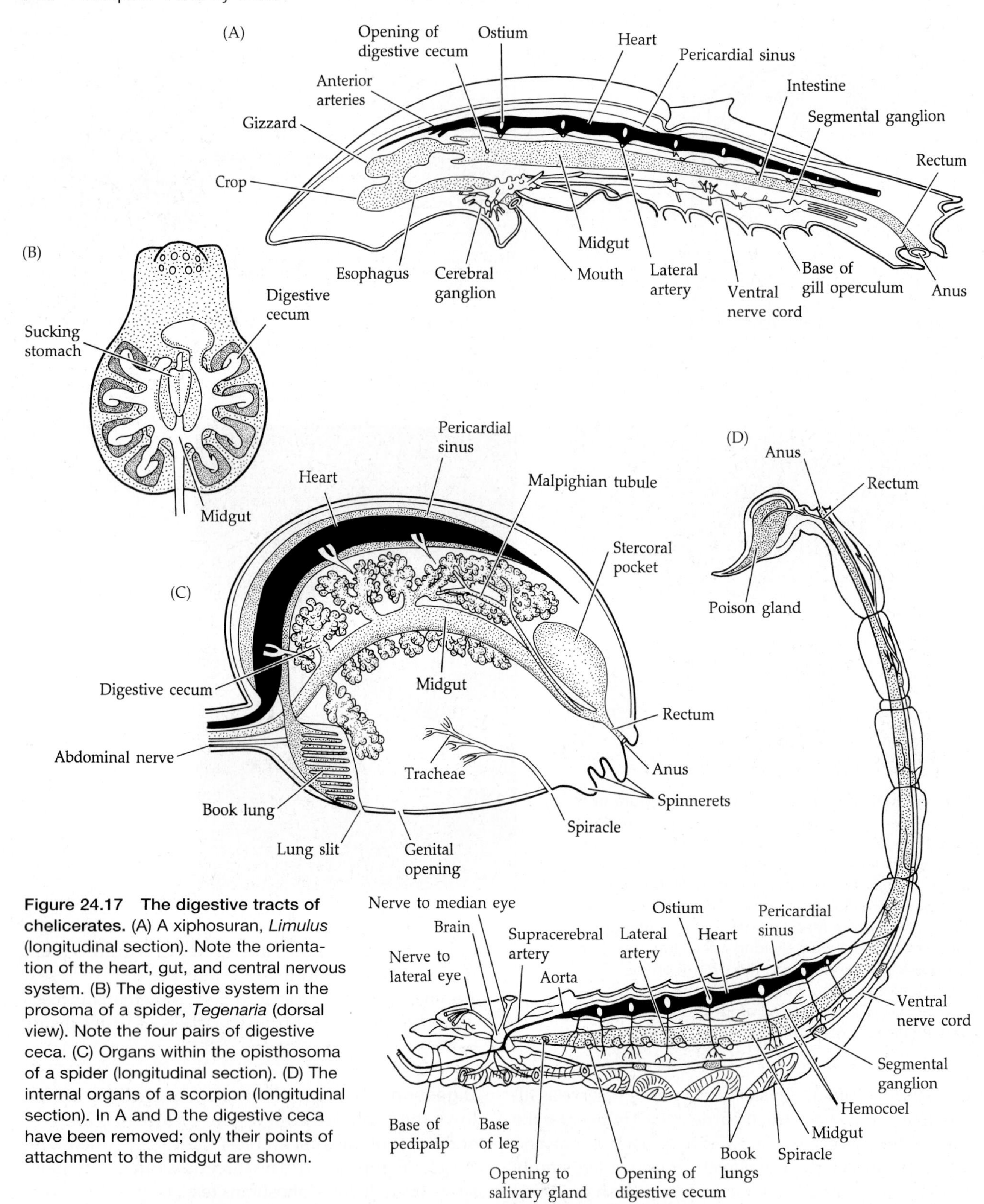

Figure 24.17 The digestive tracts of chelicerates. (A) A xiphosuran, *Limulus* (longitudinal section). Note the orientation of the heart, gut, and central nervous system. (B) The digestive system in the prosoma of a spider, *Tegenaria* (dorsal view). Note the four pairs of digestive ceca. (C) Organs within the opisthosoma of a spider (longitudinal section). (D) The internal organs of a scorpion (longitudinal section). In A and D the digestive ceca have been removed; only their points of attachment to the midgut are shown.

The euchelicerate midgut bears paired digestive ceca and is the site of final chemical digestion and absorption (Figure 24.17C). Xiphosurans have two pairs of caeca arising from the anterior part of the midgut, followed by an intestine, a short rectum (the hindgut), and the anus on the posterior margin of the opisthosoma. In *Limulus*, enzymes are produced by the midgut wall and secreted into the lumen. Apparently only preliminary protein digestion occurs extracellularly, and final breakdown takes place in the cells of the digestive ceca after absorption.

Most spiders possess four pairs of digestive ceca in the prosoma and frequently additional branched ceca in the opisthosoma (Figure 24.17B,C). Near its junction with a short rectum, the midgut expands as a spacious "mixing chamber," called the stercoral pocket.

Malpighian tubules arise from the midgut wall near the origin of the stercoral pocket. The anus is located on the opisthosoma near the spinnerets.

The midgut of scorpions bears six pairs of digestive ceca (Figure 24.17D). The first pair, the salivary glands, lies within the prosoma and produces much of the digestive juice used in preliminary external digestion. The remaining five pairs are highly convoluted and lie in the opisthosoma. These ceca produce the enzymes for final digestion and are the site of absorption of the digestive products. Two pairs of Malpighian tubules arise from the posterior region of the midgut, just in front of the short rectum. The anus is on the last opisthosomal segment.

Circulation and Gas Exchange

The euchelicerate circulatory system, like that of other arthropods, consists of a dorsal ostiate heart situated within a pericardial sinus and giving rise to various open-ended vessels (Figures 24.17 and 24.18). Blood leaves these vessels and enters the hemocoel, where it bathes the organs and supplies the structures of gas exchange before returning to the heart. The complexity of the system is primarily a function of body size; some very tiny euchelicerates (e.g., palpigrades and some mites) have lost much or all of their circulatory structures—for them, gas exchange is cutaneous. On the other hand, the xiphosurans are large animals, and their body plan demands a substantial circulatory mechanism to move the blood around inside the rigid body covering. The large tubular xiphosuran heart bears eight pairs of ostia, and it is attached to the body wall by nine pairs of ligaments that extend through the pericardium (Figure 24.17A). The organs of these big creatures are supplied with blood through an extensive arterial system arising from the heart and opening into the hemocoel close to the organs themselves. In the opisthosoma, a major ventral vessel gives rise to a series of afferent branchial vessels to the book gills. Efferent vessels carry oxygenated blood to a large branchiopericardial vessel leading back toward the heart.

The gas exchange organs of xiphosurans are unique among the chelicerates. The presence of gills is, of course, associated with their aquatic lifestyle. The structure of these opisthosomal book gills provides an extremely large surface area, a necessity for adequate gas exchange by these large animals (Figures 24.2B and 24.18C). Each gill bears hundreds of thin lamellae, like pages in a book. The blood within the lamellae is separated from the surrounding seawater by only a thin wall. Water is moved over the lamellae by metachronal beating of the gills; these movements also cause blood flow into (on the forward strokes) and out of (on the backward strokes) the gill sinuses.

(A) Gill sinus
Gill
Operculum

(B)
Heart
Pericardial sinus
Ostium
Midgut
Aorta
Abdominal artery
Stercoral pocket
Lung vein
Book lung

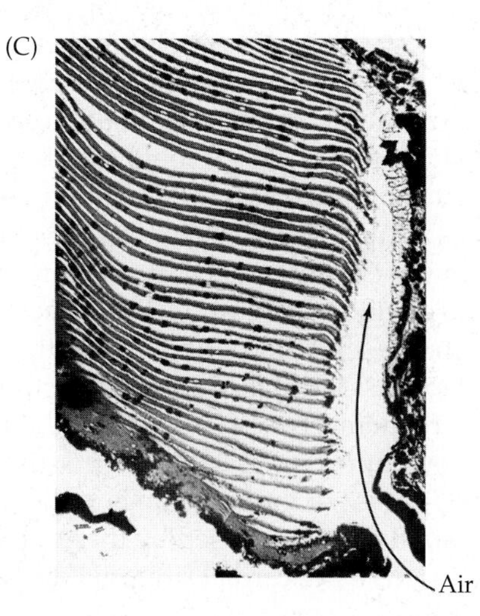

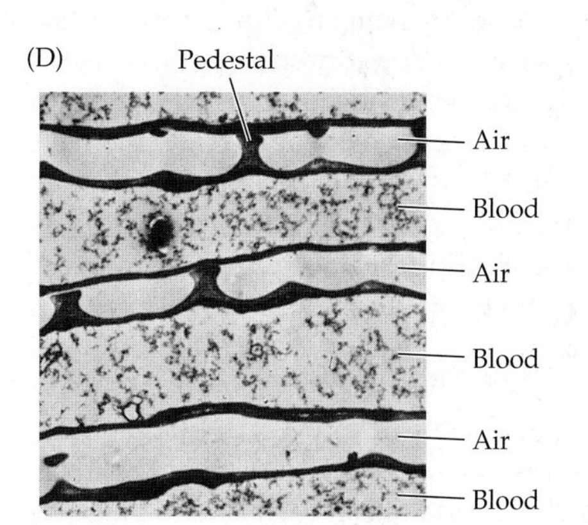

Figure 24.18 Circulatory and gas exchange structures in some chelicerates. (A) An opisthosomal appendage of *Limulus* (posterior view). Note the gill and opercular parts. (B) Major elements of the circulatory system in a spider. Note the direct route of blood from the book lung to the heart. (C) The book lung of the spider, *Lycosa* sp. (section). (D) The "leaves" of a book lung (section). Note the separation of air spaces and blood-filled leaves. Cuticular pedestals keep the air spaces from collapsing.

The heart of a spider lies within the opisthosoma and bears two to five pairs of ostia (Figure 24.18B). It is suspended within the pericardial sinus by several ligaments attached to the inside of the exoskeleton. These suspensory ligaments are stretched during systole as blood is pumped from the heart into arteries. The elasticity of the ligaments then effects diastole, expanding the heart and drawing blood into it from the pericardial sinus. The routes of the major arteries ensure that an ample supply of oxygenated blood reaches the major organs, particularly the central nervous system, muscles, and the hemocoelic spaces in the limbs in which blood pressure aids in leg extension. From the hemocoel, blood is channeled back to the pericardial sinus and the heart.

Aerial gas exchange structures in spiders include book lungs and tracheae. Generally, liphistiomorphs, mygalomorphs, and some basal araneomorphs have two pairs of book lungs but no tracheae, whereas members of Araneomorphae usually have one pair of book lungs and a system of tracheal tubes. In some groups of very small sized araneomorphs (such as the minute Synaphridae) the anterior book lungs are modified into tubular tracheae. Since the tracheae occur only in more distal groups among spiders, it is likely that they evolved separately from those of other arachnids, and of course independently from those of insects and myriapods. Phylogenetic analyses, both morphological and molecular, support this contention, and place spiders with the other pulmonate arachnid orders, sometimes even including scorpions.

Spider book lungs are located in the second, or the second and third segments of the opisthosoma. They open to the outside via spiracles, or lung slits, near the epigastric furrow (Figure 24.5C). Since they are localized and do not extend far into the body, book lungs must receive sufficient circulating blood to ensure adequate distribution of oxygen throughout the body and removal of carbon dioxide from the internal organs. Just inside each lung slit is an expanded chamber, the atrium, from which numerous flattened air spaces extend into the hemocoel (Figure 24.18D). These leaflike pockets of air are separated from one another by thin blood-filled extensions of the hemocoel. Although the book lungs are themselves relatively small, this structural arrangement provides a very large surface area between the "pages" of the book lungs and the circulatory fluid. Blood that has passed by these surfaces returns directly to the pericardial sinus via the lung vein (Figure 24.18B).

Spider tracheae open to the outside through one or two spiracles located posteriorly on the third opisthosomal segment (Figures 24.5B,C and 24.17C). These tracheae probably evolved from the book lungs and muscle apodemes on this segment in more primitive spiders. The spiracles open inward to simple or branched tubes. In spiders, the open inner ends of the tracheae do not bring the oxygen supply into direct contact with tissues, as they do in many insects; rather, a small amount of blood is necessary as a diffusion medium. When the tracheal system is extensive, there is usually a reduction in the structural components of the circulatory system.

The circulatory system of scorpions is very similar to that of spiders, except that it is constructed to accommodate an elongate and generally larger body. The tubular heart bears seven pairs of ostia and extends through most of the mesosoma (Figure 24.17D). An extensive set of arteries delivers blood to the hemocoel throughout the body and to four pairs of mesosomal book lungs.

Except where reduced or absent, the circulatory system of other euchelicerates is built along the same general plan as described above. Gas exchange is cutaneous in palpigrades and some mites, but in other euchelicerate groups it occurs through book lungs (Uropygi, Schizomida, and Amblypygi) or tracheae (Ricinulei, Pseudoscorpiones, Solifugae, Opiliones, and many mites).

The blood of chelicerates has been most extensively studied in spiders and in xiphosurans. Because *Limulus* is large, its blood chemistry has been especially well studied, and many horseshoe crabs have made the supreme sacrifice for science at the hands of laboratory physiologists—so many, in fact, that many populations of the American species (*L. polyphemus*) have become threatened. Hemocyanin, the common respiratory pigment in chelicerates, is dissolved in the blood plasma. At least in spiders, hemocyanin serves primarily for oxygen storage rather than for immediate transport and delivery of oxygen to the tissues. Hemocyanin has a very high affinity for oxygen, releasing it only when surrounding oxygen levels are quite low. Some spiders are able to survive for days after their air supply has been cut off experimentally by covering the spiracles. Apparently they obtain sufficient oxygen from their hemocyanin-bound stores and some cutaneous exchange.

Chelicerate blood also contains various cellular inclusions, but the functions of most of them are not well understood. The blood of *Limulus* includes amebocytes that may provide clotting agents. Several kinds of blood cells occur in spiders. Interestingly, it seems that all of them originate as undifferentiated cells from the muscular portion of the heart wall itself. These cells are released into the blood, where they mature and differentiate. Functions that are attributed to chelicerate blood cells include clotting, storage, combating infections, and aiding in sclerotization of the cuticle.

Excretion and Osmoregulation

Xiphosurans have two sets of four coxal glands each, arranged along each side of the prosoma near the coxae of the walking legs. The glands on each side of the body converge to a coelomic sac, from which a long, convoluted duct arises. The duct leads to a bladder-like enlargement that connects to an excretory pore at the base of the last walking legs. Surprisingly little is known about excretory physiology in xiphosurans. Apparently the coxal glands extract nitrogenous wastes from the surrounding hemocoelic sinuses and carry them to the outside. The coxal glands and their associated tubule system also function in osmoregulation, as evidenced by the production of a dilute urine when the animal is in a hypotonic medium. The digestive ceca probably aid in excretion of excess calcium by removing it from the blood and releasing it into the gut lumen.

The problems of excretion and water balance are obviously much more critical to terrestrial euchelicerates than to horseshoe crabs, and land-dwelling arachnids display a variety of structural, physiological, and behavioral adaptations to cope with them. The main excretory structures of arachnids are coxal glands and Malpighian tubules, although many groups possess additional supplementary nitrogenous waste removal mechanisms. Coxal glands persist in many arachnids (spiders, scorpions, palpigrades); in these animals, the glands lie within the prosoma and open on the coxae of some walking legs. The degree to which coxal glands function in excretion and osmoregulation varies among arachnids, but they are considered far less important than the Malpighian tubules.

The Malpighian tubules of arachnids arise from the posterior midgut. They are not homologous to the Malpighian tubules of insects or myriapods, which arise from the hindgut and are thus of ectodermal origin. The tubules branch within the hemocoel of the opisthosoma, where they actively accumulate nitrogenous waste products and release them into the gut for elimination along with the feces (Figure 24.17C). In spiders, the wastes from the tubules and the gut are mixed in the stercoral pocket prior to release from the anus. The excretory action of the Malpighian tubules is often supplemented by other mechanisms, such as the coxal glands. Nitrogenous wastes also are accumulated in the cells of the midgut wall and released into the lumen. In addition, waste products are picked up and stored by special cells called nephrocytes, which form distinct clumps in various parts of the prosoma.

Terrestrial arachnids produce complex, insoluble, nitrogen-containing excretory compounds. The major excretory product is guanine, although uric acid and other compounds also occur. Because these compounds are of low toxicity, they can be stored and excreted from the body in semisolid form, thus conserving water.

Terrestrial arachnids also display various behavioral adaptations to avoid desiccation. Most arachnids are nocturnal, remaining in cooler or more humid, protected places during the daytime. Some spiders actively drink water during dry periods or when they lose blood through injury. Desert scorpions must tolerate not only low humidities but also very high daytime temperatures. They typically bury themselves in sand or soil, or hide under rocks or tree bark during the day. In addition, some species exhibit an adaptive behavior called stilting, wherein the body is raised off the substratum to allow air to circulate underneath. While thought to be mainly a cooling strategy, this behavior, by lowering the body temperature, probably also slows the rate of evaporative desiccation. Some scorpions are also able to withstand large losses of body water—as much as 40 percent of their body weight—with no ill effects.

Nervous System and Sense Organs

As in all arthropods, the external body form of euchelicerates is generally reflected in the structure of the central nervous system. These animals show various degrees of compaction and fusion of the body somites and the associated nervous system components while still conforming to the basic arthropod plan. The cerebral ganglia, or brain, includes the protocerebrum (which innervates the eyes), the deutocerebrum (which innervates the chelicerae), and the tritocerebrum (which innervates the pedipalps). The deutocerebrum was traditionally thought to be absent in euchelicerates until Hox gene expression data showed homology of the chelicerae of euchelicerates with the deutocerebral appendages of mandibulates in the late 1990s. The tritocerebrum generally contributes to the circumenteric connectives, which unite ventrally with a large ganglionic mass formed in part by the fusion of paired anterior ganglia of the ventral nerve cord. In xiphosurans and scorpions this subenteric neuronal mass includes all of the prosomal ganglia, whereas in spiders even the opisthosomal ganglia fuse anteriorly. Thus in most spiders the adult nervous system is no longer obviously segmented (except in some members of the suborder Mesothelae), although a chain of ventral ganglia is evident during early embryonic development. The ventral nerve cord persists in the opisthosoma of xiphosurans and has five segmental ganglia; in scorpions it has seven ganglia (Figures 24.17A,D and 24.19).

The protocerebrum and deutocerebrum give rise to nerves to the eyes and chelicerae, respectively. The ventral (subenteric) ganglionic mass, which includes the fused segmental prosomal ganglia, gives rise to nerve tracts to the pedipalps and walking legs and, in spiders, bears a pair of abdominal ganglia from which arise branching nerves to the opisthosoma. Segmental ganglia on the ventral nerve cord in xiphosurans and scorpions serve the opisthosomal appendages, muscles, and sense organs.

Chapter 20 discussed some of the qualities of arthropod sense organs in terms of the imposition of the exoskeleton. Xiphosuran sense organs include tactile mechanoreceptors in the form of various spines and

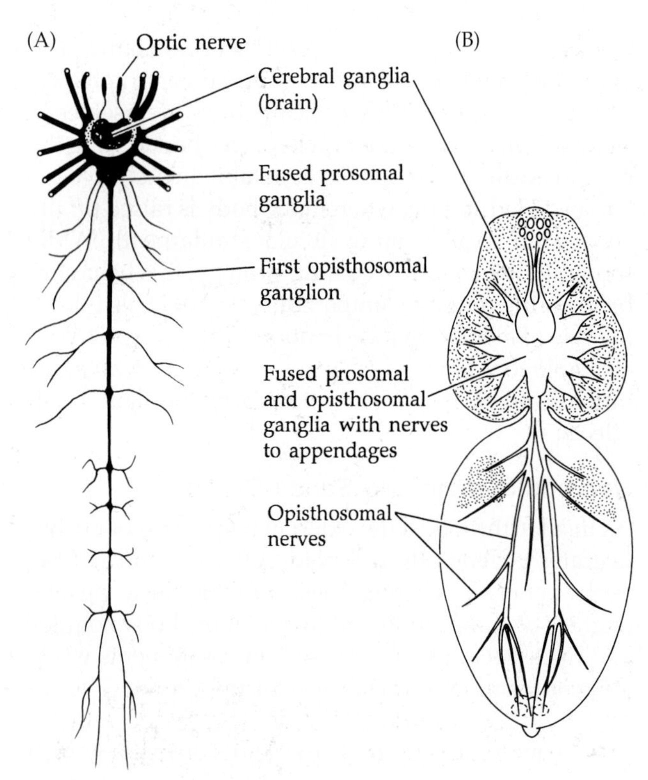

Figure 24.19 The central nervous system of (A) a scorpion and (B) a spider.

bristles, proprioceptors in the joints, chemoreceptors, and photoreceptors. The prosoma bears two simple eyes near the dorsal midline and two lateral compound eyes (Figure 24.2A). The median eyes are of the pigment cup type, but each contains a distinct cuticular lens. The lateral eyes are compound rhabdomeric units—structures not found in any other chelicerate. The thousand or so ommatidia per eye are very large and rather loosely packed together. While the resolving power of xiphosuran eyes has long been debated, certainly *Limulus* can detect movement and changes in light intensity and direction. It has been demonstrated that the light sensitivity of these receptors is regulated on a 24-hour cycle by instructions from the brain. At night, these signals increase the eyes' sensitivity to light up to a million times over daytime levels. Thus, *Limulus* may see as well at night as it does during daylight hours. It may also be capable of perceiving clear images, as male horseshoe crabs have been experimentally shown to be attracted to "models" of females.

Arachnids possess well-developed sense organs upon which much of their complex behavior depends. Most of the "hairs" on a spider or scorpion body are mechanoreceptors, collectively called hair sensilla. Simple tactile hairs (or setae) cover much of the body surface and respond to direct physical contact. A second type of setae, called trichobothria, is found on the appendages of many arachnids. They are less abundant and much thinner than simple tactile setae and are extremely sensitive (Figure 24.20A,B). *Cupiennius*, a tropical wandering spider (Ctenidae), has close to one thousand trichobothria in its legs and pedipalps. Trichobothria are stimulated by airborne vibrations, such as those caused by beating insect wings, natural air currents, and probably some sound frequencies.

Additional mechanoreceptors of arachnids include slit sense organs (Figure 24.20C). These structures may occur as single slit sense organs (or slit sensilla) or in groups of parallel slits called lyriform organs. Slit sensilla are deep grooves in the cuticle associated with sensory neurons. They detect a variety of mechanical stimuli that impose physical deformation on the cuticle around the slit. Depending on their location and orientation, spider slit sensilla serve as proprioceptors (by sensing leg movements and position), as georeceptors (by measuring bending of the pedicel under the weight of the opisthosoma), as direct mechanoreceptors (by sensing direct external pressure on the cuticle), as vibration sensors, and even as phonoreceptors.

Not all of the setae of spiders are sensory in function. We have already described the calamistrum, a row of modified macrosetae in the fourth metatarsus of cribellate spiders used to comb the silk produced by the cribellum. Another interesting type of specialized setae is found in the New World tarantulas (Theraphosidae). These mygalomorphs have modified abdominal barbed setae than can be released with their hind legs. These setae become airborne and are highly urticating, providing a defensive mechanism against some predators. Efficient proprioceptors occur in the walking legs and pedipalps of all chelicerates and are particularly well developed in arachnids. By virtue of their position and number in different joints, they convey information about the direction and velocity of appendage motion as well as the position of the limbs relative to the body and to one another. These "true" proprioceptors appear to work in concert with the lyriform organs.

Chemoreception in arachnids involves sensing both liquid and airborne chemicals that contact the body. This dual ability can be likened to capacities for both taste (contact chemosensitivity) and smell, or olfaction (distance chemosensitivity). The olfactory sense plays major roles in prey location and, for those species in which females release sex pheromones, in mating. The most important chemoreceptors are probably the hundreds of erect, hollow setae (called sensillae chaetica) with open tips that are present on the pedipalps and other areas around the mouth and are most abundant on the tips of limbs that contact the substratum. Dendrites of sensory neurons extend through the hollow hair shaft to the open tip, where they are directly stimulated by chemicals (Figure 24.20D). Spiders also bear humidity detectors called tarsal organs (Figure 24.20E).

Scorpions have a pair of large, unique, comblike structures, the pectines, on the ventral surface of the

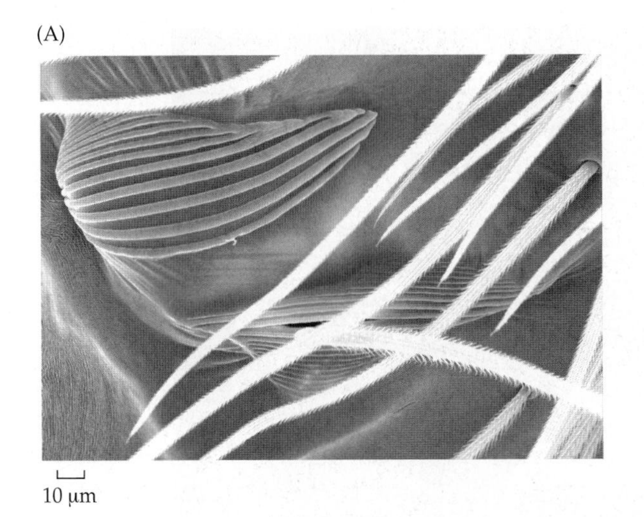

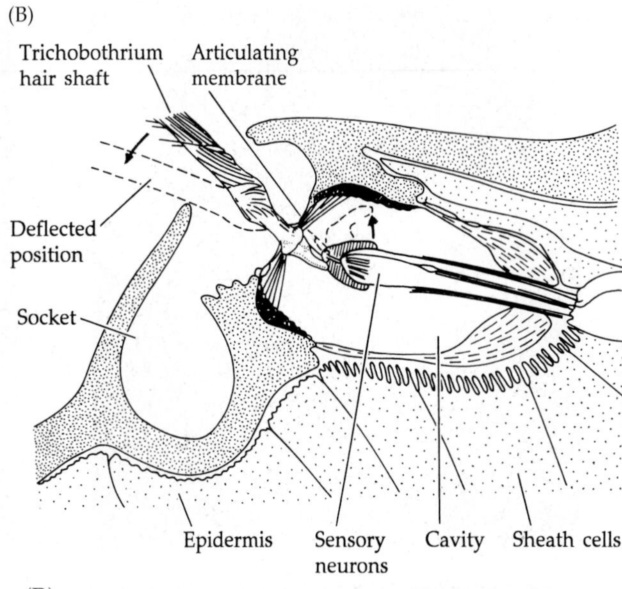

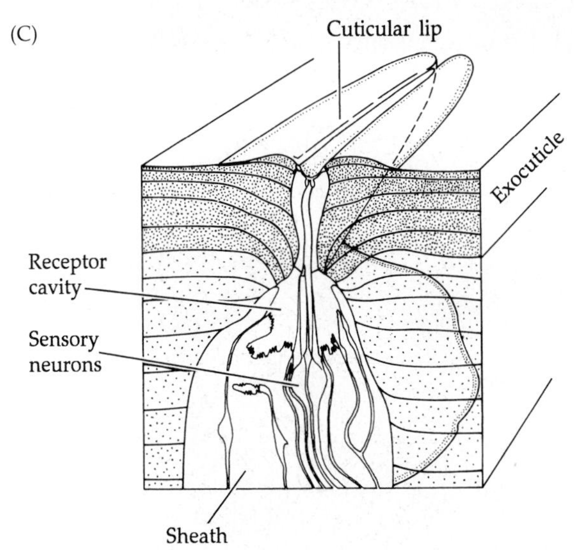

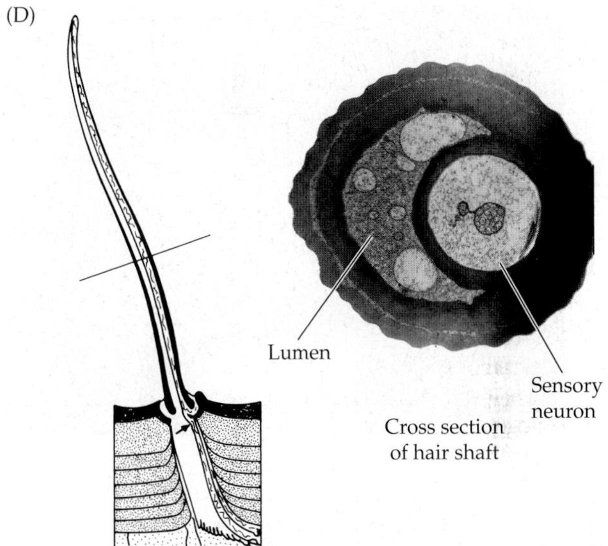

Figure 24.20 Mechanoreceptors and chemoreceptors in spiders. (A) Lyriform organs (groups of slit sensilla) on the leg patella of *Amaurobius similis* (SEM, 1.76K×). (B) The organization of a trichobothrium in its socket; from *Tegenaria*. (C) A slit sense organ from the leg of *Cupiennius* (cutaway view). (D) A chemosensitive hair. The mechanoreceptor neurons terminate at the base of the hair (arrow in illustration), whereas the chemoreceptor neurons extend through the hollow shaft to the tip (photo). (E) A humidity receptor, or tarsal organ, of the orb-weaver *Araneus*.

mesosoma (Figure 24.6C). After detailed studies of their innervation, Foelix and Schabronath (1983) suggested that the pectines act as both mechanoreceptors and chemoreceptors. Other work has shown them to be capable of detecting subtle differences in sand grain size. These versatile structures are usually held laterally erect, free to swing back and forth while the scorpion is actively moving about. Solifuges have another mysterious sensory structure called malleoli (or racket organs) on the fourth coxae and trochanters.

The importance of vision varies greatly among the arachnids. Most species possess some sort of photoreceptors, although Ricinulei and Palpigradi are entirely blind. At least some species in the other groups are also blind (e.g., some members of the orders Schizomida, Pseudoscorpiones, and different mite orders), and within the Opiliones suborder Cyphophthalmi, four of the six families are entirely

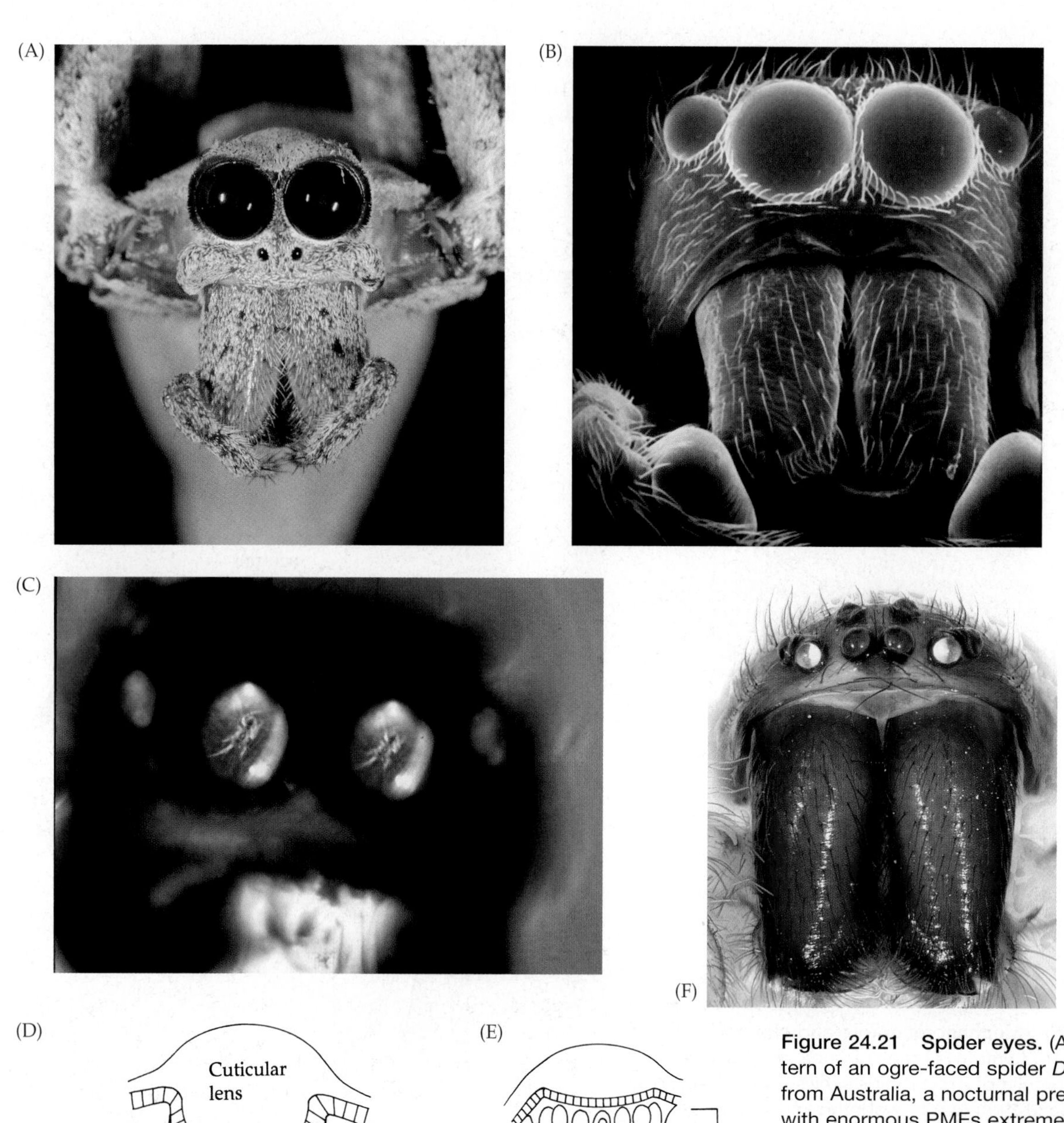

Figure 24.21 Spider eyes. (A) Eye pattern of an ogre-faced spider *Deinopis* from Australia, a nocturnal predator with enormous PMEs extremely sensitive to light. (B) "Looking into the eyes" of *Heliophanus* (family Salticidae). (C) Image of another spider as "seen through the eyes" of *Portia*, a jumping spider. (D) Section through a main eye. (E) Section through a secondary eye, showing inverted arrangement of the retinular cells and tapetum. (F) Head-on view of *Amaurobius similis* (Amaurobiidae). Note the highly reflective tapetum in the anterior lateral eyes (secondary eyes). The median eyes are the main eyes and lack a tapetum.

blind. Certain spiders depend on photoreception for prey and mate location, particularly the errant hunters (e.g., some lycosids and most salticids) (Figure 24.21B,C). Vision is relatively unimportant to many sedentary species, such as many web builders that depend more on tactile cues, vibrations, and chemosensitivity. Web builders are not blind, however, and many respond behaviorally to variations in light intensity, while some exhibit distinct escape responses when they visually detect potential predators. Nevertheless, these spiders are generally able to build their webs, capture prey, and mate with little or no visual input. Some spiders are capable of perceiving polarized light, presumably as a means of spatial orientation.

Spiders possess single-lensed rhabdomeric eyes, but the sensory units of each are simple ocelli in a cluster. Thus they are quite different from the compound eyes of xiphosurans and most other arthropods. Each eye includes a thickened cuticular lens over a vitreous body, which is a layer of cells derived from the epidermis and covering the retina (Figure 24.21D). The retina

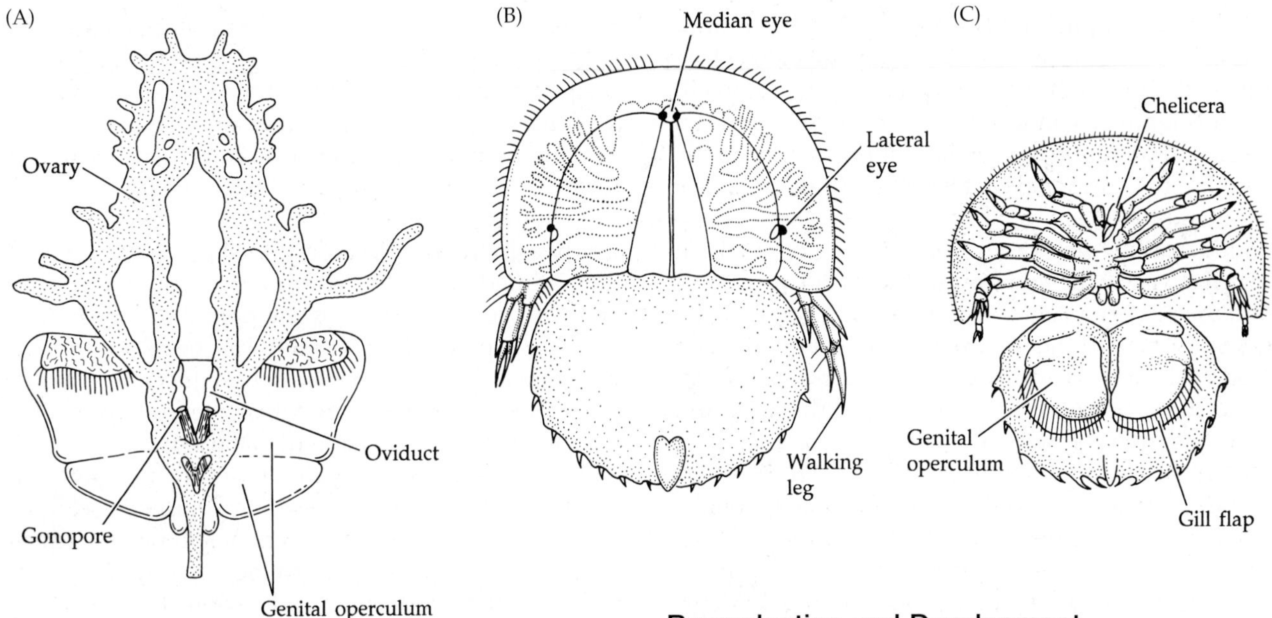

Figure 24.22 (A) Female reproductive system of *Limulus* (order Xiphosura). (B,C) Euproöps larva of *Limulus* (dorsal and ventral views).

is composed of the sensory (receptor) cells and the pigment cells. The membranes of the sensory cells bear interdigitating microvilli, confirming the rhabdomeric nature of the eyes.

Two forms of this basic eye structure occur among spiders. The anterior median eyes—the main eyes—have the light-sensitive portions of the sensory cells directed toward the lens, whereas other eyes—secondary eyes—are inverted, with the light receptor elements directed away from the lens (Figure 24.21D,E). Most of these secondary eyes contain a crystalline reflective layer called the tapetum, which may serve to collect and concentrate light in poorly lit conditions (e.g., during night hunting). The reflective nature of the tapetum produces the effect in some spiders of eyes that "shine in the dark" (Figure 24.21F).

Scorpion eyes are of the direct type and differ from spider eyes in having the retinal layer external to the epidermis. Most work suggests that scorpions depend much more on mechanoreception and chemoreception than on visual input.

There is no question that spiders, and perhaps other arachnids, are capable of modifying their behavior based on experience—that is, they can learn. We have seen some examples of such activity in the preceding section. Memory and association centers in the protocerebrum are responsible for much of this integrative activity. No doubt, at least in spiders, the ability to remember, learn, and make appropriate adaptations in behavior has played an important role in their evolutionary success. A recent collection of papers on spider behavior can be found in Herberstein (2011).

Reproduction and Development

Euchelicerates are gonochoristic and generally engage in complex mating behaviors that ensure fertilization. A few are known to be parthenogenetic (e.g., some scorpions, harvestmen, and schizomids). Males with a penis occur only in harvestmen and some mites. Free spawning never occurs, and fertilization takes place either internally or as the eggs leave the body of the female. The eggs are generally very yolky except in xiphosurans, the developmental strategy is direct in spite of the various juvenile stages through which most euchelicerates pass. We first present a summary of reproduction and development in *Limulus*, then turn to the arachnids, concentrating again on spiders (the best known arachnid developmental model) and scorpions. More recently, development in Opiliones has received attention.

The reproductive system of xiphosurans is similar in males and females. In both sexes the gonad is a single, irregularly branched mass of tissue (Figure 24.22A). Paired gonoducts lead from the gonad to a pair of pores on the ventral midline. The first pair of opisthosomal appendages lies over the gonopores, forming a genital operculum.

At the onset of the breeding season, horseshoe crabs migrate into the shallow waters of protected bays and estuaries. On the east coast of North America, this migration takes place in the spring and summer, and huge numbers of *Limulus* can be seen gathering near the shore in preparation for mating (Figure 24.2C). Mating is initiated when the male climbs onto the back of the female and grasps her with his modified first walking legs. The clasped pair moves to the shallow water, usually at a high spring tide, and the female excavates one or more shallow depressions in the sand and deposits her eggs (2,000 to 30,000 eggs per mating). The male releases sperm directly onto the eggs as they are deposited. Then the mates separate, and the female covers the fertilized eggs with sand.

Early development takes place in the sand or mud "nest." Cleavage is holoblastic, producing a stereoblastula with most of the yolk contained in the inner cells. As development continues, the surface cells at the anterior and posterior ends of the embryo divide rapidly, forming two germinal centers. Some of these rapidly proliferating cells migrate inward as the presumptive endoderm and mesoderm. The anterior germinal center gives rise to the first four segments of the prosoma and the posterior center to the rest of the body. All of the prosomal segments fuse and are eventually covered by the developing carapace-like dorsal shield.

As the yolk reserves become depleted, the embryo emerges from the sediment as a **euproöps larva** (or "trilobite larva"), so named because of its resemblance to the fossil Carboniferous xiphosuran *Euproöps* (which superficially resembled trilobites; Figure 24.22B,C). The larvae swim about and periodically burrow in the sand. Segments are formed and appendages added through a series of molts until the adult form is reached, a typical mode of development observed in other arthropod groups. The early developmental stages are supplied with an investment of yolk by the female and are protected by the nest she constructs, but the young emerge as independent feeding larvae prior to maturation.

The reproductive biology of arachnids is directly related to their success on land. They have evolved a variety of sophisticated mating behaviors, clever methods of sperm transfer, and various devices for protecting developing embryos, thereby ensuring successful procreation in terrestrial habitats. A comparison of the functional anatomy of the reproductive systems of spiders and scorpions provides a background for discussing arachnid courtship behavior and developmental patterns. But many other interesting behaviors are known from groups like amblypigids, uropygids, ricinuleids, and harvestmen, among others.

The male reproductive system of spiders consists of a pair of straight or convoluted tubular testes in the opisthosoma, which lead to a common sperm duct that opens into the epigastric furrow through a gonopore (Figure 24.23A). Each developing sperm usually bears a distinct flagellum with an unusual 9+3 axonemal pattern of microtubules (Figure 24.23B) (members of the sister families Pimoidae and Linyphiidae share a unique 9+0 pattern). The spermatozoa of spiders exhibit remarkably high structural diversity. Prior to copulation, the flagellum wraps around the head of the sperm, and a protein capsule forms around the gamete (Figure 24.23C). The sperm remains in this nonmotile state until after mating.

Although male spiders lack a penis (like almost all other arachnids), the pedipalps of the adults are modified for sperm storage and transfer and serve as copulatory organs. Sperm released from the male gonopore are placed on a specially constructed silken sperm web (Figure 24.23D). From here the sperm are picked up by the pedipalps, where they are held in special pouches or chambers and eventually transferred to the female. The pedipalps of adult male spiders vary greatly in form and complexity, generally being simple in certain mygalomorphs and more complex in most araneomorphs. In the simplest form, each pedipalp bears on its modified tarsus (called the **cymbium**) a teardrop-shaped process known as the palpal organ (Figure 24.23E). A pointed tip, or **embolus**, bears a pore that leads inward to a blind coiled sperm duct (the spermophor). Sperm are drawn into this tube from the sperm web and held until transferred to the female.

The more complex male copulatory organs are formed of structures with varying degrees of sclerotization (sclerites) and a diversity of soft parts, such as sacs that can expand through increased hemolymph pressure (hematodochae) and various type of membranes. Often some of these cymbial sclerites, as well as the cymbium itself, and other pedipalpal articles bear processes called apophyses of wildly diverse morphology (Figure 24.23F), such as the "retrolateral tibial apophysis" which characterizes a large lineage of araneomorph families (the "RTA Clade") that includes among others the salticids and the wolf spiders. Although often difficult, assessing the homology of the highly variable structures across species is an important part of morphological phylogenetic inference, particularly in araneomorphs. Because genitalia vary across species, and because in closely related species somatic morphological features often vary little, genitalia, both male and female, are extensively used in spider taxonomy. Although male genitalia need to fit with the mating organs of conspecific females, the key-and-lock hypothesis to explain the evolution and huge interspecific diversity of genitalic structures of spiders has been empirically falsified. This variation is probably part of a series of mechanisms to prevent interspecific mating, although nonproductive interspecific encounters are prevented primarily by species-specific courtship behaviors (discussed below). Following the uptake of sperm through the embolus and insertion of the palpal organ into the female's receiving structure (the epigynum), the soft hematodochae are inflated with hemocoelic fluid, thereby causing an erection of the sclerites within the female parts. Once the partners are thus coupled, sperm are injected into the female's copulatory openings and ducts. In some species, sperm transfer is preceded by copulation before the palps are charged with sperm ("pseudo-copulations").

Female spiders have a pair of ovaries in the opisthosoma. The lumen of each ovary leads to an oviduct, and the two oviducts unite to form a uterus (also called the vagina), which opens to the outside in the epigastric furrow (Figure 24.23G). Eggs are produced mainly

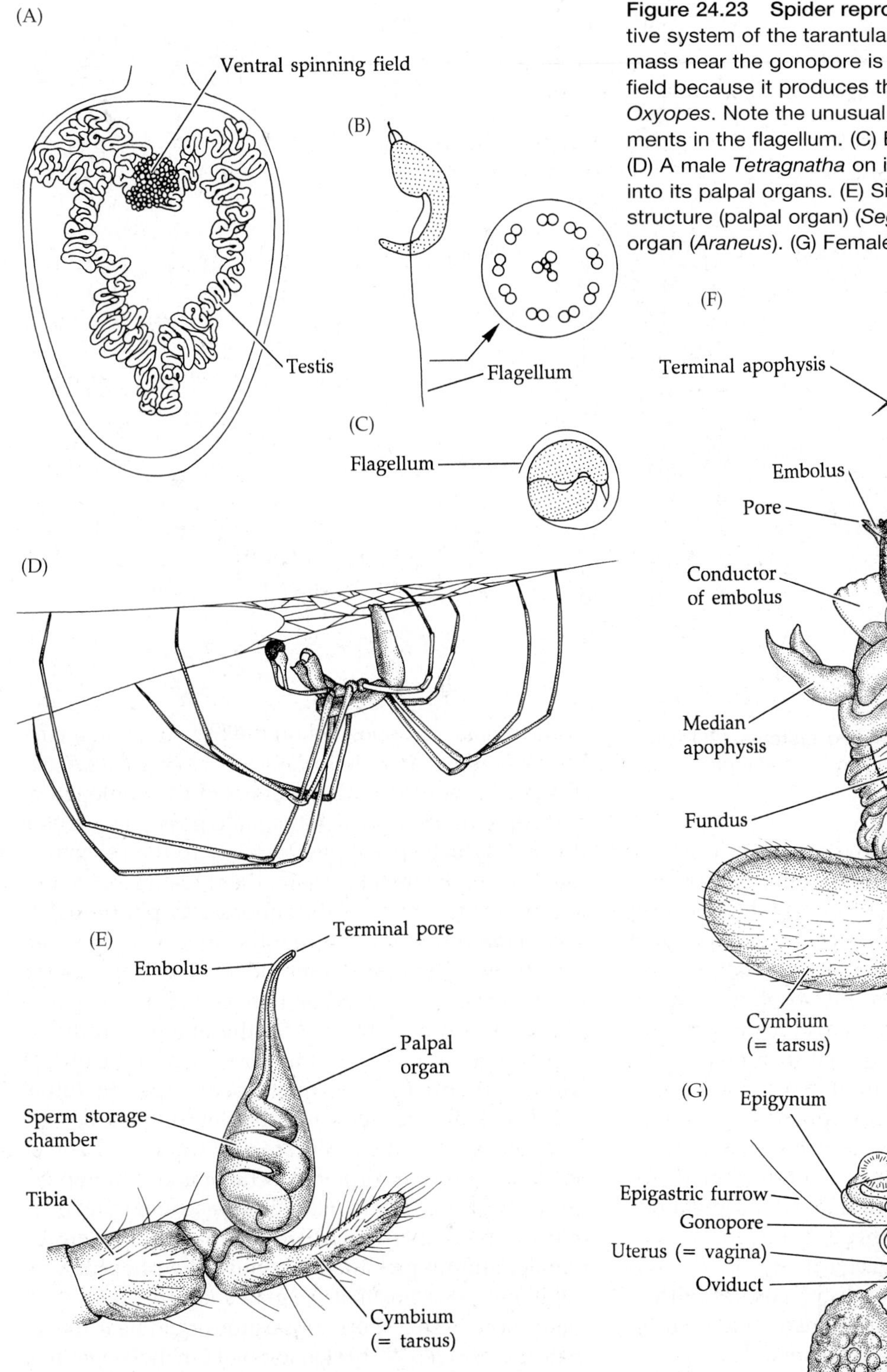

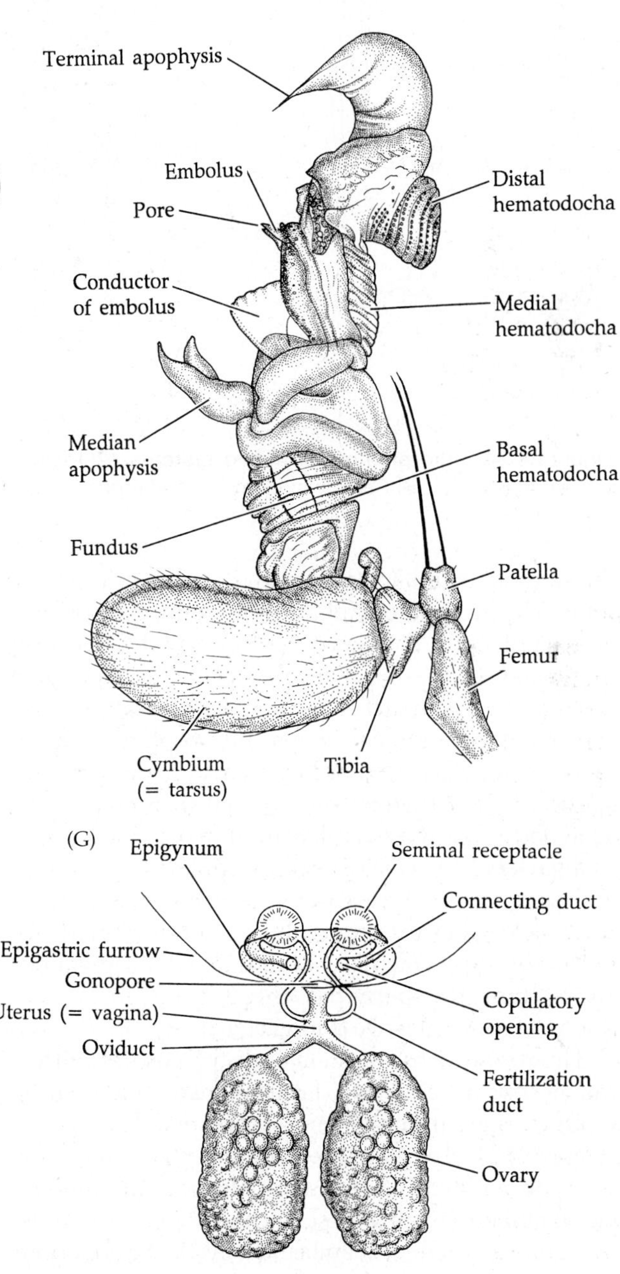

Figure 24.23 Spider reproduction. (A) Male reproductive system of the tarantula *Grammostola*. The glandular mass near the gonopore is called the ventral spinning field because it produces the sperm web. (B) Sperm of *Oxyopes*. Note the unusual 9+3 arrangement of axial filaments in the flagellum. (C) Encapsulated form of sperm. (D) A male *Tetragnatha* on its sperm web, drawing sperm into its palpal organs. (E) Simple male pedipalp copulatory structure (palpal organ) (*Segestria*). (F) Complex palpal organ (*Araneus*). (G) Female spider reproductive system.

on the exterior of the ovaries, giving them a bubbled texture; how they move to the internal ovarian lumen is not well understood.

Just inside, or lateral to, the female gonopore, there is usually a pair of copulatory openings that lead through coiled connecting ducts to paired seminal receptacles (or spermathecae). In entelegyne spiders a second pair of tubes, called the fertilization ducts, connects the seminal receptacles to the uterus (Figure 24.23G). Many spiders have a complexly structured sclerotized plate just in front of the epigastric furrow. This plate, called the epigynum, extends over the genital pore and bears the copulatory openings to

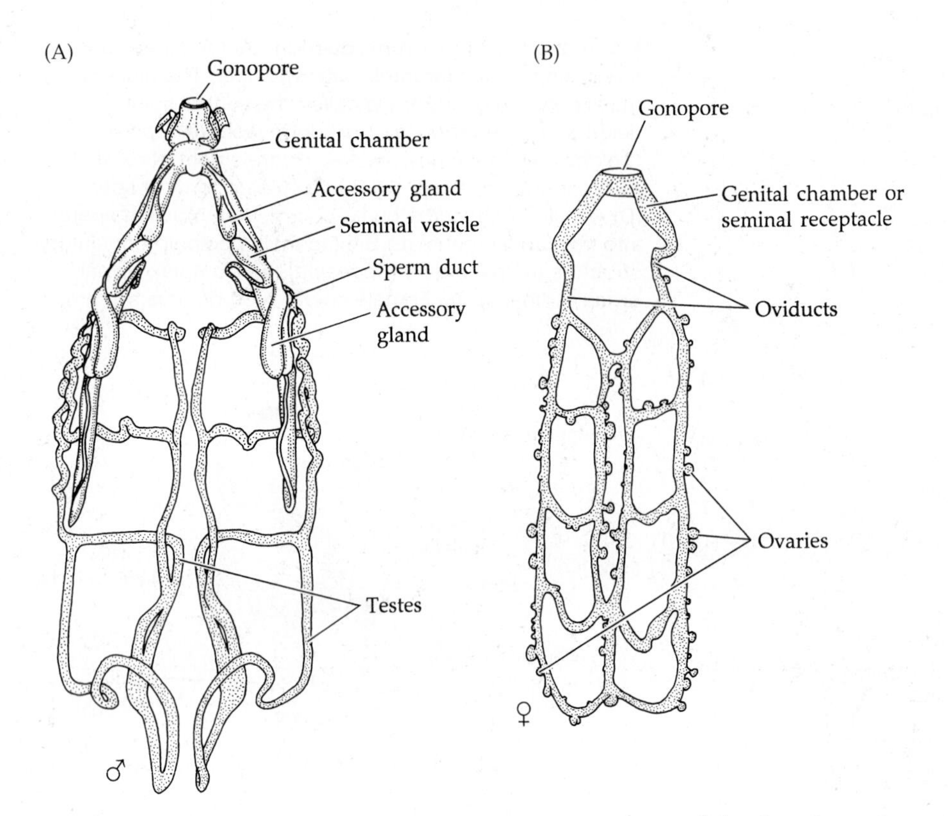

Figure 24.24 Scorpion reproductive systems. (A) Male system in *Buthus*. (B) Female system in *Parabuthus*.

the seminal receptacles through copulatory ducts or folds. The morphology of the epigynum, the position and length of the copulatory openings and connecting ducts, and other external features provide a particular topography that matches the palpal organs of conspecific males. As with male palps, the morphology of the female genitalia is used by taxonomists to discriminate species. These differences in external anatomy, as well as in body size and overall courtship behavior, result in a variety of species-specific copulatory positions among spiders. Once sperm are inside the seminal receptacles, they are stored there until the female deposits her eggs, which may be months after copulation. At that time, the sperm pass through the fertilization ducts to fertilize the ova during egg laying.

The reproductive system of scorpions lies within the mesosoma, and both, testes in males and ovaries in females are in the form of interconnected tubules (Figure 24.24). The gonads are drained by lateral sperm ducts or oviducts. The sperm ducts bear various storage chambers (seminal vesicles) and accessory glands, and unite as a genital chamber just inside the gonopore on the first segment of the mesosoma. Some accessory glands are responsible for the production of spermatophores. Each oviduct is enlarged as a genital chamber, or seminal receptacle, near its union with the gonopore.

With few exceptions, sperm transfer in arachnids is indirect. That is, the sperm leave the body of the male and are then somehow manipulated into the body of the female or deposited on the eggs outside the female's body. The only exceptions to this rule occur in most harvestmen (all members of the Phalangida) and some mites, in which the male possesses a penis through which sperm pass directly to the reproductive tract of the female. In all other arachnids, including mite harvestmen in the suborder Cyphophthalmi and many other mites, either the sperm are inserted into the female's body by modified appendages of the male, or they are placed on the ground in spermatophores and then retrieved by the female. Modified appendages are used to transfer sperm in the orders Araneae (pedipalps), Uropygi (pedipalps), Ricinulei (third walking legs), some members of Solifugae (chelicerae), and some mites (chelicerae or third legs). In other arachnids (orders Scorpiones, Schizomida, Amblypygi, Pseudoscorpiones, and many Solifugae and mites), the males deposit spermatophores on the ground and the females simply pick them up, but in Cyphophthalmi (Opiliones) it seems that the male may deposit a spermatophore with a spermatopositor organ near the female gonostome. We hasten to point out, however, that the reproductive biology of many species in these taxa, and of some entire groups (e.g., Palpigradi), has been little studied.

The events leading up to insemination often include species-specific courtship behaviors that serve as cues for species recognition. These behaviors must, of course, be compatible with the particular method of sperm transfer (Figures 24.25 and 24.26). Again, we can look to spiders and scorpions for examples. Among spiders, courtship behaviors not only ensure conspecific copulation, but also prevent the usually smaller

Figure 24.25 Courtship and mating in spiders. (A) Mating position in tarantulas. (B) Mating position in linyphiids. (C) Mating position in *Xysticus* (family Thomisidae). Mating occurs after the male has placed a series of threads over the female's body. (D) Mating position in *Araneus diadematus* (family Araneidae). (E) Use of a mating thread by some orb-weavers. In this general example, the male has spun a mating thread from an object to the female's web. When properly plucked, the thread transmits vibrations to the web, and the female responds by approaching the male and assuming a mating posture. (F) The courtship behavior of a male lycosid (*Lycosa rabida*) includes movements of the anterior legs, abdomen bobbing, and pedipalp movements. (G) The pedipalp of *Tangaroa tahitiensis* (family Uloboridae) bears organs of stridulation. The spines are scraped against the filelike serrations on the coxa. (H) The zigzag approach and courtship display of a male jumping spider (family Salticidae).

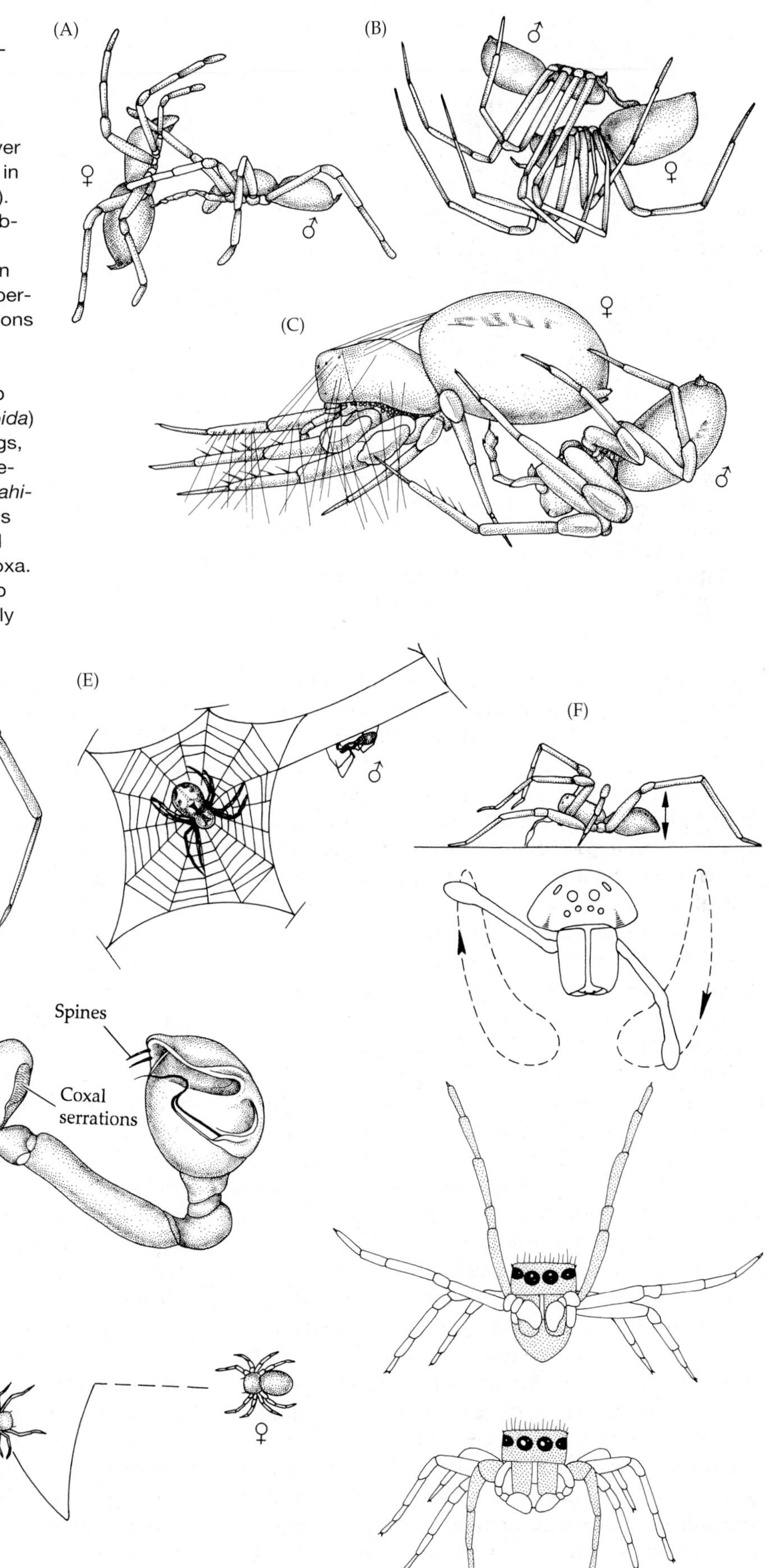

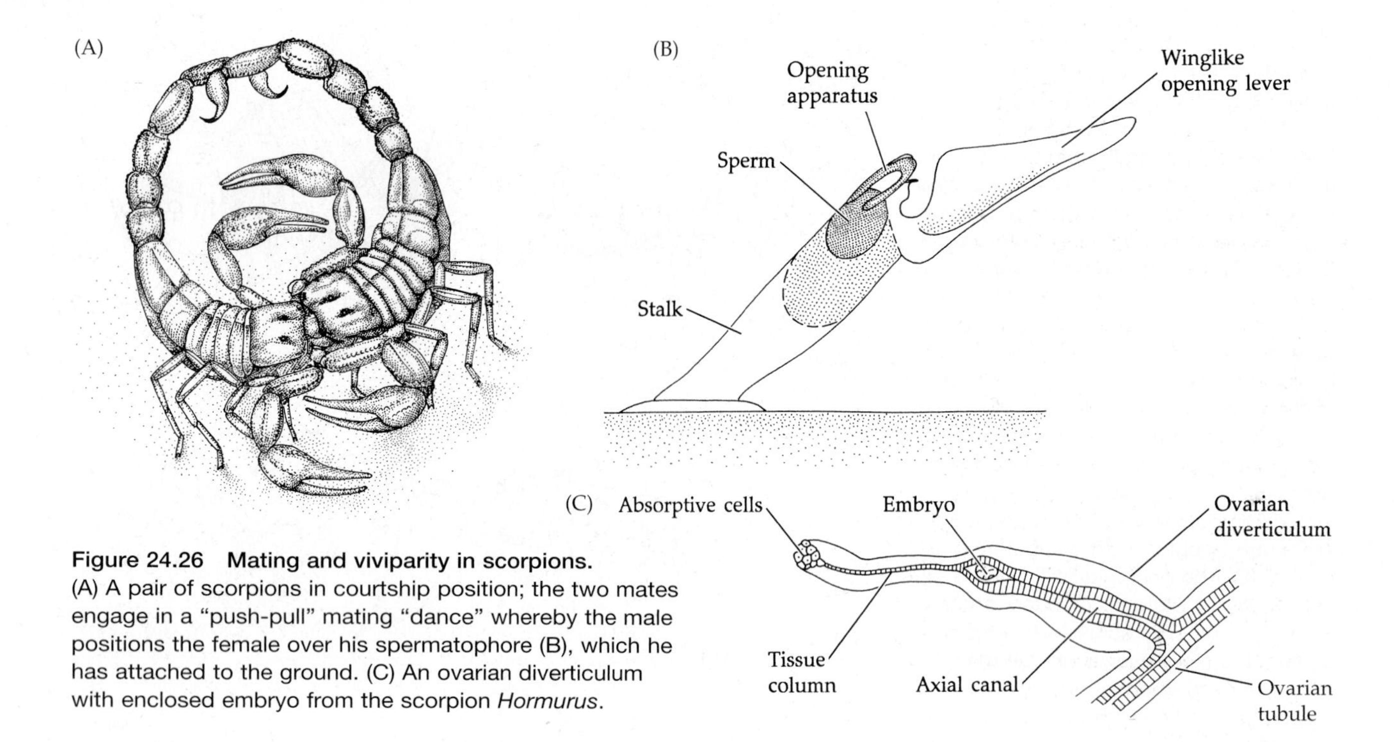

Figure 24.26 Mating and viviparity in scorpions. (A) A pair of scorpions in courtship position; the two mates engage in a "push-pull" mating "dance" whereby the male positions the female over his spermatophore (B), which he has attached to the ground. (C) An ovarian diverticulum with enclosed embryo from the scorpion *Hormurus*.

males from being mistaken for prey by the females as well as having an arousal function. Platnick (1971) classified spider courtship behaviors into three general levels. First-level courtship involves necessary contact between male and female. Among many thomisids and clubionids, mating involves the male simply climbing over the female, positioning her abdomen, and inserting a palpal organ. The males of some thomisids (e.g., *Xysticus*) and at least some species of one genus of orb-weaver (*Nephila*) place silk threads over the bodies or legs of the female preparatory to copulation. These threads are apparently only part of the recognition ritual because they are not strong enough to actually restrain the female (Figure 24.25C). A few other spiders, including certain tarantulas (Theraphosidae), also use body contact and leg touching as courtship behavior.

Second-level courtship behaviors involve the release of sex pheromones by the female spider. Some of the most complex behavior patterns occur in male spiders that detect females by olfaction, although other recognition devices may also be involved. Some male araneids are apparently led to the female's orb web by her pheromones, and the web is then recognized by contact chemoreception. Once in touch with the edge of the web, the male announces his presence to the female by plucking the threads of her orb itself, or by attaching a special mating thread to her orb, which he then plucks (Figure 24.25E). If properly orchestrated, the male's "tune" eventually attracts the female, and contact is made.

Males of some species of wolf spiders (Lycosidae) respond to pheromones emitted along with the female's dragline. When he detects a female visually, the male begins a specific set of actions in an effort to win her favor. These male behaviors involve abdomen bobbing and pedipalp waving, coupled with drumming of the pedipalps on the substratum and stridulation (Figure 24.25F). If attracted by these cues, the female responds by slowly approaching the male and sending out signals of her own in the form of particular leg movements. Stridulation, using modified pedipalps, also occurs in some uloborids and many other families (Figure 24.25G).

Among the most interesting second-level courtship behaviors are those of certain nursery-web spiders (*Pisaura*; Pisauridae). After locating a female emitting pheromones, the male captures an insect (usually a fly), spins a silk wrapping around it, and offers it to the female. Acceptance of the gift and of the male are one and the same, for the successful male copulates with the female while she devours the insect. Alas, unsuccessful males are eaten along with the offering. Postcopulatory cannibalism by females is not uncommon in certain groups of spiders. Contrary to popular belief, the males of the North American black widow (*Latrodectus mactans*) are often unharmed after copulation. In other widow species, such as *L. geometricus* and *L. hasselti*, males are cannibalized by the female.

Another interesting second-level courtship behavior occurs in the Sierra dome spider (*Neriene litigiosa*) of western North America. Upon encountering a mature virgin female, a male attacks her pheromone-laden web and packs it into a small, tight mass. This behavior hinders the evaporation and dispersal of the male-attracting pheromone, thereby reducing the likelihood of additional males locating the female and competing for her favors.

Third-level courtship behaviors depend primarily on visual recognition of prospective mates and are best known in jumping spiders (Salticidae), which often exhibit conspicuous sexual dimorphism. A male locates a female and then begins a series of behaviors that identify him as a conspecific individual. Usually the male approaches the female along a zigzag path and then performs specific movements of the opisthosoma, pedipalps, and front walking legs (Figure 24.25H). The female signals her approval and receptiveness by sitting still in a visually recognizable position. The male eventually contacts her, caresses her briefly, mounts her, and copulates.

Sex-related behaviors among spiders are not restricted to male–female encounters. Conspecific males frequently exhibit agonistic behavior in competition for a mate. When males encounter one another in the presence of a female, or even in a mating "territory" (such as on the web of a female), they may assume various threatening postures and in some cases actually engage in combat. Usually, however, one of the males retreats before any real damage is done, leaving the dominant male free to pursue his sexual interests.

The courtship behavior of scorpions does not involve copulation in any form, but instead involves deposition of spermatophores onto the ground. Courtship behaviors appear to be relatively similar among those species that have been studied, although subtle species-specific differences allowing species recognition undoubtedly occur. In a typical case, the male initiates the ritual by grasping the female's pedipalps in his, and in this face-to-face position dances her around in a series of back-and-forth steps (Figure 24.26A). Eventually the male releases a spermatophore and cements it to the ground. He then continues to move the female around until she is precisely positioned with her genital operculum over the packet of sperm. The spermatophore is a complex, species-specific structure, and it bears a special process called an opening lever (Figure 24.26B). The pressure of the female's body on this lever causes the spermatophore to burst, releasing the sperm, which can then enter her gonopore.

Arachnids undergo a variety of developmental patterns, all of which may be considered direct in terms of their life history strategies. Most species produce very yolky eggs, providing the embryos with nourishment through much of their development. By the time of hatching, many resemble miniature adults, or still contain enough yolk to carry them through subsequent development to the juvenile stage. In some acarines and all ricinuleids, the young hatch as six-legged "hexapod larvae." These immature individuals add the last pair of legs later through molting. In most arachnids the developing embryos are protected by some sort of egg case or cocoon, or they are brooded in or on the body of the female.

Nearly all spiders cement their eggs together in clusters and wrap them in silken eggsacs, the sizes and shapes of which vary among species (Figure 24.27). The eggsac provides physical protection for the embryos and also insulates them from fluctuations in environmental conditions, particularly changes in humidity. Additional protection results from placing the eggsac underground, in a nest, or in other secluded spots. Some species camouflage their cocoons with bits of detritus; others guard their cocoons or actually carry them on their bodies.

The early development of spiders includes intralecithal nuclear divisions followed by migration of the nuclei to the periphery of the embryo. The nuclei are then isolated by cytoplasmic partitioning, a process that produces a periblastula around an inner yolky mass. Gastrulation follows by formation of a germinal center of presumptive endodermal and mesodermal cells that migrate inward. Additional germinal centers produce the precursors of segments and limbs. The immature individuals of different spider species hatch from their egg membranes at different stages, but they always remain inside the cocoon and utilize their yolk reserves until they are able to feed. Most workers recognize three postembryonic, preadult stages in spider development (Figure 24.27I,J,K). The majority of spiders hatch from their egg membranes as immobile "prelarvae," characterized by incomplete segmentation and poorly developed appendages. The "prelarva" matures into a "larva" and then into a "nymph," or juvenile, which physically resembles the adult. In some spiders these early developmental changes take place in a special molting chamber inside the cocoon (Figure 24.27B). The emergence from the cocoon usually occurs at an early "nymphal" stage, when the young are fully formed spiderlings. Many female spiders even engage in postnatal care by carrying their young on their bodies or by feeding them (Figure 24.27D,E,F). (It is important to note that the terms prelarva, larva, and nymph, as used here, do not carry the same meanings as they do when used to describe indirect or mixed development, in which the larva is an independent, free-living individual, as in insects or crustaceans.)

Scorpion development is direct, and may be either ovoviviparous or viviparous. Viviparity is perhaps best studied in the Asian scorpion *Hormurus australasiae*. In this species the zygotes lie in tiny diverticula on the walls of the ovarian tubules (Figure 24.26C). Certain cells of the tubule wall absorb nutrients from the adjacent digestive ceca and supply them to the developing embryos. The eggs of *Hormurus* contain very little yolk and undergo holoblastic, equal cleavage. Ovoviviparous scorpions, on the other hand, produce yolky eggs that cleave meroblastically. The embryos of these species are brooded in the ovarian tubules, but depend on their yolk supplies for nutrients. The young

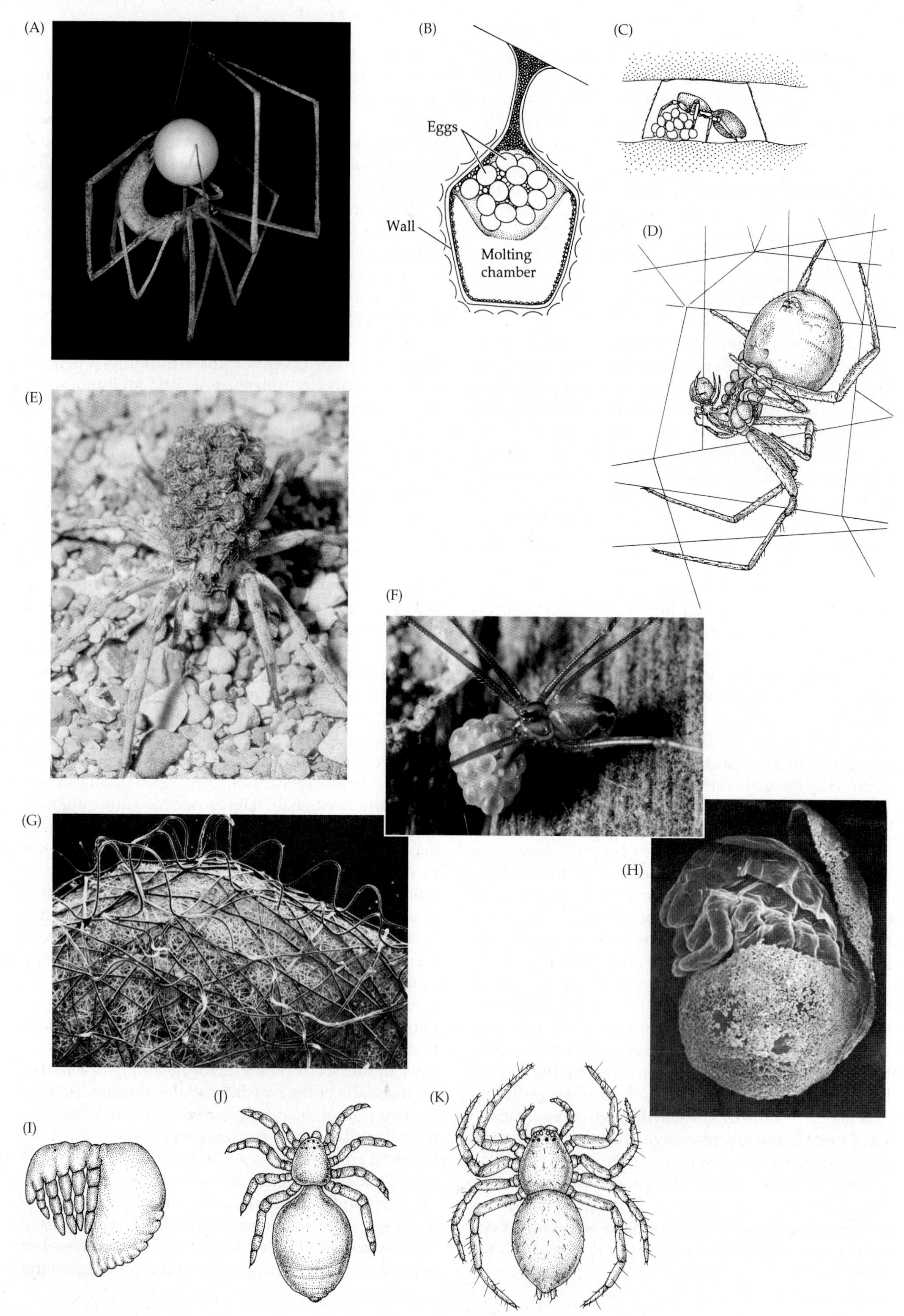
(A)
(B)
Eggs
Wall
Molting chamber
(C)
(D)
(E)
(F)
(G)
(H)
(I)
(J)
(K)

(L)

(M)

Figure 24.27 Egg cases, hatching, and parental care. (A) The ogre-faced spider *Deinopis* building the eggsac in Costa Rica. (B) The eggsac (section) of *Agroeca brunnea* (Liocranidae); the female constructs the eggsac and coats it with bits of soil. Note the eggs above the molting chamber. (C) The subterranean, silk-lined nest of a jumping spider (*Heliophanus cupreus*). (D) Female *Theridion* tending a spiderling and feeding it with regurgitated food. (E) A female wolf spider, *Lycosa*, carrying young on her back. (F) Female of *Chibchea mapuche* (Pholcidae) from Chile carrying her eggcase. (G) Surface of the eggsac of the pirate spider *Ero furcata* (Mimetidae). (H) A spider hatchling emerging from its egg case. (I,J,K) Prelarval, larval, and nymphal forms of a spider. (L) Hexapod larva of *Cryptocellus* from Panama. (M) A female armored harvestman, *Phareicranaus manauara*, tending her eggs.

eventually emerge from the female's gonopore and crawl onto her back. Here they stay until they are old enough to make periodic excursions away from their parent and eventually assume an independent life. Juvenile scorpions molt through several instars until they mature, about a year after birth.

Many other arachnids also brood their embryos, usually externally on the mother's body, as in Amblypygi, Uropygi, Ricinulei, and Schizomida. Members of these groups carry their young in some sort of sac near the female gonopore. Pseudoscorpions spin cocoons from their cheliceral silk glands. Solifuges and harvestmen are oviparous and some deposit their eggs in the soil or in leaves or logs. In all cases the young hatch and pass through a few or many instars before they mature, and the maturation process may take several years in groups like Ricinulei. Again, except for the "hexapod larvae" of mites and ricinuleids (Figure 24.27L), arachnids hatch as small immature adults, and although various names are given to these immature stages, development is strategically direct. Some harvestmen are known for their parental care, which can be in the form of maternal (Figure 24.27M), paternal or biparental. Paternal care is extremely rare in other groups of animals.

The Class Pycnogonida

Pycnogonids or pantopods (Greek *pyc*, "thick," "knobby"; *gonida*, "knees") are usually called "sea spiders" because of their superficial similarity to true, terrestrial spiders (Figures 24.28 and 24.29). Pycnogonids had been problematic in terms of their placement among the other arthropod taxa for decades. Since the turn of the twentieth century they have been associated at one time or another with virtually every major group of arthropods, as well as with the onychophorans and polychaetes. The principal problem has been uncertainty concerning homologies of the various body regions and appendages. The unique pycnogonid "proboscis," for example, has been homologized with everything from the prostomium of polychaete worms to the lips of onychophorans to various anterior regions of other arthropods. However, recent anatomical and molecular studies place the pycnogonids solidly within arthropods, most probably as the sister group to all other chelicerates. Most specialists have concluded that the pycnogonids probably arose as an early offshoot of the line leading to the modern euchelicerates, although some have argued that they constitute the sister group to all other extant arthropods, a phylogenetic hypothesis that has lost support in recent years.

Several characters are apparent shared synapomorphies between the euchelicerates and pycnogonids, including the first appendages (chelicerae/chelifores) and second appendages (pedipalps/palp)—based on the assumption that these appendage pairs, and their somites, are indeed homologous, and nervous innervation tends to support this contention. Two other apparent synapomorphies are the stenopodous uniramous legs (with certain functional similarities) and a feeding method that is largely liquid/suctorial. Pycnogonids also possess several strikingly unique features, synapomorphies not found in any euchelicerate, or in any other arthropod group, such as the odd anterior "proboscis," **ovigers** (specialized appendages between the pedipalps and first walking legs, used for a variety of purposes but most notably for brooding in males), multiple gonopores (on the second coxal segment of some or all of the walking legs), and the unique body form discussed below with a reduced abdomen.

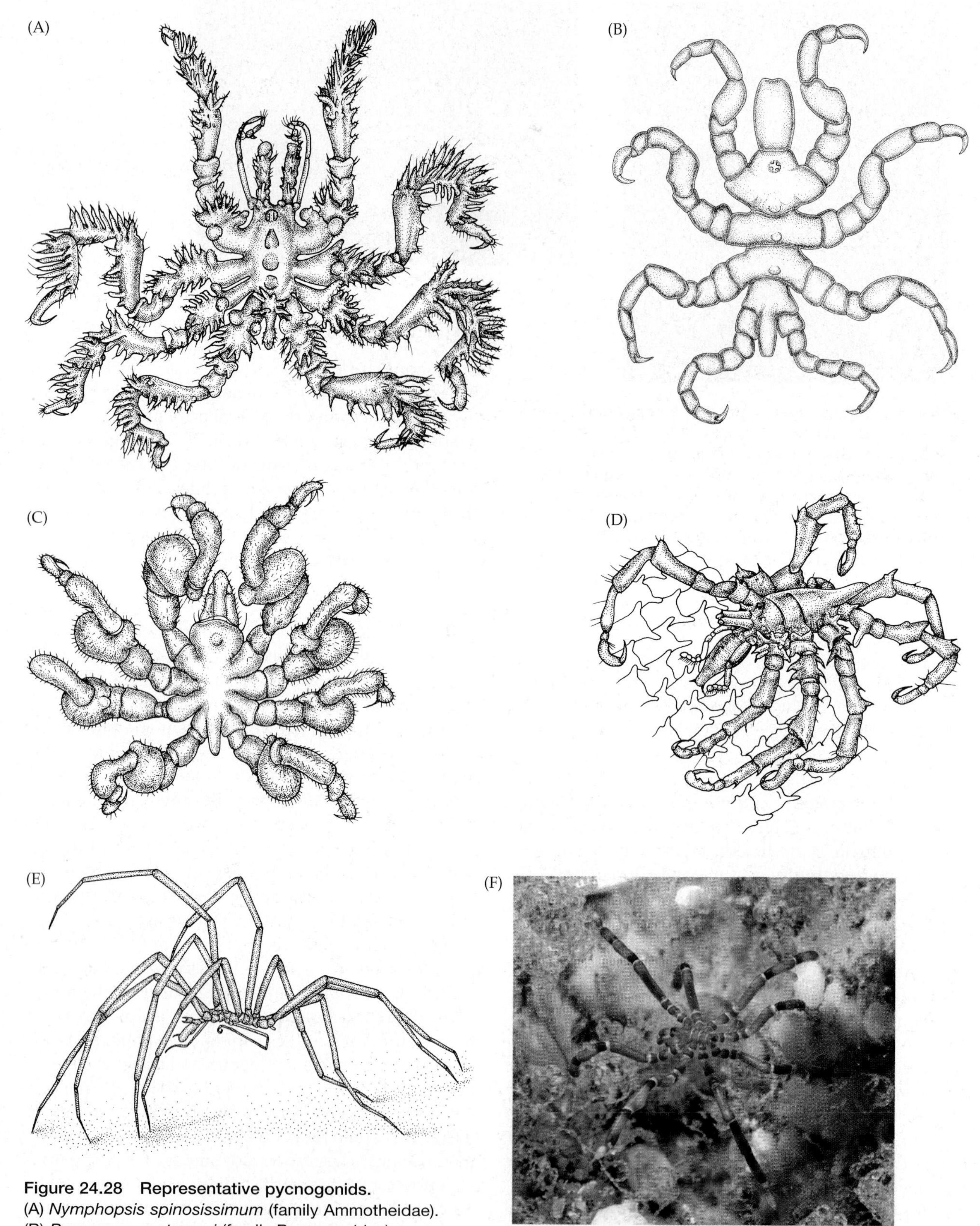

Figure 24.28 Representative pycnogonids. (A) *Nymphopsis spinosissimum* (family Ammotheidae). (B) *Pycnogonum stearnsi* (family Pycnogonidae). (C) *Tanystylum grossifemorum* (family Ammotheidae). (D) *Achelia echinata* (family Ammotheidae) feeding on an ectoproct colony, one zooid at a time. (E) The ten-legged *Decolopoda australis* (family Colossendeidae) (side view of animal walking). (F) *Anoplodactylus evansi* from Australia.

There are about 1,330 described extant species of pycnogonids, in nearly a hundred genera, with many more yet to be discovered. Pycnogonids are strictly marine; they occur intertidally and to depths of nearly 7,000 m and are distributed worldwide. Most are small,

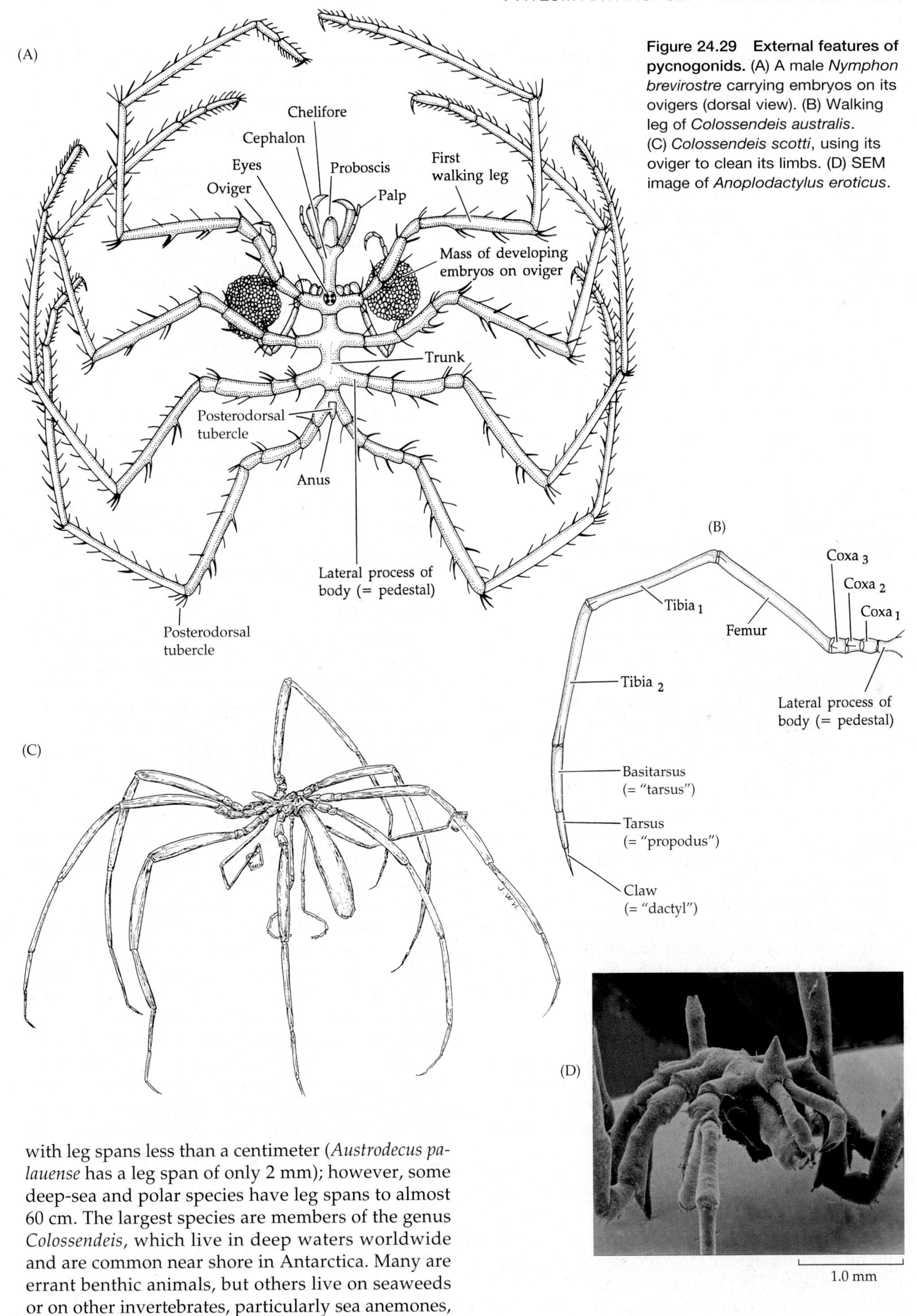

Figure 24.29 External features of pycnogonids. (A) A male *Nymphon brevirostre* carrying embryos on its ovigers (dorsal view). (B) Walking leg of *Colossendeis australis*. (C) *Colossendeis scotti*, using its oviger to clean its limbs. (D) SEM image of *Anoplodactylus eroticus*.

with leg spans less than a centimeter (*Austrodecus palauense* has a leg span of only 2 mm); however, some deep-sea and polar species have leg spans to almost 60 cm. The largest species are members of the genus *Colossendeis*, which live in deep waters worldwide and are common near shore in Antarctica. Many are errant benthic animals, but others live on seaweeds or on other invertebrates, particularly sea anemones, hydroids, ectoprocts, and tunicates. One or two species live on the bells of pelagic medusae, and six species have been collected in hydrothermal vents often associated or on the huge vestimentiferan tube worms. A review of the biology of sea spiders can be found in Arnaud and Bamber (1987).

Pycnogonid systematics is somewhat unsettled, and only recently a comprehensive hypothesis including morphology and molecules has been proposed (Arango and Wheeler 2007). Historically, the few known fossil species (which date back to the Upper Cambrian) were "dumped" into the single order Palaeopantopoda, and the living forms into the order Pantopoda. The most commonly used classification is that of Hedgpeth (1982) with subsequent modifications, which recognized just 8 families, based largely on appendage structure (particularly the reduction or loss of various appendages), but many of these are not monophyletic. Several updates of this system have been proposed in the past two decades.

The Pycnogonid Body Plan

External Anatomy

The bodies of pycnogonids are not as clearly divided into recognizable tagmata as are those of other arthropods (Figures 24.28 and 24.29), especially since they have a miniaturized abdomen. The first body "region" bears an anteriorly directed proboscis with a terminal mouth with three jaws, which varies in size and shape among species. The proboscis contains a chamber and bears an opening at its distal end (Figure 24.29A); although there is some uncertainty on this matter, most researchers now consider the mouth to be at the end (tip) of the proboscis. This anteriormost body "region" also bears paired appendages in the form of **chelifores**, palps, first walking legs, and, when present, ovigers (Figure 24.29A). The chelifores may be chelate or achelate, or absent altogether; they receive a different name from similar appendages in other chelicerates, the **cheliceres**, but it is now clear that these appendages are homologous and innervated by the deutocerebrum. The ovigers are modified legs that serve a variety of functions, including grooming (Figure 24.29C), food handling in some species, courtship, mating, and egg transfer (female to male) in many species, and brooding of the embryos by males in most species. Also located on the first segment of most species is a tubercle with four simple median eyes.

The following body segments form the "trunk" and may be variably fused, but each bears a pair of lateral processes, called **pedestals**, on which are borne the walking legs. Because of the orientation of the pedestals, the walking legs are arranged somewhat radially around the body. The posteriormost body segment carries a dorsally inserted posterodorsal tubercle, which may be a vestigial abdomen, and which bears the anus. Perhaps the most distinctive synapomorphy of the pycnogonids, compared with other arthropods, is the presence of multiple gonopores, found on some or all of the walking legs.

One of the most unusual aspects of pycnogonid morphology is the existence of polymerous species, which have more than four pairs of walking legs. This phenomenon is unique to pycnogonids and occurs in the genera *Pentanymphon*, *Pentapycnon*, and *Decolopoda* (with five pairs of legs; decapodous), and *Sexanymphon* and *Dodecolopoda* (with six pairs of legs; dodecapodous). *Callipallene brevirostris* typically bears four pairs of legs, but one specimen has been found with only three pairs. This polymery is poorly understood phylogenetically and developmentally.

Variation among different pycnogonids also occurs in appendage shape and length, spination, proboscis structure, reduction or loss of chelifores and palps, and many other external features. Several examples are shown in Figures 24.28 and 24.29 to illustrate this diversity. In all, however, the body is remarkably reduced and narrow, a feature compensated for by extensions of the gut ceca and gonads into the legs.

Locomotion

The walking legs of pycnogonids are typically nine-segmented. The junction of the first coxal segment and the pedestal is a more or less immovable joint and does not contribute to the action of the leg. The joint between the first and second coxae is hinged to provide promotion and remotion, and the rest of the joints provide the usual flexion and extension. However, the coxal joints also allow a certain amount of "twisting" and thus accentuate the anterior–posterior swing of the appendage tips during the power and recovery strokes. Some joints lack extensor muscles, and limb extension is effected by hydrostatic pressure, as in many arachnids. Note that pycnogonid specialists have given names to the leg articles that do not parallel those used for any other arthropod group (Figure 24.29B).

Most of the commonly encountered intertidal pycnogonids are quite sedentary and move very slowly. These small forms have short, thick legs that are somewhat prehensile and serve more for clinging to other invertebrates or algae than for rapid locomotion. Deep-water benthic pycnogonids tend to be more active, and these errant pycnogonids have longer and thinner legs than the sedentary forms, and they tend to walk on the tips of the legs (Figures 24.28E and 24.29C). However, some of the very large deep-sea forms (e.g., *Colossendeis*) may depend more on slow deep-ocean currents to roll them around on the bottom than on their own locomotory powers.

Many pycnogonids are also known to swim periodically by employing leg motions similar to those used in walking. Some species are known to "hang" from the water's surface, utilizing a combination of small body mass and surface tension. Several studies also report a characteristic "sinking behavior" in a number of species. When dropping to the bottom, these pycnogonids elevate all of the appendages over the dorsal surface of the body in a basket configuration. This behavior eliminates much of the frictional resistance to sinking and allows the animal to drop quickly through the water column (presumably to avoid predation), a behavior also commonly found in some ophiuroids.

Feeding and Digestion

In most species of pycnogonids, feeding habits are dictated by the form of the proboscis and food is limited to material that can be sucked into the gut. Even with this basic structural constraint, pycnogonids feed on a variety of organisms.

A few pycnogonids feed on algae, but most are carnivorous, many being generalized predators on hydroids, polychaetes, nudibranchs, and other small invertebrates. Some, perhaps many, also scavenge. Species that consume other animals usually use three cuticular teeth at the tip of the proboscis to pierce the body of their prey, and then they suck out body fluids and tissue fragments. Some pycnogonids that live on hydroids use the chelifores to pick off pieces of the host and pass them to the proboscis opening. In most species, however, the chelifores cannot reach the tip of the proboscis, and their function in feeding is questionable. Some species (e.g., *Achelia echinata*) feed on ectoprocts by inserting the proboscis into the chamber housing an individual and sucking out the zooid (Figure 24.28D). Others (e.g., *Pycnogonum littorale*, *P. rickettsi*) feed on sea anemones in a similar fashion, but rarely kill them due to the large difference in size between predator and prey.

Very little is known about the feeding habits of the deep-sea benthic forms. Undersea photographs and direct observations of aquarium specimens of the giant *Colossendeis colossea* indicate that it may walk slowly along the bottom, sweeping its palps across the substratum to sense prey that might be sucked from the mud.

The digestive tract extends from the mouth at the tip of the proboscis to the anus, which opens on the posterodorsal tubercle on the peg-like abdomen (Figure 24.30A). A chamber within the proboscis bears dense bristles that strain and mechanically mix ingested food, which continues to the pharynx. The pharyngeal and the esophageal regions have a Y-shaped foregut lumen in transverse section. The muscles of the foregut supply the suction for ingestion. A short esophagus connects with the elongate midgut or intestine, from which digestive ceca extend into the bases of each leg, providing a high surface area for digestion and absorption. A short proctodeal rectum leads to the anus.

Digestion is predominantly, if not exclusively, intracellular. The cells of the midgut and cecal walls differentiate from the basal layer and contain a bewildering array of secretion droplets, lysosomes and phagosomes, and include phagocytes that engulf ingested food materials. Some of these cells actually break free from the gut lining and phagocytize food particles while drifting in the gut lumen. Apparently these loose cells reattach to the gut wall after they have "fed." It has been suggested that upon reattachment these errant cells first pass their digested food contents to the fixed cells of the gut wall, then assume an excretory function by picking up metabolic wastes, detaching again and being eliminated via the anus. In the absence of a hepatopancreas, the midgut serves both digestive and absorptive functions.

Circulation, Gas Exchange, and Excretion

Pycnogonids lack special organs for gas exchange. The digestive ceca and the overall body plan together present a very high surface area-to-volume ratio, and exchange of gases probably occur largely by diffusion across the body and gut wall.

Traditionally, pycnogonids have been claimed to have no excretory organs at all, and to use the high surface area-to-volume ratio or the wandering cells of the midgut for excretion. Recently an excretory structure has been located in at least one ammotheid, *Nymphopsis spinosissima*, in which a simple, but standard, excretory gland has been found in the scape of the chelifore. It consists of an end sac, a straight proximal tubule, a short distal tubule, and a raised nephropore.

The circulatory system includes an elongate heart with incurrent ostia, but no blood vessels. As in other arthropods, the heart is located dorsally, within a pericardial chamber separated from the ventral hemocoel by a perforated membrane. Blood leaves the heart anteriorly and flows through the hemocoelic spaces of the body and appendages. Contraction of the heart causes a lowered pressure within the dorsal pericardial body chamber, and blood is thus drawn through the perforations in the membrane and toward the heart. Upon relaxation, the blood flows through the ostia into the heart lumen.

Nervous System and Sense Organs

The central nervous system of pycnogonids includes cerebral ganglia above the esophagus, circumenteric connectives, a subenteric ganglion, and a ganglionated ventral nerve cord (Figure 24.30A,B). The nerve cord bears a ganglion for each pair of walking legs, with additional ganglia present in the polymerous species.

The cerebral ganglia, as in most extant arthropods, include protocerebrum, deutocerebrum, and the subesophageal tritocerebrum. The protocerebrum innervates the eyes and the deutocerebrum innervates the chelifores, an arrangement similar to that in euchelicerates. Recently some authors postulated that the

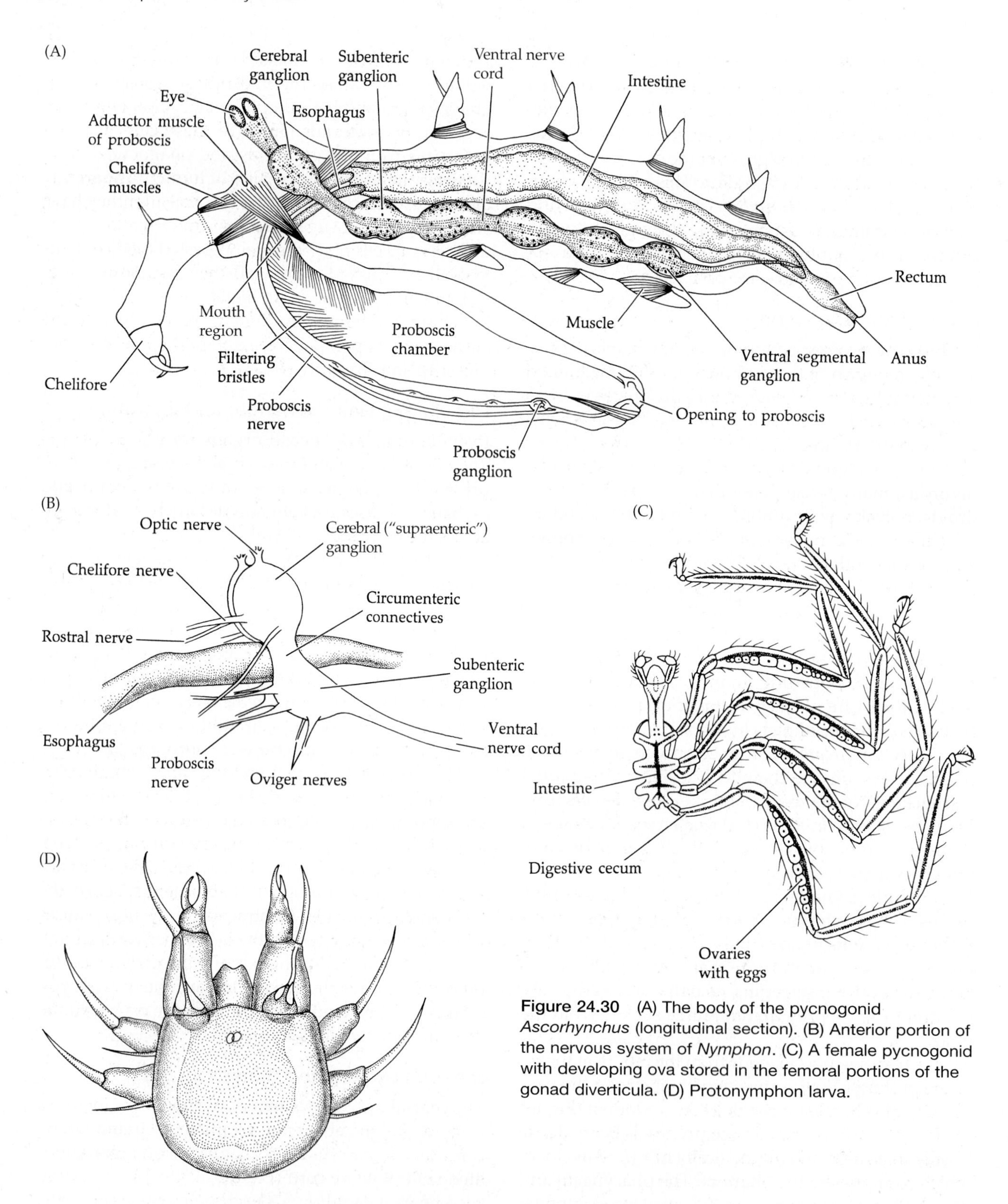

Figure 24.30 (A) The body of the pycnogonid *Ascorhynchus* (longitudinal section). (B) Anterior portion of the nervous system of *Nymphon*. (C) A female pycnogonid with developing ova stored in the femoral portions of the gonad diverticula. (D) Protonymphon larva.

chelifores were innervated by the protocerebrum, as in the "great appendages" of some Cambrian stem-group arthropods, but this was later shown to be due to the fusion of protocerebrum and deutocerebrum in the second embryonic stage. The cerebral ganglia also give rise to a well-developed ganglionated proboscis nerve.

Little work has been done on pycnogonid sense organs. Tactile reception is provided by touch-sensitive setae and probably by the palps. On the body surface, just dorsal to the cerebral ganglia, is a central tubercle with four simple eyes that provide 360° vision. Some deep-sea species lack eyes entirely.

Reproduction and Development

Pycnogonids are gonochoristic. Mating is typically followed by a period of brooding, during which the embryos are held by the male's ventrally articulated ovigers, then by the release of unique **protonymphon larvae** (Figure 24.30D). The protonymphon is six-legged, usually living in a symbiotic relationship with cnidarians, molluscs, or echinoderms. These relationships are poorly understood, but in some cases they appear to be parasitic or commensalistic, encysting in the cnidarian host. This life cycle strategy has been replaced by direct development in some species wherein the larval stage is passed within an egg case.

Sexual dimorphism is common among pycnogonids. The males bear the unique ovigers associated with the first body segment; these appendages are absent in females of some families (e.g., Phoxicilidiidae, Endeidae, and Pycnogonidae), and reduced in the females of other families. Female pycnogonids usually have enlarged limb femora.

Internally, the reproductive systems of males and females are similar and relatively simple. In both, the gonad is single and U-shaped, with extensions into the legs where gametes are produced and stored. The expanded femora in females provide space for storing unfertilized eggs (Figure 24.30C). The multiple gonopores are usually located on the ventral surface of the second coxae of two or all pairs of legs and are thus close to the regions of gamete storage. During mating, the male typically hangs beneath the female or assumes a stance over her back. As the female releases her eggs, the male fertilizes them. Following fertilization, the male gathers the eggs, either one by one or as a single mass, and glues them to his ovigers using a sticky secretion from special femoral glands (Figure 24.29A). Pycnogonids are one of the few groups of animals in which the males exclusively brood the developing embryos and, in some species, the young.

Knowledge of pycnogonid development has increased considerably in the past few years, with several studies focusing on the identity of the pycnogonid appendages with those of other arthropods using neurogenesis and Hox gene expression. Earlier in embryogenesis, there is some variation among members of different genera, but cleavage is usually holoblastic and leads to the formation of a stereoblastula. The inward movement of the presumptive endodermal and mesodermal cells is frequently accompanied by a disappearance of some cell membranes, a process leading to the formation of some syncytial tissues in the gastrula. Germinal centers become apparent as appendage buds form. The most common hatching stage is a free-swimming protonymphon larva. Through a series of molts, the protonymphon adds segments and appendages to produce a juvenile. In some species, the developing larva becomes encysted in a hydroid or stylasterine coral, emerging later with three pairs of legs. A few species are known to produce all of the appendages at once, with subsequent molts simply increasing the size and segment number of the legs (the "atypical protonymphon larva"). In *Pycnogonum littorale*, at the fifth larval molt, the three pairs of larval legs and the larval proboscis are lost; the adult proboscis appears and, at later molts, the adult limbs develop.

Chelicerate Phylogeny

Although over 150 years have passed since the discovery of pycnogonids, arguments still ensue as to their phylogenetic relationships. However, as we mentioned earlier, recent morphological, paleontological, and molecular studies point to a chelicerate ancestry. The chelifores and palps of pycnogonids are probably homologous to the chelicerae and pedipalps of the other chelicerates. We thus recognize Pycnogonida as the sister group to the remaining Chelicerata, or Euchelicerata.

Figure 24.31 is a cladogram depicting a working hypothesis of the relationships among the major chelicerate clades. However, we caution readers that the phylogenetic relationships among the Chelicerata are still debatable. The subclass Arachnida is probably a monophyletic group, although some workers have suggested that it might be diphyletic, having arisen from two separate invasions of land (one leading to the scorpions, the second to all other orders), a view that no longer has much support. Phylogenetic relationships among the 12 or 13 orders of Arachnida remain controversial; see Sharma et al. (2014) for a recent review and analysis using transcriptomic data and Shultz (2007) for a recent analysis based on morphological data.

The earliest xiphosuran, eurypterid, and chasmataspid fossils are Ordovician, but putative chasmataspid impressions are found in the late Cambrian. Like the myriapods (and perhaps the insects), arachnids probably invaded land early in the Silurian, at the same time the land plants were becoming well established, although some molecular calibrations suggest an earlier origin of these groups, probably Ordovician. Whether the earliest Paleozoic scorpions (Silurian) were aquatic remains contentious, but fully articulated fossil scorpions have been found in the marine Devonian Hunsrück Slate environment.

The oldest putative pycnogonids are a series of phosphatized larval instars from the late Cambrian Orsten of Sweden, but some authors exclude the Orsten fossils from Pycnogonida. The second oldest sea spiders come from the Silurian Herefordshire Konservat-Lagerstätte in England and are followed by a suite of taxa from the Early Devonian Hunsrück Slate of Germany.

Figure 24.31 A cladogram depicting a working hypothesis for the relationships among Chelicerata. Black dots indicate well-supported nodes. Gray dots indicate poorly supported nodes. Dashed lines indicate alternative positions for Pseudoscorpiones. Note the unresolved positions for the Palpigradi and two mite lineages (Acariformes and Parasitiformes).

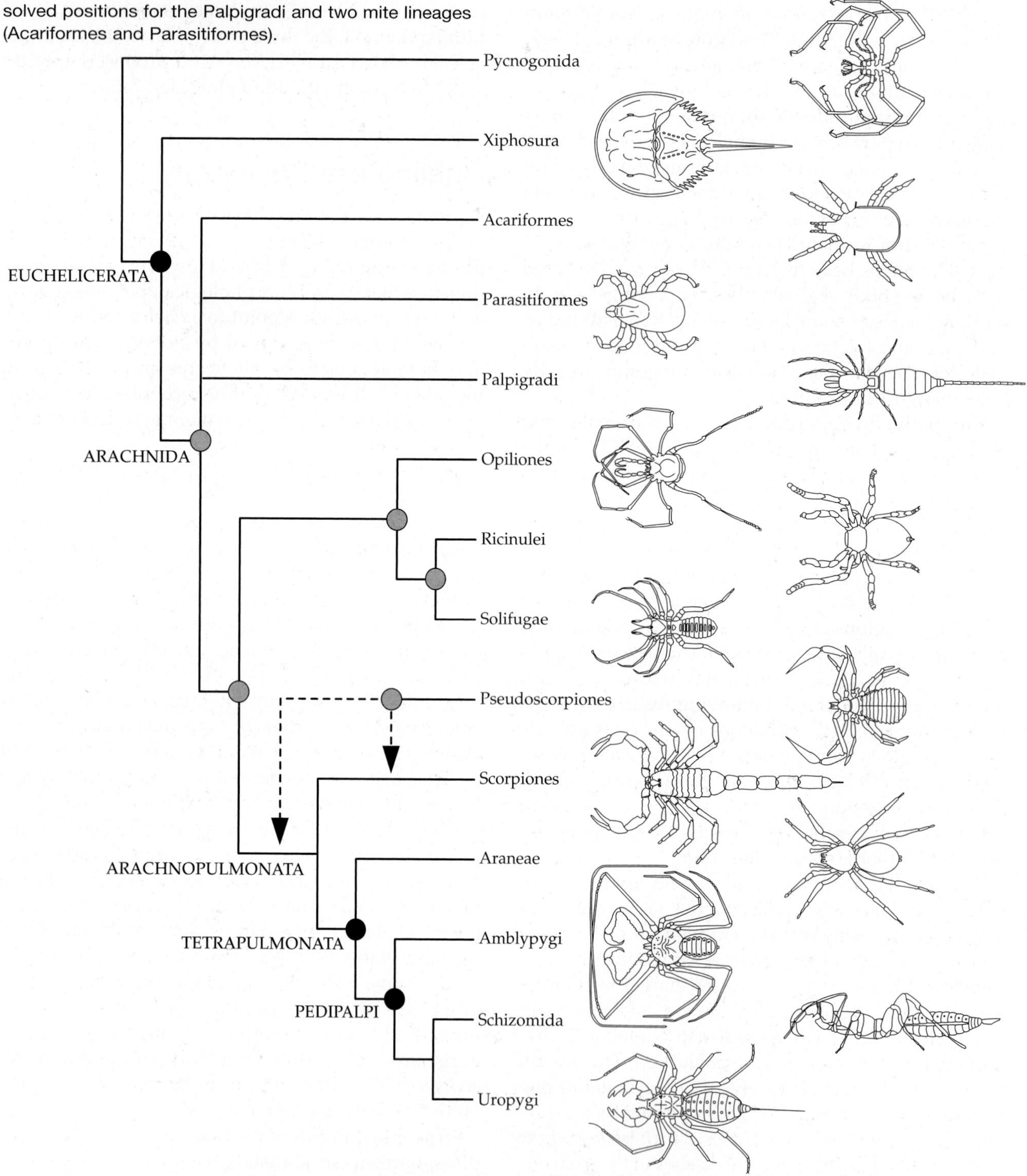

Selected References

General References

Grassé, P. 1949. *Traité de Zoologie. Vol. 6, Onychophores–Tardigrades–Arthropodes–Trilobitomorphes–Chelicerataes.* Masson et Cie, Paris.

Harrison, F. W. and R. F. Foelix. 1999. *Microscopic Anatomy of Invertebrates. Volume 8: Chelicerate Arthropoda.* Wiley-Liss, New York.

Sherman, R. G. 1981. Chelicerates. In N. A. Ratcliffe and A. F. Rowley (eds.), *Invertebrate Blood Cells, Vol. 2.* Academic Press, New York.

Strausfeld, N. J. 2012. *Arthropod Brains: Evolution, Functional Elegance, and Historical Significance.* The Balknap Press of Harvard University Press, Cambridge.

Merostomata

Fahrenbach, W. H. 1979. The brain of the horseshoe crab (*Limulus polyphemus*). III. Cellular and synaptic organization of the corpora pedunculata. Tissue Cell 11: 163–200.

Riska, B. 1981. Morphological variation in the horseshoe crab (*Limulus polyphemus*), a "phylogenetic relic." Evolution 35: 647–658.

Sekiguchi, K. and H. Sugita. 1980. Systematics and hybridization in the four living species of horseshoe crabs. Evolution 34: 712–718.

Shuster, C. N., R. B. Barlow and H. J. Brockmann (eds.) 2003. *The American Horseshoe Crab.* Harvard University Press, Cambridge.

Arachnida

Alberti, G. 2000. Chelicerata. Pp. 311–388 in K. G. Adiyodi, R. G. Adiyodi and B. G. M. Jamieson (eds.), *Reproductive Biology of Invertebrates. Volume IX, Part B. Progress in male gamete ultrastructure and phylogeny.* Oxford & IBH Publishing Co., New Delhi-Calcutta.

Avilés, L. 1997. Causes and consequences of cooperation and permanent-sociality in spiders. Pp. 476–498 in J. C. Choe and B. J. Crespi (eds.), *Social Behavior in Insects and Arachnids.* Cambridge University Press, New York.

Barth, F. G. (ed.). 1985. *Neurobiology of Arachnids.* Springer-Verlag, New York.

Barth, F. G. 2002. *A spider's World: Senses and Behavior.* Springer Verlag, Berlin.

Beccaloni, J. 2009. *Arachnids.* The Natural History Museum, London.

Blackledge, T.A. 2012. Spider silk: a brief review and prospectus on research linking biomechanics and ecology in draglines and orb webs. J. Arachnol. 40: 1–12.

Bradley, R. 2012. *Common Spiders of North America.* University of California Press.

Brownell, P. H. 1984. Prey detection by the sand scorpion. Sci. Am. 251: 86–97.

Brownell, P. H. and R. D. Farley. 1979. Detection of vibrations in sand by the tarsal organs of the nocturnal scorpion *Paruroctonus mesaensis.* J. Comp. Physiol. 131: 23–30.

Brownell, P. and G. Polis (eds.) 2001. *Scorpion Biology and Research.* Oxford University Press, Oxford.

Brunetta, L. and C. L. Craig. 2010. *Spider Silk: Evolution and 400 Million Years of Spinning, Waiting, Snagging, and Mating.* Yale University Press, New Haven.

Coddington, J. A., G. Giribet, M. S. Harvey, L. Prendini and D. E. Walter. 2004. Arachnida. Pp. 296–318 in J. Cracraft and M. J. Donoghue (eds.), *Assembling the Tree of Life.* Oxford University Press, New York.

Cooke, J. A. L. 1967. The biology of Ricinulei. Zoologica 151: 31–42.

Craig, C.L. 2003. *Spiderwebs and Silk: Tracing Evolution from Molecules to Genes to Phenotypes.* Oxford University Press, New York.

Dacke, M., T. A. Doan, D. C. O'Carroll. 2001. Polarized light detection in spiders. J. Exp. Biol. 204: 2481–2490.

Dunlop, J. A. and G. Alberti. 2008. The affinities of mites and ticks: a review. J. Zool. Syst. Evol. Res. 46: 1–18.

Dunlop, J. A. and D. Penney. 2012. *Fossil Arachnids.* Siri Scientific Press, Manchester.

Eberhard, W. G. and Huber, B. A. 2010. Spider Genitalia. Pp. 249–284 in J. Leonard and A. Cordoba (eds.), *The Evolution of Primary Sexual Characters in Animals.* Oxford University Press, Oxford, UK.

Fernández, R., G. Hormiga and G. Giribet. 2014. Phylogenomic analysis of spiders reveals nonmonophyly of orb-weavers. Curr. Biol. 24: 1772–1777.

Foelix, R. F. 1970. Structure and function of tarsal sensilla in the spider *Araneus diadematus.* J. Exp. Zool. 175: 99–124.

Foelix, R. F. 2011. *Biology of Spiders, 3rd Ed.* Oxford University Press, Oxford.

Foelix, R. F. and I.-W. Chu-Wang. 1973. The morphology of spider sensilla. II. Chemoreceptors. Tissue Cell 5(3): 461–478.

Foelix, R. F. and J. Schabronath. 1983. The fine structure of scorpion sensory organs. I. Tarsal sensilla. II. Pecten sensilla. Bull. Br. Arachnol. Soc. 6(2): 53–74.

Foelix, R. F. and D. Troyer. 1980. Giant neurons and associated synapses in the peripheral nervous system of whip spiders. J. Neurocytol. 9: 517–535.

Foil, L. D., L. B. Coons, and B. R. Norment. 1979. Ultrastructure of the venom gland of the brown recluse spider, *Loxosceles reclusa.* Int. J. Insect Morphol. Embryol. 8: 325–334.

Forster, L. 1982. Vision and prey-catching strategies in jumping spiders. Am. Sci. 70: 165–175.

Forster, R. and L. Forster. 1999. *Spiders of New Zealand and their worldwide kin.* University of Otago Press, New Zealand.

Garwood, R. J., J. A. Dunlop, G. Giribet and M. D. Sutton. 2011. Anatomically modern Carboniferous harvestmen demonstrate early cladogenesis and stasis in Opiliones. Nature Comm. 2: 444.

Garwood, R. J., P. P. Sharma, J. A. Dunlop and G. Giribet. 2014. A new stem-group Palaeozoic harvestman revealed through integration of phylogenetics and development. Curr. Biol. 24: 1–7.

Giribet, G. and 11 others. 2014. The first phylogenetic analysis of Palpigradi (Arachnida)—the most enigmatic arthropod order. Invert. Syst. 28: 350–360.

Giribet, G. and P. P. Sharma. 2015. Evolutionary biology of harvestmen (Arachnida, Opiliones). Ann. Rev. Entomol. 60: 157–175.

Gonzaga, M. O., A. J. Santos, and H. F. Japyassú. (2007). *Ecologia e comportamento de aranhas.* Interciência, Rio de Janeiro.

Gopalakrishnakone, P. Corzo, G. A. Diego-Garcia, E. de Lima, M. E. (eds.). 2015. *Spider Venoms.* Dordrecht. Springer Netherlands.

Griffiths, D. A. and C. E. Bowman (eds.). 1983. *Acarology VI, Vols. 1–2.* Wiley, New York.

Griswold, C. E., M. J. Ramírez, J. A. Coddington and N. I. Platnick. 2005. Atlas of phylogenetic data for entelegyne spiders (Araneae: Araneomorphae: Entelegynae) with comments on their phylogeny. Proc. Calif. Acad. Sci. 56: 1–324.

Haupt, J. 2003. The Mesothelae—a monograph of an exceptional group of spiders (Araneae: Mesothelae). Zoologica 154: 1–102.

Harvey, M. S, 1992. The phylogeny and classification of the Pseudoscorpionida (Chelicerata: Arachnida). Invert. Taxon. 6: 1373–1435.

Hayashi, C. Y. and R. V. Lewis. 2000. Molecular architecture and evolution of a modular spider silk protein gene. Science 287: 1477–1479.

Hayashi, C. Y, N. H. Shipley and R. V. Lewis. 1999. Hypotheses that correlate the sequence, structure, and mechanical properties of spider silk proteins. Int. J. Biol. Macromol. 24: 271–275.

Herberstein, M. E. (ed.) 2011. *Spider behaviour: flexibility and versatility*. Cambridge University Press, Cambridge.

Homann, H. 1971. Die Augen der Araneae. Anatomie, Ontogenie und Bedeutung für die Systematik (Chelicerata, Arachnida). Z. Morphol. Tiere 69: 201–268.

Hormiga, G. and C. E. Griswold. 2014. Systematics, phylogeny and evolution of orb-weaving spiders. Ann. Rev. Entomol. 59: 487–512.

Huber, B. A. 2004. Evolutionary transformation from muscular to hydraulic movements in spider (Arachnida, Araneae) genitalia: a study based on histological serial sections. J. Morphology, 261(3): 364–376.

Jackson, R. R. and R. S. Wilcox. 1998. Spider-eating spiders. Am. Sci. 86: 350–357.

Jackson, R. R., and F. R. Cross. 2011. Spider cognition. Adv. Insect Physiol. 41: 115.

Jackson, R. R. and F. R. Cross. 2015. Mosquito-terminator spiders and the meaning of predatory specialization. J. Arachnol. 43: 123–142.

Jocqué, R., and A. S. Dippenaar-Schoeman. 2006. *Spider families of the world*. Musée royal de l'Afrique centrale.

Kaston, B. J. 1970. The comparative biology of American black widow spiders. Trans. San Diego Soc. Nat. Hist. 16: 33–82.

Keegan, H. L. 1980. *Scorpions of Medical Importance*. University Press of Mississippi, Jackson.

Kovoor, J. 1987 Comparative structure and histochemistry of silk producing organs in arachnids. Pp. 160–186 in W. Nentwig (ed.), *Ecophysiology of Spiders*. Springer Verlag, Berlin.

Krantz, G. W. 1978. *A Manual of Acarology, 2nd Ed*. Oregon State University, Corvallis.

Lamy, E. 1902. Recherches anatomiques sur les trachées des Araignées. Ann. Sci. Nat. Zool. 15: 149–281.

Land, M. F. 1985. The morphology and optics of spider eyes. Pp. 53–78 in F. G. Barth (ed.), *Neurobiology of Arachnids*. Springer Verlag, Berlin.

Levi, H. W. 1967. Adaptations of respiratory systems of spiders. Evolution 21: 571–583.

Lubin, Y. and T. Bilde. 2007. The evolution of sociality in spiders. Adv. Stud. Behav. 37: 83–145.

Ludwig, M. and G. Alberti. 1990. Pecularities of arachnid midgut glands. Acta Zool. Fenn. 190: 255–259.

Michalik, P., and M. J. Ramirez, M. J. 2014. Evolutionary morphology of the male reproductive system, spermatozoa and seminal fluid of spiders (Araneae, Arachnida) – current knowledge and future directions. Arthropod Struct. Dev. 43: 291–322.

Murienne, J., M. S. Harvey and G. Giribet. 2008. First molecular phylogeny of the major clades of Pseudoscorpiones (Arthropoda: Chelicerata). Mol. Phylogenet. Evol. 49: 170–184.

Nentwig, W. 2013. *Spider Ecophysiology*. Berlin/Heidelberg: Springer.

Opell, B. D. (1998). The respiratory complementarity of spider book lung and tracheal systems. J. Morphol. 236(1): 57–64.

Palmgren, P. 1978. On the muscular anatomy of spiders. Acta Zool. Fenn. 155: 1–41.

Parry, D. A. and R. H. J. Brown. 1959. The jumping mechanism of salticid spiders. J. Exp. Biol. 36: 654.

Paul, R. J., S. Bihlmayer, M. Colmorgen and S. Zahler. 1994. The open circulatory system of spiders (*Eurypelma californicum*, *Pholcus phalangioides*): a survey of functional morphology and physiology. Physiol. Zool. 67: 1360–1382.

Paulus, H. F. 1979. Eye structure and the monophyly of Arthropoda. Pp. 299–377 in A. P. Gupta (ed.), *Arthropod Phylogeny*. Van Nostrand Reinhold, New York.

Penney, D. and P. A. Selden. 2011. *Fossil Spiders: the evolutionary history of a mega-diverse order*. Siri Scientific Press, Manchester,

Petrunkevitch, A. 1955. Arachnida. Pp. 42–162 in R. C. Moore (ed.), *Treatise on Invertebrate Paleontology, Vol. 2*. Geological Society of America, New York.

Pinto-da-Rocha, R., G. Machado and G. Giribet (eds.) 2007. *Harvestmen: The Biology of Opiliones*. Harvard University Press, Cambridge, MA.

Pittard, K. and R. W. Mitchell. 1972. Comparative morphology of the life stages of *Cryptocellus pelaezi* (Arachnida, Ricinulei). Graduate Studies Texas Tech University 1: 1–77.

Platnick, N. I. 1971. The evolution of courtship behavior in spiders. Bull. Br. Arachnol. Soc. 2: 40–47.

Platnick, N. I. and W. J. Gertsch. 1976. The suborders of spiders: A cladistic analysis. Am. Mus. Novit. 2607: 1–15.

Polis, G. A. (ed.). 1990. *The Biology of Scorpions*. Stanford University Press, Stanford, CA.

Polis, G. A. and R. D. Farley. 1980. Population biology of a desert scorpion (*Paruroctonus mesanensis*): Survivorship, microhabitat, and the evolution of life history strategy. Ecology 61: 620–629.

Punzo, B. 1998. *The biology of camel-spiders (Arachnida, Solifugae)*. Kluwer Academic Publishers, Boston.

Purcell, E. F. 1909. Development and origin of the respiratory organs of Araneae. Quart. J. Microsc. Sci. 54: 1–110.

Richter, S., M. Stein, T. Frase and N. U. Szucsich. 2013. The Arthropod Head. Pp. 223–240 in A. Minelli, G. Boxshall and G. Fusco (eds.), *Arthropod Biology and Evolution: Molecules, Development, Morphology*. Springer, Heidelberg.

Robinson, M. H. and B. Robinson. 1980. *Comparative Studies of the Courtship and Mating Behavior of Tropical Araneid Spiders. Pacific Insects Monograph, No. 36*. Department of Entomology, Bishop Museum, Honolulu, HI.

Rovner, J. S., G. A. Higashi and R. F. Foelix. 1973. Maternal behavior in wolf spiders: The role of abdominal hairs. Science 182: 1153–1155.

Sabu, L. S. 1965. Anatomy of the central nervous system of arachnids. Zool. Jahrb. Abt. Anat. Ontog. Tiere 82: 1–154.

Sauer, J. R. and J. A. Hair (eds.). 1986. *Morphology, Physiology, and Behavioral Biology of Ticks*. Eillis Horwood, Chichester.

Savory, T. H. 1977. *Arachnida*, 2nd Ed. Academic Press, New York.

Schwager, E. E., A. Schönauer, D. J. Leite, P. P. Sharma and A. P. McGregor. 2015. Chelicerata. In A. Wanninger (ed.) *Evolutionary Developmental Biology of Invertebrates 3: Ecdysozoa I: Non-Tetraconata*. Springer-Verlag, Wien.

Selden, P. A. and D. Penney. 2010. Fossil spiders. Biol. Rev. 85: 171–206.

Selden, P. A., W. A. Shear and M. D. Sutton. 2008. Fossil evidence for the origin of spider spinnerets, and a proposed arachnid order. Proc. Natl. Acad. Sci. USA 105: 20781–20785.

Sharma, P. P., R. Fernández, L. A. Esposito, E. González-Santillán and L. Monod. 2015. Phylogenomic resolution of scorpions reveals multilevel discordance with morphological phylogenetic signal. Proc. R. Soc. B Biol. Sci. 282: 20142953.

Sharma, P. P. and G. Giribet. 2014. A revised dated phylogeny of the arachnid order Opiliones. Front. Genetics 5: 255.

Sharma, P. P., S. Kaluziak, A. R. Pérez-Porro, V. L. González, G. Hormiga, W. C. Wheeler and G. Giribet. 2014. Phylogenomic interrogation of Arachnida reveals systemic conflicts in phylogenetic signal. Mol. Biol. Evol. 31: 2963–2984.

Sharma, P. P., E. E. Schwager, C. G. Extavour and W. C. Wheeler. 2014. Hox gene duplications correlate with posterior heteronomy in scorpions. Proc. R. Soc. B Biol. Sci. 281: 20140661.

Sharma, P. P., E. E. Schwager, G. Giribet, E. L. Jockusch and C. G. Extavour. 2013. *Distal-less* and *dachshund* pattern both plesiomorphic and apomorphic structures in chelicerates: RNA interference in the harvestman *Phalangium opilio* (Opiliones). Evol. Dev. 15: 228–242.

Shear, W. A. (ed.). 1981. *Spiders: Webs, Behavior and Evolution.* Stanford University Press, Stanford, CA.

Shultz, J. W. 2007. A phylogenetic analysis of the arachnid orders based on morphological characters. Zool. J. Linn. Soc. 150: 221–265.

Smrž, J., Ľ. Kováč, J. Mikeš and A. Lukešová. 2013. Micro-whip scorpions (Palpigradi) feed on heterotrophic Cyanobacteria in Slovak caves—A curiosity among Arachnida. PLoS ONE 8: e75989.

Stollewerk, A. and A. D. Chipman. 2006. Neurogenesis in myriapods and chelicerates and its importance for understanding arthropod relationships. Integr. Comp. Biol. 46: 195–206.

Tanaka, G., X. Hou, X. Ma, G. D. Edgecombe and N. J. Strausfeld. 2013. Chelicerate neural ground pattern in a Cambrian great appendage arthropod. Nature 502: 364–367.

Walter, D. E., and H. C. Proctor. 2013. *Mites: Ecology, Evolution & Behaviour. Life at a Microscale, Second Edition.* Springer, Dordrescht.

Ubick, D. and N. Dupérré. 2005. *Spiders of North America: An Identification Manual.* American Arachnological Society, Keene, New Hampshire.

Uhl, G. 2000. Female genital morphology and sperm priority patterns in spiders (Araneae). Pp. 145–156 in S. Toft and N. Scharff (eds.), *European Arachnology 2000.* Aarhus University Press, Aarhus.

Vollrath, F. and P. Selden. 2007. The role of behavior in the evolution of spiders, silks, and webs. Ann. Rev. Ecol. Evol. S. 819–846.

Watson, P. J. 1986. Transmission of a female sex pheromone thwarted by males in the spider *Linyphia litigiosa* (Linyphiidae). Science 233: 219–221.

Weygoldt, P. 1969. *The Biology of Pseudoscorpions.* Harvard University Press, Cambridge, MA.

Weygoldt, P. 2000. *Whip spiders (Chelicerata: Amblypygi). Their Biology, Morphology and Systematics.* Apollo Books, Stenstrup.

Wirkner, C. S. and L. Prendini. 2007. Comparative morphology of the hemolymph vascular system in scorpions–a survey using corrosion casting, microCT, and 3D-reconstruction. J. Morphol. 268: 401–413.

Wirkner, C.S., Tögel, M., Pass, G., 2013. The arthropod circulatory system. Pp. 343–391 in A. Minelli, G. Boxshall and G. Fusco (eds.), *Arthropod Biology and Evolution: Molecules, Development, Morphology.* Springer, Heidelberg.

Wise, D. H. 1993. *Spiders in Ecological Webs.* Cambridge University Press, Cambridge.

Witt, P. N., C. F. Reed and D. B. Peakall. 1968. *A Spider's Web.* Springer Verlag, Berlin.

Witt, P. N. and J. S. Rovner (eds.). 1982. *Spider Communication: Mechanisms and Ecological Significance.* Princeton University Press, Princeton, NJ.

World Spider Catalog. 2015. *World Spider Catalog.* Natural History Museum Bern, online at wsc.nmbe.ch

Pycnogonida

Arango, C. P. and W. C. Wheeler. 2007. Phylogeny of the sea spiders (Arthropoda, Pycnogonida) based on direct optimization of six loci and morphology. Cladistics 23: 255–293.

Arnaud, F. and R. N. Bamber. 1987. The biology of the Pycnogonida. Pp. 1–96 in J. H. S. Blaxter and A. J. Southward (eds.), *Advances in Marine Biology, Vol. 24.* Academic Press, New York.

Bain, B. 1991. Some observations on biology and feeding behavior in two southern California pycnogonids. Bijd. Dierkunde 61: 63–64.

Behrens, W. 1984. Larvenentwicklung und Metamorphose von *Pycnogonum litorale* (Chelicerata, Pantopoda). Zoomorphologie 104: 266–279.

Bergström, J., W. Stürmer and G. Winter. 1980. *Palaeoisopus, Palaeopantopus* and *Palaeothea*, pycnogonid arthropods from the Lower Devonian Hunsrück Slate, West Germany. Palaeont. Zh. 54: 7–5

Brenneis, G., A. Stollewerk and G. Scholtz. 2013. Embryonic neurogenesis in *Pseudopallene* sp. (Arthropoda, Pycnogonida) includes two subsequent phases with similarities to different arthropod groups. EvoDevo 4: 32.

Brenneis, G., P. Ungerer and G. Scholtz. 2008. The chelifores of sea spiders (Arthropoda, Pycnogonida) are the appendages of the deutocerebral segment. Evol. Dev. 10: 717–724.

Child, C. A. 1986. A parasitic association between a pycnogonid and a scyphomedusa in midwater. J. Mar. Biol. Assoc. U.K. 66: 113–117.

Cole, L. J. 1905. Ten-legged pycnogonids, with remarks on the classification of the Pycnogonida. Ann. Mag. Nat. Hist. 15: 405–415.

Fahrenbach, W. H. and C. P. Arango. 2007. Microscopic anatomy of Pycnogonida: II. Digestive system. III. Excretory system. J. Morphol. 268: 917–935.

Fry, W. G. 1965. The feeding mechanisms and preferred foods of three species of Pycnogonida. Bull. Br. Mus. Nat. Hist. Zool. 12: 195–223.

Fry, W. G. 1978. A classification within the Pycnogonida. Zool. J. Linn. Soc. 63: 35–78.

Fry, W. G. (ed.). 1978. Sea Spiders (Pycnogonida). Zool. J. Linn. Soc. 63: 1–238.

Hedgpeth, J. W. 1954. On the phylogeny of the Pycnogonida. Acta Zool. 35: 193–213.

Hedgpeth, J. W. 1978. A reappraisal of the Palaeopantopoda with description of a species from the Jurassic. Zool. J. Linn. Soc. 63: 23–34.

Hedgpeth, J. W. 1982. Pycnogonida. Pp. 169–173 in S. P. Parker (ed.), *Synopsis and Classification of Living Organisms, Vol. 2.* McGraw-Hill, New York.

Henry, L. M. 1953. The nervous system of the Pycnogonids. Microentomology 18: 16–36.

King, P. E. 1973. *Pycnogonids.* Hutchinson University Library, London.

Lehmann, T., M. Heβ and R. R. Melzer. 2012. Wiring a periscope—ocelli, retinula axons, visual neuropils and the ancestrality of sea spiders. PLoS ONE 7: e30474.

Machner, J. and G. Scholtz. 2010. A scanning electron microscopy study of the embryonic development of *Pycnogonum litorale* (Arthropoda, Pycnogonida). J. Morphol. 271: 1306–1318.

Maxmen, A., W. E. Browne, M. Q. Martindale and G. Giribet. 2005. Neuroanatomy of sea spiders implies an appendicular origin of the protocerebral segment. Nature 437: 1144–1148.

Miyazaki, K. 2002. On the shape of foregut lumen in sea spiders (Arthropoda: Pycnogonida). J. Mar. Biol. Assoc. U.K. 82: 1037–1038.

Morgan, E. 1972. The swimming of *Nymphon gracile* (Pycnogonida): The swimming gait. J. Exp. Biol. 56: 421–432.

Nakamura, K. 1981. Postembryonic development of a pycnogonid *Propallene longiceps.* J. Nat. Hist. 15: 49–62.

Nakamura, K. and K. Sekiguchi. 1980. Mating behavior and oviposition in the pycnogonid *Propallene longiceps.* Mar. Ecol. Prog. Ser. 2: 163–168.

Siveter, D. J., M. D. Sutton, D. E. G. Briggs and D. J. Siveter. 2004. A Silurian sea spider. Nature 431: 978–980.

Tomaschko, K. H., E. Wilhelm and D. Bückmann. 1997. Growth and reproduction of *Pycnogonum litorale* (Pycnogonida) under laboratory conditions. Mar. Biol. 129: 595–600.

Vilpoux, K. and D. Waloszek. 2003. Larval development and morphogenesis of the sea spider *Pycnogonum litorale* (Ström, 1762) and the tagmosis of the body of Pantopoda. Arthropod Struct. Dev. 32: 349–383.

Waloszek, D. and J. A. Dunlop. 2002. A larval sea spider (Arthropoda: Pycnogonida) from the Upper Cambrian "Orsten" of Sweden, and the phylogenetic position of pycnogonids. Palaeontology 45: 421–446.

Wyer, D. W. and P. E. King. 1974. Relationships between some British littoral and sublittoral bryozoans and pycnogonids. Estuarine Coastal Mar. Sci. 2: 177–184.

CHAPTER 25

Introduction to the Deuterostomes and the Phylum Echinodermata

With this chapter, we begin discussion of the small but important clade known as Deuterostomia. Although comprising only three phyla (Echinodermata, Hemichordata, and Chordata) and about 60,000 described living species (only about 10,500 of which are invertebrates), the Deuterostomia has figured prominently in zoological history because it is the clade to which the vertebrates belong and, thus, humans. The oldest fossils thought to be bilaterians are from the Ediacaran period, embryos found in the Doushantuo deposits of China dating 600–580 million years ago. And the oldest deuterostome fossil is a 530-million-year-old creature called *Yunnanozoon*, from the lower Cambrian Chengjiang biota of Yunnan Province, China. But molecular dating techniques suggest that the origin of the Bilateria was probably earlier, perhaps 630–600 million years ago or before.

Early in the evolution of bilaterians, there was a split into two major lineages that have long been called Protostomia and Deuterostomia. As described in Chapter 9, these groups were named over 100 years ago, and they were long defined on the basis of embryological principles. In protostomes the blastopore (the position in the embryo that typically gives rise to endodermal tissues) was said to give rise to the mouth ("protostome" = mouth first). Typically in deuterostomes, the blastopore gave rise to the adult anus, the mouth thus forming secondarily at a different location ("deuterostome" = mouth second). In both lineages, the blastopore sits at the vegetal pole of the embryo when gastrulation beings. It was long thought that deuterostomous development was a synapomorphy for the clade Deuterostomia. However, this is no longer a tenable hypothesis because some phyla with deuterostomous development (e.g., Brachiopoda) are now known to belong to the Protostome clade.

The advent of molecular phylogeny, which supports the monophyly of the clades called Protostomia and Deuterostomia, has resulted in some rearrangements, and four phyla formerly assigned to the deuterostome clade are now seen to belong

Classification of The Animal Kingdom (Metazoa)

Non-Bilateria*
(a.k.a. the diploblasts)
- PHYLUM PORIFERA
- PHYLUM PLACOZOA
- PHYLUM CNIDARIA
- PHYLUM CTENOPHORA

Bilateria
(a.k.a. the triploblasts)
- PHYLUM XENACOELOMORPHA

Protostomia
- PHYLUM CHAETOGNATHA

SPIRALIA
- PHYLUM PLATYHELMINTHES
- PHYLUM GASTROTRICHA
- PHYLUM RHOMBOZOA
- PHYLUM ORTHONECTIDA
- PHYLUM NEMERTEA
- PHYLUM MOLLUSCA
- PHYLUM ANNELIDA
- PHYLUM ENTOPROCTA
- PHYLUM CYCLIOPHORA

Gnathifera
- PHYLUM GNATHOSTOMULIDA
- PHYLUM MICROGNATHOZOA
- PHYLUM ROTIFERA

Lophophorata
- PHYLUM PHORONIDA
- PHYLUM BRYOZOA
- PHYLUM BRACHIOPODA

ECDYSOZOA

Nematoida
- PHYLUM NEMATODA
- PHYLUM NEMATOMORPHA

Scalidophora
- PHYLUM KINORHYNCHA
- PHYLUM PRIAPULA
- PHYLUM LORICIFERA

Panarthropoda
- PHYLUM TARDIGRADA
- PHYLUM ONYCHOPHORA
- PHYLUM ARTHROPODA
  - SUBPHYLUM CRUSTACEA*
  - SUBPHYLUM HEXAPODA
  - SUBPHYLUM MYRIAPODA
  - SUBPHYLUM CHELICERATA

Deuterostomia
- **PHYLUM ECHINODERMATA**
- PHYLUM HEMICHORDATA
- PHYLUM CHORDATA

*Paraphyletic group

The section on Echinodermata has been revised by Rich Mooi.

to the Protostomia. Furthermore, new embryological patterns have begun to emerge. In the Deuterostomia, the blastopore *does* consistently give rise to the anus, and the mouth forms secondarily. But among the Protostomia, gastrulation is now known to be much more variable and some protostome phyla display deuterostomous development (i.e., the blastopore becomes the anus). Evidence is accumulating that mouth formation from oral ectoderm (in the animal hemisphere), development of the anus from the blastopore, and radial cleavage, all typical of deuterostomous embryogeny, may be ancestral to both protostomes and deuterostomes. Consequently, the embryological features that once defined these two great clades of bilaterians no longer unambiguously do so, and thus our current phylogenetic/taxonomic view of Protostomia and Deuterostomia is based solely on molecular phylogenetic evidence. However, the legacy names Protostomia and Deuterostomia continue to be used for these two animal lineages. Morphological or developmental synapomorphies probably exist that define these two clades, but these have not yet been unambiguously identified—although a possible synapomorphy of the Deuterostomia is a trimeric body coelom condition, at least primitively (this feature is lacking in the phylum Chordata). A genetic synapomorphy of Deuterostomia may be the presence of the important developmental gene *Nodal*, which appears to be unique to that clade.

Within the deuterostome clade, molecular, morphological and paleontological studies now indicate that echinoderms and hemichordates comprise a sister group, a clade called Ambulacraria, and this is the sister group to Chordata. The Ambulacraria are also supported by shared Hox gene motifs and gene arrangements. These relationships imply that the features shared between chordates and hemichordates (previously thought to comprise a sister group), such as gill slits, must have been ancestral within Deuterostomia but lost in the echinoderm line (and also in pterobranch hemichordates). Gill slits in Deuterostomia have been shown to be homologous based on their gene expression patterns. Another feature shared between the Hemichordata and Chordata is the stomochord/notochord, long viewed as homologues. However, it is now thought that a group of vacuolated cells in ancestral Deuterostomia gave rise to these structures independently in hemichordates and chordates. Within Chordata, Urochordata is sister to Vertebrata (a group known as Olfactores), and Cephalochordata is sister to those. There is some evidence that the enigmatic worm genus *Xenoturbella* might be near the base of the deuterostome line, but opposing evidence suggests it is allied with the Acoelomorpha as an ancestral bilaterian clade (called Xenocoelomorpha); this is the view followed in this book.

Phylum Echinodermata

Members of the phylum Echinodermata (Greek *echinos*, "spiny"; *derma*, "skin") are familiar to many as iconic sea animals, in part because they are strictly marine. The phylum contains about 7,300 known living species, and includes the sea lilies, feather stars, starfish,[1] brittle stars, sea urchins, sand dollars, heart urchins, and sea cucumbers (Figures 25.1, and 25.2). Another 15,000 or so species are known from a rich fossil record dating back at least to early Cambrian times. Both extant and fossil forms display an enormous breadth of morphological disparity that makes the phylum a particularly interesting clade in which to explore the origins of evolutionary novelty.

Echinoderms range in size from minute sea urchins, sea cucumbers, and brittle stars less than 1 cm across, to starfish that exceed 1 m from arm tip to arm tip, and sea cucumbers that exceed 3 m in length. Although some species tolerate brackish water, echinoderms have not evolved excretory, respiratory, or osmoregulatory mechanisms allowing them to exploit terrestrial or freshwater habitats. However, in the ocean, they can be found worldwide at all depths. It is said that echinoderms are largely bottom-dwellers (i.e., benthic), and it is true that in some regions of the deep sea the only identifiable organisms in bottom photos are echinoderms. However, the statement that echinoderms are mostly benthic is true only for the adults (with the exception of a few pelagic and benthopelagic species, see Figure 25.1P,Q), because larval echinoderms are important members of pelagic communities. As larvae, echinoderms are vitally important to the ecology of planktonic communities, and as adults they perform key roles in benthic marine ecosystems in deposit and suspension feeding (notably sea cucumbers and sea lilies), predation (especially certain starfish), herbivorous grazing (many sea urchins), and in sediment processing and turnover (such as heart urchins).

Echinoderms might be the strangest metazoans on Earth. Animals with radial or bilateral symmetry are intuitively easy to understand, but the secondarily derived pentaradiality of adult echinoderms challenges our views of life. They have body systems unlike those of any other creatures, and they are missing body parts

[1]There has been some effort of late to change this familiar and earliest common name for the Asteroidea to "sea star," perhaps in an effort to avoid confusion of these animals with fish. If, upon meeting a starfish, such confusion persists, then there are bigger problems at hand than can be solved by such a name change. And what of terms such as cuttlefish, in which potential confusion could conceivably be worse? "Sea stars" are not fish, but neither are they flaming balls of fusioning hydrogen in outer space. Sea cucumbers are not vegetables, sea lilies are not flowers, and sea horses are not mammals. Venerable terms such as "starfish" can accentuate teachable moments about scientific nomenclature and relationships among organisms.

that it seems they should have. Add in the huge and bizarre fossil catalog, and this phylum can be uniquely challenging to understand. They might seem like mutated leftovers from a Precambrian alien's picnic. In fact, a somewhat complex and unique jargon has evolved just to describe the unlikely anatomy of echinoderms, although we will do our best to keep this to a minimum.

All living echinoderms possess some form of a well-developed coelom, a mesodermally-derived endoskeleton composed of unique calcareous elements, and 5-part radial symmetry (**pentaradiality**). The endoskeleton consists of ossicles and plates formed by microscopic, secreted crystals of calcium carbonate oriented along the same crystallographic axis. These microcrystals are associated with small amounts of organic tissue, and come together to form a unique, porous skeletal material called **stereom**. The stereom consists of anastomosing struts called **trabeculae** that give it a sponge-like appearance under high magnification. The hollow spaces in the stereom are occupied by living tissue (the **stroma**). In life, the ossicles and plates are attached to one another by connective tissues. Fossil echinoderm skeletons (e.g., carpoids) also have been shown to have this unique form of mineralization in their endoskeleton.

As adults, echinoderms are the only fundamentally 5-part, radially symmetrical animals, though this symmetry is modified in some lineages. Being deuterostome bilaterians, the embryology of echinoderms is frequently studied as exemplary of that clade. However, these basic embryological patterns are true only of early, bilaterally symmetrical stages of echinoderm development. The remarkable pentaradial body plan is superimposed on earlier larval bilaterality, and echinoderm adult pentaradiality is evolutionarily and developmentally derived relative to all other animals. Echinoderm adult pentaradiality is coupled with another unique body feature known as the **water vascular system**, a complex system of tubes and canals derived from a specific part of the coelom. The water vascular system serves a variety of functions, notably in giving rise to tube feet (also known as podia), which are crucial to nearly every interaction between echinoderms and their environment. These features are described in detail below. Because adult echinoderms are secondarily radially symmetrical, the terms "dorsal" and "ventral" do not have the same anatomical meaning as in other bilaterians, although these terms are still often used for convenience (and refer only to the animal's orientation on the substratum). The mouth is used as a reference point. Features on the mouth side of the adult are said to be on the **oral surface**, whereas those on the side opposite the mouth are **aboral**. For example, in a starfish, the oral region (bottom, or "ventral") faces the substratum, and the aboral region is therefore at the "top" of the animal ("dorsal").

Taxonomic History and Classification

Echinoderms have been known since ancient times. Early Europeans collected fossil heart urchins and regarded them as divine "thunder stones," burying them with their dead, and likenesses of starfish appear in 4,000-year-old Minoan frescoes. Jacob Klein is credited with coining the name Echinodermata in about 1734 in reference to sea urchins. Linnaeus placed the echinoderms among the molluscs, alongside a mixed bag of other animals in his class Vermes, which held just about every invertebrate that was not an arthropod. For nearly a century thereafter, echinoderms were allied with various other groups, including the cnidarians in Lamarck's Radiata. It was not until 1847 that Frey and Leuckart recognized the echinoderms as a distinct taxon. Recent molecular and embryological evidence places Echinodermata as a sister phylum to the Hemichordata in a clade known as the Ambulacraria.

At one time, the holothuroids, echinoids, and ophiuroids were placed in a group called the Cryptosyringida. However, molecular, developmental, and fossil evidence support sister group placement of the ophiuroids with the asteroids in a clade known as Asterozoa, and the echinoids and holothuroids in a clade called Echinozoa, with the crinoids as the earliest group to branch off (Figure 25.20). The proposed new class, Concentricycloidea, established by Baker et al. (1986) is now generally regarded to comprise highly modified asteroids adapted to life on sunken wood.

The large number of enigmatic echinoderm fossils is a rich source of argument because of varying opinions on the significance of characteristics such as feeding appendages, attachment structures, and even overall symmetry. Some have suggested that certain superficially bilaterally symmetrical fossils can be interpreted as representing the first echinoderms. Others find it more reasonable and congruent with the fossil record to suggest that these oddities exhibit secondary modification of earlier pentaradial morphology—indicating loss of rays as commonly seen in other clades. The disparate morphologies of fossil echinoderms have caused some workers to emphasize differences, recognizing upwards of 25 separate classes. Others seek fundamental similarities with overarching homologies, and recognize far fewer major clades. Some of the important extinct clades are introduced in the phylogeny section at the end of the chapter.

The classification scheme below draws from various authors and recognizes five major clades (usually referred to as classes) of living echinoderms. But readers should be aware that other classification schemes exist.

(A)
(B)
(C)
(D)
(E)
(F)
(G)
(H)
(I)
(J)
(K)
(L)

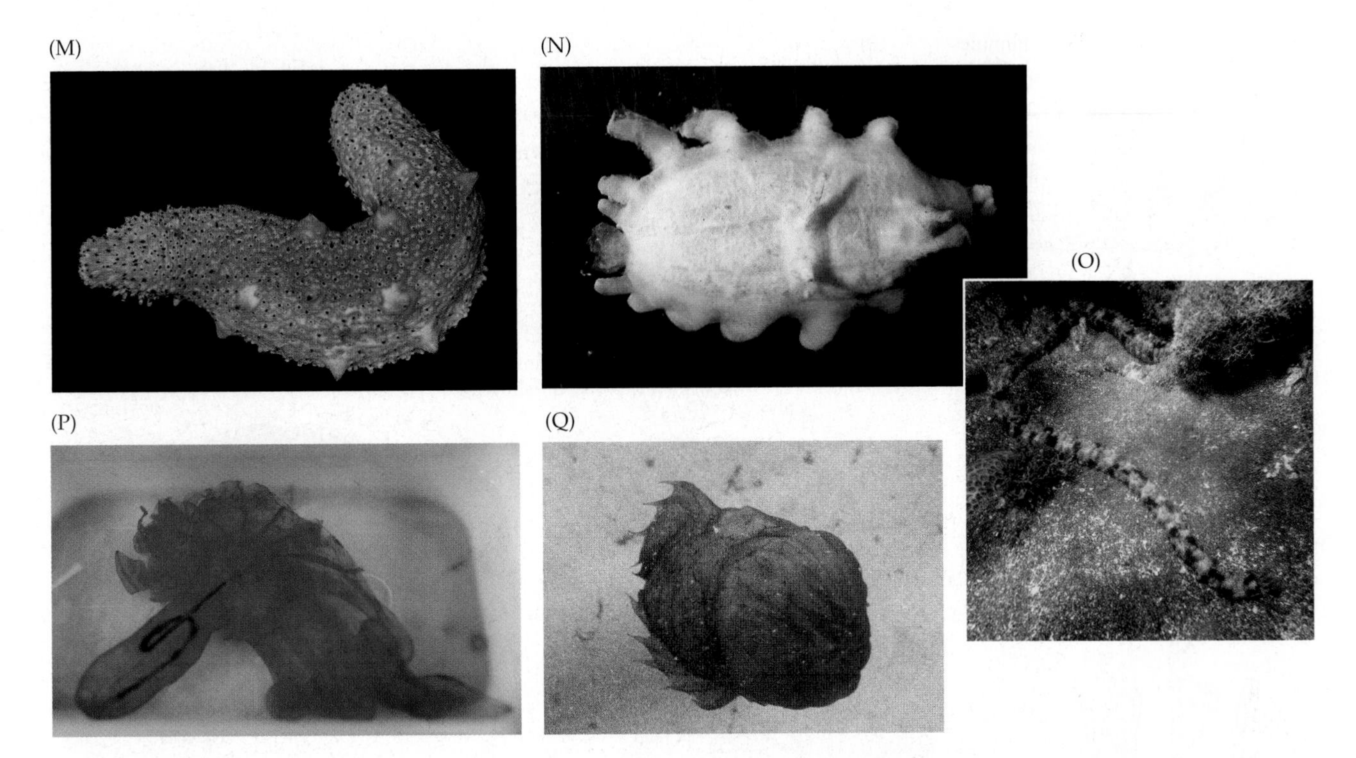

Figure 25.1 Representative echinoderms. (A) Feather star (Crinoidea). (B) A tropical starfish, *Linckia laevigata* (Asteroidea). (C) Sand-burrowing starfish, *Astropecten polyacanthus* (Asteroidea). (D) Slime starfish, *Pteraster tesselatus* (Asteroidea). (E) Indo-Pacific cushion starfish, *Culcita novaeguineae* (Asteroidea). (F) Indo-Pacific crown-of-thorns starfish, *Acanthaster planci* (Asteroidea). (G) Oral view of caymanostellid "wood star" (Asteroidea). (H) Brittle star, *Ophiopholis aculeata* (Ophiuroidea). (I) Basket star, oral view (Ophiuroidea). (J) Purple sea urchin, *Strongylocentrotus purpuratus* (Echinoidea). (K) Sand dollars, *Dendraster excentricus* (Echinoidea). (L) Heart urchin, *Lovenia elongata* (Echinoidea). (M) Sea cucumber, *Parastichopus* (Holothuroidea). (N) Deep-sea sea cucumber, *Scotoplanes* (Holothuroidea). (O) Apodid sea cucumber, *Synapta* (Holothuroidea). (P) Pelagic holothuroid, *Pelagothuria* (Holothuroidea). (Q) Abyssal swimming holothuroid, *Enypniastes* (Holothuroidea).

BOX 25A Characteristics of the Phylum Echinodermata

1. Calcareous, mesodermally derived endoskeleton composed of separate ossicles and plates made of an open meshwork structure called stereom, with the interstices being filled with living tissue (the stroma)
2. Adults with basic pentaradial symmetry; five rays extend outward from the mouth
3. Embryogeny fundamentally deuterostomous, with radial cleavage, endodermally derived mesoderm, enterocoely, and mouth not derived from the blastopore
4. Body divided into axial, perforate extraxial, and imperforate extraxial regions; axial region expresses radial symmetry derived from the rudiment; extraxial elements are derived from the bilateral, larval body
5. Water vascular system derived from hydrocoel on left side of larva; composed of complex series of fluid-filled canals expressed externally as tube feet in ambulacra; oldest parts of the ray near the mouth, youngest at ray tips
6. Body wall containing mutable collagenous tissues
7. Gut complete, anus sometimes absent
8. No specialized, distinct excretory organs
9. Circulatory structures, when present, comprise hemal system derived from coelomic cavities and sinuses.
10. Nervous system diffuse, decentralized, consisting of subepithelial nerve net, nerve ring, and radial nerves
11. Mostly gonochoristic; development direct or indirect

PHYLUM ECHINODERMATA

SUBPHYLUM CRINOZOA One living class, the Crinoidea, with the mouth facing upward and surrounded by branching arms.

CLASS CRINOIDEA Sea lilies and feather stars (Figures 25.1A and 25.2A,B). Arms radiate from central cup (calyx), with mouth and oral surface directed upward; aboral stem, when present, arising from aboral side of calyx; ambulacra on arms that bear pinnules; arms composed of confluence of axial and extraxial elements; both left and right somatocoels continuing from calyx into arms; ambulacra not calcified, forming soft tissue closure of the coelomic cavities in arms, with external radial water vessel; each side of ambulacrum with palisade of cover plates; tube feet minute, lacking suckers; ambulacra can branch equally and repeatedly; madrepor

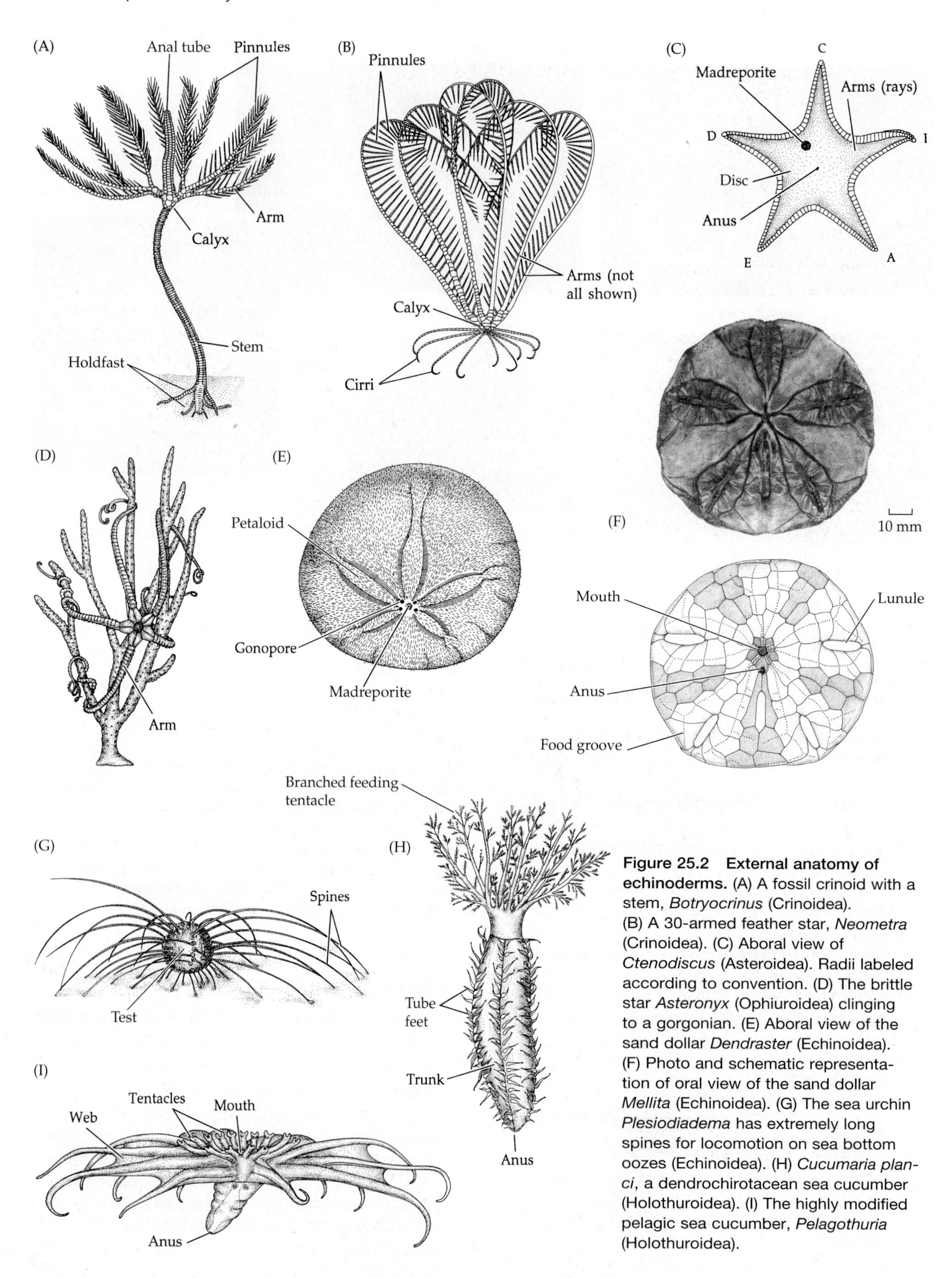

Figure 25.2 External anatomy of echinoderms. (A) A fossil crinoid with a stem, *Botryocrinus* (Crinoidea). (B) A 30-armed feather star, *Neometra* (Crinoidea). (C) Aboral view of *Ctenodiscus* (Asteroidea). Radii labeled according to convention. (D) The brittle star *Asteronyx* (Ophiuroidea) clinging to a gorgonian. (E) Aboral view of the sand dollar *Dendraster* (Echinoidea). (F) Photo and schematic representation of oral view of the sand dollar *Mellita* (Echinoidea). (G) The sea urchin *Plesiodiadema* has extremely long spines for locomotion on sea bottom oozes (Echinoidea). (H) *Cucumaria planci*, a dendrochirotacean sea cucumber (Holothuroidea). (I) The highly modified pelagic sea cucumber, *Pelagothuria* (Holothuroidea).

mouth, and anus on oral surface of calyx. About 625 living species. Sea lilies have a stem throughout their lives, but feather stars lack a stem as adults.

ORDER COMATULIDA Feather stars. Stem lost following postlarval stage; most proximal stem element in feather stars retained at base of cup to support cirri that act as attachment appendages. (e.g., *Antedon*, *Bourgeticrinus*, *Comatula*, *Mariametra*, *Tropiometra*)

ORDER CYRTOCRINIDA Aberrant, stemless sea lilies. Calyx sometimes elongate to resemble a stem; no cirri; calyx directly attached to bottom, or attached by a disk. (e.g., *Pilocrinus*, *Neogymnocrinus*, *Cyathidium*, *Holopus*)

ORDER HYOCRINIDA Sea lilies with well-developed stem attached to bottom by expanded terminal disk (holdfast); stem lacking cirri and without nodals and internodals, i.e., elements making up stem are identical. (e.g., *Hyocrinus*, *Calamocrinus*, *Ptilocrinus*, *Laubiericrinus*, *Thalassocrinus*)

ORDER ISOCRINIDA Sea lilies with calyx of very thick stereom elements; with well-developed stem composed of internodials (without cirri) and nodals (with cirri). (e.g., *Isocrinus*, *Hypalocrinus*, *Neocrinus*, *Teliocrinus*, *Endoxocrinus*, *Metacrinus*, *Saracrinus*)

SUBPHYLUM ASTEROZOA Contains the stellate forms, Asteroidea and Ophiuroidea, in which the mouth faces the substratum.

CLASS ASTEROIDEA Starfish (Figures 25.1B–G and 25.2C). Body stellate with five or more arms; arms not sharply demarcated from central disk by distinct articulations; arms consisting of coelomic extensions enclosed by both axial and extraxial body wall; anus (when present) on aboral surface; mouth directed toward substratum; ambulacral grooves with external radial water vessel; tube feet with internal ampullae, with or without suckers; madreporite aboral on CD interambulacrum. About 1,900 extant species. For an alternative classification, see Blake 1987. (In the most recent classifications, the Concentricycloidea is treated as an infraclass of the Asteroidea.)

ORDER FORCIPULATIDA With 5–50 arms; tube feet with suckers; anus present; with pincer-like pedicellariae. Widely distributed starfish, including most familiar intertidal species. Several hundred species. (e.g., *Asterias*, *Pisaster*, *Heliaster*, *Labidiaster*, *Pedicellaster*, *Pycnopodia*, *Stichaster*, *Zoroaster*)

ORDER BRISINGIDA Between 6 and 20 long, attenuated arms; fused ring of disk plates; crossed, pincer-like pedicellariae. Most are deep-sea suspension-feeders. (e.g., *Brisinga*, *Odinia*, *Freyella*, *Astrocles*)

ORDER SPINULOSIDA With 5–18 arms; tube feet with suckers; anus present; generally lacking pedicellariae. With a few hundred species. (e.g., *Echinaster*, *Henricia*, *Acanthaster*, *Mithrodia*)

ORDER NOTOMYOTIDA Usually 5 long, straight-sided arms; small disk; prominent muscle bands running inside arm coeloms from disk to arm tip. (e.g., *Benthopecten, Cheiraster*)

ORDER PAXILLOSIDA Upper surface with umbrella-like clusters of ossicles (paxillae); tube feet without suckers; anus present or absent. Epibenthic or shallow burrowers. (e.g., *Astropecten*, *Luidia*, *Platasterias*, *Ctenodiscus*)

ORDER VALVATIDA Tube feet with suckers; anus present; some possess paxillae; some possess clamp-like pedicellariae recessed into plates. Widely distributed, with several hundred species. (e.g., *Archaster*, *Asterina*, *Patiria*, *Goniaster*, *Fromia*, *Linckia*, *Solaster*, *Ophidiaster*, *Odontaster*, *Oreaster, Valvaster*)

ORDER VELATIDA Usually 5-15 arms; body relatively thick; disk large; stereom ossicles sometimes poorly developed, particularly aborally. Primarily cold-water, deep-sea forms. (e.g., *Pteraster*, *Hymenaster*, *Korethraster*, *Myxaster*, *Caymanostella*)

CLASS OPHIUROIDEA Brittle stars and basket stars (Figures 25.1H,I and 25.2D). Body with five unbranched or branched articulated arms, clearly demarcated from a central disk; oral arm plates cover ambulacral groove; coelom in arms greatly reduced; tube feet without internal ampullae or suckers; anus lacking; madreporite on CD interambulacral plate on oral surface, often reduced. About 2,000 extant species in two orders.

ORDER OPHIURIDA Unbranched arms capable of horizontal coiling, but only limited vertical movement; aboral and oral arm plates present and usually well developed; digestive glands within central disk. Includes vast majority of living brittle stars. (e.g., *Ophiura*, *Amphipholis*, *Amphiura*, *Ophiactis*, *Ophiocoma*, *Ophioderma*, *Ophiolepis*, *Ophiomusium*, *Ophionereis*, *Ophiomyxa*)

ORDER EURYALIDA Arms often highly branched, branching very extensive in basket stars; arms capable of coiling vertically as well as horizontally; dorsal arm plates absent; central disk often covered in thick dermis. Many use outstretched arms to form basketlike filter for suspension feeding. Others live with arms tightly coiled around coral branches. (e.g., *Asteronyx*, *Astrodia*, *Astrodictyum*, *Gorgonocephalus*, *Euryale*, *Asteroschema*)

SUBPHYLUM ECHINOZOA Contains the Echinoidea and Holothuroidea. Species in this subphylum lack distinct arms, but possess a specialized subdivision of the left somatocoel in which a jaw apparatus develops. This jaw apparatus is lacking in the holothuroids, but the subdivision of the somatocoel is still expressed.

CLASS ECHINOIDEA Sea urchins, pencil urchins, heart urchins, and sand dollars (Figures 25.1J–L and 25.2E–G). Body globose to flattened and discoidal, often secondarily bilateral; stereom plates sutured together by connective tissue and calcite interdigitations to form usually rigid test; test divided into coronal region derived from axial skeletal elements and apical system of extraxial elements; pedicellariae of varying types always present; movable spines mounted on tubercles; radial water vessels wholly enclosed within test; internal jaw apparatus (Aristotle's lantern) present in most groups. About 1,000 extant species in two subclasses.

SUBCLASS CIDAROIDEA Pencil urchins. Test high and globular, ambulacral plates not compound, each with a pore pair supporting a single tube foot; primary spines large, pencil-like, without epidermal covering; anus at aboral pole; ambulacral plates continuing onto peristomial membrane. About 150 living species in one order (Cidaroida). (e.g., *Cidaris*, *Histocidaris*, *Stylocidaris*, *Eucidaris*, *Phyllacanthus*, *Psychocidaris*)

SUBCLASS EUECHINOIDEA Sea urchins, heart urchins, lamp urchins, sea biscuits, sand dollars. Test globular to discoidal, with pronounced bilateral symmetry in some forms (Irregularia); numbers of tube feet and spines per plate vary; anal position varies from aboral to "posterior." Aristotle's lantern absent in heart and lamp urchins. Contains one order (Echinothurioida) that is sister to the infraclass Acroechinoidea. About 850 living species.

INFRACLASS CARINACEA Sea urchins with keeled teeth, can be globose (as in calycinoids) or extremely flattened (as in irregulars such as sand dollars). Some Irregularia lack the Aristotle's lantern, and therefore the teeth that would show the keel, but many other characters place these lanternless forms in the clade.

SUPERORDER CALYCINOIDA Globose sea urchins; anus aboral, within apical system; spines usually not hollow; with compound ambulacral plates; five pairs of gills arranged in circle on peristomial membrane. Contains the orders Echinacea. (e.g., *Arbacia*, *Stomopneustes*, *Echinometra*, *Echinus*, *Heterocentrotus*, *Paracentrotus*, *Strongylocentrotus*, *Toxopneustes*, *Tripneustes*) and Calycina (e.g., *Salenia*)

SUPERORDER IRREGULARIA Heart urchins, lamp urchins, sand dollars, sea biscuits. Body globular or discoidal, always with some bilateral symmetry; anus never wholly within apical system, shifted to "posterior" (even oral) position; spines small, forming dense covering; Aristotle's lantern absent in heart urchins and adult lamp urchins and echinoneoids. Six extant orders: Echinoneoida (living holectypoids), Holasteroida (deep-sea heart urchins), Spatangoida (heart urchins), Cassiduloida (basal lamp urchins), and Echinolampadoida (lamp urchins that form a sister group to the sand dollars, and Clypeasteroida (sand dollars and sea biscuits). (e.g., *Echinoneus*, *Urechinus*, *Pourtalesia*, *Spatangus*, *Echinocardium*, *Maretia*, *Meoma*, *Metalia*, *Lovenia*, *Cassidulus*, *Echinolampas*, *Clypeaster*, *Laganum*, *Echinocyamus*, *Echinarachnius*, *Dendraster*, *Mellita*)

A PARAPHYLETIC GROUP OF 5 ORDERS THAT CONSTITUTES A SERIES OF A SISTER GROUPS TO THE CARINACEA This group includes the order Echinothurioida, plus the "diadematacean" orders Micropygoida, Aspidodiadematoida, Diadematoida, Pedinoida, placed in a non-monophyletic, informal group.

ECHINOTHURIOIDS Among largest of echinoids, with test up to 30 cm in diameter; test flexible, collapsing upon removal from water; plates usually small and thin, joined with large amounts of collagen; anus aboral; subset of spines with poison glands; some species with club-shaped spines for walking on ooze. Usually deep-water (1,000–4,000 m). Sister group to the rest of the Euechinoidea. (e.g., *Araeosoma*, *Asthenosoma*, *Phormosoma*, *Sperosoma*)

"DIADEMATACEANS" Globose sea urchins; anus aboral; compound ambulacral plates arranged in triads; spines usually hollow; five pairs of gills on peristomial membrane. (e.g., *Astropyga*, *Aspidodiadema*, *Caenopedina*, *Diadema*, *Micropyga*)

CLASS HOLOTHUROIDEA Sea cucumbers (Figures 25.1M–Q and 25.2H,I). Body fleshy, worm-like, elongated along oral–aboral axis; skeleton usually reduced to isolated ossicles embedded in body wall; pentaradial symmetry manifested in circumoral feeding tentacles, which are derived directly from radial vessel; feeding tentacles can be highly branched; tube feet along body variably expressed, derived from embryologically interradial longitudinal canals that are not homologous with radial canals of other echinoderms (Figure 25.3A); madreporite suspended from stone canal within coelom; with circlet of feeding tentacles around mouth. About 1,700 extant species of holothuroids (= holothurians) in five orders.

ORDER APODIDA Body wormlike or snakelike, often very elongate, contractile; lacking tube feet along body; distinctive and abundant wheel or anchor ossicles in body wall; 10–25 branched feeding tentacles; respiratory trees absent. Some burrow through peristaltic action of body wall muscles, others are epifaunal at varying depths. (e.g., *Chirodota*, *Myriotrochus*, *Synapta*, *Euapta*, *Leptosynapta*)

ORDER ASPIDOCHIROTIDA Body sausage-like; tube feet present along body, often arranged to make sole-like flattened area on lower surface; oral tentacles flattened, even leaflike; oral region lacks retractor muscles; respiratory trees present. Includes largest holothuroids, and many familiar, shallow-water forms. (e.g., *Holothuria*, *Stichopus*, *Parastichopus*, *Thelenota*, *Actinopyga*, *Bohadschia*, *Mesothuria*, *Bathyplotes*, *Synallactes*)

ORDER DENDROCHIROTIDA Body sausage-like, sometimes U-shaped; body wall with small ossicles, but partially enclosed in plates in certain genera (e.g., *Psolus*); feeding tentacles typically branched; retractor muscles present in tentacles and oral region; tube feet present along body. Includes many common intertidal cucumbers, most are suspension-feeders. (e.g., *Cucumaria*, *Eupentacta*, *Phyllophorus*, *Psolus*, *Thyone*, *Ypsilothuria*, *Vaneyella*)

ORDER ELASIPODIDA Body wall often thin with elongated papillae, sails or fins; respiratory trees absent. Typically deep-sea cucumbers, some with strange body forms adapted for swimming (some are benthopelagic, and at least one species is fully pelagic). (e.g., *Deima*, *Elpidia*, *Peniagone*, *Scotoplanes*, *Laetmogone*, *Enypniastes*, *Pelagothuria*, *Benthodytes*)

ORDER MOLPADIDA Body stout, narrowed posteriorly to a distinct tail; with 10 or 15 digitate tentacles; lacking tube feet along body. Found from shallow water to over 2,000 m. (e.g., *Caudina*, *Molpadia*, *Gephyrothuria*)

The Echinoderm Body Plan

Radial symmetry is usually linked to sessile or planktonic habits of animals that face their environments on all sides as suspension feeders (e.g., feeding structures of tube-dwelling polychaetes, bryozoans, phoronids, and others) or as passive predators (e.g., cnidarians). However, echinoderms have managed to combine motility with radial symmetry, and they display many different feeding strategies and lifestyles. Much of the biology of echinoderms is associated with their unique **water vascular system** (Figure 25.3). A complex of fluid-filled canals and reservoirs, the water vascular system supports the function of the **tube feet** (or **podia**)—hydraulically operated, fleshy projections that serve in locomotion, gas exchange, feeding, attachment, and sensory perception. Much of the evolutionary story of the echinoderms is associated with elaboration of the **ambulacra**, that is, the region along the rays in which the water vascular system and skeletal plates that support the tube feet are situated.

The water vascular system is derived relatively late in embryogeny, after development of an initial, bilaterally symmetrical larva. The coelomic cavities of adult echinoderms are not expressed in the typical three body regions usually associated with deuterostomes. All the elements of the water vascular system come from a specialized part of the left mesocoelic portion of the embryonic tripartite coelom, called the **hydrocoel** (see below). The interaction of the hydrocoel with the left somatocoel leads to formation of a region on the left side of the echinoderm larva called the **rudiment**—and the rudiment represents the first glimmerings of pentaradial symmetry. In fact, it represents the first glimmerings of the adult animal itself because the adult form is produced by elaboration of the rudiment, with accompanying loss or incorporation of the bilateral parts of the old larva into the newly emerging adult. This patterning is the key to understanding the uniqueness of echinoderms and almost all of the morphologies seen throughout the phylum—even the most bizarre of the Paleozoic forms.

Developmental Roots of the Echinoderm Body Plan

Axial and extraxial body regions Echinoderms are weird animals. They look like nothing else in the animal kingdom, and even more, the echinoderm classes look nothing like one another. How has this strange anatomy come about? Part of the explanation lies in the unique transformation of the echinoderm's bilaterally symmetrical larva into a pentaradial adult. A theory that describes this complex process and body plan is called the **Extraxial/Axial Theory**, or **EAT**. The theory draws from embryology, larval development and metamorphosis, and body wall morphology. The EAT describes adult body regions and skeletal homologies in the various echinoderm lineages based on how different larval regions develop into adult body regions.

There are two main body wall regions in larval echinoderms, called the **axial** and the **extraxial regions** (Figure 25.4A). These two regions produce very different parts of the adult echinoderm, which is thus also composed of axial and extraxial regions. Each region produces skeletal material, but develops following different patterns.

During the gastrula stage, echinoderms form the three right and left pairs of coelomic compartments that characterize all archimeric deuterostomes—the prosome, mesosome, and metasome. However, in echinoderms the mesosome comes to be dominated by the left compartment, which is called the **hydrocoel**. The compartments of the metasome are called the **left and right somatocoels**. One of the most unusual features of echinoderms is the expansion of the left side of the larva as the adult begins to form. During larval development, the hydrocoel and left somatocoel become a single functional unit called the **rudiment**. The rudiment becomes the axial region, and it develops a ring around the larval esophagus. From this ring are sent out the five radial canals that characterize adult echinoderms. The adult axial region thus includes the mouth, water vascular system, tube feet, and ambulacral skeleton (which grows according to the Ocular Plate Rule, described below).

In contrast, the extraxial region is the non-rudiment part of the larva, and it makes up most of the

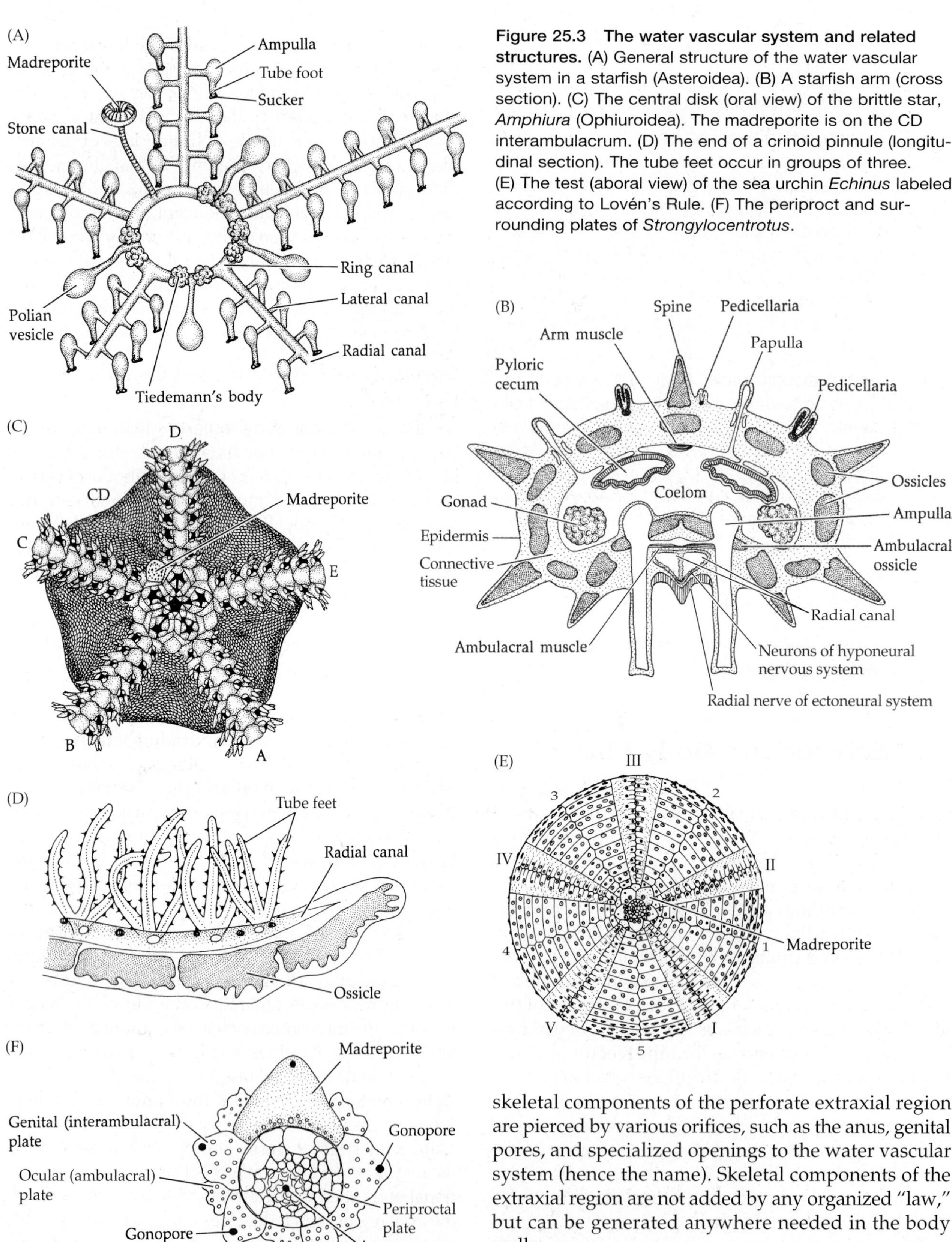

Figure 25.3 The water vascular system and related structures. (A) General structure of the water vascular system in a starfish (Asteroidea). (B) A starfish arm (cross section). (C) The central disk (oral view) of the brittle star, *Amphiura* (Ophiuroidea). The madreporite is on the CD interambulacrum. (D) The end of a crinoid pinnule (longitudinal section). The tube feet occur in groups of three. (E) The test (aboral view) of the sea urchin *Echinus* labeled according to Lovén's Rule. (F) The periproct and surrounding plates of *Strongylocentrotus*.

central disc of adult echinoderms (except in echinoids, in which it is almost entirely lost). The extraxial region consists of two parts, known as the **perforate** and **imperforate extraxial regions**. In adult echinoderms, the skeletal components of the perforate extraxial region are pierced by various orifices, such as the anus, genital pores, and specialized openings to the water vascular system (hence the name). Skeletal components of the extraxial region are not added by any organized "law," but can be generated anywhere needed in the body wall.

During evolution in the Echinodermata, alterations in relative amounts of each of the major body regions gave rise to the diverse forms seen throughout the phylum's half-billion year history (Figure 25.4B). Most significantly, in the living classes Asteroidea, Ophiuroidea, Echinoidea, and Holothuroidea, the imperforate body

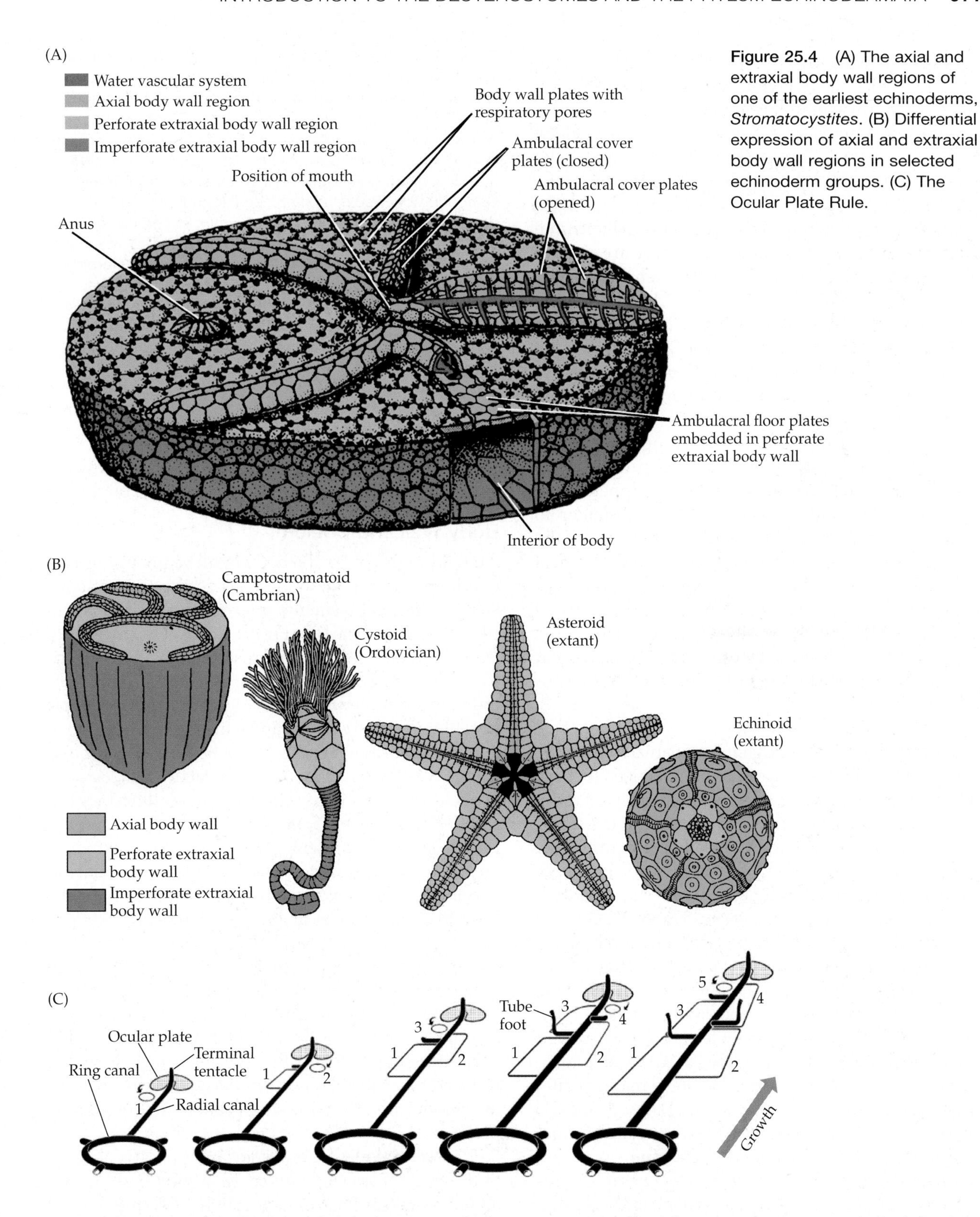

Figure 25.4 (A) The axial and extraxial body wall regions of one of the earliest echinoderms, *Stromatocystites*. (B) Differential expression of axial and extraxial body wall regions in selected echinoderm groups. (C) The Ocular Plate Rule.

wall region is greatly reduced and contributes very little to the adult body wall (Figure 25.4B). In crinoids, on the other hand, much of the main body wall below the arms and the stem is made up of imperforate extraxial body wall. Regulatory genes that help control timing and rate of development of major regions of echinoderms, and their associated coelomic structures also have distinctive expression patterns in these regions. From these patterns, researchers are starting to gain insights into the mechanisms by which the unique pentaradial symmetry of echinoderms evolved.

During development, as the axial region gives rise to the ambulacral plates and their associated tube feet, growth tightly adheres to a set of empirical "rules." Within the axial region, the tip of the growing hydrocoel farthest from the mouth is expressed as a terminal "tentacle" (similar to a tube foot) in starfish. In sea urchins, for which the dynamics of this system were first understood, this small terminal extension protrudes through a distinctive plate known as the **ocular plate**. The term "ocular" derives from an old idea that the pore through which the terminal tentacle pierces the plate in echinoids was a kind of eye. Although light sensitive tissues can be associated with the terminal tentacle itself, it is now known that there are no true eyes among echinoderms. The ontogenetic pattern by which new tube feet are laid down in the axial region is known as the **Ocular Plate Rule (OPR)**, whether or not the ocular plate itself (Figure 25.4C) is actually calcified. In all echinoderms, as the **radial canals** (hydrocoelar extensions) of the developing rudiment expand outward along the rays from the **ring canal** (the hydrocoelar ring that encircles the esophagus), new tube feet appear just proximal to the ocular tentacle (the **terminal tentacle**). A new tube foot appears first on one side of the radial canal, and then on the other, to form a zigzag pattern of tube foot additions along the growing biserial ambulacrum (Figure 25.4C). In most echinoderms, as each new tube foot appears, so does a new ambulacral plate. Therefore, a basic property of the OPR is that the oldest plates and tube feet in the ambulacrum are near the mouth, and the youngest at the tip of the ambulacrum, just proximal to the terminal tentacle.

Ray homologies Although it is not always obvious, each ray of an echinoderm can be homologized to those of any other in terms of reference to particular radii. For example the position of the opening to the water vascular system is marked by one or more **hydropores**. These minute pores, which are often borne on a special plate called the **madreporite**, lead via an internal canal to the water vascular system (Figure 25.3A). The madreporite gives a clue to body orientation because it lies between specific rays. A system of lettering has been developed in which the ambulacrum opposite the madreporite is coded A; the others are then coded B through E in a counterclockwise fashion as viewed from the aboral surface (Figure 25.2C). The madreporite (or hydropore, in forms lacking a madreporite) always lies between ambulacra C and D (i.e., on the CD interradius). Ambulacra C and D comprise the **bivium**, while radii A, B, and E comprise the **trivium**.

An alternative to this lettering system has its origins in the OPR. Called the Lovénian system, after its originator Sven Lovén (1809–1895), it is rooted in the fact that there is a specific pattern by which the first tube foot appears in a particular half-ambulacrum. Using the Lovénian system, the ambulacra are identified by Roman numerals, so that A = V, B = I, C = II, D = III, E = IV. In other words, the trivium comprises II, III, and IV and the bivium I and V (Figure 25.3E). Thus, if certain landmarks can be observed, even individual rays of echinoderms can be homologized across the phylum.

Figure 25.5 Echinoderm body wall and some skeletal elements. (A) Composite section of a sea urchin's body wall. (B) Scanning electron micrographs (SEM) of spines on the sand dollar *Echinarachnius parma* (Echinoidea). Arrows point to ciliary tracts. (C) SEM of skeletal ossicles from the central disks of ophiuroids, shown in top, side, and basal views. (D) SEM of skeletal ossicles from the sea cucumber *Psolus chitinoides* (Holothuroidea), with detail of stereom. (E) Diversity of echinoid pedicellariae surrounding the base of a large spine. (F,G) Prey-catching pedicellariae of the starfish *Stylasterias forreri* (Asteroidea): (F) pedicellariae open and extended; (G) pedicellariae retracted and closed. (H) Details of a generalized starfish pedicellaria. (I) Muscle systems in starfish pedicellariae. (J) A movable echinoid spine (section). Note the position of muscles relative to the body wall layers.

Body Wall and Coelom

An epidermis covers the body of all echinoderms and overlies a mesodermally-derived dermis that contains stereom skeletal elements referred to as **ossicles** and **plates**, as noted earlier (Figure 25.5A–D). Internal to the dermis and ossicles are muscle fibers and the peritoneum of the coelom. The degree of development of the skeleton and muscles varies greatly among groups. In echinoids, the ossicles can be firmly attached to one another to form a rigid test, and body wall musculature is weakly developed. In sea cucumbers, ossicles are seldom contiguous, and usually scattered within the fleshy dermis (Figure 25.5D), in which distinct muscle layers are present. Between these two extremes are cases in which adjacent skeletal plates articulate to varying degrees. In the arms of starfish and brittle stars, the body wall muscles are arranged in bands between the plates, providing various degrees of arm motion. In some groups, skeletal plates are developed to such a degree that they nearly obliterate internal cavities. In brittle stars, for example, each arm is made up of a column of fused, pairs of ambulacral plates called **vertebrae** (Figure 25.9A,B), and the arm coeloms are reduced to small channels. Similarly, the arm coeloms in crinoids are relatively small canals within skeletal plates.

The endoskeleton is calcareous, mostly $CaCO_3$ in the form of calcite, with small amounts of $MgCO_3$ added to increase hardness through "dolomitization." Each ossicle begins as a separate spicule-like element to which additional material is added in precise patterns to make specific elements in the skeleton. Within each ossicle, an anastomosing network of minute struts (the trabeculae) forms, creating labyrinth-like spaces (Figure 25.5D) so that the ossicle is made of porous, spongelike stereom. The spaces contain dermal cells

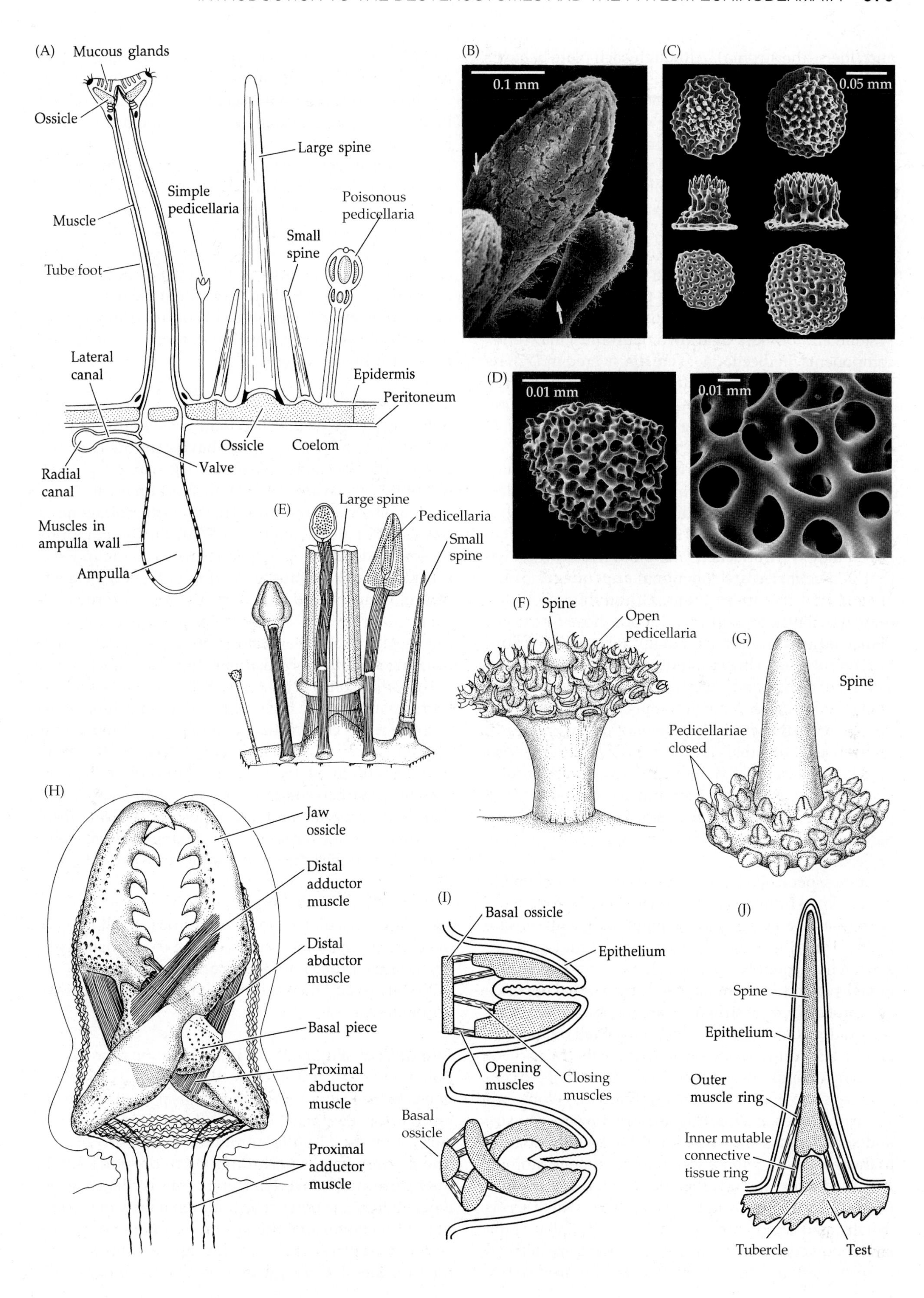

(A)
Mucous glands
Ossicle
Large spine
Simple pedicellaria
Poisonous pedicellaria
Muscle
Small spine
Tube foot
Lateral canal
Epidermis
Peritoneum
Ossicle
Coelom
Radial canal
Valve
Muscles in ampulla wall
Ampulla
(B)
0.1 mm
(C)
0.05 mm
(D)
0.01 mm
0.01 mm
(E)
Large spine
Pedicellaria
Small spine
(F)
Spine
Open pedicellaria
(G)
Spine
Pedicellariae closed
(H)
Jaw ossicle
Distal adductor muscle
Distal abductor muscle
Basal piece
Proximal abductor muscle
Proximal adductor muscle
(I)
Basal ossicle
Epithelium
Opening muscles
Closing muscles
Basal ossicle
(J)
Spine
Epithelium
Outer muscle ring
Inner mutable connective tissue ring
Tubercle
Test

and fibers (the stroma). Although each plate behaves as a single crystal, the stereom is polycrystalline, with aligned microcrystals intermixed with small amounts of organic matrix.

Plates can remain single (simple plates) or, more rarely, fuse to form compound plates. External plates can have bumps and knobs called tubercles, some of which support a variety of articulated, external projections such as spines (Figure 25.5A,E). In asteroids and echinoids, unique pincerlike structures called **pedicellariae** (Figure 25.5E–I) can also be mounted on these tubercles. Movable spines and pedicellariae respond to external stimuli independently of the main nervous system, and possess their own neuromuscular reflex components. Pedicellariae were discovered in 1778 by O. F. Müller, who described them as parasitic polyps and gave them the genus name "*Pedicellaria.*" He recorded three species of these "parasites" (*P. globifera*, *P. triphylla*, and *P. tridens*); variations on these names are still used to describe different types of pedicellariae. Nearly a century later, it was realized that pedicellariae are produced by the echinoderms themselves, but their exact nature remained elusive. Louis Agassiz believed they were attached young but it is now known that pedicellariae are functional appendages of the animal that produces them, although even today there are competing opinions about those functions. Pedicellariae vary in morphology, size, and distribution on the body. In echinoids, most are elevated on stalks, but on asteroids, they usually lie nestled directly on the body surface, sometimes in clusters. Some help keep debris and settling larvae of other invertebrates off the body, and others are used in defense. Some echinoid pedicellariae are venomous to discourage would-be predators, and others are pincers that grasp and hold objects for camouflage and protection. A few starfish even use their pedicellariae to capture prey (Figure 25.5F,G).

Spines, pedicellariae, and the epidermis contain tissues isolated from the coelom and digestive system because they are external to the main body wall (Figure 25.5H–J). Being isolated from the gut and coelom, in which nutrient supply is aided by circulation and hemal systems, nutritional needs of external appendages and epidermal structures are met by nutrients absorbed directly from the water, or by small trapped organic particulates or organisms that are then absorbed.

As in all deuterostomes (except Chordata), the coelomic system of echinoderms typically develops as a tripartite series, originating as paired **proto-**, **meso-**, and **somatocoels** that are thought to be homologous to the tripartite coeloms seen in other deuterostomes. As noted above, in adult echinoderms, these coelomic cavities do not form the usual three body regions usually associated with deuterostomes. Recall that the main body coeloms of adult echinoderms are derived from the left and right somatocoels, and are usually well developed. However, the water vascular system forms from the left mesocoel—called the hydrocoel (as described earlier in the section on The Echinoderm Body Plan). Other coelom derivatives include gonadal linings and certain neural sinuses. The main body cavities, or **perivisceral coeloms**, are lined with ciliated peritoneum, and their coelomic fluid plays major circulatory roles. A variety of coelomocytes, many phagocytic, are present in the body fluid and water vascular system. Hemoglobins occur in the coelomocytes of certain holothuroids and brittle stars. Echinoderm coelomic fluid is isosmotic with ambient seawater, and no mechanisms have evolved to control osmolarity of these fluids—a major reason that there are no terrestrial or freshwater echinoderms.

Mutable Collagenous Tissues

Between many of their skeletal ossicles, and in the body wall itself, echinoderms have a uniquely elaborated system of connective tissue. The material properties of this tissue are mutable on short timescales, and are under neuronal control. This **mutable collagenous tissue** (**MCT**) is normally rigid, but can be temporarily softened, allowing echinoderms to maintain fixed postures with zero muscular effort and energy expenditure. Long viewed as a curiosity, MCT is crucial to understanding how echinoderms maintain relatively low metabolic budgets in situations causing nutritional challenges to other phyla. In holothuroids, much of the body wall is constructed of MCT. In echinoids, muscular rings around the bases of the spines position them within a rocky crevice, then a similarly-positioned ring of MCT (so-called "catch collagen") "freezes" the spine into position to wedge the animal into place. During crawling, starfish soften their body, then use MCT in their body wall to adopt a new posture securing their position over uneven substrata, making them virtually impossible to dislodge by predators or wave action.

Water Vascular System

The water vascular system is so critical to all aspects of echinoderm biology, a discussion of its anatomy is a necessary preface to other considerations, starting with starfish and then comparing their system to other echinoderms.

Asteroidea Figure 25.3A is a schematic representation of the starfish water vascular system. The system opens to the exterior through a skeletal plate, the madreporite (or sieve plate) located off-center on the aboral surface on the CD interambulacrum (Figure 25.2C). The madreporite is perforated by hydropores set in deep furrows. The overlying epidermis is ciliated, even where it lines the pores. The traditional view of madreporite function is that it permits seawater to enter the system to replace water lost through the thin walls of the tube feet. J. C. Ferguson was able to demonstrate this process by using radioactive tracers. However, we still lack a clear understanding of how the madreporite functions.

Internally, the madreporite forms a cuplike lumen called the ampulla (distinct from the tube foot ampulla) that communicates with other coelomic derivatives of the water vascular and hemal systems. The lower end of the ampulla connects with the **stone canal**, so named because of hard skeletal deposits in its wall. A portion of the hemal system called the **axial sinus** is often intimately associated with the stone canal. The stone canal proceeds orally to join with a circular ring canal encircling the esophagus. The ring canal gives rise to radial canals extending into each arm. Interradial blind pouches called **Tiedemann's bodies** and **Polian vesicles** are attached to the ring canal (Figure 25.3A,B). There is uncertainty about the function of these pouches, but the former are suspected to produce certain coelomocytes and the latter to help regulate internal pressure within the water vascular system.

The liquid in the water vascular system consists of ambient seawater to which the animal adds specific components such as coelomocytes, organic compounds such as proteins, and a relatively high concentration of potassium ions. The fluid is moved through the system largely by cilia that line the internal epithelia of the canals, some of which, especially the stone and ring canals, contain partition-like inner extensions that could help direct the flow. The stone canal, at least in some starfish, has been described as a "ciliary pump" that draws fluid into the water vascular system from both the madreporite and the axial sinus of the hemal system.

In each arm, the radial canal gives rise to numerous **lateral canals**, each of which terminates in a tube foot–ampulla system. All tube feet form according to the Ocular Plate Rule (Figure 25.4C), with oldest ones near the mouth, and youngest at the arm tips. The tube foot is a contractile, muscular tube typically ending in a sucker, and extends externally from a bulb-like, internal ampulla (Figure 25.3B). Longitudinal and circular muscles attached to a collagenous sheath surrounding a coelomic lumen, and the tip and shaft are well supplied with nerves. Tube feet function as primary tactile, locomotory, attachment, and food manipulating organs. Each radial canal ends in an unsuckered, tentacle-like, terminal tube foot that is most likely photo receptive.

The system of tube feet and ampullae comprise a sophisticated hydrostatic system. Two sets of muscles operate around the fluid in the lumens of tube feet and ampullae. A valve in the lateral canal can effectively isolate the tube foot and ampulla from the rest of the canal. With the valve closed, muscles in the ampullar wall contract, forcing fluid into the lumen of the extending tube foot, in which the longitudinal muscles are relaxed. The sucker is then pressed against the substratum and held there by adhesive secretions of the epidermis and by the contraction of tip muscles that raise the center of the sucker and create a suction cuplike vacuum. When the longitudinal muscles of the tube foot contract, it shortens, decreasing the lumen volume and forcing fluid back into the now relaxed ampulla. When the tip muscles relax, the suction is broken. Chemicals secreted by cells in the sucker reverse the adhesive action of substances secreted by the tube foot. Tube feet bend by differential contraction of the longitudinal muscles.

Ophiuroidea The water vascular system of brittle stars is similar to that of asteroids. However, the madreporite is on the oral surface of the central disk (Figure 25.3C), and the internal plumbing is modified accordingly. In some ophiuroids the madreporite is reduced to just two hydropores.

The ring canal bears Polian vesicles, but apparently lacks Tiedemann's bodies. The ring canal gives off the usual five radial canals and also bears a wreath of buccal tube feet around the mouth. In basket stars the arms and the radial canals are branched. The suckerless tube feet are highly flexible, being fingerlike structures that can secrete copious amounts of sticky mucus. They function in feeding, digging, locomotion, and sensation.

Crinoidea The water vascular system of crinoids operates entirely on coelomic fluid. There is no centralized madreporite in modern forms. Multiple stone canals arise from the ring canal and open to coelomic channels. There can be hundreds of such canals. The main perivisceral coeloms bear ciliated funnels to the exterior through which water enters the body cavities, perhaps to regulate hydraulic pressure in the water vascular system.

From the ring canal, radial canals bifurcate into all the branches of each arm, which can have as many as 200 termini. Crinoid arms also bear side branches called **pinnules** (Figure 25.2B), into which extensions of the radial canals also reach. Topologically with respect to the coeloms, radial canals are positioned just as they are in starfish. However, ambulacral ossicles are lacking, and the canals are embedded in a soft tissue shelf, reducing weight and metabolic expense otherwise needed to produce a skeleton in the long arms. The arms are well supported by brachial ossicles, relatively large, articulated stereom elements that form the aboral surface of all arms and pinnules (Figure 25.7A,B). Suckerless tube feet occur along pinnules, often in clusters of three (Figure 25.3D), and each cluster is served by a branch from the radial canal. Equipped with adhesive papillae, the tube feet are highly mobile, functioning primarily as feeding and sensory organs.

Echinoidea Echinoids bear a special set of skeletal plates, called the **apical system**, around the aboral pole, one of these plates being a madreporite (Figure 25.3E,F) that leads to an ampulla–axial gland complex. A stone canal extends to a ring canal atop the jaw apparatus (Aristotle's lantern, see below). The ring canal gives rise to a radial canal beneath each ambulacrum, as well as to Tiedemann's bodies and Polian vesicles. Each radial canal gives off lateral canals leading to tube feet and ends with a terminal tentacle protruding through an ocular plate in the apical system. The tube feet of echinoids may be suckered or unsuckered. They serve a variety of functions including attachment, locomotion, feeding, and gas exchange.

To understand the water vascular system of sea urchins, picture a starfish in which the aboral region (the perforate extraxial region) has been reduced to a tiny button (the apical system). The result is a "pulling" of the tips of the radial canals together to converge on the apical system, leaving the rows of tube feet extending around the sides of the body like five lines of longitude on a globe (Figures 25.3E and 25.4B). However, unlike any other echinoderm, stereom plates of the test are laid down external to the water vascular components, rendering the radial canals truly internal. Pores through the plates forming the ambulacra allow communication between the tube feet and ampullae. This is very different from ophiuroids, in which internalization of the radial canal is produced by overgrowth of the vertebral ossicles, and from the holothuroids in which the longitudinal canals are not homologous to radial canals of other echinoderms (see below).

Holothuroidea The story of the holothuroid water vascular system is one of reduction, or truncation of growth through pronounced paedomorphosis. In a sense, holothuroids can be viewed as "giant larvae"—metamorphosis is suppressed, producing an adult animal in whose construction the rudiment plays only a small part. Holothuroids probably arose through paedomorphosis.

In the Holothuroidea, the stone canal usually extends from beneath the pharynx in an interradial position, giving rise to a madreporite opening to the coelom. A ring canal encircles the esophagus and bears from 1 to 50 Polian vesicles. Five radial canals arise from the ring canal to give off extensions to the **oral tentacles**, which are actually highly modified terminal tentacles and the first few tube feet. These represent the only elaborations of that part of the radial canal homologous to the radial canal in other echinoderms. In the least derived holothuroids, such as the Apodida, these tentacles, restricted to the circumoral region, are all that exist of the water vascular system, there being no canals or tube feet along the vermiform body. Early in postlarval growth of more derived forms, thin interradial extensions from the ring canal form to produce interradial, **longitudinal canals** that are later twisted in a kind of "torsion," pulling them into alignment with the radial canals leading to the feeding tentacles. The longitudinal canals lie along the inside of the larval body wall, inducing formation of tube footlike structures without respect for the Ocular Plate Rule. These tube footlike extensions can be scattered over the body wall, restricted to ambulacrum-like lines, or missing in certain radii entirely, usually along the "dorsal" or upper surface. The tube feet around the mouth serve in food gathering, whereas the tube footlike appendages along the body assist in locomotion, attachment, and touch.

Support and Locomotion

Except in holothuroids, structural support in echinoderms is maintained primarily by skeletal elements and the mutable collagenous tissue. Tube feet and gill-like extensions of the body wall (such as papulae) are supported mostly by hydrostatic pressure. Most sea cucumbers lack skeletal plates, which are reduced to minute, separated ossicles, but body wall muscles form thick sheets, adding integrity to the body and working antagonistically on coelomic fluids to function in a hydrostatic skeleton. The mutable collagenous tissues of echinoderms also contribute to overall support and aspects of locomotion (see earlier section on MCT).

Apart from sessile sea lilies (e.g., *Ptilocrinus*), most extant crinoids are capable of crawling and even swimming. Aboral cirri of feather stars (and along the stem of isocrinids) are used for temporary attachment (Figure 25.6A). During isocrinid crawling the animal lies prone and with the stem trailing, the arms pull the main body region, or **calyx** (the central cup at which the arms converge), along the bottom (Figure 25.6B) to reposition the animal. Isocrinids also "run away" from deep-sea cidaroid echinoids while casting off a piece of the stem to preoccupy the echinoid predator. Swimming among the Crinoidea is accomplished by up-and-down sweeps of the arms, which are divided into functional sets that can move alternately.

Asteroids exemplify locomotion using suckered tube feet. Starfish arms are held more or less stationary relative to the central disk, even in species with a flexible skeletal framework (e.g., *Pycnopodia*). Movement is accomplished by upwards of thousands of tube feet on the oral surface. The starfish appears to glide smoothly because of the large number of tube feet, though at any given instant they are in different phases of the power and recovery strokes (Figure 25.6C). Although control of tube foot action is not fully understood (even isolated arms crawl about normally), tube feet are coordinated to produce movement in a particular direction. There are no metachronal waves of motion as seen in many other multi-legged creatures. Most starfish move slowly, but a few (e.g., *Pycnopodia*) are relative speedsters, and otherwise sedentary asteroids become rapid "runners" upon encountering a potential predator. Species that cannot "run" have other defense mechanisms. The slow-moving Pacific cushion star, *Pteraster*

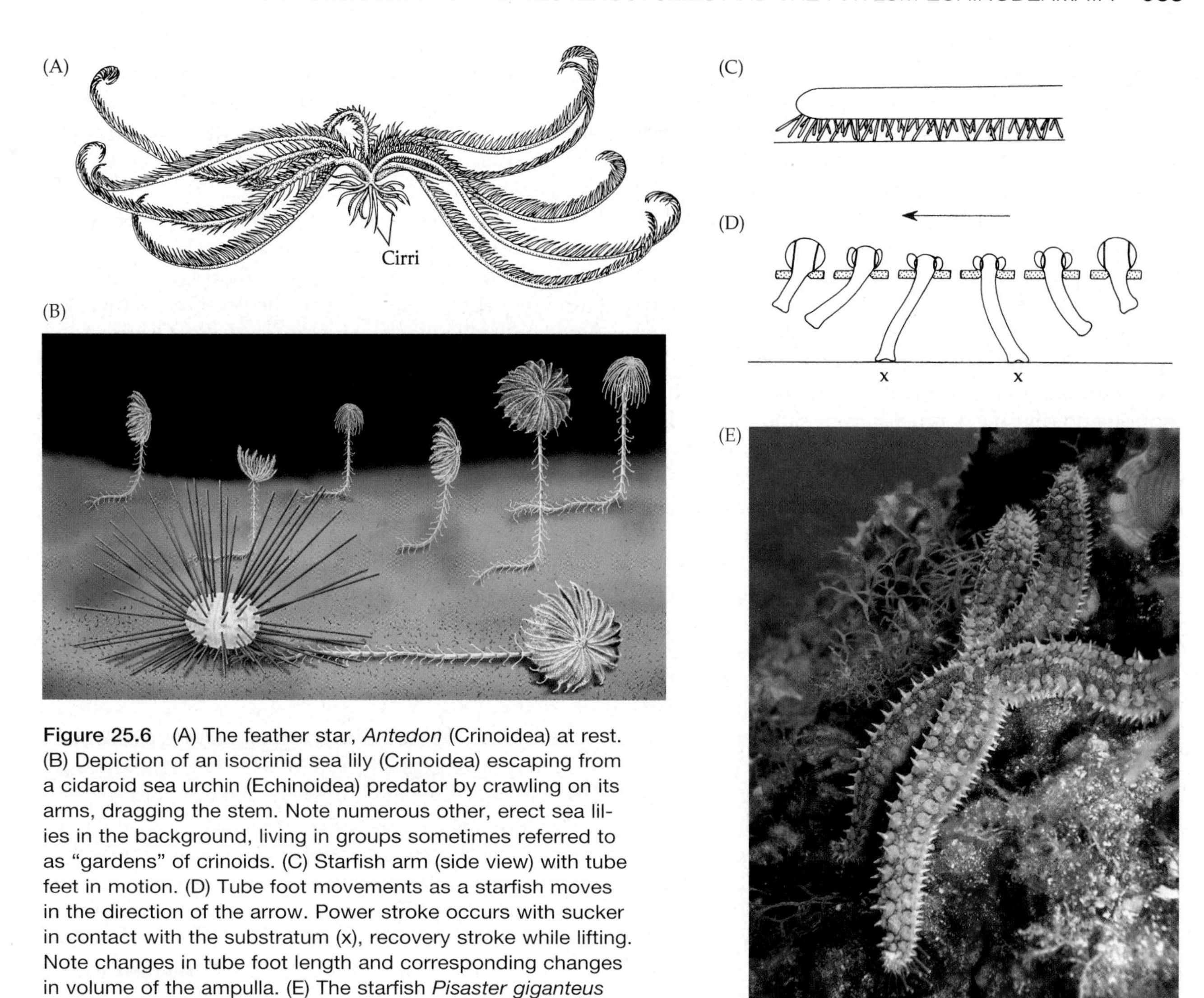

Figure 25.6 (A) The feather star, *Antedon* (Crinoidea) at rest. (B) Depiction of an isocrinid sea lily (Crinoidea) escaping from a cidaroid sea urchin (Echinoidea) predator by crawling on its arms, dragging the stem. Note numerous other, erect sea lilies in the background, living in groups sometimes referred to as "gardens" of crinoids. (C) Starfish arm (side view) with tube feet in motion. (D) Tube foot movements as a starfish moves in the direction of the arrow. Power stroke occurs with sucker in contact with the substratum (x), recovery stroke while lifting. Note changes in tube foot length and corresponding changes in volume of the ampulla. (E) The starfish *Pisaster giganteus* crawling.

tesselatus (Figure 25.1D), secretes copious amounts of mucus that discourages predators such as *Solaster* and *Pycnopodia*. Starfish on sandy bottoms (e.g., *Astropecten*) can burrow rapidly by using tube feet to move sand from below the animal to the top.

Following the action of a single tube foot during movement reveals the sources of the locomotory forces (Figure 25.6D). At the end of a recovery stroke, the tube foot extends in the direction of movement and attaches to the substratum. The sucker remains attached during the power stroke as the longitudinal muscles in the wall of the tube foot shaft contract, shortening the tube foot and pulling the body forward. At the end of the power stroke, the tube foot detaches and swings forward again. The great number of tube feet and the general flexibility of the body allow starfish to move easily over even the most irregular surfaces (Figure 25.6E).

Ophiuroids primarily use their flexible, articulated arms for crawling or clinging (Figure 25.2D). In most forms, the skeletal arrangement within the arms allows for rapid lateral, snakelike movements on a plane parallel to the disk (hence the Greek prefix *ophio*, "snake"). Only in taxa such as basket stars do the arms have full range of movement in all planes. In most species, the articulated arms are fragile and break easily when seized (hence "brittle stars"). The tube feet lack suckers and ampullae, but a well-developed lattice of muscles in their walls allows protraction, retraction, and swinging motions. Combined action of arms and tube feet allows ophiuroids to burrow into sediments, or move swiftly when escaping from predators or chasing prey.

Sea urchins move by the use of tube feet and movable spines. Strong, suckered podia are capable of a wide range of extension and motion, while spines provide stiltlike support and movement. Some urchins can excavate stone by scraping with their spines and chewing with their teeth. Some populations of *Strongylocentrotus purpuratus*, a common northeast Pacific urchin, form pockets in intertidal rocks, and *Echinometra vanbrunti* also does this on tropical eastern Pacific shores. Individuals are sometimes trapped in their self-made homes, but are protected from predation and wave action. The common Indo-Pacific reef urchin, *Echinometra mathaei*, excavates elongate channels in rock largely by the action of teeth in the jaw apparatus; in these species the test is slightly elongated to allow movement of the urchin along the channel as it harvests a "garden" of algae growing in the channel. Some irregular urchins burrow below the sediment surface. Sedentary species (e.g., *Echinocardium*) maintain an open chimney to the overlying water (Figure 25.11G). Soft-sediment burrowers often have specialized spatulate spines along the body that aid in locomotion and digging. Clypeasteroid sand dollars live on sandy bottoms, and all species lie flat with the oral surface toward the bottom except *Dendraster excentricus*, an abundant North American west coast species that lives embedded upright in the substratum (Figure 25.11F). Clypeasteroid burrowing and crawling are accomplished by the action of spines that move in metachronal waves. Sand dollar morphology is strongly tied to hydrodynamic forces. Deep marginal notches and holes (**lunules**) in the tests of some sand dollars (Figure 25.2F) are not for feeding, but act as lift spoilers to help the animals maintain stability in strong currents characterizing their coastal environments. Lift forces are also mitigated by flow along **pressure drainage channels** from the center of the body to the lunules.

Holothuroids inhabit various substrata, coral crevices, spaces under rocks, or soft sediments. They crawl or burrow in sand or mud using tube feet or peristaltic actions of body wall muscles. In epibenthic species that lodge themselves in rock crevices, the tube feet can be used for anchorage and to hold bits of shell and stone for camouflage and protection. In deep-sea forms such as *Scotoplanes* (Figure 25.1N), some tube feet are elongated for walking like small mammals, hence the nickname "sea pig." In psolids, the lower surface is modified into a footlike sole upon which the animal creeps. Among the apodids are the bizarre synaptids, with unique anchor ossicles in densities up to 1,500 per cm^2 of body wall. In lieu of tube feet, these provide a grip akin to Velcro® by protruding and retracting into the skin in peristaltic waves along the body wall. Some live completely buried, whereas others form U-shaped burrows. A few holothuroids are truly pelagic and capable of swimming (Figure 25.1P,Q).

Feeding and Digestion

Crinoidea Sea lilies and feather stars sit with their oral sides up and suspension feed, usually clinging to elevated areas exposed to currents (see chapter opener photo of the yellow feather star *Comanthina*). The arms and pinnules are held outstretched into ambient water flow, presenting a large food-trapping surface. Some forms emerge from concealment to feed only at night. Deep-water species hold their arms upward and outward, forming a funnel with which they capture detrital rain, and stalked forms raise a crown of arms like an umbrella into the current; the stalk is bent with the current flowing into the underside of the umbrella and particle capture occurs downstream, on the oral surfaces of the arms.

Food particles, including plankton and organic particulates, contact tube feet along the ambulacral grooves in arms and pinnules (Figure 25.7B). These tube feet flick the food into the grooves, which are lined with cilia that beat toward the calyx, driving food to the mouth where it is ingested. This use of the tube feet and ambulacral grooves for feeding likely represents an original function of the early water vascular system in Echinodermata.

The mouth opens to a short esophagus leading to a long intestine (Figure 25.7A,C) that loops inside the calyx, then straightens to a short rectum terminating at the anus, which is borne on an anal cone near the arm bases. The intestine can bear diverticula, some of which are branched. The histology of the crinoid gut has been described, but little is known about the digestive physiology of these animals.

Asteroidea Most starfish are opportunistic deposit feeders, scavengers, or predators, feeding on nearly any organic matter or small prey. Many species are generalists in terms of food preference and play important roles as high-level benthic predators in intertidal and subtidal communities. Others, such as *Solaster stimpsoni* from the northeastern Pacific, are specialists feeding on holothuroids, while a related species (*S. dawsoni*) preys almost exclusively on *S. stimpsoni*! The tropical crown-of-thorns starfish, *Acanthaster planci*, feeds on coral polyps, receiving great notoriety because of its implied destruction of Indo–West Pacific coral reefs. There remains disagreement concerning causes for increases in *Acanthaster* populations, and even the level of harm they actually perpetrate on the reefs, but most agree that human interference in predator–prey balance in reef communities is more to blame than *Acanthaster* itself. A major predator of *Acanthaster* is the giant triton snail *Charonia*, which is collected in high numbers for its handsome shell. In undisturbed environments, *Acanthaster* likely plays a role in maintaining coral diversity by acting as forest fires often do, opening new habitats in climax ecosystems for the establishment of other species.

Except for a few suspension feeders, starfish use an eversible portion of the stomach to obtain food. Some forms, including *Culcita* (a cushion star, Figure 25.1E), *Acanthaster* (Figure 25.1F), and *Patiria* (a bat star), spread their stomach over the surface of a food source, secrete digestive enzymes, and suck in the partially digested soup. For *Culcita* and *Patiria*, food can include encrusting sponges, algal mats, or accumulated organic detritus over which the starfish crawl. *Oreaster*, also a cushion star, feeds similarly, but switches to a predatory or scavenging mode when appropriate food

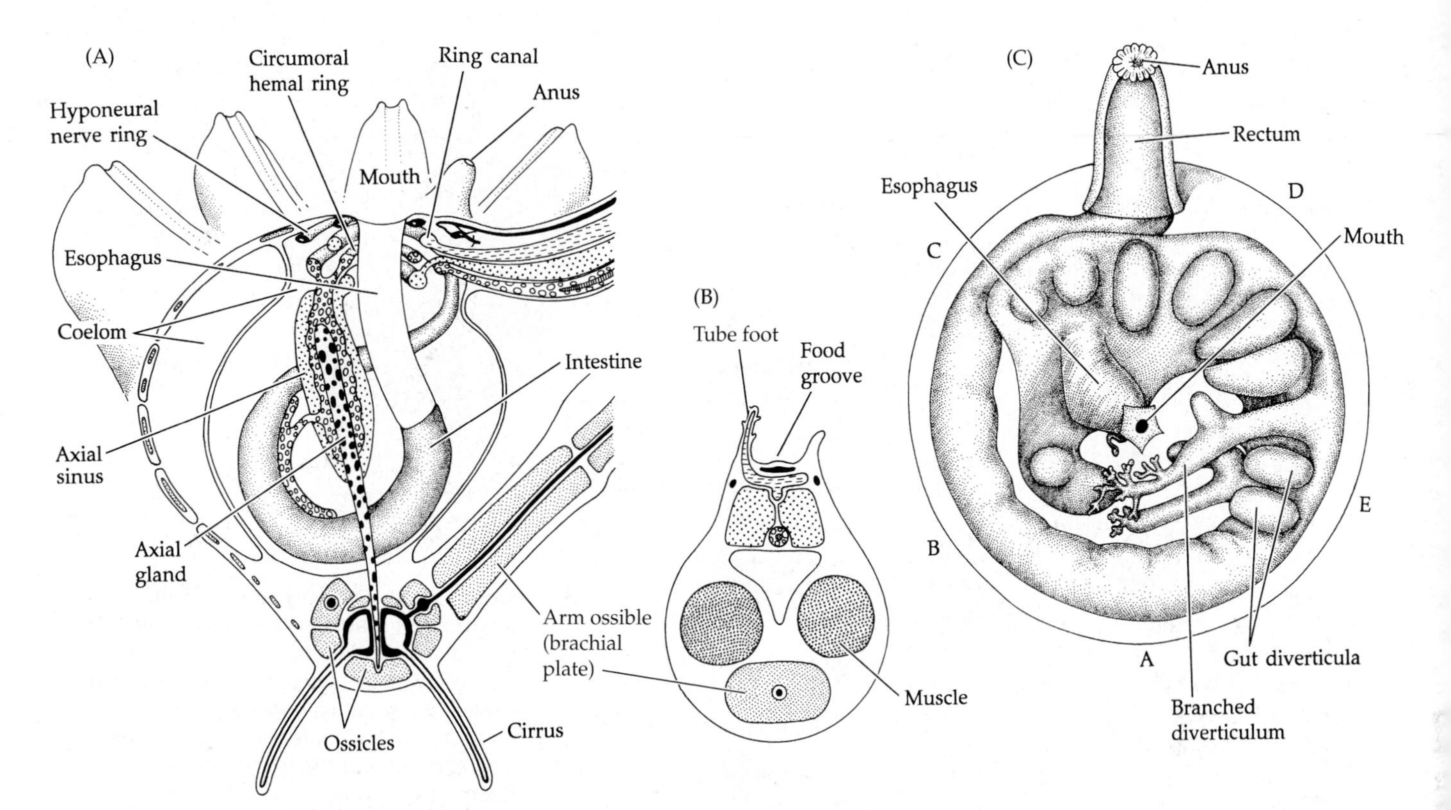

Figure 25.7 Internal anatomy of sea lilies and feather stars (Crinoidea). (A) Central disk and base of one arm (vertical section). (B) Arm with open ambulacral groove (cross section). (C) Oral surface of the feather star *Antedon* (cutaway view). Positions of radii indicated by letters.

sources are encountered. Sedentary or sessile prey, such as gastropods, bivalves, and barnacles, are eaten by asteroid predators, especially members of the Forcipulatida. The voracious Pacific ochre star *Pisaster ochraceus* (Figure 25.8D) hunches over its prey with the oral area pressed against the potential victim, holding itself in position with its tube feet. The stomach is very thin and flexible and can be slid between even the tightly clamped valves of mussels and clams to liquefy the prey's body inside its own shell so that nutrients are drawn in with the retracting stomach.

Some starfish are suspension feeders, consuming plankton and organic detritus. *Henricia*, *Porania*, and a few others are typically full-time suspension feeders, and some predators, such as *Astropecten*, are facultative suspension feeders to supplement their usual diet. Usually, particulate food material that contacts the body surface is trapped by mucus and moved by ciliary action to the ambulacral grooves and ultimately to the mouth. *Novodinia* extends its arms upward into water currents. The dozen or so arms form a large feeding surface used to capture planktonic crustaceans; the prey are grasped by clusters of pincerlike pedicellariae. A few forms, including *Stylasterias, Pycnopodia,* and *Labidiaster*, possess wreathlike circlets of pedicellariae (Figure 25.5F,G) used to capture a variety of animals, even fishes.

The starfish gut typically extends from the mouth at the junction of the arms to the aboral anus (Figure 25.8A). The mouth is surrounded by a leathery **peristomial membrane**, which is flexible to allow eversion of the stomach, and contains a sphincter muscle that closes the mouth orifice. Internal to the mouth is a very short esophagus leading to the **cardiac stomach**—the portion that is everted during feeding. Radially arranged retractor muscles retract the stomach. Aboral to the cardiac stomach is a flat **pyloric stomach**, from which arises a pair of pyloric ducts extending into each arm and leading to paired digestive glands (**pyloric ceca**) (Figure 25.8A,C). A short intestine, often bearing outpocketings (rectal glands or rectal sacs) leads from the pyloric stomach to the anus. The pyloric ceca and cardiac stomach are the main producers of enzymes (mostly proteases) carried by ciliary action through the everted stomach and released onto the food material. Digestion is completed internally, but extracellularly, after ingestion of the liquefied food. Digested products are moved through the pyloric ducts to the pyloric ceca, where they are absorbed and stored. The intestine does not seem to function in actual digestion, but the rectal sacs are known to absorb nutrients.

Many starfish harbor commensals that derive nutrition from their host's leftovers. A polynoid scale worm, *Arctonoe vittata*, is an obligate symbiont living with several species of asteroids, including the Pacific leather star, *Dermasterias*. The worm spends most of its life cruising and feeding in the host's ambulacral grooves. Not only is the polychaete chemically attracted to its host, but studies indicate that *Dermasterias* is also attracted to *Arctonoe*, suggesting that the starfish might

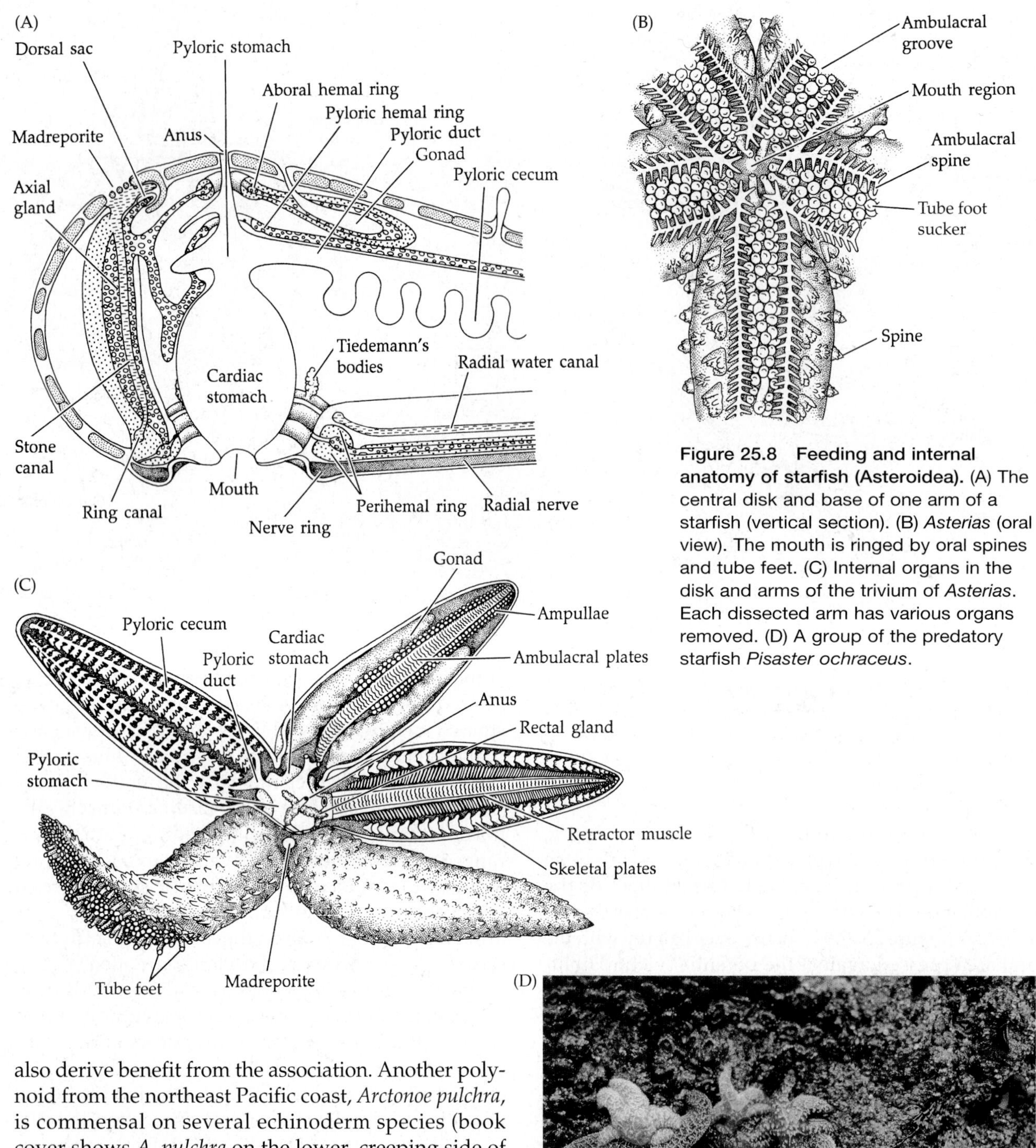

Figure 25.8 Feeding and internal anatomy of starfish (Asteroidea). (A) The central disk and base of one arm of a starfish (vertical section). (B) *Asterias* (oral view). The mouth is ringed by oral spines and tube feet. (C) Internal organs in the disk and arms of the trivium of *Asterias*. Each dissected arm has various organs removed. (D) A group of the predatory starfish *Pisaster ochraceus*.

also derive benefit from the association. Another polynoid from the northeast Pacific coast, *Arctonoe pulchra*, is commensal on several echinoderm species (book cover shows *A. pulchra* on the lower, creeping side of the sea cucumber *Parastichopus parvimensis*). Forms such as *Xyloplax* and caymanostellid "wood stars" (Figure 25.1G) that live only on sunken wood seem to be xylophagous, harboring cellulose-digesting gut bacteria that help the starfish derive nutrients from wood that starfish are otherwise incapable of digesting.

Ophiuroidea Brittle stars exhibit predation, deposit feeding, scavenging, and suspension feeding, and some species are capable of more than one feeding method. Basket stars (Figure 25.9F) are actually predators that utilize suspension-feeding behaviors to capture relatively large swimming prey (up to about 3 cm long).

Selective deposit feeding is accomplished in ophiuroids by tube feet and sometimes by spines along the arms. The arm spines and tube feet produce mucus to which organic material adheres. The tube feet roll the mucus and food into a bolus. Near the base of each

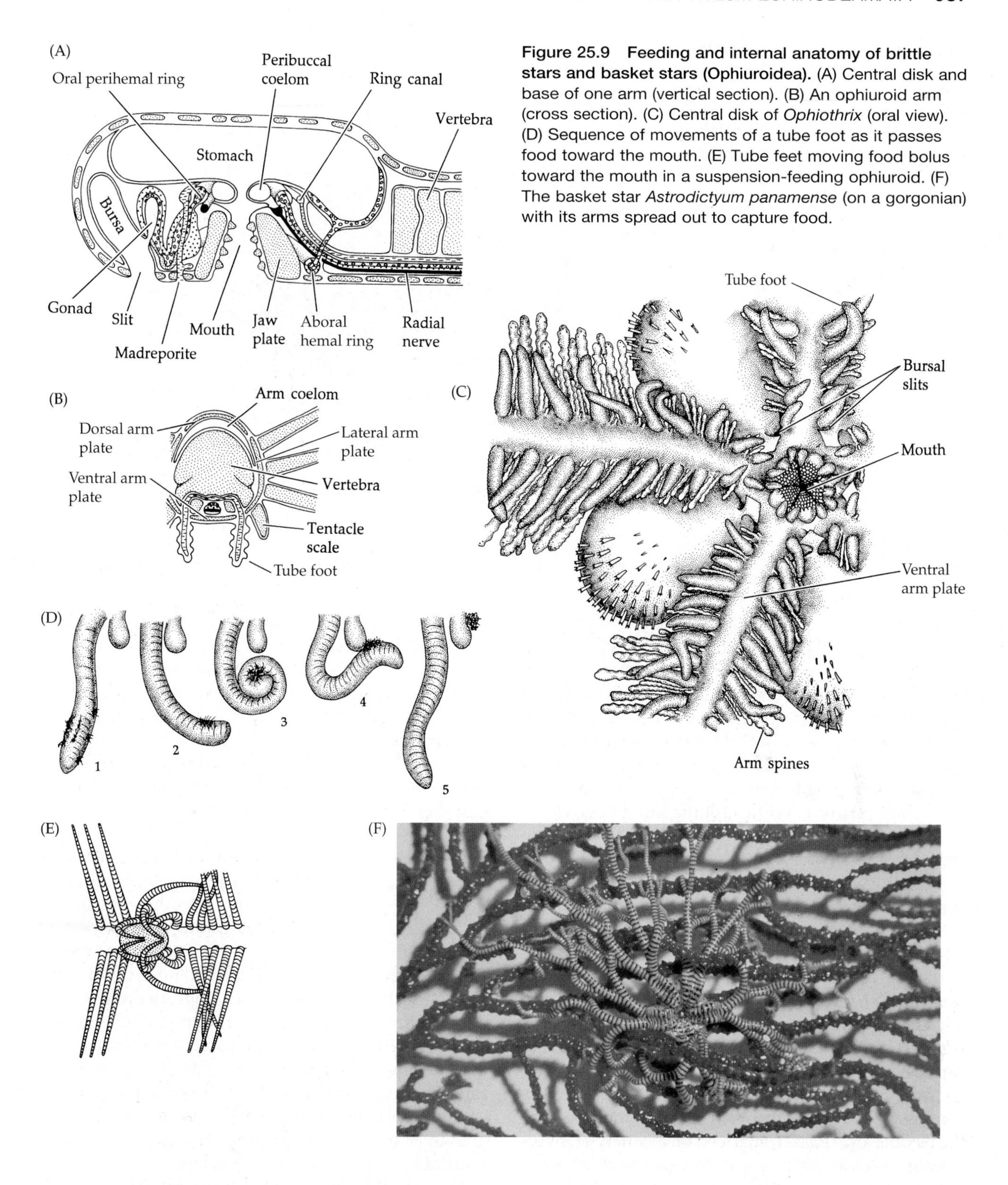

Figure 25.9 Feeding and internal anatomy of brittle stars and basket stars (Ophiuroidea). (A) Central disk and base of one arm (vertical section). (B) An ophiuroid arm (cross section). (C) Central disk of *Ophiothrix* (oral view). (D) Sequence of movements of a tube foot as it passes food toward the mouth. (E) Tube feet moving food bolus toward the mouth in a suspension-feeding ophiuroid. (F) The basket star *Astrodictyum panamense* (on a gorgonian) with its arms spread out to capture food.

tube foot is a flaplike projection called a **tentacle scale** (Figure 25.9B,D). The bolus is transferred from a tube foot onto its adjacent scale, picked up by the next tube foot, and so on, transporting the food along the arm to the mouth. Suspension feeding usually involves a similar method of transport once food is trapped. Food capture is sometimes aided by secretion of mucous threads among the arm spines and holding the arms into currents to trap plankton and organic detritus. Brittle stars using this technique typically have very long arm spines (e.g., *Ophiocoma*, *Ophiothrix*). *Astrotoma* (a euryalid with unbranched arms) extends its arms (up to 70 cm long) into the overlying water to capture planktonic copepods. Other brittle stars suspension feed by using extended tube feet to form a trap, passing clumps of food to the mouth (Figure 25.9E). Several

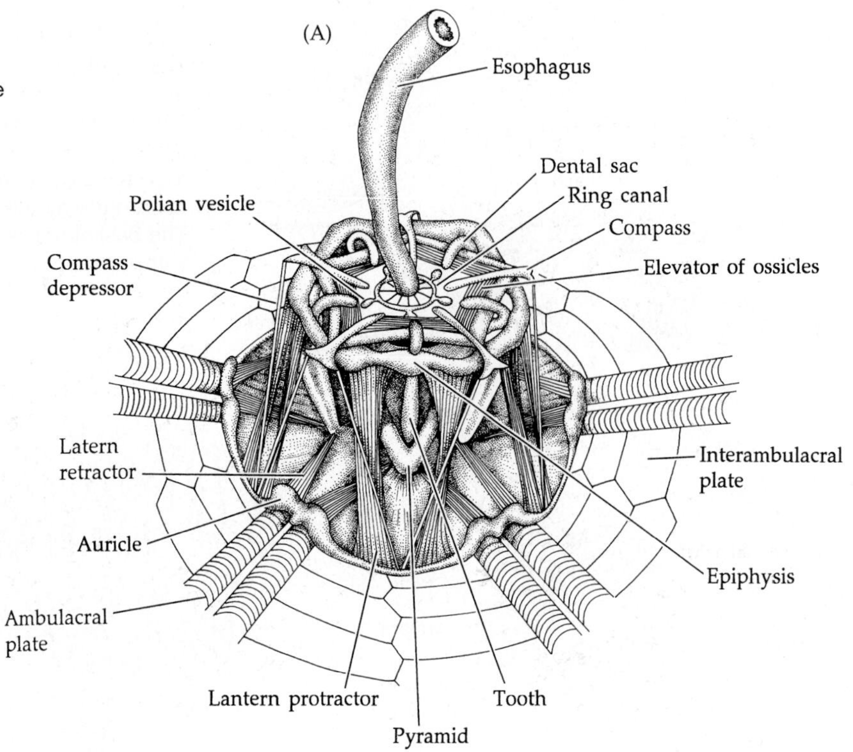

Figure 25.10 The Aristotle's lantern in sea urchins (Echinoidea). (A) The lantern as seen from inside the test. (B) The lantern of *Paracentrotus* (vertical section). (C) The lantern of *Cidaris* (aboral view), compasses removed to expose the rotules.

species burrow to form semipermanent, mucus-lined burrows. The arms extend to the surface and help maintain ventilation currents within the burrows. Such species are able to extract food from the sediment, the substratum surface, and the overlying water (Woodley 1975). The commensal brittle star *Ophiothrix lineata* lives in the atrium of the large sponge *Callyspongia vaginalis*, emerging to feed on detritus adhering to its host's outer surface. While keeping the sponge clean, the ophiuroid is supplied with food and afforded protection from predators.

Predatory suspension feeding by basket stars occurs mostly at night. At dusk the animals emerge from their hiding places and to hold their branched arms fanlike into the prevailing current, in a manner similar to that of feeding crinoids. *Astrophyton* changes position with the tidal direction, always orienting its arms into the current; it stops feeding at slack tide. When a small animal contacts an arm, the appendage curls to capture it by impaling it on microscopic hooks along the arm. Ingestion is often postponed until darkness has passed; the prey is then transferred to the mouth by the flexible arm. Basket stars feed on a variety of invertebrates, such as swimming crustaceans and demersal polychaetes.

Some brittle stars are active predators, capturing benthic organisms by curling an arm into a loop around the prey, then pulling it to the mouth. Species feeding this way usually have short arm spines that lie flat against the arm (e.g., *Ophioderma*). *Ophiura sarsii*, an ophiuroid from northern temperate seas, acts in groups to subdue and tear apart active swimming prey, such as fish and crustaceans.

In ophiuroids, the intestine and anus have been lost, and the digestive system is confined entirely to the central disk (Figure 25.9A). The mouth, surrounded by a jawlike set of modified ambulacral ossicles, leads to a short esophagus and large stomach in which digestion and absorption occurs. Digested waste material is ejected back through the mouth. The stomach fills most of the interior of the disk, reducing the coelom to a thin chamber.

Echinoidea Echinoids engage in various kinds of herbivory, suspension feeding, detritivory, and rarely, predation. In typical globose urchins, feeding is facilitated by an intricate masticatory apparatus, the **Aristotle's lantern**, that lies just inside the mouth and bears five calcareous protractible teeth (Figures 25.10 and 25.11A–D). It is an impressive example of evolutionary engineering: a complex of hard stereom elements and muscles that control protraction, retraction, and grasping movements of the teeth. Usually, the entire apparatus can be rocked so that the teeth protrude at different angles. There is great variation in lantern structure among echinoids, but the following applies to most conditions in which it is well developed.

The main structural part of the lantern (Figure 25.10) consists of five pairs of vertically oriented, calcareous, trapezoidal elements connected to form five triangular pyramids, each of which supports a single tooth in a groove called the tooth slide. The interambulacrally positioned pyramids are attached to one another by muscles that pull the pyramids (and the teeth) together to close the jaws. Along the aboral edge of each pyramid is a thickened bar, the epiphysis. Tooth honing occurs

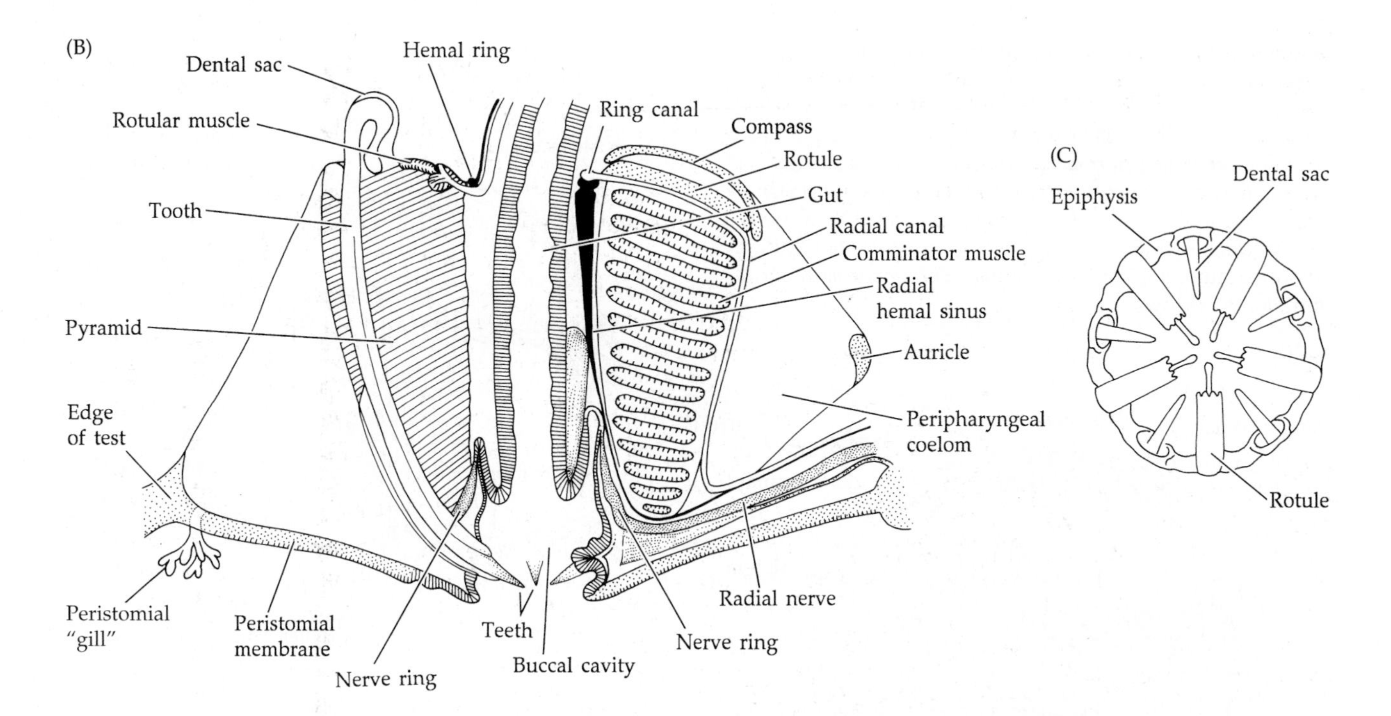

by the manner in which small platelets that make up the tooth are worn off, like a self-sharpening chisel. The tooth grows at the proximal end through the constant addition of new platelets where the tooth emerges from the top of the pyramid. This region is covered by a soft dental sac of coelomic origin that encloses the unhardened proximal end of each tooth. With normal wear, the teeth usually grow about 1 mm each week. Atop the main structure of the lantern are five, small, Y-shaped skeletal ossicles known as **compasses**, one along each ambulacral radius. The compasses and associated muscles appear to regulate hydrostatic pressure within the peripharyngeal coelom.

Protractor muscles originate from an interambulacral "lantern girdle" around the mouth and insert small skeletal elements (**rotules**) on the **epiphyses** that are attached to the proximal (upper) end of each pyramid. Contraction of these protractors pulls the entire lantern orally through the mouth, spreading the teeth apart (Figure 25.10B,C). Retractor muscles originate on thick ambulacral girdle extensions (auricles), inserting on the distal end of the lantern. Additional muscles associated with the pyramids and the rotules produce a variety of tooth movements.

Most urchins with well-developed lanterns use their teeth to scrape algae from the substratum and to bite off chunks of macroalgae. Some species feed on animal matter by similar actions. Urchins that excavate burrows in hard substrata feed on the algal film that develops on the burrow wall. Others ingest suspended particles or drift algae that enter the chamber, or organic material trapped by spines, tube feet, and pedicellariae. Some irregular urchins (e.g., cassiduloids and echinoneoids) lose the lantern before adulthood, and spatangoid heart urchins never develop one. Spatangoids burrow and swallow the sediment to digest the contained organic material (Figure 25.11G). Tube feet can be used to sort food material from the mud or sand and pass it to the mouth. The superimposed bilateral symmetry derived from evolutionary movement of the anus from the apical system to a functional "posterior" in irregular echinoids is an adaptation to sediment-swallowing habits, much like an earthworm.

Most sand dollars are "podial particle pickers," possessing a highly modified lantern with nonprotractable teeth that act as a crushing mill. As they move over the sediment, sand dollars select organic-rich material using thousands of miniaturized tube feet distributed in fields over the oral surface, add mucus to the particles, and pass them into food grooves. Specialized tube feet act as a "bucket brigade" to shunt mucus-coated particle trains to the mouth (Figure 25.11E). Related to the sand dollars, sea biscuits feed on larger particles, including bits of seagrass.

Dendraster excentricus buries the anterior portion of its body in the sand so that the posterior part extends upwards above the sediment (Figure 25.1K and 25.11F), trapping diatoms and other particulate food in the water with its tube feet, and ingesting it as described above. Tiny crustaceans can be captured by the pedicellariae. Members of most populations of *Dendraster* position themselves in groups oriented edgewise into the current, exploiting hydrodynamic properties of ambient flow to enhance each other's food collection process. Some sand dollars have high-density sand grains (especially those containing iron oxides) stored in a diverticulum of the gut as ballast to help stabilize their position on the sea bottom. Micro-echinoids (e.g., *Echinocyamus*) nestle among coarse sand grains, which are brought to the mouth by tube feet, then rotated by

Figure 25.11 Feeding and internal anatomy of sea urchins (Echinoidea). (A) A typical globose sea urchin (vertical section). (B) Internal anatomy of *Arbacia*. (C) *Arbacia* (oral view). (D) Digestive system of the sand dollar *Echinarachnius parma* (aboral view, test dissected away). (E) Food groove on oral surface of a sand dollar (cutaway view), tube feet moving food toward the mouth. (F) *Dendraster excentricus* in feeding position, half-buried in substratum. (G) The heart urchin, *Echinocardium cordatum*, in its burrow.

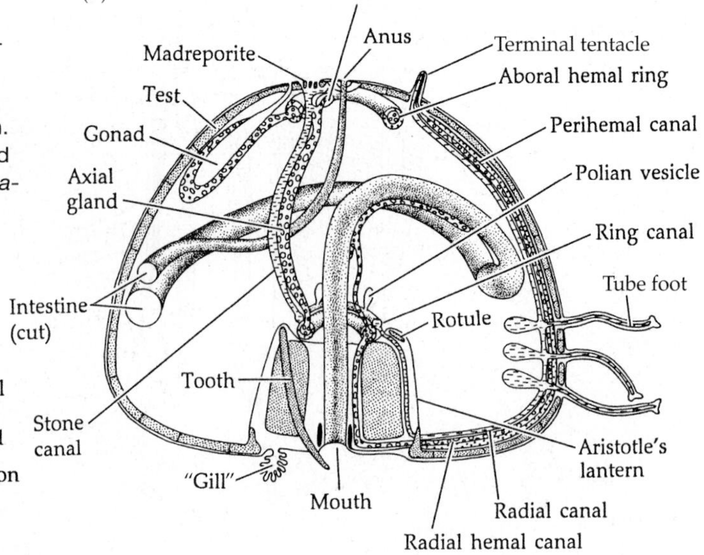

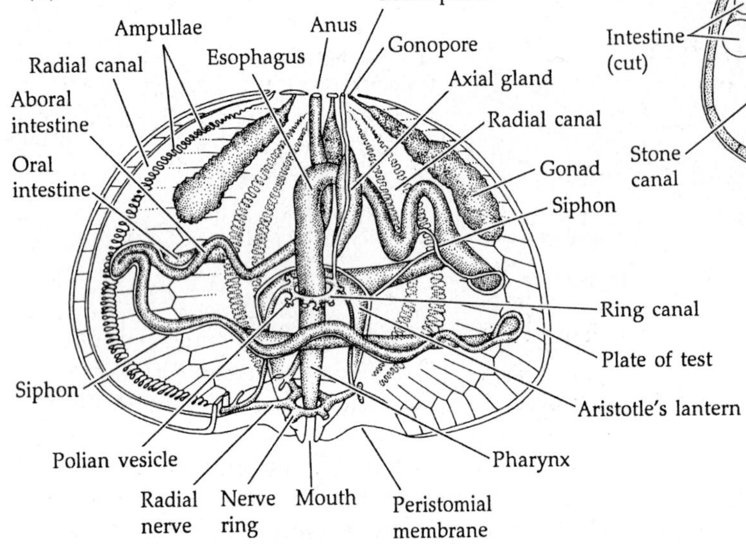

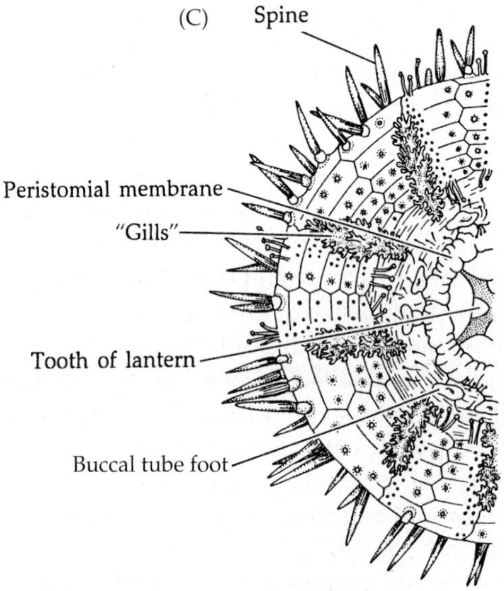

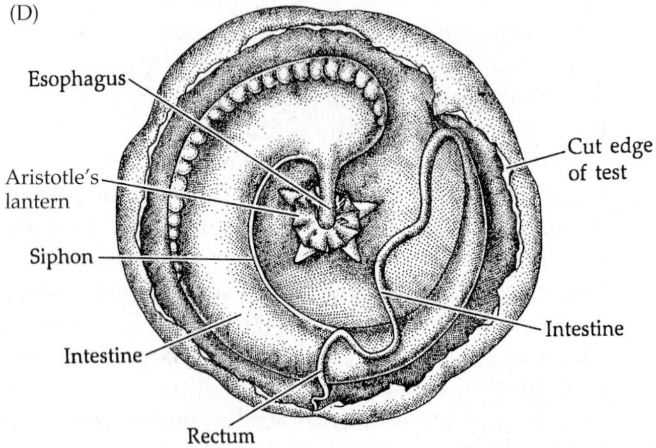

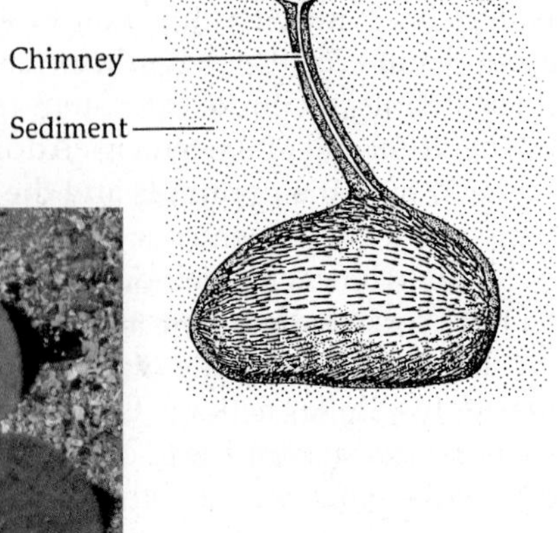

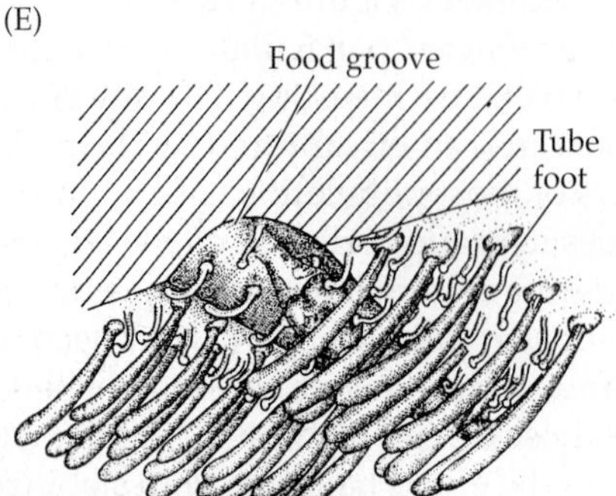

the peristomial membrane and tube feet as the teeth scrape off attached diatoms and organic detritus.

The digestive system of echinoids is essentially a simple tube from mouth to anus. The mouth is at the center of the oral surface or shifted slightly anteriorly in some irregular urchins; the anus is on the aboral surface, located centrally or shifted posteriorly. An esophagus extends through the vertical axis of the lantern (when present) to join an elongate stomach–intestine (Figure 25.11A,B,D). In most echinoids, a narrow duct (the **siphon**) parallels the intestinal tract for part of its length. Both ends of this siphon open to the intestine, possibly providing a shunt for excess water taken in with the food. In many species, blind ceca arise from the gut near the junction of the esophagus and stomach–intestine. The intestine narrows into a short rectum leading to the anus, located either centrally in the apical system, or at various points along the posterior interambulacrum. Digestive enzymes are produced by the intestinal and cecal walls, and breakdown is largely extracellular.

Holothuroidea Most sea cucumbers are suspension or deposit feeders. Many of the sedentary epibenthic or nestling forms (e.g., *Eupentacta, Psolus, Cucumaria*) extend their branched, mucus-covered tentacles (Figure 25.12D,E) into the water to trap suspended material, including live plankton and organic detritus. The tentacles are then pushed into the mouth one at a time and the food ingested (Figure 25.12F). A supply of mucus is provided by secretory cells in papillae on the tentacles and gland cells of the foregut.

More active epibenthic types (e.g., *Stichopus*) crawl across the substratum and use their tentacles to collect and ingest sediment and organic detritus (Figure 25.12C). Studies indicate that some holothuroids (e.g., *Stichopus, Holothuria*) are highly selective deposit feeders, preferentially collecting sediments rich in organic content. *Holothuria tubulosa* is so adept at selective feeding that even its fecal pellets have a higher organic content than the environmental sediments. Many apodid holothuroids ingest sediment as they burrow through the substratum by peristaltic movements.

The anterior mouth is surrounded by a whorl of buccal tentacles. The esophagus (or pharynx) leads inward and passes through a ring of calcareous plates supporting the foregut and ring canal of the water vascular system. The esophagus joins an elongate intestine, the anterior of which is often enlarged as a stomach. The intestine, along which digestion and nutrient absorption occurs, extends posteriorly, loops forward, and then posteriorly again, sometimes in coils (Figure 25.12A,B). The intestine terminates in an expanded rectum leading to the posterior anus. The rectal area is attached to the body wall by a series of suspensor muscles and often bears highly branched outgrowths, the **respiratory trees**, that extend anteriorly within the body cavity. Water is pumped via the anus into the trees for gas exchange (Figure 25.12A,B,H).

The digestive system of sea cucumbers exhibits two fascinating phenomena: (1) auto-evisceration and (2) the discharge of structures called Cuvierian tubules (Figure 25.12H,I). Evisceration is the expulsion by autotomy and muscular action of part or all of the muscles, digestive tract, and sometimes other organs, including the respiratory trees and gonads. These lost parts are usually regenerated. In some forms, these structures are expelled following rupture of the hindgut region. In others, rupture occurs anteriorly and the feeding tentacles and foregut are lost. Evisceration can be induced in the lab by chemical stress, physical manipulation, and crowding, but it also occurs in nature in some species, although its significance remains unclear. It is viewed by some as a seasonal event associated with adverse conditions, and by others as a defense mechanism wherein eviscerated parts serve as sacrificed decoys.

Cuvierian tubules are clusters of sticky, blind tubules arising from the base of the respiratory trees in certain genera (e.g., *Actinopyga, Holothuria*) (Figure 25.12A,G,H). When threatened, these holothuroids aim the anus at the potential predator and contract the body wall to discharge the tubules by rupturing the hindgut, everting the tubules onto the predator and entangling it in a mass made extremely sticky by secretions from the tubules. The Cuvierian tubules are regenerated. Such elaborate defense mechanisms are important, since holothuroids can fall prey to various fishes, starfish, gastropods, crustaceans, and even humans.

Circulation and Gas Exchange

Circulation Internal transport in echinoderms is accomplished largely by the main perivisceral coeloms, variously augmented by hemal and water vascular systems (Figure 25.13), both of which are derived from the coelom. Fluids move through these systems primarily by ciliary action and in some cases by muscular pumping. In at least one species of sea urchin (*Lytechinus variegatus*), coelomic fluid is known to be driven by movements of Aristotle's lantern.

The hemal system is a complex array of canals and spaces, mostly enclosed within coelomic channels called **perihemal sinuses**. The system is best developed in holothuroids, in which it is bilaterally arranged, and in crinoids, in which the channels can form netlike plexi. In other groups the system is radially arranged and generally parallels the elements of the water vascular system consisting of oral and aboral **hemal rings**, each with radial extensions. The two rings are connected by an axial sinus (Figure 25.13A) that lies against the stone canal. Within the axial sinus there is a core of spongy tissue called the **axial gland**, thought to be responsible for producing certain coelomocytes.

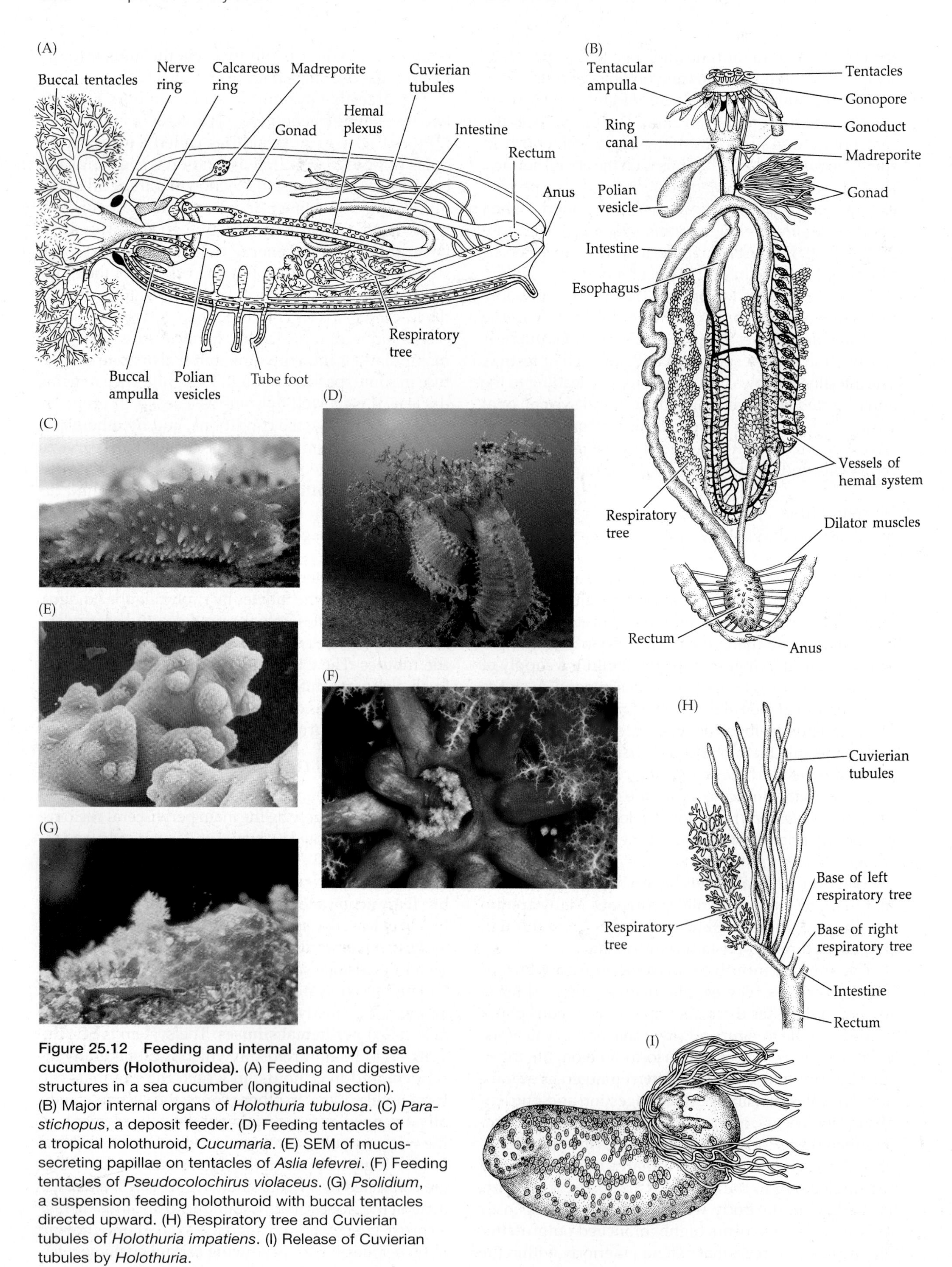

Figure 25.12 Feeding and internal anatomy of sea cucumbers (Holothuroidea). (A) Feeding and digestive structures in a sea cucumber (longitudinal section). (B) Major internal organs of *Holothuria tubulosa*. (C) *Parastichopus*, a deposit feeder. (D) Feeding tentacles of a tropical holothuroid, *Cucumaria*. (E) SEM of mucus-secreting papillae on tentacles of *Aslia lefevrei*. (F) Feeding tentacles of *Pseudocolochirus violaceus*. (G) *Psolidium*, a suspension feeding holothuroid with buccal tentacles directed upward. (H) Respiratory tree and Cuvierian tubules of *Holothuria impatiens*. (I) Release of Cuvierian tubules by *Holothuria*.

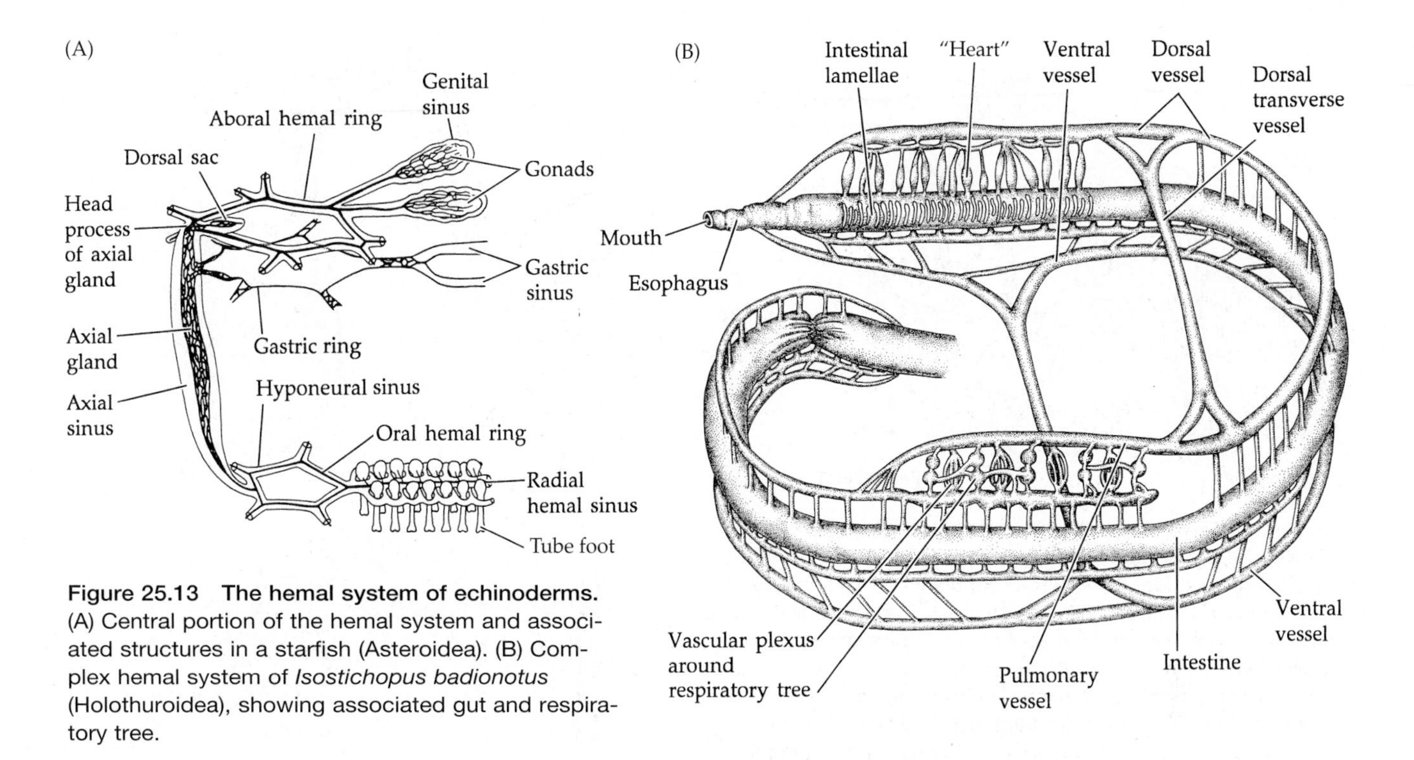

Figure 25.13 The hemal system of echinoderms. (A) Central portion of the hemal system and associated structures in a starfish (Asteroidea). (B) Complex hemal system of *Isostichopus badionotus* (Holothuroidea), showing associated gut and respiratory tree.

Radial hemal channels extend from the aboral ring to the gonads. Other radial channels arise from the oral hemal ring and are associated with the rows of tube feet; these channels are housed within a perihemal space called the hyponeural sinus (Figures 25.3B and 25.13A). A third hemal ring, the gastric ring, occurs in many echinoderms including most asteroids, and is associated with the digestive system.

Cilia move fluid through the hemal system. In asteroids and most echinoids the axial sinus bears a **dorsal sac** near its junction with the aboral hemal ring. This sac pulsates, likely to aid fluid movement within hemal spaces. The hemal system of holothuroids comprises an elaborate set of vessels intimately associated with the digestive tract (Figure 25.13B) and, when present, the respiratory trees. In many holothuroids the hemal system can include a multitude of "hearts" or circulatory pumps.

The function of the hemal system is not fully understood, but it probably helps distribute nutrients absorbed from the digestive tract. Experiments on the starfish *Echinaster graminicolus* show that absorbed nutrients appear in the hemal system within a few hours after feeding and eventually concentrate in gonads and tube feet. In sea cucumbers the hemal system probably also plays a role in gas exchange because some of the vessels are in contact with the respiratory trees.

Gas exchange Most echinoderms rely on thin-walled external processes as gas exchange surfaces. Crinoids apparently exchange oxygen and carbon dioxide between the coeloms and the ambient seawater across all exposed thin parts of the body wall, especially the tube feet. Only ophiuroids and holothuroids have specific internal organs for respiration. Given the relatively large coelomic volumes of many echinoderms, and the lack of circulatory connections from the inside to the outside of the body wall, coelomic fluid transport mechanisms figure prominently in internal gas exchange, whereas external tissues exchange directly with the ambient seawater.

Gas exchange in asteroids occurs across tube feet and special outpocketings of the body wall called **papulae** (Figure 25.14A,B), which are evaginations of the epidermis and peritoneum. Both tissues are ciliated to produce currents in both the coelomic fluid and the overlying water. The two currents move in opposite directions, setting up countercurrents that maximize concentration gradients across the surfaces of papulae.

Ophiuroids possess ten invaginations of the body wall, called **bursae**, that open to the outside via ciliated slits (Figure 25.9A,C). Water circulates through the bursae by cilia and, in some species, by muscular pumping of the bursal walls. Gases are exchanged between flowing water and body fluids. Hemoglobin occurs in the coelomocytes of a few species of ophiuroids.

Typical sea urchins possess five pairs of lobate gills around the peristome (Figure 25.11B,C). These were long viewed as important gas exchange organs. However, various authors have provided evidence of a different function. The pressure within these gills changes by manipulation of the compasses of Aristotle's lantern, so gills probably function to accommodate pressure changes in the peripharyngeal coelom during movements of the lantern, and perhaps to enhance oxygen supply to associated muscles. The main

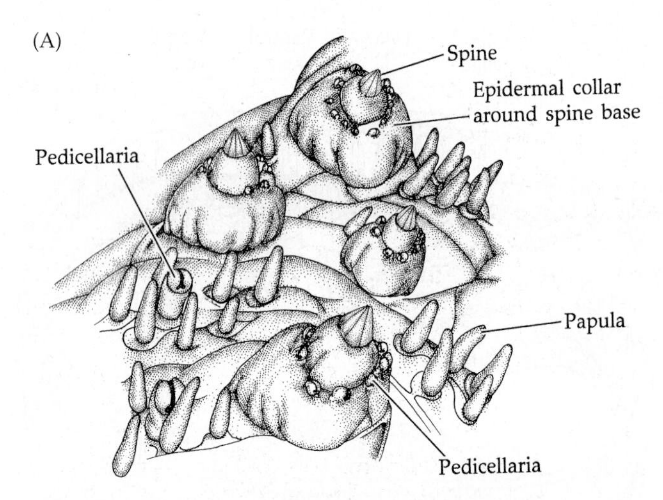

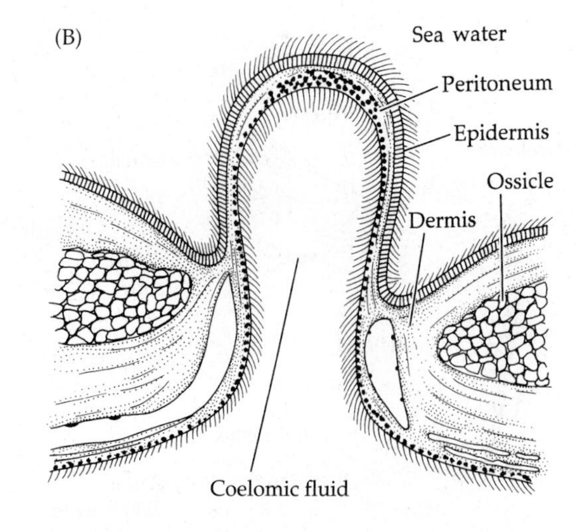

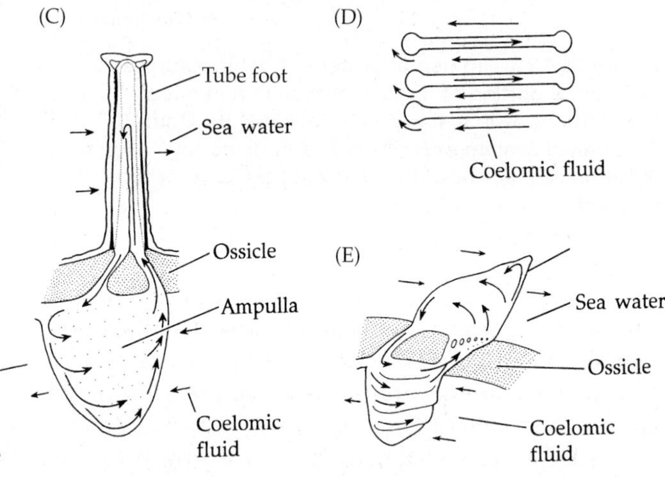

Figure 25.14 Gas exchange in echinoderms. (A) Portion of the aboral surface of *Asterias* (Asteroidea) with digitiform papulae. (B) Asteroid papula (section) lined by peritoneum and filled with coelomic fluid. (C) Ampulla and tube foot (longitudinal section) of *Strongylocentrotus purpuratus* (Echinoidea). Arrows represent countercurrents among ambient seawater, fluid of the water vascular system, and coelomic fluid. (D) Three lamelliform ampullae from *Strongylocentrotus*. Gases are exchanged between fluids of water vascular system and coelom. (E) Respiratory tube foot and ampulla (section) of the heart urchin *Echinocardium* (Echinoidea). Arrows represent countercurrents.

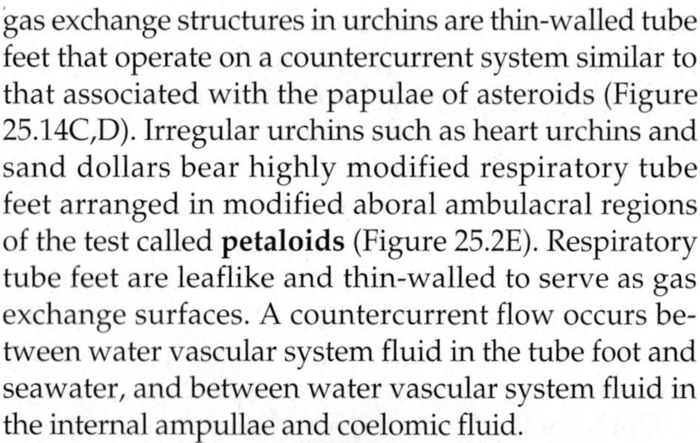

gas exchange structures in urchins are thin-walled tube feet that operate on a countercurrent system similar to that associated with the papulae of asteroids (Figure 25.14C,D). Irregular urchins such as heart urchins and sand dollars bear highly modified respiratory tube feet arranged in modified aboral ambulacral regions of the test called **petaloids** (Figure 25.2E). Respiratory tube feet are leaflike and thin-walled to serve as gas exchange surfaces. A countercurrent flow occurs between water vascular system fluid in the tube foot and seawater, and between water vascular system fluid in the internal ampullae and coelomic fluid.

In the respiratory trees of certain holothuroids, water is pumped in and out of the hindgut and branches of the respiratory trees, and gases are exchanged between the water and the coelomic and hemal systems. This process is augmented by exchange across the tube feet, which is facilitated by a countercurrent system. Hemoglobin occurs in the coelomocytes of many holothuroids.

Excretion and Osmoregulation

Excretion In most echinoderms, dissolved nitrogenous wastes (ammonia) diffuse across body surfaces to the outside. This type of excretion occurs across the tube feet and papulae in asteroids and is suspected to occur across the paired respiratory trees in holothuroids.

Precipitated nitrogenous material and other particulate wastes are phagocytized by specialized coelomocytes in body fluids and then discharged by various methods. In asteroids, waste-laden coelomocytes can accumulate in papulae, which then pinch off their distal ends, expelling the cells and waste material. Some studies suggest that rectal glands might also be involved in excretion. In ophiuroids it is suspected that coelomocytes deliver wastes to the bursae, where they are released. Phagocytic coelomocytes in echinoids accumulate wastes and transport them to the tube feet and gills for release. In holothuroids, particulate wastes are carried by coelomocytes to the respiratory trees, gut, and even the gonads, and released to the outside. Crinoid coelomocytes deposit wastes in tiny pockets along the sides of the ambulacral grooves.

Osmoregulation Echinoderms are generally strictly marine stenohaline creatures that do not have to deal with severe osmotic and ionic regulation. However, some species have been reported from brackish water. For example, *Asterias rubens* (Asteroidea) has been collected from the Baltic Sea (8% salinity), *Ophiophragmus filograneus* (Ophiuroidea) from Cedar Key, Florida (7.7% salinity), and various holothuroids from the

Black Sea (18% salinity). Some mechanism must allow these taxa to survive in such low salinities, but it is, as yet, unknown. Evidence suggests that echinoderms are osmoconformers. Both water and ions pass relatively freely across thin body surfaces, and the tonicity of the body fluids varies with environmental fluctuations. There appears to be some ionic regulation through active transport, but it is minimal.

Nervous System and Sense Organs

The derived pentaradiality of echinoderms is clearly reflected in the anatomy of their nervous systems and the distribution of their sense organs. The nervous system is decentralized, somewhat diffuse, and without significant ganglia. There are three main neuronal "systems," integrated with one another and developed to varying degrees among the clades: ectoneural (oral); hyponeural (deep oral); and entoneural (aboral). The **ectoneural system** is predominately sensory, although motor fibers do occur; the **hyponeural system** is largely motor in function. The **entoneural system** is absent from holothuroids and reduced to different degrees in other groups—except the crinoids, in which it is the primary nervous system, serving both motor and sensory functions. Note that these terms refer only to the *position* of the neuronal systems, not to their embryological origins.

The three systems are interconnected by a nerve net derived primarily from ectoneural and entoneural components. The net is described as a subepidermal plexus, but it gives rise to intraepidermal neurons and clearly has an intimate association with the epithelium both internal and external to the body wall. Except for crinoids, in which the entoneural component predominates, the most obvious nerves in echinoderms are derived from the ectoneural system. A circumoral nerve ring just beneath the oral epithelium encircles the esophagus. From this ring arise radial nerves extending along each ambulacrum. In starfish, these radial nerves appear as a distinct V-shaped, epidermal thickening in each ambulacral groove (Figure 25.3B). The entoneural nerve plexus also produces radial cords, such as those along the margins of asteroid arms. The hyponeural system generally parallels the nerves of the ectoneural system. Hyponeural neurons are subepidermal and lie near the hyponeural sinus of each ambulacral area (Figure 25.3B), giving rise to motor fibers and ganglia in the tube feet.

Sensory receptors are largely restricted to relatively simple epithelial structures innervated by a plexus of the ectoneural system. Frequently associated with outgrowths of the body wall, such as spines and pedicellariae, sensory neurons in the epidermis respond to touch, dissolved chemicals, water currents, and light. Photoreceptors occur in asteroids as light-sensitive regions comprising pigment-cup ocelli at the tips of the arms. Some brittle stars have aboral photosensitive tissues, even using calcitic microlenses in the stereom skeletal elements in the arms to enhance light detection. Echinoids seem to use light sensitivity in tube feet, mediated by the spacing of spines over their entire bodies, for spatial resolution of objects. Statocysts are known in some holothuroids, while in certain echinoids there are structures called **sphaeridia** that are thought to function in georeception. Chemoreception is understudied in echinoderms, but some evidence suggests that buccal tentacles of holothuroids and oral tube feet of echinoids are sensitive to dissolved chemicals. Chemoreception in asteroids appears to depend largely on direct contact, although distance chemoreception is reported in some species. Larvae are reported to respond to pheromones in the water column in order to metamorphose.

In spite of their rather simple nervous system and lack of specialized sense organs, many echinoderms engage in complex behaviors. Time-lapse videos of echinoderms reveal that they have complex interactions similar to any invertebrate—they just do nearly everything at such a slow pace that it is difficult to notice. There is still much to be learned about functional mediation between circuitry of the nervous system and observed behavioral responses, as in coordination of tube feet and spines during locomotion. Most echinoderms exhibit distinct righting behaviors when overturned—actions involving touch, georeception, and perhaps photoreception. Orientation to currents is known in some sand dollars and in many ophiuroids and crinoids, and responses to potential predators often include coordination of spine movements, pedicellariae, and locomotion.

Reproduction and Development

Regeneration and asexual reproduction Echinoderms have remarkable powers of regeneration. Tide pool observers often encounter starfish regenerating a new arm, or notice tube foot suckers left on a rock from which a starfish or urchin has been pulled free. Lost suckers are quickly replaced by regeneration. During the processes of evisceration and expulsion of Cuvierian tubules, multiple lost organs are replaced. Damage to the test of echinoids, even perforation and loss of plates, is usually followed by healing and regeneration of the plates and associated appendages. Studies have put to rest tales of oystermen claiming that chopping starfish into small pieces resulted in regeneration of an entire new animal from each part. While it is true that a damaged animal can grow new arms if a substantial portion of the central disk remains intact, an isolated arm soon dies. However, an exception to this is *Linckia*, which can regenerate an entire individual from a single arm, the regenerating stage being appropriately called a **comet** (Figure 25.15). Ophiuroids and crinoids frequently cast off arms or arm fragments when disturbed, then regenerate the

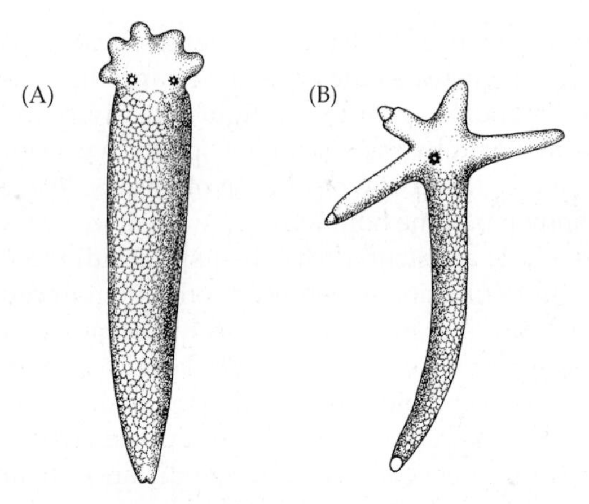

Figure 25.15 Regeneration in the starfish *Linckia* (Asteroidea). (A) Initial regeneration from a single arm, yielding a central disk with dual madreporites and five new rays. (B) Later stage, with a single madreporite and normal number of rays.

lost part. Such autotomy is also documented for certain asteroids. The Pacific coast ochre star, *Pisaster ochraceus*, autotomizes arms at their junction with the central disk when confronted by predators. Deep-sea isocrinid sea lilies autotomize distal portions of their stems to distract echinoid predators.

Asexual reproduction occurs in some asteroids, ophiuroids and holothuroids by **fissiparity**, wherein the central disk divides in two and each half forms a complete animal by regeneration. When the small 6-rayed brittle star *Ophiactis* divides, each half retains 2 to 4 arms. Asexual fission also occurs in some holothuroids, but the process is not well understood.

Sexual reproduction The majority of echinoderms are gonochoristic, but hermaphroditic species are known among asteroids, holothuroids, and especially ophiuroids. The reproductive system is relatively simple and intimately associated with derivatives of the coelom. Gonads are usually housed within genital sinuses lined with peritoneum. Holothuroids are unique among extant echinoderms in possessing a single gonad, which lies dorsally (Figure 25.12B). A single gonoduct opens between the bases of two dorsal buccal tentacles or just posterior to the tentacular whorl.

Crinoids lack distinct gonads. Gametes arise from the peritoneum of special coelomic extensions called genital canals in the pinnules extending from the proximal portion of each arm. There are no gonoducts; gametes are released by rupture of pinnule walls. Ophiuroids possess from one to many gonads attached to the peritoneal side of each bursa adjacent to the bursal slits (Figure 25.9A). Gametes are released into the bursae and expelled through the slits.

Asteroids and echinoids possess multiple gonads (Figures 25.8C and 25.11B) with gonoducts leading to interambulacral gonopores (Figure 25.8A). Typical sea urchins contain five gonads, one in each interradius. Gonopores are located on the five interambulacral genital plates of the apical system (Figures 25.3F and 25.11A). The periproct and anus have migrated posteriorly in irregular urchins, but the genital plates remain more or less centrally located on the aboral surface. In most irregular echinoids there are only four (and sometimes fewer) gonads, one being lost along the line of migration of the anus, with corresponding reduction in number of gonopores. In all urchins one of the genital plates is perforated and functions as a madreporite.

Echinoderm life history strategies vary from free spawning followed by external fertilization and indirect development, to various forms of direct development and brooding. Studies indicate that spawning is mostly a nocturnal event, wherein the animals assume characteristic postures with their bodies elevated off the substratum. In at least some asteroids and echinoids, gametogenesis is regulated by photoperiod, which ensures more or less synchronous spawning among members of the same population. In some species of free-spawning asteroids, females release pheromones that stimulate sperm release from nearby conspecific males, and many species aggregate before spawning to enhance fertilization success.

Brooding is especially common among boreal and polar species (in all groups of echinoderms) and in certain deep-sea asteroids, whose environments might be unfavorable for larval life. Brooding species produce fewer but larger and yolkier eggs than do their free-spawning counterparts. Among the crinoids, *Antedon* and a few others cement their eggs to the epidermis of the pinnules from which they emerge (Figure 25.16A,B); once eggs are fertilized by free sperm, embryos are held by the parent until hatching. Most brooding asteroids hold embryos on the body surface. One species (*Asterina gibbosa*) cements its eggs to the substratum, and another (*Leptasterias tenera*) broods its early embryos in the pyloric stomach before moving them to the outer body surface. *Xyloplax* and caymanostellid wood stars (Figure 25.1G) seem to brood within the gonads, releasing juveniles that drift or crawl away. Brooding is common among ophiuroids. Sperm enter the bursae and fertilize the eggs, and embryos are held within these sacs during development. Some echinoids brood embryos and young urchins among clusters of spines on the body, around the peristome, in depressions caused by the sunken petaloids of females or, remarkably, in body wall invaginations at the apical system through which the urchins are "born." Brooding holothuroids usually carry their embryos externally (Figure 25.16C), but some species of *Thyone* and *Leptosynapta* brood inside the coelom.

Development The tremendous numbers of eggs produced by many echinoderms, and the ease with which they can be reared in the laboratory, have made many

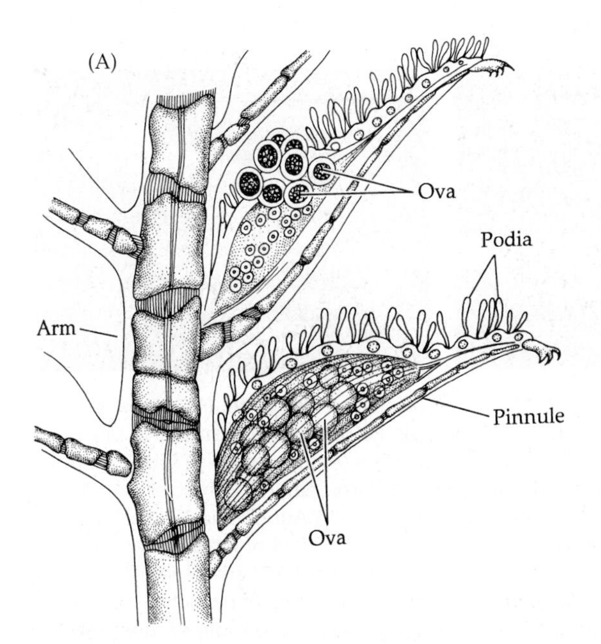

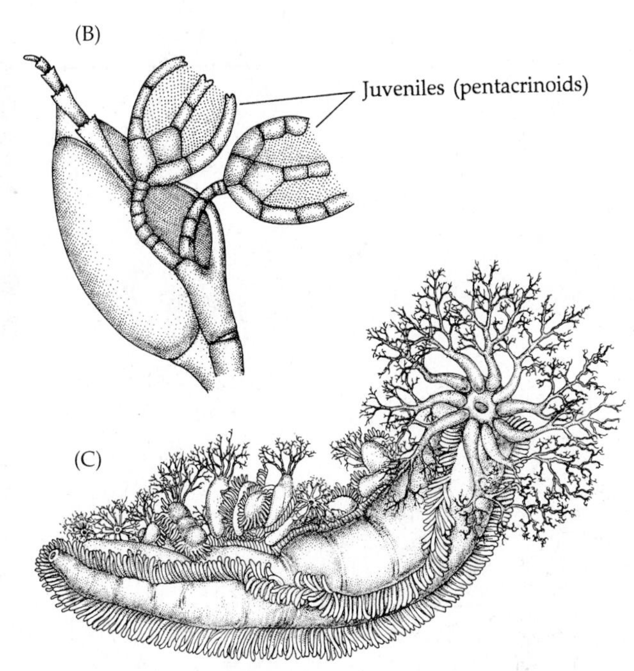

Figure 25.16 Brooding in echinoderms. (A) Portion of an arm of the feather star, *Antedon* (Crinoidea). Ova are housed within a pinnule and released to exterior. (B) Part of a pinnule of the crinoid *Phrixometra* (Crinoidea) with developing young. (C) *Cucumaria crocea* (Holothuroidea), brooding young.

of these animals embryological model organisms. Much of our information about biology of animal fertilization and early development comes from over a century of work on urchins and starfish, and early ontogeny of some echinoderms serves as a model of deuterostome development. Except in brooding species, in which development is modified by large amounts of yolk, the sequence of ontogenetic events is remarkably similar throughout the phylum. It is impossible to completely cover the vast amount of information on this subject; what is provided here is only a brief overview of indirect development based largely on urchins and asteroids, including some comparative comments on other taxa.

In free-spawning echinoderms, after sperm enters the vitelline layer on the egg, a fertilization membrane forms. The ova are usually isolecithal with relatively small amounts of yolk. Cleavage is radial, holoblastic, and initially equal or subequal, leading to a spacious coeloblastula. In groups such as echinoids, cleavage preceding the blastula becomes unequal, resulting in blastomere tiers of vegetal **mesomeres** underlain by slightly larger **macromeres** and a cluster of **micromeres** at the animal pole. (These terms refer only to the relative sizes of the cells, and are not to be confused with the same terms used in describing spiral cleavage.) The coeloblastula usually becomes ciliated and breaks free of the fertilization membrane as a swimming embryo. As in many other phyla, this stage of development is known as the **primary larva**, characterized in part by the **apical organ**—a cluster of ciliated cells with sensory ability thought to aid in orientation of the larva during settlement. Although there is evidence that the apical organ is homologous among all invertebrate primary larvae, the number of cells in, and overall shape of, the apical organ varies among the five echinoderm classes.

The blastula flattens slightly at the animal pole, forming the gastral plate from which some cells proliferate into the blastocoel as primary or larval mesoderm (sometimes called "mesenchyme"), which is destined to become larval muscles, skeletal elements (Figure 25.17G), and coelomic pouches. In most echinoids these cells are micromeres. The surrounding macromeres are presumptive endoderm and adult mesoderm; the vegetal mesomeres are presumptive ectoderm. A coelogastrula is produced by invagination of the animal pole cells. The blastopore typically forms the anus; the archenteron grows to connect with a stomodeal inpocketing that forms the mouth. Before the gut is complete, the inner end of the archenteron proliferates, as well as one or two evaginations of mesoderm, into the blastocoel. Thus, coelom formation is by archenteric pouching (enterocoely).

Some types of echinoderm larvae also develop a **larval skeleton**, comprising spicules and ossicles. The functions of the larval skeleton include defense, physical support, sites of muscle attachment, and passive orientation of larvae in the form of "arms" (not homologous to arms of starfish and other echinoderm adults) that act like the feathers on a badminton birdie. The larvae of some echinoderms have distinctive reddish pigment spots that contain caroteno-proteins, red pigments that can react to light and change the characteristics of cell membranes. The specific functions of these pigments in echinoderm larvae remain unknown.

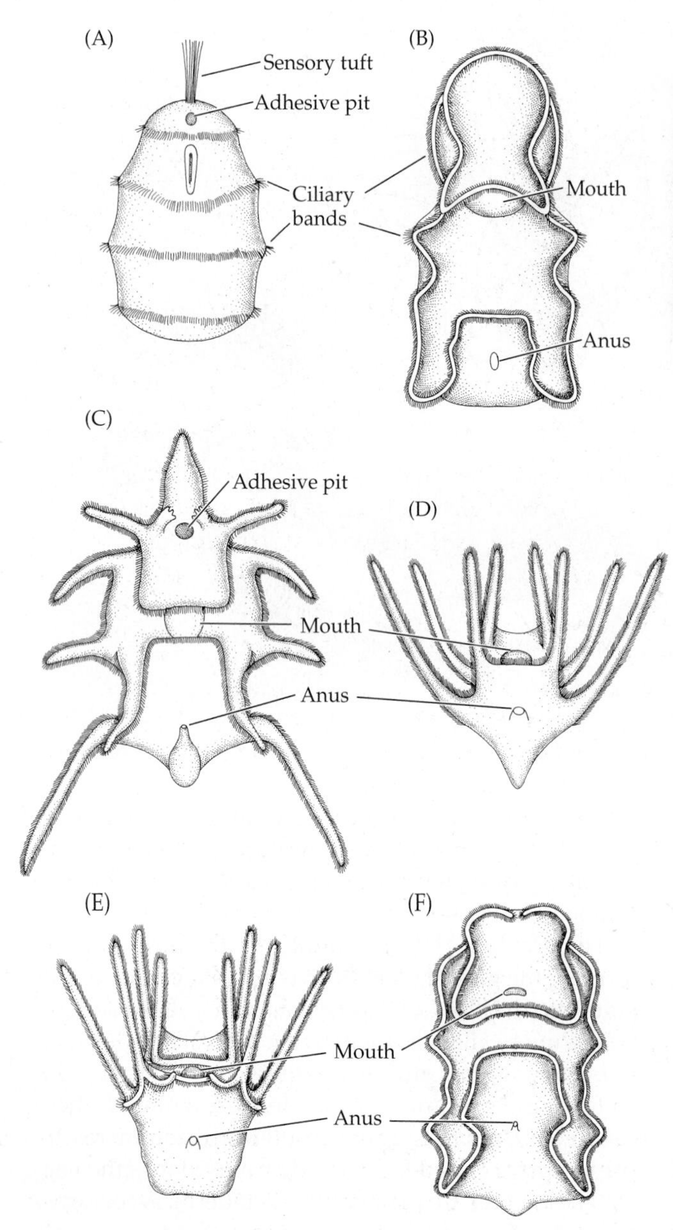

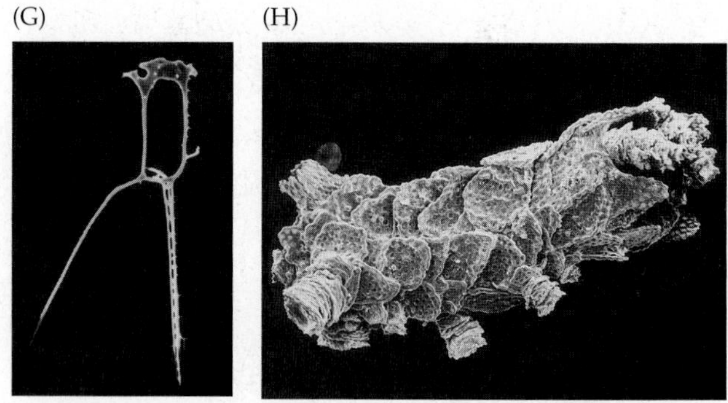

Figure 25.17 Echinoderm larval types. (A) Vitellaria larva of a sea lily (Crinoidea). (B,C) Bipinnaria and later brachiolaria larvae of a starfish (Asteroidea). (D) Ophiopluteus larva of a brittle star (Ophiuroidea). (E) Echinopluteus larva of a sea urchin (Echinoidea). (F) Auricularia larva of a sea cucumber (Holothuroidea). (G) SEM of isolated larval spicule from the sand dollar *Dendraster* (Echinoidea). (H) SEM of late pentacula stage (postlarva) of a sea cucumber (Holothuroidea); notice the rudimentary tube feet already formed.

Planktotrophic echinoderm larvae use bands of cilia along the "arms" or on the body itself to swim and generate feeding currents. However, lecithotrophism is common and apparently evolved numerous times within some echinoderm classes.

In order to understand the development of echinoderm larvae and their eventual metamorphosis to radially symmetrical adults, it is necessary to examine the embryogeny and fates of the coeloms. There are some differences in details among groups, but there is enough similarity to allow a generalized summary. The initial archenteric pouching occurs from the blind end of the developing gut, either as a pair of coeloms or as one cavity that divides into two. These coeloms pinch off another pair of cavities posteriorly, and then a third pair between the anterior and posterior ones

TABLE 25.1 Adult Fates of the Major Coelomic Derivatives in Generalized Echinoderm Development

Embryonic coelomic structure	Adult fate
Right somatocoel	Aboral perivisceral coelom; imperforate extraxial body wall
Left somatocoel	Oral perivisceral coelom; genital sinuses; most of the hyponeural sinus; perforate extraxial body wall
Right axocoel	Largely lost
Hydropore	Incorporated into madreporite
Hydrotube and dorsal sac	Parts of madreporite vesicle and ampulla
Stone canal	Stone canal
Left hydrocoel	Axial region; ring canal; radial canals; lining of tube feet lumina, plus other components of the water vascular system, including Tiedemann's bodies and Polian vesicles

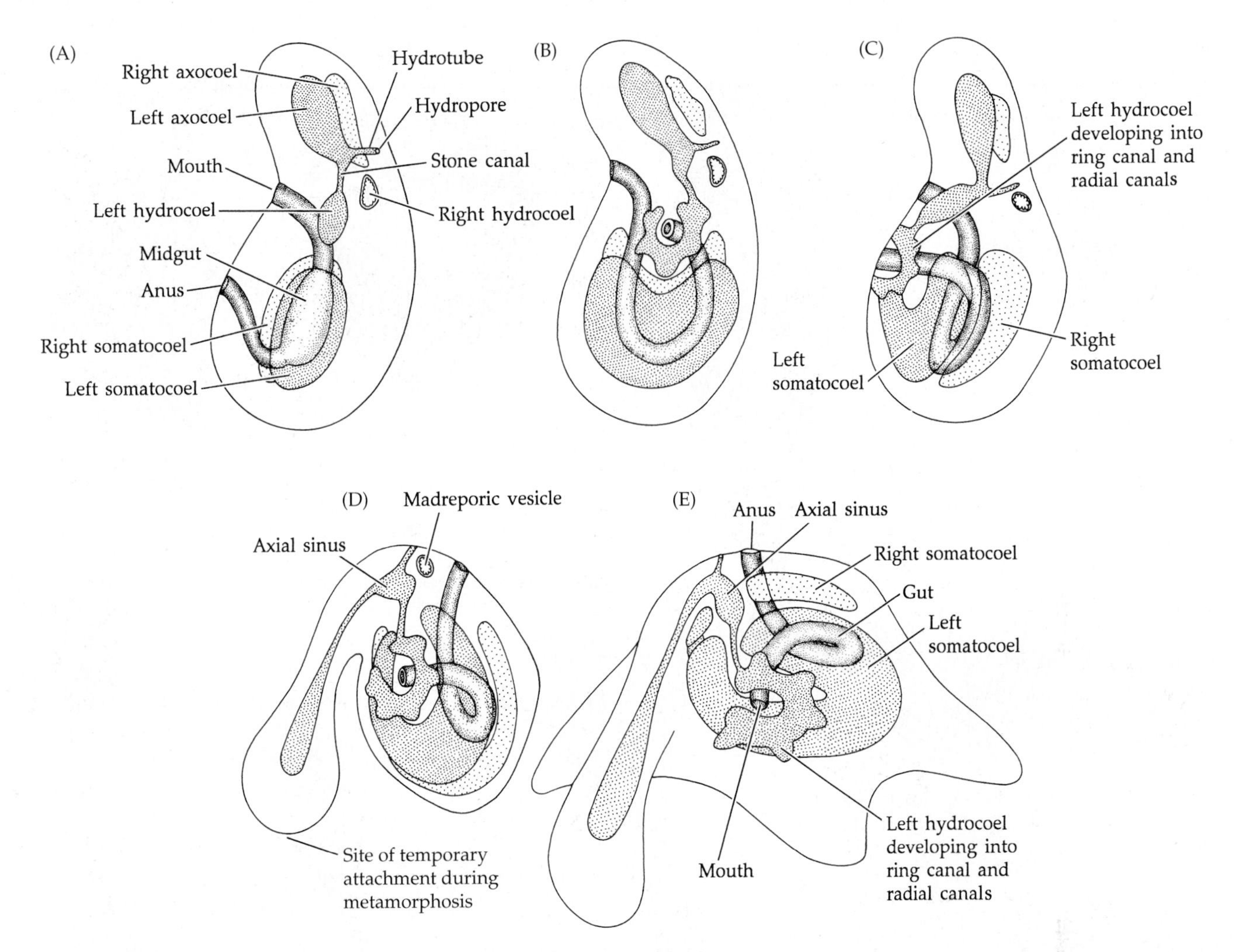

Figure 25.18 (A–E) Development of the coelom and its derivatives in a starfish (Asteroidea).

(Figure 25.18A). From front to back, these pairs of coelomic spaces are the left and right **axocoels** (known in other deuterostomes as the protocoels), mesocoels (hydrocoel + "dorsal sac"), and somatocoels (left and right somatocoels, sometimes called left and right metacoels), as seen in other trimeric deuterostomes. The left axocoel and hydrocoel remain connected to one another. From the left axocoel–hydrocoel complex, a "hydrotube" extends to the dorsal surface of the larva, opening to the outside via a hydropore. The dorsal sac becomes associated with the hydrotube leading to the hydropore. Usually, the right axocoel disappears.

The fates of coelomic derivatives are outlined in Table 25.1. As the time for metamorphosis approaches, the larva swims to the bottom, selecting and attaching to an appropriate substratum. In general, the larval left and right sides become the adult oral and aboral surfaces, respectively, although this pattern does not always reflect a precise 90° reorientation. The change from bilateral to radial symmetry involves shifts in the positions of mouth and anus, which both close during metamorphosis. Typically, embryonic openings disappear, and the stump of the foregut swings from its larval anteroventral location to the left side within the developing rudiment. The hindgut moves anteriorly and to the right (Figure 25.18B,C). Once in the adult positions, the mouth and anus reopen. Aborally, the madreporite complex arises from various parts of the left axocoel and its derivatives plus the dorsal sac. The axial sinus arises from an outpocketing of the left axocoel. The hydrocoel forms a crescent around the foregut as precursor to the ring canal, which grows radial extensions (Figure 25.18D,E), the primary lobes, destined to become the radial canals and all their derivatives. The ring canal closes to form a torus. Outgrowths of the left somatocoel produce the hyponeural sinus, interacting with the primary lobes to elaborate the structure of the rudiment. The left somatocoel comes to lie just internal to the hydrocoel, interacting with it to pattern the body cavities of the adult. The right somatocoel "stacks" just internal to the left somatocoel, forming a quasi-linear anterior–posterior axis during a process called "coelomic stacking." As these transformations take place, much of the original larval body is lost or resorbed, and the juvenile assumes benthic life. Although echinoderms are noted for this remarkable

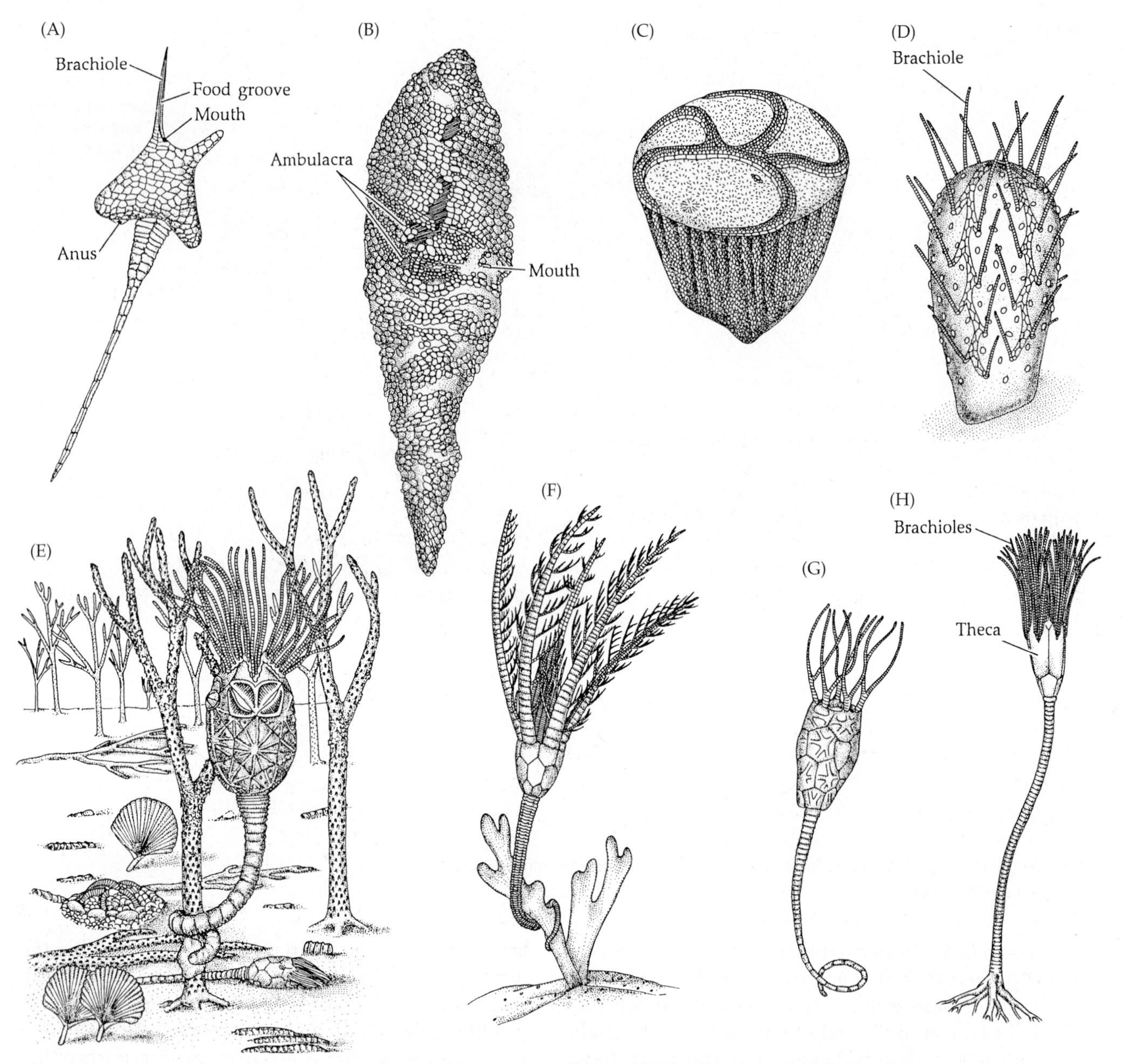

Figure 25.19 Fossil echinoderms. (A) *Dendrocystites*, an early Cambrian carpoid with a single brachiole. (B) *Helicoplacus*, a triradiate echinoderm from the Lower Cambrian. (C) *Camptostroma*, an early Cambrian, pentaradial form suggestive of the ancestral echinoderm. (D) A generalized cystoid. (E) *Lepadocystis*, a stalked Ordovician cystoid. Note the edrioasteroid in the background, attached to the substratum. (F) *Eifelocrinus*, an extinct crinoid. (G) The eocrinoid *Macrocystella*. (H) A generalized blastoid from the Carboniferous.

metamorphosis, it is expressed to varying degrees in the different classes because of differential expression of the axial and extraxial components that are related to the rudiment and larval body respectively.

Echinoderm Phylogeny

Living echinoderms are so highly derived compared with other deuterostomes that meaningful comparisons of adult body plans seem almost futile. In spite of a rich fossil record and many decades of work, the origin and subsequent evolution of echinoderms remain controversial, stemming from differing interpretations of echinoderm homologies originating from various sources in paleontology, morphology, and embryology. Almost all data support the idea that each extant class comprises a monophyletic clade, but there is much more contention concerning the relationships among fossil taxa.

First Echinoderms

Echinoderms likely originated with the Precambrian invasion of epibenthic habitats by some form of infaunal bilaterian deuterostome with similarities to the sister group of the echinoderms, the hemichordates. Echinoderms diversified rapidly, and fundamental body plans of almost all groups have roots in the Cambrian. By the end of the Ordovician, all the major clades had been established. Diversity and disparity among what seem now to be bizarre forms reached their zenith during the early to mid-Paleozoic (Figure 25.19). By the beginning of the Mesozoic, this disparity among major taxa had declined greatly, leaving the five extant classes persisting to the present.

Echinoderm origins involved modification of the bilaterian, trimeric larval trajectory so that the hydrocoel could initiate the appearance of the pentaradial adult with its unique water vascular system. Coupled with evolution of the stereom endoskeleton, these traits make it easy to distinguish echinoderms from all other early phyla (Figure 25.20). All echinoderm larvae retain their ancestral bilateral, trimeric condition. The expression of genes that dictate radiality must occur relatively late in development and must have involved major modifications of the roles of certain homeobox genes.

There are divergences of opinion concerning the significance of forms such as the extinct carpoid, helicoplacoid, and ctenocystoid homalozoans. Some members of these groups do not have the full complement of five rays (Figure 25.19A,B), and they are considered by some to be fossilized representatives of the transition to pentaradiality. This hypothesis makes certain assumptions about the homologies of various feeding appendages, structures supposed to be gill slits, and a superficial bilateral morphology in the skeleton.

Other workers have applied models of echinoderm homology and embryology to argue that the first echinoderms were likely already pentaradial. If this were true, forms such as *Stromatocystites* and *Camptostroma* (Figure 25.19C) are representative of the earliest echinoderms, with the pentaradial axial region expressed as narrow ambulacra. These thin rays remain embedded in the oral, perforate extraxial region marked by the hydropore, anus, and other small body openings (Figure 25.4A). The imperforate extraxial region is a large component devoid of orifices that aborally sealed the body wall and likely contained most of the viscera. Under this view, nonpentaradial forms, such as the aforementioned ctenocystoids and carpoids (Figure 25.19A), or the bizarre helicoplacoids (Figure 25.19B), are not representatives of basal echinoderms, but offshoots from the main lineages leading to other major echinoderm clades—lineages that were ancestrally pentaradial (Figure 25.20).

Helicoplacoids that appeared in the early Cambrian and died out soon thereafter were cigar shaped, with only three ambulacra arising from a lateral mouth, and spiraling around the body among interradial plating (Figure 25.19B). How these echinoderms made their living remains a matter of conjecture. It is tempting to consider the 3-ray pattern of helicoplacoids to be a precursor to the 2-1-2 pattern seen in the ambulacral branches of other fossil genera, such as *Stromatocystites* or *Camptostroma*, in which there seem to be two bifurcating branches (the "2"), and one nonbifurcating branch (the "1") to make a total of five rays. However, there are no parsimonious arguments against the idea that helicoplacoids (or any of the echinoderms with fewer than 5 rays) represent a reduction in ray number. The 2-1-2 pattern is also expressed in modern crinoids, but the five radial canals come off the ring canal in perfect pentaradiality, suggesting that the 2-1-2 pattern does not reflect embryological patterning of the water vascular system itself.

The one-armed carpoids seem to be derived members of the Blastozoa, a large, extinct group that includes cystoids, blastoids, and rhombiferans (Figures 25.19A,D,E,G,H and 25.20). All these taxa possess unique "free ambulacra" composed only of axial elements called **brachioles**. Basalmost blastozoans were plesiomorphically pentaradial, and derived from fossils such as *Lepidocystis* that illustrate origins of brachioles, as well as origins of the multiplated stalk derived independently from the stem seen in crinoids. Brachioles are not homologous to arms, as they do not include any coelomic elements other than the hydrocoel. "True" arms such as those seen in crinoids comprise both axial and extraxial elements, incorporating major coelomic outpocketings of both the left and right somatocoels.

Modern Echinoderms

The oldest extant echinoderm group are the crinozoans, and among living taxa the crinoids constitute the most basal clade. Crinozoa include extinct forms, the protocrinoids, as well as Crinoidea that proliferated during the Paleozoic. Crinoids bear attachment stems (distinct from the stalks of blastozoans in the nature of their ontogeny, as well as overall anatomy) arising from the aboral surface (Figure 25.19F). Protocrinoids possessed calcified ambulacral plates in the arms, but these were lost in more highly-derived forms, perhaps because the load-bearing of these profusely branched feeding structures was taken on by the elaborate, aboral brachials.

The lineage including the more "errant" asteroids, ophiuroids, echinoids, and holothuroids evolved motile feeding modes in which the mouth was directed towards the substratum (except in the unusual, strongly paedomorphic holothuroids). The tube feet became more robust and deeply involved in all aspects of life, including locomotion and active hunting of prey. Asteroids, ophiuroids, and echinoids did not appear until later (Ordovician), and the holothuroids appeared in the Silurian, perhaps by paedomorphic reduction from strange extinct forms, the ophiocistioids. At one time, asteroids were placed as sister group to a clade containing ophiuroids, echinoids, and holothuroids;

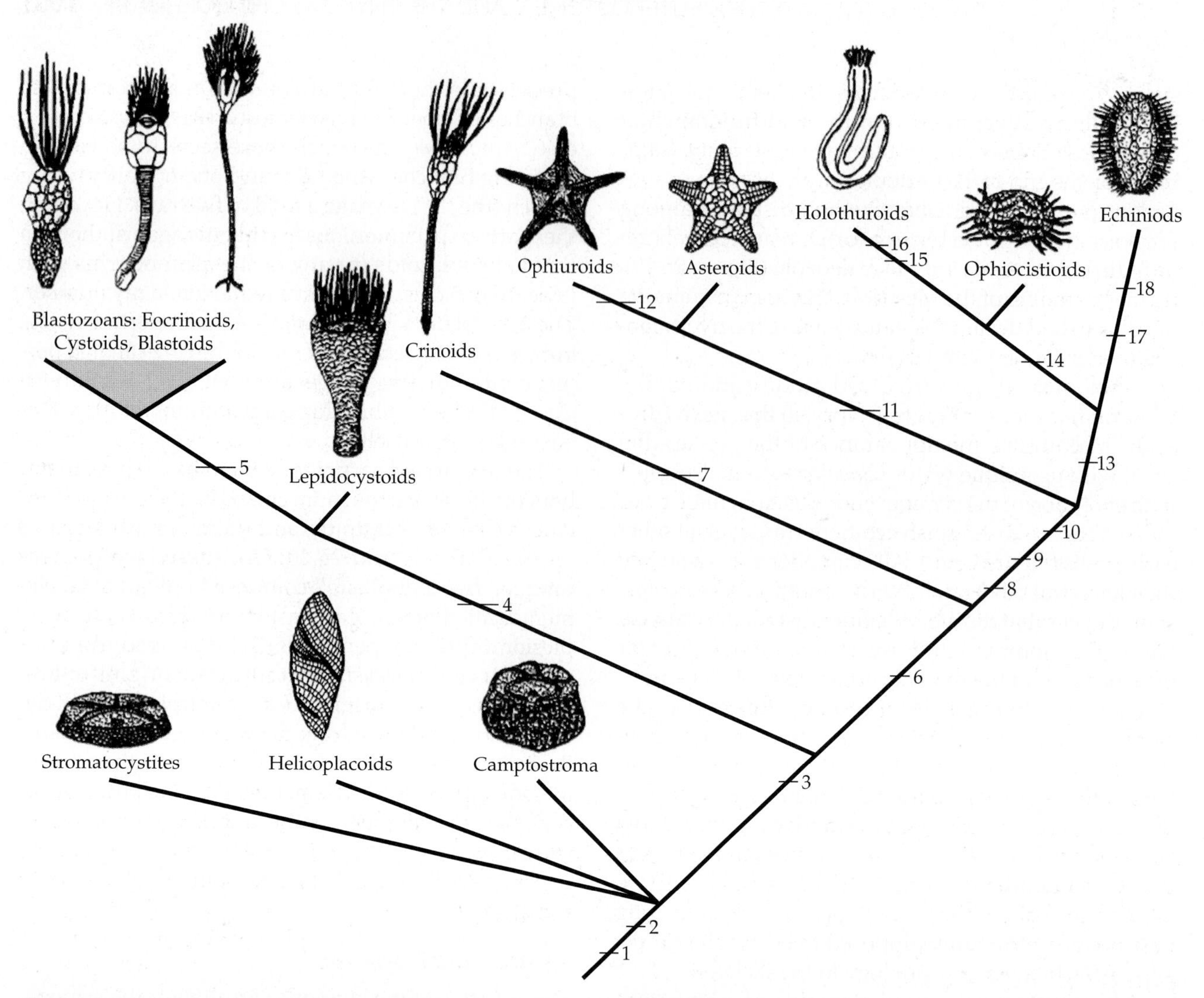

Figure 25.20 A phylogenetic hypothesis for the main groups of echinoderms, with suggested major synapomorphies. Echinoderms evolved from an infaunal ancestor similar to the sister group Hemichordata by inducing pentaradiality through a rudiment produced from the hydrocoel (1), and the evolution of a stereom endoskeleton (2). *Stromatocystites* and *Camptostroma* are edrioasteroid-like taxa likely representing the morphology of the first echinoderms—sessile animals with narrow rays (axial region), anus, and hydropore set in an upward-facing perforate extraxial region. Enigmatic derivations from this morphology had already appeared in triradial helicoplacoids. The imperforate extraxial region became an elongated sac enclosing aboral portions of the viscera (3). A diverse, entirely extinct clade, the blastozoans, are distinguished by possession of brachioles (4), feeding structures distinct from arms and made entirely of axial body wall, representing "naked" ambulacra unaccompanied by left or right coelom elements. Elaborations from the earliest blastozoans (*Lepidocystis*), included development of a stalk fashioned from imperforate extraxial region (5). Eocrinoids show the earliest morphology, with development into a demarcated stalk in cystoids and blastoids. Early, derived echinoderms such as carpoids fall into this clade by possession of a brachiole and highly defined stalk. The clade with the first extant forms is characterized by true arms containing extensions of left and right somatocoels (6). Crinoids independently evolved a stem inserted as a module within the imperforate extraxial region (7). Starfish, brittle stars, sea urchins, and sea cucumbers possess a madreporic sieve plate (8), robust tube feet for locomotion and food gathering in ambulacra lacking cover plates (9), a downward-directed mouth with the anus on the opposite (aboral) surface in which the imperforate extraxial region is virtually lost (10). Ophiuroids and asteroids are characterized by specialized plate systems such as lateral arm plates (11). Ophiuroids evolved fused ambulacral plates (vertebrae) that demarcate arms from the central disk (12). The ancestor of echinoids and holothuroids possessed an outpocketing of the left somatocoel (peripharyngeal coelom) in which the jaw apparatus develops (13). Aristotle's lantern is expressed in echinoids and forms basal to holothuroids (ophiocistioids), which begin shortening the rays and enlarging tube feet that become restricted to the oral region (14). Holothuroids take this to the extreme, suppressing the lantern as paedomorphic "giant larvae" (15). Earliest holothuroids were like modern apodids, lacking tube feet along the body. Only in more recent taxa do secondary, interradial canals (not homologous to the radial canals of other echinoderms) induce formation of tube foot-like structures along the body (16). Echinoids lost nearly all of the extraxial region, restricting it to a few plates in the unique apical system (17). Echinoids also uniquely internalized the radial water vessel (18). The earliest echinoids did not actually have a rigid test (the plates did not tightly suture together to make the test until later).

the group labeled the "cryptosyringids" due to a supposed common possession of internal radial canals—a feature now known to have arisen independently via very different, nonhomologous pathways. Recent evidence, including molecular phylogenetics, strongly supports joining of asteroids and ophiuroids in one clade, and echinoids and holothuroids in another, sister clade. Asteroids and ophiuroids are constructed along very much the same body plan—so close in fact that there remain arguments over which fossils should be placed in which of the two groups.

With support from molecular evidence, close relationship between echinoids and holothuroids is accepted by most specialists. However, holothuroids are not simply decalcified echinoids, as once supposed. With the realization that holothuroids are paedomorphic, largely lacking features that could be used to relate to those of other extant groups, morphological synapomorphies linking holothuroids with echinoids are difficult to identify. These require examination of ontogenetic patterns such as those involved in formation of the peripharyngeal coelom and Aristotle's lantern contained within.

In summary, the ancient origin of the stereom and pentaradial hydrocoel patterning to induce adult pentaradiality are keys to understanding echinoderm evolution. These features allowed departure from infaunal creatures like hemichordates to epibenthic surface-dwelling animals. The water vascular system probably originally served for suspension and perhaps detritus feeding, facilitated by small tube feet in narrow ambulacra oriented around an upward-facing mouth. Minimal participation of the ambulacra (i.e., axial region) in the body plan meant that early echinoderms likely underwent minimal metamorphosis. There was no requirement to "throw away" the larval body, which was preserved in the adult in predominant extraxial regions. As tube feet and the attendant water vascular system became increasingly significant in echinoderm lifestyles, pentaradial expression in the body plan also increased, leading to more pronounced metamorphosis to suppress the extraxial larval body in the adult. Concomitant with the more errant, active feeding lifestyle of modern taxa, specialized feeding structures such as enlarged suckered tube feet, buccal feeding tentacles, and even jaws and teeth were elaborated.

The proliferation of the echinoderm body plan since the Cambrian might seem surprising given the unique adoption of radial morphology among the deuterostomes. But it is clear that the combined qualities of supportive endoskeleton, coelomic water vascular system, and pentaradial symmetry have provided the basis for diversification and extremely disparate variations on this basic theme among the echinoderm lineages.

Selected References

General

Ameye, L., G. De Becker, C. Killian, F. Wilt, R. Kemps, S. Kuypers and P. Dubois. 2001. Proteins and saccharides of the sea urchin organic matrix of mineralization: Characterization and localization in the spine skeleton. J. Struct. Biol. 134: 56–66.

Ausich, W. I. and G. D. Webster (eds.). 2008. *Echinoderm Paleobiology*. Indiana University Press, Bloomington, Indiana.

Boolootian, R. A. (ed.). 1966. *Physiology of Echinodermata*. Interscience, New York.

Cannon, J. T., K. M. Kocot, D. S. Waits, D. A. Weese, B. J. Swalla, S. R. Santos and K. M. Halanych. 2014. Phylogenomic resolution of the hemichordate and echinoderm clade. Curr. Biol. 24(23): 2827–2832.

David, B., B. Lefebvre, R. Mooi and R. Parsley. 2000. Are homalozoans echinoderms? An answer from the extraxial-axial theory. Paleobiology 25: 529–555.

David, B. and R. Mooi. 2014. How Hox genes can shed light on the place of echinoderms among the deuterostomes. EvoDevo 5: 22–41.

David, B., R. Mooi and M. Telford. 1995. The ontogenetic basis of Lovén's Rule clarifies homologies of the echinoid peristome. Pp. 155–164 in R. Mooi and M. Telford (eds.), *Proceedings of the Ninth International Echinoderm Conference, San Francisco*. A. A. Balkema, Rotterdam.

Emlet, R. B. 1982. Echinoderm calcite: A mechanical analysis from larval spicules. Biol. Bull. 163: 264–275.

Emlet, R. B. 1983. Locomotion, drag, and the rigid skeleton of larval echinoderms. Biol. Bull. 164: 433–445.

Giese, A. C., J. S. Pearse and V. B. Pearse. 1991. *Reproduction of Marine Invertebrates. VI. Echinoderms and Lophophorates*. Boxwood Press, Pacific Grove, CA.

Grande, C., J. M. Martín-Durán, N. J. Kenny, M. Truchado-García and A. Hejnol. 2014. Evolution, divergence and loss of the Nodal signaling pathway: New data and a synthesis across Bilateria. Int. J. Dev. Biol. 58: 521–532.

Hejnol, A. and J. M. Martín-Durán. 2015 Getting to the bottom of anal evolution. Zoologischer Anzeiger, doi: 10.1016/j.jcz.2015.02.006

Hejnol, A. and 16 others. 2009. Assessing the root of bilaterian animals with scalable phylogenomic methods. Proc. R. Soc. B 276: 4261–4270.

Hendler, G., J. E. Miller, D. L. Pawson and P. M. Kier. 1995. *Echinoderms of Florida and the Caribbean: Sea stars, sea urchins, and allies*. Smithsonian Institution Press.

Hyman, L. H. 1955. *The Invertebrates. Vol. 4, Echinodermata. The Coelomate Bilateria*. McGraw-Hill, New York.

Jangoux, M. (ed.). 1980. *Echinoderms: Present and Past*. A. A. Balkema, Rotterdam.

Jangoux, M. and J. M. Lawrence (eds.). 1982. *Echinoderm Nutrition*. A. A. Balkema, Rotterdam.

Jangoux, M. and J. M. Lawrence (eds.). 1983, 1987. *Echinoderm Studies. Vols. 1 and 2*. A. A. Balkema, Rotterdam.

Jefferies, R. P. S. 1986. *The Ancestry of the Vertebrates*. British Museum of Natural History, London.

Kroh, A. and A. B. Smith. 2010. The phylogeny and classification of post-Palaeozoic echinoids. J. Syst. Palaeo. 8: 147–212.

Lawrence, J. M. 1987. *A Functional Biology of Echinoderms*. Johns Hopkins University Press, Baltimore.

Lowe, C. J. and G. A. Wray. 1997. Radical alterations in the roles of homeobox genes during echinoderm evolution. Nature 389: 718–721. [See also the correction in Nature 392: 105.]

Martín-Durán, J. M., R. Janssen, S. Wennberg, G. E. Budd and A. Hejnol. 2012. Deuterostomic development in the protostome *Priapulus caudatus*. Curr. Biol. 22: 2161–2166.

Martindale, M. Q. and A. Hejnol. 2009. A developmental perspective: Changes in the position of the blastopore during bilaterian evolution. Dev. Cell 17: 162–174.

Mooi, R. 2000. Not all written in stone: Interdisciplinary syntheses in echinoderm paleontology. Can. J. Zool. 79: 1209–1231.

Mooi, R. and B. David. 1997. Skeletal homologies of echinoderms. Paleo. Soc. Papers 1997 3: 305–335.

Mooi, R. and B. David. 1998. Evolution within a bizarre phylum: Homologies of the first echinoderms. Amer. Zool. 38: 965-974.

Mooi, R. and B. David. 2008. Radial symmetry, the anterior/posterior axis, and echinoderm Hox genes. Annu. Rev. Ecol. Evol. Syst. 39: 43–62.

Mooi, R., B. David and G. Wray. 2005. Arrays in rays: Terminal addition in echinoderms and its correlation with gene expression. Evol. Dev. 7: 542–555.

Moore, R. C. (ed.). 1966–1978. *Treatise on Invertebrate Paleontology. Parts S–U, Echinodermata*. Geological Society of America and University of Kansas Press.

Motokawa, T. 1984. Connective tissue catch in echinoderms. Biol. Rev. 59: 255–270.

Nichols, D. 1972. The water-vascular system in living and fossil echinoderms. Paleontology 15: 519–538.

Paul, C. R. C. and A. B. Smith (eds.). 1988. *Echinoderm Ontogeny and Evolutionary Biology*. Clarendon Press, Oxford.

Pearse, J. S., R. Mooi, S. J. Lockhart, and A. Brandt. 2009. Brooding and species diversity in the Southern Ocean: Selection for brooders or speciation within brooding clades? Pp. 182–196 in I. M. Krupnik, M. A. Lang and S. E. Miller (eds.), *Smithsonian at the Poles: Contributions to International Polar Year Science*. Smithsonian Institution Scholarly Press, Washington, D.C.

Pennington, J. T. and R. R. Strathmann. 1990. Consequences of the calcite skeletons of planktonic echinoderm larvae for orientation, swimming, and shape. Biol. Bull. 179: 121–133.

Peterson, K. J., C. Arenas-Mena, and E. H. Davidson. 2000. The A/P axis in echinoderm ontogeny and evolution: Evidence from fossils and molecules. Evol. Dev. 2: 93–101.

Philip, G. M. 1979. Carpoids: Echinoderms or chordates? Biol. Rev. 54: 439–471.

Sprinkle, J. 1973. *Morphology and Evolution of Blastozoan Echinoderms*. Special Publication. Museum of Comparative Zoology, Harvard University, Cambridge, MA.

Sumrall, C. D. 1997. The role of fossils in the phylogenetic reconstruction of Echinodermata. Paleo. Soc. Papers 1997 3: 267–288.

Sumrall, C. D. and G. Wray. 2007. Ontogeny in the fossil record: Diversification of body plans and the evolution of "aberrant" symmetry in Paleozoic echinoderms. Paleobiology 33: 149–163.

Swalla, B. J. and A. B. Smith. 2008. Deciphering deuterostome phylogeny: Molecular, morphological and palaeontological perspectives. Phil. Trans. Roy. Soc. B, Biol. Sci. 363: 1557–1568.

Ubaghs, G. 1975. Early Paleozoic echinoderms. Ann. Rev. Earth Planet. Sci. 3: 79–98.

Wilkie, I. C. 2005. Mutable collagenous tissue: Overview and biotechnological perspective. Prog. Molec. Subcell. Biol. 39: 221–250.

Wray, G. A. 1997. Echinoderms. Pp. 309–329 in S. F. Gilbert and A. M. Raunio (eds.), *Embryology: Constructing the Organism*. Sinauer Associates, Sunderland, MA.

Young, C. M. (ed.). 2002. *Atlas of Marine Invertebrate Larvae*. Academic Press, London.

Zamora, S., I. A. Rahman and A. B. Smith. 2012. Plated Cambrian bilaterians reveal the earliest stages of echinoderm evolution. PLoS ONE 7(6): e38296.

Crinoidea

Baumiller, T. K., R. Mooi and C. G. Messing. 2008. Urchins in the meadow: Paleobiological and evolutionary implications of cidaroid predation on crinoids. Paleobiology 34: 22–34.

Grimmer, J. C., N. D. Holland and C. G. Messing. 1984. Fine structure of the stalk of the bourgueticrinid sea lily *Democrinus conifer* (Echinodermata: Crinoidea). Mar. Biol. 81: 163–176.

Guensburg, T. E. 2012. Phylogenetic implications of the oldest crinoids. J. Paleo. 86: 455–461.

Guensburg, T. E., R. Mooi, J. Sprinkle, B. David and B. Lefebvre. 2010. Pelmatozoan arms from the Middle Cambrian of Australia: Bridging the gap between brachioles and brachials? Comment: There is no bridge. Lethaia 45: 432–440.

Heinzeller, T. and U. Welsch. 1994: Crinoidea. Pp. 9–149 in F. W. Harrison (ed.), *Microscopic Anatomy of Invertebrates*. Wiley-Liss, New York, NY.

Hemery, L. G., M. Roux, N. Ameziane and M. Eleaume. 2013. High-resolution crinoid phyletic inter-relationships derived from molecular data. Cah. Biol. Mar. 54: 511–523.

Rouse, G. W. and 11 others. 2013. Fixed, free, and fixed: The fickle phylogeny of extant Crinoidea (Echinodermata) and their Permian–Triassic origin. Molec. Phyl. Evol. 66: 161–181.

Asteroidea

Baker, A. N., F. W. E. Rowe and H. E. S. Clark. 1986. A new class of Echinodermata from New Zealand. Nature 321: 862–864.

Blake, D. B. 1987. A classification and phylogeny of post-Paleozoic sea stars (Asteroidea: Echinodermata). J. Nat. Hist. 21: 481–528.

Blake, D. B. 2013. Early asterozoan (Echinodermata) diversification: A paleontologic quandary. J. Paleo. 87: 353–372.

Emson, R. H. and C. M. Young. 1994. Feeding mechanisms of the brisingid starfish *Novodinia antillensis*. Mar. Biol. 118: 433–442.

Ferguson, J. C. and C. W. Walker. 1991. Cytology and function of the madreporite systems of the starfish *Henricia sanguinolenta* and *Asterias vulgaris*. J. Morphol. 210: 1–11.

Gale, A. S. 2011. *The phylogeny of post-Palaeozoic Asteroidea (Neoasteroidea, Echinodermata)*. Palaeontological Association, London.

Glynn, P. W. 1974. The impact of *Acanthaster* on corals and coral reefs in the Eastern Pacific. Environ. Conserv. 1(4): 295–304.

Pearse, J. S., D. J. Eernisse, V. B. Pearse and K. A. Beauchamp. 1986. Photoperiodic regulation of gametogenesis in sea stars, with evidence for an annual calendar independent of fixed day length. Am. Zool. 26: 417–431.

Smith, A. B. 1988. To group or not to group: The taxonomic position of *Xyloplax*. Pp. 17–23 in R. D. Burke et al. (eds.), *Echinoderm Biology*. A. A. Balkema, Rotterdam.

Ophiuroidea

Aizenberg, J., A. Tkachenko, S. Weiner, L. Addadi and G. Hendler. 2001. Calcitic microlenses as part of the photoreceptor system in brittlestars. Nature 412(6849): 819–822.

Ferguson, J. C. 1995. The structure and mode of function of the water vascular system of a brittlestar, *Ophioderma appressum*. Biol. Bull. 188: 98–110.

Hendler, G. 1978. Development of *Amphioplus abditus* (Verrill) (Echinodermata: Ophiuroidea). II. Description and discussion of ophiuroid skeletal ontogeny and homologies. Biol. Bull. 154: 79–95.

Hendler, G. 1982. Slow flicks show star tricks: Elapsed-time analysis of basketstar (*Astrophyton muricatum*) feeding behavior. Bull. Mar. Sci. 32: 909–918.

Hendler, G. 1983. The association of *Ophiothrix lineata* and *Callyspongia vaginalis*: A brittlestar-sponge cleaning symbiosis? Mar. Ecol. 5(1): 9–27.

Stancyk, S. E., T. Fujita and C. Muir. 1998. Predation behavior on swimming organisms by *Ophiura sarsii*. Pp. 425–429 in R. Mooi and M. Telford (eds.), *Proceedings of the Ninth International Echinoderm Conference, San Francisco.* A. A. Balkema, Rotterdam.

Woodley, J. D. 1975 The behavior of some amphiurid brittlestars. J. Exp. Mar. Biol. Ecol. 18: 29–46.

Echinoidea

Armstrong, N. and D. R. McClay. 1994. Skeletal pattern is specified autonomously by the primary mesenchyme cells in sea urchin development. Dev. Biol. 162: 329–338.

Burke, R. D. 1984. Pheromonal control of metamorphosis in the Pacific sand dollar, *Dendraster excentricus*. Science 225: 223–224.

Coppard, S. E., A. Kroh and A. B. Smith. 2012. The evolution of pedicellariae in echinoids: An arms race against pests and parasites. Acta Zool. 93: 125–148.

Ellers, O. and M. Telford. 1984. Collection of food by oral surface podia in the sand dollar, *Echinarachnius parma* (Lamarck). Biol. Bull. 166: 574–582.

Emlet, R. B. 2010. Morphological evolution of newly metamorphosed sea urchins: A phylogenetic and functional analysis. Integr. Comp. Biol. 50: 571–588.

Lewis, J. B. 1968. The function of sphaeridia of sea urchins. Can. J. Zool. 46: 1135–1138.

Mooi, R. 1986. Non-respiratory podia of clypeasteroids (Echinodermata, Echinoidea): I. Functional anatomy. Zoomorphologie 106: 21–30.

Mooi, R. 1990. Paedomorphosis, Aristotle's lantern, and the origin of the sand dollars (Echinodermata: Clypeasteroida). Paleobiology 16 (1): 25–48.

Smith, A. B. 1984. *Echinoid Paleobiology*. Allen and Unwin, London.

Smith, A. B. and C. H. Jeffery. 1988. Selectivity of extinction among sea urchins at the end of the Cretaceous period. Nature 392: 69–71.

Strathmann, R. R., L. Fenaux and M. F. Strathmann. 1992. Heterochronic developmental plasticity in larval sea urchins and its implications for evolution of nonfeeding larvae. Evolution 46: 972–986.

Telford, M. 1981. A hydrodynamic interpretation of sand dollar morphology. Bull. Mar. Sci. 31: 605–622.

Telford, M. 1983. An experimental analysis of lunule function in the sand dollar *Mellita quinquiesperforata*. Mar. Biol. 76: 125–134.

Telford, M. and R. Mooi. 1986. Resource partitioning by sand dollars in carbonate and siliceous sediments: Evidence from podial and particle dimensions. Biol. Bull. 171: 197–207.

Telford, M., R. Mooi and O. Ellers. 1985. A new model of podial deposit feeding in the sand dollar, *Mellita quinquiesperforata* (Leske): The sieve hypothesis challenged. Biol. Bull. 169: 431–448.

Telford, M., R. Mooi and A. S. Harold. 1987. Feeding activities of five species of *Clypeaster* (Echinoides, Clypeasteroida): Further evidence of clypeasteroid resource partitioning. Biol. Bull. 172: 324–336.

Timko, P. L. 1976. Sand dollars as suspension feeders: a new description of feeding in *Dendraster excentricus*. Biol. Bull. 151: 247–259.

Wray, G. A. 1992. The evolution of larval morphology during the post-Paleozoic radiation of echinoids. Paleobiol. 18: 258–287.

Wray, G. A. 1996. Parallel evolution of nonfeeding larvae in echinoids. Syst. Biol. 45 (3): 308–322.

Yerramilli, D. and S. Johnsen. 2010. Spatial vision in the purple sea urchin *Strongylocentrotus purpuratus* (Echinoidea). J. Exp. Biol. 213: 249–255.

Ziegler, A., R. Mooi, G. Rolet and C. De Ridder. 2010. Origin and evolutionary plasticity of the gastric caecum of sea urchins (Echinodermata: Echinoidea). BMC Evol. Biol. 10: 313–345.

Holothuroidea

Costelloe, J. and B. Keegan. 1984. Feeding and related morphological structures in the dendrochirote *Aslia lefevrei* (Holothuroidea: Echinodermata). Mar. Biol. 84: 135–142.

Francour, P. 1997. Predation on holothurians: A literature review. Invert. Biol. 116 (1): 53–60.

Hamel, J. F., C. Conand, D. L. Pawson and A. Mercier. 2001. The sea cucumber *Holothuria scabra* (Holothuroidea: Echinodermata): Its biology and exploitation as Beche-de-Mer. Adv. Mar. Biol. 41: 129-232.

Herreid, C. F., V. F. LaRussa and C. R. DeFesi. 1976. Blood vascular system of the sea cucumber *Stichopus moebii*. J. Morphol. 150: 423–451.

Kerr, A. M. and J. Kim. 2001. Phylogeny of Holothuroidea (Echinodermata) inferred from morphology. Zool. J. Linn. Soc. 133: 63–81.

Martin, W. E. 1969. *Rynkatorpa pawsoni* n. sp. (Echinodermata: Holothuroidea), a commensal sea cucumber. Biol. Bull. 137: 332–337.

Moriarty, D. J. W. 1982. Feeding of *Holothuria atra* and *Stichopus chloronotus* on bacteria, organic carbon and organic nitrogen in sediments of the Great Barrier Reef. Aust. J. Mar. Freshwater Res. 33: 255–263.

Pawson, D. L. and A. M. Kerr. 2001. Chitin in echinoderms? Tentacle sheaths in the deep-sea holothurian *Ceraplectana trachyderma* (Hoilothuroidea: Molpadiida). Gulf Mex. Sci. 19(2): 192

Vanden Spiegel, D. and M. Jangoux. 1987. Cuvierian tubules of the holothuroid *Holothuria forskali* (Echinodermata): A morphofunctional study. Mar. Biol. 96: 263–275.

CHAPTER 26

Phylum Hemichordata

Acorn Worms and Pterobranchs

The hemichordates constitute a phylum of exclusively marine, benthic, bilaterally-symmetrical deuterostome invertebrates (see box on this page and Box 26A). They inhabit all sea bottoms of the world, from intertidal environments to abyssal depths. The phylum includes about 135 species, most (121) of which are benthic burrowers known as acorn worms (or tongue worms) grouped in the class Enteropneusta (Figures 26.1A and 26.2A–D). The remaining species belong to the class Pterobranchia—largely colonial forms of minute zooids superficially resembling bryozoans (Figures 26.1B and 26.2E–H).

The first known hemichordate, a specimen of *Ptychodera flava* (an enteropneust), was discovered in 1821 and recorded by Eschscholtz in 1825 as an aberrant holothurian. Müller discovered in 1850 the first specimen of a tornaria larva, although it was considered to be an echinoderm larva until Metschnikoff recognized it as an enteropneust in 1869. The first pterobranch specimens described were individuals of *Rhabdopleura* dredged by G. O. Sars in 1866, but misidentified as bryozoans. S. F. Harmer later proposed their relationship with *Cephalodiscus* specimens recovered by the Challenger expedition and the relation of both with enteropneusts. Harmer also adopted the name Hemichordata for the whole group as proposed by William Bateson in 1885. Most relevant contributions to the knowledge of hemichordate biology in late nineteenth and early twentieth centuries are from J. W. Spengel, C. Dawydoff, and C. van der Horst. Nowadays hemichordates continue to provide exciting discoveries, such as the genus *Saxipendium* from hydrothermal vents, and the bizarre deep-sea enteropneust family Torquaratoridae (Figure 26.2D). Recently, a minute acorn worm, *Meioglossus psammophilus,* has been reported from Belize and Bermuda, living permanently in coral sand as part of meiofauna communities.

Enteropneust worms generally live buried in soft sediments, among algal holdfasts, or under

Classification of The Animal Kingdom (Metazoa)

Non-Bilateria*
(a.k.a. the diploblasts)
PHYLUM PORIFERA
PHYLUM PLACOZOA
PHYLUM CNIDARIA
PHYLUM CTENOPHORA

Bilateria
(a.k.a. the triploblasts)
PHYLUM XENACOELOMORPHA

Protostomia
PHYLUM CHAETOGNATHA

SPIRALIA
PHYLUM PLATYHELMINTHES
PHYLUM GASTROTRICHA
PHYLUM RHOMBOZOA
PHYLUM ORTHONECTIDA
PHYLUM NEMERTEA
PHYLUM MOLLUSCA
PHYLUM ANNELIDA
PHYLUM ENTOPROCTA
PHYLUM CYCLIOPHORA

Gnathifera
PHYLUM GNATHOSTOMULIDA
PHYLUM MICROGNATHOZOA
PHYLUM ROTIFERA

Lophophorata
PHYLUM PHORONIDA
PHYLUM BRYOZOA
PHYLUM BRACHIOPODA

ECDYSOZOA
Nematoida
PHYLUM NEMATODA
PHYLUM NEMATOMORPHA

Scalidophora
PHYLUM KINORHYNCHA
PHYLUM PRIAPULA
PHYLUM LORICIFERA

Panarthropoda
PHYLUM TARDIGRADA
PHYLUM ONYCHOPHORA
PHYLUM ARTHROPODA
SUBPHYLUM CRUSTACEA*
SUBPHYLUM HEXAPODA
SUBPHYLUM MYRIAPODA
SUBPHYLUM CHELICERATA

Deuterostomia
PHYLUM ECHINODERMATA
PHYLUM HEMICHORDATA
PHYLUM CHORDATA

*Paraphyletic group

This chapter has been revised by Fernando Pardos and Jesús Benito.

BOX 26A Characteristics of the Phylum Hemichordata

1. Bilaterally symmetrical deuterostomes, body vermiform or saccate and fundamentally trimeric, with prosome, mesosome, and metasome, each with coelomic compartments; solitary or colonial; pterobranchs with mesocoelic extensions into the arms and tentacles
2. With mesocoelic ducts and pores
3. With pharyngotremy (communication of the gut to the exterior via ciliated pharyngeal gill slits and pores)
4. Well developed, open circulatory system
5. Unique excretory structure, the glomerulus
6. Gonads extracoelic, in metasome
7. Complete gut. Deposit or suspension feeders
8. With buccal diverticulum or stomochord as supporting structure for the prosome; stomochord not homologous with the chordate notochord
9. Circular and longitudinal muscles present in body wall of proboscis and collar of enteropneusts; pterobranchs with longitudinal muscles only
10. Collagenous skeletal structures in the proboscis and gill bars of enteropneusts derived from basement membranes of adjacent epithelia
11. Short, dorsal, mesosomal, occasionally hollow nerve cord (neurochord), probably homologous with chordate nerve cord
12. Gonochoristic, with external fertilization and indirect development through a unique tornaria larva; some species with direct development; asexual reproduction common
13. Cleavage radial, holoblastic, more or less equal. Although blastopore denotes posterior end of body, both the mouth and the anus form secondarily, subsequent to closure of the blastopore. Mesoderm and body cavities form by enterocoely.
14. Strictly marine and benthic

rocks; they have been considered largely intertidal, but recent discoveries point to a substantial and undocumented enteropneust diversity occurring in the deep sea. Of the few deep-water acorn worms known, one species, *Saxipendium coronatum* (the "spaghetti worm"), is a member of hydrothermal vent communities; some other deep-water forms construct highly branched burrow systems, and recent videotaping records confirm that some deep-sea enteropneusts are epibenthic (Figure 26.2D) and use the water column to move between feeding sites. As we have seen in so many other benthic, sessile, and sedentary invertebrates, the hemichordates include a planktonic dispersal phase in their life history strategies. The life cycle of acorn worms includes a free-swimming tornaria larva stage, although direct development occurs in some species. Even in those enteropneusts with technically direct development, the pattern is strategically indirect, the embryos being at least planktonic free-living animals, even if not full-fledged larvae.

Pterobranchs are colonial, tube-dwelling hemichordates that feed with ciliated arms and tentacles (Figure 26.2E–H). Many species have been found in Antarctic waters, although some colonies are found in shallow tropical waters attached to the underside of rocks, bivalve mollusc shells, or coral rubble. Pterobranchs typically live in colonies formed by extensive aggregations of tubes called **coenecia** with one individual in each tube. Zooids of *Rhabdopleura* are connected via a stalk or stolon of living tissue, while zooids of *Cephalodiscus* are connected to a basal disk. Colonies of pterobranchs grow through asexual reproduction by budding. Sexual reproduction involves a swimming, ciliated larva that settles to form a new colony. This pattern is common among small sessile animals, which cannot afford to produce huge numbers of eggs but depend on at least a short-lived dispersal phase.

HEMICHORDATE CLASSIFICATION

The classification presented herein follows the traditional division of the phylum into two classes, Enteropneusta and Pterobranchia, and adopts the general opinion that *Planctosphaera pelagica* is the hypertrophied larva of a yet unknown enteropneust. However, there is currently growing evidence from molecular studies that pterobranchs are the sister taxon of the enteropneust family Harrimaniidae, making the Enteropneusta a paraphyletic taxon (see the phylogeny discussion later in this chapter).

CLASS ENTEROPNEUSTA Acorn, or tongue worms. Vermiform, with three body regions as proboscis, collar, and trunk; coeloms reduced by muscle development; gut elongate, straight; mouth ventral at anterior end of collar; long dorsolateral series of gill slits; dorsal, hollow nerve cord in the collar; anus posterior, terminal; marine, burrow in soft sediments or nestle under rocks or in algal holdfasts; largely intertidal although deep-water species are increasingly being discovered. About 121 described species in four families: Harrimaniidae (*Harrimania, Protoglossus, Horstia, Meioglossus, Mesoglossus, Ritteria, Saccoglossus, Saxipendium, Stereobalanus, Xenopleura*); Spengelidae (*Glandiceps, Schizocardium, Spengelia, Willeyia*); Ptychoderidae (*Balanoglossus, Glossobalanus, Ptychodera*); and Torquaratoridae (*Allapasus, Tergivellum, Torquarator, Yoda*). The former class Planctosphaeroidea was erected to include *Planctosphaera pelagica*, discovered in 1932, and viewed by most contemporary authorities as the larva of a yet unknown enteropneust.

CLASS PTEROBRANCHIA Pterobranchs. Body sacciform; with three body regions as preoral disc (= cephalic shield), tentaculate mesosome, and metasome subdivided as trunk and stalk; neurochord lacking; pharynx with one pair

(A) 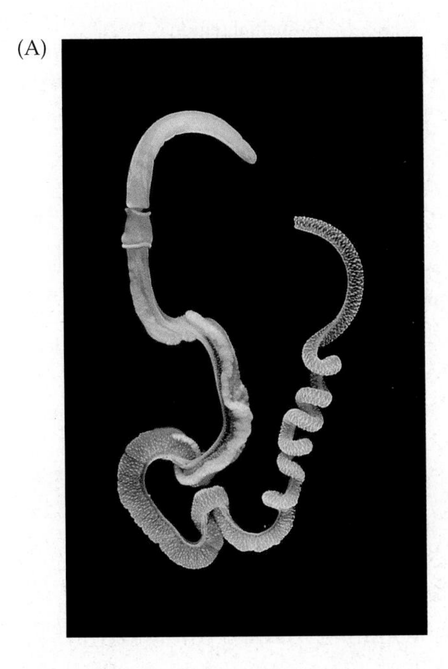

(B) 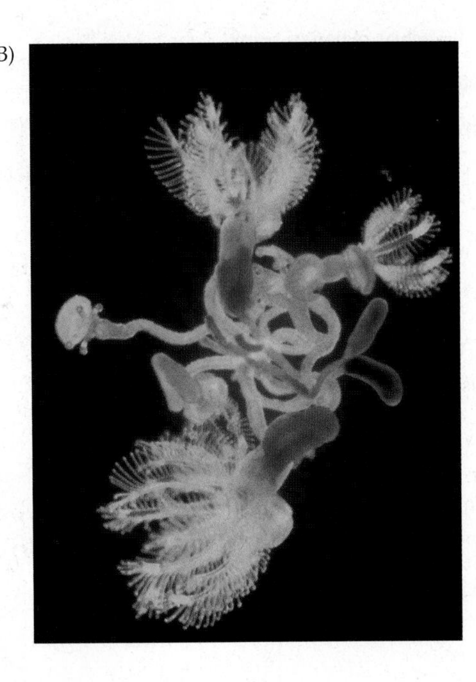

Figure 26.1 Representative hemichordates. (A) The acorn worm *Saccoglossus* (phylum Hemichordata, class Enteropneusta). (B) Portion of a *Cephalodiscus* colony (phylum Hemichordata, class Pterobranchia), showing several individuals at different stages of development.

of gill slits or none; gut U-shaped; marine; generally small (less than 1 cm), aggregating, or colonial. About 14 species, in two families: Rhabdopleuridae (*Rhabdopleura*) and Cephalodiscidae (*Cephalodiscus* and *Atubaria*). The genus *Atubaria* is represented by a single species (*A. heterolopha*), known only from 43 specimens collected in 1935 in Japan. These animals are nearly identical to *Cephalodiscus* but were found as solitary forms, without secreted casings, hence their name. However, they have never been found again and individuals of *Cephalodiscus* are known to abandon their casing when stressed, making *Atubaria* doubtful as a valid genus.

The Hemichordate Body Plan

The two classes of Hemichordata have very different appearances and ways of life: enteropneusts are burrowing, vermiform, solitary animals, whereas pterobranchs are sessile, tube dwelling, and colonial—although both share a common trimeric body plan, attesting to the evolutionary plasticity of their basic deuterostome architecture. The hemichordate body is divided into prosome, mesosome, and metasome. The prosome—the enteropneust proboscis and the pterobranch preoral disc or cephalic shield—is joined by a slender peduncle to the mesosome or collar. In pterobranchs, the latter bears two to several arms on which tentacles are located. The metasome or trunk is elongated in enteropneusts and globular to pyriform in pterobranchs, differentiated in the former by the specializations of the gut into branchiogenital, hepatic, and posterior regions, and extended posteriorly in the latter to form a tail or stolon where new individuals form by budding. Diagnostic characters of the phylum also include the presence of a dorsal buccal diverticulum called the **stomochord**, paired gill slits that connect the pharynx to the exterior, a proboscis excretory complex formed by heart vesicle, heart sinus, glomerulus, and proboscis pore, and a dorsal nerve cord in the mesosome. Enteropneusts and pterobranchs look so different and have such different ways of making a living that we treat them separately here.

Enteropneusta (Acorn Worms)

External anatomy

Enteropneusts are solitary, elongate, vermiform animals, ranging in length from a few centimeters to over 2 meters. Their bodies are clearly divided into three regions, named **proboscis**, **collar**, and **trunk**, (Figure 26.2) homologous with the prosome, mesosome, and metasome of other deuterostomes. The proboscis is short and pyriform to conical. A short, thin proboscis stalk connects the proboscis dorsally to the collar, the latter bearing the ventral mouth at its anterior end. The anus terminates at the posterior end of the long trunk. The trunk bears middorsal and midventral longitudinal ridges that correspond to the location of certain longitudinal nerves and blood vessels. In addition, the trunk is variably differentiated into regions along its length mostly by specializations of the gut (Figure 26.2A,B). Most species bear a clear, anterior, branchiogenital region characterized by the presence of two dorsolateral series of gill (= branchial) pores. Gonads also occur along this region; species from at least one family (Ptychoderidae) develop external longitudinal genital wings containing the gonads (Figure 26.2B). The anterior portion of the intestine shows a dorsolateral series of outpocketings that protrude visibly to

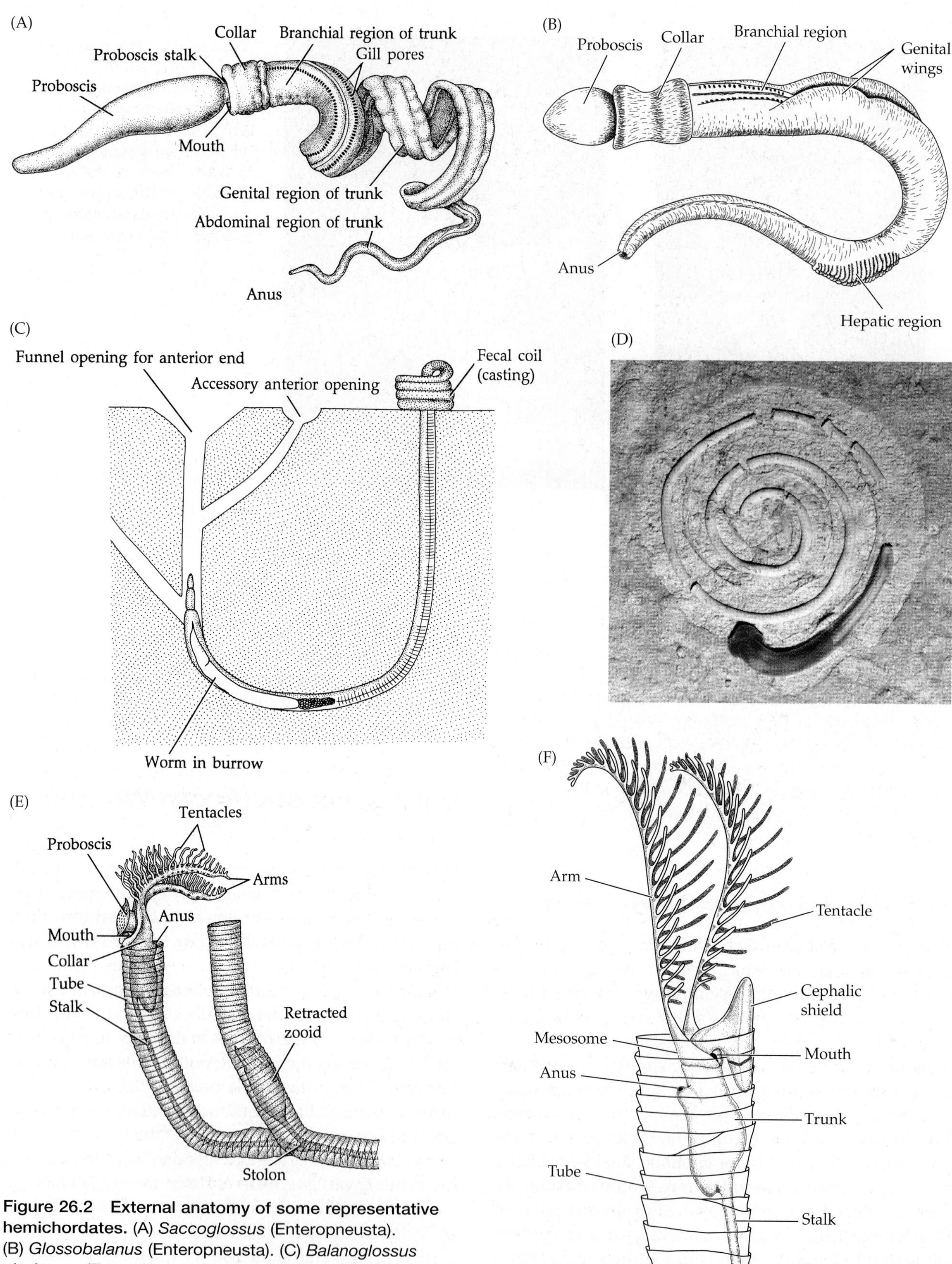

Figure 26.2 External anatomy of some representative hemichordates. (A) *Saccoglossus* (Enteropneusta). (B) *Glossobalanus* (Enteropneusta). (C) *Balanoglossus clavigerus* (Enteropneusta) in its burrow system Also see chapter opener photo of *B. sarniensis*, out of its burrow. (D) The epibenthic, deep-sea *Tergivellum cinnabarinum* (Entero-pneusta: Torquaratoridae) with its spiral cast. (E) Portion of a colony of *Rhabdopleura* (Pterobranchia). (F) Individual zooid of *Rhabdopleura*. (G) Portion of the colony of *Cephalodiscus* (Pterobranchia). (H) Individual zooid of *Cephalodiscus*. Feeding and rejection currents are shown by arrows. Water is drawn in between the tentacles and moves on a rejection current upward and away from the animal. Food particles are moved proximally along the tentacles to a food canal indicated by the arrowheads.

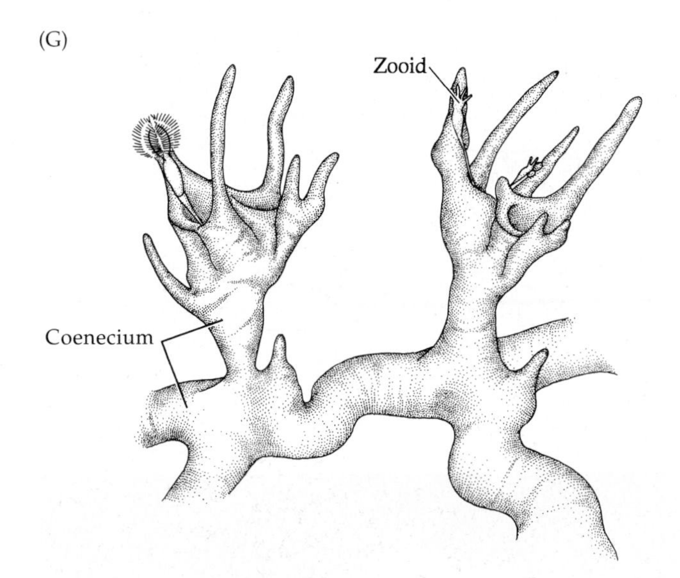

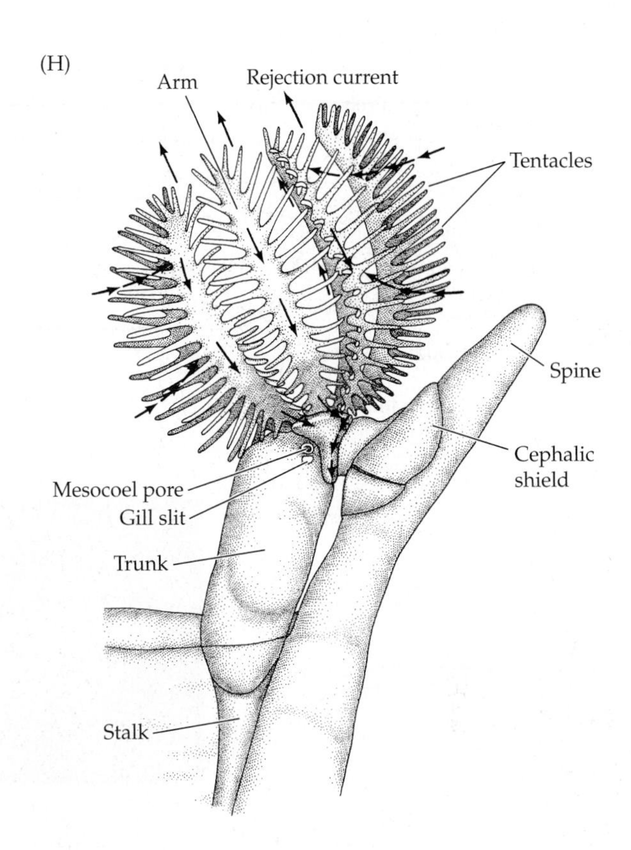

the exterior, generally exhibiting a dark greenish color and are referred to as the hepatic region. In a number of genera, no clear body subdivisions occur, except for the branchial region.

The epidermis of enteropneusts is formed by tall multiciliated cells, with a basiepithelial nerve plexus. It is usually richly supplied with gland cells, many of which are involved in mucus production—a very characteristic and abundant secretory product, with a distinctive iodine-like odor. Mucus secretion is particularly abundant on the proboscis and collar, where it is involved in trapping food particles for feeding. In some enteropneusts, certain epithelial cells produce noxious mucopolysaccharide compounds that may repel predators.

The epidermis, together with its associated nerve plexus, rests on a well-developed extracellular matrix formed by collagen fibers embedded in a ground substance. The extracellular matrix of the epidermis faces the extracellular matrix of the underlying mesothelium that lines the coelomic cavities. Both extracellular matrices fuse extensively except in places where blood lacunae occur, occupying spaces from blastocoelic origin.

Support Structures

The extracellular matrix is hypertrophied at some locations, forming skeletal supporting structures, namely the **proboscis skeleton** and the **gill bar supporting rods**. The proboscis skeleton is a single piece of cartilage shaped as an inverted Y, ventrally juxtaposed to the buccal diverticulum and with the two "horns" or **furcae** directed backwards and ventrally to the collar (Figure 26.3A,B). The gill apparatus—bars and slits—are supported by two dorsolateral sets of skeletal, collagenous pieces, each shaped as an elongated trident or "m." The central rod of the "m" lies inside primary gill bars, while secondary, or "tongue" bars bear two rods from adjacent pieces (Figure 26.3D,E).

In adult acorn worms, the buccal diverticulum, or stomochord, is a hollow and blind stiff rod formed by a highly vacuolated epithelium. It has a structural, supportive function working to antagonize the contractile pericardium, which pumps blood through the heart and into the glomerulus (Figure 26.3A). Nevertheless, body support is primarily a function of the hydrostatic nature of the body cavities, supplemented by the structural integrity of the body wall, connective tissues, and skeletal structures described above.

Coelomic Cavities

The arrangement of coelomic spaces follows the deuterostome trimeric body plan as a single protocoel in the proboscis and paired mesocoels and metacoels in the collar and trunk. The protocoel is open to the exterior through a single, left pore at the proboscis stalk in most enteropneusts; the mesocoel connects to the first, anterior-most branchial sac through two ciliated ducts, and the metacoel is blind. The metacoel sends two pairs of coelomic extensions with strong musculature into the collar, the so-called **perihemal** and **peribuccal coeloms**, to reinforce support and movement. Contraction of the longitudinal muscle fibers lining these cavities helps to retract the proboscis into the collar, thereby closing the mouth, like inserting a cork in a bottle, and may also help to pump blood into the heart.

Musculature and Locomotion

The musculature of enteropneusts derives from mesothelial linings of coelomic spaces. The arrangement of myofibrils is mostly of smooth type. Both circular and longitudinal muscles are present in the wall of the proboscis and anterior collar of acorn worms accounting

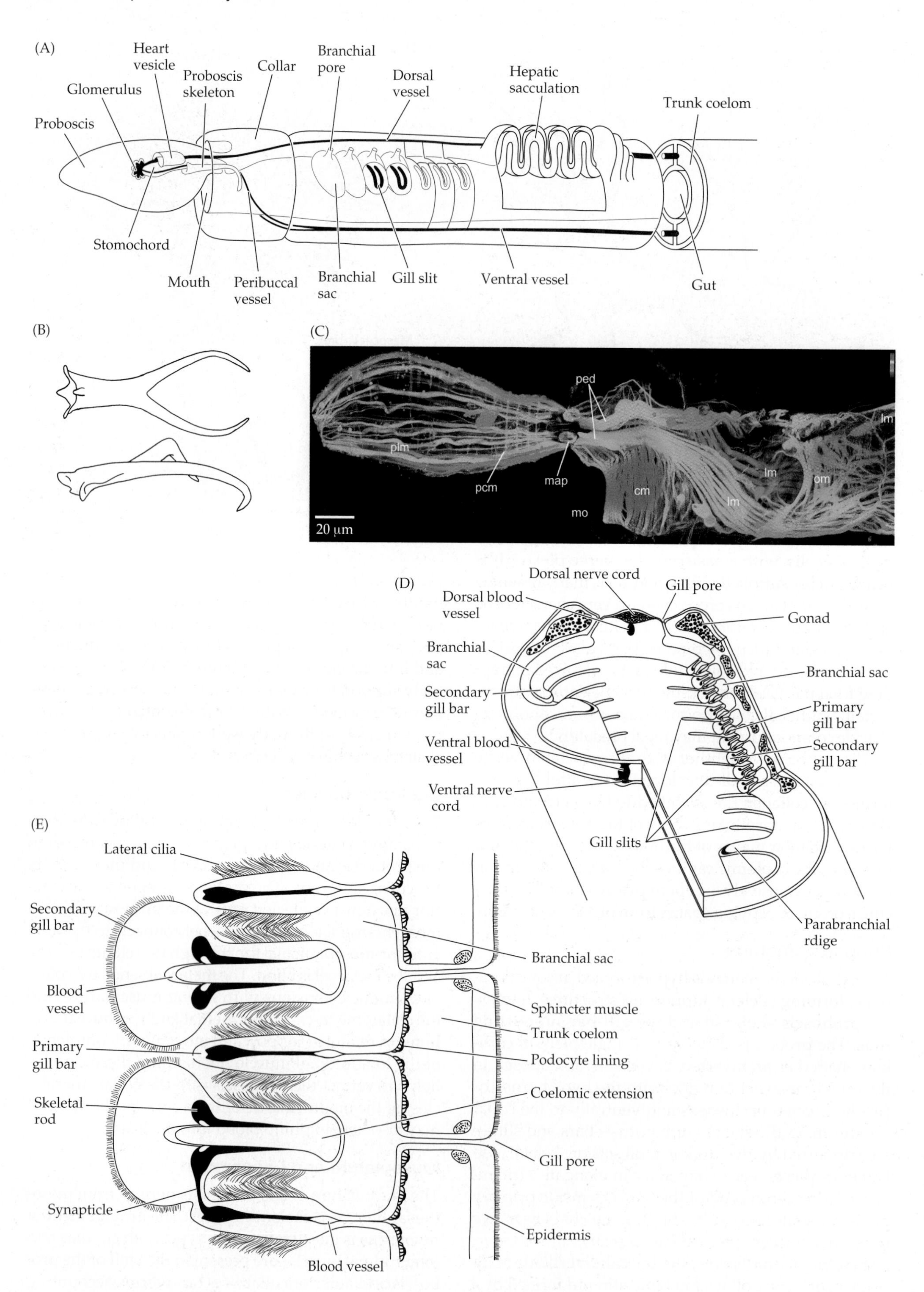
(A)
Heart vesicle
Glomerulus
Proboscis skeleton
Collar
Branchial pore
Dorsal vessel
Hepatic sacculation
Trunk coelom
Proboscis
Stomochord
Mouth
Peribuccal vessel
Branchial sac
Gill slit
Ventral vessel
Gut
(B)
(C)
ped
plm
pcm
map
mo
cm
lm
om
20 μm
(D)
Dorsal nerve cord
Gill pore
Dorsal blood vessel
Gonad
Branchial sac
Secondary gill bar
Primary gill bar
Ventral blood vessel
Ventral nerve cord
Gill slits
Parabranchial rdige
(E)
Lateral cilia
Secondary gill bar
Blood vessel
Primary gill bar
Skeletal rod
Synapticle
Branchial sac
Sphincter muscle
Trunk coelom
Podocyte lining
Coelomic extension
Gill pore
Epidermis
Blood vessel

Figure 26.3 Internal anatomy of enteropneusts. (A) Internal organization of an idealized acorn worm showing the proboscis complex, gill slits, sacs and pores, hepatic sacculations, and simplified circulatory system. (B) Proboscis skeleton of *Saccoglossus mereschkowskii* in ventral and lateral views. (C) Musculature of proboscis, collar, and anterior trunk of *Meioglossus psammophilus* revealed by Confocal Laser Scanning Microscopy (CLSM); phalloidin staining, depth coded. Depth scale follows the colors of spectral light (blue to red); cm: circular muscles; lm: longitudinal muscles; map: muscular anchoring point; mo: mouth opening; om: oblique muscles; pcm: proboscis circular muscles; ped: perihaemal diverticula. (D) 3-D reconstruction of the pharynx system of a ptychoderid enteropneust. (E) Detail of the organization of primary and secondary gill bars, branchial sacs, and pores.

for peristaltic movements in crawling or burrowing, but elsewhere only longitudinal fibers predominate (Figure 26.3C). Although most are probably strictly benthic and largely sedentary, at least one species (*Glandiceps hacksii*) sometimes swarms at the surface in shallow water, where it feeds on phytoplankton. Recent videotaping of torquaratorid enteropneusts from the deep sea show the animals detaching from the sea floor and entering the water column, presumably to be transported by currents to a new location.

Digestive System

The digestive tract of enteropneusts is a straight, regionally specialized tube, extending from the mouth to the anus. Gut musculature is scant, and the food is moved along largely by cilia. The mouth leads to a buccal cavity inside the collar, and can be closed by retraction of the proboscis against the latter. A buccal diverticulum or stomochord projects forward from its roof into the proboscis stalk. The pharynx follows the buccal cavity in the anterior part of the trunk. The pharynx bears two dorsolateral series of gill slits separated by a low, middorsal epibranchial ridge (Figure 26.3D). Two lateral parabranchial ridges divide the pharynx of ptychoderids into a dorsal branchial chamber and a ventral, digestive chamber. The gill slits number from a few to over 100 pairs. Each is a U-shaped opening in the wall of the pharynx that leads to a branchial sac and then to a dorsolateral gill pore through which water exits to the outside (Figure 26.3A,D). The septum between adjacent gill slits is a **primary gill bar** and the partition between the arms of the U of each slit is a **secondary gill bar** or tongue bar. Primary bars are solid whereas tongue bars contain an extension of the trunk coelom (Figure 26.3E). Both primary and secondary bars are supported by skeletal elements derived from the basement membrane of the gut lining and are supplied with a complex arrangement of blood vessels and sinuses. Short bridges called **synapticles** join the pharyngeal edges of primary and secondary gill bars in the family Ptychoderidae, undoubtedly contributing to the structural stability of the entire gills (Figure 26.3E).

Behind the pharynx is an esophagus, which at least in some (e.g., *Saccoglossus*) bears openings to the outside through the dorsal body wall, probably to eliminate excess water. In most species the wall of the anterior intestine folds deeply, forming two series of dorsolateral outpocketings that protrude visibly to the exterior in *Schizocardium* and the family Ptychoderidae (Figures 26.2B and 26.3A); the gut epithelium is full of dense green or brown inclusions that give a characteristic aspect to this hepatic region of the trunk. The intestine extends, more or less undifferentiated, to a short rectum terminating in the anus. The anus is dorsal in juveniles of Harrimaniids, leaving behind it a postanal region.

Feeding and Digestion

Enteropneusts are generally assumed to be deposit feeders, in some cases sedimentivorous (ingesting the substratum [*Balanoglossus*]) or detritivorous (trapping food particles from the sediment surface [*Saccoglossus*]). Others tend to be suspension or filter feeders that create ciliary currents to drive water with suspended particles into the gut. In fact, both classical and recent studies show that most species are probably capable of feeding by both methods—deposit and filtering—even simultaneously. Food material, including detritus and live organisms, is trapped in mucus secreted over the surface of the proboscis and moved posteriorly by ciliary currents (Figure 26.4). The sorting occurs at the proximal end of the proboscis and the stalk passes most large particles over the lip of the collar; these particles are then removed by special rejection currents. Most of the food is moved ventrally around the proboscis stalk, over a horseshoe-shaped structure called the **preoral ciliary organ**, and condensed into a mucous cord that is then passed into the inhalant flow of the mouth. The preoral ciliary organ includes a concentration of sensory neurons and probably also functions in chemoreception. The inhalant water current is powered by the long lateral cilia of the gill bars, leaving the body

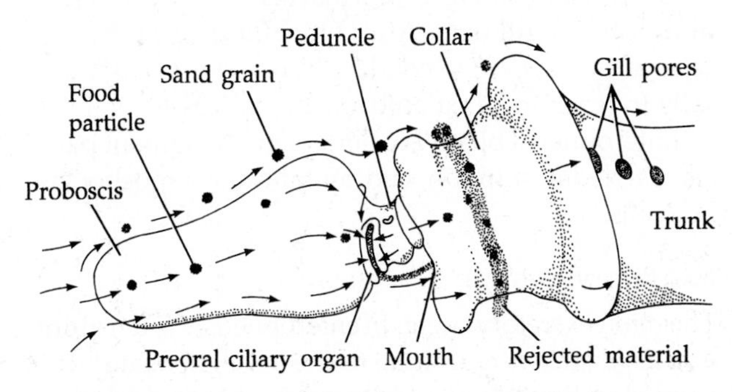

Figure 26.4 Food sorting and rejection currents (arrows) on the proboscis and collar region of the enteropneust *Protoglossus kohleri*.

through the branchial sacs and gill pores. Particles of adequate size are retained by pharyngeal mucus and transported ventrally to form a mucus-food cord that is directed posteriorly. The esophagus squeezes excess water from the food. Digestion and absorption of nutrients take place at the hepatic sacculations, first extracellularly in the most anterior diverticuli and then intracellularly in the posterior ones, whose cells store wastes and also transport nutrients to the subjacent blood sinuses. Sedimentivorous species leave characteristic sediment casts at the outside of their U-shaped burrows (Figure 26.2C,D).

Circulatory System

Enteropneusts possess a well-developed, partly open circulatory system of blood vessels and sinuses. The blood is pumped by contractile cells in the walls of blood spaces by the muscular action of the perihemal coeloms in the collar, and by a contractile heart vesicle located in the proboscis (Figure 26.3A). Two main longitudinal vessels lie in the dorsal and ventral mesenteries along the length of the trunk and in the collar (Figure 26.3A,D). Blood flows anteriorly in the **dorsal vessel** and posteriorly in the **ventral vessel**. The dorsal vessel expands in the collar as the venous sinus, which also receives blood anteriorly from a pair of lateral proboscis veins. The venous sinus leads to a larger, elongate, central sinus or heart in the proboscis, lying between the buccal diverticulum and the dorsal **heart vesicle**. The heart vesicle is a closed cavity of ectodermal origin with myoepithelial walls, whose contraction presses the central sinus against the stomochord below. This sends the blood to the excretory glomerulus in the protocoel with an increased pressure to facilitate ultrafiltration of metabolic waste. The glomerulus, buccal diverticulum, heart, and heart vesicle comprise what is often referred to as the **proboscis complex** of enteropneusts (Figure 26.3A). Blood leaves the glomerulus through two peribuccal vessels that join midventrally to form the ventral longitudinal vessel. Along the trunk, blood passes from the ventral vessel into networks of sinuses supplying both the gut and the body wall. In the pharynx, gut ventral sinuses drain into two lateral parabranchial vessels that send vessels dorsally into primary gill bars. After branching ventrally into secondary bars, the secondary bars send a vessel dorsally to reach the median trunk dorsal vessel.

Enteropneust blood is colorless, with pigment particles carried in solution, and contains few amoebocyte-like cells.

Excretory System

The main excretory organ in enteropneusts is the **glomerulus**, a structure unique to the hemichordates. It is formed of fingerlike outpocketings of protocoelic mesothelium associated with blood sinuses. Mesothelial cells are differentiated as podocytes, a cell type specialized for ultrafiltration. Pressed by the contraction of the pericardium or heart vesicle, the blood leaving the heart or central sinus is forced to pass through the glomerular sinuses, where metabolic wastes are filtered through the podocyte lining to the protocoel cavity. The primary urine is then presumably evacuated through the protocoel pore, located on the proboscis stalk in most species. Podocyte cells also line the walls of branchial sacs, suggesting a supplemental excretory role for the pharynx.

Gas Exchange

Gill bars have long been thought of as the place of gas exchange, a hypothesis undoubtedly biased by our understanding of chordate gills. However, the epithelial structure of gill bars makes this unlikely. Little work has been done on the matter of gas exchange, and it may be that other structures, such as the branchial sacs or the body surface, are also involved in this activity.

Nervous System

Most of the nervous system of all hemichordates consists of a netlike nerve plexus lying among the bases of the epithelial cells, both of ectodermal (epidermis) and endodermal (gut) origin. The plexus is thickened in enteropneusts as longitudinal tracts of neurons along the middorsal and midventral lines of the trunk (hence both dorsal and ventral longitudinal nerve cords are present). Additionally, a subepidermal (not intraepidermal) dorsal nerve cord, or **collar neurochord**, is present in the mesosome of enteropneusts, whose morphology, cellular organization, and ontogenetic neurulation process might be homologous with those of the chordate neural tube.

Acorn worms possess sensory cells over most of the body, probably serving as touch- and chemoreceptors that give these cryptic animals some information about their surroundings. As mentioned earlier, they also bear a preoral ciliary organ with a presumptive chemoreception function for feeding.

Reproduction and Development

Asexual reproduction Asexual reproduction in enteropneusts is directly related to their well-known ability to regenerate tissues and body parts when lost or damaged. They are very fragile worms and easily break when handled. Acorn worms have been described to undergo multiple transverse fissions of the body (**architomy**), leading to several regenerating individuals. Recently, **paratomy** has been described in the interstitial species *Meioglossus psammophilus*, in which new individuals develop aligned to the body axis through tissue regeneration and subsequent transverse fission, a process well known in Annelida and Acoela.

Sexual reproduction Acorn worms are gonochoristic but possess no outward evidence of sexual differences. Numerous sacciform gonads lie in the trunk, outside the mesothelium (retroperitoneal), from the branchial region to the hepatic one; they are often very elongate and in the family Ptychoderidae occupy two dorsolateral, flaplike extensions of the body called **genital wings** (Figure 26.2B). Each gonadal sac opens to the exterior through a duct and gonopore located dorsolaterally on the anterior trunk.

Fertilization is always external. Spawning in enteropneusts involves the release of mucoid egg masses by the females, followed by shedding of sperm by neighboring males. Once the eggs are fertilized, the mucous coat breaks down, thereby freeing the eggs into the seawater, where all subsequent development occurs.

Development Cleavage is holoblastic, radial, and more or less equal, attesting to the deuterostome nature of acorn worms. A coeloblastula forms, which gastrulates by invagination. The blastopore is at the presumptive posterior end, but it closes and the anus and mouth form later. By late gastrula, the embryo has acquired cilia and breaks free from the egg membrane as a free-floating plankter. Coelomic cavities (the single protocoel plus paired meso- and metacoels) derive from the archenteric endoderm as independent pouching through enterocoely. Other morphogenetic modes, including schizocoely, have been doubtfully reported in some species.

Some species (Harrimaniidae, e.g., *Saccoglossus*) have direct development from yolky eggs to juvenile worms, without an intervening larval phase (Figure 26.5A–F). Juvenile worms show a postanal tail that can not be homologized with the chordate postanal tail but most probably with the pterobranch stalk. In those that shed nonyolky eggs (Ptychoderidae, e.g., *Balanoglossus*), the hatching stage develops quickly to a characteristic, planktotrophic, **tornaria larva** with complex ciliary bands reminiscent of dipleurula larvae of echinoderms (Figure 26.5G,I). This free-swimming tornaria soon elongates and sinks, differentiating the three typical body regions and organs (Figure 26.5J–L). Gill slits develop asymmetrically—left side first, from anterior to posterior—as outpocketings of the gut that eventually open to the exterior via dorsolateral gill pores (Figure 26.5M).

Pterobranchs

The bodies of pterobranchs are small, usually pyriform or globular, but still retaining the ancestral tripartite regional division. The prosome in these odd creatures, called the **preoral disc** or **cephalic shield**, is a creeping sole that folds ventrally over the mouth. The mesosome forms a collar that bears the anteroventral mouth and extends dorsally as two to several tentaculated arms. The gut is U-shaped and the anus opens anterodorsally (Figures 26.2G,H and 26.6). The metasome is subdivided into a trunk and posterior stalk.

Pterobranchs live in colonies (*Rhabdopleura*) or aggregations (*Cephalodiscus*) consisting of zooids housed within tubular secreted casings called **coenecia** (Figure 26.2E,G). In colonies of *Rhabdopleura*, the zooids are connected to one another by tissue extensions called **stolons** (Figure 26.2E), but no such interzooidal links occur in *Cephalodiscus*. The overall forms of the aggregations and colonies vary among species. In all cases, the associated zooids are products of asexual reproduction initiated by a single sexually produced individual.

Body Wall and Cavities

The pterobranch epidermis is very simple, composed of squamous cells. Only at the cephalic shield does the epidermis become high and glandular, with several types of secretory cells that produce materials for the coenecium and mucus for gliding. Coelomic spaces follow the trimeric body pattern, with a single protocoel in the cephalic shield, and paired meso- and metacoels in the collar and trunk. The protocoel opens to the exterior via two coelomic ducts and pores at the bases of the arms. The mesocoel penetrates into the arms and tentacles, and in *Cephalodiscus*, sends two collar canals posteriorly that open to the exterior close to the single gill pore. The metacoel has no openings.

Support, Muscles, and Movement

The small size of pterobranchs correlates with a dramatic reduction of supporting structures. The proboscis skeleton is lacking and the single gill opening of *Cephalodiscus* has no supporting rods. Moreover, the pterobranch stomochord is nearly solid and lacks vacuolated cells. Most body support and protection are provided by the secreted outer casing of the colony or aggregation. Pterobranchs possess only longitudinal muscles, well developed in the cephalic shield for crawling within their tubes, and in the ventral side of the trunk and stalk. The protraction and retraction of individuals within their tubular houses are accomplished by hydrostatic pressure and contraction of longitudinal muscles, respectively.

Gut and Feeding

The mouth is located under the anteroventral edge of the collar. The gut is U-shaped, beginning with a buccal tube from which a buccal diverticulum arises dorsally. The pharynx bears one pair of gill slits in *Cephalodiscus*, but none in *Rhabdopleura*. When present, the gill apparatus is much simpler than that of enteropneusts, without secondary bars nor supporting structures and less defined branchial sacs. The slits open through pores on the exterior. A short esophagus connects the pharynx

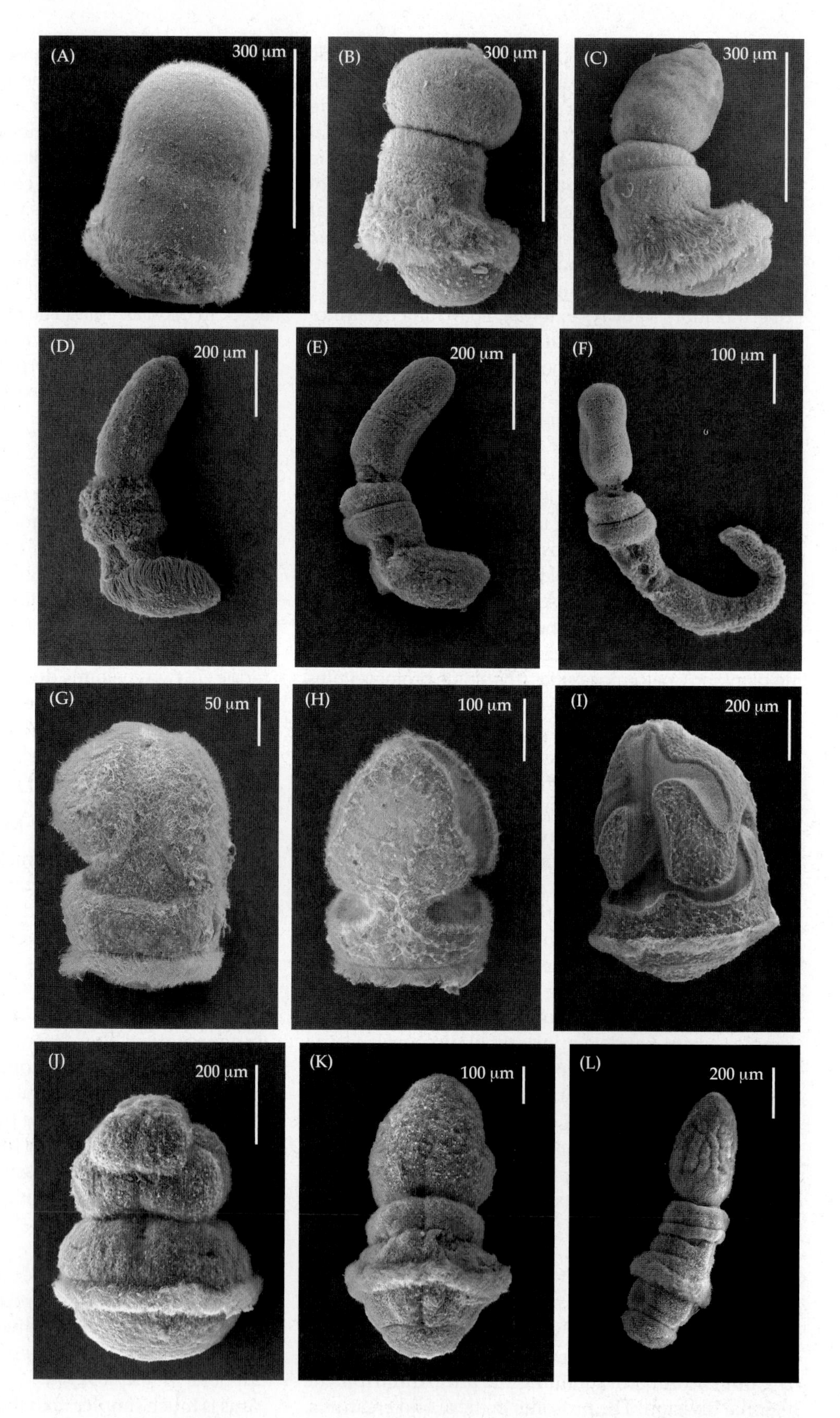

to a sacciform stomach that occupies most of the space within the trunk (Figure 26.6). The ascending portion of the gut is the intestine, which leads anteriorly to the dorsal anus.

Feeding in pterobranchs is accomplished by the mesosomal arms and tentacles. *Rhabdopleura* bears one pair of arms, whereas *Cephalodiscus* bears from five to nine pairs of arms, depending on the species (Figure

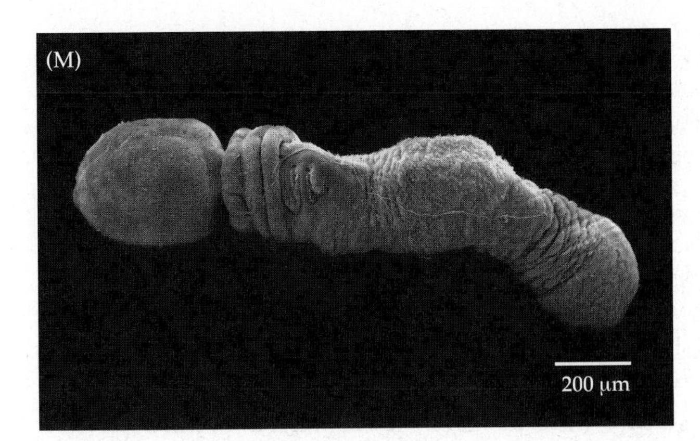

Figure 26.5 Enteropneust development. (A–F) Direct development of *Saccoglossus kowalevskii*. (A) Early embryo/larva, 36 hours after fertilization. (B) Differentiation of proboscis and collar, 56 hours after fertilization. (C) Formation of the dorsal neural fold (neurulation), 132 hours after fertilization. (D–F) Juveniles, 156, 340, and 750 hours after fertilization. Note dorsolateral gill pores behind the collar in F. (G–M) Indirect development of *Balanoglossus misakiensis*. (G) Early embryo/larva, 30 hours after fertilization. (H) Early tornaria larva, lateral view, 120 hours after fertilization. (I) Fully developed tornaria larva, ventral view, 10 days after fertilization. (J,K) Metamorphosis of tornaria, 13 and 14 days after fertilization. Proboscis and collar are formed, and larval telotroch persists. (L) Early juvenile, 12 hours after settlement. (M) Juvenile, 3 days after settlement. The telotroch has disappeared and the first two U-shaped gill slits are formed.

26.2 F,H). Each arm bears two rows of ciliated tentacles. Pterobranchs are ciliary mucus suspension feeders. During feeding, they assume a position near an opening in their tubular cases and extend their arms and tentacles into the water; *Rhabdopleura* simply spreads its two featherlike arms out into the water current, whereas *Cephalodiscus* form a so-called "feeding sphere" with its 8–10 arms and interdigitating tentacles (Figure 26.2F,H). Suspended particles and diatoms are trapped in the tentacular mucus and moved to the mouth by the action of cilia on the tentacles and arms. The single gill openings of *Cephalodiscus* eliminate the excess water entering the mouth. Digestion occurs in the stomach and intestine, well supplied by blood sinuses.

Circulation and Gas Exchange

The circulatory system of pterobranchs is a reduced and simplified version of that of enteropneusts. There is a heart or central sinus and a heart vesicle associated with the buccal diverticulum and a glomerular vessel, with the same functions of the proboscis complex in enteropneusts (Figure 26.6). Dorsal and ventral vessels are present in the trunk, continuing into the stalk. Otherwise, the circulatory system is reduced to a net of basiepithelial lacunae.

It is quite improbable that the single pair of gill slits in *Cephalodiscus* actually aid in gas exchange; instead, the small diffusion distances throughout the body probably allow general cutaneous exchange, especially over the high surface areas of the tentacles exposed to water currents.

Nervous System

Little is known about the nervous system of pterobranchs. A neuronal concentration in the mesosome named the **collar ganglion** is the putative homolog of the collar neurochord of enteropneusts, giving off branches to the cephalic shield, gill opening and tentacular system. Otherwise there is a simple intraepidermal nerve net all along the body. Touch receptors are presumably present in the tentacles.

Reproduction and Development

As in most colonial invertebrates, asexual reproduction by budding is an integral part of the life history of aggregating and colonial pterobranchs. In *Cephalodiscus*, the buds arise from near the base of the stalk of adult individuals and may eventually detach from one another when mature, living as an aggregation of individuals (Figure 26.7). Budding in *Rhabdopleura* occurs along the stolons that grow from the tips of the stalks of adult zooids. However, in this genus, buds never detach from the mother zooid, thereby forming a colony. In both genera, the aggregations or colonies arise by budding after the formation of a single sexually produced individual.

Sexes are usually separate, although hermaphroditism has been reported in *Cephalodiscus*. Gonads are

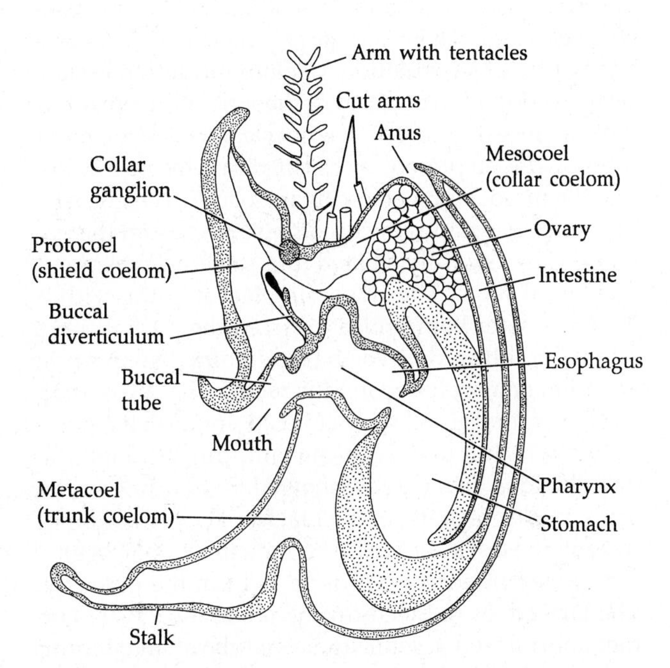

Figure 26.6 Internal anatomy of the pterobranch hemichordate *Cephalodiscus* (sagittal section).

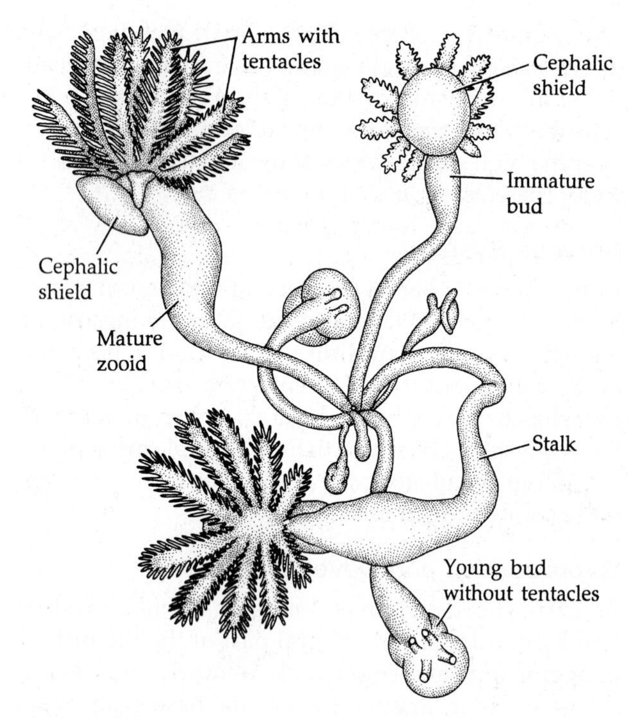

Figure 26.7 Budding in the pterobranch hemichordate *Cephalodiscus*, showing zooids at different stages of development.

single in *Rhabdopleura* and paired in *Cephalodiscus*. In *Rhabdopleura* a single pore opens on the right side of the trunk; the gonopores of *Cephalodiscus* are at the base of the arms. The eggs are shed into the tubes of the colony or aggregation, where they are fertilized. Developing embryos are brooded within the coenecium. Among the pterobranchs, only *Rhabdopleura* has been studied embryologically, and even here the details are scanty. The large yolky eggs undergo radial, holoblastic, subequal cleavage. There is some argument about the precise nature of gastrulation. Coelom formation is classically reported to occur by archenteric pouching, but the sequence of production is not clear. However, more recent work on pterobranchs has raised the possibility that the meso- and metacoel may arise via schizocoely. The embryo escapes the brooding area within the parental tube as a fully ciliated (but unnamed) planula-like larval stage. The swimming larvae settle within 24 hours after being released from the parent tube, although cultured larvae have remained swimming and unmetamorphosed for several months. The settled larva surrounds itself with a translucent, sealed cocoon within which it undergoes metamorphosis. First the oral shield is developed, followed by budding of the arms, posterior stalk, and tentacles. The formation of the gut, the last structure to become functional, begins with invagination of ectoderm to form the pharynx. The larva does not feed during metamorphosis. The metamorphosed juvenile builds a tube at an opening in the cocoon and begins to feed and secrete the coenecium for a new colony.

Hemichordate Fossil Record and Phylogeny

The hemichordate fossil record is almost entirely dominated by pterobranchs (notably the graptolites) whose tube skeletons have been well preserved. In fact, a recent study considers *Rhabdopleura* as just an extant graptolite, these being the sister taxon of cephalodiscids. Owing to their low preservational potentials, fossil enteropneusts are exceedingly rare, although some fossil genera have been described (*Mazoglossus, Megaderaion, Mesobalanoglossus*) as well as several other genera from putative trace fossils. Very recently the fossil enteropneust *Spartobranchus tenuis* has been described from middle Cambrian Burgess Shale, surprisingly associated with tubular casings. This finding compliments the very recent discovery of deep-sea torquaratorid enteropneusts that build mucous tubes.

Two very different views have existed for the relationship between enteropneusts and pterobranchs. The classical one, based on morphology and paleontology, considered these taxa as sister groups, usually ranked as classes. However, early 18S ribosomal DNA analyses suggested that pterobranchs might be an in-group of the enteropneusts, which would make Enteropneusta paraphyletic and Pterobranchia a sister taxon to the family Harrimaniidae. But in 2015, Simakov and colleagues published a large genomic analysis supporting the classical view that these two classes do comprise a sister group. Their study also supports the view that gill slits arose in the stem lineage that gave rise to the Deuterostomia.

Since their discovery, hemichordates have played a pivotal and controversial role in attempting to understand relationships among the three major deuterostome phyla (echinoderms, hemichordates, and chordates), in the nature of the deuterostome ancestor, and in the search for the evolutionary origin of chordates. Bateson coined the present phylum name after recognizing similarities with chordate embryogeny, and for many years the hemichordates were actually ranked as a subphylum of Chordata. Hemichordate traits, such as the presence of stomochord, collar neurochord, and gill slits also reinforced their phylogenetic proximity to chordates. However, current thinking is that only the gill slits and (probably) the neurochord can be homologized with those of chordates. Pharyngeal gill slits are used for filter feeding in the same way as protochordates, and moreover, gene expression data supports the hypothesis that the gill slits were present in the putative common ancestor to hemichordates and chordates. On the other hand, a close relationship with echinoderms has also been proposed through characters such as the excretory heart–kidney system and the tornaria–dipleurula larvae. Currently, a sister-group relationship between hemichordates and echinoderms (a clade called Ambulacraria) is generally favored and fairly consistently supported by molecular phylogenetic analyses.

The search for a deuterostome ancestor classically invoked an organism reminiscent of a solitary pterobranch zooid, influenced by the assumed position of pterobranchs as basal hemichordates. However, modern analyses, both morphological and molecular, and in good agreement with fossil findings, suggest that the deuterostome ancestor was a solitary enteropneust-like worm. Considerations of larval forms also agree with this hypothesis. The tornaria–dipleurula larva is shared by echinoderms and the now widely-recognized basal hemichordates, the enteropneust ptychoderids. Accordingly, the more recently derived groups, harrimaniids and pterobranchs, have a very short-lived, nonfeeding larva and are usually considered as direct developers. It may be hypothesized that the shift to direct development in solitary hemichordates allowed the transition to coloniality. Recent findings of miniaturization (*Meioglossus*) and tube dwelling (*Spartobranchus* and Torquaratoridae) in enteropneusts illustrate the evolutionary plasticity of the enteropneust body plan.

Selected References

Balser, E. J. and E. E. Ruppert. 1990. Structure, ultrastructure, and function of the preoral heart-kidney in *Saccoglossus kowalevskii* (Hemichordata, Enteropneusta) including new data on the stomatochord. Acta Zoologica 71: 235–249.

Benito, J. and F. Pardos. 1997. Hemichordata. Pp. 15–102 in F. W. Harrison and E. E. Ruppert, (eds.). *Microscopic Anatomy of Invertebrates*, Vol. 15. Wiley–Liss, Inc.

Bromham, L. D. and B. M. Degnan. 1999. Hemichordates and deuterostome evolution: robust molecular phylogenetic support for a hemichordate + echinoderm clade. Evol. Dev. 1 (3), 166–171.

Burdon-Jones, C. 1952. Development and biology of the larva of *Saccoglossus horsti* (Enteropneusta). Phil. Trans. R. Soc. Lond. B, Biol. Sci. 236: 553–590.

Cameron, C. B. 2005. A phylogeny of the hemichordates based on morphological characters. Can. J. Zoolog. 83(1): 196–215.

Cannon, J. T., A. L. Rychel, H. Eccleston, K. M. Halanych and B. J. Swalla. 2009. Molecular phylogeny of Hemichordata, with updated status of deep-sea enteropneusts. Mol. Phylogenet. Evol. 52: 17–24.

Caron, J. B., S. Conway Morris and C. B. Cameron. 2013. Tubicolous enteropneusts from the Cambrian period. Nature. doi: 10.1038/nature12017

Cedhagen, T. and H. G. Hansson. 2013. Biology and distribution of hemichordates (Enteropneusta) with emphasis on Harrimaniidae and description of *Protoglossus bocki* sp. nov. from Scandinavia. Helgoland Mar. Res. 67: 251-265.

Deland, C., C. B. Cameron, K. P. Rao, W. E. Ritter and T. H. Bullock. 2010. A taxonomic revision of the family Harrimaniidae (Hemichordata: Enteropneusta) with descriptions of seven species from the Eastern Pacific. Zootaxa 2408: 1–30.

Dilly, P. N. 1985. The habitat and behavior of *Cephalodiscus gracilis* (Pterobranchia, Hemichordata) from Bermuda. J. Zool. 207: 223–239.

Ezhova, O. V. and V. V. Malakhov. 2009. Three-dimensional structure of the skeleton and buccal diverticulum of an acorn worm *Saccoglossus mereschkowskii* Wagner, 1885 (Hemichordata: Enteropneusta). Invert. Biol. 6(2): 103-116.

Gilmour, T. H. J. 1982. Feeding in tornaria larvae and the development of gill slits in enteropneust hemichordates. Can. J. Zoolog. 60: 3010–3020.

Gonzalez, P. and C. B. Cameron. 2009. The gill slits and preoral ciliary organ of *Protoglossus* (Hemichordata, Enteropneusta) are filter-feeding structures. Biol. J. Linn. Soc. 98: 898-906.

Gonzalez, P. and C. B. Cameron. 2012. Ultrastructure of the coenecium of *Cephalodiscus* (Hemichordata: Pterobranchia). Can. J. Zoolog. 90: 1261-1269.

Hadfield, M. G. 1975. Hemichordata. Pp. 185–240 in A. C. Giese and J. S. Pearse (eds.), *Reproduction of Marine Invertebrates, Vol.* 2. Academic Press, New York.

Hadfield, M. G. 2002. Phylum Hemichordata. Pp. 553–564 in R. E. Young (ed.) *Atlas of Marine Invertebrate Larvae*. Academic Press, London.

Halanych, K.M., 1995. The phylogenetic position of the pterobranch hemichordates based on 18S rDNA sequence data. Mol. Phylog. Evol. 4: 72–76.

Halanych, K. M., J. T. Cannon, A. R. Mahon, B. J. Swalla, C. R. Smith. 2013. Tubicolous acorn worms from Antarctica. Nature. doi: 10.1038/ncomms3738

Holland, N. D., W. J. Jones, J. Ellena, H. A. Ruhl and K. L. Smith Jr. 2009. A new deep-sea species of epibenthic acorn worm (Hemichordata, Enteropneusta). Zoosystema 31 (2): 333–346.

Jones, D. O. B. and 8 others. 2013. Deep-sea surface-dwelling enteropneusts from the Mid-Atlantic Ridge: their ecology, distribution and mode of life. Deep Sea Res. Part II: Topical Studies in Oceanography 98(B): 374–387.

Kaul, S. and T. Stach. 2010. Ontogeny of the collar cord: neurulation in the hemichordate *Saccoglossus kowalevskii*. J. Morph. 271: 1240–1259.

Lester, S. M. 1985. *Cephalodiscus* sp. (Hemichordata: Pterobranchia): Observations of functional morphology, behavior and occurrence in shallow water around Bermuda Mar. Biol. 85: 263–268.

Lester, S. M. 1988. Ultrastructure of adult gonads and development and structure of the larva of *Rhabdopleura normani* (Hemichordata: Pterobranchia). Acta Zoologica 69: 95–109.

Mayer, G. and T. Bartolomaeus. 2003. Ultrastructure of the stomochord and the heart-glomerulus complex in *Rhabdopleura compacta* (Pterobranchia): phylogenetic implications. Zoomorphology 122: 125–133.

Mitchell, C. E., M. J. Melchin, C. B. Cameron and J. Maletz. 2013. Phylogenetic analysis reveals that *Rhabdopleura* is an extant graptolite. Lethaia 46: 34–56.

M'Intosh, W. C. 1887. Report on *Cephalodiscus dodecalophus*, M'Intosh, a new type of the Polyzoa, procured on the voyage of H.M.S. *Challenger* during the years 1873–76. Challenger Reports, Zoology, 20, pp. 1–37 (7 plates, Appendix pp. 38–47 by S. Harmer).

Nielsen, C., and A. Hay-Schmidt. 2007. Development of the enteropneust *Ptychodera flava*: ciliary bands and nervous system. J. Morph. 268: 551–570.

Pardos, F. and J. Benito. 1988. Blood vessels and related structures in the gill bars of *Glossobalanus minutus* (Enteropneusta). Acta Zoologica 69: 87–94.

Petersen, J. A. 1994. Hemichordates. Pp. 385–395 in Adiyodi K. G. and R. G. Adiyodi (eds.), *Reproductive Biology of Invertebrates. Asexual Propagation and Reproductive Strategies, Vol. 6, Part B.* Wiley, New York.

Simakov, O. and 63 others. 2015. Hemichordate genomes and deuterostome origins. Nature. doi: 10.1038/nature16150

Spengel, J. W. 1893. Die Enteropneusten des Golfes von Neapel. Vol. 18 of *Fauna und Flora des Golfes von Neapel.*

Stach, T. and S. Kaul. 2011. The postanal tail of the enteropneust *Saccoglossus kowalevskii* is a ciliarly creeping organ without distinct similarities to the chordate tail. Acta Zoologica 92, 150–160.

Strathmann, R. and D. Bonar. 1976. Ciliary feeding of tornaria larvae of *Ptychodera flava*. Mar. Biol. 34: 317–324.

Van der Horst, C. J. 1939. Hemichordata. In Bronn, H.G. (ed.) *Klassen und Ordnungen des Tierreichs wissenschaftlich dargestellt in Wort und Bild*. Leipzig, Akademische Verlagsgesellschaft. Vol. 4 (4), Buch 2, Tiel 2.

Welsch, U., P. N. Dilly and G. Rehkämper. 1987. Fine structure of the stomochord in *Cephalodiscus gracilis* M'Intosh 1882 (Hemichordata, Pterobranchia). Zoologischer Anzeiger 218: 3/4S: 209–218.

Woodwick, K. H. and T. Sensenbaugh. 1985. *Saxipendium coronatum*, new genus, new species (Hemichordata: Enteropneusta): the unusual spaghetti worms of the Galápagos rift hydrothermal vents. Proc. Biol. Soc. Wash. 98: 351 365.

Worsaae, K., W. Sterrer, S. Kaul-Strehlow, A. Hay-Schmidt and G. Giribet. 2012. An anatomical description of a miniaturized acorn worm (Hemichordata, Enteropneusta) with asexual reproduction by paratomy. PLoS ONE 7 (11): 1–19.

CHAPTER 27

Phylum Chordata

Cephalochordata and Urochordata

We are chordates (Latin, *chorda*, "cord"). So are cats and dogs, lemurs and anteaters, birds and fishes, frogs and snakes, whales and elephants—all conspicuous by their size and familiarity. All of these animals have a vertebral column (backbone) housing the dorsal nerve cord, and this is the key feature defining the subphylum Vertebrata.[1] But, besides the vertebrates there are two other subphyla in the Chordata, and they both lack backbones. These are the invertebrate chordates—the subphyla Cephalochordata and Urochordata. The Cephalochordata comprise 30 or so species of small, fishlike animals called lancelets, or amphioxus (Figures 27.1 and 27.2). The Urochordata (= Tunicata) are composed of about 3,000 species in four classes: the sessile filter-feeding sea squirts, or ascidians (class Ascidiacea); the pelagic tunicates, or salps (class Thaliacea); planktonic larva-like tunicates called appendicularians, or larvaceans (class Appendicularia); and the abyssal ascidian-like sorberaceans (class Sorberacea) (Figures 27.3–27.8).

Vertebrates, chordates with a backbone, account for only about 4% (about 58,000 species) of all described animal species. Those chordates that do not possess a backbone (plus the other 31 animal phyla previously described in this book) constitute the invertebrates. So we see that the division of animals into invertebrates and vertebrates is based more on tradition and convenience, reflecting a dichotomy of interests among zoologists, than it is on the recognition of natural biological groupings. "Invertebrates" is a paraphyletic group, because it excludes the clade Vertebrata. The vertebrates, on the other hand, are a monophyletic group.

The monophyly of the phylum Chordata has (in modern times) never been seriously questioned. One of the key synapomorphies defining the phylum is the **notochord**, a dorsal, elastic rod

Classification of The Animal Kingdom (Metazoa)

Non-Bilateria*
(a.k.a. the diploblasts)
- PHYLUM PORIFERA
- PHYLUM PLACOZOA
- PHYLUM CNIDARIA
- PHYLUM CTENOPHORA

Bilateria
(a.k.a. the triploblasts)
- PHYLUM XENACOELOMORPHA

Protostomia
- PHYLUM CHAETOGNATHA

SPIRALIA
- PHYLUM PLATYHELMINTHES
- PHYLUM GASTROTRICHA
- PHYLUM RHOMBOZOA
- PHYLUM ORTHONECTIDA
- PHYLUM NEMERTEA
- PHYLUM MOLLUSCA
- PHYLUM ANNELIDA
- PHYLUM ENTOPROCTA
- PHYLUM CYCLIOPHORA

Gnathifera
- PHYLUM GNATHOSTOMULIDA
- PHYLUM MICROGNATHOZOA
- PHYLUM ROTIFERA

Lophophorata
- PHYLUM PHORONIDA
- PHYLUM BRYOZOA
- PHYLUM BRACHIOPODA

ECDYSOZOA

Nematoida
- PHYLUM NEMATODA
- PHYLUM NEMATOMORPHA

Scalidophora
- PHYLUM KINORHYNCHA
- PHYLUM PRIAPULA
- PHYLUM LORICIFERA

Panarthropoda
- PHYLUM TARDIGRADA
- PHYLUM ONYCHOPHORA
- PHYLUM ARTHROPODA
  - SUBPHYLUM CRUSTACEA*
  - SUBPHYLUM HEXAPODA
  - SUBPHYLUM MYRIAPODA
  - SUBPHYLUM CHELICERATA

Deuterostomia
- PHYLUM ECHINODERMATA
- PHYLUM HEMICHORDATA
- **PHYLUM CHORDATA**

*Paraphyletic group

The section on Urochordata has been revised by C. Sarah Cohen.

[1]The subphylum Vertebrata (= Craniata) contains two groups: the jawless vertebrates, or Agnatha (hagfish and lampreys) and the jawed vertebrates, or Gnathostomata (the rest of the vertebrates). These are briefly discussed in the Chordate Phylogeny section of this chapter.

BOX 27A Characteristics of the Phylum Chordata

1. Bilaterally symmetrical, coelomate deuterostomes (coelom greatly reduced in some groups)
2. Pharyngeal gill slits present at some stage in development; used for feeding in Cephalochordata and Urochordata
3. Mesodermally-derived, dorsal notochord present at some stage in development; notochord lies dorsal to the digestive tube and ventral to the neural tube
4. Dorsal, hollow nerve cord, at least in some stage of life history; compartmentalized anteriorly and, in Vertebrata, giving rise to the brain
5. Muscular, locomotor, postanal tail at some stage in development
6. With an endodermal, pharygneal endostyle (Urochordata, Cephalochordata) or thyroid gland (Vertebrata)
7. Gut complete, usually regionally specialized
8. Circulatory system with a contractile blood vessel (or heart). Gas exchange occurs across the body wall or across epithelial tissues.
9. Gonochoristic or hermaphroditic; development variable. Cleavage radial, holoblastic, subequal, or slightly unequal. A tadpole stage is expressed at some point in the life history of all taxa.

derived from a middorsal strip of embryonic (archenteric) mesoderm, that provides structural and locomotory support in the body of larval or adult chordates. In addition, chordates have (at some point in their life history) a hollow **dorsal neural tube**, **pharyngeal gill slits**, and a **postanal tail** (a tail that extends beyond the anus) (Box 27A). They are an ancient group, with a fossil record dating to the early Cambrian, at least 530 million years ago, and with dated molecular phylogenies suggesting an age of origin at 700 to 800 million years ago. Recent molecular phylogenetic analyses have overturned the older view that Cephalochordata are the sister group to vertebrates, and instead the evidence today suggests that Urochordata is the sister group to Vertebrata (a clade known as Olfactores), and Cephalochordata is the sister group to those (see Figure 27.9 and the phylogeny section at the end of this chapter).

CHORDATE CLASSIFICATION

SUBPHYLUM CEPHALOCHORDATA The lancelets. Small (to 7 cm), fishlike chordates with notochord, gill slits, dorsal nerve cord, unique wheel organ, and postanal tail present in adults, but without vertebral column or cranial skeleton structure; gonads numerous (25–38) and serially arranged. Found in marine and brackish water environments, usually associated with clean sand or gravel sediments in which they burrow. (e.g., *Asymmetron, Branchiostoma, Epigonichthyes*)

SUBPHYLUM UROCHORDATA (= TUNICATA) The tunicates. Adult body form varies, but typically covered by thick or thin tunic (test) of a cellulose-like polysaccharide; without bony tissues; notochord restricted to tail and usually found only in larval stage (and in adult appendicularians); gut U-shaped, pharynx (branchial chamber) typically with numerous gill slits (stigmata); coelom not substantively developed; dorsal nerve cord present in larval stages. Strictly marine; 4 classes.

CLASS ASCIDIACEA Ascidians, or sea squirts. Benthic, solitary or colonial, sessile tunicates; incurrent and excurrent siphons generally directed upwards, away from the substratum; without dorsal nerve cord in adult stages; occurring at all depths. About 13 families and many genera. (e.g., *Ascidia, Botryllus, Chelyosoma, Ciona, Clavelina, Corella, Diazona, Didemnum, Diplosoma, Lissoclinum, Megalodicopia, Molgula, Psammascidia, Pyura, Styela, Trididemnum*). The relationships of three traditional ascidian orders (Aplousobranchia, Phlebobranchia, and Stolidobranchia) are in flux, as is the relationship of this class to the others.

CLASS THALIACEA Pelagic tunicates, or salps. Solitary or colonial; incurrent and excurrent siphons at opposite ends, providing locomotory current; adults without a tail; gill clefts not subdivided by gill bars; 3 orders: Pyrosomida, Salpida, and Doliolida. (e.g., *Dolioletta, Doliolum, Pyrosoma, Salpa, Thetys*)

CLASS APPENDICULARIA (= LARVACEA) Appendicularians, or larvaceans. Solitary planktonic tunicates; probably neotenic; adults retain larval characteristics, including notochord and muscular tail; body enclosed in a complex gelatinous "house" involved in feeding. (e.g., *Fritillaria, Oikopleura, Stegasoma*)

CLASS SORBERACEA Benthic, deep water, ascidian-like urochordates retaining the dorsal nerve cord in adult stages and with a greatly reduced pharynx lacking a perforated branchial chamber. These little-studied tunicates are apparently all carnivores; they range in size from 2 to 6 cm. (e.g., *Octacnemus*)

SUBPHYLUM VERTEBRATA (= CRANIATA) Craniates. Chordates with a skeletal brain case and, except for members of the Agnatha, jaws and a vertebral column (backbone) that forms the axis of the body skeleton; most with paired appendages. Several classes are generally recognized, although not all are strictly monophyletic: classes Myxini and Cephalaspidomorphi are the hagfishes and lampreys, respectively (united as the jawless or agnathan fishes); Chondrichthyes are the sharks, skates, and rays; Osteichthyes are the bony fishes (e.g., trout, tuna, perch); Amphibia includes the salamanders, frogs, toads, caecilians; Reptilia (a paraphyletic group) traditionally included the turtles, snakes, lizards, and crocodilians, but modern classifications place birds and reptiles together as Reptilomorpha (or Sauropsida). Mammalia comprises the mammals.

Phylum Chordata, Subphylum Cephalochordata: The Lancelets (Amphioxus)

Like all chordates, cephalochordates—the lancelets—possess a notochord, dorsal hollow nerve cord, pharyngeal gill slits, post-anal tail (that persists into adulthood), and true coelom (Figure 27.1 and 27.2). The name (Greek *cephala*, "head"; *chordata*, "chord") derives from the unique extension of the notochord beyond the nerve cord into the animal's head. Lancelets are small (to 7 cm), fishlike animals (e.g., *Asymmetron*, *Branchiostoma*, *Epigonichthyes*). The vernacular name amphioxus is frequently used for the common species *Branchiostoma lanceolatum* (and sometimes for others as well). Cephalochordates are cosmopolitan in shallow marine and brackish waters at tropical and warm-temperate latitudes, where they lie burrowed in clean sands with only the head protruding above the sediment. Generally, they reside semivertically in burrows, or rest on their sides on surfaces. They can and do swim, and locomotion is important to their dispersal and mating habits. However, they tend to swim erratically and without clear dorsoventral orientation. Although around 50 species of cephalochordates have been described in the literature, only 20–30 of these probably remain valid (although some species distinctions and the phylogenetic relationships among the species are still unsettled). A few fossil lancelets have been described, as old as the early Cambrian.

Surprisingly, most biologists have never seen a living lancelet in the field. They can be somewhat common, but only in the right environment and with a lot of time on one's knees in tidal flats. Most species seem to prefer clean, fine-grained sediments, but even when they are present it takes a keen eye to spot them. Occasionally they are seen swimming about in very shallow tidal pools on sand flats, darting in and out of the sand. During spawning season, the sex of an individual is quickly revealed by white testes of the males and yellowish ovaries of the females, both visible through the translucent body wall. In some areas of Asia they are common enough to be harvested for human and animal food.

The Cephalochordate Body Plan

Body Wall, Support, and Locomotion

The body of cephalochordates is completely covered by an epidermis of simple columnar epithelium, underlain by a thin connective tissue dermis. The body wall muscles are vertebrate-like and occur as chevron-shaped blocks called **myotomes** arranged longitudinally along the dorsolateral aspects of the body. These muscle blocks are large and occupy much of the interior of the body, thereby reducing the coelom to relatively small spaces. The notochord persists in adults and provides the major structural support for the body (Figure 27.1A–C).

The notochord also plays a major role in locomotion in lancelets. As a result of the action of segmental myotomes, swimming in cephalochordates is much like that of fishes, consisting largely of lateral body undulations that drive water posteriorly to provide a forward thrust. The propulsive action of these body movements is enhanced by a vertically aligned caudal fin. Unlike the vertebral column and its articulating bones, however, the notochord is a flexible elastic rod. It prevents the body from shortening when the muscles contract, causing lateral bending instead. Its elasticity tends to straighten the body, and thus assists the antagonistic action of paired myotomes. The notochord extends beyond the myotomes both anteriorly and posteriorly, providing support beyond those muscles and apparently aiding in holding the body stiff during burrowing.

Although the notochord of cephalochordates is homologous with the same structure in other chordates, it displays some unique and rather remarkable structural and functional characteristics associated with its persistence into the adult stage. The notochord of lancelets is built of discoidal lamellae that are stacked like poker chips along its length and surrounded by a sheath of collagenous connective tissue. The lamellae are composed of muscle cells whose fibers are oriented transversely. A significant amount of extracellular fluid exists in spaces and channels around and between the lamellae within the collagenous sheath. These muscle cells are innervated by motor neurons from the dorsal nerve cord. Upon contraction, the hydrostatic pressure in the extracellular spaces increases, thereby resulting in increased stiffness of the whole notochord complex. It is suspected that this action may facilitate certain kinds of movement patterns, especially burrowing.

Although commonly called "fins," the dorsal and ventral finlike structures are perhaps more appropriately called dorsal and ventral storage organs (Figure 27.1A,C). They are not homologous to the fins of fishes, and their function appears to be housing a build-up of nutritional reserves for gamete formation.

Feeding and Digestion

Cephalochordates are ciliary–mucous suspension feeders, and they employ a food-gathering mechanism very similar to that of tunicates. Water is driven into the mouth and pharynx via cilia in the **branchial chamber**, which creates the so-called **branchial pump**, and out through the pharyngeal gill slits into a surrounding **atrium**; it exits the body through a ventral **atriopore** (Figure 27.1A). Unlike the gill ventilation currents of aquatic vertebrates that are generated by muscular action, the feeding currents of lancelets (the branchial pump) are driven largely by pharyngeal cilia, as in

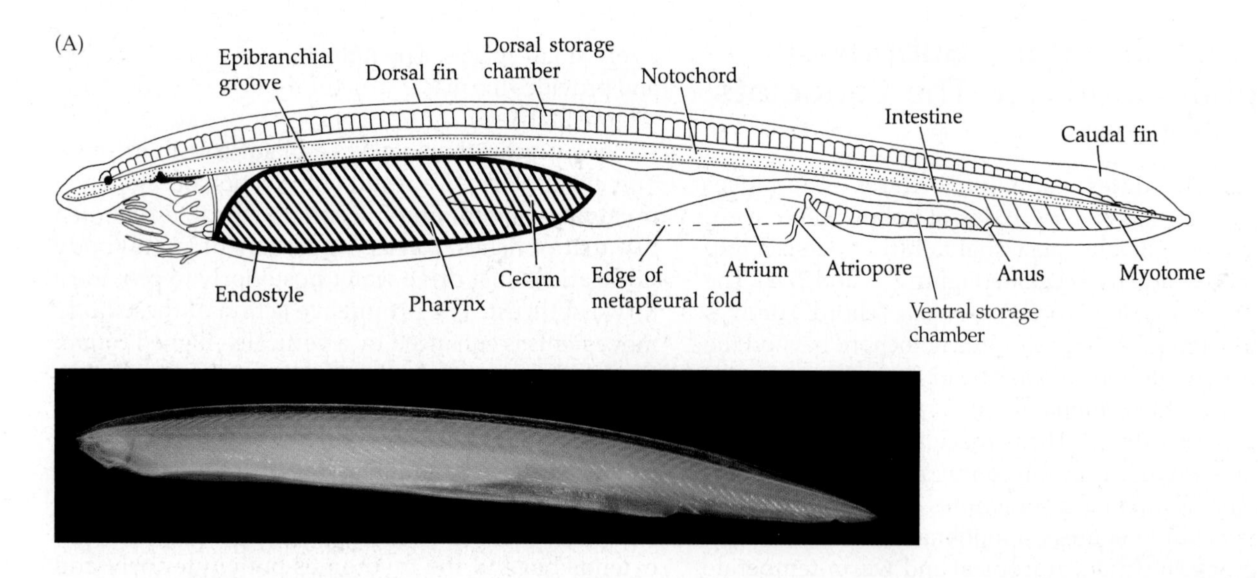

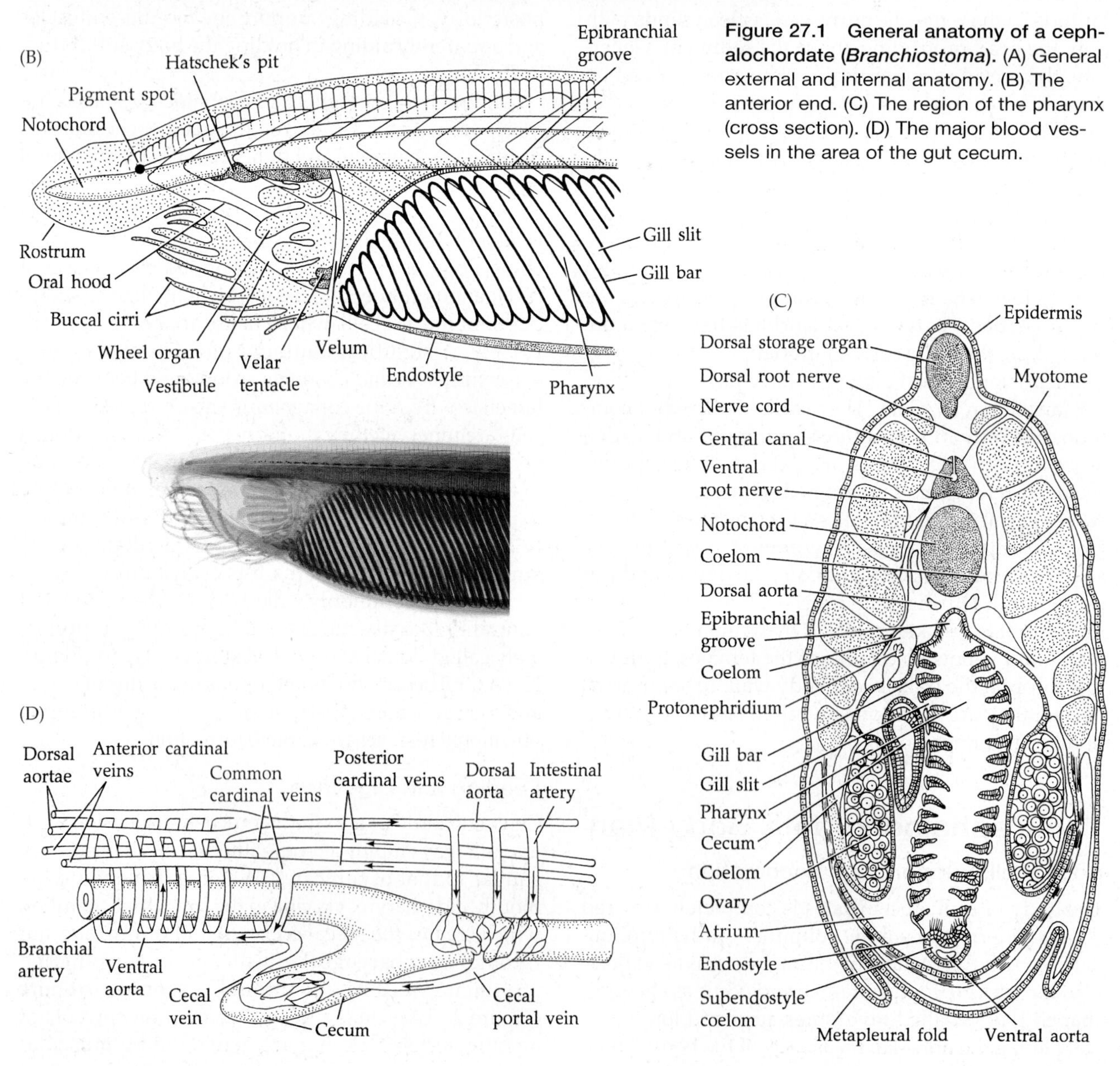

Figure 27.1 General anatomy of a cephalochordate (*Branchiostoma*). (A) General external and internal anatomy. (B) The anterior end. (C) The region of the pharynx (cross section). (D) The major blood vessels in the area of the gut cecum.

tunicates. The pump is driven by metachronal waves of lateral cilia on the branchial bars. The gill slits in lancelets function largely for feeding and have little to do with gas exchange. There are up to 200 gill slits, separated from one another by gill bars, which are supported by cartilaginous rods.

Filter feeding in lancelets involves the trapping of food, primarily phytoplankton, from the inflowing water, combined with complex handling and sorting activities that occur before the water enters the mouth. The mouth is housed within a depression called the **vestibule**, which is formed by an anterior extension of the body called the **oral hood** (Figure 27.1B). The oral hood is supported by the notochord and bears finger-like projections called **oral cirri** (= **buccal cirri**). As water enters the vestibule, the cirri act as a coarse filter to prevent sediments and other large particles from reaching the mouth. The mouth itself is a perforation in a membranous velum, which bears a set of **velar tentacles**; the tentacles provide a second screen, preventing large material from entering the mouth. The lateral walls of the vestibule bear complex ciliary bands that collectively constitute the wheel organ. The **wheel organ** appears as brown, thickened, folded epithelium in the roof and sides of the vestibule, visible through the translucent skin of living animals. Cilia of the wheel organ drive food particles to the mouth and give the impression of a rotation, hence the name. Lying on the roof of the vestibule is a mucus-secreting structure called **Hatschek's pit** (Figure 27.1B). Mucus from this pit flows over the wheel organ, and food particles trapped by the mucus are carried to the mouth along the ciliary tracts. This material is then incorporated into the general water current moving through the mouth and into the pharynx. Thread-like feces are ejected from the anus.

When feeding, most lancelets lie completely or partially buried in the bottom sand with the ventral side turned upward and the mouth opening free from interference of sand or sediment. Experiments have shown that, in *Branchiostoma lanceolatum*, particles greater than or equal to 4 μm are retained by the mucus filter with 100% efficiency. Both the oral cirri and velar tentacles bear numerous sensory cells that respond to mechanical and chemical stimulation. If the velar tentacles are stimulated, contraction of the velar muscles flick them forward and a "cough" reaction is evoked. Clogging of the oral tentacles by detritus also results in a rapid body contraction, forcing water out the mouth such that the buccal region is flushed and cleaned.

The ventral surface of the pharynx bears the **endostyle**, or hypobranchial groove (Figure 27.1A–C). An endostyle is found in the Cephalochordata and Urochordata (the so-called **protochordates**) and chordates that filter feed using a pharynx and gill slits, including primitive vertebrates (e.g., lampreys), and it is generally viewed as a synapomorphy of the Chordata. It functions by secreting mucus on which suspended particles are trapped and conveyed into the midgut. As in tunicates, the endostyle of lancelets binds iodine, although the specific cells involved differ between the two lineages. Many workers view the endostyle of urochordates and cephalochordates as homologous and a possible precursor to the vertebrate thyroid gland. As in tunicates, the endostyle produces strings of mucus that trap food from the water as it passes through the gill slits and into the atrium. The food-laden mucus is then passed dorsally along the walls of the pharynx by ciliary tracts and into the epibranchial groove, which carries the material by cilia along the dorsal midline of the pharynx and into a short esophagus. The gut extends posteriorly as an elongate intestine and opens through the anus just in front of the caudal fin. Near the junction of the pharynx and esophagus, there arises an anteriorly projecting pouch called the **digestive cecum** (sometimes called the hepatic cecum, or hepatic diverticulum). Studies indicate that this organ functions in lipid and glycogen storage and protein synthesis, and it is regarded by most workers as the evolutionary precursor of the vertebrate liver and perhaps the pancreas. Just as the liver does in vertebrates, the digestive cecum also synthesizes vitellogenin, a yolk protein precursor that is carried in the blood to the ovary where it is used in the production of yolk proteins in eggs. Digestion initially is extracellular in the gut lumen and is completed intracellularly in the walls of the intestine and especially in the cecum. In addition to storage in the cecum, food reserves accumulate in longitudinally arranged dorsal and ventral storage chambers (the "fins") along the dorsal midline and along the ventral body margin posterior to the atriopore (Figure 27.1A).

Circulation, Gas Exchange, and Excretion

The circulatory system of lancelets comprises a set of closed vessels through which blood flows in a pattern similar to that in primitive vertebrates (e.g., fishes), although there is no heart. Blood flows posteriorly along the pharyngeal region in a pair of **dorsal aortae**. Just posterior to the pharynx, these vessels merge into a single median dorsal aorta that extends into the region of the caudal fin (Figure 27.1D). Blood is supplied to the myotomes and notochord via a series of short segmental arteries and to the intestine through intestinal arteries. A capillary network in the intestinal wall collects the nutrient-laden blood and leads to a series of intestinal veins that join a large subintestinal vein called the cecal portal vein, which carries blood forward beneath the gut to another capillary bed in the digestive cecum. As in vertebrates, the vein that connects two capillary beds is called a portal vein (e.g., the hepatic portal vein and renal portal veins in fishes). Indeed, the **cecal portal vein** of cephalochordates is probably the homologue of the hepatic portal vein of vertebrates. In the digestive cecum, the nutrient and chemical composition of the blood is regulated somewhat before being distributed to the body tissues. (The vertebrate liver serves the same function via the hepatic portal system.)

Leaving the cecal capillaries is a cecal vein, which is joined by a pair of common **cardinal veins** formed by the union of paired anterior and posterior cardinal veins returning from the body tissues. These vessels merge to form the **ventral aorta** beneath the pharynx. From here blood is carried through the gill bars via afferent and efferent branchial arteries to the paired dorsal aortae, thus completing the circulatory cycle. Blood is moved through this system by peristaltic contractions of the major longitudinal vessels and by pulsating areas at the bases of the afferent branchial arteries.

The blood contains no oxygen-carrying pigments or cells and is thought to function largely in nutrient distribution rather than in gas exchange and transport. Although some diffusion of oxygen and carbon dioxide may occur across the gills, most of the gas exchange probably takes place across the walls of the **metapleural folds**, thin flaps off the body wall that lie just anterior to the atriopore (Figure 27.1C).

The excretory units in cephalochordates are protonephridia similar to the solenocytes of some other invertebrates. Numerous clusters of protonephridia accumulate nitrogenous wastes, which are carried by a nephridioduct to a pore in the atrium. However, despite the structural similarities, the homology of lancelet protonephridia with those of other invertebrates is uncertain and some specialists regard this as a case of convergent evolution.

Nervous System and Sense Organs

The central nervous system of cephalochordates is very simple. A dorsal nerve cord extends most of the length of the body and is generally expanded slightly as a cerebral vesicle in the base of the oral hood. Segmentally arranged nerves arise from the cord along the body in the typical vertebrate pattern of dorsal and ventral roots. The epidermis is rich in sensory nerve endings, most of which are probably tactile and important in burrowing. Some lancelets have a single simple eyespot near the anterior end of the dorsal nerve cord.

Reproduction and Development

Cephalochordates are gonochoristic, but the sexes are structurally very similar. Rows of 25 to 38 pairs of gonads are arranged serially along the body on each side of the atrium. The volume of gonadal tissue varies seasonally, and during the reproductive period it may occupy so much of the body as to interfere with feeding. Spawning typically occurs at dusk. The atrial wall ruptures, and eggs or sperm are released into the excurrent flow of water from the atrium; external fertilization follows.

The ova are isolecithal, with very little yolk. Cleavage is radial, holoblastic, and subequal, and leads to a coeloblastula that gastrulates by invagination (Figure 27.2). The roof of the archenteron eventually produces first a solid middorsal strip of mesoderm destined to become the notochord and then, sequentially, an anterior-to-posterior series of paired archenteric pouches along each side of the notochord. These enterocoelic pouches form the coelom and the other mesodermally derived structures such as the muscle bundles. The roof of the archenteron closes following mesoderm proliferation.

Dorsally the ectoderm differentiates into a **neural plate**. The neural plate eventually rolls inward, separating from the bordering cells, and then sinks inward as a neural tube, which forms the dorsal hollow nerve

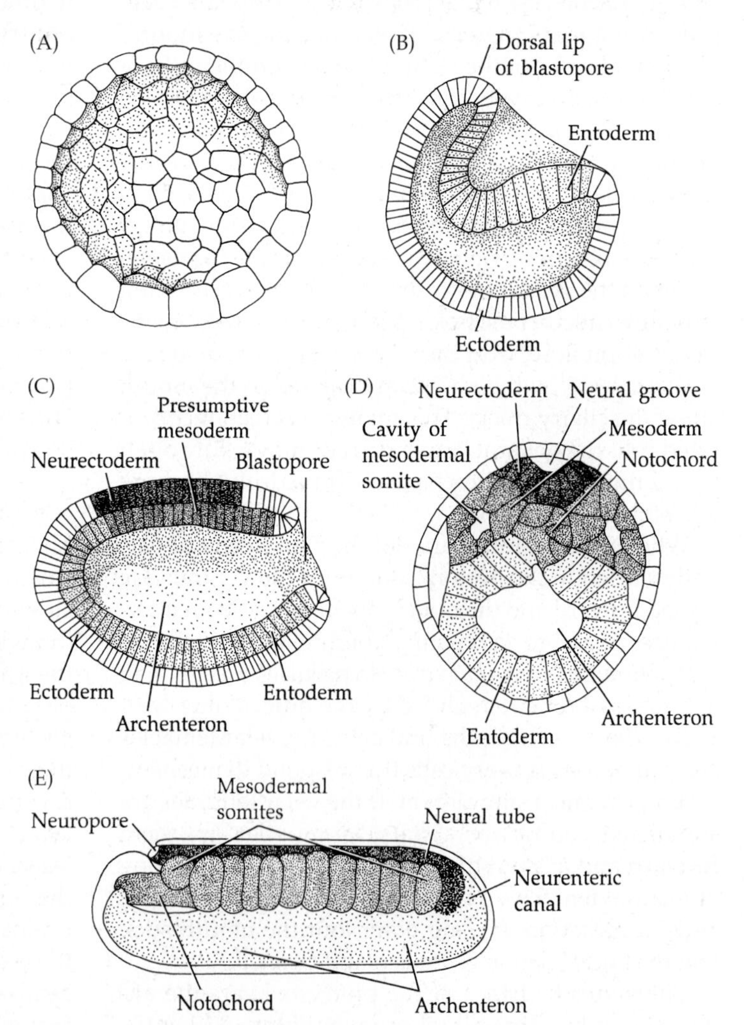

Figure 27.2 Development in a cephalochordate. (A) Coeloblastula. (B,C) Early and late gastrulae. (D) The neural groove stage (section). Note the proliferation of the mesoderm as central notochord and later coelomic cavities. (E) Lateral view showing major structures and a temporary confluence of the gut and neural tube (neurenteric canal).

cord. As this process occurs at the posterior end of the embryo, the developing nervous tissue contacts the blastopore, which remains open temporarily and connects the archenteron to the lumen of the nerve cord as a **neuropore**. Later the two structures separate, and the blastopore opens to the exterior as the anus. The mouth breaks through as a secondarily produced opening at the front end of the developing gut.

In the absence of abundant yolk reserves, development to a free-swimming larva takes place rapidly. As soon as they are able, the planktotrophic larvae swim upward in the water column, where they remain planktonic for 75 to 200 days. They alternately swim upward and then passively sink with the body held horizontally and the mouth directed downward, feeding on plankton and other suspended matter as they gradually develop into juveniles.

Phylum Chordata, Subphylum Urochordata: The Tunicates

Once relegated to near obscurity, ascidians are getting more attention lately, including as model organisms whose genomes are being studied to better understand chordate development and gene networks. The genomes of tunicates are non-duplicated, providing a simplified system closely related to the more complex genomes of vertebrates. Ascidians are also getting substantial ecological attention as some of the most dangerous marine invasive species, covering and smothering thousands of square meters of benthic substrate and causing economic losses in aquaculture worldwide. The developmental and ecological attention focused on tunicates has in turn highlighted the surprising gaps in knowledge about their systematics, distribution, reproductive and larval biology, ecology, evolutionary relationships, and regeneration abilities.

Members of the four urochordate classes are almost all marine suspension feeders, but they conduct their lives in very different ways (Figures 27.3 and 27.4). The ascidians (class Ascidiacea) include both solitary and colonial forms, with individuals ranging in size from less than 1 mm to 60 cm, and some colonies measuring several meters across or more. Ascidians are found worldwide and at all ocean depths, attached to nearly any substratum. They are most abundant and diverse in rocky littoral habitats and on deep-sea muds. Salps (class Thaliacea) are pelagic tunicates that float about singly or in cylindrical or chainlike colonies sometimes several meters long. They are known from all oceans but are especially abundant in tropical and subtropical waters. Salps occur from the surface to depths of about 1,500 meters. The larvaceans (class Appendicularia), or appendicularians, are solitary bioluminescent planktonic creatures generally not more than about 5 mm long. They resemble in certain ways the larval stages of some other tunicates, hence the name "larvaceans." Their retention of larval features, including a notochord and a nerve cord, suggests that they arose by paedomorphosis. The feces of salps and larvaceans, and the abandoned houses of the latter, constitute important sources of food and particulate organic carbon in the open sea. In fact, the filtering of submicron particles by pelagic tunicates, and subsequent repackaging as denser fecal pellets that sink more quickly, is considered to be an important aspect of biogeochemical cycling towards the deep ocean affecting carbon budgets and nutrient supplies. Finally, the unusual sorberaceans (class Sorberacea) are benthic, abyssal, ascidian-like urochordates possessing dorsal nerve cords in adult stages. They are carnivorous, lacking the typical perforated branchial chamber of other tunicates.

With the increasing application of DNA barcoding to identify cryptic species, the numbers of extant species of ascidians appears to be growing substantively. Ascidians, or their ancestors, probably date to the Precambrian. However, morphological change is so slow among living ascidian species that it is often impossible to distinguish sister species without genetic analysis, despite large genetic distances. Rapid genetic change has been documented in both mitochondrial and nuclear genomes of tunciates. Further, recent studies of mitochondrial gene order reveal rearrangements, even among species in the same genus. And nuclear genomes show dramatic changes in gene content (e.g., loss of a Hox gene cluster) and dramatic changes in intron organization. Appendicularians have been shown to have some of the fastest evolving metazoan genomes known.

Biogeographical distributions of tunicates are not well known in many cases, due to several things: a historic lack of attention, compared to many other groups; habitats that are challenging to survey in some cases; difficult morphological characters for taxonomic identification; anthropogenic effects on distributions; and challenges in working with preserved specimens. More recently, combined technological advances in morphological and genetic analysis, along with targeted collection efforts in areas of high biodiversity, are beginning to result in rapid advances in our understanding of tunicate distributions and evolutionary relationships. Nonetheless, a well-supported phylogeny of the Urochordata has not yet been achieved.

The Tunicate Body Plan

Tunicates are bilaterally symmetrical, at least during early developmental stages. Most utilize mucus-covered **pharyngeal gill slits** (= **stigmata**) for suspension feeding (Figure 27.5A,B,H). Although modified in the appendicularians, water flows into the mouth and

(A)
(B)
(C)
(D)
(E)
(F)
(G)
(H)

(I)

(J)

(K)

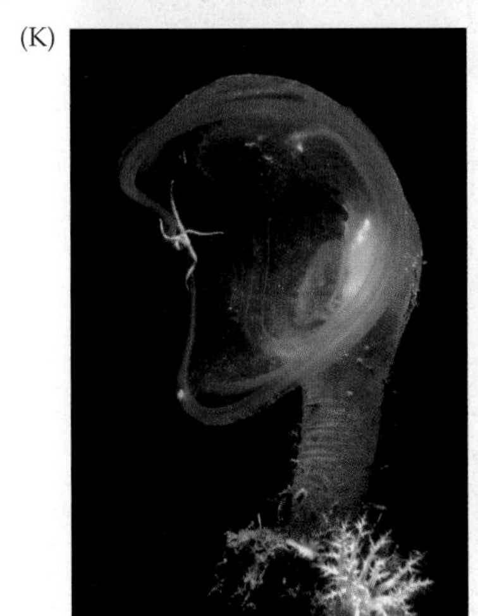

(L)

Figure 27.3 Representative Ascidians (Urochordata). (A) *Cnemidocarpa*, a solitary ascidian. (B) An intertidal compound ascidian, *Polyclinum*. (C) The solitary ascidian, *Styela*. (D) The social ascidian *Pycnoclavelina diminuta* (Sulawesi, Indonesia). (E) Colony of *Clavelina lepadiformis* from the Adriatic Sea (Croatia). (F) The common solitary ascidian *Ciona intestinalis*. (G) Colorful clavelinid ascidians. (H,I) The compound ascidian *Aplidum* sp. Note the external appearance of the colony. In I, the colony is sliced open to expose the clusters of zooids. (J) Portion of a colony of *Perophora*, with zooids arising from stolons. (K) *Megalodicopia hians*, a predatory ascidian from California with its "trap" spread open (note the small brittle star on the ascidian). (L) *Octacnemus*, a predatory deep-sea sorberacean.

pharynx by way of an incurrent **oral** (**branchial**) **siphon**, passes through the pharyngeal gill slits into a spacious water-filled atrium (**cloaca**) and exits through an excurrent **atrial siphon**. The gut is simple and U-shaped, with the anus emptying into the excurrent flow of water as it leaves the body. Because of great difference in body form relative to that of more familiar chordates, the general orientation of the bodies of urochordates is not immediately apparent and can only be fully understood by examining the events of metamorphosis, as will be described. The oral siphon is generally anterior, and the atrial siphon is either anterodorsal (in ascidians) or posterior (in thaliaceans) (Figures 27.5 and 27.6). In any case, the dorsoventral orientation of the body can be determined internally by the locations of the dorsal ganglion and a thickened ciliated groove, called the endostyle, that runs along the ventral side of the pharynx or branchial chamber (Figures 27.5A–E and 27.6C,D,H). The oral and atrial siphons are generally directed away from the substratum and set at an angle to one another, thereby reducing the potential for recycling wastewater.

The Ascidiacea is the largest and most diverse class of tunicates. Some interstitial forms are known and a few live anchored in soft sediments, but the majority of ascidians are attached to hard substrata (Figure 27.3). Three general morphological body plans of ascidians are usually recognized, although these categories do not relate directly to formal taxa since they have evolved multiple times in separate orders. Many larger species (up to 60 cm long) are called solitary ascidians because each individual develops from a single larva and is genetically unique (e.g., *Ciona, Molgula, Styela*; Figure 27.3A,C). Some solitary ascidians are found in dense aggregations, particularly on artificial substrata such as docks, pilings, and submerged ropes, where their tunics may even appear glued together, although they are still physiologically distinct. These aggregations may result from gregarious settlement, differential survival, or simply due to limited dispersal in brooding species. Social ascidians tend to live in clumps of individuals that are at least initially vascularly attached to one another at their bases (e.g., *Clavelina*; Figure 27.3D,E and chapter opener photo), although these connections may degenerate over time in some taxa (e.g., *Metandrocarpa*). Finally, a great number of species are colonial and compound ascidians and are characterized by many small feeding and reproductive units (zooids) held together in a common gelatinous matrix (e.g., *Aplidium, Botryllus*; Figures 27.3B,H,I and 27.4) and proliferating via asexual reproduction following establishment by a single sexually produced colonizing larva. In extreme cases, colonies may measure several meters across or potentially in some taxa (e.g., didemnid colonial ascidians) much more via fusion of distinct genetic individuals to form chimeric

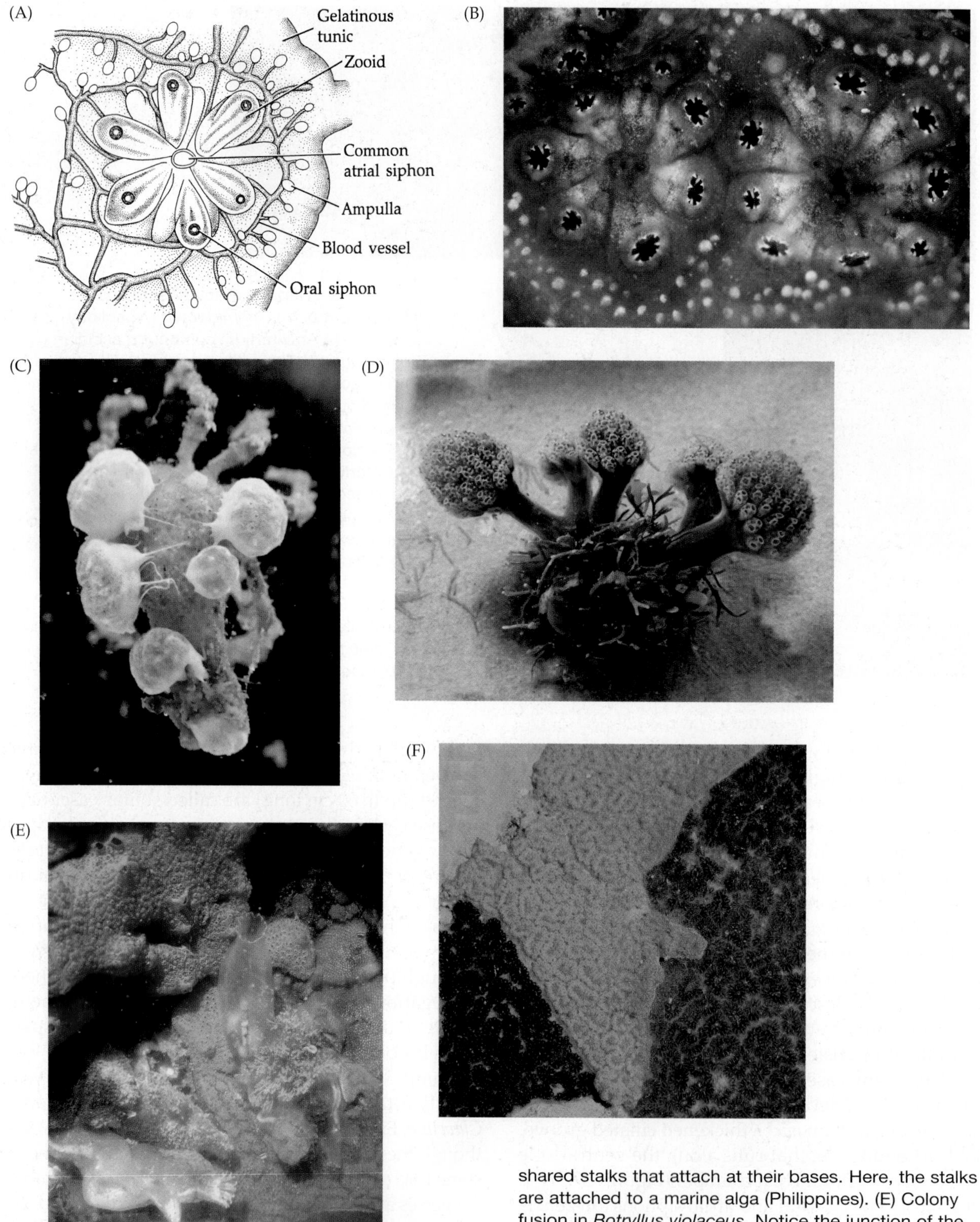

Figure 27.4 Some social, colonial, and compound ascidians. (A,B) The compound ascidian *Botryllus*. (A) Anatomical drawing; (B) living specimen. (C) Individual didemnid colonies, containing green symbionts, "walking" over the substratum (which, in this case, is the back of a small decapod crustacean in the Philippines). (D) The social ascidian *Oxycorynia*. Individual zooids are born on shared stalks that attach at their bases. Here, the stalks are attached to a marine alga (Philippines). (E) Colony fusion in *Botryllus violaceus*. Notice the junction of the salmon-colored colony in the lower right with the orange colony above; a beginning fusion event that will continue as the colonies merge. (F) Colonies of *Botryllus schlosseri*. Pigment cells in the blood produce colonies of different colors, which can be used to track the somatic outcome of fusion events. In this photos, colonies are not obviously fused, although shading differences in one colony (upper right) is indicative of incomplete somatic mixing in a past fusion event.

colonies of potentially much larger size. Colonial ascidians may regress or fragment as well, leaving a pattern of physically isolated colonies that are genetic clones distributed across the substratum.

The bodies of some solitary, social, and colonial forms are set on stalks and elevated off the substratum, although the compound forms typically grow as thin or thick sheets generally conforming to the topography of the surface on which they live. In some of these compound ascidians, zooids are arranged in regular rosettes and share a common atrial chamber that leads to a shared excurrent siphon within the outer body wall, or tunic (Figure 27.4A,B,E,F), while in other colonial species, each zooid has a distinct atrial siphon associated with each single branchial siphon. Zooids in some taxa are arranged seemingly haphazardly in the tunic, while in other taxa the zooids are arranged in regular patterns called systems. The tunic varies from thick to thin and from smooth and slick to wrinkled and leathery. In some families, the tunics can be reinforced with calcareous spicules—as in the Didemnidae. Many ascidean species are brightly colored, in some cases due to pigmented cells associated with the blood and in some tropical taxa due to symbionts. Compound ascideans frequently are some of the most colorful animals living on lower intertidal rocks and in the subtidal environment. Some solitary and social tunicates have clear tunics so that most parts of the internal anatomy are visible, sometimes highlighted with dramatic seemingly florescent coloring (e.g., the pink endostyles of "light bulb" tunicates, such as *Clavelina*). Interestingly, the Yamauchi Company in Japan has recently begun brewing beer using the ascidian *Halocynthia roretzi*, a species that has been eaten raw or cooked for many years in Japan and Korea (with a flavor likened to rubber dipped in ammonia, by western palates).

Ascidians have become so highly modified during their evolution as sessile suspension feeders that adults are not easily recognizable as chordates. The larval pharyngeal gill slits persist into adulthood. However, the dorsal hollow nerve cord and the notochord, present in the larval stage, are lost in adults, and additional neural differentiation occurs that provides sensory abilities more relevant to a sessile adult lifestyle. This neural differentiation is currently being studied with regard to relationships to vertebrate neural crest differentiation and neural placodes (embryonic ectodermal thickenings on the edge of the neural crest that give rise to neurons and sensory structures). In vertebrates, cells generated from the neural crest migrate throughout the body to give rise to diverse cell lineages producing nervous system elements and many other structures.

Thaliaceans are pelagic, ascidian-like urochordates called salps (Figure 27.6A–F). They are constructed much like their sessile counterparts except that the oral and atrial siphons are at opposite ends of the body, and in many forms the pharyngeal filtering basket is modified to accommodate the linear flow of water through the animal. The exiting water provides a means of "jet propulsion." Most are highly gelatinous and transparent. The class Thaliacea comprises three orders. Members of the order Pyrosomida are may be the most primitive thaliaceans. Pyrosomes are remarkable colonies of tiny ascidian-like zooids embedded in a dense gelatinous matrix and arranged around a long central, tubular chamber called a common cloaca (Figure 27.6A,B). The cloaca receives exhalant water from the inwardly directed atrial siphons of all the zooids; the water then exits through a single large aperture, thereby propelling the barrel-shaped colony slowly through the water. As in the ascidians, water movement in pyrosomes is generated entirely by ciliary action of the individual zooids. The remarkable light displays of large pyrosomes have long attracted interest and awe as the size of the entire colony can be several meters in length. On viewing a smaller captured specimen in a bucket on board a ship, the famed naturalist Thomas Huxley commented in 1851, "The phosphorescence was intermittent, periods of darkness alternating with periods of brilliancy. The light commenced in one spot, apparently on the body of one of the 'zooids,' and gradually spread from this as a centre in all directions; then the whole was lighted up; it remained brilliant for a few seconds, and then gradually faded and died away…". Huxley made these observations on his single specimen after noting that the entire sea was lighted intermittently by "miniature pillars of fire" producing "an ever-waning, ever-brightening soft bluish light as far as the eye could reach on every side."

The orders Doliolida and Salpida include thaliaceans that alternate between solitary sexual forms and colonial asexual stages (Figure 27.6C–E). Doliolid individuals are generally small, less than 1 cm long, whereas a single salpid may be 15 to 20 cm long but form a chainlike colony several meters in length. The members of these two orders move water through their bodies but propel themselves partially (salpids) or wholly (doliolids) by muscular action contracting their circular muscles to jet propel themselves through the water.

Thaliaceans are predominately warm-water creatures, although certain species are found in temperate and even polar seas. They are particularly abundant over the continental shelf and are frequently captured in surface waters or seen stranded on wave-swept sandy beaches after storms. Some, however, have been recorded from depths to 1,500 meters. It is thought that the very fine mesh filters of thaliaceans are easily clogged in nutrient-rich waters, potentially explaining their abundance in offshore and tropical waters.

Appendicularians are among the strangest of all urochordates and are characterized by the retention of larval features. These solitary animals live in a gelatinous casing, or **house**, that they secrete around their body (Figures 27.6G–I). The bulbous trunk of the body

Figure 27.5 General anatomy of ascidians. (A) A solitary ascidian (cutaway view). (B) The pharyngeal region of the same animal (cross section). White arrows indicate water flow; black arrows indicate path of food material. (C) A single zooid isolated from a colony of the compound ascidian *Diazona*. (D) A colony of the compound ascidian *Aplidium* (section). Two zooids are sharing a common atrial siphon. (E) A single *Aplidium* zooid brooding a larva in the atrium. (F) The endostyle of *Ciona* (cross section). The shaded cells are largely responsible for secretion of the mucus used in filter feeding. (G) The elaborate oral tentacles of the solitary ascidian *Herdmania momus* (Philippines). (H) Interior view of a bisected solitary ascidian (*Herdmania momus*) showing the elaborate folds of the branchial chamber, oral siphon bisected (whitish fingers near base of siphon are oral tentacles), larger atrial siphon intact, reddish gonads, and thick tunic. (I) A colony of *Botryllus* sp. showing three systems with the arrangements of zooids around a shared atrial siphon (in the center of each system). Fringing the entire colony are the terminal ends of the shared blood-vascular system, embedded in the tunic, that unites all the systems in the colony. The sausage-like endings are stacked several layers deep in places, as this is a very actively growing colony (orange and white blood cells also can be see).

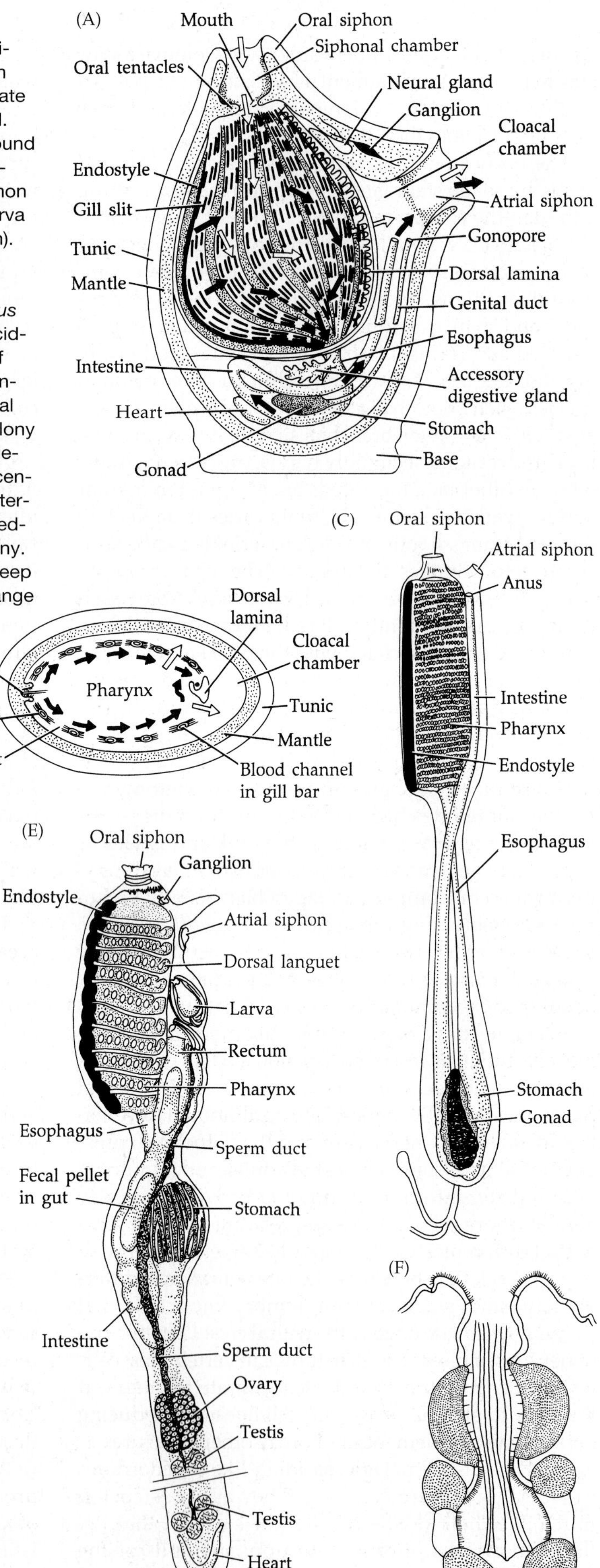

(G)

(H)

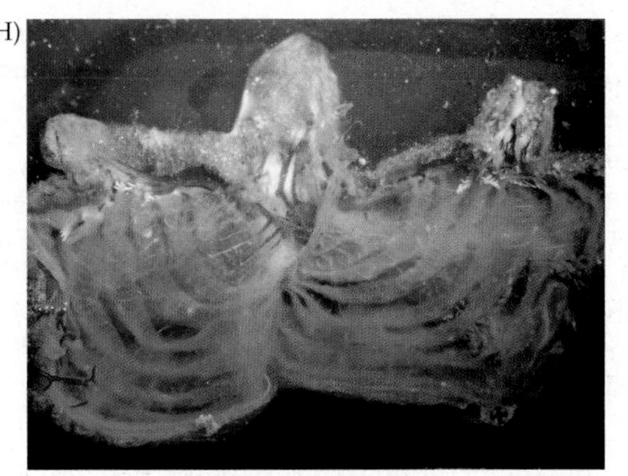

(I)

contains the major organs, including the gut, and bears a muscular tail in which the notochord is retained. The dorsal nerve cord, though reduced, extends partway along the length of the tail. Thus, we see clearly the evidence of the chordate nature of these animals retained through paedomorphosis. In other urochordates, such evidence is only obviously present during early developmental stages. (The phenomenon of paedomorphosis is discussed in Chapter 5.)

The pharynx of appendicularians is reduced and bears only two clefts. When positioned within its house, the animal produces a complex water current by beating its tail. Filtering is accomplished by meshes in the house wall and by mucous nets secreted by the animal (Figure 27.6I). The mesh filters are complex and vary among species, often being constructed of more than one size of interlaced fibers. The exiting water provides the locomotor force. Appendicularians are found in the surface waters of all oceans and are sometimes extremely abundant.

The class Sorberacea comprises a poorly-understood group of benthic, deep-sea and abyssal, ascidian-like urochordates that have a dorsal nerve cord in their adult stage. They are carnivorous, lacking the perforated branchial chamber of other urochordates (e.g., *Octacnemus*).

Body Wall, Support, and Locomotion

The body wall of tunicates includes a simple epithelium overlain by a secreted tunic of varying thickness and consistency (Figure 27.5A,D,H). The tunic is most well developed in the ascidians and some thaliaceans. It varies from soft and gelatinous to tough and leathery, and it sometimes includes calcareous spicules. The matrix of the tunic contains fibers and is composed largely of a cellulose-polymer called tunicin produced by cellulose synthetase genes that are thought to have been acquired via horizontal gene transfer. The tunic is not a simple, secreted, nonliving cuticle, however, since it also contains amebocytes and, in some cases, blood cells and even blood vessels. The tunic is an exoskeleton providing support and protection. Some ascidians harbor symbiotic bacteria in their tunics. These bacteria include a variety of species from several major groups, including taxa unique to tunicates as discussed below. It has been suggested that the ability to construct protective coverings with a cellulose material may have been a key innovation providing protection to these soft-bodied organisms allowing diversification among the protochordates.

Beneath the epidermis are muscle bands. In many species, especially among ascidians, these muscles lie within an ectodermally produced mesenchyme called the mantle (Figures 27.5A,B). Ascidians possess longitudinal muscles extending along the body wall that serve to pull the flared siphons down against the body. Circular sphincter muscles close the siphonal openings. Fine neural control of these muscles is part of the development that occurs during the transition from the motile larval to the sessile adult stage. This control allows the filter feeders to respond quickly to unwanted particles or predators near the siphonal openings. Doliolids and salps have well-developed bands of circular muscles that pump water through the body for feeding and locomotion. When they contract, water within the body is forced out the atrial siphon, thereby propelling the animal forward. When they relax, the body expands because of the resilience of the tunic, and water is drawn in through the oral siphon. As noted above, the tail muscles of appendicularians provide the action for moving water through the houses of these animals.

Tunicates have an extremely reduced coelom, considered absent by some researchers; the body cavity has been lost in concert with the evolution of the large pharynx and cloacal chamber. The cloaca is a saclike, ectodermally derived structure continuous with the epidermis of the atrial siphon (Figure 27.5A,B). The inner wall lies against the pharynx and is perforated over the gill slits, or stigmata. Thus, water that enters the pharynx via the oral siphon, flows through the stigmata, into the cloacal chamber, and out the atrial siphon.

Feeding and Digestion

We have hinted at various aspects of the feeding biology of tunicates in our comments above. Most of these animals are suspension feeders and use various kinds of mucous nets to filter plankton and organic detritus from the sea water. A few ascidians live partially embedded in soft sediments and feed on organic material in the substratum, and certain bizarre deep-sea species actually prey on small invertebrates by grasping them with the lips of the oral siphon. Below we provide a more detailed description of feeding and digestion in a suspension-feeding ascidian and then compare this with feeding in thaliaceans and appendicularians.

Water is moved through the body of an ascidian largely by the coordinated action of cilia lining the expanded pharynx, called the **pharyngeal chamber** (also known as the branchial chamber, branchial basket, or branchial sac). Water enters the oral siphon and passes through a short siphonal chamber at the inner end of which is the mouth. A ring of fleshy tentacles encircles the mouth and prevents the entrance of large particles (Figure 27.5A,G). Food-laden water passes into the pharynx, which bears a ventral longitudinal groove, the endostyle. The bottom of the groove is lined with mucus-secreting cells and bears a longitudinal row of flagella; the sides of the groove bear cilia (Figure 27.5F). The mucus, a complex mucoprotein containing iodine, is moved to the sides of the endostyle by the basal flagella and then outward by the lateral cilia. Cells near the opening of the endostyle are responsible for binding environmental iodine and incorporating it into the mucus. Sheets of mucus then move dorsally along the inner wall of the pharynx and pass over the **stigmata**. The slitlike stigmata are arranged in rows and bear lateral cilia that drive water from the pharynx into the surrounding atrial water chamber (Figure 27.5A–D,H). Thus, water passing through the stigmata also passes through the mucous sheets, on which food particles are retained. On the dorsal surface of the pharynx is a longitudinal curved ridge called the **dorsal lamina**, or a row of ciliated projections called **languets**, or both (Figure 27.5A–E). These structures serve to roll the mucous sheets into cords, which are then passed posteriorly to a short esophagus and then to a stomach. Attached to the stomach is a small pyloric gland, which extends around the intestine as a network of small tubes. Some species also bear an accessory digestive gland (Figure 27.5A). Digestive enzymes are secreted into the stomach lumen by secretory cells of the gut wall and perhaps by the associated glands, and digestion is largely extracellular. From the stomach, the gut loops forward as an intestine, through which undigested material passes to the anus, which opens into the atrium near the excurrent siphon.

Figure 27.6 Thaliaceans and appendicularians (larvaceans). (A,B) External and sectional views of the colonial thaliacean *Pyrosoma*. (C) A solitary thaliacean (*Salpa*). (D) The solitary thaliacean *Doliolum*. (E) A colonial form of *Salpa*. (F) A pelagic tunicate (Thaliacea). Note the circular bands of muscles in the body wall. (G) The appendicularian *Stegasoma* in its gelatinous house. (H) *Oikopleura*, an appendicularian removed from its gelatinous house. (I) *Oikopleura* in its house. The arrows represent water currents.

The ascidian family Didemnidae comprises some of the colonial forms in which the cloacal systems are shared and the colonies are usually hardened with aragonitic "spiculospheres." Among certain tropical genera (e.g., *Didemnum, Diplosoma, Lissoclinum, Trididemnum*) are species that maintain symbiotic bacteria in the test, the pharyngeal chamber, or the cloacal system. Three broadly different types of bacteria are found in apparent symbiotic association with tropical ascidians, cyanobacteria (including the genus *Prochloron*), alphaproteobacteria, and gammaproteobacteria. These bacteria produce a variety of secondary metabolites, including some unusual compounds found at startlingly high concentrations in the ascidians. Unusual oxidation states of vanadium are found at high levels in some taxa and some secondary metabolites are highly toxic biomolecules currently used in cancer treatment or in clinical trials. The role of these compounds in the ascidians is not well known, but it is assumed there are anti-predation benefits to the toxic compounds and nutritive benefits to hosting photosynthetic organisms such as is found in corals and sponges. Some didemnid ascidians housing such symbiotic bacteria are also remarkable for their powers of locomotion, limited though it is—colonies have been clocked at speeds of 4.7 mm per 12-hour period. Such movement may allow these ascidians to position themselves in light conditions favorable to their bacterial symbionts. Thaliaceans feed in much the same way as ascidians, except that the siphons are at opposite ends of the body. The number of pharyngeal stigmata is usually reduced, especially in doliolids, and restricted to the posterior portion of the pharynx (Figure 27.6D). The intestine extends posteriorly from the pharynx and opens into the enlarged cloacal chamber.

Ecologically, ascidians have long been known to be highly efficient filtration engineers on reefs and in other habitats including manmade structures such as marinas, pilings, and aquaculture-associated gear. The high filtration and growth rates of some solitary and

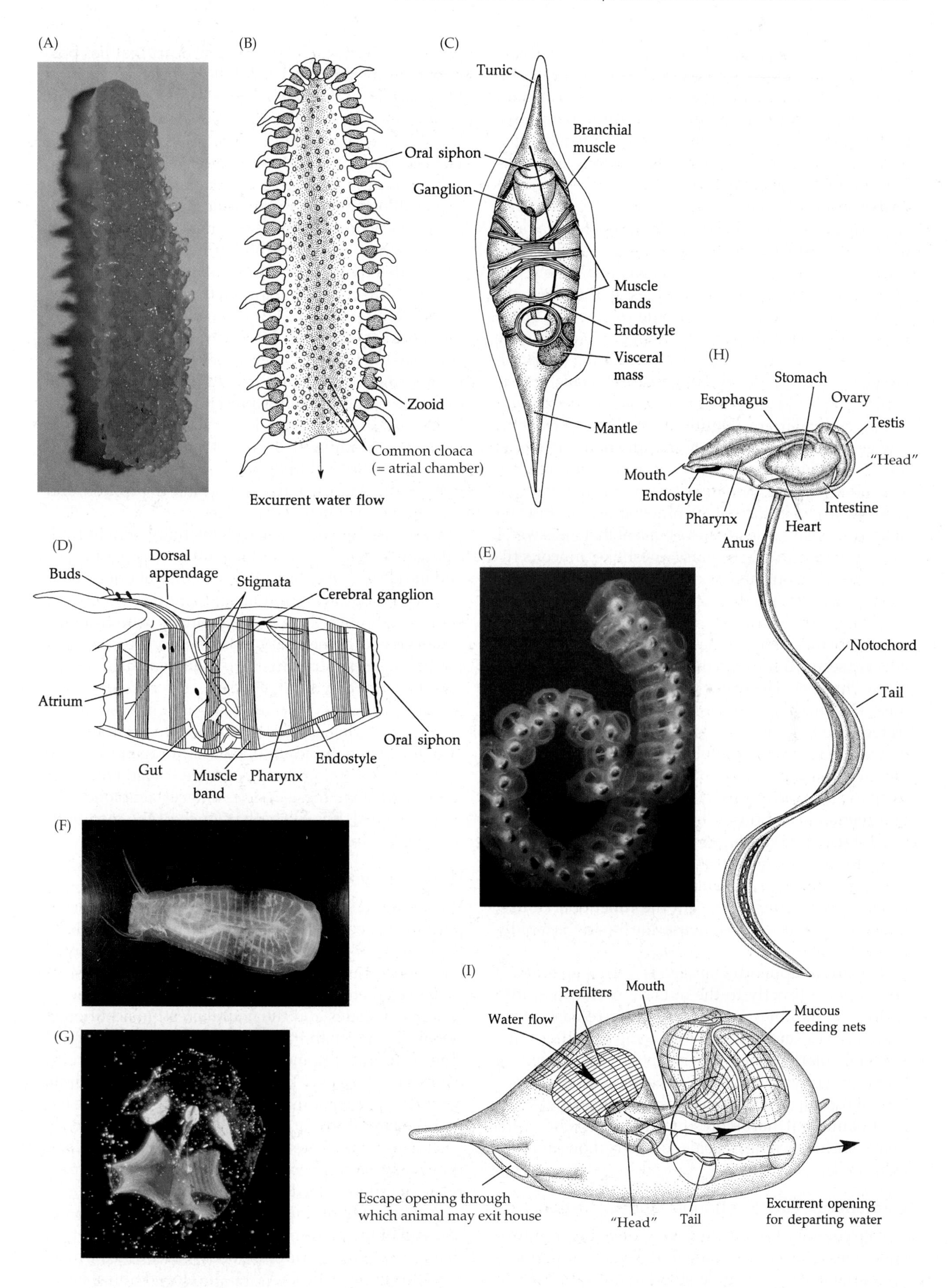
(A)
(B)
(C)
Tunic
Branchial muscle
Oral siphon
Ganglion
Muscle bands
Endostyle
Visceral mass
Zooid
Mantle
Common cloaca (= atrial chamber)
Excurrent water flow
(H)
Stomach
Esophagus
Ovary
Testis
"Head"
Mouth
Endostyle
Intestine
Pharynx
Heart
Anus
(D)
Dorsal appendage
Buds
Stigmata
Cerebral ganglion
(E)
Notochord
Tail
Atrium
Oral siphon
Endostyle
Gut
Muscle band
Pharynx
(F)
(G)
(I)
Prefilters
Mouth
Mucous feeding nets
Water flow
Escape opening through which animal may exit house
"Head"
Tail
Excurrent opening for departing water

colonial ascidians allow them to compete effectively with the aquaculture species of choice, generally bivalves. The removal of these ascidians is a costly and unwelcome addition to bivalve culture operations. Unfortunately, the rate of human-accelerated ascidian range spread is increasing globally. Recently, ascidians have become particularly infamous and menacing invasive species, in the case of *Didemnum vexillum* even covering acres of benthic habitat in the Gulf of Maine in the Northwest Atlantic. In some cases, ascidians have been observed in invasions of natural habitat, outside of harbors and bays where they have long been observed as suspiciously "cosmopolitan" species. These incursions may reflect broad changes in community structure in marine environments including the elimination of many predators, an increase in available particulate food, and increasing sea surface temperatures.

Appendicularians (Figures 27.6F–H) secrete a hollow gelatinous (mucopolysaccharide) house in which they reside and upon which they depend for feeding. The tail is directed through a tube in the house structure toward an excurrent opening. Sinusoidal beating of the muscular tail generates a current that pulls water into the house through coarse, meshlike, mucous filters that screen out large particles; eventually the water leaves the house via the excurrent opening. The pharynx of appendicularians bears only two small gill slits, which open directly to the exterior. Mucous feeding nets are secreted through the mouth and lie within the house chamber. The water current is directed through these fine-mesh nets, where food particles are concentrated. The food, net and all, is periodically ingested by way of a short buccal tube. The houses of most appendicularians bear an additional opening that serves as an escape hole through which the animal can leave and reenter. The houses are fragile and easily damaged. Damaged or clogged houses are abandoned, and new ones are manufactured rapidly, in a matter of seconds or minutes. In some species a new or "spare" house may be found beneath the functional house; after escaping the clogged house, the "spare" is rapidly inflated.

The gut of appendicularians is U-shaped and the anus opens directly to the outside rather than into a cloacal chamber. Fecal material is released into the path of excurrent water leaving the filter nets. Appendicularians are primarily herbivorous, feeding on minute phytoplankton and bacteria even below a size of 0.1 μm. They sometimes constitute the dominant planktonic herbivores in waters over the continental shelf, reaching densities of many thousands per cubic meter.

Circulation, Gas Exchange, and Excretion

The circulatory system is weakly developed in urochordates, especially in the thaliaceans and appendicularians. It is best understood in ascidians, which possess a short, valveless, tubular heart that lies posteroventrally in the body near the stomach and behind the pharyngeal chamber (Figure 27.5A,D,E). The heart is surrounded by a pericardial sac, sometimes referred to as the vestiges of the coelom (in ascidians). Blood vessels extend anteriorly and posteriorly, opening into spaces around the internal organs and also providing the blood supply to the tunic. The heartbeat is by peristaltic action driven by two myogenic pacemakers, one at either end of the tubular heart, and the direction of this motion is periodically reversed, flushing the blood first one way through the heart and then the other. Blood physiology and function are still largely matters of speculation. Curiously, many ascidians accumulate high concentrations of certain heavy metals in their blood, especially vanadium and iron. Some evidence suggests that the presence of high vanadium levels in at least some species serves to deter would-be predators. In addition, the blood includes a large variety of cell types, including amebocytes that are thought to function in nutrient transport, tunic deposition, and accumulation of metabolic wastes.

Gas exchange occurs across the body wall in tunicates, and especially across the linings of the pharynx and the cloacal chamber. Little is known about respiratory physiology in these animals. In most ascidians and some other tunicates, two evaginations arise from the posterior wall of the pharynx and lie along each side of the heart. These structures are called **epicardial sacs**, and they are thought to represent coelomic remnants. In some species the epicardial sacs are involved in bud formation during asexual reproduction, and they may also function in the accumulation of nitrogenous waste products by forming storage capsules called **renal vesicles**. Other than these vesicles and certain blood cells (nephrocytes), it is likely that much of the metabolic waste is lost from the body by simple diffusion.

Nervous System and Sense Organs

While the nervous system of tunicates is reduced, perhaps reflecting their relatively inactive sessile or floating planktonic lifestyles, it has recently received attention as possibly containing precursors of nervous system elements thought to be vertebrate innovations, such as placodes and the neural crest. In addition, a secondary mechanoreceptor organ, the **circumoral ring**, is found in some ascidian and thaliacean tunicates. This ring may be a precursor to the vertebrate secondary mechanoreceptor hair cells found in the ear and the lateral line organ. In ascidians the secondary mechanoreceptors are located around the siphons, and they show varying complexity among taxa that may relate to different feeding strategies.

A small cerebral ganglion lies just dorsal to the anterior end of the pharynx and gives rise to a few nerves that go to various parts of the body, especially the muscles and siphonal areas. A well-developed dorsal nerve cord is present in the tails of tunicate larvae, but this structure is lost during metamorphosis, except in the appendicularians. Most tunicates possess a **neural**, or **subneural, gland** located between the cerebral ganglion and the anterodorsal portion of the pharynx (Figure 27.5A,D). This gland opens to the pharynx through a small duct, but its function is unknown. Some workers have suggested that it may be the precursor of the pituitary gland of vertebrates. Sensory receptors are poorly known in tunicates, although touch-sensitive neurons are prevalent around the siphons. Ascidian larvae are well known to be responsive to light and gravity, often with a general pattern of initially swimming upwards, toward the light, but then eventually settling in darker, shaded areas, though the pattern varies among species.

Reproduction and Development

Asexual reproduction While appendicularians are probably entirely sexual in their reproductive habits, thaliaceans and many ascidians include asexual processes in their life history strategies. In social and especially compound ascidians, asexual budding allows rapid exploitation of available substrata and recovery from partial predation, as we have seen in other sessile colonial invertebrates such as sponges and bryozoans. Remarkably, some colonial ascidians can regenerate their entire bodies from a small amount of vascular tissue, an impressive feat for a chordate, and subject of ongoing investigation into the nature of regeneration pathways that differ to some extent from asexual proliferation. In fact, regeneration studies in the solitary ascidian *Ciona intestinalis* and colonial botryllid ascidians have been going on for over 100 years.

Budding in tunicates occurs in a great variety of ways and from different organs and germinative tissues (Figure 27.7). In general, initial buds are formed by a sexually produced individual (oozooid), then the asexually produced individuals (blastozooids) produce additional buds. The simplest and perhaps most primitive budding process occurs in certain social ascidians, including species of *Perophora* and *Clavelina*, where blastozooids arise from the body wall of stolons. In more complicated budding processes, the germinal tissues include various combinations of the epidermis, gonads, epicardial sacs, and gut. Doliolid thaliaceans often produce chains of buds. The chains are sometimes released intact, but eventually each blastozooid breaks loose as a separate individual. Budding in pyrosomes results in the characteristic floating colonies seen in species of this family (Figures 27.6A,B).

Some workers have divided the various types of budding among ascidians into two categories on the basis of their functional significance. Propagative budding generally occurs during favorable environmental conditions and serves to increase colony size and exploit available resources. On the other hand, survival budding tends to take place during the onset

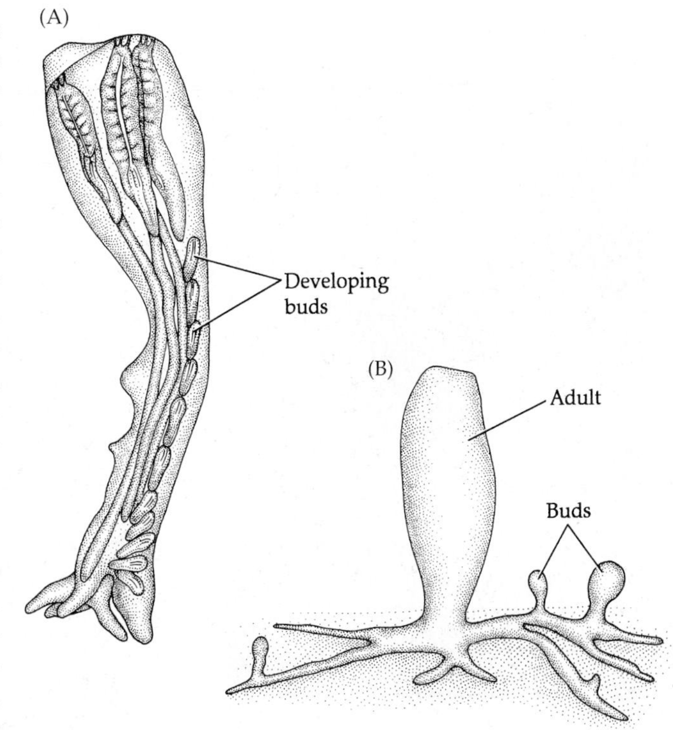

Figure 27.7 Asexual reproduction in ascidians. (A) Formation of buds in the colonial ascidian *Circinalium*. (B) General pattern of stoloniferous budding in an ascidian.

of adverse conditions and may be viewed as an overwintering or other survival device. With the return of more favorable growing conditions, these "pre-buds" quickly develop as new blastozooids. More recently, particular attention to aspects of asexual reproduction has focused on the model colonial species, *Botryllus schlosseri*, where a regular weekly cycle of zooid degeneration and replacement by developing buds (**palleal budding**) has served as a platform to investigate the relationship between developmental processes and regeneration. Recent discoveries have uncovered the presence of stem cell niches associated with the ascidian endostyle as well as stem cells circulating in the vascular system that allow whole body regeneration from isolated fragments of vascular tissue (**vascular budding**).

In botryllid ascidians, including *Botryllus schlosseri*, studies have shown that variation in blood pigments is genetically determined. The blood also contains several vertebrate-like hormones, including thyroxine, oxytocins, and vasoconstrictors. Compound species, such as *Botryllus* and *Botrylloides*, have special blood components that play a vital role in rejecting adjacent conspecific colonies unless they are closely related (e.g., close relatives and clonemates), in which case the interaction may lead to colony fusion. This system is of great interest to comparative immunologists. Through breeding studies and genetic analyses, the "fusibility system"

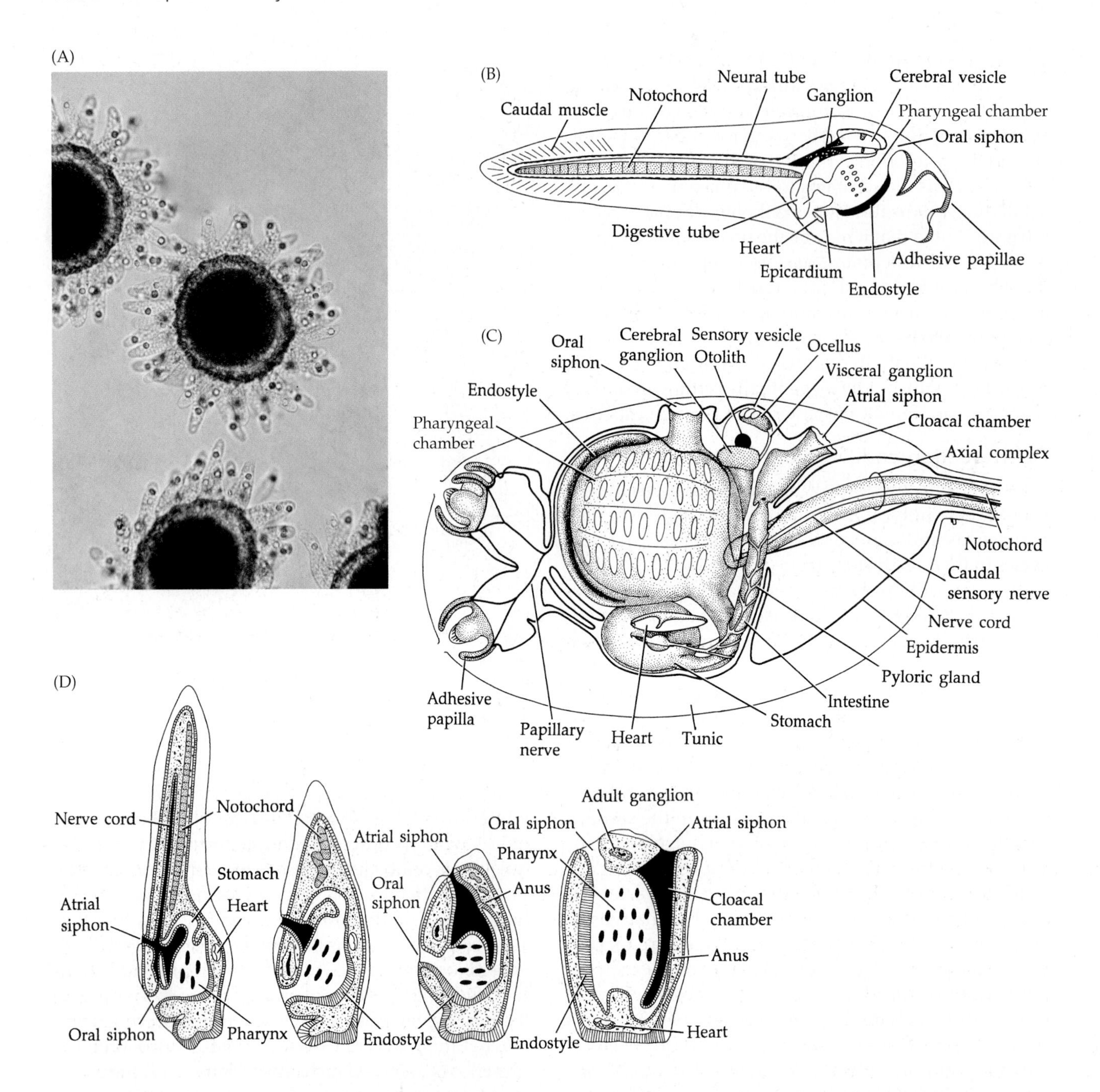

continues to be elucidated in this species. Populations of botryllids are highly genetically polymorphic at fusibility loci with hundreds of alleles present even in invasive populations worldwide. Colonies that share at least one fusibility allele can fuse to form chimeras, while non-allele-sharing colonies undergo cytotoxic rejection processes and form permanent borders. The fate of the chimeras after fusion is less well known and also a subject of ongoing inquiry. Ascidean species vary in the intensity of the physiological engagement and subsequent rejection cytotoxicity when it occurs, and additional genes (and environmental variation) are thought to also play a role in determining the outcome of these immune-like interactions. This system is also being studied as a model for the evolution of kin recognition interactions in a medical context using natural transplantation phenomena related to host–graft acceptance or rejection.

Sexual reproduction Tunicates are hermaphroditic, with relatively simple reproductive systems. Generally a single ovary and a single testis lie near the loop of the digestive tract in the posterior part of the body and, in most cases, connect through a separate sperm duct and oviduct to the cloacal chamber near the anus (Figures 27.5. However, some species have a single gonad (ovotestis), with one gonoduct, and members of a few families (e.g., Pyuridae, Styelidae) have multiple gonads. In some species, the ovaries contain high amounts of silica, but the significance of this condition is unknown.

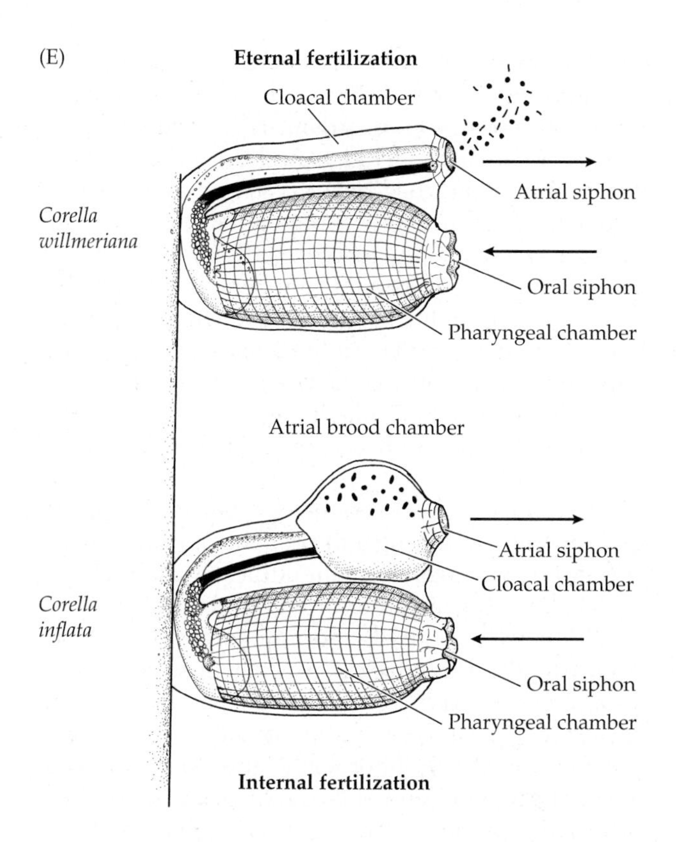

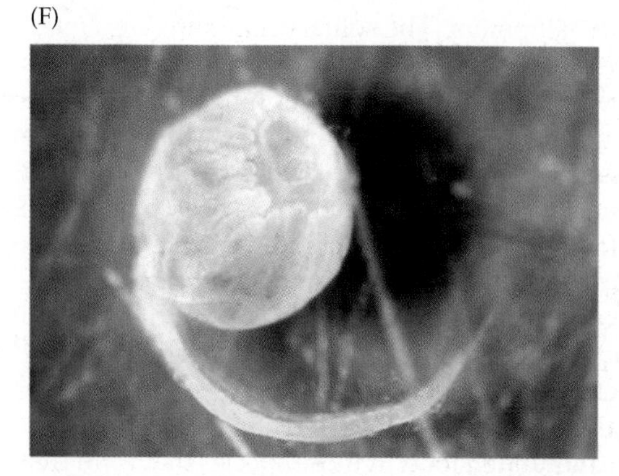

Figure 27.8 Sexual reproduction, larvae, and metamorphosis. (A) *Ciona intestinalis* egg surrounded by follicle cells. (B) An ascidian tadpole larva has many chordate features. (C) The anterior end of the larva of *Distaplia occidentalis*. (D) Metamorphosis of a settled tadpole larva. Tail resorption is followed by a reorientation of the body to bring the siphons to the adult positions. (E) Internal and external fertilization in the solitary ascidian *Corella*. Arrows show direction of water flow, in the branchial siphon and out the atrial siphon. In *C. willmeriana* eggs are being released from the atrial siphon, to be fertilized in the water (self fertilization in this species is rare). In *C. inflata*, embryos and larvae are found within the atrial brood chamber, resulting from internal fertilization (which may be self fertilization under certain conditions). (F) Tadpole larva of *Botryllus violaceous* (with ampulla).

There is considerable variation in the overall reproductive strategies among tunicates. Most large solitary ascidians produce high numbers of weakly-yolked ova, which are shed to the sea coincident with the release of sperm from other individuals. External fertilization is followed by the development of a free-swimming **tadpole larva**, which eventually settles and metamorphoses to an oozooid (Figure 27.8). In contrast to this fully indirect life-history pattern, many compound ascidians composed of tiny zooids produce relatively few eggs, but each egg has a high yolk content. These eggs are fertilized and subsequently brooded while attached to a parent zooid or detached and located in various places within the tunic or within the cloacal chamber; they are not released until the swimming tadpole larvae develop. Some compound ascidian larvae are smaller with less yolk, but they retain a nutritive connection to the parent zooid, called a placental developmental mode. And, various degrees of larval suppression also occur among some ascidians with this mixed life history strategy, with some species undergoing fully direct development. One species, *Protostyela longicauda*, produces nonswimming larvae, which are brooded. In this species, the larval "tail" is simply an extension of the tunic and contains no cellular material. The larva is sticky, and when it is released quickly adheres to any object it contacts, later making a permanent attachment. At least some of these tailless species have been shown to be derived from tailed species, and genes controlling tail development are present but suppressed in expression.

Although all thaliaceans lack a free-swimming larva, they differ markedly in their approaches to direct development. In pyrosomes, each zygote develops directly to an oozooid, with no evidence of a larval stage. The oozooid then buds to produce a colony. Doliolids produce tailed larvae, but each is encased in a cuticular capsule and does not swim. The larva metamorphoses to an oozooid. Salps undergo internal fertilization in the oviduct. The zygotes implant and form a placenta-like association with the parent in a uterine chamber in the oviduct. Here the embryos develop directly to the adult form.

Appendicularians free-spawn, and fertilization occurs externally. They develop to a tadpole-like stage and then mature by protandry into the characteristic larva-like adults.

In most tunicate species studied, cleavage is radial, holoblastic, and slightly unequal; it leads to the formation of a coeloblastula, which undergoes gastrulation by invagination. The blastopore lies at the presumptive posterior end of the body but closes as development proceeds. There is a long history of experimental work on developmental stages of ascidians leading to insights in areas such as cytoplasmic regulation within the egg and the relationship of egg and larval body axes. More recently, developmental studies in basal chordates including urochordates have focused intensely on understanding the derivation of vertebrate neural development. The solitary ascidian *Ciona intestinalis* has been the subject of a complete cell fate mapping project, including neural mapping of the larva allowing extensive comparison of the transition from the larval sensory architecture to the very different adult architecture. Particular attention has been paid to the search for homologies to vertebrates, and the emphasis has changed from marveling at the lack of vertebrate-like structures in adult ascidians to a more nuanced approach considering where antecedents to the vertebrate neural crest may lie in urochordates.

The development of chordate features is most easily seen and understood in those species that form free tadpole larvae, such as most Ascidiacea. In fact, the larvae of most ascidians consist of countable numbers of cells, making these animals excellent models for studying cellular events early in chordate development; gastrulation in ascidians starts as early as the 110-cell stage. As the embryo elongates, the gut proliferates three longitudinal strips of mesoderm—a middorsal strip that becomes the notochord, and lateral strips that form the mesenchyme and body musculature. Thus, even though the mesoderm arises from the archenteron (endoderm), it does not pouch substantively from the gut wall. A middorsal strip of ectoderm differentiates as a neural plate, which sinks inward and curls to produce the dorsal hollow nerve cord. The epidermis secretes a larval tunic, which often develops dorsal and ventral tail fins. The anterior part of the gut differentiates as the pharyngeal chamber (or basket) during larval life, and the rudiment of a cloacal water cavity forms by an ectodermal invagination producing the atrial siphon. However, these ascidian larvae are all lecithotrophic, and the gut and filtering devices do not become functional until metamorphosis. Social and colonial tunicate larvae are usually much larger than solitary larvae. The greater size includes a more highly developed adult rudiment in the trunk of the tadpole along with additional yolk, generally rendering the larval features more difficult to visualize. However, some colonial larvae are smaller and less yolky, but maintain a nutritive connection to an adult zooid with a longer development time in comparison to solitary larvae. Some species with symbiotic photosynthetic partners show a variety of structures associated with the larval and adult forms for housing the symbionts. Similarity in such structures does not necessarily correlate to phylogenetic affinities; instead, it suggests that the partnerships have evolved multiple times.

Ascidian larvae outside of the parent are short-lived. When development is fully indirect, the larvae are planktonic for only about two days or less. In some forms with a mixed life history pattern (e.g., *Botryllus*), the free larval life lasts only a few minutes and larvae may even settle and metamorphose upon the releasing parent colony, which then resorbs them into the parent colony. While the larval body is no longer visible, the germline of the resorbed larva may actually take over the germline of the resorbing adult in a phenomenon known as **germ cell parasitism**. This phenomenon is thought to be driven by kin selection and potentially by advantages accrued to both partners. Larvae may escape the high mortality common for young juvenile stages and find a ready place to settle in space-limited environments. Adults may gain benefits in avoiding senescence through an infusion of juvenile stem cells. Even though a short larval life allows dispersal over only small distances, the larvae probably play a key role in the selection of suitable substrata. The events of settling and metamorphosis of ascidian larvae are complex and varied.

Ascidian larvae possess several sensory receptors that function in settling and probably substratum selection, but are absent from the adults. A small **sensory vesicle** lies near the anterior end of the dorsal nerve cord adjacent to the developing cerebral ganglion (Figure 27.8C). This vesicle houses a light-sensitive ocellus and a statocyst (called an otolith). For most, though not all ascidian larvae, at the time of settlement the larva becomes negatively phototactic and positively geotactic. The anterior end of the larva bears two or three **adhesive papillae**, each of which is supplied with two sets of neurons. A primary, more exposed set may serve in determining an appropriate settlement location via chemosensory abilities, while the secondary set of more protected neurons may be more involved in aspects of metamorphosis related to mechanoreception following attachment. These phenomena are summarized briefly below, and are the subject of ongoing, detailed neural mapping, genetic analysis, and histological studies, particularly in the model basal chordate *Ciona*.

The settling larva of an ascidian contacts a substratum with its anterior end and may secrete mucus and undergo searching motions before releasing an adhesive from the papillae. In the larvae of many compound ascidians, the papillae evert during this process. The secretion of the adhesive apparently triggers an irreversible sequence of metamorphic events. Within minutes after attachment, resorption of the larval tail commences by one of several methods involving various contractile elements in the tail region. The animal's viscera and siphons then undergo a remarkable 90° rotation that brings these organs to their adult positions. The outer layer of the cuticle is shed, removing the larval fins from the settled juvenile (Figure 27.8D). The pharynx enlarges and the filtering mechanisms become functional as the siphons open to the environment. During all of these processes secondary attachment organs, called **ampullae**, extend from the body and permanently affix the animal to the substratum. Finally, various transient larval organs are lost, such as most of the larval nervous system and sense organs.

Larvae of some species have been reported to practice aggregated settlement. And in at least one species the larva may have the ability to distinguish close relatives from conspecific nonrelatives. Experimental studies suggest early juvenile mortality rates may be very high. Colonial species may have several advantages at this stage as they start life at a larger size and are thus able to filter water more efficiently. Tiny early settlers may suffer from a low profile in the hydrodynamic boundary layer where flow rates are near zero, making food capture challenging. Some newly-settled juvenile ascidians appear to have adaptations to increase their ability to filter feed near the boundary layer, including temporary growth of a stalk, enlarged siphon diameters relative to body size, and specific orientation relative to flow. As ascidians increase in size and escape the boundary layer, their growth rates sometimes increase dramatically. Adult filtration rates for solitary species are extremely high, on par with the bivalve filter feeders such as mussels and oysters with which they are commonly found, especially in fouling communities. Intriguingly, a large solitary *Pyura* species, invasive to Chile, has taken on the ecological role of a mussel bed, with masses of these large solitary ascidians packed together in a broad swath of the intertidal.

Chordate Phylogeny

Recent phylogenomic analyses have largely, though not entirely resolved the broad structure of chordate relationships (Figure 27.9). Both molecular and morphological analyses agree that the phylum Chordata is monophyletic. Its anatomical synapomorphies include the notochord, the endostyle (thought to be represented by the thyroid gland in vertebrates), and the muscular postanal tail. All chordates also have pharyngeal gill slits, epithelial tissue that binds iodine and secretes iodothyrosine, and a dorsal nerve cord but these three features also occur in hemichordates. However, homology of the solid dorsal nerve cord of hemichordates and the hollow cord of chordates has been challenged by recent work and needs further investigation. The notochord plays an important embryological role in inducing the chordate central nervous system, so lack of a notochord in hemichordates is evidence in support of the idea that the hemichordate dorsal nerve cord may not be homologous to the chordate dorsal nerve cord. Gill slit endoderm in both hemichordates and chordates expresses *pax1/9* and *six1* genes, supporting their homology in these two phyla. Some juvenile enteropneusts have a postanal tail that has been homologized with the pterobranch stalk. However, recent work has shown that the enteropneust tail expresses the *Hox11/13* gene, whereas the vertebrate postanal tail expresses the orthologous *Hox10–13* gene, suggesting possible homology in the postanal body regions, though not necessarily in the tail itself (the functions of these structures are altogether different in the two phyla).

The central nervous system of chordates shows a wide variety of forms, from a few small ganglia with a tail nerve cord in urochordates, to a swimming spinal cord with a hardly recognizable cerebral vesicle in cephalochordates, to a fully developed brain and spinal cord in vertebrates. The nerve cord of chordates develops dorsally as a hollow neural tube above the notochord. In the vertebrates, the nerve cord differentiates during embryogeny into a large brain and usually a well-developed spinal cord protected by the backbone. The notochord is a rodlike, skeletal structure dorsal to the gut and ventral to the nerve cord. It should not be confused with the backbone, or vertebral column, of adult vertebrates. The notochord appears early in embryogeny and plays an important role in promoting or organizing the chordate embryonic development of nearby structures. In most adult chordates, the notochord disappears or becomes highly modified. In some protochordates, and some fishes, it persists as a flexible but incompressible skeletal rod that supports the body during swimming. The embryonic development of the

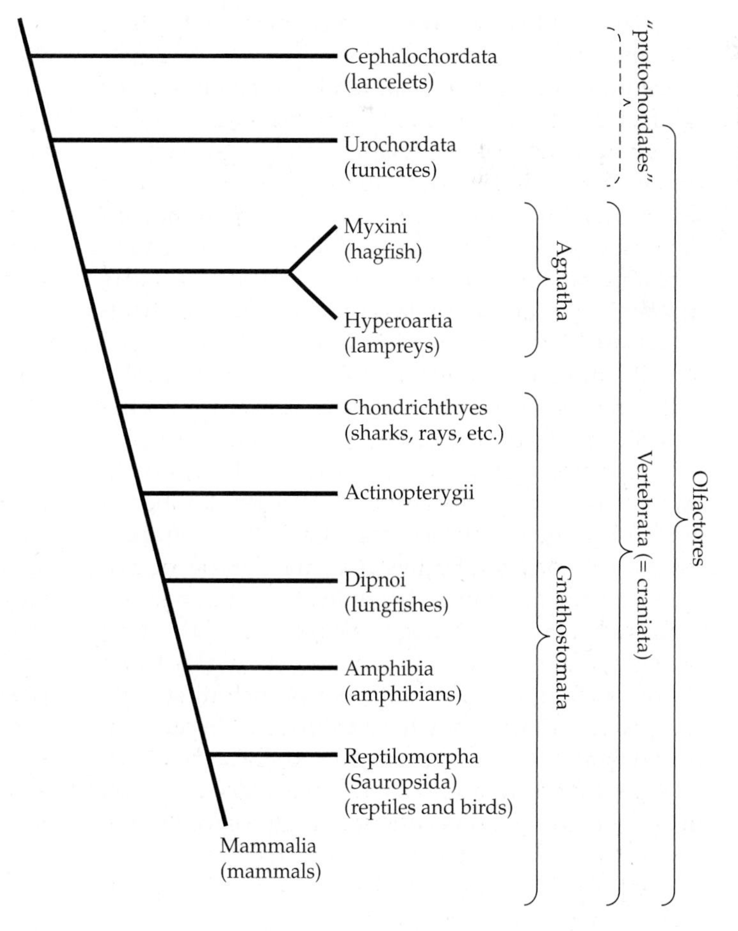

Figure 27.9 Phylogeny of the Chordata. See text for details.

notochord in ascidians has been well studied. In *Ciona intestinalis* (Figure 27.3F) it is composed of only 40 cells, which align to form a single row by an event called convergent extension. The cells undergo vacuolation to transform the notochord into a single hollowed tube. The restricted number of notochord cells is achieved by the regulated number of cell divisions under the control of a key transcription factor, Brachyury. The stomochord (buccal diverticulum) of hemichordates, once thought to be homologous with the chordate notochord, is now viewed as convergence or parallelism.

The vertebrate thyroid gland sequesters iodine and secretes T3 and T4 hormones, which regulate metabolism, post-embryonic development, and metamorphosis. The lancelet and urochordate thyroid homolog, the endostyle, is primarily involved in mucus production for feeding, but it also incorporates iodine. It has been suggested that the endostyle may even synthesize thyroid hormones, and cephalochordates have been shown to have a thyroid hormone receptor (*TR*) gene and several genes for components of the thyroid hormone production system, though not for the pituitary thyroid-stimulating hormone. Specialized secretory cells in hemichordates also bind iodine, but these are found throughout the pharynx so a direct homology with the endostyle may not be appropriate, although these cells could have been the precursors to the chordate endostyle.

Deuterostomes may be a small branch on the tree of life, but there are no stranger animals than echinoderms and chordates (especially, vertebrates). And this brings up one of the most profound synapomorphies of Chordata—the dorsoventral axis—which is thought to be inverted in comparison to the hemichordate (and probably all other Bilateria) dorsoventral axis. Early in their evolution, chordates apparently turned themselves upside-down! Hemichordate gill bars and gill slits are located dorsally (opposite the ventral mouth), while in cephalochordates and craniates the gill bars and slits are located ventrally. Gill openings are also dorsal in stem-group echinoderms. And, the dorsal epibranchial ridge of enteropneusts appears to be the anatomical and histological homolog of the ventral endostyle of chordates. Furthermore, the dorsal side of a juvenile hemichordate is determined by an expression stripe of the bone morphogenetic protein (*BMP*) gene, which is expressed dorsally, and its ventral side by an expression stripe of *BMP* antagonists, including chordin, just as has been shown for arthropods. In contrast, in vertebrates, *BMP* is expressed ventrally and chordin is expressed dorsally. However, as the mouth is ventral in both chordates and nonchordates, the evolution of chordates must have also entailed moving the mouth from the opposite side of the gill slits to the same side as the gill slits, while presumably also evolving a notochord and dorsal central nervous system from a ventral enteropneust-like nerve net. The hypothesis of chordate dorsoventral axis inversion has been around since the early nineteenth century, and it has not been without its detractors (e.g., van den Biggelaar et al. 2002). However, current EvoDevo research tends to support the idea.

But deuterostome (and chordate) oddness doesn't stop there. One of the other trends in deuterostome evolution is gene duplication, which appears to be widespread. The cephalochordates had long been thought to have the basic deuterostome Hox cluster of 14 genes, but in 2008 a fifteenth Hox gene was identified from the lancelet *Branchiostoma floridae*. In addition, in cephalochordates and vertebrates, *Hox9–14* show an independent duplication compared with protostomes. And a duplication of the posterior Hox genes into three genes (named *Hox11/13a*, *Hox11/13b*, and *Hox11/13c*) characterizes both echinoderms and hemichordates. In contrast, *Hox14* has been lost from all tetrapods and teleost fishes so far examined. Many other genes have been identified in amphioxus that show two to four paralogs in vertebrates, derived from two whole-genome duplication events. However, urochordate genomes seem to have lost many genes. Hox gene duplication, of course, is especially important for opening up new evolutionary–developmental avenues, since these are the "master genes" that are involved in anterior–posterior patterning in bilaterians.

Recent molecular phylogenetic studies indicate that Cephalochordata (lancelets) are the most basal living chordates, and that they are the sister group to the Urochordata + Vertebrata. The urochordate + vertebrate clade has been named Olfactores. The relationships of the basal vertebrates, however, are still unsettled. Although some workers view lampreys as the sister group of the gnathostomes, recent molecular trees place lampreys as the sister group of hagfishes, as a jawless clade sometimes called Agnatha. Hagfish lack a backbone (i.e., they are not true vertebrates), but they have a skull (cranium) like other craniates. Lampreys (Cephalaspidomorphia or Petromyzontidae), on the other hand, have a cartilaginous "backbone" but lack a skull! Thus, some workers define the taxon Vertebrata as Agnatha + Gnathostomata, while others view Vertebrata as Lampreys + Gnathostomata (excluding the hagfish). In the first case, the name Vertebrata is not fully descriptive, but retained out of tradition (i.e., it is a legacy name). However, if a backbone were

Rychel, A. L., S. E. Smith, H. T. Shimamoto and B. J. Swalla. 2005. Evolution and development of the chordates: collagen and pharyngeal cartilage. Mol. Biol. Evol. 23: 541–549.

Ruppert, E. E. 1994. Evolutionary origin of the vertebrate nephron. Am. Zool. 34: 542–553.

Satoh N., D. Rokhsar and T. Nishikawa. 2014. Chordate evolution and the three-phylum system. Proc. Royal Soc. B 281: 17–29.

Shu, D.-G., S. Conway Morris and X. L. Zhang. 1996. A *Pikaia*-like chordate from the lower Cambrian of China. Nature 384: 157–158.

Sobral, D., O. Tassy and P. Lemaire. 2009. Highly divergent gene expression programs can lead to similar chordate larval body plans. Curr. Biol. 19 (23): 2014–2019.

Stokes, M. D. and N. D. Holland. 1998. The lancelet. Amer. Sci. 86: 552–560.

Swalla, B. J. and A. B. Smith. 2008. Deciphering deuterostome phylogeny: molecular, morphological and paleontological perspectives. Phil. Trans. Roy. Soc. B. 363: 1557–1568.

Tiozzo, S. and R. R. Copley. 2015. Reconsidering regeneration in metazoans: an evo-devo approach. Front. Ecol. Evol. 3: 67.

Tsagkogeorga, G. and 9 others. 2009. An updated 18S rRNA phylogeny of tunicates based on mixture and secondary structure models. BMC Evol. Biol. 9: 187. doi: 10.1186/1471-2148-9-187

Turbeville, J. M., J. R. Schultz and R. A. Raff. 1994. Deuterostome phylogeny and the sister group of the chordates: Evidence from molecules and morphology. Mol. Biol. Evol. 11: 648–655.

van den Biggelaar, J. A. M., E. Edsinger-Gonzales and F. R. Schram. 2002. The improbability of dorso-ventral axis inversion during animal evolution, as presumed by Geoffroy Saint Hilaire. Contrib. Zool. 71: 29–36.

Vassalli, Q. A., E. Anishchenko, L. Caputi, P. Sordino, S. D'Aniello and A. Locascio. 2015. Regulatory elements retained during chordate evolution: Coming across tunicates. Genesis 53: 66–81.

Wada, H. and N. Satoh. 1994. Details of the evolutionary history from invertebrates to vertebrates, as deduced from the sequences of 18S rDNA. Proc. Natl. Acad. Sci. U.S.A. 91: 1801–1804.

Cephalochordata

Hirakow, R. and N. Kajita. 1994. Electron microscopic study of the development of amphioxus, *Branchiostoma belcheri tsingtauense*: The neurula and larva. Acta Anat. Nippon 69: 1–13.

Holland, N. D. and L. Z. Holland. 1990. Fine structure of the mesothelia and extracellular materials in the coelomic fluid of the fin boxes and sclerocoels of a lancelet, *Branchiostoma floridae* (Cephalochordata = Acrania). Acta Zool. 71(4) 225–234.

Holland, N. D. and L. Z. Holland. 1991. The histochemistry and fine structure of the nutritional reserves in the fin rays of a lancelet, *Branchiostoma lanceolatum* (Cephalochordata = Acrania). Acta Zool. 72(4): 203–207.

Poss, S. G. and H. T. Boschung. 1996. Lancelets (Cephalochordata: Branchiostomatidae): how many species are valid? Israel J. Zool. 42: S.13–S.66.

Ruppert, E. E. 1997. Cephalochordata. Pp. 349–504 in F. W. Harrison and E. E. Ruppert, (eds.). *Microscopic Anatomy of Invertebrates, Vol. 15.* Wiley-Liss, New York.

Ruppert, E. E., C. B. Cameron and J. E. Frick. 1999. Endostyle-like features of the dorsal epibranchial ridge of an enteropneust and the hypothesis of dorsal–ventral axis inversion in chordates. Invert. Biol. 118: 202–212.

Whittaker, J. R. 1997. Cephalochordates, the Lancelets. Pp. 365–381 in S. F. Gilbert and A. M. Raunio (eds.), *Embryology: Constructing the Organism.* Sinauer Associates, Sunderland, MA.

Urochordata

Abitua, P. B., E. Wagner, I. A. Navarrete and M. Levine, M. 2012. Identification of a rudimentary neural crest in a non-vertebrate chordate. Nature 492: 104–107.

Barham, E. 1979. Giant larvacean houses: Observations from deep submersibles. Science 205: 1129–1131.

Bates, W. R. 2007. HSP90 and MAPK activation are required for ampulla development in the direct-developing ascidian *Molgula pacifica*. Invertebr. Biol. 126: 90–98.

Berná, L. and F. Alvarez-Valin. 2014. Evolutionary genomics of fast evolving tunicates. Genome Biol. Evol. 8: 1724–1738.

Berrill, N. J. 1961. Salpa. Sci. Amer. 204: 150–160.

Berrill, N. J. 1975. Chordata: Tunicata. Pp. 241–282 in A. C. Geise and J. S. Pearse (eds.), *Reproduction of Marine Invertebrates, Vol. 2.* Academic Press, New York.

Birkeland, C., L. Cheng and R. A. Lewis. 1981. Mobility of didemnid ascidean colonies. Bull. Mar. Sci. 31: 170–173.

Bone, Q. 1989. On the muscle fibres and locomotor activity of doliolids (Tunicata: Thaliacea). J. Mar. Biol. Assoc. U.K. 69: 587–607.

Bullard, S. G. and 9 others. 2007. The colonial ascidian *Didemnum* sp. A: Current distribution, basic biology and potential threat to marine communities of the northeast and west coasts of North America. J. Exp. Mar. Biol. Ecol. 342: 99–108.

Burighel, P. and R. A. Cloney. 1997. Urochordata: Asidiacea. Pp. 221–348 in F. W. Harrison and E. E. Ruppert (eds.), *Microscopic Anatomy of Invertebrates. Vol. 15.* Wiley-Liss, New York.

Burighel, P., M. Sorrentino, G. Zaniolo, M. C. Thorndyke and L. Manni. 2001. The peripheral nervous system of an ascidian, *Botryllus schlosseri*, as revealed by cholinesterase activity. Invertebr. Biol. 120: 185–198.

Cohen, S., Y. Saito and I. Weissman. 1998. Evolution of allorecognition in botryllid ascidians inferred from a molecular phylogeny. Evolution 52(3): 746–756.

Cloney, R. A. 1978. Ascidian metamorphosis review and analysis. Pp. 225–282 in F. S. Chia and M. E. Rice (eds.), *Settlement and Metamorphosis of Marine Invertebrate Larvae*. Elsevier North-Holland, New York.

Cloney, R. A. 1990. Urochordata-Ascidiacea. Pp. 361–451 in K. G. Adiyodi and R. G. Adiyodi (eds.), *Reproductive Biology of Invertebrates.* Oxford and IBH, New Delhi.

Cloney, R. A. and S. A. Torrence. 1982. Ascidian larvae: Structure and settlement. In J. D. Costlow (ed.), *Biodeterioration.* U. S. Naval Institute, Annapolis, MD.

Cohen, S. 1996. The effects of contrasting modes of fertilization on levels of inbreeding in the marine invertebrate genus *Corella*. Evolution 50(5): 1896–1907.

Cox, G. 1983. Engulfment of *Prochloron* cells by cells of the ascidean, *Lissoclinum*. J. Mar. Biol. Assoc. U.K. 63: 195–198.

Davidson B. and B. J. Swalla. 2002. A molecular analysis of ascidian metamorphosis reveals activation of an innate immune response. Development 129: 4739–4751.

Deibel, D., M. L. Dickson, and C. Powell. 1985. Ultrastructure of the mucus feeding filters of the house of the appendicularians *Oikopleura vanhoeffeni*. Mar. Ecol. Prog. Ser. 27: 79–86.

Dolcemascolo, G., R. Pennati, F. De Bernardi, F. Damiani and M. Gianguzza. 2009. Ultrastructural comparative analysis on the adhesive papillae of the swimming larvae of three ascidian species. Invertebrate Surviv. J. 6: S77–S86.

Flood, P. R. 1991. Architecture of, and water circulation and flow rate in, the house of the planktonic tunicate *Oikopleura labradorensis*. Mar. Biol. 111: 95–111.

Gasparini, F. and 10 others. 2015. Sexual and asexual reproduction in the colonial ascidian *Botryllus schlosseri*. Genesis 53: 105–120.

Hirose, E. T. Maruyama, L. Cheng and R. W. Lewin. 1996. Intracellular symbiosis of a photosynthetic prokaryote,

Prochloron sp., in a colonial ascidian. Invert. Biol. 115(4): 343–348.

Hopcroft, R. R. and J. C. Roff. 1995. Zooplankton growth rates: Extraordinary production by the larvacean *Oikopleura dioica* in tropical waters. J. Plankton Res. 17(2): 205–220.

Huxley, T. H. 1851. Observations upon the anatomy and physiology of *Salpa* and *Pyrosoma*. Phil. Trans. R. Soc. Lond. B 141: 567–593.

Iannelli, F., F. Griggio, G. Pesole and C. Gissi. 2007. The mitochondrial genome of *Phallusia mammillata* and *Phallusia fumigata* (Tunicata, Ascidiacea): High genome plasticity at intragenus level. BMC Evol Biol 7: 155.

Jacobs, M. W., B. M. Degnan, J. D. D. Bishop and R. R. Strathmann. 2008. Early activation of adult organ differentiation during delay of metamorphosis in solitary ascidians, and consequences for juvenile growth. Invertebr. Biol. 127: 217–236.

Jeffery, W. R. 1994. A model for ascidian development and developmental modifications during evolution. J. Mar. Biol. Assoc. U.K. 74: 35–48.

Jeffery, W. R. 2015. Closing the wounds: one hundred and twenty five years of regenerative biology in the ascidian *Ciona intestinalis*. Genesis 53: 48–65.

Jeffery, W. R. and B. J. Swalla. 1997. Tunicates. Pp. 331–364 in S. F. Gilbert and A. M. Raunio (eds.), *Embryology: Constructing the Organism*. Sinauer Associates, Sunderland, MA.

Juliano, C. E., S. Zachary Swartz and G. W. Wessel. 2010. A conserved germline multipotency program. Development 137: 4113–4126.

Karaiskou, A., B. J. Swalla, Y. Sasakura and J.-P. Chambon. 2015. Metamorphosis in solitary ascidians. Genesis 53: 34–47.

Katz, M. J. 1983. Comparative anatomy of the tunicate tadpole, *Ciona intestinalis*. Biol. Bull. 164: 1–27.

Lacalli, T. C., N. D. Holland and J. E. West. 1994. Landmarks in the anterior central nervous system of amphioxus larvae. Philos. Trans. R. Soc. Lond. B Biol. Sci. 344: 165–185.

Lemaire, P. and J. Piette. 2015. Tunicates: exploring the sea shores and roaming the open ocean. A tribute to Thomas Huxley. Open Biol. 5: 150053.

Lambert, C. C. and G. Lambert (eds.). 1982. The developmental biology of the ascidians. Am. Zool. 22: 751–849. [Results of a 1981 symposium at the annual meeting of the American Society of Zoologists; nine papers, plus introductory remarks by C. Lambert.]

Lambert, G., C. C. Lambert and J. R. Waaland. 1996. Algal symbionts in the tunics of six New Zealand ascidians (Chordata, Ascidiacea). Invert. Biol. 115(1): 67–78.

Ma, L. and 7 others. 1996. Expression of an *Msx* homeobox gene in ascidians: Insights into the archetypal chordate expression in pattern. Dev. Dynam. 205: 308–318.

Mackie, G., D. H. Paul, C. M. Singla, M. A. Sleigh and D. E. Williams. 1974. Branchial innervation and ciliary control in the ascidian *Corella*. Proc. Roy. Soc. B, 187: 1–35.

Manni, L., G. Zaniolo, F. Cima, P. Burighel and L. Ballarin. 2007. *Botryllus schlosseri*: a model ascidian for the study of asexual reproduction. Dev. Dyn. 236: 335–352.

Monniot, F., R. Martoja, M. Truchet and F. Fröhlich. 1992. Opal in ascidians: a curious bioaccumulation in the ovary. Mar. Biol. 112: 283–292.

Nishida, H. 1994. Localization of determinants for formation of the anterior–posterior axis in eggs of the ascidian *Halocynthia roretzi*. Development 120: 3093–3104.

Pennati, R. and U. Rothbächer. 2014. Bioadhesion in ascidians: a developmental and functional genomics perspective. Interface Focus 5: 20140061. doi: 10.1098/rsfs.2014.0061

Piette, J. and P. Lemaire. 2015. Thaliaceans, the neglected pelagic relatives of ascidians: A developmental and evolutionary enigma. Q. Rev. Biol. 90: 117–145.

Rigon, F. 2013. Evolutionary diversification of secondary mechanoreceptor cells in tunicata. BMC Evol. Biol. 13: 112.

Ruppert, E. E. 1994. Evolutionary origin of the vertebrate nephron. Am. Zool. 34: 542–553.

Sasakura, Y., K. Nakashima, S. Awazu, T. Matsuoka, A. Nakayama and J. Azuma. 2005. Transposon-mediated insertional mutagenesis revealed the functions of animal cellulose synthase in the ascidian *Ciona intestinalis*. Proc. Natl. Acad. Sci. 102: 15134–15139.

Satoh, N. 1994. *Developmental Biology of Ascidians*. Cambridge Univ. Press, Cambridge.

Schmidt, G. H. 1982. Aggregation and fusion between conspecifics of a solitary ascidean. Biol. Bull. 162: 195–201.

Sherrard, K. and M. LaBarbera. 2005. Form and function in juvenile ascidians. II. Ontogenetic scaling of volumetric flow rates. Mar. Ecol. Progr. Ser. 287: 139–148.

Stoeker, D. 1980. Chemical defenses of ascidians against predators. Ecology 61: 1327–1334.

Svane, I. and C. M. Young. 1989. The ecology and behavior of ascidian larvae. Oceanogr. Mar. Biol. Rev. 27: 45–90.

Swalla, B. J. 1993. Mechanisms of gastrulation and tail formation in ascidians. Microsc. Res. Tech. 26: 274–284.

Torrence, S. A. and R. A. Cloney. 1982. The nervous system of ascidian larvae: Primary sensory neurons in the tail. Zoomorphologie 99: 103–115.

Tsagkogeorga, G. and 9 others. 2009. An updated 18S rRNA phylogeny of tunicates based on mixture and secondary structure models. BMC Evol. Biol. 9: 187.

Worcester, S. E. 1994. Adult rafting *versus* larval swimming: Dispersal and recruitment of a botryllid ascidian on eelgrass. Mar. Biol. 121: 309–317.

Young, C. M. and F. Chia. 1985. An experimental test of shadow response function in ascidian tadpoles. J. Exp. Mar. Biol. Ecol. 85: 165–175.

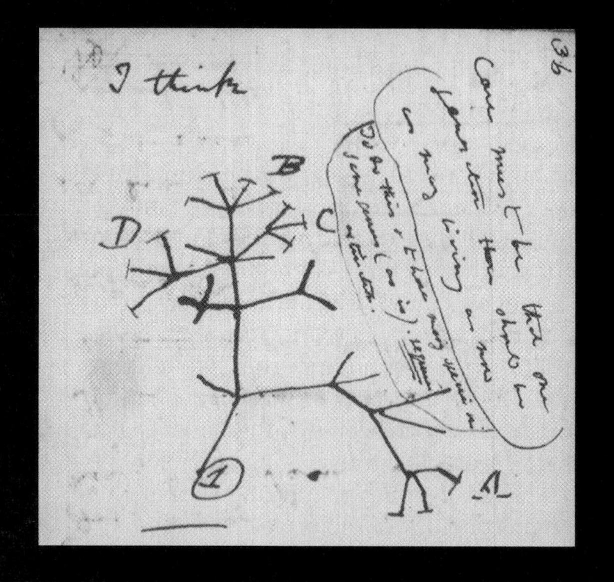

CHAPTER 28

Perspectives on Invertebrate Phylogeny

By the time you have gotten to this chapter, you will have "toured" almost *all* of the Animal Kingdom, given that invertebrates comprise roughly 1,324,402 (96%) of the approximately 1,382,402 described, living animal species. What an incredible diversity of form and function you have seen! Isn't it curious how unevenly these animals are distributed across the 32 metazoan phyla? Vastly unevenly distributed—81.5% of all described animals are arthropods, 5.8% are molluscs, and only 4.4% are chordates. The other 29 phyla make up the remaining 7%! Eight phyla have fewer than 100 described species in them, and half of those have fewer than a dozen known species—what are we to make of that? In fact, only 6 phyla comprise more than one percent each of the described animal species on Earth (Arthropoda, Mollusca, Annelida, Platyhelminthes, Nematoda, and Chordata). It would seem that the world belongs to insects, crustaceans, spiders, molluscs, worms, and vertebrates. Indeed, these are the only animals most humans ever see in their lifetimes (unless they happen to go diving on a tropical coral reef, in which case they are confronted by myriad sponges, cnidarians, and echinoderms). How fortunate that you have now been introduced to all 32 animal phyla!

In reading the "animal chapters" of this text, you have learned a good deal about the evolution of these phyla, how they are related to one another, and what their internal relationships are. You have learned that phylogenies can be constructed from a variety of different kinds of data, and that most recently the field of molecular phylogenetics has proliferated, giving us new ideas about how Earth's creatures are related to one another. The idea of phylogenetic trees should now be quite familiar to you. But let's take one last look, a "higher view" of animal phylogeny, before we close the cover on this edition of *Invertebrates*.

Molecular phylogenetics had only just begun to bear fruit in 2002, when the second edition of this book went to press. While some tantalizing suggestions of new and very different kinds of relationships among the animal phyla had been suggested, these new hypotheses had not yet been broadly tested. Since then, molecular phylogenetics has exploded onto the scene at a pace never imagined, and it now plays the primary role in reconstructing animal phylogeny. In just the past 15 years we have come from analyzing a few

This chapter has been revised by Richard C. Brusca and Gonzalo Giribet.

ribosomal genes to the point where new genome-level analyses are being published almost daily. Molecular phylogenetics is now being incorporated into studies of ecology, oceanography, biogeography, conservation biology, medicine, archeology, and anthropology. And it is building a new framework for the tree of life.

Unraveling the phylogenetic history of the Animal Kingdom has been one of biology's great challenges. The "new phylogeny," built with molecular data, has many similarities to older trees built on morphological and developmental data, but some big surprises have emerged and many uncertainties remain. The biggest challenge lies in the fact that life's deep lineages arose and began to diverge from one another so long ago, over half a billion years ago for most phyla. So the traits of animals, whether anatomical or genetic, that might be useful in revealing relationships among these ancient lineages are obscured by hundreds of millions of years of evolutionary change. But, despite this, many relationships are now well resolved.

Early molecular phylogenetic studies relied heavily on the 18S ribosomal RNA gene (also known as the nuclear small-subunit ribosomal RNA gene, or SSU rRNA). However, it quickly became apparent that understanding deep-level metazoan phylogeny required analysis of additional genes, particularly nuclear protein-coding genes, and this led to an era of multigene trees that continues today as phylogeneticists add more and more genes (and taxa) to their datasets for analysis (Chapter 2). Most recently phylogenomic studies, using large parts of the genome (often using the transcriptome[1] as a proxy for the whole genome) and analyzing hundreds or even thousands of genes, have begun to expand the scope of data available for phylogenetic analysis, and these have added increased stability to the structure of our phylogenetic framework. The advent of EvoDevo, or evolutionary developmental biology, has also begun to significantly impact our understanding of how genes relate to specific morphologies, how they work, and what roles they might have played in the unfolding of animal radiations. These new techniques have provided answers to fundamental questions, such as the identity of arthropod appendages and the nature of segmentation in animals.

Estimates of divergence times of the animal phyla, based on molecular clock calculations, suggest that the origin of the Metazoa was 875 to 650 million years ago. Some trace fossils put the emergence of the bilaterians at around a billion years ago, although the nature of those specimens has been disputed. Some recent datings that had assumed the presence of sponges in Ediacaran (and even Cryogenian) rocks due to the presence of high concentrations of the compound 24-isopropyl cholestane are now disputed. Some of the oldest metazoan fossils are from the Doushantuo Formation of southern China, dating to 600 million years ago. Sponges, cnidarians, and other apparent diploblastic animals have been reported from these deposits, as well as embryos ranging from two to thousands of cells. However, many of these Doushantuo metazoan fossils also have been disputed in one fashion or another. It seems that interpreting ancient fossils in rocks can be as challenging as inferring ancient phylogenies with morphology or molecules!

We discussed the origin of the Metazoa in several previous chapters. To reiterate, a large body of evidence, anatomical and molecular, has accumulated since the 1960s that supports the view of multicellular animals sharing a common ancestor with the protist group Choanoflagellata. The flagellated collar cells of sponges and choanoflagellates have been viewed as nearly identical and unique to these two groups. A few interesting differences between them have been noted (e.g., Mah et al. 2014), but some divergence over a half billion years is to be expected. The Metazoa are defined by a number of synapomorphies, the most obvious being multicellularity arising through the embryonic layering process called gastrulation. Also, unlike coloniality, as seen in many protist groups (including choanoflagellates), in animals the epithelial cells are in contact with each other through unique junction structures and molecules, some of which make transport of nutrients between cells possible (e.g., septate or tight junctions, desmosomes, zonula adherens). Additional metazoan synapomorphies include: striate myofibrils, actin–myosin contractile elements, the possession of animal (type IV) collagen (although collagen, or a collagen homologue also occurs in some fungi), and a basal lamina beneath the epidermis. In addition, sexual reproduction in animals involves a distinct pattern of egg development from one of the four cells of meiosis, whereas the other three cells degenerate.

Figure 28.1 presents a consensus tree of metazoan phylogeny, based primarily upon the most recent molecular phylogenetic research. Phylogenetic studies have largely been in agreement that the oldest living animal phylum is Porifera, the sponges. However, some recent molecular phylogenies have suggested Ctenophora might be the basalmost metazoans. The draft genome of *Pleurobrachia bachei*, together with other ctenophore transcriptomes, suggest that ctenophores may be rather distinct from other animal genomes in their content of neurogenic, immune, and developmental genes. However, a number of putative synapomorphies link Cnidaria and Ctenophora with the Bilateria, a clade that has been called Neuralia (Figure 28.1). The "basal Ctenophora" hypothesis was challenged by Jékely et al. (2015) and Pisani et al. (2015), both research groups suggesting it was an

[1]Whereas a genome is the complete set of genes present in a cell (or organism) and is sequenced from DNA, a transcriptome is a subset of those genes that are transcribed (or expressed) in a cell at any given time and is sequenced from RNA.

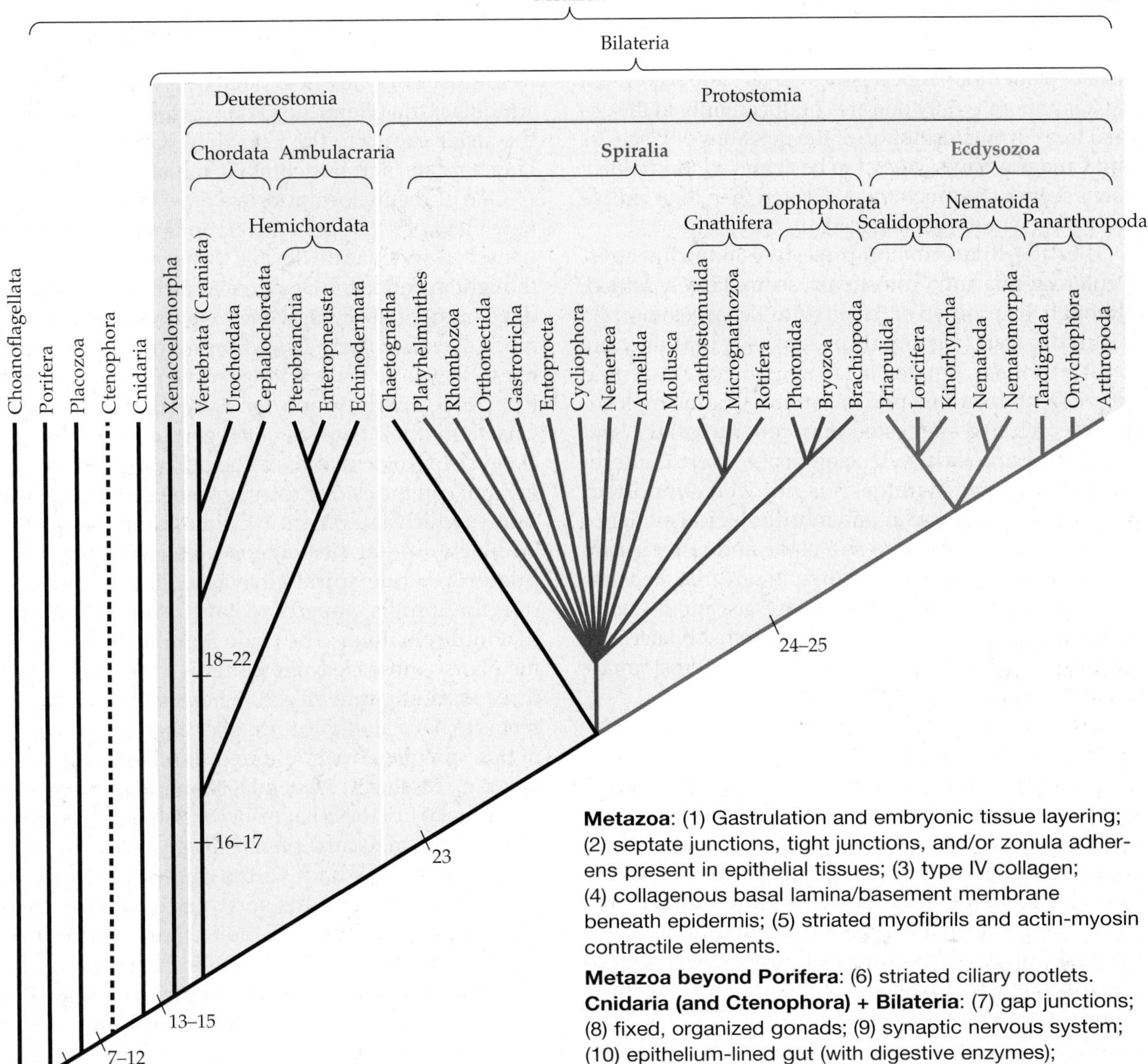

Figure 28.1 A phylogeny of Metazoa. This tree reflects a consensus view based primarily on recent molecular phylogenetic analyses. Uncertainty still exists in several regions. Strong resolution of the relative placement of Placozoa, Cnidaria, and especially Ctenophora is yet to be achieved. The Bilateria comprise two lineages, Deuterostomia and Protostomia, although the position of Xenacoelomorpha is still being debated (a general consensus is that they are basal bilaterians, not falling unambiguously within either Deuterostomia or Protostomia). Similarly, Protostomia has a difficult-to-place lineage (Chaetognatha) and two large clades—Spiralia and Ecdysozoa. Limited developmental work suggests Chaetognatha might have a form of spiral cleavage, but molecular phylogenies rarely place this genus among the Spiralia. Other than the clades Gnathifera and Lophophorata, the Spiralia have proven difficult to sort out, and these phyla are represented as a starburst on the tree. The Ecdysozoa (the "molting Protostomia") are partly resolved, although the relationships between the three major clades (Scalidophora, Nematoida, Panarthropoda) remain unclear. Putative morphological synapomorphies defining the major animal clades are indicated by numbers on the tree, and are as follows.

Metazoa: (1) Gastrulation and embryonic tissue layering; (2) septate junctions, tight junctions, and/or zonula adherens present in epithelial tissues; (3) type IV collagen; (4) collagenous basal lamina/basement membrane beneath epidermis; (5) striated myofibrils and actin-myosin contractile elements.

Metazoa beyond Porifera: (6) striated ciliary rootlets.

Cnidaria (and Ctenophora) + Bilateria: (7) gap junctions; (8) fixed, organized gonads; (9) synaptic nervous system; (10) epithelium-lined gut (with digestive enzymes); (11) with primary larva, bearing apical organ (lost in Ecdysozoa); (12) presence of opsins.

Bilateria: (13) primary symmetry bilateral; (14) cephalization, with concentration of neural cell bodies in head as a cerebral ganglion (or rudiments thereof); (15) with third germ layer, the mesoderm, formed primarily by embryonic endoderm. The Deuterostomia and Protostomia together comprise a clade called Nephrozoa, united by their possession of discrete excretory structures (i.e., bilaterians to the exclusion of Xenacoelomorpha).

Deuterostomia: (16) pharyngeal gill slits; (17) trimeric body coelom.

Chordata: (18) notochord; (19) dorsal hollow nerve cord; (20) endostyle (or its derivative); (21) tadpole larva; (22) postanal tail.

Protostomia: (23) dorsal cerebral ganglia form circumesophageal connectives to ventral nerve cords.

Ecdysozoa: (24) multilayered cuticle is shed/molted one or more times during life history, regulated by ecdysteroid hormones; (25) loss of primary larva.

Needless to say, in many cases these synapomorphies have been secondarily altered or lost in some phyla. Synapomorphies of other clades and phyla depicted on this tree have already been described in previous chapters.

artifact of methodological issues. In fact, the placement of Ctenophora is only one area of uncertainty in the animal tree. Strong resolution of the positions of Placozoa and Cnidaria have also yet to be achieved (and placozoans seem to be much more derived than their simple body plan would first suggest).

That the Bilateria comprise two main lineages, Deuterostomia and Protostomia, seems largely settled, although the position of the phylum Xenacoelomorpha is still debated (a general consensus is that they are basal bilaterians, but do not fall unambiguously within either Deuterostomia or Protostomia). Some molecular research had suggested that Xenoturbellida (and even Xenoturbellida + Acoelomorpha) were deuterostomes, but that hypothesis is not well supported. The weight of molecular and morphological evidence places Xenacoelomorpha as a clade and sister group to all other bilaterians, as shown in our tree. Because xenacoelomorphs lack excretory organs, and these (in their many forms) are present in most other bilaterians, the sister group to Xenacoelomorpha (Deuterostomia + Protostomia) has been called Nephrozoa.

The Protostomia can be divided into three main lineages, one containing the sole phylum Chaetognatha, and two large clades—Spiralia and Ecdysozoa. But neither molecular nor morphological data have convincingly resolved relationships among these three clades. Most chaetognath morphological traits have no clear homologies in other lineages, and molecularly they form a long branch that separates them from the early diversification of protostomes. Limited developmental work suggests Chaetognatha might have a form of spiral cleavage, but molecular phylogenies rarely place this genus among the Spiralia. However, their cerebral ganglia resemble those of the spiralian phyla.

Relationships within Spiralia need further refinement, but recent phylogenomic data have suggested three clades—Gnathifera, Rouphozoa (Gastrotricha + Platyhelminthes), and Lophophorata. The earlier named clade, Lophotrochozoa, which includes both the lophophorates and some of the larger groups of spiralians (annelids, mollusc, and nemerteans) has also often been recovered. Many other suggestions for relationships among the spiralian lineages have been published, but little consensus has so far been reached. Within Spiralia, the concept of Annelida including the former phyla Sipuncula and Echiura, as well as Pogonophora and the mysterious meiofaunal worms *Lobatocerebrum* and *Diurodrilus*, is strongly supported.

The Ecdysozoa (the "molting Protostomia") are partly resolved, although the sister group relationships among the three major clades (Scalidophora, Nematoida, Panarthropoda) remain poorly studied due to lack of appropriate data sets including genomic data for Kinorhyncha and Loricifera. In fact, no phylogenomic analysis has yet included representatives of all the ecdysozoan phyla. The relationships of the Deuterostomia, however, are largely settled, with the Ambulacraria (Hemichordata + Echinodermata) being the sister group to the Chordata (Cephalochordata, Urochordata, and Vertebrata/Craniata).

One of the major things we have learned from this new view of phylogenetic relationships is that developmental systems are flexible. For example, a trait long thought to define the Deuterostomia is deuterostomous development itself. However, we now know that this mode of embryogeny also occurs in some phyla belonging to the clade Protostomia (e.g., in Nematomorpha, Priapula, Brachiopoda, many Crustacea, perhaps Chaetognatha). Thus, as noted earlier in this book, the clade Deuterostomia cannot be defined by deuterostomy, nor can the clade Protostomia be defined by protostomy (which also occurs in Cnidaria and Ctenophora). Similarly, radial cleavage is plesiomorphic among Bilateria, while spiral cleavage, which is apomorphic for Spiralia, appears to have reverted back to radial multiple times. The clade Spiralia in our tree has members whose cleavage patterns are radial (e.g., the three lophophorate phyla), whereas the Gastrotricha seems to have a unique form of cleavage. Several phyla of Spiralia have trochophore or trochophore-like larvae (e.g., Mollusca, Annelida, Nemertea, and possibly Entoprocta) but these have not consistently grouped together in the molecular phylogenies.

This brings us back to the question of why animal species diversity is so strongly dominated by arthropods, molluscs, annelids, and vertebrates. Unfortunately, there is no easy explanation for why these four groups have enjoyed such expansive radiations. Could it be that this is an artifact of time—perhaps during earlier eras, the Mesozoic or Paleozoic, other phyla dominated Earth's biosphere, but simply left us little fossil evidence of their prior dominance due to their poorly fossilizable bodies? If this were true, many of those small phyla would be viewed today as "living relicts" (like horseshoe crabs). Did ctenophores, placozoans, acoel worms, gastrotrichs, nemerteans, chaetognaths, entoprocts, gnathostomulids, and other phyla that today seem almost inconsequential once dominate the world? Or perhaps the "big four" taxa have always been dominant due to something intrinsic in their biology that remains to be discovered. Another possibility is that we simply haven't looked hard enough, spent enough effort combing through interstitial habitats, deep-sea sediments, and the parasitic world—there might well be an enormous biological diversity still waiting to be discovered in such poorly sampled environments (certainly we know this is true for the roundworms, Nematoda). Stay tuned for the Fourth Edition of *Invertebrates*, to see what we have learned about the tree of life in a few years.

We finish this chapter with a note of caution and one of hope. Molecular phylogenies are resolving the puzzle of the animal tree of life one piece at a time, but morphology still has its role in phylogenetics. Gnathifera, for example, was first recognized by analyzing morphological traits, and has subsequently been consistently recovered by both morphological and molecular analyses. The monophyly of other groups, such as Scalidophora, is supported more strongly by morphological data than by molecular data but this clade is broadly accepted by invertebrate biologists. Many groups have been shuffled around the tree, mostly because they were once positioned based on single character systems (e.g., radial versus spiral cleavage; acoelomate versus blastocoelomate versus coelomate), and many of these character systems conflict with one other. Molecular phylogenetics has given us a new way to look at the animal tree, and robust and stable results have made us rethink certain long-held paradigms (e.g., Articulata versus Ecdysozoa; Atelocerata versus Tetraconata). Morphology can be misleading in some cases, as molecular analyses can also be affected by many kinds of biases or be unable to reconstruct the interrelationships of some animal groups. Thus, only by combining the information derived from the careful study of both disciplines will we be able to generate a sound animal phylogeny, and be in a position to comprehend what that phylogeny means in terms of the evolution of animal body plans.

Selected References

Aguinaldo, A. M. A., J. M. Turbeville, L. S. Linford, M. C. Rivera, J. R. Garey, R. A. Raff and J. A. Lake. 1997. Evidence for a clade of nematodes, arthropods, and other moulting animals. Nature 387: 489–494.

Antcliffe, J. B., R. H. Callow and M. D. Brasier. 2014. Giving the early fossil record of sponges a squeeze. Biol. Rev. 89: 972–1004.

Borner, J., P. Rehm, R. O. Schill, I. Ebersberger and T. Burmester. 2014. A transcriptome approach to ecdysozoan phylogeny. Mol. Phylogenet. Evol. 80: 79–87.

Borowiec, M. L., E. K. Lee, J. C. Chiu and D. C. Plachetzki. 2015. Extracting phylogenetic signal and accounting for bias in whole-genome data sets supports the Ctenophora as sister to remaining Metazoa. BMC Genomics 16, doi:10.1186/s12864-015-2146-4

Burleigh, J. G. and 27 others. 2013. Next-generation phenomics for the Tree of Life. PLoS Currents Tree of Life, 5. doi: 10.1371/currents.tol.085c713acafc8711b2ff7010a4b03733

Cannon, J. T., K. M. Kocot, D. S. Waits, D. A. Weese, B. J. Swalla, S. R. Santos and K. M. Halanych. 2014. Phylogenomic resolution of the hemichordate an echinoderm clade. Curr. Biol. 24: 1–6.

Chen, L., S. Xiao, K. Pang, C. Zhou and X. Yuan. 2014. Cell differentiation and germ-soma separation in Ediacaran animal embryo-like fossils. Nature. doi: 10.1038/nature13766

Cunningham, J. A. and 7 others. 2014. Distinguishing geology from biology in the Ediacaran Doushantuo biota relaxes constraints on the timing of the origin of bilaterians. Proc. R. Soc. B. 279: 2369–2376.

Delsuc, F., H. Brinkmann, D. Chourrout and H. Philippe. 2006. Tunicates and not cephalochordates are the closest living relatives of vertebrates. Nature 439: 965–968.

Donoghue, M. J., J. J. Doyle, J. Gauthier, A. G. Kluge and T. Rowe. 1989. The importance of fossils in phylogeny reconstruction. Ann. Rev. Ecol. Syst. 20: 431–460.

Dunn, C. W., G. Giribet, G. D. Edgecombe and A. Hejnol. 2014. Animal phylogeny and its evolutionary implications. Ann. Rev. Ecol. Evol. Syst. 45: 371–395.

Dunn, C. W. and 17 others. 2008. Broad phylogenomic sampling improves resolution of the animal tree of life. Nature 452: 745–749.

Dunn, C. W., S. P. Leys and S. H. D. Haddock. 2015. The hidden biology of sponges and ctenophores. Trends Ecol. Evol. 30(5): 282–291.

Giribet, G. 2010. A new dimension in combining data? The use of morphology and phylogenomic data in metazoan systematics. Acta Zool. 91: 11–19.

Giribet, G. 2015. Morphology should not be forgotten in the era of genomics—a phylogenetic perspective. Zool. Anz. 256: 96–103.

Halanych, K. M. 2015. The ctenophore lineage is older than sponges? That cannot be right! Or can it? J. Exp. Biol. 218: 592–597.

Hejnol, A. and J. M. Martín-Durán. 2015. Getting to the bottom of anal evolution. Zool. Anz. doi: 10.1016/j.jcz.2015.02.006

Hejnol, A., and 16 others. 2009. Assessing the root of bilaterian animals with scalable phylogenomic methods. Proc. R. Soc. B: Biol. Sci. 276: 4261–4270.

Jékely, G., J. Paps and C. Nielsen. 2015. The phylogenetic position of ctenophores and the origin(s) of nervous systems. EvoDevo 6. doi:10.1186/2041-9139-6-1

Jondelius, U. I. Ruiz-Trillo, J. Baguñà and M. Riutort. 2002. The Nemertodermatida are basal bilaterians and not members of the Platyhelminthes. Zool. Scr. 31: 201–215.

Laumer, C. E. and 10 others. 2015. Spiralian phylogeny informs the evolution of microscopic lineages. Curr. Biol. 25: 2000–2006.

Lee, M. S. Y., J. Soubrier and G. D. Edgecombe. 2013. Rates of phenotypic and genomic evolution during the Cambrian Explosion. Curr. Biol. 23: 1889–1895.

Mah, J. L., K. K. Christensen-Dalsgaard and S. P. Leys. 2014. Choanoflagellate and choanocyte collar-flagellar systems and the assumption of homology. Evol. Dev. 16: 25–37.

Marlétaz, F., and 11 others. 2006. Chaetognath phylogenomics: A protostome with deuterostome-like development. Curr. Biol. 16: R577-R578.

Marlow, H., M. A. Tosches, R. Tomer, P. R. Steinmetz, A. Lauri, T. Larsson and D. Arendt. 2014. Larval body patterning and apical organs are conserved in animal evolution. BMC Biol. 12(7): 1–17.

Moroz, L. L. and 38 others. 2014. The ctenophore genome and the evolutionary origins of neural systems. Nature 510: 109–114.

Moroz, L. L. 2015 Convergent evolution of neural systems in ctenophorans. J. Exp. Biol. 218: 598–611.

Nesnidal, M. P. and 9 others. 2013. New phylogenomic data support the monophyly of Lophophorata and an Ectoproct-Phoronid clade and indicate that Polyzoa and Kryptrochozoa are caused by systematic bias. BMC Evol. Bio. 13: 253. doi: 10.1186/1471-2148-13-253

Nielsen, C. 2015. Larval nervous systems: true larval and precocious adult. J. Exp. Biol. 218: 629–636.

Nosenko, T., and 12 others. 2013. Deep metazoan phylogeny: when different genes tell different stories. Mol. Phylog. Evol. 67: 223–233.

Novacek, M. J. and Q. D. Wheeler (eds.). 1992. *Extinction and Phylogeny*. Columbia Univ. Press, New York.

Osigus, H.-J., M. Eitel, M. Bernt, A. Donath and B. Schierwater. 2013. Mitogenomics at the base of Metazoa. Mol. Phylogenet. Evol. 69: 339–351.

Philippe, H., H. Brinkmann, P. Martinez, M. Riutort and J. Baguñà. 2007. Acoel flatworms are not Platyhelminthes: Evidence from phylogenomics. PLoS ONE 2: e717.

Philippe, H. and 8 others. 2011. Acoelomorph flatworms are deuterostomes related to *Xenoturbella*. Nature 470: 255–258.

Philippe, H. and 19 others. 2009. Phylogenomics revives traditional views on deep animal relationships. Curr. Biol. 19: 1–17.

Pick, K. S. and 10 others. 2010. Improved phylogenomic taxon sampling noticeably affects nonbilaterian relationships. Mol. Biol. Evol. 27(9): 1983–1987.

Pisani , D. and 7 others. 2015. Genomic data do not support comb jellies as the sister group to all other animals. PNAS. doi: 10.1073/pnas.1518127112

Rota-Stabelli, O. and 8 others. 2010. Ecdysozoan mitogenomics: Evidence for a common origin of the legged invertebrates, the Panarthropoda. Genome Biol. Evol. 2: 425–440.

Ryan, J. F. and 17 others. 2013. The genome of the ctenophore *Mnemiopsis leidyi* and its implications for cell type evolution. Science 342: 1242592. doi: 10.1126/science.1242592

Shen, X. and 7 others. 2015. Phylomitogenomic analyses strongly support the sister relationship of Chaetognatha and Protostomia. Zool. Scr. doi: 10.1111/zsc.12140

Strausfeld, N. J. and F. Hirth. 2013. Deep homology of arthropod central complex and vertebrate basal ganglia. Science 340: 157–161.

Vargas, P. and R. Zardoya (eds.). 2014. *The Tree of Life. Evolution and Classification of Living Organisms*. Sinauer Associates, Sunderland, MA.

Wägele, J. W. and T. Bartolomaeus (eds.). 2014. *Deep Metazoan Phylogeny: The Backbone of the Tree of Life. New Insights from Analyses of Molecules, Morphology, and Theory of Data Analysis.* De Gruyter, Berlin.

Whelan, N. V., K. M. Kocot, L. L. Moroz and K M. Halanych. 2015. Error, signal, and the placement of Ctenophora sister to all other animals. PNAS 112: 5773–5778.

Illustration Credits

Chapter Openers

Chapter 1 The Eastern Pacific Spanish shawl nudibranch, *Flabellina iodinea.* © Larry Jon Friesen.

Chapter 2 Two hairy squat lobsters (*Lauriea siagiani*) on a sponge, Cabilao Island, Philippines. © Flonline digitale Bildagentur GmbH/Alamy Stock Photo.

Chapter 3 SEM of the diatom *Rutilaria* sp. © Science Photo Library/Alamy Stock Photo.

Chapter 4 The bell medusa (*Polyorchis penicillatus*) with a hitch-hiking rock crab (*Metacarcinus gracilis*). © Larry Jon Friesen.

Chapter 5 A sea urchin (*Lytechinus pictus*) embryo at the 16-cell stage. The four large macromeres are behind the four small micromeres, and the eight mesomeres are behind the macromeres. Remnants of the fertilization envelope can be seen attached to some of the cells. © Kallista Images/Visuals Unlimited, Inc.

Chapter 6 The unusual sponge *Clathrina clathrus* (Mediterranean Sea). © Wolfgang Pölzer/Alamy Stock Photo.

Chapter 7 The widespread "crawling" hydromedusan (*Gonionemus vertens*). © Larry Jon Friesen.

Chapter 8 A ctenophore eating a ctenophore: *Beroe cucumis* about to engulf *Mnemiopsis leidyi.* © Larry Jon Friesen.

Chapter 9 It turns on a dime; a live *Xenoturbella bocki* (or possibly an undescribed species) collected by MBARI (Monterey Bay Aquarium Research Institute) from ~600 m depth, near a whalefall in Monterey Submarine Canyon, California. Courtesy of Greg Rouse.

Chapter 10 The Persian carpet flatworm, *Pseudobiceros bedfordi.* © Hans Gert Broeder/Shutterstock.

Chapter 11 The sexual vermiform stage of a rhombozoan. © Larry Jon Friesen.

Chapter 12 The eastern Pacific ribbon worm, *Baseodiscus punnetti.* © Larry Jon Friesen.

Chapter 13 The eastern Pacific nudibranch, *Melibe leonine.* © Larry Jon Friesen.

Chapter 14 *Myrianida pachycera,* a syllid annelid from Australia, showing the remarkable phenomenon found in some annelids known as epitoky. Courtesy of Greg Rouse.

Chapter 15 Unidentified entoprocts. © Robert Brons/Biological Photo Service.

Chapter 16 Colonial rotifers (*Conochilus unicornis*) that cling to the mucus mass they secrete. © Wim van Egmond/Visuals Unlimited, Inc.

Chapter 17 A subtidal, gelatinous bryozoan from southern California (*Alcyonidium* sp., Ctenostomata). © Larry Jon Friesen.

Chapter 18 The free-living soil nematode *Caenorhabditis elegans*, a model organism in many fields of biology. © Larry Jon Friesen.

Chapter 19 *Priapulus caudatus* a large, widespread priapulan in the North Atlantic Ocean. © Andreas Altenburger/Alamy Stock Photo.

Chapter 20 The giant hairy scorpion (*Hadrurus arizonensis*) fluorescing under UV light. © Rick & Nora Bowers/Alamy Stock Photo.

Chapter 21 *Polydectus cupulifer,* the teddy-bear crab, with its symbiotic partner anemones (one on each cheliped). © Larry Jon Friesen.

Chapter 22 A buckeye butterfly, *Junonia coenia.* © Larry Jon Friesen.

Chapter 23 The North American giant millipede, *Narceus americanus* (Spirobolidae). Courtesy of David McIntyre.

Chapter 24 A female orb-web spider of the genus *Eriophora* (Araneidae) from the Brazilian Amazon. Courtesy of G. Hormiga.

Chapter 25 The yellow feather star, *Comanthina* sp., Komodo Island, Indonesia. © Wolfgang Pölzer/Alamy Stock Photo.

Chapter 26 The acorn worm *Glossobalanus sarniensis* (Hemichordata: Enteropneusta). © Nature Photographers Ltd/Alamy Stock Photo.

Chapter 27 The social ascidian *Clavelina hunstmani.* © Larry Jon Friesen.

Chapter 28 A page from Darwin's notebook from around July 1837 shows an early sketch of a phylogenetic tree.

Contents

Page viii, slime mold (*Leocarpus fragilis*): Courtesy of David McIntyre. **Page ix,** acoel (*Waminoa* sp.): Courtesy of R. Groeneveld. **Page x,** annelid (*Paralvinella fijiensis*): Courtesy of Greg Rouse. **Page xi,** rotifer (*Philodina roseola*): © Wim van Egmond/Visuals Unlimited, Inc. **Page xii,** onychophoran (*Peripatoides aurorbis*): Courtesy of G. Giribet. **Page xiii,** detail of a sea star (*Astrometis sertulifera*): © Larry Jon Friesen.

Chapter 1

1.2: After Jenkins, in Lipps and Signor 1992. **1.3A,D–H:** By Marianne Collins, from S. J. Gould, *Wonderful Life*, W. W. Norton, 1989. **1.4:** "Fauna of the Burgess Shale," Carel Brest von Kempen. Acrylic on illustration board. **1.6A,C–F:** Courtesy of R. Brusca. **1.6B:** Courtesy of David McIntyre.

Chapter 2

2.6: From Misof et al. 2014.

Chapter 3

3.1A: © P. W. Johnson/Biological Photo Service. **3.1B:** © iStock.com/micro_photo. **3.1C:** © J. Solliday/Biological Photo Service. **3.1D:** © M. Kreutz/micro*scope. **3.1E:** © D. Patterson and M. Farmer/micro*scope. **3.1F:** Courtesy of M. G. Schultz/Centers for Disease Control. **3.1G:** © Wim van Egmond/Visuals Unlimited, Inc. **3.1H:** Courtesy D. Lipscomb and K.

Kivimaki. **3.1I,K:** © Larry Jon Friesen. **3.1J:** © O. R. Anderson/micro*scope. **3.2:** Modified from Archibald 2007. **3.3:** Modified from Sleigh 1989. **3.4E:** Courtesy of D. Lipscomb. **3.5A:** Courtesy of B. S. C. Leadbeater. **3.5B:** © G. F. Leedale/Biophoto Associates. **3.5C:** Courtesy of G. Brugerolle. **3.5D:** Courtesy of K. Vickerman. **3.6:** After Raikov 1994. **3.7C:** Courtesy of D. Lipscomb and K. Kivimaki. **3.7D:** © D. Patterson/micro*scope. **3.7E:** © Wim van Egmond/Visuals Unlimited, Inc. **3.7F:** © Larry Jon Friesen. **3.8A:** Courtesy of Scott Soloman. **3.8B,D–F:** © Larry Jon Friesen. **3.8C:** Courtesy of R. Brusca. **3.9:** After Grell 1973. **3.10C:** © Robert Brons/Biological Photo Service. **3.11B:** Redrawn from Grell 1973, from film E-1643 by Netzel and Heunert 1971. **3.12A** ***left*:** After Mackinnon and Hawes 1961. **3.12A** ***right*:** © Larry Jon Friesen. **3.12B:** © Dr. Peter Siver/Visuals Unlimited, Inc. **3.12C:** © Albert Lleal/Minden Pictures/Corbis. **3.12D:** Courtesy of Brian Leander. **3.12E:** © Larry Jon Friesen. **3.12F:** Courtesy of Jacques Descloitres, MODIS Land Rapid Response Team at NASA GSFC. **3.13A:** After Grell 1973. **3.14B:** After Marquardt and Demaree 1985. **3.14C:** Courtesy of D. Lipscomb and K. Kivimaki. **3.14D:** Courtesy of Brian Leander. **3.15:** After Grell 1973. **3.16:** After Miller et al. 1985. **3.17A** ***right*, E,H,I:** © Larry Jon Friesen. **3.17B–D:** Courtesy of D. Lipscomb and K. Kivimaki. **3.17F:** Courtesy D. Lipscomb. **3.17G:** © M. Kreutz/micro*scope. **3.19A,B:** After Sleigh 1973. **3.19C:** Redrawn from Grell 1973, after Parducz 1954. **3.19H:** After Grell 1973. **3.21A:** © Greg Antipa/Science Source. **3.21B:** © M. Kreutz/micro*scope. **3.22:** After Sleigh 1973. **3.23B,C,D:** Redrawn from Grell 1983, after Bardele and Grell 1967. **3.24A:** Courtesy of D. Lipscomb and K. Kivimaki. **3.24B:** From Lynn and Didier 1978, courtesy D. Lynn. **3.24C:** © M. Kreutz/micro*scope. **3.26:** © Michael Abbey/Visuals Unlimited, Inc. **3.27A,B:** Redrawn from Grell 1973, after Grell 1953. **3.27C,D:** Redrawn from Grell 1973, after Mugge 1957. **3.27E:** After Grell 1973. **3.28A,B:** Courtesy of D. Lipscomb and K. Kivimaki. **3.28C–E,G:** © Larry Jon Friesen. **3.28F:** Courtesy of D. Lipscomb and K. Kivimaki. **3.28H:** Courtesy of Claire Fackler, CINMS, NOAA. **3.29:** After Sleigh 1989. **3.30A,B:** © Science Photo Library/Alamy Stock Photo. **3.30C:** © Dr. Peter Siver/Visuals Unlimited, Inc. **3.30C:** Courtesy of Jeff Schmaltz, LANCE/EOSDIS MODIS Rapid Response Team at NASA GSFC. **3.31:** © D. Patterson/micro*scope. **3.32:** © The Natural History Museum/Alamy Stock Photo. **3.33:** © Tom Adams/Visuals Unlimited, Inc. **3.34A:** Courtesy of Patrick Keeling. **3.34B:** Courtesy of Gary Meyer. **3.35A,B:** © Robert Brons/Biological Photo Service. **3.36:** © D. Patterson/micro*scope. **3.37A:** Redrawn from Grell 1973, after Myers 1943. **3.37B:** © Michael Patrick O'Neill/Alamy Stock Photo. **3.38A:** © Wim van Egmond/Visuals Unlimited, Inc. **3.38B:** © Robert Brons/Biological Photo Service. **3.38C:** From Grell 1973, after Haeckel. **3.39A:** After Margulis and Schwartz 1988, from a drawing by L. Meszoly. **3.39B:** © M. Schliwa/Visuals Unlimited, Inc. **3.40:** Redrawn from Grell 1973, after Hollande and Enjumet 1953. **3.41:** After Sleigh 1989. **3.42A,C:** After Grell 1973. **3.42B:** After Sleigh 1989. **3.43A:** Courtesy of Dr. Stan Erlandsen/Centers for Disease Control. **3.43B:** After Grell 1980. **3.43C:** After Schmidt and Roberts 1989, from a drawing by William Ober. **3.43D:** Courtesy of Brian Leander. **3.44B:** © M. Bahr and D. Patterson/micro*scope. **3.44C:** Courtesy of Brian Leander. **3.46:** From Bricheux and Brugerolle 1987. Courtesy of G. Brugerolle. **3.47A:** After Chen 1950. **3.47B:** © D. Patterson/micro*scope. **3.49B:** After Brugerolle et al. 1979. **3.49C:** Courtesy of L. Tetley and K. Vickerman. **3.51:** © D. Patterson and M. Farmer/micro*scope. **3.52:** After Sleigh in House 1979, modified from Margulis 1970. **3.53:** Adapted from Lukes et al. 2009.

Chapter 4

4.1: Courtesy of J. DeMartini. **4.1B:** © Lebendkulturen.de/Shutterstock. **4.3B:** © Ethan Daniels/Shutterstock. **4.3C–E:** Courtesy of G. McDonald. **4.3F:** Courtesy of R. Brusca. **4.4B,C:** Courtesy of O. Feuerbacher. **4.7G:** After Brusca and Brusca 1978. **4.8F,G:** From Brusca and Brusca 1978. **4.9A:** Courtesy of R. Emlet. **4.9B:** Courtesy of P. Bergquist. **4.9C:** Courtesy of R. Brusca. **4.9D:** Courtesy of G. McDonald. **4.10A:** From Brusca and Brusca 1978. **4.10B:** After Sherman and Sherman 1976. **4.10C:** Courtesy of Dr. John R. Dolan, Laboratoire d'Oceanographie de Villefranche; Observatoire Oceanologique de Villefranche-sur-Mer. **4.10D:** Courtesy of G. McDonald. **4.10E:** © iStock.com/Velvetfish. **4.13A,B:** Courtesy of G. McDonald. **4.13C:** Courtesy of J. Haig. **4.13D:** From Sanders 1963. **4.14A:** After Fauchald and Jumars 1979. **4.14B:** Courtesy of G. McDonald. **4.14C:** Courtesy of K. Banse. **4.15:** Courtesy of G. G. Warner. **4.16A,B:** After Fauchald and Jumars 1979. **4.16C:** © Stephan Kerkhofs/Shutterstock. **4.17A,B:** Courtesy of G. McDonald. **4.17C:** Courtesy of C. DiGiorgio. **4.17E:** Courtesy of R. Brusca. **4.18:** Courtesy of P. Fankboner. **4.19A:** © Larry Jon Friesen. **4.19B:** Courtesy of David Burdick/NOAA. **4.19C:** Courtesy of P. Fankboner. **4.19D,G:** Courtesy of A. Kerstitch. **4.19E:** After Caldwell and Dingle. **4.19F:** Courtesy of T. Case. **4.19H:** © National Geographic Image Collection/Alamy Stock Photo. **4.22B:** After Mercer 1959. **4.22D:** From Jurand and Selman 1969. **4.23A:** After Wilson and Webster 1974. **4.23B:** After Goodrich 1945. **4.23D:** After Snodgrass 1952. **4.26A:** Courtesy of G. McDonald. **4.26B:** © iStock.com/Brett Petrillo. **4.26C:** Courtesy of P. Fankboner. **4.26D–G:** After Barnes 1980. **4.26H:** © Larry Jon Friesen. **4.28A:** After Kuhl 1938. **4.28B:** After Gibson 1972. **4.28C:** Courtesy of S. Riseman. **4.30:** © Larry Jon Friesen. **4.31B:** After Prosser and Brown 1961. **4.31C** ***top*:** Courtesy of T. and M. Eisner. **4.31C** ***bottom*:** © P. J. Bryant/Biological Photo Service. **4.32A:** After Wells 1968. **4.34A:** © J. Morin. **4.34B:** Courtesy of M. K. Wicksten. **4.34C:** Courtesy of R. Brusca. **4.34D:** Courtesy of G. McDonald.

Chapter 5

5.1: After Duboule, 2007. **5.7:** After Martindale, 2005. **5.15C:** After Hyman 1940. **5.18:** Courtesy of C. Nielsen.

Chapter 6

6.1A: After McConnaughey 1963. **6.1B,C:** © Robert Brons/Biological Photo Service. **6.2A,B,G,H,P:** © Larry Jon Friesen. **6.2C:** © Wolfgang Pölzer/Alamy Stock Photo. **6.2D:** © Marli Wakeling/Alamy Stock Photo. **6.2E:** © Dennis Sabo/Shutterstock. **6.2F:** © Amar and Isabelle Guillen/Guillen Photo LLC/Alamy Stock Photo. **6.2I:** © P. Petry. **6.2J:** © C. Robertson. **6.2K:** © Rick & Nora Bowers/Alamy Stock Photo. **6.2L:** From Bergquist 1978. **6.2M:** Courtesy D. Freeman. **6.2N:** Courtesy of R. Brusca. **6.2O:** Courtesy of David McIntyre. **6.3A:** After Hartman 1963. **6.3B:** After Bergquist 1978. **6.3C,D:** After Reiswig 1975. **6.4:** From Bayer and Owre 1968. **6.5A:** From Bayer and Owre 1968. **6.5B:** After Sherman and Sherman 1976. **6.5C:** © Carolina Biological Supply Co./Visuals Unlimited, Inc. **6.6:** From Bergquist 1978. **6.7A:** With permission from S. Leys (after Reiswig and Mehl 1991). **6.7B:** After Leys 1999. **6.7C:** From Leys et al. 2007. **6.8A–D,I:** After Connes et al. 1971. **6.8E,G:** From Bayer and Owre 1968. **6.8F:** After Barnes 1980. **6.8H:** After Brill 1973. **6.9A–C:** After Bergquist 1978. **6.9D:** After Hyman 1940. **6.10A:** After Bergquist 1978. **6.10B,C:** From Bergquist 1978. **6.11:** Courtesy of R. Brusca. **6.12A:** From Bergquist 1978; courtesy of P. Bergquist. **6.12B–D:** After Hyman 1940. **6.12F:** After Hartman 1969. **6.12G:** From Bergquist 1978; photo of asterose microscleres courtesy of B. Beaumont, others courtesy of P. Bergquist. **6.13:** Courtesy of J. Vacelet. **6.14A,C:** From Kelly and Vacelet 2011. **6.14B:** © 2012 MBARI. **6.14D:** From Vacelet 2007. **6.15A:** Courtesy of J. Williams. **6.15B:** Courtesy of C. Middleton. **6.15C:** Courtesy of I. Middleton. **6.16A:** From Bayer and Owre 1968. **6.16B,C:** After Hyman 1940. **6.16D,E:** From Manconi and Pronzato 2007. **6.17A,B:** After Brien and Meewis 1938. **6.17C:** Courtesy E. Bautista-Guerrero. **6.17D,E:** © Michael Patrick O'Neill/Alamy Stock Photo. **6.18:** After Hyman 1940. **6.19:** From Leys and Ereskovsky 2006. **6.20:** All images courtesy of Alexander Ereskovsky. **6.21B–F:** From Bayer and Owre 1968. **6.22:** © David Wrobel/Visuals Unlimited, Inc. **6.23A:** © Nuno Vasco

Rodrigues/Shutterstock. **6.23B:** From Rützler and Reiger 1973. **6.24:** Courtesy of R. Brusca. **6.26:** After Bergquist 1978. **6.27:** After Gazave et al. 2012.

Chapter 7

7.1A,C,K,L: © Larry Jon Friesen. **7.1B:** Courtesy of R. Brusca. **7.1D:** © WaterFrame/Alamy Stock Photo. **7.1E:** Courtesy of G. McDonald. **7.1F:** © Andrey Nekrasov/Alamy Stock Photo. **7.1G:** © Michael Patrick O'Neill/Alamy Stock Photo. **7.1H:** © Larry Jon Friesen. **7.1I:** © iStock.com/micro_photo. **7.1J:** © Robert Brons/Biological Photo Service. **7.1M:** Courtesy I. Fiala. **7.2A–C:** After Bayer and Owre 1968. **7.2D–F:** After Shimizu and Nimikawa 2009. **7.3, 7.4, 7.5A–E:** From Bayer and Owre 1968. **7.6A:** After Bayer and Owre 1968, with modification to mouth area. **7.8A–C,E:** From Bayer and Owre 1968. **7.8D,G:** © J. Morin. **7.8H:** From Alvariño 1983. **7.8I:** © D. J. Wrobel/Biological Photo Service. **7.9:** From Bayer and Owre 1968. **7.10A,B:** From Fields and Mackie 1971. **7.10C:** Courtesy of R. Brusca. **7.11A–E:** From Bayer and Owre 1968. **7.11F:** Courtesy of A. Kerstitch. **7.11G,H:** © J. Morin. **7.12A,B,D:** From Bayer and Owre 1968. **7.12C:** Courtesy of R. Brusca. **7.12E:** Courtesy of G. McDonald. **7.13A,B,D,F:** © Larry Jon Friesen. **7.14A:** From Bayer and Owre 1968. **7.14B:** After Barnes 1987. **7.15:** After Larson 1976. **7.16:** Courtesy of F. Bayer and W. R. Brown, Smithsonian Institution. **7.17:** After Cairns 1981. **7.18:** From Bayer and Owre 1968. **7.19:** After Mackie and Passano 1968. **7.20:** From Bayer and Owre 1968. **7.21B:** After Hyman 1940. **7.21D:** Courtesy of Ed Bowlby/NOAA. **7.21E:** Courtesy of C. Birkeland. **7.22A,B:** © Larry Jon Friesen. **7.23:** After Sherman and Sherman 1976. **7.24A:** From Holstein and Tardent 1983. **7.24B–E:** From Mariscal 1974. **7.25:** After Mariscal, in Muscatine and Lenhoff 1974. **7.26A–D:** From Hamner and Dunn 1980. **7.26E:** Courtesy of C. Birkeland. **7.27A:** © D. J. Wrobel/Biological Photo Service. **7.27B:** © Larry Jon Friesen. **7.27C:** Courtesy of C. Birkeland. **7.28:** After Russell-Hunter 1979. **7.29:** After Kent et al. 2001. **7.30:** Courtesy of D. Fautin. **7.31:** From Cairns and Barnard 1984. **7.32A:** After Bayer and Owre 1968. **7.32B:** Courtesy of J. Smith. **7.32C:** © Dobermaraner/Shutterstock. **7.33A:** After Wells 1968. **7.33B:** After Barnes 1987. **7.33C:** From Bayer and Owre 1968. **7.34:** From Mariscal 1974. **7.35A,B:** After Hyman 1940. **7.35C,D,F:** From Bayer and Owre 1968. **7.35E:** After Conant 1900. **7.36A:** © Larry Jon Friesen. **7.36B:** From Bayer and Owre 1968. **7.37A,C:** © Robert Brons/Biological Photo Service. **7.37B,D:** Courtesy of S. Keen and B. Cameron. **7.38A,B:** From Bayer and Owre 1968. **7.38C:** After Calder 1982. **7.39:** From Bayer and Owre 1968. **7.40:** From Stretch and King 1980. **7.41, 7.43:** From Bayer and Owre 1968. **7.44, 7.45:** Courtesy of J. Just. **7.46:** After Collins 2009.

Chapter 8

8.1A: © Larry Jon Friesen. **8.1B,F:** © D. J. Wrobel/Biological Photo Service. **8.1C,D:** G. Matsumoto/© MBARI. **8.1E:** © Reinhard Dirscherl/Corbis. **8.1G:** © Andrey Nekrasov/imageBROKER/Corbis. **8.1H:** © J. Morin. **8.1I:** Courtesy of B. Shepherd. **8.2B,C,F,L:** After Harbison and Madin 1983. **8.2E,K:** After Mayer 1912. **8.2G:** After Komai 1934. **8.2H:** After Bayer and Owre 1968. **8.3A:** From Bayer and Owre 1968. **8.4:** Courtesy of L. Madin, Woods Hole Oceanographic Institution, www.cmarz.org. **8.6:** After Hyman 1940. **8.7A:** From Bayer and Owre 1968. **8.7B:** After Franc 1978. **8.7C–E:** Courtesy of P. Fankboner. **8.8:** From Mills and Miller 1984. **8.9A:** After Komai 1922. **8.9B:** After Hyman 1940. **8.10A:** After Hyman 1940. **8.10C:** From Bayer and Owre 1968. **8.11, 8.12:** After Hyman 1940. **8.13A,B:** From Tang et al. 2011. **8.13C:** © Javier Herboso.

Chapter 9

9.1A–C,F,G: Courtesy of M. Hooge. **9.1D,H:** Courtesy of U. Jondelius. **9.1E:** Courtesy of R. Groeneveld. **9.2:** After Yamasu 1991. **9.3:** Courtesy of E. Hirose. **9.4:** After Kotikova and Raikova 2008. **9.5:** Courtesy of E. Hirose. **9.6:** After Gschwenter et al. 2002. **9.7:** Courtesy of M. Hooge. **9.8:** Courtesy of C. Todt. **9.9:** Courtesy of J. G. Achatz. **9.10:** After Reuter and Kreschenko 2004. **9.11:** After ETI World biodiversity database (original drawing by Steven Francis). **9.12:** After Achatz et al. 2010. **9.13A:** Courtesy of U. Jondelius. **9.13B:** Courtesy of M. Hooge. **9.14:** After Sterrer 1998. **9.15:** Courtesy of K. Lundin. **9.16:** After K. Lundin 1998. **9.17:** Courtesy of I. Meyer-Wachsmuth. **9.18:** After Sterrer 1998. **9.19:** After K. Lundin et al. 1998. **9.20:** After Jondelius et al. 2004. **9.21:** Courtesy of G. Rouse. **9.22:** After Ehlers and Sopott Ehlers 1997. **9.23:** Courtesy of K. Lundin (from Lundin 1998). **9.24:** After Ehlers and Sopott Ehlers 1997. **9.25:** After Telford 2008.

Chapter 10

10.1A: Courtesy of C. Laumer. **10.1B:** Courtesy of J. Smith III. **10.1C,H,J,K:** © Larry Jon Friesen. **10.1D:** © Gabbro/Alamy Stock Photo. **10.1E:** © M. Hooge. **10.1F:** © S. K. Webster/Biological Photo Service. **10.1G:** From the photo collection of Dr. James P. McVey, NOAA Sea Grant Program. **10.1I,N:** © J. Morin. **10.1L:** Courtesy of Kevin Kocot. **10.1M:** © R. Brons/Biological Photo Service. **10.3A,C,F:** © Larry Jon Friesen. **10.3E:** After Hyman 1951. **10.4B:** After Brown 1950. **10.4C:** After Hyman 1951. **10.5A:** After C. Bedini and F. Papi, in Riser and Morse 1974. **10.5B:** After Bayer and Owre 1968. **10.6A:** After Sherman and Sherman. **10.7A:** After L. T. Threadgold, 1963, *Q. J. Microsc. Sci.* 104. **10.7B:** After Barth and Broshears 1982. **10.8A:** After Schell 1982. **10.8B,C:** After Marquardt and Demaree 1985. **10.8D:** Courtesy of J. DeMartini. **10.9A,B:** After Hyman 1951. **10.9C–E:** Courtesy of J. DeMartini. **10.9F:** © Larry Jon Friesen. **10.10:** After Hyman 1951. **10.11A–C:** After Russell-Hunter 1979. **10.11D:** After Hyman 1951. **10.11E:** © M. Hooge. **10.12:** After Bayer and Owre 1968. **10.13, 10.14, 10.15:** After Hyman 1951. **10.16:** From Bayer and Owre 1968. **10.18A–E:** From Bayer and Owre 1968. **10.18F:** Courtesy of R. Hochberg. **10.19, 10.20:** After Hyman 1951. **10.21A:** From Bayer and Owre, after Ivanov 1955. **10.21C:** After Hyman 1951. **10.22A,B:** From Bayer and Owre 1968. **10.22C:** After Noreña 2015. **10.23A:** After Boolootian and Stiles 1981. **10.23B:** After Hyman 1951, drawn by Hyman from a photograph in Kato 1940. **10.23C,D:** After Scharer et al. 2011. **10.24, 10.25:** From Bayer and Owre 1968, after Kato 1940. **10.27B:** After Barnes 1980, from Smyth and Clegg 1959. **10.27C:** After Noble and Noble 1982. **10.28B *right*:** Courtesy of J. DeMartini. **10.29A:** After Olsen 1974. **10.29B:** Adapted from Smyth 1977. **10.30C:** Courtesy of J. DeMartini. **10.30D:** © Larry Jon Friesen. **10.31A:** After Marquardt and Demaree 1985. **10.31B:** After Noble and Noble 1982. **10.33A:** After Ax 1963. **10.33B:** After Karling 1974.

Chapter 11

11.1A: After Lapan and Morowitz 1972. **11.1B,C:** After Atkins 1933. **11.3A:** After Hyman 1940. **11.3B,C:** After Lapan and Morowitz 1972. **11.3D:** After from McConnaughey 1963; after Nouvel 1948. **11.5, 11.6, 11.7:** After Hyman 1940. **11.9:** Courtesy of E. Thuesen and S. Haddock. **11.10A,B:** After Shinn 1997. **11.10C:** After Hyman 1959. **11.10D:** After Feigenbaum 1978. **11.10E:** After Meglitsch 1972. **11.10F:** After Shinn 1994. **11.10G:** After Shin 1997. **11.10H:** After Burfield 1927. **11.11A–C:** Courtesy of E. Theusen and R. Bieri. **11.11D:** After Shinn 1994. **11.12:** Courtesy of M. Terazaki and C. B. Miller. **11.13A:** After Shimotori and Goto 2001. **11.13B–G:** After Hyman 1959. **11.14:** Courtesy of R. Hochberg. **11.15C,D:** After Kieneke et al. 2008. **11.15E,F:** Courtesy of R. Hochberg.

Chapter 12

12.1A,B,D,F: Courtesy of Gonzalo Giribet. **12.1C:** Courtesy of G. McDonald. **12.1E,H:** From Bayer and Owre 1968. **12.1G:** After Gibson 1982b. **12.2A,B:** After Hyman 1951. **12.2C:** From Bayer and Owre 1968; after Coe 1943. **12.3:** After Gibson 1982. **12.4:** Courtesy of S. Stricker. **12.5A–D:** After Russell-Hunter 1979, based on papers by R. Gibson. **12.5E:** After Gibson 1982b. **12.5F–H:** Courtesy of S. Stricker. **12.6:** After Gibson 1982. **12.7, 12.8, 12.9:** After Hyman 1951. **12.10B,C:** From Bayer and Owre

1968. **12.10D, 12.11, 12.12, 12.13:** After Hyman 1951. **12.14A,C:** After Gibson 1972. **12.14B:** After Gontcharoff 1961.

Chapter 13

13.1A–C: Courtesy of D. Lindberg. **13.1D–F,K:** Courtesy of G. McDonald. **13.1G–J:** © Larry Jon Friesen. **13.1L:** Courtesy of P. Fankboner. **13.1N:** © Animal Stock/Alamy Stock Photo. **13.1O:** © age fotostock/Alamy Stock Photo. **13.1P:** © Christopher Crowley/Visuals Unlimited, Inc. **13.1Q:** Courtesy of R. Brusca. **13.2A,D:** Courtesy of Christiane Todt. **13.2B,E–H:** After Hyman 1967. **13.2C:** After Scheltema and Morse 1984. **13.2E–H:** After Hyman 1967. **13.2I:** Courtesy Kevin Kocot and Jeremy Shaw. **13.2J,K:** Courtesy Kevin Kocot. **13.2L:** After Scheltema and Morse 1984. **13.3A,B,D,E:** After Lemche and Wingstrand 1959. **13.3C:** Courtesy D. Lindberg. **13.4C:** Courtesy of G. McDonald. **13.5A:** After Hyman 1967. **13.5B,C:** After Fretter and Graham 1962. **13.6B:** After Fretter and Graham 1962. **13.7A:** Courtesy of R. Brusca. **13.7B:** After Hyman 1967. **13.7E:** Courtesy of J. King. **13.7F,G:** From Brusca and Brusca 1978. **13.7H:** © WaterFrame/Alamy Stock Photo. **13.7I:** Courtesy of R. Brusca. **13.7J:** © imageBROKER/Alamy Stock Photo. **13.8A:** From Brusca and Brusca 1978. **13.11D:** © Reuters/Corbis. **13.12B:** After Lane 1960. **13.12C:** After Winkler and Ashley 1954. **13.12E–G:** Courtesy of A. Kerstitch. **13.16A,H,I:** From Brusca and Brusca 1978. **13.16B:** Courtesy of G. McDonald. **13.17:** © Larry Jon Friesen. **13.18A–D:** After Lang 1900, *Lehrbuch der vergleichenden Anatomie der wirbellosen Thiere* 3: 1–509. **13.18E:** After Morton 1979. **13.18F:** After Barnes 1980. **13.19A,B:** Modified after Miller 1974. **13.19C:** Courtesy of G. McDonald. **13.19D:** © Larry Jon Friesen. **13.19E:** After Hyman 1967. **13.20A–D:** After Trueman 1966. **13.21A:** © D. J. Wrobel/Biological Photo Service. **13.21E,F:** Courtesy of R. Brusca. **13.22A:** © Gergo Orban/Shutterstock. **13.22B:** Courtesy of R. Brusca. **13.22C:** Courtesy of G. McDonald. **13.24A:** After Solem 1974. **13.24B:** After Hyman 1967. **13.24C:** After McLean 1962, *Proc. Malac. Soc. London* 35: 23–26. **13.25A,C,E,F:** After Fretter and Graham 1962. **13.25B,D:** After Hyman 1967. **13.26:** Courtesy of C. DiGiorgio. **13.27:** © Alex Kerstitch/Visuals Unlimited, Inc. **13.28:** After Barnes 1980. **13.29:** After Fretter and Graham 1962. **13.30G–I:** Courtesy of S. Hendrixson. **13.31A:** After Yonge and Thompson 1976. **13.31B–D:** After Reid and Reid 1974. **13.32A–E:** After Hyman 1967. **13.32F:** After Bullough 1958. **13.33B:** After Hyman 1967. **13.35A:** After Cox in Moore (ed.) 1960. **13.35B:** After Pearse et al. 1987. **13.36:** After Fretter and Graham 1962. **13.37A:** © Elenarts/Shutterstock. **13.37B:** Courtesy of R. Brusca. **13.38:** Modified from Fretter and Graham 1962. **13.39:** After Hyman 1967. **13.40A,C:** After Cox, in Moore (ed.) 1960. **13.40B:** After Fretter and Graham 1962. **13.42A:** After Wells 1963. **13.42B:** After Winkler and Ashley 1954. **13.42D:** After Russell-Hunter 1979. **13.43:** After Hyman 1967. **13.44F:** © Premaphotos/Alamy Stock Photo. **13.45:** After Hadfield, in Giese and Pearse 1979. **13.46:** After Yonge, in Moore (ed.) 1960; after Hyman 1967. **13.47A:** After Fretter and Graham 1962. **13.47B,C:** After Hyman 1967. **13.47D:** After Cameron and Redfern 1976. **13.48B:** Courtesy of David McIntyre. **13.50:** After Scheltema 1993. **13.51A:** After Sherman and Sherman 1976. **13.51B:** After Hadfield, in Giese and Pearse 1979. **13.51C:** After Hyman 1967. **13.52A,B:** Redrawn from Hyman 1967, after Werner 1955, *Helg. Wissensch. Meeresuntersuchungen*, 5. **13.52C:** Redrawn from Hyman 1967, after Dawydoff 1940. **13.52E:** After Brusca 1975. **13.53:** After Fretter and Graham 1962.

Chapter 14

14.1, 14.2: Courtesy of G. Rouse. **14.3A–C:** Courtesy of G. Rouse. **14.4A:** After Meglitsch 1972. **14.4D:** After Sherman and Sherman 1976, from Storer and Usinger 1957. **14.5I–Q:** After Smith and Carlton 1975. **14.6A–F,K:** After Russell-Hunter 1979. **14.7A,D–I:** Courtesy of G. Rouse. **14.7J:** © Larry Jon Friesen. **14.8A:** From Brusca and Brusca 1978. **14.8B,C,E,F,G:** Courtesy of G. Rouse. **14.8D:** After Barnes 1980. **14.9A,B:** After Barnes 1980; inset photos courtesy of R. Brusca (A) and P. Smith (B). **14.9C–E:** After Dales 1955. **14.9F:** After Carlton and Smith. **14.9G:** After Kaestner 1967. **14.10A,B:** After Newell 1970. **14.10C:** After Borradaile et al. **14.10D,E:** Courtesy of G. Rouse. **14.11A:** After Brusca and Brusca 1978. **14.11B,D:** Courtesy of G. Rouse. **14.11C:** After Eisig 1906. **14.13E:** After Edwards and Lofty 1972. **14.13F:** Redrawn from Edwards and Lofty 1972, after Grove and Newell 1962. **14.14A,B:** After Goodrich 1946. **14.14C:** After Thomas 1940. **14.14D:** After Edwards and Lofty 1972. **14.15D,E:** After Meglitsch 1972. **14.16A,B:** After Fauvel et al. 1959. **14.16C:** After Hermans and Eakin 1974. **14.16D,E,G:** Courtesy of G. Rouse. **14.16F:** After Barnes 1980. **14.17A,C:** After Meglitsch 1972. **14.17B,E,G:** Courtesy of G. Rouse. **14.17D,F,H:** After Barnes 1980 and Fauvel et al. 1959. **14.18A,B:** After Edwards and Lofty 1972. **14.18C,D:** After Barnes. **14.18E–G:** After Edwards and Lofty 1972, after Tembe and Dubash 1961. **14.18H:** After Brinkhurst and Jamieson 1972. **14.18I:** © R. K. Burnhard/Biological Photo Service. **14.19:** After Anderson 1973. **14.20:** Courtesy of G. Rouse. **14.21B,C:** After Blake 1975. **14.22A,G:** © Larry Jon Friesen. **14.22B:** Courtesy of G. Rouse. **14.22C–F:** Courtesy of G. Giribet. **14.23:** After Kawauchi et al. 2012 and Lemer et al. 2014. **14.24A:** After Hyman 1959. **14.24B:** After Fischer 1952. **14.25A–D,F:** After Hyman 1959. **14.25E:** After Stehle 1953. **14.26, 14.27A–D:** After Hyman 1959. **14.27E–K:** Courtesy of M. Rice. **14.28A:** After Barnes 1980. **14.28B:** After Fischer 1946. **14.28D:** After MacGinitie and MacGinitie 1968. **14.28C,E–G:** Courtesy of G. Rouse. **14.28H:** Courtesy of Arthur Anker. **14.29A:** After Barnes 1980. **14.29B:** Courtesy of G. Rouse. **14.29C:** After a drawing by W. K. Fischer. **14.29D,E:** Courtesy of Ohta, from Ohta 1984. **14.30A:** After Barnes 1980. **14.30C:** After Meglitsch 1972. **14.30D:** Courtesy of M. Apley. **14.31A,B:** Courtesy of G. Rouse. **14.31C–E:** Courtesy of M. Apley. **14.32A:** After Southward 1984. **14.32C:** After Ivanov 1952. **14.32F:** After Southward 1969. **14.32G:** Courtesy of R. Hessler. **14.32D,H–L:** Courtesy of G. Rouse. **14.33A,B:** After Ivanov 1962. **14.33C–E:** Courtesy of G. Rouse. **14.34A:** After Mann 1962. **14.34B,C:** After Stuart 1982. **14.34D:** Courtesy of G. Rouse. **14.35:** After Kaestner 1967. **14.36:** After Russell-Hunter 1979, adapted from Gary and Lissmann 1938. **14.37A,B:** After Barnes 1980. **14.37C:** After Mann 1962. **14.37D:** Courtesy of G. Rouse. **14.38A:** After Barnes 1980. **14.38B, 14.39:** After Mann 1962. **14.40B:** After Barnes 1980. **14.40C:** After Barnes 1980, from Nagao 1957. **14.40D:** Courtesy of G. Rouse. **14.41A:** After Andrade et al 2015 and Struck et al. 2015. **14.41B:** After Siddall et al 2001, Rousset et al 2008, and Marotta et al. 2008.

Chapter 15

15.1A,B: Courtesy of C. Nielsen. **15.1C:** Courtesy of G. Paulay. **15.1D:** Courtesy of M. Faasse. **15.1E:** Courtesy of K. Kocot. **15.1F:** Courtesy of J. Merkel. **15.1G:** Courtesy of R. Rundell. **15.2A:** Courtesy of C. Nielsen. **15.2B:** Modified by C. Nielsen from Riisgaard et al. 2000. **15.2C:** After Hyman 1951. **15.3:** Courtesy of C. Nielsen. **15.4:** From Funch and Kristensen 1995; photos © R. Kristensen. **15.5:** Courtesy of Ricardo Neves.

Chapter 16

16.1: Courtesy of M. Sørensen. **16.2A,B:** After Sterrer 1982. **16.2C:** Courtesy of M. Sørensen. **16.3A,B,D,E:** After Nogrady 1982. **16.3C:** Courtesy of Giulio Melone. **16.3F:** © R. Brons/Biological Photo Service. **16.6:** Courtesy of M. Sørensen. **16.8:** After Hyman 1951. **16.10B:** © R. Hochberg. **16.11B,C:** After Hyman 1951. **16.11D:** After Noble and Noble 1982, from Cable and Dill. **16.11E:** After Yamaguti 1963. **16.13:** Courtesy of R. M. Kristensen. **16.14:** Courtesy of K. Worsaae and R. M. Kristensen. **16.15, 16.16, 16.17:** Courtesy of K. Worsaae and N. Bekkouche. **16.18:** Courtesy of R. M. Kristensen and K. Worsaae.

Chapter 17

17.1A–C: After Hyman 1959. **17.1D–F:** Courtesy Scott Santagata. **17.1G:** Courtesy of Kevin Lee. **17.2, 17.3A:** After Hyman

1959. **17.3B–D:** After Zimmer 1967. **17.4A:** After Hyman 1959. **17.4B,C:** After Dawydoff and Grassé 1959. **17.4D–G:** Courtesy Scott Santagata. **17.5A:** Courtesy of Bert Pijs. **17.5B,E:** Courtesy of Marco Faasse, www.acteon.nl. **17.5C:** © seasurvey.co.uk. **17.5D:** Courtesy of Hans de Blauwe. **17.5F:** Courtesy of Bernard Picton. **17.5G:** © Kåre Telnes. **17.5H:** Courtesy of Mat Vestjens and Anne Frijsinger. **17.5I:** © Neil McDaniel. **17.5J:** © Andreas Werth. **17.5K:** © Jan Hamrsky. **17.5L:** Courtesy of Daniel Nardin. **17.6A:** After Barnes 1980. **17.6B–F:** After Meglitsch 1972. **17.7A,B:** Courtesy of Claus Nielsen. **17.7C:** © J. Bailey-Brock. **17.8A,B:** After Ryland 1970. **17.8C:** Courtesy of J. Winston. **17.8D:** Courtesy of Claus Nielsen. **17.8E:** Courtesy of E. Håkansson. **17.9:** Based on Hayward and Ryland 1999 and Cheetham and Cook 1983. **17.10A–C:** Courtesy of Claus Nielsen. **17.10D:** After Ryland 1970. **17.10E:** Courtesy of H. U. Riisgaard. **17.11A:** Courtesy of Mick Otten, micksmarinebiology.blogspot.nl. **17.12:** After Gordon 1975. **17.13A–D,F:** Courtesy Claus Nielsen. **17.13E:** Courtesy of Kerstin Wasson. **17.14A:** After Woollacott and Zimmer 1975. **17.14B–E:** After Reed and Woollacott 1983. **17.15A:** Courtesy of A. Slotwinski. **17.15B:** Courtesy Claus Nielsen. **17.15C:** After Atkins 1955. **17.16A:** From www2u.biglobe.ne.jp/~gen-yu/pectinatella.html. **17.16B:** Courtesy of Hanna Hartikainen, Natural History Museum, London. **17.17A:** Courtesy Verena Haeussermann. **17.17B:** Courtesy Dirk Schories. **17.17C:** Courtesy Nina Furchheim and Anne Kaulfuss. **17.17D:** Courtesy Dave Meyer and Ben Datillo. **17.17E,F:** Courtesy Dirk Schories. **17.17G:** Courtesy A. Anker. **17.17H,I:** After Rudwick 1970. **17.17J:** Courtesy of A. Anker. **17.17K:** © Larry Jon Friesen. **17.18A:** After Williams and Rowell, in Moore 1965. **17.18B:** After Anderson 1996. **17.18C:** Redrawn from Hyman 1959, after Williams 1956. **17.18D:** After Rudwick 1970. **17.19A–C:** After Rudwick 1970. **17.19D:** Redrawn from Rudwick 1970, after Hancock 1859. **17.20:** After Rudwick 1970. **17.21A,B:** © Svetlana Maslakova. **17.21C:** © Terra Hiebert. **17.21D:** © Anne Zakrzewski. **17.21E:** © Nina Furchheim. **17.21F,G:** © Timothy Pennington. **17.21H:** After Hyman 1959.

Chapter 18

18.1: Courtesy of S. P. Stock. **18.2:** After Blaxter et al. 1998, and De Ley and Blaxter 2002. **18.3D:** After Hyman 1951. **18.3E:** Adapted from Lee and Atkinson 1977. **18.4:** Courtesy of S. P. Stock. **18.5A:** After Hyman 1951. **18.5B–H:** Courtesy of S. P. Stock. **18.6A–G:** Based on Sassar and Jenkins 1960. **18.6H–J,L:** Courtesy of S. P. Stock. **18.7A–D:** After Hyman 1951. **18.7E,F:** Courtesy of S. P. Stock. **18.8A:** After Meglitsch 1972. **18.8B,C:** After Hyman 1951. **18.8D:** After Noble and Noble 1982. **18.9A,B:** After Sherman and Sherman 1976. **18.9C–G:** Courtesy of S. P. Stock. **18.9H:** Courtesy of M. Mundo-Ocampo. **18.10A–D:** Modified from Boveri 1899. **18.10E,F:** Courtesy of S. P. Stock. **18.11B *inset*:** Courtesy of S. P. Stock. **18.13A–C:** © Larry Jon Friesen. **18.14A–F:** After Hyman 1951. **18.14G:** After Meglitsch 1972. **18.15:** After Hyman 1951.

Chapter 19

19.1A,B: After Higgins 1951. **19.1C:** After Hyman 1951. **19.1D:** Courtesy of Martin V. Sørensen. **19.2, 19.3:** Courtesy of Martin V. Sørensen. **19.4, 19.5, 19.6A:** After Hyman 1951. **19.6B:** After Storch et al. 1995. **19.6C:** After Meglitsch 1972. **19.6D,E:** Redrawn from photographs in Calloway 1982. **19.7B:** After Hyman 1951. **19.7C:** Redrawn from Hyman 1951, after Lang. **19.8A,B,E:** Courtesy of Reinhart Møbjerg Kristensen. **19.8C,D:** From Higgins and Kristensen 1986. **19.9:** After a sketch supplied by Robert Higgins. **19.10A:** From Higgins and Kristensen 1986. **19.10B:** Courtesy of Reinhart Møbjerg Kristensen. **19.11:** Modified from Pardos and Kristensen 2013 by Reinhart Møbjerg Kristensen.

Chapter 20

20.1A,B: Courtesy of R. Brusca. **20.1C,F–I:** © Larry Jon Friesen. **20.1D:** © blickwinkel/Alamy Stock Photo. **20.1E:** © J. N. A. Lott/Biological Photo Service. **20.1J:** Courtesy of R. Brusca. **20.2A:** From Kristensen 1982. **20.2B:** From Kristensen and Hallas 1980. **20.2C:** From Kristensen 1984. **20.2D,E:** From Kristensen and Higgins 1984. **20.2F, 20.3, 20.4:** Courtesy of R. M. Kristensen. **20.5A:** From Kristensen 1984. **20.5B:** Courtesy of R. M. Kristensen. **20.5C:** After Morgan and King 1976. **20.6:** Courtesy of R. M. Kristensen. **20.7:** After Morgan and King 1976. **20.8:** After Kristensen 1981. **20.9:** After Morgan 1982. **20.10A–C:** Courtesy of R. M. Kristensen. **20.11:** Courtesy of G. Giribet. **20.12A,B:** After Ramsköld and Hou 1991. **20.12C:** After Gould 1989. **20.12D *photo*:** Courtesy of Jie Yang and Xi-guang Zhang. **20.12D *illustration*:** Courtesy of Xi-guang Zhang. **20.13A:** After Manton 1977. **20.14:** Courtesy of G. Giribet. **20.16A,B:** After Borradaile and Potts 1961. **20.16C:** After Manton 1977. **20.16D:** After Barth and Broshears 1982. **20.18:** Courtesy of Georg Harzsch. **20.19A:** After Borradaile and Potts 1961. **20.19B–H:** After Anderson 1973. **20.25:** After Manton 1977. **20.27C:** Courtesy of R. Brusca. **20.27D:** © iStock.com/johnaudrey. **20.31B,C:** After Barnes 1980. **20.31D:** Courtesy of J. DeMartini. **20.32B:** After Parry, in Waterman 1960. **20.33D:** From Derby 1982. **20.33F:** After Foelix 1982. **20.34A:** After Pearse et al 1987. **20.36A,B:** After Snodgrass 1952. **20.36C,E:** After Stormer 1949. **20.36D,F:** After Bergstrom 1973. **20.36G:** © Corbin17/Alamy Stock Photo. **20.36H:** © Wim van Egmond/Visuals Unlimited, Inc.

Chapter 21

21.1A: Photo by D. Williams, courtesy of J. Yager. **21.1B:** Courtesy of J. Olesen. **21.1C:** Courtesy of G. McDonald. **21.1D:** © Larry Jon Friesen. **21.1E:** Courtesy of A. Kerstitch. **21.1F:** © Vladimir Levantovsky/Alamy Stock Photo. **21.1G:** Courtesy of L. Alberga. **21.1H:** Courtesy of P. Fankboner. **21.1I:** Courtesy of A. Anker. **21.1J:** © Larry Jon Friesen. **21.1K–M:** Courtesy of A. Kerstitch. **21.1N,O:** Courtesy of G. McDonald. **21.1P:** Courtesy of C. Holliday. **21.1Q:** © Larry Jon Friesen. **21.1R:** Courtesy of A. Anker. **21.1S:** Courtesy of David McIntyre. **21.1T:** Courtesy of A. Kerstitch. **21.1U:** © Nature Picture Library/Alamy Stock Photo. **21.1V:** From Boxshall and Lincoln 1987. **21.3B:** Courtesy of J. Olesen. **21.3C:** After Heard and Gocke 1982, *Gulf Res. Rpts.* 7: 157–162. **21.3D,F:** Courtesy of F. Schram. **21.4B:** Courtesy of N. Rabet. **21.4J–L:** From Martin et al. 1986, Zool. Scripta 15: 221–232. **21.4M:** Courtesy of J. Olesen. **21.5C,D:** Courtesy of T. Haney. **21.7A:** Courtesy of A. Kerstitch. **21.7C:** Courtesy of R. Caldwell. **21.8C, 21.9A,B,E:** After Abele and Felgenhauer, in Parker 1982. **21.10L:** Courtesy of E. Spivak. **21.12A,G:** Courtesy of E. Peebles. **21.12F:** Courtesy of J. Corbera. **21.13A,B:** After McLaughlin 1980. **21.13C:** Courtesy of J. Olesen. **21.13D:** Courtesy of M. Spindler. **21.13E:** After Bowman and Iliffe 1985. **21.15A,B:** After Bowman and Gruner 1973. **21.15C:** After a drawing by T. Haney. **21.15D,E:** Courtesy of E. Peebles. **21.15F:** Courtesy of T. Haney. **21.15J:** After Laval 1972. **21.15K:** Courtesy of E. Peebles. **21.16A:** © Larry Jon Friesen. **21.16D:** After Zullo, in Parker 1982. **21.16I:** Courtesy of J. Høeg. **21.17A,B:** After Boxshall and Lincoln 1983. **21.17C,D:** From Boxshall and Lincoln 1987, courtesy of G. Boxshall. **21.18A:** Courtesy of J. Olesen. **21.18H:** Courtesy of E. Peebles. **21.18I:** Courtesy of M. Dojiri. **21.18K:** After McLaughlin 1980. **21.18L:** Courtesy of G. McDonald. **21.20A,B,E(*b–s*):** Courtesy of D. J. Horne. **21.20C,D,E(*a*):** Courtesy of A. Cohen. **21.22F:** Photo by D. Williams, courtesy of J. Yager. **21.23B:** © K. Sandred/Visuals Unlimited, Inc. **21.24B,C:** From Schembri 1982. **21.24D:** From Abele 1985. **21.25:** After Høeg and Lutzen 1985, and Oeksnebjerg 2000; courtesy of J. T. Høeg. **21.26A–F:** From Høeg 1985. **21.26G,I:** Courtesy of J. T Høeg. **21.26H:** From Glenner and Høeg 1995. **21.27A,D–F:** After McLaughlin 1980. **21.27G:** After Kaestner 1970. **21.27H:** After Warner 1977. **21.28E–I:** After Kaestner 1970. **21.30A:** After Laverack 1964, *Comp. Biochem. and Physiol.* 13: 301–321. **21.30C:** After Cohen 1955, *J. Physiol.* 130: 9. **21.30D:** After Kaestner 1970. **21.31:** Micrographs courtesy of B. Felgenhauer. **21.32D:** © Ivan Kuzmin/Alamy Stock Photo. **21.32F:** Courtesy of C. Holliday. **21.32G:** Courtesy of C. McLay.

21.32H: © Solvin Zankl/Alamy Stock Photo. **21.33E:** © Larry Jon Friesen. **21.33H:** After Cameron 1985. **21.33J:** From Harvey et al. 2002. **21.33N:** © Larry Jon Friesen. **21.35A–E:** Courtesy of D. Waloszek. **21.35F:** Courtesy of Jean Vannier.

Chapter 22

22.1: Courtesy of Andy Murray. **22.2, 22.3:** © Larry Jon Friesen. **22.4A,H:** Courtesy of Katja Schulz. **22.4B:** Courtesy of Brian Gratwicke, image unchanged, licensed with a Creative Commons Attribution 2.0 license, creativecommons.org/licenses/by/2.0/. **22.4C–E,G:** © Larry Jon Friesen. **22.4F:** Courtesy of Sean Schoville. **22.4I:** Courtesy of Monika Eberhard. **22.4J:** Courtesy of Pisit Poolprasert. **22.4K:** Courtesy of Graham Montgomery. **22.5A–C:** © Larry Jon Friesen. **22.5D:** Courtesy of Patrick Marquez, USDA APHIS PPQ, Bugwood.org. **22.6A,E,F,I,K:** © Larry Jon Friesen. **22.6B,J:** Courtesy of Katja Schulz. **22.6C:** Courtesy of Vida van der Walt. **22.6D:** © iStock.com/Henrik L. **22.6G:** © Andre Goncalves/Shutterstock. **22.6H:** Courtesy of John Montenieri, Centers for Disease Control. **22.6I:** Courtesy of David McIntyre. **22.7A:** © Dr. Morley Read/Shutterstock. **22.7B,D,E:** © Larry Jon Friesen. **22.7C:** Courtesy of Charles Hedgcock. **22.10:** After Lawrence et al., in CSIRO 1991, based on Snodgrass 1935. **22.13:** After Chapman 1982. **22.14:** After Snodgrass 1952. **22.16C,F:** © Larry Jon Friesen. **22.20:** After Anderson and Weis-Fogh 1964. **22.21C:** © Larry Jon Friesen. **22.23A:** After B. Rodendorf, ed., 1962, Arthropoda—Tracheata and Chelicerata, in *Textbook of Paleontology*, Academy of Sciences, U.S.S.R. **22.23B:** After Snodgrass 1952. **22.23C:** After R. J. Wooten 1972, *Paleontology* 15: 662. **22.24:** © Larry Jon Friesen. **22.24C:** © Olga Bogatyrenko/Shutterstock. **22.25A:** © Larry Jon Friesen. **22.25B:** After Snodgrass 1944. **22.26B:** After Wigglesworth 1965. **22.27B:** After Sherman and Sherman 1976. **22.28B:** After Wigglesworth 1965. **22.28C:** After Chapman 1971. **22.29A:** After Snodgrass 1935. **22.29B:** After Clarke 1973. **22.29D:** Photo by S. E. Hendrixson. **22.29E:** Courtesy of Andrea Di Giulio. **22.29F, 22.30:** © Larry Jon Friesen. **22.31:** After Fretter and Graham 1976. **22.32:** After Clarke 1973. **22.34A:** Courtesy of David McIntyre. **22.35A:** Courtesy of Andrea Di Giulio. **22.35B:** After Slifer et al. 1959, *J. Morphol.* 105: 145–191. **22.36A–C:** After Michelsen 1979. **22.36D:** © Larry Jon Friesen. **22.37:** After Blaney 1976. **22.38:** After Snodgrass 1935. **22.39:** After Anderson 1973. **22.40C:** After Ross 1965. **22.41:** After Chapman 1971, after Southwood and Leston, *Land and Water Bugs of the British Isles*, 1959, Warne and Co., London. **22.41 *photo*:** © Larry Jon Friesen. **22.42:** After Chapman 1971, after Urquhart 1960. **22.43A,C:** Courtesy of David McIntyre. **22.43B,D:** © Larry Jon Friesen. **22.44:** After Misof et al. 2014.

Chapter 23

23.1A: Courtesy Alessandro Minelli. **23.1B:** Courtesy of Bikebot/Wikimedia. **23.1C,E:** Courtesy of S. Prchal. **23.1D:** © R. K. Burnard/Biological Photo Service. **23.1F,G:** Courtesy of Andy Murray. **23.1H:** Courtesy Elliot Lowndes. **23.2B,C:** After Beck and Braitwaite 1968. **23.2D,E:** After Snodgrass 1952. **23.2F,H:** After Manton 1965. **23.2G:** After Anderson 1996, *Atlas of Invertebrate Anatomy*, Univ. New South Wales Press. **23.2I:** After Lewis 1981. **23.3A,B,D:** After Russell-Hunter 1969. **23.3C:** Courtesy of David McIntyre. **23.3E:** After Manton 1965. **23.3F:** After Barth and Broshears 1982. **23.4A:** After Kaestner 1969. **23.4B,C, 23.5, 23.6:** After Lewis 1981. **23.7:** After Grenacher 1880, *Arch. Mikrosk. Anta. Entwmech.* 18: 415–467. **23.8:** After Lewis 1981, based on Tichy 1973, *Zool. Jahrb. Anat.* 91: 93–139. **23.9:** After Rilling 1968, in *Grosses Zoologisches Praktikum*, Part 13b, Fischer, Stuttgart. **23.10:** Courtesy of S. Prchal.

Chapter 24

24.1C: © Ken Lucas/Visuals Unlimited, Inc. **24.2C:** © Joe McDonald/Corbis. **24.3H–K:** Courtesy of G. Giribet. **24.4A,D,G,H,J,M,O,Q:** Courtesy of G. Giribet. **24.4P:** Courtesy of ⊠. Kovà⊠. **24.5D–F,I–O:** Courtesy of Gustavo Hormiga. **24.6A:** Courtesy of S. Prchal. **24.7D,E:** Courtesy of R. Foelix. **24.7F,G:** Courtesy of G. Hormiga. **24.8:** After Foelix 1982, Gertsch 1979, and others. **24.9A:** Courtesy of R. F. Foelix. **24.9B:** © Elliotte Rusty Harold/Shutterstock. **24.9C:** © iStock.com/BizarrePhotos. **24.10:** After Root and Bowerman 1978. **24.11:** After Foelix 1982. **24.12A:** From Foelix 1982, after slow-motion pictures by Parry and Brown 1959. **24.12C:** From Foelix 1982, after Frank 1957. **24.12D:** From Foelix 1982, after Foelix 1970. **24.12E:** Photo by R. F. Foelix. **24.13A:** Courtesy of David McIntyre. **24.13B:** © WILDLIFE GmbH/Alamy Stock Photo. **24.15:** Courtesy of G. Hormiga. **24.16A:** From Foelix 1982, after Millot et al. 1949. **24.16B:** After Foelix 1982. **24.16C,D:** Courtesy of G. Hormiga. **24.17B,C:** After Foelix 1982. **24.18B:** After Foelix 1982. **24.18C,D:** Photos by R. F. Foelix. **24.20A:** SEM by G. Hormiga. **24.20B:** From Foelix 1982, after Gauorner 1965. **24.20C:** From Foelix 1982, after Barth 1971. **24.20D:** After Foelix and Chu-Wang 1973, photo by R. F. Foelix. **24.20E:** Photo by R. F. Foelix. **24.21A,F:** Courtesy of G. Hormiga. **24.21B,C:** Photos by R. F. Foelix. **24.21D,E:** From Foelix 1982 after Hoffman 1971. **24.22A:** After Fage 1949. **24.23A:** From Foelix 1982, after Melchers 1964. **24.23B:** From Foelix 1982, after Osaki 1969. **24.23C,E,F:** After Foelix 1982. **24.25A,B:** From Foelix 1982, after von Helversen 1976. **24.25C:** From Foelix 1982, after Bristowe 1958. **24.25D,F,H:** After Foelix 1982. **24.25G:** After Bristowe 1958. **24.27A:** Courtesy of G. Hormiga. **24.27B,C:** From Foelix 1982, after Holm 1940. **24.27D:** From Meglitsch 1972, after Bristowe 1958. **24.27E:** Photo by S. Prchal. **24.27F,L,M:** Courtesy of G. Giribet. **24.27G,H:** From Foelix 1982, reprinted with permission of Harvard University Press. **24.27I,J,K:** From Foelix 1982, after Vachon 1957. **24.28A,C:** After Hedgpeth 1982. **24.28B:** After a sketch by J. W. Hedgpeth. **24.28D:** After Wyer and King 1974. **24.28E:** After Schram and Hedgpeth 1978. **24.28F:** Courtesy of M. Harros and C. Arango. **24.29A:** After Fage 1949 and other sources. **24.29B:** After Schram and Hedgpeth 1978. **24.29C:** Original drawing by J.W. Hedgpeth, from a photo of a live specimen in an aquarium. **24.29D:** Courtesy of G. Giribet. **24.30A,C:** After Fage 1949. **24.30B:** After Schram and Hedgpeth 1978. **24.30D:** After Hedgpeth 1982. **24.31:** Cladogram courtesy of G. Giribet and G. Hormiga. Based on Sharma et al. 2014 and modified from Schwager et al. 2005.

Chapter 25

25.1A,–C,E,F,I,O: Courtesy of R. Mooi. **25.1D:** Courtesy of P. Fankboner. **25.1G:** Courtesy of T. Gosliner. **25.1H:** Courtesy of P. Fankboner. **25.1J:** Courtesy of A. Kerstitch. **25.1K,L,M:** © Larry Jon Friesen. **25.1N:** Courtesy of G. McDonald. **25.1Q:** Courtesy of S. Ohta. **25.2B,C,E,I:** After Hyman 1955. **25.2F:** Courtesy of R. Mooi. **25.3C,E:** After Clark 1977. **25.3D,F:** After Hyman 1955. **25.4:** Courtesy of Rich Mooi. **25.5A:** After Barnes 1980, modified from Nichols 1962. **25.5B:** From Ellers and Telford 1984, courtesy of M. Telford. **25.5C:** From Turner 1984, photograph courtesy of R. Turner. **25.5D:** From Emlet 1982, photographs courtesy of R. Emlet. **25.5E:** After Campbell 1983. **25.5F:** After Chia and Amerongen 1975. **25.5G,H:** After Barnes 1980. **25.5I,J:** Modified from Russell-Hunter 1979. **25.6A:** After Hyman 1955. **25.6B:** Illustration by Laura Garrison. **25.6C,D:** After Clark 1977. **25.6E:** © blickwinkel/Alamy Stock Photo. **25.7A,B:** After Nichols 1962. **25.7C:** After Cuénot 1948. **25.8A:** After Nichols 1962. **25.8D:** Courtesy of R. Brusca. **25.9A,B:** After Nichols 1962. **25.9C:** After Barnes 1980. **25.9D:** After Pentreath 1970. **25.9E:** After Warner and Woodley 1975. **25.9F:** Courtesy of G. McDonald. **25.11A:** After Nichols 1962. **25.11B,C,E:** After Barnes 1980. **25.11D:** After Hyman 1955. **25.11F:** © Larry Jon Friesen. **25.11G:** After Cuénot 1948. **25.12A:** After Nichols 1962. **25.12B:** After Cuénot 1948. **25.12C:** Courtesy of P. Fankboner. **25.12D:** © WaterFrame/Alamy Stock Photo. **25.12E:** From Costelloe and Keegan 1984, courtesy of J. Costelloe, used with permission of Springer-Verlag. **25.12F:** © Stephen Frink Collection/Alamy Stock Photo. **25.12G:** Courtesy of A. Kerstitch. **25.12H:** After Barnes 1980, from a photograph by I. Bennett. **25.13A:** After Ubaghs 1967. **25.13B:**

After Herreid et al. 1976. **25.14A:** After Barnes 1980. **25.14B:** After Cuénot 1948. **25.14C,D,E:** Redrawn from Shick 1983 after the following original sources: C, after Phelan 1977 and Smith 1978; D, after Fenner 1973; E, after Smith 1980. **25.15, 25.16A,C:** After Cuénot 1948. **25.16B:** After Hyman 1955. **25.17G:** From Emlet 1982, photo courtesy of R. Emlet. **25.17H:** Courtesy of and © M. Apley. **25.18:** Redrawn from Meglitsch 1972, after Dawydoff 1948. **25.19A:** After Barnes 1980. **25.19B,C:** After Paul and Smith 1984. **25.19D,G,H:** After Nichols 1962. **25.19E:** Redrawn from Barnes 1980, from Kesling in Moore (ed.) 1967. **25.19F:** After Cuénot 1948.

Chapter 26

26.1A: © C. R. Wyttenbach/Biological Photo Service. **26.1B:** Courtesy of K. M. Halanych. **26.2A:** After Sherman and Sherman 1976. **26.2B:** Courtesy of Fernando Pardos. **26.2C:** After Hyman 1959. **26.2D:** From Jones et al. 2013. Courtesy of Daniel Jones. **26.2E:** After Pechenik 1985. **26.2F:** Courtesy of Fernando Pardos. **26.2G,H:** After Lester 1985. **26.3A,D,E:** Courtesy of Fernando Pardos and Jesús Benito. **26.3C:** From Worsaae et al. 2012. **26.4:** After Burdon-Jones 1956. **26.5A–C:** Courtesy of Sabrina Kaul-Strehlow, from Kaul and Stach 2010. **26.5D–M:** Courtesy of Sabrina Kaul-Strehlow. **26.6:** After Hyman 1959. **26.7:** After Lester 1985.

Chapter 27

27.1A,B: © Larry Jon Friesen. **27.3A:** © Daniel Gotshall/Visuals Unlimited, Inc. **27.3B:** Courtesy of R. Brusca. **27.3C:** © G. Corsi and B. Corsi/Visuals Unlimited, Inc. **27.3D,E:** © WaterFrame/Alamy Stock Photo. **27.3F:** © Larry Jon Friesen. **27.3G:** © L. S. Roberts/Visuals Unlimited, Inc. **27.3H,I:** Courtesy of R. Brusca. **27.3J:** © Scubazoo/SuperStock/Corbis. **27.3K:** © David Wrobel/Visuals Unlimited, Inc. **27.3L:** After Barnes, 1980, after photographs by Monniot and Monniot 1975. **27.4A:** After Romer 1956. **27.4B:** © Wim van Egmond/Visuals Unlimited, Inc. **27.4C,D:** Courtesy of C. S. Cohen. **27.4E:** Courtesy of J. Spaulding. **27.4F:** Courtesy of Berta Colom Sanmarti. **27.5C:** After Berrill 1935. **27.5D:** After Grassé 1948. **27.5E:** After Van Name 1945. **27.5F:** After Thorne and Thorndyke 1975. **27.5G,H:** Courtesy of C. S. Cohen. **27.5I:** Courtesy of Christopher Rieken. **27.6A:** Courtesy of R. Brusca. **27.6E:** © Gavin Newman/Alamy Stock Photo. **27.6F:** Courtesy of G. McDonald. **27.6G:** Courtesy of J. King. **27.6H:** After Pechenik 1985. **27.6I:** After Alldredge 1976. **27.7A:** After Brien, in Grassé 1948. **27.8A:** Courtesy of Vanessa Guerra. **27.8B,D:** After Seeliger, in Grassé 1948. **27.8C:** After Cloney and Torrence 1982. **27.8F:** Courtesy C. S. Cohen.

Index

Page numbers in **boldface** indicate definitions of terms; page numbers followed by f indicate figures; page numbers followed by t indicate tables; and those followed by n indicate footnotes.

A

Abarenicola pacifica, 535f
abdA gene, 30, 735
AbdB gene, 735
abdomen
  of Crustacea, **801**
  of Hexapoda, 866
Abdopus horridus, 475f
abductor muscles, **146**
aboral canal of Ctenophora, 333f, **337**
aboral region of Echinodermata, **969,** 982
abyssal plain, **18**
Abyssocladia carcharias, 238f
Acaenoplax, 523
Acalephae (akalephe), 268
Acanthamoeba, 70, 74
Acantharia, 108, 109, 110, 111
Acanthaster, 156f, 984
  planci, 970f–971f, 984
acanthella of Acanthocephala, **626,** 627f
Acanthina, 156f
Acanthobdella peledina, 541, 591, 592f, 593
Acanthobdellida, 541, 591, 592, 592f, 593
Acanthobothrium, 386
Acanthocardia tuberculata, 456f–457f
Acanthocephala, 613–615, 614f, 618, 624–626
  characteristics of, 625
  lacunar system of, 624–625, 625f
  lemniscus of, 625f, 625–626
  life cycle of, 626, 627f
  phylogeny of, 25
  reproduction of, 625f, 626
Acanthocephalus, 624, 625f
Acanthochitona, 508f
Acanthochitonida, 462
Acanthocythereis, 800f
Acanthoecidae, 125
Acanthomacrostomum, 375f
Acanthopleura, 511f
acanthopods, **74**
Acanthopriapulus, 698
acanthor larva of Acanthocephala, **626,** 627f
Acari, 916f, 917–919
Acarida, 790
Acariformes, 914, 916f, 917, 918, 962f
Acentrosomata, 377, 408, 408f, 409
Acercaria, 846, 847, 853, 854f, 889f, 890
acetabulum of Platyhelminthes, **386**
Acetes, 778
Achaearanea, 936f–937f
Achatinidae, 467
Achelata, 762f–763f, 766, 782f, 782–783
Achelia echinata, 956f, 959
aciculae of Annelida, **545**
Aciculata, 535, 545, 549, 598f
acidiphiles, **9,** 9n
Acineta, **92f**
Acmaeidae, 462
Acochlidioidea, 467
Acoela, 346, 349, 350, 351–360
  body wall and external appearance of, 352f, 353–355
  musculature, support, and movement of, 349–350, 355–356, 356f, 357f
  nervous system of, 351, 354f, 357–358, 358f
  nutrition, excretion, and gas exchange of, 356–357
  phylogeny of, 25
  pulsatile bodies of, 349, 350, 351, 353
  reproduction and development of, 358f, 358–360, 1014
  sagittocysts of, 353f, **354,** 355f
acoelomate body plan, 140, 140f
Acoelomorpha, 346, 349, 350, 351, 362, 376, 968
  phylogeny of, 407, 1050
Acoetidae, 537
acontia of Cnidaria, 278f, **279**
Acraspeda, 319
acraspedote, **285**
Acrididae, 852–853
Acrocirridae, 538
Acropora, 387
  palmata, 266f–267f
acrorhagi, **298,** 299f
Acrothoracica, 767, 791
Acteonidae, 467
Actinia, 284f
Actiniaria, 270, 284f, 290f, 292, 301f
Actiniidae, 270
Actinocyclus, 100f
Actinophrys, 98f
Actinoposthia beklemishevi, 354f
Actinopterygii, 1041f
Actinopyga, 991
Actinosphaerium, 110f
actinospore larvae of Cnidaria, 299, 300f
Actinostola, 292
Actinotrocha, 637
  brachiata, 637
actinotroch larva of Phoronida, **643,** 643f, **644**
actinula larva of Hydrozoa, 314f–315f, 316, 319
Actinulida, 316
Aculifera, 454, 522f
Adamsia, 301
adanal sensory structures of Mollusca, **513**
adductor muscles, **146,** 148
adelphophagy of Mollusca, **521**
adhesive papillae of Urochordata, **1040**
adhesive sac of Bryozoa, **656,** 656f
adhesive tubes of Gastrotricha, **429**
Adiaphanida, 377, 408f, 409
adoral ciliary zone, **208**
adoral zone of membranelles (AZM) of Ciliata, **92**
adradial canals of Ctenophora, 333f, **337**
Adula, 490
advanced character states, **44**
aedeagus of Hexapoda, **866, 883**
Aedes, 846
  aegypti, 846
  albopictus, 846
Aegidae, 47, 47f
Aegina, 335, 337f
  citrea, 337f
Aeginidae, 273
Aegiretidae, 467
Aeolidiidae, 467
Aequipecten opercularis, 512f
Aequorea, 297f, 308, 313f
  macrodactyla, 311, 313f
  victoria, 286f–287f
aequorin, **308**
aesthetascs of Arthropoda, **822,** 823f
aesthetes of Mollusca, 511f, **513**
aestivation, **22**
afferent (sensory) nerves, **168**
African relapsing fever, 918
African sleeping sickness, 123
Agariciidae, 297
Agassiz, L., 980
Agauopsis auzendei, 918
Agelas, 218f
Agelasida, 222
Agelenidae, 913, 936
Aglajidae, 467
Aglaophenia, 177f
Aglaura, 314f–315f
Agnatha, 1022, 1041f, 1042, 1043
Agnathiella, 616
Agroeca brunnea, 954f–955f
Aguinaldo, A. M., 51, 728, 751
Ahlrichs, W. H., 613

Ahyong, S. T., 765
Aiptasia, 296
Akashiwo sanguinea, 79
Akentrogonida, 815–816
Akera, 509f
Alatinidae, 272
Alaurina, 395f
Albumares, 317
Alciopinae, 537–538, 545
Alcippe, 792f–793f
Alcyonacea, 271, 284f, 285f, 288
Alcyonaria, 271
Alcyonaria (Octocorallia), 269, 285f, 320f
Alcyonidium, 645f, 655
Alcyoniina, 271
Alcyonium, 284f, 309
Alexander the Great, 858
Alexandrium, 78
Aleyrodidae, 854
algae, 55, 57n, 59–60
  blue-green, 56
  brown, 56, 57, 97, 98f, 99, 101
  chlorarachniophyte, 104–105
  chromophtye, **56**
  golden, 97, 98f, 99–100, 101
  green, 55, 56, 57, 59f, 126, 127, 128, 129
  red, 55, 56, 57, 59f, 103, 125, 126, 127, 129
  symbiotic, 57, 302–304, 353, 354f
algin, **99**
Alicella gigantea, 783, 790
alkaliphiles, **9,** 9n
Allantonema, 670f
allelochemicals, 250
allogamy of Parabasilida, **114**
Allogastropoda, 466
Allogramia, 107
Alloioplana californica, 375f
Allomalorhagida, 696, 696f
Allopora, 280f–281f, 290f
  californica, 554
allosperm of Nemertodermatida, **365**
Alma emini, 559
Alpheidae, 779f, 811–812
Alpheus, 779f, 811
Althalamida, 106
Alveolata, 62–63, 127
alveoli, **62**
  of Apicomplexa, 83
  of Dinoflagellata, **79**
Alvinella, 554
  pompejana, 554, 554f
Alvinellidae, 539
Amakusaplana acroporae, 387
Amaurobiidae, 946f
Amaurobius similis, 928f–929f, 945f, 946f
Amblypygi, 912, 919, 926, 927, 942, 950, 955, 962f
Amblypygida, 919f
ambulacra of Echinodermata, **975,** 982, 1001, 1003
Ambulacraria, 348, 350, 367, 968, 969, 1019, 1043
  phylogeny of, 25, 1049f, 1050
ambulacrum of Echinodermata, 978
amebiasis, 72
amebic dysentery, 71, 72
amebic meningoencephalitis, 64
ameboid locomotion, 141–142, 142f
ametabolous development of Hexapoda, 885f, **885**–886
amictic ova of Rotifera, 623, 623f
amino acids, and nitrogenous waste products, 158, 158f
Amiskwia, 15f
ammonia, 158
ammonotelic animals, **158**
Ammotheidae, 956f
amnesic shellfish poisoning, 99
Amoeba, 74, 75f, 76f
  proteus, 58f–59f, 70, 74, 74f
Amoebozoa, 56, 58f–59f, 60–61, 70–76, 71f
  characteristics of, 70
  nutrition of, 75, 75f
  overview of, 62
  phylogeny of, 127, 128
  reproduction of, 75–76, 76f
  support and locomotion of, 70, 74f, 74–75
amorphic development of Arthropoda, **750n**
Ampharetidae, 535f, 539, 554f
Amphibia, 1022, 1041f
amphiblastula larva, 244, 245f, 247, 247f, 249, 258
Amphibolidae, 467
Amphidinium, 80
  klebsii, 353
Amphidiscophora, 221, 257–258
amphidisc spicules of Porifera, **240,** 241f
amphids of Nematoda, 680f, **681**
Amphiesmenoptera, 846, 847, 858–859
Amphiglena, 565
Amphilinidea, 379, 382, 386, 403
Amphilonche heteracantha, 58f–59f
Amphimedon queenslandica, 216, 244
Amphinomida, 533, 542, 545
Amphinomidae, 533, 534f–535f, 565
Amphionidacea, 766, 776, 777, 806t–807t
Amphionides, 777
  reynaudii, 777, 777f
Amphioxus, 186f
Amphipholis, 419
  squamata, 419
Amphipoda, 763f, 767, 789–790, 832
Amphiporus, 441, 444f, 445f
  bimaculatus, 442f
  formidabilis, 442f
Amphiroa, 251
Amphitrite, 550
  kerguelensis, 535f
Amphiura, 976f
Amplexidiscus, 296–297
  senestrafer, 297f
Amplimatricata, 377, 408, 408f
ampullae, **290**
  of Cnidaria, 290, 290f
  of Echinodermata, 981
  of Urochordata, **1040**
Ampullariidae, 465
anabiosis of Tardigrada, **713**
anagenesis, **27,** 30, 31, 31f
Anakinetica cumingi, 659
Analges, 916f
anal pores of Ctenophora, **332,** 333f, **337**
anal somite of Crustacea, **801**
anal vesicles of Echiuridae, **584**
anamorphic development, **750n,** 830, **831,** 907
Anaphe, 743
Anapidae, 913
Anaplectus, 676f–677f
Anaspidacea, 765, 776
Anaspidea, 467
Anaspides, 776
Anatoma, 505f
Anatomidae, 463, 504
ancestral character states, **44,** 45
ancestrula of Bryozoa, **644,** 654f, 656, 657
Anchistropus, 773
Ancylostoma caninum, 676f–677f
Andrade, S. C., 532, 597, 598f
Androctonus, 927f
  australis, 936
androgonochorism of Crustacea, **825**
anecdysis of Arthropoda, **741**
Anelassorhynchus porcellus, 581f
Anelosimus, 932
  eximius, 932, 936f–937f
Anfesta, 317
Anguinella, 645f
Anheteromeyenia argyrosperma, 241f
anhydrobiosis of Tardigrada, **713**
animal architecture, 135–180
Animalia, 4, 8, 9
animal pole, **187,** 188f
animal–vegetal axis, **187,** 188f
anisogamy of Protista, **69**
anisomyarian condition of Mollusca, **490**
Annelida, 11, 317, 348, 384, 450, 531–598, 695, 751
  body forms of, 542–543
  body plan of, 541–542, 542f–543f, 574
  body wall and coelomic arrangement of, 543–545
  characteristics of, 532
  circulation and gas exchange of, 555–559, 558f–559f
  Echiuridae, 579–584
  excretion and osmoregulation of, 560–561, 561f
  feeding and digestion of, 549–555, 551f, 552f, 556f–557f
  Hirudinoidea, 541, 591–597
  nervous system and sense organs of, 172, 561–565, 562f–563f, 564f
  number of species, 2t, 1047
  phylogeny of, 25, 26, 542, 597, 598f, 1049f, 1050
  reproduction and development of, 565–572, 1014
  respiratory pigments of, 558–559
  Siboglinidae, 539, 584–591
  Sipuncula, 533, 535, 572–579
  support and locomotion of, 545–549, 546f
  symbiotic relationships of, 552–554, 554f
  taxonomic history and classification of, 532–541
  tube-dwelling, 547, 548f–549f, 552, 563
annelid cross, 190f, **191**
annules of Nematoda, **674,** 675f
Annulonemertes, 442
Anodonta, 501f
anogenital tagmata of Hexapoda, **866**
Anomalocaris, 13, 14f, 15f
  canadensis, 14f
  nathorsti, 14f
Anomalodesmata, 470, 497, 499f
Anomiidae, 469, 489
Anomma wilverthi, 844
Anomura, 762f–763f, 766, 781, 782f, 783
  feeding of, 810
  life cycle of, 814f–815f
  reproduction and development of, 831

Anopheles, 82, 84, 86, 86f, 939
gambiae, 82, 84
Anopla, 436, 450
Anoplocotyloides papillata, 385f
Anoplodactylus
eroticus, 957f
evansi, 956f
Anostraca, 737f, 765, 772
distinguishing features of, 802t–803t
nervous system and sense organs of, 823
reproduction and development of, 806t–807t, 831
antagonistic muscles, 146, 147f
Antedon, 983f, 985f, 996, 997f
antennae
of Crustacea, **801,** 803t
of Hexapoda, 860, 862f
Antennapedia (ANTP) genes, 254
antennules of Crustacea, **801,** 803t
Anthoathecata, 273, 292
Anthocerotophyta, 4n, 55n, 129
Antho karykina, 251
Anthomedusae, 273, 286f–287f
Anthopleura, 298, 303, 308, 309
elegantissima, 298, 299f, 303, 309f
xanthogrammica, 303, 303f
Anthothoe, 296
Anthozoa, 12f, 216, 266f–267f, 269, 270–271, 301f
cnidae of, 293, 295f
feeding and digestion of, 296
movement of, 290–291, 292
nerve net of, 305f
origins of, 318, 319
phylogeny of, 318, 319, 320, 320f, 321
polypoid stage of, 277, 278f, 279, 283, 284f, 285f, 319
reproduction and development of, 308–310, 309f
support of, 287, 288
Anthropocene, 6
Anthuridea, 766, 787f
antigenic variation, **84,** 116
Antigonaria arenaria, 359f
Antillesoma, 575
Antillesomatidae, 575
Antipatharia, 270
antizoea larvae, **831**
Antliophora, 846, 847, 856–858
ANTP genes, 254, 255
Anurida granaria, 845f
aorta of Cephalochordata, **1025, 1026**
Aphanolaimus, 676f–677f
Aphelenchus, 678f–679f
Aphidae, 854
aphotic zone, 19f, **19**–20
Aphragmophora, 423
Aphrocallistes vastus, 228f
Aphrodita, 556f–557f
Aphroditidae, 537
Aphroditiformia, 537
apical complex of Apicomplexa, **81,** 82f
apical organ, 208f, **346**–347, 608f, 611, 669, 1049f
of Ctenophora, **332,** 333f, 338, 339, 339f, 341
of Echinodermata, **997**
of Lophophorata, 654, 656, 656f
apical system of Echinodermata, **981,** 982
Apicomplexa, 61, 62, 81–87, 126
characteristics of, 83
nutrition of, 83
phylogeny of, 127
reproduction and life cycle of, 81–82, 83–87, 85f, 86f
support and locomotion of, 83
apicoplast of Apicomplexa, **81,** 126
Apidae, 855
Apis
dorsata, 845n
mellifera, 844, 876f
Aplacophora, 454, 459
development of, 518f, 519f
feeding of, 491
phylogeny of, 522f, 523, 524
Aplanulata, 274
Aplidium, 1028f–1029f, 1029, 1032f
Aplousobranchia, 1022
Aplysia, 456f, 502, 509, 515f
Aplysidae, 467, 487
Aplysiinae, 482
Aplysina, 252
archeri, 218f
Aplysiomorpha, 467
Apochela, 715
Apocrita, 855
apodemes of Arthropoda, **731**
Apodida, 974, 982
apomorphy, **44,** 45
Aponomma ecinctum, 918
apopyle, **226**
Appeltans, W., 6
appendage hypothesis on insect flight, 869
Appendicularia, 1021, 1022, 1027, 1031–1033, 1034f–1035f
body wall, support, and locomotion of, 1033
circulation, gas exchange, and excretion of, 1036
feeding and digestion of, 1036
nervous system and sense organs of, 1037
reproduction and development of, 1037, 1039
Appendiculata, 728
appositional eyes of Arthropoda, **749,** 749f
apterous gene, 870
Apusomonas proboscidea, 68f
aquiferous system of Porifera, **222**–227, 234, 238, 239, 254
arabinosides, 250
Arachnida, 710f–711f, 728, 911, 912, 917–927, 927f
phylogeny of, 961, 962f
Arachnocampa luminosa, 873
Arachnopulmonata, 962f
aragonite, 232, 234
Araneae, 912, 922–924, 926, 927, 928, 950, 962f
Araneidae, 913, 922f–923f, 931, 931f, 937f, 938, 951f
Araneoidea, 932
Araneomorphae, 912, 924, 930, 942
Araneus, 928, 928f–929f, 931, 934f, 945f, 949f
diadematus, 951f
Arango, C. P., 958
Arbacia, 990f
punctulata, 149
Arca vivipara, 519
Arcella, 74, 75, 76, 76f
Archaea, 3, 4, **7,** 8–9, 28, 64, 217, 221
Archaebacteria, 7, 8
Archaeidae, 913, 939f
Archaeocyatha, 256, 256f
archaeocytes of Porifera, **230**–231, 232f, 236, 240, 241f, 252
Archaeognatha, 752, 846, 847, 848, 849f
evolution and phylogeny of, 48f, 888, 889f
thorax of, 864
Archaeopteryx, 16
Archaeornithura meemannae, 16
archaeotroch, **208**
Archaeplastida, 8, 55, 57, 60, 61, 127f, 129
archenteric pouching, **196,** 196f
archenteron, **193,** 194f, 196
Archeterokrohnia, 422
Archiacanthocephala, 624, 626
Archiannelida, 535
archicoelomate hypothesis, 346, 635, 636
archicoely, **635**
Archiheterodonta, 455, 470
Architaenioglossa, 455, 465
Architectonicidae, 466
Architeuthidae, 472
Architeuthis, 472
dux, 9, 473f
architomy of Enteropneusta, **1014**
Archoophora, 376
archoöphorans, **395,** 397, 450
Archypolypoda, 909
Arcida, 469
Arcidae, 469, 488
Arcovestia ivanovi, 533f
Arctiidae, 859
Arctonoe, 553, 554f, 985–986
pulchra, 537, 986
vittata, 985
Arenicola, 546f, 547, 549, 590
excretion and osmoregulation of, 560
feeding and digestion of, 551f, 555
Arenicolidae, 535f, 539, 547, 549
Areolaimida, 672
areoles of Nematomorpha, **687,** 688f
Argasidae, 918
Argas persicus, 918
Argiope, 931, 931f
Argonauta, 481f, 518
Argonautidae, 471
Argulus, 796
foliaceus, 794f–795f
Argyroneta, 934, 935f
aquatica, 934
Argyrotheca, 663
Arhynchobdellida, 541, 592f, 593, 594f, 595, 595f, 596
Arionidae, 467
Arion lusitanicus, 505f
Aristotle, 202, 268, 454
Aristotle's lantern, 974, **988,** 988f–989f, 991, 993, 1002f, 1003
Arminidae, 467
Armohydridae, 273
Armorloricus elegans, 704f
Arnaud, F., 958
arolium of Hexapoda, **864**
artemisinin, 84
arthrodial membranes of Arthropoda, **736**
Arthropleura, 909
Arthropleurida, 909

Arthropoda, 11, 347, 348, 728–962
appendages of, 733–736
body plan of, 730–731
body wall of, 731–733
characteristics of, 729
Chelicerata, 911–962
circulation and gas exchange of, 730, 743–744, 744f, 745f
Crustacea, 761–836
digestive system of, 742–743, 743f
evolution and phylogeny of, 25, 26, 733, 734–736, 744, 751–755, 1049f
excretion and osmoregulation of, 745–746, 746f
growth of, 739f, 739–742, 740f–741f
Hexapoda, 843–890
Hox genes in, 30, 734–735, 754
joints of, 736, 736f, 738
Myriapoda, 895–909
nervous system and sense organs of, 172, 746–749, 747f, 755
number of species, 2t, 1047
origin of, 751
reproduction and development of, 750f, 750–751
support and locomotion of, 736–739, 737f, 738f
taxonomic history and classification of, 728–729
arthropodization, 730
Arthrotardigrada, 712f, 715, 716
articles of Arthropoda, **733**
articulamentum of Mollusca, **479**
articular membranes of Arthropoda, **736,** 736f
Articulata, 41, 728, 1051
Artomonema, 677
Asaphiscus, 15f
Asaphus, 753f
Asbestopluma, 236f–237f
agglutinans, 238f
desmophora, 238f
hypogea, 236
occidentalis, 236
Ascalaphidae, 856
Ascaris, 674, 674f, 678f–679f, 680f, 682f
Ascetospora, 112
Aschelminthes, 613
Ascidiacea, 1021, 1022, 1027, 1028f–1029f, 1029, 1030f, 1031
body wall, support, and locomotion of, 1033
circulation, gas exchange, and excretion of, 1036
feeding and digestion of, 1034–1036
general anatomy of, 1032f–1033f
nervous system and sense organs of, 1036
reproduction and development of, 1037, 1038, 1038f–1039f, 1039–1040, 1041
Ascoglena, 119
Ascomycota, 129
asconoid sponges, **224,** 225, 225f, 226f, 227, 236
Ascoparia, 364f
neglecta, 365
Ascorhynchus, 960f
Ascothoracida, 767, 791
Ascothorax ophiocentenis, 792f–793f
ascus, **650,** 650f
Asellota, 766
asexual reproduction, 176f, 176–177
of Acoela, 358f, 358–359
of Annelida, 566f, 567, 584
of Bryozoa, 656–657
of Cnidaria, 265, 308, 311, 313f, 319
of Ctenophora, 339
of Echinodermata, 995–996, 996f
of Enteropneusta, 1014
of Nemertea, 447
of Phoronida, 642
of Platyhelminthes, 394–395
of Porifera, 239–241, 252
of Protista, 68, 75–76, 76f, 80, 83, 93–94, 94f, 107, 111, 114, 117, 124
of Pterobranchia, 1017
of Urochordata, 1037f, 1037–1038
Asilidae, 858
Aslia lefevrei, 992f
Aspidochirotida, 974–975
Aspidodiadematoida, 974
Aspidogastrea, 378
Aspidogastrida, 378
Aspidosiphon
cristatus, 573f
parvulus, 579
Aspidosiphonidae, 575
Asplanchna, 621, 622f, 624
Assimineidae, 466
Astacidea, 765f, 766, 782, 783
Astartidae, 470
Asterias, 986f, 994f
rubens, 994
Asterina gibbosa, 996
Asteroidea, 968n, 970f–971f, 973, 976, 978f–979f, 980
circulation and gas exchange of, 993, 993f, 994f
excretion and osmoregulation of, 994
external anatomy of, 972f
feeding and digestion of, 984–986, 986f
nervous system and sense organs of, 995
phylogeny of, 1001, 1002f, 1003
reproduction and development of, 996, 996f, 998f, 999f
support and locomotion of, 982
water vascular system of, 976f, 980–981
asteroid impact, extinction event associated with, 16
Asteronyx, 972f
Asterozoa, 969, 973
Astigmata, 914
Astrodictyum panamense, 987f
Astropecten, 983, 985
polyacanthus, 970f–971f
Astrophyton, 988
astropyle of Radiolaria, **110,** 111
Astrorhiza, 107
Astrotoma, 987
asulcal side, **279**
asymmetrical body plan, **136,** 136f
Asymmetron, 1023
Atelocerata, 752, 754, 908, 1051
Athalamida, 105
Athecata, 273–274, 302f
athecate dinoflagellates, **79**
athecate hydroids, **280**
Athous haemorrhoidalis, 867
atokous annelids, **567**
Atolla, 307f, 310
atractophore of Parabasalida, **113**
Atrax robustus, 938
atrial siphon of Urochordata, **1029,** 1031, 1033, 1034
Atrina, 488f
atriopore of Cephalochordata, **1023,** 1026
atrium
of Cephalochordata, **1023**
of Entoprocta, **604,** 605f, 606, 607
of Platyhelminthes, **395**
of Porifera, **225**
Atubaria, 1009
Atya, 824f
Atypidae, 912, 937f, 938
Atypus, 938
auditory receptors, 171, 171f, 881, 881f
Augier, A., 4
Aulacoseira italica, 100f
Aulactinia incubans, 309
Aurelia, 292, 293f, 297, 307f, 310, 311f, 312f
aurita, 287, 293f
auricles
of Ctenophora, **334,** 335
of Platyhelminthes, **392,** 393f
auricular groove of Ctenophora, **334, 335**
Austeruseus faeroensis, 719f
Australeremus cooki, 811f
Austrochilidae, 912
Austrodecus palauense, 957
Austrognatharia, 616
kirsteueri, 615f
autapomorphy, **45**
Autobranchia, 455, 469, 497
circulation and gas exchange of, 506
nervous system of, 509f
auto-evisceration of Holothuroidea, 991
autogamy
of Parabasilida, **114**
of Protista, **69**
autogenesis, 125
Autolamellibranchiata, 469
autosperm of Nemertodermatida, **364,** 364f, 365
autozooids, **283,** 284f, **648,** 650, 651
avicularia of Bryozoa, **648,** 649f
Ax, P., 376, 407, 407f, 615
axial cells
of Dicyemida, **414,** 415f, 417f
of Heterocyemida, 417, 418, 418f
axial gland of Echinodermata, **991**
axial region of Echinodermata, **975,** 977f, 977–978
axial rods of Cnidaria, 285f, **288**
axial sinus of Echinodermata, **981**
Axiidea, 766, 783, 810
Axinellida, 222
Axiothella rubrocincta, 77f, 82f, 548f–549f
Axius vivesi, 762f–763f
axoblasts
of Dicyemida, **415,** 415f, 416f
of Heterocyemida, 417, 418f
axocoel of Echinodermata, 998, **999**
axoplast of Radiolaria, **109**
axopods, **65**
of Radiolaria, 109, 110f, 111
axostyle of Parabasalida, **113**
Aysheaia, 15f
pedunculata, 718, 721f
prolata, 718

B

Bacteria, 3, 4, **7,** 8, 64
bacteria, symbiotic, 23, 28
of Annelida, 554, 554f, 590

of Mollusca, 499
of Nematoda, 677, 685f, 686
of Nemertodermatida, 361, 361f, 362
of Porifera, 217, 249, 251–252
of Tardigrada, 717
of Urochordata, 1033, 1034
Baghdad boil, 121
Baker, A. N., 969
balancers of Ctenophora, **338,** 339f
Balanoglossus, 1013, 1015
clavigerus, 1010f
misakiensis, 1016f–1017f
Balanophyllia elegans, 295f, 309
Balantidium coli, 87, 93
Balanus galeatus, 825
Bamber, R. N., 958
Bangia, 129
Bangiomorpha pubescens, 129
Bankia, 490
Bankivia, 496
Barentsia, 604f–605f, 606f–607f
Barentsiidae, 606
Barnea, 490
barophiles, **9**
baroreceptors, 171
basal foot of Xenoturbellida, **367,** 367f
basal plate, **288**
basement membrane
of Cnidaria, **274**–275
of Platyhelminthes, 382
Baseodiscus, 437f
punnetti, 435
Basidiomycota, 129
Basipodella, 793f
atlantica, 793f
basopinacocytes of Porifera, **230**
Batesian mimicry, 896
Bateson, W., 1007, 1018
Bathybelidae, 423
Bathynella, 775f
Bathynellacea, 765, 776
Bathynomus, 787
giganteus, 783
Bathyteuthidae, 472
Batillipes, 715f
noerrevangi, 716f
bauplan concept, 135
Bayesian methods, **51**
Bdelloidea, 614f, 617f, 618, 620f, 621, 623, 624
Bdellonemertea, 436, 450, 451
Bdelloura, 383, 387
candida, 383f
Beklem, W. N., 832
Belostomatidae, 854
benthic boundary layer, **18**
benthic organisms, **18,** 19
Beorn leggi, 711
Berghia, 500
Beröe, 328f–329f, 330f–331f, 332, 335
ovata, 328f–329f, 337
Beroida, 328, 328f–329f, 330f–331f, 332, 334
phylogeny of, 341
reproduction and development of, 340
Bicosta, 125
Biemnida, 222
Bigelowiella natans, 103
bilateral symmetry, **138,** 138f, 345
of Bilateria, 345–359
nervous system in, 175
origin of, 206
Bilateria, 12, 216, 345–359
Deuterostomia. *See* Deuterostomia
nervous system of, 305
origin of, 10, 345, 346, 967
phylogeny of, 25, 1049f, 1050
Protostomia. *See* Protostomia
bilaterogastrea theory, 206f
Biliphyta, 4n, 55n, 129
Billingsella, 15f
binary fission, 176f
of Protista, **68,** 75–76, 76f, 81f, 83, 93–94, 94f, 95
binomen, **37**
binomial nomenclature, **37**
binominal nomenclature, **37**
biodiversity, **24**–25
humped-back model of, 24
latitudinal gradient in, **24**
bioerosion from Porifera, 252–254
biogenetic law, **183,** 184, **202**
biological classification, **4,** 26, **36,** 44–51
fully ranked, 49–50
nomenclature in, 36–39
prediction in, 36
similarity and relatedness in, 36
tree diagrams in, 36
unranked, **49,** 50
biological nomenclature, **36**–39
biological species definition, **38**
bioluminescence, **174**
of Chaetognatha, 422, 425
of Cnidaria, 308
of Crustacea, 825, 826
of Ctenophora, 328
of Dinoflagellata, 78
of Hexapoda, 873, 882
of Mollusca, 513–514
of Myriapoda, 896f, 897
of Urochordata, 1031
Biomphalaria glabrata, 255n
biotoxins. *See* toxins
Bipalium, 374f–375f, 387
kewense, 374f–375f
bipectinate condition of Mollusca, **503,** 504
biradial symmetry, **137,** 137f
biramous limbs of Arthropoda, **733,** 734f
Birgus, 820
latro, 761, 762f–763f
biting–chewing insects, 870, 871f
Bittacidae, 857
Bivalvia, 453, 454, 455, 456f–457f, 467–470, 476f, 478f
body plan of, 473–474
circulation and gas exchange of, 506
excretion and osmoregulation of, 507
feeding and digestion of, 496, 497, 497f, 498f, 499f, 502, 502f
general anatomy of, 468f–469f
locomotion of, 487, 488f, 490
mantle and mantle cavity of, 476
nervous system and sense organs of, 508, 509f, 512–513
phylogeny of, 522f, 523
reproduction and development of, 517, 519, 520f
shell of, 477, 479–481, 480f, 490
symbiotic relationships of, 499–501
Bivalvulida, 274
bivium of Echinodermata, **978**
Black Sea, 18, 18n
Ctenophora in, 336–337
salinity of, 995
blackwater fever, 84
Blaniulus guttulatus, 901
Blaps mortisaga, 85f
blastea in colonial theory, 205, 205f, **318**
blastocoel, **193**
blastocoelom, **140,** 140f, **195**
blastocoelomate body plan, 140, 140f
circulatory system in, 163, 163f
Blastocystida, 100
Blastocystis, 100
blastocystosis, 100
blastoderm of Hexapoda, 884, 885f
blastomeres, **187**
subequal, 188
blastopore, **193,** 194f, 347, 967–968
blastostyle of Cnidaria, 277f, **280**
Blastozoa, 1001, 1002f
blastula, **192**–193, 193f
of Hexapoda, 884, 885f
Blattodea, 48f, 846, 847, 850–851, 851f, 861f, 889f
Blaxter, M. L., 672
bleaching events
coral, 252, 253, **304**
sponge, 252
Bleidorn, C., 541
blind gut (incomplete gut), **149**
blood pressure and flow velocity, 164
bmp and *BMP,* 192f, 1042
Boccardia proboscidea, 533f
Bochusacea, 785
Bock, S., 366
Bodo caudatus, 121f
body plans, 135–180
asymmetry in, 136, 136f
circulation and gas exchange in, 162–168
excretion and osmoregulation in, 157–162
feeding and digestion in, 148–157
nervous system and sense organs in, 168–175
size of body in, 139
support and locomotion in, 140–148
symmetry in, 136–138
body size, 139
and surface-to-volume dilemma, 139
Boedaspis ensifer, 753f
Bolbosoma, 625f
Bolitopsis infundibulum, 335
Boloceroidae, 308
Boloceroides, 292
Bombus, 845n
Bombyliidae, 858
Bombyx, 743
mori, 888f
Bonamia, 111, 112
Bonellia, 414, 580, 584
viridis, 581f, 583f, 584
bone morphogenetic protein, 192f, 1042
book gills of Chelicerata, **928**
book lungs, **165, 744,** 745f
Boophilus annulatus, 918
Boothby, T., 28
Bopyridae, 787
Boreidae, 857
boring gland of Mollusca, **493**
Borlase, W., 436
bothria of Platyhelminthes, **386,** 386f
Bothriocephalidea, 379, 386
Bothrioplana, 393f

Bothrioplanata, 378
Bothrioplanida, 378, 408f, 409
Bothriuroidea, 914
Botrylloides, 1037
Botryllus, 177f, 1029, 1030f, 1032f–1033f, 1037, 1040
  schlosseri, 1030f, 1037
  violaceous, 1039f
Botryocrinus, 972f
Bouganvilliidae, 274
Bowen, I. D., 389
Bowerbankia, 647f, 648, 651, 655, 656
Bowmanella braziliensis, 784f
Boxshall, G., 793
Brachininae, 883
Brachinus, 859f
brachioles of Echinodermata, **1001**
Brachionus
  calyciflorus, 620f
  plicatilis, 616
Brachiopoda, 348, 543, 635, 636, 657–665, 967
  body plan of, 637–638
  body wall and support of, 660–661
  characteristics of, 657
  circulation, gas exchange, and excretion of, 662–663
  classification of, 659–660
  coelom of, 661, 663
  feeding and digestion of, 661–662, 662f, 663f
  lophophore of, 658f, 659, 661–662, 663f, 665
  marine habitat of, 17
  nervous system and sense organs of, 662f, 663
  number of species, 2t
  phylogeny of, 25, 1049f, 1050
  reproduction and development of, 663–665
  taxonomic history of, 637
Brachyura, 762f–763f, 766, 778, 779–781, 780f–781f, 831
Brachyury (bra) gene, 185, 192, 192f
  in Acoela, 351
  transcription products, 185, 1042
Braconidae, 855
Branchellion, 592
  parkeri, 592f
branchial chamber
  of Cephalochordata, **1023**
  of Urochordata, 1029, 1043
branchial pump of Cephalochordata, **1023**
branchial siphon of Urochordata, **1029**, 1031, 1033, 1034
branchial theory on insect flight, 869
Branchinecta, 770f–771f
  gigas, 772
  raptor, 772
Branchiobdellida, 541, 591, 592, 592f, 593
Branchiobdellidae, 541
Branchiocaris, 15f
Branchiopoda, 734, 735, 762f–763f, 765, 769–772, 801, 803
  distinguishing features of, 802t–803t
  excretion and osmoregulation of, 821
  gene duplication in, 29
  phylogeny of, 832, 833, 833f
  reproduction and development of, 806t–807t, 831
  thoracic appendages of, 804f–805f
Branchiostoma, 675, 1023, 1024f
  floridae, 1042
  lanceolatum, 1025
Branchipus schaefferi, 770f–771f
Branchiura, 558, 767, 790, 793–796, 794f
  distinguishing features of, 802t–803t
  nervous system and sense organs of, 825
  phylogeny of, 833f
  reproductive features of, 806t–807t
Brechites, 456f–457f
Bredocaris, 834f–835f
Brentidae, 856
brevetoxins, **79**
Briareum, 309
Brinckmannia hexactinellidophila, 301
Brisingida, 973
broadcast spawning, **178, 198**
brooding of Echinodermata, 996, 997, 997f
broom organ in Nemertodermatida, 361, 361f, **364**
brown algae, 56, 57, 97, 98f, 99, 101
brown bodies of Bryozoa, **653**
Brownell, P., 935
Brumeister, C., 454
Bryophyta, 4n, 55n, 129
Bryozoa, 348, 603, 605, 635, 636, 644–657
  body plan of, 637–638, 646–649
  body wall, muscles, and movement of, 649–650
  characteristics of, 644
  circulation, gas exchange, and excretion of, 652–653
  classification of, 646
  coelom of, 649, 650, 651, 652, 655
  colony forms of, 648–649
  feeding and digestion of, 651–652, 652f
  nervous system and sense organs of, 653–654
  number of species, 2t
  phylogeny of, 25, 1049f
  reproduction and development of, 654f, 654–657
  taxonomic history of, 636–637
  tentacle crown of, **646**, 648, 649, 650, 651
  zooid interconnections, 650–651
buccal cavity of Ciliata, **92**
buccal cirri of Cephalochordata, **1025**, 1043
Buccinidae, 466
Buccinum, 493f, 504f
Buchnera, 127
  amphidicola, 126–127
Buddenbrockia, 276, 318
  pluma, 269
budding, 176, 176f, 177
  of Bryozoa, 656–657
  palleal, **1037**
  of Porifera, 239, 240f
  of Protista, 68, 94, 94f
  of Pterobranchia, 1017, 1018f
  of Urochordata, 1037, 1037f
  vascular, **1037**
Bugula, 649f, 655, 656
  neritina, 651, 655f
Bulimulidae, 467
Bullidae, 467
Buprestidae, 855
Burgessia, 15f
Burgessochaeta, 597
Burgess Shale deposits, 13, 14f, 15f
  Arthropoda in, 751, 835
  Cephalochordata in, 1043
  Hemichordata in, 1018
  Onychophora in, 718, 721f
Burns, J., 26
bursae of Echinodermata, **993**
bursal nozzle of Acoela, 359f, **359**–360
Bursaphelenchus xylophilus, 683
Bursovaginoidea, 616
Buss, L., 215
Buthoidea, 914
Buthus, 936, 950f
  martensi, 927f
Bütschli, O., 60, 205
byssal threads of Mollusca, **487**–489, 488f, 489f
byssus gland of Mollusca, **489**
Bythotrephes, 773

C

cadherins, 255
Caecosagitta macrocephala, 422
Caenogastropoda, 455, 463–465, 464f, 478f
  circulation and gas exchange of, 504, 505
  excretion and osmoregulation of, 507
  feeding of, 491, 493, 493f, 494f
  locomotion of, 487, 490
  nervous system and sense organs of, 509f, 511, 511f
  reproduction and development of, 515, 516f, 517f, 520f, 521
Caenorhabditis elegans, 671, 825
Caerostris, 922f–923f
Calabozoidea, 766
Calamyzinae, 537
Calanoida, 767, 797
Calanus, 744f
Calappidae, 780f–781f
Calcarea, 218f, 220, 221, 222, 224, 225f, 255
  evolution and phylogeny of, 255, 256, 257f, 258
  reproduction and development of, 242, 244, 246f, 246–248
Calcaronea, 221, 243, 256f
  development of, 244, 246, 249
  phylogeny of, 258
Calcaxonia, 271
calciblastula larva, 244, 245f, 246f
calciferous glands of Annelida, **555**
Calcifibrospongia, 222
Calcinea, 221, 256f
  development of, 244, 246, 246f
  phylogeny of, 258
Calcispongia, 257
calcite, 232, 234
Caldwell, O., 637
Caligus, 794f–795f
Calkinsia, 119
Calliactis, 296, 302
Callianassa, 782f, 783, 810
Callinectes, 741
  sapidus, 828f, 831
Calliostoma, 486f
Calliostomatidae, 463
Callipallene brevirostris, 958
Calliphoridae, 858, 873
Callipodida, 897
Callistochiton viviparous, 514

Calloria inconspicua, 662f, 663f
Callyspongia vaginalis, 988
Callyspongiidae, 234f–235f
Calman, W. T., 774
calotte
of Dicyemida, **414**
of Nematomorpha, **689**
Calycina, 974
Calycinoida, 974
Calycophora, 274, 282
Calycophorae, 316
calymma of Radiolaria, **110**
Calyphora, 282
Calyptoblastea, 273
Calyptogena magnifica, 126
calyptopis, **831**
Calyptraeidae, 465, 496
calyx
of Echinodermata, **982**
of Entoprocta, **604,** 605f, 606
Camallanus, 679f
Cambrian explosion, 10, 13–15, 141
Cambropachycope clarksoni, 834f–835f
cameral fluid of Mollusca, **490**–491
campaniform sensilla of Arthropoda, **747**
Campanilidae, 465
Camptostroma, 1000f, 1001, 1002f
Canadaspis, 14f, 15f
Canadia, 15f, 597
Cancellaria cooperi, 494
Cancellariidae, 466
Cancer, 780f
gracilis, 781
magister, 441, 826f–827f
Cancridae, 780f
Candona, 800f
cannibalism, 155, 157, 521, 952
Cannon, H. G., 809
Cantharidae, 855
Capilloventer, 540
Capilloventridae, 540
Capitata, 273, 282, 283f
capitate tentacles of Cnidaria, **279**
Capitella, 534f–535f, 540
Capitellida, 540
Capitellidae, 534f–535f, 540, 557
capitulum of Chelicerata, **917**
Caprella, 788f–789f
Caprellidea, 767, 790
Capsalidea, 378
Capsaspora owczaraki, 255n
captacula of Mollusca, **499,** 512
Capulidae, 465
Carabidae, 855, 859f, 866
carapace of Crustacea, **801,** 802t, 819
carbon, in dissolved organic matter, 157
carbon dioxide
atmospheric, 7, 10, 15, 16
in body fluids, 168
in marine habitat, 17–18
as waste product of respiration, 164
carcinoecium, **302**
Carcinonemertes, 441
errans, 441
Carcinonemertidae, 441
Carcinoscorpius, 916–917
rotundicauda, 916–917
Carcinus, 792f–793f
maenas, 112, 791n
cardiac stomach of Echinodermata, **985**
cardia of Bryozoa, **651,** 652
Cardiida, 456f–457f, 470
Cardiidae, 470, 472, 489
cardinal veins of Cephalochordata, **1026**
Carditida, 470
Carditidae, 470, 519
Caridea, 762f–763f, 766, 779, 779f
feeding of, 811f
reproduction and development of, 831
thoracic appendages of, 804f–805f
caridoid escape reaction, 805f, **807**
Carinacea, 974
Carinaria, 464f
Carinariidae, 465–466
Carinina, 445f, 448f
Carinoma, 443
tremaphoros, 449
carnivores, 149, 154–157
Cnidaria as, 296, 297–298
Nemertea as, 443
Platyhelminthes as, 386–387
Porifera as, 217, 236, 237f, 238f
carotenoids, 57n
carrier cells of Porifera, **243**
Carukiidae, 272
caruncle of Annelida, **565**
Carybdea, 307f
branchi, 293
marsupialis, 267f
rastonii, 296
Carybdeida, 272
Carybdeidae, 272
Caryophyllidea, 379
Cassididae, 466
Cassiduloida, 974
Cassiopea, 303, 306
Catalogue of Life (CoL) project, 6
catch collagen, 980
categories, **38**
ß catenin, 255
Catenula lemnae, 374f–375f
Catenulida, 376, 382, 387, 392
phylogeny of, 407, 408, 408f
reproduction of, 396f
Catenulidea, 376, 384, 407
Cateria, 696
caudal organ of Gastrotricha, **431**
caudal rami of Crustacea, **801**
Caudofoveata, 453, 454, 455, 458, 458f–459f, 459
body plan of, 472–473
circulation and gas exchange of, 504, 506
locomotion of, 486
mantle and mantle cavity of, 474
phylogeny of, 522f
reproduction and development of, 514, 519
shell of, 477
Caullery, M., 588
Cavalier-Smith, T., 56, 128
Cavanaugh, C., 590
Cavoliniidae, 467
cecal portal vein of Cephalochordata, **1025**
cecum
of Arthropoda, **743**
of Cephalochordata, **1025**
of Platyhelminthes, **390**
cell fates, 191–192
cell signaling pathways, **184,** 191
cells of insect wings, **865**
Cellularia, 228, 257
cellulose, 65, 65n
Celtia, 800f
cement glands of Platyhelminthes, **395**
Cenozoic era, 16–17
centric diatoms, **100**
Centroderes, 696f
centrolecithal ova, **187,** 187f, 750, 750f
Centropyxis, 74
Centruroides exilicauda, 936
Cephalaspidea, 467
Cephalaspidomorphia, 1022, 1042
cephalic probolae of Nematoda, **677**
cephalic shield
of Crustacea, **801,** 802t
of Pterobranchia, **1015**
cephalic slits of Nemertea, **447,** 450
cephalization, **138, 175, 345**
Cephallobellus, 680f
Cephalobaena tetrapoda, 796f
Cephalobina, 672
Cephalobothrium, 386
Cephalocarida, 734, 754, 765, 768–769, 769f, 801, 803
distinguishing features of, 802t–803t
nervous system and sense organs of, 825
phylogeny of, 832, 833f, 834
reproduction and development of, 806t–807t, 831
thoracic appendages of, 804f–805f
Cephalochordata, 349, 968, 1021, 1022, 1023–1027
body wall, support, and locomotion of, 1023
circulation, gas exchange, and excretion of, 1025–1026
feeding and digestion of, 1023–1025
general anatomy of, 1024f
nervous system and sense organs of, 1026
number of species, 3t
phylogeny of, 25, 1041, 1041f, 1042, 1043, 1049f, 1050
reproduction and development of, 1026f, 1026–1027
Cephalodiscidae, 1009
Cephalodiscus, 1007, 1008, 1009, 1009f, 1010f–1011f, 1015
circulation and gas exchange of, 1017
feeding and digestion of, 1015, 1016, 1017
internal anatomy of, 1017f
reproduction and development of, 1017, 1018
cephalon of Crustacea, **798**
Cephalopoda, 453, 455, 456f–457f, 471–472, 476f
body plan of, 474
circulation and gas exchange of, 503, 503f, 506
coloration and ink of, 513–514
excretion and osmoregulation of, 507
feeding and digestion of, 491, 495–496, 502
locomotion of, 490, 491f
mantle and mantle cavity of, 477
nervous system and sense organs of, 508, 510f, 513
phylogeny of, 522f, 523–524
reproduction and development of, 517f, 517–518, 521, 521f
shell of, 481, 481f, 482
cephalothorax of Crustacea, **801**

Cephalothrix, 444f
Ceractinomorpha, 220, 221
Cerambycidae, 856, 888f
Cerastidae, 467
cerata of Mollusca, **495, 505**
Ceratium, 77f, 79, 80, 81f
hirudinella, 58f–59f
cercaria of Platyhelminthes, **401**
Cercopidae, 854
Cercopoidea, 859f
Cercozoa, 56, 63, 108, 112, 128
cerebral ganglion, **175, 746,** 755
of Annelida, 562f–563f, 563, 578, 578f, 595
of Euchelicerata, 943
of Hexapoda, 879f
of Mollusca, 507–508
of Onychophora, 725
of Pycnogonida, 959–960
of Tardigrada, 717
cerebral organs
of Nemertea, **447**
of Sipuncula, 579
Cerebratulus, 437f, 440, 444, 449
lacteus, 449
leucopsis, 437f
Ceriantharia, 270, 284f
Cerionidae, 467
Cerithideoposis californica, 401, 404f
Cerithiidae, 465
Cerithiomorpha, 455, 465
Cerithiopsidae, 466
Cervidellus, 676f–677f
Cestida, 328f–329f, 331f, 332, 334, 335, 341
Cestoda, 373, 375f, 376, 378–380, 382, 382f, 384
excretion and osmoregulation of, 391–392
feeding and digestion of, 390
nervous system of, 394, 394f
phylogeny of, 407, 408f, 409
reproduction of, 403–405
Cestum, 331f, 332
veneris, 328f–329f
Chactoidea, 914
Chaeriloidea, 914
chaetae of Annelida, 542, **543,** 543f, 544f, 545, 547, 552, 592
Chaetoceros, 97
Chaetoderma
loveni, 458f–459f
productum, 458f–459f
Chaetodermomorpha, 453, 458
Chaetognatha, 347–348, 413, 420f, 420–428, 421f, 636
body wall, support, and movement of, 423–424
characteristics of, 422
circulation, gas exchange, and excretion of, 425
feeding and digestion of, 424–425
marine habitat of, 17
nervous system and sense organs of, 425–426
number of species, 2t
phylogeny of, 25, 1049f, 1050
reproduction and development of, 426–428
taxonomic history, classification, and phylogenic relationships of, 422–423
Chaetonotida, 428f, 429, 430, 430f, 431, 432
Chaetonotus, 428f
Chaetopteridae, 532–533, 533f, 542, 552, 552f–553f
Chaetopterus, 533, 552, 552f–553f, 559
feeding and digestion of, 151, 152f, 583
reproduction and development of, 565
Chaetostephanidae, 699
Chagas disease, 122–123, 846, 854, 873
Chalcidoidea, 855
Chalinula, 241
Challengeron wyvillei, 109f
Chaos, 74
diffluens, 58f–59f
chaotic cleavage of Porifera, **244**
Chara, 129
characters, **41**–43
convergent, 42
parallel, 42
character states, **42,** 42n, **44,** 45
polarity analysis, 45
Charales, 129
Charnia, 12f
Charniodiscus, 12f
Charonia, 984
Charophyta, 4n, 55, 55n, 129
Chasmataspidida, 912
Cheilostomata, 645f, 646, 647f, 648, 648f, 651
body wall, muscles, and movement of, 650
reproduction and development of, 655
chelicerae of Chelicerata, **928**
Chelicerata, 710f–711f, 711t, 725, 728, 730, 735t, 908, 911–962
characteristics of, 912
Euchelicerata, 911, 912–955
Hox genes in, 30
number of species, 2t, 711t
phylogeny of, 48f, 752, 755, 755f, 961, 962f
Pycnogonida, 911, 912, 955–961
cheliceres, **958**
Chelifer cancroides, 919f–920f, 925
chelifores of Pycnogonida, **958,** 959, 960
Cheliplana, 374f–375f, 388f
Chelonibia, 791
chemoautotrophy, **157**
chemolithotrophs, 9n
chemoreceptors, 171–172
of Annelida, 565
of Arthropoda, 746–747, 822
of carnivores, 155
of Echinodermata, 995
of Enteropneusta, 1014
of Euchelicerata, 944, 945, 945f
of Mollusca, 511
of Nematoda, 680–681
of Nemertea, 447
of Platyhelminthes, 392, 393f, 394
of Protista, 67
chemotherapy drugs, Porifera as source of, 250
Chen, J.-Y., 3
Chengjiang deposits, 13
Arthropoda in, 751, 835
Deuterostomia in, 349, 967
Onychophora in, 718, 721f
Tardigrada in, 711
Chia, F., 484
Chibchea mapuche, 954f–955f
chiclero ulcer, 122
Chicxulub Crater, 16
chiggers, 918
Chilinidae, 467
Chilognatha, 897, 898
Chilomonas, 103
Chilopoda, 895–909
circulation and gas exchange of, 903, 904f
classification of, 897
excretion and gas exchange of, 904
feeding and digestion of, 901
locomotion of, 901, 902f
nervous system and sense organs of, 904–905
phylogeny of, 908, 909
reproduction and development of, 905–906, 906f, 908
Chimaericolidea, 378
Chirodropida, 272
Chironex, 295–296
fleckeri, 296
Chironomidae, 858
Chiroteuthidae, 472
chitin, 11n, 146n, 147f
chitin synthase gene, 254
Chitonida, 460–462
Chitonidae, 513
Chlamydomonas pulsatilla, 160
Chlamydophorus, 74f
chloragogen cells of Annelida, **555**
chloragogenous tissue of Annelida, **555**
Chlorarachnion reptans, 104f, 105
Chlorarachniophyta, 59, 59f, 103, 104f, 104–105
characteristics of, 104
classification of, 61, 63
phylogeny of, 128
Chlorella, 78, 93, 215
chlorocruorin, **168,** 168t, 559
Chlorohydra, 302
chlorophyll, 56, 57n, 66f
Chlorophyta, 55, 55n, 66f, 104, 128, 129, 302
chloroplasts, 56, 57, 57n, 66f, 100
chloroquine, 84
choanocyte chambers, **226, 227,** 229f
choanocytes, **222, 223,** 227, 228f–229f, 234–236, 239, 254
development of, 223, 247, 247f, 248
functions of, 217, 222, 223, 230
migratory, 243
in reproduction, 239, 242, 243, 243f, 248
in syconoid condition, 226
choanoderm, **223,** 227
Choanoflagellata, 3, 60, 61, 64, 124f, 124–125, 206, 254, 255
mitochondria of, 68f
phylogeny of, 128–129, 1048, 1049f
structural organization of, 184
choanosome of Hexactinellida, **227**
Choanozoa, 56
Chondrichthyes, 1022, 1041f
Chondrocladia lyra, 236, 238f
Chondrosia, 233
Chonopeltis, 796
Chonotrichida, 93
Chordata, 9, 347, 348, 349, 967, 968, 1018, 1021–1043
archicoelomate condition of, 635
brachyury gene expression in, 185
Cephalochordata, 1023–1027
characteristics of, 1022
number of species, 3t, 1047

photoreceptors of, 172
phylogeny of, 25, 1041–1043, 1049f, 1050
Urochordata, 1027–1041
Chordeuma, 907
Chordeumatida, 897
chordin (chrd) gene, 192f
chordoid larva of Cycliophora, **611,** 611f
Chordotes, 688f
chordotonal organs of Hexapoda, **881**
chrd (chordin) gene, 192f
Chromadorea, 670f, 672, 675f–679f, 681, 683f, 685f
Chromadoria, 672, 678f–679f
Chromadorida, 672
Chromalveolata, 29, 60, 61–62, 76–104
Apicomplexa, 81–87
Ciliata, 87–96
Dinoflagellata, 76–80
Haptophyta, 101–103
overview of, 62–63
phylogeny of, 127–128
Stramenopila, 97–101
chromatophores of Cephalopoda, 513
Chromera, 63
Chromista, 56, 128
Chromodorididae, 467
Chromodoris geminus, 464f–465f
chromophtye algae, **56**
Chronogaster, 675f
Chrysaora, 297, 310
Chrysomelidae, 856
Chrysopetalidae, 534f–535f, 537
Chrysopetalum, 534f–535f
Chrysopidae, 856
chrysoplasts, **100**
Chydoridae, 773
Chytridiomycetes, 129
cibarium of Hexapoda, **872**
Cicadellidae, 854
Cicadidae, 854
Cidaris, 988f–989f
Cidaroidea, 974
cilia, 89, 142–143, 143f
xenacoelomorphan, **350**
ciliary eyes, **172**
ciliary fence receptors of Chaetognatha, **423,** 425
ciliary-mucous suspension feeding, 151, 152f
Ciliata, 58f–59f, 60, 61, 62, 87–96
characteristics of, 87
nutrition of, 91f, 91–93, 92f
phylogeny of, 127
reproduction of, 93–96, 94f, 95f, 96f
support and locomotion of, 87–90, 90f
ciliate–acoel hypothesis, 346, 407
ciliated furrows or grooves of Ctenophora, **338,** 339f
Ciliophora, 56, 61
ciliophores of Micrognathozoa, **628,** 628f
Cimex
hemipterus, 873
lectularius, 873
Cimicidae, 854, 873, 883
cinclides, **296**
cinctoblastula larva, 244, 245f, 246f, **249**
cingulum of Rotifera, 619, 619f
Ciona, 1029, 1032f, 1040
intestinalis, 1028f–1029f, 1037, 1038f–1039f, 1040, 1042
Circinalium, 1037f
circulatory system, 139, 162–164, 163f
closed, 163f, **163**–164
open, 163, 163f, **164**
pumping mechanisms in, 164
circumoral ring of Urochordata, **1036**
Cirolanidae, 47, 47f, 787
Cirrata, 471
Cirratulidae, 535f, 538, 547, 565, 566f
Cirratuliformia, 538, 588, 590
cirri
of Cephalochordata, 1025
of Crustacea, **810**
of Mollusca, 497
of Tardigrada, **717**
Cirriformia, 535f
Cirripedia, 728, 763f, 767, 791, 793, 819, 825
Cirroteuthidae, 471
cirrus sac of Platyhelminthes, **399,** 403
cirumcapitular furrow of Chelicerata, **917**
clades, **4, 43,** 43f, 44
cladistics, 45
Cladocera, 765, 819
Cladoceromorpha, 765, 773
cladogenesis, 27, 31, 31f, **45,** 46
cladograms, 30, 46, 46f, 47, 49f, 50
Cladorhiza methanophila, 236
Cladorhizidae, 217, 236, 236f–237f, 238f
Clark, R. B., 206, 207, 209, 604
classification, biological. *See* biological classification
Clathria, 227f
prolifera, 224f, 231
Clathrina, 224, 225
clathrus, 218f
contorta, 246f
claustrum of Cnidaria, **279**
clavae of Tardigrada, 714f, **717**
Clavagellidae, 470
Clavelina, 1029, 1031, 1037
lepadiformis, 1028f–1029f
Clavella adunca, 794f–795f
clavoscalids of Loricifera, **703**
claw glands of Tardigrada, 715f, **716**
cleavage patterns, **187**–191, 750, 750f
of Acoela, 360
of Annelida, 572, 579, 584, 589f, 591
cell fates in, 191
of Chaetognatha, 427
of Cnidaria, 309, 310, 316, 319
coding system for, 189–191, 190f
of Crustacea, 828f, 829–830
of Ctenophora, 340
determinate, **191**
of Entoprocta, 607
equal and unequal cell divisions in, 188, 188f
of Euchelicerata, 948
of Gastrotricha, **432**
of Hexapoda, 884
holoblastic, **187,** 188f, 192–193, 828f, 829, 830
indeterminate, **191**
longitudinal, **188,** 188f
meroblastic, **187,** 188f, 830
of Mollusca, 518
of Myriapoda, 908
of Nematoda, 681
of Nemertea, 448
in Nemertodermatida, 365f, 365–366
of Onychophora, 726, 727f
of Phoronida, 644
of Platyhelminthes, 397–398
of Porifera, 243, 244, 245f, 246, 248, 249
radial, **188,** 189f, 191
spiral. *See* spiral cleavage
spiral duet, **360**
subequal blastomeres in, 188
transverse, **188,** 188f
Cleidothaeridae, 470
Clément, P., 622
ClEvx gene in Acoela, 351
Clibinarius, 302
climate change, global, 7, 10, 15–17
Clinocardium, 488f
Clio, 464f
Cliona, 252, 253f
californiana, 252–253, 253f
delitrix, 252
vermifera, 242
Clionaidae, 252
Clionidae, 467
Clitellata, 532, 540–541, 543, 597, 598f
body forms of, 542
circulation and gas exchange of, 558, 558f
excretion and osmoregulation of, 560, 561, 561f
feeding of, 549
nervous system and sense organs of, 562f–563f, 563
reproduction and development of, 567, 568f, 569, 572, 596
support and locomotion of, 547
clitellum of Clitellata, **569**
cloaca of Urochordata, **1029,** 1031, 1034
Clonorchis sinensis, 381f, 401
closed circulatory system, 163f, **163**–164
Cloudina, 11
Clubionidae, 913
Clymenella, 548f–549f
Clypeaster, 137f
Clypeasteroida, 974
clypeolabrum of Hexapoda, **862**–863
Cnemidocarpa, 1028f–1029f
cnidae, **265, 293**–296, 294f, 295f, 296f, 298
Cnidae (knide), 268
Cnidaria, 213, 214, 215, 216, 255, 265–321, 346, 347
body wall of, 275–287
cell types of, 276f
characteristics of, 269
circulation, gas exchange, excretion, and osmoregulation of, 304
cnidae of, **293**–296
compared to Ctenophora, 327, 341
defense mechanisms of, 298, 299f, 302
in Ediacaran Period, 11, 12f
feeding and digestion of, 149, 296–298, 297f
germ layers of, 139
life cycle of, 299, 300f, 308, 310, 312f, 314f, 319
luminescence of, 308
medusoid form of, **265,** 283–287
movement of, 290–293, 291f, 293f
mutualism of, 301–302
nervous system and sense organs of, 172, 304–308, 305f, 306f, 307f
number of species, 2t
origins of, 318–319
phylogeny of, 25, 26, 317–321, 320f, 341, 1048, 1049f, 1050
polypoid form of, **265,** 276–283, 277f

radial symmetry of, 275, 275f, 277
reproduction and development of, 139, 265, 274, 276, 308–316
support of, 287–290, 289f, 290f
symbiotic relationships of, 268, 292, 298, 302–304, 321
taxonomic history and classification of, 268–274
tissue layer homologies in, 268f, 276
cnidoblasts, **293**
cnidocil, **293,** 294f, 305
cnidocytes, 276, **293,** 294f
cnidoglandular band, **296**
coastal wetlands habitat, 20–22
Cobb, N., 671
coccidiosis, 82
Coccinellidae, 855–856
Coccoidea, 854
coccoliths of Haptophyta, **102**
coccosphere of Haptophyta, **102**
Cocculinida, 463
Cocculinidae, 463
Cocculiniformes, 463
cocoons
of Clitellata, 540, 568f, 569, 572
of Euchelicerata, 953
of Hexapoda, **886,** 888f
of Hirudinoidea, 596f, 597
Codonophilus, 787f
Codonophora, 274, 282, 316
Codonosiga, 125
Codosigidae, 125
Coe, W., 436
Coelenterata, 220, 269, 328
coelenteron of Cnidaria, 278f, **279,** 296, 302, 304
coeloblastula, **193,** 193f, 194f, 246, 247f, 248, 249
coelogastrula, **193**
coelom, **140,** 140f, 195f, **195**–196, 196f
of Annelida, 543–545, 557, 573, 574, 575–577, 582, 592
of Brachiopoda, 661, 663
of Bryozoa, 649, 650, 651, 652, 655
of Echinodermata, 969, 978–980, 998–999, 999f
evolutionary origin of, 206–207
formation of, 195f, 195–196, 196f
coelomic channels, **575,** 593f, 595f
coelomoducts, **162,** 560
Coelopidae, 858
Coeloplana, 331f, 338f, 340, 341
Coelosphaera hatchi, 224f
coenecia of Pterobranchia, **1008, 1015**
coenenchyme, **266n**
Coenobita, 820
coenosarc of Cnidaria, 277f, **280**
coenosteum, **289**
Coenothecalia, 271
coevolution, 15n
Colacium, 118, 119
Coleochaetales, 129
Coleoidea, 455
Coleoidia, 471
Coleopetrida, 846
Coleoptera, 844, 846, 847, 855–856, 857f, 859f
development of, 888f
feeding of, 873
locomotion of, 867
phylogeny of, 48f, 889f, 890
Coleopterida, 847, 855–856
Coleps, 87
collagen, 230, 232–233, 249
collar
of Cnidaria, **291**
of Enteropneusta, **1009,** 1010f, 1011, 1012f–1013f, 1014, 1016f–1017f
of Porifera, **223,** 254
of Pterobranchia, 1015
collar ganglion of Pterobranchia, **1017**
collar neurochord of Enteropneusta, **1014**
collar receptor of Micrognathozoa, **629**
collar tentacles of Choanoflagellata, 124
Collembola, 752, 843, 845f, 847–848
circulation and gas exchange of, 877
evolution and phylogeny of, 48f, 888, 889f
locomotion of, 867
thorax of, 864
collenchyma, 266n
collenchyme, **266n**
collencytes of Porifera, **230**
Collin, R., 485
Collinia, 87
Collinsium ciliosum, 721f
Collisella scabra, 493
colloblasts of Ctenophora, **332,** 335, 335f, 336f
collum of Myriapoda, **899**
Colobognatha, 897, 898–899
colonial theory, **204**–205, 205f, **318**
colonies, **177,** 177f
lophopodid, **649**
plumatellid, **648**–649
of Urochordata, 1037–1038
Colony Collapse Disorder (CCD), 24
Colossendeidae, 956f
Colossendeis, 957, 958
australis, 957f
colossea, 959
scotti, 957f
Colpidium, 87, 93f
colulus of Euchelicerata, **930**
Columbellidae, 466, 493
columella of Cnidaria, **288,** 289f
Comatulida, 973
comb plates of Ctenophora, 332
comb rows of Ctenophora, **332,** 333f, 334–335, 338, 339f, 341
comet of Echinodermata, **995**
commensalism, **24**
communication of insects, 881–882
comparative biology, **35**
comparative developmental genetics, 184
compass of Echinodermata, **989**
compensation sacs of Sipuncula, **577**
competent larvae, **199**
complete gut (through gut), **149, 345**
of Arthropoda, 742
of Mollusca, 472
of Nemertea, 435, 442
of Platyhelminthes, 390
complex eyes, **172**
complex jaws of Gnathifera, 613, 614, 614f
in Gnatostomulida, 615, 616
in Micrognathozoa, 614f, 630, 631f
in Rotifera, 620
compound ciliature of Ciliata, **89**
compound cirri of Mollusca, **497**
compound eyes, **172,** 173f, **748,** 749, 749f
of Crustacea, 748n, 803t, 823–825
of Hexapoda, 879–880, 880f
Concentricycloidea, 969, 973
Conchifera, 522f
Conchoderma, 792f–793f
condyles of Arthropoda, **736,** 736f
Condylocardiidae, 473–474
cone cells, crystalline, **748,** 755
congery, **248**
Conidae, 466
conjugants, **94**–95, 96f
conjugation, **69,** 94, 95–96
Conochilus, 617
Conocyema, 417, 418f
polymorpha, 417
contact suspension feeding, 150
continental edge, **18**
continental shelf, **18**
continental slopes, **18**
contractile hypothesis on exocytosis, 294
contractile vacuoles, **160,** 160f
Conus, 156f, 456f, 494, 495f
convergent evolution, 41, **42,** 43, 43f, 185
Convoluta
convoluta, 356
henseni, 356f
longifissura, 358f
Convolutidae, 352f–353f, 353, 359f, 360
Convolutriloba, 357
longifissura, 354f, 355f
Copepoda, 764f, 767, 790, 793, 794f, 797–798
distinguishing features of, 802t–803t
nervous system and sense organs of, 825
phylogeny of, 833f
reproductive features of, 806t–807t
copepodite, **831**
coprophagy, **153**
Copula sivickisi, 311
copulatory bursa of Hexapoda, **883**
coracidium of Platyhelminthes, **405**
coral, 20, 24
bleaching of, 252, 253, **304**
coralla, **290**
Corallanidae, 47, 47f, 787f
Corallimorpharia, 270, 295f, 296
Coralliochama, 489f
Coralliophilidae, 466
corallite, **288**
Corallium, 288f
corallum, **288**
Corbulidae, 470
Corella, 1039f
inflata, 1039f
willmeriana, 1039f
cor frontale of Crustacea, **818**
Corinnidae, 922f–923f
Corixidae, 854
Corliss, J. O., 89
cormidia, 281f, **282, 311**
cornea, 172, **748**
corneagen cells of Arthropoda, **748,** 755
Cornirostridae, 466
Cornu, 154f, 500f–501f, 512f, 515f, 517
aspersum, 456f, 516f
Cornulariidae, 271
Corolla, 464f
corona, of Rotifera, **617,** 617f, **619,** 619f
corona ciliata of Chaetognatha, **423**
coronal muscles of Cnidaria, **292**
Coronatae, 272, 288f, 320f
coronate larvae of Bryozoa, **656**
Coronula, 791, 792f–793f

corpora allata of Hexapoda, **879**
corpora cardiaca of Hexapoda, **879**
Corticium candelabrum, 233
Corymorpha, 282
Corynactis californica, 295f, 306f
Corynida, 273–274
Corynosoma, 624, 625f
Coscinoderma mathewsi, 239
Coscinodiscus, 98f
Cosmognathia, 616
Cossidae, 859
Costellariidae, 466
Cotylaspis, 381f
courtship behavior
of Crustacea, 826
of Euchelicerata, 950–953, 951f
of Hexapoda, 883
of Mollusca, 517
of Tardigrada, 718
coxa, of Arthropoda, **733**
coxal glands of Euchelicerata, 942–943
Craniata, 25, 1022, 1043, 1049f, 1050
Craniiformea, 658f, 659, 660, 662, 665
Cranioidea, 660
Craspedacusta sowerbyi, 286f–287f, 298
craspedote, **285**
Crassatellidae, 470
Crassiclitellata, 540, 569
Crassostrea
gigas, 104f
virginica, 497
Cratenemertea, 450, 450f
Craterostigmomorpha, 897, 904, 906
Crenobia, 393f
Crepidula, 496, 516, 516f
fornicata, 517f
Cretaceous period, 16
cribellum of Euchelicerata, **930,** 944
Cribrilaria, 648f
cribrimorphs, **650,** 650f
Criconema, 675f
Crinoidea, 970f–971f, 971–973
circulation and gas exchange of, 991, 993
excretion and osmoregulation of, 994
external anatomy of, 972f
feeding and digestion of, 984
internal anatomy of, 985f
nervous system and sense organs of, 995
phylogeny of, 1001, 1002f
reproduction and development of, 995, 996, 997f, 998f
support and locomotion of, 982, 983f
water vascular system of, 981
Crinozoa, 971–973, 1001
Crisia, 645f, 647f
cristae of Protista, **66,** 68f
Cristatella, 645f, 647f, 649
Crithidia, 124f
crural gland of Onychophora, **726**
Crustacea, 6, 710f–711f, 729, 731, 749, 761–836
body plan of, 798–802
body wall of, 731
characteristics of, 711t, 735t, 764
circulation and gas exchange of, 816–821, 820f
classification of, 4, 764–798
distinguishing features of, 802t–803t
escape reaction of, 805f, 807
excretion and osmoregulation of, 821
feeding and digestion of, 150, 809–816, 811f
gene duplication in, 29
internal anatomy of, 817f–819f
locomotion of, 802–809, 805f
migratory behavior of, 808
morphological diversity of, 762f–764f, 764
nervous system and sense organs of, 821–825
number of species, 2t, 711t
paddle-legged, 832, 833f
in Paleozoic Era, 13
phylogeny of, 25, 48f, 751, 752, 754, 755, 831–836, 908, 1050
reproduction and development of, 806t–807t, 825–831, 826f–829f
thoracic appendages of, 804f–805f
cryptobiosis of Tardigrada, **713,** 713f
Cryptocellus, 925, 955f
Cryptocercus, 850
Cryptochiton stelleri, 166f–167f, 479, 480f
Cryptochlora perforans, 105
cryptochromes, **310**
cryptocyst, **650,** 650f
Cryptodonta, 469
Cryptolithodes, 782f
Cryptomonada, 62, 63, 103, 103f, 128
characteristics of, 103
chloroplast of, 66f, 101
nucleomorph of, 59, 59f, 103, 105
phylogeny of, 127–128
Cryptomonas, 103, 103f, 128, 215
Cryptops, 902f, 903f
Cryptosporidium, 81, 83
Cryptosula, 653f
Cryptosyringida, 969, 1003
crystalline cone cells of Arthropoda, **748,** 755
crystalline style of Mollusca, **502**
ctenes of Ctenophora, **332,** 333f, 334, 338
Ctenidae, 913, 922f–923f, 938, 944
ctenidia of Mollusca, **472,** 476f, 478f, 496, 496f, 497, 498f, 503, 503f, 504, 506
Ctenizidae, 912
Ctenocephalides
canis, 858
felis, 858
Ctenocheilocaris, 797
Ctenodiscus, 972f
Ctenodrilus, 565, 566f
Ctenophora, 213, 214, 215, 216, 255, 269, 327–342, 346, 347
body plan of, 139, 332–334
characteristics of, 328
cilia of, 143, 334, 335
circulation, excretion, gas exchange, and osmoregulation of, 338
feeding and digestion of, 149, 335f, 335–337, 336f, 337f, 338f
luminescence of, 328
marine habitat of, 17
nervous system and sense organs of, 332, 338–339, 339f
number of species, 2t
phylogeny of, 25, 328, 341, 1048, 1049f, 1050
reproduction and development of, 139, 339–340
support and locomotion of, 334f, 334–335
taxonomic history and classification of, 328–332
ctenophore–polyclad theory, 407
Ctenoplana, 330f–331f, 341
Ctenostomata, 645f, 646, 647f, 648, 653
body wall, muscles, and movement of, 649–650
reproduction and development of, 655
Cubomedusae, 296, 306
Cubozoa, 216, 269, 272, 292
cnidae of, 293
feeding and digestion of, 296
nervous system and sense organs of, 305, 306–307
phylogeny of, 319, 320f
reproduction and development of, 277, 308, 310–311
Cucullanus, 674f
cucullus of Chelicerata, **925**
Cucumaria, 991, 992f
crocea, 997f
minata, 992f
planci, 972f
Cuénot, C., 28
Culcita, 984
novaeguineae, 970f–971f
Culex, 846
Culicidae, 858, 873, 888f
Cumacea, 763f, 766, 785, 819, 823
Cuninidae, 273
Cupiennius, 944, 945f
coccineus, 922f–923f
Cupuladria, 648
Curculionidae, 844, 856
Cuspidaria, 499f
rostrata, 499f
Cuspidariidae, 470
cuticle, 146
of Arthropoda, **731,** 732f, 878
of Nematoda, 673–674, 674f, 675, 675f
of Nematomorpha, 687, 688f
cuticular plaques of Siboglinidae, **589**
cuticular teeth of Chaetognatha, **423**
cuticulin layer of Arthropoda, **732**
Cutler, E., 574
Cuvier, G., 220, 328, 375, 436, 454, 574, 636, 728
Cuvierian tubules of Holothuroidea, **991,** 992f
Cyamidae, 790
Cyamus
erraticus, 788f
monodontis, 788f
scammoni, 763f
Cyanea, 297, 310
capillata, 267f
Cyanobacteria, 7, 8, 10, 56, 57, 57n, 125, 126, 252, 925, 1034
Cycadophyta, 129
Cyclestheria, 771f, 773
hislopi, 773
Cyclestherida, 765, 771f, 772, 773
Cycliophora, 347, 348, 603, 605, 609–611, 637
acoelomate body plan of, 140
characteristics of, 610
marine habitat of, 17
number of species, 2t
phylogeny of, 25, 1049f
cyclomorphosis, 773
of Tardigrada, **713**–714, 714f
Cycloneuralia, 669, 693, 694

Cyclophoridae, 465
Cyclophyllidea, 375f, 379, 382f, 386, 405, 406f
Cyclopoida, 767, 797
Cyclops vernalis, 390
Cyclorbiculina, 147f
Cyclorhagida, 696, 696f
Cyclosa, 931
Cyclostomata, 645f, 647f, 648, 648f, 651
  body wall, muscles, and movement of, 650
  circulation, gas exchange, and excretion of, 652
  reproduction and development of, 654, 656, 657
Cydippida, 328f–329f, 330f–331f, 332, 334f
  phylogeny of, 341
  reproduction and development of, 340
cydippid larva of Ctenophora, **328, 340,** 340f
cymbium of Euchelicerata, **948**
Cymothoa exigua, 816
Cymothoida, 766
Cymothoidae, 47, 47f, 787, 787f, 816
cyphonautes larva, **656,** 656f
Cyphophthalmi, 914, 924, 925, 945, 950
cyphopods of Diplopoda, **898**
Cypraeidae, 466
Cypridinacea, 825
Cypridinidae, 798, 826
Cyprinotus, 800f
cypris larvae, 791, **831**
Cyrenidae, 470
Cyrtocrinida, 973
Cyrtophora, 932
cyst, of Tardigrada, 713
cystacanth of Acanthocephala, **626,** 627f
cysticercus, **405**
cystid of Bryozoa, **646**
Cystisoma, 763f, 783
Cystonecta, 274, 282
Cytherelloidea, 800f
cytokinesis of Protista, **68**
cytoproct of Protista, **66**
cytostome of Protista, **65,** 80, 91, 117, 123

D

Dactlogyridea, 378
Dactylogyrus vastator, 402f–403f
Dactylopiidae, 854
dactylopores of Cnidaria, **289,** 290f
dactylostyles of Cnidaria, **289,** 290f
dactylozooids, **280,** 280f, 281f, 298
Dahl, E., 832
Dahlella caldariensis, 775
Dajidae, 763f
Daku riegeri, 352f–353f
Dalytyphloplanida, 377, 387
Danaus plexippus, 883, 888f
Daphnia, 29, 770f–771f, 773
Daphnidae, 773
dark-adapted eyes of Arthropoda, 749, 749f
Darwin, C., 4, 13, 27, 37, 44, 51, 55, 648
  On the Origin of Species, 4, 30
Darwinella, 221
Darwinellidae, 221
Darwin's dilemma, 13
Darwinula, 800f
Dasydytidae, 431
Davis, G. E., 765, 777
Dawydoff, C., 1007
DDT, 84
dead zones in marine habitat, 18
DeBary, H. A., 23
de Blainville, H. M., 454, 574, 636–637
Decapoda, 776, 777–778
  anatomy of, 779f, 780f–781f
  circulation and gas exchange of, 819
  classification of, 471–472, 766
  digestive system of, 816
  diversity of, 762f–763f, 779f, 780f–781f
  nervous system and sense organs of, 822
  reproduction and development of, 801, 806t–807t, 826, 831
  shrimps, 761n, 779f
  thoracic appendages of, 804f–805f
Decapodiformes, 455, 471–472, 508
Deccan Traps, 16
Decolopoda, 958
  australis, 956f
defense mechanisms
  of Chaetognatha, 423
  of Cnidaria, 298, 299f, 302
  of Enteropneusta, 1011
  of Myriapoda, 896–897
  of Platyhelminthes, 384
  of Porifera, 249, 250, 251
definitive hosts, **23, 201**
degenerative evolution of Cnidaria, 318–319
Deinopidae, 913, 938
Deinopis, 938, 946f, 954f–955f
Deinopoidea, 932
Deiopea, 331f
de Jussieu, B., 636
delamination, **193,** 194f
de Laubenfels, M. W., 231, 257
De Ley, P., 672
demersal organisms, **18**
Demodex, 916f
  brevis, 919
  canis, 919
  folliculorum, 919
Demodicidae, 919
Demospongiae, 216, 218f, 220, 221, 222
  distribution of, 249
  evolution and phylogeny of, 255, 256, 257, 257f, 258
  feeding of, 236
  fossil record of, 256f
  growth rate of, 251
  reproduction and development of, 242, 242f, 243, 244, 246f, 248
  support of, 231f, 233
  symbiotic relationships of, 251, 252
Dendraster, 972f, 989, 998f
  excentricus, 970f–971f, 984, 989, 990f
Dendrobranchiata, 761n, 766, 778, 779f, 826
dendrobranchiate gills, 778, 820f
Dendroceratida, 221
Dendrochirotida, 975
Dendrocystites, 1000f
Dendrodorididae, 467
Dendrogramma, 317–318, 318f
  discoides, 317
  enigmatica, 317, 318f
Dendrogrammatidae, 318
dendrograms, 44
Dendronotidae, 467
Dendrophiliina, 321
Dendy, A., 216
dengue, 846
Dentaliida, 470
Dentaliidae, 470
Dentalium, 520f
Deoterthron, 764f
deposit feeding, 152–153, 153f
  selective, **153**
de Quatrefages, J. L. A., 574
derived character states, **44,** 45
  shared, 51
Dermacentor, 916f
  andersoni, 918
dermal membrane of Hexactinellida, **227,** 228f
dermal pores or ostia of Porifera, **223, 226**
Dermaptera, 48f, 846, 847, 851f, 852, 866, 889f
Dermasterias, 985
Dero, 558
Derocheilocaris, 794f, 797
  ingens, 797
Desmocolecida, 672
Desmodorida, 672
Desor's larva, **450,** 451
determinate cleavage, **191**
determinate growth of Crustacea, **831**
detorted gastropods, **484**
Deuterostomia, 319, 347, 348–349, 588, 635, 636
  Chordata, 1021–1043
  coelom of, 207
  Echinodermata, 967–1003
  evolution of, 208
  Hemichordata, 1007–1019
  as legacy name, 26, 50, 347
  Nielsen theory on, 208, 209
  origin of, 1043
  Pax-6 gene of, 173
  phylogeny of, 25, 967–968, 1049f, 1050
deuterostomy, 347, 348
deutocerebrum, 746
  of Crustacea, 821
  of Euchelicerata, 943
  of Pycnogonida, 959–960
developmental polymorphisms, **200**
developmental system drift (DSD), **185**
developmental tool kits, **184**–185
Devescovina, 114f
Devonobius, 909
dextrotropic direction, **188,** 189
Diadematoida, 974
Diadumene, 292, 296, 308
Diamus montanus, 858
diapause, **22**
Diaplauxis hatti, 83
Diastylis, 784f
diatoms, 56, 57, 63, 97, 98f, 100f, 126
  invasive populations of, 99
  reproduction of, 101
  support and locomotion of, 100
  toxins of, 99
Diaulula, 464f
Diazona, 1032f
Dibranchiata, 471
dichlorodiphenyltrichloroethane (DDT), 84
Dickinsonia, 11, 12f, 317
Diclybothriidea, 378
Dicondylia, 847, 848–849, 889, 889f
Dictyodendrillidae, 221
Dictyoptera, 851

Dictyostelida, 72, 73f
Dictyostelium, 62
discoideum, 72, 73f
Dictyota, 809f
Dicyema, 415f, 416f
japonicum, 416
Dicyemida, 414–417, 417f
Didemnidae, 1031, 1034
Didemnum, 1034
vexillum, 1036
Didinium, 89, 91
nasutum, 91f
Didymophyes gigantea, 83
Didymosphenia geminata, 99
Dientamoeba fragilis, 112, 113, 114
Difflugia, 71f, 74, 74f, 75
diffusion, 164–165
Digenea, 375f, 378
digestion, 139, 148–149
digestive cecum
of Arthropoda, **743**
of Cephalochordata, **1025**
digestive gland of Crustacea, **816**
Dignatha, 908
Dimastigella, 123
Dinobryon, 97, 100
Dinoflagellata, 56, 62–63, 76–80, 127
characteristics of, 78
chloroplast of, 66f
classification of, 61, 62
diversity of, 58f–59f
nutrition of, 80
osmoregulation of, 80
overview of, 62, 63
phylogeny of, 127
plastids of, 56n, 59f
reproduction of, 80, 81f
support and locomotion of, 79–80
toxins of, 78–79
Dinomischus, 14f, 15f
Dinozoa, 61
Dioctophyma renale, 671
Dioctophymatida, 673
Diopatra, 549
cuprea, 534f–535f
Diopisthoporus lofolitus, 352f–353f
Diphyllidea, 379
Diphyllobothriidea, 379, 386, 405, 406f
Diphyllobothrium, 386
latum, 405, 406f
dipleurula larva, 644, 1019
diploblastic metazoans, **139**, 140, **265**
Diplogaster, 678f–679f
Diplogasterina, 672
Diplommatinidae, 465
Diplomonada, 63, 64, 115, 128
Diplomonadida, 61, 62, 115f, 115–117
Diplopoda, 895–909
circulation and gas exchange of, 903
classification of, 897
excretion and gas exchange of, 904
feeding and digestion of, 901–902
head and mouth appendages of, 901
locomotion of, 901, 902f
nervous system and sense organs of, 904–905
phylogeny of, 908, 909
reproduction and development of, 906–907
diplosegments of Myriapoda, **895**, 897, 898, 900f–901f, 901, 902f
Diplosoma, 1034
Diplostomida, 378, 385, 401, 404f
Diplostraca, 710f–711f, 762f–763f, 765, 770f–771f, 772–773
distinguishing features of, 802t–803t
gene duplication in, 29
locomotion of, 808
reproductive features of, 806t–807t
Diplura, 843, 845f, 847, 848
evolution and phylogeny of, 48f, 888, 889f
thorax of, 864
Dipluridae, 912, 931f
Dipnoi, 1041f
Diptera, 846, 847, 855, 856, 857f, 858, 863
development of, 888f
feeding of, 864f, 872, 873
nervous system and sense organs of, 878, 880f, 881f
phylogeny of, 48f, 889f, 890
Dipteropeltis, 796
Dipylidium, 386
direct development, 197, 198f, 199–200
of Arthropoda, 750n, 830, 885
of Enteropneusta, 1008, 1016f–1017f
of Porifera, 244, 245f
direct waves in Mollusca locomotion, 486f, 487
Discinisca lamellosa, 658f
Discinoidea, 658f, 660, 664f–665f
discoblastula, **193**, 193f
Discodorididae, 467
Discomedusae, 320f
Discophorae, 328
disjugal furrow of Chelicerata, **917**
disphaerula larva, 244, 245f, 246f, 248
disphotic zone, **20**
Disporella, 658f–659f
dissogeny of Ctenophora, **340**
dissolved organic matter, as nutrition, 65, 149, **157**
of Amoebozoa, 75
of Annelida, 577, 590
of Dinoflagellata, 80
of Euglenida, 120
of Mollusca, 501
of Porifera, 236
Dissotrocha aculeata, 620f
Distaplia occidentalis, 1038f–1039f
Distromatonemertea, 450, 450f
Ditrupa, 548f–549f
Diurodrilus, 628, 629, 1050
divergence, 43f, 52
Dll gene, 735, 863, 863n, 865
DNA, 3, 6, 51n, 51–52
apicoplast, 81
chloroplast, 57n
genome sequenced from, 1048n
kinetoplast, 123
leucoplast, 57n
long branch attraction (LBA), **51**
in micronucleus, 93
mitochondrial, 2, 123, 128, 216, 258, 319, 368, 407
in molecular phylogenetics, 51–52
next generation sequencing, **51**
nucleomorph, 105
ribosomal, 257f, 258, 319, 419, 672, 1018
transposable elements of, 28
docoglossate radula, **493**, 494f
Dodecaceria, 565, 567
Dodecolopoda, 958
Dohle, W., 755
Doliolida, 1022, 1031, 1033, 1037, 1039
Doliolum, 1034f–1035f
Dolomedes, 934
Dolops, 796
dome of Ctenophora, **338**, 339f
domoic acid poisoning, 99
Donacidae, 470
Dorididae, 467
dorsal aorta of Cephalochordata, **1025**
dorsal lamina of Urochordata, **1034**
dorsal neural tube of Chordata, **1022**, 1023
dorsal sac of Echinodermata, **993**, 998, 999
dorsal vessels
of Enteropneusta, **1014**
of Pterobranchia, 1017
dorsoventral axis inversion of Chordata, 1042
Dorvillea, 533f
Dorvilleidae, 533f, 536, 550f
Dorylaimia, 672, 673
Dorylaimida, 673
Dorylaimus, 676f–677f, 678f–679f
Dotilla, 821
Dotillidae, 820
Doushantuo Formation, 3, 1048
Bilateria in, 11, 345, 967
Ctenophora in, 342f
Porifera in, 256
Draconema, 670f, 676f–677f
Drake Passage, 16–17
Dreissena, 488
Dreissenidae, 470
Drepanophorus, 445f
Driesch, H., 183
Dromiidae, 780f–781f
Drosophila, 28, 30, 186f, 751, 881
melanogaster, 28
sensory organs of, 881f
Drosophilidae, 858
drugs, Porifera as source of, 250
Drulia, 218f–219f
duet cleavage in Nemertodermatida, 360, 365f, 365–366
Dugesia, 392, 394
tigrina, 374f–375f
Dunaliella salina, 9n
duo-gland adhesion
of Gastrotricha, **429**, 430, 430f
of Platyhelminthes, **383**
Dynamics in Metazoan Evolution (Clark), 206
dynein arms, **142**
Dysderidae, 912
dysentery, amebic, 71, 72
Dyspnoi, 914, 924
Dytiscidae, 772, 855

E

Early Devonian, 15
ecdysiotropin, **742**
ecdysis of Arthropoda, **730**, **739**
ecdysone of Arthropoda, **739**, 887
Ecdysozoa, 13, 347, 348, 669, 709, 730, **751**
cuticle of, 722, 731
molting of, 739
Nematoida, 669–690
Panarthropoda. *See* Panarthropoda
phylogeny of, 25, 26, 51, 751, 829, 1049f, 1050, 1051
Scalidophora, 693–705

Echinacea, 974
Echinarachnius parma, 978f–979f, 990f
Echinaster graminicolus, 993
Echiniscoidea, 715
Echiniscoides, 713
sigismundi, 712f
Echiniscus, 714f, 715f
Echinocardium, 984, 994f
cordatum, 990f
Echinococcus, 382f
granulosus, 382f
multilocularis, 407
Echinocyamus, 989
Echinoderes, 694f
Echinodermata, 38, 269, 319, 328, 347, 348, 967–1003
archicoelomate condition of, 635
axial and extraxial body regions of, 975–978, 977f
body plan of, 975–978
body wall of, 978–980
brooding of, 996, 997, 997f
characteristics of, 971
circulation and gas exchange of, 991–994
coelom of, 969, 978–980, 998–999, 999f
in Ediacaran Period, 11
excretion and osmoregulation of, 994–995
external anatomy of, 972f
feeding and digestion of, 980, 984–991, 1003
marine habitat of, 17
mutable collagenous tissues of, 980, 982
nervous system and sense organs of, 995
number of species, 3t
phylogeny of, 25, 1000–1003, 1002f, 1049f, 1050
ray homologies of, 978
reproduction and development of, 969, 975, 995–1000
support and locomotion of, 146f, 982–984
taxonomic history and classification of, 969–975
water vascular system of, 969, **975,** 976f, 978, 980–982, 991, 1001, 1003
Echinofabricia alata, 570f
Echinoidea, 970f–971f, 974, 976, 978f–979f, 980
circulation and gas exchange of, 993, 994f
external anatomy of, 972f
feeding and digestion of, 988f–989f, 988–991, 990f
internal anatomy of, 990f
mutable collagenous tissue of, 980
nervous system and sense organs of, 995
phylogeny of, 1001
reproduction and development of, 996, 998f
water vascular system of, 981–982
Echinolampadoida, 974
Echinometra
mathaei, 984
vanbrunti, 983
Echinoneoida, 974
Echinosquilla guerinii, 775f
Echinothurioida, 974
Echinozoa, 969, 973–975
Echinus, 976f
Echiura, 41, 531, 532, 538, 540, 572, 597
phylogeny of, 25, 1050
Echiuridae, 531, 533f, 540, 542, 543, 544, 553, 579–584
body wall and coelom of, 582
circulation and gas exchange of, 583–584
excretion and osmoregulation of, 584
feeding and digestion of, 582f, 582–583
nervous system and sense organs of, 584
reproduction and development of, 584, 585f
support and locomotion of, 582
Echiurus, 581f, 583f, 584
echiurus, 441
echolocation, 881
ecosystems, 21f
Ecteinascidia turbinata, 1043
Ectocarpus, 98f
ectoderm, **139,** 140, **191, 193,** 194f, 345
of Hexapoda, 884, 885, 885f
Ectognatha, 847
ectognathous insects, **862**
ectolecithal ova, 376, **395**
ectomesoderm, **195,** 348
ectoneural system of Echinodermata, **995**
ectoparasites, **23, 201**
ectoplasm, **141**–142, 142f
Ectoprocta, 635, 637, 644. *See also* Bryozoa
ectosome, **226**
Ediacara, 12f
Ediacara Hills, 317
Ediacaran Period, 10–12, 12f, 13, 317–318
Edmonds, S., 574
Edriopthalma, 790
Edwardsia, 309
edwardsia stage, **309**
Edwardsiella lineata, 308
Eeckhaut, I., 541
efferent (motor) nerves, **168**
egg nauplius, **830**
eggs. *See* ova
Egregia, 63
Ehlers, U., 362, 376, 384, 407
Ehrenberg, C. G., 637
Eichhorn, J. C., 711
Eifelocrinus, 1000f
Eimeria, 82
Eirenidae, 301
Eisenia, 568f
foetida, 556f–557f
Elasipodida, 975
Elateridae, 855, 867
Eldredge, N., 31
Electra, 647f, 652f, 655
Eledone, 156f, 501f
elephantiasis, **683,** 846, 873
Eleutheria, 291f, 292
Ellipura, 847, 888
Ellobiacea, 467
elongate plexes of Ctenophora, 338
Elphidium, 107
Elrathia, 15f
elytra, **867**
Embata, 621
Embiidina, 847, 853
Embioptera, 48f, 846, 847, 851f, 853, 889f
Embletonia, 500f–501f
Embletoniidae, 467
embolus of Euchelicerata, **948**
embryogenesis, **183**
embryonic development, 187–197
blastula types in, 192–193, 193f
cell fates in, 191–192
cleavage patterns in, 187–191
founder regions in, 191
gastrulation in, **193**–194, 194f
germ layers in, 183, 191, 193–194
and law of recapitulation, 202
embryophore of Platyhelminthes, **405**
Embryophyta, 129
Emerita, 151f, 782f, 810, 811f
Emiliania huxleyi, 102, 102f, 103
Encentrum astridae, 620f
Enchytraedidae, 540–541
Encyclopedia of Life (EOL) project, 6
Endeavouria, 387
Endeidae, 961
endites of Arthropoda, **733,** 735, **801**
endocrine hormones, **175**
endocuticle of Arthropoda, **732,** 732f
endocytosis, 75, **148**
endoderm, **139,** 140, **191, 193,** 194f, 345
of Hexapoda, 884, 885, 885f
endolecithal ova, **376, 395**
Endolimax nana, 70
endomesoderm, **195,** 348
endoparasites, **23, 201**
endopinacocytes of Porifera, **223, 230**
endopinacoderm, **223**
endoplasm, **141**–142, 142f
endopod of Arthropoda, **733**
endopolygeny, **68, 83**
Endopterygota, 890
endoskeleton, 145, 146f, 148
endostyle of Chordata, 1041, 1042
in Cephalochordata, **1025**
in Urochordata, 1029, 1031
endosymbiosis, 29, 125–127
primary, **57,** 59f
secondary, **57,** 59f, 79, 105, **126,** 127
endosymbiotic theory, **29,** 125–126, 126f
endotokia matricida in Nematoda, **681**
engrailed (en) gene, 731, 750
Enopla, 438
Enoplea, 670f, 672, 673, 675f–679f, 683, 684f
Enoplia, 672, 673, 678f–679f, 680
Enoplida, 673
Enoplometopus, 762f–763f
Entamoeba, 76, 117
coli, 70, 71
gingivalis, 70
hartmanni, 72
histolytica, 70, 71, 72, 76
moshkovskii, 72
enterocoel theory, **206**–207
enterocoely, **196,** 196f, 206–207
enteron of Platyhelminthes, **387**
Enteropneusta, 1007, 1008, 1009f, 1009–1015
circulation, excretion, and gas exchange of, 1014
classification of, 1008
coelomic cavities of, 1011, 1015
external anatomy of, 1009–1011, 1010f
feeding and digestion of, 1013f, 1013–1014
fossil record and phylogeny of, 1018, 1019, 1049f
internal anatomy of, 1012f–1013f

life cycle of, 1008
nervous system of, 1014
reproduction and development of, 1014–1015, 1016f–1017f
support and locomotion of, 1011–1013
entocodon, **308**
Entognatha, 846, 847–848
entognathous hexapods, **860**
entoneural system of Echinodermata, **995**
Entonomenia tricarinata, 459f
Entoprocta, 348, 450, 603–609, 604f–605f, 636, 637
acoelomate body plan of, 140
body wall, support, and movement of, 606
characteristics of, 605
circulation, gas exchange, and excretion of, 607
feeding and digestion of, 606f–607f, 607
nervous system of, 607
number of species, 2t
phylogeny of, 25, 604, 1049f, 1050
reproduction and development of, 607–609, 608f
Entosiphon, 118f, 119, 120, 120f
Enypniastes, 971f
Eoacanthocephala, 624, 625f, 626
Eoandromeda, 11, 341
octobrachiata, 342f
Eoarthropleurida, 909
Eocyathispongia qiania, 3, 256
Eognathacantha ercainella, 422
Ephelota
gemmipara, 96f
gigantea, 94f
ephemeral pools, **22**
Ephemeroptera, 846, 847, 849f, 849–850, 866, 877f
evolution and phylogeny of, 48f, 889f, 890
flight of, 868
ephippium of Arthropoda, **773**
Ephydatia fluviatilis, 236, 242f
Ephydridae, 858
Ephyra, 269
ephyrae, **308, 310,** 311f, 312f
Epiactis, 137f, 292
prolifora, 297f
epibenthic animals, **19**
epiboly, **193,** 194f
epicardial sacs of Urochordata, **1036**
Epicaridea, 766
epicone of Dinoflagellata, **79**
epicuticle of Arthropoda, **731,** 732, 732f
epidermal plate of Micrognathozoa, **628,** 628f
epifauna, **19**
epigastric furrow of Chelicerata, **924**
Epigonichthyes, 1023
epigynum
of Chelicerata, **924**
of Euchelicerata, 949–950
Epimenia, 518f
australis, 456f
verrucosa, 458f–459f
epimorphic development of Crustacea, **830**
Epiperipatus, 723f
biolleyi, 723
imthurni, 726
trinidadensis, 727f
epiphyses of Echinodermata, **989**
epiplasm of Ciliata, **87**
epipods of Arthropoda, **733, 801,** 804f–805f, 821
episphere, **209, 571**
epistome of Lophophorata, **635,** 637, 640, 661
epitheca of Dinoflagellata, **79**
epitheliomuscular cells of Cnidaria, **275,** 276, 276f
Epitheliozoa, 258
epitokous annelids, **567**
epitoky, **567**
Epitoniidae, 466, 495
Epizoanthus, 302
Eratoidae, 466
Ercaia minuscula, 835, 835f
Eremobates, 919f–921f
Erenna, 298
Eresidae, 913
Ergasilus pitalicus, 794f–795f
Eriauchenius, 939f
Erinettomorpha, 317
Eriocheir sinensis, 808
Eriophyoidea, 919
Ero furcata, 954f–955f
Erpobdella, 596f
errant animals, **19**
Errantia, 535–538, 549, 598f
Erwin, T., 844
escape reaction of Crustacea, 805f, **807**
estuary habitat, 20–22
etching cells, **252**
Eteone, 572f
Ethmoliamus, 678f–679f
euanamorphosis, 907
Eubacteria, 8, 28
Eubothrium salvelini, 390
Eucalanus, 823f
Eucarida, 749, 766, 776, 777f, 782f
phylogeny of, 832
reproduction and development of, 806t–807t, 830, 831
Euceramus praelongus, 762f–763f
Euchelicerata, 911, 912–955
body plan of, 927–928
circulation and gas exchange of, 941f, 941–942
excretion and osmoregulation of, 942–943
feeding and digestion of, 935–941, 936f–937f, 939f, 940f
locomotion of, 932–934, 933f, 934f, 935f
nervous system and sense organs of, 943–947, 944f
phylogeny of, 961, 962f
reproduction and development of, 947–955
silk and webs of, 928–932
Euchlanis, 622f
Euchlora rubra, 337f
eucoelomate body plan, 140, 140f
Eucomonympha, 114f
eucone eyes of Hexapoda, **880**
Eucopiidae, 783
Euctenizidae, 912
Eudentridae, 274
Eudistylia, 166f–167f
vancouveri, 548f–549f
eudoxids, **282**
Euechinoidea, 974
Euglena, 118, 118f, 119, 119f, 120f, 173f
spirogyra, 68f
Euglenida, 64, 117, 118–120
anatomy of, 118f
characteristics of, 118
chloroplast of, 66f
classification of, 61, 62
diversity of, 58f–59f
nutrition of, 119–120, 120f
phylogeny of, 128
reproduction of, 120, 120f
support and locomotion of, 119, 119f
Euglenozoa, 56, 62, 63, 64, 119, 128
Eugnatha, 897, 899
Eugymnanthea, 301
Euheterodonta, 455, 470
Euhyperamoeba, 74
Eukaryota, 2, **7,** 8, 9, 57, 59, 59f, 64, 213
Animalia/Metazoa, 4, 8, 9, 55, 213
bacterial genes in, 28
in Black Sea, 18n
classification of, 4, 8, 61
evolution and phylogeny of, 10, 29, 45, 127f, 184, 345, 843
Fungi, 4, 8, 55
life cycles of, 197
locomotion and support of, 140, 142
multicellularity of, 184, 205
number of species, 6, 36
origin of, 3, 7, 10, 29, 125–126, 126f, 345
Plantae, 4, 4n, 8, 55
Protista, 4, 8, 55–130
Eukoenenia
draco, 925
grassii, 925
spelaea, 919f–921f
Eukoeneniidae, 925
Eukrohnia, 426, 426f, 427
bathypelagica, 426f
fowleri, 420f, 422, 426f
Eukrohniidae, 423
Eulagisca gigantea, 537
eulamellibranch ctenidia of Mollusca, **497,** 498f
Eulecithophora, 408f
Eulepethidae, 537
Eulimidae, 466, 494
Eulimnadia texana, 825
eulittoral zone, **18**
Eumalacostraca, 765, 773, 774, 774f, 775f, 776, 782f, 784f, 801
distinguishing features of, 802t–803t
phylogeny of, 832, 833f
reproductive features of, 806t–807t
Eumecynostomum evelinae, 352f–353f
Eumetazoa, 214, 258, 317
Eumida, 550f
Eumycota, 60, 101
Euneoophora, 377, 408, 408f
Eunice, 562f
aphradiotois, 549
aphroditois, 550f
Eunicella cavolini, 266f–267f
Eunicida, 535, 536–537
Eunicidae, 536, 547, 566f, 570f
Euopisthobranchia, 455, 456f, 464f, 467
nervous system of, 509f
reproduction of, 515f
Eupentacta, 991
Euphausiacea, 766, 776–777
reproductive features of, 806t–807t
thoracic appendages of, 804f–805f
Euphausia superba, 764, 776, 777f
Euphrosinidae, 533

Euplectella, 218f–219f, 233, 251, 779
aspergillum, 218f–219f
Euplotes, 88f–89f, 90, 92, 92f, 96
Eupnoi, 914, 924
Euproöps, 948
euproöps larva, 947f, **948**
Eupulmonata, 467, 505f, 515f, 516–517
Eurotatoria, 618
Euryalida, 973
Eurycopidae, 807
euryhaline animals, **159**
Eurylepta californica, 375f
Eurypterida, 912, 913f, 915
Eurypterus, 913f, 915
remipes, 913f
Eurystomata, 646, 653, 655–656, 657
Eusarsiella, 800f
Eutardigrada, 714, 715, 719f
Euthyneura, 454, 455, 466–467, 494
euthyneury of Gastropoda, **484,** 508
Euzonus, 559
Evans, H. E., 844
Evarcha culicivora, 939
EvoDevo, 30, **184,** 192, 1048
evolution, 1–3, 10–17
of Arthropoda, 733, 734–736, 744, 751–755
bilateral condition in, 206
body size changes in, 139
of Cnidaria, 317–321
coelom origin in, 206–207
coevolution in, 15n
colonial theory of, **204**–205, 205f, **318**
convergent, 41, **42,** 43, 43f, 185
endosymbiotic theory of, **29,** 125–126, 126f
estimation of divergence times in, 1048
first life forms in, 1n, 1–2, 10
of Hexapoda, 843, 846, 869–870, 887–890
of insect flight, 869–870
and law of recapitulation, 202
long branch attraction in, 51
macroevolution in, 27–31, 31f
mesoderm in, 139–140
microevolution in, 27, 31f
of Mollusca, 521–524
of multicellularity, 2, 10, 184, 205–206, 1048
natural selection theory of, 27
of novel gene function, 185
origin of Metazoa in, 203–209
parallel, 42, 42n, 43, 43f
of Platyhelminthes, 380, 405–409, 408f
of Porifera, 254–258
reversal in, **42**
saltation theory of, 28
of triploblastic Metazoa, 139–140
evolutionary developmental biology, 30, 35, 184, 192, 346, 1048
evolutionary reversal, **42**
evolutionary species concept, **38**
Evolution of the Metazoan Life Cycle (Jägersten), 202
Excavata, 60, 61, 62, 112–124
Diplomonadida, 115–117
Euglenida, 118–120
Heterolobosea, 117–118
Kinetoplastida, 120–124
overview of, 63–64
Parabasalida, 112–115
phylogeny of, 128
exconjugants of Ciliata, **94**
Excorallana, 787f
excretion, 157–158, 160–162
nephridia in, 161–162
water expulsion vesicles in, 160, 160f
excurrent canals, 227, 227f
excurrent chimneys, **651,** 653f
Exd gene, 735, 865
Exechocentrus lancearius, 937f
exites of Arthropoda, **733, 801,** 869
exite theory on insect flight, 869
exocuticle of Arthropoda, **732,** 732f
exocytosis, **149, 294**
exopinacocytes of Porifera, **223, 230**
exopinacoderm, **223,** 228
exopod of Arthropoda, **733**
exoskeleton, 145–148, 147f
explicit phylogenetic hypotheses, **46**
extensor muscles, **146,** 147f
externa, **815**
extinction, **27**
current rate of, 6–7
in Ediacaran period, 11, 11n
of Hexapoda, 845–846, 887
macroevolution in, 27
in Mesozoic era, 16
in Paleozoic era, 15
extracellular digestion, 148
extracorporeal digestion, 148
extranuclear mitosis of Protista, 69f, **70**
extraxial/axial theory, 975
extraxial region of Echinodermata, **975**–976, 977f, 982, 1001
extrusomes of Protista, **67,** 120
exumbrella, **285**
exuvium of Arthropoda, **739**
eyes
ciliary, **172**
complex, **172**
compound, **172, 748**
rhabdomeric, **172**
eyespots, **172**
of Annelida, 565
of Protista, 66f, 68f, 80, 101
of Tardigrada, 717
eyestalks of Arthropoda, 749, 823, 825

F

Fabricia, 565
Fabriciidae, 570f
Facelinidae, 467
Facetotecta, 767, 791, 832
facultative symbionts, **23**
Farrea occa, 228f
Fasciola hepatica, 375f, 384f, 400f
Fasciolariidae, 466
fat body in Hexapoda, **875**
fate maps, **191**
Fecampia, 387
Fecampiida, 377–378
Fecampiidae, 387
feeding strategies, 149–157
deposit, 152–153
dissolved organic matter in, 157
suspension, 149–152
female heterogamety, **197**
femur of Hexapoda, **864**
Fenestrulina, 655
malusii, 648f
Ferguson, J. C., 980–981
Fermeuse Formation, 3
Ferosagitta hispida, 421f, 427
fertilization, 178–179
fibularium of Micrognathozoa, **630,** 631f
Ficidae, 466
Field, K., 51
filariasis, 683, 846, 873
filibranch ctenidia of Mollusca, **497,** 498f
Filibranchia, 469
Filifera, 274
filiform tentacles of Cnidaria, **279**
Filistatidae, 912
filopods (filopodia), **65, 74,** 74f, **142**
Filospermoidea, 615–616
Fionidae, 467
fissiparity of Echinodermata, **996**
Fissurella, 462f, 478f
Fissurellidae, 462f, 463, 504
Flabelligeridae, 535f, 538
Flabellina, 464f–465f
iodinea, 6
Flabellinidae, 467
Flaccisagitta hexaptera, 424f
flagella, 142–143, 143f
flagellar pocket of Euglenida, **119**
flagellomeres of Hexapoda, **860**
Flagellophora, 364
apelti, 360f, 361f
flagellum of Hexapoda, **860**
flame bulbs, **161**
of Nemertea, 444, 445f
of Platyhelminthes, 391, 391f
Flavella, 88f–89f
flexor muscles, **146,** 147f
flight of Hexapoda, 867–870, 868f, 869f
Florarctus heimi, 712f
Floscularia, 617f, 619f
flosculi of Priapula, **701**
Flustrellidra, 652f
Foelix, R. F., 945
Folliculina, 88f–89f
Folliculinidae, 88f–89f
foot of Mollusca, **472,** 476f, 486f, 486–487, 488f
foramen
of Brachiopoda, **659**
of Foraminifera, **106**
Foraminifera, 105–106, 106f, 107, 108f, 127
foraminiferan ooze, 107
force-sensitive organs of Arthropoda, **747**
Forcipulatida, 973, 985
forcipules of Myriapoda, **897,** 898, **901**
Formica exsectoides, 844
Formicidae, 855
FOXP2, 52
fragmentation, 176, 176f
of Nemertea, **447**
Franciscideres, 696
Fredericella, 651
Frenulata, 584–585, 591
Frenulina sanguinolenta, 658f–659f
frenulum
of Hexapoda, **866**
of Siboglinidae, **588**
Frenzel, J., 204, 214
freshwater habitat, 22
gas exchange structures in, 165
life cycle adaptations for, 22, 200
osmoregulation in, 158–159, 159f
Frey, H., 969
frontal glands of Nemertea, 446f, **447,** 450
frontal membrane of Bryozoa, 650, 650f

frontal organ
of Acoela, **354**
of Entoprocta, 608f, 608–609
of Gastrotricha, **431,** 432
frontal processes of Crustacea, **822**
Frontinella, 936f–937f
fruiting body of Protista, 72, 73f
frustules of Stramenopila, **97,** 99, 100
Fucus, 58f–59f
Fulgoridae, 854
Fuligo, 62, 72
septica, 73, 73f
Funch, P., 611, 614, 626
Fungi, 8, 55, 60, 62, 64, 125, 129
classification of, 4
nutrition of, 101
origins of, 129
funiculus of Bryozoa, **651,** 654
funnel of Mollusca, **477, 490**
furcae of Enteropneusta, **1011**
furcilia, **831**
fusibility system of Urochordata, 1037–1038
Fustiaria, 456f–457f
fusules of Radiolaria, **110**
Fuxianhuia protensa, 751

G

Gadilida, 470–471
Gadilidae, 471
Galathea Expedition (1952), 459
Galeodes arabs, 919f–921f
galls, 873
gametes, **178**
gametic nuclei of Ciliata, **94,** 95
gametocyst of Apicomplexa, **85**
Gammaridea, 767, 790, 808
Gammarus, 390
gamontogony, **83**
gamonts of Protista, **69,** 82f, 83, **85,** 85f, 107
Ganesha, 330f–331f, 332
Ganeshida, 330f–331f, 332
Garstang, W., 183, 202, 484–485, 571
Garypus, 925
titanius, 925
gas exchange and transport, 164–168
gastraea, **244n**
gastrea, **205,** 205f, **318**
gastric pouches, **285**
gastric shield of Mollusca, **502**
Gastrodes, 335
parasiticum, 340
gastroneuron, **208,** 209
Gastrophilus, 168
Gastropoda, 453, 454, 455, 456f, 462–467, 476f
body plan of, 473
circulation and gas exchange of, 503, 504, 505, 506
coiled, 463f, 482
development of, 518f, 520, 521
digestion of, 500f–501f, 502
evolution of, 482, 484, 485
excretion and osmoregulation of, 506–507
feeding of, 492, 492f, 494, 494f, 496
general anatomy of, 462f–465f
limpet-like, 462f, 482
locomotion of, 486f, 486–487
mantle and mantle cavity of, 474, 485
nervous system of, 508, 509f
phylogeny of, 522f, 523
reproduction of, 514–517, 515f, 516f
sensory organs of, 511, 512f
shell of, 477, 480f, 482
symbiotic relationships of, 499
torsion of, 483f, 483–485, 508, 521, 521f
torted and detorted, **484**
gastropores of Cnidaria, **289,** 290f
gastrostyles of Cnidaria, **289,** 290f
Gastrotricha, 347, 348, 384, 413, 428f, 428–432
acoelomate body plan of, 140
body wall of, 429–431, 430f
characteristics of, 429
circulation, gas exchange, excretion, and osmoregulation of, 431
feeding and digestion of, 431
as meiobenthic, 19
nervous system and sense organs of, 431
number of species, 2t
phylogeny of, 25, 1049f, 1050
reproduction and development of, 431–432
support and locomotion of, 431
gastrovascular canal of Ctenophora, 338, 338f
gastrovascular cavity
of Cnidaria, **265,** 296
of Platyhelminthes, **387**
gastrozooid, **280,** 280f, 281f
gastrula, **193, 244n**
gastrulation, **9, 193**–194, 194f, 213, 1048
by invagination, **193,** 194f, 244n
by involution, **194,** 194f
location and pattern of gene expression in, 192f
of Platyhelminthes, 398
of Porifera, 242, 243–244, 244n, 246
Gaussia, 764f
Gebiidea, 766, 782f, 783, 810
Gecarcinidae, 781
Gegania valkyrie, 515
Gegenbaur, K., 375, 637
Gelanor latus, 937f
Gelyelloida, 767, 798
gemmules of Porifera, **240,** 241f
gene regulatory networks, 30, **185**–187
genes, 28–30, 178, 184–187
developmental, 29–30, 185
and developmental system drift, 185
and developmental tool kits, 184–185
duplication of, 28–29, **185**
and evolutionary developmental biology, 184
evolution of novel function, 185
expression of, 30, 41, 192f
and molecular phylogenetics, 51–52
orthologous, **41, 184**–185, 191
paralogous, **41**
recruitment phenomenon, 41
regulatory, 29–30, 41, 185–187
gene transfer, 4n, **28,** 57, 59, 103, 126
to *Giardia,* 117
to Tardigrada, 713
to Urochordata, 1027
genital wings of Enteropneusta, **1015**
genotype, 41, 185
Geodia
media, 252
mesotriaena, 251
Geometridae, 859
Geonemertes, 445
Geophilomorpha, 897, 898, 901, 904, 906
Geoplana, 383f
mexicana, 392
georeceptors, 170f, 170–171
Gephyrea, 41, 574
Gephyreaster swifti, 291f
germarium of Platyhelminthes, **395**
germ cell parasitism of Urochordata, **1040**
germinal disc of Hexapoda, 884, 885f
germ layers, embryonic, **139, 183,** 191
formation of, 193–194, 194f
germovitellarium of Platyhelminthes, **395**
Gerridae, 854
Geryoniidae, 273
Ghiselin, M. T., 521
ghost larva of Loricifera, 704, 705f
ghost locus hypothesis, 255
giant cells
of Nematoda, 685f
of Nematomorpha, 689
Giardia, 115f, 116–117
agilis, 116
ardeae, 116
duodenalis, 116
intestinalis, 115, 116–117
muris, 115f, 116
psittaci, 116
Gibbifer californicus, 857f
Gibbs, P., 574
Gievzstoria expedita, 354f
Gigantocypris, 798
Gigantowales chisholmi, 900f–901f
gill bars
of Cephalochordata, 1025
of Enteropneusta, 1011, 1012f–1013f, 1013, 1014
gill bar supporting rods of Enteropneusta, **1011**
gills, 165, 166f–167f
of Hexapoda, **877,** 877f
gill slits, 202, 1018
of Chordata, **1022,** 1025, 1031, 1034, 1036, 1041, 1042
of Deuterostomia, 348–349, 968
of Enteropneusta, 1011, 1012f–1013f, 1013, 1015, 1017f
of Pterobranchia, 1015, 1017
gill theory on insect flight, 869
Ginkgophyta, 129
girdle
of Dinoflagellata, **80**
of Mollusca, **479**
Giribet, G., 469, 627, 728
gizzard of Mollusca, **502**
Glabratella, 107
Glacidorbidae, 467
gland cells
of Cnidaria, 276, 276f
of Nemertea, 442
of Platyhelminthes, 382–383
Glandiceps hacksii, 1013
gland of Deshayes, **501**
Glanduloderma, 387
Glaucidae, 467
Glaucilla, 490
Glaucophyta, 4n, 55, 55n, 59f, 129
glaucothoe, **831**
Glaucus, 490
Globigerina, 105
Globigerinella, 58f–59f, 106

Globodera, 683
glochidia of Mollusca, **519**–520, 520f
Glomerida, 897
Glomeridesmida, 897
Glomeris, 896, 907
glomerulus of Enteropneusta, **1014**
Glossina, 122
Glossinidae, 858
Glossobalanus, 1010f
Glossoscolecidae, 540
Glottidia, 664f–665f
Glycera, 549, 550f, 556f–557f
Glyceridae, 537, 550f, 557
glycocalyx, **74**
  of Porifera, **223,** 254
Glycymerididae, 469
Glypheidea, 766, 783
Gnaphosidae, 914
Gnathia, 787f
Gnathifera, 348, 613–632
  complex jaw of, 613, 614, 614f, 615, 616, 620, 630, 631f
  Gnathostomulida, 613–614, 615–616
  Micrognathozoa, 613, 614, 626–632
  phylogeny of, 25, 1049f, 1050, 1051
  Rotifera, 613–615, 616–626
Gnathiidae, 787f, 789
Gnathiidea, 766, 786
gnathobases of Arthropoda, **733**
Gnathophausia, 784f
  ingens, 785, 803
Gnathorhynchus, 388f
Gnathostomata, 1041f, 1042–1043
Gnathostomula, 616
  armata, 615f
Gnathostomulida, 429, 613–614, 614f, 615f, 615–616
  acoelomate body plan of, 140
  characteristics of, 615
  marine habitat of, 17, 19
  number of species, 2t
  phylogeny of, 25, 1049f
Gnetophyta, 129
Gnorimosphaeroma orgonense, 825
Gnosonesimida, 377
Gnosonesimora, 377, 408
Gochliomyia americana, 873
Gogia, 15f
golden algae, 97, 98f, 99–100, 101
Goldfuss, G. A., 59
Goldschmidt, R., 28, 680
Golfingia, 576f, 578f, 579, 580f
  minuta, 579
  vulgaris, 576f
Golfingiidae, 574–575
Gonactinia, 308
gonangia of Cnidaria, 277f, **280**
gonapophyses of Diplopoda, **906**
Gonatidae, 472
Gondwana, 16
Goniodorididae, 467
Gonionemus, 292
  vertens, 286f–287f
gonochory, **179, 197**
gonocoel theory, **207**, 207f
gonophores of Cnidaria, 277f, **280,** 281f
gonopores
  of Crustacea, 803t
  of Diplopoda, **898**
gonotheca of Cnidaria, 277f, **280**
Gonothyrae, 317f
Gonothyrea, 267f
gonozooids, **280,** 281f, 648
Gonyaulax, 77f, 78, 147f
Goodrich, E. S., 161
goosecoid (gsc) genes, 192f, 351
Gordioidea, 687, 689f
Gordius, 689f
  difficilis, 687
  robustus, 686f
Gorgas, W., 84
Gorgonia, 285f
gorgonin, **288**
Gorgonorhynchus, 441
Götte's larva, **397**
Gould, S. J., 31, 201, 202, 203
grades, 43f, **43–44**
Gradungulidae, 912
Graffilla, 387
Grammostola, 949f
grana, **57**
Grandiderella, 809f
Grant, R. E., 220
Grantia, 226
Granuloreticulosa, 63, 105–107, 106f
  characteristics of, 105
  classification of, 61, 62
  diversity of, 58f–59f
  nutrition of, 107
  phylogeny of, 128
  reproduction and life cycle of, 107, 108f
  support and locomotion of, 106–107
Grapsidae, 780f, 781
grasping spines of Chaetognatha, **423,** 425
grazing carnivores, 155
great oxidation event, 7
Greeffiella, 670f
  minutum, 9, 671
green algae, 55, 56, 57, 59f, 126, 127, 128, 129
greenhouse gases, 7
Greenwood, A. D., 59
Gregarina, 81
Grell, K., 215
growth zone of Annelida, 565, 571
Gryllacridadae, 853
Gryllidae, 853
Grylloblattodea, 48f, 846, 847, 851f, 852, 853, 889f
Gryllotalpidae, 853
gubernaculum of Nematoda, **681,** 682f
Guilding, L., 718
Gurney, R., 832
gut
  complete or through, **149, 345**. *See also* complete gut
  incomplete or blind, **149**
Gwynia capsula, 658f
Gwynioidea, 658f
Gymnoblastea, 273
Gymnodinium
  breve, 79
  catenatum, 78
Gymnolaemata, 646, 648, 650, 651, 656
Gymnoplea, 767
Gymnosomata, 467
Gymnostreptus, 902f
Gyrinidae, 855
Gyrocotyle fimbriata, 382f
Gyrocotylidea, 379, 382, 382f, 403
Gyrodactylidea, 378
Gyrodactylus, 381f

H

Habellia, 15f
habitat, 17–23
  estuaries and coastal wetlands, 20–22
  freshwater, 22
  loss of, 7, 22
  marine, 17–20
  and osmoregulation, 158–159
  terrestrial, 22–23
Hadromerida, 222
Hadzi, J., 205, 407
Haeckel, E., 59, 60, 183, 184, 244n, 247, 247n, 728
  colonial theory of, 204–205
  recapitulation law of, 183, 202
  tree of life, 4, 5f
Haeckelia, 335–336, 337f
  rubra, 302, 337f
Haementeria ghilianii, 591
Haemoproteus, 82, 83
Halechiniscus, 715f
Halicephalobus mesphisto, 671
Halichondria, 250
  moorei, 250
  okadai, 250
  panicea, 236, 251
  poa, 252
Halichondrida, 222
Haliclona, 218f, 224f, 241
  cratera, 252
  sonorensis, 252
Haliclystus, 292
Halicreatidae, 273
Halicryptus, 698, 699f, 700f
  spinulosus, 698
Halictidae, 855, 858
Haliotidae, 463, 504
Haliotis, 493f, 521f
  rufescens, 154f, 456f, 494f
Haliplanella, 308
Halisarca, 220, 228f–229f, 233, 243, 244, 248
  dujardini, 246f
  harmelini, 251
Halisarcida, 243
Halisarcidae, 244
Hall, B. K., 201
Hallucigenia, 15f, 718–719, 721f
  sparsa, 721f
Halobacterium, 9n
  salinarum, 9n
Halobiotus, 717
  crispae, 712f, 713, 714f
Halocynthia roretzi, 1031
Halocyprida, 767, 798, 800f
Halocyprididae, 826
halophiles, **9**
halophytes, 22
Halosydna, 553
Halotydeus destructor, 919
halteres, **867**
Haminoeidae, 467
Hammer, E., 244n
hamuli of Hexapoda, **866**
Hancock, A., 637
Hansen, E. D., 205, 407
Hanseniella magna, 899
Hansenocaris, 791
Hanson, E. D., 205
Haootia, 3
  quadriformis, 3, 317
Hapalogastridae, 781

haplodiploidy, **197**
Haplognathia, 616
  simplex, 615f
Haplogonaria amarilla, 356f
Haplopharyngida, 376
Haplopharynx rostratus, 390
Haplosclerida, 222, 243
Haploscleromorpha, 221, 222, 258
Haplosporidia, 61, 62, 63, 104f, 111–112, 128
Haplosporidium, 112
  littoralis, 112
  nelsoni, 104f
Haplotrematidae, 467
Haplozoon, 79
  axiothellae, 77f
haptocysts of Ciliata, **92**
haptonema of Haptophyta, **102**
Haptophyta, 56, 61, 62, 101–103, 102f, 126
  characteristics of, 102
  nutrition of, 101
  overview of, 62, 63
  phylogeny of, 127–128
  support and locomotion of, 100
Haptopoda, 914
HAR1 (human accelerated region 1), 52
Harbison, R., 328, 340
Harmer, S. F., 1007
Harmothoe, 533f
Harpacticoida, 767, 797
Harpidae, 466
Harrimania kupfferi, 254
Harrimaniidae, 1008, 1013, 1015, 1018
Hartmanella, 71f, 74
Hatschek, B., 269, 328, 604, 637
Hatschek's pit of Cephalochordata, **1025**
head
  of Crustacea, **798**
  of Hexapoda, 860–863
head shield
  of Crustacea, **801,** 802t
  of Pterobranchia, **1015**
heart, 164
heart vesicle of Enteropneusta, **1014**
hectocotyli of Mollusca, **517**–518
Hectocotylus, 518n
Hedgpeth, J. W., 958
Helicidae, 467, 505f
Helicinidae, 463
Heliconius, 29
Helicoplacus, 1000f
Helicoprorodon, 67f, 93f
Helicotylenchus, 678f–679f
Heliophanus, 946f
  cupreus, 954f–955f
Helioporacea, 271
Heliopora coerulea, 309
Helioporidae, 271
Helix, 154f, 517
Helminthomorpha, 897, 898, 907, 909
hemal channels of Onychophora, 723f, **724,** 724f
hemal plexus of Phoronida, 640f, 642
hemal rings of Echinodermata, 991
hemal system
  of Chaetognatha, 425
  of Echinodermata, 981, 991, 993, 993f
hemerythrin, **168,** 168t
hemianamorphosis, 907
Hemichordata, 347, 348, 349, 636, 967, 968, 969, 1007–1019
  archicoelomate condition in, 635
  characteristics of, 1008
  classification of, 1008–1009
  Enteropneusta, 1009–1015
  marine habitat of, 17
  number of species, 3t
  phylogeny of, 25, 1002f, 1018–1019, 1041, 1042, 1049f, 1050
  Pterobranchia, 1015–1018
Hemicytherura, 800f
Hemigrapsus sexdentatus, 780f–781f, 826f–827f
hemimetabolous development of Hexapoda, **885,** 885f, 886, 886f
Hemiptera, 846, 847, 853–854, 854f, 863
  defense mechanisms of, 859f
  feeding of, 871, 872f, 873
  leg modifications of, 865f
  locomotion of, 867, 869f
  phylogeny of, 48f, 889f, 890
  reproduction and development of, 883, 886f
  sensory organs of, 881f
  sound production structure of, 882
Hemirotatoria, 618
Hemithiridoidea, 660f
Hemithiris psittacea, 663
hemocoel, **164,** 730n
  of Arthropoda, **730,** 731, 751, 816
hemocyanin, **167**–168, 168t, 819, 903, 942
hemoglobin, **167,** 168t
  of Annelida, 559
  of Crustacea, 819
hemolymph, **164**
  of Arthropoda, **743**
  of Mollusca, 503f, 503–504
Hennig, W., 45
Henricia, 985
Hepatus, 780f–781f
Herberstein, M. E., 947
herbivores, 149, **153**–154, 154f
Herdmania momus, 1032f–1033f
hermaphroditism, 179, **197**
  of Annelida, 567
  of Bryozoa, 654
  of Chaetognatha, 426
  of Cnidaria, 309, 316
  of Crustacea, 825
  of Ctenophora, 328, 339
  of Echinodermata, 996
  of Entoprocta, 607
  of Gastrotricha, 431, 432
  of Mollusca, 514, 515f, 516–517, 517f
  of Nematoda, 681
  of Phoronida, 642
  of Platyhelminthes, 395, 399
  of Porifera, 241
  protandric, **179**
  protogynic, **179**
  of Tardigrada, 717
  of Urochordata, 1038
hermatypic corals, **303**
Hermodice carunculata, 534f–535f
Herpetomonas, 124f
Hesiolyra bergi, 554, 554f
Hesionidae, 537
Hesperiidae, 859
Hesperonoe adventor, 553
Hesse, R., 680
Heteractis, 284f
Heteranomia squamula, 419
Heterobranchia, 455, 464f, 466
  circulation and gas exchange of, 505, 505f
  locomotion of, 490
  lower, 455, 466
  reproduction of, 515
Heterochaerus langerhansi, 353
heterochrony, **203**
heterocoely of Chaetognatha, **428**
Heteroconchia, 455, 469
Heterocorallia, 320
Heterocyemida, 414, 417–418, 418f
Heterodera, 683
Heterodonta, 455, 456f–457f, 469, 470
heterogamety, **197**
heterokont flagellation of Stramenopila, **101**
Heterokrohnia, 422
  involucrum, 421f
Heterokrohniidae, 423
Heterolobosea, 61, 62, 63, 64, 115f, 117–118, 128
Heteronemertea, 436, 437f, 438, 440f, 441
  body wall of, 439
  feeding and digestion of, 442f, 443
  osmoregulation of, 445
  phylogeny of, 450, 450f
  reproduction and development of, 448f, 449
heteronomous annelids, **542,** 543, 552, 557
heteronomy, **542**
Heterophlias, 788f–789f
Heterophrynus batesii, 919f
Heteropora, 645f
Heterordera glycines, 670f
Heterorhabditidae, 677
Heterorhabditis, 675f
  bacteriophora, 670f, 682f
Heteroscleromorpha, 221, 222, 258
Heteroserolis, 787f
Heterostropha, 466
Heterotardigrada, 714, 715
Heteroteuthis, 514
heterotrophs, 9n, **56,** 65, **149**
heteroxenous life cycle of Apicomplexa, **85**–86
heterozooids, **648**
Hexabranchus sanguineus, 495
hexacanth larva of Platyhelminthes, **405**
Hexacorallia, 269, 270, 293, 320, 320f
Hexactinellida, 218f–219f, 221, 223, 227–228, 239
  development of, 244, 246f, 248
  distribution of, 249
  evolution of, 255, 257
  fossil record of, 256f, 257–258
  internal anatomy of, 228f
  phylogeny of, 257, 257f
  skeleton of, 146f, 220, 221, 233, 234
  taxonomic history and classification of, 220
Hexadella, 220, 233
Hexamermis, 670f, 682f
Hexamita, 115f, 116, 117
  meleagridis, 116
  salmonis, 116
Hexanauplia, 832
Hexapoda, 710f–711f, 728, 729, 731, 843–890, 895, 904
  abdomen of, 866
  body regions of, 860f
  characteristics of, 711t, 735t, 844

circulation and gas exchange of, 875f, 875–877
classification of, 4, 846–859
evolution and phylogeny of, 25, 48f, 754, 755, 755f, 832, 833, 834, 843, 846, 869–870, 887–890, 889f, 908
excretion and osmoregulation of, 877–878
feeding and digestion of, 864f, 870–875, 871f, 874f
general morphology of, 859–866
head of, 735t, 860–863
locomotion of, 866–870
nervous system and sense organs of, 878–883, 879f, 880f, 881f
number of species, 2t, 711t, 844
reproduction and development of, 860, 864–865, 883–887, 884f, 885f, 886f
thorax of, 863–866
Hexasterophora, 221, 258
Hexathelidae, 912, 924
Hexibranchidae, 467
hibernation, **22**
Higgins larvae, **704,** 704f, 705f
Himantarium gabrielis, 896f
Himasthla, 403
hindgut irrigation, **165,** 166f–167f
Hippodiplosia, 653f, 654f, 655
Hippolytidae, 779f
Hipponicidae, 465
Hippoporida calcarea, 302, 302f
Hippoporidridae, 646
Hippospongia, 233
Hirsutia, 785
Hirsutiidae, 785
Hirtopelta, 499
Hirudinea, 532, 542, 543, 597
Hirudinida, 541, 592, 593
Hirudinoidea, 532, 541, 591–597
body wall and coelom of, 592, 593f
circulation and gas exchange of, 595, 595f
excretion and osmoregulation of, 595
feeding and digestion of, 593–594, 594f
nervous system and sense organs of, 595f, 595–596
phylogeny of, 593
reproduction and development of, 596f, 596–597
support and locomotion of, 592–593, 593f
Hirudo, 593f, 594f, 595f, 597
medicinalis, 592f, 594
verbana, 594
Hislopia, 655
Histeridae, 872
Histioteuthis, 456f–457f
Histomonas meleagridis, 113, 114
Histoteuthidae, 472
Hochberg, Rick, 413
Hofstenia, 353
miamia, 352f–353f, 357f
Holarchaeidae, 922, 938
Holasteroida, 974
Holaxonia, 271
holoblastic cleavage, **187,** 188f, 192–193, 828f, 829, 830
Holometabola, 846, 847, 855, 857f, 889f, 890
holometabolous development of Hexapoda, **885,** 885f, 886–887, 887f
holoplanktonic species, **20, 199,** 200
Holothuria, 991, 992f
impatiens, 992f
tubulosa, 991, 992f
Holothuroidea, 971f, 974–975, 976
circulation and gas exchange of, 991, 993, 993f, 994
excretion and osmoregulation of, 994
external anatomy of, 972f
feeding and digestion of, 991, 992f
mutable collagenous tissue of, 980
nervous system and sense organs of, 995
phylogeny of, 1001, 1002f, 1003
reproduction and development of, 996, 997f, 998f
support and locomotion of, 978f–979f, 982, 984
symbiotic relationships of, 554f
water vascular system of, 982
Holothyrida, 914, 917
Holozoa, 64, 129, 255n
Holterman, M., 672
Homalorhagida, 696
Homarus, 747f, 828
americanus, 761, 782
homeobox genes, 10, 29–30, 216
of Ctenophora, 341
of Echinodermata, 1001
Hox. *See* Hox genes
of Porifera, 244
homologues, **41,** 42, 43, 44, 50
homology, **41**–43, 185
homonomous annelids, **541,** 542, 542f–543f, 545, 555, 557
homoplasy, **42,** 46, 185
Homoscleromorpha, 222
cell types in, 228, 230, 249
collagen of, 232
distribution of, 249
evolution and phylogeny of, 257, 257f, 258
homeobox genes of, 216
reproduction and development of, 239, 243, 244, 246, 246f, 248–249
skeleton of, 220, 233
taxonomic history and classification of, 220
honeybees, 24, 844–845, 845n, 865f
navigation of, 882–883
parthenogenesis of, 180
hood of Chaetognatha, **423**
Hoplocarida, 765, 773, 775–776
distinguishing features of, 802t–803t
phylogeny of, 832
reproductive features of, 806t–807t
Hoplonemertea, 436, 438, 440f, 441f
body wall of, 439f
circulation and gas exchange of, 444f
excretion and osmoregulation of, 445, 445f
feeding and digestion of, 441, 442f, 443f
nervous system and sense organs of, 446f
phylogeny of, 450, 450f, 451
reproduction and development of, 449, 449f
horizontal furrows of Xenoturbellida, **366,** 366f, 367
horizontal gene transfer. *See gene transfer*
hormones, **175**–176
endocrine, 175
neurohormones, 175
Hormurus, 952f, 953
australasiae, 953
host in symbiotic relationships, 23
definitive, **23, 201**
intermediate, **23, 201**
house of Appendicularia, **1031,** 1034f–1035f, 1036
Hox genes, 10, 29–30, **185,** 186f, 254, 348, 730, 968, 1027
in Arthropoda, 30, 734–735, 754
in Chordata, 1041, 1042
clusters of, **185,** 186f
in Cnidaria, 299, 310
evolutionary potential of, 30
in Rhombozoa, 414
in *Xenoturbella,* 350
H-system of collecting canals in Nematoda, 679f, 680
Hubrechtida, 450, 450f
Hubrechtidae, 436, 438
hull cells of Platyhelminthes, **398**
hull membrane of Platyhelminthes, **398**
humped-back model (HBM) of biodiversity, 24
Hutchinsoniella, 769f
Huxley, T. H., 521, 1031
Hyale, 788f–789f
Hyalophora euryalus, 859f
Hydra, 265, 267f, 277f, 302
cnidae of, 295f
life cycle of, 314f
movement of, 291, 292
nematocysts of, 293
nerve net of, 305f
reproduction and development of, 176f, 313–315
Hydractinia, 280f–281f, 302
Hydractiniidae, 274
hydranth of Cnidaria, 277f, 279f, **280**
hydric habitat, **23**
Hydrobiidae, 466
hydrocaulus of Cnidaria, 277f, **279,** 279f, **280**
Hydrocenidae, 463
hydrocoel of Echinodermata, **975,** 999, 1001
Hydroctena, 341, 341f
hydrogenosomes of Parabasalida, **113**
Hydroidolina, 270, 273, 286f–287f, 320f
Hydromedusae, 297
luminescence of, 308
nervous system and sensory organs of, 305–306
reproduction of, 311, 313f, 316
Hydrophilidae, 855
hydropores of Echinodermata, **978,** 998, 999
hydrorhiza of Cnidaria, 277f, 279f, **280,** 280f
hydrostatic skeleton, 144f, 144–145
hydrotheca of Cnidaria, 277f, **280**
hydrothermal vents, 1n, 9, 9n, 19, 20
hydrotube of Echinodermata, 998, 999
Hydrozoa, 267f, 269, 270, 272–273, 276
body wall of, 276, 276f
cnidae of, 293, 294, 295f
coloniality of, 319
diversity of, 280f–281f
DNA of, 216
evolution and phylogeny of, 302, 302f, 319, 320f
feeding and digestion of, 296, 297–298

growth forms of, 279f, 279–280
life cycle of, 314f, 316
medusoid stage of, 283, 285, 286f–287f, 311, 316
movement of, 291, 292
nematocyst of, 295f
nervous system and sensory organs of, 305, 305f, 306, 307f
polypoid stage of, 277f, 279, 279f, 279–280, 311, 316
radial symmetry of, 275f
reproduction and development of, 308, 311–316
support of, 287, 288, 290f
symbiotic relationships of, 251
taxonomic history and classification of, 269–270
tissue layers of, 268f
Hygrophila, 467
Hyman, L. H., 149, 202, 205, 360, 376, 574, 636, 637, 646
Hymeniacidon sanguinea, 252
Hymenolepis diminuta, 385, 390
Hymenoptera, 847, 855, 857f, 866
feeding of, 864f, 872, 873
leg modifications of, 865f
phylogeny of, 48f, 889f, 890
reproduction and development of, 884, 885
Hyocrinida, 973
Hyolithes, 15f
Hyperia medusarum, 790
Hyperiidea, 302, 767, 789, 790
Hyperoartia, 1041f
hyperparasitic parasites, **23**
Hypochilidae, 912, 924
Hypochilus, 928f–929f
Hypoconcha, 780f–781f
hypocone of Dinoflagellata, **79**
hypodermis of Arthropoda, **731**
hypognathous condition, **862,** 863f
Hypomzyostoma dodecephalis, 536f
hyponeural system of Echinodermata, **995**
hypostome of Cnidaria, 277f, **278**
hypostracum of Mollusca, **477**
hypotheca of Dinoflagellata, **79**
Hypsibius annulatus, 718f
Hypsogastropoda, 455, 465
Hyptiotes, 937
Hyriidae, 470

I

Ichneumonidae, 855
Ichthyosporea, 62, 255n
Ichthyotomus sanguinarius, 554, 554f
idiosoma of Chelicerata, **917**
Idmidronea, 648f
Idotea, 787f
Idoteidae, 787f
Ikeda, 580, 582f, 584
Ilbiidae, 467
Illacme plenipes, 896
Ilyocypris, 800f
imaginal discs of Hexapoda, **864**
imagos, **883**
imperforate extraxial region of Echinodermata, **976**
Incirrata, 471
incomplete gut, **149**
incurrent pores of Porifera, 227f, 234
independent effectors, **173**–174
indeterminate cleavage, **191**
indeterminate growth of Crustacea, **831**
indirect development, 197, 198f, 198–199, 750n, 885
induction process, **196**
infauna, **19**
infraciliature of Ciliata, **89**
infundibulum of Ctenophora, 331, 333, **337**
Infusoria, 59
infusoriform larvae
of Dicyemida, **414,** 415, 416, 417f
of Heterocyemida, 417, 418f
infusorigens of Dicyemida, **415**–416, 417f
Ingolfiella, 788f–789f
Ingolfiellidea, 767, 790
ingression, **193,** 194f
ink sac of Cephalopoda, 514
inquilism, **24**
Insecta, 4, 728, 843, 846, 847, 848–859
evolution and phylogeny of, 48f, 888–890
instars of Arthropoda, **739,** 741, 887
Integrated Taxonomic Information System (ITIS), 6
intermediate hosts, **23, 201**
intermolts of Arthropoda, **739**
interna, **815**
internal transport, 162–163
International Code of Zoological Nomenclature (ICZN), **37,** 38
interplate ciliated groove of Ctenophora, **338**–339
interradial canals of Ctenophora, 333f, **337**
intersex characteristics of Crustacea, **825**–826
interstitial cells of Cnidaria, 276, 276f
interstitial organisms, **19**
intertentacular organ of Bryozoa, **655**
Intestina, 375
Intoshia
linei, 419
variabi, 419
intracellular digestion, 148, 149
intranuclear mitosis of Protista, 69f, **70**
involution, **194,** 194f
Iodamoeba buetschlii, 70
Iphionidae, 537
Ircinia oros, 252
Iridia, 107
iris cells of Arthropoda, 748
Irish potato blight, 99
Irregularia, 974
Ischnochitonida, 495
Isocrinida, 973, 982, 996
Isodiametra
divae, 356f
earnhardti, 356f
pulchra, 358f
isogamy of Protista, **69**
Isolaimida, 673
isolecithal ovum, **187,** 187f, 198, 198f
Isopoda, 763f, 766, 786–789, 787f
phylogeny of, 832
thoracic appendages of, 804f–805f
Isoptera, 48f, 850
Isostichopus badionotus, 993f
Iuroidea, 914
Ivanov, A. V., 588
Iwata's larva, **450**
Ixodes pacificus, 918
Ixodida, 914, 917
Ixodidae, 918

J

Jackson, A., 215
Jägersten, G., 202, 206f
Jakobida, 63, 64, 117, 128
Janaria mirabilis, 302, 302f
Janthina, 490, 494
Janthinidae, 466
Jassa, 828f–829f
jaws, complex, of Gnathifera. *See* complex jaws of Gnathifera
Jeannel, R., 28
Jefferies, R. P. S., 1043
Jékely, G., 1048
Jennings, J. B., 387
jet propulsion, **292,** 335
Joeropsididae, 787f
Joeropsis, 787f
Johanssonia, 594f
Jones, M. L., 588
Jonston, J., 454
Julida, 897, 899
Juliidae, 467
Jurassic, 16

K

kala-azar, 121, 122
Kalyptorhynchia, 377, 388f, 389
Kamptozoa, 603, 637
Kapp, H., 428
Karenia, 79
brevis, 79
Karling, T., 362, 376, 407, 407f
karyomastigont systems of Diplomonadida, 116, 117
karyotype, 41
kelps, 97, 98f, **99**
kenozooids, **648,** 654f
kentrogon, **812,** 814f–815f
Kentrogonida, 812
Keratella slacki, 624
Keratosa, 221–222
kernels in gene regulatory networks, 185–187
Kerriidae, 854
Khan, Genghis, 858
Kimberella, 11, 524
quadrata, 524
kineties of Ciliata, **87**
kinetocysts of Radiolaria, **111**
kinetodesmal fiber of Ciliata, **89**
Kinetoplastida, 64, 119, 120–124, 121f
characteristics of, 122
classification of, 61, 62
diversity of, 58f–59f
nutrition of, 123
phylogeny of, 128
reproduction and life cycle of, 124
support and locomotion of, 123
kinetoplast of Kinetoplastida, **123**
kinetosome, **87, 142**
Kinorhyncha, 687, 693, 694, 694f, 695–698
blastocoelom of, 140
body wall of, 696
characteristics of, 695
circulation, gas exchange, excretion, and osmoregulation of, 697
classification of, 696
feeding and digestion of, 697, 697f
marine habitat of, 17, 19

nervous system and sense organs of, 697–698
number of species, 2t
phylogeny of, 25, 1049f, 1050
reproduction and development of, 697f, 698
segmentation of, 695, 696
support and locomotion of, 697

Kiwa, 781
hirsuta, 781
Klein, J., 969
kleptocnidae
of Ctenophora, **336**
of Mollusca, **495**
Kocot, K. M., 523
Koenenia, 919f–920f
Kofoidinium, 80
Komai, T., 340
Kowalevsky, A., 454, 637
krill, **776,** 777n, 810
Kristensen, R. M., 614, 626, 701, 716
Krohn, A., 414
Krohnitta subtilis, 421f
Krohnittellidae, 423
Krohnittidae, 423
Kronborgia, 387
Kruppomenia minima, 458f–459f
K-selected species, 187, 200
Kukalová-Peck, J., 733n, 864n, 869, 870, 888
Kulindroplax, 523

L

labellum of Hexapoda, **872**
labial palps of Mollusca, **496**
labial probolae of Nematoda, **677**
Labidiaster, 985
labidognathous chelicerae, 922f–923f, **924**
Laboea, 93
labyrinthulids, 97, 101
lacinia mobilis, **783**
Lacrymaria, 90
lacunae
of Nemertea, 443, 444f
of Platyhelminthes, **373**
lacunar system of Acanthocephala, **624–625,** 625f
Laevicaudata, 765, 772
Laevidentaliidae, 470
Laevipilina
antartica, 499
hyalina, 456f
Lagerstätten, 751
Lamarck, J.-B., 4, 220, 269, 375, 532, 574, 728, 765, 969
Lameere, A., 414
Lamellibranchiata, 454, 467–470
Lamellisabella, 585, 586f, 588, 589f
zachsi, 588
Lampea, 335
Lampyridae, 855, 882
land, adaptations to, 22–23
gas exchange structures in, 165
life cycle in, 200
osmoregulation in, 159, 159f
Lang's vesicle, **396**
languets of Urochordata, **1034**
Laniatores, 914, 919f–920f, 924
Lankester, E. R., 207, 728
lappets, **306,** 307f
Laqueoidea, 664f–665f
Laqueus erythraeus, 664f–665f

larva
acanthor, of Acanthocephala, 626, 627f
actinotroch, **643,** 643f, **644**
of Annelida, 570f–571f, 571, 572f, 579, 584
of Brachiopoda, 664f–665f, 665
of Bryozoa, 655, 655f, 656, 656f
chordoid, **611,** 611f
of Cnidaria, 265, 299, 300f, 308, 309, 309f, 311, 314f, 316, 316f, 317f, 319
competent, **199**
coronate, **656**
of Crustacea, 812f–815f, 823, 826, 830, 831
of Ctenophora, 328, 340, 340f
of Cycliophora, 609f, 610–611, 611f
cyphonautes, **656,** 656f
Desor's, **450,** 451
of Dicyemida, 414–415, 416, 417f
dipleurula, 644, 1019
of Echinodermata, 975, 997–998, 998f, 1001
of Enteropneusta, 1008, 1015, 1016f–1017f
of Entoprocta, 608, 608f, 609
episphere of, **209**
of Euchelicerata, 947f, 948, 953
feeding and nonfeeding types, 194, 197
of Heterocyemida, 417, 418, 418f
hexacanth, **405**
of Hexapoda, 886–887, 887f
Higgins, **704,** 704f, 705f
infusoriform, **414,** 415, 416, 417, 417f, 418f
Iwata's, **450**
lecithotrophic, **194, 197,** 198f, 198–199
of Loricifera, 704, 704f, 705f
metamorphosis of, 199
in mixed development, 200
of Mollusca, 519f, 519–521, 520f, 521f
of Nematoda, 671, 681, 683
of Nematomorpha, 671, 687, 688f, 690
of Nemertea, 449f, 449–450
Pandora, **610,** 611
of Phoronida, 643f, 643–644
planktotrophic, **194, 197,** 198, 198f
planula. *See* planula larva
of Platyhelminthes, 392, 397, 397f, 398f, 401, 401f, 402f–403f, 405
of Porifera, 239, 243–244, 245f, 246f, 247, 247f, 248, 249
of Priapula, 700f, 701
primary, **346,** 347, 348, 796f, 826, **997,** 1049f
Prometheus, 609f, **610,** 611
of Pterobranchia, 1018, 1019
of Pycnogonida, 960f, 961
Schmidt's, **450**
settlement process, 199
tadpole, 1038f–1039f, **1039,** 1040, 1043
tornaria, **1015,** 1016f–1017f, 1019
in trochaea theory, 208, 208f
trochophore, 208, 208f, 348
of Urochordata, 1038f–1039f, 1039, 1040–1041
vermiform, **414,** 415
Wagener's, **418,** 418f
Larvacea, 1022
larval skeleton of Echinodermata, **997**
Lasionectes, 769f, 805f
lasso cells of Ctenophora, **335**
lateral canals of Echinodermata, **981,** 982

lateral fins of Chaetognatha, **423**
lateral gene transfer. *See gene transfer*
Laternulidae, 470
latitudinal biodiversity gradient, **24**
Latrodectus, 922f–923f, 938
geometricus, 952
hasselti, 952
mactans, 952
Laurasia, 16
Laurentaglyphea, 783
Laurer's canal of Platyhelminthes, **400**
Lecanicephalidea, 379
Lecithoepitheliata, 377, 396f
lecithotrophy, **194, 197,** 198f, 198–199
Lecudina tuzetae, 83
legacy names, **26, 50, 347,** 348
leg theory on insect flight, 869
Leishmania, 120, 121, 122, 124f
braziliensis, 122
donovani, 122
infantum, 122
major, 122
mexicana, 122
leishmaniasis, 122
Lejopyge, 15f
Lemmatophora typa, 870f
lemniscus of Acanthocephala, 625f, **625**–626
Leocinclis, 119
Lepadocystis, 1000f
Lepas anatifera, 763f
Lepetellidae, 463
Lepetidae, 462
Lepetodrilidae, 463
Lepidocystis, 1001, 1002f
Lepidopleurida, 460
Lepidoptera, 16, 857f, 858–859, 859f, 866, 932
abdominal limbs of, 30, 735
classification of, 846, 847
development of, 887f, 888f
dispersal of, 869
feeding of, 863, 864f, 871, 872
phylogeny of, 48f, 889f, 890
sensory organs of, 881f
silk production of, 869, 932
wing-flapping rates of, 868
Lepidurus, 772
arcticus, 772
Lepocinclis, 58f–59f
Leptasterias tenera, 996
Leptocottis, 788f–789f
Leptodora, 770f–771f, 773
kindtii, 773
Leptodoridae, 773
Leptogorgia, 288f
Leptomedusae, 273, 286f–287f
Leptomitus, 15f
Leptomonas, 120, 121, 124f
Leptostraca, 765, 774–775, 803, 819
phylogeny of, 832
thoracic appendages of, 804f–805f
Leptosynapta, 996
Leptothecata, 273, 292
Lernaeodiscus porcellanae, 814f–815f
Leucauge, 710f–711f
Leucilla, 227
nuttingi, 218f
Leuckart, R., 269, 328, 422, 728, 969
Leucocytozoon, 82
Leuconia, 223
leuconoid sponges, **224,** 225f, 227, 236

leucoplasts, 57n
Leucosolenia, 224, 228f–229f, 234f–235f, 246
Leucothea, 328f–329f, 335
leukorrhea, **113**
Lévi, C., 220
levotropic direction, **188,** 189
Leys, S., 244n
life, first appearance of, 1n, 1–2, 10
life cycles, 178f, 197–201
  adaptations to land and fresh water in, 200
  classification of, 197
  direct development in, 197, 198f, 199–200
  indirect development in, 197, 198f, 198–199
  meroplanktonic, **198**
  mixed development in, 197, 198f, 200
  of parasites, 201
  settlement and metamorphosis in, 199
ligaments of Bivalvia, **479**
light-adapted eyes of Arthropoda, 749, 749f
Lightiella, 769f
Ligia, 763f, 787f
  pacifica, 763f
Lima, 456f–457f, 512
Limacidae, 467, 505f
limax amoebozoans, **74**
Limidae, 469, 490
Limifossor, 458f–459f
Limnocnida, 314f–315f
Limnocythere, 800f
Limnognathia maerski, 614, 626–632
  body wall and muscles of, 629, 630f, 631f
  jaw structures of, 630, 631f
  sculptured winter egg of, 632, 632f
Limnomedusae, 273, 286f–287f
Limnoridea, 766
Limoida, 469
Limulidae, 916
Limulus, 710f–711f, 915f, 916
  circulation and gas exchange of, 941f, 942
  digestion of, 939, 940, 940f
  ectocommensal on, 383f, 387
  nervous system and sense organs of, 173, 944
  polyphemus, 915f, 916, 942
  reproduction and development of, 947, 947f
Linckia, 995, 996f
  laevigata, 970f–971f
Lincoln, R., 793
Lindberg, D. R., 521, 522f
Linderiella, 770f–771f
Lineus, 447
  longissimus, 435
  ruber, 443, 448f, 450
  viridis, 450
Linguatula serrata, 796f
Lingula, 658f–659f, 659, 660f, 663, 665
  anatina, 664f–665f
Linguliformea, 658f–659f, 659–660, 661, 662, 664f–665f, 665
Linguloidea, 658f–659f, 659, 664f–665f
Linnaeus, C., 39n, 671, 728
  binomial nomenclature system of, 37
  on echinoderms, 969
  on Mollusca classification, 454
  on priapulans, 698
  on sipunculans, 574
  on sponge classification, 220
  Systema Naturae, 37, 374–375, 574, 698, 844
  Vermes phylum of, 374–375, 532, 574, 728, 969
  Zoophyta category, 268, 328, 636
Linuche, 303
Linyphiidae, 913, 936, 936f–937f, 948
Liocranidae, 954f–955f
Liphistiidae, 924
Liphistiomorpha, 928
Liponema brevicornis, 291f, 292
Lipostraca, 772
lips of Nematoda, **677**
Liriope tetraphylla, 286f–287f
Lissoclinum, 1034
Listriolobus, 581f
Lithistida, 234
Lithobiomorpha, 897, 898, 903, 905, 906
Lithobius, 900f–901f, 903f
  forficatus, 905f, 906f
Lithodidae, 781
Lithophaga, 490
Lithophora, 377
Lithotelestidae, 271
Litobothridea, 379
littoral region, 18
Littorina, 463f, 493f, 507f, 511f, 512f, 516f
Littorinimorpha, 455, 465–466, 509f
Lobata, 328f–329f, 332, 333–334, 341
  feeding and digestion of, 335
  general anatomy of, 331f
  reproduction and development of, 340
Lobatocerebrum, 1050
lobopods (lobopodia), **65, 74,** 74f, **142**
  of Onychophora, **722,** 723
locomotion, 140–148
  ameboid, 141–142, 142f
  cilia and flagella in, 142–143, 143f
  Reynolds numbers in, 141, 142
Loliginidae, 472
Loligo, 473f, 501f, 504f, 509, 510f, 517f, 518
long branch attraction (LBA), **51**
longitudinal canals of Echinodermata, **982**
longitudinal cleavage, **188,** 188f
longitudinal fission of Cnidaria, 308, 309f, 311
Lopadorhynchidae, 534f–535f, 538
Lopadorhynchus, 534f–535f
lophocytes of Porifera, **230,** 231f
Lophogastrida, 766, 783–785, 804f–805f
Lophogorgia, 177f
Lophophorata, 348, 635–665
  Brachiopoda, 657–665
  Bryozoa, 644–657
  Phoronida, 638–644
  phylogeny of, 1049f, 1050
  taxonomic history of, 636–637
lophophore of Lophophorata, **635,** 637–638
  in Brachiopoda, 658f, 659, 661–662, 663f, 665
  in Bryozoa, 646, 651
  in Phoronida, 638, 639f, 640f, **641,** 642, 644
Lophophus, 649
lophopodid colonies of Bryozoa, **649**
Lophotrochozoa, 25, 348, 1050
lorica, 701
  of Loricifera, 702, 703
  of Priapula larva, 700f, 701
  of Protista, 57, 87, 88f–89f, 100, 119, 125
  of Rotifera, 617f, **618**
Loricifera, 687, 693, 694, 701–704
  blastocoelom of, 140
  characteristics of, 703
  life cycle of, 704, 705f
  marine habitat of, 17, 19
  number of species, 2t
  phylogeny of, 25, 1049f, 1050
Lottia, 462f, 478f
  gigantea, 493
Lottiidae, 462, 462f
Lovén, S., 454, 978
Lovenia elongata, 970f–971f
Lovén's Rule, 976f, 978
Loxokalypodidae, 605
Loxophyllum, 88f–89f
Loxorhynchus, 780f
Loxosceles reclusa, 938
Loxosoma pectinaricola, 608f
Loxosomatidae, 605
Loxosomatoides, 603
Loxosomella, 604f–605f, 606f–607f, 608–609
  harmeri, 608f
  leptoclini, 606f–607f, 608f
  vancouverensis, 605f
  vivipara, 608f
Lubomirskiidae, 240, 248
Lucernaria, 267f, 291f
Lucifer, 778
luciferase, **174,** 882
luciferin, **174,** 308, 882
Lucinida, 470
Lucinidae, 499
Luffariella variabilis, 250
Lumbricidae, 533f, 540, 556f–557f
Lumbriculidae, 532, 541, 549, 591
Lumbricus
  circulation and gas exchange of, 557, 558f
  digestive system of, 557f
  excretion and osmoregulation of, 561f
  nervous system and sense organs of, 562f–563f
  reproduction and development of, 568f
  terrestris, 533f, 540
Lumbrinereis, 534f–535f
Lumbrineridae, 534f–535f, 536
luminescence. *See* bioluminescence
Lunataspis, 917
Luncinidae, 470
lung of Mollusca, **505,** 505f
lunules of Echinodermata, **984**
Lurifax vitreus, 499
lurking predators, 155
Lycaenidae, 859
Lycosa, 941f, 954f–955f
  erythrognatha, 938
  rabida, 951f
Lycosidae, 913, 936, 952
Lycoteuthidae, 472
Lycoteuthis, 514
Lyme disease, 918
Lymnaea, 518f
Lymnaeidae, 467
lymphatic channels of Platyhelminthes, **385,** 390
Lynceidae, 772
Lynceus, 770f–771f

Lyrocteis, 330f–331f
  imperatoris, 328f–329f
Lysmata californica, 762f–763f, 779f
Lytechinus
  pictus, 188
  variegatus, 991
Lytocarpus, 295

M

Macandrevia cranium, 664f–665f
Maccabeidae, 699
Maccabeus, 698, 700
  tentaculatus, 700, 701
Macellicephala, 554f
Macellomenia
  morseae, 459f
  schanderi, 459f
Macracanthorynchus hirudinaceus, 625f
Macrobdella decora, 594
Macrobiotus, 715f
Macrocheira kaempferi, 709, 761
macroconjugant of Ciliata, **95,** 96f
Macrocypris, 800f
Macrocystella, 1000f
Macrocystis, 63, 98f
Macrodasyida, 428f, 429, 430f, 431, 432
macroevolution, 27–31
macroherbivory, 153–154
macromeres, **188,** 189–190
  of Chaetognatha, 427
  of Echinodermata, **997**
  of Porifera, 247, 247f, 248
  quartet of, **189**
macronucleus of Protista, **70, 93,** 94, 94f, 95, 96f
Macropodus opercularis, 401
macroseptum of Annelida, **567**
macrospheric generation of Granuloreticulosa, **107**
Macrosteles quadrilineatus, 29
Macrostomida, 376, 382, 387, 395
Macrostomorpha, 376, 388f, 395, 396f, 407
  phylogeny of, 408, 408f
Macrostomum, 374f–375f, 388f, 396
  hystrix, 397f
  lignano, 397f, 407
Mactridae, 468f–469f, 470
Madin, L., 328
Madreporaria, 270–271
madreporite of Echinodermata, **978,** 980–981, 996f
MADS box transcription factor genes, 10
Magellania
  flavescens, 662f
  venosa, 658f
Magelona pitelkai, 534f–535f
Magelonidae, 532, 534f–535f, 542
  circulation and gas exchange of, 559
  nervous system and sense organs of, 561
Magiciada, 854
magnetotactic capabilities, 883n
Magnoliophyta, 129
Mah, J. L., 1048
Majidae, 780f
Majoidea, 778
major claw of Crustacea, **826**
Malachia, 454
Malacobdella, 440, 442, 442f, 443, 447, 449
  grossa, 437f
Malacosporea, 274
Malacostraca, 725, 765, 768, 773–774
  anatomy of, 774f, 775f, 777f
  distinguishing features of, 802t–803t
  diversity of, 762f–763f
  external morphology of, 765f
  nervous system and sense organs of, 822, 823
  phylogeny of, 832, 833f
  reproductive features of, 806t–807t
Malacovalvulida, 274
Malacozoa, 454
malaria, 81, 82, 84, 86, 86f, 846, 872, 873
Malawispongiidae, 240
Maldanidae, 539, 548f–549f, 549
Maldanomorpha, 539
male atrium of Platyhelminthes, **395**
male heterogamety, **197**
Malldida, 469
Malleidae, 469
Malletiidae, 469
Mallicephala, 553
Malpighian tubules, 161f, 162, **743, 745,** 746f, 908
  of Euchelicerata, 941, 943
  of Hexapoda, 878, 878f
  of Myriapoda, 903f, 904
  of Tardigrada, 714, 717
Mammalia, 1022, 1041f
Manayunkia aestuarina, 153f
mancas, **783**
mandible
  of Crustacea, **801,** 810
  of Hexapoda, 863, 870, 871, 872
  of Myriapoda, 895, 897, 901, 906
Mandibulata, 731, 752, 754, 908
Mangelia, 493f
Mantispidae, 856
mantle
  of Brachiopoda, **660**–661, 663, 665
  of Mollusca, **472,** 474–477, 479f, 481, 505
  of Urochordata, 1033
mantle canals of Brachiopoda, **661,** 661f
mantle cavity of Mollusca, **472,** 474–477, 476f, 478f, 485, 505, 506
Mantodea, 846, 847, 851f, 851–852
  leg modifications of, 865f
  phylogeny of, 48f, 889f
Mantophasmatodea, 48f, 846, 847, 851f, 853, 889f
manubrium of Cnidaria, **278, 285,** 286f
Marchantiophyta, 4n, 55n, 129
Margaritidae, 463
Margaritiferidae, 470
Margaritiflabellum, 317
Marginellidae, 466
Marginifera, 658f–659f
Margulis, L., 29, 125
Marimermithida, 673
marine habitat, 17–20
  gas exchange structures in, 165
  osmoregulation in, 158, 159f
  pH of, 17–18
  regions of, 19f
Markuelia, 694
Marphysa, 570f
Marrella, 15f
Marteilia, 112
Martesia, 490
Martin, J. W., 765, 777, 801
Martinssonia elongata, 835f
mastax of Rotifera, 621, 621f
Masterman, T., 637
Mastigias, 303, 303f
mastigonemes, **143**
Mastigophora, 60
Mastigoproctus, 919f–921f, 927
Mastophora, 938
maxillae
  of Crustacea, **801,** 810, 820f
  of Hexapoda, 863, 870, 871
  of Myriapoda, 895, 897, 898, 901
maxillipeds of Crustacea, **801,** 802t
Maxillopoda, 790, 832, 833f
maxillules of Crustacea, **801,** 810
maximum likelihood algorithm, **51**
Mayorella, 71f, 74
Mayr, E., 38, 39–40
Mazocraeidea, 378
Mazoglossus, 1018
McClintock, B., 28
McHugh, D., 540
McKinney, M. L., 203
McNamara, K. J., 203
Meara, 360–361, 362
  stichopi, 361f, 362, 363, 363f, 364, 365f
mechanoreceptors, 169, 746–747
  of Crustacea, 822, 824f
  of Euchelicerata, 944, 945, 945f
Mecoptera, 48f, 846, 847, 856–857, 857f, 889f
medicinal leeches, 594
medusoid form of Cnidaria, **265,** 283–287, 308, 319
  feeding and digestion of, 296
  movement of, 292–293, 293f
  nervous system and sense organs of, 305–308, 307f
  support of, 287
medusoid hypothesis on Cnidaria, **319**
Medusozoa, 267f, 269, 271–274, 283, 299
  feeding and digestion of, 296
  phylogeny of, 319, 320, 320f, 321
  polypoid stage of, 277
Megadasys, 428f, 431
Megaderaion, 1018
megaesthetes of Mollusca, **479**
Megalaspis acuticauda, 753f
Megalodicopia hians, 1029f
Megalomma, 564f
megalopa of Crustacea, **831**
Megaloptera, 48f, 846, 847, 856, 857f, 889f
Meganyctiphanes, 777f
megascleres, **234**
Megascolecidae, 540
Megascolides australis, 540
megaspheric generation of Granuloreticulosa, **107**
Mehlis's glands of Platyhelminthes, **400**
meiofauna, **19**
Meioglossus, 1019
  psammophilus, 1007, 1012f–1013f, 1014
Meiopriapulus, 698
  fijiensis, 700, 701
meiosis, 178
Melanopsidae, 465
Meldal, B. H., 672
Melibe, 494, 500
Melita, 788f
Mellita, 972f
Meloidae, 856
Meloidogyne, 683
  incognita, 685f
Melongenidae, 466
Melosira, 98f

Membracidae, 854
Membranipora, 645f, 648, 649, 651, 653f, 654, 655, 656f
meningoencephalitis, amebic, 64
Menippe adina, 828f–829f
Menodium, 119
Mercenaria, 468f–469f
Mereschkowsky, K., 29, 125
meridional canals of Ctenophora, 333f, 334f, **337,** 338f
meridional rows of Ctenophora, **332**
Meristoderes, 696f
Merlia normani, 224f
Mermithida, 670, 673
meroblastic cleavage, **187,** 188f, 830
merogony of Apicomplexa, **83**
meroplanktonic animals, **20, 198**
Merostomata, 710f–711f, 912, 913f, 915f, 915–917
Mertensia ovum, 335
Mertensiidae, 341
Mesanthura, 787f
mesenchyma, 266n
mesenchymal cells, 266n
mesenchyme, **266n**
  of Cnidaria, 265
  of Ctenophora, **334,** 338
  of Platyhelminthes, 373, 380, 382, 383, 385, 390
Mesenchytraeus, 541
mesenterial filament of Cnidaria, 278f, **279,** 290f, 296
mesentoblast, **195,** 195f
mesic environment, **23**
Mesobalanoglossus, 1018
mesocoel, **196,** 196f, **636,** 638, 640, 661
  of Echinodermata, **980,** 999
  of Enteropneusta, 1011
  of Pterobranchia, 1015
mesoderm, **139, 191, 193,** 195–196, **345,** 348
  evolution of, 139–140
  formation of, 195, 195f, 196
  of Hexapoda, 884–885, 885f
  of Platyhelminthes, 380
mesoglea, **195,** 265, **266n**
mesohyl, **223,** 227, 230, **266n**
mesomeres of Echinodermata, **997**
Mesomycetozoea, 62
mesonotum of Hexapoda, **863**
mesoparasites, **23, 201**
mesopeltidium of Chelicerata, **928**
mesosoma, **636**
  of Chelicerata, **928**
  of Echinodermata, 975
  of Hemichordata, 1009
Mesostigmata, 914, 917
Mesostoma, 156f, 393f
Mesotardigrada, 714
Mesothelae, 912, 922f, 924, 937, 943
mesothorax of Hexapoda, **863**
Mesozoa, 414
Mesozoic era, 16
metaboly of Euglenida, **119**
Metabonellia, 580, 584, 585f
  haswelli, 581f, 585f
metacercaria of Platyhelminthes, **401,** 403
metachronal model of arthropod locomotion, 739
metacoel, **196,** 196f, **636,** 638, 661, 663
  of Enteropneusta, 1011
  of Pterobranchia, 1015
metamerism
  of Annelida, **531**
  of Mollusca, 524
metamorphosis, 199
  of Crustacea, 812f–813f, **830**–831
  of Hexapoda, 887
  of Mollusca, 519, 519f
metanauplius, **830,** 835
Metandrocarpa, 1029
metanephridia, 161f, **162**
  of Annelida, 560, 561f, 573, 578, 584, 595
  of Brachiopoda, 663
  compared to protonephridia, 162
metanephromixium, **162**
metanotum of Hexapoda, **863**
metapeltidium of Chelicerata, **928**
metapleural folds of Cephalochordata, **1026**
metapodium of Mollusca, **487**
Metapolycope, 800f
metasoma, **636**
  of Chelicerata, **928**
  of Echinodermata, 975
  of Hemichordata, 1009
metathorax of Hexapoda, **863**
metatroch, **208,** 571
Metazoa, 55, 56, 60, 61, 62, 64, 125
  Bilateria. *See* Bilateria
  classification of, 4, 8, 25
  definition of, **9**
  non-Bilateria. *See* non-Bilateria Metazoa
  origin of, 10–17, 129, 203–209, 204f
Metchnikoff, E, 148
Metridium, 296, 298, 308
  giganteum, 277
  senile, 266f–267f
Metschinkowiidae, 240
Metschnikoff, E., 205, 205f, 340, 1007
microaesthetes of Mollusca, **479**
Microcerberidea, 766
microconjugant of Ciliata, **95,** 96f
Microcyema, 417–418, 418f
  gracile, 417
Microdajus pectinatus, 793f
Microdecemplicida, 909
microevolution, 27, 28, 30, 31f
Micrognathozoa, 347, 348, 429, 613, 614, 626–632
  acoelomate body plan of, 140
  body wall and muscles of, 629, 630f, 631f
  characteristics of, 629
  ciliophores of, 628, 628f
  collar receptor of, 629
  complex jaws of, 614f, 630, 631f
  epidermal plates of, 628, 628f
  feeding, and digestion of, 630–631, 631f
  locomotion of, 629–630
  as meiobenthic, 19
  number of species, 2t
  phylogeny of, 25, 1049f
  sculptured winter egg of, 632, 632f
  ventral muscle plate of, 631, 631f
  zip-junctions of, 628
micromeres, **188,** 189–191
  of Chaetognatha, 427
  of Echinodermata, **997**
  of Porifera, 247f, 248
micronucleus of Protista, **70, 93,** 94, 94f, 95, 96f
Microphallus, 390f
  papillorobustus, 390
Micropilina arntzi, 514
Microplana, 387
micropores of Apicomplexa, **83**
micropredators, **23**
Micropygoida, 974
microscleres, **234**
microspheric generation of Granuloreticulosa, **107**
Microsporidia, 60, 64, 129
Microstoma caudatum, 302
microtriches of Platyhelminthes, **384,** 384f
microtubules, **142**
Micrura, 450
  verrilli, 437f
Mictacea, 766, 785
mictic ova of Rotifera, 623, 623f
Mictocaris, 785, 786f
  halope, 785
midsagittal plane, **138,** 138f
migratory behavior of Crustacea, 808
Millepora, 280f–281f, 290, 295
Milleporidae, 288, 320
Miller, S., 1n
Millitauropus, 902
Milne-Edwards, H., 454, 637
Milne-Edwards openings of Crustacea, **820**
Milnesium
  swolenskyi, 711
  tardigradum, 713
Mimetidae, 913, 937f, 938, 954f–955f
Ministeria vibrans, 255n
Minot, C., 376, 436
Minyadidae, 292
miracidia of Platyhelminthes, **401**
Miracrustacea, 834
Misof, B., 832, 887, 888, 890
Misophrioida, 767, 798
mitochondria of Protista, 66, 68f
mitosis, 69f, 69–70, 76, 80, 83, 94–95, 101, 124
mitosomes of Diplomonadida, **116**
Mitridae, 466
mixed development, 197, 198f, 200, 830
mixonephridium, **162**
mixotrophs, **56**
Mnemiopsis, 331f, 337
  leidyi, 328f–329f, 335, 336–337, 340
Modocia, 15f
Moinidae, 773
molecular developmental biology, 27, 184, 191
molecular genetics, 184
molecular phylogenetics, 26, 47, **51**–52, 1047–1051
Molgula, 1029
Mollusca, 347, 348, 450, 453–524, 579
  body plan of, 472–474
  body wall of, 474, 479f
  characteristics of, 454
  circulation and gas exchange of, 478f, 503f, 503–506, 504f
  coloration and ink of cephalopods, 513–514
  development of, 472, 518–521
  digestion of, 500f–501f, 501–503, 502f
  in Ediacaran Period, 11

excretion and osmoregulation of, 472, 478f, 506–507
feeding of, 155f, 491–501, 492f, 493f, 494f
hypothetical ancestral, 521
locomotion of, 486f, 486–491, 488f
mantle and mantle cavity of, **472,** 474–477, 476f, 478f, 479f, 481, 485, 505, 506
morphological diversity of, 456f–457f
nervous system and sense organs of, 172, 472, 507–513
number of species, 2t, 1047
phylogeny of, 25, 521–524, 522f, 1049f, 1050
reproduction of, 472, 478f, 514–518
shell of, 476f, 477–483, 479f, 480f, 481f, 488f, 490
symbiotic relationships of, 499–501
taxonomic history and classification of, 454–472
torsion of, 483f, 483–485
molluscan cross, 190f, **191**
Molluscoides, 637
Molpadida, 975
molting of Arthropoda, 730, **739,** 739f, 740f–741f, 741, 742f
in Crustacea, 808, 831
in Hexapoda, 887
molt-inhibiting hormone (MIH) of Arthropoda, **741**
Monhysterida, 672
Moniezia, 394f
Moniliformis, 624
moniliformis, 390
Monoblastozoa, 204, 214, 214f
Monobrachiidae, 273
Monobrachium, 280f
Monocotylidea, 378
Monocystis, 83
Monodella, 786f
monodisc strobilation, **310**
Monogenea, 373, 378, 381f, 384, 401, 401f
phylogeny of, 407, 408f, 409
Monogononta, 614f, 617f, 618, 620f, 622, 623
monomyarian condition of Mollusca, **490**
Mononchida, 673
Mononchus, 678f–679f
monopectinate condition of Mollusca, **505**
monophyly, **4, 40,** 40f
Monopisthocotylea, 378
Monoplacophora, 453, 455, 456f, 459–460, 473, 478f
circulation and gas exchange of, 503, 506
excretion and osmoregulation of, 507
feeding of, 495
locomotion of, 486
mantle and mantle cavity of, 477
phylogeny of, 522f, 523, 524
reproduction of, 514
shell of, 479
symbiotic relationships of, 499
monopodial growth of Cnidaria, **279,** 279f
Monorhaphis chuni, 251
Monosigidae, 125
Monostilifera, 436, 437f, 438, 442f
phylogeny of, 450, 450f, 451
monoxenous life cycle
of Apicomplexa, **85**
of Kinetoplastida, 124
Monstrilloida, 767, 798
Montchadskyellidea, 378
Montipora, 387
Montucaris, 785
Mopalia, 495
muscosa, 456f
Mopaliidae, 495
Mora, C., 6
Mormonilloida, 767, 798
morphogenesis, **196**–197
morphogens, **196**
mosaic ova, **191**
motor (efferent) nerves, **168**
Motyxia, 896f, 897
mouth cone of Loricifera, **702**
mucocysts
of Ciliata, **92,** 93f
of Dinoflagellata, **80**
of Euglenida, **119**
mucous-net suspension feeding, 151, 152f
mucrones of Loricifera, **704**
mud mounds, 257
Müller, J., 637, 869, 1007
Müller, K. J., 701, 752, 835
Müller, O. F., 113, 980
Müller, P., 84
Müller's larvae, **397,** 397f, 398f
multicellularity, 139
evolution of, 2, 10, 184, 205–206, 1048
multiple fission of Protista, **68,** 76, 83, 94
multipolar delamination, 248
multipolar egression, **249**
multipolar ingression, **193**
Multitubulatina, 429
Multivalvulida, 274
Murex brandaris, 453n
Muricea
californica, 285f
fruticosa, 289f
Muricidae, 466, 493, 500f–501f, 515f
Mus, 186f
Musca domestica, 72
Muscidae, 858
muscle arms of Nematoda, 674f, **675**
muscles, 144–148
antagonistic, 146, 147f
contraction of, 148
in hydrostatic skeleton, 144f, 144–145
Muspiceida, 673
mutable collagenous tissue (MCT) of Echinodermata, **980,** 982
mutualism, **23**–24, 301
Mycale, 231f
vansoesti, 251
Mygalomorphae, 912, 924
Myida, 470
Myidae, 470
Myochamidae, 470
myocytes of Porifera, 228f–229f, **230,** 239, 275
Myodocopa, 767, 798, 799f, 800f
Myodocopida, 767, 798, 800f
myoepithelial cells of Cnidaria, **275,** 276f, 290, 290f
Myolaimina, 672
myonemes
of Ciliata, **90**
of Cnidaria, **275,** 276f, 290f
Myopea callaeum, 357f
Myopsida, 472
myotomes of Cephalochordata, **1023**
Myrianida gidholmi, 566f
Myriapoda, 710f–711f, 728, 729–730, 731, 895–909
characteristics of, 711t, 735t, 899
circulation and gas exchange of, 903–904, 904f
classification of, 897–899
excretion and osmoregulation of, 904
feeding and digestion of, 901–903, 903f
head and mouth appendages of, 735t, 899–901
locomotion of, 901, 902f
nervous system and sense organs of, 904–905, 905f
number of species, 2t, 711t
phylogeny of, 48f, 754, 755, 755f, 908–909
reproduction and development of, 905–908
Myriochelata, 908
Myrmecodesmus adisi, 904
Myrmecotypus, 922f–923f
Myrmeleontidae, 856
Mysida, 763f, 766, 783, 784f, 819
Mysidacea, 783
mysis larva, **831**
Mystacocarida, 767, 790, 794f, 797
distinguishing features of, 802t–803t
nervous system and sense organs of, 825
phylogeny of, 833f
reproduction and development of, 806t–807t, 831
Mytilida, 469
Mytilidae, 468f–469f, 469, 488, 490
Mytilus, 468f–469f, 498f
californianus, 489f, 497, 498f
edulis, 497
Mytridium, 93
Myxini, 1022, 1041f
Myxobolus cerebralis, 267f, 299, 300f, 301
Myxogastrida, 71, 72, 73, 73f
Myxospongiae, 221, 222
Myxosporea, 274
myxospores, **299**
Myxosporidium bryozoides, 269
Myxozoa, 55, 60, 64, 265, 267f, 274, 277, 293
body wall of, 276
classification of, 42, 55, 269
life cycle of, 301, 316
as parasitic, 299, 299f, 316, 321
phylogeny of, 25, 319, 320f, 321
reproduction of, 316
Myzophyllobothrium, 386f
myzorhynchus of Platyhelminthes, **386,** 386f
Myzostoma, 536f
cirriferum, 536f
Myzostomida, 541, 553

N

Nabokov, V., 859n
Nacellidae, 462
Naegleria fowleri, 64, 118
nagana, 122
naiads, **886**
Naididae, 540, 554
naked amebas, **74,** 75–76
Namanereidinae, 537

Nanaloricida, 701
Nanaloricidae, 701, 703, 704
Nanaloricus, 702f, 704
mysticus, 701, 702f, 703f, 704
Nanosella fungi, 855
Naraoia, 15f
Narceus americanus, 902f
Narcomedusae, 273, 292
Nasanov gland, **882**
Nassariidae, 466
Nassula, 88f–89f
Nassulopsis, 91f
elegans, 93f
Nasuia deltocephalinicola, 29
Natantia, 778
Natica, 492f
Naticidae, 466, 486f, 493
Natronobacterium gregoryi, 9n
natural selection, 27, 31
naupliar eye of Crustacea, **823**
nauplius of Crustacea, **801,** 830
Nausithoe, 310
Nautilida, 471
Nautilidia, 455, 471
Nautiliniellidae, 537
Nautilus, 455n, 456f, 471, 471f, 477, 523, 524
circulation and gas exchange of, 503
digestion of, 502
locomotion of, 490–491, 491f
reproduction of, 518
sensory organs of, 513
shell of, 481, 481f, 482
Navanax, 511
Navicula, 100f
navigation mechanisms of insects, 882–883
Neanthes succinea, 189
Nebalia, 774f
Nebaliopsis typica, 774, 775
Nebela collaris, 71f
neck of Platyhelminthes, **382**
Nectocaris, 14f
Nectocarmen antonioi, 280f–281f
Nectonema, 671, 687, 688f, 689, 689f, 690
Nectonematoidea, 687
nectophores, **280,** 282
Needham's sac of Cephalopoda, **517,** 517f
Neis, 332
nekton, **20**
Nemata, 671
nematocyst batteries, **293**
nematocysts, **80, 293,** 294f, 302
Nematoda, 669, 670, 670f, 671–686, 693
blastocoelom of, 140
body wall, support, and locomotion of, 673–675
characteristics of, 672
circulation, gas exchange, excretion, and osmoregulation of, 677–680, 679f
classification of, 672–673, 673f
cuticle of, 673–674, 674f, 675, 675f
feeding and digestion of, 677, 678f–679f
life cycles of, 682–686, 685f
nervous system and sense organs of, 680f, 680–681
number of species, 2t, 1047
phylogeny of, 25, 673f, 1049f, 1050
reproduction and development of, 671, 681–686, 682f, 683f
nematodesmata of Ciliata, **91**
Nematodinium, 80
nematogens
of Dicyemida, **414,** 415, 415f, 416, 417f
of Heterocyemida, 418, 418f
Nematoida, 348, 429, 669–690, 693, 694, 751
Nematoda, 671–686
Nematomorpha, 686–690
phylogeny of, 25, 1049f, 1050
Nematomorpha, 669, 670, 671, 686–690, 693, 694
acoelomate body plan of, 140
body wall, support, and locomotion of, 687, 688f
characteristics of, 687
circulation, gas exchange, excretion, and osmoregulation of, 689
classification of, 687
feeding and digestion of, 687–689
nervous system and sense organs of, 688f, 689
number of species, 2t
phylogeny of, 25, 1049f, 1050
reproduction and development of, 671, 688f, 689
Nematopsides, 80
Nematostella vectensis, 308, 310
Nemertea, 348, 376, 384, 435–451
body plan of, 139
body wall of, 439, 439f, 440f
characteristics of, 436
circulation and gas exchange of, 438, 443–444, 444f
excretion and osmoregulation of, 444–445, 445f
feeding and digestion of, 438, 440–443, 441f
nervous system and sense organs of, 445–447, 446f
number of species, 2t
phylogeny of, 25, 450f, 450–451, 1049f, 1050
reproduction and development of, 447–450, 448f, 449f
support and locomotion of, 439–440
taxonomic history and classification of, 436–438
Nemertes, 436
Nemertina, 436
Nemertoderma, 362
westbladi, 363, 363f, 364, 365, 365f, 366
Nemertodermatida, 346, 349, 350–351, 360–366
body structure of, 362
cell and tissue organization of, 362
ciliary rootlet structure of, 362
development of, 365f, 365–366
muscles of, 349–350, 362–363, 363f
nervous system of, 364
nutrition, excretion, and gas exchange of, 364
phylogeny of, 25
proboscis of, 361, 361f
reproduction of, 361–362, 364f, 364–365, 365f
support and movement of, 361, 362–364
symbiotic bacteria associated with, 361, 361f, 362
zonula adherens of, 362, 363f
Nemertoscolex parasiticus, 441
Neoacarus, 916f–917f
Neocephalopoda, 455, 471
Neocopepoda, 767
Neodasys, 428f, 429, 431, 432
Neodermata, 378, 384, 403
phylogeny of, 407, 408f, 409
neodermis of Platyhelminthes, **384**–385
Neoechinorhynchus, 624
Neofibularia, 250
Neogastropoda, 455, 465, 466, 493
circulation and gas exchange of, 504f
development of, 521
digestion of, 500f–501f
feeding of, 494
nervous system of, 509f
reproduction of, 515f
Neoglyphea, 783
Neolepetopsidae, 462
Neoloricata, 460
Neomenia carinata, 458f–459f
Neomeniomorpha, 453, 458–459
Neometra, 972f
Neomphalida, 463
Neomphalidae, 463
Neomphalus, 496
Neomus, 302
Neonemertea, 436, 438, 440f
Neonesidea, 800f
Neoophora, 376
neoöphorans, **395,** 396f, 398, 399f
Neopilina, 453, 460f
galatheae, 524
Neoproterozoic era, 10
Neoptera, 846, 847, 850, 889f, 890
Neosabellaria cementarium, 535f
Neoscona arabesca, 939f
Neotenotrocha, 628
neoteny, **203**
Neothyris lenticularis, 659
Nephila, 928f–929f, 931, 932, 952
nephridia, 161–162
of Annelida, 560, 561f, 578, 584
of Arthropoda, 745, 821
of Mollusca, 506–507, 507f
of Onychophora, 725
nephridioduct, 161, 560, 561, 561f
nephridiopores, 161
of Annelida, 560, 573, 578
of Gastrotricha, 431
of Platyhelminthes, 391, 391f
nephrocoel theory, **207**
nephromixia, **162**
Nephrops norvegicus, 609f
nephrostomes of Phoronida, 641f, 642
Nephrozoa, 1049f, 1050
Nephtyidae, 537, 545
Nephtys, 545
Nepidae, 854
Nereididae, 164, 534f–535f, 537, 545, 550f
Nereis, 550f
circulation and gas exchange of, 555, 557
fucata, 553
irrorata, 566f
nervous system and sense organs of, 562f
support and locomotion of, 545, 546f, 547
virens, 547, 555
Nereocystis, 63
Neriene litigiosa, 952
neritic zone, **19,** 19f
Neritidae, 463

Neritiliidae, 463
Neritimorpha, 455, 463
Neritopsidae, 463
nerve nets, 174f, **174**–175
of Cnidaria, **305,** 305f
of Ctenophora, 332, 338
of Echinodermata, 995
nervous system, 168–175
cerebral ganglion in, 175
impulse conduction in, 168–169, 169f
nerve nets in, 174f, 174–175
neural crest, of Chordata, **1043**
neural crest cells of Chordata, **1043**
neural gland of Urochordata, **1037**
Neuralia, 347, 1048
neural plate of Cephalochordata, **1026**
neural tube, dorsal, **1022**
neurohormones, **175**
neurons, **168**
neuropodium of Annelida, **543,** 544f
neuropore of Cephalochordata, **1027**
Neuroptera, 48f, 846, 847, 856, 857f, 889f
Neuropterida, 846, 856
neurosecretory cells of Platyhelminthes, 392
neurotoxic shellfish poisoning (NSP), 79
neurotroch of Annelida, 571
Newby, W. W., 540
next generation sequencing, **51**
nidamental glands
of Mollusca, **518**
of Phoronida, **642,** 643
Nielsen, C., 208f, 208–209
Niobia, 313f
Niphargus, 790
nitrogenous wastes, 157–158, 158f
Nitsche, H., 637
NK genes, 254, 255
Noctiluca, 79, 80
Noctuidae, 859
Nodal gene, 348, 968
nomenclature, biological, **36**–39
binomial, **37**
binominal, **37**
principles of, 37–38
non-Bilateria Metazoa, 139
Cnidaria, 213, 265–321
Ctenophora, 213, 327–342
Placozoa, 213, 214, 215–216
Porifera, 213, 214, 216–258
notochord of Chordata, 968, **1021**–1022, 1041–1042
in Cephalochordata, 1023, 1025, 1026
in Urochordata, 1031
Notodelphys, 794f–795f
Notomastus, 564f
Notomyotida, 973
Notonecta, 690
Notonectidae, 854
notoneuron, **208**
notopodium of Annelida, **543,** 544f
Notosaria nigricans, 660f
Notostigmophora, 897, 898
Notostraca, 762f–763f, 765, 772
distinguishing features of, 802t–803t
reproductive features of, 806t–807t
Novocrania, 659, 662
lecointei, 658f
Nucella, 494f, 515f, 521
emarginata, 494f
nuchal organs of Annelida, 564f, **565,** 578f, 579
nuchal tentacles of Sipuncula, **572**
Nucleariida, 62, 64
nuclear ß-catenin, 192f
nucleomorph, **59,** 126
of Chlorarachniophyta, 105
of Cryptomonada, **103**
Nucula, 496, 497f
Nuculana, 496
Nuculanidae, 469
Nuculaniformii, 469
Nuculida, 469
Nuculidae, 469
Nuculiformii, 469
Nuda, 328
Nudibranchia, 466, 482
circulation and gas exchange of, 505
digestion of, 500f–501f
feeding of, 495
locomotion of, 490
Nudipleura, 455, 464f, 466–467
Nummulites, 107
gizehensis, 107
nuptial net of Myriapoda, 906, 907f
nurse cells of Porifera, 240, 242, 248
nutritive-muscular cells of Cnidaria, **275,** 276, 276f
Nyctalops crassicaudatus, 919f–921f
Nymphalidae, 859, 888f
Nymphon, 960f
brevirostre, 957f
Nymphopsis spinosissima, 956f, 959
nymphs, **886**

O

Oakley, T. H., 832
Obelia, 314f–315f
geniculata, 316
oblate bells of Cnidaria, **292**
obligatory symbionts, **23**
oceanic zone, **19,** 19f
Oceanospiralleles, 590
ocelli, **172**
of Hexapoda, **860,** 879, 883
Ochetostoma, 581f
Ochromonas, 99f
Ockham's razor, 46n
Octacnemus, 1029f, 1033
Octocorallia, 269, 271, 285f, 320f
Octomitis, 117
Octopoda, 471
Octopodidae, 471
Octopodiformes, 455, 471
Octopoteuthidae, 472
Octopus, 495
anatomy of, 475f
bimaculoides, 456f
chierchiae, 475f
dofleini, 472
sensory organs of, 512f, 513
Octospiniferoides, 624
ocular plate of Echinodermata, **978**
Ocular Plate Rule, 975, 977f, **978,** 981, 982
Ocypodidae, 781, 826
Ocyropsis, 339
Odaraia, 15f
Odonata, 846, 847, 849f, 850
evolution and phylogeny of, 48f, 889f, 889–890
flight of, 868
nervous system and sense organs of, 879
odontoblasts of Mollusca, **492**
Odontogriphus, 14f
odontophore of Mollusca, **472, 492**
oecial vesicle of Bryozoa, **655**
Oegopsida, 472
Oenonidae, 536–537
Oesophagostomum, 679f
Oikopleura, 1034f–1035f
ojassie, 645f, 646
Olenellus gilberti, 753f
Olenidae, 729
Olenoides, 15f
Olfactores, 968, 1022
phylogeny of, 1041f, 1042, 1043
Oligochaeta, 532, 597
Oligostraca, 832
Olindiasidae, 273
Olivellidae, 466
Olividae, 466
olynthus of Porifera, 226f, **248**
Omalogyra, 494, 515
Omalogyridae, 466
Ommastrephidae, 472
ommatidia, **172,** 173f, **748,** 749, 749f, 755n
of Crustacea, 823–825
of Hexapoda, 879, 880, 880f
omnivores, 149
Onchidiidae, 467
Onchidorididae, 467
Onchnesoma, 578
Onchocerca volvulus, 672, 683, 685–686, 781
Onchoproteocephalidea, 379, 380
oncomiracidium of Platyhelminthes, **401,** 401f
oncosphere of Platyhelminthes, **405**
Onega, 11
Oniscidea, 766, 786, 789
On the Origin of Species (Darwin), 4, 30
ontogenetic analysis, 45
ontogeny, **183**
and phylogeny, 201–203
Onuphidae, 534f–535f, 537
Onychocaudata, 765, 772, 773
Onychodictyon ferox, 721f
Onychognathia, 616
Onychophora, 348, 709, 710f–711f, 711, 718–727, 730, 733
body plan of, 722–723
characteristics of, 711t, 721, 735t
circulation and gas exchange of, 724–725
in Ediacaran Period, 11
excretion and osmoregulation of, 725
feeding and digestion of, 724
locomotion of, 723–724
nervous system and sense organs of, 722, 725–726, 746
number of species, 2t, 711t
phylogeny of, 25, 26, 755f, 1049f
reproduction and development of, 726–727, 727f
systematics and biogeography of, 727
Oomyceta, 97, 99, 101
Oomycota, 63
Oonopidae, 912
Ooperipatus, 726
Oopsacas minuta, 246f
oostegites of Crustacea, **828**
ootype of Platyhelminthes, **399**
Opabinia, 13, 14f, 15f
Opalina, 58f–59f, 101, 102f
open circulatory system, 163, 163f, **164**

Opheliidae, 534f–535f, 538, 549
Ophiactis, 996
Ophiocoma, 987
Ophioderma, 988
Ophiopholis aculeata, 970f–971f
Ophiophragmus filograneus, 994
Ophiothrix, 987, 987f
  fragilis, 152f
  lineata, 988
Ophiura sarsii, 988
Ophiurida, 973
Ophiuroidea, 970f–971f, 973, 976, 976f
  circulation and gas exchange of, 993
  excretion and osmoregulation of, 994
  external anatomy of, 972f
  feeding and digestion of, 986–988, 987f
  phylogeny of, 1001, 1002f, 1003
  reproduction and development of, 995, 996, 998f
  support and locomotion of, 983
  water vascular system of, 981
Ophryotrocha, 536, 550f
Opilioacarida, 917, 918
Opilioacariformes, 914, 916f–917f, 917
Opiliones, 914, 919f–920f, 924–925, 935, 942, 945
  feeding of, 873
  phylogeny of, 962f
  reproduction and development of, 947, 950
Opiloacarida, 914
opisthaptor of Platyhelminthes, **385,** 385f
Opisthobranchia, 454, 466, 494
opisthognathous condition, **862,** 863f
Opisthokonta, 3, 60, 61, 62, 64, 124–125, 255n
  Choanoflagellata, 124–125
  phylogeny of, 128–129
Opisthorchis
  sinensis, 381f, 401
  viverrini, 401
opisthosoma, 728, 911, 912, 922, **927**–928, 928f–929f
  of Amblypygi, 919
  of Arachnida, **917,** 918
  of Araneae, 922, 924
  in circulation and gas exchange, 941, 941f, 942
  of Euchelicerata, 915
  in excretion and osmoregulation, 943
  in feeding and digestion, 935, 940f, 940–941
  nervous system in, 943
  of Opiliones, 924
  of Palpigradi, 925
  of Pseudoscorpiones, 925
  in reproduction, 948, 953
  of Ricinulei, 925
  of Schizomida, 926
  of Scorpiones, 926
  of Siboglinidae, 586f, **588,** 589f, 590
  of Solifugae, 926
  of Uropygi, 926, 927
Opisthoteuthidae, 471
Opisthothelae, 912, 924
opportunistic symbionts, 23
oral ciliature of Ciliata, **89**
oral cirri of Cephalochordata, **1025,** 1043
oral disc of Cnidaria, **278,** 278f
oral grooves of Ctenophora, **335**
oral hood of Cephalochordata, **1025,** 1043
oral lobes of Ctenophora, **334,** 334f, 335
oral papillae of Onychophora, **722,** 722f
oral ridges of Loricifera, **703**
oral siphon of Urochordata, **1029,** 1031, 1033, 1034
oral styles of Kinorhyncha, **695,** 695f
oral stylets
  of Loricifera, **702**
  of Tardigrada, **716–717**
oral sucker of Platyhelminthes, **385**–386
oral surface of Echinodermata, **969**
oral tentacles of Echinodermata, **982**
Orbiculariae, 932
Orbiniidae, 534f–535f, 538, 545, 564f, 569f
orb webs, 928–932, 936–938
Orchestia, 808
Orchestoidea, 808
Oreaster, 984
Oreohelicidae, 467
organic matter, dissolved, as nutrition, 65, 149, **157**
  of Amoebozoa, 75
  of Annelida, 577, 590
  of Dinoflagellata, 80
  of Euglenida, 120
  of Mollusca, 501
  of Porifera, 236
organ of Berlese, **883**
organ of Tömösváry, **171, 905,** 905f
Oribatida, 914
Ornithonodorus moubata, 918
orogenital pores of Platyhelminthes, **395**
Oropsylla montana, 857f
Orsten deposits, 13, 13n, 751, 752, 755
  Chelicerata in, 961
  Crustacea in, 797, 801, 834, 834f–835f, 835
  Onychophora in, 718
  Tardigrada in, 711
Orthalicidae, 467
orthognathous chelicerae, 922f–923f, **924**
orthologous genes, **41, 184**–185, 191
orthomitosis of Protista, 69f, **70,** 76, 83
Orthonectida, 415f, 418–420, 419f, 420f
  body organization and life cycle of, 204
  cilia of, 143
  classification of, 347, 348, 413, 414
  marine habitat of, 17
  number of species, 2t
  in origin of metazoan condition, 204, 204f
  phylogeny of, 25, 1049f
Orthoporus ornatus, 896f, 904
Orthoptera, 846, 847, 851f, 852–853, 859f, 861f, 862f
  feeding of, 870
  forewings of, 866
  leg modifications of, 865f
  nervous system and sense organs of, 879f
  phylogeny of, 48f, 889f
Orzeliscus, 715f
Oscarella, 220, 233
  carmela, 216
Oscarellidae, 222
Oscillatoria spongeliae, 252
osculum of Porifera, **225,** 227, 238, 247
Osedax, 20, 539, 585, 586f–587f, 588, 589f, 590, 591
  frankpressi, 586f–587f
  rubiplumus, 586f–587f, 589f
osmoconformers, **159**
osmoregulation, 158–162
  and habitat, 158–159
  nephridia in, 161–162
osmoregulators, **159**
osmotic hypothesis on exocytosis, 294
osphradia of Mollusca, **511,** 511f, 513
ossicles of Echinodermata, **978,** 978f–979f
Osteichthyes, 1022
ostia of Calcarea, **225**
Ostrachodermata, 454
Ostracoda, 767, 790, 798, 799f–800f
  distinguishing features of, 802t–803t
  nervous system and sense organs of, 825
  phylogeny of, 832, 833f
  reproductive features of, 806t–807t
Ostrea, 517
Ostreida, 469
Ostreidae, 469, 488
Ostreobium, 252
Otinoidea, 467
Otitidae, 858
Ottoia, 15f, 694
  prolifica, 698
  tricuspida, 698
ova, 179, **187,** 187f
  centrolecithal, **187,** 187f
  ectolecithal, 376, **395**
  endolecithal, **376, 395**
  isolecithal, **187,** 187f, 198, 198f
  mosaic, **191**
  regulative, **191**
  telolecithal, **187,** 187f, 198f
ovaries, **178,** 178f
ovicell of Bryozoa, **655,** 655f
Ovicides, 441
oviducts, **179**
ovigers of Pycnogonida, **955,** 961
oviparity, **199,** 726
ovipositor of Hexapoda, **864**
ovoviviparity, **199,** 448
ovular nuclei of Protista, **70**
Ovulidae, 155, 466
Owen, R., 55
Owenia, 532, 533f, 556f–557f, 570f
Oweniidae, 532, 533f, 542, 563, 570f
Oxidus gracilis, 896
Oxycorynia, 1030f
oxygen
  atmospheric, 7–8, 10, 13
  in estuary habitat, 20
  in marine habitat, 18
  respiratory pigments carrying, 167–168, 168t
  uptake and transport of, 164–168, 165f
oxygen minimum zones (OMZs), 18
Oxymonada, 63, 64, 128
Oxyopes, 949f
Oxyopidae, 914, 936
Oxyurida, 672
Ozobranchidae, 594, 595
ozopores of Myriapoda, **896**

P

Pachygrapsus, 780f
paedomorphosis, **203,** 1033
Paguristes, 782f
  pilosus, 811f
Pagurus, 302, 782f
Palaeacanthocephala, 624, 625f
Palaeocopida, 767, 798
Palaeoheterodonta, 455, 469–470

Palaeonemertea, 436–438, 437f, 440f
body wall of, 439, 439f
circulation and gas exchange of, 443, 444f
excretory system of, 445f
feeding and digestion of, 442f
nervous system and sense organs of, 446f
phylogeny of, 450, 450f, 451
reproduction and development of, 448f, 449
Palaeopantopoda, 958
Palaeoptera, 846, 847, 849f, 849–850
evolution and phylogeny of, 889–890
flight of, 868
Palcephalopoda, 455, 471
Paleoloricata, 473
Paleozoic era, 12–16
Palinuridae, 782
palleal budding of Urochordata, **1037**
pallial line of Mollusca, **481**
Pallisentis fractus, 625f
Palola, 567
viridis, 566f
Palpigradi, 914, 919f–920f, 925, 945, 950, 962f
palp proboscis of Mollusca, **496**
Palythoa toxica, 268
palytoxin, **268**
Pamphagus, 76, 76f
Panagrellus redivivus, 671
Panama Seaway, 17
Panarthropoda, 348, 669, 693, 695, **709,** 710f–711f, 751
Arthropoda, 728–962
comparison of phyla, 711t
Onychophora, 718–727
phylogeny of, 25, 26, 755f, 1049f, 1050
Tardigrada, 711–718
Pancrustacea, 4, 754, 755, 755f, 833, 889f, 908
Pandalus, 820f
Pandaros, 250
Pandeidae, 274
Panderia, 753f
Pandora larvae of Cycliophora, **610,** 611
Pandoridae, 470
Pangaea, 10, 15, 16
Pannychia, 554f
Panorpidae, 857
Panpulmonata, 455, 467
pansporocysts, **301**
Pantopoda, 958
Panulirus, 782f
argus, 805f, 808
Papilio
appalachiensis, 29
canadensis, 29
glaucus, 29
Papillionidae, 859
papulae of Echinodermata, **993,** 994f
parabasal fiber of Parabasalida, **113**
Parabasalida, 63, 64, 112f, 112–115
characteristics of, 113
classification of, 61, 62
nutrition of, 114
phylogeny of, 128
reproduction of, 114f, 114–115
support and locomotion of, 113–114
Parabuthus, 936, 950f
Paracanthobdella livanowi, 541, 593
Paracentrotus, 988f–989f
Paracerceis, 787f
Parachela, 715
paracrine system of Porifera, **239**
Paracyrba wanlessi, 939
Paradoxopoda, 908
Parafossarulus, 401
paragastric canals of Ctenophora, 333f, **337**
Paragiopagurus fasciatus, 762f–763f
Paragonimus, 781
westermani, 404f
Paragordius, 688f, 689, 689f
ParaHox genes, 254, 255
ParaHoxozoa, 254
Paraleptophlebia, 877f
parallelism, **42,** 42n, 43, 43f
paralogous genes, **41**
Paralvinella fijiensis, 535f
paralytic shellfish poisoning (PSP), 78–79
Paramecium, 88f–89f, 93, 149
activity and sensitivity of, 67
bursaria, 93
caudatum, 91f, 94, 95f
ingestion by *Didinium,* 91, 91f
locomotion of, 89, 90.90f
in origin of metazoan condition, 204f
reproduction of, 94, 94f, 95, 95f, 96
water expulsion vesicles of, 160f
Paramecynostomum diversicolor, 352f–353f
paramylon of Euglenida, **120**
Paramyxea, 112
Paranema, 120f
Paranemertes, 441
peregrina, 441, 441f, 442f
Paraneoptera, 847, 853, 890
paranotal lobe hypothesis on insect flight, 869, 870f
Parapagurus, 302
dofleini, 301f
parapet of Cnidaria, **291**
paraphyly, **4, 40,** 40f, 44–45
Parapinnixa, 780f
parapodia of Annelida, 543, 544f, 544–545, 547, 558f
parapyles of Radiolaria, **110,** 111
Parasagitta elegans, 424f
Parascaris equorum, 683f
Paraseison, 621
annulatus, 617f
kisfaludyi, 620f
parasites, 23, 201
ectoparasites, **23, 201**
endoparasites, **23, 201**
mesoparasites, **23, 201**
parthenogenetic, **201**
primary and intermediate hosts of, **201**
Parasitiformes, 914, 916f, 917, 918, 962f
parasitism, **23**
parasitoids, **23**
Paraspadella gotoi, 421f, 427, 427f
Parastichopus, 971f, 992f
parvimensis, 986
tremulus, 364
Parastomonema, 677
Paratomella
rubra, 352f–353f, 357
unichaeta, 357f
paratomy of Enteropneusta, **1014**
Paravortex, 387
paraxonemal rod
of Dinoflagellata, **80,** 80n
of Euglenida, 119
of Kinetoplastida, 123
parazoan grade of body construction, 217
parenchyma, 266n
parenchyme, **266n,** 373, 382
parenchymella larvae, 244, 245f, 246f, 248
Paroctopus digueti, 475f
paroral membrane of Ciliata, **92**
parsimony, 46, 46n
parthenogenesis, 179n, **179**–180, **201**
of Crustacea, 825, 831
of Euchelicerata, 947
of Hexapoda, 884
of Micrognathozoa, 632
of Rotifera, 622, 623
of Tardigrada, 717
Paruroctonus, 935
mesaensis, 935
Patella, 504
Patellidae, 462
Patellogastropoda, 455, 462, 478f
circulation and gas exchange of, 504
development of, 520
excretion and osmoregulation of, 507
feeding of, 492, 493
sensory organs of, 511
shell of, 480f, 482
torsion of, 484
Patiria, 137f, 984
Patu, 924, 937f
Paucijaculum samamithion, 422
Paucitubulatina, 429, 430f
Pauropoda, 895, 896f, 897, 899
feeding and digestion of, 902
head and mouth appendages of, 901
locomotion of, 901
nervous system and sense organs of, 904
phylogeny of, 908
reproduction and development of, 907, 908
Pauropus silvaticus, 896f
Paussinae, 883
pavement cells, 223
pax1/9 genes, 1041
Pax-6 gene, 30, 173, 185, 754
Paxillosida, 973
pdm gene, 870
pearls, 481, 481n
Pearse, V. B., 215
Pecten, 512, 512f, 513
Pectinaria, 548f–549f, 551f, 552
Pectinariidae, 539–540, 548f–549f, 550
Pectinatella, 645f, 649
magnifica, 646
Pectinidae, 456f–457f, 469, 488, 490
Pectinidina, 469
pectinobranch condition of Mollusca, **505**
pedal cords of Mollusca, **507**
pedal disc of Cnidaria, **277,** 278f, 284f, 292
pedal ganglion of Annelida, **563**
pedestals of Pycnogonida, **958**
Pedicellaria, 980
globifera, 980
tridens, 980
triphylla, 980
pedicellariae of Echinodermata, 978f–979f, **980**
Pedicellina, 604f–605f
cernua, 604f–605f, 608f
Pedicellinidae, 605–606
pedicel of Hexapoda antennae, **860**

pedicle of Brachiopoda, **659,** 661
pediculosis, **873**
Pediculus, 873
humanus, 52
Pedinoida, 974
Pedipalpi, 962f
pedipalps of Chelicerata, **928,** 948, 950
peduncle, **283,** 801n
peg organs of Hexapoda, 880f, **881**
Pelagia, 296, 297, 310, 312f
noctiluca, 296
Pelagica, 450, 450f
pelagic organisms, **18**
pelagic zone, **19,** 20
pelagosphera larva of Sipuncula, **579,** 580f
Pelagothuria, 971f, 972f
Pelecypoda, 454, 467–470
pellicle of Protista, **65**
Pelomyxa, 74, 75
pelta of Parabasalida, **113**–114
Peltogaster paguri, 812f–813f
Peltospiridae, 463
Penaeus, 778, 828f
setiferus, 779f
Penicillata, 896, 897, 898, 909
penis papilla of Platyhelminthes, **399**
Pennaria, 316
pennate diatoms, **100**
Pennatula, 284f
Pennatulacea, 12f, 266f–267f, 271, 280, 283, 284f, 292
Pennington, J. T., 484
Pentanymphon, 958
Pentapora, 645f, 651
Pentapycnon, 958
pentaradial symmetry, **137,** 137f
of Echinodermata, 968, **969,** 977, 995, 1001, 1003
Pentastomida, 767, 790, 796f, 796–797
distinguishing features of, 802t–803t
nervous system and sense organs of, 825
phylogeny of, 833f
Pentastomum, 796f
Pentatrichomonas homonis, 112
Pentazonia, 897, 898, 907, 909
Penthaleidae, 919
Penthaleus major, 919
Peracarida, 749, 763f, 766, 783, 784f, 801
phylogeny of, 832
reproduction and development of, 806t–807t, 831
thoracic appendages of, 804f–805f
Peranema, 120
Perenereis, 550f
pereon, 768n, **801**
pereonite of Crustacea, **801**
pereopods of Crustacea, **801**
perforate extraxial region of Echinodermata, **976,** 982
periblastula, **193,** 193f, 750f
peribuccal coelom of Enteropneusta, **1011**
pericardial chamber of Mollusca, **472, 503**
Periclimenes brevicarpalis, 302
peridinin of Dinoflagellata, **80,** 126
Peridinium, 58f–59f, 77f
volzii, 81f
perihemal coelom of Enteropneusta, **1011,** 1014
perihemal sinuses of Echinodermata, 991
periodomorphosis, 907
periostracum
of Brachiopoda, **660**–661
of Mollusca, **477,** 482
Peripatidae, 719, 727
Peripatoides, 720f
aurorbis, 720f
Peripatopsidae, 719, 727
Peripatopsis, 723f, 726
alba, 720f
capensis, 720f, 724f
moseleyi, 720f, 723f, 724f, 727f
sedgwicki, 722f
Peripatus, 709, 725
peripheral tentacles of Sipuncula, **572**
Periphylla, 310
Periplaneta americana, 390
periplast of Cryptomonada, **103**
Periplomatidae, 470
perisarc of Cnidaria, 277f, **280**
peristomial membrane of Echinodermata, **985**
peristomium of Annelida, **541**
in Hirudinoidea, 591
in Siboglinidae, **588**
peritoneum, **140**
peritrema of Myriapoda, **903**
Peritrichida, 95
peritrophic membrane of Arthropoda, **743, 874**
perivisceral coeloms of Echinodermata, **980,** 991
Permian period, 15
Perophora, 1029f, 1037
petaloids of Echinodermata, **994**
Petasidae, 273
petasma of Crustacea, **826**
Petrochirus diogenes, 762f–763f
petroleum, **56**
Petrolisthes, 782f
armatus, 824f
cabriolli, 814f–815f
elegans, 151f
Petromyzontidae, 1042
Petta, 556f–557f
Peyssonal, J.-A., 636
Pfiesteria, 79
piscicida, 79
shumwayae, 79
Phacellophora, 292, 297
camtschatica, 781, 790
Phacops rana, 753f
Phacus, 119
Phaeodaria, 108, 109, 110, 111
phaeodium of Radiolaria, **110**
Phaeophyta, 56, 61, 63, 97, 101
Phagocata, 394
phagocytosis, 148f, **148**–149
of Amoebozoa, 75, 75f
of Dinoflagellata, 80
of Diplomonadida, 117
of Granuloreticulosa, 107
of Parabasilida, 114
Phagodrilus, 549
Phaladidae, 490
Phalangida, 924, 925, 950
Phalangiotarbida, 914
Phallonemertes murrayi, 437f
Phareicranaus manauara, 919f–920f, 955f
pharyngeal canals of Ctenophora, 333f, **337**
pharyngeal chamber of Urochordata, **1034**
pharyngeal gill slits of Chordata, **1022,** 1041
in Cephalochordata, 1023
in Urochordata, **1027,** 1031, 1034, 1036
pharyngeal glands of Platyhelminthes, **387**
pharyngeal pump of Hexapoda, **875**
pharynx
of Acoela, 356
of Ctenophora, 333f, 334, 337
of Gastrotricha, 431
of Nematoda, 677, 678f–679f
of Platyhelminthes, 387–388, 388f, 389f, 398, 398f
simplex, **356, 387**
Phascolion, 533f, 573f, 578
crypta, 579
strombi, 579
Phascolopsis, 575, 576
Phascolosoma, 573f, 575, 576, 576f, 579, 580f
agassizii, 578
Phascolosomatidae, 575, 577
Phasmatodea, 48f, 847
Phasmida, 846, 847, 851f, 852, 889f
phasmids of Nematoda, 680f, **681**
Phenacolepadidae, 463
Phengodidae, 855
phenotype, 185
plasticity of, **200**
Pheretima, 568f
pheromones, **176,** 883
Pherusa, 535f
Phidiana, 464f
Philactinoposthia novaecaledoniae, 352f–353f
Philodina roseola, 617f
Philonexis, 518
Philoponella republicana, 932
Phlebobranchia, 1022
Pholadida, 470
Pholadidae, 470
Pholadomyidae, 470
Pholas, 490
Pholcidae, 912, 954f–955f
Pholoidae, 537
Phoneutria, 922f
fera, 938
Phonognatha, 931
phonoreceptors, 171, 171f, 881, 881f
Phoratopidea, 766
phoresis, **24**
Phoronida, 635, 636, 638–644
body plan of, 637–638
body wall, body cavity, and support of, 638–641
characteristics of, 638
circulation, gas exchange, and excretion of, 640f, 641–642
feeding and digestion of, 640f, 641
marine habitat of, 17
nervous system of, 642
number of species, 2t
phylogeny of, 25, 1049f
reproduction and development of, 642–644
taxonomic history of, 637
Phoronis, 637, 638, 644
architecta, 639f
australis, 641f
hippocrepia, 641, 643f
ijimai, 639f, 641f, 642
ovalis, 642, 643, 644

pallida, 639f, 643f
psammophila, 639f
vancouverensis, 639f, 641f, 642
Phoronopsis, 638
californica, 639f
harmeri, 638, 640, 641f, 642, 644
viridis, 642
photic zone, **19,** 19f, 20
Photis conchicola, 808, 809f
photoautotrophy, **56, 157**
photocytes of Hexapoda, 882
photophores of Mollusca, 513–514
photoreceptors, 169, 172–173, 173f, 747–748
of Annelida, 564f, 564–565
of Chaetognatha, 423, 426
of Crustacea, 823
of Echinodermata, 995
of Euchelicerata, 945–946
of Gastrotricha, 431
of Mollusca, 513
of Platyhelminthes, 392
photosynthesis, 7, 8, 56, 57n, 149
of Protista, 65, 68
photosystems, **57n**
Phoxicilidiidae, 961
Phragmatopoma californica, 548f–549f
Phragmophora, 420f, 423
Phreatoicidea, 766
Phreatoicopis terricola, 383f
Phrixometra, 997f
Phthipodochiton, 523
Phthiraptera, 854
Phthirus, 873
phycobilins, 57n
phycobiliproteins, 57n
Phylactolaemata, 645f, 646, 647f, 648, 651
body wall, muscles, and movement of, 649
circulation, gas exchange, and excretion of, 652
feeding and digestion of, 651, 652
reproduction and development of, 655, 657
Phyllangia, 284f
Phyllidiidae, 467
Phyllobothriidea, 379, 380
Phyllobothrium, 386
phyllobranchiate gills, 778, 779, 820f
Phyllocarida, 734, 765, 773, 774f, 774–775
distinguishing features of, 802t–803t
phylogeny of, 833f
reproductive features of, 806t–807t
thoracic appendages of, 804f–805f
Phyllodocida, 535, 537–538, 545, 560
Phyllodocidae, 537–538, 545, 550f, 557, 572f
phyllopodia of Arthropoda, **734, 803**
Phyllorhiza, 297
phylogenetic fuse hypothesis, 12
phylogenetic hypotheses, explicit, **46**
phylogenetics, molecular, 26, 47, **51**–52, 1047–1051
phylogenetic systematics, **45,** 50–51
similarity concept in, 36, 50–51
phylogenetic trees, **43**–44, 46, 47, 47f, 48f–49f
new species in, 50
phylogeny, 25–31, 44–52, 1047–1051
construction of, 44–51
heterochrony in, 203
and ontogeny, 201–203
origin of term, 4
physa, **277**–278
Physa, 515f
Physalia, 280, 280f–281f, 282, 295, 298
physalis, 280
symbiotic relationships of, 302
Physarum, 72
polycephalum, 73f
Physidae, 467
Physonecta, 274, 282, 316
Phytophthora infestans, 99
phytoplankton, **20**
Picrophilus oshimae, 9n
Pieridae, 859
pigment cells of Arthropoda, 748
pigments
in coloration and ink of cephalopods, 513–514
light-sensitive, 172
respiratory, 167–168, 168t, 558–559
in shells of Mollusca, 477
Pikaia, 15f
Pilidiophora, 436, 438, 449, 450f, 451
pilidium of Nemertea, 449f, **449**–450
Pimm, S., 6
Pimoidae, 948
pinacocytes, **223,** 228, 229f, 236, 246
pinacoderm of Porifera, **223,** 228, 229f, 230
Pinnidae, 469, 488
Pinnotheridae, 780f
pinnules
of Echinodermata, **981**
of Siboglinidae, **588**
pinocytosis, **148,** 148f, 149
in Amoebozoa, 75, 75f
in Apicomplexa, 83
in Kinetoplastida, 123
in Parabasilida, 114
Pinophyta, 129
piroplasms, 81–87
Pirsig, R. M., 36
Pisani, D., 25, 1048
Pisaster
giganteus, 37, 38, 983f
giganteus capitatus, 37
ochraceus, 985, 986f, 996
Pisaura, 952
Pisauridae, 913, 952
Piscicola, 596f
Piscicolidae, 592f, 594, 595, 596f
Pista, 550
placenta of Onychophora, 726, 727f
Placentonema gigantisima, 9, 671
placids of Kinorhyncha, 695f, **696**
Placiphorella, 495
velata, 492f
Placobdella, 595f
Placobranchus, 501
placoids of Tardigrada, **717**
Placozoa, 205, 214f, 215–216
absence of nervous system in, 238
characteristics of, 215
classification of, 213–214, 255, 341, 347
feeding and digestion of, 149
marine habitat of, 17
mitochondrial DNA of, 319
number of species, 2t
phylogeny of, 25, 255, 341, 346, 1049f, 1050
plagiognath chelicerae, **924**
Plagiorchiida, 378, 385, 390f, 401, 404f
Plagiorhynchus, 624
Plakina trilopha, 246f
Plakinidae, 222
plakula, **205**
Planctosphaera pelagica, 1008
Planctosphaeroidea, 1008
plankton, **20**
planktotrophy, **194, 197,** 198, 198f
Planocera, 393f, 397f
Planorbidae, 467
Plantae, 4, 4n, 8, 55, 55n, 56, 57, 60, 61, 125, 129
planula larva, **265,** 308–310, **316,** 317f, 320f
anthozoan, 309, 309f
cubozoan, 306–307, 308, 311
hydrozoan, 314f–315f, 316, 316f
photoreceptors of, 306–307
scyphozoan, 312f
staurozoan, 311
planuloid–acoeloid hypothesis, 346
plasmagel, 74
plasmasol, 74
plasmodium, **299,** 300f
Plasmodium, 81, 82, 83, 85, 872, 873
berghei, 82
chabaudi, 82
falciparum, 82, 84, 86
life cycle of, 86f, 86–87
reichenowi, 84
vivax, 84
yoelii, 82
plasmotomy of Protista, **68, 83**
plasticity, phenotypic, **200**
plastids, 57n, 79
origin and spread of, 59f
primary, **57,** 59f
secondary, 59f
Platnick, N. I., 952
Platycopida, 767, 798, 800f
Platycopioida, 767, 798
Platyctenida, 330f–331f, 332, 334, 341
Platydesmida, 897
Platydesmus, 387
manokwari, 387
Platydorididae, 467
Platyelmia, 375
Platyelminthes, 375
Platyhelminthes, 349, 350, 351, 360, 362, 373–409, 414
acoelomate body plan of, 140
body plan of, 139
body wall of, 382–385, 383f, 384f
characteristics of, 376
cilia of, 143
circulation and gas exchange of, 380, 390–391
defense mechanisms of, 384
excretion of, 391–392
feeding and digestion of, 149, 374, 386–390
life cycle of, 400–405, 402f–403f, 404f, 406f
nervous system and sense organs of, 373, 380, 392–394
number of species, 2t, 1047
osmoregulation of, 380, 391–392
phylogeny of, 25, 341, 347, 380, 405–409, 408f, 1049f, 1050
reproduction and development of, 374, 380, 394–405
support, locomotion, and attachment of, 380, 383, 383f, 385f, 385–386, 386f

symbiotic relationships of, 387
taxonomic history and classification of, 317, 374–376
Platylepas, 791
Platynereis, 545
dumerilii, 534f–535f
Plecoptera, 846, 847, 850, 851f
phylogeny of, 48f, 889f
Plectida, 672
Plectonchus, 682f
Plenocaris, 15f
Pleocyemata, 761n, 766, 778
pleonites of Crustacea, **801**
pleon of Crustacea, **801**
pleopods of Crustacea, **801,** 819
plerocercoid stage of Platyhelminthes, **405**
Plesiodiadema, 972f
plesiomorphy, **44,** 45
pleural folds of Crustacea, **801**
pleura of Arthropoda, **731**
Pleurobrachia, 328f–329f, 330f–331f, 334f, 336f, 339
bachei, 143f, 327, 328, 340, 1048
Pleurobranchidae, 466, 482
pleurobranchs of Decapoda, **778**
Pleuroceridae, 465
Pleurodasys helgolandicus, 431
pleuromitosis of Protista, 69f, **70**
Pleuroncodes planipes, 762f–763f
Pleurostigmophora, 897, 898
Pleurotomariidae, 463, 504
Pleurotomarioidea, 485
Pliciloricidae, 701, 703, 704, 704f
Pliciloricus, 703, 704
enigmaticus, 702f
gracilis, 702f, 704f
hadalis, 701
Plocamia karykina, 251
Plumatella, 645f, 649
plumatellid colonies of Bryozoa, **648–649**
pneumatophores, **280,** 282
pneumostome, **165,** 166f–167f
of Mollusca, **505,** 505f
Pocillopora damicornis, 308
podia of Echinodermata, 969, **975**
Podocopa, 767, 798, 799f, 800f
Podocopida, 767, 798, 800f
Podocoryne, 302
carnea, 215
Podoplea, 767
podosoma of Chelicerata, **917**
Poecilosclerida, 222
Poecilostomatoida, 767, 798
Pogonophora, 532, 538, 539, 540, 585, 588, 597
phylogeny of, 1050
polar cap of Dicyemida, **414,** 415f
polar capsules of Cnidaria, **269, 294**
polar fields of Ctenophora, 339, 339f
polar filaments of Cnidaria, **299,** 299f, 301
polarity analysis, character state, 45
Polaromonas vacuolata, 9n
Polian vesicles of Echinodermata, **981,** 982
Polinices, 156f, 486f
Polis, G., 155
Politolana, 170f
Pollicipes, 151f
polymerus, 792f–793f
pollination, Hexapoda in, 844–846
Polyartemiidae, 772
polyaxial cleavage of Porifera, **244**
Polybrachia, 585, 586f, 588
Polycelis tenuis, 389
Polyceridae, 467
Polychaeta, 6, 532, 597
Polychelida, 766, 783
Polychoerus gordoni, 359f
Polycladida, 377, 389, 389f, 395, 396f
phylogeny of, 408, 408f
Polycladidea, 408f
Polyclinum, 1028f–1029f
Polycope, 800f
Polycystina, 108, 109, 110, 111
Polydectus cupulifer, 781
Polydesmida, 897, 899, 900f–901f, 907
polydisc strobilation, **310**
Polydora, 549, 551f
alloporis, 554
polyembryony, **656**
of Hexapoda, 885
polygastry, **311**
Polykrikos, 80
Polymitarcyidae, 849
polymorphism
developmental, **200**
of Rotifera, 624
Polymorphus, 624
Polyneoptera, 846, 847, 850, 851f, 889f, 890
Polynoidae, 533f, 537, 554f
feeding of, 549
support and locomotion of, 545
Polyommatini, 859n
Polyophthalmus, 565
Polyopisthocotylea, 378
Polyorchis, 267f, 286f–287f
Polypes, 220
Polyphemidae, 773
polyphyly, **4,** 6, 40f, **40**–41
polypides, **646,** 651, 653, 656
Polyplacophora, 453, 454, 455, 456f, 460–462, 473, 476f, 478f
development of, 518f, 519f
feeding of, 495
generalized anatomy of, 461f
nervous system of, 508f
phylogeny of, 522f, 523, 524
sensory organs of, 513
shell of, 479, 480f
Polypodium hydriforme, 282–283
polypoid form of Cnidaria, **265,** 276–283, 277f, 308, 319
movement of, 290–292, 291f
nervous system of, 305
support of, 287
polypoid hypothesis on Cnidaria, **319**
Polypoidozoa, 269, 272, 282, 320f, 321
Polystilifera, 436, 437f, 438, 450, 450f
Polystoma integerrimum, 402f–403f
Polystomatidea, 378
polythridium of Priapula, **700**
polytroch larvae of Annelida, **571**
Polyxenida, 897
Polyzoa, 603, 637
Polyzoniida, 897
Pomatias, 509f
Pomatiasidae, 466
Pomatiopsidae, 466
Pontodoridae, 538
Porania, 985
Porcellanidae, 826
Porcellio, 745f, 820f, 821
Porifera, 213, 214, 216–258, 269
activity and sensitivity of, 238–239
aquiferous system of, **222**–227, 234, 238, 239, 254
biochemical agents of, 250
bioerosion from, 252–254
bleaching of, 252
body forms of, 216, 218f–219f, 224f
body structure of, 223–228
cell reaggregation in, 231–232
cell types of, 228–231, 229f, 231f, 232f
defense mechanisms of, 249, 250
development of, 243–249, 252
distribution and ecology of, 249–250
in Ediacaran Period, 11
evolution of, 254–258
excretion of, 236–238
feeding and digestion of, 149
gas exchange of, 238
growth rates of, 250–251
harvest and sale of, 233, 233f
number of species, 2t
nutrition of, 234–236
origin of, 254–255
phylogeny of, 25, 254–258, 257f, 1048, 1049f
reproduction of, 239–243, 240f, 241f, 242f, 243f
support of, 232–234, 234f–235f
symbiotic relationships of, 217, 236, 251–254
taxonomic history and classification of, 220–222
Poritiina, 321
porocytes, **225**
Poromyata, 456f–457f, 470, 497
Poromyidae, 470
Porpita, 280f–281f, 298
portal vein, cecal, of Cephalochordata, **1025**
Portia, 938, 946f
Portunidae, 780f, 807
postanal tail of Chordata, **1022,** 1023, 1041
postecdysis of Arthropoda, **739**
Postelisa, 98f
postlarva phase of Crustacea, **830,** 831
postmolt period of Arthropoda, **741**
post-parthenogenetic hermaphroditism of Gastrotricha, **432**
Potamocypris, 800f
potato blight, 99
Praecambridium, 11
Praesagittifera shikoki, 353f
Praxillura maculata, 152f
predation, 154–157, 156f
prehensors of Myriapoda, **897,** 898, **901**
premolt state of Arthropoda, **739**
preoral ciliary organ of Enteropneusta, **1013,** 1014
preoral disc of Pterobranchia, **1015**
pressure drainage channels of Echinodermata, **984**
pretarsus of Hexapoda, **864**
preventral organs of Onychophora, **722**
Priapula, 41, 636, 687, 693, 694, 698–701
blastocoelom of, 140
body wall, support, and locomotion of, 699–700
characteristics of, 698

circulation, gas exchange, and osmoregulation of, 700–701
classification of, 699
feeding and digestion of, 700
marine habitat of, 17
nervous system and sense organs of, 700f, 701
number of species, 2t
phylogeny of, 25, 1050
reproduction and development of, 701
Priapulida, 1049f
Priapulidae, 699, 699f
Priapulites, 694
Priapulopsis, 698
australis, 699f
Priapulus, 698, 700f, 701
bicaudatus, 699f
caudatus, 347, 700f, 701
Priapus humanus, 698
primary endosymbiosis, **57,** 59f
primary gill bar of Enteropneusta, **1013**
primary hosts, **201**
primary larva, **346,** 347, 348, 1049f
of Echinodermata, **997**
of Pentastomida, 796f, 826
primary pigment cells of Arthropoda, **748,** 755
primary plastids, **57,** 59f
Primavelans, 653f, 655
insculpta, 654f
primeval soup theory, 1n
primitive character states, **44**
Primno, 788f–789f
Proarticulata, 11, 317
probolae of Nematoda, **677**
Probopyrus bithynis, 828f–829f
Proboscidactyla, 280f, 301
proboscides of Mollusca, **496**
proboscis
of Enteropneusta, **1009,** 1010f, 1011, 1012f–1013f, 1013, 1014, 1016f–1017f
of Pycnogonida, 955, 958, 959, 961
proboscis apparatus of Nemertea, 438, 441f, 441–442, 442f, 443f, 445, 449
proboscis complex of Enteropneusta, **1014**
proboscis skeleton
of Enteropneusta, **1011,** 1012f–1013f
of Pterobranchia, 1015
Procarididea, 778, 779f
Procaridoida, 766
Procaris, 778, 811
ascensionis, 779f, 811f
procercoid stage of Platyhelminthes, **405**
Prochloron, 1034
Procryptobia, 123
proctodeal ectoderm, **193**
Proctonotidae, 467
procuticle of Arthropoda, **732,** 732f
proecdysis of Arthropoda, **739,** 741
progenesis, **203**
proglottids of Platyhelminthes, **382,** 391, 391f, 392, 403, 405f
prognathous condition, **862,** 863f
Progoneata, 908
Progradungula otwayensis, 936f–937f
Progymnoplea, 767
prohaptor of Platyhelminthes, **385,** 385f
Prokaryota, 3, 4, **7**–9, 56
Archaea, 4, 7, 8, 64
Bacteria, 4, 7, 8, 64
body plan of, 125
chemoautotrophy of, 157
classification of, 8, 36, 45
evolution of, 7, 10, 28, 116
first appearance of, 1, 1n
number of species, 3, 6, 36
photosynthesis of, 8
symbiotic relationships of, 23, 29, 57, 59f, 125, 126, 126f
transposable elements of, 28
Prokoeneniidae, 925
Prolecithophora, 377, 396f
proloculum of Granuloreticulosa, **107**
Prometheus larva of Cycliophora, 609f, **610,** 611
Proneomenia, 508f
antarctica, 458f–459f
pronotum of Hexapoda, **863**
propodium of Mollusca, **487**
Propontocypris, 800f
Proporus bermudensis, 357f
proprioceptors, **171,** 747, 747f
of Crustacea, 822–823
of Euchelicerata, 944
of Hexapoda, 881
Prorhynchida, 377, 408, 408f
Proscoloplos, 564f
Proseriata, 377, 389, 392, 396f
phylogeny of, 408f, 409
Proserpinellidae, 463
Proserpinidae, 463
Prosobranchia, 454
prosoma, **636,** 911
of Echinodermata, 975
of Euchelicerata, 915, 917, 918, 922, 924, 925–926, **927**–928, 940, 944, 948
of Hemichordata, 1009
prosopyles, **226**
Prosorhochmus, 449f
Prosthiostomum, 387
Prostigmata, 914
Prostoma, 448
prostomium of Annelida, **541**
in Hirudinoidea, 591
in Siboglinidae, **588**
protandry, **179**
protein kinase C in Porifera, 257
Proteobacteria, 20
Proteocephalidea, 379
proterosoma of Chelicerata, **928**
Proterospongia, 204f, 206
Proterozoic eon, 10
prothorax of Hexapoda, **863**
Protista, 8, 42, 55–130, 217
activity and sensitivity of, 67–68
Amoebozoa, 60–61, 70–76
Apicomplexa, 61, 81–87
body plan of, 57, 64–70
body structure of, 64–65
Chlorarachniophyta, 62, 104–105
Choanoflagellata, 62, 124–125
Ciliata, 61, 87–96
circulation of, 65
classification of, 4, 4n, 60–64
Cryptomonada, 62, 103
Dinoflagellata, 61, 76–80
Diplomonadida, 62, 115–117
Euglenida, 62, 118–120
excretion of, 65
gas exchange of, 65
Granuloreticulosa, 62, 105–107
Haplosporidia, 62, 111–112
Haptophyta, 62, 101–103
Heterolobosea, 62, 117–118
Kinetoplastida, 62, 120–124
mitosis in, 69f, 69–70
nutrition of, 65–66, 66f, 67f
origin of, 125–127
Parabasalida, 62, 112–115
phylogeny of, 125–130, 127f
Radiolaria, 62, 108–111
relationships among, 127–130
reproduction of, 68–70
single-celledness of, 64–65
Stramenopila, 61, 97–101
support and locomotion of, 60, 65
taxonomic history and classification of, 59–60
Protoalcyonaria, 271
Protobonellia, 533f, 580
Protobranchia, 455, 469, 496, 497f, 502
protocerebrum, 746
of Crustacea, 821
of Euchelicerata, 943
of Pycnogonida, 959–960
protochordates, **1025,** 1041, 1041f, 1043
protocoel, **196,** 196f, **636,** 637, 638, 640
of Echinodermata, **980**
of Enteropneusta, 1011
of Pterobranchia, 1015
protoconch of Mollusca, **482**
Protoctista, 4n
Protodrilida, 535
Protodrilidae, 535
Protodriloididae, 535
Protoglossus kohleri, 1013f
protogyny, **179**
protonephridia, **161,** 161f
of Annelida, 560, 561f, 571
of Cephalochordata, 1026
compared to metanephridia, 162
of Entoprocta, 607
of Gastrotricha, 431
of Nemertea, 438, 444–445, 445f
of Platyhelminthes, 373, 380, 391f, 391–392
of Priapula, 701
protonephromixium, **162**
protonymphon larvae, 960f, **961**
Protopalina, 101
Protoperidinium, 79
depressum, 77f
protopod of Arthropoda, **733**
in Crustacea, **801**
in Hexapoda, **860,** 863–864
Protostomia, 319, 346, 347–348, 413–432, 632, 636
Chaetognatha, 420–428
Ecdysozoa. *See* Ecdysozoa
as legacy name, 26, 50, 347
origin of coelom in, 207
Pax-6 gene in, 173
phylogeny of, 25, 51, 829, 967–968, 1049f, 1050
Spiralia. *See* Spiralia
trochaea theory on, 208, 208f, 209
Protostyela longicauda, 1039
protostyle of Mollusca, **502**
prototroch, **208, 571**
Protozoa, 55, 56, 60, 128
protozoea, **830**
protractor muscles, **146**
Protura, 843, 845f, 847, 848
evolution and phylogeny of, 48f, 888, 889f

thorax of, 864
proventriculus of Hexapoda, **875**
Pruvot, G., 523
Pruvotina impexa, 458f–459f
Prymnesiophyta, 61, 62, 63, 101
Psammogorgea, 285f
Pseudamphithoides incurvaria, x808, 809f
Pseudoarctolepis, 15f
pseudoblastomeres of Onychophora, **726,** 727f
Pseudocarcinus gigas, 761
Pseudocellus, 925
Pseudoceros
bajae, 374f–375f
canadensis, 392
ferrugineus, 374f–375f
Pseudochactoidea, 914
pseudocoelom, 140, 140f
pseudofeces of Mollusca, **498**
pseudogamy, **179n**
Pseudonitzschia, 99
australis, 99
pseudophalangia of Micrognathozoa, **630**
Pseudoplexaura, 285f
pseudopods (pseudopodia), **141**
in ameboid locomotion, 141–142, 142f
of Protista, 65, 74–75, 114
Pseudoscorpiones, 914, 919f–920f, 925, 942, 945, 950, 962f
pseudosimplex stage of Tardigrada, **713**–714, 714f
Pseudosquilla bigelowi, 605f
pseudotrachea, **165, 744,** 745f
of Crustacea, **821**
pseudozoea larvae, **831**
Psocodea, 846, 847, 854, 854f, 872–873, 890
phylogeny of, 48f, 889f
Psocoptera, 854
Psolidium, 992f
Psolus, 991
chitinoides, 978f–979f
Psychidae, 859
psychrophiles, **9**
ptenoglossate radula, **494**
Ptenothrix, 845f
Pteraeolidia, 500
Pteraster tesselatus, 970f–971f, 982–983
Pteriida, 469
Pteriidae, 469, 488
Pterioida, 469
Pteriomorphia, 455, 456f–457f, 469, 499
Pterobranchia, 1007, 1009f, 1010f–1011f, 1015–1018
body wall and cavities of, 1015
circulation and gas exchange of, 1017
classification of, 1008–1009
feeding and digestion of, 1008, 1010f–1011f, 1015–1017
internal anatomy of, 1017f
nervous system of, 1017
phylogeny of, 1018, 1019, 1049f
reproduction and development of, 1008, 1017–1018, 1018f, 1019
support, muscles, and movement of, 1015
Pterognathia, 616
Pterokrohniidae, 423
Pteromonas lacerata, 68f
Pterosagitta draco, 427
Pterosagittidae, 423
Pterospora floridiensis, 82f
pterothorax of Hexapoda, **863**
Pterotrachea, 464f, 493f
Pterygota, 843, 846, 847, 848, 849
evolution and phylogeny of, 888, 889f, 889–890
thorax of, 863, 864, 865
Pterygotus, 915
buffaloensis, 913f
Ptiliidae, 856
Ptilocerembia, 853
Ptilocrinus, 982
Ptilosarcus gurneyi, 266f–267f
ptychocysts, 294f
Ptychodera flava, 1007
Ptychoderidae, 1008, 1009, 1013, 1015
Ptychogastriidae, 273
puerulus larvae, **831**
Pugettia, 780f
Pulex irritans, 858, 873
Pulmonata, 454, 466
pulsatile bodies, **349, 353**
of Acoela, 349, 350, 351, 353
of Nemertodermatida, 349, 350, 361
pulsatile organs of Hexapoda, **875**
Pulsellidae, 471
pumping mechanisms in circulatory system, 164
Punciidae, 798
Puncturella, 462f
pupae of Hexapoda, **886,** 887, 888f
pupation of Hexapoda, **886**
Pupillidae, 467
pusules of Dinoflagellata, **80**
Pycnoclavelina diminuta, 1028f–1029f
Pycnogonida, 710f–711f, 735t, 911, 912, 955–961
circulation, gas exchange, and excretion of, 959
external anatomy of, 957f, 958
feeding and digestion of, 959
fossil record of, 751
Hox genes of, 30
locomotion of, 958–959
nervous system and sense organs of, 959–960
phylogeny of, 962f
reproduction and development of, 961
Pycnogonidae, 956f, 961
Pycnogonum
littorale, 959, 961
rickettsi, 959
stearnsi, 956f
Pycnophyes, 696f, 697f
Pycnopodia, 982, 983, 985
pygidium
of Annelida, **541**
of Arthropoda, **750**
Pygocephalomorpha, 783
pyloric ceca of Echinodermata, **985**
pyloric stomach of Echinodermata, **985**
pylorus of Bryozoa, **652**
Pyralidae, 859
Pyramidellidae, 467, 494
pyrenoids, 104, 104n
pyriform organ of Bryozoa, **654**
Pyrodinium bahamense, 78
Pyrolobus fumarii, 9n
Pyrosoma, 1034f–1035f
Pyrosomida, 1022, 1031, 1039
Pyura, 1041
Pyuridae, 1038

Q

quadriradial symmetry, **137,** 137f
Quammen, D., 846
quartet of macromeres, **189**

R

rachiglossate radulae, **493,** 493f, 494f
rachis, **283**
radial canals, **285**
of Cnidaria, 285, 286f
of Echinodermata, **978,** 981, 982, 999, 1001
radial cleavage, **188,** 189f
indeterminate, 191
radial spokes, **142**
radial symmetry, **136**–138, 137f
biradial, **137,** 137f
of Cnidaria, 275, 275f, 277, 304
of Ctenophora, 327, 332, 333
nerve nets in, 174, 174f
pentaradial, **137,** 137f, 968, **969,** 977, 995, 1001, 1003
quadriradial, **137,** 137f
Radiaspis radiata, 753f
Radiata, 41, 220, 269, 969
radiations, 43f
explosive adaptive, 50f
Radiolaria, 58f–59f, 108–111, 109f–111f
characteristics of, 108
classification of, 61, 62, 63
phylogeny of, 128
reproduction of, 111
support and locomotion of, 109–111
radioles of Annelida, **552,** 552f–553f
radula, **153**–154, 154f, **472,** 491, 492f, 492–494, 496, 499
docoglossate, 493, 494f
ptenoglossate, 494
rachiglossate, 493, 493f, 494f
rhipidoglossate, 492, 493f, 494f
taenioglossate, 493, 493f
toxoglossate, 493f, 494, 495f
radular membrane of Mollusca, **492**
radular sac of Mollusca, **492**
Raff, R., 51
Ramisyllis, 566f
Ranella, 511f
Rangeomorpha, 317
Rapana, 493
Raphidioptera, 48f, 847, 856, 857f, 889f
Rathkea, 313f
Rathkeidae, 274
Reaka-Kudla, M., 20
recapitulation concept, **183, 202**
rectal accessory gills of Hexapoda, **877**
red algae, 55, 56, 57, 59f, 103, 125, 126, 127, 129
redia of Platyhelminthes, **401**
red queen hypothesis, 178
red tides, 77f, 78, 79
reduction bodies of Porifera, **241,** 241f
Reduviidae, 846, 854, 873
reef mounds, 257
regeneration, 176, 191
of Annelida, 565–567, 566f, 579, 584
of Cnidaria, 308, 313
of Ctenophora, 339
of Echinodermata, 995–996, 996f
of Enteropneusta, 1014
of Nemertea, 447
of Phoronida, 642
of Platyhelminthes, 394–395

of Porifera, 231, 239
of Urochordata, 1037
Regier, J. C., 832
regulative ova, **191**
Rehbachiella, 772, 834f–835f
Remipedia, 729, 754, 762f–763f, 765, 768, 769f, 801
distinguishing features of, 802t–803t
nervous system and sense organs of, 825
phylogeny of, 832, 833f, 834
reproduction and development of, 806t–807t, 831
thoracic appendages of, 804f–805f
renal appendages of Mollusca, **507**
renal sacs of Mollusca, **507**
renal vesicles of Urochordata, **1036**
renette cells of Nematoda, **679,** 679f, 680
Reniera, 239, 243
Renilla, 266f–267f, 308
reproduction, 176–179
asexual, 176f, 176–177
sexual, 177–179
Reptantia, 450, 450f, 778
Reptilia, 1022
Reptilomorpha, 1022, 1041f
repugnatorial glands of Arthropoda, **743**
reservoir of Euglenida, **119**
respiration, 164, 165f
respiratory pigments, 167–168, 168t, 558–559
respiratory tree of Echinodermata, **991,** 992f, 993, 993f, 994
Resticula nyssa, 620f
Retaria, 112
reticulopods (reticulopodia), **65,** 106, 106f, 107, **142**
retinula of Arthropoda, **748**
Retortamonada, 63, 128
retractor muscles, **146**
retrocerebral organ of Rotifera, **622,** 622f
retrograde waves in Mollusca locomotion, 486f, 487
retrolateral tibial apophysis of Euchelicerata, 948
Retusidae, 467
reversal, evolutionary, **42**
Reynolds, O., 141
Reynolds numbers, 141, 142, 807, 810, 867
in suspension feeding, 150
rhabdites, 354
of Platyhelminthes, 383f, 384
Rhabditida, 672, 674f, 679f, 680f, 685f
Rhabditina, 672
Rhabditis, 678f–679f, 680f
Rhabditophora, 376, 384
phylogeny of, 407, 408, 408f
Rhabdocalyptus dawsoni, 228f
Rhabdocoela, 377, 387, 389, 389f, 395, 396f
phylogeny of, 408, 408f, 409
rhabdoids
of Acoela, **354**
of Platyhelminthes, 383f, 383–384
rhabdome of Arthropoda, **748**
rhabdomere of Arthropoda, **748**
rhabdomeric eyes, **172**
Rhabdopleura, 1007, 1008, 1010f, 1015
digestion and feeding of, 1015, 1016, 1017
fossil record and phylogeny of, 1018
reproduction and development of, 1017, 1018
Rhabdopleuridae, 1009
Rhagoletis, 29
Rhampsinitus transvaalicus, 916f–917f
Rhapidioptera, 846
rheoreceptors, 393f
Rhinebothriidea, 379
rhinophores of Mollusca, **511**
rhipidoglossate radulae, **492,** 493f, 494f
Rhizaria, 60, 61, 62, 104–112
Chlorarachniophyta, 104–105
Granuloreticulosa, 105–107
Haplosporidia, 111–112
overview of, 63
phylogeny of, 128
Radiolaria, 107–111
Rhizocephala, 767, 791, 791n
life cycle of, 812, 812f–815f, 815
Rhizophora, 812
rhizopods, 74
Rhizostomae, 288f
Rhizostomeae, 272
Rhobdomonas, 119
Rhodactis howesii, 268
Rhodophyta, 4n, 55, 55n, 57, 125, 129
Rhodopidae, 466
rhombogens of Dicyemida, **414,** 415
Rhombozoa, 204, 204f, 347, 413, 414–418, 415f, 416f
cilia of, 143
Dicyemida, 414–417
Heterocyemida, 417–418
marine habitat of, 17
number of species, 2t
phylogeny of, 25, 1049f
rhopalia (rhopalium), **306,** 307, 307f
Rhopalonematidae, 273, 316
Rhopalura, 415f
granosa, 419
littoralis, 419
ophiocomae, 419, 419f, 420
Rhynchelmis, 568f
Rhynchobdellida, 541, 593, 594, 594f, 595, 595f, 596
rhynchocoel, of Nemertea, 435, 441, 442, 449
Rhynchocoela, 435, 436
Rhynchodemus, 387
rhynchodeum of Nemertea, 441
Rhynchonellida, 660
Rhynchonelliformea, 658f, 659, 660, 660f
feeding and digestion of, 661, 662, 662f, 663f
nervous system and sense organs of, 663
reproduction and development of, 664f–665f, 665
Rhynchoscolex, 397
Rhyncobdellida, 592f
Rhyniella praecursor, 848
Rhyniognatha, 887
Rice, M., 574
Richtersius coronifer, 713f
Ricinoides, 925
atewa, 919f–920f
crassipalpe, 919f–920f
Ricinulei, 919f–920f, 924, 925
circulation and gas exchange of, 942
classification of, 914
nervous system and sense organs of, 945
phylogeny of, 962f
reproduction and development of, 912, 948, 950, 953, 955
Riedl, R., 615
Rieger, R. M., 613
Riftia, 20, 585, 590
pachyptila, 20, 585, 586f–587f, 590
rigid skeleton, 144, 145–148
ring canal of Echinodermata, **978,** 981, 982, 999
ring furrows of Xenoturbellida, **366,** 366f, 367
Riodinidae, 859
Ripistes, 552
Rissoellidae, 466
Rissoidae, 465
Ristau, D. A., 220n
river blindness, 672, 683, 685–686
RNA, 51n
microRNA, 341
ribosomal, 51, 116, 341, 367, 407, 414, 694, 751, 1043, 1048
transcriptome sequenced from, 1048n
RNA polymerase, 8
Rochdalia parkeri, 870f
rock snot, 99
Rocky Mountain spotted fever, 918
Rodinia supercontinent, 10
Romaña's Sign, 123
rosette cells, 190f, **191, 337,** 338, 338f
Ross, R., 84
Rossella, 251
Rostanga pulchra, 251, 482
rostellum of Platyhelminthes, **386,** 386f
Rotaliella, 107
Rotaria, 619f
Rota-Stabelli, O., 13
Rotifera, 429, 613–615, 616–626
blastocoelom of, 140
body cavity of, 619
body forms of, 617, 617f
body wall and external anatomy of, 618–619
characteristics of, 617–618
cilia of, 143
circulation, gas exchange, excretion, and osmoregulation of, 621–622
feeding and digestion of, 620–621, 621f
nervous system and sense organs of, 622, 622f
number of species, 2t
phylogeny of, 25, 1049f
polymorphism of, 624
reproduction and development of, 622–624, 623f
summer and autumn cycles of, 623, 623f
support and locomotion of, 619f, 619–620
Rotiferophthora angustispora, 624
rotules of Echinodermata, **989**
Rouphozoa, 1050
Roux, W., 183
rowing propulsion of Cnidaria, **292**
r-selected species, 187, 198
rudiment of Echinodermata, **975**
Rudman, W. B., 483
Ruggiero, M. A., 6
Rugiloricus, 703, 704
cauliculus, 704
manuelae, 704
renaudae, 704

Rugoconites, 317
Rugosa, 320
Runcinidae, 467
Runcinoidea, 467
Runga, 936f–937f

S

Sabella, 547, 552
Sabellariidae, 535f, 539, 548f–549f
Sabellastarte magnifica, 535f
Sabellida, 539, 547, 588, 590
Sabellidae, 535f, 539, 548f–549f, 552, 564f, 588
Saccharomycetes, 129
Saccocirridae, 534f–535f, 535
Saccocirrus, 534f–535f
Saccoglossus, 1009f, 1010f, 1013, 1015
  kowalevskii, 1016f–1017f
  mereschkowskii, 1012f–1013f
Saccosporidae, 274
Sacculina carcini, 792f–793f
Sacculinidae, 791n
sacculus of Onychophora, **725**
Sacoglossa, 467, 501
Sadocus polyacanthus, 916f–917f
Sagartia, 296
  troglodyte, 310
Sagartiomorpha, 302
Sagitta, 421f
  bipunctata, 170f
Sagittidae, 423
Sagittiferidae, 354
sagittocysts in Acoela, 353f, **354,** 355f
Sahnicythere, 800f
Saipanetta, 800f
Saldidae, 854
Salinella, 204, 204f
  salve, 214, 214f
Salpa, 335, 1034f–1035f
Salpida, 1022, 1031, 1033, 1039
Salpingoeca, 124f, 125
Salpingoecidae, 125
saltation theory, 28
Salticidae, 914, 933, 936, 939, 946f, 951f, 953
Salvelinus frontinalis, 390
Sapphirina, 826f–827f
Sarcodina, 60
Sarcophagidae, 858
Sarcoptes scabiei, 916f
Sarcoptiformes, 914, 917
Sareptidae, 469
Sarotrocercus, 15f
Sars, G. O., 1007
Sars, M., 269
Sarsia, 313f
Sarsostraca, 772
Saturniidae, 859, 859f, 888f
Sauropsida, 1022
Saxipendium, 1007
  coronatum, 1008
saxitoxins, **78,** 79n
Scalidophora, 348, 429, 669, 693–704, 751
  Kinorhyncha, 695–698
  Loricifera, 701–704
  phylogeny of, 25, 1049f, 1050, 1051
  Priapula, 698–701
scalids of Scalidophora, **693,** 694, 699, 700, 701
  in Loricifera, **703**
Scalpellum, 791
  scalpellum, 825
scan-and-trap method of suspension feeding, 150
scape of Hexapoda antennae, **860**
Scaphandridae, 467
Scapholeberis, 773
Scaphopoda, 453, 455, 456f–457f, 470–471, 476f
  body plan of, 474
  circulation and gas exchange of, 506
  feeding of, 499
  general anatomy of, 470f
  locomotion of, 490
  mantle and mantle cavity of, 476–477
  phylogeny of, 522f, 523
  sensory organs of, 512
  shell of, 481
Scarabaeidae, 856, 872
Scathophaga stercoraria, 857f
Scatophagidae, 858
scavenging, 154–157
Schabronath, J., 945
Scheltema, A. H., 454, 524
Schinderhannes bartelsi, 751
Schindewolf, O., 28
Schistosoma, 407–408
  haematobium, 407
  mansoni, 381f, 404f, 407
schistosomiasis, 404f
Schizocardium, 1013
schizocoely, 195f, **196,** 206, 207
schizogony, **83, 301**
Schizomida, 919f–921f, 925–926, 927
  circulation and gas exchange of, 942
  classification of, 914
  nervous system and sense organs of, 945
  phylogeny of, 962f
  reproduction and development of, 950, 955
Schleip, W., 26
Schmidt, I., 236
Schmidtea mediterranea, 407
Schmidt's larva, **450**
Schmitt, W., 785
Schneider, A., 637
Schrödl, M., 214
Schultze, M., 436
Schulze, F. E., 215
Scissurellidae, 463, 504
Scleractinia, 270–271, 304, 320–321
  support of, 288, 289, 289f
Scleraxonia, 271
sclerites
  of Arthropoda, **731**
  of Mollusca, **454,** 477
scleroblasts, **288**
Sclerocypris, 799f
sclerocytes of Porifera, **230,** 231f, 233
Sclerolinum, 585
Sclerospongiae, 220, 222
sclerotization of Arthropoda, 732
Scolecida, 604
scolex of Platyhelminthes, **382, 386,** 386f
Scolopendra, 897, 901, 902f
  cingulata, 904f, 907f
  heros, 896f, 907f
Scolopendromorpha, 897, 898, 900f–901f, 903, 903f, 904, 904f, 906
Scolopendropsis duplicata, 898
Scoloplos, 569f
  armiger, 534f–535f
Scopimera, 820
Scorpiones, 914, 926, 927f, 950, 962f
Scorpionoidea, 914
Scotoplanes, 971f, 984
Scrippsia, 137f
Scrobicularia, 488f
Scruparia, 645f
Scudderia furcata, 171f
Scutigera, 900f–901f, 902f
  coleoptrata, 896f, 898
Scutigerella, 896f
  immaculata, 902, 908
Scutigeromorpha, 729, 897, 900f–901f, 903, 904, 905, 906
Scyllaeidae, 467
Scyllaridae, 782
Scypha, 218f, 221, 227, 247f
scyphistoma, **310,** 311f
Scyphistoma, 269
Scyphomedusae, 276, 306, 310, 320
scyphopolyp, **310,** 311f
Scyphozoa, 216, 269, 272
  cnidae of, 293
  life cycle of, 312f
  medusoid form of, 287f, 307–308
  nervous system and sense organs of, 307–308
  phylogeny of, 319, 320, 320f
  polypoid stage of, 277, 279
  reproduction and development of, 308, 310, 311f, 312f
Scytodes, 938, 939f
Scytodidae, 912, 938
secondary endosymbiosis, **57,** 59f, 105, **126,** 127
secondary gill bar of Enteropneusta, **1013**
secondary pigment cells of Arthropoda, 748
secondary plastids, 59f
Sedentaria, 538–539, 598f
sedentary animals, **19**
Segestria, 949f
segmental gland of Onychophora, **725**
segmentation genes, 346
Seilacher, A., 317
Seison, 621
Seisonidea, 614f, 617f, 618, 620f, 620–621, 622
selective deposit feeders, **153**
Selenella, 648, 649f
self-fertilization, 179
Selkirkia, 694
Semaeostomae, 288f
Semaeostomeae, 272
Semelidae, 470, 497
Semibalanus balanoides, 763f
seminal bursa of Acoela, 359f, **359**–360
seminal receptacle, **179**
seminal vesicle, **178**
sense organs, 169–173
sensilla, **747**
  of Chelicerata, 944
  of Crustacea, **822,** 824f
  of Hexapoda, **881**
sensory (afferent) nerves, **168**
sensory vesicle of Urochordata, **1040**
Sepia, 417, 491f, 513
Sepiida, 471–472, 482
Sepiidae, 472
Sepiolidae, 472
Sepioteuthis lessoniana, 456f–457f
septa of Cnidaria, **288,** 289f
Sergestes, 778

serial endosymbiotic theory, **125–126**, 126f
serial homology of Annelida, **531**
Serolidae, 787f
Serpulidae, 539, 548f–549f, 552, 570f
sessile animals, **19**
as predators, 155
setal-net suspension feeding, 150, 151f
settlement process, 199
Sexanymphon, 958
sex determination, 197
sexual reproduction, 177–179
Sharma, P. P., 961
Sheina orri, 798
shell, of Mollusca, 476f, 477–483, 479f, 480f, 481f, 488f, 490
shell eyes of Mollusca, **479**
shellfish poisoning
amnesic, 99
neurotoxic, 79
paralytic, 78–79
shell glands of Mollusca, **474,** 477
Shinn, G., 413
shiny spheres of Placozoa, **215**
shrimp, use of term, 761n
Shultz, J. W., 961
Siberian Traps, 16
Siboglinidae, 532, 533f, 539, 584–591
body plan of, 541, 542, 588–590
circulation, gas exchange, excretion, and osmoregulation of, 590
nervous system and sense organs of, 590
nutrition of, 590
reproduction of, 590–591
taxonomic history and classification of, 532, 540, 588
Siboglinum, 585, 586f, 588, 589f
fiordicum, 590
veleronis, 586f
weberi, 588
Sicariidae, 912
Sicyonia, 778
ingentis, 710f–711f
Sigalionidae, 537
Sigwart, J. D., 522f
Silene, 2
Silicea, 257
Silicispongia, 257
Siliquariidae, 465, 486f, 487
silk, spider, 928–932, 934, 936f–937f, 936–938
silk glands, **928,** 928f–929f, 932, 934
Silurian deposits, 523, 843
similarity, concept of, 36, 50–51
simple cilia of Ciliata, **89**
simple eyes of Hexapoda, **860,** 879
simplex stage of Tardigrada, **716**
Simpson, G. G., 38, 39
Simuliidae, 858
Simulium, 683, 781
siphon
of Echinodermata, **991**
of Echiuridae, **583**
of Mollusca, **477,** 482, 488f, **490,** 499, 499f, 505
of Urochordata, **1029,** 1031, 1033, 1034
siphonal canal of Mollusca, **482**
Siphonaptera, 856, 857f, 857–858, 872, 873
classification of, 846, 847
feeding of, 871
phylogeny of, 48f, 889f
Siphonariidae, 467
Siphoniulida, 897
Siphonocryptida, 897
Siphonodictyon, 250
coralliphagum, 250f
siphonoglyphs of Cnidaria, 278f, **279**
Siphonomecus, 575, 576
Siphonophora, 274, 280f–281f, 282, 282f
classification of, 273
locomotion of, 292
nervous system and sense organs of, 306
phylogeny of, 320
polymorphism of, 280
reproduction of, 311, 316
Siphonophorae, 328
Siphonophorida, 897, 902
Siphonosoma, 575, 576, 580f
Siphonosomatidae, 575
Siphonostomatoida, 767, 798
siphonozooids, **283**
siphuncle of Mollusca, **481**
Sipuncula, 532, 533, 533f, 535, 572–579, 597
body plan of, 541, 542, 543, 544, 574, 575f, 575–577
circulation and gas exchange of, 163, 168, 557
classification of, 81, 524, 531, 532
excretion and osmoregulation of, 578
feeding and digestion of, 153f, 577–578
nervous system and sense organs of, 578f, 578–579
phylogeny of, 25, 41, 524, 531, 574f, 1050
reproduction and development of, 106, 579, 580f
support and locomotion of, 144f, 577
taxonomic history and classification of, 436, 532, 574f, 574–575
Sipunculidae, 574, 576, 577
Sipunculus, 535, 574, 576, 576f, 577
norvegicus, 573f
nudus, 573f, 575f, 576f, 579
Siriella, 763f
Sirilorica, 694
sister taxa, **46,** 46f
six1 genes, 1041
Skara, 834f–835f, 835
skeleton, 144–148
endoskeleton, 145, 146f, 148
exoskeleton, 145–148, 147f
hydrostatic, 144f, 144–145
rigid, 144, 145–148
Slabber, M., 422
sleeping sickness, 122, 123
sleeze of sponges, 220n
slime glands of Onychophora, **724,** 725f
slime nets, 97, 101
slime papillae of Onychophora, **722,** 722f
slit sense organs, **171**
slit sensilla
of Arthropoda, **747**
of Euchelicerata, 944
Smittina cervicornis, 251
Snodgrass, R., 752, 754
Snowball Earth events, 10, 11
Solaster, 983
dawsoni, 984
stimpsoni, 984
sole, of Mollusca foot, **472,** 486
Solemyida, 469
Solemyidae, 469, 499
solenia of Cnidaria, **279**
Solenidae, 470
solenocytes, **161**
Solenofilomorpha crezeei, 357f
Solenogastres, 456f, 458f–459f, 458–459
body plan of, 472–473
circulation and gas exchange of, 506
classification of, 453, 454, 455
development of, 519
digestion of, 502
locomotion of, 486
mantle and mantle cavity of, 474
nervous system of, 508f
phylogeny of, 522f, 523
reproduction of, 514, 514f
shell of, 477
Solifugae, 914, 919f–921f, 924, 926, 942, 950
phylogeny of, 962f
Solmarisidae, 273
somatic ciliature of Ciliata, **89**
somatocoels of Echinodermata, **975, 980,** 998, 999, 1001
somites of Arthropoda, **731,** 733
Sorbeoconcha, 455, 465
Sorberacea, 1021, 1022, 1027, 1033
sorocarp of Protista, **72**
sorting area of Mollusca, **502**
Spadella, 422, 424, 425, 427
celphaloptera, 427
Spadellidae, 423
spadix of Mollusca, **518**
Sparassidae, 914, 922f–923f, 924
Spartobranchus, 1019
tenuis, 1018
Spatangoida, 974
Spathebothriidea, 379
spat of Mollusca, **488**
spawning, broadcast, **178, 198**
speciation, **27,** 28, 31, 31f
common models of, 50f
in hybridization, 29
phylogenetic tree representing, 50
species
classification of, 4
number of, 2t–3t, 3, 6, 1047
Species Plantarum (Linnaeus), 37
specific epithet, **37**
Speciospongia confoederata, 218f–219f
Spelaeogriphacea, 766, 785, 804f–805f
Spelaeogriphus, 786f
Speleonectes, 769f
ondinae, 762f–763f
tulumensis, 824f
Spemann, H., 183
Spengel, J. W., 454, 1007
Spengelidae, 1008
Spengelomenia bathybia, 459f
Speonebalia cannoni, 775
sperm, 178–179
spermatic cysts of Porifera, **242,** 242f
spermatophores, **178**
Spermatophyta, 129
sperm duct, **178**
sperm follicles of Porifera, **242,** 242f
sphaeridia of Echinodermata, **995**
Sphaeriidae, 470, 519
Sphaerodoridae, 538
Sphaeroeca, 125, 206
volvox, 204f
Sphaerolichida, 914

Sphaeromatidae, 787f, 812
Sphaeromatidea, 766
Sphaerotheriida, 897
Sphaerotheriidae, 907
Sphaerotherium, 907
Sphaerulariopsis, 670f
Sphecidae, 855
Spheciospongia, 252
  vesparia, 251
spherical symmetry, **136,** 136f, 138
spherulous cells of Porifera, **231**
Sphingidae, 859
Sphodros, 937f, 938
spicules
  of Nematoda, **681,** 682f
  of Porifera, 233–234, 234f–235f, 240, 241f, 251, 257
spider silk and webs, 928–932, 934, 936f–937f, 936–938, 946
spidroins, **928**
Spinicaudata, 765, 772, 773
spinnerets, **928,** 928f–929f
Spinoloricus cinziae, 702, 702f
spinoscalids
  of Kinorhyncha, 695f, **695**–696
  of Loricifera, **703**
Spinulosida, 973
Spiochaetopterus, 533f
Spionida, 538–539
Spionidae, 533f, 538, 547, 549, 550
spiracles, **165, 744**
  of Hexapoda, 877
  of Myriapoda, 903
spiral cleavage, **188**–189, 319, 347–348
  of Acoela, **360**
  of Chaetognatha, 427
  coding system for, 189–191, 190f
  of Crustacea, 829
  determinate, 191
  germ layers in, 191
  of Platyhelminthes, 397–398
  radial cleavage compared to, 189f
Spiralia, 346, 347–348, 398, 413, 604–605, 719, 728, 751
  Annelida, 531–598
  cerebral ganglia of, 346
  Cycliophora, 609–611
  Entoprocta, 603–609
  Gastrotricha, 428–432
  gastrulation and gene expression in, 192f
  Gnathifera, 613–632
  as legacy name, 26, 348
  Lophophorata, 635–665
  Mollusca, 453–524
  Nemertea, 435–451
  Orthonectida, 418–420
  phylogeny of, 25, 51, 348, 407, 1049f, 1050
  Platyhelminthes, 373–409
  Rhombozoa, 414–418
  spiral cleavage of, 319, 347–348, 360, 408
Spirobolida, 897, 898, 899
Spirobranchus, 539
  giganteus, 548f–549f, 570f
spirocysts, **293,** 298
Spirolina, 106f
Spironoura, 680f
Spironucleus, 116, 117
Spirorbinae, 539
Spirostreptida, 897, 899
Spirostreptus, 902f
Spirula, 482
Spirulida, 471
Spirulidae, 471
Spirurina, 672
Spodoptera, 938
Spondylidae, 469, 487
Spondylus, 513
sponge bleaching, 252
Spongia, 233
  officinalis, 233f
Spongiae, 220
Spongiaria, 220
Spongicola, 251
  venusta, 779
Spongida, 220
Spongilla, 218f–219f, 241f
Spongillida, 222
Spongillidae, 241f
spongin, 217, **230,** 232–233
sponging insects, 871f, 872
spongocoel, **225**
spongocytes of Porifera, **230,** 231f
spore valves of Cnidaria, **299**
sporocysts
  of Apicomplexa, **85**
  of Platyhelminthes, **401**
sporogony of Apicomplexa, **83**
sporosacs, **315**
Sporozoa, 60, 112
Spriggina, 11
Squilla, 744f, 775f
stabilimenta of Euchelicerata, **931,** 931f
stalking predators, 155
Staphylinidae, 855
star-cell organ of Entoprocta, **607**
statoblasts of Phylactolaemata, **657,** 657f
statoconia of Mollusca, **511**
statocysts, 170f, **170**–171
  of Acoela, 357–358
  of Annelida, 565
  of Brachiopoda, 663
  of Cnidaria, 306, 308
  of Crustacea, 822–823, 823f
  of Echinodermata, 995
  of Mollusca, 511, 512
  of Nemertodermatida, 363f, 363–364
  of Platyhelminthes, 392
  of Urochordata, 1040
  of Xenoturbellida, 369, 369f
statoliths, **170,** 170f
Stauroteuthidae, 471
Staurozoa, 3, 269, 271, 288f, 292
  phylogeny of, 319, 320f
  polypoid stage of, 277, 279
  reproduction of, 311
Stegasoma, 1034f–1035f
Stegodyphus, 932
Stegophrynus dammermani, 919f
Steinböck, O., 362
Steinernema, 678f–679f, 682f, 685f
  scapterisci, 671
Steinernematidae, 677
Stelexomonas, 125
Stemmiulida, 897
stem nematogens
  of Dicyemida, 416, 417f
  of Heterocyemida, 418, 418f
Stemonitis, 73f
Stenodictya lobata, 870f
stenohaline animals, **159**
Stenolaemata, 646
Stenoplax, 518f
Stenopodidea, 766, 778, 779, 779f
stenopods of Arthropoda, **734,** 734f
Stenopus, 779, 779f
  hispidus, 824f
Stenorhynchus, 780f
Stentor, 58f–59f, 88f–89f, 90, 92, 92f
Stephanocea, 68f
Stephanoceros, 617f
Stephanodrilus, 592f
Stephanoeca, 125
Stephanopogon minuta, 115f
Stephanoscyphus, 310
Stephen, A., 574
stercoral pocket of Euchelicerata, 940–941
stereoblastula, **193,** 193f, 248
stereogastrula, **193**
stereom of Echinodermata, **969**
sternite of Arthropoda, **731**
Sterreria, 360f
  psammicola, 357f
Stiasny-Wijnhoff, G., 436
Stichocotylida, 378
Stichodactyla mertensii, 277
Sticholonche, 111
Stichopus, 991
stichosome of Nematoda, **677**
stigmata, **172**
  of Protista, **68,** 80
  of Urochordata, **1027, 1034**
Stilbonematidae, 677
Stoecharthrum, 419
Štolc, A., 269
Stolidobranchia, 1022
stolonal funiculus of Bryozoa, **651**
Stolonifera, 271
stolons of Pterobranchia, **1015**
stomachic plexus of Phoronida, 640f, 642
Stomatolepas, 791
Stomatopoda, 765, 775–776
stomoblastula of Porifera, **246,** 247f
stomochord of Hemichordata, 968, **1009**
  in Enteropneusta, 1011, 1013
  in Pterobranchia, 1015
stomodeal ectoderm, **193**
stomodeal plane of Ctenophora, **333,** 333f, 334
stomodeum of Ctenophora, **333,** 333f
Stomolophus meleagris, 312f
Stomphia, 291f, 292
stone canal of Echinodermata, **981,** 982, 991, 998
Stramenopila, 58f–59f, 61, 66f, 97–101, 128
  characteristics of, 97
  diversity of, 98f
  nutrition of, 99, 101
  overview of, 62, 63
  phylogeny of, 97, 127–128
  reproduction of, 99, 101
  support and locomotion of, 100–101
Strausfeld, N. J., 755
Strepsiptera, 48f, 846, 847, 855, 856, 857f, 889f
Streptomyces avermectinius, 686
Streptoneura, 454
streptoneury of Gastropoda, **484,** 508
Strickland code, 37
Strigamia, 900f–901f
strobila
  of Cnidaria, **310,** 311f, 312f
  of Platyhelminthes, **382**

Strobila, 269
strobilation, **308, 310,** 311
  monodisc, **310**
  polydisc, **310**
stroma of Echinodermata, **969**
Stromatocystites, 977f, 1001, 1002f
Stromatoporoida, 256f
Strombidae, 465
Strombidium, 93
Strombomonas, 119
Strombus, 487, 511
Strongylocentrotus, 153, 976f, 994f
Strongylocentrotus purpuratus, 970f–971f, 983, 994f
Struck, T., 538
Stunkard, H. W., 414
Styela, 152f, 1028f–1029f, 1029
Styelidae, 1038
stygobionts, **22**
Stygobromus, 790
Stygocarella, 775f
Stylasterias, 985
  forreri, 978f–979f
Stylasteridae, 274, 288, 320
style sac of Mollusca, **502,** 502f
stylet
  of Crustacea, **812**
  of Hexapoda, 871
  of Loricifera, **702**
  of Nematoda, **677,** 678f–679f
  of Nemertea, 441, 442f
  of Platyhelminthes, **389, 395**
  of Tardigrada, 714f, 715f, **716, 717**
stylet glands of Tardigrada, 714f, 715f, **717**
Stylobates, 301
  aenus, 301f
Stylocephalus, 83, 85
  longicollis, 85, 85f
Stylochus, 397, 397f
Stylommatophora, 467, 505f, 516f
Stylonychia, 90f
Styraconyx qivitoq, 712f
subepidermal plexus of Ctenophora, 338
Suberites, 251
sublittoral zone, **18**
subneural gland of Urochordata, **1037**
subumbrella, **285**
Succineidae, 467
sucking insects, 871f, 871–872
sulcal side, **279**
sulcus of Dinoflagellata, **80**
superpositional eyes, of Arthropoda, **749,** 749f
support structures, 140–148
  in hydrostatic skeleton, 144f, 144–145
  in rigid skeleton, 144, 145–148
supralittoral zone, **18**
surface-to-volume dilemma, **139**
suspension feeding, 149–152
  ciliary-mucous mechanism in, 151, 152f
  contact methods of, 150
  mucous-net, 151, 152f
  scan-and-trap method, 150
  setal-net, 150, 151f
  tentacle or tube feet, 151–152, 152f
sutures of Mollusca, **481**
swimming bells, 280, 281f, 282
swimming motion, 736–737, 737f
  of Crustacea, 803, 805f, 807, 808
  of Mollusca, 490, 491f
  Reynolds numbers in, 141
Sybota, 937f
Sycon, 218f, 227, 235f
  development of, 247, 247f, 247n
  distribution of, 249
syconoid sponges, **224,** 225, 225f, 226–227, 236
Syllidae, 534f–535f, 538
Syllis ramosa, 566f
Symbiodinium, 78, 302–303, 499, 500
symbiogenesis, 2, 29, **125**
Symbion, 609f
  americanus, 609, 610
  pandora, 609, 609f, 610, 611f
symbiosis, **23**
symbiotic relationships, **23**–24, 125–127
  of Amoebozoa, 70
  of Annelida, 552–554, 554f
  of Chlorarachniophyta, 105
  of Ciliata, 87, 93
  of Cnidaria, 268, 292, 298, 302–304, 321
  of Dinoflagellata, 78, 79
  endosymbiotic theory of, **29,** 125–126, 126f
  of Granuloreticulosa, 107
  of Hexapoda, 872–873
  macroevolution in, 29
  of Mollusca, 499–501
  of Nematoda, 677, 685f, 686
  of Parabasalida, 112, 113
  parasitic life cycle in, 201
  photosynthetic by-products as food source in, 149
  of Platyhelminthes, 387
  of Porifera, 217, 236, 249, 251–254
  primary, **57,** 59f
  of Radiolaria, 111
  secondary, **57,** 59f, 79, 105, **126,** 127
  of Tardigrada, 717
  of Urochordata, 1033, 1034, 1040
symmetry, 136–138
  bilateral, **138.** *See also* bilateral symmetry
  radial, **136.** *See also* radial symmetry
  spherical, **136,** 136f, 138
Symphyla, 895, 896f, 897, 899
  circulation and gas exchange of, 903
  feeding and digestion of, 902
  head and mouth appendages of, 901
  locomotion of, 901
  nervous system and sense organs of, 904
  phylogeny of, 908
  reproduction and development of, 907–908
Symphyta, 855
Symphytognathidae, 924, 937f
Symplasma, 228, 257
symplesiomorphy, **44,** 45
sympodial growth of Cnidaria, 279f, **279**–280
Symsagittifera, 351
Synagoga mira, 791
Synalpheus, 811
  regalis, 812
synapomorphy, **44,** 45, 46, 47, 1048, 1049f
Synapta, 153f, 971f
synapticles of Enteropneusta, **1013**
Syncarida, 765, 776, 806t–807t
synchronous broadcast spawning, **178, 198**
syncytial theory, **205,** 407
syncytial tissues, 227–228, 228n
Syndermata, 616
Syndesmis, 387, 387f
Synechococcus, 252
syngamy, **68, 178**
synonyms, **37**
Synotaxidae, 936f–937f
Synura, 98f, 100
Syrphidae, 858
Systellommatophora, 467
Systema Naturae (Linnaeus), 37, 374–375, 574, 698, 844
systematics, **35,** 39–40
  phylogenetic, **45**
syzygy, **85**

T

Tabanidae, 858
Tabaota, 390
table palintomy of Porifera, **244,** 246
tabulae, **288**
Tabulata, 320
Tachinidae, 858
Tachypleus, 916
  gigas, 916
  tridendatus, 916
tactile receptors, 170, 170f
tadpole larva, 1043
  of Urochordata, 1038f–1039f, **1039,** 1040
Taenia, 375f, 386
  saginata, 382f, 405, 406f
  solium, 386f, 405, 405f, 407
taenidia of Hexapoda, **877**
taenioglossate radula, **493,** 493f
Taeniopoda eques, 710f–711f
Tagelus, 488f
tagmata of Arthropoda, **730,** 802t
tagmosis of Arthropoda, **730**–731, 746
tail fan of Crustacea, **801,** 807
tail fin of Chaetognatha, **423**
tail-flip of Crustacea, **807**
Tainisopidea, 766
Talitrus
  pacificus, 790
  sylvaticus, 790
Tanaidacea, 766, 785
Tangaroa tahitiensis, 951f
tanning, **146, 732**
Tantulocarida, 764f, 767, 790, 791–793, 793f
  distinguishing features of, 802t–803t
  nervous system and sense organs of, 825
  reproductive features of, 806t–807t
Tanystylum grossifemorum, 956f
tapeum of Mollusca, **513**
Tardigrada, 348, 709, 710f–711f, 711–718, 730, 733
  blastocoelom of, 140
  body plan of, 715f, 715–716
  characteristics of, 711, 711t
  circulation and gas exchange of, 717
  feeding, digestion, and excretion of, 716–717
  locomotion of, 716
  as meiobenthic, 19
  nervous system and sense organs of, 717
  number of species, 2t, 711t
  phylogeny of, 25, 26, 755f, 1049f

reproduction and development of, 717–718
tarsus of Hexapoda, **864**
Tatjanellia grandis, 582f
taxis, **169**
taxon, **38**
taxonomy, **40**
Technitella, 107
Tecthya crypta, 250
Tectitethya crypta, 250
Tedania, 250
Tegella, 655
Tegenaria, 928f–929f, 940f, 945f
Tegiticula, 845
tegmentum of Mollusca, **479**
tegmina of Hexapoda, **866**
Tegulidae, 463
tegument of Platyhelminthes, 384f, **384**–385
Telestacea, 284f
Telesto, 284f
Tellinidae, 470, 497
Tellinoidea, 497
teloanamorphosis, 907
teloblast, **750**
teloblastic development
of Annelida, **531, 571**
of Arthropoda, **750**
of Tardigrada, 718
telolecithal ova, **187,** 187f, 198f
telopod of Arthropoda, **733,** 735, 864
telotroch, **208**
telson of Arthropoda, **750, 801**
Temnocephala caeca, 383f
Temnocephalidae, 387
Tenagodus, 486f
Tenebrionidae, 856, 878, 883
tension hypothesis on exocytosis, 294
tentacle
of Ctenophora, **332,** 333, 333f, 335, 335f, 341
in deposit feeding, 153
of Echinodermata, 978, 987
of Sipuncula, **572,** 577
in suspension feeding, 151–152, 152f
tentacle canal of Ctenophora, 333f, **337**
tentacle crown of Bryozoa, **646,** 648, 649, 650, 651
tentacle scale of Echinodermata, **987**
tentacle sheaths of Ctenophora, **332,** 333f
tentacle shedding or budding of Anthozoa, 308
tentacular plane of Ctenophora, **333,** 333f, 334
Tentaculata, 328, 637
tentillae of Ctenophora, **332,** 333, 333f, 336f
tentorium of Hexapoda, **860**
Tephritidae, 858
Terebella, 550
Terebellidae, 535f, 540, 550
Terebelliformia, 539–540, 547, 560
Terebrasabella, 547
Terebratalia transversa, 664f–665f
Terebratella sanguinea, 659
Terebratelloidea, 658f, 662f, 663f
Terebratulida, 660
Terebratulina, 660f
Terebridae, 466
Teredinidae, 470, 490, 501
Teredo, 489f, 490
navalis, 155f
Tergipedidae, 467
tergite of Arthropoda, **731**
Tergivellum cinnabarinum, 1010f
terminal tentacle of Echinodermata, **978**
Terpios, 251
terrestrial habitat, 22–23
gas exchange structures in, 165
life cycle adaptations in, 200
osmoregulation in, 159, 159f
Terrestricytheridae, 798
Testacea, 454
Testacella, 482, 494
testate amebas, **74**
reproduction of, 76
testes, **178,** 178f
test of Protista, **65**
of Amoebozoa, 74
of Dinoflagellata, 77f
of Foraminifera, 106, 107
Tethya
aurantia, 218f–219f
aurantium, 252
orphei, 252
seychellensis, 252
Tethydidae, 467
Tethys Sea, **16,** 17
Tetilla, 244
Tetrabothriidea, 379
Tetrabranchiata, 471
Tetracapsula
bryosalmonae, 269
bryozoides, 269
Tetracapsuloides bryosalmonae, 269
Tetraclita, 828f
Tetraconata, 833, 1051
Tetractinellida, 222, 234, 244
Tetractinomorpha, 220, 221
Tetragnatha, 949f
Tetragnathidae, 913, 931
Tetrahymena, 87
pyriformis, 89f
tetrapartite ommatidium of Arthropoda, **748,** 755n
Tetraphyllidea, 379–380, 386
Tetraplatiidae, 273
Tetrapulmonata, 926, 927, 962f
Tetrigidae, 853
tetrodotoxin of Chaetognatha, 425
Tetrophthalmi, 924
Tettigoniidae, 853, 859f
Tevnia, 585
jerichonana, 586f–587f
Thalassia testudinum, 108f
Thalassiosira, 97
Thalassisobates litoralis, 904
Thalassocalyce inconstans, 332, 334, 334f
Thalassocalycida, 332, 334
Thalassophysa, 111f
Thaliacea, 1021, 1022, 1027, 1029, 1031, 1034f–1035f
body wall, support, and locomotion of, 1033
circulation, gas exchange, and excretion of, 1036
feeding and digestion of, 1034
nervous system and sense organs of, 1036
reproduction and development of, 1037, 1039
Thaumastodermatidae, 431
Thaumatoconcha, 799f
Thayer, C., 659
theca
of Cnidaria, **288,** 289f
of dinoflagellates, **79**
of hydroids, **280**
Thecamoeba, 74
Thecata, 273
Thecideida, 660, 661, 663
Thecidellina meyeri, 658f
Thecideoidea, 658f
Thecosomata, 467
Thecostraca, 763f, 767, 790–791, 792f–793f
distinguishing features of, 802t–803t
phylogeny of, 832, 833f
reproductive features of, 806t–807t
Thectardis, 11
Thelyphonellus amazonicus, 919f–921f
Thelyphonida, 926, 927
Themiste, 577, 579
dyscrita, 573f
lageniformis, 579
pyroides, 579
Theraphosa blondi, 922f–923f, 924
Theraphosidae, 912, 922f–923f, 944, 952
Theridiidae, 913, 922f–923f, 936f–937f, 939
Theridion, 954f–955f
thermophiles, **9,** 9n
thermoreceptors, 67, 173
Thermosbaenacea, 766, 783, 785
Thermosbaena mirabilis, 785
Thermozodium esakii, 714
Thetispelecaris, 785
thigmotaxis, 170
Tholoarctus natans, 716
Thomas, L., 413
Thomisidae, 914, 936, 939, 951f
Thompson, D., 139
Thompson, J. V., 454, 637, 830n
Thompsonia, 815
Thoosa, 242f
Thoracica, 763f, 767, 791
thoracomeres, 768n, **801**
Thoracophelia mucronata, 534f–535f
thoracopods of Crustacea, 802t
thorax
of Crustacea, **801**
of Hexapoda, 863–866
Thorectidae, 221
Thraciidae, 470
threatened species, 7
through gut, **149, 345**. *See also* complete gut
Thylacodes
arenarius, 496
squamigerus, 496
thylakoids, **56**–57, 66f, 80
Thyone, 996
Thysanocardia nigra, 573f
Thysanoptera, 48f, 846, 847, 853, 854f, 889f, 890
Thysanozoon, 375f, 384
brocchii, 383f
Thysanura, 847, 848, 849f, 887
evolution and phylogeny of, 888, 889f
Thysiridae, 470
tibia of Hexapoda, **864**
Tichenor, J., 179
tick fever, 918
Tiedemann's bodies of Echinodermata, **981,** 982
Tinerfe, 330f–331f
Tipulidae, 858

tissues, 138–139
  in circulatory system, 162
Titaniloricus, 703
  inexpectatovus, 702
Titanus giganteus, 855
Titiscania, 482
Titiscaniidae, 463
Tityus, 936
Tjalfiella, 340
toes of Loricifera, **704,** 704f
Tomopteridae, 538
Tomopteris, 545
Tonicella lineata, 461f
Tonicia, 511f
Tonna, 500f–501f
Tonnidae, 466
tornaria larva, **1015,** 1016f–1017f, 1019
Torquaratoridae, 1007, 1008, 1010f, 1019
torsion of Gastropoda, 483f, 483–485, 508, 521, 521f
torted gastropods, **484**
touch receptors, 170, 170f
toxicysts of Ciliata, **92,** 93f
toxins
  of Chaetognatha, 425
  of Cnidaria, 268, 295–296
  of Dinoflagellata, 78–79
  of Euchelicerata, 935–936, 938
  of Mollusca, 494, 495f
  of Nemertea, 441
  of Platyhelminthes, 389
  of Porifera, 250
  of Stramenopila, 99
toxoglossate radula, 493f, 494, 495f
Toxoplasma, 81, 83
  gondii, 83
toxoplasmosis, 82
Toxopneustes roseus, 154f
trabeculae of Echinodermata, **969**
trabecular syncytium of Hexactinellida, **227**
tracheae, **165, 877**
tracheal system, **165,** 166f–167f, **744,** 745f
  of Hexapoda, 876f, 877
  of Myriapoda, 903–904, 904f
Tracheata, 752
Trachelomonas, 119
tracheoles of Hexapoda, **877**
Tracheophyta, 4n, 55n, 129
Trachylina, 270, 273, 286f–287f, 316, 320f
Trachymedusa, 273
Trachymedusae, 273, 292
transcription factors, 10, **184,** 191
transfer choanocytes of Porifera, **243,** 243f
transport, internal, 162–163
  of oxygen, 167–168
transposable elements, 28
transposase, 28
transverse canals of Ctenophora, 333f, **337**
transverse cleavage, **188,** 188f
transverse fission
  of Cnidaria, 308
  of Enteropneusta, 1014
  of Nemertea, 447
  of Platyhelminthes, 395f
transverse plane, **138,** 138f
Trapezia, 780f
Trebius heterodont, 794f–795f
tree diagrams, **36,** 43–44
Trematoda, 373, 375f, 376, 378, 384, 385
  life cycle of, 404f
  phylogeny of, 407, 408f, 409
Trembley, A., 313
Trepaxonemata, 376, 408, 408f
Treptoplax reptans, 215
Tresus, 468f–469f, 488f
Tretomphalus, 107
  bulloides, 108f
Triassic, 16
Triatoma, 123
Triatominae, 846, 854, 873
Tribolium confusum, 390
Tribrachidium, 317
Tricellaria, 649f
  occidentalis, 648f
trichimella larva, 244, 245f, 246f, **248**
Trichinella, 683
  spiralis, 671, 683, 684f
trichobothria, **170,** 944, 945f
trichobranchiate gills, 778, 779, 782, 783, 820f
Trichocephalida, 673
trichocyst, **67**
  of Ciliata, **92**
  of Dinoflagellata, **80**
trichogon, **815,** 815f
trichomes of Hexapoda, **877**
Trichomonas, 117
  murius, 112f
  tenax, 112, 113
  vaginalis, 112, 113, 114
Trichonympha, 112, 113, 114f, 115
Trichoplax, 204f, 205
  adhaerens, 205, 214f, 215–216
Trichoptera, 48f, 846, 847, 857f, 858, 889f
Trichurida, 673
Trichuris trichiura, 683, 684f
Tricladida, 377, 380f, 389, 389f, 396f
Tricolia, 478f
Tridachia crispata, 464f–465f
Tridacna, 9, 147f, 474, 499
  gigas, 474
  maxima, 456f–457f
Tridactylidae, 853
Trididemnum, 1034
Triebelina, 800f
Trigoniida, 470
Trigoniidae, 470
Trigonotarbida, 914
Trilobita, 710f–711f, 711t, 728–729, 752, 753f, 754f
  phylogeny of, 755f
Trilobitomorpha, 728
Trilobozoa, 317
trimeric body plan, **636**
Trimusculidae, 496
Trimusculoidea, 467
Trinchesia, 500f–501f
trinomen, **37**
Triopha, 494f
Triops, 771f, 772, 824f
Triopsidae, 772
Tripedalia cystophora, 296
Tripedaliidae, 272
Triphoridae, 466
triploblasty, **139,** 140, 345
  body plan in, 140, 140f
  and turbellarian theory on Cnidaria, **318**–319
Triplonchida, 673
Triticella, 645f, 648, 655
tritocerebrum
  of Crustacea, 821
  of Euchelicerata, 943
Tritonia diomedea, 511
Tritrichomonas, 114
  foetus, 113
trivium of Echinodermata, **978**
trochaea, **208,** 208f
Trochida, 463
Trochidae, 463, 515f
trochophore larva, 208, 208f, 348, 644
  of Annelida, 570f, 571, 572f, 579, 580f, 585f, 589f
  of Mollusca, 519, 519f, 520
Trochozoa, 348, 449
trochus of Rotifera, 619, 619f, 620, 620f
Trombicula, 916f, 918
Trombidiformes, 914, 917
trophocytes of Porifera, 240, **242,** 242f
trophosome
  of Nematoda, **677**
  of Siboglinidae, **590**
Truncularioopsis trunculus, 453n
trunk
  of Crustacea, **798**
  of Enteropneusta, **1009,** 1010f, 1012f–1013f
  of Siboglinidae, **588**
Trypanorhycha, 380
Trypanosoma, 58f–59f, 84, 120, 121, 122, 124f
  brucei, 121f, 122, 123
  cruzi, 122–123, 873
  gambiense, 122
  rhodesiense, 122
trypanosomes, 120–124
trypanosomiasis, 846, 873
Trypanosyllis californiensis, 534f–535f
tube feet
  of Asteroidea, 980, 981, 982, 983, 993
  of Crinoidea, 981
  development of, 977–978, 981
  Dll expression in, 735
  of Echinoidea, 982, 989, 990f, 991, 995
  evolution of, 1001, 1002f
  in excretion, 994
  in feeding, 151–152, 152f, 984, 986, 986f, 987, 987f, 989
  in gas exchange, 993, 994
  in hemal system, 993
  of Holothuroidea, 982, 984
  in hydrostatic skeleton, 144
  in locomotion, 982, 983, 983f, 984
  nervous system and sense organs in, 995
  of Ophiuroidea, 981, 983, 986, 987, 987f
  in water vascular system, 969, **975,** 980
Tubifex, 301
Tubificidae, 540
Tubiluchidae, 699, 699f
Tubiluchus, 698, 700
  corallicola, 699f, 700
Tubipora, 288f
Tubulanus, 441, 444f, 446f, 448
  sexlineatus, 437f
Tubularia, 277f, 282, 314f
Tubulipora, 645f
Tullberg, T., 454
Tunicata, 1021, 1022
tunicin, 1033, 1043
tun stage of Tardigrada, **713**
Turbellaria, 112, 376
turbellarian theory, **318**–319, 524
Turbeville, J. M., 436

Turbinidae, 463
Turridae, 466
Turritella, 496
Turritellidae, 465, 496
Tydeus starri, 916f
Tylenchina, 672
Tylenchulus semipenetrans, 670f
Tyler, S., 613
Tylodinidae, 467
tympanic organs, **171, 881,** 881f
tympanum, 171, 171f, 881, 881f
Typhlatya, 811f
Typhloscolecidae, 538
typhlosole of Annelida, **555**
Typosyllis, 566f

U

Ubx gene, 30, 735
Uca, 826, 826f–827f
  princeps, 762f–763f
Uloboridae, 913, 922, 930, 937, 938, 951f
Umbellula, 283
Umbonium, 496
Umbrachulida, 467
Umbraculidae, 467
Unguiphora, 377
unguitractor plate of Hexapoda, **864**
Unionida, 470, 519
Unionidae, 470
unipolar ingression, **193,** 194f
Uniramia, 754
uniramous limbs of Arthropoda, **734,** 734f
United States Geological Service (USGS), 6
unranked classification, **49,** 50
Upogebia, 783, 810
Uraraneida, 914
Urbilateria, **346**
urea, **158,** 158f
Urechis, 582–583, 583f, 584, 585f
  caupo, 540, 553, 581f, 582f, 584
ureotelic animals, **158**
uric acid, **158,** 158f
uricotelic animals, **158**
Urnaloricidae, 701
Urnatella, 603
  gracilis, 607
urns of Sipuncula, **577,** 578
Urochordata, 349, 968, 1021, 1022, 1025, 1027–1041
  biogeographical distribution of, 1027
  body wall, support, and locomotion of, 1033–1034
  circulation, gas exchange, and excretion of, 1036
  feeding and digestion of, 1034–1036
  nervous system and sense organs of, 1036–1037
  number of species, 3t
  phylogeny of, 25, 1041f, 1042, 1043, 1049f, 1050
  reproduction and development of, 1037f, 1037–1041, 1038f–1039f
Urodasys, 428f
uropods of Crustacea, **801**
uropolar cells of Dicyemida, **414,** 415f
Uropygi, 914, 919f–921f, 926–927, 942, 950, 955
  phylogeny of, 962f
Urosalpinx, 493
Urosporidium, 112
Uschakov, P. V., 588
uterine bell of Acanthocephala, 625f, 626
uterus of Platyhelminthes, **395,** 403, 405f

V

Vaejovis spinigerus, 936
vagabond's disease, **873**
vagina, of Platyhelminthes, **399,** 403
Vaginacola, 88f–89f
Vahlkampfia, 75f
Valvatida, 973
Valvatidae, 466
valves of Bivalvia, **473, 479,** 481
Valvifera, 766
Vampyromorpha, 471
Vampyroteuthis, 491f
Vanadis, 564f
van Beneden, E., 414
van den Biggelaar, J. A. M., 1042
van der Horst, C., 1007
vanes of Porifera, **223,** 254
van Leeuwenhoek, A., 59, 115, 116, 616
Vannella, 71f, 74
Van Valen, L., 178
Vargula, 800f
vascular budding of Urochordata, **1037**
vasoperitoneal tissue of Phoronida, 642
Vaucheria, 127
Vauxia, 15f
vegetal pole, **187,** 187f, 188f
veins of insect wings, **865,** 866f, 870
Velamen, 332
velarium of Cnidaria, **285,** 292
velar tentacles of Cephalochordata, **1025**
Velatida, 973
Velella, 267f, 282, 283f, 298
veliger of Mollusca, **472,** 488, **519,** 520, 520f, 521
velum, **285**
  of Cnidaria, 285, 286f, 292
  of Mollusca, **519,** 520f
Velutinidae, 482
Vendia, 11
Vendobionta, 317
Vendozoa, 317
Venerida, 470
Veneridae, 468f–469f, 470
venom
  of Euchelicerata, 935–936, 938
  of Mollusca, 494, 495f
ventral aorta of Cephalochordata, **1026**
ventral muscle plate of Micrognathozoa, **631,** 631f
ventral organs of Onychophora, **722**
ventral vessels
  of Enteropneusta, **1014**
  of Pterobranchia, 1017
vents, hydrothermal, 1n, 9, 9n, 19, 20
Vericrustacea, 832
Vermes, 374–375, 532, 728, 969
Vermetidae, 465
  feeding of, 151, 496
  locomotion of, 487
Vermetus, 496
vermiform larvae of Dicyemida, **414,** 415
vermigon, **812,** 815f
vernal pools, 22
Verongida, 252
Verongimorpha, 222
Veronicellidae, 467
Verruca, 792f–793f
vertebrae of Echinodermata, **978**
Vertebrata, 9, 968, 1021, 1022
  as legacy name, 1043
  number of species, 3t
  phylogeny of, 25, 1041f, 1042–1043, 1049f, 1050
Verticillitidae, 222
Vertiginidae, 467
vesicular nucleus of Protista, **70**
Vespidae, 855
vestibule
  of Cephalochordata, **1025**
  of Chaetognatha, **423**
Vestimentifera, 532, 538, 539, 540, 585, 588, 590, 591, 597
vestimentum of Siboglinidae, **588**
Vetericaris, 778
Vetigastropoda, 455, 463, 478f
  circulation and gas exchange of, 504
  development of, 520
  excretion and osmoregulation of, 507
  feeding of, 492, 493f, 494f
  reproduction of, 515f
  sensory organs of, 511
  shell of, 482
  torsion of, 484, 485
vibraculum of Bryozoa, **648,** 649f
Vibrio, 425
Victorella, 655
vinegaroons, 926, 927
Viridiplantae, 4n, 55n, 129
viruses, 6, 8
visceral cords of Mollusca, **507**
vitellarium of Platyhelminthes, **395**
vitellogenesis, **187**
Viviparidae, 465, 496
Viviparus, 493f, 496
Vogt, C., 375
Voight, O., 215
volcanism, 16, 17
Voltzow, J., 485
Volutidae, 466
Volvox, 204, 204f, 244, 246, 621
von Baer, K., 183
von Eschscholtz, J. F., 328, 1007
von Frisch, K., 882
von Graff, L., 454
von Ihering, H., 454
von Siebold, K., 59, 728
Vorticella, 90, 92, 94f
  campanula, 96f
Vosmaer, R., 328

W

Waddycephalus teretiusculus, 796f
Wagener's larvae of Heterocyemida, **418,** 418f
walking legs, 734f
  of Hexapoda, 866–867
  of Onychophora, **722,** 723
Wallace, A. R., 27, 37, 55
Wallin, I., 125
Waloszek, D., 701, 752, 835
Waminoa, 352f–353f, 357
water conservation, 158
water expulsion vesicles, **160,** 160f
water vascular system of Echinodermata, **969, 975,** 976f, 978, 991, 1001, 1003
webs, spider, 928–932, 936f–937f, 936–938, 946
Weigert, A., 532, 597
Wells, M., 4
Westblad, E., 365, 366

whale falls, 20
Wheeler, W. C., 469, 958
wheel organ of Cephalochordata, **1025,** 1043
Wigglesworth, V. B., 869
Wiley, E. O., 38
Wilkinson, C. R., 252
Wilson, E. B., 183
  spiral cleavage coding system of, 189–191, 190f, 828f
Wilson, E. O., 6, 844
Wilson, H. V., 231
wings, insect, 867–870, 868f, 869f
  veins in, **865,** 866f, 870
Wingstrandarctus, 717
  corallinus, 712f, 714f
Wiwaxia, 14f, 15f
Woese, C., 7
Wolbachia, 686, 884, 884n
  pipientis, 884n
Woodley, J. D., 988
World Register of Marine Species (WoRMS), 6
Wright, T. S., 637
Wuchereria, 683
  bancrofti, 683, 846

X

Xanthidae, 780f
xanthophylls
  of Dinoflagellata, 80
  of Stramenopila, 101
Xenacoelomorpha, 206, 345, 347, 349–369, 380, 384
  acoelomate body plan of, 140
  characteristics of, 346
  feeding and digestion of, 149
  nervous system of, 392
  number of species, 2t
  phylogeny of, 25, 407, 1049f, 1050
Xenobalanus, 791
Xenocarida, 832, 833f
Xenocoelomorpha, 968
Xenodasys, 428f
Xenophoridae, 465
Xenosiphon, 574, 576
Xenotrichula, 428f
Xenoturbella, 25, 346, 349, 350, 366–369, 968
  bocki, 350, 366, 366f, 367, 367f, 368, 368f, 369f
  westbladi, 366
Xenoturbellida, 349, 351, 366–369
  body structure of, 366f, 367f, 367–368, 368f
  development of, 369
  nervous system and sense organs of, 366, 367, 368–369
  nutrition, excretion, and gas exchange of, 368
  phylogeny of, 1050
  reproduction of, 366, 369
  support and movement of, 368, 369f
xeric habitats, **23**
Xestospongia, 137f
  muta, 242f, 251, 252
Xiaoheiqingella, 694
Xiphinema index, 683
Xiphosura, 911, 912, 915f, 915–917
  body plan of, 927
  circulation and gas exchange of, 941, 942
  digestion of, 939, 940, 940f
  excretion and osmoregulation of, 942–943
  locomotion of, 932
  nervous system and sense organs of, 943, 944, 946
  phylogeny of, 962f
  reproduction and development of, 947f, 947–948
X-organ of Gastrotricha, **432**
Xylophaga, 490
Xylophagidae, 490
Xyloplax, 986, 996
Xylotrechus nauticus, 888f
Xysticus, 951f, 952
Xystodesmidae, 897

Y

Yager, J., 768
Y-cells of Gastrotricha, **431**
yeast, 4
y-larva, 790, 791, 792f–793f, 832
Yohoia, 14f
Yoldia, 496
yolk, **187,** 187f
Yunnanozoon, 349, 967
Yunnanpriapulus, 694

Z

Zanclea, 301
Zancleida, 273
Zeillerioidea, 664f–665f
Zeroseius ometes, 916f
Zimmer, R., 642–643
zip-junctions of Micrognathozoa, **628**
Zoantharia, 270
Zoanthidea, 270
zoea, **830,** 831
zoecium of Bryozoa, **648,** 648f, 649
Zonosagitta
  bedoti, 424f
  pulchra, 424f
zonula adherens, **362**
  of Nemertodermatida, 362, 363f
  of Xenoturbellida, 368
Zoochlorella, 78
zoochlorellae, **302,** 303
zoonotic pathogens, **23**
Zoophyta, 59, 328, 636
Zoophytes, 220
zooplankton, **20**
zoospores, **255**
Zooxanthella, 78, 302–303
zooxanthellae, 78, **302**–303, 303f, 304, 306, 309, 321
Zoraptera, 846, 851f, 853
  phylogeny of, 48f, 889f
Zorotypus, 853
Zygentoma, 48f, 847
Zygiella, 934f
zygote, 178, **187**

About the Book:
Editor: Andrew D. Sinauer
Project Editor: Martha Lorantos
Proofreader: Rebecca J. Rundell
Production Manager: Christopher Small
Photo Researcher: David McIntyre
Book Design and Layout: Janice Holabird
Cover photo and concept: Larry Jon Friesen
Illustrations: Nancy Haver and Elizabeth Morales
Indexer: Linda Hallinger, Herr's Indexing Service
Cover and Book Manufacturer: RR Donnelley